THE ENCYCLOPEDIA OF
PSYCHOACTIVE PLANTS

THE ENCYCLOPEDIA OF
PSYCHOACTIVE
PLANTS

ETHNOPHARMACOLOGY and Its APPLICATIONS

CHRISTIAN RÄTSCH

FOREWORD BY ALBERT HOFMANN

Translated by John R. Baker
with assistance from
Annabel Lee and Cornelia Ballent

Park Street Press
Rochester, Vermont

Park Street Press
One Park Street
Rochester, Vermont 05767
www.InnerTraditions.com

Park Street Press is a division of Inner Traditions International

Originally published in German under the title *Enzyklopädie der psychoaktiven Pflanzen*
First U.S. edition published in 2005 by Park Street Press

Library of Congress Cataloging-in-Publication Data

Rätsch, Christian, 1957-
[Enzyklopädie der psychoaktiven Pflanzen. English]
The encyclopedia of psychoactive plants : ethnopharmacology and its applications / Christian Rätsch ; foreword by Albert Hofmann ; translated by John R. Baker with assistance from Annabel Lee and Cornelia Ballent.
p. cm.
Summary: "The most comprehensive guide to the botany, history, distribution, and cultivation of all known psychoactive plants"—Provided by publisher.
Includes bibliographical references and index.
ISBN 0-89281-978-2 (hardcover)
1. Psychotropic plants--Encyclopedias. I. Title.

QK99.A1R2813 2005
615'.788—dc22

2004027227

Printed and bound in China by Regent Publishing

10 9 8 7 6 5 4 3 2 1

Text layout by Priscilla Baker
This book was typeset in Minion

Contents

Foreword

In this world, the point at which something happens is determined by the circumstances that call for it to happen. This *Encyclopedia of Psychoactive Plants* had to appear at just this time, for our contemporary society has need of such a work.

This need is connected with the spiritual and material dilemma of our times. It is not necessary to list all of the things that are no longer right in our world. But we can mention some: in the spiritual domain, materialism, egoism, isolation, and the absence of any religious foundation; on the material level, environmental destruction as a result of technological development and over-industrialization, the ongoing depletion of natural resources, and the accumulation of immense fortunes by a few people while the majority become increasingly destitute.

These ominous developments have their spiritual roots in a dualistic worldview, a consciousness that splits our experience of the world into subject and object.

This dualistic experience of the world first emerged in Europe. But it had already been at work in the Judeo-Christian worldview, with its god that sits enthroned above creation and humankind, and his admonition to "subdue . . . and have dominion . . . over every living thing that moveth upon the earth."

This is now occurring at a terrifying rate.

A change for the better will come about only when a general shift in consciousness takes place. Our fractured consciousness, which Gottfried Benn characterized as a "fateful European neurosis," must be replaced by a consciousness in which creator, creation, and created are experienced as a unity.

All means and all ways that will help lead to a new and universal spirituality are worthy of support. Chief among these is meditation, which can be enhanced and intensified through a variety of methods, including yogic practices, breathing exercises, and fasting, and through the appropriate use of certain drugs as pharmacological aids.

The drugs I am referring to belong to a special group of psychoactive substances that have been characterized as psychedelics and, more recently, as entheogens (psychedelic sacraments). These effect an enormous stimulation of sensory perceptions, a decrease or even neutralization of the I–Thou boundary, and alterations in consciousness in the form of both sensitization and expansion.

The use of such psychedelic drugs within a religio-ceremonial framework was discovered among Indian tribes in Mexico at the beginning and in the middle of the twentieth century.

This sensational discovery led to ethnobotanical investigations to remote areas around the world to search for psychoactive plants, the results of which were documented in numerous publications and pictures. The encyclopedic compilation of ancient knowledge and new discoveries about psychoactive plants that is in your hands was produced by a well-qualified author who has contributed important new insights on the basis of his own fieldwork. It is an undertaking of great value.

Disseminating knowledge about psychoactive plants, together with the proper ways to use them, represents a valuable contribution within the context of the many and growing attempts to bring about a new, holistic consciousness. Transpersonal psychology, which is becoming ever more important in psychiatry, pursues the same goal within a therapeutic framework.

The holistic perspective is more easily practiced on living nature than on the inanimate objects created by humans. Let us look into a living mandala instead, such as that found in the calyx of a blue morning glory, which is a thousand times more perfect and beautiful than anything produced by human hand, for it is filled with life, that universal life in which both the observer and the observed find their own individual places as manifestations of the same creative spirit.

Albert Hofmann, Ph.D.
summer 1997

Preface

My grandmother taught me many wise things that I have followed my entire life. In particular, her saying "An ounce of practice is worth a pound of theory" has had a considerable influence upon me and made it much easier to follow the path that has led me to the psychoactive plants.

It was during the 1967 "Summer of Love," when I was ten years old, that I first heard of hashish. I was listening to the radio. A menacing voice spoke of the "horrible dangers" that were descending upon our imperiled youth with the "new wave of drugs" from the United States. The picture that continues to dominate our drug policies was sketched out in dramatic fashion: Hashish was a gateway drug that inevitably, even compellingly, led to death from a golden shot of heroin. This was terrible news! But by that time I had already learned that I should not trust my teachers or the conservative politicians. I instinctively felt that the voice on the radio was lying. As a result of that broadcast, I yearned for nothing more than finally to try hashish myself (my experiments with cigarettes were already behind me, and I had noticed that I could not find any use or enjoyment there). At that time, it was not as easy to obtain hashish as it is today. Two years went by before I had my first opportunity. Up until then, I had only smoked dried banana peels and inhaled chloroform that I had synthesized myself. One morning, on the school bus, an older student walked down the aisle and whispered, "Hash, hash, anyone want hash?" "I do!" I cried, barely able to contain my joy and excitement. Back then, one gram cost about 3.50 marks (roughly one U.S. dollar), my entire allowance. But what does money mean when we're talking about the fulfillment of a two-year-old dream?

With the hashish in my pocket, I sat through my classes, bored to death as usual, waiting for the time when I would finally make it back home. After the ordeal of school, the time had finally come. I stood at home with my precious stash and pondered the best way to smoke it. Tobacco was not an option, for I genuinely disliked it. I went into the kitchen, saw a small bag of dried peppermint leaves, and immediately knew that I had found the appropriate admixture. I pedaled my bike into the nearby forest, stuffed a pipe full of mint and hashish, and lit it up. I immediately sensed that this mixture was easy to inhale, a wonderful contrast to those disgusting cigarettes.

Although the effects were mild, they were enough to make me want to continue my experiments. The next time, I went into the forest with a friend and we smoked the pipe together. An incredible sense of mirth overwhelmed us, and we almost split our sides with laughter.

I now know that my quest to obtain hashish and my deliberations as to what it could best be combined with marked the beginnings of my ethnopharmacological research. Today, I still search for psychoactive plants in all corners of the world and experiment with them until I have had meaningful experiences with them and learned what I could from them. And I still feel that I am being lied to when the media and the politicians talk about "drugs" and "narcotics," and I think to myself, "Oh, if only you too had smoked a nice pipe of hashish when you were twelve; so many problems could have been avoided!"

During my fieldwork in Nepal, I learned that the three fundamental evils of existence are hate, envy, and ignorance. The tantric doctrine has developed a number of methods for becoming aware of these evils and overcoming them by means of altered states of consciousness. It is my hope that all people—especially the politicians and the psychiatrists of Western countries—will one day understand that ignorance is one of the main reasons behind the catastrophic condition of our Mother Earth!

During my extended journeys to the various continents, I have seen time and again how people in all cultures, and of all social strata, religions, and skin colors, consume psychoactive plants and psychoactive products. Why do people ingest psychoactive substances? Because a fundamental drive for inebriation, ecstasy, blissful sleep, knowledge, and enlightenment is written right into our genes.

While working on the manuscript of this book, I realized that it would be my first "life work." The results of twenty years of research and experience are compiled in this work. I have collected information all over the world, assembled a large and specialized library, attended countless meetings and symposia, photographed my way through the plant world, and experimented with as many psychoactive plants as I could. The knowledge I have gained has now been distilled and organized into this encyclopedia.

"Thoughts are free, . . .
for my thoughts tear the fences and walls asunder. . . ."
GERMAN FOLK SONG

Introduction

Every day, most people in most cultures, whether Amazonian Indians or western Europeans, ingest the products of one or more psychoactive plants. Even the Mormons, who claim that they do not use "drugs," have a psychoactive stimulant: Mormon tea (*Ephedra nevadensis*), which contains the very potent alkaloid ephedrine, the model substance for amphetamine.

The use of psychoactive substances is extraordinarily common in the countries of South America. After rising, a typical Amazonian Indian will drink guaraná, cacao, or maté (and sometimes all three together). After breakfast, he will place the first pinch of coca in his mouth, where, periodically renewed, it will remain until evening. In the afternoon, he will shift to a fermented beverage made of maize or manioc. In late afternoon, some powder that contains tryptamines may be snuffed into the nose. Ayahuasca is often used in the evening. It goes without saying that every free minute is filled with the smoking, chewing, sniffing, or licking of tobacco.

Among the Tukano Indians, the use of psychoactive plants is mythologically associated with the origins of the world. The Sun Father was a *payé*, a shaman, who gave the shamans of our time all of their knowledge and abilities. At the beginning of the world, he carried in his navel *vihó*, a snuff obtained from the bark of the parica tree (*Virola* spp.). The ayahuasca vine came into the world through his daughter. As she was lying in labor, one of her fingers broke off. The midwife at her side took the finger and guarded it in the *maloca*, the cosmic roundhouse. A young man who saw this stole the finger. He buried it, and the ayahuasca liana grew from that spot. Another daughter of the Sun Father was also heavy with child. As she writhed about in the pains of labor, one of her fingers broke off as well. This time, the midwife took the finger and buried it herself. This gave rise to the first coca plant. Because these plants are associated with the origins of the world, they are considered sacred.

In the modern Western world, the use of psychoactive plant products is very widespread, but their sacredness has been profaned. How many of us today, when we are sipping our morning coffee, are aware that the Sufis venerated the coffee bush as a plant of the gods and interpreted the stimulating effects of caffeine as a sign of God's favor? Who of us, lying in bed and smoking the first cigarette of the day, knows that tobacco is regarded as a gift of the gods that aids shamans in journeying into other realities? How many recall the frenzied Bacchanalia in honor of Dionysos as they drink a glass of wine with their lunch? And the evening beer in front of the television is downed without any knowledge of the sacred origin of this barley drink. Our ancestors, however, the Germanic peoples and the Celts, knew this, and they venerated such drinks and immortalized them in their poetry:

> It is certain that the Celts knew of alcohol. The Greek and Roman authors of antiquity regarded them as passionate lovers of inebriating beverages. Drunkenness is a common theme in the epics, especially in Ireland. Gods and heroes competed with one another in their sheer unquenchable thirst for alcohol, whether in the form of wine, beer, or hydromel, the Celtic mead we still remember today. No religious festival was celebrated without an uninhibited drinking bout, a tradition which survives in our time in the form of (supposedly) folk customs. The most important aspect of such rituals is the *lifting off*, the *unleashing*, by means of which one forgets that man is an earthbound being. (Markale 1989, 203)

Indeed, it is this lifting off, this fact of getting "high," the unleashing, the ecstasy, that is at the heart of the use of psychoactive plants and psychoactive products. This encyclopedia is a testimony to the wealth of knowledge that humans have acquired about these substances. Through proper use and proper knowledge, we too—like our ancestors—may learn to once again recognize the sacred nature of inebriants and utilize these to have profound experiences of the sacredness of nature.

What Are Psychoactive Plants?

Psychoactive plants are plants that people ingest in the form of simple or complex preparations in order to affect the mind or alter the state of consciousness.

Consciousness is an energy field that can expand, shift amorphously like an amoeba through the hidden corners of the world, dissolve in the ocean of desire, or crystallize in geometric clarity. Through the use of psychoactive plants and products, consciousness can be paralyzed, subdued, and contained; it can also be animated, stimulated, and expanded. Because psychoactive plants affect the mind, they have been characterized as *mind-moving substances*. The renowned Berlin toxicologist Louis Lewin (1850–1929) referred to

"The peculiar, mysterious longing and desire for stimulants that is common to almost all peoples has, to the extent that we are aware of historical traditions, always prevailed and been satisfied in the most varied of ways. Inducing a happy mood in which the emotions, sorrows, and everything else that may weigh upon the soul can be forgotten, shifting into a state of partial or completely absent consciousness in which the individual, detached from the present, surrounded by the glowing and shining images of an excessively amplified imagination, becomes free from the misery of his every day life or from bodily pains, artificially inducing peace and sleep for the fatigued body and mind in all cases where these necessary requirements for life cannot be brought about in the normal manner, and finally the wish to gain creative strength, both physically and mentally, by means of these stimulantsthese are the primary reasons why these agents are used."

LOUIS LEWIN
ÜBER PIPER METHYSTICUM
(1885, 1)

"Only plants had consciousness. Animals got it from them."

DALE PENDELL
PHARMAKO/POEIA
(1995)

"Every life's heart and desire
Burns with greater rapture, flickers
    more colorfully,
I welcome every inebriation,
I stand open to all torments,
Praying to the currents, taken with
    them
Into the heart of the world."

HERMANN HESSE
VERZÜCKUNG [RAPTURE]
(1919)

all those substances that produce some sort of psychoactive effects as *phantastica.* Carl Hartwich (1851–1917), a pharmacist, described them as "human means of pleasure." Timothy Leary (1920–1996) preferred to speak of them as *neurobotanical substances.* Today, the terms *psychotropic* ("influencing the psyche") and *psychopharmacologic* ("affecting the mind") are often used to refer to these substances.

In the pharmacological literature, which commonly refers to them as *mind-altering substances,* psychoactive substances are clearly and systematically classified by precise scientific definitions (cf. Inaba and Cohen 1994; Seymour and Smith 1987; Wagner 1985):

🌿 Stimulants ("uppers")

This category comprises substances that wake one up, stimulate the mind, incite the initiative, and may even cause euphoria but that do not effect any changes in perception. Among the most important plants in this category are coffee, tea, cacao, guaraná, maté, ephedra, khat, and coca.

🌿 Sedatives, Hypnotics, Narcotics ("downers")

This category includes all of the calmative, sleep-inducing, anxiety-reducing, anesthetizing substances, which sometimes induce perceptual changes, such as dream images, and also often evoke feelings of euphoria. The most important psychoactive plants and products in this category are poppy, opium, valerian, and hops.

🌿 Hallucinogens ("all-arounders")

This category encompasses all those substances that produce distinct alterations in perception, sensations of space and time, and emotional states. Most of the plants discussed in this encyclopedia fall into this category. Over the course of time, these substances have been referred to under a variety of names:

–Psychotomimetics ("imitating psychoses")
–Psychotica ("inducing psychoses")
–Hallucinogens (Johnson; "causing hallucinations")
–Psychedelics (Osmund; "mind manifesting")
–Entheogens (Ruck et al.; "evoking the divine within")
–Entactogens (Nichols; "promoting self-knowledge")
–Empathogens (Metzner; "stimulating empathy")
–Eidetics ("giving rise to ideas")
–Psychotogenics ("affecting the mind")
–Psychodysleptics ("softening the mind")

Today, the most commonly used term is still *hallucinogen.* By definition, a hallucinogen is a substance that evokes hallucinations (Siegel 1995b), which are now medically defined as "sensory delusions that may involve several (to all) senses (= complexes) and are not the result of corresponding external sensory stimuli but possess a reality for the affected person; also occur in schizophrenia, stimulated brain states (e.g., due to poisoning, epilepsy, brain injuries, the effects of hallucinogens)" (*Roche Lexikon Medizin* 1987, 725).

Because the term *hallucination* now has a psychopathological tinge to it, nonmedical circles and publications usually prefer the terms *psychedelic, entheogen,* or *visionary substance* and accordingly speak of psychedelic, entheogenic, or visionary experiences:

The awakening of the senses is the most basic aspect of the psychedelic experience. The open eye, the naked touch, the intensification and vivification of ear and nose and taste. This is the Zen moment of satori, the nature mystic's high, the sudden centering of consciousness on the sense organ, the real-eye-zation that this is it! I am eye! I am hear! I knose! I am in contact! (Leary 1998, 34)

Shamans, of course, the traditional specialists in psychoactive substances, do not speak of psychoactive drugs, psychotropics, or hallucinogens—not to mention narcotics—but of plant teachers, magical plants,[1] plants of the gods, sacred beverages, et cetera. They revere these mind-altering plants and make them offerings; they use them not as recreational drugs or as something to get "high" with in the evening but as sacraments in their rituals. The shamans regard these plants as sacred because they make it possible for them to contact the true reality and the gods, spirits, and demons. They are sacred because within them dwell plant spirits, plant gods, or devas that one can ally oneself with and that are esteemed as the teachers, mothers, ambassadors, and *doctores* (physicians) of other realities. In addition, these sacred plants have the power to heal. They can liberate the ill from their afflictions and drive out harmful, disease-causing spirits. They also can bring spiritual awakening to healthy people and make possible mystical experiences. With the aid of these plants, one does not lose control, for control is ultimately an illusion.[2] And they are used not to escape from reality but to recognize true reality:

We can see that these plants do more than simply maintain our body. They also promote and nourish our souls and make possible the enlightenment of our mind. Their existence is offering, sacrifice, and selfless love. The earth on which they grow is itself a sacrificial altar—

1   It should be noted, however, that many plants that are characterized as "magical plants" in folklore or literature are not psychoactive (cf. Schöpf 1986; Storl 1996c; Weustenfeld 1995).

2   "The fear of the psychedelic experience is quite literally the fear of losing control. Dominator types today don't understand that it's not important to maintain control if you are not in control in the first place" (Terence McKenna, in Sheldrake et al. 2001, 50).

and we who receive their blessings are the sacrificial priests. Through plants, the outer light of the sun and the stars becomes the inner light which reflects back from the foundations of our soul. This is the reason why plants have always and everywhere been considered sacred, divine. (Storl 1997, 20)

The Use of Psychoactive Plants

Humans have a natural drive to pursue ecstatic experiences (Weil 1976; Siegel 1995a). The experience of ecstasy is just as much a part of being human and leading a fulfilling and happy life as is the experience of orgasm. In fact, many cultures use the same words to refer to ecstasy and to orgasm.[3] The possibility of having ecstatic experiences is one of the fundamental conditions of human consciousness. All archaic and ethnographic cultures developed methods for inducing such experiences (Bourguignon 1973; Dittrich 1996). Some of these methods are more efficacious than others. The most effective method of all is to ingest psychoactive plants or substances.

These methods, however, require certain skills, for there are many factors that play a role in shaping the effects and the contents of the experiences. The most important is proper use— that is, a responsible and goal-oriented use.

Fitz Hugh Ludlow (1836–1870), whose book *The Hasheesh Eater* (published in 1857) was the first American literary work on the effects of hashish, has given us an amazing description of the proper way to use hashish:

There is a fact which can be given as a justification for the craving for drugs without coming close to dubious secondary motives, namely, that drugs are able to bring humans into the neighborhood of divine experience and can thus carry us up from our personal fate and the everyday circumstances of our life into a higher form of reality. It is, however, necessary to understand precisely what is meant by the use of drugs.

We do not mean the purely physical craving. . . . That of which we speak is something much higher, namely the knowledge of the possibility of the soul to enter into a lighter being, and to catch a glimpse of deeper insights and more magnificent visions of the beauty, truth, and the divine than we are normally able to spy through the cracks in our prison cell. But there are not many drugs which have the power of stilling such craving. The entire catalog, at least to the extent that research has thus far written it, may include only opium, hashish, and in rarer cases alcohol, which has enlightening effects only upon very particular characters. (Ludlow 1981, 181)

There are many different ways to use psychoactive plants. The reasons they are consumed range from relaxation, recreation, and pleasure (hedonism) to medical and therapeutic treatments and to ritual and religious ceremonies and spiritual growth. It is the task of culture and society to provide the individual with patterns for using them that serve these purposes.

Drug Culture

Both experience and research have very clearly demonstrated that every culture in the world either has used or still does use psychoactive substances in traditional contexts:

Every society, every time has its drug culture. Corresponding to the complexity of society, its drug culture may also be more or less complex, oriented, for example, around just one central drug or encompassing a number of drugs. It can also be subdivided into internal cultures that can contradict one another. (Marzahn 1994, 82)

These "internal cultures" are often referred to as "subcultures" or "scenes." Within these cultural structures, cultural patterns often form that seem to be archetypical for human existence. Marzahn analyzed traditional rituals that employ psychoactive substances—he uses the term *drug*, most likely as a provocation—and from these developed a model that suggests that, throughout the world, common drug cultures continually emerge and establish themselves:

Yet the deepest meaning of the common drug culture appears to lie in the fact that this *internal order* is required because of this exiting, this stepping over boundaries; it is precisely what a culture of border crossers needs. In the context of the common drug culture, the use of drugs is not banished out of time and space. Rather, it has a clear and circumscribed place within both. People gather at a special place and surround themselves with the proper space and with beautiful devices. The communal use of the drug has a beginning and an end. It takes place according to an internal order, which has been derived from experience, and does not simply allow whatever anyone might want. With time, it has become condensed into a ceremony, a rite. This internal order and its outer form, the ritual, are what make it possible for the drug to be used properly and protect people from harm and destruction. In all normal drug cultures, it is a duty of those who have already had the experience to introduce the inexperienced to this order. (Marzahn 1994, 45)

"I allow dew drops to fall from the flowers onto the fields, which inebriate my soul."

SONG OF NEZAHUALCOYOTL IN *ANCIENT NAHUATL POETRY* (BRINTON 1887)

"Most people never realize that the purpose of intoxication is to sharpen the mind."

ROBERT E. SVOBODA *AGHORA* (1993, 175)

"Religions are false means for satisfying genuine needs."

KARLHEINZ DESCHNER *BISSIGE APHORISMEN* [BITING APHORISMS]

3 E.g., among the Tukano and Desana Indians (Reichel-Dolmatoff 1971).

Ritual Uses of Psychoactive Substances

It is possible to classify many psychoactive plants and the products made from them into different types of rituals that reflect how they are used. These include:

Shamanic initiation
  Fly agaric mushroom (*Amanita muscaria*)
  Cigar tobacco (*Nicotiana rustica*)

Shamanic healing rituals
  Ayahuasca
  Snuffs
  Cebil (*Anadenanthera colubrina*)
  Mushrooms
  San Pedro (*Trichocereus pachanoi*)

Shamanic ritual circles
  Ayahuasca
  Mushrooms (*Panaeolus* spp., *Psilocybe* spp.)
  Hemps (*Cannabis* spp.)
  San Pedro (*Trichocereus pachanoi*) and cimora

Vision quests
  Kinnikinnick
  Thorn apples (*Datura* spp.)
  Tobaccos (*Nicotiana* spp.)

Rites of passage
  Balche'
  Iboga (*Tabernanthe iboga*)
  Hemp (*Cannabis indica* in Jamaica, among the Rastafarians)

Rituals of greeting
  Kava-kava (*Piper methysticum*)
  Cola nuts (*Cola* spp.)
  Pituri

Burial rituals
  Cola nuts (*Cola* spp.)
  Alcohol

Divination
  Henbanes (*Hyoscyamus* spp.)
  Angel's trumpets (*Brugmansia* spp.)
  Ololiuqui (*Turbina corymbosa*)
  Salvia divinorum
  Thorn apples (*Datura* spp.)
  Incense

Rain magic and rain ceremonies
  Henbane (*Hyoscyamus niger*)
  Saguaro (*Carnegia gigantea*) wine

Healing rituals within a religious cult
  Nutmeg (*Myristica fragrans*)

Harmful magic
  Yagé, ayahuasca
  Datura, Brugmansia, Solandra

Purification rituals
  Cassine, yaupon, black drink (*Ilex cassine, Ilex vomitoria*)
  Guayusa (*Ilex guayusa*)
  Enemas
  Incense

Sexual magic rituals (Tantra, Taoism, cult of Aphrodite)
  Mandrakes (*Mandragora* spp.)
  Damiana (*Turnera diffusa*)
  Oriental joy pills
  Yohimbé (*Pausinystalia yohimba*)

Initiation into secret societies and cults or cultic communities
  Iboga (*Tabernanthe iboga*)
  Madzoka medicine

Religious ceremonies led by priests
  Incense
  Wine (libations)
  Spiced coffee

Mystery cults
  Haoma
  Kykeon
  Wine

Socially integrative ritual circles
  Plants: cocas (*Erythroxylum* spp.), hemps (*Cannabis* spp.), khat (*Catha edulis*), cola nuts (*Cola* spp.)
  Products: balche', beer, chicha, palm wine, wine

Perceptual training
  Tea (*Camellia sinensis*) ceremony
  Kodoh (incense)

Meditation
  Hemp (*Cannabis indica*)
  Coffee (*Coffea arabica*)
  Khat (*Catha edulis*)
  Tea (*Camellia sinensis*)
  Soma

In many cultures, the experts are the shamans, or sometimes the priests, diviners, or medicine people. However, in our culture there is a deep chasm, a wound, for the people who preserved our own traditional knowledge have disappeared as a result of forced Christianization, imperialism, the Inquisition, the persecutions of witches, the Enlightenment, and positivism. And yet in spite of this, the psychoactive life continues to pulse in the inner cultures. And the archaic patterns continue to remain relevant, so appropriate uses of psychoactive substances continue to emerge. This has produced what may be called "underground experts" in the proper use of psychoactive substances:

> Through rhythm, internal order, and ritual, the common drug culture provides an orientation and a foothold in dealing with drugs: for our aspiration, because it embeds drug use within an understanding about the proper way to live, about the goals and forms of life, and about the role that befits drugs within these; for our knowledge, because it provides information about the mechanisms of actions, benefits, and drawbacks of drugs that is based upon experience and traditional knowledge; for our feelings, because it provides us with security in the simultaneously affirming and shy respect for drugs, thereby protecting us from ill-conceived fear and fascination, from both a demonizing worship and a demonization; and finally for our actions, because it develops and passes down rules that are recognized and respected because experience and validation have shown them to be meaningful and because they tell us which drugs, in which dosage, when, where, and with whom are beneficial and which are not. (Marzahn 1994, 47)

The Most Important Considerations: The Theory of Dosage, Set, and Setting

The theory of dosage, set, and setting provides a useful model for better understanding the effects of psychoactive plants. Dr. Timothy Leary (1920–1996), a Harvard professor, conducted scientific experiments with psychedelic substances (LSD and psilocybin) in the early 1960s. On the basis of his own experiences and his systematic observations, he and his colleagues Ralph Metzner and Richard Alpert (Ram Dass) developed this theory (Leary et al. 1964), which states that there are three main factors responsible for the experiences induced by psychedelics. The first factor is the dosage—a truism since ancient times, or at least since Paracelsus. The set is the internal attitudes and constitution of the person, including his expectations, his wishes, his fears. The third aspect is the setting, which pertains to the

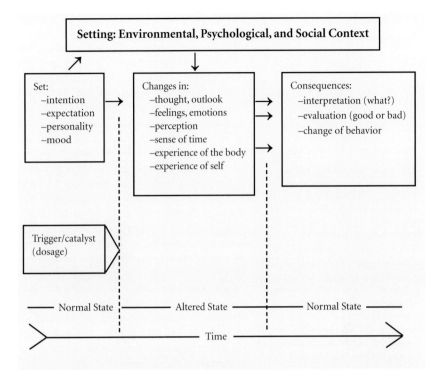

surroundings, the place, and the time—in short, the space in which the experiences transpire. This theory clearly states that the effects are equally the result of chemical, pharmacological, psychological, and physical influences.

The model that Timothy Leary proposed for the psychedelics also applies to experiences with other psychoactive plants (including the stimulants and narcotics). All three factors must be carefully considered when one wishes to have experiences with and understand these plants. Even in the same person, the same plant can evoke very different effects if the dosage, set, and setting are changed.

The first factor, of course, pertains to the choice of plant. Then the proper dosage must be consumed. But what is a "proper" dosage? It is the amount that will produce the desired effects. But since the effects are not solely the result of the dosage, the proper dosage can be determined only by taking the other factors into consideration as well. As the saying goes, "An ounce of practice is worth a pound of theory." This is especially true in the present context. When experimenting, one should always begin with low dosages. It is better to use too little than too much. You can always use more on the next occasion. If one rashly takes too much, the result may be unpleasant or even dangerous. When ingesting strychnine, for example, the dosage is extremely crucial. A small dosage can produce wonderful sensations and sexual vigor, whereas a large dosage can be lethal.

American Indians, for example, recognize three dosage levels for magic mushrooms: a medicinal, an aphrodisiac, and a shamanic. For the

"In different cultures, drugs are often used in completely different manners. This demonstrates that the consumption of drugs is culturally shaped to a very large extent. Which substances are used, when, by whom, how, how often, and in which dosage, where, with whom, and why, and also which conceptions are related to this are largely dependent upon the cultural membership of a user. Because of these influences, inebriation is experienced and lived out in very different ways, and a drug may be used for different purposes, may be assigned different functions."

BLÄTTER
(1994, 123)

medicinal dosage, a quantity is administered that does not produce any psychoactive effects but that can heal certain ailments. The aphrodisiac dosage is higher: The mind is activated but not over-powered by visions or hallucinations; perception and sensitivity are heightened and the body is aroused and invigorated. The shamanic dosage catapults consciousness into an entirely different reality that is flooded with cosmic visions and enables a person to peer into worlds that are beyond the normal experience of space and time.

Set is perhaps the most important factor for becoming aware of the efficaciousness of a psychoactive plant, especially when a hallucin-ogenic substance is involved. These substances have the ability to activate, potentiate, and sometimes mercilessly expose everything that a person has in his or her consciousness or buried beneath it. People who were raised under the repressive conceptions of the Catholic religion, for example, may need to struggle with the original sin that was laid upon them in the cradle, whereas a nature-venerating pagan may perceive his or her partner as a temple of divine desire.

In traditional cultures, set is shaped primarily by the worldview that all individuals share and is especially expressed in a tribe's mythology. The mythology provides a kind of cartography of the visionary worlds and other realities. Using this cartography, an explorer of consciousness can reach the desired goal. And he can always count on the help of the shaman who accompanies him, for the shaman is the best cartographer of the other, visionary reality. Even when a person gets lost in that world, the shaman can bring him back. The contents of the visions, in other words, are shaped by culture.

Psychoactive Plants and Shamanic Consciousness

The shaman is not only a hunter, warrior, healer, diviner, and entertainer, but also a natural scientist and thinker. Anthropologist Elizabeth Reichel-Dolmatoff reports that among the Tanimuka, a Tukano Indian group, shamanism is consequently referred to as "thinking." The shaman, first and foremost, is a visionary who has genuine visions:

> A shaman is one who has attained a vision of the beginnings and the endings of all things *and who can communicate that vision.* To the rational thinker, this is inconceivable, yet the techniques of shamanism are directed toward this end and this is the source of their power. Preeminent among the shaman's techniques is the use of the plant hallucinogens, repositories of living vegetable gnosis that lie, now nearly forgotten, in our ancient past. (McKenna 1992, 7)

"Shamanism is the door to the real world." The ethnopsychologist Holger Kalweit spoke these words at a symposium in September 1996 entitled "The Shamanic Universe." What he meant is that shamanic consciousness is the real world or, as the Indians say, the "true reality."

For many Indians in the Central and South American rain forests, the everyday world is an illusion, a superficial necessity.[4] "To those who know, this appears as the world of effects, whereas the world of myths is the world of causes" (Deltgen 1993, 125). Ayahuasca or yagé, the "drink of true reality," helps people pierce through the illusion that is the everyday world and penetrate into the heart of reality. The reality that is

Left: This beaded head of a jaguar bears witness to the great importance that psychoactive substances play in shamanism. The jaguar is a symbol of the shaman—for he can transform himself into this powerful animal—but he is also his power animal and ally. The shaman uses a psychoactive plant, the peyote cactus (*Lophophora williamsii*), to establish contact with this ally. The fantastic world he enters is echoed in the artistry of the beadwork. The visionary cactus itself is portrayed on the animal's cheek. (By an unknown Huichol artist, ca. 1996)

Center: Shamans throughout the world use psychoactive substances in order to penetrate into the other world, the other reality. Drumming helps ensure a safe journey during the visionary adventure. (Nepali shaman at Kalinchok, 1993)

Right: During a dramatic visionary experience, the shaman obtains his special abilities and powers by dying as a person and being reborn as a shaman. Psychoactive substances often produce experiences of death and rebirth, as well as near-death experiences. (Huichol yarn painting, ca. 1995)

4 This is very similar to the Indian concept of *maya*, "the illusion that conceals reality as a result of ignorance" (Zimmer 1973, 31*).

Name	Active Substance	Dosage		Duration of Effects
		Alkaloid	Drugs (dried)	
Mushrooms	psilocybin psilocin	20 mg	3–5 g 2–8 mushrooms	4 hours
Peyote	mescaline phenethylamine	0.5–1.0 g	4–14 "buttons"	4–8 hours
Vines	lysergic acid amide	1–5 mg	5–40 seeds (up to 300!)	4–12 hours
Datura	tropanes (scopolamine/atropine)	5 mg	12–60 seeds 3–5 flowers	3–6 hours up to 24 hours

experienced under the influence of ayahuasca is the reality of the myths, which appears to be more real and more meaningful. "The drug is a medium, a vehicle between this reality and that. It is the gateway to knowledge. The *kumú* [shaman], however, is the mediator between the two worlds, and may be more passive or more active, depending upon his power and his talents" (Deltgen 1993, 141). The effects of the "ingestion of these hallucinogens are not understood as an action produced by a special, that is, active chemical substance, but as a contact with spirit beings (owners, 'mothers,' species spirits), who control the corresponding plant and embody its 'essence'" (Baer 1987, 1). The spirits of the plant are the same spirits who aid the shaman in the healing process: "The hallucinogenic plants, or the spirits that dwell within them, open the eyes of those who take them; they enable them to recognize the nonordinary reality, which is considered to be reality per se, and it is ultimately they, and not the shaman, who free the patient from his affliction" (Baer 1987, 79). Not everyone, however, can control the spirit helpers: "The *caji* [ayahuasca] thus does not make the shaman. To the contrary: he who is called to be a shaman, the spiritually gifted, is able to make something out of the drugs and their effects" (Deltgen 1993, 200).

Like their shamans, most of the Indians of the Amazon base their lives upon the visions they receive through ayahuasca: "Our ancestors oriented the entire rhythm of their lives around the ayahuasca visions; whether it had to do with the making of weapons, drawings, art, colors, clothing, medicine, or something else, or it involved determining the most favorable time for a journey or to till the fields. They used the ayahuasca visions in their attempts to better organize themselves" (Rivas 1989, 182).

Shamans throughout the world consume psychoactive plants and products so that they may be able to enter the shamanic state of consciousness and travel to the visionary world, the other reality. The substances shamans use are very diverse both chemically and pharmacologically. The active substances they contain belong to different classes that are analogs of or related to different endogenous neurotransmitters (see the box above).[5] Nevertheless, they are all pharmacological stimuli for achieving the selfsame purpose: to produce the shamanic state of consciousness.[6]

This fact was verified through the research of Adolf Dittrich, who demonstrated that experiences in altered states of consciousness—and compared to everyday reality, the shamanic state of consciousness is very altered—are identical at the core, no matter which pharmacological and/or psychological stimuli elicited them (Dittrich 1996).

On the basis of my own experiences with a variety of psychoactive plants, I can attest that different active substances can evoke the same state of consciousness, e.g., trance, but will not always do so, for the same drug can produce totally different effects in different people. In particular, the drugs found in datura exhibit striking differences (cf. Siegel 1981). Even in the same person, the same substance can induce very different effects depending upon the dosage, set, and setting. In order to produce the same state, i.e., the shamanic state of consciousness, more than just a psychoactive substance is needed. The user must also have the appropriate intention and the appropriate external conditions.[7] The drug experience is heavily influenced by the mythological and cosmological matrix of the user and by the ritual that is taking place in the external world. Mythology and cosmology provide the topography or cartography of the shamanic world and show the ways into it and back out. The ritual provides the outer framework that facilitates the user's transition from everyday reality to shamanic reality and back.

The reasons why a plant is being used will strongly affect the content of the experiences. If it is being used to perform shamanic tasks, then it will tend to evoke shamanic realities. As with all human abilities, however, this talent is not the same in everyone. Only the most talented can become shamans. In the same way, humans all differ with regard to our boldness and courage. Only the most courageous of us can become shamans. Fearful people should not confront the

"Emissaries of the plant kingdom merge with human bodies and aid people in attaining other states of consciousness. Only the gods know which powers of nature are here at work. Possessed people give themselves in to sexual activities and join in the cosmic dance of joy. They celebrate festivals in the truest sense of the word. These festivities are an expression of that fundamental and timeless form of religious ceremony which is an invitation to the gods. Through this adoration, man makes a request, he offers the gods his body and soul, so that they will 'take over' these. Enlightenment."

TIMOTHY LEARY
ON THE CRIMINALIZATION OF THE NATURAL
(NO DATE)

5　Psilocybin is an analog of serotonin, mescaline of dopamine, LSA of the tryptamine-like endopsychedelics, and scopolamine of acetylcholine (Rätsch 1993c, 42).

6　It should be noted that Western psychiatry has or still does use the active substances in these four drugs for psychotherapeutic purposes. The discovery of mescaline led to a revolution in European psychiatry (Hermle et al. 1993; La Barre 1960). Psilocybin and its derivatives were used extensively in psycholytic therapy (Leuner 1981). Because of their hypnotic properties, the alkaloids in ololiuqui were tested in experimental psychiatry (Heim et al. 1968; Isbell and Gorodetzky 1966; Osmund 1955). Scopolamine was used as a truth serum and in narcoanalysis, and in psychiatry it is still used as a "chemical straitjacket" (Rätsch 1991b).

7　Typically, the altered state of consciousness produced by alcohol is viewed medically as being fundamentally different from the state produced by hallucinogens (Winkelman 1992, 186). This notwithstanding, Mexican shamans use both hallucinogens and alcohol to induce a shamanic trance. The effects of alcohol that we are familiar with (ranging from being slightly tipsy to delirium tremens) are thus at least partially conditioned by culture. After all, the ancient descriptions of the Dionysian frenzy (cf. Emboden 1977) are fundamentally different from the descriptions of alcohol intoxication in modern industrialized societies.

gods and demons. It is for these reasons that in most societies that have institutionalized shamans, the use of plants with visionary effects is embedded within an exclusively ritual context. The visionary experiences take place against the familiar background of one's own culture.

The shamanic use of psychoactive plants follows a specific basic pattern, whereby it is relatively unimportant which substance is being used. First and foremost are the form, meaning, and purpose (function) of a ritual. The structure of the ritual follows a pattern that I have termed the "psychedelic ritual of knowledge" (cf. Rätsch 1991b):

Preparation
Collecting and preparing the drugs
Sexual abstinence, fasting
Practical purification (bathing, sweat house,[8] enemas)
Symbolic purification

Utilization
Offerings to the gods (e.g., incense)
Prayers to the gods and/or plant spirits
Ingestion of the drug
Soul journey during trance
Communication with the plant spirits/gods/animal spirits

Integration
Diagnosis/prophecy
Instructions for how to behave
Offerings of thanks

The Fear of Psychoactive Plants

The fear of consciousness-expanding plants is at least as old as the Bible. In Genesis, this fear is thematically expressed as the Fall. The fruit of the tree of knowledge transforms a person into a god. But since we are allowed to worship only one god, no one else can stand on the same level as him (or her?).

In many hierarchical cultures with an imperialistic orientation (emphasizing power instead of knowledge), immediate mystical, ecstatic, or religious experience is heavily regulated and is usually even forbidden. The direct experience of the world has been replaced by an elaborate, theologically driven religion and is monopolized by the state. Paradise, that other reality, is administered by bureaucrats who have not personally experienced it and who sell it to the needy and those who crave ecstasy. Jonathan Ott has referred to this mechanism as the "pharmacratic inquisition" (1993). The Mexican inquisition provides the best historical example of the suppression of personal experience and its replacement by a state monopoly for administering the divine.

As the Europeans pushed into the New World, they encountered for the first time shamans, whom they contemptuously labeled "magicians" and "black artists." The shamans' gods and helping spirits were degraded as false gods, idols, and the devil's work; their sacred drinks were defamed as witches' brews. An Inquisition report from the colonial period written by D. Pedro Nabarre de Isla (issued on June 29, 1620) notes:

As for the introduction of the use of a plant or root named peyote . . . for the purpose of uncovering thievery, divinations about other occurrences, and prophesizing future events, this is a superstition which is to be condemned because it is directed against the purity and integrity of our sacred Catholic faith. This is certain, for neither this named plant nor any other possesses the power or intrinsic property of being able to bring about the alleged effects, nor can anything produce the mental images, fantasies, or hallucinations that are the basis of the mentioned divinations. In the latter, the influences and workings of the devil, the real cause of this vice, are clear, who first makes use of the innate gullibility of the Indians and their idolatrous tendencies and then strikes down many other people who do not sufficiently fear God and do not possess enough faith.

Even today, the sacred plants of the Indians and/or their active constituents are forbidden throughout the world. While the use of peyote, mescaline, psilocybin (the active principle of Mexican magic mushrooms), DMT, et cetera, is in principle exempt from punishment, the possession of or trafficking in these is nevertheless illegal (Körner 1994). The drug laws of our time, in other words, are rooted in the spirit of the Catholic Inquisition. As long as the sacred plants and substances of the Indians remain illegal, the war against the indigenous peoples of the Americas will not be over. Generally speaking, the U.S. "War on Drugs" is a continuation of European colonialism and an instrument for criminalizing the Indians and their spiritual kin.

This phobia about drugs is nothing new, for drugs have been viewed as wild and reprehensible since ancient times (think of the persecution of the mystical followers of Dionysos, as well as of the witches, alchemists, and hippies). The fear of drugs and the experiences associated with them is found even throughout the various camps of shamanism fans and in academic circles. Mircea Eliade, for example, discounted the use of drugs to produce trance and (archaic) ecstasy as "degenerate shamanism" (1975, 382). Many members of the New Age movement have claimed that they

8   The ritual and hygienic use of sweat houses (temazcal in Aztec) was widespread in Mexico before the arrival of the Spanish. Today, it has almost vanished (cf. Cresson 1938).

can attain "it" without drugs. There are also anthropologists who argue that just because "their" shaman apparently enters into trance without any pharmacological support, other shamans—about whom they know nothing—also should not need drugs. It appears, however, that almost all traditional shamans prefer pharmacological stimuli (Furst 1972a; Harner 1973; Ripinsky-Naxon 1993; Rosenbohm 1991; Vitebsky 1995). As one source puts it, "The Indians view the drugs as nourishment for the soul and venerate them because of their wondrous properties" (Diguet in Wagner, 1932, 67).

When the Christian Europeans encountered their first shamans, they saw them as black magicians, master witches who had allied themselves with the devil and who, with his help, were leading the other members of their tribe down the road to ruin. In the early ethnographic literature, they are referred to as magicians, witch doctors, medicine men, weather makers, mediums, and the like. A large portion of the literature on shamanism specialized in demonstrating that shamans are con artists who use sleight of hand to trick the other members of their tribes and that, in the best of cases, they are charlatans whose methods are irrational and superstitious.

In traditional psychiatry and psychoanalytical-oriented anthropology, shamans were regarded as schizophrenics, psychopaths, and sufferers of arctic hysteria, that is, as people who are ill. Strange indeed that these ill people are the very ones who concern themselves with the task of healing. In contrast, shamans have been glorified and proclaimed as saviors in antipsychiatric circles. This attitude gave rise to images of "psychiatric utopias in which the shaman was the leader" (Kakar 1984, 95). In the more recent ethnographic literature, especially that based upon the approach known as cognitive anthropology, shamans are looked at from the perspective of that which they represent for their communities: people who, because of their calling and their special gift of trance, are able to divine, diagnose, and heal. In doing so, they maintain harmony in the community, preserve the tribal myths and traditions, and ensure the survival of their people.[9]

The interdiction of psychoactive plants and their effects is not bolstered solely by questionable politically based laws, but also receives support from the side of established science. Here, two concepts from psychiatry have played key roles: *psychotomimetic* and *model psychosis*. The first is a term for a substance that is said to mimic a psychosis; the second is a term used to characterize the experience. As a result, these plants and the effects they produce are viewed not as something sacred or mystical but as something pathological. This is reminiscent of such anthro-pologists and religious scholars as George Devereux and Mircea Eliade, who regarded shamans as psychopaths or people suffering from hysteria.

Since the end of the nineteenth century, Western psychiatry has known of and used drugs that alter consciousness (Grob 1995; Strassman 1995). The first such substance to be tested and used in psychiatry was mescaline. Mescaline was first extracted, chemically identified, and synthesized from the Mexican peyote cactus at the end of the nineteenth century. At that time, the effects of mescaline upon healthy subjects were thought to be the same as those that were otherwise known only from psychiatric patients. This led to the idea of the pharmacologically induced "model psychosis" (cf. Leuner 1962; Hermle et al. 1988). During the twentieth century, other substances with similar effects were discovered in the plant world, synthesized in the laboratory, and tested on patients and even on prisoners (Hermle et al. 1993).

The concept of the model psychosis is simply another form of ethnocentrism. Whereas the Inquisition saw the workings of the devil in these psychoactive substances, psychiatrists interpreted the sacred visions as psychotic-like states, that is, as "artificially" induced mental illnesses. Today, however, the model psychosis concept has itself landed on the rubbish heap of modern high-technology science. Recent research into the brain activity of true psychotics and of healthy users of psychedelics, using PET scans, has demonstrated that very different regions of the brain are active in each (Hermle et al. 1992).

Another opinion prevalent in our world holds that "drugs" cannot be used intelligently but will automatically be "misused" (cf. Dobkin de Rios and Smith 1976). In our culture, it is commonly argued that narcotic drugs lead to "addiction" or "dependency." Here, the views vary widely. In addition, the addictive potential of a substance is often used as the only definition of an inebriant (also frequently referred to as an "addictive drug"). Since addictive behaviors can arise with respect to almost every substance, many foods, luxury goods, and numerous medicines should also be seen as addictive substances. Many people, for example, are "addicted" to chocolate (cf. Ott 1985). Some have even argued that sugar is a drug, and an addictive one at that (McKenna and Pieper 1993). So, are chocolate and sugar invigorating foodstuffs, delicious luxury goods, or addictive drugs?

Since ancient times, psychoactive substances have been used by athletes as doping agents (cf. *Mammillaria* **spp.**). In the modern world of competitive sports, the plant substance **ephedrine** and its derivatives (amphetamines), camphor (cf. *Cinnamomum camphora*), **strychnine**, and **cocaine**

9 See Jilek (1971) for an extensive study of the contradictions and changes that have taken place in the images of shamans.

"The world is as one perceives it and what one perceives of it."

Albert Hofmann
Lob des Schauens [In Praise of Looking]
(1996)

have all been used. Of course, the use of doping agents is condemned, regarded as unsportsmanlike, forbidden, and strongly proscribed (Berendonk 1992). Many athletes, however, are like the "closet shamans" (see p. 20), constantly on the lookout for new ways to augment their performance. Recently, preparations of the ascomycete *Cordyceps* were successfully used for doping purposes. The athlete involved could not have her victory disallowed, however, for this was a dietary supplement, not a forbidden substance.

The Study of Psychoactive Plants

Science begins by collecting data, facts, and objects and ends with systematic knowledge. This process characterized all of the early works of science, which condensed and concentrated the knowledge of their time and their world. There is also the human desire to experiment. We learn by trying things out, and we change our behavior as a result of our experiences. It is striking that all of the great plant researchers have been avid collectors of both information and materials and have also tested the effects of as many plants as possible on themselves, for how can one evaluate the effects of a plant if one has never seen or touched it, not to mention ingested it?

The study of psychoactive plants began with the beginnings of botany. Theophrastus (ca. 370–287 B.C.E.), the "father of botany," has given us descriptions of numerous psychoactive plants and substances. Systematic science, which some trace back to the poet Homer (ninth–eighth centuries B.C.E.), was already being practiced in ancient times:

But Homer, who was the forefather of the sciences and of the history of ancient times and was a great admirer of Circe, attributed Egypt with the fame of its valuable herbs. . . . At least he described a great number of Egyptian herbs which were given to his Helen by the Pharaoh's wife, and spoke of that renowned *nepenthes*, which induced one to forget sorrow and forgive and which Helen should have had all the mortals drink. But the first of whom we still have knowledge was Orpheus, who reported some interesting things about herbs. We have already mentioned the admiration which Musaios and Hesiod, following his lead, had for *polium*. Orpheus and Hesiod recommended the burning of incense. . . . After him, Pythagoras, the first person known for his knowledge, wrote a book about the effects of plants in which he attributed their discovery and origin to Apollo, Asclepius, and all of the immortal Gods in general. Democritus also produced such a compilation; both visited the magicians in Persia, Arabia, Ethiopia, and Egypt. (Pliny, *Natural History* 25.12–3)

In late ancient times, other books of herbal lore joined Pliny's (23–79 C.E.) *Natural History*. The most important of these was Dioscorides's (ca. first century) *Teachings on Medicines*, which is still important in our time. This work provides information about numerous psychoactive plants, including their various names, preparations, and uses (cf. Rätsch 1995a).

In the Middle Ages, descriptions of psychoactive plants were found especially in the writings of Arabic and Indian authors, such as Avicenna (980–1037). In Germany, many plants (including hemp, henbane, and deadly nightshade) were described by the abbess Hildegard von Bingen (1098–1179) (Müller 1982).

The great period of the "fathers of botany" dawned at the beginning of the modern era. This period witnessed the publication of voluminous herbals full of information about psychoactive plants. Among their authors are Leonhart Fuchs (1501–1566), Jacobus Theodorus Tabernaemontanus (1522–1590), Hieronymus Bock (1498–1554), Otto Brunfels (ca. 1490–1534), and Pierandrea Matthiolus (1500–1577).

During the colonization of the New World, the Spanish king sent physicians and botanists to Mexico and Peru. Their task was to investigate the indigenous flora to determine potential medicinal uses. The results were published in a number of compendia dedicated to American flora and its healing effects. All of these works contain numerous references to psychoactive plants and their medicinal and psychoactive uses (Pozo 1965, 1967).

Left: The first botanical and chemical investigations of the peyote cactus (*Lophophora williamsii*) were conducted at the end of the nineteenth century. The extraordinary psychedelic effects of mescaline, an alkaloid isolated from this cactus, have influenced the history of European psychiatry.

Right: Two psychoactive plants in an intimate embrace: an Amazonian ayahuasca vine (*Banisteriopsis caapi*) winding itself up the trunk of a coral tree (*Erythrina mexicana*).

The systematic study of psychoactive plants first began in the nineteenth century. Dr. Ernst von Bibra (1806–1878), a baron from Lower Franconia, was a private scholar typical of his time. He was wealthy by birth, achieved academic distinction, and dedicated his life to his studies, which he preferred to carry out within his own four walls. He studied medicine and philosophy in Würzburg and later lived in Nuremberg. When he was not traveling, he spent most of his time at his estate at Schwebheim. Bibra held liberal political views and was actively involved in the Revolution of 1848. Because of this, he was forced to leave the country for a time, during which he journeyed through South America (1849–50). While there, he became acquainted not only with many exotic cultures but also with a number of South American inebriants, especially coca and guaraná.

Just one year after Bibra published the remarkable journal of his travels,[10] his groundbreaking book *Die narkotischen Genußmittel und der Mensch* [The Narcotic Agents of Pleasure and Man] (Nuremberg 1855; published in English as *Plant Intoxicants* in 1995) appeared. A unique work, it became a true literary sensation, providing the first detailed descriptions of the psychoactive drugs that were known at the time and their effects. The author's own experiences, as well as his liberal disposition, were very discernible:

> One could not name a single country in the whole wide world in which the inhabitants are not using some sort of narcotic. Indeed, almost all use several of them. Although perhaps only a few tribes use certain of these substances, millions of people employ the vast majority of them. (Bibra 1995, 218)

In his book, Bibra reported at great length about coffee, tea, maté, guaraná, cacao, fahan tea, fly agaric, thorn apple, coca, opium, lactucarium, hashish, tobacco, betel, and arsenic. The conclusions of his discussion have a very modern ring:

> We have learned from experience that man can live without narcotics or without alcoholic drinks, which we wish to include here because of their similar effects. By taking these substances, however, man's life becomes brighter and therefore they ought to be approved. (221)

Clearly, the notion that we have a right to inebriation was already current at that time!

In German-speaking countries, Bibra's work launched a wave of interdisciplinary drug research that has continued into the present day. He was the chief source of inspiration for pharmacist Carl Hartwich (1851–1917), who compiled the most voluminous work on psychoactive plants to date

(Hartwich 1911), as well as for toxicologist Louis Lewin (1850–1929). Even Albert Hofmann (b. 1906), a modern-day Swiss chemist specializing in the investigation of naturally occurring substances, feels a kinship with the baron, for Bibra called upon the chemists who would come after him to dedicate themselves to the study of psychoactive plants.

Arthur Heffter (1860–1925) took Bibra at his word. He was the first person to test an isolated plant component, in his case mescaline, by trying it out on himself. It is for this reason that we still refer to this method of conducting research through self-experimentation as the Heffter technique.

At about the same time as Bibra, the American Mordecai Cubitt Cooke (1825–1913) also was studying human inebriants, which he poetically described as the "Seven Sisters of Sleep" (Cooke 1860, reprinted in 1989). Paralleling Cooke's work, the Scotsman James F. Johnston was investigating the chemistry of everyday life and the substances that humans ingest for their pleasure. He published his work in 1855, the same year as Bibra.

In Italy, Paolo Mantegazza (1831–1910) is regarded as a pioneer of drug research (Samorini 1995b). In 1871, Mantegazza published in Milan his 1,200-page main work *Quadri della natura umana: Feste ed ebbrezze* [Pictures of Human Nature: Festivals and Inebriations].

Mantegazza was partial to coca, and in 1858 he published a sensational work entitled *Sulle virtù igieniche e medicinali della coca e sugli alimenti nervosi in generale* [On the Hygienic and Medicinal Virtues of Coca and Nerve Nourishment in General]. Like Bibra and Hartwich, Mantegazza was interested in all agents of inebriation and pleasure and was guided and inspired by these his entire life. Since most of his writings have appeared only in Italian, they have not attracted as much international attention as the publications of Bibra, Johnston, and Cooke.

Mantegazza's classification of inebriants is especially interesting. He divided the "nerve nourishment" into three families: 1. alcoholic nerve nourishment, with the two branches fermented and distilled beverages; 2. alkaloid nerve nourishment, with the branches caffeine and narcotics (among the narcotics, he included opium, hashish, kava-kava, betel, fly agaric, coca, ayahuasca, and tobacco); and 3. aromatic nerve nourishment (sage, oregano, rosemary, cinnamon, pepper, chili, etc.).

The psychiatrist Emil Kraepelin (1856–1926), who in 1882 published his medical and psychologically oriented book *Über die Beeinflussung einfacher psychologischer Vorgänge durch einige Arzneimittel* [On the Influencing of Simple Psychological Processes by Some Medicines], followed a path different from Bibra's. That same

Dr. Ernst Freiherr von Bibra (1806–1878) was a pioneer in the ethnopharmacological study of psychoactive substances. His work *Die narkotischen Genußmittel und der Mensch* [The Narcotic Agents of Pleasure and Man] (1855) was the first comprehensive book on the topic and is still an important reference.

Shen-Nung, the legendary Red Emperor, is regarded as the founder of Chinese herbal medicine. He personally tried each herb, including the poisonous and inebriating ones, before recommending their use for healing purposes. Shen-Nung was thus the founder of the ethnopharmacological method of conducting bioessays, also known as the Heffter technique. (Ancient Chinese

10 *Reise in Süd-Amerika* [Journey in South America] (Mannheim 1854).

"Oh! Joy! Joy! I have seen the birth of life, the beginnings of movement. The blood pounds in my veins as if they would burst. I want to fly, swim, bark, bleat, roar, would that I had wings, a carapace, a rind, could fume smoke, would have a trunk, could coil my body, divide myself and enter into everything, could effuse myself into scents, unfurl myself like a plant, flow like water, vibrate like sound, shimmer like the light, assume every form, penetrate into every atom, sink down to the foundation of matter—be matter!"

GUSTAVE FLAUBERT
LA TENTATION DE SAINT ANTOINE
[THE TEMPTATION OF ST. ANTHONY]
(1979, 189)

11 In the older ethnographic literature, the key role that psychoactive substances play in shamanism was typically suppressed. Some authors even went so far as to suggest that those shamans who "required" drugs to produce their trance state (which was referred to as toxic ecstasy) were "degenerate" and no longer capable of "true ecstasy." In the years following Eliade's "classic" publication (1951), the more recent ethnological and cultural anthropological literature on shamanism has essentially corrected this earlier view, for it has now been determined that almost all shamans ingest drugs (cf. Rosenbohm 1991) and that pharmacological techniques for inducing altered states of consciousness are much more important than all the others.

year witnessed the publication of the revised second edition of *Die Schlaf- und Traumzustände der menschlichen Seele mit besonderer Berücksichtigung ihres Verhältnisses zu den psychischen Alienationen* [The Sleeping and Dreaming States of the Human Mind, with Special Emphasis Upon Their Relationship to the Psychic Alienations], by Heinrich Spitta, a dream researcher and professor of philosophy. Both books, each in its own way, dealt with the chemical agents that can be used to induce altered states. Shortly thereafter, the neurologist and "father of dream theory" Sigmund Freud (1856–1939) published his work *Ueber Coca* [On Coca], which helped make the use of cocaine fashionable. These pioneering works led to the development of psychopharmacology or pharmacopsychology, a field that has attracted psychiatrists, pharmacologists, pharmacognosists, and chemists. *Pharmacopsychology* has been defined as "the doctrine of influencing the mental life by means of chemically effective substances introduced into the body" (Lippert 1972, 10).

The most important chemist in this history of research is the Swiss Albert Hofmann. Not only did he invent LSD while investigating the ergot alkaloids, but also he discovered the active principles in the magic mushrooms of Mexico as well as other American Indian magical drugs. Also of note is Alexander T. Shulgin, an American chemist of Russian descent who has played an especially significant role in the area of structure-effect relationships.

In anthropology or ethnology, investigations into the use of psychoactive plants did not begin until the twentieth century.[11] Among the pioneers of psychoactive ethnology are Pablo Blas Reko, Weston La Barre, Johannes Wilbert, Peter Furst, and Michael Harner. Today, the role played by Carlos Castaneda is a source of considerable controversy.

Toward the end of the nineteenth century, ethnobotany began to emerge as a specialized branch of science. The term was introduced in 1895 by John W. Harshberger (1869–1929). Both ethnologists and botanists have specialized in ethnobotany. A British scholar, Richard Spruce (1817–1893), was one of the pioneers of ethnobotany. Richard Evans Schultes (1915–2001), a former professor and former director of the Botanical Museum at Harvard University, is universally regarded as the "father of psychoactive ethnobotany." His investigations in Mexico and South America have led to the discovery of numerous psychoactive plants (Davis 1996). Many of Schultes's students have themselves become renowned ethnobotanists or ethnopharmacologists, including Timothy Plowman (1944–1989), Wade Davis, Mark J. Plotkin, and Tom Lockwood. The American botanist William Emboden is noted for making a creative leap to art history and has

published many important works in this area.

Ethnomycology, the study of the cultural uses of fungi, was founded by the banker R. Gordon Wasson (1898–1986). In some ways, Jonathan Ott, a chemist who investigates natural substances, has become Wasson's successor. Many other discoveries in ethnomycology have been made by Paul Stamets, Gastón Guzmán, and Jochen Gartz.

During the past thirty years, ethnopharmacology, the study of the cultural uses of pharmacologically active substances and their cognitive interpretations, has developed into a specialized field within the disciplines of ethnobotany and ethnomedicine. It is a young field that is very interdisciplinary in nature. This encyclopedia is a work of this nature.

Finally, we should also mention *closet shamans*. This term has come to be used for amateurs and hobbyists who experiment at home with psychoactive plants and preparations, occasionally making astonishing discoveries that are then eagerly taken up and pursued by scientists. Almost all of the research into ayahuasca analogs has been conducted by these closet shamans.

Most of the important discoveries in the field of psychoactive plants, including those having to do with their chemistry and pharmaceutical uses, have been made by German-speaking scientists. Is this an expression of some need of the German "soul"? Why this concentration on German soil? Is the Germanic god Wotan still at work? Wotan is both the god of knowledge and the restless shaman who will do whatever he can to satisfy his immeasurable thirst for knowledge. It was he who stole the Mead of Inspiration and brought it to us humans (Metzner 1994b).

Psychoactive Plants as Factors in the Development of Culture

The use of—and the need for—psychoactive plants is very ancient. Some authors have suggested that the roots lie somewhere in the Paleolithic period (Ripinsky-Naxon 1989; Westermeyer 1988). It appears that the connection to shamanism was already present at an early date (La Barre 1972). Although I personally do not believe that shamanism was one of the very first religions, I do think that the altered states of consciousness and visions induced by psychoactive plants have led to significant cultural innovations.

A person ingests a substance obtained from the environment and sinks into a flood of pictures, visions, and hallucinations. He is confronted by a previously unimagined quantity of images—images that seem somehow familiar, or archetypically known, we might say. Moreover, these images are complex and intricate, following one another in incredible sequences, and they have such detail that you cannot shake the feeling that you have somehow landed on the molecular level or are somewhere far away, in the depths of infinite space. Where do these pictures come from? Do they arise in the human brain as a result of the material interaction between molecules from without and the brain stem? Can we, using these substances that come from outside of ourselves, look into realities that truly are outside and for which we normally have no perception? No matter the answer, the wonder or the mystery remains the same! Wherever the images come from, they are present, they can be perceived, they are a reality that can be experienced.

Many cultures and many researchers have concerned themselves with these questions. Although no one has been able to provide a definitive answer, the hypotheses and positions that have been put forth can be divided into two camps. One assumes that all reality is merely a projection of our own consciousness; the other holds that there are numerous or even infinitely many different realities in the external world.

We can take shamanism seriously only if we follow the second view, for if we assume that the shaman is only flying around within his own skull, then he would not be able to recover, liberate, and bring back stolen souls.

The internal images and visions induced by psychoactive plants appear to have influenced human art since the Stone Age (Biedermann 1984; Braem 1994). African rock art has been interpreted as an expression of altered states of consciousness, most likely induced by mushrooms or similar substances (Lewis-Williams and Dowson 1988, 1993). American Indian rock art has also been inspired by experiences with psychoactive plants (Wellmann 1978, 1981).

The images in the world of Hieronymus Bosch have been interpreted as the product of drug experiences as well. Nineteenth-century art would have been inconceivable without psychoactive plants (Kupfer 1996a, 1996b). To the observer, many of the pictures of the surrealists, especially those of Max Ernst, René Magritte, and Salvador Dalí, appear to be "drug pictures" or remind one of one's own experiences in altered states. Hashish appears to have played a role in the development of surrealism, the philosophy of which was set forth in the *Surrealist Manifesto* of 1924: "Surrealism rests upon the belief in the greater reality of certain forms of associations that have been neglected until now, in the omnipotence of dreams, in the non-utilitarian play of thought" (Breton 1968, 26 f.).

The founder of surrealism compared this art form with the effects of psychoactive substances:

Surrealism does not permit those who follow it to abandon it whenever they will. Everything points to the fact that it affects the mind in the same way as stimulants do; like these, it produces a certain condition of need and is able to drive a person to terrible revolts. Once again, if you will, we stand before a very artificial paradise, and our penchant to enter into it falls with the same rights under the same Baudelairean criticism as all the others. Thus, the analysis of the mysterious effects and special pleasures which it can impart—in some ways, Surrealism appears to be a *new vice*, which is not suitable for just a few; like hashish, it is able to satisfy all those who are particular—thus, such an analysis must be undertaken within this investigation.

Surrealistic images are like those pictures from opium inebriation, which a person is no longer evoking, but which are "spontaneously and tyrannically presented to him. He is incapable of fending them off; for the will has lost its power and no longer controls his abilities." (Baudelaire) The question remains as to whether one ever "evoked" the images at all. (Breton 1968, 34)

It appears that experimentation with psychoactive substances provided an important impetus to the art scene surrounding fantastic realism. Only a few artists, however, have publicly admitted as much. Ernst Fuchs even denied his drug experiences in one of his early biographies (Müller-Ebeling 1992). For most artists, it appears that the use of hashish and marijuana does not necessarily affect the creative process but functions instead as a way to focus concentration, in the way that some Indians use hashish in their meditation practice (e.g., Gustav Klimt). Albert Paris Gütersloh, an admitted cannabis user, provided a realistic assessment of the situation:

Every [artist] of my generation has made the acquaintance of hashish, and when I walk through the academy and sniff, I am certain: everyone in my class, at least, has as well. Does this mean that we are all hash artists? (cited in Behr 1995)

The discipline of anthropology provides us with many examples of cultural goods or artifacts that are the direct result of visionary experiences with psychoactive plants and products (Andritsky 1995). The yarn paintings of the Huichol are representations of their experiences with peyote, and the visions induced by ayahuasca have been the subject of numerous ayahuasca paintings.

The most important magical plants of Mexico were first described in the Aztec-language work of Fray Bernardino de Sahagun, the Florentine Codex: 515 Tlapatl (*Datura* spp.); 516 Nanacatl (*Psilocybe* spp.); 517 Peyotl (*Lophophora williamsii*); 518 Toloa (*Datura innoxia*). (Paso y Troncoso edition)

THE PSYCHOACTIVE PLANTS

"My heart wears flowers and fruits in
the midst of the night . . .
I, Cinteotl [=Xochipilli], was born in
paradise. I come from the land of
flowers.
I am the new, the glorious, the
unequaled flower.
Cinteotl was born of the water; as a
mortal, as a young man he was
born
from the heavenly blue House of the
Fishes. A new, victorious god. He
shines like the sun. His mother
lived in the House of the Twilight,
as colorful as a Quetzal, a new,
delightful flower."

AZTEC HYMN
IN *MEXIKANISCHE MYTHOLOGIE*
[MEXICAN MYTHOLOGY]
(NICHOLSON 1967, 115 f)

Depicted here in a state of ecstasy,
Xochipilli, the flower prince, was the
Aztec god of psychoactive plants,
eroticism, spring, inspiration, and
music. This pre-Columbian statue
clearly illustrates the great importance
that people have placed upon visions
as well as the manners in which such
experiences have found expression
in art.

12 Many psychopharmacological drugs
and medicines used in psychiatry exert
a palliative effect upon sick people but
have no effects upon healthy
individuals (cf. V. Faust 1994).

Which plants have been included in this
encyclopedia? I considered all of those plants that
my own research and experience have indicated to
be psychoactive as well as those plants that other
researchers or the literature have reported to be
psychoactive. Here, we must keep in mind that
there are plants for which a majority of subjects
have reported no psychoactive effects. There also
are plants that have a reputation of being
hallucinogenic, but which no one has yet tried. To
date, many of these plants have been the object of
only cursory investigation. There also are a
number of plants that have not yet been botanically determined or identified. The situation is
complicated by the fact that the botanical data
contained in the ethnographic literature are often
incorrect, or at least very imprecise. Sometimes, it
was difficult to decide whether a particular plant
should be included in this work or not. One such
case is St. John's wort (*Hypericum perforatum* L.),
which the ancient Germans used as a sedative and
in modern phytotherapy is generally regarded as
a natural tranquilizer (Becker 1994). St. John's
wort and the oil it yields do indeed exert a psychoactive
effect, but only upon patients suffering from
mental or emotional afflictions. As a rule, healthy
individuals do not notice any psychotropic effects,
even after ingesting large amounts.[12] Such uncertain cases—to the extent that they are known—
have not been included in these monographs.

Reflecting our current state of knowledge, I
have treated the various psychoactive plants discussed in this book in several different ways. Well-
known plants that have been investigated in some
detail are examined in a very systematic fashion in
the major monographs. Plants that have been little
studied or about which very little is known are

discussed in informal minor monographs. A
number of very well-known and well-researched
plants that are purported to produce psychoactive
effects and are sometimes referred to as "legal
highs" are considered in their own small section of
informal monographs. This is followed by another
section that focuses on a number of psychoactive
plants whose botanical identity is unknown.

Because they are not plants in the strict sense,
psychoactive fungi are presented in a section of
their own. The section on psychoactive fungi is
followed by another that focuses on psychoactive
products that are obtained through often intricate
procedures and/or from combinations of plants.
Finally, there is a short section that examines the
psychoactive constituents of plants. This section
also serves as an aid in locating the plants discussed in the monographs.

On the Structure of the Major Monographs

The monographs are arranged alphabetically
according to botanical names. Below the scientific
name may been found the most common English
name(s) or, when none is known, a common
international name.

Some of the monographs treat not just one
species but, rather, a number of species of the same
genus. This is done either because the traditional
users make no distinction between the different
species or because the species all contain the same
active constituents and/or are sources of the same
products.

Family

Here, information is provided about the botanical
family to which the plant belongs, along with
additional details about taxonomy.

Forms and Subspecies

Any known forms, varieties, cultivars, or subspecies
of the plant are listed here.

Synonyms

Most plants have been described in the botanical
literature under more than one name. Under this
heading may be found these nonvalid botanical
names (including misspellings in the literature).

Folk Names

Folk and popular names are given here. Often,
information is also furnished about the particular
language a name is taken from, and translations of
many of the terms are provided. Please note that
the names of indigenous tribes and tribal languages referenced in these sections will be spelled
in a variety of ways, rather than uniformly. The
variant spellings reflect the spellings found in the
literature referenced for each plant.

History

Here may be found the most important information about the history of the plant, including its discovery, botanical description, and historical uses.

Distribution

Under this heading is provided information concerning the range as well as the natural occurrence of the plant in question.

Cultivation

Information about the more simple and successful methods of growing and cultivating a plant is provided under this heading. It should be noted, however, that more is required to successfully grow these plants than simply reading this information; it is also helpful to have a "green thumb," experience, skill, and a deep love of the plant world.

Appearance

Here is provided a brief description of the plant. Other plants that might possibly be mistaken for it are mentioned, and the plant's distinguishing features are emphasized. It should be noted that the information contained under this heading does not always conform to the standardized botanical descriptions (which may be found in the botanical literature).

Psychoactive Material

Under this heading, information is provided about the parts of the plant that are utilized, as well as products obtained from them (where appropriate, the pharmaceutical names of the raw drugs are also given).

Preparation and Dosage

Here may be found information for preparing and dosing the various raw drugs. While every attempt has been made to ensure that this information is as accurate as possible, it must be explicitly stated that this information should not be regarded as definitive. Identical dosages can produce very different responses in different individuals.

Ritual Use

Information concerning the traditional uses of the plant in shamanic rituals, priestly ceremonies, domestic festivals, and other experiences may be found here. As in the Folk Names sections, the spelling of indigenous tribal names will reflect the variant spellings found in the literature referenced.

Artifacts

Where possible, reference is made to three types of artifacts associated with the plant:

— Artifacts composed of the plant or manufactured from it
— Artistic representations of the plant (in paintings, architecture, etc.)
— Art works (paintings, poetry, music, theater pieces, etc.) whose inspiration has come from the use of the plant.

Medicinal Use

Many psychoactive plants are also of medicinal and therapeutic significance. Sometimes a plant's medicinal applications are much more important than its psychoactive uses. For these reasons, as much information as possible is provided about the medicinal uses of the plant under discussion. This includes ethnomedical, folk medical, biomedical, and homeopathic uses.

Constituents

Under this heading may be found a comprehensive listing of the known constituents of the plant.

Effects

Here, the effect or the pattern of effects of the plant is described. Once again, it should be kept in mind that different individuals can have very different experiences with the same plant.

Commercial Forms and Regulations

Many plants and/or the raw drugs obtained from them are available through commercial sources. Some plants are subject to particular regulations or laws. Pertinent information is included under this heading.

Literature

Here, references are provided to specialized literature on the plant under discussion.

Symbol Key

Those sources marked in the monographs with a * are listed in the general bibliography located at the end of this book (pages 878–907). Those marked with a ** may be found in the general literature on psychoactive fungi (pages 689–693), which follows the section devoted to them.

Terms in bold print within the running text refer to other entries in this book.

Where a question mark has been inserted in a table column, it indicates that the missing information poses an important question for further research.

The Most Important Genera and Species from A to Z

Major Monographs

The following genera are discussed in the major monographs:

Acacia, Aconitum, Acorus, Agave, Alstonia, Anadenanthera, Areca, Argemone, Argyreia, Ariocarpus, Artemisia, Arundo, Atropa

Banisteriopsis, Boswellia, Brugmansia, Brunfelsia

Calea, Calliandra, Camellia, Cannabis, Carnegia, Catha, Cestrum, Cinnamomum, Coffea, Cola, Coleus, Convolvulus, Corynanthe, Coryphantha, Cytisus

Datura, Desfontainia, Diplopterys, Duboisia

Echinops, Ephedra, Erythrina, Erythroxylum, Escholzia

Heimia, Humulus, Hyoscyamus

Ilex, Iochroma, Ipomoea

Juniperus, Justicia

Lactuca, Latua, Ledum, Leonurus, Lolium, Lonchocarpus, Lophophora

Mammillaria, Mandragora, Mesembryanthemum, Mimosa, Mitragyna, Mucuna, Myristica

Nicotiana, Nuphar, Nymphaea

Pachycereus, Papaver, Passiflora, Paullinia, Pausinystalia, Peganum, Pelecyphora, Petroselinum, Phalaris, Phragmites, Phytolacca, Piper, Psidium, Psychotria

Rhynchosia

Salvia, Sassafras, Sceletium, Scopolia, Solandra, Solanum, Sophora, Strychnos

Tabernaemontana, Tabernanthe, Tagetes, Tanaecium, Theobroma, Trichocereus, Turbina, Turnera

Vaccinium, Veratrum, Virola, Vitis

Withania

Left: *Latua pubiflora*, known as the tree of the magicians, is one of the world's rarest shamanic plants. The flower of this nightshade is 3 to 4 cm in length. (Photographed near Osorno, in southern Chile)

Acacia Spp.
Acacia Species

Numerous acacias have played a role in ethnopharmacology and medical history. Some species (such as gum arabic) are used as sources for an excipient and incense, some are used as beer additives, and others provide DMT and other tryptamines. (Woodcut from Tabernaemontanus, *Neu Vollkommen Kräuter-Buch*, 1731)

Left: Many Australian acacias contain high concentrations of *N,N*-DMT and are thus suitable for the production of psychedelic ayahuasca analogs. Although our studies of Australia's psychoactive flora have only just begun, they already have demonstrated great promise. (*Acacia* spp., photographed in southeastern Australia)

Right: Catechu, the resin of the catechu tree (*Acacia catechu*), is one of the main ingredients in betel quids. (Photograph: Karl-Christian Lyncker)

13 Seeds of *Datura stramonium* are also added to dolo beer (Voltz 1981, 176).

Family
Leguminosae: Mimosaceae (Fabaceae) (Legume Family)

Synonyms
Many species of the genus *Acacia* were formerly assigned to the genera *Mimosa, Pithecolobium, Senegalia,* and *Racosperma.* In addition, some species previously described under the genus name *Acacia* have now been reclassified as *Anadenanthera* (see **Anadenanthera colubrina**) and *Mimosa* (see **Mimosa tenuiflora, Mimosa spp.**).

General
The genus *Acacia* encompasses 750 to 800 species (other sources list only approximately 130) found in tropical and subtropical regions throughout the world (Harnischfeger 1992). Most are medium-sized trees, the leaves of which are usually pinnate but sometimes edentate. The flowers appear in clusters and produce podlike fruits.

Some species are sold as cut flowers under the name "mimosa." *Acacia farnesiana* (L.) Willd. yields an **essential oil** that is used as an aromatic substance in aromatherapy and in the manufacture of perfumes (Bärtels 1993, 89*). Some acacias (such as gum arabic) have been used since ancient times as excipients for compound medicines and **incense**. Some species find use as additives in psychoactive products (**betel quid, beer, balche', pituri**; for pulque, cf. **Agave spp.**). Many species are suitable for producing **ayahuasca analogs**. The bark and/or leaves of numerous Australian acacia species (*A. maidenii, A. phlebophylla, A. simplicifolia*) contain high concentrations of *N,N*-DMT (Fitzgerald and Sioumis 1965; Ott 1994, 85 f.*; Rovelli and Vaughan 1967).

Acacia angustifolia (Mill.) Kuntze [syn. *Acacia angustissima* (Mill.) Kuntze, *Acacia filiciana* Willd.]—pulque tree, timbre
The root of this Mexican acacia provides an additive to pulque (a fermented beverage made from *Agave* spp.) that may have psychoactive effects. The Aztecs called this small tree *ocpatl*, "pulque drug"; in contemporary Mexican Spanish, it is known as *palo de pulque*, "tree of pulque." *Acacia albicans* Kunth [syn. *Pithecolobium albicans* (Kunth) Benth.] was also used as a pulque additive.

Acacia baileyana F. von Muell.
This Australian acacia is found in New South Wales. It contains psychoactive β-**phenethylamines**, including tetrahydroharman, and may be suitable as an MAO-inhibiting additive in the preparation of **ayahuasca analogs**.

Acacia campylacantha Hochst. ex A. Rich [syn. *Acacia polyacantha* Willd. ssp. *campylacantha*]
The leaves of this Old World species contain *N,N*-DMT and other tryptamines (Wahba Khalil and Elkeir 1975). In West Africa, the bark is traditionally used as a psychoactive additive to a type of **beer** known as *dolo*[13] that is brewed from certain cereal grains (*Sorghum* spp., *Pennisetum* spp.), sometimes with the addition of **honey**. The alcohol content normally ranges between 2 and 4%, and from 8 to 10% when honey has been added (Voltz 1981, 176). Dolo is consumed as a libation during offering ceremonies and other rites as well as in daily life. Its properties are held in high regard: "*Dolo* imparts strength and courage and brings a joy of living. It is customary to drink *dolo* when performing strenuous work. The farmer who is making a piece of wilderness cultivable, the smith who is working hard at the anvil, the warrior who is preparing himself for battle, the woman who is in labor, the dancer who will be wearing the heavy, sacred mask . . . , all of them receive strength and courage from *dolo*, which is offered to them by their mother, wife, or sister" (Voltz 1981, 178).

Acacia catechu (L. f.) Willd.—catechu tree

This acacia species, found in India, Indonesia, and Malaysia, can grow as tall as 20 meters. It is also known as cutch tree, khair, kath, katha, khadira, and ercha. Its heartwood is boiled in water for twelve hours to concentrate the extract, which is known as catechu, katechu, catechu nigrum, extractum catechu, succus catechu, terra catechu, terra japonica, pegu, black catechu, cutch, cachou, katha, khair, terra giapponica, khadira, and cato de pegú. Essentially four types are found in trade: Pegu catechu (= Bombay catechu), the most common type; Bengali catechu; Malaccan catechu; and Camou catechu (Harnischfeger 1992, 31). Catechu is an ancient Indian drug and is still officinal in Germany as well *(DAB6).** In Vedic times, the bark of *Acacia catechu* was known as *somatvak* and was associated with **soma**.

Catechu is odorless and has a bitter, astringent taste that slowly turns sweet. It is largely water soluble and can be crystallized back out again. It is composed of flavonols and glycosides (fisetin, quercetin [cf. *Psidium guajava, Vaccinium uliginosum*], quercitrin), as well as flavonoids (catechine, catechin tanning agents) and red pigments (Harnischfeger 1992, 31). Catechu is thus responsible for the reddish coloration of the saliva that occurs when **betel quids** are chewed (Atkinson 1989, 775*). In India and Nepal, catechu is used in dyeing and tanning. In the local ethnomedicine, it is employed as a tonic and for digestive ailments and skin diseases. However, the greatest economic significance of catechu is as a (coloring) additive to **betel quids** (Storrs 1990, 5*). In Indian medicine, catechu is an ingredient in recipes for treating ulcers on the mucous membranes of the mouth, inflamed throats, and toothaches (Harnischfeger 1992, 32). Catechu is a definite tannin drug that is suitable for treating inflammations of the mucous membranes and diarrhea (Pahlow 1993, 453*). Catechu has no psychoactive effects of its own but is simply an important component of a psychoactive product; however, it may have synergistic effects in this.

Acacia confusa Merr.

This acacia species contains *N,N*-DMT and is usable as an additive in **ayahuasca analogs**.

Acacia cornigera (L.) Willd. [syn. *Acacia spadicigera* Cham. et Schlechtend.]—horned acacia

The large binate thorns of this striking acacia are hollow and provide a home for ants. In Mayan, the small tree (also known as *akunte'*) is called *subin*, "dragon." It plays an important role in the magical preparation of the ritual drink known as **balche'**. It is possible that parts of the tree were formerly added to the drink. The bark may contain *N,N*-DMT. The Maya of San Antonio (Belize) use its roots and bark to treat snakebites. The root is made into a tea that is also consumed as an aphrodisiac and as a remedy for impotence. Other preparations are used to treat asthma and headaches (Arvigo and Balick 1994, 81*).

Acacia maidenii F. von Muell.—maiden's wattle

All parts of this beautiful, upright, silvery tree contain tryptamines. The bark contains 0.36% *N,N*-DMT (Fitzgerald and Sioumis 1967). The leaves are usable in **ayahuasca analogs** as a source

The *Acacia catechu* tree, indigenous to South Asia, produces a substance known as catechu, which is an important ingredient in betel quids. (Engraving from Pereira, *De Beginselen der Materia Medica en der Therapie,* 1849)

Top left: The horned acacia (*Acacia cornigera*) is an important magical plant for the Lacandon Maya. (Photographed in the Selva Lacandona, Chiapas, Mexico, 1996)

Below left: The bark of *Acacia maidenii*, collected in New South Wales, contains high concentrations of *N,N*-DMT.

Right: Flowers and leaves of the Australian *Acacia maidenii*.

* Editor's note: DAB6 refers to the sixth edition of a German pharmacopoeia entitled *Deutsches Apotheker Buch.*

Above: The Australian *Acacia phlebophylla* is apparently the rarest species of acacia in the world. Its leaves contain large quantities of *N,N*-DMT.

Below: Seedpods of *Acacia phlebophylla.*

"In Canaan the prime oracular tree was the acacia—the 'burning bush.' . . . The acacia is still a sacred tree in Arabia Deserta and anyone who even breaks off a twig is expected to die within the year."

Robert Graves
The White Goddess
(1948, 440, 441*)

of DMT (Ott 1993, 246*). This acacia is easily cultivated in temperate zones (e.g., in California and southern Europe).

Acacia nubica Bentham—Nubian acacia
The leaves of this African acacia contain *N,N*-DMT and other constituents (Wahba Khalil and Elkeir 1975). However, the concentrations do not appear to be sufficient for producing **ayahuasca analogs.**

Acacia phlebophylla F. von Muell.—buffalo sallow wattle
This Australian species is rich in *N,N*-DMT. The leaves contain 0.3% *N,N*-DMT (Rovelli and Vaughan 1967), and are usable as a source of DMT for **ayahuasca analogs** (Ott 1993, 246*). This acacia may be the rarest species of the genus and is found only on Mount Buffalo.

Acacia polyantha Willd. [syn. *Acacia suma* (Roxb.) Buch.-Ham.]—white catechu tree
The resin of this Indian acacia is sometimes used as catechu or as a catechu substitute in **betel quids** (see above). The leaves apparently contain *N,N*-DMT. Interestingly, the Sanskrit name of this plant is *somavalkah*, which associates it with the divine drink **soma.** This is also suggested by the Malayalam name *somarayattoli* (Warrier et al. 1993, 26*).

Acacia retinodes Schlechtend.—swamp wattle
This Australian acacia is found primarily in swampy and humid areas. It contains **nicotine** (Bock 1994, 93*). No traditional use of this plant is known.

Acacia senegal (L.) Willd. [syn. *Acacia verek* Guill. et Perrott, *Senegalia senegal* (L.) Britt.]—gum arabic tree
This African acacia is chiefly significant as the source of gum arabic, which is used as a binding agent in **incense** and for other purposes. The leaves contain *N,N*-DMT (Wahba Khalil and Elkeir 1975), although the concentration is very low. It is apparently not particularly suitable for producing **ayahuasca analogs.**

Acacia simplicifolia Druce
The bark of the trunk of this acacia, which is found in Australia and New Caledonia, is said to contain up to 3.6% alkaloids; 40% of these are MMT, 22.5% *N,N*-DMT (= 0.81% DMT total concentration), and 12.7% 2-methyl-1,2,3,4-

tetrahydro-ß-carboline. The leaves contain up to 1% *N,N*-DMT, along with MMT, N-formyl-MMT, and 2-methyl-1,2,3,4-tetrahydro-ß-carboline (Poupat et al. 1976). Both the bark and the leaves are suitable for the production of **ayahuasca analogs.**

Acacia spp.—wattle
According to the reports of some closet shamans, the bark and leaves of many species of wattles (as acacias are known in Australia) clearly contain *N,N*-DMT. It is said that these can be used to make smokable extracts that produce definite tryptamine hallucinations. The Aborigines burned some species of *Acacia* to ashes, which they then added to **pituri.**

Commercial Forms and Regulations
Acacia seeds are occasionally sold in ethnobotanical specialty shops. Gum arabic is available without restriction and may be purchased in pharmacies in Germany. It is readily available in the United States as well.

Literature
See also the entry for **ayahuasca analogs.**

Clarce-Lewis, J. W., and L. J. Porter. 1972. Phytochemical survey of the heartwood flavonoids of *Acacia* species from arid zones of Australia. *Australia Journal of Chemistry* 25:1943–55.

Fitzgerald, J. S., and A. A. Sioumis. 1965. Alkaloids of the Australian Leguminosae, V: the occurrence of methylated tryptamines in *Acacia maidenii* F. von Muell. *Australian Journal of Chemistry* 18:433–34.

Harnischfeger, Götz. 1992. Acacia. In *Hagers handbuch der pharmazeutischen praxis.* 5th ed. Vol. 4:26–43. Berlin: Springer.

Poupat, Christiane, Alain Ahond, and Thierry Sévenet. 1976. Alcaloïdes de *Acacia simplicifolia. Phytochemistry* 15:2019–20.

Rovelli, B., and G. N. Vaughan. 1967. Alkaloids of *Acacia*, I: *N,N*-dimethyltryptamine in *Acacia phlebophylla* F. von Muell. *Australian Journal of Chemistry* 20:1299–1300.

Voltz, Michel. 1981. Hirsebier in Westafrika. In *Rausch und Realität*, edited by G. Völger. Vol. 1:174–81. Cologne: Rautenstrauch-Joest-Museum.

Wahba Khalil, S. K., and Y. M. Elkheir. 1975. Dimethyltryptamine from the leaves of certain *Acacia* species of northern Sudan. *Lloydia* 38(2): 176–77.

Aconitum ferox Wallich ex Seringe

Blue Aconite

Family

Ranunculaceae (Buttercup Family); Helleboreae Tribe

Forms and Subspecies

Aconitum ferox may be a subspecies or variety of **Aconitum napellus.** In Tibetan medicine, several forms of *Aconitum ferox* are distinguished from one another on the basis of their pharmacological properties (Aris 1992, 233*).

Synonyms

Aconitum ferox L.
Aconitum napellus var. *ferox*
Aconitum virorum Don
Delphinium ferox Baill.

Folk Names

Aconite, atis, ativish (Nepali, "very poisonous"), ativisha (Sanskrit, "poison"), bachnag (Persian), bachnâg (Hindi), bikh, bis, bis-h, bish (Arabic), black aconite, blue aconite, bong-nag, bong nga, gsang-dzim, Himalayan monkshood, Indian aconite, jádwár, kalakuta, mithavis (Hindi), monk's hood, nang-dzim, nilo bikh, phyi-dzim, singya, sman-chen (Tibetan, "great medicine"), valsa-nabhi (Malay), vasanavi (Tamil), vatsamabhah (Sanskrit), vatsanabha, vatsanabhi (Malayalam), visha (Sanskrit, "poison"), wolfbane

History

Vedic and later Sanskrit texts indicate that the root of this *Aconitum* species was already being used as an arrow poison in ancient India in early times (cf. **Aconitum spp.**). In contrast to their original use, these poison arrows were used not in the hunt but in warfare (Bisset and Mazars 1984, 19). In the *Shushrutasamhita,* the Ayurvedic writings of Shushruta (ca. 300 C.E.), *Aconitum ferox* is referred to as "vatsanabha." Today, *Aconitum chasmanthum* is usually sold under the name "vatsanabha" (13). In the tenth century, the Persian physician Alheroo described the plant under the name *bish.* Europeans first became aware of *Aconitum ferox* in the nineteenth century during journeys to Nepal. During the nineteenth century, there was a thriving trade in the root tubers of *Aconitum ferox,* which were brought from Lhasa via Le (Mustang) to Ladakh (Laufer 1991, 57).

Distribution

Blue aconite is found in Nepal, Kashmir (northern India), Garhwal, Sikkim, and Bhutan at altitudes of 2,000 to 3,000 meters (Manandhar 1980, 7*). It is a typical Himalayan plant and has even been observed growing at 3,600 meters (Polunin and Stainton 1985, 5*). It is said to grow at altitudes as high as 4,500 meters (Pabst 1887, 7*).

Cultivation

Propagation occurs via seeds, which can be simply strewn about or grown in beds. Blue aconite prefers to grow over a stony or rocky substrate and also thrives in crevices and the hollow spaces between stones.

Appearance

This perennial plant produces tuberous roots and can grow up to 1 meter in height. The lower, long-stemmed leaves are pinnate and deeply retuse. Toward the top of the plant, the leaves become smaller and their stalks shorter. The helmet-shaped, blue-violet flowers are located in clusters at the end of the smooth, erect stem. The flower stalks grow directly out of the leaf axils. The fruit is a five-cusped, funnel-shaped capsule that opens at the top. In the Himalayas, blue aconite blooms during the monsoon season (from July to September, or until October in higher elevations). The root tubers, which have a dark brown cortex and are yellowish inside, regenerate annually.

Aconitum ferox is very similar to **Aconitum napellus** but is somewhat smaller and more stocky. It also bears fewer flowers, and these are spaced farther apart.

Aconitum ferox is easily confused with *Aconitum*

The flowers of the blue aconite (*Aconitum ferox*). Tantrists who follow the left-handed path smoke the herbage and roots of the plant as a potent inebriant.

"The man who is struck by an arrow that has been smeared with the seeds of *shalmali* [*Bombax ceiba* L.] and *vidari* [*Ipomoea digitata* L.], together with *mula* [*Raphanus sativa* L.] and *vatsanabha* [*Aconitum ferox*] and the blood of the muskrat, will bite ten people, each of which will then bite ten other people in turn."

KAUTILIYA ARTHASHASTRA
(14, 1: SUTRA 29)

Aconite is one of the most dangerous of all poisonous plants. However, like all poisons, it is also a valuable medicine. For this reason, aconite was formerly known as healing poison or poisonous remedy. (Woodcut from Tabernaemontanus, *Neu Vollkommen Kräuter-Buch*, 1731)

heterophyllum Wall. ex Royle, known as *bachnak*, *atis*, or *prativisa* (Bisset and Mazars 1984, 15). However, *Aconitum heterophyllum* has cordate leaves with serrate edges, whereas *Aconitum ferox* has the same deeply retuse and pinnate leaves as *Aconitum napellus*. Blue aconite can also be mistaken for another Himalayan species, *Aconitum spicatum* (Brühl) Stapf, which bears blue flowers as well (Polunin and Stainton 1985, 6*).

Psychoactive Material
—Root tuber (tubera aconiti ferocis, bish root)
—Herbage

Preparation and Dosage
When used in Ayurvedic medicine, the tubers are steeped in the milk or urine of sacred cows after harvesting to "purify" them. This removes the potent toxins from the root. Milk is said to more effectively detoxify the tubers (Warrier et al. 1993, 44*).[14] The root tuber is ground into a paste for external application to treat neuralgia.

For tantric and psychoactive purposes, of course, the root is not detoxified. It is simply dried, minced, and consumed in **smoking mixtures**, normally with ganja (flowers of *Cannabis indica*). The leaves are dried and smoked.

Aconitum ferox is the most poisonous plant of the Himalayas and can very easily prove lethal! As little as 3 to 6 mg of aconitine, corresponding to only a few grams of dried or even fresh plant material, is sufficient to kill an adult.

Ritual Use
Among Indian Tantrists is an extreme sect known as the Aghoris. They follow the "left-handed" path, which regards sexuality and drugs as important methods for expanding consciousness. The Aghoris ingest plants associated with Shiva (hemp, *Datura metel*, opium from *Papaver somniferum*) and poisons (cobra venom, mercury, arsenic) so that they may experience the divine consciousness of their master. Aghoris produce mixtures of various plants for their large smoking tubes (*chilam*). One mixture for "advanced" individuals consists of ganja and *Aconitum ferox* roots (Svoboda 1993, 175).

Shiva is the Hindu god of inebriants and poisons. Myths relate how he personally tried all poisons at the beginning of the world. This caused him to turn blue, as blue as the flowers of blue aconite. Similarly, a Tantrist can assimilate himself with the god by ingesting every poison and surviving (according to the motto "That which does not kill me only makes me stronger"). In another version of this story, the beating of the primeval ocean, or the churning of the milk ocean (*samudramathana*), not only brought forth the sacred cow but also caused the essences of all poisons to swirl up. Petrified with fear, the gods

hurried to Kailash, where Shiva sat in meditation. They bade him help. Shiva took the poison in his hand and drank it. His wife Parvati became afraid for him and squeezed her husband's throat, which caused the poison to remain there and turned him completely blue. It is for this reason that Shiva is also known as Nilakanta, "blue throat." By performing this act, Shiva saved all creatures from death by poisoning. But a little of the poison slipped out of his hand and onto the Himalayas. Today, it still flows in the veins of blue aconite and other poisonous plants.

Artifacts
Hindu art contains numerous images of Shiva, many of which depict him with a blue skin color. Sometimes only his throat is blue. The *Sarada-tilaka Tantra* describes Shiva in his form as a "blue throat": He shines like a myriad of rising suns and has a glowing crescent moon in his long, matted hair. His four arms are adorned with snakes. He has five heads, each of which has three eyes, and is clad in only a tiger skin and is armed with his trident. It is possible that the plant spirit of *Aconitum ferox* has the same appearance.

Aconitum ferox, along with other species (including *Aconitum napellus*), is portrayed on Tibetan medical thangkas (paintings). One leaf of the Tibetan medical tree is dedicated to the plant, and this depicts how the "great medicine" can be used to make a medicinal butter (Aris 1992, 179, 233*).

Medicinal Use
In Ayurvedic medicine, the "purified" root tubers are used to treat neuralgia, painful inflammations, coughs, asthma, bronchitis, digestive problems, colic, weak hearts, leprosy, skin afflictions, paralysis, gout, diabetes, fever, and exhaustion (Warrier et al. 1993, 41ff.*).

These and other Himalayan species of aconite (*Aconitum heterophyllum*, *Aconitum balfourii* Stapf; cf. **Aconitum spp.**) find many uses in Tibetan medicine. The roots are regarded as a remedy for colds and "cold"; the herbage is used to treat ailments resulting from "heat." In Tibetan medicine, *Aconitum ferox* is also known as *sman-chen*, "great medicine"; the crushed roots, mixed with bezoar stones, are used as a universal antidote. The root is also used to treat malignant tumors (Laufer 1991, 57). The great medicine is also esteemed as a remedy for demonic possession (Aris 1992, 77*). In Nepalese folk medicine, blue aconite is used to treat leprosy, cholera, and rheumatism (Manandhar 1980, 7*).

Constituents
The entire plant contains the diterpenoid alkaloids aconitine and pseudoaconitine[15] (Mehra and Puri 1970). The root tuber contains the greatest

14 "An external change of the drug is also effected by boiling the tuber in cow urine, a common custom, which apparently protects them against the attacks of insects, which they are otherwise very prone to. In this condition, it largely loses its color and also yields a dark brown solution in water after just a short period of time. For medicinal purposes, this latter form of the drug is completely useless; it can be used only as a poison for killing wild animals, as is often done in India" (Pabst 1887, 8*).

15 Pseudoaconitine has the same properties as aconitine and is chemically related to veratrum acid (cf. **Veratrum album**).

concentrations of these constituents and is thus the most dangerous part of the plant (cf. *Aconitum napellus*).

Effects

In Ayurvedic medicine, the root is attributed with sweet, narcotic, sedative, anti-inflammatory, diuretic, nervine, appetite-stimulating, digestion-promoting, stimulant, anaphrodisiac, calming, and antipyretic effects (Warrier et al. 1993, 41*).

The effects of a tantric smoking mixture containing aconite are said to be extreme. Even experienced Tantrists emphatically warn against its use (cf. *Aconitum napellus*).

Commercial Forms and Regulations

The seeds may sometimes be purchased in nurseries.

Literature

See also the entries for *Aconitum napellus*, *Aconitum* spp., and **witches' ointments**.

Bisset, N. G., and G. Mazars. 1984. Arrow poisons in South Asia, part I: Arrow poisons in ancient India. *Journal of Ethnopharmacology* 12:1–24.

Mehra, P. N., and H. S. Puri. 1970. Pharmacognostic investigations on aconites of "*ferox*" group. *Research Bulletin of the Punjab University* 21:473–93.

Laufer, Heinrich. 1991. *Tibetische Medizin*. Ulm: Fabri Verlag. (Orig. pub. 1900.)

Rau, Wilhelm. 1994. *Altindisches Pfeilgift*. Stuttgart: Franz Steiner Verlag.

Svoboda, Robert E. 1993. *Aghora: At the left hand of God*. New Delhi: Rupa.

Aconitum napellus Linnaeus

Monkshood, Blue Rocket

Family

Ranunculaceae (Buttercup Family); Helleboreae Tribe

Forms and Subspecies

Monkshood is a polymorphous species with many subspecies and cultivated forms; it is regarded as taxonomically complex (Colombo and Tomè 1993):

Aconitum napellus ssp. *compactum* (Rchb.) Gayer
Aconitum napellus ssp. *napellus*
Aconitum napellus ssp. *neomontanum* (Wulfen) Gayer
Aconitum napellus ssp. *pyramidale* (Mill.) Rouy et Fouc.
Aconitum napellus ssp. *tauricum*
Aconitum napellus ssp. *vulgare* Rouy et Fouc.

It is possible that *Aconitum ferox* may be a synonym or a subspecies or variety of *Aconitum napellus* (cf. Warrier et al. 1993, 41*).

Synonyms

Aconitum compactum (Rchb.) Gayer
Aconitum neomontanum Wulfen
Aconitum pyramidale Mill.

Folk Names

Abnehmkraut, aconit, aconite, aconit napel, aconito napello, akonit, akoniton, altweiberkappe, apolloniabraut, apolloniakraut,[16] apolloniawurz, arche noah, blauelsterkraut, blauer akonit, blaukappen, blaumützen, blue aconite, blue rocket, casque-de-Jupiter (cap of Jupiter), eisenhut, eisenhütlein, eisenkappe, eliaswagen, eysenhütlein, fischerkiep, fliegenkraut, franzosenkapp, fuchskraut, fuchsschwanz, fuchswurz, giftkraut, goatsbane, goekschl, groß eysenhütlein, gupfhauben, hamburger mützen, härrgottslotscha, helm, helmblume, herrgottslatsche, herrnhut, heuhütli, hex, holtschoe, hummelkraut, isenhübli, jakobsleiter, judenkappe, jungfernschuh, kalessen, kappenblume, kapuzinerchäppli, kapuzinerkappe, königsblume, kutscherblume, marienscheusäken, mönchskappe, mönchswurz, monkshood, münchskapffen, muttergottesschühlein, napellus major, narrenkappe, noarnkopp, nonnenhaube, Odins hut, pantöffelchen, pantöffelken, papucha, paterskappe, pfaffenhütchen, pferdchen, poutsche, ra-duggam'dzim-pa (Tibetan), ranunculus montana, reiterkapp, reiter-zu-pferd, rössel, satanskraut, schawwerhaube, schlotfegerskappen, schneppekapp, steinkraut, sturmhut, tauben, taubenschnabel, teufelswurz, thora quasi phtora interitus (Latin, "doom"), totenblume, trollhat (Nordic, "troll's cap"), tübeli, tuifelkappe, venuskutschen, venuswägelchen, venuswagen, wolfgift, wolfkraut, wolfskraut, wolfswurz, würgling, ziegenschuh, ziegentod

History

Theophrastus (ca. 370–287 B.C.E.) already provided a very precise description of the plant and its effects and origin. In ancient times, monkshood, or aconite, was a feared poison associated

16 This name is also given to, and in fact is primarily used for, *Hyoscyamus niger*.

Monkshood (*Aconitum napellus*) was once a dreaded toxic and witches' plant. Today, it is a popular garden ornamental. (Woodcut from Tabernaemontanus, *Neu Vollkommen Kräuter-Buch*, 1731)

In ancient times, monkshood (*Aconitum napellus*) was feared as a poison; in the Middle Ages and the early modern period, it was purportedly used in the manufacture of witches' and flying ointments.

17 According to legend, Claudius was poisoned with a mushroom (*Amanita phalloides*) (Deltgen and Kauer 1973; Wasson 1972). But this did not prove powerful enough, so *Aconitum* was used as well.

with the legendary Colchian "witch" Medea (who was probably a Scythian shaman; cf. *Cannabis ruderalis*) and the gloomy underworld. Like henbane (*Hyoscyamus albus*), the plant was said to have sprung from the slaver of Cerberus, the hound of hell, and both plants were known as *apollinaris* ("Apollo's plant"). Another legend states that monkshood rose from the blood of Prometheus, which dripped onto the rocks whenever the eagle came and ate his liver (Gallwitz 1992, 111).

Monkshood was an important "battle drug" in Roman politics. Emperor Claudius died in 54 C.E. from aconite poisoning (Schöpf 1986, 77*).[17] The ancient Germans may have used the plant in their magical rituals, such as when the Berserkers were transformed into wolves. In the fourteenth century, Konrad von Megenberg described monkshood and its poisonous effects in his *Buch der Natur* [Book of Nature]. Monkshood was and still is regarded as the most poisonous and most dangerous plant in Europe (Roth et al. 1994, 89*).

Distribution

Monkshood occurs from Italy to Ireland and from Spain to the Himalayas. It is often found in subalpine zones. It belongs to the typical flora of the Alps and is (still) common in Switzerland.

Cultivation

Monkshood can be propagated by seeds or from separated tubers. Handling fresh root tubers can have dangerous toxic effects! The seeds are either sown in the spring by pressing them directly into the ground or raised in seedbeds. Monkshood prefers nutrient-rich soils and good, humus-rich earth. It also thrives in moist soils.

Appearance

This perennial herbaceous plant can grow as tall as 150 cm. The palmate leaves are deeply divided into five to seven lobes. The luxuriant, dark blue, helmet-shaped flowers form at the end of the stalk (terminal racemes). The sepal has the exact shape of a bumblebee, and this insect is also the plant's most important (and perhaps even its only) pollinator. The follicular fruits contain multiple seeds. The blooming period is from June to August. The plant develops a new tuberous root each year, while the root from the previous year dies.

Aconitum napellus can be very easily confused with *Aconitum ferox* and many other *Aconitum* species (*Aconitum* spp.). This does not present a problem from a pharmacological perspective, as most *Aconitum* species contain very similar active constituents. Some individuals have also confused monkshood with larkspur (*Delphinium* spp.; cf. *Delphinium consolida*).

Psychoactive Material

—Root (tubera aconiti, radix aconiti, aconiti tuber)
—Herbage (herba aconiti, aconiti herba)

Corresponding to the plant's growth cycle, the drug should not be stored and used for periods in excess of one year (Roth et al. 1994, 88*).

Preparation and Dosage

The dried herbage can be smoked (see *Aconitum ferox*). However, no information is available concerning dosage. People must be cautioned against improper use of this plant. Even harvesting the leaves can cause the active constituents to enter the body and produce unintentional symptoms of poisoning (Roth et al. 1994, 89*). As little as 3 to 6 mg of aconitine, which often corresponds to only a few grams of dried or even fresh plant material, can be lethal for adults. Ingested orally, as little as 0.2 mg of aconitine can produce toxic symptoms.

Tinctures were formerly used to treat migraine headaches and neuralgia. Up to five drops per day were ingested (Vonarburg 1997a, 65).

The roots were purportedly used in the manufacture of **witches' ointments**. They were also added to **wine** (cf. *Vitis vinifera*), which was drunk both for healing and for inebriating purposes (Pahlow 1993, 117*).

Although the plant is considered extremely toxic, children in Iceland have been reported to eat the flowers because of their honeylike sweetness (Olafsson and Ingolfsdottir 1994).

Ritual Use

In ancient times, monkshood was definitely used as a ritual poison:

> Bent on his destruction, Medea mixed in a cup a poison, which she had brought long ago from the Scythian shores. This poison, they say, came from the mouth of the Echidnean dog. There is a cavern with a dark, yawning throat and a way down-sloping, along which Hercules, the hero of Tiryns, dragged Cerberus with chains wrought of adamant, while the great dog fought and turned away his eyes from the bright light of day. He, goaded on to mad frenzy, filled all the air with his threefold howls, and sprinkled the green fields with white foam. Men think that these flecks of foam grew; and, drawing nourishment from the rich, rank soil, they gained power to hurt; and because they spring up and flourish on hard rocks, the country folk call them aconite. (Ovid, *Metamorphoses* 7.406 ff.)

Monkshood was presumably also used in other Scythian preparations and in shamanic-magical rituals, e.g., to transform oneself into a wolf. It may already have been used in antiquity to prepare flying ointments. Since the early modern period, monkshood has been one of the chief ingredients in **witches' ointments**. Many of its folk names suggest that the plant was used for both ritual and psychoactive purposes: *hut des Jupiter* (hat of Jupiter), *venuswagen* (Venus's wagon), *wolfskraut* (wolf's plant), *hut des trolls* (hat of the troll), *Odins hut* (Odin's hat), *hex* (witch), et cetera.

Artifacts

In Christian art, the plant appears in paintings as a symbol of death (e.g., in *Maria Lactans* by the Master of Flémalle and in *The Lamentation of Christ*) (Gallwitz 1992, 113 f.). In Europe, the plant was used as a symbol of the venomousness of nature. Monkshood is portrayed on Tibetan medicine thangkas alongside *Aconitum ferox* and *Aconitum* spp. (Aris 1992, 233*).

Gustav Meyrink (1868–1932), an author who was experienced in alchemy and the occult and who wrote about numerous psychoactive plants (cf. *Cannabis indica, Lophophora williamsii, Veratrum album, Amanita muscaria*), composed a very insightful story about monkshood, "Der Kardinal Napellus" [Cardinal Napellus]. In this story, he describes a sect "known as the 'blue brothers,' the followers of which have themselves buried alive when they sense that they are approaching their end." The founder of the order, Cardinal Napellus, transformed himself into the first monkshood plant after his death. All of the plants were said to be derived from him. The sign of the order, of course, is the flower of *Aconitum*

napellus, and a field of aconite grows in the cloister garden. The novitiates start the plants when they are accepted into the order, and they baptize these with blood and sprinkle them with blood shed from the wounds produced by flagellation. "The symbolic meaning of this strange ceremony of blood christening is that the person should magically plant his soul in the garden of paradise and nourish its growth with the blood of his desires." The brothers in the order use the plant in a psychoactive manner: "When the flowers withered in the fall, we collected their poisonous seeds, which resemble small human hearts. In the secret tradition of the blue brothers, these represent the 'mustard seed' of faith, of which it is written that he who has it can move mountains, and we ate of these. Just as their terrible poison changes the heart and brings a person into a condition between life and death, so shall the essence of faith transform our blood—into miraculous power in the hours between the gnawing pain of death and ecstatic rapture" (Meyrink 1984). The story is reminiscent of the tantric use of *Aconitum ferox*.

Medicinal Use

Because it has long been feared as a potent poison, monkshood never attained any great significance in folk medicine. In Western phytotherapy, tinctures of monkshood are used externally to ease the pains of gout, sciatica, and neuralgia and to treat feverish colds in their onset. They are less frequently used internally (Pahlow 1993, 116*).

In homeopathy, Aconitum napellus hom. is used in dilutions of D3 and greater in accordance with the medical descriptions to treat nervous and psychic ailments, e.g., as a result of anger, fright, agitation, or neuralgia (Pahlow 1993, 116*; Roth et al. 1994, 89*). Hahnemann had high praise for the agent, for "its powers to help are miraculous" (Buchmann 1983, 29*). It is still used for numerous purposes today (Vonarburg 1997a, 1997b).

Constituents

The entire plant contains the alkaloid aconitine (= acetylbenzoylaconine) and aconitine acid. The highest concentrations are in the root, which is thus the most dangerous part of the plant. The root tubers contain large amounts of diterpenoid alkaloids of the so-called aconitine type (0.3–2.0%). The structures of some of these have not yet been determined (Bugatti et al. 1992). Aconitine is the primary alkaloid; mesaconitine, hypaconitine, napelline, and *N*-diethylaconitine are also present. In some subspecies, the primary alkaloid is mesaconitine (Olafsson and Ingolfsdottir, 1994). Aconitine is also present in all the other parts of the plant, although typically in only low concentrations. Aconitine has even been detected in the nectaries. As a result, it may be

"So is the man who has given aconite to three uncles to ride by on swaying feathers and to look down on us from there? Keep quiet when he comes by; if you so much as say 'That's the man who . . .' you'll be treated as if you'd accused him in court."

JUVENAL
SATIRES
(I.158–61)

"Monkshood is associated with the Nordic god Odin and the goddess Hel. Earlier tales referred to it as 'Odin's helm,' [and] it was supposedly used as an ingredient in 'lycoanthropic transformation ointments.'"

MAGISTER BOTANICUS
MAGISCHES KREUTHERKOMPENDIUM
[COMPENDIUM OF MAGICAL HERBS]
(1995, 194*)

> "It is established that of all poisons the quickest to act is aconite, and that death occurs on the same day if the genitals of a female creature are but touched by it. . . . Fable has it that aconite sprang out of the foam of the dog Cerberus when Hercules dragged him from the underworld, and that this is why it grows around Heraclea in Pontus, where is pointed out the entrance to the underworld used by Hercules. Yet even aconite the ancients have turned to the benefit of human health, by finding out by experience that administered in warm wine it neutralizes the sting of scorpions. It is its nature to kill a human being unless in that being it finds something else to destroy. Against this alone it struggles, regarding it as more pressing than the find. What a marvel! Although by themselves both are deadly, yet the two poisons in a human being perish together so that the human survives."
>
> PLINY
> *NATURAL HISTORY*
> (27.4f.)

Aconitine

Mesaconitine

possible to produce a psychoactive **honey** from the plant.

Effects

When applied to the skin, monkshood is said to provoke sensations of tingling and hallucinations. Because of this, monkshood was purported to have been an important ingredient in **witches' ointments**. It is said to produce the sensation of wearing a garment made of fur or feathers. In the Rhineland, it is said that "[s]imply smelling the plant will make the nose swell up" (Gallwitz 1992, 113). Monkshood has a strong stimulating or inebriating effect upon horses. They become "foamy," that is, fiery; for this reason, horse dealers at one time fed their animals monkshood before offering them for sale.

The description of the course of effects of monkshood poisoning is not exactly enticing: "The longer the time that the alkaloid and the drug are in contact with the mouth, the more pronounced will the local, sensory nerve effects in the mouth and throat be observable following acute aconitine or aconite poisoning. The tingling and burning may be followed by a loss of speech and a sensation of paralysis in the tongue and in the area around the mouth, so that speaking becomes difficult. When absorbed, feelings of tingling and formication [the sensation of being crawled upon by insects] in the fingers, hands, and feet very characteristically soon appear, sometimes with twitching of the face, followed by paralysis of the facial muscles. The person who has been poisoned is also disturbed by an unbearable sensation of coldness (the feeling of 'ice water in the veins') with hypothermia, caused by stimulation of the cold centers. This is followed by numbness, symptoms of paralysis in the arms and legs, and difficulties in breathing. Seeing green, dizziness, buzzing in the ears, and trigeminal pain have also been observed. Nausea and vomiting can occur, but can also be absent, as can diarrhea and increased urination. Respiratory difficulties and especially peculiar heart disturbances . . . can result in loss of consciousness and death due to heart or breathing problems. But consciousness may also

be retained until death, which under these circumstances can occur within in the first hour" (Fühner 1943, 217f.*).

Commercial Forms and Regulations

In Europe, the wild plant—like all *Aconitum* species—is protected (Roth et al. 1994, 89*). The seeds of various subspecies, varieties, and cultivars are available in nurseries.

Literature

See also the entries for **Aconitum ferox**, **Aconitum spp.**, and **witches' ointments**.

Bauerreiss, Erwin. 1994. *Blauer Eisenhut.* Bad Windsheim: Wurzel-Verlag.

Bugatti, C., M. L. Colombo, and F. Tomè. 1992. Extraction and purification of lipolalkaloids from *Aconitum napellus* roots and leaves. *Planta Medica* 58 suppl. (1): A695.

Colombo, M. L., and F. Tomè. 1993. Nuclear DNA amount and aconitine content in *Aconitum napellus* subspecies. *Planta Medica* 59 suppl.: A696.

Deltgen, Florian, and Hans Gerd Kauer. 1973. The Claudius case. *Botanical Museum Leaflets* 23 (5): 213–44.

Gallwitz, Esther. 1992. *Kleiner Kräutergarten: Kräuter und Blumen bei den Alten Meistern im Städel.* Frankfurt/M.: Insel TB.

Meyrink, Gustav. 1994. Der Kardinal Napellus. In *Fledermäuse,* 1:101–13. Berlin: Moewig.

Olafsson, Kjartan, and Kristin Ingolfsdottir. 1994. Aconitine in nectaries and other organs from Icelandic populations of *Aconitum napellus* ssp. *vulgare. Planta Medica* 60:285–86.

Vonarburg, Bruno. 1997a. Blauer Eisenhut (1. Teil). *Natürlich* 17 (1): 64–67.

———. 1997b. Blauer Eisenhut (2. Teil). *Natürlich* 17 (2): 64–67.

Wasson, R. Gordon. 1972. The death of Claudius or mushrooms for murderers. *Botanical Museum Leaflets* 23 (3): 101–28.

Aconitum spp.

Aconite Species

Family

Ranunculaceae (Buttercup Family); Helleboreae Tribe

Many *Aconite* species have worldwide ethno-botanical significance as medicines, psychoactive products, and arrow poisons.

Uses as Medicine

The following species of aconites (*chuan wu tou*) are used in traditional Chinese medicine and in Japanese kampo medicine (as cited in Wee and Keng 1992, 16 f.*; Schneebeli-Graf 1992, 55*):

Aconitum carmichaelii Debeaux (*chuan wu tou* or *bushi*); also:

var. *wilsonii* (Stapf ex Moltet) Munz (*tsao wu tou*)

Aconitum chinense Sieb. et Zucc.

Aconitum hemsleyanum E. Pritz

Aconitum transsectum Diels

Aconitum vulparia Rchb. ex Spreng. [syn. *Aconitum lycoctonum* auct. non. L.]

Only the dried rhizomes are used; they lose their potent toxicity during the drying process. In traditional Chinese medicine, aconite roots are characterized as stimulating, cardiotonic, analgesic, narcotic, and locally anesthetic. They stimulate the *yang* energy and are used for all *yang* ailments. The dosage lies between 3 and 8 g (Reid 1988, 115*).

The species *Aconitum carmichaelii* is common throughout southern China. The folk medicine of the region uses its roots for headaches, paralysis of one side of the body (hemiplegia), overheating of the body, rheumatism, arthritis, contusions, bruises, and broken bones. Pharmacological studies in China have demonstrated that this drug stimulates the body's own immune system. However, it has not yet been possible to isolate from the root a substance that might be responsible for these effects. It is possible that there may be a synergistic effect of several or even all of the active constituents (Chang et al. 1994). The Chinese medicinal drug (*fu tzu*) has the highest concentration of alkaloids (Bisset 1981).

In Japanese kampo medicine, which is based on Chinese herbalism, the roots of the species *Aconitum carmichaelii* are known as *bushi* and are used for weak digestion (cf. Murayama and Hikino 1984). Pharmacological investigations have determined that the so-called aconitans A, B, C, and D have hypoglycemic effects, i.e., they lower blood sugar levels (Hikino et al. 1989, 1983).

Psychoactive Products

Aconite (*Aconitum napellus*) is said to have been an important ingredient in **witches' ointments**.

Aconitum ferox, a species found throughout the Himalayas, is a component of tantric **smoking mixtures** with drastic effects. Some Chinese species, the identity of which are unfortunately uncertain, but whose root drugs are known by the name "fu tzu" (including *Aconitum carmichaelii*), provided one of the main ingredients in **han-shih powder**.

Many Taoist elixirs of immortality contain large quantities of aconite, along with ominous fungi (*Psilocybe* **spp.**), arsenic, mercury, hemp (***Cannabis sativa***), and *Digitalis* spp. (cf. ***Digitalis purpurea***) (Cooper 1984, 54*).

Use as Arrow Poison

Aconitum was used as an arrow poison in ancient Europe and in Asia and North America (Alaska) (Bisset 1989). In ancient China, the most important source of arrow poison was the root of *Aconitum carmichaelii* (*wu tou, fu tzu, tsao wu*) (Bisset 1979, 1981). Many northern Asian hunting peoples used the toxic root tubers of the following species of aconite to make their arrow poisons:

Aconitum delphinifolium DC.

    ssp. *chamissonianum* (Reichb.)

    ssp. *paradoxum* (Reichb.) Hult.

Aconitum fischeri Reichb.

Aconitum japonicum Thunb.

Aconitum kamtschaticum Reichb.

Aconitum maximum Pall. ex DC.

Aconitum napellus Thunb. non L.

The Asian species *Aconitum carmichaelii* is used in traditional Chinese medicine. Its roots were once used to produce elixirs for prolonging life.

Herb Paris (*Paris quadrifolia* L.; Liliaceae), once regarded as a poisonous plant but apparently only slightly toxic, was formerly known as *dollwurz* (crazy plant). It was included among the aconite plants under the name *Aconitum pardalianches*. Because the folk name *dollwurz* was primarily used to refer to the root of the deadly nightshade (*Atropa belladonna*) and obviously refers to its hallucinogenic effects, it is possible that herb Paris may be a psychoactive plant that has been forgotten. (Woodcut from Fuchs, *Läebliche abbildung und contrafaytung aller kreüter*, 1545)

White monkshood (*Aconitum septentrionale*).

This ancient Chinese representation of various aconite species is from the *Ch'ung-hsiu cheng-ho pen-ts'ao*.

"Arrows are prepared with the juice of aconite. They very quickly kill that which they hit."

Avicenna
Canon Medic
(1608)

Aconitum sachalinense Fr. Schmidt
Aconitum subcuneatum Nakai
Aconitum yezoense Nakai

The harvesting of the roots is often accompanied by magical rites. The arrow poisons are usually produced using other substances as well. The Ainu, the original inhabitants of Japan, combined the principal ingredient with the leaves of *Artemisia vulgaris*, the toxin of the Japanese puffer fish (*Dasyatis akajei* Müller et Henle), and even ***Nicotiana tabacum*** (Bisset 1976). The notorious fugu fish provided an additional ingredient for an especially powerful arrow poison (Bisset 1976, 91; cf. **zombie poison**). The manner in which the Ainu tested whether the poison was usable and potent enough provides an interesting fact for explaining the effects of *Aconitum* in **witches' ointments**. They made a small cut in the thenar below the thumb and held a freshly cut root tuber against this. The poison caused the thumb to become numb and (temporarily) paralyzed. An experienced poison maker could evaluate the effectiveness of the root by the duration of its effects (Bisset 1976, 91).

Constituents
Most aconite species contain the very toxic aconitine alkaloids as well as the slightly toxic alkamines. Those species that are used for medicinal purposes have higher amounts of alkamines, whereas the species used to produce arrow poisons contain higher amounts of aconitines (Bisset 1976).

In China, the roots of *feng-feng*, or the plant *Siler divaricatum* (Turcz.) Benth. et Hook. f. (Umbelliferae), were once used as an antidote for aconite poisoning. But it has been said that the root of this plant "produces madness" (Schultes and Hofmann 1992, 56*). Although it has sometimes been claimed that *Siler divaricatum* is psychoactive, there is no evidence supporting this assertion.

Literature
See also the entries for **Aconitum ferox, Aconitum napellus**, and **witches' ointments**.

Bisset, N. G. 1976. Hunting poisons of the North Pacific region. *Lloydia* 39 (2/3): 87–124. (Includes a very detailed bibliography.)

———. 1979. Arrow poisons in China. Part I. *Journal of Ethnopharmacology* 1:325–84.

———. 1981. Arrow poisons in China. Part II: *Aconitum*—botany, chemistry, and pharmacology. *Journal of Ethnopharmacology* 4 (3): 247–336.

———. 1989. Arrow and dart poisons. *Journal of Ethnopharmacology* 25:1–41.

Chang, Jan-Gowth, Pei-Pei Shih, Chih-Peng Chang, Jan-Yi Chang, Fang-Yu Wang, and Jerming Tseng. 1994. The stimulating effect of radix aconiti extract on cytokines secretion by human mononuclear cells. *Planta Medica* 60:576–78.

Hikino, Hiroshi, Masako Kobayashi, Yukata Suzuki, and Chohachi Konno. 1989. Mechanisms of hypoglycemic activity of aconitan A, a glycan from *Aconitum carmichaelii* roots. *Journal of Ethnopharmacology* 25:295–304.

Hikino, Hiroshi, Hiroshi Takata, and Chohachi Konno. 1983. Anabolic principles of *Aconitum* roots. *Journal of Ethnopharmacology* 7:277–86.

Murayama, Mitsuo, and Hiroshi Hikino. 1984. Stimulating actions on ribonucleic acid biosynthesis of aconitines, diterpenic alkaloids of *Aconitum* roots. *Journal of Ethnopharmacology* 12:25–33.

Murayama, M., T. Mori, H. Bando, and T. Amiya. 1991. Studies on the constituents of *Aconitum* species. *Journal of Ethnopharmacology* 35 (2): 159–64.

Rätsch, Christian. 1996. Das 'Heilgift' Aconit. *Dao* 4/96:68.

Acorus calamus Linnaeus

Calamus, Sweet Flag

Family

Araceae (Arum Family)[18]

Forms and Subspecies

Several varieties have been described, reflecting differences in the genomes and geographical distribution (Motley 1994, 397):

Acorus calamus var. *americanus* (Raf.) Wulff (North America, Siberia)

Acorus calamus var. *angustatus* Bess. (southeast Asia, Japan, Taiwan)

Acorus calamus var. *calamus* L. (Eurasia)

Acorus calamus var. *verus* L. (tetraploid form)

Acorus calamus var. *vulgaris* L. (Europe, India, Himalayas)

Synonyms

Acorus aromaticus Gilb.

Acorus odoratus Lam.

Acorus vulgaris L.

Acorus vulgaris (Willd.) Kerner

Folk Names

Ackermagen, ackerwurtz, ackerwurz, acore, acore aromatique, acore odorant, acore vrai, acori, acoro, acoro verdadero, acrois, ajîl-i-turkî (Persian), akoron (Greek), aksir-i-turki, a-notion ao-titara, bach, bacha, bajegida (Kannada), beewort, belle angélique, bhadra (Sanskrit), bhuta-nashini (Sanskrit), boja, bojho (Nepali), bueng, calamo aromatico, calamus, canna cheirosa, chalmis, ch'ang (Chinese), ch'ang-jung, ch'ang-p'u, cinnamon sedge, dálau, dárau, déngau, deutscher ingwer, deutscher zittwer, erba cannella, erba di Venere (Italian, "plant of Venus"), flagroot, galanga des marais, ganghilovaj (Gujarati), gewürzkalmus, ghorabach, gladdon, gora vatch (Hindi), iggur, ighir jammu, jerangau, kahtsha itu (Pawnee, "medicine that lies in the water"), kalmoes, kalmuß, karmes, karmsen, kaumeles, ki we swask, kni (Egyptian), kolmas, kolmes, lubigan (Tagalog), magenwurz, Mongolian poison, moskwas'wask, muskrat root, muskwe s uwesk, musquash, myrtle flag, myrtle grass, myrtle sedge, nabuguck (Chippewa), nagenwurz, pai-ch'ang, peze boao ka (Osage, "flat plant"), pine root, pow-e-men-artic ("fire root"), rat root, reed acorus, roseau aromatique, roseau odorant, safed-bach (Hindi), schiemen, schiemenwurz, schwertenwurzel, sete, shui-ch'ang-p'u, shyobu (Japanese), sih kpetawote, sinkpe tawote (Lakota, "food of the muskrat"), sunkae (Lakota, "dog penis"), sweet calomel, sweet cane, sweet cinnamon, sweet flag, sweet flagroot, sweet grass, sweet myrtle, sweet rush, sweet segg, tatar, themeprü (Assamese), ugragandha (northern India), vaambu, vacha, vaj, vasa (Telugu), vasambu (Tamil), vash (Arabic), vashampe (Malayalam), vekhand (Marathi), venerea (Roman), venuspflanze, Venus plant, wada-kaha, warch, watchuske mitsu in, water flag, wechel, weekas, wee-kees, wehkes ("muskrat root"), wekas, wika, wike, wiken, wye (Kashmiri), yellow flag, zehrwurzrhizome, zwanenbrood (Dutch, "swan bread")

History

The history of calamus is still largely unknown. It is more than questionable whether the *akoron* of Dioscorides was actually calamus (Schneider 1974, 1:42*). In ancient times, it was believed that *akoron* was indigenous to the legendary gardens of Colchis (on the Balkan Peninsula on the Black Sea). Whether calamus was used as an aphrodisiac in ancient times, as it is in modern Egypt, cannot be determined with certainty. But if the ancient names do in fact refer to calamus, then it is likely that it was used for this purpose (cf. Pliny, *Natural History* 25.157).[19] In Italy, it is still regarded as a "plant of Venus" (Samorini and Festi 1995, 33). The "calamus" of the Bible is now interpreted as *Andropogon aromaticus* L. or *Cymbopogon* spp. (cf. **Cymbopogon densiflorum**). Remnants of *Acorus calamus* were reportedly found in the tomb of Tutankhamun (Motley 1994, 400; cf. also Germer 1985, 238 f.*). It has also been suggested that calamus was an ingredient in some **witches' ointments.**

Chinese sources contain what may be the oldest reference to sweet flag. The related but smaller species *Acorus gramineus* Soland. (*p'u*) was mentioned in the ancient Chinese *Shih Ching*, or the Book of Songs (ca. 1000–500 B.C.E.) (Keng 1974, 403*).

In Europe, sweet flag was well known during the late Middle Ages and has been esteemed as a medicinal plant since that time. It is not known whether it was present in pre-Columbian America. In any case, North American Indians became aware of its hallucinogenic effects as a result of ethnobotanical research (Motley 1994). The notion that calamus can have hallucinogenic effects was first published by Hoffer and Osmond (1967, 55 f.*).

Distribution

Sweet flag is apparently indigenous to Central Asia or India (Motley 1994) and is common on Sri Lanka and in the Himalayas. It has spread throughout the world as a result of cultivation (Hooper 1937, 80*). The plant was not introduced

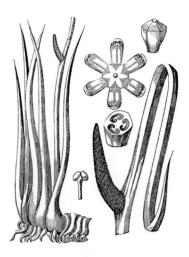

The botany of *Acorus calamus* was not clarified until quite recently. Its most characteristic feature is the almost phalluslike inflorescence. (Engraving from Pereira, *De Beginselen der Materia Medica en der Therapie,* 1849)

18 Botanists have recently questioned whether sweet flag does indeed belong in the Araceae Family (cf. Grayum 1987).

19 "*Akoron*, some call it *choros aphrodisias* [= dance of Venus's plant], the Romans *venerea* [= Venus plant], also *Radix nautica* [= ship's root], the Gauls *peperacium* [= water pepper], has leaves that resemble those of the sword lily but are narrower, and not dissimilar roots, although these are not intertwined with one another and do not grow straight, but grow toward the light at an angle and are interrupted at intervals, whitish, with a pungent taste and not unpleasant scent. The thick and white ones, which have not been eaten and are full of aroma, are superior. Such a one is that which in Colchis and Galatia is called *splenion* [= agent against spleen disorders]" (Dioscorides, *De materia medica* 1.2).

"A Penobscot Indian had the following dream: A muskrat told him that it was a root and where it could be found. When the man awoke, he went in search of the muskrat root and made it into a medicine. In this way, he healed his people of the plague."

FRANK G. SPECK
MEDICINE PRACTICES OF THE NORTHEASTERN ALGONQUIANS (1917)

Above: The small Chinese relative of calamus (*Acorus gramineus*).

Right: The characteristic inflorescence of calamus (*Acorus calamus*).

into central Europe until the sixteenth century; since then, it has established itself along creeks and slow-moving bodies of water and in lakes.

Cultivation

Calamus is propagated vegetatively by planting divided pieces of the rhizomes or scions with shoots. Calamus requires a marshy or very moist location. It can also survive in still water and does particularly well along the moist margins of ponds.

In North America, the muskrat (*Ondatra zibethica*) appears to have played a substantial role in increasing the range and occurrence of the plant. The animal is "magically" attracted to the rhizome and not only eats the rhizomes of the fresh plant but also collects parts of these and stores them for future use. Under the proper conditions, these pieces may then produce new roots. It is possible that the muskrat's characteristic scent may be due in no small part to its consumption of calamus (Morgan 1980, 237).

Appearance

Calamus is a perennial plant that may grow as tall as 120 cm. The rootstock (rhizome) spreads by creeping. The light to lushly green leaves are gladiate (like a sword blade) and distichous (in two rows). Rubbing them releases the typical calamus scent. The tiny, inconspicuous yellow-green flowers are attached to a spadix 5 to 8 cm in length. In its area of origin (India), sweet flag blossoms from April to June; in central Europe, from June to July.

The very similar but considerably smaller

species *Acorus gramineus* Soland. is found throughout Asia. It is easily recognized by its very small leaves (10 to 20 cm in length), which also exude the typical calamus aroma when rubbed.

In North America, calamus is often confused with *Iris pseudacorus* L., commonly known as yellow flag, and *Iris versicolor* L., known as blue flag (Motley 1994, 400).

Psychoactive Material

—Rhizome (rhizoma calami, calami rhizoma, calamus root)
—Calamus oil (calami aetheroleum, oleum calami)

Preparation and Dosage

Calamus oil is used as an aromatic additive to **snuff powders** and snuffing tobacco (see *Nicotiana tabacum*) (Hooper 1937, 80*) and in alcoholic beverages (spirits, **alcohol**, **beer**) (Motley 1994, 398).

A tea (infusion or decoction) from chopped rootstock (1 teaspoon per cup) can be drunk to treat feelings of weakness, nervousness, and stomach and intestinal cramps and as a nervine or aphrodisiac (Frohne 1989). A strong decoction can also be used as a bath additive. Calamus is an ingredient in many bitter cordials (cf. **theriac**).

According to some North American Indians, an amount of calamus equivalent in size to a finger is sufficient to produce psychoactive effects. However, very high dosages (200 to 300 g of dried roots) have also been tested.

Ritual Use

In ancient China, calamus was clearly used in shamanism. However, this may have been the smaller species (*Acorus gramineus* Soland. or *Acorus gramineus* Soland. var. *pusillus* (Sieb.) Engl.) known as *ch'ang-p'u* (also *shi chang pu*). Mêng Shen wrote:

> Those who wish to see spirits use the raw ma fruits [**Cannabis sativa**], ch'ang-p'u [*Acorus gramineus*], and k'uei-chiu [*Podophyllum pleianthum* Hance, syn. *Dysosma pleiantha* (Hance) Woods.; cf. *Podophyllum peltatum*], ground in equal amounts, and make these into pills the size of a marble and take these every day when they look into the sun. After a hundred days, they will be able to see spirits. (Li 1978, 23*)

In China, calamus is one of the oldest auspicious plants. It is said that the Taoist An-ch'i-sheng used wild calamus as an elixir, which caused him to become not only immortal, but invisible as well. Unfortunately, the methods of preparing and ingesting calamus for this purpose have not been passed down to us. Bundles of calamus leaves, together with *Artemisia vulgaris* (cf. **Artemisia**

spp.), are still used as talismans during the dragon boat festival, and they are hung over the house door to protect against evil spirits (Motley 1994, 402).

In Kashmir, the root is regarded as auspicious and should be the first thing a person looks upon on the morning of the traditional new year's festival (*navroj*) (Shah 1982, 299*). Indian snake charmers use pieces of calamus root to charm cobras (Motley 1994, 403).

Many North American Indians regard calamus as a panacea and tonic. The Iroquois used the root to detect witches and evil magic. Many Indians of the Northeast woodlands believe that the root has apotropaic effects and consequently hang it in the house or sew it into their children's clothing. The "spirits of the night" (nightmares) then stay away. The Winnebago, Ponca, Pawnee, Omaha, and Dakota make garlands of calamus grass that are used in secret rites (*wakan wacipi*, "sacred dance") and as hunting talismans (Howard 1953; Morgan 1980, 235). The Chippewa combine calamus with *Aralia nudicaulis* L. and then boil a decoction in which they soak their fishing nets to ensure a good catch or to chase away rattlesnakes (Motley 1994, 404).

The Cheyenne use calamus roots as **incense** in their sweat lodge ceremonies. They simply toss pieces of the root onto the glowing stones in the sweat lodge. The smoke is said to be cleansing and beneficial to health. Pieces of calamus root as well as calamus leaves are also occasionally added to **smoking mixtures** or mixed with tobacco (*Nicotiana* spp.) (cf. **kinnikinnick**).

The Cree reputedly used calamus roots as a hallucinogen. It is said that they chewed a piece of root as long as a finger. The accuracy of this information, which has been repeated often in the psychedelic literature, is somewhat in doubt (cf. Morgan 1980; Ott 1993*; Schultes and Hofmann 1995*), as all experiments with American calamus—even those involving very high dosages (up to 300 g of rhizomes!)—have been completely unsuccessful. If the Cree did indeed possess a hallucinogen, it probably was not *Acorus calamus*. One Cree name for calamus—or, as the source notes, a very *similar* plant—is *pow-e-men-artic*, "fiery pepper root." The Cree frequently placed pieces of calamus root, which they called *wee-kees* (muskrat root), in their medicine bundles (Johnston 1970, 308*).

Amazingly, several evangelical churches in Lutheran parishes burned calamus in the 1950s as an incense during Easter (Motley 1994, 402).

Artifacts

A section of naturalist North American poet Walt Whitman's (1819–1892) renowned collection of poems *Leaves of Grass* bears the heading "Calamus." It is possible that the poems contained within this section were inspired by calamus or its effects (Morgan 1980, 235 f.).

Medicinal Use

In the Ayurvedic system of medicine, calamus is used to treat sleeplessness, melancholy, neuroses, epilepsy, hysteria, memory loss, and fever (Vohora et al. 1990, 53). Calamus is used together with saffron (see *Crocus sativus*) and milk to induce labor (Motley 1994, 403). Nepali Sherpas use a paste made from fresh rootstock as an antiseptic agent to treat wounds in animals (Bhattarai 1989, 47*). The Nepalis use the root for colds and coughs (Manandhar 1980, 9*) and as a tonic for the nerves (Singh et al. 1979, 188*). In Ayurvedic and Tibetan medicine, calamus is an important psychoactive plant: "*Vacha* literally means 'speak' and describes the power of the word, the intelligence, or the self-expression that is stimulated by this medicinal plant" (Lad and Frawley 1987, 175*).

It is for this reason that calamus root, when used as **incense**, has the effect of illuminating and strengthening the mind. It is often found in Tibetan incense mixtures that are burned to strengthen the nerves and to increase meditative concentration. It is also regarded as a rejuvenation tonic and as "nourishment for the Kundalini serpent" (Lad and Frawley 1987, 176*).

In the forest regions and neighboring plains of North America, calamus is a medicine that the Indians use for a great number of purposes. Decoctions of the root are used to treat stomach and intestinal ailments, digestive difficulties, and cramps. Fresh pieces of the root are chewed for headaches, colds, sore throats, and bronchitis. Dried, the root is also used to prepare a medicinal and ritual **snuff** (Morgan 1980).

Calamus is smoked or burned as incense to treat headaches, coughs, and colds (Motley 1994, 404). The Blackfeet, who obtained calamus roots via long trade routes, used it as an abortifacient. The root was chewed as a cure-all. To treat headaches, the Blackfeet would burn a mixture of ground root and tobacco (*Nicotiana* spp.) and inhale the smoke (Johnston 1970, 307 f.*). The Chippewa manufactured a medicine from calamus root, the bark of *Xanthoxylum americanum* Mill., the root cortex of *Sassafras albidum*, and the root of *Asarum canadense* L.[20] to treat colds and bronchitis (Morgan 1980, 240).

In traditional Chinese medicine, the rootstock of *Acorus gramineus* is used to treat forgetfulness, lack of concentration, hearing difficulties, buzzing in the ears, epilepsy, mental illnesses, sensations of being full in the stomach, and gastritis (Paulus and Ding 1987, 128*).

Constituents

Calamus root contains high levels of **essential oil** with decadienal, caryophyllene, humulene,

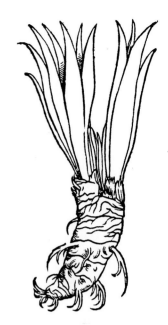

Calamus is an exotic plant that is also commonly referred to as sweet flag. For centuries, it was known in the West only in the form of its rhizome. Herbals listed it under the name "Acorus of the apothecaries." The illustrations in these books were based primarily upon the artists' imaginations. (Woodcut from Lonicerus, *Kreuterbuch*, 1679)

Calamus
"In paths untrodden,
In the growth by margins of pond-waters,
Escaped from the life that exhibits itself,
From all the standards hitherto publish'd, from the pleasures, profits, conformities,
Which too long I was offering to feed my soul,
Clear to me now standards not yet publish'd, clear to me that my soul,
That the soul of the man I speak for rejoices in comrades,
Here by myself away from the clank of the world,
Tallying and talk'd to here by tongues aromatic. . . ."

WALT WHITMAN
LEAVES OF GRASS
(1900)

20 Like the European hazelwort (*Asarum europaeum* L.), *Asarum canadense* L. also contains asarone (the name of which was derived from the genus name).

The yellow gladiolus was once known as *Acorus vulgaris,* or common calamus. This was intended to distinguish it from the "calamus of the apothecaries," the true calamus. (Woodcut from Fuchs, *Läebliche abbildung und contrafaytung aller kreüter,* 1545)

Asarone

Eugenol

curcumene, and β-asarone as well as the bitter principles acorone, neoacorone, and acorine, tanning agents, and mucilage (Chinese calamus contains α-asarone and β-asarone as well as eugenol, safrole, α-humulene, sekishone, etc.). The essential oil of *Acorus calamus* var. *americanus* is devoid of β-asarone (Motley 1994, 407). Plants from India contain especially high concentrations of asarone (Baxter et al. 1960; Vohora et al. 1990). Indian plants have also been reported to produce psychotropic effects (Motley 1994, 405).

The rhizomes of *Acorus gramineus* contain high amounts of essential oil, consisting of α-asarone, β-asarone, eugenol, safrole, α-humulene, sekishone, and other constituents (Paulus and Ding 1987, 128*).

Effects

Asarone is regarded as the inebriating principle in the raw drug[21] (Baxter et al. 1960; Motley 1994, 399). Laboratory experiments have verified its effects upon the central nervous system (Vohora et al. 1990). It also has inebriating effects, which are presumably due to a metabolite, TMA or **trimethyl-methamphetamine** (cf. *Myristica fragrans*). The essential oil has tonic effects, strengthens the stomach, and relieves cramps. It also has anti-bacterial effects. β-asarone is also reputed to have toxic and carcinogenic effects. Pharmacologically, asarone is said to act in a manner similar to **papaverine** (Motley 1994, 405).

The assertion that calamus is hallucinogenic appears to be due more to wishful thinking than to actual experiences with the plant. Even with very high dosages (up to 100 g of decocted, dried rhizomes), I have been unable to detect any type of hallucinogenic, psychedelic, entheogenic, or other visionary effect. Instead, the effects of asarone appear to be more sedative in nature. I also am unaware of any experimentally inclined psycho-nauts who have been able to report successful experiments with calamus. In my opinion, calamus can be stricken from the list of so-called legal highs, at least until new evidence of its psychoactivity appears.

Commercial Forms and Regulations

Calamus root (*rhizoma calami*) can be obtained in herb shops and pharmacies. Calamus oil has been taken off the market because of its (doubtful)

carcinogenic effects (Motley 1994, 407). In Germany, calamus is allowed to be used as an aromatic agent for schnapps and similar items as long as the amount of asarone per liter of the beverage it is added to does not exceed 1 mg (Roth et al. 1994, 92*).

Literature

See also the entry for **essential oil**.

Abel, Gudrun. 1987. "Chromosomenschädigende Wirkung von β-asaron in menschlichen Lymphocyten." *Planta Medica* 53:251–53.

Baxter, R. M., P. C. Dandiya, S. I. Kandel, A. Okany, and G. C. Walker. 1960. Separating of hypnotic potentiating principles from the essential oil of *Acorus calamus* L. of Indian origin by gas-liquid chromatography. *Nature* 185:466–67.

Frohne, Dietrich. 1989. Kalmuswurzelstock. In *Teedrogen,* ed. M. Wichtl, 260–62. Stuttgart: WVG.

Grayum, M. H. 1987. A summary of evidence and arguments supporting the removal of *Acorus* from the Araceae. *Taxon* 36:723–29.

Howard, James. 1953. Notes on two Dakota "holy dance" medicines and their uses. *American Anthropologist* 55:608–9.

Morgan, George R. 1980. The ethnobotany of sweet flag among North American Indians. *Botanical Museum Leaflets* 28 (3): 235–46.

Motley, Timothy J. 1994. The ethnobotany of sweet flag, *Acorus calamus* (Araceae). *Economic Botany* 48 (4): 397–412. (Very good bibliography.)

Samorini, Giorgio, and Francesco Festi. 1995. *Acorus calamus* L. (calamon aromatico). *Eleusis* 1:33–36.

Speck, Frank G. 1917. Medicine practices of the northeastern Algonquians. In *Proceedings of the Nineteenth International Congress of Americanists* (Washington, D.C.): 303–21.

Vohora, S. B., Shaukat A. Shah, and P. C. Dandiya. 1990. Central nervous system studies on an ethanol extract of *Acorus calamus* rhizomes. *Journal of Ethnopharmacology* 28:53–62.

Whitman, Walt. 1900. *Leaves of grass.* Philadelphia: David McKay.

21 "In high doses, α- and β-asarone can induce visual hallucinations and LSD-like states of inebriation" (Roth et al. 1994, 92*).

Agave spp.

Agaves, Mezcal Plants

Family

Agavaceae (Century Plant Family; Zander 1994, 95*), previously Liliaceae (Lily Family)

Forms and Subspecies

There are approximately 136 species in the genus *Agave* in Mexico and neighboring regions (Gentry 1982). Many of the larger species are ethnobotanically and ethnopharmacologically significant.

Species used in the manufacture of fermented beverages (pulque, suguí, tesgüino, tizwin, mesagoli) and distilled spirits (tequila, mescal, pisto):

Agave americana L. (century plant, teometl, mescale)
Agave americana L. var. *expansa* (Jacobi) Gentry (mescal maguey)
Agave asperimma Jacobi
Agave atrovirens Karw. ex Salm. (maguey, metl, tlacametl)
Agave bocicornuta Gentry (mescal luchuguilla, sa'pulí)
Agave cerulata Trel. ssp. *dentiens* (Trel.) Gentry
Agave durangensis Gentry
Agave ferox Koch (maguey)
Agave hookeri Jacobi
Agave latissima Jacobi [syn. *Agave macroculmis* Tod., *A. coccinea* hort. non Roezl ex Jacobi]
Agave mapisaga Trel. (maguey manso, maguey mapisaga)
Agave mescal Koch (mescal agave)
Agave multifilifera Gentry (chahuí)
Agave pacifica Trel. (mescal del monte, mescal casero, gusime)
Agave palmeri Engelm.
Agave parryi Engelm.
Agave polianthiflora Gentry (ri'yéchili)
Agave potatorum Zucc. [syn. *Agave scolymus* Karw.] (tlacametl)
Agave potatorum Zucc. var. *verschaffeltii* (Lem.) Berger [syn. *Agave verschaffeltii* Lem.] (tlacametl)
Agave rhodacantha Trel.
Agave salmiana Otto ex Salm-Dyck [syn. *Agave atrovirens* Karw. var. *salmiana* (Otto ex SD.) Trel., *Agave atrovirens* Trel. and "of authors" (Gentry 1982, 13)] (maguey de pulque, tlacametl)
Agave shrevei Gentry (mescal blanco, o'tosá)
Agave tequilana Weber (tequila agave, maguey, blue agave)
Agave tequilana Weber cv. Azul ("blue variety")
Agave vivipara L. [syn. *Agave angustifolia* Haw.] (babki, mescal de maguey)
Agave weberi Cels
Agave wocomahi Gentry (mescal verde, ojcome, pine maguey)

Agave zebra Gentry

For fibers, medicines, and sacrificial thorns (*pencas*):

Agave americana L.
Agave sisalana Perrine (henequen, sisal agave, kih)
Agave fourcroydes Lem. [syn. *Agave ixtlioides* Lem.] (henequen agave)

Folk Names

Agave, century plant, chupalla, henequen, maguei, maguey, mescal plant, meskalpflanze, metl, mezcal plant, pita

For the fermented juice:

Agave wine, iztac octli, mesagoli, mescal beer, metl, octli, pulque, suguí, tesgüino, tizwin, vino mescal, wine

For the distilled liquor:

Agave schnapps, mescal, mezcal, pisto, tequila, tuché (Huichol), vino mescal

History

Roasted remains of agaves, dated to approximately 8,000 years ago, have been recovered from the caves of Tehuacán (Mexico) (Wolters 1996, 28*). In prehistoric times, agave was already playing an important role as a source of food, inebriants, and materials in Mexico and the American Southwest. Some agaves were even used as poisons to stun fish in isolated bodies of water (Bye et al. 1975). The Mexican agaves were first described around 1577

Above left: The Mexican *Agave salmiana* is the most important species used to produce pulque and tequila.

Below left: In Yucatán (Mexico), great quantities of fibers obtained from the leaves of the sisal agave (*Agave sisalana*) are used to produce a variety of fiber products. This species can also be used to produce inebriating beverages. (Plantation at San Antonio Tehuitz, Yucatán)

Right: The inflorescence of the century plant (*Agave americana*) develops when the plant is about fourteen years old. Afterward, the plant dies.

The blue agave (*Agave tequilana*) is regarded as the best plant for the production of tequila.

Mayahuel, the goddess of agave, the source of the inebriating pulque drink. (Codex Laud, 9r.)

Above: Eating the *gusano de mescal*, an insect larva that lives in the mescal agave, will allegedly produce psychoactive effects.

Below: Because the mescal worm is purportedly psychoactive and is, moreover, (still) legal, a California manufacturer had the idea of putting this unappetizing little creature into a lollipop.

What Is Mescal?

The name *mescal* has created a great deal of terminological confusion about both psychoactive plants and their products.

The term is used to refer to a type of agave (the mescal agave), while the alcoholic spirits that are distilled from this species are also known as mescal or mezcal.

In southern California, *Yucca whipplei* Torr. is known as maguey, but also as mescal (Timbrook 1990, 247*).

Even the peyote cactus (**Lophophora williamsii**) is called mescal or mescalito, while the buttons that are sliced from this cactus are called mescal buttons or mescal heads. In addition, the seeds of **Sophora secundiflora** are known as mescal beans.

In the "drug scene," mescaline trips are often known as mescalitos.

In North America, the species *Agave felgeri* Gentry is known as *mescalito.*

With so many associations with the word *mescal,* it is not surprising that some people are firmly convinced that mescal spirits contain mescaline and are able to induce psychedelic effects.

In addition, it is rumored that the worm (actually a larva) that is contained in *mescal con gusano* contains special constituents and will induce hallucinogenic effects when eaten. Some people maintain that several worms must be eaten to obtain a sufficient dosage.

by Francisco Hernández. The Conquistadors also remarked about the use of the fermented juice (pulque) (Gentry 1982).

According to Aztec historical records, pulque was "invented" in Central America between 1172 and 1291, after the Aztecs had migrated from the north (Gentry 1982, 8). Pulque has probably been used for a considerably longer period of time and was known to many peoples and tribes. Pulque and similar alcoholic beverages also played a role among the tribes of northern Mexico and in the American Southwest (cf. **beer, chicha**). The Apaches, for example, used agave to make fermented drinks (*tizwin*) that were ritually consumed during tribal festivals (Barrows 1967, 75*).

Today, the Mexican agaves are important primarily because they are used in the production of tequila, and they are popular throughout the world as ornamental plants.

Distribution

The genus *Agave* is indigenous to Mexico and the American Southwest. Numerous species of the genus are from Mexico and have been cultivated for various purposes since prehistoric times (Dressler 1953, 120f.*).

Cultivation

Agaves are propagated using bulblets (offshoots), which are planted in growing fields shortly before the beginning of the rainy season. After twelve to eighteen months, the plants are transplanted to production fields. When this occurs, all of the roots are cut away from the rootstock (Rehm and Espig 1996, 328*). Agaves are succulents (photosynthesis occurs according to the crassulacean pattern) and can easily survive long periods without water. Some species thrive in the desert, others in tropical rain forests. Although the quality of the soil is not important, it should be well drained.

Appearance

Most agaves, and especially those species that are used in the production of pulque and spirits, are quite similar and rather uniform in appearance. They are hardy plants with thick, fleshy roots from which the fleshy leaf rosette grows. The lanceolate, cultrate, or hastate leaves are sharply pointed at the ends, and most have serrated edges and a very sharp, hard, woody tip. At the end of their lifetimes, the plants produce a panicled inflorescence on a straight, smooth stalk. The bulblets form in the axils of the flower bracts. From 1,000 to 4,000 bulblets may grow on one inflorescence (Rehm and Espig 1996, 327*).

Psychoactive Material

— Aguamiel (Spanish, "honey water"), the sugary juice that collects in the interior (the shaft) of the plant.

Shortly before the plant is ready to develop its inflorescence, a sap (*aguamiel, metl*) that is very high in sugar accumulates in the shaft of the plant underneath the leaf crown. This sap apparently ferments as a result of microbial (*Pseudomonas lindneri*), wild yeast, or fungal activity (Gonçalves 1956). The plant produces the fermented drink known as pulque (also *mezcal* or *vino mezcal*) on its own. This process can be artificially controlled by removing a portion of the leaf crown. When this is done, the plant will produce a much greater quantity of the inebriating juice (around 2 liters per day). The plant will produce new pulque daily for up to one month (Bye 1979a, 152f.*).
— Mescal worm (*gusano de mescal*)

Preparation and Dosage

The plant juice can either be tapped while it is fermenting or fermented in a covered, but not tightly closed, vat.[22] Pulque contains 3% to 4% alcohol (Havard 1896, 34*). Various plants have been and still are added to the pulque to improve it and to modify its psychoactive effects (see table).

The Serí Indians of northern Mexico boil the narrow leaves of *Agave cerulata* Trel. ssp. *dentiens* (Trel.) Gentry, a plant they call *heme*, chop them into small pieces, and place them into the carapace of a sea turtle. They then press them with a stone, so that the juice collects within the carapace. The juice, which ferments within just a few days, is diluted with water prior to consumption (Felger and Moser 1991, 223*).

The Tarahumara manufacture suguí or tesgüino from a number of agaves. They boil the leaves in water, press the agave hearts (mescal hearts), or make an extract from the chopped leaves. Fermentation occurs on its own (Bye et al. 1975, 88).

The Indians of Arizona prepared their mescal beer from the inflorescences of *Agave parryi* and *Agave palmeri* (Havard 1896, 34*).

Alcoholic spirits (tequila, mescal) are distilled either from the plant juice (aguamiel) or from boiled and mashed leaves. The Yaqui Indians fortify their mescal liquor with the leaves of **Datura innoxia**. In Mexico, it is also common to use marijuana flowers (cf. **Cannabis sativa**), sugar, and chili pods (see **Capsicum** spp.) as additives to mescal (Reko 1936, 64*). Damiana (**Turnera diffusa**) is also a good tequila additive. In fact, there are a great number of recipes for tequila (Walker and Walker 1994).

It is said that the mescal worm (a larva approximately 5 cm in length) that is added to some mescal spirits should be eaten whole if a psychoactive effect is desired. Two or three worms is considered an effective dosage. Recently, a California manufacturer began producing sugar-free lollipops, each containing a mescal worm.

Pulque Additives

(Adapted with modifications from Bye 1979a, 153*; Bye 1979b, 38*; Bye et al. 1975; Furst 1974, 71*; Havard 1896, 39*; Marino Ambrosio 1966; Kuehne Heyder 1995.)

Anacardiaceae	
Rhus schinoides Willd. ex Schult.	fruits[23]
[syn. *Schinus terebinthifolius* Raddi]	
Burseraceae	
Bursera bipinnata Engl.	bark, resin
Cactaceae	
Lophophora williamsii	cactus flesh
"root"	
Convolvulaceae	
Turbina corymbosa	seeds
Ipomoea violacea [?]	
Gramineae	
Triticum aestivum L.	wheat flour
Leguminosae	
Acacia angustifolia (Mill.) Kuntze	root[24]
[syn. *Acacia angustissima* (Mill.) Kuntze]	
(*palo de pulque*, tree of pulque, Ocpatl, pulque drug; cf. **Acacia** spp.)	
Acacia albicans Kunth	
[syn. *Pithecolobium albicans* (Kunth) Benth.]	
Calliandra anomala (Kunth) McBride	
Mimosa spp.	root
Phaseolus spp. (*frijolillo*; a wild species of bean)	root
Prosopis juliflora DC.[25] (mesquite)	fruit pods
Sophora secundiflora (Ortega) Lag. ex DC.	seeds
Solanaceae	
Datura innoxia	root
Datura lanosa (cf. **Datura** spp.)	root
Strophariaceae	
Psilocybe spp.	fructification

In the Yucatán, the roots of one maguey agave (perhaps *Agave americana* var. *expansa*) were used as an additive to **balche'**.

Ritual Use

Among the Aztecs, pulque was a drink sacred to the gods that could be drunk only on ritual occasions. The dosage was limited to four bowls. Men over the age of seventy, however, were allowed to drink until they were inebriated. Sacrificial celebrations were followed by ritual drinking bouts:

And on the following day, wine [= pulque] was drunk and the after-celebration of the festival took place. The wine that was drunk is

22 The siphonlike cucurbit vessels used to tap pulque are known as *acacote*. The pulque sack itself (*bota*) is made from an animal skin and a cow horn.

23 The so-called Brazilian pepper is regarded as a poisonous plant (Morton 1978).

24 The leaves have been shown to contain the alkaloid *N*-methyl-β-phenethylamine (Argueta et al. 1994, 1338*); it may also be present in the root.

25 The closely related species *Prosopis nigra* (Grisebach) Hieron. contains β-carbolines (Ott 1993, 263*).

Above: The Brazilian pepper tree (*Schinus terebinthifolius*), known as a toxic plant, is used to fortify pulque obtained from agaves.

Below: The seedpods of the mesquite tree (*Prosopis juliflora* DC.) contain 25 to 30% sugar and are thus ideal for use as a fermentation agent. Boiled and crushed in water, they yield a fresh, sweet beverage known in northern Mexico as *atole*. If allowed to stand, it quickly ferments. After one to two days, a beerlike drink results (**chicha**). The plant is also added to Mexican pulque.

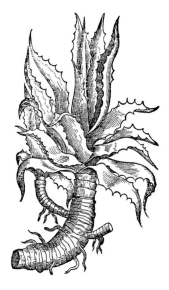

An early European illustration of the American agave, which was interpreted as a relative of the aloe. (Woodcut from Gerard, *The Herball or General History of Plants,* 1633*)

The mescal agave (*Agave horrida* Lem.); its name is derived from the Aztec word *mexcalmetl.* (From Hernández, 1942/46 [Orig. pub. 1615]*)

called blue wine. Everyone drank wine, the old men, the old women, and the chiefs of the nobility, the married, the adults, and the princes of blood and the leaders of the adults. And the first among the ranks of the young who were already strong, they too drank wine, but they drank in secret, they did not show themselves, they used the night as protection, they hid themselves under the grass so that they would not be seen. But if someone discovered them and it became known that they had drunk wine, then they would hit them with a jaw club so that their flesh would swell up, and they would shave their heads as slaves, drag them, kick them, throw them to the ground, and do everything evil to them until it sometimes occurred that they killed them. And after they had quieted their lust, they would throw them down, throw them out of the house. (Sahagun, Florentine Codex 2:34*)

The inebriating beverage was used as a ritual offering and libation for the gods and also was needed for human sacrifices. Before the ceremony, those whom the Aztecs were to sacrifice were required to drink four bowls of pulque, to which *Datura innoxia* or a decoction of the bark of the incense tree *Bursera bipinnata* had apparently been added. Once inebriated in this manner, they were allowed to have the priests rip the hearts out of their still-living bodies on the sacrificial altar.

In his 1541 work, *Historia de los indios de la Nueva España,* Motolinia had already alluded to the addition of *ocpatli,* apparently *Acacia angustifolia* (cf. **Acacia spp.**), and this addition was for-

bidden in the colonial period. The additive as well as the resulting drink were both known as *teoctli,* "divine pulque," or *xochioctli,* "flower pulque" (Ott 1996d, 428*).

The Huaxtec, who live on the Gulf of Mexico, used pulque in all their rituals and glorified the state of inebriation it produced. Pulque was also used during the sexual magical rites held to honor the erotic images of their gods. Men and women would copulate before the statues as the priests administered pulque **enemas** to them (even today, pulque is still regarded as an aphrodisiac). Following this, the men and women would perform ritual anal coitus. It appears that the pulque that was used for these purposes was fortified with thorn apple roots (*Datura innoxia*) (Kuehne Heyder 1995).

Large quantities of spirits distilled from agaves were also consumed in shamanic rituals, especially during the peyote festivals of the Huichols (cf. *Lophophora williamsii*):

> The shaman took a few swigs from a bottle of a potent liquor made from the agave plant, then passed it to me. I matched him swig for swig. Then he picked up the bowl of peyote gruel and took a long drink. I counted the gulps and took the same amount. This continued throughout the night. (Siegel 1992, 28*)

It should be noted that the **mescaline** contained in the peyote cactus strongly suppresses the effects of **alcohol**.

Pieces of agave are also used in ritual healing and fertility ceremonies, usually as amulets (Bye et al. 1975, 91). In Aztec sacrificial ceremonies, the tips of the leaves (*pencas*) were driven as thorns into the skin of the victims. They were also used in raising boys to be noblemen. Anyone who behaved inappropriately was punished using agave thorns (Gentry 1982, 10).

Artifacts

Aztec manuscripts contain numerous depictions of the pulque goddess Mayahuel and the foaming drink itself, as well as of drinking rituals, drinking sacrifices, and libations (Gonçalves 1956). Pulque also appears in Aztec songs and poems (Guerrero 1985).

In Cholula (Puebla), pre-Columbian wall paintings have been discovered that depict the ritual drinking of pulque. Peter Furst has identified the blossoms portrayed in one of the paintings as those of *Turbina corymbosa.* He has suggested that the seeds of this plant, which produce psychedelic effects, were used as an additive to the pulque (Furst 1974, 71*).

Agaves, tequila bottles, and states of drunkenness resulting from tequila are common elements in the paintings of such Mexican artists as Eugenia

Marcos, Elena Climent, Joel Renón, and Ricardo Martínez. Tequila is praised in many Mexican poems and songs (Artes de México 1994).

Medicinal Use

Different agaves find numerous uses in folk medicine. They are used to treat wounds, snakebites, skin diseases, foot fungus, venereal diseases, toothaches, rheumatism, diarrhea, et cetera (Wolters 1996, 31f.*).

In Mexico, it is widely believed that *mezcal con gusano* has aphrodisiac effects, as the worm is thought to contain active constituents. And in general, tequila and mescal are popularly associated with sex and eroticism.

Preparations of *Agave americana* are also used in homeopathy (Wolters 1996, 35*).

Constituents

Agaves contain saponins, steroid saponins, hecogenin glycosides, large amounts of sugar (up to 8%), vitamin C, polysaccharides, and minerals (Wolters 1996, 34*). *Agave americana* contains saponine, a pungent **essential oil**, from 0.4 to 3% hecogenin, and oxalic acid (Roth et al. 1994, 103*). Agave juice contains 8% sugar (agavose), essential oil, and some papain. Pulque contains 2 to 4% alcohol as well as large amounts of vitamin C and has 204 calories per liter.

Effects

The effects of pure pulque are similar to those of **balche'**, **chicha**, and **palm wine**. However, there is also a noticeably refreshing component. In a pulque inebriation, one remains clearer than in a beer inebriation. Pulque that has been fortified with *Psilocybe* **spp.** is not merely inebriating, but also visionary. Visions of snakes are said to appear with some regularity (Havard 1896, 39*).

Commercial Forms and Regulations

A number of agave species are available in nurseries throughout the world as ornamentals. Pulque is found only in Mexico. The corresponding distilled spirits (tequila, mescal) are sold around the world and are subject to local regulations regarding alcoholic beverages. The best-quality tequila, manufactured using the blue agave (*Agave tequilana* cv. Azul), is only infrequently found outside of Mexico. Similarly, tequilas that have been stored and aged for longer periods are not easily obtained outside of Mexico.

Literature

See also the entries for **alcohol**, **balche'**, **beer**, and **chicha**.

Artes de México. 1984. El tequila. *Arte Tradicional de México.* 27.

Barrios, Virginia B. de. 1984. *A guide to tequila, mezcal and pulque.* Mexico: Minutae Mexicana.

Benitez, Fernando. 1973. *Ki: El drama de un pueblo y de una planta.* Mexico City: Fondo de Cultura Econlanta.

Bye, Robert A., Don Burgess, and Albino Mares Trias. 1975. Ethnobotany of the western Tarahumara of Chihuahua, Mexico. 1: Notes on the genus *Agave. Botanical Museum Leaflets* 24 (5): 85–112.

Castetter, E. F., W. H. Bell, and A. R. Grove. 1938. The early utilization and the distribution of agave in the American Southwest. *University of New Mexico Bulletin* (Biological Series) 5 (4).

Gentry, Howard Scott. 1982. *Agaves of continental North America.* Tucson: University of Arizona Press.

Gonçalves de Lima, Oswaldo. 1956. *El maguey y el pulque en los codices Mexicanos.* Mexico City: Fondo de Cultura Economica.

Guerrero, Raúl. 1985. *El pulque.* Mexico: INAH.

Kuehne Heyder, Nicola. 1995. "Uso de alucinogenos de la huaxteca: La probable utilización de la datura en una cultura prehispanica." *Integration* 5:63–71.

Marino Ambrosio, A. 1966. The pulque agaves of Mexico. PhD thesis, Department of Biology, Harvard University, Cambridge, Mass.

Morton, Julia F. 1978. Brazilian pepper—its impact on people, animals and the environment. *Economic Botany* 32 (4): 353–59.

Nandra, K. S., and I. S. Bhatia. 1980. *In vivo* biosynthesis of glucofructosans in *Agave americana. Phytochemistry* 19:965–66.

Walker, Ann, and Larry Walker. 1994. *Tequila.* San Francisco: Chronicle Books.

"All inebriating beverages, including the hallucinogenic ones, stood under the protection of the goddess Mayahuel, who was originally only a simple farmer's wife. The myth reports how she wanted to kill a mouse in the fields one day. But the animal escaped, danced fearlessly around her, and laughed at her. Dumbfounded, Mayahuel ultimately noticed that the mouse had nipped on a maguey plant, from which a milky juice was dripping. She collected this juice and took it with her to her house so that her husband could try it. After drinking it, the two of them became cheerful and completely relaxed, and life appeared to them to be pure joy. Because they consecrated the drink to the gods, these thanked Mayahuel by naming her the goddess of pulque, while her husband became Xochipilli ('flower prince'), the lord of flowers and games."

AZTEC MYTH
IN *MEXIKANISCHE MYTHOLOGIE*
[MEXICAN MYTHOLOGY]
(NICHOLSON 1967, 69 f.*)

The pulque agave, or maguey (*Agave atrovirens* Karw.), is known in Aztec as *metl.* (From Hernández, 1942/46 [Orig. pub. 1615]*)

Alstonia scholaris (Linnaeus) R. Brown
Dita Tree

Family
Apocynaceae (Dogbane Family)

Forms and Subspecies
None

Synonyms
Echites malabarica Lam.
Echites scholaris L.

Folk Names
Chatian (Hindi), chatiun, chattiyan, chhatim (Bengali), chhation, daivappala, devil tree, devil's tree, dirita, dita (Tagalog), dita tree, ditta, elilampala, elilappalai, maddale (Kannada), milky pine (in Australia), nandani, pala (Malayalam, Tamil), palai, palimara, pulai, saittan ka jat, saptachadah, saptaparna (Sanskrit, "seven-leaved"), saptaparnah, saptaparni, satvin (Marathi, "seven-leaved"), schulholzbaum, shaitan (Arabic, "devil"), shaitan wood, tanitan, weißquirlbaum, yaksippala

History
The dita tree has been used in South Asia to manufacture writing parchment since ancient times (Miller 1988, 20*). The wood was formerly used to make writing tablets for schoolchildren (Gandhi and Singh 1991, 89*). The related species

Alstonia venenata R. Br. [syn. *Echites venenata* Roxb.] was used for similar purposes.

Although the seeds were used in the tantric cult, no traditional use of this plant as a hallucinogen is known (Scholz and Eigner 1983, 77*).

The tree is named after a professor from Edinburgh, Scotland, C. Alston (1685–1760). In Europe, the bark was once sold as a febrifuge and tonic (Schneider 1974, 1:77*).

Distribution
Although the dita tree is from India, it is now found throughout all of Southeast Asia (and Myanmar, the Philippines, and Thailand) (Padua et al. 1987, 14). It also occurs in the tropical rain forests of the east coast of Australia and on the Solomon Islands.

Cultivation
It may be possible to propagate the plant using seeds. The most successful method is to transplant young trees.

Appearance
This evergreen tree can grow to a height of 30 meters. Its bark is rough and gray. The branches are arranged circumambiently around the trunk, so that the crown looks like a parasol. The large lanceolate leaves, which can grow as long as 25 cm, are arranged in clusters of seven. The greenish yellow flowers are small and inconspicuous; the fruits hang in pairs and form thin, slightly undulating or curved pods that can grow to 20 to 45 cm in length. A sticky and bitter sap flows through the bark.

The genus *Alstonia* encompasses some forty-three species that are found in all tropical zones. Some of these are difficult to distinguish from *Alstonia scholaris*, and they are presumably often confused with one another.

Psychoactive Material
— Bark, root cortex
— Leaves
— Latex

Preparation and Dosage
In India, when the intended use is medicinal, the bark (which possesses no aphrodisiac properties) is boiled together with rice.

The seeds are preferred when the proposed use is aphrodisiac or psychoactive in nature. Two grams of the seeds are crushed and allowed to steep in water overnight. The next day, the liquid is filtered and drunk. The dosage for aphrodisiac

Above: Leaves and pseudoflowers of the dita tree (*Alstonia scholaris*).

Below: The bark of the dita tree (*Alstonia scholaris*) is rich in alkaloids.

Right: The dita tree (*Alstonia scholaris*) is both revered and feared in India and Nepal.

purposes can vary considerably among individuals. It is best to begin with 3 g per person and then slowly increase the dosage (Gottlieb 1974, 33*; Miller 1985, 11*).

The leaves of the related species *Alstonia theaeformis* (Bogotá tea) are brewed to make a tea that is consumed for its stimulating effects (Lewin 1980 [orig. pub. 1929], 352*).

Ritual Use

The tree is considered "evil" in India. Some tribal peoples do not merely fear the tree but avoid it altogether. They believe that an evil spirit dwells within the tree that can possess any person who walks underneath the tree or sleeps in its shadow. Some also believe that the guardian of the tree can bring death to those who sleep under its branches. These conceptions may be based upon the fact that the tree can induce visions. Because of this negative folklore, however, the tree has also been spared from the destruction being visited upon other tropical trees (Gandhi and Singh 1991, 89*).

The seeds of the tree play a significant role in the sexual magical practices of the Indian tantric cult. Unfortunately, little is known about this use (Miller 1988, 21 f.*).

The Australian Aborigines used the latex as an adhesive for attaching ceremonial decorations (such as feathers) to their skins for rituals (Pearson 1992, 25*). It is possible that they also knew of and utilized the psychoactive properties of the dita tree. Apart from this, we know of no traditional usages for psychoactive purposes.

Artifacts

In tantric magic, mantras (magical formulae) were written on pieces of bark parchment that were then used as amulets.

Medicinal Use

The bark is generally regarded as a tonic (Wright et al. 1993, 41). In Ayurvedic medicine, it is also used to treat fever, malaria, lower abdominal ailments, diarrhea, dysentery, digestive problems, leprosy, skin diseases, pruritus, tumors, chronic ulcers, asthma, bronchitis, and frailty. Both the latex and the tender leaves are applied externally to tumors (Sala 1993, 1:97*). In India, the bark and root cortex are boiled with rice and ingested by girls daily for one to two weeks to treat leukorrhea (Bhandary et al. 1995, 152*). In the regions of Ganjam and Godawari, it is used to treat insanity and epilepsy (Scholz and Eigner 1983, 77*), while in Nepal it is used as a febrifuge and to treat malaria (Manandhar 1980, 15*). In Assam, a cold-water extract is drunk to treat malaria (Boissya et al. 1981, 221*). In the Philippines, the bark is used as a tonic and to treat diarrhea disorders of all types. A decoction of the young leaves is drunk for beri-beri (Padua et al. 1987, 14).

The bark of the Southeast Asian species *Alstonia angustifolia* Wall., *Alstonia macrophylla* Wall. ex G. Don, and *Alstonia spathulata* Bl. is also used as a traditional treatment for malaria and as a tonic (Padua et al. 1987, 13). In Africa, the species *Alstonia congensis* Engl. and *Alstonia boonei* De Wild. are also made into medicines for treating malaria (Wright et al. 1993, 41 f.).

Constituents

The seeds contain hallucinogenic **indole alkaloids** (alstovenine, venenatine, chlorogenine, reserpine) as well as chlorogenic acid (Miller 1985, 10*). The bark through which the latex flows contains the alkaloids ditamine, echitamine (= ditaine), and echitenines (Miller 1985, 10*; Rätsch 1992, 73*). Ditamine, echitamine, alstovenine, and venenatine occur in all parts of the plant (Scholz and Eigner 1983, 77*).

Most *Alstonia* species contain **indole alkaloids** (Majumder and Dinda 1974; Mamatas-Kalamaras et al. 1975). The New Caledonian *Alstonia coriacea* Pancher ex S. Moore even contains a yohimbine derivative (Cherif et al. 1989). The Malaysian species *Alstonia angustifolia* Wall. contains thirty-one alkaloids, **yohimbine** being the primary one (Ghedira et al. 1988). The Australian species *Alstonia muelleriana* Domin contains a complex mixture of indole alkaloids (Burke et al. 1973).

Effects

The bark is alleged to have aphrodisiac and, as a result of MAO inhibition (see **ayahuasca**), psychoactive effects. The primary constituent "alstovenine demonstrates MAO-inhibiting effects in low doses and in higher doses CNS-stimulating effects, stereotypy, and spasms. In contrast, the effects of venenatine are reserpine-like [cf. *Rauvolfia* spp.]" (Scholz and Eigner 1983, 77*). Alstonia "helps retain erection and delays orgasm during intercourse" (Miller 1985, 9*).

The alkaloid echitamine is said to kill the malaria pathogen; however, it is some ten times less effective than quinine. The effects upon malaria have not yet been clearly demonstrated pharmacologically (Wright et al. 1993).

Commercial Forms and Regulations

None

Literature

See also the entries for ***Mitragyna speciosa*** and **yohimbine**.

Burke, David E., Gloria A. Cook, James M. Cook, Kathleen G. Haller, Harvey A. Lazar, and Philip W. le Quesne. 1973. Further alkaloids of *Alstonia muelleriana*. *Phytochemistry* 12:1467–74.

Cherif, Abdallah, Georges Massiot, Louisette Le Men-Olivier, Jacques Pusset, and Stéphane

"Once upon a time, in the Western Ghats or hills of India lived a shepherd called Ramu who played the flute beautifully. Every day while his goats grazed in the mountains, Ramu sat under the Chatian tree and played his flute.

Now, in this Chatian tree lived a fierce spirit. When Ramu first came to sit under the tree he was just about to strike him dead when he heard the boy's flute and was charmed by the melody.

The spirit danced among the leaves and branches. Soon, when he was used to Ramu coming every day, he ventured down from the tree and introduced himself. From then on Ramu would play and the spirit would dance in great happiness. The two became good friends."

MANEKA GANDHI AND YASMIN SINGH
A FOLKTALE FROM MADHYA PRADESH
(1989, 89–90*)

"It was in *Tantric* India that the seed of the dita tree was first used as an aphrodisiac. Use of the drug was accompanied by an exercise that prolonged erection and delayed orgasm by control of specific genital muscles."

RICHARD ALAN MILLER
THE MAGICAL & RITUAL USE OF
APHRODISIACS
(1985, 11*)

Labarre. 1989. Alkaloids of *Alstonia coriacea*. *Phytochemistry* 28 (2): 667–70.

Gandhi, Manoj, and Virender Kumar Vinayak. 1990. Preliminary evaluation of extracts of *Alstonia scholaris* bark for *in vitro* antimalarial activity in mice. *Journal of Ethnopharmacology* 29 (1): 51–57.

Ghedira, K., M. Zeches-Hanrot, B. Richard, G. Massiot, L. Le Men-Olivier, T. Sevenet, and S. H. Goh. 1988. Alkaloids of *Alstonia angustifolia*. *Phytochemistry* 27 (12): 3955–62.

Hawkins, W. L., and R. C. Elderfield. 1942. Alstonia alkaloids. II. A new alkaloid, alstoniline from *A. constricta*. *Journal of Organic Chemistry* 7:573–80.

Hu, W., J. Zhu, and M. Hesse. 1989. Indole alkaloids from *Alstonia angustifolia*. *Planta Medica* 55:463–66.

Majumder, Priya L., and Biswanath N. Dinda. 1974. Echinoserpidine: A new alkaloid of the fruits of *Alstonia venenata*. *Phytochemistry* 13:645–48.

Mamatas-Kalamaras, Stylianos, Thierry Sévenet, Claude Thal, and Pierre Potier. 1975. Alcaloïdes d'*Alstonia vitiensis* var. *novo ebudica monachino*. *Phytochemistry* 14:1637–39.

Padua, Ludivina S. de, Gregorio C. Lugod, and Juan V. Pancho. 1987. *Handbook of Philippine medicinal plants*. Vol. 1. Laguna, Luzon: University of the Philippines at Los Baños.

Wright, Colin W., David Allen, J. David Phillipson, Geoffrey C. Kirby, David C. Warhurst, Georges Massiot, and Louisette Le Men-Olivier. 1993. *Alstonia* species: Are they effective in malaria treatment? *Journal of Ethnopharmacology* 40:41–45.

Anadenanthera colubrina (Vellozo) Brennan

Cebíl, Villca

Family

Leguminosae (Legume Family); Mimosoideae Section: Eumimoseae

Forms and Subspecies

There are two geographically isolated varieties or subspecies (von Reis Altschul 1964):

Anadenanthera colubrina var. *colubrina* Altschul: only in eastern Brazil[26]

Anadenanthera colubrina var. *cebil* (Grisebach) Altschul: in the southern Andes region and neighboring areas (Argentina, Bolivia, Paraguay, Peru, southeastern Brazil)

Synonyms

Acacia cebil Grisebach
Anadenanthera excelsa Grisebach[27]
Anadenanthera macrocarpa (Benth.) Spegazzini
Piptadenia cebil Grisebach
Piptadenia colubrina Benth.
Piptadenia grata (Willd.) Macbr.
Piptadenia macrocarpa Bentham = *A. colubrina* var. *cébil*

Folk Names

Aimpä, aimpä-kid, algarobo, angico, angico do cerrado, cabuim, cebil, cébil, cebíl, cebil blanco, cebil colorado, cebilo, cevil, cevil blanco, cevil colorado, cibil, curubu'y, curupai, curupai-curú, curupaí, curupaù blanca, curupaú barcino, curupay,[28] curupáy, curupaytí, guayacán,[29] hataj (Wichi name for the **snuff**), hatax, huilca, huillca, jataj, kurupá, kurupaí, kurupaîraî, kurupayara,

quebracho,[30] sebil, sébil, sevil, tara huillca, tèék, tek (Wichi), uataj, uillca, uña de gato (Spanish, "cat claw"),[31] vilca, vilcas, villca, wilka, wil'ka, willca,[32] willka, xatax

The names used for the tree are usually the same as the names given to the **snuff** that is prepared from it.

History

The seeds of the variety known as *cebíl* were being smoked in pipes over 4,500 years ago in the Puna region of northwestern Argentina (Fernández Distel 1980).[33] Its use appears to have had a particularly profound effect upon the culture of Tiahuanaco (literally, "dwelling of the god").

Cebíl's usage as a snuff in the southern Andes is first mentioned in the *Relación* of Cristobal de Albornoz. Use as an additive to maize beer (**chicha**) was first described by Polo de Ondegardo in 1571. The Mataco Indians are said to have brewed a *vino de cebil* (cebíl **wine**) even during the twentieth century.

It is uncertain whether the reports about the use of villca seeds that have come to us from the colonial period actually do refer to the seeds of *Anadenanthera colubrina*. Even today, other trees are also referred to as *vilca* (*Acacia visco* Lorentz ex Griseb., *Aspidosperma quebracho-blanco*).

Distribution

See "Forms and Subspecies" (above). In the region of Salta (northwestern Argentina), entire forests of cebíl trees stretch across the mountains and slopes.

26 This variety does not appear to have been used for psychoactive purposes (C. Manuel Torres, pers. comm.).

27 Some authors regard this name not as a synonym but as a name for a separate species—*Parapiptadenia excelsa* (Griseb.) Burk.—popularly known as *cebil, cebil blanco, sacha cebil*, or *horco-cebil* (Santos Biloni 1990, 18*).

28 This name may be linguistically related to *curupira*, a term for a mythical protective spirit of the forest (cf. Pavia 1995, 90*).

29 In Argentina, the folk name *guayacán* is also used to refer to other hardwood trees, e.g., *Caesalpinia paraguariensis* (D. Parodi) Burkart (Santos Biloni 1990, 100*).

30 In Chile, the name *quebracho* is also used for *Cassia closiana* Phil. (= *Senna candoleana*) (Donoso Zegers and Ramírez García 1994, 38*).

31 The name *uals ria to* is used in Peru primarily to refer to the claw thorn (*Uncaria tomentosa*); one of its uses is as an *ayahuasca* additive.

32 In Peru, the white quebracho tree (*Aspidosperma quebracho-blanco*) is also called *willca* (Santos Biloni 1990, 118*).

33 During excavations in the area of Jujuy (Argentina), C. Manuel Torres found a chilamlike pipe that was some five thousand years old. The pipe still contained clearly identifiable remains of the seeds. Unfortunately, the object was lost when it was sent to Sweden for chemical analysis.

Cultivation

The dried seeds can be germinated and then planted. The tree is relatively fast growing and can be cultivated in both tropical and subtropical climate zones.

Appearance

The tree, which grows to a height of only 3 to 18 meters, has a bark that is almost black and often features conical thorns or knotty constrictions. The leaves are finely pinnate and up to 30 cm in length. The whitish yellow flowers are globose. The leathery, dark brown seedpods can grow as long as 35 cm; they contain reddish brown seeds that are 1 to 2 cm wide, very flat, and roundish to rectangular. The tree is very difficult to distinguish from the closely related *Anadenanthera peregrina* (von Reis Altschul 1964).

In the twilight of evening, the tree "goes to sleep," i.e., the pinnate leaves fold together, opening again the following morning. The stems of the leaves contain small glands that exude a sweet liquid. Certain types of ants are attracted to this and consume the nectar. At the same time, the ants destroy other pests that might pose a threat to the tree.

The tree is often confused with other species from the same family. For example, according to an oral communication from C. M. Torres, even professional botanists have incorrectly identified one tree found in the San Pedro de Atacama (northern Chile) that is also known as *vilca, Acacia visco* Lorentz ex Griseb. [syn. *Acacia visite* Griseb., *A. platensis* A. Manganaro, *Manganaroa platensis* (Mang.) Speg.], as *A. colubrina*.

The botanical identification is not always easy, as the species can exhibit considerable variation. The variety *colubrina*, for example, can form seedpods that are exactly like those of the genus *Prosopis* (von Reis Altschul 1964, 11).

Psychoactive Material

—Seeds (semen anadenanthera colubrina)

Preparation and Dosage

The ripe seeds are dried and may be lightly roasted, after which they are ground as fine as possible. As little as 150 mg to 0.5 g of this powder is effective when ingested nasally. One gram (which roughly corresponds to the weight of a large seed) represents a potent visionary dosage.

For smoking, the ripe, dried seeds are lightly roasted and then coarsely crushed. Some five to eight seeds are mixed with cut tobacco (*Nicotiana tabacum*) and occasionally with the leaves of aromo (*Amaranthus* spp., *Acacia caven* (Mol.) Molina, or *Acacia farnesiana*; cf. **Acacia spp.**) and rolled into a cigarette. An effective dosage is said to be half a cigarette per person.

When intended for oral ingestion, the seeds or the juice that is pressed from them is mixed with **chicha** and drunk. Two or three seeds can be boiled in water with the root of *Polypodium* spp. and drunk. Boiled, the seeds can also be mixed with **honey** and eaten. Another recipe calls for ingesting six ground seeds with some liquid (von Reis Altschul 1972, 38).

Ritual Use

Before the arrival of the Spanish, villca seeds must have had great ritual and religious importance in Peru, as high-ranking Andean priests and certain soothsayers *(umu)* were also known as *villca* or *vilca camayo* (Cobo 1990, 267*; Salomon and Urioste 1991, 256*; *villac* [sic] in Arriaga 1992, 31*; von Reis Altschul 1967). One Indian shrine *(huaca)* was also referred to as *villca, vilcacona,* or *vilcabamba,* "place of the villca trees" or "villca forest," and an especially sacred mountain is known as Villca Coto. The primeval survivors of a great flood retreated to the peak of this mountain (Ibid., 51*). There are numerous other examples of this kind (cf. von Reis Altschul 1972). Moreover, *villca* also appears to have been a name for **enemas**.

Villca seeds had great ritual significance as a **beer** additive in **chicha** intended for ceremonial consumption. Here, the "juice" of villca was trickled into the fermented beverage and consumed by soothsayers *(umu)* or "magicians" (= shamans) so that they could peer into the future (Cobo 1990*).

The ritual or shamanic use of snuffs made from this species of *Anadenanthera* has been documented for the following tribes: Quechua, Piro, Chiriguano, Yabuti, Atacama (Kunza), Comechingón, Diaguita, Allentiac, Millcayac, Humahuaca (Omoguaca), Ocloya, Mataco (Mataguayo, Nocten), Vilela, and Guaraní (von Reis Altschul 1972).

The oldest archaeological evidence for a ritual or shamanic use of cebíl seeds comes from the Puna region of northwestern Argentina (Fernandez Distel 1980).

The shamans of the Wichi (= Mataco), who live in northwestern Argentina, still use a snuff they call *hataj* (Califano 1975). The Mataco shamans prefer smoking the dried and roasted seeds in pipes or cigarettes over sniffing the powder. They believe that it is only through *hataj* that they can penetrate into the other reality and have an effect upon it (Arenas 1992; Califano 1975; Domínguez and Pardal 1938). In recent years, some of the Mataco have become converts to Christianity. When this occurred, they immediately equated the biblical tree of knowledge with cebíl (Arenas 1992). The Mataco, however, regard this not as a "forbidden fruit" but as the fruit of a sacred tree that the shamans use to perform healings. The shaman Fortunato Ruíz has described cebíl seeds as a "door into the other

From top to bottom:
The cebíl tree (*Anadenanthera colubrina* var. *cebil*) develops long seedpods that open in August and cover the ground with cebíl seeds. (Photographed in the cebíl forests of Salta, northwestern Argentina)

An opened seedpod of *Anadenanthera colubrina* var. *cebil*, showing the bufotenine-rich seeds.

The seeds of the southern Brazilian *Anadenanthera colubrina* var. *colubrina*.

The false villca tree, often mistaken for *Anadenanthera colubrina*, is actually an acacia (*Acacia visco*). (Photographed in San Pedro de Atacama, northern Chile)

Representation of a shamanic or ritual hunt on a Mochican ceramic vessel (ca. 500 C.E.). The stag is "hanging" in a villca tree (*Anadenanthera colubrina*), which can be clearly recognized by its seedpods and pinnate leaves.

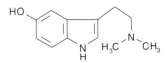

Bufotenine

34 The author believes that the seedpods are those of *Anadenanthera peregrina*; however, in pre-Columbian Peru, it is much more likely that *A. colubrina* was being used.

35 While the algorrobo (*Prosopis chilensis*) does indeed produce seedpods, neither are these segmented (as are those of *Anadenanthera*), nor do their ends taper to a small, fine point. In Peru, Chile, and Argentina, algorrobo is used as a fermenting agent for **beer** and **chicha**.

world." He smokes the seeds with tobacco and aromo—just as his ancestors did five thousand years before. Northwestern Argentina is thus the site of the longest uninterrupted tradition of ritual/shamanic use of a psychoactive or psychedelic substance on the planet.

Artifacts

Numerous pre-Columbian objects associated with snuff use (snuff trays, snuff tubes) have been found in northwestern Argentina (Puna) and northern Chile (Atacama Desert). The iconography of these objects was influenced by the visions produced by the cebíl seeds (see **snuffs**). A number of pipes made of clay have also been recovered from the region; the heads of some still contained cebíl seeds.

The petroglyphs and geoglyphs in the Atacama Desert as well as the images depicted on the ceramics of the Argentinian Puna region are clearly reminiscent of cebíl visions.

The hallucinations cebíl can induce appear to have exerted a considerable influence upon the iconography of the so-called Tiahuanaco style. The iconography of Chavín de Huantar is interwoven with similar motifs. The intertwined and entangled snakes that come out of the head of the oracle god can, for example, be interpreted as cebíl-induced hallucinations.

A two-thousand-year-old shamanic textile from the Chavín culture features depictions of seedpods that can easily be interpreted as those of *Anadenanthera colubrina* (Cordy-Collins 1982*).[34] In fact, a variety of iconographic elements in the Chavín culture have been interpreted as representations of *Anadenanthera* spp. (Mulvany de Peñaloza 1984*).

Several paintings on ceramics from the pre-Columbian Moche or Chimu include depictions of trees. The iconographic contexts and the botanical representations of these trees indicate that they may very well be construed as *Anadenanthera colubrina* (archaeologists typically interpret these as "algarrobo trees"[35] [Kutschner 1977, 14*; Lieske 1992, 155]).

In 1996, the German artist Nana Nauwald produced a painting about an experience with cebíl seeds. Entitled *Nothing Is Separate from Me,* the painting depicts the typical "wormlike" visions.

The novel *The Inca* includes a number of descriptions of psychoactive villca use (Peters 1991*).

The Mataco make bags, nets, et cetera, from agave fibers, some of which are dyed using extracts of cebíl bark. The seeds were also formerly used to make armbands.

Medicinal Use

A tea made from cebíl seeds and the root of *Polypodium* **spp.** is consumed for digestive

problems. The seeds are drunk in **chicha** as a remedy for fever, melancholy, and other mysterious afflictions. In **honey,** they are used as a diuretic or to promote female fertility (von Reis Altschul 1972, 38). At the same time, cebíl is also regarded as an abortifacient (78). The resin of the variety *colubrina* is used like gum arabic (see *Acacia* **spp.**) and is said to be effective in the treatment of coughs. Sundried seeds of the variety *colubrina* are ingested in **snuff** form to treat constipation, chronic influenza, and headaches (78).

The Mataco use a decoction of the fresh (i.e., still green) cebíl pods as a wash for the head to treat headaches.

Constituents

The seeds contain tryptamines, primarily **bufotenine**. Some varieties contain only bufotenine (Pachter et al. 1959*). One species described for Argentina, so-called *Piptadenia macrocarpa* (= cebíl), contains bufotenine (Fish and Horning 1956). Other analyses found that samples of seeds from *Piptadenia macrocarpa* contained 5-MeO-MMT, DMT, DMT-*N*-oxide, bufotenine, and 5-OH-DMT-*N*-oxide; seeds from "*Piptadenia excelsa*" contained DMT, bufotenine, and bufotenine-*N*-oxide, while seeds from "*Piptadenia colubrina*" contained only bufotenine (Farnsworth 1968, 1088*). Old samples of seeds were found to contain only 15 mg/g of bufotenine (de Smet and Rivier 1987).

According to an as yet unpublished analysis by Dave Repke, freshly harvested and quickly dried seeds from trees in northeastern Argentina (Salta) contain primarily bufotenine (over 4%) and an additional alkaloid (perhaps serotonin), but no other tryptamines or alkaloids. The same chemist found 12% bufotenine (!) in one of the samples (per oral communication from C. M. Torres).

Whether the seedpods or root cortex contains tryptamines has not yet been investigated. The ripe seedpods do contain some bufotenine.

Effects

The effects of cebíl snuff last for some twenty minutes and consist of profound hallucinations that are often in black and white, less frequently in color. These are not, or are only rarely, geometrical but are, rather, very flowing and decentralized. They are clearly reminiscent of the depictions of the Tiahuanaco culture.

When smoked, cebíl seeds also produce hallucinogenic effects that are very pronounced for approximately thirty minutes and that disappear completely within two hours. Because of the short duration of these effects, cebíl is an ideal drug for shamanic diagnoses. The effects begin with a sensation of bodily heaviness. After some five to ten minutes, visual hallucinations begin to appear when the eyes are closed. These either appear as phosphenes (entoptic or endogenous images of light that the "inner eye" sees in the form of characteristic patterns) or flow together in worm- and snakelike manners. Symmetrical and crystallographic hallucinations are less common. In rare cases, there may be strong visions having the character of reality (experiences of flying, journeys into other worlds, transformations into animals).

Experience has shown that before smoking or sniffing cebíl, it is useful to chew coca (*Erythroxylum coca*) (or sniff some **cocaine**). This helps the visions become clearer and also obviates possible side effects.

Commercial Forms and Regulations
None

Literature

See also the entries for ***Anadenanthera peregrina***, **bufotenine**, and **snuff**.

Altschul. *See* von Reis Altschul.

Arenas, Pastor. 1992. El 'cebil' o el 'árbol de la ciencia del bien y del mal.' *Parodiana* 7 (1–2): 101–14.

Brazier, J. D. 1958. The anatomy of some timbers formerly included in Piptadenia. *Tropical Woods* 108:46–64.

Califano, Mario. 1975. El chamanismo Mataco. *Scripta Ethnologica* 3 (2): 7–60.

Dasso, María Cristina. 1985. El shamanismo de los Mataco de la margen derecha del Río Bermejo (Provincia del Chaco, Republica Argentina). *Scripta Ethnologica* suppl. (5): 9–35.

de Smet, Peter A. G. M., and Laurent Rivier. 1987. Intoxicating paricá seeds of the Brazilian Maué Indians. *Economic Botany* 41(1): 12–16.

Domínguez, J. A., and R. Pardal. 1938. El hataj, droga ritual de los indios Matako: Historia su empleo en América. Ministerio del Interior, Comisión Honoraria de Reducciones de Indios (Buenos Aires), *Publicación* no. 6:35–48.

Fernández Distel, Alicia A. 1980. Hallazgo de pipas en complejos precerámicos del borde de la Puna Jujeña (Republica Argentina) y el empleo de alucinógenos por parte de las mismas cultura.

Universidad de Chile *Estudios Arqueológicos* 5:55–79.

Fish, M. S., and E. C. Horning. 1956. Studies on hallucinogenic snuffs. *The Journal of Nervous and Mental Disease* 124 (1): 33–37.

Flury, Lázaro. 1958. El Caá-pí y el Hataj, dos poderosos ilusiógenos indígenos. *América Indígena* 18 (4): 293–98.

Giesbrecht, A. M. 1960. Sobre a ocorrência de bufotenina em semente de *Piptadenia falcata* Benth. *Anais da Associação Brasileira de Química* 19:117–19.

Granier-Doyeux, Marcel. 1965. Native hallucinogenic drugs *Piptadenias*. *Bulletin on Narcotics* 17 (2): 29–38.

Lieske, Bärbel. 1992. *Mythische Bilderzählungen in den Gefäßmalereien der altperuanischen Moche-Kultur*. Bonn: Holos Verlag.

Rendon, P., and J. Willy. 1985. Isolation of bufotenine from seeds of the *Piptadenia macrocarpa* Benth. *Revista Boliviana de Química* 5:39–43.

Torres, Constantino Manuel, and David Repke. *Anandenanthera* (monograph in preparation).

———. 1998. The use of *Anadenanthera colubrina* var. *cebil* by Wichi (Mataco) shamans of the Chaco Central, Argentina. *Yearbook for Ethnomedicine and the Study of Consciousness,* 1996 (5): 41–58. Berlin: VWB.

von Reis Altschul, Siri. 1964. A taxonomic study of the genus *Anadenanthera*. *Contributions from the Gray Herbarium of Harvard University* 193:3–65.

———. 1967. Vilca and its uses. In *Ethnopharmacologic search for psychoactive drugs,* ed. Daniel H. Efron, 307–14. Washington, D.C.: U.S. Dept. of Health, Education, and Welfare.

———. 1972. *The genus* Anadenanthera *in Amerindian cultures*. Cambridge: Botanical Museum, Harvard University.

Wassén, S. Henry, and Bo Holmstedt. 1963. The use of paricá: An ethnological and pharmacological review. *Ethnos* 28 (1): 5–45.

An experience with cebíl:

"We darkened the room of our bungalow in the rainforest. The powder was relatively simple and unproblematic to sniff into the nose. It does not burn like others (e.g., *Anadenanthera peregrina*). The slight prickling in the nose is tolerable.

"At first I noticed how my body, especially the arms and legs, became heavy like lead; but the body sensation was warm and very pleasant (it was somewhat reminiscent of the initial effects of ketamine). I closed my eyes and waited for the coming effects with anticipation. After about five minutes, dancing phosphenes swirled before my eyes. The hopping and jumping points of light joined together into flowing forms and structures. It was as if the floodgates of the universe had been opened: Flowing patterns crashed into my field of view. From every point flowed streams and rivers of threads of light that quickly intertwined in and throughout one another, always in and throughout one another. And all this with incredible speed.

"Flowing patterns, yes, exactly the patterns that shoot out of the head of the god of Tiahuanaco! I then knew that it must have been this exact same snuff that had provided the inspiration for the artists of Tiahuanaco.

"The quickly changing patterns turned into a chaotic river of spermatozoa. They twisted and darted and shot in every direction, as if they—almost aggressively—wanted to fertilize the entire universe. After this appeared geometric figures that came forth from the depths of space and fell tunnel-like into my field of view.

"Up until now, I had not seen any colors. But now I had visions in pale color. The speed of the visions decreased, and suddenly they were over.

"As I opened my eyes in the darkened room, the brightness around me suddenly changed. For a moment I felt a trace of nausea. I had to burp, and then everything was wonderful. It was a truly new visionary experience. The effects lasted for a total of about 25 minutes."

Anadenanthera peregrina (Linnaeus) Spegazzini

Cohoba, Yopo

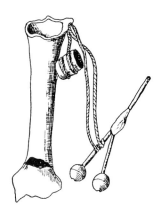

A device for sniffing niopo powder from *Anadenanthera peregrina*, used by the Guajibo Indians of the Upper Orinoco, Venezuela. (From Hartwich, *Die menschlichen Genußmittel*, 1911)

Left: The bark of *Anadenanthera peregrina* is often warty in nature. This provides an easy way to distinguish it from the closely related and very similar *Anadenanthera colubrina*.

Center: The typical finely pinnate leaves of *Anadenanthera peregrina*.

Right: Pods and seeds of *Anadenanthera peregrina*, collected in Guyana.

Family
Leguminosae (Legume Family); Mimosoideae
Section: Eumimoseae

Forms and Subspecies
There are two geographically isolated varieties:
Anadenanthera peregrina var. *peregrina* Altschul: northern Brazil to the Antilles
Anadenanthera peregrina var. *falcata* (Benth.) Altschul: South America (in Brazil, only in the east)

Synonyms
Acacia angustiloba DC.
Acacia microphylla Willd.
Acacia niopa (Kunth) Humb.
Acacia niopo Humb. et Bonpl.
Acacia paniculata Willd.
Acacia peregrina Willd.
Inga niopa Willd.
Mimosa (?) *acacioides* Benth.
Mimosa acacioides Schombrugk
Mimosa niopo Poir.
Mimosa peregrina L.
Piptadenia falcata Spegazzini
Piptadenia niopo Spruce
Piptadenia peregrina (L.) Benth.

Folk Names
Acuja, ai'yuku, akúa, a'ku:duwha, algarroba de yupa, angíco, angico rosa, angico vermelho, anjico, black parica, bois écorce, bois rouge, cahoba, cajoba, candelón, caobo, cehobbâ, cogiba, cogioba, cohaba, cohiba, cohoba, cohobba, cohobbû coiba, cojiba, cojobilla, curuba, curupa, curupá, dópa, ebãnã, ebena, hakúdufha, hisioma, iopo, jop, khoba, kohobba, niopa, niopo, niupo, noopa, nopa, nopo, nupa, ñiopo, ñope, ñopo, ñupa, parica, paricá, paricachí, paricarama, parica rana, paricauva, savanna yoke, tabaco-rapé, tan bark, yacoana, yarupi, yarupio, yoco, yop, yopa, yopo, yópo, yoto, yu'a', yu'ä, yupa, yuuba, zumaque

History
Archaeological remains of seeds that were definitely used in ritual contexts have been found in Brazil, Chile, Colombia, Peru, Haiti, the Dominican Republic, and Puerto Rico (Ott 1996).

The **snuff** known as *cohoba*, which is made from *Anadenanthera peregrina*, was mentioned several times in early colonial sources, e.g., by Fra Bartolomé de las Casas (Safford 1916). In the early sixteenth century, Gonzalo Fernández de Oviedo y Valdés was the first to note that the powder was obtained from the seeds of a tree belonging to the legume family (Torres 1988). The island of Cuba was apparently named after *cohoba*.

The first botanical description of the tree was provided by Linnaeus in 1753.

Distribution
The tree thrives only in the tropics, where it prefers drier locations such as savanna-like regions (grasslands), open plains, and fallow lands. It grows best in sandy and/or clay soils. In South America, it occurs naturally in Brazil, Guyana, Colombia, and Venezuela. The tree had already been planted on some Caribbean islands prior to the arrival of the Spanish, and it now grows wild in these areas. The relatively rare variety *falcata* occurs only in southern Brazil and Paraguay.

It is possible that these *Anadenanthera* grow even in Belize, in Central America (per oral communication from Rob Montgomery).

Cultivation
The ripe, dried seeds are easy to germinate and plant. The tree requires poor and relatively dry soil. It can be started in the moist tropics but quickly dies.

Appearance
This tree grows only to a height of 3 to 18 meters. It has a gray to black bark that is often covered with conical thorns. The leaves are finely pinnate

and up to 30 cm long. The flowers are small and globose. The leathery, dark brown seedpods, which can grow as long as 35 cm, contain very flat and roundish seeds that are reddish brown in color and 1 to 2 cm across.

The tree is easily confused with *Anadenanthera colubrina*.

Psychoactive Material
—Seeds
—Seedpods (with seeds)
—Bark (used by the Yecuana; von Reis 1991)

Preparation and Dosage
The ripe, dry seeds are usually lightly roasted and then ground into a fine, grayish green powder that is often mixed with an alkaline plant ash or ground snail shells and other additives (e.g., tobacco). The addition of the alkaline substances liberates the alkaloids (Brenneisen n.d.).

The Otomac collect the seedpods, which they then break, moisten, and allow to ferment. These are then mixed with manioc flour (*Manihot esculenta* Crantz) and slaked lime from various species of land snails, kneaded to a paste, and heated over a fire. The dried product is ground into a fine powder before being used as a snuff.

The Maué produce their snuff, which they call *paricá*, from the seeds of the variety *peregrina*, the ashes of an unidentified vine, and the leaves of an *Abuta* (*Abuta* is an **ayahuasca** additive) or *Cocculus* species.

The dosage is usually determined by the sniffing instrument that is used.

> The indigenous peoples of the Amazon region knew of the technique of caoutchouc production [from the latex of *Hevea* spp.] long before the arrival of the Conquistadors. The Omogua, for example, used vessels of caoutchouc that they would fill with an inebriating agent [*Anadenanthera peregrina* powder]. A hole was drilled in the bottom, through which they introduced a tube for removing the inebriant and blowing it in one another's nostrils. (Pavia 1995, 137*)

The minimum dosage is approximately 1 g of seeds (when applied nasally). The snuff can be ingested in a series of dosages. The ground seeds are also administered in the form of an **enema**.

Ritual Use
Many tribes use the roasted seeds to manufacture **snuffs** that are used for shamanic purposes and that hunters also ingest to locate their prey. The Taino made great use of this powder during their healing rituals and tribal celebrations (Rouse 1992; Torres 1988). The shamanic use of both varieties of this species has been documented for

the following tribes: Arawak, Guajibo, Cuiva-Guajibo, Maipure, Otomaco, Taino, Tukano, Yanomamö/Waika, Yecuana, Ciguayo, Igneri, Chibcha, Muisca, Guane, Lache, Morcote, Tecua, Tunebo (= Tama), Achagua, Caberre (Cabre), Cocaima, Piapoco, Arekana, Avane, Bainwa, Bare, Carutana, Catapolitani, Caua, Huhuteni, Ipeca, Maipure, Siusi, Tariana, Airico, Betoi, Jirara (Girara), Lucalia, Situfa (Citufa), Otomac, Pao, Saruro, Sáliva, Yaruro, Chiricoa, Puinave, Guaipunavo, Macú, Guaharibo, Shirianá, Yecuana, Omagua, Mura, Maué, Mundurucú, and various tribes in Paraguay.

Artifacts
The Caribbean Taino carved figures of gods from the hard and durable wood of *Anadenanthera* (von Reis 1991). Numerous objects of snuff paraphernalia have been discovered in the Dominican Republic (Alcina Franch 1982). One of these is a sniffing tube in the form of a naked woman who is spreading her legs and wearing a death's-head. In order to use this tube, you must place the skull against your nose. The other end, the opening of the vagina, is used to take in the snuff (Rouse 1992).

A recording of a snuff ritual with *epená* was published under the name *Hekura—Yanomamö Shamanism from Southern Venezuela* (London: Quartz Publications, !QUARTZ004, 1980).

Donna Torres has produced a painting of *Anadenanthera peregrina* (published on the book cover of Ott 1995*).

A science-fiction story by Reinmar Cunis (1979) entitled *Zeitsturm* [Time Storm] involves journeying between realities, made possible by tryptamine derivatives from "*Piptadenia peregrina*."

Medicinal Use
Both varieties produce a resin that resembles gum arabic (see *Acacia* spp.) in appearance and is used in the same manner. A decoction of the bark of the variety *peregrina* is used to treat dysentery and gonorrhea. The variety *falcata* is used to treat pneumonia.

Constituents
The seeds of both varieties contain the tryptamines *N,N*-DMT, 5-MeO-DMT, and 5-OH-DMT (= **bufotenine**) as well as their *N*-oxides. Traces of β-**carbolines** have also been detected (Ott 1996).

The presence of appreciable quantities of bufotenine is characteristic of this species (Stromberg 1954). Only bufotenine could be detected in old seed material (from Spruce's collection) (Schultes et al. 1977). It is possible that bufotenine may accumulate through the hydrolysis of *N,N*-DMT and 5-MeO-DMT when the seeds are stored.

> "'Piptadenia peregrina,' he said in a monotone voice, 'that is the key.'..."
>
> REINMAR CUNIS
> *ZEITSTURM* [TIME STORM]
> (1979, 205)

5-MeO-DMT

β-carboline

The cover of a German science-fiction novel, in which a drug obtained from *Anadenanthera peregrina* plays a central role.

The bark also contains *N*-methyltryptamine, 5-methoxy-*N*-methyltryptamine, and 5-methoxy-*N,N*-dimethyltryptamine (Legler and Tschesche 1963). Another analysis found that the bark contains MMT, 5-MeO-MMT, DMT, and 5-MeO-DMT (Farnsworth 1968, 1088*). The seedpods also contain DMT.

Effects

When ingested nasally, the snuff induces psychedelic effects and produces multidimensional visions. Experiences of ego dissolution, death and rebirth, transformations into animals, and flying are common. The effects of the snuff last for some ten to fifteen minutes, although aftereffects may be noticeable for up to an hour.

During medicinal and pharmacological experiments, it was difficult to recognize the psychoactive effects (Turner and Merlis 1959).

Commercial Forms and Regulations

None

Literature

See also the entries for **Anadenanthera colubrina** and **snuff**.

Alcina Franch, José. 1982. Religiosidad, alucinogenos y patrones artisticos tainos. *Boletin de Museo del Hombre Dominicano* 10 (17): 103–17.

Brenneisen, Rudolf. n.d. Anadenanthera. In *Hagers Handbuch der pharmazeutischen Praxis,* 5th ed. Suppl. vol. Berlin: Springer. (in press).

Coppens, Walter, and Jorge Cato-David. 1971. Aspectos etnograficos y farmacologicos el yopo entre los Cuiva-Guajibo. *Antropología* 28: 3–24.

Cunis, Reinmar. 1979. *Zeitsturm.* Munich: Heyne.

Fish, M. S., N. M. Johnson, and E. C. Horning. 1955. *Piptadenia* alkaloids: Indole bases of *P. peregrina* (L.) Benth. and related species. *Journal of the American Chemical Society* 77:5892–95.

Legler, Günter, and Rudolf Tschesche. 1963. Die Isolierung von *N*-Methyltryptamin, 5-Methoxy-*N*-methyltryptamin und 5-Methoxyl-*N,N*-dimethyltryptamin aus der Rinde von *Piptadenia peregrina* Benth. *Die Naturwissenschaften* 50:94–95.

Ott, Jonathan. 1996. *Anadenanthera peregrina* (Linnaeus) Spagazzini. Unpublished file from electronic database. Jalapa, Veracruz. Cited 1998.

Rouse, Irving. 1992. *The Taínos: Rise and Decline of the people who greeted Columbus.* New Haven and London: Yale University Press.

Safford, William E. 1916. Identity of *cohoba,* the narcotic snuff of ancient Haiti. *Journal of the Washington Academy of Sciences* 6:547–62.

Schultes, Richard Evans, Bo Holmstedt, Jan-Erik Lindgren, and Laurent Rivier. 1977. De Plantis Toxicariis e Mundo Novo Tropicale Commentationes XVIII: Phytochemical examination of Spruce's ethnobotanical collection of *Anadenanthera peregrina. Botanical Museum Leaflets* 25 (10): 273–87.

Stromberg, Verner L. 1954. The isolation of bufotenine from *Piptadenia peregrina. Journal of the American Chemical Society* 76:1707.

Torres, Constantino Manuel. 1988. El arte de los Taíno. In *Taíno: Los descubridores de Colón,* ed. C. M. Torres, 9–22. Santiago: Museo Chileno de Arte Precolombino.

Turner, William J., and Sidney Merlis. 1959. Effect of some indolealkylamines on man. *A.M.A. Archives of Neurology and Psychiatry* 81:121–29.

von Reis, Siri. 1991. *Mimosa peregrina* Linnaeus, species plantarum 520. 1753. *Integration* 1:7–9.

Areca catechu Linnaeus

Betel Palm

Family

Arecaceae, Palmae (Palm Family); Subfamily Ceroxylinae-Arecineae, Areceae Tribe

Forms and Subspecies

Numerous forms and varieties have been described, although these may represent only local races (cf. Raghavan and Baruah 1958):

Areca catechu f. *communis* (Philippines)
Areca catechu var. *alba* (Sri Lanka)
Areca catechu var. *batanensis* (Philippines)
Areca catechu var. *deliciosa* (India)
Areca catechu var. *longicarpa* (Philippines)
Areca catechu var. *nigra* (Java)
Areca catechu var. *silvatica* (may be the wild form)

Often, the local people give their own names to the "varieties." These are usually based upon the appearance and size of the seeds and appear to have no botanical relevance. The cultivated palm is likely derived from *Areca catechu* var. *silvatica*.

The people of Sri Lanka make a distinction between the varieties *hamban-puwak,* which has long oval nuts, and *rata-puwak* or *Batavia-puwak,* which produces large round nuts (Macmillan 1991, 427*).

Synonyms

Areca guavaia nom. nud.

Folk Names

Adike, arbor areka, areca, areca nut palm, arecanut tree, arecapalme, arecca, arekapalme, arekpalme, arequero (Portuguese), aréquier, aréquir, arreck, ataykkamaram, avellana d'India, betelnußpalme, betelnut tree, betel palm, buoga, bynaubaum, catechupalme, fobal, fufal (Arabic), fûfal, ghowa, gooroaka, goorrecanut palm, gouvaka (Sanskrit), gurvaca, kamuku, kamunnu, kavunnu (Malayalam), mak, noix d'arec, paan supari, pak-ku, pakkumaram (Tamil), pan of India, papal (Persian), pinang (Malay), pinangpalme, ping-lang, pinlang, puga, pugah (Sanskrit), puwak, pynan, pynanbaum, sopari (Hindi), supari, surattu supary, tambul, tuuffel (Arabic)

The palm *Chrysalidocarpus lutescens* H. Wendl. [syn. *Areca lutescens* hort. non Bory] is often sold as an ornamental under the name "areca palm" (Bärtels 1993, 39*).

History

The name *areca,* which means "cavalier," may be derived from the Kanarese word *adeke* or the Malayalam *adakka.* In early Sanskrit works, the palm is referred to as *gouvaka.* It was already mentioned in Jataka and Pali writings. The first description of the palm, however, is purportedly that of Herodotus (ca. 340 B.C.E.). Later, both the palm and the chewing of betel were more or less precisely discussed by many Arabic and European travelers (e.g., Abd Allah Ibn Ahmad, Marco Polo, Vasco da Gama, Garcia da Orta, Abul Fazal, Jacobus Bontius) in their travel reports. The British traveler R. Knox, in his *Historical Relation of the Island of Ceylon* (London, 1681), was obviously impressed, and he described both the use of the betel nut and its economic significance. The first European pictorial representation of the betel nut is a copperplate engraving by Carolus Clusius in *Aromatum et simplicium aliquot medicamentorum . . . historia* (Antwerp, 1605).

Distribution

Almost all betel palms have been planted by humans. The origin of the assumed wild form has not been fully ascertained, although it may have come from the Sunda Isles or the Philippines (cf. Raghavan and Baruah, 1958). Since it can thrive only in regions with tropical rain forests, it is limited to such areas in Hindustan, Indochina, Pakistan, Sri Lanka (Ceylon), the Maldives, Madagascar, Egypt, East Africa, Arabia, southern China, Taiwan, Indonesia, Malaysia, Fiji, and Melanesia (Stewart 1994, 39*). Betel palms grow wild in Malabar (India).

Cultivation

The betel palm is grown primarily for its seeds (betel nuts), although it is also planted as an ornamental. Avenues lined with betel palms are typical features of most palaces and parks in India.

The betel palm can be grown in a variety of soil types. Cultivation is performed using pregerminated seeds. The saplings need to grow in the shade, as they may otherwise fall victim to the intense tropical sun. It is for this reason that trees that grow quickly and provide shade (e.g., *Erythrina indica* Lam.; see **Erythrina spp.**) are first planted in betel palm plantations.

The palms bear fruit when they are ten to fifteen years of age. Typically, only the ripe fruits are harvested. One palm can bear fruit for forty-five to seventy years (Raghavan and Baruah 1958, 328). Cultivated betel palms are often infected by fungi, especially *Ganoderma lucidum* (Leys.) Karst. (see **"Polyporus mysticus"**) (Raghavan and Baruah 1958, 330f.).

Appearance

This fan palm can grow as tall as 25 meters and develop a trunk between 30 and 50 cm in

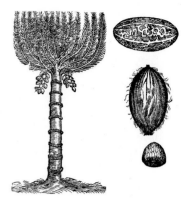

In Europe, the betel plant was once regarded as a species of date palm that is able to make a person "drunk." Although this botanical illustration is inaccurate, the betel nut itself is depicted as it occurs in nature. (Woodcut from Gerard, *The Herball or General History of Plants,* 1633*)

Betel palms (*Areca catechu*) can attain a stately height.

Left: The tuftlike inflorescence of the betel palm (*Areca catechu*).

Above center: Betel "nuts" are actually the seeds of the areca fruit (*Areca catechu*).

Below center: Fermented and colored betel nuts from Varanasi (India).

Above right: Slicing betel nuts reveals the astonishing, fractal-like structure of their natural inner world.

Below right: A typical leaf of *Areca triandra*, the nuts of which can be used as a betel substitute.

36 "In addition, the following substitutes were also mentioned: among the Weddas on Ceylon, the bark of the mora tree (*Nephelium longana* Camb.); on the Philippines, the bark of **Psidium guajava** Raddi; in Cochin, China, the poisonous roots of *Derris elliptica* Lour., which are otherwise used as an arrow poison. Finally, Ibn Baithar (thirteenth century) names red sandalwood and coriander as substitutes for areca nut, although he does not specifically refer to their use in betel chewing" (Hartwich 1911, 529*).

diameter. The loculate fronds grow to some 2 meters in length. The male and female flowers are found in spadices located below the leaves. The palm can produce up to three such spadices, each of which yields 150 to 200 fruits. The ovoid fruits, which can be as long as 7 cm in length, contain one brown, reticulate seed (the endosperm, or actual betel nut) that can weigh from 3 to 10 g.

The betel palm is easily confused with the Caribbean king palm (*Roystonea regia*; cf. Anzeneder et al. 1993, 33*) and with some species of the genus *Veitchia*, found in the Philippines and Oceania (Stewart 1994, 196*). It is difficult to distinguish from the closely related species *Areca triandra* Roxb. (India) and *Areca vestiaria*.

Psychoactive Material
—Areca nuts (arecae semen, formerly semen arecae, nuces arecae); also known as betel nut, areca nut, noix d'arec, puwag

In Sri Lanka (Ceylon), the seeds of the closely related species *Areca concinna* Thwaites are sometimes chewed as a substitute for the true betel nut (Raghavan and Baruah 1958, 318). In the Philippines, the seeds of another related species, *Areca ipot* (known as *bungang-ipot*), are used as a substitute (Stewart 1994, 40*). The seeds of the palm *Areca laxa* Ham. serve as a substitute on the Andaman Islands, while *Areca nagensis* Griff is used in Bengal and *Areca glandiformis* Lam. and *Calyptrocalyx spicatus* Blume are used on the Moluccas for the same purpose (Hartwich 1911, 529*). In Assam, the seeds of *Gnetum montanum* Mark. [syn. *G. scandens* Roxb. (Gnetaceae)], known locally as *jagingriube*, are chewed as a substitute for areca nuts (Jain and Dam 1979,

54*). In India, the bark of *Loranthus falcatus* L. (Loranthaceae) is used as a substitute for areca nuts and has narcotic effects. The fruits of *Pinanga dicksonii* Blume are also used as an areca substitute in India, while the fruits of *Pinanga kuhlii* Blume are used in the Malay Archipelago for the same purpose (Hartwich 1911, 529*).[36]

In many areas of India, freshly harvested betel nuts are preferred. In order to maintain their freshness, these may be stored for several months in a vessel full of water. When the nuts dry, they become very hard and can then be chewed only with difficulty. Sometimes, however, even dried betel nuts can be found in the market. These are dried in the sun for six to seven weeks before sale (as so-called *chali* nuts). In Malaysia, cracked betel nuts are smoked with gum benzoin, which imparts to them a pleasant aroma; these are sold in the markets as *pinang ukup* (see **incense**). In addition, whole, ripe, dried nuts (*pinang kossi*); halved, dried nuts (*pinang blah*); smoked nuts (*pinang salai*); and semi-ripened, salted nuts (*pinang asin*) are also sold in the markets.

Sometimes, nearly ripe betel nuts are harvested and boiled in a decoction of betel leaves (**Piper betle** L.); pieces of bark from *Syzygium jambolanum* DC., *Pterocarpus santalinus* L., *Adenanthera pavonia* L., and *Ficus religiosa* L.; and some slaked lime and oils. This lends them a reddish color (from the red sandalwood) and a beautiful luster. Such nuts have a more aromatic taste and remain soft for a longer period of time (Raghavan and Baruah 1958, 332 f.).

Occasionally, freshly harvested, tender, unripe nuts are boiled in a solution of lime, dried, and exported. Cut into slices, these nuts are sold under the name *kali* (Macmillan 1991, 427*).

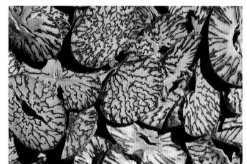

Preparation and Dosage

Betel nuts have their greatest ethnopharmacological significance as the primary ingredient in **betel quids**.

Fermenting the fruits can even produce an areca **wine** (Raghavan and Baruah 1958, 316). Leaves that have been inoculated with beer yeast (*Saccharomyces cerevisiae*) are employed for alcoholic fermentation.

One betel quid contains approximately one-quarter to one-half nut. According to Roth et al. (1994, 141*), the maximal individual dosage is 4 g. Eight to 10 g of powdered seed can be sufficient to produce lethal effects.

A dosage of 2 mg of the isolated main alkaloid, arecoline, produces strong stimulating effects. The individual dosage should not exceed 5 mg.

Ritual Use

The most important ritual use of the betel nut occurs in ceremonies involving **betel quids** (cf. also *Piper betle*).

In Melanesia, betel nuts are considered magical once a magician has uttered an appropriate formula over them. They then carry the magical power of the words in themselves and can transfer this to a goal (a person, an action, an object). They are often used as carriers of love magic.

In India, the flowers of the betel palm are one of the flowers used as ceremonial offerings. The tree itself is symbolically venerated as Ganesha (Gupta 1991, 79*).

The leaves of the betel palm also have ritual significance. They are used in Buddhist ceremonies and during initiations. On Sri Lanka, watertight bowls are woven from the leaves, and newborn boys are ritually bathed in these.

In Southeast Asia (Indonesia), betel palm leaves are placed before the door of a newlywed couple and attached to their house as a sign of honor (Meister 1677, 57*).

Artifacts

In India, the hard areca nuts are carved into small bottles or containers for storing **incense**.

The palm is occasionally found depicted in Indian and Thai art.

Medicinal Use

In India, betel nuts are administered primarily to dispel tapeworms (Raghavan and Baruah 1958, 338). Betel nuts were also once a popular anthelmintic in Europe, especially in veterinary medicine (Macmillan 1991, 426*; Pahlow 1993, 430*). They also found use in folk medicine for diarrhea and similar ailments.

Betel nuts are used for a variety of purposes in Ayurveda and Unani, the two traditional medical systems of India and its neighboring regions. They are administered to treat digestive problems and

nervous disorders. A decoction of them is also esteemed as a tonic and aphrodisiac (especially in combination with other substances) (Raghavan and Baruah 1958, 338). Similar uses of betel nuts can be found in traditional Chinese medicine and in Cambodia. Malay magicians and poisoners use a mixture of betel nuts and opium (see *Papaver somniferum*) to poison and rob their victims.

In Iran, areca nuts are mixed with sugar and coriander and administered to induce labor (Hooper 1937, 86*).

Constituents

The seeds contain various alkaloids (0.3 to 0.6%) of a relatively simple chemical structure: 0.1 to 0.5% arecoline (primary alkaloid), as well as arecaine, arecaidine, arecolidine, guvacoline, isoguvacine, and guvacine. Tanning agents (tannins: galotannic acid, gallic acid, D-catechol, phlobatannin), mucilage, resin, carbohydrates (saccharose, galactan, mannan), proteins, saponines, carotene, minerals (calcium, phosphorus, iron), and fat (sitosterol) are also present (Raghavan and Baruah 1958, 335 ff.). When betel nuts are chewed in combination with slaked lime, the alkaloid arecoline is transformed into arecaidine.

Recently, *Areca* seeds were discovered to contain four new polyphenolic substances (NPF-86IA, NPF-86IB, NPF-86IIA, NPF-86IIB) that may be able to inhibit a membrane-bound enzyme (5'-nucleotidase) (Uchino et al. 1988).

Effects

Arecoline, the primary alkaloid, is a parasympathomimetic. It has stimulating effects, strongly promotes salivation, and has anthelmintic

An inebriating beverage can be produced by allowing the fresh fruits of the betel palm to ferment.

"Its fruits are like nutmegs, when this fruit areca is broken out of its red-yellow mold. It is universally esteemed by the inhabitants of India, from kings to the lowest beggar, because they chew this fruit, both green and dry, smeared with betel flowers and a little lime from shells, more because it is a custom passed down from their ancestors than from necessity.

"And it is true that it imparts a well-scented breath and purple-red lips. It is for this reason that the Portuguese ladies do not wish to kiss any European man, regardless of how disgusting they might not otherwise be, before they have chewed this fruit, or one generally known as betel, claiming that the Dutch or the Germans stink from their throats when they do not chew this fruit."

GEORGE MEISTER
DER ORIENTALISCH-INDIANISCHE KUNST- UND LUSTGÄRTNER [THE ORIENTAL-INDIAN ART AND PLEASURE GARDENER] (1677, CH. 8, 1*)

Areca vestiaria is easily mistaken for the betel palm. Its fruits and seed may also contain the stimulating substance arecoline.

An ancient Chinese illustration of pin-lang, the betel palm, together with its inflorescence. (From the *Nan-fang ts'ao-mu chuang* [Plants of the Southern Regions], early fourth century C.E.)

Arecoline

"After a few days, a great festival was to take place. The people came to the festival from near and far. Mongumér-anim [a primeval being, a culture hero] was supposed to kill the pig, but during the night before the festival Mana seduced Mongumér-anim's wife. For this reason, he was afraid of Mongumér-anim and would not let his club leave his hands. During the night . . . , when the singing was well under way, Mana used the opportunity to kill Mongumér-anim. He gave him a blow to the head with a club and then fled. . . . The people mourned Mongumér-anim. His Nakari [the unmarried girls of his totemic group] wrapped him in eucalyptus bark and placed him in his grave. The next morning, an areca palm had grown from the grave, a beautiful, slender tree that already bore ripe fruit and that had previously been unknown. All of the people came by and admired the tree and tried its nuts. . . . From that time on, it has been customary to chew betel nuts."

New Guinea origin myth in *Die Marind-anim von Holländisch-Süd-New-Guinea* [The Marind-Anim of Dutch Southern New Guinea] (Wirz 1922, 10:126)

(worm-killing) properties; it can also induce brachycardia (deceleration of the heartbeat) and tremors. Eight to 10 g of the seed can be lethal, death resulting from cardiac or respiratory paralysis (Roth et al. 1994, 140*). The poly-phenolic substances (NPF-86IA, NPF-86IB, NPF-86IIA, NPF-86IIB) have tumor-inhibiting and immune-strengthening effects (Uchino et al. 1988). The oil of areca nuts has antifertility properties (Roth et al. 1994, 140*). An aqueous extract strengthens the body's own immune system (Raghavan and Baruah 1958, 339). As for the psychoactivity of the pure areca nut:

> The effects of the common areca nut are only slight, resulting at most in a sense of dizziness that is short in duration. However, there are some forms that can have strong toxic effects. The seed of *Areca catechu* L. var. *nigra* from Java (*akar pining hitam*) produces narcolepsy and sedation and can cause death. Other forms have inebriating effects: such as one from Burma known as "*toung-noo*," one from the Moluccas known as "*pining-mabok*," and another from Ceylon. (Hartwich 1911, 528 f.*)

Commercial Forms and Regulations

"Since betel is nonaddictive, it does not appear on any of the international lists of addictive drugs" (Roth et al. 1994, 141*). Betel nuts are freely sold and easily available in all the countries of Asia. In Europe, they are occasionally available in pharmacies.

Literature

See also the entry for **betel quids** as well as Balick and Beck 1990*; there is also a specialized journal entitled *Arecanut and Spices Bulletin*.

Bavappa, K. V. A., ed. 1982. *The areca nut palm.* Kasaragod: Central Plant Crop Research Institute Publication.

Chang, C. S. C., and C. E. De Vol. 1973. The effects of chewing betel nuts in the mouth. *Taiwania* 18 (2): 123–41.

Chaudhuri, S. K., and D. K. Ganguly. 1974. Neuromuscular pharmacology of harmine and arecoline. *Indian Journal of Medical Research* 62 (3): 362–66.

Johnston, G. A. R., P. Krogsgaard-Larsen, and A. Stephanson. 1975. Betel nut constituents as inhibitors of γ-aminobutyric acid uptake. *Nature* 258:627–28.

Raghavan, V., and H. K. Baruah. 1958. Arecanut: India's popular masticatory—history, chemistry, and utilization. *Economic Botany* 12: 315–45. (Contains an excellent bibliography of older works.)

Rätsch, Christian. 1996. Betel, die Palme mit der erregenden Frucht. *Dao* 5/96:68.

Uchino, Keijiro, Toshiharu Matsuo, Masaya Iwamoto, Yasuhiro Tonosaki, and Akira Fukuchi. 1988. New 5'-nucleotidase inhibitors, NPF-86IA, NPF-86IB, NPF-86IIA, and NPF-86IIB from *Areca catechu*. Part I. Isolation and biological properties. *Planta Medica* 54:419–25.

Wirz, Paul. 1922. *Die Marind-anim von Holländisch-Süd-New-Guinea.* Vols. 10 and 16. Hamburg: Abhandlungen aus dem Gebiet der Auslandskunde, Völkerkunde, Kulturgeschichte und Sprachen.

Argemone mexicana Linnaeus
Mexican Prickly Poppy

Family
Papaveraceae (Poppy Family)

Forms and Subspecies
In addition to the common yellow-blooming *Argemone mexicana* L. var. *typica* Prain, there is a white-blooming form that is known as *chicalote* in Mexico and is usually referred to as *Argemone mexicana* L. var. *ochroleuca* Sweet (Martínez 1987, 1050*). Another form that is almost thornless has been described under the name *Argemone mexicana* L. f. *leiocarpa* (Greene) G.B. Ownb. (Lucas 1962, 3; Grey-Wilson 1995, 74*).

There is only one named cultivar, notable for its very large and beautiful flowers (Grey-Wilson 1995, 74*): *Argemone mexicana* L. cv. Yellow Lustre.

Three previously described varieties are now recognized as species in their own rights (Grey-Wilson 1995, 75, 78*):
Argemone mexicana var. *hispida* Wats. = *Argemone munita* Dur. et Hilg.
Argemone mexicana var. *rosea* (Hook.) Reiche = *Argemone rosea* Hook.
Argemone mexicana var. *rosea* Coulter ex Greene = *Argemone sanguinea* Greene

Synonyms
Argemone alba var. *leiocarpa* Fedde
Argemone leiocarpa Greene
Argemone mexicana L. var. *leiocarpa* Prain
Argemone mexicana var. *ochroleuca* Britton
Argemone mucronata Dum.
Argemone ochroleuca Sweet[37]
Argemone ochroleuca L. var. *barclayana* Prain
Argemone spinosa Moench
Argemone sulphurea Sweet ex London
Argemone versicolor Salisb.
Ectrus mexicanus Nieuwland
Papaver spinosum Bauhin

Folk Names
Amapolas del campo (Spanish, "field poppy"), Bermuda thistle, bhatbhant (Hindi), bird-in-the-bush, brahmadanti (Sanskrit), carbincho, cardo, cardo lechero, cardo santo (Spanish, "sacred thistle"), cardosanto, cardui flava, carhuinchu, carhuinchunca, carquincho, caruancho, chadron béni, chadron mabré, chicallotl, chicalote,[38] chichicallotl, chichilotl (Aztec), chillazotl, donkey thistle, fischgemüse, fischkraut, flowering thistle, gailshe, gamboge thistle, gold thistle of Peru, guechinichi (Zapotec), h-am (Maya), hierba loca[39] (Spanish, "crazy herb"), infernal fig, ixkanlol (Maya, "yellow flower"), Jamaican thistle, kantankattiri (Malayalam), kawinchu (Quechua), k'i'ix k'an lòl (modern Maya, "prickly yellow flower"), k'i'ix sák lòl (modern Maya, "prickly white flower"), kutiyotti (Tamil), Mexican poppy, Mexican prickly poppy, Mexican thistle, Mexican thorn poppy, mexikanischer stachelmohn, mihca:da:c (Mixe), mizquitl, pavero messicano (Italian), pavot du mexique, pavot espineux (French), pharamgi dhattura (Hindi), pili katili (Hindi), ponnummattai (Tamil), ponnummattu (Malayalam), prickly pepper, prickly poppy, queen thistle, satayanasi, shate (Zapotec), stachelmohn, stinking thistle, svarnasiri (Sanskrit), teufelsfeige, thistle, thistley-bush, tlamexaltzin (Nahuatl), tsolich (Huastec, "lost"), XaSáokS (Serí), xaté (Tarascan), xicólotl, yellow thistle, zèbe dragon (Creole, "dragon herb")

History
During the time of the Aztecs, the prickly poppy was known as the nourishment of the dead; souls would refresh themselves on it in the realm of the dead and in the rain-rich paradise (Rätsch 1985). Prickly poppy is mentioned in numerous documents from the colonial period (Sahagun, Hernández, *Yerbas y hechizerias,* etc.) and in Europe was already well known by 1597, when it was described by John Gerard. At the beginning of the twentieth century, Chinese residents of Mexico were said to produce a kind of opium from the prickly poppy that they used as a legal substitute for *Papaver somniferum* (Reko 1938, 94f.*). Today, the dried plant is smoked as a marijuana substitute (see *Cannabis indica*) and aphrodisiac. In India, the plant is called *pharamgi dhattura* because of its psychoactive properties, and it is regarded as a sister of *Datura metel* (Warrier et al. 1993, 169*).

Distribution
The plant is from the American tropics but is now found throughout the world (Franquemont et al. 1990, 89*). It is common in tropical Africa (Lucas 1962) as well as India and Nepal.

Cultivation
Prickly poppy is very easily grown from seed. The seeds may be either simply dispersed in spring or planted in seedbeds. The plant prefers light, sandy soils, but with sufficient sunlight it can adapt to any type of soil (Grubber 1991, 23*). It can tolerate climates that are tropically moist, hot and dry, subtropical, or moderate. Under cultivation, it can thrive for two or more years.

Appearance
This annual plant, which can grow up to 1 meter in height, has several branches and produces a

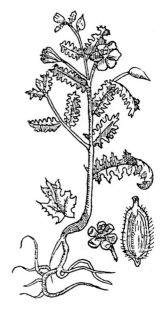

The Aztec name for the Mexican prickly poppy (*Argemone mexicana* L.) is *chicallotl,* "thorn." (From Hernández, 1942/46 [Orig. pub. 1615]*)

"And all poisonous herbs are eaten in the Underworld.
And all who go to the Underworld eat prickly poppy [*Argemone mexicana*].
And all that is not eaten here on the Earth
is eaten in the Underworld.
And it is said that nothing else is eaten."

BERNARDINO DE SAHAGUN
IN *EINIGE AUSGEWÄHLTE KAPITEL AUS DEM GESCHICHTSWERKE DES FRAY BERNARDINO DE SAHAGUN* [SEVERAL SELECTED CHAPTERS FROM THE HISTORICAL WORKS OF FRA BERNARDINO DE SAHAGUN] (SELER 1927, 302 f.*)

---

37 Some authors regard this taxon as a separate species (Grey-Wilson 1995, 74, 75*).

38 This name is also used to refer to the white-blooming Mexican *Argemone platyceras* Link et Otto (Grey-Wilson 1995, 76f.*). Another Mexican member of the poppy family, *Bocconia arborea* Wats., is also called chicalote or chicalote de árbol (Martínez 1987, 1058*).

39 Other psychoactive plants, such as *Datura* spp., are also known by this name.

The Mexican prickly poppy (*Argemone mexicana*) was one of the sacred plants of the Aztec rain god, Tláloc. (Codex Vaticanus 3773, fol. 23)

Left: The white-blossomed variety of the Mexican prickly poppy (*Argemone mexicana* var. *ochroleuca*).

Above right: The typical yellow flower of the Mexican prickly poppy (*Argemone mexicana*) reveals its affinity with the opium poppy.

Below right: The North American prickly poppy (*Argemone albiflora*) produces narcotic effects similar to those of its Mexican relative. (Photographed in the Badlands, South Dakota)

yellowish latex. The bluish leaves are compound and have thorny ends; some are deeply retuse. The flowers, which appear singly, can grow 4 to 6 cm across and have six yellow petals. The four- or six-chambered fruits are heavily thorned capsules that stand erect and are filled with small black seeds. The plants often bear flowers and fruits at the same time. In the tropics, prickly poppy can flower throughout the year.

The plant is easily confused with the closely related *Argemone platyceras* Link et Otto (also found in Mexico) and with the North American species *Argemone albiflora* Hornemann and *Argemone polyanthemos* (Fedde) G. Ownb. [syn. *Argemone alba* James]. It is also very similar to the South American (Argentinean) species *Argemone subfusiformis* Ownb. ssp. *subfusiformis*, which in the local Spanish is also known *cardo santo* or *cardo amarillo* (Bandoni et al. 1972). The blue prickly poppy of Hawaii, *Argemone glauca* (Prain) Pope, is also very similar and is practically indistinguishable from the white-flowered *Argemone mexicana* var. *ochroleuca*. The former, however, has leaves that are somewhat bluer in color.

Occasionally, the prickly poppy may be confused with Mary's thistle, *Silybum marianum* (L.) Gaertn. (Grey-Wilson 1995, 74*).

Psychoactive Material
—Leaves
—Flowers
—Capsules
—Latex, dried

Preparation and Dosage
The dried herbage can be smoked alone or in **smoking blends**. The latex that is tapped from the

capsule can be dried and smoked. No information is available about dosages (Gottlieb 1973, 9*). In Urubamba (Peru), *gringos* smoke the dried flowers as a marijuana substitute (Franquemont et al. 1990, 89*). Further research is needed to determine the appropriate dosages.

Mexican Opium?
Chicalote, el opio mexicano, or chicalote opium, allegedly results when *Argemone mexicana* is pollinated by **Papaver somniferum**. This "produces capsules which, in an unripe state, do indeed allow one to obtain a product which, like opium, induces self-forgetfulness and total contentment" (Reko 1938, 94*). Botanical experiments have demonstrated that this is not possible and that the idea appears to have sprung from the author's imagination (Emboden 1972, 63 f.*; Tyler 1966, 278*).

Ritual Use
It is not entirely clear whether the Aztecs or any other Mesoamerican peoples used the prickly poppy for psychoactive purposes. Since it was regarded as a nourishment of the dead, it is possible that its consumption or use may have been controlled or prevented; in any case, its use was limited to the priests. It may have been utilized for shamanic journeys into the worlds beyond (Rätsch 1985).

The prickly poppy was a sacred plant of the Aztec rain god Tláloc, who reigned in Tlálocan, the "kingdom of dreams" (Knab 1995, 67*):

> The rain was attributed to the rain god, the rain priest. He created, allowed to fall, scattered the rain and the hail, enabled the trees, the grass, the maize to blossom, sprout up, become green, burst open, grow. Moreover, he was also said to be responsible when people drowned in water or were killed by lightning.
>
> And he was adorned in the following manner: a thick mask of soot over his face, his face painted with liquid cautschuk, he is smeared with soot; his face is spotted with a paste from the seeds of the prickly poppy, he wears the raiment of the dew, he wears the garb of the fog, he bears a crown of heron feathers, a neckband of green gems, he wears sandals of foam, and bells, he has white rushes for hair. (Sahagun, Florentine Codex 1: 4*)

Tláloc was also associated with two other psychoactive plants: iztauhiatl (**Artemisia mexicana**) and yauhtli (*Tagetes lucida*; see **Tagetes** spp.) (Ortiz de Montellano 1980).

Sacrificial foods that included prickly poppy seeds were prepared for a variety of ceremonies (Sahagun, Florentine Codex 2:21*). The Aztecs used prickly poppy seeds to make a dough that was ground so fine that it became a kind of tar. They used this tar to form an image of their (highest) god Huitzilopochtli. During celebrations in honor of the god, the priest would "kill" this image with a spear. Its "flesh," which was called "god food," was distributed among the worshippers (Sahagun, Florentine Codex 3:1, 2*).

Artifacts

Numerous pre-Columbian sculptures, wall paintings, frescoes, ceramics, and illuminated manuscripts depict the rain god Tláloc (García Ramos 1994). The prickly poppy, however, does not appear to have been portrayed in any of these contexts (cf. *Turbina corymbosa*).

The flower painter Hans Simon Holtzbecker (from Hamburg) painted a botanically correct portrait of the plant for the Gottorfer Codex (ca. 1650) (de Cuveland 1989, table 52*).

Medicinal Use

The medicinal use of prickly poppy juice to treat eye ailments is common and is found, for example, among the Mixe and the Maya (Lipp 1991, 187*; Roys 1976, 94*). The Serí Indians of northern Mexico prepare a tea from leaves wrapped in linen that they drink for kidney pains. The tea is also said to dispel the "bad" blood that accumulates during birth (Felger and Moser 1974, 427*). The Pima Indians of northern Mexico also use the leaves to treat kidney ailments (Pennington 1973, 221*); a decoction is drunk for difficulties with urination (Eldridge 1975, 316*). The Yucatec Maya utilize the plant for gallbladder disorders (Pulido and Serralta 1993, 47*).

In Peru, a plaster made of prickly poppy is used to treat muscle pains (Chavez 1977, 192*). The inhabitants of many Caribbean islands apply the latex to remove warts and use a decoction for sleeplessness and other sleep disorders. A tea from the leaves is used for asthma (Seaworth 1991, 128*).

In Ladakh, an aqueous extract of crushed leaves is used externally to treat eye diseases and eczema (Navchoo and Buth 1989, 141*). In Uttar Pradesh (India), the latex is combined with oil and cumin powder (*Cuminum cyminum* L.) to make a paste that is applied externally as a treatment for skin diseases, eczema, and flesh worms (Siddiqui et al. 1989, 484*). In Nigeria and Senegal, the prickly poppy is esteemed for its sedative effects. Use of the leaves as a sedative was known even in Europe (Schneider 1974, 1:123*; Watt 1967).

In Hawaii, the yellowish latex of *Argemone glauca* is used to treat toothaches, neuralgia, and ulcers (Krauss 1981, 44*).

Constituents

Although it has often been claimed that **morphine** is present in the prickly poppy, this information is strongly contested (Blohm 1962, 25*). Nevertheless, the entire plant is rich in alkaloids, with a concentration of 0.125% in the roots and stalk (Roth et al. 1994, 142*). The leaves, stalks, and seeds contain the alkaloids berberine and protopine (fumarine, macleyine) (Oliver-Bever 1982, 30). The roots also contain coptisine, up to 0.099% α-allocryptopine (= α-fagarine), chelerythrine, and dihydrochelerythrine. The rather toxic sanguinarine and dihydrosanguinarine are also present in the seeds (Bose et al. 1963). Argemonine was isolated from the leaves and capsules and identified as *N*-methylpavine (Martell et al. 1963). The entire plant contains the isoquinoline alkaloids (–)-canadanine, queilantifoline, queleritrine, allocryptatopine, (–)-tetrahydropalmatine, reticuline, sanguinarine, esculerine, and meta-hydroxy-(–)-estilopine (Lara Ochoa and Marquez Alonso 1996, 37*).

Effects

Little is known about the plant's psychoactive effects: "The seeds have a cannabis-like effect and the herb, juice and flowers are reputed to be narcotic in many countries" (Oliver-Bever 1982, 30). There are increasing reports from Mexico of aphrodisiac and euphoriant effects after smoking the dried herbage. The thickened latex has induced potent narcotic effects and delirium.

Commercial Forms and Regulations

The seeds are occasionally available in nurseries or ethnobotanical specialty shops. The plant is not subject to any regulations or legal restrictions.

Literature

See also the entries for *Papaver somniferum* and *Papaver* spp.

Bandoni, A. L., R. V. D. Rondina, and J. D. Coussio. 1972. Alkaloids of *Argemone subfusiformis*. *Phytochemistry* 11:3547–48.

Bose, B. C., R. Vijayvargiya, A. Q. Saifi, and S. K. Sharma. 1963. Chemical and pharmacological studies of *Argemone mexicana*. *Journal of Pharmaceutical Sciences* 52:1172.

García Ramos, Salvador. 1994. *Tláloc: El dios de la lluvia*. México City: GV Editores.

Lucas, G. Lloyd. 1962. Papaveraceae. In *Flora of tropical East Africa*. London: The Secretary for Technical Cooperation.

Martell, M. J., T. O. Soine, and L. B. Kier. 1963. The structure of argemonine, identification as (–)-methylpavine. *Journal of the American Chemical Society* 85:1022–23.

The Aztecs used the seeds of *Argemone mexicana* for ritual purposes and associated them with the underworld. (Photograph: Karl-Christian Lyncker)

Berberine

Protopine

Allocryptopine

"The prickly poppy is so full of sharp and poisonous thorns that a person who has one of these stick in his throat will doubtlessly go directly to Heaven or to Hell."

JOHN GERARD
THE HERBALL
(1597)

"The four of us smoked [the prickly poppy] and did more than just good to ourselves. As the stick was making its second round, an agreeable state of inebriation began in me. My head was blown free, my body was pleasurably warm, and I could feel how my blood whipped through its canals. The circle of friends gave me additional comfort, in particular as they appeared in a special glow in the evening sun. I found myself among beloved people. This feeling did not search long for an expression, but found one with gentle and yet rapid speed. My eyes lost their focus, and all of my other senses were stimulated in the most delicious manner. Even after the time of the bodies, the senses long remained in that fantastic state in which they are forbidden to perceive all those obscenities, to name all those realities, that we normally do. I found it difficult to steer my steps through the streets, to use fork and knife appropriately at the table, to enjoy the wine from a glass. The shortly measured sleep of this night—not much more than four hours—allowed us to experience the morning in complete and rested freshness."

Ossi Urchs
"Ein ganz besonders Rausch"
[A Very Special Inebriation]
in *Isoldens Liebestrank* [Isolden's Love Drink]
(Müller-Ebeling and Rätsch 1986, 142 f.*)

Oliver-Bever, B. 1982. Medicinal plants in tropical West Africa. *Journal of Ethnopharmacology* 5 (1): 1–71.

Ortiz de Montellano, Bernardo. 1980. Las hierbas de Tláloc. *Estudios de Cultura Náhuatl* 14: 287–314.

Ownbey, G. 1961. The genus *Argemone* in South America and Hawaii. *Brittonia* 13: 91–109.

Rätsch, Christian. 1985. *Argemone mexicana*—food of the dead. Unpublished lecture manuscript.

Stermitz, F. R., D. K. Kim, and K. A. Larson. 1973. Alkaloids *of Argemone albiflora, Argemone brevicornuta* and *Argemone turnerae*. *Phytochemistry* 12:1355–57.

Watt, J. M. 1967. African plants potentially useful in mental health. *Lloydia* 30:1–22.

Argyreia nervosa (Burman f.) Bojer

Baby Hawaiian Wood Rose, Silver Morning Glory

Family
Convolvulaceae (Morning Glory Family)

Forms and Subspecies
There may be an African variety.

Synonyms
Argyreia speciosa (L. f.) Sweet
Convolvulus speciosus L. f.

Folk Names
Baby Hawaiian woodrose, bastantri (Sanskrit), chamang-pins-dansaw, elefantenwinde, elephant creeper, Hawaiian baby woodrose, hawaiianische holzrose, Hawaiian woodrose, holzrose, jamang-pi-danok, jatapmasi, marikkunni, marututari, mile-a-minute, miniature wood-rose, monkey rose, samandar-ka-pat (Hindi), samudrappacca, samudrasos, samuttirappaccai (Tamil), samuttirap-palai, silberkraut, silver morning glory, soh-ring-kang, vrddhadarukah (Sanskrit), woodrose, woolly morning glory

Argyreia nervosa is often confused with *Ipomoea tuberosa* L.[40] [= *Merremia tuberosa* (L.) Rendle; syn. *Operculina tuberosa* (L.) Meissn.], which is also known and sold under the name "Hawaiian wood rose." Its Hawaiian name is *pili-kai*.

History
The plant is originally from India, where it has been used medicinally since ancient times. It must have been introduced into Hawaii at a very early date, for its "home" now lies in the Pacific Islands. We know of no traditional use as an entheogen. The discovery that the wood rose is a potent psychedelic is a result of phytochemical research (Shawcross 1983).

Distribution
The baby Hawaiian wood rose is found throughout India and on Sri Lanka at altitudes of up to 900 meters. It is common in Uttar Pradesh (India), both in the wild and in cultivation. The plant is part of the indigenous flora of Australia and also occurs in Africa. It is now planted in all tropical regions as an ornamental or an inebriant (Bärtels 1993, 214*).

Cultivation
The plant is easily grown from seed. These are either planted after having been germinated or placed in germinating pots. The baby wood rose requires a great deal of water and a warm, preferably tropical climate. Unfortunately, when grown as an indoor plant, it almost never develops flowers (and therewith no seeds). It can also be propagated through cuttings (Grubber 1991, 33*).

Appearance
This vigorous perennial vine, which can climb as high as 10 meters, produces a latexlike sap in its cells. The opposite petiolate leaves are cordate and can grow up to 27 cm in length. Their undersides are covered with hairs and have a silvery appearance (hence the name silver morning glory). The violet- or lavender-colored flowers are funnel shaped and attached to cymes. The sepals are also covered with hairs. The roundish fruits are berry shaped and contain smooth brown seeds. Each seed capsule contains from one to four seeds (one dosage).

The genus *Argyreia* consists of some ninety species (Bärtels 1993, 214*), many of which are easily confused with *Argyreia nervosa*. It is also easily mistaken for the vine *Calystegia sepium* (L.)

---

40 *Ipomoea tuberosa* L. does not contain any lysergic acid derivatives and also has no known psychoactive effects (Ott 1993, 140*).

Brown. It is sometimes even confused with the Hawaiian wood rose *Merremia tuberosa*.

Psychoactive Material
— Seeds
— Roots

Preparation and Dosage
Four to 5 g represent a good starting dosage (Ott 1993, 140*). Generally, four to eight seeds (corresponding to approximately 2 g) are considered sufficient to produce an LSD-like experience (Gottlieb 1973, 17*). Thirteen or fourteen seeds are given as a maximum dose. The seeds should be ground before use (Ott 1979, 58*) and can be washed down with water. The seeds can also be chewed thoroughly (Jackes 1992, 13*). The highest dosage that has been reported in the literature is fifteen seeds (Smith 1985).

The seeds are also used in a preparation known as Utopian bliss balls. These consist of five *Argyreia* seeds, damiana herbage (*Turnera diffusa*), ginseng root (*Panax ginseng*), fo-ti-tieng (*Centella asiatica*; cf. **herbal ecstasy**), and bee pollen.

The dosage for *Merremia tuberosa* is also given as four to eight seeds (Gottlieb 1973, 18*); the psychoactivity of this plant, however, is uncertain (Schuldes 1995; cf. Grierson 1996, 88).

Ritual Use
To date, we know of no traditional use of this psychoactive plant (Brown and Malone 1978, 14*). The baby Hawaiian wood rose is a possible candidate for the **soma** plant, which was described as a vine.

It is unknown whether the shamanic Huna religion used the seeds as enthogenic, magical, or medicinal agents, although this is possible. In Hawaii, poor individuals who were unwilling or unable to pay the exaggerated black market prices for Hawaiian marijuana (*Cannabis indica*) used and still use the seeds as an inebriant (Brown and Malone 1978, 15*; Emboden 1972*). In contrast, the plant does not appear in the traditional ethnobotany of Hawaii (cf. Krauss 1993).

Today, the seeds are used in the white Australian drug scene as psychedelic agents. It is not known whether the Aborigines ever used them. In the Californian subculture, the seeds as well as preparations made with them are used in sexual magical rituals à la Crowley.

Artifacts
None

Medicinal Use
The plant has been used in Ayurvedic medicine since ancient times. The root is regarded as a tonic for the nerves and brain and is ingested as a rejuvenation tonic and aphrodisiac and to increase intelligence. It is also prescribed for bronchitis, cough, "seminal weakness," nervousness, syphilis, diabetes, tuberculosis, arthritis, and general debility (Warrier et al. 1993, 1:173*). The baby Hawaiian wood rose is also used in the folk medicine of Assam (Jain and Dam 1979, 53*). Many *Argyreia* species, e.g., *Argyreia pilosa* Arn., also find use in Indian folk medicine as febrifuges (Bhandary et al. 1995, 153*).

"It struck me that I had remained in the real world during the wood rose session [14 seeds], whereby I understood it much better. As a result, while many of the interesting aspects of a regular trip had remained out of reach (strange worlds, adventures . . .), the thing had not been nearly as strenuous. If you do not leave the real world, then you will not have any difficulties integrating yourself back into it."

KRIK, DESCRIBING AN *ARGYREIA NERVOSA* EXPERIENCE IN *PSYCHOAKTIVE PFLANZEN* [PSYCHOACTIVE PLANTS] (SCHULDES 1995, 98*)

Left: The tropical climber *Calystegia sepium* is often mistaken for the wood rose (*Argyreia*); the seeds of *Calystegia* also appear to contain psychoactive substances. (Photographed in Palenque, Mexico)

Above center: The inflorescence of the baby Hawaiian wood rose (*Argyreia nervosa*). (Photographed on Oahu, Hawaii)

Below center: The effects of the Hawaiian wood rose (*Meremmia tuberosa*) are said to be similar to those of *Argyreia nervosa*. (Photographed on Oahu, Hawaii)

Right: The seeds of *Argyreia nervosa* are rich in psychoactive ergot alkaloids. (Photograph: Karl-Christian Lyncker)

Ergine

Chanoclavine

Lysergol

Ergometrine

Many species of *Argyreia* contain psychoactive constituents. The silvery leaf is typical of the genus. (*Argyreia* spp., photographed in Varanasi, India)

Constituents

The seeds contain 0.3% **ergot alkaloids** and are thus the most potent of all vine drugs (Hylin and Watson 1965). The ergot alkaloids agroclavine, ergine, isoergine (= isolysergic acid amide), chanoclavine-I and -II, racemic chanoclavine-II, elymoclavine, festuclavine, lysergene, lysergol, isolysergol, molliclavine, penniclavine, stetoclavine, isosetoclavine, ergometrinine, lysergic acid-α-hydroxyethylamide, isolysergic acid-α-hydroxy-ethylamide, and ergonovine (ergometrine) have been demonstrated to be present (Brown and Malone 1978, 15*; Chao and Der Marderosian 1973b, 2436 f.). Chanoclavine-I is one of the principal constituents not just in *Argyreia nervosa* but also in most species of *Argyreia* as well as in other representatives of the Family Convolvulaceae. The overall alkaloid composition is reminiscent of that of *Turbina corymbosa*. The related vine *Stictocardia tiliafolia* (Desr.) Hallier f. from Panama also contains large quantities of ergot alkaloids (ergine, chanoclavine-I, chanoclavine-II, festuclavine, lysergol, ergometrinine, lysergic acid-α-hydroxy-

Argyreia Species Containing Significant Concentrations of Psychoactive Ergot Alkaloids (Ergolines)

(From Chao and Der Marderosian 1973b; Hylin and Watson 1965; Ott 1993, 158 f.*)

Name	Occurrence
Argyreia acuta	Asia
Argyreia barnesii (Merr.) Oostr.	Philippines
Argyreia cuneata (Willd.) Ker-Gawl	South India
Argyreia hainanensis	China
Argyreia luzonensis (Hall. f.) Oostr.	Philippines
Argyreia mollis (Burm. f.) Choisy	Sumatra
Argyreia nervosa (Burm. f.) Bojer	Pacific, Asia
Argyreia obtusifolia Loureiro	China
Argyreia philippinensis (Merrill) Oostr.	Philippines
Argyreia speciosa (L. f.) Sweet	Africa[41]
Argyreia splendens (Hornem) Sweet	China
Argyreia wallichi Choisy	Asia

ethylamide, and ergonovine [ergometrine]) (Chao and Der Marderosian 1973b, 2437).

Effects

Most psychonauts characterize the effects of four to eight seeds as very similar to those of LSD (Smith 1985), that is, entailing typical psychedelic patterns and sensations. Reports describe colorful visions of a mystical nature. The effects typically last for six to eight hours or even longer (Ott 1979, 58*). *Argyreia* is also regarded as an aphrodisiac: "Following ingestion, the user attains a euphoric state which is soon followed by a pleasant tingling throughout the body that can last for several hours" (Stark 1984, 28*). There may also be mild side effects, including nausea, exhaustion, and subsequent constipation (Jackes 1992, 13*). When taken in high doses, the trip will sometimes begin with intense nausea (Smith 1985).

Commercial Forms and Regulations

The seeds are available in nurseries and are not subject to any additional regulations.

Literature

See also the entries for *Ipomoea violacea* and *Turbina corymbosa*.

Chao, Jew-Ming, and Ara H. Der Marderosian. 1973a. Ergoline alkaloidal constituents of Hawaiian baby wood rose, *Argyreia nervosa* (Burm. f.) Bojer. *Journal of Pharmaceutical Sciences* 62 (4): 588–91.

———. 1973b. Identification of ergoline alkaloids in the genus *Argyreia* and related genera and their chemotaxonomic implications in the Convolvulaceae. *Phytochemistry* 12:2435–40.

Grierson, Mary, and Peter S. Green. 1996. *A Hawaiian florilegium: Botanical portraits from paradise.* Lawai, Kaui, Hawaii: National Tropical Botanical Garden.

Hylin, John W., and Donald P. Watson. 1965. Ergoline alkaloids in tropical wood roses. *Science* 148:499–500.

Shawcross, W. E. 1983. Recreational use of ergoline alkaloids from *Argyreia nervosa*. *Journal of Psychoactive Drugs* 15 (4): 251–59.

Smith, Elvin D. 1985. Notes on the proposed experiment with *Argyreia nervosa*. *Psychedelic Monographs and Essays* 1:30–37 (pagination lacking).

Z[ubke], A[chim]. 1997. *Argyreia nervosa*: Viel Wind um eine psychedelische Winde. *HanfBlatt* 4 (35): 18–21.

41 The entheogenic effects of this species, which is probably only an African variety or race of *Argyreia nervosa*, have been demonstrated; chemical studies, however, are lacking (Ott 1993*).

Ariocarpus fissuratus (Engelm.) K. Schum.

False Peyote, Living Rock

Family
Cactaceae (Cactus Family)

Forms and Subspecies
This variable species is divided into two varieties:
Ariocarpus fissuratus var. *fissuratus* (Engelm.) K. Schum.
Ariocarpus fissuratus var. *lloydii* (Rose) Anderson

Synonyms
Anhalonium engelmanni Lem.
Anhalonium fissuratum (Engelm.) Engelm.
Ariocarpus intermedius
Ariocarpus lloydii Rose
Mammillaria fissurata Engelm.
Roseocactus fissuratus (Engelm.) Berger
Roseocactus intermedius
Roseocactus lloydii (Rose) Berger

Folk Names
Chaute, chautle, dry whiskey, falscher peyote, false peyote, falso peyote, hikuli sunamí (Tarahumara, "false peyote"), lebender stein, living rock, living star, pata de venoda (Spanish, "deer paw"), peyote, peyote cimarrón (Spanish, "wild peyote"), pezuña de venado, star cactus, star rock, sternenkaktus, sunami, tsuwíri (Huichol), wollfruchtkaktus

History
This cactus, which is usually referred to as false peyote or dangerous peyote (see *Lophophora williamsii*), was certainly already well known in pre-Spanish times. Colonial sources, however, make no mention of it. Today, it is a sought-after species for many cactus enthusiasts and breeders.

Distribution
This species is found only in southwestern Texas, New Mexico, and northern Mexico.

Cultivation
The plant can be grown from seed; it requires well-draining cactus soil (otherwise, like *Lophophora williamsii*).

Appearance
Ariocarpus fissuratus is a small tuberous cactus that grows only a few centimeters tall. Its nodes end in pointed triangles, which give the plant a starlike appearance. The flower is pink-violet. The furrows of the variety *lloydii* are considerably smaller, so it does not have such a jagged appearance (Preston-Mafham 1995, 15*).
Ariocarpus fissuratus is easily confused with the closely related *Ariocarpus retusus* Scheidw. The

Huichol Indians also refer to the latter species as *tsuwíri*, "bad peyote"; it is known in Spanish as *falso peyote*, "false peyote," and may have been used as a peyote substitute. Also very similar, with violet or white flowers, is *Ariocarpus kotschoubeyanus* (Lem.) K. Schum., which is found in the Mexican states of Durango, Nuevo León, and San Luis Potosí (Preston Mafham 1995, 16*). This species is also known as false peyote or deer paw (Bravo Hollis and Scheinvar 1995, 63*).

Psychoactive Material
—Buttons (aboveground cactus flesh)

Preparation and Dosage
Unknown; it is apparently eaten while fresh or dried until its effects become noticeable.

It is said that the cactus was formerly used by the inhabitants along the Texas-Mexico border to fortify the maize beer (**chicha**) they called *tizwin*; it purportedly made them "temporarily crazy and uncontrollable" (Havard 1896, 38*).

Ritual Use
If this cactus has any ritual use at all, it is only as a peyote substitute (see *Lophophora williamsii*). The Huichol Indians strongly warn against eating this cactus, for it has the reputation of being associated with sorcery (Furst 1971).

Artifacts
A related species of *Ariocarpus* is depicted on a Laotian stamp.

Medicinal Use
Unknown

Constituents
Both varieties have been found to contain the β-**phenethylamines** hordenine and *N*-methyltyramine. The var. *fissuratus* also yielded *N*-methyl-3, 4-dimethoxy-phenethylamine (McLaughlin 1969; Mata and McLaughlin 1982, 95*). *Ariocarpus*

The blossoming *Ariocarpus trigonus* of Mexico, on a stamp from the Southeast Asian country of Laos.

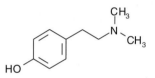

Hordenine

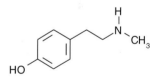

N-methyltyramine

Above: An *Ariocarpus* button, used as a peyote substitute.

Left: The relatively rare northern Mexican *Ariocarpus fissuratus* is known as false peyote or dangerous peyote.

From top to bottom:
The rare *Ariocarpus kotschoubeyanus*.

Ariocarpus retusus, also known as false peyote.

A rare variety, *Ariocarpus retusus* var. *furfuraceus*.

Ariocarpus trigonus, which resembles an agave or an aloe and also contains psychoactive substances.

42 *Ariocarpus denegrii* (Fric) W.T. Marsh. is now known as *Obregonia denegrii* Fric (see **Lophophora williamsii**); *Ariocarpus strobiliformis* Werderm. is now *Pelecyphora strobiliformis* (Werderm.) Kreuzgr. (cf. **Pelecyphora aselliformis**).

retusus contains hordenine, *N*-methyltyramine, *N*-methyl-3,4-dimethoxy-β-phenethylamine, and *N*-methyl-4-methoxy-β-phenethylamine (Braga and McLaughlin 1969; Neal and McLaughlin 1970). Other species of *Ariocarpus* have also yielded hordenine and methyltyramine (Bruhn 1975; Mata and McLaughlin 1982, 95*; Speir et al. 1970).

Effects

The renowned Huichol shaman Ramón Media Silva described the effects as contrasting with the pleasant effects of peyote: "When you eat it, you become crazy; you fall into the canyons, you see scorpions, snakes, dangerous animals, you are unable to walk, you fall, you often fall to your death by falling from the cliffs."

The effects of *Ariocarpus* are said to be very dangerous, particularly for those who do not possess a strong "Huichol heart" (Furst 1971, 183).

Commercial Forms and Regulations

The cactus (as well as other species of the genus) is available in cactus nurseries. Often, however, it is sold for astronomical prices.

Literature

See also the entries for **Lophophora williamsii**, **β-phenethylamines**, and **mescaline**.

Braga, D. L., and J. L. McLaughlin. 1969. Cactus alkaloids. V: Isolation of hordenine and *N*-methyltyramine from *Ariocarpus retusus*. *Planta Medica* 17:87.

Bruhn, Jan G. 1975. Phenethylamines of *Ariocarpus scapharostus*. *Phytochemistry* 14:2509–10.

Furst, Peter T. 1971. *Ariocarpus retusus*, the "'false peyote" of Huichol tradition. *Economic Botany* 25:182–87.

McLaughlin, J. L. 1969. Cactus alkaloids. VI: Identification of hordenine and *N*-methyl-tyramine in *Ariocarpus fissuratus* varieties *fissuratus* and *lloydii*. *Lloydia* 32:392.

Neal, J. M., and J. L. McLaughlin. 1970. Cactus alkaloids. IX: Isolation of *N*-methyl-3,4-dimethoxy-β-phenethylamin and *N*-methyl-4-methoxy-β-phenethylamin from *Ariocarpus retusus*. *Lloydia* 33 (3): 395–96.

Speir, W. W., V. Mihranian, and J. L. McLaughlin. 1970. Cactus alkaloids. VII: Isolation of hordenin and *N*-methyl-3,4-dimethoxy-β-phenethylamin from *Ariocarpus trigonus*. *Lloydia* 33 (1): 15–18.

The Mexican Species of the Genus *Ariocarpus* Scheidw. and Their Distribution

(From McLaughlin 1969; Zander 1994, 121*)

Name and Synonyms[42]	State (Mexico)
Ariocarpus agavoides (Castañeda) Anderson [syn. *Neogamesia agavoides* Castañeda]	Tamaulipas
Ariocarpus fissuratus (Engelm.) K. Schum. [syn. see above]	
A. *fissuratus* var. *fissuratus* (Engelm.) K. Schum.	Southwest Texas, Coahuila
A. *fissuratus* var. *lloydii* (Rose) Anderson	Coahuila, Durango, Zacatecas
Ariocarpus kotschoubeyanus (Lem.) K. Schum. [syn. *Anhalonium kotschoubeyanus* Lem., *Roseocactus kotschoubeyanus* (Lem.) Berger]	Nuevo León, Durango, San Luis Potosí
Ariocarpus retusus Scheidw. [syn. *Anhalonium furfuraceum* (S. Wats.) Coult., *Anhalonium retusum* (Scheidw.) Salm-Dyck, *Ariocarpus furfuraceus* (S. Wats.) H.J. Thomps.] A. *retusus* Scheidw. var. *furfuraceus*	Coahuila, Zacatecas, San Luis Potosí
Ariocarpus scaphorostrus Böd	Nuevo León
Ariocarpus trigonus (F.A.C. Web.) K. Schum. [syn. *Anhalonium trigonum* F.A.C. Web.]	Nuevo León, Tamaulipas

Artemisia absinthium Linnaeus

Absinthe, Wormwood

Family
Compositae: Asteraceae (Aster Family); Antemideae Tribe

Forms and Subspecies
The wild form occasionally differs from the cultivated form. There also are several chemotypes (see "Constituents").

Synonyms
Absinthium majus Geoffr.
Absinthium officinale Lam.
Absinthium vulgare Lam.

Folk Names
Absint-alsem (Dutch), absinth, absinthe, absinthium vulgare, absinthkraut, agenco, ajenjo, ajenjo común, ambrosia (ancient Greek), apsinthos, artenheil, assenzio vero (Italian), bitterer beifuß, botrys, common wormwood, eberreis, echter wermut, gengibre verde (Spanish, "green ginger"), grande absinthe, green muse, grüne fee (German, "green fairy"), heilbitter, hierba santa (Spanish, "sacred herb"), la fée verte, magenkraut, ölde, rîhân (Arabic), sage of the glaciers, schweizertee, wermôd (Saxon), wermut, wermutkraut, wermutpflanze, wor-mod (Old English), wormod, wormwood, wurmkraut

History
Wormwood and its qualities were already well known in ancient times. This and other species of *Artemisia* were sacred to the Greek goddess Artemis—hence their name (Vernant 1988). However, it is uncertain whether the early sources used the Greek name *absinthion* as a catch-all term for a number of *Artemisia* spp. or even other plants (asters) (Schneider 1974, 1:136 ff.*).

In medieval times, the powers of wormwood were praised in Latin hexameter in the *Hortulus* of Walahfried Strabo (ninth century) (Stoffler 1978). Hildegard von Bingen euphorically praised it as "the most important master against all exhaustions" (*Physica* 1.109).

In the sixteenth century, Spanish Jesuits brought the Old World plant, which was known as *hierba santa*, "sacred herb," to the entire world, particularly Central and South America (Hoffmann et al. 1992, 37*).

In central Europe, the **essential oil** (also known as absinthe oil) was distilled from the plant and mixed with **alcohol**. This drink, known as absinthe, became a fashionable drug, especially in nineteenth-century artistic circles. Chronic use, however, had terrible side effects (brain damage; so-called absinthism) (Schmidt 1915). It is still unclear whether absinthism was due to the thujone or to other ingredients (e.g., heavy metal salts) (Proksch and Wissinger-Gräfenhahn 1992, 363). Because wormwood was an inebriating drug, and because it was also used as an illegal abortifacient (by quack physicians), it was soon banned as a result of "increasing misuse" (Vogt 1981) in France in 1922 (Arnold 1988, 3043) and in Germany in 1923. At about the same time, the "green fairy" (as the psychedelic drink was called) was also made illegal in Switzerland under threat of severe fines and imprisonment (Rätsch 1996). Today, absinthe cannot be (officially) obtained anywhere.

Since the early 1990s, many Swiss "scene" bars have been selling beverages known as *die grüne fee* ("the green fairy"). These drinks do not contain any genuine and illegal absinthe but consist of other commercial alcoholic beverages. The true green fairy is available only through private channels. No one has been able to explain to me why absinthe became known as the green fairy. One woman conjectured that it might have something to do with the effects, for absinthe is said to carry people away as if they had been enchanted by a fairy. Others suggested that it refers to the often greenish color of the absinthe. One Swiss man informed me that absinthe is the "most psychedelic alcohol there is."

Distribution
Wormwood is found throughout Europe, North Africa, Asia, and North and South America. Large numbers of the plant grow wild in Valais (Switzerland).

Cultivation
Wormwood is quite easy to grow from its very small seeds. The best method is to sow the seeds in a bed sheltered from the rain and press them a little into the ground. The seeds should be watered with care so that they are not constantly shifted around and their germination disturbed (Grubber 1991, 67*). Wormwood prefers dry soils; it also thrives well on rocky subsoil. Most of the areas in which wormwood is grown for pharmaceutical use are in eastern Europe (Proksch and Wissinger-Gräfenhahn 1992, 360).

Appearance
The perennial, upright, somewhat branched shrubby herb grows to a height of 50 to 100 cm. The finely pinnate, whitish gray leaves are covered on both sides with fine hairs and have a feltlike,

Wormwood (*Artemisia absinthium*) can differ in appearance from one location to another. (Woodcut from Tabernaemontanus, *Neu Vollkommen Kräuter-Buch*, 1731)

Found seldom in the wild, wormwood (*Artemisia absinthium*) is an ancient European medicinal and inebriating plant. (Photographed in Valais, Switzerland)

silky surface. When crushed, they immediately exude the characteristic aromatic-bitter scent of the essential oil. The spherical, clustery yellow flowers are attached paniculate to the ends of the branches. The flowering lasts from July to September. The stalks wilt in the fall. The root-stock produces new shoots in the spring.

Artemisia absinthium is easily confused with other members of the genus, including mugwort (*Artemisia vulgaris*) (see *Artemisia* **spp.**). Wormwood is almost indistinguishable from *Artemisia mexicana.*

Psychoactive Material
— Aboveground herbage (absinthii herba, herba absinthii, absinthii cacumina florentia, summitates absinthii, wermutkraut)

The percentages of constituents in the plant are highest when it is harvested during the flowering season. The dried herbage should be stored away from light.

Preparation and Dosage
The fresh or dried herbage (it is best to use leaves only from the ends of the twigs) are added to boiling water and allowed to steep for five minutes. One g of dried leaves in 1 cup of hot water represents a single medicinal dosage (Roth et al. 1994, 146*).

Wormwood herbage can also be smoked alone or as an ingredient in **smoking mixtures**; it is also used as an **incense**, e.g., in smudge bundles (cf. *Artemisia* **spp.**).

In ancient times, the plant was already being used to produce medicinal wines:

A wine, the so-called wormwood wine, is also made from it, especially in Propontis and in Thrace, where it . . . is used when fever is

lacking. They also drink it in the summer before, for they believe that is wholesome to the health. . . . But the juice of absinthe appears to exert the same effects, except that we do not consider it good to drink, for it is contrary to the stomach and causes headaches. (Dioscorides 3.23)

In ancient China, wormwood was used as an additive to rice wine (cf. **sake**).

In 1797, M. Pernod, a Frenchman who was living in Switzerland at the time, developed the emerald green drink known as absinthe by distilling a preparation of an herb mash of wormwood, anise (*Pimpinella anisum* L. [syn. *Anisum vulgare* Gaertn.]), fennel, lemon balm (*Melissa officinalis* L.), hyssop, and other herbs (Arnold 1988, 3043). Absinthe definitely has a much more pleasant taste when only the distilled oil of *Artemisia absinthum* is used. Herbal extracts can impart an unpleasantly bitter taste to the liquor.

Absinthe was also produced by macerating the following herbs in a high-proof **alcohol** (brandy or similar spirits, with up to 85% ethanol content) (Albert-Puleo 1978, 69):

Wormwood leaves	*Artemisia absinthium*
Angelica root	*Angelica archangelica* L. (cf. **theriac**) [syn. *Archangelica officinalis* Hoffm.]
Calamus root	*Acorus calamus*
Dictamnus leaves	*Origanum dictamnus* L. [syn. *Amaracus dictamnus* (L.) Benth.]
Star anise fruits	*Illicium verum* Hook. f.
Cinnamon bark	*Cinnamomum verum* Presl.
Peppermint	*Mentha piperita* L., *Mentha* spp. (cf. **Mentha pulegium**)
Hyssop herbage	*Hyssopus officinalis* L.
Fennel seeds	*Foeniculum vulgare*

To prepare absinthe, coriander (*Coriandrum sativum* L.), sweet marjoram (*Majorana hortensis* Moench [syn. *Origanum majorana* Boiss.]), nutmeg (*Myristica fragrans*), marjoram (*Origanum vulgare* L., *Origanum* spp.), chamomile (*Chamomilla recutita* (L.) Rauschert [syn. *Matricaria chamomilla* L.]), parsley (*Petroselinum crispum*), juniper (*Juniperus communis* L.; cf. *Juniperus recurva*), and spinach (*Spinacia oleracea* L.) have also been used (Pendell 1995, 103*).

Dale Pendell, one of the last of the Beat poets, developed a recipe of his own that induces profound psychoactive effects:

30 g	Wormwood leaves	(*Artemisia absinthium*)
8.5 g	Hyssop herbage	(*Hyssopus officinalis*)
1.8 g	Calamus root	(*Acorus calamus*)
6.0 g	Lemon balm leaves	(*Melissa officinalis*)
30 g	Anise seed	(*Pimpinella anisum*)
25 g	Fennel seed	(*Foeniculum vulgare*)
10 g	Star anise fruits	(*Illicium verum*)
3.2 g	Coriander seeds	(*Coriandrum sativum*)

Lightly crush the herbs and place them in a vessel that can be sealed. Add 800 ml of 85 to 95% **alcohol**. Allow the vessel to stand closed for a week, shaking it slightly from time to time. Then add 600 ml of water and allow the entire contents to macerate for an additional day. Pour off the liquid and squeeze the remaining fluid out of the herbs and into the extract. The herbs can then be added to vodka or another type of spirits and squeezed once more (Pendell 1995, 112*).

Contemporary (Swiss) absinthe recipes are kept secret. The wormwood is distilled together with other herbs. The color may be clear, greenish, or yellowish. The taste is strongly reminiscent of anisette or Pernod. Absinthe is diluted with water prior to consumption (approximately 1:1). The resulting mixture is milky-cloudy.

An absinthelike drink named *yolixpa* (Nahuatl, "in the view of the heart") is distilled in Puebla (Mexico) and drunk ritually (Knab 1995, 219*). It is produced from *aguardiente* (sugarcane spirits; cf. **alcohol**) to which such herbs as *Artemisia mexicana* have been added. Absinthelike love drinks made of alcoholic spirits and the appropriate herbs were once produced in Switzerland as well (Lussi 1997).

German wormwood wine[43] contains only trace amounts of the essential oil (Fühner 1943, 239*).

Ritual Use

In ancient times, the name *artemisia* (which was derived from that of the goddess Artemis, the sister of Apollo and the god of healing) was primarily used to refer to wormwood, mugwort, and related species (cf. *Artemisia* spp.).[44] Unfortunately, very few ancient texts have come down to us that are able to cast light upon the connection between these plants and the virgin goddess. The Greek word *artemisia* means "intactness," a clear reference to the chasteness of the goddess, who, as the mistress of wild animals, functions as a mixture of Amazon, witch, and shamaness. In ancient Greece, Artemis was revered as the patron goddess of virgins. In the ancient Orient, she was regarded as the ruler of the Amazons. During the Italian Renaissance, she became Diana, the witch goddess. In spring, during the time of the full moon, ecstatic and orgiastic Artemis festivals were held to honor the goddess. As part of these festivities, the goddess was symbolically consumed in the form of wormwood and mugwort. In Laconia, boisterous Artemis festivals were held that featured obscene activities, wild dances, travesties, and masks. The men would wear women's masks and the women would strap on phalluses (Giani 1994, 89*). It appears that these festivals were actually mystery rites and fertility rituals.

Artifacts

Absinthe was a legendary drug among artists and Bohemians at the end of the nineteenth century (Conrad 1988). It was popularized primarily through the absinthe pictures of the Parisian painters Henri de Toulouse-Lautrec (1864–1901) and Édouard Manet (1832–1883). The manic-depressive painter Vincent van Gogh (1853–1890) appears to have been addicted to absinthe. His paintings, especially those in which brilliant yellow tones predominate (the renowned "Van Gogh yellow"), are good representations of the perceptual changes caused by thujone (Arnold 1988). Pablo Picasso also helped immortalize absinthe (Adams 1980). Paul Gauguin even took an ample supply of absinthe with him when he traveled to Tahiti. Alfred Jarry referred to absinthe as "holy water" (Pendell 1995, 110*).

Absinthe was also a source of literary inspiration for such writers as Arthur Rimbaud, Ernest Dowson, Charles Cros, H. P. Lovecraft, Charles Baudelaire, Oscar Wilde, Jack London, Ernest Hemingway, Gustave Kahn, Victor Hugo, Alfred de Musset, and Paul Verlaine (Conrad 1988; Pendell 1995: 103 ff.*). These authors have left us with a number of poems praising absinthe.

Medicinal Use

In ancient Egypt, wormwood was commonly used as a remedy, an aromatic substance, and an additive to **wine** (cf. *Vitis vinifera*) and **beer** and to dispel worms and to treat pains in the anal region. Today, wormwood is still used in Yemen to alleviate the pains associated with parturition (Fleurentin and Pelt 1982, 102 f.*).

In European folk medicine, wormwood is one of the most important gynecological agents for abortion and to induce menstruation and labor. In tea form, it is consumed primarily for stomach pains, lack of appetite, feelings of fullness, gallbladder problems, vomiting, and diarrhea (Pahlow 1993, 339*).

In homeopathy, absinthium is used in accordance with the medical descriptions to treat such ailments as epilepsy and nervous and hysterical spasms (Pahlow 1993, 340*).

Constituents

Wormwood contains large quantities of bitter substances (absinthine) and an **essential oil** that is rich in thujone. The four primary components of the essential oil are (+)-thujone (= α-thujone), cis-epoxyocimene, trans-sabinylacetate, and chrysanthenylacetate. Wormwood develops a variety of chemotypes; for this reason, the composition of the essential oil can vary considerably. Any one of the four primary components can dominate, depending upon the location where the herb originated. For example, (+)-thujone dominates in altitudes of up to 1,000 meters (Proksch and

"Use it to help yourself,
and boil the bitter green of the
  woody wormwood;
then pour the juice from spacious
  bowls
And wash the highest part of the
  head with it.
When you have washed the fine hairs
  with this brew,
then remember to lay thereon,
bundles of leaves tied together,
and wrap a snug bandage around the
  hair after the bath.
Before too many hours have passed
in the course of time,
you will be amazed by this agent
and by all of its other powers."

WALAHFRIED STRABO
HORTULUS 9

"[T]he use of psychedelic *Artemisia* preparations combined synergistically with the lunar effect would have facilitated the ecstatic and orgiastic rites of Artemisia."

MICHAEL ALBERT-PULEO
"MYTHOBOTANY, PHARMACOLOGY, AND CHEMISTRY OF THUJONE-CONTAINING PLANTS AND DERIVATIVES"
(1978, 68)

43 In Turin [Italy], wormwood wine was made from *Artemisia absinthium* as well as *Artemisia pontica* L. and *Artemisia abrotanum* L. (Hartwich 1911, 772*).

44 According to Pliny (*Natural History* 25.73), the name is derived from that of a certain Artemisia, the wife of King Mausolus of Caria.

Thujone

Wissinger-Gräfenhahn 1992, 360). Thujone has a molecular symmetry not unlike that of **THC** (Castillo et al. 1975).

In addition to the essential oil, the herbage also contains sesquiterpene lactones, glycosides of camphor oil, tanning agents, and quercetin (cf. *Acacia* spp., *Psidium guajava*, *Vaccinium uliginosum*, **kinnikinnick**) (Proksch and Wissinger-Gräfenhahn 1992, 361).

Effects

The extremely bitter wormwood tea has been demonstrated to soothe the stomach (Hoffmann et al. 1992, 37*). The pharmacological effects of thujone, which is chemically related to camphor (see *Cinnamomum camphora*) and pinene, are very similar to those of **THC** (Castillo et al. 1975). The literature contains frequent reports of hallucinations as well as of spasms and epileptic-like seizures following consumption of absinthe (Arnold 1988, 3043; Schmidt 1915; Walker 1906).

Because of the presence of thujone, a potent psychoactive substance, absinthe liquor is much stronger than other types of alcoholic beverages and produces different effects (cf. **alcohol**):

> The absinthe did indeed have inebriating effects upon me, but these were quite different than with "normal" schnapps. The stimulant effects of absinthe were quite strong, it woke me up and also kept me awake for a long time. I was partially bathed in aphrodisiac sensations, and I partially flowed in that direction. As the effects increased, I had the sensation that I was floating away. It was like the kiss of the green fairy. — Unfortunately, the next day the head was in as much pain as the inebriation had been delightful during the previous evening. I had never before experienced such a brutal hangover. (Rätsch 1996, 286)

A line of **cocaine** is said to be a very effective treatment for the torments of an absinthe headache.

In comparison to absinthe, the effects of the herbage when smoked are quite mild, producing only a slight euphoria.

Commercial Forms and Regulations

In central Europe, wormwood herbage is officinal (*DAB10, Helv. VII, ÖAB90, BHP83*); the minimum amount of essential oil must be 0.2% (Proksch and Wissinger-Gräfenhahn 1992, 362). The herbage is sold without restriction; only absinthe is illegal. However, we find the same thing occurring here as with other instances of legal proscriptions, for the illegal substance continues to be distilled in underground circles. Today, absinthe is banned around the world, but it is still being illegally produced in some of the German-speaking regions of Switzerland according to old, traditional recipes. Absinthe connoisseurs pay precious little attention to the fact that the substance is severely restricted. In Switzerland, absinthe was made illegal primarily because it was being used (or abused) to terminate pregnancies. Today, anyone caught distilling absinthe illegally faces a fine of 100,000 Swiss francs (Rätsch 1996).

Literature

See also the entries for **Artemisia mexicana**, **Artemisia spp.**, **essential oil**, and **THC**.

Adams, B. 1980. Picasso's absinth glasses: Six drinks to the end of the era. *Artforum* 18 (8): 30–33.

Albert-Puleo, Michael. 1978. Mythobotany, pharmacology, and chemistry of thujone-containing plants and derivatives. *Economic Botany* 32:65–74.

Arnold, Wilfred Niels. 1988. Vincent van Gogh and the thujone connection. *Journal of the American Medical Association* 260 (20): 3042–44.

————. 1989. Absinthe. *Scientific American.* June:113–17.

Castillo, J. D., M. Anderson, and G. M. Rubboton. 1975. Marijuana, absinthe and the central nervous system. *Nature* 253:365–66.

Conrad, Barnaby, III. 1988. *Absinthe: History in a bottle.* San Francisco: Chronicle Books.

Lussi, Kurt. 1998. Der Liebestrank der Aphrodite: Eine Rezeptsammlung aus der Innerschweiz. *Yearbook for Ethnomedicine and the Study of Consciousness* 1996 (5): 79–97. Berlin: VWB.

Proksch, Peter, and Ulrike Wissinger-Gräfenhahn. 1992. *Artemisia.* In *Hagers Handbuch der pharmazeutischen Praxis,* 5th ed., 4:357–77. Berlin: Springer.

Rätsch, Christian. 1996. "Die Grüne Fee": Absinth in der Schweiz. *Yearbook for Ethnomedicine and the Study of Consciousness* 1995 (4): 285–87. Berlin: VWB.

Schmidt, H. 1915. L'Absinthe, l'aliénation mentale et la criminalité. *Annales d'Hygiène Publique et Médecine Légale* 23 (4th series): 121–33.

Vernant, Jean-Pierre. 1988. *Tod in den Augen— Figuren des Anderen im griechischen Altertum: Artemis und Gorgo.* Frankfurt/M.: Fischer.

Vogt, Donald D. 1981. Absinthium: a nineteenth-century drug of abuse. *Journal of Ethnopharmacology* 4 (3): 337–42.

Vogt, Donald D., and Michael Montagne. 1982. Absinthe: Behind the emerald mask. *The International Journal of Addictions* 17 (6): 1015–29.

Walker, E. E. 1906. The effects of absinthe. *Medical Record* 70:568–72.

Zafar, M. M., M. E. Hamdard, and A. Hameed. 1990. Screen of *Artemisia absinthium* for antimalarial effects on *Plasmodium berghei* in mice: a preliminary report. *Journal of Ethnopharmacology* 30:223–26.

Artemisia mexicana Willdenow et Spreng.

Mexican Wormwood

Family
Compositae: Asteraceae (Aster Family); Antemideae Tribe; Abrotanum Section

Forms and Subspecies
Today, *Artemisia mexicana* is usually regarded as a subspecies of the North American western mugwort (Argueta et al. 1994, 628*; Lee and Geissman 1970; Ohno et al. 1980, 104; Pulido Salas and Serralta Peraza 1993, 16*): *Artemisia ludoviciana* Nutt. ssp. *mexicana* (Willd.) Keck (cf. **Artemisia spp.**). The plant also has one variety: *Artemisia mexicana* var. *angustifolia* (Mata et al. 1984).

Synonyms
Artemisia ludoviciana ssp. *mexicana* (Willd.) Keck
Artemisia vulgaris ssp. *mexicana* (Hall.) Clem.

Folk Names
Agenjo del país, ajenjo, ajenjo del país, altamisa, altamiza, altaniza, ambfe (Otomí), artemisia, azumate de puebla, cola de zorillo, ("little tail of the fox"), ensencio de mata verde ("incense of the green bush"), epazote de castilla, estafiate,[45] estaphiate, estomiate, green wormwood, guietee, guitee (Zapotec), haway, hierba de San Juan (Spanish, "Saint John's herb"), hierba maestra (Spanish, "master herb"), hierba maistra, incienso verde (Spanish, "green incense"), istafiate, istafiatl, ixtauhyatl (Aztec), iztauhiatl, iztauhyatl (Nahuatl), kamaistra (Popoluca), kaway si'isim, Mexican wormwood, mexikanischer beifuß, mexmitzi (Otomí), osomiate, quije-tes (Zapotec), ros'sabl'i (Rarámuri), si'isim (Maya), te ts'ojol (Huastec), tlalpoyomatli (Aztec), tsakam ten huitz (Huastec), tsi'tsim (Yucatec), xun, zizim

History
The Aztecs and other Indians of Mesoamerica were already using Mexican wormwood for ritual and medicinal purposes in pre-Columbian times. Today, the prime significance of the plant is in folk medicine. In Mexico, the herbage is often smoked as a marijuana substitute (cf. **Cannabis indica**).

The first European to describe Mexican wormwood and compare it to its European counterpart was the Franciscan priest and book burner Diego de Landa (1524–1579).

Distribution
The plant occurs in both the dry and warm regions of Mexico (the Valley of Mexico, San Luis Potosí, Veracruz, Chihuahua) and the Yucatán Peninsula (Martínez 1994, 134*). It is also said to occur in Arizona and New Mexico (Ohno et al. 1980, 104).

Cultivation
See *Artemisia absinthium.*

Appearance
Mexican wormwood, which can grow as tall as 1 meter in height, is so similar to the European species that even experts can have difficulty distinguishing the two. Some botanists and ethnobotanists believe that it is a variety or subspecies of **Artemisia absinthium**.

Psychoactive Material
—Herbage without the roots
—Roots

Preparation and Dosage
The fresh herbage can be added to aguardiente, mescal, tequila (cf. **Agave spp.**), or any other distilled spirits (cf. **alcohol**) for optimal extraction (Martínez 1994, 134*). Mexican wormwood is one of the herbs used to manufacture the absinthelike herbal liquors of Central Mexico known as *yolixpa.*

The dried herbage can be smoked. One to 3 g produces mild psychoactive effects. Three to 4 g of the dried herbage, taken internally, has strong anthelmintic effects (Martínez 1994, 135*). Higher dosages can induce abortions.

Ritual Use
The Aztecs were already using *Artemisia mexicana* as a ritual **incense** in pre-Columbian times:

> Tlalpoyomatli, its leaves are smoky, gray, soft; it has many flowers. Incense is made from this plant: it produces an agreeable scent; it produces a perfume. This incense spreads, it is distributed over the entire country. (Sahagun, Florentine Codex 11:6*)

Mexican wormwood (*Artemisia mexicana*) is almost indistinguishable from European wormwood. The Mexican plant, however, is more potently psychoactive. (Photographed in Veracruz, Mexico)

45 *Artemisia ludoviciana* Nutt. and *Artemisia klotzchiana* Basser are also known as *estafiate* (cf. **Artemisia spp.**).

"[In the Yucatán], There is wormwood, much leafier and more aromatic than what we have here, and with longer and narrower leaves; the Indians grow it for the scent and for pleasure, and I have noticed that the plants grow better when the Indian women put ashes at the foot."

FRAY DIEGO DE LANDA
AN ACCOUNT OF THE THINGS OF YUCATÁN, XLIX
(2000:158*)

Coumarin

The aromatic plant was sacred to Uixtociuatl, the Aztec goddess of salt and salt makers. The Aztec name for Mexican wormwood, *itztauhyatl*, is sometimes translated as "water of the goddess of salt" (Argueta et al. 1994, 628*). During her festival, which occurred in the seventh month (Tecuilhuitontli), the goddess was portrayed by a priestess who carried a staff that was used in a dance:

While dancing, she swings her shield around in a circle, makes movements with it. And she carries a rush staff, decorated with papers and sprinkled with caoutchouc, and furnished on three sides with shells. And where the staff bears the chalice-shaped enlargements, there too is wormwood herbage. Crossed feathers are on it, it bears crossed feathers. When dancing, she supports herself on this, places it firmly into the ground and circles around it, making movements towards the four directions. And ten days long they sang and danced for her in the manner of women; everyone was occupied with this, the salt people, the salt makers, the old women and the women in middle age and the maidens and the girls who had just grown to be maidens. While the sun is still there, still shines, they begin to dance. They are arranged in rows, they arrange themselves in rows. Using a rope, which they call "flower rope," they take hold of one another, forming a long row. They wear a wormwood flower on their heads. And they sing, they scream loudly, sing with a very high voice, their song is just as the centzontle sings somewhere in the forest, like a clear little bell are their voices. (Sahagun, Florentine Codex 2:26*)

Mexican wormwood is one of the plants sacred to Tláloc, the rain god, who was also associated with *Argemone mexicana* and *Tagetes lucida* (see *Tagetes* spp.).

Documents from the colonial period make no mention of any use of this plant as a psychoactive plant. However, Jacinto de la Serna did refer to Mexican wormwood in the same breath as peyote (*Lophophora williamsii*) and ololiuqui (*Turbina corymbosa*) (Garza 1990; Ott 1993, 393*). In modern Mexico, the leaves are smoked as a marijuana substitute. It is possible that ritual forms for using the plant have developed in conjunction with this use.

Artifacts
The plant was sometimes depicted in connection with the Aztec goddess Uixtociuatl (= Huixtocihuatl) and her festival.

Medicinal Use
The herbage is used as an antispasmodic in Mexican folk medicine (Cerna 1932, 303*). An extract obtained with a mixture of alcohol and water is medicinally drunk for stomach ailments and digestive problems (Martínez 1994, 134*). The plant is listed in the Mexican pharmacopoeia as an anthelmintic and a stomachic (Dibble 1966, 66*; Lara Ochoa and Marquez Alonso 1996, 55*). In modern folk medicine (which was influenced by the Aztecs), the roots and herbage are used to treat epilepsy and rheumatism and to induce menstruation and abortion, and are also drunk as a tonic (Reza D. 1994). Teas made from the plant are drunk to treat lack of appetite. Alcoholic extracts with *albahaca* (see *Ocimum micranthum*) are said to heal diseases caused by "bad winds" (Argeuta et al. 1994, 628 f.*). The Yucatec Maya use the herbage as a fumigant for treating headaches (Pulido Salas and Serralta Peraza 1993, 16*). Decoctions are drunk for coughs, asthma, and diarrhea (Roys 1976, 310*). Both the Yucatec Maya and other Indians also use the plant for birth control (to induce menstruation and abortion).

Constituents
In addition to the **essential oil**, which is composed in part of the terpenes borneol, alcafor, limonene, α-phellandrene, and β-phellandrene, the primary active component is santonin. An alkaloid of unknown structure is also said to be present (Martínez 1994, 134*). The herbage contains azulene, butenolide, **coumarins**, flavones, polyacetylenes, lactones, and sesquiterpenes (armefolin, 8-α-acetoxyarmexifolin, artemexifolin) (Argueta et al. 1994, 628*; Dibble 1966, 66*; Lara Ochoa and Marquez Alonso 1996, 55*). Although it is likely present, thujone has not yet been detected.

A sample from Arizona was found to contain the eudesmanolides (sesquiterpene lactones) douglanin, ludovicin-A, ludovicin-B, and ludovicin-C. Mexican plants contain the sesquiterpene lactones arglanin, douglanin, armexin, estafiatin, chrysartemin-A,[46] and artemolin (Lee and Geissman 1970; Ohno et al. 1980, 104; Romo et al. 1970).

Effects
Smoking the dried herbage initially produces a mild, pleasant stimulation that can—depending upon dosage and sensitivity—increase to a euphoric state, very much like the effects of marijuana.

Taken internally, the herbage and the oil it yields have anthelmintic and abortive effects. Overall, the plant is said to be less toxic than *Artemisia absinthium* and therefore more easily tolerated (Martínez 1994, 134*).

Commercial Forms and Regulations
In Mexico, the dried herbage is available in markets and herbal shops.

46 *Chrysartemins* also occur in *Chrysanthemum* spp., which are used as additives in tea (*Camellia sinensis*) and **sake** (Romo et al. 1970).

Literature

See also the entries for ***Artemisia absinthium,
Artemisia* spp.,** and **essential oil.**

Lee, K. H., and T. A. Geissman. 1970. Sesquiterpene
lactones of *Artemisia* constituents of *A. ludoviciana* ssp. *mexicana. Phytochemistry* 9:403–8.

Mata, Rachel, Guillermo Delgado, and Alfonso
Romo de Vivar. 1984. Sesquiterpene lactones of
Artemisia mexicana var. *angustifolia.
Phytochemistry* 23 (8): 1665–68.

Ohno, Nobuo, Jonathan Gershenzon, Catherine
Roane, and Tom J. Mabry. 1980. 11,13-dehydro-

desacetylmatricarin and other sesquiterpene
lactones from *Artemisia ludoviciana* var.
ludoviciana and the identity of artecanin and
chrysartemin B. *Phytochemistry* 19:103–6.

Reza D., Miguel. 1994. *Herbolaria azteca.* México,
D.F.: Instituto Mexiquense de Cultura.

Romo, J., A. Romo de Vivar, R. Treviño, P. Joseph-
Nathan, and E. Díaz. 1970. Constituents of
Artemisia and *Chrysanthemum* species: the
structures of chrysartemins A and B.
Phytochemistry 9:1615–21.

Artemisia spp.

Artemisia Species

Family

Compositae: Asteraceae (Aster Family); Antemideae Tribe

To date, a number of species of this genus have
been described that display interesting pharmacological properties that can be characterized as
stimulating, tonic, and antispasmodic (Morán et
al. 1989a). In all of the places where species of
Artemisia are found—and they are found almost
worldwide—they are used for ethnomedicinal
purposes. For example, *Artemisia herba alba* L. is
used in Arabic folk medicine to treat diabetes. Its
abilities to lower blood sugar levels have been
experimentally verified (Twaij and Al-Badr 1988).
The Nepalese Sherpas use the juice of freshly
pressed leaves of *Artemisia dubia* Wall. ex Besser
(*titepati, kemba girbu*) as an antiseptic and a
decoction for fevers (Bhattarai 1989, 47*). The
malarial agent artemisinin (= quinghaosu) was
discovered in the Asian *Artemisia annua* L. (El-
Feraly et al. 1986).

Many *Artemisia* species are used ritually as
incense, in the peyote cult (see ***Lophophora
williamsii***), and as medicines. Even mugwort
(*Artemisia vulgaris*), which was introduced from
Europe, is used as "sage" (*tägyi*). Some species that
have gynecological effects are sacred to the Greek
goddess Artemis (Brøndegaard 1972).

Artemisia frigida contains camphor (the plant
is even regarded as a source of camphor; cf.
Cinnamomum camphora). Some species of
Artemisia contain the psychoactive substance
thujone (see **essential oils**). Methoxylated
flavonoids are common in the genus (Rodríguez et
al. 1972). Many species of *Artemisia* have muscle-

relaxing and antiasthmatic effects (Morán et al.
1989c) and are thus suitable for use in **smoking
blends.** They include:

Artemisia scoparia Waldst. et Kit.
Artemisia sieversiana (Ehrh.) Willd.
Artemisia argyi Leveille et Vaniot
Artemisia caerulescens ssp. *gallica* (Willd.) K. Pers.

The West European *Artemisia caerulescens* ssp.
gallica is rich in an essential oil with a high thujone
content (Morán et al. 1989b).

Artemisia copa **Phil.—copa-copa, copa tola**
This species is found in northern Chile. The
inhabitants of the Toconse oasis (Atacama Desert)
claim that this plant has the power to induce
dreams (Aldunate et al. 1981, 205*). It apparently
even has hallucinogenic properties (Aldunate et al.
1983*).

Artemisia ludoviciana **Nutt.—prairie sagebrush,
western mugwort, white sage, präriebeifuß**
This variable species is divided into the following
varieties and subspecies (Ohno et al. 1980, 104):

Artemisia ludoviciana Nutt. [syn. *Artemisia
gnaphalodes, Artemisia purshiana* Bess.]
Artemisia ludoviciana var. *ludoviciana* Nutt.
Artemisia ludoviciana ssp. *albula* (Woot.) Keck.
Artemisia ludoviciana ssp. *mexicana* (Willd.)
Keck. [syn. ***Artemisia mexicana*** Willd.]

The subspecies differ with regard to the
composition of the sesquiterpene lactones that are
present (Ohno et al. 1980).

Ethnobotanical research now suggests that

Pontic wormwood (*Artemisia pontica* L.) contains thujone. In former times, it was used together with wormwood to manufacture absinthe. (Woodcut from Fuchs, *Läebliche abbildung und contrafaytung aller kreüter,* 1545)

Above: The North American prairie sagebrush (*Artemisia ludoviciana*) is the most important ritual incense of the Plains Indians. It also contains an essential oil with stimulating effects.

Right: Pati, a Himalayan species of mugwort (*Artemisia* spp.), is used as an incense to support meditation and as a psychoactive beer additive. (Photographed in Langtang, Nepal)

"Artemis—
Queen, hear me,
Much beckoned daughter of Zeus,
Thundering, highly praised Titaness,
Exalted archeress!
All-illuminating, torch bearing,
Goddess Diktynna, who smiles upon
    the childbed;
Helper in labor,
But who herself knows the childbed
    not.
She who unties the girdle, Friend of
    madness,
Dispeller of Troubles, Huntress,
Runner, Hurler of arrows,
Friend of the hunt, who storms
    through the night."

ORPHIC HYMN

Paleo-Indians brought the use of mugwort as incense with them into the New World from Asia some 30,000 years ago (Storl 1995).

There is almost no ritual among the Plains Indians that does not include smudging with *Artemisia ludoviciana*. The ascending aromatic smoke is a prayer. It links together Maká, the Mother Earth, with Wakan Tanka, the Great Spirit, who is active in all creatures. The Plains Indians use western mugwort primarily for spiritual purification, to dispel disease spirits and negative powers, to treat possession, and to protect the home. The herbage is also used as an **incense** in peyote ceremonies, as pillows (support) for "Father Peyote" (cf. *Lophophora williamsii*), and as an altar covering. The herbage as well as the leaves are suitable for use as a tobacco (*Nicotiana tabacum*) substitute and are a component of ritual and medicinal **smoking blends** and **kinnikinnick**.

The aboveground portion of the plant contains an **essential oil** with thujone as well as the lactone glycosides santonin and artemisin, which are responsible for the anthelmintic effects. The sesquiterpene lactone anthemidin has been found in *Artemisia ludoviciana* (Epstein et al. 1979). Four santanolides (ludovicin-A, -B, -C, and luboldin) as well as camphor have also been detected (Domínguez and Cárdenas 1975). The essential oil has antibacterial properties (Overfield et al. 1980, 99). A variety of guaianolides have been discovered in *Artemisia ludoviciana* var. *ludoviciana* (Ohno et al. 1980). Occasionally, mild psychoactive effects (euphoria, sensations of being "high") have been reported following deep inhalation.

The species *Artemisia tridentata* Nutt. is used in the Great Plains as an alternative to *Artemisia ludoviciana*. *Artemisia tridentata* also contains sesquiterpene lactones (Asplund et al. 1972). Sagebrush (*Artemisia arbuscula arbuscula*) is used as an incense as well. It contains an essential oil with cineole, camphor (cf. *Cinnamomum camphora*), camphene, *p*-cymene, and other compounds (Epstein and Gaudioso 1984). A number of Plains Indians also use *Artemisia cana* Pursh and the subspecies *cana* as a ritual **incense**. This plant is also rich is sesquiterpene lactones (Bhadane and Shafizadeh 1975; Lee et al. 1969).

Artemisia nilagirica (Clarke) Pamp.

The Lodha, a tribe from West Bengal (India), call this species *ote-paladu*. Tribe members inhale the smoke of the burning herbage as a sedative. This effect is also widely known in Southeast Asia. The Santal use an oil pressed from the leaves as a local anesthetic. The Oraon smoke the dried leaves to induce hallucinations (Pal and Jain 1989, 466).

Artemisia tilesii Ledeb.

The Yupik Eskimos live in southwest Alaska. Because of the paucity of flora in the tundra, they know of only a very few medicinal plants. The fresh or dried herbage of this small *Artemisia* is used to treat skin diseases, painful joints, and chest colds. A decoction is made from the herbage that is said to be strong enough once it has turned green. It is adminstered externally and internally. The ample **essential oil** consists almost entirely of thujone and isothujone, whereby thujone predominates. Thujone has potent psychoactive powers, while the effects of isothujone are similar to those of **codeine** (Overfield et al. 1980).

Artemisia tournefortiana Reichenb.—burnak

This species is indigenous to the Himalayas. In Ladakh, it is used as a psychoactive additive to **beer** (Navchoo and Buth 1990, 319*).

Literature

See also the entries for **Artemisia absinthium**, **Artemisia mexicana**, and **essential oils**.

Aldunate, Carlos, Juan J. Armesto, Victoria Castro, and Carolina Villagrán. 1983. Ethnobotany of pre-altiplanic community in the Andes of northern Chile. *Economic Botany* 37 (1): 120–35.

Asplund, R. O., Margaret McKee, and Padma Balasubramaniyan. 1972. Artevasin: A new sesquiterpene lactone from *Artemisia tridentata*. *Phytochemistry* 11:3542–44.

Bhadane, Nageshvar R., and Fred Shafizadeh. 1975. Sequiterpene lactones of sagebrush: The structure of artecanin. *Phytochemistry* 14:2651–53.

Bohlmann, Ferdinand, and Christa Zdero. 1980. Neue Sesquiterpene aus *Artemisia koidzumii*. *Phytochemistry* 19:149–51.

Brøndegaard, V. J. 1972. Artemisia in der gynäkologischen Volksmedizin. *Ethnomedizin* 2 (1/2): 3–16.

Domínguez, Xorge Alejandro, and Enrique Cárdenas G. 1975. Achillin and deacetylmatricarin from two *Artemisia* species. *Phytochemistry* 14:2511–12.

Epstein, William W., and Ellen E. Ubben Jenkins. 1979. Anthemidin, a new sesquiterpene lactone from *Artemisia ludoviciana*. *Journal of Natural Products* 42 (3): 279–81.

Epstein, William W., and Larry A. Gaudioso. 1984. Volatile oil constituents of sagebrush. *Phytochemistry* 23 (10): 2257–62.

Feraly, Farouk el-, Ibrahim A. Al-Meshal, Mohammed A. Al-Yahya, and Mohammed S. Hifnawy. 1986. On the possible role of qinghao acid in the biosynthesis of artemisinin. *Phytochemistry* 25 (11): 2777–78.

Lame Deer, Archie Fire, and Richard Erdoes. 1992. *Gift of power: The life and teachings of a Lakota medicine man.* Rochester, Vt.: Bear and Co.

Lee, K. H., R. F. Simpson, and T. A. Geissman. 1969. Sesquiterpenoid lactones of *Artemisia*, constituents of *Artemisia cana* ssp. *cana*, the structure of canin. *Phytochemistry* 8:1515–21.

Morán, A., M. J. Montero, M. L. Martín, and L. San Román. 1989a. Pharmacological screening and antimicrobial activity of the essential oil of *Artemisia caerulescens* subsp. *gallica. Journal of Ethnopharmacology* 26:197–203.

Morán, A., M. L. Martín, M. J. Montero, A. V. Ortiz de Urbina, M. A. Sevilla, and L. San Román. 1989b. Analgesic, antipyretic and anti-inflammatory activity of the essential oil of *Artemisia caerulescens* subsp. *gallica. Journal of Ethnopharmacology* 27:307–17.

Morán, A., R. Carrón, M. L. Martín, and L. San Román. 1989c. Antiasthmatic activity of *Artemisia caerulescens* subsp. *gallica. Planta Medica* 55:351–53.

Ohno, Nobuo, Jonathan Gershenzon, Catherine Roane, and Tom J. Mabry. 1980. 11,13-dehydrodesacetylmatricarin and other sesquiterpene lactones from *Artemisia ludoviciana* var. *ludoviciana* and the identity of artecanin and chrysartemin B. *Phytochemistry* 19:103–6.

Overfield, Theresa, William W. Epstein, and Larry A. Gaudioso. 1980. Eskimo uses of *Artemisia tilesii* (Compositae). *Economic Botany* 34 (2): 97–100.

Pal, D. C., and S. K. Jain. 1989. Notes on Lodha medicine in Midnapur District, West Bengal, India. *Economic Botany* 43 (4): 464–70.

Rodríguez, E., N. J. Carman, G. Vander Velde, J. H. McReynolds, T. J. Mabry, M. A. Irwin, and T. A. Geissman. 1972. Methoxylated flavonoids from *Artemisia. Phytochemistry* 11:3509–14.

Storl, Wolf-Dieter. 1995. Das esoterische Pflanzen-Lexikon: Beifuß. *Esotera* 11/95:137–39.

Twaij, Husni A. A., and Ammar A. Al-Badr. 1988. Hypoglycemic activity of *Artemisia herba alba. Journal of Ethnopharmacology* 24:123–26.

Arundo donax Linnaeus

Giant Reed

Family
Gramineae: Poaceae (Grass Family); Festuceae Tribe

Forms and Subspecies
There is a small form that has striped leaves and is frequently cultivated as an ornamental: *Arundo donax* L. cv. Variegata.

Synonyms
Arundo bambusifolia Hkr.
Arundo bengalensis Retz.
Arundo glauca Bub.
Arundo sativa nom. nud.

Folk Names
Arundo cypria, arundo tibialis, auleticon, barinari (Hindi), calamia, calamus, calamus cyprius, cana, cane of Spayne, cane sticks, canna, canna hispanica, caña brava, carizzo, carizzo de castilla, casab (Arabic), donax, flötenrohr, giant reed, great reed, guna pipi (Siona, "rock reed"), harundo, hasab (Arabic), hispanischried, italienisches rohr, juco, juinanashu(p)jua (Kamsá), kalamos (Greek), kinapipi (Secoya, "rock reed"), kyprisches rohr, nalaka (Sanskrit), navadna trstenika (Slowenic), nbj.t (ancient Egyptian), pfahlrohr, pfeilrohr, pilco, rede, rede of Spayne, ried, riesenschilf, riet, rohr, rohr aus syrien, roseau, shaq (Chumash), spanisches rohr, Spanish cane, Spanish reed, tubito, uenyinanashuf, xapij, xapij-aacöl (Serí, "great reed grass"), yuntu (Mapuche), zahm rohr

History
Archaeological finds, e.g., of flutes made from the stalks, demonstrate that *Arundo donax* was used widely in ancient Egypt since at least the time of the New Kingdom (Germer 1985, 204*). The stems have been used around the world to make shafts for arrows (Timbrook 1990, 246*). The plant has long been associated with the pastoral god Pan, in part because its shafts were used to make pipes of Pan. *Arundo donax* may have been the wondrous "twelve gods' plant" of late antiquity

The reed (*Arundo donax*) is the largest species of true grass known in Europe. (Woodcut from Tabernaemontanus, *Neu Vollkommen Kräuter-Buch,* 1731)

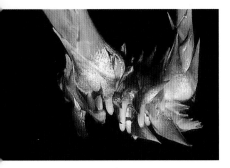

Above: The phalluslike shoots on the roots of *Arundo donax* may explain why the grass is sacred to the lusty god Pan.

Center: The giant reed (*Arundo donax*), with its typical spikes. (Photographed on Naxos, Greece)

Right: One of the many cultivated forms of *Arundo donax* (cv. Variegata) bred for ornamental purposes.

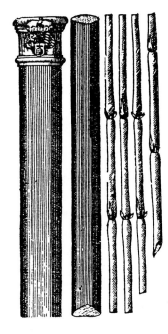

It was once thought that the Greco-Roman columns of ancient times had been inspired by the stalks of the reed (*Arundo donax*). (Woodcut from Tabernaemontanus, *Neu Vollkommen Kräuter-Buch*, 1731)

(see **dodecatheon**). Because it appears in the legend of the Buddha, the reed is also sacred to Buddhists (Gupta 1991, 18f.*). It is only recently that the psychoactive properties of the reed have become known (Ott 1993, 245*).

Distribution
The giant reed is originally from the Mediterranean region, but it spread quickly throughout the world. It has been present in the New World since the sixteenth century.

Cultivation
The simplest method is to plant root segments that have been dug up and separated from the main root or to take scions with young shoots. The scion, with its small piece of root, can be placed in water before transplanting. Young, phalluslike roots (which may help to explain the association with the phallic god Pan) will form almost overnight.

Appearance
The stalks, which grow in bundles from the nodular rhizomes, can grow 4 to 6 meters tall. The lanceolate leaves are 3 to 5 cm wide and over 50 cm in length. The symmetrical panicles can grow as long as 70 cm. In the tropics, the grass can grow over 10 meters tall. The striped form that is grown as an ornamental reaches a height of only about 3 meters.

Arundo donax is easily confused with *Phragmites australis*.

Psychoactive Material
— Rhizome (rhizoma arundinis donacis)

Preparation and Dosage
The fresh rhizome is cleaned, cut into small pieces, and macerated in an alcohol-water mixture (1:1). The maceration can be concentrated by evaporation. The residue, which is rich in alkaloids, can then be prepared in a manner appropriate for **ayahuasca analogs**.

The Shipibo shamans from Caimito use the giant reed as an **ayahuasca** additive. Northern Peruvian folk healers (*curanderos*) sometimes set up crosses of reeds when making the San Pedro drink (see *Trichocereus pachanoi*) so that the brew will not boil over. Otherwise, it will not bring good fortune (Giese 1989, 229*).

Little is known about dosages. Fifty mg of the extract (in combination with 3 g of *Peganum harmala* seeds) does not appear to produce any psychedelic effects. Unfortunately, little is also known about toxic dosages. Great care should be exercised when experimenting with *Arundo donax* (cf. *Phragmites australis*).

Ritual Use
In ancient times, the reed not only was consecrated to the nature god Pan but also was sacred to Silvanus and Priapus. It is not known whether the reed was used as a psychoactive agent in the cult of Pan. On the other hand, the syrinx, the pipes of Pan, are made from reeds, and these not only produce beautiful melodies but also can spread a "Panic terror" (Borgeaud 1988). This story

may be a metaphor for the great psychoactive power of the root (for most people, DMT experiences are profoundly terrifying).

Apart from this, there are only a few rumors of a ritual use as a psychoactive plant that can be taken seriously:

> There are statements about a secret Sufi tradition in which *Arundo donax* and *Peganum harmala* have been linked with mystical initiation. If this is in fact true, then this would be evidence for the use of a reliable ayahuasca analog in the ancient Near East—the celebrated **soma** of the Arians. (DeKorne 1995, 28)

Artifacts

Several ancient Egyptian paintings depict grasses and thickets of grasses that can be interpreted as either *Arundo donax* or **Phragmites australis** (Germer 1985, 204*). The stems were made into panpipes. They also appear to have been used as a model for the design of certain columns.

In the New World, the shafts of *Arundo donax* were used not only in the manufacture of arrows but also as ritual objects. The poles for the prayer flags of the Huichols (cf. *Lophophora williamsii*) are made from *Arundo donax* stems (per oral communication from Stacy Schaeffer). Today, Ecuadoran Indians still make panpipes from the stalks (Vickers and Plowman 1984, 13*). In Colombia, shamans wear the fronds as ear ornaments (Bristol 1965, 103*).

Medicinal Use

The rootstock was used in folk medicine primarily as a diuretic, i.e., an agent that promotes urination (Wassel and Ammar 1984).

In homeopathy, an essence of fresh root shoots called "Arundo mauritanica—water reed" was an important remedy around 1863 (Schneider 1974, 1:144 f.*).

Constituents

The rhizome contains at least five tryptamines: *N,N*-DMT, 5-MeO-DMT, **bufotenine**, dehydrobufotenine, and bufotenidine (DeKorne 1995, 27; Ghosal et al. 1969; Wassel and Ammar 1984). Little is known about other constituents.

Effects

According to Dioscorides, the flower tufts of *Arundo donax*—just like those of **Phragmites**

australis—induce deafness if they get into the ear (1:114).

The reports about the effects of an **ayahuasca analog** made with *Arundo donax* are not very promising and do not encourage others to experiment:

> For example, I once ingested one gram of *Peganum harmala* extract with 50 mg of an *Arundo donax* extraction. There was no psychoactivity at all, but I did suffer a modest allergic reaction. Within an hour I noticed that my vision was impaired—there was some difficulty in focusing on the print in a magazine. Later, my eyes felt watery and slightly swollen. The next day, I had a medium conjunctivitis with occasional hives appearing on my body. It took three days for these symptoms to subside. Obviously, one should take extreme care when experimenting with any new plant species, especially those which have no known history of shamanic usage. (DeKorne 1994, 97*)

Commercial Forms and Regulations

None

Literature

See also the entries for *Phalaris arundinacea*, *Phragmites australis*, ayahuasca analogs, *N,N*-DMT, and 5-MeO-DMT.

Borgeaud, Philippe. 1988. *The cult of Pan in ancient Greece.* Chicago and London: The University of Chicago Press.

DeKorne, Jim. 1995. Arundo donax. *Entheogene* 4:27–28.

Ghosal, S., et al. 1969. *Arundo donax* L. (Graminae): Phytochemical and pharmacological evaluation. *Journal of Medicinal Chemistry* 12:480–83.

Machen, Arthur. 1994. *Der Große Pan.* Munich: Piper.

Valenčič, Ivan. 1994. Ali vsebuje navadna trstenika (*Arundo donax*) psihedelik DMT? *Proteus* 56:258–61.

Wassel, G. M., and N. M. Ammar. 1984. Isolation of the alkaloids and evaluation of the diuretic activity of *Arundo donax. Fitoterapia* 15 (6): 357–58.

> "Pan, the Mighty, I call to you
> the god of the shepherds, the totality
>   of the universe—
> Heaven, Ocean, Earth, the queen of
>   all,
> and the immortal fire,
> for all are the limbs of Pan.
> Come, blessed one, Jumper,
> running in a circle,
> He who rules with the Horae,
> Goat-footed god:
> Friend of souls ardent for god,
> Ecstatic, cave-dweller—
> You play the world's harmony
> with merry flute tones,
> . . . .
> Off then, Blessed one, Ecstatic one,
> to the libations of sacred virtue!
> A blessed end shall join life;
> To the marrow of the earth
> Enthrall the Panic terrors
> Power!"
>
> ORPHIC HYMN

> "There *is* a real world, but it lies behind this luster and this illusion, behind all of the 'hunting for Gobelin tapestries, dreams at full speed'! Behind it, as if behind a veil. . . . The old ones knew what it meant to lift the veil. They called it: beholding the god Pan."
>
> ARTHUR MACHEN
> DER GROSSE PAN [THE GREAT PAN]
> (1994, 10)

Atropa belladonna Linnaeus

Belladonna, Deadly Nightshade

The deadly nightshade (*Atropa belladonna*) is one of the most significant medicinal plants in the history of pharmacy. (Woodcut from Tabernaemontanus, *Neu Vollkommen Kräuter-Buch*, 1731)

"When we encounter the deadly nightshade during our excursion through the columned halls of the forest, a strange feeling comes over us, as if a secretive being with fixed, staring eyes were standing behind the mysterious plant. Its sparkling, shiny black fruits reflect back to us in the dark light of the forest. The deadly nightshade has an aura of danger about it, and we can feel as we look at it that caution is advised."

BRUNO VONARBURG
DIE TOLLKIRSCHE [THE DEADLY NIGHTSHADE]
(1996, 61)

Family

Solanaceae (Nightshade Family); Atropoideae (= Solanoideae), Atropeae (= Solaneae) Tribe

Forms and Subspecies

Two varieties are distinguished on the basis of the color of their flowers and ripe fruits (Lindequist 1992, 423):

Atropa belladonna var. *belladonna*: violet flowers, black fruits

Atropa belladonna var. *lutea* Döll.[syn. *Atropa lutescens* Jacq. ex C.B. Clarke, *Atropa pallida* Bornm., *Atropa belladonna* L. var. *flava*, perhaps *Atropa acuminata* Royle ex Lindl.]: pure yellow flowers, yellow fruits

Synonyms

Atropa belladonna L. ssp. *gallica* Pascher
Atropa belladonna L. ssp. *grandiflora* Pascher
Atropa belladonna L. ssp. *minor* Pascher
Atropa lethalis Salisb.
Atropa lutescens Jacq. ex C.B. Clarke
Atropa pallida Bornm.
Belladonna baccifera Lam.
Belladonna trichotoma Scop.

Folk Names

Banewort, beilwurz, belladonna, belladonne, belledame, bennedonne, bockwurz, bollwurz, bouton noir, bullkraut, cerabella, chrottebeeri, chrotteblueme "toad flower"), deadly nightshade, deiweilskersche, dol, dollkraut, dolo, dolone, dolwurtz, dulcruyt, dwale, dway berry, English belladonna, great morel, groote nachtschaed, große graswurzel, hexenbeere, hexenkraut, höllenkraut, irrbeere, jijibe laidour (Moroccan), judenkernlein, judenkirsche, lickwetssn, mandragora theophrasti, mörderbeere, morel, morelle furieuse, poison black cherry, pollwurz, rasewurz, rattenbeere, satanskraut, saukraut, schlafapfel, schlafbeere, schlafkirsche, schlafkraut, schwarzber, schwindelbeere, sleeping nightshade, solanum bacca nigra, solanum lethale, solatrum mortale, strignus, teufelsauge, teufelsbeere, teufelsbeeri, teufelsgäggele, teufelsgückle, teufelskirsche, tintenbeere, todeskraut, tollbeere, tolle tüfus-beeri, tollkraut, tollkirsche, tüfus-beeri, waldnachtschaden, waldnachtschatt, waldnachtschatten, uva lupina "wolf's berry"), uva versa, walkerbaum, walkerbeere, wolfsauge, wolfsbeere, wolfskirsche, wutbeere, wuth-beer, yerva mora

History

Since ancient times, belladonna has been feared as a poisonous plant and demonized as a plant of witches. It has even been suggested that the plant was responsible for the extinction of the dinosaurs. These mighty lizards may have poisoned themselves on the plant or caused their own demise through hallucinations.

It is possible that Dioscorides described the deadly nightshade under the name *stychnos manikos* (Schneider 1974, 1:160*). However, this name has caused great confusion and continues to pose an ethnobotanical puzzle (cf. **Datura stramonium, Solanum spp., Strychnos nux-vomica**).

Belladonna may be identical to the *morion*, the "other, growing near caves," "male" mandrake (**Mandragora officinarum**). *Morion* literally means "male member" and refers to the plant's use as a *tollkraut* (in Middle High German, *toll* = mad, as in "crazy," and *kraut* = plant). The deadly nightshade has been used as an aphrodisiac since antiquity.

The genus name is derived from that of Atropos (= "the terrible, merciless"), one of the Three Fates or Goddesses of Destiny, who determine life and death. It is Atropos who cuts the thread of life.

In the ancient Orient, belladonna was used as an additive to **beer** and **palm wine**. The Sumerians appear to have used it to treat numerous ailments caused by demons.

Little is known about the history of the deadly nightshade during the early Middle Ages. In the eleventh century, the plant was used as a "chemical weapon" in the war between the Scots and the invading Danes. The Scots added juice from the berries to their dark beer and gave this to the thirsty Danes, who were subsequently overpowered as they lay in a delirious stupor (Schleiffer 1979, 143 ff.*; Vonarburg 1996, 62).

The demonization and denouncement of the plant, which was utilized in pagan rituals, had already begun by the time of Hildegard von Bingen:

> The deadly nightshade has coldness in it, and yet holds disgust and paralysis in this coldness, and in the earth, and at the place where it grows, the devilish prompting has a certain part and a role in its arts. And it is dangerous for a man to eat and to drink, for it destroys his spirit, as if he were dead. (*Physica* 1:52)

This demonization continued into later times (*teufelsbeere* = "devil's berry," *teufelsgäggele* = "little devil's berry," *teufelskirsche* = "Devil's cherry"), as the plant was linked to the **witches' ointments** and regarded as a dangerous and demonic plant. Because belladonna can easily produce toxic states that can prove lethal, its role as a magical plant has never been significant.

The Italian herbalist Matthiolus was the first to mention the name *belladonna*, "beautiful woman," explaining that Italian women would drip juice pressed from the plant into their eyes in order to make themselves more beautiful. The juice contains **atropine**, which effects a temporary dilation of the pupils (mydriasis). At the time, large black pupils were the epitome of beauty. Because of this dilatory effect, belladonna juice also gained great significance in eye medicine. Ophthalmologists still use **atropine**, named after *Atropa*, to achieve the same effect. Atropine, the active principle of the plant, was first isolated from the deadly nightshade in 1833 by the German apothecary Mein (Vonarburg 1996, 62).

Distribution

Belladonna is indigenous to central and southern Europe and Asia Minor. From there, it spread throughout Western Europe, as far as Iran, and all across North Africa. It is rare in Greece, where it is found only in mountainous regions. In the Alps, it grows at altitudes of up to 1,700 meters (Kruedener et al. 1993, 128*). It prefers shady locations and requires chalky soils (Vonarburg 1996, 61).

Cultivation

The simplest and most successful method of cultivating belladonna is to take cuttings from newly formed shoots or layers of the rootstock. This must be done in the spring. Cultivation from seed is rather difficult, as less than 60% of the seeds are viable. This notwithstanding, the use of seeds is important in commercial production (Morton 1977, 284*). Belladonna is cultivated on a large scale in southern and eastern Europe, Pakistan, North America, and Brazil.

Appearance

This herbaceous perennial, which can grow as tall as 1.5 meters, develops straight, ramified stalks, oblong leaves, and bell-shaped, brownish violet flowers that are enclosed in a five-cusped green calyx. The fruit, which is initially green but then turns shiny black, is roughly the size of a cherry and sits upon the five-pointed calyx. Belladonna blooms between June and August and often already bears fruits at this time. The variety *lutea* has yellow flowers, yellow fruits, and a green stem.

The deadly nightshade produces an attractive nectar that bees and bumblebees enthusiastically collect and transform into psychoactive **honey** (Hazlinsky 1956). The plant is also pollinated by these insects (Vonarburg 1996, 62).

Although belladonna can be mistaken only for other *Atropa* species (see page 85), the *Scopolia* species (see *Scopolia carniolica*) are an occasional source of confusion.

Historical sources often confused the deadly

nightshade with the bittersweet nightshade (*Solanum dulcamara*) and the black nightshade (*Solanum nigrum*; cf. **Solanum spp.**) (Schneider 1974, 1:160*) and occasionally with the sleeping berry (**Withania somnifera**). The herb Paris (*Paris quadrifolia* L.; cf. **Aconitum spp.**) has also been regarded as a form of belladonna (Schwamm 1988, 133).

Psychoactive Material

—Leaves (belladonnae folium, belladonnae herba, folia belladonnae, herba belladonnae, solani furiosi, belladonnablätter). The pharmaceutical raw drug is sometimes falsified with leaves of tree-of-heaven (*Ailanthus altissima* L.), pokeberry (*Phytolacca americana* L., **Phytolacca acinosa**), *Hyoscyamus muticus*, *Physalis alkekengi* L. (cf. **Physalis spp.**), and *Scopolia carniolica*.

—Roots (belladonnae radix, radix belladonnae, belladonnawurzel). The pharmaceutical raw drug is sometimes falsified with the roots of pokeberry (*Phytolacca americana* L., **Phytolacca acinosa**) or *Scopolia carniolica*.

—Fresh or dried fruits (belladonnae fructus, fructus belladonnae).

Preparation and Dosage

The leaves of the wild form should be harvested in May or June, as their alkaloid content is greatest at this time. They are dried in the shade and must then be stored away from light in an airtight container. It is best to harvest the fruits when they are almost ripe. These must be dried in a dry, ventilated location. Both the leaves and the fruits are suitable for use in **smoking blends** and can be combined with dried fly agaric mushrooms

Left: The typical violet-tinged flower of the deadly nightshade (*Atropa belladonna*).

Right, from top to bottom:
The rare yellow-blooming variety of the deadly nightshade (*Atropa belladonna* var. *lutea*).

The shiny black fruit of the deadly nightshade has a seductively sweet taste.

The ripe fruit of the yellow-blooming *Atropa belladonna* var. *lutea* is also yellow and is easily mistaken for the fruits of the mandrake (*Mandragora*).

Belladonna, the goddess of the deadly nightshade, is depicted with a wreath made from the leaves and fruits of *Atropa belladonna*. She appears to be deep in a nightshade dream. (*Belladonna*; engraving after a painting by Gabriel Max, printed in the *Jugendstil* journal *Gartenlaube*, 1902)

Atropos, the goddess of destiny who severs the thread of life, provided the inspiration for the botanical name of the deadly nightshade. (Floor mosaic, Roman period, Cyprus)

(*Amanita muscaria*) and hemp (*Cannabis indica*). Pharmaceutical cigarettes made of belladonna leaves soaked in a tincture of opium (cf. *Papaver somniferum*) were still being prescribed as recently as 1930 (Schneider 1974, 1:162*).

Ingestion of one or two fresh berries produces mild perceptual changes approximately one to two hours after consumption. Three or four fresh berries is regarded as a psychoactive aphrodisiac; three to a maximum of ten berries represents a hallucinogenic dosage. Ten to twenty berries is said to be lethal; among children, as few as two or three berries can cause death (Vonarburg 1996, 62). Extreme care should be exercised with *Atropa belladonna*! With some individuals, even the smallest amounts can have devastating results (delirious states). Using the plant as a fumigant or as an ingredient in **smoking blends** is the least dangerous method of consumption.

When used (internally) for medicinal purposes, 0.05 to 0.1 g of dried and powdered leaves is regarded as an average individual dosage (Lindequist 1992, 429). The therapeutic dosage of **atropine** is listed as 0.5 to 2 mg. A pleasant psychoactive dosage is 30 to 200 mg of dried leaves or 30 to 120 mg of dried roots, either smoked or ingested internally (Gottlieb 1973, 5*).

Belladonna is reputed to have been an ingredient in **witches' ointments** and was used as a magical fumigant. One traditional "oracular incense" had the deadly nightshade as its chief component and active ingredient (cf. **incense**). It contained:

leaves of fool's parsley [*Aethusa* spp., *Apium* spp., or *Sium* spp.], harvested during the new moon,
acorns [*Quercus* spp.], plucked during the full moon while naked,
leaves and flowers of the deadly nightshade, harvested at midday,
leaves of vervain [*Verbena officinalis*], plucked by hand in the afternoon,
leaves of wild peppermint [*Mentha* spp.], picked in the morning,
leaves of the thistle [*Viscum album*], from the previous year, cut at midnight.
(Magister Botanicus 1995, 185*)

Unfortunately, no precise details about the amounts to use have come down to us.

Belladonna berries can also be mashed, fermented, and distilled into **alcohol** and were formerly used as a psychoactive additive to **beer**, **mead**, **palm wine**, and **wine**. They are also an ingredient in the Moroccan spice mixture known as *ras el hanout* (Norman 1993, 96 f.*).

Ritual Use
Since ancient times, deadly nightshade has probably been used in the same or a very similar manner as mandrake (see **Mandragora officinarum**). It is possible that its root was also used as a substitute or an alternative to that of the mandrake. Folktales have preserved the remnants of a belladonna cult that suggest that this may have been the case. In Hungary, for example, the root "is dug up on the night of St. George while naked, and a bread offering is made as if to an elfish monster" (Höfler 1990, 90*). In Romania, the deadly nightshade is also known as the wolf cherry, the flower of the forest, the lady of the forest, and the empress of herbs.

Belladonna is common in some regions of southern Germany. It is uncertain whether the plant is part of the indigenous flora or was introduced during the early Middle Ages. The German names for the plant are suggestive of its psychoactive effects (*schlafbeere* = "sleeping berry," *rasewurz* = "mad root," *tollkirsche* = "crazy cherry") and contain pagan references (*wolfsauge* = "wolf's eye," *wutbeere* = "rage berry"); the wolf is the animal of Wotan and *wut* (= fury, ecstasy) is his characteristic (*wuotan*, "the furious"). The deadly nightshade is also associated with the daughters of Wotan: "On the Lower Rhine, its fruits are known as *Walkerbeeren* ["Valkyrie berries"], and the plant itself is known as the *Walkerbaum* ["Valkyrie tree"], and anyone who ate of the berries would fall victim to the Valkyries" (Perger 1864, 182 f.*). The Valkyries are the daughters of Heaven and Earth (Wotan and Erda) who accompany the souls of heroes who have fallen in battle to Valhalla, where they are entertained with inebriating **mead** while they await the Götterdämmerung, the Twilight of the

Gods (i.e., the rebirth of the world). Since Wotan is the lord of the Wild Chase as well as of the hunt and the forest, he was also closely associated with hunters. Consequently, as late as the nineteenth century, southern German hunters would still often consume three or four belladonna berries before going out hunting, a practice that was said to sharpen their perception and make them better hunters.[47]

Although the deadly nightshade is regarded as a classic witches' plant, only a very few details have come down to us concerning its magical use in witches' rituals. In his work *Magiae naturalis* [Natural Magic], Giovanni Battista della Porta (ca. 1535–1615) wrote that a person could use an arcanum (secret means) to transform himself into a bird, fish, or goose—the sacred animal that was sacrificed to Wotan/Odin at the time of the winter solstice—and thereby have much amusement. He listed belladonna first among the agents that could be used for these purposes (Schleiffer 1979, 139 f.*).

In Celtic rituals and in the neo-pagan rituals of certain modern "witches cults" (Wicca) that are based upon these, following a fasting period of fourteen days[48] (one fortnight), the oracular fumigation noted above was and is still carried out during the night of the full moon preceding the festival of Samhain (November 1). A tea made of *Amanita muscaria* was also consumed on that day:

> The members of a band of people knowledgeable about herbs would then gather in the "sacred" night and select one of their own, who would then sit before the incense vessel as the oracular priest/ess and inhale the toxic fumes.
>
> The resulting toxic effects of the smoke would place the priest/ess into a trance state in which he or she would then serve as oracle and answer the questions of the others or establish contact to the spirits or gods. It is also interesting that a priest or priestess was never allowed to function as the oracle on two consecutive occasions. (Magister Botanicus 1995, 185 f.*)

Artifacts

In the nineteenth century and in the 1920s, both the berry of the deadly nightshade and belladonna in its anthropomorphic form were frequently portrayed in printed material (e.g., by Erich Brukal, Paul Wending; cf. Rätsch 1995, 138*). It is not known whether these pictures were the result of the artists' personal experiences with belladonna preparations. It is possible that the legends concerning the witches' plant and the spirit that lives within it provided all the inspiration they needed.

The short film *Belladonna*, by herman de vries and others, depicts a magical ritual in which witches use the deadly nightshade. It shows a young woman walking through the forest in search of the plant; when she finds it, she disrobes and rubs its fruits all over her body. The film then attempts to portray the resulting psychoactive effects. An experimental film entitled Atropa belladonna—*Die Farbe der Zeit* [Atropa belladonna—The Color of Time] was also inspired by the myth of the beautiful woman and the spirit of the plant (Friel and Bohn 1995).

Belladonna has also left its marks on both psychedelic and heavy metal music (e.g., Ian Carr, *Belladonna*, as well as the band Belladonna) and in the music of Andreas Vollenweider.

The book *Right Where You Are Sitting Now: Further Tales of the Illuminati*, by Robert Anton Wilson, provides a turbulent literary version of belladonna inebriation (1982, 13–26).

Medicinal Use

The deadly nightshade has been used for medicinal purposes since ancient times, e.g., as an analgesic (cf. **soporific sponge**). It was often administered to "dispel demons" (in other words, it was likely used in therapies to treat depressions, psychoses, and other mental illnesses). Rudiments of such psychiatric usage have been preserved in northern Africa.

In Morocco, the dried berries, together with a small amount of water and some sugar, are made into a tea that can "help produce a good mental condition." This tea is also an aphrodisiac for men. It is also said "that a small dose of belladonna clears the mind and enables one to do intellectual tasks" (Venzlaff 1977, 82*). A couple of fresh berries are also said to increase the ability to remember.

In the nineteenth century, extracts of the root and herbage were used to treat jaundice, dropsy, whooping cough, convulsive cough, nervous ailments, scarlet fever, epilepsy, neuroses, renal colic, various skin diseases, eye inflammations, and diseases of the urinary and respiratory tracts, the throat, and the esophagus (Schneider 1974, 1:161*).

A mother tincture obtained from the fresh plant together with the rootstock at the end of the flowering season (Atropa belladonna hom. *PFX* and *RhHAB1*, Belladonna hom. *HAB1*) as well as various dilutions (normally D4 and above) are used in homeopathy for numerous purposes, depending upon the medical description (Vonarburg 1996, 63).

Constituents

The entire plant contains between 0.272 and 0.511% **tropane alkaloids**; the variety *lutea* contains only 0.295% (Lindequist 1992, 424). The

The psychedelic wave of the late 1960s also swept over many jazz musicians in the early 1970s. In 1972, the trumpeter Ian Carr dedicated an entire album (*Belladonna*) to the hallucinogenic deadly nightshade. In this cover photograph, however, the musician is actually shown standing among field poppies (*Papaver rhoeas*). (CD cover 1990, Linam Records.)

47 As Elisabeth Blackwell wrote in her *Herbal*, "The berries, which look so lovely, almost always bring either deadly vomiting or a deadly rage, as tragic experiences have confirmed" (Heilmann 1984, 96*).

48 During the first seven days of the fasting period, an apple could be eaten at midnight on every second day; otherwise, only water could be consumed (Magister Botanicus 1995, 185*).

stalk can contain up to 0.9% alkaloids, the unripe fruits up to 0.8%, the ripe fruits from 0.1 to 9.6%, and the seeds approximately 0.4%. In the living plant, (–)-hyoscyamine predominates; following harvest, during the drying and storage process, this is converted to **atropine**. The dried leaves contain 0.2 to 2% alkaloids, the dried root 0.3 to 1.2%. Hyoscyamine is the primary component (68.7%), apoatropine is the secondary alkaloid (17.9%), and many other tropane alkaloids are also present (433).

The alkaloids in the plant apparently pass into the tissues of animals that have eaten the foliage, the fruits, or the roots. In one eighteenth-century case, an entire family experienced hallucinations after eating a rabbit. Rabbits appear to be fond of the plant and do not display any toxic reactions (Ruspini 1995).

Effects

The clinical picture of the effects of belladonna is rather homogeneous (and reminiscent of the effects of **Datura** and **Brugmansia**):

> Within a quarter of an hour, the following toxic symptoms appear: psychomotor disquiet and general arousal, not infrequently of an erotic nature, urge to speak, great euphoria (cheerfulness, urge to laugh), but also fits of crying, strong desire for movement, which may be manifested as an urge to dance, disturbances of intention, manneristic and stereotypic movements, choreatic states, ataxia, disturbances of thought, sensations of befuddlement, confused speech, screaming, hallucinations of a diverse nature; increasing states of excitation culminating in frenzy, rage, madness, with complete lack of ability to recognize the surroundings. (Roth et al. 1994, 158*)

Death can result from respiratory paralysis. The (main) effects last for three to four hours; the effects on vision may continue for three to four days.

Belladonna-induced hallucinations are typically described as threatening, dark, demonic, devilish, hellish, very frightening, and profoundly terrifying. Many users have compared the effects to those of a "Hieronymus Bosch trip" and have indicated that they have no intention of repeating the experiment (Gabel 1968; Illmaier 1996; Pestolozzi and Caduff 1986).

The alkaloids also cause the mucous membranes to become very dry and the face to turn red, while the pulse rate accelerates and the pupils dilate.

Commercial Forms and Regulations

In Germany, belladonna leaves and roots are available in pharmacies and require a physician's prescription (Lindequist 1992, 431).

Literature

See also the entry for **atropine**.

Dräger, B., and A. Schaal. 1991. Isolation of pseudotropine-forming tropinone reductase from *Atropa belladonna* root cultures. *Planta Medica* 57 suppl. (2): A99–100.

Friel, Gunnar, and Ralf Bohn. 1995. Atropa belladonna—*Arbeiten am Film*. Vienna: Passagen Verlag.

Gabel, M. C. 1968. Purposeful ingestion of belladonna for hallucinatory effects. *Journal of Pediatrics* 76:864–66.

Hazlinsky, B. 1956. Poisonous honey from deadly nightshade. *Zeitschrift für Bienenforschung* 3:93–96.

Heltmann, H. 1979. Morphological and phytochemical studies in *Atropa* species. *Planta Medica* 36:230–31.

Illmaier, Thomas. 1996. Die unerbittlich schöne Frau. *Grow!* 5/96:20–23.

Kessel, Joseph. 1929. *Belladonna*. Munich: Piper. (A novel.)

Lindequist, Ulrike. 1992. Atropa. In *Hagers Handbuch der pharmazeutischen Praxis*, 5th ed., 4:423–37. Berlin: Springer.

Münch, Burchard Friedrich. 1785. *Practische Abhandlung von der Belladonna und ihrer Anwendung*. Göttingen, Germany: Diederich.

Pestolozzi, B. C., and F. Caduff. 1986. Gruppenvergiftung mit Tollkirschentee. *Schweizerische medizinische Wochenschrift* 116:924–26.

Rowson, J. M. 1950. The pharmacognosy of *Atropa belladonna* L. *Journal of Pharmacy and Pharmacology* 2:201–16.

Ruspini, G. 1995. Belladonna e conigli. *Eleusis* 3:29–30.

Schwamm, Brigitte. 1988. *Atropa belladonna, eine antike Heilpflanze im modernen Arzneischatz*. Stuttgart: Deutscher Apotheker Verlag. (Quellen und Studien zur Geschichte der Pharmazie, vol. 49.) (Excellent bibliography.)

Vonarburg, Bruno. 1996. Die Tollkirsche (1. Teil). *Natürlich* 10/96:61–64.

Wilson, Robert Anton. 1984. *Right Where You Are Sitting Now: Further Tales of the Illuminati*. Berkeley, Calif.: And/Or Press.

Other Species of Belladonna

The genus *Atropa* is composed of four to six species, all of which occur only in Eurasia (D'Arcy 1991, 79*; Symon 1991, 147*). The genus is treated inconsistently in the taxonomic literature. The only species that has attained any ethnopharmacological significance as a psychoactive plant is *Atropa belladonna*. Only the Indian belladonna has acquired a certain ethnomedicinal usage. All species contain **tropane alkaloids**, primarily hyoscyamine and **atropine**, along with apoatropine, belladonnine, and cuscohygrine (in the roots). The leaves also contain quercetin and camphor oil derivatives and **coumarins (scopoletin)** (Lindequist 1992, 423*).

Atropa aborescens [nom. nud.?]

This species, collected in Martinique, "contains narcotic poisonous substances" (von Reis and Lipp 1982, 266*).

Atropa acaulis L.

Synonym for *Mandragora officinarum*

Atropa acuminata Royle ex Lindl. [syn. *Atropa belladonna* var. *acuminata*, *Atropa belladonna* C.B. Clarke non. L.]—Indian belladonna, sagangur

This species occurs only in the Indian districts of Barmula, Kinnaur, Simla, and Nainital. It is very similar to *Atropa belladonna*, in particular the yellow-flowered subspecies. But this species produces yellow flowers and black fruits (Morgan 1977, 289*). For this reason, it has recently been regarded more as a synonym, although it may represent a local variety or subspecies. Indian belladonna has almost the same alkaloid content as the European species, but it has a higher concentration of scopolamine (approximately 30% of the total alkaloids). It is cultivated in Afghanistan and Kashmir for pharmaceutical purposes. The Indian raw drug is very often falsified with the roots of *Phytolacca acinosa* (Morton 1977, 290*). *Atropa acuminata* is also regarded as a synonym for *Atropa baetica*.

Atropa baetica Willk.—Iberian belladonna

This species is found in Spain and possesses a higher alkaloid content (with only a little **scopolamine**) than *Atropa belladonna*.

Atropa caucasica Kreyer—Caucasian belladonna

The only other Asian species besides *Atropa acuminata*.

Atropa cordata—heart-leaved belladonna

This species is apparently a broad-leaved European form of *Atropa belladonna*.

Atropa digitaloides—finger-leaved belladonna

This species is apparently a small-leaved European form of *Atropa belladonna*.

Atropa komarovii Blin. et Shal—Turkmenic belladonna

Found only in Turkmenistan, this species is being tested for commerical cultivation as a source of alkaloids.

Atropa mandragora (L.) Woodville

Synonym for *Mandragora officinarum*.

Atropa x *martiana* Font Quer

Hybrid from *Atropa baetica* and *Atropa belladonna*.

Atropa pallidiflora Schönb.-Tem.

Has an alkaloid concentration that is approximately as high as that of *Atropa belladonna*, although the mixture contains approximately 30% scopolamine.

Atropa rhomboidea Gill. et Hook

Now known as *Salpichroa origanifolia* (Lam.) Baill., this plant occurs in the southern portion of South America as far as Tierra del Fuego (Zander 1994, 496*).

The fruit of the deadly nightshade (*Atropa belladonna*), whole and in cross section.

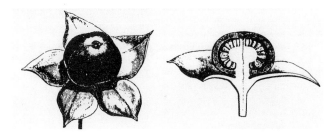

Banisteriopsis caapi (Spruce ex Grisebach) Morton

Ayahuasca Vine

Inflorescence and fruit of the ayahuasca vine (*Banisteriopsis caapi*). Under cultivation, the vine rarely develops flowers. (Drawing by Sebastian Rätsch)

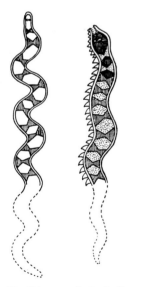

The Tukanos and other Indians of the Amazon regard the ayahuasca vine as a snake that can bear humans into the world of the spirits. (Traditional representation, from Koch-Grünberg, *Zwei Jahre bei den Indianern Nordwest-Brasiliens,* 1921*)

49 The Makú Indians use this name (which is applied throughout Amazonia to *Banisteriopsis caapi*) to refer to another vine from the same family: *Tetrapteris methystica.*

Family

Malpighiaceae (Barbados Cherry Family); Pyramidotorae, Banisteriae Tribe

Forms and Subspecies

Two varieties have been distinguished (D. McKenna 1996):

Banisteriopsis caapi var. *caupari*
Banisteriopsis caapi var. *tukonaka*

The first form has a knotty stem and is considered to be more potent; the second has a completely smooth stem.

The Andoques Indians distinguish among three forms of the vine, depending upon the types of effects each has upon the shamans: *iñotaino'* (transformation into a jaguar), *hapataino'* (transformation into an anaconda), and *kadanytaino'* (transformation into a goshawk) (Schultes 1985, 62). The Siona make a distinction among the following cultivated forms: *wa'i yahé* ("flesh yahé," with green leaves), *ya'wi yahé* ("pekari yahé," with yellow-striped leaves), *naso ānya yahé* ("monkey snake yahé"), *naso yahé* ("monkey yahé," with striped leaves), *yahé repa* ("proper yahé"), *tara yahé* ("bone yahé," with knotty stems), *'aíro yahé* ("forest yahé"), *bi'ā yahé* ("bird yahé," with small leaves), *sia sewi yahé* ("egg *sewi* yahé," with yellowish leaves), *sēsé yahé* ("white-lipped peccary yahé"), *wēki yahé* ("tapir yahé," of large size), *yaí yahé* ("jaguar yahé"), *nea yahé* ("black yahé," with dark stems), *horo yahé* ("flower yahé"), and *sisé yahé* (Vickers and Plowman 1984, 18 f.*).

Synonyms

Banisteria caapi Spruce ex Griseb.
Banisteria quitensis Niedenzu
Banisteriopsis inebrians Morton
Banisteriopsis quitensis (Niedenzu) Morton

Folk Names

Amarón wáska, "boa vine"), ambi-huasca (Inga, "medicine vine"), ambiwáska, ayahuasca amarilla, ayahuascaliane, ayahuasca negra, ayahuasca vine, ayawasca, ayawáska, bejuco de oro ("gold vine"), bejuco de yagé, biaj (Kamsá, "vine"), biáxa, biaxíi, bichémia, caapi,[49] caapí, camárambi (Piro), cauupuri mariri, cielo ayahuasca, cuchi-ayahuasca, cushi rao (Shipibo, "strong medicinal plant"), doctor, hi(d)-yati (d)yahe, iáhi', kaapi, kaapistrauch, kaheé, kahi, kalí, kamarampi (Matsigenka), máo de onça, maridi, natem, natema, nepe, nepi, nishi (Shipibo, "vine"), oo'-na-oo (Witoto), purga-huasca, purga-huasca de los perros, rao (Shipibo, "medicinal plant"), reéma (Makuna), sacawáska, sacha-huasca (Inga, "wild vine"), seelenliane, seelenranke, shuri-fisopa, tiwaco-mariri, totenliane, vine of the dead, vine of the soul, yagé, yagé cultivado, yagé del monte, yagé sembrado, yahe, yaje, yáje, yajé, yajén, yaji, yaxé (Tukano, "sorcerer's plant")

History

The word *ayahuasca* is Quechuan and means "vine of the soul" or "vine of the spirits" (Bennett 1992, 492*). The plant has apparently been used in South America for centuries or even millennia to manufacture psychoactive drinks (**ayahuasca**, *natema, yahé,* etc.). Richard Spruce (1817–1893) collected the first botanical samples of the vine between 1851 and 1854 (Schultes 1957, 1983c*). The original voucher specimens have even been tested for alkaloids (Schultes et al. 1969).

The German ethnographer Theodor Koch-Grünberg (1872–1924) was one of the first to observe and describe the manufacture of the caapi drink from *Banisteriopsis caapi* (1921, 190 ff.). The pharmacology was first elucidated in the mid-twentieth century (see **ayahuasca**).

Distribution

It is not certain where the plant originated, as it is now cultivated in Peru, Ecuador, Colombia, and Brazil, that is, throughout the entire Amazon basin. Wild plants appear to be chiefly stands that have become wild (Gates 1982, 113).

Cultivation

The plant is cultivated almost exclusively through cuttings, as most cultivated plants are infertile (Bristol 1965, 207*). A young shoot or the tip of a branch is allowed to stand in water until it forms roots, after which it is transplanted or simply placed into humus-rich, moist soil and watered profusely. The fast-growing plant thrives only in moist tropical climates and does not typically tolerate any frost.

Appearance

This giant vine forms very long and very woody stems that branch repeatedly. The large, green leaves are round-ovate in shape, pointed at the end (8 to 18 cm long, 3.5 to 8 cm in width), and opposite. The inflorescences grow from axillary panicles and have four umbels. The flowers are 12 to 14 mm in size and have five white or pale pink sepals. The plant flowers only rarely (Schultes 1957, 32); in the tropics, the flowering period is in

January (although it can also occur between December and August). The winged fruits appear between March and August (Ott 1996) and resemble the fruits of the maple (*Acer* spp.). The plant is quite variable, which is why it has been described under several different names (see "Synonyms").

The vine is closely related to *Banisteriopsis membranifolia* and *Banisteriopsis muricata* (see *Banisteriopsis* spp.) and can easily be confused with these (Gates 1982, 113). It is also quite similar to *Diplopterys cabrerana*.

Psychoactive Material
— Stems, fresh or dried (banisteriae lignum)
— Bark, fresh or dried, of the trunk (banisteriae cortex)
— Leaves, dried

Preparation and Dosage
In Amazonia, dried pieces of the bark and the leaves are smoked. The Witotos powder dried leaves so that they can smoke them as a hallucinogen (Schultes 1985).

The vine is rarely used alone to produce **ayahuasca** or yagé:

> The Tukano prepare the yajé by dissolving it in cold water, not, as is done by other tribes to the south, by boiling. Short pieces of the liana are macerated in a wooden mortar, unmixed with the leaves or with other ingredients. Cold water is added, and the liquid is passed through a sieve and placed in a special ceramic vessel. This solution is prepared two or three hours before its proposed ceremonial use, and it is later drunk by the group from small cups. These drinking vessels hold 70 cubic cm and between drinks, six or seven in number, intervals of about an hour elapse. (Reichel-Dolmatoff 1970, 32).

In between they drink **chicha**, a slightly fermented **beer**, and smoke copious amounts of tobacco (*Nicotiana rustica, Nicotiana tabacum*).

The vine is usually prepared together with one or more additives so that it develops either psychedelic (using plants that contain DMT, primarily *Psychotria viridis*) or healing (e.g., with *Ilex guayusa*) powers (see list at **ayahuasca**).

Small baskets made of strips of ayahuasca bark 4 to 6 mm in thickness (total dry weight = 13 to 14 g) are now being produced in Ecuador; each basket corresponds to the dosage for one person. These little baskets are stuffed with leaves of *Psychotria viridis* (approximately 20 g) and boiled to prepare a psychedelic drink.

Ritual Use
The Desana, a Colombian Tukano tribe, drink pure **ayahuasca** only on ritual occasions, although these do not have to be associated with any particular purpose, such as healing or divination. Only men may consume the drink, although the women are involved as dancers (i.e., as entertainment). The ritual begins with the recitation of creation myths and is accompanied by songs. It lasts for eight to ten hours. Very large amounts of **chicha** are also consumed while the ritual is in progress (Reichel-Dolmatoff 1970, 32).

For more on ritual use, see **ayahuasca**.

Artifacts
See **ayahuasca**.

Medicinal Use
In some areas of the Amazon, and among the followers of the Brazilian Umbanda cult, a tea made from the ayahuasca vine is drunk as a remedy for a wide variety of diseases and may also be used externally for massaging into the skin (Luis Eduardo Luna, pers. comm.).

Constituents
The entire plant contains alkaloids of the β-carboline type. The principal alkaloids are **harmine**, **harmaline**, and tetrahydroharmine. Also present are the related alkaloids harmine-*N*-oxide, harmic acid methylester (= methyl-7-methoxy-β-carboline-1-carboxylate), harmalinic acid (= 7-methoxy-3,4-dihydro-β-carboline-1-carboxyl acid), harmic amide (= 1-carbamoyl-7-methoxy-β-carboline), acethylnorharmine (= 1-acetyl-7-methoxy-β-carboline), and ketotetrahydronorharmine (= 7-methoxy-1,2,3,4-tetrahydro-1-oxo-β-carboline) (Hashimoto and Kawanishi 1975, 1976). Also present are shihuninine and dihydroshihunine (Kawanishi et al. 1982).

The stems contain 0.11 to 0.83% alkaloids, the branches 0.28 to 0.37%, the leaves 0.28 to 0.7%, and the roots between 0.64 and 1.95%. Of these, 40 to 96% is harmine. Harmaline is completely absent in some samples, while in others it can comprise as much as 15% of the total alkaloid content (Brenneisen 1992, 458). The stems and bark also contain large quantities of tanning agents.

It has also been reported that the vine contains **caffeine**. This information is probably due to confusion with *Paullinia yoco* (cf. *Paulinia* spp.) (Brenneisen 1992, 458).

Effects
The vine functions as a potent MAO inhibitor, whereby only the endogenous enzyme MAO-A is inhibited (see **ayahuasca**). As a result, both endogenous and externally introduced tryptamines, such as *N,N*-DMT, are not broken down and are thus able to pass across the blood-brain barrier.

The ayahuasca vine (*Banisteriopsis caapi*) blooms in January. The plant only flowers in the tropics.

Harmine

Harmaline

"Caapi is a decoction of a Malpighiaceae shrub (*Banisteria*) and is prepared in the following way by the men only, for the women do not drink any Caapi. The roots, stems, and leaves are pounded in a wide, trough-shaped mortar into a greenish brown mass that is washed in a pot with a little water, squeezed thoroughly, and then pounded in the mortar and washed again. The resulting mush, which bears something of a resemblance to cow dung, is strained through two fine sieves placed on one another into the Caapi vessel, whereby this is aided by hitting against the edge of the sieve. The pot with the unappetizing drink is covered carefully with leaves and placed in front of the house for a time. The Caapi vessel always has the same bulging urn shape and is always painted with the same yellow pattern against a dark red background. Remarkably, these are very similar to the patterns that are painted on the round exterior of the signal drums. At the upper edge, the vessel has two leaf-shaped handles that protrude out horizontally and are used to carry it, and two holes, in which a string for hanging it is attached. It is never washed but is newly painted from time to time.

"The effects of Caapi resemble hashish inebriation. One can see how the Indians say that everything is much larger and more beautiful that it really is. The house is enormously large and magnificent. They see many, many people, especially many women.—The erotic appears to play a central role in this inebriation. — Large, colorful snakes wind their way up and down the house posts. All of the colors are garishly colorful. Some who drink Caapi suddenly fall into a deep unconsciousness state and then have the most beautiful dreams, and admittedly also the most beautiful headache when they awake—hangover."

THEODOR KOCH-GRÜNBERG
ZWEI JAHRE BEI DEN INDIANERN NORDWEST-BRASILIENS [TWO YEARS AMONG THE INDIANS OF NORTHWEST BRAZIL]
(1921, 119f.*)

When the vine is used alone, it produces mood-enhancing and sedative properties. In higher dosages, the **harmine** present in the plant (above 150 to 200 mg) can induce nausea, vomiting, and shivering (Brenneisen 1992, 460).

In the 1960s, Reichel-Dolmatoff was able to take part in numerous ayahuasca rituals among the Desana. He wrote the following about his experience with a repeated administration of a drink that was said to have been made only from *Banisteriopsis caapi*:

My own experience was as follows: first draft, pulse 100, a sense of euphoria, followed by a passing drowsiness; second draft, pulse 84; fourth drink, pulse 82 and strong vomiting; sixth draft, pulse 82, severe diarrhea. Almost immediately there appeared to me spectacular visions in color of a multitude of intricate designs of marked bilateral symmetry, which passed slowly in oblique bands before my range of vision, my eyes being half closed. The visions continued, becoming modified, for more than twenty minutes, during which time I was entirely conscious and able to describe my experience very clearly on the tape recorder. There were no acoustical phenomena and no figures represented. (Reichel-Dolmatoff 1970, 33)

Commercial Forms and Regulations

Pieces of the vine are only rarely offered in ethnobotanical specialty stores. There are no regulations concerning the plant.

Literature

See also the entries for *Banisteriopsis* **spp.**, *Diplopterys cabrerana*, and **ayahuasca**.

Brenneisen, Rudolf. 1992. *Banisteriopsis*. In *Hagers Handbuch der pharmazeutischen Praxis,* 5th ed., 4:457–61. Berlin: Springer.

Elger, F. 1928. Über das Vorkommen von Harmin in einer südamerikanischen Liane (Yagé). *Helvetica Chimica Acta* 11:162–66.

Friedberg, C. 1965. Des *Banisteriopsis* utilisés comme drogue en Amerique du Sud. *Journal d'Agriculture Tropicale et de Botanique Appliquée* 12:1–139.

Gates, Brownwen. 1982. A monograph of *Banisteriopsis* and *Diplopterys*, Malpighiaceae. *Flora Neotropica*, no. 30, The New York Botanical Garden.

———. 1986. La taxonomía de las *malpigiáceas* utilizadas en el brebaje del ayahuasca. *América Indígena* 46 (1): 49–72.

Hashimoto, Yohei, and Kazuko Kawanishi. 1975. New organic bases from Amazonian *Banisteriopsis caapi*. *Phytochemistry* 14:1633–35.

———. 1976. New alkaloids from *Banisteriopsis caapi*. *Phytochemistry* 15:1559–60.

Hochstein, F. A., and A. M. Paradies. 1957. Alkaloids of *Banisteria caapi* and *Prestonia amazonicum*. *Journal of the American Chemical Society* 79:5735–36.

Lewin, Louis. 1928. Untersuchungen über *Banisteria caapi* Spruce (ein südamerikanisches Rausch-mittel). *Naunyn Schmiedeberg's Archiv für Experimentelle Pathologie und Pharmakologie* 129:133–49.

———. 1986. Banisteria caapi, *ein neues Rauschgift und Heilmittel.* Berlin: EXpress Edition, Reihe Ethnomedizin und Bewußtseinsforschung—Historische Materialien 1. (Orig. pub. 1929.)

Kawanishi, K., et al. 1982. Shihuninine and dihydroshihunine from *Banisteriopsis caapi*. *Journal of Natural Products* 45:637–39.

McKenna, Dennis. 1996. Lecture given at Ethnobotany Conference, San Francisco.

Mors, W. B., and P. Zaltzman. 1954. Sôbre o alcaloide de *Banisteria caapi* Spruce e do *Cabi paraensis* Ducke. *Boletím do Instituto de Quimica Agricola* 34:17–27.

Morton, Conrad V. 1931. Notes on *yagé*, a drug-plant of southeastern Colombia. *Journal of the Washington Academy of Sciences* 21:485–88.

Ott, Jonathan. 1996. *Banisteriopsis caapi*. Unpublished electronic file. Cited 1998.

Reichel-Dolmatoff, Gerardo. 1969. El contexto cultural de un alucinogeno aborigen: *Banisteriopsis caapi*. *Revista de la Academia Colombiana de Ciencias Exactas, Físicas y Naturales* 13 (51): 327–45.

———. 1970. Notes on the cultural context of the use of yagé (*Banisteriopsis caapí*) among the Indians of the Vaupés, Colombia. *Economic Botany* 24 (1): 32–33.

Schultes, Richard Evans. 1985. De Plantis Toxicariis e Mundo Novo Tropicale: Commentationes XXXVI: A novel method of utilizing the hallucinogenic *Banisteriopsis*. *Botanical Museum Leaflets* 30 (3): 61–63.

Schultes, Richard Evans, et al. 1969. De Plantis Toxicariis e Mundo Novo Tropicale: Commentationes III: Phytochemical examination of Spruce's original collection of *Banisteriopsis caapi*. *Botanical Museum Leaflets* 22 (4): 121–32.

Banisteriopsis spp.

Banisteriopsis Species

Family

Malpighiaceae (Barbados Cherry Family); Banisteriae Tribe

Today, some ninety-two species of the genus *Banisteriopsis* are recognized. Most species occur in the tropical lowlands of Central and South America. A few species are also found in Asia.

Banisteriopsis argentea (Spreng. ex A. Juss.) Morton

A native of India, this species contains tetrahydroharman, 5-methoxytetrahydroharman, **harmine**, **harmaline**, and the **β-carboline** leptaflorin (Ghosal et al., 1971). The leaves contain only 0.02% alkaloids [(+)-N_b-methyltetrahydroharmane, **N,N-DMT**, N,N-DMT-N_b-oxide, (+)-tetrahydroharmine, harmaline, choline, betaine, (+)-5-methoxytetrahydroharmane] (Ghosal and Mazumder 1971). We know, however, of no traditional use as a psychoactive plant (Schultes and Farnsworth 1982, 147*). *Banisteriopsis argentea* may be synonymous with *Banisteriopsis muricata* (see below).

Banisteriopsis inebrians Morton

In the Amazonian lowlands of Ecuador, *Banisteriopsis inebrians* is known as *barbasco*. In South America, the word *barbasco* is used primarily to refer to fishing trees (*Piscidia* spp.) and other plants that can be used to poison fish (e.g., *Clibadium* spp.). The Indians pound the fresh roots of *Banisteriopsis inebrians*, place the result into a coarse-meshed basket, and put this in the water. The fish poison then spreads as a milky exudation (Patzelt 1996, 261*).

In southern Colombia (in the Vaupés and Río Piraparaná region), this ayahuasca species is used ritually to prepare yagé or kahi primarily by the Barasana (see **ayahuasca**). In the language of the Barasana, this species is known as *kahi-ukó*, "yagé catalyst," *yaiya-sūava-kahi-ma*, "red jaguar yagé," and *kumua-basere-kahi-ma*, "yagé for shamanizing." Under the influence of this vine, it is said that one sees thing in shades of red, dances, and is able to see people who are normally invisible. According to the mythology of the Barasana, this vine was brought to people in the yuruparí trumpet; for this reason, it is also known as *h̄e-kahi-ma*, "yuruparí yagé" (Hugh-Jones 1977, 1979; Schultes 1972, 142 f.*). Today, it is regarded as a synonym for *Banisteriopsis caapi*. It contains the same alkaloids (O'Connell and Lynn 1953).

Banisteriopsis maritiniana (Juss.) Cuatrecasas var. *laevis* Cuatrecasas

This species is found in the Amazon region of Colombia. The Makuna Indians purportedly use it to manufacture yajé (Schultes 1975, 123).

Banisteriopsis muricata (Cavanilles) Cuatrecasas [syn. *Banisteria acanthocarpa* Juss., *Banisteria muricata* Cav., *Banisteriopsis argentea* (H.B.K.) Robinson in Small, *Heterpterys argentea* H.B.K., and others]

In Ecuador, where this species is known as *mii*, the Waorani use it as the basis for **ayahuasca**. The shaman *(ido)* prepares the drink from bark scrapings that are slowly boiled. He can use the drink to heal a person as well as to send him a disease or even death. A disease can be healed only when the person who has caused the illness also brews the healing drink (Davis and Yost 1983, 190 f.*).

The Witoto from Puca Urquillo on the Rió Ampiyacu (Peru) call this vine *sacha ayahuasca*, "wild vine of the soul," and say that it can be used in place of *Banisteriopsis caapi* (Davis and Yost 1983, 190 f.*). In Peru, this plant is also known as *ayahuasca de los brujos* ("ayahuasca of the sorcerers"); in Bolivia it is called *bejuco hoja de plata* ("silver leaf vine"); in Argentina, *sombra de tora* ("shadow of the steer"); and in El Salvador, *bejuco de casa* ("vine of the house"), *pastora* ("shepherdess"; cf. **Salvia divinorum**, **Turnera diffusa**), and *ala de zompopo*. Of all the species of *Banisteriopsis*, this plant is the most widely distributed.

The vine is also found in the lowlands of southern Mexico (Selva Lacandona) and in Petén (Guatemala) (per oral communication from Rob Montgomery). It is possible that the ancient Maya may have used it to produce a kind of "mayahuasca" (see **ayahuasca analogs**).

The plant contains both **β-carbolines** (**harmine**, etc.) and **N,N-DMT**. DMT is present not in the vine itself (i.e., the stems) but in the leaves.

Intertwining stems of *Banisteriopsis muricata*, found in Petén (Guatemala) and Chiapas (Mexico), recall numerous illustrations of cosmic umbilical cords from the Classic and post-Classic Mayan period. Some people believe that the Maya used this vine to brew a type of "mayahuasca." (Photographed in Tikal)

This yellow-blossomed vine was published under the name *Banisteria tomentosa*. (Copperplate engraving, colorized, nineteenth century)

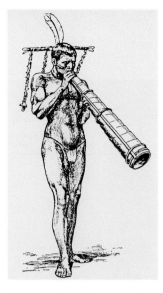

An Indian plays on the yuruparí trumpet; according to mythical tradition, the trumpet came from the heavens filled with *Banisteriopsis* spp. (From Koch-Grünberg, *Zwei Jahre bei den Indianern Nordwest-Brasiliens*, 1921*)

"Ayahuasca [from *Banisteriopsis* spp.] is drunk among the Cashinahua to obtain information which would otherwise remain concealed. The hallucinations are regarded as the experiences of one's own dream spirit; they are clues about the future and memories of the past, and with them the drinker can learn about things, people, and events that are far removed."

ARA H. DER MARDEROSIAN, ET AL. "THE USE AND HALLUCINATORY PRINCIPLE OF A PSYCHOACTIVE BEVERAGE OF THE CASHINAHUA TRIBE (AMAZON BASIN)" (1970, 7)

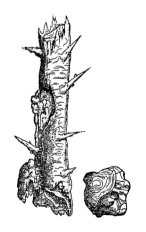

Early illustration of the frankincense tree, which was long unknown in Europe. (Woodcut from Gerard, *The Herball or General History of Plants*, 1633*)

This American species may be identical to the Indian *Banisteriopsis argentea* (see above).

Banisteriopsis quitensis (Niedenzu) Morton
This species is purported to have hallucinogenic effects (Schultes and Farnsworth 1982, 188*). Today, it is regarded as a synonym for *Banisteriopsis caapi*.

Banisteriopsis rusbyana (Niedenzu) Morton
This name is now regarded as a synonym for *Diplopterys cabrerana*.

Literature
See also the entries for *Banisteriopsis caapi*, *Diplopterys cabrerana*, **ayahuasca**, and **ayahuasca analogs**.

Der Marderosian, Ara H., Kenneth M. Kensinger, Jew-Ming Chao, and Frederick J. Goldstein. 1970. The use and hallucinatory principle of a psychoactive beverage of the Cashinahua tribe (Amazon Basin). *Drug Dependence* 5:7–14.

Ghosal, S., and U. K. Mazumder. 1971. Malpighiaceae: Alkaloids of the leaves of *Banisteriopsis argentea*. *Phytochemistry* 10:2840–41.

Ghosal, S., U. K. Mazumder, and S. K. Bhattacharya. 1971. Chemical and pharmacological evaluation of *Banisteriopsis argentea* Spreng. ex Juss. *Journal of Pharmaceutical Science* 60:1209–12.

Hugh-Jones, Stephen. 1977. Like the leaves on the forest floor . . . space and time in Barasana ritual. In *Actes du XLIIᵉ Congrès International des Américanistes* (Paris). 2:205–15.

———. 1979. *The palm and the Pleiades: Initiation and cosmology in Northwest Amazon.* New York: Cambridge University Press.

O'Connell, F. D., and E. V. Lynn. 1953. The alkaloids of *Banisteriopsis inebrians* Morton. *Journal of the American Pharmaceutical Association* 42:753–54.

Schultes, Richard Evans. 1975. De Plantis Toxicariis e Mundo Novo Tropicale: Commentationes XIII: Notes on poisonous or medicinal Malpighiaceous species of the Amazon. *Botanical Museum Leaflets* 24 (6): 121–31.

Boswellia sacra Flückiger

Frankincense Tree

Family
Burseraceae (Bursera Family)

Forms and Subspecies
The true frankincense tree can exhibit considerable variation, depending upon its place of origin. For this reason, it has been described under numerous names that refer solely to local varieties, forms, or races. The taxonomy of the genus *Boswellia* is represented very irregularly in the literature, particularly the nonbotanical literature. An additional difficulty lies in the fact that many species of the genus produce resins that are sold under the name frankincense (Watt and Sellar 1996, 22 f.).

Synonyms
Boswellia bhau-dajiana Birdwood
Boswellia carteri Birdwood
Boswellia thurifera sensu Carter

Folk Names
Ana, bayu, beyo, djau der, echter weihrauchbaum, encens, frankincense, frankincense tree, incense tree, kundara (Persian), kundur (Persian), lebona (Hebrew), libanotis (Greek), lubân, luban-tree, maghrayt d'scheehaz (Arabic), mohr (Somali), mohr madow, mohr meddu, neter sonter (Egyptian), oliban, olibanum (Roman), olibanumbaum, seta kundura (Hindi), weihrauchbaum, weihrauchstrauch, weyrauch, wicbaum, wichboum

History
Frankincense, the true incense, is the golden yellow, pleasantly aromatic resin of the bushlike frankincense tree, great forests ("balsam gardens") of which thrive near the Red Sea, especially in Arabia (the ancient land of incense known as Sa'kalan) and Somalia (the legendary land of Punt) (Wissman 1977). For at least four thousand years, frankincense has been obtained in these regions from incisions made in the bark of the tree (Howes 1950). In ancient times, this was the most coveted of all resins used for incense, and it was transported along the famed Incense Road—probably the most important trade route of antiquity—between Egypt and India (Groom 1981; Kaster 1986).

Since ancient times, frankincense resin has been used to manufacture **incense**, cosmetics, and perfumes. Arabian women still burn frankincense to perfume their bodies, in particular the vulva (Martinetz et al. 1989). This not only lends them a more pleasant scent but also is said to make them more erotic.

Frankincense has been attributed with psychoactive powers since the early modern period (Lonicerus 1679, 738*; Menon and Kar 1971; Farnsworth 1972, 68*). As a result, frankincense was swallowed, burned, or smoked in the Ottoman Empire, in Arabia, and even in Europe (often in combination with opium; cf. *Papaver somniferum*). Other incense plants have also been claimed to induce hallucinogenic effects (cf. *Bursera bipinnata*).

The history of frankincense and the tree it is obtained from is simultaneously a history of mistaken identities and confusion, as all drop-shaped aromatic resins were once referred to as "frankincense" (Schneider 1974, 1:185f.*). Since all species of *Boswellia*—and their resins—display considerable variation, the botanical classification is often a question of chance (cf. Hepper 1969). The botanical identity of the stock plant was not determined until the middle of the nineteenth century (Carter 1848; Hepper 1969).

Distribution
The frankincense tree occurs in Somalia and southern Arabia. In Somalia, it is primarily found in areas between 100 and 1,800 meters above sea level (Pabst 1887, 1:54*).

Cultivation
Cultivation methods (assuming any are actually known) are a well-protected secret of the peoples who live from the collection of frankincense. The ancient Egyptians attempted to plant frankincense trees in Egypt but were unsuccessful in spite of their great knowledge of gardening (Dixon 1969). They excavated small trees together with soil that they then shipped back to Egypt in tubs. The trees died shortly thereafter.

Appearance
This small, rather graceful tree grows from 4 to 5 meters in height, and in rare cases to 6 meters. It has a robust trunk and a dark brown, papery bark that sheds repeatedly and immediately regrows. Each year, new shoots form that are thickly adorned with short yellow trichomes. The pinnate leaves grow in clusters on the ends of the branches. The small, petiolate flowers grow from the leaf axils and are arranged in paniculate racemes. The whitish flowers have five petals and ten red stamens. The small fruits form light brown capsules that have three lobes, in which the angular stones sit individually with the small seeds. Flowering occurs in April.

The true frankincense tree is very easily confused with *Boswellia serrata* Roxb., the Indian incense tree, especially because this species is also a source of incense (Indian frankincense) (Schneider 1974, 1:187*). The very similar *Boswellia papyrifera* (Del.) Hochst. is easily distinguished on the basis

of its height; it grows much taller and is more ramified than the other species.

Psychoactive Material
— Resin (olibanum, Somalian olibanum, Aden olibanum, Bible incense, Arabian olibanum)

In Persia (Iran), a distinction is made between two types of olibanum: *kundara zakara*, "male incense," is dark yellow to reddish in color and comes in the form of round drops; *kundara unsa*, "female incense," is yellowish-whitish, pale, and transparent and usually comes in the form of oblong drops (Hooper 1937, 92*).

Resins used to counterfeit frankincense are false incense (spruce resin; *Picea* spp.), gum arabic (cf. *Acacia* spp.), fir resin (*Abies* spp.), mastic (*Pistacia lentiscus* L.), sandarac (the resin of *Tetraclinis articulatia* (Vahl) Mast or *Callitris quadrivalvis* Vent.), and calcite crystals (Pabst 1887, 1:56*).

Preparation and Dosage
The resin is obtained by making long (4 to 8 cm), deep incisions into the bark. A special scalpel-like instrument known as a *mengaff* is used for this purpose. According to Theophrastus, the resin should be collected during the dog days, i.e., the hottest time of the year. Pliny also noted that the first incision into the bark of the stock plant should be made around the time of the rising of the Dog Star (Sirius).

Frankincense is an important component in many recipes for psychoactive **incense**. It is also an ingredient in the **Oriental joy pills** and was formerly used as a spice in wine (cf. *Vitis vinifera*).

Above: The true frankincense tree (*Boswellia sacra*) in its natural habitat. (Photograph: Walter Hess)

Below: Two qualities of olibanum, the resin of the frankincense tree (*Boswellia sacra*). Left, high-quality ware from Eritrea; right, the coarser, impure product from Ogaden.

"[Amun] transformed herself into the form of the majesty of her spouse, the king of Upper and Lower Egypt; they [Amun and Thoth] found her as she rested in the splendor of her palace.—She awakened from the scents of the god; she laughed at his majesty. . . . She was filled with joy to see his beauty, his body went into her body, [the palace] was flooded with the scents of the god, all of his scents were [scents] from Punt [the land of incense]."

Temple inscription at Der el-Bahari (ca. 1550 b.c.e.) cited in *Weihrauch und Myrrh* [Frankincense and Myrrh] (Martinetz et al. 1989, 103)

"God does not permit himself to be held, nor heard, nor seen, but in the scent of the divine flower or the taste of the sacred juice, it is most likely that the conception of him will be awakened. Using incense, balsam, and sacred oils, through tonic and heart-strengthening plant extracts, the Egyptians speak directly to the gods, they begin to breathe, to speak, to open their mouths and live."

PAUL FAURE
MAGIE DER DÜFTE [MAGIC OF
SCENTS]
(1990, 23)

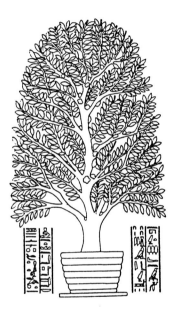

An ancient Egyptian illustration of the frankincense tree as a potted plant, in the grave of Hatshepsut (1504–1483 B.C.E.). The inscription reads: "Greening incense tree, 31 pieces; brought here among the delicacies for the majesty of the god Amun, the lord of the earthly throne. Never has its like been seen since the creation of the universe." (From Engel)

50 *Boswellia serrata* is sometimes alleged to be hallucinogenic (Schultes and Hofmann 1980, 367*).

51 H 15 (also known as H15-ayurmedica) is a standardized extract of the gum resin of *Boswellia serrata*. One tablet contains 400 mg of the dried lipophilic extract (Etzel 1996, 91).

Other Types of Olibanum

The following species of *Boswellia*, which occur in East Africa and India, also yield resins that are referred to and marketed as olibanum:

Boswellia frereana Birdw. gekar, elemi olibanum,	African elemi, yegaar
Boswellia papyrifera (Del.) Hochst. [syn. *Amyris papyrifera* Gaill. ex Del.] olibanum	Ethiopian olibanum, Eritrean
Boswellia serrata Roxb.[50] [syn. *Boswellia thurifera* Colebr., *Boswellia glabra* Roxb., *Canarium balsamiferum* Willd.]	Indian incense, salakhi, lobhan, thus indica

Ritual Use

For both cultic and economic reasons, frankincense was the most important **incense** of the ancient Assyrians, Hebrews, Arabs, Egyptians, and Greeks. The resin was used as a fumigant and offered to the gods at every ceremony. The Assyrians burned it especially for Ishtar, the queen of the heavens, for Adonis, the god of resurrecting nature, and for Bel, the Assyrian high god. The Assyrian kings, who were also high priests, offered frankincense to the tree of life, which was sprinkled with wine (cf. *Vitis vinifera*) as it was bathed in smoke. The pagan, pre-Islamic Arabs consecrated frankincense to Sabis, their sun god, and the entire supply was required to be stored in the temple of the sun. Among the Hebrews, frankincense was one of the ingredients of the sacred incense and a symbol of divinity. The Bible refers to it as a sacred incense and an article of tribute and trade. It later became the most important incense of the Catholic Church. In central Europe, the resin of the frankincense tree became known primarily through the Catholic Church. At the time of Charlemagne, it was burned not only during masses but also during the "trials by ordeal" that were common at the time.

Egyptian and Greek magicians of the late ancient period used the smoke to conjure demons, the intermediary beings that they wished to put to use. In Egypt, the frankincense tree was consecrated to Amun of Thebes. Incense was also sacred to Hathor, the goddess of drunkenness (cf. *Mandragora officinarum*). For the Romans as well, there was no ceremony, no triumphal procession, no public or private celebration that did not include the use of this aromatic resin. It was said that frankincense "enabled one to recognize god." Frankincense manna was sacred to the sun and the oracular god Apollo (cf.

Hyoscyamus albus). Frankincense was also important in the cult of Aphrodite. Offering the goddess incense was a way to ensure that the hetaerae, or temple servant girls, would be provided with sufficient clientele. In Ethiopia, frankincense is still burned to "control evil spirits" (Wilson and Mariam 1979, 30*). Similar practices have been preserved in certain Swiss folk customs (Vonarburg 1993).

Artifacts

Both frankincense and the tree from which it comes are frequent subjects in ancient Egyptian works of art (wall paintings, poetry). There are a great many incense vessels and other devices for burning incense (cf. Martinetz et al. 1989).

Medicinal Use

Frankincense was used for numerous medicinal purposes in ancient times and was praised by Hippocrates, Celsus, Dioscorides, Galen, Marcellus, and Serenus Sammonicus. It was used to produce oils to treat colds, **enemas to treat constipation**, agents for cleaning wounds, bandages to treat the "sacred fire" (cf. *Claviceps purpurea*), and ointments against frostbite, burns, skin nodules, rashes, scabies, warts, psoriasis, inflammations, excrescence, watery eyes, scars, ear inflammations, boils, rheumatism, and gout. More recently, an extract of *Boswellia serrata* (H 15)[51] has been used with success in Western medicine and phytotherapy to treat rheumatoid arthritis (Etzel 1996). **Essential oils** distilled from a number of different types of frankincense are also becoming increasingly important in aromatherapy (Watt and Sellar 1996).

In traditional Chinese medicine, frankincense is generally regarded as a stimulant, and it is used to treat leprosy, skin diseases, menstrual cramps, coughs, and lower abdominal pains. The smoke or the **essential oil** is inhaled for coughs.

In the early modern period, frankincense was even used for "psychiatric" purposes as a mood-enhancer:

The smoke of Olibani is good for heavy eyes / when taken in there. It removes sorrow / increases reason / strengthens the heart / and makes one of cheerful blood. (Lonicerus 1679, 738*)

In Ethiopia, frankincense is burned as a fumigant to treat fever and as a tranquilizer (Wilson and Mariam 1979, 30*).

Constituents

All types of frankincense are composed of 53% resin ($C_{30}H_{32}O_4$), gum, essential oil, boswellic acids, bitter principles, and mucilage. Frankincense contains 5 to 10% **essential oil**, consisting of pinene, limonene, cadinene, camphene, π-cymene, borneol, verbenone, verbenol, dipentene, phellandrene, olibanol, and other substances. The composition of the essential oils of the different species varies somewhat (Tucker 1986). The essential oil from bejo (olibanum from Somalia) contains 19% α-thujene and 75% α-pinene, as well as sabinene, α-cymene, limonene, β-caryophyllene, α-muurolene, caryophyllene oxide, and other, unknown substances. The oil of olibanum Eritrea consists of approximately 52% octyl acetate; the oil of olibanum Aden contains approximately 43% α-pinene (Watt and Sellar 1996, 28).

For years, both the literature and the media have reported claims that pyrochemical modifications and reactions produce **THC** when frankincense is burned (Martinetz et al. 1989, 138; Faure 1990, 30).[52] To date, however, THC has not been detected in any genus other than *Cannabis*. The results of recent studies at the Pharmaceutical Institute of the University of Bern have demonstrated that no THC is produced when frankincense resin is burned; not even one nanogram was detected (Kessler 1991). But since there are numerous types of olibanum, it may be that some of these do contain THC or produce it when burned while others do not. However, the smoke was not "investigated with respect to other psychotropic substances, so that its last secrets are still preserved" (Hess 1993, 11).

Effects

Both olibanum and the incense of the church have long been attributed with inebriating, euphoriant, and mood-improving effects (Menon and Kar 1971). The *Universallexikon* of 1733 to 1754 states:

> It strengthens the head, reason, and sense, but when it is used needlessly, it awakens painful days in the head and is damaging to the reason, otherwise it cleanses the blood, strengthens the heart, takes away sorrow, and makes the blood cheerful.

Cases of "olibanum addiction" are still observed and noted in the toxicological literature (Martinetz et al. 1989, 138). It is not unlikely that many people attended church services in the past because of the inebriating effects of olibanum.

Commercial Forms and Regulations

The olibanum that is now found in trade comes primarily from *Boswellia sacra*, the true frankincense tree, which is indigenous to Somalia, Iran, and Iraq. Olibanum is traded in different qualities and is named after its place of origin (Aden, Eritrea, Beyo). The unrefined drops (olibanum electum) are the best. Olibanum is available without restriction and can be obtained from sources specializing in devotional articles and incenses.

Literature

See also the entry for **incense**.

Carter, H. J. 1848. A description of the frankincense tree of Arabia. *Journal of the Royal Asiatic Society* (Bombay Branch) 2:380–90.

Dixon, D. M. 1969. The transplantation of punt incense trees in Egypt. *The Journal of Egyptian Archaeology* 55:55–65.

Etzel, R. 1996. Special extract of *Boswellia serrata* (H 15)* in the treatment of rheumatoid arthritis. *Phytomedicine* 3 (1): 91–94.

Faure, Paul. 1990. *Magie der Düfte*. Munich and Zurich: Artemis.

Groom, N. St. J. 1981. *Frankincense and myrrh*. London: Longman.

Hepper, F. Nigel. 1969. Arabian and African frankincense trees. *Journal of Egyptian Archaeology* (London) 55:66–72.

Hess, Walter. 1993. Weihrauch-Beweihräucherung, Harze und Balsame. *Natürlich* 13 (12): 6–17.

Howes, F. N. 1949. *Vegetable gums and resins*. Waltham, Mass.: Chronica Botanica.

———. 1950. Age-old resins of the Mediterranean region. *Economic Botany* 1:307–16.

Kaster, Heinrich L. 1986. *Die Weihrauchstraße: Handelswege im alten Orient*. Frankfurt/M.: Umschau.

Kessler, Michael. 1991. Zur Frage nach psychotropen Stoffen im Rauch von brennendem Gummiharz der *Boswellia sacra*. Inaugural diss., Basel, Switzerland.

Martinetz, Dieter, Karlheinz Lohs, and Jörg Janzen. 1989. *Weihrauch und Myrrhe*. Stuttgart: WVG.

Menon, M. K., and A. Kar. 1971. Analgesic and psychopharmacological effects of the gum resin of *Boswellia serrata*. *Planta Medica* 19:333–41.

Tucker, Arthur O. 1986. Frankincense and myrrh. *Economic Botany* 40 (4): 425–33.

Vonarburg, Bruno. 1993. Wie die Innerrhoder "räuchelen." *Natürlich* 13 (12): 13.

Watt, Martin, and Wanda Sellar. 1996. *Frankincense and myrrh*. Saffron Walden, U.K.: The C. W. Daniel Co. Ltd.

Wissmann, Herman v. 1977. *Das Weihrauchland Sa'kalan, Samarum und Moscha*. Vienna: Verlag der Österreichischen Akademie der Wissenschaften (Philosophisch-Historische Klasse, 324).

52 "Also brought to our attention as a result of repeated observations of addiction [!], we began to consider which constituents could produce these effects. We became aware of the fact that there is a possibility for synthesizing the hashish constituent Δ^9-tetrahydrocannabinol through the conversion of olivetol (5-pentyl-resorcin) with verbenol. . . . Verbenol, like phenols and phenol ether, has frequently been described as a component in incenses; in addition, other phenolic structures might be formed in the course of the combustion process, so that we are of the opinion that the formation of a basic tetrahydrocannabinol structure . . . is entirely possible. . . . It is also entirely possible that such inebriating and stimulating substances may be produced during the chewing process or in the digestive tract as a result of enzyme activity, whereby the essential oils are of course also of some importance" (Martinetz et al. 1989, 138).

Brugmansia arborea (Linnaeus) Lagerheim

Angel's Trumpet Tree

Depiction of an angel's trumpet or tree datura on an Incan drinking vessel. (Copy by C. Rätsch)

The treelike angel's trumpet (*Brugmansia arborea*) is a common ornamental throughout the world. Especially in subtropical zones, it can attain a stately height and develop woody stems. (Photographed in Peru)

53 In folk taxonomy, the name *borrachera* is also used to refer to and classify other plants, e.g., *Iresine celosia* L. and *Iresine herbstii* Hook. f. (cf. **Iresine spp.**, **cimora**), as well as all other species of **Brugmansia**, but also **Iochroma fuchsioides** (Bristol 1969, 184).

54 In Indonesia, *kecubong* is normally used to refer to **Datura metel**.

55 The name *tecomaxochitl* is normally used for **Solandra spp.** (cf. Díaz, 1979: 84*).

Family
Solanaceae (Nightshade Family); Subfamily Solanoideae, Datureae Tribe, Brugmansia Section

Forms and Subspecies
There are presumably different cultivars.

Because of their variability and the many hybrids, angel's trumpets are often very difficult to classify (Preissel and Preissel 1997). The botanical literature is also quite chaotic concerning the taxonomy of these plants (Bristol 1966 and 1969; Lockwood 1973).

Synonyms
Datura arborea L.
Datura arborea Ruíz et Pav.
Datura cornigera Hook.
Brugmansia candida Pers. sensu latu

Folk Names
Almizclillo, angel's trumpet tree, baumdatura, baumstechapfel, borrachera, (Spanish, "inebriator"),[53] campachu, campanilla, chamico, cimora, cojones del diablo, engelstrompetenbaum, floripondio, großer stechapfel, guarguar, hierba de los compañones, huántac (Zaparo-Quechua), huanto, huánto (Quijo), huántuc (Quechua), huarhuar, isshiona (Zaporo), kecubong (Bali),[54] maícoma, mai ko, mai ko' mo, mataperro (Spanish, "dog killer"), misha huarhuar, misha rastrera blanca, qotu (Quechua), saharo, tecomaxochitl (Nahuatl),[55] toé, tree stramonium, trombeteiro (Brazil)

History
All of the angel's trumpets are from South America. They are now known only as cultigens and not as wild plants. It has still not been determined which wild plants yielded the known species and hybrids. This implies that the plant must have long played a role in human cultures. For this reason, it is very likely that angel's trumpets were already being used for ritual and

psychoactive purposes in prehistoric times. *Brugmansia arborea* is from the Andes region. The earliest description of the Indian use of this potent hallucinogen is probably that of Bernabé Cobo (1653) (Bastien 1987, 115*). The species "was first described in 1714 by Louis Feullée, and Linnaeus's description is based upon his illustration" (Stary 1983, 96*).

Distribution
This relatively rare species has an extensive range, from Ecuador to Peru and Bolivia and into northern Chile. It grows wild in the Bolivian province of Bautista Saavedra, in the lower valleys of Camata (Bastien 1987, 114*).

Cultivation
This angel's trumpet, like all other *Brugmansia* species, is most easily propagated by cuttings. A sharp knife is used to cut off the end of a branch, some 20 cm in length, which is then stripped of all but the newest leaf buds. The cutting is then placed in water. Roots appear after two to three weeks. Soon thereafter, it can be planted in nutrient-rich soil. Because the plant does not tolerate frost, in central Europe it can be grown only in pots.

The plant is grown for pharmaceutical purposes (**scopolamine** production) in the Andes region, Brazil, the southern United States, and India (Lindequist 1992*). The angel's trumpet also enjoys a wide distribution as an ornamental.

Appearance
The treelike perennial bush can grow as tall as 5 meters in height. It produces trumpet-shaped (sometimes double), five-pointed flowers that are white or cream-white in color and hang slightly to the side. At night, these exude an exhilaratingly sweet scent. The long calyx is single and deeply incised (an important diagnostic feature). The smooth fruits, when they do develop, are berrylike and contain large brown seeds. They are almost spherical (another important diagnostic feature). Most angel's trumpets only rarely or never produce fruits. The leaves, which frequently grow along the same side of the stalk, are oblong-elliptic and pointed at the ends and can be of varying lengths. In the tropics and temperate zones, angel's trumpets blossom throughout the year. The flowers wilt after about five days.

Brugmansia arborea is easily confused with the white-blossomed **Brugmansia aurea** and **Brugmansia candida**.

Many plant lovers, gardeners, anthropologists,

and even botanists confuse all *Brugmansia* species with the thorn apple (*Datura*). The two genera, however, can be easily distinguished by the position of their flowers. The flowers of all species of *Brugmansia* hang more or less straight down, while the flowers of the *Datura* species point more or less upward, often steeply. In addition, no *Brugmansia* is known to produce a thorny fruit.

Psychoactive Material
— Leaves
— Fresh flowers (used to produce the homeopathic mother tincture)
— Seeds

Preparation and Dosage
The leaves are extracted in cold water or steeped in hot water. A psychoactive dosage is usually given as four leaves or one flower brewed into a tea. The crushed seeds may be added to **chicha** (Bastien 1987, 114f.*). The leaves are also used as one of the main ingredients of the **cimora** drink and as an additive to preparations of San Pedro (*Trichocereus pachanoi*). Dried leaves are smoked alone or in **smoking blends** together with other ingredients, e.g., *Cannabis indica.*

Extreme caution should be exercised when ingesting any species of *Brugmansia*. Angel's trumpets are the most potent hallucinogens in the plant kingdom, producing hallucinations that are no longer recognized as such. South American shamans urgently warn unknowledgeable people against using these plants. Angel's trumpets are used for psychoactive purposes almost exclusively by experienced shamans. Overdoses can result in states of delirium that can last for days and have aftereffects that persist for weeks. Proper dosage presents an additional problem. People react very differently to the **tropane alkaloids**. In other words, the same dosage can produce completely different effects in different people. In the toxicological literature, one can read that heavy overdoses can be fatal; such cases, however, are only poorly documented (cf. **Brugmansia suaveolens**).

In contrast to internal ingestion, smoking the dried leaves is relatively harmless. Smoking an amount that corresponds to one to two cigarettes produces only subtle effects. The *Brugmansia* effects become more obvious when the leaves are combined with hemp products (*Cannabis indica, Cannabis sativa*).

Ritual Use
The Indians regard angel's trumpet as sacred. The priests of the Andean peoples smoked the leaves so that they could make prophecies, divine, and diagnose. Many Andean peoples use the seeds as an additive to the **chicha** (maize beer) that is drunk at village festivals and religious rituals.

Artifacts
It is astonishing that relatively few artifacts or artistic renditions are associated with angel's trumpet. Where such objects do exist, it is usually almost impossible to determine which species is being represented (cf. **Brugmansia candida**). Angel's trumpets are frequently depicted in the paintings of the American artist Donna Torres. Jürgen Mick has masterfully portrayed *Brugmansia arborea* in his comic story *Träume* [Dreams] (Carlsen, Hamburg, 1993).

Medicinal Use
In Peru, the leaves of this and other angel's trumpets are used in the treatment of tumors (Chavez V. 1977, 231*). It is possible that the seeds may have been used for anesthetic purposes in pre-Columbian times, perhaps in combination with coca leaves (*Erythroxlum coca*) (Bastien 1987, 115*).

A number of potencies of Datura arborea hom. *HAB34* and Datura arborea hom. *HPUS78* are used in homeopathy in accordance with the medical description. The mother tinctures are produced by extracting the flowers in strong spirits (Lindequist 1992*).

Constituents
All parts of the plant contain **tropane alkaloids**. The leaves contain 0.2 to 0.4% total alkaloids, of which 0.01% is hyoscyamine, 0.13% is **scopolamine**, and 0.07% is **atropine**. The stems contain only 0.16% total alkaloids; the seeds contain chiefly hyoscyamine. The roots also contain the alkaloids (−)-3,6-ditigloyloxytropane, 7-hydroxy-3,6-ditigloyloxytropane, tropine, and pseudotropine. **Coumarins** and **scopoletin** are both present in all parts of the plant (Lindequist 1992, 1140).

Effects
Brugmanisa arborea induces strong parasympatholytic effects (Jacinto et al. 1988). Characteristic symptoms include mydriasis (dilation of the pupils), often persisting for days, along with an extreme dryness of the mucous membranes. Depending upon the dosage and individual reactions, there can also be profound hallucinations with a complete loss of reality, delirium, coma, and death through respiratory paralysis (Lindequist 1992*).

Angel's trumpets are said to have potent narcotic effects. In Peru, intoxicating a person against his or her will is known as *chamicado*, which means "touched by the angel's trumpet" (Bastien 1987, 114*).

Commercial Forms and Regulations
The seeds and plants of all *Brugmansia* species are available without restriction and can be purchased

"The Indians use [the angel's trumpet] to get drunk, and when they take too much, they completely lose their senses, so that they cannot see or hear with open eyes. They are accustomed to exploit this for evil purposes. Not long ago, it happened that one of my friends was given *chamico* so that he could be robbed. When he awoke, he was so angry that he ran around naked, with only his shirt, and fell into a river. They seized him and kept him locked up until, after two days, he reawakened from his condition. Juice from the leaves, mixed with vinegar and applied above the liver, mitigates mild fever and is very good for high fever. A *mate* [= tea] of this solution heals chronic fever."

Bernabé Cobo
Historia del Nuevo Mundo
[History of the New World]
(1653; cited in Bastien 1987, 115*)

Scopolamine

Hyoscyamine

Scopoletin

"His body was burning with heat, his throat parched with dryness, and for the first time, he recalled the cup of Huacacachu [= *Brugmansia*] that he had ingested. Then he knew that he had journeyed through the land of visions, and not through the desert, as he had thought . . ."

DANIEL PETERS
THE INCA
(1991, 299*)

in many nurseries and garden shops. The mother tincture requires a prescription (Lindequist 1992*).

It is strange that this most potent and most dangerous of all plant hallucinogens is not included on any list of illegal drugs, while such plants as **Cannabis** and **Erythroxylum**, which by comparison are almost completely harmless, are prohibited. This situation is a strong indication that most current drug laws are not founded upon scientific knowledge (cf. Körner 1994*).

Literature
See also the entries for the other *Brugmansia* species, **scopolamine**, and **tropane alkaloids**.

Bristol, Melvin L. 1966. Notes on the species of tree daturas. *Botanical Museum Leaflets* 21 (8): 229–48.

———. 1969. Tree datura drugs of the Colombian Sibundoy. *Botanical Museum Leaflets* 22 (5): 165–227.

Jacinto, José Maria Serejo S., José Antonio Lapa, and Souccar Caden. 1988. Estudio farmacológico do extrato bruto do *Datura arborea* L. *Acta Amazônica*, Supplement 18 (1–2): 135–43.

Lindequist, Ulrike. 1992. Datura. In *Hagers Handbuch der pharmazeutischen Praxis*, 5th ed., 4:1138–54. Berlin: Springer.

Lockwood, Tommie E. (See obituary in *Economic Botany* 29 [1975]: 4–5.) 1973. Generic recognition of *Brugmansia*. *Botanical Museum Leaflets* 23:273–84.

———. 1979. The ethnobotany of *Brugmansia*. *Journal of Ethnopharmacology* 1:147–64.

Mick, Jürgens. 1993. *Träume* [Dreams]. Hamburg: Carlsen.

Preissel, Ulrike, and Hans Georg Preissel. 1997. *Engelstrompeten: Brugmansia und Datura*. 2nd edition. Stuttgart: Verlag Eugen Ulmer.

Shah, C. S., and A. N. Saoji. 1966. Alkaloidal estimation of aerial parts of *Datura arborea* L. *Planta Medica* 14:465–67.

Brugmansia aurea Lagerheim
Golden Angel's Trumpet

"The aim and purpose of the Brugmansia inebriation was—as a rule—to establish contact with the gods or the spirits of the ancestors. With their help, one could try to positively influence one's own future and that of the tribe. In the inebriated state, one saw oneself transported into a different level of consciousness that made it possible to communicate with the supernatural powers, to ask them for help, and to receive their teachings. Brugmansias were the keys that opened the door to this other world."

ULRIKE AND HANS-GEORG PREISSEL
ENGELSTROMPETEN [ANGEL'S TRUMPET]
(1997, 14 f.)

Family
Solanaceae (Nightshade Family); Subfamily Solanoideae, Datureae Tribe, Brugmansia Section

Forms and Subspecies
There is a yellow- and a white-blooming form. Several mutations have also been observed: "The clones of *Brugmansia aurea* are often of bizarre appearance. They are frequently infected by viruses" (Plowman 1981, 441).

Synonyms
Datura aurea (Lagerh.) Saff.

Schultes and Raffauf have recently suggested that **Brugmansia candida** is a synonym for *Brugmansia aurea* (1990, 421*). On the other hand, *Brugmansia aurea* is also regarded as a synonym for *Brugmansia candida*.

Folk Names
Borrachero, floripondio, gelbe baumdatura, golden angel's trumpet, goldene baumdatura, goldene engelstrompete, golden tree datura, guantu, huandauj, kiéri (Huichol),[56] kiéri-nánari (Huichol, "root of the kiéri"), yellow tree datura

History
Golden angel's trumpet was first discovered and described at the end of the nineteenth century by the Swedish botanist Nils Gustaf von Lagerheim (1860–1926) (Lagerheim 1893). In South America, its ethnobotanical significance is similar to that of *Brugmansia candida* (Plowman 1981).

Distribution
The original range of *Brugmansia aurea* extended from Colombia into southern Ecuador. It is not known when the species was introduced into Mexico. In the Andes, it is primarily found in altitudes between 2,000 and 3,000 meters (Plowman 1981).

Cultivation
See *Brugmansia arborea*.

Appearance
This perennial, treelike shrub has a woody stem and is usually heavily branched. The smooth, marginated leaves are oval-cuspidate. The calyx is simple and only slightly incised. The long, funnel-shaped, five-pointed, and normally luminously yellow flowers hang down at an angle. They are larger than the flowers of **Brugmansia arborea** and more stocky than the blossoms of **Brugmansia candida**. The smooth fruits are somewhat fatter and shorter than those of **Brugmansia candida**.

The white-blossomed form is very easily mistaken for **Brugmansia candida**.

56 The name *kiéri* is used usually for *Solandra* spp. and less frequently for *Datura innoxia*.

Psychoactive Material
— Stems and stem pith
— Leaves
— Flowers
— Seeds

Preparation and Dosage
The Canelos scrape the green pith from split stems, press this, and ingest it when it is swollen with water (Whitten 1985, 155).

In Ecuador, the juice pressed from the pith of a 5 cm long, finger-thick piece of stem is used as a "prophetic" dose (Metzner 1992). The juice is drunk with some water.

The dried leaves and flowers can be smoked alone or in **smoking blends** (cf. *Brugmansia arborea, Brugmansia suaveolens*).

Ritual Use
The shamans of the Canelo Indians use angel's trumpet to establish contact with their spirit helpers and animal spirits. With their aid, they are able to detect sorcerers who carry out harmful magic in secret and magically send "worms" and other diseases into their victims' bodies (Whitten 1985). In Ecuador, the juice of the plant is ingested to induce prophetic dreams that can then be interpreted as portents concerning the next phase of one's life (Metzner 1992).

The seeds are used as an inebriating additive to **chicha** (maize beer), which is consumed at village festivals and religious rituals.

In Mexico, the Huichol apparently use this angel's trumpet in a manner similar to *Solandra* spp.

Artifacts
Kiéri plants are sometimes depicted in the visionary art of the Huichol (see *Solandra* spp.).

Medicinal Use
Identical to the use of *Brugmansia candida*

Constituents
Golden angel's trumpet contains large amounts of **tropane alkaloids**. The total alkaloid content was

The luminous yellow-gold blossom of the relatively rare yellow angel's trumpet (*Brugmansia aurea*).

found to be 0.9%, with the main alkaloid **scopolamine** (hyoscine) making up some 80% of the mix (Plowman 1981, 440). In addition, apo-atropine, 3α-tigloyloxyltropane-6β-ol, tigloidine, 6β-acetoxy-3α-tigloyloxyltropane, apohyoscine, hyoscyamine/atropine, norhyoscyamine/noratro-pine, 6β-hydroxyhyoscyamine, and tropane-3α-ol are also present (el Imam and Evans 1990, 149*).

Effects
This species has been reported to produce intense dreams of a prophetic nature (Metzner 1992). Apart from this, the general effects are similar to those of *Brugmansia candida.*

Commercial Forms and Regulations
See *Brugmansia arborea.*

Literature
See also the entries for the other *Brugmansia* species, **scopolamine**, and **tropane alkaloids**.

Lagerheim, Gustav. 1893. Eine neue, goldgelbe *Brugmansia. Gartenflora* 42:33–35.

Metzner, Ralph. 1992. Divinatory dreams induced by tree datura. In *Yearbook for ethnomedicine and the study of consciousness* 1:193–98. Berlin: VWB.

Plowman, Timothy. 1981. Brugmansia (Baum-Datura) in Südamerika. In *Rausch und Realität*, ed. G. Völger, 2:436–43. Cologne: Rautenstrauch-Joest-Museum.

Whitten, Norman. 1985. *Sicuanga runa.* Urbana: University of Illinois Press.

"Most of the ethnographic reports about tree daturas list the species used under the name *Brugmansia* (*Datura*) *arborea* (L.) Lagerh. This name is assigned without distinction to every white-flowered *Datura*. While this species is indeed used with great pleasure, it is one of the rarer species. It usually grows in the higher regions of Ecuador and south toward Bolivia. *Brugmansia aurea* Lagerh. is used much more commonly. This plant grows at altitudes between 2,000 and 3,000 meters, primarily in the northern Andes of Venezuela and in the south toward Ecuador, where it is frequently found along roadsides, rivers, and drainage ditches. The flowers are also sometimes golden yellow."

TIMOTHY PLOWMAN
"BRUGMANSIA (BAUM-*DATURA*) IN SÜDAMERIKA" [BRUGMANSIA (DATURA TREE) IN SOUTH AMERICA] (1981, 439)

Brugmansia x *candida* Persoon

White Angel's Trumpet

Family
Solanaceae (Nightshade Family); Subfamily Solanoideae, Datureae Tribe, Brugmansia Section

Forms and Subspecies
This quite variable angel's trumpet may be a natural hybrid between **Brugmansia aurea** and **Brugmansia versicolor** (Giulietti et al. 1993). There is a white form and a form that produces peach-colored flowers.

A form cultivated by the Sibundoy Indians that has tiny, stunted leaves has been described under the name *Datura candida* (Pers.) Saff. cv. Munchira; the "normal" form was referred to as *Datura candida* (Pers.) Saff. cv. Biangan (Schultes 1979b, 147 f.*). Additional forms include *Datura candida* cv. Quinde, *Datura candida* cv. Andres, *D. candida* cv. Ocre, *D. candida* cv. Amaron, and *D. candida* cv. Salaman (Bristol et al. 1969).

A form used commercially and ornamentally is known by the name *Datura candida* cv. Flintham Hall (Imam and Evans 1990*).

Methysticodendron amesianum, which Richard Evans Schultes (1955) described as a new genus and species, is now known to be nothing more than a "monstrous" cultivar (Bristol 1965, 272*). It may be best described as *Brugmansia* x *candida* f. Culebra.

From Peru, the following forms, all of which are used for psychoactive purposes (cf. **Trichocereus pachanoi, cimora**), have been described:
Brugmansia x *candida* f. Cimora oso
Brugmansia x *candida* f. Cimora galga
Brugmansia x *candida* f. Cimora toto curandera

Synonyms
Brugmansia candida Pers.
Datura affinis Saff.
Datura arborea Ruíz et Pav. non L.
Datura aurea x *D. versicolor*
Datura candida (Pers.) Saff.
Datura x *candida* (Pers.) Saff.
Datura pittieri Saff.
Methysticodendron amesianum Schultes

Folk Names
Almizclillo, amarón, andaqui, biangán, biangán borrachera, borachero, borrachera, borrachera de agua, borrachero (Spanish, "inebriator"),[57] borracherushe, buyés, buyés borrachera, buyés borracherushe, cacao sabanero, cambanda, campana (Spanish, "bell"), campanilla (Spanish, "little bell"),[58] cari, chamico,[59] chontaruco, chontaruco borrachera, cimora,[60] cucu, culebra, culebra-borrachero (Spanish, "snake inebriant"), danta ("tapir"), danta borrachera, flor de campana (Spanish, "bell flower"), floripondio, floripondio blanco, goon'-ssi-an borrachero (Kamsá), guamuco blanco, guamuco floripondio, huama, kampaana wits (Huastec, "bell of the mountain"), kampachu (Quechua), kampána nichim (Tzeltal, "bell flower"), kin-de-borrachero (Inga), lengua de tigre ("tongue of the jaguar"), lipa-ca-tu-ue (Chontal), maikoa, mets-kwai borrachero (Masá, "jaguar inebriant"), misha, mitskway borrachero, munchira, mutscuai, ngunsiana, nitkwai boracero (Kamsá), nit-waí-boracero (Inga), palpanichium, po:bpihy (Mixe), queen of the night, quinchora borrachera, quinde, quinde borrachero, reinweißer stechapfel, Sta. Maria wits (Huastec, "St. Mary's flower"), salamán, salamanga, salvanje, tecomaxochit (Náhuatl), trombita (Spanish, "little trumpet"), ts'ak tsimin (Lacandon, "horse medicine"), tu:tk-hiks (Mixe), white angel's trumpet

History
In 1935, H. García-Barriga collected the first specimen of this species in the Sibundoy Valley of Colombia (located at an altitude of 2,200 meters). Later, numerous forms were described for the Sibundoy region on the basis of the Indian ethnobotanical classification of the angel's trumpets. *Datura* (*Brugmansia*) *candida* cv. Culebra was originally thought to represent a different genus and was described by Richard Evans Schultes under the name *Methysticodendron amesianum* (Schultes 1955). This form has very long, thin leaves that look like snakes; for this reason, the Sibundoy Indians call this form *culebra-borrachero* and the Kamsá call it *mitskway-borrachero*, both of which mean "snake inebriant" (Schultes 1979b, 148 f.*). It is not known when this ethnobotanically significant species spread into Central America.

Distribution
The plant is originally from Colombia or Ecuador (Franquemont et al. 1990, 99*) and is still common in these areas. It is usually found at altitudes between 1,500 and 2,500 meters. It was apparently introduced into Mexico in pre-Columbian times (Berlin et al. 1974, 280*).

Cultivation
Propagation can take place only by cuttings, although this is very simple. The stem is simply placed in the ground and watered (Bristol 1965, 276*). Apart from this, cultivation is the same as with **Brugmansia arborea.** Nitrogen-rich soils

57 In South America, many inebriating plants are known as *borrachero* or *borrachera*. Some of these (e.g., **Brugmansia spp.**, **Iochroma fuchsioides**) have psychoactive properties. But the name is also applied to some plants that are not known to exert any psychoactive effects; **Pilocarpus alvaradoi** Pittier (Rutaceae), for example, is known as *borrachero* in Venezuela (Blohm 1962, 45*).

58 In Spanish, the European snowdrop (*Galanthus nivalis*) is referred to as *campanilla blanca*.

59 This name is used in South America (and especially in Chile) for **Datura stramonium.**

60 In Peru, this name is also applied to other plants that are used as additives to cimora and San Pedro drinks: *Iresine* spp., *Pedilanthus tithymaloides*, and *Hippobroma longiflorum* (see **Trichocereus pachanoi**).

have been found to increase alkaloid production in the plant.

Appearance

Growing to heights of up to 8 meters, this treelike shrub always produces flowers but only very rarely produces fruits. The smooth fruits are slender, spindle-shaped, and pointed at the end. They are somewhat more slender and longer than the fruits of *Brugmansia aurea* (which is how the two species may be distinguished). This angel's trumpet typically bears snow-white flowers that hang almost straight down, are often double, and can reach a length of more than 30 cm. In southern Mexico, the flowers of this species sometimes have a pink margin. The form of the flowers is so variable that it is often very difficult to make a definite species identification.

Brugmansia x *candida* is very often confused with *Brugmansia aurea* and has even been regarded as a synonym for the latter. It can also be mistaken for *Brugmansia arborea.*

Psychoactive Material

— Leaves
— Flowers

Preparation and Dosage

Shamans in Colombia use primarily cold-water extracts of the leaves. Normally, only pairs of leaves—and even numbers of pairs—are used. Depending upon the size of the leaves, the Sibundoy Indians regard two to twenty-four (= twelve pairs) leaves per person as a shamanic dosage. For "normal" people, this dosage would likely result in extreme delirium and dangerous toxic symptoms.

In the Kamsá tradition, the "jaguar inebriant," which is made from the fresh leaves of *Brugmansia* x *candida* f. Culebra (= *Methysticodendron amesianum*), can be produced and drunk only during a waning moon. No more than one hour before they are to be ingested, the leaves are plucked from the shrub, crushed, and placed in cold water for about half an hour. Directly before the extract is consumed, it is warmed a little and stirred, but never boiled. The liquid is then strained off. The shamans never drink all of the liquid at once but, rather, take a few sips at a time over a period of some three hours. This procedure evidently enables them to achieve the dosage that is appropriate for them. If the shaman has not fallen into trance after three hours, a helper will prepare another drink and give small sips of it to him until the desired state of consciousness has been achieved (Schultes 1955, 9).

Northern Peruvian *curanderos* (folk healers) drink a decoction of the leaves to induce a clairvoyant trance. The freshly pressed juice of the leaves and/or flowers is also ingested alone or

mixed with **alcohol** and sugar (Bristol 1965, 285*).

In Peru, at least three cultivated forms (see above) are used as one of the main ingredients in the psychoactive drink known as **cimora** and as additives to preparations of San Pedro (*Trichocereus pachanoi*).

The dried leaves and flowers can be smoked alone or mixed with other plants such as *Cannabis indica* and *Nicotiana rustica* to produce **smoking blends** (cf. *Brugmansia arborea*).

Ritual Use

In Colombia (Sibundoy), extracts of the leaves are drunk at shamanic and religious ceremonies of the Kamsá and Inga Indians, primarily to learn about methods for performing witchcraft, divination, prophecy, and shamanic therapy.

Among the Kamsá, the form described as *Methysticodendron amesianum* is known as *metskwai borrachero* or *mitskway borrachero*, the "jaguar inebriant" (Schultes 1955, 10). Corresponding to the strongest of all shamanic animals, this plant thus represents a very potent shamanic vehicle (cf. *Nymphaea ampla, Solanum* spp.). The shamans of the Kamsá use this agent almost exclusively for divination and prophecy. They normally turn to it only when faced with a truly difficult case, for it can occur that the body of the shaman who has used it lies in a comatose or delirious state for two to three days while his soul explores the secret corners of nonordinary reality. During this procedure, an assistant is always present. He does not merely watch over the shaman's body but also pays heed to any messages the shaman may utter (Schultes 1955, 8 f.).

In modern Mexico, angel's trumpets are used as alternatives to the herbaceous thorn apples (*Datura innoxia, Datura stramonium*) (Heffern 1974, 100*). In some Mixe settlements, angel's trumpets are used for divination and diagnosis (Lipp 1991, 187*). Three flowers is the suggested effective dosage, although six flowers may be administered if the desired effects are not achieved with three. The fresh flowers are macerated in hot water and then pressed with a cloth (190).

The Huastec, who live on the Gulf of Mexico, believe that a person who has eaten *Brugmansia candida* leaves "sees" reality (Alcorn 1984, 624*).

Cultivated plants of *Brugmansia* x *candida* often develop double flowers.

"Our ancestors were accustomed to it, so it is said,
to drink many medicines, even more than we do.
By drinking these medicines they could see, so it is said,
the forms in which things appear.
And once they drank yagé
and angel's trumpets,
and suddenly, so it is said,
a falcon flew past,
and the bird fell dead in the courtyard.
And then they said: O God, what is going to pass?
And as they asked this,
the best among the drinkers of medicine answered:
People from another world will come,
flying, and they come to drive us from our land."

FRANCISCO TANDIOY
IN *VON DEN WURZELN DER KULTUR*
[FROM THE ROOTS OF CULTURE]
(RÄTSCH 1991b, 161*)

The leaves of the cultivated "snake plant" angel's trumpet (cv. Culebra) are often grossly deformed.

The Tzeltal, who live in the Selva Lacandona (Chiapas, Mexico), smoke the dried leaves (with or without tobacco, *Nicotiana rustica*) when they wish to perform divinations (Rätsch 1994c*).

Artifacts

The Mexican art nouveau artist Saturnino Herrán (1887–1919) painted a fresco, *Nuestros dioses*, in the center of which Coatlicue, the Aztec earth goddess, is portrayed adorned with angel's trumpets (López Velarde 1988, 113). During the art nouveau movement, glass lamp shades were manufactured in the shape of angel's trumpet flowers (cf. *Brugmansia arborea*). In a cloth print (Paris, 1896) based upon a design by Alphonse Mucha, *Brugmansia x candida* appears as a floral element playfully surrounding a young woman.

The prophetic powers of the angel's trumpet are portrayed in a theater piece by Francisco Tandioy (McDowell 1989, 139).

Medicinal Use

In the Sibundoy Valley, the fresh flowers and leaves of *Methysticodendron amesianum* are heated in water and applied as a plaster to treat tumors, swelling, swollen knees, et cetera. Medicine men sometimes bathe patients suffering from fever and chills in a warm decoction of the leaves and flowers (Schultes 1955, 9 f.).

In Colombia, preparations of *Brugmansia candida* are used to treat muscle cramps, erysipelas (a skin inflammation), swollen inflammations, and colds. The Tzeltal Indians of Chiapas (Mexico) use the leaves to treat diseases that are caused by "winds" in the body.

Constituents

All forms of *Brugmansia x candida* contain **tropane alkaloids**. The primary constituent is **scopolamine** (hyoscine); also present are meteloidine and hyoscyamine. The Culebra form (= *Methysticodendron amesianum* Schultes) contains hyoscyamine, scopolamine, and **atropine**; L-scopolamine makes up 80% of the plant's total alkaloid content (Bristol 1965, 286*). The young leaves have the highest concentration of alkaloids, up to 0.56% total content (Griffin 1976). The cultivar Flintham Hall was found to contain 0.55% total alkaloids, with scopolamine the primary constituent. Also present are 6β-acetoxy-3α-tigloyloxyltropane, tigloidine, 6β-tigloyloxyltropane-3α-ol, 3α-tigloyloxyltropane-6β-ol, hyoscyamine/atropine, norhyoscyamine/noratropine, 6β-hydroxy-hyoscyamine, and tropane-3α-ol (el Imam and Evans 1990, 149*).

Effects

The Sibundoy Indians say that they encounter numerous giant snakes in the visions they receive while under the influence of this powerful magical plant. One Sibundoy provided the following description of his first encounter with the "snake plant":

> The first time, I drank six leaves [of the Culebra form] at night. I became drunk. I saw forests full of trees, people from other places, animals, tree stumps, meadows full of all kinds of snakes which came towards me at the edge of the pasture—all in green—to bite me. As the inebriation became stronger, the house began to lean against the rest of the world, as did the things in the house. . . . But the snakes still wanted to kill me. (Bristol 1965, 283*)

Apart from this, the effects of this plant should not differ much from those of the other species of *Brugmansia* (see *Brugmansia arborea*).

Commercial Forms and Regulations

See *Brugmansia arborea*.

Literature

See also the entries for the other *Brugmansia* species, **Trichocereus pachanoi**, **cimora**, **scopolamine**, and **tropane alkaloids**.

Bristol, Melvin L., W. C. Evans, and J. F. Lampard. 1969. The alkaloids of the genus *Datura*, section *Brugmansia*. Part VI: Tree datura drugs (*Datura candida* cvs.) of the Colombian Sibundoy. *Lloydia* 32 (2): 123–30. (Includes a listing of additional literature.)

Giulietti, A. M., A. J. Parr, and M. J. C. Rhodes. 1993. Tropane alkaloid production in transformed root cultures of *Brugmansia candida*. *Planta Medica* 59:428–31.

Griffin, W. J. 1966. Alkaloids in *Datura*, section *Brugmansia*: The peach flowered form of *Datura candida* sens. lat.. *Planta Medica* 14:468–74.

———. 1976. Agronomic evaluation of *Datura candida*—a new source of hyoscine. *Economic Botany* 30:361–69.

López Velarde, Ramón. 1988. *Saturnino Herrán*. Mexico City: Fondo Editorial de la Plastica Mexicana.

McDowell, John Holmes. 1989. *Sayings of the ancestors: The spiritual life of the Sibundoy Indians*. Lexington: The University Press of Kentucky.

Pachter, I. J., and A. F. Hopkinson. 1960. Note on the alkaloids of *Methysticodendron amesianum*. *Journal of the American Pharmaceutical Association, Science Ed.* 49:621–22.

Schultes, Richard Evans. 1955. A new narcotic genus from the Amazon slopes of the Colombian Andes. *Botanical Museum Leaflets* 17:1–11.

Brugmansia x *insignis* (Barbosa Rodrigues) Lockwood ex Schultes

Magnificent Angel's Trumpet

Family
Solanaceae (Nightshade Family); Subfamily Solanoideae, Datureae Tribe, Brugmansia Section

Forms and Subspecies
The Siona make a distinction among at least four "species" of this plant, the names of which refer to totemic and shamanic elements (Vickers and Plowman 1984, 29*):

muhu pehí—"thunder angel's trumpet"
semé pehí—"Paca angel's trumpet"
sesé pehí—"white-lipped peccary angel's trumpet"
tãkiyaí pehí—"*tãki*-cats angel's trumpet"

Synonyms
Datura insignis Barb. Rodr. in Vellosia
Datura x *insignis* Barb. Rodr.
Datura suaveolens x *D. versicolor*

Folk Names
Ain, ain-va-i (Kofán), angel's trumpet, danta borrachera, floripondio, guando, hayapa, huanduj, jayapa, ku-a-va-u, ku-wá-oo (Inga, "pink angel's trumpet"), magnificent angel's trumpet, maricaua, muhu pehí, pehí (Secoya), pimpinella borrachera, saaro (Matsigenka), sacha-toé, toa-toé, tree-datura, ts'ak tsimin (Lacandon, "tapir medicine"), wandú (Quechua), xayápa (Mashco)

History
The Amazon Indians of Ecuador use the stem of this angel's trumpet as a hallucinogen. The Mashco, who live in the southwestern region of the Amazon (Peru), are composed of two tribes (Huachipaire and Zapiteri). Their most important shamanic plant is the magnificent angel's trumpet, which they call *xayápa*.

Distribution
This hybrid of *Brugmansia suaveolens* and *Brugmansia versicolor* was likely the result of cultivation. The plant is from the West Amazon, and many Indians plant it in their house gardens (Vickers and Plowman 1984, 29*). The species has spread into other tropical areas. It is frequently found growing wild in the Selva Lacandona (Chiapas, Mexico).

Cultivation
In Amazonia, this angel's trumpet is propagated with cuttings. The Indians take a piece of stem or branch approximately 50 cm long and simply stick this into the moist ground (Califano and Fernández Distel 1982, 131).

Appearance
This species is most likely the result of a cross between *Brugmansia suaveolens* and *Brugmansia versicolor* (Schultes 1977b, 124*) and looks exactly like an intermediate stage between the two species. It is most easily recognized by its flowers. They are convex, like those of *Brugmansia suaveolens*, but not as obese, and they hang almost straight down, although not as steeply as those of *Brugmansia versicolor*.

In the tropics, *Brugmansia* x *insignis* can grow into a proper and heavily branching tree that can reach a height of over 5 meters. In Amazonia, it blossoms between November and April. The flowers exude a potent perfume in the evening. This cultivar almost never develops fruits (Califano and Fernández Distel 1982, 131).

In addition to the yellowish-reddish blooming form, there is also a form with luminously yellow blossoms that is easily mistaken for *Brugmansia aurea*. This species is also easily confused with *Brugmansia suaveolens* and *Brugmansia versicolor*.

Psychoactive Material
— Stems
— Leaves
— Flowers

Preparation and Dosage
The Secoya grate the stem and boil it for an entire day. They then pour off the decoction and boil it down some more. Unfortunately, no precise information about dosages is known, as use of the plant is restricted to knowledgeable shamans (Vickers and Plowman 1984, 29*).

The Siona and Secoya also use this angel's trumpet as an **ayahuasca** additive. The leaves are burned to ashes in a pot and powdered. This powder is mixed into the finished ayahuasca to potentiate the visions (Vickers and Plowman 1984, 29*). The leaves are also used as an

Left: The cultivar *Brugmansia* x *insignis* cv. Orange produces beautiful trumpet-shaped yellow flowers.

Below: *Brugmansia* x *insignis* is a highly significant shamanic plant among the peoples of the rain forests of northern South America and southern Mexico. (Wild plant, photographed in the Selva Lacandona, Mexico)

ayahuasca additive in the area of Loreto (Peru) (Schultes and Raffauf 1990, 422*).

The Kofán drink an infusion of the leaves for psychoactive purposes. An infusion of six leaves in 200 ml of water is sufficient to induce a hypnotic state (Schultes and Raffauf 1990, 422*). The Kofán also drip juice pressed from fresh flowers into the nostrils of their hunting dogs "so that they can hunt better" (421).

The Mashco prepare a hallucinogenic drink from this plant. Both the drink and the plant itself are known as *xayápa*. The Mashco take stalks of various thicknesses, cut these into pieces about 70 cm in length, and carry them into the ritual house, which is located outside of the settlement in the jungle. There, the bark is peeled from the pieces of stalk, pounded, and boiled in water for several hours. The long period of boiling yields a thick concentrate that possesses "enough hallucinogenic power." The preparation of *xayápa* is usually carried out by a knowledgeable—typically older—person who also assists the *xayápa* drinker during his journey (Califano and Fernández Distel 1982, 135). The shamans of the Huachipaire Mashco also ingest the drink in the form of an **enema** (140).

Ritual Use

The Huachipaire Mashco have several rules associated with the ritual consumption of the *xayápa* drink that must be observed without exception: Ingestion must take place at night; the drinker must lie on the ground or a platform, uncovered, with open arms, and be able to observe the nocturnal sky above; the liquid must be drunk with the lips directly from the pot without touching the pot; the assistant or assistants may not speak to the drinker, even when the latter encourages them to do so; when the sun rises, the drinker must be dipped completely naked into the water of a nearby stream or river so that the last effects of the drink dissipate. In the weeks following ingestion, the drinker must adhere to a specific diet. In no case may he consume certain fishes and birds, bananas, and sugarcane, lest he fall victim to fever, skin spots, or stomach ailments. The drink is customarily ingested to localize a lost or stolen object, discern the future, heal illnesses, or renew the body. The Mashco believe that under the influence of *Brugmansia*, the body is renewed or rejuvenated and thereby healed of all diseases (Califano and Fernández Distel 1982, 135f.). A longer life can thus be expected.

In Colombia and Peru, shamans also ingest preparations of angel's trumpet for diagnostic purposes (Schultes and Raffauf 1990, 422*).

Artifacts
See *Brugmansia arborea.*

Medicinal Use
The fresh leaves are tied to inflamed or painful areas. The freshly pressed plant juice is also used to treat pains. An infusion of the leaves is drunk as a sedative (Schultes and Raffauf 1990, 421 f.*).

Constituents
This angel's trumpet contains the **tropane alkaloids atropine, scopolamine**, and hyoscine. The bark appears to be particularly rich in alkaloids (Califano and Fernández Distel 1982, 134).

Effects
The Mashco granted permission to anthropologists Mario Califano and Alicia Fernández Distel to try the *xayápa* drink several times under their instruction and supervision. They drank approximately one quarter of a liter of the bitter, almost viscous drink. This resulted in a series of hallucinations pertaining to the "social life that we had experienced a few days earlier," and they saw family members and friends approach them as if from a different world. The effects lasted a total of twelve hours and were characterized by visual hallucinations, illusory feelings, acoustic and olfactory hallucinations, and a profound dryness of the mouth. They occasionally fell into periods of sleep from sixty to ninety minutes long, with prophetic dreams, but they also experienced nervous discomfort and euphoria (Califano and Fernández Distel 1982, 137f.).

The Lacandon (Chiapas, Mexico) say that horses who have eaten the leaves of this angel's trumpet become inebriated "as if they were drunk."

Commercial Forms and Regulations
See *Brugmansia arborea.*

Literature

See also the entries for the other **Brugmansia** species, **scopolamine**, and **tropane alkaloids.**

Califano, Mario, and A. Fernández Distel. 1982. The use of a hallucinogenous plant among the Mashco (southwestern Amazonia, Peru). *Zeitschrift für Ethnologie* 107:129–43.

Brugmansia sanguinea (Ruíz et Pavón) D. Don

Bloodred Angel's Trumpet

Family

Solanaceae (Nightshade Family); Subfamily Solanoideae, Datureae Tribe, Brugmansia Section

Forms and Subspecies

One form from Sibundoy that has heavily serrated leaves has been described under the name *Datura sanguinea* Ruíz et Pav. cv. Guamuco (Schultes 1979b, 148).

Datura vulcanicola [syn. *Brugmansia vulcanicola* (Barclay) Lockw.], a species originally described by A. S. Barclay (1959), is now regarded as a subspecies: *Brugmansia sanguinea* ssp. *vulcanicola* (Riviera et al. 1989). The variety [or cultivar β] *flava* Dunal is a yellow-blooming variety (= *Brugmansia lutea* = *Datura rosei*) cultivated primarily in Colombia. A form with pure red flowers occurs in the highlands of southern Colombia and northern Ecuador and is referred to as *Brugmansia sanguinea* cv. Sangre. In Sibundoy, there is a cultivar *Brugmansia sanguinea* cv. Guamuco.

A form recently discovered in eastern Ecuador (Pelileo, Napo Province) at an altitude of approximately 2,500 meters appears to be an intermediate between *Brugmansia sanguinea* and *Brugmansia sanguinea* ssp. *vulcanicola* (possibly it is a hybrid between the species and the subspecies).

In southern Chile, there is also a form whose flowers are almost entirely green; the outer margin of the calyx is slightly red, and occasionally almost purple.

Synonyms

Brugmansia bicolor Pers.
Brugmansia lutea Hort. ex Gardeners
Brugmansia vulcanicola (Barclay) Lockwood
Datura (*Brugmansia*) *rosei* Saff.
Datura sanguinea Ruíz et Pav.
Datura vulcanicola Barclay

Folk Names

Belladonna tree, bloodred angel's trumpet, blutroter stechapfel, borrachero, borrachero rojo, bovachero, campanilla encarnada, chamico, el guantug (Ecuador), floripondio, floripondio boliviano, floripondio encarnado, guamuco (Kamsá, Inga),[61] guamuco floripondio, guamucu borrachera (Inga), guando, guantug, guántug, huaca (Quechua, "grave"), huacacachu, huántug, humoco, koo-wá-oo, misha colorada, misha curandera, misha huarhuar, misha rastrera, perecillo, poroporo, puca champancho (Quechua, "red Brugmansia"), puca-campanilla, qotu (Quechua), tonca, tonga, yerba de huaca

History

In Colombia, this sacred plant was used in pre-Columbian times in ritual contexts in the cult of the sun. It was apparently this species of angel's trumpet that José de Acosta mentioned in 1590 under the name *floripondio*. The incredible effects of the *tonga* drink, prepared from *Brugmansia sanguinea*, were first described in 1846 by the Swiss Johann J. von Tschudi (Hartwich 1911, 519*). Today, shamans in Ecuador continue to utilize the plant as a hallucinogen. In Peru, the seeds are still used as popular additives to **beer, chicha,** and coffee (cf. ***Coffea arabica***).

This angel's trumpet is now also known as *floripondio boliviana* because its flowers have the same colors as the Bolivian flag: red, yellow, and green (Bastien 1987, 114*).

Distribution

This rather cold-resistant species is distributed throughout the Andes, from Colombia to Ecuador, Peru, and Bolivia and into southern Chile. It is usually found around 2,000 meters above sea level. On Chiloé, an island off the southern coast of Chile, it is found at altitudes as low as sea level. In Charazani, Cochabamba, and the area of La Paz (Bolivia), it is frequently cultivated as an ornamental (Bastein 1987, 114*).

The subspecies *Brugmansi sanguinea* ssp. *vulcanicola* is found only in the mountainous region around the Puracé volcano in Colombia at altitudes above 3,000 meters (Rivera et al. 1989).

Cultivation

This species is propagated through seeds or cuttings. Of all species of *Brugmansia*, this species is most easily grown from seed. It is best to pre-germinate the seeds, e.g., in moist cloths or in thoroughly moistened soil in seedbeds or greenhouses. The seedlings should be transplanted with care (into pots where the climate is too cold for outdoor planting).

This angel's trumpet is cultivated commercially in Ecuador to produce **scopolamine** for the pharmaceutical industry and thus represents one of the world's primary sources of this compound (Rivera et al. 1989).

Appearance

This perennial, heavily branching angel's trumpet forms a woody stem and can grow from 2 to 5 meters in height. The margins of the gray-green, hairy leaves are coarsely serrated and are usually smaller than those of other *Brugmansia* species. The bloodred angel's trumpet does not exude a scent at night, an important criterion for

Iconographic element from the pre-Columbian Tello obelisks; possibly a representation of the bloodred angel's trumpet (*Brugmansia sanguinea*) with its characteristically shaped fruits.

61 In Sibundoy (Colombia), the name *guamuco* also refers to *Spigelia pedunculata* H.B.K.

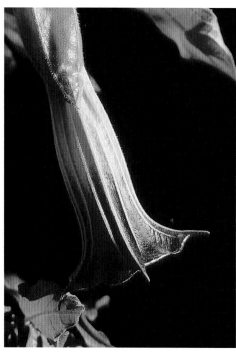

Top left: The typical flower shape and color of *Brugmansia sanguinea*

Bottom left: The seeds of *Brugmansia sanguinea* are similar to the seeds of *Datura innoxia* but are considerably larger (two to three times as large).

Right: The relatively rare *Brugmansia sanguinea* ssp. *vulcanicola* of Ecuador

A shaman with a lance and animal spirit (bird) receives a branch with angel's trumpet flowers and fruits from a woman. (Taken from a Keru lacquer picture of the colonial period, late sixteenth century, South America)

recognition. The blossoms do not produce any perfume, which enables the species to be very clearly identified. The flowers are normally greenish at their base, yellow in the middle, and red at their margin. But there are also green-red, pure yellow, yellow-red, and almost entirely red varieties.

The oval-obese, pointed fruits have a smooth surface and are usually half-covered by the dried calyx. This species produces fruits more regularly than does any other species of *Brugmansia*. In contrast to the normal form, the subspecies *vulcanicola* produces smooth seeds.

This angel's trumpet is the most easily identifiable of all *Brugmansia* species. Nevertheless, it is still occasionally confused with *Brugmansia aurea* and *Brugmansia suaveolens*. It has even been mistaken for *Iochroma fuchsioides*.

Psychoactive Material
—Leaves
—Fruits/seeds

Indians of the Ecuadoran highlands believe that this angel's trumpet provides a toxic or inebriating honey when bees collect its nectar.

Preparation and Dosage
The seeds are used as an additive in preparations of *Trichocereus pachanoi* (cf. **cimora**) and to fortify **chicha**. The fruits or seeds are boiled to produce a decoction known as *tonga*. Only shamans are allowed to drink *tonga*, for it is said that it would cause normal people to lose their mind. For information about dosages, see *Brugmansia arborea*.

The folk healers (*curanderos*) of northern Peru add the leaves and flowers to their San Pedro drinks (cf. *Trichocereus pachanoi*) so that they may "see" better. The woody stems are used to produce magic wands for *mesa* (or table; a temporary altar used by curanderos) rituals (Giese 1989, 251*).

The dried leaves may be smoked alone or in **smoking blends**. They are also a component of South American asthma cigarettes, which are smoked to relieve the symptoms of asthma.

Ritual Use
In pre-Spanish and late colonial times, the priests in the sun temple of Sogamoza (north of Bogotá, Colombia) ingested *tonga* during religious rituals (Lockwood 1979, 149). During the pre-Spanish period, the Chibchas gave the widows and slaves of deceased rulers a mixture of *Brugmansia*, **chicha** (maize beer), and tobacco extract (*Nicotiana tabacum*) so that they would be sedated but still alive as they were buried along with the deceased (Lockwood 1979, 150). Contemporary shamans and diviners still use *tonga* to induce a prophetic trance, to diagnose diseases, and to localize lost objects as well as to divine the future.

In the Darién and Chocó regions, the seeds were boiled to produce a decoction that was administered with **chicha** to children so that they could enter a clairvoyant state in which they would receive the power to "see" gold and treasures (Lewin 1980 [orig. pub. 1929], 182*).

In many parts of South America (e.g., southern Chile), the seeds may be secretly mixed into a person's coffee (cf. *Coffee arabica*) to harm, induce aphrodisiac effects in, or make fun of him or her. Depending upon the dosage, the victim may fall into a coma, become sexually aroused, or carry out comical stereotypical acts (cf. *Scopolia carniolica*).

Artifacts
This or other species (cf. *Brugmansia arborea*) appear to be represented on various objects of the pre-Columbian Chavín culture (Mulvany de Peñaloza 1984*).

An Indian drawing of a woman under a *borrachero* tree has been incorrectly interpreted as a depiction of *Brugmansia vulcanicola* (Schultes and Hofmann 1992, 128*). The species shown is in fact *Iochroma fuchsioides*.

In Sri Lanka, the beautiful flowers are sometimes depicted on batiks.

Medicinal Use
In the Colombian Sibundoy Valley, flowers of the bloodred angel's trumpet, leaves of the Culebra form of *Brugmansia* x *candida*, and the stems and leaves of *Phenax integrifolius* Webb. are macerated in water and made into a plaster for treating rheumatism. Heated leaves are also bound over

swollen infections, and an infusion of the leaves is used to wash inflamed areas (Schultes and Raffauf 1990, 422*). The leaves are also used in Peruvian folk medicine for treating inflammations (Chavez V. 1977, 189*). The healers of the Callaway use the leaves externally to treat rheumatism and arthritis (Bastien 1987, 114*).

Constituents

The entire plant contains **tropane alkaloids**. The flowers contain chiefly **atropine** and only traces of **scopolamine** (hyoscine). The seeds contain approximately 0.17% total alkaloids; of this, 78% is scopolamine. The alkaloids apohyoscine, hyoscyamine, choline, tropine, and pseudotropine and two unknown alkaloids have also been detected (Leary 1970). The roots contain the highest alkaloid concentration as well as 0.08% littorine (Evans and Woolley 1969). This angel's trumpet produces a psychoactive or toxic **honey**.

The subspecies *Brugmansia sanguinea* ssp. *vulcanicola*, which is native to Colombia, is especially rich in scopolamine and atropine. The flowers contain the highest concentrations of alkaloids (0.83%), followed by the fruits (0.74%), while the leaves contain only 0.4% (Rivera et al. 1989). This is probably the most potent of all *Brugmansia* species.

Effects

All parts of the plant produce strong hallucinations and delirium. The overall effects of this species are the same as those of the other *Brugmansia* species (cf. ***Brugmansia arborea***).

Commercial Forms and Regulations

See ***Brugmansia arborea***.

Literature

See also the entries for the other *Brugmansia* species, **scopolamine**, and **tropane alkaloids**.

Evans, W. C., V. A. Major, and M. Pethan. 1965. The alkaloids of the genus *Datura*, section *Brugmansia III: Datura sanguinea* R. and P. *Planta Medica* 13:353–58.

Evans, W. C., and Valerie A. Woolley. 1969. Biosynthesis of the (+)-2-hydroxy-3-phenylpropionic acid moiety of littorine in *Datura sanguinea* and *Anthocercis littorea*. *Phytochemistry* 8:2183–87.

Leary, John D. 1970. Alkaloids of the seeds of *Datura sanguinea*. *Lloydia* 33 (2): 264–66.

Rivera, A., E. Calderon, M. A. Gonzalez, S. Valbuena, and P. Joseph-Nathan. 1989. *Brugmansia sanguinea* subsp. *vulcanicola*, a good source of scopolamine. *Fitoterapía* 60 (6): 542–44.

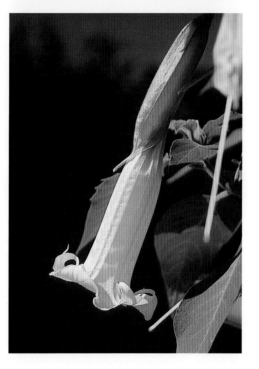

Top left: The ripe fruit of *Brugmansia sanguinea*.

Bottom left: The yellow-blooming variety *Brugmansia sanguinea* var. *flava* is also known by the names *Brugmansia lutea* and *Datura chlorantha*.

Top right: A green-blooming variety of *Brugmansia sanguinea*. (Photographed in southern Chile)

Bottom right: The fruit of *Brugmansia sanguinea* produces a large number of seeds. In South America, these are often used to fortify maize beer.

Brugmansia suaveolens (H.B.K.) Berchtold et Presl

Aromatic Angel's Trumpet

"The natives call it *Huacacachu, yerba de Huaca,* or *Borrachero* and prepare from its fruits a potent narcotic drink known as *tonga.* Its effects are terrible. I once had the opportunity of seeing them in an Indian who wanted to contact the spirits of the ancestors. The horrible sight of this scene has burned itself so deeply into my memory that I will never forget it. Shortly after consuming the tonga, the man fell into a dark brooding, his eyes stared dully at the ground, his mouth was tightly, almost convulsively shut, his nostrils gaped widely, cold sweat covered his forehead and earthly pallid face, the jugular veins of his throat were swollen as thick as a finger, his chest rose slowly and gaspingly, his arms hung rigidly alongside his body. Then his eyes moistened and filled with great tears, his lips twitched slightly and convulsively, his carotid pounded visibly, his respiration accelerated, and his extremities made repeated automatic movements. This state may have lasted for a quarter of an hour, and then all of these manifestations increased in intensity. His now dry but completely reddened eyes rolled wildly in their sockets. All of his facial muscles were contorted in the most horrible fashion. A thick white foam emerged from between his half-opened lips. The pulse on his forehead and throat beat with terrible rapidity. His breath was shallow and unusually accelerated and was no longer able to lift his chest, which exhibited only a slight vibration. Copious amounts of sticky sweat covered his entire body, which was continuously shaken by the most horrible convulsions. His limbs were twisted in the most terrible way. A low, unintelligible murmuring alternated with a piercing and heart-rending scream, a muffled crying, and a deep groaning or moaning. Long did this terrible state last, until the severity of the symptoms diminished and calmness appeared. Immediately, women hurried over, washed the Indian's entire body with cold water, and laid him comfortably onto several fleeces.

"There followed a peaceful sleep that lasted for several hours. That evening, I saw the man again as he was just describing his visions and his

Family
Solanaceae (Nightshade Family); Subfamily Solanoideae, Datureae Tribe, Brugmansia Section

Forms and Subspecies
One form with a very large calyx has been described under the name *Datura suaveolens* β *macrocalyx* Sendtner. The Shuar and Achuar know of several "species" of this *Brugmansia*; these cannot, however, be botanically distinguished (Bennett 1992, 493*, Descola 1996, 88*).

Synonyms
Datura gardneri Hooker
Datura suaveolens Humb. et Bonpl. ex Willd.

Folk Names
Ain-vai (Kofán), almizclillo, angel's trumpet, aromatic angel's trumpet, baikua, bikut, borrachero, campana, canachiari (Shipibo), chinki tukutai maikiua (Achuar, "angel's trumpet to blow on small birds"), datura d'Egitti, datura d'Egypt, duftende engelstrompete, fleur trompette, flor de campana, floripondia, floripondio blanco, guando, huanduc (Quechua), ishauna (Zapara), juunt maikiua (Achuar, "large angel's trumpet"), maikiua (Achuar), maikiuwa (Achuar/Shuar), maikoa (Jíbaro), maikua, maikuna, ohuetagi (Huaorani), peji (Secoya), sprengels engelstrompete, toá, toé, toé canachiari (Shipibo), trompeta del juicio, ts'ak tsimin (Lacandon, "horse medicine"), tsuaak, tsuak, tu-to-a-vá-a (Kofán, "white angel's trumpet"), vau (Kofán), wahashupa (Sharanahua), weiße engelstrompete, wohlriechender stechapfel, yawa maikiua (Achuar, "dog's angel's trumpet"), yumi maikiua (Achuar, "heaven's water angel's trumpet")

History
In South America, aromatic angel's trumpet has apparently been used for ritual and medicinal purposes since pre-Columbian times. It is possible that this species may have been known even in pre-Spanish Mexico; there, it continues to possess a certain significance as a hallucinogenic shamanic plant. This angel's trumpet was first described by Alexander von Humboldt (1769–1859). Because of its beauty and its bewitchingly delicious scent, it is now the most commonly cultivated species of *Brugmansia*. Among the Jíbaro Indians, it has important meaning as a ritual inebriant (Descola 1996*).

Distribution
Aromatic angel's trumpet is found throughout the Andes and Cordilleras as well as Central America. Through cultivation, it has spread into other regions of the world. It is now part of the flora of Nepal and can be found in the Himalayas at altitudes of up to 1,700 meters (Polunin and Stainton 1985, 289*).

Cultivation
The simplest method of cultivation is through cuttings (see **Brugmansia arborea**). This angel's trumpet can also be grown from seed. In northern climates, sowing (possible throughout the year) should be done in pots on a window ledge at temperatures of 20 to 25°C (time to germination is two to three weeks). A sterile, porous substrate, e.g., sandy, loose soil, works best. The soil must be kept well moistened. The plant should be transplanted while still small into a large pot filled with peat-rich soil or into the garden. Prune back in late fall and allow to overwinter in the cellar. Water the plant thoroughly in the spring. Leaves will appear again quite quickly. The plant requires much water and thrives best in semi-shaded areas.

Appearance
This large perennial bush with woody stems is often heavily branched and can grow as tall as 5 meters. It has very large, usually smooth-margined leaves that are oval and pointed at the ends. The flowers, which can grow as long as 30 cm, hang down at an angle and are usually pink. The calyx and corolla each have five points (an important point for classification). During the evening and at night, the flowers exude a bewitching and inebriating scent. The fruits, which form only very rarely, are short and spindle-shaped with an irregularly gibbous surface and contain large (approximately 1 cm) light brown seeds. This species also occurs in a form with pure white blooms (e.g., in Argentina). In the Himalayas, only the white-blooming form is found (Polunin and Stainton 1985, 289*).

This angel's trumpet is easily mistaken for *Brugmansia* x *insignis.*

Psychoactive Material
— Leaves
— Flowers
— Stems
— Juice pressed from the fresh stems
— Seeds

Left: The typical flower of *Brugmansia suaveolens*. (Photographed in Chiapas, Mexico)

Right: A rare pure-white-blooming form of *Brugmansia suaveolens*. (Photographed in northwest Argentina)

Preparation and Dosage

The fresh leaves, seeds, and flowers may be eaten fresh or drunk in the form of an infusion. This tea is sometimes mixed with alcoholic beverages. The fresh flowers are also added to milk and drunk (Hall et al. 1978, 251). To produce an aphrodisiac tea, pour hot water over one fresh flower and allow this to steep for ten minutes. The fresh leaves can be added to white rum, tequila, or other types of spirits (**alcohol**). The leaves also can be made into a decoction or used as an **ayahuasca** additive (Schultes and Raffauf 1990, 422*). For information about dosages and dangers, see *Brugmansia arborea*.

In the Himalayas, the dried leaves are used in tantric **smoking blends** in the same manner as those of *Datura metel.*

Ritual Use

This species is the most commonly used *Brugmansia* in the upper Amazon region. The Jíbaro or Shuar and Achuar drink a tea of the plant they call *maikuna* to obtain a vision that can help them acquire an *arutam wakani,* "visionary soul" (cf. *Nicotiana tabacum*). Once acquired, this soul will be sent forth to make inquiries in the "other world" (Bennett 1992, 493*). Among the Achuar, the visions of the *arutam* are especially important because they restore to a warrior (the former hunter of shrunken heads) the power that he has lost as a result of ritual war killing. To do this, the warrior will go to a secluded placed deep in the forest and consume angel's trumpet juice as well as tobacco juice by himself, away from all others. The effects allow him to see an *arutam* soon thereafter:

Arutam is initially a vision, the fruit of a change in consciousness induced by fasting, the repeated ingestion of tobacco juice, and especially the high doses of scopolamine that are liberated in the thorn apple [sic] [207] preparation. . . . The circumstances under which *arutam* appear [are] exceptionally stereotypical. Exhausted by the inebriation, physically weakened by a lack of food, the senses focused completely on the desired encounter, the supplicant waits alongside the path until he suddenly perceives the rustling of a distant wind, which swells into a hurricane and descends with all its power over the clearing, while a strange figure or a monster slowly approaches: Perhaps a gigantic jaguar whose eyes spout fire, it might also be two intertwined giant anacondas, an overpowering harpy, a sneering bunch of armed strangers, a dismembered human body whose limbs crawl along the ground, or a flaming head that falls from the skies and rolls around twitching. . . . The wind calms down as quickly as it had arisen, and out of the sudden quiet steps an old man. It is *arutam.* . . . (Descola 1996, 318f.*)

The Jíbaro drink the freshly pressed juice of the stems to become brave and to peer into the future. Unruly children are given some of this drink so that they will learn how to behave properly while delirious (Harner 1984, 143ff.*). The Kofán and Achuar give the plant to their dogs to improve their hunting abilities (Descola 1996, 88*; Schultes 1981, 34*).

The shamans of the Tzeltal of southern Mexico smoke the leaves "in order to see things," i.e., for divination and divinatory diagnosis of the causes of illness. But they warn: People who smoke too much will see demons and ultimately "go crazy."

In Nepal, the leaves of this angel's trumpet together with those of *Cannabis indica* are smoked by sadhus and tantric practitioners for meditation or for yoga exercises (cf. also *Aconitum ferox*).

Artifacts

A white-blooming *Brugmansia suaveolens* is portrayed on a still life with flowers (1833) by Johan Laurentz Jensen (1800–1856). (See also *Brugmansia arborea* and *Brugmansia candida.*)

Medicinal Use

In Latin America, the leaves of this *Brugmansia* are very commonly used in folk medicine as an external treatment for wounds, rashes, and ulcers (Berlin et al. 1990, 33ff.*). The Achuar also use the conversations with the spirits of his ancestors to a circle of attentive listeners. He appeared to be very exhausted and worn out, his eyes were glazed, his body slack, and his movements sluggish.

"In earlier times (and still today among the wild tribes), only the physicians and magicians made use of the thorn apple (Peruvian thorn apple = angel's trumpet) to induce ecstasy when performing their conjurations, in which they pretended that this had enabled them to enter into a closer relationship to the gods and, as they put it, 'to speak confidentially with the powerful spirits.' But after Christianity suppressed the magicians and the belief in one god had spread widely, at least in appearance, the Indians themselves stated that they used the tonga to establish contact with the gods of their ancestors and to obtain from them insights into the treasures that were hidden in the graves (huacas). Hence the name: huacacachu (grave plant). The Mestizos use the plant for this purpose much more frequently than do the Indians, who have an unlimited awe of and veneration for the graves of their ancestors. The Cholos (mixtures of Indians with Chinos, Chinos are a mix of mulattos and Mestizos, Mestizos are a mix of whites and Indian women) very often give the pressed juice of the fruits of the thorn apple, mixed with chicha (an alcoholic drink, usually from maize), to the women as an aphrodisiac."

JOHANN VON TSCHUDI
REIZESKIZZEN [TRAVEL SKETCHES]
(1846, 2:21f.)

Cuscohygrine

leaves to treat battle wounds and snakebites (Descola 1996, 88*). Use of the flowers and leaves, and sometimes also the seeds, as aphrodisiacs is found throughout the world. Even the scent is regarded as an aphrodisiac.[62]

Some Lacandon Mayans use angel's trumpet as a remedy to treat their animals. As one tribe member said, "That is a medicine for the chickens. I treat my chickens with it when they have a rash on the eyes. I take the stem and rub it over it, then they quickly get healthy" (Rätsch 1994b, 60*).

Constituents

The **tropane alkaloid** constituents of this *Brugmansia* species have a characteristic composition that chemically distinguishes this plant from all other *Brugmansia* species. The above-ground herbage contains **scopolamine** (hyoscine), apohyoscine, norhyoscine, **atropine**, and noratropine as well as high concentrations of the tigloy-esters of these substances. The roots contain scopolamine, meteloidine, atropine, littorine, 3α-acetoxytropane, 6β-(α-methybutyryloxy)-3α-tigloyloxytropane, 3α,6β-ditigloyloxytropane-7β-ol, 3-α-tigloyoxytropane-6β-ol, tropine, and cuscohygrine. The principal alkaloid in the corolla of the flowers is norhyoscine (Evans and Lampard 1972). The leaves contain 0.09 to 0.16% alkaloids. Some of the esters also occur in the genera *Solandra* and *Datura* (Evans and Lampard 1972). The alkaloid content is highest during the flowering period (Roth et al. 1994, 294*).

Effects

In Colombia, it is commonly believed that the scent of aromatic angel's trumpet induces sleep and intense dreams often with erotic overtones. In southern Colombia, where there are entire boulevards of aromatic angel's trumpet trees, people suffering from sleep disorders walk past the scented trees in the evenings. In Peru, it is thought that those who sleep under an aromatic angel's trumpet will become permanently insane (Schultes 1980, 115*). "Even the scent of the flowers is said to possess narcotic properties and induce headaches as well as nausea" (Roth et al. 1994, 294*).

The hallucinations evoked by aromatic angel's trumpet can last for up to three days (Bennett 1992, 493*). Overdoses can result in anti-cholinergic delirium (Hall et al. 1978). The toxicological literature contains reports of five deaths alleged to have resulted from an overdose of *Brugmansia suaveolens* (Roth et al. 1994, 294*).

Commercial Forms and Regulations

See *Brugmansia arborea*.

Literature

See also the entries for the other *Brugmansia* species, **scopolamine**, and **tropane alkaloids**.

Evans, W. C., and J. F. Lampard. 1972. Alkaloids of *Datura suaveolens*. *Phytochemistry* 11:3293–98.

Hall, Richard C. W., Betty Pfefferbaum, Earl R. Gardner, Sondra K. Stickney, and Mark Perl. 1978. Intoxication with angel's trumpet: Anticholinergic delirium and hallucinosis. *Journal of Psychedelic Drugs* 10 (3): 251–53. (About *Datura suaveolens*.)

---

62 Some Lacandon Mayans see angel's trumpets as relatives (*u bäho'*) of the plant they call *k'äni bäkel*, the "yellow scented" (*Solandra* spp.). And indeed, the plants are members of the same family (Solanaceae). The scent of *Solandra* is quite similar to that of *Brugmansia suaveolens*.

Brugmansia versicolor Lagerheim

Amazonian Tree Datura

Family

Solanaceae (Nightshade Family); Subfamily Solanoideae, Datureae Tribe, Brugmansia Section

Forms and Subspecies

Presumably none

Synonyms

Datura versicolor (Lagerh.) Saff.

Folk Names

Amazonian datura, Amazonia tree datura, bunte engelstrompete, canachiari (Shipibo), sacha-toé, toé, tree datura

History

Although this angel's trumpet appears to be an important Amazonian shamanic plant, almost no ethnobotanical or ethnopharmacological studies of it have been carried out to date. This may be due at least in part to the fact that the uses of the plant that have been noted in ethnographic reports may have been listed under incorrect botanical names. It is very likely that a great deal of the information recorded for *Brugmansia suaveolens* and *Brugmansia* x *insignis* actually pertains to *Brugmansia versicolor*.

This plant first became known to botanists when the Swedish botanist Nils Gustaf von Lagerheim (1860–1926), who was also the first to describe *Brugmansia aurea*, discovered it in Ecuador in 1895.

Distribution

This angel's trumpet is from the northwestern Amazon region (the basin of Guyaquil) and is adapted to the tropical climate. It is found primarily in Ecuador (Zander 1994, 226*) but is also common in northern Peru (Schultes and Raffauf 1990, 424*).

Cultivation

Propagation is performed through cuttings (as with *Brugmansia arborea* and *Brugmansia* x *insignis*).

Appearance

This perennial plant grows into a treelike shrub that can attain a height of 3 meters. The large, trumpet-shaped flowers have smooth corollas and hang straight down (an important characteristic for identification). The flowers are usually various shades of pink and yellow (hence the name *versicolor*). The calyx is simply dentate. The smooth fruit capsule is thin, spindle-shaped, and approximately 15 cm in length and, like the flower

itself, hangs straight down. The leaves have a smooth margin and are oval with a pointed end.

Amazonian tree datura is easily confused with *Brugmansia* x *candida* and *Brugmansia* x *insignis*. Crossing *Brugmansia aurea* with *Brugmansia versicolor* yielded the hybrid *Brugmansia* x *candida* (Schultes and Hofmann 1980, 267*).

Psychoactive Material

— Fresh stalks
— Leaves

Preparation and Dosage

A shamanic dosage consists of 1 to 2 ml of the juice pressed from fresh stalks. The dried leaves and flowers can be smoked alone or in **smoking blends**. For more about dosages and dangers, see *Brugmansia arborea*.

Ritual Use

This species is one of the most important shamanic plants in the Amazonian regions of Ecuador and Peru. In spite of this, almost nothing is known about its usage, which is presumably quite similar to the usage of *Brugmansia aurea*, *Brugmansia* x *insignis*, and *Brugmansia suaveolens*. In the Peruvian Amazon, *Brugmansia versicolor* is used as an **ayahuasca** additive and is cultivated in home gardens specifically for this purpose (Ott 1993, 222*).

Artifacts

See *Brugmansia arborea*.

Medicinal Use

It is possible that this plant is used in folk medicine as a means of birth control (Schultes and Raffauf 1990, 424*).

Constituents

The entire plant contains **tropane alkaloids**. Chemical studies are lacking.

Effects

It is said that the scent of this species not only induces sedative effects but also can, at high dosages (e.g., when a person sleeps underneath this angel's trumpet at night), result in temporary or permanent insanity. A myth of the Juruna tribe tells how under certain circumstances the scent can also lead one to become a shaman:

> One day Uaiçá went hunting. In the forest, he
> saw many, yes, very many dead animals lying
> beneath a tree. Uaiçá stood and looked, unable
> to comprehend how this could happen. As he

Top: The tropical *Brugmansia versicolor* can be recognized by its vertically hanging flowers.

Bottom: A cultivated form of *Brugmansia versicolor* with a double trumpet flower

"Perhaps the Brugmansias would have become extinct long ago if they had not impressed so many people with their ornamental value. These decorative and yet easily cultivatable plants are making their ways into ornamental gardens with increasing frequency—both planted in the ground and as potted plants, depending upon the geographical situation. It would be wonderful if the survival of an interesting genus of plant could be ensured at least in this manner."

ULRIKE AND HANS-GEORG PREISSEL
ENGELSTROMPETEN [ANGEL'S TRUMPETS]
(1997, 17)

was thinking about it, he walked around the tree. No sooner had he walked beneath the tree than he felt dazed, and he immediately fell down and slept. He had many dreams. He dreamt of singing people, of the tapir and all the other animals. In a dream, he also saw Sinaá, an ancestor of the Juruna. He spoke long with him. When Uaiçá awoke, he immediately went back home, for it was late and the sun was already setting. The following day, he returned to the tree, where he again fell down and went to sleep. He dreamed of the same things: of Sinaá, singing people, animals, and his people. For several days, Uaiçá came to the tree, under which he always had the same dreams after he had fallen asleep. He had fasted since the first day. He ate nothing. During the last visit, Sinaá told Uaiçá in a dream: 'Do not come under this tree any more. This is enough.'

"After Uaiçá woke up, he scraped off a little of the bark of the tree and went to the riverbank. There he made a tea from the bark and drank of it. Then he was inebriated, and he jumped into the water and caught fish with his hands. . . . Uaiçá never went back to the tree. He now drank the tea he had brewed from the bark scrapings, and in this way he acquired many abilities." (Karlinger and Zacherl 1976, 172 f.*)

Apart from this, the effects of this plant should not differ from those of other *Brugmansia* species.

Commercial Forms and Regulations
See *Brugmansia arborea.*

Literature
See also the entries for the other **Brugmansia** species, **scopolamine**, and **tropane alkaloids**.

Lagerheim, G. 1895. Monographie der ecuadorianischen Arten der Gattung *Brugmansia* Pers. *Engler's Botanisches Jahrbuch* 20:655–68.

Brugmansia spp. and Hybrids
Angel's Trumpets and Hybrids

Family
Solanaceae (Nightshade Family); Subfamily Solanoideae, Datureae Tribe, Brugmansia Section

Because angel's trumpets are such beautiful plants, they have attracted the enthusiastic attention of gardeners all around the world. There is hardly a tropical or subtropical region in which one will not find angel's trumpets being used as ornamentals. As a result of the now global distribution of these plants, even specialized botanists (including myself) are no longer able to keep track of the genus and the hybrids that have been produced from it (cf. Lockwood 1973).

It is a difficult enough task to distinguish among the species discussed above, not to mention the local varieties. The situation is made more confusing by the utter taxonomic confusion and the multitude of popular names. The many commercial names of plants and seeds that can be obtained from nurseries and seed dealers are more the result of the seller's imagination than botanical nomenclature.

To truly untangle the taxonomy of angel's trumpets, extensive comparative genetic studies would be necessary. These, however, are both expensive and time-consuming and would presumably not justify the economic benefits that would result.

Below may be found some of the names that have appeared in the literature. These refer either to very little known species, subspecies, or varieties or to cultivated forms and hybrids. Only three types can in fact be distinguished on the basis of the actual flower shape: *Brugmansia* x *candida* (= *B. aurea*), *B. sanguinea*, and *B. suaveolens* (cf. Schultes 1979b*). For this reason, the following taxa have been assigned to these types (most species and hybrids are sterile, so the shapes of the fruits cannot be used for classification purposes):

Brugmansia x *candida* type:
— *Brugmansia dolichocarpa* Lagerh. [syn. *Datura dolichocarpa* (Lagerh.) Saff., *Datura carpa*]
  This form is very similar to *B. versicolor.*
— *Datura* (*Brugmansia*) *cornigera* (Hook.) Lagerh.
  A form with very large flowers; described for the Valley of Mexico (Safford 1921, 183).
— *Datura* (*Brugmansia*) *mollis* Saff.
  A yellow-flowering form from Ecuador; apparently synonymous with *B.* x *candida.*

— *Datura rubella* Saff.

Described only for an herbarium specimen from Ecuador (Safford 1921, 185).

Brugmansia sanguinea type:
— *Datura* (*Brugmansia*) *chlorantha*

Yellow-blooming form; presumably identical to *B. sanguinea.*
— *Datura pittieri* Saff.

A form of *B. sanguinea* that produces light-colored blossoms.
— *Datura* (*Brugmansia*) *rosei* Saff.

Reddish-blooming form of *B. sanguinea* from Ecuador; also used as a name for a cross between *Datura innoxia* and *B. aurea* (Lockwood 1973, 280).
— *Brugmansia vulcanicola* (Barclay) Lockw. [syn. *Datura vulcanicola* A.S. Barclay]

See *B. sanguinea.*

Brugmansia suaveolens type:
— *Datura affinis* Saff.

Nonsterile form that produces an oval fruit, from the area of Quito, Ecuador; apparently synonymous with *B. arborea* or *B. suaveolens.*
— *Datura suaveolens* x *Datura candida* cv. Flintham Hall
— *Brugmansia longifolia* Lagerh. [syn. *Datura longifolia* (Lagerh.) Saff.]

Presumably a long-leafed, white-blooming form of *B. suaveolens.*

Today, most botanists accept four species of angel's trumpets: *B. arborea*, *B. aurea*, *B. sanguinea*, *B. suaveolens*. All the other names refer to forms, subspecies, hybrids, and races (D'Arcy 1991, 94; Schultes 1979b, 141*). It may be that only *B. aurea*, *B. sanguinea*, and *B. suaveolens* are true, independent species.

Crosses of *Brugmansia suaveolens* and *Brugmansia versicolor* are frequently encountered. These often yield spectacularly beautiful flowers in various colors (especially white and yellow). Several of the crosses and hybrids have been affected by certain viruses that do not kill the plant but simply alter the form of its flowers. Some of the cultivars are not amenable to more precise specification.

Crosses with Other Genera
Some botanists have succeeded in producing crosses between the species *Datura* and *Brugmansia*. The following hybrids have been successfully produced (Lockwood 1973, 280):

Datura innoxia (fem.) x *Brugmansia aurea*
Datura innoxia (fem.) x *Brugmansia suaveolens*

Synonyms with Other Species
Some of the nightshades that have been described under the name *Brugmansia* are now assigned to the genus *Juanulloa*:
— *Brugmansia aurantiaca* Hort. ex Walpers is an outdated synonym for the nightshade *Juanulloa parasitica* Ruíz et Pav.
— *Brugmansia coccinea* Hort. ex Siebert et Voss. is a synonym for *Juanulloa aurantiaca* Otto et A. Dietr.
— *Brugmansia floribunda* Paxton (= *Brugmansia parviflora* Paxton) is a synonym for a *Juanulloa* species.

Some species of the genus *Juanulloa* are used as **ayahuasca** additives.

Literature
See also the entries for the other **Brugmansia** species, **scopolamine**, and **tropane alkaloids**.

D'Arcy, William G. 1991. The Solanaceae since 1976, with a review of its bibliography. In *Solanaceae III: Taxonomy, chemistry, evolution*, ed. Hawkes, Lester, Nee, and Estrada, 75–138. London: Royal Botanic Gardens Kew and Linnean Society.

Lagerheim, G. 1895. Monographie der ecuadorianischen Arten der Gattung *Brugmansia* Pers. *Engler's Botanisches Jahrbuch* 20:655–68.

Lockwood, Tom E. 1973. Generic recognition of *Brugmansia. Botanical Museum Leaflets* 23 (6): 273–84.

Safford, William E. 1921. Synopsis of the genus *Datura. Journal of the Washington Academy of Sciences* 11 (8): 173–89.

Top left: This flower of *Brugmansia suaveolens* x *versicolor* has been deformed by a virus.

Bottom left: As a result of the long history of cultivation and the numerous hybrids that have been produced, many angel's trumpets have developed into forms that are impossible to assign to a specific species.

Top right: The hybrid *Brugmansia suaveolens* x *versicolor* is a popular garden ornamental.

Bottom right: A virus has produced this mutated form of a *Brugmansia* species cultivar, which develops "glovelike" flowers.

Brunfelsia spp.

Manaca, Brunfelsia

The genus *Brunfelsia* is named after the German physician, botanist, and theologist Otto Brunfels (ca. 1489–1534), one of the fathers of botany. Brunfels produced an important work on herbalism, *Contrafayt Kreuterbuch*, in 1532. (Contemporary woodcut)

Family
Solanaceae (Nightshade Family); Subfamily Cestroideae, Salpiglossidae Tribe

Forms and Subspecies
Today, forty to forty-five species of *Brunfelsia* are botanically accepted (D'Arcy 1991, 78*; Schultes and Raffauf 1991, 34*). Several of these have importance as medicinal or ornamental plants or as additives to psychoactive preparations (Plowman 1977).

Species used for psychoactive purposes:
— *Brunfelsia chiricaspi* Plowman
   Borrachero, chiricaspi, chiric-caspi,[63] chirisanango, covi-tsontinba-ko (Kofán), sanango, yaí uhahai (Siona, "jaguar Brunfelsia")
— *Brunfelsia grandiflora* D. Don ssp. *grandiflora*[64] [syn. *B. calycina* Benth. var. *macrantha* Bailey, *B. tastevinii* Benoist]
   Borrachera, chinikiasip (Shuar), chiricaspi, chiric sanango, keya-honi, mucapari (Shipibo-Conibo)
— *Brunfelsia grandiflora* D. Don ssp. *schultesii* Plowman
   Bella unión, borrachero, chipiritsontinbaka (Kofán), chiricaspi chacruco (Quechua), chiricaspi picudo, chiricaspi salvaje, chiricsanango, huha hay (Siona), sanango, uhahai
— *Brunfelsia uniflora* (Pohl) Benth.[65] [syn. *Brunfelsia hopeana* (Hook.) Benth., *Franciscea uniflora* Pohl]
   Bloom of the lent, boas noites, ("good nights"), camgaba, camgamba, ("tree of the gambá-opossum"), Christmas bloom, flor de natal, ("Christmas tree"), gerataca, good night, jerataca, jeratacaca ("snake bite medicine"), manaca, manacá, mercurio dos pobres ("poor man's quicksilver"), Paraguay jasmine, Santa María, umburapuama ("medicine tree"), vegetable mercury, white tree
— *Brunfelsia maritima* Benth.
   Borrachera ("inebriator")[66]
— *Brunfelsia mire* Plowman
   Borrachera

History
The genus *Brunfelsia* was named for the German physician, botanist, and theologist Otto Brunfels (1489–1543). When the Portuguese arrived in northern Brazil, they were able to observe the use of *Brunfelsia uniflora* among the Indians. The inhabitants of the Amazon manufactured arrow poisons from extracts of the root. The *payés*, or shamans, used the root for healing and in magical activities (Plowman 1977, 290f.). The first description of the plant (*Brunfelsia uniflora*) was published in 1648 in Piso's *De Medicina Brasiliensi*.

Today, *Brunfelsia uniflora* enjoys the greatest phytomedicinal and pharmaceutical importance in Brazil and is grown in plantations as the stock plant for the manaca root drug. The word *manaca* is derived from *manacán*, which means "the most beautiful woman of the tribe" and alludes to the bush's beauty (Plowman 1977, 290). Because of their attractive flowers and colors, several *Brunfelsia* species (*B. americana*, *B. australis*, *B. uniflora*, *B. pilosa*) are now grown in tropical gardens throughout the world or raised as potted ornamentals.

Distribution
The genus *Brunfelsia* is originally from northern (tropical) Brazil and the Caribbean Islands. Because most of its species are so beautiful, the genus has spread as an ornamental into all of the tropical zones of the world. It has also been successfully propagated in the frost-free areas of the Mediterranean region (Bärtels 1993, 180*).

The species with ethnomedical significance are all from the Amazon, where they are planted by many of the Indians. *Brunfelsia chiricaspi* is found only in primary forest (Plowman 1973a, 258f.; 1977, 305).

Cultivation
Most species of *Brunfelsia* are propagated by cuttings, root pieces, or scions. In cultivation, they rarely produce fruits. Brunfelsias require a tropical climate and thrive best in loose soil. *Brunfelsia chiricaspi* is not cultivated (Plowman 1977, 305).

Indoor plants (*B. uniflora*, *B. pauciflora*) must be watered regularly with water that has been allowed to stand. Between April and August, they should be fertilized every fourteen days.

Appearance
The species discussed here are very similar in appearance and are all easily mistaken for one another. They usually form evergreen shrubs that can grow as tall as 3 meters. The leaves are alternate and elliptical in shape, and they taper to a point. Their upper side is leathery and dark green, while their underside is pale green. The flowers, which are typically borne on short stalks, are almost always violet but are sometimes white or in rare cases yellow (*Brunfelsia americana*) or creamy white (*Brunfelsia undulata*). Often, a plant will bear both white and violet flowers simultaneously. The fruits, which only rarely develop, are round green berries. The seeds are relatively large.

The flowers fade after only a few days. The

---

63 On the Putumayo (Colombia), this name is also used to refer to the fever tree (*Stephanopodium peruvianum* Poeppig et Endlicher; Dichapetalaceae) (Schultes 1983a, 262*).

64 Possibly a synonym for *Brunfelsia pauciflora* var. *calycina* (Benth.) J.A. Schmidt (Roth et al. 1994, 174*).

65 The common garden plant *Brunfelsia uniflora* (Pohl) D. Don is very easily confused with the very similar *Brunfelsia australis* Benth. (Plowman 1977, 290). The constituents of *Brunfelsia australis* are practically unknown. Little is known about any possible psychoactive effects; reports of any psychoactive use are lacking. The plant is, however, regarded as poisonous.

66 In South America, the name *borrachera* or *borrachero*, "inebriator," is given to almost all inebriating nightshades (cf. *Brugmansia*, *Datura*, *Iochroma*).

flowers of the species *Brunfelsia pauciflora* are dark violet the first day, light lilac the second, and almost white on the third day. As a result, this popular ornamental species is also popularly known by the name *yesterday, today, and tomorrow.*

At night some species (e.g., *Brunfelsia americana*) exude a sweet scent that has inebriating effects and is reminiscent of the scent of **Brugmansia suaveolens**. In tropical regions, brunfelsias bloom throughout the year. As potted plants in temperate zones (central Europe), the flowering period lies between spring and late summer.

Those species that are cultivated as ornamentals are quite similar in appearance to and are easily confused with those used for psychoactive purposes. Even a trained botanist can have difficulty identifying the species. The species *Brunfelsia hopeana* (= *B. uniflora*) and *Brunfelsia pilosa* Plowman, for example, are almost always regarded as one and the same (Plowman 1975, 47). For this reason, it can be assumed that the species' identifications provided in the ethnobotanical literature are not reliable. Accordingly, as a rule this monograph will not differentiate among species that are used for the same purpose (unless absolutely accurate data are available).

Brunfelsia maritima is deceptively similar to *B. grandiflora* and has even been confused with the latter in herbarium specimens. *B. grandiflora* is also frequently confused with *Brunfelsia latifolia* (Pohl) Benth. and *Brunfelsia bonodora* (Vell.) Macbr. (Plowman 1977, 298).

Brunfelsia grandiflora ssp. *schultesii* Plowman is distinguished from *B. grandiflora* ssp. *grandiflora* solely on the basis of its much smaller flowers and fruits. There is no ethnobotanical distinction made between the two subspecies or forms; both are known as *chiricaspi*, "cold tree," and each is used in the same manner (Plowman 1973a; 1977, 299).

Psychoactive Material
—Leaves
—Stems
—Roots (manaca roots, *manacá*, radix manaca, radix brunfelsiae)

In Brazil, several species are used as sources for manaca roots: *Brunfelsia uniflora, Brunfelsia australis,* and *Brunfelsia* spp.

Preparation and Dosage
There are a variety of traditional and pharmaceutical preparations of the raw drug. Leaves can be steeped in hot water (Schultes 1966, 303*). Leaves and stalks can also be decocted in boiling water. As little as 100 mg/kg of an extract of manaca root (*B. uniflora*) is sufficient to produce pharmacological effects (Iyer et al. 1977, 358).

For medicinal purposes, *Brunfelsia grandiflora* can be prepared in several ways. Scrapings of the

bark can be added to cold water or **chicha** (maize beer). The bark of other trees (*remo caspi: Pithecellobium laetum* Benth.; *chuchuhuasi: Heisteria pallida* Engl.; *huacapurana: Campsiandra laurifolia* Benth.) can be added to potentiate the dosage. Unfortunately, the amount of bark used to make the extract is not known. The root can also be added to **alcohol**. Here, 50 g of the root cortex is added to 1 liter of *aguardiente* (cane sugar spirits). One shot glass of this is taken before every meal (Plowman 1977, 300).

The Jíbaro produce a type of **ayahuasca** using **Banisteriopsis spp.**, *Brunfelsia grandiflora*, and a botanically unidentified vine known as *hiaji*. First the *Banisteriopsis* pieces are boiled for fourteen hours, after which the other ingredients are added and the entire mixture boiled down until a thick solution results (Plowman 1977, 303).

When used for psychoactive and magical purposes, the wild-growing *Brunfelsia chiricaspi* is preferred over the cultivated varieties of *Brunfelsia grandiflora* (Plowman 1973a, 259).

Brunfelsia can also be smoked. Men and women of the Yabarana roll cigarettes out of manaca bark and tobacco (**Nicotiana tabacum**) (Wilbert 1959, 26 f.*).

Ritual Use
In Ecuador, Amazonian Indians use *Brunfelsia grandiflora* as a hallucinogen. The shamans of the Shuar drink a tea of the leaves and stems in order to induce "strong feelings" that they can then use for healing purposes (Bennett 1992, 493*). The Siona scrape the bark of *B. grandiflora* ssp. *schultesii* and drink a cold-water extract of this. Two mouthfuls are said to be an effective dosage

Left: The typical yellow flowers of *Brunfelsia americana*.

Top right: *Brunfelsia australis* can have both white and violet flowers simultaneously.

Center right: The South American *Brunfelsia grandiflora* ssp. *grandiflora* is one of the shamanic plants known as *chiricaspi*.

Bottom right: *Brunfelsia grandiflora* ssp. *schultesii*, named for the botanist Richard Evans Schultes, is a rare shamanic plant.

Top left: The beautiful *Brunfelsia maliformis* from Jamaica is one of the rarest members of the genus.

Bottom left: Fruits and seeds of *Brunfelsia grandiflora* ssp. *schultesii.*

Right, from top to bottom:
The prostrate shamanic plant *Brunfelsia mire* is almost unknown and has been little studied.

Brunfelsia pauciflora var. *calycina* is a popular ornamental from Brazil.

Brunfelsia pauciflora cv. Floribunda compacta, a plant found in tropical gardens.

The green fruit of *Brunfelsia plicata*, from Jamaica.

Brunfelsia uniflora is the source of the manaca root drug.

(Vickers and Plowman 1984, 29f.*). They drink the extract "to obtain visions and alleviate pains." The *Brunfelsia* extract is often drunk prior to the ingestion of **ayahuasca** or combined with yoco (cf. ***Paullinia* spp.**) (Plowman 1977, 305). The shamans of the Kofán drink *Brunfelsia grandiflora* to diagnose diseases. The shamans of the Lama Indians, who live in northern Peru, regard *B. grandiflora* as a spiritual leader. They consume it during their initiation and receive from it special powers that they can use to heal as well as to cause diseases (Plowman 1977, 303).

Both subspecies of *Brunfelsia grandiflora* are used as **ayahuasca** additives and are said to potentiate its effects (Schultes and Raffauf 1991, 34*). In Iquitos, urban *ayahuasqueros* say that *Brunfelsia grandiflora* makes ayahuasca more powerful and induces acoustical effects "like rain in the ear." During a new moon, the Witoto on the Río Ampiyaco (Peru) add brunfelsia to ayahuasca (they add pieces of the bark to cold ayahuasca) to obtain power (Plowman 1977, 303).

Artifacts
Apparently none; cf. **ayahuasca**.

Medicinal Use
In Brazil, the manaca root is used as a remedy for syphilis and as an abortifacient (Bärtels 1993, 180*). It is used in folk medicine to treat rheumatism, syphilis, yellow fever, snakebites, and skin diseases (Iyer et al. 1977, 356). It is a very important fever medicine; *chiricaspi* means "cold tree" and refers to the plant's ability to lower body temperature (Schultes and Raffauf 1991, 34*).

The stalks of *Brunfelsia grandiflora* are grated and added to cold water. The resulting solution is rubbed over or massaged into areas affected by rheumatism. A cold-water extract is also drunk to treat arthritis and rheumatism (Plowman 1977, 300).

A homeopathic preparation made from manaca roots was introduced around 1862. Franciscea uniflora (essence of fresh root), as it was known, was an important agent for a time (Schneider 1974, 1:198*).

Constituents
In the older literature, the constituents were listed as brunfelsia alkaloids with such names as franciscaine, manacine, brunfelsine (Brandl 1885), and even mandragorine (cf. ***Mandragora officinarum***)—all obsolete names for the "only little understood chemical components" of the roots (Schultes 1979b, 154*).

The species *Brunfelsia uniflora, B. pauciflora,* and *B. brasiliensis* contain the non-nitrogenous compound **scopoletin** (= 6-methoxy-7-hydroxy-coumarin). The alkaloid cuscohygrine, which also occurs in ***Atropa belladonna*** and ***Erythroxylum*** *coca*, has been isolated from an unidentified species of the genus (Mors and Ribeiro 1957; Schultes 1979b, 155*).

Brunfelsia uniflora and *B. pauciflora* contain the alkaloids manacine and manaceine as well as aesculetine. The concentration of manacine is highest in the bark (of *B. uniflora*), reaching 0.08% (Roth et al. 1994, 175*).

Effects
The peculiar effects of the manaca root were described at an early date: heavy salivation, slackness, general sedation, partial paralysis of the face, swollen tongue, and blurred vision. There were also more drastic reports: "wild deliria and persistent feeblemindedness." "One kind of *manacá* has the property of causing intoxication, blindness, and the retention of urine during the day; but after having drunk the infusion of the root or bark of this tree, a man is always happy in his hunting and fishing" (Plowman 1977, 292).

Laboratory studies of the **scopoletin** extracted from *Brunfelsia uniflora* (= *B. hopeana*) have demonstrated clear depressive effects upon the central nervous system (Iyer et al. 1977, 359). "Manacine stimulates glandular secretion and kills by respiratory paralysis. Manacein has similar effects" (Roth et al. 1994, 175*).

Brunfelsia chiricaspi is said to be the most psychoactively potent of all brunfelsias. However, the effects do not seem particularly alluring. They begin within just a few minutes and are first manifested as tingling, numbness, et cetera (similar to when an arm or leg has fallen asleep). These are followed by a profound sensation of coldness and an inability to move, foaming at the mouth, shaking, and nausea. Aftereffects include dizziness and exhaustion. The feelings of dizziness and weakness persist into the following day (Plowman 1977, 306 f.). Plowman, one of the few researchers to have actually tried the drink, compared the overall effects to those of **nicotine** (on nonsmokers). He assumes that brunfelsia is added to **ayahuasca** in order to achieve a higher concentration in the body or a stronger effect upon bodily processes. The shamans could then use the resulting condition to heal specific ailments.

Jonathan Ott (per oral communication with the author) notes that he almost died as a result of a self-experimentation with brunfelsia. To date, there are no reports of pleasant visionary experiences. However, for understandable reasons, few psychonauts have dared to explore the depths of the brunfelsia state.

Commercial Forms and Regulations
Some brunfelsias (usually *Brunfelsia pauciflora* and *Brunfelsia uniflora*) are sold in nurseries as ornamentals. Manaca root is officinal in Brazil and is listed in the Brazilian pharmacopoeia. In principle, manaca roots are available without restriction.

Literature

Beckurts, H. 1895. Chemische und pharmakologische Untersuchung der Manacá-Wurzel. *Apotheker Zeitung* 72:622–23.

Brandl, J. 1885. Chemisch-pharmakologische Untersuchung über die Manacá-Wurzel. *Zeitschrift für Biologie* 31:251–92.

Brewer, E. P. 1882. On the physiological action of manacá. *The Therapeutic Gazette*, n.s., 3 (9): 326–30.

de Almeida Costa, O. 1935. Estudio farmacognóstico de Manacá. *Revista da Flora Medicinal* 1 (7): 345–60.

Erwin, J. L. 1880. Manacá—proximate properties of the plant. *The Therapeutic Gazette*, n.s., 1 (7): 222–23.

Hahmann, C. 1920. Beiträge zur anatomischen Kenntnis der *Brunfelsia hopeana* Benth., im Besonderen deren Wurzel, Radix Manaca. *Angewandte Botanik* 2:113–33, 179–91.

Iyer, Radhakrishnan P., John K. Brown, Madhukar G. Chaubal, and Marvin H. Malone. 1977. *Brunfelsia hopeana*. I. Hippocratic screening and antiinflammatory evaluation. *Lloydia* 40:356–60.

Mors, Walter B., and Oscar Ribeiro. 1957. Occurrence of scopoletin in the genus *Brunfelsia*. *Journal of Organic Chemistry* 22:978–79.

Plowman, Timothy. 1973a. Four new brunfelsias from northeastern South America. *Botanical Museum Leaflets* 23 (6): 245–72.

———. 1973b. The South American species of *Brunfelsia* (Solanaceae). PhD diss., Harvard University.

———. 1975. Two new Brazilian species of Brunfelsia. *Botanical Museum Leaflets* 24 (2): 37–48.

———. 1977. *Brunfelsia* in ethnomedicine. *Botanical Museum Leaflets* 25 (10): 289–320.

———. 1979. The genus *Brunfelsia*: A conspectus of the taxonomy and biogeography. In *The biology and taxonomy of the Solanaceae*, ed. J. G. Hawkes et al., 475–91. London: Academic Press.

"The juice of this plant [*Brunfelsia grandiflora*] plunges them [the Indians] into a kind of intoxication or stupefaction which lasts a little more than a quarter of an hour and from which they acquire magical powers, enabling them to heal all sorts of diseases through incantations. While the effects of the drink act on their brains, they are unable to fall asleep. They believe they see all kinds of fantastic animals: dragons, tigers [= jaguars], wild boars, which attack them and tear them to bits, etc. This action of *honi* lasts four or five hours depending on the quantity ingested."

PATER C. TASTEVIN
IN "*BRUNFELSIA IN ETHNOMEDICINE*"
(PLOWMAN 1977, 302)

The first illustration of the Brazilian manaca root *Brunfelsia uniflora* (= *Brunfelsia hopeana*). (Engraving from Piso, *De Medicina Brasiliensi*, 1648)

Calea zacatechichi Schlechtendal

Aztec Dream Grass, Zacatechichi

Zacachichic (*Conyza filaginoides* Hieron.), the false zacatechichi, was probably used as dream grass. (From Hernández, 1942/46 [Orig. pub. 1615]*)

Left: Inflorescence of the Aztec dream grass (*Calea zacatechichi*)

Right: The entire aboveground herbage provides the dream-inducing raw drug of *Calea zacatechichi*.

67 This name is also given to the tree *Lucuma salicifolia* H.B.K., which has psychoactive properties and is known in modern Mexico as *zapote borrachero*.

68 In the Yucatán, this Mayan name is also used for the closely related species *Calea urticifolia* (Mill.) DC. as well as its subspecies *Calea urticifolia* var. *axillaris* (DC.) Blake (Barrera Marin et al. 1976, 214; Martínez 1987, 1069).

Family
Compositae (Aster Family); Subfamily Heliantheae, Galinsoginae Tribe/Subtribe

Forms and Subspecies
A number of varieties have been described (Flores 1977, 12 ff.):

Calea zacatechichi var. *calyculata* Robinson
Calea zacatechichi var. *laevigata* Standley
Calea zacatechichi var. *macrophylla* Robinson et Greenman
Calea zacatechichi var. *rugosa* (DC.) Robinson et Greenman
Calea zacatechichi var. *xanthina* Standley et L.O. Williams
Calea zacatechichi var. *zacatechichi*

There is also said to be a form that occurs only in Guadalajara (Flores 1977, 15).

Synonyms
Aschenbornia heteropoda Schauer
Calea rugosa Hemsley
Calea ternifolia Kunth var. *ternifolia*
Calydermos rugosus DC.

Folk Names
Ahuapatli, amula, atanasia amarga, Aztec dream grass, bejuco chismuyo, betónica, chapote,[67] chichicxihuitl (Nahuatl, "bitter plant"), chichixihuitl, cochitzapotl, dream herb, falso simonillo, hierba amarga, hoja madre ("leaf of the mother"), iztactzapotl, jaral, jaralillo, juralillo, mala hierba, matasano, oaxaqueña ("the one from Oaxaca"), paiston, poop taam ujts, prodigiosa, pux lat'em (Huastec), sacachcichic, sacachichic, sacatechichi, simonillo, techichic, tepetlachichixihuitl (Nahuatl, "bitter plant of the mountains"), thle-pelacano, thle-pela-kano (Chontal, "leaf of god"), tsuleek' ethem ("racoon's trachea"), tzicinil, tzikin, xikin (Maya, "dove's plant"),[68] xtsikinil, x-tzicinil, yerba amarga ("bitter plant"), zacachichi, zacachichic, zacate amargo (Mexican, "bitter grass"), zacatechi, zacatechichi, zacate de perro (Mexican, "dog grass")

History
This Compositae was used for magical and medicinal purposes in pre-Columbian times. It is possible that *Calea zacatechichi* helped Aztec magicians (*nagualli*) travel deeper into Tlálocan, the realm of dreams.

The Aztec name *zacatechichi* is translated as "bitter grass." The first botanical description of the plant comes from the nineteenth century (1834). Its psychoactive use was first described by Thomas MacDougall (1968). Research into its pharmacology and phytochemistry did not begin until relatively recently (Flores 1977).

Distribution
Aztec dream grass grows primarily in the highlands of central Mexico (1,500 to 1,800 meters in altitude), in the mountainous regions of Oaxaca, Veracruz, and Chiapas, in Jalisco and Morelos, and in the lowlands of Yucatán (Barrera Marin et al. 1976*; Martínez 1987*). The plant also occurs in Costa Rica in association with pines (*Pinus* spp.) and oaks (*Quercus* spp.) (Schuldes 1995, 23*). It is more easily found in pure pine forests (Flores 1977, 12).

In Mexico, the closely related species *Conyza filaginoides* DC. is also known as *zacatechichi* (Schultes 1970, 48*).

Cultivation
Dream grass can be grown from germinated seeds. The dried fruit husks should be removed prior to planting. It is best planted in good topsoil and watered thoroughly.

Appearance
The herbaceous, branched plant grows to a height of approximately 1.5 meters and in rare cases to 3 meters. It has small, oval leaves that are crispate (curled) on the edges and forms small yellow or

occasionally whitish flowers. The undersides of very young leaves are violet. The plant is difficult to recognize and is easily mistaken for a number of other plants. Its most noticeable feature is its intense green color. It sometimes occurs in small fields that distinguish themselves from the surrounding vegetation through their green luminescence.

Dream grass is very easily confused with the closely related *Calea cordifolia*, which also bears yellow flowers.

Psychoactive Material
—Leaves and stems, before the fruits ripen

Preparation and Dosage
The dried drug is used to prepare a tea, either an infusion or a decoction. The dried leaves and stems can also be smoked in a pipe or in the form of a cigarette (MacDougall 1968, 105).

When used for folk medicinal purposes—e.g., to treat malaria—a total of 10 g of dried herbage is made into a tea and drunk three times a day (Schultes 1970, 49*).

In Mexico, an alcohol extract of the leaves of the closely related *Calea urticifolia* (Mill.) DC. var. *axillaris* (DC.) Blake was formerly drunk as an inebriant (von Reis Altschul 1975, 324*).

Ritual Use
Although the Aztecs and other Mesoamerican peoples were almost certainly using the plant ritually in pre-Columbian times, very little information is contained in the sources. Dream grass is probably identical with *chichixihuitl*, "bitter herb," an inebriating plant mentioned in sources from the colonial period.

The Chontal Indians of Oaxaca, whose language is related to Mayan, call the plant *thle-pela-kano*, "leaf of god," and venerate it as a plant of the gods. The *curanderos* (healers) of the Chontal boil crushed fresh leaves to produce a powerful and astringent brew that they drink in order to produce visions and clairvoyant, dreamlike states. They also lie down in a partially or fully darkened room and smoke a cigarette of dried leaves. The *curanderos* report having altered, dreamlike states in which they can hear the voices of the gods and spirits, recognize the causes of illnesses, look into the future, and locate lost or stolen objects. This form of divination is known as oneiromancy (divination through dreams).[69] The Chontal healers consider a handful of dried herbage (approximately 60 g) an effective dosage.

The fresh herbage is sometimes placed under the pillow to induce dreams.

Artifacts
As of this writing, no artifacts are known.

Medicinal Use
Colonial medical texts from Yucatán indicate that crushed fresh leaves were used to prepare an herbal plaster to treat a swollen scalp. Steamed leaves were applied to treat skin diseases (Roys 1976, 290, 295*). Today, the Yucatán Maya still use dream grass as an herbal medicine (Barrera M. et al. 1976). During the Aztec period, the plant was also used to treat "cold stomach" (Flores 1977, 8).

The herbage is used in Mexican folk medicine as a laxative and febrifuge. Teas made from the plant are used as appetite stimulants (once the bitter taste has disappeared from the mouth) and stomach tonics and are also beneficial for diarrheal diseases (Mayagoitia et al. 1986, 230). The herbage also finds folk medicinal use in the treatment of headaches and diabetes, as a stimulant, and for menstrual complaints (Argueta V. et al. 1994, 1407*; Jiu 1966, 252*).

Constituents
The herbage contains a complex of horrible-tasting bitter principles consisting of several sesquiterpene lactones: germacranolide[70] (1β-acetoxy-zacatechinolide, 1-oxo-zacatechinolide), germacrene 7, caleicine I and II, caleocromene A and B, caleine A and B, zexbrevine and analogs, and budleine A and analogs (Argueta V. et al 1994, 251*; Bohlmann and Zdero 1977; Herz and Kumar 1980; Lara Ochoa and Marquez Alonso 1996, 123 f.*; Mayagoitia et al. 1986, 231; Quijano et al. 1979). The flavones acacetin and *O*-methyl-acacetin have also been found (Herz and Kumar 1980). Several studies have indicated the presence of an alkaloid (?) of unknown structure that has mild psychoactive and central sedative properties. According to Díaz (1979, 79*), there are different chemical races of the plant, of which one is psychoactive while the other(s) are not. This would explain why Chontal healers distinguish between "good" and "bad" specimens of the plant.

The active ingredients are water soluble. They may also be alcohol soluble, as tinctures are also used (cf. Schuldes 1995, 23*).

Effects
The subtle psychoactive effects on humans are best described as dream-inducing or oneirogenic. *Calea* also appears to promote sleep. Animal studies have demonstrated that cats quickly fall asleep when administered a dosage equivalent to that which induces dreaming in humans (Mayagoitia et al. 1986, 230).

A group of Mexican researchers headed by José Díaz conducted a double-blind experiment using a placebo and a preparation of *Calea zacatechichi* and registered a significant increase in the number of meaningful dreams in the subjects who had ingested *C. zacatechichi* (Mayagiotia et al. 1986). The geomancy researcher Paul Devereux, whose

Germacranolide

69 The Melanesian dream fish (*Kyphosus fuseus*) is said to produce oneirogenic effects like those of *Calea zacatechichi* (cf. Ott 1993, 410).

70 Germacranolides are apparently of chemotaxonomic significance in the genus *Calea* (Ferreira et al. 1980).

"I played tamboura in a concert of classical Indian music after I had drunk the *Calea zacatechichi* tea. As I did, I dove into the sound of the tamboura. My teacher describes the tamboura as a 'river,' the water, the bearer of the music (the tablas are the currents, the voices are the spirit). I had always regarded the river as only a surface, and that with my tamboura playing I would form the surface upon which the others could flow. But the *Calea* literally brought a new depth into this picture. I was not only the carrying wave upon which the other musicians could drift along . . . I saw that we all played under water this time, all dove down to the depths of the ocean. . . ."

DAWN DELO
(PERSONAL COMMUNICATION,
29 JULY 1996)

Dragon Project investigates dream activity in ancient cultic sites, is planning to conduct an additional study of the induction of waking dreams by *Calea zacatechichi*.

A decoction made with a heaping tablespoon (approximately 25 g) of dried, chopped herbage, together with one standard joint, is regarded as an effective dosage for producing oneirogenic effects. After consuming the two, the person should lie down in a darkened room or go to bed:

> After some 30 minutes, sensations of relaxation and calmness begin. The heartbeat is perceived more consciously. The stated amount of 25 grams clears the thoughts and the senses. (Schuldes 1995, 23*)

Some subjects have reported experiencing marijuana-like effects (cf. *Cannabis*) after smoking a cigarette made with *Calea*. I personally cannot (yet) confirm such an effect. The only effects I have noticed are an increase of blood flow to the head and mild sensations of being "high."

The effects described in the literature are not reliable (cf. Ott 1993, 422*). To date, no side effects have been reported.

Commercial Forms and Regulations

In Mexico, the dried herbage is occasionally found in markets or herb shops. It is more rarely found in international specialty stores. There are no regulations concerning its use.

Literature

Bohlmann, Ferdinand, and Christa Zgero. 1977. Neue Germacrolide aus *Calea zacatechichi*. *Phytochemistry* 16:1065–68.

Ferreira, Zenaide S., Nídia F. Roque, Otto R. Gottlieb, Fernando Oliveira, and Hugo E. Gottlieb. 1980. Structural clarification of germacronolides from *Calea* species. *Phytochemistry* 19:1481–84.

Flores, Manuel. 1977. *An ethnobotanical investigation of* Calea zacatechichi. Senior honors thesis, Harvard University.

Giral, Francisco, and Samuel Ladabaum. 1959. Principio amargo del zacate chichi. *Ciencia* 19 (11–12): 243.

Herz, Werner, and Narendra Kumar. 1980. Sesquiterpene lactones of *Calea zacatechichi* and *C. urticifolia*. *Phytochemistry* 19:593–97.

Lourenço, Tânia O., Gokithi Akisue, and Nídia F. Roque. 1981. Reduced acetophenone derivatives from *Calea cuneifolia*. *Phytochemistry* 20 (4): 773–76.

MacDougall, Thomas. 1968. *Calea zacatechichi*: A composite with psychic properties? *Garden Journal* 18:105.

Martínez, Mariano, Baldomero Esquivel, and Alfredo Ortega. 1987. Two caleines from *Calea zacatechichi*. *Phytochemistry* 26 (7): 2104–6.

Martínez, Mariano, Antonio Sánchez F., and Pedro Joseph-Nathan. 1987. Thymol derivatives from *Calea nelsonii*. *Phytochemistry* 26 (9): 2577–79.

Mayagoitia, Lílian, José Díaz, and Carlos M. Contreras. 1986. Psychopharmacologic analysis of an alleged oneirogenic plant: *Calea zacatechichi*. *Journal of Ethnopharmacology* 18 (3): 229–43.

Quijano, L., A. Romo de Vivar, and Tirso Rios. 1979. Revision of the structures of caleine A and B, germacranolide sesquiterpenes from *Calea zacatechichi*. *Phytochemistry* 18:1745–47.

Calliandra anomala (Kunth) McBride

Red Powder Puff

Family
Leguminosae (Legume Family): Mimosaceae

Forms and Subspecies
None

Synonyms
Calliandra grandiflora (L'Hér.) Benth.

Folk Names
Cabellito, cabellitos de ángel, cabellitos de una vara, cabello de angel, cabellos de ángel, cabeza de angel (Spanish, "angel's head"), canela, chak me'ex k'in (Lacandon, "the red beard of the sun/of the sun god"), ch'ich' ni' (Tzotzil, "bloody nose"), clagot, coquito, engelshaupt, hierba de canela, lele, meexk'in, pambonato, pombotano, red powder puff, saqaqa (Totonac), tabardillo, tepachera, tepexiloxóchitl, texoxóchitl, timbre, timbrillo,

tlacoxilohxochitl, tlacoxiloxochitl (Aztec), tlama-catzcatzotl, tzonxóchitl, u me'ex k'in, xiloxóchitl

History

The spectacular red powder puff is originally from Mexico, where it was already being used for medicinal purposes in pre-Spanish times. The first reports about the plant are from Francisco Hernández, made in the fifteenth century. The Aztecs are said to have used the plant as a narcotic (Emboden 1979, 4*).

Both *Calliandra anomala* and the genus *Calliandra* as a whole have been little studied, even though the genus encompasses several interesting medicinal plants and very attractive and beautiful ornamental shrubs.

Distribution

Calliandra anomala occurs in the tropical zones of Central and South America. In Mexico, it is primarily found in Chiapas, Veracruz, Oaxaca, Morelos, Chihuahua, and Sinaloa (Martínez 1994, 319*).

Cultivation

The shrub can be propagated from seeds or cuttings. The seeds must be pre-germinated to ensure success. The plant requires a warm to moist/hot climate; it does not tolerate cold or frost (Grunner 1991, 19*).

Appearance

This branched shrub can grow as tall as 6 meters, although it usually attains a height of only 3 to 4 meters. The finely pinnate leaves are opposite. The bark is thick and covered with short hairs and has an olive-colored sheen. The characteristic inflorescences develop at the tips of the branches. The actual white flowers are inconspicuous and arranged in rings around the branch. From these sprout the enormously long, luminously red filaments that give the inflorescence the appearance of a powder puff. In the tropics, the bush blooms throughout the year. The fruits, which usually appear in February, are long, flat pods that contain several flat seeds.

The genus encompasses some 110 species that are found primarily in the tropical zones of the Americas (Anzeneder et al. 1993, 53*; Bärtels 1993, 144*). The species *Calliandra fulgens* Hook. and *Calliandra tweedi* Benth. also produce red filaments and can thus have a similar appearance.

Psychoactive Material

— Bark (cortex calliandrae)
— Resin (sap)
— Root
— Buds/flowers (cabellitos)

Preparation and Dosage

Calliandra anomala was used as a pulque additive (see *Agave* spp.) and may have been used as an additive to cacao (*Theobroma cacao*).

The plant can allegedly be used to manufacture a **snuff**: "Several days after several incisions were made into the bark, the resin that had appeared was collected, dried, powdered, mixed with ashes, and sniffed" (Schuldes 1995, 24*). The powdered root has an irritating effect upon the mucous membranes of the nose (it is a sneezing powder, similar to *Veratrum album*). To date, no other effects have been reported.

The total daily dosage should not exceed 120 g; in one known case of overdose, a dog died following a dosage of 90 g (Martínez 1994, 320*).

The closely related species *Calliandra angustifolia* and *Calliandra pentandra* are used in South America as **ayahuasca** additives.

Ritual Use

In Aztec mythology and cosmology, *Calliandra* is associated with the heavenly realm of the dead (the House of the Sun in Heaven) and with the nourishment for reborn souls:

> The third place to which one went was in the House of the Sun in Heaven. Those who had fallen in battle went there, those that had either died right in battle, so that they were carried away on the battlefield, that their breath ceased there, that fate found them there, or those that were brought home so that they could be sacrificed later, whether in the *Sacrificio gladiatorio* or by being thrown alive into the fire, or stabbed to death, or thrown onto the cactus [*Coryphantha* spp.], or in battle, or bound with pine chips—all of these go to the house of the sun. . . . And where those who had fallen in battle dwell, there are wild agaves [*Agave* spp.], thorny plants, and groves of acacias [*Acacia* spp.]. And he can see all of the offerings that are brought to them, that can make it to him. And after they have spent four years in this manner, they change into birds with bright feathers: hummingbirds, flower birds, into yellow birds with black, hollow depressions around their eyes; into chalk-white butterflies, into downy butterflies, into butterflies (as large) as drinking vessels, which suck **honey** from all types of flowers, the flowers of the *equimitl* [*Erythrina* spp.], the tzompantli tree [*Erythrina americana*], the *xiloxochitl* [*Pseudobombax ellipticum* H.B.K.; cf. **amapola**], and the *tlacoxilohxochitl* [*Calliandra anomala*]. (Sahagun, in Seler 1927, 301f.*)

It is possible that the shrub may have had a ritual significance among the Maya, for the Lacandon of Chiapas still refer to it as *chäk me'ex k'in*, "the red beard of the sun god."

Top: The astonishing flower of the tropical *Calliandra anomala*. (Photographed in Palenque, Chiapas, Mexico)

Bottom: Various *Calliandra* species are used as ayahuasca additives in shamanic contexts.

Artifacts
None known

Medicinal Use
The Aztecs trickled the sap of the plant into the nose in order to induce a hypnotic sleep (Argueta V. et al. 1994, 251*; Emmart 1937*). To treat coughs, the root was chewed or peeled, ground, and taken in water with **honey** (Emboden 1979, 4*). The root is still used in folk medicine today to treat diarrhea, fevers, and malaria. A cold-water extract of the root is used as an eyewash (Martínez 1994, 320*). In Mexico, the shrub is becoming increasingly important in the treatment of diabetes (Argueta V. et al. 1994, 251*).

The Tzotzil Indians (Chiapas, Mexico) use this and other species of *Calliandra* to treat severe diarrhea. They macerate the root in water, boil the result, and drink three to five cups of this extract daily (Berlin and Berlin 1996, 212).

Around 1900, the bark of two Mexican species (cortex calliandrae, cortex pambotani) was used in Europe to treat marsh fever (Schneider 1974, 1:215*).

Constituents
The root drug contains quantities of tannins, fat, a resin called glucoresina, a glycoside called calliandreine, an **essential oil**, and minerals (Martínez 1994, 319f.*). The bark is said to contain harmane (per oral communication from Rob Montgomery). It is rumored that the bark also contains *N,N*-DMT. Felix Hasler and David Volanthen did not find any DMT in an analysis of *Calliandra* stem cortex material from southern Mexico. If DMT is in fact present in the stem cortex, it would have to be in amounts less than 0.1%. The root cortex has not yet been studied.

Calliandra angustifolia and *Calliandra pentandra* have been found to contain harmane and *N,N*-DMT. The closely related *Calliandra houstoniana* contains an alkaloid; this species is also the source of a gum resin that has industrial use (Cioro 1982, 74*). The leaves of *Calliandra portoricensis* Benth. contain saponines, tannins, flavonoids, and glycosides (Aguwa and Lawal 1988). Rare derivatives of pipecolic acid as well as derivatives of piperidine also occur in the genus (Marlier et al. 1979; Romero et al. 1983).

Effects
The effects of the resin have been characterized as hypnotic and sleep-inducing (Emboden 1979, 4*). It is unknown whether anyone has had psychoactive experiences with the plant.

The related species *Calliandra portoricensis* has sedative effects upon the nervous system (Adesina 1982; Berlin and Berlin 1996, 213).

Commercial Forms and Regulations
None

Literature

Adesina, S. K. 1982. Studies on some plants used as anticonvulsants in Amerindian and African traditional medicine. *Fitoterapia* 53:147–62.

Aguwa, C. N., and A. M. Lawal. 1988. Pharmacologic studies on the active principles of *Calliandra portoricensis* leaf extracts. *Journal of Ethnopharmacology* 22:63–71.

Berlin, Elois Ann, and Brent Berlin. 1996. *Medical ethnobiology of the Highland Maya of Chiapas, Mexico.* Princeton, N.J.: Princeton University Press.

Marlier, Michel, Gaston Dardenne, and Jean Casimir. 1979. 2S,4R-Carboxy-2-Acetylamino-4-Piperidine dans les feuilles de *Calliandra haematocephala. Phytochemistry* 1979:479–81.

Romeo, John T. 1984. Insecticidal aminoacids in leaves of *Calliandra. Biochemistry and Systematic Ecology* 12 (3): 293–97.

Romeo, John T., Lee A. Swain, and Anthony B. Bleecker. 1983. *Cis*-4-hydroxypipecolic acid and 2,4-*cis*-4,5-*trans*-4,5-dihydroxypipecolic acid from *Calliandra. Phytochemistry* 22 (7): 1615–17.

Calonyction muricatum (L.) G. Don
[*syn. Ipomoea muricata (L.) Jacq.*]

See under *Ipomoea* spp.

Camellia sinensis (Linnaeus) O. Kuntze

Tea Plant

Family

Theaceae (outdated: Ternstroemiaceae; Camelliaceae) (Camellia Family); Subfamily Theoideae (Camellioideae), Theeae (Camellieae) Tribe

Forms and Subspecies

The two basic forms (or races?) differ ecologically and especially economically. Assam tea is the source of black tea, China tea of green and brown tea. Whether the Assam tea plant is a variety (*Camellia sinensis* var. *assamica*), a subspecies (*Camellia sinensis* ssp. *assamica*), or a species in its own right (*Camellia assamica*) has still not been definitively clarified. Most authors presume that there are two varieties: *Camellia sinensis* var. *sinensis* and *Camellia sinensis* var. *assamica* (Teuscher 1992, 629). Numerous hybrids have been produced from both sorts; crosses have also been undertaken to produce higher yields.

Various species from the same genus are occasionally used as tea surrogates (i.e., *Camellia kissi* in Tibet and Nepal, *Camellia japonica* in Japan).

Synonyms

Camellia assamica (J.W. Masters) W. Wight
    (Assam tea plant)
Camellia assamica ssp. *lasiocalyx* (Watt) Wight
Camellia bohea (L.) Sweet
Camellia chinensis (Sims) Kuntze
Camellia oleosa (Lour.) Rehder
Camellia thea Link
Camellia thea var. *lasiosalyx* Watt
Camellia viridis (L.) Sweet
Thea bohea L.
Thea cantonensis Lour.
Thea chinensis Sims
Thea cochinchinensis Makino
Thea grandiflora Salisb.
Thea oleosa Lour.
Thea parviflora Salisb.
Thea sinensis L. (China tea plant)
Thea stricta Hayne
Thea viridis L. (green tea)
Theaphylla assamica J.W. Masters
Theaphylla cantonensis (Lour.) Raf.
Theaphylla lanceolata Raf.
Theaphylla laxa Raf.
Theaphylla viridis Raf.

Folk Names

Arbre à thé, caha (Sanskrit), cajnoe derevo (Russian), cay (Hindi), cha, châ (Hindi), ch'a, chai, châ'î sabz (Persian), charil, gur gur cha, herba thee, kaiser-thee, ojandonnassame tzshe, syamaparni (Sanskrit), tè, tea plant, tea-shrub, teebaum, teepflanze, teyila (Malayalam), têyilai (Tamil), théier, tzshe noky

History

The earliest written reference to the tea plant is contained in a document from 221 B.C.E., according to which the Chinese emperor Tsching-schi-huang-ti had introduced a tax on tea (Temming 1985, 9).

Legend has it that Bodhidharma, a disciple of the Buddha, brought tea from India to China together with the Buddhist teachings (ca. 519 C.E.). There, it was enthusiastically received and passed on to Southeast and East Asia. The first handbook on tea was written by the Chinese Lu-Yu (740–804).

In 801, the Buddhist monk Saichô brought the tea plant to Japan (Okakura 1979, 34). The Zen monk Esai wrote the first Japanese book on tea (and its healing properties) during the early thirteenth century (Iguchi 1991).

The European Engelbert Kämpfer provided the first botanical description of the tea plant after visiting Japan in 1712. Tea arrived in Europe in 1610, when Dutch merchants brought it from Japan to Amsterdam (Gilbert 1981). The very first European description of the beverage, in Johan Neuhof's *Reisebericht* [Travel Report] (1655–1657), praised its psychoactive effects:

> The power and effect of this drink is / that it dispels immoderate sleep; but afterward those in particular feel very good / who have overburdened their stomachs with food / and have loaded the brain with strong beverages: for it dries and removes all other moisture / and dispels the rising vapors or fog / which provoke sleep; it fortifies the memory / and sharpens the mind. (In Temming 1985, 14)

Distribution

The tea plant is originally from the triangle of countries formed by South China, Assam (northeastern India), and Cambodia. Today, it is planted in almost all tropical and subtropical regions of the world. The economically most important areas in which the plant is cultivated are in China, Japan, India, Sri Lanka, and Indonesia. Growing areas in Australia (North Queensland), KwaZulu-Natal, East Africa (Kenya), southern Brazil, the Caucasus Mountains, and the Seychelles (Mahé) are also gaining in importance. The place that is most renowned for its tea is Darjeeling, a small country in the Himalayas that culturally is a part of Nepal but politically is a protectorate of India (Vollers 1981).

An oversized tea plant in front of a tea field and a typical Chinese pagoda. (1669 copperplate engraving, printed in Amsterdam)

Botanical illustration of the Chinese tea plant. (Engraving from Pereira, *De Beginselen der Materia Medica en der Therapie*, 1849)

"The spirit of tea is like the spirit of the Tao; it flows spontaneously, wanders here and there, and resists every compulsion."

JOHN BLOFELD
DAS TAO DES TEETRINKENS [THE TAO OF TEA DRINKING]
(1986, 9)

Cultivation

The tea plant is usually propagated from cuttings, although it can also be grown from seed. The plant requires an average annual temperature of 20°C and a minimum of 1,300 mm of precipitation. The plant does not require any particular type of soil (for more on cultivation, see Franke 1994, 85–94). The first harvest can be had three years after propagation, but large harvests will not be produced until after six to seven years. The plant is harvested throughout the year, sometimes at short intervals (ten to fourteen days).

Appearance

This evergreen tree can grow as tall as 10 meters; in cultivation, it is maintained as a bush some 1.5 meters in height. It has elliptical, dentate, and leathery leaves that can grow as long as 10 cm. The flowers have five white petals and yellow pistils. The fruit is in capsules that may be monolocular, bilocular, or trilocular.

Psychoactive Material

— The young leaves (folia theae, thea folium); the best qualities are from young, small leaves from sorts that are planted in favorable altitudes (Darjeeling).

The processing method determines the type of tea. Green tea consists of unfermented, dried leaves (thea viridis folium), black tea of fermented leaves (thea nigrae folium), and oolong (also known as white or brown tea) of semi-fermented leaves.

Steps in processing include plucking, drying using hot steam or wilting, rolling of the wilted leaves, fermenting, and firing or roasting.

Preparation and Dosage

Tea is prepared by brewing the leaves in boiling or hot water, resulting in a simple infusion. Steeping time varies by sort. Darjeeling tea should not be steeped for longer than one minute, while heavily fermented black teas can be steeped for up to three minutes and oolong teas up to ten. With green teas, the amount of time is dependent upon the quality. The best sorts (e.g., Japanese gyokuro) require only thirty seconds, and they can be reinfused several times. Black tea should always be

prepared with water that has reached a rolling boil, fine green teas should be brewed with hot water that is only between 60 and 70°C. Steeping tea for too long a time releases the bitter tannins.

Dosages of tea vary among individuals. Some people can tolerate up to thirty-five cups of tea per day, while others can handle little more than one cup at breakfast. One tea bag per cup yields approximately 60 mg of caffeine. The yield is lower with loose tea (only about 40 mg of caffeine is obtained from the same weight).

The renowned Tibetan butter tea, which is also made in Mongolia, is prepared from brick tea (pressed black tea leaves bound together with ox blood). Shavings from the brick are boiled in a mixture of milk and water (1:2) and flavored with rice, ginger (see **Zingiber officinale**), orange peel, various spices, and salt. Finally, a piece of yak butter (not rancid butter, as is often incorrectly stated) is added to the souplike tea. The entire mixture must then be churned in a special tubular vessel until an emulsion results.

Tea is sometimes combined with other plants to alter its aroma. Moroccan tea, a mixture of green Chinese tea and the North African nana mint (*Mentha* x *nana*), is quite typical. This tea is brewed strong and heavily sweetened (in Morocco, it is drunk primarily during usage of kif; cf. **Cannabis sativa**). In Yemen, tea is aromaticized with twigs of **Catha edulis**. In eastern Asia, oolong tea is often mixed with the flowers of *Chrysanthemum* spp.

A number of plants have been or are utilized as stimulating tea substitutes; mate (**Ilex paraguariensis**) is an especially popular alternative. **Ilex cassine, Ilex guayusa, Ilex vomitoria**, other *Ilex*

Top right: One of the many cultivated sorts of the tea plant that was bred to produce Japanese green tea (*Camellia sinensis* cv. Yutaka midori)

Bottom left: A Tibetan woman making the notorious butter tea

Bottom right: The tea plant (*Camellia sinensis*) with its typical fruit capsules.

species, coca (*Erythroxylum coca*), and *Ephedra* **spp.** have also been used. The African rooibos tea consists of the leaves of the leguminous *Aspalathus linearis* (Burm. f.) R. Dahlgr. ssp. *linearis*, which is devoid of **caffeine** and other stimulating constituents (Rehm and Espig 1996, 257*).

Ritual Use

The legend of the origin of tea explains both its stimulating effects and its ritual significance: A pious monk—according to some versions, Bodhidharma, a disciple of the Buddha—was constantly falling asleep while meditating in the cloister. Angered by the fact that he could not keep his eyes open, he abruptly cut off his eyelids and cast them away. The first tea plant grew from the ground where they had fallen, its leaves resembling the eyelids. Other monks witnessed this miracle, collected some of the leaves, and poured water over them. They immediately noticed the animating power of the new beverage, and from that time forward, they always drank tea before meditating (Temming 1985, 9).

Customs surrounding the use of tea, some of which exhibit marked cultic or ceremonial qualities, have developed all across the world (Goetz 1989). In China, tea was initially drunk by Taoists and Buddhists to aid them in their meditations and sexual practices. This tradition evolved into the Chinese tea ceremony (Blofeld 1986), which culminated in the Japanese tea cult:

> For us, the tea cult became more than simply an idealized form of drinking; it is a religion of the art of living. Tea drinking gradually became a pretext for venerating purity and refinement, it became a sacred act through which host and guest came together to create the greatest bliss. (Okakura 1979, 35)

The tea path (*cha-no-yu*) is a true entheogenic ritual in which a ceremonial master not only prepares the substance but also determines the spiritual direction of the circle. At the beginning of the ritual, which is conducted in a special house (teahouse) or a room furnished especially for the occasion, incense sticks (joss sticks based on aloe wood, *Aquilaria agallocha*) or special mixtures of various fumigatory substances (see **incense**) are burned. The tea is prepared in a ritual fashion: Powdered green tea (*macha*) is added to hot water (approximately 60°C) and whipped with a tea whisk in a tea vessel made of stone (*chawan*) until it froths. The dosage per person is "three and one-half sips." The guests must ritually wash themselves before the ceremony (ablutions) and be prepared for philosophical discussion if the occasion should arise (Ehmcke 1991; Hammitzsch 1977; Iguchi 1991; Sadler 1992; Soshitsu Sen XV 1991; Staufelbiel 1981):

Certainly, the tea path is not the path for many, even though many follow the path. Only a few knowers attain its ultimate goal—finding in the tea path a path to their true selves. They are liberated from their concerns about the transitoriness of all that is earthly, they take part in the eternal, they find their way back to nature, because they are in harmony with all living beings. (Hammitzsch 1977, 125)

Similarly to the manner in which **wine** has shaped Occidental philosophy, Eastern philosophy has been borne on the wings of the spirit of tea:

> Teaism is the art of shrouding beauty in order to discover it, and to suggest something which one does not dare reveal. It is the delicate secret of laughing softly and yet inscrutably about one's self, and is thus the good mood itself—the smile of philosophy. (Okakura 1979:19)

Tea has long been prepared as an aphrodisiac (cf. Stark 1984, 109*) and plays an important role in the Chinese and Japanese arts of love (Soulié 1983).

The Japanese name for a tea mortar is *cha-usu*. This word also refers to a particular aspect of erotic play: The man lies on his back, and the woman squats over him and places his "tea pestle" (*kine*) into her "tea mortar" (Heilmann 1991, 46). In many Taoist and similarly erotic rituals, drinking tea is a required practice.

Tea leaves are an ingredient in the initiatory drink of the Afro-American Candomblé cult (see **Madzoka medicine**).

Artifacts

Tea has influenced not only the Taoist and Zen Buddhist philosophies but also the associated arts (Soshitsu Sen XV 1991). For example, there are numerous depictions of Taoist saints drinking tea.

There are also many Chinese and Japanese marriage pictures and other erotic representations (*shunga*) that depict often intimately intertwined lovers drinking tea as they are having sex (Heilmann 1991; Marhenke and May 1995[N]; Soulié 1983). These erotic interludes are often shown taking place in the teahouse (following the tea ceremony).

The Japanese tea path has produced countless artifacts, especially those intended for use in conducting the ceremony (Ehmcke 1991). In 1989, the Japanese director Hiroshi Teshigahara produced a motion picture entitled *Rikyu, the Tea Master*. The film clearly portrays both the subtleties and the difficulties of the tea path (the film music was composed by the Japanese avant-gardist Toru Takemitsu).

A Taoist saint drinking tea while riding a dragon through the fields of bliss. (Chinese woodcut)

Although this American band calls itself The Tea Party, it is likely they are referring not to true tea but, rather, to a cozy *Cannabis* session. It is difficult to imagine that this cover illustration is meant to represent the effects of a tea party. (CD cover 1993, EMI Records)

"Tea is better than wine, for you
drink it without inebriation."

CHINESE SAYING (FIFTH CENTURY)

Medicinal Use

Before tea began its triumphant march through the world as an agent of pleasure, it was used primarily for medicinal purposes. In traditional Chinese medicine, the "foam of liquid jade" is regarded as an excellent panacea. It was first mentioned as a medicine in a Chinese herbal from the sixth century and was recommended particularly for people who slept too much (Leung 1995, 241 f.*). In the Chinese literature, tea was attributed with the following properties:

promotes the circulation of blood into all parts of the body; aids clear thinking and mental wakefulness; promotes the excretion of alcohol and other harmful substances (fats and nicotine) from the bodily organs; strengthens the body's resistance to a broad spectrum of diseases; accelerates the metabolism and the absorption of oxygen by the organs; prevents loss of teeth; cleans and invigorates the skin, which contributes to the maintenance of a youthful appearance; prevents or slows down anemia; purifies the urine and promotes its excretion; resists the effects of the summer heat; is good for the eyes and makes them more shiny; promotes digestion; soothes discomfort in limbs and joints; prevents harmful mucous secretions; alleviates thirst; combats tiredness or attacks of depression; enlivens the spirit and brings about a general feeling of well-being; increases life expectancy. (Blofeld 1986, 209)

In Japan, "newborn tea"—*gyokuro*, literally "precious dew," referring to the first harvest of the year—is generally attributed with potent healing properties and is regarded as a rejuvenant. Many Japanese drink green tea together with a shot of **sake** or whiskey (**alcohol**) when they have a cold.

Strong infusions of tea are suitable for external application in treating skin ailments (athlete's foot, skin eruptions, inflamed abrasions).

Constituents

Depending upon their source and fermentation process, tea leaves contain 0.9 to 5% **caffeine** (previously: theine or teine), which occurs freely or bound with glycosides; 0.05% theobromine; some theophylline ($C_7H_8N_4O_2$); the purine derivatives xanthine, methylxanthine, and adenine; 5 to 27% tanning agents (tannin, polyphenoles, gallic acid, and catechin derivatives); and chlorophyll (only in fresh or unfermented leaves). Also present are vitamins (A, B_2, C, D, P, nicotinic acid), minerals (e.g., manganese), and carbohydrates (dextrin, pectin), as well as traces of **essential oils**, which are responsible for the aroma (the fresh leaves contain some four to five times as much essential oil as dried or fermented leaves; Aleíjos 1977, 103). The

greatest amounts of essential oil are found in the so-called flying tea from Darjeeling (the first harvest of the year, which is exported by air freight; cf. Vollers 1981).

Effects

Because of the often high amounts of **caffeine** it can contain (up to 4.5%), tea has strong excitant and stimulant effects. The tanning agents are strongly astringent and "tanning" and are used as dyes in tanning hides. The stimulating effects of tea manifest themselves more slowly than those of coffee (see ***Coffea arabica***) but also persist longer, as the caffeine often must first be liberated from the bond to the tanning agents and the glycosidic compounds. The tannic acids form toxic alkaloids and stimulate the digestion of fats. The essential oil has euphoriant as well as calming effects upon the nerves (Aleíjos 1977, 106; Blofeld 1986, 212). The essential oil as such has stimulating effects very much like those of caffeine.

Japanese studies on the pharmacology and pharmacokinetics of green tea have demonstrated that the national drink of Japan has anticarcinogenic effects, lowers cholesterol levels, and has hypoglycemic properties. It also hinders the development of arteriosclerosis. Numerous longitudinal studies in Japan have shown that drinkers of Japanese green tea develop cancer significantly less frequently than those who do not drink tea (Blofeld 1986, 214; cf. also Scholz and Bertram 1995).

The relatively high amounts of vitamin P in tea have positive effects upon high blood pressure and heart diseases.

A recent study of the medicinal effects of black tea revealed that the hot-water extract (what is normally drunk as "tea") has antiulcerogenic effects (Maity et al. 1995). Theaflavine has bactericidal properties (Vijaya et al. 1995).

Strong tea has general detoxifying properties and is a useful antidote for alcohol poisoning, overdoses of hashish and opium, and nicotine or heroin withdrawal (Blofeld 1986, 211).

Tea is also used in homeopathy, both as a mother tincture and in various dilutions (Thea chinensis hom. *HAB34*, Thea sinensis hom. *HPUS78*). According to the homeopathic medical description, tea is used among other things to treat stomach weakness, headaches, circulatory problems, states of excitation, and ill feelings (Teuscher 1992, 638 f.).

Commercial Forms and Regulations

Tea is sold in various forms on the international market. Different qualities of black tea and green tea (*sencha*), including oolong from specified areas, are available. Mixtures (e.g., English tea, East Friesian tea; cf. Haddinga 1977) as well as perfumed or aromatized teas (e.g., vanilla, Earl Grey,

"One drinks tea to forget the noise of the world."

LIN YUTANG
WEISHEIT DES LÄCHELNDEN LEBENS
[WISDOM OF THE SMILING LIFE]
(1960)

cinnamon) are also available. The most commonly sold type of tea in the world is bag tea (black tea). There are also such specialized teas as Japanese powder tea (*macha*), Tibetan brick tea, Chinese cake tea, rice tea (*genmaicha*), et cetera (Adrian et al. 1983, Maronde 1973).

Tea is an agent of pleasure that is sold freely throughout the world and is usually classified as a foodstuff.[71]

Literature

See also the entry for **caffeine**.

Adrian, Hans G., Rolf L. Temming, and Arend Vollers. 1983. *Das Teebuch*. Munich and Lucerne: C. J. Bucher.

Aleíjos. 1977. *T'u Ch'uan—grüne Wunderdroge Tee*. Vienna: Universitätsbuchhandlung W. Braumüller.

Blofeld, John. 1986. *Das Tao des Teetrinkens*. Bern, Munich, and Vienna: O. W. Barth.

Burgess, Anthony, Alain Stella, Nadine Beauthéac, Gilles Brochard, and Catherine Donzel. 1992. *Das Buch vom Tee*. Munich: Heyne.

Das, Minati, Joseph Rajan Vedasiromoni, Saran Pal Singh Chauhan, and Dilip Kumar Ganguly. 1994. Effects of the hot-water extract of black tea (*Camellia sinensis*) on the rat diaphragm. *Planta Medica* 60:470–71.

Ehmcke, Franziska. 1991. *Der japanische Tee-Weg: Bewußtseinsschulung und Gesamtkunstwerk*. Cologne: DuMont.

Gilbert, Richard M. 1981. Einführung des Tees in Europa. In *Rausch und Realität*, ed. G. Völger, 1:386–89. Cologne: Rautenstrauch-Joest-Museum für Völkerkunde.

Goetz, Adolf. 1989. *Teegebräuche in China, Japan, England, Rußland und Deutschland*. Berlin: VWB. (With an essay "Der Schaum von flüssiger Jade" by C. Rätsch.)

Haddinga, Johann. 1977. *Das Buch vom ostfriesischen Tee*. Leer: Schuster.

Hammitzsch, Horst. 1977. *Zen in der Kunst der Tee-Weges*. Bern, Munich, and Vienna: Scherz. (Formerly titled *Zen in der Kunst des Tee-Zeremonie*.)

Heilmann, Werner, ed. 1991. *Japanische Liebeskunst—Das japanische Kopfkissenbuch*. Munich: Heyne.

Iguchi, Kaisen. 1991. *Tea ceremony*. Osaka: Hoikusha.

Kaufmann, Gerhard, ed. 1977. *Tee: Zur Kulturgeschichte eines Getränkes*. Hamburg: Altonaer Museum. (Exhibition catalog.)

Maity, S., J. R. Vedasiromoni, and D. K. Ganguly. 1995. Anti-ulcer effect of the hot-water extract of black tea (*Camellia sinensis*). *Journal of Ethnopharmacology* 46:167–74. (Includes an excellent bibliography.)

Maronde, Curt. 1973. *Rund um den Tee*. Frankfurt/M.: Fischer TB.

Marquis, F., and Fr. W. Westphal. 1836. *Taschenbuch für Theetrinker oder der Thee in naturhistorischer, culturlicher, merkantilischer, medicinisch-diätetischer und luxuriöser Hinsicht*. Weimar: Voigt.

Okakura, Kakuzo. 1979. *Das Buch vom Tee*. Frankfurt/M.: Insel.

Oppliger, Peter. 1996. *Der Grüne Tee: Genuß und Heilkraft aus der Teepflanze*. Küttigen/Aarau: Midena Verlag.

Sadler, A. L. 1992. Cha-no-yu: *The Japanese tea ceremony*. Rutland, Vt., and Tokyo: Charles E. Tuttle Co.

Scholz, E., and B. Bertram. 1995. *Camellia sinensis* (L.) O. Kuntze: Der Teestrauch. *Zeitschrift für Phytotherapie* 17:231–46. (Very good bibliography.)

Soshitsu Sen XV. 1991. *Ein Leben auf dem Teeweg*. Zurich: Theseus Verlag.

Soulié, Bernard. 1983. *Japanische Erotik*. Fribourg and Geneva: Liber.

Staufelbiel, Gerhardt. 1981. Die Teezeremonie in Japan. In *Rausch und Realität*, ed. G. Völger, 2:576–81. Cologne: Rautenstrauch-Joest-Museum für Völkerkunde.

Temming, Rolf L. 1985. *Vom Geheimnis des Tees*. Dortmund: Harenberg.

Teuscher, Eberhard. 1992. *Camellia*. In *Hagers Handbuch der pharmazeutischen Praxis*, 5th ed., 4:628–40. Berlin: Springer.

Vijaya, K., S. Ananthan, and R. Nalini. 1995. Antibacterial effect of theaflavin, polyphenon 60 (*Camellia sinensis*) and *Euphorbia hirta* on *Shigella* spp.—a cell culture study. *Journal of Ethnopharmacology* 49:115–18.

Vollers, Arend. 1981. *Darjeeling: Land des Tees am Rande der Welt*. Braunschweig: Verlagsservice.

Yutang, Lin. 1960. *Weisheit des lächelnden Lebens*. Reinbek: Rowohlt.

"With its aroma and clear foam
Tea resembles the nectar of the immortals.
The first cup swept the cobwebs from my thoughts,
the entire world appeared in a gleaming light.
The second freed the spirit like purifying rain,
the third made me one with the immortals. . . ."

CHIAO-JÊN (A POET OF THE T'ANG DYNASTY)
"THE TAO OF TEA"

---

71 "In Tunisia, the medical establishment regards the black tea that is used there 'almost as a dangerous drug' and regards its consumption as 'a widespread toximania'" (Aléijos 1977, 109).

Cannabis indica Lamarck

Indian Hemp

Illustration of *bangue*, the Indian hemp plant (*Cannabis indica*). (Woodcut from Garcia da Orta, *Colloquies on the Simples and Drugs of India*, 1987)

Family
Cannabaceae [= Cannabinaceae; also Cannabiaceae, Cannabidaceae] (Hemp Family); *Cannabis* is sometimes classified within the Moraceae Family (cf. Zander 1994, 165*)

A Preliminary Note on the Botany of *Cannabis* spp.
Contemporary botanists are of two minds regarding the genus *Cannabis* (Clarke 1981; Schmidt 1992; Small et al. 1975). One school regards the genus as monotypic and suggests that there is only one species, *Cannabis sativa*, which can be divided into several varieties and numerous sorts (Anderson 1980; Small and Cronquist 1976; Stearn 1974). The other group adheres to the concept of three species (Emboden 1974a, 1974b, 1981b, and 1996; Schultes et al. 1974).

This encyclopedia subscribes to the division of the genus into three species.

Forms and Subspecies
Wild or feral Indian hemp is sometimes referred to as *Cannabis indica* Lam. var. *spontanea* Vavilov (Schmidt 1992, 641).

Synonyms
Cannabis foetens Gilibert
Cannabis macrosperma Stokes
Cannabis orientalis Lam.
Cannabis sativa α-*kif* DC.
Cannabis sativa ssp. *indica* (Lam.) E. Small et Cronq.
Cannabis sativa var. *indica* Lam.

Folk Names
Azalla, azallû (Assyrian), bandsch, bang, banj, bengali, bengué, bhamgi (Tamil), bhang, bhanga, black prince, bota, cánamo de India (Spanish), canapem indiana (Italian), canhamo, canhamo da India, cannabis, can xa, caras, charas, charras, churrus, doña Juanita, gai ando (Vietnamese), gañajâ, gañca, gangué, ganja, gánzigùnu (Assyrian), garda (Kashmiri), ghariga, ghee ("clarified butter"), gunjah, haschischpflanze, hemp, hierba santa (Spanish, "sacred herb"), Indian hemp, juanita, jvalana rasa, kamashwar modak, kañcavu, kancha, keralagras, kerala grass, kimbis (Mesopotamia), konopie indyjskie, kumari asava, la amarilla, lai chourna, la mona, la santa rosa (Spanish, "the sacred rose"), liamba, madi,

maguoon, manali, maría rosa, marihuana, marijuana, mariquita, mazar-i-sharif, menali, misarai, mustang gold, parvati, qunnab, qunubu (Assyrian), ramras, rosamaría, santa rosa, shivamuli, siddhi (Bengali, "miraculous ability"), soft hemp, tarakola, the herb, true hemp, utter, vijaya (Sanskrit, "the victor"), yaa seep tit (Thai, "drug"), zacate chino

Many of these names are also used for *Cannabis sativa* and hemp hybrids (see *Cannabis* **x and hybrids**).

History
We do not know with certitude when Indian hemp was first cultivated, when it was first used as a medicinal and pleasure plant, or where its ritual use began (Abel 1980; Merlin 1972; Schultes 1973). It is very likely that it was used in prehistoric times in the Indus Valley and in Mesopotamia. Its psychoactive effects were obvious from the beginning and were utilized for both ritual and medicinal purposes. Some authors are of the opinion that *Cannabis indica* was the miraculous Arian drug **soma** (Behr 1995). It is certain that hemp was used as a soma substitute during the post-Vedic period. In India, its use as a medicine has been documented as far back as 1400 B.C.E. In northern India and the Himalayas, hemp has been utilized since prehistoric times in shamanism (cf. ***Cannabis ruderalis***, ***Cannabis sativa***), the tantric cult, yoga, and magical contexts. Many of these uses have continued into the present day (Chopra and Chopra 1957; Sharma 1977).

The story of the Assassins, those "fanatic treacherous murderers," has been reworked often to demonstrate the "horrible effects" of hashish (e.g., Meck 1981; Nahas 1982). It has been argued that *assassins* itself means "hashish people" or "hashish eaters." The leader of this group supposedly used hashish, produced from the resin of hemp plants, to made his followers pliable, so that they would blindly carry out any murderous order. However, "[n]owhere, whether in any of the Oriental or any of the Occidental sources, is there even a suggestion that a captive Assassin made mention of the use of hashish or any other drugs" (Gelpke 1967, 274).

Indian hemp first came to the attention of Europe in the nineteenth century (Martius 1855). In 1811, an illustrated book was published in Paris that depicted the customs of the Hindus. It contained numerous scenes of Indians enjoying hemp from various water pipes and other smoking devices (Solvyns 1811). Indian hemp and the hashish it provides were immediately adopted for medicinal use; artists also discovered that hemp could be a source of inspiration, while occult circles tested it for use in inducing clairvoyant states (Hoye 1974; Meyrink 1984). The studies of the French psychiatrist Moreau de Tours (1804–1884) proved to be very influential, from both a medical and a cultural perspective (Scharfetter 1992). As a result of his published work, a number of artists, poets, and Bohemians were inspired to found the Club de Hashishins in Paris (Haining 1975; Müller-Ebeling 1992b). The renowned **Oriental joy pills** were also circulating in Marseilles at this time.

The systematic demonization of what is actually the most harmless inebriant and agent of pleasure known is the work of drug policies promulgated in and by the United States (cf. Herer and Bröckers 1993). The illegal status of hemp is a recent phenomenon and is based not upon scientific data but upon sociopolitical goals and economic structures (Hess 1996). In recent years, even judges have begun to push for the decontrol of *Cannabis* products on the basis of the individual's "right to inebriation" (Neskovic 1995).

Today, hemp is the most commonly consumed of all illegal drugs in the world, although its users typically classify hemp products not as drugs but as agents of pleasure (Drake 1971; Haag 1995). Wherever hemp is used, a hemp culture has developed along with it (Giger 1995; Novak 1980; Rätsch 1996a; Vries 1993). In the 1990s, hemp experienced a renaissance when its potential as a useful economic crop with outstanding ecological qualities was rediscovered (Galland 1994; Herer and Bröckers 1993; Hesch et al. 1996; Rätsch 1995b; Robinson 1996; Rosenthal 1994; Sagunski et al. 1995; Wasco 1995).

Distribution

The range of *Cannabis indica* is limited to northern India, Afghanistan, Pakistan, and the Himalayas (Macmillan 1991, 421*). It is difficult to assess whether the hemp that was used in ancient Mesopotamia was in fact *Cannabis indica*. Only in the Himalayas has it been observed growing wild. A large occurrence of wild plants can be found in Taratal, near the Dhaulagiri massif; the wild plant is known as *tara khola* (Haag 1995, 75). On the other hand, *Cannabis indica* and hybrids that have been produced from it are now grown throughout the world (cf. *Cannabis* x **and hybrids**).

Cultivation

All species of *Cannabis* can be grown from both seeds and cuttings (clones). Propagation from cuttings requires some skill, a green thumb, and considerable luck, but it does ensure purely female descendants (see *Cannabis* x **and hybrids**).

The seeds can be germinated either in seedbeds or in germinating pots. They can be presprouted

by placing them in moist and warm (21°C) paper towels placed in a dish and kept in a dark location. This method provides the clearest indication as to which of the seeds are the most vigorous. The seed coat will open in just a few days, after which the seed can be placed into the soil (0.5 cm deep). The young seedlings do not tolerate direct sunlight and must not be allowed to dry out. A seedling can be transplanted as soon as it has sprouted its first pair of leaves. In central Europe, it is best to begin germination in April (at home or in a greenhouse). The young plants should not be placed outdoors (i.e., on a balcony or in a garden) until the middle of May. It is also possible to sow or scatter the seeds directly onto the ground in May, although the success rate of such a method is considerably lower. In the Himalayas, *Cannabis indica* is self-sowing.

Cannabis plants need relatively copious amounts of water to grow. For this reason, they must be watered regularly. Branching can be induced by occasionally pinching off the new leaves at the ends of the branches. Flower formation can be promoted by partially defoliating the plant from time to time. As soon as hemp begins to flower, it should no longer be heavily watered. Large amounts of light and only a little water will help ensure that the inflorescences are rich with resins. Opinions about fertilizer use vary considerably.

Appearance

Indian hemp typically grows to a height of only about 1.2 meters. It is heavily branched, which gives it a conelike appearance not unlike that of a Christmas tree. Because of its many oblique side branches, this species forms many more (female) flowers than the other hemp species do, which makes it particularly attractive for the production of psychoactive products. The aril is heavily articulated, in contrast to its more smooth appearance in **Cannabis sativa** (cf. Clarke 1981, 158). The seeds are somewhat darker and smaller than those of *Cannabis sativa*. Apart from its size and the heavily branched appearance, the main distinguishing feature of the species is the shape of its leaves, which are usually much broader and more oval than those of the other two species.

Top: A typical leaf of Indian hemp (*Cannabis indica*).

Bottom: Produced from *Cannabis* leaves, bhang is sold openly in Varanasi, the sacred city of Shiva, the god of hemp and hemp use. One ball represents a weak dose, two produce moderate effects, and three induce profound psychoactive effects.

Left: Indian hemp (*Cannabis indica*) is distinguishable primarily by its short stature and Christmas-tree-like appearance. (Female wild plants, photographed in the Himalayas, Nepal)

From top to bottom:
The two most important products of *Cannabis indica* are the dark resin rubbed from its female flowers (charas) and the dried female inflorescences (ganja).

The hashish produced in Lebanon is known as red Lebanese.

Moroccan hashish is also known as green hashish. It is produced by pressing finely chopped and sifted female flowers.

Indian hemp is almost always dioecious. The male plants are somewhat more slender and taller than the female plants.

This plant is very easily confused not only with the other hemp species, but also with other plants, such as false hemp (*Datisca cannabina* L.), which is remarkably similar and has even been mistaken for Indian hemp in herbariums (Small 1975).

Psychoactive Material
— Female flowers/inflorescences (ganja)
— Leaves (bhang)
— Herbage of flowers and leaves (cannabis indicae herba, herba cannabis indicae, summitates cannabis)
— Seeds
— Resin (resina cannabis indicae, charas = churrus, hashish)
— Oil from the resin (hashish oil)
— Oil from the seed (hemp oil)

Preparation and Dosage
There are many ways to prepare *Cannabis indica*. For psychoactive use, the most popular parts are the female flowers and the resin. The leaves of the female plant are also used. Male plants are practically useless. All products can be either smoked or eaten (drunk) (Rippchen 1995). The most common method of use is to smoke the dried flowers of the female plant, which should be harvested before the seeds form and slowly dried in the shade.

The most valuable products of the plant are the resin and the resin glands that are rubbed off the female flowers. The resin can be harvested or obtained in a number of ways (Gold 1994). The most valuable resin is obtained by rubbing the female inflorescences with the hands. The resin and some of the resin glands stick to the surface of the hands and collect there as more flowers are rubbed. The result can be scratched or scraped from the hand. Kneading the collected material produces a soft, aromatic, black or deeply dark olive green mass known in the Himalayas as charas (= charras, chura, churrus). Charas can be mixed either into various foods (pudding, cakes, cookies, etc.) or with other herbs to produce **smoking blends**.

In India, Pakistan, Afghanistan, and Nepal, the resin is graded into different sorts, depending upon the source and the intended use: Kashmiri or Dark Brown Kashmiri, Manali or Finger Hashish, Rajasthani (resin mixed with other pieces of the plant), Indian Gold or Black Gold (high-grade resin gilded with gold leaf), Black (soft, pure resin), Bombay Black (resin to which opium, *Papaver somniferum*, or **morphine** has been added), Parvati, (hand-rubbed resin), Pakistani or Brown Pakistani (brown resin), Afghani or Black Afghani (hand-rubbed resin), and Moldy Afghani (inferior quality).

After charas, the most potent product is the nonrubbed, deleaved dried female inflorescence. This product is usually called ganja and is either smoked alone or mixed with other herbs (e.g., **Datura metel**, **Turnera diffusa**, **Brugmansia suaveolens**, **Amanita muscaria**, **Nicotiana rustica**, **Aconitum ferox**). Ganja can also be eaten or drunk.

A third psychoactive *Cannabis indica* product is bhang. The term *bhang* refers to the small, resinous leaves of the plant as well as to drinks that are prepared from them. Bhang is prepared from watery hemp leaves, i.e., leaves that have been soaked in water, ground, and mixed with sugar and molasses (this method is typical for the region around Varanasi/Benares). But bhang can also be made with milk products:

> The drink called *bhang lassie* (*thandai, poust, siddhi, ramras*), which is made of yogurt, water, **honey**, pepper [cf. *Piper* spp.], and hemp flowers, symbolizes the sacred Ganges and can be obtained for pennies throughout India even today. It is equally venerated by pilgrims and by participants in marriage ceremonies and temple festivals. When **alcohol** is added to bhang, it is called *loutki*; if opium tincture [see *Papaver somniferum*] is also added to the preparation, then the Indians call the drink *mourra*. Bhang mixed with ice cream produces *gulfi*, also called *hari gulfi* (green ice), which is particularly popular in northern India. (Haag 1995, 78)

On occasion, the leaves will be drunk only with water or milk; such drinks, known as *thandai*, are

Bhang Recipe (Nepal)

Required ingredients:
— hemp flowers (ganja)
— spices (e.g., cardamom, turmeric, nutmeg [**Myristica fragrans**], cloves, pepper [**Piper spp.**], cinnamon)
— sugar or **honey**
— milk (water buffalo)

Optional ingredients:
— poison nut (*Strychnos nux-vomica*)
— opium (*Papaver somniferum*)
— thorn apple seeds (*Datura metel*)
— ground nuts (e.g., almonds)
— ghee (clarified butter)

Finely chop the hemp flowers and mix with the spices (and optional ingredients).
Dissolve sugar or honey in the milk, then mix in the hemp and the spices.

consumed for refreshment (Morningstar 1985). Ganja can also be used to brew **beer** (Rosenthal 1996).

For tantric **smoking blends**, hemp flowers (ganja) are sometimes mixed with cobra venom—the cobra is a sacred animal and a symbol of Shiva. Crystallized cobra venom[72] is mixed with chopped hemp flowers or hashish and smoked in a chillum. Other tantric mixtures contain *Aconitum ferox*, *Datura metel*, *Brugmansia arborea*, opium (see *Papaver somniferum*), tobacco (*Nicotiana tabacum*), or henbane (*Hyoscyamus niger*).

Around 1870, Indian cigarettes were made in Paris from the following ingredients:

0.3 g belladonna leaves (*Atropa belladonna*)
0.15 g henbane leaves (*Hyoscyamus niger*)
0.15 g thorn apple leaves (*Datura stramonium*)
0.5 g Indian hemp leaves, impregnated with opium extract and cherry laurel infusion (*Prunus laurocerasus* L.)

This recipe is reminiscent of **witches' ointments** as well as modern **smoking blends**. Another recipe for Indian cigarettes mentions paper impregnated with a tincture of *Cannabis indica*, opium (see *Papaver somniferum*), and *Lobelia inflata*.

In Cambodia, the wood of the botanically unidentified **shlain** tree is added to hemp flowers and leaves in order to increase their effects when smoked (cf. Rätsch 2001, 51*).

The psychoactive dose when smoking *Cannabis indica* is approximately twice as high as when eating it; around 50% of the THC is taken into the smoke. A normal dosage is between 5 and 10 mg of THC. This corresponds to about 0.25 g of smoked flowers or 0.1 g of smoked charas (resin). These guidelines should be used with care, however, because THC content can vary considerably (Schmidt 1992, 650). The products of *Cannabis indica* are generally more potent than those of *Cannabis sativa*.

Ritual Use

Hemp has been a drug of shamans since ancient times (Eliade 1975, 376 ff.*; Knoll-Greiling 1950; Sebode and Pfeiffer 1988, 16). Shamans are generally attributed with the discovery of pharmacologically efficacious plants, including the discovery of hemp and its multitude of uses (Merlin 1972). In central and east Asia, hemp was already in use during the Neolithic period. Our word *shaman* originated in the same area. In the Tungusic language, *shaman* refers to the healing and prophesizing master of consciousness (Sebode and Pfeiffer 1988, 7). The earliest literary and ethnohistorical evidence for hemp is contained in shamanic texts from ancient China (Li 1974a, 1974b).

In Nepal, shamanism has been and continues

The Chillum Cult

The word *chillum* (pronounced *tschillum*, and sometimes spelled *chilam*) refers to a conical tube for smoking hemp. The smoking of chillums is an ancient tradition that is still alive in the Himalayas and India (Knecht 1971; Morningstar 1985). Generally speaking, the Himalayan region is the most tradition-rich area of the world as far as hemp is concerned (Fisher 1975; Sharma 1972, 1977). It is not known how long chillums have been in use. It is also not clear whether the chillum is an ancient invention of the Himalayan peoples or is derived from the head (upper part) of the Muslim hookah (the traditional Oriental water pipe) (Morningstar 1985, 150).

The chillum is the typical smoking device of sadhus and yogis, who use it constantly in their rituals of worship, meditation, and yogic practice (Bedi 1991; Gross 1992; Hartsuiker 1993).

When Western travelers ("hippies") journeyed to India and Nepal in the 1960s, they quickly learned about the indigenous use of the chillum and hemp by sadhus. They brought back to the West both the smoking device and the knowledge of its proper use. Soon, large numbers of Indian and Nepalese chillums were being imported by shops specializing in Indian articles and by "head shops." Many Western hashish smokers possess one or more chillums and know how to use them in a traditional manner.

A chillum is smoked not alone, by one person, but in a smoking circle (*chilam chakri*). One person fills the chillum with the smoking mixture (e.g., hashish and tobacco, hashish and marijuana, or hashish and *Datura metel*) and then hands it to the next person in the circle for lighting. The chillum is lit with two matches (which represent the masculine and the feminine poles of the universe). Before lighting the mixture in the chillum, the person holds the chillum against his or her forehead (the third eye) and utters a short formula (*japa*), usually "Bum Shankar!" This consecrates the smoke to the Hindu god Shiva. Hemp smokers regard both Shiva and his son Ganesha as the gods of hemp smoking. After the chillum has been started, it is passed around the circle, usually in a clockwise direction. When the chillum is "through," its owner taps out the remnants of the smoking mixture and carefully cleans it with a piece of cloth.

Chillum smoking is a relatively elaborate process that demonstrates the profound respect the consumer has for the plant as well as for the Asian tradition, and often reveals a religious attitude toward hemp smoking. Most Europeans who use chillums today learned about its use not while in India or Nepal but from other hemp smokers. The European tradition of chillum smoking can now look back upon more than thirty years of tradition, and chillum use is being passed on from one generation to the next (Rätsch 1996a).

to be of great importance for many indigenous peoples who have had only limited contact with Western medicine. Most of the peoples of Nepal practice a religion that draws from various sources. Here, elements from the Vedic period, the ancient Bön religion of Tibet, Tibetan Lamaism, and a variety of schools of Hinduism have blended into a harmonious whole. Shamans can be found in almost every village. They are usually called *jakri*, a word that means "magician" (both male and female). These shamans inhabit a polytheistic cosmos in which the Buddha is as much at home as are the ancient Bön demons and the Vedic and Hindu gods:

> According to shamanic tradition, Indra, the original Vedic god, discovered Cannabis and sowed it in the Himalayas so that it would always be available to humans, who could

72 Cobra venom can also be sniffed and is said to induce extreme psychedelic effects (per oral communication from Ossi Urchs). In the Himalayas, it is believed that hemp that grows over a buried cobra will become especially potent as a result of its venom and have extremely strong inebriating effects (Sharma 1972, 208).

A traditional hookah, a simple water pipe for consuming hashish or marijuana. (Drawing by C. Rätsch)

attain joy, courage, and greater sexual desire though the herb. (Haag 1995, 78)[73]

The shamans venerate Shiva, whose roots can be traced back to the Vedic Rudra. They regard him as the primordial first shaman, who had a perfect understanding of the shamanic arts and who taught this to certain chosen people. One Nepalese name for Shiva is *vijaya*, "the victorious"; in the Vedic scriptures, hemp is referred to by the same name. Shiva is also known as Bhangeri Baba, "the lord of hemp" (Storl 1988, 83, 198, 201). According to shamanic tradition, it was he who discovered hemp and sowed it in the Himalayas so that humans would always have access to it. Shiva also gave people the various recipes for its use: "In Nepal, ascetics, shamans, and magicians have been consuming small amounts of this agent since ancient times in order to induce trance states" (Gruber 1991, 144).

Smoking is the most common method of consuming the various hemp products (Knecht 1971). Hemp leaves, the female flowers (ganja), or bits of the sticky, aromatic resin (charas) are stuffed into a chillum, either alone or mixed with thorn apple leaves (*Datura metel*), henbane (*Hyoscyamus niger*), aconite (*Aconitum ferox, Aconitum* **spp.**), or tobacco (*Nicotiana tabacum*). The chillum, a symbol and an attribute of Shiva, is then held to the forehead and consecrated to the god with the words *Bum Shankar*, "hail to the benefactor" (Morningstar 1985).

In Nepal, hemp is often drunk in the form of bhang (Müller-Ebeling and Rätsch 1986, 20*). The shamans of the Himalayas drink bhang in order to induce the states of trance or ecstasy that they require for their healing rituals. They offer bhang at the phallus-shaped shrines of Shiva (sacred stones, lingams). The offerings actuate the healing powers of the god, for no one loves hemp and the state it produces as much as Shiva himself. The inebriated god sends forth his healing power, which the shaman channels and transmits to the patient. Although it is usually only the shaman who smokes ganja or drinks bhang during the shamanic healing sessions, hemp is also used as a medicine. By smoking hemp, and by virtue of his gifts, a shaman who is devoted to Shiva can produce an especially efficacious medicine:

Smoking is an unbecoming, a dissolution, a process of death. In this small, spinning pyre, the husks of delusion that entwine us burn to ash. The rotting corpses of our transgressions, the cadavers of old karma, roast therein and are transformed to snow white ash. . . . The bolt to the door of the "transcendent" is shattered; the demonic hordes of Shiva, the ethereal images of natural forces and the shapes of souls, dance before the eyes of the

initiate. The dead and the gods appear! In an even deeper samadhi, all manifestations, all appearances, cease, and it simply *is*. In total absorption, Shiva sits on Kailash, the holy mountain, the mountain of snow, the mountain of ash. . . . After the chillum has been smoked all the way to the end and the meditation is over, the shaman takes the ashes and rubs them onto his forehead or he places them on his tongue as prasad, for the sacred white powder is regarded as the best medicine. (Storl 1988, 204, 205*)

Hemp is the most important ritual drug of Indian and Nepalese Tantrists. They call it *vijaya*, "the victorious," and regard it as "the only true aphrodisiac" (Bharati 1977, 209). For this reason, hemp preparations are used in the erotic rituals for couples, during which the two lovers are transformed into the gods Shiva and Parvati (Aldrich 1977). The *sadhaka* (or Tantrist) places a bowl with a hemp preparation in front of himself on a mandala and asks the tantric "goddess of the divine nectar" to consecrate the hemp. After this, he carries out several ritual gestures (*mudras*) over the vessel. He then speaks a mantra to the guru, the teacher, in order to offer him the libation. Finally, he touches his heart and drinks the drink to honor the god, usually Shiva, he has chosen for this purpose (Bharati 1977, 207 f.).

During one tantric ritual that is still conducted in northern India today, hemp (bhang) is transformed into *amrita*, the drink of the gods (cf. **soma**):

1. As an act of preparation and ritual purification, the leaves of the Cannabis plant are rubbed with black pepper [cf. *Piper* **spp.**], water is added, and the mixture is filled into a stone vessel.
2. A *yantra* (ritual diagram) of a circle, square, and triangle is drawn. The primordial feminine energy *ardhar shakti* is venerated in this *yantra.*
3. The vessel with *bhang* is placed upon the *yantra.* This is followed by meditation and recitation.
4. Using a *mantra* (a kind of magical formula), *vijaya* (the name of the goddess) is called into the *bhang* container and welcomed.
5. Using a particular *mantra* (magical formula), *bhang* is transformed into *amrita* (a drink of the gods).
6. With a ritual gesture of veneration [*mudra*], the vessel full of *bhang* is raised to the forehead and a prayer in honor of the *guru* (religious teacher) is uttered.
7. The *bhang* preparation is ingested.

The activity described is accompanied by recitations and ritual gestures (*mudras*). (Moser-Schmitt 1981, 545)

73 According to a different Indian myth, the first hemp plant arose at the spot where a drop of juice pressed from *Datura metel* fell upon the earth (Schleiffer 1979, 60*).

Since the post-Vedic period, Brahmans have been using hemp as an adjunct to meditation and to promote concentration as well as to deepen their understanding of the sacred texts (Rig Veda, Atharva Veda, Puranas, etc.). Orthodox Brahmans from the area of Varanasi (= Benares) and Allahabad (Uttar Pradesh) still regularly ingest bhang every Friday (Bharati 1977, 207).

In Mesopotamia, and especially among the Assyrians, hemp was burned as a sacred **incense** (Bennett et al. 1995, 15, 19). The Scythian hemp ritual is discussed under *Cannabis ruderalis* (cf. also "**Trees with Special Fruits**"). In the Occult movement, hemp was used as a visionary incense (Bennett et al. 1995, 280 ff.; Meyrink 1984).

On the Caribbean island of Jamaica, Indian hemp is at the center of the Rastafari cult, a movement that arose in the twentieth century and is said to have its roots in Ethiopia (see Rätsch 2001, 137–42). The ritual music of the Rastas is reggae, and their sacrament is hemp (ganja). One leader of the Rastas summarized the cultural meaning of hemp in the following manner:

> We use this herb as medicine and for spiritual experiences. It helps us to overcome illness, suffering, and death. . . . We use our herb in our church—as incense for God, just as the Roman Catholics use incense in their church. We burn our incense in order to venerate our God through spiritual experience. . . . It gives us spiritual comfort, we praise God in peace and love, without force. . . . When we are depressed, when we are hungry, we smoke our little herb and we meditate on our God. The herb is a true comfort to us. (In Kitzinger 1971, 581)

In the Rastafarian community, the first inebriation produced by smoking ganja has the character of an initiation. The young smoker is supposed to receive a vision that will mark him as a full member of the community and reveal his path through life (Rubin and Comitas 1976). "Ganja is the most strongly shared experience among the brothers" (Gebre-Selassie 1989, 156). The Rastas, it should be noted, eschew **alcohol**, which may be used only as a solvent for ganja and be consumed in medicines. Alcohol inebriation is viewed as reprehensible, harmful, aggression promoting, and asocial (Blätter 1990, 1993).

In Mexico, an Indian cult calls hemp *la santa rosa*, "the sacred rose," and venerates it as a sacred plant. The cult members chew hemp flowers at their meetings and use the psychoactive effects to intuitively speak sacred words, for divination, and as an expression of the divine (Williams-Garcia 1975). This hemp cult may have its roots in a pre-Columbian use of other psychoactive plants (possibly *Salvia divinorum*).

Artifacts

A Sumerian necklace from Ur incorporates a number of elements that are strongly reminiscent of *Cannabis* leaves (Emboden 1995, 99*). An ancient depiction of bull killing suggests that hemp may very well have played a role in the Mithraic mysteries. From the wounds of the bull that Mithras killed as a sacrifice to create the world, blood is shown flowing in the shape of a hemp leaf (Bennett et al. 1995, 146; cf. *Peganum harmala*, **haoma**).

Evidence of the effects of hemp consumption on art (painting) is not as obvious as with other psychoactive plants. This is certainly due to the fact that the effects of hemp are only rarely visionary in nature. With many artists, it is also impossible to state whether their works were influenced by hemp or other psychoactive substances because the artists themselves refuse to discuss the matter (Müller-Ebeling 1992b).

Hemp provided Aubrey Beardsley, one of the greatest artists of the art deco movement, with inspiration during his short life (1872–1898). He described Warden's Extract of Cannabis Indica, available at pharmacies, as "my mental nourishment" (Behr 1995, 185). It is very likely that other art deco artists created their works while under the influence of hemp; information about this, however, is scant (Müller-Ebeling 1994). Consequently, it is not surprising that elements of the art deco style reemerged in the psychedelic art of the 1960s.

Hashish had a significant influence upon surrealism (Breton 1968). Other artists were also inspired by hemp. Picasso (cf. *Artemisia absinthium*) was quite familiar with hashish and was of the opinion that it made one happy and stimulated the imagination. In contrast, Alfred Kubin experienced its effects on more of an existential level and felt compelled to transform his hashish visions into art (Behr 1995, 208 f., 244 f.). A recent work by the American artist Alex Grey, known for his psychedelic visions in *Sacred Mirrors*, has hemp as its theme, featuring a hemp goddess for the "Cannabis Cup" (Rätsch 1995d, 306).

Since the 1960s, hemp, hemp leaves, hemp consumption, smoking paraphernalia, and caricatures of hemp smokers and of persecution by the police have all been the subjects of posters and postcards.

In the art of the Rastafarian movement, the hemp plant is sometimes depicted as a sacred tree. Many Rasta pictures clearly have been inspired or influenced by the heavy hemp consumption of the painter (e.g., Ivan Henry Baugh, Jah Wise) (Haus der Kulturen der Welt 1992).

It is possible that numerous Sanskrit texts were inspired by hemp consumption. It is certain that hashish inebriation had an enormous influence on the *Tales of a Thousand and One Nights* (cf. *Papaver somniferum*).

> "The gods gave people hemp out of compassion so that they could attain enlightenment, lose their fear, and retain their sexual desire."
>
> RAJA VALABHA (SEVENTEENTH-CENTURY SANSKRIT TEXT)

> "To the Hindu the hemp plant is holy. A guardian lives in the bhang leaf. . . . To see in a dream the leaves, plant, or water of bhang is lucky. . . . A longing for bhang foretells happiness. . . . It cures dysentery and sunstroke, clears phlegm, quickens digestion, sharpens appetite, makes the tongue of the lisper plain, freshens the intellect, and gives alertness to the body and gaiety to the mind. Such are the useful and needful ends for which in his goodness the Almighty [Shiva] made bhang. . . . [T]he quickening spirit of bhang is the spirit of freedom and knowledge. In the ecstasy of bhang the spark of the Eternal in man turns into light the murkiness of matter. . . . Bhang is the Joy-giver, the Sky-flier, the Heavenly-guide, the Poor Man's Heaven, the Soother of Grief. . . . No god or man is as good as the religious drinker of bhang."
>
> HEMP DRUG COMMISSION REPORT (1884)
> IN *THE BOOK OF GRASS* (ANDREWS AND VINKENOOG 1968, 145)

At the dawn of the twentieth century, hashish was a well-known drug of fashion, as were opium and cocaine. ("Hashish," fantasy from Schlosser and Wenisch; illustration from the periodical *Das Magazin,* January 1930)

Hashish inebriation inspired the literary efforts of many nineteenth-century authors. The works of Charles Baudelaire (*Paradis artificiels*), Fitz Hugh Ludlow (*The Hasheesh Eater*), Maurice Magre (*La nuit de haschish et d'opium*), Walter Benjamin (*Über Haschisch*), Leo Perutz (*Der Meister des letzten Tages*), and Ernst Jünger (*Annäherungen*) all rank among the classics of world literature (Kimmens 1977).

The poets of the Beat generation—Jack Kerouac, Gary Snyder, Allen Ginsberg, Paul Bowles—regarded hashish use as an important source of inspiration, and their work has provided us with numerous examples attesting to this fact. For the authors of the psychedelic generation—Robert Anton Wilson, Robert Shea, Tom Robbins, Mohammed Mrabet, Stephen Gaskin, Hunter S. Thompson, Tom Wolfe—smoking hashish was an obvious source of inspiration. The smug novel *Budding Prospects: A Pastoral,* from the best-selling author T. Coraghessan Boyle (1984), relates the turbulent story of the attempts of several hippies to grow hemp and the paranoia that accompanies their efforts. The Rastafarian movement and its hemp consumption, and the inseparably related reggae, has also been the subject of literary treatments (e.g., Thelwell 1980, Zahl 1995).

Even more than such literature, underground comics have clearly been inspired by hemp and drawn for readers who are under the plant's influence while they are reading them. One series of comic books, featuring the stories of various authors and illustrators, was even called *Dope Comix.* The works of artists Robert Crumb and Gilbert Shelton (*The Fabulous Furry Freak Brothers*) are among the classics of this genre. The tales of the Freak Brothers were quickly translated into German, and they became a true underground hit (1975). The motto of the three permanently stoned Freak Brothers expresses the sentiments of many hemp users: "As we all know, dope will get you through times of no money better than money will get you through times of no dope!"

The German counterpart of the American Gilbert Shelton is Gerhard Seyfried. His comics and caricatures (*Wo soll das alles enden* [Where Will It All End?] and *Freakadellen und Bulletten* [Freak Sandwiches and Pig Burgers]) provide a clear and amusing document of the German underground during the 1970s and 1980s. Seyfried has recently produced a poster about hemp (1994). During the 1990s, Walter Moers's comics and caricatures about *das kleine Arschloch* ("the little asshole") enjoyed considerable popularity. In one volume titled *Sex, Drogen und Alkohol,* Moers provides a bitingly satirical characterization of the effects of a variety of psychoactive substances, including hashish. While the underground comics primarily give expression to typical "stoner"

humor, the more "artistically serious" comics (e.g., Edition Comic Art) reveal a different side of the hashish state. The French comic artist Moebius, who gained renown primarily as a result of his collaboration with the filmmaker Alejandro Jodorowsky (*Montana Sacra*) and their joint opus *John Difool* (better known as *Der Incal*), has created worlds that his readers perceive as "extremely psychedelic." The artist himself has stated that marijuana provided him with a source of inspiration and that he learned things from his experiences with inebriated states, but also that he has now grown out of these (Moebius 1983).

Some comics focus exclusively on the topic of hemp, e.g., the works of Pete Loveday (*Highter Breiter: Der definitive Hanf Comix,* Edition Rauschkunde, 1995; *Russell' Big Strip Stupormarket,* John Brown Publishing, 1995). The collected works of Harold Holmes bear the seal "Cannabis-Friendly Comicx"—this pertains especially to *Der Abenteuer von Harolod Hedd* (Raymond Martin Verlag, 1995). From time to time, hemp also appears in children's comics, albeit in disguise. In Bud Sagendorf's *Popeye,* the wondrous plant is camouflaged as spinach. The favorite food of the Frenchman Peyo's Smurfs (who live in fly agaric mushrooms) is "sarsaparilla" (cf. **Veratrum album**). In Voss's *Der Drogenbaum* [The Drug Tree], the plant turns into an autonomous thinking being (Volksverlag, 1984).

Hemp has also left its mark upon movies. Many movies from the first half of the twentieth century, such as *Reefer Madness,* were intended as deterrents but are unintentionally comical. Other "stoner" films that have appeared since the 1960s are intentionally humorous. Like the Freak Brothers of the comic world, the films of Cheech and Chong push marijuana use to satirical heights. The film *Up in Smoke* (1978) is a true cult classic of the hemp culture. The film *Rembetiko* (1985), which brings the Greek hemp and music scene of the 1930s and 1940s to life with much music and hashish, is well worth watching. *The Harder They Come,* a film based upon Michael Thelwell's novel of the same name (1980), provides deep insight into the world of Rasta and reggae.

The music that is associated with smoking hemp and with marijuana and hashish is as varied as the cultural landscape of our planet (Rätsch 1995a). Although there is no hemp music per se, much traditional and ethnic music has been related to the ritual ingestion of hemp products for centuries. This includes both classical Indian music and Jajouka, the ecstatic music of Morocco (Welte 1990). Rembetiko, also known as rebetiko, is sometimes called the Greek blues, but it is actually a folk music from the 1930s and 1940s that was influenced by Oriental elements and was played primarily in Greek cafés by those under the influence of hashish (Dietrich 1987; cf. also Rätsch 2001, 92–93*).

Some music has been inspired and composed as a result of smoking hemp, while other music is played while directly under its influence. Some music draws upon texts from the hemp culture, and some is played for an audience whose members are inebriated as they listen. The musical excursions into the world of hemp-influenced consciousness are as multifarious as the possibilities of using the hemp plant. Indeed, the perception of music is profoundly affected by *Cannabis* (Fachner et al. 1995). This new listening experience has also produced new music (Mezzrow 1995; Shapiro 1988). Jazz, for example, has been profoundly affected by these new listening experiences. And the reggae that has developed in Jamaica is a "pure stoner music" (Epp 1984).

Today, an increasing number of hemp leaves are gracing the covers of CDs as well as the rainbow-colored CDs themselves. The hemp leaf has become a political symbol of the underground and the counterculture. The leaf signals both a disaffection with current political and social systems and a peaceful way of achieving inebriation and enjoying music. Some bands even name themselves after the plant and its products: Bongwater, Gunjah, Hash, The Smoke (Calm 1995).

For information about smoking devices and other paraphernalia, see *Cannabis sativa*.

Medicinal Use

In ancient times, the Assyrians were especially known for using Indian hemp (*azallu, qannapu, ganzigunnu*) and hashish (*martakal*) in their medicine (Thompson 1949, 220ff.*). Numerous cuneiform texts attest to this fact. Hemp root was prescribed for difficult births. For abdominal pains, the entire plant was boiled and administered as an **enema**. Hemp oil or hemp in petroleum was rubbed onto swollen stomachs. The roasted seeds were administered for treating *arimtu*, a disease causing a type of shaking in the extremities. Crushed hemp seeds, mixed with the seeds of a **Mesembryanthemum** species, were given to "suppress the spirits" that were responsible for causing (probably) some type of depression. A mixture of hemp and cereal flour was used as an antidote. Mixed with other plants and with "pig oil," hemp was applied as a small anal compress. Hemp was also used in **beer** (*kurunnu*); this brew was drunk to treat diseases that had been caused by witchcraft (Thompson 1949, 221f.*). It is possible that the Assyrians adopted the practice of inhaling hemp smoke from the Scythians (cf. *Cannabis ruderalis*), who long had trade relationships with the Assyrians before eventually contributing to their destruction. The Assyrians

Discography of Hemp Music (a small selection)

Traditional and Ethnic Hemp Music

Jilala and Gnaoua—Moroccan Trance Music. SUB CD013-36. Sub Rosa Records, 1990. (Recorded by Paul Bowles.)

L'Ensemble Traditionnel de l'Orissa. *L'Inde—Musique traditionnelle de danse Odissi.* ARN 64045. Arion Records, 1975.

Maroc—Festival de Marrakech. PS 65041. Playasound Records, 1989.

The Master Musicians of Jajouka, featuring Bachir Attar. *Apocalypse Across the Sky.* 314-510857-2. Axiom Records, 1992. (Includes accompanying text by William S. Burroughs.)

Rembetica: Historic Urban Folk Songs from Greece. CD 1079. Rounder Records, 1992. (Historical original recordings [from the 1930s] from the legendary *tekedes* [hashish cafés])

Rembetiko—Original Filmmusik. CD CMC 013009. PROTON/Videorent, 1985.

Songs of the Underground. The Greek Archives. Vol. 5. F.M. Records 631.

Reggae

Big Blunts—Smokin' Reggae Hits. Vols. 1, 2, and 3. Tommy Boy Records, 1995 ff.

Culture. *International Herb.* 44006. Shanachie Records, 1992.

Dub Syndicate. *Stoned Immaculate.* ON-U LP56. On-U Sound Records, 1991.

Inner Circle. *The Best of Inner Circle.* 74321 12734 2. Island Records, 1992.

Peter Tosh. *Bush Doctor.* 1C 064-61 708. EMI Electrola Records, 1978.

Peter Tosh. *Legalize It.* CDV 2061. CBS/Virgin Music, 1976.

Tougher Than Tough: The Story of Jamaican Music. 4 CDs. Island Records, 1993.

Zion Train. *Natural Wonders of the World in Dub.* WWLP/CD5. Zion Records, 1994.

Jazz, Pop, Rock, Metal, Ambient/Techno/Trance, etc.

Alex Oriental Experience. *Studio Tapes 1976–78.* WR 08517122. Wiska Records, 1996.

Black Crowes. *The Southern Harmony and Musical Companion.* 512 263-2. Def American Records, 1992.

Blue Cheer. *Oh! Pleasant Hope.* 1971/LMCD 9.51080 Z. Line Records, 1991.

Cannabis Weekend. Dope Records, 1995.

Children of the Bong. *Sirius Sounds.* 540394-2. Ultimate Records, 1995.

Cypress Hill. *Black Sunday.* CK 53931. Ruffhouse/Columbia Records, 1993.

Dr. John, The Night Tripper. *Remedies.* AMCY-231. Atlantic, 1970.

Dope on Plastic. Vols. 1, 2, and 3. React CDs, 1994 ff.

Embryo. *Turn Peace.* EFA 01045-26. Schneeball Records, 1990.

Freaky Fuckin Weirdoz. *Senseless Wonder.* PD 75331. RCA Records, 1992.

Give 'em Enough Dope. Vols. 1, 2, and 3. CD 001/310. Wall of Sound, ca. 1995 ff.

Godfathers. *Dope, Rock 'n' Roll and Fucking in the Streets.* GFTR CD 020. Corporate Image, 1992.

The Golden Dawn. *"Power Plant."* LIK 24. Reissue Charly Records, 1988.

Gong. *Camembert Electrique.* CD LIK 64. Charly Records, 1990.

Gong. *Flying Teapot.* CD LIK67. Charly Records, 1973, 1990.

Green Piece. *Northern Herbalism.* CD 003. Kiff Records, 1996.

Hasch stoppt Hass—Alkohol killt. Vince Records 019, ca. 1995.

Hans Hass Jr. *Magic Ganja.* AIM0085. Aquarius Records, 1996.

Hempilation: Freedom Is Normal. Capricorn Records, 1995.

Highzung. LC-8248. Rockwerk Records, 1992.

Idjo. *Argile.* 3055-2. Schneeball/Indigo, 1995.

Jefferson Airplane. *Long John Silver.* NL89133. RCA Records, 1978.

Joint Venture. *Dinger.* Fun Beethoven Records, ca. 1994.

Marijuana's Greatest Hits Revisited. 7-5042-2. Rehash Records, 1992.

MC5. *High Time.* R2 71034. Orig. Atlantic, 1971; reissue Rhino Records, 1992.

New Riders of the Purple Sage. *Adventures of Panama Red.* CK 32450. Columbia Records, 1973.

David Peel and The Lower East Side. *Have a Marijuana.* LECD 9.01050. Elektra, 1968; Line Records, 1991.

Pro Cannabis—Tranceformed Ambient Collection. DO CD 01. Dope Records, 1994; distributed by EFA. (Featuring Robert Anton Wilson.)

Rausch. *Glad.* 848546-2. Vertigo, 1991.

Reefer Songs—23 Original Jazz and Blues Vocals. Jass CD-7. Jass Records, 1989.

The Sky Is High . . . 25 jazzige Reefer Songs der 30er und 40er Jahre. LC 4590. Transmitter, 1995.

Snow Bud and The Flower People. *Green Thing.* FH-339D. Flying Heart Records, 1991.

Sweet Smoke. *Just a Poke.* LC 0162 EMI. Electrola Records, 1970.

Tad. *Inhaler.* 74321 16570 2. Giant Mechanic Records, 1993.

Ten Years After. *Stonedhenge.* 820 534-2. Decca, 1969; reissue Dream, 1989.

U.S. Homegrown. COA 70003-2. City of Angels, 1995.

Witthüser und Westrupp. *Der Jesus Pilz—Musik vom Evangelium.* 2021098-7. Pilz Records, 1971.

Zentralpark. *Haschisch in Marseille.* Peace Records, 1995.

Spoken Word (and related)
Cheech and Chong. 9 3250-2. Warner Bros. Records, 1972.
Cheech and Chong. *Greatest Hit.* WB K 56 961. Warner Bros. Records, 1981.
Cheech and Chong. *Up In Smoke.* 7599-27367-2. Warner Bros. Records, 1978. (Soundtrack from the film of the same name.)
Mick Farren's Tijuana Bible. *Gringo Madness.* CDWIK 117. Ace Records, 1993.
Mohammed M'Rabet. *The Storyteller and The Fisherman.* SUB CD015-38. Psalmodia Sub Rosa Records, 1990. (Translated and read by Paul Bowles [cf. Mrabet 1995].)

Numerous record and CD covers feature hemp leaves (e.g., *ProCannabis*) or suggest *Cannabis* use (*Marijuana's Greatest Hits*). Music and texts make frequent references to the plant. (CD covers: 1995, Dope Records; 1992, Rehash Records)

74 Comparable homeopathic agents (relations) include *Belladonna* (deadly nightshade), *Hyoscyamus* (henbane), and *Stramonium* (thorn apple), the "witches' herbs" of old. *Anhalonium lewinii,* the extract of *Lophophora williamsii,* is a substitute (Boericke 1992*).

inhaled hemp smoke to relieve worries, cares, and sadness (Thompson 1949, 220*). Since these afflictions often hide behind the masks of demons, it is very likely that hemp was also used in exorcisms.

Since the beginnings of Ayurvedic medicine, *Cannabis* products have been an indispensable part of that tradition's medicinal trove. The leaves (bhang) are ingested for cramps, earaches (otalgia), lower abdominal complaints, diarrhea (including bloody dysentery), body pains, and hemorrhaging. The crushed leaves are used as **snuffs** to treat headaches and other ailments. The resin (charas) is used especially as an aphrodisiac, usually combined with opium (*Papaver somniferum*), poison nut (*Strychnos nux-vomica*), thorn apple seeds (*Datura metel*), and spices (cf. **Oriental joy pills**). In Nepal, hemp is used as a tonic, stomach medicine, and pain and sleeping agent. Sick people are prescribed hemp drinks for a variety of ailments, including depression, lack of appetite, inconstancy, and altitude sickness, a frequent occurrence in the Himalayas (Morningstar 1985). In Kashmir, the roasted leaves and flowers of the female plant are mixed with **honey** and used as sleeping pills (Shah 1982, 298*).

Immigrants from India introduced the plant into the Caribbean and taught the peoples there about its many uses. As a result, in Jamaica ganja has become an important part of bush medicine and Rasta medicine. It is used as a general remedy and restorative (Witt 1995, 80 ff.), as an efficacious means of relaxing, and also as an analgesic in the same manner that aspirin is used in Germany or the United States (Kitzinger 1971, 581). The Zionist-Coptic Church of Ethiopia encourages the Jamaican Rastas in such use, declaring that "the *Herb* may definitely be grown for its use as an asthma medicine, as a remedy for glaucoma, and for joint inflammations; also to aid in the treatment of cancer, as well as for economic use in the clothing industry and for producing paper to use, for example, in the manufacture of Bibles" (Gebre-Selassie 1989, 161). Ointments produced from crushed leaves and fat are applied externally

to treat pain. A poultice is used to treat open wounds and internal pains. Newborns are sometimes rubbed with a mush made of hemp. Hemp tea is a popular prophylactic as well as therapeutic drink for practically all complaints. It is especially effective in treating eye weakness and night blindness (West 1991).

In the nineteenth century, Europeans discovered the analgesic properties of Indian hemp (Martius 1855; O'Shaughnessy 1839). This led to the development of a number of anodynes that were made from *Cannabis indica* and marketed both in Europe and in the United States (Edes 1893; Mattison 1891). In central Europe, the seeds were mixed with an extract of henbane (see *Hyoscyamus niger*) and used to treat gonorrhea (V. Robinson 1930, 39). Around the beginning of the twentieth century, numerous cigarettes and medicinal smoke powders based on *Cannabis indica* were utilized to treat asthma, lung ailments, neuralgia, and sleep disorders (cf. **incense, smoking blends**).

Homeopathic medicine uses *Cannabis indica* (Cannabis indica hom. *HAB34,* Cannabis indica hom. *HPUS78*) in accordance with the medical description to treat a wide variety of ailments, including asthma, impotence, lack of appetite, sexual exhaustion, nightmares, and nervous disorders (Boericke 1992, 187*; Schmidt 1992, 644).[74]

The American physician Lester Grinspoon has noted that *Cannabis* has shown great promise as a medicine to treat the following ailments: depression, pain, headaches, migraines, menstrual cramps, paralysis, traumatic injuries, spasms, epilepsy, asthma, glaucoma, the side effects of cancer therapy, and AIDS (Grinspoon 1996; Grinspoon and Bakalar 1995; cf. also Roffman 1982). Overall, more and more physicians are expressing the wish that hemp products be made available for therapeutic purposes so that they can prescribe them for their patients (Clarke and Pate 1994; Grotenhermen and Karus 1995; Iversen 1993). Even psychiatry is beginning to revise its opinion of the plant (Baumann 1989; Hess 1996). But it is especially those patients who have had very good experiences with illegal self-medication who are demanding the (long overdue) legalization of *Cannabis* products (Corral 1994; Rathbun and Peron 1993). Studies have been designed to assess the medicinal use of *Cannabis* as an adjunct in AIDS therapy (Doblin 1994), and voters in California, Arizona, and other states have passed initiatives supporting the legalization of medical marijuana (ADH 1997).

Constituents
The resin, the female inflorescences, and the leaves of hemp all contain an **essential oil** and other substances, primarily cannabinoids, more than sixty of which are already known structurally and

pharmacologically (Brenneisen 1986; Clarke 1981; Hollister 1986; Mechoulam 1970; Schmidt 1992). The main active constituent is delta-9-tetra-hydrocannabinol (Δ^9-THC, corresponding to Δ^1-THC, abbreviated as **THC**). The resin (hashish) contains the four primary components, the so-called cannabinoids: Δ^1-tetrahydrocannabinol (THC) with three variants, two of which—cannabidiol (CBD) and cannabinol (CBN)—result as artifacts only when the resin is stored. These substances are responsible for the psycho-active effects of hemp. The structures of some thirty other cannabinoids with only mild or no psychoactive effects have been described. The resin also contains various sugars, flavonoids, alkaloids (choline, trigonelline, piperidine, betaine, proline, neurine, hordenine, cannabissativine), and chlorophyll. The THC content can exhibit considerable variation. Some plants (fiber hemp) contain very little or none, while in others it can constitute as much as 25% of the resin. Four to 8 mg is regarded as an efficacious psychoactive or analgesic dosage (Schmidt 1992).

The characteristically scented **essential oil**, which imparts hemp with its typical bouquet, contains eugenol, guaiacol, sesquiterpenes, caryo-phyllene, humulene, farnesene, selinene, phellan-drene, limonene, and other substances.

The constituents of the seeds, lignans, et cetera, are similar to those of *Cannabis sativa*.

Effects

The primary effect of consuming hemp is a mild to profound sense of euphoria accompanied by rich associative and imaginative abilities, a stimulated imagination, and a sense of physical well-being. Very often, the effects of hemp are also perceived as aphrodisiac or erotic (Amendt 1974; Blätter 1992; Cohen 1982; Lewis 1970).[75] When cannabis is smoked, these effects are manifested within ten minutes; when eaten or drunk, in forty-five minutes to two hours. The euphoric phase lasts for one to two hours, after which a calming effect becomes predominant. The effects often culminate in a more or less dream-rich sleep. Hemp products can also potentiate the effects of other substances (e.g., of such nightshades as *Atropa belladonna*, *Brugmansia* spp., *Datura* spp., *Hyoscyamus niger*, and, of **cocaine, nicotine**, opium [*Papaver somniferum*], ayahuasca, **ayahuasca analogs**, and *Piper methysticum*). The effects of cannabis are generally contrary to those of tobacco (*Nicotiana tabacum*). Nicotine suppresses the effects of THC, while THC potentiates the effects of nicotine (cf. **smoking blends**).

When larger quantities of hashish are eaten or drunk, visionary states, enlivened imagination, hallucinations, and even near-death experiences can be the result (Baudelaire 1971; Benjamin 1972; Cohen 1966; Haining 1975; Hofmann 1996; Kimmins 1977; Ludlow 1981; Robinson 1930; Tart 1971). Overdoses can lead to circulatory problems, anxiety, and vomiting. In Nepal, strongly brewed tea (cf. *Camellia sinensis*) is recommended for overdoses. In the European drug scene, a high dose of vitamin C is a recommended first-aid step. Dangerous symptoms, not to mention deaths from *Cannabis* overdoses, are unknown (Grinspoon and Bakalar 1994; Hess 1996; Hollister 1986; Mikuriya 1973; Schmidt 1992).

The effects of *Cannabis* products are primarily the result of the principal constituent, **THC**. THC has euphoric, stimulant, muscle-relaxing, anti-epileptic, antiemetic, appetite-stimulating, bron-chodilating, hypotensive, antidepressant, and analgesic effects. Cannabidiol (CBD) has no psychoactive effects but does have sedative and analgesic effects. Although cannabinol (CBN) is mildly psychoactive, it serves primarily to lower intraocular pressure and as an antiepileptic. Cannabigerol (CBG) is not psychoactive but does have sedative and antibiotic effects and also lowers intraocular pressure. Cannabichromene (CBC) has sedative effects and promotes the analgesic effects of THC (Grotenhermen and Karus 1995, 7). The lignans contained in the seeds suppress allergies.

The official, state-sanctioned and -supported psychiatry is dominated by the strangest notions and preconceptions about the long-term effects of frequent or chronic *Cannabis* use; for example, it is hypothesized that hemp is a "gateway drug" and that it contributes to a so-called amotivational syndrome (Täschner 1981). These "psychiatric symptoms" are pure invention and have no empirical basis (cf. Hess 1996). A politically independent, sociological study of the long-term effects of chronic hemp use yielded an interesting picture: "The chances that a person will think and work creatively and productively while under the influence of hemp increase with increasing experience with hemp" (Arbeitsgruppe Hanf and Fuss 1994, 103). Many studies of long-term use have demonstrated that *Cannabis* products are among the most harmless psychoactive agents of pleasure that humans have thus far discovered (cf. Blätter 1992; Grinspoon 1971; Hess 1996; Michka and Verlomme 1993; Schneider 1995).

Recent discussions have focused on the influ-ence of *Cannabis* upon driving behavior. Lawma-kers have based their legislation on the erroneous assumption that the effects of hemp are more dangerous than those of **alcohol**—even though a number of studies have shown that drivers under the influence of hashish drive considerably slower and with greater care than either sober or drunk drivers (Karrer 1995; Robbe 1994, 1996).

Commercial Forms and Regulations

Hemp products containing THC were banned in most countries as a result of the Single Convention

"The great disappointment of that which we know is what YOU encounter, terrible Bhairava, the heads of hope hang bloody around YOUR loins. I greet YOU, YOU who first placed into the earth the seeds of ganja, from which my knowledge of YOU grows. The lord of the burning pyres, may my smoke be as limitless as YOU, Uma [=Unmatta Bhairava]. Lowering my hemp-red eyes toward within, I experience YOU while inebriated, and I have left the world behind me. Bom Shankar! In YOUR honor do I raise my dschillum [= chillum] to my brow, so that I may merge with YOU. Om nama Shiva!"

NEPALESE HYMN TO SHIVA
(FIFTEENTH CENTURY)
IN VON HANF IST DIE REDE
[SPEAKING OF HEMP]
(BEHR 1995, 44)

THC

---

75 The homeopathic essence (Tct. Cannabis, homoeopath.) was once even called Aphrodisiacum (Arends 1935, 15*).

"*Ong Hrîn*. Ambrosia which is derived from Ambrosia, you who causes Ambrosia to rain down, make Ambrosia for me again and again. Bring Kâlikâ under my control. Give me *siddhi*, the miraculous abilities [= hemp]; *svâhâ*."

Mantra for consecrating *vijaya* (= hemp)
in *Tantra of the Great Liberation* (Mahanirvana Tantra)
(Avalon 1972, 73)

on Narcotic Drugs (1961) and are therefore not allowed to be marketed. There are only a few exceptions: "When signing the Single Convention, Bangladesh, India, and Pakistan retained the right to permit the non-medical use of opium and cannabis" (Haag 1995, 174).

In Germany, the *Betäubungsmittelgesetz* (Law on Narcotic Drugs) does not allow *Cannabis* to be used as medicine (Körner 1994, 56).[76] Thus prohibition applies even to hemp preparations devoid of active constituents: "Homoeopathic drugs and preparations are subject to the stipulations of the *Betäubungsmittelgesetz* and thus are not open to trafficking" (Schmidt 1992, 653).

Only the seeds are explicitly allowed to be sold without restriction (Körner 1994, 38, 56*). Many countries now allow the cultivation of fiber hemp (see *Cannabis sativa*) or varieties with very low THC contents for industrial use.

However, many sorts of hashish from around the world, numerous sorts of marijuana (especially the potent hybrids from Holland, Acapulco gold, Thai sticks, etc.; cf. *Cannabis* x **and hybrids**), and, more rarely, hash oil are available on the black market. Holland has its "coffee shops" (cf. *Coffea arabica*) and bars where the police tolerate the sale of small quantities of hemp preparations (cf. Haag 1995). The legal situation for hemp consumers varies greatly from one country to another, and even within countries. While much of Europe now regards hemp consumption as a minor offense (Bührer n.d.), some Southeast Asian countries (Malaysia, Singapore, the Philippines) may impose the death penalty on convicted hemp consumers.

Literature

See also the entries for *Cannabis ruderalis, Cannabis sativa*, and **THC**, as well as the Italian bibliography (SISSC 1994*).

Abel, Ernest L. 1980. *Marihuana: The first twelve thousand years*. New York: Plenum Press.

ADH [abbreviation]. 1997. Die Wende in Amerika? *Hanfblatt* 4 (26): 24–26.

Andrews, George, and Simon Vinkenoog, eds. 1968. *The book of grass: An anthology of Indian hemp*. New York: Grove Press.

Aldrich, Michael R. 1977. Tantric cannabis use in India. *Journal of Psychedelic Drugs* 9 (3): 227–33.

Aldrich, Michael R., ed. 1988. Marijuana—an update. *Journal of Psychoactive Drugs* 20 (1): 1–138.

Amendt, Günter. 1974. *Haschisch und Sexualität*. Stuttgart: Enke.

Anderson, Loran C. 1980. Leaf variation among *Cannabis* species from a controlled garden. *Botanical Museum Leaflets* 28 (1): 61–69.

Anonymous. 1994. *Marihuana für DOS—Was Sie schon immer über Hanf fragen wollten, aber nie zu wissen wagten!* Mannheim: TopWare PD-Service GmbH (TopWare 539).

Arbeitsgruppe Hanf und Fuss, ed. 1994. *Unser gutes Kraut: Das Porträt der Hanfkultur*. Solothurn: Nachtschatten Verlag; Löhrbach: Werner Pieper's MedienXperimente.

Avalon, Arthur [= Sir John Woodroffe]. 1972. *Tantra of the great liberation (Mahanirvana Tantra.)* New York: Dover.

Baudelaire, Charles. 1971. *Artificial paradise; on hashish and wine and means of expanding individuality*. New York: Herder and Herder.

Barber, Theodore X. 1970. *LSD, marihuana, yoga and hypnosis*. Chicago: Aldine.

Baumann, Peter. 1989. Hanf heute—Wert und Unwert. *Schweizerische Ärztezeitung* 70 (4): 134–40.

Bedi, Rajesh. 1991. *Sadhus: The holy men of India*. New Delhi: Brijbasi Private Limited.

Behr, Hans-Georg. 1995. *Von Hanf ist die Rede: Kultur und Politik einer Pflanze*. Frankfurt/M.: Zweitausendeins. (Orig. pub. Basel: Sphinx Verlag, 1982.)

Benjamin, Walter. 1972. *Über Haschisch*. Frankfurt/M.: Suhrkamp.

Bennett, Chris, Lynn Osburn, and Judy Osburn. 1995. *Green gold—the tree of life: Marijuana in magic and religion*. Frazier Park, Calif.: Access Unlimited.

Bharati, Agehananda. 1977. *Die Tantra-Tradition*. Freiburg i. Br.: Aurum.

Blätter, Andrea. 1990. *Kulturelle Ausprägungen und die Funktionen des Drogengebrauchs*. Hamburg: Wayasbah.

———. 1992. Das Vergnügen, die Sucht und das Bewußtsein—Einstellungen zum Cannabiskonsum. In *Yearbook for Ethnomedicine and the Study of Consciousness*, 1992 (1):117–32. Berlin: VWB.

———. 1993. Der erlernte Rausch—Die Funktionen des Cannabiskonsums auf Jamaika und in Deutschland. In *Yearbook for Ethnomedicine and the Study of Consciousness*, 1993 (2):119–45. Berlin: VWB.

Bowles, Paul, and Mohammed Mrabet. 1992. *El Limón*. Munich: Goldmann.

Boyle, T. Coraghessan. 1984. *Budding prospects: A pastoral*. New York: Viking.

Brenneisen, Rudolf. 1986. Hanf-Dampf in allen Gassen. *Uni-Press*, no. 51:7–9.

Bührer, Tony. n.d. *Haschisch Studie: Zur Klassifizierung von Cannabis (Konsum, Anbau, Kleinhandel) als Bagatelldelikt*. Der Grüne Zweig 125. Löhrbach: Werner Pieper's MedienXperimente.

Calm, Sven F. 1995. Music like Gunjah. *HanfBlatt*, no. 7:25–26.

Cherniak, Laurence. 1995. *Das große Haschisch-Buch. Teil 1: Marokko, Libanon, Afghanistan und der Himalaya*. Markt Erlbach: Raymond Martin Verlag.

---

76 The *Kommentar zum* [Commentary on the] *Betäubungsmittelgesetz* (BtMG 1994) states: "The use of Cannabis plants to manufacture Cannabis cigarettes, to manufacture medicaments and Cannabis tinctures (cough agents, sleeping agents, asthma and migraine agents) is forbidden."

Chopra, I. C., and R. N. Chopra. 1957. Use of cannabis drugs in India. *Bulletin on Narcotics* 9:4–29.

Clarke, Robert C. 1997. *Hanf—Botanik, Anbau, Vermehrung und Züchtung.*, Aarau: AT Verlag. (Orig. pub. Berkeley, Calif.: Ronin Publ., 1981 [under the title *Marijuana Botany*]).

Clarke, Robert C., and David W. Pate. 1994. Medical marijuana. *Journal of the International Hemp Association* 1 (1): 9–12.

Cohen, Sidney. 1966. *The beyond within.* New York: Atheneum.

———. 1982. Cannabis and sex: Multifaceted paradoxes. *Journal of Psychoactive Drugs* 14 (1–2): 55–58.

Corral, Valerie. 1994. A patient's story: Medical marijuana. *MAPS Bulletin* 4 (4): 26–29.

Cosack, Ralph, and Roberto Wenzel, eds. 1995. *Das Hanf-Tage-Buch; Neue Beiträge zur Diskussion über Hanf, Cannabis, Marihuana.* Hamburg: Wendepunkt Verlag.

De Leeuw, Hendrik. 1939. *Flower of joy.* New York: Lee Furman.

Dietrich, Eberhard. 1987. *Das Rebetiko: Eine Studie zur städtischen Musik Griechenlands.* 2 parts. Hamburg: Karl Dieter Wagner.

Doblin, Rick. 1994. A comprehensive clinical plan for the investigation of marijuana's medical use in the treatment of the HIV-related wasting syndrome. *MAPS Bulletin* 5 (1): 16–18.

Drake, William Daniel, Jr. 1971. *The connoisseur's handbook of marijuana.* San Francisco: Straight Arrow.

Edes, R. T. 1893. *Cannabis indica. Boston Medical and Surgical Journal* 129 (11): 273.

Emboden, William A. 1974a. Cannabis—a polytypic genus. *Economic Botany* 28:304–10.

———. 1974b. Species concepts and plant nomenclature. *California Attorneys for Criminal Justice Forum*, no. 5 (Aug./Sept.): 2–4.

———. 1981a. Cannabis in Ostasien—Ursprung, Wanderung und Gebrauch. In *Rausch und Realität*, ed. G. Völger, 1:324–29. Cologne: Rautenstrauch-Joest-Museum für Völkerkunde.

———. 1981b. The genus *Cannabis* and the correct use of taxonomic categories. *Journal of Psychoactive Drugs* 13 (1): 15–21.

———. 1996. Cannabis: The generation and proliferation of mythologies placed before U.S. courts. In *Yearbook for Ethnomedicine and the Study of Consciousness*, 1995 (4): 143–52. Berlin: VWB.

Epp, Rainer. 1984. The king's music: Über die Musik der Rastafaris. In *Rastafari-Kunst aus Jamaika*, ed. Wolfgang Bender, 49–56. Bremen: edition CON.

Fachner, Jörg, E. David, and M. Pfotenhauer. 1995. EEG-Brainmapping in veränderten Bewußtseinszuständen unter Cannabiseinwirkung beim Hören ausgewählter Musikstücke—ein Fallbeispiel. *Curare* 18 (2): 331–58.

Fisher, James. 1975. Cannabis in Nepal: An overview. In *Cannabis and culture*, ed. V. Rubin, 247–55. The Hague: Mouton.

Galland, Jean-Pierre, ed. 1993. *Première journée internationale du cannabis.* Paris: Éditions du Lézard.

Galland, Jean-Pierre. 1994. *Fumée clandestine il était une fois le cannabis.* Paris: Éditions du Lézard.

Gebre-Salassie, Girma. 1989. *Babylon muß fallen: Die Rasta-Bewegung in Jamaika.* N.p.: Raymond Martin Verlag.

Gelpke, Rudolf. 1967. Der Geheimbund von Alamut—Legende und Wirklichkeit. *Antaios* 8:269–93.

Giger, Andreas. 1995. Bewußtseins-Design mit Cannabis: Das Portrait der Hanfkultur. *Curare* 18 (2): 325–29.

Grinspoon, Lester. 1971. *Marihuana reconsidered.* Cambridge, Mass.: Harvard University Press. 1996. Cannabis als Arznei. In *Cannabis*, ed. Jürgen Neumeyer, 43–55. [Munich]: Packeispresse Verlag Hans Schickert.

Grinspoon, Lester, and James B. Bakalar. 1997. *Marihuana, the forbidden medicine.* Rev. ed. New Haven, Conn.: Yale University Press.

Gross, Robert Lewis. 1992. *The sadhus of India: A study of Hindu asceticism.* Jaipur and New Delhi: Rawat Publications.

Grotenhermen, Franjo, and Michael Karus. 1995. *Cannabis als Heilmittel: Eine Patientenbroschüre.* 2nd ed. Cologne: Nova-Institut.

Gruber, Ulrich. 1991. *Nepal.* Munich: Prestel.

Haag, Stefan. 1995. *Hanfkultur weltweit. Über die Hanfsituation in fast 100 Ländern rund um den Äquator.* Rev. ed. Löhrbach/Solothurn: Edition Rauschkunde.

Hager, Steven, ed. 1994. *High times—greatest hits: Twenty years of smoke in your face.* New York: St. Martin's Press.

Hai, Hainer. 1981. *Das definitive Deutsche Hanf-Handbuch.* Der Grüne Zweig 73. Löhrbach: Die Grüne Kraft.

Haining, Peter, ed. 1975. *The hashish club: An anthology of drug literature.* 2 vols. London: Peter Owen.

Hartsuiker, Dolf. 1993. *Sadhus: Holy men of India.* London: Thames and Hudson.

Hartwich, C. 1911. *Haschisch Anno 1911.* Repr. Löhrbach: Werner Pieper's MedienXperimente (Edition Rauschkunde), 1997. (The reprint is of a chapter from *Die menschlichen Genußmittel.*)

Hasan, Khwaja A. 1975. Social aspects of the use of cannabis in India. In *Cannabis and Culture*, ed. V. Rubin, 235–46. The Hague: Mouton.

Haus der Kulturen der Welt, ed. 1992. *Rastafari-Kunst aus Jamaika.* Berlin: CON Verlag.

MAURICE MAGRE

LA NUIT DE HASCHICH ET D'OPIUM

"I found myself in a wondrous inner world. I existed alternately in different places and in different conditions. Once I steered my gondola through a moonlit Venetian canal, and then mountain upon mountain rose before my eyes, and the magnificence of the rising sun bathed the icy peaks in purple light. And then I spread my feathered leaves out like a giant fern in the primordial silence of some untouched tropical jungle and swayed and nodded softly in the scented breeze above a riverbed, upon whose waves equally thick clouds of music and perfume arose. My soul transformed itself into a plant being and was strangely and unimaginably enraptured."

FITZ HUGH LUDLOW
THE HASHEESH EATER
(1857)

Herer, Jack, and Mathias Bröckers. 1993. *Die Wieder-entdeckung der Nutzpflanze Hanf Cannabis Marihuana*. Frankfurt/M.: Zweitausendeins.

———. 1996. *Die Wiederentdeckung der Nutzpflanze Hanf Cannabis Marihuana*. Abridged and revised edition. Munich: Heyne.

Hesch, R., A. Meyer, F. Beckmann, and K. Hesch. 1996. *Hanf: Perspektiven für eine ökologische Zukunft: Eine realistische Betrachtung*. Lemgo: Taoasis Verlag.

Hess, Peter. 1996. Medizinische und psychiatrische Aspekte von Cannabis. *Jahrbuch des Europäischen Collegiums für Bewußtseinsstudien* (1995): 157–77.

Hofmann, Albert. 1996. Rudolf Gelpke und der Hanfrausch. *Jahrbuch des Europaischen Collegiums für Bewußtseinsstudien* (1995): 103–12.

Hollister, Leo E. 1986. Health aspects of cannabis. *Pharmacological Review* 38 (1): 1–20.

Hoye, David. 1974. *Hasheesh: The herb dangerous*. San Francisco: Level Press.

Iversen, Leslie L. 1993. Medical uses of marijuana? *Nature* 365:12–13.

Jaque, Axel, et al. 1996. *Hanf CD-ROM*. Gelsenkirchen: Media Factory.

Karrer, Barbara. 1995. *Cannabis im Straßenverkehr*. Aachen: Verlag Shaker.

Kimmens, Andrew C., ed.. 1977. *Tales of hashish: A literary look at the hashish experience*. New York: William Morrow.

Kitzinger, Sheila. 1971. The Rastafarian brethren of Jamaica. In *Peoples and cultures of the Caribbean*, ed. Michael M. Horowitz, 580–88. Garden City, N.Y.: The Natural History Press.

Knecht, Sigrid. 1971. Rauchen und Räuchern in Nepal. *Ethnomedizin* 1 (2): 209–22.

Knoll-Greiling, Ursula. 1950. Die sozial-psychologische Funktion des Schamanen. In *Beitrage zur Gesellungs- und Völkerwissenschaft (Festschrift Richard Thurnwald)*, 102–24. Berlin: Gebr. Mann.

La Valle, Suomi. 1984. *Hashish*. London: Quartet Book.

Lewis, Barbara. 1970. *The sexual power of marijuana*. New York: Wyden.

Li, Hui-Lin. 1974a. An archaeological and historical account of *Cannabis* in China. *Economic Botany* 28:437–48.

———. 1974b. The origin and use of *Cannabis* in eastern Asia: Linguistic-cultural implications. *Economic Botany* 28:293–301.

Ludlow, Fitz Hugh. 1857. *The hasheesh eater: Being passages from the life of a pythagorean . . .* New York: Harper & Brothers.

Martius, Georg. 1855. *Pharmakologisch-medicinische Studien über den Hanf*. Erlangen: Junge und Sohn, Repr. Berlin: VWB, 1996.

Mattison, J. B. 1891. *Cannabis indica* as an anodyne and hypnotic. *The St. Louis Medical and Surgical Journal* 56 (Nov.): 265–71.

Mechoulam, Raphael. 1970. Marijuana chemistry. *Science* 168 (3936): 1159–66.

Meck, Bruno. 1981. *Die Assassinen: Die Mördersekte der Haschisch Esser*. Düsseldorf and Vienna: Econ.

Merlin, Mark D. 1972. *Man and marijuana*. Rutherford, N.J.: Fairleigh Dickinson University Press.

Meyrink, Gustav. 1984. Haschisch und Hellsehen. In *Das Haus zur letzten Latern*, 2:28–35. Berlin: Moewig.

Mezzrow, Mezz. 1995. *Die Tüte und die Tröte—Kiffen und Jazz: Really the Blues*. Löhrbach: Werner Pieper's MedienXperimente (Edition Rauschkunde).

Michka and Hugo Verlomme. 1993. *Le Cannabis est-il une drogue?* Geneva: Georg Éditeur.

Mikuriya, Tod H., ed. 1973. *Marijuana: Medical papers 1839–1972*. Oakland, Calif.: Medi-Comp Press. (Contains all important medical and pharmacological articles in the designated time period.)

Moebius. 1983. *Reisen der Erinnerung*. Cologne: Taschen.

Moreau de Tours, J. J. 1973. *Hashish and mental illness*. New York: Raven Press.

Morningstar, Patricia J. 1985. *Thandai* and *chilam*: Traditional Hindu beliefs about the proper uses of *Cannabis. Journal of Psychoactive Drugs* 17 (3): 141–65.

Moser-Schmitt, Erika. 1981. Sozioritueller Gebrauch von Cannabis in Indien. In *Rausch und Realität*, ed. G. Völger, 1:542–45. Cologne: Rautenstrauch-Joest-Museum für Völkerkunde.

Mrabet, Mohammed. 1995. *M'hashish: Kiff-Stories aus Marokko*. Der Grüne Zweig 49. Recorded by Paul Bowles, with an afterword by Werner Pieper. Löhrbach: Werner Pieper's MedienXperimente.

Müller-Ebeling, Claudia. 1992a. Die frühe französische Haschisch- und Opiumforschung und ihr Einfluß auf die Kunst des 19. Jahrhunderts. *Jahrbuch des Europäischen Collegiums für Bewußtseinsstudien* (1992): 9–19. Berlin: VWB.

———. 1992b. Visionäre und psychedelische Malerei. In *Das Tor zu inneren Räumen*, ed. C. Rätsch, 183–96. Südergellersen: Verlag Bruno Martin.

———. 1994. Kunst im Rausch. *Esotera* 4/94:90–95.

Nahas, Gabriel G. 1982. Hashish in Islam 9th to 18th century. *Bulletin of the New York Academy of Medicine* 58 (9): 814–31.

Neskovic, Wolfgang. 1995. Das Recht auf Rausch—Vom Elend der Drogenpolitik. In *Das Hanf-Tage-*

Buch, ed. Ralph Cosack and Roberto Wenzel, 141–64. Hamburg: Wendepunkt Verlag.

Neumeyer, Jürgen, ed. 1996. *Cannabis.* [Munich]: Packeispresse Verlag Hans Schickert.

Novak, William. 1980. *High culture: Marijuana in the lives of Americans.* New York: Alfred A. Knopf.

O'Shaughnessy, W. B. 1839. On the preparation of the Indian hemp or gunja. Reprint in *Marijuana: Medical Papers 1839–1972*, ed. T. H. Mikuriya, 3–30. Oakland, Calif.: Medi-Comp Press.

Rätsch, Christian. 1992. *Hanf als Heilmittel. Eine ethnomedizinische Bestandsaufnahme.* Löhrbach: Werner Pieper's MedienXperimente; Solothurn: Nachtschatten Verlag.

———. 1994. Der Nektar der Heilung. *Dao* 4/94:44–46.

———. 1995a. Biorohstoff Hanf 1995: Internationales Technisch-wissenschaftliches Symposium und Produkt- und Technologieschau 2.–5. Marz, Frankfurt a.M./Messe. *Curare* 18 (1): 231–33.

———. 1995b. Get high beyond style! Hanf, Musik und Kultur. In *Hanfkultur weltweit. Über die Hanfsituation in fast 100 Ländern rund um den Äquator*, by Stefan Haag, 179–89. Rev. ed. Löhrbach/Solothurn: Edition Rauschkunde.

———. 1996a. Die Hanfkultur—Eine kulturanthropologische Betrachtung. In *Jahrbuch des Europäischen Collegiums für Bewußtseinsstudien* (1995): 113–46.

———. 1996b. Die Pflanze der Götter. *Esotera* 6/96:52–57.

———. 1996c. Hanf als Heilmittel: Ethnomedizinische Befunde. In *Cannabis*, ed. Jürgen Neumeyer, 72–87. [Munich]: Packeispresse Verlag Hans Schickert.

Rathbun, Mary, and Dennis Peron. 1993. *Brownie Mary's marijuana cookbook and Dennis Peron's recipe for social change.* San Francisco: Trail of Smoke Publishing Co.

Rippchen, Ronald. 1995. *Die Hanfküche: Gesund, traditionell, exotisch, psychoaktiv.* Löhrbach/Solothurn: Edition Rauschkunde.

Robbe, H. W. J. 1994. *Influence of marijuana on driving.* Maastricht, Netherlands: Institute for Human Psychopharmacology, University of Limburg.

———. 1996. Influence of marijuana on driving. *Jahrbuch des Europaischen Collegiums für Bewußtseinsstudien* (1995): 179–89.

Robinson, Rowan. 1996. *The great book of hemp.* Rochester, Vt.: Park Street Press.

Robinson, Victor. 1930. *An essay on hasheesh.* New York: Dingwall-Rock.

Roffman, Roger A. 1982. *Marijuana as medicine.* Seattle: Madrona Publishers. (Foreword by Sidney Cohen.)

Rosenthal, Ed. 1996. *Marijuana beer.* Oakland, Calif.: Quick American Archives.

Rosenthal, Ed, ed. 1994. *Hemp today.* Oakland, Calif.: Quick American Archives.

Rubin, Vera, ed. 1975. *Cannabis and culture.* The Hague: Mouton.

Rubin, Vera, and Lambros Comitas. 1976. *Ganja in Jamaica: The effects of marijuana use.* Garden City, N.Y.: Anchor Press/Doubleday.

Sagunski, Horst, Eva-Susanne Lichtner, and Corinna Hembd. 1996. *Hanf: Das Praxisbuch.* Munich: Ludwig Verlag.

Sauer, J., and L. Kaplan. 1969. *Canavalia* beans in American prehistory. *American Antiquity* 34 (4): 417–24.

Scharfetter, Christian. 1992. Jacques-Joseph Moreau de Tours (1804–1884)—Haschisch-induzierte Phänomene als Psychosenmodell. *Jahrbuch des Europäischen Collegiums für Bewußtseinsstudien* (1992): 1–8. Berlin: VWB.

Schmidt, Stephan. 1992. Cannabis. In *Hagers Handbuch der pharmazeutischen Praxis*, 5th ed., 4:640–55. Berlin: Springer.

Schneider, Wolfgang. 1995. *Risiko Cannabis? Bedingungen und Auswirkungen eines kontrollierten, sozial-integrierten Gebrauchs von Haschisch und Marihuana.* Berlin: VWB. (Written with the assistance of Wolfgang Haves.)

Schultes, Richard Evans. 1973. Man and marijuana. *Natural History* 82 (7; Aug./Sept.): 58–64.

Schultes, Richard E., William M. Klein, Timothy Plowman, and Tom E. Lockwood. 1974. Cannabis: An example of taxonomic neglect. *Botanical Museum Leaflets* 23 (9): 337–67.

Sebode, Christina, and Rolf Pfeiffer. 1988. Schamanismus. *Salix* 1.87:7–33.

Shapiro, Harry. 1988. *Waiting for the man: The story of drugs and popular music.* London and New York: Quartet Books.

Sharma, G. K. 1972. Cannabis folklore in the Himalayas. *Botanical Museum Leaflets* 25 (7): 203–15.

———. 1977. Ethnobotany and its significance for cannabis studies in the Himalayas. *Journal of Psychedelic Drugs* 9 (4): 337–39.

Siegel, Ronald K. 1976. Herbal intoxication: Psychoactive effects from herbal cigarettes, tea and capsules. *Journal of the American Medical Association* 236 (5): 473–76.

Simmons, J. L., ed. 1967. *Marihuana: Myths and realities.* North Hollywood, Calif.: Brandan House.

Small, Ernest. 1975. The case of the curious "*Cannabis.*" *Economic Botany* 29:254.

———. 1978. *The species problem in cannabis: science and semantics.* 2 vols. Toronto: Corpus.

"Core, throw hemp into the wine! Let us drink of the juice of rapture!"

Ursula Haas, libretto to the opera by Rolf Liebermann *Freispruch für Medea* [Acquittal for Medea] (1994/95)

Dried parsley leaves are sometimes smoked as a hemp substitute. Whether this produces psychoactive effects is an open question. (Woodcut from Lonicerus, *Kreuterbuch*, 1679)

Small, Ernest, H. D. Beckstead, and Allan Chan. 1975. The evolution of cannabinoid phenotypes in *Cannabis. Economic Botany* 29:219–32.

Small, Ernest, and A. Cronquist. 1976. A practical and natural taxonomy for *Cannabis. Taxon* 25 (4): 405–35.

Solvyns, Baltazard. 1811. *Les hindous.* Paris: Mame Frères.

Stearn, William T. 1974. Typification of *Cannabis sativa* L. *Botanical Museum Leaflets* 23 (9): 325–36.

Storl, Wolf-Dieter. 1988. *Feuer und Asche—Dunkel und Licht: Shiva—Urbild des Menschen.* Freiburg i. Br.: Bauer.

Täschner, Karl-Ludwig. 1981. *Haschisch: Traum und Wirklichkeit.* Wiesbaden: Akademische Verlagsanstalt.

Tart, Charles. 1971. *On being stoned: A psychological study of marijuana intoxication.* Palo Alto, Calif.: Science and Behavior Books.

Thelwell, Michael. 1980. *The harder they come.* New York: Grove Press.

Touw, Mia. 1981. The religious and medicinal uses of *Cannabis* in China, India and Tibet. *Journal of Psychoactive Drugs* 13 (1): 23–34.

Vries, Herman de. 1993. Ein versandkatalog als kulturdokument: udopea. *Integration* 4:66–67.

Waskow, Frank. 1995. *Hanf und Co.—Die Renaissance der heimischen Faserpflanzen.* Göttingen: Verlag Die Werkstatt/AOL-Verlag. (Published by the Katalyse-Institut.)

Welte, Frank Maurice. 1990. *Der Gnawa-Kult: Trancespiele, Geisterbeschwörung und Besessenheit in Marokko.* Frankfurt/M.: Peter Lang.

West, M. E. 1991. Cannabis and night vision. *Nature* 351 (27.6.91): 703–4.

Williams-Garcia, Roberto. 1975. The ritual use of cannabis in Mexico. In *Cannabis and culture,* ed. Vera Rubin, 133–45. The Hague: Mouton.

Witt, Konrad. 1995. Die Bedeutung der Pflanze Cannabis in der Rastafari-Bewegung. Master's thesis, Tübingen.

Wolke, William, ed. 1995. *Cannabis-Handbuch.* Rev. ed. Markt Erlbach: Raymond Martin Verlag.

Zahl, Peter-Paul. 1995. *Teufelsdroge Cannabis.* Berlin: Verlag Das Neue Berlin.

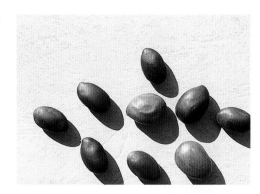

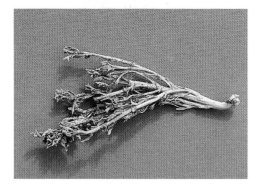

From top to bottom:

The seeds of *Canavalia maritima,* a plant used as a marijuana substitute.

In California, the dried herbage of wild dagga (*Leonotis leonurus*) is smoked as a marijuana substitute. Careful cultivation may be able to increase the psychoactivity of the stock plant.

The inhabitants of the Atacama Desert (northern Chile) smoke the dried, resinous herbage of *pupusa* or *chachalana,* a desert plant that has not yet been botanically identified, as a marijuana substitute.

Marijuana Substitutes

These are plant drugs that are smoked in place of *Cannabis* flowers in order to induce an identical or similar effect. (After Ott 1993* and Schultes and Hofmann 1995*; modified and expanded)

Botanical Name	Popular Name	Drug	Place/Culture
Alchornea floribunda M.-A.	niando	root	Africa
Anethum graveolens	dill	herbage	U.S.A.
Argemone mexicana	prickly poppy	leaves	Mexico
Artemisia mexicana	estafiate	herbage	Mexico
Calea zacatechichi	zacatechichi	herbage	Mexico, U.S.A.
Canavalia maritima (Aubl.) Thouars[77] [syn. *Canavalia obtusifolia*] (Leguminosae)	frijolillo	leaves	Mexico
Capsicum fructescens (cf. *Capsicum* spp.)	paprika	rotten fruits	U.S.A.
Catharanthus roseus	periwinkle	leaves	Florida
Cecropia mexicana Hemsl.[78] [syn. *Cecropia obtusifolia* Bert.]	chancarro	leaves	Mexico (Veracruz)
Cestrum laevigatum Schlecht. (cf. *Cestrum parqui*)	maconha	leaves	Brazil
Cymbopogon densiflorus	lemongrass	flower extract	Tanzania
Daucus carota	carrot	herbage	U.S.A.
Helichrysum spp.	strawflower	herbage	
Helichrysum foetidum (L.) Moench		herbage	Zulu/Africa
Helichrysum stenopterum DC.		herbage	Africa
Hieracium pilosella	håret høgeurt	herbage	Denmark
Hydrangea paniculata	hydrangea	leaves	U.S.A.
Hydrangea spp.	hydrangea	flowers, leaves	U.S.A.
Lactuca sativa L.	lettuce	leaves	U.S.A.
Lactuca serriola	wild lettuce	leaves	U.S.A.
Lactuca virosa	wild lettuce (thickened juice)	lactucarium	U.S.A.
Leonotis leonurus	wild dagga	herbage	Hottentots
Leonurus sibiricus	marijuanillo	herbage	Mexico (Chiapas)
Mimosa spp.[79]	dormilona	herbage	San Salvador
Musa x *sapientum*	banana	inner peel	cosmopolitan
Myristica fragrans	nutmeg	seed, aril	U.S.A., Europe
Nepeta cataria	catnip	herbage	cosmopolitan
Nepeta spp.	catnip	herbage	cosmopolitan
Petroselinum crispum	parsley	flowering herbage	U.S.A., Europe
Piper auritum	gold pepper	leaves	Belize
Sceletium tortuosum	kougoed	herbage, roots	South Africa
Sida acuta Burm. chichibe	malva amarilla,	herbage	Mexico, Belize
Sida rhombifolia L.	escobilla	herbage	Mexico[80]
Turnera diffusa	damiana	herbage	cosmopolitan
Zornia latifolia DC. (Leguminosae) [81]	maconha brava	desiccated leaves	Brazil
Zornia diphylla (L.) Pers. yerba de la víbora[82]	maconha brava,	leaves	Brazil
Unidentified	*Pupusa/Chachalana*[83]	herbage	Atacama/Chile
	canna	herbage	South Africa

It is thought that the fruits of the bay bean (*Canavalia maritima*) had a ritual or magical significance in ancient Peru. This painting on a Moche vessel may be a depiction of these fruits. It is uncertain whether they actually produce psychoactive effects.

77 Various no more precisely identified species of the genus *Canavalia* have been found in archaeological contexts dating to the prehistoric period in Peru and northern Mexico (Dressler 1953, 126*). *Canavalia maritima* is native to Polynesia. It is highly questionable whether the plant truly does have psychoactive effects. The presence of L-betonicine has been demonstrated (Schultes and Hofmann 1995, 37*).

78 The primary ethnobotanical importance of the genus *Cecropia* in South America is as a source of ash for coca chewing (cf. *Erythroxylum coca*).

79 It is possible that the plant referred to is *Mimosa pudica*, which is known in Guatemala and Mexico by the same name (*dormilona*, "sleeping plant") (Ott 1993, 400*). In Mexico, the following plants are also known as *dormilona*: *Mimosa albida* Humb. et Bonpl., *Mimosa pigra* L., *Phyllanthus lathyroides* H.B.K., *Neptunia oleracea* Bour., and *Bellis perennis* L. Interestingly, *Psilocybe aztecorum* is known as *dormilón* (Martínez 1987, 317*).

80 This plant, which contains ephedrine, is also used in the Brazilian Candomblé cult as a "liturgical" (i.e., sacred) plant (Voeks 1989, 127*).

81 To date, no biodynamic or psychoactive compounds have been found in the tropical genus *Zornia* (Schultes 1981, 20*; Schultes and Hofmann 1995, 59*).

82 The Pima of northern Mexico drink a tea made from the leaves of the "snake plant" to treat shivering (Pennington 1973, 223*).

83 Possibly *Senecio eriophyton* Remy (cf. *Senecio* spp.), an aromatic Compositae known in the Atacama Desert as *chachacoma* and used as a medicinal incense (Hofmann et al. 1992, 83*).

Cannabis ruderalis Janischewsky

Weedy Hemp

Family
Cannabaceae (= Cannabinaceae) (Hemp Family)

Forms and Subspecies
None

Synonyms
Cannabis intersita Sojak
Cannabis sativa L. ssp. *spontanea* Serebr. ex Serebr. et Sizov
Cannabis sativa L. var. *ruderalis* (Janisch.)
Cannabis sativa L. var. *spontanea* Mansfield
Cannabis spontanea Mansfield

Folk Names
Anascha, konopli, mimea, momea, mumeea, penka, penscha, ruderalhanf, russischer hanf, weedy hemp, wilder hanf, wild hemp

History
Weedy hemp was already being used for shamanic and ritual purposes in central Asia in prehistoric times. The description that Herodotus (ca. 500–424 B.C.E.) provided us about the ancient Scythian[84] use of hemp in purification and burial rituals has been confirmed by archaeological findings in the Altai Mountains of Mongolia. The Scythians also smoked hemp for pleasure (Rocker 1995). This small, wild hemp is still used in Mongolia for shamanic and medicinal purposes. Recently, a Scythian shaman was found in an undisturbed and frozen grave in the Altai Mountains. Among the goods found with her were hashish and other hemp products (*Stern* no. 18, 1994, 194 ff.).

This hemp species was first described in 1924 by the Russian Janischewsky. Today, it is primarily used to breed low-growing varieties of hemp that contain THC (see *Cannabis* x **and hybrids**).

Distribution
Cannabis ruderalis now grows wild from the Caucasus Mountains to China. Its species name expresses its preference for so-called ruderal places, i.e., rocky locations, scree, and rubbish sites. *Cannabis ruderalis* originally occurred in the wild only in southeastern Russia (Emboden 1979, 172*). It is likely that the Scythians introduced it into Mongolia, where it then became wild.

Cultivation
See *Cannabis indica, Cannabis* x **and hybrids.**

Appearance
This hemp species grows to a height of only 30 to 60 cm. It has few if any branches and rather small leaves. The inflorescence is small and forms only on the end of the stalk. The seed coat has a fleshy base.

Psychoactive Material
— Female flower
— Seeds
— Resin

Preparation and Dosage
The female inflorescences are dried and smoked or inhaled as a kind of **incense**. The flowers are also suitable for use as a fumigant in sweat lodge rituals (cf. Bruchac 1993), both alone and in combination with **Artemisia absinthium**, **Artemisia mexicana**, or other *Artemisia* species.

A shamanic incense with psychoactive effects can be mixed using equal parts of hemp flowers, the tips of juniper branches (*Juniperus communis* L., **Juniperus recurva**, *Juniperus* spp.), thyme (*Thymus* spp.), and wild rosemary (**Ledum palustre**).

In Russia, sedative, aphrodisiac, and analgesic foods are prepared from hemp, saffron (**Crocus sativus**), nutmeg (**Myristica fragrans**), cardamom, **honey**, and other ingredients (cf. **Oriental joy pills**).

Ritual Use
The oldest known literary evidence of the use of hemp is from Herodotus. In a comprehensive chapter of his *History*, he describes the social structure, religion, mythology, and customs of the Scythians. Their burial or death ritual is particularly significant:

> When they have buried the dead, the relatives purify themselves as follows: they anoint and wash their heads; as to their bodies, they set up three sticks, leaning them against one another, and stretch, over these, woolen mats; and, having barricaded off this place as best they can, they make a pit in the center of the sticks and the mats and into it throw red-hot stones.
>
> Now, they have hemp growing in that country that is very like flax, except that it is thicker and taller. This plant grows both wild and under cultivation, and from it the Thracians make garments very like linen. Unless someone is very expert, he could not tell the garment made of linen from the hempen one. Someone who has never yet seen hemp would certainly judge the garment to be linen.

---

[84] In ancient times, the name Scythian was a catchall phrase for nomadic horse-riders who lived on the Black Sea, along the Danube, and in southern Russia and who spoke several Indo-Iranian languages or dialects. Many of the Scythian tribes developed extensive trade relationships with the Pontic Greeks. They were feared as brave and wild warriors and were consequently a respected people (Pavlinskaya 1989).

The Scythians take the seed of this hemp and, creeping under the mats, throw the seed onto the stones as they glow with heat. The seed so cast on the stone gives off smoke and a vapor; no Greek steam bath could be stronger. The Scythians in their delight at the steam bath howl loudly. This indeed serves them instead of a bath, as they never let water near their bodies at all. But their women pound to bits cypress and cedar and frankincense wood on a rough rock and mix water with it.[85] When they have made of the wood and the water a thick paste, they smear it all over their bodies and faces. A wonderful scent pervades them from this; a day later they take off the plaster, and they have become shining clean. (Herodotus 4.73–75; 1987, 307*)

It seems obvious that the hemp seeds were still attached to the inflorescences, for how else could a "smoke and a vapor" have resulted that would make the Scythians "howl loudly" in their delight? Herodotus is describing a cult activity in which the relatives of the deceased would accompany the soul of the dead to the next world while in a shamanic trance. The ritual was intended to benefit both the soul of the deceased and the souls of those remaining behind. Hemp loosened the barriers of death and enabled people to take part in the immortality of the soul; it allowed a collective way to overcome grief. Meuli (1935) has characterized this ritual use of weedy hemp as a "family shamanism" devoid of any pronounced specialization (Jettmar 1981, 310). Similar rituals are known to have occurred among other peoples (e.g., the Assyrians; cf. *Cannabis indica*) and tribes of antiquity (the Thracians and Massagets). The Massagets, a nomadic tribe from central Asia, camped together around fires into which certain "fruits" had been thrown. After inhaling the smoke, the participants sprang up elatedly (312).

Artifacts

The deeply frozen Scythian burial mounds of Pazyryk Kurgan (Altai Mountains, Mongolia) have yielded leather bags containing hemp seeds together with vessels for burning incense (cf. **incense**). The bags are 2,400 years old. The rather small seeds indicate that they came from wild-growing plants—most likely *Cannabis ruderalis* (Clarke 1996, 104). The Russian archaeologist S. I. Rudenko excavated a number of bronze incense vessels, over which a felt-covered frame still stood (Rudenko 1970). The report on the excavation noted:

In the southwest corner of the grave chamber of Parzyryk Kurgan II, a bundle of six staffs was found. Below this stood a rectangular bronze vessel on four legs, filled with broken stones. The lengths of the staffs was 122.5 cm, their diameter some 2 cm, at the lower end about 3 cm. A small strap that held the staffs together had been pulled through openings 2 cm below the upper end of each staff. Glued to all of the staffs was a narrow spiral strip of birch bast. North of this, in the western half of the chamber, a second bronze vessel was found, of the type of a Scythian kettle. It too was full of stones. Over this were spread six of the same type of staff, partially broken and thrown about when robbers had broken in. These, together with the incense vessel, were covered by a large leather wrap.

In addition to the stones already mentioned, both vessels were found to contain a large quantity of hemp seeds (*Cannabis sativa* L. of the variety *C. ruderalis* Janisch). Hemp seeds were also in one of the leather bottles described previously, which was attached to one of the poles of the hexapod that stood over the vessel in the form of a Scythian kettle. The stones in the incense vessels had been scorched and a portion of the hemp seeds blackened. In addition, the handles of the kettle that had been used as an incense vessel were wrapped around with birch bast. Clearly, the vessel had been so heated by the glowing stones that it could not be picked up with bare hands. . . . Consequently, we have here complete sets of those utensils that were necessary to carry out the purification ritual that Herodotus so precisely recorded in reference to the Pontic Scythians. Sets of utensils for inhaling hemp were present in all of the Pazyryk kurgans, without exception. Even though the vessels and the wraps had been stolen by plunderers, with the exception of Kurgan II, the staffs remained in all of the kurgans. It follows that the smoking of hemp was practiced not only during the purification rituals but also in daily life. . . . Here, both men and women smoked. (In Jettmar 1981, 311)

Medicinal Use

A Mongolian medicine known as *bagaschun*, said to be a kind of cure-all and apparently made from hemp, juniper (cf. *Juniperus recurva*), and bat guano, is known from the Altai region. This preparation is also known as *mumio* and is highly esteemed as a tonic in Russian folk medicine (Rätsch 1991).

Today, *Cannabis ruderalis* grows throughout the entire region once inhabited by the Scythians. It is still used in Russian and Mongolian folk medicine to treat depression. Recently, the Mongolian Academy of Sciences sponsored a project to document the shamanic, folk medical, and Lamaistic knowledge of medicinal plants. It

Scythian incense vessels for inhaling hemp smoke, found in prehistoric graves in the High Altai (Mongolia). (From Rudenko, *Frozen Tombs of Siberia*, 1970)

85 The "cedar wood" actually comes from a species of juniper (*Juniperus* spp.; cf. *Juniperus recurva*). The descendants of the Scythians in the Hindu Kush still inhale juniper smoke to induce a shamanic trance (Jettmar 1981, 312).

"A colleague of mine at work is a 'Russian-German' from Tadjikistan. As we were first smoking something together a few days ago during lunch, he told me the following: 'Where I come from, no one buys hashish or grass. It grows wild everywhere. To harvest, we only needed to ride through the pastures of hemp on our horses. After just a half hour, the legs of the horses were covered with resinous plant parts. All we needed to do was scrape it off.'"

DIETMAR. B.
IN *HANFKULTUR WELTWEIT* [HEMP CULTURE WORLDWIDE]
(HAAG 1995, 51)

was found that in the Mongolian tradition *Cannabis sativa* and *Cannabis ruderalis* are used for different medicinal purposes. *Cannabis sativa* is usually used as a source of oil, whereas *Cannabis ruderalis* is esteemed more for its psychoactive properties (Mr. Günther, Ulaanbaatar, pers. comm.). It is very likely that Mongolian shamans in the Altai use *Cannabis ruderalis* as well as juniper to induce shamanic trances (Jettmar 1981).

Constituents

This hemp species contains more or less the same cannabinoids as are found in *Cannabis indica* and *Cannabis sativa*. The amount of **THC**, however, is considerably lower. Only 40% or less of the cannabinoids that were measured could be identified as THC; in *Cannabis sativa*, the THC content is around 70% (Beutler and Der Marderosian 1978, 390).

Effects

See *Cannabis indica*.

Commercial Forms and Regulations

See *Cannabis indica*.

Literature

See also the entries for *Cannabis indica*, *Cannabis sativa*, and **THC**.

Benet, Sula. 1975. Early diffusion and folk uses of hemp. In *Cannabis and culture*, ed. V. Rubin, 39–49. The Hague: Mouton.

Beutler, John A., and Ara H. Der Marderosian. 1978. Chemotaxonomy of cannabis I. Crossbreeding between *Cannabis sativa* and *C. ruderalis,* with analysis of cannabinoid content. *Economic Botany* 32 (4): 387–94.

Bruchac, Joseph. 1993. *The Native American sweat lodge.* Freedom, Calif.: The Crossing Press.

Brunner, Theodore F. 1977. Marijuana in ancient Greece and Rome? The literary evidence. *Journal of Psychedelic Drugs* 9 (3): 221–25.

Clarke, Robert C. 1995. Scythian *Cannabis* verification project. *Journal of the International Hemp Association* 2 (2): 194.

Haag, Stefan. 1995. *Hanfkulture weltweit. Uber die Hanfsituation in fast 100 Ländern rund um den Äquator.* Der Grüne Zweig 73. Rev. ed. Lörbach: Die Grüne Kraft.

Janischewsky. 1924. *Cannabis ruderalis. Proceedings Saratov* 2 (2): 14–15.

Jettmar, Karl. 1981. Skythen und Haschisch. In *Rausch und Realität,* ed. G. Völger, 1:310–13. Cologne: Rautenstrauch-Joest-Museum für Völkerkunde.

Meuli, K. 1935. Scythia. *Hermes* 70/1. Berlin.

Pavlinskaya, Larisa. 1989. The Scythians and Sakians, eighth to third centuries B.C. In *Nomads of Eurasia*, ed. Vladimir Basilov, 19–39. Los Angeles: Natural History Museum; Seattle: University of Washington Press.

Rätsch, Christian. 1991. Neues aus der Dreckapotheke: Mumio. Unpublished manuscript, Hamburg.

Rocker, Tom. 1995. Hanfkonsum im Altertum: Die Skythen. *Hanfblatt* 2 (11): 19.

Rudenko, S. I. 1970. *Frozen tombs of Siberia: The Pazyryk burials of Iron Age horsemen.* Berkeley: University of California Press.

Cannabis sativa Linnaeus

Fiber Hemp

Family

Cannabaceae (= Cannabinaceae) (Hemp Family)

Forms and Subspecies

In the mid-nineteenth century, the renowned botanist Alphonse-Louis-Pierre Pyramus de Candolle (1806–1893) attempted to standardize the taxonomy of *Cannabis* and proposed the following varieties:

Cannabis sativa var. α *Kif* DC. (Moroccan hemp)
Cannabis sativa var. β *vulgaris* DC. (fiber hemp)
Cannabis sativa var. γ *pedemontana* DC. (wild hemp)
Cannabis sativa var. δ *chinensis* DC. (Chinese hemp, giant hemp) [= *C. chinensis* (Del.) A. DC., *C. gigantea* Del. ex Vilm. = *C. sativa* cv. Gigantea]

According to Clarke (1981, 159), this species can be subdivided into the following subspecies and varieties (although it is certainly not a good idea to propose a subspecies *indica* as well as a variety *indica*):

Cannabis sativa var. *sativa* (the common, cultivated fiber hemp)
Cannabis sativa var. *spontanea* (has smaller seeds, occurs wild)
Cannabis sativa ssp. *indica* (very rich in cannabinoids) [= ***Cannabis indica***]
Cannabis sativa var. *indica* (fruits smaller than 3.8 mm)
Cannabis sativa var. *kafiristanica* (short fruits)

It has also been suggested that the species can be divided into four phenotypes (chemotypes) (cf. Clarke 1981, 160). In my opinion, however, there is no justification for such a division, as any one population can exhibit considerable variation in cannabinoid content (Hemphill et al. 1978; Latta and Eaton 1975). Two chemotypes have been described for Africa (Boucher et al. 1977).

Synonyms

Cannabis americana Houghton
Cannabis chinensis Delile
Cannabis culta Mansfield
Cannabis erratica Sievers
Cannabis generalis Kraus
Cannabis gigantea Crevost
Cannabis intersita Sojak
Cannabis lupulus Scopoli
Cannabis macrosperma Stokes
Cannabis pedemontana Camp
Cannabis sativa monoica Holuby
Cannabis sativa ssp. *culta* Sereb. ex Sereb. et Sizov

Folk Names

Agra, al-haschisch, anascha, asa, atchi e erva, bang, bangi, banj, baretta, bästling, bengi, beyama, bhamgi, bhang, bhanga, bhangalu, bhangaw, bhangi, birra, bota (Spanish), bushman grass, cabeça de negro, canamo, cáñamo, canape (Italian), canep (Albanian), cangonha, canhamo, cannabis, cannabus, cannacoro, ceviche, cha de birra, chamba, chanvre, charas, chira, chrütli (Swiss German, "little herb"), daboa, dacha, dagga, da hola herb, dakka, damó (Tagalog, "grass"), darakte-bang, dendromalache, deutscher hanf, dhagga, diamba, dirijo, djamba, dumo, doña juanita, donna juanita, durban poison, el-keif (Lebanese), entorpecente ("sedative agent"), epangwe, erva, esra (Turkish, "the secret one"), faserhanf, fêmea, femmel, fimmel, fumo brabo, fumo d'angola, fumo de caboclo, füve (Hungarian), gallow grass, gañca, ganja, gemeiner hanf, gnaoui, gongo, gosale (Persian), gras, graspflanze, grass, green goddess, grifa, habibabli, hafion, hajfu (Turkish), hamp (Swedish), hampa (Danish), hanaf, hanf, hanif, hapis ciel (Serí, "green tobacco"), hapis-coil (Serí), happy smoke, haschisch, haschischpflanze, hashisch, hashîsh (Arabic), hasisi (Greek), hasjet, hemp, henep, hennup (Dutch), hierba santa ("sacred herb"), hierba verde ("green herb"), huntul k'uts (Lacandon, "a different tobacco"), indracense, injaga, kabak, kamanin (Japanese), kamonga, kamugo, kanab, kañcaru, kancha, kannabion, kannabis, kansa, kemp (Flemish), kenvir (Bulgarian), kif, knaster, konopie, konopli, kraut, lopito, lubange, ma, maconha, maconha di pernambuco, maconha negra, macusi (Huichol), makhlif, malak, mala vida ("bad life"), malva, mapouchari, mara-ran (Ka'apor, "false malaria"), maria-johanna, maria juana, maricas, mariguana, marihuana, marijuana,[86] marimba, mariquita, masho, masmach, mästel, mavron, mbange, mbanji, mbanzhe, mfanga, mmoana (Lesotho), moconha, morrao, mota (Mexican), mulatinha, muto kwane, myan rtsi spras, nasha, nederwiet, njemu, nsandu, ntsangu, nutzhanf, opio do pobre (Portuguese, "opium of the poor"), Panama red, panga, planta da felicidade (Portuguese, "plant of happiness"), penek, pot, potagua ya, pungo, rafi, rauschgiftpflanze, riamba, rosamaria, rosa maría, sadda, samenhanf, sangu, santa rosa (Mexican, "sacred rose"), shivamuli, siddhi, siyas (Turkish, "the black one"), ssruma, starker tobak, swazi, taima, tedrika, tiquira, trava (Croatian), tujtu (Cuicatleca), ugwayi abadala ("smoke of the ancestors"), uhtererê, uluwangula, umbaru, umburu, wacky weed, weed, wee-wee, whee, wiet, yama, yesil (Turkish,

Fiber hemp (*Cannabis sativa*), also known as *ta-ma*, is one of the oldest food, ritual, and medicinal plants of Chinese culture. (Illustration from the *Chih-wu-ming-shih-t'u k'ao*)

86 In Mexico, tree tobacco (*Nicotiana glauca*; see *Nicotiana* spp.) is also known by the name *marijuana* (V. Reko 1936, 62*).

The (ancient) Chinese character (*ma*) for *Cannabis sativa*.

Woodcut of a male hemp plant from the herbal of John Gerard (1633). At that time, the botanically male plant was thought to be the female.

"the green one"), zahret-el-assa, zerouali, zhara, ziele konopi

History

The oldest archaeological evidence for the cultural use of hemp points to it originally having been used in shamanic contexts (cf. ***Cannabis indica, Cannabis ruderalis***). Hemp seeds, which could be identified as those of *Cannabis sativa,* were recovered in the Neolithic linear band ceramic (*linearbandkeramik,* or LBK) layers of Eisenberg in Thuringia, Germany (Renfrew 1973, 163*; Willerding 1970, 358*). The layers were dated to around 5500 B.C.E. Hemp seeds have also been found in the excavations of other, somewhat more recent Neolithic layers, such as those in Thainigen (Switzerland), Voslau (Austria), and Frumusica (Romania) (Renfrew 1973, 163*). These finds date from a period of peaceful, horticultural, pre-Indo-European cultures who venerated the Great Goddess (Gimbutas 1989) and knew of shamanism (Probst 1991, 239). The linear band ceramics that lend their name to this Stone Age cultural epoch are decorated with graphics representing the archetypical motifs and patterns of hallucinatory or psychedelic themes (Stahl 1989).

In Bavaria, finds of clay pipe bowls with wooden stems, discovered during excavations of the barrows of Bad Abbach-Heidfeld, indicate that cannabis or its products were being smoked more than 3,500 years ago, possibly together with sleeping poppy or opium (***Papaver somniferum***) (Probst 1996, 174). There is also evidence from the early Germanic period:

> Hemp remnants from the prehistoric period of northern Europe were uncovered in 1896 when the German archaeologist Hermann Busse opened an urn grave in Wilmersdorf (Brandenburg). The vessel that was found dates from the fifth century B.C. and contained sand mixed with plant remains. The botanist Ludwig Wittmaack (1839–1929) was able to identify fruits and fragments of the seed coats of *Cannabis sativa* L. among these. (Reininger 1941, 2791)

Among the Germanic peoples, hemp was sacred to Freya, the goddess of love, and apparently was used as an inebriant in ritual and aphrodisiac contexts. Like Indian hemp (***Cannabis indica***), the German fiber hemp did and does have inebriating effects:

> But here as well, the fresh plant also possesses an extremely potent, unpleasant, often intoxicating scent, and it is known that dizziness, headache, and even a kind of drunkenness frequently result if a person spends too much time in a blooming hemp field. It has also been observed that roasting

hemp, as it is called, produces a similarly intoxicating scent. (Martius 1996, 31)

Fiber hemp was mentioned as a source of food several times in the ancient Chinese *Shih Ching,* the Book of Songs (ca. 1000–500 B.C.E.) (Keng 1974, 399 f.*). The Egyptians probably learned of hemp at about the same time.

The useful and medicinal hemp plant was very well and very widely known in ancient times. Theophrastus provided a botanically correct description of the plant under the name *dendromalache.* As many ancient authors (e.g., Varro, Columbarius, and Gellus) attest, hemp was known and esteemed in antiquity as a good source of fiber, and it was planted widely. Pliny wrote extensively about hemp, which he called *cannabis.*

Concerning the origin of the term *cannabis,* we know that there was a classical Greek expression *cannabeizein,* meaning "to inhale hemp smoke." Another word from that period is *methyskesthai,* "to become inebriated through drug use"; Herodotus used this word to describe the inebriation that the inhabitants of an island in the Araxes (Araks) produced using smoke (cf. ***Cannabis ruderalis*, trees with special fruits**). Hemp's ability to improve mood did not escape Democritus (460–371 B.C.E.), the "laughing philosopher," who called the plant *potamaugis.* He noted that when this plant is drunk in wine (cf. ***Vitis vinifera***) together with myrrh (*Commiphora molmol* Engl.), it produces delirium and visions. He was especially struck by the immoderate laughter that invariably followed such a drink. Galen (ca. 130–199 C.E.) wrote that in Italy it was customary to serve small cakes containing hemp for dessert. These would increase the desire to drink, although eating too many had stupefying effects (6:549 f.). It was considered a sign of good manners to offer guests hemp, for it was regarded as a "promoter of high spirits" (cf. **Oriental joy pills**).

It is likely that hemp spread from Arabia and Egypt into the rest of Africa at an early date. Numerous pipes and smoking devices have been found in archaeological contexts, some of which still contain remnants of **THC** (Van der Merve 1975). It appears that the introduced hemp, with its better effects, supplanted the use of indigenous smoking herbs (***Leonotis leonurus, Sceletium tortuosum***) (Du Toit 1981, 511).

As human culture spread around the world, hemp went along (see Rätsch 2001*). In many places, e.g., Morocco and Trinidad, hemp cultivation has come to have an irreplaceable economic importance for the indigenous peoples (Joseph 1973; Lieber 1974; Mikuriya 1967).

Distribution

Cannabis sativa is from either central Europe or central Asia. But as an anthropophilous species, it already had become widespread during the Neolithic period. Today, it is found almost everywhere in the world and has adapted to very different soil types and climate zones. It is unknown as a wild plant.

Cultivation
See *Cannais indica* and *Cannabis* x **and hybrids**.

Appearance
Fiber hemp can vary considerably in appearance. Like the other *Cannabis* species, it usually occurs in two sexes, but it can also be hermaphroditic in cultivation. It has few if any branches and has the largest leaves of the three species. The individual "fingers" of the leaves are long, lanceolate, and very slender (an important feature for recognizing the plant).

Cannabis sativa is sometimes confused with the anaphrodisiac monk's pepper (*Vitex agnus-castus* L.; Verbenaceae), the leaves of which are remarkably similar to those of *Cannabis sativa*. An illustration from the Viennese Dioscorides that has appeared in many publications as the "oldest representation of the hemp plant" (e.g., Fankhauser 1996) is in fact *Vitex agnus-castus*.

Psychoactive Material
— Female flowers
— Resin glands
— Resin
— Red hemp oil (hash oil, cannabis resinoid)
— Seeds (cannabis sativae fructus, fructus cannabis, semen cannabis, hemp fruits, hemp grains, hemp seeds)
— Leaves

Preparation and Dosage
It is primarily the dried female inflorescences and the resin or resin-rich preparations that are used for psychoactive purposes; these are smoked or ingested (cf. *Cannabis indica*).

The inflorescences are usually referred to as marijuana (= marihuana) or "grass." Well-known sorts of Colombian marijuana from *Cannabis sativa* include Santa Marta Gold (= *Muños de oro*; yellow-brown color), Blue Sky Blonde (yellowish color), Red Dot (= *Punto rojo*; yellow color with reddish specks), and Mangoviche Grass. Panama Red, from Panama, and Maui Waui, from Hawaii, are legendary.

Cannabis sativa is just as well suited as **Cannabis indica** and **Cannabis ruderalis** for producing hashish. In Mexico, hashish, the pressed resin, is known as *marijuana pura* ("pure marijuana") and is obtained in the following manner: "All you need is to walk wearing the heavy leather pants typical of the rancheros through the field with the diabolical flora and a knife, which is used to scrape off the resin that has adhered to the

pants, and then roll it into little balls" (V. Reko 1936, 65*). Numerous kinds of hashish are obtained from *Cannabis sativa*: Green Turkish (sometimes adulterated with henna, *Lawsonia inermis* L. [syn. *Lawsonia alba* Lam.]), Yellow (from Syria), Yellow Lebanese, Red Lebanese, Zero-Zero (pure resin gland powder, pressed), Black Moroccan (hand-rubbed resin), Green Moroccan (pressed resin glands and flowers),[87] and Pollen (unpressed resin glands; has nothing to do with flower pollen).

It is also possible to make hashish at home. The female inflorescences should be chopped into large pieces and then rubbed on gauze over a bowl. A fine dust consisting of the valuable resin glands and fine resinous leaf tips will collect in the bowl. This powder should be dried and pressed, and then the hashish is ready. Using this method, between 30 and 50 g of hashish can be obtained from 1 kilo of plant material (plant tips with inflorescences) (Haller 1996).

Red hemp oil (= hashish oil) is produced as a resinoid by extracting the female inflorescences and then evaporating the solvent (ethanol; cf. **alcohol**). The essential oil of hemp, which smells slightly of fresh hemp flowers, is obtained by steam distillation.

Cannabis sativa is often used as an additive in alcoholic beverages. In former times, it was used in place of **Humulus lupulus** as an additive in **beer** (cf. *Cannabis indica*). Since 1996, a hemp beer is again being produced in Switzerland that is freely available (at least in that country). In South America, hemp flowers are added to drinks made with **Trichocereus pachanoi** (cf. **cimora**). Democritus's famous recipe for a hemp **wine** is suitable for internal use: Macerate 1 teaspoon of myrrh (*Commiphora molmol*, cf. **incense**) and a handful of hemp flowers in 1 liter of retsina or dry Greek white wine (cf. *Vitis vinifera*). Strain before drinking. Hemp can also be used to make liquors. Mexicans "chop up the flowers and the top parts of the stalks, rub these with sugar and chili [cf. *Capsicum* spp.], and mix everything in a glass of milk or mescal (agave liquor) [see *Agave* spp.]" (V. Reko 1936, 64*).

Left: The female flower of *Cannabis sativa* forms the THC-rich resin on its hairs.

Top right: The male flower of fiber hemp (*Cannabis sativa*).

Bottom right: A rare African form of *Cannabis sativa*, with red stalks.

87 "The African aphrodisiac khala-khif consists of normal marijuana mixed with hemp resin; a blue mold is then allowed to develop on the mixture for about one month, after which this potent product is dried and smoked" (Stark 1984, 60*).

Ritual Use

Our present state of knowledge does not allow us to state with certainty when and where the ritual use of *Cannabis sativa* began (cf. **Cannabis indica**). In central Europe, it may already have been in use in shamanic contexts during the Neolithic period (Probst 1991; Stahl 1989). It is certain that shamans in ancient China knew of hemp, which they used to produce a shamanic state of consciousness so they could divine and heal. Ancient Chinese literature is filled with information about hemp's medicinal use. In the earliest sources on the Chinese use of herbs, it is said that chronic use of *ma-fen* ("hemp fruits") will enable one to "see devils" that can then be pressed into service. Unfortunately, these sources do not say how the hemp was ingested, i.e., whether it was eaten, drunk, or burned as incense (Li 1975*).

Hemp products had a cultic significance among the ancient Greeks. The Greek archaeologist Sotiris Dakaris, who has been investigating the oracle of the dead at Acheron since 1959, discovered "bags full of black clumps of hashish" in Ephyra (Vandenberg 1979, 24*). It is entirely possible that the temple sleepers at Acheron were administered a hemp preparation so that their dreams would be especially vivid. It is also possible that hemp, as "Scythian fire" (cf. **Cannabis ruderalis**), was used as an **incense** in the cult of Asclepius, the god of healing.

Remnants of hemp were recovered from the ancient Egyptian grave of Amenophis IV (Akhenaten; 1550–1070 B.C.E.) in Tell el-'Amârna. Hemp pollen has been identified on the mummy of Ramses II. Egyptian mummies were stuffed with hashish (Balabanova et al. 1992). Thus, the ritual use (death cult) of hemp during the dynastic period in Egypt (New Kingdom) has been demonstrated for the second millennium B.C.E. (Manniche 1989, 82f.*). This has also made it possible to identify the ancient Egyptian word *šmšmt* as "hemp." In Egypt, hashish continues to possess a ritual significance as a socio-integrative element at social events. People smoke out of the water pipe together after meals and at concerts and dance performances (Sami-Ali 1971).

In medieval Islamic society, hemp products were used primarily as sacred plants to support meditation in various Sufi and dervish orders. The plant became so closely identified with its mystical use by the Sufis that it became known as the "hashish of the poor" [= Sufis] (Rosenthal 1971, 13).

In South Africa, where the plant is known as *dagga*, hemp is now usually smoked for hedonistic purposes. In times past, however, it played an important role in numerous tribal rituals (Du Toit 1958, 1975, 1980; Morley and Bensusan 1971; Watt 1961). Its smoke was inhaled for divination, and it was sometimes smoked collectively for healing dances (cf. **Ferraria glutinosa**, **kanna**). *Dagga* was often used ritually together with other psychoactive plants (see **Mesembryanthemum** spp., **Sceletium tortuosum**, **Tabernanthe iboga**).

In Switzerland, the hemp fields in the *allmend* (the collective fields of a community) were once the site for various pagan and erotic rituals that the authorities interpreted as "witches' dances" or the "witches' sabbath" (Lussi 1996).

In modern Germany, ritualized hemp use based upon traditional shamanic patterns (cf. **Cannabis indica**) is becoming increasingly common. Because of the legal situation, however, these so-called hemp healing circles have not yet been described in any detail.

Artifacts

Pipes intended for hemp smoking have been found in Gallo-Romanic graves (Brosse 1992, 181*). Celtic and Germanic graves have yielded inflorescences of *Cannabis sativa* (cf. **Papaver somniferum**).

A large number of smoking devices have been invented in Africa. In addition to water pipes with hoses (so-called *argile*), these include horn pipes, earth pipes, and gourd pipes (Du Toit 1981, 518 ff.).

When it comes to smoking devices, creativity knows no limits. Numerous pipes have been devised and used to smoke hemp. In addition to standard tobacco pipes and Oriental water pipes (hookahs), devices have been developed specifically for smoking hemp. There are pure pipes, bongs (water pipes, made out of laboratory glass, plastic, or ceramics), ka-booms (smoking tubes with extra airflow), and others in a wide variety of designs. One astonishing invention comes from California, where an ocean creature, the sea urchin (*Clypeaster rosacea*), produces a shell that can be made into a natural and ideal pure pipe. Apart from the shell, all that is needed is a small screen, which is placed in the sea urchin's oral cavity. The smoke is drawn through its anal opening. Consequently, in the local counterculture it has become customary to speak of "ritual analingus." Recently, Nick Montefiore and James Hassal developed a high-tech pipe the size of a credit card and made entirely of metal (for pure smoking). When it came on the market, it was immediately awarded the BBC Designer Prize.

Most of the time, however, hemp products and the **smoking blends** prepared from them are smoked in the form of a self-rolled cigarette, a so-called joint (called also spliff, doobie, reefer, number, etc.). A joint is made using either the same commercial rolling papers that are used to roll cigarettes or special commercial rolling papers that differ from normal rolling papers principally in that they are larger. In 1986, the cigarette company BAT sponsored an exhibit in Paris entitled *Les papiers du paradis* [The Papers of Paradise]. The catalogue for the exhibit made it very clear that most of the rolling papers were intended for use in rolling joints.

For modern treatments of *Cannabis sativa* in painting, music, literature, comics, and films, see *Cannabis indica*. In art, no distinction is made among the different *Cannabis* species.

Medicinal Use

For additional information on medicinal use, see also *Cannabis ruderalis* (and see Rätsch 2001*).

The medical pyramid inscriptions and papyri of the ancient Egyptians indicate that hemp was used as a medicine in myriad ways. One translation reads, "A treatment for the eyes: celery; hemp; is ground and left in the dew overnight. Both eyes of the patient are to be washed with it early in the morning" (P. Ramesseum II, 1700 B.C.E.). This recipe has been interpreted as a treatment for glaucoma, which was common in ancient Egypt. This interpretation is very revealing, for ophthalmologists still have not developed a medicine better than hemp for treating glaucoma (cf. *Cannabis indica*).

Hemp was introduced into New Spain (Mexico, Peru) in the early colonial period. Since that time, it has been prized as a stimulant. Hemp, mixed with *aguardiente* (alcohol made of sugarcane; cf. **alcohol**), is used both internally and externally as a remedy for scorpion stings and tarantula bites (Bye 1979a, 145*).

At the beginning of the early modern period, all of the "fathers of botany" were in agreement that hemp possessed a "warm and dry nature" and could therefore dispel gas and flatulence. They wrote that it provided a good medicine for ear ailments. The use of the boiled root as a wrap for painful limbs is also frequently mentioned. The most important information about this early medicinal use comes from Tabernaemontanus, whose *Kräuterbuch* [Book of Herbs] is one of the most comprehensive works of its kind. It says, "Those women who have cramps in the womb / for them hemp should be burned / and held to the nose" (1731, 937*). This is apparently the first written reference to medicinal hemp smoking (to treat uterine cramps) in the German-language literature.

In the nineteenth century, so-called Indian cigarettes were sold in European pharmacies. These were smoked to treat asthma, lung and larynx ailments, neuralgia, sleeplessness, et cetera (cf. *Cannabis indica*). They were made from hemp leaves that had been soaked in an extract of opium (*Papaver somniferum*), together with belladonna leaves (*Atropa belladonna*), henbane leaves (*Hyoscyamus niger*), thorn apple leaves (*Datura stramonium*), and sometimes *Lobelia inflata* or cherry laurel schnapps (*Prunus laurocerasus* L.).

Such mixtures are reminiscent of the recipes for both **witches' ointments** and **smoking blends** (including **kinnikinnick**). The recommended dosage was one cigarette as needed (Fankhauser 1996, 156f.).

Hemp has been a part of the homeopathic materia medica since its inception. Homeopathy was established as a medical method by the physician Samuel Hahnemann (1755–1843), who wrote the following about hemp (*Cannabis sativa*):

> Until now, hemp has been usefully applied for acute gonorrhea and several types of jaundice. This organotropic tendency is found again in testing the symptoms of the urinary organs. In Persian inns, the herb is used to alleviate tiredness among those who are traveling on foot. Here, too, there are suitable symptoms for testing. For a long time, I administered hemp juice in the mother tincture, in the dosage of the smallest portion of a drop. But now I find that the dilution C30 is able to more highly develop these medicinal powers. (Buchmann 1983, 19f.*)

In the homeopathic doctrine of medicine, it has become customary to distinguish between *Cannabis sativa* and **Cannabis indica**, as the medical descriptions and symptom pictures of the two species vary considerably. *Cannabis sativa* (Cannabis sativa hom. *HPUS78*, Cannabis hom. *HAB34*) is prescribed primarily for urine retention as well as diseases of the urinary tract (gonorrhea, inflammation of the penis) and the respiratory organs. A substitute is *Hedysarum ildefonsianum*, a Brazilian species of sweet clover (Boericke 1992, 190*).

Hemp seed oil is now used in folk medicine to treat neurodermatitis. It is applied to the affected areas of the skin (this treatment is said to be amazingly successful).

Constituents

The chemistry of *Cannabis sativa* is very complex but is now quite well understood (Lehmann 1995). The main psychoactive constituent is **THC** (cf. *Cannabis indica*), which is contained primarily in the resin and the female flowers and, in lower concentrations, in the leaves. The most highly concentrated product is hashish oil, which contains approximately 70% THC. The resin contains up to 25% THC. Studies of older material have demonstrated that, even when stored for long periods, THC only very slowly oxidizes into the much less active CBN (Harvey 1990).

The **essential oil** contained in the plant, and especially the hashish, contains caryophyllene oxide. This odoriferous substance has been used to train police dogs to detect the drug (Martin et al.

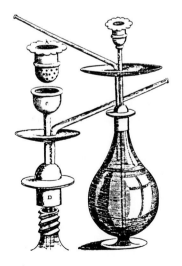

Persian water pipe for using hashish. (From Neander, *Tabacologia*, 1626)

The male hemp plant in blossom. (Woodcut from Tabernaemontanus, *Neu Vollkommen Kräuter-Buch,* 1731)

Hordenine

1961; Nigam et al. 1965). Hemp's essential oil usually either is devoid of THC or contains only trace amounts.

The seeds contain an oil rich in lignan, proteins, and the enzyme edestinase (St. Angelo and Ory 1970). The growth hormone zeatin has been found in immature seeds (Rybicka and Engelbrecht 1974). The seeds also contain the alkaloids cannabamine A, B, C, and D, piperidine, trigonelline, and L-(+)-isoleucine-betaine (Bercht et al. 1973). Hemp seed oil, which is obtained by cold pressing the seeds, is very rich in unsaturated fatty acids (vitamin F).

The pollen contains Δ^9-THC as well as THCA, an alkaloid-like substance, flavones, and phenolic compounds (Paris et al. 1975).

The leaves of *Cannabis sativa* have be shown to contain choline, trigonelline, muscarine, an unidentified betaine, and, astonishingly, hordenine, a β-**phenethylamine** alkaloid present in many cacti (El-Feraly and Turner 1975). The leaves of Thai and African populations have yielded water-soluble glycoproteins, serine-*O*-galactoside, and hydroxyproline (Hillestad and Wold 1977; Hillestad et al. 1977).

The roots of *Cannabis sativa* have been found to contain friedelin, epifriedelinol, N-(*p*-hydroxy-β-phenethyl)-*p*-hydroxy-*trans*-cinnamamide, choline, and neurine as well as the steroids stigmast-5-en-3β-ol-7-on (= 7-keto-β-sitosterol), campest-5-en-3β-ol-7-on, and stigmast-5,22-dien-3β-ol-7-on (Slatkin et al. 1975).

Effects
See *Cannabis indica.*

Commercial Forms and Regulations
See *Cannabis indica.*

Literature
See also the literature lists for **Cannabis indica** and **THC.**

Bercht, C. A. Ludwig, Robert J. J. Ch. Lousberg, Frans J. E. M. Küppers, and Cornelis A. Salemink. 1973. L-(+)-isoleucine betaine in *Cannabis* seeds. *Phytochemistry* 12:2457–59.

Boucher, Françoise, Michel Paris, and Louis Cosson. 1977. Mise en évidence de deux types chimiques chez le *Cannabis sativa* originaire d'Afrique du Sud. *Phytochemistry* 16:1445–48.

Brenneisen, Rudolf. 1996. *Cannabis sativa*—Aktuelle Pharmakologie und Klinik. *Jahrbuch des Europäischen Collegiums für Bewußtseinsstudien* (1995): 191–98.

Clarke, Robert C. 1995. Hemp (*Cannabis sativa* L.) cultivation in the Tai'an District of Shandong Province, People's Republic of China. *Journal of the International Hemp Association* 2 (2): 57,60–65.

Dayanandan, P., and J. P. B. Kaufman. 1975. *Trichomes of Cannabis sativa.* Ann Arbor: University of Michigan Press.

Du Toit, Brian M. 1958. Dagga (*Cannabis sativa*) smoking in Southern Rhodesia. *The Central African Journal of Medicine* 4:500–1.

———. 1975. Dagga: The history and ethnographic setting of *Cannabis sativa* in southern Africa. In *Cannabis and culture*, ed. V. Rubin, 81–116. The Hague: Mouton.

———. 1980. *Cannabis in Africa.* Rotterdam: A. A. Balkema.

———. 1981. *Cannabis* in Afrika. In *Rausch und Realität*, ed. G. Völger, 1:508–21. Cologne: Rautenstrauch-Joest-Museum für Völkerkunde.

Emboden, William A. 1990. Ritual use of *Cannabis Sativa* L.: A historical-ethnographic survey. In *Flesh of the Gods*, ed. P. Furst, 214–36. Prospect Heights, Ill.: Waveland Press.

Fankhauser, Manfred. 1996. *Haschisch als Medikament: Zur Bedeutung von Cannabis sativa in der westlichen Medizin.* Unpublished inaugural diss., Bern.

Feraly, Farouk el-, and Carlton E. Turner. 1975. Alkaloids of *Cannabis sativa* leaves. *Phytochemistry* 14:2304.

Gimbutas, Marija. 1989. *The language of the goddess.* New York: Harper and Row.

Grotenhermen, Franjo, and Renate Huppertz. 1997. *Hanf als Medizin: Wiederentdeckung einer Heilpflanze.* Heidelberg: Haug.

Haller, Andi. 1996. *Hausgemachtes Haschisch und andere Methoden zur Cannabis-Verarbeitung.* Markt Erlbach: Raymond Martin Verlag.

Harvey, D. J. 1990. Stability of cannabinoids in dried samples of cannabis dating from around 1896–1905. *Journal of Ethnopharmacology* 28:117–28.

Hemphill, John K., Jocelyn C. Turner, and Paul G. Mahlberg. 1978. Studies on growth and cannabinoid composition of callus derived from different strains of *Cannabis sativa*. *Lloydia* 41 (5): 453–62.

Hillestad, Agnes, and Jens K. Wold. 1977. Water-soluble glycoproteins from *Cannabis sativa* (South Africa). *Phytochemistry* 16:1947–51.

Hillestad, Agnes, Jens K. Wold, and Thor Engen. 1977. Water-soluble glycoproteins from *Cannabis sativa* (Thailand). *Phytochemistry* 16:1953–56.

James, Theodore. 1970. Dagga: A review of fact and fancy. *South African Medical Journal* 44:575–80.

Joseph, Roger. 1973. The economic significance of *Cannabis sativa* in the Moroccan Rif. *Economic Botany* 27:235–40.

Latta, R. P., and B. J. Eaton. 1975. Seasonal fluctuations in cannabinoid content of Kansas marijuana. *Economic Botany* 29:153–63.

Lehmann, Thomas. 1995. *Chemische Profilierung von Cannabis sativa* L. Master's dissertation, Bern.

Lieber, Michael. 1974. The economics and distribution of *Cannabis sativa* in urban Trinidad. *Economic Botany* 29:164–70.

Liggenstorfer, Roger. 1996. Hanf in der Schweiz. *Jahrbuch des Europäischen Collegiums für Bewußtseinsstudien* (1995): 147–56.

Lussi, Kurt. 1996. Verbotene Lust: Nächtliche Tänze und blühende Hanffelder im Luzerner Hexenwesen. *Yearbook for Ethnomedicine and the Study of Consciousness* 1995 (4): 115–42. Berlin: VWB.

Martin, L., D. Smith, and C. G. Farmilo. 1961. Essential oil from fresh *Cannabis sativa* and its use in identification. *Nature* 191 (4790): 774–76.

Martius, Georg. 1996. *Pharmakologisch-medicinische Studien über den Hanf.* 1855. Reprint, Berlin: VWB.

Meijer, Etienne de. 1994. Diversity in cannabis. PhD diss., Wageningen, Netherlands. (Distributed by the International Hemp Association IHA, Amsterdam.)

Mikuriya, Tod H. 1967. Kif cultivation in the Rif mountains. *Economic Botany* 21 (3): 231–34.

Morley, J. E., and A. D. Bensusan. 1971. Dagga: Tribal uses and customs. *Medical Proceedings* 17:409–12.

Nigam, M. C., K. L. Handa, I. C. Nigam, and L. Levi. 1965. Essential oils and their constituents XXIX. The essential oil of marihuana: Composition of genuine Indian *Cannabis sativa* L. *Canadian Journal of Chemistry* 43:3372–76.

Paris, M., F. Boucher, and L. Cosson. 1975. The constituents of *Cannabis sativa* pollen. *Economic Botany* 29:245–53.

Probst, Ernst. 1991. *Deutschland in der Steinzeit.* Munich: C. Bertelsmann.

———. 1996. *Deutschland in der Bronzezeit.* Munich: C. Bertelsmann.

Reininger, W. 1941. Haschisch. *Ciba-Zeitschrift* 7 (80): 2765–95.

Rosenthal, Franz. 1971. *The Herb: Hashish versus medieval Muslim society.* Leiden: E. J. Brill.

Rybicka, Hanna, and Lisabeth Engelbrecht. 1974. Zeatin in *Cannabis* fruit. *Phytochemistry* 13:282–83.

St. Angelo, Allen J., Robert L. Ory, and Hans J. Hansen. 1970. Properties of a purified proteinase from hempseed. *Phytochemistry* 9:1933–38.

Sami-Ali. 1971. *Le haschisch en Égypte.* Paris: Payot.

Segelman, Alvin, R. Duane Sofia, and Florence H. Segelman. 1975. *Cannabis sativa* L. (marihuana): VI. Variations in marihuana preparations and usage—chemical and pharmacological consequences. In *Cannabis and culture,* ed. V. Rubin, 269–91. The Hague: Mouton.

Slatkin, David J., Joseph E. Knapp, and Paul L. Schiff Jr. 1975. Steroids of *Cannabis sativa* root. *Phytochemistry* 14:580–81.

Smith, R. Martin, and Kenneth D. Kempfert. 1977. Δ^1-3,4-*cis*-tetrahydrocannabinol in *Cannabis sativa. Phytochemistry* 16:1088–89.

Spinger, Alfred. 1980. Zur Kulturgeschichte des Cannabis in Europa. *Kriminalsoziologische Bibliographie*: 26–27.

———. 1982. Zur Kultur und Zeitgeschichte des Cannabis. In *Haschisch: Prohibition oder Legalisierung,* ed. W. Burian and I. Eisenbach-Stangl, 34–43. Weinheim and Basel: Beltz.

Stahl, Peter W. 1989. Identification of hallucinatory themes in the Late Neolithic art of Hungary. *Journal of Psychoactive Drugs* 21 (1): 101–12.

Sterly, Joachim. 1979. Cannabis am oberen Chimbu, Papua New Guinea. *Ethnomedizin* 5 (1/2): 175–78.

Taura, Futoshi, Satoshi Morimoto, and Yukihiro Shoyama. 1995. Cannabinerolic acid, a cannabinoid from *Cannabis sativa. Phytochemistry* 39 (2): 457–58.

Tobler, Friedrich. 1938. *Deutsche Faserpflanzen und Pflanzenfasern.* Munich and Berlin: Lehmanns Verlag.

Van der Merwe, Nikolaas. 1975. Cannabis smoking in thirteenth–fourteenth-century Ethiopia: Chemical evidence. In *Cannabis and culture,* ed. V. Rubin, 77–80. The Hague: Mouton.

Van der Werf, Hayo. 1994. Crop physiology of fiber hemp (*Cannabis sativa* L.). Dissertation, Wageningen Agricultural University, Wageningen, Netherlands. (Distributed by the International Hemp Association IHA, Amsterdam.)

Watt, J. M. 1961. Dagga in South Africa. *Bulletin on Narcotics* 13:9–14.

"In Bohemia, hemp is a febrifuge. To cure lumbago in France (Côte d'Or), you should tie a thread from the male hemp plant around the hips."

SIEGFRIED SELIGMANN
DIE MAGISCHEN HEIL- UND SCHUTZMITTEL AUS DER BELEBTEN NATUR [THE MAGICAL HEALING AND PROTECTIVE AGENTS FROM THE ANIMATED NATURE]
(1996, 121*)

Cannabis x and Hybrids

Hemp Hybrids

As the very wide leaves indicate, this *Cannabis* sort, bred for the (still illegal) production of high-quality marijuana, was derived primarily from *Cannabis indica.*

"Since every cannabis sort is genetically unique and exhibits at least a few genes that cannot be found in other sorts, these unique genes are lost forever when a sort dies out. If genetic problems should appear as a result of excessive inbreeding of commercial sorts, then these new sorts might not be as resistant to previously unknown threats from the environment. Then, for example, it would be possible for a plant disease to spread with great speed and simultaneously befall and destroy several fields. For a farmer, the loss of a large part of the crop means immense financial losses. In this way, entire sorts can disappear forever."

ROBERT C. CLARKE
HANF [HEMP]
(1997, 14*)

Family

Cannabaceae (= Cannabinaceae) (Hemp Family)

Hemp connoisseurs regard any marijuana (= female hemp flowers) that contains more than just a few seeds as inferior in quality. They prefer those psychoactive, **THC**-rich sorts that form no or few seeds. These are known as *sinsemilla*, literally "without seeds" (Mountain Girl 1995). When producing *Cannabis* crosses or sorts, a fundamental distinction is made between hybrids that can be grown outdoors and those that thrive only under conditions of artificial light (so-called indoor sorts).

Crosses between *Cannabis indica* and *Cannabis ruderalis* are popular, as they are very small as well as very potent. Crosses with *Cannabis ruderalis* are well suited for growing outdoors, as they flower early regardless of the length of the day.

Because of pressure from law enforcement, cannabis intended for smoking is being grown in closed rooms with ever greater frequency. The cultivation of highly potent sorts in greenhouses is now especially common in the Netherlands (Jansen 1991).

Most marijuana growers no longer use seed to produce new plants; they use cuttings (clones) of female plants instead. To take a cutting, a sharp knife is used to separate a vigorous 8 to 10 cm long shoot from the mother plant. The leaves are removed and the shoot is immediately placed in a container of lukewarm water. Then the shoot is placed in a watered piece of rock wool that is full of holes. To promote root formation, a root hormone may be added to the water. Shoots will most readily develop roots if they are kept in a warm room (soil temperature between 21 and 24°C) with very high (at least 80%) air humidity (e.g., in a small, heated greenhouse). Once the shoots have formed roots, they may be planted in soil in pots.

The most important factor affecting the formation of THC-rich flowers is the amount of light the plants are exposed to: "When two clones of a female hemp plant grow in two totally different surroundings, i.e., one perhaps in shade and the other in full sun, their genotypes remain identical. But the clone grown in the shade will grow tall and slender and mature late, whereas the clone exposed to sunlight will remain small and bushy and mature much sooner" (Clarke 1997, 28 f.).

One important goal of cultivation is shortening the time it takes for the THC-rich inflorescences to develop without producing seeds. For this reason, many sorts or hybrids are

evaluated by the amount of time between seed germination and the full development of the resinous flowers. For example:

Skunk Special	flowers after 9 weeks
Super Skunk	flowers after 7 weeks
Big Bud	flowers after 9 weeks
California Orange Bud	flowers after 9 weeks
California Indica	flowers after 7 weeks
Misty	flowers after 10 weeks
NL Shiva	flowers after 9 weeks
Shiva Shanti	flowers after 7–8 weeks
NL Masterkush	flowers after 10 weeks
Haze	flowers after 11 weeks
Afghaan	flowers after 8 weeks
Durban Poison	flowers after 9 weeks
Hindu Kush	flowers after 6–7 weeks
Northern Lights	flowers after 7–8 weeks
Jack Herer	flowers after 10 weeks

Spectacular results were achieved in experiments in which *Cannabis sativa* was grafted onto **Humulus lupulus** and *Humulus japonicus*. Here, four-week-old hops seedlings were cut straight across. The stems were split, and a cannabis stalk that had also been split was placed into each hops seedling's stem, after which the two were tied together with cellulose. Over 30% of these grafted plants survived and developed into large plants. When THC-rich hemp is grafted onto *Humulus*, it continues to produce high quantities of constituents. Unfortunately, this does not occur when hops is grafted onto cannabis (Crombie and Crombie 1975).

There is a very rich literature on methods for cultivating all sorts and hybrids of hemp, including Behrens 1996, Frank and Rosenthal 1980, Starks 1981, and Stevens 1980. High-tech methods have been developed to provide optimal watering to hemp fields in dry or very dry areas (prairies, deserts). Special hydroponic techniques have been developed for indoor growing (Storm 1994).

There is now a vigorous trade in legal seeds (cf. ***Cannabis indica***) of particular sorts and crosses for growing both indoors and outdoors.

Literature

See also the entries for the other *Cannabis* species.

Behrens, Katja. 1996. *Leitfaden zum Hanfanbau in Haus, Hof und Garten.* Frankfurt/M.: Eichborn.

Coffman, C. B., and W. A. Gentner. 1979. Greenhouse propagation of *Cannabis sativa* L. by vegetative cuttings. *Economic Botany* 33 (2): 124–27.

Crombie, Leslie, and W. Mary L. Crombie. 1975. Cannabinoid formation in *Cannabis sativa* grafted inter-racially, and with two *Humulus* species. *Phytochemistry* 14:409–12.

Frank, Mel, and Ed Rosenthal. 1980. *Das Handbuch für die Marihuana-Zucht in Haus und Garten*. Linden: Volksverlag.

Jansen, A. C. M. 1991. *Cannabis in Amsterdam: A geography of hashish and marihuana*. Muiderberg, Netherlands: Dick Coutinho.

Mann, Peggy. 1987. *Pot safari: A visit to the top marijuana researchers in the U.S.* New York: Woodmere Press.

Margolis, Jack S., and Richard Clorfene. 1979. *Der Grassgarten*. Linden: Volksverlag.

Mountain Girl. 1995. *Sinsemilla: Königin des Cannabis*. Markt Erlbach: Raymond Martin Verlag.

Starks, Michael. 1981. *Marihuana-Potenz*. Linden: Volksverlag.

Stevens, Murphy. 1980. *Marihuana-Anbau in der Wohnung*. Linden: Volksverlag.

Storm, Daniel. 1994. *Marijuana hydroponics: High-tech water culture*. Berkeley, Calif.: Ronin.

Carnegia gigantea (Engelmann) Britton et Rose

Saguaro, Giant Cactus

Family
Cactaceae (Cactus Family); Cereeae Tribe, Cereanae Subtribe

Forms and Subspecies
None

Synonyms
Cereus giganteus Engelm.

Folk Names
Cardón grande, giant cactus, great thistle, ha'rsany (Pima), harsee, hoshan (Papago, Pima), mojepe, mojépe, moxéppe (Serí), pitahaya, riesenkaktus, saguaro, saguarokaktus, sahuaro, sahuro, sah-wáh-ro, sajuaro, sauguo (Mayo), suhuara

History
Archaeological discoveries indicate that the prehistoric Hohokam (1150–1350 C.E.) used the saguaro for a variety of purposes (Hodge 1991, 48; Nabhan 1986, 32). The cactus has continued to play a central role in the cultures of the Southwest into the present day. In 1540, Spanish conquistadors, marching to the north under the command of Coronado, first mentioned the cactus as well as the wine produced from it under the name *pitahaya* (Bruhn 1971, 324). It was first described in a botanical publication in 1848, under the name *Cereus giganteus*. The genus name used today was coined to honor Andrew Carnegie, a passionate desert researcher (Hodge 1991, 6).

Distribution
The giant saguaro is native to Arizona, Southern California, Baja California, and northern Sonora (Mexico).

Cultivation
Propagation by seed is possible, but it is extremely difficult and rarely successful. Because of this, most attempts to restore the saguaro forests of Arizona have met with failure (Hodge 1991, 35 ff.). The fruits cannot be picked by hand but must be collected using a long pole (2 to 5 meters long), to the end of which is attached another pole (*kuibit*) (Bruhn 1971, 325). The cactus requires an extreme desert climate with very high summer temperatures. It tolerates frost and snow in winter (Nabhan 1986, 16 f.).

Appearance
The cactus can grow to a height of more than 12 meters. It has a main trunk and eight to twelve side branches that rise vertically. The skeleton has twelve to twenty-four ribs. The white flowers emerge from green, scaly buds on the tips of the trunk and branches. The flowers have luminously yellow stamens and pistils. The cactus does not flower until it is between fifty and seventy-five years old (Bruhn 1971, 323). The fruits are 6 to 9 cm long and contain a crimson flesh, in which some 2,200 seeds are found.

The cactus occasionally takes on a deformed appearance. Such specimens are known popularly as "monarchs with crowns" (Hodge 1991, 31 ff.).

The saguaro (*Carnegia gigantea*) is the largest of all column cacti. In Arizona, entire saguaro forests can be seen. (Photographed in its natural habitat)

The cactus can live 150 to 175 years and attain a weight of six to ten tons. Its high water content (80 to 95%) enables the cactus to flower and fruit regularly, even during yearlong droughts (Bruhn 1971, 323). Flowering time is normally in the spring. Pollination occurs through bats and birds as well as other agents (Hodge 1991, 16). The **honey** that is collected from the flowers has no psychoactive effects. In Arizona, it is regarded as a culinary specialty.

Psychoactive Material
— Fruit (pitahaya, tjúni, a-a, a-ag, nol-bia-ga)

Preparation and Dosage
In the area in which the cactus is found, fermented drinks (beerlike or **wine**) made from its fruits are known as *tiswin, sawado, saguaro, haren, ha'-san na'vai* ("saguaro drink"), and *na'vait*. Among the O'odham (= Papago), the wine is known as *nawait*.

Boiling the fruit flesh yields a sweet brown syrup (*sítoli*) that can be either eaten as is or fermented.[88] When a fermented beverage is made from this syrup or from fresh fruits with water, the **alcohol** content is only 5% or less (Hodge 1991, 47f.). Thus, the beverage is not a wine but a rather beerlike drink (very similar to the South American **chicha**). Fermentation takes about seventy-two hours. Possible additives are unknown (Bruhn 1971, 326). The Serí Indians of northern Mexico also brewed a fermented drink from saguaro fruits. Known as *imám hamáax*, "fruit wine," it was made by crushing the fruits in a basket and mixing the result with water. Fermentation was complete after a few days. They more rarely produced a true wine without water (Felger and Moser 1991, 247*).

Ritual Use
The Tohono O'odham (= Papago) venerate the saguaro as a sacred tree. They explain that it arose from drops of sweat that fell from the eyebrows of I'itoi, the older brother of the tribal pantheon, in the morning dew and condensed into pearls. According to a different origin myth, the cactus was once a boy. When his mother was not watching, he became lost in the desert and fell into a tarantula hole. He reemerged as a cactus. This may explain why, after a child is born, the O'odham bury the placenta next to a saguaro. Doing so is said to secure the child a long life. On the vernal equinox, the O'odham sing special songs throughout the night to aid the cactus fruits in their development (Hodge 1991, 47).

The O'odham make cactus wine in July (= *harsany paihitak marsat*, "saguaro harvest month") for their annual rain ceremony, which was established by I'itoi, the Older Brother (Bruhn and Lundström 1976, 197). The wine consumed in the ceremony is made from fruits or syrup contributed by all the families (Bruhn 1971, 326). The ritual is both a conjuration of rain—an extremely important ceremony in the desert—and a socially integrative tribal celebration and harvest festival. In a kind of sympathetic magic, all the members of the tribe drink copious amounts of *nawait*. They do this in imitation of nature, for "the earth drinks water" so that the plants, and especially the cacti, can thrive. During the festival, songs and texts are presented that describe the life cycle of the cactus, the proper ways of harvesting the fruits, and the influence the cactus spirit has upon the "rain house" in which the weather is made (Underhill 1993, 21ff.). The elders of the tribe pray to the four directions. During the festival, a person is not allowed to ask for a drink but must wait until one is offered (Hodge 1991, 48).

As is the case in the ceremonies of many tribes of the Southwest, a ceremonial clown appears at the festival and makes fun of the ritual. The O'odham ceremonial clown (Naviju dancer) is regarded as a personification of the saguaro. And in general, the giant cactus is construed as an "Indian" (Bruhn 1971, 327).

The Serí, who live in the Sonoran Desert of Mexico, share the O'odham belief that the cactus was originally a person. For this reason, they also bury the placenta of a newborn at its roots so that the child will enjoy a long life (Felger and Moser 1991, 248*; Lindig 1963).

To date, we know of no psychoactive use of the cactus flesh or an alkaloid-rich preparation made from it. It is possible that there might have once been such a use, for the saguaro is also regarded as a peyote substitute (see ***Lophophora williamsii***).

Artifacts
Representations of the giant cactus in varying degrees of abstraction are found as graphic elements in the baskets that are woven out of yucca (*Yucca* spp.), catclaw (*Acacia greggii*), and other desert plants (Hodge 1991, 47). A figure of the Naviju dancer, the personification of the cactus, is on display at the Arizona State Museum in Tucson.

The saguaro cactus is depicted on numerous Western paintings and has become something of a symbol of the Wild West.

The O'odham artist Leonard F. Chana has produced an acrylic painting, *When the Clouds Come*, that depicts the harvest of the saguaro fruit. (It was published as a postcard by Indigena Fine Art Publishers in 1995.) The Luiseno–Hunkpapa Sioux painter Robert Freeman has immortalized the cactus in his painting *Lady in Waiting* (1990).

The O'odham and other tribes have a number of songs that praise the cactus; some of these have been recorded, translated, and published. Some songs, especially the dream songs, are said to have

88 In Sonora and Baja California, the somewhat larger and sweeter fruit of the cactus *Cereus thurberi* Engelm., known as *pitahaya dulce*, is used in a very similar manner (Havard 1896, 36*).

been inspired by the effects of the wine (Bruhn 1971, 327; Densmore 1929; Underhill 1993).

The skeletons of decayed cacti are used as raw materials for numerous products. They also serve as fence posts and are now used throughout the world as window decorations (to suggest a Wild West ambience).

Medicinal Use

The Mexican Serí Indians cut a piece from the trunk of the living cactus, remove the thorns, and heat the cactus flesh over hot wooden coals. The flesh is then wrapped in a cloth and applied to rheumatic or painful areas (Bruhn and Lundström 1976, 197; Felger and Moser 1974, 421*). Apart from this, no ethnomedical or folk medicinal uses have been recorded.

Constituents

Saguaro flesh has been found to contain the β-phenethylamines carnegine, gigantine, salsolidine, 3-methoxytyramine, 3,4-dimethoxyphenethylamine, arizonine, and dopamine (Bruhn and Lundström 1976; Mata and McLaughlin 1982, 96*). The alkaloids carnegine, gigantine, and salsolidine are closely related to the constituents of peyote (*Lophophora williamsii*) (Bruhn 1971, 323). The main alkaloid is salsolidine (= norcarnegine), which makes up some 50% of the total alkaloid content. This alkaloid was first discovered in a *Salsola* species (Chenopodiaceae) and also occurs in *Pachycereus pecten-aboriginum* (Bruhn and Lundström 1976, 199). Altogether, the cactus contains 0.7% alkaloids (Bruhn 1971, 323).

The entire air-dried fruit contains approximately 7% sugar and 13% protein. The fruit syrup consists of up to 63% sugar. The seeds contain high amounts of tannin and are approximately 16% protein (Bruhn 1971, 324 f.).

Effects

The sap, which flows from the cactus when it has been wounded, is very bitter. When ingested, it typically produces nausea and dizziness (Bruhn and Lundström 1976, 197).

In laboratory tests with monkeys and cats, the alkaloid gigantine was found to induce hallucinations (Bruhn and Lundström 1976, 197). It is interesting to ponder, however, how we are able to recognize that animals that are incapable of speech are having hallucinations.

All that has been reported about the effects of saguaro wine is that it produces "good feelings" (Bruhn 1971, 327).

Commercial Forms and Regulations

The cactus is listed as an endangered species and is protected. In Arizona, only saguaro honey is available.

Literature

Bruhn, Jan G. 1971. *Carnegiea gigantea*: The saguaro and its uses. *Economic Botany* 25 (3): 320–29.

Bruhn, Jan G., and Jan Lundström. 1976. Alkaloids of *Carnegiea gigantea*. Arizonine, a new tetrahydroisoquinoline alkaloid. *Lloydia* 39 (4): 197–203. (Additional literature.)

Densmore, Francis. 1929. Papago music. *Bureau of American Ethnology Bulletin* 90.

Hodge, Carle. 1991. *All about saguaros*. Phoenix: Arizona Highways Books.

Lindig, Wolfgang. 1963. Der Riesenkaktus in Wirtschaft und Mythologie der sonorischen Wüstenstämme. *Paideuma* 9:27–62.

Nabhan, Gary Paul. 1982. *The desert smells like rain: A naturalist in Papago Indian country*. San Francisco: North Point Press.

———. 1985. *Gathering the desert*. Tucson: The University of Arizona Press.

———. 1986. *Saguaro*. Tucson, Ariz.: SPMA. (Includes an excellent bibliography.)

Underhill, Ruth Murray. 1993. *Singing for power: The song magic of the Papago Indians of southern Arizona*. Tucson and London: The University of Arizona Press.

Wild, Peter. 1986. *The saguaro forest*. Flagstaff, Ariz.: Northland Press.

"The dreams and feelings that are experienced in inebriation are generally attributed to a supernatural origin and are considered essential for certain undertakings. Among the Pima and the Papago, drunkenness in the context of the annual rain dance held great significance. A fermented drink was obtained from the juice of saguaro, pitahaya, or nopal cactus fruits. In a kind of sympathetic magic, they believed that drinking alcohol would induce clouds to form, which would soon burst and satiate the world with water."

Serge Bramley
Im Reiche des Wakan [In the Realm of the Wakan]
(1977, 82*)

Carnegine

Salsolidine

Catha edulis (Vahl) Forsskål ex Endlicher

Khat

Family

Celastraceae[89] (Bittersweet Family); Subfamily Celastroideae, Celastreae Tribe

Forms and Subspecies

Khat farmers of Ethiopia make a distinction between two varieties: *ahde*, "white," and *dimma*, "red," in reference to the color of the leaves. The red leaves are said to be more potent (Getahun and Krikorian 1973, 359ff.). Apart from this, no varieties or forms have been described botanically (Brenneisen and Mathys 1992, 730).

Synonyms

Catha edulis Forsk.[90]
Catha forskalii A. Rich
Catha inermis G.F. Gmel.
Celastrus edulis Vahl
Dillonia abyssinica Sacleux
Trigonotheca serrata Hochst.

Folk Names

Abessinischer tee, Abyssinian tea, al-qât, Arabian tea, arabischer tee, Arab tea, bushman's tea, cat, cath, chat, chat tree, flower of paradise, gat, jaad (Somali), jât, kafta (Arabic, "leaf"), kat, kât, kath, kathbaum, katstrauch, khat, khatstrauch, miraa, mirra, mirungi, muhulo (Tanzania), muirungi (Kenya), musitate (Uganda), qaad (Somali), qat, qât, qatbaum, qatstrauch, Somali tea, somalitee, thé des abyssins, tschat

History

The use of psychoactive khat leaves is very old, with roots that definitely predate coffee (*Coffea arabica*) drinking. It is very likely that khat was first chewed as an agent of pleasure and a stimulant[91] in Ethiopia. The plant was first included in a list of medicines in 1222; it is also mentioned in the book *The Wars of 'Âmda Syon I* ('Âmda Syon I was a Christian king who ruled in the early fourteenth century) (Getahun and Krikorian 1973, 356). A history book by Al-Maqrîzî (1364–1442) notes the following about Abyssinian plants: "Among them is a tree that is called *gât*. It does not bear fruit, people eat its leaves, and these resemble the small leaves of the orange tree. They expand the memory and in doing so call the forgotten back into mind. They give pleasure and diminish the desire for food, sexuality, and sleep. For the inhabitants of that land, not to mention the educated, the consumption of this tree is associated with great longing" (Schopen 1978, 46f.).

The Sufis and the wandering dervishes played a great role in the early spread of khat use (Schopen 1981). They regarded the ingestion of the leaves as a sacred activity and used khat to achieve mystical experiences, believing that "[i]n doing so, you see things of rare knowledge that belong to God's magnificence" (Schopen 1978, 52).

The name *khat* is apparently derived from the Arabic *kut*, "sustenance or driving principle," or from the place name Kafa (in Ethiopia), which is also thought to be the source of the word *coffee*. Most folklore suggests that both the khat bush and the practice of chewing khat are from Yemen. It is said that the goatherd Awzulkernayien observed how his goats ate the leaves of a shrub, after which they behaved in a frisky manner. The goatherd then tried the fresh leaves, whereupon he immediately felt more awake and stronger than he ever had before. Before turning in to sleep that night, he chewed a few more of the leaves he had brought with him. He was unable to sleep the entire night and spent it instead in prayer and meditation. Following this, khat was proclaimed to be a sacred tree and regarded as a wondrous medicine (Getahun and Krikorian 1973, 353f.).

According to a different legend, two saints who often prayed throughout the night were constantly falling asleep or fighting to avoid sleep. They prayed to God that he would give them a means to prevent them from falling asleep. An angel appeared to them and showed them a plant to eat that would help them remain awake and pray through the night (Getahun and Krikorian 1973, 356).

Charles Musès has proposed the theory that khat was already regarded as a "food of the gods," "divine food," or "food of existence" in ancient Egypt and was used for magical purposes. The plant is thought to have been known as *kht* in Egyptian (Musès 1989). Others have suggested that the Homeric **nepenthes** was in fact khat. It has also been suggested that khat was the magical medicine Alexander the Great used to miraculously heal his troops. Even the smoke of Delphi (cf. *Hyoscyamus albus*) was said to have come from khat leaves and to have been inhaled as a psychoactive **incense** (Elmi 1983, 164).

The plant was described in 1775 by the Swedish botanist Pehr Forsskål (1732–1763), who lived in Yemen for many years and eventually died there. Pharmacognostic and chemical studies of the khat bush began in the German-language areas at the end of the nineteenth century (cf. Beitter 1900 and 1901). In the 1920s and 1930s, a variety of pharmaceuticals and agents of pleasure made from khat (e.g., Catha-Cocoa Milk; cf. *Theobroma cacao*) were sold in London (Brenneisen and Mathys

---

89 In the literature, this family name is often written as *Celestraceae* (e.g., Elmi 1983, 164); according to Zander (1994, 171*), the correct spelling is *Celastraceae* (cf. also Frohne and Jensen 1992, 175*).

90 The describing author was Pehr Forsskål (Krikorian 1985, 515; Zander 1994, 710*). The correct abbreviation is actually Forssk., but Forsk. has become accepted internationally.

91 Louis Lewin (1980*) placed khat in the group of the Excitantia, together with camphor (*Cinnamomum camphora*), tea (*Camellia sinensis*), coffee (*Coffea arabica*), and **betel**.

1992, 735). The beginnings of ethnographic khat research (in Yemen) began in the 1970s with the groundbreaking work of Armin Schopen (1978). It was only during the early 1980s that Swiss scientists discovered the actual psychoactive constituent, the amphetamine-like cathinone (Kalix 1981).

Distribution

The bush is very likely from the area around Lake Tana (Harar) in Ethiopia. From there, it spread to East Africa (via Kenya), Tanzania, Aden, Arabia, and Yemen (Getahun and Krikorian 1973). The khat bush can thrive in quite different ecological zones and can be found in both tropical and cooler mountain regions. The wild khat bush grows in the tropical rain forest of the Gurage country in Shewa, Ethiopia. It is cultivated in Arabia, Zambia, and Somalia, and even as far away as Afghanistan (Getahun and Krikorian 1973, 357).

Cultivation

Because the plant only rarely produces seeds in cultivation, khat is best propagated by cuttings (approximately 35 cm long) from the young branches. Propagation is best performed in a dry, hot climate (Grubber 1991, 43*). The cuttings—usually two—are placed in a hole filled with water. Khat can be planted at any time of the year, so long as the young plants can be provided with sufficient water. The bushes are planted in rows about 1 meter apart. Sorghum is often planted between the rows.

Propagation can also occur using seeds, but this is never done in the areas under cultivation (Getahun and Krikorian 1973:364).

Khat requires a climate that is the same as or similar to the climate required for growing coffee (*Coffea arabica*), i.e., approximately 1,200 mm of precipitation. As a mountain plant, the bush can tolerate a mild frost. The first harvest may be taken from the bush when it is three years old, although typically it is not taken until after five to eight years. Khat planting is done primarily by males (Getahun and Krikorian 1973, 365; R. Schröder 1991, 126*). Khat bushes are often inhabited by an insect of the genus *Empoasca*, although this does not cause damage. In fact, a greater number of young shoots (the best merchandise) are formed as a result of the insect eating the plant (Getahun and Krikorian 1973, 367).

Important areas of cultivation are found especially in Ethiopia and Yemen and now also in northern Madagascar, Afghanistan, Turkistan, and even Israel. Some 60% of the fertile areas of Yemen are used for planting khat (Brenneisen and Mathys 1992, 732).

Left: Every day, large segments of the population of Yemen chew the light green leaves of the khat bush (*Catha edulis*) as a stimulant and inebriant.

Right: The small fruits of *Catha edulis*.

Appearance

This evergreen, fast-growing bush can grow into a tree as large as an oak (15 to 20 meters tall); under cultivation, it usually is kept to a height of 3 to 5 meters or, in rare cases, 7 meters (Getahun and Krikorian 1973, 356). The more the bush is trimmed, the more rapidly young shoots appear.

On blooming branches, the leaves are always opposite, but they can be alternate on young branches and plants (Brennesien and Mathys 1992, 730; Krikorian 1985). The leaves have a serrated margin and a shiny, slightly leathery upper surface. Young leaves at the ends of branches are light green; older leaves are dark green. The leaves sometimes take on a red hue. The small, star-shaped flowers are white and are borne in clusters in axillary cymes. The fruit capsules are 7 to 8 mm long and have four chambers. When they mature, they open up like small flowers (Krikorian 1985).

The genus *Catha* consists of only a few species (Wang 1936), probably a maximum of three. The two others are *Catha transvaalensis* Codd. [syn. *Catha cassinoides* N.K.B. Robson] and *Catha abbottii* Van Wyk et Prins;[92] *Catha spinosa* Forssk. nows bears the botanically valid name *Maytenus parviflora* (Vahl) Sebsebe (Brenneisen and Mathys 1992, 730). These African bushes can all be mistaken for khat, although they themselves have no ethnopharmacological significance.

Psychoactive Material

— Leaves (Catha-edulis leaves, khat leaves)
— Fresh leaves and twigs, and also the leaf buds
— Dried leaves (khat tea)

92 *Catha transvaalensis* has been found to contain sesquiterpenes but no other psychoactive alkaloids (Mathys 1993, 15). To date, no studies of *Catha abbottii* have been carried out (Brenneisen and Mathys 1992, 732).

Preparation and Dosage

The fresh leaves should be chewed as soon after harvesting as possible. They should be no more than two days old. They require no further treatment and do not need to be mixed with any other substances. A person simply takes as many of the leaves into the mouth as possible, then chews the leaves for some ten minutes before either spitting them out or swallowing them (Getahun and Krikorian 1973, 371). The juice of the chewed leaves is swallowed periodically (Schopen 1978, 85). The longer the constituent-rich juice is retained in the mouth, the more pronounced the effects. In Yemen, fresh leaves are also pounded in mortars.

The fresh leaves and branch tips are (more rarely) brewed or boiled into a tea. In South Africa, infusions of khat are known as bushman's tea. In Yemen, roasted khat leaves were once used to prepare "coffee" (Schopen 1978, 86). They also can be ground, mixed with honey or sugar, and made into candies (Getahun and Krikorian 1973, 357). In Somalia, the leaves are sometimes dried in the sun and then crushed. Cardamom, cloves, and water are then mixed with the powder to produce a paste that is taken as a quid. Fresh or dried khat branches are added to tea (*Camellia sinensis*) for flavor. In Ethiopia, khat is even used to make **mead**: "The qât infusion is fermented with **honey**. This produces a brown, bitter, meadlike drink with mild inebriating effects" (Schopen 1978, 85).

In Arabia (Yemen), dried leaves are smoked both alone and with other substances, especially hashish (*Cannabis indica*; cf. also **smoking blends**) (Getahun and Krikorian 1973, 357). Dried leaves that are still green are used as a (medicinal) **incense**. The dried leaves may also be ground and formed into balls using a binding agent; when ingested, such preparations give strength to pilgrims bound for Mecca. Dried leaves can also be mixed with water to prepare a paste for ingestion by old people without teeth (Getahun and Krikorian 1973, 366).

Leaves that have been damaged by frost take on a dull ashen color and should not be used, as they are said to cause headaches (Getahun and Krikorian 1973, 367).

It is generally said that tobacco (*Nicotiana tabacum*) should be smoked along with khat, as it will potentiate the effects of the khat (Schopen 1978, 86).

Only the leaf buds, the young leaves, and the ends of the branches contain sufficient constituents. The main psychoactive component breaks down rapidly when the material is dried. In contrast, it can remain unchanged for months if the fresh leaves are frozen (Brenneisen and Mathys 1992, 732).

Between 100 and 200 g of leaves are typically consumed during a khat session (R. Schröder 1991, 127*). Ethiopian khat farmers eat between $1/4$ and $3/4$ kg of khat leaves in the morning—of the finest quality, to be sure (Getahun and Krikorian 1973, 374).

Cathinone, the primary active constituent, is some three times less toxic than amphetamine. An alcoholic extract of khat, in a dosage of 2 g per kilogram of body weight, has been demonstrated to have lethal effects on mice (Brenneisen and Mathys 1992, 738). One gram of khat leaves contain 3.27 mg of cathinone/cathine (Ahmed and El-Qirib 1993, 214).

Ritual Use

Most Muslims who live in areas where khat is grown regard both the bush and its leaves as sacred and utter a prayer of thanks before they use it (Getahun and Krikorian 1973, 356). In Ethiopia, traditionally only the older men chewed khat, and then only in conjunction with religious rites. They chewed the leaves and drank coffee so that they could remain awake for the long prayers. Often, they also smoked hashish on these occasions. Over the course of time, people began to chew khat leaves while watching over the sick, at marriages and funerals, and during business negotiations. Today, khat leaves are chewed by men and women of all ages, including students and children (Getahun and Krikorian 1973, 371 f.).

Ethiopian dervishes use khat as part of their religious healing ceremonies. They chew consecrated leaves and then spit upon the ill before pronouncing prayers and magical formulae over them (Schopen 1978, 87).

In Yemen, the ritual use of khat is widespread for certain festivals and religious events: engagements, marriages, burials. Most Yemenites chew khat every day in a social round whose structure adheres to precisely defined ritual forms. Because of their socially integrative nature, these rounds play a central role in Yemenite society (Schopen 1978). The participants come together in the afternoon, during the "blue hour." Most are men, but women occasionally take part as well. These daily khat rounds take place in the main rooms of private houses as well as in special khat rooms in the offices of the government, large companies, et cetera. The participants pluck fresh leaves from the twigs and stuff these into their mouths. The leaves are then moistened with saliva and chewed thoroughly. A pitcher of water is passed around continuously, "for the alkaloids work only when the cell juice of the leaves that have been mixed with saliva gets to the stomach by drinking" (R. Schröder 1991, 127*). Because tobacco (or, less often, hashish) smoking is regarded as an absolutely essential part of chewing khat, cigarettes, pipes, or the hoses of large water pipes are also passed around. The participants often sing and make music together. The activities change as the effects of the khat run their course. At first, the members of the round converse excitedly with one another about contemporary political topics, current events, gossip, and Islam. As the effects begin to diminish some two hours later, the participants weary and the conversations dwindle. At this time, the circle breaks up (Schopen 1978, 1981).

Artifacts

In Yemen, numerous Arabic-language poems both glorify and criticize khat use (Schopen 1978). It is possible that many aspects of Arabic art have been influenced by the use of khat. The samar music of Yemen is composed specifically for the afternoon khat rounds and is played and sung during these social gatherings. At least one album of samar music that was recorded on-site has been released internationally: *Music from Yemen Arabia: Samar* (Lyrichord Discs, LLST 7284).

In Tanzania, the wood of the khat bush is used to manufacture spoons and combs (Schopen 1978, 86).

Medicinal Use

Khat is generally used only infrequently as a medicine. The leaves are mentioned in only two Arabic pharmacopoeias. Khat is said to calm the stomach and cool the intestines and is recommended for the treatment of depression and melancholy (Schopen 1978, 87). In Yemen, it is used also as an appetite suppressant (Fleurentin and Pelt 1982, 96f.*). More rarely, the fumes of burning khat are inhaled for the treatment of headaches (Schopen 1978, 88).

In Africa, khat root is used to treat influenza, stomach problems, and diseases of the chest (Getahun and Krikorian 1973, 357).

In Ethiopia, it is believed that khat can cure 501 diseases and afflictions, as the numerical value of its Arabic name, *ga-a-t*, is 400 + 100 + 1 (Getahun and Krikorian 1973, 370). Khat is also used there as an aphrodisiac (Krikorian 1984), as well as to treat depression and melancholy. A khat leaf is applied to the forehead as a remedy for headaches. Among the Masai and Kipsigi tribes, the leaves are used to treat gonorrhea. It is also said that regular consumption of khat provides protection against malaria. In Saudi Arabia, khat is used to treat asthma and fever (Brenneisen and Mathys 1992, 735).

Constituents

During the early phase of khat research, it was thought that the leaves contained **caffeine**, an assumption that none of the studies was able to verify. Later, katin (= cathine) or "celastrina" was regarded as the active principle (Krikorian and Getahun 1973, 379). Soon thereafter, it was suggested that **ephedrine** was responsible for the plant's effects. Some investigators claimed that *d*-norpseudoephedrine was also present (Krikorian and Getahun 1973, 387). The constituents primarily responsible for the stimulating effects upon the central nervous system (CNS) are the khat phenylalkylamines or khatamines (phenyl-propylamines) cathinone and cathine (= *S,S*-[+]-norpseudoephedrine) (Brenneisen and Geisshüsler 1985). Other CNS stimulants, the phenylpentyl-amines merucathine, pseudomerucathine, and merucathinone, are present in small amounts, and there is also some *R,S*-(−)-norephedrine (Brenneisen and Geisshüsler 1985, 293; Brenneisen et al. 1984). The actual primary psychoactive and stimulating constituent is cathinone (= *S*-(−)-cathinone or *S*-(−)-[alpha]-amino-propiophenone) (Brenneisen and Mathys 1992, 731; Kalix 1992).

The amount of constituents present in the fresh leaves can vary considerably, depending upon provenance, place of cultivation, age, and quality (Geisshüsler and Brenneisen 1987). The alkaloid content varies between 0.034% (leaves from Harar, Ethiopia) and 0.076% in leaves from Aden. Surprisingly, leaves from khat bushes grown in the United States and Europe have been found to contain little or almost no alkaloids (Krikorian and Getahun 1973, 379, 388). In leaves from Ethiopia, the cathinone content was measured as approximately 0.9 mg per leaf (fresh weight) (Halket et al. 1995, 111).

The flavonoid glycosides myricetin-3-*O*-β-D-galactoside, dihydromyricetin-3-*O*-rhamnoside, myricetin-3-*O*-rhamnoside, and quercetin-3-*O*-β-D-galactoside have been found in air-dried leaves and branch tips (Al-Meshal et al. 1986). These substances are similar to those contained in *Psidium guajava*.

Fresh leaves contain several polyphenolics (El Sissi and Abd Alla 1966). They are also rich in vitamins (especially vitamin C, but also thiamine, niacin, riboflavin, and beta-carotene) and minerals (Mg, Fe, Ca) as well as tannins, catechol tannins, sugars (mannitol, glucose, fructose, rhamnose, galactose, xylose), flavonoids, glycosides, amino acids (phenylalanine, choline, etc.), and proteins (Krikorian and Getahun 1973). There are also reports of an **essential oil** (Qédan 1972).

Effects

The primary effect of khat is an increase in energy and wakefulness (Widler et al. 1994). Khat chewing initially induces a cheerful mood, gaiety, and euphoria, together with a certain talkativeness. This state of arousal diminishes after about two hours. The stimulating effects usually begin with a tingling sensation on the head. It is said that khat "produces a social delirium" (Remann 1995, 79). The effect of the leaves is often compared to that of a "combination of **caffeine** and **morphine**" (R. Schröder 1991, 125*).

Sufis and dervishes use khat to produce an

"Let the jewels of qât go around,
emerald leaves of the small leaves.
Its ingestion sweetens my heart,
its sight my eye,
my condition and my times
are pleasant because of it.
Its hearts [i.e., the leaves]
bear the secrets that they place in our hearts.
They then flow into the most secret of thoughts.
[Qât is] the Burâq of the ascension of my heart,
as soon as it needs it.
Gabriel is my heart,
who travels to the highest heaven.
. . . .
Its use, say the Mursidûn [Sufis],
is like the enlightenment of the mysteries,
the seclusion of the forty days.
I have never wanted to ascend into the heaven of
my view in the universe, unless
I can make qât my ladder"

Abdallâh ibn Al-Imâm Saraf ad-Dîn
Qasîde (sixteenth century)
(in Schopen 1978, 85)

Cathinone

Cathine

"For centuries, khat has been used traditionally in Islamic cultures and tolerated by the Koran as a part of religious and social life. Consuming khat in groups is especially popular, as this is said to increase the ability to communicate and stimulate the fantasy and power of imagination. When working, khat is consumed individually, primarily for its performance-enhancing and hunger-suppressing effects."

Rudolf Brennesien and
Karoline Mathys
Catha
(1992, 735)

"At exactly two o'clock in the afternoon, the proud Yemenite people roll down their store blinds and leave work to go straight to their inebriation, the qat inebriation. I have rarely seen a people who are as full of themselves as the Yemenite people are with qat. I have seldom encountered a people that could puff up its cheeks so mightily. With qat. Show me your cheeks, and I will tell you if you are from Yemen. Fat cheeks appear to be the primary physical effect of this people's drug."

Micky Remann
Der Globaltrottel [The Globe-Trotting Moron]
(1995, 79)

† Translator's note: Effective July 1, 1992, khat leaves were added to Appendix 2 of the Swiss Federal Drug List, thereby making them illegal to import, possess, or use in Switzerland. Cathinone was already listed in the same appendix as a hallucinogen.

ecstatic state, but khat "will not induce this if the greatest intention is not present. If it does not appear, then you are negligent" (Schopen 1978, 200). In other words, ecstatic effects occur only when set and setting are taken into consideration.

Cathinone, the main active constituent, has been described as a "natural amphetamine" and has correspondingly similar effects (Kalix 1992). Cathinone interacts with the neurochemistry of dopamine (Pehek et al. 1990) and frees catecholamine at the synapses (Kalix 1992). It has the identical or at least very similar pharmacological properties and the same sympathomimetic effects as amphetamine (Kalix 1992; Widler et al. 1994). However, the effects of the leaves appear to be conditioned by the synergistic effects of cathinone and the other constituents (Krikorian and Getahun 1973, 278). Khat, or the mixture of substances contained in the leaves, also has interesting cholesterol-lowering effects (Ahmed and El-Qirib 1993, 215).

Apart from its psychoactive effects, khat also has an antidiabetic effect. Long-term chronic use can cause stomach problems, undernourishment, and nervousness. Ethiopian Christians claim that "insanity" is prevalent among Muslims because of their constant khat use (Krikorian and Getahun 1973, 378). A World Health Organization (WHO) document from 1964 notes, "Physical dependence (in the sense in which this is understood for morphine and substances with morphine-like effects or of the barbiturate type) does not occur [with khat], even when some tolerance to the effects has been acquired" (cited in Getahun and Krikorian 1973, 375).

Commercial Forms and Regulations

In Ethiopia, khat is divided into three commercial grades, depending upon the size and age of the leaves as well as their taste and tenderness: *kudda* (first quality), *uretta* (second quality), and *kerti* (third quality). In Kenya, a distinction is made between the qualities *giza* (best) and *kangeta* (lower quality). The best quality, *giza-bomu*, does not even make it to market because the plantation owners themselves consume it (Geisshüsler and Brenneisen 1987, 276). Some two hundred different sorts are recognized in Yemen (Schopen 1978, 66ff.). All the attempts to suppress khat use in Yemen or even to replace it with chewing gum (!) have—quite rightly—failed (Schopen 1978, 11).

Today, khat leaves are used in all those areas of the world in which ethnic groups from the traditional khat countries have settled. To serve them, shipments are sent by air freight daily to France, Italy, England, Switzerland,† and even the United States. Around the world, some two to eight million portions of khat are chewed every day. The average price for a bundle of 50 g is approximately ten dollars (Brenneisen and ElSohly 1992, 99, 109).

In Arabia, the dried leaves are sold in supermarkets for use as tea (R. Schröder 1991, 127*). In contrast, the fresh leaves are forbidden, as they are in Djibouti (Brenneisen and ElSohly 1992, 111).

Upon the recommendation of WHO, cathinone in its pure form was made an internationally controlled substance listed in Schedule I of the U.N. Convention on Psychotropic Substances (Brenneisen and ElSohly 1992, 109).

On the black market, one can occasionally find so-called khat pills (Nexus). Although the label often indicates that these pills contain extracts of *Catha edulis*, they actually consist of pure 2-CB, a synthetic phenethylamine with empathogenic effects (Schulgin and Schulgin 1991, 503ff.*).

Literature
See also the entry for **ephedrine**.

Ahmed, M. B., and A. B. El-Qirbi. 1993. Biochemical effects of *Catha edulis*, cathine, and cathinone on adrenocortical functions. *Journal of Ethnopharmacology* 39:213–16.

Beitter, A. 1900. *Pharmacognostisch-chemische Untersuchung der Catha edulis*. Strassburg: Schlesier und Schweikhardt.

———. 1901. Pharmakognostisch-chemische Untersuchung der *Catha edulis*. *Archiv der Pharmazie* 239:17–33.

Brenneisen, Rudolf, and Mahmoud A. ElSohly. 1992. Socio-economic poisons: Khat, the natural amphetamine. In *Phytochemical resources for medicine and agriculture*, ed. H. N. Nigg and D. Seigler, 97–116. New York: Plenum Press.

Brenneisen, Rudolf, and S. Geisshüsler. 1985. Psychotropic drugs. III: Analytical and chemical aspects of *Catha edulis* Forssk. *Pharm. Acta Helvetica* 60 (11): 290–301.

Brenneisen, Rudolf, S. Geisshüsler, and X. Schorno. 1984. Merucathine, a new phenylalkylamine from *Catha edulis*. *Planta Medica* 50:531.

Brenneisen, Rudolf, and Karoline Mathys. 1992. Catha. In *Hagers Handbuch der pharmazeutischen Praxis*, 5th ed., 4:730–40. Berlin: Springer.

Brilla, R. 1962. *Über den zentralerregenden Wirkstoff der frischen Blätter von Catha edulis* Forsskal. Dissertation, Bonn.

Brücke, Franz Th. von. 1941. Über die zentrale Wirkung des Alkaloids Cathin. *Naunyn-Schmiedeberg's Archiv für Experimentelle Pathologie und Pharmakologie* 198:100–6.

Elmi, Abdullahi S. 1983. The chewing of khat in Somalia. *Journal of Ethnopharmacology* 8:163–76.

El Sissi, H. I., and M. F. Abd Alla. 1966. Polyphenolics of the leaves of *Catha edulis*. *Planta Medica* 14:76–83.

Friebel, H., and R. Brilla. 1963. Über den zentralerregenden Wirkstoff der frischen Blätter und Zweigspitzen von *Catha edulis* Forssk. *Naturwissenschaften* 50:354–55.

Geisshüsler, S. 1988. Zur Chemie, Analytik und Pharmakologie von Phenylalkylaminen aus *Catha edulis* Forssk. (Celastraceae). Dissertation, Bern.

Geisshüsler, S., and Rudolf Brenneisen. 1987. The content of psychoactive phenylpropyl and phenylpentenyl khatamines in *Catha edulis* Forssk. of different origin. *Journal of Ethnopharmacology* 19:269–77.

Getahun, Amare, and A. D. Krikorian. 1971. Chat: Coffee's rival from Harar, Ethiopia. I: Botany, cultivation and use. *Economic Botany* 25:353–77.

Giannini, A., H. Bunge, J. Shasheen, and W. Price. 1986. Khat: Another drug of abuse? *Journal of Psychoactive Drugs* 18:155–58.

Halket, J. M., Z. Karasu, and I. M. Murray-Lyon. 1995. Plasma cathinone levels following chewing khat leaves (*Catha edulis* Forssk.). *Journal of Ethnopharmacology* 49:111–13.

Kalix, Peter. 1981. Cathinone, an alkaloid from khat leaves with amphetamine-like releasing effect. *Psychopharmacology* 74:269–79.

———. 1988. Khat: A plant with amphetamine effects. *Journal of Substance Abuse Treatment* 5:163–69.

———. 1990. Pharmacological properties of the stimulant khat. *Pharmacology and Therapeutics* 48:397–416.

———. 1992. Cathinone, a natural amphetamine. *Pharmacology und Toxicology* 70:77–86. (Excellent bibliography.)

Kennedy, John G. 1987. *The flower of paradise: The institutionalized use of the drug qat in North Yemen.* Dordrecht: D. Reidel Publishing.

Kennedy, John G., J. Teague, and L. Fairbanks. 1980. Qat use in North Yemen and the problem of addiction: A study in medical anthropology. *Culture, Medicine and Psychiatry* 4:311–44.

Krikorian, Abraham D. 1984. Kat and its use: A historical perspective. *Journal of Ethnopharmacology* 12:115–78.

———. 1985. Growth mode and leaf arrangement in *Catha edulis* (kat). *Economic Botany* 39 (4): 514–21.

Krikorian, A. D., and Amare Getahun. 1973. Chat: Coffee's rival from Harar, Ethiopia. II: Chemical composition. *Economic Botany* 25:378–89.

Margetts, E. L. 1967. Miraa and myrrh in East Africa: Clinical notes about *Catha edulis. Economic Botany* 21:358–62.

Mathys, Karoline. 1993. Untersuchung der pharmakologischen Wirkung von *Catha edulis* Forssk. (Khat) im Menschen. Dissertation, Bern.

Meshal, Ibrahim A. al-, Mohamed S. Hifnawy, and Mohammad Nasir. 1986. Myricetin, dihydromyricetin, and quercetin glycosides from *Catha edulis. Journal of Natural Products* 49 (1): 172.

Musès, Charles. 1989. The sacred plant of ancient Egypt. In *Gateway to inner space*, ed. C. Rätsch, 143–58. Bridport, England: Prism Press.

Pehek, E., M. Schlechter, and B. Yamamoto. 1990. Effects of cathinone and amphetamine on the neurochemistry of dopamine in vivo. *Neuropharmacology* 29:1171–76.

Qédan, S. 1972. Über das ätherische Öl von *Catha edulis. Planta Medica* 21:410–15.

Remann, Micky. 1995. *Der Globaltrottel.* 2nd ed. Der Grüne Zweig 177. Löhrbach: Werner Pieper's MedienXperimente.

Schopen, Armin. 1978. *Das Qât: Geschichte und Gebrauch des Genußmittels Catha edulis Forssk. in der Arabischen Republik Jemen.* Wiesbaden: Franz Steiner.

———. 1981. Das Qât in Jemen. In *Rausch und Realität*, ed. G. Völger, 1:496–501. Cologne: Rautenstrauch-Joest-Museum für Völkerkunde.

Special issue devoted to *Catha edulis* (khat). 1980. *Bulletin of Narcotics* 32 (2): 1–99.

Van Wyk, A. E., and M. Prins. 1987. A new species of *Catha* (Celastraceae) from southern Natal and Pondoland. *South African Journal of Botany* 53:202–5.

Wang, Chen-hwa. 1936. The studies of Chinese Celastraceae. I. *The Chinese Journal of Botany* 1:35–68.

Weir, S. 1985. *Qat in Yemen: Consumption and social change.* Dorset: British Museum Publications.

Widler, Peter, Karoline Mathys, Rudolf Brenneisen, Peter Kalix, and Hans-Ulrich Fisch. 1994. Pharmacodynamics and pharmacokinetics of khat: A controlled study. *Clinical Pharmacology und Therapeutics* 55 (5): 556–62.

"The most common method of consuming khat is to chew it. As a part of the religious social life, khat was traditionally used in the Islamic cultures. Today, khat sessions are still subjected in part to strict, ritualized customs. . . .

"Khat use induces a general state of well-being. The consumers become cheerful, excited, and talkative. Problems appear to be more easily dealt with, the sense of space and time partially disappears, without hallucinations being produced. All tiredness vanishes and, as a result of the anorexic effect, all feelings of hunger as well."

KAROLINE MATHYS
"UNTERSUCHUNGEN DER PHARMAKOLOGISCHEN WIRKUNG VON *CATHA EDULIS*" [RESEARCH INTO THE PHARMACOLOGICAL EFFECT OF *CATHA EDULIS*]
(1993, 6)

Cestrum nocturnum Linnaeus

Night-Blooming Jessamine

Family
Solanaceae (Nightshade Family); Subfamily Cestroideae, Cestreae Tribe

Forms and Subspecies
One variety has been described for Mexico: *Cestrum nocturnum* L. var. *mexicanus.*

Synonyms
Cestrum hirtellum Schlechtendal
Chiococca nocturna Moc. et Sessé

Folk Names
Akab-xiu (Mayan, "night plant"), ak'ab-yom, äk'a'yo'om (Lacandon, "night foam"), arum ndalu (Javanese), dama de noche, ejek tsabalte', galán de noche, galán de tarde, hammerstrauch, hedeondilla, hedioncilla, hediondilla, hierba de zorillo,[93] hierba hedionda, huele de noche, ijyocxibitl, iscahuico (Totonac), ishcahuico'ko, it'ib to'ol (Huastec), lady of the night,[94] mach-choch, minoche, mocxus, nachtschaum, nachtschaumbaum, night-blooming jasmine, night-blooming jessamine, orquajuda negro, palo huele de noche, parqui, pipiloxihuitl, pipiloxohuitl (Náhuatl), putanoche ("whore's night"), scauilojo (Totonac), tzisni sanat, tzisnutuwan, tzon tzko kindi t oan (Amuzgo), zitza kiwi (Totonac)

In Peru, one *Cestrum* species that has not been botanically identified is known locally as *hierba santa*, "sacred herb."

History
Most *Cestrum* species are indigenous to the Amazon basin, and many occur in the Andes (Hunziker 1979, 70). It is unknown whether these psychoactive plants were used for ritual or medicinal purposes during pre-Spanish times, but it is possible. To date, no traditional psychoactive use has been documented. Overall, little ethnobotanical and ethnopharmacological research has been conducted into the genus *Cestrum* (cf. *Cestrum parqui*).

Distribution
The shrub is indigenous to the West Indies, Central America, and South America; in Mexico, it occurs in Coahuila, Guerrero, Oaxaca, Veracruz, and Chiapas (Martínez 1994, 437*). As a result of cultivation, it is also found in Southern California (Enari n.d., 22).

Cultivation
Propagation can occur with seeds or cuttings. The seeds are either pre-germinated or sowed in seedbeds. The cuttings (approximately 20 cm long) are separated from the branch tips and placed in water until roots develop. They can then be planted in soil. The plant does not tolerate frost or a cold climate and requires considerable water. In central Europe, it can be grown only as a houseplant or in a greenhouse. In tropical regions, the bush often is planted for the scent it produces at night (Morton 1995, 130*).

Appearance
This perennial bush can grow as tall as 4 meters. It bears shiny leaves. The 2 to 3 cm long, funnel-shaped flowers are greenish white and grow in clusters. At night, they open and exude a sweet, delicious scent that is very intense and penetrating. The white fruits are round but slightly oval and grow to 2 cm in length. The bush can bloom three or four times a year (Morton 1995, 130*). When rubbed, the fresh leaves release a scent similar to that of the fresh leaves of *Datura innoxia* and *Datura stramonium.*

Today, 175 to 250 species are recognized in the genus *Cestrum* (D'Arcy 1991, 78*; Hunziker 1979, 70). Many species are easily confused with one another. *Cestrum nocturnum* is easily mistaken for *Cestrum diurnum* L. (dama de noche, day jessamine), a species originally from the Antilles. It can also be confused with the Guatemalan species *Cestrum aurantiacum* Lindl., which develops magnificent yellow flowers.

Cestrum nocturnum is occasionally crossed with *Cestrum diurnum* L., as the hybrid (*Cestrum*

Left: Inflorescence of the tropical night-blooming jessamine *Cestrum nocturnum*, a typical night-scented plant.

Right: Many *Cestrum* species are easily confused with one another. (*Cestrum aurantiacum*, from southern Mexico)

93   This name is also given to *Artemisia mexicana.*

94   A Caribbean relative, *Cestrum latifolium* Lam., is also known by this name (Wong 1976, 37*).

nocturnum x *diurnum*) is more easily adaptable to nontropical climates. *Cestrum nocturnum* can be confused with many other yellow-blooming species of the genus (cf. *Cestrum parqui*).

Psychoactive Material
— Leaves
— Flowers

Preparation and Dosage
When dried, the leaves may be smoked alone or in **smoking blends** (cf. *Cestrum parqui*). Fresh or dried leaves can be decocted into a tea (Argueta et al. 1994, 830*). No information concerning dosages is available.

Ritual Use
In the mythology of the Lacandon of Naha', who have preserved the pre-Hispanic cosmology of the Maya (cf. **balche'**) into the present day, the lord of death (Kisin) was born from a flower of *Cestrum nocturnum*. It is possible that the ancient Maya may have used the plant in necromantic rituals. Apart from this, we do not yet know of any traditional use of the plant for psychoactive purposes.

Artifacts
None

Medicinal Use
The Yucatec Maya use decoctions of the plant as medicinal baths to treat cold sweats as well as a curious illness known as *ak'ahkilka* ("night sweats") (Pulido S. and Serralta P. 1993, 61*).

In Mexican folk medicine, an extract of the leaves is used as an antispasmodic, especially in the treatment of epilepsy (Martínez 1994, 438*). It is frequently used to treat headaches and illnesses resulting from *susto*, "fright" (Argueta et al. 1994, 830*).

Constituents
The composition of the powerful scent is as little understood as are most of the constituents (Morton 1995, 130*). Chemical studies of *Cestrum nocturnum* are lacking (Aguilar Contreras and Zolla 1982, 56*). The sapogenin steroids tigogenine, smilagenine, and yuccagenine have been found only in the leaves (Arbain et al. 1989, 76; Argueta et al. 1994, 830*).

The characteristic constituents of the genus *Cestrum*, i.e., those that are chemotaxonomically relevant, are saponines (Schultes 1979b, 151*). Alkaloids, tanning agents, and glycosides are also present in the genus (Wong 1976, 137*). Many species contain alkaloids of the **nicotine** type (Schultes and Raffauf 1991, 36*). *Cestrum diurnum* contains a principle that behaves like **atropine** and produces the same effects (Morton 1995, 24*). The saponines yuccagenine (0.5%) and tigogenine (0.04%) are found in the entire plant (Frerichs, Arends, and Zörnig, *Hagers Handbuch* 1980, 821*).

Effects
Simply inhaling the scent of this plant deeply is said to be sufficient to induce psychoactive effects (Argueta et al. 1994, 830f.*). The berries as well as the leaves are also reputedly able to induce hallucinations (Aguilar Contreras and Zolla 1982, 56*; Enari n.d., 22).

One child reportedly experienced profound hallucinations after consuming several fruits of *Cestrum diurnum* (Morton 1995, 24*).

Commercial Forms and Regulations
In the tropical regions of the Americas, young bushes are sold in tree farms.

Literature
See also the entry for *Cestrum parqui*.

Arbain, Dayar, Jack R. Cannon, Afriastini et al. 1989. Survey of some West Sumatran plants for alkaloids. *Economic Botany* 43 (1): 73–78.

Enari, Leonid. n.d. *Poisonous plants of Southern California*. Arcadia, Calif.: Dept. of Arboreta and Botanic Gardens.

Halim, A. F., R. P. Collins, and M. S. Berigare. 1971. Isolation and characterization of the alkaloids of *Cestrum nocturnum* and *Cestrum diurnum*. Analysis of the essential oil of *Comptania peregrina*. *Planta Medica* 20:44.

Hunziker, Armando T. 1979. South American Solanaceae: A synoptic survey. In *The biology and taxonomy of the Solanaceae*, ed. J. G. Hawkes, R. N. Lester, and A. D. Skelding, 49–85. London: Academic Press.

Karawya, M. S., A. M. Rizk, et al. 1971. Phytochemical investigation of certain *Cestrum* species: General analysis, lipids, and triterpenoids. *Planta Medica* 20:363.

Ma'ax, K'ayum, and Christian Rätsch. 1994. *Ein Kosmos im Regenwald*. 2nd ed. Munich: Diederichs.

"In the beginning Hachäkyum, our true lord, created the night foam tree, for Hachäkyum was to create Kisin, the lord of death. He created Kisin.

"He planted the night foam tree, for when it was night, the flowers of the night foam tree were to open and Kisin was to blossom out of these. Kisin, the lord of death, was born there of the night foam tree."

<small>FROM THE LACANDON CREATION MYTH</small>
<small>(IN MA'AX AND RÄTSCH 1994, 41)</small>

A Mexican species of the genus *Cestrum* is known in Aztec as *iyacxihuitl*, "stink tree." The name refers not to the scent of the flowers but to the typical inebriating nightshade scent of its leaves. (From Hernández, 1942/46 [Orig. pub. 1615*]

Cestrum parqui L'Héritier

Willow-Leafed Jessamine, Palqui

Above: The willow-leafed jessamine (*Cestrum parqui*) in blossom.

Right: This delicate jessamine (*Cestrum elegans* (Brongn. ex Neum.) Schlechtend. [syn. *C. purpureum* (Lindl.) Standl.]) is from Mexico. Like many other *Cestrum* species, it contains psychoactive and toxic constituents.

95 In the Andes, the closely related species *Cestrum matthewsii* Dun. (which may possibly also produce narcotic or other psychoactive effects) is also known by the name *hediondilla* (Bastien 1987, 116 f.*).

Family
Solanaceae (Nightshade Family); Subfamily Cestroideae, Cestreae Tribe

Forms and Subspecies
None

Synonyms
Cestrum salicifolium H. et B.
Cestrum virgatum Ruíz et Pavón

Folk Names
Alhuelahuen, chilenischer hammerstrauch, duraznillo negro, green cestrum, hediondilla ("stinking"),[95] paipalquen, paique, palguin, palki, palqui, palqui blanco, palquin, parqui, parquistrauch, willow-leafed jessamine, yerba santa

History
Since pre-Columbian times, the Mapuche of southern Chile have used this plant for medicinal and probably ritual purposes. The Spanish missionary Bernabé Cobo described the medicinal use of a plant known as *hediondilla* in his *Historia del Nuevo Mundo* (1653) (Bastien 1987, 117*). Louis Lewin provided an early description of the use of the wood and leaves as a tobacco substitute (cf. *Nicotiana tabacum*) among the Cholos Indians (Lewin 1980 [orig. pub. 1929], 411*). Palqui generally appears to have been smoked prior to the introduction of tobacco (Hartwich 1911, 48, 523*):

According to Ochsenius [1884], the Chonos (sic) Indians on the island of Chiloe smoke the herbage of another nightshade known as *palguin* (*Cestrum parqui* L'Hérit.) when tobacco is lacking. It is possible that this is a remnant of a smoking custom that is older than [that of] tobacco. (Hartwich 1911, 48 f.*)

Distribution
The plant is from central Chile but had spread as far as Peru, Argentina, Uruguay, and Brazil at an early date (von Reis and Lipp 1982, 267*). In Chile, it occurs as far south as Osorno and Chiloé (Hartwich 1911, 523*; Montes and Wilkomirsky 1987, 164*). It has now been introduced into the Mediterranean region as well as California (Zander 1994, 179*).

Cultivation
Propagation is best achieved using seed. The plant is sometimes grown as an ornamental.

Appearance
The bush, which can grow as tall as 1.5 meters, has narrow, lanceolate, pale green leaves. The yellow, tubular, five-pointed flowers are located in panicles or clusters at the ends of the branches. In South America, the flowers bloom in October and November, and they exude a strong, inebriating scent. The plant produces small, oval-round berries (approximately 5 mm in length) that take on a shiny black color as they mature.

The palqui bush is easily confused with *Cestrum aurantiacum* Lindl. Other similar species include *Cestrum elegans* (Brongn. ex Neum.) Schlecht., *Cestrum ochraceum*, and *Cestrum laevigatum* Schlecht. (Roth et al. 1994, 209*).

Psychoactive Material
— Leaves
— Bark
— Wood

Preparation and Dosage
The leaves of *Cestrum parqui* are dried, chopped, and smoked alone or in **smoking blends**, e.g., with **Cannabis sativa**. Three or four leaves per person is a good starting dosage. The leaves are an ingredient in psychoactive fumigations using **Latua pubiflora** (cf. also **incense**).

A decoction of leaves and bark or a bark tea (infusion) is drunk for folk medicinal purposes.

In Brazil, the dried leaves of the closely related species *Cestrum laevigatum* Schlecht. are known as *maconha* and are smoked as a marijuana substitute (Schultes and Hofmann 1992, 38*).

Ritual Use

In southern Chile, this sacred plant is used in shamanic healing activities. The plant possesses the virtue or power known as *contra*, which resists attacks by sorcerers or black shamans (*tué-tué* or *chonchones*). Since illnesses are often produced by other shamans, they can best be healed by a shaman with the aid of palqui. The stems are used to make wooden crosses that are attached to the windows or outer walls of houses as a magical protection against disease demons. A tea also protects from *susto* ("fright") and *mal de ojo* ("evil eye") and is drunk during purification ceremonies (*limpia*) (Hoffmann et al. 1992, 172*).

The shamans of the Kamsá (Sibundoy, Colombia) refer to one *Cestrum* species as *borrachero andoke*. They press the leaves of the plant in water and drink the resulting solution so that they can see things as if under the influence of **ayahuasca** (Schultes and Raffauf 1991, 36*).

Artifacts

The stems are made into wooden crosses and amulets.

Medicinal Use

The Mapuche of southern Chile drink an infusion of the leaves to treat smallpox, tuberculosis, and leprosy; to treat herpes, to wash out wounds (Houghton and Manby 1985, 99f.*); and also to treat fever (Montes and Wilkomirsky 1987, 164*; Schultes 1980, 114*). A tea or decoction of the bark is ingested as a powerful analgesic and sleeping agent (Hoffmann et al. 1992, 171f.*). The leaves and the freshly pressed juice of the plant are used especially in the treatment of ant bites. In Chile, it is said that "wherever the devil has put ants, there God has planted a palqui tree" (Mösbach 1992, 105*). In the Andes, the leaves are used primarily to treat wounds (Bastien 1987, 116f.*).

The Colombian Sibundoy Indians drink a tea prepared from the closely related species *Cestrum ochraceum* Francey [syn. *Cestrum ochraceum* var. *macrophyllum* Francey] to treat headaches, pain, swelling, fever, and rheumatism (Bristol 1965, 267*). It is said that the patient will fall into a mild state of delirium if he drinks (too much) of the tea (Schultes 1981, 34*; Schultes and Raffauf 1991, 36*). In Brazil, *Cestrum laevigatum* is used as a sedative (Frerichs, Arends, Zörnig, *Hagers Handbuch* 1980, 820*). On the Brazilian coast, the leaves are smoked as a marijuana substitute (cf. **Cannabis indica**) (Schultes 1979b, 151*).

Constituents

Cestrum parqui contains solasonine, a glycosidic steroidal alkaloid, and solasonidine (Montes and Wilkomirsky 1987, 164*; Schultes 1979b, 151*). The bitter alkaloid parquine has the empirical formula $C_{21}H_{39}NO_8$ and produces effects like those of **strychnine** or **atropine** (Roth et al. 1994, 209*). A triterpene and fitoesterol are also present. The leaves and fruits contain tigogenin, digallogenin, digitogenin, and ursolic acid (Montes and Wilkomirsky 1987, 164*). The fruits contain at least three alkaloids. Solasonine is the primary active constituent (Hoffmann et al. 1992, 172*). The alkaloid is found in the leaves as well as the wood (Hartwich 1911, 523*).

Cestrum parqui and *Cestrum laevigatum* contain gitogenin and digitogenin.

Effects

Pharmacologically, the extract has an atropine-like effect (Montes and Wilkomirsky 1987, 164*; cf. **atropine**).

Smoking *Cestrum parqui* leaves produces effects that are clearly psychoactive and are reminiscent of the effects of smoking **Brugmansia** leaves. However, no dryness of the mouth occurs. The effects are relatively weak and are perceived as a mild euphoria and physical relaxation.

Commercial Forms and Regulations

In Chile, the dried leaves are available at most herb stands and in shops that sell natural medicines. Apart from this, the plant is not sold.

Literature

See also the entry for *Cestrum nocturnum*.

Silva, M., and P. Mancinell. 1959. Chemical study of *Cestrum parqui*. *Boletin de la Sociedad Chilena de Química* 9:49–50.

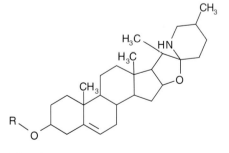

Solasonine

From top to bottom:
The dried leaves of *Cestrum parqui* can be smoked as a mild inebriant.

A cultivated variety of jessamine, known by the name *Cestrum rubrum*. It also contains psychoactive constituents.

The fruits of *Cestrum rubrum*.

Cinnamomum camphora (Linnaeus) Siebold

Camphor Tree

Botanical illustration of the Southeast Asian camphor tree *Cinnamomum camphora*, previously known as *Camphora officinarum*. (Engraving from Pereira, *De Beginselen der Materia Medica en der Therapie*, 1849)

Family
Lauraceae (Laurel Family); Subfamily Lauroideae, Cinnamomeae Tribe, Cinnamominae Subtribe

Forms and Subspecies
Distinctions were previously made among different forms, varieties, and even species that are now understood simply as chemical races (Morton 1977, 103f.*). The variety *Cinnamomum camphora* var. *linaloolifera*, which is especially rich in sesquiterpenes, is still important. Most of the distinctions are geographical in nature (Chaurasia 1992, 896):

Cinnamomum camphora ssp. *formosana* (Taiwanese camphor)

Cinnamomum camphora ssp. *japonicum* (Japanese camphor)

Cinnamomum camphora ssp. *newzealanda* (New Zealand camphor)

Synonyms
Camphora camphora Karst.
Camphora officinarum Nees
Cinnamomum camphora Fries
Cinnamomum camphora (L.) Nees et Eberm.
Cinnamomum camphora Presl et Eberm.
Cinnamomum camphoriferum St. Lag.
Laurus camphora L.
Laurus camphorifera Salisb.
Persea camphora Spr.

Folk Names
Alcanfor (Spanish), baum-camphera, borneo-campher, borneo-kampfer, camfora (Italian), campherbaum, camphero, camphor laurel, camphor tree, camphre, camphrier du japon, chang (Chinese), chang-shu, cusnocy (Old Japanese), cutakkarpuram (Malay), gaara-boon (Tai), ga bur (Tibetan), gaburi (Mongolian), gum camphor, japaansche kamferboom (Dutch), kafr (Czech), kamfer, kamferboom (Dutch), kámforfa (Hungarian), kampferbaum, kampferlorbeer, kanfur (Arabic), kapor, kapur, kâpûr, karpura, karpurah (Sanskrit), karpuram (Tamil), kuso-noki (Japanese), laure à camphre, laurocanfora (Italian), re

History
In China and Japan, camphor has been obtained from the camphor tree since at least the ninth century C.E. (Morton 1977, 105*). In Asia, camphor has been a much praised aphrodisiac[96] and remedy since ancient times (Warrier et al. 1994, 2, 81*).

Arabs were using camphor as early as the eleventh century for all types of medical purposes

(Bärtels 1993, 123*). The first camphor tree was brought to Europe in 1676 and planted in Holland (Morton 1977, 103*). Since 1910, camphor has been produced synthetically in Germany from α-pinene (turpentine). In the Roaring Twenties, camphor often was used as an inebriant.

Distribution
The tree is indigenous to India, China, and Formosa (Taiwan). From there, it spread throughout the tropical zones of Southeast Asia. In the Mediterranean region, it is even grown as an ornamental (Bärtels 1993, 123*).

Cultivation
The camphor tree can be propagated from seed, cutting, scion, or rootstock. Cuttings that contain high amounts of camphor rarely develop roots on their own. The tree is usually grown from the seeds of twenty- to twenty-three-year-old mother trees. The seeds of younger trees are infertile. Seeds will germinate only when fresh, and only a very few of the planted seeds do so. The germination period is approximately ninety days. When the seedlings are six months old, they are trimmed for the first time and transplanted (Morton 1977, 104*). Trees that are older than thirty years yield the most camphor.

In the tropics (Sri Lanka, India), the tree thrives best at altitudes between 1,220 and 1,800 meters when precipitation is between 114 and 368 cm per annum.

Most commercial camphor tree plantations are on Taiwan, but there are others in India and in the Republic of Georgia (Morton 1977, 103*).

Appearance
This evergreen tree can grow as tall as 50 meters. It develops a gnarled stem (up to 5 meters in diameter) and a projecting crown. It has long-stemmed, leathery, smooth, oblong leaves that are shiny green on their upper surface and dull blue-green on their underside. When young, the leaves often have a reddish color. When rubbed, the leaves smell strongly of camphor; this is the most reliable method for identifying the tree. The greenish white flowers are small and rather inconspicuous. They form axillary panicles 5 to 7 cm in length. The fruits are small, one-seeded berries surrounded by a calyx (Chaurasia 1992, 896).

Because of its habitus, the tree is easily confused with the true or Ceylon cinnamon tree (*Cinnamomum verum* Presl [syn. *Cinnamomum ceylandicum* Bl.]). However, the leaves of the cinnamon tree smell (almost exaggeratedly) of cinnamon when they are rubbed. The genus

96 In contrast, camphor is regarded as an anaphrodisiac in Cuba and is used in medicine as such (Morton 1977, 106*).

Cinnamomum encompasses approximately 150 to 200 species, most of which occur in eastern Asia. Many of these resemble the camphor tree (Bärtels 1993, 123*; Chaurasia 1992, 884).

Psychoactive Material
— Leaves
— Fruits (fructus camphora)
— Camphor (camphora; depositum in the oil cells, Japanese camphor)
— Camphor tree oil (cinnamomi camphorae aetheroleum, oleum camphorae, oleum cinnamomi camphorae, camphor oil, huile de camphre)

The so-called Borneo camphor (also known as *kapur*) is from the stock plant *Dryobalanops aromatica* Gaertn., a member of the resin-producing family Dipterocarpaceae. The odoriferous substance borneol is distilled from its wood. Crystals of pure camphor sometimes crystallize out of its trunk (Martin 1905).

There is also a safrole camphor (cf. *Sassafras albidum*) and a parsley camphor (= apiol; see *Petroselinum crispum*).

Preparation and Dosage
Actual camphor is obtained by careful distillation from small pieces of the wood. It crystallizes out and is ready to be used.

The information concerning dosages for internal use varies. Up to 10 g is said to produce pleasant inebriating effects. However, reactions differ from person to person[97]: "Serious toxic effects occurred as a result of ingestion of 10 to 20 g of camphor; lethal poisoning from 6 g subcutaneous" (Fühner 1943, 237*).

In India and Nepal, camphor (*kapur*) is used primarily as a stimulating additive to **betel quids** and as an ingredient in **incense**.

The most important Japanese incense mixture for Buddhist services and ceremonies consists of five or seven coarsely chopped ingredients. The proportions can be varied as desired, so new scent compositions are constantly being produced (cf. **incense**). The *shokoh-5* mixture is a combination of:

Aloe wood	*Aquilaria agallocha*
White sandalwood	*Santalum album*
Cloves	*Syzygium aromaticum*
Cassia cinnamon	*Cinnamomum aromaticum*
Camphor	*Cinnamomum camphora* ssp. *japonicum*

The *shokoh-7* mixture consists of the same five substances plus ginger (*Zingiber officinale*) and amber (Morita 1992).

The leaves of the Cambodian camphor tree (*Cinnamomum tetragonum*) are made into a stimulating drink (von Reis Altschul 1975, 78*).

Ritual Use
In Japan, camphor is an important ingredient in ritual **incenses**. It is also one of the most important incense materials in the traditional Tibetan Tantra cult (Yeshe Tsogyal 1996) and has enormous ritual significance, especially in southern India. In the region of Nordarcot is a sacred mountain known as Arunachala, "red mountain," which is said to be hollow inside and inhabited by beings with extraordinary spiritual abilities. A large temple at the mountain is dedicated to a goddess of the same name:

> Once a year, the priests celebrate her great festival. As soon as it begins in the temple, a gigantic flame is lit on the peak of the mountain [and] is fed by great quantities of butter and camphor. It burns for days and is visible for miles. (Brunton 1983, 153)

This cult is closely associated with Shiva, the god of ecstasy and inebriants, to whom camphor is also sacred:

> According to our sacred legends, the god Shiva once appeared as a fiery flame at the peak of the sacred red mountain. For this reason, once a year the priests of the temple light the great fire to commemorate this event, which must have occurred thousands of years ago. I assume that the temple was built for this festival, as Shiva still protects the mountain. (Brunton 1983, 165)

In Varanasi (= Benares), Shiva's sacred city, there is a shrine to Krishna in which a golden statue of the young god and lover is venerated. Offerings include flowers (e.g., *Cestrum nocturnum*), fruits (thorn apples; cf. *Datura metel*), and dyes. The incense that is burned at this site is camphor (Brunton 1983, 217).

In Malaysia, Borneo camphor had a ritual and magical significance for the indigenous Malay people:

> Along with the Hantu belief and the conception that things in nature could be charmed was a peculiar custom which, however, was found only among the Jakun and was known only by the name camphor language (Bhasa Kapor). The natives used the term "pantang kâpûr" (Malay, "pantang" = forbidden) for this, which expresses the fact that it is forbidden to use the normal Malay language while they are searching for camphor. . . . In fact, the Jakun believe that a "bisân" ["woman"] or spirit watches over the camphor trees [*Dryobalanops aromatica*] and that it is impossible to obtain camphor before a person has made her kindly disposed to

Top: Leaves and buds of the camphor tree (*Cinnamomum camphora*)

Bottom: Natural camphor from India. (Photograph: Karl-Christian Lyncker)

97 "A man who accidentally ingested 3.7 g of camphor at one time became dizzy, felt cold in his extremities, and experienced great anxiety, cold sweats, mild delirium, somnolence, and weak pulse, soon thereafter extreme heat, rapid pulse, and red urine" (Roth et al. 1994, 232*).

"Because the [camphor] tree can become several hundred years old and develop into an immense tree, it enjoys special veneration in China and Japan."

ANDREAS BÄRTELS
FARBATLAS TROPENPFLANZEN [COLOR ATLAS OF TROPICAL PLANTS] (1993, 123*)

"The king washed the jewel in purifying water, attached it to the top of a victory banner, enveloped it in the aromatic smoke of camphor and sandalwood, and spread an immeasurable series of offerings out in front of it. He then bathed, put on clean garments, and, after paying his respects to the gods of the four directions, spoke the following prayer: If this incomparable jewel that I have found truly is the perfect and precious wish-fulfilling jewel, so may everything humans and other beings wish for fall like a blessing rain!"

YESHE TSOGYAL
DER LOTUSGEBORENE IM LAND DES SCHNEES [THE LOTUS-BORN IN THE LAND OF SNOW] (1996, 39 f.)

Camphor

98 In China, recent or petrified alga colonies (*Collenia sinensis*) were also thought to be "dragon brains" and used for medicinal purposes (Read 1977, 9*).

168

them. During the night, it gives off shrill sounds . . . , and this is evidence that camphor trees are near. To placate the camphor spirit, the Jakun give it a part of their food before they themselves eat . . . , eat some soil, and utilize the special language. . . . (Martin 1905, 972 f.)

Since the beginning of the twentieth century, reports of the psychoactive use of camphor have increased:

Indeed, since about two decades one encounters in the upper circles of English society camphor eating men and women, who ingest the agent in milk, alcohol, pills, etc. The same can be found in the United States and in Slovakia. Women maintain that it gives them a fresh complexion. But the true motive appears to be that they achieve a certain state of excitation or inebriation that admittedly, it seems to me, requires a special deposition. (Lewin 1980 [orig. pub. 1929], 302*)

Today, camphor is used in Amazonia by mestizo shamans in connection with **ayahuasca**.

Artifacts

In Japan, the aromatic camphor wood is carved into ritual masks, e.g., of *tengu* (see **Amanita muscaria**, **ibotenic acid**) for the *gagaku* dance festivals (since the second century).

Medicinal Use

Since very early times, the camphor tree has been one of the most important medicinal plants in the Chinese materia medica. In Chinese, the white, aromatic camphor resin is known as *long nao xiang*, "dragon brain."[98] The Yellow Emperor used it as a remedy for headaches and hemorrhoids:

We do not know whether the congealed camphor reminded them of a brain and was ascribed to the king of animals because it was so rare and precious, or whether the name is derived from the fact that camphor was reserved for the emperor, the "dragon." (Fazzioli 1989, 23)

In China and Tibet, the camphor tree was long regarded as the "king of far eastern medicinal plants," for "camphor is comparable to a 'wild man' (Yeti, the snow man of the Himalayas)" (Kaufmann 1985, 106*). In Nepal, camphor is used as a stimulant, vermifuge, and digestive agent (Singh et al. 1979, 188*).

In Ayurvedic medicine, camphor is prescribed for inflammations, heart weakness, coughs, asthma, spasms, flatulence, diarrhea, and dysentery (Warrier et al. 194, 2:81*). Camphor is often

administered as a sedative, to cool off, so to speak, and for hysteria and nervousness:

Camphor increases *prana*, opens the senses, imparts clarity to the mind. . . . A pinch of camphor powder is sniffed when the nose is congested, for headaches, and to increase perception. During a *puja*, a religious service, camphor is burned as incense in order to purify the atmosphere and promote meditation. . . . To treat the respiratory tract, an infusion of camphor can also be boiled and the fumes inhaled. For internal use, only raw camphor should be used, not the synthetic camphor that is frequently sold in stores. (Lad and Frawley 1987, 179 f.*)

In Western medicine, camphor is highly important in the treatment of coughs and colds and well as fits of shivering (Morton 1977, 106*; Pahlow 1993, 388*). In homeopathy, camphora is used in accordance with the medical description for such ailments as colic and spasms (Roth et al. 1994, 233*).

Constituents

All parts of the plant contain camphor oil and **essential oils** with sesquiterpenes (campherenone, campherenol, camphor derivatives); the white substance camphor (empirical formula $C_{10}H_{16}O$) is precipitated from this. The amount of camphor can vary considerably. The leaves of Indian camphor trees contain 22.2% camphor.

The composition of the essential oil is complex and varies according to location, climate, et cetera; azulene, bisabolone, cadinene, camphene, α-camphorene, carvacrol, cineole (main component), π-cymol, eugenol, laurolitsine, Δ-limonene, orthodene, α-pinene, reticuline, safranal, safrole, salvene, and terpineol are some of the substances that have been identified. Safrole is often present in great quantities, and large amounts are contained in the wood. The highest concentrations of safrole are found in the roots (Morton 1977, 104*). The leaves also contain large amounts of safrole (cf. *Sassafras albidum*) (Chaurasia 1992, 896).

The heartwood of the stem contains sesquiterpenes and cyclopentenones (Takaoka et al. 1979). The alkaloids laurolitsine and reticuline are present in the roots (Chaurasia 1992, 896). The seeds contain primarily laurine and an oil whose composition is the same as that of coconut oil (cf. *Cocos nucifera*). The entire plant contains traces of caffeic acid, quercetin, camphor oil, and leucocyanidin (Chaurasia 1992, 896).

Effects

In the medical and toxicological literature, one repeatedly reads that high dosages of camphor can induce hallucinations (Morton 1977, 107*):

After ingestion of app. 1.2 g, the following can appear: a pleasant feeling of warmth of the skin and a general stimulation of the nerves, a need for movement, a tingling in the skin, and a peculiar, inebriation-like, ecstatic, mental excitement. "According to one such self-experimenter, the effects were clear and obvious with tendencies of the most beautiful kind." This condition lasted for one and a half hours. After ingestion of 2.4 g, a need for movement was felt. All movements were easier. When walking, the thighs were raised higher than usual. Mental work was impossible. A flood of thoughts occurred, one idea wildly followed another, quickly, without one persisting. *The consciousness of the personality was lost.* (Lewin 1980 [orig. pub. 1929], 302 f.*)

The inebriating effects of camphor are often compared to those of **alcohol**:

Following ingestion of larger amounts of camphor, nausea and vomiting can quickly remove the greatest part of the substance. When resorbed, mild toxic effects include central stimulation, dizziness, headache, a state of inebriation like that produced by alcohol, with sensory delusions and delusional ideas; kidney irritation occurs, rarely hematuria. With frequent use of camphor, "camphor addiction" can develop. (Fühner 1943, 237*)

Commercial Forms and Regulations

Since camphor is relatively easy to synthesize, pharmacies now offer almost only synthetic camphor (*Camphora synthetica* DAB 8). It is an open question as to whether this has the fine qualities of the natural product. In spite of its name, the so-called camphor oil available in pharmacies has had all of its camphor removed.

Literature

See also the entries for **incense** and **essential oil**.

Brunton, Paul. 1983. *Von Yogis, Magiern und Fakiren: Begegnungen in Indien.* Munich: Knaur.

Chaurasia, Neera. 1992. *Cinnamomum.* In *Hagers Handbuch der pharmazeutischen Praxis*, 5th ed., 4:884–911. Berlin: Springer.

Fazzioli, Edoardo. 1989. *Des Kaisers Apotheke.* Bergisch-Gladbach: Gustav Lübbe.

Martin, Rudolf. 1905. *Die Inlandstämme der malayischen Halbinsel.* Jena: Gustav Fischer.

Morita, Kiyoko. 1992. *The book of incense: Enjoying the traditional art of Japanese scents.* Tokyo: Kodansha International.

Takaoka, Daisuke, Minoru Imooka, and Mitsuru Hiroi. 1979. A novel cyclopentenone, 5-dodecanyl-4-hydroxy-4-methyl-2-cyclopentenone from *Cinnamomum camphora.* *Phytochemistry* 18:488–89.

Yeshe Tsogyal. 1996. *Der Lotusgeborene im Land des Schnees: Wie Padmasambhava den Buddhismus nach Tibet brachte.* Frankfurt/M.: Fischer.

Shiva is the god of ecstasy and psychoactive plants. Here, he is shown inhaling the smoke of a sacred incense. Shiva is particularly fond of the psychoactive camphor. (Hindu devotional picture, detail; India)

"In Spain, the children carry a small linen sack containing camphor to ward off the evil eye. . . . In India, the child is bathed in smoke to protect against the evil eye. When a newly married Tamil couple returns from their procession through the streets, during which they are exposed to the looks of the crowd, a vessel is filled with camphor and pepper [**Piper spp.**], the camphor is lit, and the vessel with the burning camphor is swung around the heads of the newlyweds. . . . If a child in India has become ill as a result of the evil eye, then a piece of burning camphor is waved before him. If a Jewish child has developed a stomachache, headache, or fever because of the evil eye, or if a singer has suddenly become hoarse, then some camphor is placed before the door of the house and ignited, if the cattle have become ill from the evil eye, then camphor is burned in front of them."

SIEGFRIED SELIGMANN
DIE MAGISCHEN HEIL- UND SCHUTZMITTEL AUS DER BELEBTEN NATUR [THE MAGICAL HEALING AND PROTECTIVE AGENTS FROM THE ANIMATED NATURE]
(1996, 146 f.*)

Cocos nucifera Linnaeus

Coconut Palm

Family
Palmae (Palm Family) (previously Arecaceae)

Forms and Subspecies
Numerous varieties and cultivars are grown in the tropics and subtropics (Stewart 1994, 88*). There are cultivars for ornamental purposes that have only a short trunk and develop inedible, small yellow fruits. One variety with green fruits is known as *Cocos nucifera* var. *viridis*. Only the tall-growing varieties (*Cocos nucifera* var. *typica* Nar.) are easily distinguishable from the dwarf forms (*Cocos nucifera* var. *nana* [Griff.] Nar.) (Franke 1994, 240*).

Synonyms
Cocos butyraceum
Cocos nana Griff.

Folk Names
Coconut, coconut palm, coco nut tree, coco palm, cocotero (Spanish), cocotier (French), cocus, dab (Bengali), green gold, ha'ari, hach kokoh, khopra (Hindi), kôkô, kokoh, kokosnußpalme, kokospalme, kuk, kuk-anâ (Ka'apor), mabang, mbang ntnag, naral (Marathi), narial (Hindi), narikela, narikelamu, narikera, nariyal (Sanskrit), narkol (Bengali), niu (Samoa), obi, ogop, palmeer-baum, palmenbaum, pol, suphala (Sanskrit), tenga, tengu (Kannada), tenkai, tennaimaram (Tamil), thengu, thenna (Malayalam)

In Europe, the coconut palm (*Cocos nucifera*) is regarded as a symbol of the tropics.

The coconut palm produces the inebriating palm wine in the interior of the young leaf shoots. (Woodcut from Tabernaemontanus, *Neu Vollkommen Kräuter-Buch,* 1731)

History

Coconut trees have been used culturally in India for three thousand to four thousand years. They first appear in European literature in the sixth century and became an officinal agent in Europe with the adoption of Arabic medicine (Schneider 1974, 1:341*). They were known by the names *nuces indicae, carya indica,* and *Indian nut.* The name *cocos* means "grimace" and was given to the palm by the Spanish because of the "eye" where the nut is attached to the fruit (Bremness 1995, 49*).

In the older literature, the coconut palm was often characterized as "the most useful of all trees" because every part of it can be utilized (Meister 1677, 43*). This palm is a source of food, medicine, fibers, copra, and other raw materials as well as various inebriating beverages. **Palm wine** was even mentioned in ancient Sanskrit literature.

Culturally and economically, the coconut palm is one of the most important plants of the tropics. Coconut oil provides 8% of the world's oil and fat supply. The oil is used to make a variety of products, including margarine (Udupa and Tripathi 1983, 64).

Distribution

Now pantropical, the coconut tree apparently came from Asia or Melanesia (Zander 1994, 194*). However, there were already coconut trees in Colima (Mexico) when the first Europeans arrived there (Dressler 1953, 129*).

Coconut trees represent the typical vegetation associated with the beaches of the islands of the Indian Ocean, India, Southeast Asia, Central and South America, the Caribbean, and Melanesia.

Cultivation

The natural propagation and distribution of coconut palms occurs through the coconuts, which fall into the water, are carried away, and are then washed up in suitable locations. The palm thrives in sand, preferably at or near the beach; it can tolerate up to 1% salt in the groundwater. For cultivation, the fruits can be laid out (in areas rich in rain, under a roof) with the narrow side facing down. Up to half of the coconut can be lightly buried in sand. After four to five months, the fruit will have developed roots and formed a shoot. After six to twelve months, the seedling can be planted in the desired location. The germination period can be shortened by wrapping the coconut in a plastic bag left somewhat open at the top (Rehm and Espig 1996, 87 f.*).

Appearance

The slender, slightly leaning coconut palm can grow up to 30 meters in height. It develops pinnate leaves that can be as long as 6 meters. It has cream-colored panicles and large, solitary fruits (coconuts) that hang in thick clusters between the leaf stalks.

The coconut palm can be easily confused with the king coconut (*Cocos butyracea*), assuming that this is in fact a separate species.

Psychoactive Material

— Coconut
— Coconut milk (coconut water)
— Bleeding sap (toddy); **palm wine** (suri, tuaco, vino de coco)

Preparation and Dosage

The first detailed discussion (1677) of the manner in which **palm wine** is obtained from the coconut palm provides a precise description of the method that is still used throughout Southeast Asia and on the islands of the Indian Ocean:

Now follows the usefulness of the noble palmeer wine. . . . This wine, which is the juice of this tree, is called major by the inhabitants of Java, tuaco by those of the Malabars, and surii by the Dutch, and is tapped from the tree in the following manner: While the flower is still growing, one cuts the same at the front with a broad knife made for this purpose and places such shortened little branches in a piece of bamboo (this bamboo is a hollow tube almost as wide as a leg, which the inhabitants of India generally use to build their houses) or in a narrow pot, open at the top, which is standing in the sun. When they are visited by their warders, also known as divitores, [they] . . . use several steps that have been chopped in to climb quickly up. They pour the sura into a pober or Indian gourd attached to their persons at least twice every 24 hours, that is, early for what was collected during the night, and evenings, for what was collected during the day. . . . This sura or juice, when drunk immediately while fresh, is delicious and good and sweet beyond measure, especially that which comes from those grown around Cannanor or in the Kingdom of Calicuth, on the coast of Cannera and Malabar, which tastes quite agreeable, almost as sweet as a newly pressed cider. But if you drink a little too much of this, you will very easily get drunk from it. (Meister 1677, 49*)

As a result of fermentation and enzymatic processes, palm wine changes considerably during the course of a single day:

The palm wine that is collected in the morning tastes like sweet cider until about 10 o'clock, however with the oily aftertaste of the coconut, it then begins to ferment and at around 12 o'clock foams over the edge of the bottle or bamboo vessel in which it is being kept in the open. In the evening, toward 3 o'clock, it is then an inebriating drink, a "fire water," as the natives call it. . . . If the palm grower wishes to hinder the fermentation, he takes some calcium from shells and mixes this into the palm juice. (Schröter, in Hartwich 1911, 627*)

If the palm wine is allowed to stand for a longer time, it ferments into palm vinegar. Production of the bleeding sap can be stimulated and increased by hitting the inflorescence with a special wooden stick or bone.

The alcohol that is distilled from the flower sap (*toddy, tonwack*) is known as *arrak* (Fernando 1970). On the Marquesas Islands, fermented coconut milk is distilled into a type of brandy (**alcohol**). On Rennel Island, one of the southernmost of the Solomon Islands, a drink obtained from coconuts is known as *kava kava ngangi*. In spite of its name, the drink contains no *Piper methysticum* (Holmes 1979).

Coconut flakes are an ingredient in **betel quids** as well as **Oriental joy pills**.

Ritual Use

In India, coconuts are thrown into the sea as offerings to placate the spirits of the monsoon. In Gujarat, the palm is venerated as a familial deity. Muslims toss pieces of coconut and limestone over the heads of newlyweds to dispel evil spirits. The Bengalis believe that coconuts have eyes and are able to see if someone is lying beneath the palm so they will not fall on that person's head (Gandhi and Singh 1991, 65*). Because coconuts are as large as a person's head, they are offered to the bloodthirsty goddess Bhadrakali ("auspicious

black [goddess]"), a terrifying manifestation of Shiva's wife Parvati, in place of real human sacrifices (Gandhi and Singh 1991, 66*).

The Yoruba of Africa believe that at the beginning of creation, the coconut was a pure, loving, and virtuous person who was later transformed into the plant. For this reason, the palm is a sacred tree that is venerated and respected.

Coconut palm wine enjoys great ritual significance, especially in western New Guinea but also in other areas: "Drinking palm wine is part of certain idolatrous ceremonies, but in private life the palm wine drinkers are despised and are not as common as the habitual drunkards are among us" (Schröter, in Hartwich 1911, 627*).

For more on ritual use, see **palm wine**.

Artifacts

In Southeast Asia, coconuts are made into boxes for tobacco snuff (Meister 1677, 48*). In Oceania, half shells were and still are used to manufacture vessels for drinking kava (cf. *Piper methysticum*).

The wooden beating sticks—called *pudscha*—that were used to stimulate juice production were regarded as idols and venerated accordingly (Hartwich 1911, 627*).

Because the coconut palm is a symbol of tropical, South Seas romance, it is depicted on numerous pictures intending to invoke such an ambience. It is possible that there are art objects that were inspired by the use of coconut palm wine, but there are no reports about this.

Medicinal Use

On Samoa, the coconut is used in a multitude of ways as a remedy for stomach problems, constipation, open wounds, puerperal fever, gonorrhea, inflammations, eye ailments, problems associated with pregnancy, and stings by the very poisonous stonefish (*Synanceja* spp. and others) (Uhe 1974, 6 f.*). It is used in similar manners in the folk medicine of other South Pacific islands. In Polynesia, coconut milk is used as a solvent for medicinal herbs (Whistler 1992, 82).

In India (Karnataka), a tea made from the tender flower buds is drunk every morning for three days to balance out all menstrual irregularities (Bhandary et al. 1995, 157*). The oily exudation of heated coconut shells is used in Ayurvedic medicine as a treatment for parasites (Venkataraman et al. 1980). Coconut milk is prescribed for gastritis, stomach ulcers, and heartburn (Udupa and Tripathi 1983, 64).

On the Malay Peninsula, the ground root is administered as an antidote for poisoning with *Datura metel* (Perry and Metzger 1980, 304*).

The Fang of Central Africa use the bark to obtain a medicine to treat toothaches (Akendengué 1992, 169*).

The use of coconut flakes and meat as aphrodisiacs[99] and as treatments for venereal

"I [must] mention a kind of bat, which can easily grow as large as a proper, complete hen in its head and hair, long of mouth and short of ears, the wings, as smooth as those of our own bats, are as long as a fathom when you stretch them apart, the Dutch call them suiri-cats, the Portuguese murchsebes, and in the Malay language eansching duack. These monstrous bats, although very rarely seen, often came and hung with their sharp claws on the coconut tree leaves, from which the sura or palm wine is tapped, and inebriate themselves on this to such an extent that they would often remain the entire night until the sun had risen in order to get enough sleep, and could thus be encountered, and I brought down such a one with a good flint rifle, and my slaves then made themselves very merry when they ate this, and I too consumed it out of curiosity, as they are not poisonous but rather taste like the best chicken meat."

GEORGE MEISTER
DER ORIENTALISCH-INDIANISCHE KUNST- UND LUSTGÄRTNER [THE ORIENTAL-INDIAN ART AND PLEASURE GARDENER]
(1677, CH. 6, P. 3*)

99 The tender fruit flesh of the Seychelles's coco-de-mer (*Lodoicea maldivica* [J.F. Gmel.] Druce [syn. *Lodoicea seychellarum* Labill]) also has the reputation of being a powerful aphrodisiac. No constituent that might produce such effects has yet been found (Müller-Ebeling and Rätsch 1989, 39 f.*).

diseases is widespread. In Indonesia, coconut shells are burned and the ashes mixed with **wine** to treat syphilis (Perry and Metzger 1980, 404*). In Indonesia, a flaccid or sick "Venus rider" would dangle his damaged member through a hole in a fresh coconut and bathe it in the coconut milk to provide it with new vigor or to cure the venereal disease he had acquired (Meister 1677, 46*). In Islamic medicine, the penis is packed in a mush made from fresh coconut meat in order to give it new energy (Moinuddin Chishti 1984, 96*). In the Bahamas, the tender coconut flesh is mixed with nutmeg (*Myristica fragrans*) and ingested to heal "weakness" (Eldridge 1975, 314*).

Coconut oil, obtained from the dried endosperm of the seed, has great significance in the cosmetic industry.

Constituents

The plant contains an **essential oil**, wax, and oil. The bleeding sap, which ferments into **palm wine**, contains proteins, ashes, 15% sugar (saccharose), and enzymes (Perry and Metzger 1980, 304*; Rehm and Espig 1996, 74, 89*).

The milk of a still-green coconut fruit has been found to contain 1,3-diphenylurea, a compound that stimulates cellular growth (Wong 1970, 110*). Coconut flakes contain proteins, carbohydrates, and vitamin B complex.

Effects

Because of its low alcohol content, the **palm wine** obtained from the bleeding sap—even when consumed in large quantities—has stimulating, almost refreshing and invigorating effects that do, however, tend toward drunkenness. The effects of

drinks fermented from the milk are different: "Fermented coconut milk has a high alcohol content: too much will result in toxic symptoms" (Udupa and Tripathi 1983, 64).

Commercial Forms and Regulations

Coconuts are available throughout the world wherever fruits and vegetables are sold. On the other hand, **palm wine** can be obtained only where it is made, as it is quite perishable. Arrak can be obtained throughout Southeast Asia but is only infrequently available in the West.

Literature

See also the entries for *Areca catechu* and **palm wine**.

Fernando, T. 1970. Arrack, toddy, and Ceylonese nationalism. *Ceylon Studies Seminar* 9:1–33, Colombo.

Guzmán-Rivas, P. 1984. Coconut and other palm use in Mexico and the Philippines. *Principes* 28 (1): 20–30.

Holmes, Lowell D. 1979. The kava complex in Oceania. *New Pacific* 4 (5): 30–33.

Udupa, K. N., and S. N. Tripathi. 1983. *Natürliche Heilkräfte*. Eltville am Rhein: Rheingauer Verlagsgesellschaft.

Venkataraman, S., T. R. Ramanujam, and V. S. Venkatasubbu. 1980. Antifungal activity of the alcoholic extract of coconut shell—*Cocas nucifera* L. *Journal of Ethnopharmacology* 2:291–93.

Whistler, Arthur. 1992. *Polynesian herbal medicine*. Lawai, Kauai, Hawaii: National Tropical Botanical Garden.

Coffea arabica Linnaeus

Coffee Bush

Branch and "beans" (= seeds) of the coffee bush. (Copperplate engraving from Peter Pomet, *Der aufrichtige Materialist und Specerey-Händler*, Leipzig 1717)

Family

Rubiaceae (Coffee Family); Subfamily Cinchonoideae, Coffeeae Tribe

Forms and Subspecies

The variety *Coffea arabica* L. var. *abyssinica* A. Chev. (wild form) occurs in the mountain forests of Ethiopia. In principle, two varieties that were derived from early Arabic plantations are now under cultivation:

Coffea arabica L. var. *arabica* (= var. *typica* Cramer)
Coffea arabica L. var. *bourbon* (B. Rodr.) Choussy

A very large number of mutants and cultivated forms have been described. The following are of economic interest:

Coffea arabica L. cv. Caturra (stocky growth, productive)
Coffea arabica L. cv. Mundo novo (very good yield)
Coffea arabica L. cv. Catuai vermelho (red fruits)
Coffea arabica L. cv. Catuai amarelo (yellow fruits)
Coffea arabica L. cv. Mragogipe (gigantic form)
Coffea arabica L. cv. Mokka (very low growing)

The last of these cultivars, which is also known by the names *mokha* and *moka*, has also been described as a variety:

Coffea arabica L. var. *mokka* Cramer

Synonyms
Coffea laurifolia Salisb.
Coffea mauritiana Host. non Lamk.
Coffea vulgaris Moench
Jasminum arabicum laurifolia de Juss.

Folk Names
Arabian coffee, Arabica coffee, arabica-kaffee, arabischer kaffee, bergkaffee, bun (Yemen), buna ("wine"), buni (Ethiopian), cabi, café, caféier, cafeiro, cafeto, chia-fei (Chinese), coffa, coffee, coffee bush, coffee tree, common coffee, kaffeebaum, kaffeepflanze, kaffeestrauch, kahawa (Swahili), kahwa (Arabic), kahwe (Turkish), kahweh, k'hoxwéeh (Navajo), koffie (Dutch), kopi, qahûa, qahwa (Arabic, "wine"), qahwe

History
Long before the first coffee was ever brewed, the berries of the coffee bush were being chewed in Africa for stimulating purposes (by about the sixth century). Coffee drinking was discovered long after khat chewing (see *Catha edulis*). The word *coffee* is sometimes derived from the Arabic word for wine,[100] *gahwe*; but the Arabic name for coffee, *kahwa*, is more likely derived from the place-name Kafa (in Ethiopia). In Ethiopia, the story told to explain the discovery of coffee is almost identical to that told in Yemen to explain the discovery of khat. A goatherd watched as his goats scampered around excitedly after they had eaten from the coffee bush. He took some of the beans and gave them to the village priest, who then experimented with them until he experienced their stimulating power and was thus better able to recite the long prayers (Mercatante 1980, 171*). The first mention of coffee use in Yemen comes from the twelfth century (Meyer 1965, 137).

Coffee is highly esteemed among African Sufis, for it enables them to take part in their mystical rituals night after night without falling asleep and makes it easier for them to attain religious ecstasy. The Sufis and the wandering dervishes played a great role in the spread and popularization of coffee.

In the sixteenth and seventeenth centuries, coffee became known both in Europe and on the African Swahili coast (Sheikh-Dilthey 1985, 253). Coffee was enthusiastically received in Europe, where it was praised as a cure-all and used as an aphrodisiac (Müller 1981). The first complete botanical description of the plant was not made until the mid-nineteenth century (Meyer 1965, 142).

Today, coffee is probably the most commonly consumed stimulating beverage in the world (Morton 1977, 356*). As a result, the coffee bush is one of the most culturally important psychoactive plants that exist.

Because of its economic significance, coffee has

often led to violent altercations and warlike actions. In the 1920s and 1930s, a veritable "witches' war" broke out in Puebla (Mexico), during which over one hundred Nahuat Indians lost their lives (Knab 1995*).

Distribution
The coffee bush apparently originated in Abyssinia, i.e., southwest Ethiopia (Schneider 1974, 1:343*). It is still indigenous to the region (Baumann and Seitz 1992, 927; Meyer 1965). Wild plants have also been observed in Sudan.

Cultivation
The coffee bush requires a tropical climate to thrive and does not tolerate frost. It must be raised in partial or complete shade. Anyone who wants to grow the plant in the climate zone of central Europe must raise it as a potted plant or in a tropical greenhouse. The seeds are placed on peaty, sandy seed soil. They should be not covered with soil but, instead, gently pushed into the soil and kept continuously moist. The germination time is quite variable but usually requires between two and four weeks (at temperatures between 25 and 30°C). The germinated seeds or seedlings can be transplanted into a suitable pot. They should be fed frequently and well watered. In principle, sowing can be performed throughout the year, but because of the plant's biorhythms, it is best done between November and January. After a growing period of about three years, the plant produces its first fruits. These contain the coffee beans.

Coffee plantations are found in many tropical countries. Coffee is an economically important source of income for many so-called Third World countries. Outside of Africa, the most important coffee-growing regions are in Mexico (Chiapas), Guatemala, Nicaragua, Colombia, and Brazil. In tropical Africa, the closely related species *Coffea liberica* Bull. is grown as a source of coffee beans. Robusta coffee (*Coffea canephora*) is also cultivated on a large scale in Africa. *Coffea canephora* provides some 20% of the world's supply of coffee beans, while about 80% comes from *Coffea arabica* (Baumann and Seitz 1992, 928).

The seeds of the ripe fruits of the coffee bush (*Coffea arabica*) are referred to as coffee beans, even though the plant is not related to the Legume Family.

100 In Arabic, **wine** is usually referred to as *khamr*, a word that means "inebriating" (cf. *Vitis vinifera*).

Appearance

The perennial coffee bush can grow to a height of about 4 meters. It is heavily foliated, with shiny leaves (6 to 20 cm long, 2.5 to 6 cm wide) that can persist for two to three years. The white, star-shaped flowers (calyx approximately 3 mm long) are in thick glomerules and exude a fine, delicious scent vaguely reminiscent of that of jasmine (*Jasminum* spp.). The green, oval fruits (berries) turn bright red when ripe (only the cultivar Catuai amarelo develops yellow berries).

The genus *Coffea* consists of some ninety species, many of which resemble the coffee bush. *Coffea arabica* is very similar to two tropical species, *Coffea congoensis* Froehn. and *Coffea eugenioides* S. Moore, and is easily confused with them (Meyer 1965, 138).

Other *Coffea* Species That Yield Coffee
(After Baumann and Seitz 1992 and Meyer 1965; amended.)

Trade Name	Stock Plant
Congo Coffee	*Coffea canephora* Pierre ex Froehner [syn. *C. arabica* L. var. *stuhlmannii* Warb., *C. bukobensis* Zimm., *C. laurentii* De Wild., *C. maclaudii* A. Chev., *C. ugandae* Cramer, *C. welwitschii* Pierre ex De Wild.]
Robusta Coffee	*Coffea canephora* var. *canephora* [syn. *Coffea robusta* Lind.]
Nganda Coffee	*Coffea canephora* var. *nganda* Haarer [syn. *Coffea kouilouensis* Pierre ex De Wild.]
Liberian Coffee	*Coffea liberica* Bull ex Hiern
Inhambane Coffee	*Coffea racemosa* Lour.
Rainforest Coffee	*Coffea dewevrei* De Wild. et Dur.

Psychoactive Material

— Seeds (coffee beans, semen coffeae, coffeae semen, green coffee)
— Roasted coffee beans (coffeae semen tostae)

The roasted coffee beans must be kept well sealed, in the dark, and away from humidity.

Preparation and Dosage

After the ripe fruits (coffee cherries, coffee berries) have been harvested by hand, they are spread out in a layer 3 to 4 cm deep to dry in the sun. The drying fruits are raked often, sometimes several times a day. After three to four weeks, the fruits are completely dry. The beans now lie loosely in the fruit coat, which is then removed by rubbing with the hand or with machines (so-called hullers). To brew coffee, the seeds must be roasted. The green coffee beans are roasted for differing lengths of time and by different methods, either on clay or metal plates above a fire or with industrial machines. The roasting process gives the beans their aroma, an important factor in determing the market quality of the beans.

The roasted beans are coarsely ground and brewed for ten minutes in boiling water or boiled for several minutes in water. These methods are common in Africa and Scandinavia. More often, the roasted beans are ground and placed in a filter or a suitable coffee maker. Boiling water is then poured slowly over them.

A normal cup of coffee, brewed from 5 g of ground coffee and 300 cm^3 of water, contains 70 to 80 mg of **caffeine** (Roth et al. 1994, 248*). Approximately 250 mg of caffeine is contained in a double espresso.[101] If the daily consumption of coffee is so great that a person is ingesting 1.5 to 1.8 g of caffeine daily, "caffeinism" may result (Baumann and Seitz 1992, 935). Still, some people are said to drink up to fifty cups of strong coffee daily. The French writer Voltaire was one such person (Huchzermeyer 1994).

In Africa, coffee is usually spiced with cardamom (*dawa ya chai*, "tea medicine"), and also with ginger roots (*Zingiber officinale*) when being made into medicinal drinks. In Africa, ten to twelve roasted coffee beans are brewed with water when the drink is intended for medicinal purposes. When the beans are chewed for medicinal purposes, children take one or two beans, while adults take from seven to fourteen beans (Sheikh-Dilthey 1985, 254).

The following ingredients are used to make a purgative that is administered on the day after giving birth:

5   cups of water
    "Very many" crushed coffee beans
2   betel leaves (*Piper betle*)
1   spoonful of dried dill herbage (**Anethum graveolens**)
1   teaspoon of Ajwan cumin (*Trachyspermum ammi* [L.] Sprague)
2   sticks of cinnamon (*Cinnamomum verum* Presl)
5   cardamom seeds (*Elettaria cardamomum* [L.] Maton)
5   cloves (*Syzygium aromaticum*)
2   teaspoons of molasses (from sugarcane)

All of the ingredients are chopped and boiled in the water. After filtering, about two cups remain (Sheikh-Dilthey 1985, 255).

In Ethiopia and other African countries, the dried and/or roasted leaves of the coffee bush may

---

101 In the underground drug scene, the term *double espresso* is also used to refer to **cocaine** (because of the similar effects).

also be chopped and boiled in water. Milk is then added, and the product is drunk either sweetened or salted. In Ethiopia, an infusion of the leaves or fruit husks is known as *hoja* and is drunk with milk (Wellman 1961).

A number of other stimulating plants are used as coffee substitutes, e.g., ***Ilex guayusa***, but also the roasted seeds of *Abrus precatorius*. The roasted root tubers of chicory (*Cichorium intybus* L. var. *sativum* Lam. et DC.), which contain no stimulating or psychoactive constituents, are the source of chicory coffee (Rehm and Espig 1996, 255*). In Yemen and the surrounding countries, an infusion of dried khat leaves is used as a coffee substitute (see ***Catha edulis***). Dandelion roots (*Taraxacum officinale* Weber), fig fruits (*Ficus carica* L.), sugar beet roots (*Beta vulgaris* L.), lupine seeds (***Lupinus* spp.**), rye grains (*Secale cereale* L.), and barley grains (*Hordeum distichon* L.) are also used as coffee substitutes or counterfeits. Some ***Psychotria*** species are known by the name *wild coffee* and are said to have formerly been used in Jamaica and on other Caribbean islands as coffee substitutes.

Ritual Use
In East Africa, it is believed that spirits live in the coffee beans and that the beans therefore possess magical powers a person can draw upon through rituals and incantations. According to an Arab legend, the archangel Gabriel presented the first coffee to the ailing Muhammad for his recuperation (Brunngraber 1952, 128*). For this reason, coffee is sacred in Islam, and it is even used as a ceremonial drink. In Swahililand, copious amounts of coffee are drunk during all religious rites, at the evening readings of the Koran, and at the midnight worship services at the mosques (presumably so that people will not fall asleep during the sermons):

> The greatest of the Islamic festivals on the Swahili coast is Maulidi al Nabi, the celebration of the prophet's birthday. . . . On this occasion, people of all ethnic groups gather in the bigger cities and take part in processions through the city that are led by groups of musicians singing religious songs in praise of Mohammed. When it gets dark, the processions meet in a great square in front of a mosque. In the light of the torches or light bulbs, wrapped in the scent of ubani (incense [cf. ***Boswellia sacra***]), all of the praying people listen attentively deep into the night as the life story of Mohammed is recited in prose or poetry. As this is going on, spiced coffee is passed out and drunk by all who are present. (Sheikh-Dilthey 1985, 255)

The use of coffee to support prayers, meditations, and secret rituals was of great importance in many Sufi orders.

The customs associated with drinking coffee in Viennese coffeehouses also have a ritual character, though the coffee drinkers themselves do not usually regard them as rituals (Thiele-Dohrmann 1997; Weigel et al. 1978). In some circles, the magical use of coffee has been preserved in the form of reading coffee grounds, a traditional folk oracular method. For many Westerners, the act of preparing coffee in the morning has become a small, personal ritual that helps them prepare for the day. Many coffee drinkers are not "officially" available before they have had their morning coffee, i.e., coffee opens a person to the world. Afternoon coffee parties and coffee breaks at work also have a ritual and socio-integrative character.

Artifacts
As a stimulating and awakening work drug, coffee has certainly made an indirect contribution to the productive abilities of creative artists. Many musicians have been inspired by coffee. If the American composer Frank Zappa (1940–1993), whom many music lovers regarded as a psychedelic musician, is to be believed, coffee and cigarettes were his "basic food groups" and the foundation of his musical productivity. The greatest musical work devoted to coffee is the very worldly *Kaffeekantate* [Coffee Cantata] of Johann Sebastian Bach (1685–1750), which was composed to be played in coffee- and teahouses as well as more traditional venues. The hymn "Cigarettes and Coffee," by the rock bard Jerry "Captain Trips" Garcia (1942–1995) (featured in the soundtrack to the film *Smoke*, 1995), is quite well known, as is the crossover ballad "Caffeine" from the heavy metal band Faith No More (on the album *Angeldust*, 1992).[102]

The recent anthology *Music for Coffeeshops* (Dreamtime Records, 1995), the maxi-single "Coffee Shop" (from the crossover band the Red Hot Chili Peppers, WEA, 1996), and the album *Locked in a Dutch Coffeeshop* (by Eugene Chadbourne and Jimmy Carl Black, ca. 1993) are referring not to true coffeehouses but to the renowned Dutch coffee shops in which hashish and other hemp products (***Cannabis indica***) are sold in a quasi-legal environment.

Medicinal Use
In Africa, roasted coffee beans are chewed to treat headaches, malaria, and general weakness (Sheikh-Dilthey 1985, 254). In Arabia, coffee grounds are eaten as a folk medicinal treatment for dysentery and applied externally to suppurating wounds and inflammations (Baumann and Seitz 1992, 930). Decoctions of roasted coffee beans are used in Haiti to treat hepatitis, liver ailments, edema, anemia, and conditions of weakness (Baumann and Seitz 1992, 934).

In the United States, paramedical circles have

Botanical illustration of the coffee bush. (Engraving from Pereira, *De Beginselen der Materia Medica en der Therapie,* 1849)

102 It is questionable whether the songwriters were actually speaking of coffee and caffeine, for in the music scene, the terms *coffee* and *double espresso* are sometimes used as cover names for cocaine.

In the seventeenth century, drinking coffee or tea was an enormously popular activity. Meetings, or lodges, clearly provided the ritual and collective space for a communal "inebriation" using stimulating beverages. (Copperplate engraving from *Die neueröffnete lustige Schaubühne menschlicher Gewohnheiten und Thorheiten*, Hamburg, 1690)

claimed that coffee **enemas**, administered every two hours, can heal cancer. This therapy is usually recommended to cancer patients by other cancer patients. This treatment is responsible for at least two deaths (Eisele and Reay 1980).

In homeopathic medicine, Coffea—Kaffee is an important agent obtained from a tincture of unroasted seeds (Schneider 1974, 1:245*). Preparations of roasted coffee beans (Coffea arabica tosta hom. *HAB1*) are also used in homeopathy, including for the treatment of neuralgia and sleep disorders (Baumann and Seitz 1992, 936).

Constituents

The green beans contain purine alkaloids. In addition to concentrations within a normal range of 0.58 to 1.7% **caffeine**, there are slight concentrations of theobromine (cf. *Theobroma cacao*), theophylline, paraxanthine, theacrine, liberine, and methylliberine. Also present are chlorogenic acids, in concentrations of 5.5 to 7.6%, of which 60 to 80% is 5-caffeoylquinic acid. A portion of the caffeine is bound to the chlorogenic acids. The beans contain approximately 16% coffee oil with diterpene alcohols. Coffee wax contains fatty acid derivatives of 5-hydroxytryptamine (Baumann and Seitz 1992, 931). Green coffee beans have also occasionally been found to contain concentrations of 3% caffeine (Roth et al. 1994, 248*).

Roasting the seeds has almost no effect upon their caffeine content, but the chlorogenic acids are reduced to about 10% of their original concentration. Roasting also creates new compounds, including nicotinic acid, 5-hydroxyindole, alkane, trigonelline, and polymer pigments, which are responsible for the brown coloration of the beans. The source of the characteristic coffee aroma, which plays such an important role in determining the commercial value, remains unknown. The average caffeine content of roasted coffee is around 1% (Baumann and Seitz 1992, 932 f.).

The red pigmentation of the fruits is the result of anthocyanins and the aglycone cyanidin. The

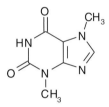

Theobromine

Caffeine

hull (pulp) of the fruit contains large amounts of tanning agents (Baumann and Seitz 1992, 928).

Whether the leaves contain caffeine, other purines, or chlorogenic acids is unknown (Roth et al. 1994, 248*).

Effects

Coffee has strong stimulating effects and induces wakefulness, accelerates the pulse rate, and promotes perspiration. At a certain dosage, which varies from person to person and also depends upon the degree to which a person has become habituated to coffee, mental abilities are improved. Coffee often improves heart activity and urinary excretion. Very high dosages can produce profound disturbances in perception, trembling, nervousness, and sleep disturbances. The discussions about the beneficial or harmful effects of coffee upon health are apparently not over and are constant subjects of the popular media and health advocates. The chlorogenic acids are responsible for coffee's "acid content"; in large quantities, they can make the stomach acidic, resulting in heartburn, stabbing pains, and, over a period of time, stomach ulcers (Roth et al. 1994, 248*).

A nutrition scientist has noted: "If we summarize the results of the rather comprehensive research into the acute effects of caffeine and the long-term effects of daily coffee, then coffee should be ranked among the most harmless of all drugs" (Huchzermeyer 1994).

Commercial Forms and Regulations

Viable seeds (in packages designed to prevent germination) are available in nurseries and seed stores. Coffee beans are subject only to the prevailing food laws. Various types of coffee are available; Colombian coffee, Turkish mocha, and Italian espresso are especially popular. Decaffeinated coffees, commercial goods that have been treated to remove the **caffeine,** are also offered.

Literature

See also the entry for **caffeine.**

Baumann, Thomas W., and Renate Seitz. 1992. *Coffea.* In *Hagers Handbuch der pharmazeutischen Praxis,* 5th ed., 4:926–40. Berlin: Springer.

Eisele, John W., and Donald T. Reay. 1980. Deaths related to coffee enemas. *Journal of the American Medical Association* 244 (14): 1608–9.

Haberland, Eike. 1981. Kaffee in Äthiopien. In *Rausch und Realität,* ed. G. Völger, 2:492–95. Cologne: Rautenstrauch-Joest-Museum für Völkerkunde.

Hentschel, Kornelius. 1997. *Geister, Magier und Muslime: Dämonenwelt und Geisteraustreibung im Islam.* Munich: Diederichs.

Huchzermeyer, Hans. 1994. Kaffee: Wirkungen einer alltäglichen "Dröhnung." In *Köstlichkeiten: Von "sinnvollen" Essen und Trinken* (Jubiläumsschrift). Minden: Institut für Ernährungsmedizin.

Jacob, Heinrich Eduard. 1934. *Sage und Siegeszug des Kaffees.* Hamburg: Rowohlt.

Meyer, Frederick G. 1965. Notes on wild *Coffea arabica* from southwestern Ethiopia, with some historical considerations. *Economic Botany* 19:136–51.

Müller, Irmgard. 1981. Einführung des Kaffees in Europa. In *Rausch und Realität*, ed. G.Völger, l:390–97. Cologne: Rautenstrauch-Joest-Museum für Völkerkunde.

Schnyder-v. Waldkirch, Antoinette. 1988. *Wie Europa den Kaffee entdeckte: Reisebericht der Barockzeit als Quellen zur Geschichte des Kaffees.* Zurich: Jacobs Suchard Museum.

Sheikh-Dilthey, Helmtraut. 1985. Kaffee, Heil- und Zeremonialtrank der Swahiliküste. *Curare*, Sonderband 3/85:253–56.

Sylvain, Pierre G. 1958. Ethiopian coffee: Its significance to world coffee problems. *Economic Botany* 12:111–30.

Thiele-Dohrmann, Klaus. 1997. *Europäische Kaffeehauskultur.* Zurich and Dusseldorf: Artemis & Winkler.

Weigel, Hans, Werner J. Schweiger, and Christian Brandstätter. 1978. *Das Wiener Kaffeehaus.* Vienna, Munich, and Zurich: Verlag Fritz Molden.

Wellman, F. L. 1961. *Coffee.* London: Leonard Hill.

Cola spp. (*Cola acuminata* and *C. nitida*)

Cola Tree

Family

Sterculiaceae (Cocoa Family); Sterculieae Tribe, Sterculiinae Subtribe

Forms and Subspecies

The two most important trees that provide cola nuts are so similar that they can be distinguished only on the basis of the structure of the nuts they produce:

Cola acuminata (P. Beauv.) Schott et Endl.—small cola tree

Characteristic:	Four- to six-part nut
Synonyms:	*Sterculia acuminata* Schott·et Endlicher
	Cola pseudoacuminata Engl.
Variety:	*Cola acuminata* var. *trichandra* K. Schum.

Cola nitida (Vent.) Schott et Endl.—large cola tree

Characteristic:	Two-part nut
Synonyms:	*Cola vera* K. Schum.
	Cola acuminata Engl.
	Cola acuminata var. *latifolia* K. Schum.
	Sterculia nitida Vent.
Varieties:	*Cola nitida* var. *alba* (white seeds/flowers)
	Cola nitida var. *mixta*
	Cola nitida var. *pallida*
	Cola nitida var. *rubra* (red seed core)
	Cola nitida var. *sublobata* (very large seeds)

Folk Names

Abata kola, abé, afata, ajauru, ajo pa, alie a uke, al mur, aloko, alou, ang-ola, apo, ashaliya, atara, ataras, atarashi, awasi, awedi, ballay cornu, b'are, 'bari, bar ni da mugu, bese, bese-fitaa ("white cola"), bese hene ("king's cola"), bese koko ("red cola"), bese kyem, bese-pa (Ghanese, "good cola"), besi, bichy nuts, bise hene, bise kyem, bise pa ("good cola"), bisi, bisihin, bisi tur, bissy, bitter cola, bobe, buesse, buessé, burduk'u, chigban, chousse, cola, colatier, cola tree, dabo, 'dan agyaragye, 'dan agyegye, 'dan badum, 'dan katahu, 'dan kataku, 'dan kwatahu, 'dan laka, 'dan richi, daushe, dibe, doe-fiah, ebe, ebi, e esele, egin-obi, ehoussé, ehuese, ereado, erhesele, eseri, evbe gabari, evbe gbanja, evbere, evbi, eve, evi, ewe, ewese, fakani, farafara, farsa, fatak, fecho, fetjo, gabanja, gandi, ganjigaga, gazari, ge, go, "nut"),

> "To the extent that they have still preserved their own natural and cultural qualities, cola fruits and cola trees occupy a high position in profane and religious ceremonies of the populace. . . . The prophet is said to have rested under a cola tree and to have distributed cola nuts to his followers. . . . The thing with Mohammed is, of course, an unmitigated swindle, for the cola tree does not occur in Arabia or in East Africa, and the prophet was never in West Africa. . . . The Islamic missionaries played their part from an early date and surrounded the cola with mystical embellishments."
>
> RUDOLF SCHRÖDER
> *KAFFEE, TEE UND KARDAMOM*
> [COFFEE, TEA, AND CARDAMOM]
> (1991, 116*)

Other *Cola* Species Used for Pleasure and Other Purposes

The genus *Cola* consists of fifty to sixty species, some of which have attained importance as agents of pleasure, medicines, or ritual drugs. In addition to *Cola acuminata* and *Cola nitida* (the most important members of the genus), the following species are also used. (From Seitz et al. 1992, 940.)

Name	Distribution	Use
Cola anomala K. Schum.	Cameroon	seeds chewed for pleasure
Cola ballayi Cornu ex Hekkel	Central and East Africa	seeds used in cult activities and for pleasure
Cola cordifolia R. Br.	Africa, Southeast Asia	seeds eaten, bark used as medicine
Cola sphaerocarpa A. Cheval.	Central and East Africa	seeds chewed for pleasure
Cola verticillata Stapf ex Cheval. (owe cola)	Congo, Ivory Coast	seeds used for pleasure

Left: The rather low-growing cola tree (*Cola acuminata*).

Right: The fruit of *Cola nitida*.

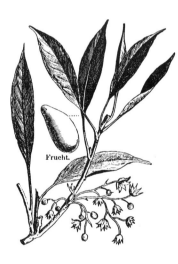

Botanical illustration of *Cola acuminata*. (From Meyers, 5th ed.)

godi ("tree"), godoti, gola, gonja, gooroo nuts, gor, gore, goriya, goro, gorohi, goron 'yan k'asa, gotu, gotu kola, guere, guéré, guiti, guli, gura, gura nuts, guresu, gurésu, guro, gwanja, gwe, gwolo, hak'orin karuwa, halon, halou, hannunruwa, hapo, hure, huré, ibe oji, ibi, ibong, ihié, inkurma, jouro, kanu, kanwaga, kobe, kola, kolabaum, kolai, kola nut tree, kolaxame, ko-tundo, kui, kuruo, k'waryar goro, k'waryar yaraba, k'yank'yambishi, k'yanshe, labuje, labure, lou, maandin, mabanga, marsa, mbuesse, mbuessé, minu, na fo ("white cola"), nafo, na he ("red cola"), nahé, nata, ngoro, ntawiyo, ntawo, obi, obí (Yoruba),[103] obi abata, obi gbanja, obi gidi, oji, oji ahia, oji aniocha, oji anwe, oji inenabo, oji odi, oji ugo, ombene, oro, oue, oué, oure, ouré, sandalu, saran-waga, siga, suture, tino uro, togo, tohn-we-eh, toli, tolo, toloi, tshere, tugule, tugure, tugwi, tui, ture, tutugi, uro, vi, wa na, we-eh, we na, wé na, wobe ihie, wore, woroe, wuro, yétou

These folk names almost always apply to both *Cola* species (Ayensu 1978, 255*).

History

Indigenous to western Africa, the cola nut (*Cola nitida, Cola acuminata*) was originally reserved for the gods. During a visit to the earth, however, one of the gods left a piece behind and humans found it. Because of their stimulating powers, cola nuts were used for magic and as amulets and aphrodisiacs. Today, they still play a central role in the religious and social life of many West and Central African cultures.

Clusius provided the first description of cola in 1605. Europe first became aware of the cola nut in the second half of the sixteenth century. The first cola plantations were established in the West Indies around 1680 (Schröder 1991, 119*). Nevertheless, the stock plant long remained unknown (Schumann 1900). In 1865, the seeds were found to contain **caffeine** (Schneider 1974, 1:346*).

The original Coca-Cola was a potent psychotropic beverage made with an extract of cola nuts and *Coca* leaves (***Erythroxylum novogranatense***).

Distribution

The genus *Cola* is originally from tropical West Africa. *Cola acuminata* is found from Togo to Angola, and *Cola nitida* from Liberia to the Ivory Coast and in Senegal and Nigeria. As a result of cultivation, both species have now spread into the tropical zones of the New World and Southeast Asia.

Cultivation

Propagation occurs using large, undamaged seeds from the center of the fruit. To germinate, the seeds are placed in well-moistened seedbeds or pressed directly into the ground; no other treatment is necessary. The seeds germinate after three to five weeks. The tree can also be propagated using cuttings taken from root shoots (Eijnatten 1981). The variety *Cola acuminata* var. *trichandra* K. Schum. is especially suitable for cultivation (Seitz et al. 1992, 941).

Cola is now also planted in the region of Bahia (Brazil) for use in the Afro-Brazilian Candomblé cult (Voeks 1989, 126*).

Cola trees require a moist, warm, tropical climate and thrive especially in rain forests. They prefer alluvial and humus soils.

Appearance

This evergreen tree, which can grow as tall as 25 meters, develops pale yellow, purple-striped flowers and star-shaped composite fruits with large, woody hulls. The alternate leaves of *Cola nitida* are shiny and light green in color, while those of *Cola acuminata* are leathery and dark green. The leathery/woody fruits (so-called follicular fruits) can weigh as much as 3 kg. Enveloped within a mucilaginous layer are the large (up to 3 cm) seeds, with two (*C. nitida*) or four to six (*C. acuminata*) seed leaves. These cola "nuts" turn reddish brown when they dry. In the tropics, *Cola* species bloom throughout the year, with the main flowering period occurring at the beginning of the rainy season.

The two species *C. acuminata* and *C. nitida* are easily confused with the tropical species *Cola quinqueloba* (K. Schum.) Garcke as well as with other *Cola* species.

Psychoactive Material

—Seeds (nuts, semen cola, semen colae, colae semen, cotyledones colae, embryo colae, nuces sterculiae, nux colae)

Preparation and Dosage

The cola nut is the dried seed heart that has had its hull, i.e., the seedling or embryo of the plant, removed. In the pharmaceutical trade, only the seeds of *Cola acuminata* and *Cola nitida* may be referred to as cola nuts (Seitz et al. 1992, 942).

The seeds are freed from the fruit by breaking

103 This Yoruba name is originally from Africa. Santería followers in South America now also use the name to refer to the coconut (*Cocos nucifera*) (González-Wippler 1981, 97*).

open the follicles by hand. The white seed hull that they are attached to can be removed in various ways. The cola nuts may be soaked overnight in water, so that the swollen hulls can be pulled off the next day, or they may be allowed to dry in large piles for five to six days. As soon as the hull turns brown, it disintegrates. After this, the nuts need only be washed. Freshly harvested cola nuts are sometimes placed in a termite mound. The termites then eat the seed hull cleanly away but do not touch the cola nuts (Schröder 1991, 123*).

Some of the bitter red and white seeds are chewed fresh (Bremness 1995, 50*), but most are placed in water (so that they will remain soft) or dried in the sun.

An average daily dosage is 2 to 6 g or 1 to 3 g three times daily (Seitz et al. 1992, 944). The nuts are also used to manufacture extracts, tinctures, and wine extracts. Depending upon the production method, these products can exhibit considerable variation in the amounts of active constituents they contain.

Ritual Use

In West Africa and the Sahel zone, all of life is heavily shaped by the cola nut (Uchendu 1964), which represents the most important socio-integrative element. Cola nuts are offered to every guest as a gesture of respect and deference, they are presented to a lover as a token of one's feelings, they are exchanged at the end of business negotiations to seal the contract, and they are offered to the ancestors, orishas, spirits, and gods. The stimulating nuts are ingested at all social and religious events. They are chewed or given to others at burials, name-giving ceremonies, baptisms, and sacrifices. In the royal courts (e.g., in northern Ghana), all political meetings and discussions begin with a communal chewing of cola. The nuts are placed at forks in the road as protective amulets, they are given to lepers and beggars as gifts, they are handed to physicians and healers as a welcoming greeting, and they are given to soothsayers as payment for divinations (Drucker-Brown 1995).

The social meetings at which cola nuts are ceremonially distributed and communally consumed are strongly reminiscent of the manners in which *Catha edulis* is used in Yemen, *Erythroxylum coca* and *Erythroxylum novogranatense* in South America, *Ilex cassine* and *Ilex vomitoria* in southeastern North America, *Ilex paraguariensis* in southern South America, *Piper methysticum* in Oceania, *Camellia sinensis* in Japan, *Cannabis sativa* in Morocco, and **betel** in Southeast Asia (cf. Graebner 1927).

Cola nuts have also attained a ritual significance in Latin America. They are one of the liturgical plants of the Candomblé cult and are an indispensable element in the initiation of new members into the cult (Voeks 1989, 126*).

Cola Counterfeits and Cola Substitutes

The drug can be counterfeited by using the seeds of lower-quality (i.e., with less caffeine) *Cola* species as well as the fruits/seeds of false colas (some of which have no caffeine at all) (Seitz et al. 1992, 943).

Lower-Quality *Cola* Species

Cola anomala K. Schum.		Cameroon
Cola astrophora Warb.	(kpadu cola)	Togo
Cola digitata Mast.		
Cola lepidata K. Schum.		
Cola pachicarpa K. Schum.		
Cola supfiana Busse	(avatimecola)	

False Colas

Coula edulis Baill.	Oleaceae	West Africa
Dimorphandra mora Schomb.	Fabaceae	Guiana, Trinidad
Garcinia cola Heckel (bitter cola)	Guttiferae	Sierra Leone
Garcinia floribunda (bitter cola)	Guttiferae	Lagos
Heritiera litoralis Dryander	Sterculiaceae	Africa, Indonesia, Antilles
Lucuma mammosa Griseb.	Sapotaceae	Southeast Asia
Napoleona imperialis Beauv.	Lecythidaceae	Benim, Nigeria
Pentadesma butyraceum G. Don	Guttiferae	Kenya, West Africa

In the Afro-American Santería cult (cf. **madzoka medicine**), a sacred liquid known as *omiero* is drunk at the initiation of a new cult member (*santero*). *Omiero* should consist of 101 herbs, representing all of the orishas (Yoruba gods).[104] Yet because it is almost impossible to collect all of these plants, the number of sacred orisha herbs has been reduced to twenty-one. *Omiero* is prepared from these twenty-one herbs as well as the following ingredients: rainwater, seawater, river water, holy water, sacrificial blood, rum, **honey**, *manteca de corojo*, cocoa butter, cascarilla, pepper (*Piper* **spp.**), and cola nuts (González-Wippler 1981, 95). Alone, the presence of the many cola nuts, the rum (see **alcohol**), and the cocoa butter (see *Theobroma cacao*) is enough to ensure that the preparation has stimulant or mild psychoactive effects. Unfortunately, the botanical identity of the twenty-one orisha herbs is not fully known. They do include *Solanum nigrum* (cf. *Solanum* **spp., witches' ointments**), lettuce (*Lactuca virosa*), cinnamon, and fern (see *Polypodium* **spp.**), all of which could contribute to the drink's psychoactivity (González-Wippler 1981, 96).

Artifacts

The cola nuts themselves represent artifacts, as they were used as currency in Africa for a time (Schröder 1991, 116*).

Medicinal Use

The fruits have many folk medicinal uses, especially in Africa (Akendengué 1992, 171*). They are used primarily as a tonic and stimulant and to treat dysentery, fever with vomiting, and exhaustion (Ayensu 1978, 257*). Many African women chew cola nuts to avoid vomiting while

The so-called cola nuts are rust-brown in color when dry.

104 "The orishas are energies that largely represent the different aspects of nature" (Neimark 1996, 23).

"Without the cola nut, for which, if circumstances called for it, the riding horse or the bed slave would be given up, there would be no courtship, no marriage contract, no dowry, no oath, no symbolic expression of friendship or enmity, and no provisions for the dead on their journey. For the people of A. E. N., the cola nut was also responsible for the Adam's apple in men's throats, as they explain in a legend. According to this, once, as the Creator was wandering on the earth looking after the people, he took a piece of cola nut that he was chewing out of his mouth and laid it on a tree drum. And because he forgot this when he departed, a man who had observed this forgetfulness was able to take possession of the delicacy. But the god remembered his absentmindedness and returned, whereupon the man tried to swallow the divinely tasting morsel. The creator, however, stopped him from doing so with a quick grab of his neck, and since that time, men bear this knotty deformation on their throats."

RUDOLF BRUNNGRABER
HEROIN: ROMAN DER RAUSCHGIFTE
[HEROIN: A NOVEL OF DRUGS]
(1952, 127 f.*)

The stimulating extract of the cola nut was mixed with an extract of coca to produce what is probably the most famous refreshing drink in the world: Coca-Cola.
(Advertisement, late nineteenth century)

pregnant and to treat or suppress emerging migraines (Seitz et al. 1992, 944). To a certain extent, cola is also regarded as an aphrodisiac (Drucker-Brown 1995, 132 f.).

In Europe, cola nuts were once utilized to treat migraine headaches, neuralgia, vomiting, seasickness, and diarrhea (Schneider 1974, 1:347*). Today, *Cola* preparations are consumed around the world for physical and mental exhaustion (see **energy drinks**). A mother tincture (Cola hom. *HAB*) is used in homeopathy (Seitz et al. 1992, 945).

Constituents

The composition of the constituents is the same in both species. The purines **caffeine** and theobromine (cf. *Theobroma cacao*) occur in all parts of the plant but are concentrated in the seeds and seedlings. Cola nuts from *Cola acuminata* contain up to 2.2% caffeine, and those from *Cola nitida* up to 3.5% caffeine. Both contain less than 1% theobromine (Brown and Malone 1978, 11*; Seitz et al. 1992, 942). Also present are the polyphenols leucoanthocyanidin and cathecine and large amounts of starch (Seitz et al. 1992, 940). Caffeine and cathecine are primarily present in the form of a caffeine-cathecine complex (especially in fresh nuts) that previously was wrongly thought to be a glycoside and named colanine (Seitz et al. 1992, 941).

Effects

Cola nuts have pronounced powers to stimulate, to wake up a person and keep him awake, as well as tonic effects, i.e., they generally invigorate a person and promote concentration. The effects of freshly chewed nuts are more pronounced, as the caffeine-cathecine complex they contain is broken down more rapidly. Since this complex decays as the seeds dry, the alkaloids are not as easily removed from the tissue and hence are more slowly absorbed. To date, no negative effects of *Cola* use during pregnancy have been observed (Seitz et al. 1992, 944).

Commercial Forms and Regulations

In Africa, numerous commercial goods are produced in different areas. Tinctures and refreshing beverages are manufactured in many countries. All *Cola* products are available around the world without restriction (Seitz et al. 1992). Only the pertinent food regulations need be taken into consideration.

Literature

See also the entry for **caffeine**.

Agiri, Babatunde A. 1975. The Yoruba and the pre-colonial kola trade. *Odu—A Journal of West African Studies* 12:55–68.

———. 1977. The introduction of nitida kola into Nigerian agriculture, 1880–1920. *African Economic History* 3:2–5.

———. 1981. Kola-Handel in Westafrika. In *Rausch und Realität*, ed. G. Völger, 2:528–32. Cologne: Rautenstrauch-Joest-Museum für Völkerkunde.

———. 1986. Trade in gbanja kola in south western Nigeria, 1900–1950. *Odu—A Journal of West African Studies* 30:25–45.

Akinbode, Ade. 1982. *Kolanut production and trade in Nigeria*. Ibadan: NISER.

Chevalier, August, and Em. Perrot. 1911. *Les kolatiers et los noix de kola*. Paris: Augustin Challamel.

Drucker-Brown, Susan. 1995. The court and the kola nut: Wooing and witnessing in northern Ghana. *The Journal of the Royal Anthropological Institute* 1 (1): 129–43.

Eijnatten, Cornelis L. M. 1981. Probleme des Kola-Anbaus. In *Rausch und Realität*, ed. G. Völger, 2:522–27. Cologne: Rautenstrauch-Joest-Museum für Völkerkunde.

Ford, Martin. 1992. Kola production and settlement mobility among the Dan of Nimba, Liberia. *African Economic History* 20:51–63.

González-Wippler, Migene. 1981. *Santería: African magic in Latin America*. Bronx, N.Y.: Original Products.

Graebner, F. 1927. Betel und Kola. *Ethnologica* 3:295–96, Leipzig.

Lovejoy, Paul E. 1970. The wholesale kola trade of Kano. *African Urban Notes* 5 (2): 141.

———. 1980a. *Caravans of kola: The Hausa kola trade 1700–1900*. Zaria and Ibadan, Nigeria: Ahmadu Bello University Press.

———. 1980b. Kola in the history of West Africa. *Cahiers d'Etudes Africaines* 20 (1/2): 97–134.

———. 1995. Kola nuts: The "coffee" of the Central Sudan. In *Consuming habits*, ed. J. Goodman et al., 103–125. London and New York: Routledge.

Neimark, Philip J. 1996. *Die Kraft der Orischa: Tradition und Rituale afrikanischer Spiritualität*. Bern, Munich, and Vienna: O. W. Barth.

Schumann, K. 1900. Die Mutterpflanze der echten Kola. *Notizblatt des Königl. botanischen Gartens und Museums zu Berlin* 3 (21): 10–18.

Seitz, Renate, Beatrice Gehrmann, and Ljubomir Kraus. 1992. Cola. In *Hagers Handbuch der pharmazeutischen Praxis*, 5th ed., 4:940–46. Berlin: Springer.

Uchendu, V. 1964. Kola hospitality and Igbo lineage structure. *Man* 64:47–50.

Coleus blumei Bentham

Coleus

Family
Labiatae (Lamiaceae) (Mint Family)

Forms and Subspecies
Numerous *Coleus blumei* hybrids are raised as indoor plants and ornamentals (Roth et al. 1994, 256f.*).

Synonyms
None

Folk Names
Buntblatt, buntnessel, coleus, coleus scutellaire, common coleus, el ahijado ("the godchild"), el nene ("the child"), la'au fai sei (Samoa), manto de la virgen (Peru), painted nettle, patharcheer, patharchur

History
Coleus is used primarily as an ornamental. Very little is known about its ethnobotany. Its psychoactive use among the Mexican Mazatecs was discovered in 1962 in connection with Gordon Wasson's early research into **Salvia divinorum** (Ott 1993, 381*) and has been only rudimentarily investigated. Phytochemical studies of the plant have increased in recent years, but these have been focused primarily on enzymatic processes (Kempin et al. 1993; Petersen 1992, 1993).

Distribution
Coleus is from Southeast Asia and was not brought to the Americas until the colonial period at the earliest (Schultes 1970, 42*). Today, it is a pantropic ornamental.

Cultivation
Propagation occurs primarily through cuttings. A young shoot some 10 cm long or a young branch is separated from the mother plant and all leaves except the last pair at the end of the stem are carefully removed. The stem is placed in a glass of

water. Within two weeks, the first roots will appear. After three to four weeks, the small plant can be transplanted into humus-rich soil. It should be watered well and not allowed to stand in direct sunlight. Because it does not tolerate any frost, in cold climates coleus can be kept only as a houseplant.

Appearance
This herbaceous or bushy plant can grow to a height of about 80 cm. The colorful green-red leaves are decussate and ovate-acuminate; they have serrated margins and a slightly sinuate upper surface. The small flowers grow in terminal racemes or panicles. The plant can bloom throughout the year in the tropics. As a houseplant, it usually blooms from June to September. The plant apparently never or only extremely rarely develops fruits.

There are a large number of *Coleus blumei* hybrids, some of which can be mistaken for other *Coleus* species. The popular cultivar Verschaffetii is especially easy to confuse with *Coleus forskohlii* (Poir.) Briq [syn. *Coleus barbatus* Benth.]. A species from Borneo, *Coleus pumilus* Blanco [syn. *Coleus rehneltianus* Berger], also has a very similar appearance.

Psychoactive Material
— Leaves

Preparation and Dosage
The leaves are dried and smoked alone or mixed with other herbs (cf. **smoking blends**).

In the tropics, the leaves dry slowly but do not grow moldy like those of other plants. Psychoactive effects can appear when smoking as few as three leaves.

Ritual Use
The Mazatecs include coleus in the same "family" as **Salvia divinorum**, whereby *Salvia* is the "female" and coleus the "male." They also make an additional distinction: *Coleus pumilis* Blanco [syn. *Coleus rehneltianus* Berger] is *el macho*, "the male," while the two forms of *Coleus blumei* are *el nene*, "the child," and *el ahijado*, "the godson" (Schultes 1970, 42*). The fresh leaves are used in exactly the same manner as those of **Salvia divinorum**, that is, they are chewed as quids. Mazatec soothsayers apparently use coleus only as a substitute for *Salvia divinorum*.

Artifacts
None

"Having magico-religious significance, *Coleus* is used as a divinatory plant. The leaves are chewed fresh or the plants are ground, then diluted with water for drinking."

RICHARD E. SCHULTES AND ALBERT HOFMANN
PLANTS OF THE GODS
(1992, 69*)

Above: In German, *Coleus blumei* is known as *buntblatt*, "colorful leaf." As this photograph demonstrates, the name is very appropriate. (Photographed in Palenque, Mexico)

Left: *Coleus pumilis*, a close relative of *Coleus blumei*, is also said to induce psychoactive effects. (Photographed in Palenque, Mexico)

Coleus forskohlii in bloom.
(Photographed on the Seychelles)

Medicinal Use

On Samoa, the herbage is used to treat elephantiasis (Uhe 1974, 15*). In Southeast Asia, it is used to treat dysentery and digestive problems (Valdés et al. 1987, 474), and in Papua New Guinea it is used to treat headaches (Ott 1993, 381*). Coleus is also used as a medicinal plant in the San Pedro cult (cf. *Trichocereus pachanoi*).

The closely related species *Coleus atropurpureus* Benth. was once used to prevent conception (Schneider 1974, 1:349*).

Constituents

Coleus was recently found to contain salvinorin-like substances (cf. **salvinorin A**) of an as yet undetermined chemical structure (cf. **diterpenes**). It is possible that these diterpenes are chemically modified by drying or burning and transformed into efficacious substances. However, additional chemical and pharmaceutical research is needed to clarify this situation.

Rosmarinic acid has been biosynthesized in cell cultures of *Coleus blumei* (Häusler et al. 1992; Meinhard et al. 1992, 1993).

A diterpene (forskolin = coleonol) that is potently bioactive has been found in the related species *Coleus forskohlii* (Poir.) Briq. [syn. *Coleus barbatus* Benth.] (Valdés et al. 1987). It is possible that *Coleus blumei* may also contain forskolin or a similar substance. However, an initial investigation of Indian plants was unable to detect any forskolin (Valdés et al. 1987, 479).

Forskolin activates the enzyme adenylate cyclase, an intracellular neurotransmitter that can bind to various receptors. This means that forskolin is able to exert strong indirect effects upon neurotransmission (D. McKenna 1995, 103*). Whether this can result in psychoactive effects is unknown.

Effects

Some 30% of subjects who smoked dried Mexican *Coleus blumei* leaves reported effects similar to those produced by smoking a small dosage of *Salvia divinorum* (increase in pulse rate, sensations of bodily heaviness, rolling sensations, lights dancing before the eyes). It may be that a particular bodily chemistry is required to react to the plant. It is also possible that the effects are perceived only after repeated attempts (as is the case with *Cannabis* and *Salvia divinorum*).

In the specialized literature, the psychoactivity of coleus is highly controversial:

> Coleus can be found in every specialized book on inebriating drugs. . . . I myself, as well as a larger number of people that I know, [have] undertaken experiments with this plant, some of them using very large amounts of leaves. In no case was there any type of effect. . . .

A communication from the ethnopharmacologist Daniel J. Seibert suggests the same. He was in the area of the Mazatecs, and wrote me that only one single Indian there maintained that coleus is psychoactive. All of the other Indians denied this. (Schuldes 1995, 78*)

Commercial Forms and Regulations

Living coleus plants can be obtained in virtually every nursery. The plant is not subject to any rules or legal regulations.

Literature

See also the entries for *Salvia divinorum*, **diterpenes**, and **salvinorin A**.

Dubey, M. P., R. C. Srimal, S. Nityanand, and B. N. Dhawan. 1981. Pharmacological studies on coleonol, a hypotensive diterpene from *Coleus forskohlii*. *Journal of Ethnopharmacology* 3 (1): 1–13.

Garcia, L. L., L. L. Cosme, H. R. Peralta, et al. 1973. Phytochemical investigation of *Coleus blumei*. I. Preliminary studies of the leaves. *Philippine Journal of Science* 102:1.

Häusler, E., M. Petersen, and A. W. Alfermann. 1992. Isolation of protoplasts and vacuoles from cell suspension cultures of *Coleus blumei*. *Planta Medica* 58 suppl. (1): A598.

Karwatzki, B., M. Petersen, and A. W. Alfermann. 1992. Properties of hydroxycinnamate: CoA ligase from rosmarinic acid–producing cell cultures of *Coleus blumei*. *Planta Medica* 58 suppl. (1): A599.

Kempin, B., M. Petersen, and A. W. Alfermann. 1993. Partial purification and characterization of tyrosine aminotransferase from cell suspension cultures of *Coleus blumei*. *Planta Medica* 59 suppl.: A648.

Lamprecht, W. O. Jr., H. Applegate, and R. D. Powell. 1975. Pigments of *Coleus blumei*. *Phyton* 33:157.

Meinhard, J., M. Petersen, and A. W. Alfermann. 1992. Purification of hydroxyphenylpyruvate reductase from cell cultures of *Coleus blumei*. *Planta Medica* 58 suppl.: A598–99.

———. 1993. Rosmarinic acid in organ cultures of *Coleus blumei*. *Planta Medica* 59 suppl.: A649.

Petersen, M. 1992. New aspects of rosmarinic acid biosynthesis in cell cultures of *Coleus blumei*. *Planta Medica* 58 suppl. (1): A578.

———. 1993. The hydroxylation reactions in the biosynthesis of rosmarinic acid in cell cultures of *Coleus blumei*. *Planta Medica* 59 suppl.: A648.

Valdés, L. J. III, S. G. Mislankar, and A. G. Paul. 1987. *Coleus barbatus* (*C. forskohlii*) (Lamiaceae) and the potential new drug forskolin (coleonol). *Economic Botany* 41 (4): 474–83.

Convolvulus tricolor Linnaeus

Dwarf Morning Glory

Family
Convolvulaceae (Morning Glory Family)

Forms and Subspecies
There are three subspecies, as well as cultivars that are distinguishable on the basis of the color of their flowers; 'Royal Ensign', for example, is characterized by a nonclimbing, bushy growth pattern and gentian blue flowers.

Synonyms
None

Folk Names
Bunte ackerwinde, dreifarbige winde, dwarf morning glory

History
Dioscorides may have known this vine by the name *helxine*; he states that "the juice of the leaves has, when drunk, power to loosen the stomach" (4.39).[105] However, the taxonomic history of this vine is by no means clear (Schneider 1974, 1:362*). It has been suggested that the plant may have been an ingredient in the Eleusinian initiatory drink (see **kykeon**). Ethnopharmacological research into the plant is needed.

Distribution
The plant is from southern Europe (Italy or Portugal) and occurs throughout the entire Mediterranean region as well as North Africa (Festi and Alliota 1990*; Schönfelder 1994, 158*). It has become naturalized in Denmark. In Germany, it is usually found only in botanical gardens.

Cultivation
Sowing is best performed between April and June. The germinated seeds (time to germination is fourteen to twenty days at 15 to 18°C) should be planted directly outdoors. This vine is also suitable for growing on a balcony. The plant loves calciferous soils and thrives best in a sunny location. Flower production can be promoted by providing only a little fertilizer. The blooming period is from July to September.

Appearance
This bushy, annual vine attains a height of only about 35 cm. The funnel-shaped, five-pointed, three-colored flowers (yellow inside, white in the middle, and blue on the margin) are solitary and long-stemmed (as are the leaves). The corolla is 1.5 to 4 cm in length. The stigma has two oblong lobes

(this characteristic distinguishes the genus *Convolvulus* from *Ipomoea*).

Convolvulus tricolor is sometimes confused with *Ipomoea violacea* (even in the specialized literature), and especially its synonym *Ipomoea tricolor* (e.g., Bauerreuss 1995*; Roth et al. 1994*).

Psychoactive Material
— Seeds (semen convulvuli, vine seeds)

Preparation and Dosage
The crushed seeds are drunk in the form of a cold-water infusion. To date, dosages have not been reported.

Ritual Use
A traditional use of *Convolvulus tricolor* as a psychoactive substance is as yet unknown, although such a use is entirely possible. Some "closet shamans" believe that the seeds of this vine may have been an ingredient in **kykeon**, the initiatory drink of the Eleusinian mysteries.

Artifacts
None known

Medicinal Use
This vine may have been used in folk medicine as a laxative, similar to the use of scammony (*Convolvulus scammonia* L.) or greater bindweed (*Calystegia sepium* (L.) Br. [syn. *Convolvulus sepium* L.]) (Pahlow 1993, 353*). Scammony has been used as an aid in childbirth and to induce contractions in both ancient and modern times (Albert-Puleo 1979).

Constituents
The seeds may contain **ergot alkaloids**, ergoline, and other lysergic acid derivatives. Trace amounts of these alkaloids (0.001% of fresh weight) have been detected in plants from Denmark (Genest and Sahasrabudhe 1966).

The dwarf morning glory (*Convolvulus tricolor*) is indigenous to the Mediterranean region. It is sometimes confused with the Mexican morning glory (*Ipomoea violacea*).

Two vines from the genus *Convolvulus* that may be identical to *Convolvulus tricolor* L. and *Convolvulus scammonia* L. Both species contain alkaloids. (Copperplate engraving from Dioscorides, 1610)

105 The *helxine* has also been interpreted as the field convolvulus (*Convolvulus arvenis* L.).

The closely related field convolvulus (*Convolvulus arvensis* L.) contains **tropane alkaloids,** including tropine, cuscohygrine, and hygrine (Todd et al. 1995).

Another relative, *Convolvulus pseudocantabricus* Schrenk., is said to contain alkaloids with analgesic effects, substances very similar to those found in *Turbina corymbosa. Convolvulus scammonia* appears to contain **ergot alkaloids** (Albert-Puleo 1979).

Effects
The seeds may possibly have a hypnotic effect.

Commercial Forms and Regulations
The seeds are available in nurseries and seed shops and are not subject to any regulations.

Literature
See also the entries for *Argyreia nervosa, Ipomoea violacea, Ipomoea* spp., *Turbina corymbosa,* and **ergot alkaloids.**

Albert-Puleo, Michael. 1979. The obstetrical use in ancient and early modern times of *Convolvulus scammonia* or scammony: Another non-fungal source of ergot alkaloids? *Journal of Ethnopharmacology* 1 (2): 193–95.

Genest, K., and M.R. Sahasrabudhe. 1966. Alkaloids and lipids of *Ipomoea, Rivea* and *Convolvulus* and their application to chemotaxonomy. *Economic Botany* 20 (4): 416–28.

Todd, Fred G., Frank R. Stermitz, Patricia Schultheis, Anthony P. Knight, and Josie Traub-Dargatz. 1995. Tropane alkaloids and toxicity of *Convolvulus arvensis. Phytochemistry* 39 (2): 301–3.

Corynanthe spp.

Pamprama

Family
Rubiaceae (Coffee Family); Subfamily Cinchonoideae, Cinchoneae Tribe

The genus *Corynanthe* is composed of five or six species. It is very closely related to *Pausinystalia yohimba* and is often confused with it. The small trees of this genus occur in the tropical rain forests of western Africa. The species that have been studied to date (*Corynanthe pachyceras* K. Schum., *Corynanthe mayumbensis* [Good] N. Hallé) contain **indole alkaloids** of the corynanthein-yohimbine group (Chaurasia 1992, 1029). The bark (pseudocinchonae africanae cortex) of *Corynanthe pachyceras*[106] contains approximately 5.8% indole alkaloids, including corynanthine (= rauhimbine), corynanthidine (= α-yohimbine), corynanthein, dihydrocorynanthein, corynantheidin, corynoxein, corynoxin, and β-**yohimbine.** In the pharmaceutical trade, the bark is often used as a counterfeit or substitute for the true yohimbé bark (from *Pausinystalia yohimba*) (Chaurasia 1992; Neuwinger 1994, 701*).

The bark of *Corynanthe pachyceras* is used in the Ivory Coast to manufacture arrow poisons (Neuwinger 1994, 701*; 1997, 780*). In the former French Equatorial Africa, the bark is used to fortify fermented beverages (**beer, palm wine**). The bark extract has weak analgesic and local anesthetic effects. In animal experiments, it decreased the toxicity of amphetamine (cf. **ephedrine**) by 100%! In West Africa, the bark is esteemed as an aphrodisiac (Chaurasia 1992, 1031; Raymond-Hamet 1937).

(Advertisement, ca. 1920)

Literature
See also the entries for *Pausinystalia yohimba* and **yohimbine.**

Chaurasia, Neera. 1992. *Corynanthe.* In *Hagers Handbuch der pharmazeutischen Praxis,* 5th ed., 4:1029–32. Berlin: Springer.

Goutarel, R., M. M. Janot, R. Mirza, and V. Prelog. 1953. Über das reine Corynanthein. *Helvetica Chimica Acta* 36:337–40.

Karrer, P., R. Schwyzer, and A. Flam. 1952. Die Konstitution des Corynantheins und Dihydrocorynantheins. *Helvetica Chimica Acta* 35:851–62.

Raymond-Hamet. 1937. Über die Wirkungen von Corynanthine auf die männlichen Genitalfunktionen. *Archiv für Pharmakologie und experimentelle Pathologie* 184:680–85.

106 Synonyms: *Corynanthe macroceras* K. Schum., *Pausinystalia pachyceras* (K. Schum.) De Wild., *Pseudocinchona africana* A. Chev. ex E. Perrot, *Pseudocinchona pachyceras* (K. Schum.) A. Chev.

Coryphantha spp.

Coryphantha Species

Family
Cactaceae (Cactus Family); Cereeae Tribe, Coryphanthanae Subtribe

Species
The following species have been found to contain β-phenethylamines (often hordenine) with (presumably) psychoactive effects (Howe et al. 1977; Kelley H. et al. 1972):

Coryphantha cornifera (DC.) Lem.
Coryphantha durangensis (Rünge) Br. et R.
Coryphantha echinus (Engelm.) Br. et R. [syn. *Coryphantha cornifera* var. *echinus*]
Coryphantha elephantidens (Lem.) Lem.
Coryphantha greenwoodii H. Bravo
Coryphantha ottonis (Pfeif.) Lem.
Coryphantha pectinata (Engelm.) Br. et R.
Coryphantha vivipara var. *arizonica*

Ethnobotanically relevant species:

Coryphantha compacta (Engelm.) Britt. et Rose (peyote substitute)
Coryphantha macromeris (Engelm.) Britt. et Rose [syn. *Lepidocoryphantha macromeris*] (peyote substitute)
Coryphantha macromeris var. *runyonii*
Coryphantha palmeri Britt. et Rose (narcotic)
Coryphantha ramillosa Cutak

Folk Names
Biznaga de piña, donana, falscher peyote, huevos de coyote (Spanish, "the eggs [= testicles] of the coyote"), mulato (for *Coryphantha macromeris*), stachelkaktus, warzenkakteen

Distribution
Most members of the genus are from Mexico, although some occur from northern Mexico to Texas.

Cultivation
Like all cacti, these species can be grown from seed. *Coryphantha* thrives best in sandy and clayey soils and requires much sun and much water during the blooming period (but do not keep wet). No water at all should be given during the winter (Hecht 1995, 26*).

Appearance
The most ethnopharmacologically interesting of these cacti, *Coryphantha compacta*, is a slightly depressed spherical cactus with a maximum diameter of 8 cm. The whitish, 1 to 2 cm long thorns are arranged in a radial pattern. Most *Coryphantha* species are spherical, heavily thorned

ball cacti (Preston-Mafham 1995*) that often develop magnificent, sun-yellow flowers. They can be confused with some species from the genera *Ferocactus* and *Echinocactus*, as well as with **Mammillaria spp.**

Psychoactive Material
— Cactus flesh, fresh or dried

Preparation and Dosage
The thorns must first be removed, after which the aboveground portion is eaten fresh. The dosage is given as eight to twelve cacti (*Coryphantha macromeris*) (Gottlieb 1973, 12*).

Ritual Use
Presumably the only ritual or shamanic use of *Coryphantha* species is (for some of them) as peyote substitutes (see *Lophophora williamsii*).

Artifacts
None

Medicinal Use
Presumably similar to that of *Lophophora williamsii*

Constituents
β-phenethylamines (hordenine, normacromerine, calipamine, methyltyramine and derivatives, synephrine, macromerine, metanephrine, tyramine) have been found in many *Coryphantha* species (Bruhn et al. 1975). Most species contain primarily hordenine (Howe et al. 1977; Mata and McLaughlin 1982, 97–100*; Ranieri et al. 1976).

Effects
Coryphantha compacta is "taken by shamans as a potent medicine and greatly feared and respected by the Indians" (Schultes and Hofmann 1992, 67*).

Commercial Forms and Regulations
Many species of the genus are available in cactus shops.

Left: Many species in the genus *Coryphantha* are spherical in shape and regarded as peyote substitutes. (*Coryphantha recurva*, photographed in Arizona)

Right: *Coryphantha echinus*, from northern Mexico, produces psychoactive effects.

Hordenine

"The goddess sits at the top of the round cactus,
our mother, butterfly of obsidian.
Look, there in the springlike fields,
nourished by the hearts of deer,
is our mother, queen of the earth,
adorned with fresh clay and new feathers.
From the four directions of heaven,
because she breaks lances:
she is transformed into a deer.
Across the stony ground
come Xiuhnelli and Mimich,
to see you."

AZTEC PRAYER
IN *LA LITERATURE DE LOS AZTECOS* (A. GARIBA)

Literature

See also the entries for ***Lophophora williamsii*** and β-phenethylamines.

Bruhn, J., S. Agurell, and J. Lindgren. 1975. Cactaceae alkaloids. XXI: Phenethylamine alkaloids of *Coryphantha* species. *Acta Pharm. Suecica* 12:199.

Howe, R. C., R. L. Ranieri, D. Statz, and J. L. McLaughlin. 1977. Cactus alkaloids. XXXIV: Hordenine HCl from *Coryphantha vivipara* var. *arizonica*. *Planta Medica* 31:294.

Keller, W. J., and J. L. McLaughlin. 1972. Cactus alkaloids. XIII: Isolation of (–)-normacromerine

from *Coryphantha macromeris* var. *runyonii*. *Journal of Pharmaceutical Science* 61:147.

Kelly Hornemann, K. M., J. M. Neal, and J. L. McLaughlin. 1972. Cactus alkaloids XII: β-phenethylamine alkaloids of the genus *Coryphantha*. *Journal of Pharmaceutical Science* 61:41–45.

Ranieri, R. L., J. L. McLaughlin, and G. K. Arp. 1976. Isolation of β-phenethylamines from *Coryphantha greenwoodii*. *Lloydia* 39 (2–3): 172–174.

Crocus sativus Linnaeus

Saffron Crocus

The saffron crocus (*Crocus sativus*) is the source of the coveted and precious saffron spice, high dosages of which produce opium-like effects. (Woodcut from Tabernaemontanus, *Neu Vollkommen Kräuter-Buch*, 1731)

True saffron consists of the stigmas of *Crocus sativus*. (Photograph: Karl-Christian Lyncker)

Family

Iridaceae (Iris Family)

Forms and Subspecies

A very late-blooming form has been described as the variety *Crocus sativus* L. var. α *autumnalis*. The saffron grown in Kashmir is referred to as *Crocus sativus* L. var. *cashmirianus* (Bowles 1952). The subspecies *Crocus sativus* ssp. *cartwrightianus* is said to be endemic in Greece (Baumann 1982, 158*).

Synonyms

Crocus autumnalis Mill.
Crocus hispanicus
Crocus luteus L.
Crocus orientalis

Folk Names

Abir (Persian), crocus (Roman), gewürzsafran, hay saffron, karcom (Hebrew), karkom, karkum (Persian), kesar (Sanskrit), kesara (Hindi), kesari, krokos (Greek), krokus, kumkumkesari, plam phool (Pakistani), saffron, saffron crocus, safrankrokus, sn-wt.t (ancient Egyptian), z'afarân (Arabic/Yemen), zafran

History

The saffron crocus is one of the very oldest of all cultivated plants. A wild form is no longer known (Czygan 1989, 413). Saffron was first mentioned in conjunction with the name of a city on the Euphrates: Azupirano, "Saffron City" (ca. 2300 B.C.E.). The plant was already being cultivated on Crete and Thera (Santorini) during the Minoan period (Basker and Negbi 1983, 228). Because of the plant's color, the Greek scholar Carl Ruck

believes that in Archaic Greece the saffron crocus was used entheogenically as a substitute for fly agaric (*Amanita muscaria*), which was originally venerated as sacred and eaten ritually (Ruck 1995, 133*). The earliest written record of saffron is presumably in the *Iliad* and in the Old Testament Song of Songs. The first documentation for Kashmir is from the fifth century B.C.E. (Basker and Negri 1983, 228).

In the eighteenth and nineteenth centuries, saffron was used as an inebriant—certainly, a very expensive one—whose effects were said to be like those of opium (cf. ***Papaver somniferum***). Although it is known that saffron has psychoactive properties, this aspect of the plant has been only little studied. The reason is the still very high price that true saffron commands (in comparison, **cocaine** is a veritable "middle-class" drug).

Because true saffron has always been extremely expensive, the coveted spice has often been counterfeited. Moreover, the name has been used for a wide variety of plants (Schneider 1974, 1:378*). In ancient times, saffron was an important source of dye, especially for coloring royal garments (Basker and Negbi 1983, 230).

Saffron also played a role in perfumery, as Aristophanes has intimated (*The Clouds*, l.51).

In the tenth century, saffron was cultivated in Spain and from there exported into all the countries of Europe (Hooper 1937, 107*). The Upper Valais (Switzerland) is a renowned and venerable area of saffron cultivation. The so-called krummenegga ("saffron fields") are located there, having been established in 1420 by knights returning from the Crusades. In 1979, following long years of neglect, a saffron guild constituted itself with the aim of reinvigorating saffron cultivation (Vonarburg 1995).

Distribution
Because the wild form is unknown, only the range of saffron culture can be given. This is established primarily in western Asia, Asia Minor, Turkey, Iran, Greece, India, and Spain.

Cultivation
Propagation occurs vegetatively by the separation of small tubers. The precise methods of cultivation are usually kept secret for economic reasons.

Saffron is the most expensive spice in the world, which is why its cultivation has such great economic significance in the areas where it is grown. Twenty thousand stigmas yield a mere 125 g; according to a different calculation, 1 kilo of dried filaments requires some 60,000 flowers or 120,000 to 150,000 stigmas (Vonarburg 1995, 75).

Appearance
This perennial tuberous plant blooms in the fall. It has very narrow and elongated leaves. The violet-veined flower sits at the end of the stalk. It has three yellow stamens, a thin yellow style, and three long, red, funnel-shaped stigmas that project from the flower.

The saffron crocus is very similar to the meadow saffron (= autumn crocus; *Colchicum autumnale* L.)[107] and is easily confused with it, especially because this plant also blooms in the fall (Bowles 1952).

Psychoactive Material
— Saffron (croci stigma, flores croci, crocus): the brick red stigma held together by a small piece of the style. The dried stigmas are approximately 20 to 40 mm long. They have a strong aromatic scent and a spicy-hot taste.

Two qualities are distinguished:

— Crocus electus (saffron tips, free of the remains of the styles)
— Crocus naturalis (with many pieces of styles)

The stigmas must be kept out of the light and stored in an airtight container, or the volatile essential oil will evaporate and the color will fade.

The entire flower is used for folk medicinal purposes.

Besides the Greek saffron, the Hippocratics mention an "Egyptian saffron" that was used externally. This likely is a reference to yellow safflower (*Carthamus tinctorius* L.), as the Egyptians themselves did not plant saffron (which they called the "blood of Hercules"). Instead, they imported it from Crete and southwest Asia. The saffron threads are often mistaken for or counterfeited by the petals of the safflower (*Carthamus tinctorius*) (Norman 1991, 33*). Curcuma (*Curcuma longa* L., Zingiberaceae) is known as saffron spice as well as Indian saffron. To add to the confusion, the autumn crocus is also known as meadow saffron (Basker and Nagbi 1983, 232).

Preparation and Dosage
In ancient times, saffron was used primarily as a wine additive (cf. *Vitis vinifera*) that provided an additional inebriating effect (Norman 1991, 33*). Saffron is an important ingredient in laudanum or tinctura opii crocata (cf. *Papaver somniferum*, soporific sponge). Saffron is also found in the so-called Swedish herb mixes (cf. **theriac**) as well as **Oriental joy pills** and other aphrodisiacs. In ancient China, saffron was used as an additive to **sake**.

A Greek papyrus from the Egyptian Arsinoites (third century B.C.E.) contains a recipe; unfortunately, there is no information about what the mixture should be used for:

The plaster of Dionysus: two drams of copper oxide, three obols of rosebud hearts (perhaps specifically *Rosa gallica*), three obols of saffron, one-half obol poppy juice (*Papaver somniferum*), three obols of white (acacia) gum (*Gummi arabicum*). Stir these (things) in wine as smoothly as possible (and) make ointments, apply. (In Hengstl et al. 1978, 272)

Perhaps this was some type of aphrodisiac ointment, for saffron has always enjoyed a reputation as an aphrodisiac and agent of love.

To date, no risks have been documented at a maximum daily dosage of 1.5 g. Twenty grams is given as a lethal dose, while 10 g can induce abortion (per *Monographie der Kommission E*; cf. Czygan 1989, 414).

Ritual Use
The roots of the ritual use of saffron, which was regarded as sacred, lie in Minoan Crete and Thera, and most likely in the entire range of the Minoan culture. As the many saffron frescoes in different shrines suggest, saffron had an important ritual significance on Crete and Thera. The saffron crocus was apparently associated with the priestly veneration of the Minoan goddess, with the

"To the Aither
—a smoking offering of saffron—
You high-reaching house of Zeus,
Indestructible in eternal power,
Bearer of the stars, the sun,
the moon,
Vanquisher of all, fire-breathing,
substance which ignites all life!
Far-illuminating ether,
most noble primeval substance of the
universe,
magnificent first seed, bearer of light,
flaming from the fire of the stars—
To you sounds my pleading call:
Oh show your cheerful visage!"

ORPHIC HYMN

"Emperor Marcus Aurelius bathed in saffron water, because it beautified the skin and supposedly also increased male potency. The celebration halls were adorned with crocuses as an auspicious sign for the orgy that the carousal would hopefully turn into, and crocus flowers were placed in the hair."

ANTHONY MERCATANTE
DER MAGISCHE GARTEN [THE MAGICAL GARDEN]
(1980, 51*)

"Saffron, which is often used in the home kitchen to color food dishes, takes the place of opium for children. It is frequently used as an analgesic and antispasmodic, as an agent to promote menstruation and uterine spasms, and externally for inflammations of the glands (breasts), panaritis, hemorrhoidal knots, some eye ailments, and facial pain. In high dosages, it induces abortion."

HAGERS HANDBUCH DER PHARMAZIEUTISCHEN PRAXIS KOMMENTAR to the *DEUTSCHE APOTHEKER BUCH*, FIRST EDITION

107 Meadow saffron is an ancient Colchic magical and witches' plant that can produce serious toxic effects if used incorrectly (Rätsch 1995a, 190ff.*).

The petals of the safflower (*Carthamus tinctorius*), which do not have any psychoactive effects, are a source of false saffron. (Woodcut from Fuchs, *Läebliche abbildung und contrafaytung aller kreüter*, 1545)

Safranal

worship of nature, and with fertility. Wall paintings from Thera make it clear that priestesses carried out the saffron harvest (Doumas 1992). It is possible that saffron was also involved in the ritual embalming and preparation of the dead in Egypt.

Saffron was sacred to the goddess Hecate, for Orphic hymns invoked the ruler of the shadows as the "sea goddess in saffron robes." In the Orphic mysteries, part of the cult of Dionysos (cf. *Vitis vinifera*), saffron was a ritual **incense** that was burned during the recitation or singing of hymns.

To date, we know of no traditional or ritual use of saffron as a psychoactive substance.

Artifacts

Both the saffron crocus and its harvest are subjects of Minoan wall paintings (Marinatos 1984). The saffron paintings from Thera (Santorini, Xestes 3, Room 3a, first floor) reflect the loving manner in which the plant was treated (Doumas 1992, 152 ff.; Douskos 1980).

Garments dyed with saffron have been preserved from antiquity, the Middle Ages, and the early modern period. In contrast, the "saffron yellow" robes that Buddhist monks in Sri Lanka wear are not dyed with true saffron, as has been wrongly assumed (Basker and Negbi 1983).

The novel *Die Safranhändlerin* [The Saffron Merchantess], by H. Glaesener (1996), provides an amusing description of the world of the medieval spice trade.

Medicinal Use

Saffron is one of the oldest and most used medicines of the Hippocratics. It was said to be an effective antidote for drunkenness (see *Vitis vinifera*) and to increase male potency. According to Pliny, saffron was a panacea and an aphrodisiac: "It induces sleep, has a gentle effect upon the head, and whets the sex drive" (21.137). For this reason, saffron was also an important ingredient in love drinks in ancient Rome (Mercatante 1980, 50*). During the Renaissance, it was said that smelling a crocus in bloom "expands the heart and the tools of the mind and stimulates to coitus."

In the mystical medicine of Islam, the following is said about saffron: "It is an excellent agent for the blood and for strengthening the soul. It assuages joint pains and strengthens the sex drive in young men" (Moinuddin 1984, 99*).

Since the Middle Ages, saffron has been used as a remedy for "St. Anthony's fire" (ergotism; cf. *Claviceps purpurea*). In Victorian England, it was used to treat constipation and found its way to the source of the problem as an **enema** (Mercatante 1980, 51*).

In Western medicine, saffron was once used as a nerve calmative and to treat spasms and asthma, but it no longer has any medical significance. In folk medicine, saffron is still used as a sedative and antispasmodic (Czygan 1989, 414). In homeopathy, the mother tincture is prepared from the dried filaments (stigmas) and is used primarily to treat women and children (Vonarburg 1995, 76).

Saffron also found its way into traditional Chinese medicine, where it is used as a psychoactive remedy:

> Among the ailments that are generally treated with saffron are depression, constricted feelings in the chest, fear, shock, confusion (mental and emotional disturbances), coughing blood, period pains and other menstrual complaints, blood congestion [accumulation of blood in the capillaries], and abdominal pains following childbirth. Long-term use of saffron can free one from depressions and feelings of anxiety and produce sensations of happiness. (Leung 1995, 186*)

In Baluchistan (Pakistan), 10 g of ground flowers (not just the pistils), which are known as *khakhobe*, are drunk mornings and evenings in a mixture of liquid yogurt as a remedy for dysentery (Goodman and Ghafoor 1992, 52*). In Yemen, saffron is still used as an aromatic stimulant (Fleurentin and Pelt 1982, 90 f.*).

Constituents

Saffron contains 8 to 13% solid oil and up to 1% **essential oil**, as well as oleanolic acid derivatives, glycosides, the bitter substance picrocrocine (which when stored transforms into safranal, the aromatic substance that gives saffron its characteristic scent), and crystalline yellow dyes (α-crocine = crocetine-di-β-D-gentiobiosylester, crocetine, and others) (Czygan 1989, 414). Saffron also contains the vitamins riboflavin (100γ/g!) and thiamine (Bhat and Broker 1953). The essential oil has a rather complex structure (Zarghami 1970): "The principal component of the essential oil is safranal, which produces the scent typical of the drug. Safranal is first produced during drying, which is why this step merits particular attention during processing" (Pahlow 1995, 78*).

Effects

The psychoactive effects of saffron have been occasionally described as "spasms of laughter" and "delirium" (Vonarburg 1995, 76); "in its effects, saffron comes close to opium [cf. *Papaver somniferum*]; in low dosages, it excites, cheers, and produces laughter . . . , in contrast, in high dosages it sedates, promotes sleep, sopor" (Most 1843, 536*). The essential oil and its vapors also produce psychoactive effects, which have been described as "a sedative effect upon the brain, sleep-inducing, produc[ing] headaches [and] cheerful delirium, and paralyz[ing] motor nerves. Blindness. Peculiar orgasm" (Roth et al. 1994, 276). Actual reports of direct experiences with the drug are not available, presumably because of its high cost.

Saffron promotes protein digestion because it stimulates enzymatic activity. It also stimulates uterine activity and can thus have abortifacient effects. Saffron has the highest riboflavin content of any plant (as a percentage of weight) and as a result appears to lower cholesterol levels (Basker and Begbi 1983). The extract has stimulating and antispasmodic properties (Hooper 1937, 107*).

Commercial Forms and Regulations

Saffron was formerly an important official drug. Today, it is listed only in the ÖAB, Ph. Eur. 1/III and Ph. Helv. VI.* Because it is classified as a spice, saffron is freely available.

Saffron is very often counterfeited for sale. Red or yellow pieces of marigold (*Calendula officinalis* L.) flowers or dyer's saffron (= safflower; *Carthamus tinctorius* L.) are often sold as saffron (even in such producer countries as Greece and Spain). The petals of **Tagetes spp.** (American saffron) have also appeared in trade. Paprika powder (*Capsicum fructescens*) and curcuma (*Curcuma longa* L.) are frequently sold as ground saffron. The red coloration is often achieved using red sandalwood (*Pterocarpus santalinus* L.f.). Saffron powder may also be made heavier by the addition of very dense additives (barium sulfate, brick dust, glycerol) (Czygan 1989, 415).

Literature

Basker, D., and M. Negbi. 1983. Uses of saffron. *Economic Botany* 37 (2): 228–36.

Bhat, J. V., and R. Broker. 1953. Riboflavine and thiamine content of saffron, *Crocus sativus* L. *Nature* 172:544.

Bowles, E. H. 1952. *A handbook of* Crocus *and* Colchicum. London: Bodley Head.

Czygan, Franz-Christian. 1989. Safran. In *Teedrogen*, ed. Max Wichtl, 413–15. Stuttgart: WVG.

Doumas, Christos. 1992. *The wall-paintings of Thera*. Athens: The Thera Foundation.

Douskos, I. 1980. The crocuses of Santorini. In *Thera and the Aegean world*, ed. C. Doumas, 2:141–46. London.

Glaesener, Helga. 1996. *Die Safranhändlerin*. Munich: List.

J. Hengstl, ed., with G. Häge and H. Kühnert. 1978. *Griechische Papyri aus Ägypten als Zeugnisse des öffentlichen und privaten Lebens*. Munich.

Madan, C. L., B. M. Kapur, and U. S. Gupta. 1966. Saffron. *Economic Botany* 20:377–85.

Marinatos, Nannto. 1984. *Art and religion in Thera: Reconstructing a Bronze Age society*. Athens: Mathioulakis.

Nauriyal, J. P., R. Gupta, and C. K. George. 1977. Saffron in India. *Arecanut Spices Bulletin* 8:59–72.

Pfander, H., and F. Wittwer. 1975. Untersuchungen zur Carotinoid-Zusammensetzung im Safran. *Helvetica Chimica Acta* 58:1608–20.

Vonarburg, Bruno. 1995. Homöopathisches Pflanzenbrevier 19: *Crocus sativus*. *Natürlich* 15 (10): 75–78.

Zarghami, N. S. 1970. *The volatile constituents of saffron* (Crocus sativus L.). PhD thesis, University of California, Davis.

The saffron crocus in bloom, photographed in its cultivation area in Valais, Switzerland. (Photograph: Walter Imber)

---

*   Editor's note: These are all European pharmacopoeias.

Cytisus canariensis (Linnaeus) O. Kuntze

Canary Island Broom

Family
Leguminosae (Legume Family); Subfamily Papilionoideae, Genisteae Tribe, Cytisinae Subtribe

Forms and Subspecies
The taxonomy of the genus *Cytisus* (= *Genista*) is rather confusing and ambiguous, especially with respect to the Canary Island species (cf. Kunkel 1993). One variety is occasionally described under the name *Cytisus canariensis* (L.) O. Kuntze var. *ramosissimus* (Poir.) Briq.

Synonyms
Cytisus attleyanus hort.
Cytisus canariensis Steud.
Cytisus ramosissimus Poir.
Genista canariensis L.

Folk Names
Canary Island broom, kanarischer ginster, kytisos, Spanish broom, spartion, spartium

History
Canary Island broom is from the island group of the same name. It may have been a ritual plant of the Guancha, the indigenous people of the islands who, in the fifteenth century, still possessed a Stone Age culture and venerated the Great Goddess (Tara) in painted ritual caves (see Braem 1995, 114–28). It was likely introduced into the New World at an early date, for many ships that were bound for New Spain stopped at the Canary Islands and, when they left, carried the islands' native plants with them. Knowledge of the use of broom as an inebriant may have traveled along as well.

Yaqui shamans from northern Mexico use Canary Island broom for ritual purposes (Fadiman 1965). In the United States, the flowers are smoked as a tobacco substitute (cf. **Nicotiana tabacum**) (Fadiman 1965).

Distribution
The bush is endemic to the Canary Islands. As a result of cultivation (as an ornamental), it can now be found in the entire Mediterranean region and in North, Central, and South America.

Cultivation
Propagation can occur via seeds as well as cuttings. The seeds should be pre-germinated in January before planting. The bush does not tolerate frost (Grubber 1991, 19*).

Appearance
This evergreen bush can grow up to 2 meters in height. The small green leaves are tripartite. The aromatic, light yellow, labiate flowers develop on the upper ends of the branches. The plant flowers between May and July. The fruits are small pods (15 to 20 mm) that contain several small, beanlike seeds.

Canary Island broom is easily mistaken for other species of the genera *Cytisus* and *Spartium* (see **Cytisus** spp.).

Psychoactive Material
— Flowers

Preparation and Dosage
The flowers are dried and chopped. They can be rolled into cigarettes (joints) or placed in a pipe and smoked by themselves or together with other herbs (cf. **smoking blends**). The flowers can also be used to prepare an aphrodisiac drink:

> The flowers of the Canary Island broom are dried over a low flame, then brewed with water, filtered, and drunk. After ingesting this liquid, a person is transported into a state of total euphoria, which includes more intense sensations of sexual arousal along with more intense perception, and a great deal of calm and quiet. (Stark 1984, 56*)

One dosage is the amount of dried leaves contained in one to three normal cigarettes (joints) (Fadiman 1965).

The bright yellow blossoms of Canary Island broom (*Cytisus canariensis*).

Ritual Use

A Yaqui shaman discovered the psychoactive use of this plant. After ingesting another psychoactive plant (most likely peyote; see *Lophophora williamsii*), he was shown in a vision that the flowers of the Canary Island broom should be smoked. Further study of the ritual use is needed.

Artifacts

None

Medicinal Use

None

Constituents

Canary Island broom contains large quantities of **cytisine** (Ott 1993, 407*) and other alkaloids. Detailed chemical studies are lacking (Schultes and Hofmann 1980, 153*).

Effects

Smoking the dried leaves produces effects that have been described as mildly psychedelic without unpleasant side effects or aftereffects (Allen and Allen 1981, 211*). A small dose (one cigarette per person) produced sensations of relaxation with positive feelings for some two hours. Higher dosages (two to three cigarettes) produced an increase in intellectual abilities (clarity, flexibility) as well as an increase in alertness. Although there have been reports of sharpened perception and greater intensity of colors, hallucinations have not been observed. Closing the eyes stimulated the imagination. The effects lasted a maximum of five hours. Apart from a (rare) slight headache the following day, no side effects or aftereffects have been reported (Fadiman 1965).

Commercial Forms and Regulations

The plant is available as an ornamental in nurseries.

Literature

See also the entries for *Cytisus* spp. and cytisine.

Braem, Harald. 1995. *Magische Riten und Kulte: Das dunkle Europa.* Stuttgart and Vienna: Weinbrecht.

Fadiman, James. 1965. *Genista canariensis*: A minor psychedelic. *Economic Botany* 19:383–84.

Kunkel, Günther. 1993. *Die Kanarischen Inseln und ihre Pflanzenwelt.* 3rd ed. Stuttgart: Gustav Fischer.

"The magical beliefs have the following to say about broom: 'Whoever has become ill as a result of enchantment with formulae, he must urinate down through an upside-down broom made of broom, and then he will become healthy.'"

G. W. GESSMANN
DIE PFLANZE IM ZAUBERGLAUBEN [THE PLANTS IN SUPERSTITION] (N.D., 43*)

Cytisus spp. and Relatives

Broom Species

Family

Leguminosae (Legume Family); Subfamily Papilionoideae, Genisteae Tribe, Genistinae (previously Cytisinae) Subtribe

The originally Old World genus *Cytisus* (= *Genista*) encompasses some fifty species, twenty-three to thirty-three of which are found in Europe (Wink 1992, 1124). Many species contain the alkaloids anagyrine, **cytisine**, lupanine, *N*-methylcytisine, and sparteine (Allen and Allen 1981, 210*). Chinolizidine alkaloids (of the sparteine type) are of chemotaxonomic significance (van Rensen et al. 1993).

Cytisus scoparius (L.) Link [syn. *Genista angulata* Poiret, *Genista glabra* Spach, *Genista hirsuta* Moench, *Genista scoparia* (L.) Lam., *Genista scoparius* DC., *Genista vulgaris* Gray, *Sarothamnus ericetorum* Gandoger, *Sarothamnus obtusatus* Gandoger, *Sarothamnus scoparius* (L.) Wimm. ex W.D.J. Koch, *Sarothamnus vulgaris* Wimm., *Spartium angulosum* Gilib., *Spartium glabrum* Mill., *Spartium scoparium* L.]—Scotch broom

The use of Scotch broom for inebriating purposes is allegedly based upon the observation that sheep behave in an excited and peculiar manner after eating the plant (Brown and Malone 1978, 8*).

For several years, dried Scotch broom flowers have been regarded as a "legal high" and used in **smoking blends**. Users have repeatedly reported mild euphoric effects and distinct synergistic effects when the flowers are mixed with other substances, especially *Cannabis sativa.*

The common name of the plant is derived from the fact that its branches were once used to make brooms. Witches are said to have made their flying brooms from *Cytisus* (Ludwig 1982, 143*). Perhaps this use, and its association with witches' flight, is what led to the plant's psychoactive use (cf. **witches' ointments**).

Broom flowers are used in folk medicine as a dehydrating agent and blood purifier. In phytotherapy, their sole use is as a decorative drug in tea mixtures (Wink 1992, 1128).

The aboveground parts of broom as well as its seeds contain the alkaloid sparteine (= lupinine),

Cytisine

Lupinine

"They joined together the flowers of oak, broom, and meadow queen and, with the help of their magic, created the most beautiful and perfect maiden in the world."

FROM THE WELSH *MABINOGION* (MARKALE 1989, 142*)

sarothamnine, and genisteine. **Cytisine** does not appear to be present (Brown and Malone 1978, 9*), although the flowers do contain phenethylamine derivatives (tyramine, etc.) (Wink 1992, 1127). For this reason, they should not be used in combination with MAO inhibitors (see *Peganum harmala*, **ayahuasca analogs**). Sparteine binds to nicotinergic acetylcholine receptors (Wink 1992, 1130). This property may account for the weak psychoactive effects.

Cytisus spp.—broom species

Some *Cytisus* species contain **cytisine** and were apparently frequently smoked as tobacco substitutes (*Nicotiana tabacum*).

In colonial Peru, a plant that was used as a medicinal **incense** may have been a *Cytisus* species. The source notes, "Another herb, *chuquicaylla*, similar to broom, is used as a fumigant for fever" (in Andritsky 1989, 267*).

Genista spp.—broom

Species of the genus *Genista* are easily confused with *Cytisus canariensis* and with *Spartium* spp. Dyer's greenweed (*Genista tinctoria* L.) contains the alkaloid **cytisine**, as does the German broom (*Genistra germanica* L.): "Some alkaloids of *Genista* species exhibit hallucinogenic effects" (Roth et al. 1994, 372*). No information is available that would suggest that these plants were used traditionally for psychoactive purposes.

Spartium junceum L. [syn. *Sarothamnus junceus* Link, *Spartianthus junceus* (L.) Link]—Spanish broom

Spanish broom is easily mistaken for Canary Island broom (*Cytisus canariensis*). Spanish broom has a rich alkaloid content and high concentrations of **cytisine**. It also appears to induce psychoactive effects:

The drug evidently has weak hallucinogenic properties: One artist repeatedly drank decoctions of Spanish broom flowers as a "cardio-

tonic," as he believed that he was using Cytisi scoparii flos. He reported that afterward he experienced very intense dreams, during which he had seen very colorful images. . . . After ingesting a tea infusion of seeds and branch tips (dosage unknown), a woman is said to have experienced vomiting, disturbances of vision, and feelings of drunkenness. (Wink 1994, 771)

In the highlands of Ecuador, this originally European plant is known as *retama*. Drunk as a tea, it is said to have abortifacient or prophylactic effects. The dried flowers are smoked there to treat asthma (Schultes 1983a, 262*). In southern Peru, the flowers are ground and added to **chicha**, (brewed from maize) in order to make it "more inebriating" (Franquemont et al. 1990, 82*). The bush is also called *retama* in Peru and is ingested together with *markhu* (*Ambrosia peruviana* Willd.), *guaco* (*Mikania scandens* Willd.; see *Mikania cordata*), *Coca* (*Erythroxylum coca*), rosemary (*Rosmarinus officinalis* L.; cf. *Fabiana imbricata*), and *nijnd* (*Myrica pubescens* H. et B. ex Willd. var. *glandulosa* Chev.) to treat rheumatism (Bastien 1987, 131*).

Literature

See also the entries for **Cytisus canariensis** and **cytisine**.

van Rensen, I., M. Veit, R. Greinwald, P. Cantó, and F.-C. Czygan. 1993. Simultaneous determination of alkaloids and flavonoids as a useful tool in chemotaxonomy of the genus *Genista*. *Planta Medica* 59 suppl.: A592.

Wichtl, Max. 1989. Besenginsterkraut. In *Teedrogen*, ed. Max Wichtl, 91–93. Stuttgart: WVG.

Wink, Michael. 1992. *Cytisus*. In *Hagers Handbuch der pharmazeutischen Praxis,* 5th ed., 4:1124–33. Berlin: Springer.

———. 1994. *Spartium*. In *Hagers Handbuch der pharmazeutischen Praxis,* 5th ed., 6:768–72. Berlin: Springer.

"Herbs on the fire: Broom,
    nightshades, thorn apple.—How
    that crackles and fumes.
The old man extinguishes the
    lantern, bends over the pan, and
    inhales the toxic smoke;
he can barely remain standing, it
    sedates him so.
And the terrible buzzing in the ears!"

GUSTAV MEYRINK
COAGULUM
(1984, 179*)

Opposite page, left column, from top to bottom:

Scotch broom (*Cytisus scoparius*), found throughout central Europe, is smoked as a tobacco substitute.

Scotch broom (*Cytisus scoparius*).

A cultivated variety of broom.

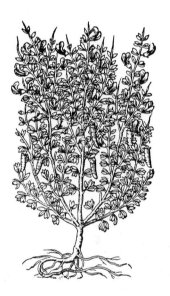

The dried flowers and leaves of two species of broom are smoked for inebriating purposes. The upper illustration shows Scotch broom (*Cytisus scoparius*), a very common plant in Germany, while the lower illustration is of Spanish broom (*Spartium junceum*), which is more common in southern Europe. (Woodcut from Tabernaemontanus, *Neu Vollkommen Kräuter-Buch*, 1731)

Right column, from top to bottom: The southern broom (*Cytisus australis*).

Spanish broom (*Spartium junceum*) occurs throughout the entire Mediterranean region. (Photographed on Crete)

Broom (*Spartium junceum*) was introduced into South America. This specimen was photographed in the Altiplano of Peru.

Datura discolor Bernhardi

Sacred Datura

This Italian techno-pop group took its name from the sacred datura plant. The album includes descriptions of hallucinations caused by "devil's weed." A path to eternity? (CD cover 1993, ZYX Music)

Family
Solanaceae (Nightshade Family); Subfamily Solanoideae, Datureae Tribe, Dutra Section

Forms and Subspecies
Presumably none

Synonyms
Datura thomasii Torr.

Folk Names
A'neglakya, desert datura, é"ee kamóstim (Serí, "plant that makes one squint"), é"ee karóokkoot (Serí, "plant that makes one crazy"), hehe camóstim, hehe carócot, heilige datura, heiliger stechapfel, holy datura of the Zuni, malykatu (Mohave), sacred datura, sacred thornapple, Thomas' thornapple, toloache

History
The history of this potently hallucinogenic species of *Datura* is shrouded in mystery. Although this genus is ethnopharmacologically highly interesting and has been the object of a great deal of research over the last century, many questions still remain (Avery 1959). There is also some taxonomic confusion (cf. *Datura* spp.). In the ethnobotanical literature, *Datura discolor* is usually listed as *Datura innoxia* or *Datura meteloides* (syn.). And indeed, the ranges of the two species overlap. Moreover, the ethnopharma-

cological uses of both are almost identical, and the two species share many folk names. In the American Southwest, however, it has become customary to refer to *Datura discolor* as sacred datura or holy datura of the Zuni and to *Datura innoxia* as toloache or devil's weed.

Distribution
The primary range of this relatively rare thorn apple species extends across the American Southwest and northern Mexico. The plant has also occasionally been reported to occur in the West Indies.

Because of its high alkaloid content, *Datura discolor* is grown commercially in Egypt as a source of pharmaceutical scopolamine (Saber et al. 1970).

Cultivation
As with all *Datura* species, *Datura discolor* is propagated from seeds. Often, the seeds must simply be scattered over the ground. Seeds also can be grown in seedbeds or germination pots. They should be gently pressed into the soil or germinating substrate (0.5 to 1 cm deep) and watered regularly. The germination period is relatively short (five to ten days). The seedlings are somewhat sensitive. They do not tolerate direct or intense sunlight or complete shade. They will not survive excessive watering, yet if the soil or substrate dries out, the seedlings will die. The seedlings quickly grow into small, robust plants, which can then be repotted or planted in the ground. At this time, the plants will tolerate more exposure to the sun.

While most *Datura* species require relatively large amounts of water, they need little other care. *Datura* is self-sowing, so once it has been in a garden, it will likely be seen again in subsequent years.

Although the daturas are originally from subtropical and tropical zones, they adapt well to the climate of central Europe. In this area, seedlings should not be transplanted into the open until mid-May. Wild, self-sowing plants quickly adapt to local ecological conditions.

Appearance
This annual plant develops a multibranched, bushy, prostrate, laterally growing herbage that is dark green in color, with soft, slightly serrated leaves. The white flowers have a striking trumpetlike shape and are sometimes tinged violet on the inside. The flowers grow from axils on the branches and point slightly sideways or almost

In the southwestern region of the United States, the thorn apple (*Datura discolor*) is known as sacred datura. (Photographed in Zion Canyon)

straight up. They blossom in the evening, exuding a sweet, delicate, delicious scent. They wither during the course of the following day. The green fruits, which have only a few long thorns, are pendulous. They contain numerous black seeds (an important feature for identification). Apart from this, the plant is very similar to **Datura innoxia**, although it is somewhat smaller in every respect.

Psychoactive Material
— Seeds
— Leaves
— Roots

Preparation and Dosage
A medicinal tea can be prepared from the dried, ground seeds of *Datura discolor*, cinnamon bark (*Cinnamomum verum*), the leaves of desert lavender (*Hyptis emoryi* Torr.), and sugar (Felger and Moser 1991, 320, 366*). Unfortunately, we have no information about dosages.

The dried leaves can be smoked alone, in **kinni-kinnick**, or in other **smoking blends**. The fresh root can be chewed. Apart from this, the plant is used in the same manner as *Datura innoxia*.

Typically, preparations of *Datura discolor* must be dosed more carefully than those of *Datura innoxia*, as the former contains higher concentrations of alkaloids.

Ritual Use
According to the mythology of the Serí Indians, *Datura discolor* was one of the first plants of creation; as a result, humans should avoid contact with the plant (Felger and Moser 1991, 366*). Because inappropriate use can be very dangerous, the plant is used only by shamans.

In the American Southwest, the ritual use of *Datura discolor* is similar to that of **Datura innoxia** (see there). *Datura discolor*, however, is used much more rarely.

Artifacts
See *Datura innoxia*.

Medicinal Use
The Serí Indians of northern Mexico drink a tea made from the seeds to treat a swollen throat (Felger and Moser 1974, 428*).

The ethnomedical use of *Datura discolor* is similar to that of **Datura innoxia**.

Constituents
The entire plant contains between 0.13 and 0.49% alkaloids (primarily **tropane alkaloids**), half of which is hyoscine (= **scopolamine**). The alkaloid concentrations of the plant can vary considerably as it grows. The highest concentrations have been found to occur in the stems during the fruiting phase (Saber et al. 1970).

The dried herbage contains 0.17% alkaloids. The principal alkaloid is hyoscine/**scopolamine** (0.08% by dry weight). Apohyoscine, norhyoscine, hyoscyamine, meteloidine, tropine, and c-tropine are also present.

The dried roots contain 0.31% alkaloids, chiefly hyoscine/scopolamine, along with norhyoscine, atropine, littorine, meteloidine, $3\alpha,6\beta$-ditigloyloxytropane, $3\alpha,6\beta$-ditigloyloxytropane-7β-ol, cuscohygrine (the primary alkaloid in the roots), tropine, and c-tropine (Evans and Somanabandhu 1974).

Effects
See *Datura innoxia*.

Commercial Forms and Regulations
Datura discolor is found only rarely in nurseries. Both the plant and the seeds are available without restriction.

Literature
See also the entries for **Datura innoxia** and **tropane alkaloids**.

Avery, A. G., ed. 1959. *Blakeslee—the genus* Datura. New York: Ronald Press.

Evans, William C., and Aim-On Somanabandhu. 1974. Alkaloids of *Datura discolor*. *Phytochemistry* 13:304–5.

Saber, A. H., S. I. Balbaa, G. A. El Hossary, and M. S. Karawya. 1970. The alkaloid content of *Datura discolor* grown in Egypt. *Lloydia* 33 (3): 401–52.

"He who consumes the thorn apple drink believes that he is consorting with spirits and demons."

Ernst Freiherr von Bibra
Die narkotischen Genussmittel [The Narcotic Agents of Pleasure and Man] (1855, 142*)

Datura fruits (thorn apples). Top, from left to right: *Datura innoxia, D. discolor, D. stramonium*; bottom, from left to right: *D. metel, D. ceratocaula, D. quercifolia*. (From Festi 1995, 118 f.*)

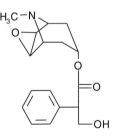

Scopolamine

Datura ferox Linnaeus

Chinese Datura

See *Datura innoxia*.

Datura innoxia Miller

Toloache, Mexican Thorn Apple

This stamp from the small Southeast Asian country of Laos depicts the Mexican thorn apple (*Datura innoxia* = *Datura meteloides*) with an unusual violet-colored flower.

A pre-Hispanic Mexican representation of the toloache fruit (*Datura innoxia*). (From Camilla 1995)

Family

Solanaceae (Nightshade Family); Subfamily Solanoideae, Datureae Tribe, Dutra Section

Forms and Subspecies

Today, two subspecies are generally accepted:
Datura innoxia Mill. ssp. *quinquecuspida* Torr.
Datura innoxia Mill. ssp. *lanosa* (Bye)

Synonyms

The taxonomy of this *Datura* species has resulted in many errors and divergent interpretations, as well as numerous synonyms (cf. Ewan 1944):
Datura guayaquilensis H.B.K.
Datura hybrida Tenore
Datura inoxia Mill.
Datura lanosa Barclay ex Bye
Datura metel Dunal non L.
Datura metel Sims non L.
Datura metel Ucria
Datura metel L. var. *quinquecuspida* Torr.
Datura meteloides DC. ex Dunal

Folk Names

A-neg-la-kia (Mazatec),[108] a'neglakya (Zuni), chamico, chànikah, ch'óhojilyééh, ch'óxojilghéí (Navajo, "crazy maker"), dekuba (Tarahumara), devil's weed, dhatura (Pakistani), dhaturo (Nepali), hehe camóstim (Serí, "plant that produces grimaces"), hehe carócot (Serí, "plant that makes crazy"), hierba del diablo, hierba hedionda, hippomanes, hoozhónee yilbéézh (Navajo, "beautyway decoction"), hyoscyamus de Peru, Indian apple, Jamestown weed, jimsonweed, kâtundami (Pima), kiéli, kielitsa (Huichol, "bad kieli"), kiéri,[109] kí-ki-sow-il (Coahuilla), kusi, loco weed,[110] menj (Arabic/Yemen), Mexican thorn apple, mexikanischer stechapfel, moapa, moip, nacazcul, nacazul, nocuana-patao (Zapotec), nohoch xtóhk'uh (Mayan, "great [plant] in the direction of the gods"), ñongué blanco, ntígíliitshoh (Navajo, "great sunflower"), ooze apple, poison lily, pomum spinosum, rauchapfel, rikúi, rikuri, sacred datura, sape enwoe be (Tewa), solanum manicum, stechapfel, tapate, tecuyaui (Garigia), telez-ku, thorn apple, tikúwari (Tarahumara), tlapa, tohk'u, tolachi, tolguacha, toloa, toloache, toloache grande, toloatzin ("nodding head"), tolochi, tolohuaxihuitl (Aztec, "nodding herb"), tolovachi, toluache, toluah ("nodding"), uchurí (Tarahumara), u'teaw ko'hanna (Zuni, "the white flower"), wichurí, xtóhk'uh (Yucatec Mayan, "in the direction of the gods"), xtoku (Mayan, "in the direction of the gods"), yerba del diablo (Spanish, "devil's herb")

History

Toloache is the most ethnopharmacologically significant of all thorn apple species in the New World. Archaeological studies of ritual rooms dating to 1200 to 1250 C.E. have demonstrated that the prehistoric Pueblo Indians of the Southwest used the seeds for ritual purposes (Litzinger 1981, 64; Yarnell 1959). Our present state of knowledge does not allow us to say with certainty how long this thorn apple has been used in Mexico, although such use certainly has its roots in the prehistoric period. Today, many Mexican Indians still frequently use *Datura innoxia* for ritual and medicinal as well as aphrodisiac purposes.

Distribution

The original range of *Datura innoxia* extended from the American Southwest and Mexico to Guatemala and Belize. From there, the species spread into the islands of the Caribbean. It was introduced into Asia at an early date. In India, it often occurs in association with *Datura metel*. It also grows wild in Greece and Israel (Dafni and Yaniv 1994*).

Cultivation

For information on cultivation, see *Datura discolor.*

Datura innoxia is grown commercially in Central America, North Africa, Ethiopia, India, and England for pharmaceutical purposes (as a source of **scopolamine**) (Gerlach 1948).

Appearance

Datura innoxia is usually a 1- to 2-meter-tall annual plant; in the tropics, where it can grow to more than 3 meters in height, it can thrive as a perennial. The root can grow up to 60 cm long. The light to dull green plant is heavily branched and develops hairy leaves with serrated margins. The white, funnel-shaped flowers grow almost perpendicularly out of the axils. The flowers bloom at night, exuding a delicious scent, and begin to wither the following day. In central Europe, the plant flowers between June and September; in protected locations, the plant can produce flowers into November.

The fruits are pendulous and covered with numerous short thorns. The seeds have an ocher color with hints of orange. They are larger than the black seeds produced by *Datura discolor* and *Datura stramonium* and are easily mistaken for those of *Datura metel* and *Datura wrightii*.

Datura innoxia is very similar to the Asian *Datura metel* and is easily confused with this plant. It is in fact questionable whether the two species

---

108 This Mazatec name (Díaz 1979, 84*) is identical to the Zuni name for *Datura* (*a'neglakya;* cf. Schultes and Hofmann 1992, 106*).

109 The name *kiéri* (or *kiéli*) is usually used for *Solandra* spp., and less often for *Brugmansia.*

110 This name is normally used for various species in the genera *Astragalus* and *Oxytropis.*

should be regarded as distinct. Recent phytochemical studies have shown that the two species are extremely similar (Mino 1994). It may be that they are actually subspecies or varieties of the same species. The two species (or forms) are most easily distinguished on the basis of their stalks. *Datura innoxia* has green stalks with soft hairs, while *Datura metel* has smooth stalks that are purple in color.

Datura innoxia is also easily confused with *Datura discolor* and *Datura wrightii*, although it occurs in a geographical area different from that of the latter two species.

Psychoactive Material
— Leaves (daturae innoxiae herba), fresh or dried
— Roots
— Flowers
— Seeds

Preparation and Dosage
The dried leaves and flowers can be smoked alone or in combination with other herbs and substances (cf. **smoking blends**).

Shamans of the Yucatec Maya (*hmenó'ob*, "the doers") use the leaves of *Datura innoxia* and tobacco (*Nicotiana tabacum, Nicotiana undulata*; cf. *Nicotiana* spp.) to roll cigars known as *chamal*. They typically use one leaf of each plant per cigar. A shaman will smoke this cigar until he has attained the altered state of consciousness he desires (which can differ considerably from one individual to the next).

The seeds and leaves may be crushed and treated with a fermenting agent to produce an alcoholic drink (Havard 1896, 39*). The roots are often used as an inebriating additive to pulque (see *Agave* spp.), **beer**, or **chicha**. The Tarahumara add the seeds to the maize **beer**, known as *tesgüino*, they brew (Bye 1979b, 35*). Both the tribes that live along the Colorado River and the Paiute fortified their beer with the seeds and leaves of *Datura innoxia* (Havard 1896, 39*).

The Yaqui Indians make an ointment by adding crushed seeds and leaves to lard; they rub this onto their abdomens to induce visions.

Fresh roots may be crushed and applied externally, chewed, or dried and powdered. Unfortunately, the literature does not provide any precise information about the amounts of roots that should be chewed or eaten.

For smoking, up to four leaves is regarded as an appropriate dosage for eliciting aphrodisiac effects. Consuming the plant in this manner effectively rules out the possibility of overdose. Teas made from the leaves must be dosed with care. As little as one large leaf can be sufficient to induce profound hallucinations. Because alkaloid concentrations can vary considerably (see "Constituents"), and because individuals react

quite differently to **tropane alkaloids**, detailed information about dosages is rarely provided. Thirty to forty seeds is regarded as a potent visionary or hallucinogenic dosage. However, as few as ten seeds can lead to profound perceptual changes. For information about lethal dosages, see ***Datura stramonium***.

In Pakistan, 150 g of leaves, fruits, or flowers is regarded as a lethal dose (Goodman and Ghafoor 1992, 40*). This amount appears to be quite high.

Ritual Use
In Aztec medical texts, toloache is cited numerous times as a remedy, especially for fever (Rätsch 1991a, 254 ff.*):

> Toloa. It is also a fever medicine. It is drunk in a weak infusion. And where gout is, it is applied to it, one is rubbed with it there. It soothes, dispels, wards off [pain]. It is not inhaled and it is not breathed in. (Sahagun, Florentine Codex 11:7*)

The written sources do not contain any clear indications that the plant was used as an inebriant. Since ritual and magical use occurs throughout modern Mexico, it can be assumed that the potent inebriating powers of *Datura innoxia* were also being utilized in pre-Hispanic times. It has even been suggested that victims destined for human sacrifice were given a *Datura* drink to prepare them for their death (Rätsch 1986a, 234*). The administration of *Datura* preparations for initiatory purposes also appears to have been known in Mesoamerica (cf. ***Datura wrightii***).

In the Yucatán (southern Mexico), *Datura innoxia*, which is known as *xtohk'ùh* ("in the direction of the gods"), is a rare plant. But it is frequently planted in house gardens as both an ornamental and a source of drugs. The *hmenó'ob* (shamans) use *Datura* not just as a medicine but first and foremost as a ritual drug. When divining with a quartz crystal (*ilmah sastun*), they either smoke cigars (*chamal*) rolled from datura leaves or eat datura seeds (Rätsch 1987*). They say of the cigars and the seeds *hach mà'lo' ta wòl*, "they are very good for your consciousness." In the resulting inebriated state, the shaman is able to see things in the crystal that can provide insights into questions posed beforehand (e.g., pertaining to stolen or lost objects, the causes of illness, sorcery). Some modern Mayan shamans also use tarot cards[111] (which were introduced into Mexico some one hundred years ago) when they perform divinations (Rätsch 1988b). Occasionally, they may smoke *Datura* before laying out the cards. Eating the seeds enables the *h-mèn* to travel to *yuntsil balam*, "the jaguar lord," when a sick person has lost his *ah-kanul*, "protector spirits." The aromatic flowers are also regarded as an excellent

Top: The typical upright, funnel-shaped flower of the Mexican thorn apple (*Datura innoxia*), also known as toloache

Bottom: The thorny fruit of *Datura innoxia* typically hangs straight down.

111 For more on the meaning European tarot cards have in Mexican *brujería*, see Devine (1982).

Jugo de toloache, a Mexican magical drink made from *Datura innoxia*, is used in love magic.

offering for the gods (Rätsch and Probst 1985, 1138). Among the Maya, use of the plant as an aphrodisiac (smoking the dried leaves) and a love magic (presenting the flowers to the desired person) is also common (Kennedy and Rätsch 1985; Rätsch and Probst 1985).

In urban *brujería*,[112] toloache plays an important role in the preparation of magic powders (*verdadero polvo de toloache*), in the manufacture of aphrodisiac ointments and bath additives, and in love magic. In some parts of central Mexico, *Datura innoxia* is venerated in churches as a quasi-Catholic healer known as Santo Toloache, who is called upon to effect love magic.

Many Mexicans look upon the plant with respect, timidity, or disdain. It has an intimate connection to dark practices that may appear eerie to the ignorant (Madsen and Madsen 1972*). The plant has a reputation for causing insanity,[113] of being toxic, and of being misused by the *brujos* ("sorcerers") for harmful magic. According to many Mexican shamans, toloache is especially dangerous because it gives its users power. The Huichol regard it as a "bad plant of the gods" and usually associate it with sorcery (cf. **Solandra spp.**). Carlos Castaneda received the following explanation of the magical properties of the "devil's weed" from his teacher, Don Juan:

> The second portion of the devil's weed is used to fly.... The unguent by itself is not enough. My benefactor said that it is the root that gives direction and wisdom, and it is the cause of flying. As you learn more, and take it often in order to fly, you will begin to see everything with great clarity. You can soar through the air for hundreds of miles to see what is happening at any place you want, or to deliver a fatal blow to your enemies far away. As you become familiar with the devil's weed, she will teach you how to do such things. (Castaneda 1968, 92*)

Datura innoxia plays an important role in Indian divination. The Náhuatl-speaking peoples, and those who belong to the language family of the Maya, use the thorn apple as a prophetic and oracular plant. The Mixtec are said to ingest *Datura innoxia* as a traditional hallucinogen for divinatory purposes (Avila B. 1992*). Many tribes of the Southwest (the Colorado River and Paiute tribes and the Coahuilla) smoked the leaves and added them to drinks (**chicha**, pulque; cf. **Agave spp.**) to induce a prophetic delirium (Barrows 1967, 75*).

Datura innoxia is sacred to the Navajo, who view it with great respect and use it because of its great potency. The Navajo have many names for datura, including *ch'óhojilyééh*, "producing madness," and *hoozhónee yilbéézh*, "Beautyway decoc-

tion" (Brugge 1982, 92). The Navajo collect the thorn apple, which is ritually addressed as "little white hair," according to a specific ritual. They begin by sprinkling maize pollen over the plant and uttering the following prayer: "Little white hair, forgive me for taking you. I do not do this out of arrogance. I would like you to heal me. I will take only as much as I need" (Abel 1983, 193).

In many Navajo healing ceremonies, visions and dreams play a central role. The medicine men or shamans learn from the visions and attain powers they can then use for healing (Haile 1940). During the ceremony known in the literature as the Beautyway, preparations of *Datura* are ingested to produce such visions (Brugge 1982, 92). The Navajo medicine men also use the thorn apple to treat hallucinations.[114] In secret ceremonies, the seeds are eaten as well.

The Navajo ingest small portions of *Datura* to protect themselves from the attacks of witches (Simmons 1980, 154). At the same time, the magical powers of the plant are also used for both positive and negative love magic (Hill 1938, 21), in which a person attempts to mix *Datura* into the food or smoking tobacco of the person he desires (Tierney 1974, 49).

The Navajo *ajilee* ceremony has been described in the ethnographic literature under the names Excess Way, Prostitution Way, and Frenzy Witch-craft. *Ajilee* is the name of a myth, a magical song, and a ritual in which the performer is transformed into a *Datura* spirit and is able to gain power over the women he desires as well as the game he wishes to hunt (Haile 1978; Luckert 1978). *Ajilee* is not one of the main healing rituals, and some Navajo (especially those who have been Christianized) regard it as witchcraft. Four magical plants, including *Datura innoxia* and probably also **Argemone mexicana** and the locoweeds (**Astragalus spp.**), play a central role in the *ajilee* myth and ritual. The ritual is intended to summon desired women (especially virgins from the Hopi and Pueblo Bonito) for sexual enjoyment. The same songs are also used to attract game. The ritual is also used to heal people who are suffering from sexual excesses as well as women who have been forced to prostitute themselves (Haile 1978). The person conducting the ritual is transformed into *Datura innoxia*, and the plant's aphrodisiac effects give him magical power over the woman he desires. As a result of the confusing and stupefying effects of *Datura*, the ritual performer also gains power over animals (Haile 1978, 26, 35 ff.).

It is said that a few medicine men who perform *Datura* divinations, also known as criminal telepathy, live in the region to the east of the Lukachukai Mountains. They use the plant to detect thieves and locate lost objects (Simmons 1980, 154).

The Apaches use the powdered root in secret

The Cultural Significance of *Datura innoxia* among the Navajo
(From Müller-Ebeling and Rätsch 1998)

— Agent for inducing visions
— Love magic
— Aphrodisiac
— Hunting magic
— Divination (criminal telepathy)
— Diagnosis (of disease causes)
— Medicine
— Magical protection
— Agent of pleasure

ceremonies as a ritual medicine. The Coahuilla utilize it to produce ritual delirium. The Costanoan smoke dried leaves as a hallucinogen. The seeds are mixed with tobacco (*Nicotiana tabacum*) to make a "love medicine" that is smoked during rituals of love magic. Hopi medicine men chew the root to induce a visionary state for diagnosing diseases (cf. *Mirabilis multiflora*). The Luiseño give their youths juice from the roots during their initiations (cf. *Datura wrightii*). The Shoshone brew a hallucinogenic tea for secret rites (Moerman 1986, 148 f.*).

The plant was introduced into Baluchistan (Pakistan) from the Americas. Now growing wild, it is a well-known inebriant that is referred to by the Sanskrit name *dhatura* (cf. *Datura metel*). The people of the region smoke a few crushed seeds or a dried leaf mixed with tobacco (*Nicotiana tabacum*) (Goodman and Gharfoor 1992, 40*). In India, *Datura innoxia* is used in the same manner as *Datura metel*.

Artifacts

Strangely, artifacts associated with toloache are relatively uncommon in Mexico. The Denver Museum has a postclassical Mayan ceramic that represents a three-dimensional head wearing ear ornaments that are clearly naturalistic representations of datura.

Numerous incense vessels have been found in western, central, and southern Mexico, the forms of which are reminiscent of the thorny fruits of *Datura innoxia* (Kan et al. 1989, 129, 201). In some Indian areas, e.g., among the Lacandon of Naha', this tradition of incense vessels still exists (Ma'ax and Rätsch 1994, 58*). Litzinger (1981) regards these vessels as true representations of the inebriating fruit. Today, *Datura* seeds are still burned for medicinal and ritual purposes. Incenses that include *Datura* seeds (copal, *pom;* cf. *Bursera bipinnata*) are definitely capable of eliciting profound psychoactive effects (cf. **incense**).

Numerous ceramic vessels ("spiked vessels") have been found in the American Southwest and in northern Mexico. These appear to be representations of thorn apple fruits and were presumably used as vessels for burning incense (Camilla 1995, 106 f.*; Litzinger 1981, 58 ff.).

In a kiva (cosmological ritual room) at Kuaua, near Bernalillo, New Mexico, a Pueblo IV wall painting (1300–1550 C.E.) depicts a figure holding a *Datura* flower in its hand (Wellmann 1981, 92*).

The Hopi women of Moki traditionally twisted their hair into two round buns that they wore on the sides of their heads. Although the two buns were thought to represent squash blossoms and were known by that name, they actually represented the sacred *Datura innoxia* (Furst and Furst 1982, 56). Several petroglyphs at Moki resemble thorn apple flowers as viewed from above. In the literature, these also have been wrongly interpreted as squash blossoms (Patterson 1992, 189).

The Zuni use a head ornament in various ritual dances that is intended to recall the headpiece of A'neglakya, the personification of *Datura innoxia*. They wrap a dried fruit of *Martynia louisiana* Mill. with colored woolen ribbons and tie it to a leather headband. Early anthropologists interpreted this headpiece as a symbol of the squash flower. The Zuni themselves approved of this error, as it helped keep their sacred *Datura* secret (Müller-Ebeling and Rätsch 1998).

Navajo jewelry features a type of chain that is known both publicly and in the popular literature as the squash blossom necklace. However, these ornaments are representations not of squash flowers but of the flowers of a plant that had much greater cultic significance: the sacred thorn apple (*Datura innoxia*, **Datura discolor**). The term *squash blossom* has come to be a widely used alias for *Datura*, which is used and venerated in secret (Müller-Ebeling and Rätsch 1998).

Some Shoshone petroglyphs depict visions that were obtained while under the influence of *Datura innoxia* (Camilla 1995, 109*).

The American artist Georgia O'Keeffe (1887–1986) produced several paintings of luxuriously beautiful *Datura innoxia* flowers, e.g., the oil paintings *White Trumpet Flower* (1932) and *Jimson Weed* (ca. 1934). These paintings, which are regarded as typical of O'Keeffe's style and expression (Castro 1985), have been reproduced on numerous calendars and postcards.

Medicinal Use

Datura innoxia plays a significant role in the ethnomedicine of the tribes of the Southwest (**Datura discolor** is used in precisely the same manner, albeit much less frequently). The Apaches use juice freshly pressed from the flowers and roots to disinfect wounds. The Coahuilla rub plants crushed in water onto their horses' saddle sores. The Costanoan smear an ointment made from the leaves onto burns. They use dew drops that have collected in the flowers as an eyewash.

"How Datura came to be:
In olden times, there lived a boy and a girl (the name of the boy was A'neglakya and the name of the girl was A'neglakyatsi'tsa), brother and sister, in the inside of the earth. But they often came up to the earth and wandered about. They observed everything very closely and reported to their mother what they had seen and heard. This constant talking did not please the divinities (the twin sons of the sun father) at all. Once, when the divinities encountered the boy and the girl, they asked them: 'How are you?' The brother and the sister answered: 'We are happy!' (Sometimes A'neglakya and A'neglakyatsi'tsa appeared on the earth as people.) They told the divinities how they had caused a person to fall asleep and see spirits and could have someone wander around who could see where a theft had occurred. After this encounter, the divinities decided that A'neglakya and A'neglakyatsi'tsa knew too much and that they had to be banished from the world. And so the divinities caused the brother and the sister to vanish from the earth forever. But on the place where they disappeared, flowers grew—exactly those kinds of flowers that they had worn on their head when they had visited the earth. The divinities named the plants A'neglakya, after the boy. The first plant had many children who spread over the entire world. Some of the flowers are colored yellow, some blue, some red, some are entirely white— the colors of the four cardinal points."

ZUNI MYTH
IN "ETHNOBOTANY OF THE ZUNI
INDIANS"
(STEVENSON 1915)

Although it is known as a "squash blossom" necklace, this Navajo design actually symbolizes the sacred *Datura*.

Heated leaves are applied to the chest to relieve difficulties in breathing. The Mahuna use the plant to treat rattlesnake and tarantula bites. The Navajo use it to treat the castration wounds of their sheep. Fresh thorn apple leaves or aqueous extracts are applied to the skin to treat wounds. The root is chewed for severe pain. The Zuni used the root as an anesthetic when performing surgery (cf. **soporific sponge**). The Tubatulabal consume the plant to relieve constipation and use it to treat inflammations, wounds, and swelling (Brugge 1982, 92; Moerman 1986, 148 f.*).

The Aztecs utilized thorn apple leaves in the treatment of broken bones (e.g., skull fractures), abscesses, and swollen knees. They usually placed leaves that they had warmed in a steam bath directly onto the affected area. The Maya use the leaves to treat rheumatism (Pulido S. and Serralta P. 1993, 61*). Smoking the dried leaves to treat asthma, bronchitis, and coughs is a very common practice.

Toloache is one of the most important aphrodisiacs[115] and sedatives of Mexican folk medicine. In rural areas, toloache brews are administered during childbirth to induce a twilight sleep and to mitigate the pains of childbirth (Heffern 1974, 98*). Ointments made with lard and extracts of *Datura innoxia* are often used to treat skin diseases as well as muscle and joint pain. Along with the plant, this use was introduced into Europe at an early date. John Gerard, in his sixteenth-century work *The Herball*, wrote:

> The juice of Thornapple, boiled with hog's grease, cureth all inflammations whatsoever, all manner of burnings and scaldings, as well of fire, water, boiling lead, gunpowder, as that which comes by lightning and that in very short time, as myself have found in daily practice, to my great credit and profit. (In Grieve 1982, 2:806*)

The plant finds use in the ethnomedicine of every region of the Old World into which it has spread. In Israeli folk medicine, a decoction of the leaves is drunk to treat diarrhea and a paste of the leaves is applied externally to treat pain (Dafni and Yaniv 1994, 13*). In Asia, this introduced species is used in the same manner as *Datura metel* and *Datura stramonium* (Shah and Joshi 1971, 420*; Singh et al. 1979, 188*).

Constituents

The entire plant is rich in **tropane alkaloids**. **Scopolamine** (the primary alkaloid) and hyoscyamine predominate in the aboveground parts, while the flowers contain large amounts of tyramine and the stems large amounts of meteloidine.

The roots contain the following alkaloids: hyoscyamine, **scopolamine**, cuscohygrine, 3-tigloyl-oxytropane, 3-hydroxy-6-tigloyloxytropane, 6-hydroxyhyoscyamine, 6-tigloyloxyhyoscyamine, and tropine (Ionkova et al. 1989). A different analysis detected tigloidine, **atropine**, pseudotropine, 7-hydroxy-3,6-ditigloyloxytropane, 3α,6β-di-tigloyl-oxytropane [428], hyoscine, and meteloidine (Evans and Wellendorf 1959).

The seeds contain a total of 0.3% alkaloids (0.09% scopolamine, 0.21% hyoscyamine).

In addition to the alkaloids, the leaves also contain phenolic compounds (caffeic and hydroxy-cinnamic acid esters).

Some plants produce considerably more **scopolamine** than others (Hérouart et al. 1988). This fact helps explain some of the difficulties in determining dosages.

Effects

The effects of *Datura innoxia*—and in fact of all *Datura* species—are heavily dependent upon dosage and can vary greatly depending upon the method of application (Weil 1977). The Indian division into three stages has particular relevance here: A mild dosage produces medicinal and healing effects, a moderate dosage produces aphrodisiac effects, and high dosages are used for shamanic purposes.

The effects of four leaves smoked together by one couple appear to be typical for *Datura innoxia*:

> The skin acquired an unimagined sensitivity. A simple, light caress became a tender, fulfilling experience. The blood collected in our lower abdomens so quickly that it demanded we join. The normal sexual functions were extremely heightened. Every form of erotic exchange and sexual activity had a special deliciousness. The time to orgasm was much longer, and the orgasm itself appeared to last for minutes. During the phase of sexual activity, we were both pleasantly free of thoughts, uninhibited, and very much focused on the moment. The effects lasted the entire night, so that there were many couplings. The next morning, after a short sleep with erotic dreams (!), we awoke with a clear consciousness, a very pleasurable warm sensation in the body, a still overly senstive skin, and a dry throat. (Rätsch and Probst 1985, 1139)

Shamanic dosages induce profound visions, strong hallucinations, and delirium. As with those produced by *Brugmansia suaveolens*, the hallucinations can have a metaphysical character or can be quite banal.

Overdoses can begin with initial excitation, the urge to dance, frenzy, and fits of laughter before leading to acute hallucinosis and finally to death through respiratory paralysis (Siegel 1981*). In Mexico, peyote (*Lophophora williamsii*) is used

115 Throughout the world, this and other *Datura* species have great importance as aphrodisiacs (Kennedy and Rätsch 1985).

as an antidote for toloache overdose (Nadler 1991, 95*).

Commercial Forms and Regulations

In Mexico, various preparations allegedly made from toloache (such as magical juices and powders) are sold at the *brujería* markets. Chemical analysis of one *legítimo polvo de toloache* ("legitimate toloache powder") revealed that the sample did not contain any alkaloids and therefore could not consist of *Datura* (Hasler 1996).

In Europe, potted plants and seeds are available at nurseries without restriction. Pharmaceutical preparations of *Datura* are made almost exclusively from **Datura stramonium** or **Datura metel** (see there.)

Literature

See also the entries for **Datura discolor, Datura stramonium, Datura wrightii,** and **tropane alkaloids.**

Abel, Friedrich. 1983. *Nur der Adler sprach zu mir.* Bern: Scherz.

Anon. 1974. Navajo witchcraft. *El Palacio* 80 (2): 38–43.

Basey, Keith, and Jack G. Woolley. 1973. Biosynthesis of the tigloyl esters in *Datura*: The role of 2-methylbutyric acid. *Phytochemistry* 12:2197–2201.

Boitel-Conti, M., E. Gontier, J. C. Laberche, C. Ducrocq, and B. S. Sangwan-Norreel. 1995. Permeabilization of *Datura innoxia* hairy roots for release of stored tropane alkaloids. *Planta Medica* 61:287–90.

Brugge, David M. 1982. Western Navajo ethnobotanical notes. In *Navajo Religion and Culture,* ed. D. M. Brugge and Ch. J. Frisbie, 89–97. Santa Fe: Museum of New Mexico Press.

Castro, Jan Garden. 1985. *The art and life of Georgia O'Keeffe.* New York: Crown Publishers.

Devine, Mary Virginia. 1982. *Brujería: A study of Mexican-American folk-magic.* St. Paul, Minn.: Llewellyn Publications.

Evans, W. C., and M. Wellendorf. 1959. The alkaloids of the roots of *Datura. Journal of the Chemical Society* 59:1406–9.

Ewan, Joseph. 1944. Taxonomic history of the perennial southwestern *Datura meteloides. Rhodora* 46 (549): 317–23.

Furst, Peter T., and Jill L. Furst. 1982. *North American Indian art.* New York: Rizzoli.

Gerlach, George H. 1948. *Datura innoxia*, a potential commercial source of scopolamine. *Economic Botany* 2:436–54.

Gontier, E., M. A. Fliniaux, J. N. Barbotin, and B. S. Sangwan-Norreel. 1993. Tropane alkaloid levels in the leaves of micropropagated *Datura innoxia* plants. *Planta Medica* 59:432–35.

Haile, Father Berard. 1940. A note on the Navaho visionary. *American Anthropologist* n.s. 42:359

———. 1978. *Love-magic and butterfly people: The Slim Curly version of the ajilee and Mothway myths.* Flagstaff: Museum of Northern Arizona Press.

Hasler, Felix. 1996. Analytisch-chemische Untersuchung von "Toloache-Pulver." Unpublished laboratory report, Bern.

Hérouart, D., R. S. Sangwan, M. A. Fliniaux, and B. S. Sangwan-Norreel. 1988. Variations in the leaf alkaloid content of androgenic diploid plants of *Datura innoxia. Planta Medica* 54:14–17.

Hill, W. W. 1938. Navajo use of jimson weed. *New Mexico Anthropologist* 3 (2): 19–21.

Hiraoka, N., M. Tabata, and M. Konoshima. 1973. Formation of acetyltropine in *Datura* callus cultures. *Phytochemistry* 12:795–99.

Ionkova, Iliana, L. Witte, and A. W. Alfermann. 1989. Production of alkaloids by transformed root cultures of *Datura innoxia. Planta Medica* 55:229–30.

Kan, Michael, Clement Meighan, and H. B. Nicholson. 1989. *Sculpture of ancient West Mexico.* Los Angeles: County Museum of Art.

Kennedy, Alison Bailey, and Christian Rätsch. 1985. *Datura:* Aphrodisiac? *High Frontiers* 2:20, 25.

Kluckhohn, Clyde. 1967. *Navaho witchcraft.* Boston: Beacon Press.

Leete, Edward. 1973. Biosynthetic conversion of α-methylbutyric acid to tiglic acid in *Datura meteloides. Phytochemistry* 12:2203–5.

Litzinger, William. 1981. Ceramic evidence for prehistoric *Datura* use in North America. *Journal of Ethnopharmacology* 4:57–74.

———. 1994. Yucateco and Lacandon Maya knowledge of *Datura* (Solanaceae). *Journal of Ethnopharmacology* 42:133–34.

Luckert, Karl W. 1978. *A Navajo bringing-home ceremony: The Claus Chee Sonny version of Deerway Ajilee.* Flagstaff: Museum of Northern Arizona Press.

Mino, Yoshiki. 1994. Identical amino acid sequence of ferredoxin from *Datura metel* and *D. innoxia. Phytochemistry* 35 (2): 385–87.

Müller-Ebeling, Claudia, and Christian Rätsch. 2003. Kürbisblüten oder Stechäpfel: Die Entschlüsselung eines indianischen Symbols. In *Stechapfel und Englestrompete,* ed. Markus Berger, 99–107. Solothurn: Nachtschatten Verlag.

Patterson, Alex. 1992. *Rock art symbols of the greater Southwest.* Boulder, Colo.: Johnson Books.

Rätsch, Christian. 1988. Tarot und die Maya. *Ethnologia Americana* 24 (1), Nr. 112:1188–90.

"This person is going to drink you.
Give him a good life.
Show him what he wants to know."
PRAYER TO TOLOACHE
(FROM SCHULTES AND HOFMANN
1992:107*)

These Shoshone petroglyphs from Wyoming show the visions that are obtained under the influence of *Datura innoxia.* (From Camilla 1995)

These Indian incense vessels clearly resemble thorn apple fruits. (From Camilla 1995)

Rätsch, Christian, and Heinz Jürgen Probst. 1985. Xtohk'uh: Zur Ethnobotanik der Datura-Arten bei den Maya in Yucatan. *Ethnologia Americana* 21 (2), Nr. 109:1137–40.

Simmons, Marc. 1980. *Witchcraft in the Southwest: Spanish and Indian supernaturalism on the Rio Grande.* Lincoln and London: University of Nebraska Press (Bison Book).

Stevenson, Matilda Coxe. 1915. Ethnobotany of the Zuñi Indians [of the Extreme Western Part of New Mexico]. Bureau of American Ethnology to the Secretary of the Smithsonian Institution, *Thirtieth Annual Report,* 1908–1909.

Tierney, Gail D. 1974. Botany and witchcraft. *El Palacio* 80 (2): 44–50.

Weil, Andrew. 1977. Some notes on *Datura. Journal of Psychedelic Drugs* 9 (2): 165–69.

Yarnell, R. A. 1959. Evidence for prehistoric use of *Datura. El Palacio* 66:176–78.

Datura metel Linnaeus
Indian Thorn Apple

This early illustration of the Indian thorn apple (*Datura metel*) is botanically quite accurate. (Woodcut from Garcia da Orta)

Family
Solanaceae (Nightshade Family); Subfamily Solanoideae, Datureae Tribe, Dutra Section

Forms and Subspecies
Because of the considerable variation within this *Datura* species, numerous forms, varieties, and subspecies have been described, and the taxonomy is confusing (Avery 1959). White-flowering varieties are now usually referred to as *Datura metel* var. *alba*, and violet-blooming varieties as *Datura metel* var. *fastuosa*. There also are a number of cultivars: *Datura metel* cv. Fastuosa (double violet flowers), cv. Chlorantha (yellow double trumpets), cv. Coerulea (blue flowers), cv. Atrocarmina, cv. Lilacina, cv. Violace (violet flowers), cv. Alboplena, cv. Flavaplena, et cetera. *Datura metel* L. f. *pleniflora* Degener has triple yellow flowers.

Synonyms
Brugmansia waymannii Paxton
Datura alba Eisenb.
Datura alba Nees
Datura bojeri Raffeneau-Delile
Datura cathaginensis Hort. ex Siebert et Voss
Datura chlorantha Hook.
Datura cornucopaea Hort. ex W.W.
Datura dubia Pers.
Datura fastuosa L.
Datura fastuosa L. var. δ *alba* Bernh.
Datura fastuosa L. var. *flaviflora* Schulz (yellow blooming)
Datura fastuosa L. var. α *glabra* Bernh.
Datura fastuosa L. var. β *parviflora* Nees
Datura fastuosa L. var. γ *rubra* Bernh.
Datura fastuosa L. var. β *tuberculosa* Bernh.
Datura huberiana Hort.
Datura humilis Desfontaines
Datura hummatu Bernh.
Datura indica nom. nud.
Datura muricata Bernh.
Datura nigra Rumph. in Hasskarl.
Datura nilhummatu Dunal
Datura pubescens Roques
Datura timoriensis Zipp. ex Spanoghe
Stramonium fastuosa (L.) Moench

Folk Names
Arhi-aba-misang, bunjdeshtee (Persian), chosen-asagau (Japanese, "Korean morning beauty"), da dhu ra (Tibetan), datula, datur-a (Mongolian), datura (Sanskrit), datura engletrompet (Danish), datura indica, datura kachubong, devil's trumpet flower of Ceylon, dhatra (Santali), dhattûra (Sanskrit), dhatura (Sanskrit, "heterogeneous"), dhatûrâ, dhatur-ma, dhaturo, dhétoora (Hindi), dhustura, dhustûra, dhutro (Bengali), dhutura (Bengali), dootura, dornäpfel, dotter (Dutch), doutro, doutry, dutra, dutro, dutroa, dutro banguini, engelstrompete, engeltrompet, ganga bang, gelber stechapfel, goozgiah (Persian), hearbe dutroa, Hindu datura, hummatoo, indischer stechapfel, insane herb, jous-mathel (Arabic), jowz massel, kachubong (Philippines), kala dahtoora, kala dhutura (Hindi, "black datura"), kalu antenna, kalu attana, karoo omatay (Tamil), kechubong, kechubong hitam ("black datura"), kechubong puteh ("white datura"), kechu-booh (Egyptian), kechubung (Malayan), kecubong (Bali), keppate jad, krishna dhattura, man-t'o-lo (Chinese),[116] menj (Arabic/Yemen), metelapfel, metelnuß, mnanaha (Swahili), mondzo (Tsonga), nao-yang-hua (Chinese), neura, neurada, ñongué morado, nucem metellam arabum, nulla oomantie, nux metel, nux-methal, paracoculi, pig-ble, rauchapfel, rauchöpfel, rotecubung, shan-ch'ieh-êrh (Chinese), Shiva's plant, stechöpfel, stramonia, talamponay, takbibug, tatorah (Arabic), thang-

phrom dkar-po (Tibetan), thorn apple, umana, unmata (Sanskrit, "divine inebriation"), unmeta, violettblauer stechapfel, violettblaue engelstrompete

History

The Indian thorn apple was first mentioned in Sanskrit literature (Vamana Purana, Garuda Purana). The Arabic physician Avicenna (Ali al-Husayn Abd Allah Ibn Sina, 980–1037) discussed its medicinal use and the importance of dosage among the Arabs, who classified the plant as one of the so-called *mokederrat*, the narcotica (Avery 1959, 3). This thorn apple also appears in very ancient Tibetan and Mongolian texts, the existence of which demonstrates that *Datura metel* was indigenous to Asia prior to the fifteenth century (Siklós 1993, 1996). It is not known when this thorn apple spread into Africa. Today, *Datura metel* is still a psychoactive plant of great ethnopharmacological significance, especially in India, Southeast Asia, and Africa.

Distribution

This species most likely originated in northern India but then spread quickly throughout Southeast Asia. It is now found in the Philippines, in Indonesia, and on the islands of the Indian Ocean (Seychelles, Mauritius, etc.). It presumably spread into Africa and the New World (Central and South America, the Caribbean) through human activity.

Cultivation

Propagation is performed with seeds (cf. *Datura discolor*). It is best to soak the seeds overnight before sowing. The next morning, they should be pressed 1 to 2 cm deep into sandy, humus-rich seedling soil and lightly covered. Do not allow the soil to dry out. The time to germination is fourteen to thirty-five days. In central Europe, the seeds should be sown (in the open) between April and July, preferably in June. The plant is sensitive to frost but can be trimmed back in late fall and allowed to overwinter in the cellar. With a little luck, the plant will develop shoots again the following spring.

Datura metel is cultivated commercially as a source of alkaloids (**scopolamine**) chiefly in subtropical and tropical regions throughout the world (especially in India and Africa).

Appearance

Datura metel is an annual or biennial plant that has an herbaceous, bushy appearance. It can grow to more than 2 meters in height and develops numerous branches. The soft leaves are light to dull green in color and have a slightly serrated margin. The plant has smooth, violet or dark purple stalks. The funnel-shaped flowers, which can be white, violet, or yellow (depending on the variety, subspecies, or cultivar), point upward at an angle. They open in the evening, exude a pleasant scent during the night, and then wither over the course of the next day or two. *Datura metel* often produces filled double or triple flowers. The variety *fastuosa* frequently develops violet double flowers. In the tropics, the plant will blossom throughout the year. In central Europe, the flowering period is from June to October.

The fruit, which hangs upward at an angle, has a few short thorns that are often only roundish bumps. The kidney-shaped seeds are yellow ocher and almost identical to those of *Datura innoxia* and *Datura wrightii*.

Datura metel, especially the variety *alba*, is very easily confused with **Datura innoxia**. It is sometimes even mistaken for certain forms of **Datura stramonium**.

Psychoactive Material

— Leaves
— Seeds
— Roots
— Flowers (these are used in Chinese medicine, where they are known as *yang jin hua*; Lu 1986, 82*)

Because the alkaloid content of the entire plant increases until the plant reaches the end of its reproductive phase, the raw drug is best collected during or after the end of fruit development (Afsharypuor et al. 1995).

Preparation and Dosage

A narcotic or inebriating drink is prepared by adding equal parts of *Datura metel* seeds and/or leaves and hemp (*Cannabis sativa*) flowers to **wine** (Perry and Metzger 1980, 392*). In Asia, the leaves are often taken together with wine or **sake** (Penzer 1924, 160). Thorn apple seeds are used to fortify *rokshi* (barley spirits; see **alcohol**) in Darjeeling and Sikkim. The seeds are also used in **betel quids**.

A unique method for preparing the plant was discovered in East India. Here, women feed datura leaves to a certain type of beetle (the exact species is unfortunately unknown) for a period of time and collect the beetle's excrement. They then mix this into an unfaithful husband's food for revenge. Overall, there are a number of traditional preparations in India:

> India has zones of datura use. For example, Bengal. The particularly passionate smoke *Cannabis indica*, ganjah, with two or three thorn apple seeds or a quantity of leaves as additives. In order to potentiate and alter the effects of alcoholic drinks upon the brain, they soften seeds in the drink, filter this, and mix it with palm wine. This is done, e.g., in Madras

Top: The flowers of the Asian *Datura metel* (= *Datura alba*) point upward.

Bottom: *Datura metel* is recognized primarily by its fruits, which have few thorns and hang at an angle, and by its smooth, violet stems. (Photographed in Uttar Pradesh, India)

Left: *Datura metel* var. *fastuosa*, common in northern India, has violet-tinged flower margins and smooth stems. (Photographed in Uttar Pradesh, India)

Center: This cultivated form of *Datura metel* var. *fastuosa* produces double flowers.

Right: *Datura metel* f. *pleniflora* develops filled yellow flowers (triple trumpet).

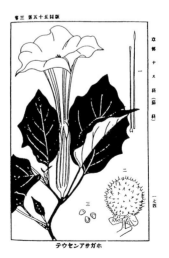

Datura metel var. *alba*, depicted in an early Japanese woodcut from Siinuma Yokusai

province. Or, as in Bombay, they allow the smoke of roasted seeds to come into contact with an alcoholic beverage for a night. It is certain that active components of the plant volatize in this way, and they can then be absorbed by the alcohol. (Lewin 1980 [orig. pub. 1929], 181*)

The dried leaves (or less often the flowers and seeds) are an important ingredient of tantric **smoking blends** (cf. *Aconitum ferox*). A mixture of equal parts of *Datura metel* leaves and hemp (*Cannabis indica*) flowers is particularly esteemed for its inebriating and aphrodisiac effects. The seeds are also added to magical or psychoactive **incenses**. In Malaysia, the seeds are mixed with aloe wood (*Aquilaria agallocha*), cat's eye resin (from *Balanocarpus maximus* King; Dipterocarpaceae), or leban resin (from *Vitex pubescens* Vahl.; Verbenaceae) and burned as an inebriating incense (Gimlette 1981, 216*).

In Malaysia, a hallucinogenic paste is mixed from opium (cf. *Papaver somniferum*), *Datura* seeds, the green shoots of a wild yam species known as *gadong* (*Dioscorea triphylla* Lam.; cf. *Dioscorea composita*), and the green inner bark of *Glycosmis citrifolia* (Rutaceae) (Gimlette 1981, 220*).

The seeds are a main ingredient of **Oriental joy pills** and other similar aphrodisiacs. In Burma (Myanmar), the seeds are added to curries to increase their aphrodisiac effects (Perry and Metzger 1980, 391*). In Oceania, they are added to kava drinks to potentiate their inebriating effects (see *Piper methysticum*). On Java, the thorn apple is prepared as an inebriant in the following manner: Fully grown, ripe, but still unopened fruits are collected and opened. The seeds are dried in the sun and then ground. They then may be mixed with tobacco (Indonesian tobacco [*Nicotiana tabacum*] perfumed with clove oil) or rolled into a tobacco leaf and smoked. In Japan as well, dried leaves were once smoked together with tobacco (*Nicotiana tabacum*) (Lewin 1980 [orig. pub. 1929], 181*).

On Mactan Island (off the coast of Cebu, in the Philippines), young flowers that have not yet unfurled are plucked and dipped briefly in boiling water. They are then laid out in the sun to dry. The dried flowers are crumbled, rolled in a cigarette paper, and smoked. The effects are said to be similar to those of marijuana, but stronger.

The homeopathic mother tincture is produced from ripe seeds with 90% ethyl alcohol (medicinal content of the tincture 1/10).

In Malaysia, fifty seeds taken internally is regarded as a hallucinogenic or (when used for criminal purposes) deliriant dosage (Gimlette 1981, 214*). One hundred seeds (= 1 g) can produce dangerous states and toxic effects. In India, 125 seeds have been reported to be lethal (Gimlette 1981, 217*).

Ritual Use

According to the Vamana Purana, the thorn apple grew from the chest of the Hindu god Shiva, the lord of inebriants (cf. *Cannabis indica*). In the Garuda Purana, it is said that *Datura* flowers should be offered to the god Yogashwara (= Shiva) on the thirteenth day of the waxing moon in January (Mehra 1979, 63 f.). In Nepal, where *Datura* is usually known as *dha-tur-ma*, the plant is considered to be sacred to Shiva. *Dhatur* is interpreted as another name for Shiva; *ma* means "plant." Thorn apple flowers and fruits are among the most important offering gifts of the Newari of Nepal. At every family *puja* (devotional service or offering ceremony), Shiva is first offered *Datura* fruits in order to please him. In Varanasi, Shiva's sacred city, metel fruits and rose flowers are made into sacrifical garlands (*malas*) for the god of inebriation and sold to pilgrims and devotees at the entryways to his temples. These *Datura* chains are then devoutly placed around the lingam, the deity's phallic-shaped image, as fresh flowers are tossed over it (cf. "Artifacts"). The lingam is normally placed in a yoni, the cosmic vulva. Fresh metel fruits are placed into it as offerings.

In Uttar Pradesh (northern India), it is common knowledge that *Datura metel* can be used for inebriating purposes. Smoking the plant is regarded as pleasurable and not dangerous, whereas eating or drinking it is considered dangerous and is generally avoided. Yogis and sadhus in particular smoke thorn apple leaves or seeds together with hemp (*Cannabis indica*) and other herbs (*Aconitum ferox, Nicotiana tabacum*).

In Tibet and Mongolia, this thorn apple is used as an **incense** in secret Vajramabhairava Tantra rituals intended to transform wealth into poverty and to dispel certain spirits or energies. The fruits or seeds are used to induce insanity (Siklós 1995, 252).

In China, the white-blossomed *Datura metel* var. *alba* was considered to be sacred because it was believed that shimmering dew drops had rained from the heavens onto its flowers while the Buddha was giving a sermon. The Chinese Buddhists called it *man-t'o-lo*, after a non-translatable passage from a sutra named *man t'o lo hua*. In ancient China, it apparently was popular to steep the aromatic flowers in **wine** or **sake** before consumption. In the sixteenth-century *Pen tsao kang mu*, Li Shih-chên wrote of the plant's properties:

> Tradition has it that if someone laughs while the flowers are being plucked for use with wine, the wine will evoke laughter in all who drink it. If the flowers are picked while someone is dancing, the wine will induce dancing.

It is possible that this information may be referring to an ancient shamanic ritual.

The Igorot, a Malayan tribal people of Luzon (Philippines), boil the leaves to make an inebriating soup that is eaten communally in a ritual circle.

In Africa, *Datura metel* is used for criminal telepathy and in initiations. The seeds are also used to poison victims so they can be robbed. The toxic and hallucinogenic properties of the plant are well known in East Africa. Seeds are added to the locally brewed **beer** to potentiate its effects (Weiss 1979, 49).

In Tsongaland, which stretches from Mozambique to the Transvaal, *Datura metel* var. *fastuosa* is utilized as a hallucinogenic ritual drug (*mondzo*) in the initiation of girls into women (similar to the use of **Datura wrightii** in the initiation of boys). At their initiation, the girls are painted with red ocher (a symbol of menstrual blood). One after another, they are made to lie down in a fetal position on a mat made from palm fronds while the others dance around them holding on to their hips. Special songs are sung. Afterward, the initiates are tied to a tree (*Euphorbia cooperi* N.E. Br.). Others beat the tree with a stick until white latex (a symbol of spermatozoa) issues from its bark.

The next stage is a water ritual, through which the initiates are cleansed and are supposed to cast aside their childish past. Before they ingest the thorn apple, the girls are required to stretch an animal skin over a vessel of water. Older women perforate the skin with sticks and stir the water.

Following this symbolic defloration, a "school-mother" covered entirely in *Datura* leaves, toad skins, and dog teeth bursts out from the bushes. The initiates are covered in blankets and laid onto palm mats as rhythmic drumming prepares them to receive the *Datura* drink. The schoolmother approaches the initiates, spits upon them, and tells them repeatedly that they will soon hear the voice of the fertility god. She then places cubes of clay, from which pieces of straw protrude, between the girls' legs (the girls' pubic hair is shaved off prior to the ceremony). The clay cubes symbolize the fact that when the pubic hair regrows, it will belong to a woman, not a girl. Then the thorn apple drink is carried around in a ceremonial seashell. It is made by boiling the herbage in water and is said to contain human fat or powdered human bones as well. The schoolmother holds the drink in her hands and sings, "We dig up the medicinal plants that are known to all. Take the medicine, about which you have already heard so much!" Now the initiates drink and listen for the voice of the fertility god. They experience certain visions that are shaped and channeled by means of music and song. At the end of the initiation, the girls are freed from their coverings, dressed in new clothes, and adorned with ornaments. Finally, they dance and sing. The young women are now able to marry (Johnston 1972).

Artifacts

Datura metel flowers are sometimes depicted in Hindu/tantric art, usually in connection with images of Shiva in his various forms. One famous eighteenth-century painting shows a lingam-yoni statue (= the cosmic union of phallus and vulva) upon which a thorn apple flower has been placed in offering (Mookerjee 1971, 49). The plant is also represented on Tibetan medicine thangkas (Aris 1992, 67*). In the Kathmandu Valley, thangkas and statues show Unmata Bhairab, the "divinely inebriated thorn apple Bhairab" (a special tantric form of Shiva), standing straight up.

Hans Simon Holtzbecker, a flower painter from Hamburg, painted a masterful portrait of the plant for the Gottorfer Codex (ca. 1650) (de Cuveland 1989, table 50*).

Numerous Oriental fairy tales mention the inebriating and aphrodisiac properties of the plant (Penzer 1924, 158–62). E. T. A. Hoffmann (1776–1822) wrote a story, "Datura fastuosa (Der schöne Stechapfel)" [The Beautiful Thorn Apple], in which he romantically describes the psychoactive effects of the plant's scent (Hoffmann 1967, 329–80).

In Tsongaland, special music as well as *Datura fastuosa* songs are used during initiations to control the visionary state (Johnston 1975). One techno-pop band, Datura, has even taken its name from the thorn apple.

"The [metel] seeds are a forbidden article among the Dutch, both in India and in Holland, inasmuch as the beer brewers use it to fortify their weak beer, and the brandy distillers do the same; but when they use a bit too much of it, it makes the people who drink of it senseless for a time, or it weakens their reason, so that they imagine some wondrous and almost laughable things, depending upon how their humors are inclined, as for example that one would be a great lord, king, or prince, or another that he wants to swim in the water, even though he is lying in a bed or is in the room, while another has foolish fantasies. Some misuse this datura to force the favors of the fairer sex, although with some this grave thing would not require such force, but when you drink a little warm milk, it is soon over, otherwise it is in 4 degrees of cold nature, it robs a person of sense and reason, and before this of the sense of feeling, because it induces a heavy sleep, blinds the vision, makes the head silly, as if drunken, so that he would pluck at his own clothes and limbs like a born idiot, and would make comical gestures like an Indian monkey when you spread Urtica Indiaca over his claws."

GEORG MEISTER
DER ORIENTALISCH-INDIANISCHE KUNST- UND LUSTGÄRTNER [THE ORIENTAL-INDIAN ART AND PLEASURE GARDENER] (1677, CH. 9, 31*)

"The smoke apples are also an exotic growth / recently come to us from the Oriental lands / it is now grown in many gardens / more for fun / than for usefulness or use / for it is beautiful and amusing to look upon / especially when it is flowering and bearing fruits."

PIERANDREA MATTHIOLUS
KRÄUTERBUCH [HERBAL] (1627, 377*)

Traditional representation of *Datura metel* on a Tibetan medicine thangka (close-up)

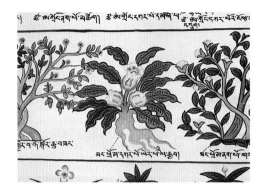

"Not very far from the gate there bloomed a Datura fastuosa (beautiful thorn apple), with its wonderfully scented, large, funnel-shaped flowers in such splendid magnificence, that Eugenius thought with shame about the wretched appearance that the same plant displayed in his own garden.... There floated, as if borne by the evening airs, the sweet accords of an unknown instrument from the remote magical bushes, and the wondrous heavenly tones of a woman's voice ascended luminously.—It was one of those melodies that can spring only from the deepest breast of the fires of love of the south; it was a Spanish romance that the hidden one sang."

E. T. A. HOFFMANN
"DATURA FASTUOSA (DER SCHÖNE STECHAPFEL)" [THE BEUATIFUL THORN APPLE]
(1967, 358)

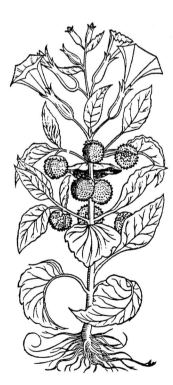

A woodcut depicting the thorn or metel apple. (From the herbal of Hieronymus Bock, 1577)

Medicinal Use

There is evidence that *Datura metel* seeds have been used in Indian folk medicine as well as Ayurveda since a very early date. In the Ayurvedic system, preparations of *Datura* are used to treat numerous illnesses and ailments: headaches, mumps, chicken pox, furuncles, wounds that will not heal, pains of all types, rheumatism, muscle tension, nervous disorders, spasms, convulsions, epilepsy, insanity, syphilis and other venereal diseases, asthma, bronchitis, and overdoses of opium (see ***Papaver somniferum***). The seeds were even once used as a substitute for opium (see **morphine**).

In the Indian medical system known as Unani, which was shaped substantially by Avicenna and is still practiced today (Chishti 1988), *Datura metel* was and is used in similar or identical manners to its use in the Ayurvedic system.

In the Indian folk medicine of the Santal, thorn apple is administered as a remedy for a large number of illnesses: headaches, otitis, wounds, mumps, pain, dropsy, insanity, rheumatism, muscle tension, epilepsy, spasms, delirium febris, pimples, smallpox, syphilis, venereal diseases, and orchitis (Jain and Tarafder 1970, 251). In Karnataka, crushed fresh leaves are applied externally to treat mumps. An infusion is applied externally to treat scorpion stings. Thorn apple is mixed with the leaves of *Solanum nigrum* (see ***Solanum* spp.**) and *Erythrina variegata* L. (see ***Erythrina* spp.**) to make a tonic (Bhardary et al. 1995, 155 f.*). In Uttar Pradesh (northern India), a paste obtained from the seeds is used to treat parasitic skin diseases (Siddiqui et al. 1989, 484*). Powdered seeds are ingested together with dried seedlings of ***Cannabis sativa***, roots of *Laportea crenulata*, and roots of ginger (*Zingiber officinale*) and used as a remedy for pain and cramping (Jain and Borthakur 1986, 579*).

In Java, the seeds are placed onto teeth, inserted into cavities, or chewed lightly to relieve dental pain. *Datura metel* var. *alba* is also used for numerous purposes in traditional Chinese medicine. Mixed with wine (see ***Vitis vinifera***) and hemp (see ***Cannabis indica***), it is used as a narcotic. The flowers and seeds are used to treat skin eruptions and other skin diseases, colds, and nervous disorders.

Datura metel is used to treat asthma in all the regions of the world in which it occurs (Perry and Metzger 1980, 391*; Baker 1995*). In East Africa, dried leaves are either smoked in the form of cigars or burned in incense vessels and inhaled for this purpose (Weiss 1979, 49). In the Philippines, the fresh herbage is placed in an open fire so that asthmatics can inhale the resulting smoke (cf. **incense**). In Europe, this species of *Datura* quickly became known as a medicinal plant under the name *rauchapfel* ("smoking apple") because its leaves can be smoked to treat asthma.

Constituents

All forms and varieties of *Datura metel* contain potently hallucinogenic **tropane alkaloids** (Afsharypuor et al. 1995). Of all the thorn apple species, *Datura metel* contains the highest concentrations of **scopolamine**. Also present are hyoscyamine, **atropine**, meteloidine, norscopolamine, norhyoscyamine, hydroxy-6-hyoscyamine, and datumetine. The entire plant also contains **withanolides**: daturiline, withameteline, daturilinole, secowithameteline, and various daturametelines (Lindequist 1992, 1142*).

The leaves have been found to contain 0.5% alkaloids, the flowers 0.1 to 0.8%, the fruits 0.12%, the roots 0.1 to 0.2 %, and the seeds 0.2 to 0.5% (Lindequist 1992, 1142*).

Effects

The effects of *Datura metel* are essentially the same as those of ***Datura innoxia*** (see there). However, some of what is known indicates that the former can produce effects specific to the species. For example, **smoking blends** made with *Datura metel* seeds and tobacco (with clove oil) have cheering effects and produce a sleep with lively dreams.

In Tsongaland, the hallucinogenic effects are controlled by music, resulting in auditory hallucinations and synesthetic perceptions in which the music is perceived as colors and stereotypical patterns. The contents of these visions include blue-green patterns, green snakes or worms, whirlpools, and sandbanks. The snakes are interpreted as ancestral gods and the auditory hallucinations as the spoken messages of the fertility god (Johnston 1977).

Overdoses will usually result in a delirious state that sometimes can last for days, after which little or nothing can be recalled. Thieves, criminals, and bands of robbers (e.g., the Thuggs) make use of this property when they wish to sedate their victims and rob or rape them without disturbance (Gimlette 1981, 204 ff.*).

In Southeast Asia, licorice (*Glycyrrhiza glabra* L.) is a recommended antidote for overly strong doses of *Datura metel* (Perry and Metzger 1980, 392*).

Commercial Forms and Regulations

The seeds and potted plants of all cultivars, forms, varieties, and subspecies are freely available. In Germany, preparations of the homeopathic mother tincture (Datura metel hom. *HAB34*) and various dilutions are available from pharmacies (Lindequist 1992, 1142*). The mother tincture as well as dilutions to D3 require a prescription (cf. *Datura stramonium*).

Literature

See also the entries for the other *Datura* species.

Afsharypuor, Suleiman, Akbar Mostajeran, and Rasool Mokhtary. 1995. Variation of scopolamine and atropine in different parts of *Datura metel* during development. *Planta Medica* 61:383–86.

Avery, A. G. 1959. Historical review. In *Blakeslee—the genus Datura*, ed. A. G. Avery, 3–15. New York: Ronald Press.

Chishti, Hakim G. M. 1988. *The traditional healer: A comprehensive guide to the principles and practice of Unani herbal medicine.* Rochester, Vt.: Healing Arts Press.

Hoffmann, E. T. A. 1967. *Meister Floh und letzte Erzählungen.* Vol. 4 of the collected works. Frankfurt/M.: Insel.

Jain, S. K., and C. R. Tarafder. 1970. Medicinal plant lore of the Santals. *Economic Botany* 24 (3): 241–78.

Johnston, Thomas F. 1972. *Datura fastuosa*: Its use in Tsonga girls' initiation. *Economic Botany* 26:340–51.

———. 1975. Power and prestige through music in Tsongaland. *Human Relations* 27 (3): 235–46.

———. 1977. Auditory driving, hallucinogens and music-color synesthesia in Tsonga ritual. In *Drugs, rituals and altered states of consciousness*, ed. B. M. du Toit, 217–36. Amsterdam: Balkema Press.

Mookerjee, Ajit. 1971. *Tantra Asana—Ein Weg zur Selbstverwirklichung.* Basel: Basilius Press.

Penzer, N. M. 1924. *The ocean of story.* London: Sawyer.

Siklós, Bulcsu. 1993. Datura rituals in the Vajramahabhairava-Tantra. *Curare* 16:1–76, 190 (addendum).

———. 1995. Flora and fauna in the Vajramahabhairava-Tantra. *Yearbook for ethnomedicine and the study of consciousness,* 1994 (3): 243–66. Berlin: VWB.

———. 1996. *The Vajrabhairava Tantras: Tibetan and Mongolian versions, English translation and annotations.* Vol. 6 of *Buddhica Britannica.* Trink, U.K.: The Institute of Buddhist Studies.

Weiss, E. A. 1979. Some indigenous plants used domestically by East African coastal fishermen. *Economic Botany* 33 (1): 35–51.

Left: The South American thorn apple (*Datura stramonium* ssp. *ferox*) is usually called *chamico* or *miyaya*. (Photographed in Ecuador)

Right, from top to bottom:
These seeds, although of different colors, all came from the same *Datura stramonium* ssp. *ferox* fruit. (Photograph: Karl-Christian Lyncker)

The black seeds of *Datura quercifolia* are comparatively large in size. (Photograph: Karl-Christian Lyncker)

The light-colored seeds of *Datura metel*. (Photograph: Karl-Christian Lyncker)

The seeds of *Datura metel* var. *fastuosa* are very difficult to distinguish from those of the variety *metel*. (Photograph: Karl-Christian Lyncker)

Datura stramonium Linnaeus

Common Thorn Apple

The stigma of the common thorn apple (*Datura stramonium*), shown here enlarged and covered with pollen, resembles a phallus. Is this an indication of the aphrodisiac effects of the plant? (From Louis Figuier, *The Vegetable World*, London 1869)

Family
Solanaceae (Nightshade Family); Subfamily Solanoideae, Datureae Tribe, Stramonium Section

Forms and Subspecies
This diverse species is now divided into four varieties:

Datura stramonium L. var. *godronii* Danert [syn. *Datura inermis*] has thornless fruits and light violet flowers.

Datura stramonium L. var. *inermis* (Juss. ex Jacq.) Timm. has smooth fruits, white flowers, and green stalks.

Datura stramonium L. var. *tatula* Torr. has thorny fruits, violet flowers, and violet-tinged shoots, leaf stalks, and leaf veins. The karyotype of *Datura stramonium* L. var. *tatula* Torr. is almost identical to that of *Datura wrightii* (Spurná et al. 1981).

Datura stramonium L. var. *stramonium* Off. has thorny fruits, white flowers, and green shoots.

The following varieties were described at an earlier date:

Datura stramonium L. var. β *canescens* Wallich in Roxburgh

Datura stramonium L. var. β *chalybea* Koch

The following subspecies are also accepted today:

Datura stramonium L. ssp. *ferox* (L.) Barclay (Franquemont et al. 1990, 99*), which is probably from South America and not from China

Datura stramonium L. ssp. *quercifolia* (H.B.K.) Bye

Datura stramonium L. ssp. [or var.] *villosa* (Fern.) Saff.

Synonyms
Datura bernhardii Lundström
Datura bertolonii Parl. ex Guss.
Datura capensis Hort. ex Bernhardi
Datura ferox L. (Estramonia de la Chino)
Datura inermis Jacq.
Datura laevis L. f.
Datura loricata Sieber
Datura lurida Salisb.
Datura parviflora Salisb.
Datura peregrinum
Datura pseudo-stramonium Sieber
Datura quercifolia H.B.K.
Datura spinosum Lam.
Datura tatula L.
Datura villosa Fernald

Datura wallichii Dunal
Stramonium ferox Boccone
Stramonium foetidum Scopoli
Stramonium spinosum Lam.
Stramonium vulgare Moench
Stramonium vulgatum Gaertner

Folk Names
Ama:ymustak, ama:y'uhc (Mixe, "dangerous plant"), aña panku (Quechua), apple of Peru, arhiaba, asthmakraut, atafaris, attana, azacapanyxhuatlazol-patli (Nahuatl), chamaka, chamico (Quechua), chasse-taupe, chililiceño tapat (corruption of *tlapatl*), cojón del diablo, common thorn apple, concombre à chien, concombre zombi (Caribic, "zombie cucumber"), devil's apple, devil's trumpet, dhatura, donnerkugel, doornappel (Dutch), dornapfel, dornkraut, dutry, el-rita (Morocco), endormeuse, estramonio, fêngch'ieh-êrh (Chinese), gemeiner stechapfel, héhe caroocot (Seri, "plant that makes crazy"), herbe aux sorciers (French, "sorcery plant"), herbe de taupes, hierba del diablo ("plant of the devil"), hierba hedionda ("stinking plant"),[117] hierba inca ("Inca plant"), higuera loca ("crazy fig"), igelkolben, ix telez ku, Jamestown weed, jimsonweed, jimson weed, jouj macel (Arabic), khishqa khishqa (Quechua, "very thorny"), kieli-sa (Huichol, "bad *kieli*"), kratzkraut, manzana del diablo ("apple of the devil"), manzana espinosa ("thorny apple"), matul (Tzeltal), mehen xtohk'u'u (Mayan, "little plant in the direction of the gods"), menj (Arabic/Yemen), mezerbae, mezzettoni, miaia, miaya (Mapuche), mixitl, miyaya, moshobaton tahui (Shipibo), muranha (Swahili), niungué, noce puzza, noce spinosa, ñongué, ñongué morada, papa espinosa (Spanish, "thorny potato"), parbutteeya, patula (Turkish), patura, pomme de diable, pomme épineuse, rurutillo (from the Quechua *ruru*, "fruit"), santos noches, schlafkraut, schwarzkümmel, semilla de la virgen ("seeds of the virgin"), shinah azqhi, simpson weed, stachelnüß, stachelnüß, stink weed, stramoine, stramoine commune, stramonio, stramonio comune, stramonium, taac-amai'ujts (Mixe), ta:g'amih (Mixe, "grandmother"), tatula (Persian, "to prick"), tc'óxwotjilyáih (Navajo), teufelsapfel, thanab (Huastec), thanab thakni' ("white thanab"), thangphrom dkar-po (Tibetan), thorn apple, tohk'u (Mayan, "the direction of the gods"), tollkraut,[118] toloache, tonco-onco, torescua (Tarascan), tukhmtâtûrâ (Persian), tzitzintlapatl (Aztec, "thorny *tlapatl*"), weißer stechapfel, wysoccan, xholo (Zapotec), yacu toé, yoshu chosen asago (Japanese, "exotic morning flower"), zigeunerapfel

---

117 This name is also given to the nightshades *Cestrum nocturnum* and *Cestrum parqui*. The ground leaves of these *Cestrum* species have the same scent as the crushed leaves of *Datura stramonium*.

118 This name is applied to a number of psychoactive nightshades: *Atropa belladonna*, *Hyoscyamus niger*, *Scopolia carniolica* (Arends 1935, 268*).

History

The origins of this potently hallucinogenic thorn apple species are unknown and have been the subject of considerable botanical debate (Symon 1991, 142*). Some authors have suggested that *Datura stramonium* is an Old World species from the region of the Caspian Sea, while others maintain that it originated in Mexico. It is less frequently assumed that the species is from the eastern coast of North America (Schultes and Hofmann 1995*). Still other authors believe that the plant is from Eurasia and did not arrive in Mexico until the colonial period (Berlin et al. 1974, 489*).

In the seventeenth century, the plant's use as an inebriant ("drunk in wine") was documented in Chile (Hoffmann et al. 1992, 145*). *Datura tatula* (= *Datura stramonium* L. var. *tatula*) has been interpreted as the "lost inebriant" of the Shawnee (Tyler 1992). In 1676, a troop of soldiers in Jamestown, Virginia, was served a salad made from thorn apple leaves by their cook. The soldiers fell into a state of delirium that got out of hand and acted like idiots (see "Effects"). As a result, the plant also became known by the name *Jamestown weed*, which eventually became *jimsonweed*. In Mexico, *Datura stramonium* is generally regarded as a "younger sister" of *Datura innoxia*, and it is used in the same ways.

The interpretation that *Datura stramonium* is the plant that Theophrastus and Dioscorides referred to as *strychnós manikós* is quite uncertain (cf. Dieckhöfer et al. 1971, 432; Marzell 1922, 170*). It is more likely that these ancient descriptions refer to the poison nut (*Strychnos nux-vomica*). The trance-inducing smoke of Delphi (cf. *Hyoscyamus albus*) has been attributed to an incense made with *Datura stramonium* (Lewin 1980 [orig. pub. 1929], 183*).

In the European literature, this and other species of thorn apple (*Datura innoxia*, *Datura metel*) were described in all the herbals compiled by the fathers of botany. The first botanically precise illustrations of *Datura stramonium* are contained in the herbals of Hieronymus Bock and Pierandrea Matthiolus. It is widely believed that Gypsies brought this thorn apple to Europe (Perger 1864, 183*). Matthiolus wrote that *Tatula Strominio altera* was an Oriental plant.

Distribution

Today, *Datura stramonium* is commonly found throughout North, Central, and South America; North Africa; central and southern Europe; the Near East; and the Himalayas. The plant is very common on the islands of the Caribbean (Concepción 1993, 554). In the Himalayas (Nepal), the violet-blooming *Datura stramonium* var. *tatula* is the most common form. The subspecies *ferox* grows primarily in Central and

South America. In Germany and Switzerland, the common thorn apple has been growing wild (usually in rubbish dumps and on roadsides) since at least the sixteenth century (Lauber and Wagner 1996, 802*). It has also spread into Israel and Greece (Dafni and Yaniv 1994*).

Cultivation

Cultivation occurs in the same manner as with *Datura discolor* (see there).

The common thorn apple is cultivated commercially for pharmaceutical purposes (as a source of raw drugs and of **scopolamine**). It has been determined that cultivated thorn apples produce considerably more scopolamine when exposed to intense light (Cosson et al. 1966). In contrast, nitrogenous fertilizers have no effect (Demeyer and Dejaegere 1991). It is likely that alkaloid production in the plant can be stimulated by the addition of sugar (saccharose) (Dupraz et al. 1993).

Appearance

This annual plant can grow to a height of about 1.2 meters. It produces many forked, bald branches. The margins of the rich green leaves are coarsely serrated. The funnel-shaped, five-pointed flowers grow from the axils and point straight up; in the common form, they are white. This species produces the smallest flowers (6 to 9 cm long) of the *Datura* species. The egg-shaped green fruits, which are densely covered with short, pointed thorns, are quadripartite and always grow straight up from the axils, a feature which makes it easy to distinguish this species (including all of its varieties and/or subspecies) from the other *Datura* species. The kidney-shaped, applanate seeds (up to 3.5 mm long) are usually black.

The subspecies *ferox*, which was previously thought to be a distinct species, bears leaves that are more clearly and more deeply serrated than those of the common form, while its fruits have longer and slightly curved thorns. The seeds are somewhat lighter in color; they can be brown or black (and may occur in different colors in the same fruit). The flowers are pure white.

The variety *tatula* produces smaller violet flowers. The variety *stramonium* has many short

The common thorn apple (*Datura stramonium*) produces small white flowers.

One of the first European illustrations of the common thorn apple (*Datura stramonium*), clearly showing the typically upright fruits. (Woodcut from Tabernaemontanus, *Neu Vollkommen Kräuter-Buch*, 1731)

Top left: The upright fruit of *Datura stramonium*

Bottom left: The fruits of the South American thorn apple (*Datura stramonium* ssp. *ferox*) are characterized by their especially long thorns. (Photographed in San Pedro de Atacama, northern Chile)

Center: The fruits of *Datura stramonium* var. *godronii* are completely smooth.

Right: *Datura stramonium* var. *tatula* is distinguished on the basis of its violet flowers. (Wild plant, photographed in northern California)

119 The name *tollkörner* is also given to the seeds of **Anarmita cocculus** (also known as fructus cocculi) (Arends 1935, 268*).

120 In Iran, the name *kachola* is also used for the poison nut (**Strychnos nux-vomica**) (Hooper 1937, 112*).

thorns; the subspecies *ferox* has only a few long, sometimes slightly curved thorns; the variety *quercifolia* has even fewer thorns that are somewhat shorter but thicker at the basis. The subspecies *villosa* (cf. **Datura spp.**) has very pileous (hairy) branches, stalks, and calyxes.

Datura stramonium can be confused with small forms of **Datura discolor**, **Datura innoxia**, and **Datura metel**.

Pyschoactive Parts
— Leaves (stramonii folium, folia stramonii, stramonium leaves, thorn apple leaves)
— Seeds (stramonii semen, semen stramonii, thorn apple seeds, tollkörner,[119] kachola[120])
— Flowers
— Roots (radix stramonii, tollwurzel)

Preparation and Dosage
The herbage is harvested shortly after the flowering phase and hung in the shade to dry. It may be smoked alone or in **smoking blends** with other herbs:

> The leaves of *Datura stramonium* are said to be smoked by the Utahs, the Indians of the Great Salt Lake, as well as the Pima and Maricopa, together with the leaves of *Arctostaphylos glauca* or alone. (Lewin 1980 [orig. pub. 1929], 183*)

Even into the twentieth century, the herbage was one of the primary ingredients in asthma cigarettes (cf. **Cannabis indica**). One gram of leaves (alkaloid content = app. 0.25%) is regarded as a therapeutically efficacious dosage for smoking (Lindequist 1992, 1148*). As with all information concerning thorn apple dosages, however, this information should be used with care: "When the drug is administered through inhalation of smoking powders and 'asthma cigarettes,' the amount of applied alkaloids is incalculable" (Roth et al. 1994, 291*).

The Huastec, who live on the Gulf Coast of Mexico, make a magical medicine from thorn apple leaves, slaked lime, and chili pods (*Capsicum annuum* var. *annuum*; cf. **Capsicum spp.**) (Alcorn 1984, 93*). In South America, a paste of freshly

ground leaves (of the subspecies *ferox*) and vinegar is prepared for external use (Schultes 1980, 115*). In the Andes region, *Datura stramonium* (usually the subspecies *ferox*) is used as an additive to San Pedro drinks (see **Trichocereus pachanoi**). The thorn apple, which is also known as the zombie cucumber, is an active ingredient in **zombie poison**.

Four to 5 g of dried leaves contains enough alkaloids to produce fatal results (Lindequist 1992, 1149*), and as little as 0.3 g can be toxic (Roth et al. 1994, 291*). In Morocco, inhaling the smoke of forty seeds that have been strewn over glowing coals is regarded as a hallucinogenic dosage (Vries 1984*). For use in psychoactive **incense**, see *Datura metel*.

In homeopathic medicine, thorn apple is also used in composite medicines, e.g., Stramonium Pentarkan, which consists of *Datura stramonium*, Ignatius beans (cf. **Strychnos spp.**), calcium phosphate, zinc, and passionflower (**Passiflora spp.**).

Ritual Use
In Mexico and neighboring regions, the psychoactive use of *Datura stramonium* is similar to that of **Datura innoxia** (see there). Among the Huastec, it is said that *Datura stramonium* leaves can kill witches and sorcerers (*brujas* and *brujos*) (Alcorn 1984, 624*). The Yucatec Maya call this plant *mehen xtoh-k'uh* ("little plant in the direction of the gods").

The Mixe of Oaxaca (Mexico) believe that *Datura stramonium* contains a plant spirit in the form of a very old woman. For this reason, one Mixe name for the plant is *ta:g'amih*, "grandmother" (cf. **Datura wrightii**). When a portion of the plant is to be harvested, the people make a small offering of three pebbles or a couple of branches. They also speak a prayer:

> Grandmother, do us a favor and cure the illness [name of person] is suffering from. Here we pay you, we carry [the plant] to see what illness [she or he] has. We are sure that you will remedy [the illness]. (Lipp 1991, 37*)

The seeds are then swallowed in a ritual context for divination (cf. **Datura innoxia**)—following the same pattern as the mushroom ritual (see *Psilocybe mexicana*—in the following dosages: Men take three times nine seeds (= twenty-seven), while women three times seven seeds (= twenty-one). In contrast to mushrooms and ololiuqui (*Turbina corymbosa*), however, *Datura* seeds can be ingested during the day (Lipp 1991, 190*).

Although preparations of *Datura stramonium* (cf. **Datura innoxia**), e.g., *jugo de toloache* and *polvo de toloache*, are sold in Mexican herb stores, such transactions usually occur under the table, as the (Catholic) population believes that this plant was created by the devil (Bye and Linares 1983, 4*).

Chamico, the common South American name for the thorn apple (ssp. *ferox*), is derived from the Aymara word *chamakani*, "soothsayer" (Guevara 1972, 160). The plant apparently has had a long tradition as a prophetic and oracular plant (similar to **Brugmansia sanguinea**). The Mapuche use a psychoactively effective brew made with seeds of *Datura stramonium* ssp. *ferox* (*miyaya*) to treat (mental) illnesses that are produced by the *wefukes* spirits and to educate their children[121] (Munizaga 1960).

In North America, the most significant use of *Datura stramonium* in ritual contexts is as an ingredient in **smoking blends** and **kinnikinnick**, which are used to aid in vision quests. If the interpretation of the term *wysoccan* as a common name for *Datura stramonium* is in fact correct, then the Algonquian used the plant as a ritual narcotic.

In Europe, the thorn apple was associated with witches' rituals and **witches' ointments** in the early modern period. In Germany, Russia, and China, the seeds were added to **beer** to lend it potent narcotic properties (Marzell 1922, 172*). In Europe, the seeds served as an incense, a custom allegedly derived from the Gypsies:

> The seeds are used in fumigations to chase away ghosts or to invoke spirits. All of the gypsies' arts are said to come primarily from a precise knowledge of the juices of the thorn apple. (Perger 1864, 183*)

The Gypsies used the thorn apple as an oracular plant in a ritual reminiscent of shamanism:

> On Andrea's night (November 30), thorn apple seeds are placed outside in the open. The next morning, they are then thrown into the fire. If the seeds burn with a loud crackle, then the winter will be dry but very cold. . . . When the tent gypsies wish to find out if a sick person is going to become healthy or not, they ask the "magic drum." An animal skin is marked with lines, each one of which has a special meaning. Nine to twenty-one thorn apple seeds are strewn across the skin, and

these are set in motion by hitting the skin with a small hammer a certain number of times (9 to 21). The position of the seeds on or between the lines then tells whether the sick person will recover or die. This same procedure is also performed for sick animals or to recover stolen objects. (Marzell 1922, 173, 174*)

Artifacts

Among the edifices fashioned from the bizarre, alchemically suggestive constructions of floral and artistic elements contained in his painting *The Garden of Desires*, the late Middle Ages painter Hieronymus Bosch (ca. 1450–1516) included several depictions of fruits that appear to be naturalistic depictions of thorn apples. The entire painting is filled with allusions to the abilities of strange fruits to alter consciousness (Beagle 1983). Perhaps some of the visions of Hieronymus Bosch were produced by *Datura stramonium* (cf. **Claviceps purpurea**). If this interpretation is correct, then the thorn apple would have been indigenous to Europe prior to its first contact with the Americas (see "Distribution").

In her novel *The Clan of the Cave Bear*, best-selling American author Jean Auel describes how sorcerers of the prehistoric Neanderthal prepared an inebriating beverage from the thorn apple and ingested it during their tribal ceremonies and dances and to induce visions (Auel 1980).

The Scandinavian death-metal band Tiamat sang of *Datura stramonium* and its effects in their song "Whatever That Hurts" (*Wildhoney*, Magic Arts 1994; Gaia Century Media, 1994).

Medicinal Use

Aztec medical texts provide the following description of *Datura stramonium*:

> Mixitl. It is of average size, round, green-leaved. It has seeds. The ground seeds are applied where there is gout. It is not edible, not drinkable. It paralyzes one, closes one's eyes, constricts one's throat, holds back the voice, makes one thirsty, deadens the testicles, splits the tongue.
>
> It cannot be noticed when it has been drunk. Those whom it paralyzes—when that person's eyes are closed, he remains behind with closed eyes for all time. That which he looks at, he looks at forever. One becomes stiff, dumb. This can be relieved a little with wine [= pulque; cf. **Agave spp**.]. I take Mixitl. I give someone Mixitl. (Sahagun, Florentine Codex 11:7*)

The Yucatec Maya roast the leaves on a clay or metal disk (*comal*) and then place them on areas affected by muscle pains and rheumatism (Pulido S. and Serralta P. 1993, 61*). Apart from this, the

"More serious however are the effects that religious fanatics, clairvoyants, miracle workers, magicians, priests, and deceivers induce in men by using datura, who inhale the smoke of the burning plant during cultic ceremonies or who are given it internally. The sorcerer's or devil's weed—*herbe aux sorciers, herbe au diable*—was used to produce fantastic hallucinations or illusions and the deceits which result. In demonology, this plant in particular has a role whose significance outsiders of course can scarcely imagine."

LOUIS LEWIN
PHANTASTIKA
(1980[ORIG. PUB. 1929], 180F.*)

"In magical beliefs, the thorn apple is used as an agent for producing ecstasy. For example, the priests of the sun temple in the Peruvian city of Sagomozzo would chew the seeds of this plant in order to achieve the inspiration that was the prerequisite for divination. Thorn apple extracts play an enormous role both in witches' ointments and in narcotic magical incenses. It is generally known that the so-called asthma cigarettes have stramonium as an additive even today. Various people have told us that those in the know smoke these asthma cigarettes because they allegedly stimulate the sex drive."

MAGNUS HIRSCHFELD AND RICHARD LINSERT
LIEBESMITTEL [APHRODISIACS]
(1930, 175*)

"Decoction of Jimsonweed
Slimy trailing plants distil
Claustrophobia and blood mixed seed
Cursed downstairs against my will."

TIAMAT
WILDHONEY
(1994)

---

121 The Jíbaro Indians (Ecuador) use *Brugmansia suaveolens* in a similar manner.

"Among the Transylvanian Gypsies, when a newlywed couple returns to the camp, water is poured over them, after which a weasel skin filled with thorn apple seeds [from *Datura stramonium*] is rubbed over them. The weasel skin protects them from misfortune and the thorn apple seeds from the evil eye. . . .

"Before the nomadic Gypsies of Hungary move into their winter caves, they never forget to light a fire of dried thorn apple bushes in front of each dwelling and shake some alum into the embers. They also carry some of this fire into the caves, for it is an excellent agent against the evil eye."

Siegfried Seligman
Die magischen Heil- und Schutzmittel aus der belebten Natur [The Magical Healing and Protective Agents from the Animated Nature]
(1996, 257*)

folk medical uses are the same as with *Datura innoxia.*

In Peru and Chile, a tea made from the leaves is drunk to alleviate pain (Schultes 1980, 115*). In Peru, a tea made with *Datura stramonium* ssp. *ferox* is drunk for stomachaches (Franquemont et al. 1990, 40*). The Mapuche use a tea of the fresh herbage of *Datura stramonium* ssp. *ferox* as a narcotic and a *Datura* ointment for toothaches. The entire plant is administered in various preparations to treat pains, inflammations, cancer, and neuritis (Houghton and Manby 1985, 100*).

In Uttar Pradesh (India), juice pressed from the fruits is massaged into the scalp to treat dandruff (Siddiqui et al. 1989, 484*). In Southeast Asia, the roots are used to treat the bites of rabid dogs and insanity, while the leaves are smoked for asthma (Macmillan 1991, 423*).

Throughout the world, *Datura stramonium* is regarded as an aphrodisiac (Guevara 1972, 160) and as an agent for treating asthma (Baker 1995*; Dafni and Yaniv 1994, 13*; Mösbach 1992, 105*; vries 1984*; Wilson and Mariam 1979, 30*). When used for asthma, either the leaves are smoked or the seeds, burned as **incense**, are inhaled. In the Canary Islands, where this species is known as *santos noches* ("holy nights"), the dried leaves are also smoked for asthma (Concepción 1993, 54).

During the early modern period, *Datura stramonium* was used to make love drinks, but it was also recommended for mental disorders and other diseases:

In some parts of France and Germany, both the herbage and the seeds of this narcotic poisonous plant are used as a home remedy for toothaches, wheezing, and other nervous afflictions of a chronic nature. The seeds are placed in the hollow, painful tooth, and a small pipe full of one part of the leaves and eight parts of tobacco [see *Nicotiana tabacum*] are smoked once daily or as often as the asthma attacks occur. Dried, it is also made into cigars that are smoked for the same purpose. In the hands of the physician, the Tinctura Seminum Stramonii, five to fifteen drops two to three times daily, is a very effective agent against a pathologically increased sexual desire, nymphomania, and satyriasis, but should never be allowed to become a folk medicine. (Most 1843, 141*)

In Peru, *chamico* leaves (*Datura stramonium* ssp. *ferox*) are applied externally as a facial wash to treat headaches and migraines. An industrially manufactured perfume called Chamico is dabbed onto the face for the same purpose. This scent is also used to promote one's own attractiveness, for love magic, and to increase male potency. It is unknown whether this perfume is made using *Datura*, but it is very unlikely. However, the

enclosed instructions include a "prayer to the chamico perfume" that makes reference to its hypnotic effects.

In Europe, the plant has been widely used for medicinal purposes since the eighteenth century. In 1747, Elisabeth Blackwell wrote the following in her *Herbal*: "Some use the leaves as a cooling agent when a person has received a burn, and to fight inflammations. The seeds have a power to make one slack and sedated" (Heilmann 1984, 82*).

Cigarettes made from *Datura stramonium* were being smoked as a treatment for asthma and mental illnesses as late as the twentieth century (Hirschfeld and Linsert 1930, 174*).

In homeopathy, Datura stramonium hom. (usually in dilutions of D3 and greater) is used in accordance with the medical description to treat such ailments as whooping cough, asthma, neuralgia, and nervous excitement (Pahlow 1993, 304*). It is used especially for disturbances of the mind, for "the entire power of this agent appears to expend itself in the brain" (Boericke 1992, 720*).

Constituents

The entire plant contains **tropane alkaloids**. The alkaloid content can vary greatly and lies between 0.25 and 0.36% (with one recorded instance of 0.5%) in the leaves and between 0.18 and 0.22% in the roots. The flowers can contain as much as 0.61% alkaloids and the seeds up to 0.66%. The main alkaloids in all parts of the plants are L-hyoscyamine and L-**scopolamine**; also present are apoatropine, tropine, belladonnine, and hyoscyamine-*N*-oxide. Dried leaves and seeds contain 0.1 to 0.6% alkaloids. Apoatropine and tropanole arise only when the raw drug is stored in an inappropriate manner or for too long a time (Roth et al. 1994, 291*). Young plants contain chiefly **scopolamine** and older ones primarily hyoscyamine.

Datura stramonium L. var. *tatula* Torr. contains primarily hyoscyamine (Spurná et al. 1981). In addition to the alkaloids that are regarded as the primary active constituents, **withanolides**, lectines, peptides, and **coumarins** are also present.

The seeds of the Argentinean *Datura stramonium* ssp. *ferox* have been found to contain 3α-tigloyloxytropane (= tigloyltropeine), 3-phenylacetoxy-6β, 7β-epoxytropane (= 3-phenylacetoxyscopine), aposcopolamine (= apohyoscine), 7β-hydroxy-6β-propenyloxy-3α-tropoyloxytropane, traces of 7β-hydroxy-6β-isovaleroyloxy-3α-tigloyoxytropane, the pyrrolidine alkaloid hygrine, and the previously unknown 3-phenylacetoxy-6β,7β-epoxytropane (= 3-phenylacetoxyscopine) and 7β-hydroxy-6β-propenyloxy-3α-tropoyloxytropane (Vitale et al. 1995).

Effects

The profile of effects of *Datura stramonium* is essentially the same as that of *Datura innoxia* and *Datura metel.* Among the characteristic effects are dryness of the mouth, difficulty in swallowing, dilation of the pupils, restlessness, confusion, and hallucinations. The effects sometimes begin after only a half an hour but may occasionally appear after four hours, and they can persist for days (Gowdy 1972; Lindequist 1992, 1148*; Roth et al. 1994, 292*).

In his *History and Present State of Virginia*, Robert Beverly described the oft-quoted effects that occurred when English soldiers at Jamestown unknowingly or accidentally ate thorn apple leaves as a salad:

The James-Town Weed (which resembles the Thorny Apple of *Peru*, and I take it to be the Plant so call'd) is supposed to be one of the greatest Coolers in the World. This being an early Plant, was gather'd very young for a boil'd salad, by some of the Soldiers sent thither, to pacifie the troubles of *Bacon*; and some of them eat plentifully of it, the Effect of which was a very pleasant Comedy; for they turn'd natural Fools upon it for several Days: One would blow up a Feather in the Air; another wou'd dart Straws at himself, and another stark naked and grinning like a monkey, sitting in the corner, tried to mow the grass; a Fourth would fondly kiss, and paw his Companions, and sneer in the Faces, with a Countenance more antick, than any in a *Dutch Droll*. In this frantick condition they were confined, lest they should in their Folly destroy themselves; though it was observed, that all their Actions were full of Innocence and good nature. Indeed, they were not very cleanly; for they would have wallow'd in their own Excrements, if they had not been prevented. A Thousand such simple Tricks they play'd, and after Eleven Days, return'd themselves again, not remembering anything that had pass'd. (In Safford 1920, 557–58)

In the Canary Islands, thorn apple grows like a weed. Many young tourists have consumed teas made from the flowers, swallowed or smoked the seeds, or eaten the fresh leaves. The majority of the experiences they have reported have been unpleasant. One man smoked thorn apple seeds and became feverish for three days. Another man who drank a tea made with the flowers collected and ate his feces for three days. Others who had eaten the seeds went swimming and decided to swim to one of the neighboring islands. Some have felt themselves transported back in time, where they conversed with the Guancha, the indigenous inhabitants of the islands, who have been "extinct for 600 years" (cf. **Cytisus canariensis**). Many simply felt ill. Reports mention nausea, headache, and confusion. Positive experiences, which do also occur, are only rarely mentioned.

Occasionally, *Datura stramonium* intoxications can also prove fatal (*MMWR* 44 [3] 1995; Roth et al. 1996, 291f.*).

People who smoked asthma cigarettes often reported "undesired" side effects, namely, "dreams with sexual overtones" (Schenk 1954, 78*; Hirschfeld and Linsert 1930, 174f.*). The medical literature also contains accounts of such erotic effects (Dieckhöfer et al. 1971, 432).

Commercial Forms and Regulations

All commercially available pharmaceutical forms (herbage, extracts, tinctures, homeopathic preparations [Datura stramonium hom. *HAB1*]) can be obtained only in a pharmacy with a prescription from a physician. *Datura stramonium* is a proscribed substance under the Cosmetic Regulations (from 19 June 1985, appendix 1, 301). In contrast, both the seeds and potted plants are freely available.

Literature

See also the entries for **Datura discolor**, **Datura innoxia**, **Datura metel**, and **tropane alkaloids**.

Auel, Jean. 1980. *The clan of the cave bear*. New York: Crown Publishers.

Beagle, Peter S. 1983. *Der Garten der Lüste: Unsere Welt in den modernen Malereien des Hieronymus Bosch*. Cologne: DuMont.

Concepción, José Luis. 1993. *Costumbres, tradiciones y remedios medicinales canarios: Plantas curativas*. La Laguna, Tenerife: ACIC.

Cosson, L., P. Chouard, and R. Paris. 1966. Influence de l'éclairement sur les variations ontogéniques des alcaloides de *Datura tatula*. *Lloydia* 29 (1): 19–25.

Demeyer, K., and R. Dejaegere. 1991. Influence of the N-form used in the mineral nutrition of *Datura stramonium* on alkaloid production. *Planta Medica* 57 suppl. (2): A27.

Dieckhöfer, K., Th. Vogel, and J. Meyer-Lindenberg. 1971. *Datura stramonium* als Rauschmittel. *Der Nervenarzt* 42 (8): 431–37.

Dupraz, Jean-Marc, Philippe Christen, and Ilias Kapetanidis. 1993. Tropane alkaloid production in *Datura quercifolia* hairy roots. *Planta Medica* 59 suppl.: A659.

———. 1994. Tropane alkaloids in transformed roots of *Datura quercifolia*. *Planta Medica* 60:158–62.

Gowdy, J. M. 1972. Stramonium intoxication: Review of symptomatology in 212 cases. *Journal of the American Medical Association* 221:585–87.

Guevara, Dario. 1972. *Un mundo mágico-mitico en la mitad del mundo: Folklore ecuatoriano*. Quito: Impr. Municipal.

"In Kenya, I was able to learn from the British Secret Service—when I was reporting on the Mau Mau uprising for *Life* magazine—that the Mau Mau secret society had collected large amounts of thorn apple seeds and leaves (*Datura stramonium* and *Datura fastuosa* L. or *Datura alba* Nees). An informant told the English that it was planned to use the black cooks and servants to add a powder of this drug to the food on a certain evening, so that they would be helpless during the massacre that was planned for the night. A person who is under the influence of this drug will allow anything to happen.... Shortly after this incident, I heard that the English had been given the order to exterminate and burn all of the thorn apple plants.... During an extended drive, I was then able to determine that the blacks had simply disregarded the order."

HANS LEUENBERGER
ZAUBERDROGEN [MAGIC DRUGS]
(1969, 184*)

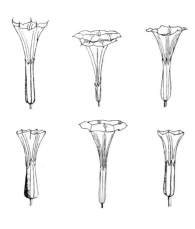

Top row, from left to right: *Datura stramonium*, *D. metel*, *D. innoxia*

Bottom row, from left to right: *D. stramonium ssp. ferox*, *D. wrightii*, *D. leichhardtii*. (From Festi 1995, 122 f.)

Hilton, M. G., and M. J. C. Rhodes. 1993. Factors affecting the growth and hyoscyamine production during batch culture of transformed roots of *Datura stramonium*. *Planta Medica* 59:340–44.

Munizaga A., Carlos. 1960. Uso actual de *miyaya* (*Datura stramonium*) por los araucanos de Chile. *Journal de la Société des Américanistes* 52:4–43.

Portsteffen, A., B. Dräger, and A. Nahrstedt. 1991. Isolation of two tropinone reductases from *Datura stramonium* root cultures. *Planta Medica* 57 suppl. (2): A107.

Spurná, Vera, Marie Sovová, Eva Jirmanová, and Alena Sustácková. 1981. Chromosomal characteristics and occurrence of main alkaloids in *Datura stramonium* and *Datura wrightii*. *Planta Medica* 41:366–73.

Tyler, Varro E. 1992. John Uri Lloyd and the lost narcotic plants of the Shawnees. *Herbalgram* 27:40–42.

Vitale, Arturo A., Andrés Acher, and Alicia B. Pomilio. 1995. Alkaloids of *Datura ferox* from Argentina. *Journal of Ethnopharmacology* 49:81–89.

Wein, Kurt. 1954. Die Geschichte von *Datura stramonium*. *Kulturpflanze* 2:18–71.

Datura wrightii Regel

Wright's Datura

Family
Solanaceae (Nightshade Family); Subfamily Solanoideae, Datureae Tribe, Dutra Section

Forms and Subspecies
The karyotype of this plant is almost identical to that of *Datura stramonium* L. var. *tatula* Torr. (see *Datura stramonium*). It is possible that *Datura wrightii* is only a local (California) variant of *Datura innoxia* (Hickman 1993, 1070).

Synonyms
Datura metel var.[?] *quinquecuspida* Torr.
Datura meteloides Dunal in DC.
Datura wrightii Bye
Datura wrightii Hort.

Folk Names
Kalifornischer stechapfel, kiksawel (Cahuilla), kusi (Diegueno), malkapit, manai (Yokut), manet, manit (Gabrieliño), manitc (Serrano), mánoyu (Miwok), momoy (Chumash), mo'moy, monayu (Miwok), nakta mush (Luiseño), naktanuuc (Cupeño), smalikapita (Yuma), tanabi, tanábi (Mono), tanai, tañai, táñai, taña'nib (Mono), thornapple, toloache, Wright's datura, Wright's stechapfel

History
This Southern California species of thorn apple has apparently been used for ritual and medicinal purposes for more than five thousand years (Grant 1993; cf. Boyd and Dering 1996, 266 f.*). During the colonial period, shamanic use of the plant was extremely common among many tribes, much to the distress of Catholic missionaries. Although it is occasionally rumored that some contemporary Chumash youth have been using the plant in an attempt to explore their cultural roots, there is as yet no evidence to support this (cf. Baker 1994).

Distribution
This *Datura* is found only in Southern California and is especially common in the territories once occupied by the Chumash (Los Angeles and Ventura Counties).

Cultivation
See *Datura discolor*.

Appearance
Datura wrightii is almost indistinguishable from *Datura innoxia*, but it grows in a prostrate and creeping manner and produces pendulous fruits with many thin thorns. *Datura wrightii* is also easily confused with *Datura discolor* and less frequently with *Datura metel*.

Psychoactive Material
— Roots, fresh or dried and powdered
— Leaves

Preparation and Dosage
The fresh root is crushed and extracted in water (Timbrook 1990, 252*). Unfortunately, the ethnographic sources provide little information about dosages (cf. *Datura innoxia*).

Left: *Datura wrightii*, the predominant Southern California species, has a creeping pattern of growth. (Wild plant, photographed near Moorpark, in the former territory of the Chumash Indians)

Right: The fruit of *Datura wrightii* hangs downward at an angle.

Several Southern California tribes used the seeds or entire fruits to produce a beerlike drink (cf. **beer**). The fresh seeds (or fruits) were ground and then added to water. It is possible that the tribes added other fermentation agents, such as manzanita fruits (*Arctostaphylos manzanita* Parry). The vessel containing the liquid was set in the sun so that fermentation (by means of wild yeasts) would begin quickly. Fermentation was complete after one to two days. The drink, which was only mildly alcoholic, must have been extremely potent (Balls 1962, 67).

As with all other species of *Datura*, the seeds and dried leaves of this species are suitable as additives to smoking blends and incense.

Ritual Use

This *Datura* species played an especially important role in the initiation rites (*chungichnich* cult, *manet, kiksawel*) of the Indians who once lived in Southern California (Gayton 1928; Jacobs 1996).

The Chumash regarded this *Datura* as a female spirit being, "the old woman Momoy" (*momoy* is the Chumash name for *Datura wrightii*; Baker 1994). They had shamans specializing in the use of *Datura*; such a shaman would be known as *alshukayayich* ("one who causes intoxication) or, in Spanish, *toloachero* ("datura giver") (Applegate 1975, 10; Walker and Hudson 1993, 43). The thorn apple was regarded as a "dream helper" that shamans used frequently to induce prophetic dreams.

The most significant use of thorn apple occurred in conjunction with the initiation of boys into men. Before ingesting the drink, which would be prepared by the initiate's grandmother, the initiate was required to fast and to avoid all consumption of meat. During the fast, he would smoke a great deal of tobacco (*Nicotiana attenuata, Nicotiana bigelovii*; cf. **Nicotiana spp.**) (Applegate 1975). Normally, the initiate was left alone, in a cave or a dwelling, after he received the drink and surrendered himself to the visions it induced. Any questions he had could be answered only by the *Datura* spirit, for it was said that "the *Datura* will teach you everything." During the visionary state, finding a spirit ally in the form of an animal (coyote, hawk, etc.) was considered an especially fortunate occurrence. The initiate usually fell into a delirious state for some twenty-four hours, from which he only gradually awoke. Afterward, a *Datura* shaman would help him interpret the visions and develop them into a plan for the rest of his adult life (Applegate 1975).

The Chumash also used *Datura* in conjunction with sweat lodge rituals. Unfortunately, the precise manners of such use have not come down to us (Timbrook 1987, 174). It is possible that the seeds may have been used as a psychoactive incense that was strewn over the glowing rocks (cf. **Artemisia spp.**).

Many other California tribes (Coahuilla, Yokut, Gabrieliño, Luiseño, Diegueno, Dumna) also used a potent *Datura* tea to initiate their youths into the mysteries of life (Beau and Siva Saubel 1972, 61 ff.). The visions and dreams they experienced were intended to be signposts for their future lives (Jacobs 1996).

Similar to the use of *Datura*, several California tribes used red ants for psychoactive purposes during their initiatory rites (Blackburn 1976*; Groark 1996; cf. **Nepeta cataria**).

The shamans of the Miwok ate the roots or drank a decoction of the fresh herbage in order to acquire supernatural powers and peer into the future (Barrett and Gifford 1933, 169). Shamans also used *Datura* for harmful purposes (Applegate 1975).

The Kawaiisu use *Datura wrightii* as a ritual medicine in the initiation of boys and to produce visions and prophetic dreams (Moerman 1986, 149*).

Artifacts

The initiatory use of *Datura* is well known from the American Southwest, e.g., among the Chumash (Timbrook 1987, 174 f.) The Chumash have a long tradition of creating ritual paintings on rocks and in caves. Some of these five-thousand-year-old paintings have been interpreted as evidence of a *Datura* cult (Grant 1993). They also incorporate what are clearly shamanic elements (Hedges 1992). *Datura* visions have apparently shaped all of the Chumash rock art. Many paintings provide symbolic representations of elements that were

"With this [*Datura*] they [the Chumash] intoxicate themselves. They take it in order to become strong, in order not to fear anyone, to prevent snakes from biting them and that darts and arrows may not pierce their bodies, etc."

Franciscan missionary in "Virtuous Herbs" (Timbrook 1987, 174)

"In California, the leaves of the stalk and sometimes also the root of the [*Datura wrightii*] plant were squeezed, softened in water, and drunk after being decocted. This drink called forth hallucinations—an Indian would speak of visions—as well as dreams which make it possible to look into the future and make supernatural beings visible. The drink also produced clairvoyance and revealed things that would not be revealed in the context of normal visions: events which had taken place hundreds of kilometers away or only in the future."

Serge Bramley in *Reiche des Wakan* [In the Realm of the Wakan] (1977, 82 f.*)

significant in the visions. In some ways, the painters translated their visions into the symbolic code of the Chumash culture (Hudson 1979; Wellmann 1981*).

Medicinal Use

The Chumash drank infusions or decoctions of the root to treat pain, especially in cases of broken bones and injuries. They also drank *Datura* for snakebites, apparently as a kind of sympathetic magic. It was said that snakes would bite into a thorn apple with their fangs in order to make their teeth poisonous before they would bite an animal or a person. In other words, the Chumash used the same toxin to combat the venom, in the manner of the basic principles of homeopathy. To treat asthma, they inhaled the smoke of the dried leaves as a medicinal **incense** (Timbrook 1987, 174).

The Kawaiisu administer the pressed root internally to alleviate strong pain and apply a paste of it externally to treat broken bones and swelling. A tea made with the roots is used as a medicinal bath for rheumatism and arthritis (Moerman 1986, 149*).

Constituents
See *Datura innoxia.*

Effects
See *Datura innoxia.*

Commercial Forms and Regulations
Datura wrightii is occasionally found in specialty nurseries in California. The plant is not subject to any regulations.

Literature
See also the entry for ***Datura innoxia.***

Applegate, Richard B. 1975. The datura cult among the Chumash. *The Journal of California Anthropology* 2 (1): 7–17.

Baker, John R. 1994. The old woman and her gifts: Pharmacological bases of the Chumash use of *Datura. Curare* 17 (2): 253–76. (Very good bibliography.)

Balls, Edward K. 1962. *Early uses of California plants.* Berkeley: University of California Press.

Barrett, S. A., and E. W. Gifford. 1933. Miwok material culture. *Bulletin of Milwaukee Public Museum* 2 (4).

Bean, Lowell John, and Katherine Siva Saubel. 1972. *Temalpakh: Cahuilla Indian knowledge and usage of plants.* Morongo Indian Reservation, Calif.: Malki Museum Press.

Blackburn, Thomas. 1977. Biopsychological aspects of Chumash rock art. *Journal of California Anthropology* 4:88–94.

Gayton, Anna Hadwick. 1928. The narcotic plant *Datura* in aboriginal American culture. PhD thesis, University of California.

Grant, Campbell. 1993. *The rock paintings of the Chumash.* Santa Barbara, Calif.: Santa Barbara Museum of Natural History.

Groark, Kevin P. 1996. Ritual and therapeutic use of "hallucinogenic" harvester ants (*Pogonomyrmex*) in native south-central California. *Journal of Ethnobiology* 16 (1): 1–29.

Hedges, Ken. 1976. Southern California rock art as shamanic art. In *American Indian rock art*, ed. Kay Sutherland, 2:126–38. El Paso, Texas: Archaeological Society.

———. 1992. Shamanistic aspects of California rock art. In *California Indian shamanism*, ed. Lowell John Bean, 67–88. Menlo Park, Calif.: Ballena Press.

Hickman, James C., ed. 1993. *The Jepson manual: Higher plants of California.* Berkeley: University of California Press.

Hudson, Travis. 1979. Chumash Indian astronomy in south coastal California. *The Masterkey* 53 (3): 84–93.

Jacobs, David. 1996. The use of *Datura* in rites of transition. *Jahrbuch für Transkulturelle Medizin und Psychotherapie* 6 (1995): 341–51.

Timbrook, Jan. 1987. Virtuous herbs: Plants in Chumash medicine. *Journal of Ethnobiology* 7 (2): 171–80.

Walker, Phillip L. and Travis Hudson. 1993. *Chumash healing.* Banning, Calif.: Malki Museum Press.

This Chumash Indian rock art image was allegedly inspired by the ritual use of *Datura wrightii.*

Datura spp.

Thorn Apple Species

Family

Solanaceae (Nightshade Family); Subfamily Solanoideae, Datureae Tribe

Eleven species of *Datura* are usually accepted today (D'Arcy 1991, 78*). Some botanists have recently suggested that the genus *Datura* is indigenous only to the New World and did not spread into Asia (*D. metel*) and Australia (*D. leichhardtii*) until the past four hundred years. I cannot accept this purist view in any way (cf. *Datura metel*). These botanists do not appear to possess any detailed ethnohistorical knowledge and appear to have overlooked the fact that the very name of the genus is Sanskrit in origin (Symon and Haegi 1991).

Datura kymatocarpa A.S. Barclay

This species (if in fact it is a distinct species and not simply one of the many varieties of *Datura innoxia*) is found only in the tropical valley of the Río Balsa (Mexico). It is recognizable by its hairy fruits (Barclay 1959, 257). To date, no ethnobotanical use has been reported.

Datura lanosa Barclay ex Bye [syn. *Datura innoxia* ssp. *lanosa*]—rikuri, rikúi

Only recently described (Bye 1986), this thorn apple species occurs exclusively in northern Mexico and may simply be a local variety of *Datura innoxia*. The name the Tarahumara use to refer to the plant (*rikuri*) is derived from *rikú*, "drunken" (Bye et al. 1991, 34). The name is linguistically related to *kiéri/kíeri*, a word the Huichol use chiefly for *Solandra* spp.

Datura leichhardtii F. Muell. ex Benth. [syn. *Datura pruinosa* Greenman]—Leichhardt's datura, Australian thorn apple

This Australian species, which is very common on the continent and is almost the only species there, is also said to occur in very remote areas of Mexico and Guatemala (Symon and Haegi 1991). It has small, round, drooping fruits with numerous short thorns. Apart from this, the plant is very similar to *Datura stramonium*. In Australia, where it is used as a **pituri** substitute, the plant is also known as "killer of sheep" (Low 1990, 187*).

Datura pruinosa Greenman—pruinose thorn apple

This Mexican species is found only in Oaxaca at altitudes between 550 and 1,550 meters. It has very small flowers and finely haired leaves that look as though they have been affected by frost. The dried herbage contains 0.16% alkaloids (the primary alkaloid is **atropine**; also present are apoatropine, noratropine, hyoscine [= **scopolamine**], norhyoscine, apohyoscine, littorine, tigloidine, 3α-tigloyloxytropane, meteloidine, tropine, and Ψ-tropine) (Evans and Treagust 1973). The chemical composition is practically identical to that of *Datura leichhardtii*. The name *Datura pruinosa* is now usually regarded as a synonym for *Datura leichhardtii* (Symon and Haegi 1991, 198).

Datura quercifolia H.B.K. [syn. *Datura stramonium* ssp. *quercifolia* (H.B.K.) Bye]—oak-leaf datura

This *Datura* is limited to Texas, Arizona, and northern Mexico. Its fruits have long thorns and its leaves resemble those of an oak (hence the name). It is probably identical to *Datura stramonium* (Safford 1921, 177) and is now best regarded as a subspecies of *Datura stramonium* (Bye 1979b, 37*).

The Pima Indians of northern Mexico roast the fruits, which they call *toloache*, and then grind them and mix them with fat to produce an ointment they apply to open wounds. Together with the leaves of a *Physalis* species known as *coronilla* or *kokovuri*, the fruits are boiled to produce a decoction for treating coughs (Pennington 1973, 228*).

Datura reburra A.S. Barclay

This species has been described for the Mexican state of Sinaloa. The plant is similar to *Datura discolor*, however, the thorns are longer and thinner (Barclay 1959, 259). It is likely only a variety of *Datura discolor*.

Datura villosa Fernald [syn. *Datura stramonium* ssp./var. *villosa* (Fern.) Saff.]—shaggy thorn apple

This species occurs in Jalisco and San Luis Potosí (Mexico); it may be identical to *Datura stramonium* (Safford 1921, 177).

Datura (*Ceratocaulis*) *ceratocaula* Ortega [syn. *Datura macrocaulis* Roth, *Apemon crassicaule* Raf., *Datura sinuata* Sessé et Moc., *Ceratocaulus daturoides* Spach.]—tlapatl

This species is found only in central Mexico (México, Querétaro, Oaxaca). It is a water plant that has the appearance of a vine instead of an herbaceous plant or bush. It has thick, forked stalks and thornless fruits that hang to the side. In Mexico, it is known as *tornaloco* ("maddening plant"), and it is apparently identical to the magical plant the Aztecs called *atlinan*, "his

"The effects of all species are similar, since their constituents are so much alike. Physiological activity begins with a feeling of lassitude and progresses into a period of hallucinations followed by deep sleep and loss of consciousness. In excessive doses, death or permanent insanity may occur. So potent is the psychoactivity of all species of *Datura* that it is patently clear why peoples in primitive cultures around the world have classed them as plants of the gods."

RICHARD EVANS SCHULTES AND
ALBERT HOFMANN
PLANTS OF THE GODS
(1992, 111*)

This ancient Mexican representation of a flower in the process of opening can be interpreted as *Datura*.

The fruit of the rare Mexican *Datura quercifolia* resembles a weapon and protrudes horizontally.

As the species name suggests, the deeply emarginated leaf of *Datura quercifolia* resembles that of an oak.

mother is water,"[122] or *tlapatl*. The Aztecs regarded it as the "Sister of Ololiuqui" (see ***Turbina corymbosa***) (Schultes and Hofmann 1992, 41, 111*). An Aztec-language source provided the following description of the plant:

> It is small and round, blue, green-skinned, broad-leaved. And it blossoms white. Its fruit is smooth, its seeds black, foul-smelling. It causes one harm, takes away the appetite, makes one mad, inebriates one.
>
> He who eats of it will not want any other food until he dies. And if he eats it regularly, he will always be confused, mad; he will always be possessed, never again calm. And where gout is present, it is applied thinly as an ointment in order to heal this. It is also said to be sniffed, for it will cause harm, it takes away a person's appetite. It causes harm, makes one mad, takes away the appetite. I take Tlapatl; I eat, I go around and eat Tlapatl.
>
> This is what is said of him who goes around and is contemptuous, who goes around with arrogance, presumptuousness, who goes around and eats the Mixitl and Tlapatl herbs; he goes around and takes Mixitl and Tlapatl. (In Sahagun, Florentine Codex 11:7*)

An Aztec magical formula from the colonial period invoked the plant spirit of this *Datura* in the following manner:

> I call to you, my mother, she who is of the beautiful water!
>
> Who is the god, or who has the power to break and consume my magic?
>
> Come here, sister of the green woman Ololiuqui, of she by means of which I go and leave the green pain, the brown pain, so that it hides itself.
>
> Go and destroy with your hands the entrails of the possessed, so that you test his power and he falls in shame.

(Jacinto de la Serna, in *Documentos Ineditos para la Historia de Espawe* 104:159–60; cf. Safford 1921, 182; also Rätsch 1988a, 142*)

This *Datura* is said to have very potent narcotic effects. Little is known about any modern use (Schultes and Hofmann 1992, 111*). In Mexico City, preparations of *Datura ceratocaula* are supposedly used as a drug of fashion in some circles. I have also heard that some Mexican psychiatrists administer combinations of ketamine and *Datura ceratocaula* to their patients for psychotherapeutic purposes.

Datura velutinosa Fuentes—silky thorn apple
This species has recently been described for Cuba. However, the name is apparently a synonym for ***Datura innoxia***.

Literature
See also the entries for the other ***Datura*** species as well as ***Brugmansia*** spp.

Barclay, Arthur S. 1959. New considerations in an old genus: *Datura*. *Botanical Museum Leaflets Harvard University* 18 (6): 245–72.

Bye, Robert A. 1986. *Datura lanosa*, a new species of datura from Mexico. *Phytologia* 61:204–6.

Bye, Robert A., Rachel Mata, and José Pimentel. 1991. Botany, ethnobotany and chemistry of *Datura lanosa* (Solanaceae) in Mexico. *Anales del Instituto Biológico de la Universidad Autónoma Nacional de Mexico*, ser. bot. 61:21–42.

Evans, William C., and Peter G. Treagust. 1973. Alkaloids of *Datura pruinosa*. *Phytochemistry* 12:2077–78.

Festi, Francesco. 1995. Le herbe del diavolo. 2. Botanica, chimica e farmacologia. *Altrove* 2:117–41.

Safford, William E. 1921. Synopsis of the genus *Datura*. *Journal of the Washington Academy of Sciences* 11 (8): 173–89.

Satina, Sophie, and A. G. Avery. 1959. A review of the taxonomic history of *Datura*. In *Blakeslee: The genus Datura*, ed. Amos G. Avery, Sophie Satina, and Jacob Rietsema, 16–47. New York: The Ronald Press Co.

Symon, David E., and Laurence A. R. Haegi. 1991. *Datura* (Solanaceae) is a New World genus. In *Solanaceae III: Taxonomy, chemistry, evolution*, ed. Hawkes, Lester, Nee, and Estrada, 197–210. London: Royal Botanic Gardens Kew and Linnaean Society.

122 Some authors (e.g., Diaz) have identified *atl inan* as the plant *Rumex pulcher*. However, this plant is not known to produce any inebriating effects.

Desfontainia spinosa Ruíz et Pavón

Latuy

Family
Desfontainiaceae (only one genus); occasionally, the genus is assigned to the family Loganiaceae (Brako and Zarucchi 1993, 618*).

Forms and Subspecies
The variety *Desfontainia spinosa* Ruíz et Pav. var. *hookeri* (Dun.) Voss ex Vilmorin occurs in Chile (Emboden 1979, 176*). A small-leaved (Andean) form has been described as *Desfontainia spinosa* var. *parvifolia* (D. Don) Hooker (Brako and Zarucchi 1993, 618*).

Synonyms
Desfontainia obovata Kraenzlin
Desfontainia parvifolia D. Don
Desfontainia spinosa var. *hookeri* (Dun.) Reiche

Folk Names
Borrachera de páramo ("inebriator of the swamps"), chapico ("chili water"), desfontainia, intoxicator, latuy, latuye, mëchai, michai, michai blanco, michay, michay blanco, muerdago, taique, trau-trau (Mapuche, "unique"), trautrau[123]

History
Richard Evans Schultes discovered the psychoactive use of this beautiful plant in the Sibundoy Valley of Colombia in 1941 (Davis 1996, 173*). Unfortunately, the plant has been little studied or tested since that time.

Distribution
The bush occurs from Colombia (Sibundoy) to southern Chile (Chiloé), as well as in Ecuador and in the higher Andean regions of Argentina (Brako and Zarucchi 1993, 618*). In southern Chile, the bush is found from the Río Maule to Magallanes, most frequently south of Valdivia, and typically in the underwood of lenga and coigüe forests. It has also been observed in Costa Rica (Zander 1994, 230*).

Cultivation
In southern Chile, *Desfontainia* is a recommended garden ornamental (Donoso Zegers and Ramírez García 1994, 49*). Methods of cultivation are still unknown. It likely can be propagated from seed or, even more easily, from cuttings. The plant requires moist to very moist soil (swampy areas, marshes).

Appearance
Desfontainia spinosa is a small evergreen bush or shrublike tree that grows up to 2 to 3 meters in height. It has thick, thorny, mid- to dark green

leaves and large, funnel-shaped flowers that are orange-red with yellow margins. The leaves resemble those of the English holly (*Ilex aquifolium* L.; cf. **Ilex cassine**). The flowers are similar to those on some nightshades, e.g., **Iochroma fuchsioides**.

The plant is easily confused with several species of the genus *Berberis*, especially *Berberis darwinii* Hook. In Chile, many species of *Berberis* (*B. actinacantha* Mart., *B. chilensis* Gill. ex Hook, *B. darwinii*, *B. serrata*, *B. dentata*) are known as *michay* (Mapuche, "yellow tree"; Mösbach 1992, 78*) and are used as sources of yellow dye (Donoso Zegers and Ramírez García 1994*). *Berberis* fruits are used to make **chicha**.

Psychoactive Material
— Leaves
— Fruits

Preparation and Dosage
The leaves can be brewed or decocted into a hallucinogenic tea. The fruits are considered to be more effective and presumably are prepared as a decoction. No information concerning dosages is available. The fruits may once have been used to prepare a potently psychoactive **chicha**.

Ritual Use
The shamans of the Kamsá in the Sibundoy Valley of Colombia drink a tea made from the leaves when they "want do dream" or receive visions to diagnose diseases (Schultes 1977, 100).

The *machis* (shamans) of the Mapuche appear to use the plant in the same manner as they use **Latua pubiflora**. This use, however, requires further research.

Artifacts
The folklore of Chiloé (an island in southern Chile) speaks of a mythical figure named El Trauco who may originally have been a plant spirit

Left: The evergreen herbage of the Chilean shamanic plant *Desfontainia spinosa*, shown in blossom

Right: El Trauco, a satyrlike forest spirit, is a popular figure in the mythology of Chiloé. It is possible that he is a representation of the plant spirit of *Desfontainia spinosa*, which is also known as *trau-trau*. (Section of a statue in Ancud, Chiloé, southern Chile)

123 This name is also used to refer to a fructiferous plant (*Ugni candollei* [Barn.] Berg), the berries of which are used to make chicha (Mösbach 1992, 95*).

of *Desfontainia*, known locally as *trau-trau*. El Trauco is a small, perverse man, a "satyr of the forest." He has a stone ax that he uses to cut down trees, and he looks like a mushroom spirit. There is a large statue of El Trauco in Ancud (Chiloé), and small replicas carved from stone are sold as souvenirs.

The Mapuche of southern Chile use the leaves to obtain a yellow dye that they use to color wool and the material they use for their traditional garments (Mösbach 1992, 101*).

Medicinal Use

The leaves are used in Chile as a folk remedy for upset stomachs. One older Chilean book about medicinal plants surprisingly makes reference to *Desfontainia* but states that it has no medicinal use (Urquieta Santander 1953, 87).

Constituents

No constituents have been identified to date (McKenna 1995, 100*). A Dragendorff test for alkaloids yielded negative results (Schultes 1977, 100).

In southern Chile, the plant is regarded as poisonous (Mösbach 1992, 101*). However, no toxic component has yet been identified. Recently, Rob Montgomery and I collected information in Chiloé (May 1995) that indicates that the plant is well known among indigenous plant specialists, who regard it as nonpoisonous but hallucinogenic.

Effects

Smoking two dried leaves produced clear psychoactive effects with perceptual changes (flickering lights, feelings of being "high").

Commercial Forms and Regulations

None

Literature

Schultes, Richard Evans. 1977. De Plantis Toxicariis e Mundo Novo Tropicale Commentationes XV: *Desfontainia*: A new Andean hallucinogen. *Botanical Museum Leaflets* 25 (3): 99–104.

Urquieta Santander, Carlos. 1953. *Diccionario de medicacitsis e Mund*. 5th ed. Santiago de Chile: Editorial Nascimento.

Diplopterys cabrerana (Cuatrecasas) B. Gates

Yahé Vine

"Like *Banisteriopsis caapi*, *Diplopterys cabrerana* grows in Amazonian lowlands, and the plant has been collected only in southern Colombia and Venezuela, eastern Ecuador, northern Perú and western Brazil. Like *B. caapi*, *D. cabrerana* rarely flowers, and is normally cultivated by shamans for use in *ayahuasca*. Both plants are commonly propagated by cuttings."

JONATHAN OTT
AYAHUASCA ANALOGUES
(1994, 24)

Family

Malpighiaceae (Barbados Cherry Family)

Forms and Subspecies

None

Synonyms

Banisteria rusbyana Niedenzu
Banisteriopsis cabrerana Cuatrecasas
Banisteriopsis rusbyana (Niedenzu) Morton
Banisteriopsis rusbyana sensu ethnobotanical, non (Niedenzu) Morton

The literature also contains the spelling *Diplopteris*.

Folk Names

Biaxíi, chagropanga, chagropanga azul pisco, chagrupanga (Inga, "*chagru* leaf"), chakruna, ka-hee-ko (Karapaná), kahi (Tukano, "that which causes vomiting"), kamárampi (Campa, "vomit"), mené kahi ma, mené kahima, nyoko-buko guda hubea ma (Barasana), nyoko-buku guda hubea ma, oco-yagé ("water yagé"), oco yáge, yaco-ayahuasco (Quechua/Peru), yagé, yage-oco, yageúco, yagéúco, yahéliane, yahé 'oko (Siona-Secoya, "*Banisteriopsis* water"), yahé-oko (Kofán), yahé vine, yajé, yajé oko, yaji, yají

History

This vine was first named *Banisteria rusbyana* in honor of Henry Hurd Rusby (1855–1940), one of the pioneers of ethnobotany (the name unfortunately fell victim to the synonym). Rusby was one of the first white people to witness an **ayahuasca** ceremony, which he actually filmed. He also was one of the first druggists and botanists to intensively investigate coca (**Erythroxylum coca**), guaraná (**Paullinia cupana**), and **Fabiana imbricata** (Rossi-Wilcox 1993*).

The confusion surrounding the botanical identity of the plant was not clarified until 1982 (Gates 1982, 214).

Distribution

This tropical vine is found only in the Amazon basin (Ecuador, Peru, Brazil, Colombia). It grows wild in the forests but is most often found in cultivation.

Cultivation
The plant is cultivated in house gardens using cuttings. A young shoot or the tip of a branch is allowed to sit in water until it develops roots; it can also be placed directly into the moist jungle soil.

Appearance
This very long vine has opposite leaves that are oblong-oval and retuse-attenuate in shape. The inflorescences, each of which bears four tiny flowers, grow from the petiolar axils. However, the plant only rarely develops flowers, and almost never under cultivation.

The closely related species *Mezia includens* (Niedenzu) Gates [syn. *Diplopterys involuta* (Turcz.) Niedenzu] is known in Peru as *ayahuasca negro*. It is possible that this species was once also used for psychoactive purposes (Schultes 1983b, 353*). The very similar species *Diplopterys mexicana* B. Gates is common in Mexico (Gates 1982, 215).

Diplopterys cabrerana is easily confused with **Banisteriopsis caapi**. The two species are most easily distinguished on the basis of their leaves. Those of *Diplopterys* are distinctly wider and larger in size.

Psychoactive Material
— Fresh or dried leaves

Preparation and Dosage
The Desana, Barasana, and other Indians in the Colombian regions of the Amazon use the leaves of this vine (which is closely related to *Banisteriopsis*) to make **ayahuasca** (Bristol 1965, 211*; Reichel-Dolmatoff 1979a, 35*). In the Colombian Sibundoy, an inebriating beverage known as *biaxíi* is boiled from **Banisteriopsis caapi** and the leaves of *Diplopterys cabrerana* (see **ayahuasca**).

The Shuar use the leaves as an ayahuasca additive (Bennett 1992*), as do the Siona-Secoya (Vickers and Plowman 1984, 19*) and the Mocoa Indians of Colombia. Unfortunately, the sources do not provide any precise information about the quantities of leaves to use (cf. Bristol 1966).

Ritual Use
The Barasana of the lower Piraparaná use the stems to make a hallucinogenic drink that they call *yagé* and use in the same manner as **ayahuasca** (Schultes 1977b, 116*). Apart from this, the primary use of the leaves is as a source of *N,N*-DMT as an ayahuasca additive (Der Marderosian et al. 1968; Gates 1982).

Artifacts
See **ayahuasca**.

Medicinal Use
None, except when used medicinally in **ayahuasca**.

Constituents
The leaves contain 0.17 to 1.75% *N,N*-DMT (Agurell et al. 1968; Der Marderosian et al. 1968; Poisson 1965). In addition to the main alkaloid, DMT, they also contain *N*-methyltryptamine, **5-MeO-DMT**, **bufotenine**, and *N*-methyltetrahydro-β-carboline (cf. **β-carbolines**). The main alkaloid in the stems is *N,N*-DMT; also present are 5-MeO-DMT and *N*-methyltetrahydro-β-carboline (Pinkley 1973, 185*).

Effects
See *Psychotria viridis* and **ayahuasca**.

Commercial Forms and Regulations
In the Colombian Sibudoy region, Indians and shamans trade in finished preparations of the plant (Bristol 1966, 123). Apart from the fact that the legal situation regarding plants and products that contain DMT is unclear, the plant is not subject to any restrictions.

Literature
See also the entries for **Banisteriopsis caapi**, **Banisteriopsis spp.**, and **ayahuasca**.

Agurell, S., B. Holmstedt, and J. E. Lindgren. 1968. Alkaloid content of *Banisteriopsis rusbyana*. *American Journal of Pharmacy* 140:148–51.

Bristol, Melvin L. 1966. The psychotropic *Banisteriopsis* among the Sibundoy of Colombia. *Botanical Museum Leaflets* 21 (5): 113–40. (Primarily discusses *Banisteriopsis rusbyana* = *Diplopterys cabrerana*.)

Cuatrecasas, José. 1965. *Banisteriopsis caapi, B. inebrians, B. rusbyana*. *Journal d'Agriculture Tropicale et de Botanique Appliqueé* 12:424–29.

Der Marderosian, Ara H., K. M. Kensinger, J. Chao, and F. J. Goldstein. 1970. The use and hallucinatory principles of a psychoactive beverage of the Cashinahua tribe (Amazon basin). *Drug Dependence* 5:7–14.

Der Marderosian, A. H., H. V. Pinkley, and M. F. Dobbins IV. 1968. Native use and occurrence of *N,N*-dimethyltryptamine in leaves of *Banisteriopsis rusbyana*. *The American Journal of Pharmacy* 140:137–47.

Gates, Bronwen. 1982. A monograph of *Banisteriopsis* and *Diplopterys*, Malpighiaceae. *Flora Neotropica* no. 30. (A publication of the Organization for Flora Neotropica.)

Poisson, J. 1965. Note sur le "Natem," boisson toxique péruvienne et ses alcaloïdes. *Annales Phärmaceutique Françaises* 23 (4): 241–44.

A South American ethnobotanist with the *Diplopterys cabrerana* vine. (Photograph: Bret Blosser)

DMT

5-MeO-DMT

Bufotenine

"Amazingly, the spirit of the *Diplopterys cabrerana* plant spoke English. When I asked him about his nature, I was answered: *Power and Beauty*."

Duboisia hopwoodii F. v. Mueller

Pituri Bush

The Australian pituri tree (*Duboisia hopwoodii*) in bloom. The tiny inflorescences are almost invisible. (Photographed in North Queensland)

Family
Solanaceae (Nightshade Family); Subfamily Cestroideae, Anthocercideae/Salpiglossideae Tribe

Forms and Subspecies
None

Synonyms
Anthoceris hopwoodii
Duboisia piturie Bancroft

Folk Names
Bedgerie, bedgery, camel poison, emu plant, pedgery, petcherie, picherie, pitchery, pitchiri, pitjuri, pitschuri, pituri,[124] pituribaum, pituribusch, pituri bush, pituristrauch, pizuri, poison bush

History
It is possible that the Aborigines have been using the psychoactive pituri bush for hedonistic and ritual purposes since the settlement of Australia. The plant and its dried, fermented leaves were a valuable article of trade and played an important role in the indigenous economy.

The plant was first described in 1878 by the great German/Australian botanist Ferdinand J. H. von Mueller (1825–1896), who also recognized it as the source of **pituri** (Hartwich 1911, 518*). In 1879, an alkaloid was isolated and was named piturine. Only in recent decades has the plant become the focus of more detailed studies.

Distribution
Duboisia hopwoodii is found primarily in the Australian interior. The plant is not found either in the Victoria Desert or in Tasmania (Barnard 1952, 5).

Cultivation
This and other *Duboisia* species are propagated by seeds or cuttings from the branch ends (Barnard 1952).

Appearance
This branched, evergreen shrub has a woody stem and can grow as tall as 2.5 to 3 meters. Its wood is yellow and has a noticeably vanilla-like scent. The green leaves are lineal/lanceolate (12 to 15 cm long, 8 mm wide), entire, and tapered at the petioles. The white, sometimes pink-spotted flowers are campanulate (up to 7 mm long) and occur in clusters at the branch ends. Flowering occurs between January and August. The fruit is a black berry (6 mm long) containing numerous tiny seeds.

Duboisia hopwoodii is easily confused with other *Duboisia* species and also can be mistaken with *Anthoceris* spp. (Solanaceae).

Psychoactive Material
— Leaves

Preparation and Dosage
The leaves are collected in August when the plant is flowering and are hung up to dry or roasted over a fire. They are either chewed as quids (cf. **pituri**) or rolled together with alkaline substances into cigars for smoking: "The Australian Aborigines sometimes smoke moistened pituri leaves mixed with plant potash" (Stark 1984, 98*).

The pituri quids consist of chopped *Duboisia hopwoodii* leaves mixed with acacia leaves (cf. **Acacia** spp.), "small, dried berries, and unopened flower buds in the form of a caper" (Maiden 1888, 370). *Duboisia hopwoodii* leaves also can be chewed by themselves, although the effects are not considered particularly strong. It is said that the addition of the plant ashes is what brings out the full stimulating effects.

Ritual Use
See **pituri**.

Artifacts
The rock art—paintings, spray paintings, and rock carvings—of the Aborigines can be traced back to the earliest times. Spiritlike *wondjinas*, Dreamtime animals, magical totems, "x-rays," and visions of the Milky Way are all among the earliest works of Aboriginal art. The abstract paintings found on bark bast (in Arnhem Land) appear to be very old.

The semi-abstract art of the rain forest peoples in the region of Cairns (North Queensland) also has a long tradition. They used natural pigments to paint especially their war shields: "All of the images represented food for daily use and medicines or antidotes that are obtained from a wide variety of trees. Each one of these drawings is associated with a story with a certain meaning" (Hollingsworth 1993, 115). It is certain that pituri and corkwood (*Duboisia* spp.) were depicted in this manner, for the rain forest peoples used them as inebriants, medicines, and fish poisons.

In the nineteenth century, some Aborigines adapted European painting techniques and began to orient themselves around European art. It was only in the mid-twentieth century that a contemporary Aboriginal style of its own developed, borrowing from the ritual sand paintings of the "outback" tribes. These contemporary Aboriginal

124 The non-Aboriginal inhabitants of Australia now refer to all chewing tobacco (*Nicotiana* spp.) as *pituri* (Peterson 1979, 178*).

paintings appear extremely psychedelic to the eyes of many Western observers. They usually portray Dreamings, and also often primordial beings associated with the clan of the painter or his family members. Many of these paintings tell the myths of the Dreamtime and show the Dream paths or "song lines" of the ancestors. They appear to show what the Dream soul of the painter sees as it flies over the Dream land. They are cartographies or topographies of the Dreamtime. The artist, e.g., Clifford Possum Tjapaltjarri (sometimes called "the Van Gogh of Aboriginal art"), sees himself as a "cartographer of the Dreaming" (Johnson 1994, 47). The Aboriginal art not only records the effects of the Dreamtime on our contemporary world but also creates them anew with every painting and every piece. The art is the reality of the Dreaming. Colin McCormick has said that some of the Aboriginal painters creating this type of art are inspired by pituri inebriation.

Sometimes Dreamings of particular plants are depicted. Some plants appear as totems or ancestor spirits, while others appear symbolically, often simply as individual points in the Dreamings. Theoretically, all plants, including pituri, are totems and can be represented. Clifford Possum has produced a painting, *Corkwood Dreaming* (1982), that has a very psychedelic appearance (Johnson 1994, ill. 34, pp. 94, 95, 165). It is the Dream of his mother. Since the term *corkwood* can refer to many different plants, and not just to species of *Duboisia* (e.g., also *Hakea* spp.[125]), it is uncertain whether this painting is a secret "pituri Dreaming."

Plants are often simply suggested by lines, points, and dabs of color. In his autobiographical novel *Songlines*, Bruce Chatwin describes how an Aborigine painted "pitjuri" ("Pitjuri is a mild narcotic which Aboriginals chew to suppress hunger") as a "squiggle" in the center of a painting (Chatwin 1988, 260). It is only through the artist's explanation that one can decipher the meaning of the squiggle.

Medicinal Use

Pituri is now regarded as a bush medicine, a wild medicinal plant that "bushwalkers" use as an analgesic (Cherikoff 1993, 171*; Lassak and McCarthy 1992, 33*).

Constituents

Duboisia hopwoodii contains various potently stimulating as well as toxic alkaloids: piturine (possibly identical to nicotine), duboisin, D-nor-nicotine, and **nicotine** (Hicks and LeMessurier 1935). The presence of nicotine is contested but possible (Peterson 1979, 178*). D-nor-nicotine is regarded as the primary constituent (Barnard 1952, 12; Bottomley et al. 1945). Dried leaves can contain between 2.4 and 5% nicotine/nornicotine.

Gas chromatography has also demonstrated the presence of myosmine, *N*-formylnornicotine, cotinine, *N*-acetylnornicotine, anabasine, anatabine, anatalline, and bipyridyl (Luanratana and Griffin 1982).

The root has been found to contain the hallucinogenic **tropane alkaloid** hyoscyamine (Kennedy 1971). Traces of **scopolamine**, **nicotine**, nornicotine, metanicotine, myosmine, and *N*-formylnornicotine have also been detected (Luanratana and Griffin 1982).

Effects

Carl Lumholz compared the effects of *Duboisia hopwoodii* to those of tobacco (*Nicotiana tabacum*) and opium (*Papaver somniferum*) (1889, 49). Carl Hartwich, who studied *Duboisia hopwoodii* intensively, wrote that the plant's effects "are inebriating, it invokes passionate dreams. It also takes away . . . feelings of hunger and thirst" (Hartwich 1911, 834*). These effects reminded him of the effects of coca (*Eythroxylum coca*).

When the leaves are smoked alone, they produce an effect similar to that of marijuana (see *Cannabis indica*). There have also been reports of "invigorating, mildly psychedelic and erotic properties of the plant" (Stark 1984, 98*).

Commercial Forms and Regulations

None

Literature

See also the entries for *Duboisia* spp., *Nicotiana* spp., *Goodenia* spp., and **pituri**.

Barnard, Colin. 1952. The Duboisias of Australia. *Economic Botany* 6:3–17.

Bottomley, W., R. A. Nolte, and D. E. White. 1945. The alkaloids of *Duboisia hopwoodii*. *Australian Journal of Science* 8:18–19.

Chatwin, Bruce. 1988. *The songlines.* New York: Penguin.

Hicks, C. S., and H. LeMessurier. 1935. Preliminary observations on the chemistry and pharmacology of the alkaloids of *D. hopwoodii*. *Australian Journal of Experimental Biology and Medical Science* 13:175–78.

Hollingsworth, Mark. 1993. Die Cape-York-Halbinsel und Nord-Queensland. In *Aratjara: Kunst der ersten Australier* (exhibition catalogue), 109–15. Cologne: DuMont.

Johnson, Vivien. 1994. *The art of Clifford Possum Tjapaltjarri.* East Roseville, New South Wales: Craftsman House (Gordon and Breach Arts International).

Kennedy, G. S. 1971. (–)-Hyoscyamine in *Duboisia hopwoodii*. *Phytochemistry* 10:1335–39.

Luanratana, O., and W. J. Griffin. 1982. Alkaloids of *Duboisia hopwoodii*. *Phytochemistry* 21:449–51.

"Pituri is an entryway to the Dreamtime."

Colin McCormick
(1994)

"In order to understand the Australian Aborigines, in order to recognize the illuminating power that radiates from the multidimensional depths of their paintings—which the Western observer usually regards as childlike and primitive—you must also know about *pituri* (*Duboisia hopwoodii*). With this nightshade, a person can find an entryway into the primordial 'Dreamtime,' the place of the totem animals, the original images that eternally exist."

Wolf-Dieter Storl
Von Heilgöttern und Pflanzengottheiten
[On Healing Gods and Plant Divinities]
(1993, 341*)

Nicotine

Nornicotine

125 There are more than 130 species of *Hakea*, many of which the Aborgines use as food, medicines, and the basis for a mildly alcoholic drink (Low 1992b, 184 f.).

Lumholz, Carl. 1889. *Among cannibals.* London: John Murray.

Maiden, Joseph Henry. 1888. Some reputed medicinal plants of New South Wales.

Proceedings (Linnean Society of New South Wales), 2nd ser., 3 (24): 367–71.

Senft, Em. 1911. Über *Duboisia hopwoodii. Pharm. Praxis* 1.

Duboisia spp. and Hybrids
Corkwood Trees

"The Australian Aborigines bored holes into the trunk of the corkwood tree, which they then poured water or another liquid into and allowed to stand overnight. The next morning, they drank the juice, which was so potent that it produced an inebriated state with noticeably erotic sensations."

RAYMOND STARK

APHRODISIAKA [APHRODISIACS] (1984, 76*)

Botanical illustration of *Duboisia myoporoides.* (From *Köhler's Medizinalpflanzen,* 1887/89)

Family

Solanaceae (Nightshade Family); Subfamily Cestroideae, Anthocercideae/Salpiglossideae Tribe

The genus *Duboisia* is composed of just three species (D'Arcy 1991, 78*), all of which are native to Australia and two of which are endemic (Haegi 1979). The genus is closely related to the genera *Anthoceris, Anthotroche* (both endemic to Australia), and *Brunfelsia* (cf. *Duboisia hopwoodii*). In Australian English, all of these are known by the name *corkwood* because of their corklike bark (Dowling and McKenzie 1993, 151 f.*).

The constituents of *Duboisia* appear to be extremely variable with regard to the concentration, distribution, and mixture ratio, and this has produced considerable confusion in the phytochemical literature. The alkaloid content depends upon a number of factors: location of the plant, time of collection, and the existence of chemical races and hybrids (Dowling and McKenzie 1993, 153*). In addition to **nicotine** and nicotine derivatives, all species of *Duboisia* have been found to contain the following **tropane alkaloids**: hyoscine (= **scopolamine**), hyoscyamine, norhyoscyamine, tigloidine, valeroidine, poroidine, isoporoidine, butropine, valtropine, 3α-tigloyloxytropane, 3α-acetoxytropane, norhyoscine, apohyoscine, tropine (= tetramethylputrescine), 6-hydroxyhyoscyamine. Most of these tropanes also occur in the genus *Datura.* Today, corkwoods (*Duboisia* spp.) are used by the international pharmaceutical industry to manufacture agents for the treatment of travel sickness (Lewington 1990, 149*).

Various (?) *Duboisia* species are also found on Papua New Guinea. The Papuas smoke and even chew these along with both indigenous and introduced tobacco species (*Nicotiana suaveolens, N. fragrans,* **Nicotiana tabacum, Nicotiana spp.**). The Papuas discovered the *Duboisia* leaves and their effects partially on their own, while their use and even the plant material was introduced into New Guinea via trade relationships with the Torres Strait Islands.

Duboisia leichhardtii F. Muell.—Leichhardt's corkwood

This bushlike tree, which has a straight trunk and can grow as tall as 7.5 meters, is the least known of all *Duboisia* species. The 0.5 to 1.5 cm long flowers are the most distinct feature of this species; the petals are long, slender, and tapered at the ends. This corkwood species thrives only in clayey and sandy soils. Its natural range is restricted to central and western Queensland and western New South Wales (Dowling and McKenzie 1993, 152*; Morton 1977, 299*). The concentration of **tropane alkaloids** contained in the species is quite high and can be increased through breeding and hybridization (Luanratana and Griffin 1980a). The dried leaves contain approximately 1.4% alkaloids, primarily **scopolamine** (Morton 1977, 299*). It is not yet known whether this species was or is used by the Aborigines.

Duboisia myoporoides R. Br. [syn. *Natalaea ligustrina* Sib.]—corkwood, onungunabie, ngmoo

This evergreen, shrublike tree can grow as tall as 15 meters. It has lanceolate leaves 10 cm long and 3 cm wide. The small flowers are white and have five pinna. The fruits are 0.5 cm long, oval, and greenish yellow in color. They turn black when ripe. Both the flowers and the fruits develop in July (the winter or rainy period). Corkwood is a typical rain forest tree of the Australian east coast (Pearson 1992, 95*). It grows in clayey and sandy soils, and sometimes even on sandy beaches near the coast (Dowling and McKenzie 1993, 152 ff.*).

The leaves of this *Duboisia* species, which is also known as eye-plant or elm, are used as an alternative for **Duboisia hopwoodii** as **pituri** or as a pituri substitute. The Aborigines obtained a "stupefying drink" from corkwood (Cribb and Cribb 1984, 222*) and used the alkaloids in other ways as well (Pearson 1992, 95*; Stark 1984, 76*).

At the beginning of the twentieth century, the Australian raw drug was used as a substitute for *Atropa belladonna.* The preparation Duboisia (an essence of the fresh leaves of *D. myoporoides*) is an important homeopathic agent (Schneider 1974, II:44*). The Aborigines of New Caledonia use the fresh leaves as an antidote for ciguatera poisoning (Bourdy et al. 1992; Dufva et al. 1976; Ott 1993, 376*).

It has long been known that *Duboisia myoporoides* contains large amounts of scopolamine (Emboden 1979, 146*); in fact, this species is now grown commercially as a source of **scopolamine** for the pharmaceutical industry (Morton 1977, 294*). The alkaloids **nicotine**, nornicotine, **atropine**, and **scopolamine** have been detected in all parts of the plant. The main alkaloids in the leaves are scopolamine and hyoscyamine (Cougoul et al. 1979). Approximately two mouthfuls of leaves contains 50 mg of nicotine and 20 mg of scopolamine (Ott 1993, 376*). Also present are tropine, 3α-acetoxytropane, α-alkylpiperidine alkaloids (e.g., pelletierine), and myrtine. The roots contain a quinolizidine alkaloid as well as β-**phenethylamine** derivatives (Bachmann et al. 1989). Providing root cultures or the plant with certain tropane precursors, e.g., putrescine, ornithine, arginine, and tropine, has been found to substantially increase the biosynthesis of scopolamine (Yoshioka et al. 1989).

The Australian mistletoe (**Benthamia alyxifolia**), which lives as a parasite on *Duboisia myoporoides*, apparently accumulates scopolamine in its leaves as a result.

Duboisia myoporoides R. Br. x *Duboisia leichhardtii* F. Muell.—hybrid corkwood

In Australia, the two treelike *Duboisia* species have been bred to produce a hybrid that is grown on large plantations as a source of alkaloids. This hybrid has been found to be especially rich in tropane alkaloids and therefore useful for commercial cultivation (Luanratana and Griffin 1980a, 1980b). The hybrid has the advantage of being almost devoid of nicotine and sometimes developing scopolamine concentrations as high as 3% (Morton 1977, 301*). Several methods for influencing and increasing the alkaloid content have been discovered and developed (Luanratana and Griffin 1982). The hybrid (presumably) plays no part in Aboriginal ethnobotany.

Literature

See also the entries for **Duboisia hopwoodii**, **pituri**, and **scopolamine**.

Bachmann, P., L. Witte, and F.-C. Czygan. 1989. The occurrence of β-phenethylamine derivatives in suspension culture of *Duboisia myoporoides*. *Planta Medica* 55:231.

Bourdy, G., et al. 1992. Traditional remedies used in the Western Pacific for the treatment of ciguatera poisoning. *Journal of Ethnopharmacology* 36 (2): 163–74.

Cougoul, N., E. Miginiac, and L. Cosson. 1979. Un gradient métabolique: Rapport Scopolamine/Hyoscyamine dans les feuilles du *Duboisia myoporoides* en fonction de leur niveau d'Insertion et du stade de croissance. *Phytochemistry* 18:949–51.

Dufva, E., et al. 1976. *Duboisia myoporoides:* Native antidote against ciguatera poisoning. *Toxicon* 14:55–64

Griffin, W. J., H. P. Brand, and J. G. Dare. 1975. Analysis of *Duboisia myoporoides* R. Br. and *Duboisia leichhardtii* F. Muell. *Journal of Pharmaceutical Science* 64 (11): 1821–25.

Haegi, L. 1979. Australian genera of the Solanaceae. In *The biology and taxonomy of the Solanaceae*, ed. J. G. Hawkes et al., 121–24. London: Academic Press.

Luanratana, O., and W. J. Griffin. 1980a. Cultivation of a *Duboisia* hybrid. Part A. Nutritional requirements and effects of growth regulators on alkaloid content. *Journal of Natural Products* 43 (5): 546–51.

———. 1980b. Cultivation of a *Duboisia* hybrid. Part B. Alkaloid variation in a commercial plantation. *Journal of Natural Products* 43 (5): 552–58.

———. 1982. The effect of a seaweed extract on the alkaloid variation in a commercial plantation of a *Duboisia* hybrid. *Journal of Natural Products* 45 (3): 270–71.

Yoshioka, Toshiro, Hikaru Yamagata, Aya Ithoh, Hiroshi Deno, Yasuhiro Fujita, and Yasuguki Yamada. 1989. Effects of exogenous polyamines on tropane alkaloid production by a root culture of *Duboisia myoporoides*. *Planta Medica* 55:523–24.

Because of the corklike bark of *Duboisia myoporoides*, this nightshade has been given the name *corkwood*. (Photographed in North Queensland)

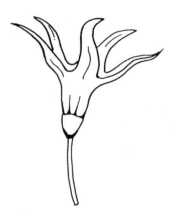

Flower of *Duboisia leichhardtii* F. v. Muell. (Drawing by C. Rätsch)

Echinopsis spp.

Sea Urchin Cactus

According to Zander (1994, 249*), *Echinopsis* Zucc. is now the taxonomically valid genus name for *Trichocereus* spp. The genera *Chamaecereus*, *Lobivia*, *Setiechinopsis*, and *Soehrensia* were all incorporated into the genus *Echinopsis*. Yet because the traditionally accepted genus name *Trichocereus* is so widely used worldwide—in the botanical as well as the ethnobotanical and ethnopharmacological literature, and in academic botanical gardens—in this encyclopedia I will continue to use the established genus name *Trichocereus*. It is questionable whether the inclusion of *Trichocerus* in the genus *Echinopsis* can be botanically justified and is sensible. For example, the most recently revised edition of *Flora del Ecuador* lists the San Pedro cactus under the botanical name *Echinopsis* (*Trichocereus*) *pachanoi* (Patzelt 1996, 108).

Literature

Patzelt, Erwin. 1996. *Flora del Ecuador*. 2nd ed. Quito: Banco Central del Ecuador.

Ephedra gerardiana Wallich ex Stapf

Somalata

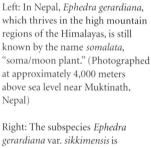

Left: In Nepal, *Ephedra gerardiana*, which thrives in the high mountain regions of the Himalayas, is still known by the name *somalata*, "soma/moon plant." (Photographed at approximately 4,000 meters above sea level near Muktinath, Nepal)

Right: The subspecies *Ephedra gerardiana* var. *sikkimensis* is prevalent in Sikkim.

Family

Ephedraceae (Ephedra Family); Monospermae Section

Forms and Subspecies

The following varieties are currently accepted:

Ephedra gerardiana Wall. var. *gerardiana*
Ephedra gerardiana Wall. var. *saxatilis*—tsafad
Ephedra gerardiana Wall. var. *sikkimensis* Stapf

Synonyms

Ephedra saxatilis Royle var. *sikkimensis* (Stapf) Flories
Ephedra vulgaris Rich.

Folk Names

Amsania, asmânia, asmani-booti, budagur, bûd-sûr, bûtsûr, chefrat, cheldumb, chewa, ehewa, khanda, khanda-phog, khanna, ma houng (Tibetan), ma-huang (Chinese), mtshe (Tibetan),

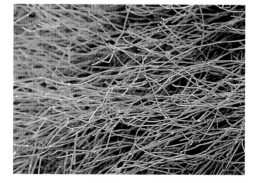

narom (Pakistani), oman (Pashto), phok, raci, sang kaba (Sherpa, "*kaba* incense"), sikkim ephedra, soma, somalata (Sanskrit, "moon plant"), somlata (Nepali), thayon (Ladakhi), tootagantha (Hindi), trano, tsafad, tsapatt-tsems, tse, tseh (Tamang), tutgantha (Hindi), uman (Pashto), uroman

History

This plant must already have been known in the Vedic or post-Vedic period, for it was used as a **soma** substitute (cf. also **haoma**). It was botanically described in the eighteenth century. The species name refers to John Gerard, the English herbal author, who published one of the earliest precise descriptions and illustrations of *Ephedra* under the name *Vua marina* (Gerard 1633, 1117).

Distribution

The species is found in the Himalayas (from Afghanistan to Bhutan) in altitudes between 2,400 and 5,600 meters (Navchoo and Buth 1989, 143*). It prefers drier Alpine regions and high mountain deserts (with less than 50 cm of precipitation annually). In Nepal, somalata is common in Langtang and the Mustang district. In Sikkim, the variety *sikkimensis* is most common.

In Nepal, this high mountain *Ephedra* species is most frequently encountered at altitudes between 3,000 and 4,000 meters, often in association with **Juniperus recurva** and **Rhododendron** spp. (Malla 1976, 34). In the high mountains, *Ephedra* herbage is an important source of nourishment for yaks and goats during the winter (Polunin and Stainton

1985, 384). It is likely that these animals also eat the plant as a stimulant.

Cultivation
The plant can be grown from seed. It requires a humus-poor, rocky soil; it can survive with very little water and is able to grow in very dry locations. This robust plant even thrives in soils that contain salt, such as in the neighborhood of salt lakes (Hemsley and Rockhill 1973, 18).

Appearance
Somalata is a perennial herbaceous plant that has practically no leaves and consists solely of fibrous, segmented stalks (in older specimens, some fifteen segments). The small and inconspicuous yellow flowers grow directly from the stalks at the segments. The small, round red fruits (6 mm in diameter) ripen in autumn (August to September). The fruits are edible. The herbage typically does not grow taller than 20 cm, although it can attain an overall height of 60 cm (Morton 1977, 33*).

As with all other species of *Ephedra*, this species is very easily confused with its relatives.

Psychoactive Material
— Dried stalks; these are collected during the monsoon season (July) when they are flowering, as the alkaloid content is greatest then (Manandhar 1980, 35*).

Preparation and Dosage
The dried herbage (stalks) is boiled in water for approximately ten minutes. Six grams of dried herbage is regarded as a medicinally efficacious individual dosage. Dosages as high as 20 g may be used for euphoriant purposes.

In the Himalayas, ashes of the plant are said to be used as **snuff** (von Reis and Lipp 1982, 6*).

Ritual Use
In the post-Vedic period, when the Aryans were no longer able to find the original psychedelic soma plant in the Indus Valley and the knowledge of the plant was being kept secret or lost, the sacred soma drink (which corresponds to the Persian **haoma**) was prepared with substitute plants, which included *Ephedra* sp. (cf. **soma**). It is for this reason that this Himalayan *Ephedra* species is still known by the name *somalata*, "plant of the moon" (Singh et al. 1979, 189*). While the effects of the plant are stimulating, they are not visionary. The closely related and very similar or synonymous Himalayan species *Ephedra saxatilis* Stapf is also known as somalata.

In contrast to the Tibetans, the Nepali Tamang cremate their dead. The cremations are carried out on small chortens (Lamaist shrines) that are erected outside of the villages specifically for this purpose. Dried bundles of *Ephedra* herbage are used as **incense** during the cremation ceremonies. The smoke has a surprisingly pleasant, fine, slightly spicy fragrance somewhat reminiscent of the scent of a forest fire.

Artifacts
A Pakistani Gandhara sculpture (first to sixth century C.E.) in the Archaeological Museum of Peshawar (India) depicts the Buddha as an herbalist. Farmers are shown offering him bundles of stalks which Mahdihassan (1963 and 1991) has interpreted as those of *Ephedra gerardiana*.

Medicinal Use
In Ayurvedic medicine, an *Ephedra* tea (6 g per dosage) is used to treat colds, coughs, wheezing, bronchitis, asthma, arthritis, and dropsy. To avoid undesirable side effects (such as tachycardia), licorice (*Glycyrrhiza glabra* L.) can be added to the *Ephedra* tea.

In Nepali folk medicine, the herbage is used as a tonic for asthma, hay fever, and diseases of the respiratory tract (Manandhar 1980, 35*; Singh et al. 1979, 189*). In Tibetan medicine, *Ephedra* is regarded as a rejuvenant. In Ladakh, the powdered plant is ingested along with water as an expectorant and to treat "blood diseases" (Navchoo and Buth 1989, 144*).

Constituents
The herbage contains 0.8 to 1.4% alkaloids, of which 50% is **ephedrine** and 50% is composed of other alkaloids, such as pseudoephedrine (Manandhar 1980, 35*; Morton 1977, 34*). Bitter and tanning agents are also present. Fertilizing the plant with amino acids increases the biosynthesis of ephedrine (Ramawat and Arya 1979).

Effects
A decoction of *Ephedra gerardiana* elevates blood pressure; constricts blood vessels; has diuretic, stimulating (natural stimulant), and euphoriant effects; and causes allergic symptoms (hay fever, asthma) to disappear. The effects last six to eight hours.

Commercial Forms and Regulations
Herbage of *Ephedra gerardiana* that is sold outside of the Himalayan region is subject to the same regulations as other *Ephedra* species (see *Ephedra sinica*).

Literature
See also the entries for *Ephedra sinica*, *Ephedra* **spp.**, and **ephedrine**.

Hemsley, W. Botting, and W. Woodville Rockhill. 1973. *Two small collections of dried plants from Tibet.* New Delhi: Pama Primlane (The Chronica Botanica).

"Today Ephedra is the vehicle of Ephedrine. Formerly it was the carrier of a large but fixed quantum of soul."

S. Mahdihassan
Indian Alchemy or Rasayana
(1991, 100)

Woodcut of a large *Ephedra* species, from the herbal of John Gerard (1633)

Mahdihassan, S. 1963. Identifying soma as ephedra. *Pakistan Journal of Forestry* (October):370ff.

———. 1991. *Indian alchemy or rasayana.* Delhi: Motilal Banarsidass Publ.

Quazilbash, N. N. 1948. Some observations on Indian ephedra. *Quarterly Journal of Pharmacy and Pharmacology* 21:502ff.

Ramawat, Kishan Gopal, and Harish Chandra Arya. 1979. Effect of amino acids on ephedrine

production in *Ephedra gerardiana* callus culture. *Phytochemistry* 18:484–85.

Rätsch, Christian. 1995. Mahuang, die Pflanze des Mondes. *Dao* 4/95:68.

Stein, Sir A. 1932. On ephedra, the hum plant and soma. *Btn. School of Oriental Studies London Institution* 6:501ff.

Ephedra sinica Stapf
Ma-huang

Family
Ephedraceae (Ephedra Family); Ephedra Section

Forms and Subspecies
None

Synonyms
None

Folk Names
Ask-for-trouble,[126] Chinese ephedra, Chinese joint fir, chinesisches meerträubel, mahuang, ma-huang, máhuáng, ts'ao ma-huang

History
Ma-huang is one of the oldest known medicinal plants of China. It has been estimated that its medicinal use may date back six thousand years (Bremness 1995, 102*; Morton 1977, 35*). Ma-huang was first mentioned in the herbal text of the legendary Shen Nung, and it has occupied a firm place in the Chinese materia medica since that time. There are several ma-huang species in China that are used for medicinal purposes, but the most important of these is *Ephedra sinica.* China was the primary source of ephedrae herbae until 1925 (Morton 1977, 35*).

Distribution
Ephedra sinica is found in northern China from Xinjiang Uygur to Hebei Province and into Outer Mongolia. It is found only at an altitude of about 1,500 meters (Morton 1977, 33*) and chiefly on steep slopes in semiarid regions (Bremness 1995, 102*).

The other *Ephedra* species are clearly isolated geographically. *Ephedra shennungiana* is found only in Fujian (Zander 1994, 256*). *Ephedra equisetina* grows only at altitudes between 1,220

Above: The typical red fruits of ma-huang, or Chinese ephedra (*Ephedra sinica*)

Top right: The herbage of *Ephedra sinica*, also known as ma-huang, as sold on the international herb market

Bottom right: Ma-huang buds

126 This American name is supposedly derived from the saying "In treating patients without much knowledge, you are asking for trouble," which a Chinese herb dealer said to a student after the student had wrongly recommended the use of the ma-huang plant (Lu 1991, 84*).

and 1,700 meters in Inner Mongolia (Morton 1977, 33*). Recently, it has been suggested that *E. shennungiana* is a synonym for *E. equisetina* (Hiller 1993, 48). *Ephedra intermedia* is found from Inner Mongolia to Pakistan, but it thrives only in lower altitudes.

Cultivation

Cultivation occurs via seeds that are sown in light, sandy soil in the spring. The plant can also be propagated using pieces of the rootstock (Morton 1977, 34*). The seeds germinate best when they are extracted by hand from ripe fruits still attached to the stalks. The plant requires a dry, warm climate.

Species of *Ephedra* are also cultivated in Australia, Kenya, the United States (South Dakota), and England (Morton 1977, 33*).

Appearance

This perennial horsetail-type plant, which can grow as tall as 75 cm, develops leafless, segmented canes that are round in cross-section. The male flowers resemble catkins. The red fruits contain several black seeds and are attached to short stalks that develop from the stem segments. The fruits develop in late autumn.

The species *sinica* is almost impossible to distinguish from other *Ephedra* species on the basis of its external appearance. The most reliable method of identification is to consider the geographical distribution, the altitude (see under "Distribution"), and the height the plant can attain: *E. sinica* grows to between 45 and 75 cm in height, *E. equisetina* to between 60 and 180 cm, and *E. intermedia* only to 30 to 60 cm (Morton 1977, 33*).

Ma-huang species are easily confused with the species **Ephedra gerardiana**, *Ephedra likiangensis* Florin, *Ephedra przewalskii* Stapf, and *Ephedra distachya* L. (the raw drug can be counterfeited with these species) (Paulus and Ding 1987, 124*).

Psychoactive Material

— Ma-huang: dried stems (ephedrae herba, herba ephedra)

Ephedrae herba is derived primarily from *Ephedra sinica* but is also available as a mixture of various species. The quality of the commercial drug can vary substantially (Liu et al. 1993, 377).

— Ma-huang gen: dried roots (radix ephedrae)

Dried roots should be stored away from light (Hiller 1993, 52).

Preparation and Dosage

A tea (made with 1 heaping teaspoon of *Ephedra* herbage boiled in ¼ liter of water for five to ten minutes) can alleviate hay fever, bronchitis, asthma, or asthmatic complaints very well. The

fresh or dried herbage can also be added to heavy wine or brandy. The astringent taste can be improved by the addition of cardamom, anise, and fennel. The daily dosage of ma-huang herbage is listed as 1.5 to 9 g by itself or in combination preparations (as a tea); the daily dosage of the root ranges from 3 to 9 g (Paulus and Ding 1987, 123*).

The Chinese preparation *mimahuang* is obtained from the chopped raw herbage and honey (10:2). The stems are roasted until the honey has been absorbed and they are no longer sticky (Hiller 1993, 53).

Ritual Use

Because traditional Chinese medicine is rooted in shamanism (Schneider 1993) and the use of ma-huang certainly dates back at least five thousand years, it can be assumed that northern Chinese and Mongolian shamans used *Ephedra* for magical, medicinal, and ritual purposes. Unfortunately, we have not yet found any sources that can confirm this assumption. It is interesting that *Ephedra*'s name (*ma-huang*) places it in the same taxonomic category (*ma*) as **Cannabis sativa** (*ma-fen*). This may be due to the fact that both plants produce euphoriant and stimulating effects and would thus have been very useful to shamans.

Because ma-huang continues to be used as an ingredient in tonics and vitalizing aphrodisiacs, it can also be assumed that Taoist alchemists utilized the plant in their quest for long life and immortality and in their magical sexual rites.

Artifacts

None known

Left: *Ephedra intermedia*, the middle ephedra, is also a source of ma-huang.

Right: In Asia, the species *Ephedra equisetina* is also known as ma-huang and is used in the same manner as *Ephedra sinica*.

Asian *Ephedra* Species That Provide the Chinese Drug Ma-huang
(From Liu et al. 1993; Morton 1977, 33 ff.*; Paulus and Ding 1987, 123*; Schneider 1974, 2:54*)

Ephedra equisetina Bunge	mu-ts'ê ma-huang
Ephedra intermedia Schrenk et Meyer	ma-huang
Ephedra shennungiana Tang	ma-huang
Ephedra sinica Stapf	ts'ao ma-huang

"The roots of Chinese medicine lie in the shamanic tradition of the Shang and Zhou periods (between the sixteenth and first centuries B.C.E.), a tradition associated with magic and geomancy that is now regarded as superstition. During the course of time, and stimulated by the rational orientation of the state philosophy of Confucianism, this increasingly crystallized into the rational and empirical beginnings of a pre-scientific medicine."

WOLFGANG SCHMIDT
DER KLASSIKER DES GELBEN KAISERS ZUR INNEREN MEDIZIN [THE CLASSIC OF THE YELLOW EMPEROR ON INTERNAL MEDICINE] (1993, 10)

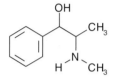

Ephedrine

麻黄

麻黄根

Both of these ma-huang preparations (the herbage and the root) are used in traditional Chinese medicine. (Ancient Chinese illustration)

Medicinal Use

In traditional Chinese medicine, ma-huang has been used with success to treat asthma for more than five thousand years (Wee and Keng 1992, 77*). Generally, both the stems and the roots are used to treat diseases of the lungs and bladder. The stems are used especially in the treatment of fever, colds, headaches, bronchial asthma, and hay fever, while the root is administered for excessive perspiration (Paulus and Ding 1987, 123*).

Constituents

Air-dried herbage contains 1 to 2.5% alkaloids (sometimes as much as 3.3%!), primarily *l*-**ephedrine**, *d*-pseudoephedrine, and *l*-norephedrine. Also present are the analogs norpseudoephedrine, methylephedrine, and methylpseudoephedrine. The alkaloid content is greatest in material collected in autumn. Analyses of different commercial ma-huang preparations from Taiwan have shown that the herbage of *Ephedra sinica* consistently exhibited the highest alkaloid content (approximately 1.1 to 2.1%), followed by *Ephedra equisetina*. *Ephedra intermedia* had the lowest amounts (0.8 to 1.5%) (Liu et al. 1993). The roots and fruits are almost completely devoid of alkaloids (Hiller 1993; Morton 1977, 34*).

Ephedroxane, an anti-inflammatory principle, was also discovered in ma-huang drugs (Konno et al. 1979). In addition to the alkaloids, which should be regarded as the primary active constituents, there are tanning agents, saponines, flavonoids (vicenine, lucenine, etc.), an **essential oil**, and dextrose (Paulus and Ding 1987, 124*).

Effects

Ephedra herbage has an arousing effect on the central nervous system that is similar to that of **ephedrine**: It stimulates, awakens, accelerates the pulse, and constricts the blood vessels. Extracts of the entire plant effect vasoconstriction, stimulate circulation, elevate blood pressure, arouse the central nervous system, are strongly diuretic, suppress the appetite, alleviate bronchial spasms, and relieve the symptoms of hay fever (for at least eight hours).

Both *Ephedra* extracts and ephedrine hydrochloride are regarded as excellent aphrodisiacs, especially for women. Because of the potent vasoconstrictive effects, high dosages of *Ephedra* can produce temporary impotence in men in spite of erotic arousal!

People with elevated blood pressure and heart problems should avoid using *Ephedra*. The drug should not be used when heart arrhythmia or high blood pressure is present (Paulus and Ding 1987, 123*).

MAO inhibitors (*Peganum harmala*, harmaline and harmine) can potentiate the effects of *Ephedra* preparations considerably (Hiller 1993, 53).

Commercial Forms and Regulations

According to the *DAB10*, ma-huang is officinal under the name *ephedrae herba* (ephedra herbage). A tincture, Tinctura Ephedrae *EB6*, is prepared from powdered *Ephedra* herbage and diluted ethyl alcohol (1:5). The pharmacopoeia of Chinese medicine states that the alkaloid content may not be less than 0.8% (Hiller 1993, 51).

In Germany, *Ephedra* herbage and especially preparations made with it are available only with a prescription. The International Olympic Committee and the German Sport Federation have included medicaments containing ephedrine in their lists of doping agents as prohibited stimulants (Hiller 1993, 54).

In the United States, both the herbage and herbal tablets and tinctures were available without restriction prior to 2004, but an FDA ban on supplements containing Ephedra went into effect on April 12, 2004 (cf. Hirschhorn 1982).

Literature

See also the entries for **Ephedra gerardiana**, **Ephedra** spp., and **ephedrine**.

Hiller, Karl. 1993. *Ephedra.* In *Hagers Handbuch der pharmazeutischen Praxis*, 5th ed., 2:46–57. Berlin: Springer.

Hirschhorn, Howard H. 1982. Natural substances in currently available Chinese herbal and patent medicines. *Journal of Ethnopharmacology* 6 (1): 109–19.

Hu, Shiu-Ying. 1969. Ephedra (ma-huang) in the new Chinese materia medica. *Economic Botany* 23:346–51.

Konno, Chohachi, Takashi Taguchi, Mitsuru Tamada, and Hiroshi Hikino. 1979. Ephedroxane, anti-inflammatory principle of *Ephedra* herbs. *Phytochemistry* 18:697–98.

Liu, Ying-Mei, Shuenn-Jyi Sheu, Shiow-Hua Chiou, Hsien-Chang Chang, and Yuh-Pan Chen. 1993. A comparative study on commercial samples of ephedrae herba. *Planta Medica* 59:376–78.

Rätsch, Christian. 1995. Mahuang, die Pflanze des Mondes. *Dao* 4/95:68.

Schmidt, Wolfgang G. A. 1993. *Der Klassiker des Gelben Kaisers zur Inneren Medizin: Das Grundbuch chinesischen Heilwissens.* Freiburg: Herder.

Ephedra spp.

Ephedra Species

Family

Ephedraceae (Ephedra Family)

Folk Names

Ephedra, ephedrakraut, joint fir, meerträubchen, meerträubel, meerträubelarten, meerträubl, uva maritima

Distribution

Ephedra species occur primarily in Eurasia and the Americas. Some three species are found in the Himalayan region (see **Ephedra gerardiana**). There are also several species in China and central Asia (see **Ephedra sinica**). Ephedra species also grow in Europe, chiefly in the eastern Mediterranean region (Greece, Turkey, Cyprus). These *Ephedra* bushes can be mistaken for an *Ephedra*-like broom (*Genista ephedroides* DC.) that is found throughout the Mediterranean.

There is some confusion regarding the taxonomy and nomenclature of the genus *Ephedra*. While as many as seventy-seven species were once described and accepted (Stapf 1889), a revision of the genus has resulted in only some forty-four well-defined species. Questions regarding synonyms, subspecies, and varieties continue to produce considerable confusion in the literature (cf. Zander 1994, 225f.*).

History

Ephedra is one of the oldest plants used by humans. The Neanderthals of Shanidar (modern Iraq) used the plant for ritual and apparently medicinal purposes. Plant remains (pollen) have been recovered from the caves of Shanidar, a Neanderthal burial site dating to approximately 30,000 B.P. (Solecki 1975). *Ephedra* herbage and other bioactive flowers (*Senecio* spp., *Achillea* sp., *Centaurea solstitialis* L., *Muscari* spp.) were placed with the deceased for his last journey. The species has been identified as *Ephedra altissima* Desf. (= *E. distachya* type, *E. fragilis* type) (Leroi-Gourhan 1975; Lietava 1992). But it is possible that it may in fact have been a different species, such as *Ephedra alata* Decne., *Ephedra foliata* Bois. et Kotschy, or *Ephedra fragilis* ssp. *campylopoda* (Solecki 1975, 881).

During archaeological excavations in the southeast of the Kara-kum Desert (Turkmenistan), a three-thousand-year-old temple site that looked exactly like a pre-Zoroastrian shrine was uncovered from under enormous walls of sand. Large clay vessels and basins in which great quantities of a presumably fermented ritual drink were obviously prepared were found on the well-preserved fire altar. Some of the remains of the brewing were able to be identified, with astonishing results: Drinks containing *Ephedra* were being brewed at the site. This temple may have been the home of Zoroaster (= Zarathustra), the founder of the religion that now bears his name. This find suggests that *Ephedra* was an ingredient in the inebriating **haoma** drink. Remains of *Papaver somniferum* were also found on associated objects (pestles, etc.) (Sarianidi 1988). Today, *Ephedra* species are still known as *hum*, *huma*, and *yahma* in the Harirud Valley (Baluchistan). The name appears to preserve a certain reminder of the ancient fire cult.

Constituents and Effects

Almost all *Ephedra* species contain the amphetamine-like **ephedrine** as well as the related alkaloids pseudoephedrine and norephedrine, along with tanning agents, saponines, flavonoids, and an **essential oil**. The Mediterranean species exhibit their highest concentrations of alkaloids in August, which is why they should be collected at that time. *Ephedra major* contains 0.69% alkaloids, *E. distachya* 0.35%, and *E. campylopoda* only 0.14% (Tanker et al. 1992). Extracts of the entire plant for all the species mentioned have the same effects as **Ephedra gerardiana** or **Ephedra sinica**.

Woodcut of a small *Ephedra* species from the herbal of John Gerard (1633)

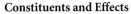

Left: The winged ephedra (*Ephedra alata*), common in North Africa and southwest Asia, was already being used for ritual purposes during the Stone Age.

Right: *Ephedra andina*, known in the Andes as *pingo-pingo*, is devoid of ephedrine.

Top left: *Ephedra breana*, known locally as *pingo-pingo*, occurs primarily in the higher regions of the Atacama Desert (northern Chile), where it can grow into proper trees with very woody trunks. (Photographed in the Atacama Desert)

Top center: Andean ephedra (*Ephedra andina*) grows at very high altitudes.

Top right: *Ephedra distachya*, found in Greece, develops red berries with a sweet taste. (Photographed on Naxos, Greece)

Bottom left: Swiss ephedra, *Ephedra helvetica*, is found in the Alps. (Photographed in Switzerland)

Ephedra Species with the Greatest Ethnobotanical Significance

Species with ethnobotanical and ethnopharmacological significance are found primarily in South America and Asia. The most important of these are discussed below.

Ephedra americana Humb. et Bonpl. ex Willd.—pinku-pinku

Known as *pinku pinku*, *naranja naranja* ("orange"), and *refresco* ("refresher"), this shrub is found from Ecuador to Argentina and also occurs in the high mountains. In Peru, a tea prepared from the plant is drunk as a tonic (Franquemont et al. 1990, 40*). The related but more prostrategrowing species *Ephedra rupestris* Benth. is known in Quechuan as *pampa pinku pinku*, and it is drunk as a tea for lung ailments (Franquemont et al. 1990, 40 f.*).

Ephedra andina Poepp. ex C.A. Mey. [syn. *Ephedra americana* var. *andina* Stapf]—pingo-pingo

This South American bush is called *pingo-pingo* (literally, "tubes") and is also known as *canotu* ("joint" or "hashish cigarette"), *solupe*, and *transmontaga*. In Chilean folklore, this plant is a symbol for "he-men" and heartbreakers—perhaps because of its purported aphrodisiac effects. The plant is used in (folk) medicine to treat bronchitis, asthma, and whooping cough (Mösbach 1992, 60*). In contrast to the other species of the genus, *Ephedra andina* does not contain any ephedrine, although vicenine-I and -II as well as flavones and camphor oil have been demonstrated to be present (Gurni and Wagner 1982; Montes and Wilkomirsky 1987, 40*). Further research is needed to determine whether this plant is psychoactive. The very similar species *Ephedra multiflora* Phil. is also known as *pingo-pingo* (Aldunate et al. 1981, 209*).

Ephedra breana Phil.—pingo-pingo

This species grows into a proper tree that develops a woody, thick trunk as large as 20 cm in diameter. It is found only in the extremely dry high desert of Atacama (northern Chile), where it is known as *pingo-pingo* or *tume*. The edible fruits are known as *granada*, "pomegranates" (Aldunate et al. 1981, 209*; 1983*). A decoction of the fresh or dried stems has very strong stimulating and mood-enhancing effects with pleasurable aphrodisiac sensations. In some ways, these effects are reminiscent of those of MDMA (cf. **herbal ecstasy**).

Ephedra campylopoda C.A. Mey [syn. *Ephedra fragilis* Desf. ssp. *campylopoda* (C.A. Mey) Aschers et Graebn., *E. fragilis* Desf. var. *campylopoda* (C.A. Mey) Stapf]—polik stap

This species is found primarily on Cyprus and the Greek isles (Sfikas 1990, 94*). In ancient times, it apparently was known as "food of Saturn." Unfortunately, almost nothing is known about its early history. The modern Greeks consider the plant to be poisonous. Decoctions have only weakly stimulating, albeit very refreshing, effects.

Ephedra distachya L. [syn. *E. maxima* Saint-Lager, *E. vulgaris* L.C. Rich, *E. distachya* L. ssp. *distachya*]—sea grape, little sea grape

This species, which can grown as tall as 50 cm, is indigenous to the Mediterranean region and is also found in flat areas from the Black Sea to Siberia. It is sometimes regarded as a synonym for **Ephedra gerardiana**, which occurs only in high mountains. This species is primarily used in homeopathy (Ephedra distachya hom. *HAB1*, Ephedra distachya spag. Zimpel hom. *HAB1*, Ephedra vulgaris hom. *HPS88*). The herbage is nearly as rich in alkaloids as that of *Ephedra sinica*.

Ephedra helvetica C.A. Mey. [syn. *Ephedra distachya* L. ssp. *helvetica* (C.A. Mey.) Aschers]—Swiss ephedra

This central European species has rather thin green stems and a creeping growth pattern, achieving a height of only 20 to 50 cm. It thrives chiefly in rocky areas. Swiss ephedra is found primarily in Ticino and Valais (Rhône Valley), where it may be collected from the wild (Lauber and Wagner 1996, 82*). Teas made from the plant have stimulating effects that are similar in potency to teas made from **Ephedra sinica**.

Ephedra intermedia Schrenk et Mey.—narom

Four geographically isolated varieties of this species are distinguished:

— var. *glauca* (Regel) Stapf (Transcaspian region, Pamir, Mongolia)
— var. *persica* Stapf (Iran, western Afghanistan)
— var. *schrenkii* Stapf (northwest Iran, Turkistan)
— var. *tibetica* Stapf (Afghanistan, Pakistan, India, Tibet)

In Baluchistan (Pakistan), the aboveground herbage of the variety *tibetica* is used for dyeing and tanning (Goodman and Gharfoor 1992, 14*). A decoction of 25 g of the stems is drunk for back pain or as a general tonic (Goodman and Gharfoor 1992, 52*). In Pakistan, the herbage is burned to ashes and mixed with **Nicotiana tabacum** to produce a chewing tobacco (Morton 1977, 36*). In China, the variety *glauca* is one of the three ma-huang species (see *Ephedra sinica*). The Persian variety is known as *hôm, hum,* or *huma* and is regarded as a substitute for or ingredient in **haoma**.

Ephedra major Host [syn. *Ephedra equisetiformis* Webb. et Berth., *E. nebrodensis* Tineo ex Guss., *E. scoparia* Lange]—large ephedra

There are two subspecies of large ephedra: *E. major* ssp. *major* and *E. major* ssp. *procera* (Fisch. et C.A. Mey.) Markg. [syn. *E. procera* Fisch. et C.A. Mey.]. The subspecies *major* is found in Spain and along the Mediterranean Sea as far as western Asia. The subspecies *procera* is indigenous to Dalmatia, Greece, southwest Asia, and the Caucasus (Zander 1994, 256*). This rather rare and fairly tall plant contains more than 2.5% alkaloids, 75% of which is ephedrine (Morton 1977, 34*). It is thus a very good source of ephedrine.

Ephedra monosperma C.A. Mey—Tibetan ephedra

This high mountain species is found almost exclusively in Tibet, where it is known by the name *mtshesdum*. It has been a part of the Tibetan pharmacopoeia since ancient times and was mentioned in the Blue Beryl Treatise of Sangye Gyamtso (1653–1705). The herbage is used to treat "liver fever" and bleeding, but it is particularly renowned for its refreshing and rejuvenating properties (Aris 1992, 1:69, 2:225*).

Ephedra nevadensis Wats—Mormon tea

This species, which can attain a height of approximately 90 cm, predominates in the American Southwest. Analyses of coprolites (fossilized feces) have demonstrated that prehistoric Indians of the Caldwell Cave Culture (1200–1450) used this species for ritual or medicinal purposes (to treat diarrhea) (Sobolik 1996, 8; Sobolik and Gerick 1992). The Coahuilla Indians of Southern California call the plant *tú-tut.* They make an infusion of the plant for use as a stimulating tea (Barrows 1967, 73f.*). Because of the aphrodisiac effects of the plant and the teas made from it, it is also known as "whorehouse tea" (Morton 1977, 36*). It is one of the favorite drinks of the Mormons, who are otherwise avowedly "anti-drug."

Ephedra torreyana S. Wats—Torrey joint fir

The Navajo call this species, which grows to a height of only 60 cm, *tl'oh azihii libáhígíí* ("gray rasp grass"). They use its stems as a diuretic to treat kidney ailments and venereal diseases and to alleviate postpartum pain. They roast the stems before brewing in order to remove their bitter taste (Mayes and Lacy 1989, 54*).

Ephedra trifurca Torr.—tlanchalahua

This Mexican species has been used medicinally since the pre-Columbian period. Teas (infusions, decoctions) made from it are used in folk medicine to achieve weight loss and to suppress the appetite (Martínez 1994, 304*). Such uses reflect the high alkaloid content of the plant.

Literature

See also the entries for **Ephedra gerardiana,** **Ephedra sinica,** and **ephedrine.**

Aldunate, Carlos, Juan J. Armesto, Victoria Castro, and Carolina Villagrán. 1983. Ethnobotany of pre-altiplanic community in the Andes of northern Chile. *Economic Botany* 37 (1): 120–35.

Bottom left: *Ephedra microsperma,* a species from central Asia.

Bottom center: Dwarf ephedra (*Ephedra minima*) is found on the Tibetan Plateau.

Top right: Large ephedra (*Ephedra major*)

Bottom right: The subspecies *Ephedra major* ssp. *procera* (Fischer et C.A. Mey.) Markg. is found in southern Greece.

The stimulating Mormon tea (*Ephedra nevadensis*) is occasionally found growing in rocks. (Photographed in Black Canyon, Colorado)

Groff, G. Weidman, and Guy W. Clark. 1928. The botany of *Ephedra* in relation to the yield of physiologically active substances. *University of California Publications in Botany* 14 (7): 247–82.

Gurni, Alberto A., and Marcello L. Wagner. 1982. Apigeninidin as a leucoderivative in *Ephedra frustillata*. *Phytochemistry* 21 (9): 2428–29.

Leroi-Gourhan, Arlette. 1975. The flowers found with Shanidar IV, a Neanderthal burial in Iraq. *Science* 190:562–64.

Lietava, Jan. 1992. Medicinal plants in a Middle Paleolithic grave Shanidar IV? *Journal of Ethnopharmacology* 35:263–66.

Nawwar, M. A. M., H. H. Barakat, J. Buddrust, and M. Linscheid. 1985. Alkaloidal, lignan and phenolic constituents of *Ephedra alata*. *Phytochemistry* 24 (4): 878–79.

Nielson, C. H., C. Causland, and H. C. Spruth. 1927. The occurrence and alkaloidal content of various ephedra species. *Journal of the American Pharmaceutical Association* 16 (4).

Sarianidi, W. 1988. Die Wiege des Propheten. *Wissenschaft in der UdSSR* 5:118–27.

Sobolik, Kirstin D. 1996. Direct evidence for prehistoric sex differences. *Anthropology Newsletter* 37 (9): 7–8.

Sobolik, Kirstin D., and Deborah J. Gerick. 1992. Prehistoric medicinal plant usage: A case study from coprolites. *Journal of Ethnobiology* 12 (2): 203–11.

Solecki, Ralph S. 1975. Shanidar IV, a Neanderthal flower burial in northern Iraq. *Science* 190:880–81.

Stapf, Otto. 1889. Die Arten der Gattung Ephedra. *Denkschrift der Kaiserlichen Akademie der Wissenschaften (Wien), Mathematisch-naturwissenschaftliche Klasse* 56:1–112.

Tanker, N., M. Coskun, and L. Altun. 1992. Investigation on the *Ephedra* species growing in Turkey. *Planta Medica* 58 suppl. (1): A695.

Wallace, James W., Pat L. Porter, Elisabeth Besson, and Jean Chopin. 1982. *C*-glycosylflavones of the Gnetopsida. *Phytochemistry* 21 (2): 482–83.

Erythrina americana Miller

American Coral Tree

Family
Leguminosae (Legume Family); Subfamily Papilionoideae

Forms and Subspecies
None

Synonyms
Corallodendron americanum Kuntze
Corallodendron triphyllum nom. nud.
Erythrina carnea Ait.
Erythrina enneandra DC.
Erythrina fulgens Loisel.

Folk Names
American coral tree, amerikanischer korallenbaum, bolita grande, cehst (Mixe), chakmolche' (Mayan, "the tree of the red puma"), chak-moolché', chocolín, chotza, colorín, cosquelite, demti (Otomí), equimite, hutukuu' (Huastec), iquemite, iquimite, jiquimiti, k'ante' (Mayan, "yellow tree"), korallenstrauch, lakatilá (Totonac), lakatili, lakatilo, lak'tanga, li-pashcua (Chontal), madre alcaparra, madre brava, madre cacao, madre chontal, ma-ja-ñú (Chinantec), palo de coral, pali de pite, parencsuni, patol, pat-olli, pichoco, pito, pito pichoco, puregue, purenchecua (Tarascan), purgne, quemite, quimiti, sompantle, sompantli, sumpantle, te'batai (Otomí), tlalni, tsejch (Mixe), tsizch, tu saba (Mixtec), tzinacancuáhuitl, tzite (Quiché), tzompancuahuitl (Aztec "tzompantli tree"),[127] tzompantli, tzompomitl (modern Nahuatl), tzon të kichilo, uhkum, xk'olok'max (Yucatec Mayan), xoyo' (Mayan), zompantli, zompantlibaum, zompantlibohne, zumpantle

The edible red flowers are known in Veracruz as *gasparitos* and in Nahuatl as *cozquelite*.

History
The red seeds have been found in prehistoric strata. The tree and its hieroglyph appear in pre-Columbian Mayan manuscripts (Codex Dresdensis) under the name *k'ante'*, "yellow tree" (Rätsch 1986, 223*). The tree is also mentioned in Aztec sources, as well as sources in other languages, from the early and late colonial period. The genus and many of its species were first described by Linnaeus.

Because of the paralyzing effects of the extract, this agent was once misused for vivisections (Roth et al. 1994, 327*).

In Veracruz, the flowers are cooked and eaten

127 The closely related species *Erythrina chiapasana* Kruk., which occurs in the highlands of Chiapas, is also known by this name (Martínez 1987, 112*).

as a vegetable. They are regarded as an aphrodisiac food (Reko 1938, 127*; Ott 1993, 423*). The seeds were once used in a kind of dice game (*patol*) (Krukoff 1939, 210).

Distribution
The tree occurs from northern Mexico to Guatemala. It prefers a dry and warm climate. Its range is concentrated around central Mexico (Morelos, Puebla, Veracruz, Colima, Guerrero, Oaxaca) (Krukoff 1939, 299).

Cultivation
Cultivation is performed simply by planting pre-germinated seeds in the soil. The seeds should be watered well, but not too much. In Mexico, the tree has been planted as a living fence since pre-Columbian times (Krukoff 1939, 210).

Appearance
The American coral tree can grow to a height of some 6 to 8 meters. It has large, wide, and attenuate leaves that are attached to the stalks in clusters of three. The luminously red flowers can grow as long as 10 cm and are arranged in upright clusters. The tree loses its foliage in the winter. The flowers begin developing while the tree is still bare (January to March). The slightly constricted pods then ripen as the leaves emerge. The pods contain two to five bright red, bean-shaped seeds.

The tree is easily confused with the very similar species *Erythrina mexicana* Kruk. (whose seeds, like those of *E. americana*, are known as *equimitl* in Aztec) as well as with other **Erythrina species**. It is almost identical to *Erythrina standleyana* Kruk. and can be distinguished from the latter species only on the basis of its geographical distribution (Krukoff 1939, 300 f.).

Psychoactive Material
— Seeds (*colorines, equimitl, tzite*)

Preparation and Dosage
When intended for internal use, the seeds must be ground. The effective dosage is given as no more than half of a seed. But this guideline should be used with great care, as no dependable information is available!

Ritual Use
The pre-Columbian Maya associated this tree with the direction south, whose symbolic color is yellow. The Mayan name *k'ante'*, "yellow tree," refers not to the color of the red flowers or the red seeds but to the yellow dye that is obtained from the root cortex (Krukoff 1939, 210). The tree is invoked in magical formulae uttered to treat possession (*tancasil*). In the prophetic texts of the shamanic jaguar priest (*chilam balam*), the plant is mentioned in an anthropomorphized form as a

divine being known as *ah kantenal*, "he of the yellow tree" (Rätsch 1986, 223*). There are only a few vague indications that contemporary Yucatec Mayan shamans use the seeds for healing rituals and divinations (Garza 1990, 188*).

The Huastec still use the wood to make ritual masks (Alcorn 1984, 640*).

The traditional divinatory priests of the Kanjobal (Guatemala) still use the ancient Indian 260-day calendar for their divinations. They use the seeds of the coral tree to count the days. This calendar divination has been maintained into the present time and has an important function for the Kanjobal in solving personal problems and social conflicts (Hinz 1984).

In the Aztec culture, the coral tree was closely related to human sacrifice. Before the victims were disemboweled and butchered for sale at the great market of Tenochtitlán (present-day Mexico City), their heads were removed. The skulls were placed on a rack of vertical poles. They were speared onto the wooden stakes, so that the growing numbers of skulls were placed one on top of the other. In Aztec, this rack was known as a *tzompantli* ("skull frame"), and it was always located in close proximity to the main temple (Krickeberg 1975, 239). Several pre-Hispanic stone sculptures depicting the *tzompantli* have been preserved. Since it is known by the name *tzompantli tree*, the coral tree was directly or symbolically associated with this practice. Unfortunately, the sources do not provide us with any insights into the actual nature of this relationship. It is possible that the seeds were administered to the selected victims to sedate them (cf. **Datura innoxia, Bursera**

One of the earliest European illustrations of a botanically correct branch of the American coral tree. (Woodcut from Gerard, *The Herball or General History of Plants*, 1633)

Botanical illustration of the American coral tree (*Erythrina americana*). (Colored copperplate engraving from Trew 1750.)

bipinnata). The wood the frame was made from was definitely not that of the coral tree, for coral tree wood is very soft and would not have provided sufficient support for the many skulls.

Artifacts

The wood was and still is used to manufacture ritual objects, including masks and figures of gods (Aguilera 1985, 128f.*). In central Mexico, small ithyphallic images of gods were still being carved from the wood in the twentieth century; these were placed in the kitchen as a magical protection against food spoilage (Reko 1938, 127 f.*).

A statue of the god Maximón made from the wood of a closely related species is still cultically venerated in the Guatemalan highlands, among the Tzutujil of Lago Atitlán. The fly agaric mushroom (**Amanita muscaria**) is associated with this sacred wooden figure of Maximón, who is said to have risen from a coral tree (*Erythrina rubrinervia*) when it was struck by lightning. According to legend, the tree had been standing in the midst of a group of fly agaric mushrooms when it was hit by the lightning bolt. A man is said to have eaten a piece of these mushrooms and become young again (Lowy 1980).

In Mexico, the red seeds are frequently used in the manufacture of amulets and necklaces:

> It is also said that the seeds can incite love. According to folk belief, when a girl is wearing a chain of such seeds around her neck, she will soon become so incapable of resisting a man's wishes that she will simply give herself to him.—Because of this tradition, it is probable that certain ladies who belong to the category of those who do not die as long as they love still like to decorate themselves with these ominous necklaces today so that they can tell the world of the living about the magic which affects them. (Reko 1938, 127*)

A beautiful portrait of the plant by the flower painter Georg Dionys Ehret (1708–1770) was published in 1750 in the *Plantae selectae* of Christoph Jakob Trew (Trew 1981, table 8*).

Medicinal Use

The modern Huastec use the leaves as a medicine for sleeplessness, restlessness, and "crying out at night" (Alcorn 1984, 640*). In Mexican folk medicine, a decoction of the flowers is ingested for chest pains. The juice tapped from the trunk is used for scorpion stings. The bark is drunk as a diuretic and purgative (Krukoff 1939, 210).

Constituents

The seeds and, to a lesser extent, the flowers and other parts of the plant contain erythrina alkaloids (erythrane, erythroidine, coralline, coralloidine, and erythro-coralloidine). The seeds have been found to contain 1.61% alkaloids (erysopine, erysovine, erybidine, erysodine, and erythrartine) (Lara Ochoa and Marquez Alonso 1996, 39*; Martínez 1994, 78*). The flowers contain 0.11% alkaloids (α-erythroidine and β-erythroidine) (Aguilar et al. 1981).

The raw drug is known as Mexican curare (Krukoff 1939, 205; Roth et al. 1994, 327*).

Effects

The seeds are said to produce a so-called women's inebriation, i.e., nymphomaniac/ecstatic states with a strong desire for sex:

> The first such intoxication is described in a report from the year 1719. An Indian woman prepared a dish from the red beans, which appeared to her to be edible, and gave this to other women. All who ate of it began to laugh without reason, babbled all kinds of nonsensical things, and talked shamelessly. Later, they reeled like drunkards and finally fell into a deep sleep, so that they had to be carried home.
>
> In September 1738, an honorable young girl accidentally ate some of the red zompantli beans and shortly thereafter lost her mind. She ran through the streets with her skirt raised, laughing horribly, much to the dismay of the women and the derision of the men. Neighbors brought her back home, where she fell into a hot fever, ripped apart all of her bedding, and died on the third day thereafter. . . . In all of these cases, the ingestion of the red beans initially produced immoderate gaiety, then confused speech, wobbling as if drunk, and a heightened libido. The intoxicated persons then fell into a deep sleep, from which they usually did not reawaken. (V. Reko 1938, 129 ff.*)

These and similar reports describe the alleged effects of the red seeds. It is difficult to determine whether these are in fact authentic or belong to the domain of legend. In any case, these reports have had such an effect that since that time no one else has attempted to try the seeds himself.

In Mexico, the boiled flowers are eaten as a vegetable and produce mild hypnotic effects (Aguilar et al. 1981).

Commercial Forms and Regulations

In Mexico, *colorines* are available in the markets and in stores selling devotional articles. Indians often sell chains of the seeds near ruins and other tourist attractions.

Literature

See also the entry for *Erythrina* spp.

Aguilar, María Isabel, Francisco Giral, and Ofelia Espejo. 1981. Alkaloids from the flowers of *Erythrina americana. Phytochemistry* 20 (8): 2061–62.

Folkers, K., and R. T. Major. 1937. Isolation of erythroidin, an alkaloid of curare action, from *Erythrina americana. Journal of the American Chemical Society* 59:1580 ff.

Hargreaves, R. T., R. D. Johnson, D. S. Millington, M. H. Mondal, W. Beavers, L. Becker, C. Young, and

K. L. Rinehart, Jr. 1974. Alkaloids of American species of *Erythrina. Lloydia* 37:569 ff.

Hinz, Eike. 1984. Kanjobal Maya divination: An outline of a native psycho-sociotherapy. *Sociologus* 34 (2): 162–84.

Krickeberg, Walter. 1975. *Altmexikanische Kulturen.* Berlin: Safari-Verlag.

Krukoff, B. A. 1939. The American species of Erythrina. *Brittonia* 3 (2): 205–337.

Lowy, Bernard. 1980. Ethnomycological inferences from mushroom stones, Maya codices, and Tzutuhil legend. *Revista/Review Interamericana* 10 (1): 94–103.

Ramirez, E., and M. D. Rivero. 1935. Contribución al estudio de la acción farmocodinámica de la *Erythrina americana. Anales del Instituto Biológico de la Universidad Nacionál de México* 6:301–5.

The rain god Chac, with his tapir nose (= God B), is shown sitting on the *k'ante'*, the "yellow tree" of the south (*Erythrina americana*). (Codex Dresdensis, 31c)

Erythrina berteroana Urban

Pito Coral Tree

Family

Leguminosae (Legume Family); Subfamily Papilionoideae

Forms and Subspecies

None

Synonyms

None

Folk Names

Aposhí, aposí, chilicote, colorín, coral bean, coralina, k'änte'(Lacandon, "yellow tree"), peonía, pioneo, pito coral tree, pito-korallenbaum, tzinacancuáhuitl (Aztec), tzompantli

History

For the history of the coral trees, see *Erythrina americana.*

In Guatemala and El Salvador, the young flowers of this species of coral tree are eaten as a vegetable (fresh or frozen). When a large amount is consumed, the meal has sedative effects and induces a long and deep sleep. The tree is frequently cultivated in Central America (as a living fence). In former times, crushed branches were used as a fish poison (Morton 1994).

Distribution

This species is found primarily in Guatemala, El Salvador, and southern Mexico. It also occurs in central Mexico and in rare cases in northern Mexico.

Cultivation

The tree is difficult to grow from pregerminated seeds. In contrast, it is easily propagated from cuttings (of trunks or branches that have already become woody). It does not place any particular demands upon the soil. It should be watered well but not overly much. It does not tolerate any cold or frost (Grubber 1991, 26*).

Appearance

This shrublike tree, which can grow as tall as 9 meters, develops thorny branches. The 6 to 9 cm long leaves are arranged on the stems in groups of three. The red flowers (3 to 6 cm long) develop in loose, many-flowered clusters. The brilliant red, bean-shaped seeds are enclosed within the seedpods (two or three per pod).

This tree is easily confused with the closely related species *Erythrina flabelliformis* (see ***Erythrina* spp.**).

Psychoactive Material

— Seeds (*colorines*)

Preparation and Dosage

One quarter or one half of a seed is chewed thoroughly and swallowed (Gottlieb 1973, 9*). Otherwise, see ***Erythrina americana.***

Botanical illustration of the pito coral tree (*Erythrina berteroana*). (Colored copperplate engraving from Trew, 1750)

Left: The large flower of the coral tree known as pito (*Erythrina berteroana*) is edible. (Photographed in Naha', Chiapas, Mexico)

Right: A traditional string of red seeds of *Erythrina berteroana*, as produced by the Lacandon

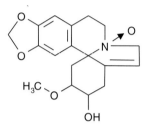

Erythratine-*N*-oxide

"Almost all of the Indian women of this region (Nayarit) know about the colorines or zompantli beans which, when ingested, induce hot dreams and cause them to concern themselves with thoughts of fleshly desire. They say that one can eat only a very little of these, as too much will produce fever, pains in the breasts and lower abdomen, and a daze akin to drunkenness. They keep this custom a secret from the men. But in confession, some of them ask whether this agent is Christian or abominable, and to justify it they cite the advantage that those who use it never defile themselves with a man. I regard the bean as diabolic, for it clearly incites lewd behavior and probably also leads to infertility. It will therefore be no great wrong if its use is forbidden at every opportunity, in particular in conversations during confession."

FATHER SALAMIELLA
FROM A LETTER IN THE ARCHIVE OF
TEPIC
IN *MAGISCHE GIFTE* [MAGICAL
POISONS]
(V. REKO 1938, 131 f.*)

Ritual Use
See *Erythrina americana.*

Artifacts
The seeds have been used to make necklaces since ancient times. They also are a component of amulets.

A botanically accurate portrait of the plant by the flower painter Georg Dionys Ehret (1708–1770) was published in 1760 in the *Plantae selectae* of Christoph Jakob Trew (Trew 1981, table 58*).

Medicinal Use
A tea made from the flowers is occasionally drunk as a "sleeping pill" (Morton 1994).

Constituents
The seeds contain erythrina alkaloids (erysodine, erysopine, erysothiopine, erysothiovine, α- and β-erythroidine, and hypaphorine), which are also present in the flowers in lesser amounts. These alkaloids are responsible for the sedative effects.

The new alkaloid erythratine-*N*-oxide was isolated from this species (Soto-Hernandez and Jackson 1994).

Effects
The psychoactive effects of the seeds have been characterized as narcotic, sedative, and mildly inebriating, and purportedly also as aphrodisiac (cf. *Erythrina americana*).

Commercial Forms and Regulations
In southern and central Mexico, the red seeds can be obtained in Indian markets and shops that sell devotional articles. Indian women sometimes offer strings of seeds for sale in tourist areas (e.g., Palenque).

Literature
See also the entries for *Erythrina americana* and *Erythrina* spp.

Morton, Julia F. 1994. Pito (*Erythrina berteroana*) and chipilin (*Crotalaria longirostrata*), (Fabaceae), two soporific vegetables of Central America. *Economic Botany* 48 (2): 130–38.

Soto-Hernandez, M., and Anthony H. Jackson. 1994. *Erythrina* alkaloids: Isolation and characterisation of alkaloids from seven *Erythrina* species. *Planta Medica* 60:175–77.

Erythrina spp.

Coral Tree Species

Family

Leguminosae (Legume Family); Subfamily Papilionoideae

Coral trees are found primarily in the tropical zones of the New and Old Worlds (Standley 1919), although there are also species in Australia. The genus is composed of some one hundred species (Bärtels 1993, 142*). The seeds contain chiefly cytisine or other erythrina and curare-like alkaloids (El-Olemy et al. 1978; Wandji et al. 1994). It is for this reason that they generally are regarded as poisonous. Only the seeds of the Andean species *Erythrina edulis* Triana [syn. *Erythrina esculenta* Sprague, *E. edulis* Posada-Arango, *E. lorenoi* F. Maebr., *E. megistophylla* Diels] can be eaten (Bärtels 1993, 68*). They are often sold as "beans" at Indian markets. The seeds of some species contain lectins (Peña et al. 1988).

Erythrina corallodendron L. [syn. *Erythrina corallodendron* var. *occidentalis* L., *E. spinosa* Mill., *E. inermis* Mill., *E. corallifera* Salisb., *Corallodendron occidentale* Kuntze]—madre del cacao

This tree is indigenous to Central America, where it is found both in cultivation and in the wild. It is an important shade tree in the tropical cacao plantations (cf. *Theobroma cacao*). Its red seeds, known as *colorines*, are strung together to make bracelets and necklaces. They allegedly contain "hallucinogenic substances" (Bärtels 1993, 68*).

Erythrina falcata Benth.—seibo

This beautiful blooming tree is known in Peru as *pisonay*. It also occurs in Bolivia, Brazil, and Paraguay, as well as in northwestern Argentina, where it is known by the folk names *seibo, ceibo, seibo del noroeste, seibo de jujuy, seibo de salta, seibo de Tucuman, seibo de la selva, seibo rosado, seiba,* and *suiñandi* (Santos Biloni 1990, 21*). According to a physician in Tartagal (pers. comm.), it may also be called *seibo silvestre*, and it is said to be the source of a hallucinogenic **snuff**. The national tree of Argentina, *Erythrina crista-galli* L., is also known as *seibo*. It plays a role in numerous legends; it is said that an ugly Guaraní maiden named Anahí was transformed into the beautiful flower (Santos Biloni 1990, 171*).

Erythrina flabelliformis Kearney [syn. *Erythrina purpusi* Brand.]—fan-shaped coral bush

The shamans of the Tarahumara formerly used the seeds in rituals, although we do not know how (Bye 1979b, 38*). The seeds probably also were added to the *tesgüino* beer that was brewed from agaves (*Agave* spp.) or maize (*Zea mays*) to potentiate the effects (Bye 1979b, 38*). The seeds are or were strung into necklaces by northern Mexican Indians (Bye 1979b, 37*). They are used as an alternative to mescal beans (*Sophora secundiflora*). The Serí Indians of northern Mexico boil the seeds to produce a decoction that they drink to treat diarrhea (Felger and Moser 1974, 425*). The Pima Indians grind the seeds and mix them with lard to produce an ointment that is applied to treat inflammations. Partially ground seeds—which are regarded as poisonous—are swallowed as a purgative (Pennington 1973, 222*). The Tarahumara used the seeds to treat toothaches and lower abdominal ailments. The Indians of the Barranca de Batopilas region applied a kind of ointment made from ground seeds to their eyelids to improve their vision (Bye 1979b, 37*).

The seeds contain numerous erythrina alkaloids, including 14% erysotrine, 45% erysodine, 40% erysovine, and approximately 1% eryspine (Bye 1979b, 38*). The Tarahumara say that this plant produces erotic dreams. The extract has effects like those of curare (Díaz 1979, 87*). To date, there is no evidence that the plant produces psychoactive effects (Schultes and Hofmann 1980, 338*).

Erythrina fusca Lour.

This coral tree, known as *amasisa* or *gachica*, is found in Amazonia, where it is used as an **ayahuasca** additive. It has been found to contain the alkaloids erythraline, erythramine, and erythratine.

Erythrina glauca Willdenow—amasisa

The "blue" coral tree is found in the Amazon basin. It is known as *amasisa* in Colombia and as *assacú-rana* in Brazil. The Tikuna Indians boil the bark to wash out wounds. In Brazil, a tea made from the roots is drunk to treat rheumatic complaints and liver ailments, while higher dosages are used as a purgative. Very strong concentrations of the root tea are said to produce narcotic effects. The chemistry of this species is unknown (Schultes and Raffauf 1990, 241*). Parts of the plant are used as **ayahuasca** additives (Ott 1993, 217*).

Erythrina indica Lamarck [syn. *Erythrina variegata* L.]—mandara

This tree is sacred in India and Nepal, where it is associated both with the production of *amrita*, the drink of immortality (cf. **soma**), and with Shiva's paradise. According to Vedic mythology, the tree

The enchanting flower of a Brazilian species of coral bush, *Erythrina crista-galli*

"The consumption of just a small amount of 'colorines' produces an extraordinary increase in blood pressure. For this reason, the Indians in the Sierra of Tlaltizepam (Michoacan) use this in the form of an extract for acts of revenge, when they wish that a person be struck by a 'blow.'"

LUTZ ROTH, MAX DAUNDERER, AND KURT KORMANN
GIFTPFLANZEN—PFLANZENGIFTE [POISONOUS PLANTS—PLANT POISONS] (1994, 327*)

arose when the milk of the primordial ocean was churned to make the drink of the gods. Indra saw the tree rise up from the depths and planted it in his pleasure garden. It is regarded as one of the five heavenly trees and is venerated for its abilities to fulfill wishes (*kalpavriksha*). Krishna stole the tree from Indra's garden and brought it to humans. The wood of the mandara tree is sacred and is burned as a sacrificial offering on the fire altar

known as *homa*. The bright red flowers are offered to Shiva, and the tree is generally closely associated with the god. The three leaves on each stalk symbolize the trinity of Hindu gods: Brahma, Vishnu, and Shiva (Gupta 1991, 39 f.*).

The seeds and bark of the species found in the Himalayas were or are used as a fish poison. It is quite possible that the tree was once used as a hallucinogen. In Sri Lanka, the trees are cultivated as a support for the cubeb pepper vine (*Piper cubeba*; cf. *Piper* spp.) to climb (Macmillan 1991, 415*). A tonic is prepared from the plant together with the leaves of *Solanum nigrum* L. (see *Solanum* spp.) and the seeds of *Datura metel* (Bhandary et al. 1995, 155 f.*).

Erythrina mulungu Mart.—mulungu

The bark of this Brazilian species was once used medicinally as a narcotic in the form of a Galenic preparation. It "contains a narcotic with opium-like effects" (Schneider 1974, 2:66*).

Top left: The Indian coral bush (*Erythrina indica*) in full bloom. This tree is venerated as sacred. (Photographed in the Arum Valley, Nepal)

Above left: The characteristically colored leaves of the cultivated *Erythrina indica* var. *variegata* cv.

Right: This traditional necklace of the Aborigines of northeastern Australia incorporates *Erythrina* seeds in its design.

Colorines

Many Mexican markets have stands that offer fresh and dried herbs, images of saints, amulets, and candles and incense. Often they sell *Erythrina* seeds under the name *colorines* as "magical seeds" or "magic beans" (Bye and Linares 1983, 6*). However, the seeds of other plants are also sold under this name and attributed with identical or similar magical significance (Martínez 1987*).

Name of the Stock Plant	Description of the Seeds
Abrus precatorius L.	red-black seeds (small, roundish)
Capparis indica (L.) Fawc. et Rendl.	red berries
Erythrina americana Mill.	red seeds (bean shaped)
Erythrina berteroana Urb.	red seeds (bean shaped)
Erythrina breviflora DC.	dark brown seeds (bean shaped)
Erythrina corallodendron L.	red seeds (bean shaped)
Erythrina coralloides DC.	scarlet red seeds with a black line
Erythrina flabelliformis Kearn.	red to yellow seeds (bean shaped)
Erythrina herbacea L.	red seeds (bean shaped)
Erythrina lanata Rose	red seeds (bean shaped)
Erythrina lepthorriza DC.	black seeds (bean shaped)
Erythrina occidentalis Standl.	red seeds (bean shaped)
Erythrina phaseloides DC.	red seeds (bean shaped)
Erythrina spp.	red seeds (bean shaped)
Hamelia xorullensis H.B.K.	?
Ormosia istmensis Standl.[128]	red seeds (round, humped)
Ormosia macrocalyx Ducke	red seeds (round, humped)
Ormosia toledana Standl.	red seeds (bean shaped)
Ormosia spp.	red-orange seeds (round, humped)
Ormosia spp.	red-black seeds (round, humped)
Piscidia americana Moc. et Sess.	?
Rhynchosia pyramidalis (Lam.) Urb.	red-black seeds (small, round)
Rivina humilis L.	red seeds
Sophora conzatti Standl.	red seeds (bean shaped)
Sophora purpusii T.S.	red seeds (bean shaped)
Sophora secundiflora (Ort.) Lag.	red to yellow seeds (bean shaped)
Sophora tomentosa L.	red or yellow seeds (bean shaped)

128 In West Africa, the related species *Pericopsis laxiflora* (Benth. ex Bak.) van Meeuwen [syn. *Aromosia laxiflora* (Benth. ex Bak.) Harms, *Ormosia laxiflora* Benth. ex Bak.] is reported to have a "kind of hypnotic or hallucinogenic effect" (Neuwinger 1994, 635*).

As the jaguar indicates, the Argentinean coral tree (*Erythrina falcata*), known locally as *seibo*, is clearly a shamanic tree that was or secretly still is used as a hallucinogen. (From Pedro de Montenegro, *Materia médica misionera*, seventeenth century)

Erythrina poeppigiana (Walpers) Cook

Parts of this coral tree, which is native to the Amazon basin, are used as **ayahuasca** additives (Ott 1993, 217, 270*). In Latin America, the flowers of this species are eaten as a vegetable or salad.

Erythrina standleyana Krukoff—chakmolche', pito del monte

In Yucatán (Mexico), the seeds are believed to offer magical protection against "evil winds" (*k'ak'as ik'o'*). The Maya place them on the altar for the rain ceremony *ch'a'chak* (cf. **balche'**) (Barrera M. et al. 1976, 303*). It is not known whether the seeds were or are used for psychoactive purposes. It is possible that this species is actually only a variety of *Erythrina americana*.

Erythrina vespertilio Benth.—batswing coral tree

This small tree is found in tropical eastern Australia, where the local Aborigines make the red, beanlike seeds into chains and (magical) ornaments. They also make shields from the wood of the tree (Pearson 1992, 106*). Whether the seeds were used for psychoactive purposes is unknown. The plant is rich in alkaloids (Collins et al. 1990, 40*).

Erythrina spp.

The seeds of many species of *Erythrina* are known as *colorines* and are used for magical as well as ethnopharmacological purposes (see the table at left). The ethnographic literature makes frequent mention of species that certainly were not compared to herbarium specimens and whose botanically identity may thus be incorrect. For this reason, it is not possible to provide a precise classification.

In Venezuela, the ashes of the wood of several *Erythrina* species known as *bucare* (also *anauco*, *ceibo*, *immortelle*) are used as additives in the tobacco mixture known as *chimó* (see **Nicotiana tabacum**).

Literature

See also the entry for *Erythrina americana*.

Amer, M. E., M. Shamma, and A. J. Freyer. 1991. The tetracyclic *Erythrina* alkaloids. *Journal of Natural Products* 54:329–63.

Games, D. E., A. H. Jackson, N. A. Khan, and D. S. Millington. 1974. Alkaloids of some African, Asian, Polynesian and Australian species of *Erythrina*. *Lloydia* 37:581 ff.

Olemy, M. M. el-, A. A. Ali, and M. A. El-Mottaleb. 1978. Erythrina alkaloids. I. The alkaloids of the flowers and seeds of *Erythrina variegata*. *Lloydia* 41:342–47.

Peña, Claudia, Fanny Villarraga, and Gerardo Pérez. 1988. A lectin from the seeds of *Erythrina rubrinervia*. *Phytochemistry* 27 (4): 1045–48.

Standley, P. C. 1919. The Mexican and Central American species of *Erythrina*. *Contributions of the U.S. Herbarium* 20:175–82.

Wandji, J., Z. Tanee Fomum, F. Tillequin, A. L. Skaltsounis, and M. Koch. 1994. Erysenegalenseine H and I: Two new isoflavones from *Erythrina senegalensis*. *Planta Medica* 60:178–80.

Right, from top to bottom:
In Mexico, the red seeds of the rain forest tree *Ormosia* are also known as *colorines* and are used to make necklaces.

The red-black seeds of an *Ormosia* species from the southern Mexican rain forest are also known as *colorines*.

A typical Mexican amulet, in the center of which is an *Erythrina* seed

In Mexico, the small black-red seeds known as rosary peas (*Abrus precatorius*) are also called *colorines*

These red mescal beans (*Sophora secundiflora*) are easily mistaken for the red seeds of the other tree legumes.

Below: The red seeds of this southern Mexican *Erythrina* species are also known as *colorines*.

Erythroxylum coca Lamarck

Coca Bush

Mama Coca presents the coca bush, her sacred plant, to the Spanish invaders. (Engraving by Robida, from the French edition of Mortimer, ca. 1904)

Family
Erythroxylaceae (Coca Family)

Forms and Subspecies
The genus *Erythroxylum* (previously *Erythroxylon*) encompasses some three hundred species. Apart from *Erythroxylum coca* and *Erythroxylum novogranatense*, however, none of these contains any significant amount of cocaine. The most commonly cultivated species is *Erythroxylum coca*, which is subdivided into two varieties that are distinguished on the basis of morphological, geographical, and ecological characteristics (Plowman 1982):

Erythroxylum coca Lam. var. *coca*—huanuco,
  Bolivian coca (humid mountain regions from
  Ecuador to Bolivia)
Erythroxylum coca var. *ipadú* Plowman—ipadú,
  Amazonian coca (tropical lowlands,
  Amazonia)

Synonyms
Erythroxylon coca Lam.[129]
Erythroxylon peruvianum Prescott (= *E. coca* var.
  coca)
Erythroxylum bolivianum Burck (= *E. coca* var. *coca*)
Erythroxylum peruvianum Prescott (= *E. coca* var.
  coca)

Folk Names
Erythroxylum coca var. *coca*:
Bolivian coca, bolivianische coca, botô, ceja de montaña coca, Ceylon huanuco, coca, coca bush, coca del Perú, cocaine plant, cocaine tree, cocamama, cocastrauch, cocca, cochua, coco, cuca, divine plant of the Incas, gran remedio, huanacoblatt, huánuco coca, koka, khoka (Aymara, "tree"), kuka (Quechua), la'wolé (Mataco), mamacoca, Peruvian coca, spadie

Erythroxylum coca var. *ipadú*:
Batú, botô (Makú), coca, coca-á (Siona), daallímü, ebee, hibi, hibia, hibio, huangana-coca (Bora), igatúa (Karijona), ipadó, ipadu (*língua-geral*), ipadú, ipatú (Yucuna), ípi (Bora), jibína (Witoto), kaheé (Makuna), majarra coca, pató (Tatuyo), patoó (Kubeo), pelejo coca, tsi-paa, ypadu, ypadú

The word *coca* is from the Aymara language and means simply "tree" (Weil 1995). This is an expression of the great cultural significance of the plant.

History
The coca bush is originally from the rain forests of the Andean foothills. For millennia, it has been cultivated and used in South America for a multitude of purposes. The oldest archaeological evidence of coca chewing has been dated to approximately 3000 B.C.E. In the dry lowlands of Peru, numerous pre-Columbian graves have yielded remains of coca leaves (cf. *Erythroxylum novogranatense*), lime, and artifacts associated with coca use (Hasdorf 1987; Martin 1969; Towle 1961, 58 ff.*). Archaeological finds of coca leaves are extremely rare in the Andean highlands, primarily because of the poor state of preservation of botanical materials and the clumsy excavation methods of past decades. Recently, *Erythroxylum coca* var. *coca* was identified in a prehistoric settlement in the Upper Mantaro Valley (Peru) (1000–1460; Late Intermediate and Late Horizon periods) (Hastorf 1987). Hairs taken from mummies found in northern Chile have been tested for cocaine and its most significant metabolite (benzylecgonine). Trace amounts were detected in almost all of the samples. The oldest of the mummies dates to some four thousand years ago (Cartmell et al. 1991).

Coca had an extremely important function in many pre-Columbian cultures as an article of economic exchange, a medicine, an aphrodisiac, a remedy, and a ritual inebriant. Without coca, the civilizations of the Andes would have been inconceivable (Mortimer 1974). The Spanish first encountered the widespread use of coca when they moved into South America to subdue and suppress the indigenous cultures. They understood this custom as little as they understood other aspects of Indian culture. The government of New Spain quickly forbade coca use in the years between 1560 and 1569, using a dubious line of reasoning that is quite reminiscent of many modern arguments in support of some drug laws: "The coca plant is nothing more than idolatry and witchcraft which only appears to strengthen evil by deceit, possesses no true virtues, but probably takes the lives of a number of Indians who in the best case only escape from the forests with damaged health" (cited in Voigt 1982, 36). In the seventeenth century, the Inquisition viewed the veneration of coca as a sign of witchcraft and magic but found it a difficult custom to overcome. For the Indians, who regarded the coca bush as sacred and not diabolic, life in the oxygen-poor high mountains was inconceivable without coca. For this reason, they held firm to their tradition and disregarded the laws of New Spain and the

Catholic Church. When the separation from the Spanish motherland occurred, the use of coca was normalized and, ultimately, legalized in Peru and Bolivia. Today, the use of coca is associated with Indian identity; coca is, so to speak, the expression of the Indian way of life and of indigenous culture (Instituto Indigenista Interamericano 1986, 1989; Lobb 1974).

In 1565, the Spanish physician Nicolas Monardes wrote that the Indians chewed coca together with tobacco. Monardes brought the first coca plant to Europe in 1569 (other sources say 1580) (Morton 1977, 180*). The first botanical illustration of the plant was made by Clusius in 1605 (Lloyd and Lloyd 1911, 3). The main constituent, **cocaine**, was first isolated from the leaves in 1859 by the German chemist Albert Niemann.

At the end of the nineteenth century, cigars and cigarettes made of coca leaves were being smoked in Philadelphia. It appears that the leaves were also smoked in England, where they were known as Peruvian tobacco (Lindequist 1993, 90).

Henry Hurd Rusby worked on behalf of the Parke and Davis Company to have coca leaves incorporated into the American pharmacopoeia (Rossi-Willcox 1993*). In 1863, the Corsican chemist Angelo Mariani created his Vin Mariani, a coca extract in sweet wine. Fans of the preparation included Queen Victoria, Pope Leo XIII (who served from 1878 to 1903), Mozaffer-et-Dine (the shah of Persia), Thomas Edison (who invented the motion picture and numerous other devices), and a great number of artists and intellectuals (Andrews and Solomon 1975, 243–46; Voigt 1982, 22).

The most important botanical and ethnobotanical studies of the past century were conducted by Timothy Plowman (1944–1989), who produced forty-six publications about coca and *Erythroxylum* (Davis 1989, 98).

In recent years, the governments of Bolivia and Peru have been working to have coca products legalized around the world. The discussions, however, continue to make a moral dichotomy between "good coca" and "bad cocaine" (Cabieses 1985; Henman 1990).

Distribution

The coca bush (*coca* variety) is originally from the rain forests on the mountain slopes of Peru and Bolivia, the so-called *yungas* (Schröder 1991, 112*). It is found at altitudes of up to 2,000 meters, although it is most often cultivated between 500 and 1,500 meters. As a result of cultivation, the coca bush is now found in many regions of the world (Indonesia, Seychelles, East Africa, India) (Potratz 1985); Ceylon huanuco, which was successfully cultivated in Sri Lanka, is now quite well known (Macmillan 1991, 415*).

Amazonian coca (*ipadú* variety) is found only in tropical lowlands (Amazon basin) (Plowman 1979b, 46).

Cultivation

The seeds of *Erythroxylum coca* var. *coca* are sown naturally by birds that eat the ripe drupes from the bush and excrete the seeds undigested. In the Andes, this variety is propagated almost exclusively from seeds (Plowman 1979b, 46). Coca seeds become infertile when they dry (normally after three days). The seeds are pressed into shaded seedbeds for germination. The seedlings are transplanted when they have grown about as large as a person's hand. The plants are placed into the ground at a distance of about 1.5 meters from one another. In South America, transplanting is carried out during the rainy season. The large fields and plantations of coca in the Andes are known as *cocal* (singular) or *cocales*.

The bush does not place any great demands upon the soil. It prefers loose, humus-rich soil, which should be fertilized often with plant compost. It thrives well in loamy soil formed from weathered slate. Chalky soils are not suitable (Bühler and Buess 1958, 3047). This skiophilous plant requires high humidity and ample precipitation (at least 2,000 mm per year) and does not tolerate frost.

From the time of planting, a period of some eighteenth months is necessary before the first leaves can be harvested. A bush produces for twenty to thirty years. During the rainy season, the bushes can be harvested every fifty to sixty days. During the dry season, harvesting can be performed only every three or four months. The plant is not disturbed by the removal of almost all of its leaves. If the leaves are not harvested, the bush will grow into a proper tree.[130] The leaves of these coca trees are almost devoid of effects.

In the Amazon region, coca growing is almost exclusively a male activity. In contrast, the planting of food crops is usually done by women. The Amazonian coca bush is pruned to a height of about 1.5 meters. Such bushes are known as *ilyimera*, "little birds." Amazonian coca is propagated solely through cuttings, as this variety does not produce viable seeds (Plowman 1979b, 46 f.).

Appearance

The coca plant usually grows as a bush. The elliptical foliage leaves, whose length varies by subspecies, are arranged spirally. The bark of young plants is reddish in color. The scaly leaves that appear at the base of young branches are a characteristic feature of the plant. The tiny, monoclinous white flowers develop from the axes of these scaly leaves. The flowers are radially symmetrical and have ten stamens that are joined

An engraving of an herbarium specimen of *Erythroxylum coca*. The two "transparent" leaves clearly show the vein pattern typical of the variety *coca*, in which two lines run parallel to the central vein. (From Mariani, *La Coca et la Cocaïne*, 1885)

130 Other species of the genus (*E. areolatum, E. tortuosum*) are sources of timber (Anzeneder et al. 1993, 65*).

Left: A branch tip of the coca bush *Erythroxylum coca* var. *coca*

Center: The tiny flowers of the coca bush *Erythroxylum coca* var. *coca*

Right: The leaves of the coca bush *Erythroxylum coca* retain their elasticity when dried.

at their bases. As they ripen, the small oval fruits (drupes) initially turn yellow and then luminously red.

Erythroxylum coca var. *coca* usually grows to only 3 to 5 meters in height, although it can grow taller. *Erythroxylum coca* var. *ipadú* grows to a height of only about 3 meters and can be recognized by its long and very thin branches. The leaves are larger, somewhat rounder, and more elliptical than those of *coca* variety, and they do not taper at the end (Plowman 1979b, 46). In Amazonia, coca bushes are often covered completely with lichens.

The coca bush is very easily confused with other species of the genus *Erythroxylum*, as many of these have a similar appearance. The most certain method of botanical identification is to chew the dried leaves together with an alkaline substance. If the mucous membranes of the mouth become numb, then the plant can only be one of the two species that contain cocaine (*Erythroxylum coca* and **Erythroxylum novogranatense**) or their varieties.

It appears that the scientific literature continues to confuse some of the species of the genus *Erythroxylum* with *E. coca* or subsume them within this species. Because of the many local varieties of the coca plant, even botanists can have difficulty identifying the species in question (Plowman et al. 1978).

Psychoactive Material
— Dried leaves (cocae folium)

The leaves must be dried (roasted) before use or they will not produce the intended effects. Freshly picked leaves can be either lightly roasted or made into a tea. The freshly harvested leaves are dried in such a manner that they retain their green color while also remaining supple and elastic. Drying can occur in the sun or by artificial means. If the leaves are dried in an oven or similar manner, the temperature should not exceed 40°C (= 104°F) lest the cocaine content be adversely affected (Schröder 1991, 114*). The taste of the dried leaves of huanuco (*coca* variety) is strongly reminiscent of that of green Chinese tea (**Camellia**

sinensis). In contrast, Amazonian coca (variety *ipadú*) is somewhat bitter in taste (Koch-Grünberg 1921, 175*).

Preparation and Dosage
Coca leaves can be chewed, smoked or otherwise burned and inhaled, or ingested in the form of an extract (tea, decoction, tincture, etc.).

By far the most common method of ingestion is chewing or, more precisely, sucking the leaves. In the Andes, coca quids are usually known as *acullico*, and coca chewing is known as *acullicar*. At the beginning of the Incan period, coca leaves were chewed together with tobacco leaves (see **Nicotiana tabacum**), a practice that was also observed during the colonial period but now appears to have largely disappeared. The Swiss naturalist Johann Jacob von Tschudi (1818–1889), who also was the first to observe and report the use of angel's trumpet (see **Brugmansia sanguinea**), provided a very thorough description of the Andean use of coca that still applics today:

At least three times, but normally four times per day, the Indians rest from their work so that they may chew coca. For this purpose, they carefully remove the individual leaves from the *Huallqui* (bag), remove the veins, place the divided leaf into their mouth, and chew this for as long as it takes for a proper ball to form under their molars, they then take a thin, moistened little stick of wood and dip this into slaked lime and then place this together with the adhering powder into the ball of coca in their mouths; they repeat this a couple of times until it has the proper spice; some of the copious amounts of saliva, which mixes with the green juice of the leaves, is spit out, but most of this is swallowed. When the ball no longer produces enough juice, they throw it away and begin with another. I have often watched how a father would pass an almost juiceless ball to his little boy, who eagerly took it into his mouth and chewed on it for a long while. (In Bühl and Buess 1958, 3052 f.)

The coca leaves must be mixed with an alkaline substance (known as "sweetening" the coca quid) for the **cocaine** to be released so that it can be absorbed through the mucous membranes of the mouth (Cruz Sánchez and Guillén 1948; Rivier 1981; Wiedemann 1979, 280). In South America, either plant ash or burned/slaked lime from various sources is used for this purpose (Gantzer et al. 1975, 10).

In the Andes, coca is chewed together with what is known as *llipta* (Quechuan), scrapings from a cake of ash. *Llipta*—also known as *chile*, *llucta*, *llinta*, *lliptu*, and *tocra*—is made from the ashes of various plants (see the table on pages 247–48). The ashes are produced not by burning the plants but by roasting them thoroughly. To do this, the plant pieces are placed in a metal or ceramic pot. The pot is kept over a fire until the plant pieces break down into an ashlike powder. The ashes are then moistened with lemon juice, boiling water, **chicha** (maize beer), sugarcane schnapps (**alcohol**), sweetened tea (*Camellia sinensis*, *Ilex paraguariensis*), salt water, or even urine and kneaded together with a carrier substance such as potato flour or starch. The mixture is formed into large disks, small pyramids, snakes, etc., and allowed to dry in the air for a day (Bühler and Buess 1958, 3054; Franquemont et al. 1990, 66 f.*). When dried, the *llipta* is as hard as a rock. Pieces are broken off and added to the coca quids.

In Bolivia and northwestern Argentina, coca leaves are now chewed with sodium bicarbonate (*bicarbonato de sodio*, *bica*, *yuspe*), which is sold in plastic bags weighing 20 g. The Mataco (Wichi, "people of the place") chew coca in the style of the Andes, whereby they "eat" the entire leaf. They stuff their mouths so full that their faces have an enormous bulge. They then simply toss some *llipta* (sodium bicarbonate) into their open mouths.

A variety of substances may be added to coca quids in order to modify their psychoactive or medicinal effects or to make the effects more specific. In the triangle of countries formed by Bolivia, Argentina, and Chile, coca quids may be chewed together with the ashes of the flowers and fruits (without seeds) of a large columnar cactus (*Trichocereus pasacana*) that is frequently confused with the San Pedro cactus (cf. *Trichocereus pachanoi*) (C. M. Torres, pers. comm.). It is possible that the (main) alkaloid of the cactus, candicine,[131] alters the effects of the coca. The Argentinean Mataco obtained their *yista* (= *llipta*) from the ashes of the cactus flesh of a *Trichocereus* species (*tso'nahlak*). This was said to potentiate the effects of the coca (cf. *Trichocereus* spp.).

There are also a number of substances used to aromatize coca quids and improve their taste. Leaves of the rosary pea (*Abrus precatorius* L.; cf. *Rhynchosia pyramidalis*), known in northern Peru as *misquina*, which are roasted beforehand so that they will not produce any toxic effects, impart a licorice-like taste to coca quids. The leaves of *Tagetes pusilla* H.B.K.[132] (cf. *Tagetes* spp.), known in southern Peru as *pampa anis* ("prairie anise"), are also used to lend the coca quids an aromatic taste (Plowman 1980, 254).

The Peruvian Campa Indians like to add the bark of the *chamairo* vine (*Mussatia hyacinthina* [Standl.] Sandw.; Bignoniaceae), which is also used for medicinal purposes, to their coca. This practice can also be observed in other parts of Peru; the bark is sold in markets for precisely this purpose (Plowman 1980, 255 f.).

The Amazonian preparation is very different from the Andean and is identical among all the tribes except one. The leaves of the Amazonian coca (*E. coca* var. *ipadú*) are plucked fresh daily from the bush and immediately roasted on a cassava baking tin. This roasting must be done gently and carefully so that the leaves do not carbonize. The men then pound the roasted leaves in large mortars made from hollowed-out trunks of the hardwood trees *Tabebuia* spp., mahogany (*Swietenia mahagoni* [L.] Jacq.), or chontaduro palm (*Guilielma speciosa* Martius). While they are pounding, the leaves of other plants are turned into ashes over a charcoal fire. The gray ash is mixed with the green coca powder in more or less equal amounts and is then ready for consumption. A person typically takes a spoonful, which he or she carefully moistens with saliva and then pushes between the cheek and the teeth with the tongue. There, the mixture slowly dissolves over a period of thirty to forty-five minutes and is gradually swallowed.

Of all the Amazonian coca additives, by far the most popular are the ashes of the large, fresh leaves of *Cecropia sciadophylla*, known variously as *pêtuy'*, *göra-ñá*, *guarumo*, or *setico* (Schultes and Raffauf 1990, 313*). Other species from the genus *Cecropia* as well as *Pourouma cecropiaefolia* are also utilized (Plowman 1979b, 47). Occasionally, other plant substances may be used as well. The Witoto sometimes add some powdered root of *Chelonanthus alatus* to the

Left: A typical cake of ash (*llipta*) from the southern Andes, scrapings of which are added to the coca leaves

Right: In northwestern Argentina, industrially manufactured baking powder (sodium bicarbonate) is usually used as an alkaline coca additive.

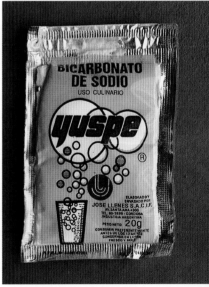

131 *Trichocereus* from a California culture was found to contain 0.075% candicine (Shulgin 1995, 26*).

132 The fresh leaves of this *Tagetes* species are also chewed alone as a cold remedy (Plowman 1980, 254).

"Coca enhances performance and suppresses both hunger and tiredness. But its primary effect is to induce the latent power of those visions which lead one closer to the 'reality of dreams.'"

Wolfgang Müller
Die Indianer Amazoniens [The Indians of Amazonia] (1995, 197*)

A branch of the coca bush (*Erythroxylum coca* var. *coca*) from the highlands of Bolivia. (From Mariani, *La Coca et la Cocaïne,* 1885)

133 The resin consists of 30% protamyrine, 25% proteleminic acid, 37.5% protele resin, and oils. In Amazonia, it is often burned in the local churches as an incense (Schultes 1957, 246). It is also the most important **incense** in the Brazilian Santo Daime cult (see **ayahuasca**).

coca/ash mixture to lend it a "bitter taste" (Schultes 1980, 57) or mix it with dried and powdered leaves of *Tachia guianesis* Aublet (Gentianaceae) to improve the taste (Schultes and Raffauf 1986, 276*).

An unusual method of preparing coca was discovered among a small group of the Tanimuka (on the Río Apaporis, Colombia). They aromatize the ashes of *Cecropia* leaves with **incense**. To do this, they make incisions in the bark of *Protium heptaphyllim* March and tap the resin, which they then allow to age for three to four months. The resin,[133] which is known in Amazonia under the names *o-mo-tá, hee-ta-ma-ká, brea, pergamín, tacamahaca,* and *breuzinho,* is broken into small pieces and rolled into a semidried leaf of an *Ischnosiphon* species to make a kind of cigarette. The men who are involved in the preparation of this coca will put this cigarette in their mouths and light it, but they will not inhale. Instead, they blow through the resin-filled tube so that the aromatic smoke flows out of the other end. When the cigarette is burning well, they place its tip into the *Cecropia* ashes for a couple of minutes to fumigate it. The scent is absorbed by the plant ash, thereby giving the finished mixture of coca and ash a resinous, incenselike aroma (Schultes 1957; Schultes and Raffauf 1990, 117*; Uscátegui M. 1959, 297*).

The Makú Indians use their coca (*ipadu, botô*) in a manner different from that of all the other Amazonian tribes. The leaves are roasted, mixed with the ashes of fresh green banana leaves (*Musa* spp.), and finely crushed in a ritual context. This powder is then mixed with flour (cassava, farina, tapioca) and made into bread. The bread is prepared fresh every evening and is regarded as food; it is not just chewed but properly eaten (Prance 1972a, 19*).

The coca/ash powder is also sniffed (cf. **snuffs**) in some regions of Colombia, although such use has been little documented (Schultes 1980, 53).

Coca can be combined with almost any other psychoactive substance. Coca will sometimes even potentiate the effects of another substance, e.g., *Anadenanthera colubrina.* Coca leaves are also suitable for use as a stimulating ingredient in **incense** and **smoking blends** and are especially well suited for smoking together with *Cannabis sativa.*

When chewing coca, one should avoid drinking maté (*Ilex paraguariensis*) entirely, not because the two substances produce a negative synergy, but because the anaesthetized mucous membranes of the mouth are unable to detect the temperature of the maté and can be scalded very easily without notice. Chronic coca chewing will occasionally result in mild inflammations of the mucous membranes. A tea made of the leaves and bark of

Pagamea macrophylla Spruce ex Benth. may be drunk to counteract such problems (Schultes 1980, 57).

The usual dosage for a medicinal tea is given as 5 g of dried coca leaves (Morton 1977, 180*). Much larger amounts are consumed, however, when the coca is chewed. With an average use of 60 g of good leaves per day, it can be assumed that 100 to 200 mg of **cocaine** are being absorbed. In some Amazon tribes (e.g., the Yucuna), it is not unusual to observe men consuming up to 1 pound of coca/ash powder daily (Schultes 1980, 51). When smoked, even small amounts (0.1 g or more) of the roasted leaves will produce stimulating effects. The Omagua smoke leaves as they chew them (Bühler and Buess 1958, 3054).

Ritual Use

The ritual uses of coca are manifold. The leaves are used as parts of offerings and for oracles, socially integrative forms of interaction, shamanic healings, initiations, and tribal festivals. The ritual use of coca is probably as old as the use of the leaves in general, at least five thousand years. Unfortunately, little is known about the use of coca in pre-Hispanic times. Grave goods clearly indicate that coca was given to the dead for their journey to the other world. The representations of coca chewing contained on pre-Columbian artifacts also point to a very ancient ritual use.

The ethnohistorical evidence from the colonial period is rather limited and was obviously filtered through the "devil's glasses" of the Catholic Spanish. In *The Naturall and Moral Historie of the West Indies* (ca. 1570), José de Acosta provides an amazingly unprejudiced account: "The Ingua [= Inca] are said to have used coca as an exquisite and regal thing which they most often used in their offerings by burning it in honor of their gods."

The significance of coca in the kingdom of the Incas has been summarized as follows:

In ancient Peru, where coca was venerated as a gift of the gods of the sun, there were few ceremonies which did not require the drug. At great festivities, coca leaves were burned as fumigants, and priests adorned with coca wreaths would use the smoke to divine. Only with a quid of coca in his mouth could one dare to approach the gods, and coca was one of the gifts offered to the priests. The coca offering had a special significance. (Bühler and Buess 1958, 3061)

Coca is sacred to the Indians because it makes possible the connection between humans and the gods (Allen 1988, 132; Lloyd and Lloyd 1911) and also deepens the contact between people, e.g., as a

Additives for Coca Chewing

(From Aldunate et al. 1981*; Fernandez Distel 1984; Plowman 1980; Prance 1972a, 1972b*; Schultes 1957, 1980, 1983b*; Schultes and Raffauf 1986,* 1990*; Wiedemann 1979; revised and expanded.)

Name	Part	Form Used
Plants		
Abrus precatorius L.	dried leaves	powder
Amaranthus sp. (*ataco, aromo*)	herbage without roots	ashes
Aristeguietia (*Eupatorium*) *discolor* (DC.) King et Robinson (*isphinhuy*)	herbage	ashes
Astrocaryum munbaca Mart. (*rui-ré-gö* palm)	leaves	ashes
Baccharis tricuneata (L. f.) Pers. (*tayanqa*)	herbage	aroma
Brugmansia spp.	fresh leaves	pieces of leaves
Cactaceae (*k'achilana*)	cactus flesh	ashes
Capsicum spp.	fruit	chili powder
Cecropia spp. (*yarumo* trees)[134]	fresh leaves	ashes
Cecropia ficifolia Warburg (*wa-kö'-bö-ta*)	leaves	ashes
Cecropia palmata Willd.	leaves	ashes
Cecropia peltata L.	leaves	ashes
Cecropia sciadophylla Martius (*guarumo, setico*)	leaves	ashes
Chelonanthus alatus (Willd.) Pulle	root	powder
Chenopodium ambrosioides L.	entire plant	ashes
Chenopodium hircinum Schrad. (*yuyo, quinoa, ch'api*)	herbage	ashes
Chenopodium pallidicaule Aell.	entire plant	ashes
Chenopodium quinoa Willd.	entire plant	ashes
Chenopodium spp. (*ajarilla, illincoma*)	herbage without roots	ashes
Cortaderia atacamensis (Phil.) Pilg. (*cortadera*)	flowers	ashes
Costus amazonicus (Loes.) Macbr. (*nã'-ka*)	leaves	ashes
Costus erythrocoryne K. Schum.	leaves	ashes
Diplotropis martiusii Benth. (*ko-ma'-ma*)	leaves	ashes
Distictella pulverulenta Sandw. ("vine for coca ashes")	leaves	ashes
Eupatorium sp. (*suytu suytu*)	herbage	aroma
Helianthus annuus L. (sunflower)	flower petals	ashes
Heliconia sp.	roots	ashes
Ipomoea batatas (L.) Lam. (sweet potato)	rhizome	flour
Iriartea exorrhiza Martius (*paxiúba* palm)	leaves	ashes
Musa x *paradisiacum* L. (banana)	roots	ashes
Musa sapientum L. (banana)	leaves	ashes
Musa spp. (banana)	leaves	ashes
Mussatia hyacinthina (Standl.) Sandw. (*chamairo*)[135]	bark	pieces of bark
Nicotiana tabacum (tobacco)	leaves	paste, powder
Octea opifera Martius	fruits	ashes,[136] powder
Octea simulans C.K. Allen	leaves	ashes
Palms, various	leaves	ashes
Plumbago coerulea H.B.K. (*asul ñuqchu*)	herbage	ashes
Portulacca oleracea L. (*verdolaga*)	herbage without roots	ashes
Pourouma cecropiaefolia Mart. (*curúra, uva de monte*)	leaves	ashes
Protium heptaphyllum Marchal (*breuzinho*)	resin	smoke (to fumigate the coca)
Puya weberbaueri Mez (*tayñu*)	flowers, dried	ashes
Schinus molle L.	fruits	red pepper powder
Senecio sp. (*chula-chula*)	leaves	pieces of leaves
Solanum topiro Humb. et Bonpl.	seeds	powder
Solanum tuberosum L. (potato)	tuber	flour

134 There are some fifteen species of the genus *Cecropia* in Amazonia, many of which have ethnobotanical uses (Berg 1978); cf. **snuffs** and marijuana substitutes (**Cannabis indica**).

135 *Chamairo* can also be chewed alone; it produces mild euphoriant effects (Plowman 1980, 258).

136 According to the Taiwano, this additive makes the coca stronger and "better" for certain ritual dances (Schultes 1983a, 258*; Schultes and Raffauf 1990, 170*).

Top: In northern Chile, the herbage of an *Amaranthus* species is added to coca quids. (Photographed in San Pedro de Atacama)

Bottom: Various species of the genus *Chenopodium* are traditionally used as coca additives.

Top: In Amazonia, the leaves of *Cecropia peltata* are one of the primary sources of the ashes used as an alkaline additive to finely ground Amazonian coca (*Erythroxylum coca* var. *ipadú*).

Bottom: The flower petals of the American sunflower (*Helianthus annuus*) not only are used as a coca additive but also are regarded as an aphrodisiac in Indian medicine.

Name	Part	Form Used
Stylogyne amplifolia Macbride	leaves	ashes (?)
Styrax anthelminticum Schultes	bark	ashes
Styrax spec. nov.	bark	ashes
Suaeda aff. *divaricata* Moq. (*jume*)	herbage without roots	ashes
Tachia guianensis Aubl.	leaves	powder
Tachigalia cavipes (Spruce ex Benth.) Macbride	bark	powder
Tachigalia paniculata Aubl. var. *comosa* Dwyer	leaves	ashes[137]
Theobroma cacao L.	fruits	ashes
Trichocereus pasacana (Webb.) Britt. et Rose	flowers, fruits without seeds	ashes
Trichocereus sp.	fruits	ashes
Vernonia sp.	stalks	ashes
Vicia faba L.	roots	ashes
Vochysia ferruginea Martius	leaves	powder[138]
Zea mays L. (corn/maize)	corn silk, stems, cobs	ashes

Animals
bones, various sources
mollusks (selection):

		ashes
Melongena melongena L.	shells	burned lime
Strombus spp.	shells	burned lime
Strombus gallus L.		
Strombus gigas L.		
Strombus pugilis L.		
Strombus raninus Gmelin		
Venus spp.	shells	burned lime

Minerals

clay		dried or burnt
limestone (*mombi*)		burned/slaked lime
soil		calciferous
stalagtites/stalagmites		burned lime

Other

baking powder		
manioc flour (*fariña*)		starch
sodium bicarbonate (*bica*)		
sugar molasses		saccharose

137 The unripe fruit husks of the closely related *Tachigalia ptychophysca* Spruce ex Bentham are esteemed as an aphrodisiac (Schultes 1978a, 184*).

138 The addition of this leaf powder is said to prevent and heal the sores that can develop in the mouth as a result of constant coca chewing (Schultes 1977, 117*).

139 In the Caribbean, the related species *Erythroxylum rotundifolia* Lunan is used as an ingredient in love drinks (McClure and Eshbaugh 1983*).

love magic and an aphrodisiac[139] (Mortimer 1974, 429).

Whenever the Indians of the Andes come together, they offer, exchange, and chew coca with one another. People invite one another to chew coca as a way of initiating a social exchange. Although the practice is essentially the same throughout the region, the actual form of the exchange varies from place to place (Allen 1988, 126 ff.). Before the leaves are placed in the mouth to be moistened, three of them are placed together like a fan and held before the brow. The person then turns to face the highest of the neighboring mountains and consecrates the leaves with the words *poporo apú*.

In the Andes, coca leaves still number among the most important of all ritual offerings. A pile of offerings known as an *apacheta* can be found at the highest location of a mountain pass. This typically consists of a pile of hand-size stones, over which coca leaves are strewn as "payment" for a safe crossing of the pass. Chewed coca quids, bottles of **beer**, *aguardiente*, and pure **alcohol** may also be placed there. The coca leaves are a gift to Pachamama, the Mother Goddess. The offering of coca leaves at sacred places gives the Andean Indians a deep connection with their world (Allen 1988, 130). The offerings also have a medicinal importance:

Coca, along with other aromatic plants, is either burned as a smoke offering or offered in the natural state in the form of especially attractive leaves. The simplest type of offering consists of six beautifully shaped coca leaves, over which some distilled spirits and llama fat are dripped. More extensive offerings are composed of one hundred and forty-four *aita* (each consisting of six coca leaves) in rows of twelve. The *curandero*—medicine man—does this when he celebrates his mass for the ill. He asks that the relatives of the ill person bring offerings of aromatic herbs, llama fat, mussels, a rosary, sweets, and a woven cloth with coca leaves (*inkuña*). The sweets are offered in the beginning, during the "sweet" mass. This is followed by the "Apostles' mass," which is named after the twelve rows of coca leaves that are offered. Only then can the healer make a diagnosis. After this, he burns the offerings to appease the gods. (Wiedemann 1992, 7)

In the Andes region, coca leaves are an absolutely indispensable part of magical and religious healing rituals. The roots of many illnesses lie in the spirit world: *puquio*, a sleeplessness that afflicts a person who has not shown respect to the sacred springs; *huari*, a disease that is produced by the spirits that reside in ancient ruins; *japipo*, a sickness caused by spirits who steal parts of the soul; *tinco* or *tasko*, which occurs when a person encounters an aggressive soul; *susto*, "fright," which is induced by great emotional burdens. To heal these peculiar afflictions, which do not fit into the Western view of symptom-oriented diagnosis, the traditional healer must locate and visit the place where the spirits live or at which the encounter with them occurred. When he has found the correct location, he will continue to offer the appropriate beings as many coca leaves as needed until they permit themselves to be asked to remove the illness from his patient (Hoffman et al. 1992, 75*).

Specialized diviners use coca leaves for casting oracles (*coca qhaway*). These oracles are consulted by people suffering from illnesses and other types of problems (Allen 1988, 133 ff.; Franquemont et al. 1990, 67*; Quijada Jara 1982, 39 ff.). This ancient technique of coca oracles is still practiced today:

The contemporary oracular use of coca leaves is but a pale reflection of such renowned state oracles of the Incan period as Pachamama ("lord of the earth") near Lima. There, in the Lurin Valley, lies one of the oldest cult centers in South America, the settlement of which began 10,000 years ago. Pilgrims came from the jungle regions and even from Central America to listen to the words of the oracle after a prolonged period of fasting. (Andritsky 1987, 52)

Through his art, the coca oracle played an important function in structuring the community (Allen 1988, 133 ff.) and therefore had enormous social responsibility. A person had to undergo a long period of training to become a coca oracle and needed a great ability to empathize with his clients. Lest the clients become suspicious of the diviner's intent, the ritual had to be carried out precisely:

Through the ritual acts of the oracle, the spiritual principle of the coca leaves, "coca mama," acquires a new quality together with the "wamanis" [local mountain spirits]: because of his abilities to mediate, and in an altered state of consciousness, the diviner enters into a dialogue with these spirit powers. His role is that of interpreter, who lays out the structure of the coca leaves and their pattern on the oracular cloth according to specific rules. . . . Because many coca diviners are also healers, diagnosing with coca leaves is an important activity. (Andritsky 1987, 52)

Sometimes, other objects such as ergot-infected grains (cf. *Claviceps purpurea*) may also be thrown and read with the coca leaves.

Peruvian shamans inhale copious amounts of coca smoke so that they can enter an ecstatic state in which they are able to travel to the world beyond. In doing so, they cross over a "bridge of coca smoke" and enter a different reality, the shamanic universe, in which they are able to heal (Martin 1969).

Most tribes in the Amazon basin use coca as a stimulant and agent of pleasure, chewing it almost daily. Besides manioc (*Manihot esculenta*), the main source of food, the ipadú bush is the most important cultigen. But it also has a ritual significance. The Tukano Indians believe that the first coca plant arose from a finger joint of a daughter of the lord of the animals. Because the ***Banisteriopsis caapi*** vine arose from the finger joint of another of his daughters, **ayahuasca** and coca are regarded as "siblings" (Schultes and Raffauf 1990, 167*). Among the Tukano tribe of the Yebámasa, ipadú has both a ritual and a hedonistic significance:

Every adult male spends some three hours a day preparing the coca powder from the roasted leaves of the coca bush. The men consume this powder throughout the day almost without interruption. It makes them more able to be productive, prevents tiredness, and suppresses their hunger. But the Yebámasa do not eat it for these reasons alone. When they ingest coca, they also partake of the magical power that lies within, which has invigorating effects upon them and thus protects their body and spirit. In addition, the coca powder has an important social function:

Top: A pre-Columbian lime or *llipta* pouch from Peru

Bottom: A pre-Columbian lime container of wood with mother-of-pearl and shell inlay, for use with coca. (Peru, Inca period)

"When the divine son of the sun, Manco Capac, climbed down from the rocks of Lake Titicaca, he gave humans light, knowledge of the gods, and knowledge of the arts and of coca, a divine plant which satiates the hungry, gives new strength to the tired and exhausted, and helps the unhappy to forget their cares."

GARCILASO DE LA VEGA
(1539–1616)

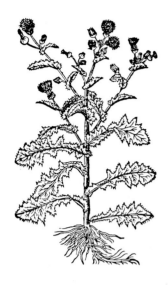

In the Atacama Desert of northern Chile, the dried leaves of the sow-thistle or hare's lettuce (*Sonchus oleraceus* L.) are chewed together with an alkaline substance as a coca substitute. The Atacameños call the wild plant, which is originally from Eurasia, *wirikocha*. This name is a reference to Viracocha, one of the most important deities of the Andes, who was regarded as the highest god in the Incan religion. This ancient Andean god of the heavens would select his priests and shamans with a lightning bolt. (Woodcut from Tabernaemontanus, *Neu Vollkommen Kräuter-Buch*, 1731)

the reciprocal offering of coca powder is a gesture of contact and friendship. (Deltgen 1979, 23*)

Artifacts

Numerous artifacts are associated with the coca bush. These include the paraphernalia for consuming the leaves, depictions of the plant and the goddess that dwells within, and many cultural products that have been inspired by its stimulating effects.

Both the Andean and the Amazonian Indians manufacture special containers for storing and transporting (carrying along) coca leaves. In the Andes, coca purses (*chuspa, pisca, mochila, guambis*) are used for storage. Over the centuries or even millennia, the production of these purses has become a great art. The woven bags are usually decorated with abstract patterns and images of symbolically important animals and gods. For example, the highest god, Viracocha (literally "father of the sun"; also spelled Huiracocha and Virakocha), is represented by a small duck (Wiedemann 1992, 17).

In the Amazonian regions, the coca and ash powder is usually stored in coca bottles (*cuya*) made from *jicaras* (fruit coats of tree gourds or calabash trees, *Crescentia cujete* L.) or coca bags (*tuturí*) produced with paper made from the bark of a *Ficus* sp. or an *Eschweilera* sp. The lime containers (*checo, iscupuru, calero mombero*) are usually made from smaller gourds. The Amazonian coca spoons were once made chiefly from jaguar bones (Schultes 1980, 51). Today, Western-style spoons are also used.

In Europe, coca has influenced or inspired numerous poets, writers, and artists. The first European poem—a hymn—that appears to have been dedicated to coca was penned by the English physician and diplomat Abraham Crowley (1618–1667).

In the late nineteenth century, a potent coca wine known as Vin Mariani played an especially significant role in stimulating, inspiring, and supporting the creative efforts of numerous artists, intellectuals, and politicians. The writers Alexandre Dumas, Henrik Ibsen, Octave Mirbeau, Sully-Prudhomme, and particularly Jules Verne and H. G. Wells all "lived" on Mariani wine and wrote their best works while under its influence. The French composers Charles Gounod (1818–1893) and Jules Massenet (1842–1912) were passionate enthusiasts of the "wondrous coca wine" and praised this "beneficent creator."

Modern music contains numerous references to cocaine but only rarely makes mention of coca. Merrell Fankhauser dedicated a hymn to the plant ("Treasure of the Inca") on his CD *Jungle Lo Lo Band* (Legend Music LM 9015, 1994).

Medicinal Use

We know with certainty that coca leaves were an important medicine in pre-Hispanic times. Unfortunately, the paucity of sources from the period means that we do not have any specific information concerning the ways in which they were utilized. For more on the use of coca in pre-Columbian trepanations, see *Erythroxylum novogranatense*.

Today, the folk medicinal uses of coca are so manifold that it has been called the "aspirin of the Andes." Coca is used for pains of all types, neuralgia, rheumatism, colds, flu, digestive problems, constipation, colic, upset stomachs, altitude sickness, exhaustion, and states of weakness and to ease labor (Quijada Jara 1982, 35 ff.). Coca leaves are burned or smoked to treat bronchitis, asthma, and coughs (Morton 1977, 180*). In the nineteenth century, the popular *Encyclopädie der medizinisch-pharmazeutischen Naturalien- und Rohwarenkunde* of Eduard Martiny (1854) listed coca smoke as a treatment for asthma. Coca leaves appear to have been smoked for this purpose quite often in England, where they were imported under the name *Peruvian tobacco.*

A brewed coca tea (*mate de coca*) is recommended in cases of diabetes and to suppress the appetite of people who are overweight, as a stomach tonic, as an aid in digestion, and to treat diarrhea, states of exhaustion, and especially altitude and travel sickness. The tea is effective both therapeutically and as a preventative for *soroche* or *la puna*, the altitude sickness that is common on the Altiplano, the high plateau of the Andes (Europeans are especially fond of the tea) (Sarpa and Aimi 1985; Schneider 1993, 19*).

In Peru, a tea made with coca leaves and *cedrón* (*Lippia citridoria* [Ort. ex Pers.] H.B.K. [syn. *Aloysia triphylla* (L'Hérit.) Britt.]; cf. **marari**] is drunk to treat stomach pains and other indispositions.

The usage information provided for an extract of *E. coca* var. *coca* produced in Bolivia and known as *Jarabe de coca* states that the product will improve physical beauty, sexual functioning, digestion, and mental activity; stimulate the appetite and the circulation; strengthen the bones; and promote liver activity.

The wandering Callawaya healers of the Andes use a mixture of coca leaves and other herbs (see *Cytisus* spp., *Mikania cordata*) to treat rheumatism (Bastien 1987, 131*).

Today, many star athletes (soccer, football, baseball, etc.) utilize pure **cocaine** as a doping agent. Such use has its roots in the use of coca by the runners who carried messages throughout the Incan empire. These "postal runners" traversed great distances in the high mountains to take the messages contained in knotted strings from one corner of the empire to another. Without coca, this pre-Hispanic postal service would certainly have collapsed.

The Kofán do not (or only rarely) chew coca for hedonistic purposes but cultivate the bush solely for use as a medicine. They and other Amazonian tribes drink a tea made from ipadú leaves to treat pains in the region of the heart. In the Vaupés region of Colombia, consumption of a tea brewed from leaves of coca and *Vochysia laxiflora* Stafleu[140] is recommended when a person is unable to urinate (Schultes 1977b, 117*; 1980, 57).

Although once officinal, coca leaves are no longer used in European medicine. The only modern use occurs in homeopathic medicine, where the agent *Erythroxylon coca hom. HPUS88* is produced by macerating fresh or dried leaves (Lindequist 1993, 96).

Constituents

Depending upon their source, coca leaves can have an alkaloid content of 0.5 to 2.5%. The primary alkaloids are **cocaine** and cuscohygrine. Among the most important secondary alkaloids are cinnamoylcocaine, α-truxilline, and β-truxilline. Peruvian and Bolivian coca leaves contain the highest amounts of cocaine, which makes up some 75% of the total alkaloid content (Morton 1977, 178*). In a dried state, these may contain as much as 2% cocaine!

The fresh leaves in particular contain an **essential oil** as well as flavonoids (rutin, quercitrin, isoquercitrin), tanning agents, vitamins (A, B, C), protein, fat, and large amounts of minerals, especially calcium and iron. Approximately 100 g of coca leaves suffice to provide the recommended daily dose of all important minerals and vitamins (Duke et al. 1975). Both fresh as well as dried leaves possess good nutritional value (305 calories per 100 g)—which is why the Indians regard coca as a food.

The essential oil of *E. coca* var. *coca*, which has a scent reminiscent of that of grass, consists of 38% α-dihydrobenzaldehyde, 16.1% *cis*-3-hexen-1-ol, 13.6% methyl salicylate, 10.4% *trans*-2-hexanal, some *N*-methylpyrrole, 1-hexanol, and *N,N*-dimethylbenzylamine,[141] and several as yet unidentified substances (Novák and Salemink 1987).

The leaves and the bark contain the **tropane alkaloids** cuscohygrine and hygrine. The seeds and the bark also contain some cocaine (Bühler and Buess 1958, 3046; Morton 1977, 178*).

Effects

The Indians classify coca as a food and emphasize the nutritional value of the leaves (Hanna 1974). The Andean Indians say that when coca is chewed properly and with respect, it can soak up sadness and pain and protect the chewer like a mother (Allen 1988, 135). Coca chewing has a regulating

effect upon blood sugar levels. Apparently, chewing coca will raise a blood sugar level that is too low and lower a blood sugar level that is too high. In other words, coca use helps maintain blood sugar concentrations at a level that the body requires (Burchard 1975). Coca chewing counteracts the stresses associated with high altitudes and appears to improve oxygen absorption in the thin mountain air (Bittmann 1983; Bolton 1979; Bray and Dollery 1983). The nutritional value of coca leaves is higher in the forms in which Amazonian coca is prepared and typically ingested in the region (where everything is swallowed) (Schultes 1980, 52). In addition to its nutritional value, coca deadens the stomach nerves, thereby suppressing sensations of hunger. In general, chewing coca has stimulating and animating effects that can range from a general improvement in mood to aphrodisiac sensations and even euphoria. One quid of coca to which *chamairo* bark has been added is said to induce a "sensation of well-being and calm" (Plowman 1980, 256). When mixed with *Trichocereus* spp., coca quids can apparently have profoundly stimulating, perhaps even mildly psychedelic, effects (Fernandez Distel 1984).

The cocaine that is released from the leaves by chewing remains in the body for some seven hours in the form of its metabolite, ecgonine, albeit only in very low concentrations. The active amount is present in the blood for one to two hours. It is this that is responsible for the stimulating effects of the coca quid (Holmstedt et al. 1978). For information on pharmacology, see **cocaine**.

When a quid of coca is placed in the mouth, moistened well with saliva, and mixed with an alkaline substance, it takes a few minutes for the cocaine to dissolve out of the leaves and spread throughout the entire mouth, mixed with the copious amounts of saliva that are also produced. The surface of the mucous membranes immediately become numb. The quality of the coca can be recognized by the rapidity with which this numbing effect takes place. After an additional five to ten minutes, the stimulating effects of the cocaine become clearly noticeable. This effect slowly increases during the following minutes, persists for about 45 minutes, and then quickly dissipates.

It has often been claimed that cocaine can destroy the nasal septum. However, the effects of *coca y bica* ("coca leaves and bicarbonate of soda") upon the mucous membranes of the cheeks are much more destructive. Longtime *coqueros* ("coca users") must develop a kind of leathery surface in their mouths. In my own experience, the effects of a coca quid upon the mouth's mucous membranes are considerably more damaging than those that cocaine produces on the mucous membranes of the nose.

Cocaine

Cuscohygrine

140 For the treatment of asthma, the bark of this tree is tossed into the fire and the resulting smoke inhaled (Schultes 1977b, 117*).

141 This substance is also present in the essential oil of black tea (see *Camellia sinensis*).

Left: The leaves of hare's lettuce (*Sonchus oleraceus*), a plant introduced from Europe, are used as a coca substitute.

Right: When coca is in short supply, the inhabitants of the oases in the Atacama Desert will chew the leaves of a plant introduced from Europe, the quince (*Cydonia oblonga* = *Cydonia vulgaris*).

Coca chewing can indeed lead to various problems in the oral cavity. In the Vaupés region of Colombia, the bark of *Tachigalia cavipes* (Spruce ex Benth.) Macbride is powdered and strewn over the ulcerous wounds that are said to result from immoderate coca chewing (Schultes 1978a, 184*).

Commercial Forms and Regulations

The cultivation, trade, and use of coca leaves is permitted in Peru, Bolivia, and northwestern Argentina (Gran Chaco). Both the dried leaves and *llipta* and other alkaline substances (sodium bicarbonate) are sold at markets, herb shops, newspaper stands, and other small shops, usually under the name *coca y bica*. Three qualities are usually available: *regular* ("normal"), *seleccionada* ("select"), and *super seleccionada* ("super select") or *sele desfolillada*. While the use of coca is

forbidden in northern Chile, it is tolerated. The Aymara Indians and the Aborigines of the Atacama coast are permitted to use coca.

Coca leaves are also processed into tea bags for brewing tea (so-called *maté de coca*) and are even sold in supermarkets. In Peru, various flavors and combinations are available, including some containing spices, chamomile (**Matricaria recutita**), *hierba luisa* ("Louisa's herb" [?]), anise (*Pimpinella anisum* L. or *Tagetes pusilla* H.B.K.; cf. **Tagetes spp.**), mint (*Mentha* spp.), *canelo* ("caneel"; probably imported cinnamon, and most likely originally *Canella winterana* [L.] Gaertn.), and *muun* (*Minthostachys andina* [Britt.] Epling, *Satureja* spp., or *Mentha virides* L.; cf. Bastien 1987, 133).

For years, the governments of Peru and Bolivia have been making efforts to have coca leaves legalized so that they can be exported throughout the world. The legalization of coca would have profoundly positive effects upon the economic situation in both countries (Weil 1995).

In Europe, coca preparations (the leaves) are regarded as obsolete, although theoretically they are available through pharmacies, as no prescription is required (Lindequist 1993, 96). However, because they contain cocaine, they are subject to various drug laws (Körner 1994, 96*). In Germany, they are listed as trafficable substances in List II of the drug laws (Körner 1994, 57*).

142 While this wild relative of cultivated coca does contain cocaine, the concentrations are significantly lower. The Barasana say that you can eat this coca and that it was "the coca of our fathers" (Schultes and Raffauf 1990, 166*).

143 According to the Mapuche of southern Chile, a tea made from the leaves of this daisy thistle (also known as *ulhuihuaca*) lowers body temperature and purifies the blood (Houghton and Manby 1985, 100*). The root of the plant, which is also referred to as *facu*, *llaupange*, *pangue*, *nalca*, and *oder*, contains large amounts of tannins (Mösbach 1992, 81*).

144 These leaves are also chewed "against the cold" (Plowman 1980, 254).

Coca Substitutes

(From Aldunate et al. 1981*; Henman 1981; Plowman 1980; Schultes 1980; Schultes and Raffauf 1990, 166 ff.*; von Reis and Lipp 1982, 233*; supplemented.)

Name	Plant Part	Location/Culture
Chenopodium arequipensis (Cuatr.) Cuatr. (*pariente de la coquilla, coquilla*)	leaves	Chile (Atacama)
Cordia nodosa L. (*tabaco chuncho*)	leaves	Andean foothills/Campa Indians
Couma macrocarpa Barb. Rodr. (*sorva, juansoco*)	leaves	Upper Amazon
Cydonia oblonga Mill. (*membrillo, quitte*)	leaves	Chile (Atacama)
Dodonea viscosa L.	leaves	High Andes
Erythroxylum acuminatum Ruiz et Pav. (*coca de mono*)	leaves	Colombia and Peru
Erythroxylum cataractarum Spruce (*coca de pescado*)[142]	leaves	Barasana (Río Piraparaná)
Erythroxylum fimbriatum Peyr. (*coca brava*)	leaves	Upper Amazon
Erythroxylum gracilipes Peyr.	leaves	Ecuador/Quechua
Erythroxylum macrophyllum Cav. (*coca brava*)	leaves	Upper Amazon
Lacmellea spp. (two species)	leaves	Upper Amazon
L. lactescens (Kuhlm.) Markgraf	leaves	
L. cf. *peruviana* (Heu. et Muell. Arg.) Markgraf	leaves	
Rosa sp. (*rosa*)	leaves	Chile (Atacama)
Sonchus oleraceus L. (*wirikocha*)[143]	leaves	Chile (Atacama)
Stylogyne amplifolia Macbride (*jipina coca, coca silvestre*)	leaves	Río Putumayo/Witoto
Urmenetea atacamensis Phil. (*coquilla, coca del suri*)	leaves	Chile (Atacama)
Werneria dactylophylla Sch. Bip.	leaves[144]	High Andes

Literature

See also the entries for *Erythroxylum novogranatense* and cocaine.

Allen, Catherine J. 1981. To be Quechua: The symbolism of coca chewing in Highland Peru. *American Ethnologist* 8:157–71.

———. 1988. *The hold life has: Coca and cultural identity in an Andean community.* Washington and London: Smithsonian Institution Press.

Andrews, George, and David Solomon, eds. 1975. *The coca leaf and cocaine papers.* New York and London: Harcourt Brace Jovanovich. (An anthology of the most important historical works.)

Andritzky, Walter. 1987. Das Koka-Orakel. *Esotera* 3/87:50–57.

Bayona Vengoa, Moisés. 1993. *La coca del Perú.* Quillabamba: self-published.

Berg, C. C. 1978. Éspecies de *Cecropia* da Amazônica Brasileira. *Acta Amazônica* 8 (2): 149–82.

Bittmann, Bente. 1983. On coca chewing and high-altitude stress. *Current Anthropology* 24 (4): 527–29.

Bohm, B. A., F. R. Ganders, and T. Plowman. 1982. Biosystematics and evolution of cultivated coca (Erythroxylaceae). *Systematic Botany* 7:121–33.

Bolton, Ralph. 1979. On coca chewing and high-altitude stress. *Current Anthropology* 20 (2): 418–20.

Bray, Warwick, and Colin Dollery. 1983. Coca chewing and high-altitude stress: A spurious correlation. *Current Anthropology* 24 (3): 269–82.

Bühler, A., and H. Buess. 1958. *Koka. Ciba-Zeitschrift* 92 (8): 3046–76.

Burchard, Roderick E. 1975. Coca chewing: A new perspective. In *Cannabis and culture,* ed. Vera Rubin, 463–84. The Hague and Paris: Mouton.

———. 1992. Coca chewing and diet. *Current Anthropology* 33 (1): 1–24.

Cabieses, Fernando. 1985. Ethnologische Betrachtungen über die Cocapflanze und das Kokain. In "Ethnobotanik," special issue, *Curare* 3/85:193–208.

Califano, Mario, and Alicia Fernández Distel. 1977. El empleo de la coca entre los Mashco de la Amazonia del Peru. *Årstryck Goteborgs Etnografiska Museum* (1977): 16–32.

Cartmell, Larry W., Arthur C. Aufderheide, Angela Springfield, Cheryl Weems, and Bernardo Arriaza. 1991. The frequency and antiquity of prehistoric coca-leaf-chewing practices in northern Chile: Radioimmunoassay of a cocaine metabolite in human-mummy hair. *Latin American Antiquity* 2 (3): 260–68.

Cruz Sánchez, G., and A. Guillén. 1948. Estudio químico de las substancias alcalinas auxiliares del cocaismo. *Revista de Farmacología y Medicina Experimental* (Lima) 1 (2): 209–15.

Davis, Wade. 1989. Obituary: Timothy Charles Plowman. *Journal of Ethnopharmacology* 26:97–100.

Duke, James A., David Aulik, and Timothy Plowman. 1975. Nutritional value of coca. *Botanical Museum Leaflets* 24 (6): 113–19.

Fernández Distel, Alicia. 1984. Contemporary and archaeological evidence of llipta elaboration from the cactus *Trichocereus pasacana* in Northwest Argentina. *Proceedings 44 International Congress of Americanists,* BAR International Series 194.

Freud, Sigmund. 1884. Ueber Coca. *Centralblatt für die gesamte Therapie* 2:289–314.

———. 1885. Beitrag zur Kenntnis der Cocawirkung. *Wiener medizinische Wochenschrift* 35:129–33.

Fuentes, Manuel A. 1866. *Mémoire sur la coca du Pérou.* Paris: Lainé et Havard.

Gagliano, Joseph. 1994. *Coca prohibition in Peru: The historical debates.* Tucson and London: The University of Arizona Press.

Gantzer, Joachim, Hartmut Kasischke, and Ricardo Losno. 1975. *Der Cocagebrauch bet den Andenindianern in Peru, unter Berücksichtigung sozialmedizinischer und ideologiekritischer Aspekte.* Hannover, Germany: ASA.

Gutiérrez-Noriega, Carlos. 1949. El hábito de la coca en el Perú. *América Indígena* 9 (2): 143–54.

Gutiérrez-Noriega, Carlos, and Viktor W. von Hagen. 1951. Coca—the mainstay of an arduous life in the Andes. *Economic Botany* 5:145–52.

Hanna, Joel M. 1974. Coca leaf use in southern Peru: Some biosocial aspects. *American Anthropologist* n.s. 76:281–96.

Hastorf, Christine A. 1987. Archaeological evidence of coca (*Erythroxylum coca,* Erythroxylaceae) in the Upper Mantaro Valley, Peru. *Economic Botany* 41 (2): 292–301.

Henman, Anthony R. [= Antonil]. 1981. *Mama coca.* Bremen, Germany: Verlag Roter Funke.

———. 1990. Coca and cocaine: Their role in "traditional" cultures in South America. *The Journal of Drug Issues* 20 (4): 577–88.

Holmstedt, Bo, E. Jaatmaa, K. Leander, and Timothy Plowman. 1977. Determination of cocaine in some South American species of *Erythroxylum* using mass fragmentography. *Phytochemistry* 16:1753–55.

Holmstedt, Bo, J.-E. Lindgren, L. Rivier, and T. Plowman. 1978. Cocaine in blood of coca chewers. *Botanical Museum Leaflets* 26 (5): 199–201.

"See how densely it is set with leaves,
every leaf is fruit, and such substantial fare,
that no fruit will dare to compete with it.
Moved by his land's coming fate
(whose soil must be exposed to plunder because of its treasures),
our Varikocha first sent coca,
furnished with leaves of wondrous nutritional power,
whose juice is sucked in and passed to the stomach,
and allows hunger and work to be endured for long periods;
And which gives more help to our weak and tired bodies
and refreshes our tired spirit more than
your Bacchus and your Ceres are capable of together.
A supply of three leaves suffices for a march of six days.
With this supply, the Quitoita can traverse
the mighty, cloud-covered Andes,
the terrible Andes, between the abundance of wind, rain, and snow of winter
and that more modest earth
which brings forth the small but powerful coca bush,
this warrior who provides cheer to the warlike Venus."

ABRAHAM CROWLEY (1618–1667)
BOOK OF PLANTS

The tiny flowers of the coca bush (*Erythroxylum coca* var. *coca*), shown in various views. (Copperplate engraving, nineteenth century)

"Among the persons to whom I gave coca, three reported profound sexual excitation which they unhesitatingly attributed to the coca. A young writer, who was able to take up his work after a long silence because of coca, renounced the use of coca because he found this side effect undesirable."

Sigmund Freud
"Ueber Coca" [On Coca]
(1884, 314)

Instituto Indigenista Interamericano. 1986. *La coca andina: Visión indígena de una planta satanizada.* Mexico City: Joan Boldó; Climent Editores and Instituto Indigenista Interamericano.

———. 1989. *La coca . . . tradición, rito, identidad.* Mexico City: Instituto Indigenista Interamericano.

Jenzer, R. 1910. *Pharmakognostische Untersuchungen über Pilocarpus pennatifolius* Lemaire *und Erythroxylon Coca* Lamarck *mit besonderer Berücksichtigung ihrer Alkaloide.* Dissertation, Bern.

Jeri, F. R., C. Sanchez, T. del Pozo, and A. Fernandez. 1978. The syndrome of coca paste. *Journal of Psychedelic Drugs* 10 (4): 361–70.

Leon, Luis A. 1952. Historia y extinction del coca-ísmo en el Ecuador. *América Indígena* 12:7–32.

Lindequist, Ulrike. 1993. *Erythroxylum.* In *Hagers Handbuch der pharmazeutischen Praxis*, fifth edition, 5:88–98. Berlin: Springer.

Lloyd, John Uri, and John Thomas Lloyd. 1911. Coca—"The divine plant" of the Incas. *Lloyd Library Bulletin* no. 18.

Lobb, C. Gary. 1974. El uso de la coca como manifestación de cultura indígena en las montañas occidentales de sud-america. *América Indígena* 34 (4): 919–38.

Mariani, Angelo. 1885. *La coca et la cocaïne*, Paris: Librairie A. Delahaye and É. Lecrosnier.

Martin, Richard T. 1969. The role of coca in the history, religion, and medicine of South American Indians. *Economic Botany* 23:422–38.

Mayer, Enrique. 1986. Coca use in the Andes. In *Drugs in Latin America*, ed. Edmundo Morales, 1–51. Publ. 37. Williamsburg, Va.: Studies in Third World Societies.

Monge, C. 1953. La necesidad de estudiar el problema de la masticación de las hojas de coca. *América Indígena* 13 (1): 47–54.

Mortimer, W. Golden. 1974. *History of coca: "The divine plant" of the Incas.* San Francisco: And/or Press. (A Fitz Hugh Ludlow Memorial Library Edition; orig. pub. 1901.)

Nachtigall, Horst. 1954. Koka und Chicha. *Kosmos* 50 (9): 423 ff.

Novák, Michael, and Cornelis A. Salemink. 1987. The essential oil of *Erythroxylum coca. Planta Medica* 53:113.

Pacini, D., and C. Franquemont, eds. 1986. *Coca and cocaine: Effects on people and policy in Latin America.* Cultural Survival Report 23. Cambridge: Cultural Survival, Inc.

Plowman, Timothy. 1967. Orthography of *Erythroxylum* (Erythroxylaceae). *Taxon* 25 (1): 141–44.

———. 1979a. Botanical perspectives on coca. *Journal of Psychedelic Drugs* 11:103–17.

———. 1979b. The identity of Amazonian and Trujillo coca. *Botanical Museum Leaflets* 27 (1–2): 45–68.

———. 1980. Chamairo: *Mussata hyacinthina*—An admixture to coca from Amazonian Peru and Bolivia. *Botanical Museum Leaflets* 28 (3): 253–61.

———.1981. Amazonian coca. *Journal of Ethnopharmacology* 3:195–225.

———. 1982. The identification of coca (*Erythroxylum* species): 1860–1910. *Botanical Journal of the Linnean Society* 84:329–53.

———. 1983. New species of *Erythroxylum* from Brazil and Venezuela. *Botanical Museum Leaflets* 29 (3): 273–90.

———. 1984a. The ethnobotany of coca (*Erythroxylum* spp., Erythroxylaceae). *Advances in Economic Botany* 1:62–111.

———. 1984b. The origin, evolution and diffusion of coca *Erythroxylum* spp., in South and Central America. In *Pre-Columbian plant migration*, ed. Doris Stone, 125–63. No. 76. Cambridge: Papers of the Peabody Museum in Archaeology and Ethnography.

———. 1987. Ten new species of *Erythroxylum* (Erythroxylaceae) from Bahia, Brazil. *Fieldiana* (Botany) n.s. 19:1–41.

Plowman, T., L. Rudenberg, and C. W. Greene. 1978. Chromosome numbers in neotropical *Erythroxylum* (Erythroxylaceae). *Botanical Museum Leaflets* 26 (5): 203–9.

Potratz, Egbert. 1985. Zur Botanik der Coca-Pflanze. In "Ethnobotanik," special issue, *Curare* 3/85:161–76.

Quijada Jara, Sergio. 1982. *La coca en las costumbres indígenas.* Huancayo, Peru: Imprenta Ríos.

Ricketts, Carlos. 1952. El cocaísmo en el Perú. *América Indígena* 12:309–22.

———. 1954. La masticación de las hojas de coca en el Perú. *América Indígena* 14:113–26.

Rivier, Laurent. 1981. Analysis of alkaloids in leaves of cultivated *Erythroxylum* and characterization of alkaline substances used during coca chewing. *Journal of Ethnopharmacology* 3:313–35.

Rocha, José A. 1996. Die Coca-Pflanze. *Infoemagazin* 11:27–28.

Rury, Phillip M. 1981. Systematic anatomy of *Erythroxylum* P. Browne: Practical and evolutionary implications for the cultivated cocas. *Journal of Ethnopharmacology* 3:229–63.

Rury, Phillip M., and Timothy Plowman. 1983. Morphological studies of archaeological and recent coca leaves (*Erythroxylum* spp.). *Botanical Museum Leaflets* 29 (4): 297–341.

Rusby, Henry Hurd. 1886. The cultivation of coca. *American Journal of Pharmacy* 58:188–95.

Scarpa, Antonio, and Antonio Aimi. 1985. An ethno-medical study of *soroche* (i.e. altitude sickness) in the Andean plateaus of Peru. In "Ethnobotanik," special issue, *Curare* 3/85:209–26.

Schatzman, Morton, Andrea Sabbadini, and Laura Forti. 1976. Coca and cocaine: A bibliography. *Journal of Psychedelic Drugs* 8 (2): 95–128.

Scheffer, Karl-Georg. 1981. Coca in Südamerika. In *Rausch and Realität*, ed. G. Völger, 2:428–35. Cologne: Rautenstrauch-Joest-Museum für Völkerkunde.

Schultes, Richard Evans. 1957. A new method of coca preparation in the Colombian Amazon. *Botanical Museum Leaflets* 17:241–64.

———. 1980. Coca in the Northwest Amazon. *Botanical Museum Leaflets* 28 (1): 47–60.

———. 1981. Coca in the Northwest Amazon. *Journal of Ethnopharmacology* 3 (2): 173–94.

———. 1987. Coca and other psychoactive plants: Magicoreligious roles in primitive societies of the New World. In *Cocaine: Clinical and biobehavioral aspects*, eds. S. Fischer, A. Raskin, and E. Uhlenhuth, 212–50. New York: Oxford University Press.

Turner, C. E., Y. Ma, and M. A. Elsohly. 1981. Gas chromatographic analysis of cocaine and other alkaloids in coca leaves. *Journal of Ethnopharmacology* 3:293–98.

Von Glascoe, Christine, Duane Metzger, Aquiles Palomino O., Ernesto Vargas P., and Carter Wilson. 1977. Are you going to learn to chew coca like us? *Journal of Psychedelic Drugs* 9 (3): 209–19.

Wagner, C. A. 1978. Coca y estructura cultural en los andes peruanos. *América Indígena* 38 (4): 877–902.

Walger, Th. 1917. Die Coca: Ihre Geschichte, geographische Verbreitung und wirtschaftliche Bedeutung. Berlin: Supplement to *Tropen-pflanzer* [Tropical Planters] 17.

Weil, Andrew. 1975. The green and the white. *Journal of Psychedelic Drugs* 7:401–13.

———. 1978. Coca leaf as a therapeutic agent. *American Journal of Drug and Alcohol Abuse* 5:75–86.

———. 1995. The new politics of coca. *The New Yorker* 71 (12): 70–80.

Wiedemann, Inga. 1979. The folklore of coca in the South-American Andes: Coca pouches, lime calabashes and rituals. *Zeitschrift fur Ethnologie* 104 (2): 278–309.

———. 1992. *Cocataschen aus den Anden*. Berlin: Haus der Kulturen der Welt.

Erythroxylum novogranatense (Morris) Hieronymus

Colombian Coca Bush

Family
Erythroxylaceae (Coca Family)

Forms and Subspecies; Synonyms
There are two regionally distinct cultivated varieties of the Colombian coca bush:

— *E. novogranatense* Morris var. *novogranatense* [syn. *Erythroxylum coca* var. *novogranatense* Morris]: Colombian coca (dry/hot regions of northern South America)

— *E. novogranatense* var. *truxillense* (Rusby) Plowman [syn. *Erythroxylum truxillense* Rusby, *Erythroxylum hardinii* E. Machado, *Erythroxylum coca* Lam. var. *spruceanum*]: Trujillo coca (coastal zone of northern Peru)

Folk Names
—var. *novogranatense*:
Coca, Colombian coca, hahio, hayo, hayu, koka, kolumbianischer kokastrauch
—var. *truxillense*:
Coca, coca de trujillo, Trujillo coca, trujillo-kokastrauch, tupa ("kingly" or "noble"), small-leaved coca, Peruvian coca, Java coca

History
Within its range, this coca species has been used for as long as has **Erythroxylum coca**. During excavations of the ancient Valdivia culture (Ecuador), lime containers were found that have been dated to 2100 B.C.E. All of the coca leaves that have been recovered from archaeological excavations in the coastal region of Peru are Trujillo

Typical leaves of the four types of coca: *Erythroxylum novogranatense* var. *truxillense*; *Erythroxylum coca* var. *coca*; *Erythroxylum coca* var. *ipadú*; *Erythroxylum novogranatense* var. *novogranatense*. (From Burck, 1892)

In the early twentieth century, coca preparations were a standard part of the home medicine cabinet; they were used as tonics and for hedonistic purposes and also as remedies. (Advertisement from a German magazine, ca. 1915)

Top: Branch tip with leaves and fruits of the Colombian coca bush *Erythroxylum novogranatense*.

Bottom: The flower of *Erythroxylum novogranatense*.

coca (Cohen 1978; Griffiths 1930; Plowman 1979, 55). The plant has been in use for at least three thousand years, primarily in the pre-Hispanic Moche and Nazca cultures.

The first European source to describe the use of coca (from *E. novogranatense* var. *novogranatense*) is the report of Amerigo Vespucci (after whom America is named) from 1499:

> We spied an island in the sea, which lay some 15 miles from the coast, and decided to go there and see if it was inhabited. We encountered the most corrupt and hideous people that we had ever seen: their faces and their expression were very hideous, and all of them had their cheeks filled with a green herb which they chewed upon the entire time like animals, so that they could hardly speak; and each of them carried two gourds around their neck, one of which was filled with the herb that they had in their mouths, and the other with a white powder that looked like ground plaster, and from time to time they dipped a stick into the powder after they had moistened it in their mouths, and placed it deep into each side of the mouth, to bring the powder to the herb which they chewed; they did this with great frequency. We were very amazed about this and could not understand their secret or why they did this.

During the early colonial period, the Spanish chronicler Pedro de Cieza de León provided a very precise description of the use of *Erythroxylum novogranatense* in Colombia and the northern Peruvian coast and of several coca substitutes that unfortunately were not identified:

> Everywhere that I have traveled in the West Indies, I have noticed that the natives take great pleasure in keeping roots, twigs, or plants in their mouths. In the area around the city of Antiocha (Antigua in Colombia), some of them chewed small coca leaves and in the province of Arma other plants, and in Quimbaya and Acerma they cut strips from a kind of small tree that has soft wood and is evergreen, and they kept this between their teeth the entire time. Among most tribes that are subject to the cities of Cali and Popayan, they keep the leaves of the small coca of which I have spoken in the mouth and dip a mixture that they prepare from small gourd bottles that they carry and place this into their mouths, and chew it all together; they do the same with a kind of earth that is like chalk. In all of Peru, it was and is custom to have this coca in the mouth, and they keep it there without removing it from the morning until they go to sleep. When I asked some of the

Indians why they always have this plant in their mouths (which they do not eat, but merely keep between their teeth), they said that because of it they feel no hunger and that it gives them great power and strength. I believe that it apparently exerts some type of effect of this nature, although I find it a disgusting practice that was to be expected of such people as these Indians. (In von Hagen 1979, 101 f.)

Neither in Colombia nor in Peru were the Spanish able to exterminate the coca use that they were so unable to comprehend. During the colonial period, the Indians put up with many things—too many things—but one thing they would not tolerate: the end of their coca use. In these areas, the situation has changed little into the present day.

The name Trujillo comes from that of a city in northwestern Peru; the plant is cultivated in a desert zone near Trujillo (Plowman 1979, 52). Trujillo coca is the type that is used even today as the source of aroma and taste for the production of Coca-Cola. During the early 1860s, before the beginning of Prohibition, Dr. John S. Pemberton (1831–1888) successfully introduced Coca-Cola, which was inspired by Vin Mariani, to the market. In addition to extracts of Trujillo coca and the West African cola nut (*Cola* spp.), the drink contained Mediterranean sweet wine and damiana extract (*Turnera diffusa*) and thus was a rather potent, psychoactive, and probably also aphrodisiac drink (Pendergrast 1996, 40). Although there has been no **cocaine** in Coca-Cola since 1903, the drink has acquired a mythological significance and has become one of the most popular soft drinks in history. Today, Coca-Cola often is regarded as a symbol of American cultural imperialism.

In Colombia, *Erythroxylum novogranatense* var. *novogranatense* increasingly is cultivated for the (illegal) production of cocaine.

Distribution

The two varieties of this coca species generally occur at lower altitudes than *Erythroxylum coca* (Towle 1961, 60*). The species prefers generally warmer climates. The variety *novogranatense* is indigenous chiefly to northern Colombia, both in the foothills of the Andes and in the Sierra Madre, and has been cultivated there by numerous Indian tribes since ancient times.

Adapted to a desert climate, Trujillo coca now has a very small range and area of cultivation that lies in the northern Peruvian coastal strip near the town of Trujillo (= Truxillo) (Plowman 1979, 52). Its range was probably much larger in pre-Columbian times, extending throughout the entire coastal strip (Rostworowski 1973). Smaller populations that are grown primarily in home gardens as medicinal plants have been found in northwestern Ecuador and neighboring Colombia (Plowman 1979, 56).

During the colonial period, the Dutch introduced *Erythroxylum novogranatense* var. *truxillense* into their colonies on Java and Sumatra. The English brought the plant to Ceylon (Schröder 1991, 113*).

Cultivation

In principle, this coca species is propagated in the same manner as *Erythroxylum coca.* This coca bush is usually trimmed back to remain small. The bush, which is pruned to a height of about 1 meter, is called a *pajarito*, "little bird."

Cultivation is occasionally subject to ritual rules. In the Sierra Madre de Santa Marta, for example, new coca fields (which are regarded as sacred) can be planted only after consultation with medicine men or priest shamans (*mames*) (Baumgartner 1994; Bühler 1958, 3059).

Because it could hardly survive otherwise, Trujillo coca planted in the desert of northern Peru is artificially irrigated (Plowman 1979, 51).

Appearance

For nonspecialized botanists, *Erythoxylum novogranatense* is difficult to distinguish from *Erythroxylum coca.* The most characteristic features are probably the size and the structure of the leaves. The leaves of the variety *novogranatense* are not as wide as those of *Erythroxylum coca* and often have a slightly yellowish tinge.

Trujillo coca can grow as tall as 3 meters and is most easily recognized by its smaller, relatively narrow, lanceolate, tapered leaves. The leaves have a characteristic scent reminiscent of that of Coca-Cola (which is how the leaves can be distinguished from those of the other types even when dry). In addition, the flowers are arranged in small clusters and are attached to longer stalks (Plowman 1979).

Psychoactive Material

—Leaves

Erythroxylum novogranatense has an aromatic taste. As a result, many users prefer it to *Erythroxylum coca,* even though it contains less **cocaine.**

Preparation and Dosage

Erythroxylum novogranatense is prepared in a fashion somewhat similar to *Erythroxylum coca.* The variety *novogranatense* is often prepared and used in precisely the same manner as *E. coca* var. *ipadú*, and the variety *truxillense* in the same manner as *E. coca* var. *coca.* In order to be able to manifest their effects, the leaves of *Erythroxylum novogranatense*, like those of *E. coca*, must be mixed with an alkaline substance (see the table on pages 247–48).

In Colombia, lime from shells, limestone mineral deposits, or plant ashes is usually used as an alkaline additive. There, the ash cakes are known as *mambe* or (in Spanish) *lejia*. The Indians of the Sierra Madre de Santa Marta (Kogi, Arhuaco, Ika), as well as their pre-Columbian and colonial-period ancestors, the Tairona, chewed or chew coca leaves with slaked lime (*yotinwe*) from seashells and saltwater snails (Nicholl 1990, 389; Reichel-Dolmatoff 1955).

The Kogi prepare their coca in the following manner: Fresh coca leaves are dried or toasted and ground. The lime additive is obtained from the shells of snails and shells (*Melongena melongena* L., *Venus* spp., *Strombus* spp., and others) from the Caribbean coast (Reichel-Dolmatoff 1955). From time to time, the Kogi make a pilgrimage to the coast to acquire new supplies of shells, which are then burned in fires of grass piled up in the shape of pyramids. The lime they obtain in this manner is transformed into slaked lime by the addition of water or through the humidity of the air, and it is stored in special gourds (*Cucurbita pepo* L.).

Among the Arhuaco Indians, the processes of harvesting and drying are quite elaborate:

To dry, [the leaves] are spread onto large stone plates near the houses. When the conditions are right, then the early morning's harvest is completely dry in the evening. Quick drying seals the stimulating alkaloids into the leaf, and the *coca del dia* ["coca of the day"] is the leaf that fetches the highest price. When first dried, the leaves are brittle and easily broken. After drying, they are piled up and allowed to sit for two to three days. This causes the leaves to perspire, which gives them back some of their suppleness and moisture. Afterward, they are very quickly dried in the sun once more and then packaged. Every one of the steps in this process is a delicate affair. If the first drying stage lasts too long, then the leaves will turn brown and become damp. If the leaves are allowed to perspire for too long a time, then they will turn gray and musty, look moldy, and have . . . a *caspa*, which means scales. To store, the coca leaves are first pressed

The shells of Caribbean snails of the genus *Strombus* provide the Indians of northern Colombia (Kogi, Ika, etc.) with a source of lime for chewing coca.

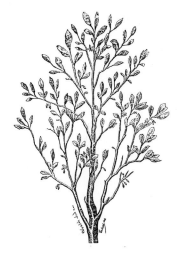

Engraving of a pressed herbarium specimen of *Erythroxylum novogranatense.* (From Mariani, *La Coca et la Cocaïne*, 1885)

"In ancient times, it is said, the coca tree was an exceptionally beautiful woman. But because she misused her body, she was killed and cut into two halves. From one of the two pieces grew a tree. And this tree was given the name *Mamacoca* or *Cocamama.* Since this event it has been used as an agent of pleasure."

RAFAEL LARCO HOYLE
ARS ET AMOR: PERU [ART AND LOVE: PERU]
(1979, 22)

"When you chew coca, it is not as though you are 'high.' You don't have any 'drug experience,' but rather the feeling that you have been freed of a burden."

CHARLIE NICHOLL
TREFFPUNKT CAFÉ "FRUCHTPALAST" [MEETING PLACE CAFÉ "FRUIT PALACE"]
(1990, 397)

"What kind of meaning did life have for the trephined? A 'normal' way of life was no longer possible. They constantly found themselves in an exceptional mental and physical state. How did they bear this? Their extraordinary experiences, inebriation, hallucinations, pains, self-denials, placed them outside of the community. They represented something like their own social status. Did they perhaps comprise an order that was obligated to perform quite specific psychological and spiritual tasks?"

Inge von Wedemeyer
" 'Cultura cefálica' in Alt-Peru"
(1969, 309)

with wooden weights, then folded in banana leaves, and finally wrapped in sackcloth or coarse woolen fabric. If the coca leaves have been dried and packaged properly and are then stored in a cool and dry place, they will retain their power for an entire year, and the *coquero* will be supplied until the next harvest. (Nicholl 1990, 395f.)

In Colombia, it is or was common to mix coca powder with a paste of ground fresh tobacco leaves (*Nicotiana tabacum*) or dried and powdered tobacco leaves for chewing (as well as smoking?). It appears that powdered coca leaves are also used as "sniffing tobacco" in Colombia (cf. **snuffs**).

Ritual Use

Archaeological finds as well as the few available ethnohistorical sources clearly demonstrate that the cultures of the Peruvian desert (Nazca, Mochica) have always used coca, more specifically the variety *truxillense*. For example, coca leaves were placed in the mouths of the dead so that they would be sufficiently stimulated for their "last journey" (Bühler 1958).

The Mochica used coca as an aphrodisiac and in erotic rituals that have been immortalized on ceramic grave goods. Unfortunately, little is known about these erotic rites, although the ceramics do provide a clear picture. The goal of the erotic activities was not reproduction (95% of the erotic images depict heterosexual anal intercourse)[145] but the production of altered states of consciousness that in turn were utilized to obtain insights into the normally invisible world (Larco Hoyle 1979, 145).

Trepanation, the act of making openings in the skulls of living people, also was widely practiced in these cultures. Such large numbers of finds have been made that it is improbable that all of these operations were carried out with the intention of removing brain tumors. It is likely that the numerous trepanations (as many as 45% of the skulls found in Paracas were trephined; von Wedemeyer 1969, 302) were carried out for ritual and religious rather than medical reasons. It appears "as if the trepanations were also intended to induce personality changes and increases. In modern medicine, the remarkable observation has been made that peculiar euphoric states often occur after serious head injuries. The sensation of weight is suspended, people think they are floating, they have a kind of 'experience of traveling to heaven' " (von Wedemeyer 1969, 307). It may be that trepanation was an important shamanic method for producing ecstatic states of consciousness.

The chewing of coca (of the variety *novogranatense*) was widespread throughout Colombia during pre-Hispanic times (Uscátegui M. 1954).

Numerous gold objects from the time of the Tairona make reference to the use of coca. As with the later Kogi and Ika, the use of coca among the Tairona was associated with divination and the shamanic priesthood:

According to Castelanos (1601, II, Santa Marta 1, Str. 16), women were also involved in divination, which allows us to assume that there was a distinctive group of religious specialists (shamans?), for the priesthood in the more narrow sense was restricted to the men. Priestly qualifications were earned during a 16- to 20-year fast in isolation and visionary experiences under the influence of aromatic woods that were marked solely for the temple fires, augmented by the use of coca. (Bischof 1986, 25)

The descendants of the Tairona have preserved both these rituals and the restrictions concerning coca use:

Almost everywhere [in northern Colombia], the use of coca is limited to the men and strictly forbidden to the women. The extent to which coca chewing is considered a prerogative of the men can be seen in a report . . . about the Ijca [= Ika]. At the beginning of a lunar eclipse, these Indians attempt to influence the astronomical events by exchanging the roles of man and woman, i.e., the men sit down upon the floor and begin to spin, while the women watch, chewing coca as they do. (Bühler 1958, 3059)

Among the Arhuaco, only married men are allowed to chew coca. When a young man wishes to marry, he is instructed in many things by the *mame*, the shaman priest, and initiated into the mysteries of adult life. At the wedding, the newlywed then receives his own lime bottle, called a *poporo* (from *Lagenaria* spp.), and from that time on, he may chew coca—as much as he wants to or can. The lime bottle is the symbol of his initiation and is regarded as an amulet (Nichol 1990, 394 f.).

Both the use of coca and the associated symbolism are very similar among the Kogi (Baumgartner 1994; Ereira 1993; Müller-Ebeling 1995). At his initiation into manhood, each young boy receives his own lime gourd (*poporo*), which he will carry with him without interruption for the rest of his life. The initiate is told that the gourd bottle symbolizes a woman whom he will ritually wed during his initiation ceremony. When the initiate introduces the wooden stick (lime spoon) into the gourd bottle for the first time during this ritual, he is said to "deflower" his "partner," thereby winning her for his "bride." The

145 Pedro de Cieza de León, a chronicler from the colonial period, provided a short description of the sexual practices of the last Mochica: "The women practiced sodomy [i.e., anal copulation] with their spouses or other men, even while they were nursing their own children. . . . In spite of the fact that there was no shortage of women, some of whom were beautiful, I was assured that most of these indulged in the abominable vice of anal copulation, about which they were very proud" (cited in von Hagen 1979, 67, 70).

stick that is introduced is understood to be the phallus, the rubbing of the stick represents coitus, and the gourd bottle is the vulva. The Kogi men are supposed to suppress all sexual activities and experience their erotic life solely through the constant use of coca (Reichel-Dolmatoff 1985, 1:87–90*; Uscátegui M. 1959, 282*; Ochiai 1978):

> The little figure-eight-shaped gourd [*poporo*] used as a lime container is an image of the cosmos, and the stick that is inserted into it is its axis. It follows, by Kogi logic, that the gourd is a womb, the stick is a phallus, the coca leaves to be chewed are female, and the powdered lime is semen. (Reichel-Dolmatoff 1987, 78)

Artifacts

Many archaeological objects associated with coca and its use are known from both the northern Peruvian desert and northern Colombia.

The spectacular gold artwork of the Tairona and related peoples contains numerous representations of people using coca. There are also gold lime containers in the form of gourd bottles as well as other shapes. The small gold lime bottles shaped like a coca-chewing Indian holding coca bags in his hands are especially impressive. The opening to the bottles is located at the crown of the head (Bray 1979a, cat. no. 143a). A number of richly decorated gold lime spoons have also been discovered (Bray 1979b).

The ritual chewing of coca is a frequent theme in the paintings and the ceramics of the Moche and the Nazca cultures (Kutscher 1977*; Naranjo 1974, 610; Towle 1961, 58 f.*). The Mochican ceramics include numerous anthropomorphic vessels depicting people using coca (Dietschy 1938, 1999). Some authors have attempted to associate the erotic art of the Mochica (Larco Hoyle 1979), which they consider "perverse," with the use of coca. They regard this art as an expression of sexual pathology resulting from a "cocaine psychosis" (cf. Kaufmann-Doig 1978, 22). I mention this hypothesis only to point out the "perversions" of some would-be scholars.

The ceramics also show the use of coca in (ritual) trephinations:

> There is a Peruvian clay sculpture which shows a surgeon holding the patient's head between his knees, chewing coca, and spitting this into the wound as a local anesthetic. The saliva is attributed with magical as well as therapeutic effects. It has, for example, a distinctive function in confession. A person would spit out his sins onto a bundle of grass that would then be thrown into the river; this is a special way to make a cathartic statement. (von Wedemeyer 1969, 306)

Artifacts were even made from the bush itself. The wood of older plants was and still is carved into images of the gods (*idolos*). Some beautiful examples of these can be seen in the Ethnology Museum in Basel (Switzerland). They are used primarily as altar objects in shamanic healing seances.

Today, the Indians still use bags for the leaves, gourds for the lime, and lime sticks. Among the Indians of the Colombian Sierra Madre, the coca bags are known as *mochilas*, *tutu*, or *kuetand diaja*. The lime gourds are called *poporo*, *yoburo*, or *kuetand-tuky*, and the lime spoons *sokane*.

In some ways, Trujillo coca also left its mark upon the early products of Coca-Cola. The ads produced around the turn of the twentieth century in particular feature young, rosy, healthy-looking women who are well dressed and are usually shown together with bouquets of roses. It almost seems as though the spirit of Mama Coca is still present in these illustrations.

Medicinal Use

The Mochica used coca for medicinal purposes; unfortunately, no documents about actual applications have come down to us (von Hagen 1979, 156 f.). In pre-Columbian times, coca apparently was commonly used as a local anesthetic for trephinations and other surgical procedures (Dietschy 1938).

The contemporary folk medical uses of *Erythroxylum novogranatense* are more or less identical to the uses of **Erythroxylum coca**.

Constituents

The constituents of the two varieties are essentially the same as those in **Erythroxylum coca**, with some slight exceptions. The greatest difference is in the concentration of **cocaine**. Both varieties

Representation of a caterpillar (?) with the typical leaves of the coca bush. (After a painting on a jar from Nazca, Peru, fifth to tenth century)

Coca-chewing shamans beneath a starry sky. The lime container, from which lime is removed with a spoon and placed in the mouth, is clearly visible. (Painting on a Moche jar, Chimu, ca. 500 C.E.)

A shaman with all of the coca paraphernalia (bag for leaves, gourd for lime) beneath a starry sky that is spanned by a double-headed snake (rainbow, a ladder of coca). (Painting on a Moche jar, Chimu, ca. 500 C.E.)

The Trujillo coca bush (*Erythroxylum novogranatense* var. *truxillense*), which is cultivated in the desert of northern Peru, is most readily recognized by its small leaves and long branches. (From Mariani, nineteenth century)

contain less than 1% cocaine and other alkaloids. The leaves and the bark contain the **tropane alkaloids** cuscohygrine and hygrine.

Of the four types of coca, the variety *truxillense* contains the highest concentration of **essential oil** and other pleasant flavor agents (Plowman 1979, 52).

Effects

The effects correspond to those of *Erythroxylum coca*, although they are somewhat milder as a result of the lower amounts of cocaine.

Commercial Forms and Regulations

The Indian use of coca is tolerated in Colombia. In Peru, the cultivation and use of Trujillo coca is legal.

Some years ago, a *maté de coca* made with Trujillo coca (in tea bags) was exported to the United States. Sales were quickly suspended, however, as the tea was found to contain small amounts of **cocaine** (Siegel et al. 1986).

The legal situation (laws regulating drugs and medicines) is identical to that pertaining to *Erythroxylum coca*.

Literature

See also the literature for *Erythroxylum coca* and **cocaine**.

Allen, Frederick. 1994. *Coca-Cola-Story: Die wahre Geschichte*. Cologne: vgs.

Baumgartner, Daniela. 1994. Das Priesterwesen der Kogi. In *Yearbook for Ethnomedicine and the Study of Consciousness*, 1994 (3):171–98. Berlin: VWB.

Bischof, Henning. 1986. Politische Strukturen und soziale Organisation im Bereich der Tairona-Kultur. In *Tairona—Goldschmiede der Sierra Nevada de Santa Marta, Kolumbien*, 22–27. Hamburg: Hamburgisches Museum fur Völkerkunde.

Bray, Warwick. 1979a. *El Dorado: Der Traum vom Gold*. Hannover: Bücher-Büchner.

———. 1979b. *Gold of El Dorado*. New York: American Museum of Natural History.

Bühler, A. 1958. Die Koka bei den Indianern Südamerikas. *Ciba Zeitschrift* 92:3052–62.

Cohen, M. N. 1978. Archaeological plant remains from the central coast of Peru. *Ñawpa Pacha* 16:23–50.

Dietschy, Hans. 1938. Die Heilkunst im Alten Peru. *Ciba Zeitschrift* 58:1990–2017.

Ereira, Alan. 1993. *Die großen Brüder: Weisheiten eines urtümlichen Indio-Volkes*. Reinbek, Germany: Rowohlt.

Griffiths, C. O. 1930. Examination of coca leaves found in a pre-Incan grave. *Quarterly Journal of Pharmacy and Pharmacology* 3:52–58.

Hamburgisches Museum für Völkerkunde. *Tairona–Goldschmiede der Sierra Nevada de Santa Marta, Kolumbien*. 1986. (An exhibition catalogue)

Harms, H. 1922. Übersicht der bisher in altperuanischen Gräbern gefundenen Pflanzenreste. In *Festschrift Eduard Seler*, 157–86. Stuttgart: Strecker und Schröder.

Kaufmann-Doig, Federico. 1978. *Sexualverhalten im Alten Peru*. Lima: Kompaktos.

Larco Hoyle, Rafael. 1979. *Ars et Amor: Peru*. Munich: Heyne.

Müller-Ebeling, Claudia. 1995. Die Botschaft der Kogi. *Esotera* 5/95:24–29.

Naranjo, Plutarco. 1974. El cocaísmo entre los aborígenes de Sud América. *América Indígena* 34 (3): 605–28.

Nicholl, Charles. 1990. *Treffpunkt Café "Fruchtpalast": Erlebnisse in Kolumbien*. Reinbek, Germany: Rowohlt.

Ochiai, Ines. 1978. El contexto cultural de la coca entre los indios kogi. *América Indígena* 37 (1): 43–50.

Pendergrast, Mark. 1996. *Fur Gott, Vaterland und Coca-Cola: Die unautorisierte Geschichte der Coca-Cola-Company*. Munich: Heyne.

Plowman, Timothy. 1979. The identity of Amazonian and Trujillo coca. *Botanical Museum Leaflets* 27 (1–2): 45–68.

Reichel-Dolmatoff, Gerardo. 1955. Conchales de la costa caribe de Colombia. *Anais Do XXXI Congreso Internacional de Americanistas* (São Paulo), 619–26.

———. 1987. The Great Mother and the Kogi universe: A concise overview. *Journal of Latin American Lore* 13:73–113.

———. 1991. *Los Ika: Sierra Nevada de Santa Marta, Colombia—Notas Etnograficas 1946-1966*. Bogotá: Universidad Nacional de Colombia.

Rostworowski de Diez Canseco, M. 1973. Plantaciones prehispánicas de coca en la vertiente del pacifico. *Revista del Museo Nacional* (Lima) 39:193–224.

Rusby, Henry H. 1900. The botanical origin of coca leaves. *Druggists Circular and Chemical Gazette*, Nov.: 220–23.

———. 1901. More concerning Truxillo coca leaves. *Druggists Circular and Chemical Gazette*, March: 47–49.

Siegel, Ronald K., Mammoud A. Elbomly, Timothy Plowman, Phillip M. Rury, and Reese T. Jones. 1986. Cocaine in herbal tea. *Journal of the American Medical Association* 255 (1): 40.

Uscátegui M., Nestor. 1954. Contribución al estudio de la masticación de las hojas de coca. *Revista Colombiana de Antropología* 3:209–89.

von Hagen, Victor W. 1979. *Die Wüstenkönigreiche Perus*. Bergisch Gladbach, Germany: Bastei-Lübbe.

von Wedemeyer, Inge. 1969. "Cultura cefálica" in Alt-Perú: Ein Beitrag zur Bedeutung der Schädelbehandlungen. *Antaios* 10:298–312.

Eschscholzia californica Chamisso

Golden Poppy, California Poppy

Family

Papaveraceae (Poppy Family); Subfamily Papaveroideae, Eschscholzieae Tribe

Forms and Subspecies

Numerous forms (over thirty) have been cultivated for ornamental purposes, especially white- and red-blooming cultivars, including some with double flowers (Grey-Wilson 1995, 55*). Several varieties or subspecies of the wild plant have been described (Kreis 1993, 111):

Eschscholzia californica Cham. var. *alba*
Eschscholzia californica Cham. var. *crocea* (Benth.) Jepson
Eschscholzia californica Cham. f. *dentata* (deeply incised leaves)
Eschscholzia californica Cham. var. *douglasii* (Benth.) Gray
Eschscholzia californica Cham. var. *maritima* (Greene) Jepson (perennial)
Eschscholzia californica Cham. ssp. *mexicana* (Greene) C. Clarke
Eschscholzia californica Cham. ssp. [or var.] *peninsularis* (Greene) Munz (Southern California, Baja California)

Synonyms

Chryseis californica Torr. et Gray
Eschscholtzia californica Cham.
Eschscholtzia douglasii Benth.
Eschscholtzia douglasii (Hook. et Arn.) Walp.
Eschscholtzia maritima Greene
Eschscholtzia mexicana Greene

Folk Names

Amapola amarilla, amapola de California, amapola de los indios ("opium of the Indians"), amapolla, California poppy, Californian poppy, copa de oro (Spanish, "cup of gold"),[146] cululuk (Rumsen), globe du soleil, golden poppy, goldmohn, indianischer mohn, kalifornischer mohn, knipmutsje (Dutch), pavot de californie, schlafmützchen, slapmutshe, yellow poppy

History

The Indians of California have used the golden poppy for medicinal and/or psychoactive purposes since prehistoric times.

The genus is named after the surgeon Dr. J. F. Eschscholtz (1793–1831), who served as naturalist in the 1816 and 1824 Russian expeditions to the coast of the Pacific Northwest (Grey-Wilson 1995, 55*).

The golden or California poppy is the official state flower of California (Bremness 1995, 250*).

The plant has been cultivated in European gardens since 1790 (Grey-Wilson 1995, 55*).

Since the 1960s, the California poppy has been regarded as a "legal high" and as a marijuana substitute (cf. *Cannabis indica*) (Kreis 1993, 113).

Distribution

The plant is indigenous to western North America (California, Oregon). It thrives at altitudes of up to 2,000 meters above sea level and requires moderately dry soil and much sun. It occurs as far south as Baja California and northern Mexico (ssp. *mexicana*).

Cultivation

Propagation occurs through the seeds. These are pressed into the ground to a depth of 0.5 to 1 cm and watered. At temperatures between 10 to 22°C, they germinate within eight to fifteen days. The young plants must be transplanted so that their later growth is not disturbed. In central Europe, it is advisable to raise the plants indoors and transplant them into a garden or planter box at the end of April.

California poppy seeds germinate so effectively that only a few seeds are needed. The plant can tolerate various soil types.

Cultivation for pharmaceutical purposes occurs primarily in southern France (Kreis 1993, 112).

Appearance

The California poppy is an annual that can grow as tall as 40 cm. The multipinnate leaves are opposite and are blue-green or even grayish in color. The silky flowers are luminously orange-yellow and sit at the end of long, slender stalks. The wild plant blossoms from June to August. The fruits are long, thin, tapered pods that point straight up and contain numerous tiny seeds.

The genus *Eschscholzia* is composed of some ten species that occur in the wild only in North America, especially California. *Eschscholzia californica* is easily confused with *Eschscholzia caespitosa*

Left: One of the many cultivated sorts of golden poppy (*Eschscholzia californica*).

Right: The flower and fruit of the golden poppy (*Eschscholzia californica*). (Wild plant, photographed in northern California)

146 This name is normally used for *Solandra* spp.

"The plant promotes sleep and alleviates pain. It should be able to substitute for opium for children."

HAGERS HANDBUCH
(1930)

Magnoflorine

Allocryptopine

Protopine

[147] Coptisine and sanguinarine are present in most members of the poppy family; however, they do not produce any noticeable psychoactive effects (Brown and Malone 1978, 8*).

Benth. [syn. *E. tenuifolia* Benth.], a species found in the Sierra Nevada of California, and with *Eschscholzia lemmonii* Greene (Grey-Wilson 1995, 60 f.*).

Psychoactive Material
— Leaves/herbage (eschscholziae herba, herba eschscholtziae, herba eschscholziae, eschscholzienkraut)
— Flowers
— Fruits

Preparation and Dosage
The leaves, flowers, and fruits can be dried and smoked, either alone or mixed with other herbs (cf. **smoking blends**). The effects are very mild. The maximum dosage appears to be rather open (Gottlieb 1973, 9*).

To make a tea with sedative effects, add one to two heaping teaspoons of dried herbage to one cup of boiling water and allow to steep for thirteen minutes.

The fresh fruits (and the fresh herbage) can be chewed if desired. Tinctures and extracts at first should be dosed according to the instructions on the packaging material. Dosages can then be increased as desired for experimentation.

Ritual Use
It is possible that the California Indians may have used the plant for ritual purposes in prehistoric times.

Artifacts
Because of its beauty, the golden poppy is often depicted in paintings of California and on postcards and posters (e.g., The Panama-Pacific International Exposition, 1915).

Medicinal Use
The California Indians use the flowers, stalks, and leaves primarily as a sedative for toothaches (Bremness 1995, 250*). They usually chew the fresh leaves. A decoction of the flowers is used to treat head lice. Two flowers may be placed under a child's pillow to aid in sleep. Indian women avoid the golden poppy while pregnant (Kreis 1993, 113).

In northern Mexico, the golden poppy is used in folk medicine in the same manner as opium (cf. *Papaver somniferum*) (Martínez 1994, 36*). In pediatric medicine, it is also popular in place of opium as a mild sedative and analgesic (Sturm et al. 1993). Tinctures made from material collected during the flowering period are preferred (Columbo and Tomé 1993). In homeopathic medicine, preparations of the fresh flowering plant (Eschscholtzia californica hom. *PFX*, Eschscholtzia californica hom. *HPUS88*) are used in accordance with the medical description to treat such ailments as sleep disturbances (Kreis 1993, 114).

Constituents
The entire plant contains alkaloids. The concentrations are highest during the flowering period. Material collected at this time can contain as much as 1.1% alkaloids by dry weight. The root contains up to 2.7% alkaloids, including 0.014% magnoflorine (= escholine; an aporphine alkaloid), 0.013% (–)-α-canadinmethohydroxide, 0.05% norargemonine, and 0.08% bisnorargemonine; these substances are present in the herbage in only trace amounts. The primary alkaloid in the root is allocryptopine (approximately 1.8%). The seeds contain protopine (= macleyine; = fumarine; = biflorine), allocryptopine, chelerythrine, and other substances (Kreis 1993, 111). The herbage contains magnoflorine as well as californidine, protopine, allocryptopine, sanguinarine (= pseudochelerythrine), coptisine,[147] chelerythrine (= toddaline), escholzine (= eschscholtzine; = californine), *N*-methyllaurotetanine (= lauroscholtzine), corydine, isocorydine, chelirubin, macarpine, chelilutine, and *O*-methylcariachine (Kreis 1993, 112). The alkaloids in the aboveground parts of the plant increase dramatically when the plant is grown under strong light (Colombo and Tomé 1991).

Effects
The psychoactive effects of the golden poppy are very subtle: "*Eschscholzia* elevates the body's oxygen supply and promotes the absorption of vitamin A. . . . When smoked, the leaves and flowers induce a mild state of euphoria—side effects are unknown!" (Bremness 1994, 250*).

In animal experiments (mice), the extract has exhibited clear sedative and anxiolytic effects. In other words, the pharmacological behavior is that of a tranquilizer (Rolland et al. 1991). The extract extends the effects of barbiturates (Kreis 1993, 113).

The two alkaloids chelerythrine and sanguinarine bind to vasopressin (V_1) receptors, a neurochemical behavior that may explain the golden poppy's psychoactivity (Granger et al. 1992).

Commercial Forms and Regulations
The seeds may be purchased in most nurseries (usually a variety of cultivars are offered). Health food stores in the United States sell liquid extracts of California poppy (fresh roots, leaves, and flowers) from guaranteed organic sources. All products are available without restriction.

Literature
See also the entries for *Papaver* spp.

Colombo, M. L., and F. Tomé. 1991. Growth and alkaloid content in *Eschscholzia californica* in controlled conditions. *Planta Medica* 57 suppl. (2): A91.

———. 1993. Nuclear DNA changes during morphogenesis in calli of *Eschscholzia californica*. *Planta Medica* 59 suppl.: A596.

Granger, I., C. Serradeil-Le Gal, J. M. Augereau, and J. Gleye. 1992. Benzophenanthrine alkaloids isolated from *Eschscholzia californica* cell suspension cultures interact with vasopressin (V_1) receptors. *Planta Medica* 58:35–38.

Kreis, Wolfgang. 1993. *Eschscholzia*. In *Hagers Handbuch der pharmazeutischen Praxis*, 5th ed., 5:110–15. Berlin: Springer.

Rodríguez, Eloy, Mann Chin Shen, Tom J. Marby, and Xorge A. Domínguez. 1973. Isorhamnetin 3-O-glucoside 7-O-arabinoside from *Eschscholzia mexicana*. *Phytochemistry* 12:2069–71.

Rolland, Alain, Jacques Fleuretin, Marie-Claire Lanhers, Chafique Younos, René Misslin, François Mortier, and Jean-Marie Pelt. 1991. Behavioural effects of the American traditional plant *Eschscholzia californica*: Sedative and anxiolytic properties. *Planta Medica* 57:212–16.

Rusby, Henry Hurd. 1889. *Eschscholzia californica* Chamisso. *Druggists Bulletin* 3 (6): 176–79.

Sturm, S., H. Stuppner, N. Mulinacci, and F. Vinceieri. 1993. Capillary zone electrophoretic analysis of the main alkaloids from *Eschscholzia californica*. *Planta Medica* 59 suppl.: A625.

Fabiana imbricata Ruíz et Pavón

Fabiana Bush, Pichi-Pichi

Family

Solanaceae (Nightshade Family); Subfamily Cestroideae, Nicotianeae Tribe

Forms and Subspecies

It is possible that one subspecies characterized by a treelike pattern of growth may exist in southern Chile.

Synonyms

None

Folk Names

Coa, fabiana bush, fabianastrauch, fabiane, fabiane imbriquée, k'oa,[148] k'oa santiago, monte derecho (Spanish, "right mountain"), monte negro (Spanish, "black mountain"), Peru false heath, peta, piche, picheng, pichi,[149] pichi-picheng, pichi-pichi, pichi-romero,[150] pichirromero, romero,[151] romero pichi, tola[152]

History

The plant was brought to Europe during the early colonial period and was propagated in the botanical garden of Madrid, which was constructed specifically for the cultivation and distribution of plants from the New World. Strangely, the literature from the colonial period makes no mention of *Fabiana* (Hoffmann et al. 1992, 184*). The plant, which has enjoyed great success in folk medical contexts, did not attract the attention of the medical community until the nineteenth century. Henry Hurd Rusby was one of the first researchers and druggists to investigate *Fabiana* in more detail (Rusby 1885; Rossi-Wilcox 1993, 5*). Because of his work, the plant was introduced into the United States as a medicine under the name *pichi-pichi* (Rusby 1890). Around the turn of the century, it was also accepted into European pharmacopoeias as a diuretic (Schneider 1974, 2:86*).

The genus is named for Francisco Fabiano y Fuero (1719–1801), the archbishop of Valencia, who was a supporter of the botanical sciences (Genaust 1996, 243*).

Distribution

The fabiana bush is indigenous to Chile, where it occurs primarily in the south between Coquimbo and Magallanes. It also is found in Patagonia, including the Argentinean portion (Hoffmann et al. 1992, 184*), and in Bolivia, Peru, and several areas of Brazil.

There are some twenty-one species in the genus *Fabiana* (D'Arcy 1991, 78*), most of which are little known ethnobotanically. It is possible that the inhabitants of the Andes use the name *pichi-pichi* for many species. In Chile, several *Fabiana* species are known as *tola*[153] (Mösbach 1992, 105*):

Fabiana barriosii Phil.
Fabiana denudata Miers.
Fabiana ericoides Dun.

Pichi-pichi is sometimes confused or counterfeited with the related species *Fabiana bryoides* Phil. and *Fabiana friesii* Dammer. These two species occur only in the high mountains (Atacama, northern Chile).

Cultivation

Propagation occurs via the tiny seeds, which like those of all nightshades should be pregerminated

148 In South America, this and similar names (*khoa, khoba*) are also used for pennyroyal (*Mentha pulegium*; cf. **kykeon**), which is also used as a ritual incense in traditional Indian ceremonies. In the Atacama Desert, a plant known as *koa*, *Lepidophyllum quadrangulare* (Mey.) Benth. et Hook, also is used as a ritual incense (Munizaga A. and Gunckel 1958, 32*).

149 The Mapuche also use this name for a plant of similar appearance from the Family Papilionaceae: *Anarthrophyllum andicola* (Gill. ex H. et A.) F. Phil. (Mösbach 1992, 84*). Further studies are needed to determine whether this plant was or is also used as a ritual **incense**. *Lepidophyllum cupressiforme* (Lam.) Cass. also is called *pichi* (Mösbach 1992, 110*).

150 This same folk name is used in Chile for *Anarthrophyllum andicola* (Gill. ex H. et A.) F. Phil., a common Andean plant that bears a distant resemblance to *Fabiana imbricata* and is also used in folk medicine as a diuretic (Montes and Wilkomirsky 1987, 106*).

151 *Romero* is actually the Spanish name for rosemary (*Rosmarinus officinalis* L.).

152 This Quechua word is also used to refer to *Baccharis boliviensis* (Wedd.) Cabr. (Mösbach 1992, 110*).

153 In Spanish, *tola* is actually used to refer to *Baccharis tola*.

The pichi-pichi plant (*Fabiana imbricata*) is from southern Chile.

"The South American inebriant 'pichi-pichi' is dried shoot apexes."

LUTZ ROTH, MAX DAUNDERER, AND KURT KORMANN

GIFTPFLANZEN—PFLANZENGIFTE [POISONOUS PLANTS—PLANT POISONS]

(1994, 347*)

Fabianine

and potted when they are seedlings. The bush thrives best in rocky and poor soils. In southern Chile, it is grown in nurseries for use in gardens and as an ornamental (Hoffmann et al. 1992, 184*). In Europe, the plant can be grown in a cold house and in areas that are largely free of frost (Spain, Ireland).

Appearance

This bushy shrub can grow as tall as 3 meters and usually has a large number of branches at the ends of its stems. Tiny, almost rod- or needle-shaped leaves arranged like scales sit on the straight stems. The small, white or violet (the color is variable), trumpet-shaped flowers grow from the tips of the branches. The fruits develop into oval capsules 5 to 6 mm in length. The flowering period is usually from November to January in South America and from May to June in Europe.

Psychoactive Material

— Herbage (herba fabianae imbricatae, summitates fabianae, pichi, pichi-pichi-kraut)
— Wood (lignum fabianae, lignum pichi-pichi)

Preparation and Dosage

For psychoactive use, the tips of the branches are dried and, if necessary, chopped into small pieces. The dried herbage can then be burned as **incense** or strewn over glowing charcoal. When burned, the aromatic twig ends give off a resinous smoke that is easy to inhale and has a sweet/acrid scent somewhat reminiscent of that of pine or fir. No information is available concerning the dosage of this type of preparation. Overdoses from breathing in such smoke do not appear to be known. *Fabiana* and *Latua pubiflora* are combined as components of a psychoactive incense.

The fresh or dried herbage as well as the bark may be used when the intended use is medicinal in nature. A decoction of the bark is a potent diuretic. A tea made by steeping a tablespoon of the fresh or dried herbage is drunk as a tonic (Hoffmann et al. 1992, 186*).

Ritual Use

Fabiana herbage is a sacred **incense** that is burned at all traditional, originally Indian ceremonies (Aldunate et al. 1983). The Aymara of northern Chile keep bundles of the dried herbage, which they light when needed. The glowing herbage gives off copious amounts of smoke. The Indians of this region use *Fabiana* as an incense at all religious ceremonies and festivals and especially at the traditional ceremonies in which they make offerings to Pachamama. Patients and rooms are fumigated with the plant to treat illness. The smoke is said to banish spirits and ward off demons. In the Atacama Desert, the burning of incense is regarded as a "payment" for the dead and for general

purification. The spirits of the dead are tamed and dispelled by the smoke (Aldunate et al. 1981, 210*). Other *Fabiana* species are used in a similar manner (see the table on page 265).

Artifacts
None

Medicinal Use

In Chilean folk medicine, an infusion of the fresh herbage has been used since ancient times to treat kidney ailments and urinary tract problems and as a diuretic (Donoso Zegers and Ramírez García 1994, 55, 104*; Houghton and Manby 1985, 100*; San Martín A. 1983). The tea is also used to promote digestion (Razmilic et al. 1994). In Peru and Chile, it is believed that the plant has strong anthelmintic effects upon sheep and goats (Schultes 1980, 115 f.*).

The use of the plant as a diuretic and a treatment for venereal diseases has become generally accepted in the folk medicine of South America. The herbage also was listed as a diuretic in international pharmacopeias. A mother tincture (fabiana imbricata) occasionally finds use in homeopathic medicine (Boericke 1992, 329*).

Constituents

The aboveground parts of *Fabiana imbricata* contain an **essential oil,** resin (fabiana resin), a bitter principle, an alkaloid of rather simple structure known as fabianine, various sugars (D-manoheptulose, D-arabitinole, D-mannitol, D-galactose, D-xylose, primaverose), a glycoside (fabiana-glycotannoid), various alkanes, fatty acids, erythroglaucin, physcion, and acetovanillone (Hoffmann et al. 1992, 186*; Knapp et al. 1972; Roth et al. 1994, 347*). Various murolanes and amorphane sesquiterpenes[154] (3,11-amorphadiene) have been found in the leaves and herbage (Brown 1994; Brown and Shill 1994). The herbage also contains the flavonoids and glycosides quercetin (cf. *Artemisia absinthium, Psidium guajava, Vaccinium uliginosum,* kinnikinnick), camphor oil, and quercetin-3-*O*-rhamnoglucoside (= rutin) (Hörhammer et al. 1973). The herbage (especially the twigs) also contains the **coumarin scopoletin,** although the concentrations can vary considerably (Knapp et al. 1972, 3092; Razmilic et al. 1994; Roth et al. 1994; 347*), as well as a substance known as fabiatrina (Montes and Wilkomirsky 1987, 166*).

To the extent that they have been chemically investigated, other species of *Fabiana* (e.g., *Fabiana denudata* and *Fabiana squamata*) appear to have similar constituents (Knapp et al. 1972).

Effects

An extract of the herbage has potent diuretic effects and is thus very successful in the treatment

---

154 Some of these sesquiterpenes also occur in *Pernettya furens* (cf. **Pernettya** spp.) and *Matricaria recutita* (Brown 1994, 432).

of kidney and urinary tract ailments (Montes and Wilkomirsky 1987, 166*). An aqueous-alcohol extract of the herbage inhibits the enzyme β-glucuronidase (Razmilic et al. 1994) and has antiseptic effects (Hoffmann et al. 1992, 184*). Teas made from the herbage have general tonic effects (Hoffmann et al. 1992, 186*). The smoke, when inhaled deeply, has euphoric and inebriating effects that are sometimes rather subtle but also can be quite pronounced in some individuals. Quercetin may be responsible for the narcotic effects (cf. **Psidium guajava**).

The aromatic and resinous smoke of *Fabiana denudata* is easily inhaled and has mild stimulating effects. The yellowish, intense smoke of *Fabiana bryoides* has a lemony, fragrant, and yet strange aroma, is not so easily inhaled, and does not produce the same stimulating effects.

Commercial Forms and Regulations

In Europe, the tips of fabiana herbage (fabiana herba, herba pichi-pichi, or summitates fabianae) are difficult to obtain. In pharmacies, fabiana usually is available only as a homeopathic mother tincture (under the name *Fabiana imbricata* or *pichi-pichi*). The living plant occasionally is sold in nurseries (as a cold-house plant; Roth et al. 1994, 347*).

Literature

See also the entries for **coumarins** and **scopoletin**.

Aldunate, Carlos, Juan J. Armesto, Victoria Castro, and Carolina Villagrán. 1983. Ethnobotany of pre-Altiplanic community in the Andes of northern Chile. *Economic Botany* 37 (1): 120–35.

Brown, Geoffrey D. 1994. The sesquiterpenes of *Fabiana imbricata*. *Phytochemistry* 35 (2): 425–33.

Brown, G. D., and Joanne Shill. 1994. Isolation of 3,11-amorphadiene from *Fabiana imbricata*. *Planta Medica* 60:495–96.

Other *Fabiana* Species Used as Incense

(From Aldunate et al. 1981, 209 f.*)

Species	Folk Names
Fabiana bryoides Phil.	k'oa, k'oa santiago, pata de loro ("parrot's claw"), pata de perdíz ("partridge's claw")
Fabiana densa Remy var. *ramulosa* Wedd.	tara ("dyer's bush"), tara macho ("male dyer's bush")
Fabiana denudata Miers.	alma tola ("soul dyer's bush"), leña de alma ("firewood of the soul"), tara hembra ("female dyer's bush"), tolilla ("little dyer's bush")
Fabiana squamata Phil.	k'oa pulika

Hörhammer, L., Hildebert Wagner, M. T. Wilkomirsky, and M. Aprameya Iyengar. 1973. Flavonoide in einigen chilenischen Heilpflanzen. *Phytochemistry* 12:2068–69.

Knapp, J. E., N. R. Farnsworth, M. Theiner, and P. L. Schiff, Jr. 1972. Anthraquinones and other constituents of *Fabiana imbricata*. *Phytochemistry* 11:3091–92.

Munizaga A., Carlos, and Hugo Gunckel. 1958. Notas etnobotanicas del pueblo atacameño de Socaire. *Universidad de Chile, Centro de Estudios Antropologicos, Publicación* 5:7–35.

Razmilic, I., G. Schmneda-Hirschmann, M. Dutra-Behrens, S. Reyes, I. López, and C. Theoduloz. 1994. Rutin and scopoletin content and micropropagation of *Fabiana imbricata*. *Planta Medica* 60:140–42.

Rusby, Henry Hurd. 1885. The new Chilean drug "pichi." *Therapeutic Gazette* 9:810–13.

———. 1890. The status of pichi as a remedy in genito-urinary disease. *Medical Record*, July 5: 5–7.

San Martín A., José. 1983. Medicinal plants in central Chile. *Economic Botany* 37 (2): 216–27.

Top: *Fabiana bryoides*, found in the Atacama Desert, is the most important ritual fumigant of the oasis inhabitants.

Bottom: The naked fabiana (*Fabiana denudata*), which is adapted to the extreme desert climate, is used as incense during shamanic healing ceremonies. (Wild plant—still living!—photographed in San Pedro de Atacama, northern Chile)

Heimia salicifolia (H.B.K.) Link et Otto

Sinicuiche

The flower of *Heimia salicifolia*. (From Wasson)

155 In Mexico, the same name is used to refer to the henna plant *Lawsonia inermis* L. (Lythraceae); in Mexican folk medicine, this plant was also used as a sedative (Díaz 1979, 77*).

156 This name is also used for *Passiflora* spp.

157 In Mexico, this name is also used for the plants *Selloa glutinosum* Spreng. and *Baccharis glutinosa* Pers. (Díaz 1979, 76*).

158 In Mexico, this name is also used for *Helenium mexicanum* H.B.K. and *Ipomoea orizabensis* Leden [syn. *Ipomoea thryanthina* Lindl.] (Díaz 1979, 76*).

Family

Lythraceae (Loosestrife Family)

Forms and Subspecies

One variety has been described for Mexico:
Heimia salicifolia var. *mexicana* Link

Synonyms

Heimia salicifolia (Kunth) Link
Heimia syphillitica DC.
Nesaea salicifolia H.B.K.
Nesaea syphilitica Steud.

Folk Names

Abre-o-sol (Portuguese, "sun opener"), anchinol, anchinoli, chapuzina, cuauxihuitl (Nahuatl, "meadow fire"),[155] escoba colorada (Spanish, "colorful broom"), escoba de arroyo (Spanish, "broom of the creek"), escobilla del río, flor de San Francisco, garañona, granadilla,[156] granadillo, grandadillo, hachinal, hanchinal, hanchinol, hanchinoli, hauchinal, hauchinol, hauchinoli, heimia, herva de la vida (Portuguese, "herb of life"), herva de vida, hierba de San Francisco (Spanish, "herb of St. Francis"), hierba jonequil, huachinal, huauchinolli (Nahuatl, "the burning of the wood"), jara, jara negra, jarilla, jarrila,[157] ko'ß̣i la'wo, maajaji lop'om, maan witsiil (Huastec, "yellow are its flowers"), penaganaq'te, quiebra arado, quiebra yugo, rosilla de Puebla (Spanish, "little rose of Puebla"), sinicuiche, sinicuichi, sinicuil, sinicuilche, sinicuitl, to: la'-gaik, witlat lek'e, xonecuili, xonecuite, xoneculli, xoneguilchi, xonochilli, xonocuili (Nahuatl, "twisted foot"), yerba de las ánimas (Spanish, "herb of the spirits")[158]

History

No information is available concerning the prehistoric use of *Heimia salicifolia*. It is possible that the plant was associated with the cult of Xochipilli, the Aztec god of spring and desire (Wasson 1974*). During the nineteenth century, the plant was a recommended treatment for syphilis (Argueta V. et al. 1994, 851*).

It often has been (incorrectly) assumed that the plant was named for the renowned Alsatian mycologist Roger Heim (cf. *Psilocybe mexicana*). However, the genus *Heimia*, which consists of just three species, was actually named for Ernst Ludwig Heim (1747–1834), a physician from Berlin who introduced Alexander von Humboldt to botany (Genaust 1996, 281*).

The name *sinicuiche* is used both for the plant and for the drink that is prepared from it. The Mexican names *sinicuiche* and the derivatives *sinicuilche* and *sinicuil* are also used for other inebriating, psychoactive, or poisonous plants: *Abrus precatorius* L., *Rhynchosia* spp., *Piscidia* spp. (cf. *Lonchocarpus violaceus*), and *Erythrina* spp. (D. McKenna 1995, 102*; Reko 1938, 145 f.*; Schultes 1970, 35*).

Calderón (1896) was the first to describe *Heimia salicifolia* as the source of a psychoactive preparation. He noted the optical effects (yellow vision) and the acoustical phenomena that occur following ingestion of the leaves. However, he did not experience any of these effects in self-experiments using 5, 10, and 15 g of leaves (Malone and Rother 1994, 141). Calderón appears to have produced a morphogenetic field that still exerts itself and continues to develop today. Victor Reko contributed greatly to this reputation with his dramatic description of a "magical drink causing oblivion" (Reko 1938, 140–47*).

Distribution

The bush occurs primarily in the highlands of Mexico but is also found in Baja California (Martínez 1994, 293*). It is found throughout South America as far south as Argentina.

Cultivation

Propagation occurs through cuttings as well as the tiny seeds. The seeds are sown in seedbeds or pots. The soil must be of a fine consistency and should be pressed down with a tile and smoothed. The seeds are broadcast onto the soil and pressed in gently with a flat object. The seeds should then be only moistened with a water sprayer, and not watered by pouring. The soil should be kept slightly moist until the seeds have germinated. The seedlings do not tolerate any direct sunlight. Only when they have developed their first true leaves can they be placed in the sun and watered thoroughly. The soil should be allowed to dry out between waterings. When transplanting, be aware that even very young plants develop a large root system (Grubber 1991, 61 f.*).

The plant prefers loose soils that dry quickly after watering. It thrives best in warm, dry zones and does not tolerate frost. In central Europe, it can be kept only as a houseplant.

Appearance

The perennial, herbaceous shrub can grow to a height of over 3 meters. It has many wooden branches and narrow, willowlike leaves 2 to 9 cm in length. The yellow flowers (2 cm in diameter) have six petals and are usually arranged in pairs on the leaf axils. The tiny seeds are contained in ribbed fruits that are chalicelike capsules (5 mm in length).

The two other *Heimia* species are almost identical in appearance and are difficult even for experts to distinguish: *Heimia myrtifolia* Cham. et Schl. (indigenous to southwest Brazil)[159] and *Heimia montana* (Griseb.) Lillo (found throughout Bolivia and Argentina) (Malone and Rother 1994, 136; Rother 1990). *Heimia myrtifolia* grows to a height of only about 1 meter and looks like a dwarf form of *Heimia salicifolia.*

Psychoactive Material
— Leaves
— Herbage/branch tips

Preparation and Dosage
Sinicuiche, the Mexican "magical drink causing oblivion," is made from the leaves:

> The preparation of the drink involves laying the slightly wilted leaves in water for a day and then pressing these thoroughly the following day. The juice obtained in this manner is allowed to ferment. In this way, one obtains a peculiar, not unpleasant-tasting drink whose effects, however, are certainly not due to the only low quantities of alcohol that are present but are derived from other substances that are produced during fermentation. (Reko 1938, 142*)

A more modern recipe calls for adding one handful of freshly crushed wilted leaves per person to water and allowing this to sit in the sun for a couple of days, whereupon the liquid will begin to ferment slightly. One cup of this is said to induce yellowish vision and mild euphoria (D. McKenna 1995, 102*). The cold-water extract of the leaves is sticky. Even with dosages as high as 15 g of dried leaves, no psychoactive effects could be observed (Martínez 1994, 295*).

The fresh or dried leaves can be brewed into a tea, both alone and in combination with other herbs.

The fresh herbage can be added to 60 to 80% ethanol to produce an alcoholic extract (tincture). Twenty to 25 g of this tincture is said to be an effective psychoactive dosage.

Ritual Use
To date, no traditional ritual use of *Heimia salicifolia* is known.

Artifacts
A floral element on the famous Aztec statue of Xochipilli appears to be a naturalistic representation of the sinicuiche flower (Wasson 1974).

The occult writer Hanns Heinz Ewers (1871–1943) immortalized the "magical drink causing oblivion" in his novel *Die blauen Indianer* [The Blue Indians] (Reko 1938, 141*).

Medicinal Use
In Mexican folk medicine, sinicuiche is regarded as a narcotic, inebriant, diuretic, and febrifuge (Díaz 1979, 77*; Jiu 1966, 254*). The Huastec use the bush as a medicinal bath additive (Alcorn 1984, 665*). In Mexican folk medicine, a tea made from the leaves is thought to promote digestion (Martínez 1994, 294*). The herbage is also used in the treatment of rabies and to counteract the "evil eye" (Argueta V. et al. 1994, 851*). The plant is widely used to treat syphilis (Malone and Rother 1994, 136).

In Mexico, *Heimia salicifolia* is used primarily for ethnogynecological purposes. Infertile women are said to be helped by a bath prepared from sinicuiche, *pericón* (*Tagetes lucida*; cf. **Tagetes** spp.), rosemary (*Rosmarinus officinalis* L.[160]; cf. **incense, essential oils**), and lavender (*Lavandula angustifolia* Mill. [syn. *Lavandula officinalis* Chaix]).[161] To promote conception, women are advised to drink a tea made from sinicuiche twigs, *dormilona* (*Mimosa pudica*; cf. **Mimosa** spp.), *gobernadora* (*Larrea tridentata* [DC.] Cav.), and *raíz de la fuerza* ("root of power"; unidentified) or *raíz hijera* (?) daily. If a woman remains childless, she should consume a tea brewed from sinicuiche twigs together with *cuatecomate* (*Crescentia alata* H.B.K.), *pericón* (*Tagetes lucida*), and maize cobs (**Zea mays**). To increase fertility or treat sexual weakness and frigidity, ovarian inflammations, and uterine ailments, the vagina should be exposed to the steam from a tea of rosemary and sinicuiche. After giving birth, and to treat the symptoms of a potential miscarriage, a drink made of sinicuiche, cinnamon (*Cinnamomum verum*), pulque (cf. **Agave** spp.), and *piloncillo* (?) should be drunk (Argueta V. et al. 1994, 851*).

The Maká Indians of Chaco, in Paraguay, use fresh *Heimia* leaves as an extracting plant paste for treating wounds caused by thorns that have remained in the body. The leaves are said to simplify the removal of the thorn and also appear to promote the healing of the wound (Arenas 1987, 290*). The Pilagá of the Argentinean Chaco place fresh leaves onto sores, drink a decoction of the root for stomachaches, and bathe in the decoction for scabies (Filipov 1994, 188*).

Constituents
The bush contains the quinolizidine alkaloids lythrine, cryogenine (= vertine), heimine, sinicuichine, anelisine, heimidine, lyfoline, dehydrodecodine, abresoline, demethyllasubine-I and -II, epidemethoxyabresoline, sinine, lythridine, vesolidine, and cryofoline. Cryogenine, the main alkaloid, has anticholinergic and antispasmodic

From top to bottom:
The sinicuiche bush (*Heimia salicifolia*), with ripe fruits.

Heimia salicifolia in bloom.

The raw drug that can be obtained in Mexico from *Heimia salicifolia* consists primarily of the thin branches.

159 This species is rumored to be used for hallucinogenic purposes (Grubber 1991, 61*).

160 Rosemary is an important ingredient in many aphrodisiac bath additives (Rätsch 1995a, 334*).

161 "Lavender flowers have sedative effects upon the central nervous system and on the nervous system of the trachea" (Pahlow 1993, 206*).

"The farmers in certain parts of Mexico use the word *sinicuichi* for a drink that is produced from a plant that bears the same name. When taken over an extended period of time, gross forgetfulness occurs. The sinicuichi drinkers usually have no clear and correct orientation concerning space and time. They forget things that took place only a few hours ago, and they tell you things as great news that they told you just minutes earlier without being aware that they are repeating themselves. In contrast, ideas and memories from years ago, which have more or less engraved themselves into their memories, are correctly reproduced. Some babble that they are able to precisely recall occurrences that took place during the first days of their lives, and others that they can even recall events that took place before they came into the world, things that their grandfathers may have experienced and that they describe as if they were their own memories from a time before they were born."

VIKTOR A. REKO
MAGISCHE GIFTE [MAGICAL POISONS]
(1938, 140 f.*)

Cryogenine

effects (Malone and Rother 1994, 137; Scholz and Eigner 1983, 75*). The four Lythraceae alkaloids that have been best studied are vertine (= cryogenine), lyfoline, lythrine, and nesodine (Malone and Rother 1994). The biological precursor of vertine is phenylalanine (Rother and Schwarting 1972).

The leaves contain 15% tannins, 9% bitter agents, and 14% resins (Martínez 1994, 294*). The roots and seeds are devoid of alkaloids (Dobberstein et al. 1975; Malone and Rother 1994, 139).

Effects

The drink brewed from *Heimia salicifolia* produces only mild psychoactive effects:

> Sinicuiche has a weak intoxicating effect. It induces a pleasant, slightly euphoric dizziness and numbness, and the surroundings are perceived as being darker. Auditory hallucinations occur as the inebriated person hears indistinct sounds from a great distance. The world around one shrinks. No unpleasant aftereffects are known. (Scholz and Eigner 1983, 75*)

There have been repeated reports of yellowish vision and mild auditory hallucinations, tunnel effects, and tunnel vision (D. McKenna 1995, 102*; Rob Montgomery, pers. comm.). Chills and shivering have also been reported (Bob Wallace, pers. comm.).

Animal experiments have demonstrated that the alkaloids have anticholinergic and antispasmodic effects (D. McKenna 1995, 102*). The pharmacology of vertine (= cryogenine) is said to be identical to that of the whole extract (Kaplan and Malone 1966). Self-experiments with the alkaloids vertine, lythrine (310 mg, corresponding to 36 to 156 g of dried branch tips), and acetylsalicylic acid did not result in any detectable psychoactivity (Malone and Rother 1994, 142).

Commercial Forms and Regulations

The seeds are occasionally available from ethnobotanical specialty sources. The plant is not subject to any restrictions.

Literature

Appel, H.-G., Ana Rother, and A. E. Schwarting. 1965. Alkaloids of *Heimia salicifolia*, 2: Isolation of nesodine and lyfoline and their correlation with other Lythraceae alkaloids. *Lloydia* 28:84–89.

Blomster, R. N., A. E. Schwarting, and J. M. Bobbitt. 1964. Alkaloids of *Heimia salicifolia*, 1: A preliminary report. *Lloydia* 27:15–24.

Calderón, J. B. 1896. Estudio sobre el arbusto llamado sinicuichi. *Anales del Instituto Médico Nacionál* 2:36–42.

Dobberstein, R. H., J. M. Edwards, and A. E. Schwarting. 1975. The sequential appearance and metabolism of alkaloids in *Heimia salicifolia*. *Phytochemistry* 14:1769–75.

Douglas, B., J. L. Kirkpatrick, R. F. Raffauf, O. Ribeiro, and J. A. Weisbach. 1964. Problems in chemotaxonomy II. The major alkaloids of the genus *Heimia*. *Lloydia* 27:25–31.

Hörhammer, R. B., A. E. Schwarting, and J. M. Edwards. 1970. The structure of sinicuichine. *Lloydia* 33 (4): 483.

Kaplan, Harvey R., and Marvin H. Malone. 1966. A pharmacologic study of nesodine, cryogenin and other alkaloids of *Heimia salicifolia*. *Lloydia* 29:348–59.

Kaplan, H. R., R. V. Robichaud, and M. H. Malone. 1965. Further pharmacologic studies of cryogenine, an alkaloid isolated from *Heimia salicifolia* Link et Otto. *The Pharmacologist* 7:154 (abstract 103).

Lema, William J., James W. Blankenship, and Marvin H. Malone. 1986. Prostaglandin synthetase inhibition by alkaloids of *Heimia salicifolia*. *Journal of Ethnopharmacology* 15:161–67.

Malone, Marvin H., and Ana Rother. 1994. *Heimia salicifolia*: A phytochemical and phytopharmacologic review. *Journal of Ethnopharmacology* 42:135–59.

Reko, Victor A. 1926. *Sinicuichi*. *La Revista Médica de Yucatán* 14:22–27.

———. 1935. *Sinicuichi*: Der vergeglich machende Zaubertrank. *Pharmaceutische Monatshefte* 16:155–57.

Rother, Ana. 1985. The phenyl- and biphenyl-quinolizidines of in vitro–grown *Heimia salicifolia*. *Journal of Natural Products* 48:33–41.

———. 1989. *Heimia salicifolia*: In vitro culture and the production of phenyl- and biphenyl-quinolizidines. In *Medicinal and aromatic plants II*, vol. 7 of *Biotechnology in agriculture and forestry*, ed. Y. P. S. Bajaj, 246–63. Heidelberg: Springer.

———. 1990. Alkaloids of *Heimia montana*. *Phytochemistry* 29:1683–86.

Rother, Ana, H.-G. Appel, J. M. Kiely, A. E. Schwarting, and J. M. Bobbitt. 1965. Alkaloids of *Heimia salicifolia*. III: Contribution to the structure of cryogenine. *Lloydia* 28:90–94.

Rother, Ana, and A. E. Schwarting. 1972. Phenylalanine as a precursor for cryogenin biosynthesis in *Heimia salicifolia*. *Phytochemistry* 11:2475–80.

Humulus lupulus Linnaeus

Hops

Family

Cannabaceae (Cannabinaceae; Hemp Family); sometimes placed in the Moraceae Family (cf. Zander 1994, 315*)

Forms and Subspecies

There is only one species,[162] a wild form with different varieties, but numerous cultivars (Edwardson 1952; Small 1978). The cultivars were formerly described as unique species (see "Synonyms"); they are now regarded as varieties. Today, morphological differences are the basis for recognizing five varieties: *Humulus lupulus* var. *cordifolius* (Miq.) Maxim in Franch. et Sav., *Humulus lupulus* var. *lupuloides* E. Small, *Humulus lupulus* var. *lupulus* L., *Humulus lupulus* var. *neomexicanus* Nelson et Cockerell, and *Humulus lupulus* var. *pubescens* E. Small. The most famous of the cultivated sorts are Saazer hops (Chechnya, Bohemia), Nittelfrüh hops (Bavaria), Tettnanger hops (Switzerland), Fuggles hops (England), Goldings hops (England), and Cascade hops (United States).

Synonyms

Cannabis lupulus (L.) Scopoli
Humulus americanus Nutt. (= *Humulus lupulus* var. *lupuloides* E. Small)
Humulus cordifolius Miq. (= *Humulus lupulus* var. *cordifolius* [Miq.] Maxim in Franch. et Sav.)
Humulus lupulus var. *brachystachyus* Zapalowics
Humulus neomexicanus (Nelson et Cockerell) Rydberg (= *Humulus lupulus* var. *neomexicanus* Nelson et Cockerell)
Humulus volubilis Salisb.
Humulus vulgaris Gilib.
Humulus yunnanensis Hu (may be a distinct species)
Lupulus communis Gaertn.
Lupulus humulus Mill.
Lupulus scandens Lam.

Folk Names

Bierhopfen, gemeiner hopfen, hop, hopf, hopfen, hoppen, hoppho, hops, houblon, hupfen, lupolo, luppolo, lupulo, vigne du nord

History

Pliny appears to have been the first to mention hops; Dioscorides was not yet aware of the plant. It is first referred to by name in medieval manuscripts, which also make mention of *homularien*, "hops gardens," a reference not to cultivated fields but to large wild occurrences of the plant (DeLyser and Kasper 1994, 166). Hildegard von Bingen was the first to more precisely describe the plant's psychoactive effects and its use as a preservative (in **beer**). Hops also appears in the works of all of the fathers of botany and was botanically described by Linnaeus.

Because no fossil precursors have been found, the phylogeny of the plant is entirely unknown. Hops is the closest relative of hemp (*Cannabis indica, Cannabis sativa*).

Today, hops is the most commonly used beer additive. This use was invented toward the end of the Middle Ages by Christian monks, who had a great interest in the anaphrodisiac qualities of hops flowers. Abbé Adalhard, in the *Statutae abbatiae corbej* (822 C.E.), was the first to document the use of hops in beer. However, the use of hops as a beer additive did not become commonplace until the sixteenth century, when the Bavarian Purity Law—the first German drug law—was promulgated (DeLyser and Kasper 1994, 168; Wilson 1975).

Distribution

Hops appears to have originated in northern Eurasia. Beer-loving humans spread the plant, which is now found throughout the world. Its presence in central Europe has been documented to as early as the eighth century. Today, hops is cultivated in all of the world's temperate zones. One of the most important areas of commercial cultivation is in Tasmania (Pearce 1976), which has the best air in the world, copious rainfall, and now—unfortunately—an overabundance of UV radiation (harmful to humans).

In central Europe, wild hops can be found growing in lowland forests and fens, on the sides of paths, and in hedges.

Cultivation

Female plants only are grown vegetatively, i.e., through clones and cuttings (Gross 1900).

Appearance

Hops is a 6- to 8-meter-long (in cultivation, it can grow up to 12 meters), perennial, dextrorotatory, twisting climber with three to five lobed, opposite leaves.

The plant is dioecious; the male flower is paniculate, the female a so-called strobile that becomes the fruit cone (hop cone or achene). The flowering period is between July and August, and the fruit cones ripen from September to October (in Australia and Tasmania, the ripening period is from April to May).

The stems are a source of fiber (similar to that

Today, the primary use of the sedative hops plant is as an additive to beer. (Woodcut from Tabernaemontanus, *Neu Vollkommen Kräuter-Buch*, 1731)

---

162 The genus *Humulus* is composed of three species: *H. lupulus* L., *H. japonicus* Sieb. et Zucc. [syn. *H. scandens* (Lour.) Merr.], and *H. yunnanensis* Hu (Wohlfahrt 1993, 447). *H. japonicus* can be used only for ornamental purposes.

A female hops plant (*Humulus lupulus*) with its characteristic flowers (hops cones).

Lupulone

"Of Hops:
The hop is warm and dry, and it has some moisture, and it is not very suitable for the use of man, for it causes melancholy to increase in man, and it makes the mind of man sorrowful, and it burdens his intestines. But with its bitterness, it keeps certain corruptions away from the beverages to which it is added, so that they are that much more imperishable."

Hildegard von Bingen
Physica 1.61

---

of hemp, but not as durable). In former times, the fiber was made into linen cloth (DeLyser and Kasper 1994).

Psychoactive Material
— Female inflorescences (strobiles), also known as hops flowers (stroboli lupuli, lupuli stroboli) or hop glands (lupuli glandula)

Because the active constituents in the dried plant material are continuously broken down by oxidation, flowers that are more than one year old should not be used. The flowers must be harvested before they lose their glandular leaves.

Preparation and Dosage
A calming hops tea can be brewed with two heaping teaspoons of hops flowers and $^1/_4$ liter of boiling water. To increase the sedative effects, valerian (*Valeriana officinalis* L.) may be added. Many breweries produce heavily hopped beer (pilseners), most of which have a very bitter taste and also are useful as soporific drinks.

Ritual Use
Hops—the "soul of Christian beer"—was first ingested "ritually" by monks to suppress their natural urges. These chaste men drank huge quantities of beer so that they could resist the temptations of the devil, i.e., their own desires.

However, hops has never played any particular role as a true ritual plant. In recent centuries, hops flowers have occasionally found use as an **incense** or an ingredient in incense blends. The plant was assigned to the planet Mars (Culpeper) and the element water.

The Omaha Indian tribe had a society of buffalo doctors (*te'ithaethe*) composed of men and women to whom the buffalo had appeared in a dream. The members of this society were specialized in the treatment of wounds. Their most important medicine consisted of wild hops, the root of the nightshade *Physalis heterophylla* Nees., and American sweet cicely, *Osmorhiza longistylis* (Torr.) DC. They chewed these three ingredients and then spat them with some water onto the wounded region (Kindscher 1992, 269*).

Artifacts
Hops and the harvesting of hops are the subjects of many paintings and illustrations from the nineteenth and early twentieth centuries (e.g., *The Hops Harvest*, by William Henry Pyne, and *The Hops Pluckers*, by H. Stelzer).

Medicinal Use
Hops and hops extracts are used as sedatives in both folk medicine and biological medicine. The German Federal Health Office has recommended the use of hops tea as a sedative for unease, anxiety, and sleep disorders.

Hops pillows are used in aromatherapy to promote calm and to relieve difficulty in sleeping. These pillows have been known since at least the eighteenth century and were used in cases in which "opium had already failed" (DeLyser and Kasper 1994, 167).

In homeopathy, the agent Humulus lupulus is used primarily as a sedative.

Constituents
Hop flowers contain 15 to 30% resin, the bitter acids humulone and lupulone (and their auto-oxidation products), and an essential oil with mono- and sesquiterpenes (2-methyl-3-buten-2-ol, β-caryophyllene, **farnesene**, humulene, 2-methyl-isobutyrate, methyl-*n*-octylketone, myrcene, post-humulene-1, posthumulene-2), along with minerals, flavonoids, chalcones, polyphenoles, and catechines.

The yellow hops granules, which contain the bitter substance lupulone, are found in hops flowers. Lupulone, which has antibiotic properties, gives beer its characteristic bitterness. It has calming effects on humans and inhibits premature ejaculation. Also present are enzyme-inhibiting polyphenoles (Williams and Menary 1988).

Numerous chemical races (chemocultivars, chemovarieties) have been described for hops. These vary both quantitatively and qualitatively in their concentrations of bitter substances and essential oils (Wohlfart 1993, 448).

The leaves contain camphor oil, quercetin and quercetin glycosides (rutin; cf. *Psidium guajava*), proanthocyanidins (procyanidin, prodelphini-dine), ascorbic acid, and quebrachitol (Wohlfart 1993, 448).

Because hops is very closely related to hemp (*Cannabis sativa*), attempts have been made to find cannabinoids (**THC**) in the former. To date, none has been detected.

Hops produces a yellow coloring agent that was once used in dyeing

Effects
Hops is a sedative and also has been characterized as a "mild hypnotic" (Roth et al. 1994, 406*; Lee et al. 1993). It has pronounced sedative effects, especially when used in combination with valerian (*Valeriana officinalis*), and is effective in treating sleep disorders and withdrawal from **diazepam** addiction (Brattström 1996).

The bitter substances have antibacterial, antimycotic, spasmolytic, and estrogenic effects. Because of the estrogenic properties of hops, chronic beer consumption can result in a feminization of the male body. This can be expressed in morphological changes, e.g., the development of what are known as "beer breasts." The effects of hops are not affected by **alcohol** (Brattström 1996). Some of the narcotic effects of the plant may be due to the presence of quercetin (cf. *Psidium guajava*).

Fresh hops cones may irritate the skin (i.e., provoke allergic reactions) and produce a dermatitis known as hops pluckers' disease. Hops eye, a kind of conjunctivitis, can also occur (Wohlfart 1993, 453 ff.)

Side effects (apart from the allergic reactions) and interactions are unknown.

Commercial Forms and Regulations
All hops drugs and preparations, including homeopathic preparations (Humulus lupulus hom. *HAB1*, Humulus lupulus hom. *PFX*, Lupulus, Lupulinum hom. *HPUS88*), are available without restriction.

Literature
See also the entries for *Cannabis* x **and hybrids** and **beer.**

Brattström, A. 1996. Wirksamkeitsnachweis von Phytopharmaka am Beispiel einer Hopfen-Baldrian-Kombination. *Forschende Komplementärmedizin* 3 (4): 188–95.

DeLyser, D. Y., and W. J. Kasper. 1994. Hopped beer: The case for cultivation. *Economic Botany* 48 (2): 166–70.

Edwardson, J. R. 1952. Hops—their botany, history, production, and utilization. *Economic Botany* 6:160–75.

Gross, E. 1900. *Hops: In their botanical, agricultural and technical aspect and as an article of commerce.* London: Scott, Greenwood and Co.

Lee, K. M., J. S. Jung, D. K. Song, M. Kräuter, and Y. H. Kim. 1993. Effects of *Humulus lupulus* extract on the central nervous system in mice. *Planta Medica* 59 suppl.: A691.

Pahlow, Mannfried. 1985. *Hopfen und Baldrian.* Stuttgart: J. F. Steinkopf.

Pearce, H. R. 1976. *The hop industry in Australia.* Melbourne: Melbourne University Press.

Simmonds, P. L. 1877. *Hops: Their cultivation, commerce, and uses in various countries.* London: E. and F. N. Spon.

Small, E. 1980. The relationships of hop cultivars and wild variants of *Humulus lupulus. Canadian Journal of Botany* 58 (6): 676–86.

Stevens, R. 1967. The chemistry of hop constituents. *Chemical Review* 67:19–71.

Williams, Elizabeth A., and Robert C. Menary. 1988. Polyphenolic inhibitors of alpha-acid oxidase activity. *Phytochemistry* 27 (1): 35–39.

Wilson, D. G. 1975. Plant remains from the Graveney boat and the early history of *Humulus lupulus* L. in Western Europe. *New Phytologist* 75:627–48.

Wohlfart, Rainer. 1993. Humulus. In *Hagers Handbuch der pharmazeutischen Praxis*, 5th ed., 5:447–58. Berlin: Springer.

Zuurbier, K. W. M., S. Y. Fung, J. J. C. Scheffer, and R. Verpoorte. 1993. Possible involvement of chalcone synthase in the biosynthesis of bitter acids in *Humulus lupulus. Planta Medica* 59 suppl.: A588.

The Japanese hops (*Humulus japonicus*) is not suited for use as a beer additive, although it may have psychoactive properties. (Photographed in Songlisan, South Korea)

Hyoscyamus albus Linnaeus
Yellow Henbane

Family
Solanaceae (Nightshade Family); Subfamily Solanoideae, Hyoscyameae Tribe, Hyoscyaminae Subtribe

Forms and Subspecies
Three varieties are usually distinguished:
Hyoscyamus albus L. var. *desertorum*
Hyoscyamus albus L. var. *canariensis*
Hyoscyamus albus L. var. *albus*

Synonyms
Hyoscyamus luteus nom. nud.

Folk Names
Altersum, apollinaris, bíly blín (Bohemia), diskíamos (modern Greek), dontochorton (Cyprus), gelbes bilsenkraut, helles bilsenkraut, hyoskyamos, obecny (Bohemia), Russian henbane, sikran (Morocco), weiß bülsen, weiß bülsenkraut, weißes bilsenkraut, yellow henbane, zam bülsenkraut[163]

History
Hyoscamus albus was the most commonly used magical and medicinal plant of European antiquity (Schneider 1974, 2:184*). Though it is commonly referred to as yellow henbane today, it was more commonly known as white henbane in the

163 Many of the folk names given to *Hyoscyamus niger* are also used for yellow henbane (Schneider 1974, 2:184*).

A botanically accurate illustration of white henbane (= yellow henbane). (Woodcut from Gerard, *The Herball or General History of Plants*, 1633*)

seventeenth and eighteenth centuries. Much earlier it was described in detail by Dioscorides (4.69), who characterized it as the species with the greatest medicinal value. Pliny had similar things to say in his *Natural History* (first century):

The plant that is known to us as *apollinarus* and by some as *altercum*, and among the Greeks as *hyoskyamos*, is also ascribed to Hercules. There are several species of this: one [**Hyoscyamus niger**] has a black seed and almost purple-red flowers, is thorny at the calyx, and grows in Galatia; while the common species [**Hyoscyamus muticus** (?)] is whiter, bushier, and taller than the poppy; the seeds of the third are similar to those of the wild radish; all three species produce frenzy [*insania*] and dizziness. The fourth species is soft, woolly, and fatter than the other species, has a white seed, and thrives in coastal areas. This [*Hyoscyamus albus*] is used by physicians, as is the one with the reddish seeds. But sometimes the white seed also becomes red when it is not yet ripe, and this is discarded. In general, the plant is not collected anywhere until it has become dry. It has the property of wine, which is why it unnerves the senses and the head. The seeds are used both by themselves and pressed as juice. This juice is pressed separately, as is that of the stems and leaves. The root is also made use of; but this, in my opinion, is a hazardous medicine. It is known that the leaves as well confuse the mind when more than four are taken in a drink; but according to the opinion of the elderly, in wine they will dispel fever. From the seeds is produced . . . an oil that will confuse the reason even if dripped into the ear, and it is remarkable that those who have drunk of it are given medicines as if for a poison, and yet it itself is used as a medicine. (Pliny 25.17, 35–37)

In England, the tobacco (*Nicotiana rustica, Nicotiana tabacum*) imported from the New World was identified as a species of henbane (*Hyoscyamus peruvianus*). In the seventeenth century, the word *tobacco* was a kind of synonym for herbs that could be smoked. The so-called little yellow henbane (*Hyoscyamus luteus*) that was grown in

Yellow henbane (*Hyoscyamus albus*) is a southern European plant.

many English gardens and appears to sow itself was known as English tobacco (Gerard 1633, 356*). This is a clear indication that henbane, a European native, was once numbered among the smoking herbs and may have been used for ritual purposes (cf. Golowin 1982*).

Today, yellow henbane has only a negligible pharmaceutical significance in the production of **tropane alkaloids** (Sauerwein and Shimomura 1991).

Distribution

Yellow henbane is found primarily in southern Europe (Spain, Italy, Greece) and in the Near East. It is very common in the Golan Heights of Israel (Dafni and Yaniv 1994, 12*). It prefers sunny locations close to the sea.

Cultivation

Hyoscyamus albus lives from one to three years and is the most easily grown of all henbane species. The seeds need only be loosely broadcast over sandy, clayey, or even poor soil. Water occasionally at first, but never overwater. This heat-resistant plant also thrives in crevices of old walls and between rocks. In the Mediterranean region, the plant flowers from April to May, while in central Europe (where it is grown as a cultivar), it flowers from June to September. The entire plant is harvested while still in bloom and hung by the roots in a well-aired location. Drying requires from three to six weeks.

Appearance

The plant grows to a height of 40 to 50 cm. Although the stems are vertical, the plant often takes on a bushy appearance. The light green stems and serrated leaves are very pileous, as are the calyxes and the fruits. On Cyprus and in Greece, the plant flowers from January to July. The flowers have a light yellow color but often are dark violet on the inside. The seeds are whitish, ocher, or, less frequently, gray in color.

Psychoactive Material
— Leaves
— Herbage
— Seeds

Preparation and Dosage

The dried herbage can be smoked as a treatment for asthma, bronchitis, and coughs (in an amount equal to that contained in one cigarette). An aphrodisiac **smoking blend** can be produced by combining equal parts of the plant, hemp flowers (*Cannabis indica, Cannabis sativa*), and dried fly agaric mushrooms (*Amanita muscaria*). A "prophetic delirium" can be produced by inhaling the smoke of burning henbane seeds. The fresh or dried herbage can be added to wine and used as a remedy for pains and cramps. For information about dosages, see *Hyoscyamus muticus* and *Hyoscyamus niger*. In Morocco, it is said that twice the amount that can be taken up by the fingertips is sufficient to produce hallucinogenic effects (Vries 1984*).

Ritual Use

Henbane, and in particular this species, was certainly the most important ancient means for producing a trance state and clearly was ingested by many oracles and soothsayers (sibyls, Pythias). It was the "dragon plant" of the ancient earth oracle of Gaia, the "madness-inducing" plant of the Colchic oracle of the witch goddess Hecate, the "Zeus bean" of the oracle of Zeus-Ammon of late ancient times and the Roman Jupiter, and the "Apollo's plant" of Delphi and other oracles of the god of "prophetic madness"[164] (Rätsch 1987).

The seeds, both alone and in combination with other substances, were usually burned as a ritual **incense** and inhaled, or the leaves were added to **wine** and drunk. When the soothsayers and prophetesses inhaled the smoke or drank the wine after their ritual ablutions, they called to the oracular deity, usually Apollo. When they had been possessed by the god, they would lose their human consciousness and proclaim the messages of Apollo through their mouths. Priests then "translated" (i.e., interpreted and proclaimed as the words of the oracle) their often unintelligible babbling, sighing, and groaning (Kerényi 1983; Maas 1993; Parke 1985, 1988; Roberts 1984).

In Morocco, the herbage or the seeds of *sikram*, "inebriant," as the plant is known (cf. *Hyoscyamus muticus*), are still burned for psychoactive purposes or used as an ingredient in psychoactive incenses, usually in combination with the seeds of *Peganum harmala* (Vries 1994*).

Artifacts

Curiously, there are no objects from ancient times that can be interpreted as representations of yellow henbane.

Medicinal Use

The legendary physician Hippocrates (ca. 460–370 B.C.E.) praised the medicinal use of henbane. His students, the Hippocratics, administered the seeds with wine to treat fever, tetanus, and gynecological ailments. Donkey's milk was listed as an antidote for overdoses.

Yellow henbane was one of the most important analgesics of antiquity. According to Galen (ca. 130–199 C.E.), it was the main ingredient in a soporific and sedative agent known as *philonion*, which consisted of five parts saffron (*Crocus sativus*), one part each of *Pyrethrum, Euphorbium,* and *Spica nardi*, twenty parts each of white pepper (*Piper album* = *Piper nigrum*) and henbane, and ten parts opium (cf. **soporific sponge**).

On Cyprus, crushed leaves are still used as an analgesic plaster. The dried leaves are smoked together with tobacco (*Nicotiana tabacum*) as a remedy for asthma (Georgiades 1987, 2:56*). In the Golan Heights (Israel), various preparations of the leaves (decoctions, pastes) are used externally to treat skin diseases, open wounds, headaches, rheumatism, inflammations of the eye, and insect stings (Dafni and Yaniv 1994, 13*).

Constituents

The entire plant contains the **tropane alkaloids** hyoscyamine and **scopolamine**, along with apo-scopolamine, norscopolamine, littorine, tropine, cuscohygrine, tigloidine, and tigloyloxytropane, in concentrations similar to those found in *Hyoscyamus niger*.

Effects

The psychoactive effects of henbane were well known in ancient times and were characterized as *mania*, or madness. It should be noted, however, that the Greeks used the term *madness* to characterize not a pathological state but, rather, a dramatic alteration of consciousness:

> Madness (*mania*): in kind there is but one madness, but in form it appears in a thousand ways. Its nature is a chronic condition of being out-of-oneself. . . . Even the inebriation of wine can heat one to insanity; even edible things produce frenzy, such as the mandrake [*Mandragora officinarum*] or henbane. But all of this does not fall under the name of madness; for it passes as quickly as it arrived. (Aretaeus, *De causis et signis morborum chronicorum* 1:6)

This "madness" was regarded as a "divine alteration of the normal, orderly condition":

> [W]e divided the divine kind of madness into four parts, each with its own deity. We attributed prophetic inspiration to Apollo, mystical inspiration to Dionysus, poetic inspiration to the Muses, and the fourth kind to Aphrodite and to Love. We said that the madness of love was the best kind. (Socrates in Plato, *Phaidros* 265)

Because henbane sedates the externally oriented consciousness, humans are opened up to the divine: "If the divine madness of prophetic enthusiasm is to come over humans, then the sun of consciousness within them must set; the human light must disappear within the divine light" (Philos of Alexandria). In the same way, the "divining madness," a kind of prophetic and clairvoyant state of consciousness (or a form of trance), was induced by the sacred "plant of Apollo" (cf. Pliny 26.140). Apart from this, the

This early modern illustration of white henbane (= yellow henbane) shows the characteristic flowers and fruit capsules. (Woodcut from Tabernaemontanus, *Neu Vollkommen Kräuter-Buch*, 1731)

164 The so-called smoke of Delphi has also been attributed to *Catha edulis, Datura stramonium, Cannabis indica,* and *Laurus nobilis*.

"Now the effects of henbane, which were well known in ancient times, correspond to that which we know about Dionysian possession. Henbane causes states of delirium that are filled with visions and hallucinations and can develop into violent attacks of insanity; after this comes an irresistible need for sleep, which leads to very deep slumber."

JACQUES BROSSE
MYTHOLOGIE DER BÄUME
[MYTHOLOGY OF TREES]
(1990, 108*)

effects of this species of henbane are similar or identical to the effects produced by the other species.

Commercial Forms and Regulations
None

Literature
See also the entries for the other *Hyoscyamus* species.

Kerényi, Karl. 1983. *Apollo*. Dallas: Spring Publications.

Maas, Michael. 1993. *Das antike Delphi*. Darmstadt, Germany: WBG.

Parke, H. W. 1985. *The oracles of Apollo in Asia Minor*. London: Croom Helm.

———. 1988. *Sibyls and sibylline prophecy in classical antiquity*. London: Routledge.

Rätsch, Christian. 1987. Der Rauch von Delphi: Eine ethnopharmakologische Annäherung. *Curare* 10 (4): 215–28.

Roberts, Deborah H. 1984. *Apollo and his oracle in the Oresteia*. Hypomnemata, no. 78. Göttingen, Germany: Vandenhoeck und Ruprecht.

Sauerwein, M., and K. Shimomura. 1991. Production of tropane alkaloids in *Hyoscyamus albus* transformed with *Agrobacterium rhizogenes*. *Planta Medica* 57 suppl. (2): A108–9.

Hyoscyamus muticus Linnaeus
Egyptian Henbane

"Among the plant remains discovered in the animal necropolis at Saqqâra, which unfortunately could not be dated, were also parts of *H. [Hyoscyamus] muticus*. It can be assumed that this poisonous plant as well was already part of the flora of ancient Egypt. Unfortunately, the medicinal texts from Pharaonic times do not contain any plant names that can be definitely construed as inebriants. Egyptian henbane was first mentioned as a medicinal plant in a papyrus written in Greek from the 1st century C.E. . . . Today, this plant continues to be used as a medicine and inebriant in Egypt."

RENATE GERMER
FLORA DES PHARAONISCHEN ÄGYPTEN [FLORA OF PHARAONIC EGYPT]
(1985, 169*)

Family
Solanaceae (Nightshade Family); Subfamily Solanoideae, Hyoscyameae Tribe, Hyoscyaminae Subtribe

Forms and Subspecies
One subspecies occurs in Morocco (Vries 1984*; 1989, 39*): *Hyoscyamus muticus* L. ssp. *falezles* (Saharan henbane).

Synonyms
Hyoscyamus betaefolius Am.
Hyoscyamus datura nom. nud.
Hyoscyamus insanus Stocks
Scopolia datora Dun.
Scopolia mutica Dun.

Folk Names
Ägyptisches bilsenkraut, bhang, Cyprus henbane, Egyptian henbane, giusquiamo egiziano, Indian henbane, jusquiame d'Egypt, kohi-bhang, kohi-bung, mountain hemp, pitonionca (Greek, "dragon plant"), sakra, sakran, sekaran (Arabic, "the inebriating"), sikrane (Morocco), sikran sahra, ssakarân, traumkraut

History
Ancient Egyptians knew the henbane that grew among them as *sakran*, "the drunken one," using a word borrowed from Aramaic (Kottek 1994, 129*). It is mentioned as a medicinal plant in a papyrus from the first century that was written in Greek. From that time until the present, the plant has been in use in Egypt as a medicinal and inebriating plant (Germer 1985, 169*).

In Arabia, the plant was used for criminal purposes. The powdered herbage, mixed with puréed dates (*Phoenix dactylifera*) or milk, was given to the intended victims, who then fell into a delirious state and were easily robbed (Morton 1977, 309*).

Today, Egyptian henbane is the most important of all *Hyoscyamus* species for pharmaceutical and economic purposes.

Distribution
The plant prefers desert regions and occurs from Egypt to as far south as Sudan, and also in Syria, Afghanistan, Pakistan, and northern India.

Cultivation
The plant thrives in dry, rocky soils and a desert climate. The seeds (when they can be obtained) are scattered over sandy clay soil and pressed down lightly. The seeds should be well watered; later, the plant often can tolerate short dry periods. Even when the leaves appear wilted because of a shortage of water, a few drops of water will quickly restore them. After it is harvested, the herbage should be hung by the roots in a shady, airy location to dry (the plant may require up to six weeks to dry thoroughly).

The plant is cultivated for pharmaceutical purposes in Egypt, the former Yugoslavia, Greece, Pakistan, and India.

Appearance
The annual to biennial plant can grow as tall as 90 cm. The almost square stem is an important characteristic. Apart from this, it is quite similar to

Hyoscyamus albus and *Hyoscyamus niger* (see there).

Psychoactive Material
— Leaves/herbage (folia hyoscyami mutici, hyoscyami mutici herba)
— Seeds

Preparation and Dosage
Either the fresh leaves are used as a poultice or the dried herbage and seeds are used internally. When taken internally, a single dosage should not exceed 0.25 g; the total daily dosage should not exceed 1.5 g. Because reactions to **tropane alkaloids** can vary significantly from one individual to another, it is difficult to provide dosage guidelines that apply to everyone. Anyone who wishes to experiment with Egyptian henbane for therapeutic or psychoactive purposes should exercise great care and begin with a very small dosage, which then can be slowly increased.

The dried leaves and seeds are suitable for use as ingredients in **incense** and **smoking blends**. The herbage also can be used to brew **beer** (see recipe under *Hyoscyamus niger*).

In Morocco, twenty seeds from the subspecies *falezles* are taken orally in a date (cf. **palm wine**), chewed thoroughly, and swallowed to induce hallucinations. They also are used as an ingredient in *majun* (see **Oriental joy pills**) (Vries 1984*; 1989, 39*).

Ritual Use
It has been suggested that the magical Homeric **nepenthes** was Egyptian henbane (Millspaugh 1974, 488*). The ancient Assyrians sometimes brewed beer with henbane (Thompson 1949, 230*). According to Aelia (ca. 170–240 C.E.), when digging up this henbane, great care had to be exercised, similar to that taken with the mandrake (see *Mandragora officinarum*). Instead of using a dog to pull out the plant, however, a bird was tied to the plant (2:251).

It is very likely that Egyptian henbane was used as a ritual inebriant in ancient Egypt. Unfortunately, very little is known about this plant from late ancient times. It appears to have played a role in the cult of the dead; remnants recovered from the animal necropolis at Saqqâra have been definitely identified as henbane (Germer 1986, 169*; Manniche 1989, 20*). However, it has not yet been possible to equate the plant with a particular hieroglyphic name. For this reason, the use of Egyptian henbane is known only from the Greek literature. A Greek papyrus from the Egyptian city of Arsinoites (third century B.C.E.) contains a quite interesting recipe, although it does not state what it was used for:

Left: Egyptian henbane (*Hyoscyamus muticus*) in bloom.

Right: A stem of *Hyoscyamus muticus*, showing the ripe seed capsules.

Other [recipe]: for the plaster he mixed three parts of white gum, one part of [copper] oxide, one-half part burnt copper, as much henbane juice as copper. Stir these [things] until smooth, dissolve in water, use.

A recipe with which one "can put a person to sleep for two days" (presumably to induce prophetic dreams) has come down to us in the late ancient Leiden magical papyrus:

Mandrake root [*Mandragora officinarum*], one ounce, licorice, one ounce, henbane, one ounce, ivy [*Hedera helix*], one ounce, you pound these together. . . . If you want to do this with skill, for every portion give the four-fold amount of wine [see *Vitis vinifera*], you moisten everything from morning until evening, you pour it off, you allow it to be drunk; very good. (Griffith and Thompson 1974, 149 f.*)

The Arabs liked to spice their coffee (see *Coffea arabica*) with crushed henbane seeds. It is possible that the plant also played a role in the secret rites of the dervishes or Sufis. In the twentieth century, the Towara Bedouins of the Sinai Peninsula were still smoking the leaves to produce "an inebriation with delirium" (Lewin 1981, 177*).

In India, *Hyoscyamus muticus* (the effects of which are more pronounced than those of *Hyoscyamus niger*) is used in place of opium (see *Papaver somniferum*) as an inebriant (Macmillan 1991, 421*). In the Punjab and in Baluchistan, the leaves are smoked together with *Cannabis indica* (Lewin 1981, 178*).

Today, Egyptian henbane is known in some counterculture circles as "dream herb" (Lindequist 1993, 463) and is smoked by itself or in combination with other substances (**smoking blends**).

Artifacts
The magical henbane has been a plant of the gods since early times. According to Josephus Flavius, the turban (*mitra*) of the highest priest of the Hebrews was adorned with a branch of henbane (Kottek 1994, 129*). Apart from this, no artifacts are known.

Medicinal Use
In Nigeria, the herbage is used as an antispasmodic, to treat asthma, and for seasickness. In Iran, the smoke produced by burning the seeds is used to treat toothaches (Morton 1977, 308*). Moroccans use the seeds of the subspecies *falezles* for the same purpose (Vries 1984*).

Constituents
Egyptian henbane has the highest alkaloid content of any species of *Hyoscyamus*. The alkaloid content can be as much as 2% dry weight. Moreover, numerous methods have been discovered for increasing alkaloid concentrations in plants that are cultivated for pharmaceutical purposes (Misra et al. 1992; Oksman-Caldentey et al. 1991; Sevón et al. 1992; Sevón et al. 1993; Vanhala et al. 1992). The plant forms chiefly hyoscyamine with only traces of hyoscine (= **scopolamine**) and **atropine** (Misra et al. 1992). Also present are the **tropane alkaloids** scopolamine, aposcopolamine, norscopolamine, littorine, tropine, cuscohygrine, tigloidine, and tigloyloxytropane (Lindequist 1993). Amazingly, the flowers contain the highest concentration of alkaloids (2%), followed by the leaves (1.4 to 1.7%) and the seeds (0.9 to 1.3%). The lowest concentrations are in the stems (0.5 to 0.6%) (Morton 1977, 308*).

Effects
Of all the *Hyoscyamus* species, Egyptian henbane has the most profound inebriating effects. In India, some users are said to have developed feeblemindedness, tarantism, and exhibitionism (Morton 1977, 308*). Other symptoms of higher dosages include hallucinations as well as possible unpleasant side effects (mydriasis, dry mouth, disturbances of coordination, flight of ideas, delirium) (cf. *Hyoscyamus niger*).

Commercial Forms and Regulations
According to *Hagers Handbuch der pharmazeutischen Praxis*, the herbage (hyoscyami mutici herba) can be obtained from a pharmacy without a prescription (in Germany) (Lindequist 1993, 464). Egypt does not permit the export of viable seeds (Morton 1977, 308*).

Literature
See also the entries for the other *Hyoscyamus* species.

Lindequist, Ulrike. 1993. *Hyoscyamus.* In *Hagers Handbuch der pharmazeutischen Praxis*, 5th ed., 5:460–74. Berlin: Springer.

Misra, H. O., J. R. Sharma, and R. K. Lal. 1992. Inheritance of biomass yield and tropane alkaloid content in *Hyoscyamus muticus. Planta Medica* 58:81–83.

Oksman-Caldentey, K.-M., M.-R. Laaksonen, and R. Hiltunen. 1989. "Hairy root" cultures of *Hyoscyamus muticus* and their hyoscyamine production. *Planta Medica* 55:229.

Oksman-Caldentey, K.-M., N. Sevón, and R. Hiltunen. 1991. Hyoscyamine accumulation in hairy roots of *Hyoscyamus muticus* in response to chitosan. *Planta Medica* 57 suppl. (2): A105.

Sevón, N., M. Suomalainen, R. Hiltunen, and K.-M. Oksman-Caldentey. 1992. Effect of sucrose, nitrogen, and copper on the growth and alkaloid production of transformed root cultures of *Hyoscyamus muticus. Planta Medica* 58 suppl. (1): A609–10.

———. 1993. The effect of fungal elicitors on hyoscyamine content in hairy root cultures of *Hyoscyamus muticus. Planta Medica* 59 suppl.: A661.

Vanhala, L., R. Hiltunen, and K.-M. Oksman-Caldentey. 1991. Virulence of different *Agrobacterium* strains on *Hyoscyamus muticus. Planta Medica* 57 suppl. (2): A109–10.

Vanhala, L., T. Seppänen-Laakso, R. Hiltunen, and K.-M. Oksman-Caldentey. 1991. Fatty acid composition in transformed root cultures of *Hyoscyamus muticus. Planta Medica* 58 suppl. (1): A616–17.

Hyoscyamus niger Linnaeus

Black Henbane

Family
Solanaceae (Nightshade Family); Subfamily Solanoideae, Hyoscyameae Tribe, Hyoscyaminae Subtribe

Forms and Subspecies
The species can be divided into a number of varieties (cf. Strauss 1989):

Hyoscyamus niger L. var. α *agrestis* Kit.—flowers usually pale yellow
Hyoscyamus niger L. var. *annuus* Sims—annual, most commonly planted variety
Hyoscyamus niger L. var. *chinensis* Makino—Chinese variety
Hyoscyamus niger L. var. *niger*—wild form
Hyoscyamus niger L. var. *pallidus* (Wadst. et Kit.) Koch [= *H. niger* L. var. β *pallidus* Kit.)—biennial

Synonyms
Hyoscarpus niger (L.) Dulac
Hyoscyamus agrestis Kit.
Hyoscyamus auriculatus Ten.
Hyoscyamus bohemicus Schmidt (cf. **Hyoscyamus spp.**)
Hyoscyamus lethalis Salisb.
Hyoscyamus officinalis Cr.
Hyoscyamus pallidus Waldst. et Kit. ex Willd.
Hyoscyamus persicus Boiss. et Buhse
Hyoscyamus pictus Roth
Hyoscyamus sinensis Makino
Hyoscyamus syspirensis Koch
Hyoscyamus verviensis Lej.
Hyoscyamus vulgaris Neck

Folk Names
Alterco, alterculum, altercum (Arabic), apollinaris (Roman, "plant of Apollo"), apolloniakraut, asharmadu (ancient Assyrian), banj (Persian), bazrul (Hindi), beléndek (Anglo-Saxon), belene, beleño (Spanish),[165] beleño negro, belinuntia (Gaelic), bendj, bengi (Arabic), bilinuntia (Celtic, "plant of Bel[enus]"), bilisa, bilsa, bilse, bilsen, bilsencruydt, bilzekruid (Dutch), bilzenkruid, black henbane, blín, blyn (Bohemian), bolmört (Swedish), bolonditó csalmatok (Anglo-Saxon), bulmeurt (Danish), caliculares, caniculata, cassilagine, caßilago, caulicula, demonaria, dens caballinus, dentaria, dente cavallino, dioskyamos (Greek, "god's bean"), dollkraut, dordillen saett, dulbillerkraut, dulldill, dull-dill, dullkraut, endromie, erba del dento, faba iouis, faba lupina, faba suilla, fabulonia, fetid nightshade, foetid nightshade, gemeines bilsenkraut, giusquiamo (Italian), giusquiamo nero, gur (ancient Assyrian), hannebane, henbain, henbane (English), henbell, herba canicularis, herba pinnula, herbe aux chevaux, herbe aux dents, hisquiamum, hogbean, hühnertod, hyoscyamus (Roman), hyoskyamos (Greek, "hog's bean"), Indian henbane, insana, iosciamo, iupiters beame, iusquiame, iusquiamo, iusquiamus, jupitersbohne, jupitersbon (Swiss, "Jupiter's bean"), jusquaime noire, jusquiamus, kariswah (Newari), khorasanijowan (Bengali), khurasani ajavayan, khurasani ajowain (Hindi), khurassani jamani, khursani ajwan (Nepali), kurasaniajowan (Hindi), lang dang, lang-tang (Chinese), lang-thang-tse (Tibetan), meimendro (Portuguese), meimendro negro, milicum, milimandrum, nicotiana minor, palladia, parasikayavani (Sanskrit), piliza, pilsener krutt, pilsenkrawt, poison tobacco, pythonion (Greek, "dragon plant" or "plant of the Pythia"), rasenwurz, rindswurz, rindswurzel, saubohnen, saukraut, säukraut, schlafkraut, schwarzes bilsenkraut, shakruna (Aramaic), sickly smelling nightshade, sikran, stinking nightshade, stinking roger, swienekruud, symphoniaca, taubenkraut, teufelsauge, teufelsaugn, tollkraut, tornabonæ congener, totenblumenkraut, veleño negro, zahnkraut, zigeunerkorn, zigeunerkraut[166]

History
The ethnobotanist Wolf-Dieter Storl has conjectured that henbane was already in use in Eurasia for ritual and shamanic purposes in the Paleolithic period. When the Paleoindians migrated from Asia into the Americas via the Bering Strait, they brought their knowledge of the use of the plant with them. But because they were unable to find the plant they knew on the American continent, they substituted the similar and related tobacco (**Nicotiana tabacum**).

Henbane was used as a ritual plant by the pre-Indo-European peoples of central Europe. In Austria, two handfuls of henbane seeds were discovered in a kind of urn along with bones and snail shells; the find dates from the early Bronze Age (Graichen 1988, 69).

Ancient authors (Dioscorides, Pliny) were familiar with black henbane (see **Hyoscyamus albus**). It has been suggested that henbane was the magical Homeric **nepenthes** (Hocking 1947, 313). Carl Ruck believes that henbane, under the name *hyoskyamos*, was sacred to the goddess Deo-Demeter-Persephone, for her sacred animal was the sow, the "mother swine" (Ruck 1995, 141*)—perhaps being a "lucky swine" was a sign that a person had been allowed to ingest the plant.

In the Celtic regions, the plant was known by the name *belinuntia*, "plant of the sun god Bel."

Illustration of *lang-tang*, or Chinese henbane (*Hyoscyamus niger* var. *chinensis* Makino), from an ancient Chinese herbal (*Chêng-lei pên-ts'ao*, 1249 C.E.).

165 According to J. M. Fericgla, the Spanish *veneno*, "poison," may have been derived from this name.

166 This name is also used for **Datura stramonium.**

Top left: The annual black henbane *Hyoscyamus niger* var. *annuus*, cultivated in northern Germany.

Bottom left: The ripe fruit panicle of black henbane (*Hyoscyamus niger*).

Bottom center: This flower, deep violet inside with fine violet veins outside, and otherwise yellow, is typical of black henbane (*Hyoscyamus niger* var. *niger*).

Bottom right: This bright yellow flower is typical of one variety of black henbane (*Hyoscyamus niger* var. α *agrestis*).

The Gauls poisoned their javelins with a decoction of henbane. The healing properties of the plant were mentioned in medieval Anglo-Saxon pharmacopoeias. The plant's name is derived from the Indo-European *bhelena* and is said to have meant "crazy plant" (Hoops 1973, 284). In Proto-Germanic, *bil* appears to have meant "vision, hallucination" or "magical power, miraculous ability" (vries 1993). There was even a goddess (Asin) known as Bil, a name interpreted as "moment" or "exhaustion." She is understood to be the image in the moon or one of the moon's phases. She may have been a henbane "fairy" or the goddess of henbane and may even have been a goddess of the rainbow: *Bil-röst* is the name of the rainbow bridge that leads to Asgard. *Bil* would then also be the original word for "heaven's bridge" (Simek 1984, 48 f.).

In the fourteenth century, Guy de Chauliac described a narcotic inhalation of henbane for medicinal purposes. A similar fumigation was mentioned in *The Thousand and One Nights* (Hocking 1947, 313, 314). But the plant was often used as a fumigant for magical purposes. Albertus Magnus, in his work *De Vegetabilibus* (ca. 1250; 6.362 f.), reported that necromancers (conjurers of the dead) used henbane to invoke the souls of the dead as well as demons.

In the infamous bathhouses of the late Middle Ages, henbane seeds were strewn over glowing coals in order to heat up the erotic atmosphere. The smoke, mixed with steam, appears to have induced strong aphrodisiac effects (Rätsch 1990, 148*).

Henbane was already being demonized during the Middle Ages and associated with alleged witchcraft (Müller-Ebeling 1991): "The witches drank the decoction of henbane and had those dreams for which they were tortured and executed. It was also used for witches' ointments and was used for making weather and conjuring spirits. If there was a great drought, then a stalk of henbane would be dipped into a spring and the sun-baked sand would be sprinkled with this" (Perger 1864, 181*). In a Pomeranian witchcraft trial in 1538, "a witch confessed" that she had given a man henbane seeds so that he would run around "crazy" (= sexually aroused). In a file from an Inquisition trial, it was noted that "a witch admits" having once strewn henbane between two lovers and uttering the following formula: "Here I sow wild seed, and the devil advised that they would hate and avoid one another until these seeds had been separated" (Marzell 1922, 169*).

Henbane was also renowned as a **beer** additive with potent effects (Marzell 1922, 170*). This use was forbidden by the Bavarian Purity Law of 1516, the first German drug law (Kotschenreuther 1978, 83*).

The ancient use of henbane has been preserved into the present day, especially on Cyprus and in North Africa, primarily in Morocco and Egypt. There, henbane, often mixed with Spanish fly (cantharis; *Lytta vesicatoria*), is used to treat diseases of the female sex organs and as an analgesic, aphrodisiac, and inebriant (when mixed with hashish; cf. **Cannabis sativa**) (Venzlaff 1977*).

Distribution

Black henbane is the most widely distributed of all henbane species. It is found from Europe to Asia (Lauber and Wagner 1996, 804*) and grows wild from the Iberian Peninsula to Scandinavia (Morton 1977, 303*). It is common in North Africa (especially Morocco). In the Himalayas (Uttar Pradesh), it thrives at altitudes of up to 3,600 meters (Jain and Borthakur 1986, 579*). It has become naturalized in North America and Australia.

Cultivation

Propagation occurs via seeds that need only be pressed into sandy and clayey soil (in March and April). However, the likelihood of germination is greater when the seeds are allowed to germinate in seedling soil and then transplanted. Henbane also sows itself. Because the plant requires a nitrogen-rich soil, a pure nitrogen or calcium cyanamide fertilizer should be used. Do not overwater. The plant must be protected from potato beetles and henbane fleas (*Psylloides hyoscyami*) (Morton 1977, 304*).

Black henbane is grown for pharmaceutical purposes in central and eastern Europe (Romania, Bulgaria, Albania) and in India, although not as frequently as **Hyoscyamus muticus**. The plant material is collected or harvested during the

middle of the flowering phase (June to August). The herbage dries very slowly.

Appearance

Depending upon location and climate, black henbane can be an annual or a biennial; the former is more common. This upright plant can grow as tall as 80 cm and possesses undivided, serrated, very pungent leaves.[167] The alternate, quinquelobate flowers are in thick panicles. This *Hyoscyamus* species has the largest flowers of all the species in the genus. They are typically pale yellow with violet veins, but there is a unusual variety that has lemon or bright yellow corollas without veins. The black seeds are very small, and many remain in the fruit. In central Europe, the flowering period is from June to October. In the Mediterranean region, it begins in May and is usually over by July or August.

Black henbane is easily confused with *Hyoscyamus muticus*, although the latter has smaller flowers that lack the violet veins and is more pilose. Black henbane is most similar to the species *Hyoscyamus reticulatus*, which has corollas that are purple-violet with reticulate veins. Also similar is yellow henbane (*Hyoscyamus albus*), which can be distinguished by its smaller, pure yellow flowers and the smaller, almost round leaves with few serrations.

The Asian variety known as *lang-tang* was first described as *Scopolia japonica* L. or *Scopolia sinensis* Hemsl. (see *Scopolia carniolica*) and is still confused with these nightshades today (Li 1978, 19*).

Psychoactive Material

— Leaves (hyoscyami folium, DAB10 [Eur.], ÖAB, Helv. VII)
— Herbage without the roots (herba hyoscyami, hyoscyami herba)
— Seeds (hyoscyami semen); the seeds are listed in Chinese herbals under the name *tian xian zi* (Lu 1986, 80*)
— Henbane oil (hyoscyamus oil)

Preparation and Dosage

The chopped, dried herbage can be used as an ingredient in **incense** and **smoking blends**, for brewing **beer**, to spice **wine**, and as a tea (infusion, decoction). The seeds are most appropriate for use in incense recipes.

Dosages must be assessed carefully no matter what type of preparation is being considered. According to *Hagers Handbuch der pharmazeutischen Praxis*, the therapeutic individual dosage of prepared *Hyoscyamus* (with a standardized alkaloid content of 0.05%) is 0.5 g, and the daily dosage 1.5 g (maximum 3 g) (Lindequist 1993, 469). Apart from this, see the guidelines given for *Hyoscyamus muticus*.

Recipe for Henbane Beer

40 g dried, chopped henbane herbage (*herbae hyoscyamus niger conc.*)
5 g bayberry or another *Myrica* species (this aromatic ingredient is optional)
approx. 23 liters of water
1 liter (approx. 1.2 kg) brewing malt (barley malt)
900 g honey (e.g., spruce or pine honey)
approx. 5 g dried top-fermenting yeast
brown sugar

Boil the henbane and bayberry (if desired) in 1 liter of water (to ensure the necessary sterility). Leave the henbane in the water until it has cooled.

Sterilize the brewing vessel (plastic bucket) with boiling water. Then add the liquid malt to the bucket along with 2 liters of hot water and the honey. Stir until the ingredients are thoroughly dissolved. Add the henbane water together with the herbage and the bayberry. Stir thoroughly. Add cold water to make a total of approx. 25 liters of liquid. Pitch the yeast into the mixture and cover.

In order for the top-fermenting yeast to be effective, the wort should be allowed to stand in a warm location (20° to 25°C.). Fermentation will begin slowly because the tropane alkaloids will initially inhibit the yeast. The main fermentation will be over in four to five days, and the after-fermentation will then begin. The yeast will slowly settle and form a layer at the bottom of the bucket.

The beer can now be poured into bottles. A heaping teaspoon of brown sugar can be added to each (0.7 liter) bottle to promote an additional after-fermentation. Henbane beer tastes best when stored before use in a cool place for two to three months.

Oleum hyoscyamin infusum (henbane oil) is obtained by boiling the leaves in oil. It can be used externally for therapeutic or erotic massage.

Ritual Use

The Assyrians called henbane by the name *sakiru*. They used the plant for medicinal purposes, as an inebriating additive to **beer**, and as an **incense** (in combination with sulfur) to protect against magic (Thompson 1949, 230*).

In ancient Persia, henbane was known as *bangha*, a name that was subsequently applied to hemp (**Cannabis sativa**) and other psychoactive plants. Like the still uncertainly unidentified **haoma**, henbane had a religious significance as a ritual drug. Many Persian sources describe journeys to other worlds and visions that were

Beer brewed with henbane acquires a red color and recalls the witches' beers (always described as red) that were consumed during the nocturnal Sabbats.

Botanical illustration of black henbane. (Engraving from Pereira, *De Beginselen der Materia Medica en der Therapie*, 1849)

167 In her *Herbal*, Elisabeth Blackwell noted, "The entire plant stinks, overpowers the head, and the smell alone betrays its powers to make one stupid" (Heilmann 1984, 124*).

"Sleeping within my orchard,
My custom always of the afternoon,
Upon my secure hour thy uncle stole,
With juice of cursed hebenon in a
    vial,
And in the porches of my ears did
    pour
The leperous distilment; whose effect
Holds such an enmity with blood of
    man
That swift as quicksilver it courses
    through
The natural gates and alleys of the
    body,
And with a sudden vigour doth
    posset
And curd, like eager droppings into
    milk,
The thin and wholesome blood."

Wɪʟʟɪᴀᴍ Sʜᴀᴋᴇsᴘᴇᴀʀᴇ
Hᴀᴍʟᴇᴛ (1.5)

"But those who were best able to
drink were aware that they should
make themselves great; finally they all
pranced around as if they had eaten
henbane seeds."

Gʀɪᴍᴍᴇʟsʜᴀᴜsᴇɴ
Sɪᴍᴘʟɪᴄɪssɪᴍᴜs (1.32)

evoked by various henbane preparations. Prince Vishtasp, who has gone down in history as the protector of Zoroaster (= Zarathustra), drank *mang*, a preparation of henbane and **wine**. After doing so, he fell into a deathlike sleep that lasted for three days and three nights. During this time, his soul journeyed to the Upper Paradise. According to a later source, he drank a mixture of *hom* (= haoma) and henbane in wine. Another Persian visionary named Viraz also made a three-day journey into other worlds with the aid of a mixture of henbane and wine. At the end of the third night, "the soul of the righteous [= Viraz] had the feeling of being in the midst of plants and inhaling scents. It sensed an intensely scented wind that blew from the south. The soul of the righteous inhaled this wind through its nose" (Couliani 1995, 141*).

The Celts knew black henbane as *beleno* and consecrated it to Belenus, the god of oracles and the sun. It was burned as a fumigant in his honor. Druids and bards who inhaled the smoke were taken to the "other world," where they could communicate with fairies and other beings.

Henbane also appears to have been one of the most important ritual plants of the Vikings. Iron Age Viking graves have yielded hundreds of henbane seeds. The grave of a woman on Fyrkat (Denmark) is widely known. The most important grave good worn by the woman was a leather bag filled with countless henbane seeds (Robinson 1994, 544, 547*).

The oldest ethnohistorical evidence of the Germanic use of henbane as a magical plant is contained in the nineteenth book of the collection of church decrees (*Deutsche Bußbuch* [German Book of Atonement]) of the Bishop Burchard von Worms (died 1025). In one confessional question, the following ritual was described in astonishing detail:

> Did you do as certain women are wont? If they need rain and have none, they gather several girls and select from these a small maiden as a kind of leader. They disrobe her and take her outside of the settlement to a place where they find hyoscyamus, which is known as *bilse* in German. They have her pull out this plant using the little finger of the right hand and tie the uprooted plant to the small toe of the right foot with any kind of string. Then the girls, each of whom is holding a rod in her hands, lead the aforementioned maiden to the next river, pulling the plant behind her. The girls then use the rods to sprinkle the young maiden with river water, and in this way they hope to cause rain through their magic. Then they take the young maiden, as naked as she is, who puts down her feet and moves herself in

the manner of a crab, by the hands and lead her from the river back to the settlement. If you have done this or have agreed to do this . . . (in Hasenfratz 1992, 87*)[168]

Such activities bring henbane into association with Donar, the Germanic god of thunder, weather, and storms. The Romans associated the plant with their god Jupiter, whom they equated with the Germanic god of thunder. In Switzerland, the folk name *jupitersbon*, "Jupiter's bean," is still used today.[169]

Of all the Germanic gods, Donar was the most enthusiastic drinker and the most able to hold his drink. Because of this, the strongly inebriating bock beers were consecrated to him. The **beer** of the thunder god was brewed with henbane, his plant. Because of the great demand for henbane, which is quite rare in Germany and northern Europe, ancient Germans planted henbane gardens specifically for the purpose of brewing beer. These fields of cultivated henbane stood under the protection of Wotan/Odin, Donar's father, and were considered sacred. This history of many of the sites where these gardens once stood lives on in their names, such as Bilsensee (Henbane Lake), Billendorf (Henbane Village), Bilsengarten (Henbane Garden), and especially the Bohemian Plzeň (= Pilsen = henbane) (Römpp 1950, 271*).[170]

During the Middle Ages and the early modern period of Europe, henbane was generally associated with witchcraft and magic, and especially with oracles and love magic. Lonicerus noted, "The old women use this plant to make magic, they say that anyone who carries the root around with them will remain invulnerable."

It was also believed that henbane smoke could make one invisible, and the leaves were smoked in a pipe for this purpose (Hinrichsen 1994, 107). If any plant was in fact a true ingredient in **witches' ointments**, it was henbane (cf. Marzell 1922, 168*):

> Henbane poison acts quickly, for it is absorbed by the skin. And something else. It works especially quickly and intensively when taken up through the mucous membranes. But since henbane was also smeared on the broomsticks that men and women rode upon nude, the effects were profound. Among the women, the effects were considerably stronger and were felt more rapidly, as the mucous membranes of the anus and vagina came into contact with the broomstick during wild movements and the effects were immediate. (Hug 1993, 140)

In modern occultism, henbane seeds were used as fumigants to conjure spirits, especially for necromancy, or the conjuration of the dead. The

---

168 In Romania, a similar ritual using the mandrake *(Mandragora officinarum)* instead of henbane was still being conducted in the twentieth century.

169 Ernst Schoen, *Nomina popularia plantarum medicinalium,* Zurich, Galenica, 1963, p. 36.

170 The city that gave its name to the modern, heavily hopped beer known as pilsener acquired its name from henbane, which thus also gave its name to the true or original pilsener beer. In Switzerland, the ancient name *pilsener krut* lives on in the name *Pilsenkraut* (henbane plant).

following recipe was used to mix a powder for use as a fumigant (cf. **incense**):

1 part fennel root/seeds (*Foeniculum vulgare*)
1 part olibanum (*Boswellia sacra*)
4 parts henbane (*Hyoscyamus niger*)
1 part coriander seeds (*Coriandrum sativum* L.)
1 part cassia bark (*Cinnamomum cassis* Presl)

It was said that one should take this incense into a ghostly, dark forest and light a black candle and the incense vessel on a tree stump. The powder was to be burned until the candle suddenly went out. Then one would see the spirits of the night appear in the darkness from out of the smoke. To dispel the spirits, a mixture of equal parts of asafetida ("devil's dirt") and olibanum should be burned (Hyslop and Ratcliffe 1989, 15*).

Henbane was and still is used for psychoactive purposes in Asia. The *Pên-ts'ao Ching*, a very ancient Chinese herbal, contains the following information about the subspecies known as *lang-tang*:

[The seeds], when [properly prepared] and ingested over a long period of time, make it possible for one to go for a very long way, are useful for the mind, and increase power. . . . Moreover, through them one can communicate with spirits and see devils. If they are taken in excess, they will make one stupid. (Li 1978, 19*)

The seeds appear to have been used together with those of *Caesalpinia decapetala* as a psychoactive **incense** (Li 1978, 20 f.*).

In southern Kashmir (on the border of the Himalayas), the leaves, mixed with tobacco (*Nicotiana tabacum),* are smoked as a hallucinogen (Shah 1982, 297*). Unfortunately, nothing is known about the ritual or hedonistic context of this use.

Artifacts

Although henbane is an unmistakable member of the plant kingdom and has played a significant role in European culture and pharmaceutical history, no artifacts are known (cf. *Hyoscyamus muticus*). Similarly, the plant does not appear in depictions of witches from the early modern period. Henbane does occasionally appear as a floral element in the utilitarian art of the art nouveau movement. It is depicted on medicine thangkas used to illustrate Tibetan medicine.

Medicinal Use

The use of henbane smoke to treat toothaches (Rowell 1978, 263*) and asthma is widespread. In Darjeeling and Sikkim, henbane is used for these purposes and to treat nervous diseases (Biswas 1956, 71). In Uttar Pradesh, the plant is used to heal bones (Jain and Borthakur 1986, 579*). In Nepal, the leaves are smoked to treat asthma and used as a sedative and narcotic (Singh 1979, 190*). In traditional Chinese medicine, the smoke of the seeds of Chinese henbane (*lang-dang-zi*) is inhaled in treatments for coughs, bronchial asthma, rheumatism, and stomach pains.

In Europe, preparations of henbane have been used for medicinal purposes since ancient times as analgesics and antispasmodics (cf. **soporific sponge**) to treat stomach cramps, whooping cough, toothaches, lower abdominal inflammations, neuralgia, and, in the form of cigarettes, asthma (Rätsch 1995a, 114–21*). Such use continued into the twentieth century.

Hildegard von Bingen recommended the psychoactive herbage as an antidote for alcohol inebriation: "But so that a drunken person will return to himself, he should lay henbane in cold water, and moisten his forehead, temples, and throat (with this), and he will fare better" (*Physica* 1.110).

A modern German folk medicinal recipe for treating fungal diseases (such as that caused by *Candida albicans*) instructs, "[P]our boiling hot meat broth over a third part stinging nettles [*Urtica dioica* L., *U. urens* L.], a sixth part henbane, a sixth part ground nutmeg [see *Myristica fragrans*], a pinch of saffron [see *Crocus sativus*], a third part balm [*Melissa officinalis* L.], and allow to steep in the refrigerator for four hours, then drink daily of this for four weeks" (*Natur*, June 1996, 60).

In homeopathy, the agent Hyoscyamus niger is used in accordance with the medical description to treat such ailments as unease, agitated states, sleep disturbances, and spasmodic digestive disorders (Lindequist 1993, 472).

Pharmaceutical adhesive bandages containing henbane extracts have been developed to treat travel sickness. They are adhered behind the ear (a traveler who is wearing such a henbane bandage is always on a trip . . .).

Constituents

The leaves and the herbage contain 0.03 to 0.28% **tropane alkaloids**. The principal alkaloids S-(–)-hyoscyamine (which is transformed into **atropine** by drying) and S-(–)-**scopolamine** are present in ratios of 2:1 to 1:1. Trace amounts of aposcopolamine, norscopolamine, littorine, tropine, cuscohygrine, tigloidine, and tigloyloxytropane are also present, as well are flavonoids (rutin) and **coumarin** derivatives (Lindequist 1993, 467).

The homeopathic mother tincture contains a minimum of 0.007 and a maximum of 0.01% alkaloids, calculated as hyoscyamine.

Effects

The parasympathicolytic effects of the drugs and of preparations of black henbane are due to the

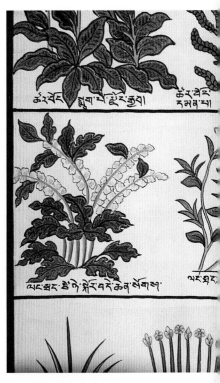

Illustration of Chinese henbane (*Hyoscyamus niger* var. *chinensis*) on a Tibetan medical thangka (detail).

"Henbane is the treasure flower of the Underworld."

MARGRET MADEJSKY (SEPTEMBER 1997)

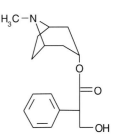

Hyoscyamine

Scopolamine

"In Prussia, [the black henbane] is placed under the roof or in the posts of the stall on the evening of Midsummer's Day to protect against the witches. . . . In Mecklenburg, a cow that has been hexed is fumigated with henbane (dulldill), which is plucked on Midsummer's Day between 11:00 and 12:00 o'clock."

SIEGFRIED SELIGMANN
DIE MAGISCHEN HEIL- UND SCHUTZMITTEL AUS DER BELEBTEN NATUR [THE MAGICAL HEALING AND PROTECTIVE AGENTS FROM THE ANIMATED NATURE]
(1996, 71*)

"Horrible, stinking smoke of burned henbane swirls from a pan and lays itself upon the senses, as heavy as the hands of torment."

GUSTAV MEYRINK
IN "DER UNTERGANG"
(IN MEYRINK 1984, 139)

principal alkaloids hyoscyamine (or **atropine**) and **scopolamine**. Peripheral inhibition with simultaneous central stimulation is characteristic. The primary effects last for three to four hours. Hallucinogenic aftereffects may persist for as long as three days. The alkaloids can cross via the blood into the placenta and have also been detected in breast milk (Lindequist 1993, 469).

Among the unpleasant side effects are severe dryness of the mouth, locomotor disturbances, and farsightedness. Overdoses can lead to delirium, coma, respiratory paralysis, and death. However, only a very few instances of lethality are reported in the toxicological literature (Lindequist 1993, 470). For this reason, the actual lethal dosage is not known precisely. The plant is also toxic to grazing cattle, deer, fish, and many species of birds. Pigs appear to be immune to the effects of the toxins (Morton 1977, 305*) and actually appear to enjoy the inebriating effects. This may be the source of the ancient name *hog's bean*.

Low dosages (0.5 to 1 liter) of beer brewed with henbane have inebriating effects, while higher dosages (1 to 1.5 liters) are aphrodisiac (henbane beer is the only beverage that makes one thirstier the more one drinks!). Very high dosages (more than 2 to 3 liters) can induce delirious and "inane" states, confusion, memory disturbances,[171] and "crazy" behaviors.

Commercial Forms and Regulations

The plant enjoys protected status and is included on the World Conservation Union's Red List of endangered plants. The herbage requires a prescription and may be purchased only in a pharmacy. Henbane oil is available without restriction and may be purchased in shops other than pharmacies. Homeopathic preparations are subject to varying regulations (Lindequist 1993, 471).

Literature

See also the entries for the other *Hyoscyamus* species.

Graichen, Gisela. 1988. *Das Kultplatzbuch*. 2nd ed. Hamburg: Hoffmann und Campe.

Hinrichsen, Torkild. 1994. *Erzgebirge: "Der Duft des Himmels."* Hamburg: Altonaer Museum.

Hocking, George M. 1947. Henbane: Healing herb of Hercules and Apollo. *Economic Botany* 1:306–16.

Hoops, Johannes. 1973. Bilsenkraut. *Reallexikon der germanischen Altertumskunde* 1:284.

Hug, Ernst. 1993. *Wolfzahn, Bilsenkraut und Dachsschmalz: Rückblick in ein Schwarzwalddorf.* St. Margen, Germany: Selbstverlag Ernst Hug.

Klein, G. 1907. Historisches zum Gebrauche des Bilsenkrautextraktes als Narkotikum. *Münchener medizinische Wochenschrift* 22:1088–89.

Lindequist, Ulrike. 1993. *Hyoscyamus*. In *Hagers Handbuch der pharmazeutischen Praxis*, 5th ed., 5:460–74. Berlin: Springer.

Makino, T. 1921. *Hyoscyamus niger* L. var. *chinensis* Makino (Solanaceae). *Journal of Japanese Botany* 2 (5): 1. (In Japanese.)

Meyrink, Gustav. 1984. *Des deutschen Spießers Wunderhorn 2: Der violette Tod.* Rastatt, Germany: Moewig.

Müller-Ebeling, Claudia. 1991. Wolf und Bilsenkraut, Himmel und Hölle: Ein Beitrag zur Dämonisierung der Natur. In *Gaia—Das Erwachen der Göttin*, ed. Susanne G. Seiler, 163–82. Brunswick, Germany: Aurum.

Schiering, Walther. 1927. Bilsenkraut: Eine okkultistisch-kulturgeschichtliche Betrachtung. In *Zentralblatt für Okkultismus*, 23-31, Leipzig. Repr. in Bauereiss 1995, 81–91.*

Simek, Rudolf. 1984. *Lexikon der germanischen Mythologie*. Stuttgart: Kröner.

Storl, Wolf-Dieter. 1996. Lecture on Bilsenkraut at the Rothenburg near Mariastein i.L. (Switzerland).

Strauss, A. 1989. *Hyoscyamus* spp.: In vitro culture and the production of tropane alkaloids. In *Medicinal and Aromatic Plants* 11, vol. 7 of *Biotechnology in Agriculture and Forestry*, ed. Y. P. S. Bajad, 286–314. Berlin: Springer. (Includes a comprehensive bibliography.)

Vries, Herman de. 1993. *Heilige bäume, bilsenkraut und bildzeitung*. In *Naturverehrung und Heilkunst*, ed. C. Rätsch, 65–83. Südergellersen, Germany: Verlag Bruno Martin.

171 The "drink of forgetfulness" that Gudrun gave to Sigurd (in the Volsunga Saga) has often been interpreted as a henbane brew.

Hyoscyamus spp.

Henbane Species

Family

Solanaceae (Nightshade Family); Subfamily Solanoideae, Hyoscyameae Tribe, Hyoscyaminae Subtribe

The genus *Hyoscyamus* Tourn. contains some twenty recognized species, all of which are indigenous only to Eurasia (D'Arcy 1991, 78*; Symon 1991, 141*). Some of these species are very rare and therefore do not play any significant ethnobotanical role. The species are similar, some very similar, in appearance, and thus sometimes can be difficult to identify (Lu and Zhang 1986, 67). All species contain the **tropane alkaloids** hyoscyamine and **scopolamine**, along with apo-scopolamine, norscopolamine, littorine, tropine, cuscohygrine, tigloidine, and tigloyloxytropane (Lindequist 1993, 461).

Hyoscyamus aureus L.—golden henbane, gold henbane

In the Bible, this henbane species (the most common of the five species that occur in Israel) is mentioned by the name *shikrona* (Zohary 1986, 187*). Today, it is quite common in the Golan Heights (Dafni and Yaniv 1994, 12*). In Israeli folk medicine, the seeds and leaves are burned and the rising smoke is inhaled to treat toothaches and tooth decay. A decoction of the leaves is used in the form of eyedrops to treat eye inflammations; the crushed fresh leaves are mixed with olive oil and applied to open wounds. The steam produced by boiling the leaves or the smoke of burning leaves is inhaled as a treatment for asthma and other diseases of the respiratory tract. A poultice made of crushed leaves mixed with flour is applied externally to treat headaches (Dafni and Yaniv 1994, 13 f.*).

Hyoscyamus bohemicus F.W. Schmidt

This "Bohemian" species is found from northern China through central Asia and into the Near East. In traditional Chinese medicine, the seeds are used in a manner similar to those of *lang-tang* (*Hyoscyamus niger* var. *chinensis*) (Lu 1986, 80*; Lu and Zhang 1986, 71). To date, we know of no ethnic psychoactive use of this species. However, it is quite possible that in central Asia the species may be smoked in place of *Hyoscyamus niger*. This species also is regarded as a synonym or variety of *Hyoscyamus niger* (Lindequist 1993, 464).

Hyoscyamus boveanus (Dun. in DC.) Asch. ex Schweinfurth

This species is found in the eastern desert of Egypt. The Bedouins call it *saykaraan*, a word borrowed from Arabic that means "to become inebriated" (see *Hyoscyamus muticus*). In former times, the leaves and flowers of this little-known henbane species were smoked as an inebriant by the Bedouins and Nubians (some of whom still continue the practice), sometimes mixed with tobacco (*Nicotiana tabacum*) or other herbs (presumably *Cannabis sativa*). The plant is still sold in herb stands in the markets of the Nile Valley (Goodman and Hobbs 1988, 84 f.). This species is apparently synonymous with *Hyoscyamus niger*.

Hyoscyamus desertorum Boiss. (= *H. albus* var. *desertorum*)—desert henbane

The inhabitants of the Negev Desert and the northern Sinai smoke the dried leaves and seeds of this plant as a treatment for toothaches, chest pains, coughs, asthma, and hysteria (Dafni and Yaniv 1994, 14*).

Hyoscyamus x *györffyi*

This is a hybrid of *Hyoscyamus niger* and *Hyoscyamus albus* (Ionkova et al. 1994). It has no ethnopharmacological significance, although it does contain high concentrations of tropane alkaloids.

Hyoscyamus pallidus Kitaib.

The Assyrians appear to have used this Near East species of henbane to treat toothaches and as an inebriant (Thompson 1949, 216*).

Hyoscyamus physaloides L.

The Tungus roasted the seeds of this central and East Asian henbane species and brewed them into a drink. They drank this special "coffee" after eating, presumably for its inebriating effects (Rowell 1978, 263*). In Siberia, the entire plant, including the root, is used as an inebriant and opium substitute (see *Papaver somniferum*). The root apparently also was used there as a potently inebriating, hallucinogenic **beer** additive (Hartwich 1911, 522*).

Hyoscyamus pusillus L.

This species is found from northern China through central Asia and into the Near East, just as *Hyoscyamus bohemicus* is (Lu and Zhang 1986, 71). Likewise, its seeds are used in traditional Chinese medicine (Lu 1986, 80*). To date, we know of no psychoactive use of this species in any Asian culture. However, in central Asia it is quite possible that this species is used as an inebriant in place of *Hyoscyamus niger*. The ancient Assyrians

"Today, the inhabitants of Siberia at Jenissei still use the plant [*Hyoscyamus physaloides*] as an agent of pleasure. They chop the leaves and the root into small pieces and add this to fermenting or already finished beer. One glass of this beer produces an utter confusion. The drinker begins to speak without knowing what. His senses are dulled, and he loses any sense of proportion. When he walks, he imagines that he is encountering insurmountable obstacles. Every moment, he sees his close and unavoidable death before him."

CARL HARTWICH
DIE MENSCHLICHEN GENUßMITTEL
[THE HUMAN AGENTS OF PLEASURE]
(1911, 522*)

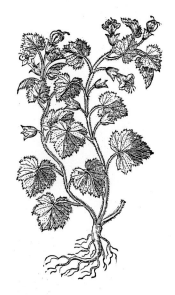

A strange and botanically unrecognizable species of Cretan henbane. (Woodcut from Gerard, *The Herball or General History of Plants*, 1633)

In former times, more species of henbane were recognized and illustrated than are botanically recognized today. This species had reddish flowers. (Woodcut from Gerard, *The Herball or General History of Plants*, 1633)

appear to have used this henbane species as an analgesic and inebriant (Thompson 1949, 216*). In Arabic, the plant is known as *sufairâ*.

Hyoscyamus reticulatus L.

Like *Hysocyamus desertorum*, this species is found only in the Negev Desert, where the Bedouins use it for folk medicinal purposes (Dafni and Yaniv 1994, 17*). The inhabitants of the Negev smoke the dried leaves to treat toothaches, as a sedative, and as an inebriant (Dafni and Yaniv 1994, 14*). In Syria and Iran, the plant is known as *bazr-i-banj*, *kohi bang*, *banj barri*, and *benj*. Physicians there use the seeds like opium (cf. **Papaver somniferum**) (Hooper 1937, 128*).

Literature

See also the entries for the other **Hyoscyamus** species.

Goodman, Steven M., and Joseph J. Hobbs. 1988. The ethnobotany of the Egyptian eastern desert: A comparison of common plant usage between two culturally distinct Bedouin groups. *Journal of Ethnopharmacology* 23:73–89.

Ionkova, Iliana, L. Witte, and A. W. Alfermann. 1994. Spectrum of tropane alkaloids in transformed roots of *Datura innoxia* and *Hyoscyamus* x *györffyi* cultivated in vitro. *Planta Medica* 60:382–84.

Lindequist, Ulrike. 1993. *Hyoscyamus*. In *Hagers Handbuch der pharmazeutischen Praxis*, 5th ed., 5:460–74. Berlin: Springer.

Lu An-ming and Zhang Zhi-yu. 1986. Studies of the subtribe Hyoscyaminae in China. In *Solanaceae: Biology and systematics*, ed. William G. D'Arcy, 56–78. New York: Columbia University Press.

Ilex cassine Walter

Cassina Tree

Family
Aquifoliaceae (Holly Family);[172] Iliceae Tribe

Forms and Subspecies
Three varieties and one form are botanically accepted (Galle 1997, 165f.):

Ilex cassine L. var. *cassine*
Ilex cassine L. var. *angustifolia* Ait.
Ilex cassine L. var. *mexicana* (Turcz.) Black
Ilex cassine L. f. *aureo-bractea*

The myrtle-leaf holly (*Ilex myrtifolia* Walt.), which is found in Florida, is sometimes regarded as a unique species, and occasionally as a variety of *Ilex cassine*: *Ilex cassine* var. *myrtifolia* Walt. (Bell and Taylor 1982, 75).

Synonyms
Ageria germinata Raf.
Ageria heterophylla Raf.
Ageria obovata Raf.
Ageria palustris Raf.
Aquifolium carolinesse Cat. et Duh.
Ilex aquifolium carolinianum Duh. (= *I. cassine* var. *angustifolia*)
Ilex cassinaefolia Loes.
Ilex cassene L.
Ilex cassine [alpha] L.
Ilex cassine corymbosia W.T. Mill.
Ilex cassine L. f. *glabra* Loes. (= *I. cassine* var. *mexicana*)

Ilex cassine L. f. *hirtella* Loes. (= *I. cassine* var. *mexicana*)
Ilex cassine var. *latifolia* Ait.
Ilex cassinoides Link (= *I. cassine* var. *angustifolia*)
Ilex cassinoides Du Mont (= *I. cassine* var. *angustifolia*)
Ilex castaneifolia Hort. ex Loes.
Ilex chinensis DC.
Ilex dahoon Walt.
Ilex dahoon var. *angustifolia* (Willd.) Torr. et Gray (= *I. cassine* var. *angustifolia*)
Ilex dahoon var. *grandiflora* Koch
Ilex dahoon var. *laurifolia* (Nutt.) Nutt.
Ilex dahoon var. *ligustrum* (Ell.) Woods (= *I. cassine* var. *angustifolia*)
Ilex lanceolata Griseb.
Ilex ligustrina Elliot (= *I. cassine* var. *angustifolia*)
Ilex mexicana (Turcz.) Black (= *I. cassine* var. *mexicana*)
Ilex phillyreifolia Hort. ex Dippel
Ilex prinoides Willd.
Ilex ramulosa Raf.
Ilex watsoniana Spach (= *I. cassine* var. *angustifolia*)
Pilostegia mexicana Turcz. (= *I. cassine* var. *mexicana*)
Prinos cassinoides Hort. ex Steudel

Folk Names
Black drink plant, cassena,[173] cassiana, cassina, cassinabaum, cassina tree, cassine, dahoon,

---

172 This family is very closely related to the Celestraceae (cf. *Catha edulis*) (Schultes and Raffauf 1990, 79*).

173 "The word is said to be derived from *assie*, 'little leaf,' and comes from the language of the also extinct Timuna Indians, who adopted it from the Creek who migrated into their territory" (Hartwich 1911, 468*).

Left: The fruits of the cassina tree (*Ilex cassine*), a native of the southern region of North America, are red berries. (Photographed in Florida)

Right: The *Ilex* species that are used in North America to prepare the "black drink" are often confused with the closely related holly (*Ilex aquifolium*).

dahoon holly, dahoon-holly, dahoon plant, holly-ilex, southern yaupon, yaupon, yupon

History

This plant was sacred to the Indians of Florida and the East Coast of North America and was used in the same manner as *Ilex vomitoria* to produce the "black drink" (Galle 1997, 165; Millspaugh 1974, 416*). The leaves were traded over long distances (Havard 1896, 40*).

Ilex cassine and *Ilex vomitoria* are still confused with one another today. Yet both can be easily distinguished from one another on the basis of their morphology and geographical distribution. The two species have sometimes been regarded as synonymous, which is inaccurate. Of course, the confusion has only been increased by the fact that Indians used the two plants in an almost identical manner, so that ethnographers did not make a distinction between the two.

Distribution

This North American *Ilex* species is indigenous to the margins of swamps and waterways. It usually is found near the coast and occurs in Virginia, Florida, and the Gulf Coast as far as the Colorado River (Texas). The variety *mexicana* is found in Mexico.

Cultivation

Unknown; presumably from seed

Appearance

This heavily branched tree can grow as tall as 8 meters and develops a projecting crown. The leaves are 6 to 10 cm long, lanceolate, tapered at the ends, and shiny on the upper surface. They are considerably longer and more narrow than the leaves of *Ilex vomitoria*. The fruits of *Ilex cassine* are luminously red and are more solitary that those of *Ilex vomitoria*, which are borne in thick clusters on the branches.

Cassina is often confused with holly (*Ilex aquifolium* L.,[174] which occurs in many varieties), especially in the ethnological literature.

Psychoactive Material

— Leaves

Preparation and Dosage

The fresh leaves are boiled in water (for at least ten minutes) until a black decoction has been produced. This tea is known as the "black drink" (cf. *Ilex vomitoria*).

A more potent preparation requires additional time: The leaves are first roasted and then boiled in water for at least half an hour. The decoction should be stirred well and/or poured into a different vessel repeatedly until it becomes foamy (Havard 1896, 40*). The taste of the black drink is slightly reminiscent of that of oolong tea (see *Camellia sinensis*).

Particularly potent versions of the drink were made for rituals and festivals, and various herbs and roots were added to potentiate and/or alter its effects. For example, plants that would induce vomiting—button snakeroot (*Eryngium aquaticum* L.), *Iris versicolor* L., and lobelia (*Lobelia inflata*)—were added. The participants in these rituals would often vomit violently, an act that was regarded as a ritual purification (Havard 1896, 41*).

Sometimes fermenting agents were added to the black drink. In such cases, the drink had inebriating as well as potent stimulant effects (Havard 1896, 41f.*). Tobacco (*Nicotiana tabacum, Nicotiana* spp.) was a less common additive (Waldman 1985, 63*).

Ritual Use

The tribes of the coastal areas from the Carolinas to Florida and Texas, as well as inland tribes on both sides of the Mississippi, used cassina in the form of the black drink in such important annual ceremonies as the Green Corn Festival (also known as the Busk Ritual). The focus of this festival was the renewal of the world during the continually recurring passage of the year (Waldman 1985, 63*). Participants drank great quantities of the drink as part of this ritual's events. Sometimes only the men were allowed to partake of the black drink (Millspaugh 1974, 416*).

"The 'black drink,' or cassina, has a stimulating effect as a result of its high caffeine content. The drink was made and used in large areas of the Southeast and beyond as far as Texas. The use occurred in a strict ritual framework and was limited to high-ranking males. Its emetic effects, which the literature often treats as a curiosity, were understood as an inner cleansing and preparation for contact with the spiritual world."

ANDREA BLÄTTER
"DROGEN IN PRÄKOLUMBISCHEN NORDAMERIKA" ["DRUGS IN PRE-COLUMBIAN NORTH AMERICA"] (1996A, 174*)

174 Although holly (*Ilex aquifolium*) does not contain any caffeine, it does contain some theobromine (Alikaridis 1987).

"The leaves of the *Ilex cassine* tree were the primary ingredient of the renowned 'black drink' of the Indians of the Southeast. This drink induced immediate vomiting. It was generally used in the context of purification rituals, especially those that took place prior to war parties. The Creek used this emetic in almost all of their religious ceremonies."

SERGE BRAMLEY
IM REICHE DES WAKAN [IN THE REALM OF THE WAKAN]
(1977, 84*)

At the tribal festivals of the Apalachicola tribe, copious amounts of black drink were offered in and consumed from large snail shells (*Busycon* spp.). Participants also smoked enormous amounts of tobacco (*Nicotiana rustica*).

For more on ritual uses, see *Ilex vomitoria*, the use of which was very similar to that of *Ilex cassine*.

Artifacts

In Florida and elsewhere, the large, sinistrorse shells of the sea snail *Busycon contrarium* Conrad were used as drinking vessels for the cassina drink (cf. Moore 1921). Some of these shells were engraved with markings representing mythical or shamanic beings.

Medicinal Use

The decoction was used primarily as an emetic (Millspaugh 1974, 416*). The Cherokee, Alabama,

Creek, and Natchez tribes used a decoction of the leaves and young shoots to induce vomiting, for urinary problems (gravel), as a sudorific, for dropsy, and for purification (including "moral") (Moerman 1986, 232*).

Constituents

The leaves contain 0.27 to 0.32% **caffeine** (Havard 1896, 40*), along with a tanning agent and perhaps other substances. More recent data indicate that the leaves contain not caffeine but, more likely, **theobromine** (an important active principle in *Theobroma cacao*). Cyanidin-3-xylosylglucoside is present in the fruits (Alikaridis 1987, 126).

Effects

Because of its caffeine content, the black drink has both stimulating and strongly diuretic effects

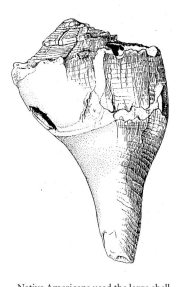

Native Americans used the large shell of the sea snail *Busycon contrarium* as a drinking vessel for consuming the cassina drink. (From Moore, "Notes on Shell Implements from Florida," 1921)

Ilex Species Used in the Production of Stimulating Drinks

The genus *Ilex* consists of some four hundred to six hundred species that are found throughout the world, although they are concentrated in South America and Asia. Caffeoylquinic acid has been found in fifteen species. Purines occur in nineteen species. Many species contain **caffeine** and are or were used to produce stimulating beverages (Hartwich 1911, 452*).

Botanical Name	Occurrence	Indigenous Name(s)
Ilex amara (Vell) Loes.	Brazil	cauna, caurina, congohinha
Ilex affinis Gard.	Brazil	congonha do campo
Ilex argentina Lillo [syn. *Ilex tucumanensis* Speg.]	Argentina	palo de yerba
Ilex brevicuspis Reiss.	South America	mate
Ilex cassine Walt.	southern North America	cassina, black drink
Ilex caroliniana (Lam.) Loes.	Carolinas (U.S.A.)	cassine
Ilex congonhinha Loes.	Brazil	congonhinha
Ilex conocarpa Reiss.	Brazil	congonha, catuaba do mato
Ilex cuyabensis Reiss.	Mato Grosso	congonha
Ilex diuretica Mart.	Brazil	congonha
Ilex dumosa Reiss.	Brazil, Uruguay, Paraguay	congonha miuda, caa-chiri
Ilex dumosa var. *guaranina* Loes.	Brazil	congonha
Ilex fertilis Reiss. ex Mart.	South America	maté
Ilex glabra A. Gray	North America	black drink
Ilex glazioviana Loes.	Brazil	congonhinha
Ilex guayusa Loes.	Ecuador	guayusa
Ilex microdonta Reiss.	South America	
Ilex paraguariensis St.-Hil.	South America	maté (true maté)
Ilex perado Ait. [syn. *Ilex quercifolia* Meerb.]	North America	black drink
Ilex pseudobuxus Reiss.	South America	maté
Ilex pseudothea Reiss.	Brazil	
Ilex tarapotina Loes.	eastern Peru	maté
Ilex theezans Mart.	Brazil, Argentina	cauna amarga, páo d'aceite, caa-na
Ilex verticillata A. Gray	North America	black drink
Ilex vitis-idaea Loes.	Peru	maté
Ilex vomitoria Ait.	southeastern U.S.	yaupon, black drink
Ilex yunnanensis Franch.	China, Tibet	
Ilex yunnanensis var. *eciliata* Hu	China	shui-cha-tze ("water tea")

(Havard 1896, 41*). Higher dosages (to which individuals have varying responses) can induce vomiting. When additional substances were added to the drink to promote fermentation, the primary effects were certainly alcoholic, although it is likely that a person would not fall asleep as quickly as with other alcoholic beverages. The addition of lobelia (cf. *Lobelia inflata*) or tobacco (cf. *Nicotiana* spp.) substantially increased the psychoactive and emetic effects.

The red fruits are said to be toxic to humans. Detailed information about such toxic effects is lacking.

Commercial Forms and Regulations
None

Literature
See also the entries for *Ilex guayusa, Ilex paraguariensis, Ilex vomitoria,* and **caffeine**.

Alikaridis, F. 1987. Natural constituents of *Ilex* species. *Journal of Ethnopharmacology* 20:121–44.

Bell, C. Ritchie, and Bryan J. Taylor. 1982. *Florida wild flowers and roadside plants.* Chapel Hill, N.C.: Laurel Hill Press.

Galle, Fred C. 1997. *Hollies. The genus* Ilex. Portland, Ore.: Timber Press.

Hale, E. M. 1891. *Ilex Cassine*, the aboriginal North American tea. USDA Division of Botany *Bulletin,* no. 14.

Hu, Shiu-Ying. 1949. The genus *Ilex* in China. *Journal of the Arnold Arboretum* 30:341 ff.

Hudson, Charles, ed. 1979. *Black drink. A Native American tea.* Athens: University of Georgia Press.

Hume, H. Harold. 1953. *Hollies.* New York: Macmillan.

Moore, Clarence B. 1921. Notes on shell implements from Florida. *American Anthropologist,* n.s., 23:12–18.

A blooming branch of the maté bush (*Ilex paraguariensis*).

Ilex guayusa Loesener

Guayusa

Family
Aquifoliaceae (Holly Family); Iliceae Tribe

Forms and Subspecies
Native Americans make a distinction between a wild form and the cultivated plant. Apart from this, no varieties or other forms have been described to date. It has been suggested that *Ilex guayusa* is a cultivated form of *Ilex paraguariensis* (Shemluck 1979, 156).

Synonyms
Ilex guayusa Loesener emend. Shemluck
Ilex guayusa var. *utilis* Moldenke

Folk Names
Aguayusa, guañusa, guayupa, guayusa,[175] guayúsa, guayusa holly, guayyusa, huayusa, kopíniak (Záparo), rainforest holly, wais (Shuar), wayus (Achuar), wayusa, weisa (Jíbaro)

In Peru, the pepper plant *Piper callosum* Ruíz et Pav., which continues to enjoy great ethnomedicinal significance, is known in the local language as *huayusa.*

History
The ritual use of guayusa (whose name is a Quechuan word) in South America is very ancient. In Niño Korin (Bolivia), carefully wrapped leaves that had served as grave goods and were radiocarbon-dated to 355 C.E. were discovered. Chemical analyses revealed that the leaves still contained **caffeine**. The guayusa leaves were found together with paraphernalia for using snuff powder and devices for administering enemas. It is uncertain, however, whether the leaves were used in the making of stimulating **snuffs** or psychoactive **enemas** (Schultes 1972, 115 f.; Wassén 1972, 29). As early as the seventeenth and eighteenth centuries, Jesuits and other missionaries were suggesting that the plant was used for psychoactive and medicinal purposes (Schultes 1972, 126 ff.) It is possible that the use of guayusa was much more widespread in pre-Columbian America than it is today. It appears as though the great demand for the plant led to its large-scale cultivation (Schultes 1979, 144 f.).

The English botanist Richard Spruce (1817–1893) made a precise record of the use of guayusa but surprisingly neglected to collect any botanical material (Schultes 1972, 120). The plant was first described in the early twentieth century (Loesener 1901). Even today, little is known about it, and much ethnopharmacological research remains to be done.

175 In Ecuador, other plants are also known by the name *guayusa*, particularly several members of the Piperaceae Family, as well as *Siparuna eggersii* Hieron. (Monimiaceae) and possibly others as well (Patiño 1968, 315). One member of the genus *Hedyosmum* (Chloranthaceae), the species identification of which is uncertain, not only is known by the name *guayusa* but also is used to produce a stimulating drink (Schultes 1972, 128).

"The jungle Indians, for example, ingested guayusa as an emetic. The pot of guayusa, in which the leaves and water were transformed into a greenish brew, was always simmering. The drink caused anyone who partook of it to vomit out all of the food that had not been digested during the night. In this manner, they practiced medicine according to a primitive Hippocratic principle."

Victor W. von Hagen
Die Wüstenkönigreiche Perus
[The Desert Kingdoms of Peru]
(1979*)

Distribution

Guayusa occurs only in the tropical rain forests of the western Amazon region, chiefly in Ecuador, but also in northeastern Peru and southwestern Colombia (Schultes 1979, 144).

Cultivation

Propagation occurs only with cuttings that are grown in house gardens or in small plantations.

Appearance

This evergreen tree grows to a height of some 15 meters and is able to develop a very thick trunk (Patiño 1968, 314). The oval leaves are alternate and attenuate and grow to a length of 2.5 to 4.5 cm. The flowers grow in clusters from the leaf stalk axils (Shemluck 1979). Cultivated plants never produce flowers. After the initial description by Richard Spruce, more than one hundred years passed before a botanist, Homer Pinkley, was able to actually observe a guayusa flower. He reported that the flowers grow directly out of the trunk of the tree, are smaller than 3 cm in size, resemble a small bowl, and are pale yellow in color (Schultes 1972, 120 f.).

Ilex guayusa is very easily confused with *Ilex paraguariensis* as well as with other *Ilex* species (Shemluck 1979, 158).

Psychoactive Material

— Leaves

A report from the colonial period describes how the Indians would drink preparations of *tripiliponi*, a similar plant with somewhat larger leaves and less stimulating effects, together with lime or orange juice as a substitute for guayusa. Unfortunately, it has not yet been possible to determine the botanical identity of the *tripiliponi* plant (Patiño 1968, 311).

Preparation and Dosage

The plucked leaves are threaded onto a string and hung to dry either in the sun or in the house. The dried leaves are then boiled over low heat for at least ten minutes but preferably for half or even an entire hour.

Five leaves is often given as the dosage for

These *Ilex guayusa* leaves from Ecuador are rich in caffeine.

one cup of tea (Schultes 1972, 132). A chemical analysis of a guayusa drink prepared by the Jíbaro (Achuar) revealed that it contained 3.3% **caffeine**. Each morning the men consume an average of 2.2 liters of the beverage. Most of them vomit half of this after some forty-five minutes. As a result, they ingest approximately 690 mg of caffeine, the equivalent of about eight cups of coffee (*Coffea arabica*) (Lewis et al. 1991, 25).

The Jíbaro often boil the leaves for hours or even overnight (Schultes 1972, 129). Although only the men of the Jíbaro are allowed to make guayusa (they call the tea *wayus*), the women and children also drink the decoction. The Jíbaro even give their dogs some of the tea before they go hunting so that they may better "see."

Ritual Use

Because there is almost no evidence apart from that which has been found in graves, we can only conjecture about the pre-Columbian use of guayusa leaves in South America. However, a find of leaves in the grave of a shaman of the Tiahuanaco culture in Bolivia demonstrates that the leaves were both known and desired; they were held in high esteem and seen as a valuable grave good (Schultes 1972; Wassén 1972).

In 1682, the Spanish Jesuit Juan Lorenzo Lucero sent a letter to the viceroy of Peru, Melchor Navarra y Rocafull, in which he described guayusa:

> They [the Jíbaro] mix together all of these diabolic herbs [*Banisteriopsis caapi, Brugmansia, Datura,* or similar plants] with *guañusa* and tobacco [*Nicotiana tabacum*], which was also invented by the devil, and allow this to boil until only a little liquid remains, the quintessence of evil, and the faith of those who drink this has been twisted by the devil through the fruit of enchantment and always to the disadvantage of all." (In Patiño 1968, 311)

Guayusa has a long tradition as a love magic. In Ecuador, it is said that when a person gives guayusa to a lover, he or she will always return, no matter how far away the journey has taken him or her. The Indians of the eastern borders of the Andes drink guayusa tea as a tonic, as a ritual emetic, and to have "hunting dreams" (Müller 1995, 196*). The shamans or medicine men of the Kamsá Indians (Colombia) also use guayusa leaves, although we do not know how (Schultes 1979, 144).

The best-known example of guayusa use is the daily ritual use among the Jíbaro (Shuar, Achuar). Every morning, over a period of about one hour, and often in the company of one another, the men of the tribe drink some 2.2 liters of a guayusa decoction that has been boiled for at least one hour. They then stick their finger down their throat or tickle their throat with a feather so that they will vomit (Patiño 1968, 312). This causes them to regurgitate about half of the liquid they have consumed. The drink wakes up the men and gives them strength; also, they say that they will not need to eat anything all day (an important effect for a hunter) because of it. The vomiting that follows the consumption of these large quantities of guayusa does not appear to be caused by any substances contained in the drink but is instead a learned and trained body practice (Lewis et al. 1991). The vomiting is said to expel the undigested food from the previous day that burdens the stomach and also helps prevent an overdose.

Large amounts of guayusa are also consumed before the most important tribal ceremonies, such as the women's tobacco ritual, the victory celebration (*tsantsa*), and the (no longer practiced) manufacture of shrunken heads (Schultes 1972, 130 f.).

The Jíbaro say that guayusa can have narcotic or hypnotic properties, as a result of which they may obtain "little dreams" in which they can see whether a hunting expedition will be successful. Seeing heavily boiling guayusa in a dream is regarded as a good omen (Karsten 1935; Patiño 1968, 312 f.; Schultes 1972, 131).

The Shuar use the leaves as an **ayahuasca** additive (Bennett 1992, 492*). They also drink guayusa tea before, during, or after ingesting **ayahuasca** "to wipe out the bitter taste" and "to prevent a hangover"; moreover, "it gives power and strength in the use of ayahuasca" (Schultes and Raffauf 1990, 80*; Shemluck 1979, 157).

In addition to the Jíbaro, the following tribes are known to use guayusa: Omagua, Kokama, Pánobo, Kaschibo, Koto, Pioché, Lamisto, Kichos, Canelo, Mocoa, Aguano, Kandoschi, Sabela, Chívaro, Mayoruna, Tschayahuita, Tschamikuro, Chebero, Omurana, Yagua, Auischiri, Ssimaku, Ikito, Záparo, Yameo, and Pintsche (Patiño 1968, 312).

Artifacts

The Indians of Ecuador and northeastern Peru produce special guayusa drinking bowls or vessels known as *guayuceros* (Schultes 1972, 126 f.)

Medicinal Use

In the Amazon, it is believed that guayusa tea is good for calming the nerves and also for pregnant women. The tea is drunk for stomach problems and as an aphrodisiac (Shemluck 1979, 157). The Mocoa Indians use guayusa to treat liver pains, malaria, syphilis, and stomachaches and to regulate menstruation (Schultes and Raffauf 1990, 80*). The ashes of guayusa leaves, mixed with **honey** and barley, are said to provide a remedy for amenorrhea (absence of menstruation). Guayusa

leaves, boiled together with the bark of *Paullinia yoco*, are drunk as a treatment for dysentery and stomach pains (Patiño 1968, 314). It is sometimes claimed that guayusa is useful for treating venereal diseases, chills, and infertility (Schultes 1972, 128).

The Jíbaro and other Indians drink guayusa as a "health tonic" (Schultes 1972, 120). It is said that a woman may become fertile if she drinks guayusa sweetened with **honey**. A woman who drinks a decoction of guayusa that has been sweetened with honey from a type of bee known as *apaté* will "immediately" become pregnant (Patiño 1968, 313). In Ecuador, the leaves are sold in herb markets as an "antispasmodic agent" (Schultes 1972, 135).

Constituents

The leaves typically contain 1.7 to 1.8% **caffeine**, and less frequently 3 to 4%. One wild plant actually contained 7.6% caffeine—a world record for a plant's caffeine concentration. Also present are low amounts of theobromine (0.003 to 0.12%) and traces of other dimethylxanthines. Emetine and other substances with emetic effects (such as those found in ipecacuanha and similar plants; cf. *Psychotria* spp.) have not yet been detected (Lewis et al. 1991, 25, 27, 28).

The caffeine contained in guayusa leaves maintains its quality for a long period of time. Samples dating to over one thousand years ago have been found to contain more than half of the concentration (1.0%) of more recent leaves (1.8%) (Holmsted and Lindgren 1972).

It is possible that triterpenes and chlorogenic acid are also present, as both are well represented in the genus *Ilex* (Schultes and Raffauf 1990, 79*).

Effects

Guayusa tea has potent stimulating effects that will wake up those who ingest it and keep them awake. The emetic (inducing vomiting) effects that have been so often described are actually culturally learned and do not have any pharmacological basis (Lewis 1991, 27). The Jíbaro do not like to use wild plants with high caffeine concentrations, as they are not interested in a caffeine overdose. Instead, their intention is to produce a very specific kind of stimulation that they do not want to exceed. The Achuar report that the typical symptoms of an overdose include severe headache, bloodshot eyes, and disturbing hallucinations (pseudohallucinations, delusions, illusions) (Lewis et al. 1991, 27).

Commercial Forms and Regulations

Living guayusa plants are sometimes available through ethnobotanical specialty sources. The dried leaves are normally available only in Ecuador. The plant is not subject to any legal restrictions.

"The *wayus* ritual was approaching its inevitable end. The properties of this morning infusion are not merely social in nature: it is primarily an emetic. Small amounts of *wayus* have no particular effects. But here it is poured down without interruption, just like the manioc beer, until the large, black pot has been drunk to the bottom. Then a persistent nausea will begin if one does not expel the great quantities of liquid from the stomach as quickly as possible. . . . [They are devoted] to the daily habit of vomiting. Without this energetic purging which gives back to the organism the innocence of an empty stomach, the men would be unable to begin the day. The Achuar see the cleansing spewing of the physiological remnants as a good way to cast off the past and to experience their return to the world each morning with an entirely new bodily feeling."

PHILIPPE DESCOLA
LEBEN UND STERBEN IN AMAZONIEN
[LIFE AND DEATH IN AMAZONIA]
(1996, 61*)

Literature

See also the entries for ***Ilex cassine, Ilex para-guariensis, Ilex vomitoria,*** and **caffeine.**

Holmstedt, Bo, and Jan-Erik Lindgren. 1972. Alkaloid analysis of botanical material more than a thousand years old. *Etnologiska Studier* 32:139–44.

Karsten, Rafael. 1935. *The head-hunters of western Amazonas: The life and culture of the Jibaro Indians of eastern Ecuador and Peru.* Commentationes Humanarum Litterarum (7, 174, 380).

Lewis, W. H., E. J. Kenelly, G. N. Bass, H. J. Wedner, M. P. Elvin-Lewis, and D. Fast W. 1991. Ritualistic use of the holly *Ilex guayusa* by Amazonian Jíbaro Indians. *Journal of Ethnopharmacology* 33:25–30.

Loesener, Theodor. 1901. *Monographia Aquifoliacearum.* Nova Acta 78. Halle, Germany: Abhandlungen der Kaiserlichen Leop.-Carol.-Deutschen Akademie der Naturforscher.

Patiño, Victor Manuel. 1968. Guayusa, a neglected stimulant from the eastern Andean foothills. *Economic Botany* 22:311–16.

Schultes, Richard Evans. 1972. *Ilex guayusa* from 500 A.D. to the present. *Etnologiska Studier* 32:115–38.

———. 1979. Discovery of an ancient guayusa plantation in Colombia. *Botanical Museum Leaflets* 27 (5–6): 143–53.

Shemluck, Melvin. 1979. The flowers of *Ilex guayusa*. *Botanical Museum Leaflets* 27 (5–6): 155–60.

Wassén, S. Henry. 1972. A medicine-man's implements and plants in a Tihuanacoid tomb in highland Bolivia. *Etnologiska Studier* 32:3–114.

Ilex paraguariensis Saint-Hilaire

Maté Bush

Family

Aquifoliaceae (Holly Family); Iliceae Tribe

Forms and Subspecies

This variable species is difficult to separate into varieties, forms, and subspecies (Hölzl and Ohem 1993, 508). Occasionally, wild maté is described as *Ilex paraguariensis* var. *genuina*. Three varieties are distinguished along the Rio Grande do Sul (Brazil): var. *talo roxo* ("red stem"), var. *talo branco* ("white stem"), and var. *piriquita*. Botanically, the species is divided into three varieties:

Ilex paraguariensis St.-Hil. var. *paraguariensis*
Ilex paraguariensis St.-Hil. var. *sincorensis* Loes.
Ilex paraguariensis St.-Hil. var. *vestita* (Reiss.) Loes.

Synonyms

Ilex bonplandiana Muenter
Ilex bonplandiana Münter
Ilex congonhas Liais
Ilex curtibensis Miers
Ilex curtibensis Miers var. *gardneriana* Miers
Ilex domestica Reiss.
Ilex gongonha Martius
Ilex mata St.-Hil.
Ilex mate St.-Hil.
Ilex paraguaiensis Lamb.
Ilex paraguaiensis Unger
Ilex paraguajensis Endlicher
Ilex paraguariensis D. Don
Ilex paraguayensis Hook.
Ilex paraguayensis Morong et Britt.
Ilex paraguayiensis Ed. Winkler
Ilex paraguayriensis Bonpl.
Ilex paraguensis D. Don
Ilex sorbilis Reiss.
Ilex theaezans Bonpl.
Ilex theezans Bonpl.
Ilex vestita Reiss.
Rhamnus quitensis Spreng.

Folk Names

Caá (Guaraní, "leaf"), caáchiri, caá-cuy, caá-cuyo, caaguagu, caá-guazú, caúna, caunina, congoin, congoinfe, congonha (Brazil), congonhas, congoni, erva mate, grünes gold, herba da Bartholomei, herva-mate, jesuitentee, jesuiten-teestrauch, Jesuit tea, kaá, kaá-maté, mate, maté, mate bush, maté-palme, matepflanze, mateteestrauch, mathee, matte, palo de yerba mate, Paraguay tea, paraguay-tee, südseetee, yerba, yerbabaum, yerba mate, yerba maté, yerva de palo

History

In South America, maté appears to have been used as an agent of pleasure and a ritual drug for millennia. Pre-Columbian graves in the Andes region of Peru have yielded maté leaves. In northern Argentina, Indian graves containing silver maté drinking utensils have been discovered.

One of the earliest European illustrations of the Paraguayan and Argentinean maté tree, together with the leaf that is the source of the stimulating tea. (From Pedro de Montenegro, *Materia médica misionera*, seventeenth century)

The Guaraní Indians used maté for shamanic purposes as well.

During the early colonial period, attempts were made to enslave the Indians. The Spanish kings, however, quickly forbade this practice, whereupon the Jesuits forced the Indians onto reservations and compelled them to establish maté plantations so that they could become a part of the cash economy. In commemoration of this "great achievement" of Christian charity, maté was first known by the name *Jesuit tea* (Schröder 1991, 102*).

In Argentina and Paraguay, non-Indians have also been cultivating maté trees since 1606 (Santos Biloni 1990, 196*). In Brazil, maté has been proclaimed the symbolic tree of Rio Grande do Sul.

The physician and botanist Aimé Bonpland (1773–1858), one of Alexander von Humboldt's traveling companions, described the plant that was the source of maté in 1821. The plant received its valid botanical name the following year.

Distribution

The true maté bush is found only between the twentieth and thirtieth parallels of South America. Its range extends over parts of Paraguay, northern Argentina, Uruguay, Brazil, and Bolivia (Hölzl and Ohem 1993, 508).

Cultivation

Maté was originally an underwood plant in the extensive araucaria forests. Today, maté plants often stand alone, because the large forests have been cut down. In Argentina, maté is grown on large plantations (*yerbatales*). The main source of maté, however, is Brazil. When grown on plantations, the bush is maintained at heights between 2 and 5 meters. The harvest takes place between May and September of every second year.

Propagation occurs via the seeds but is rather complicated, as it entails a six-month stratification:

The distribution is performed by birds (pheasant species). They peck up the seeds, and as these pass through the stomach and intestinal tract, the seed's hard outer shell is destroyed to such an extent that it can germinate after being eliminated. If the seeds are placed in the soil without any preparation, then the seed is unable to break through the shell and it will rot in the ground. The Jesuits solved the problem in their own way: they mixed the seeds into chicken feed! (Schröder 1991, 104*)

The maté tree prefers alluvial soils and does not tolerate clayey or calcareous soils. The young trees grow rapidly, and the first harvest occurs when they are three to six years old. During the harvest, up to 95% of the leaves (including branches) may be removed from the trees. Trees remain productive for fifty to sixty years.

Appearance

This evergreen tree typically has a light-colored bark. It can attain heights of 15 to 20 meters and has an oblong-oval crown. The alternate leaves, which can grow from 6 to 20 cm in length, have a serrate-crenate margin and a leathery upper surface. They are dark green on the upper side and light green below. The closely clustered axillary inflorescences have forty to fifty flowers with tetraphyllous or pentaphyllous calyxes. The reddish stone fruit is round and contains four to eight seeds. The trees blossom in the (South American) spring, i.e., from October to November. Although most plants are male, there are also female and even dioecious flowers (Schröder 1991, 103*).

Maté is sometimes confused with *Ilex aquifolium* L., as both species display considerable variation and can thus exhibit an astonishing degree of similarity. *Ilex aquifolium*, however, does not contain any **caffeine** and is therefore unsuitable for use as a maté substitute (Hölzl and Ohem 1993).

The very similar *palo de yerba* (*Ilex argentina* Lillo [syn. *Ilex tucumanensis* Speg.]) occurs in Argentina. The leaves of this plant were once used as a maté substitute (Santos Biloni 1990, 37*). In Chile, one plant (*Citronella mucronata* [R. et P.] D. Don) from the Family Icacinaceae is known as *yerba mate de Chile* (Mösbach 1992, 90*).

Psychoactive Material
— Leaves (mate folium, folia mate, mate)

The tea that is made from the leaves is known variously as maté tea, Jesuit tea, mission tea, Paraguay tea, parana tea, St. Bartholomew's tea, maté, thé du Paraguay, chimarrão, erva maté, and yerba maté.

In South America, the following species are used as substitutes for or additives to the true maté: *Ilex brevicuspis* Reiss., *Ilex conocarpa* Reiss., *Ilex dumosa* Reiss., *Ilex microdonta* Reiss., *Ilex pseudobuxus* Reiss., and *Ilex theezans* Mart. (Hölzl and Ohem 1993, 508). In Bolivia, *Coussarea hydrangeaefolia* Benth. et Hook. (Rubiaceae) is regarded as the true maté (Hartwich 1911, 452*).

The leaves of the maté bush (*Ilex paraguariensis*) are quite large.

Botanical illustration of *Ilex paraguariensis*. (From *Köhler's Medizinalpflanzen*, 1887/89)

"Maté is an agent of pleasure for invigorating one's powers and to increase the sense of well-being. It is the national drink of southern Brazil, northern Argentina, Paraguay, Uruguay, and Chile. The use of maté is part of the sociocultural environment; the guest is greeted with the drink; it is seen as the equivalent of the North American peace pipe. Maté is derived from the Guarani word *mati*, which originally referred to the drinking vessel; the term was applied to the drink, the drug, and the plant."

Josef Hölzl and Norbert Ohem
"Ilex"
(1993, 511)

"Yari's slender body transformed itself into a tree trunk and the dainty arms and delicate fingers became the small yerba leaves. 'The seeds of your blue fruits will open only if a bird has eaten and then eliminated them. You will give yourself to all and be desired by all, you shall calm, console, enliven.' . . . Since that time there has been yerba, whose tea calms, consoles, enlivens. It is also called maté. Every year, the indios travel to Mbaracuyurú to harvest its leaves. Yet every year the bravest of them are torn apart there by jaguars and strangled by Sucuriús. Meanwhile, a gentle boy died of confusion. Caa Yarí will also make jokes and caress, and she lives with the glowing love of a woman who must wait for a year. The young man who experiences such a thing will not want to eat again. Crazed with love, he will wander confused in the forest until he dies."

A GUARANÍ TALE
IN *DSCHUNGELMÄRCHEN* [JUNGLE
FAIRY TALES]
(MELZER 1987, 57F.)

176 The use of penis bones as aphrodisiacs is a worldwide phenomenon. The Maya of southern Mexico also use coati penis bones to manufacture love drinks. The Chol, who live in the region of Palenque, drink penis bone scrapings in alcohol three times daily (Helfrich 1972, 153*).

177 Other *Tabebuia* species are used in South America as **ayahuasca** additives. The Caribbean *Tabebuia bahamensis* (Northrop) Britt. is an important ingredient in love drinks (McClure and Eshbaugh 1983*).

Preparation and Dosage

The freshly harvested leaves and twigs are quickly heated at a high temperature so that they will retain their green color (if dried slowly, they turn black). They are then dried, slowly burned, or roasted over a wood fire or reheated in a metal cylinder (a process known as toasting). Today, a variety of industrial processing techniques are used. The dried, roasted leaves are powdered or finely chopped and sold commercially. In Europe, a distinction is made between green and brown maté (also known as toasted maté leaves or mate folium tostum).

Maté has an acrid, smoky, mildly astringent taste. The normal individual dosage is 2 g of dried leaves to one cup of water. Hot, but not vigorously boiling, water is poured over the leaves, and the mixture is allowed to steep for five to ten minutes. The effects of infusions that are allowed to steep for only a short period are more stimulating, but even cold-water infusions are pleasant tasting and stimulating (Schröder 1991, 103*).

In South America, maté almost always is drunk through a straw (*bombilla*) from gourd containers or bottles known as *cuias*. The maté powder (*chimarrón*) is placed in a bottle and hot water is then added. After the liquid has been sucked out, new water is poured over the used leaves. This process can be repeated several times and has led to a ritualized use of maté. Maté is usually consumed unsweetened. Sometimes, lime or lemon juice may be added.

In Paraguay, maté was typically sweetened before drinking. The Guaraní Indians sweetened the tea with the dried, crumbled leaves of the sweet-tasting *Stevia rebaudiana* (Bertoni) Hemsl. (Compositae), which contains not sugar but, rather, sweet-tasting **diterpenes** (König and Goez 1994, 791; Soejarto et al. 1983, 9). The Guaraní call the *Stevia* plant *kaá hêê*, "sweet herb." *Kaá* is also their name for the maté tree (Schröder 1991, 102*).

Ritual Use

Many Indian legends portray the tree as one of the most important plants made by the god of creation. During the colonial period, the Spanish were of the opinion that Santo Tomé (Saint Thomas) had brought the tree and the drink to the Indians. The Guaraní and Caingang venerated maté as a magical plant that enabled them to establish contact with the supernatural world. They believe(d) that a spirit named Ka'a Yary lives in the maté tree and protects good, industrious workers but punishes those who do not believe in plant souls (Cadogan 1950; Schaden 1948).

The stimulating effects of maté were discovered by Indian shamans in southern South America. They drank potent decoctions for stimulation and to produce the wakefulness they needed

for their nocturnal rituals. The Indians developed their ritual consumption of maté in the pre-Columbian period. They sat together in circles and passed around a vessel filled with maté leaves and hot water. The stimulation was enjoyed communally and was used for telling one another stories.

Today, just as England has its five o'clock tea (cf. *Camellia sinensis*) and Yemen its afternoon khat chewing (cf. *Catha edulis*), the communal drinking of maté is a part of the daily life of all social circles in contemporary southern Brazil, Paraguay, and northern Argentina (Chaco).

Artifacts

Almost all of the artifacts associated with maté are objects used in its consumption, especially the *cuia* (gourd bottle) and the *bombilla* (sucking straw). Nowadays, *cuias* are often made of pure silver, although they are still often made in the shape of a bottle gourd. The sucking tubes are usually simple in appearance, and the best are also made of silver.

Medicinal Use

Among the Indians of Argentina, it is customary to ingest almost all of their medicinal herbs in maté tea (Filipov 1994, 182*). In the countries in which it is produced, maté is generally thought to strengthen the stomach and is used to treat rheumatism and fever and as a plaster for sores (Hölzl and Ohem 1993, 511).

The men of the Maká Indians (Chaco, Paraguay) produce an aphrodisiac by pouring hot water over maté and the penis bone of a coati (*Nasua nasua* [Procynidae]) (Arenas 1987, 285 f.*).[176] When made as a cold-water extract, the drink is known as *tereré*. To treat stomach ailments, they make a tea of maté and the bark of *Tabebuia caraiba* (Mart.) Bur. (293).[177]

Homeopathic medicine uses an essence of the bark of the closely related species *Ilex verticillata* (L.) A. Gray under the name *Prinos verticillatus* (Schneider 1974, 2:193*). Maté hom. *HAB34* or Ilex paraguariensis hom. *HPUS88* is used in accordance with the medical description to treat digestive weakness and other problems (Hölzl and Ohem 1993, 511).

In Europe, maté is used primarily as an aid to dieting and fasting. Because of its stimulating effects and relatively high vitamin content, it is an ideal beverage for use during fasts.

Constituents

The leaves contain 0.4 to 1.6% **caffeine**, 0.3 to 0.45% theobromine, and traces of theophylline (cf. *Theobroma cacao*). In addition to these purines, the leaves also contain vitamin C, 0.01 to 0.78% **essential oil**, an enzymatic substance, caffeoylquinic acid (chlorogenic acids 3,5-, 4,5-, and 3,4-dicaffeoylquinic acid, neochlorogenic acid, cryptochlorogenic acid), flavonoids (isoquercetin, camphor oil glycosides, rutoside), saponines, menisdaurine, and several phenols (Hölzl and Ohem 1993).

Effects

Maté has a stimulating and invigorating effect that refreshes both the body and the mind. High dosages can produce euphoric feelings with clear wakefulness. The appetite is usually suppressed. Because these effects are induced by at least three different substances (caffeine, theobromine, and chlorogenic acid), the effects of maté are not identical to those of caffeine alone. Side effects or unwanted effects are unknown (Hölzl and Ohem 1993, 511).

According to reports from Guaraní shamans, the consumption of large quantities of maté enables them to enter clairvoyant trance states.

Commercial Forms and Regulations

Maté is available throughout the world without restriction. The form typically sold in Argentina (*la hoja elaborada con palo*) consists of crushed leaves together with pieces of stems. In Germany, maté is also sold in tea bags. This material usually consists of toasted maté to which aromatic substances have been added. Recently, tea bags containing both maté and guaraná (*Paullinia cupana*) have appeared on the market.

According to the *DAB86*, maté leaves should have a caffeine content of no less than 0.6% (Hölzl and Ohem 1993, 510). Commercially available maté tea is occasionally adulterated with material from related species (*Ilex brevicuspis* Reiss., *Ilex dumosa* Reiss. var. *guaranina* Loes.) (Santos Bilonio 1990, 196*).

Literature

See also the entries for **Ilex cassine, Ilex guayusa, Ilex vomitoria,** and **caffeine.**

Baltassat, F., N. Darbour, and. S. Ferry. 1984. Étude du contenu purique des drogues à caféine: I.—Le maté: *Ilex paraguariensis* Lamb. *Plantes Médicinales et Phytothérapie* 18:195–203.

Cadogan, León. 1950. El culto al árbol y a los animales sagrados en el folklore y las tradiciones guaraníes. *América Indígena* 10 (1): 327–33.

Graham, Harold N. 1984. Maté. In *The methylxanthine beverages and foods: Chemistry,*

A *cuia*, the traditional maté drinking vessel (made from a tree gourd), together with a *bombilla* tube.

consumption, and health effects, 179–83. New York: Alan R. Liss, Inc.

Hölzl, Josef, and Norbert Ohem. 1993. *Ilex.* In *Hagers Handbuch der pharmazeutischen Praxis,* 5th ed., 5:506–12. Berlin: Springer.

König, Gabriele, and Christiane Goez. 1994. *Stevia.* In *Hagers Handbuch der pharmazeutischen Praxis,* 5th ed., 6:788–92. Berlin: Springer.

Melzer, Dietmar H. 1987. *Dschungelmärchen.* Friedrichshafen, Germany: Idime.

Schaden, Egon. 1948. A Erva do Diablo. *América Indígena* 8:165–69.

Schmidt, M. 1988. Mate—eine vergessene Heilpflanze? *PTA heute* 2 (1): 10–11.

Scutellá, Francisco N. 1993. *El mate: Bebida national argentina.* Buenos Aires: Editorial Plus Ultra.

Soejarto, Djaja D., César M. Compadre, and A. Douglas Kinghorn. 1983. Ethnobotanical notes on stevia. *Botanical Museum Leaflets* 29 (1): 1–25.

Vázquez, Alvaro, and Patrick Moyna. 1986. Studies on mate drinking. *Journal of Ethnopharmacology* 18:267–72.

Chlorogenic acid

Caffeine

Ilex vomitoria [Solander in] Aiton

Yaupon

Family
Aquifoliaceae (Holly Family); Iliceae Tribe

Forms and Subspecies
Two subspecies and one form are now botanically accepted:

Ilex vomitoria Ait. ssp. *vomitoria*
Ilex vomitoria Ait. ssp. *chiapensis* (Sharp) E. Murr.
Ilex vomitoria Ait. f. *pendula* Foret et Solymosy

A variety with pileous leaves and longer hairs on the branches has been described for Mexico (Chiapas) (Sharp 1950): *Ilex vomitoria* [Soland. in] Aiton var. *chiapensis* A.J. Sharp. Another variety has become known under the name *Ilex vomitoria* [Soland. in] Aiton var. *yawkeyii* Tarbox (Schultes 1950, 99).

Synonyms
Ageria cassena (L.) Raf.
Ageria cassena (Michx.) Raf.
Casine yapon Bartram [nom. nud.]
Cassine amulosa Raf. [nom. sphalm.]
Cassine caroliniana Lam.
Cassine paragua (L.) Mill.
Cassine paragua Mill. [non *Cassine peragua* L.]
Cassine peragua L.
Cassine peragua Mill.
Cassine ramulosa Raf.
Cassine vomitoria Swanton [nom. nud.]
Cassine yaupon Gatschert [nom. nud.]
Emetila ramulosa Raf.
Hierophyllus cassine (L.) Raf.
Hierophyllus cassine (Walt.) Raf.
Ilex atramentaria Bart.
Ilex caroliniana (Lam.) Loes.
Ilex carolinianum (Lam.) Loes.
Ilex cassena Michx.
Ilex cassine L.
Ilex cassine β L.
Ilex cassine (L.) Walt.
Ilex cassine Walt.
Ilex floridana Lam.
Ilex floridiana Lam.
Ilex ligustrina Jacquin
Ilex opaca Soland. in Ait.
Ilex peragua (L.) Trel.
Ilex religiosa Bart.
Ilex vomitoria Soland. in Ait.
Ilex vomitoria var. *chiapensis* Sharp (= *I. vomitoria* ssp. *chiapensis*)
Oreophila myrtifolia Schelle
Prinos glaber L.

Ilex vomitoria, also known as yaupon, has relatively small, round leaves.

Folk Names
Black drink tree, cassena, cassena vera floridanorum, cassiana, cassina, cassine, holly, holly-ilex, Virginia yaupon, yap (Waccon, "wood"), yapon, yaupon, yaupon holly, yop

History
We do not know with certainty how long *Ilex vomitoria* has been used in southeastern North America. It was certainly already known in prehistoric times, for it was described in the first sources from the colonial period (cf. *Ilex cassine*).

Perhaps the earliest description of the yaupon plant in the botanical literature is contained in Bauhin and Cherler's great encyclopedia, *Historia plantarum universalis* [Universal History of the Plant World] (1651). In contrast, the "black drink" (and its stimulating and appetite-suppressing effects) had already been described in 1542 by Nuñez Cabeça de Vaca in his *Relación y comentarios*.

During the American Civil War, toasted yaupon leaves were used as a substitute for tea (*Camellia sinensis*) and coffee (*Coffea arabica*). Because of its low **caffeine** concentration, the plant is only rarely used today, as many other sources of caffeine are now widely available. Yaupon was once thought to be the source of maté (Schneider 1974, 2:192*; cf. *Ilex paraguariensis*).

The Cherokee are said to have used a "southern yaupon" form as a "hallucinogen" to "evoke ecstasies" (Moerman 1986, 232*). Unfortunately, we have no pertinent recipes or other information about special additives that may have been used to make the mildly caffeinated black drink into such a potent beverage.

Distribution
The tree is indigenous to the southeastern regions of North America and to areas in the proximity of the North American Caribbean coast. One variety or subspecies occurs in Mexico (Chiapas) (Sharp 1950).

Cultivation
The tree requires moist to moderately dry soil. The plant presumably is propagated from seeds. Because the Indians collected only wild material, they did not develop any methods of cultivation.

Appearance
This evergreen tree can grow to a height of 6 meters. It has multiple stems and many branches and produces white flowers and scarlet berries. The shiny, emarginated leaves are alternate and are similar in appearance to those of *Ilex para-*

guariensis, although they are usually smaller. The more rounded leaves on the upper branches are some 4 cm in length and are distinctly smaller than those of *Ilex cassine*. The fruits ripen in October.

Psychoactive Material
— Leaves
— Fruits

Preparation and Dosage
Hartwich mentioned three methods for making the "black drink" or "yaupon holly tea": "1. by decocting the fresh leaves, 2. the dried leaves, 3. a decoction which must ferment and which is then said to be inebriating" (Hartwich 1911, 468*). Unfortunately, we have no information about the potential fermenting agent for this drink. Saw palmetto fruits may have been used as an additive (cf. **palm wine, wine**).

The Mikasuki Indians, who now live in the Florida Everglades (to where they were deported), still use a ceremonial drink called black drink at certain tribal rituals. It is obtained or fermented from one or more Everglades plants. It is possible that it is made not from an *Ilex* species but from the fruits of the saw palmetto (*Serenoa repens* [Bartr.] Small [syn. *Serenoa serrulata* (Michx.) Nichols]; cf. **palm wine**). Neither the drink nor its associated ritual has been studied to date.

Ritual Use
The Indians of the southern coastal region of North America generally believed that *Ilex vomitoria* and the black drink prepared from it enabled them to attain a state of ceremonial purity, by means of which they were ideally prepared for all rituals and ceremonies (Moerman 1986, 232*). Among the Cherokee, only those warriors who had already proved their courage were allowed to partake of the black drink. It is said to have induced ecstatic states (Hamel and Chiltoskey 1975, 62).

In Oklahoma, the black drink was consumed by socially high-ranking or important personages to cleanse themselves for their public duties. Otherwise, the drink brewed from *Ilex vomitoria* was used in precisely the same manner as that prepared from *Ilex cassine*.

Artifacts
In Oklahoma, the large shells of the sea snail *Busycon contrarium* Conrad or *Busycon perversum* L.[178] were decorated and used for drinking yaupon ritually; they resemble Mesoamerican objects (cf. *Ilex cassine*).

Mark Catesby's comprehensive work *The Natural History of Carolina, Florida and the*

Bahama Islands (1754) contains an illustration (fig. 25) of a yaupon branch with fruits, around which a snake is coiled. It is possible that the artist's intent was to express the medicinal or sacred quality of the plant.

Medicinal Use
The primary ethnomedicinal use of the plant was as an emetic, an agent for inducing vomiting (Schultes 1950). The Cherokee used yaupon in the same manner as *Ilex cassine*, i.e., to treat dropsy and gravel (Hamel and Chiltoskey 1975, 62).

Constituents
The leaves contain relatively low amounts of **caffeine**, typically between 0.27 and 0.32%, but approximately 7% tanning agents (Hartwich 1911, 468*; Power and Chestnut 1919). More recent studies (Stone County, Mississippi) found only 0.09% caffeine, along with 0.04% theobromine and no theophylline. Substances with actual emetic effects have not yet been reported (cf. *Ilex guayusa*). The fruits contain cyanidin-3-xylosylglucoside.

Effects
The black drink appears to have produced only mild stimulating effects. In the literature, it is occasionally claimed that a more potent decoction is able to "induce hallucinations" (Turner and Szczawinski 1992, 156*). The emetic effects are also questionable. The "vomiting rituals" of the Indians that were previously described presumably entailed ritual methods (e.g., putting one's finger in one's throat; cf. *Ilex guayusa*) that did not depend upon any pharmacological effects.

Commercial Forms and Regulations
None

Literature
See also the entries for *Ilex cassine, Ilex guayusa, Ilex paraguariensis*, and **caffeine**.

Hamel, Paul B., and Mary U. Chiltoskey. 1975. *Cherokee plants and their uses: A 400 year history*. Sylva, N.C.: Herald Publishing Co.

Power, F. B., and V. K. Chestnut. 1919. *Ilex vomitoria* as a native source of caffeine. *Journal of the American Chemical Society* 41:1307–12.

Schultes, Richard Evans. 1950. The correct name of the yaupon. *Botanical Museum Leaflets* 14 (4): 97–105.

Sharp, A. J. 1950. A new variety of *Ilex vomitoria* from southern Mexico. *Botanical Museum Leaflets* 14 (4): 107–8.

"[Yaupon] is highly regarded by the Indians and used for many purposes, which gives it an even greater character. They say that the virtues of the shrub have been known since the earliest times and that for their use, it has always been prepared in the same manner as today. After drying or roasting the leaves in a pot over the fire, they are stored for use. From these they make their favorite beverage by preparing a potent decoction which they drink in great quantities, both for their health and also with great desire and good pleasure, without any kind of sugar or other additives. They drink great amounts of this, and time and again, and thus swallow many quarts. They have an annual custom in the spring in which the drinking takes place in a ceremony. The inhabitants of the village gather in the communal house after they have cleaned their own houses by burning all of their furniture and replacing this anew. The king is first served the drink in a great bowl or snail shell which has never been used before. After this, the other eminent personages receive the drink according to their rank. The women and the children are the last to receive the brew. They say that it makes lost appetites return, that it strengthens the stomach, and that it gives them power and courage for war."

MARK CATESBY
THE NATURAL HISTORY OF CAROLINA, FLORIDA AND THE BAHAMA ISLANDS
(1754, VOL. 2, CH. 57)

178 Today, in Florida, these snail shells are used primarily as ritual objects by members of the Afro-Cuban Santería cult. The shells of Caribbean snails are exported even as far away as the Himalayas, where they are used ritually as sacred snails.

Iochroma fuchsioides (Bentham) Miers

Yas

Right, from top to bottom:
The red-flowered yas bush, *Iochroma fuchsioides*.

The Ecuadoran species *Iochroma grandiflorum* bears large flowers.

Iochroma grandiflorum, photographed in Salala near Las Huaringas, in northern Peru.

179 The Kamsá use the same name to refer to *Cestrum* spp. (Schultes and Raffauf 1991, 37*).

180 The closely related species *Iochroma lanceolatum* Miers is also known by this name (Bristol 1965, 290*).

181 In California, the species *Iochroma coccineum* Scheid., *Iochroma lanceolatum* Miers, and *Iochroma tubulosum* Benth. also have been successfully cultivated as ornamentals (Grubber 1991, 40 f.*).

Family
Solanaceae (Nightshade Family); Subfamily Solanoideae, Solaneae Tribe

Forms and Subspecies
Twelve to fifteen species are currently accepted in the genus *Iochroma* (D'Arcy 1991, 79*; Schultes and Raffauf 1991, 37*). To date, no subspecies or varieties have been described for the species *Iochroma fuchsioides*.

Synonyms
Iochroma fuchsioides (Humb. et Bonpl.) Miers
Lycium fuchsioides H.B.K.

Folk Names
Árbol de campanilla, borrachera, borrachera andoke,[179] borrachero, campanitas (Spanish, "little bell"), dotajuanseshe (Kamsá), flor de quinde (Spanish, "flower of the hummingbird"),[180] guatillo, hacadero, hummingbird's flower, iochroma, isug yas gyeta, paguando, paguano, tatujansuche, tetajuanse, totubjansush, totubjansushe, totufjansush, totujanshve, yas

The Kamsá Indians (Colombia) also call the closely related species *Iochroma gesnerioides* (H.B.K.) Miers *borrachera*. It is not known whether this plant has psychoactive effects or was ever used for such purposes (Schultes and Raffauf 1991, 38*). *Iochroma umbrosa* Miers has been listed as an inebriating plant, but this information appears to be incorrect (Bristol 1965:290 f.*).

History
To date, we know of no pre-Columbian use of the yas bush. Richard Evans Schultes discovered the psychoactive use of this beautiful and rare plant in 1941 in the Sibundoy Valley of Colombia (Davis 1996, 173*). Detailed ethnobotanical and ethnopharmacological studies are lacking.

Distribution
The yas bush is found in the high Andes mountains of Colombia and Ecuador at altitudes around 2,200 meters.

Cultivation
Propagation occurs with seeds or cuttings. The cuttings are best taken in February or early March and placed in water. They require considerable time to develop roots (Grubber 1991, 41*).

In subtropical regions, yas is cultivated as an ornamental (Bristol 1965, 290*). The plant has been successfully grown in northern California,[181] e.g., in the botanical gardens of Berkeley and San Francisco (Strybing Arboretum). The plant does not tolerate any frost.

Appearance
This perennial bush can attain a height of 3 to 4 meters. It has woody stems and reddish brown branches and develops lanceolate, light green leaves that are up to 10 cm in length. The deep red, trumpet-shaped flowers are 2.5 to 4 cm long and hang in umbelliform clusters. The berries are fruits with a diameter of approximately 2 cm that remain partially enclosed by the wilted calyxes.

Yas is very similar to the Colombian species *Iochroma cyaneum* (Lindl.) M.L. Green [syn. *Iochroma lanceolatum* (Miers) Miers, *Iochroma tubulosum* Benth.], which is commonly cultivated as an ornamental in tropical and subtropical regions. The bush is also easily confused with the related Central American species *Iochroma*

coccineum Scheid. (Bärtels 1993, 155*). In addition, the plant is astonishingly similar to the Mexican *Fuchsia fulgens* Moç. et Sessé ex DC. (Onagraceae).

Psychoactive Material
— Leaves
— Flowers

Preparation and Dosage
The leaves can be harvested and dried at any time of the year except the winter months. The flowers are collected as soon as they exhibit the first signs of wilting (Grubber 1991, 41*). The dried leaves are smoked or brewed into a tea. No precise information concerning dosages is available (Gottlieb 1973, 21*). It is likely that the leaves and flowers can be used in **smoking blends**, in combination with other plants.

In Colombia, the (fresh) leaves are drunk in the form of an infusion or a decoction.

Ritual Use
The shamans of the Colombian Kamsá Indians ingest preparations of this nightshade when confronted with cases that are difficult to diagnose.

Iochroma fuchsioides is also used as an **ayahuasca** additive.

Artifacts
Despite what has been previously published, an old Indian drawing (*A woman under a* borrachero *tree*) by Francisco Tumiña Pillimue of Colombia portrays not *Brugmansia sanguinea* ssp. *vulcanicola* but, rather, *Iochroma fuchsioides* (Schultes and Raffauf 1977; Schultes and Raffauf 1991, 38*). Above the bush is a bird, most likely a transformed shaman or the vision-bringing spirit of the plant (Hernández de Alba 1949). The drawing, however, is much more reminiscent of the larger-flowered *Iochroma grandiflorum* Benth., which occurs in the Ecuadoran and Peruvian Andes.

Medicinal Use
Iochroma fuchsioides is used ethnomedicinally as a narcotic for difficult births and digestive disorders (Schultes and Hofmann 1995, 46*).

The folk healers (*curanderos*) of northern Peru use an *Iochroma* species known as *contrahechizo* ("anti-magic") as an additive to the San Pedro drink (cf. **Trichocereus pachanoi**) and as a purgative remedy for treating diseases caused by harmful magic. The plant is said to induce vomiting and diarrhea, thereby cleansing the body of all poisons and negative influences (cf. Giese 1989, 229, 250*).

In northern Peru, *Iochroma grandiflorum* is regarded as a typical medicinal plant from the region of Las Huaringas. The *curanderos* believe that the plant is especially powerful in this area because it absorbs the water of the sacred lake. It is known in the area as *campanitas* ("little bells")[182] or *yerba para mal hechizo* ("herb against bad magic"). It is used primarily as a bath additive for removing magic. It is sometimes combined with *Fuchsia* spp. for this purpose. It is possible that the *curanderos* may also use the plant for psychoactive purposes.

Constituents
To date, no alkaloids have been detected in *Iochroma fuchsioides*, although **withanolides** have been found (Raffauf et al. 1991).

The closely related *Iochroma coccineum* Scheid. has also been found to contain **withanolides** (Alfonso and Bernardinelli 1991; Alfonso et al. 1992).

Left: The lush inflorescence of the blue *Iochroma cyaneum*.

Center: An unidentified *Iochroma* species.

Right: *Fuchsia fulgens* is astonishingly similar to the red-flowered yas bush (*Iochroma fuchsioides*).

The "inebriating" nightshade *Iochroma fuchsioides* is shown on this old Indian drawing (*A woman under a* borrachero *tree*) from Colombia. The bird shown hovering above the tree is likely a transformed shaman or the vision-inducing spirit of the plant.

(FROM GREGORIO HERNÁNDEZ DE ALBA, *NUESTRA GENTE—NAMUY MISAG*, 1949)

182 This name is a folk taxonomic indication of the plant's (botanical) relationship to the angel's trumpets (*Brugmansia* spp.), which are known in Latin America as *campana*.

Effects
The inebriation produced by this plant is said to last for days or have aftereffects. No reports of self-experiments are available.

Commercial Forms and Regulations
None.

Literature
See also the entry for **withanolides**.

Alfonso, D., and G. Bernardinelli. 1991. New withanolides from *Iochroma coccineum*. *Planta Medica* 57 suppl. (2): A67.

Alfonso, D., G. Bernardinelli, and I. Kapetanidis. 1992. Four new withanolides from *Iochroma coccineum*. *Planta Medica* 58 suppl. (1): A712–13.

Hernández de Alba, Gregorio. 1949. *Nuestra gente—namuy misag*. Popayán, Colombia: Editorial Universidad del Cauca.

Raffauf, Robert F., Melvin J. Shemluck, and Philip W. Le Quesne. 1991. The withanolides of *Iochroma fuchsioides*. *Journal of Natural Products* 54 (6): 1601–6.

Schultes, Richard Evans. 1977. A new hallucinogen from Andean Colombia: *Iochroma fuchsioides*. *Journal of Psychedelic Drugs* 9 (1): 45–49.

Schultes, Richard Evans, and Alec Bright. 1977. A native drawing of an hallucinogenic plant from Colombia. *Botanical Museum Leaflets* 25 (6): 151–59.

Ipomoea violacea Linnaeus
Morning Glory Vine

The purple- or violet-flowered vine known in Aztec as *mecapatli*, "cord medicine," can be interpreted as *Ipomoea violacea* L. or *Ipomoea purpurea* (L.) Lam. (From Hernández, 1942/46 [Orig. pub. 1615]*)

183 Now regarded as a synonym (Wasson 1971, 340).

Family
Convolvulaceae (Morning Glory Family); Subfamily Convolvuloideae, Ipomoeae Tribe

Forms and Subspecies
At least two varieties have been described:

Ipomoea violacea var. *rubrocaerulea* Hook.
Ipomoea violacea var. *tricolor* Cav.[183]

The following sorts (cultivars), which are grown throughout the world as ornamentals, are known under their own names: 'Blaustern' 'Blue morning glory', 'Crimson rambler', 'Darling', 'Heavenly Blue' (= 'Kaiserwinde', 'Blaue Trichterwinde'), 'Klimmende blauwe winde', (Dutch) 'Morning glory', 'Pearly Gates', 'Summer Skies', 'Blue Star', 'Flying Saucers', and 'Wedding Bells'.

Synonyms
Ipomoea tricolor Cav.
Ipomoea rubrocaerulea Hook.
Ipomoea violacea Lunan et auct. mult., non L.
Pharbitis rubrocaerulea (Hook.) Planch.

Folk Names
Badoh negro (Zapotec, "black badoh"), badolngás ("black badoh"), badungás ("black badoh"), bejucillo (Spanish, "little tendril"), blaue trichterwinde, coatlxoxouhqui, dreifarbige prunkwinde, ipomée, kaiserwinde, la'aja shnash (Zapotec, "seeds of the Virgin"), mantos de cielo ("coat of heaven"), ma:sung pahk (Mixe, "bones of the children"), mehen tu'xikin (Lacandon, "little stink ear"), michdoh, morning glory, morning glory vine, pih pu'ucte:sh (Mixe, "flower of the broken plates"), pihyupu"ctesy (Mixe, "flower of the broken plate"), prachtwinde, prunkwinde, purpurwinde, quiebraplato (Mexico, "breaker of plates"), tlitliltzin (Aztec, "black divine"), trichterwinde, xha'il (Mayan, "that from the water"), ya'axhe'bil, yaxce'lil

The seeds, which are used primarily in ritual contexts, are known among the Chinantec and Mazatec as *piule*, the same name given to the seeds of **Turbina corymbosa**. The Zapotec call them *badoh negro*, the same name they give to the plant itself (cf. **Rhynchosia pyramidalis**). The Aztecs are said to have called them *tlililtzin* or *tlitliltzin*, "the very black" (Schultes and Hoffmann 1992, 46*). In both the pharmaceutical and the ethnobotanical literature, and in other sources, the seeds of *Ipomoea violacea* are confusingly referred to by the name *ololiuqui*.

History
It has not yet been possible to confirm with certainty whether *Ipomoea violacea* is identical to the Aztec entheogenic plant known as *tlitliltzin*, but it is very likely. The psychoactive use of this plant in divinations and healing rituals has been documented since the late colonial period.

This vine was first botanically described by Linnaeus. Since his time, it has been repeatedly described anew under other names. At the same time, this beautiful plant has been changed through cultivation and breeding to such an extent that many of the new sorts must sometimes appear to be new, previously undescribed species. Closet shamans, gardeners, and breeders often confuse this plant with ololiuqui (*Turbina corymbosa*).

Distribution

The plant is originally from the tropical regions of Mexico but is now found as a garden ornamental (with numerous sorts) in tropical and subtropical areas around the world. In temperate zones, the plant occurs only as an ornamental annual.

Cultivation

This fast-growing *Ipomoea* species does well in topsoil, although it prefers slightly alkaline soils. It can be grown in protected locations in the open, e.g., in a flower box on a balcony facing south. It requires copious amounts of water but can survive for several days without watering. Even plants that appear to have dried out can recover if they are watered well.

To grow the plant, place four or five seeds in a flowerpot during March or April. If the seeds are kept at a temperature between 18 and 20°C, germination will take place in ten to twenty days (or sooner). The plant can be placed in the open or transplanted after mid-May.

Appearance

This vine branches early and can grow as long as 3 meters. The leaves are cordate. The first two leaves that develop from the young shoots exhibit a characteristic fissuring (a sure way to identify the seedlings).

In the Mexican tropics, *Ipomoea* can flower throughout the year. It is very frequently seen in full bloom in February and March. In temperate zones, where it grows only as an annual, the flowering period is between June and October (depending upon the cultivar or sort). The flowers of the wild form are usually up to 8 cm wide and luminously violet; the flowers of the cultivated sorts can grow as large as 10 cm and may have a blue, pink, or white color. The flowers unfurl in the morning and close before the onset of twilight (often around four o'clock in the afternoon). The stigma is mono- to tricephalous. The seeds are black, elongate-triangular, some 7 to 8 mm long, and approximately 4 mm wide.

Ipomoea violacea is very frequently confused with *Ipomoea purpurea* (even in the specialized literature) as well as with other *Ipomoea* species.

Psychoactive Material

— Seeds (semen ipomoeae violaceae, ipomoeae violaceae semen)

Preparation and Dosage

Among the Mixe, a dosage consists of twenty-six seeds. The seeds should be ground by a ten- to fifteen-year-old virgin and mixed with water; otherwise, the seeds will not "speak" (Lipp 1991, 190*). The methods of preparing and administering the seeds are more or less the same for all the peoples of Oaxaca.

According to Schuldes, a low dosage consists of 20 to 50 seeds, a moderate dosage of 50 to 150, and a high dosage of 300 seeds or more; only with moderate to high dosages was he able to observe LSD-like effects (Schuldes 1993, 86 f.*). Other sources suggest chewing and swallowing 5 to 19 g of seeds or grinding them and allowing them to sit in water for thirty minutes (Gottlieb 1973, 37*).

A cold-water extract of three hundred crushed or ground seeds is said to correspond to approximately 200 to 300 μg of LSD (Rob Montgomery, pers. comm.; Veit 1993, 548). The LD_{50} of the isolated alkaloids is said to lie between 164 and 214 mg/kg in rats and between 1 and 2 g for humans.

Ritual Use

It is very likely that the Nahua-speaking peoples (e.g., the Aztecs) were ritually using the seeds of *Ipomoea violacea* during pre-Hispanic times. Evidence of the ritual use of *tlitliltzin* in the colonial period is contained in the report of Pedro Ponce (*Breve Relacíon de los Dioses y Ritos de la Gentilidad*, par. 46):

> On the ways in which one finds lost objects and other things that people want to know: They drink *ololiuhque* [sic; cf. *Turbina corymbosa*], peyote [cf. *Lophophora williamsii*], and a seed which they call *tlitlitzin*. These are so strong that they sedate the senses [of the natives] and that—so they say—little black men appear before them which tell them what they want to know about. Others say that Our Lord appears to them, while still others [say] that it is angels. And when they do this, they enter a room, close themselves in, and have someone watch so that they can hear what they say. And it is not allowed for people to speak to them before they have reawakened from their delirium, lest they go insane. And then they ask what they have said, and that is so. (Cf. Andrews and Hassig 1984, 218*)

The Zapotec use *Ipomoea* seeds in the same manner as *Turbina* seeds (MacDougall 1960). The black *Ipomoea* seeds are often referred to as *macho*, "male," while the light-colored *Turbina* seeds are *hembra*, "female" (Wasson 1971, 340). The Mixe regard *Turbina corymbosa* and *Ipomoea violacea* as siblings. Mixe shamans, however, consider *Ipomoea* to be more effective as well as more powerful

From top to bottom:
The wild form of the morning glory (*Ipomoea violacea*) is indigenous to tropical Mexico. (Photographed in Naha', Chiapas, Mexico)

The most commonly sold cultivar: *Ipomoea violacea* cv. Heavenly Blue.

The species *Ipomoea violacea* is rather easily identified by its first set of leaves. The Indians equate these with snake tongues.

Mexican shamans ingest the seeds of *Ipomoea violacea*, known as *badoh negro*, during healing rituals.

The Lacandon regard one
Aristolochia species as a "sibling" of
the morning glory (*Ipomoea
violacea*). Whether this plant is
psychoactive is a subject for future
research. (Photographed in the
Yucatán, Mexico)

184 It is doubtful whether Reichel-
    Dolmatoff's botanical identification is
    correct. If it is, then *Ipomoea violacea*
    must have already made its way to
    Colombia in pre-Columbian times,
    perhaps through trade. The plant in
    question may be *Ipomoea carnea* Jacq.,
    which is indigenous to South America
    and is actually more potent. In
    another article (1978), Reichel-
    Dolmatoff refers only to a
    Convolvulaceae.

185 In Yucatec Mayan, the closely related
    species *Ipomoea cornicalyx* Moor and
    Ipomoea seleri Millsp. are also known
    as *tu'xikin* (Barrera M. et al. 1976,
    249*; Martínez 1987, 1137*).

186 Very often, the noninitiated populace
    regards ritual plants and fungi with
    psychoactive effects (e.g., fly agaric
    mushrooms, thorn apple, henbane) as
    poisonous, or at least very dangerous.

(Lipp 1991, 187*). And indeed, the alkaloid content of *Ipomoea* seeds is considerably higher than that of *Turbina* seeds (Hofmann 1971, 354).

The *mamas* (shaman priests) of the Kogi of the Colombian Sierra Madre are said to use *Ipomoea violacea* in ritual contexts (Baumgartner 1994; Reichel-Dolmatoff 1987):

All this astronomical imagery is related to certain present-day rituals in which a high-ranking priest, wearing a jaguar mask impersonating Sun, cohabits in the main temple with his sister-wife who wears a mask made from the wide gaping maw of a black jaguar, impersonating Moon. The ritual takes place during the dark of the moon, and under the influence of a hallucinogen, probably Morning Glory (*Ipomea violacea*).[184] Black jaguars (*nébbi abáxsë*) are said to be fairly frequent in Colombia and can interbreed with normally colored jaguars; both jaguars, also the black one, have dark pelt markings which, in Kogi terms, are sun-spots and moon-spots and which, according to the priests, are the marks of incest. We would call them *maculae*. The gaping gullet of Moon's mask might be interpreted as another representation of the devouring womb of the Great Mother. . . . In any event, the ritual here described is one of recreation, as stated clearly by the Kogi; it is the conquest of darkness. . . .

The principal mythical jaguar priest was *Kasindúkua*, a lord or máma with very ambivalent attributes. He was an expert curer of human illness, and his principal power objects were a jaguar mask and a bluish pebble or seed called *nebbis kwái*/jaguar's testicle; the Great Mother herself had given him these objects, and from a number of texts it appears that they suggest the use of hallucinogenic drugs. But putting on the mask and, simultaneously, swallowing the "testicle," ordinary reality began to change; illnesses became visible to the eye in the form of black beetles and thus could be destroyed; women turned into luscious pineapples, and maize stalks were transformed into armed men. (Reichel-Dolmatoff 1987, 101, 105)

The plant in question may be *Ipomoea carnea* Jacq. (see *Ipomoea* spp.). The plant is not described in any more detail in another source:

The Kogi compare the brilliant reflection on the water with certain luminous sensations a person might perceive under the influence of a drug. A priest or any other devout believer who tries to establish contact with the supernatural sphere by fasting, meditation, and the use of hallucinogenic drugs such as morning glory (Convolvulaceae) will sometimes perceive quite suddenly a brilliant light which he believes to be a direct manifestation of a divine being. Since these trance states are often accompanied by horrifying visions, these sudden flashes of light are greatly feared by the common people who will associate them with a ghostly and dangerous dimension. (Reichel-Dolmatoff 1978, 24–25)

Artifacts

A watercolor of a morning glory appeared in *Types of Floral Mechanism*, by Arthur Harry Church (1865–1937). Flowering vines also appear as floral elements in American art deco windows (cf. *Turbina corymbosa*).

Medicinal Use

The Lacandon know *Ipomoea violacea* as *äh tu'xikin*, literally "the feces/rot of the ear."[185] In the Selva Lacandona, this violet-blooming vine thrives along paths, in clearings, and around ancient ruins and overgrown milpas (*pakche'kol*). The Lacandon view it as a relative of *äh mehen tu'xikin*, "little feces/rot of the ear" (*Aristolochia foetida* H.B.K.; Aristolochiaceae), and of *nukuch tu'xikin*, "big feces/rot of the ear" (*Aristolochia grandiflora* Swartz [syn. *Aristolochia gigas*]). This *Ipomoea* is a remedy for a disease that the plant itself causes: "If you play with the flowers of the purple vine, you get earaches, your ear rots. The purple vine has a disease. If you pick the flower or play with the purple vine, your ear rots. But I mean that it is a different ear [scrotum, genital labia]. It is the same whether women or men play with it, their ear will rot." According to information provided by several of the Lacandon, this "ear disease" can be healed by roasting an *Ipomoea* flower and placing it (for how long?) into the diseased "ear" (Rätsch 1994c, 77*). I believe that this Lacandon statement is a rudiment from the time in which the psychedelic use of the vine was restricted to religious specialists. The warnings not to play with the flowers, i.e., not to be careless with them or not respect them, is evidence of a protective attitude toward the plant.[186]

Constituents

The leaves contain the **ergot alkaloids** (ergoline derivatives) ergometrine, isolysergic acid amide, and lysergic acid amide (LSA). The biosynthesis of these **indole alkaloids** occurs in the leaves, while the subsequent translocalization leads to an accumulation of the alkaloids in the seeds. The seeds contain a variety of lysergic acid derivatives: (+)-lysergic acid amide, (+)-isolysergic acid amide, lysergic acid hydroxyethylamide, chanoclavine, elymoclavine, and ergometrine. The concentrations of these active compounds can vary considerably, depending upon the location of the plant and the sort (Gröger 1963a; Roth et al. 1994, 428*[187]). In young seeds, the chanoclavine content is very high, but it declines as they mature and the lysergic acid amide levels increase (Gröger 1963b).

The cultivars 'Heavenly Blue', 'Pearly Gates', 'Summer Skies', 'Blue Star', 'Flying Saucers', and 'Wedding Bells' all contain psychoactive alkaloids (Der Marderosian and Youngken 1966).

Effects

Eating the seeds can induce profound side effects (nausea, vomiting, indisposition, lassitude), which probably result from non-water-soluble alkaloids and other substances. The fewest side effects result from ingesting a cold-water extract of the ground or crushed seeds. Cold-water extracts have distinct hallucinogenic effects that are, however, not exactly the same as those of LSD. Visions of "small people" are typical (Turner and Szczawinski 1992, 178*). The effects have narcotic and hypnotic components that Indian shamans utilize for their soul journeys. Most Westerners who have experimented with morning glory seeds typically have little desire to repeat the initial experiment.

The seeds also have a stimulating effect upon the uterus, probably because of the presence of ergonovine (Der Marderosian et al. 1964).

Commercial Forms and Regulations

In nurseries and flower shops, the seeds are usually sold under the name *Ipomoea tricolor*. The plant is not subject to any legal restrictions.

Literature

See also the entries for ***Ipomoea* spp., *Turbina corymbosa*, indole alkaloids,** and **ergot alkaloids.**

Baumgartner, Daniela. 1994. Das Priesterwesen der Kogi. In *Yearbook for ethnomedicine and the study of consciousness*, 1994 (3):171–98. Berlin: VWB.

Der Marderosian, Ara H. 1965. Nomenclatural history of the morning glory, *Ipomoea violocea*. *Taxon* 14:234–40.

———. 1967. Psychotomimetic indole compounds from higher plants. *Lloydia* 30:23–38.

Der Marderosian, Ara H., Anthony M. Guarino, John J. DeFeo, and Heber W. Youngken, Jr. 1964. A uterine stimulant effect of extracts of morning glory seeds. *The Psychedelic Review* 1 (3): 317–23.

Der Marderosian, Ara, and Heber W. Youngken, Jr. 1966. The distribution of indole alkaloids among certain species and varieties of *Ipomoea*, *Rivea* and *Convolvulus* (Convolvulaceae). *Lloydia* 29 (1): 35–42.

Gröger, D. 1963a. Über das Vorkommen von Ergolinderivaten in *Ipomoea*-Arten. *Flora* 153:373–82.

———. 1963b. Über die Umwandlung von Elymoclavin in *Ipomoea*-Blättern. *Planta Medica* 4:444–49.

Hofmann, Albert. 1963. The active principles of the seeds of *Rivea corymbosa* and *Ipomoea violacea*. *Botanical Museum Leaflets, Harvard University* 20:194–212.

———. 1971. The active principles of the seeds of *Rivea corymbosa* (L.) Hall f. (ololiuhqui, badoh) and *Ipomoea tricolor* Cav. (badoh negro). In *Homenaje a Roberto J. Weitlaner*, 349–57. Mexico City: UNAM.

MacDougall, Thomas. 1960. *Ipomoea tricolor*: A hallucinogenic plant of the Zapotecs. *Boletín del Centro de Investigaciones Antropológicas de México* 6:61–65.

Reichel-Dolmatoff, Gerardo. 1978. The loom of life: A Kogi principle of integration. *Journal of Latin American Lore* 4 (1): 5–27.

———. 1987. The Great Mother and the Kogi universe: A concise overview. *Journal of Latin American Lore* 13:73–113.

Veit, Markus. 1993. *Ipomoea*. In *Hagers Handbuch der pharmazeutischen Praxis*, 5th ed., 5:534–50. Berlin: Springer.

Wasson, R. Gordon. 1966. Ololiuqui and the other hallucinogens of Mexico. In *Summa Antropologica en Homenaje a Roberto J. Weitlaner*, 329–48. Mexico City: UNAM.

Perhaps inspired by a good dosage of seeds from *Ipomoea violacea*, one American wave band took its name, Morning Glories, from that of the psychedelic vine (CD cover 1994, Cargo Records). The British band Oasis recently raised some eyebrows with its album *(What's the Story) Morning Glory?* (1995).

Ergometrine

Lysergic acid amide

Chanoclavine

187 Unfortunately, these authors erroneously illustrated their monograph with a photograph of *Convolvulus tricolor*.

Ipomoea spp.

Ipomoea *Species*

Family

Convolvulaceae (Morning Glory Family); Subfamily Convolvuloideae, Ipomoeae Tribe

A variety of both wild and cultivated vines in the Family Convolvulaceae can be found in all the vegetation zones of Mexico. One very common plant in the Indian areas is the sweet potato (*Ipomoea batatas* L.), which is grown for food. Today, many species of these vines are grown throughout the world as ornamentals for their colorful, funnel-shaped flowers. The number of species has been estimated to exceed five hundred (Schultes and Farnsworth 1982, 151*). Some species are esteemed for their medicinal properties, e.g., jalap bindweed (*Exogonium purga* (Wender.) Benth. [syn. *Ipomoea purga* (Wender.) Hayne, *Convolvulus jalapa* Schiede non L.]; cf. Veit 1993, 543 ff.). But of all the species of Mexican vines, only two have been ritually ingested in shamanic contexts: *Turbina corymbosa* and *Ipomoea violacea,* in all known variations.

Ipomoea batatas (L.) Poir.—sweet potato, batate

This commonly grown *Ipomoea* species has small pink flowers and develops thick tubers containing starches and sugars that many Indians use for food. Although it has sometimes been suggested that the seeds may contain alkaloids (cf. Schultes and Hofmann 1980, 367*), there is no evidence to support this assertion. In the culture of inebriants, the sweet potato has played a role solely as a fermenting agent for making **beer** and as an additive to coca quids (cf. *Erythroxylum coca*). It likely originated in Mexico but had already spread far into South America and even to Polynesia in pre-Columbian times. *Ipomoea tiliacea* (Willd.) Choisy [syn. *Ipomoea fastigata* (Roxb.) Sweet] is presumably the wild form from which the sweet potato was cultivated (Dressler 1953, 135*). Tubers infected with *Ceratocystis fimbriata* contain sesquiterpenes (Inoue et al. 1977).

Ipomoea sp. aff. *calobra*—weir vine

This vine, found in only a limited area of southern Queensland (Australia), apparently contains LSD-like indole alkaloids. In addition to their psychoactive effects, these alkaloids also have potent toxic properties (at least on sheep and cattle) (Dowling and McKenzie 1993, 117 f.*).

Ipomoea carnea Jacq. [syn. *Ipomoea fistulosa* Mart. ex Choisy, *Ipomoea carnea* ssp. *fistulosa* (Mart. ex Choisy) D. Austin; cf. Hubinger T. et al. 1979, 33*]—manjorana, canudo, toé

This vine species produces flesh-colored flowers. It is found throughout the entire Amazon basin as well as in adjacent regions (Almeida Falcão 1971). In Ecuador, it is known as *florón* and *borrachera,* "inebriator" (Patzelt 1996, 178*), the same names used for many **Brugmansia** species, and as *matacabra* ("goat killer"). The concentrations of ergot alkaloids appear to be very high. To date, the ergoline alkaloids agroclavine and α-dihydrolysergol have been detected in the plant (Asolkar et al. 1992, 371*; Schultes and Farnsworth 1982, 151*). This makes the seeds of this vine, the effects of which are often stronger than those of *Ipomoea violacea,* a potent precursor substance. In Ecuador, the seeds (which contain **ergot alkaloids**) are said to be still used as a shamanic inebriant (Lascano et al. 1967; Ott 1993, 127*). In the Ucayali region of Peru, the vine is known as *toé* ("inebriant"). In the Pucallpa area, it is used as an additional inebriating ingredient in **ayahuasca** (Mellington Curichimba Marín, pers. comm.). In Amazonia, where the plant is feared as a poison, it is known by the names *manjorana, canudo,* and *algodão bravo*.

The plant is known as *chok'obkat* in Yucatec Mayan. There, the flowers are considered a source of honey (Téllez V. et al. 1989, 67*; cf. **honey**).

In the older literature, this South American plant can be found under the synonym *Ipomoea fistulosa,* which is now regarded as the subspecies *Ipomoea carnea* ssp. *fistulosa* (Mart. ex Choisy)

Above: *Ipomoea carnea* (= *Ipomoea fistulosa*) is regarded as one of the most potent morning glories.

Center: The botanical identity of many morning glories (*Ipomoea* spp.) is difficult to ascertain. (Photographed in the Kathmandu Valley, Nepal)

Right: The flower of the sweet potato (*Ipomoea batatas*).

D. Austin (Austin 1977; Hubinger T. et al 1979, 33*). In Argentina, the Pilagá Indians use the ashes of this subspecies (known locally as *we'daGaik'gel'ta*) to treat burns and boils (Filipov 1994, 186*).

Ipomoea crassicaulis Robinson [syn. *Ipomoea fistulosa* Mart.]

This plant, which develops large white flowers, is known in Mexico as *palo santo de castilla*, "sacred tree of castilla" (Martínez 1987, 1137*). This name may indicate a former use for psychoactive purposes. The seeds contain **ergot alkaloids** (Ott 1993, 127*).

Ipomoea hederacea Jacquin [syn. *Convolvulus hederaceus* var. *beta* L., *C. trilobus* Mach., *Ipomoea barbigera* Sweet, *I. coerulea* Roxb., *I. desertorum* House, *I. punctata* Pers., *I. scabra* Gmel., *I. triloba* Thunb., *Pharbitis hederacea* (L.) Choisy]—Japanese morning glory

This annual vine, which can grow as long as 2 to 3 meters, is known as *asagau*, "morning flower,"[188] in Japanese. It is regarded as an aphrodisiac in Asia and for this reason is sometimes seen in Japanese erotic art (Marhenke and May 1995, 49*). The beautiful, blue-flowered plant can be found in the Himalayas at altitudes of up to 2,000 meters and has also gone wild in the American tropics. It even occurs (introduced) in the Selva Lacandona, where it is easily mistaken for *Ipomoea violacea*. In Mexico, it is known by the name *manto de la virgen*, "coat of the Virgin" (Martínez 1987, 1137*). The seeds are regarded as poisonous in Iran (Hooper 1937, 130*).

Ipomoea hederacea possesses pubescent seminal leaves and forms capsules with four to six seeds. The seeds have a certain importance as a pharmaceutical raw drug (pharbitidis semen, kaladana) for kalana resin. **Ergot alkaloids** have been detected in the seeds (Abou-Char 1970; Veit 1993, 535). Seeds from Pakistan have been found to contain the alkaloids lysergol, chanoclavine, penniclavine, isopenniclavine, and elymoclavine (Asolka et al. 1992, 372*).

Ipomoea involucrata P. Beauv.

Among the Central African Fang, this African vine is known as *nguenga*. The medicine men of the Fang (cf. *Tabernanthe iboga*) use the fresh, whole plant to make a magical medicine with stimulating effects that is used to treated victims of magic (Akendengué 1992*). Whether this medicine does in fact have psychoactive properties is a subject for future research.

Ipomoea muricata (L.) Jacq. [syn. *Calonyction muricatum* (L.) G. Don]—lakshmana

In India, this white-blossomed vine with heart-shaped leaves is known as *lakshmana*[189] and is associated with Lakshmi, the goddess of good fortune. It is said that the root provides nourishment for the immaterial kundalini serpent (cf. *Cannabis indica*), which rests in the pelvic region of the lower abdomen and represents the creative sexual energy. In Ayurvedic medicine, *lakshmana* is one of the most important *vajikarana* (aphrodisiacs). According to one tantric recipe, the plant is mixed with bezoars and ground to produce an ointment that is applied to the forehead. This ointment is said to make love magic and mystical experiences possible. The root is attributed with healing powers for snakebites. Snake charmers use the root as a magical protection against cobras (Kumaraswamy 1985; Rätsch 1990, 51*). It has even been suggested that this climber is identical to the Vedic **soma**.

All parts of the plant contain up to 3.7% behenic acid, which has stimulating effects upon the central nervous system and also appears to have psychoactive and aphrodisiac effects. The seeds have been found to contain **ergot alkaloids** (Veit 1993, 535) as well as the alkaloid ipomine

The Aztecs knew one morning glory species (possibly *Ipomoea heterophylla* Ort.) as *totoycxitl* ("bird claw"). (From Hernández, 1942/46 [Orig. pub. 1615]*)

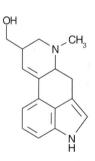

Lysergol

Top: The deep violet flowers of *Ipomoea hederacea*. (Photographed in Kathmandu, Nepal)

Bottom: *Ipomoea nil* is common throughout Asia and Oceania.

188 In Japan, this vine is regarded as a "sibling" of the thorn apple (see *Datura metel*).

189 The pink-blossomed morning glory *Ipomoea sepiaria* Roxb. [syn. *Ipomoea maxima* auct. non. (L. f.) Sweet] is also known in Sanskrit as *lakshmana* and bears the same name in many other Indian languages. It is also regarded as a tonic, aphrodisiac, and rejuvenant (Warrier et al. 1995, 3:237).

Left: The beach morning glory *Ipomoea pes-caprae.* (Photographed in the Seychelles)

Center: The purple morning glory, *Ipomoea purpurea*, is commonly found in cultivation.

Right: A color variant of *Ipomoea purpurea*.

(Asolkar et al. 1992, 372*). In the pharmaceutical trade, the seeds are available under the name *kaladana*.

Ipomoea murucoides Roem. et Schult.

This white-blooming *Ipomoea* species grows as a shrub or tree. In Mexico, where it is known as *palo bobo*, "inebriating tree," or *arbol del muerto*, "tree of the dead," it is considered to be a poisonous plant and is said to cause paralysis (Jiu 1966, 252*). The Aztecs knew the plant as *micacuahuitl*. In Sonora, it is still known today as *palo santo*, "sacred tree" (Martínez 1987, 1137*). Whether the plant was or is used for psychoactive purposes is unknown.

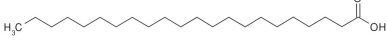

Behenic acid

Damascenone

In the Yucatán, the seeds of one *Ipomoea* species known locally as *xtontikin* are used for ethnogynecological purposes as a postpartum treatment. (Photographed in the Yucatán)

Ipomoea nil (L.) Roth [syn. *Convolvulus hederaceus* L., *C. hederaceus* var. *zeta* L., *C. nil* L., *C. tomentosus* Velloso, *Ipomoea cuspidata* Ruíz et Pav., *I. githaginea* A. Richard, *I. scabra* Forssk., *Pharbitis nil* (L.) Choisy, *Ipomoea hederacea* auct. non. Jacq.]

This blue-blossomed vine (there are also violet-blue and purple-red forms) is almost indistinguishable from *Ipomoea hederacea* and is often confused with that species (both have pubescent seminal leaves). It is found in tropical regions throughout the world. In Japan, it has been cultivated as an ornamental (*Ipomoea nil* cv. Imperialis) since the fifteenth century. In Mayan, it is known as *tsotsk'abil*, "hairy arm," and is probably one of the sources of nectar for producing a **honey** that is used ritually. The seeds, which are listed in the pharmaceutical literature under the name *pharbitidis semen*, contain glycosides and approximately 0.5% **ergot alkaloids** (consisting of lysergol, chanoclavine, penniclavine, isopennyclavine, and elymoclavine, but no ergometrine) (Veit 1993, 535 f.). The alkaloids lysergol, chanoclavine, penniclavine, isopenniclavine, and elymoclavine have been extracted from seeds of a Pakistani sort (Asolka 1992, 372*). Gebberellin is also present in the seeds (Koshioka et al. 1985). The seeds are the source of the pharmaceutically important kaladana resin.

Ipomoea pes-caprae (L.) Brown [*Ipomoea pes-caprae* (L.) Brown var. *brasiliensis* Ooststroom; syn. *Ipomoea biloba*]—beach vine

This prostrate species grows on sandy beaches. In Mexico, it is known by many names, inluding *hierba de la raya*, "plant of the stingray" (Martínez 1987, 1138*). This is an interesting association, as stingray spines were used in the ancient Mayan culture for ritual bloodletting, a part of the vision quest (Furst 1976c*). The leaves of this vine find ethnomedicinal use throughout the world to treat ailments such as rheumatism. In Thailand, extracts of the leaves are used to treat inflammations and jellyfish stings. The leaves contain β-damascenone and *E*-phytol, both of which have antispasmodic effects very similar to those of **papaverine** (Pongprayoon et al. 1992). The seeds contain

ergot alkaloids. Whether the seeds are suitable for psychoactive use is unknown, although they are used as an ingredient in the initiatory drink of the Afro-American Candomblé cult (see **madzoka medicine**).

Ipomoea purpurea (L.) Roth [syn. *Pharbitis purpurea* (L.) Voigt; *Ipomoea purpurea* var. *diversifolia* (Lindl.) O'Donell [syn. *Ipomoea mexicana* Gray]—early-blooming morning glory (numerous cultivars)

This climber, which can grow as long as 3 meters, is native to the American tropics but is now found as an ornamental around the world (Tykac and Severa 1985, 128). The plant as well as its seeds are very often confused with *Ipomoea violacea* and its seeds (Ott 1993, 162*). While it is often claimed in the literature that *I. purpurea* seeds contain ergoline and other ergot alkaloids, most studies (except one: Wilkonson et al. 1986) have reported an absence of alkaloids. The flowers may be purple-red, white, pink, light blue, or deep violet. The seeds are available in nurseries (usually under the name *purple morning glory*). The seeds are considerably smaller (3 mm by 4 mm) and rounder than those of *Ipomoea violacea.*

The following morning glory species contain **ergot alkaloids** and/or **indole alkaloids** and may possibly be used for psychoactive purposes (Ott 1993, 127*; Schultes and Farnsworth 1982, 187*):

Ipomoea argyrophylla Vatke
Ipomoea coccinea L.[190] [*Ipomoea coccinea* L. var. *hederifolia* House; syn. *Quamoclit coccinea* Moench]
Ipomoea leptophylla Torr.
Ipomoea littoralis Blume
Ipomoea medium Choisy
Ipomoea muelleri Benth.

Further studies of the genus *Ipomoea* certainly represent a highly interesting field of ethnopharmacology and will require extensive pharmacological studies with human subjects (Heffter technique!).

Two vines that are related to *Ipomoea* are sometimes claimed to be psychoactive or hallucinogenic: *Merremia tuberosa* (cf. **Argyreia nervosa**) and *Stictocardia titiaefolia* (Choisy) Hall f. (Schultes and Farnsworth 1982, 187*; Schultes and Hofmann 1980, 367*).

Literature

See also the entries for *Ipomoea violacea*, *Turbina corymbosa*, **ergot alkaloids**, and **indole alkaloids.**

Abou-Char, C. I. 1970. Alkaloids of an *Ipomoea* seed known as kaladana in Pakistan. *Nature* 225:663.

Almeida Falcão, Joaquim Inácio de. 1971. Convolvulaceae do Amazonas. *Acta Amazônica* 1 (1): 15–20.

Austin, Daniel F. 1977. *Ipomoea carnea* Jacq. vs. *Ipomoea fistulosa* Mart. ex. Choisy. *Taxon* 26 (2/3): 235–38.

———. 1991. *Ipomoea littoralis* (Convolvulaceae): Taxonomy, distribution, and ethnobotany. *Economic Botany* 45 (2): 251–56.

Der Marderosian, Ara, and Heber W. Youngken, Jr. 1966. The distribution of indole alkaloids among certain species and varieties of *Ipomoea*, *Rivea* and *Convolvulus* (Convolvulaceae). *Lloydia* 29 (1): 35–42.

Gröger, D. 1963. Über das Vorkommen von Ergolinderivaten in *Ipomoea*-Arten. *Flora* 153:373–82.

Inoue, Hiromasa, Natsuki Kato, and Ikuzo Uritani. 1977. 4-hydroxydehydromyopororone from infected *Ipomoea batatas* root tissue. *Phytochemistry* 16:1063–65.

Koshika, Masaji, Richard P. Pharis, Rod W. King, Noboru Murofushi, and Richard C. Durley. 1985. Metabolism of [^{3}H]gebberellin A_5 in developing *Pharbitis nil* seeds. *Phytochemistry* 24 (4): 663–71.

Kumaraswamy, R. 1985. Ethnopharmacognostical studies of the Vedic jangida and the Siddha kattuchooti as the Indian mandrake of the ancient past. In "Ethnobotanik," special issue, *Curare* 3/85:109–20.

Lascano, C. et al. 1967. Estudio fitoquímico de la especie psicotomimética *Ipomoea carnea*. *Ciencias Naturales* 10:3–15.

Pongprayoon, U., P. Baeckström, U. Jacobsson, M. Lindström, and L. Bohlin. 1992. Antispasmodic activity of β-damascenone and *E*-phytol isolated from *Ipomoea pes-caprae*. *Planta Medica* 58:19–21.

Tykac, Jan, and Frantisek Severa. 1985. *Kletterpflanzen und rankende Pflanzen*. Hanau, Germany: Dausien.

Veit, Markus. 1993. *Ipomoea*. In *Hagers Handbuch der pharmazeutischen Praxis*, 5th ed., 5:534–50. Berlin: Springer.

Wilkinson, R. E. et al. 1986. Ergot alkaloid contents of *Ipomoea lacunosa*, *I. hederacea*, *I. trichocarpa*, and *I. purpurea* seeds. *Canadian Journal of Plant Science* 66:339–43.

Botanical illustration of a morning glory from the tropical genus *Ipomoea*. (Engraving from Pereira, *De Beginselen der Materia Medica en der Therapie*, 1849)

An early illustration of a morning glory (*Ipomoea* spp.) with cordate leaves. (Woodcut from Gerard, *The Herball or General History of Plants*, 1633)

190 The seeds contain alkaloids, including elymoclavine (Asolka et al. 1992, 371; Gröger 1963). This species is found throughout the Amazon basin (Almeida Falcão 1971).

Juniperus recurva Buchanan-Hamilton ex D. Don

Drooping Juniper

Family
Cupressaceae (Juniper Family)

Forms and Subspecies
There is a dwarf form that does not have its own botanical name. *Juniperus recurva* var. *squamata* (Don) is the variety that grows in Kashmir (Weyerstahl et al. 1988).

Synonyms
Juniperus macropoda Auct. non Boiss.[191]
Juniperus squamata D. Don

Folk Names
Apurs (Pakistan), bsang (Tibetan, "incense"), dhupi (Nepali, "incense tree"),[192] drooping juniper, hapusha (Sanskrit), hochgebirgswacholder, shang-shing (Tamang, "incense tree"), weeping blue juniper

History
Species of juniper are found throughout the world and are especially common in Europe, Asia, and North America. In almost every region, they are used for ritual, magical, and medicinal purposes. In most of the cultures in which shamanism is found, juniper is an **incense** of the shamans. It is likely one of the oldest fumigants of humankind. The earliest written documents of ancient times (e.g., from Dioscorides and Pliny) provide evidence of numerous juniper species. Alexander the Great was presumably acquainted with the Himalayan juniper. We do not know how long this juniper has been used in the Himalayas for ritual purposes. The first botanical description of the plant was made in the nineteenth century.

Distribution
Drooping juniper occurs from Pakistan to southwestern China and is particularly common in Nepal (Langtang and Helumbu), where there are entire forests of the tree (Shrestha 1989). It grows at altitudes of at least 3,000 meters and up to 4,500 meters. In the subalpine zones, it sometimes forms large forests ("incense forests").

In northern India, it is said that the juniper forests of the Himalayas are the abode of the gods. Certain sacred sites (e.g., in Muktinath) have solitary juniper trees, some of which are very old, that are venerated as sacred trees.

Cultivation
Unknown; this species presumably is propagated and cultivated in the same manner as common juniper (*Juniperus communis* L.) and Chinese juniper (*Juniperus chinensis* L.)

Appearance
Drooping juniper can grow as tall as 12 meters but often attains a height of only 3 to 5 meters. It usually grows in a stocky or prostrate fashion. The species name is derived from the fact that the tips of the branches curve downward. The relatively soft needles are 6 to 8 mm long. The large, oval fruits (8 to 13 mm) contain just one seed and are violet-brown to black in color (Polunin and Stainton 1985, 390).

Drooping juniper is very easily confused with the closely related species *Juniperus excelsa* M. Bieb. (Goodman and Gharfoor 1992, 52*).

Top: Drooping juniper (*Juniperus recurva*) can be recognized by its downward-curving branch tips. (Photographed in Langtang, Nepal)

Bottom: An incense powder (*dhup*) for use in rituals is produced simply by rubbing the branches of *Juniperus recurva*.

191 *Juniperus macropoda* Boiss. may possibly be synonymous with *Juniperus excelsa* M. Bieb. (Goodman and Gharfoor 1992, 14*).

192 In Nepali, the same name is also given to other *Juniperus* species (e.g., *Juniperus communis* L. var. *saxatilis* Pallas [Indian juniper], *Juniperus excelsa* M. Bieb. [Himalayan pencil cedar, *shukpa*], and *Juniperus wallichiana* Hook [*techokpo*]), and to both cultivated and introduced forms and cultivars (e.g., *Juniperus chinensis* L., *J. polycarpa* Koch); it is also used for the conifer *Cryptomeria japonica* (L. f.) D. Don (= *dhupi salla*; cf. Storrs 1988, 88 f.) and the Himalayan cypress *Cupressus torulosa* D. Don, which is also used as a temple incense (Polunin and Stainton 1985, 389).

Psychoactive Material
— Fresh or dried branch tips
— Heartwood
— Resin

Preparation and Dosage
Drooping juniper is one of the most important materials for the manufacture of Tibetan incense and is made into incense powders and sticks. The branch tips are used for the former, while both the tips and the (heart)wood may be used for the latter. Most Tibetan incense blends based upon drooping juniper also contain other *Juniperus* species, branch tips of the Himalayan cypress (*Cupressus torulosa* D. Don), and various *Artemisia* species (cf. **Artemisia spp.** and **incense**).

Nepali shamans use freshly ground twigs, white sandalwood (*Santalum album* L.), and *nepali kagas* or daphne paper (*Daphne papyracea* Wall. ex Steud. [syn. *D. cannabina* Lour.]) to make incense cones approximately 10 cm long that they use in rituals. Normally, however, the fresh, half-dried, or dried branch tips are simply strewn onto glowing coals and the resulting smoke is deeply inhaled. Concrete information about dosages is almost impossible to ascertain, as the amounts used can range from small branch tips to entire bundles or even large branches.

Ritual Use
Drooping juniper is sacred to almost every Himalayan people. Branches of the plant are often attached to the ends of the masts holding prayer flags (as a symbol of the World Tree or shamanic tree). Wanderers in the mountains also use the branches as amulets to protect them from falling or being hit by rocks; the branches are offered to the mountain spirits at the piles of stones typically found at passes. But the primary ritual use of the plant is as an incense.

The Tibetans characterize the incense they make from drooping juniper as a "food of the gods." The incense is associated with a special ritual of purification they call *bsang* (pronounced "shang"), which is the same name they give to the tree. In this ritual, branch tips are burned while special mantras (conjuration formulas) are recited. The shamans of western Tibet also use juniper as a trance-inducing incense (Schenk 1994*).

The Buddhist peoples of the Himalayas (Tibetans, Bhotyas, Tamang, Sherpas) use fresh or dried juniper branches as an incense during their morning prayers (*puja*) to the Buddha Shakyamuni. The Sherpas fumigate with juniper when conjuring spirits and cremating bodies.

Nepali shamans (*jhākri*) use a variety of incenses, the most important of which is juniper. They use either the branches or the resin. Drooping juniper has definite psychoactive effects upon these shamans:

It was striking that the shaman (*jhākri*) began the exorcising activities by bending over an incense bowl filled with glowing juniper needles of *recurva* (other species are not used) and clumps of resin and deeply inhaling the smoke before drumming himself into trance with the aid of a large lama drum. (Knecht 1971, 218)

In Pakistan, juniper is venerated as a sacred tree that the shamans of the Dards use as an incense. The shamans and trance dancers (*bitaiyo*) of the neighboring Hunza also inhale juniper smoke:

The state of true trance is attained by inhaling the smoke of a juniper fire, by biting on juniper branches, and by drinking the blood from the severed head of a male kid goat. . . . The trance dancer (bitan) runs around jumping wildly. . . . He repeatedly interrupts his fast run, . . . to listen to the musical instruments and, finally, breaking down over a drum, he sings his prophecy in an ancient language that he is incapable of speaking or understanding when he is awake. (In Knecht 1971, 219)

The Hunza attribute shamanic abilities directly to the effects of juniper smoke:

In Hunza, the bitaiyo are regarded as persons with supernatural powers whose services as prophets, magicians, and healers were called upon. They manifest their abilities only after inhaling the smoke of burning juniper branches and drinking warm goat's blood. After this, they danced to rhythmic drum beats until they had attained the trance state. When asked about the future, they passed on the messages of the fairies in the form of songs. (Felmy 1986, 19)

To increase the psychotropic effects, juniper needles may be mixed with the seeds of Syrian rue (**Peganum harmala**) before being cast onto a fire.

Artifacts
Branches and trunks of drooping juniper may be erected at cloisters and shrines as symbols of the World Tree or shaman's tree. Juniper (and other materials) is burned as incense in numerous incense vessels. The vessels usually are made of bronze or brasslike alloys and typically are decorated with Buddhist symbols (the eight auspicious symbols, dragons, et cetera).

Medicinal Use
In Darjeeling, an area that is under Indian jurisdiction but culturally is part of Nepal, the branches

"Everything has come together in the kitchen, the center of the house, and while those in search of healing are still conversing, the shaman is already beginning to sing. Behind him, juniper is being burned. He groans, for he is soon singing so rapidly that he has difficulty taking a breath. It appears to be a mild hyperventilation, for he can fight for air only once in a while. While he recites in song, he arranges the table before which he kneels as an altar. Now, completely surrounded by billows of juniper smoke, he blows, utters inarticulate sounds, and breathes more and more heavily until high, sharp screams pass over his lips—signs of the approaching deity."

AMÉLIE SCHENK
WAS IST SCHAMANENTUM? [WHAT IS SHAMANISM?]
(1996, 23*)

Although *Juniperus* species are used throughout the world for both ritual and medicinal purposes, only the species *Juniperus recurva* has a reputation for being psychoactive. (Woodcut from Gerard, *The Herball or General History of Plants*, 1633.)

Traditional representation of a high mountain species of Juniper (*Juniperus pseudosabina*) on a Tibetan medical thangka.

"Whoever carries a Kranewitt bush [Wotan's belt, *Juniperus communis*] on his hat is protected from dizzy spells and becoming tired. The smoke of the 'Martin's belt' dispels snakes, worms, and spirits. The drink made from its berries enables one to see the future."

HANS SCHÖPF
ZAUBERKRÄUTER [MAGICAL PLANTS] (1986, 151 f.*)

are burned to dispel insects and mosquitoes.

In Darjeeling and in Sikkim (formerly a tiny Himalayan kingdom, now a part of India), the ripe fruits are added to the locally brewed millet and rice **beer** for flavoring (Biswas 1965). In other regions, they are used to perfume homemade millet schnapps (*rokshi*). Juniper berries are said to have a purifying effect upon the aura and the subtle body and therefore find use as incense in both Ayurvedic and tantric medicine. In Ayurveda, the berries are ingested as a diuretic and to improve digestion. They also are made into a paste that is applied externally to treat arthritis, swelling, and pain (Lad and Frawley 1987, 214*).

Constituents

Drooping juniper is rich in an **essential oil** whose composition presumably is similar to that of the essential oil of *Juniperus communis* L.[193] The essential oil is most heavily concentrated in the tips of the branches, the fruits (0.46 to 0.88%), and the heartwood. Isocedrolic acid and 4-ketocedrole have been found in the entire plant. The needles contain biflavones, cupressoflavones, and their derivatives (Asolkar et al. 1992, 380*). The composition of the essential oil of the Kashmiri variety *squamata* has been worked out; it consists of 23.6% limonene, 16.3% sabinene, 14.6% α-pinene, and some α-thujone, myrcene, terpinene, and Δ-cadinene, as well as trace amounts of other substances (Weyerstahl et al. 1988, 260). The dried leaves have yielded biflavones (amentoflavone, hinokiflavone, isocryptomerin) and flavonol-*O*-glycosides (quercetin-*O*-α-L-rhamnosides and kaempferol-3-*O*-β-D-glucoside) (Ilyas et al. 1977).

The smoke of *Juniperus recurva* has been chemically investigated for psychoactive components. The gaseous phase contains over forty substances, the primary components of which have been identified as acetone, benzole, toluole, ethylbenzole, o-xylole, m-xylole, and most likely limonene (Knecht 1971, 220). To date, no actual psychoactive component has been discovered or isolated and pharmacologically studied.

Some sources in the literature have asserted that norpseudoephedrine is present in this species of Himalayan juniper (Schuldes 1995, 45*). Such claims most likely are incorrect (cf. **ephedrine**).

Effects

Indian and Nepali scientists have determined that the smoke from the fresh wood has emetic effects causing prolonged vomiting (Suwal et al. 1993, 72*); the extract of the branch tips appears to have anticarcinogenic effects.

Drooping juniper often has been attributed with the ability to produce psychoactive effects. However, it is uncertain whether the scent has a psychological effect or the smoke has a pharmacological effect. It is possible that only certain specially endowed individuals are able to experience the psychoactive effects of the smoke.

In my own experience, the smoke promoted the recall of memories. I found myself immediately transported to the Himalayas, where I mentally participated in archaic shamanic rituals.

Commercial Forms and Regulations

In Nepal and India, the dried branch tips are sold in herb markets and by sellers of incense (e.g., around the stupa of Bodnath, near Kathmandu). Drooping juniper is not subject to any restrictions.

Literature

See also the entries for **essential oils** and **incense**.

Biswas, K. 1956. *Common medicinal plants of Darjeeling and the Sikkim Himalayas*. Alipore, India: West Bengal Government Press.

Felmy, Sabine. 1986. *Märchen und Sagen aus Hunza.* Cologne: Diederichs.

Ilyas, Mohammad, Najma Ilyas, and Hildebert Wagner. 1977. Biflavones and flavonol-*O*-glycosides from *Juniperus macropoda*. *Phytochemistry* 16:1456–57.

Knecht, Sigrid. 1971. Rauchen und Räuchern in Nepal. *Ethnomedizin* 1 (2): 209–22.

Malla, S. B., et al., eds. 1976. *Flora of Langtang and cross section vegetation survey (central zone)*. Kathmandu: His Majesty's Government, Dept. of Medicinal Plants.

Rätsch, Christian. 1995. Einige Räucherstoffe der Tamang. In *Yearbook for Ethnomedicine and the Study of Consciousness,* 1995 (4):153–61. Berlin: VWB.

Shrestha, Bom Prasad. 1989. *Forest plants of Nepal.* Lalitpur, Nepal: Educational Enterprise.

Weyerstahl, P. H. Marschall-Weyerstahl, E. Manteuffel, and V. K. Kaul. 1988. Constituents of *Juniperus recurva* var. *squamata* oil. *Planta Medica* 54:259–61.

---

193 This consists of α-pinene, sabinene, camphene, cadinene, juniperol, juniperine, junen, and terpineol-4.

Justicia pectoralis Jacquin

Justicia

Family
Acanthaceae (Acanthus Family)

Forms and Subspecies
There is one variety that occurs chiefly in Venezuela and Ecuador and is of ethnopharmacological importance: *Justicia pectoralis* Jacq. var. *stenophylla* Leonard.

Synonyms
Dianthera pectoralis (Jacq.) Murr.
Eclobium pectorale (Jacq.) Kuntze
Psacadocalymma pectorale (Jacq.) Bremek
Rhytiglossa pectoralis (Jacq.) Nees
Stethoma pectoralis (Jacq.) Raf.

Folk Names
Boo-hanak, buhenak, carpenter bush, carpenter grass, curía,[194] fresh cut, garden balsam, herbe à charpentier, justicia, justizia, kokoime, kumaru-ka'a (Ka'apor, "tonka bean plant"), mahfarahenak (Maitá), marica (Shipibo-Conibo), masci-hiri, masha-hara-hanak, mashahari, masha-hiri (Waika), mashi-hiri, mashihíri, paxararok (Ninam), pira-pishí-ka'a (Ka'apor, "fish(?)-plant"), sua-ka-henako (Yanomamö, "leaves to use on women"), tilo (Cuba), tilo casero, tilo criollo, tilo de jardín, tilo natural, toyeau, trebo, yacu piri-piri, ya-ko-yoó (Puinave), zeb shêpantyê

History
The first report of the Venezuelan Indian use of justicia as a **snuff** was made in 1953 (Schultes 1990, 61). The ethnopharmacology and chemistry of the plant are still essentially unknown.

Distribution
The plant grows along waterways in the tropical rain forests of Mexico and Central America, on the Caribbean islands (Cuba), and in northern South America. The variety *stenophylla* occurs only in South America.

Cultivation
Propagation occurs via the seeds or the planting of rootstocks that have been separated from another plant. The simplest method is to use cuttings that have begun to develop roots or scions (stems that have developed roots on the lower nodes). In South America, the plant is cultivated as an ornamental. The Yanomamö Indians cultivate it for the manufacture of psychoactive **snuffs**. They grow the plant in the semishaded areas between banana trees. The plant does not tolerate frost.

Appearance
The plant, which can grow as tall as 70 to 80 cm, develops vertical stalks that lean at the tops and sometimes develop roots on their lower nodes. The numerous light green, somewhat rough leaves are narrow and lanceolate, 2 to 5 cm long and 2 to 3 cm wide. The flowers, which are typical of those of the family, develop at the tips of the stalks. The calyxes are only 5 mm long and are usually white or light violet in color. In the tropics, the flowering period is from November to April. The fruits, which contain the flat, reddish brown seeds, develop from December to March.

The variety *stenophylla* is chiefly distinguishable by a more stocky pattern of growth (up to 30 cm tall) and narrower leaves (1 to 2 cm wide).

The plant is very easily confused with other *Justicia* species, some eighty of which occur in Mexico alone. Worldwide, there are approximately four hundred species in the genus (cf. Daniel 1995).

Psychoactive Material
— Leaves, fresh or dried

Preparation and Dosage
A sedative tea can be made by pouring hot water over a handful of fresh leaves. Allow to steep for five to ten minutes, and sweeten with **honey** as desired.

In Guadeloupe (Caribbean), the fresh herbage is steeped in **wine**, sweetened with **honey**, and used as a love drink (Müller-Ebeling and Rätsch 1986, 126*).

Only leaves that have been dried in the shade are used for psychoactive purposes. These are ground into a fine powder and used primarily as an additive to the **snuff** known as *epená*. Justicia powder often is mixed with the dried resin of *Virola* spp. (Prance 1972a, 17*).

Today, the dried leaves are often mixed with

The leaves of the South American *Justicia pectoralis* var. *stenophylla* are used primarily as an aromatic additive to psychoactive snuffs.

194 This name is also used to refer to the Venezuelan *Justicia caracasana*, which is used as an additive in *chimó* (cf. *Nicotiana tabacum*) (Ott 1993, 410*).

"*Chonó-Rau, Chonó Ininti*, or *Rimon Ininti* (*Justicia* sp.) [is one] master plant that smells like lemons; it is ingested especially by the students of the shamans [of the Shipibo-Conibo Indians]. During a two-week long dietary period (abstinence), the student drinks the initiatory water of the leaves. Afterward, he takes ayahuasca and encounters the lord of the plants either in the resulting visions or in his nocturnal dreams."

ANGELIKA GEBHART-SAYER
DIE SPITZE DES BEWUßTSEINS [THE PINNACLE OF CONSCIOUSNESS] (1987, 337)

Betaine

Umbelliferone

195 *Tilo* is actually a Spanish term for the linden tree (which is not indigenous to the Americas). Today, Florida is home to numerous Cubans who have preserved their own businesses and relationship structures. They trade in justicia leaves and flowers, which are then (as if there were not already enough confusion) sold under the American name *linden flowers with leaves.*

marijuana (*Cannabis indica*) for smoking (see **smoking blends**); the mixture has a pleasant aroma. *Justicia pectoralis* also appears to be used as an ingredient in the tobacco preparation known as *chimó* (see **Nicotiana tabacum**).

Ritual Use

The most important use of the leaves of the variety *stenophylla* is as an additive to psychedelic **snuffs** that are based on the dried resin of DMT-containing species of the genera *Anadenanthera* and *Virola*. The dried leaves acquire an aromatic scent. They are used in this way by various tribes in the Amazon region. The Waika or Yanomamö use *Justicia* leaves and *Virola* resin to manufacture a snuff they call *machohara*. They say that while each of the two ingredients can be snuffed by itself in order to induce mild visions, the combination of the two has better effects and is more potent (Schultes 1990, 68).

It is possible that *Justicia pectoralis* may have been used as a **snuff** in prehistoric Mexico.

The Shipibo say that the plant awakens the spirit of work in humans and brings good fortune in fishing. To achieve these benefits, a person should drink a decoction of the leaves (Arévalo V. 1994, 185*).

Artifacts

Yanomamö women place bundles of the leaves into holes in their earlobes for decorative purposes.

Medicinal Use

The Yanomamö use *Justicia pectoralis* var. *stenophylla* as an aphrodisiac for women (Schultes 1990, 64f.) The Kofán Indians of Colombia make a decoction of a related species, *Justicia ideogenes* Leonard, which they use to treat the symptoms of old age (Schultes 1993, 131*).

In Cuba, *Justicia pectoralis* is known as *tilo*,[195] more rarely as *tila*, and is drunk as a mild nerve tea (sedative) that has an aromatic/sweet taste. Moreover, the plant is used in Cuban folk medicine as a remedy for heartburn, epilepsy, arteriosclerosis, baldness, runny nose, blindness, colic, lack of appetite, weakness, sleeplessness, headache, scurf, coughing, and depression (Seoane Gallo 1984, 876*). In the Caribbean, teas made from the plant are used chiefly for coughs and colds. The freshly pressed juice of the leaves is dripped onto bleeding wounds (Seaworth 1991, 70*). In Trinidad, a decoction is drunk to treat flu, fever, cold chest, coughing, pneumonia, and vomiting (Wong 1976, 139*). The plant is used as an aphrodisiac in Guadeloupe (Müller-Ebeling and Rätsch 1986, 126*).

In Mexican folk medicine, the plant known locally as *trebo* is administered to treat elevated body temperatures (Argueta V. et al 1994, 1519*).

Constituents

The leaves were at one time determined to contain *N,N*-DMT, a finding that later was discovered to be incorrect (Ott 1993, 410*). However, this possibility has again been raised (Schultes 1990).

Known to be present are betaine, umbelliferone, an **essential oil**, various **coumarins** (**scopoletin** and others), benzopyrane, and justicidine B (Macrae and Towers 1984; Seaworth 1991, 70*). Small amounts of vasicine and traces of tryptamines also have been detected (Schultes 1990, 66).

Large quantities of **coumarins** are produced as the leaves dry, and these give the raw plant material its characteristic scent (Schultes 1990, 68). The genus *Justicia* is also known to contain lignans (Ghosal et al. 1979).

Effects

The plant is sometimes described as hallucinogenic (Daniel 1995, 75). Apart from its mild sedative effects, however, little is known about the psychoactive properties of the plant. There are some reports of hypnotic and sedative effects, which can be attributed to the **coumarin** the plant contains (Macrae and Towers 1984).

Commercial Forms and Regulations

The seeds are occasionally available through sources specializing in ethnobotanical plants.

Literature

See also the entry for **snuffs**.

Daniel, Thomas F. 1995. *Flora of Chiapas. Part 4: Acanthaceae.* San Francisco: Dept. of Botany, California Academy of Sciences (pages 75f.).

Ghosal, Shibnath, Shanta Banerjee, and Radhey S. Srivastava. 1979. Simplexion, a new lignan from *Justicia simplex. Phytochemistry* 18:503–5.

Macrae, W. Donald, and G. H. Neil Towers. 1984. *Justicia pectoralis*: A study of the basis for its use as a hallucinogenic snuff ingredient. *Journal of Ethnopharmacology* 12:93–111.

Schultes, Richard Evans. 1990. De Plantis Toxicariis e Mundo Novo Tropicale Commentationes XXXVI: *Justicia* (Acanthaceae) as a source of an hallucinogenic snuff. *Economic Botany* 44 (1): 61–70.

Lactuca virosa Linnaeus

Wild Lettuce

Family
Compositae: Asteraceae (Aster Family); Subfamily Cichorioideae, Lactuceae Tribe

Forms and Subspecies
The taxonomy of the wild *Lactuca* species has not been completely clarified. It is possible that there may be several varieties of *Lactuca virosa*.

Synonyms
Lactuca agrestis nom. nud.
Lactuca sylvestris nom. nud.

Folk Names
Bitter lettuce, German lactucarium, giftlattich, giftsalat, kompaßpflanze, lactuca agresti, lactucke, laitue vireuse (French), lattichopium, lattig, latuga velenosa (Italian), leberdistel, lettuce, lettuce opium, lopium, prickly lettuce, stinksalat, totenkraut, wild lettuce, wilder lattich

History
Egyptian grave paintings dated to approximately 4500 B.C.E. depict plants whose appearance is strongly reminiscent of a wild lettuce species or a cultivated form bred from it (Whitaker 1969, 261). There was already some confusion about the name and source of *Lactuca* pharmacologic material in ancient times (Schneider 1974, 2:222 ff.*). Thus, we still do not know with certainty which lettuce species was known to and used by the ancient Egyptians. It may have been *Lactuca scariola* L. [syn. *Lactuca serriola* Torner], *Lactuca virosa*, or a form of *Lactuca sativa* (Harlan 1986, 7). The garden salad *Lactuca sativa* presumably was derived from *Lactuca scariola* in ancient Egypt (Lindqvist 1960).

Wild lettuce clearly had a ritual (divination) and medicinal significance in ancient times:

> The wild lettuce, which the prophets knew as "Titan's blood," Zoroaster as *pherumbros*, the Romans as *Lactuca silvatica*, is similar to the garden lettuce but has a sturdier stalk, whiter, thinner, coarser, and bitter-tasting leaves.
> Overall, it is similar in its effects to the poppy, which is why some also mix its juice into opium. . . . It generally is able to make one sleepy and relieve pain. Furthermore, it promotes the catamenia and is also drunk for scorpion and spider stings. The seed is taken like that of the garden lettuce and inhibits pollution and coitus. After it has been dried in the sun, the juice that is pressed from it is stored in earthen vessels like the other juices. (Dioscorides 2.165)

It may be that the "twelve gods' herb" that Pliny praised as a panacea was a species of *Lactuca*, perhaps even wild lettuce itself (see **Dodecatheon**). The Arabic physician Avicenna (= Ibn Sina, 980–1036), who established the use of opium (cf. *Papaver somniferum*) in Islamic medicine, wrote, "Opium is sometimes also produced from the seeds of *Lactuca agrestis* (lettuce), this is only mildly sedating" (*Qanun*, 5.526).

Hildegard von Bingen certainly played a role in establishing the psychoactive reputation of the plant:

> The lettuces, which can be eaten, are very cold, and when eaten without spice they make the brain of man empty with their useless juice. . . . But the wild lettuces have almost the same nature. For anyone who would eat lettuces, which are useless and are called weeds, either raw or cooked, would become mad, that is insane, and he will become empty in the core, because those are neither warm nor cold, but simply a useless wind that dries out the fruit of the earth and brings forth no fruit. And those lettuces grow from the foam of the earth's sweat and are therefore useless. (*Physica* 1.90/91)

In 1543, Leonard Fuchs depicted a lettuce plant in his *Kräuterbuch* [Herbal] under the name *Lactuca capitata*, a wild or already cultivated species of *Lactuca* (Whitaker 1969, 262).

Wild lettuce was once an important opium

The stalks and leaves of the forest lettuce, regarded as the "fourth race of incense spice," are also said to contain an opium-like juice. Unfortunately, it is no longer possible to botanically identify the plant in question. (Woodcut from Tabernaemontanus, *Neu Vollkommen Kräuter-Buch*, 1731)

Left: Wild lettuce (*Lactuca virosa*) in bloom.

Right: Lactucarium, obtained from the latex of wild lettuce, is regarded as an opium substitute.

"There is also a form of white lettuce which the Greeks called opium lettuce because a juice with sleep-promoting powers flows richly within it. It is thought, however, that all lettuce species bring sleep. This was the only species of lettuce that grew in Italy in ancient times and is recalled in the Latin name for lettuce [*lactuca*], which is derived from the word for milk."

PLINY
NATURAL HISTORY
(19.38)

"The salad or lettuce (*Lactuca sativa*) increases the milk of the sucklers and lessens the fire of love. The sharp eye of the eagle is said to be due to the fact that it eats lettuce from time to time."

K. RITTER VON PERGER
DEUTSCHE PFLANZENSAGEN
[GERMAN PLANT LEGENDS]
(1864, 201*)

"Among the most important attributes of the fertility god Min is lettuce. At the festival of the god, a small bed with this plant was included in the procession. Lettuce was depicted in numerous relief images of Min (and borrowing from him, in a special form of Amun as well, e.g., in the temple of Luxor). The plant was regarded as an aphrodisiac, which also explains its popularity as a cultic gift of offering (preserving the procreative power = life)."

MANFRED LURKER
LEXIKON DER GÖTTER UND SYMBOLE DER ALTEN ÄGYPTER [LEXICON OF THE GODS AND SYMBOLS OF THE ANCIENT EGYPTIANS]
(1987, 124)

196 In former times, lactucarium was also obtained from blooming *Lactuca sativa* L., the cultivated garden salad (Schneider 1974, 2:226*).

197 Even chlorinated *Lactuca sativa* leaves have been smoked as a marijuana substitute (cf. *Cannabis indica*).

substitute (Coxe 1799; Schneider 1974, 2:226*). It has been claimed that some North American Indians smoked lettuce (Miller 1993, 48*). In 1792, a physician in Philadelphia named Kore invented lactucarium, the thickened latex (Bibra 1855, 254*).

Distribution

Lactuca virosa's original range was in southern Europe. It now grows wild in many parts of central Europe, where it was once cultivated as a source of lactucarium. Wild lettuce is also widespread in southern North America, where it was introduced from Europe.

Cultivation

Propagation occurs via the seeds (just as with garden salad). The seeds are simply broadcast over the ground in spring (Grubber 1991, 66*). Lettuce prefers loose, well-drained humus soil (topsoil).

Appearance

Lactuca virosa is an annual or sometimes biennial plant that can grow from 60 to 150 cm tall. It has a round stalk that branches into panicles at the top. The leaves are spinose and toothed, and their midveins have spines on the lower side. The flowers are light yellow and basket-shaped. The blackish fruits are narrow and resemble wings. The plant is most easily recognized by the white latex that flows through all of its parts and issues when the plant is injured.

Lactuca virosa is easily confused with prickly lettuce, *Lactuca scariola* [syn. *Lactuca serriola* L.]. The leaves of the latter, however, are deeply incised/retuse. Lettuce is also sometimes confused with various species of the genus *Sonchus* (cf. *Erythroxylum coca*).

Psychoactive Material

— Thickened juice (latex): lactucarium,[196] lactucarium germanicum
— Dried leaves[197]

Preparation and Dosage

The dried leaves can be smoked alone or in **smoking blends** with other herbs.

Lactucarium is obtained in a variety of ways:

The juice of the plant can be obtained by using an electric juicer and then drunk. More frequently, the upper part of the plant is cut repeatedly and the milk that emerges collected. This is allowed to dry and can then be smoked. The entire plant can also be dried and smoked. The largest single dosage of lactucarium was 0.3 g, the maximum daily dosage 1 gram. (Schuldes 1995, 46*)

Lactucarium can be obtained simply by col-

lecting the latex and allowing it to dry. The dried latex then can be either dissolved in **alcohol** and drunk or mixed together with other herbs (mint, hemp, thorn apple; cf. **smoking blends**) and smoked:

Lettuce opium was often used by North American Indians who smoked the dried resin or sap obtained from the plant. They cut the flower heads off, gathered the sap that drained, and then let it air dry. This process was repeated over a two-week period by cutting just a little bit off the top each time. (Miller 1985, 96)

The leaves and lactucarium were also ingredients in the so-called **witches' ointments**.

A psychoactive dosage of wild lettuce leaves is 28 g (Miller 1985, 96*). When used for medicinal purposes, the largest single dosage of lactucarium is 0.3 g, while the total daily dosage is 1.0 g (Roth et al. 1994, 444*).

Ritual Use

Lettuce was sacred to the popular Egyptian god Min (Keimer 1924), who was already being worshipped in the time of the Old Kingdom. The Greeks saw Min as a form of their lusty god Pan (cf. *Arundo donax*). Min was usually portrayed with an erect penis. He was the god of the desert, the lightning, and the sandstorm as well as a god of fertility and procreation. He wore a headdress made of ostrich feathers. His symbols were the phallus and the lettuce. His festival, which was a kind of harvest celebration, was celebrated during the first month of summer. At this festival, a statue of Min was carried in a sacred procession on a bed of lettuce (Harlan 1986, 6).

Artifacts

Lettuce was frequently depicted in Egyptian works of art and is often found in association with representations of the god Min (Lurker 1987, 124). Min's ceremonial procession is depicted in many wall paintings in buildings from the time of Ramses II, Ramses III, Herihor, Seti I, Amenhotep III, Sosestris, and Thutmosis III (Harlan 1986, 6).

Medicinal Use

Aphrodisiacs were once made from lettuce and the lactucarium it yields. The ancient Egyptians possessed a book of love agents, which has unfortunately been lost. For this reason, the Egyptians' recipes for aphrodisiacs based upon lettuce are unknown. Oddly enough, the ancient Greeks attributed lettuce with the opposite effect, namely that of an anaphrodisiac (Harlan 1986, 8).

In the modern period, lactucarium was used as a sedative and as a substitute for opium (cf. *Papaver somniferum*).

In homeopathy, Lactuca virosa is used according to the medical description (which differs from that for Lactuca sativa) to treat sleeplessness, dry cough, and other ailments (Roth et al. 1994, 444*):

This agent primarily affects the brain and the circulatory system. Delirium tremens with sleeplessness, cold and tremor. Hydrothorax and aszites. Impotence. Feelings of lightness and tightness that affect the entire body, especially the chest. It appears to be a true lactagogue. Pronounced affects upon the extremities. (Boericke 1992, 460*)

Constituents

Lactucarium contains the sedative sesquiterpene lactone/bitter substance (guaianolide) lactucine ($C_{15}H_{16}O_5$) and its p-hydroxyphenylacetic acid ester, lactupicrine (= lactucopicrine),[198] along with triterpene alcohols (lactucerol), a melampol-glycoside (lactuside A), and other guaianolides (11β,13-dihydrolactucine, 8-deoxylactucine, jac-quineline, zaluzanin derivatives) (Stojakowska et al. 1993, 1994). The older literature speaks of the presence of a "hyoscyamine-like alkaloid" (cf. **tropane alkaloids**) (Frohne and Pfänder 1983, 67 f.*).

It should be noted that opium-like alkaloids are also present in garden salad or lettuce (*Lactuca sativa* L.) (Bibra 1855, 259*). When the cabbage lettuce forms its stalk, a white, milky sap develops that contains alkaloids with sedative effects (Rätsch 1995a*, 1995c*). The variety *Lactuca sativa* var. *capitata* L., which is grown for use in salads, is also known as lettuce opium and French lactucarium (Brown and Malone 1978, 23*).

Effects

Lactucarium has analgesic, sedative, and cough-suppressing effects (Stojakowska et al. 1993). The effects were once compared to those of the deadly nightshade (**Atropa belladonna**) (Harlan 1986, 10). They also have been used to explain the effects of kava (see **Piper methysticum**). The effects have been described as a "languid dream state" (Miller 1985, 97*) or an aphrodisiac high.

The pioneer Freiheer Ernst von Bibra (1806–1878), who experimented with lactucarium a great deal, came to the following conclusion: "Quite similar to opium, lactucarium from various species and varieties of Lactuca species also possesses somewhat diverging properties, but in its main effects, just as with opium, it is the same" (Bibra 1855, 255*).

The sedative effects are attributed to the sesquiterpene lactones of the guaiane type that occur in the latex and often are present in the form of glycosides (Stojakowska et al. 1994, 93).

Commercial Forms and Regulations

All preparations of *Lactuca virosa* may be sold without restriction.

Recently, lettuce preparations (such as Lettucene) have been sold as "extra strong" hashish substitutes under such provocative names as *hash oil* and *hashish* (see **Cannabis indica**). These consist of damiana (**Turnera diffusa**) and extracts of *Lactuca sativa* and sometimes also contain yohimbé bark (**Pausinystalia yohimba**).

Literature

See also the entry for **Papaver somniferum**.

Coxe, John Redman. 1799. Inquiry into the comparative effects of the opium officinarum, extracted from the *Papaver somniferum* or white poppy of Linnaeus; and of that procured from *Lactuca sativa*, or common cultivated lettuce of the same author. *Transactions of the American Philosophical Society*, o.s., 4:387–414.

Harlan, Jack R. 1986. Lettuce and the sycomore: Sex and romance in ancient Egypt. *Economic Botany* 40 (1): 4–15.

Helm, J. 1954. *Lactuca sativa* L. in morphologisch-systematischer Sicht. *Kulturpflanze* 2:72–129.

Keimer, L. 1924. Die Pflanze des Gottes Min. *Zeitschrift für Altägyptische Sprache und Altertumskunde* 59:140–43.

Lindqvist, K. 1960. On the origin of cultivated lettuce. *Hereditas* 46:319–50.

Lurker, Manfred. 1987. *Lexikon der Götter und Symbole der alten Ägypter*. Bern, Munich, and Vienna: Scherz.

Stojakowska, A., J. Malarz, and W. Kisiel. 1994. Sesquiterpene lactones in tissue culture of *Lactuca virosa*. *Planta Medica* 60:93–94.

Stojakowska, A., J. Malarz, W. Kisiel, and S. Kohlmünzer. 1993. Callus and hairy root cultures of *Lactuca virosa*. *Planta Medica* 59 suppl.: A658.

Whitaker, Thomas W. 1969. Salads for everyone—a look at the lettuce plant. *Economic Botany* 23:261–64.

Zubke, Achim. 1998. Lactucarium. *HanfBlatt* 5 (41): 12–15.

Top: The flower of *Lactuca sativa* var. *sativa*.

Bottom: A cultivated form of the garden salad (*Lactuca sativa* cv.), here shown in flower, during which phase it produces a narcotic juice.

Lactucine

198 Lactucine and lactupicrine also occur in the dandelion (*Taraxacum officinale* Web.) (Frohne and Pfänder 1983, 68*).

Latua pubiflora (Grisebach) Baillon
Latúe, Tree of the Magicians

Family
Solanaceae (Nightshade Family); Subfamily Cestroideae, Nicotianeae Tribe

Forms and Subspecies
The genus is composed of just one species (D'Arcy 1991, 78*); it is possible that a distinction may be made between two forms on the basis of pronounced or almost absent thorns. The Mapuche Indians make a distinction between a male (thornless) and a female (thorny) form.

Synonyms
Latua pubiflora (Griseb.) Phil.
Latua venenata Phil. (misspelling in the
   literature)
Latua venenosa Phil.
Lycioplesium pubiflorum Griseb.

Folk Names
Árbol de los brujos (Spanish, "tree of the magicians"), baum der zauberer, latua, latue, latué (Mapuche, "he who kills"), latúe, latue-hue, latuhue, latuy,[199] latuyé, palo de bruja ("tree of the witch"), palo de brujos ("tree of the magicians"), palo mato ("tree of death"), tayu, tree of the magicians, witches tree

History
It appears that this plant was already being used for shamanic purposes in southern Chile in pre-Spanish times. Sources from the colonial period do not mention any kind of use. The genus and species were first described by the German botanist Rudolph A. Philippi (1858). The name *latúe* is from the Mapudungun (Mapuche) language and means "that which causes death." In Chile, many rumors circulate concerning the lethal effects of the tree, which is generally so feared that no one will speak of it (which does not make research any easier!).

In earlier times, Chilean fishers used the shrub as a fish poison. They combined the juice of the plant with bark from the sacred shamanic tree known as *canelo* (*Drimys winteri* Forst; see **incense**) (Plowman et al. 1971, 74). The not-unpleasant-tasting plant juice was or is used to poison food (Houghton and Manby 1985, 100*). Such use occurs primarily as a method of taking revenge on people who are the objects of desire but do not reciprocate the feeling.

Distribution
Latua pubiflora is endemic to Chile (Hoffmann 1994, 222) but is found only in very clearly delimited areas, e.g., in the Cordillera Pelada near Osorno (Mösbach 1992, 104*). The shrub is somewhat more common on the mountain ridges of the Cordillera de San Juan la Costa, which is in the center of the Mapuche region. *Latua* is also said to be indigenous to the island of Chiloé, but it is extremely rare there (Plowman et al. 1971).

Latua pubiflora is one of the rarest of all psychoactive plants. To date, it has never been cultivated anywhere else, nor has it been spread by humans.

Cultivation
Propagation can occur with pregerminated seeds (experiments are presently being carried out). The Mapuche propagate the beautiful plant using cuttings taken from green branches. *Latua* requires a mild, frost-free climate with much rain. However, the soil should not become too moist.

The leaves of the latúe plant grow considerably larger, and its appearance is (even more) beautiful when it is grown in the shade.

Appearance
While this perennial shrub can attain heights between 2 and 10 meters, it usually grows to 3 to 4 meters. The plant has one or more main stems that become woody and can grow up to 25 cm in diameter. The gray branches grow curiously in all directions. On some plants, these branches are thickly covered with long, sharp, hard thorns, while on other individuals the thorns are almost entirely absent. The gray-green, lanceolate leaves are alternate and can grow as long as 8 cm (although they usually reach only 3 cm). The violet, bell-shaped flowers hang from the thorny branches as if in panicles and attain a length of some 4 cm. The fruits are small, round, yellow-green berries that contain a great number of tiny seeds. Flowering occurs between October and March, although individual shrubs may flower more than once a year. The fruits typically ripen in March. Pollination occurs via hummingbirds (Plowman et al. 1971, 68).

Latua is easily mistaken for *Dasyphyllum diacanthoides* (Less.) Cabr. [syn. *Flotowia diacanthoides* Less.], known as *tayo* or *palo santo* ("sacred tree"), which is used in folk medicine (Plowman et al. 1971, 70).

Psychoactive Material
— Leaves, fresh or dried, and the juice obtained by
   pressing them
— Stem/wood
— Bark, fresh or dried

Latua pubiflora, known as the tree of the magicians, is one of the rarest of all shamanic plants. Its range is restricted to a small area in southern Chile. (Photographed near Osorno)

199 This same name is used in Chiloé
   (southern Chile) to refer to
   Desfontainia spinosa.

Left: The raw plant material from *Latua pubiflora* is used for folk medicinal purposes.

Right: *Rhaphithamnus spinosus*, which is used as an antidote for *Latua* overdoses, has a similar appearance to the tree of the magicians, especially in its long thorns. (Photographed in Chiloé)

— The entire plant except for the root (tot.)
— Flowers, fresh or dried

Preparation and Dosage

The shamans of the southern Chilean Mapuche (so-called *machi*; cf. Bacigalupo 1995) normally use a juice obtained by pressing freshly harvested leaves, which they dilute with water. The freshly pressed plant juice is also mixed with **wine**; occasionally, a decoction made from the bark of young shoots is preferred. Dried parts of the plant are used to make a tea (infusion) (Houghton and Manby 1985, 100*). Unfortunately, no information is available concerning the quantities that are used.

The Mapuche shamans ingest some of the tea every twenty to thirty minutes. In this way, the effects gradually become evident and overdoses can be avoided (Plowman et al. 1971, 81).

There are many recipes for (psychoactive) latúe incense (*saumerio de latúe*). Usually, equal parts of *canelo* bark (*Drimys winteri*), *romero* herb (**Fabiana imbricata**), and palqui leaves (**Cestrum parqui**) are mixed with *Latua* leaves. Another recipe calls for equal parts of *Fabiana*, *Latua*, and *Cestrum parqui*. A third recipe lists a mixture of *Latua*, wheat, maté (**Ilex paraguariensis**), and *Fabiana*, to which horse bones are sometimes added (Nakashima Degarrod n.d., 11 f.).

The flowers and leaves are dried in the shade, after which they can be smoked. Initial experiments have indicated that the dosage should not exceed 1 g per person.

For external treatments, alcohol extracts (*licor*) are prepared from the leaves, often in combination with other herbs. A handful of the dried raw plant (tot.), boiled in water, is used as a bath additive to treat rheumatism (*baño de reumatismo*).

The traditional antidote for *Latua* overdose is a decoction of *hierba mora* (*Solanum nigrum*; see **Solanum** spp.) and the herbage of an *Oxalis* species or the fruit of *Rhaphithamnus spinosus* (A. Juss.) Moldenke (Plowman et al. 1971, 75).

Ritual Use

This plant formerly was used by the shamans (*machi*) of the Mapuche in the region of Valdívia.[200] Most Mapuche shamans are female; only a very few men hold this position (Philip 1994).

One Mapuche group, the Huilliche, still revere the plant as a shamanic tree, for it brings power, knowledge, and realization; offers magical protection; and can heal (Nakashima Degarrod n.d., 8).

An offering (bread, cooked chicken, wheat porridge, tobacco) must be made to the plant before it is harvested. Then a short but important prayer is said to the spirit of the plant: "Little plant, come to me, I will take something of you so that you will give me health."

Many shamans claim that they acquired their powers and abilities by consuming latúe preparations during their initiation (Nakashima Degarrod n.d., 11).

For the Mapuche shamans, latúe is the most important **incense** for dispelling evil spirits, bad moods, worries, and grief. To this end, the herb, which is always mixed with other substances (see above), is scattered over an open fire.

The plant is also used by black shamans (*kalku*) for their nefarious deeds (witchcraft, death magic).

Artifacts

No artifacts are known. However, it is possible that the Mapuche shamans do manufacture or use some.

Medicinal Use

Latúe is considered to be an aphrodisiac and was utilized as an ingredient in love drinks (Bodendorf and Kummer 1962). The Mapuche believe that small doses of the plant help develop physical strength and that children should be given some *Latua* so that they become "big and strong" (Nakashima Degarrod n.d., 13). Mapuche shamans administer weak infusions of the leaves and fruits internally to alleviate pain (Nakashima Degarrod n.d., 15).

In Chile, the bark of the bush, known as *tayu*, is used in folk medicine as a decoction for treating contusions and bruises (Schultes 1970, 48*). Shamans recommend a decoction of latúe, **Fabiana imbricata**, and palqui (**Cestrum parqui**) as a medicinal bath additive. A *licor de latúe* is applied externally to treat rheumatism, arthritis, coughing, pain, et cetera.

200 Not to be confused with the archaeological culture of Valdivia, Ecuador (Baumann 1981). The Chilean Valdivia lies in the heart of Araucana. This ostensibly European city was founded in 1552 by the conquistador and general Pedro de Valdivia (1505–1553).

"The Huilliche [a Mapuche tribe] believe that *Latua pubiflora* is inhabited by a spirit and is thus sacred. It is thought that the plant can bring physical strength, healing, death, and good luck and contains shamanic powers. These powers, however, are only associated with the female plant. The female plant is considered to be that one which produces fruits and has thorns, which are lacking in the male plant. Only the female plant is used by the Huilliche. It is thought that the plant is jealous and possessive. For this reason, the people keep the rituals associated with it secret. In addition, they show the plant a deep respect, for they believe that they will otherwise be severely punished. Several conceptions about the plant exhibit syncretism with Christianity. For some people, the plant is sacred because it is associated with Jesus: the thorns are said to have been used to crown Christ before the crucifixion."

LYDIA NAKASHIMA DEGARROD
CONTEMPORARY USES OF THE LATUA RUBIFLORA AMONG THE HUILLICHE OF CHILE
(N.D.)

A statue of El Brujo, "the magician," lord of *Latua pubiflora*. (Sculpture in the Museum of Ancud, Chiloé)

Constituents

The entire plant contains the **tropane alkaloids atropine** and **scopolamine** (Schultes and Farnsworth 1982, 166*); the concentration of atropine (0.18%) is greater than that of scopolamine (0.08%) (Plowman et al. 1971, 86). The leaves contain 0.18% hyoscyamine (= atropine) and lesser amounts of scopolamine (Bodendorf and Kummer 1962). Earlier analyses suggested that the leaves contain the highest concentrations of active constituents, while the stem contains less and the fruits are devoid of alkaloids. Yet other analyses have indicated that the stem contains the highest alkaloid content and the seeds and leaves only slight amounts (Plowman et al. 1971, 86).

Effects

Latúe is said to cause severe delirium and visual hallucinations and induces pronounced dryness of the mouth, pupillary dilation, headaches, and confusion. The effects last up to three days; after-effects (similar to those of *Brugmansia*) can linger for weeks. Even a tea made from the leaves is capable of inducing hallucinations and cramps (Houghton and Manby 1985, 100*). It is also said that *Latua* can produce permanent "imbecility" (Murillo 1889). The shamans are apparently not harmed by the plant; on the contrary, it helps them learn about things that are otherwise hidden.

Smoking the dried leaves produced in me very pleasant bodily effects with aphrodisiac sensations and great mental relaxation with associative thoughts (very similar to the effects of various species of the genus *Brugmansia*).

Commercial Forms and Regulations

None

Literature

See also the entries for **atropine, tropane alkaloids,** and **scopolamine.**

Bacigalupo, Ana Mariella. 1995. Renouncing shamanistic practice: The conflict of individual and culture experienced by a Mapuche *machi*. *Anthropology of Consciousness* 6 (3): 1–16.

Baumann, Peter. 1981. *Valdivia: Die Entdeckung der ältesten Kultur Amerikas.* Frankfurt/M.: Fischer.

Bodendorf, K., and H. Kummer. 1962. Über die Alkaloide in *Latua venenosa. Pharmazeutische Zentralhalle Deutschlands* 101:620–22.

Hoffmann, J. Adriana E. 1994. *Flora silvestre de Chile: Zona araucana.* Santiago: Ediciones Fundación Claudio Gay.

Miranda, J. B. 1918. Estudio químico, fisiológico y terapéutico de *Latua venenosa* (Palo de Bruja). *Actes de la Société Scientifique du Chile* 23 (3): 10–26.

Murillo, A. 1889. *Plantes Médicinales de Chile.* Paris: Imprimerie de Lagny.

Nakashima Degarrod, Lydia. n.d. *Contemporary uses of the* Latua pubiflora *among the Huilliche of Chile.* Unpublished manuscript, written ca. 1988.

Philip, Arturo. 1994. *La curación chamánica: Experiencias de un psiquiatra con la medicina aborigen americana.* Buenos Aires: Editorial Planeta.

Philippi, Rudolph A. 1858. *Latua* Ph., ein neues Genus der Solanaceen. *Botanische Zeitung* 33 (Aug.): 241–42.

Plowman, Timothy, Lars Olof Gyllenhaal, and Jan Erik Lindgren. 1971. *Latua pubiflora*—magic plant from southern Chile. *Botanical Museum Leaflets* 23 (2): 61–92.

Silva, M., and P. Mancinelli. 1959. Atropina en *Latua pubiflora* (Griseb.) Phil. *Boletín de la Sociedad Chilena de Química* 9:49–50.

Vásquez, A. 1864. Substancias del *Latua venenosa* de Chiloé, Latue o árbol de los brujos. *Anales de la Sociedad de Farmacia de Santiago* 2 (3): 71–75.

Ledum palustre Linnaeus

Wild Rosemary

Family

Ericaceae (Heath Family); Subfamily Rhododendroideae

Forms and Subspecies

Two Eurasian subspecies are distinguished: *Ledum palustre* L. ssp. *palustre* (European wild rosemary) and *Ledum palustre* ssp. *sibiricus* (Siberian wild rosemary). The Greenland wild rosemary has recently been recognized as a subspecies: *Ledum palustre* ssp. *groenlandicum* (Oed.) Hult. (Zander 1994, 341*).

The genus itself consists of only a few species.[201]

Synonyms

Ledum groenlandicum Oed.

In the summer of 1996, the American Horticultural Association incorporated the genus *Ledum* into the genus *Rhododendron*. Whether this taxonomic innovation will gain acceptance remains to be seen. Whether it is botanically justified is also an open question.

Folk Names

Altseim, baganz, bagen, bagulnik, bieneheide, bienenscheide, böhmischer rosmarin, borse, brauerkraut, bûpesbupt (Makah, "cranberry"), cistus ledonfoliis rosmarini ferugineis, einheimischer lorbeer, flohkraut, getpors (Swedish, "goat porsch"), gichttanne, gräntze, gruitkraut, gruiz, grund, gruut, hartheide, heidenbienenkraut, Hudson's Bay tea, kiefernporst, kienporst, kienrost, klopovnik, kühnrost, Labrador tea, labradortee, ledo, ledón des marais, ledumporst, ledum silesiacum, lunner, marsh tea, mirtus, moerasrozenmarijn, moor-rosmarin, morose, mottenkraut, mutterkraut, myrto, nûwaqwa'ntî (Quinault), pors, porsch, porskraut, porst, post, postkraut, rausch, rosmarinkraut, rosmarinus sylvestris, roßkraut, sautanne, schweineposse, sqattram, sumpfporst, tannenporst, ti:mapt ("tea plant"), waldrosmarin, wanzenkraut, weiße heide, wilder rosmarin, wild rosemary, zeitheide, zeitheil

History

It appears that the authors of antiquity were not aware of wild rosemary. Pliny, however, does mention a plant named *Ledum* that had narcotic effects (Rowell 1978, 271*). In the later herbals of the fathers of botany, the plant appears under the name *Ledum silesiacum* (Clusius) or *wild rosemary* (Tabernaemontanus). Although the woodcuts that were included in the works of such authors as Tabernaemontanus and Gerard are botanically accurate and clearly identifiable, the first scientific description of the plant was made by Linnaeus, who even wrote a medicinal work about the plant in 1775 (Vonarburg 1995, 78).

From top to bottom:
Wild rosemary (*Ledum palustre* ssp. *palustre*) in bloom.

Greenland tea (*Ledum palustre* ssp. *groenlandicum*) is a North American species.

The fruits of wild rosemary (*Ledum palustre* ssp. *palustre*).

201 The genus *Ledum* is composed of some six species. The closely related Labrador tea bush (*Ledum latifolium* Jacq.) is very similar in appearance to though somewhat larger than wild rosemary; it grows in Greenland, Labrador, and northern North America. The Indians and the Inuit (Eskimos) use its aromatic leaves as a medicinal tea that they drink for asthma and colds and also as incense. The Kwakiutl Indians of the Canadian Pacific coast obtained a narcotic from the Greenland porsch (*Ledum groenlandicum* Oed. [syn. *Ledum latifolium* Ait. non Jacq.]) that they used in shamanic healing. The Inuit used the dried tips of the twigs of *Ledum decumbens* (Ait.) Lodd. (possibly a synonym for *Ledum latifolium*) as a medicinal incense for treating children's ailments (cf. Rätsch 1996).

Left: The North American wild rosemary species *Ledum glandulosum* ssp. *columbianum*.

Right: The leaves of an as yet unidentified species of wild rosemary (*Ledum* spp.) collected in North America are almost twice as long as the leaves of related species.

"The shamans of the Giljak and the other Amur natives bend over a thick, strongly scented smoke that develops when dried leaves of wild rosemary (*Ledum palustre*) are laid onto glowing coals. Wold. Aerseniew was able to determine that every true shamanic activity of the Orotsha (as compared to the shamanic games) is begun by rubbing the knee with heated leaves of *Ledum palustre*."

GEORG TSCHUBINOW
BEITRÄGE ZUM PSYCHOLOGISCHEN VERSTÄNDNIS DES SIBIRISCHEN ZAUBERERS [CONTRIBUTIONS TO A PSYCHOLOGICAL UNDERSTANDING OF THE SIBERIAN MAGICIAN] (1914, 44*)

Distribution

Wild rosemary occurs almost exclusively in high and transitional bogs, often growing in association with pine and birch trees. Although very widespread, it is only rarely found in the wild. It grows in the region of the Alps and in northern Europe and in central, eastern, and northern Asia (Siberia), as well as Japan. A closely related species occurs in North America. In Germany, this rare plant is protected.

Wild rosemary is regarded as a relic from the ice age, as it is adapted to a cool and moist climate.

Cultivation

Propagation is possible from seed. However, there are few reports concerning experiences with cultivation.

Appearance

Wild rosemary is a shrublike evergreen plant that can grow to a height of 1.5 meters. It has lanceolate leaves that are hairy on their undersides and white flowers in terminal corymbs. The capsule fruits are ovate and hang down. The plant flowers in May and June.

Wild rosemary is very easily confused with several closely related American species. The glandular wild rosemary (*Ledum glandulosum* Nutt. ssp. *columbianum* [Piper] C.L. Hichc.) is especially similar and also has a similar scent.

The name *wild rosemary* (*Rosmarinum sylvestre*) originally was applied because the plant's morphology, particularly the structure of the resinous leaves, was reminiscent of that of the kitchen spice rosemary. The two plants, however, are not related (Greve 1938a, 1938b).

Psychoactive Material

— Blooming wild rosemary herbage (herba ledi palustris, ledi palustris herba), dried or fresh; also the young shoots (used to produce alcoholic extracts and the homeopathic mother tincture)

Preparation and Dosage

The blooming herbage is air-dried in the shade. It then can be used by itself as an **incense**. When dried, wild rosemary is easily ignited and burns with a bright and crackling flame. When the burning branch tips are blown out, the leaves and stalks will continue to glow and give off a white, aromatic smoke. The scent is resinous, fragrant, pleasant, and somewhat reminiscent of that of juniper and fir. A delicate, resinous scent with a slightly acrid character lingers in the room. When used for psychoactive purposes, large amounts must be burned and inhaled.

Little information is available concerning dosages for oral use. When the herbage is prepared as a tea, psychoactive or toxic effects are rare. In contrast, alcohol extracts, which are rich in the essential oil, can have pronounced effects.

To produce Labrador tea, the leaves are collected in May (before the plant blooms) and roasted in an oven. The tea is made simply by brewing the roasted leaves with boiling water (Turner and Efrat 1982, 65).

Ritual Use

In addition to juniper (cf. ***Juniperus recurva***), the shamans of the Tungus (whose language has given us the word *shaman*) and the neighboring Gilyak used especially wild rosemary as a ritual and trance-inducing **incense**. They inhaled deep breaths of the smoke in order to enter a shamanic state of consciousness. Sometimes they would chew the root as well as inhale the smoke. The shamans of the Ainu, the aboriginal inhabitants of northern Japan, also esteemed wild rosemary incense. Ainu shamanesses would prepare a potent tea of wild rosemary to treat menstrual pains and coliclike lower abdominal cramps (Mitsuhashi 1976).

While wild rosemary was used in Europe for medicinal purposes, its primary significance was as an inebriating additive to **beer** (*grutbier*) and as a ritual plant. Alongside henbane (*Hyoscyamus niger*) and thorn apple (*Datura stramonium*), wild rosemary was the most important psychoactive additive to the Germanic beers (Wirth 1995, 146) that were brewed prior to the Bavarian Purity Law of 1516 (cf. **beer**, *Humulus lupulus*). Because wild rosemary oil can provoke aggressive behavior, it has been suggested that the berserkers obtained their legendary "berserker rage" by drinking porst (wild rosemary) beer, which was common in Scandinavia (Sandermann 1980; Seidemann 1993).[202]

Artifacts
No artifacts have become known to date.

Medicinal Use
In Siberia, wild rosemary has a long history of use as a folk remedy. Bone and joint pains were treated by rubbing the joints with the fresh plant, and partially or completely dried herbage was burned to ward off insects. In Russia, wild rosemary was given to those who were as "drunk as a fish" (Rowell 1978, 21*).

Because of its narcotic effects, wild rosemary is used in folk medicine to treat whooping cough; patients both ingest it internally and inhale its smoke as a medicinal incense. The Sami (Lapps) inhale the steam of decoctions of the plant for colds. They also use a decoction to bathe painful joints and areas affected by frostbite. They drink a tea made from the plant for coughs. The Poles use wild rosemary as a fumigant to treat all manner of lung ailments (Greve 1938b, 76 f.).

Many Indian tribes of northern North America drank Labrador tea as a tonic (Gunther 1988, 43*).

Because ledol, one of wild rosemary's constituents, has abortifacient effects, preparations of wild rosemary were once used to induce abortions. This was often accompanied by strong toxic reactions.

Various potencies are used in homeopathy to treat articular rheumatism, sciatica, rheumatism of the shoulder, contusions, et cetera (Vonarburg 1995).

Constituents
The entire plant contains an **essential oil** (0.5 to 1% in the leaves) composed of ledol (= ledum camphor), palustrol, myrcene, ericoline, and other substances. It also contains the glycoside arbutine, the flavonoids hyperoside and quercetin (cf. *Artemisia absinthium, Fabiana imbricata, Humulus lupulus, Psidium guajava, Vaccinium uliginosum*), resins, and traces of alkaloids (Roth et al. 1994, 452; Tattje and Bos 1981; Vonarburg 1995, 78).

Effects
The **essential oil** can induce states of inebriation and spasms, but also abortions. Ledol has potent inebriating and narcotic effects that definitely can assume an aggressive nature. The effects of the alcoholic extract are very similar to those of **alcohol**. **Beer** that has been brewed with wild rosemary or to which it has been added (*grutbier*) has much more potently inebriating effects as a result. Generally speaking, ledol potentiates the effects of alcohol (similar to *Piper methysticum*).

The isolated resin exhibits potent anesthetizing effects that may possibly account for the plant's cough-suppressing properties (Greve 1938b, 79).

Commercial Forms and Regulations
Wild rosemary can be obtained from pharmacies without a physician's prescription as a mother tincture under the name *Ledum (D1)* (Hahnemann himself introduced wild rosemary as a homeopathic agent). This preparation is produced from the dried, resinous leaves (*HAB4a*).

Wild rosemary is a protected plant in many areas and tends to grow in regions that are protected.

Literature

Brekhman, I. I., and Y. A. Sam. 1967. Ethnopharmacological investigation of some psychoactive drugs used by Siberian and Far-Eastern minor nationalities of U.S.S.R. In *Ethnopharmacologic search for psychoactive drugs*, ed. D. Efron, 415. Washington, D.C.: U.S. Dept. of Health, Education, and Welfare.

Greve, Paul. 1938a. *Der Sumpfporst*. Hamburg: Hansischer Gildenverlag. (Almost identical with Greve 1938b.)

———. 1938b. *Ledum palustre* L.—Monographie einer alten Heilpflanze. Dissertation, Hamburg.

Mitsuhashi, Hiroshi. 1976. Medicinal plants of the Ainu. *Economic Botany* 30:209–17.

Rätsch, Christian. 1996. Sumpfporst: Eine archaische Schamanenpflanze. *Dao* 2/96:68.

Sandermann, W. 1980. Berserkerwut durch Sumpfporst-Bier. *Brauwelt* 120 (50): 1870–72.

Seidemann, Johannes. 1993. Sumpfporstkraut als Hopfenersatz. *Naturwissenschaftliche Rundschau* 46 (11): 448–49.

Tattje, D. H. E., and R. Bos. 1981. Composition of essential oil of *Ledum palustre*. *Planta Medica* 41:303–7.

Turner, Nancy J., and Barbara S. Efrat. 1982. *Ethnobotany of the Hesquiat Indians of Vancouver Island*. Cultural Recovery Paper no. 2. Victoria: British Columbia Provincial Museum.

Vonarburg, Bruno. 1995. Homöopathisches Pflanzenbrevier. 16: Sumpfporst. *Natürlich* 15 (6): 77–80.

Wirth, H. 1995. Der Sumpfporst. In *Heimische Pflanzen der Götter: Ein Handbuch für Hexen und Zauberer*, ed. Erwin Bauereiss, 145–46. Markt Erlbach: Raymond Martin Verlag.

"When administered in excess, porsch irritates the mucous membranes of the stomach and intestines. Muscle and joint pains also appear, heart palpitations, cold shivers, persistent sensations of coldness, shortness of breath, sweating, sudden states of excitation, depressed moods, appearance of a sensation of sedation and drunkenness . . ."

PAUL GREVE
LEDUM PALUSTRE L.—
MONOGRAPHIE EINER ALTEN
HEILPFLANZE [*LEDUM PALUSTRE* L.—
MONOGRAPH ON AN OLD MEDICINAL
PLANT]
(1938a, 80)

An early illustration of wild rosemary, which was placed near the aromatic cistus (*Cistus ladaniferus* L.) because of its resinous scent. (Woodcut from Gerard, *The Herball or General History of Plants*, 1633)

202 Many authors continue to believe that the berserkers used fly agaric mushrooms (cf. *Amanita muscaria*) to induce their battle ecstasy. The fly agaric, however, is completely unsuited for this purpose, because its effects tend to be more sedative and opium-like, conducive to dreaming and listening to music.

Leonurus sibiricus Linnaeus

Siberian Motherwort

Top: In Mexico, Siberian motherwort (*Leonurus sibiricus*) is known as *marijuanillo*, "little hemp." (Photographed in Chiapas)

Bottom: The Asian species *Leonurus heterophyllus* is very similar in appearance to Siberian motherwort. The herbage is used in traditional Chinese medicine. (Photographed in Songlisan, South Korea)

203 This name is also used to refer to the very similar *Leonurus japonicus* Houtt. (Karting et al. 1993, 648).

Family
Labiatae (Lamiaceae) (Mint Family); Subfamily Lamioideae (= Stachyoideae), Lamieae (= Stachydeae) Tribe, Lamiinae Subtribe

Forms and Subspecies
Two varieties are said to be distinguished in East and Southeast Asia (Hasler n.d., 1).

Synonyms
Leonurus artemisia

Folk Names
Altamisa, amor mío (Spanish, "my love"), chinesischer löwenschwanz, chinesisches mutterkraut, coda di leone, gras zum segen der mutter, ich-mau-thao (Vietnamese), i-mu-tsao (Chinese), mahjiki (Japanese),[203] marihuanilla (Spanish, "little marijuana plant"), marijuanillo (Spanish, "little hemp"), mehajiki (Japanese), motherwort, rangadoronphul, Siberian motherwort, sibirischer löwenschwanz, sibirisches herzgespann, sibirisches mutterkraut, t'uei, yakumosos (Japanese)

History
Siberian motherwort appears in the ancient Chinese *Shih Ching*, the Book of Songs (ca. 1000–500 B.C.E.), under the name *t'uei* (Keng 1974, 402*). It was sometimes praised as a medicinal plant in later ancient Chinese herbals.

It is not known when the plant spread into the New World, nor is it known when it was first smoked as an inebriant.

Distribution
Siberian motherwort is found in southern Siberia, China, Korea, Japan, Vietnam, and Southeast Asia. It now grows wild in Brazil (coastal regions) and Mexico (Chiapas).

Cultivation
Propagation occurs through the seeds. These are lightly covered with soil, moistened well, and exposed to sunlight. The seeds germinate and the plant grows surprisingly quickly. The seeds also can also be sown directly in the garden. Seedlings can be planted in pots or garden beds. The plant does not tolerate frost and should overwinter indoors in a pot. In areas free of frost, it can develop into a perennial bush. The plant can be given much water and fertilizer but will also thrive under less ideal conditions.

Appearance
The plant, which grows straight up and usually has a single stem, can attain a height of over 2 meters. The stem branches in a maxilliform manner. The dark green leaves are finely pinnate. The violet flower spikes develop at the ends of all the branches and can develop into long, attractive inflorescences.

Leonurus sibiricus is easily confused with motherwort (*Leonurus cardiaca* L.), *Leonotis quinquelobatus* Gilib. [syn. *Leonurus villosus* Desf. ex Spreng.], and especially the East Asia and Siberian species *Leonurus japonicus* Houtt. [syn. *Leonurus artemisia* (Lour.) S.Y. Hu, *Leonurus heterophyllus* Sweet, *Leonurus sibiricus* auct. non L.] (cf. Kartnig et al. 1993, 648) and *Leonurus lanatus* (L.) Pers.

Psychoactive Material
— Flowering herbage
— Dried leaves
— Root

Preparation and Dosage
The leaves, collected while the plant is in bloom, are dried and smoked as a marijuana substitute (cf. *Cannabis indica*). Typically, 1 to 2 g of dried leaves is sufficient for one joint. No truly toxic dosage is known. In laboratory experiments with rats, even very high dosages (750 to 3,000 mg *Leonurus sibiricus* extract per kg body weight *per os*) did not result in the death of the animals (Hasler n.d., 4).

Because the effects of the pure herbage are not especially pronounced, they can be synergistically potentiated by mixing the herbage with *Cannabis indica* or *Cannabis sativa* (cf. **smoking blends**).

Ritual Use
The flowers are used in the *pujas* (devotional and offering activities) of Hindus in Assam (Boissya et al. 1981, 221*). No traditional or ritual use for psychoactive purposes is known.

In Veracruz, Mexico, the plant is used in folk magic intended to make the "groom return" (Argueta V. et al 1994, 114*).

Artifacts
None

Medicinal Use
The seeds and fruits are thought to have medicinal value. The dried herbage can be found in any Chinese herb pharmacy (Keng 1974, 402*). It is used to treat loss of potency, overly heavy menstruation, postpartum bleeding, and painful menstruation (Stark 1984, 81*). It also is used as a diuretic (Ott 1993, 411*). North American Indians are said to have used the herbage as an aid in labor

(Kartnig et al. 1993, 653). In southern Mexico, the root is drunk in the form of a tea to treat women's ailments and to induce menstruation. Leaves macerated in **alcohol** are applied externally to treat rheumatism (Argueta V. et al. 1994, 114*).

Constituents

Most of the active components of the plant appear to be unstable and occur in the herbage in only low concentrations. The plant has been found to contain 0.1% of the flavone glycoside rutin (cf. *Psidium guajava*) (Hayashi 1963). The alkaloid leonuridine has been detected in the seeds. The herbage contains 0.02 to 0.04% of the alkaloid leonurine (chemical formula $C_{14}H_{21}O_5N_3$), which apparently is a guanidine derivative (Hayashi 1962). A total of four guanidine derivatives (4-guanidinobutanol, arginine, 4-guanidine butyric acid, and leonurine) have been described (Reuter and Diehl 1970, 1971). Chinese researchers have described an alkaloid "A" (with the chemical formula $C_{20}H_{30(32)}O_{16}N_6$) for the leaves. Also present is L-stachydrine, which is characteristic for the Family Labiatae (Reuter and Diehl 1970), as well as glycosidic bitter substances, syringic acid, rosmarinic acid, and caffeic acid depsides. Of particular interest for the discussion of the psychoactive effects was the discovery that the **essential oil** contains three new **diterpenes**: leosibiricin, leosiberin, and the isomer isoleosiberine (Savona et al. 1982). These diterpenes may have effects similar to those of **salvinorin A**. Leosibiricine and leosibirine also occur in motherwort (*Leonurus cardiaca*) and presumably in many other species of the genus *Leonurus* (Knöss and Glombitza 1993). Leonurine is also present in *Leonurus japonicus* (Kartnig et al. 1993, 649).

Hasler has argued that it is likely that *Leonurus sibiricus* also contains other substances that have not yet been detected (methoxylated phenyl bodies, amides) that may be biologically transformed into psychoactive methoxyphenylalkylamines (Hasler n.d., 8).

Effects

The effects are occasionally described as mildly narcotic or cannabis-like. They are by no means spectacular, unless the plant is combined with other substances.

In animal experiments, the administration of an extract produced distinct CNS activity. The alkaloid "A" elevates the tone of the uterus (Hasler n.d., 3).

Animal studies using an extract of the related species *Leonurus quinquelobatus* Gilib. [syn. *Leonurus cardiaca* L. ssp. *villosis* (Desf. ex Urv.) Hyl.] observed clear sedative and narcotic effects (Rács and Rács-Kotilla 1989).

Commercial Forms and Regulations

Both the herbage and the live plant are freely available (but usually difficult to obtain).

Literature

See also the entries for **diterpenes** and **salvinorin A**.

Hasler, Felix n.d. *Leonurus sibiricus*. Unpublished manuscript, Basel. (13 pp., ca. 1994.)

Hayashi, Y. 1962. Studies on the ingredients of *Leonurus sibiricus* L. (I). *Yakugaku Zasshi* 82:1020–25.

———. 1963. Studies on the ingredients of *Leonurus sibiricus* L. (II). *Yakugaku Zasshi* 83:271–74.

Kartnig, Theodor, Kerstin Hoffmann-Bohm, and Renate Seitz. 1993. Leonurus. In *Hagers Handbuch der pharmazeutischen Praxis*, 5th ed., 5:645–54. Berlin: Springer.

Knöss, W., and K.-W. Glombitza. 1993. Diterpenes in cultures of *Leonurus cardiaca*. *Planta Medica* 59 suppl.: A655–56.

Rács, G., and E. Rács-Kotilla. 1989. Sedative and antihypertensive activity of *Leonurus quinquelobatus*. *Planta Medica* 55:97.

Reuter, G., and H.-J. Diehl. 1970. Arzneipflanzen der Gattung *Leonurus* und ihre Wirkstoffe. *Die Pharmazie* 25 (10): 586–89.

———. 1971. Guanidinderivate in *Leonurus sibiricus* L. *Die Pharmazie* 26 (12): 777.

Savona, Giuseppe, Franco Piozzi, Maurizio Bruno, and Benjamin Rodriguez. 1982. Diterpenoids from *Leonurus sibiricus*. *Phytochemistry* 21 (11): 2699–701.

Serradell, M. N., and P. Blancafort. 1977. Leonurine. *Drugs of the Future* 2 (9): 597–99.

"In the first period after the influx of L.s. [= *Leonurus sibiricus*, 2 x 1 gram, smoked], little happened apart from the fact that I got into a basic meditative state in which all material things appeared to increase in depth and importance. Cognitive abilities and emotional experience remained largely unchanged. During the hour after the smoking, no additional increase in the effects occurred."

FELIX HASLER
LEONURUS SIBIRICUS
(N.D., 6)

Isoleosiberine

Leosibiricine

Leosibirine

The motherwort found in Germany (*Leonurus cardiaca* L.) is an old folk remedy for nervousness, sleeplessness, ill humor, and anxiety states. (Woodcut from Brunfels, *Novi Herbarii Tomus Secundus*, 1532)

Lolium temulentum Linnaeus

Bearded Darnel

Left: The species *Lolium multiflorum* is a close relative of bearded darnel.

Right: A ripe spike of bearded darnel (*Lolium temulentum*).

Family
Gramineae: Poaceae (Grass Family); Hodeae Tribe

Forms and Subspecies
One variety has been found throughout Egypt since ancient times, i.e., for at least five thousand years (Germer 1985, 215*): *Lolium temulentum* var. *macrochaeton* A. Br.

Synonyms
Lolium maximum Willd.

Folk Names
Bearded darnel, borrachera (Canary Islands, "inebriator"), cizaña, darnel, darnel grass, delirium grass, dolik (Dutch), dower, drunken lolium, hammerl (Austria), hierba loca (Spanish, "crazy herb"), huedhuedcachu (Mapuche, "crazy plant" or "plant that makes crazy"), ivraie ("inebriating"), ivraie enivrante (French), jamdar (Persian), jollo, joyo, loglio ubriacante (Italian), lolch, ollo, rauschgras, schwindelhaber, schwindelhafer, schwindelweizen, tares, taumellolch, tobgerste, tollkorn, tollkraut (Bahamas)

History
Bearded darnel has been found in Stone Age deposits in Europe and was well known among the ancient Egyptians (Christiansen and Hancke 1993, 118*). It may even have been mentioned in the Bible, which speaks of separating the "wheat from the chaff," an indication that the wheat fields may have been infested with *Lolium* (Germer 1985, 215).

Because it produces an inebriating cereal grain, bearded darnel may have been associated with the cult of the goddess Demeter (Ruck 1995, 141*; cf. **kykeon**). On the Canary Islands, the plant apparently was used as the source of an inebriant (Daria et al. 1986, 188). The Gauls are also said to have taken advantage of the "staggering effect" of the plant and used it to brew **beer**. During the Middle Ages, bearded darnel frequently found its way into cereal grain plantings and was

consequently harvested and baked into bread, which induced psychoactive effects when ingested and caused staggering (hence the many folk names referring to this effect). Incidents of mass cases of toxic symptoms—similar to those resulting from ergot (*Claviceps purpurea*)—are said to have resulted (Roth et al. 1994, 465*). Bearded darnel also may have been one of the components of the **witches' ointments**.

Distribution
The grass may have originated in the Near East; it has been found in association with humans since the early Egyptian period (Aichele and Hofmann 1991, 142*). It is especially common in central Europe, where it prefers grain fields, roadsides, and wastelands.

Cultivation
Lolium temulentum is easily propagated from seed, which normally needs only to be broadcast onto the ground. Bearded darnel prefers calciferous soils and grows well between oats and barley.

Appearance
This annual grass, which is green to bluish green in color, has stiff, upright, coarse stalks that can attain a height of up to 75 cm (more rarely 100 cm). The stalks are often branched at the base but do not exhibit any leaf shoots. The ears can be as long as 20 cm and possess long awns (beards). The glumes are two to four times longer than the lemmas. The flowering period is from June to August. The brown fruits are elongated.

It was once thought that the bearded darnel was infested with ergot (*Claviceps purpurea*) (Hooper 1937, 137*). But the grass is actually infected by one or more fungi: *Endoconidium temulentum* Prillieux et Delacroix, *Chaetonium kunzeanum* Zopf, and/or *Gibberella subinetii* (Mont.) Saccardo (Blohm 1962, 11*; Christiansen and Hancke 1993, 118*). As many as 80% of bearded darnel seeds may be infected with a fungus (Germer 1982, 215*):

> When the seeds germinate, the fungus grows up through the entire plant to the new grains and is thereby transmitted to the next generation. This kind of long-lasting relationship between host plant and fungus is known as a cyclic symbiosis. As far as we are aware, it is only the fungus which receives an advantage from this coexistence. The fungus does not harm the grass, but the grass does not benefit from it. (Christiansen and Hancke 1993, 118*)

Bearded darnel is easily confused with Italian ray grass (*Lolium multiflorum* Lamk. [syn. *Lolium italicum* A. Br.]), which is very similar in appearance (Aichele and Hofmann 1991, 142*). Bearded darnel also resembles the common ray or rye grass (*Lolium perenne* L.), which contains loliolide (cf. *Salvia divinorum*).

Psychoactive Material
— Spikes (fructus lolii temulenti, lolii temulenti fructus, bearded darnel fruits, seeds, darnel fruits)

Preparation and Dosage
The seeds of bearded darnel were used as an ingredient in **witches' ointments**, as a fermenting agent for producing fermented or distilled drinks (**alcohol**), and as an additive to **beer** (Europe) or **chicha** (Peru): "The seed was also sometimes purposely mixed into the grain in order to make the beer more inebriating" (Aichele and Hofmann 1991, 142*).

In an animal (mice) experiment with the primary active constituent, loline (= temuline), dosages as high as 200 mg/kg injected intraperitoneally did not produce any toxic effects (Ott 1993, 159f.*).

Ritual Use
It has been rumored that a secretive mystery cult in Lebanon macerates the seeds of bearded darnel in water. They drink the extract to induce a state of religious ecstasy (Ott 1993, 155*).

Artifacts
None

Medicinal Use
"The bearded darnel is little used in the healing arts; only in homeopathy is it used for rheumatic and gout ailments, stomachaches, dizziness, and trembling limbs" (Wirth in Bauereiss 1995, 144*). The plant was or is used in folk medicine as an abortifacient (Blohm 1962, 11*). In the Canary Islands, the seeds were used in folk medicine as a tranquilizer (Ott 1993, 155*).

Constituents
The spikes contain up to 0.06% of the pyridine base temuline (Roth et al. 1994, 465*). Two as yet unidentified alkaloids are also present. Perloline has been detected in the stalks (Dannhardt and Steindl 1985). The narcotic or inebriating alkaloid

tenuline, also known as loliin (= loline), is a metabolic product of the parasitic rust fungus *Endoconidium temulentum*, which almost always grows on the grains (Christiansen and Hancke 1993, 118*).

Effects
Bearded darnel is said to induce profound perceptual changes (Blohm 1962, 11*): "Drunkenness, staggering, headaches, clouding of the thought process, disturbances of vision, violent vomiting, colic, sleepiness or sleeping sickness, death resulting from respiratory paralysis. Lethal outcomes are rare, but the central effects can persist for days" (Roth et al. 1994, 465*).

"The constituent temuline induces disturbances in the coordination of movement, motor paralysis, and can provoke a spontaneous respiratory paralysis. Its atropine-like effects are manifested as a dilation of the pupils" (Wirth in Bauereiss 1995, 143*).

When darnel seeds are mixed into cereal grains, the bread or beer that is made from them can make one "crazy" (Mösbach 1992, 64*).

The characteristic "staggering" effects can also be induced by *Chaerophyllum temulentum* L. (Apiaceae), which contains a volatile alkaloid (chaerophylline?) (Roth et al 1994, 210*).

Commercial Forms and Regulations
Lolium temulentum is listed as an endangered plant (Roth et al. 1994, 465*). Apart from this, there are no regulations concerning the plant.

The mother tincture (= TM), Lolium temulentum, can be obtained in German pharmacies.

Literature
See also the entry for *Claviceps purpurea*.

Dannhardt, G., and L. Steindl. 1985. Alkaloids of *Lolium temulentum*: Isolation, identification and pharmacological activity. *Planta Medica* 51:212–14.

Darias, V., L. Bravo, E. Barquin, D. Martin Herrera, and C. Fraile. 1986. Contribution to the ethnopharmacological study of the Canary Islands. *Journal of Ethnopharmacology* 15 (2): 169–93.

Katz, I. 1949. Contribution à l'Étude de l'Ivraie Envivrante (*Lolium temulentum* L.). Thesis, École Polytechnique Fédérale, Zurich.

"Home brewers often used cheap surrogates for their top-fermenting beer instead of the expensive hops; the old beer brewers of Gotha, for example, used 'dower,' the seeds of the annual spike grass 'Taumel-Lolch' (*Lolium temulentum* L.), which caused one to 'dower,' or stagger after drinking such a beer, thus feigning the effects of a potent drink. It was easy to collect this weed in the summer grain."

OTTO LUDWIG
IM THÜRINGER KRÄUTERGARTEN [IN THE THURINGIAN HERB GARDEN] (1982, 180*)

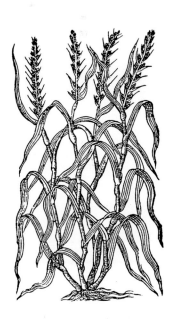

The inebriating properties of bearded darnel have been known since antiquity. This botanically correct illustration is from the early modern period. (Woodcut from Tabernaemontanus, *Neu Vollkommen Kräuter-Buch*, 1731)

Lonchocarpus violaceus (Jacquin) DC.

Balche' Tree

Family
Leguminosae: Papilionaceae (Fabaceae) (Legume Family); Subfamily Papilionoideae: Dalbergieae, Lonchocarpinae Tribe

Forms and Subspecies
It is possible that the tree the Lacandon cultivate is a variety or an as yet undescribed subspecies.

Synonyms
Lonchocarpus longistylus Pittier
Lonchocarpus maculatus DC.[204]
Lonchocarpus punctatus
Lonchocarpus violaceus H.B.K.
Lonchocarpus yucatanensis Pittier
Lonchocarpus violaceus Rth. (Pool 1898;
    Neuwinger 1994, 623*); described for India
    but is likely a different species, and the name is
    probably obsolete

Folk Names
Baälche, bá'alché', ba'che', balche (Mayan),[205] balche' (Lacandon, "thing of wood/essence of the forest"), balché, balche'baum, lancepod, lance-pod (Trinidad), palo de patlaches (Mexico), patachcuahuitl (Nahuatl), pitarilla (Spanish), saayab, sakiab, samea (Chiapas), sayab (Yucatán),[206] violet lancepot, xbalché (colonial-period Mayan)

History
The tree was known to the pre-Columbian Maya, who used it for ritual purposes. It was first mentioned in early colonial sources from the Yucatán (Diego de Landa's *Relación de las cosas de Yucatán*, Mayan dictionaries). In 1665, de Rochefort described one member of the genus for the Antilles (Allen and Allen 1981, 396*). Balche' was first scientifically described in the nineteenth century.

Distribution
The main range is the tropical rain forests of southern Mexico (Chiapas, Yucatán; cf. Steggerda 1943, 209*) and the neighboring Petén region (Guatemala). The tree also grows on Trinidad and Tobago (Graf 1992, 558, 1033*) and has been described as native to Puerto Rico, the Lesser Antilles, Isla Margarita, Venezuela, and Colombia (Morton 1995, 44*).

Several species of the genus grow wild in the forest[207]; they are called *matabuey* in the local Spanish (Berlin et al. 1974, 277*). The very similar species *Lonchocarpus santarosanus* Donn., which is known in Huastec as *ehtiil i thal te'* ("like the *thal* tree"), is found on the Gulf Coast (Alcorn 1984, 691*).

Cultivation
The Lacandon (Chiapas, Mexico) know the tree only in its cultivated form. They propagate the tree primarily through cuttings, which are started from thin branches (approximately 30 cm long) that are set in the ground. The seeds can be germinated and planted as well. The tree requires a moist, tropical climate. It grows very quickly and will even grow back after the trunk or branches have been removed.

Appearance
Balche' is a midsized tree that can attain a height of up to 10 meters in cultivation. It has a smooth, light bark that is often covered with lichens (Rätsch 1994). The leaves are lanceolate. The violet flowers are in panicles. The fruits, which contain only one or two seeds, are flat pods 8 to 12 cm in length. The plant flowers in May, and the fruits ripen in January and February.

The tree is easily confused with the common fish-catcher (*Piscidia piscipula* [L.] Sargent [syn. *Piscidia erythrina* (Loefl.) L.]), which is frequently encountered in the rain forests of Mexico. In Lacandon, the fish-catcher is known as *ya'ax balche'* ("first balche' tree") and is regarded as a very toxic wild relative of the true balche' tree.

Top: The wonderfully scented panicle of the balche' tree, *Lonchocarpus violaceus*. (Photographed in Naha', Chiapas, Mexico)

Bottom: The typical spearlike fruits of *Lonchocarpus violaceus*. (Photographed in Naha', Chiapas, Mexico)

204 This name is now regarded as an old synonym for the species *Gliricidia sepium* (Jacq.) Steud.

205 In Quintana Roo (Mexico), the closely related species *Lonchocarpus castilloi* Standl. is also known as *balche* in Mayan (Cioro 1982, 123*).

206 This Mayan name is also applied to other trees from the Legume Family, including *Gliricidia sepium* (Jacq.) Steud. (Arvigo and Balick 1994, 140).

207 The genus *Lonchocarpus* H.B.K. consists of some 175 species, most of which are found in the tropics of the New World (Allen and Allen 1981, 395*).

Psychoactive Material
— Fresh bark, more rarely the fresh flowers

Preparation and Dosage
The bark is knocked off the freshly cut trunk by beating it with a piece of wood. To prepare the ritual drink that is also known as **balche'**, the Lacandon add two to ten pieces of bark (1 meter long, 10 to 20 cm wide) to approximately 180 liters of water. The more fresh bark that is added to the balche' drink, the more potent are its psychoactive effects. Fresh flowers are sometimes added to the finished balche' drink.

Ritual Use
The drink that is prepared from the bark is consumed communally only in ritual contexts (offering ceremonies, harvest festivals, initiations, the treatment of illnesses) (see **balche'**).

Artifacts
Strangely, the tree has not been identified in pre-Columbian Mayan art. Only drinking scenes, drinking vessels, and ceremonial offerings of the drink were depicted (e.g., in the *Codex Dresdensis*). It is quite possible that some Mayan art was inspired by experiences with balche'.

The Lacandon use a multitude of ritual and drinking vessels, some of which are carved with decorations and other ornaments.

Medicinal Use
Mayan recipe books from the colonial period contain a number of recipes using balche'. Crushed fresh leaves were rubbed onto lesions caused by smallpox, a disease introduced from Europe. A tea made from the leaves was drunk for "loss of speech" (Roys 1976, 216*).

Constituents
The bark and the seeds contain rotenone[208] (which can be clearly recognized by its scent; cf. Morton 1995, 44*) as well as several rotenoides or saponines, flavonoids, and tannins (Delle Monache et al. 1978; Menichini et al. 1982; Neuwinger 1994, 623*). Also present in the bark are the prenylated stilbenes A-, B-, C-, and D-longistyline (Delle Monache et al. 1977; de Smet 1983, 140*). The fruits (pods and seeds) appear to contain the highest concentrations of rotenone (Morton 1995, 44*). It is sometimes thought that an alkaloid is also present. An initial analysis of a balche' drink made with *Lonchocarpus violaceus* did not detect any alkaloids (Hartmut Laatsch, pers. comm.).

The seeds of the closely related *Lonchocarpus sericeus* (Poir.) H.B.K., which are thought to be poisonous, have been found to contain 0.5% enduracidine and a related acid (Fellows et al. 1977). Whether these substances also occur in *Lonchocarpus violaceus* is unknown.

Effects
The longistylines are chemically related to kavains and kavapyrones (cf. *Piper methysticum*) and to hispidine (see *"Polyporus mysticus"*) and likely elicit similar effects. Rotenone is found in a number of plants, including the tuba root (*Derris elliptica* [Sweet] Benth.[209]; Fabaceae); it is regarded as an abortifacient and is thus dangerous to ingest during pregnancy (Roth et al. 1994, 298*). The effects of the balche' drink are likely due to the presence of the longistylines:

Chemically, the structures of the longistylines leave some room for interpretation. Thus, for example, it may be that an aminisation in the body (similar to that of myristicine from nutmeg [cf. *Myristica fragrans*]) into mescaline-like alkaloids is responsible for the effects. But the structural similarity with the styryl pyrones of *Piper methysticum* suggests that the drink is more likely on the same level as kava-kava, which also agrees quite well with the effects that you have described. (Hartmut Laatsch, pers. comm., July 1, 1987)

Rotenone (chemical structure $C_{23}H_{22}O_6$; synonyms derrin, tubatoxin, 1,2,12,12α-tetrahydro-2α-isopropenyl-8,9-dimethoxy[1]-benzopyrano-(3,4-β)furo (2,3H)[1]benzopyran-6(6αH)-on) belongs to the group of pyrano derivatives (Roth et al. 1994, 912) and is regarded as a potent fish poison. In humans, the lethal dose (LD_{50}) is estimated to lie around 0.3 to 0.5 g/kg (912*). The toxic effects are said to be more pronounced when the constituents are inhaled rather than ingested. Symptoms include numbness of the mucous membranes, nausea, vomiting, tremors, tachypnoea, and respiratory paralysis; "rotenone excites the nerves; spasms occur, which result in nerve paralysis, finally leading to exitus" (299, 912*). Rotenone does not appear to be responsible for the psychoactive effects (at least not by itself).

In South America, *Lonchocarpus* species with high rotenone contents (*L. rariflorus, L. floribundus*, known as *timbó* or *barbasco*) are used as fish poisons (im Thurn 1967, 234; Heizer 1958; Schultes and Raffauf 1990, 244f.*). One species (*Lonchocarpus utilis*) is even used as a curare substitute (Schultes and Raffauf 1990, 308*). Rotenone is the primary active component of the common fish-catcher *Piscidia piscipula* (L.) Sargent [syn. *Piscidia erythrina* (Loefl.) L.; Fabaceae] (Roth et al. 1994, 573*), known in Lacandon as *ya'ax ba'che'* ("first balche' tree"):

"Toxic symptoms: Following consumption of too large a quantity of the alcoholic [rotenone] tincture, vomiting, salivation, sweating, numbness, and shaking were observed" (Roth et al. 1994, 573*).

Longistyline

Rotenone

208 Geoffroy first isolated rotenone from *Lonchocarpus nicou* (Aubl.) DC. [syn. *Robinia nicou*] between 1892 and 1895 and presented it under the name *nicouline* (Allen and Allen 1981, 396*).

209 The root of this plant is also used as a substitute for the areca nut (*Areca catechu*)

"And so it shall remain for the creatures of the earth. For always shall it be so for them. They shall receive the seeds of the tree and plant them. Oh, it is true, only when it is planted does the tree grow, the true balche' tree. Not all balche' trees are good, only this one here. With it, the creatures of the earth will not die. They will plant it always. The creatures of the earth shall plant it....

"And so is it until today. They make the balche' drink. They drink it. As my father told me, if the gods had not shown us the proper tree, we would have died. But so, as Hachäkyum the original ancestor showed, so it has remained until today. We make the balche' drink for our lords."

<small>FROM THE LACANDON CREATION MYTH
IN *EIN KOSMOS IM REGENWALD*
[A COSMOS IN THE RAINFOREST]
(MA'AX AND RÄTSCH 1984, 72)</small>

Commercial Forms and Regulations
None

Literature
See also the entry for **balche'**.

Delle Monache, F., F. Marletti, G. B. Marini-Bettolo, J. F. de Mello, and O. Gonçalves de Lima. 1977. Isolation and structure of longistylines, A, B, C, and D, new prenylated stilbenes from *Lonchocarpus violaceus*. *Lloydia* 40:201–8.

Delle Monache, F., L. E. C. Suarez, and G. B. Marini-Bettolo. 1978. Flavonoids from the seeds of six *Lonchocarpus* species. *Phytochemistry* 17:1812–13.

Fellows, Linda E., Robert C. Hider, and Arthur Bell. 1977. 3-[2-amino-2-imidazolin-4(5)-yl]alanine (enduracididine) and 2-[amino-2-imidazolin-4(5)-yl] acetic acid in seeds of *Lonchocarpus sericeus*. *Phytochemistry* 16:1957–59.

Gonçalves de Lima, O., J. F. Mello, Franco Delle Monache, I. L. d'Albuquerque, and G. B. Marini-Bettolo. 1975. Sustancias antimicrobianas de plantas superiores. Communicaçao XLVI: primeras observaçoes sobre os efeitos biologicos de extractos de cortex do caule e raizes de balche.

L. violaceus (syn. *L. longistylus* Pittier) a planta mitica dos maias do Mexico e da Guatemala e Honduras Britanica. *Rev. Inst. Antib. Recife.*

Heizer, Robert F. 1958. Aboriginal fish poisons. *Anthropological Papers*, no. 38, BAE, bull. 151.

im Thurn, Everard F. 1967. *Among the Indians of Guiana.* New York: Dover.

Kamen-Kaye, Dorothy. 1977. Ichthyotoxic plants and the term "Barbasco." *Botanical Museum Leaflets, Harvard University* 25 (2): 71–90.

Ma'ax, K'ayum, and Christian Rätsch. 1984. *Ein Kosmos im Regenwald: Mythen und Visionen der Lakandonen-Indianer.* Cologne: Diederichs. Repr. (revised edition) Munich, 1994.

Menichini, F., F. Delle Monache, and G. B. Marini-Bettolo. 1982. Flavonoides and rotenoides from Thephrosiae and related tribes of Leguminosae. *Planta Medica* 45:2434.

Pool, J. F. 1898. Nekoe, ein indisches Fischgift. *Chemisches Zentralblatt* 1:520.

Rätsch, Christian. 1994. Lichens in northern Lacandon culture. In *Yearbook for Ethnomedicine and the Study of Consciousness* 1994 (2):95–98. Berlin: VWB.

Lophophora williamsii (Lemaire ex Salm-Dyck) Coulter

Peyote Cactus

The flower of the peyote cactus, which the Huichol lovingly call *tútu*. (Botanical illustration from 1888)

Family
Cactaceae (Cactus Family); Subfamily Cactoideae, Cereeae (= Cacteae) Tribe, Echinocacteinae (= Echinocactinae) Subtribe

Forms and Subspecies
The Huichol differentiate among several forms of the peyote cactus on the basis of their graphic structure. One form resembles the flower of a *Solandra* species and is consequently called *kieri*; another is reminiscent of maize (**Zea mays**). Both of these forms are especially useful for the shamans. Another form is referred to as "doors to the other world," while the Native American Church refers to a form with twelve or fourteen segments as "chief." A cactus afflicted by a virus (as sometimes happens) is known as *culebra*, "snake."

One form with deep violet-pink flowers has been described botanically: *Lophophora williamsii* (Lem.) Coult. f. *jordanniana*.

Synonyms
Anhalonium lewinii Hennings
Anhalonium williamsii (Lem.) Lem.
Anhalonium williamsii (Lem.) Rümpler
Ariocarpus williamsii Voss
Echinocactus lewinii (Hennings) K. Schum.
Echinocactus williamsii Lem. ex Salm-Dyck
Lophophora echinata Croizat
Lophophora echinata var. *lutea* Croizat
Lophophora fricii Habermann
Lophophora lewinii (Hennings) Rusby
Lophophora lewinii (Hennings) Thompson
Lophophora lutea Backeb.
Lophophora williamsii var. *decipiens*
Lophophora williamsii var. *lewinii* (Hennings) Cult.
Mammillaria lewinii Karsten
Mammillaria williamsii Coult.

Folk Names
Azee (Navajo), bacánoc, bad seed, beyo (Otomí), biisung (Delaware), biote, biznaga, camaba (Tepehuano), challote, chaute, chiee (Cora), ciguri,

devil's root, diabolic root, divine herb, dry whiskey, dumpling cactus, hicouri, hículi, hikuli, híkuli (Tarahumara), híkuli walula saelíami, híkuli wanamé, hikuri, hikúri, ho (Mescalero), huatari (Cora), hunka (Winnebago), icuri, Indian dope, jícori, jicule (Huichol), jículi, jícuri, jicurite, kamaba, kamba, makan (Omaha), medicine of god, medizin, mescal,[210] mescalito, mezcal buttons, moon, muscale, nezats (Wichita), P, pee-yot (Kickapoo), peiotl, pejori (Opata), pejote, pejuta (Dakota, "medicine"), pellote, peotl, peyote, peyote cactus, peyotekaktus, peyotl, péyotl (Aztec, "root that excites" [?]),[211] peyotle zacatecensis, peyotl-kaktus, piule,[212] raíz diabólica ("devil's root"), rauschgiftkaktus, schnapskopf, seni (Kiowa), señi, tuna de tierra ("earth cactus"), turnip cactus, uocoui, walena (Taos), white mule, wohoki, wokowi (Comanche), xicori

History

In the region of Trans-Pecos, Texas, peyote buttons have been discovered in archaeological contexts approximately six thousand years old (Boyd and Dering 1996, 259*; Furst 1966*). In archaeological excavations in northeastern Mexico, remains of peyote have been recovered that have been dated to approximately 2500 to 3000 B.P. (Adovasio and Fry 1976*; Schultes and Hofmann 1995, 132*). A cave burial in Coahuila (810–1070 C.E.) yielded pieces of peyote that still contained alkaloids (Bruhn et al. 1978).

In Mexico, peyote was already being used as a ritual entheogen during the prehistoric period. In colonial times, Indians were forbidden to use peyote, and the Inquisition punished its use severely (Leonard 1942). The Huichol peyote cult, which has largely survived since the pre-Hispanic period and has been preserved in a relatively pure form, has been well studied (Schaefer and Furst 1996).

The ritual use of the peyote cactus presumably was spread into North America by the Mescalero Apache and apparently also by the Lipan of Mexico (Opler 1938). The first description of the use of peyote in the area that now is part of the United States dates to around 1760. During the American Civil War, peyote use was already established among many tribes of the Plains (Schultes 1970, 30*). The peyote cult is quite widespread in North America; its members can be found in almost every North American tribe. Most, however, are members of the tribes of the Southwest. The number of peyotists has been estimated to be over 250,000 (Evans 1989, 20). During the past one hundred years, the North American Indian peyote cult has been very well documented and the subject of numerous studies (Gerber 1980; Goggin 1938; Gusinde 1939; Hayes 1940; La Barre 1989; Opler 1940; Slotkin 1956; Stewart 1987; Wagner 1932).

The taxonomic history of this magical plant is rather confusing and reflects the scientific vanity often found among botanists (Schultes 1937a, 1937b; 1970, 32*). The first botanically accurate description comes from Hernández (1615), who used the name *peyote zacatecensis* (Schultes 1970, 30*). He also described a *peyote xochimilcensis*, which is likely *Cacalia cordifolia* (cf. **Calea zacatechichi**), an herbaceous plant that is still sold in Jalisco today as an aphrodisiac (Schultes 1966, 296*). The cactus was first described in terms of modern botanical taxonomy in the mid-nineteenth century by the French botanist and cactus aficionado Antoine Charles Lemaire (1800–1871), who gave it the name *Echinocactus williamsii* (1839). Later, the German botanist Paul Christoph Hennings (1841–1908) also described the cactus, which he named *Anhalonium lewinii* in honor of Louis Lewin (1850–1929) (Hennings 1888). It was long thought that these two names referred to different species or at least subspecies or varieties. Today they are regarded as synonyms for *Lophophora williamsii*. In 1894, the North American botanist John Merle Coulter (1851–1928) assigned the cactus to the genus *Lophophora*. The first chemical analyses were published by Louis Lewin (1888) and Arthur Heffter (1894).

Peyote is one of the best studied of all psychoactive plants (Bruhn and Holmstedt 1974). The **mescaline** that was isolated from it at the end of the nineteenth century revolutionized European psychiatry. At the beginning of the twentieth century, peyote became a drug of fashion in artistic and occult circles (Rouhier 1996). Tinctures of peyote were formerly available without restriction (Gartz 1995).

Distribution

The peyote cactus occurs in desert areas from Texas to central Mexico (Tamaulipas, Coahuila, Nuevo León, San Luis Potosí, Zacatecas). In Pecos (Texas), the peyote grows naturally under mesquite trees (*Prosopis juliflora* [SW.] DC.). North of Mexico, the most important natural occurrence of the cactus is in the Mustang Plains of Texas (Morgan 1983a, 1983b).

The peyote cactus (*Lophophora williamsii*), one of the classic plants of the gods, in flower.

210 See *Agave* spp. for a discussion of the confusion surrounding the term *mescal* (cf. also Schultes 1937b).

211 The etymology is not entirely certain; the name *peyotl* may be derived from the Aztec *pepeyoni*, "excite," or *peyona-nic*, "stimulate" (cf. Anderson 1980, 139).

212 Although *piule* appears to be a corruption of *peyote*, in Mexico the word is used as a catchall term for psychoactive plants that are used by the so-called *piuleros*, or traveling diviners.

Cultivation

Propagation occurs chiefly through the seeds. The seeds need only be pressed lightly into sandy or clayey cactus soil and moistened a little each day. Germination may take several weeks. Because the seedlings are very small, they are easily overlooked. Overly heavy watering may wash them away and consequently hamper their growth. The seeds can be planted throughout the year. The cactus does not tolerate any frost.

Peyote requires a porous, clayey soil rich in minerals and nutrients. Apart from this, it is not demanding and will survive even some withering. In the summer, it should be exposed to the sun and watered in moderation, but never kept moist. It requires almost no water in the winter (Hecht 1995, 60*).

After a five-year growing period, a peyote cactus is large enough to be harvested for use. In time, the stump will grow new heads. Peyote is one of the slowest growing of all cacti. There is, however, a gardening technique for accelerating its growth:

> Since peyote grows so slowly, one can quadruple the growth rate by splicing a button to a similar diameter limb of a *T[richocerus] pachanoi* or any other *Trichocereus*. This is done by carefully cutting each surface perfectly flat and smooth before grafting them. Until the graft takes, the peyote button may be held to the surface of the *Trichocereus* by spreading multiple strings with small weights attached to them across the top of the button. A light ring of petroleum jelly should be painted around the cut to prevent desiccation of the contacting surfaces.
>
> In four years the button will be very large. It may then be cut off, re-rooted and returned to the soil. (In DeKorne 1994, 87*)

Appearance

The fleshy, thornless cactus can grow up to 20 cm in height. Although it usually appears as a single head, it can have numerous ribs and take on different shapes. Clusters of fine hairs grow along the ribs. The carrot-shaped root grows from 8 to 11 cm in length. The light pink flowers develop from the center of the head, reaching a diameter of up to 2.2 cm. They fade after a few days. The flowering period is from March to September. The fruit is a club-shaped pink berry that contains the black, coarse, 1 to 1.5 mm long seeds.

Lophophora diffusa (Croiz.) Bravo [syn. *Lophophora echinata* var. *diffusa*] is the only other species in the genus; in Mexican Spanish, it is known as *peyote de Querétaro*, "peyote from Queretaro" (the only place where it occurs) (Diaz 1979, 88*). It contains the alkaloid *O*-methylpellotine (Bruhn and Agurell 1975).

The peyote cactus has been repeatedly confused with other cacti (Schultes 1937b; see table on pages 337 and 338). Most similar is *Turbinicarpus lophophoroides* (Werderm.) Buxb. et Backeb. (Brenneisen and Helmlin 1993, 707). Peyote also can be confused with the African *Euphorbia obesa* Hook.

Psychoactive Material

— Buttons (*Lophophora williamsii* shoot, *Lophophora williamsii* head, *Lophophora williamsii* crown, mescal buttons, peyote buttons)

The potency of the buttons decreases very little even when they are stored for long periods (Schultes 1970, 31*).

Preparation and Dosage

Peyote buttons are the heads of the cactus cut off above the root. The buttons can be either eaten fresh or dried and then chopped or powdered for later use. The fresh or dried buttons can be boiled or decocted in water. Both the buttons and the tea produced from them have an extremely bitter taste.

Dosages vary considerably, both between individuals and in rituals. Dosages ranging from four to thirty buttons may be ingested (Schultes 1970, 33*). Among the Kiowa, ten to twelve buttons are usually consumed per person per night (Havard 1896, 39*). Strong psychedelic effects and visions appear only when an amount corresponding to 200 to 500 mg **mescaline** is ingested (Ott 1996). Approximately 27 g dry weight corresponds to some 300 mg mescaline (DeKorne 1994, 88*). Antonin Artaud (1896–1948) made an interesting observation about dosages among the Tarahumara (1936):

> It is with the peyotl as it is with all that is human. It is a wonderful magnetic and alchemical principle as long as one knows how it should be taken, that is, in the prescribed and then eventually larger dosages. . . . Anyone who has *truly* drunk ciguri [= peyote], the proper amount of ciguri, of the PEOPLE, not of the undetermined GHOST, he knows how the

Left: This peyote cactus has been grafted onto a stalk of *Trichocereus pachanoi*.

Right: The African *Euphorbia obesa* is a perfect example of plant symmetry. This spurge is sometimes mistaken for the true peyote cactus.

things are made, and can no longer lose his mind, because God is in his nerves and controls him from there. But ciguri drinking means in particular not to exceed the dosage, for ciguri is the infinite, and the secret of the therapeutic effects of medicines is tied to the extent to which our organism takes them in. To exceed what is necessary means to DISTURB the effects. (Artaud 1975, 19f.)

The Tarahumara formerly produced an **incense**, "supposedly a resin [of *Pinus* sp.?] to which pieces of pejote had been mixed," for use in their rituals (e.g., the Chumari Dance) (Zabel 1928, 264). Dried peyote pieces may be smoked, both by themselves and in **smoking blends** with other plants. Peyote powder is used as an additive in alcoholic beverages (**beer, chicha, balche'** [?], pulque, and mescal [cf. *Agave* spp.]):

In Narárachi, dried *jíkuri*, i.e., peyote and other cactus species [*Ariocarpus fissuratus*, *Cory-phantha* spp., *Echinocereus* spp., *Mammillaria* spp., *Pachycereus pecten-aboriginum*], are ground, mixed with water to make *tesgüino* [maize beer], and drunk; it was also formerly added as an additive to *tesgüino* so that it would be "more enjoyable." (Deimel 1980, 78)

A *tinctura de peyotl* can be produced using 50 g of dried ground peyote. The powder is first moistened with a little water and then 100 ml of a spirit with a high percentage of **alcohol** is added. The mixture is allowed to stand in a sealed vessel for two days. The extract is then filtered. A medicinal dosage (e.g., for heart ailments) consists of thirty drops taken three times a day (Deimel 1986, 87).

The homeopathic mother tincture is obtained from fresh peyote buttons with **alcohol**:

Manufacture [of Lophophora williamsii hom. *HPUS78*]: To prepare 1000 ml of the mother tincture, add 754 ml of 94.9% alcohol (*V/V*) to a moist plant mass consisting of 50 g solid material and 283 ml of supplemented plant juice. . . . Medicinal content 1/20. (Brenneisen and Helmlin 1993, 711)

Approximately 3¹/₂ tablespoons of the mother tincture will produce psychedelic effects.

Ritual Use

The ritual use of peyote has its roots deep in prehistoric times and can be demonstrated for both Pecos (Texas) and Mexico (Furst 1965, 1996*). The Aztecs were well acquainted with the cactus and its effects. In the chapter "in which the names of the many plants that make one confused, crazy are named" (Sahagun), the following is said about peyote:

This peyote is white and grows only in the northern region known as Mictlan. It exerts an effect like a mushroom upon those who eat or drink of it. That person will also see many things that make him afraid or cause him to laugh. It influences one for perhaps a day, perhaps two days, but then it leaves in the very same way. And yet it still causes harm to one, stirs one up, inebriates one, exerts an influence upon one. I take peyote, I am stirred up. (Sahagun, Florentine Codex 11:7*)

In his *Natural History of New Spain* (1615), the Spanish physician Francisco Hernández wrote the following about peyote, which he called *peyotl zacatecensis:*

This root is attributed with wondrous properties, if you can believe the things that are said about it. Those who take it receive the divine gift of clairvoyance and are able to know future things, like prophets. . . . The Chichimec believe that the power of this root makes it possible.

The Aztecs generally referred to the nomadic tribes of the north as the Chichimec. Among these peoples were most likely the ancestors of the Huichol, who now live in the Sierra Madre, and the Tarahumara, from the high north. Both the Huichol and the Tarahumara (along with other cultures, such as the Yaqui and the Cora[213]) have preserved the peyote cult, which can be traced back to prehistoric times (Benzi 1972; Deimel 1980; Lumholtz 1902; Rouhier 1927). As a result of contact with the Spanish conquerors, however, some Catholic elements have made their way into the Huichol ceremonies (Zingg 1982).

The desert region the Aztecs called *mictlan*, "realm of the dead," is the Huichol paradise that they know as *wirikuta*. In the Third Song of the Peyote, it is said, "In Wirikuta grows a flower that speaks, and you understand it" (Benítez 1975, 78).

Once a year, the *mara'akame*, the Huichol shamans, make a pilgrimage to *wirikuta*, traveling to their mythic origin and to the venerated peyote grounds. They undertake this journey "to find their lives" and to hunt the peyote they require for this purpose. For them, the peyote is fundamentally feminine. The peyote cactus is both the origin and the center of the universe. In the cosmology of the Huichol, it is symbolized as Grandfather Fire (= the sun), as a deer, and as a maize plant. This trinity reflects the foundation (diversity and integrity) of the Huichol culture: Gathering (peyote), hunting (deer), and agriculture (maize) form an inseparable mythical and practical unity (cf. Hell 1988).

The journey to *wirikuta* is a journey to the origins of the world and of culture. Here in

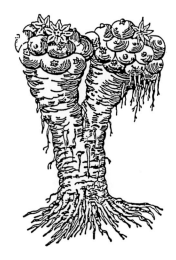

An early colonial illustration of the Mexican peyote cactus (*Lophophora williamsii*), showing the root and flowers.

213 One seventeenth-century source (Ortego, *Historia del Nayarit*) provides a description of the peyote dance festival of the Cora (cf. Schultes 1938; Schultes and Hofmann 1987, 134).

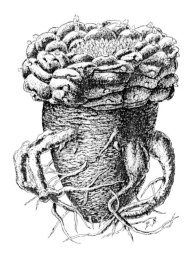

The earliest botanical illustration of the peyote cactus, included in the article "Eine giftige Kaktee, *Anhalonium Lewinii* n. sp." (Henning 1888)

The peyote altar in the form of a thunderbird. The flute at the top represents the beak, under which is Father Peyote, placed precisely in the middle of the half moon, symbolizing the peyote way. The long fire for lighting cigarettes (at the bottom) forms the tail of the thunderbird.

214 Dream catchers are small, flat, woven objects resembling spiderwebs. It is believed that they capture dreams in their webs and can thus be associated with a particular place.

wirikuta, the gods created the world in all its manifestations; here is where time began, with all its consequences; here is where the peyote grows, which keeps alive the memory of creation by leading humans back to the source of all being, thereby enabling them to take part in the divine. Following the ritual preparations, during which the shamans consecrate the peyote fields with a *muvieri* ("fan," a wooden staff with feathers), the peyote cactus is hunted with a small bow and arrows. The cactus is the blue deer of creation, and when the "flesh that has been hunted" is eaten, it opens the pilgrims' eyes to the beginning of the world. The pilgrims travel together to the source and return from it like newborns. They collect large numbers of peyote buttons that they take back home with them (Myerhoff 1980). They need the peyote buttons for festivals and shamanic journeys, and to treat the sick, most of whom will take peyote either by themselves while under the supervision of the shaman or together with him or her (Berlin 1978).

At the great peyote festivals (*híkuri neirra*), all Huichol, whether young or old—including even the elderly, small children (!), and pregnant women—ingest the sacred cactus (Haan 1988, 128 ff.; Schaeffer 1997). While they consume the peyote, they also smoke great quantities of a **smoking blend** made of *Nicotiana rustica* and *Tagetes lucida* (see *Tagetes* **spp.**) (Schaefer and Furst 1996, 154).

The Tarahumara, who live in northern Mexico, once had an extensive peyote cult that apparently has disappeared in recent years (Deimel 1996). Peyote was consumed communally, especially at the tribal festivals (dances) (Artaud 1975; Zabel 1928). The legendary long-distance runners of the Tarahumara used peyote or *chilitos* (fruits of *Mammillaria* **spp.**, *Epithelantha micromeris*) as a (ritual?) doping agent. In former times, the Tarahumara used peyote for divination, diagnosis, and treating the ill and as a protection from harmful magic, thieves, and enemies (Deimel 1980, 80 ff.). The use of peyote among the Yaqui Indians of Sonora and Arizona was similar (e.g., in the Pascola or Deer Dance). The uses of peyote and the visions of mescalito described in Carlos

Huichol Peyote Symbolism

Peyote

Deer   Maize

Peyote is food for the soul.
Deer and maize are food for the body.

Castaneda's books (1973, 1975) have no connection to the Yaqui tradition, and anthropologists have raised serious doubts about their veracity (cf. Fikes 1993; La Barre 1989, 271–75, 307 f.).

The Blackfeet, who did not become acquainted with the peyote cactus until the twentieth century, used it to support their vision quests (Johnston 1970*).

In North America, the most important use of peyote is as a sacrament in peyote meetings—ritual circles—of the Native American Church. The history and ritual forms of the Native American Church have been very well documented (Brave Bird 1993; Dustin 1962; Smith and Snake 1996; Wagner 1932). Today, most people call the Native American Church the peyote religion or the peyote way (Aberle 1991; Gerber 1980; Stewart 1987). The peyote religion, which transcends individual tribes, is a syncretic form of organized spirituality that combines elements of Indian traditions and ideas from Christianity (Steimmetz 1990).

North American peyotists offer so-called dream catchers before gathering the buttons in Texas peyote fields.214 The peyote ritual itself begins with the smoking of cigarettes made with homemade tobacco (preferred brand: Bull Durham; cf. *Nicotiana tabacum*) or sometimes with smooth sumac leaves (*Rhus glabra* L.; cf. **kinnikinnick**) that have been rolled in cigarette papers or corn husks (Schultes 1937a, 138 f.). The branch tips of red cedar (*Juniperus virginiana* L.; cf. *Juniperus recurva*) are the preferred ritual **incense** at peyote meetings. A variety of *Artemisia* species (see **Artemisia spp.**) may also be used for fumigation (Schultes 1937a). The Kiowa and neighboring tribes often wore necklaces of mescal beans (*Sophora secundiflora*) at the peyote rituals (Schultes 1937a, 148), a practice that has led some researchers to suggest that the North American peyote cult may have grown out of a prehistoric psychoactive mescal bean cult (cf. La Barre 1957, 1989).

The typical peyote meeting usually takes place in a tepee, less often in a hogan (Navajo round building), but always at night. The tepee symbolizes the entire universe; in its middle burns a fire, which is surrounded by the circle of people. In front of the fire, which must be maintained throughout the night, an altar in the shape of a half-moon is built. Upon this altar are placed the ritual objects: a talking stick, a gourd rattle, drums, and flutes. The talking stick symbolizes the connection between heaven and earth, between the Great Spirit and humans; the rattle or rattles are seen as direct prayers to God; and the drum is the heart and the drumbeat the heartbeat. A large, living peyote cactus is placed between the fire and the altar. This is addressed as Chief Peyote or

Grandfather Peyote and is venerated as an incarnation of the Great Spirit. The participants (men, women, young people, children) sit in a circle around the fire and the altar. The leader of the ritual, who is usually known as a roadman (Brito 1989), first smokes and purifies the tepee with **incense**, offers prayers, may read something from the Bible, and sings songs. He distributes the fresh peyote buttons, the powder, or the tea. When possible, each participant determines his or her own dosage. The participants should not speak to one another and should not leave the circle until the following morning. They may step outside only if the roadman has given them permission to do so. As the cactus manifests its effects, the talking stick, rattles, and drums are passed around in a clockwise direction. Whoever holds the talking stick sings his own peyote songs. He accompanies himself with the rattle or the drum. Sometimes a special drummer accompanies the participants. The objects go around several times, interspersed with periods of silence. From time to time, more peyote may be ingested, incense burned, and water drunk. Sometimes, tobacco is also smoked and fanned with the peyote fan. The Sioux medicine man Leonard Crow Dog has described the meaning and purpose of the ritual:

> Grandfather Peyote unites us all in love, but first he must separate us, cut us off from the outer world in order to bring us to look into ourselves. . . . A new understanding dawns within you—joyful and hot like the fire, or bitter like the peyote. . . . You will see people that bend themselves into a ball, as if they were still in the belly of their mother, remember things that happened before you were born. Time and space grow and shrink in an inexplicable manner—an entire lifetime of being, learning, understanding, compressed into a few seconds of insight, or time stands still, does not move at all, a minute becomes an entire lifetime. (Lame Deer and Erdoes 1979, 247 f.)

The peyote meetings do not take place according to any calendrical schedule but rather at agreed-upon times.

Westerners have drawn upon the North American peyote meeting as a model for modern ritual circles in which a variety of psychoactive substances (MDMA [cf. **herbal ecstasy**]; *Psilocybe cyanescens, Psilocybe semilanceata*, LSD [cf. **ergot alkaloids**]) may be used (Müller-Ebeling and Rätsch n.d.; Rätsch 1995d).

Artifacts

Numerous artifacts, especially several clay figurines, discovered in western Mexico (Colima, Nayarit, Guerrero) are clearly related to the pre-Columbian shamanic peyote cult. Many clay figurines from shaft-and-chamber tombs (200 B.C.E. to 500 C.E.) in northwestern Mexico depict shamans in ritual positions and gestures (Furst 1965, 1969). One of these figurines shows a man putting a peyote-like object into his mouth (Gerard 1986, fig. 78). A ceramic object from Colima (first century C.E.) shows a child holding a peyote cactus in each hand (cf. Guerra 1990, 125*). A vessel decorated with plants, including what are clearly ten peyote roots, was recovered in the same area (Gerard 1986, fig. 88; a similar piece can be seen in Kan et al. 1989, 163). A ceramic vessel (ca. fifth century B.C.E.) found in Monte Albán (Oaxaca) apparently was used as a pipe for ingesting **snuff**; it represents a little deer with a peyote button in its mouth (Furst 1976b, 155*; Schultes and Hofmann 1992, 133*).

In Aztec poetry, peyote cacti are metaphorically referred to as "flowers" (Brinton 1887*), and their praises are sung often (Quezada 1989*). In Aztec magical formulas, the cactus is invoked as the "green woman" (Ruiz de Alarcón 1984*).

Most of the modern artifacts that have been inspired by peyote experiences or allude to the cactus and its magical effects are from the Huichol Indians of Mexico.

After the peyote festival, and after peyote experiences in general, the Huichol manufacture offering gifts as thanks for the visions they received and the healings those visions brought. The gifts are offered at mountain shrines (piles of rocks), at sacred places, on the "tree of the wind" (*Solandra* spp.), et cetera. The most important offerings are *nierika* (votive offerings), *rukuri* (prayer cups, votive snakes), and *tsikuri* ("god's eyes"). The word *nierika* (= *nearika*) refers to "the entrance to the other world, where after the dark passage it becomes light again" (Bollhardt 1985, 30). *Nierika* originally were simple disks of wood with a hole (shields) or carved wooden objects, e.g., animals (snakes) or humans (such as the earth goddess Nakawé or the peyote goddess Wuili Uvi), covered with glue and decorated with colorful threads (wool yarn) or glass beads (Lumholtz 1989). The most common ornamentations are representations of peyote in various

Bottom left: Representation of peyote cacti, arranged like the four cardinal points. (Huichol yarn painting)

Top right: A wooden Huichol mask, decorated with colored woolen yarn. The mouth is a peyote cactus. Beneath the eyes are peyote symbols, which are worn as face paintings during the peyote hunt and at tribal festivals.

Bottom right: Representation of a *peyotero* or peyote pilgrim. (Huichol yarn painting)

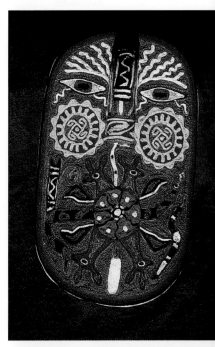

The peyote bird has become the symbol of the peyote religion or cult. Its appearance is taken from the water bird *Anhinga anhinga*. The peyote bird is regarded as the bringer of visions.

degrees of abstraction or symbols of the peyote cactus (stylized deer, maize plants, birds, et cetera). Like the visions (which because of their personal nature should not be discussed), the *nierika* often were very personal and were utilized only as votive offerings.

> In their ritual importance, nierikas are faces as well as mirrors with two sides. They represent the external appearance of humans, things, or even elements. The hole that every nierika has in its center symbolizes a magical eye through which gods and humans look at one another. It also enables one to see things from a great distance. Nierikas can be found in temples, caves, and at springs. Like all other articles of offering, these too are carried to wirikuta. (Haan 1988, 162)

The *rukuri* are made from the lower portion of gourds and are decorated in the same very colorful manner as the *nierika*. The *tsikuri* are yarn crosses that symbolize mystical or visionary sight and are used as magical protection for the house and farm, for example to ward off the "evil eye" (Haan 1988, 160 ff.). Similar offerings to the peyote deity were being made during the colonial period, as Ruiz de Alarcón has described (1984, 50*).

Among the Huichol, these offering goods have developed into a highly unusual art form. Visions perceived while under the influence of peyote are represented by colorful woolen yarn glued onto large, flat pieces of wood (Berrin 1978; Straatman 1988; Valadez 1992). These yarn paintings, which are now sold throughout the world under the name *nierika*, can be characterized as true psychedelic art, for they are the products of psychedelic experiences. "The yarn paintings reproduce the visions in both color and idea. Looking at them takes one into the other world" (Müller-Ebeling 1986, 290*). According to the Huichol, there are two types of yarn paintings being produced today: those that "are produced for purely commercial reasons and have only a decorative value, and those that express the personal spiritual experience of the invisible made visible and are of profound magico-religious significance" (Haan 1988, 169).

The Huichol also manufacture a variety of jewelry pieces (armbands, necklaces, chains for amulet bags, earrings) from woven yarn or glass beads. These pieces are largely and sometimes exclusively decorated with more or less abstract representations or symbols (deer) of peyote. The Huichol also weave images of peyote in varying degrees of abstraction into their festive and ritual clothing (Schaefer 1989, 1993a, 1993b).

The silver jewelry of the tribes of the Southwest shows the peyote bird in a variety of forms, styles, and interpretations. This long-necked bird is known as *anhinga* or, in the regional English, as snake bird, water bird, or water turkey. It is a cormorant-like bird (*Anhinga anhinga*; Anhingidae) from the Gulf Coast whose feathers are very desired for the production of peyote fans (Bahti 1974, 61), which apparently are used to induce hallucinations or visions. The first peyote birds appeared in Navajo silver work around 1940 (Bahti 1974, 61).

A great deal of paraphernalia is used in the North American peyote cult, including staffs, fans, rattles, and drums. Over the years, a particular, recognizable (i.e., standardized) style has evolved in the production of these artistic objects that immediately reveals their association with the peyote cult (Wiedman 1985).

Toward the end of the nineteenth century, North American Indians began to paint pictures of peyote ceremonies and the visions they experienced there. One of the first to do so was the Kiowa Silverhorn, also known as Haungooah (Wiedman and Greene 1988). The peyote cult is an important theme in the paintings of many modern Indian artists (e.g., Archie Blackowl, *Peyote Mother*; Woody Crumbo, *Religious Peyote Ceremony*; Jerry Ingram, *Peyote Dream*; Al Momaday, *The Peyote Dreamers*; Alfred Whiteman, *Peyote Chief*).

Peyote and experiences with peyote have been worked into numerous comic stories. Magical rituals with the peyote cactus were featured in the issue "Jikuri" (no. 3 in the series *Julian B.*) by Plessix and Dieter (1993). Jeronaton used an Aztec story as the basis of his comic *Im Reich Peyotls* [In the Kingdom of Peyotl] (1982). Gilbert Shelton (born 1940) has his Freak Brothers encounter Carlos Castaneda's Don Juan in the parody *Die mexikanische Odyssee* [The Mexican Odyssey] (Berlin: Rotbuch Verlag, 1990). The Freak Brothers take peyote with Don Juan and see Mescalito, the peyote spirit. In Shelton's comic short "Ein Indianer kommt zu Essen" ["An Indian Comes to Dinner"] (in *U-Comix* no. 45 (1984): 58-59), an Indian visits an average American family. The television is broken—the picture is only in black and white—and needs repair. The Indian pulls a couple of peyote buttons from his pocket. After he and the two white people have eaten four buttons each, the pictures appear in color once more. The whites, filled with enthusiasm, decide to quit their jobs.

The picture story *Im Reiche des Mescal* [In the Kingdom of Mescal] describes an Indian boy's first experience in the visionary world of peyote (Schäfer and Cuz 1968).

The Sirens of Titan (1959), the lovingly sarcastic novel of the future by American cult author Kurt Vonnegut, appears to have been influenced by peyote experiences. In the book, the hero repeatedly states that he thinks he is on a peyote trip. Richard Wilson wrote a science-fiction book about peyote, *Der Sonnentanz* [The Sun Dance], in which one can read, "Stick to the peyote" (Wilson 1981, 10). The Beat poet Allen Ginsberg wrote numerous poems while under the influence of peyote and referred to the cactus itself in some of them: "Peyote is certainly one of the world's great drugs" (Ginsberg 1982, 43). Peyote generally appears to have been one of the more important sources of inspiration for the poets of the Beat generation (e.g., Michael McClure, Ken Kesey). The best-selling Indian author Natachee Scott Momaday (born 1934) incorporated a peyote experience into his novel *House Made of Dawn* (Momaday 1968). The peyote cactus also peers from between the lines in some of his other works (Momaday 1969).

Peyote has also provided a great deal of musical inspiration (see the box on page 334). Mexican Indians have traditionally used and passed down tribal and healing peyote songs. And singing plays a central role in the modern North American peyote cult. Songs are the primary means by which participants communicate with one another during the ceremonies so that they can bond together and open themselves to the world of visions. These ritual peyote songs are not folk songs in our sense of the phrase but are individual works of art that serve to control altered states of consciousness. Many Indians who follow the peyote road say that the cactus teaches each person his or her own songs. These songs have power and function as "medicine." Only rarely are the songs composed of text (words); instead, the songs usually consist of a series of sounds and notes to which certain meanings are attributed (Merriam and d'Azevedo 1957). In most cases, the songs are accompanied by drums and/or rattles. The monotonous rhythm usually has about two hundred beats per minute.[215] A comparative ethnomusical study containing transcriptions of numerous peyote songs was produced in the late 1940s (McAllester 1949).

Peyote has appeared in the music world many times, often in very different forms. The contemporary composer Györgi Ligeti (born 1923) wrote the opera *Le Grand Macabre* during the years 1974 to 1977. The main role is sung by a female character named Mescalina. Unfortunately, the composer has not let it be known whether he was influenced by the soul of the cactus.

The Seattle grunge band Pearl Jam, the most successful "alternative music" group of the mid-1990s, took its unusual name from the jam that lead singer Eddie Vedder's grandmother (Pearl) used to make from peyote and other magical plants. During the mid-1990s, a psychedelic band named The Wild Peyotes appeared in San Francisco. The country rock singer Calvin Russell issued an album entitled *Dream of the Dog*, the cover of which is decorated with a picture of

peyote in the style of the Huichol. The singer, who is of Indian descent and whose grandmother was a Comanche peyote woman, makes direct reference to the Indian "medicine." Other musicians and bands have also used Huichol peyote images as cover art, both to demonstrate their ties to Indian culture and spirituality and to give expression to their own psychedelic experiences (e.g., Santana, *Shangó*, 1982 CBS).

Medicinal Use

To the Indians, peyote is *the* medicine per se, a kind of all-purpose remedy, especially for the body and spirit (Crow Dog 1993). The Indian uses of the cactus for medicinal purposes are correspondingly numerous (Anderson 1996a; Deimel 1985; Schultes 1938):

> Without a doubt, peyote is the most important medicine now being used by the North American Indians. It has replaced the older, less spectacular plant medicines [e.g., *Sophora secundiflora*]. It is used in daily life as a home remedy. And all peyote ceremonies, whether by Mexican or by North American Indians, are definitively healing rituals in which the sick are given high dosages of the cactus. (Schultes 1938)

The cactus was already being used for medicinal purposes during Aztec times: "Peyotl. It is a fever medicine. It is eaten, it is moderately drunk, only some" (Sahagun, Florentine Codex 11:5, 30*). Decoctions of peyote were administered as **enemas** to treat high fever, a method that is still in use among the Huichol today. They utilize a cold-water decoction made from dried, powdered peyote.

The Kickapoo Indians of northern Mexico use freshly cut slices of peyote to treat headaches or sunstroke. They use a linen cloth to tie the cactus slice to the head of the afflicted person. A decoction of peyote is drunk for arthritis (Latorre and Latorre 1977, 350*). In Mexico, peyote also is used to treat overdoses of toloache (*Datura innoxia*) (Nadler 1991, 95*). It also finds use in Mexican folk medicine. The local Spanish term *empeyotizarse* means "to treat oneself with peyote."

The Native American Church has used peyote with success to treat alcoholism (cf. **alcohol**) (Albaugh and Anderson 1974; Pascarosa and Futterman 1976).

In homeopathy, Anhalonium (Anhalonium lewinii hom. *HAB34*, Lophophora williamsii hom. *HPUS78*) is used in the dilutions D3 to 6 to treat depression and other disorders (Boericke 1992, 62 f.*; Brenneisen and Helmlin 1993, 710 f.):

> In homeopathic therapy, it has been in use for several decades. . . . In addition to circulatory

"The gods were already present in the head of the *mara'akame* when he took the peyote, but the drug caused their voices to appear to sing down directly from heaven."

RONALD SIEGEL
HALLUCINATIONS
(1995b, 39*)

"The peyotl leads the self back to its true source. When someone has experienced such a visionary state, it is impossible for one to confuse the lie with the truth as before."

ANTONIN ARTAUD
DIE TARAHUMARAS [THE TARAHUMARAS]
(1975, 28)

"This small, wooly rascal turns on like a lamp. Daddy Peyote is the plant representation of the sun."

NATACHEE SCOTT MOMADAY
HOUSE MADE OF DAWN
(1968)

Mescaline

---

215 Rhythms of 200 to 220 bpm are regarded as trance-inducing (Goodman 1992*). Modern techno music (cf. **herbal ecstasy**) purposefully uses this range to produce altered and extraordinary states of consciousness (Rätsch 1995d, 316 ff.*).

Discography: Peyote Music

Indian Music

Denny, Bill, Jr.
  Intertribal Peyote Chants (Canyon Records, 1984)
Guy & Allen
  Peyote Canyon (Soar Sound of America Records, 1991)
  Peyote Brothers (Soar Sound of America Records, 1993)
  Peyote Strength (Soar Sound of America Records, 1994)
Mother Eagle Kaili
  Huichol Sacred Music/Musica y Canto Ceremonial Huichol (Paraiso, 1995)
Nez, Billie
  Peyote Songs from Navaholand (Soar Sound of America Records, 1992; Spalax Music, 1993)
Primeaux and Mike
  Walk in Beauty: Healing Songs of the Native American Church (Canyon Records, 1995)
Primeaux, Mike, and Attson
  Healing and Peyote Songs in Sioux and Navajo (Canyon Records, 1994)
Turtle, Grover, and Sam Sweezy
  32 Cheyenne Peyote Songs (Indian Records, 1979)
Various Singers
  Cheyenne Peyote Songs (Indian House, 1975)
  Indiens Yaquis: Musique et dances rituelles (Arion, 1978)
  The Kiowa Peyote Meeting (Ethnic Folkways Records, 1973)
  Musical Atlas: Mexico (EMI Records, 1982)
  Music of the Plains: Apache (Asch Records, 1969)
  Musiques Mexicaines (Ocora Disques, n.d.)
  Navajo Peyote Ceremonial Songs, vol. 1 (Indian House, 1981)
  Peyote Songs from Rocky Boy (Montana), vols. 1–3 (Canyon Records, 1978 ff.)
  Yankton Sioux Peyote Songs (Indian House, 1976)

Non-Indian Music

Ligeti, György
  Le Grand Macabre (Wergo Schallplatten, 1991)
Peyote
  Alcatraz/I Will Fight No More (RundS Records, ca. 1994)
Russell, Calvin
  Dream of the Dog (SPV Recordings, 1995)

The peyote cult, which has spread like wildfire among the Indians of North America, has also generated numerous recordings of peyote songs. This movement can be seen as a Western, high-technology manifestation of Indian medicine. (CD cover, 1994, Soar Sound of America Records)

problems, the following therapeutic indications are given for Anhalonium: clouded consciousness, headaches, migraines, hallucinations, sleeplessness, emotional disease states, weak nerves, and brain exhaustion. . . . Yet how modest these areas of application are in contrast to the transdimensional and exotic otherworldliness of the peyotl inebriation. A little bottle of the agent is a part of that power that can take the earth out of its course and lift our spirit to the heavens, and also a part of that guiding hand without which life and law are impossible. (Gäbler 1965, 199, 204 f.)

Constituents

To date, more than fifty alkaloids have been isolated from the peyote cactus and described. A fresh cactus that has been well watered has a total alkaloid content of approximately 0.4%, while specimens that live under extremely dry conditions have as much as 2.74% and dried buttons up to 3.7% (Brenneisen and Helmlin 1993, 708 f.). The alkaloid concentration can exhibit considerable variation (Todd 1969).

In addition to **mescaline**, the primary alkaloid, peyote also contains the β-**phenethylamines** tyramine, *N*-methyltyramine, hordenine, candicine, anhalamine, lophophorine, pellotine, *O*-methylpellotine, *N,N*-dimethyl-3-methoxytyramine, dopamine, epinine, 3-methoxytyramine, *N*-methylmescaline, *N*-formylmescaline, *N*-acetylmescaline, *N*-formylanhalamine, *N*-acetylanahalamine, isoanhalamine, anhalinine, anhalidine, anhalotine, isoanhalidine, anhalonidine, and various derivatives (Mata and McLaughlin 1982, 105 f.*). Apparently only mescaline has definite psychoactive effects (McLaughlin and Paul 1966). In spite of the popular misconception, the cactus's hairs do *not* contain **strychine**.

Lophophora diffusa contains pellotine, lophophorine, anhalamine, anhalonidine, **mescaline**, and *O*-methylpellotine (Mata and McLaughlin 1982, 105 f.*).

Effects

Depending upon the dosage used, peyote can have healing, aphrodisiac, or psychedelic/visionary effects. The psychedelic effects typically begin some 45 to 120 minutes after ingestion. Nausea and vomiting commonly occur prior to the onset of the visions; for this reason it is said that with peyote, the hangover comes before the effects. The psychedelic effects last for six to nine hours. Aftereffects are rare; there are a few reports of headaches the following day. The visionary world opened by peyote is very similar to the "other reality" induced by **psilocybin**, *Psilocybe* spp., LSD (cf. **ergot alkaloids**), and **mescaline**.

Among the Huichol, the visions peyote induces appear to exhibit many constants and are all of a mystical nature (Myerhoff 1975). The late renowned Huichol shaman Ramón Media Silva gave a very precise description of the effects of peyote and the special nature of the effects upon shamans:

The first time one puts the peyote into one's mouth, one feels it going down into the stomach. It feels very cold, like ice. And the inside of one's mouth becomes dry, very dry. And then it becomes wet, very wet. One has much saliva then. And then, a while later one feels as if one were fainting. The body begins to feel weak. It begins to feel faint. And one begins to yawn, to feel very tired. And after a while one feels very light. The whole body begins to feel light, without sleep, without anything.

And then, when one takes enough of this,

one looks upward and what does one see? One sees darkness. Only darkness. It is very dark, very black. And one feels drunk with the peyote. And when one looks up again it is total darkness except for a little bit of light, a tiny bit of light, brilliant yellow. It comes there, a brilliant yellow. And one looks into the fire. One sits there, looking into the fire which is Tatewarí. One sees the fire in colors, very many colors, five colors, different colors. The flames divide—it is all brilliant, very brilliant and very beautiful. The beauty is very great, very great. It is a beauty such as one never sees without the peyote. The flames come up, they shoot up, and each flame divides into those colors and each color is multicolored—blue, green, yellow, all those colors. The yellow appears on the tip of the flames as the flame shoots upward. And on the tips you can see little sparks in many colors coming out. And the smoke which rises from the fire, it also looks more and more yellow, more and more brilliant.

Then one sees the fire, very bright, one sees the offerings there, many arrows with feathers and they are full of color, shimmering, shimmering. That is what one sees.

But the mara'akame [= shaman], what does he see? He sees Tatewarí, if he is chief of those who go to hunt the peyote. And he sees the Sun. He sees the mara'akame venerating the fire and he hears those prayers, like music. He hears praying and singing.

All this is necessary to understand, to comprehend, to have one's life. This we must do so that we can see what Tatewarí lets go from his heart for us. One goes understanding all that which Tatewarí has given one. That is when we understand all that, when we find our life over there. But many do not take good care. That is why they know nothing. That is why they do not understand anything. One must be attentive so that one understands that which is the Fire and the Sun. That is why one sits like that, to listen and see all of that, to understand. (Myerhoff 1974, 219–220)

For many North American Indians, it is particularly important that they receive a vision through the peyote that will provide them with direction for their lives (Anderson 1996b, 92 ff.). Navajos have reported that their dream experiences are disturbed by frequent peyote use (Dittmann and Moore 1957). There are many reports of mystical experiences, which are valued even more highly than visions:

Peyotists seldom have visions and regard them as mere distractions from what is important.

It can perhaps be said that mystical experience consists in the harmony of all immediate experience with that which the individual regards as the highest good. Peyote has the remarkable ability to provide a person with a mystical experience of unlimited duration. (Tedlock and Tedlock 1978, 107)

Western peyote users have compared the visionary or psychedelic effects of peyote with the higher yogic states of consciousness described in the literature of India (e.g., James 1964) but have also disparaged it as "drunkenness" or characterized it as an artificial psychosis (cf. **mescaline**). It has repeatedly been reported that the psychoactive effects of peyote are much more significant, more profound, and more spiritual when they are experienced as part of an Indian ritual (Ammon and Patterson 1971).

An extract of the cactus has antibiotic properties (McCleary et al. 1960). Huichol women claim that peyote stimulates lactation.

Commercial Forms and Regulations

During the colonial period (seventeenth century), the Inquisition forbade the Indians of Mexico to use peyote for ritual and religious purposes and threatened them with severe penalties if they were caught using it (Leonard 1942). The Indians of North America were also long forbidden to use peyote (Bullis 1990; Camino 1992). In the United States, the members of the Native American Church have been allowed to use peyote for religious and sacramental purposes since 1995. They are permitted to collect the cactus in the peyote fields of Texas and to buy peyote powder in prepackaged tea bags. The cactus, which is becoming increasingly rare, is covered by the Washington treaty on endangered species (Deimel 1996, 24).

In Germany, live peyote cacti are not covered by the current drug laws, so it is permitted to buy and sell the cactus (Körner 1994*). But the main active constituent, **mescaline** (as well as preparations containing mescaline), is included among the "narcotic drugs for which trafficking is not allowed" in Appendix I of the German drug laws. If the cactus was not subject to the drug laws because of its mescaline content, then it would be a medicine requiring a prescription (Brenneisen and Helmlin 1993, 710). Only those homeopathic preparations diluted at D4 or greater are allowed to be sold. Recently, the mother tincture has become available through pharmacies, but it is difficult to obtain.

Nevertheless, living peyote plants do occasionally make their way into cactus and flower shops (sometimes under the name *living rocks*). Seeds are available from ethnobotanical specialty sources.

This richly illustrated children's book, *Im Reiche des Mescál* [In the Realm of Mescál], is based upon an Indian story in which a child has a peyote experience that teaches him about the other reality, the "realm of mescál." (Title page, Synthesis Verlag, 1968)

Left, from top to bottom: Cactus
aficionados consider the North
American sand dollar cactus
(*Astrophytum asterias*) to be the most
beautiful star cactus. Members of the
Native American Church regard it as
a symbol for the peyote cactus.

Astrophytum myriostigma, known in
Germany as *bischofsmütze* ("bishop's
miter"), is called *peyote cimarrón*
("feral peyote" or "wild peyote") in
northern Mexico.

The very rare ball cactus *Aztekium
riterii* occurs only in the Mexican
state of Nuevo León.

The red fruit of the small ball cactus
Epithelantha micromeris is used as a
doping agent.

Right, from top to bottom:
In Mexico, some species of the genus
Ferocactus are called *biznaga*, a name
also given to the peyote cactus.
Several species contain β-phenethyl-
amines.

Mammillaria heyderi, known by the
name *híkuli* or *peyote*, produces
edible red fruits (so-called *chilitos*)
that have a wonderful taste.
(Photographed in Teotihuacan)

The rare cactus *Obregonia denegrii* is
from Tamaulipas.

Peyote Substitutes and Plants with the Name *Peyote*

(From Anderson 1996b, 162f; Bruhn and Bruhn 1973; Bye 1979b*; Deimel 1996, 22; Díaz 1979*; Martínez 1987*; Ott 1993*; Schultes 1937a, 1937b, 1966*; Shulgin 1995*; expanded.)

Botanical Name	Indian Name	Constituent(s)
BROMELIACEAE[216]		
Tillandsia mooreana L.B. Smith	waráruwi	?
CACTACEAE		
Ariocarpus fissuratus (Engelm.) K. Schum. [=*Roseocactus fissuratus* (Engelm.) Bg.]	peyote cimarrón, híkuli sunami	hordenine, tyramine [β-**phenethylamines**]
Ariocarpus kotschoubeyanus (Lem.)	peyote cimarrón	hordenine, et cetera
Ariocarpus retusus Scheidw.	chaute	hordenine, β-**phenethylamines**
Astrophytum asterias (Zucc.) Lem.	peyote	alkaloids (?)
Astrophytum myriostigma Lem. [several varieties var. *columnare* (Sch.) Tsuda var. *nudum* (R. Mey.) Backeb. var. *quadricostatum* (Moell.) Baum]	peyote cimarrón	alkaloids
Aztekium riterii Boedeker	péyotl	?
Carnegia gigantea (Engelm.) Britt. et Rose	saguaro	arizonine, carnegine
Coryphantha compacta (Engelm.) Britt. et Rose	bakana, wichurí, santa poli	dimethoxyphenethylamine (DMPEA)
Coryphantha macromeris (Engelm.) Lem.	mulato	DMPEA, macromerine
Echinocereus triglochidiatus Engelm. [syn. *E. salm-dyckianus* Scheer]	híkuli, pitallita[217]	3-hydroxy-4-methoxyphen-ethylamine, alkaloids
Epithelantha micromeris (Engelm.) Web. [syn. *E. polycephala, E. rufispina, Mammillaria micromeris* Engelm.]	híkuli mulato, chilito, híkuli rosapara	alkaloids (?), triterpenes
Lophophora diffusa (Croizat) Bravo	peyote	pellotine, lophophorine, anhalidine, anhalonidine, hordenine, **mescaline**
Lophophora fricii Habermann [syn. *Lophophora williamsii*]	chiculi hualala	pellotine, lophophorine, **mescaline**
Mammillaria craigii Lindsay [= *M. standleyi* (Britt. et Rose) Orcutt]	peyote de San Pedro, wichurí, witculiki	?
Mammillaria grahamii Engelm.	híkuli, peyote	?
M. grahamii var. *oliviae* (Orcutt) L.	híkuli, peyote	?
Mammillaria heyderi (Mu.) Britt. et Rose	híkuli, biznaga de chilitos	DMPEA
Mammillaria longimamma DC. [syn. *Dolichothele longimamma* (DC.) Britt. et Rose]		alkaloids
Mammillopsis senilis (Lodd.) Weber [syn. *Mammillaria senilis* Lodd.] cabeza de viejo	peyote cristiano, híkuli dewéame,	alkaloids (?)
Obregonia denegrii Fric	hikuli sunami, peyoti, peyotillo	β-**phenethylamines**, hordenine, tyramine
Pachycereus pecten-aboriginium (Engelm.) Britt. et Rose	chawe', cardillo, wichowaka	β-**phenethylamines**
Pelecyphora aselliformis Ehr.	peyote meco, peyotillo	anhalidine, hordenine, pellotine
Pelecyphora pseudopectinata Backeb.	peyote	hordenine
Solisia pectinata Britt. et Rose	peyotillo	hordenine, methyltyramine
Strombocactus disciformis (DC.) Britt. et Rose	peyote, peyotillo	alkaloids
Turbinicactus pseudomacrochele (Bckbg.) Ruxb. et Backeb.	peyotillo	hordenine

216 Jonathan Ott has suggested that the Bromeliaceae Family, and perhaps even the genus *Tillandsia*, may contain potent entheogenic species (cf. Ott 1993, 108, 420*; cf. also **shahuán-peco**).

217 This species is quite variable, with numerous forms, varieties, and subspecies already described (cf. Preston-Mafhan 1995, 44*).

Botanical Name	Indian Name	Constituent(s)
COMPOSITAE		
Cacalia cordifolia L. f.	peyote, peyote xochimilcensis	alkaloids (?)[218]
Cacalia decomposita Gray	peyote, maturi, matarique, hongo de los pinos ("mushroom of the pines")	sesquiterpene lactones, alkaloids (?)
Cacalia spp. (cf. *Calea zacatechichí*)	peyote	?
Senecio canicida Moc. et Sess.	clarincillo, itzcuinpatli, hierba del perro	
Senecio cardiophyllus Hemsl.	peyote, palo bobo, piote	sesquiterpene lactones
Senecio grayanus Hemsl.	palo loco	pyrrolizidine
Senecio hartwegii Benth.	peyote de Tepíc	pyrrolizidine
Senecio praecox (Cav.) Gray	quantlapatzinzintli, palo bobo, palo loco, candelero, texcapatli	
Senecio tolucanus DC.	guantlapazinzintli, peyote	?
CRASSULACEAE		
Cotyledon sp.	peyote	glycosides
CYPERACEAE		
Scirpus spp.	bakana, bakánowa	alkaloids
LYCOPERDACEAE		
Lycoperdon spp.	kalamota, pedo del diablo	?
LEGUMINOSAE		
Rhynchosia longeracemosa Mart. et Gal.	peyote	alkaloids
ORCHIDACEAE		
Oncidium cebolleta (Jacq.) Sw.[219] [syn. *O. ascendens* Lindl., *O. longifolium* Lindl.]	híkuli	alkaloids (?)

Left: An *Oncidium* species from northern Mexico. Several Mexican species of *Oncidium* bear a strong resemblance to *Oncidium cebolleta*, known as *híkuli* ("peyote"). Whether this or a closely related species does in fact have psychoactive effects is a subject for future research.

Right: In the future, bromeliads of the genus *Tillandsia*, which live on trees and branches, may become important as psychoactive plants. One species in northern Mexico is regarded as a peyote substitute and is said to produce psychoactive effects.

218 In Mexico, this species is attributed with aphrodisiac powers. To date, no psychoactive effects or constituents have been identified (Schultes and Hofmann 1992, 37*).

219 This orchid, which occurs from Mexico to Paraguay, possesses a strong rhizome and blooms in late winter (Wiard 1987, 109). It is occasionally available through the international orchid trade (Rysy 1992, 140). The genus *Oncidium* is one of the most species-rich genera in the New World (Stacy 1975).

Literature

See also the entries for *Ariocarpus fissuratus* and **mescaline**.

Since the summer of 1996, The Peyote Foundation has published a periodical entitled *The Peyote Awareness Journal* (Kearny, Arizona).

Aberle, David F. 1982. *The peyote religion among the Navaho.* 2nd ed. Chicago and London: The University of Chicago Press.

Albaugh, B. J., and P. O. Anderson. 1974. Peyote in the treatment of alcoholism among American Indians. *American Journal of Psychiatry* 131:1247–50.

Ammon, Günter, and Paul G. R. Patterson. 1971. Peyote: Zwei verschiedene Ich-Erfahrungen. In Bewußtseinserweiternde Drogen aus psychoanalytischer Sicht, special issue, *Dynamische Psychiatrie*: 47–71. Berlin.

Anderson, Edward F. 1980. *Peyote: The divine cactus.* Tucson: The University of Arizona Press.

———. 1996a. Peyote and its derivatives as medicine. *Jahrbuch für Transkulturelle Medizin und Psychotherapie* 6 (1995): 369–79.

———. 1996b. *Peyote: The divine cactus.* 2nd ed. Tucson: The University of Arizona Press. (Very good, up-to-date bibliography.)

Artaud, Antonin. 1975. *Die Tarahumaras.* Hamburg: Rogner und Bernhard.

Bahti, Tom. 1974. *Southwestern Indian ceremonials.* Las Vegas: KC Publications.

Benítez, Fernando. 1975. *In the magic land of peyote.* Austin and London: University of Texas Press.

Benzi, Marino. 1972. *Les derniers adorateurs du peyotl.* Paris: Gallimard.

Berrin, Kathleen, ed. 1978. *Art of the Huichol Indians.* San Francisco: The Fine Arts Museum.

Blanco Labra, Víctor. 1992. *Wirikuta: La tierra sagrada de los huicholes.* Mexico City: Daimon.

Bollhardt, Thomas Benno. 1985. Nearika: Visionen der Huichol. In *Umgarnte Mythen*, 9–75. Freiburg: Völkerkundemuseum Freiburg. (An exhibition catalogue.)

Brave Bird, Mary. 1993. *Ohitika woman.* New York: Grove Press.

Brenneisen, Rudolf, and Hans-Jörg Helmlin. 1993. *Lophophora.* In *Hagers Handbuch der pharmazeutischen Praxis*, 5th ed., 5:707–12. Berlin: Springer.

Brito, Silvester J. 1989. *The way of a peyote roadman.* New York: Peter Lang.

Bruhn, Jan G., and Stig Agurell. 1975. *O-methylpellotine, a new peyote alkaloid from Lophophora diffusa. Phytochemistry* 14:1442–43.

Bruhn, Jan G., and Catarina Bruhn. 1973. Alkaloids and ethnobotany of Mexican peyote cacti and related species. *Economic Botany* 27:241–51.

Bruhn, Jan G., and Bo Holmstedt. 1974. Early peyote research: An interdisciplinary study. *Economic Botany* 28:353–90.

Bruhn, Jan G., J.-E. Lindgren, and Bo Holmstedt. 1978. Peyote alkaloids: Identification in a prehistoric specimen of *Lophophora* from Coahuila, Mexico. *Science* 199:1437–38.

Bullis, Ronald K. 1990. Swallowing the scroll: Legal implications of the recent Supreme Court peyote cases. *Journal of Psychoactive Drugs* 22 (3): 325–32.

Camino, Alejandro. 1992. El peyote: Derecho histórico de los pueblos indios. *Takiwasi* 1 (1): 99–109.

Casillas Romo, Armando. 1990. *Nosología mítica de un pueblo: Medicina tradicional huichola.* Guadalajara: Editorial Universidad de Guadalajara.

Crow Dog, Mary (= Mary Brave Bird). 1994. *Lakota woman: Die Geschichte einer Sioux-Frau.* Munich: dtv.

d'Azevedo, Warren L. 1985. *Straight with the medicine: Narratives of Washo followers of the Tipi Way.* Berkeley, Calif.: Heyday Books.

Deimel, Claus. 1980. *Tarahumara.* Frankfurt/M.: Syndikat.

———. 1985. Die Peyoteheilung der Tarahumara. *Schreibheft* 25:155–63.

———. 1986. Der heilsame Rausch. In "Mexiko," special issue, *Geo Special*, no. 2:86–87.

———. 1996. *Híkuri ba—Peyoteriten der Tarahumara.* Ansichten der Ethnologie 1. Hannover: Niedersächsisches Landesmuseum.

Dittmann, Allen T., and Harvey C. Moore. 1957. Disturbance in dreams as related to peyotism among the Navaho. *American Anthropologist* 59:642–49.

Dustin, C. Burton. 1962. *Peyotism and New Mexico.* Albuquerque, N.M.: self-published.

Ellis, Havelock. 1897. A note on the phenomenon of mescal intoxication. *The Lancet* 75 (1): 1540–42.

———. 1902. Mescal—a study of a divine plant. *Popular Science Monthly* 61:52–71.

Evans, A. Don. 1989. The purpose and meaning of peyote as a sacred material for Native Americans. In *The concept of sacred materials and their place in the world*, ed. George P. Horse Capture, 20–35. Cody, Wyo.: The Plains Indian Museum.

Fikes, Jay Courtney. 1993. *Carlos Castaneda, academic opportunism and the psychedelic sixties.* Victoria, B.C.: Millennia Press.

Furst, Peter T. 1965. West Mexican tomb art as evidence for shamanism in pre-Hispanic Mesoamerica. *Antropológica* 15:29–80.

———. 1969. A possible symbolic manifestation of funerary endo-cannibalism in Mexico. In *Verhandlungen des XXXVIII. Internationalen Amerikanistenkongresses*, 2:385–99. Munich: Klaus Renner.

———. 1981. Peyote und die Huichol-Indianer in Mexiko. In *Rausch und Realität*, ed. G. Völger, 2:468–75. Cologne: Rautenstrauch-Joest-Museum für Völkerkunde.

Furst, Peter T., and M. Anguiano. 1977. "To fly as birds": Myth and ritual as agents of enculturation

"The ocean drew itself out of the ocean,
and after the ocean
came the gods.
The gods stepped forth
like flowers.
In the way of the flowers
they followed the ocean
and they came to the womb
to the place of origin
to the place of birth."

HUICHOL PEYOTE SONG
IN *RAUSCH UND ERKENNTNIS*
[INEBRIATION AND KNOWLEDGE]
(HÖHLE ET AL. 1986, 53*)

"I speak to the peyote as if to a mother. She praises me and she rebukes me. She strengthens me and gives me good advice."

A Navajo
in *Nur der Adler sprach zu mir* [Only the Eagle Spoke to Me] (F. Abel 1983, 85)

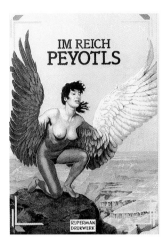

The Mexican peyote cactus has inspired numerous comic book artists such as Jeronaton and stimulated them to develop fantastic stories about the magical effects of this plant of the gods. (Title page, Edition Becker & Knigge, 1982)

among Huichol Indians of Mexico. In *Enculturation in Latin America: An anthology*, ed. Johannes Wilbert, 95–181. Los Angeles: UCLA Latin American Center Publications.

Furst, Peter T., and Salomón Nahmad. 1972. *Mitos y arte huicholes.* Mexico City: SepSetentas.

Gäbler, Hartwig. 1965. *Aus dem Heilschatz der Natur.* Stuttgart: Paracelsus Verlag.

Gartz, Jochen. 1995. Ein früher Versuch der Kommerzialisierung von Peyotl in Deutschland. *Integration* 6:45.

Gerber, Peter. 1980. *Die Peyote-Religion.* Zurich: Völkerkundemuseum der Universität.

Ginsberg, Allen. 1982. *Notizbücher 1952–1962.* Reinbek, Germany: Rowohlt.

Gerard, John. *Glanz und Untergang des Alten Mexiko.* 1986. Mainz: Philipp von Zabern. (An exhibition catalogue.)

Goggin, John M. 1938. A note on Cheyenne peyote. *New Mexico Anthropologist* 3 (2): 26–32.

Gusinde, Martin. 1939. *Der Peyote-Kult: Entstehung und Verbreitung.* Vienna-Mödling: Missionsdruckerei. [Cf. the book review by Marvin K. Opler in *American Anthropologist* 42 (1940): 667–69.]

Haan, Prem Lélia de. 1988. *Bei Schamanen.* Frankfurt/M.: Ullstein.

Hayes, Alden. 1940. Peyote cult on the Goshiute Reservation at Deep Creek, Utah. *New Mexico Anthropologist* 4(2): 34–36.

Heffter, Arthur. 1894. Über Pellote: Ein Beitrag zur pharmakologischen Kenntnis der Kakteen. *Naunyn-Schmiedebergs Archiv für experimentelle Pathologie und Pharmakologie* 34:65.

Hell, Christina. 1988. *Hirsch, Mais, Peyote in der Konzeption der Huichol.* Hohenschäftlarn, Germany: Klaus Renner Verlag.

Hennings, Paul. 1888. Eine giftige Kaktee, *Anhalonium lewinii* n. spp.. *Gartenflora* 37:410–12.

James, Joyce. 1964. Shouted from the housetops: A peyote awakening. *Psychedelic Review* 1 (4): 459–83.

Kan, Michael, Clement Meighan, and H. B. Nicholson. 1989. *Sculpture of ancient West Mexico.* Los Angeles: County Museum of Art.

La Barre, Weston. 1957. Mescalism and peyotism. *American Anthropologist* 59:708–11.

———. 1960. Twenty years of peyote studies. *Current Anthropology* 1 (1): 45–60.

———. 1979. *The peyote cult.* 5th ed. Norman: University of Oklahoma Press.

———. 1981. Peyotegebrauch bei nordamerikanischen Indianern. In *Rausch und Realität*, ed. G. Völger, 2:476–78. Cologne: Rautenstrauch-Joest-Museum für Völkerkunde.

Lame Deer and Richard Erdoes. 1979. *Tahca Ushte— Medizinmann der Sioux.* Munich: List.

Leonard, Irving A. 1942. Peyote and the Mexican Inquisition, 1620. *American Anthropologist*, n.s., 44:324–26.

Lewin, Louis. 1888. Ueber *Anhalonium lewinii.* *Archiv für experimentelle Pathologie und Pharmakologie* 24:401–11.

Lumholtz, Carl. 1902. *Unknown Mexico.* 2 vols. New York: Charles Scribner's Sons.

———. 1989. *A nation of shamans: The Huichols of the Sierra Madre.* The Shamanic Library, no. 1. Oakland, Calif.: Bruce I. Finson. (Orig. pub. 1900 as *Symbolism of the Huichol Indians.*)

Marriott, Alice, and Carol K. Rachlin. 1972. *Peyote.* New York and Scarborough, Ontario: Mentor Book.

McAllester, David P. 1949. *Peyote music.* Viking Fund Publications in Anthropology, no. 13. New York: Viking Fund.

McLaughlin, J. L., and A. G. Paul. 1966. The cactus alkaloids. I: Identification of *N*-methylated tyramine derivatives in *Lophophora williamsii.* *Lloydia* 29 (4): 315–27.

McLeary, James A., Paul S. Sypherd, and David L. Walkington. 1960. Antibiotic activity of an extract of peyote [*Lophophora williamsii* (Lemaire) Coulter]. *Economic Botany* 14:247–49.

Mellen, Chase, III. 1963. Reflections of a peyote eater. *The Harvard Review* 1 (4): 63–67.

Merriam, Alan P., and Warren L. d'Azevedo. 1957. Washo peyote songs. *American Anthropologist* 59:615–41.

Momaday, N. Scott. 1968. *House made of dawn.* New York: Harper & Row.

———. 1969. *The way to Rainy Mountain.* Albuquerque: University of New Mexico Press.

Morgan, George Robert. 1976. *Man, plant, and religion: Peyote trade on the Mustang Plains of Texas.* Dissertation, University of Colorado (microfilm no. 76–23, 637).

———. 1983a. The biogeography of peyote in South Texas. *Botanical Museum Leaflets* 29 (2): 73–86.

———. 1983b. Hispano-Indian trade of an Indian ceremonial plant, peyote (*Lophophora williamsii*), on the Mustang Plains of Texas. *Journal of Ethnopharmacology* 9:319–21.

Mount, Guy, ed. 1988. *The peyote book: A study of Native Medicine.* Arcata, Calif.: Sweetlights Books.

Müller-Ebeling, Claudia, and Christian Rätsch. n.d. Kreisrituale mit Peyote und MDMA. In *MDMA: Die psychoaktive Substanz für Therapie, Ritual und Rekreation*, rev. ed., ed. Constanze Weigle and Ronald Rippchen. Der Grüne Zweig 103. Löhrbach: Werner Pieper's MedienXperimente.

Myerhoff, Barbara G. 1973. *Organization and ecstasy: Peyote and the Huichol case.* Unpublished manuscript.

———. 1974. *Peyote hunt: The sacred journey of the Huichol Indians.* Ithaca, N.Y.: Cornell University.

———. 1975. Peyote and Huichol worldview: The structure of a mystic vision. In *Cannabis and culture*, ed. Vera Rubin, 417–38. The Hague and Paris: Mouton.

———. 1980. *Der Peyote Kult.* Munich: Trikont.

Nahmad Sittri, Salomón, Otto Klineberg, Peter T. Furst, and Barbara G. Myerhoff. 1972. *El peyote y los huicholes.* Mexico City: SepSetentas.

Opler, Marvin Kaufman. 1940. The character and history of the southern Ute peyote rite. *American Anthropologist*, n.s., 42:463–78.

Opler, Morris E. 1938. The use of peyote by the Carrizo and Lipan Apache tribes. *American Anthropologist*, n.s., 40:271–85.

Ott, Jonathan. 1996. *Lophophora williamsii* (Lemaire) Coulter. Unpublished electronic file. (Cited 1998.)

Pascarosa, Paul, and Sanford Futterman. 1976. Ethnopsychedelic therapy for alcoholics: Observations in the peyote ritual of the Native American Church. *Journal of Psychedelic Drugs* 8 (3): 215–21.

Pinkson, Tom Soloway. 1995. *Flowers of wirikuta: A gringo's journey to shamanic power.* Mill Valley, Calif.: Wakan Press.

Roseman, Bernard. 1966. *The peyote story: The Indian mind drug.* Hollywood, Calif.: Wilshire Book Co.

Rouhier, Alexandre. 1927. *Le Peyotl* (Echinocactus williamsii). Paris: Gaston Doin et Cie.

———. 1996. *Die Hellsehen hervorrufenden Pflanzen.* Berlin: VWB. (Reprint of 1927 Altmann edition published in Leipzig, translated from the original French by E. Stöber.)

Rysy, Wolfgang. 1992. *Orchideen: Tropische Orchideen für Zimmer und Gewächshaus.* 4th ed. Munich: BLV.

Schaefer, Stacy. 1989. The loom and time in the Huichol world. *Journal of Latin American Lore* 15 (2): 179–94.

———. 1993a. Huichol Indian costumes: A transforming tradition. *Latin American Art* Spring 93:70–73.

———. 1993b. The loom as a sacred power object in Huichol culture. In *Art in small scale societies*, ed. R. Anderson and K. Field, 118–30. New York: Prentice Hall.

———. 1995. The crossing of the souls: Peyote, perception and meaning among the Huichol Indians of Mexico. *Integration* 5:35–49.

———. 1997. Peyote and pregnancy. *Yearbook for Ethnomedicine and the Study of Culture,* 1996 (5): 67–78. Berlin: VWB.

Schaefer, Stacy, and Peter T. Furst, eds. 1996. *People of the peyote: Huichol Indian history, religion, and survival.* Albuquerque: University of New Mexico Press.

Schäfer, Georg, and Nan Cuz. 1968. *Im Reiche des Mescal.* Essen: Synthesis Verlag.

Schultes, Richard E. 1937a. Peyote and plants used in the peyote ceremony. *Botanical Museum Leaflets* 4 (8): 129–52.

———. 1937b. Peyote (*Lophophora williamsii*) and plants confused with it. *Botanical Museum Leaflets* 5 (5): 61–88.

———. 1938. The appeal of peyote (*Lophophora williamsii*) as a medicine. *American Anthropologist*, n.s., 40:698–715.

Slotkin, J. S. 1956. *The peyote religion: A study in Indian-White relations.* Glencoe, Ill.: The Free Press. [Cf. the book review by Weston La Barre in *American Anthropologist* 59 (1957): 359 f.]

Smith, Huston, and Reuben Snake. eds. 1996. *One nation under God: The triumph of the Native American Church.* Santa Fe, N.M.: Clear Light Publishers.

Stacy, John E. 1975. Studies in the genus *Oncidium. Botanical Museum Leaflets* 24 (7): 133–67.

Steinmetz, Paul B., Jr. 1990. *Pipe, Bible, and peyote among the Oglala Lakota.* Knoxville: The University of Tennessee Press.

Stewart, Omer C. 1987. *Peyote religion: A history.* Norman and London: University of Oklahoma Press.

Stewart, Omer C., and David F. Aberle. 1984. *Peyotism in the West.* University of Utah Anthropological Papers no. 108. Salt Lake City: University of Utah Press.

Straatman, Silke. 1988. *Die Wollbilder der Huichol-Indianer.* Vol. 6 of Marburger Studien zur Völkerkunde. Marburg/Lahn: Marburger Studien zur Völkerkunde.

Tedlock, Dennis, and Barbara Tedlock, eds. 1978. *Über den Rand des tiefen Canyon.* Cologne: Diederichs.

Todd, James S. 1969. Thin-layer chromatography analysis of Mexican populations of *Lophophora* (Cactaceae). *Lloydia* 32 (3): 395–98.

Valadez, Mariano, and Susana Valadez. 1992. *Huichol Indian sacred rituals.* Oakland, Calif.: Dharma Enterprises.

Vonnegut, Kurt. 1959. *The sirens of Titan.* New York: Dell.

Wagner, Günter. 1932. Entwicklung und Verbreitung des Peyote-Kultes. *Baessler-Archiv* 15:59–144.

Wiard, Leon A. 1987. *An introduction to the orchids of Mexico.* Ithaca, N.Y., and London: Comstock/Cornell University Press.

Wiedman, Dennis. 1985. Staff, fan, rattle and drum: Spiritual and artistic expressions of Oklahoma peyotists. *American Indian Art Magazine* 10 (3): 38–45.

Wiedman, Dennis, and Candace Greene. 1988. Early Kiowa peyote ritual and symbolism: The 1891 drawing books of Silverhorn (Haungooah). *American Indian Art Magazine* 13 (4): 32–41.

Wilson, Richard. 1981. *Der Sonnentanz und andere Storys.* Hamburg: Xenos.

Zabel, Rudolf. 1928. *Das heimliche Volk.* Berlin: Deutsche Buch-Gemeinschaft.

Zingg, Robert M. 1982. *Los huicholes.* 2 vols. Mexico City: INI.

Facsimile title page of Rouhier 1927.

Mammillaria spp.

Mammillaria Cacti

Family
Cactaceae (Cactus Family); Cacteae Tribe, Echino-cactinae Subtribe

Species
The genus *Mammillaria* is composed of 150 to 200 species. It thus has more species than any other genus in the Cactus Family except *Opuntia*. Species of ethnobotanical relevance:

Mammillaria craigii Lindsay [syn. *Mammillaria standleyi* (Britt. et Rose) Orcutt] (a peyote substitute)
Mammillaria grahamii Engelm. (a peyote substitute)
M. grahamii var. *oliviae* (Orcutt) L. (a peyote substitute)
Mammillaria heyderi (Mu.) Britt. et Rose (a peyote substitute)

Top: The Mexican ball cactus *Mammillaria compressa*.

Bottom: Because it flowers frequently, *Mammillaria bocasana* is a popular specimen for the home.

M. heyderi var. *coahuilensis* (Boedeker) J. Lüthy comb. nov.
M. heyderi var. *gummifera* (Engelm.) L. Benson
M. heyderi var. *hemispaerica* (Engelm.) Engelm.
M. heyderi var. *macdougalii* (Rose) L. Benson
M. heyderi var. *meiacantha* (Engelm.) L. Benson
Mammillaria longimamma DC. [syn. *Dolichothele longimamma* (DC.) Britt. et Rose, *Dolichothele uberiformis* (Zucc.) Britt. et Rose, *Mammillaria uberiformis* Zucc.] (a peyote substitute)
Mammillopsis senilis (Lodd.) Weber [syn. *Mammillaria senilis* Lodd.] (a peyote substitute)

Among the immediate relatives of *Mammillaria crinita* (order Stylothelae), the following species have been shown to contain an alkaloid that may have psychoactive properties (the so-called *M. wildii* profile) (Lüthy 1995):

Mammillaria anniana Glass et Foster
Mammillaria aurihamata (after Reppenhagen 1991)
Mammillaria bocasana Poselger [syn. *M. eschauzieri* (Coult.) Craig]
Mammillaria brevicrinata Boedeker (after Reppenhagen 1991)
Mammillaria crinita DC. [syn. *Mammillaria zeilmanniana* Boedeker (Mother's Day cactus), *Mammillaria gilensis* Boedeker]
Mammillaria duwei Rogozinski et Braun (after Reppenhagen 1991)
Mammillaria erythrosperma Boedeker
Mammillaria fittkaui Glass et Foster
Mammillaria limonensis Reppenhagen [= *Mammillaria fittkaui* ssp. *limonensis* (Reppenhagen) Lüthy comb. nov.]
Mammillaria mathildae Kraehnbuehl et Kraiz [= *Mammillaria fittkaui* ssp. *mathildae* (K. et K.) Lüthy comb. nov.]
Mammillaria monancistracantha Reppenhagen
Mammillaria ojuelensis n.n.
Mammillaria puberula Reppenhagen
Mammillaria pygmaea (Britt. et Rose) Berg
Mammillaria schwarzii Shurly
Mammillaria variabilis Reppenhagen
Mammillaria wildii A. Dietrich [= *Mammillaria crinita*, which is propagated and sold on the international market under the name *Mammillaria wildii*]

Synonyms
See above.
Mammillaria senilis Lodd. = *Mammillopsis senilis* (Lodd.) Weber

Folk Names

Biznaga de chilillos, biznaga de chilitos, falscher peyote, false peyote, híkuli, jículi, jícuri, mammillaria cactus, mammillarienarte, Mother's Day cactus, muttertagskaktus, peyote, warzenkaktus, wichuriki

History

The genus and several of its many species were described at the beginning of the nineteenth century. Today, the genus is well studied and has been the subject of several taxonomic revisions (Lüthy 1995). Many members of the genus are now in international demand as ornamentals.

Distribution

The genus *Mammillaria* is American in origin and is most heavily concentrated in northern Mexico. Most species live in hot and dry regions (deserts), some on rocky and craggy soils as well as on sandy and volcanic soils.

Cultivation

Many mammillaria cacti can be easily grown from seed. The tiny seeds need only be lightly pressed into porous soil and covered with pure sand. Keep very moist and in the sun at first. The time to germination is two to six weeks (at 20 to 25°C). In central Europe, the plant can be grown only as a houseplant or in a greenhouse.

Appearance

Those species of ethnobotanical and chemotaxonomic interest are all rather similar in appearance. They are small, spherical, heavily thorned, piliferous woolly cacti with flowers approximately 1.5 cm in length. The flowers, which may be white, pink, red, or violet, appear between March and May. The fruits typically are small red pods reminiscent of chili pods (*Capsicum* spp.). For this reason, they are known as *chilitos* in Mexico.

Psychoactive Material

— Fruits (*chilitos*)
— Cactus flesh

Preparation and Dosage

For use as a doping agent, a handful of the fresh fruit is recommended. The cactus flesh, dried and powdered, was drunk in maize beer (**chicha**) as a peyote substitute (see *Lophophora williamsii*).

Ritual Use

The Tarahumara of northern Mexico used some species of *Mammillaria* as peyote substitutes (see *Lophophora williamsii*) (Bye 1979b*; Deimel 1996, 22; Díaz 1979, 80*). *Mammillaria craigii* purportedly was formerly eaten by Tarahumara shamans wanting to obtain "clear vision" so that they could "see" witches and sorcerers (*brujos*)

(Bye 1979b, 30*). The flesh of *Mammillaria grahamii* var. *oliviae* was consumed by shamans and participants in secret ceremonies so that they could undertake a journey into a realm of brilliant colors. It is said that a person will go insane if he or she consumes this cactus immoderately (Bye 1979b, 31*).

Artifacts

Several illustrations can be found on postage stamps.

Medicinal Use

It is possible that the Tarahumara used mammillaria cacti for folk medicinal purposes similar to the ways in which they used the true peyote.

Roasted pieces or heart pieces of *Mammillaria heyderi* are placed in the ear canal to relieve headaches. The long-distance runners of the Tarahumara eat its fruits (*chilitos*) for doping purposes. This cactus is also thought to prolong life (Bruhn and Bruhn 1973, 244).

Mammillaria species containing latex are sold at Mexican markets as folk remedies and to counteract witchcraft.

Left, from top to bottom: *Mammillaria heyderi* is alleged to have potent psychoactive effects; however, no reports of experiences with the cactus are available.

Mammillaria zeilmanniana is known as Mother's Day cactus because its luminous flowers appear around May, the month in which Mother's Day falls.

Mammillaria wildii, a cactus that contains an active alkaloid.

Top right: The fruits of several mammillaria cacti are known as *chilitos* ("little chilis"). In Mexico, *chilitos* are regarded as doping agents, and the long-distance runners of the Tarahumara eat them to maintain their strength.

These Laotian postage stamps show two Mexican mammillaria.

"Among the most important 'false Peyotes' of the Tarahumara Indians are several species of *Mammillaria*, all of them round and stout-spined plants."

RICHARD EVANS SCHULTES AND
ALBERT HOFMANN
PLANTS OF THE GODS
(1992, 48)

Hordenine

Constituents

Several species contain the alkaloid hordenine (Howe et al. 1977). Other β-phenethylamines have been more frequently reported (Knox et al. 1983; West and McLaughlin 1973). To date, **mescaline** has not been detected (Shulgin 1995*).

The flowers and fruits of those species exhibiting the *M. wildii* profile have yielded a new alkaloid whose probable chemical formula is $C_{13}H_{13}NO_3$ (Lüthy 1995, 58 f.). Most mammillaria contain a latex composed of terpenes.

Effects

Hordenine and other β-**phenethylamines** may have psychoactive effects. Whether the newly discovered alkaloid is indeed psychoactive must be determined by further human pharmacological research.

Commercial Forms and Regulations

Many species of the genus *Mammillaria* can be found in cactus and flower shops. Because the genus is very popular among cactus aficionados and collectors, the best shops will often have a large selection. However, care must be exercised as far as the botanical information is concerned. Numerous species are sold under the name *Mammillaria zeilmanniana*.

Literature

See also the entries for **Lophophora williamsii** and β-**phenethylamines**.

Bruhn, Jan G., and Catarina Bruhn. 1973. Alkaloids and ethnobotany of Mexican peyote cacti and related species. *Economic Botany* 27:241–51.

Howe, Roberta C., Jerry L. McLaughlin, and Duwayne Stantz. 1977. *N*-methytyramine and hordenine from *Mammillaria microcarpa*. *Phytochemistry* 16:151.

Knox, M. J., W. D. Clark, and S. O. Link. 1983. Quantitative analysis of β-phenethylamines in two *Mammillaria* species (Cactaceae). *Journal of Chromatography* 265:362–75.

Lüthy, Jonas M. 1995. *Taxonomische Untersuchung der Gattung* Mammillaria HAW. (*Cactaceae*). Bern: Arbeitskreis für Mammillarienfreunde und J. Lüthy.

Reppenhagen, W. 1991–1992. *Die Gattung Mammillaria*. 2 vols. Titisee-Neustadt, Germany: Druckerei Steinhart.

West, L. G., and J. L. McLaughlin. 1973. Cactus alkaloids XVIII: Phenolic β-phenethylamines from *Mammillaria elongata. Lloydia* 36 (3): 346–48.

Mandragora officinarum Linnaeus
Mandrake

Family

Solanaceae (Nightshade Family); Subfamily Solanoideae, Solaneae Tribe, Mandragorinae Subtribe; chemotaxonomic subgroup composed of the genera *Mandragora* and *Scopolia* (cf. **Scopolia carniolica**) (Jackson and Berry 1979, 511)

Forms and Subspecies

Mandragora officinarum probably occurs in several varieties that were originally described as separate species (Jackson and Berry 1979):

Mandragora officinarum L. var. *haussknechtii*
Mandragora officinarum L. var. *hybrida*
Mandragora officinarum L. var. *officinarum*
Mandragora officinarum L. var. *vernalis* (very early-blooming form)

Synonyms

Atropa acaulis L. 1762
Atropa mandragora L.
Atropa mandragora (L.) Woodville 1794[220]
Mandragora acaulis Gaertn.
Mandragora haussknechtii Heldr.

Mandragora hispanica Vierhapper
Mandragora hybrida Hausskn. et Heldr.
Mandragora mas Gersault
Mandragora neglecta G. Don
Mandragora officinalis Bertoloni
Mandragora officinalis Mill.
Mandragora praecox Sweet
Mandragora vernalis Bertolini

Folk Names

Abu'l-ruh (Old Arabic, "master of the life breath"), abu-roh, adam koku, adam-kökü (Turkish, "man root"), adamova golowa (Russian, "Adam's head"), alrauinwortel (Dutch), alraun, alraune, alraunmännchen, alraunwurzel, alrune (Swedish), alrüneken, althergis, antimelon ("in the apple's place"), antimenion (Greek, "counter rage"), apemum (Egyptian/Coptic), archine, armesünderblume, astrang-dastam harysh, atzmann, Ανδρωπομορφος (ancient Greek, according to Pythagoras "in human form"), baaras (Hebrew, "the fire"), bayd al-jinn (modern Arabic, "testes of the demon"), bhagner, bid-l-gul, bombochylos

220 This old synonym has caused much confusion in the botanical, ethnobotanical, and history of medicine literature (e.g., in Eliade 1982, 215–34).

344

(Greek, "a juice that produces dull sounds"), ciceron (Roman, "plant of Circe"), Circe's plant, diamonon, dirkaia, dollwurz, drachenpuppe, dudaim, dûdâ'îm (Hebrew), dukkeurt (Danish, "mad root"), erdmännchen, erdmännlein, folter-knechtwurzel, galgenmännlein, geldmännlein, giatya bruz, gonogeonas, hausväterchen, hemionus, henkerswurzel, hundsapfel, hunguruk koku, jebrûah (Syrian/Aramaic, "manlike plant"), kam-maros (Greek, "subject to fate"), kindleinkraut, kirkaia ("plant of Circe"), καλανθροποζ (Cyp-riot, "good man"), lakhashmana, lakmuni, lebruj, liebesapfel, liebeswurzel, love apple, lufahat, luffah manganin (Arabic, "mad apple"), luffat, main de gloire (French), mala canina (Roman, "dog apple"), mala terrestria (Roman, "earth apple"), mandraghorah, mandragora, mandragóra, man-dragore, mandrake, männlicher alraun, mannikin (Belgian, "little man"), mannträgerin, mano di gloria, mardami, mardom ghiah (Persian, "man's plant"), mardum-gia (ancient Persian, "man plant"), matragun (Romanian, "witch's drink"),[221] matraguna, matryguna (Galician), m^{c}ntr^{c}gwrw (Egyptian), mehr-egiah (Persian, "love plant"), mela canina (Italian, "dog apple"), menschen-wurzel, minos, Μανδραγοραζ (ancient Greek), namtar ira (Assyrian, "the male [plant] of the god of the plagues"), natragulya (Hungarian), Oriental mandrake, pevenka trava (Russian, "the plant that screams"), pisdiefje (Dutch), planta semihominis (Roman, "half-man plant"), pomo di cane (Italian, "dog apple"), putrada, rakta vindu, rrm.t (Egyptian), Satan's apple, siradsch elkutrhrub (Anda-lusian Arabic, "root of the demon El-sherif"), sirag al qutr (Arabic), sirag el-kotrub (Arabic/Palestine, "devil's lamp"), taraiba, taraila (Morocco), tepilla-lilonipatli,[222] thjofarót (Icelandic, "thieve's root"), thridakias, tufah al-jinn (modern Arabic, "apple of the demon"), tufah al-Majnun (Arabic, "[love] apple of Majnun"),[223] tufhac el sheitan (Arabic, "apple of the devil"), womandrake (English), yabrough (Syrian Arabic, "life giver"), yabruh (Arabic), ya pu lu (Chinese), yavruchin (Ara-maic), yubru-jussanam, zauberwurzel

History

The mysterious mandrake or *Mandragora*—the queen of all magical plants—is not a character from a fairy tale but a real plant that is found especially in the eastern Mediterranean region. There are only two European species, the botanical identity of which was long uncertain (cf. ***Mandragora* spp.**). Mandrake has quite correctly been described as "the most famous magical plant in history" (Heiser 1987*). Its medicinal and magical uses, its aphrodisiac and psychoactive effects, and its mythology and the legends surrounding the plant all raise it above the level of any other magical plant (Schlosser 1987; Schöpf 1986*; Starck 1986). There is probably no other

The legendary mandrake (*Mandragora officinarum*), shown here in bloom. (Wild plant, photographed in Cyprus)

plant about which such a rich and varied literature has been produced (cf. Hansen 1981*).

Probably the oldest written mention of the mandrake occurs in the cuneiform tablets of the Assyrians and the Old Testament; these refer primarily to the area of Babylon. In Assyrian, the mandrake was known as *nam-tar-gir(a)* (^{is}nam-tar-*gir$_{12}$).[224] Nam Tar was the god of plagues; *(g)ira* means "male." An Ugaritic cuneiform text from Ras Shamra (fifteenth to fourteenth century B.C.E.) appears to refer to a ritual; the text reads, "[P]lant mandragoras in the ground . . ." (Schmidbauer 1968, 276). Mesopotamian cunei-form texts make frequent mention of a wine known as cow's eye, which was purportedly a wine mixed with mandrake. "The effects of the man-drake upon the pupils could thus be the reason for the strange name 'cow's eye'" (Hirschfeld and Linsert 1930, 162*).

In ancient times, mandrake was an enormously important ritual, inebriating, and medicinal plant. The German name *alraune* suggests an Old Germanic use of the plant: "Alraun comes from Alrun, and originally meant 'he who knows the runes' or the 'all knowing'" (Schmidbauer 1969, 281). The Germanic seeresses (*seidkona*, *wölwas*), who by late ancient times were known far beyond Europe's borders for their miraculous abilities (e.g., Albruna and Weleda; one was even active in Egypt!), would enter a prophetic ecstasy with the aid of such magical agents and shamanic tech-niques (Derolez 1963, 240*). With the Christian-ization of Germania, *Mandragora* (as an ancient pagan ritual plant) was demonized. Hildegard von Bingen was the first to denounce the mandrake:

The Alraun is warm and somewhat watery and is spread by that ground from which Adam was made; it somewhat resembles a person. With this plant, however, also because of its similarity to a person, there is more diabolical whispering than with other plants and it lays snares for him. For this reason, a person is driven by his desires, whether they are good or bad, as he also once did with the idols. . . . It is harmful through much that is corruptive of the magicians and phantoms, as

221 This name is also used for ***Atropa belladonna*** and ***Scopolia carniolica***.

222 The ancient Aztecs are said to have known the mandrake by this name (Cerna 1932, 304*). However, *Mandragora* was not part of the pre-Columbian flora.

223 A play on the Arabic love story "Majnun and Layla."

224 According to Thompson (1949, 217*), the Assyrian *namtargira* is strikingly similar phonetically to the Greek *mandragora*.

many bad things were once caused by the idols. (*Physica* 1.56)

Although *Mandragora* is numbered among the witches' plants (cf. **witches' ointments**), it was often counterfeited during the Middle Ages because it was also valued as a talisman and bringer of luck. Surrogates were sold in pharmacies even as late as the twentieth century. Because of the difficulty in obtaining actual plant material, mandrake has never attained much significance as a psychoactive plant in the hippie subculture or among the modern closet shamans. Surprisingly, the psychoactivity of the root has never been the object of any systematic study.[225]

Distribution

Mandragora officinarum is found in southern Europe from Portugal to Greece; it is quite common in Greece and Italy (Festi and Aliotta 1990*; Viola 1979, 175*). It never occurs in the wild north of the Alps (Beckmann 1990, 129*). But the root is winter-hardy and can be grown in central and northern Europe. It is also found in North Africa, Asia Minor, and the Middle East and on most of the Mediterranean islands (Cyprus, Crete, Sicily) (Georgiades 1987, 50*; Sfikas 1990, 246*). It often thrives in dry, sunny locales, usually along paths and around ancient temples. However, it is one of Europe's rarest plants.

Cultivation

Propagation occurs through the seeds (which are very similar in appearance to the seeds of *Datura innoxia*). It is best to pregerminate the seeds (as

Top: The ripe fruits of the mandrake (*Mandragora officinarum*) exude an exhilarating scent that in ancient times was thought to incite love. (Wild plant, photographed in Cyprus)

Bottom: The seeds of *Mandragora officinarum* are very similar to those of some species of *Datura*. (Photograph: Karl-Christian Lyncker)

225 The Solanaceae are typically given a great deal of consideration in the modern toxicological literature. Remarkably, mandrake does not appear in most compendiums on poisonous plants. In my opinion, this is due to the fact that its alleged toxicity has been greatly exaggerated. Moreover, the rarity of the plant means that almost no incidence of hospitalized cases of poisoning have occurred. The experientially oriented literature on psychedelics also makes little mention of the plant (it is not mentioned in Hartwich 1911* or in Lewin 1980 [orig. pub. 1929]*!) or speaks of it only in passing. There are no firsthand reports of experiences with the plant. It appears as though the magical aura that has surrounded the root since ancient times has frightened away psychedelic experimenters, many of whom are otherwise so ready to try anything new.

with those of *Datura discolor*). The seedlings should be transplanted into very large pots, as the plant will develop a very large root over the years. The first flowers will develop when the plant is in its fourth year. The plant can be grown in topsoil to which a small amount of sand has been added. The plant should never be overwatered, especially when it is in its dormant stage.

Although the plant does not actually tolerate frost, it can be maintained as a perennial in central Europe. To achieve this, the plant or its location should be covered with a pile of leaves in the fall. The leaves should then be removed in the spring. In central Europe, the plant does not develop leaves until early summer.

Appearance

The mandrake is a stemless perennial plant whose fleshy root can grow as long as 100 cm, sometimes taking on a bizarre or anthropomorphic shape. Most of the year, the plant is hidden in the ground.

Once every year, the long, wide leaves, which form a characteristic rosette, grow directly out of the roots. The bluish or violet, bell-shaped, quinquelobate flowers grow on short stalks from the center of the rosette. The leaves wither as the yellow berries (fruits) mature. But the root remains alive and will develop leaves and flowers again the following spring. The golden yellow fruits have a fruity scent (similar to that of the fruits of *Physalis* spp.) but taste more like tomatoes (fellow members of the Nightshade Family). The leaves have a scent somewhat reminiscent of that of fresh tobacco (*Nicotiana tabacum*).

Mandragora officinarum can be easily confused with the autumn mandrake. Both European species are very similar in their anatomy. However, the rhizome of *M. officinarum* is larger than that of *M. autumnalis*. The primary difference between the two is their flowering period. *M. officinarum* blossoms in May, while *M. autumnalis* blooms in the fall (September to November) (cf. **Mandragora** spp.).

In the ethnographic literature and in literature pertaining to the history of medicine, the true mandrake is frequently confused with the mayapple (*Podophyllum peltatum* L.) and other plants (see the table on page 347).

Psychoactive Material
— Root (mandragorae radix, mandrake root)
— Root cortex
— Leaves
— Fruits (mandrake fruits, love apples, dudaim, Arabic lofah)

Preparation and Dosage
The fruits should be consumed only while fresh. No overdoses have been observed, even with quantities as high as ten fruits.

The leaves can be either chewed while fresh or dried for later use. They are best collected before the fruiting period and dried in the shade. They can be smoked alone (as a tobacco substitute; cf. *Nicotiana tabacum*) or mixed with other herbs into **smoking blends**. They also can be used as **incense**.

The root can be burned as incense as well. While the burning root pieces give off a rather unpleasant aroma reminiscent of that of burned food, the smoke itself is rather easy to inhale. When used as an incense, mandrake can be combined with the pleasant-smelling olibanum (cf. *Boswellia sacra*). When smoked or used as an incense, the psychoactive effects of mandrake are very subtle.

The most common use involves tinctures of the dried root. It is only seldom eaten. The alkaloids present within the plant are quite water soluble, which is why tinctures are derived from an aqueous whole extract.

The mandrake is very suitable for making or improving **beer** and **wine**. Mandrake beer is brewed in the same manner as henbane beer (see *Hyoscyamus niger*). Fifty grams of dried root should be used for 20 liters of liquid. Cinnamon sticks and/or saffron (*Crocus sativa*) can be added to improve the taste of mandrake beer. As little as 1/2 to 1 liter of mandrake beer can produce very noticeable effects. Care should be exercised when determining dosages!

The ancient Greeks frequently added the fresh or dried root to wine (cf. *Vitis vinifera*) for use as a love drink. Dioscorides has passed on a complete recipe for making mandrake wine (περι μανδραγοριτοω):

Mandrake wine. Cut the cortex of the root and add 1/2 mine [= 8 ounces], wrapped in linen, to 1 metrete [= 36.4 liters] of must and allow to sit for three months, then pour into another container. The average dosage is 1/2 cotyle [= 5 ounces]. It is drunk with the addition of twice the amount of must. It is said that 1 hemine [= 10 ounces] of this, mixed with 1 chus [= 10 pounds = 120 ounces], will cause sleep and sedate; 1 cup, with xestes [= 1 pound, 8 ounces] of wine, can kill. When used correctly, it has analgesic effects and thickens the discharges. Whether used as a fumigant, as an enema, or as a drink, it has the same effects. (5.81).

When I wish to make mandrake wine, I add a handful (approximately 23 g) of chopped mandrake root (mandragorae radix conc.) to a bottle of retsina (0.7 liter). The mixture is then allowed to steep for a week. Do not filter out the root pieces; allow them to remain in the wine until it has been drunk. A few (two or three) cinnamon

Plants That Were Used as a Substitute for or to Counterfeit the True Mandrake

(From Brøndegaard and Dilg 1985; Dahl 1985; Emboden 1974*; Rätsch 1986, 1987, 1994; Wlislocki 1891, 90*; modified and supplemented.)

American mandrake[226] (= mandrake root)	*Podophyllum peltatum* L.
Belladonna root	*Atropa belladonna*
Calamus root	*Acorus calamus*
Canna	*Canna edulis* Ker-Gawl. (cf. **canna**), *Aureliana canadensis*[227]
Carrot root	*Daucus carota*
Cimbola, cimitrk root[228] (= common celandine)	*Chelidonium majus* L.
Common tormentil, five fingers	*Potentilla erecta* (L.) Räuschel [syn. *Tormentilla erecta* L.]
False mandrake	*Allium victorialis* L.
Galangal	*Alpinia officinarum* Hance, *A. galanga* (cf. *Kaempferia galanga*)
Ginseng (sometimes called the mandrake of the East)	*Panax ginseng*, *Panax pseudochinseng* Wall., *Panax* spp.
Iris root	*Iris pseudacorus* L.
Jangida root	*Withania somnifera*
Karengro root	*Orchis mascula* (L.) L.
Kougoed root (sometimes called the mandrake of the South)	*Sceletium tortuosum*
Orchids (sometimes called the mandrakes of the North)	*Orchis* spp.
Scopolia root	*Scopolia carniolica*
Shang-luh	*Phytolacca acinosa*
White bryony	*Bryonia cretica* L. ssp. *dioica* (Jacq.) Tutin[229] [syn. *Bryonia dioica* Jacq.], *Bryonia alba* L.

Left: The root of the true mandrake (*Mandragora officinarum*) is the most renowned magical agent in European history.

Right: In the early modern period, the rhizome of the Asian (= East Indian) *Alpinia galanga*, a member of the ginger family, was fraudulently sold as mandrake root.

226 The root of the American mandrake (= mayapple) is sold in North American "voodoo drugstores" as *mandrake root*. It is esteemed as a talisman. The root is said to bring luck in love (it makes the bearer of the talisman attractive and lovable in the eyes of the other sex), attract wealth (= money), protect from magical spells and the evil eye, and ward off demons. In general, the root of the American mandrake is attributed with the same properties as the true mandrake.

227 This taxon cannot be identified from the literature.

228 In northern Dalmatia, where chelidonium root is known as *cimitrk*, it is used in love magic in a very similar manner to the true mandrake (Mitrovic 1907, 233).

229 See Hylands and Masour (1982) for information on the chemistry of white bryony.

Because the true mandrake is not indigenous to central Europe, the fleshy root of the native bryony (*Bryonia* sp.) was once used to carve counterfeit "little mandrake men."

230 Women accused of being witches, including Joan of Arc, the Virgin of Orleans, were charged with having worn a mandrake on their chest (Schmidbauer 1969, 282).

231 The Biblical *dûdâ'îm* has been construed as *Cucumis dudaim* L., as *Citrus medica* L., and even as the common edible field mushroom (*Agaricus campestris* L.) and as jasmine (*Jasminum* spp.) (Moldenke and Moldenke 1986, 138*). Others have interpreted *dûdâ'îm* as flowerpots, cherries, jujube fruits (*Zizyphus*), blackberries, bananas (*Musa x sapientum*), and melons (*Cucumis aegypticus reticulatus*) (Friedreich 1966, 159 f.). "Luther, who used the translation 'lilies,' in particular got onto the wrong track, for these are the plant symbols of chastity, the greatest of all repressive achievements" (Müller-Ebeling n.d., 97).

sticks and 1 tablespoon of saffron (cf. *Crocus sativus*) can be added if desired; this will considerably improve the earthy, slightly bitter taste. One liqueur glass (40 to 60 ml of wine) is an effective dosage.

An aphrodisiac drink can also be made using the following recipe (after Miller 1988, 51*; modified):

1 bottle	white wine (variety as desired)
28 g	vanilla pods (*Vanilla planifolia* Andr.)
28 g	cinnamon sticks (*Cinnamomum verum* J.S. Presl)
28 g	rhubarb root (*Rheum officinale* Baill. or *R. palmatum* L.)
28 g	mandrake root (*Mandragora officinarum*)

Coarsely chop all of the ingredients and allow to steep in the wine for two weeks. Shake once a day if possible. Then filter the liquid through a sieve and, if desired, color with St. John's wort (*Hypericum perforatum* L.) or saffron (*Crocus sativus*). **Honey** (preferably in combination with royal jelly) may be added to sweeten the drink. Experiment to obtain the desired dosage.

The root pieces can also be added to any type of spirits (**alcohol**). Alcoholic drinks containing mandrake roots are still prepared in Romania, where it is said, "A few mandrake fibers in wine or schnapps keeps the bartender's customers" (Eliade 1982, 226).

In ancient times, the very thin root cortex of the mandrake was used in a variety of ways:

A juice is prepared from the cortex of the bark by crushing this while fresh and pressing this; it must then be placed in the sun and stored in an earthen vessel after it has thickened. The juice of the apples is prepared in a similar manner, but this yields a less potent juice. The cortex of the root that is pulled off all the way around is put on a string and hung up to store. Some boil the roots with wine until only a third part remains, clarify this and then put it away, so that they may use a cup of this for sleeplessness and immoderate pain, and also to induce lack of sensation in those who need to be cut or burned themselves. The juice, drunk in a weight of two obols with honey mead [cf. **mead**], brings up the mucus and the black bile like hellebore [*Veratrum album*]; the consumption of more will take life away. (Dioscorides 4.76)

The literature very rarely includes information about dosages. According to *Hagers Handbuch*, the therapeutic dosage is fifteen to thirty drops of the tincture of an aqueous extract of the root in alcohol (Roth et al. 1994, 485*). Thirty to fifty drops of the mother tincture will induce aphrodisiac/psychoactive effects.

Mandrake is purported to have been an ingredient in **witches' ointments**.[230]

Ritual Use

In ancient times, the primary ritual significance of the mandrake was in erotic cults. Because of the poor quality of the sources that have come down to us, however, only rudimentary information about these practices is available. The most important source about the use of mandrake in the Orient is the Old Testament, where the fruits (love apples) are mentioned numerous times under the Old Hebrew name *dûdâ'îm*, and namely as an aphrodisiac (not all interpreters of the Bible recognize the identification of *dûdâ'îm* with *Mandragora*).[231] According to Rabbi Jacob ben Asher (1269–1343), the name *dûdâ'îm* has its basis in numerology. The numerical value of the word is identical to that of the Hebrew word *ke'adam*, "like a human," and is an allusion to the anthropomorphic shape (Rosner 1993, 8). It is possible that the mandrake, which according to kabbalistic principles is a symbol for becoming one, may have been used in secret mystical rites in ancient Israel (Weinreb 1994, 252 –67).

The aphrodisiac quality was attributed primarily to the scent of the ripe, golden yellow fruits (Fleisher and Fleisher 1994). The Book of Genesis contains an account of what may have been an archaic magical ritual:

And Reuben went in the days of wheat harvest [May], and found mandrakes [*dûdâ'îm*] in the field, and brought them unto his mother Leah. Then Rachel said to [her sister] Leah, Give me, I pray thee, of thy son's mandrakes. And she said unto her, Is it a small matter that thou hast taken my husband? and wouldest thou take away my son's mandrakes also? And Rachel said, Therefore he shall lie with thee tonight for thy son's mandrakes. And Jacob came out of the field in the evening, and Leah went out to meet him, and said, Thou must come in unto me; for surely I have hired thee with my son's mandrakes. And he lay with her that night. And God hearkened unto Leah, and she conceived, and bare Jacob the fifth son. (Genesis 30:14–17)

A similar ritual with the magical fruit appears to be the basis of the much cited text of the erotic Song of Solomon: "[T]here will I give thee my loves, The mandrakes give a smell." (Song of Solomon 7:12, 13)

In the Near East today, the aromatic fruits of the mandrake are still regarded as aphrodisiacs (Fleisher and Fleisher 1994; Moldenke and Moldenke 1986, 137ff.*) and used in love magic (Rosner 1993, 7).

In another legend from the postbiblical period, the creation of the mandrake is attributed to Adam himself:

> When Adam was separated from his wife Eve for a long period, the long abstinence played a trick on him. He fantasized about her presence with such ardor that his semen, which shot forth as a result of the loving embrace and sprayed onto the floor, gave rise to a plant which took on human form, the *Caiumarath*, the mandragora. (Müller-Ebeling n.d., 97; Starck 1986, 21)

The most extensive description of this magical and erotic root and the ritual surrounding its harvest is from Flavius Josephus (first century), who wrote in Greek so that he could make the customs of the people of Judea more comprehensible to the Greeks. It is possible that he obtained his magical and botanical knowledge from the Essenes, with whom he lived for some time (Kottek 1994, 163*):

> In the valley that is on the north side of the city (Machairos)[232] is a special place known as Baaras, where a root of the same name grows. Every evening, it shines with a fire red glow[233]: But if someone wishes to approach it to pull it out, then it is difficult to take hold of, it pulls away from the hands and cannot be held until one has poured menstrual blood or urine onto it. But even then, any direct contact with the root means sudden death, unless one carries it in the hand so that the tip of the root faces down. Alone, one can also take possession of the root without any danger in the following way: You dig the earth up all around it, so that only a small piece of the root remains covered by the soil. Then a dog is tied to it. When this tries to follow the person who has tied it, it will of course quite easily pull the root from the ground. But in that moment the dog will die, as if to atone for he who has really taken away the plant. From then on, one can take hold of the root without fear. The reason that this root is so sought after in spite of its dangerous nature is to be found in its peculiar effects: for it has the power to dispel the so-called demons, that is the spirits of evil deceased persons who enter into the living and kill them if they are not aided, simply by coming near the afflicted. (Flavius Josephus, *History of the Judean War* 7.6, 3)

In ancient Egypt, mandrake fruits were used as gifts of love during courtship and probably were eaten as aphrodisiacs. The love plant appears to have been associated with Hathor, the goddess of love. The mandrake **beer** that was consecrated to

her played an important role in the famous myth describing the destruction of the human race and the creation of heaven (Brunner-Traut 1991, 101–6).

The sun god Ra was angry with humans because they had contrived to attack him. In his anger, he created the terrible lion-headed goddess Sekmet (an early form of Hathor) to punish the race of humans. She raged among the people for an entire day and was not yet finished when the sun set, for she wanted to utterly extinguish humanity. But Ra did not want this, and he thought of a trick to end the goddess's deadly rampage. He had mandrake fruits brought to him from Elephantine, an island in the Nile (Brugsch 1918, 31; Tercinet 1950, 17; Thompson 1968, 43); according to other versions and/or translations, hematite[234] or "red ocher" was brought as well (Brunner-Traut 1991, 103). At the same time, he ordered the production of a huge amount of barley beer (seven thousand jugs). He mixed the mandrakes (and the hematite or red ocher) into this and had the fields covered in the bloodred beer (the "sleeping drink"). The following sunrise, when the goddess saw the beer, she first perceived her reflection and thus recognized herself. Because of its red color, she thought the beer was human blood, and she eagerly drank it to the last drop. "Her countenance became gentle as a result, and she drank; this did her heart well. Drunken did she return, without having recognized the people" (Brunner-Traut 1991, 104).

Out of thankfulness, humans never again rose up against Ra. Sakmet transformed herself into the cow Hathor and carried Ra into the heavens.[235]

In commemoration of these dramatic events, which took place at the beginning of time, Ra established the Hathor festival (literally, "festival of drunkenness"), during which young maidens consecrated to the goddess would make a beer known as *sdr.t* (= "sleeping drink" [?]) using a similar recipe. The Hathor festivals were ecstatic orgies with obscene performances, sacrificial activities, and wild music (Cranach 1981*). Later, Hathor was celebrated as the inventor of beer and the "mistress of drunkenness without end" (Thompson 1968, 46).

The mandrake was also a sacred love magic in ancient Greece. Even the plant's collection took place under the auspices of the love goddess: "One should, it is said, make three circles around the mandrake with a sword and cut it while facing west.[236] And when cutting the second piece, one should dance around the plant and speak as much as possible about the mysteries of love" (Theophrastus, *History of Plants*, 9.8).

The Cypriot cult of Aphrodite developed directly out of the Oriental cult of the love goddess Ishtar, Astarte, Asherot, et cetera. J. Rendel Harris proposed the theory that the Greek cult of

232 In this city, located by the Red Sea, stood a giant rue tree (*peganon*, presumably *Ruta montana* [Kottek 1994, 130 f.] or *Peganum harmala*). Here too is where the lost cities of Sodom and Gomorrah were said to have been located, and John the Baptist was supposedly beheaded here (cf. Schlosser 1987, 88). The magical root, in other words, was at home in a completely "sinful" land.

233 This luminous plant (*baaras* is derived from the Hebrew *ba'ar*, "to burn") has been the subject of considerable discussion. To date, botanists know of no plant that gives off light during the night. In moist areas and rain forests, however, it has been observed that rotting roots or tree trunks phosphoresce and glow in an almost otherworldly manner at night. Rahner has proposed the following interpretation of the luminous mandrake, which I find very compelling: "The glow worms in Palestine like to rest upon the beautiful leaf rosettes of the spring mandrake, so that the plant looks like a glowing lamp. When persons looking for the plant approach, the light goes out, to reappear once more on another mandrake. Today, the Arabs in the Holy Land still refer to the mandrake as 'Sirag el-Kotrub,' 'devil's lamp'" (Rahner 1957, 210*).

234 Hematite, or bloodstone, is a common iron oxide mineral. It usually has a black or reddish luster. When hematite is ground or cut, the powder takes on a bloodred color.

235 "In an ancient Egyptian myth of the gods, it is told how the mandrake fruits came from Nubia, from whence they were brought to the temples and royal palaces of Egypt. There they were given to a goddess in a vessel filled with beer. This magic drink made her eyes shiny and caused her to enter into such an inebriated state 'that after sunrise she could no longer see'" (Kreuter 1982, 19*).

236 "Folk belief regarded the mandrake as a chthonic plant, one that was classed with the dark demons. For the place of these spirits is in the West, and so it was believed that a look to the West would banish the spirits of darkness. The Greeks directed their sacrifices for the dead and curses to the West. . . . One had to protect oneself from the wind [from the West], for otherwise the scent of the still unrisen plants could transmit the evil influence of the plant demon. But the look to the West can also mean that the rhizotome wished to insure himself of the powers of the nocturnal spirits that are thought to be present in the root and for this reason asked with a look for 'permission' as it were to remove it without harm" (Rahner 1957, 205*).

Aphrodite could be traced back to the Greek assimilation of the Oriental conceptions about the mandrake (Harris 1917). Aphrodite was also known as Mandragoritis ("she of the *Mandragora*"; Greek μανδραγοριτιζ η Αχροδιτη), a name passed down to us by Heschius (*Lexicon*; cf. Rahner 1957, 201, 364, note 21*; Schlosser 1987, 22; Thompson 1968, 55). The *Mandragora* thus had an intimate connection to the love goddess and was sacred to her (cf. *Papaver somniferum*). In the late ancient Mysteries of the Great Goddess, Aphrodite was identified with Hecate (Apuleius, *Metamorphoses*). Thus, the "mandrake of Hecate" was nothing other than the sacred plant of the love goddess.

The mandrake, and especially its root, was the plant of Hecate. The chthonic goddess (also known as Enodia or Trivia) was from Kairen (Asia Minor) and bore many Asian attributes. As the goddess of the three paths, she had three forms, three heads, and six arms. She was simultaneously rooted in heaven, on earth, and in the underworld. As the goddess of the nocturnal spooks (visual hallucinations), she was accompanied in her wild activities by barking dogs and noisy specters (auditory hallucinations). Hecate was both the harmful witch goddess (poison foods) and the goddess of birth (aphrodisiacs, erotic hallucinations). A description of Hecate's magical garden is contained in the saga of the Argonauts: "Many mandrakes grow within" (*Orph. Argonaut.* 922 f.).[237]

Eusebios (ca. 260–339 C.E.), the Christian and Greek church author, held Hecate, the goddess of the underworld, to be the "ruler of all evil demons," the "black one," or a "demon of love madness" that was the equivalent of Aphrodite. She is the mother of Italy's Circe and Colchis's Medea, the "cosmic super witch" (Luck 1962, 61*).[238] She sends to humans dampening sleep and heavy dreams and causes epilepsy (the "sacred disease") and madness (*mania*); in other words, she is capable of inducing altered states of consciousness. It almost appears as though the dark goddess would reveal herself only through the effects of the mandrake juice. As Democritus (ca. 470–380 B.C.E.), the "laughing philosopher," described in his lost work *Cheirokmeta* ("things made by hand"), one could invoke the goddess with the mandrake:

> Now it is known that the book *Cheirokmeta* is from Democritus. And how many adventurous things is this man able to report, who after Pythagoras was the most dedicated student of magic! He reports about the plant *aglaophotis*, which received its name because of the wonder of humans for its special color, and which thrives in the marble quarries of Arabia on the Persian side, which is why it is also called *marmaritis* [= marble plant]; the

magicians use it when they wish to invoke the gods. (Pliny 24.160)

Many late ancient conjurations (magic papyri) appeal to Hecate as the most important goddess. She was usually invoked in love magic, and then often in association with dogs, and even with Cerberus (Luck 1990, 129 ff.*). In the magic papyri, Medea was sometimes invoked in place of Hecate (Luck 1990, 50*).

The mandrake was also the plant of Circe (Dierbach 1833, 204*). The daughter of Helios, she was knowledgeable about magic and lived on the Italian coast above Sicily (Pliny 25.10 f.). Today, Monte Cicero remains as the sacred mountain of Circe.

It is possible that the mandrake was already identified with **moly** at an early date.

We can only speculate about the ritual use of mandrake in Germania. The mandrake apparently was associated primarily with love magic and divination but also with magical and ritual healing. Hildegard von Bingen described a small ritual to influence the psyche (a kind of ritual healing of depression):

> And when a person is so confused in his nature that he is always sad and always in distress, so that he often has weakness and pain in his heart, he should take mandrake after it has already been pulled with its root from the ground, and he should place it for a day and a night in a spring . . . and then place it, washed by the spring, next to him in his bed. The plant will become warm from his sweat, and he should say: "God, who makes humans from the dirt of the earth without pain. Now I place this earth, which has never been stepped over, beside me, so that my earth will also feel that peace which you have created." (*Physica* 1.56)

In Renaissance magic and in early modern occultism, mandrake was utilized as an **incense** that was under the influence of the moon. In Mecklenburg, mandrakes formerly were placed under the pillow so that one would have prophetic dreams (Schmidbauer 1969, 281 f.).

In Romania, an erotic mandrake cult existed into the early twentieth century (Eliade 1942). In this cult, the mandrake was regarded as the plant of life and death and was seen as an aphrodisiac and magical love agent. The plant was harvested during the full moon between Easter and Ascension Day. The harvest was conducted according to ritual guidelines:

> The plant should be collected without others knowing; . . . women and girls dance naked around the mandrake, sometimes they are

---

237 Dierbach (1833, 195*) considers the mandragora of Hecate to be the mandragora of Theophrastus and regards it as the deadly nightshade, *Atropa belladonna*. I am doubtful about this interpretation, as the deadly nightshade is not indigenous to the land of Hecate, whereas the mandrake is.

238 "In the myths and poetry of the Greeks, we learn about the great witches such as Circe and Medea. But perhaps these were not originally witches, but goddesses or priestesses of deities of a long-lost religion. Their knowledge of herbs, roots, and mushrooms represents primeval experiences whose secrets were guarded and which gave them special powers. They were priestesses in their own culture; subsequent generations made them into dangerous magicians" (Luck 1990, 46*). Similar things can be said about Roman poetry (cf. Luck 1962, 60 f.*).

content with simply letting down their hair. . . .
The couples caress and embrace one another.
To collect the leaves of the mandrake, the girls
lie on one another in the manner of the sex
act. . . .

Four young girls pluck the mandrake and
speak magical formulas over it; they bury it in
the middle of the street, where they then
dance completely naked. During the dance,
four young boys remain close by to watch over
them; they repeat:

"Mandrake, good mother,
marry me in this month,
if not in this, then in the next,
but make it so that I remain a girl no
longer." (Eliade 1982, 223, 219)

Offerings (salt, bread, sugar, **wine**, **alcohol**, eggs,
et cetera) were placed into the hole from which the
mandrake was taken. The root was carried home,
carefully washed, and worn as a talisman.

Artifacts

Since ancient times, mandrake root has been used
as an amulet and has been prepared for this
purpose (Scanziani 1972). The root was carved
into so-called mandrake men (*atzmann*, gallow's
man) or fashioned into dolls. These dolls needed
to be magically animated. For example, "an Italian
charlatan animated a human figure that had been
carved from a mandrake root using a hemp seed
[*Cannabis sativa*] placed in the pudenda" (von
Luschan 1891, 742). In southern Tirol, this custom
is still alive today:

The so-called "Galgenmandl" [little man of
the gallows], the root of the mandrake in
anthropomorphic from, could often be found
in the pantries and entryways of the houses of
the old farms. The "Galgenmandl" was
regarded as a good house spirit, and the
"Weibele" [little woman] as the protectress of
the grain chest. . . . In December 1968, a little
mandrake coffin was discovered during the
"Rauter over Francis' Festival." It had been
walled in over the door of the kitchen. It was
estimated to be 300 years old. (Fink 1983, 74)

There are even crucifixes make of mandrake
roots (root crosses) that are preserved in churches
as "miraculous objects" (Bauer 1993).

The mandrake apparently was brought from
Palestine to Egypt during the Eighteenth Dynasty
(New Kingdom; 1551–1305 B.C.E.); it became a
houseplant in Egypt (Germer 1985, 170*;
Manniche 1989, 117*). Gardens were sacred to the
goddess Hathor, which is why mandrakes were
also grown there (Hugonot 1992). A collar
containing mandrake fruits that were cut in half
was discovered in the grave of Tutankhamen

(Germer 1985, 171*). The yellow fruits (love
apples) appear often in Pharaonic art (Scanziani
1972, 50 f.). In the love songs of the New
Kingdom, the fruits (*rrm.t*) are often mentioned
in conjunction with lotus flowers (*Nymphaea
caerulea*) (Emboden 1989)[239]:

Celebrate a beautiful day! . . .
Give balsam and pleasant scents to your nose,
garlands of lotuses and love apples on your
chest,
while your wife, who is in your heart, sits by
your side.

The mandrake also appears in Greek poetry.
Lucian (app. 120–180 C.E.) states that a person
under the influence of *Mandragora* will fall asleep
(*Timon* 2). The fictional island of Hypnos
("sleep"), the island of the darkly rising dreams, is
the place "where only high shooting poppies
[*Papaver somniferum*] grow rampant and
Mandragora blooms, while silent butterflies, the
only birds of this land, flutter around" (Lucian,
Verae historiae 2.33).

A tiny fragment (thirty-two verses) from a
comedy by the Attic writer Alexis (app. 372–270
B.C.E.) has come down to us. It bears the name η
Μανδραγοριζομεψν, "the woman sedated by
Mandragora." This fragment demonstrates the use
of mandrake for purposes of love (Schlosser 1987,
46; Starck 1986, 8, 15).

The unscrupulous Renaissance politician and
writer Niccolò Machiavelli (1469–1527) wrote a
comedy, *La Mandragora*, about the power of the
mandrake to promote fertility (Schmidbauer
1969, 277; Tercinet 1950, 105). The plant also
appears in many other works of world literature,
e.g., by Apuleius (*Metamorphoses*, *The Golden
Ass*), Shakespeare (*Romeo and Juliet*, *Macbeth*, and
other works), E. T. A. Hoffmann (*The Little Zack*),
Goethe (*Faust*), Gustave Flaubert (*Salammbô*, *The
Temptation of Saint Anthony*), Marcel Schwob (*Le
Roi au Masque d'Or*), Gustav Meyrink, and many
others (Peters 1886; Tabor 1970*; Tercinet 1950).
In 1810, Friedrich Baron de la Motte Fouqué
(1777–1849) published *History of the Little Man of
the Gallows*, which influenced other literary works
(Fouqué 1983). The "little man of the gallows," the
mandrake produced by the last discharge of a
hanged man, has been the subject of numerous
literary works (Schlosser 1987). In 1940, John
Palmer published *Mandragora*, an English novel
that takes place in India and involves a number of
psychoactive drugs.

The occult author Hanns Heinz Ewers (1871–
1943) immortalized the magical root and the
female being that arises from it in his novel
Alraune (1911). There have been several film
versions of this book, the first in 1918 (director
unknown). The second, made in 1927 by director

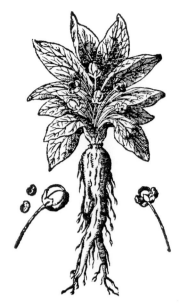

In former times, the true mandrake
(*Mandragora officinarum*) was
construed as a little mandrake man.
(Woodcut from the herbal of
Matthiolus, 1627)

239 Emboden (1989) has hypothesized that
both *Nymphaea caerulea* and
mandrake were used for shamanic
experiences or practices in dynastic
Egypt.

"The magic of roots is as old as humankind. Roots were at home in those places where the dead were buried and where one conceived of the underworld, the realm of shadows. And yet roots are the intermediaries of life. In a way that is difficult to comprehend, both realms seem to be near and familiar—to life and to death. It is in such magical domains that the spirits and enchanted ones live, the dwarves and the goblins. The mandrake is also a rare and fabulous being between the worlds—powerful and dangerous at the same time."

Marie-Luise Kreuter
Wunderkräfte der Natur
[Miraculous Powers of Nature]
(1982, 204*)

"On the Babylonian dedaim trees, human heads hang like fruits; mandrakes sing, the root Baaras slips through the grass."

Gustave Flaubert
The Temptation of Saint Anthony
(1979, 188*)

Clothed and "animated" little mandrake men. (Dutch copperplate engraving, eighteenth century)

240 In ancient times, representations of plants were used as illustrations in medicinal, pharmaceutical, and botanical works. Unfortunately, few of these pictures have been preserved; cf. Stückelberge 1994, 79.

and screenwriter Henrik Galeen, was also titled *Alraune.* The first sound version, directed by Richard Oswald, appeared in 1930 (Seesslen 1980, 93, 99; Seesslen and Weil 1980, 139). In the film, the mandrake is

the product of an act of artificial insemination in which a scientist uses a murderer's semen to impregnate a prostitute. He raises the child, primarily because he wishes to prove his theory that the character of a person is shaped much more by the environment and education than by nature. But even with the best intentions, Alraune can not be freed of her criminal disposition. But the film suggests that it is not just the genetic makeup that is responsible, but also the "soulless" act of its procreation, which makes her into a human monster, a mechanistic vamp whose sole purpose in life is to drive men to ruin, destruction, and suicide. (Seesslen 1980, 93)

A mandrake appears in a television version of Hoffmann's *The Little Zack.* A woman named Mandrake/Mandrax appears in the fantasy film *The Magic Bow* (USA 1981; directed by Nicholas Corea).

In his story *Alräunchen* [Little Alracine], esoteric author Manfred Kyber (1880–1933) tells of a child of the same name who is a "changeling, a little root man, who is rooted deep in the earth and becomes a changeling when the roots are ripped from the ground" (Kyber 1985).

In his 1990 novel *Another Roadside Attraction*, inspired in part by *Psilocybe* spp., American cult author Tom Robbins (born 1936) describes the effects of mandrake in a dialogue between Jesus and Tarzan: "John the Baptist turned me on to mandrake roots. It was a great experience, but once was enough." He [Jesus] shielded his eyes from the brilliant memory of the visions. "These days I am *stoned* on my own, one could say naturally" (Robbins 1987, 335).

The mandrake, the little mandrake man or woman, the little man of the gallows, and the themes related to this group of myths have been featured in illustrations since ancient times. In ancient Hellas, the mandrake became an important symbol of pharmacology and the healing arts. The oldest preserved manuscript of *De materia medica* (= Περι ψληζ ιατρικηζ), written about 68 C.E. by Dioscorides Pedanius of Anarzaba (Cilicia), is the *Codex Vindobonensis medicus graecus 1* (also known as the Viennese Dioscorides, ca. 512 C.E.). It contains an illustration that demonstrates the central role mandrake played in Greek pharmacy and medicine: Folio 5v shows an ancient studio in which book illustrations were made. The goddess Epinoia ("power of thought") is in the room,

holding a mandrake plant in her hands. To the left of her sits the illustrator,[240] who is capturing a little mandrake man on the canvas. Dioscorides is visible on the right, studying a book (on medicinal plants) and communicating information to the illustrator (Stückelberger 1994, 82, table 17). Dioscorides is also shown sitting in Folio 4 of the *Codex medicus graecae*, no. 5. Here, Hereusis, the goddess of scientific research, is handing him an anthropomorphic mandrake (Krug 1993, 107, fig. 42*). At the bottom of the illustration is a dying dog, the means by which it was possible to obtain the mandrake.

Since the dawn of the Middle Ages, the magical hunt for the mandrake has been a popular subject of illustrations in herbals and books on health (Heilmann 1973). In 1974, there was even an entire exhibit in Prague about mandrake illustrations from the fifteenth to the seventeenth centuries (Vrchotka 1974).

The mandrake has rarely been a subject of paintings. One work from Otto Boyer (nineteenth century), called *Alraun*, depicts an old witch offering Victorian ladies a little mandrake man.

In contrast, the mandrake has figured often in comic book art. Stimulated by John Donne's poem "Mandrake Root," Lee Falk began producing the comic series *Mandrake* in 1934. In this series, Mandrake is a mysterious magician from the Himalayas who is capable of overcoming both time and space. In the French comic *The Smurfs*, by Peyo, an "evil magician" is constantly using mandrakes to enchant the little blue mushroom dwellers (cf. **Veratrum album**). In the story "Blue Smurfs and Black Smurfs" (Carlsen Comics 1979), the magician uses mandrake roots and snake venom to make a little smurf (i.e., a homunculus). In the story "An Alchemist Awakes," author Alexis portrays his hero hallucinating the life of a normal citizen while under the influence of a "mandrake baked with cheese" (in *Einsame Phantasien*, 37–38; Linden: Volksverlag, 1983). The psychedelic comic artist Caza illustrated a phantasy story titled "Mandragore," in which a magical ritual transforms a mandrake dug out from under a gallows into a seductive but ominous woman (in *Gesammelte Werke* 4; Linden: Volksverlag, 1980). The Italian comic artist Paolo Eleuteri Serpieri published a volume titled *Mandragora* (1995) as part of his extremely bizarre and erotic science-fiction series *Morbus Gravis*, which is heavily influenced by psychedelic experiences. The story is actually about a "miracle flower." One comic artist of the psychedelic avant garde has even identified himself with the plant and now draws under the pseudonym Mandryko (Rätsch 1986, 97–99).

The mandrake has left few traces in music. The magical root was the subject of the rock band Deep Purple's "Mandrake Root" (Oh-Boy 1-9048, bootleg; original on the album *Shades of Deep*

Purple, EMI, 1968). The psychedelic band Gong released a song entitled "Mandrake" on their album *Shamal* (Virgin Records 1975/1989). The acid-rock band Mandrake Paddle Steamer, active from 1969 to 1971, took their name from the plant (Forgotten Jewels Records FJ 001, 1989). In the mid-1990s, a neopsychedelic underground band named Mandragora formed in England and released an album titled *Over the Moon* (Delec CD 027, ca. 1995).

Medicinal Use

The ancient Assyrians used mandrake as an analgesic and anesthetic. It was administered to treat toothaches, complications associated with childbirth, and hemorrhoids. The root was powdered and given in **beer** to treat stomach ailments and burned as a fumigant to dispel "poison from the flesh" (exorcism) (Thompson 1949, 218f.*). In Egypt, mandrake has certainly been used for medicinal purposes since the beginning of the New Kingdom. The identification of mandrake in the even older *Papyrus Ebers* (ca. 1600 B.C.E.) is contested (Heide 1921). If it is correct, however, then the *Papyrus Ebers* contains seven recipes with mandrake (even "mandrakes from Elephantine"). These include preparations to treat *pend* worms, pain (or "pain demons"), inflammations of the skin, and bone pain; to "make the skin smooth"; "to soften hardening of the joints"; and to treat a "sick tongue."

Few plants in ancient times had such a wide spectrum of uses as the mandrake. It was used as a sleeping agent, analgesic and anesthetic, antidote, abortifacient, aphrodisiac, and inebriant and also in love magic. The medical indications for which it was used were correspondingly numerous and included the following ailments: abscesses, arthritis, inflammations and diseases of the eyes, discharge, anxiety, possession, depression, swollen glands, inflammation, uterine inflammation, complications during labor, painful joints, tumors, ulcers, gout, hemorrhoids, skin inflammations, hip pains, hysteria, impotence, bone pains, headaches, cramps, liver pains, stomach ailments, melancholy, menstrual problems, amenorrhea, spleen pains, sleeplessness, snakebite, pain, side pains, scrofula, tubercles, infertility, poisoning, callosities, loss of speech, worms, wounds, erysipelas, and toothaches (Rätsch 1994).

The use of mandrake root as a sleeping agent was a widespread practice in early ancient times (Valette 1990, 468*). In fact, the term *hypò mandragóra katheúdein* (literally, "sleep under the mandrake") was synonymous with *sleepy*. Two recipes for such use have been preserved in the late ancient Leiden papyrus (Griffith and Thompson 1974*):

Another [agent] when you wish to have a man sleep for two days: mandrake root [μανδρακοροσ ριζα], one ounce; licorice [?], one ounce; henbane [*Hyoscyamus muticus*], one ounce; ivy [*Hedera helix*], one ounce; you crush these together. . . . If you wish to do this skillfully, give to each part the four-fold amount of **wine**, you moisten everything from the morning until the evening, you shake it off, you have it drunk; very good. (*Col.* 24.6–14)

The *Corpus Hippocraticum* notes that mandrake root was used as a sedative and anesthetic agent, as well as a remedy for psychological anxiety and depression. The Hippocratics used mandrake to treat melancholy (*Corp. Hippocrat.* 420, 19) and for severe spasms (Berendes 1891, 223*).

According to Aristotle (384–322 B.C.E.), the mandrake, along with opium (cf. *Papaver somniferum*), wine (cf. *Vitis vinifera*), and bearded darnel (*Lolium temulentum*), belonged to the Hypnotica class of plants; in his work *On Sleep and Wakefulness*, he listed mandrake as a sleeping agent (Kreuter 1982, 24*; Starck 1986, 8). In *The Republic*, the philosopher Plato (427–347 B.C.E.) described *Mandragora* as an anesthetic comparable to **mead** (488c). The Greek physician Aretaios (second century C.E.) referred to *Mandragora* as an anesthetic for use in surgical procedures. Generally speaking, mandrake was the most important narcotic or anesthetic agent of ancient and late ancient times and in the Middle Ages (cf. **soporific sponge**).

The physician and natural scientist Aulus Cornelius Celsus, who was active in the time of Tiberius (14–37 C.E.), mentioned mandrake fruits as a sedative; he used the root to treat the flow of mucus in the eyes and a decoction of the root as a remedy for toothache (121, III, Ch. 18). He wrote:

There is another, more effective method of inducing sleep. Crush mandrake with opium and henbane seeds [*Hyoscyamus niger*] in wine.

For headaches, ulcers, inflammations of the uterus, hip pains, liver, spleen, or side pains, or for all cases of female hysteria and loss of language, a bolus of the following recipe, supported by rest, will heal the affliction. A drachma each of silica, acorns, Syrian rue [*Peganum harmala*]. Two drachmas each of rhizinus and cinnamon; three drachmas each of opium, panacea root, dried mandrake fruits, flowers of the round cyperus grass [cf. *Cyperus* spp.], and 56 peppercorns [cf. *Piper* spp.]. Each must be crushed alone and then everything should be mixed together. Passum should be added from time to time, so that it acquires a certain consistency.

The root of the greater celandine (*Chelidonium majus* L.) was used in magical contexts as a substitute for the true mandrake. Although the greater celandine is a member of the Poppy Family (Papaveraceae), it is not known to induce any psychoactive effects. (Woodcut from Fuchs, *Läebliche Abbildung und Contrafaytung aller Kreüter*, 1545)

"If you so much as turn the black of your eye, then a little mandrake will slip on by!"

APULEIUS
METAMORPHOSES
(2.22)

"The true mandragoras is the 'tree of knowledge' and the love that arises from consuming it is the source of the human race."

HUGO RAHNER
GRIECHSICHE MYTHEN IN CHRISTLICHER DEUTUNG [GREEK MYTHS IN CHRISTIAN INTERPRETATION]
(1957, 221*)

These little mandrake men from southwest Asia were carved from roots of *Mandragora officinarum*. They were used to ensure good fortune and as love magic. (From von Luschan 1891)

A small amount is given in the form of a little ball or, when dissolved in water, as an enema. (In Thompson 1968, 101 f.)

Rufus of Ephesus (first century C.E.) mixed a decoction of mandrake root with poppy (*Papaver somniferum*) and chamomile (*Matricaria recutita*) (Tercinet 1950, 24). The renowned physician Galen (131–210 C.E.) has passed down an interesting recipe, calling for a preparation of mandrake root mixed with myrrh (*Commiphora* sp.), cassia (*Cinnamomum cassia* Blume), cedar (*Cedrus libani* Rich.), pepper (*Piper* spp.), saffron (*Crocus sativus*), and henbane seeds (*Hyoscyamus niger*) as an application to painful areas of the body (13.92; cf. Tercinet 1950, 24). The important role mandrake played as an analgesic is confirmed in a variety of sources, including Serenus Samonicus (first century C.E.). Even into the early Renaissance, mandrake preparations were the sole anesthetic (cf. **soporific sponge**).

In an early medieval Persian manuscript that presumably drew upon much older original sources, *Mandragora* is listed along with opium, *Datura metel*, and hemp (*Cannabis indica*) as agents for inducing sleep (Berendes 1891, 43*).

In his book *The Passover Plot*, the Englishman Hugh J. Schonfield has argued that "Jesus was given a sponge dipped in vinegar while on the cross, which is a third, albeit very hidden indication of *Mandragora* in the gospels." Schonfield believes that the vinegar contained mandrake juice, which induced a deathlike state in Christ. This was done so that he could be removed from the cross as quickly as possible and be brought back to life with the aid of a physician. "The plan failed when one of the soldiers—unexpectedly and entirely in violation of the rules—pierced Christ in the side with his lance" (Hansen 1981, 27 f.*). It is not possible to determine whether this story is true or even has an element of truth to it. However, it was customary among the Romans to administer to those who were being crucified a mandrake wine, which the literature of the early Middle Ages (fifth century) referred to as *morion*, "death" drink (Thompson 1968, 225). In general, even into the early modern period, it was a common practice to give mandrake preparations to condemned people before their sentences (torture, execution) were carried out. The story of the little man of the gallows becomes understandable in this light (see Schlosser 1987; cf. Beckmann 1990, 130*):

And thus the hangman dripped the juice of crushed seeds [of henbane, deadly nightshade, and mandrake fruits, or fly agaric mushrooms] into the water which they, only appearing to be heartless, would use to refresh the unconscious victim during their tortures and awaken them to new torments. (Golowin 1970, 30)

The Ancient Meaning of the Mandrake (*Mandragora*)

There are a number of related roots and magical plants that are regarded as anthropomorphic and have magical effects.

There are the (three) species of *Mandragora*, two of which are similar and are construed as male and female.

The fruits are the love apples; these are female.

The root is a male phallus.

The mandrake is a plant of the gods:
— The love apples are sacred to the love goddess (Astarte, Aphrodite, Hathor, etc.).
— The root is consecrated to the chthonic deity of the underworld (Hecate).
— It is a phallic plant of the god of the heavens and lightning (Ra, Zeus).

The plant may be harvested only ritually (through magical acts, conjurations, sacrifices).

The plant, and especially the root, is an amulet.

The plant is a medicine and provides a typical *pharmakon*:
— It is a poison and can kill.
— It stimulates fertility and gives life.

The plant is an aphrodisiac:
— The fruits are the "love" apples.
— The scent of the fruits incites desire.
— The root ensures the readiness of the opposite sex for love.
— The root products stimulate potency.

The root and its juice are the source of a medicinally valuable narcotic; it is:
— analgesic
— sedative
— anesthetic

The mandrake was the connection between
— heaven and earth
— divine favor and human art

The mandrake was added to alcoholic drinks (beer, wine, etc.) to improve their psychoactive effects.

In Romania, mandrake has been used for a variety of folk medicinal purposes. A decoction of mandrake was applied externally and/or ingested for pains in the limbs, the sacrum, and the back and for fevers. Fresh leaves were chewed to treat toothaches. Mandrake leaves were burned and the resulting smoke inhaled for coughs (Eliade 1982, 227). Mandrake also was burned to treat headaches (cf. **incense**). For this purpose, pieces of mandrake roots were combined with mugwort (*Artemisia* spp.), mint (*Mentha* spp., *Mentha pulegium*), and cloves.

The root was used in Russian folk medicine in similar ways (Rowell 1978, 269*). The root also found use in European folk medicine. In her

Herbal, Elisabeth Blackwell wrote: "This plant is used externally for all inflammations, acute ulcers, and enlarged and hardened glands. Some drip the juice into the eyes to treat heat and redness in the same. Because this plant is rarely found in this land, henbane [*Hyoscyamus niger*] is commonly used in its place (for example in the unguent Populeon [cf. **witches' ointments**])" (Heilmann 1984, 94*). In southern Tirol, mandrake juice was rubbed onto women in labor who were giving birth at home to alleviate their pains (Fink 1983, 238).

In homeopathy, preparations of mandrake made from the root (Mandragora hom. *HAB34*, Mandragora officinarum hom. *HPUS88*, Mandragora e radice siccato hom. *HAB1*, Mandragora, ethanol Decoctum hom. *HAB1*) are administered in accordance with the medical description for such ailments as headaches.

Constituents

Mandrake, especially its root (0.3 to 0.4%), but also its leaves, contains the psychoactive and anticholinergic **tropane alkaloids scopolamine** ([L]-scopolamine/[D,L]-scopolamine; Roth et al. 1994*), atropine, apoatropine, L-hyoscyamine, mandragorine, cuscohygrine (= bellaradin),[241] nor-hyoscyamine (= solandrine), 3α-tigloyloxytropane, and 3,6-ditigloyloxytropane (Jackson and Berry 1973 and 1979; Maugini 1959; Staub 1962). This alkaloid mixture was previously described under the name *mandragorine* (Ahrens 1889; Hesse 1901).

The dried root contains between 0.2 and 0.6% alkaloids. The tropane alkaloid belladonnine occurs only in the dried roots (Jackson and Berry 1973). In addition to the alkaloids, the root also contains coumarins (scopoline, **scopoletin**), sitosterol, sugars (rhamnose, glucose, fructose, saccharose), and starches (Müller 1982; Tercinet 1950).

It was once thought that the fruits were poisonous and hence inedible; they are actually quite harmless, containing only trace amounts of alkaloids (Germer 1985, 170*). The fruits also contain β-methylesculetin.

The aromatic components of the scent of the mandrake fruit have recently been chemically identified. The composition, and especially the high concentrations of sulfuric chemicals, is unusual for an aromatic substance. The **essential oil** is composed primarily of ethyl acetate, ethyl butyrate, butyl acetate, butanol, butyl butyrate, hexyl acetate, hexanol, ethyl octanoate, ethyl-3-hydroxy butyrate, 3-methyl thiopropanol, 3-phenylpropanol, and eugenol. Also present are methyl butyrate, ethyl-2-methyl butyrate, hexanal, propyl butyrate, limonene, (E)-2-hexanal, ethyl hexanoate, amyl alcohol, 3-hydroxy-2-butanone, isopropyl benzole, propyl hexanoate, hexyl butyrate, octyl acetate, benzaldehyde, indanone, linalool, octanol, ethyl 3-methyl-thiobutyrate, ethyl decanoate, ethyl benzoate, α-terpinol, γ-hexa

lactone, benzyl acetate, carvon, decanol, isobutyl decanoate, β-phenethyl isobutyrate, ethyl laurate, benzyl alcohol, henylethyl alcohol, 3-phenylpropyl acetate, methyl eugenol, γ-octalactone, 2-ethyl-4-hydroxy-5-methyl-3(2H)-furanone, ethyl cinnamate, γ-decalactone, (E)-cinnamyl acetate, cinnamoyl alcohol, (E)-isoeugenol, γ-dodecalatone, and vanillin (Fleisher and Fleisher 1992 and 1994).

Effects

Although the mandrake has been one of the most renowned of all psychoactive medicinal plants for millennia and has provided inspiration for a great number of authors, the enormous amount of literature it has generated contains very few experiential reports. One of the earliest is from the church patriarch Augustine (354–430 C.E.), who reports in his own words that he bit into a root but only "found a disgusting bitter taste" (Rahner 1957, 201*). The late ancient lexicographer Suidas stated that the mandrake has "a fruit which has hypnotic effects and allows everything to sink into forgetfulness" (*Lexicon* 136; *Lexicographi Graeci* 3.317). The folk saint and seeress Hildegard von Bingen (1098–1179 C.E.) noted that the mandrake produced "illusions" (*Physica* 1, 56).

Schenk (1954, 36*) claimed that the root produces "inebriation, narcosis, hallucinations, visions"; he provided only a single example:

> Here is also the place to mention the peculiar case of a 40-year-old painter who was the victim of an unusual *Mandragora* poisoning. Since his childhood, he had suffered from headaches during the foehn.[242] An acquaintance advised him to treat his affliction with a tea cure, specifically with a "mandrake tea." He procured three mandrake roots, boiled them, and drank several cups of the decoction. The next day, his pupils were extremely dilated, his mouth was dry, and apart from a slight sensation of dizziness, he was otherwise without complaints. During the next three days, he repeatedly drank of this tea. In the meantime, however, the alkaloids had certainly become more concentrated, for the roots had been left in the teapot. On the fourth day, the painter was in an utterly confused state, his face had become very red, and he was racing around in his dwelling. He carried his bed into the stairwell, and tried to throw furniture and paintings from the window. The horrified landlady called the doctor, who had him taken to the hospital. His face remained red for some time, the pupils were immoderately dilated. He fumbled with the bed covers with his hands. He had lost any sense of orientation. But after two days, he was able to be released without any complaints. (Schenk 1954, 37 f.*)

"The Mandragorae is a plant which causes such a deep sleep that one could cut a person and they would not feel the pain. For the mandragoras symbolize aspiration through contemplation. This tranquility makes it possible for a person to fall into a sleep of such delightful sweetness that he will no longer feel anything of the cutting which his earthly enemies visit upon him, that he no longer pays attention to any worldly things. For the soul has now closed its eyes to all that is outside—it lies in the good sleep of the internal."

THOMAS CISTERCIENSIS
HOHELIED-EXEGESE [SONG OF SOLOMON EXEGESIS]
(CITED IN MÜLLER-EBELING 1987, 145)

"The symbolism of its form and its hallucinatory effects made the mandrake into a mythical plant that was a part of two worlds, the worldly and the underworldly."

JACQUES GÉLIS
DIE GEBURT [THE BIRTH]
(1989, 63)

In acid rock music, *Mandragora* has often been used as a symbol for psychedelic and heavenly experiences. (Record cover, ca. 1969)

241 According to Schultes and Hofmann (1980, 298*), cuscohygrine and mandragorine are identical. The empirical formula of the alkaloid "mandragorine" is $C_{15}H_{19}NO_2$ (Roth et al. 1994: 485*).

242 *Translator's note*: The foehn is a high-pressure weather pattern common in southern Germany.

"There is an animal that is called the elephant. No sexual drive dwells within this animal. When it wishes to produce children, it goes back to the East, close to paradise. It is there that the so-called mandragora tree grows. That is where the female and the male go. The female takes the fruit from the tree first, offers it to her spouse as well, and plays with him until he too takes of it, and when he has eaten, he unites with the female from behind, because of the fact that they do not have any harmony with one another. Only once does he have coitus, and she immediately becomes pregnant."

PHYSIOLOGUS (43)[243]

The mandrake has been the subject of numerous literary treatments. In the age of comic books, it became a magician with mysterious supernatural powers. (Title page of an Italian edition; the original appeared in 1941)

243 As late as the early Middle Ages, the story of the male elephant eating mandrake leaves that grow near the Garden of Eden as an aphrodisiac before mating was often found in books about animals (Hansen 1981, 32*).

Roth et al. (1994, 485*) maintain that the effects of mandrake are similar to those of *Atropa belladonna*. Typical clinical symptoms include dryness of the mouth and other mucous membranes, dilated pupils, farsightedness, and muscular atony, as well as, according to Roth et al. (1994, 485), increase in the frequency of the pulse. All of these symptoms of "mandrake intoxication" are very similar to the homeopathic medical description (cf. Mandl 1985, 133*).

On several afternoons, I have consumed a glass of wine in which mandrake root had been steeped. The effects become apparent in some fifteen to twenty minutes. These are associated with a slight sense of euphoria. Pleasant, sometimes lusty sensations run through the body. Visual perception is only mildly affected; a slight farsightedness manifests. During the nights that have followed the consumption of mandrake wine, I have always experienced increased dream activity, often with erotic content. After consuming 0.5 liter of mandrake beer, I noted the following:

The effects of the alcohol are not apparent. I notice a slight pressure in my head, as also occurs with henbane or thorn apple. It is more fun to dance than to sit at the computer. It is a desire to fall into the rhythm of the music. . . . Obliviousness to self, pleasurable bodily sensations . . . pleasant tingling on the scalp. Slightly dry lips, distinct changes in the visual field, as if the perspective had shifted somewhat. (Protocol, December 28, 1994)

When I was in Cyprus, I ate all of the ripe mandrake fruits I could find in order to test their aphrodisiac or mind-altering properties. I did not notice any direct psychoactive effects. But I had an increased number of erotic dreams during those nights.

Commercial Forms and Regulations

When *Mandragora officinarum* makes it into the pharmaceutical trade, it is normally in the form of the chopped root (mandragorae radix conc.). In Germany, Mandragora conc. can be purchased only from a pharmacy, and because of recent regulatory changes, a prescription is now needed. Even the homeopathic preparations now require a prescription. The seeds are only very rarely available through flower or ethnobotanical sources.

Literature

See also the entries for *Atropa belladonna*, *Mandragora* spp., witches' ointments, tropane alkaloids, and scopolamine.

Ahrens, F. B. 1889. Über das Mandragorin. *Berichte der Deutschen Chemischen Gesellschaft* 22:2159.

Bauer, Wolfgang. 1993. Das wundertätige Wurzelkreuz in der Kirche von Maria Straßenengel. *Integration* 4:39–43.

Brøndegarrd, V. J., and Peter Dilg. 1985. Orchideen als Aphrodisiaca. In *Ethnobotanik*, ed. V. J. Brøndegarrd, 135–57. Berlin: Mensch und Leben.

Brugsch, Heinrich. 1918. Die Alraune als ägyptische Zauberpflanze. *Zeitschrift für ägyptische Sprache und Altertumskunde* 29:31–33.

Brunner-Traut, Emma. 1991. *Altägyptische Märchen.* Munich: Diederichs.

Dahl, Jürgen. 1985. Die Zauberwurzel der kleinen Leute . . . *Natur* 6/85:83–84.

Eliade, Mircea. 1942. Le Mandragore et le mythe de la "naissance miraculeuse." *Zalmoxis* 3:3–48.

———. 1982. *Von Zalmoxis zu Dschingis-Khan.* Cologne: Hohenheim.

Emboden, William. 1989. The sacred journey in dynastic Egypt: Shamanistic trance in the context of the narcotic water lily and the mandrake. *Journal of Psychoactive Drugs* 21 (1): 61–75.

Ewers, Hanns Heinz. 1911. *Alraune—Die Geschichte eines lebenden Wesens.* Munich: Georg Müller Verlag.

Fink, Hans. 1983. *Verzaubertes Land: Volkskult und Ahnenbrauch in Südtirol.* Innsbruck and Vienna: Tyrolia.

Fleisher, Alexander, and Zhenia Fleisher. 1992. The odoriferous principle of mandrake, *Mandragora officinarum* L. Aromatic plants of Holy Land and the Sinai. Part IX. *Journal of Essential Oil Research* 4:187–88.

———. 1994. The fragrance of biblical mandrake. *Economic Botany* 48 (3): 243–51.

Fouqué, Friedrich de la Motte. 1983. Eine Geschichte vom Galgenmännlein. In *Teufelsträume— phantastische Geschichten des 19. Jahrhunderts,* ed. Horst Heidtmann, 7–33. Munich: dtv.

Frazer, J. 1917. Jacob and the mandrakes. *Proceedings of the British Academy* 8:346 ff.

Gélis, Jacques. 1989. *Die Geburt.* Munich: Diederichs.

Golowin, Sergius. 1970. *Hexer und Henker im Galgenfeld.* Bern: Benteli.

Harris, J. Rendel. 1917. The origin of the cult of Aphrodite. John Rylands Library *Bulletin* 3:354–81. (Published in Manchester, England.)

Hartwich, Carl. 1911. Die Mandragorawurzel. *Schweizerische Wochenschrift für Chemie und Pharmazie,* no. 20. (Published in Zurich.)

Heide, Frits. 1921. Alrunen i det gamle Ägypten. *Tidsskrift for Historisk Botanik* 1:21.

Heilmann, Karl Eugen. 1973. *Kräuterbücher in Bild und Geschichte.* Munich: Kölbl. (Contains numerous illustrations of mandrakes from various works.)

Hesse, O. 1901. Über die Alkaloide der Mandragorawurzel. *Journal für praktische Chemie* 172:274–86.

Hugonot, J.-C. 1992. Ägyptische Gärten. In *Der*

Garten von der Antike bis zum Mittelalter, ed. M. Carroll-Spillecke, 9–44. Mainz, Germany: Philipp von Zabern.

Hylands, Peter J., and El-Sayed S. Mansour. 1982. A revision of the structure of cucurbitacin S from *Bryonia dioica*. *Phytochemistry* 21 (11): 2703–7.

Jackson, Betty P., and Michael I. Berry. 1973. Hydroxytropane tiglates in the roots of *Mandragora* species. *Phytochemistry* 12:1165–66.

———. 1979. *Mandragora*—taxonomy and chemistry of the European species. In *The biology and taxonomy of the Solanaceae*, ed. J. G. Hawkes et al., 505–12. London: Academic Press.

Killermann, H. 1917. Der Alraun (Mandragora). *Naturwissenschaftliche Wochenschrift*, n.s., 16:137–44.

Krauss, Friedrich S. 1913. Ein Altwiener Alraunmännchen. *Anthropophyteia* 10:29–33.

Kyber, Manfred. 1985. *Das Manfred Kyber Buch*. Reinbek, Germany: Rowohlt.

Marzell, Heinrich. 1927. Alraun. In *Handwörterbuch des Deutschen Aberglaubens*, 1:311–23. Berlin: de Gruyter.

Maugini, E. 1959. Ricerce sul Genere *Mandragora*. *Nuovo Giornale Botanico Italiano e Bolletino della Societa Botanica Italiana*, n.s., 66 (1–2): 34–60.

Mitrovic, Alexander. 1907. Mein Besuch bei einer Zauberfrau in Norddalmatien. *Anthropophyteia* 4:227–36.

Müller-Ebeling, Claudia. 1987. Die Alraune in der Bibel. In *Die Sage vom Galgenmännlein im Volksglauben und in der Literatur*, by Alfred Schlosser (orig. pub. 1912), 141–49. Berlin: Express Edition.

———. n.d. Die Alraune in der Bibel. In *Das Böse Bibel Buch*, by Roland Ranke Rippchen, 97–100. Löhrbach: Werner Pieper's MedienXperimente.

Palmer, John. 1940. *Mandragora*. London: Victor Gollancz.

Peters, Hermann. 1886. Alraune. *Mitteilungen aus dem germanischen Nationalmuseum* 1 (1884–86): 243–46.

Randolph, Ch. Brewster. 1905. The mandragora of the ancients in folklore and medicine. *Proceedings of the American Academy of Arts and Sciences* 40:487–537.

Rätsch, Christian. 1986. Die Alraune heute. In *Der Alraun: Ein Beitrag zur Pflanzensagenkunde*, by Adolf Taylor Starck (orig. pub. 1917), 87–109. Berlin: Express Edition.

———. 1987. Einleitung. In *Die Sage vom Galgenmännlein im Volksglauben und in der Literatur*, by Alfred Schlosser (orig. pub. 1912), vii–xxiv. Berlin: Express Edition.

———. 1994. Die Alraune in der Antike. *Annali dei Musei Civici dei Rovereto* 10:249–96.

Robbins, Tom. 1971. *Another roadside attraction*. Garden City, N.Y.: Doubleday.

Rosner, Fred. 1980. Mandrakes and other aphrodisiacs in the Bible and Talmud. *Koroth* 7. (Published in Jerusalem.)

———. 1993. Pharmacology and dietics in the Bible and Talmud. In *The healing past: Pharmaceuticals in the biblical and rabbinic world*, ed. Irene and Walter Jacob, 1–26. Leiden: Brill.

Scanziani, Piero. 1972. *Amuleti e Talismani*. Chiasso, Switzerland: Elvetica Edizioni SA.

Schlosser, Alfred. 1987. *Die Sage vom Galgenmännlein im Volksglauben und in der Literatur*. Berlin: Express Edition. (Orig. pub. 1912.)

Schmidbauer, Wolfgang. 1969. Die magische Mandragora. *Antaios* 10:274–86.

Scholz, E. 1995. Alraunenfrüchte—ein biblisches Aphrodisiakum. *Zeitschrift für Phytotherapie* 16:109–10.

Seesslen, Georg. 1980. *Kino des Utopischen*. Reinbek, Germany: Rowohlt.

Seesslen, Georg, and Claudius Weil. 1980. *Kino des Phantastischen*. Reinbek, Germany: Rowohlt.

Starck, Adolf Taylor. 1986. *Der Alraun: Ein Beitrag zur Pflanzensagenkunde*. Berlin: Express Edition. (Orig. pub. 1917.)

Staub, H. 1942. Non-alkaloid constituents of mandrake root. *Helvetica Chimica Acta* 25:649–83.

———. 1962. The alkaloid constituents of mandragora root. *Helvetica Chimica Acta* 45:2297.

Stückelberger, Alfred. 1994. *Bild und Wort*. Mainz, Germany: Philipp von Zabern.

Tercinet, Louis. 1950. *Mandragore, qui es-tu?* Paris: Édité par l'Auteur.

Thompson, C. J. S. 1968. *The mystic mandrake*. New York: University Books.

Vaccari, A. 1955. La Mandragora, erba magica. *Fitoterapia* 26:553–59.

von Luschan, F. 1981. [Untitled]. In *Verhandlungen der Berliner Gesellschaft für Anthropologie, Ethnologie und Urgeschichte* 1891:726–46.

Vrchotka, Jaroslav. 1974. *Mandragora: Illustrovaná Kniha Vĕdecká 15.–17. Století*. Prague: Nationalmuseum.

Weinreb, Friedrich. 1994. *Schöpfung im Wort: Die Struktur der Bibel in jüdischer Überlieferung*. Weiler im Allgäu, Germany: Thauros Verlag. (Pages 252–67 are on the mandrake in the Bible.)

Winter, Gayan S. 1997. *Die Nacht der Mandragora*. Munich: Heyne.

"There is a trick in which an egg that is placed under a hen produces a humanlike figure, as I myself have seen and am also able to perform. The magicians attribute miraculous powers to one such figure, which they call mandrake."

Heinrich Cornelius Agrippa von Nettesheim
Die magische Werke [The Magical Works]
(ca. 1510)

Mandragora spp.

Mandrake Species

The autumn mandrake (*Mandragora autumnalis*) was once regarded as the "little mandrake woman." (Woodcut from the herbal of Matthiolus, 1627)

The leaf crown of *Mandragora autumnalis*.

Family

Solanaceae (Nightshade Family); Subfamily Solanoideae, Solaneae Tribe, Mandragorinae Subtribe

Currently, four to six species are botanically accepted in the genus *Mandragora*. All occur only in Eurasia and northern Africa (D'Arcy 1991, 78 f.*; Symon 1991, 147*).

Mandragora autumnalis Spreng. [syn. *Mandragora autumnalis* Bertol., *Mandragora microcarpa* Bertol., *Mandragora foemina* Gersault, *Mandragora foemina* Thell., *Mandragora haussknechtii* Heldr., *Mandragora officinalis* Moris ex Miller, *Mandragora officinarum* Bertol. non Linnaeus]—autumn mandrake

Since ancient times, the autumn mandrake has been known as the "female" mandrake, the counterpart to the "male" *Mandragora officinarum*. It was said:

> One kind [of *Mandragora*] is female, black, called *thridakias*, it has narrower and smaller leaves with an ugly and pungent scent, is spread over the earth, and has apples like the fruits of the rowan tree [*Sorbus domestica* L.], yellow, pleasant scented, among them a fruit like the pear, the roots are very large, two or three, grown together, black on the outside, white within, and with a thick rind. It does not develop a stalk. (Dioscorides 4.76)

Pliny (first century) reported on the psychoactive and medicinal properties of the plant:

> It is not the mandrake of all countries which produces a juice; but if it does provide one, then it is collected at the time of the wine harvest [cf. *Vitis vinifera*]. It then has a potent aroma, that of the root, but mostly that of the fruit. The fruit is collected when it is ripe; it is dried in the shade, and the juice is thickened by the sun after it has been extracted. The same is done with the juice of the roots, which is obtained either by pressing or by boiling in

red wine until it has been reduced to a third. The leaves are best stored in a strong brine (salt water); their juice is a poison that cannot be healed; this harmful property is not completely removed by the brine when the leaves are stored therein. Its specific scent is oppressing to the head, but there are countries in which the fruits are eaten. Persons that are unfamiliar with its properties are convinced that the scent of this plant would make them mute and that too high a dosage of the juice is deadly. If a dosage is given that is appropriate to the strength of the patient, the juice has an anesthetic effect; a medium dosage is a cyathus. It is also given for wounds from snakes and before the body is cut or pierced to lower the sensitivity to pain. (Pliny 25)

Mandragora autumnalis is easily confused with **Mandragora officinarum** (Berry and Jackson 1976). The primary distinction between the two is the time at which they flower. The autumn mandrake blooms in the fall (September to November).[244] Both species are found in southern Europe, from Portugal to Greece (Festi and Aliotta 1990*; Viola 1979, 175*). Autumn mandrake is also common in northern Africa (Morocco).

In Morocco, where the autumn mandrake is known variously as *taraiba*, *taraila*, *bid l'gul*, and *bioe al ghorl*, a finger-sized piece of the root is taken together with a nutmeg (**Myristica fragrans**) in order to produce a "good head" (Vries 1984*; 1989, 39, 40, 44*), and the root is still used there for hunting treasures (Vries 1989, 39*). Apparently it is also used in a magical **incense**.

Mandragora autumnalis contains the same **tropane alkaoids** as **Mandragora officinarum**. Little is known about its effects:

> After consuming vegetables that had been contaminated with *Mandragora autumnalis* leaves, fifteen people developed symptoms of poisoning. The latent period was 1 to 4 hours, with a mean of 2.7 hours. No association between the latent period and the degree of intoxication was observed. All patients exhibited disturbances of vision, dry mouth, tachycardia, mydriasis, and reddening of the skin. Also observed were, in 14 of the 15 patients, dry skin and mucosa, hallucinations, and overactivity; in 9 of 15 patients, agitation/delirium, confusion, headaches, and problems with micturition; in 8 patients, difficulties swallowing and stomachaches. One patient developed an acute psychosis of short duration. (Mechler 1993, 765)

Mandragora caulescens C.B. Clarke [syn. *Anisodus humilis* (Hook. f.)]—Himalayan mandrake

Four subspecies of Himalayan mandrake have been described to date (Mechler 1993, 765):

Mandragora caulescens ssp. *brevicalyx* Grierson et Long
Mandragora caulescens ssp. *caulescens*
Mandragora caulescens ssp. *flavida* Grierson et Long
Mandragora caulescens ssp. *purpurascens* Grierson et Long

This yellow-blooming species, known as *kattuchooti* or *chi'ieh shen*, is found in the high Himalayas, primarily at altitudes between 3,000 and 4,000 meters (Polunin and Stainton 1985, 287, plate 93*). Its range extends through Tibet and into western China (Deb 1979, 94). It is common in Sikkim and Darjeeling and grows in western Sichuan, northwestern Yunnan, and eastern Xizang (Tibet) at altitudes between 2,200 and 4,200 meters (Lu 1986, 81f.). In Sikkim, this mandrake species is used for magical rituals (Mehra 1979, 162*) and sometimes as an alternative to *Withania somnifera*; it also has been interpreted as the Vedic *jangida*. It is used in traditional Chinese and Tibetan medicine to treat stomach ailments (Mechler 1993, 765). The root contains 0.13% hysocyamine and may possibly also contain mandragorine. **Scopolamine** and cuscohygrine have not been detected (Mechler 1993, 765).

Mandragora chinghaiensis Kuang et A.M. Lu—Chinese mandrake

Also known as *chinghai chi'eh shen*, this recently described mandrake species is endemic to the Qinghai-Xizang Plateau of western China, where its root is used in local folk medicine (Lu 1986, 82). In Tibet, it is used to treat pain and as a substitute for *Mandragora caulescens*. The entire plant contains 0.19% hyoscyamine and 0.12% **scopolamine**; the root contains 0.21% hyoscyamine and 0.48% scopolamine (Mechler 1993, 765).

Mandragora morion nom. nud.

The plant that is listed in the older literature under the name *Mandragora morion* is identical to a *Mandragora* species, to *Atropa belladonna*, or to some other species of nightshade (cf. *Solanum* spp., *Withania somnifera*). The ancient literature has the following to say about *Mandragora morion*:

It has been reported that there is also another species [perhaps *Mandragora turcomanica* or *M. caulescens*] called *morion* [from *moria*, "dullness of the senses," or *morion*, "male member"] that grows in shady places and

around rocky caves; it has leaves like those of the white *Mandragora*, but smaller and about as long as a span, white, and arranged in a circle around the root, which is tender, white, a little larger than a span, and as thick as a thumb. This, when drunk in the amount of a drachma [ca. 3.8 g] or eaten with pearl barley in bread or a side dish, is said to induce a deep sleep; a person will namely sleep in the same position as they were in when they partook of it, without any sensation, for three to four hours from the time at which they ingested it. The physicans also use this when they wish to cut or burn. The root is also said to be an antidote [*antidoton*] when taken with the so-called *Strychnos manikos*.[245] (Dioscorides 4.76)

Mandragora shebbearei Fischer—Tibetan mandrake

This species or variety of mandrake is said to occur only in Tibet. It may be identical to *M. caulescens*.

Mandragora turcomanica Mizgireva—Turkoman mandrake

This rare species, which grows only in Turkmenistan, produces violet flowers and has been used as a medicine by the people of the Sumbar Valley since antiquity. It appears that Asian authors of the Middle Ages, such as Abu-Reichan Beruni (973–1048), identified this Asian species with the European *Mandragora* of the ancient literature, and according to Khlopin (1980, 227), it is identical to the "male" *Mandragora* of Dioscorides (cf. ***Mandragora officinarum***). Its large, juicy, golden yellow fruits were regarded as edible (when eaten in moderation). The Turkoman mandrake thrives only in clayey soils at altitudes of 600 meters (Khlopin 1980).

The Parsis had a sacred plant with inebriating or entheogenic qualities that was known as haoma and was mentioned often in the Avesta. In our time, the botanical identity of **haoma** is as uncertain as that of the Indian **soma** or the Greek ambrosia. It is possible that the word *haoma* was used to refer to a number of plants (cf. ***Peganum harmala***). It has long been suggested that *haoma* and the mandrake were one and the same. The discovery and description of the Turkoman mandrake has provided this hypothesis with renewed support:

When one compares the description of the haoma of the Avesta with the white male *Mandragora* of the ancient and medieval scholars, then it can be seen that these are possibly the same plant. Otherwise, this white male *Mandragora* would have to be identified with the Turkoman *Mandragora*. In other words, the Avestan Aryans [the ancient Parsis]

The plant that Leonhard Fuchs depicted under the name *Mandragora morion* appears to be the deadly nightshade (*Atropa belladonna*). (Woodcut, 1545)

245 This plant has not been identified with certainty, although many authors have interpreted it as the common thorn apple (*Datura stramonium*). The plant, which presumably came from India or the Caspian Sea, has been construed as the trance-inducing inebriant used by the Pythia on several occasions (Mehra 1979, 167*). Pliny reported that the "one that makes mad" (*manikon*) produces effects in which "one is led to perceive apparitions and hallucinations" (21.178).

"There are a few more wondrous stories about these plants [mandrakes]. It is said of the root that it has a very similar appearance to the organs of procreation of both sexes. Although it is only rarely found, when a root looks like the male organ and comes into a man's possession, then it will secure for him the love of a woman. In this way, the lesbian Phaeon was loved in such a passionate manner by Sappho. Much has been said about this, not only by the magicians, but also by the Pythagorean philosophers."

PLINY
NATURAL HISTORY
(25.147–150)

used the Turkoman species of *Mandragora* to produce the divine drink and called it haoma. . . . When the Indian Aryans penetrated into northern India from the west after the fall of the Indo-Iranian union, they found the Himalayan species of *Mandragora*, which received the name **soma**. (Khlopin 1980, 230 f.)

Literature

See also the entries for **Atropa belladonna**, **Mandragora officinarum**, **tropane alkaloids**, and **scopolamine**.

Berry, Michael I., and Betty P. Jackson. 1976. European mandrake (*Mandragora officinarum* L.

and *M. autumnalis* Bertol.): The structure of the rhizome and root. *Planta Medica* 30:281–90.

Deb, D. B. 1979. Solanaceae in India. In *The biology and taxonomy of the Solanaceae*, ed. J. G. Hawkes et al., 87–112. London: Academic Press.

Khlopin, Igor N. 1980. *Mandragora turcomanica* in der Geschichte der Orientalvölker. *Orientalia Lovaniensia Periodica* 11:223–31.

Lu, An-ming. 1986. Solanaceae in China. In *Solanaceae: Biology and systematics*, ed. William G. D'Arcy, 79–85. New York: Columbia University Press.

Mechler, Ernst. 1993. *Mandragora*. In *Hagers Handbuch der pharmazeutischen Praxis*, 5th ed., 5:762–67. Berlin: Springer.

Mesembryanthemum spp.
Ice Plants

Left: The ice plant, *Mesembryanthemum crystallinum*, received its name because its leaves appear as though they are covered with ice.

Right: South African species of *Mesembryanthemum* are sometimes called midday flowers because their flowers are often open only during that time.

246 German-language literature continues to use the family name Aizoaceae (Frohne and Jensen 1992, 125*; Zander 1994*). The name Mesembryanthemaceae is preferred in English-language literature (Herre 1971) (cf. *Sceletium tortuosum*).

Family

Aizoaceae (Ice Plant Family)[246]

It is likely that the ancient Assyrians used a species of the genus *Mesembryanthemum* (which they called *dilbat*) in combination with Indian hemp (**Cannabis indica**) as a medicine for "suppressing the spirits" (Thompson 1949, 222*).

In northern Peru, one *Mesembryanthemum* species is known in the local Spanish as *hierba de la señorita*, "plant of the maiden." Folk healers of the San Pedro cult (see **Trichocereus pachanoi**) use its fleshy leaves in herbal amulets (known as *seguros*, "insurances"), especially for love magic (Giese 1989, 252*).

Today, two or more species are known in South Africa as **kanna** (also *channa* or *kougoed*). It has been suggested that the *kanna* or *channa* that was described some 250 years ago as a plant that the Hottentots chewed for its inebriating effects was a *Mesembryanthemum* species. More recently, a plant used as kanna was identified as **Sceletium tortuosum**.

Mesembryanthemum crystallinum L. [syn. *Cryophytum crystallinum* (L.) N.E. Br., *Gasoul crystallinum* (L.) Rothm.], an ice plant from the foothills of the Cape of Good Hope, is cultivated for use as a vegetable and in salads. In 1785, Liebig suggested it for use as a medicine in Germany, where the species became known by the name *Hottentot fig.* Around 1900, it was said that the flowers of the plant (flores candiae) were being used for "superstitious purposes" (Schneider 1974, 2:322*). It has been claimed that the Hottentots of the Kalahari Desert chewed or sniffed the roots (**snuffs**), and there have been reports that this ice plant, which contains mesembrine and oxalic acid, has induced (psychoactive) intoxications (Festi and Samorini 1995, 32 f.).

The Bantu use the roots of *Mesembryanthemum mahonii* N.E. Br. to brew a potently inebriating **beer**. In South Africa, the closely related species *Trichodiadema stellatum* (Miller) Schwantes is used as a yeast substitute for baking bread and brewing beer. The roots purportedly

possess potent inebriating properties (Festi and Samorini 1995, 31).

It is very likely that mesembrine is present in another closely related species that is also known as Hottentot fig, *Carpobrotus edulis* (L.) N.E. Br. [syn. *Mesembryanthemum edule* L.]. The plant has sedative properties and may be psychoactive (Festi and Samorini 1995).

Tribal peoples of South Africa used another species in the same family, *Rabaiea albinota* (Haw.) N.E. Br. [syn. *Nananthus albinotus* N.E. Br.], as an additive to smoking and sniffing tobacco (*Nicotiana tabacum*, **snuffs**) (D. McKenna 1995, 101*).

The alkaloids mesembrine and mesembrinine and six additional derivatives are known to occur in the genus (Frohne and Jensen 1992, 125*; D. McKenna 1995, 101*). They have analgesic effects and stimulate circulation (Tyler 1966, 280*). Mesembrine has both sedative and **cocaine**like properties (Scholz and Eigner 1983, 75*). According to information provided by the chemist K. Trout, some members of the genus contain *N,N*-DMT.

The alkaloid mesembrine has also been detected in several other species of the Family Aizoaceae: *Drosanthemum hispidum* (L.) Schwantes [syn. *Mesembryanthemum hispidum* L.], *Sceletium anatomicum* (Haw.) L. Bolus [syn. *Mesembryanthemum anatomicum* Haw.], **Sceletium tortuosum** (L.) N.E. Br. [syn. *Mesembryanthemum tortuosum* L.], *Trichodiadema barbatum* (L.) Schwantes [syn. *Mesembryanthemum barbatum* L.], *Trichodiadema bulbosum* (Miller) Schwantes [syn. *Mesembryanthemum stellatum* Miller], and *Trichodiadema intonsum* (Haw.) Schwantes [syn. *Mesembryanthemum intonsum* Haw.] (Festi and Samorini 1995, 32).

Literature

See also the entries for **Sceletium tortuosum** and **kanna.**

Bodendorf, K., and P. Kloss. 1961. Über Abbau und Biogenese der Alkaloide Mesembrin und Mesembrenin. *Archiv der Pharmazie* 66:654–61.

Festi, Francesco, and Giorgio Samorini. 1995. *Carpobrotus edulis* (L.) N.E. Brown in Phillips (Fico degli Ottentotto/Hottentots Fig). *Eleusis* 2:28–34.

Herre, H. 1971. *The genera of the Mesembryanthemaceae.* Cape Town: Tafelberg Publishers.

Popelak, A., E. Haack, G. Lettenbauer, and H. Spingler. 1960. Zur Konstitution des Mesembrins. *Naturwissenschaften* 47:156.

Popelak, A., G. Lettenbauer, E. Haack, and H. Spingler. 1960. Die Struktur des Mesembrins und Mesembrenins. *Naturwissenschaften* 47:231–32.

Some *Mesembryanthemum* species may once have been used for psychoactive purposes.

The inebriating ice plant *Mesembryanthemum expansum* is now known by the botanical name *Sceletium expansum.* (Woodcut, eighteenth century)

Mesembrine

Methysticodendron amesianum Schultes

See *Brugmansia* x *candida*

Mimosa tenuiflora (Willdenow) Poiret

Jurema, Tepescohuite

Family
Leguminosae: Mimosaceae-Fabaceae (Mimosa-like); Subfamily Mimosoideae

Forms and Subspecies
A variety from the Guianas that was recently described under the name *Acacia tenuiflora* var. *producta* Grimes may be a form of this taxonomically uncertain species (Grimes 1992).

Synonyms
Acacia tenuiflora Willd.
Mimosa cabrera Karst.
Mimosa hostilis (Mart.) Benth.
Mimosa jurema nom. nud.
Mimosa nigra Huber nom. nud.
Mimosa tenuefolia L. (misspelling in the literature)
Mimosa tenuiflora Karst.

Folk Names
Ajucá, cabrero ("goatherd"), carbón ("charcoal"), carbonal, espineiro, jurema, jurema negro,[247] jurema preta, jurema prêta, tepescahuite, tepescohuite, veuêka, vinho da jurema

History
The Aztecs already knew of the mimosa tree during pre-Columbian times. The name *tepescohuite*, which is now common in Mexico, is derived from the Aztec *tepus-cuahuitl*, "metal tree," a reference to the tree's extremely hard wood. The Mexican tree was botanically described in 1810. Only in the past few years was it recognized that this tree is the same species as the Brazilian *Mimosa hostilis* (Ott 1996b, 11*). The jurema cult, in which drinks made from this *Mimosa* are consumed, was first described in 1788. Until recently, it was thought that the cult had died out, but it is now experiencing a great renaissance.

Distribution
The tree grows wild in southern Mexico (it is common in Oaxaca and on the Pacific coast of Chiapas), Central America, Venezuela, and Brazil (especially in the northeast: Minas Gerais, Bahia, Pernambuco). It thrives best in tropical lowlands but can grow at altitudes of up to 1,000 meters (Sánchez León 1987).

Cultivation
Until recently, little was known about methods for cultivating the plant. Initial experiments have demonstrated that it is likely possible to propagate the tree through cuttings.

Appearance
This mimosa is a bushy tree that can grow as tall as 8 meters. It has pinnate leaves and short, sharp thorns along its branches. The white flowers occur in clusters, and the fruits are small and lanceolate (2 to 4.5 mm wide, 5 to 7 mm long). The pods each contain three or four of the fruits (Sánchez León 1987).

Psychoactive Material
— Dried trunk cortex
— Dried root cortex

Preparation and Dosage
In Brazil, *vinho do jurema* is sometimes made with passionfruit juice (cf. *Passiflora* spp.). An **ayahuasca analog** can be made by combining 9 to 12 g of the dried root cortex with 3 g of *Peganum harmala*.

Ritual Use
In former times, the Pancarú Indians and many Indians of the eastern Amazon region (such as the Karirí, Tusha, and Fulnio) used the root to make jurema drinks (*ajucá* or *veuêka*), which could induce shamanic states of consciousness (Gonçalves de Lima 1946). Unfortunately, only very rudimentary information about the exact preparations and the rituals is available. According to the older ethnographic literature (summarized in Schultes and Hofmann 1980, 153 ff.*), the "mysterious drink" gave the shamans fantastic and

Left: Known in Mexico as *tepescohuite*, *Mimosa tenuiflora* is an excellent source of DMT. (Photographed in Chiapas, Mexico)

Right: In Mexico, *tepescohuite* bark is used to produce numerous preparations for treating wounds and as general tonics.

247 This name is also used for *Mimosa nigra*, which contains DMT and was used as a substitute (albeit only rarely) for *Mimosa tenuiflora* to produce jurema drinks (Schultes and Hofmann 1980, 155*).

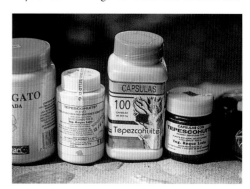

meaningful dreams and brought on "an enchantment, transporting them to heaven."

> An old master of ceremonies, wielding a dance rattle decorated with a feather mosaic, would serve a bowlful of the infusion made from jurema roots to all celebrants, who would then see glorious visions of the spirit land, with flowers and birds. They might catch a glimpse of the clashing rocks that destroy souls of the dead journeying to their goal or see the Thunderbird shooting lightning from a huge tuft on his head and producing claps of thunder by running about. (R. H. Lowie, in Schultes and Hofmann 1980, 154*)

The Indian use of jurema decreased significantly during the twentieth century, partly as a result of the destruction of indigenous cultures and partly due to the increasing popularization of **ayahuasca** and the ayahuasca churches.

Since the beginning of the twentieth century, the ritual use of *vinho do jurema* has been integrated into the Afro-Brazilian Candomblé and macumba cults. It is likely, however, that *Mimosa tenuiflora* has rarely been used to prepare the *vinho do jurema*; the more likely source is ***Pithecolobium diversifolium.***

In contemporary Brazil, jurema is ritually consumed in a variety of circles of different ethnic origin (Novaes da Mota 1987).

To date, no evidence of any ritual use of *Mimosa tenuiflora* in Mexico is known.

Artifacts

Some Afro-Brazilian ayahuasca cults venerate Indian spirits (*caboclos*) as saints. Among these saints is Cabocla Jurema, who is regarded as the goddess of the forest. She is presumably a personification of *Mimosa tenuiflora*.

Medicinal Use

In Mexican folk medicine, the powdered bark of the trunk is used—apparently with great success—to treat burns, inflammations, and wounds (Grether 1988; Sánchez León 1987). The analgesic effects of the powdered bark for the treatment of burns became known throughout the world when the international press reported its successful use in the treatment of victims of two catastrophes, a natural gas explosion in 1982 and an earthquake in 1985, as a result of which the death rate of the burn victims declined significantly (Anton et al. 1993).

In Mexico, the powdered bark, contained in gelatin capsules, is taken as a tonic, often in combination with the ground bark of *uña de gato* (*Uncaria tomentosa*). Among the rural Brazilian population, the bark of the trunk also is used as a home remedy for exhaustion and debility.

Indian women in Brazil use the fresh root cortex as an aphrodisiac love magic by rubbing it onto the soles of the feet of the men they desire. Whether this occurs with the knowledge of the men is an open question.

Constituents

The bark of the trunk has been found to contain several triterpene saponins (mimonosides A, B, and C) as well as steroid saponins (3-*O*-β-D-glucopyranosyl-campesterol, 3-*O*-β-D-glucopyranosyl-stigmasterol, 3-*O*-β-D-glucopyranosyl-β-sitosterol) that are clearly bioactive (Jiang, Beck, et al. 1991; Jiang, Weniger, et al. 1991b; Lara Ochoa and Marquez Alonso 1996, 99*). Also present are lupeol, campesterol, stigmasterol, and β-sitosterol. The bark contains large amounts of calcium oxalate crystals and a great deal of starch and tannins (Anton et al. 1993), as well as small quantities of alkaloids, of which *N,N*-DMT, 5-hydroxytryptamine, and β-**carbolines** have been identified (Lara Ochoa and Marquez Alonso 1996, 99*; Meckes-Lozoya et al. 1990). New chalcones have been detected in the bark and were named kukulkanins, after the Mayan deity Kukulcan ("feathered serpent") (Dominguez et al. 1989).

Recent studies of the Mexican root cortex, which is rich in alkaloids, have yielded sensational results. The dried root cortex contains approximately 1% *N,N*-DMT. The root bark of Brazilian plants has been found to contain 0.57% *N,N*-DMT (Farnsworth 1968, 1088*; Pachter et al. 1959*; Schultes and Hofmann 1980*).

Effects

When spread onto burns, the powdered trunk bark produces analgesic effects that last for two to three hours and clearly shortens the regeneration period of the epidermis. The bark appears to have a stimulating effect on the immune system (Anton et al. 1993).

If the ethnographic literature is accurate, then a decoction of the root produces psychedelic effects when ingested orally. No information is available as to whether the Indians also use MAO-inhibiting additives in such cases. But if the root cortex does contain β-carbolines, then the tea could indeed be orally efficacious. In modern contexts, the only additive that has been observed being used is passionfruit juice (see ***Passiflora* spp.**). The oral efficacy of a decoction of the root cortex presumably is increased by the addition of the passionfruit juice, which allegedly has MAO-inhibiting properties.

I smoked 1 g of dried, coarsely chopped Mexican root cortex (corresponding to approximately 100 mg of *N,N*-DMT) in a pipe. As it was being lit, the smoke immediately emitted an almost overly obvious characteristic DMT scent. However, I felt only very mild DMT effects. It is

"*Mimosa tenuiflora* has become the most important source of DMT for anahuasca [= ayahuasca analogs]."

JONATHAN OTT
PHARMACOTHEON
(1996d, 254*)

possible that a bark extract (cold-water decoction!) concentrated by evaporation may be sufficiently concentrated to produce good DMT effects.

Commercial Forms and Regulations

In Mexico, tepescohuite is available as dried bark and as powdered bark in markets, drugstores, and natural foods stores. It is unclear whether tepescohuite may be imported into Europe. Because of the presence of the alkaloids, it was not allowed to be approved for medicinal use in Europe (cf. Anton et al. 1993, 156).

Literature

See also the entries for *Mimosa* spp., **ayahuasca analogs**, and *N,N*-DMT.

Anton, R., Y. Jiang, B. Weniger, J. P. Beck, and L. Rivier. 1993. Pharmacognosy of *Mimosa tenuiflora* (Willd.) Poiret. *Journal of Ethnopharmacology* 38:153–57.

Dominguez, Xorge A., Sergio Garcia G., Howard J. Williams, Claudio Ortiz, A. Ian Scott, and Joseph H. Reibenspies. 1989. Kukulkanins A and B, new chalcones from *Mimosa tenuefolia*. *Journal of Natural Products* 52 (4): 864–67.

Gonçalves de Lima, Oswaldo. 1946. Observações sôbre o 'vinho da Jurema' utilizado pelos índios Pancurú de Tacaratú (Pernambuco). *Arquivos do Instituto de Pesquisas Agronomicas* 4:45–80.

Grether, R. 1988. Note on the identity of tepescohuite in Mexico. *Boletín de la Sociedad Botanica de México* 48:151.

Grimes, James W. 1992. Description of *Acacia tenuifolia* var. *producta* (Leguminosae, Mimosoideae), a new variety from the Guianas, and discussion of the typification of the species. *Brittonia* 44 (2): 266–69.

Jiang, Y., J. P. Beck, L. Italiano, M. Haag, and R. Anton. 1991. Biological effects of the saponins from *Mimosa tenuiflora* on fibroblast cells in culture. *Planta Medica* 57 suppl. (2): A38.

Jiang, Y., B. Weniger, G. Massiot, C. Lavaud, and R. Anton. 1991. Saponins from the bark of *Mimosa tenuiflora*. *Planta Medica* 57 suppl. (2): A38–39.

Meckes-Lozoya, M., et al. 1990. Dimethyltryptamine alkaloids in *Mimosa tenuiflora* bark (tepescohuite). *Arch. Invest. Med.* 1990:175–77.

Novaes da Mota, Clairice. 1987. *As* jurema *told us:* Kariki shoko *and* shoco modo *of utilization of medicinal plants in the context of modern northeastern Brazil.* Ann Arbor, Mich.: University of Michigan Press. (UMI microfilm order no. 8717395.)

Sánchez León, Victor. 1987. *El tepescohuite.* Tuxtla Gutiérrez, Chiapas: Instituto de Historia Natural. (Plantas de Chiapas—Yashté-1.)

Mimosa spp.

Mimosa Species

Many mimosas in Central and South America, Australia, and Oceania contain DMT and other tryptamines. This copperplate engraving from Sibly's appendix to Culpeper's herbal associates the (psychoactive) mimosas with the (psychoactive) mandrake (bottom left).

Family

Leguminosae: Mimosaceae-Fabaceae (Mimosa-like); Subfamily Mimosoideae

The family is composed of some five hundred species, the majority of which occur in South America. They require a tropical or subtropical climate (Schultes and Raffauf 1990, 246*). Mimosa species are often confused with acacias (see *Acacia* **spp.**) and with *Anadenanthera peregrina* and *Anadenanthera colubrina*.

Mimosa pudica L.—sensitive mimosa

It is possible that the well-known sensitive mimosa, whose leaves immediately fold together when touched, has a certain importance as a psychoactive substance. In Amazonia, where the plant is known as *chami*, it is made into a tea for treating sleep disorders (Duke and Vasquez 1994*). In Belize (Arvigo and Balick 1994, 215*) and on the Caribbean island of La Réunion, the stalks, leaves, and roots are used as sedative and sleeping agents. In Brazil, the plant is called *jurema*, while the variety *acerba* Benth. is known as *jurema branca* (cf. **Mimosa tenuiflora**, **Pithecellobium** **spp.**). Both forms are used as ingredients in the initiatory drink of the Afro-American Candomblé cult (see **madzoka medicine**).

The plant is known as *punyo-sisa* in Quechuan. Its leaves are placed in the pillows of old people and children so that they will sleep better (Schultes 1983, 261*). In the Amazon region, women soak the leaves in the juice pressed from the roots and smear the resulting juice between their breasts and on the soles of their feet. They claim that this gives them "increased sexual power" (Gottlieb 1974, 66*).

In the Philippines, *Mimosa pudica* is regarded as an aphrodisiac for frigid women. They pick and boil the leaves. The leaves fold together when picked and open up again when boiled. The opened leaf is a symbol for the vagina when it is open for sexual activity.

In India, the leaves are chewed and the resulting mush is spread onto fresh wounds to stop bleeding (Bhandary et al. 1995, 154*).

The plant contains norepinephrine (Schultes and Raffauf 1990, 246*). The narcotic effects are thought to be due to the alkaloid mimosine (Wong 1976, 123*). The aerial parts of the plant contain two *C*-glycosylflavones: 2"-*O*-rhamnosyl-orientine and 2"-*O*-rhamnosylisoorientine (Englert et al. 1994). The root contains tannin (Wong 1976, 123*).

Mimosa scabrella Benth.—bracaatinga

The bark of this small tree contains *N,N*-DMT, MMT, *N*-formyl-MMT, and 2-methyl-1,2,3,4-tetrahydro-β-carboline—i.e., both psychedelic tryptamines and MAO-inhibiting β-**carbolines**. The bark presumably is suitable for making **ayahuasca analogs**. No traditional psychoactive use is known.

Mimosa verrucosa Benth.

This species is rumored to be psychoactive or hallucinogenic (Schultes and Farnsworth 1982, 188*). However, there are no chemical analyses of or detailed reports about any possible use of the plant.

Mimosa spp.

There appears to be a number of mimosas that may be of chemical interest and may possibly be suitable as sources of *N,N*-DMT for other **ayahuasca analogs**. There also appears to be a variety of species that are smoked as marijuana substitutes (cf. **Cannabis indica**) in Central America.

Literature

See also the entries for ***Mimosa tenuiflora***, **ayahuasca analogs**, β-**carbolines**, and *N,N*-DMT.

Englert, Jürgen, Yulin Jiang, Pierre Cabalion, Ali Oulad-Ali, and Robert Anton. 1994. *C*-glycosylflavones from aerial parts of *Mimosa pudica*. *Planta Medica* 60:194.

Top: The delicate and modest sensitive mimosa (*Mimosa pudica*) folds its pinnate leaves together at the slightest touch.

Bottom: The seeds of *Mimosa scabrella*, a DMT-containing species.

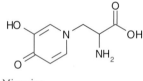

Mimosine

Mitragyna speciosa Korthals

Kratom

Branch tip and flower of the kratom tree (*Mitragyna speciosa*).

The typical leaf arrangement of the genus *Mitragyna*.

Family
Rubiaceae (Coffee Family)

Forms and Subspecies
None

Synonyms
Mitragyna religiosa nom. nud.
Mitragyne speciosa (Korth.) (misspelling in the literature)

Folk Names
Biak, biak-biak, gra-tom, katawn, kratom, kraton, kutum, mabog, mambog, mitragyne

History
In the nineteenth century, it was reported that kratom was being used in Malaysia as an opium substitute and to heal "opium addiction" (Beckett, Shellard, and Tackie 1965, 241; Tyler 1966, 285*; Wray 1907a, 1907b). Phytochemical research into the plant began around 1920 (Field 1921). Pharmacological studies of the main active constituent began soon thereafter (Grewal 1932a, 1932b).

Distribution
The tree is indigenous to Thailand and from the northern Malay Peninsula to Borneo and New Guinea (Macmillan 1991, 416*).

Cultivation
The tree grows in marshy regions. No information about propagation is available.

Appearance
This tropical tree or shrub often grows to a height of only 3 to 4 meters, although it will sometimes grow as tall as 12 to 16 meters. It has a straight trunk with forked branches that grow upward obliquely. The oval, green leaves have a very large surface area (being 8 to 12 cm long) and are tapered at the ends. The deep yellow flowers grow in globular clusters attached to the leaf axils on long stalks. The seeds are winged (Emboden 1979, 184*).

Kratom is easily confused with other members of the genus *Mitragyna*, such as the African species *Mitragyna brunonis* (Wall. ex G. Don) Craib.

Psychoactive Material
— Leaves (kratom)

Preparation and Dosage
The dried leaves can be smoked, chewed, or made into the extract known as *kratom* or *mambog* (Wray 1907b). They also can be powdered, brewed with hot water, and drunk as a tea; 8.8 g has been given as a dosage (Macmillan 1991, 416*). Another method of preparation involves powdering the dried leaves and boiling them in water until a syrup results (which is easy to preserve); a dosage of this is 0.38 g. The syrup can be mixed with finely chopped leaves of the palas palm (*Lincuala paludosa*) and made into pills. This product is known as *madat* in Malaysia and is smoked in long bamboo pipes (Macmillan 1991, 416*).

The fresh leaves can be chewed together with betel nuts (**Areca catechu**) (Scholz and Eigner 1983, 75*). Salt is often added to prevent constipation. A typical user will chew three to ten mouthfuls of leaves through the day (Suwanlert 1975).

The main active constituent, mitragynine, appears to be well tolerated and has few toxic effects even when taken in high dosages. In studies with mice, even extreme dosages of 920 mg/kg body weight did not produce any toxic effects (Jansen and Prast 1988, 117).

Ritual Use
In Thailand, kratom is used primarily as an opium substitute. It is possible that some type of ritual method of use similar to opium smoking may have developed (see **Papaver somniferum**). Unfortunately, this area has not been the subject of ethnographic research.

Artifacts
None

Medicinal Use
In Thai medicine, kratom is used to treat diarrhea (Ott 1993, 413*). Drivers of *tuk-tuks* (three-wheeled motorized "taxis") in Bangkok consume kratom as an amphetamine substitute (Schuldes 1995, 52*). In Malaysia, the leaves are used as a folk medicinal treatment for worms (Said et al. 1991).

In West Africa, the related species *Mitragyna stipulosa* (DC.) O. Kuntze is used in folk medicine as a local anesthetic. The bark is drunk in **palm wine** (cf. *Cocos nucifera*) to counteract poisoning and as a diuretic (Ayensu 1978, 222*).

Constituents
The plant contains numerous **indole alkaloids**: mitragynine, ajmalicine, corynanthedin, isomitraphylline, mitraphylline, mitraversine, paynantheine, speciogynine, speciofoline, speciophylline, stipulatine (= rotundifoline), rhynchophylline, mitragynaline, corynantheidinaline, mitragynalinic acid, and corynantheidinalinic acid (Beckett, Shellard, Philipson, and Lee 1965; Beckett, Shellard, and Tackie 1965; Houghton et al. 1991; Tyler 1966, 286*).

The main active constituent, mitragynine (66% of the total alkaloid mixture), is present especially in the leaves. Young leaves of plants of Thai origin contain 7α-hydroxy-*7H*-mitragynine (1.6% of the total alkaloid mixture) (Ponglux et al. 1994). A total of approximately 0.5% alkaloids is present in the dried leaves (Beckett, Shellard, and Tackie 1965, 242). A new indole alkaloid, 3-dehydromitragynine, was discovered in the fresh leaves (Houghton and Said 1986).

Mitragynine is chemically related to **psilocybin** and other **ergot alkaloids** (D. McKenna 1995, 102*), e.g., alstovenine (cf. *Alstonia scholaris*). Mitraphylline and isomitraphylline belong to the **yohimbine** type (Ponglux et al. 1994).

The fresh leaves also contain (–)-epicatechine. Several flavonoids are present in the dried leaves. Both the dried and the fresh leaves have been found to contain ursolic acid (Said et al. 1991).

The alkaloid mitraspecine is present in the wood and bark (Beckett, Shellard, and Tackie 1965).

Several of these alkaloids also occur in other species (e.g., *Mitragyna parvifolia*) (Jansen and Prast 1988, 115; Shellard 1974, 1983).

Effects
Self-experiments, the descriptions contained in the literature, and the pharmacological properties of the constituents indicate that the effects of kratom are simultaneously stimulating like those of coca (*Erythroxylum coca*) and sedating like those of opium (see *Papaver somniferum*)—in other words, paradoxical (Ponglux et al. 1994). Kratom's effect is as if one were chewing coca while smoking opium (Jansen and Prast 1988). When fresh leaves are chewed, the stimulating effects can begin in as soon as five to ten minutes (Suwanlert 1975).

The pure alkaloid mitragynine has the following primary effects: "a) increase in the excitability of the cranio-sacral and sympathetic portion of the involuntary nervous system, b) increase in the excitability of the medulla and the motor centers of the CNS" (Scholz and Eigner 1983, 75*). These are indeed indications of a paradoxical substance (cf. Grewal 1932a, 1932b; Jansen and Prast 1988). The effects of mitragynine have even been compared to those of **codeine** (Macko et al. 1972).

The alleged kratom addiction is said to be a Thai cultural phenomenon (Jansen and Prast 1988, 117).

Commercial Forms and Regulations
Although the plant does not produce addiction per se, it does alter behavior, and it is now illegal in Thailand (D. McKenna 1995, 102*; Said et al. 1991; Schuldes 1995, 52*). Apart from this, the plant is not subject to any regulations. Unfortunately, no commercial forms are available.

Literature
See also the entries for *Alstonia scholaris* and **indole alkaloids**.

Beckett, A. H., E. J. Shellard, J. D. Philipson, and M. L. Calvin. 1966. The *Mitragyna* species of Asia Part IV: Oxindole alkaloids from the leaves of *Mitragyna speciosa* Korth. *Planta Medica* 14:266–76.

Beckett, A. H., E. J. Shellard, J. D. Philipson, and C. M. Lee. 1965. Alkaloids from *Mitragyne speciosa* (Korth.). *Journal of Pharmaceutical Pharmacology* 17:753–55.

Beckett, A. H., E. J. Shellard, and A. N. Tackie. 1965. The *Mitragyna* species of Asia—the alkaloids of the leaves of *Mitragyna speciosa* Korth.: Isolation of mitragynine and speciofoline. *Planta Medica* 13 (2): 241–46.

Field, E. J. 1921. Mitragynine and mitraversine, two new alkaloids from species of *Mitragyne*. *Transactions of the Chemical Society* 119:887–91.

Grewal, K. S. 1932a. The effect of mitragynine on man. *British Journal of Medical Psychology* 12:41–58.

———. 1932b. Observations on the pharmacology of mitragynine. *The Journal of Pharmacology and Experimental Therapeutics* 46:251–71.

Houghton, Peter J., Aishah Latiff, and Ikram M. Said. 1991. Alkaloids of *Mitragyna speciosa*. *Phytochemistry* 30 (1): 347–50.

"*Mitragyna speciosa*. . . . Its leaves have narcotic properties like opium. Just as with opium, regular use of the plant can lead to addiction."

H. F. MACMILLAN
TROPICAL PLANTING AND GARDENING
(1991, 416*)

Mitragynine

Houghton, Peter J., and Ikram M. Said. 1986. 3-dehydromitragynine: An alkaloid from *Mitragyna speciosa*. *Phytochemistry* 25: 2910–2912.

Jansen, Karl L. R., and Colin J. Prast. 1988. Ethnopharmacology of kratom and the *Mitragyna* alkaloids. *Journal of Ethnopharmacology* 23:115–119.

Macko, E., J. A. Weisbach, and B. Douglas. 1972. Some observations on the pharmacology of mitragyne. *Archive International de Pharmacodynamie* 198:145–61.

McMakin, Patrick D. 1993. *Flowering plants of Thailand: A field guide*. Bangkok: White Lotus.

Ponglux, Dhavadee, Sumphan Wongseripipatana, Hiromitsu Takayama, Masae Kikuchi, Mika Kurihara, Mariko Kitajima, Norio Aimi, and Shin-ichiro Sakai. 1994. A new indole alkaloid, 7α-hydroxy-*7H*-mitragynine, from *Mitragyna speciosa* in Thailand. *Planta Medica* 60:580–81.

Said, Ikram M., Ng Chee Chun, and Peter J. Houghton. 1991. Ursolic acid from *Mitragyna speciosa*. *Planta Medica* 57:398.

Shellard, E. J. 1974. The alkaloids of *Mitragyna* with special reference to those of *M. speciosa* Korth. *Bulletin of Narcotics* 26:41–54.

———. 1983. *Mitragyna*: A note on the alkaloids of African species. *Journal of Ethnopharmacology* 8:345–47.

Suwanlert, S. 1975. A study of kratom eaters in Thailand. *Bulletin of Narcotics* 27:21–27.

Wray, L. 1907a. "Biak": An opium substitute. *Journal of the Federated Malay States Museum* 2:53.

———. 1907b. Notes on the anti-opium remedy. *The Pharmaceutical Journal* 78:453.

Mucuna pruriens (Linnaeus) DC.

Cowhage

"The total indole alkylamine content [of cowhage] was studied from the point of view of its hallucinogenic activity. It was found that marked behavioral changes occurred which could be equated with hallucinogenic activity."

RICHARD SCHULTES AND ALBERT HOFMANN
PLANTS OF THE GODS
(1992, 50*)

Family
Leguminosae-Papilionaceae (Legume Family): Subfamily Papilionoideae: Phaseoleae, Erythrininae Tribe

Forms and Subspecies
The species can be divided into at least two or three subspecies (Zander 1994, 385*; Lassak and McCarthy 1992, 66*):

Mucuna pruriens (L.) DC. ssp. *deeringiana* (Bort) Hanelt
Mucuna pruriens (L.) DC. ssp. *pruriens*
Mucuna pruriens (L.) DC. ssp. *gigantea* (15 cm long fruits)

Mucuna utilis Wall. ex Wight, once considered a species in its own right, is now regarded as a variety (Allen and Allen 1981, 448*): *Mucuna pruriens* (L.) DC. var. *utilis* (Wall. ex Wight) Backer.

Synonyms
Dolichos pruriens L.
Mucuna deeringianum (Bort) Merr.
Mucuna prurita Hook. f.
Mucuna prurita Wight
Mucuna utilis Wall. ex Wight
Mucuna utilis Wall. ex Wight var. *utilis* Backer ex Burck
Stizolobium deeringianum Bort
Stizolobium pruriens (L.) Medik.
Stizolobium pruritum Piper

Folk Names
Acharriya-pala, afrikanische juckbohne, akushi (Bengali), baidhok, balagana, chiikan (Mayan), chipororo, chiporro, cowhage, cowhage-winde, cowitch, cow itch, demar pirkok (Cuna), haba, huacawuru (Shipibo), itchweed, jeukboontje (Dutch), juckbohne, juckende fasel, juckfasel, kachaguli, kawanch, kiwach (Hindi, "bad to rub"), korodu, kuhkrätze, mucunán, ojo de vaca (Spanish, "eye of the cow"), ojo de venado (Spanish, "eye of the deer"), ojo de zamuro, oyobe, pica pica, pois à gratter, pois pouillieux, pwa gwatê, shabun baranti (Shipibo), siliqua hirsuta, velvet bean, wich yuk (Lacandon, "deer eye"), wodza, zizi, zootie

History
Almost nothing is known about the early history of the plant. In India, it has long been used for ethnomedicinal purposes. The genus name *Mucuna* is derived from the Tupí word *mucunán*, which is used in Amazonia to refer to several members of the genus (Allen and Allen 1981, 446*). In 1688, Hans Sloane brought to London a collection of cowhage seeds that were exhibited there as "itch powder" (Allen and Allen 1981, 447*). One substance obtained from the plant, L-dopa, has become rather well known and has revolutionized the treatment of Parkinson's disease (Remmen and Ellis 1980).

Distribution

It is no longer possible to determine where cowhage originated, although it may have come from tropical Asia (Zander 1994, 385*). It is now found in tropical regions in both hemispheres (the Americas, Africa, and Asia). It often grows in cultivated areas and at the borders of forests, near the ocean, and in sandy soils.

Cultivation

The plant can be grown from pregerminated seeds and also propagated vegetatively. Little is known about its cultivation.

Appearance

Cowhage is a vigorous climber that produces clusters of luxuriant violet inflorescences. The seedpods are approximately 8 to 10 cm long and covered with fine hairs. The pods contain the large, round, flat seeds, which are dark brown with a darker stripe.

Cowhage is easily confused with a wisteria species (*Wisteria sinensis*). The genus *Mucuna* is composed of some 150 species, many of which are very similar to one another (Allen and Allen 1981, 446*).

Psychoactive Material

— Seeds
— Leaves

Preparation and Dosage

For aphrodisiac and ethnomedicinal purposes, the dried seeds are ground and ingested with liquid. They also may be suitable for use in producing **ayahuasca analogs**. An aphrodisiac dosage (for men) is regarded as 15 g (Argueta V. et al. 1994, 1151*).

For medicinal purposes, the fruits are boiled (decoction).

The dried leaves—which dry in one day even in the humid tropics—can be smoked; one to two cigarettes is regarded as an effective dosage (Anonymous 1995).

Ritual Use

In India (Karnataka), the seeds are used to produce an aphrodisiac; two ground seeds are taken in cow's milk in the evening (Bhandary et al. 1995, 155*). It is possible that aphrodisiacs made from cowhage seeds are also used in sexual rituals of the tantric cult (cf. **Alstonia scholaris**). A folk love magic for strengthening the procreative powers suggests that this might be so:

> Here, two plants are used: *Mucuna pruritus* [sic] and *Feronia elephantum* [= *Limonia acidissima* L.; Rutaceae]. These are dug up with the following words: "Oh plant, you have been uprooted by bulls. You are the bull who

foams over from lusty strength: And now you are being dug up by me for a bull of this kind!" ... After they have been chopped and softened in water, decoctions of [the plants] are mixed with some milk. The patient, who sits upon a club or an arrow, drinks the mixture while reciting the magical formula for procreative power ...: "Oh Indra, give power to this agent; its heat is like that of the fire. Like the male antelope, you, oh plant, possess all the power there is, you brother of the great soma." (I. Shah 1994, 198*)

Artifacts

The large seeds are made into amulets wherever they are found, including Mexico, Guatemala, the Caribbean, tropical Africa (Ghana), and India. The seeds are often used in necklaces or as pendants (Madsen 1965, 110).

Medicinal Use

In Mexico, the powdered seeds are regarded as a potent aphrodisiac (Martínez 1994, 255*). A Mexican folk medicinal practice involves washing the eyes of newborns with a decoction of the seeds to prevent eye inflammation (Patten 1932, 210). In Puebla, a decoction of the fruits is drunk as an anthelmintic (Argueta V. et al. 1994, 1151*). In Brazil, the plant is used as an aphrodisiac and nerve tonic. The Cuna Indians also use it as an aphrodisiac (Duke 1975, 290*). In Trinidad, crushed seeds are ingested with sugarcane juice to treat intestinal worms (Wong 1976, 126*).

The seeds have long been used in Ayurvedic medicine and the Unani system as an aphrodisiac (Bhattacharya et al. 1971, 53). The tribal peoples of Bastar use the seeds to increase semen production and to cure "nocturnal dreams" (wet dreams?) (Jain 1965, 241*).

In the folk medicine of Nepal, the seeds are prescribed as a treatment for nervous disorders. In India, four or five hairs from the seedpod are taken with milk or buttermilk as an anthelmintic (Bhandary et al. 1995, 155*).

A common belief in Southeast Asia is that the seeds of *Mucuna pruriens* var. *utilis*, which are known as *achariya-pala*, can suck out the venom of a scorpion simply by being placed on the wound (Macmillan 1991, 424*). In the countries of the region, "herbal medicine [uses] a decoction of the root and the hulls as a diuretic and to alleviate inflammations of the nasal cavities" (Stark 1984, 69*).

The seed is also used as an anthelmintic in West African folk medicine (Ott 1993, 400 f.*).

Homeopathic medicine uses a tincture made from the hairs of the seedpods known as Dolichus pruriens—Cowhage (Schneider 1974, 2:334*). Extracts of the seeds may be suitable for the

Top: The fruits of *Mucuna pruriens* ssp. *gigantea*. (Photographed in Palenque, Mexico)

Bottom: This cowhage (*Mucuna pruriens*) seed, which contains DMT, is being used as the centerpiece of a traditional necklace made by the Lacandon of the rain forest of southern Mexico.

Flower and seed pod of an as yet undescribed *Mucuna* species from Panama, which may be a potent source of DMT.

5-MeO-DMT

DMT

treatment of Parkinson's disease (Hussain and Manyam 1997).

Constituents

The seeds contain *N,N*-DMT, DMT, DMT-*N*-oxide, **5-MeO-DMT**, and **bufotenine** along with two **β-carbolines** (Bhattacharya et al. 1971). Serotonin (= 5-hydroxytryptamine) and L-dopa have also been detected (Argueta V. et al. 1994, 1151*; Ott 1993, 400*), as have the alkaloids mucunine, mucunadine, prurienene, prurieninine, mucuadine, mucuadinine, mucuadininine, prurienidine, and **nicotine** (Allen and Allen 1981, 447*). Tryptamines are present in the leaves.

The substances aproteine and mucunaine occur in the hairs on the seedpods and are responsible for the skin irritation (Seaworth 1991, 142*).

The cells contain a phenoloxidase, tyrosine, that is transformed into L-dopa when given to the cells as a substrate (Woerdenbag et al. 1989; cf. also Remmen and Ellis 1980).

It is likely that other members of the genus *Mucuna* also contain appreciable amounts of psychoactive tryptamines. L-dopa has been detected in several *Mucuna* species (Remmen and Ellis 1980; Yoshida 1976).

Effects

Animal experiments (with rats) have demonstrated that an extract of the seeds probably has hallucinogenic effects (Bhattacharya et al. 1971). Very few pharmacological experiments with humans have been conducted. There is one very interesting report describing the effects of smoking the leaves:

> Smoking a joint the size of a cigarette will produce a general CNS stimulation (a pounding "tryptamine high"). The ingestion of 3 grams of harmala [seeds of **Peganum harmala**] and two smoked mucuna joints led to a pounding in my head, accompanied by colorful geometric patterns. A slight irritation developed within an hour to a delicate feeling (pulsating colorful patterns moved around me in a spiral fashion, I felt a strong need to lie down. Very delicate and detached.). (Anonymous 1995, 33)

Commercial Forms and Regulations

The seeds are sometimes sold as beads in shops that trade in artwork and antiquities from Africa and other overseas locations. The plant is not subject to any specific regulations.

Literature

See also the entries for *N,N*-DMT and **ayahuasca analogs**.

Anonymous. 1995. *Mucuna pruriens. Entheogene* 5:33.

Bhattacharya, S. K., A. K. Sanyal, and S. Ghosal. 1971. Investigations on the hallucinogenic activity of indole alkylamines isolated from *Mucuna pruriens* DC. *Indian Journal of Physiology* 25 (2): 53–56.

Hussain, Ghazala, and Bala V. Manyam. 1997. *Mucuna pruriens* proves more effective than L-DOPA in Parkinson's disease animal model. *Phytotherapy Research* 11:419–23.

Madsen, Claudia. 1965. *A study of change in Mexican folk medicine.* New Orleans: Middle American Research Institute.

Patten, Nathan van. 1932. Obstetrics in Mexico prior to 1600. *Annals of Medical History*, n.s., 4 (2): 203–12.

Remmen, Shirley F. A., and Brian E. Ellis. 1980. DOPA synthesis in non-producer cultures of *Mucuna deeringiana. Phytochemistry* 19:1421–23.

Woerdenbag, H. J., N. Pras, H. W. Frijlink, C. F. Lerk, and Th. M. Malingre. 1989. Cyclodextrin-facilitated bioconversion of β-estradiol by cultured cells of *Mucuna pruriens* and derived phenoloxidase preparations. *Planta Medica* 55:681.

Yoshida, Takeo. 1976. A new amine, stizolamine, from *Stizolobium hassjoo. Phytochemistry* 15:1723–25.

Myristica fragrans Houttuyn

Nutmeg

Family
Myristicaceae (Nutmeg Family)

Forms and Subspecies
It is likely that there are a number of varieties and cultivars that differ especially in regard to their psychoactive effects.

Synonyms
Myristica amboinensis Gandoger
Myristica americana Rottb.
Myristica aromatica Lamk.
Myristica aromatica Swartz
Myristica moschata Thunberg
Myristica officinalis L. f.
Myristica philippensis Gandoger

Folk Names
Almendra de la semilla, balla (Banda), Banda nutmeg, bazbaz (Persian), bisbâsa al-hindî (Arabic/Yemen), buah pala (Malayalam), bush-apal, chan-thet (Laotian), hindî, jaephal (Hindi), jan-thet (Tahi), jauz-i-bûyâ (Arabic, "fragrant nut"), ju-tou-k'ou, juz, mada shaunda, massa, miskad, moscada, moscata miristica (Italian), moschocaria, moschocarydia, muscade, musca-dier, muscadier cultivé, muscatennußbaum, muschatennuß, muskach'u (Callawaya), muskat-nußbaum, musque, myristica moschata, noix muscade, nootmuskaat (Dutch), noz moscada, nuce muscata, nuez moscada, nutmeg, nutmeg tree, pala banda, roudoukou (Chinese)

History
It is very likely that the nutmeg tree originated on the island of Banda (Meister 1677, 57*; Van Gils and Cox 1994, 118), where it was derived from a wild form through cultivation. Nutmeg appears to have first arrived in Europe with the Crusaders (Norman 1991, 46*). During the seventeenth century, the trade in nutmeg seeds flourished. The Dutch Vereenigde Oostindische Companie (V.O.C.), which held the monopoly on the Spice Islands, controlled the trade.

In England, abortions were once induced by grinding nutmeg and adding it to **beer** (Fühner 1943, 240*). During the 1950s and 1960s, great quantities of nutmeg powder were ingested in the United States as a marijuana substitute. Today, the plant is regarded as hallucinogenic (Bastien 1987, 138*). In Indonesia, it appears that a psychoactive use was (formerly) unknown (Van Gils and Cox 1994, 124). However, Rumphius (1741–1755) does report of an incident on Banda Island in which two soldiers slept beneath a nutmeg tree and woke up the next morning completely drunk (Van Gils and Cox 1994, 123).

Distribution
The nutmeg tree is endemic to the Indonesian province of the Moluccas, formerly known as the Spice Islands (Van Gils and Cox 1994, 117). It is now planted in numerous tropical areas.

Cultivation
Propagation is via seeds, which must be pregerminated with care. The seedlings can then be planted in the desired location.

The tree requires a tropical climate with heavy precipitation (2,210 to 3,667 mm per year). It especially prefers rich, volcanic soil and actually thrives only in a maritime environment (Van Gils and Cox 1994, 118). In cultivation, the tree first produces fruits when it is seven or eight years old and then continues to produce for many years (twenty to thirty) (Pahlow 1995, 72*). Although it bears fruits throughout the year, April and November are the main harvest periods (Van Gils and Cox 1994, 120).

Appearance
The tree, which can attain a height of 20 meters, bears evergreen leaves (approximately 8 cm in length) on short stalks. The inconspicuous,

One of the earliest European illustrations of the nutmeg tree, this depiction is botanically quite correct. (Woodcut from Garcia da Orta, *Colloquies on the Simples and Drugs of India,* 1987)

Left: The nutmeg tree (*Myristica fragrans*), with a ripe fruit that has opened to offer a glimpse of the "nut" and the surrounding mace "flower."

Right: One type of *Myristica fragrans* from the Seychelles has very darkly colored "nuts" and deep orange arils.

An ancient Chinese illustration of the nutmeg, known as *ju-tou-k'ou*.

whitish flowers are diclinous and hang in loose clusters. The tree is dioecious, although some plants have male and female flowers. The pale yellow fruits are reminiscent of apricots but are somewhat longer. When they ripen, they split open from top to bottom, exposing the dark brown seed and the surrounding red arils (Isaac 1993).

The true nutmeg tree is very easily confused with *Myristica argentea* Warb. (New Guinea), *M. malabarica* Lam. (India), *M. speciosa* Warb., and a species indigenous to the Moluccas, *Myristica fatua* Houtt. The seeds and arils of these and other species (see table below) were sometimes sold as nutmeg counterfeits (Schneider 1974, 2:339*).

Two species of nutmeg trees indigenous to Australia (*Myristica insipida* and *Horsefieldia australiana*) develop fruits and "nuts" that are very similar to those of the true nutmeg tree. The Aborigines use all parts of the tree, including the so-called nuts, as sources of materials, food, and medicines (Wightman and Andrews 1991, 14*). It is entirely conceivable that they also discovered the psychoactive effects of nutmeg oil.

Psychoactive Material

— Nutmeg (myristicae seed, myristicae nux, nuces aromaticae, nuces nucistae, nuclei myristici, nux moschata, semen myristicae)
— Arils (macis, mace, gul-i-jauz, flower of nutmeg, nutmeg flower, arillus myristicae)
— Nutmeg oil (myristicae aetheroleum, aetheroleum myristicae, macidis aetheroleum, myristici essentia, oleum macidis, oleum myristicae, oleum myristicae aethereum, oleum nucis

moschati, oleum nucis moschati aethereum, essential oil of nutmeg, mace oil)

Preparation and Dosage

Seeds obtained from the ripe fruits are dried in the sun, over a charcoal fire, or in drying houses and then limed. Originally the seeds were covered with lime to inhibit their ability to germinate; today, the lime serves to protect the seeds from insect infestation and damage (Pahlow 1995, 72*).

The nutmeg flowers (arillus, macis) are separated from the seeds after the fruits have ripened. They are then pressed flat and dried in drying houses or in the sun. As they dry, their brilliant red color typically transforms into a warm yellow.

The light yellow essential oil of nutmeg is obtained by steam distillation of the seeds and seed coats (Isaac 1993, 868; Van Gils and Cox 1994, 120). Steam distillation can also be used to extract an essential oil from the green leaves. The essential oil of the leaves is used chiefly to adulterate or counterfeit the true nutmeg oil (Isaac 1993, 869).

The following recipe is used to produce a mixture known as *obat penenang*, "sedative medicine" (this amount corresponds to the dosage for a child; an adult should use 1.5 times this amount):

9 leaves of *violtjes* (*Viola odorata* L.)
3 leaves of *daun seribu* (*Achillea millefolium* L.)
2 arils of mace (*Myristica fragrans*)
1 piece of rhizome (app. 4 cm long) of *jahe merah* (*Zingiber officinale*)
1 piece of rhizome (app. 4 cm long) of *dringo* (*Acorus calamus*)

The calamus and ginger roots are chopped into small pieces, mixed with the other ingredients, placed in a pot with three glasses of water, and boiled for fifteen minutes. The liquid is then strained off and sweetened with **honey** to taste. A glass of this is drunk two or three times daily for a period of one to two weeks (Van Gils and Cox 1994, 123).

In the Moluccas, nutmeg seeds are combined with *jahe merah* (**Zingiber officinale**), *sereh* (lemongrass, *Cymbopogon nardus* [L.] Rendle; cf. **Cymbopogon densiflorus**), cloves (*Syzygium aromaticum*), and soaked raw rice to produce an ointment for medicinal purposes (Van Gils and Cox 1994, 122).

Powdered nutmeg also can be used as an ingredient in **smoking blends** and **snuffs** (Weil 1965).

The information regarding dosages for psychoactive purposes varies considerably. In 1576, Lobelius reported that a pregnant English woman fell into delirium after having eaten ten or twelve nutmeg seeds (Van Gils and Cox 1994:123). Malcolm X, alias Malcolm Little (1925–1965),

Other Species of the Genus *Myristica* That Are Sources of Nutmeg and Mace
(From Isaac 1993; Pahlow 1995, 73*)

Stock Plant	Trade Name	Source
Myristica argentea Warb. [syn. *Myristica finschii* Warb.]	Horse nutmeg, Macassar nuts, Papua nuts, Papuan nutmeg/mace, akum, gagom, heen, Macassar mace	Indonesia, New Guinea
Myristica impressinerva J. Sinclair		
Myristica iners Bl.		
Myristica malabarica Lam. [syn. *Myristica fatua* Houtt., *M. dactyloides* Wall. non Gaertn., *M. notha* Wall., *M. tomentosa* Graham non Thunb.]	Malabar nuts, Bombay mace	India
Myristica malaccensis Kh.	Malacca nuts	Indonesia
Myristica succedanea Reinw. ex Bl. [syn. *Myristica speciosa* Warb., *M. radja* Miq., *M. resinosa* Warb., *M. schefferi* Warb.]	Batjang nutmeg, pala maba, onem, tidore, gosara onim	Moluccas
Myristica umbellata Elmer		

Fruits Referred to as "Nutmeg" and Used as Substitutes or Counterfeits

(From Isaac 1993, 881)

None of these surrogates contains myristicin, and only a few contain safrol.

Name	Stock Plant	Family
Brazilian nutmeg	*Cryptocarya moschata* Nees et Mart.	Lauraceae
Calabash nutmeg	*Monodora myristica*	Annonaceae
California nutmeg	*Torreya californica* Torr.	Taxaceae
Chilean nutmeg	*Laurelia sempervirens* (R. et P.) Tulasne	Monimiaceae
Kua kung	*Laurelia sempervirens* (R. et P.) Tulasne	Monimiaceae
Large mace bean	*Acrodiclidium puchurymajor* (Mart.) Mez.	?
Mace bean	*Monodora myristica*	Annonaceae
Madagascan nutmeg	*Ravensara aromatica* Sonn.	Lauraceae
Nuces caryophyllatae	*Ravensara aromatica* Sonn.	Lauraceae
Otoba nutmeg	*Dialyanthera otoba* (H. et B.) Warb.	Myristicaceae
Owere seed	*Monodora myristica*	Annonaceae
Pichurim nut	*Acrodiclidium puchurymajor* (Mart.) Mez.	?
Plum nutmeg	*Atherosperma moschatum*	Monimiaceae

noted in his autobiography that a matchbox full of nutmeg powder produced a "high" corresponding to that produced by three to four joints of marijuana (Schleiffer 1979, 100*). Usually, two or three nutmeg seeds is regarded as a hallucinogenic dosage (Sherry et al. 1982, 61). Controlled experiments have determined that a dosage of up to 15 g of seed powder will produce unequivocal psychoactive effects (Isaac 1993, 884). According to Leung (1995, 157*), 7 to 8 g will induce hallucinations and euphoria.

When using the essential oil, as little as a few drops can be sufficient to produce clear psychoactive effects. Nutmeg oil is applied sublingually, i.e., under the tongue, from whence it slowly disperses. The medicinal and toxicological literature does not contain any definitive information concerning toxic amounts. In one source, it was noted that rabbits die within thirteen hours and five days of ingesting 8 to 21 g of nutmeg oil (Isaac 1993, 871).

Ritual Use

In ancient India, the nutmeg seed was known as *mada shaunda*, "stupefying fruit," and was utilized as an aphrodisiac, as an additive to **betel quids**, and as an important ingredient in curries. It was used not only in the kitchen but also in medicine (Ayurveda) and magic. In Malaya, nutmeg seeds were eaten to treat possession; in other words, a psychoactive substance was used to treat a mental illness. During the Middle Ages, it was regarded as an agent that incited to the "traffic of Venus." An unusual love magic practice has continued into modern times (see margin text on the next page).

Smoking mace flowers in order to get "high" is common practice among students in Papua New Guinea (David Orr, pers. comm.). Such use is likely more hedonistic than ritual in nature. It has been observed that finely ground nutmeg seeds are sniffed during the Indonesian shadow theater (*wayang*) (Weil 1965) (cf. **snuffs**).

In the Moluccas, nutmeg seeds are used in religious healing rituals. When all other methods have failed to help a seriously ill child, then nutmeg seeds will be placed around his or her neck. Prayers are spoken that ask God to heal the child and reveal his or her fate (Van Gils and Cox 1994, 123).

Artifacts

None

Medicinal Use

The nutmeg seed plays an important role in traditional Indonesian medicine. It is used in treatments for stomachaches, stomach cramps, kidney problems, rheumatism, nervousness, vomiting, whooping cough, and other ailments. On the Moluccas, nutmegs are used primarily as sedatives for children and for those suffering from sleep disorders. The powdered seeds are ingested with milk or a banana drink. The mixture *obat penenang*, "calming medicine" (see above), is said to be especially efficacious; however, indigenous healers are of the opinion that a person can become somewhat "dependent" on this drink (Van Gils and Cox 1994, 123). A mixture of mace, *Viola odorata* leaves, red ginger (*Zingiber officinale*), and *Phaseolus radiatus* L. beans is said to improve concentration.

In Malaysian medicine (and in the medicine of the Malaysian Muslims), nutmeg seeds and flowers are used as stimulants, digestives, aphrodisiacs, and tonics. They are even ingested in treatments for malaria and "imbecility" (Van Gils and Cox 1994, 122). To relieve headaches, oil of nutmeg is applied to the temples or ingested in tea (one drop). An ointment known as *param* is applied externally for rheumatism and limb pains.

In the Moluccas, the oil of the related species *Myristica malabarica* Lam. is also used to treat headaches (Van Gils and Cox 1994, 122).

"This massa or balla in the Bandamic language, known as the nutmeg tree in German, grows, as is now thought, on the island of Banda, and as the coconut tree is the most useful in the world, this may be the rarest, because it does not grow in any place in all of India other than here. . . . The fruit or green nut, when preserved in the same manner as our Welsh nuts, and eaten early on an empty stomach with its outer green shell, is a superb preservative from unhealthy air. Indeed, it can greatly refresh even the ill and cheer them up with fresh spirits. . . . The good doctors can judge how the usefulness of this noble fruit and its true heart can be made use of in oil, water, and other things suitable for medicine, but this is a little different than the Indians [= Asian Indians/Indonesians] believe that it is good to use. The nut is said to strengthen the brain, sharpen the memory, warm and fortify the upset stomach, dispel flatulence or wind, impart good breath, dispel urine, constipate the red dysentery, in sum, for all types of ailments of the head, brain, stomach, liver, and mother's complaints, while the oil alone is said to be better and more powerful than the seed for all ailments now conceived. But the flower is said to help a cold, upset stomach and dispel all evil moistures and uprising winds beyond all measure. . . . The seeds are of two kinds, the male and the female, the male is blackish brown and twice as long as the female, but this last type genus feminini is the best and most powerful."

GEORGE MEISTER
DER ORIENTALISCH-INDIANISCHE KUNST- UND LUST-GÄRTNER [THE ORIENTAL-INDIAN ART AND PLEASURE GARDENER] (1677, CH. 8, 3*)

"Like peppercorns, the nutmeg seed is used for an unappetizing love magic: The girl swallows it, and when the seed has come out again, she powders it and mixes it into the food of her beloved (Franconia). As a result of the physical admixing, the aphrodisiac steers the boy's kindled desire for love solely toward this girl."

AIGREMONT
VOLKSEROTIK UND PFLANZENWELT
[FOLK EROTICISM AND THE PLANT WORLD]
(1986, 2:83*)

Because they are sold as nutmeg "flowers," the arils (mace) that envelop the nutmeg seed were once thought to be the actual flowers of the nutmeg tree. (Woodcut from Tabernaemontanus, *Neu Vollkommen Kräuter-Buch*, 1731)

In India, preparations of nutmeg are used in place of opium (see *Papaver somniferum*) when opium is contraindicated for a patient. A tonic is obtained using brandy (**alcohol**) and salt (Isaac 1993, 884). In Yemen, the seeds are used as tranquilizers and the "flowers" (arils) to treat headaches (Fleurentin and Pelt 1982, 92 f.*). Use of nutmeg as an aphrodisiac was and still is very widespread (from India to Arabia and into Europe; Weil 1965).

In homeopathic medicine, tinctures of nutmeg seeds and mace flowers (Myristica fragrans hom. *HAB1*, Nux moschata hom. *PFX*, Nux moschata hom. *HPUS88*) are often used in accordance with the medical description to treat nervous complaints and perceptual disturbances (Isaac 1993, 886).

Constituents

The **essential oil** of the nutmeg seed consists of approximately 4% myristicin, 39% sabinene, 13% α-pinene, 9% β-pinene, 4% α-phellandrene, 4% limonene, 1% γ-terpinene, 1% π-cymene, 1% terpinolene, and traces of other substances (safrole, eugenol, isoeugenol). The composition can vary considerably (Janssens et al. 1990). The essential oil obtained from the leaves contains 80% α-pinene and 10% myristicin (Bastien 1987, 138; Isaac 1993, 869).

Myristicin, elemicine, and safrole appear to be responsible for the psychoactive effects. Apparently an amination takes place during metabolism that transforms these substances into centrally active amphetamine derivatives (Isaac 1993, 883; Shulgin and Naranjo 1967; Weil 1965, 1967). Amination of myristicin yields MMDA (= 3-methoxy-4,5-methylendioxyamphetamine), a known entactogenic compound (Shulgin and Shulgin 1991*). Elemicine is transformed into TMA (3,4,5-trimethoxyamphetamine), a substance related to mescaline. Aminization of safrole results in MDMA (3,4-methylendioxymethamphetamine), which is now known in counterculture circles as ecstasy or the "love drug" (cf. **herbal ecstasy**). Although myristicin has been pharmacologically demonstrated to have a mild MAO-inhibiting effect (Isaac 1993, 883), it is likely not suitable for use as an **ayahuasca analog**. The trimyristine that is present in the extract has a sedative effect upon chickens (Sherry et al. 1982).

Effects

Hildegard von Bingen provided an early description of the psychoactive powers of nutmeg seeds and their MDMA-like, empathogenic effects (cf. **herbal ecstasy**):

The nutmeg seed has great warmth and a good mixture in its powers. And when a person eats the nutmeg seed, it opens his heart and cleanses his mind and brings him to a good understanding. Take, as always, nutmeg and the same weight of cinnamon and some cloves and powder these. And then make some little cakes of this powder and bread crumbs and some water and, and eat these often, and it will subdue the bitterness of the heart and of your mind, and it opens your heart and your dull mind, and it makes your spirit happy and cleanses your mind, and it diminishes all of the harmful juices in you, and it imparts a good juice to your blood, and it makes you strong. (*Physica* 1.21)

The pharmacologist Johann E. Purkyne provided a very detailed description of the psychoactive effects of nutmeg in his work *Einige Beiträge zur physiologischen Pharmacologie* (1829):

On the narcotic effects of nutmeg. . . . I first ingested an entire seed one morning, in pieces with sugar, which was not unpleasant. The effects that I felt were unimportant; some indolence in the outer senses and in the locomotor system, rather persistent in that they lasted the entire day, but not disturbing, whether on the thinking or on the other faculties; only I did notice that a small glass of wine after eating affected me in a disproportionately strong manner. One afternoon, after a moderate meal, I ingested three nutmeg seeds. The effects soon manifested themselves: an irresistible sleepiness overcame me, and I spent the afternoon slumbering on a small sofa, in what was otherwise an uncomfortable position, reveling in pleasant, peaceful dreams that were occasionally interrupted by external disturbances. . . . After these effects had completely subsided, I made another experiment in which I grated some two drams [= 8.74 g] of nutmeg into pure brandy and drank this. Here, too, I found the effects to be significantly different, in that instead of a calm sleepiness, I was affected by a general unease in the muscular system and dizziness. (In Sajner 1965, 16 ff.)

Prison inmates who have used nutmeg as a substitute drug (for *Cannabis indica*) have reported visual and auditory hallucinations, sensations of floating, and disturbances of the body schemata (Van Gils and Cox 1994, 123).

Nutmeg oil has induced out-of-body, shamanic experiences (Devereux 1992).

Commercial Forms and Regulations

Nutmeg seeds and flowers are classified internationally as spices, i.e., foodstuffs, and are subject only to the various regulations pertaining to food. Grades of differing quality are available. The legal situation pertaining to the essential oil is not entirely clear. The essential oil is sometimes available from sources dealing in aromatic substances.

Literature
See also the entry for **essential oil**.

Devereux, Paul. 1992. An apparently nutmeg-induced experience of magical flight. In *Yearbook for Ethnomedicine and the Study of Consciousness*, 1992 (1):189–91. Berlin: VWB.

Forrest, J. E., and R. A. Heacock. 1972. Nutmeg and mace, the psychotropic spices from *Myristica fragrans*. *Lloydia* 35:440–49.

Greenberg, S., and E. L. Ortiz. 1983. *The spice of life*. New York: Amaryllis Press.

Isaac, Otto. 1993. *Myristica*. In *Hagers Handbuch der pharmazeutischen Praxis*, 5th ed., 5:863–94. Berlin: Springer.

Janssens, Jos, Gert M. Laekeman, Lug A. C. Pieters, Jozef Totte, Arnold G. Herman, and Arnold J. Vlietinck. 1990. Nutmeg oil: Identification and quantitation of platelet aggregation. *Journal of Ethnopharmacology* 29:179–88.

Payne, R. B. 1963. Nutmeg intoxication. *New England Journal of Medicine* 269:36–38.

Sajner, Josef. 1965. Joh. Ev. Purkynes Beschreibung der pharmakologischen Wirkung der Muskatnuß. *Die Medizinische Welt* 46:2613–15.

Sherry, C. J., L. E. Ray, and R. E. Herron. 1982. The pharmacological effects of a ligroin extract of nutmeg (*Myristica fragrans*). *Journal of Ethnopharmacology* 6 (1): 61–66.

Shulgin, Alexander T. 1963. Composition of the myristicin fraction from oil of nutmeg. *Nature* 197:379.

Shulgin, Alexander T., Thornton Sargent, and Claudia [*sic*] Naranjo. 1967. The chemistry and psychopharmacology of nutmeg and of several related phenylisopropylamines. In *Ethnopharmacologic search for psychoactive drugs*, ed. D. Efron, 202–14. Washington, D.C.: U.S. Dept. of Health, Education, and Welfare.

Truitt, Edward B., Jr. 1967. The pharmacology of myristicin and nutmeg. In *Ethnopharmacologic search for psychoactive drugs*, ed. D. Efron, 215–22. Washington, D.C.: U.S. Dept. of Health, Education, and Welfare.

Van Gils, Carl, and Paul Alan Cox. 1994. Ethnobotany of nutmeg in the Spice Islands. *Journal of Ethnopharmacology* 42:117–24.

Weil, Andrew. 1965. Nutmeg as a narcotic. *Economic Botany* 19:194–217.

———. 1967. Nutmeg as a psychotropic drug. In *Ethnopharmacologic search for psychoactive drugs*, ed. D. Efron, 188–201. Washington, D.C.: U.S. Dept. of Health, Education, and Welfare.

"Because mescaline-like metabolites are produced from myristicin and elemicine, the nutmeg seed has also been tested as an inebriating drug. The danger of addiction, however, can be ruled out completely, for anyone who has willingly ingested an overdose of nutmeg in order to experience the hallucinogenic effects acquires such an aversion to this spice that he can no longer take it."

MANNFRIED PAHLOW
GEWÜRZE: GENUß UND ARZNEI
[SPICES: PLEASURE AND MEDICINE]
(1995, 73*)

Myristicin

Nicotiana rustica Linnaeus

Turkish Tobacco, Wild Tobacco

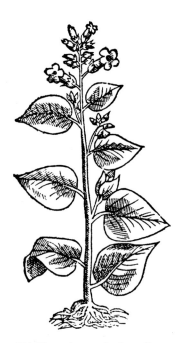

This illustration may be the earliest non-Indian representation of the nicotine-rich wild tobacco (*Nicotiana rustica*). (From Francisco Hernández, *Rerum Medicarum Novae Hispaniae thesaurus*, Rome, 1651)

Family

Solanaceae (Nightshade Family); Subfamily Cestroideae, Nicotianeae Tribe, Rubiflorae Subtribe

Forms and Subspecies

A number of varieties have been named that are distinguishable from one another primarily on the basis of their phytogeography (Hartwich 1911, 29*):

Nicotiana rustica L. var. *rustica* (indigenous to Texas and Mexico, also occurs in Brazil)
Nicotiana rustica L. var. *texana* Comes (indigenous to northern Mexico, Sonora, and Texas)
Nicotiana rustica L. var. *jamaicensis* Comes (Mexico, Guatemala, Jamaica)
Nicotiana rustica L. var. *brasilia* Schrank (indigenous to Brazil, cultivated in Hungary)
Nicotiana rustica L. var. *asiatica* Schrank (cultivated in Syria, Arabia, Persia, and Abyssinia)
Nicotiana rustica L. var. *humilis* Schrank (cultivated in Peru)

There are apparently also several cultivars, as well as a hybridogenic cultivated race known as Machorka.

Synonyms

Hyoscyamus luteus nom. nud.

Folk Names

Andumucua (Tarascan), Aztec tobacco, bauerntabak, cathérinaire, ch'aque khuri (Quechua), c'jama saire (Aymara), gelbbilsenkraut, herba legati, herba medicea, herba prioris, herba reginae, herbe divine, herbe sacrée, herbe sainte, huaña, indianisch bilsenkraut, Indian tobacco, klein nicotianskraut, kraut der ambassadoren, k'ta tobaco (Quechua), k'uru (Aymara), latakia, machene, macuche, mahorka, makucho (Huichol), nicotiana media, nicotiane, noholki'k'uuts (Mayan, "south tobacco"), nohol xi k'uts (modern Mayan, "southern tobacco"), panacea (Latin, "cure-all"), pesietl, petum, petún, piciete, picietl (Nahuatl), piciétl, piciyetl ("little tobacco"), pycielt, qonta saire (Aymara), sana sancta indorum, San Pedro,[248] sayre (Quechua), sero (Susu), tabaco blanco (Spanish, "white tobacco"), tabaco macuche, tabaco rupestris (Spanish, "rural tobacco"), tabaquillo (Spanish, "little tobacco"), tangoro, tawa, tenapete, teneshil (modern Nahuatl), tobaco cimarrón (Spanish, "wild tobacco"), toeback, tombac, tönbeki, toutoune estamboule, türkentabak, türkischer tabak, Turkish tobacco, turkomani tambaku (Afghani), tûtûn, um-wéh (Paez), upawoc, veilchentabak, warimba, wilder tabak, wild tobacco,[249] ya, yé, yellow henbane, yetl, yétl

History

The genus *Nicotiana* is named after the French envoy Jean Nicot, who in 1560 sent seeds of *Nicotiana rustica* from Portugal, where he had grown the plant in his garden, to Paris, thereby promoting awareness of the plant (Schneider 1974, 2:359*).

It is very likely that wild tobacco was being cultivated in Mexico in pre-Columbian times (Dressler 1953, 138*). Apparently, it was not derived from a wild form but arose through hybridization and further cultivation—possibly from *Nicotiana paniculata* L. and *Nicotiana undulata* Ruíz et Pav. (Schultes and Raffauf 1991, 41*; Wilbert 1987, 6; cf. *Nicotiana* spp.).

Analyses of plant material left as grave goods have revealed that *Nicotiana rustica* was in use in ritual contexts in the Andes at the time of the flourishing of the Tiahuanaco culture (Bondeson 1972). This tobacco species was first described by Francisco Hernández (1651). In Europe, it initially was known under the name *Hyoscyamus peruvianus*, or Peruvian henbane (cf. *Hyoscyamus* spp.) (Schneider 1974, 2:360*). Sahagun (11.7) documented the plant's psychoactive powers. However, as an agent of pleasure, wild tobacco never achieved the same significance as *Nicotiana tabacum*.

Many Egyptian mummies have been found to contain **nicotine**, but it is not known how the nicotine got into the bodies. The simplest answer would be through smoking. But what might the ancient Egyptians have been smoking? It is generally accepted that the genus *Nicotiana* originated in the New World. Does this mean that the Egyptians had established trade relationships with pre-Columbian peoples there? The chemist Svetlana Balabanova, who took part in the investigations of the mummies, has hypothesized that wild tobacco, which is also widely known under the name *yellow henbane*, is an Old World plant that the Egyptians used as an **incense** and which was used even later as the European "strong tobacco"[250] (Pahl 1996). But there is no evidence to suggest that the nicotine might not have come from another source or have been the product of thousands of years of deposition and storage. Nicotine is also present in some species of the genus *Datura*. Investigating whether the mummies also contain **tropane alkaloids** may yield interesting insights into these questions.

---

248 In South America, this name is also given to the psychedelic cactus *Trichocereus pachanoi* as well as to other *Trichocereus* species (*Trichocereus* spp.).

249 *Lobelia nicotianaefolia* is also known as wild tobacco (cf. *Lobelia inflata*).

250 In the German-speaking regions, the name *strong tobacco* (*starker tobak*) was once also given to hemp (*Cannabis sativa*).

The yellow flowers of wild tobacco (*Nicotiana rustica*).

Distribution

Today, wild tobacco is found throughout the world. It originated in either Mexico or northern South America. It grows wild in Nayarit, Jalisco (Mexico), and the Andes, where it is found even at altitudes above 3,400 meters (Bastien 1987, 153*). In pre-Columbian times, it is said to have occurred as far north as Canada (Hartwich 1911, 32*).

Cultivation

The plant is propagated from seeds. It usually is sufficient simply to scatter the seeds over loose soil. Like the seeds of *Nicotiana tabacum*, they also can be pregerminated. In temperate zones (central Europe), the seeds should be sown between March and May. The plant does well in normal soil. The Huichol prefer to cultivate the plant in soil that has been fertilized with the ashes of burned trees.

In pre-Columbian times, wild tobacco was grown in the Andes, Mexico, and what is now the southwestern and eastern United States. It is now cultivated on a large scale as a source of **nicotine** for the manufacture of insecticides (Rehm and Espig 1996, 252*).

Appearance

This annual herbaceous plant grows to a height of 60 to 80 cm. Its leaves are smaller and more rounded than those of *Nicotiana tabacum*, and its yellow flowers are somewhat shorter and smaller. The flowering period is in June and July. The fruits are round capsules that contain numerous tiny, reddish brown seeds.

Wild tobacco can be confused with other *Nicotiana* species (e.g., with *Nicotiana langsdorffii* Weinm.; see *Nicotiana* **spp.**).

Psychoactive Material

— Leaves

Preparation and Dosage

The leaves are dried in the sun in a well-aired location. They are then usually powdered and mixed with other substances (lime or *Tagetes lucida*; cf. *Tagetes* **spp.**). This method of preparation was recorded in the early colonial period:

> Those who sell *piciete* crush it with lime and then powder both between the hands. Some do this with the incense of the earth, then take it in their hands and their mouth to soothe headaches or induce drunkenness. (Sahagun 11)

The Aztecs mixed the leaves with various unidentified plants and the gum (used as an **incense**) of the American storax, *Liquidambar styraciflua* L. (Emboden 1979, 4*). The Warao (Venezuela) mixed tobacco leaves with the resin of *Protium heptaphyllum* (Aubl.) March., known as

caraña, curucay, or *tacamahaca,* which normally was used as an **incense** in **ayahuasca** rituals (Wilbert 1991, 183). The Huichol use the powdered leaves as a ritual **incense**. The Mazatec use the powder they call San Pedro (cf. *Trichocereus pachanoi*) in all of their rituals. They also ingest it.

The juice pressed from the fresh leaves is used for shamanic purposes. The fresh or dried leaves can be prepared as a cold-water extract, infusion, or decoction. Such extracts can be drunk or administered as an **enema**.

In North America, the leaves are added to **kinnikinnick** and other **smoking blends** (Hartwich 1911, 32*). The Warao roll cigarettes up to 90 cm in length for shamanic initiations.

In Iran and Iraq, the leaves are made into sniffing tobacco (cf. **snuffs**). The leaves are placed in *arrak* (palm sugar/palm alcohol; cf. **alcohol, palm wine**), dried, and then mixed with the ashes of *Ephedra pachyclada* Boiss (cf. *Ephedra* **spp.**), known locally as *huma,* or they are perfumed with jasmine oil (cf. *Jasminum* **spp.**) (Hooper 1937, 143 f.*).

In the Himalayan region, a smoking blend known as *khamera* is made from leaves of wild tobacco that have been perfumed with *keora* (*Pandanus tectorius* Parkins. ex Du Roi [syn. *Pandanus odoratissimus* L. f.]; cf. *Pandanus* **spp.**), leaves of the moschatel *Delphinium brunonianum* Royle (cf. *Delphinium consolida*), powdered sandalwood (*Santalum album* L.), rose petals (known as *gúl-kand*), the fruits of *Zizyphus jujuba* Mill. [syn. *Zizyphus vulgaris* Lam.], cardamom, and the wilted leaves of the betel palm (*Areca catechu*). The tobacco dealers do not reveal the relative amounts of the different ingredients they use (Atkinson 1989, 756 f.*).

Wild tobacco has an extremely high nicotine content and is considerably more potent than *Nicotiana tabacum* and all other *Nicotiana* species. For this reason, it should be used with great care. Dosages can vary so greatly from one individual to the next that it is not possible to provide any specific guidelines (cf. **nicotine**).

Ritual Use

Wild tobacco (*picietl*) was sacred to the Aztecs, who used it like peyote (*Lophophora williamsii*) or

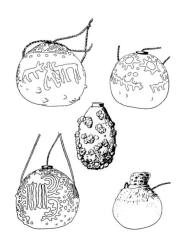

The tobacco gourds of the Huichol are used to store and transport the powder they produce from *Nicotiana rustica*. (From Lumholtz, *Symbolism of the Huichol Indians*, 1900)

Clay figure of the Aztec goddess Cihuacoatl, the "soul of wild tobacco." (From Krickeberg)

Left: Preparation of wild tobacco for a *singando.*

Right: Folk healers (*curanderos*) drinking a wild tobacco extract through their noses (a practice known as *singando*). (Photographed at the Laguna Shimbe, Las Huaringas, northern Peru)

ololiuqui (*Turbina corymbosa*) for magical healing with incantations and in divination (Ruiz de Alarcón 1984):

> When they smoked it and became inebriated from it, they called to the demons to find out about future events and to ask for advice for the requests of others who had retained them for this purpose. (Fuentes y Guzmán, in Maurer 1981, 347)

The colonial chronicler Jéronimo Mendieta wrote in his *Historia eclesiástica indiana*:

> Others say that some see the plant called *picietl*, which the Spanish call tobacco, as the body of the goddess Ciuacoatl. And that for this reason it has several medicinal effects. It must be smoked with great care, for it is very dangerous, as it takes away the minds of those who partake of it and makes them behave crazy and insane. (In Maurer 1981, 347 f.)

Among the Aztec, Ciuacoatl or Cihuacoatl, "woman-snake," was a mother and earth goddess. She was the patroness of midwives and watched over the sweat bath. She—the soul of wild tobacco—was described in the following manner:

> In this garb does she allow herself to be seen by people—adorned with lime, like a lady from the palace—: she wears ear pegs of obsidian, she appears in a white garment, she is dressed in a white costume, is completely white; she wears her woman's coiffure put up. During the night, she cries, she screams, she is also a portent of war. Her image is adorned in the following manner: her face is half red [and] half black, she wears a crown of eagle feathers, she wears a golden ear peg, she wears a collarlike outer garment, she carries a blue weaver's knife. (Sahagun 1.6)

The fact that the goddess is adorned with lime may be an indication of wild tobacco being prepared with lime.

The Huichol regard this tobacco species as a

manifestation of the fire god Tatewari, who was originally a falcon but was transformed into a plant (Siegel et al. 1977, 16). Wild tobacco is sacred to the Huichol and is a part of all of their ceremonial activities (peyote rituals, drinking festivals, peyote pilgrimages; cf. *Lophophora williamsii*). It sometimes is smoked in combination with *Tagetes lucida* (see *Tagetes* spp.).

The Mazatec refer to wild tobacco as San Pedro (St. Peter[251]), thereby associating the plant with the saint that holds the key to heaven (cf. *Trichocereus pachanoi*). Tobacco, powdered and mixed with lime, is exchanged or offered both as an agent of pleasure for establishing social structures and in magico-religious contexts at all ceremonies (shamanic healings, divinations, mushroom circles; cf. *Psilocybe* spp.). The Mazatec regard the smoke as a magical protection against rattlesnakes, scorpions, and giant centipedes.

In South America (Tiahuanaco culture), powdered wild tobacco was used as a ritual **snuff** (Bondeson 1972). Unfortunately, no details are known about such use. This use could still be observed during the colonial period, with the note that the Indians used *sayre*, as they called wild tobacco, for many things and would also sniff the powder "to cleanse their heads" (Bastien 1987, 153*).

The Warao of Venezuela have developed an extremely complex mythology and cosmology on the basis of their tobacco experiences. The anthropologist Johannes Wilbert has been working for decades to decipher and comprehend this complex structure. The Warao use the leaves to make cigars some 90 cm in length that can be smoked only by shamans and initiates. A man must fast and may drink only water during the seven days that precede the use of one of these cigars. Only a few inhalations are taken from the cigar, as most people fall to the ground after their first deep inhalation and enter an extremely altered state of consciousness (Wilbert 1996).

Many shamans acquire their abilities to travel into other realities through the help of tobacco. They learn to enter the house of tobacco, to use the tobacco smoke to ascend into the heavens, and to communicate with the plant spirits of tobacco,

251 The name Peter is derived from the Latin word *petrus*, "rock." This implies that the saint was a "sacred stone."

which often appear in the form of snakes. When performing healings, shamans often blow tobacco smoke onto the ill person to free him or her of disease spirits or to protect the patient from them. For the initiated shamans, tobacco smoke is a door into another world, the world of visions, the world beyond space and time.

The Cariña Indians use a mixture of wild tobacco and ginger (*Zingiber officinale*) to promote the ability to see at night among those who are being trained to become shamans. The juice of both plants is dripped into the eyes of the candidates so that they later will be able to see and recognize both good and evil spirits (Wilbert 1987, 166).

Peruvian folk healers practice an extreme form of ingesting wild tobacco during their San Pedro rituals (cf. *Trichocereus pachanoi*). The process, which is known as *singando*, consists of the ritual drinking of a decoction of *tabaco blanco* (wild tobacco) through the nostrils. Today, *singando* is still frequently practiced by the *curanderos* of northwestern Peru. The healers consecrate themselves to the mountain gods and fall into a trance-like altered state of consciousness as a result of the potent effects of the **nicotine**. *Singando* is an important part of the *mesa* rituals, during which San Pedro or, less frequently, *floripondio* (*Brugmansia sanguinea*, *Brugmansia* spp.) is drunk. The tobacco decoction is produced by macerating the tobacco leaves, sometimes mixed with other plants, in water, **alcohol**, and scented waters (e.g., eau de cologne, agua florida). The drinking vessels are usually made from the shells of ocean mollusks. Shells of the pearl oyster (*Pteria sterna* [Gould, 1851] [syn. *Pteria peruviana* (Reeve, 1857)]) are preferred, as they are pointed at the ends.

Artifacts

The Huichol Indians make tobacco bottles (*yékwe*) from tree gourds (*Crescentia cujete* L.) for ceremonial purposes. Some of these bottles are decorated with visionary elements or images from peyote experiences (see *Lophophora williamsii*). They also are used as offerings to *Solandra* spp., the magical "tree of the wind." The Tzeltal Indians (Chiapas, Mexico) sometimes make a tobacco pouch for **smoking blends** (*bankilal*) containing wild tobacco from the penis (called *yat kohtom*) and/or scrotum of the coati (*Nasua nasua, N. narica*).

Many archaeological objects from the Mesoamerican region are related to tobacco, although it cannot be determined which tobacco species they are associated with (see *Nicotiana tabacum*). In South America, a variety of types of tobacco pipes and forked cigar holders are carved from wood (Wilbert 1987).

Medicinal Use

The Callawaya wandering healers of South America recommend wild tobacco as a treatment for swollen muscles. The fresh leaves are warmed in the sun for a half hour and then massaged into the painful areas (Bastien 1987, 153*). In Peru, a decoction of the leaves is drunk to treat dysentery.

In Aztec medicine, wild tobacco was placed on the lower abdomen to treat swollen stomachs, smoked for asthma, and used to treat uterine ailments, sleeplessness, headaches, inflammations of the spleen, toothaches, syphilis, snakebites, and arrow wounds (Hernández 1959, 81f., 376*; Sahagun 11.7).

In modern Mexico, wild tobacco is smoked together with *Ephedra nevadensis* (see *Ephedra* spp.) as a treatment for headaches.

Constituents

Nicotiana rustica is very rich in **nicotine** (3.9 to 8.6%) and other pyrrolidine alkaloids (nornicotine, anabasine). It also contains traces of harmala alkaloids and tobacco camphor (Bastein 1987, 153*; Díaz 1979, 85*). The dried leaves may contain as much as 16% nicotine. Tobacco smoke has been found to contain more than nine hundred substances (Siegel et al. 1977, 18).

Effects

The Italian Girolamo Benzoni described the potent psychoactive effects of *Nicotiana rustica* in his early colonial work *Historia del Mondo Nuovo* (1568):

> They light one end of the cigar, place the other end in the mouth, breathe through this, and fill themselves up with the dreadful smoke, so that they lose their minds. Some take so much of it that they fall over as if they were dead and remain unconscious for the greater part of the day or night. (In Maurer 1981, 348)

Wild tobacco can induce hallucinations that shamans are able to utilize. Wilbert (1996) has distinguished the hallucinations wild tobacco produces among the Warao Indians on the basis of their phenomenological effects:

— dreamlike and chromatic
— multisensory perceptions
— brilliant occurrences of light
— intuitive knowledge and spontaneous insights
— soul-escort by a psychopomp
— tunnel experiences

However, such phenomena appear only among initiated shamans. Nonshamans who consume the same amounts as shamans do may experience toxic effects that can be life-threatening (cf. Wilbert 1991).

"The great magician, who as physician, spirit conjurer, and preserver of the old tribal messages is the most important personage of a Taulipáng community, knows how to enter a trance state by means of excessive smoking and drinking a strong tobacco brew. In this state, visions come to him that he feels he has truly experienced when he awakes.... The magic physician must drink a great deal of tobacco juice when curing the ill so that he can send his soul to the mountain spirits, the Mauarí. He searches among them for those Mauarí who know how the sick person can be helped."

JOSEFINE HUPPERTZ
IN *GEISTER AM RORAIMA* [SPIRITS ON THE RORAIMA]
(KOCH-GRÜNBERG 1956, 14 f.*)

"South American Indians consider tobacco as food that can be eaten or drunk, and in many societies shamans are referred to as 'tobacco eaters.'"

JOHANNES WILBERT
"DOES PHARMACOLOGY CORROBORATE THE NICOTINE THERAPY AND PRACTICES OF SOUTH AMERICAN SHAMANISM?"
(1991, 182)

Nicotine

Nornicotine

Anabasine

Commercial Forms and Regulations

None

Literature

See also the entries for *Nicotiana tabacum*, *Nicotiana* spp., and **nicotine**.

Bondeson, Wolmar E. 1972. Tobacco from a Tiahuanacoid culture period. *Etnologiska Studier* 32:177–84.

Maurer, Ingeborg. 1981. Die Rauchenden Götter—Tabak in Kunst, Geschichte und Religion der Maya. In *Rausch und Realität*, ed. G. Völger, 1:346–50. Cologne: Rautenstrauch-Joest-Museum für Völkerkunde.

Pahl, Carola. 1996. Schon die alten Ägypter frönten der Drogensucht. *Frankfurter Rundschau* 20 (July).

Siegel, Ron K., P. R. Collings, and José L. Diaz. 1977. On the use of *Tagetes lucida* and *Nicotiana rustica* as a Huichol smoking mixture. *Economic Botany* 31:16–23.

Wilbert, Johannes. 1972. Tobacco and shamanistic ecstasy among the Warao Indians of Venezuela. In *Flesh of the gods*, ed. Peter Furst, 55–83. New York: Praeger.

———. 1975. Magico-religious use of tobacco among South American Indians. In *Cannabis and culture*, ed. Vera Rubin, 439–61. The Hague: Mouton.

———. 1979. Magico-religious use of tobacco among South American Indians. In *Spirits, shamans, and stars*, ed. David Browman and Ronald A. Schwarz, 13–38. The Hague: Mouton.

———. 1987. *Tobacco and shamanism in South America*. New Haven and London: Yale University Press. (Excellent bibliography.)

———. 1991. Does pharmacology corroborate the nicotine therapy and practices of South American shamanism? *Journal of Ethnopharmacology* 32:179–86.

———. 1996. Illuminative serpents: Tobacco hallucinations of the Warao. Lecture presented at the Entheobotany Conference, San Francisco, October 18–20.

Nicotiana tabacum Linnaeus

True Tobacco

> "In many ways, the history of tobacco is as gripping as any novel. The detectives of science had to make use of all of their acumen to illuminate all of the botanical, national, economic, and even linguistic problems that surround this mystical plant, which has completed its triumphant march across the entire globe in the course of the past four centuries."
>
> ØVRE RICHTER FRICH
> *VITAMIN DER SEELE* [VITAMIN OF THE SOUL]
> (1936, 9)

Family

Solanaceae (Nightshade Family); Subfamily Cestroideae, Nicotianeae Tribe

Forms and Subspecies

Numerous varieties have been described, most of which can be distinguished on the basis of their phytogeography (Hartwich 1911, 27 f.*):

Nicotiana tabacum L. var. *brasiliensis* Comes (indigenous to Brazil and northern South America)

Nicotiana tabacum L. var. *fruticosa* Hook. f. (indigenous to Mexico and Brazil; the most commonly cultivated variety)

Nicotiana tabacum L. var. *havanensis* Comes (indigenous to Mexico, and introduced from there to Cuba and Manila)

Nicotiana tabacum L. var. *lancifolia* Comes (indigenous to Ecuador and Central America)

Nicotiana tabacum L. var. *macrophylla* Schrank (Maryland tobacco; indigenous to Mexico)

Nicotiana tabacum L. var. *virginica* Comes (indigenous to the Orinoco, and introduced from there to Virginia)

All of these varieties have been used to produce numerous hybrids, cultivars, and sorts. The most important market forms are Virgin and Burley.

Synonyms

Nicotiana chinensis Fisch.
Nicotiana fruticosa L.
Nicotiana lancifolia Willd. ex Lehm.
Nicotiana latissima Mill.
Nicotiana loxensis H.B.K.
Nicotiana macrophylla Lehm.
Nicotiana mexicana Schlechtend.
Nicotiana nepalensis Lk. et Otto
Nicotiana pilosa Dun.
Nicotiana ybarrensis H.B.K.
Nicotiana tabacum L. var. *subcordata* Sendtner
Nicotiana tabacum L. var. *macrophyllum* Dun.

Folk Names

Ægte tobaksplante, alee (Bara), a'-li, anjel, apagu (Cuicatleca), ascut, a'xcu't (Totonac), ayic (Popoluca), bujjerbhang (Arabic), bunco (Malabar), buncus, chimó, ch'ul winik ("human piss"), cocorote, cuauhyetl, cultivated tobacco, cutz, de-oo-we (Witoto), dê'-oo-wê, dhum-kola, dhuum-rapatra (Sanskrit), doonkola, duma, dumkola (Singhalese), dunkala, echter rauchtabak, echter tabak, elee (Baniwa), e'-li, finak, gemeiner tabak, guácharo, guexa (Zapotec), gueza, hach k'uts (Lacandon), hapis copxot (Seri), hepeaca (Tarahumara), herba sancta, herbe petum, huepaca, huipá (Tarahumara), indianisch wundkraut, iyatl, jaari, jacha, jakhon, jakhu, ju'uikill (Mixe), kapada, kherm'-ba (Kofán), kuanmat, kultur-tabak, kuts, k'uts (Maya), k'úts, kutz, kuutz, ku'utz, lixcule, lixculi, lu-kux-ree (Yucuna), lukux-rí, maay (Huastec), majoris peti, may (Tzeltal), may wamal, me-e (Chontal), mitó (Siona), moo-loo (Desana), moy (Tzotzil), mulú (Tukano), mu-lu', nát'oohlijiníh, nát'oohntl'ízíkíih (Navajo), nát'oohxiit'aalíh, nicotiana maior, nicotiane, nicotianskraut, otzi (Zoque), pa-ga-ree-moo-le (Desana), pagári-mulé (Desana), pahu"ky (Mixe), pëtrem (Mapuche, "that which is smoked"), petum, petun, piciete, picietl, poga, poghako, poghéi elley (Tamil), pop siwa, poyile, püchrem (Mapuche), puthem, quahyetl, quaryetl, quauyétl (Nahuatl), rauchtabak, ro-hú (Chinantec), rome (Shipibo-Conibo), salóm, sana sancta, sang-yen (Chinese), sayri, sidí, suma, symphytum indicum, tabac, tabacco, tabacco vero, tabaci, taback, tabaco, tabaco cimarrón, tabaco de la montaña, tabaco huitl, tabaku, tägyi (Comanche), takap, tamaku, tambaku (Hindi), tambracoo, tamer, tenejiete, tenexiet, thnam, thuok, tobacco, toback, tobak, tombeki, tosu, toutoune kordestani, tranco corto, true tobacco, tsaank (Shuar), tsank, tuma, tumak, tumbaku, uar (Cuna), uipa (Guarigia), uxkut (Tepehuano), vesciakola (Veddah, "cola leaf"), virgineischer tabak, Virginian tobacco, virginiatabak, ya, yaná (Cora), ye'-ma (Tariana), yen (Chinese), yerba santa (Spanish, "sacred herb"),[252] yetl (Aztec), yinheu, youly, ysé, yuyi (Otomí)

History

True tobacco was first cultivated in either Mexico or Peru. Whichever is correct, it spread into the other region at a very early date (Dressler 1953, 138 f.*). It apparently was not derived from a wild form but was produced by hybridization (Schultes and Raffauf 1991, 41*). *Nicotiana sylvestris* Spegazz. et Comes may have been a precursor form. In Central and South America, tobacco is the most important and most commonly used shamanic plant.

In ancient Mesoamerica, tobacco was religiously venerated and was regarded as a plant of the gods. With its aid, priests induced an inebriation that opened contact to the world of the gods (Elferink 1983; Robicsek 1978). The "smoking god" of Palenque (= God K)—which the Lacandon know well and refer to as *k'uh ku ts'uts'*, "the god that smokes"—bears witness to this ritual. Tobacco played a significant role in the pre-Columbian culture of the Maya. In addition to its numerous social meanings, tobacco also is regarded by the Indians as a universal antidote for all types of animal bites and poisonings.

The earliest report about tobacco came from the feather of the monk Romano Pane, a companion of Christopher Columbus (1451–1506),

who sent tobacco seeds to Charles V in 1518. The first botanical description is from Hernández (1525), who compared tobacco to henbane (*Hyoscyamus niger, Hyoscyamus* spp.), which was well known in Europe (cf. *Nicotiana rustica*). In Europe, tobacco was received as a panacea and miracle agent and used in folk medicine in numerous ways (Kell 1965).

Tobacco appears to have been introduced to the South Pacific during pre-Columbian times (Feinhandler et al. 1979). The Portuguese brought it to India in the fifteenth or sixteenth century (Gupta 1991, 62*). By the sixteenth century, it was apparently well known in India and Nepal. Tobacco, and the practices of smoking and chewing it, was introduced to southern and Southeast Asia in the seventeenth century by the Dutch. Tobacco finally conquered all of Asia in the nineteenth century. Today, tobacco is one of the most frequently used psychoactive agents of pleasure in the world.

Distribution

True tobacco is a pure cultigen that was originally cultivated in either Central America or Amazonia and the neighboring regions. It is now grown throughout the world. Turkey has become renowned in the tobacco industry for its tobacco, and tobacco plants for commercial production are grown even in Germany. Although it is originally from the tropics, the tobacco plant has adapted very well to subtropical, dry/warm, and temperate climates.

Cultivation

Tobacco is propagated from seeds. In the tropics, the seeds need only be scattered over the earth, which is usually fertilized with ashes. In central Europe, the seeds must be sown between the middle and the end of March in greenhouses or in windowboxes in porous, sandy soil. The seeds should be gently pressed into the soil and will germinate in ten to twenty days when kept at 18 to 20°C. The young plants should then be transplanted into larger pots or beds. Tobacco requires much sun, a great deal of fertilizer, and copious amounts of water. It thrives best in sheltered areas.

If a tobacco plant develops flowers too quickly,

When true tobacco (*Nicotiana tabacum*) was first introduced to Europe from the Americas, it was used as a medicinal plant and was listed in the early modern herbals under the name *Indian beinwell*. (Woodcut from Tabernaemontanus, *Neu Vollkommen Kräuter-Buch*, 1731)

True tobacco (*Nicotiana tabacum*), with flowers and fruits.

252 The same name is used for *Piper auritum*.

"Many Amazonian Indians believe that there is a mighty life-giving and life-preserving power in tobacco. It is also believed that [tobacco] can strengthen the ability to resist harmful influences, purify, and enlighten. As a vehicle of the shaman, tobacco promotes contact with supernatural beings: the expelled smoke forms a kind of ladder to the heavens and [is] the medium through which religious authorities receive their energy. Few healing ceremonies occur in which tobacco smoke is not blown or tobacco leaves laid down. An ailing person must also ingest great quantities of tobacco so that the power of such an immunization can defend against the renewed attacks of disease demons."

Wolfgang Müller
Die Indianer Amazoniens [The Indians of Amazonia]
(1995, 197*)

"Never has a plant that has entered into the circle of culture been laden with such curses or been persecuted with such severe laws as tobacco, and never has one passed so triumphantly across the entire earth—evidence as to how much humans have always been and still are inclined to pursue the enticements of pleasure more quickly than the demands of social welfare."

Julius Lippert
Die Kulturgeschichte in einzelnen Hauptstücken [The History of Culture in Individual Main Sections]
(1885, 127)

The Lacandon still enjoy smoking cigars rolled from unfermented tobacco leaves.

they should be removed immediately so that the plant will continue to grow and develop more and larger leaves.

Appearance

This annual herbaceous plant can attain a height of 2 to 3 meters. Its large leaves (30 to 40 cm long) are oblong-elliptic. The campanulate, funnel-shaped, five-pointed flowers grow in panicles and have a light green calyx and pink petals. In North America and Europe, the plant flowers between July and September. The capsule-shaped fruits contain numerous tiny brown seeds.

True tobacco is occasionally confused with other *Nicotiana* species (*Nicotiana* spp.).

Psychoactive Material

— Leaves (dried and/or fermented; folia nicotianae, nicotianae folium, herba tabaci, herba nicotianae virginianae, tobacco leaves)
— Herbage (nicotianae virginianae herba)

Preparation and Dosage

The leaves can be dried in a variety of ways. For the smoking industry, tobacco leaves are "fermented" in a manner that varies depending upon their intended use. Cigarette tobacco is dried slowly in a moist environment. Cigar tobacco is air-dried, chewing tobacco is dried over a fire, and Turkish tobacco is dried in the sun (Macmillan 1991, 419*). Fruit juices are added to some pipe tobaccos. The drying process ("fermenting") produces the special, preferred aroma. The leaves also can be laid on top of one another to ferment, a process that can require several months.

The Indians hang the leaves in the shade or, rarely, spread them out in the sun to dry.

Smoking tobacco should be yellowish or brown. The leaves are sometimes bleached with sulfur to give them a light yellow color. The taste is improved by adding sugar solutions, spices, salts, and coloring agents to the leaves. Chewing tobacco is produced from tobacco leaves that have been allowed to sit in a tobacco solution. Industrial tobacco snuffs may be aromatized with extracts of juniper berries (*Juniperus communis* L.; cf. *Juniperus recurva*), calamus root (*Acorus calamus*), sassafras wood (*Sassafras albidum*), and spices

(Wagner 1985, 172*). In Burma (Myanmar), "the tobacco is doused with urine to improve the flavor" (Hartwich 1911, 113*).

Tobacco leaves are smoked by themselves or in **smoking blends** with other herbs (e.g., *Cannabis indica, Datura innoxia*) (cf. **kinnikinnick**). The Siona roll tobacco leaves in dried banana leaves (*Musa* x *sapientum*) for smoking (Vickers and Plowman 1984, 31*). In Mexico, tobacco is usually rolled in corn husks (*Zea mays*). The dry yet still elastic tobacco leaves also can be rolled into cigars by themselves, without any additions. Shamans of the Yucatán (Mexico) make cigars from one leaf of *Nicotiana tabacum* and one leaf of **Datura innoxia**. Birch bark is sometimes added to the tobacco to dilute the effects (Hartwich 1911, 91*).

In Siberia, smoking tobacco is mixed with spruce bark (*Picea omorika* [Panc.] Purkyne), scrapings of birch wood (*Betula* spp.), fir wood (*Abies* spp.), and moss (*Polytrichum*) (Hartwich 1911, 110*).

The men and women of Burma (Myanmar) roll different kinds of cigarettes. Men's cigarettes consist of finely cut tobacco that is wrapped in the foliage leaves of *Ficus* spp., *Cordia dichotoma* Forst. f. [syn. *Cordia myxa* Roxb., *Cordia obliqua* Willd.] (Cordiaceae), *Careya arborea* Roxb. (Barringtoniaceae), or *Tectona grandis* L. f. (Verbenaceae). Women's cigarettes are wrapped in the epidermis of the spathes of **Areca catechu**, corn husks, or bamboo skin that has been flattened on a heated stone. The filling is a mixture of chopped tobacco leaves, stems, and roots; the roots of a *Euphorbia* species; the pith of a plant known as *oh'ne* (*Streblus asper* Lour.; Moraceae); sometimes palm sugar (cf. *Cocos nucifera*, **palm wine**); and finely chopped banana leaves (Hartwich 1911, 113*).

In Europe, tobacco is often rolled into joints together with hashish (see **Cannabis indica, Cannabis sativa**). From a pharmacological perspective, however, this combination is rather pointless, as the combination of the two substances results in a negative synergy. The tobacco suppresses the effects of the hashish, and the hashish potentiates the effects of the nicotine.

The western Amazon region and Venezuela are home to a practice known as tobacco licking or sucking. Here, tobacco is boiled to produce a kind of syrup called *ambíl* or *chimó* (*chimú*). The syrup is collected on a stick that is then dipped in coca powder (*Erythroxylum coca* var. *ipadú*) or plant ashes (cf. **Erythroxylum coca**) and licked (Kamen-Kaye 1971, 1975). This method of preparation and use has its roots in the pre-Hispanic period. Varying the additives and the quality of the tobacco leaves yields a *chimó manso* (mild), *chimó dulce* (sweet), *chimó bravo* (courageous), or *chimó fuerte* (strong). The *chimó* paste is placed between the lip and the lower front teeth, where it slowly dissolves. The black saliva is spat out. Habitual

 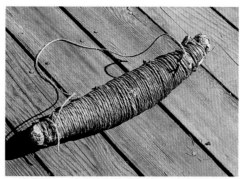

Left: Indians of the North American prairie make fresh tobacco leaves into braids that are later used as chewing tobacco. (Lakota tobacco braid, from South Dakota)

Right: A *mapacho*, a roll of tightly glued tobacco leaves from the Amazon.

chimó users (whether men, women, or children) use the paste from the morning until the evening (Kamen-Kaye 1971, 17). As with cigarette smoking, tobacco licking has no predetermined dosages.

The Siona make their *ambíl* from tobacco leaves that are boiled, pressed, and then boiled again together with the pressed juice until a dark brown syrup results. The ashes of the fruit husks of *cacao colorado de monte* (a *Herrania* species, possibly *Herrania breviligulata*), banana peels (see *Musa* x *sapientum*), and the bark of *Paullinia yoco* are also added to the syrup. The thick mixture is stored in gourds and can be sucked or even swallowed for use (Kamen-Kaye 1971, 53). The Witoto add avocado seeds (*Persea americana* Mill.) to the boiling tobacco decoction and sweeten their *ambíl* with cane sugar. They usually add a salty plant ash from the wood of rain forest trees of the genus *Lecythis* or palm wood from the genera *Bactris* and *Chamaedorea* (Kamen-Kaye 1971, 36). They often store *ambíl* in the fruit shells of a wild cacao species (*Theobroma glaucum* Karsten; see *Theobroma* spp.); they believe that this improves the taste considerably. The Kogi use manioc flour (*Manihot esculenta* Crantz) or *sagú* (*Maranta arundinacea* L.) to thicken the tobacco decoction. Other Indians use *sugii* (*Sorghum* spp.), a type of millet (Kamen-Kaye 1971, 33; 1975, 58). In Venezuela, the ashes from the wood of an *Erythrina* species (*Erythrina* spp.) are used as well (Plotkin et al. 1980, 295).

Amazonian **snuffs** are made by mixing finely ground, dried green tobacco leaves in equal parts with the ashes of a wild cacao species (*Theobroma subincanum* Mart.; see *Theobroma* spp.). Sometimes, a pinch of chili (*Capsicum* spp.) or some coca powder (*Erythroxylum coca* var. *ipadú*) may also be added (Schultes and Raffauf 1991, 42*). In Europe, Schneeberger Schnupftabak, a type of sniffing tobacco, was made with *Veratrum album* as an additive.

In French Guiana and Suriname, tobacco leaves and an alkaline additive, the ashes of the stem wood of a tree known as *mahot cochon* or *okro-oedoe* (*Sterculia excelsa, S. pruriens* [Aubl.] K. Schum.[253]), are used to produce a liquid that is sucked through the nose. The ashes are sprinkled over the fresh tobacco leaves, which are then moistened with water. After being allowed to sit for a period of time, the treated leaves are then pressed. The juice is sucked into the nose, immediately producing pronounced psychoactive effects characterized as an "overwhelming sensation of ecstasy." The effects persist for some twenty to thirty minutes (Plotkin et al. 1980).

Peruvian shamans use fresh tobacco leaves to prepare cold-water extracts, often adding other plants, aromas, or **alcohol** (Vickers and Plowman 1984, 31*). This preparation usually is taken nasally before the ingestion of *Trichocereus pachanoi* or **cimora**.

An astonishingly similar use occurs (occurred?) in Africa:

> Wherever they [the Wadchichi, Lake Tanganyika] go and stand, they carry a gourd of

253 *Sterculia pruriens* contains a cyanogenic substance that is considered toxic (Schultes and Raffauf 1990, 447*).

254 In the Putumayo region, the dried leaves of the closely related species *Cephalis williamsii* Standley are smoked together with tobacco (Schultes and Raffauf 1990, 379*).

255 This may be an incorrect botanical identification from the older literature. Kamen-Kaye believes that the leaves of a *Psychotria* species (*Psychotria* spp.), a genus that is very closely related to and easily confused with the genus *Palicourea*, may have been or are used as an additive containing tryptamines (1971, 47).

Chimó Additives

Venezuelan *chimó* preparations are made from tobacco leaves and a number of additives (Kamen-Kaye 1971, 46 f.):

amapolo (cf. **amapola**)	opium	*Papaver somniferum*
anís	anise seeds	*Pimpinella anisum* L.
cafecito	leaves	*Cephalis tinctoria*[254]
cafecito blanco	*chimó* leaves	*Palicourea chimó*[255]
chivata/cervata		unidentified
clavo de olor	cloves	*Syzygium aromaticum*
cocuy	cocui liquor	*Agave cocui*
curía	justicia leaves	*Justicia caracasana* [syn. *Rhytiglossa caracasana, Ecbolium caracasana*], ***Justicia pectoralis***
nuez moscada	nutmeg	***Myristica fragrans***
panela/papelón	brown sugar	*Saccharum officinarum* L.
quina negra	leaves	*Guettarda sabiceoides*
sarrapia	tonka beans	*Dipteryx odorata* (Aubl.) Willd. [syn. *Coumarouna odorata* Aubl.]
tamo de caraota	bean straw	Leguminosae spp.
vainilla	vanilla	*Vanilla planifolia* Andr. [syn. *V. fragrans* (Salisb.) Ames]
Alkaline additives:		
cernada	plant ashes	***Erythrina* spp., *Musa* spp.**
uroa	sodium carbonate, sodium bicarbonate	

"Hachäkyum, our true lord, made tobacco. He planted it, he tried it out: 'Ah, it is very delicious. That is very good for my creatures, for its smoke dispels the *us* flies, its juice kills the ticks and the flesh worms [Spanish: *colmoyotes*].' Hachäkyum gave tobacco to the ancestors. But they did not need to die, for they did not inhale the smoke. When you inhale the smoke, your consciousness turns, the heart beats faster, the stomach aches. When you inhale the smoke, you quickly become inebriated, you must vomit, the muscles ache. Then you must drink a lot of water or pour it over the head, then you will become healthy once more."

A LACANDON MYTH
IN "*TS'AK*"
(RÄTSCH 1994B, 55*)

256 An alkaline substance is not necessary to liberate the alkaloid (nicotine) from the plant material (unlike *Erythroxylum coca* and *Areca catechu*). However, laboratory studies have demonstrated that an alkaline additive potentiates the effects of the nicotine (Kamen-Kaye 1971, 41).

257 The same bark ash is also added to **snuffs** (made from *Virola* spp.).

tobacco and wear a metal or wooden clamp that hangs from a rope around their necks. From time to time, they fill the gourd with water and press the juice of the moistened tobacco into the hollow of their hand. They sniff this from their hand into their nose, and then the clamp is carefully set in place so that nothing flows out. (Lippert 1885, 28)

In Amazonia, a popular pastime involves sprinkling a partially dried tobacco leaf with coca powder (*Erythroxylum coca* var. *ipadú*), rolling it up, and chewing it as a quid (Schultes and Raffauf 1991, 42*). Ingesting the tobacco orally together with slaked lime increases the effects noticably.[256]

The Jíbaro mix tobacco juice with *Banisteriopsis caapi* and *piripiri* (probably a *Cyperus* species; see *Cyperus* spp.). The mestizo *ayahuasqueros* of Iquitos mix tobacco juice with **ayahuasca**. Dried tobacco leaves are moistened with saliva and allowed to remain overnight in a hollow that has been cut into the trunk of a *lupuna* tree (*Trichilia tocacheana* C. DC.; Meliaceae). This makes them thoroughly wet, and the leaves also absorb the toxic juice of the tree (Schultes and Raffauf 1991, 42*).

In Mexico, a decoction of tobacco leaves, stems of *capulín agarroso* (*Conostegia xalapensis* [Bonpl.] Don), guava leaves (*Psidium guajava*), avocado leaves (*Persea americana* Mill.), *muicle* herbage (*Justicia spicigera* Schlechtend. or *Justicia mexicana* Rose; cf. *Justicia pectoralis*), *Dyssodia porophylla* (Cav.) Cav., and garlic (*Allium sativum* L.) is prepared as an external remedy used to treat edema of the legs (Argueta V. et al. 1994, 1301*).

In India, fresh or dried green tobacco leaves are chewed or used as additives to betel quids (Jain and Borthakur 1986, 579*). Indian bidis are sometimes made from unfermented tobacco and thorn apple leaves (*Datura metel*). In the Near East, a smoking blend known as *guracco* is made primarily from hemp (*Cannabis indica*), sometimes opium (cf. *Papaver somniferum*), raw sugar, fruits, the leaves of a *Eugenia* species, the leaves of *Rhododendron campanulatum* D. Don (cf. *Rhododendron caucasicum*), *Marrubium candidissimum* L., and tobacco. A mixture of tobacco and rhododendron leaves is used as snuff. In the Himalayan region, tobacco is added to homemade **alcohol** (Hartwich 1911, 92*).

Tobacco is added to numerous psychoactive products, including **ayahuasca, balche', beer, betel quids, enemas, incense, smoking blends, snuffs,** and **witches' ointments**.

Forty to 60 mg of **nicotine** represents a lethal dosage for adults (Roth et al. 1994, 517*). Depending upon the type of tobacco and the manner of preparation, this dosage can correspond to very different amounts of tobacco leaves. It is possible for a person to die after consuming just one

commercially manufactured cigarette (cf. *Nicotiana rustica*). One "normal" cigarette contains approximately 1 g of tobacco, which usually corresponds to a concentration of 5 to 10 mg of nicotine (calculated as a salt) (Wagner 1985, 172*).

Ritual Use

In Mesoamerica, tobacco has a long history as a plant of the gods. The plant not only was venerated but also was offered to the gods and was always smoked at rituals and ceremonies and during shamanic healings (cf. *Nicotiana rustica*).

Today, tobacco continues to play a central role in Mexican shamanism, although often it is only in the form of industrially manufactured, commercial cigarettes. The Nahuat offer tobacco leaves or cigarettes during shamanic rituals to treat soul loss (Knab 1995, 160*). Tobacco also is used as a fumigant for magical protection from evil sorcerers, spirits, and snakes. The Mayan shamans of the Yucatán smoke tobacco (usually in combination with *Datura innoxia*) to diagnose diseases and to dispel the spirits that cause them. The Lacandon smoke an especially large number of cigars during their communal **balche'** drinking rituals. They also use cigars as gifts when courting a bride. In general, tobacco and cigars are their most important traditional gifts for establishing or strengthening social relationships.

The priestly and shamanic use of tobacco has also been documented for pre-Columbian Central America and the Caribbean islands. There, tobacco often was mixed with other substances, including the balsam of *Liquidambar styraciflua* L., which was used an as **incense** (Elferink 1983).

In Colombia, the Indian use of tobacco is widespread and has its roots in pre-Columbian times (Uscátegui M. 1956). Tobacco has great ritual, medicinal, and magical significance for almost every tribe in the Amazon region. It is smoked, chewed, sniffed (cf. **snuffs**), drunk in the form of pressed juice or brew (decoction, coldwater extract), administered rectally as an **enema**, or added to **ayahuasca** (Schultes and Raffauf 1990, 432 ff.*).

Freshly pressed tobacco juice (from the leaves and stems) or an aqueous decoction of tobacco is often added to **ayahuasca** to potentiate its hallucinogenic effects. While under the influence of ayahuasca, most Amazonian Indians also smoke thick cigars or snuff tobacco practically without interruption (Schultes and Raffauf 1991, 42*).

In Amazonia, tobacco sniffing is usually hedonistic in nature, although it can also occur in ritual contexts. During **ayahuasca** ceremonies, many Indians in the western Amazon region sniff great quantities of tobacco powder mixed with the bark of a wild cacao species (*Theobroma subincanum* Mart.; cf. *Theobroma* spp.),[257] sometimes even mixed with ground chili pods (*Capsicum* spp.) (Reichel-Dolmatoff 1971*; Schultes and Rffauf 1990, 433*).[258]

During shamanic initiations, the Tukanos frequently administer great quantities of tobacco juice to novices so that they will vomit and pass out. Only a person who has survived this "chemical torture"—the amounts administered would quickly kill a "normal" person—is able to become a proper shaman (Schultes and Raffauf 1990, 435*).

The Aguaruna (an Ecuadoran Jíbaro tribe) mix tobacco juice with **ayahuasca** and administer the result as a ritual **enema**. Before receiving the enema, they drink ayahuasca and tobacco water alternately until they have to vomit. The enema is given after this (Schultes and Raffauf 1991, 43*).

Children of the Shuar sometimes receive tobacco water in place of *Brugmansia suaveolens* so that they can find their dream soul (*arutam*) (Bennett 1992, 493*).

The Mapuche burn tobacco leaves to fumigate rooms (cf. **incense**) in which sick people are staying or have stayed in order to dispel the spirits and causes of disease (Houghton and Manby 1985, 100*). Mapuche shamans smoke a great deal of tobacco, often in combination with other plants and tobacco species (*Nicotiana acuminata* [Grah.] Hook., *Nicotiana* spp., *Nicotiana rustica*), to attain an ecstatic or trancelike state (Houghton and Manby 1985, 100*). They also blow tobacco smoke onto the ill (Mösbach 1992, 105*).

In former times, initiates among the Ayoreo Indians of the Paraguayan Chaco had to drink a cold-water extract (maceration) of tobacco leaves as part of the process of becoming a shaman (*naijna*). Each novice was required to fast for two days before receiving the drink. If he did not have to vomit after ingesting the drink, he could become a shaman. Afterward, he had to fast for another two days and then survive a drink of tobacco leaves and a caper plant known as *najnur* (*Capparis speciosa* Griseb.). Finally, he had to smoke the dried roots of *Jatropha grossidentata* and *Manihot anomala* (Schmeda-Hirschmann 1993, 109*).

True tobacco, although it did not originate in North America, was cultivated at an early date by many North America tribes, especially in Montana and Virginia (hence the name *Virginia tobacco*) (cf. *Nicotiana* spp.). The Crow Indians, a typical Plains tribe, even had a secret society that was dedicated to the growing, care, and use of tobacco (Lowie 1975). The modern peyote ceremony of the Native American Church is opened with tobacco smoking (Bull Durham sort) (see *Lophophora williamsii*). Today, *Nicotiana tabacum* is a common ingredient in ritual **smoking blends** and in **kinnikinnick**.

The shamanic use of tobacco has spread outside of the New World (Tschubinow 1914, 45*). "Tungus and Sojot shamans smoke imported tobacco to achieve ecstasy. Manchu shamans smoke tobacco and blow it onto sick persons. Tobacco is new to Asia, but inhaling the smoke of various plants in order to enter trance was an ancient practice of the local shamans" (Ruben 1952, 239*) (see *Juniperus recurva*, *Ledum palustre*). In eighteenth-century Siberia, "priests who were to divine would rub tobacco leaves between their hands and scatter these in brandy, which they then drank so that they could attain the necessary rapture" (Hartwich 1911, 109*). In Nepal, tobacco is venerated as a sacred plant of Shiva and is used by the shamans of the country (cf. *Aconitum ferox*, *Cannabis indica*). Tobacco was also used as a snuff in Nepal.

In Papua New Guinea, shamanism and sorcery are closely linked to the smoking and chewing of both the indigenous and the introduced tobacco species (cf. *Nicotiana* spp.). Certain aspects of the mythology and the culture of the Papuas are reminiscent of characteristics of the Australian Aborigines (cf. **pituri**):

> Among the Fore of the eastern highlands, the medicine men, who are also called *dream men* or *smoke men*, obtain their great knowledge from a knowing that is provided to them in dreams that are induced by psychotropic plants and the inhalation of tobacco smoke. (Michel 1981, 261)

According to the Eipo tribe (New Guinea), tobacco originated at the beginning of time from the excrement of birds (260).[259]

Around 1851, an article about witches' flights (cf. **witches' ointments**) appeared in the *Rheinischen Antiquarius*, in which a kind of "tobacco shamanism" was described:

> But around midnight, when he had lit a tobacco pipe and had smoked a little, he would have fallen into a swoon or sleep as a result of constant tobacco drinking, after which he would have only just dreamed or fancied that he had crept over a deep well or a cistern, constantly in danger of falling into this deep well. But when he awoke before the sun rose, he found himself next to or among his comrades [who sat imprisoned twenty-seven German miles away], in great amazement, and had the tobacco pipe, half-full, in his hands, which he then relit and smoked to the end. He did not know whether he had driven there on a goat or a coat or a poker or in some other manner, but had only been always fancying or dreaming, as if he had dreamed while asleep, how he crept over a deep well. (Stramberg 1986, 50)

Many people see the psychoactive tobacco as a brain food, a "vitamin of the soul." (Book cover, 1936)

258 In Colombia, the dried roots and leaves of *Piper interitum* Trelease (cf. *Piper* spp.) are used as a substitute for sniffing tobacco (Davis and Yost 1983, 182*; Schultes 1978b, 226f.*).

259 "This notion corresponds to deliberations as to whether birds had already carried tobacco seeds from the Americas to Polynesia and then to New Guinea [and perhaps Australia] during prehistoric times" (Michel 1981, 260) (cf. also Feinhandler et al. 1979).

The smoking god of Palenque demonstrates that the smoking of tobacco and other plants was of tremendous ritual and religious significance in pre-Hispanic times. (Relief, ca. eighth century)

260 The word *cigar* is derived from the Quiché term *siq'ar*, which refers to tobacco that has been rolled without a cover leaf (Richter 1921). Quiché is a Mayan language spoken in the highlands of Guatemala.

261 In this context, it should be noted that it cannot be assumed that the cigars of the smoking god were made only of tobacco. There is a great deal of ethnographic evidence attesting to the fact that the shamans of the various Mayan peoples smoke other plants (**Brugmansia suaveolens, Datura innoxia, Datura stramonium, Amanita muscaria**) besides tobacco.

262 The Yucatec Maya believe that burning tobacco will keep snakes at bay (Redfield 1950, 124).

In Europe, the rise of the smoking salon led to the development of ritual forms of communal smoking (smoking clubs), rudiments of which have been preserved in modern times. This includes the act of offering cigarettes in social gatherings.

Artifacts

Ancient Mexican art, especially the art of the Maya, is filled with references to tobacco (cf. **Nicotiana rustica**). There are frequent depictions of smoking gods. In particular, the Mayan god known in the literature as God K has cigars,[260] cigarettes, or smoking tubes (*chamal*) as its attributes (Robicsek 1978, 59 ff.).[261] God K appears often in Mayan manuscripts and reliefs from the Classic Mayan period (300 to 900 C.E.). The best-known image of God K is as the smoking god of Palenque. It is possible that tobacco visions provided the inspiration for some of the representations of other worlds (Robicsek 1978; cf. Dobkin de Rios 1974a*).

The use of tobacco has generated countless paraphernalia. The most important, of course, are the pipes, which have become objects of an almost cultic fixation among Western smokers in particular. The invention of the pipe is described in a whimsical story:

> In South Africa, people once allowed tobacco to glow in a hole in the earth, and several people would suck the smoke from this hole using tubes. It was a small step to attach a small coal pot and portable smoking altar to the end of the tube itself—and so our tobacco pipe was born. (Lippert 1885, 127)

Containers for storing tobacco preparations (snuff boxes, cigar boxes, tobacco tins) also come in many forms. These objects are often decorated with iconographic elements of cultural significance. One cigar box, for example, featured depictions of mandrakes (**Mandragora officinarum**).

In Venezuela, traditional *chimó* containers (*cajeta, cuca, chimoera, cachita*) are made from cow horns (*cacho, cuerna de res*), to which a spatula (*paletica, pajuela*) of horn, wood, bone, or silver is usually attached (Kamen-Kaye 1971, 20 ff.).

Medicinal Use

Tobacco is used throughout Central and South America as a remedy for snakebites (Schultes and Raffauf 1991, 42*). Today, the Tzeltal still use tobacco plasters as a treatment for this purpose (Berlin et al. 1974, 445*).

In Mexico, the use of tobacco (juice) as a pesticide is found among almost all Indians as well as the mestizos. Tobacco juice (*u yits k'uts*) is applied to insect stings and bites. The term *u yits*

k'uts, "the juice of the tobacco," is used for the tobacco condensate that can be seen in the saliva after a cigar has been smoked. The Lacandon believe that tobacco is poisonous, but that it is precisely for this reason that it is also a general antidote that has the power to neutralize even potent poisons. A ritual still exists in which snake venom is neutralized with tobacco.[262] When a venomous snake such as a nauyaca (*Bothrops atrox*) or a rattlesnake (*Crotalus terrificus*) is found in a settlement or milpa, it must immediately be killed. A cigar stub is pushed into the throat of the dead snake to neutralize its venom. The snake is then buried. When spending the night in the jungle, making a circle around the camp with cigar ashes is said to keep away the poisonous snakes. It is possible that the Lacandon may also have once treated snakebites with tobacco (Rätsch 1994b, 55*).

In Venezuela, *chimó* is used in folk medicine as a remedy for scorpion stings, centipede bites, insect stings (wasps, bees, et cetera), and snakebites. The flesh worms (*Dermatobia hominis*) that live under the skin are also treated and expelled with *chimó*. *Chimó* users believe that the preparation is good for the teeth, for warding off the "evil eye," for dispelling hunger and exhaustion as well as evil spirits, and for healing a variety of illnesses (coughs, headaches, dysentery, toothaches, asthma, influenza, stomachaches) (Kamen-Kaye 1971, 23 ff.).

Among the Maká Indians (Paraguay), only men and those women who are not nursing will smoke or chew tobacco (it is said that tobacco use will cause nursing women to have "bad milk"). The Maká used the resinous tars that remain in the pipe as a remedy for treating wounds (Arenas 1987, 291*).

The Shipibo use combinations of tobacco leaves and the pith of the stems of **Brugmansia suaveolens** as a plaster to threat aching wisdom teeth (Arévalo V. 1994, 259*).

In India, tobacco leaves are crushed together with the leaves of *Erythrina stricta* Roxb. (cf. **Erythrina spp.**) and *Desmodium caudatum* (cf. **ayahuasca analogs, soma**) to make a paste that is applied to the skin to treat ulcers (Jain and Borthakur 1986, 579*).

In German folk medicine, tobacco was smoked, burned, or chewed to treat toothaches (cf. **incense**). Decoctions of the leaves were administered as **enemas** in the treatment of broken bones (Pabst 1887, 2:140*).

In homeopathy, Nicotiana or Tabacum is usually used in higher potencies in accordance with the medical description to treat such ailments as angina (Roth el al. 1994, 518*):

> The symptomatics of Tabacum are extremely striking. The nausea, dizziness, deathly paleness, the vomiting, the icy cold and sweating

with intermittent pulse, are all highly characteristic. (Boericke 1992, 741*)

Constituents

As of 1989, 2,549 different substances had been detected and described in tobacco! The entire plant contains **nicotine** (primary constituent) as well as nornicotine and other pyridine alkaloids (anabasine, nicotyrine). The alkaloid content can vary considerably, ranging from 0.05 to 4% (Roth et al. 1994, 516*). In addition to the alkaloids, amines, flavones, **coumarins**, pyrrolidine, and piperidine are also present. Fermented tobacco contains up to 0.4% free nicotinic acid (Wagner 1985, 173*).

Sun-dried leaves contain a mixture of aromatics consisting of hundreds of substances (primarily volatile acids) (Kimland et al. 1973). The substance known as tobacco camphor (= nicotianine) has been characterized as a volatile substance (cf. *Cinnamomum camphora*).

The smoke of commercial cigarettes has been found to contain myristicin (cf. *Myristica fragrans*, **essential oils**). The hallucinogenic effects of tobacco that are sometimes reported may perhaps be due (at least in part) to this component of the smoke (Schmeltz et al. 1966).

The study of the condensate of cigarette smoke for alkaloids has led to astonishing discoveries about its constituents. The principal alkaloid in the analyzed mass is the β-**carboline** harmane, followed by *N*-methylanabasine, nicotinamide, and anabasine. Lower concentrations of 2,2'-bipyridyl, β-nicotyrine, 2,6-dimethylquinoline, and myosmine have also been detected (Brown

Plants That Are Smoked (Sniffed or Chewed) as Tobacco Substitutes

(From Hartwich 1911*; Low 1990*; Ott 1993*; Schultes and Raffauf 1986, 275*; Vickers and Plowman 1984, 13*; modified and supplemented.)

Plant Source	Plant Part	Active Constituent
Adriana glabrata Gaudich	leaves	?
Asperula spp.	herbage	coumarin
Cestrum parqui	leaves, wood	parquine
Chelonanthus alatus (Aubl.) Pulle (*tabaco bravo*, wild tobacco)	leaves	?
Clerodendrum floribundum R. Br.	leaves	alkaloids
Cytisus spp.	leaves	**cytisine**
Dalbergaria picta (Karsten) Wiehler (*soma m?tó*, *secoya*, cooking tobacco)	leaves	?
Desmodium lasiocarpum (Beauv.) DC.	leaves (?)	DMT (?), β-**carbolines**
Lobelia inflata L. (Indian tobacco)	herbage (without roots)	α-lobeline
Lobelia tupa	leaves	?
Mandragora officinarum L. (mandrake)	leaves	**tropane alkaloids**
Nicotiana glauca (see **Nicotiana** spp.)	leaves	anabasine
Notholeana nivea (*inca sayre*; Polypodiaceae)[263]	?	?
Piper interitum Trelease (cf. **Piper** spp.)	leaves	**essential oil** and other substances
Rheum palmatum L. (cf. **soma**)	leaves	anthrachinone
Rhododendron sp. (cf. **Rhododendron** *caucasicum*)	leaves, bark	arbutine and other substances
Stemodia lythrifolia (bush tobacco)	leaves	?
Syzygium spp. (lilly pilly leaves)	leaves	**essential oil**
Trichodesma zeylanicum	leaves	alkaloids
Tussilago farfara L. (coltsfoot)	leaves	senkirkin
Typha latifolia L. (cattail)	spadix	?

Left: The spadixes of the cattail *Typha latifolia* are often smoked as substitutes for real cigars.

Right: Dried coltsfoot leaves (*Tussilago farfara*) are a popular tobacco substitute.

263 This plant, whose name means "Inca tobacco," may have been smoked or even used as a **snuff** (Alvear 1971, 22*). It is doubtful, however, that the botanical name is correct.

"The religious leader of his group, who heals illnesses but primarily establishes the connection to the nonhuman persons and powers, also uses tobacco—often to a very considerable degree—the ingestion of which enables him to attain the state that the now almost endless literature refers to as 'inebriation' and 'ecstasy' but often calls 'trance,' 'possession,' 'entrancement,' 'rapture,' or 'passion' as well, in which he can directly contact the deities and spirits, have them manifest themselves in his hut or let them enter his body, or search out their location by sending out his spirit double or through an ecstatic journey that the body and the soul undertake together."

GERHARD BAER
DER VON TABAK BERAUSCHTE [HE WHO IS INEBRIATED ON TOBACCO] (1986, 70*)

and Ahmad 1972, 3486; Janiger and Dobkin de Rios 1973, 1976; Poindexter and Carpenter 1962).

Effects

The effects of tobacco are primarily the result of the **nicotine**. Low dosages of tobacco produce invigorating and stimulating effects that suppress feelings of hunger. Moderate dosages can easily result in nausea, vomiting, diarrhea, anemia, and dizziness. High dosages can led to delirium with hallucinations (cf. *Nicotiana rustica*) and to death due to respiratory paralysis (Wagner 1985, 173*). Reactions to tobacco and to different dosages are greatly dependent upon familiarity with the drug. Chronic smokers can survive sublethal dosages without problem.

Chronic use of tobacco can lead to serious health problems (cancer, lung disorders, throat problems, smoker's leg). In Mexico, "smoker's cough" (bronchial catarrh) is treated with damiana tea (*Turnera diffusa*).

Combining tobacco with other substances, e.g., in **betel quids**, can have synergistic effects, about which very little pharmacological data is available.

Commercial Forms and Regulations

Today, tobacco is the only sacred plant of ancient times that is still legally available throughout the world (in the form of cigarettes or rolling tobacco)! In Germany, up to one hundred plants may be privately cultivated without taxation. Any person growing more than one hundred plants must report this to the customs authorities.

In Europe, any tobacco product intended for sale must include a warning about the potential for health problems. In Belgium, smoking has been prohibited in public buildings since 1987. In the United States, smoking in public is increasingly being restricted. The World Health Organization (WHO) has recently adopted an international framework for restricting tobacco advertising.

Literature

See also the entries for *Nicotiana rustica, Nicotiana* spp., and **nicotine**.

There are numerous journals that are sponsored by the cigarette industry and that continually publish the most up-to-date chemical studies. The most important of these journals is *Beiträge zur Tabakforschung International* (since 1978).

Brown, E. V., and I. Ahmad. 1972. Alkaloids of cigarette smoke condensate. *Phytochemistry* 11:3485–90.

Califano, Mario, and Alicia Fernández Distel. 1978. L'emploi du tabac chez les Mashco de l'Amazonie sud-occidentale du Pérou. *Bulletin de la Société Suisse des Américanistes* 42:5–14.

Elferink, Jan G. R. 1983. The narcotic and hallucinogenic use of tobacco in pre-Columbian Central America. *Journal of Ethnopharmacology* 7:111–22.

Feinhandler, Sherwin J., Harold C. Fleming, and Joan M. Mohahon. 1979. Pre-Columbian tobaccos in the Pacific. *Economic Botany* 33 (2): 213–26.

Hartmann, Günther. 1981. Tabak bei den südamerikanischen Indianern. In *Rausch und Realität*, ed. G. Völger, 1:224–35. Cologne: Rautenstrauch-Joest-Museum für Völkerkunde.

Heimann, Robert K. 1960. *Tobacco and Americans.* New York, Toronto, and London: McGraw-Hill.

Janiger, Oscar, and Marlene Dobkin de Rios. 1973. Suggestive hallucinogenic properties of tobacco. *Medical Anthropology Newsletter* 4 (4): 6–10.

———. 1976. *Nicotiana* an hallucinogen? *Economic Botany* 30:295–97.

Kamen-Kaye, Dorothy. 1971. *Chimó:* An unusual form of tobacco in Venezuela. *Botanical Museum Leaflets* 23 (1): 1–59.

———. 1975. *Chimó*—why not? A primitive form of tobacco still in use in Venezuela. *Economic Botany* 29:47–68.

Kell, Katharine T. 1965. Tobacco in folk cures in Western society. *Journal of American Folklore* 78 (308): 99–114.

Kimland, B., A. J. Aasen, S.-O. Almqvist, P. Arpino, and C. R. Enzell. 1973. Volatile acids of sun-cured Greek *Nicotiana tabacum. Phytochemistry* 12:835–47.

Lewis, Albert B. 1924. *Use of tobacco in New Guinea and neighboring regions.* Anthropology Leaflet 17. Chicago: Field Museum of Natural History.

Lippert, Julius. 1885. *Die Kulturgeschichte in einzelnen Hauptstücken.* Leipzig: G. Freytag.

Lowie, Robert H. 1975. *The tobacco society of the Crow Indians.* New York AMS Press. (Orig. pub. 1919.)

Michel, Thomas. 1981. Tabak in Neuguinea. In *Rausch und Realität*, ed. G. Völger, 1:258–62. Cologne: Rautenstrauch-Joest-Museum für Völkerkunde.

Plotkin, Mark J., Russell A. Mittermeier, and Isabel Constable. 1980. Psychotomimetic use of tobacco in Surinam and French Guiana. *Journal of Ethnopharmacology* 2:295–97.

Poindexter, E. H., and R. D. Carpenter. 1962. Isolation of harman and norharman from tobacco and cigarette smoke. *Phytochemistry* 1:215–21.

Redfield, Robert. 1950. *A village that chose progress: Chan Kom revisited.* Chicago and London: University of Chicago Press.

Richter, Elsie. 1926. Zigarre und andere Rauchwörter. In *Congresso intern. degli Americanisti* XXII (Rome): 2:296–306.

Richter Frich, Øvre. 1936. *Vitamin der Seele: Eine unterhaltsame Kulturgeschichte um den Tabak.* Hamburg: Paul Zsolnay Verlag.

Robicsek, Francis. 1978. *The smoking gods: Tobacco in Maya art, history, and religion.* Norman: University of Oklahoma Press.

Schivelbusch, Wolfgang. 1981. Die trockene Trunkenheit des Tabaks. In *Rausch und Realität,* ed. G. Völger, 1:216–23. Cologne: Rautenstrauch-Joest-Museum für Völkerkunde.

Schmeltz, Irwin, R. L. Stedman, J. S. Ard, and W. J. Chamberlain. 1966. Myristicin in cigarette smoke. *Science* 151:96–97.

Schopen, Armin. 1981. Tabak in Jemen. In *Rausch und Realität,* ed. G. Völger, 1:244–47. Cologne: Rautenstrauch-Joest-Museum für Völkerkunde.

Stramberg, Chr. von, et al. 1986. Hexenfahrten. In *Spuk- und Hexengeschichten,* ed. Hermann Hesse, 26–62. Frankfurt/M.: Insel. (Article orig. pub. ca. 1851 in *Rheinischen Antiquarius* 2 (4): 334–61.)

Tiedemann, Friedrich. 1854. *Geschichte des Tabaks und anderer ähnlicher Genußmittel.* Frankfurt/M.: H. L. Brönner.

Uscátegui M., Nestor. 1956. El tabaco entre las tribus indígenas de Colombia. *Revista Colombiana de Antropología* 5:12–52.

Volprecht, Klaus. 1981. Tabak und sein Gebrauch in Afrika. In *Rausch und Realität,* ed., G. Völger, 1:248–57. Cologne: Rautenstrauch-Joest-Museum für Völkerkunde.

Walther, Elisabeth. 1981. Kulturhistorisch-ethnologischer Abriß über den Gebrauch von Tabak. In *Rausch und Realität,* ed. G. Völger, 1:208–15. Cologne: Rautenstrauch-Joest-Museum für Völkerkunde.

Wynder, E. L., and D. Hoffmann. 1967. *Tobacco and tobacco smoke: Studies on experimental carcinogenesis.* New York: Academic Press.

In Europe, tobacco (*Nicotiana tabacum*) was originally known as the henbane of Peru or *Hyoscyamus peruvianus*. (Woodcut from Gerard, *The Herball or General History of Plants,* 1633)

Nicotiana spp.

(Wild) Tobacco Species

Family

Solanaceae (Nightshade Family); Subfamily Cestroideae, Nicotianeae Tribe

Today, ninety-five species are botanically recognized in the genus *Nicotiana* (D'Arcy 1991, 78*). Schultes and Raffauf (1990, 432*), however, recognize only sixty-six species, albeit with numerous subspecies. Of these, some forty-five species are indigenous to the Andes region (Schultes and Raffauf 1991, 41*). Sixteen to twenty species are endemic to Australia.

Many *Nicotiana* species are called wild tobacco or Indian tobacco; *Lobelia nicotianaefolia* is also known as wild tobacco and **Lobelia inflata** as Indian tobacco, and **Lobelia tupa** is known as devil's tobacco.

During prehistoric times, only wild tobacco was smoked in North America (Dixon 1921; Setchell 1921).

Nicotiana acuminata (Grah.) Hook.

This wild tobacco occurs in southern Chile as the varieties *multiflora* (Phil.) Reiche and *acuminata*. The Mapuche have been smoking the leaves of this plant, which they call *pëtrem*, since pre-Columbian times. Mapuche shamans (*machi*) smoke a great deal at their healing ceremonies (cf. **Latua pubiflora**); it is said to enable them to enter an ecstatic state (Mösbach 1992, 105 f.*).

Nicotiana attenuata Torr. ex Wats.—coyote tobacco

This wild North American species was smoked by the tribes of the Pacific Northwest (Heizer 1940). The leaves usually were mixed with bearberry leaves (*Arctostaphylos uva-ursi* [L.] Spreng.; cf. **kinnikinnick**) or other unidentified native plants (French 1965, 380 f.). The Blackfeet cultivated this tobacco, which they called *pistacan* or *mah-wat-osis* ("hard tobacco"), and developed a tobacco plant ceremony in its honor (Johnston 1970, 319*).

Remains of this tobacco species have been identified in 1,300-year-old layers (Basket Maker III period) in the American Southwest (Jones and Morris 1960). The Hopi Indians know this wild tobacco by the name *pí-bú* and use it in various manners in their rituals. Because of this, the plant is also known as Hopi tobacco. The Navajo call it *tzilnát'ooh,* "mountain tobacco," and use it in many ceremonies and healing rituals, in which it is placed in their prayer sticks (see **Phragmites australis**).

The Coahuilla Indians of Southern California refer to this wild tobacco as *pivat-isil,* "coyote tobacco." They would grind the leaves with a stone, mix it with some water, and chew the result (Barrows 1967, 75*). The neighboring Chumash mixed the powdered leaves with lime obtained by burning shells. The mixture, known as *pespibata,* could be either sucked on as a quid or steeped in

In Europe, many American tobacco species were once known as Peruvian henbane. (Woodcut from Tabernaemontanus, *Neu Vollkommen Kräuter-Buch,* 1731)

Left: Blue tobacco (*Nicotiana glauca*), originally from northwestern Argentina, is now grown throughout the world as an ornamental. (Wild plant, photographed near Salta, northwest Argentina)

Right: *Nicotiana langsdorffii* Schrank, one of the many South American tobacco species from Brazil.

"The old woman also tried to use plant leaves cured over a fire for smoking in a clay pipe. Everybody watched her as she became dizzy and collapsed, and they said: 'This time she is dying; just let her die.' But the woman came to again, and the first thing she did was have another smoke. It did her no harm and so everybody learned how to smoke."

A MATACO INDIAN MYTH
IN *FOLK LITERATURE OF THE MATACO INDIANS*
(WILBERT AND SIMONEAU 1982, 116*)

"In Mexico, a very strong tobacco species whose leaves are smoked, *Nicotiana glauca* Grah. (Solanc.), is also known as marihuana. Of course, this marihuana (also called macuchi) does not do anything other than that which any other strong tobacco does."

VICTOR A. REKO
MAGISCHE GIFTE [MAGICAL POISONS]
(1936, 62*)

264 "Ground to a fine powder and mixed with some **chicha**, this plant can invoke almost the same symptoms as *Datura stramonium*" (Hargous 1976, 158*).

water and drunk. The effects were described as profoundly euphoric. This use apparently was also related to the initiation with *Datura wrightii*.

Nicotiana bigelovii (Torr.) Wats. [syn. *Nicotiana plumbaginifolia* var. *bigelovii* Torr.]—pespibata

This tobacco species was sacred to the Southern Californian Chumash. It was used as an offering and was smoked by special tobacco shamans, known as smoke healers, for ritual healings or was mixed with lime and ingested (Timbrook 1990, 252*). Other Indians of the Southwest also smoked or smoke the leaves (Barrows 1967, 74*; Setchell 1921, 404).

The smoke of this plant is so potent that it can cause a person to lose consciousness or fall into a deep trance. The Indians of California made use of this effect in their initiations and healing rituals. Today, this sacred plant is considered to be very rare.

Nicotiana clevelandii (Gray)—Cleveland's tobacco, pavivut

The Serí Indians of northern Mexico call this wild tobacco *xeezej islítx*, "the inner ear of the badger." They dry and smoke the small, wide leaves in clay pipes, in the shell of a worm snail (*Tripsycha tripsycha* in the literature; probably *Serpulorbis* spp.), or in pieces of the canes of *Phragmites australis*. The Serí appear to be the only people who still smoke this tobacco species today (Felger and Moser 1991, 369*). In earlier times, it was also used in Southern California (Setchell 1921, 412 f.).

Nicotiana glauca Grah. [syn *Nicotidendron glauca* (Grah.) Griseb., *Siphaulax glabra* Raf.]—blue tobacco, macuchi

This shrublike plant, which has bluish leaves and yellow flowers, was first described for the province of Salta (northwestern Argentina), where it occurs at altitudes up to about 3,700 meters. It quickly spread throughout all of the Americas and is now found even in Europe and Asia, where it is grown as an ornamental and has also become wild. Remains of this plant have been found in archaeological layers of the Nazca culture (Bruhn et al. 1976, 45).

In the southern Andes region of Peru, the plant is known by the Quechuan name *supay kayku* (*supay* essentially means "devil"). The plant is made into a medicinal bath to treat diseases that are caused by *suq'a* ("evil spirits"). Traditional healers known as *p'aqus* rub the leaves and add them to **chicha**. They drink this secretly to induce an inebriated state and so that they can diagnose and heal diseases (Franquemont et al. 1990, 100*).264 The plant is known as *belen-belen* or *huelen-huelen* ("very left") in the Andes and as *palqui extranjero*, "foreign palqui bush" (*Cestrum parqui*), in Chile (Mösbach 1992, 105*).

The Pilagá Indians of Argentina call this shrub *konyel'kaik*. They use its fresh leaves as a plaster for treating headaches (Filipov 1994, 190*).

The Serí Indians of northern Mexico call the plant *noj-oopis caacöl*, "hummingbird that sucks out the large justicia" (Felger and Moser 1991, 369*; cf. *Justicia pectoralis*).

The plant's main alkaloid is anabasine (Leete 1982). Nornicotine is also present (Argueta V. et al. 1994, 1306*). Although it has sometimes been asserted that the plant contains **nicotine**, this is doubtful (Roth et al. 1994, 515*). The pharmacological effects of anabasine, however, are very similar to those of nicotine (Blohm 1962, 94*).

Nicotiana langsdorffii Weinm.

This wild tobacco species occurs in southern Brazil and Chile (Hartwich 1911, 29*). The shamans of the Mapuche (*machi*) and other peoples formerly smoked it during their healing rituals (cf. *Latua pubiflora*).

Nicotiana palmeri A. Gray

The Navajo call this wild tobacco species *tipénát'ooh*, "sleeping tobacco," and use it in numerous ceremonies and healing rituals. They also place this tobacco in their prayer sticks (see *Phragmites australis*).

Nicotiana plumbaginifolia Torr.

California Indians collected and smoked this wild tobacco (Barrows 1967, 74*). The plant also occurs in South America, where it was also smoked (Hartwich 1911, 29*).

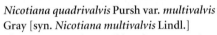

Top left: This ornamental tobacco plant develops flowers of different colors.

Bottom left: *Nicotiana alata* is an important starting plant for breeding ornamental tobaccos.

Top center: Red ornamental tobacco (*Nicotiana* x) is sometimes used as a tobacco substitute.

Bottom center: A wild tobacco species (*Nicotiana* sp.), photographed in the ruined city of Yagul (Oaxaca, Mexico).

Top right: The wild tobacco (*Nicotiana* sp.) of Altiplano, Peru, is both smoked and used as a ritual incense. (Photographed at Lake Titicaca)

Nicotiana quadrivalvis Pursh var. *multivalvis* Gray [syn. *Nicotiana multivalvis* Lindl.]

This variety of a wild tobacco species was cultivated by the Indians on the Columbia River (Pacific Northwest) and was smoked for hedonistic or ritual purposes (Barrows 1967, 74*). The flowers were smoked as well as the leaves (Hartwich 1911, 31*).

Nicotiana sylvestris Spegazz. et Comes—wild tobacco

This persistent tobacco species blooms throughout the entire year and even tolerates light frost. It is originally from the Bolivian highlands and may have been a precursor of *Nicotiana tabacum*. The sticky leaves can be dried and smoked.

Nicotiana trigonophylla Dunal—desert tobacco

Desert tobacco is used by the Indians of the Southwest (Barrows 1967, 74*). The Tarahumara call it *bawaráka* or *wipake* and use it as an analgesic (Díaz 1979, 85*). It is also known as *tabaco del coyote*, "tobacco of the coyote." The Serí Indians of northern Mexico call this wild tobacco species *hapis casa*, "that which is smoked, rots," and believe that the plant has magical powers. The small, narrow leaves were apparently ritually collected and then slowly dried. The effects are said to be very strong (Felger and Moser 1991, 369*).

Nicotiana undulata Ruíz et Pav. [syn. *Nicotiana tabacum* var. *undulata* Sendtner]—Yaqui tobacco

The Quechua know this Central and South American tobacco species as *kamasayri*. Applied externally, it is used to treat stomachaches (Franquemont et al. 1990, 100*). It is thought that the Maya used a preparation of the leaves as a medicinal or ritual psychoactive **enema**. The hybridization of this species with *Nicotiana paniculata* may have given rise to *Nicotiana rustica* (Emboden 1979, 42*).

Nicotiana x—ornamental tobacco

There have been few chemical studies of the cultivated ornamental tobaccos, which produce a variety of flower colors. Some psychonauts are now investigating and testing these plants for possible psychoactive effects.

Anabasine

Top: The rare wild tobacco species *Nicotiana cordifolia* develops very large leaves that are well suited for rolling cigars.

Bottom: An unidentified *Nicotiana* species.

A healing ceremony on Hispaniola, showing a shaman smoking some type of tobacco and falling into a trance. (From Benzoni, *La Historia del Mondo Nuevo,* 1568)

Nicotiana spp.—various species (wild tobacco)
There is evidence suggesting that many species of wild tobacco were and still are used in South America for ethnomedical, hedonistic, or ritual purposes. The Indians regard such wild tobaccos, most of which have not yet been identified, as more potent than the cultivated species (*Nicotiana rustica, Nicotiana tabacum*). The following species have been smoked (Hartwich 1911, 29 ff.*):

Nicotiana alata Link et Otto (Brazil, Uruguay, Paraguay)

Nicotiana alata var. *persica* Comes (Brazil)
Nicotiana angustifolia Ruíz et Pav. (Chile, Brazil)
Nicotiana glutinosa L. (Peru, Chile)
Nicotiana mexicana Schlecht. (Mexico, Guatemala, Bolivia; syn. or var. of **Nicotiana tabacum**)
Nicotiana paniculata L. (Peru)
Nicotiana pusilla L. (Mexico, Cuba)
Nicotiana repanda Willd. (Mexico)

Nicotiana species have been identified in archaeological excavations, including in a grave of the Tiahuanaco culture (in association with **Ilex guayusa**) (Bruhn et al. 1976). Their use appears to be very ancient.

Wild Tobacco in Australia

Some twenty species of the genus *Nicotiana* occur in Australia (Bahadur and Farooqui 1986; Burbidge 1960; Haegi 1979). All of the Australian *Nicotiana* species—usually referred to as wild tobacco[265]—are small annuals that seldom grow taller than 1 meter in height. All of the species are regarded as poisonous, especially for grazing cattle and sheep (Dowling and McKenzie 1993, 91*). Among the most poisonous species are *Nicotiana megalosiphon* and *Nicotiana velutina*, both of which are used as chewing tobacco.

All of the Australian *Nicotiana* species contain pyridine alkaloids, primarily **nicotine** and nornicotine (Dowling and McKenzie 1993, 92*). The Aborigines used and still use at least six of the indigenous species for psychoactive purposes, usually as chewing tobacco or **pituri** (cf. *Duboisia hopwoodii*). **Goodenia** spp., **Trichodesma zeylanicum**, and other species are used as substitutes (see **Nicotiana tabacum**) (O'Connell et al. 1983, 97 f.*).

Nicotiana gossei is regarded as the most potently effective wild tobacco species, followed by *Nicotiana ingulba* and *Nicotiana benthamiana*. The species *Nicotiana velutina* and *Nicotiana megalosiphon* are considered only weakly effective and are consumed only in extreme emergencies (O'Connell et al. 1983, 98*). The wild tobacco leaves are dried and made into rolls. For use, they are first moistened with saliva, then dipped in plant ashes (from **Acacia** spp., *Eucalyptus microtheca, Eucalyptus* sp., *Ventilago viminalis,* and others), and then chewed (cf. **pituri**).

Until recently, the Alyawara, who live in central Australia, also used wild species of tobacco to produce hunting poisons. The leaves were macerated in water and the resulting extract was poured into the watering sites of emus. When emus drank this toxic water, they became sedated and were easy prey for hunters (O'Connell et al. 1983, 98*).

Wild tobacco was also used in the rituals of the rainmakers.[266] It was said that when mixed with saliva, the chewed leaves of *Nicotiana* species smelled like rain. Chewing made it possible for the

---

265 *Solanum mauritianum,* which is also known as wild tobacco or wild tobacco tree, was not smoked as tobacco. The fruits of this plant are edible. The Aborigines used the soft leaves as toilet paper (Low 1992a, 78*). However, this plant is usually regarded as poisonous (Jackes 1992, 40*).

266 Ancient Germans also used a nightshade (*Hyoscyamus niger*) for rain magic. Some authors have conjectured that nightshades can induce auditory hallucinations that sound like rain or flowing water. For this reason, such plants were utilized in sympathetic magic.

Nicotiana Species Used as Chewing Tobacco and Pituri Quids (Australia)

Species	Alyawara Name (from O'Connell et al. 1983, 108*)
Nicotiana benthamiana Domin.	*ngkulpa putura*
Nicotiana gossei Domin.	*ngkulpa inpiynpa*
Nicotiana ingulba J.M. Black	*ngkulpa nguninga*
Nicotiana megalosiphon Heurck et J. Muell.	*ngkulpa ntarrilpa*
Nicotiana stimulans Burbidge	
Nicotiana velutina Wheeler	*ngkulpa ntarrilpa*

"life essence" to leave the leaves. To make rain, the rainmaker and his assistants spat the chewed masses into the sky. The rainmaker could also spit the chewed mass onto a smooth stone. He would then take a pearl oyster (*Pinctada margaritifera* L.), which he would wear on a string around his neck at other ceremonies and which had great magical significance and great value to him, and rub it over the mass of leaves and saliva for a while, singing various magical songs as he did so. The pearl oyster was then smeared with blood, placed in mulga leaves, and covered with grass, after which it was hung from a mulga tree (*Acacia aneura* F. Muell. ex Benth.; cf. *Acacia* **spp.**). The oyster would then attract the rain (Mathews 1994, 26).

Literature

See also the entries for *Duboisia hopwoodii*, *Nicotiana rustica*, *Nicotiana tabacum*, **pituri**, and **nicotine**.

Bahadur, Bir, and S. M. Farooqui. 1986. Seed and seed coat characters in Australian nicotiana. In *Solanaceae: Biology and systematics*, ed. William G. D'Arcy, 114–37. New York: Columbia University Press.

Bruhn, Jan G., Bo Holmstedt, Jan-Erik Lindgren, and S. Henry Wassén. 1976. The tobacco from Niño Korin: Identification of nicotine in a Bolivian archaeological collection. *Göteborgs Etnografiska Museum Årstryck* 1976:45–48.

Burbidge, N. T. 1960. The Australian species of *Nicotiana* L. (Solanaceae). *Australian Journal of Botany* 8:342–80.

Dixon, Roland B. 1921. Words for tobacco in American Indian languages. *American Anthropologist*, n.s., 23:19–49.

French, David H. 1965. Ethnobotany of the Pacific Northwest Indians. *Economic Botany* 19:378–82.

Goodspeed, Thomas H. 1954. *The genus* Nicotiana. Waltham, Mass.: Chronica Botanica Company.

Haegi, L. 1979. Australian genera of the Solanaceae. In *The biology and taxonomy of the Solanaceae*, ed. J. G. Hawkes et al., 121–24. London: Academic Press.

Heizer, Robert F. 1940. The botanical identification of northwest coast tobacco. *American Anthropologist* 42:704–6.

Jones, V. H., and E. A. Morris. 1960. A seventh-century record of tobacco utilization in Arizona. *El Palacio* 67 (4): 115–17.

Leete, E. 1982. Tobacco alkaloids and related compounds. 46: Biosynthesis of anabasine from DL-(4,5-13C,2,6-14C)-lysine in *Nicotiana glauca* examined by 13C-NMR. *Journal of Natural Products* 45:197–205.

Mathews, Janet. 1994. *Opal that turned into fire.* Broome, Wash.: Magabala Books.

Setchell, William Albert. 1921. Aboriginal tobaccos. *American Anthropologist*, n.s., 23 (4): 397–414.

Tso, T. C. 1972. *Physiology and biochemistry of tobacco plants.* Stroudsburg, Pa.: Dowden, Hutchinson & Ross.

"One plant person I know likes tree tobacco [*Nicotiana glauca*] and smokes it, but I've always found it rather harsh. Eaten, of course, it can kill just as quickly as does nicotine, and has. Of the wild tobaccos of western North America, *Nicotiana attenuata*, coyote tobacco, is said to be particularly fine. But *Nicotiana acuminata* should also be tried...."

DALE PENDELL
PHARMAKO/POEIA
(1995, 33*)

Nuphar lutea (Linnaeus) Sibthorp et Smith

Yellow Water Lily

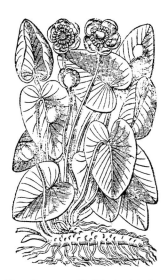

The yellow water lily (*Nuphar lutea*), which is indigenous to Europe, produces opium-like effects. (Woodcut from Tabernaemontanus, *Neu Vollkommen Kräuter-Buch*, 1731)

Family
Nymphaeaceae (Water Lily Family)

Forms and Subspecies
It was formerly assumed that the genus *Nuphar* was composed of twenty-five or twenty-six species. Currently, only two species are recognized, and *Nuphar lutea* has been divided into nine subspecies (Beal 1956; Slocum et al. 1996, 165*):

Nuphar lutea ssp. *lutea* Beal
Nuphar lutea ssp. *macrophylla* (Small) Beal [syn. *N. advena* (Ait.) Ait. f.]
Nuphar lutea ssp. *orbiculata* (Small) Beal
Nuphar lutea ssp. *ozarkana* (Miller et Standley) Beal
Nuphar lutea ssp. *polysepala* (Engelm.) Beal
Nuphar lutea ssp. *pumila* (Timm) Beal [syn. *N. microphylla* Beal]
Nuphar lutea ssp. *sagittifolia* (Walter) Beal
Nuphar lutea ssp. *ulvacea* (Miller et Standley) Beal
Nuphar lutea ssp. *variegata* (Engelm.) Beal

Synonyms
Nenuphar luteum nom. nud.
Nuphar advena (Aiton) Aiton f.
Nuphar luteum Smith
Nuphar microphylla Beal
Nymphaea lutea L.
Nymphaea luteum S. Sm.

Folk Names
Amello, American spatterdock, andere seerose, Cape Fear spatterdock, carfano maschio, gael seebluomen, geel seeblume, geelseeblumen, gelber mummel, gelbe seerose, gelbe teichrose, gelbe wasserlilie, gele plomp (Dutch), lake rose, madonaïs, mummel, nailufar (Arabic, "water lily"), naunufero, nenuphar, ninfea, ninfea gialla (Italian), ninupharo, nuphar, nuphara, nuphar jaune, nymphe minor, nymphon, pond lily, seeblume, seekandel, spatterdock, teichrose, yellow water lily, yellow water-lily

History
The yellow water lily or lake rose was described by Dioscorides:[267]

> Other lake rose. There is yet another Nymphaia (some call it Nymphon, its flower is called Nuphar) with leaves similar to that mentioned before [*Nymphaea alba* L.]; it has a large and rough root and a yellow, shiny flower like the rose. The root and the seed, when drunk in dark wine, have a good effect against the flow of women. But it grows in the region of Thessaly on the Peneus River. (3.139 [149])

The root was often used medicinally as an alternative to *Nymphaea alba* L. (Schneider 1974, 2:365*). No traditional use as a psychoactive substance is known.

Distribution
Varieties of the yellow water lily occur in Europe, North America, and Asia. The plant thrives in standing and slowly flowing waters at depths of up to about 1.5 meters.

Cultivation
The yellow water lily is easily propagated through scions taken from the creeping rhizome (Slocum et al. 1996).

Appearance
This perennial plant has oval, heart-shaped, long-stemmed, floating leaves and large flowers the color of egg yolks that tower above its leaves. The capsule fruit is bottle-shaped. The plant flowers between June and August. The rhizome can grow to a length of several meters and a width of up to 10 cm.

The yellow water lily is easily confused with the Japanese water lily (*Nuphar japonica* DC.). However, the latter has slightly reddish leaves and develops orange-red flowers (Slocum et al. 1996, 165*).

Psychoactive Material
— Roots (rhizoma nupharis lutei, nupharis lutei rhizoma, water lily root)
— Seeds

Preparation and Dosage
The fresh root is chopped and macerated in red wine (cf. **Vitis vinifera**). To date, no information regarding psychoactive dosages has become available. The seeds are edible.

Ritual Use
According to Dioscorides, the yellow water lily grew especially in Thessaly. It is possible that the plant may have played a role in the ancient witch cult of the area. The water lily has also been named as an ingredient in the early modern **witches' ointments**.

An old recipe for a magical use has come down to us from central Europe; it appears to suggest a continuation of the ancient witches' use:

> The "water lily," collected at the moment when the sun enters the sign of Cancer, and dried in

---

267 The *Nymphaia* of Theophrastus should probably also be construed as *Nuphar lutea*.

the air of midnight, is an agent against dizziness if this plant is hung on the wall and merely looked upon. (Schöpf 1986, 141*)

Artifacts
None

Medicinal Use
In the medicine of antiquity, the yellow water lily was used in ethnogynecology to treat such conditions as vaginal discharge.

The Japanese water lily (*Nuphar japonica* DC.) is used in Japan as a sleeping agent and sedative (Meister 1677, 114*) and to treat syphilis, circulatory problems, and postpartum complications (Tsumura 1991, 175*).

In homeopathy, the essence of the fresh root is known as Nuphar luteum—Yellow Water Lily (Schneider 1974, 2:365*). The agent is used as an alternative to Yohimbinum (see *Pausinystalia yohimba*), as some parts of the medical descriptions overlap considerably:

Nuphar luteum. Induces nervous weakness with pronounced symptoms in the sexual domain. . . . Male.—Complete lack of sexual desire; sexual organs are flaccid; the penis withdrawn. Impotence, with involuntary ejaculations while defecating and urinating, spermatorrhea, pain in the testicles and penis. (Boericke 1992, 556 f.*)

Constituents
The entire plant, and especially the rootstock (rhizome), contains up to 0.4% nupharine (empirical formula $C_{18}H_{24}O_2N_2$) and β-nupharidine. But the primary alkaloid is desoxynupharidine (= α-nupharidine; empirical formula $C_{15}H_{23}NO$) (Reichert et al. 1949, 3:839*; Roth et al. 1994, 520*). The root also contains 5.9% tannic acid, dextrose, 1.2% saccharose, 18.7% starch, metarabinic acid, and fat (Reichert et al. 1949, 3:839*).

The rhizome of the Japanese water lily contains the alkaloids nupharidine, desoxynupharidine, and nupharamine. Also present are β-sitosterol, oleic acid, palminic acid, nicotinic acid, and tannins (nupharine-A, -B, and –C) (Tsumura 1991, 175*).

Effects
Nupharine is said to have opium-like effects and to induce trancelike states (Goris and Crete 1919). Atropine- and papaverine-like effects have been observed in animal studies (cf. **atropine**, **papaverine**). Desoxynupharidine has tonic effects and raises the blood pressure (Roth et al. 1994, 520*).

Commercial Forms and Regulations
The plant is protected (as are all indigenous Nymphaeaceae). Apart from this, the plant is a "legal high."

Literature
See also the literature for the entries *Nymphaea ampla*, *Nymphaea caerulea*, and **papaverine**.

Beal, Ernest O. 1956. Taxonomic revision of the genus *Nuphar* of North America and Europe. *Journal of the Elisha Mitchell Society.*

Goris, A., and L. Crete. 1919. Sur la nupharine. *Bulletin de Science et pharmacologie* 17:13–15.

The yellow water lily (*Nuphar lutea*) harbors narcotic powers. (Wild plant, photographed in northern Germany)

"Coebe. Is a type of the Nymphe minor, found frequently in Jappon in watery locations. Its leaves, of the shape of a heart, float upon the water, filled with small veins; the flowers are blue, in shape like Consolita Regalis or larkspur. When its flowers wilt, little balls like onions appear in their place, which are quite watery in taste, and which the Japponese consider to be cold in nature. They, the Japponese, when their sick are unable to sleep, give this onion to the patient in their food or drink, otherwise it is not used in any other way as a medicine."

GEORGE MEISTER
DER ORIENTALISCH-INDIANISCHE KUNST- UND LUSTGÄRTNE R [THE ORIENTAL-INDIAN ART AND PLEASURE GARDENER] (1677, CH. 10, 31*)

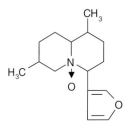

Nupharidine

Nymphaea ampla (Salisbury) DC.

White Water Lily

"The coolness of my foot,
the coolness of my hand,
when I cooled the eruption! [lit.:
    "fire"]
Five are my white hail-stones,
my black hail-stones,
[my] yellow hail-stones.
Then I cooled the eruption.
Thirteen are the layers of my red
    dressing,
my white dressing,
[my] black dressing,
[my] yellow dressing,
when I received the force of the
    eruption.
A black fan was my symbol,
when I received the force of the
    eruption.
With me comes the white [aquatic]
    ixim-ha-plant,
when I received the force of the
    eruption.
With me comes the white *nab*-water-
    lily,
when I received the force of the
    eruption.
Shortly ago I applied [to it]
the coolness of my foot,
the coolness of my hand.
Amen."

MAGICAL FORMULA OF THE MAYA
FOR TREATING ERUPTIONS
(ROYS, *RITUAL OF THE BACABS* MS
114 f.)

Family
Nymphaeaceae (Water Lily Family); Apocarpiae Group, Subgenus *Brachyceras*

Forms and Subspecies
The species is divided into two varieties:

Nymphaea ampla var. *pulchella* (DC.) Caspary
Nymphaea ampla var. *speciosa* (Martius et
    Zuccarini) Caspary

Moreover, in Mexico (and elsewhere) there are a number of crosses and hybrids of different *Nymphaea* species, including *Nymphaea ampla* cultivars, that have acquired a certain importance as ornamentals (Slocum et al. 1996*).

Synonyms
None

Folk Names
Apepe, japepe, lolha' (Mayan, "flower of the water"), mexikanische seerose, naab[268] (Mayan), nab, nikte'ha' (Mayan, "flower/vulva of the water"), ninfa, ninfea, nukuchnaab (Mayan, "large water lily"),[269] pan de manteca, quetzalxochiatl (Aztec, "Quetzal feather flower"), saaknaab, saknaab (Mayan, "white water lily"), sol de agua (Spanish, "sun of the water"), u k'omin (Lacandon), u k'omin ha', water-lily, white water lily, xikinchaak (Mayan, "the ear of the rain god"), zac-nab

History
The white water lily is represented in both pre-Columbian art and in Mayan manuscripts (Emboden 1983). It was first described scientifically in the nineteenth century. The first reports of its psychoactive use were published in the 1970s.

Distribution
In southern Mexico, the plant occurs throughout the entire Mayan lowlands (Emboden 1979a). It is also found on the higher elevated lake plateaus of Chiapas (Lagunas de Montebello, Yahaw Petha' Lago Metzabok, Laguna de Najá, et cetera) It is common in the cenotes (natural wells, limestone caverns) near Mérida in the northern Yucatán (Roys 1976, 267*) and in Lago Petén Itzá in Guatemala. It is also said to occur in Brazil.

Cultivation
The rhizome can be multiplied and will thrive when placed in a pond with drainage or in slow-moving water.

Appearance
The plant develops a thick rhizome and has long-stemmed cordate leaves. The white flowers protrude some 20 to 30 cm above the floating leaves.

Nymphaea ampla is easily confused with the very similar European *Nymphaea alba* L.

Psychoactive Material
— Buds or flowers
— Rhizome

Preparation and Dosage
The white water lily can be prepared as a tea or decoction. However, no information concerning dosages is currently known (cf. ***Nymphaea caerulea***).

The dried buds and flowers can be smoked alone or in **smoking blends**. One or two buds is said to be a psychoactive dosage. The fresh rhizome can be eaten raw or cooked. Eating an entire rhizome produces a mild sensation of being high (Brett Blosser, pers. comm.).

Ritual Use
The water lily appears to have been used as an additive to the **balche'** drink.

It has been said that Brazilians used the flowers as a narcotic inebriant with opium-like effects during the first half of the twentieth century (Emboden 1979a, 51). In the 1960s, "hippies" in Chiapas allegedly used the flowers as a recreational drug.

Artifacts
The water lily is a very common subject in Mesoamerican art. The rain god of Teotihuacan (Tláloc; cf. ***Argemone mexicana***) is often depicted with water lily leaves, buds, and flowers. Sometimes he even has the buds in his mouth (Pasztory 1974). Curiously, the renowned ethnomycologist R. Gordon Wasson interpreted these very images of water lilies as representations of entheogenic mushrooms (see ***Psilocybe mexicana***). (See Emboden 1981 and 1982 for more on this subject.)

The water lily was portrayed especially often in the art of the Classic Mayan period in iconographic contexts (Rands 1953) that can be interpreted in a variety of ways. There are essentially three motifs: water lilies sprouting from the backs of crocodiles swimming in the water; the head of the "earth monster," around which water lilies are entwined; jaguars that are either wearing the stalks and buds of water lilies as a head ornament or dancing with water lilies. The association between the jaguar and the water lily is especially dominant (Rands 1953, 88; cf. Emboden and Dobkin de Rios 1981).

---

268 In the Yucatán, this name is also used for *Centella asiatica* Urban (Barrera M. 1976, 115*). *Naab* is both a name for the water lily and the measure between the thumb and little finger (app. 20 cm).

269 *Mehen-naab*, "little water lily," is a name used in the Yucatán for *Hydrocotyle umbellata* L., thereby establishing a folk taxonomic relationship between the plants (Barrera M. et al. 1976, 112*).

The water lily is frequently seen on ceramic vessels that appear to depict primarily visionary scenes from the underworld or other worlds (Coe 1973; Emboden 1979b):

> In association with the transforming tadpoles and balche' vessels, the water lily appears to invoke shamanic ecstasy. The tadpole changes its form and becomes a toad; the shaman undergoes a similar transformation and manifests himself in his alter ego. In some cases, the toad is seen in a human form which is offering libations; these may consist of balche' or of balche' to which water lilies have been added, so that the transformation is easier. (Emboden 1979a, 51)

There is even a Mayan hieroglyph known by the name *jaguar–water lily,* which has played an important role in the deciphering of the Mayan writing system (Coe 1993, 257). In all likelihood, the jaguar–water lily is a transformed shaman. In the American tropics, the jaguar is the most important shamanic animal and/or is the animal that is identical to the shaman and whose shape he can assume (Reichel-Dolmatoff 1975*; Walter 1956*).

The *uay* glyph (see illustration on right) may indicate that the shaman was transformed into a jaguar by means of a potent balche' drink and is traveling in another reality. The iconographic element of the water lily may possibly be a symbol for the balche' drink, the water in which the inebriated jaguar swims, the inebriation itself, or the other reality. The water lily also appears as a ritual scepter and is depicted over **balche'** vessels (Emboden 1979a, 50; 1992, 81).

Medicinal Use

The water lily was invoked in numerous Mayan magical conjurations from the colonial period to heal ulcers and skin diseases (Roys 1965, 39 f., 123). It has been said that the plant is still used for ethnomedicinal purposes in the Yucatán today (Barrera M. et al. 1976*).

Constituents

The flowers have been found to contain aporphine, a substance that is closely related to the **opiate** apomorphine (Tamminga et al. 1978). It differs only in the lack of two hydroxyl groups. It is possible that aporphine can be transformed into apomorphine through processing, storage, or metabolism (Emboden 1979a, 50). Aporphine is also found in the poppy species *Papaver fugax* Poir. (Phillipson et al. 1973; cf. **Papaver spp.**). Alkaloids of the aporphine type also occur in the Family Lauraceae (e.g., *Litsea sebifera* Pers., *Litsea wightiana* Hook. f., *Actinodaphne obovata* Bl.) (Uprety et al. 1972). The backbone chain of aporphine is boldine, a substance also contained in the leaves of the boldo tree (*Peumus boldus*), which have traditionally been used as **incense**.

In addition to aporphine, *Nymphaea ampla* appears to contain quinolizidine alkaloids (Emboden 1983).

Effects

A tea made from the buds is said to have psychoactive effects (Dobkin de Rios 1978). It can also have psychodysleptic effects and can induce vomiting, but it does not have any toxic aftereffects (Emboden 1979a, 51). Further human pharmacological experiments are required.

Commercial Forms and Regulations

None

Literature

See also the entries for **Nymphaea caerulea** and **balche'**.

Coe, Michael D. 1973. *The Maya scribe and his world.* New York: The Grolier Club.

———. 1993. *Breaking the Maya code.* London: Thames and Hudson.

Conrad, H. S. 1905. *The waterlilies: A monograph of the genus* Nymphaea. Washington, D.C.: Carnegie Institution.

Dobkin de Rios, Marlene. 1978. The Maya and the water lily. *The New Scholar* 5 (2): 299–307.

Emboden, William A. 1979a. *Nymphaea ampla* and other Mayan narcotic plants. *Mexicon* 1:50–52.

———. 1979b. The water lily and the Maya scribe. *The New Scholar* 8 (2): 103–27.

———. 1981. Pilz oder Seerose—literarische und bildliche Zeugnisse von *Nymphaea* als rituellem Psychotogen in Mesoamerika. In *Rausch und Realität,* ed. G. Völger, 1:352–57. Cologne: Rautenstrauch-Joest-Museum. für Völkerkunde.

———. 1982. The mushroom and the water lily. *Journal of Ethnopharmacology* 5:139–48.

———. 1983. The ethnobotany of the Dresden Codex with special reference to the narcotic

The bud and open flower of the white water lily (*Nymphaea ampla*), found throughout southern Mexico and Guatemala. (Wild plant, photographed in Lago Petén Itzá, Guatemala)

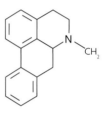

Aporphine

The *uay* glyph from the Classic Mayan period (from a vase painting). The Mayan word *uay* (literally "transformation," "magic") refers to either the nagual (an animal-shaped alter ego) or a shaman who is able to change his shape as he wishes. The picture shows a jaguar (inebriated?) swimming in a lake. The glyph text reads (from top to bottom): Water lily—jaguar—his nagual/his animal transformation—Seibal [place glyph]—*ahau* ("lord"); freely translated: "The lord of the city of Seibal has the water lily–jaguar as his nagual (animal spirit/animal form)."

A dancing jaguar (= shaman) with water lilies (*Nymphaea ampla*) in a vase painting from the Classic Mayan period.

Nymphaea ampla. Botanical Museum Leaflets 29 (2): 87–132.

———. 1992. Medicinal water lilies. *Yearbook for Ethnomedicine and the Study of Consciousness* 1992 (1):71–88. Berlin: VWB.

Emboden, William A., and Marlene Dobkin de Rios. 1981. Narcotic ritual use of water lilies among ancient Egyptians and the Maya. In *Folk healing and herbal medicine*, ed. G. G. Meyer, Karl Blum, and J. G. Cull. Springfield, Ill.: Charles C. Thomas Publishers.

Pasztory, Esther. 1974. *The iconography of the Teotihuacan Tlaloc*. Washington, D.C.: Dumbarton Oaks.

Phillipson, J. David, Günay Sariyar, and Turhan Baytop. 1973. Alkaloids from *Papaver fugax* of Turkish origin. *Phytochemistry* 12:2431–34.

Rands, Robert L. 1953. The water lily in Maya art: A complex of alleged Asiatic origin. *Anthropological Papers*, no. 34. Smithsonian Institution BAE Bulletin 151:75–153.

Roys, Ralph L. 1965. *Ritual of the Bacabs*. Norman: University of Oklahoma Press.

Tamminga, C. A., et al. 1978. Schizophrenia symptoms improve with apomorphine. *Science* 200 (5): 567–68.

Uprety, Hema, D. S. Bhakuni, and M. M. Dhar. 1972. Aporphine alkaloids of *Litsea sebifera, L. wightiana* and *Actinodaphne obovata*. *Phytochemistry* 11:3057–59.

Walter, Heinz. 1956. Der Jaguar in der Vorstellungswelt der südamerikanischen Naturvölker. MS diss., Hamburg.

Nymphaea caerulea Savigny

Blue Lotus Flower

Family
Nymphaeaceae (Water Lily Family); Apocarpiae Group, Subgenus *Brachyceras*

Forms and Subspecies
None (cf. Slocum et al. 1996, 164*)

Synonyms
Nymphaea coerulea (misspelling!)

Folk Names
Blaue lotusblume, blauer lotus, blaue seerose, blue lotus flower, himmelblaue seerose, λωτοσ (ancient Greek), lotus (Roman), νψμφαιασ (ancient Greek), nymphaea (Roman), ssn (Egyptian), utpala (Sanskrit)

History
Blue and white lotuses were the most important cultivated (ritual) plants of ancient Egypt. They grew wild in ponds and in the lowlands of the Nile and were planted in all natural and artificial (built) bodies of water (Hugonot 1992). They were esteemed for their beauty, their enchanting hyacinth-like scent, their symbolism, and probably also their inebriating effects. The buds and flowers were popular head and hair ornaments. Both the living and the dead were festooned with garlands made from the plant. The garlands in the grave of the great pharaoh Ramses II (1290–1223 B.C.E.) were made almost entirely of white and blue lotus leaves (Germer 1988*). The Egyptian lotus (*Nymphaea lotus* [L.] Willd.) was described by Dioscorides, who was certainly aware of the blue lotus as well.

The blue lotus was first mentioned in the Egyptian Book of the Dead (Hornung 1993, 167, 364). The ancient Egyptians appear to have eaten the rhizome.

In ancient Egypt, the beautiful blue lotus of the Nile (*Nymphaea caerulea*) was cultivated as a sacred plant. It has now disappeared almost completely from the Nile region. (Copperplate engraving, colorized, nineteenth century)

Left: The blue lotus (*Nymphaea caerulea*) is now rare.

Right: This hybrid water lily was derived in part from *Nymphaea caerulea*.

Distribution

The blue lotus is found only in the Nile delta, the lowlands of the Nile, and, less frequently, Palestine (Zander 1994, 397*). Today, it has almost completely disappeared from around the Nile and is seriously endangered.

Cultivation

The plant can be propagated by placing pieces of the rhizome in still bodies of water.

Appearance

The blue lotus has blue, sky blue, or sometimes slightly violet-tinged flowers that sit on long stems some 20 to 30 cm above the water's surface. The long-stemmed, floating leaves are round.

The blue lotus is easily confused with the violet-blooming *Nymphaea nouchali* Burman f. [syn. *Nymphaea stellata* Willdenow].[270]

Psychoactive Material

— Buds or flowers

Preparation and Dosage

Six buds or flowers that have already opened and closed again should be boiled in water. The flowers should be squeezed in a linen cloth so that their greenish brown juice runs into the water.

Ritual Use

In ancient Egypt, the blue lotus was closely linked to the detailed and visionary concepts of the afterlife and rebirth. Numerous buds, petals, and garlands have been found as mummy decorations and grave goods. The flower stands for the enlightened and reawakened consciousness of the deceased; it is "that lotus flower which shines in the earth" (Book of the Dead, chapter 174, line 30; cf. Dassow 1994). In the story of the battle between Horus and Seth, the lotus flower appears as a symbol of the divine, all-seeing eye. When Seth tracks down the resting Horus beneath a tree in an oasis, he rips both eyes from the sleeper and buries them in the sand, whereupon they are transformed into lotus flowers.

Because of the mythological, cosmological, symbolic, and artistic significance of the water lily, William Emboden (1978) has suggested that the ancient Egyptians used the blue lotus for its narcotic effects to produce a shamanic ecstasy among an elite priesthood. Since the blue lotus is usually portrayed in association with mandrakes (*Mandragora officinarum*) and poppy flowers (from *Papaver somniferum* or *Papaver rhoeas*; cf. *Papaver* spp.), it is highly probable that these images represent an "iconographic recipe." A psychoactive ritual drink consisting of lotus buds, mandrake fruits, and poppy capsules is entirely conceivable (Emboden 1989).

Artifacts

A portrait head of Tutankhamen showing his head emerging from a lotus flower was discovered. The water lily was associated with the sun god Ra as the bringer of light. Usually, blue lotus flowers are portrayed together with yellow mandrake fruits and red poppy capsules (see above). They very often appear in scenes that have a shamanic, visionary, or initiatory character. Stylized lotus flowers were an important ornamental element in the art (container shapes, capitals of columns) of ancient Egypt (Emboden 1989). The blue lotus flower was also a symbol of Osiris, the god of the mystery cult, who was also regarded as the lord of **beer** and wine (see *Vitis vinifera*).

Medicinal Use

In ancient Egypt, water lilies were prescribed to treat the liver, to remedy constipation, to counteract poisons, and to regulate the urine. The petals were used both externally and internally, the latter being primarily in the form of **enemas** (Rätsch 1995, 351*).

Constituents

"No pharmaceutical properties of *Nymphea lotus* and *Nymphea coreulea* [sic!] are known" (Germer 1979, 28*). Yet the leaves and flowers are alleged to have narcotic properties. According to Emboden (1978), *Nymphaea caerulea* presumably contains alkaloids. The flowers produce an exquisite **essential oil** that is said to have aphrodisiac properties.

Effects

Three to six buds, drunk as a tea, are said to induce hypnotic effects. The effects of the decoction become apparent some twenty minutes after ingestion. The initial symptoms include muscle tremors and nausea. Then comes a sensation of calm with alterations of color perception, auditory hallucinations, and other changes in auditory perception. The effects dissipate quickly after some two hours.

Commercial Forms and Regulations

Today, the blue lotus is a rare plant. It is doubtful whether it is sold in any form. The essential oil is only rarely available. When found, it is usually dissolved in sandalwood oil and is extremely expensive (John Steele, pers. comm.).

Literature

See also the entry for *Nymphaea ampla*.

Dassow, Eva von, ed. 1994. *The Egyptian book of the dead*. San Francisco: Chronicle Books.

Emboden, William A. 1978. The sacred narcotic lily of the Nile: *Nymphaea caerulea*. *Economic Botany* 32 (4): 395–407.

The white-blossomed Egyptian lotus (*Nymphaea lotus*) had a symbolic meaning in ancient Egypt that was very similar to that of the blue lotus. However, the white lotus does not appear to be psychoactive. (Woodcut from *The Penny Magazine*, July 7, 1834)

Ancient Egyptian representations of the sacred lotus as cut flowers for sacrifice and decoration. (From Engel)

270 There are uncertain reports suggesting that the flowers of *Nymphaea stellata* also have narcotic effects (Emboden 1978, 401).

"O you two fighters, tell the Noble One, whoever he may be, that I am this lotus-flower which sprang from the earth. Pure is he who received me and prepared my place at the nostril of the Great Power [Re]. I have come into the Island of Fire, I have set right [*maat*] in the place of wrong. . . . I have appeared as Nefertum, the lotus at the nostril of Re."

FROM *THE EGYPTIAN BOOK OF THE DEAD* (DASSOW 1994, CH. 174)

———. 1989. The sacred journey in dynastic Egypt: Shamanistic trance in the context of the narcotic water lily and the mandrake. *Journal of Psychoactive Drugs* 21 (1): 61–75.

Hornung, Erik. 1993. *Das Totenbuch der Ägypter.* Munich: Goldmann.

Hugonot, J. C. 1992. Ägyptische Gärten. In *Der Garten von der Antike bis zum Mittelalter*, ed. M. Carroll-Spillecke, 9–44. Mainz: Philipp von Zabern.

Pachycereus pecten-aboriginum (A. Berger) Britton et Rose

Pitayo

Family
Cactaceae (Cactus Family)

Forms and Subspecies
None

Synonyms
Cereus pecten-aboriginum A. Berger
Pachycereus pectenaboriginum

Folk Names
Bigi-tope (Zapotec), bitaya mawali (Tarahumara), cardón, cardón barbón, cardón espinoso, cardón pelo, carve, cawé, chave, chawé, echo, hecho, kammbaumkaktus, órgano, pitahayo, pitayo, sagüera, sahueso, shawé (Tarahumara), wichowaka, xáasx (Serí)

History
This striking cactus was certainly already known in prehistoric times and was being used for ethnobotanical purposes similar to those for which it is used today (Strombom and Bruhn 1978). Mexican Indians use the thorny fruit as a hairbrush and plant the cactus as a living fence. In former times, the wood of the cactus was used as a building material. The Tarahumara collect the fruits and seeds for food (Bruhn and Lindgren 1976, 175).

Distribution
The cactus is found only in Mexico: Baja California, Chihuahua, Sonora, Colima, and also near Tehuantepec.

Cultivation
In Mexico, the cactus is propagated by planting pieces that have been cut from young shoots. The cactus is often planted in dense rows to protect houses and gardens as a living fence.

Appearance
This cactus develops long, straight stems that grow upward, parallel to one another, and recall the pipes of an organ. The cactus can grow to a height of 5 to 6 meters, is somewhat thorny, and bears thorny fruits.

This species is easily confused with the closely related *Pachycereus pringlei* (S. Wats) Britt. et Rose [syn. *Cereus pringlei* S. Wats], which is known in Sonora by the names *cardón*, *cardón pelo*, and *sagüera* (Martínez 1987, 1176*). A similar species (*Pachycereus emarginatus* [DC.] Britt. et Rose) is known in Mexico by the names *órgano* and *pitayo* but is cultivated primarily as a living fence (Dressler 1953, 140*). The alkaloid-containing *Pachycereus weberi* (Coult.) Backeb. also has a similar appearance (Mata and McLaughlin 1980).

Psychoactive Material
— Young branches (stalks)

Preparation and Dosage
The fresh cactus flesh is pounded in a hollowed-out stone. The resulting juice is mixed with water (approximately one part cactus juice to three parts water). Unfortunately, there is no information describing the amounts of this solution that must be drunk in order to induce psychoactive effects (Bruhn and Lindgren 1976, 175).

The Tarahumara press the juice from the fresh stalks and drink it to induce visions. The juice is either mixed with *tesgüino* (maize beer, **chicha**) or boiled and fermented (Bye 1979b, 34*).

Pachycereus pecten-aboriginum in its natural environment in Baja California (Mexico).

Ritual Use

In former times, the Tarahumara Indians of northern Mexico drank the cactus juice during secret ceremonies in the western canyons (Bruhn and Lindgren 1976, 175; Pennington 1963, 166 f.). Apart from this, the cactus is used as a peyote substitute (see *Lophophora williamsii*).

Artifacts

A pre-Hispanic clay vessel from Colima has the shape of four stalks of a column cactus as a decorative element. It may be a representation of this cactus.

Medicinal Use

The cactus is used in Mexican folk medicine to treat stomach ulcers and cancer (Bruhn and Lindgren 1976, 175). Heated, the fresh cactus flesh of *Pachycereus pringlei* is applied externally to treat rheumatism (Felger and Moser 1974, 421*).

Constituents

The cactus flesh contains the β-phenethylamines carnegine (= pectenine), 3-hydroxy-4-methoxy-phenethylamine, salsolidine, 3,4-dimethoxyphen-ethylamine, heliamine, 3-methoxytyramine, and arizonine (Bye 1979b, 35*; Bruhn and Lindgren 1976; Mata and McLaughlin 1982, 109*).

Effects

Freshly pressed cactus juice, mixed with water, is said to produce effects similar to those of peyote (see *Lophophora williamsii*), including dizziness and visions (Bruhn and Lindgren 1976, 175; Pennington 1963, 167). The fermented juice also has strong purgative effects (Bye 1979b, 34*).

Commercial Forms and Regulations

None

Literature

See also the entry for *Lophophora williamsii*.

Bruhn, Jan G., and Jan-Erik Lindgren. 1976. Cactaceae alkaloids. XXIII: Alkaloids of *Pachycereus pecten-aboriginum* and *Cereus jamacaru. Lloydia* 39 (2–3):175–77.

Mata, Rachel, and Jerry L. McLaughlin. 1980. Tetrahydroisoquinoline alkaloids of the Mexican columnar cactus *Pachycereus weberi. Phytochemistry* 19:673–78.

Pennington, C. W. 1963. *The Tarahumara of Mexico: Their environment and material culture.* Salt Lake City: University of Utah Press.

Strombom, J., and J. G. Bruhn. 1978. Alkaloids of *Pachycereus pectenaboriginum*, a Mexican cactus of ethnopharmacologic interest. *Acta Pharm. Suecica* 15:127.

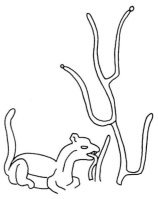

This ancient Mexican petroglyph (Olmec horizon) shows a wild cat as it licks a cactuslike plant. This may be a reference to the psychoactive powers of a column cactus of the genus *Pachycereus*. The cat may be a shaman who has transformed himself with the aid of the cactus juice.

Carnegine

Salsolidine

Papaver somniferum Linnaeus

Opium Poppy

Family

Papaveraceae (Poppy Family)

Forms and Subspecies

Over the course of time, numerous sorts, forms, varieties, and subspecies have been described for this highly variable plant. The most important are:

Papaver somniferum var. *album* DC. (white blooms)
Papaver somniferum var. *apodocarpum* Huss. (acaulescent seed capsule)
Papaver somniferum var. *glabra* (produces an especially high-quality opium; Macmillan 1991, 417*)
Papaver somniferum var. *hortense* Huss. (flat stigmal disk)
Papaver somniferum var. *nigrum* DC.
Papaver somniferum ssp. *setigerum* (DC.) Corbière (this may be the wild form; cf. Grey-Wilson 1995, 172*; Macmillan 1991, 417*; Zander 1994, 417)
Papaver somniferum ssp. *somniferum*

An unusual form, called hens and chicks, has the capsule surrounded by small secondary capsules (Grey-Wilson 1995, 173*). In addition, a number of cultivars have been produced as ornamentals, e.g., 'Black Peony', 'Golden Peony', 'Pink Chiffon', 'White Cloud', et cetera (Grey-Wilson 1995, 172*).

Papaver somniferum var. *glaucum* (Boiss. et Hausskn.) O. Kuntze is now regarded as a separate species: *Papaver glaucum* Boiss. et Hausskn. (the so-called tulip poppy) (Zander 1994, 416*).

Synonyms

Papaver glaucum Boiss. et Hausskn.
Papaver nigrum DC.
Papaver officinale Gmel.
Papaver setigerum DC.

Folk Names

Adormidera, aguna (Lithuanian), amapola, amapola de opio, biligasgase (Kannada), black poppy, bloed-zuipers-bloem (Flanders), calocatanos (Gallic), feldmohn, garden poppy, gartenmohn,

This early illustration of the white-blooming variety of the opium poppy is botanically accurate. (Woodcut from Gerard, *The Herball or General History of Plants*, 1633)

guia-guiña (Zapotec), kasa-kasa (Tamil), kasha-khasa (Malayam), kavl-a-kůknâr (Persian), kish-kâsh (Arabic/Yemen), koknâr (Persian), koquenar, madi-huada (Mapuche, "lovely gourd"), magan, magen, mâgen, mâgenkraut, magesamo, maggona (Estonian), maggons (Latvian), mago, magsat, magsomkraut, mâhan, mahonnus (Vulgar Latin), mahunus, mak (Slavic), manus, mechones, meconium, mekon, miconium, mohn, namtilla (ancient Assyrian, "plant of life"), nocuana-bizuono-huse-achoga-becala (Zapotec), oehlmagen, oehlsaamen, opium poppy, papaver (Latin), papâver, papaver album, papavero indiano (Italian), papavero somnolente, papœg, papula, pavot des jardins, pavot somnifère, popig, popœg (Anglo-Saxon), poppy, post (Hindi), pôst-a-kůknâr (Afghani), posto (Bengali), schlafmohn, schwarzer magsaamen, slaapbol (Dutch), white poppy, ying su ke (Chinese)

Folk Names for Opium

Affion (Arabic), affium, afin (Kannada), afion, a-fu-yung (Chinese), afyun (Arabic), ahiphena, amapola (Spanish), amfion (Portuguese), amphion, amsion, aphenam (Sanskrit, "foamless"), aphim (Hindi), aphu (Marathi), arfiun, chandu, maslach (Turkish), meconium, meseri, milk of poppies, misri (Egyptian), nagaphena, O, offion, ofium, opio (Latin), opion (Greek), opium, poust, tschandu, tschibuk

History

The opium poppy is known only as a cultivar. Although it is often thought that the plant was first cultivated in Asia, its home actually lies in central and/or southern Europe (Grey-Wilson 1995, 169*). Poppy was being cultivated in northern Italy, Switzerland, and southern Germany as early as the Neolithic period. It probably was used both as a source of food and as an inebriating plant: "[T]he inebriating and sedative effects of the seeds and the oil obtained from them may not have escaped the lake dwellers. In any case, both the type of use and the frequency of its occurrence and the quantities of poppy seeds that have been found demonstrate that we are looking at an important cultivar of the lake dwellers" (Hoops 1973, 233; cf. Hartwich 1899). Although it has not yet been possible to determine exactly when humans began growing poppies in the southern and northern Germanic territories, the practice certainly dates to a very early time. The Germanic peoples planted poppies (Proto-Germanic *magan*) in poppy or *magan* fields that were known as *odâinsackr* and were regarded as convalescent sites at which Odin/Wotan would effect healing miracles.[271]

The poppy is one of the most important medicinal plants in the history of pharmaceuticals. It contains a latex—the "juice of the

plant of forgetting" (Ovid)—known in ancient times by such names as *tears of the moon* and *tears of Aphrodite*. When the latex is exposed to air, it coagulates into a brown mass known as raw opium or simply opium. The methods for obtaining opium were discovered not in Southeast Asia—as is so often assumed—but in Stone Age central Europe, in the area of Lake Constance or in Provence (Hartwich 1899; Seefelder 1996, 11).

The earliest mention of the poppy, however, is on a Sumerian tablet (ca. 3000 B.C.E.), where it is described as the "plant of happiness." The first literary mention is in Homer's *Iliad* (cf. **nepenthes**). Our word *opium* is derived from the Greek *opion*, "latex [of the poppy]," which in turn is derived from *opos*, "plant juice." Roman reports demonstrate that the Gauls were well aware of the opium poppy and its properties (Höfler 1990, 93*). Walahfried Strabo (809–849), in his hexametric *Hortulus*, praised the psychoactive effects of the German opium poppy as "sacred to Ceres" (Schmitz 1981, 380; Stoffler 1978, 91). The Vikings are known to have used opium for medicinal purposes, and probably also as an inebriant (Robinson 1994, 547*).

At the present time, we still do not know when the Egyptians first began to use opium. Some authors have conjectured that the poppy was already known and being used in the time of the Old Kingdom, while other authors have suggested that the Egyptians did not become acquainted with and come to appreciate the plant and its opium until the time of the New Kingdom or even late antiquity (Bisset et al. 1994; Merrillees 1962).

In China, opium has been documented as far back as the third century. The Chinese physician Hua To used narcotics of opium and *Cannabis indica* in surgical procedures (Geddes 1976, 201). The Arab scholar Avicenna (= Abu Ali al-Hosein ben Abdallah Ibn Sina, 980–1036), who is considered to be the most important physician of the Middle Ages, is known as the "father of sleep" because he introduced the use of opium into Islamic medicine (Seefelder 1996, 52ff.).

In fifteenth-century Beijing, opium was celebrated as the best of all aphrodisiacs and apparently was used in great quantities (Duke 1973, 393). In Siam (Thailand), opium was highly regarded by the kings of Ayutthaya since at least the fourteenth century. At that time, production was probably already in the hands of the mountain tribes, who still number among its most important producers (Geddes 1976, 208). During the Middle Ages, opium was listed in all European pharmacopoeias (Schneider 1974, 3:20*), and opium preparations were used as anesthesias (cf. **soporific sponge**).

In 1670, the English physician Thomas Sydenham invented laudanum, a tincture of opium, saffron (*Crocus sativus*), cinnamon (*Cinnamomum*

271 Höfler 1990, pp. 92 f.*; cf. *Hyoscyamus niger*.

verum), powdered cloves (*Syzygium aromaticum*), and Spanish **wine** that was to have a far-reaching impact. Until the nineteenth century, laudanum was one of the most effective of all universal remedies and was also drunk for its inebriating effects.

In the seventeenth century, opium was one of the most important trade articles of the Dutch East India Company (Meister 1677, 93*). The Württemberg pharmacopoeia of 1741 characterized opium (= meconium thebaicum) as a "divine medicine" (Schneider 1974, 3:21*). Goethe described it as the "quintessence of the sweet slumber juices" (*Faust* 1). **Morphine**, isolated from opium by the German pharmacist Sertürner (1805), was the first pure active plant substance ever to be extracted and made available; the event revolutionized the pharmaceutical industry.

Opium became famous through the works of many nineteenth-century authors. Thomas de Quincy (1785–1859) saw in opium a manna, an ambrosia, a panacea, a universal remedy, a "mysterious balm to fulfill all human wishes" (De Quincey 1985, 183).

The Opium War of 1840–1842, which the English initiated purely from economic motives, led to far-reaching changes in world politics and the shape of international trade (Behr 1980; cf. Geddes 1976, 202 f.; Solomon 1978). In the 1920s, the use of opium took on social forms that appeared threatening to the ruling class and ultimately led to global prohibition (Johnson 1981; Kohn 1992*). Today, opium is important primarily as the starting point for the illegal production of heroin.

Distribution

This cultivar has spread into all regions of the world. Large areas of cultivation—either for the pharmaceutical industry or for illegal heroin production—are found in the Golden Triangle, northern Thailand, central Asia, Turkey, Mexico, Tasmania, and Austria.

In Switzerland, the opium poppy now occurs as a wild or feral plant (Lauber and Wagner 1996, 144*).

Cultivation

The plant is easily propagated from seeds, which should be broadcast in spring. Some of the seeds will germinate after ten to fifteen days. The seedlings do not like to be transplanted. Once poppies have appeared in the garden, it is relatively certain that they will always reappear, for the plant sows itself quite readily. It also can become wild and begin to appear in neighboring gardens. When harvesting the capsules, care should be taken to leave the heads of some of the plants untouched so that they can ripen and produce seeds for the following year. Poppies will also spread if the dried plants are composted, as the seeds will be distributed with the compost.

Poppies thrive best in warmer soils that are rich in nutrients, contain a great deal of humus, and are well tended. The plant requires a great deal of lime and consequently prefers lime-rich soils (Heeger and Poethke 1947, 236). For more on commercial methods of cultivation, see Griffith (1993).

Appearance

This annual plant possesses a distinct taproot, from which the perpendicular, simple or only slightly branching stem (as tall as 175 cm) develops. The gray-blue or, more rarely, greenish leaves are ovate-oblong with a more or less serrated margin or irregular lobes. The long flower stalk, which can be either hairless or only slightly pileous, bears a single flower with four petals that can vary in color (white, pink, violet, bluish, purple, light red, luminous red, dark red, almost black). The pistil itself already bears a resemblance to the fruit capsule. The smooth, round, capsule-shaped fruit has a corona and ranges in size from 2 to 6 cm, depending upon the location, sort, variety, and subspecies. It can be rather slender in appearance or very obese-gibbous. One capsule may contain up to two thousand of the tiny, kidney-shaped seeds. The seeds can be creamy yellow, brown, blue-green, or black in color. In central Europe, the plant flowers in June and July. The fruits are mature by August at the latest. A white, milky sap (latex) flows throughout the plant.

The opium poppy is easily confused with the

Just as the flowers of *Papaver somniferum* can vary in color and size, the seeds also appear in a variety of colors. (Photograph: Karl-Christian Lyncker)

Left: The opium poppy (*Papaver somniferum*) is a popular ornamental.

Right: A dark violet–blooming variety of the opium poppy (*Papaver somniferum* var. *nigrum*).

Top right: In Southeast Asia, the dried leaves of the kratom tree (*Mitragyna* sp.) are smoked as an opium substitute.

Bottom right: The arils of the Californian chestnut (*Aesculus californicus*) were formerly used as an opium substitute.

tulip poppy, *Papaver glaucum* Boiss. et Hausskn. [syn. *Papaver somniferum* var. *glaucum* (Boiss. et Hausskn.) O. Kuntze], which does not contain any psychoactive alkaloids. It is also occasionally confused with the Oriental poppy (see *Papaver* spp.).

Psychoactive Material

— Fruit capsules (fructus papaveris immaturi, capita papaveris immaturi, pericarpium papaveris, opium capsules, poppy heads)
— Opium (latex)
— Seeds (semen papaveris, poppy seeds)
— Leaves (folia papaveris, poppy leaves)
— Roots

Opium Substitutes

(From Emboden 1979; Emboden 1986, 165*; Low 1990, 199*; Ludwig 1982, 134 f.*; Millspaugh 1974, 168*; Seefelder 1996; supplemented.)

The following plants and products are or have been used as opium substitutes:

Name	Stock Plant	Notes
Amapola	various	
Amapola silvestre (wild opium)	***Bernoullia flammea*** Oliv.	seeds
Asafoetida	*Euphorbia* spp.	latex used as medicinal **incense**
Black tar	*Papaver somniferum*	raw opium that has been enriched with heroin through diacetylization
California buckeye (fruit shell)	*Aesculus californicus* (Hippocastanaceae)	$^1/_8$ of the potency of true opium
California poppy	***Eschscholzia californica***	tinctures
Chicalote	***Argemone mexicana***	latex
	Argemone platyceras Lk. et Otto	latex
Flanders/Corn poppy	*Papaver rhoeas* L. (see *Papaver* spp.)	
	Papaver somniferum	
	Papaver bracteatum (cf. *Papaver* spp.)	
Heroin		synthesized from **morphine**
Indian pipe	*Monotropa uniflora* L. (Monotropaceae)	dried herbage
Kratom	***Mitragyna speciosa***	leaves
Lactucarium (latex)	***Lactuca virosa***	"Lactuca agrestis"
	Lactuca sativa L.	
	Lactuca serriola L. [syn. *Lactuca scariola* L.]	
	Lactuca quercina L.	
Morphine		synthesized from thebaine
Ohio buckeye	*Aesculus glabra* Willd.	hypnotic component aesculine
Pituri	***Duboisia hopwoodii***	nornicotine
Red buckeye	*Aesculus pavia* L.	hypnotic component aesculine

Preparation and Dosage

The leaves are collected during the period in which the fruit capsules are maturing, and are dried in the shade. They can be smoked alone or in **smoking blends**. A rather subtle opium effect will become apparent after consuming several joints. The dried leaves also can be brewed or boiled to make a tea.

The two most important products are the capsules (poppy heads) and the milky sap (latex). The plant contains the greatest amount of latex at the end of the flowering phase and while the fruit is first ripening. The latex level declines again as the capsule matures. For this reason, both the capsules and the latex are harvested shortly after the flowers have wilted. The capsules are broken off where they connect to the stalk and are used fresh or set out to dry. To dry, they should be spread out in a single layer (perhaps in the sun), as they may otherwise become moldy. The seeds continue to mature as the heads ripen and will be usable for sowing the following year (although they will not be as vital as those of a completely matured plant).

The latex, which oozes out of incisions made in the ripe capsules, dries to a brown mass known as raw opium. The highest yield can be obtained when

the capsule wall is cut with a small, special knife perpendicularly or obliquely to the longitudinal axis in the afternoon or evening hours between the 8th and the 10th day after the petals have fallen off. The white latex that emerges quickly hardens and turns brown. The sticky mass can be scraped off and collected the next day. Each capsule can yield app. 20–50 mg of raw opium. At least 20,000 poppy capsules are required to obtain 1 kg of opium. This corresponds to a poppy field approximately 400 m² in size. (Wagner 1985, 162*)

The brown raw opium is then pressed into balls or flat cakes that will slowly dry into a hard, crumbly, solid mass. The balls or cakes should be stored in a dark, airtight location.

Opium is consumed in numerous ways: orally (opium eating, opium drinking), rectally (as a suppository or **enema**), smoked, or, when sterilized and dissolved in a saline solution, injected. Opium has a very bitter and characteristic taste (earthy-herby) that, once tasted, will not be forgotten.

Opium usually contains about 10% **morphine** (although concentrations can vary considerably). A moderate psychoactive dosage consists of the amount of opium that contains about 30 mg of morphine, i.e., around 300 mg (0.3 g) of opium.

In Rome, "slumber drinks," whose main ingredient was opium, were very popular during the time of the Caesars (cf. **soporific sponge**). These drinks were also known as "pain-relieving *catapotium*." One recipe that has come down to us lists the following ingredients: *sili* (presumably *Chaerophyllum temulentum* L.[272]), *Acorus calamus*, *ruta* (the seeds of either *Ruta graveolens* or *Peganum harmala*), castoreum (a glandular secretion of beavers), *cinnamomum* (cinnamon, probably *Cinnamomum verum*), "tears of the poppy" (opium), panax root (undeterminable), *mandragora* (mandrake root, *Mandragora officinarum*), dried "apples" (presumably mandrake fruits), *Lolium temulentum*, and peppercorns (*Piper nigrum*; cf. *Piper* **spp.**). The ingredients were chopped; raisin wine (cf. *Vitis vinifera*), a very sweet and heavy **wine**, was dripped in; and the entire concoction was then rubbed into a mass (Schmitz 1981, 380; Seefelder 1996, 36). This mixture is strongly reminiscent of the later, opium-laced **soporific sponges** as well as **theriac**.

In China, aphrodisiac "spring agents" were mixed from opium, ginseng roots (*Panax ginseng*), and musk.

Above: A normal and a mutated form of the poppy capsule (*Papaver somniferum*), known as hens and chicks.

Left: Opium is obtained by making incisions in the unripe fruit capsules of the opium poppy.

There are a number of methods for preparing an opium or poppy tea. The freshly harvested capsules can be boiled in water for fifteen to twenty minutes (until they look like well-cooked vegetables). After the liquid has cooled, strain and drink. A clearly effective dosage consists of two handfuls of capsules per person. The tea tastes like artichoke water. The pods can be boiled together with the juice of half a lemon (which will apparently have a favorable effect upon the solubility of the alkaloids). The dried capsules can be prepared in a similar manner. These, however, are best when ground (e.g., in a coffee grinder), thoroughly moistened with ample lemon juice, and then boiled in water for a short time. Allow to sit for ten to thirty minutes, then strain and drink. This preparation has a slightly unpleasant taste.

The fresh, not fully ripened capsules can be made into an opium and rum pot. A container that can be completely sealed is filled to the brim with opium capsules, after which rum is added until all the fruits are covered. To improve the bitter taste and to potentiate the effects, several female inflorescences of *Cannabis indica* or *Cannabis sativa* and flowers of *Datura metel* or another *Datura* species (*Datura* **spp.**) may be added. Seal the container and allow the mixture to sit in a relatively warm location for six months. Then pour off the liquid and vigorously squeeze the opium capsules in a strainer. As little as one shot glass of the liquid will induce clearly perceptible opium effects.

In ancient and late ancient times, opium was usually dissolved in wine (*Vitis vinifera*) for consumption (Krug 1993, 14*). Both the poppies and opium were also added to **beer** and **mead**. In

272 This umbelliferous plant has the same effect on grazing cattle as the bearded darnel (*Lolium temulentum*) (Roth et al. 1994, 210*).

"In the name of Allah, the Compassionate, the Merciful. Praise be to Allah, whose might created opium and whose power allows it to heal diseases."

Abu'l Qâsem Yazdi
"Tract for Opium Smokers"
in *Vom Rausch im Orient und Okzident*
(Gelpke 1995, 51*)

"The Indians [= from India/Indonesia] use this afion or amfion [= opium], before the Javanese and Malay, also the Malabarans, Ceylonese, as well as the Moors from the Arabic Coast, as well as the Persians and Turks, usually to increase their sensual pleasures."

George Meister
Der Orientalisch-Indianische Kunst- und Lustgärtner [The Oriental-Indian Art and Pleasure Gardener]
(1677, ch. 9, 29*)

In twentieth-century Europe, it was believed that opium was a Chinese invention as well as a typical Chinese vice. This led to the image of the Chinese opium den, which continues to etch its way into many people's minds. (Magazine illustration, Germany, ca. 1920.)

his sixteenth-century herbal, Tabernaemontanus provided a recipe for barley mead whose active ingredient was *magsaamen* (= *Papaver somniferum*)! In India, a drink was made from **wine**, hemp seeds (*Cannabis indica*), poppy seeds, and opium (Duke 1973, 392). Opium is also an ingredient in bhang drinks (cf. *Cannabis indica*). In ancient India, opium was swallowed mixed with *araq* (**alcohol**).

Opium is one of the key ingredients in the **Oriental joy pills**. In the Orient, opium, either alone or mixed with such other substances as hashish (cf. *Cannabis indica*), spices, ambergris, musk, olibanum (cf. *Boswellia sacra*), and powdered pearls and precious stones (lapis lazuli, rubies, emeralds), was made into little balls—which were sometimes even gilded—that were swallowed or administered anally (Croutier 1989, 55). A ball 0.5 to 0.7 mm in diameter was the dosage for a rectal suppository, which was pushed as deeply into the rectum as possible. There it would quickly dissolve, so that the effects would become apparent after as little as ten to fifteen minutes. When administered in this manner, great care needed to be exercised with dosage.

The use of poppy capsules, poppy seeds, and opium as **incense** is very ancient. Opium and bitumen were inhaled as fumigants for treating toothaches (Schmitz 1981, 380). During the Middle Ages, a medicinal and psychoactive incense was made using opium, *Mandragora officinarum*, and arsenic (Seefelder 1996, 200).

In China, smoking opium (*chandu*) is produced by dissolving raw opium in water and bringing it to a boil. The resulting mass, which remains moist, is then allowed to ferment for several days or weeks. Fermentation is complete when an elastic, kneadable mass results. The *chandu* is now ready for use. When used alone, *chandu* is not "smoked" but, rather, heated in the bowl of a pipe and vaporized. The vapor is inhaled deeply (Hogshire 1994, 86). A dosage for an opium pipe is a ball of opium the size of a pea. For noticeable visionary effects, several pipes should be smoked at short intervals. The desired effects typically begin after the fifth pipe.

In China and Laos, opium was and is mixed with tobacco (*Nicotiana tabacum*) for smoking (Geddes 1976, 202; Westermeyer 1982, 56). In Laos, the resinous remains (a condensate of the smoke or vapor) are scratched out of frequently used opium pipes, mixed with raw opium, and sold under the name *khe dya-feen*, "little opium tails" (Westermeyer 1982, 56). In Sumatra, opium is mixed with the leaves of *Ficus hypogaea* (von Reis Altschul 1975, 53*). In India, opium is smoked with hemp (*Cannabis indica*) or wild tobacco (*Nicotiana rustica*). In Morocco, dried poppy capsules are smoked to induce sleep (Vries 1984*), and sometimes chicken dung is added to

smoking opium to extend it (Bourke 1996, 161*).

Laudanum is a tincture that was originally made of opium, saffron (*Crocus sativus*), cinnamon (*Cinnamomum verum*), clove powder (*Syzygium aromaticum*), and Spanish wine (cf. *Vitis vinifera*). It was later produced using just opium, saffron, and high-proof **alcohol** (70% ethanol) (opii tinctura). The pharmaceutically standardized tincture should contain about 1% morphine. The largest single therapeutic dose of opium tincture has been listed as 1.5 g (Wagner 1985, 165*).

Ritual Use

Opium was ingested in the Minoan culture to produce the ecstatic states needed for religious ceremonies (Kritikos 1960). On Crete, seeresses would give oracles and divine the future while under the influence of opium:

> Around 1300–1250 B.C.E., in the land of the goddesses of health and healing [Crete], opium was inhaled or used as an incense; this is evidenced by an ash heap and a tube-shaped vase with an opening on the side that was found during excavations . . . of the divine idols of Gazi. . . . The same effects were expected from opium smoke as were later of tobacco smoke: cheerfulness, forgetfulness, or ecstasy. (Faure 1990, 123*)

Demeter, the earth goddess and mother of grain, who was adorned in wreaths of poppies, was originally venerated on Minoan Crete. From there, her cult spread to the other Greek islands and the mainland. One of her sacred plants was the poppy (cf. **kykeon**):

> The inebriating plant can be demonstrated to have been everywhere in the Demeter cult, which was not limited to Eleusis, but was found throughout the entire settlement area of Magna Graecia, for example, in Enna on Sicily, upon whose sacred mountain a Demeter shrine was enthroned in which quite riotous initiation mysteries were also held. Similar rites and festivals were associated with the goddesses corresponding to Demeter, for example, Cybele in Asia Minor and later Ceres in Rome, as the influence of the Greek cults of the gods manifested themselves in the new centers of power. (Seefelder 1987, 19)

In the ancient world, the poppy was regarded as the nourishment of divining dragons, as a mysterious magical plant, and as a sleeping and dreaming agent. According to Theocritus, the poppy grew from the tears that Aphrodite shed as she mourned her youthful lover Adonis. The plant was sacred to many gods and goddesses: The Great

Mother goddess Cybele was depicted holding poppy capsules in her hand, as was Hypnos, the god of sleep, the "resolver of cares." Hermes/Mercury carried the plant in his left hand (cf. **moly**). Thanatos, or Death, was decorated with garlands of poppies, while Nyx, the goddess of the night, was portrayed with poppies wrapped around her temples.

In late antiquity, poppy seeds were an important ritual smoke offering to Hypnos, the god of sleep, at the Orphic mysteries (a cult of Dionysos). The poppy also symbolized the prophetic dream. Opium appears to have been used in the sacred **incense** of Epidauros and in preparations for inducing the healing and vision-giving temple sleep. Poppies and opium occupied a firm place in the religious healing cult of ancient times (Krug 1993*).

The opium poppy was also a magical and ritual plant among the Germanic tribes. It was sacred to the southern Germanic (Frankish) god Lollus. Ludwig Bechstein described an amazingly long-lived pagan custom in *Der Sagenschatz des Frankenlandes* [Treasury of Tales from Franconia] (1842):

One can still read of a purported pagan idol whose type and name belongs to Franconia quite alone. This is Lollus, Löllus, or Lullus, whose special veneration is said to have taken place on the Main River (near the later city of Schweinfurt). The bronze image of the idol, in the form of a youth with curly golden hair, was found. From around his neck, a garland of *magsamenköpfen* (poppy seed capsules) hung down over his chest. The right hand of the image is reaching for its mouth and grasping the tongue with its thumb and index finger; in the left it holds a cup of wine, in which stand ears of grain. Apart from a loincloth, the body was completely naked. The image is said to have stood in a sacred, enclosed grove by the banks of the Main, and the people are said to have brought it offerings of grapes and ears of grain at certain times. (In Hasenfratz 1992, 109 f.*)

The name Loll(us) suggests the German word *lallen* ("slur"). In other words, Lollus was an oracular god who, so to speak, was inebriated from opium and/or **wine** and slurred his words, perhaps "speaking in tongues." The name Lull(us) is suggestive of *lullaby*, "to put to sleep." After all, even in the present day one can sometimes hear of the rural custom of administering some poppy juice (raw opium) to restless or crying children to put them to sleep. Perhaps the image of Lollus represents an iconographic recipe: poppy heads (opium) are added to wine along with grain (perhaps ergot; cf. *Claviceps purpurea*). If a person partakes of this drink, then he or she will "speak in tongues."[273] Speaking in tongues, also known as glossalalia, is a type of unconscious flow of speech that has been known since ancient times and appears both in shamanic rituals and in religious cults (Goodman 1974).

Opium played a role in the meditations and mystical rituals of several Islamic sects and secret societies (Sufis, dervish orders). Because of the secrecy of their traditions, however, no details are known about such activities (Seefelder 1996, 56).

Women in Oriental harems were quite fond of using opium and developed certain rituals in their dreary solitude:

The nights in the harem swelled with *keyf* (ultimate fulfillment) induced by opium pills and the drowsy peace of sated senses. The women indulged in drawn-out opium rituals, spending the evenings inhaling hookahs or eating opium, the "elixir of the night," dreaming of faraway lands beyond the latticed windows. Mostly, they preferred eating rather than smoking opium because the effect lasted longer, dreams lingering until the rising of the sun. Amnesia followed; night after night of this induced chronic insomnia. The women began forgetting their distant homes, their lives before the seraglio. In order to remember, they told stories to one another. A thousand stories of faraway lands, stories told in the night. At first it was a thousand nights of stories, but even numbers brought bad luck, so they added one. (Croutier 1989, 56)

In Asia, opium was often used as an aphrodisiac in the erotic rituals of the Taoists and Tantrists (cf. *Camellia sinensis*, **Oriental joy pills**).

In Asia, opium is still used today by fakirs, yogis, sadhus, and shamans (cf. *Aconitum ferox*, *Cannabis indica*). The shamans of the Miao, a mountain tribe in northern Thailand, smoke opium before a healing ceremony in order to enter the trance that is needed to heal (Geddes 1976, 218 f.). In this condition, they can travel to heaven and act on behalf of the ill person while they are there. In Thailand, opium is also used as an offering to sacred trees and rocks. According to a legend of the Akha tribe (Thailand), the first poppy plant arose from the heart of a beautiful woman who was killed because she had given herself in love to all the men (Anderson 1993, 117*).

Artifacts

A statue of Tammuz (sacred ram) standing on a plant that was found in Ur and dates to the time of the Sumerians may be one of the oldest poppy artifacts. The flowers depicted on the statue are strongly reminiscent of those of *Papaver somniferum* (Emboden 1995, 100*).

"I dropped some more laudanum into the glass. The night was a coat that provided warmth and security, I pulled it more closely around my body. Time became space, quite thick, like a small chamber that lay no longer inside the pyramid but far below it. Nothing occurred any more, only peaceful quiet, unassailable solitude."

ERNST JÜNGER
ANNÄHERUNGEN [APPROACHES]
(1980, 203*)

Inebriated
Your sun is stronger than the sun of Africa.
I have no parasol against it.
If it is true that I inebriate myself, then it is more on you than on opium.

JEAN COCTEAU
(1988, 57)

The poppy flower and the opium that is obtained from its fruit capsule have long been used as seductive names for sensuous perfumes. (Advertisement in a German magazine, ca. 1915)

273 A similar iconographic recipe can be found on an image of Demeter.

Top: In ancient times, the poppy was sacred to the goddess Demeter. Opium, the juice it yields, was known as the tears of Demeter. (Greek relief of Demeter with poppy capsules, Corinth, first century)

Bottom: A set of Chinese utensils for smoking opium. (San Francisco, Chinatown, ca. 1920)

One of the most spectacular artifacts of ancient times is the Cretan "poppy goddess." This terracotta piece, which dates to the late Minoan period (1400–1100 B.C.E.) and was in the shrine at Gazi, depicts a half-naked woman with raised hands who looks, ecstatically enraptured, into the distance and wears a headband into which three incised opium capsules have been placed. This poppy goddess has been interpreted as a "personification of the goddess of sleep or of death" (Sakellarakis 1990, 91).

A number of incised poppy capsules are depicted on the portal of the former Eleusinion in Athens. A golden tablet from Mycenae shows Demeter giving three poppy capsules to Perseus, the founder of the "mushroom city." A Boeotian plate shows the same goddess with a torch, two ears of grain, and two opium capsules. A coin of King Pyrrhus of Epirus depicts the goddess as the earth mother with ears of grain and opium capsules. In a terra-cotta relief from the Campani Collection, Demeter peers ecstatically into the infinite while holding spikes of grain and incised opium capsules in her two snake-entwined hands. A bouquet of ears of grain and poppy flowers is clearly recognizable on the mystical chest (*kiste mystica*) of the Eleusinian cult. Poppies and poppy capsules were portrayed on numerous ancient coins dating from the seventh century B.C.E. on (Seefelder 1996, 15).

Ancient Egyptian frescoes frequently depict the poppy, usually in association with mandrake fruits (***Mandragora officinarum***) and lotus flowers (***Nymphaea caerulea***). While some authors interpret the poppy that is shown as the opium poppy (Emboden 1995*), others usually see it as the Flanders poppy (*Papaver rhoeas*; Germer 1985*) or, more rarely, as the Oriental poppy (*Papaver orientale*; cf. ***Papaver*** spp.) (Seefelder 1996, 13). The Egyptian Museum in Berlin houses a (New Kingdom) statue of Isis as a cobra, surrounded by ears of grain and opium capsules.

In Ayutthaya, the ancient Siamese capital, incised poppy capsules were carved into quartz crystals during the fourteenth century. In Thailand (Siam) and Burma, "opium weights"—which were used to weigh other things besides opium—were made of metal alloys fashioned into the shape of animals (ducks, lions, birds, elephants). Regarded as symbols of luck, they were used even as a means of payment (Braun 1983; Greifenstein n.d., 55 ff.). Today, they can be found only in antique shops. In the nineteenth century, numerous miniatures (book paintings) with erotic contents were painted in Thailand. Many of these depicted the so-called opium dens. Not only was a great deal of opium smoked in such places, but all forms of erotic play were indulged in as well (Haack 1984, 55, 121).

Over the course of time, numerous devices for smoking as well as inhaling opium have been developed (Hartwich 1911*). An ivory opium pipe found at the shrine of Astarte (a predecessor of Aphrodite) at Kition (Cyprus) was dated to the twelfth century B.C.E. (Karageorghis 1976); it may be the oldest archaeological evidence of opium smoking in Europe. In Thailand, water pipes are usually fashioned from bamboo tubes and round gourds or coconut shells (*Cocos nucifera*). In China, water pipes were made from brass following the same principle. In China, Korea, and Japan, long, thin pipes with small bowls—which held exactly one dose of opium—were especially common. These pipes are frequently depicted in Chinese and Japanese art.

In the Middle Ages, opium was the subject of countless Arabic poems, stories, and novels (Gelpke 1995*). Many of the stories in *The Thousand and One Nights* either were inspired by opium or are direct descriptions of opium and its effects (Croutier 1989, 56).

In the nineteenth century, opium was both a very widespread people's drug, e.g., as "poppy tea" in England (London et al. 1990), and the drug of choice for many artists (Berridge and Edwards 1987; Kramer 1981). It was supremely important to many poets and writers, who immortalized it in their works (Hayter 1988). Novalis (1772–1801) sang the praises of opium in his *Hymnen an die Nacht* ("Hymns to the Night"), Edgar Allan Poe (1809–1849) wrote most of his work while under the influence of opium, and E. T. A. Hoffmann (1776–1822) also knew of and utilized the effects of opium. Charles Baudelaire (1821–1867) wrote about opium in *The Artificial Paradise* (cf. ***Cannabis indica***) and worked his experiences into the collection of poems known as *The Flowers of Evil*.

From a literary point of view, Thomas De Quincey (1785–1859) had an especially important influence and helped set the tone for later authors and imitators. His book *The Confessions of an English Opium-Eater* appeared in London in 1822 and has served as a kind of template for later literary treatments of the effects of opium and its associated dependency.

Descriptions of the joys and sorrow of opium use have made their way into many novels and literary reports of self-experiences (such as diaries) (Cobbe 1895; Cocteau 1948; Detzer 1988; Ekert-Rotholz 1995; Jones 1700; Magre 1929; Schweriner 1910). Even the Opium War has become the subject of literary treatments (e.g., Fraser 1987; Thompson 1984).

In the nineteenth century, opium was the subject of many paintings. Carl Spitzweg (1808–1885) produced a painting in 1856 (*Chibuk Smoking Oriental on a Divan*) that portrays Oriental opium smoking (Seefelder 1996, 61). In his oil painting *The Siesta* (1876), John Frederick Lewis depicts a beautiful woman in Oriental garb, inebriated on opium. His painting *In the Bey's Garden, Asia Minor* (1865) portrays a lady of the harem as she picks flowers next to a large opium poppy plant. Eugène Delacroix (1798–1863) used the same motif as the subject of his oil painting *Odalisque* (1845). The oil painting *Odalisque and Slave,* (1842), by Jean-Auguste-Dominique Ingres (1780–1867), shows a woman, inebriated on opium and almost naked, as her servant plays music for her. The eunuch in Jean-Léon Gérôme's (1824–1904) oil painting *The Guard of the Harem* (1859) is shown holding a meter-long opium pipe in his hand (Croutier 1989, 31, 45, 55f., 124). *The Lascar's Room* (1873), a picture by Gustave Doré (1832–1883), has become a kind of archetype for the opium dens. In the art nouveau movement, the opium poppy was often used as a floral element or was placed at the center of focus.

Many nineteenth-century Japanese woodcuts depict erotic scenes in which lovers are shown smoking opium from long, thin pipes before, during, or after coitus and while drinking tea (*Camellia sinensis*) (Marhenke and May 1995*). The effects of opium also inspired several colored woodcuts by Katsushika Hokusai (1760–1849) (e.g., *Smoking Ghost*).

During the Golden Twenties, opium smoking was often portrayed in pictures and illustrations of Berlin society and other circles in Germany and in San Francisco. The drawings of Paul Kamm, Max Brüning, and D. Fenneker were especially popular and were published in several magazines (*Berliner Leben, Der Junggeselle,* and others). These illustrations played a great role in helping to develop the stereotype of the opium den (cf. **morphine**). A German emergency currency note (value two marks) of the time shows a physician armed with an **enema** syringe as he utters the words "I am Doctor Eisenbarth, I cure the people with my own art. The night watchman Didelum, I gave him 10 pounds of opium."

In several volumes of the popular children's book series Mecki, the hero, an anthropomorphic hedgehog, is shown smoking opium conspicuously often. His visions are shown in words and pictures (*Mecki among the Chinese,* 1955; *Mecki and Prince Aladdin,* 1958; *Mecki on the Moon,* 1959).

Opium is also a subject of several comic books (e.g., Hergé, *Tim and Struppi, The Blue Lotus*; Francis Leroi and Marcelino Truong, *The Bamboo Dragon*; Daniel Torres, *Opium*).

In music, the most conspicuous traces of opium can be found in Hector Berlioz's (1803–1869) *Symphonie Fantastique,* which serves as a kind of soundtrack for imagining an opium experience. Apart from this, almost nothing is known about the influence opium may have had on the music of the nineteenth century.

The cover of the album *Spitfire* (1976), by the psychedelic band Jefferson Starship, shows a Chinese woman riding on a dragon growing out of the smoke of her opium pipe. A short-lived British neopsychedelic band of the mid-1990s was called Opium Den. The California avant-garde metal band Tool titled its first album *Opiate* (BMG 1992).

In the 1960s, a film version of De Quincey's *The Confessions of an English Opium-Eater* was made in England.

Medicinal Use

As early as the time of the Middle Kingdom, the Egyptians were aware of the sleep-promoting effects of the poppy: "A remedy for too much crying in a child: špn [poppy] seeds; fly dung from the wall; is made to a paste, strained and drunk for four days. The crying will cease instantly" (Papyrus Ebers, 782, in Manniche 1989, 131*).

Such uses of the poppy and poppy juice have been preserved even into our day.[274] In modern Egypt, opium is said to incite men to war and to love and to produce spectacular dreams. It is usually eaten mixed with spices, or it may be smoked. It is a popular aphrodisiac, especially in **Oriental joy pills**. The ancient Assyrians valued even the root as an aphrodisiac (Thompson 1949, 227*).

Opium was one of the most important medicines of the ancient Hippocratics, who used it to treat almost all illnesses, especially dropsy, diarrhea, uterine disorders, inflammations of rectal fistulas, hysterical complaints, and, of course, sleep disorders (Krug 1993*; Rätsch 1995a, 240–249*).

In Germanic folk medicine, poppy juice (opium) was taken internally to protect against nocturnal pests, bloodsucking vampires, nightmares, and nickel goblins (Höfler 1990, 94*). As late as the twentieth century, pharmacists were still making poppy pacifiers to calm small children (Nadler 1991, 58*). "The dried poppy leaves are only occasionally used as sedatives in folk medicine" (Heeger and Poethke 1947, 235).

Opium is used throughout the world as a folk medical treatment for coughing (cf. **codeine**) and diarrhea (Fleurentin and Pelt 1982, 92f.*; Paulus and Ding 1987, 394*).

The opium poppy was also an important ingredient in poplar ointment (unguentum populeum; cf. **witches' ointments**). In her *Kräuterbuch* ("Herbal"), Elisabeth Blackwell wrote, "The leaves are taken among cooling ointments, they are considered useful for burned areas, inflammations, acute swelling, and come to the unguent, Popul." (Heilmann 1984, 106*).

Opium is used in homeopathy (usually in higher potencies) according to the medical description for such conditions as agitated states (Bomhardt 1994):

"Opium will expand beyond all measures,
Stretch out the limitless,
Will deepen time, make rapture bottomless,
With dismal pleasures
Surfeit the soul to point of helplessness."

CHARLES BAUDELAIRE
"POISON"

"Baudelaire went
to a baseball game
and bought a hot dog
and lit up a pipe
of opium."

RICHARD BRAUTIGAN
(1968, 58*)

The art and literature of the Golden Twenties were shaped primarily by the use of opium and other psychoactive substances (morphine, cocaine). (Book title, 1920)

274 According to Langham (1579), children should be given the ground seeds together with hemp (*Cannabis sativa*) and almonds in milk or **beer** (ale).

"To Hypnos—
a smoke offering of poppy seeds—
Hypnos, who rules the departed
and mortal humans without
        exception.
As well as the living beings,
as many as the broad earth
        nurtures—
For you alone govern them all,
you approach them all,
with softly wrought chains
binding the bodies, a remover
        of cares,
giving comfort to the exertions
and healing consolation to all
        afflictions.
You turn away the worries of death
and preserve the souls."

ORPHIC HYMN

An opium smoker and paraphernalia:
A) spoons for cleaning the pipe bowl;
B) needle for picking up the opium;
C) longitudinal view of the pipe bowl.
(From Hartwich 1911)

The resurrection of Kore (Persephone) in spring. The goddess of rebirth, shown here in Minoan garb, bears three poppy capsules in her hand. The two plants to the sides of the goddess may be mandrakes. A man (rhizotome) is shown serving as a birth-helper. (Seal, Boeotia, fourth century B.C.E.)

Hahnemann said that the effects of poppy juice are more difficult to evaluate than almost any other medicine. The effects of opium, as they are expressed in the insensibility of the nervous system, the suppression of bodily functions, the sleepy dazed feeling, the lack of pain, the inactivity, the general sluggishness, and the lack of vital reactions, represent the main indications for the homeopathic use of this drug. (Boericke 1992, 571*)

Constituents

The entire plant (except for the roots and flower petals) contains a latex that coagulates into opium. Opium contains some forty alkaloids, known collectively as **opium alkaloids**. Opium can contain 3 to 23% **morphine**, 0.1 to 2% **papaverine**, 0.1 to 4% **codeine**, 1 to 11% narcotine, and 0.1 to 4% thebaine; the other alkaloids are present in only trace amounts (Heeger and Poetke 1947). The composition of the alkaloids, and in particular the morphine concentration, can vary greatly. Modern techniques make it possible to determine quickly the morphine content of a particular sample (Hsu et al. 1983).

Callus tissue has been found to contain the alkaloids sanguinarine, norsanguinarine, dihydrosanguinarine, oxysanguinarine, protopine, cryptopine, magnoflorine, and choline (Furuya et al. 1972).

The characteristic scent of opium is the result of some seventy substances, of which pyrazine appears to play an especially important role (Buchbauer et al. 1994).

Poppy seeds contain practically no or only slight traces of alkaloids (Norman 1991, 49*). They are rich in oil, carbohydrates, calcium, amino acids (apart from tryptophan), and proteins. However, **codeine** can be produced from the seeds through the process of digestion with pepsin.

Effects

In the older literature, the effects of opium are compared to those of wine (cf. *Vitis vinifera*) remarkably often (Schmitz 1981, 384). All opium users make a clear distinction between the effects of smoked opium and those of ingested opium (Cocteau 1957, 86). When eaten or drunk, opium usually has stronger physical effects that are perceived as a paradisiacal state and blissfulness:

Opium . . . loosens the soul from its entanglements with everyday things and the outer world. . . . Opium makes one silent and gentle. It inspires and gives flight to the imagination, including the erotic, increases sensitivity and the sense of tenderness, while simultaneously diminishing the need to move and be active, the need to communicate,

ambition, sexual potency, emotions, and aggressiveness in general. (Gelpke 1995, 42*)

Descriptions of the effects of opium tend to place great emphasis on the cheerfulness that develops and the soothing effect this has upon the mood:

Opium . . . communicates serenity and equipoise to all the faculties, active or passive: and with respect to the temper and moral feelings in general, it gives simply that sort of vital warmth that is approved by the judgment, and that would probably always accompany a bodily constitution of primeval or antediluvian health. Thus, for instance, opium, like wine, gives an expansion to the heart and the benevolent affections: but then, with this remarkable difference, that in the sudden development of kindheartedness that accompanies inebriation, there is always more or less of a maudlin character, which exposes it to the contempt of the bystander. . . . But the expansion of the more benign feelings incident to opium is no febrile access but a healthy restoration to that state the mind would naturally recover upon the removal of any deep-seated irritation or pain that had disturbed and quarrelled with the impulses of a heart originally just and good. (De Quincey 1998, 41)

Although the contents of the visions may be shaped substantially by the individual, there are many reports of their vegetative nature: "Opium is the only vegetable substance that communicates the vegetable state to us. Through it, we get an idea of that other speed of plants" (Cocteau 1957, 91f.).

Opium users describe encounters with the soul of the plant, often in the form of an enchantingly beautiful and loving woman or goddess (Schwob 1969). Telepathic and clairvoyant states (e.g., seeing through walls) are often said to be characteristic of the effects of opium (Arsan 1974).

With irregular or occasional use of opium, the "ecstasy of the opiophage is strongly permeated with sexual conceptions, while parallel to this a strong arousal of the sexual apparatus [has been] observed" (Hirschfeld and Linsert 1930, 252*).

Opium's effects are the result of a synergy between the primary alkaloids: The main active constituent, **morphine**, has sedative-hypnotic, narcotic, antitussive, respiratory-suppressing, and constipating effects. **Papaverine** increases the flow of blood into the corpus cavernosum of the penis; **codeine** is the best cough medicine known. The effects of opium are manifested quite rapidly and persist for six to eight hours at an almost constant strength.

Among the undesirable side effects of opium use are constipation, nausea, and vomiting (which usually occurs the following day). Metoclopramide (e.g., Paspertin) is an effective antidote for these symptoms. Chronic use can lead to dependency structures with "addictive behavior." However, the so-called addictive potential of opium is not as great as portrayed by the sensationalistic press and uninformed politicians.

Commercial Forms and Regulations

The opium poppy is classified as a drug, trafficking in which is not allowed (German BtMG Anlage II). In Germany, only 10 m² of one's own garden may be planted with opium poppies. Since 1984, only "detoxified" poppy capsules, from which the **morphine** has been removed, may be sold in German flower shops. In Denmark, poppy capsules have been illegal for decorative purposes since 1986 (Roth et al. 1994, 536*).

Opium is subject to numerous drug laws throughout the world and may be prescribed only on special forms. Opium is available for sale today only in India and Pakistan (see *Cannabis indica*). It appears as though both the pharmaceutical lobby and organized criminals (Mafia) have a great interest in the difficulties associated with prescribing opium. As a result of these difficulties, the former is able to market its very expensive synthetic opiates and the latter are in a better position to sell illegal heroin.

Literature

See also the entries for *Papaver* spp., **Oriental joy pills**, **soporific sponge**, **codeine**, **morphine**, **opium alkaloids**, and **papaverine**.

Arsan, Emanuelle. 1974. *Emmanuelle oder Die Schule der Lust.* Reinbek, Germany: Rowohlt.

Behr, Hans-Georg. 1980. *Weltmacht Droge.* Vienna and Dusseldorf: Econ.

Berridge, V., and G. Edwards. 1987. *Opium and the people: Opiate use in nineteenth century England.* London: Yale University Press.

Bisset, Norman G., Jan G. Bruhn, Silvio Curto, Bo Holmstedt, Ulf Nyman, and Meinhart H. Zenk. 1994. Was opium known in 18th dynasty ancient Egypt? An examination of materials from the tomb of the chief royal architect Kha. *Journal of Ethnopharmacology* 41:99–114.

Bomhardt, Martin. 1994. Opium. *Homöopathische Einblicke* 20:5–22.

Boyes, Jon, and S. Piraban. 1991. *Opium fields.* Bangkok: Silkworm Books.

Braun, Ilse, and Rolf Braun. 1983. *Opiumgewichte.* Landau, Germany: Selbstverlag.

Buchbauer, Gerhard, Alexej Nikiforov, and Barbara Remberg. 1994. Headspace constituents of opium. *Planta Medica* 60:181–83.

Cobbe, William Rosser. 1895. *Doctor Judas: A portrayal of the opium habit.* Chicago: S. C. Griggs.

Cocteau, Jean. 1948. *Opium—Journal d'une desintoxication.* Paris: Libraire Stock.

———. 1957. *Opium: The diary of a cure.* Trans. Margaret Crosland and Sinclair Road. London: Peter Owen Limited.

———. 1988. *Ich war im Paradies.* Bielefeld, Germany: Pendragon-Verlag.

Croutier, Alev Lytle. 1989. *Harem: The world behind the veil.* New York: Abbeville Press.

De Quincey, Thomas. 1998. *Confessions of an English opium-eater.* Oxford: Oxford University Press.

Detzer, Eric. 1988. *Poppies: Odyssey of an opium eater.* San Francisco: Mercury House.

Duke, James A. 1973. Utilization of *Papaver. Economic Botany* 27:390–400.

Durresi, S., and P. Rizo. 1991. Determination of the contents of morphine in the capsules of two varieties of poppies grown in Albania. *Planta Medica* 57 suppl. (2): A100–1.

Ekert-Rotholz, Alice. 1995. *Mohn in den Bergen: Eine junge Frau verfällt dem Opium.* Frankfurt/M.: Fischer TB.

Farrère, Claude. 1920. *Opium.* Munich: Thespis-Verlag.

Fraser, George Macdonald. 1987. *Flashman—Der chinesische Drache.* Frankfurt/M. and Berlin: Ullstein.

Furuya, T., A. Ikuta, and K. Syono. 1972. Alkaloids from callus tissue of *Papaver somniferum. Phytochemistry* 11:3041–44.

Geddes, William Robert. 1976. *Migrants of the mountains: The cultural ecology of the Blue Miao (Hmong njua) of Thailand.* Oxford: Clarendon Press.

Goodman, Felicitas D. 1974. *Speaking in tongues: A cross-cultural study of glossolalia.* Chicago and London: The University of Chicago Press.

Greifenstein, Ute I., n.d. *Fremdes Geld.* Frankfurt/M.: Commerzbank und Museum für Völkerkunde.

Griffith, William. 1993. *Opium poppy garden.* Berkeley, Calif.: Ronin.

Haack, Harald. 1984. *Der Liebe zur Freude: Erotische Buchmalerei aus Thailand.* Dortmund: Harenberg.

Hartwich, Carl. 1899. Über *Papaver somniferum* und speziell dessen in den Pfahlbauten vorkommende Reste. *Apothekerzeitung*: 39–41.

Hayter, Alathea. 1988. *Opium and the romantic imagination: Addiction and creativity in De Quincey, Coleridge, Baudelaire and others.* Wellingborough, U.K.: Crucible.

Heeger, E. F., and W. Poethke. 1947. *Papaver*

The German band Die Toten Hosen seems not to be propagating a new religion with its album *Opium fürs Volk* ("Opium for the People") but to be intent on distributing the narcotic. (CD cover, 1996, JKP Records.)

"Here it pleases me well, in
the circle of my light little poems
to now make mention of
the field poppy [*Papaver*], which the
mother Latona
partook of in sorrow for the rape of
her daughter, so it is said,
that longed-for forgetfulness frees
the breast
of its cares."

Walahfried Strabo
in Hans-Dieter Stoffler's Der
Hortulus des Walahfried
Strabo, 1978, 16

"When I am under the spell of opium and the present has pulled away from my mental eye, then Ot-Chen and Chen-Hoa appear to me like two fairy-tale princesses, and it pleases me to rock in old, wonderful dreams. The smoking chamber Chen-Tas expands and becomes a magnificent marble palace in which I care for the divine peace as a sovereign prince. The tumult of Fouchow Road is submerged, and I feel myself surrounded by the majestic calm of the old forests in which my imperial ancestors once walked."

CLAUDE FARRÈRE
OPIUM
(1920, 151 f.)

"The Slovakian mother who wishes to heal her epileptic child walks fully disrobed three times around it without anyone seeing her, and uttering a spell, she shakes from a poppy head the seeds into her hand and shakes these around the child; then the poppy is swept together and, with the cut-off nails of the fingers of one hand and the toes of one foot, but crosswise, from the right hand and the left foot or vice versa, she stuffs these into an infernal tube and buries it where the sun will never shine."

SIEGFRIED SELIGMANN
DIE MAGISCHEN HEIL- UND SCHUTZMITTEL AUS DER BELEBTEN NATUR [THE MAGICAL HEALING AND PROTECTIVE AGENTS FROM THE ANIMATED NATURE]
(1996, 219*)

somniferum L., der Mohn: Anbau, Chemie, Verwendung. *Die Pharmazie* suppl. 4 (1): 235–340.

Hillestad, Agnes. 1980. Glycoproteins of the opium poppy. *Phytochemistry* 19:1711–15.

Hogshire, Jim. 1994. *Opium for the masses*. Port Townsend, Ore.: Loompanics.

Hoops, Johannes. 1973. *Mohn*. In *Reallexikon der germanischen Altertumskunde*, 3:233–34.

Hsu, An-Fei, Dorothy Brower, Ronald B. Etskovitz, Peter K. Chen, and Donald D. Bills. 1983. Radioimmunoassay for quantitative determination of morphine in capsules of *Papaver somniferum*. *Phytochemistry* 22 (7): 1665–69.

Husain, Akhtar, and J. R. Sharma. 1983. *The opium poppy*. Lucknow, India: Central Institute of Medicinal and Aromatic Plants.

Jevons, F. R. 1965. Was Plotinus influenced by opium? *Medical History* 9.

Johnson, Bruce D. 1981. Die englische und amerikanische Opiumpolitik im 19. und 20. Jahrhundert: Konflikte, Unterschiede und Gemeinsamkeiten. In *Rausch und Realität*, ed. G. Völger, 2:656–61. Cologne: Rautenstrauch-Joest-Museum für Völkerkunde.

Jones, John. 1700. *The mysteries of opium reveald (. . .)*. London: Printed for Richard Smith.

Kapoor, L. D. 1995. *Opium poppy: Botany, chemistry, and pharmacology*. Binghamton, N.Y.: The Haworth Press.

Karageorghis, Vasso. 1976. A twelfth-century BC opium pipe from Kition. *Antiquity* 50:125–29.

Kramer, John C. 1981. The metapsychology of opium. *Journal of Psychoactive Drugs* 13 (1): 71–79.

Krikorian, Abraham D. 1975. Were the opium poppy and opium known in the ancient Near East? *Journal of the History of Biology* 8 (1).

Kritikos, Pan. G. 1960. *Der Mohn, das Opium und ihr Gebrauch im Spätminoicum*. Athens: Archives of the Academy of Athens III.

Kritikos, Pan. G., and S. N. Papadaki. 1967. The history of the poppy and of opium and their expansion in antiquity in the eastern Mediterranean. *Bull. Narcotics* 19 (3): 17 ff.; (4): 5 ff.

Lamour, Catherine, and Michel R. Lamberti. 1972. *Les grandes manœuvres de l'opium*. [Paris]: Éditions du Seuil.

Latimer, Dean, and Jeff Goldberg. 1981. *Flowers in the blood: The story of opium*. New York: Franklin Watts.

London, M., T. O'Regan, P. Aust, and A. Stockford. 1990. Poppy tea drinking in East Anglia. *British Journal of Addiction* 85:1345–47.

Magre, Maurice. 1929. *La nuit de haschich et d'opium*. Paris: Flammarion.

Merlin, Mark D. 1984. *On the trail of the ancient opium poppy*. Madison, N.J.: Fairleigh Dickinson University Press.

Merrillees, R. S. 1962. Opium trade in the Bronze Age Levant. *Antiquity* 36:287–92.

Moorcock, Michael. 1988. *Der Opium-General*. Bergisch Gladbach, Germany: Bastei-Lübbe.

Rush, James R. 1981. "Opiumfarmen" auf Java in der Kolonialzeit. In *Rausch und Realität*, ed. G. Völger, 2:568–71. Cologne: Rautenstrauch-Joest-Museum für Völkerkunde.

Sakellarakis, J. A. 1990. *Heraklion—Das archäologische Museum*. Athens: Ekdotike Athenon.

Schmitz, Rudolf. 1981. Opium als Heilmittel. In *Rausch und Realität*, ed. G. Völger, 1:380–85. Cologne: Rautenstrauch-Joest-Museum für Völkerkunde.

Schweriner, Oskar T. 1910. *Opium*. Berlin: Carl Duncker.

Schwob, Marcel. 1969. Das Opiumhaus. In *Das Spiegelkabinett*, ed. W. Pehnt. Munich: dtv.

Seefelder, Matthias. 1987. *Opium—eine Kulturgeschichte*. Frankfurt/M.: Athenäum.

———. 1996. *Opium: Eine Kulturgeschichte*. 3rd ed. Landsberg, Germany: ecomed. (Very good bibliography.)

Solomon, Robert. 1978. The evolution of opiate use in China: The origins of the illicit international trade. *Journal of Psychedelic Drugs* 10 (1): 43–49.

Spence, Jonathan. 1972. Das Opiumrauchen im China der Ch'ing-Zeit (1644–1911). *Saeculum* 23 (4).

Thompson, E. V. 1984. *Opium und Mandelaugen*. Munich: Knaur.

Westermeyer, Joseph. 1982. *Poppies, pipes, and people: Opium and its use in Laos*. Berkeley: University of California Press.

Papaver spp.

Poppy Species

Family
Papaveraceae (Poppy Family)

The genus *Papaver*, to which **Papaver somniferum** belongs, has some seventy species, more than any other member of its family (Grey-Wilson 1995, 128*). While the genus *Papaver* is very rich in alkaloids, psychoactive species are few (Sariyar et al. 1994). Prickly poppy (**Argemone mexicana**), California poppy (**Eschscholzia californica**), and greater celandine (*Chelidonium majus* L.; cf. **Mandragora officinarum**) are all closely related to *Papaver*.

Papaver bracteatum Lindl. [syn. *Papaver orientale* var. *bracteatum* Ledeb.]—Oriental poppy
This Oriental poppy is endemic to the Caucasus, eastern Turkey (Anatolia), and northwestern Iran. Its taxonomic history is more than confusing. There are many similar types, i.e., ornamental forms that have been produced by breeding this species with other species (Grey-Wilson 1995, 138*). While this species does not contain any opium, **morphine**, or **codeine**, it does contain a large amount of thebaine ($C_{19}H_{21}NO_3$), the precursor substance for the pharmaceutical synthesis of morphine, codeine, and Nalaxon (Morton 1977, 124*). Some chemotypes exhibit concentrations of up to 98% thebaine in the total alkaloid content (Brenneisen and Borner 1985, 303; Kettenes et al. 1981). It is hoped that increasing the cultivation of *Papaver bracteatum* for pharmaceutical purposes will help limit the cultivation of *Papaver somniferum* (Kettenes-van de Bosch et al. 1981).

Papaver orientale L. [syn. *Papaver pollakii* A. Kerner]—Oriental poppy
This Oriental poppy is endemic to Anatolia (Turkey) (Grey-Wilson 1995, 156*). It develops into an herbaceous perennial that can grow as tall as 80 cm. As a result of its distinctive beauty, it is an extremely popular garden ornamental. There are numerous cultivars of *Papaver orientale* (e.g., 'Cedric's Pink', 'Mrs. Perry', 'Goliath') that are widespread ornamentals (Grey-Wilson 1995, 157*). The plant, which is occasionally characterized as psychoactive, is difficult to distinguish from *Papaver bracteatum* and is sometimes mistaken for *Papaver somniferum*. It does not contain any significant **opium alkaloids**. The main alkaloid is oripavine or isothebaine, which is of interest only to the pharmaceutical industry (Brenneisen and Borner 1985, 303).

Papaver rhoeas L. [syn. *Papaver strigosum* (Boenn.) Schur]—Flanders poppy, Shirley poppy
The annual Flanders poppy—also called the wild field poppy (or corn poppy, wild poppy, *pavot coquelicot*)—is a wild European and Mediterranean plant that is sometimes confused with *Papaver somniferum*. The variety *Papaver rhoeas* var. *oblongatum* occurs on Cyprus and is one of the sacred flowers of Aphrodite (Georgiades 1987, 2:64*; Rätsch 1995a, 245*).

The ancient Assyrians called the plant *ukushrim* or *irrû* and apparently used it in a manner similar to that of true opium (Thompson 1949, 225*). The plant was formerly used as a children's sedative and sleeping agent (Schneider 1974, 3:26*). The psychoactive, opium-like effects that are sometimes attributed to the plant are doubtful.

The main alkaloid in the aboveground parts of *Papaver rhoeas* is (+)-rhoeadine (0.06%). The total alkaloid content is 0.11 to 0.12% (Roth et al. 1994, 534*). Also present are the alkaloids allocryptopine, protopine, coulteropine, berberine, coptisine, sinactine, (+)-isocorydine, (+)-roemerine, and (+)-rhoeagenine (Kalav and Sariyar 1989). Several of these alkaloids are also present in *Eschscholzia californica*. In contrast, no **opium alkaloids** are present.

Top: Many poppy species (*Papaver* spp.) have extremely colorful flowers.

Bottom: The Oriental poppy (*Papaver orientale*).

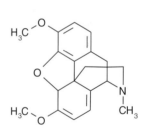

Thebaine

Oripavine

From top to bottom:
In the wild, the Flanders poppy
(*Papaver rhoeas*) can cover entire
fields.

This variety of Flanders poppy
(*Papaver rhoeas* var. *oblongatum*) is
found in Cyprus. (Wild plant,
photographed in Cyprus)

The yellow horned poppy (*Glaucium
flavum*) is indigenous to the
Mediterranean region. (Wild plant,
photographed on Naxos, Greece)

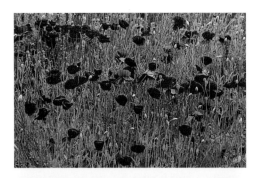

The Oriental poppy (*Papaver orientale*
var. *bracteatum*), cultivated in many
gardens as an ornamental, also
contains opiates. (Woodcut from
Tabernaemontanus, *Neu Vollkommen
Kräuter-Buch*, 1731)

The wild Flanders poppy (*Papaver rhoeas*)
has only weak sedative properties.
(Woodcut from Tabernaemontanus, *Neu
Vollkommen Kräuter-Buch*, 1731)

Glaucium flavum Crantz [syn. *Papaver cornutum*
nom. nud., *Glaucium luteum* Scop.]—yellow
horned poppy

The horned poppy is indigenous to the Mediterranean region (Grey-Wilson 1995, 41*). Dioscorides described it under the name *mekon keratitis*:

> After eating or drinking this mekon keratitis,
> the same manifestations appear as with poppy
> juice [= opium; cf. *Papaver somniferum*]. One
> also counters it with the same agents. The fruit
> is collected in the summer, when it is dry. The
> decoction of the root is taken with **wine** [see
> *Vitis vinifera*], it helps against dysentery.
> (Dioscorides 4.68)

The plant was formerly used in folk medicine together with **honey** to treat ulcers (Grey-Wilson 1995, 36*). The horned poppy is sometimes characterized as psychoactive. The plant does contain glaucine, which has the same effects as **codeine** (Roth et al. 1994, 374*).

Literature
See also the entries for *Papaver somniferum* and **morphine**.

Brenneisen, R., and S. Borner. 1985. Psychotrope
Drogen. IV. Zur Morphinalkaloidführung von
Papaver somniferum und *Papaver bracteatum*.
Pharm. Acta Helvetica 60 (11): 302–10.

Kalav, Y. N., and G. Sariyar. 1989. Alkaloids from
Turkish *Papaver rhoeas*. *Planta Medica* 55:488.

Kettenes-van de Bosch, J. J., C. A. Salemink, and I.
Khan. 1981. Biological activity of the alkaloids of
Papaver bracteatum Lindl. *Journal of
Ethnopharmacology* 3 (1): 21–38. (Includes
literature useful for a deeper understanding.)

Sriyar, Günay, Aynur Sari, and Afife Mat. 1994.
Quarternary alkaloids of *Papaver spicatum*.
Planta Medica 60:293.

Passiflora spp.

Passionflowers, Passion Fruits

Family

Passifloraceae (Passionflower Family); Subfamily Passifloreae

Forms and Subspecies

The genus *Passiflora* encompasses some four hundred, but no more than five hundred, species (Meier et al. 1994, 34; Vanderplank 1991). The following species and subspecies have ethnopharmacological significance:

Passiflora caerulea L. cv. Constance Elliot—blue crowned passionflower

Passiflora edulis Sims var. *edulis* Sims—passion fruit, maracuja, purple passion fruit, purpurgranadille

Passiflora edulis Sims var. *flavicarpa* Degener—granadilla, granadille, yellow passion fruit

Passiflora foetida L.—**amapola**, t'usku', tsyquitieco, pok'pok'

Passiflora incarnata L. [syn. *Granadilla incarnata* Medik., *Grenadilla incarnata* Medik., *Passiflora edulis* Sims var. *kerii* Masters, *Passiflora kerii* Spreng.]—true passionflower, passionflower, wild passion vine, maypop

Passiflora involucrata (Masters) A. Gentry [syn. *Passiflora quadriglandulosa* var. *involucrata* (Masters) Killip, *Passiflora vitifolia* var. *involucrata* Masters]—chontay huasca

Passiflora jorullensis H.B.K.—coanenepilli

Passiflora laurifolia L.

Passiflora quadrangularis L. [syn. *Passiflora macrocarpa* Masters]—tumbo, bate

Passiflora rubra L.—liane zombie (cf. **zombie poison**)

History

Most passionflowers are tropical plants native to Central and South America. In pre-Columbian times, many Indians used some species as a source of food (twelve to sixty species are edible) and as sedatives and medicines. When Spanish missionaries invaded the New World, they took *Passiflora* as a sign from God, seeing the unusual flowers as a symbol of the mystery and the passion of their savior. Among those who played an important role in promoting this view were the cloister student and artist Jacomo Bosio, the Jesuit J. B. Ferrari in Siena, Father Simone Parlasca, and Pope Paul V (Klock 1996, 13).

The English herbalist John Gerard may have been the first to report on the mysterious new plant (Meier 1995b, 116; Rätsch 1991a, 203*). In the eighteenth and nineteenth centuries, botanists helped spread many passionflower species across much of the globe (Meier 1995b, 115). Most species were described in the nineteenth century (Schneider 1974, 3:31*).

Today, passion fruits are regarded as one of the most prized exotic fruits in the world (Mollenhauer 1962). The genus is still awaiting comprehensive ethnopharmacological research, particularly with regard to its psychoactive usefulness (cf. Meier 1995b, 119).

Distribution

Almost all species of the genus *Passiflora* are indigenous to the tropical rain forests of the Americas, especially South America. *Passiflora incarnata* was originally found in the Caribbean region as well as in southeastern North America (Meier et al. 1994, 35). Only a few cold-resistant species (*P. caerulea* L., *P. incarnata* L., *P. lutea* L.) can survive outdoors in temperate zones. *P. caerulea* now grows wild in southern Europe (Italy and Greece) (Meier 1995b). Passion fruits are grown in Portugal and Spain. Many *Passiflora* species, such as the beautiful *P. amethystina*, have become established in Southeast Asia as ornamentals.

Cultivation

The passionflowers are becoming increasingly important as houseplants and ornamentals. All species can be grown from seed (see Klock 1996), which can be sown in very loose, airy potting soil

Top right: The color of the enchanting flower of *Passiflora amethystina* Mikan led to its being named after the amethyst, the sacred stone of Dionysos. (Photographed in Japan)

Bottom right: The grenadilla, or passion fruit (*Passiflora edulis*).

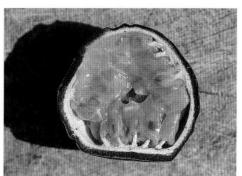

Bottom left: One of the few passionflowers that can be cultivated in Europe: *Passiflora caerulea*. (Wild plant, photographed in Cyprus)

Bottom center: The juicy and vitamin-rich fruit of the maracuja (*Passiflora edulis*). In Brazil, the juice pressed from this fruit is used in the production of the psychoactive *vinho do jurema*.

415

Top left: *Passiflora rosea*, known as tumbo, photographed in the Andes of northern Peru.

Bottom left: *Passiflora* aff. *foetida* is commonly called *amapola*, "opium." (Photographed at Tikal, Guatemala)

Right: One of the many as yet undescribed tropical *Passiflora* species in the Amazonian rain forests.

275 "The quantities of harmane in *Passiflora incarnata* are very slight and lie according to the state of knowledge at the time in the range of 50 to 300 mg in 100 g of dried drug. This is far removed from the effective dosage, which was investigated in the 1950s and was said to lie at 10 to 35 mg daily for a sedative effect in humans" (Meier 1995b, 119).

throughout the year; in central Europe, however, the period from November to April is the best time for planting (time to germination: two to six weeks at 20 to 25°C). As houseplants, passion-flowers should be well watered between April and September and given fertilizer every fourteen days. In spring, the shoots should be cut back to a length of 10 to 15 cm (or pot up as needed).

Appearance

All *Passiflora* species are evergreen climbing vines or bushes with many-lobed leaves and unmis-takable, bizarre flowers with three styles, some seventy-two filaments, and five anthers. The fruits are usually oval and, in many species, edible.

Psychoactive Material

— Passionflower herbage (leaves and stalks; passiflorae herba, herba passiflorae) from the stock plant *Passiflora incarnata* L.; *Passiflora caerulea* L. was formerly used as well but is now regarded as a counterfeit of the drug (Meier 1995b)
— Roots of *Passiflora involucrata*
— Fruit juice of *Passiflora edulis*
— Calyxes of *Passiflora foetida*

Preparation and Dosage

To make calmative teas, the dried herbage of *Passiflora incarnata* is best combined with valerian roots (*Valeriana officinalis*), hop cones (*Humulus lupulus*), and St. John's wort (*Hypericum perfora-tum* L.) or with valerian roots, balm (*Melissa officinalis* L.), anise (*Pimpinella anisum* L.), and mint (*Mentha* sp.) (Meier 1995b, 124 f.). The recommended daily dosage of the dried herbage of *Passiflora incarnata* is 4 to 8 g; as a tea, 2.5 g per cup, taken three or four times daily (Meier 1995b, 122; Wichtl 1989). Tea can also be prepared using 15 g of passionflower herbage and 150 g of boiling water. To date, no interactions with other substances (negative synergies) are known (Meier et al. 1994, 46).

The herbage can be smoked alone or in **smoking blends** (overdoses are unknown).

In Mexico, the flowers of *Passiflora foetida* are

known as **amapola**, "opium," and are made into a tea that is used as an opium substitute (Argueta V. et al. 1994, 119*).

The roots of *Passiflora involucrata* are suitable for use in preparing **ayahuasca analogs.**

Passion-fruit juice is used together with **Mimosa tenuiflora** and probably **Pithecellobium spp.** to produce the ayahuasca-like drink known as *jurema*. Further study of this use is needed.

Ritual Use

In the region of Iquitos, the roots of the Ama-zonian species *Passiflora involucrata* are used as an **ayahuasca** additive "so that the visions become more intense" (Rob Montgomery, pers. comm.). Maracuja juice (*P. edulis* var. *edulis*) plays an imprecisely understood role in the little-studied Brazilian *jurema* cult, whose practices are similar to those surrounding ayahuasca.

Artifacts

The Brazilian *jurema* cult may possess artifacts that make reference to the passionflower.

Medicinal Use

In Amazonia, a tea of maracuja leaves (*Passiflora edulis*) is drunk as a sedative (Duke and Vasquez 1994, 130*). A tea made from the leaves of tumbo (*Passiflora quadrangularis*) is used as a narcotic and sedative. The Kubeo Indians say that a decoction of the leaves of *Passiflora laurifolia* has sleep-inducing effects (Schultes and Raffauf 1986, 269*). The Indians of the Caribbean and Central America also know of *Passiflora* species that they use as sedatives and sleeping agents.

In European folk medicine and phytotherapy, *Passiflora incarnata* is taken as a tea or as part of a combination preparation for states of nervous unrest (Meier 1995b, 122; Wichtl 1989). In home-opathy, a mother tincture (Passiflora incarnata hom. *HAB1*, *PFX*, *HPUS88*) is used for such purposes as calming and to promote sleep (Meier et al. 1995, 47).

Constituents

It was once thought that harmane alkaloids were the active constituents in *Passiflora incarnata* and other species (Löhdefink and Kating 1974; cf. β-carbolines, harmine and harmaline).[275] One can sometimes read in the literature that 100 g of dried *Passiflora incarnata* herbage contains approxi-mately 10 mg of harmane alkaloids, but this amount is highly questionable (Meier 1995b, 120). It is possible that cinnamic acid derivatives and **coumarins** were mistaken for harmanes during the analysis (Meier et al. 1994, 38). Maltol (a γ-pyrone), once thought to be the main active constituent, is actually a by-product produced when the raw plant material is heated and cannot be responsible for the effects. The most recent research indicates that the *C*-glycosylflavones apigenine and luteoline are the main active constituents (Meier 1995b, 120; Meier 1995a; Meier et al. 1994).

The following compounds are present in *Passiflora incarnata*: vicenine-2, isoorientine-2"-*O*-glucoside, schaftoside, isoschaftoside, isoorientine, isovitexine-2"-*O*-glucoside, isovitexine, and swertisine. Orientine and vitexine are present only in trace amounts. Saponarine, which was thought to be a component as well, is absent (Meier 1995a). *Passiflora jorullensis* contains passicol, harmol, harmane, harmine, harmalol, and harmaline (Emboden 1979, 187*).

The mucilaginous pulp (mesocarp) of passion fruit (*Passiflora edulis*) consists primarily of 2 to 4% citric acid, relatively little ascorbic acid (only 20 to 50 mg per 100 g of pulp), carotenoids (0.5 to 2.5 mg per 100 g of pulp), starch, and more than two hundred aromatic substances (Meier 1995b, 116 ff.). There is no evidence indicating whether harmanes occur in the fruit.

The roots of *Passiflora involucrata* appear to be rich in β-**carbolines** with MAO-inhibiting properties. The chemistry of the flowers of *Passiflora foetida* has not yet been clarified (Argueta V. et al. 1994, 119*).

Effects

Maracuja juice increases the efficacy of *vinho do jurema* (see **Mimosa tenuiflora**), as it allegedly possesses MAO-inhibiting properties (cf. **ayahuasca analogs**).

Animal experiments have demonstrated that an aqueous extract of *Passiflora incarnata* both deepens and prolongs sleep. The neuropharmacological effects have been compared to those of *Cannabis sativa* (Speroni and Minghetti 1988). Although calmative effects are often mentioned, there is no pharmacological evidence to support them. The effects appear to be more anxiolytic (Meier 1995b, 123).

When smoked, the herbage of *Passiflora incarnata* induces a marijuana-like "high" (Brown and Malone 1978, 11*). The effects are very subtle. It has been claimed that smoking *Passiflora* inhibits MAO, so that orally administered *N,N*-**DMT** can be effective.

Smoking *Passiflora jorullensis* induces a state of euphoria that is said to be like that produced by *Cannabis sativa* (Emboden 1979, 187*). Whether the *Passiflora rubra* of the Dominican Republic is able to produce a zombielike state is unknown.

Commercial Forms and Regulations

The seeds of various species can be obtained in nurseries and flower shops (the blue crowned passionflower is frequently sold under the name *Passiflora caerulea*). Passion fruits are now found on fruit stands in much of the world. Tea mixtures and herbal tablets based upon *Passiflora incarnata* are available without restriction and can be purchased in pharmacies, herb shops, health food stores, et cetera.

Literature

See also the entries for **ayahuasca**, **ayahuasca analogs**, β-**carbolines**, and **harmaline and harmine**.

Killip, Ellsworth P. 1938. The American species of Passifloraceae. *Publications of the Field Museum of Natural History*, botanical series, 19:1–613.

Klock, Peter. 1996. *Das große Buch der Passionsblumen*. Hamburg: Lagerstroemia Verlag.

Löhdefink, J., and H. Kasting. 1974. Zur Frage des Vorkommens von Harmanalkaloiden in Passiflora-Arten. *Planta Medica* 25:101–4.

Martin, F. W., and H. Y. Nakasone. 1970. The edible species of *Passiflora*. *Economic Botany* 24:333–34.

Meier, Beat. 1995a. *Passiflora herba*—pharmazeutische Qualität. *Zeitschrift für Phytotherapie* 16 (2): 90–99.

———. 1995b. *Passiflora incarnata* L.—Passionsblume: Portrait einer Arzneipflanze. *Zeitschrift für Phytotherapie* 16 (2): 115–26.

Meier, Beat, Anne Rehwald, and Marianne Meier-Liebi. 1994. *Passiflora*. In *Hagers Handbuch der pharmazeutischen Praxis*, 5th ed., 6:34–49. Berlin: Springer.

Mollenhauer, H. P. 1962. Die Grenadilla (*Passiflora edulis* Sims). *Deutsche Apotheker-Zeitung* 102:1097–1100.

Speroni, E., and A. Minghetti. 1988. Neuropharmacological activity of extracts from *Passiflora incarnata*. *Planta Medica* 54:488–91.

Vanderplank, John. 1991. *Passion flowers and passion fruit*. London: Cassell Publishers Limited.

Wichtl, Max. 1989. Passionsblumenkraut. In *Teedrogen*, ed. M. Wichtl, 362–64. Stuttgart: WVG.

The bizarre blossom of the passionflower (*Passiflora incarnata*) has long stimulated the human imagination. (Drawing by Sebastian Rätsch)

Luteoline

Apigenine

"This plant, which the Spaniards in the West Indies named *Granadilla* because of its similarity to the fruit of the pomegranate, is the same that the Virginians called *Maracoc*. The Spanish Friars first called it *Flos Passionis*, passion flower, because of an imagined conception, and reported that it was epitome of the passion story of Our Savior. The fruit, which thrives in the West Indies, is large and red and reminiscent of a pomegranate. But the husk is thinner, and although the flesh of the fruit is without taste, the juice is sour. The Indians as well as the Spaniards open the fruit in the manner of eggs and eagerly suck out the juice. The fruit is a mild laxative."

JOHN GERARD
THE HERBALL
(1591)

Paullinia cupana Kunth ex H. B. K.

Guaraná Vine

Family
Sapindaceae (Soapberry Family); Subfamily Sapindoideae, Paullinieae Tribe

Forms and Subspecies
The wild form is *Paullinia cupana* H.B.K. var. *typica* (Seitz 1994, 53). The cultivated form, *Paullinia cupana* H.B.K. ssp. [or var.] *sorbilis* (Mart.) Ducke, is often regarded as a subspecies or variety (Erickson et al. 1984, 273).

Synonyms
Paullinia cupana H.B.K.
Paullinia sorbilis Mart.
Paullinia sorbilis (L.) Mart.

Folk Names
Brasilianischer kakaobaum, Brazilian cocoa, camú-camu (Shipibo), cipo-guaraná, cupana, cupána, dschungeltee, guarana, guaraná, guaraña, guaranáliane, guaranáranke, guaranastrauch, guaranáuva, guaraná vine, guaranazeiro, naraná, naranajeiro, uabano, uaraná, uraná

History
The use of guaraná—which the Indians regard as a gift from the gods—is said to have been discovered by the Amazonian Satéré-Mawé tribe and to have a tradition dating back thousands of years (Carneiro M. 1989, 60f.*; Pavia 1995, 137*; Straten 1996, 62). A drink made of water and the ground fruits of the guaraná vine was originally consumed by shamans so that they could acquire secret knowledge. Many Indians use guaraná drinks, which they call "elixirs of eternal youth," as a hunting drug. The Amazonian Indians have known of the plant and its stimulating products for centuries.

In Europe, however, reports of the jungle vine did not appear until the mid-seventeenth century (Straten 1996, 60f.). The plant became more well known through Alexander von Humboldt, who became acquainted with both the plant and the drink it yields during his travels from the Orinoco to the Rio Negro. The high caffeine content was detected in 1840 and has been confirmed numerous times (Berrédo Carneiro 1931). Today, guaraná is increasingly being exported to Europe, both as an agent of pleasure and for pharmaceutical processing (Schröder 1991, 108f.*). Guaraná has become a popular coffee substitute (see *Coffea arabica*), and is now also known as a "techno drug" (cf. **herbal ecstasy**). It has also played a significant role in the development of **energy drinks**.

Distribution
The natural range is in central Amazonia from the Rio Madeira to the Rio Tapajos and on the Rio Negro and the Orinoco. Wild stands occur only in these areas. Amazingly, these regions are also the only places where the plant has been successfully cultivated (Schöder 1991, 109*).

Cultivation
Guaraná can be grown either from viable seeds or from cuttings taken from prunings. The method of pollination, which has a very strong influence on the yield, was long unknown. It now appears that hundreds of insects are responsible (Esteves Gondim 1984).

Guaraná has now been cultivated successfully for years in the central Amazon region (Schultes 1979, 259). The primary area of cultivation, consisting of some fifteen thousand acres (six thousand hectares), is near Manaus (Erickson et al. 1984). One plant produces about 1 kg of seeds per harvest. It is astonishing that guaraná cultivation has not been commercially successful in any other part of the world. Apart from the Central Amazon region, the plant has been cultivated only in Sri Lanka, Uruguay, and Central America (Seitz 1994, 54).

Appearance
This woody, vinelike climber can grow over 12 meters long. It climbs up supporting trees by means of its spreading branches. The perennial underwood plant has a smooth, vertical stem and large, long leaves consisting of five ovate-oblong individual leaves. The white, scentless flowers are attached to short stalks and grow from the branch axis in bushy panicles. The three-chambered ovary usually contains a single developed seed that is about 1 cm in size, round, and embedded in a thin, dark brown seed coat. The seed is filled with two starchy cotyledons (Tschirch 1918). The flowers begin to blossom when the rainy period ends. The

Left: The young guaraná vine (*Paullinia cupana*) in the Amazonian rain forest.

Right: Around the world, stimulating drinks are now manufactured from guaraná seeds (*Paullinia cupana*).

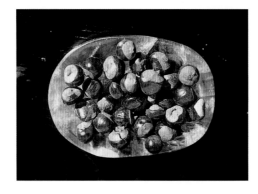

fruits (guaraná clusters) mature some three months later (Seitz 1994, 54).

The guaraná vine can be easily confused with other species of the genus *Paullinia* (see ***Paullinia spp.***).

Psychoactive Material
— Seeds
— Pasta guarana (pasta guaraná, pasta seminum paulliniae, massa guaranae, guarana paste, guaraná paste)

Preparation and Dosage
The traditional method of preparing the seeds is to manufacture guaraná breads (also known as *bastão*):

The fruits are placed in water, allowing the outer hull to swell so that it can be more easily removed. The seeds are then lightly roasted so that the starch adheres the cotyledons together. The seeds that have been prepared in this way are then ground with the shells or pounded with a pestle. Mixed with manioc starch flour [*Manihot esculenta* Crantz] and perhaps also cacao [***Theobroma cacao***], it is kneaded with water to produce a doughy paste. This guarana paste, whose caffeine content can vary considerably depending upon the amounts of the admixtures, is dried in the sun or over a low fire for the domestic market in the form of rolls about a foot long and the width of a wrist. (Schröder 1991, 110*)

These loaves are easily transported. The Amazonian Indians use the ossified tongue (hyoid bone) of an Amazonian fish called *pirarucú* (*Arapaima gigas* Cuvier) to rasp off the amount they wish to use and drink the powder suspended in water.

For pharmaceutical and industrial processing, guaraná is prepared by mixing the dried, shelled, more or less strongly roasted and powdered seeds with water to produce a paste (Seitz 1994, 54).

One-half teaspoon of the prepared powder is often given as an average daily dosage. One gram of pharmaceutical guaraná paste represents a single dose (Seitz 1994, 56).

Ritual Use
According to the mythology of the Brazilian Tupí Indians, the guaraná plant had a shamanic origin. The shamaness Omniamasabé, who possessed a very comprehensive knowledge of the "real world that is hidden to humans," was impregnated in the forest solitude by the snake god Mboy. Shortly thereafter, she gave birth to a son. Her jealous brothers enlisted a hostile shaman to kill the son. This shaman drank **ayahuasca** and transformed himself into an ara parrot. In this form, he

searched out and killed the boy. As the tears of the mother poured over the body, it was transformed into the guaraná vine. Since that time, shamans eat the guaraná fruits so that they may be initiated into the secrets of the knowledgeable shamaness Omniamasabé.

Most Amazonian Indians, including the Mundurucú, tell some version of the following origin myth:

A son born to a couple of the Maué tribe, so the story goes, was an exceptional child who spread happiness and good will wherever he went, a veritable angel. A jealous evil spirit resolved to eliminate the youngster. Despite close supervision by the tribe the child slipped out alone one day to collect fruit in the forest. The evil spirit, Iurupari, transformed himself into a snake that attacked and killed the child. When rescuers found him he was lying facing the sky, bearing a benevolent expression in death, eyes opened wide.

Soon thereafter a shattering bolt of lightning shook the earth, halting the lamentations of the assembled tribe. Enter the mother who gave a lengthy discourse on how she had received divine instructions to bury the excised eyes of the child. No one wanted to accept the gruesome task so a lottery was conducted, and the interment performed by the loser. Later a shrubby plant sprouted from the buried eyes. This was the first guaraná and its origin accounts for ripe fruit having the appearance of living eyes. (In Erickson et al. 1984, 280)

For the Indians, this story also explains the plant's abilities to help a person stay awake. The fruit's eyelike shape is interpreted as a kind of signature for mystical vision. Because of this, the plant has a certain significance as a shamanic plant and is ingested when diagnosing diseases (cf. Karlinger 1983, 128–32). For this reason, the Indians do not pick the guaraná berries until the first "eye" has opened (Seitz 1994, 54). Some Amazonian Indians also use guaraná for ritual fasts (Straten 1996, 143).

In recent years, guaraná has acquired a reputation in Europe as a "techno drug" (Walder 1995). Using copious amounts of guaraná makes it easier to dance through the night at techno parties, which have a distinct ritual character. In certain circles, guaraná is also used as a substitute for **cocaine**.

Artifacts
The Indians also use the guaraná paste to manufacture decorative articles. They shape it into figures of humans, animals, and plants, and even into relief pictures with depictions of

"The first description of guaraná was written in the year 1669 by the Jesuit missionary J. F. Betendorf, who spoke about the diuretic effects of the drink and its use for headaches, fever, and cramps. In the eighteenth century, interest in guaraná was the reason for searching for a trade route across central Brazil, from Mato Grosso across the source rivers of the Rio Tapajós to the Rio Maués. The merchants needed 9–10 months to journey back and forth by boat, which had to be pulled over land in some places, but the effort was obviously worthwhile. The Maué intensified the cultivation to meet the demand, and the Brazilians themselves built plantations."

Bruno Wolters
Drogen, Pfeilgift und Indianermedizin [Drugs, Arrow Poisons, and Indian Medicine] (1994, 154 f.*)

"On this occasion, I noticed that a missionary seldom undertakes a journey without taking along the prepared seeds of the cupana vine. This preparation requires great care. The Indians grind the seeds, mix them with manioc flour, wrap the mass in banana leaves, and allow this to ferment in the water until the mass has turned saffron yellow. This dough is dried in the sun, and after water has been poured over it, it is consumed in the morning instead of tea. The drink is bitter and strengthens the stomach, but I found the taste to be very unpleasant."

ALEXANDER VON HUMBOLDT
REISE IN DIE AEQUINOCTIAL-
GEGENDEN DES NEUEN CONTINENTS
[TRAVELS TO THE EQUINOCTIAL
REGIONS OF THE NEW CONTINENT]
(STUTTGART 1862, VOL. 5, P. 113)

To the Indians, the medicinal value of this plant [*Paullinia cupana*] and its ability to stimulate the brain and keep the body active and vital is nothing short of a miracle."

MICHAEL VAN STRATEN
GUARANA
(1996, 62)

village life. These sculptures are often decorated with colors that are produced from indigenous natural dyes [e.g., *Bixa orellana*] and minerals. The traditional value of these objects far exceeds their simple beauty. Because they are made from one of the Indians' most valuable medicines and foods, they acquire a spiritual, almost religious significance. (Straten 1996, 64)

Medicinal Use

Many Indians of the Amazon esteem guaraná as an aphrodisiac. In the region of the Peruvian Amazon, the subspecies *sorbilis* especially is regarded as an aphrodisiac (Rutter 1990*). Apart from this, the plant is used primarily as a remedy for intestinal diseases (Schröder 1991, 110*). Guaraná also finds use in folk medicine for treating menstrual pains, difficulties with digestion, conditions of weakness, diarrhea, and fever (Seitz 1994, 56).

In phytotherapy and alternative medicine, guaraná has been found to be effective primarily as an antidepressant, to treat "coffee addiction," for migraines, and for chronic fatigue syndrome (CFS) (Straten 1996, 70, 155).

In homeopathy, a tincture of pure seeds, Guarana hom. *HAB34* or Paullinia sorbilis hom. *HPUS88*, is used to treat headaches and other afflictions (Schneider 1974, 3:33*; Seitz 1994, 57).

Constituents

Guaraná is the strongest of all **caffeine** drugs. It is some three times stronger than coffee (*Coffea arabica*) and eight times as potent as maté (*Ilex paraguariensis*).

The seeds contain approximately 5% caffeine, 3% fatty oil, 9% tannic acids, 8% resin, 10% starch, 50% fiber, minerals (potassium, sodium, magnesium, calcium), some protein, sugar, and water. The oil is sometimes attributed with "hallucinogenic" properties, for example, in the package inserts of commercial guaraná products that are sold on the techno scene.

The leaves contain 0.38% caffeine and up to 1.2% theobromine (cf. *Theobroma cacao*). All the other parts of the plant also contain traces of caffeine and other purines (Seitz 1994, 54).

The active constituents and concentrations change as the seeds are made into the paste. Cyanogenic glycosides are presumably produced during this process. The paste contains 3.6 to 5.8% caffeine, 0.03 to 0.17% theobromine, 0.02 to 0.06% theophylline, up to 12% tanning agents (proanthocyanidin, [+]-catechin, [−]-epicatechin), saponins, seed fat, minerals, and water. A portion of the caffeine occurs in a complex bond with the tanning agents. This complex was formerly referred to as guaranine (Seitz 1994, 55).

Effects

Guaraná has potent stimulating effects due to the **caffeine** and the "guaranine." However, the overall effects are distinctly different from those produced by coffee and other purine drugs. It is thought that the long duration of the stimulant effects (in contrast to the relatively short duration of the effects of coffee) is due to the complex binding of caffeine to the tanning agents. Guaraná also has an inhibiting effect upon sensations of hunger and thirst. It has been characterized as "a harmless, mild antidepressant" (Straten 1996, 11). There have been many reports of aphrodisiac effects (Miller 1988, 57*; Straten 1996, 61). Curiously, some people do not experience any effects or paradoxically become tired, even with high dosages.

Commercial Forms and Regulations

To protect its own economy and its monopoly status, Brazil does not allow the export of viable seeds (Schröder 1991, 109*). Apart from this, guaraná is subject to the various laws regulating foods.

Guaraná is the main ingredient in numerous manufactured products that are freely available (supermarkets, body shops, health food stores, gas stations, et cetera). Buzz Gum, a sugar-free chewing gum made with guaraná extract, is sold in the United States. Gumdrops containing guaraná extract as well as **caffeine** and taurine are sold in Switzerland. In Germany, tea bags containing guaraná and maté (*Ilex paraguariensis*) have recently become available in stores.

Literature

See also the entries for *Paullinia* spp. and **caffeine**.

Berrédo Carneiro, Paulo E. de. 1931. *Le Guarana et Paullinia Cupana H.B.K.: Contribution à l'etude des plantes à caféine.* Paris: Jouve et Cie. Editeurs.

Erickson, H. T., Maria Pinheiro F. Corréa, and José Ricardo Escobar. 1984. Guaraná (*Paullinia cupana*) as a commercial crop in Brazilian Amazonia. *Economic Botany* 38 (3): 273–86.

Esteves Gondim, Carlos José. 1984. Alguns aspectos da biologia reprodutiva do guaranazeiro (*Paullinia cupana* var. *sorbilis* (Mart.) Ducke—Sapindaceae. *Acta Amâzonica* 14 (1–2): 9–38.

Henman, Anthony Richard. 1982. Guaraná (*Paullinia cupana* var. *sorbilis*): Ecological and social perspectives on an economic plant of the central Amazon basin. *Journal of Ethnopharmacology* 6:311–38.

Karlinger, Felix, and Elisabeth Zacherl. 1976. *Südamerikanische Indianermärchen*, Cologne: Diederichs.

Schultes, Richard Evans. 1979. The Amazonia as a source of new economic plants. *Economic Botany* 33:259–66.

Seitz, Renate. 1994. *Paullinia*. In *Hagers Handbuch der pharmazeutischen Praxis*, 5th ed., 6:52–59. Berlin: Springer.

Straten, Michael van. 1996. *Guarana: Energiespendende und heilkräftige Samen aus dem Amazonas-Regenwald.* Aarau and Stuttgart: AT Verlag.

Tschirch, A. 1918. Über den Bau der Samenschale von *Paullinia cupana* Kunth. *Schweizerische Apotheker-Zeitung* 56 (35): 445–47.

Walder, Patrick. 1995. Technodrogen. In *Techno*, ed. Philipp Anz and Patrick Walder, 192–97. Zurich: Verlag Ricco Bilger.

Paullinia spp.

Paullinia Species

Family

Sapindaceae (Soapberry Family); Subfamily Sapindoideae, Paullinieae Tribe

Several species of the genus *Paullinia* occur in the tropics of Central and South America. The genus encompasses some 150 to 200 species in all, some of which contain large quantities of **caffeine** (see *Paullina cupana*), while others (e.g., *Paullinia curruru* L.) have very toxic properties and have been used as fish poisons and to produce arrow poisons with curare-like effects (Blohm 1962, 67*; Millspaugh 1974, 167*; Schultes 1942, 314). Species in the very closely related genus *Serjania* also contain powerful fish poisons (Schultes 1942, 314).

Paullinia australis St.-Hil.

Indigenous to the forests of Brazil, this species is said to produce a toxic or psychoactive **honey** (Millspaugh 1974, 167*). The leaves and roots of Argentinean plants contain an alkaloid with sedative and narcotic properties (Schultes 1942, 314).

Paullinia carpopodea Camb.

In Brazil, the leaves of this species are used in folk medicine to treat pain. They may contain a narcotic or analgesic principle.

Paullinia emetica Schultes

The Karijona Indians of Colombia used a tea made from the leaves of this species medicinally and ritually as a potent emetic agent (Schultes 1977b, 119*). At the present, we can only conjecture as to whether the species also contains **caffeine** or has psychoactive properties.

Paullinia pinnata L. [syn. *Paullinia angusta* N.E. Br., *Paullinia nitida* Steud., *Paullinia pinnata* DC., *Serjania curassavica* Radlk.]—timbosipo, cururu apé, guaratimbo

In Amazonia, this evergreen climber is known as *sapo huasca* (*sapo* means "toad"!) and is pur-portedly used as an inebriant. The root cortex is said to contain a toxin ("timboine" or "timbol") with narcotic properties (Duke and Vasquez 1994, 132*; Millspaugh 1974, 167*; Schultes 1942, 314).

In Paraguay, where the plant is known as *erejna*, the Ayoré Indians use *Paullinia pinnata* L. to treat rheumatism (Schmeda-Hirschmann 1993, 107*). The vine is called *bejuco de zarcillo* ("earring vine") in Venezuela. The stalks and roots were used as fish poison and to commit suicide (Blohm 1962, 67*).

Few chemical studies of the plant have been conducted to date.

Paullinia yoco Schultes et Killip ex Schultes—yoco

This woody vine, which can grow to a very long length and can have a diameter of up to 12 cm, is found in southern Colombia (Putumayo) and Ecuador, where it is known as *yoco* or *yoko* (Schultes 1942). The Indians make a distinction among a number of forms: *yoco blanco*, *yoco colorado*, *huarmy yoco*, *taruco yoco*, *yagé yoco*, *canaguche yoco*, *verde yoco*, and others; however, these are not distinguished botanically (Schultes 1942; 1987, 527). The name *yagé yoco* suggests an association with *Banisteriopsis caapi* or a use of yoco as an additive in **ayahuasca**. The name *canaguche yoco* indicates the addition of yoco to a type of **chicha** known as *chicha de canaguche*, which is fermented from the fruits of *Mauritia minor* Burrett (cf. **palm wine**) (Schultes 1942, 312). The Kofán differentiate between two ecological forms with the names *to-to-oa-yoko*, "white yoco," and *cu-i-yoko*. The first form is preferred because it contains more latex (Schultes 1981, 23*). The use of yoco was discovered only in the 1920s (Schultes 1942, 309). The Inga, Kofán, and Coreguaje use yoco as a daily morning stimulant. These Indians never go on hunting expeditions or journeys without taking a supply of pieces of the vine (Schultes 1942, 322).

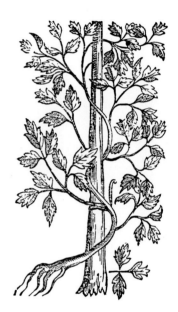

The Mexican plant *aquiztli* (*Paullinia fuscescens* H.B.K.) was the first species in the caffeine-producing genus *Paullinia* to be described and illustrated. (From Hernández, 1942/46 [Orig. pub. 1615]*)

"Yoco is undoubtedly the most curious caffeine-rich plant that people have bent to their use. A forest liana of the westernmost Amazon of Columbia, Ecuador, and Peru, it is the only species the bark of which is employed in the preparation of a stimulant drink. The liana is the most important non-food plant in the life of numerous tribes of Indians; when a local supply of the wild source is exhausted, the natives find it necessary to abandon their home-site and re-locate in another area where the plant is found in greater abundance. It appears that the liana is rarely or never cultivated, probably because it is extremely slow growing."

RICHARD EVANS SCHULTES
"A CAFFEINE DRINK PREPARED FROM BARK"
(1987, 527)

A milky latex that contains a very high level of **caffeine** flows through the bark. Yoco bark is drunk only in the form of a cold-water extract. The vine (epidermis, bark) is scraped, and the bark pieces and the caffeine-rich latex yield a mass that is then pressed in cold water. One dosage is prepared from 15 to 28 ounces (= 420 to 790 g) of scraped bark and a tree gourd (*jicara*) full of water (Schultes 1987). The effects are a powerful stimulation with tingling in the fingers. A general sensation of well-being and clear wakefulness manifests within a few minutes of consuming the drink. The appetite is profoundly and persistently suppressed. Most yoco users drink two *jicaras* in the morning right after rising and do not eat until the late afternoon (Schultes 1942, 323).

The bark is also added to the tobacco preparation known as *ambíl* (see **Nicotiana tabacum**) and is mixed with *Ilex guayusa.*

Yoco bark contains 2.73% **caffeine** (Rouhier and Perrot 1926). No other active constituents have been detected. The buds have also been found to contain caffeine (Schultes 1942, 313).

Literature

See also the entries for *Paullinia cupana* and **caffeine.**

Perrot, E., and Alexandre Rouhier. 1926. Le yocco, nouvelle drogue simple à caféine. *Comptes Rendus Hebdomadaires des Séances Acad. de Science* 182:1494–96.

Rouhier, A., and E. Perrot. 1926. Le "yocco," nouvelle drogue simple à caféine. *Bulletin de Science Pharmacologique* 33:537–39.

Schultes, Richard Evans. 1942. Plantae Colombianae II—Yoco: A stimulant of southern Colombia. *Botanical Museum Leaflets* 10 (10): 301–24.

———. 1987. A caffeine drink prepared from bark. *Economic Botany* 41:526–27.

Pausinystalia yohimba (K. Schum.) Pierre ex Beille

Yohimbé Tree

At the beginning of the twentieth century, yohimbé bark and yohimbine enjoyed great popularity in Germany as psychoactive aphrodisiacs. (Magazine advertisement, 1915)

Family
Rubiaceae (Coffee Family)

Forms and Subspecies
Presumably none

Synonyms
Corynanthe johimbe K. Schum.
Corynanthe yohimba K. Schum.
Corynanthe yohimbe K. Schum.
Pausinystalia macroceras Kennedy (non [Schum.] Pierre)
Pausinystalia johimbe (K. Schum.) Pierre ex Beille
Pausinystalia yohimbe Pierre

Folk Names
Johimbe, liebesbaum, lustholz, pau de cabinda (Portuguese), potenzbaum, potenzholz, potenzrinde, yohambine (Arabic), yohimba, yohimbe, yohimbé, yohimbébaum, yohimbehe, yohimbéhé (French), yohimbehon, yohimbene, yohimbe tree, yohimbé tree, yohumbe, yumbehoa

History
Since ancient times, the bark of this tree has been used in Africa as an aphrodisiac, especially among the Bantu peoples (Miller 1993, 70*). It is possible that the ancient Egyptians may have known about and even imported the bark through trade relations with western Africa. In Cameroon, this jungle tree has long been esteemed as an aphrodisiac and stimulant (Dalziel 1937).

The German chemist Spiegel isolated the alkaloid **yohimbine** from the bark in 1896, and the compound subsequently found use in Western medicine as a treatment for impotence and as a local anesthetic (Brown and Malone 1978, 20*; Schneider 1974, 3:34*). The tree was correctly botanically described in 1901 (Gilg and Schumann 1901, 94 f.). The bark is a source of alkaloids of pharmaceutical value (Oliver-Bever 1982, 39).

Distribution
The tree occurs in the tropical forests of Nigeria and Cameroon and in the Congo (Hutchinson and Dalziel 1963, 112).

Cultivation
The plant can be propagated either from seeds or from cuttings. Details, however, are lacking.

Appearance
This evergreen tree, which can grow to a height of 30 meters, somewhat resembles an oak. It has oval, attenuated leaves (7 to 13 cm long) and bushy inflorescences and produces winged seeds. The

light or gray-brown bark is 4 to 8 mm thick with both longitudinal and transverse fissures and is usually heavily overgrown with lichens.

The yohimbé tree is easily confused with *Pausinystalia macroceras* (K. Schum.) Pierre ex Beille and with some members of the genus *Corynanthe* (*Corynanthe* spp.).

Psychoactive Material

— Bark (cortex yohimbe, yohimbe cortex, yohimbehe cortex, yohimberinde, yohimbeherinde, potenzrinde, yohimbé bark)

The raw drug is apparently counterfeited with the bark of other *Pausinystalia* species and *Corynanthe* spp.:

It is interesting, by the way, that the natives in the French Congo used the bark of a tree that they called "endun" and that Pierre named "Pausinystalia trillesii" as an aphrodisiac. This bark also contains yohimbine; the tree itself probably belongs to the genus *Corynanthe*. (Hirschfeld and Linsert 1930, 172*)

Preparation and Dosage

Only the dried bark is used. It can be prepared as an extract in **alcohol** (tincture) or as a tea (cf. Geschwinde 1996, 146*).

To make tea, 6 teaspoons of yohimbe bark per person should be boiled with 500 mg of vitamin C for ten minutes and then sipped slowly (from Gottlieb 1974, 76*; Miller 1985, 117*).

The following ingredients can be used to decoct a tea for a "firm erection" (from Gottlieb 1974, 81*):

1 tablespoon yohimbe bark (cortex yohimbe)
1 teaspoon dita seeds (*Alstonia scholaris*), crushed
1 tablespoon cola nuts (*Cola* spp.), broken
1 tablespoon sarsaparilla bark (*Smilax* spp.)

The ingredients should be boiled in water for ten minutes and then sipped slowly.

The pharmaceutical industry uses yohimbé extracts to manufacture aphrodisiacs and medicines to treat impotence. These extracts are usually combined with **atropine**, *Turnera diffusa*, *Strychnos nux-vomica*, *Strychnos* spp., *Liriosma ovata*, or other substances.

The bark is also used in aphrodisiac **smoking blends** (Brown and Malone 1978, 20*). In western Africa, it was or is still used as an additive to iboga (see *Tabernanthe iboga*).

A "stimulant for the sexual organs" consists of ten drops of a 1% solution (Boericke 1992, 803*). For more on dosages, see **yohimbine**.

Ritual Use

It is likely that yohimbé was once used in western Africa as an initiatory drink in fetish and ancestor cults and in initiations into secret societies. It has been conjectured that yohimbé was used together with *Tabernanthe iboga* in the *bieri* cults of the Fang. Unfortunately, no detailed evidence has come down to us (cf. *Alchornea* spp.).

Today, yohimbé is used chiefly in North America but also in Europe for sexual magical rituals that borrow from Indian Tantra and the techniques of various occultists (Aleister Crowley). Miller (1985, 117 ff.*) has recommended using yohimbé as a sacrament for pagan wedding ceremonies.

Artifacts

See *Tabernanthe iboga*.

Medicinal Use

In Cameroon, yohimbé is used in folk medicine to treat impotence resulting from witchcraft (cf. Amrain 1907).

Preparations containing yohimbé are used in modern phytotherapy and in Western medicine to treat frigidity and impotence and are also used in veterinary medicine (Pahlow 1993, 484*). In homeopathy, Yohimbinum is regarded as an alternative for *Nuphar lutea*. It

arouses the sexual organs, affects the central nervous system and the respiratory center. Is an aphrodisiac in physiological dosages, but is contraindicated for all acute and chronic inflammations of the abdominal organs. Homeopathically, it is said to be able to help with congestive conditions of the sexual organs. Causes hyperemia of the mammary glands and stimulates milk production. (Boericke 1992, 802*)

Constituents

The bark of the trunk of trees that are older than fifteen to twenty years contains 2 to 15% indole alkaloids: yohimbine (= corynine, quebrachine), β-yohimbine (= corynanthidine, isoyohimbine, mesoyohimbine, rauwolscine), β-yohimbine (= amsonine), yohimbinine, corynanthine (= rauhimbine), corynanthein, dihydrocorynanthein, alloyohimbine (= dihydroyohimbine), pseudoyohimbine,

The coarsely chopped bark (cortex yohimbe) of the West African yohimbé tree (*Pausinystalia yohimba*) contains the potent aphrodisiac yohimbine as well as numerous other indole alkaloids.

"In Africa, the black sorcerers have their disciples drink johimbe and iboga to produce the inebriation required to prepare them for the great fetishistic initiations. The candidate ingests a great quantity of iboga, whether in the natural state or as a decoction. Shortly thereafter, all of his nerves tense up in an extraordinary manner; an epileptic fit comes over him, while he unconsciously utters words that, when they are picked up by the initiated, have a prophetic meaning and demonstrate that the fetish dwells within them."

ALEXANDRE ROUHIER
DIE HELLSEHEN HERVORRUFENDEN PFLANZEN [THE PLANTS THAT INDUCE CLAIRVOYANCE]
(1927 = 1986*)

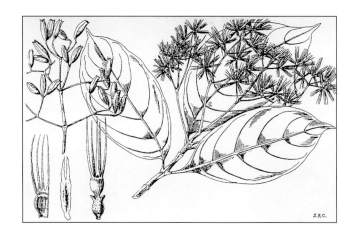

Yohimbine

"The warlike Masai of East Africa call their ultra-hard ritual drug 'motoriki' or simply 'ol motori,' the soup. It is cooked from the bark of the yohimbé tree together with the roots of *Acokanthera* [sp.; an Apocynaceae]—which also provide an arrow poison.

"Archaic drug rituals almost always include an animal sacrifice. Among the Masai, a bull is killed on such occasions; they catch its blood in the opened dewlap and mix it into the finished brew of bark and root pieces. The motoriki drink produces an epilepsy-like tetanus in which the Morani—the young Masai warriors—are visited by terrible sights in which they fight with demons and wild animals. The horror visions are so strong that the inebriated men must be watched over and held on to so that they will not injure themselves or others. And yet there are repeatedly deaths from people running amok or respiratory paralysis of the intoxicated. Whoever survives this inebriation will no longer fear anything."

PETER LEIPPE
GEGENWELT RAUSCHGIFT: KULTUREN UND IHRE DROGEN [COUNTER-WORLD INEBRIANT: CULTURES AND THEIR DRUGS]
(1997, 21 f.*)

Botanical illustration of the kinkele tree (*Pausinystalia macroceras*). In the Congo and Gabon, its bark is also called yohimbé, and it is used for aphrodisiac purposes.

tetrahydromethylcorynanthein, and ajmalicine (Oliver-Bever 1982, 39; Paris and Letouzey 1960; Poisson 1964; Roth et al. 1994, 544*). The average yohimbine content in commercial material (cortex yohimbe) ranges from 1.67 to 3.4% (Roth et al. 1994, 545*). Apart from the alkaloids, the bark also contains tannic acid and a coloring agent (Pahlow 1993, 484*).

Effects

Yohimbé bark is reputed to be hallucinogenic (Schultes and Farnsworth 1982, 189*). The psychoactive effects are due primarily to the main active constituent, **yohimbine** (Roth et al. 1994, 545*). Yohimbine has sympatholytic effects and local anesthetic effects like those of **cocaine**; it also has vasodilating effects, particularly upon the sex organs (Oliver-Bever 1982, 40). Yohimbine interacts with other psychopharmacological agents (Roth et al. 1994, 544*). Overdoses can be very unpleasant (Sacha Runa and Lady Sanna 1995). For information about psychoactivity, see **yohimbine**.

Preparations of the bark usually produce only mild or subtle effects.

Commercial Forms and Regulations

The bark (cortex yohimbe) is available without restriction, while the pure alkaloid requires a prescription. In the United States, pharmaceutical extracts of yohimbé bark, often in a tincture together with saw palmetto fruits (*Serenoa repens*; cf. *Turnera diffusa*, **palm wine**, **wine**), are offered in health food stores and are not subject to any restrictions.

Literature

See also the entries for *Corynanthe* spp., *Alchornea* spp., and **yohimbine**.

Amrain, Karl. 1907. Die Stärkung männlicher Kraft. *Anthropophyteia* 4:291–93.

Chaurasia, Neera. 1992. Corynanthe. In *Hagers Handbuch der pharmazeutischen Praxis*, 5th ed., 4:1029–32. Berlin: Springer.

Dalziel, J. M. 1937. *The useful plants of west tropical Africa*. London: Crown Agents.

Gilg, E., and K. Schumann. 1901. Über die Stammpflanze der Johimberinde. *Notizblatt des Königl. botanischen Gartens und Museums zu Berlin* 3 (25): 92–97.

Hutchinson, J., and J. M. Dalziel. 1963. *Flora of west tropical Africa*. 2nd ed. Vol. 2. London: Crown Agents for Oversea Governments and Administrations.

Oliver-Bever, B. 1982. Medicinal plants in tropical West Africa. *Journal of Ethnopharmacology* 5 (1): 1–71.

Paris, R., and R. Letouzey. 1960. Répartition des alcaloïdes chez le *Johimbe. Journal d' Agriculture Tropicale et de Botanique Appliquée* 7:256.

Poisson, J. 1964. Recherches récentes sur les alcaloïdes de Pseudocinchona et du *Yohimbe. Annales de Chimie (Paris)* 9:99–121.

Sacha Runa and Lady Sanna. 1995. Yohimbe-Rinde—Überdosis. *Entheogene* 5:12–13.

Peganum harmala Linnaeus

Syrian Rue, Harmel

Family
Zygophyllaceae[276] (Caltrop Family)

Forms and Subspecies
None

Synonyms
None

Folk Names
Aspand (Kurdish), besasa (Egypt, "plant of Bes"), churma, epnubu (Egypt), gandaku, haoma (Persian), harmal, harmale, harmalkraut, harmal rutbah (Arabic/Iraq), harmel, harmelkraut, harmelraute, hermel, hermelkraut, hermelraute, hom (Persian), kisankur, moly, mountain rue, pegano, pégano, peganon, sipand (Persian), steppenraute, Syrian rue, syrische raute, techepak (Ladakhi), tukhm-i-isfand, uzarih (Turkish), wilde raute

History
Syrian rue is likely an ancient ritual plant that was used for psychoactive purposes from an early date. It may have been the **haoma** of ancient Persia and was in any case later used as a substitute for it. The plant appears in the ancient literature (Dioscorides) under the name *peganon*. The name *peganon* (also *peganum*) may have been derived from that of Pegasus, the winged horse of ancient mythology that was begotten by Poseidon, the god of the sea, and the dying Medusa. The plant has also been interpreted as the legendary magical plant known as **moly**.

The seeds, which are used both medicinally and ritually, were imported from Persia to India by the Muslims at an early date (Hooper 1937, 148*). The plant was present in central Europe by the fifteenth century at the latest and was portrayed by the "fathers of botany." In the Near East and North Africa, Syrian rue has retained its great significance as a ritual **incense** into the present day. In North America and central Europe, it is

increasingly being used to produce **ayahuasca analogs**.

Distribution
Syrian rue is distributed from the eastern Mediterranean region over northern India (Rajasthan) and into Mongolia and Manchuria (Schultes 1970, 25*). It is common in Yemen and the Negev Desert. In southern Europe, the plant can occasionally be found on Cyprus and less frequently in Greece (Sfikas 1990, 140*).

Cultivation
Although Syrian rue thrives without problem in extreme desert climates, cultivation of the plant is often difficult. The plant almost never grows in central Europe. It is more likely that suitable soils and climates are found in southern Europe. In California, cultivation is relatively simple. There, the seeds are simply broadcast onto normal, moistened potting soil and gently pressed in. When there is sufficient—but not too much—sunlight and the air is warm, some of the seeds will germinate. Each can be repotted as a seedling as soon as its root has become strong enough.

Growing larger quantities of Syrian rue requires considerable experience in and knowledge of gardening.

Appearance
This herbaceous perennial, which can grow as tall as 50 to 100 cm, has ramified roots and many thin stalks. The multifid leaves are opposite and have a fragile appearance. The white flowers have five petals and ten yellow pistils. The delicate, solitary flowers are at the ends of long stalks attached to the leaf axils. The trilocular fruit is round and acquires a reddish color as it matures. It contains numerous gray or almost black triangular seeds (up to 3 mm long).

The herbage begins to develop new shoots in February, and flowers appear in April. The plant is green in early summer and develops fruits in July.

An old illustration of Syrian rue. (Woodcut from the herbal of Matthiolus, 1627)

Left: Syrian rue (*Peganum harmala*), displaying a mature fruit.

Right: The seeds of *Peganum harmala*, known as harmel or hermel, contain large amounts of **harmaline and harmine** and are thus very well suited for making ayahuasca analogs. (Photograph: Karl-Christian Lyncker)

276 The plant was formerly assigned to the Family Rutaceae.

The herbage has a characteristic scent that some people find unpleasantly obtrusive.

Psychoactive Material

— Seeds (semen harmalae, semen rutae silvestris, harmalae semen, harmalasamen, harmala seeds, harmel seeds)

Preparation and Dosage

Burning the dried seeds as a fumigant is by far the most common use of *Peganum harmala*. The seeds are simply thrown onto glowing wood or charcoals, sometimes in combination with other substances (cf. **incense**). The seeds also are used as an ingredient in **smoking blends**, e.g., with hashish (cf. *Cannabis indica*). In Ladakh, the seeds are toasted on a red-hot iron plate, very finely ground, and then smoked, either alone or mixed with tobacco (*Nicotiana tabacum*) (Navchoo and Buth 1990, 320*). An especially effective smoking mixture can be obtained from 15 g of seeds and the juice of a lemon. The ground seeds are boiled carefully in some water and the lemon juice until a paste results. Mixed with tobacco and smoked, this is said to produce inebriating and aphrodisiac effects.

In Pakistan, 5 to 10 g of unground seeds are ingested with water after a meal for internal purposes (Goodman and Ghafoor 1992, 25*). In Morocco, the seeds are added to wine (*Vitis vinifera*) to make harmel wine. The powdered seeds are taken as a **snuff** to produce a "clear mind" (Vries 1984*).

Up to 20 g of powdered seeds have been ingested for psychoactive purposes, but such a dosage can have severe toxic effects. "Over 4mg/kg (oral) of the two substances [**harmine** and **harmaline**] have hallucinogenic effects on humans" (Roth et al. 1994, 548*). Some 3 to 4 g of crushed seeds (approximately one teaspoonful) is effective as an MAO inhibitor for activating the DMT in **ayahuasca analogs**. (By the way, it is not necessary to swallow the seeds or, in gelatin capsules, the powder. This amount can be extracted in cold water, and the solution can be used.)

In recent years, many psychonauts have experimented with adventurous combinations of harmel seeds and other psychoactive substances (*Lophophora williamsii*, *Phalaris arundinacea*, *Psilocybe* spp., *Trichocereus pachanoi*, mescaline, LSD, et cetera) (DeKorne 1995*; Turner 1994*). Great care should be exercised when conducting such self-experiments. It is better to use too little than too much!

Ritual Use

If Syrian rue was in fact the original **haoma** plant, then its greatest ritual significance would have been in pre-Zoroastrian Persia. It is possible that the plant may also have been the secret inebriant used in the ancient mystery cult of Mithra.

Dioscorides noted that the late ancient Syrians knew the plant as *besasa*, a name that is usually interpreted to mean "plant of Bes." Bes was a misshapen god of dwarf form and with the face of an old man who was especially beloved among the simple Egyptian people, for he was a protector spirit who turned away all evil. His image, in the form of small amulets, was affixed to head rests, beds, mirrors, and cosmetic jars. The small figures of Bes were fumigated with Syrian rue seeds to promote their apotropaic powers.

Syrian rue was a sacred plant in the ancient Orient. The Koran states, "Every root, every leaf of harmel, is watched over by an angel who waits for a person to come in search of healing." For this reason, it is said that dervishes in Buchara also esteem and ritually utilize harmel seeds for their inebriating effects.

Syrian rue seeds, in the form of small incense balls (*sepetan*), are still offered by burning great quantities during Nouruz (New Day), the ancient Iranian and now Islamized spring and new year's festival (on March 21, the vernal equinox). The ascending smoke is distributed throughout the entire house to keep away all misfortune (Schlegel 1987). In Persia (Iran, Iraq), the seeds are scattered over glowing coals at weddings to ward off evil spirits and the evil eye. It is said that the smoke is also capable of dispelling epidemic diseases (Hooper 1937, 148*).

In Baluchistan (Pakistan), the seeds are used to neutralize the enchantments of a *jin* (jinn) and to banish all evil spirits in general. A person who has fallen under the spell of or has been possessed by a *jin* is urged to inhale as much as possible of the smoke rising from the crackling seeds on the charcoals. It is said that such a treatment usually results in a rapid improvement (Goodman and Ghafoor 1992, 25*). Harmel is also used as a fumigant in Turkey to counteract the effects of the evil eye.

In North Africa, Syrian rue has been regarded as a magical and medicinal panacea since ancient times. The seeds are used as **incense**, both alone and in combination with other substances. The seeds are scattered over charcoal to dispel evil spirits. The smoke is inhaled to treat headaches, the consequences of the evil eye, and venereal diseases. In Morocco, an incense of Syrian rue seeds, alum, and olibanum (*Boswellia sacra*) is burned during the wedding night to fan the flames of desire (vries 1985). Fumigations of harmel seeds and alum, performed to defend against demons, are said to produce inebriating effects.

In the Himalayas and neighboring regions, shamans use the seeds as a magical incense. The shamans of the Hunza, who live in what is now Pakistan, inhale the smoke to enter a clairvoyant trance. The shamans (*bitaiyo*) then enter into a close, lusty sexual contact with the divining fairies, who give them important information and the ability to heal (cf. *Juniperus recurva*).

Artifacts

It is possible that Syrian rue may have played a role in the development of the floral patterns and elements of Islamic art. If the plant was indeed used in the haoma cult or the Mithraic mysteries, it likely would have inspired numerous cult images.

In modern Iran, fruit capsules of *Peganum harmala* are strung together and attached to articles of clothing as protective amulets (*panja*).

Medicinal Use

Traditionally, Syrian rue seeds have been used primarily for gynecological purposes. In India, the seeds (*hurmur, lahouri, marmara*) are burned as an **incense** to ease the birthing process. In Indian folk medicine, the seeds (*harmal, is-band*) are regarded as an aphrodisiac, an asthma remedy, and an agent for treating menstrual difficulties. In Pakistan, infertile women and women with severe labor or uterine pains are fumigated with the seeds; in addition, special pipes are used to blow the smoke directly into the vagina (Goodman and Ghafoor 1992, 24f.*; Hassan 1967). Ill people generally inhale the smoke for all manner of afflictions. A tea made with 5 to 10 g of seeds is drunk after a meal to control flatulence (Goodman and Ghafoor 1992, 54*). In Rajasthan, the smoke of glowing seeds is used as an antiseptic agent to fumigate wounds (Shah 1982, 301*).

In the folk medicine of Asia Minor and central Asia, preparations of Syrian rue are used as aphrodisiacs. In Persia, the seeds were regarded as a purifying medicine and an aphrodisiac (Hooper 1937, 148*), and are used in Iran today. The Bedouins of the Negev Desert use the plant to promote menstruation and to induce abortions (Bailey and Danin 1981; Shapira et al. 1989). The herbage is traditionally used to treat unusual skin disorders (El-Rifaie 1980). A decoction of the seeds is drunk for stomach ailments, heart problems, and sciatica. A strong decoction can act as a tranquilizer.

Constituents

The herbage and seeds contain the β-**carbolines** harmine, harmaline, and related bases, e.g., harmalol and harmidine (Al-Shamma et al. 1981; Degtyarev et al. 1984). Also present are quinazoline alkaloids with a similar structure: (−)-vasicine, (±)-vasicine, vasicinone, pegaline, tetrahydroharman, and desoxyvasicinon. The alkaloid content of the seeds can vary between 2 and 6% (Roth et al. 1994, 548*). In addition to the β-carbolines, Syrian rue herbage contains a pleasantly scented **essential oil** that, when used in massage oils, has a relaxing effect upon the musculature. Vitamin C and fatty acids are also present.

Effects

The seeds have antidepressive effects and stimulate the imagination. There are reports of dreamlike states following the ingestion of larger quantities. The alkaloids as well as the total extract of the seeds act as MAO inhibitors, i.e., they suppress the excretion of the endogenous enzyme monoamine oxydase (= MAO), which metabolizes certain endogenous neurotransmitters (serotonin) as well as foreign toxins. This makes it possible for certain substances (*N,N*-DMT, 5-MeO-DMT, β-phenethylamines) to be orally efficacious (cf. *Banisteriopsis caapi*, **ayahuasca**, **ayahuasca analogs**).

Studies conducted at the University of Kansas (Lawrence) have shown that the harmine present in Syrian rue seeds has antibiotic effects on microorganisms (microbes) (Al-Shamma et al. 1981; cf. also Harsh and Nag 1984). In animal studies, an extract of the stems and leaves has been demonstrated to have abortifacient and antifertility properties (Shapira et al. 1989).

It is possible that the β-carbolines may pass into the smoke and can thus potentiate the effects of other smoked substances (e.g., **THC**).

Commercial Forms and Regulations

The seeds are sold freely and can be obtained from nurseries and sources specializing in ethnobotanical supplies. In California, Syrian rue is considered a noxious weed and may not be imported or sold. Apart from this, the plant is not subject to any restrictions.

Literature

See also the entries for **ayahuasca analogs** and **harmine** and **harmaline**.

Bailey, C., and A. Danin. 1981. Bedouin plant utilization in Sinai and the Negev. *Economic Botany* 35:145–62.

Degtyarev, V. A., Y. D. Sadykov, and V. S. Aksenov. 1984. Alkaloids of *Peganum harmala*. *Chemistry of Natural Compounds* 20 (2): 240–41.

Fritzsche, J. 1847. Bestandtheile der Samen von *Peganum harmala*. *Justus Liebig's Annalen der Chemie* 64:360–64.

Haas, Volkert. 1977. *Magie und Mythen im Reich der Hethiter: 1. Vegetationskulte und Pflanzenmagie.* Hamburg: Merlin Verlag.

Harsh, M. I., and T. N. Nag. 1984. Anti-microbial principles from *in vitro* tissue culture of *Peganum harmala*. *Journal of Natural Products* 47 (2): 365–67.

Hassan, I. 1967. Some folk uses of *Peganum harmala* in India and Pakistan. *Economic Botany* 21:384.

Kashimov, H. N., M. V. Teclezhenetskaya, N. N. Sharakhimar, and S. Y. Yunasov. 1971. The dynamics of the accumulation of alkaloids in

Harmine

Harmaline

"Harmel rue, Syrian rue, Peganum harmala. A plant of southern Europe and the Orient. All parts of it have a strong and unpleasant scent. The seed is three-sided and tastes very bitter. . . . According to Pseudo-Aristotle, this rue species was eaten in ancient times as a remedy against fascination. In modern Greece, women hang Peganum on their little ones' heads as an amulet. On Cephalonia, little pieces of it are placed in children's swaddling clothes.—In Morocco, the seeds are carried in amulet bags to protect against the evil eye and the djnûn. It is also used as a fumigant.—In Punjab, the seeds are mixed with bran and salt and burned to ward off the evil eye, the djnûn, and such things."

SIEGFRIED SELIGMANN
DIE MAGISCHEN HEIL- UND SCHUTZMITTEL AUS DER BELEBTEN NATUR [THE MAGICAL HEALING AND PROTECTIVE AGENTS FROM THE ANIMATED NATURE]
(1996, 121f.*)

Peganum harmala. Chemistry of Natural Compounds 3:364–65.

Müller, K. O. 1932. Über die Verbreitung der Harmelstaude in Anatolien und ihre Bindung an die menschlichen Wohnstätten. *Berichte der Deutschen botanischen Gesellschaft* 274 (Berlin).

Munir, C., M. I. Zaidi, A. Nasir, and Atta-Ur-Rahman. 1995. An easy rapid metal mediated method of isolation of harmine and harmaline from *Peganum harmala*. *Fitoterapia* 66:73–76.

Quedenfeld, M. 1887. Nahrungs- , Reiz- und kosmetische Mittel bei den Marokkanern. *Zeitschrift für Ethnologie* 19:241–84.

Rifaie, M. el-. 1980. *Peganum harmala*: Its use in certain dermatoses. *International Journal of Dermatology* 19 (4): 221–22.

Schipper, A., and O. H. Volk. 1960. Beiträge zur Kenntnis der Alkaloide von *Peganum harmala*. *Deutsche Apotheker-Zeitung* 100:255–59.

Schlegel, Christiane. 1987. Nouruz: Das Neujahrs- und Frühlingsfest der Iraner. Unpublished seminar paper, Universität Bremen.

Shamma, A. al-, S. Drake, D. L. Flynn, L. A. Mitscher, Y. H. Park, G. S. R. Rao, A. Simpson, J. K. Swayze, T. Veysoglu, and S. T.-S. Wu. 1981. Antimicrobial agents from higher plants: Antimicrobial agents from *Peganum harmala* seeds. *Journal of Natural Products* 44 (6): 745–47.

Shapira, Zvia, J. Terkel, Y. Egozi, A. Nyska, and J. Friedman. 1989. Abortifacient potential for the epigeal parts of *Peganum harmala*. *Journal of Ethnopharmacology* 27:319–25.

vries, herman de. 1985. hermel, harmel, harmal, peganum harmala, die steppenraute, ihr gebrauch in marokko als heilpflanze und psychotherapeutikum. *Salix* 1 (1): 36–40.

"The alkaloid-rich seeds (harmine and harmaline) [of Syrian rue] are used because of their sudoriferous effect to heal febrile diseases of a general nature. The seeds are roasted and the fumes are inhaled. They also serve as an anthelmintic. In addition, they are used to spicen foods. It is said that they are esteemed among the dervishes in Buchara for their inebriating effect."

K. O. MÜLLER
(1932)

Pelecyphora aselliformis Ehrenberg

False Peyote, Peyotillo

The rare Mexican hatchet cactus (*Pelecyphora aselliformis*) produces peyote-like effects.

Family
Cactaceae (Cactus Family)

Forms and Subspecies
None

Synonyms
None

Folk Names
Asselkaktus, falscher peyote, false peyote, hatchet cactus, peotillo, peyote (see *Lophophora williamsii*), peyote meco, peyotillo, piote

History
Indians of northern Mexico once used this relatively rare cactus in a similar manner to or as a substitute for peyote (see *Lophophora williamsii*).

The first botanical description of the psychoactive cactus was made by the Berlin physician and botanist Christian Gottfried Ehrenberg (1795–1876). A powder of the cactus was formerly sold in Paris under the name *poudre de peyote*, "peyote powder."

Distribution
This cactus occurs only in northern Mexico (San Luis Potosí) (Preston-Mafham 1995, 167*; Zander 1994, 422*).

Cultivation
The plant is propagated from seeds, which are planted in the same manner as those of *Lophophora williamsii*.

Appearance
This solitary cactus can grow to a height of 10 cm. It has a round form with lateral, flattened tubercles that are arranged in a spiral fashion and have scalelike pectinate spines. Because of this, the cactus sometimes resembles a deeply convoluted brain. The flowers are up to 3 cm across and are bright violet. The fruits are red pods.

Peyotillo can easily be confused with the closely related species *Pelecyphora strobiliformis* (Werderm.) Kreuz. [syn. *Ariocarpus strobiliformis* Werderm., *Encephalocarpus strobiliformis* (Werderm.) Berger; cf. *Ariocarpus fissuratus*], which is found in Nuevo León (Mexico) (Preston-Mafham 1995, 167*). Another very similar species is *Pelecyphora pseudopectinata* Backeb. [syn. *Neolloydia pseudopectinata* (Backeb.) Anderson, *Turbinicarpus pseudopectinatus* (Backeb.) Glass et Foster]; in Tamaulipas, this cactus is also called peyote (Díaz 1979, 90*). *Turbinicarpus valdezianus* (Moell.) Glass et Foster [syn. *Pelecyphora valdezianus* Moell.] is also quite similar, but it is smaller (growing to a height of only 2.5 cm) and occurs in Coahuila (Preston-Mafham 1995, 194*).

Psychoactive Material
— Fresh or dried cactus flesh (buttons)

Preparation and Dosage
The flesh of the cactus (the aboveground portion or the head) can be eaten fresh or dried. No information concerning dosages is known.

Ritual Use
Only as a peyote substitute (see *Lophophora williamsii*)

Artifacts
See *Lophophora williamsii*.

Medicinal Use
See *Lophophora williamsii*.

Constituents
The cactus contains hordenine, anhalidine, pellotine, 3-dimethyltrichocerine, some **mescaline**, *N*-methylmescaline, and other β-**phenethylamines**

(Mata and McLaughlin 1982, 110*; Neal et al. 1972).

Effects
One cactus, eaten fresh, is said to produce peyote-like effects (cf. *Lophophora williamsii*). Although the effects are not quite as dramatic, they do include the typical visual changes and phenomena (William Emboden, pers. comm.).

Commercial Forms and Regulations
This rare cactus is almost never found in the international cactus trade. It may be possible to obtain seeds from ethnobotanical mail-order suppliers.

Literature
See also the entry for *Lophophora williamsii*.

Neal, J. M., P. T. Sato, W. N. Howald, and J. L. McLaughlin. 1972. Peyote alkaloids: Identification in the Mexican cactus *Pelecyphora aselliformis* Ehrenberg. *Science* 176:1131–33.

The rare and beautiful *Pelecyphora aselliformis* cactus was used as a peyote substitute in Mexico. (Copperplate engraving, colorized, from Jacques Brosse, *Great Voyage of Discovery*, 1764–1843)

Petroselinum crispum (Miller) Nyman ex A.W. Hill

Parsley

Family
Apiaceae (Umbelliferae) (Carrot Family); Subfamily Apioideae, Amminae Tribe

Forms and Subspecies
The species is divided into two subspecies (Frank 1994, 105):

Petroselinum crispum ssp. *crispum* (leaf parsley; has a smooth-leaved and a crisp-leaved form, as well as three chemotypes [see "Constituents"])
Petroselinum crispum ssp. *tuberosum* (Bernh. ex Rchb.) So (root parsley, parsley root)

Synonyms
Apium hortense E.H.L. Krause
Apium laetum Salisb.
Apium petroselinum L.
Apium romanum Zuccagni
Apium vulgare Druce
Carum petroselinum Benth. et Hook.
Helosciadium oppositifolium Reuss
Ligusticum levisticum Elsmann
Petroselinum hortense Hoffm.
Petroselinum macedonicum (Lonitzer) Bubani
Petroselinum petroselinum Karst.
Petroselinum sativum auct. non. Hoffm.

Petroselinum sativum Hoffm.
Petroselinum vulgare Kirschl.
Selinum petroselinum E.H.L. Krause
Sium oppositifolium Kit.
Wydleria portoricensis DC.

Folk Names
Apio ortense (Italian), apium, bittersilche, elixanter, gartenpetersilie, jaubert, maghdunes (Iraq), oxillatrum, parsley, perejil (Spanish), persil, peterchen, peterlein, peterling, peterselie (Dutch), petershiljen, petersilie, petersilienkraut, petersill, petersillig, petroselino, petrosella, pitar saleri (Hindi), prezzemolo, silk, tukhm-i-kalam (Persian)

History
It is possible that Dioscorides described parsley under the name *sison* as a seed that was savored in Syria (3.57). Whether the ancient Egyptians used the plant is a subject of debate (Germer 1985, 144 f.*). One of the earliest descriptions of parsley mentions a psychoactive property: "It produces seriousness in the mind of a person" (Hildegard von Bingen, *Physica* 1.68). It has been listed as a medicine in all pharmacopoeias since the Middle Ages (Schneider 1974, 3:43*).

The chief significance of parsley is culinary; it is

"Following the ingestion of a larger quantity of the essential oil [of parsley], first a central state of arousal, then inebriated states are possible (an effect of the hallucinogenic myristicin?)."

Franz-Christian Czygan
Petersilienfrüchte [Parsley Fruits]
(1989, 369)

"Parsley was already being planted in the gardens of German farmers during the early Middle Ages. . . . It plays a great, often shrouded role in folk eroticism. First, it is regarded as an aphrodisiac for the man because of its strong aroma. It is . . . attributed with powerful arousing effects."

MAGNUS HIRSCHFELD AND RICHARD LINSERT
LIEBESMITTEL [APHRODISIACS] (1930, 156*)

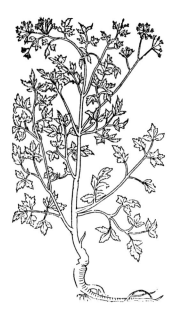

Wild parsley is said to be more psychoactive than garden parsley. (Woodcut from Fuchs, *Läebliche abbildung und contrafaytung aller kreüter*, 1545)

used as a kitchen spice, soup seasoning, and aromatic substance (including for alcoholic beverages; cf. **alcohol**). In the history of psychoactive substances, the plant is of only minor importance. It may have been an ingredient in **witches' ointments** and **theriac**. It was often used as a **beer** additive. Since the 1960s, the dried herbage has been smoked as a marijuana substitute (cf. *Cannabis indica*). The root is sometimes used as an ingredient in **incense**, while parsley oil is used in the (illegal) manufacture of psychoactive **phenethylamines** of the MDA or MDMA type (see *Myristica fragrans*, **herbal ecstasy**; Shulgin and Shulgin 1991*).

Distribution
Parsley is thought to have originated in the Mediterranean region. As a result of cultivation, it is now found throughout the world and has become wild in some areas.

Cultivation
Parsley is very easily grown from seed. The seeds need only be broadcast onto a bed of good topsoil and watered.

Appearance
This biennial fragrant plant has pinnate, incised leaves, a smooth stalk, and a spindle- or turnip-shaped vertical root. The root of the subspecies *tuberosum* is substantially thicker and more bulbous than that of the rest of the species. The white umbel, which grows from the center of the branching stalk, does not appear until the second year. For this reason, most hobby gardeners are unfamiliar with flowering parsley. The flowering period is from June to July. The gray-brown, 2 to 3 mm long fruits mature on the ten- to twenty-flower pedicels, which are arranged on the umbel in a radial manner.

Parsley can be confused with the only other member of the genus, *Petroselinum segetum* (L.) Koch. It is also very similar to the toxic dog parsley (*Aethusa cynapium* L.) and poison hemlock (*Conium maculatum*) (Frank 1994, 106).

Psychoactive Material
— Herbage (petroselini herba, folia petroselini, herba petroselini, parsley leaves), fresh or dried

Parsley (*Petroselinum crispum*) is not normally associated with psychoactive substances. But it does contain powerful active constituents.

— Seeds (semen petroselini, petroselini fructus)
— Parsley fruit oil (petroselinum aetheroleum e fructibus, oleum petroselini, parsley seed oil, grünes apiol, apiolum)
— Root (petroselini radix, radix petroselini, parsley root)

Preparation and Dosage
The subspecies *crispum* is used primarily for its herbage, while the subspecies *tuberosum* is used chiefly for its root.

A daily medicinal dose is regarded as 6 g of the dried herbage (Frank 1994, 115). For the ingestion of powdered parsley fruits, a therapeutic single dose is 1 g. For a cold- or hot-water extract, a daily dose is listed as 1 to 3 g of seeds crushed shortly before being steeped (112). A hot-water extract or infusion should be allowed to steep for five to ten minutes.

Parsley fruit oil is obtained by distilling the mature fruits. The composition of the oil varies depending on the chemical race (see "Constituents"). As a result, the different oils have correspondingly different applications and dosages. The oil of the apiol race is used to induce abortions. For this purpose, either a single dose of up to 10.8 g or a daily dose of 1 g for one to two weeks is ingested (Frank 1994, 109). Only the oil of the myristicin race can be used for psychoactive purposes (cf. *Myristica fragrans*). Unfortunately, no reliable information regarding dosages is available.

Ritual Use
Parsley herbage played a magical and apotropaic role in the customs of central Europe:

> In Moravia, the plant makes the influence of witches upon cows ineffective if it is sown between the 24th and the 26th of June. In many communities, a wreath of parsley is placed on a child's head on its first birthday, for it has then survived the most dangerous time. According to a widely held superstition, pulling a parsley root from the ground will bring death to that person who was thought of when it was planted. In Galacia, the Ruthenian bride carries bread and parsley on the way to the church so as to ward off evil spirits. Garlic and parsley are tied to the linen cloth under which a woman in labor lies in order to protect her from magic. (Schöpf 1986, 124*)

Artifacts
None

Medicinal Use
Parsley herbage is used in folk medicine to purify the blood and to treat diseases of the urinary tract.

In homeopathy, both an essence of the fresh

herbage—Petroselinum–Petersilie (Petroselinum crispum hom. *HAB1*, Petroselinum sativum hom. *HPUS88*)—and a tincture made from the mature fruits—Petroselinum e seminibus—are used (Schneider 1974, 3:43*).

Constituents

The entire plant contains an **essential oil** consisting of myristicin, *p*-apiol (= parsley camphor), monoterpenes, and sesquiterpenes. The seeds contain the highest concentration of essential oil (2 to 6%; average 2.7%) (Czygan 1989, 268; Fühner 1943, 240*; Roth et al. 1994, 552*). Three chemotypes (chemical races) have been distinguished on the basis of the principal constituents of the essential oil of the mature fruits (Frank 1994, 106; Warncke 1992):

— Myristicin race, with 49 to 77% myristicin, 0 to 3% apiol, and 1 to 23% allyltetramethoxybenzol
— Apiol race, with 58 to 80% apiol, 9 to 30% myristicin, and up to 6% allyltetramethoxybenzol
— Allyltetramethoxybenzol race, with 50 to 60% allyltetramethoxybenzol, 26 to 37% myristicin, and traces of apiol

The essential oil of the root of the subspecies *tuberosum* is composed chiefly of apiol (principal constituent), β-pinene, and myristicin but has traces of elemicine, limonene, bisabolene, sesquiphellandrene, and germacrene-A (Czygan 1989, 370 f.; Frank 1994, 116). The herbage contains flavones (apiine) and furanocoumarin (cf. **coumarins**). The fruits are rich in a fatty oil (petroselinic acid). The roots contain polyacetylene and furanocoumarin. Parsley herbage has a high vitamin C content (165 mg per 100 g) and also contains nicotine amide and considerable potassium (1%).

Effects

The essential oil of the apiol race has powerful abortive effects (Fühner 1943, 240*) and also can induce coma (Frank 1994, 109). The essential oil of the myristicin race has primarily psychoactive and inebriating effects comparable to those of *Myristica fragrans* (Czygan 1989, 369).

Commercial Forms and Regulations

Fresh parsley is one of the most commonly sold herb seasonings. The dried herbage, the seeds, and the dried root (chopped drug) can be procured in herb shops and pharmacies (without restriction). The seeds can also be obtained in flower shops.

Literature

See also the entries for **witches' ointments** and **essential oils**.

Czygan, Franz-Christian. 1989. Petersilienfrüchte [and] Petersilienwurzel. In *Teedrogen*, ed. M. Wichtl, 368–69 and 370–71. Stuttgart: WVG. (Two separate articles.)

Frank, Bruno. 1994. *Petroselinum*. In *Hagers Handbuch der pharmazeutischen Praxis*, 5th ed., 6:105–19. Berlin: Springer.

Warncke, D. 1992. Untersuchungen über die Zusammensetzung der ätherischen Öle von *Petroselinum crispum* (Mill.) A.W. Hill und *Petroselinum segetum* (L.) Koch unter besonderer Berücksichtigung von Handelsdrogen und Handelsölen. Diss. (biology), Würzburg.

In the drug culture, the word *parsley* is a pseudonym for marijuana. This is not inappropriate, as the plant does contain inebriating essential oils. (Woodcut from Tabernaemontanus, *Neu Vollkommen Kräuter-Buch*, 1731)

Myristicin

Apiol

Allyltetramethoxybenzol

Phalaris arundinacea Linnaeus

Reed Canary Grass

Family
Gramineae: Poaceae (Grass Family)

Forms and Subspecies
Several varieties and cultivars have been described, including *Phalaris arundinacea* var. β *picta* L. (from North America), known as bent grass. The widespread cultivar *Phalaris arundinacea* cv. Turkey Red produces primarily **5-MeO-DMT** (Appleseed 1995, 37).

Synonyms
Baldingera arundinacea (L.) Dumort.
Phalaroides arundinacea (L.) Rauschert
Typhoides arundinacea (L.) Moench

Folk Names
Bentgrass, canarygrass, canary grass, glanzgras, militz, phalaridos, randgräs, reed canarygrass, reed canary grass, reed grass, rohrglanzgras

History
Reed canary grass was known even in ancient times. It cannot be determined whether this or another species (e.g., *Phalaris aquatica* L. or *Phalaris canariensis* L.) is the *phalaridos* described by Dioscorides. A number of grasses appear in herbals from the early modern period. The fact that *Phalaris* is psychoactive was discovered during phytochemical studies of the grasses for agricultural purposes. For several years, closet shamans have been experimenting with possibilities for using this and other grasses for psychoactive purposes (cf. **Arundo donax**, **Phalaris spp.**, **Phragmites australis**).

Distribution
The grass is found in Eurasia, North Africa, and North America. Thick stands grow on the banks of rivers and lakes and in wet meadows, often in reed fields and large sedge swamps (so-called phragmitetea).

Cultivation
The grass can be grown from seed or propagated by root cuttings (Appleseed 1995). The seeds need only be broadcast onto the ground. The grass prefers nutrient-rich, acid soils and must be near water or watered frequently.

Appearance
The perennial grass develops gray-green stalks that can grow up to 2 meters in height and can branch. The long, wide leaves have rough edges and are attached to the stalks. The panicle can take on a light green or red-violet hue. The spikelets bear a single flower. The flowering period is from June to August (Christiansen and Hancke 1993, 74 f.*). Large specimens can be confused with small forms of **Phragmites australis**.

Psychoactive Material
— Leaves

Preparation and Dosage
While the dried grass can be smoked, smoking almost never yields any effects. An extract obtained from the leaves is more suitable for smoking. It can be produced in the following manner: The dried leaves are finely chopped or powdered and, preferably, freeze-dried (or frozen and unfrozen several times). The material prepared in this fashion is placed in a blender with water and minced into a mush that is made acidic by the addition of an acid (e.g., acetic acid) and lightly simmered. The material is then boiled down until a tarlike mass remains. This mass can then be dissolved in alcohol (or a mixture of ethanol and water). The resulting solution is then impregnated into material suitable for smoking (e.g., damiana herbage; cf. **Turnera diffusa**). After being dried, the preparation should be quite potent (cf. DeKorne 1994, 127 ff.*).

Reed canary grass is increasingly being used to produce **ayahuasca analogs**. To date, however, there are very few detailed reports about optimal dosages, and definitive information about the races or strains of the grass to use is also lacking (Appleseed 1993).

A combination of 125 mg of an extract of **Peganum harmala** seeds and 50 mg of *Phalaris* extract produced unequivocal psychedelic effects "accompanied by strong waves of nausea" (DeKorne 1994, 98*). A combination of 60 g fresh weight of *Phalaris* and 3 g of **Peganum harmala** produced strong toxic effects (Festi and Samorini 1994).

Ritual Use

To date, we know of no traditional use of *Phalaris arundinacea* as a psychoactive substance. However, the Roman poet Ovid (43 B.C.E.–17 C.E.) described a shamanic transformation that was induced by (an unfortunately unidentified) "grass." In the story of Glaucus, a fisher from Anthedon in Boeotia, Glaucus himself described his wondrous metamorphosis into a sea god:

> I sought the cause if any God had brought
> this same abowt,
> Or else sum jewce of herb. And as I so did
> musing stand,
> What herb (quoth I) hath such a powre? And
> gathering with my hand
> The grasse, I bote it with my toothe. My
> throte had scarcely yit
> Well swallowed downe the uncouth jewce,
> when like an agew fit
> I felt myne inwards soodeinly to shake, and
> with the same,
> A love of other nature in my brest with
> violence came.
> And long I could it not resist, but sayd: Deere
> land, adeew,
> For never shall I haunt thee more. And with
> that woord I threw
> My bodye in the sea. The Goddes thereof
> receyving mee,
> Vouchsaved in theyr order mee installed for
> too bee,
> Desyring old Oceanus and Thetis for theyr
> sake,
> The rest of my mortalitie away from mee to
> take.
> They hallowed mee, and having sayd nyne
> tymes the holy ryme
> That purgeth all prophanednesse, they
> charged mee that tyme
> To put my brestbulk underneathe a hunred
> streames. Anon
> The brookes from sundry coastes and all the
> Seas did ryde uppon
> My head. From whence as soone as I
> returned, by and by
> I felt my self farre otherwyse through all my
> limbes, than I
> Had beene before. And in my mynd I was
> another man.
> Thus farre of all that mee befell make just
> report I can.
> Thus farre I beare in mynd. The rest my
> mynd perceyved not.
> Then first of all this hory greene gray grisild
> beard I got,
> And this same bush of heare which all along
> the seas I sweepe,
> And theis same myghty shoulders, and theis
> grayish armes, and feete
> Confounded into finned fish.

(Ovid, *Metamorphoses* 13.1099–24; in Nims 1965, 348–49)

Perhaps the "grass" was *Phalaris arundinacea*, and a preparation was known in ancient times that would have been suitable in rituals for animal transformation.

Artifacts

None

Medicinal Use

Dioscorides noted that the "crushed plant, treated with water or wine to make a juice, has the power to have good effects on bladder disorders" (3.149).

Constituents

The entire grass contains **indole alkaloids**, the composition of which can vary greatly depending upon race, strain, location, time of collection, et cetera (Marten 1973; Ostrem 1987). *N,N*-DMT, MMT, and **5-MeO-DMT** are usually present (Matum et al. 1979). The grass also can have high concentrations of gramine, a very toxic alkaloid (Appleseed 1995).

Effects

Smoking a suitable preparation can produce effects like those produced by *N,N*-DMT. While some of the **ayahuasca analogs** that have been tested to date have indeed yielded ayahuasca-like effects, many of the reports describe unpleasant experiences (cf. Festi and Samorini 1994).

Commercial Forms and Regulations

The seeds are available through ethnobotanical specialty sources.

Literature

See also the entries for ***Arundo donax, Phragmites australis,* ayahuasca analogs**, *N,N*-**DMT**, and **5-MeO-DMT.**

Appleseed, Johnny. 1993. Ayahuasca analog plant complexes of the temperate zone: *Phalaris arundinacea* and the *Desmanthus* spec. *Integration* 4:59–62.

————. 1995. *Phalaris in großen Mengen. Entheogene* 4:36-37.

Festi, Francesco, and Giorgio Samorini. 1994. Alcaloidi indolici psicoattivi nei generi *Phalaris* e *Arundo* (*Graminaceae*): Una rassegna. *Annali dei Musei Civici di Rovereto* 9 (1993): 239–88. (Very good bibliography.)

Marten, G. C. 1973. Alkaloids and palatability of *Phalaris arundinacea* grown in diverse environments. *Agronomy Journal* 165:199–201.

Marum, P., A. W. Hovin, and G. C. Marten. 1979. Inheritance of three groups of indole alkaloids in reed canarygrass. *Crop Science* 19:539–44.

Nims, John Frederick, ed. 1965. *Ovid's* Metamorphoses, *the Arthur Golding translation 1567.* New York: Macmillan.

Ostrem, L. 1987. Studies on genetic variation in reed canarygrass, *Phalaris arundinacea* I: Alkaloid type and concentration. *Hereditas* 107:235–48.

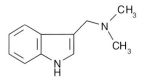

Gramine

"An extremely potent smokable form of DMT can be extracted from the reed canary grass (*Phalaris arundinacea*). . . . *Phalaris* DMT is something brand new—derived from one of the ayahuasca analog plants, it is a natural form of DMT and 5-MeO-DMT which can be grown by anyone anywhere on the planet outside of the polar regions. It has no somatic side effects (nausea, vomiting), nor is it dependent for its extraction on complicated laboratory procedures, equipment or knowledge; hence it isn't necessary to rely upon a profit-oriented monopoly of dealers to obtain. It comes on fast, is too intense, and subsides rapidly: just like the way we live our lives."

JIM DEKORNE
PSYCHEDELIC SHAMANISM
(1994, 103, 105)

A copperplate engraving of *Phalaris* grass, from the German edition of Dioscorides (1610).

Phalaris spp.

Canary Grasses

Family

Gramineae: Poaceae (Grass Family)

Following a comprehensive revision of the genus *Phalaris*, a total of twenty-two species is now accepted. The greatest number of species (eleven) are found in the Mediterranean region, where they are part of the indigenous flora. Four species are native to the American Southwest (Baldini 1995). Like *Phalaris arundinacea*, many species exhibit considerable variability. The various species appear to have different chemotypes and chemical races. For this reason, experimenting with unknown types of *Phalaris* without previously analyzing their constituents can be extremely dangerous. Many grasses contain gramine, a very toxic alkaloid.

Phalaris aquatica L. [syn. *Phalaris bulbosa* auct. non. L., *Phalaris commutata* Roem. et Schult., *Phalaris nodosa* Murray, *Phalaris tuberosa* L.]— water canary grass

Originally from the Mediterranean region, this species is now found throughout the world. *Phalaris aquatica* is very common in Australia, where it is despised in sheep pastures as a poisonous grass (McBarron 1991, 17). This species is thought to contain the highest concentrations of *N,N,*-DMT in the genus (Baxter and Slaytor 1972; Mack et al. 1988). Whether the Aborigines used this grass in any way is unknown. There also is no evidence to determine whether the grass was present in Australia before the arrival of the Europeans or whether it was introduced along with the cattle and sheep. There are several varieties (e.g., var. *australia*, var. *uneta*), some of which represent chemical races. Some sorts or strains contain primarily *N,N,*-DMT, while in others **5-MeO-DMT** predominates (Mack and Slaytor 1979; Mulvena and Slaytor 1982, 1983). This grass is being increasingly tested for use in developing **ayahuasca analogs**.

Phalaris spp.—cane canary grass

Ancient Egyptian graves have yielded grave garlands into which pieces or entire stalks (including panicles) of *Phalaris* species were worked (Germer 1985, 219*). It is possible that psychoactive tryptamines may be present in a number of *Phalaris* species.

Literature

See also *Phalaris arundinacea* and **ayahuasca analogs**.

Anonymous. 1995. Phalaris special. *Eleusis* 49–51.

Baldini, Riccardo M. 1993. The genus *Phalaris* L. (Gramineae) in Italy. *Webbia* 47:1–53.

———. 1995. Revision of the genus *Phalaris* L. (Gramineae). *Webbia* 49:265–329.

Baxter, C., and M. Slaytor. 1972. Biosynthesis and turnover of *N,N*-dimethyltryptamine and 5-methoxy-*N,N*-dimethyltryptamine in *Phalaris tuberosa*. *Phytochemistry* 11:2767–73.

Mack, J. P. G., et al. 1988. *N,N*-dimethyltryptamine production in *Phalaris aquatica* seedlings: A mathematical model for its synthesis. *Plant Physiology* 88:315–20.

Mack, J. P. G., and M. Slaytor. 1979. Indolethylamine *N*-methyltransferase of *Phalaris tuberosa*. *Phytochemistry* 18:1921–25.

McBarron, E. J. 1991. *Poisonous plants.* Melbourne, Sydney, and London: Inkata Press.

Mulvena, D. P., and M. Slaytor. 1982. Separation of tryptophan derivatives in *Phalaris aquatica* by thin layer chromatography. *Journal of Chromatography* 245:155–57.

———. 1983. *N*-methyltransferase activities in *Phalaris aquatica*. *Phytochemistry* 22 (1): 47–48.

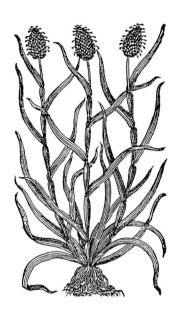

This early modern illustration of a canary grass may represent the DMT-rich species *Phalaris aquatica*. (Woodcut from Tabernaemontanus, *Neu Vollkommen Kräuter-Buch*, 1731)

Above: The grass *Phalaris aquatica* (=*Phalaris tuberosa*) is rich in DMT and other tryptamines.

Top right: Many *Phalaris* species have not been chemically investigated to date. There are strong indications, however, that many members of the genus contain psychoactive constituents. (A *Phalaris* species, photographed in Crete)

Bottom right: The common canary grass (*Phalaris canariensis*) is known primarily as a source of bird food. This species also contains alkaloids.

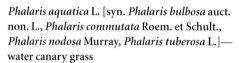

Phragmites australis (Cav.) Trinius ex Steudel

Reed

Family
Gramineae: Poaceae (Grass Family)

Forms and Subspecies
At least two subspecies have been described (Germer 1985, 205*):

Phragmites australis ssp. *altissimus* (Benth.) Clayton
Phragmites australis ssp. *stenophyllus* (Boiss.) Bor.

Synonyms
Arundo isiaca Del.
Arundo phragmites L.
Arundo vulgaris Lam.
Phragmites communis L.
Phragmites communis Trin.
Phragmites communis var. *isiacus* (Del.) Coss. et DR.

Folk Names
Calamus vallaris, canna sepiaria, carrizo, carrizo de panocha, common reed, 'eqpe'w (Chumash), gemeines rohr, gemeines schilfrohr, harundo, 'iqpew, lók'aa' (Navajo, "tube"), kalamos, phragmites (Greek), rancül, reed, reedgrass, ried, rohr, schelef, schilf, schilfrohr, topo, xapij

History
In ancient Egypt, reeds were used for a number of purposes, especially as a source of materials (Germer 1985, 205*). The plant was described by Theophrastus, Dioscorides, and Pliny. Apart from its use as a fermenting agent, no traditional psychoactive use of the plant has been documented to date.

Over time, reeds have been used for a wide variety of purposes, including as roofing material, as a source of cellulose, and in the manufacture of arrows, cane mats, and musical instruments (Aichele and Hofmann 1991, 120*; cf. ***Arundo donax***). The plant has even been used as a source of nourishment. The seeds have been made into porridge, the young shoots are a good vegetable, and the sweet pith can be used to make fermented beverages (**beer**) (Bremness 1995, 202*; Timbrook 1990, 246*).

Distribution
The reed is the largest grass in central Europe, where it is often encountered along the shores of lakes (in the water) in so-called reed fields. The grass can grow on land, but only where the water table is close to the surface and does not subside for any length of time, e.g., in sedge meadows and fens (Christiansen and Hancke 1993, 89*). The common reed is now found throughout the world.

Cultivation
The plant is propagated primarily vegetatively. The grass can be easily grown from a piece of the root (rhizome). Reeds prefer marshy soil and require a great deal of nutrient-rich water. They are well suited for use as ornamentals in garden ponds. However, they do not tolerate acidic water (Christiansen and Hancke 1993, 89*).

Appearance
This perennial marsh grass develops a thick, creeping, branching rhizome from which runners grow into the swampy subsurface. The canes can grow from 1 to 3 meters tall. The leaves have rough margins and can attain a length of 40 to 50 cm and a width of 1 to 2 cm. The very large, 15 to 40 cm long panicle is multiflorous and develops four- to six-flowered dark violet spikelets. The flowering period is from July to September (Christiansen and Hancke 1993, 88*). The seeds do not ripen until the winter, when the plant also loses its leaves. The panicle then usually turns light white in color. The new shoots begin to appear in early summer and grow rather slowly. The subspecies *altissimus* can grow to a height of at least 5 meters. In the tropics, the reed can attain a height of up to 10 meters and is then easily confused with ***Arundo donax*** (Aichele and Hofmann 1991, 120*). The reed can be easily distinguished from *Arundo donax* by the fact that its panicle hangs only to one side (Germer 1985, 205*).

Psychoactive Material
— Root (reed root, radix arundinis vulgaris)

Preparation and Dosage
The fresh or dried rootstock (20 to 50 g) is boiled for at least fifteen minutes, combined with 3 g of ***Peganum harmala*** seeds, and drunk as an **ayahuasca analog**. Exercise care when determining dosages!

Ritual Use
To the Navajo, the reed is a sacred plant of ritual significance. According to the Navajo creation myth, the reed saved humanity (i.e., the Navajo) during the Great Flood. The Navajo received the reed from a holy person. Humans, animals, and insects climbed into the magical reed, which immediately grew up to the sky. So that it would be able to grow straight up, a holy person took a feather and attached it to the ascending reed, like the feathers on an arrow. For this reason, the reed still has a flower that flutters like a feather in the wind. The shaft of the reed is used to make prayer

The cosmopolitan reed (*Phragmites australis*) contains potent psychoactive substances.

"I boiled 45 g of roots of *Phragmites australis* for 15 minutes to make a tea. I then ingested a normal dose of *Peganum harmala*, 3 g. It was the most exceptional and pristine experience of my life and definitely the most powerful trip with an ayahuasca analog that I had had to that point. Very visual, with awe-inspiring insights into myself and the world. God, what a day! Six hours full of mind-shattering insights and revelations. Unbelievable sensations of intense beauty. Visions of golden worlds beyond any conception. . . . I was emotionally deeply moved by the exquisiteness and beauty of the experience. . . . There was no nausea or other side effects."

ANONYMOUS
"PHRAGMITES AUSTRALIS"
(1995, 39)

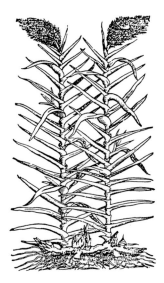

The common reed (*Phragmites australis*) is the largest naturally occurring true grass in central Europe. (Woodcut from Tabernaemontanus, *Neu Vollkommen Kräuter-Buch*, 1731)

poles for all ceremonies and healing rituals (Mayes and Lacy 1989, 101 f.*).

The Serí Indians of northern Mexico used fragments of the reed to smoke wild tobacco species (see *Nicotiana* spp.).

Artifacts

The reed is depicted in numerous works of art from ancient Egypt, e.g., in the wall paintings at Medinet Habu and Amarna. A hieroglyph (j) was derived from the characterisic flower panicle (Germer 1985, 205 f.*).

The Navajo make the stalks into prayer poles, and many cultures use them to make arrow shafts.

Medicinal Use

In late ancient times, the finely ground root was mixed with onions to prepare a wrap or plant poultice for removing thorns and splinters. "Mixed [with] vinegar, it soothes dislocations and hip pains" (Dioscorides 1.114). In Europe, the herbage was once used as a diuretic (Schneider 1974, 3:54*). An infusion of the roots is used in folk medicine for the same purpose (Aichele and Hofmann 1991, 120*); it can also be used to treat mucus obstruction, coughing, lung pains, and hiccups (Bremness 1995, 202*).

The Navajo use a tea as an emetic agent to treat certain stomach and skin problems (Mayes and Lacy 1989, 101*).

Constituents

The rootstock contains *N,N*-DMT, 5-MeO-DMT, **bufotenine**, and gramine (Wassel et al. 1985).

Effects

Dioscorides stated that the flower tufts of *Phragmites australis*—like those of *Arundo donax*—cause deafness if they get into the ear (1.114).

Reports about the psychoactive effects of *Phragmites australis* are based almost exclusively on experiences with **ayahuasca analogs** that are composed of the root extract, lemon juice, and *Peganum harmala* seeds. Unpleasant side effects (nausea, vomiting, diarrhea) are usually mentioned (Eros 1995).

Commercial Forms and Regulations

None

Literature

See also the entries for *Arundo donax*, *Phalaris arundinacea*, and **ayahuasca analogs**.

Anonymous. 1995. *Phragmites Australis*—Eine weitere Pflanze zur Ayahuasca-Bereitung. *Entheogene* 4:39–40.

Eros. 1995. *Phragmites australis:* positiv. *Entheogene* 5:43.

Wassel, G. M., S. M. El-Difrawy, and A. A. Saeed. 1985. Alkaloids from the rhizomes of *Phragmites australis* Cav. *Scientia Pharmaceutica* 53:169–70.

Phytolacca acinosa Roxburgh

Pokeweed, Shang Lu

Family

Phytolaccaceae (Pokeweed Family)

Forms and Subspecies

The Chinese make a distinction between a form with white flowers and a white root, which they regard as harmless and edible, and a form with red flowers and a reddish root, which is considered dangerous, toxic, and hallucinogenic (Li 1978, 21*). The edible type is presumably the variety *esculenta* Maxim., which formerly was regarded as a distinct species (see "Synonyms").

Synonyms

Phytolacca esculenta Van Houtte

Folk Names

Cancer-root, Chinese pokeweed, chinesische kermesbeere, dpa'-bo dkar-po, dpa'-bo ser-po (Tibe-tan), fu, Indian poke, jaringo, jaringo sag (Nepali), juniper, kermesbeerspinat, pokeweed, shang lu, shang-lu, sweet belladonna, tibetische kermesbeere, white pokeberry, yellow pokeberry

History

The edible variety (var. *esculenta*) is mentioned in the ancient *Shih Ching*, the Book of Songs (ca. 1000–500 B.C.E.), under the name *fu* (Keng 1974, 402*). The leaves have long been eaten as a vegetable (Li 1978, 21*). The plant is still in use in traditional Chinese and Tibetan medicine. The genus *Phytolacca* is now quite well known pharmacologically and chemically (Woo 1978). The psychoactivity of the plant is debated.

Distribution

In the Himalayas, the plant occurs at altitudes between 2,000 and 3,000 meters (Polunin and

Stainton 1985, 342*). It is found in Tibet, China, Korea, Japan, and India and has become naturalized in some parts of Europe (e.g., Greece). The plant also may be found in many European botanical gardens.

Cultivation

The plant is propagated from seeds, which should be pregerminated and planted in good topsoil. This perennial is quite easy to grow in central Europe. The aboveground herbage dies back after the fruiting period. The root sends forth new shoots the following spring.

Appearance

This bushy, heavily ramified plant can grow to a height of about 1 meter. It has large, oblong, attenuated leaves that can grow as long as 26 cm. The stems are normally light green in color but may also be violet. The terminal flowers grow in clusters. The flowers are whitish, while the ripe berries are dark violet to black. The plant flowers in June and the fruits mature by August. The racemes sometimes bear fruits and flowers simultaneously. The plant has a turnip-shaped root tuber.

Shang lu is easily confused with American pokeberry (*Phytolacca americana* L. [syn. *Phytolacca decandra* L.]). North American Indians of the Pacific Southwest allegedly used this species as a narcotic (Emboden 1986, 164*).[277]

In contrast to *Phytolacca americana*, *Phytolacca acinosa* has vertical upright inflorescences and infructescences; in the American relative, both of these lean to the side.

The closely related species *Rivina humilis* L. (Phytolaccaceae), which is known as *coralillo*, *colorines*, or *hierba mora*,[278] is said to be identical to the Aztec narcotic *amatlaxiotl* (Díaz 1979, 93*).

Psychoactive Material

— Root

Preparation and Dosage

Both the manner(s) in which the root should be prepared for psychoactive use and the dosage that should be used have not come down to us. It is possible that the root was used as an additive in the production of **sake**, for the few sources do mention a "brewed" preparation.

In Nepal, the young, tender leaves and stalks are cooked and eaten as vegetables (Malla 1982, 193). This use is the source of the German name for the leaves, *kermesbeerspinat* ("kermes berry spinach").

Ritual Use

In ancient China, the root was placed into the same category as ginseng (**Panax ginseng**) and mandrake (**Mandragora officinarum**). The root

also was used as a substitute for belladonna root (**Atropa belladonna**) (Emboden 1986, 164*).

T'ao Hung-ching reported that the plant "is used by the Taoists. When it is boiled or brewed and consumed, it is good for lower abdominal parasites and to see spirits" (Li 1979, 22*). Su Sung wrote, "In olden times, it was used a great deal by the magicians [= shamans]" (Li 1979, 22*). Su Ching provides more precise information:

> There are two forms of this medicine, one red and one white. The white kind is used in the healing arts. The red kind can be used to conjure spirits; it is very toxic. Otherwise, it can be used only externally for inflammations. If eaten, it is very terrible: it causes bloody stools. It can be lethal. It causes one to see spirits. (Li 1979, 22*)

Unfortunately, nothing more is known about any shamanic or alchemical use of the plant.

Artifacts

The plant is depicted on Tibetan medicine thangkas (Aris 1992, 79, 235*).

Medicinal Use

In traditional Chinese medicine, the roots of shang lu (*Phytolacca acinosa* Roxb. var. *esculenta* Maxim.) are used to treat tumors, edema, and bronchitis (Yeung et al. 1987; Yi 1992). In Asian folk medicine, the roots are especially esteemed for their anti-inflammatory and antirheumatic effects.

In Tibetan medicine, the roots are attributed with cooling qualities. They are utilized as an antidote, to treat chronic fever, and to treat the

Left: The typical infructescence of pokeweed (*Phytolacca acinosa*), the root of which was used in ancient China for psychoactive purposes.

Right: The flower of American pokeweed (*Phytolacca americana*), a plant that Indians used for many purposes, including as a narcotic.

"Plant [*Phytolacca acinosa*] said to have narcotic properties, and produces a bitter toxic substance. The leaves make an excellent potherb if well-boiled."

OLEG POLUNIN AND ADAM STAINTON
FLOWERS OF THE HIMALAYA
(1985, 342*)

277 The German name for American pokeberry (*Phytolacca americana* L.) is *amerikanischer nachtschatten* ("American nightshade") (Schneider 1974, 2:59*).

278 This name is also used for black nightshade (*Solanum nigrum*; see *Solanum* spp.).

Shang lu (*Phytolacca acinosa*), a Chinese magical and medicinal plant that is related to pokeweed (*Phytolacca americana*), as depicted in the herbal *Chêng-lei pên-ts'ao* (1249 C.E.).

"The [shang lu] flowers—Ch'ang-hau'—are esteemed for treating apoplexy. The root is so poisonous that it is normally used only externally."

RICHARD SCHULTES AND ALBERT HOFMANN

PLANTS OF THE GODS

(1992, 54*)

pain of wounds. Nepalese Sherpas use a paste made from the roots as a potent purgative to treat food poisoning (Bhattarai 1989, 51*).

Constituents

The roots of *Phytolacca acinosa* Roxb. var. *esculenta* Maxim. have been found to contain various saponins (esculentoside-I, esculentoside-N, phytolaccagenin derivatives) (Yi 1992). The triterpenes phytolaccagenin and acinospesigenin have been detected in the leaves (Spengel and Schaffner 1990). The fruits have yielded acids of the 28,30-dicarboxy-oleanene type and its ester (Spengel et al. 1992). The triterpenoids acinosolic acid, phytolaccagenin (empirical formula $C_{31}H_{40}O_7$), phytolaccagenic acid, esculentic acid, jaligonic acid, phytolaccagenin-A, and acinosolic acids A and B have also been identified (Harkar et al. 1984).

Proteins with abortifacient effects occur in the roots, leaves, and seeds (Yeung et al. 1987).

The species *Phytolacca bogotensis* H.B.K., which is toxic to cattle, has been found to contain cyanoglycosides (Schultes 1977b, 111*).

Effects

Apart from the ancient Chinese sources, according to which the use of shang lu "enables one to see spirits" (i.e., is hallucinogenic), there are no reports of psychoactive experiences.

Sedative effects are possible, as other species in the genus are used for narcotic purposes. The Kofán Indians of Colombia produce a fish poison from the leaves of *Phytolacca rivinoides* Kunth et Bouché and the leaves of a *Phyllanthus* species (Schultes 1977b, 112*).

The saponins (triterpene aglycones) that are present in the roots of the genus have immune-enhancing, anti-inflammatory, and molluscicide effects (Parkhurst et al. 1990; Yi 1992).

Commercial Forms and Regulations

Because the bush is highly regarded as a beautiful ornamental, its seeds are occasionally available from seed suppliers and flower shops.

Literature

Barbieri, L., G. M. Aron, J. D. Irvin, and F. Stirpe. 1982. Purification and partial characterization of another form of the antiviral protein from the seeds of *Phytolacca americana* L. (Pokeweed). *Biochemical Journal* 203:55–59.

Harkar, S., T. K. Razdan, and E. S. Waight. 1984. Further triterpenoids and ^{13}C NMR spectra of oleanane derivatives from *Phytolacca acinosa*. *Phytochemistry* 23 (12): 2893–98.

Malla, Samar Bahadur, ed. 1982. *Wild edible plants of Nepal*. Bulletin no. 9. Kathmandu: Department of Medicinal Plants.

Parkhurst, Robert M., David W. Thomas, Robert P. Adams, Lydia P. Makhubu, Brian M. Mthupha, L. Wolde-Yohannes, Ephraim Mamo, George E. Heath, Janeen K. Strobaeus, and William O. Jones. 1990. Triterpene aglycones from various *Phytolacca dodecandra* populations. *Phytochemistry* 29 (4): 1171–74.

Spengel, Sigrid, St. Luterbacher, and Willi Schaffner. 1992. Phytolaccagenin and phytolaccagenic acid from berries, roots, leaves, and calli of *Phytolacca dodecandra*. *Planta Medica* 58 suppl. (1): A684.

Spengel, Sigrid, and Willi Schaffner. 1990. Acinospesigenin—ein neues Triterpen aus den Blättern von *Phytolacca acinosa*. *Planta Medica* 56:284–86.

Woo, W. S. 1978. *The chemistry and pharmacology of Phytolacca plants*. Seoul: Natural Product Research Institute, Seoul Natural University.

Yeung, H. W., Z. Feng, W. W. Li, W. K. Cheung, and T. B. Ng. 1987. Abortifacient activity in leaves, roots and seeds of *Phytolacca acinosa*. *Journal of Ethnopharmacology* 21:31–35.

Yi, Yang-Hua. 1992. Two new saponins from the roots of *Phytolacca esculenta*. *Planta Medica* 58:99–101.

Piper auritum H.B.K.

Gold Pepper

Family
Piperaceae (Pepper Family); Pipereae Tribe

Forms and Subspecies
It is possible that there are varieties, forms, or subspecies that may be distinguished on the basis of their leaves. However, the taxonomy of the neotropical *Piper* species is quite confusing.

Synonyms
Piper auritum Kunth
Piper umbellatum L.

    Piper sanctum (Miq.) Schl. may also be a synonym; Martínez 1987 (page 1188) lists practically the same Mexican names for the two species.

Folk Names
Acoyo, acuya, acuyo, aguiyu, alahan, bakanil a iits' (Huastec), cordoncillo, cordoncillo blanco, corriemineto, coyoquelite, gold pepper, goldpfeffer, hierba anís ("anise herb"), hierba de Santa María (Spanish, "the herb of Saint Mary),[279] hierba santa, hinojo sabalero, ho'ben (Lacandon, "the herb of the five"), hoja de anís, hoja de cáncer, hoja santa (Mexico, "sacred leaf"), homequelite, ixmaculan, jaco, jinan (Totonac), maculan, ma'haw, ma'jóo, mak'ulan, mecaxóchitl (Nahuatl), momo, mumun, mumun te' (Tzeltal), omequelite, omequilit-dos quelite, Santa María,[280] tlampa, tlanepa, tlanepaquelite, tlanipa, totzoay, tzon tzko ntko, wo, woo, xalcuahuitl, xmaculan (Mayan/Quintana Roo), x-mak-ulam, xmak'ulan, x'obel (Mayan/San Antonio [Belize]), yerba santa

History
Gold pepper is an ancient traditional Mayan remedy that was mentioned as a medicinal plant in the few sources from the colonial period (e.g., the Motul dictionary and the *Relación de las cosas de Yucatán*) (Roys 1976, 263*). In contemporary Mexico, the primary use of the plant is as a seasoning; fish and other seafood are wrapped in the large, aromatic leaves and braised (Bye and Lianres 1983, 6*; Cioro 1982, 143*).

    In Panama, the leaves were or still are used to catch fish. Apparently, their scent attracts a food fish known as *sábalo pipwu* (Gupta et al. 1985).

    In Brazil, the leaves were used in the industrial production of raw safrole for the international market (Rob Montgomery, pers. comm.).

Distribution
Gold pepper is found from Mexico through Central America and into South America. It is very common among the tropical flora of Mexico (Chiapas), Belize, Panama, and Brazil and has been carried into other tropical areas.

Cultivation
The plant is most easily propagated through cuttings (approximately 15 to 20 cm long) taken from the lower stems. In tropical areas, it can very easily go wild and can displace other pepper plants (e.g., *Piper methysticum*), thereby causing some ecological damage (e.g., in Hawaii).

Appearance
This evergreen perennial bush, which can grow to a height of 4 to 5 meters, develops branched green stems that do not lignify on their lower ends until quite late. The leaves are opposite, oval, and tapered at the end and project straight out from the stem or droop slightly. The green-white, very thin inflorescences extend straight up and can attain a length of more than 10 cm.

    Gold pepper is easily confused with the very similar species *Piper sanctum* (possibly a synonym), which is also known as *hoja santa* and is also rich in safrole (Martínez 1994, 185*). However, *Piper sanctum* grows to a height of only 1.5 meters and does not occur in the southeastern lowlands (Argueta et al. 1994, 813*).

    The closely related and similar, but generally smaller, species *Piper amalago* L. (see *Piper* spp.) also contains safrole and is used ethnobotanically in very similar ways (Arvigo and Balick 1994, 64 f.*). Some Maya regard this species as the "female" counterpart of the "male" gold pepper.

    Gold pepper is almost identical in appearance to *Piper methysticum*; most laymen can distinguish the two species only by the scent of the leaves.

Psychoactive Material
— Fresh leaves
— Dried leaves
— Essential oil

Preparation and Dosage
Shade-dried leaves may be smoked by themselves or in combination with other herbs (see **smoking blends**). Fresh leaves are added to **alcohol** (*aguardiente* = sugarcane alcohol, mescal; see *Agave* spp.) (Argueta et al. 1994, 49*).

    The **essential oil**, which is easily obtained through steam distillation (Gupta et al. 1985), is suitable as a precursor for the synthesis of amphetamine derivatives (e.g., MDMA; cf. **herbal ecstasy**).

    An orally administered dose of 9 g/kg of plant extract did not have any lethal effects upon rats.

The tropical gold pepper (*Piper auritum*) can be recognized by its very large leaves, its perpendicular flowers, and the strong safrole scent of its leaves and young shoots. (Wild plant, photographed in Belize)

279 In South America, another herbaceous pepper plant (*Pothomorphe peltata* [L.] Miquel) is also known by this name (Vickers and Plowman 1984, 26*).

280 This same name is used in the Santo Daime cult to refer to marijuana (*Cannabis sativa*) (cf. ayahuasca).

When administered via injection, the LD_{50} is calculated as 2g/kg (Argueta et al. 1994, 50*).

Ritual Use

Today in Belize, the large leaves are smoked, most likely as a marijuana substitute (cf. **Cannabis indica**) and for hedonistic purposes. To date, we know of no traditional rituals in which gold pepper has been used for its psychoactive properties.

The natives of the West Indies (or Mexico) are said to have once used *Piper plantagineum* Schlecht., a species found throughout the Caribbean region, as a narcotic in a manner similar to the way kava-kava (**Piper methysticum**) is used. It is possible that this species is synonymous with *Piper auritum*. Unfortunately, almost nothing is known about it.

Artifacts

None

Medicinal Use

In Belize (San Antonio, Cayo District), the large, fresh leaves are heated over a wood fire and laid over painful areas of the back, especially around the small of the back. The Yucatec Maya of Quintana Roo use the leaves as a stimulant, as an analgesic, and to treat asthma, bronchitis, dyspnea, weak digestion, stomachaches, head colds, erysipelas, fever, gout, rheumatism, and wounds (Cioro 1982, 143*; Roys 1976, 263*). In Mexican folk medicine, the leaves are used for ethno-gynecological purposes. A tea made from the leaves and mixed with **honey** is used to treat scorpion stings. Juice pressed from the leaves is ingested to relieve asthma, coughing, and bronchitis (Argueta et al. 1994, 49*).

The fresh leaf buds and young shoots can be eaten as mild stimulants. When eaten, a mild numbness is produced in the mouth that feels very similar to the anesthesia of the mucous membranes that is caused by **Piper methysticum**.

Constituents

The leaves contain 0.47 to 0.58% **essential oil** (Martínez 1994, 185*). The essential oil is also present in the stalks, although in much lower concentrations (Oscar and Poveda 1983).

The essential oil has a characteristic safrole or sassafras scent and consists of up to 70% safrole; also present are some forty other substances, including α-thujene, α-pinene, camphene, sabinene, β-pinene, myrcene, β-phellandrene, carene, α-terpinene, limonene, 1,8-cineole, γ-terpinene, β-phellandrene, *cis*-sabinene hydrate, nonanon-2, r-cymene, terpinolene, linalool, camphor, borneol, r-cymene-8-ol, bornylacetate, eugenol, D-

elemene, α-cubenene, muurolene, α-copaene, β-bourbonene, paraffin, β-caryophyllene, humulene, myristicin, β-bisabolene, elemicine, D-cadinene, cadina-1,4-dien, spathulenole, β-caryophyllene oxide, and *n*-hexadecane (Gupta et al. 1985; Argueta et al. 1994, 49*).

The leaves have been found to also contain the flavonoid 3'-hydroxy-4',7-dimethoxyflavone, β-sitosterol, and the **diterpene** *trans*-phytol. Various phenoles are also present in the leaves (Ampofo et al. 1987). The roots contain isoquinoline alkaloids, phenylpropenoids, and safrole (Argueta et al. 1994, 49*; Hansel et al. 1975; Nair et al. 1989).

Effects

The pharmacological effects of the leaves are clearly the result of their high safrole content (cf. *Sassafras albidum*).

Commercial Forms and Regulations

Although the plant is not subject to any regulations, it is not available as a living plant or as dried raw plant material. Because it is a precursor for the synthesis of MDMA and closely related amphetamine derivatives, safrole is subject to registration (cf. **herbal ecstasy**). In some areas, trade in safrole or in preparations with a high safrole content is regulated or even prohibited.

Literature

See also the entries for **Piper betle, Piper methysticum, Piper spp.**, and **essential oils**.

Ampofo, Stephen A., Vassilios Roussis, and David F. Wiemer. 1987. New prenylated phenolics from *Piper auritum*. *Phytochemistry* 26 (8): 2367–70.

Collera Zúñiga, Ofelia. 1956. *Contribución al estudio del* Piper auritum. Mexico City: Tesis, Facultad de Ciencias Químicas.

Gupta, Mahabir P., Tomás D. Arias, Norris H. Williams, R. Bos, and D. H. E. Tattje. 1985. Safrole, the main component of the essential oil from *Piper auritum* of Panama. *Journal of Natural Products* 48 (2): 330.

Hänsel, Rudolf, Anneliese Leuschke, and Arturo Gomez-Pompa. 1975. Aporthine-type alkaloids from *Piper auritum*. *Lloydia* 38:529–30.

Nair, Muraleedharan G., John Sommerville, and Basil A. Burke. 1989. Phenyl propenoids from roots of *Piper auritum*. *Phytochemistry* 28 (2): 654–55.

Oscar, C. C., and A. L. J. Poveda. 1983. *Piper auritum* (H.B.K.), Piperaceae Family: Preliminary study of the essential oil from its leaves. *Ing. Ciencias Químicas* 7 (1/2): 24–25.

Safrole

Piper betle Linnaeus

Betel Pepper

Family
Piperaceae (Pepper Family); Pipereae Tribe

Forms and Subspecies
The two most frequently cultivated and utilized varieties differ from one another primarily in their concentrations of **essential oil** and oleoresins:

Piper betle L. var. *bangla*: 5.9% oleoresin, 1.6% essential oil

Piper betle L. var. *metha-thakpala*: 4.9% oleoresin, 2.4% essential oil

Numerous cultivars are distinguished in Sri Lanka: 'Rata Bulath-vel', 'Siribo Bulath', 'Naga Walli-Bulath' (with spotted leaves), 'Getatodu-Bulath', 'Mala-Bulath', 'Gal-Bulath', and 'Dalu-Kotu-Bulath' (Macmillan 1991, 427*).

Synonyms
Chavica auriculata Miq.
Chavica betle (L.) Miq.
Chavica chuvya Miq.
Chavica densa Miq. l.c.
Chavica sibirica (L.) Miq. l.c.
Piper malamiris L. l.c.p.p.
Piper pinguispicum C. DC. et Koord.
Piper siriboa L.

Folk Names
Beatelvine, betel, bétel, betele, betel-leaf, betel pepper, betelpfeffer, betel vine, betle, betre (Malay, "single leaf"), bettele, bettele-pfeffer, bu, buio, bulath (Singalese), bulath-vel, buru, daun syry (Malay), fu-liu, fu-liu-t'êng (ancient Chinese), ikmo (Philippines), liu, mô-lû, nagavalli (Sanskrit), paan, pan, pelu (Thai), pu, sirih, tambul (Sanskrit), tambula (Sanskrit), tembul, veth-thile

History
In Southeast Asia and India, the use of betel leaves must be very ancient (cf. *Areca catechu*, **betel quids**). The plant is mentioned in early Sanskrit texts.

The first European representation of the betel leaf (although entirely inaccurate) can be found on a copperplate engraving from the *Delle navigationi e viaggi* of Giovanni Battista Ramusio (1485–1557), published in Venice in 1553. The first botanically correct representation was published in Paris in 1758 in *Histoire générale des voyages*, by Antoine-François Prévost.

Today, betel pepper (fresh betel leaves) is one of the most important articles of trade in Southeast Asia and in all areas in which large numbers of Indians or Tamils have settled.

Distribution
Betel pepper is indigenous to the Indo-Malayan region but is now grown in all of southern and Southeast Asia and even in the Seychelles and in Mauritius, Madagascar, and eastern Africa. It appears to have originated in central or eastern Malaysia. Some authors have suggested that the plant is originally from Java (Gupta 1991, 79*).

Cultivation
Propagation is performed almost exclusively with cuttings taken from the stem (10 to 20 cm in length). They are either placed in water until they develop roots or placed in moist cultivation beds. The plant requires moist and humus-rich soil and a semishaded location (Macmillan 1991, 427*).

In the tropics, the leaves of this evergreen plant can be harvested throughout the year. They normally are picked in the early morning.

Appearance
Betel pepper is a climbing half-shrub that bears shiny, light green, heart-shaped leaves (up to 18 cm in length). The sheen of the leaves is a reliable characteristic for distinguishing this species from other species of *Piper*, with which it can easily be confused (cf. **Piper spp.**). The "buds" (spikes) hang on the leafstalks like long, light-colored threads. The male spikes are cylindrical; the female grow to a length of only 4 cm. The fruit is a spherical drupe about 6 mm in diameter.

Psychoactive Material
— Betel leaves (folia piperis betle, piperis betle folium, betel pepper leaves)

Only fresh leaves are suitable for making **betel quids**; dried leaves can be used for medicinal purposes. The leaves are pressed after they are collected.

Occasionally the "buds" (= spikes) are also used for betel quids.

Preparation and Dosage
Fresh, undamaged leaves that have not begun to dry are used almost exclusively for psychoactive preparations. A normal dosage is one leaf per **betel quid**. A tea can be brewed from fresh or dried leaves. Again, one leaf is used per dosage.

Ritual Use
In India, all of life is ritually associated with the betel pepper. When a parcel of land is being prepared for cultivating betel pepper, a goat is first sacrificed while special mantras are recited. The head of the goat is buried in one corner of the

The betel pepper (*Piper betle*) develops aesthetically perfect heart-shaped leaves, which are used to make betel quids.

One of the first botanical illustrations of the betel pepper. (Copperplate engraving from Antoine-François Prévost, *Histoire générale des voyages*, Paris 1758)

future betel field (*paan mara*), the four hoofs are buried in the four cardinal directions, and the blood, mixed with earth, is distributed along the borders of the field as a landmark. Then a number of *shobhanjana* (*Moringa oleifera*) trees are planted. The betel vines will later grow up the branches of these fast-growing trees. Rows of mandara trees (*Erythrina indica*; see **Erythrina spp.**) are planted along the margins of the field as windbreaks. Anyone who enters the field must perform a gesture of veneration, for the field is regarded as a temple and is revered accordingly (Gupta 1991, 77 f.*).

Betel leaves are regarded as sacred and are among the more important offerings that can be made to Shiva, to whom all inebriating plants are sacred (cf. *Aconitum ferox, Cannabis indica, Datura metel, Strychnos nux-vomica*). Myths describe how the betel vine first grew only in heaven. Shiva asked the plant to go to the people on the earth. At first the vine refused because it was afraid that it would not be sufficiently respected and venerated. Shiva promised the plant that its leaves would be used with respect in all ceremonies. When he had convinced the plant, it came down from heaven to the earth. For this reason, it is considered good manners to offer guests a few betel leaves (with or without areca nuts; cf. *Areca catechu*). Betel leaves are also used for sprinkling sacred water during every ceremony. When the leaves are combined with cloves, castoreum, salt, and red, black, white, and yellow colors, they are considered a sure agent for banishing demons (Gupta 1991, 78 f.*).

For more on ritual uses, see **betel quids**.

Artifacts

The heart-shaped leaves have been depicted in Indian art since ancient times and are often used as an ornamental decoration on objects for making or consuming **betel quids**.

Medicinal Use

In the folk medicine of southern and Southeast Asia, betel leaves are chewed or eaten to treat coughing, inflammations of the mucous membranes, diphtheria, inflammations of the middle ear, and all types of stomach ailments. In India, the leaves are also used to treat snakebites and as an aphrodisiac (Gupta 1991, 79*).

In Southeast Asia, the roots and inflorescences are used in cases of weak digestion (Macmillan 1991, 424*); the same custom is found in the Seychelles and in other places with an Indian population. In the Seychelles, the leaves are "chewed in order to stay healthy. Seven leaves, finely chopped and placed on wounds, promote healing. A compress is also said to be effective against varicose veins" (Müller-Ebeling and Rätsch 1989, 29*).

Eugenol

"I would like to add that the scent of this [betel] leaf, when chewed, is reminiscent of that of our mugwort. It produces a breath that incites sensuous pleasure to a high degree, and the number of those who chew the leaf is very large. It simultaneously restores and strengthens, so that it incites anew to the joys of Venus."

Francesco Carletti (sixteenth century)

in *Kaffee, Tee und Kardamom* [Coffee, Tea and Cardamom] (R. Schröder 1991, 129*)

Constituents

The leaves contain 0.2 to 2.6% **essential oil** with phenolic constituents (eugenol, isoeugenol, allyl-pyrocatechol, chavicol, carvacrol) as well as nonphenolic substances (cineole, cadinene, and α-caryophyllene) (Roth et al. 1994, 569*). Also present are safrole, anethol, hentricontane, pentatriacontane, β- and γ-sitosterol, stearic acid, and triacontol. The pungent substance piperine, present in most *Piper* species, has not been detected in the betel pepper.

A team of Chinese researchers isolated and clarified neolignans (methylpiperbetol, piperol A, piperol B, crotepoxide) from the stems (and leaves) (Yin et al. 1991). Betel pepper flowers contain large amounts of essential oil, primarily with eugenol and isoeugenol.

Effects

The leaves have stimulant, antibiotic, digestion-promoting, and antiflatulence effects (Roth et al. 1994, 569*). They have a clear stimulating and awakening effect and open the perception. The effects appear to be synergistically potentiated by the other ingredients in **betel quids**.

The essential oil has anthelmintic properties (Ali and Mehta 1970) and appears to have anti-mutagenic and cancer-inhibiting effects. As a result, the betel leaf is an important health-promoting component of the betel quid. Pharmacological investigations of aqueous leaf extracts of Indonesian plants carried out at the Center for Research for Traditional Medicine (Airlangga University, Surabaya) have demonstrated that they stimulate phagocytosis, thereby strengthening the body's immune system (Sutarjadi et al. 1991). On the other hand, the neolignan crotepoxide is said to have pronounced cytotoxic effects (Yin et al. 1991).

Commercial Forms and Regulations

Because betel leaves are internationally recognized as not being an "addictive drug" or "narcotic," the plant is not subject to any laws regarding medicines or similar regulations but is classified as a foodstuff (the laws regulating such products may apply). In Switzerland, the fresh leaves are available in shops selling Indian articles.

Literature

See also the entries for *Areca catechu, Piper auritum, Piper methysticum, Macropiper excelsum,* and **betel quids**.

Ali, S. M., and R. K. Mehta. 1970. Preliminary pharmacological and anthelmintic studies of the essential oil of *Piper betle. Indian Journal of Pharmacy* 32:132–33.

Patel, R. S., and G. S. Rajorhia. 1979. Antioxidative role of curry (*Murray koenigi*) and betel (*Piper* betle) leaves in ghee. *Journal of Food Science and Technology* 16:158–60.

Sen, Soumitra. 1987. Cytotoxic and histopathological effects of *Piper betle* L. varieties with betel nut, lime, and tobacco. PhD thesis, University of Calcutta.

Sutarjadi, M., H. Santosa, S. Bendryman, and W. Dyatmiko. 1991. Immunomodulatory activity of *Piper betle*, *Zingiber aromatica*, *Andrographis paniculata*, *Allium sativum*, and *Oldenlandia corymbosa* grown in Indonesia. *Planta Medica* 57 suppl. (2): A136.

Yin, M.-L., J. Liu, Z.-L. Chen, K. Long, and H.-W. Zeng. 1991. Some new PAF antagonistic neolignans from *Piper Betle*. *Planta Medica* 57 suppl. (2): A66.

"This folia bettele, which grows throughout India and snakes up both wild and good trees like the pepper or cubeb fruit, is smeared with the already frequently mentioned areca fruit or pynan, together with a little shell lime, and is chewed and eaten by both the great and the common man as a necessity of all necessities; it produces a pleasantly scented breath and cleanses the mouth and the gums of all scorbutic blood. Likewise, it is also laid onto injuries like our plantain or plantago, which makes them thoroughly healthy and they quickly heal."

GEORGE MEISTER
DER ORIENTALISCH-INDIANISCHE KUNST- UND LUSTGÄRTNER [THE ORIENTAL-INDIAN ART AND PLEASURE GARDENER]
(1677)

Piper methysticum Forster f.

Kava

Family
Piperaceae (Pepper Family)

Forms and Subspecies
There are numerous cultivars that can be distinguished on the basis of morphological and chemical differences. Botanically, however, few of these have been described as varieties (Hölzl et al. 1993, 201).

In contrast, the Polynesians differentiate among a large number of varieties. In Fiji, six are counted; they differ from one another in the height, length, and thickness of the knots on their stems and the color (from green to purple). *Yagona leka*, which is stocky but develops the best aroma, is particularly esteemed. On the island of Tahiti, fourteen varieties were once recognized and differed from one another solely in their inebriating qualities (Lewin 1886, 6). In Hawaii, a particular distinction is made with regard to the variety known by the name *black awa*, the stems of which are nearly black; in addition, the following forms are also named: *apu, kau la'au, ke'oke'o, kuaea* (= *nene*), *kumakua, liwa, makea, mamaka, mamienie, mo'i, mokilana, papa, papa ele'ele*, and *papa kea* (Singh 1992, 20). Twenty-one varieties are recognized on the Marquesas Islands and five in Papua New Guinea, and Vanuatu has been reported to have seventy-two different cultivars (Lebot and Cabalion 1988). The existence of such variety may be the reason for the rather different experiences with kava drinks in the different regions. Recently, completely new and previously unknown varieties are said to have been discovered on Vanuatu (Kilham 1996).

Synonyms
Macropiper latifolium Miq.
Macropiper methysticum (G. Forst.) Hook. et Arnott
Macropiper methysticum Miq.
Piper decumanum Opitz
Piper inebrians Bertero
Piper inebrians Soland.

Folk Names
Agona, angona, angooner, ava, ava-ava, awa, 'awa (Hawaiian), awa-awa, cáva, gea, gi, intoxicating pepper, kava, kava-kava, kawa, kawa-kawa, kawa pepper, kawapfeffer, malohu, maluk, meruk, milik, poivre enivrant, rauschpfeffer, sakau, wati, yagona, yakona, yangona, yaona, yaqona, yaquona

The Polynesian word *awa* or *kava* means "bitter," "pungent," "sour," or "sourish"; *yangona* (and its derivatives) means "drink" as well as "bitter" and, thus, "bitter drink" (Singh 1992, 15). In most cases, the names given to the plant and to the drink prepared from it are identical. *Piper methysticum* does not grow on Rennel Island (southern Solomon Islands), and no drink made from the plant is used there. However, a drink made there from coconuts (*Cocos nucifera*) is called, strangely enough, *kava kava ngangi* (Singh 1992, 16).

History
Kava is the most important psychoactive agent in Oceania (Lebot et al. 1992). On most of the islands of Polynesia, the use and cultivation of the plant appears to have spread along with the settlement of the islands. Both the plant and kava drinking have also spread into many of the islands of Melanesia (Singh 1992, 15). It has been conjectured that Polynesians colonized Easter Island (Rapa Nui) in the third or fourth century because a chief was "led" there by a kava-induced vision (Ripinsky-Naxon 1989, 221*).

The ethnologist R. W. Williamson has worked out strong resemblances between the Vedic **soma** ritual and the Polynesian kava ceremonies and has conjectured that at least the ritual of kava spread from India to Oceania. There, the kava pepper was used as a substitute for the Indian soma plant (Williamson 1939). Another ethnologist has

Botanical illustration of the inebriating pepper (*Piper methysticum*). (From De Lessert, *Icones selectae plantarum*, 3:89 [1837])

argued that Polynesia was originally settled by two cultures, which on the basis of their "drug" consumption he called the betel people and the kava people. Even today, the areas in which betel is chewed and those in which kava is preferred can be geographically clearly distinguished (Churchill 1916). The custom of chewing **betel quids** only rarely overlaps with that of drinking kava.

The first Europeans to become acquainted with kava were Captain James Cook (1727–1779) and his fellow travelers. In 1777, Johann Georg Forster (1754–1794), who accompanied Cook, provided the first botanical description of the plant and the associated ceremony (Vonarburg 1996, 57). The report of Cook's journey (1784) noted that "when several of the members of the ship's crew partook of the drink, it was observed that it induced an effect like that of a strong dose of an alcoholic drink or even more a stupefaction such as produced by opium [cf. *Papaver somniferum*]. The effects of kava have also been compared to those of wild lettuce [cf. *Lactuca virosa*] and those of hashish [cf. *Cannabis indica*]" (Lewin 1886, 44).

Many islanders used or use kava as an everyday beverage, just as tea (*Camellia sinensis*) or coffee (*Coffea arabica*) is consumed in other parts of the world (Gajdusek 1967; Lewin 1886, 18). There are official kava bars in Fiji and on other islands.

On many South Sea islands, the **alcohol** that was introduced by missionaries has supplanted the use of kava and caused substantial devastation to the indigenous cultures. Fortunately, this situation has seen some reversal in recent decades, as an increase in ethnic identity has given new life to traditional values. As a result, large amounts of kava are once again being consumed in many places, and this has helped to successfully counteract the growth of alcoholism.

Of all of the psychoactive plants that have been introduced into Australia, kava appears to have acquired the greatest significance among the Aborigines. Since 1980, kava drinking has been part of the culture of the Northern Territory (Lebot et al. 1992, 72, 199–202). Some Aborigines use it to treat alcoholism, while others drink such high overdoses of kava that new problems have arisen (Prescott and McCall 1988; Singh 1992, 17).

Kava was first used therapeutically in Europe around 1820. It was initially used primarily in the treatment of venereal diseases (Lewin 1886, 17).

The first pharmacognostic and pharmacological studies were carried out at the close of the nineteenth and the beginning of the twentieth centuries (Lewin 1886; Penaud 1908). Today, kava is a popular "natural tranquilizer" (Vonarburg 1996, 61).

Distribution

The original home of kava is unknown; it is occasionally found in New Guinea and on the New Hebrides. Wild plants are unknown, although stands of plants that have become wild are encountered from place to place. Since all of the cultivars are sterile, the plant can have spread only through human activity. It may have developed from *Piper wichmannii* C. DC.

Prehistoric Polynesians brought the plant to Hawaii (= Sandwich Islands) at a very early date. Once there, it spread quickly (Krauss 1981, 2*).

The plant does not occur in New Zealand (cf. *Macropiper excelsum*) or on Easter Island (Whistler 1992a, 185).

Cultivation

The plant is propagated from cuttings (approximately 15 to 20 cm long) taken from the lower stems or from young stems separated from the rootstock (also called a stump) when the root is harvested. The new plant develops shoots after a short growth period. The plant grows into a substantial shrub and is ready to be harvested after five to six years at the most. Kava plantations are fertilized almost exclusively with ash from wood and are well tended:

> The cultivation of kava requires great care, skill, and diligence. The soil is often subjected to treatment with the rake for this purpose, freed of weeds and fertilized with lime from shells and coral. . . . In areas where the plant is still cultivated, it is a question of honor for every family to grow good kava. Before the arrival of the missionaries, the kava fields were divided into three parts. The best was given to the gods that can cause harm—it was taboo, i.e., sacrosanct, the second to the atuas, the gods of sleep, and the third was the family's portion. . . . . It was preferred to locate the plantations in places that were raised, on cliffs, and dry. But when there was no other way, one could also find the plants in lower and wetter valleys at the margins of rivers. The plants that develop here do not taste as well and are less aromatic than the former. The plantations are reminiscent of young fig plantations. (Lewin 1886, 13)

The most important commercial areas for growing kava are now found in Samoa, Fiji, and Vanuatu.

Appearance

This bushy, evergreen shrub usually grows to a height of about 2 meters, although it can grow to a height of more than 5 meters. The light green, alternate, heart-shaped leaves can grow as long as 30 cm. The greenish white male inflorescences can attain a length of up to 6 cm and form in spikes attached at the leaf axils; female flowers are unknown (Whistler 1992a, 185). The fruits are

Left: A variety of inebriating pepper (*Piper methysticum*) that has dark stems is known as the black awa form. (Photographed on Oahu, Hawaii)

Center: The large rootstocks of the Oceanian inebriating pepper (*Piper methysticum*) are used in both fresh and dried form to prepare the tonic, stimulating, and inebriating kava drink.

Right: Kava, the inebriating pepper (*Piper methysticum*), is Oceania's most important psychoactive plant. (Photographed in Hawaii)

said to form one-seeded berries (Lewin 1886). The juicy root (stump) can grow very large, develop multiple branches, and weigh from 2 to 10 kilos.

Piper methysticum is easily confused with similar *Piper* species (e.g., *Piper tutuilae* C. DC.), which are also called *kava* or *ava* (Uhe 1974, 23*).

The closely related species *Piper puberulum* (Benth.) Benth. var. *glabrum* (C. DC.) A.C. Sm. [syn. *Macropiper puberulum* Benth., *Piper macgillivrayi* C. DC. ex Seem.], which is very common in Tonga, is similar in appearance (although it has red inflorescences) and is known as *kavakava'uli* or *kavakava'ulie*, and on Niue even as *kavakava*, but it is not used for psychoactive purposes (Weiner 1971, 443; Whistler 1992b, 73 f.; Whistler 1992a, 169). Another quite similar species is *Piper latifolium* Forst. (also known as bastard kava; Lewin 1886, 8), which grows on the Marquesas. On the Society Islands, *P. latifolium* is known as *avavahai*. Any psychoactive use of this plant is unknown (Steinmetz 1973, 6).

The kava plant is so similar to the American species **Piper auritum** that almost the only way to distinguish the two species is by the scent of their leaves. Steinmetz reported a Caribbean species that is also very similar (*Piper plantagineum* Schlecht.) and that the natives of the West Indies or Mexico allegedly once used in a similar manner to kava (Steinmetz 1973, 6).

Psychoactive Material

— Root (rhizome, kava-kava rhizome, kava-kava rootstock, kava pepper root, piperis methystici rhizoma, radix kava-kava, rhizoma kava-kava, rhizoma kavae, waka); usually the peeled stump that has been freed of small roots

The dried plant material must be stored away from light. The stump loses some 60% of its moisture as it dries. The kava from Vanuatu is especially high in quality.
— Fresh leaves
— Fresh or dried stems (lewana)

Preparation and Dosage

The freshly dug root is freed of its small secondary roots, peeled and chopped, and then prepared while either fresh or dried. Kavains (kavapyrones) are not easily soluble in water but do dissolve well in alcohol. For this reason, it is best to prepare an alcohol tincture of the stump. In the pharmaceutical industry, the dried root is used to obtain alcohol/water or acetone extracts with 94% ethanol and 1% ethylmethylketone. The yield, or kavapyrone content, is greatest in a pure alcohol extract (31.6 to 35.4%) and makes up some 30% in alcohol/water mixtures (cf. Hölzl et al. 1993, 203). Sixty to 120 mg of kavapyrones is listed as a medicinal dosage (the amount can vary considerably depending upon the preparation); in clinical studies, 200 to 300 mg were administered daily for a period of several days. In spite of the daily use by countless numbers of Polynesians, the pharmaceutical literature warns against using the plant for a period exceeding three months. Pregnant women and people with endogenous psychoses should also avoid kava (Hölzl et al. 1993, 210).

The traditional production of the refreshing and inebriating kava drink (also known as *ava*, *kavakava*, *sakau*, *wati*, *viti grog*, and *fiji grog*) is identical on almost all of the islands. Normally, the fresh roots are peeled and then chewed by young men (less frequently by girls or young women) for about ten minutes and insalivated. This process can increase the volume of the root pieces considerably. The chewed material is then mixed with water in special sacred vessels (kava bowls, *tanoo*, *kanoa*) made from the hard wood of *vesi* (*Intsia bijuga* [Colebr.] O. Ktze. [syn. *Afzelia bijuga* A. Gray]; Leguminosae [Caesalpiniaceae]) and "fermented" shortly before use. (In the early literature, one could occasionally read that the drink was allowed to "ferment"; this information, however, appears to be based on an error; Lewin 1886, 24.) The resulting milky drink is filtered through a sieve made from the inner bark of *Hibiscus tiliaceus* L. (*vau*, *fau*) or from coconut fibers (*Cocos nucifera*) and poured into drinking bowls. The drink is consumed only while fresh, as it becomes flat and unappetizing if allowed to stand for too long (Steinmetz 1973, 13 ff.).

The finished kava drink has a dark, sometimes brown, yellow, or gray cloudy color and a characteristic taste that can differ in aroma but may also be soaplike, very bitter, or astringent. The drink induces an anesthesia on the surface of the mouth similar to that produced by coca (cf. **Erythroxylum coca**).

In Fiji, the kava drink was once prepared not

by chewing (*mama*) the root but by grating it with large mushroom corals (a practice that presumably was also found in other places) (Ford 1967, 165). In Hawaii, kava was made using coconut milk (cf. *Cocos nucifera*) instead of water (Krauss 1981, 2*). In addition, Hawaiian Huna sorcerers (*kahunas*) would boil a poisonous drink from roots collected on days of heavy rain together with the leaves of *Tephrosia piscatoria* [syn. *Theophrosia purpurea*] *Daphne indica*, and a *Lagenaria* species (Kepler 1983; McBride 1988; cf. also Singh 1992, 15).

Typically, each person drinks one to four coconut shells' worth of kava drink (= 0.5 to 2.0 liters) at the kava ceremonies. Many Polynesians drink a couple of bowls of freshly prepared kava every day. Some "enthusiastic kava drinkers consume the drink 6 to 8 times a day" (Lewin 1886, 19).

The old notion that kava acquires its inebriating or psychoactive effects only after it has been "fermented" (insalivated) has been clearly refuted (Schmidt 1994, 376 f.). However, the insalivation does appear to enable the kavapyrones (which do not easily dissolve in water) to release in the emulsion and thus be absorbed when the fresh beverage is consumed.

The inebriating (psychoactive) effects become apparent only after the consumption of several liters: "A certain numbness appears only after the ingestion of some 9 liters of the kava drink" (Vonarburg 1996, 58). Chronic consumption of very high doses (13 liters per day, corresponding to approximately 310 to 440 g of dried rootstock) can lead to toxic effects (rash, hair loss, yellow coloration of the skin, reddening of the eyes, loss of appetite, et cetera) (Hölzl et al. 1993, 211). Daily dosages of 4 liters or less will not induce these symptoms or will do so only extremely rarely.

The traditional methods of preparation use some 100 g of dried plant material per 100 ml of water, corresponding to about 70 mg of kavapyrones, oftentimes more (Hölzl et al. 1993, 203). The lethal dosage for humans is unknown. In mice, the LD_{50} is 1,500 mg of kavapyrones per kilogram of body weight (Hölzl et al. 1993, 212).

The inebriating effects of kava can also result from or be potentiated by various additives:

Kava Additives
Other substances are occasionally added to the kava drink (Holmes 1967, 107; Lewin 1886, 23; Singh 1992, 23):

Chili pods	*Capsicum* spp.	Polynesia
Kava leaves	*Piper methysticum*	New Guinea
Coconut milk	*Cocos nucifera*	Hawaii
Thorn apple seeds	*Datura metel*	Fiji
Yagoyagona extract	*Piper puberulum*	Fiji

But kava can also be used alone and without any preparation. A piece of the fresh rootstock about as long and as thick as a finger is a good dosage for inducing psychoactive effects. It should be chewed well and then swallowed. The effects of kava appear to be potentiated by the addition of *Cannabis*.

A tonic can be prepared by emulsifying equal parts of ground kava root and lecithin in a blender. Kava roots are sometimes used as an ingredient in **betel quids**. It is possible that kava roots and honey may be used to brew a **mead** whose effects are more inebriating than those of a cold-water extract of chewed roots. Whether the inebriating beverage known as **keu** was indeed made from *Piper methysticum*, as has been suggested, is unknown.

In the Society Islands, juice from the root of *Piper tristachyon* was formerly used to "ferment" an inebriating beverage known as *ava ava* (von Reis Altschul 1975, 45*).

Ritual Use
The traditional ritual uses of kava include the kava ceremonies as well as the use of the plant for magical purposes. The more original kava ceremonies are especially well documented in the ethnographic literature and still exist, in the same or at least a similar form, in Fiji, Samoa, and Vanuatu (Lebot et al. 1992; Singh 1992).

Kava ceremonies range from formal to informal in nature. They can function as a greeting for guests, as a part of tribal deliberations, and as a part of the relaxing, social drinking rounds that take place in the evenings. The basic pattern of the ceremonies is always the same. First, the drink is prepared, accompanied by prayers and songs. Then the participants sit either in two groups, one facing the other, or in a circle. The priest, chief, politician, or host distributes equal portions of the drink to all of the participants. The ceremony, which is usually accompanied by collective singing, ends after a number of rounds. At the conclusion, the location at which the ceremony has taken place, the temple, and the ceremonial objects are all cleansed. Sometimes dancing accompanies the ceremony (Singh 1992).

In some places, only men are allowed to take part in the kava ceremonies, while on other islands everyone can drink. The women of Tonga once had their own drinking societies (Lewin 1886, 20). Some initiation ceremonies, such as the initiation of girls into the sacred hula dances, also involve kava. On Niue, it was once only the priests who drank kava, which they did to obtain visions (Singh 1992, 16).

Any person who saw the shark-shaped sea god Sekatoa in the water would have to ceremonially purify himself with a kava drink (Singh 1992, 28).

At their ceremonies or libations, the

Samoans—through their chief—ask the gods for health, long life, a good harvest, and success in war. In Samoa, the largest roots are called *lupesina* ("great respect"); they are presented as gifts to people of respect but are not consumed (Cox and O'Rourke 1987, 454).

Kava roots were or are placed as offerings in temples and shrines or hung together with small branches of *Waltheria americana*. Kava roots also are placed on the graves of deceased family members as a last farewell. Perhaps this should be seen in the context of certain mythological traditions, according to which the first kava plant grew upon the grave of a Tongan leper. On the Marquesas Islands, it is believed that the plant was born as a child of the god Atea, who provides food, sends the rain, is the lord of the farmers, and was transformed into the inebriating plant. One story told in Tonga describes how the cooked daughter of the host was placed before the great chief Loua during a feast. When he smelled the roast, he had the well-done flesh buried. The first kava plant grew from the grave. In Vanuatu, it is said that an old man observed a rabbit chewing on a kava root. After watching this on several occasions, he tried the root himself and invented the kava drink (Singh 1992, 18 f.).

On the islands of Vanuatu and other islands of the South Pacific, kava is used in magic, especially magic intended to harm others (Singh 1992, 29). The practice is known as *elioro* in Vanuatu and is used to send out disease or death to a specific person. The sorcerer buries a "deadly object"— usually a kava root upon which incantations have been uttered or a blood-filled bamboo tube—at a spot where it is assumed that the intended victim will pass by. By passing or, even better, walking over the spot, the unsuspecting victim assimilates the harmful magic and then becomes ill or dies (Ludvigson 1985, 56). In contrast, in Hawaii kava is regarded as a means for removing magic (Singh 1992, 15).

Artifacts

The majority of the artifacts associated with kava are those used in its preparation and consumption (shells, bowls, mortars, drinking vessels).

The large, round wooden bowls used in preparing kava frequently feature carved legs (often depictions of people). Strings made of coconut fiber are used to attach cowrie shells (*Cypraea moneta* L., *Cypraea annulus* L.) to the kava bowls of the chiefs for magical protection. In Samoa, the *wa ni tanoa*, "king's vessel," was sometimes decorated with the renowned gold cowrie (*Cypraea aurantium* Gmelin), the symbol of the ruler's office (Ford 1967, 166, 167).

The drinking vessels of Fiji (*m'bilo, bilo ni yagona, ipu'ava, 'apu 'awa*) are made from halves of coconut shells (*Cocos nucifera*) to which strings of coconut fibers are sometimes attached. The resinous remnants of the drink impart a glasslike finish to these coconut shells after they have been used enough times. This layer is sometimes scraped off and ingested as an especially potent form of kava (Lewin 1886, 27; Singh 1992, 26). In Tonga, banana leaves are woven together to make single-use kava cups. On the Hawaiian and other Polynesian islands, ritual kava-drinking vessels are made from calabash gourds (*Lagenaria* spp.) (Dodge 1995).

In Fiji and Samoa, there are numerous kava songs that are sung at ceremonies, when greeting people, when making kava, and on other occasions. Some of these songs have been published in ethnomusical recordings (e.g., *Unique Fiji: The Nakamakama Villagers in Mekes and Songs*, Olympic Records no. OL-6159, 1979). One psychedelic rock band from England took its name—Kava Kava—from that of the inebriating plant. The plant can also be seen in the paintings of some Hawaiian and Polynesian artists.

Medicinal Use

In Samoa, kava is regarded as an aphrodisiac, tonic, and stimulant. The rootstock is used to treat gonorrhea and elephantiasis (Uhe 1974, 23*; Weiner 1971, 443). The plant is widely used as an internal and external analgesic (Whistler 1992a, 186).

In Hawaii, restless and feverish children are given in the morning and in the evening kava roots that their mothers have prechewed (Krauss 1981, 2*). In Tonga, an infusion of crushed yellow (semiwilted) leaves is administered to crying children as a calmative (Weiner 1971, 443). In New Caledonia, the fresh leaves are chewed for bronchitis (Weiner 1971, 443); in Tonga, the fresh leaves are rubbed onto the stings of giant centipedes, insects, and poisonous fish (Whistler 1992b, 73). In Oceania, kava is used as an antidote for poisoning by **strychnine** or *Strychnos nux-vomica* (Pfeiffer et al. 1967, 155; Schmidt 1994, 474), a traditional use whose effectiveness has been pharmacologically verified (Singh 1992, 39).

In Papua New Guinea, great quantities of kava are chewed and swallowed to induce a kind of numbness for painful tattooing procedures (Steinmetz 1973, 23).

In Western phytotherapy, kava preparations are used to treat states of nervous anxiety, tension, and restlessness (Hölzl et al. 1993, 210; Schmidt 1994) and—according to the claims of certain pill manufacturers—to increase concentration and performance (Hänsel and Woelck 1995). Preparations in which kava is combined with St. John's wort (*Hypericum perforatum* L.) are used as mild antidepressants (cf. Becker 1994, 3*). The essence or mother tincture (Piper methysticum hom. *HAB34*, Piper methysticum hom. *HPUS88*)

One English band from the early 1990s took its name, Kava Kava, from that of the inebriating pepper of the South Seas. The band's music incorporated both psychedelic and funk influences. However, the CD booklet does not tell us whether the band members were inspired by kava or by more traditional European psychedelics. (CD cover 1995, Delerium Records)

is used in homeopathy for such conditions as states of excitation and exhaustion (Hölzl et al. 1993, 212).

Constituents

Kavalactones (= kava pyrones, kavapyrones, α-pyrones, kavains) occur in all parts of the plant, usually totaling a concentration of over 5%, with 1.8% kavain, 1.2% methysticin (= kavahine, kavakin, kavatin, kanakin), 1% demethoxy-yangonin, 1% yangonin, 0.6% dihydrokavain, 0.5% dihydromethysticin, and traces of dihydrokavain-5-ol, 11,12-dimethoxyhydrokavain, 11-hydroxy-12-methoxykavain, 11-methoxy-nor-yangonin, 11-methoxy-yangonin, and the two ethylketones cinnamoylacetone and methylendioxy-3,4-cinnamoylidenacetone (Schulgin 1973; Young et al. 1966). The plant has been found to contain amides (2-methoxy cinnamic acid pyrrolidide, cinnamic acid pyrrolidide), chalcones (flavokavin A and B), and free and aromatic acids (anisic acid, benzoic acid, capronic acid, hydroxy cinnamic acid, and derivatives) (Hölzl et al. 1993, 202; Klohs 1967). A pale yellow **essential oil** has also been described (Lewin 1886, 30).

The leaves contain 0.71% transient piper-methysticin (an alkaloid); this compound is found in the stems in lower concentrations but not in the roots (Cox and O'Rourke 1987, 454). Dihydro-kavain, dihydromethysticin, and yangonin are present in the stems. Trace amounts of the substance cepharadione A were discovered in the roots (according to the DAB supplemental volume 6). This substance is also found in other *Piper* species (*Piper* spp.) (Jaggy and Achenbach 1992).

Kavapyrones are chemically related to longi-stylines (cf. *Lonchocarpus violaceus*, **balche'**).

Effects

Potent psychoactive effects of the local drink have been reported particularly for Pohnpei (Ponape) (Hambruch 1917; Thurnwald 1908). It is said that after several rounds, the participants in the drinking ritual leave their bodies and are able to glide over the tropical island world in a disembodied state and journey to the heavens, to the home of the kava plant. They experience sensations of fraternization and unity with their environment as well as erotic visions. These and similar statements in the older literature, according to which kava may have hallucinogenic effects, have been cast in serious doubt by many authors who have had numerous experiences of their own (Cox and O'Rourke 1987, 454). The legendary hallucinogenic effect has occasionally been attributed to the additives that may be used (in particular *Datura metel*; see above).

Frequent mention is made of euphoric effects that begin shortly after the consumption of larger amounts and subside some two to three hours

later (Roth et al. 1994, 572*). There is general agreement among both the authors and the kava consumers alike that the drink quenches thirst better than **beer**, has mild stimulating and invigorating effects that revitalize the body after strenuous exertion, clears the head, and stimulates the appetite. In contrast, the aphrodisiac or anaphrodisiac effects are the subject of debate (Lewin 1886; Steinmetz 1973). "Too, *kava* is a means of maintaining or enhancing intimacy" (Gregory 1995, 44). Louis Lewin summarized the reported psychoactive effects in the following way:

> Following not too large amounts, a sensation of happy lightheartedness, comfort, and satisfaction appears without any physical or mental excitation. At first, speaking is easy and free and the vision and hearing are more acute for finer impressions. The agent reveals a calming power. The drinkers never become angry, mad, quarrelsome, or paralyzed as with alcohol, which the Fiji Islanders also especially esteem as an advantage of this beverage. The natives and the whites regard it as a sedative in cases of accidents. Both consciousness and the rational faculties remain intact. When somewhat larger quantities are consumed, then the limbs become limp; the muscle power no longer appears to be under the jurisdiction and control of the will; walking becomes slower and more unsteady; the people appear as if half-drunk; one feels the need to lie down. The eye sees objects that are present but does not want to and cannot fix upon them on command, just as the ear perceives without being able or willing to give an account of that which is being heard. An overpowering tiredness and a need to sleep that controls every sensation becomes apparent in the drinker; he becomes somnolent and finally falls asleep. Some Europeans have observed this power of kava to lame the senses and ultimately lead to sleep, which is like magic, on their own selves. Often, it merely produces a torpid/somnolent state accompanied by disconnected dreams and, according to some reports, by erotic visions as well. (Lewin 1886, 44 f.)

Numerous pharmacological studies have demonstrated that the psychoactive effects of kava are due to the kavapyrones; moreover, they are not caused by one isolated substance but instead appear to be due to the mixture (Meyer 1967, 140). In experiments with mice, extracts have produced strong sedative effects (Hölzl et al. 1993, 203):

> Like meprobamat or benzodiazepine [cf. **diazepam**], the kavapyrones are capable of lowering the excitability of the limbic system,

Kavain

Methysticin

whereby the inhibition of the activity of the limbic system is regarded as an expression of a suppression of emotional excitability and an improvement in the mood. (Hölzl et al. 1993, 204)

Muscle-relaxing, antispasmodic, analgesic, local anesthetic, and nerve-protecting effects have all been pharmacologically demonstrated. The kavapyrones also cause a prolongation or deepening of anesthesia (induced, e.g., by chloroform, ether, laughing gas, or barbiturates), for which methysticin has the strongest synergistic effects. Kava extracts have antagonistic effects on dopamine, apomorphine, and amphetamine (cf. **ephedrine**) (Hölzl et al. 1993, 205; Meyer 1976). Kava also potentiates the effects of **alcohol** (e.g., the duration of sleep following inebriation; cf. Zubke 1997). The local anesthetic effects are very similar to those of **cocaine**, procaine, and lidocaine, and the duration of effects is similar (Hölzl et al. 1993, 206; Meyer and May 1964; Singh 1992, 40). There is some evidence suggesting that the kavapyrones bind to the GABA and/or benzodiazepine receptors ([^3H]-GABA bond, [^3H]-diazepam bond), thereby exhibiting an affinity similar to that of **muscimol** and **diazepam** (Hölzl et al. 1992). Human pharmacological studies on healthy subjects using 210 mg or even 300 to 600 mg of kavapyrones per day have demonstrated that the quality of sleep is improved, anxiety states are dissipated, and information processing in the brain is improved, while reaction times are unaffected (Hölzl et al. 1993, 207; Hänsel and Kammerer 1996). Often the desired effects do not become apparent until after several days of regular consumption (Schmidt 1994, 376). In rare instances, kava use may result in mild allergic reactions. However, "there are no indications of physical and/or psychological dependency" (Hölzl et al. 1993, 210).

It has frequently been reported that kava can induce marijuana-like effects (cf. *Cannabis indica*), but that these effects are very subtle and are perceived only following repeated ingestion of the substance (Miller 1985, 59*; Zubke 1997).

Commercial Forms and Regulations

Kava, both raw and in its various preparations, is available without restriction throughout the world (even in herb shops, health food stores, supermarkets, et cetera). Many South Pacific islands have bars in which no alcohol is served but various preparations of kava are.

Numerous preparations and products (capsules, tablets, coated tablets, solutions, tinctures) are available in European and Western markets, including capsules containing kava extracts and the oil of St. John's wort (*Hypericum perforatum* L.; cf. Becker 1994*) for treating stress and capsules with extracts of kava and valerian (*Valeriana officinalis*) for relaxation. Each of the Antares® 120 tablets contains 120 mg of kavapyrones; these thus have one of the highest concentrations of all commercial forms (Schmidt 1994, 376). Each capsule of the psychopharmacological agent known as Neuronika contains 200 mg of kavain (cf. Kretschmer 1970). Many products contain only 10 mg of kavapyrones in each pill.

Literature

See also the entries for *Piper auritum*, *Piper betle*, *Piper* spp., *Macropiper excelsum*, **keu**, and **betel quids**.

Brunton, R. 1989. *The abandoned narcotic: Kava and cultural instability in Melanesia.* Cambridge: Cambridge University Press.

Buckley, Joseph P., Angelo R. Furgiuele, and Maureen J. O'Hara. 1967. Pharmacology of kava. In *Ethnopharmacologic search for psychoactive drugs*, ed. D. Efron, 141–51. Washington, D.C.: U.S. Dept. of Health, Education, and Welfare.

Churchill, W. 1916. *Sissano: Movements of migration within and through Melanesia.* Washington, D.C.: Carnegie Institution. (*See* pages 124–44.)

Cox, Paul Alan, and Lisa O'Rourke. 1987. Kava (*Piper methysticum*, Piperaceae). *Economic Botany* 41:452–54.

Dodge, Ernest S. 1995. *Hawaiian and other Polynesian gourds.* Honolulu: Ku Pa'a Publishing.

Ford, Clellan S. 1967. Ethnographical aspects of kava. In *Ethnopharmacologic search for psychoactive drugs*, ed. D. Efron, 162–73. Washington, D.C.: U.S. Dept. of Health, Education, and Welfare.

Gajdusek, D. Carleton. 1967. Recent observations on the use of kava in the New Hebrides. In *Ethnopharmacologic search for psychoactive drugs*, ed. D. Efron, 119–25. Washington, D.C.: U.S. Dept. of Health, Education, and Welfare.

Garner, Leon F., and Jeremy D. Klinger. 1985. Some visual effects caused by the beverage kava. *Journal of Ethnopharmacology* 13 (3): 307–11.

Gregory, Robert J. 1995. Reflections on the kava (*Piper methysticum*, Forst.) experience. *Integration* 6:41–44.

Hambruch, P. 1917. Die Kawa auf Ponape. *Studien und Forschungen zur Menschen- und Völkerkunde* 14:107–15.

Hänsel, R. and H. U. Beiersdorff. 1959. Zur Kenntnis der sedativen Prinzipien des Kava-Rhizoms. *Arzneimittel-Forschung* 9:581–85.

Hänsel, Rudolf, and Susanne Kammerer. 1996. *Kava-kava*, Basel: Aesopus.

Hänsel, Rudolf, and Helmut Woelck. 1995. *Spektrum Kava-Kava.* 2nd ed. Arzneimitteltherapie heute. Basel: Aesopus.

A very early illustration of the production of kava. (Woodcut from *Captain Cook's Three Famous Voyages around the World*)

"The wonder drug of Polynesia represents an enrichment of our treasure of phytotherapeutic medicines. Its use for difficult states of unrest and anxiety is not a task for self-medication. For many of modern man's disturbances of health ('nervous sleep disturbance, stress-related tension'), it is certainly worth trying this medicinal plant, which is to be preferred over the quick use of the benzodiazepines [cf. **diazepam**]."

MICHAEL SCHMIDT
KAVA-KAVA
(1994, 377)

"During a time of hunger and need, the Tui Tonga (king of Tonga) wanted to visit his subjects on the island of Eua and sent his messengers to the chief of the island to announce the impending visit. The chief and his wife were put in a difficult position because they lacked all that they needed for the festive greeting in honor of the divine Tui Tonga. After grappling between their hearts and their duty for days and nights, the no-longer-young couple decided to offer Tui Tonga their only child, a beautiful little girl. The beautiful child was dead when Tui Tonga arrived, the parents were grieving beyond all measure, and the entire island was grieving with them. The king was overcome by the events and deeply moved by the faithfulness of his subjects. Tui Tonga bade the mourners to place the dead child in the ground, and as he departed he promised that he would provide them with consolation during the year. Exactly one year to the day later, a plant that no one had ever seen before grew above the head of the victim. The chief called it 'kava,' and a new plant shot forth from the foot end and was given the name *do* (sugarcane). So did kava come to the earth as a mild relief for grieving souls, and *do*, the sweet sugarcane, was given to us to counteract the bitterness of memory that can poison and exhaust the body."

WALTER HURNI
"KAVA—GESCHENK DER GÖTTER"
[KAVA—GIFT OF THE GODS]
(1997, 66)

Holmes, Lowell D. 1967. The function of kava in modern Samoan culture. In *Ethnopharmacologic search for psychoactive drugs*, ed. D. Efron, 107–18. Washington, D.C.: U.S. Dept. of Health, Education, and Welfare.

Hölzl, Josef, S. Wiltrud Juretzek, and Elisabeth Stahl-Biskup. 1993. *Piper*. In *Hagers Handbuch der pharmazeutischen Praxis*, 5th ed., 5:52–59. Berlin: Springer.

Hurni, Walter. 1997. Kava—Geschenk der Götter. *Natürlich* 17 (11): 65–68.

Jaggy, H., and H. Achenbach. 1992. Cepharadione A from *Piper methysticum*. *Planta Medica* 58:111.

Keller, F., and Murle W. Klohs. 1963. A review of the chemistry and pharmacology of the constituents of *Piper methysticum*. *Lloydia* 26:1–15.

Kepler, Angela Kay. 1983. *Hawaiian heritage plants*. Honolulu: Oriental.

Kilham, Chris. 1996. *Kava: Medicine hunting in paradise*. Rochester, Vt.: Park Street Press.

Klohs, Murle W. 1967. Chemistry of kava. In *Ethnopharmacologic search for psychoactive drugs*, ed. D. Efron, 126–32. Washington, D.C.: U.S. Dept. of Health, Education, and Welfare.

Koch, Gerd. 1981. Kawa in Polynesien. In *Rausch und Realität*, ed. G. Völger, 1:194–99. Cologne: Rautenstrauch-Joest-Museum für Völkerkunde.

Kretschmer, Wolfgang. 1970. Kavain als Psychopharmakon. *Münchener Medizinische Wochenschrift* 112 (4): 154–58.

Lebot, Vincent, and P. Cabalion. 1988. *Kavas of Vanuatu: Cultivars of* Piper methysticum *Forst*. Technical Paper no. 195. Nouméa, New Caledonia: South Pacific Commission.

Lebot, Vincent, Mark Merlin, and Lamont Lindstrom. 1992. *Kava: The Pacific drug*. New Haven, Conn., and London: Yale University Press. (Cf. book review by John Baker in *Yearbook for Ethnomedicine and the Study of Consciousness*, 1994 (3): 355 f. Berlin: VWB.)

Lewin, Louis. 1886. *Über* Piper methysticum *(Kawa)*. Berlin: August Hirschfeld.

Ludvigson, Tomas. 1985. Healing in central Espíritu Santo, Vanuatu. In *Healing practices in the South Pacific*, ed. Claire D. F. Parsons, 51–64. Honolulu: University of Hawaii Press (The Institute for Polynesian Studies).

McBride, L. R. 1988. *Practical folk medicine of Hawaii*. Hilo: Petroglyph.

Meyer, Hans J. 1967. Pharmacology of kava. In *Ethnopharmacologic search for psychoactive drugs*, ed. D. Efron, 133–40. Washington, D.C.: U.S. Dept. of Health, Education, and Welfare.

Meyer, Hans J., and H. U. May. 1964. Lokalanästhetische Eigenschaften natürlicher Kawa-Pyrone. *Klinische Wochenschrift* 42:407.

Penaud, A. 1908. *Le kawa-kawa*. Bordeaux: Thèse de doctorat.

Pfeiffer, Carl C., Henry B. Murphree, and Leonide Goldstein. 1967. Effect of kava in normal subjects and patients. In *Ethnopharmacologic search for psychoactive drugs*, ed. D. Efron, 155–61. Washington, D.C.: U.S. Dept. of Health, Education, and Welfare.

Prescott, J., and G. McCall, eds. 1988. *Kava: Use and abuse in Australia and the South Pacific*. Monograph no. 5. Sydney: University of New South Wales, National Drug and Alcohol Research Center.

Schmidt, Michael. 1994. Kava-Kava: Heilpflanze aus der Südsee. *PTA heute* 8 (5): 374–78.

Shulgin, Alexander T. 1973. The narcotic pepper: The chemistry and pharmacology of *Piper methysticum* and related species. *Bulletin of Narcotics* 25:59–74.

Singh, Yadhu N. 1983. Effects of kava on neuromuscular transmission and muscle contractility. *Journal of Ethnopharmacology* 7:267–76.

———. 1986. *Kava: A bibliography*. Suva, Fiji: University of the South Pacific, Pacific Information Centre.

———. 1992. Kava: An overview. *Journal of Ethnopharmacology* 37:13–45. (Contains an excellent bibliography that provides a good basis for further research.)

Steinmetz, E. F. 1973. *Kava-kava: Famous drug plant of the South Sea islands*. San Francisco: Level Press.

Thurnwald, Richard. 1908. Nachrichten aus Nissau und von den Karolinen. *Zeitschrift für Ethnologie* 40:106–15.

Vonarburg, Bruno. 1996. Kava-Kava stellt sie wieder auf die Beine. *Natürlich* 3/96:57–61.

Weiner, Michael A. 1971. Ethnomedicine in Tonga. *Economic Botany* 25:423–50.

Whistler, W. Arthur. 1992a. *Polynesian herbal medicine*. Hawaii: National Tropical Botanical Gardens.

———. 1992b. *Tongan herbal medicine*. Honolulu: University of Hawaii Press.

Williamson, R. W. 1939. *Essays in Polynesian ethnology*. Cambridge, U.K.: Cambridge University Press. (See pages 51–112, 274–75.)

Young, Richard L., John W. Hylin, Donald L. Plucknett, Y. Kawano, and Roy T. Nakayama. 1966. Analysis for kavapyrones in extracts of *Piper methysticum*. *Phytochemistry* 5:795–98.

Z[ubke], A[chim]. 1997. Kava: Die Südseedroge. *Hanfblatt* 4 (28): 29–31.

Piper spp.

Pepper Species

Family
Piperaceae (Pepper Family); Pipereae Tribe

The genus *Piper* includes some 1,000 to 1,200 species, many of which are ethnobotanically significant (Hölzl et al. 1993, 191; Schultes and Raffauf 1990, 364*). Half of all *Piper* species occur in the American tropics. These include epiphytic plants, climbers, half-shrubs, and small trees. A large number of **essential oils** occur in the genus, so many leaves, inflorescences, and fruits are highly aromatic and have therefore attracted cultural attention. Some *Piper* species are said to have psychoactive, and others aphrodisiac, effects. Safrole and asarone have been identified in various species (such as *Piper divaricatum* Meyer, *P. manassausense*, *P. futokadsura*, and *P. sarmentosum*) (Avella et al. 1994). *Piper abutiloides* Kunth, *Piper cincinnatoris* Yuncker, and *Piper lindbergii* C. DC., which are used in Brazilian folk medicine as analgesics, are pharmacologically active (Costa et al. 1989). It has even been suggested that the common black pepper (*Piper nigrum* L.) is capable of inducing hallucinogenic effects (Schultes and Hofmann 1980, 368*).

The so-called red pepper comes not from a *Piper* species but from the Peruvian pepper tree (*Schinus molle* L.; cf. Norman 1991, 53*). In South America, it is used to aid in the fermentation of **chicha** and also as a **beer** additive.

Piper amalago L. [syn. Piper medium Jacq.]—amalago pepper
The leaves of this bush, which is indigenous to Central America (southern Mexico, Belize), are smaller and narrower than those of *Piper auritum*, but the plant is otherwise quite similar in appearance. When rubbed, its leaves smell strongly of the **essential oil** safrole. It may be possible to use this pepper species for psychoactive purposes. The Maya, who call the plant *yaaxpehelche'*, regard it as the "younger sibling" or "female" counterpart of *Piper auritum*.

Piper angustifolium Ruíz et Pavón—matico pepper
It is not known whether this American pepper species has psychoactive effects by itself. Because of the disinfectant properties of its fresh leaves, the plant is also known as soldier's herb. Its leaves and inflorescences are an ingredient in various Aztec cacao recipes (see **Theobroma cacao**) and have a mild stimulating effect because of the **essential oil** that is present (Rätsch 1991a, 185*). Some authors regard *Piper angustifolium* as a synonym for *Piper elongatum*, which is also known as matico pepper.

Piper cubeba L. [syn. Cubeba officinalis Miq. (or Raf.)]—cubeb pepper
This climbing shrub, which is indigenous to the Sunda Islands and eastern Asia, grows preferentially on *Erythrina indica* [syn. *E. variegata*] (cf. **Erythrina spp.**) and is the source of the fruit that is sold under the names *cubeb, kubeb, cubeb pepper, pimenta cubeba*, and *fructus cubebae* (Macmillan 1991, 415*; Norman 1991, 54*). The fruits contain 10 to 20% **essential oil**, 2.5% cubebin ($C_{20}H_{20}O_6$), and amorphous cubeb acid. Large doses of the essential oil can induce irritation in the urinary tract as well as headaches, which is why one of the fruit's folk names is dizzy corns. Such typical CNS symptoms as anxiety states and delirium have also been reported. Two grams has been given as a well-tolerated single dosage, while the daily dosage should not exceed 10 g (Roth et al. 1994, 570*). Hildegard von Bingen described the psychoactive effects as well as an anaphrodisiac effect that is difficult to understand:

> The cubeb is warm, and this warmth in itself is of the proper mixture, and it is also dry. And when someone eats cubeb, then any unseemly desires that are within him are moderated. But it also makes his spirits cheerful and his reason and knowledge pure, for the useful and moderate warmth of the cubeb extinguishes the unseemly flames of desire in which the stinking and slimy liquids are hidden, and it makes the spirit of man and his reason illuminatingly clear. (*Physica* 1.26)

Cubeb is used in folk medicine in cases of weakness of memory and to increase the sexual appetite (aphrodisiac) (Gottlieb 1974, 26 f.*; Hölzl et al. 1993, 196). In Yemen, where they are known as *kebâb*, the fruits are regarded as an aphrodisiac and nerve tonic (Fleurentin and Pelt 1982, 92 f.*). In former times, cubeb was often used as a spice. Today, it is used only in Asian cusine (e.g., as an

Many wild pepper species, such as the Central American *Piper amalago*, contain stimulating and inebriating essential oils. (Wild plant, photographed in Belize)

Red pepper is obtained not from a *Piper* species but from the American tree *Schinus molle* L. (From Hernández, 1942/46 [Orig. pub. 1615]*)

451

Left: The fruits of the aphrodisiac cubeb pepper (*Piper cubeba*).

Right: In India, the long pepper (*Piper longum*) is renowned as an aphrodisiac.

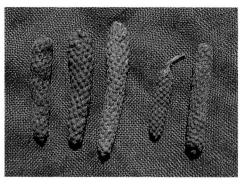

Botanical illustration of the aphrodisiac cubeb pepper. (Engraving from Pereira *De Beginselen der Materia Medica en der Therapie*, 1849)

"On Cubeb: The Javanese, because they attribute many good virtues to this fruit, are protective of it, and so that it cannot be reproduced by any nation, it is said that they lay it in warm water before they trade it, which makes it incapable of developing. Apart from this, they swallow these and eat them for all kinds of ailments, like fever, cold stomach, and such things. In addition, when soaked in wine and then drunk, it is said to help strengthen the work of Venus."

GEORGE MEISTER
DER ORIENTALISCH-INDIANISCHE KUNST- UND LUSTGÄRTNER [THE ORIENTAL-INDIAN ART AND PLEASURE GARDENER] (1677, CH. 9, P. 18*)

ingredient in curries). It is one of the primary ingredients in the Moroccan spice mixture *ras el hanout*, which also contains cardamom (*Elettaria cardamomum*), nutmeg fruits and flowers (*Myristica fragrans*), galanga (*Alpinia* sp.; cf. *Kaempferia galanga*), long pepper (*Piper longum*), cinnamon (*Cinnamomum verum*), cloves (*Syzygium aromaticum*), ginger (*Zingiber officinale*), rose buds (*Rosa* sp.), lavender flowers (*Lavandula angustifolia* Mill.), Spanish fly (*Cantharides*), ash berries (*Fraxinus* sp.?), paradise corns (*Amomum melegueta*), black pepper (*Piper nigrum*), peanuts (*Arachis hypogaea* L.), turmeric (*Curcuma longa*), cassia (*Cinnamomum cassia*), fennel seeds (*Nigella sativa*), monk's pepper (*Vitex agnus-castus*), belladonna (*Atropa belladonna*), and violet root (*Viola odorata* L.) (Norman 1991, 96 f.*). The consumption of large quantities of this spice mixture is said to produce psychoactive and aphrodisiac effects. Cubeb pepper is also an ingredient in **Oriental joy pills** and was once used as an additive to wine (see *Vitis vinifera*).

Piper elongatum Vahl [syn. *Artanthe elongata* (Vahl) Miq., *Piper angustifolium* Ruíz et Pavón, *Piper purpurascens* D. Dietr., *Steggensia elongata* (Vahl) Kunth]—matico pepper

The matico or soldiers' pepper comes from the Central and South American tropics and has a long history of use as a medicine and as an agent of pleasure. The leaves contain 0.3 to 6% **essential oil**, in which asarone and parsley apiol are present alongside the primary component, dillapiol (cf. *Acorus calamus*, *Petroselinum crispum*). Matico pepper is used in Panama as an aphrodisiac and stimulant (Hölzl et al. 1993, 198). In Mexico, it is one of the traditional spices for cacao (see *Theobroma cacao*). It is possible that mild psychoactive effects can result from the consumption of high doses of the leaves.

Piper interitum Trelease—tetsi pepper

The Kulina Indians of Peru use the leaves and roots of *Piper interitum*, which they call *tetsi*, to produce a **snuff** used as a substitute for tobacco snuff (cf. *Nicotiana tabacum*) that is alleged to have psychoactive properties (Schultes 1978b, 227*; Schultes and Raffauf 1990, 365 f.*).

Piper longum L. [syn. *Chavica roxbhurgii* Miq., *Chavica sarmentosa* (Roxb.) Miq., *Piper latifolium* Hunter, *Piper sarmentosum* Roxb.]— long pepper, pippali

In Asia and Arabia, the unripe fruits of the long pepper are used as a spice, an aphrodisiac, and a medicine (Fleurentin and Pelt 1982, 92 f.*; Rätsch 1995). They contain approximately 1% **essential oil** with sesquiterpene hydrocarbons and *p*-cymene, dihydrocarveol, terpinoles, and α-thujene as well as amides (piperidine and others). The drug has vasodilatory properties (Hölzl et al. 1993, 200). In Asia, long pepper has been used as a spice for much longer than black pepper (Norman 1991, 52*). While black pepper has been regarded as an aphrodisiac in Europe since ancient times, long pepper has an even greater reputation. Long pepper is a principal ingredient in numerous recipes for the aphrodisiac preparations used in tantric rituals (cf. **Oriental joy pills**). It is regarded as an "inciter" in Ayurvedic medicine. Its qualities are pungent, heating, and sweet, which is why it strengthens the functions of the genital system and is said to provide the organs of desire with a warming energy (Lad and Frawley 1987, 249*). The Ananga-Ranga, an ancient Indian book on the art of the love, lists a tantric "secret agent"— possibly with psychoactive effects—that awakens the lingam (= phallus) to life:

> Take a few corns of black pepper [*Piper nigrum*], seeds of the thorn apple [*Datura metel*], one pod of pinpalli (*Piper longum*, which yields the pepper that works slowly, or betel powder [*Areca catechu*]) with lodhra peel or *Morinda citrifolia*, which is used for dyeing; rub this with light honey and [rub it on the lingam]. This agent is unsurpassable. (Johann Wolfgang von Goethes, Ananga-Ranga 1985, 65)

The spice mixture *trikatu*, "three spices," which is widely known in India, consists of equal parts of long pepper, black pepper, and dried pieces of gingerroot (*Zingiber officinale*). This mixture is considered to be the most important Ayurvedic stimulant. *Trikatu* is a rejuvenator for *agni*, the inner fire. At the same time, it is important as an

agent that is taken together with other medicines; its stimulating effects potentiate or improve the assimilation of all kinds of active substances.

Piper plantagineum Schlecht.

This Caribbean species was once allegedly used in the West Indies (Mexico) in a similar manner to *Piper methysticum*; it may be identical to *Piper auritum*.

Piper sp.—syryboa

In his book *Der Orientalisch-Indianische Kunst-und Lustgärtner* [The Oriental-Indian Art and Pleasure Gardener] (1677), George Meister, who traveled to the East Indies, described a species of *Piper* that was used in a similar manner to or as a substitute for betel pepper (*Piper betle*):

> On Foliis Syryboae. These run lengthwise up the trees in the same way as folia bettele or pepper. The fruit is almost that of a long pepper species, pungent taste, looking like the so-called aments that hang on the hazel nuts in the spring, but somewhat thicker and longer, almost a span in length. These are cut from one another and eaten along with filled bettele leaves and the fruit areca [cf. *Areca catechu*]. In addition, they also take the flower, known as canange, which has yellow petals, with this, so that it has not just a pleasant scent but also a good taste. (Ch. 9, 20)

Unfortunately, the species of pepper described here as an additive to **betel quids** cannot be determined with certainty. The "canange flower" is very likely the blossom of the ylang-ylang tree (*Cananga odorata*; cf. **essential oils**).

Piper spp.—masho-hara

The Tanimuka and Yucuna Indians of the Río Miritiparaná (Amazonia) boil the very aromatic leaves of one *Piper* species to prepare a drink that is said to invigorate the elderly (Schultes 1993, 135*). Other species of *Piper* that are also known as *masho-hara* or *yauardi-hena* are used as ritual **snuffs** in Amazonia. The Muinane from the region of La Pedrera make a snuff from the dried leaves of a *Piper* species and tobacco (*Nicotiana tabacum*). Shamans chew or smoke various *Piper* species to track down cases of witchcraft. The Canelo use a *Piper* species that they call *guayusa* (cf. *Ilex guayusa*) as a stimulant (Schultes and Raffauf 1990, 367 f.*). One *Piper* species endemic to Papua New Guinea that has not yet been botanically described contains kavalactones (cf. **keu**).

Literature

See also the entries for ***Piper auritum*, *Piper betle*, *Piper methysticum*,** and ***Macropiper excelsum*.**

Atal, C. K., K. L. Dhar, and J. Singh. 1975. The chemistry of Indian *Piper* species. *Lloydia* 38:256–64.

Avella, Eliseo, Pedro P. Díaz, and Aura M. P. de Díaz. 1994. Constituents from *Piper divaricatum*. *Planta Medica* 60:195.

Costa, Mirtes, Luiz C. di Stasi, Mizue Kirizawa, Sigrid L.J. Mendaçolli, Cecilia Gomes, and Gustaf Trolin. 1989. Screening in mice of some medicinal plants used for analgesic purposes in the state of São Paulo. *Journal of Ethnopharmacology* 27:25–33.

Goethes, Johann Wolfgang v., ed. Ananga-Ranga. 1985. *Orientalische Liebeslehre*. Munich: Goldmann.

Hölzl, Josef, S. Wiltrud Juretzek, and Elisabeth Stahl-Biskup. 1993. *Piper*. In *Hagers Handbuch der pharmazeutischen Praxis*, 5th ed., 5:52–59. Berlin: Springer.

Ilyas, M. 1976. Spices in India. *Economic Botany* 30:273–80.

Rätsch, Christian. 1995. *Piper longum*, der ayurvedische Scharfmacher. *Dao* 6/95:68.

The common pepper (*Piper nigrum*, fresh fruits) has also been attributed with psychoactive and even hallucinogenic effects.

The pepper species (*Piper* sp.) known in Aztec as *mecaxochitl*, "cord flower," is used as an additive to Indian cacao drinks. (From Hernández, 1942/46 [Orig. pub. 1615]*)

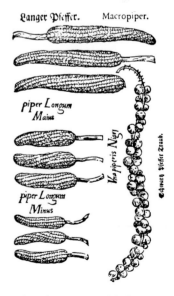

The infructescences of the long pepper (*Piper longum*) and black pepper (*Piper nigrum*) have been used as spices, stimulants, and aphrodisiacs. (Woodcut from Tabernaemontanus, *Neu Vollkommen Kräuter-Buch*, 1731)

Psidium guajava Linnaeus

Guajava Tree, Guava

Family
Myrtaceae (Myrtle Family)

Forms and Subspecies
There are a number of wild forms and cultivars, which differ primarily in the size of their fruits (Lutterodt and Maleque 1988, 219).

Synonyms
Psidium guajava Raddi
Psidium pomiferum L.
Psidium pyriferum L.

Folk Names
Aci'huit, äh pichib, al-pil-ca (Chontal), arazá, a'sihui't (Totonac), asiuit, asiwit, bec (Huastec), bek, bijui (Zapotec), bimpish (Shipibo-Conibo), bui, ca'aru (Coya), carú, chak-pichi (Mayan, "red guava"), chalxócotl (Aztec), coloc, cuympatan, djambubaum, enandi (Tarascan), gouyave, guabes-baum, guáibasim (Mayo), guajavabaum, guajava tree, guajave, guajave-apfel, guava,[281] guave, guavenbaum, guayaba, guayaba dulce, guayaba manzana, guayaba perulera, guayabilla, guayabilla cimarrona, guayabo, guayabo colorado, guayabo del monte, guayabo de venado, guayabo morada, guayavabaum, guyav, huayabo, jaljocote pichi, jalocote, jukoin papoxtiks, julú, kautonga, kolok, kuava, ku'ava, kuawa (Hawaiian), kuma (Siona), lacow (Huave), mo'eyi (Cuitlatec), ngoaba (Fang), ñi-joh (Chinantec), nulu (Cuna), pachi', palo de guayabo blanco, pata (Tzotzil), patan, pehui (Zapotec), pichi, pichi' (Mayan), pichib, pici, pitchcuy, pocs-cuy (Zoque), pojosh (Popoluca), posh-keip (Mixe), potoj, potos, pox (Mixe), poxr, puitá, sacpichi, sahuintu, saiyú, sumbadán (Zoque), tuava, tzon t kichi (Amuzgo), ushca-aru (Tepe-huano), vayevavaxi-te (Huichol), vî papalagi, xalá-cotl (Nahuatl), xalcolotl, xalxócotl, xaxokotl, xaxucotl (Náhutl), xoxococuabitl (Aztec), yaga-huií (Zapotec)

History
The plant was originally native from Mexico to Brazil but is now grown throughout the world as a tropical economic plant (Anzeneder et al. 1993, 59*). In Peru, it was already in cultivation by the eighth century B.C.E. (Root 1996, 105*). The first report about the guava tree is contained in *Relación de las cosas de Yucatán*, written by the Franciscan monk Diego de Landa (1524–1579). One of the earliest botanical descriptions of the tree as well as a copperplate engraving of the fruits were provided by the East Indies traveler George Meister (1677). The chewing of the leaves as a

narcotic and diarrhea medicine is well known in the tropics. In contrast, the psychoactive use of the leaves was only recently discovered in Ghana (Lutterodt and Maleque 1988, 220).

Distribution
The guava tree is apparently originally from Mexico but is now found in all of the tropical zones from Mexico to Peru (Dressler 1953, 154*). It also is planted in other parts of South America (Brazil, Paraguay, Argentina), where it sometimes occurs as a wild or feral plant (Santos Biloni 1990, 222*). It prefers a distinctly tropical climate and has spread throughout the world (Africa, Oceania, Southeast Asia, India).

Cultivation
Propagation occurs through cuttings or seeds. In nature, the seeds are spread by birds, bats, rodents, and humans. The seeds contained in the fruits pass through the digestive tract without harm and are excreted "well fertilized" (Lutterodt and Maleque 1988, 219).

Appearance
The small, gnarled evergreen tree, which does not grow much taller than 10 meters, has squamous bark and opposite elliptical leaves (5 to 15 cm long, 3 to 6 cm wide). The large white flowers have five petals and have a large number of stamens (up to 275). The fruits (approximately 7.5 cm in length) are initially green but turn yellow and exude a fruity aroma as they mature. The fruits of the wild form contain a great number of seeds and only a little pulp. This situation is reversed in cultivated fruit trees, which develop only a few seeds while producing a great deal of pulp (Lutterodt and Maleque 1988, 219 f.).

The guava tree is easily confused with *Psidium acutangulum* DC., which also is called *guayaba* and cultivated for its edible fruits (Vickers and Plowman 1984, 24*).

Psychoactive Material
— Leaves (djambu leaves, djambu folium, folia djambu)
— Bark
— Root cortex

Preparation and Dosage
The fresh leaves are chewed or drunk as a decoction as needed. Overdoses do not appear to occur.

In Southeast Asia, and especially in China, a narcotic *Psidium* drug is obtained in a very unusual manner. The fresh leaves are fed to insects

The fruits of the guava tree are rich in vitamins. In contrast, the narcotic principle is found in the leaves. (Copperplate engraving from Meister, *Der Orientalisch-Indianische Kunst- und Lustgärtner* [The Oriental-Indian Art and Pleasure Gardener], 1677)

281 The South American *Ugni molinae* Turcz. [syn. *Myrtus ugni* Mol., *Eugenia ugni* (Mol.) Hook. et Arn.] (Myrtaceae) is known in German as *chilenische guava* ("Chilean guava") (Zander 1994, 556*; see **chicha**).

(walking sticks, praying mantises, especially *Hepteropteryx dilata*) as their exclusive food. The dung the insects excrete is collected, kneaded into small balls, dried, and stored in an airtight container. When needed, some of these balls are dissolved in hot water and drunk. The wine-colored drink is said to have a "pleasant taste" (Lutterodt 1992, 156).

Ritual Use

Use of the leaves as a psychoactive substance was first observed among the Ga tribes who live near the coast in Ghana. They chew the fresh leaves. It has not been reported whether ritual customs (e.g., communal chewing as a socially integrative element, magical acts, healing ceremonies) are associated with this use. The Ga say that the chewed leaves exert a depressing effect upon the central nervous system that is useful in cases of sleeplessness, and that the leaves also suppress the effects of **alcohol** (Lutterodt and Maleque 1988, 220).

In the Philippines, the bark is chewed in **betel quids** as a substitute for areca nuts (*Areca catechu*) (Hartwich 1911, 529*).

Artifacts
None

Medicinal Use

In many traditional medicine systems, the leaves are used as an analgesic, a neuroleptic, and an agent for treating diarrhea (including that caused by cholera) (Lutterodt 1992, 151). However, folk medical knowledge of the beneficial effects of the leaves as a diarrhea medicine (which have been pharmacologically confirmed) is not as widespread as one might assume (Lutterodt 1992, 155).

In Hawaii, the fresh young leaves are chewed and swallowed to treat diarrhea (Krauss 1981, 24*). An infusion of the leaves is used for the same purpose in Trinidad (Wong 1976, 133*).

The Yucatec Maya drink a decoction made from the bark or the leaves to treat diarrhea (Pulido S. and Serralta P. 1993, 46*). In Belize, a tea made from the leaves is gargled to treat mouth sores and bleeding gums. A decoction of nine

leaves and nine young fruits (boiled for twenty minutes) is drunk three times daily before meals in cases of diarrhea, dysentery, upset stomach, and colds (Arvigo and Balick 1994, 121*).

In South America, teas made from the leaves are drunk to treat digestive disorders (Anzeneder et al. 1993, 59*). In Chile and Peru, the leaves are chewed to strengthen the teeth (Schultes 1980, 110*). In Panama, the leaves are chewed for toothaches (Lutterodt and Maleque 1988, 220).

The Fang of central Africa use the leaves to make an anthelmintic juice (Akendengué 1992, 169*). In Samoa, the leaves are used as a cough medicine and as an antidote for all types of poisonings (Uhe 1974, 22*).

Constituents

The leaves contain some 10% tannin, β-sitosterol, maslenic acid, guaijavolic acid, essential oil (chiefly caryophyllene, along with β-bisabolene, aromadendrene, β-selinene, nerolidiol, caryophyllene oxide, sel-11-en-4a-ol, and eugenol), triterpenoids (oleanolic, ursolic, crategolic, and guaijavolic acids),[282] a quercetin derivative, guaijaverin (= 3-α-l-arabopyranoside), and several unidentified substances (Argueta et al. 1994, 711*; Lutterodt and Maleque 1988, 220; Wong 1976, 133*). The glycoside quercetin[283] and its derivative (quercetin-3-arabinoside) are regarded as the primary active constituents (responsible for the narcotic effects) (Lutterodt and Maleque 1988, 229).

In an earlier study, guava leaves were found to contain the polyphenols quercetin, guaijaverin, leucocyanidin, and amritsoside (Seshadri and Vashishta 1965). Opiates (**opium alkaloids**) and cannabinoids (cf. **THC**) have not been detected (Lutterodt 1992, 152).

The fruit contains large amounts of vitamins (A, B, C), about two to three times as much as an orange (Arvigo and Balick 1994, 121*).

Effects

Animal experiments using a leaf extract demonstrated a distinct morphinelike effect as a result of inhibition of acetylcholine release (cf. **morphine**). This effect was likely produced by the quercetin contained in the leaves (Lutterodt 1989 and 1992, 152). The active constituent does not appear to bind to the opioid receptors and is not thought to have any "addictive potential" (Lutterodt and Maleque 1988, 225). Toxic effects and overdoses are unknown (Argueta et al. 1994, 711*).

A hot-water extract of dried leaves has antibacterial effects upon *Sarcina lutea*, *Staphylococcus aureus*, and *Mycobacterium phlei*. An aqueous extract of the fresh leaves has fungicidal effects (Arvigo and Balick 1994, 121*).

The tropical fruit tree *Psidium guajava* has leaves that produce narcotic effects.

Quercetin

282 The crategolic acid that has been found in the leaves is also found in clove (*Syzygium aromaticum*; cf. **essential oils**) and appears to be at least partially responsible for the analgesic effects (Brieskorn et al. 1975).

283 Quercetin (= cyanidanol, cyanidenolen 1522, 3,3',4',5,7-pentahydroxyflavone, meletin, sophorin, ericin) is very common in the plant kingdom, especially in the bark of trees. Significant quantities are found in the Douglas fir (*Pseudotsuga menziesii* [Mirb.] Franco), pansies (*Viola tricolor* L.), the false hellebore (*Adonis vernalis* L.), the horse chestnut (*Aesculus hippocastanum* L.), the true chamomile (*Matricaria recutita*), the English hawthorn (*Crataegus laevigata* [Poir.] DC.), hops (*Humulus lupulus*), oaks (*Quercus* spp.), the apple tree (*Malus sylvestris* Mill., *Malus* spp.), catechu resin (cf. *Acacia* spp.), *Fabiana imbricata*, wormwood (*Artemisia absinthium*), and many members of the Family Ericaceae (*Ledum palustre*, *Vaccinium uliginosum*, *Arctostaphylos* spp.; cf. kinnikinnick) (Römpp 1995, 3746*).

Whether the psychedelic underground duo Ween named their debut album *Pure Guava* for the narcotic effects of guava leaves or used the fruits as a symbol for other substances is something the listener must decide. (CD cover 1992, Electra Records)

"Xalxocotl—Guajava-apple
The Xalxocotl tree has thin foliage, sparse foliage. Its wood is brown-red on the surface. Its branches fall off. Its fruit is pale, dark yellow, delicate, very delicate, grainy. Its heart is like sand. They are round; it is sweet-sour. It makes one belch, makes one's teeth dull, increases the flow of saliva. My teeth are dull. My saliva is flowing. I am becoming sour. My teeth are dull."

BERNARDINO DE SAHAGUN
FLORENTINE CODEX (11.6)

Commercial Forms and Regulations

Guava leaves are sometimes found as ingredients in tea mixtures (stomach teas) sold in pharmacies (Pahlow 1993, 437*).

Literature

Brieskorn, Carl Heinz, Klaus Münzhuber, and Gerhard Unger. 1975. Crataegolsäure und Steroidglukoside aus Blütenknospen von *Syzygium aromaticum*. *Phytochemistry* 14:2308–9.

Cheng, J. T., and R. S. Yang. 1983. Hypoglycemic effects of guava juice in mice and human subjects. *American Journal of Chinese Medicine* 11 (1–4): 74–76.

Khadem, H. el-, and Y. S. Mohammed. 1958. Constituents of the leaves of *Psidium guajava*. II: Quercetin, avicularin and guaijaverin. *Journal of the Chemical Society* (London): 3320–23.

Lutterodt, George D. 1989. Inhibition of gastrointestinal release of acetylcholine by quercetin as a possible mode of action of *Psidium guajava* leaf extracts in the treatment of acute diarrhoeal disease. *Journal of Ethnopharmacology* 25:235–47.

———. 1992. Inhibition of Microlax*-induced experimental diarrhoea with narcotic-like extracts of *Psidium guajava* leaf in rats. *Journal of Ethnopharmacology* 37:151–57.

Lutterodt, George D., and Abdul Maleque. 1988. Effects on mice locomotor activity of a narcotic-like principle from *Psidium guajava* leaves. *Journal of Ethnopharmacology* 24:219–31.

Osman, A. M., M. E. Younes, and A. E. Sheta. 1974. Triterpenoids of the leaves of *Psidium guajava*. *Phytochemistry* 13:2015–16.

Seshadri, T. R., and K. Vasishta. 1965. Polyphenols of the leaves of *Psidium guajava*—quercetin, guaijaverin, leucocyanidin and amritsoside. *Phytochemistry* 4:989–92.

Psychotria viridis Ruíz et Pavón

Chacruna

Family

Rubiaceae (Coffee Family)

Forms and Subspecies

It is possible for white thorns (domatia) to develop along the central nerve on the underside of some chacruna leaves. South American *ayahuasqueros* distinguish different forms of the plant on the basis of the number of these thorns. Plants with three thorns per leaf are considered to be particularly potent, medicinal, and well suited for the production of **ayahuasca**. A form with nine thorns is regarded as the highest quality.

Synonyms

Psychotria psychotriaefolia (Seem.) Standley may be a synonym (cf. ***Psychotria* spp.**).

Folk Names

Amirucapanga, cahua (Shipibo-Conibo), chacrona, chacruna, chagropanga, chalipanga, horóva (Campa), kawa (Cashinahua/Sharanahua), oprito (Kofán, "heavenly people"), sami ruca

History

It is not known when the use of chacruna in Amazonia first began. It is presumably as old as the use of ***Banisteriopsis caapi*** and **ayahuasca**. But it was only in the 1960s that the American ethno-botanist Homer Pinkley (a student of Schultes) first observed and described the psychoactive use of the plant among the Kofán Indians of Colombia, who use it as an ayahuasca additive (Pinkley 1969). Linnaeus, who provided the first botanical description of the genus *Psychotria*, derived the name of the genus from *Psychotrophum* (Patrick Browne), a term that had already been circulating in the literature. Unfortunately, he did not provide any reason for this action. It is quite possible that the genus name means that it "influences the psyche" (cf. Pinkley 1969).

Distribution

The tropical bush is at home primarily in the undisturbed forests of the Amazon lowlands but has spread from Colombia to Bolivia and into eastern Brazil as a result of extensive cultivation. It is said to occur also north of the Amazon region and into Central America (Pinkley 1969, 535). Today, there are also plantations in Hawaii and northern California.

Cultivation

The plant is difficult to propagate from seed. The seeds can require sixty days to germinate. Sometimes, only one seed in a hundred will germinate. In contrast, cultivation from cuttings is much easier and more successful. A small branch needs only to be set in the ground and watered thoroughly. Plants can be grown even from a branch piece having only two leaves, and it is possible for individual leaves and leaf pieces to develop into plants. It has been claimed that a young plant once developed from a piece of leaf that was accidentally covered with soil. The plant requires moist, humus-rich soil. It can survive an occasional flooding of its location, as occurs in Amazonia (Pinkley 1969).

Appearance

The evergreen bush can grow into a small tree with a very woody trunk, but in cultivation it is usually maintained at a height of 2 to 3 meters. It has long, narrow, ovate leaves that are light green to dark green in color and whose upper side is shiny. The flowers have greenish white petals and are attached to long stalks. The red berry fruits contain several small ovate-oval retuse seeds (approximately 4 mm in length). The convex side is streaked with three parallel grooves with irregular edges.

Psychotria viridis is easily confused with other *Psychotria* species. *Psychotria psychotriaefolia* in particular is very similar in appearance and may in fact be a synonym (see *Psychotria spp.*).

Psychoactive Material

— Leaves

Preparation and Dosage

The leaves must be collected in the morning and are used both fresh and dried to manufacture **ayahuasca** and **ayahuasca analogs**. The dried leaves are coffee brown in color. The leaves also can be used to produce an extract that thickens to a tarlike mass and can be smoked.

As little as 1 ml of the juice pressed from the fresh leaves is said to contain some 100 mg of *N,N*-DMT (cf. Russo 1997, 6).

Ritual Use

See **ayahuasca**.

Artifacts

See **ayahuasca** ("Ayahuasca Music—A Discography," on page 711).

Medicinal Use

The Machiguenga use juice that has been freshly pressed from the leaves of *Psychotria viridis* or another *Psychotria* species (**Psychotria spp.**) as eyedrops for treating migraine headaches (Russo 1997, 5). While *Psychotria viridis* does have a reputation as a medicinal plant, such use has been little studied to date (see also **ayahuasca**).

Constituents

The leaves contain 0.1 to 0.61% *N,N*-DMT along with traces of MMT and MTHC (= 2-methyl-tetrahydro-β-carboline). The DMT content is typically around 0.3%. *Psychotria* leaves appear to contain the highest concentrations of DMT in the early morning, which is why they should be collected at that time (Dennis McKenna, pers. comm.).

Effects

The Kofán Indians say that by mixing *Psychotria viridis* leaves into their yagé (= ayahuasca; cf. **Banisteriopsis caapi**), they are able to see the *oprito*, the small "heavenly people" that bear the same name as the plant (Pinkley 1969, 535). When used as an ayahuasca additive, the leaves manifest typical DMT effects (see **ayahuasca**).

Commercial Forms and Regulations

The dried leaves are occasionally available from sources specializing in ethnobotanical products. The legal situation with respect to the raw plant material has not been clarified.

Literature

See also the entries for *Psychotria* **spp.**, **ayahuasca**, and **ayahuasca analogs**.

Der Marderosian, Ara H., et al. 1970. The use and hallucinatory principles of a psychoactive beverage of the Cashinahua tribe (Amazonia basin). *Drug Dependence* 5:7–14.

Pinkley, Homer V. 1969. Etymology of *Psychotria* in view of a new use of the genus. *Rhodora* 71:535–40.

Prance, G. T., and A. E. Prance. 1970. Hallucinations in Amazonia. *Garden Journal* 20:102–7.

Russo, Ethan B. 1992. Headache treatments by native peoples of the Ecuadorian Amazon: A preliminary cross-disciplinary assessement. *Journal of Ethnopharmacology* 36:192–206.

———. 1997. An investigation of psychedelic plants and compounds for activity in serotonin receptor assays for headache treatment and prophylaxis. *MAPS* 7 (1): 4–8.

Left: A large-leaved variety of the DMT-containing plant *Psychotria viridis*.

Right: A small-leaved type of *Psychotria viridis*.

DMT

"*Psychotria* L. (Rubiaceae), derived from χψξη (soul, life) and τρεωειν (nourish, maintain); according to P. Browne, the seeds of Ps. herbacea are used on Jamaica to produce a pleasant, coffeelike drink. Linnaeus contracted the name Psychotrophum that Browne had originally coined."

G. C. WITTSTEIN
ETYMOLOGISCH-BOTANISCHES HANDWÖRTERBUCH [ETYMOLOGICAL BOTANICAL HAND DICTIONARY] (ANSBACH, 1852)

Psychotria spp.

Wild Coffee, Psychotria Species

Family

Rubiaceae (Coffee Family)

Most of the approximately 1,200 to 1,400 *Psychotria* species that have been described are found in the tropical zones of Central and South America, although a few species occur in the rain forests of Malaysia and in New Caledonia (Standley 1930). In the Caribbean, the seeds of some species, e.g. *Psychotria nervosa*, are referred to as wild coffee and drunk as a coffee substitute (cf. *Coffea arabica*). The fruits of many *Psychotria* species (*P. involucrata* Swartz, *P. nudiceps* Standley) are regarded as poisonous (Schultes 1969, 158; 1985). *N,N*-DMT has been demonstrated to be present in several species. Some contain the alkaloid psychotridine, and others indoles (Lajis et al. 1993). Some species (*Psychotria poeppigiana* Muell. Arg., *Psychotria ulviformes* Sterm.) appear to contain opium-like constituents (Elisabetsky et al. 1995, 78). The Yucatec Maya regard the Central American species *Psychotria acuminata* Benth. (*ix-anal*) and *Psychotria tenuifolia* Sw. (*x'anal*) as "male" and "female" counterparts and use them to treat nervousness and sleeplessness (Arvigo and Balick 1994, 45, 105*). In Europe, *Psychotria emetica* (L. fil.) Mutis, the Peruvian vomit plant, was known in particular as a counterfeit for ipecac (*Cephaelis ipecacuanha* [Brot.] Tussac [syn. *Psychotria ipecacuanha* (Brot.) Stokes]) (Rätsch 1991a, 136 f.*; Schneider 1974, 3:135 f.*). The vomit-inducing substance emetine is said to occur in numerous *Psychotria* species (Fisher 1973, 231).

Psychotria brachypoda (Muell. Arg.) Britton

This *Psychotria* is used traditionally as a pain medicine. The species contains active constituents with opium-like, analgesic effects (Elisabetsky et al. 1995).

Psychotria carthaginensis Jacquin—sameruca

According to information provided by the Colombian Makuna Indians, eating the fruit of this bush will induce perceptual alterations that can persist for days, nausea, weakness, and fever (Schultes 1969, 158). The leaves, which contain some *N,N*-DMT, are used as an ayahuasca additive (Schultes 1985, 118).

Psychotria colorata (Willd. ex R. et S.) Muell. Arg.

This bush is known as *perpétua do mato* in the Brazilian Amazon, where it is used in folk medicine to treat ear and lower abdominal pain. The Caboclos produce eardrops by heating the flowers in banana leaves on hot ashes. A decoction of the roots and fruits is drunk to treat abdominal pains. The leaves and flowers have been found to contain alkaloids with opium-like effects whose structures have not yet been determined (Elisabetsky et al. 1995).

Psychotria poeppigiana Muell. Arg.—oreja del diablo (Spanish, "devil's ear")

In Amazonia (Ecuador), the nectar of this species is used as a traditional ear medicine. The leaves are rich in *N,N*-DMT and are evidently well suited for use as an ayahuasca additive (**ayahuasca analogs**) (Rob Montgomery, pers. comm.). In the Putumayo region of Colombia, the roots are used to treat lung ailments (Schultes 1985, 119; Schultes and Raffauf 1990, 395*).

Among the Ka'apor, *Psychotria poeppigiana* Muell. Arg. is called *yawaru-ka'a*, "black jaguar plant," or *tapi'i-ka'a*, "tapir plant" (Balée 1994, 303*). These names suggest that the plant may be used for shamanic purposes (animal transformation).

Psychotria psychotriaefolia (Seem.) Standley

In the Colombian Putumayo region, the leaves of this species are used together with **Banisteriopsis caapi** to produce **ayahuasca**. In Ecuador, both the

Left: *Psychotria nervosa* is also known as wild coffee. However, nothing is known about any possible psychoactivity.

Right: Some *Psychotria* species develop very striking flowers and fruits. (A *Psychotria* species, photographed in Chiapas, Mexico.)

leaves and the fruits are used for this purpose (Schultes 1969, 158). The addition of this plant to the mixture is said to deepen and prolong the visions. The leaves contain *N,N*-DMT. The Kofán Indians call the plant *oprito*. They use this same name to refer to the "heavenly people" that they contact while under the influence of **ayahuasca** (164). This species may be synonymous with *Psychotria viridis*.

Psychotria spp.

Among the many members of the genus *Psychotria*, there are certainly other species that contain *N,N*-DMT and may be suitable for use as ayahuasca additives. We already know of some as yet undescribed members of the genus that are used to make **ayahuasca** and are often called by the name *chacruna*.

Literature

See also the entries for *Psychotria viridis*, **ayahuasca**, and *N,N*-DMT.

Elisabetsky, Elaine, Tânia A. Amador, Ruti R. Albuquerque, Domingos S. Nunes, and Ana do C. T. Carvalho. 1995. Analgesic activity of *Psychotria colorata* (Willd. ex R. et S.) Muell. Arg. alkaloids. *Journal of Ethnopharmacology* 48:77–83.

Fisher, H. H. 1973. Origin and uses of ipecac. *Economic Botany* 27:231–34.

Lajis, Nordin H., Zurinah Mahmud, and R. F. Toia. 1993. The alkaloids of *Psychotria rostrata*. *Planta Medica* 59:383–84.

Schultes, Richard Evans. 1969. De Plantis Toxicariis e Mundo Novo Tropicale Commentationes IV. *Botanical Museum Leaflets* 22 (4): 133–64.

———. 1985. De Plantis Toxicariis e Mundo Novo Tropicale Commentationes XXXIV: Biodynamic Rubiaceous plants of the Northwest Amazon. *Journal of Ethnopharmacology* 14:105–24.

Small, John K. 1928. *Psychotria sulzneri*. *Addisonia* 13:47–48.

Standley, Paul C. 1930. *The Rubiaceae of Colombia*. Botanical Series, vol. 8, no. 1. Chicago: Field Museum of Natural History.

While the leaves of *Psychotria poeppigiana* are not traditionally used in the production of ayahuasca, they do contain high concentrations of DMT.

"*Psychotria carthaginensis* has also been reportedly used as an *ayahuasca* admixture, and preliminary studies likewise detected DMT in leaves of this species. . . . Various unidentified species of *Psychotria* are also used as *ayahuasca* admixtures by Peruvian Sharanahua and Cashinahua Indians. *Psychotria* leaves called *pishikawa* and *batsikawa* were added to *ayahuasca* by Sharanahua of the upper Río Purús region . . . , and the latter was said to be inferior. . . . *Nai kawa* was thought to be *P. alba*, *P. carthaginensis*, *P. horizontalis* or *P. marginata*. Clearly, more detailed taxonomic and ethnobotanical studies are needed to clarify the identity of these *Psychotria* species. A number of species of the genus are used ethnomedicinally throughout Amazonia."

JONATHAN OTT
AYAHUASCA ANALOGUES
(1994, 25*)

Rhynchosia pyramidalis (Lam.) Urban

Bird's Eyes

Family

Leguminosae: Papilionideae (Legume Family); Subfamily Fabodeae

Forms and Subspecies

The genus consists of some three hundred species that are found in the tropical and subtropical regions of both hemispheres (Schultes and Hofmann 1980, 338*).

Synonyms

Dolicholus phaseoloides Sw.
Rhynchosia phaseoloides (Sw.) DC.

Folk Names

Äh mo' ak' (Lacandon, "ara parrot vine"), antipusi, atecuixtle, atecuxtli, bejuco culebra, bird's eyes, casanpulgas, chanate pusi, cha'pak' (Mayan), colorín chiquito, colorincito, colorines (cf. *Erythrina americana*), coralito, frijol de chintlatla-hua, frijolillo, guarecitas, gun-ma-muy-tio-ña (Chinantec), krebsaugenbohne, liucai-nofal (Chontal), negritos, ojitos de picho (Spanish, "little eyes of the dove"), ojo de cangrejo (Spanish, "crab's eye"), ojo de chanate (Mexico, "eye of the thrush [*Cassidix mexicanus*]"), ojo de culebra (Spanish, "eye of the snake"), ojo de pajarito (Spanish, "eye of the little bird"), ojo de zanate (Mexico, "eye of the thrush [*Cassidix mexicanus*]"), pega palo, peonía, perico, peyote (see *Lophophora williamsii*), pipilzíntli, piule, pulguitas, puren-sapicho, salti-pús, senecuilche (see *Heimia salicifolia*), shasham wupu'är (Pima), sinicuiche, xenecuilche

Plants and Fungi Known in Mexico as *Piule*
(from Martínez 1987, 757*; Ott 1993, 419*; Santesson 1938; supplemented)

RHYNCHOSIA SPP.

Rhynchosia longeracemosa (Mart. et Gal.) Rose	*piule, peyote* (cha'pak)
Rhynchosia minima (L.) DC.	*piule*
Rhynchosia pyramidalis (Lam.) Urban	*piule*
Rhynchosia spp.	*piule*

CACTI

Lophophora williamsii	*piule, peyote*

VINES (CONVOLVULACEAE)

Ipomoea violacea	*piule*
Turbina corymbosa (L.) Raff.	*piule*

FUNGI

Psathyrella sepulchralis Sing., Sm. et Guz.	*piule de barda*
Psilocybe mexicana Heim	*piule de churis*
Psilocybe zapotecorum Heim	*piule de barda*

History

The Aztecs may have used the striking seeds of this plant for ritual purposes (Schultes and Hofmann 1980, 340*). The red-black seeds, which are known by the name *piule* (Santesson 1938), were or are used ritually in the village of San Pedro Nexapa, on the slopes of Popocatepetl (Mexico) (Wasson and Wasson 1957, 306 f.). In Mexico, the name *piule* has been used as a catchall term for psycho-active plants since the twentieth century (Martínez 1987, 757*; cf. *Psilocybe mexicana, Turbina corymbosa*). The word *piule* may have been derived from the Nahuatl *peyotl* (= *Lophophora williamsii*). Accordingly, *piuleros* are those people who use a psychoactive substance (*piule*) to divine and/or heal (Santesson 1937a, 1937b). Some species, e.g., *Rhynchosia longeracemosa* Mart. et Gal., are now also known by the name *peyote* (Schultes 1966, 296*).

Distribution

This climber is found throughout the tropical and warm regions of Mexico and on many islands of the Caribbean (Cuba) (von Reis and Lipp 1982, 139*). It usually grows at the edge of forests and in clearings. It is frequently found in fallow milpas (slash-and-burn gardens).

Cultivation

The seeds are best pregerminated in a mixture of soil and moss. The seedlings must be planted in topsoil and watered well as soon as the seeds have opened and the young shoots have become visible (Grubber 1991, 56*). The plant requires a moist, warm climate and in northern zones can thus be grown only as a houseplant.

Appearance

The vine, which can grow to a length of several meters, has the typical leaves of the Legume Family, in which three leaves sit upon each stalk. The greenish flowers are arranged in long racemes. The bean-shaped seedpods are constricted between the two small, red-black, almost spherical hard seeds (4 to 6 mm long).

The kidney-shaped seeds of the closely related *Rhynchosia longeracemosa* are "mottled light- and dark-brown" (Schultes and Hofmann 1992, 55*).

Rhynchosia pyramidalis is often confused with *Abrus precatorius* L. (jequirity, rosary pea), which is widely feared as a poisonous plant. It too produces red-black seeds, although they are somewhat larger (6 to 7 mm long). Jequirity can be recognized by its smaller, pinnate leaves. The seeds of *Abrus precatorius* contain abrin, a lectin mixture that is unstable when heated and one of the most potent of all known toxins, along with several alkaloids (Ghosal and Dutta 1971; Nwodo 1991; Nwodo and Alumanah 1991; Roth et al. 1994, 83 f.*). In Mexico, the seeds of *Abrus precatorius* are known as *colorines* (see *Erythrina* spp.). They are associated with the mescal bean cult (see *Sophora secundiflora*); ashes from the leaves are used as a coca additive (see *Erythroxylum coca*).

Psychoactive Material
— Seeds (semina rhynchosiae phaseoloides, bird's eyes, *colorines*)
— Stalks

Preparation and Dosage
In entheogenic rituals in the high valleys of Mexico, twelve untreated seeds were ingested with six pairs of *Psilocybe aztecorum* per person (Wasson and Wasson 1957, 306).

Ritual Use
To date, the only description that is available pertains to the ritual use of the seeds in connection with the ingestion of mushrooms. The ingestion of the seeds is presumably more symbolic in meaning, for the red-black seeds represent bodiless, free-floating eyes, a symbol of psychedelic and prophetic vision.

The Zapotec of Miahuatlan are said to have used the seeds of the closely related species *Rhynchosia minima* (L.) DC. [syn. *Dolicholus minimus*] in magical rituals (Díaz 1979, 87*).

Artifacts
The small, durable seeds are made into amulets and chains (cf. *Erythrina americana*, *Erythrina* spp., *Sophora secundiflora*).

Wall paintings at Teopantitla (near Teotihuacán) allegedly show the seeds falling out of the hand of the rain god Tláloc (D. McKenna 1995, 102*). The red-black coloration is said to be an indication of the seeds' hallucinogenic use (Schultes 1970c; Schultes and Hofmann 1980, 340*).

Medicinal Use
The seeds are regarded as a narcotic and poison in Mexican folk medicine (Jiu 1996, 254*). The Yucatec Maya use the root along with other herbs to produce a medicine to treat pellagra[284] (Pullido S. and Serralta P. 1993, 37*). The Pima of northern Mexico grind the seeds on a mortar and strew the powder into the eyes of those who are suffering from the "evil eye" (Pennington 1973, 223*).

In the Dominican Republic, the stalks are used to prepare an aphrodisiac drink (Díaz 1979, 87*).

Constituents
The chemistry of the constituents has not yet been clarified. Reports about the alkaloids are contradictory (Santesson 1937a). The seeds apparently contain alkaloids similar to those in *Sophora secundiflora* and *Erythrina* spp. (D. McKenna 1995, 102*). The root may possibly contain niacin or nicotine amide, for it is used in the Yucatán as a folk medicine to treat pellagra (maidism). Whether the flavonol rhynchosin (Adinarayana et al. 1980) occurs in the plant is unknown.

Effects
In Mexico, it is commonly believed that the seeds cause "imbecility" or "madness" (Díaz 1979, 87*; Jiu 1996, 254*). There are as yet no reports of actual psychoactive effects. An extract of the seeds is said to have curare-like activity (Schultes and Hofmann 1980, 340*).

Commercial Forms and Regulations
The seeds are sometimes available through the international seed trade. Mexican Indians sometimes sell necklaces with beads of *Rhynchosia* seeds.

Literature
See also the entries for *Erythrina* spp. and *Sophora secundiflora*.

Adinarayana, Dama, Duvvuru Gunasekar, Otto Seligmann, and Hildebert Wagner. 1980. Rhynchosin, a new 5-deoxyflavonol from *Rhynchosia beddomei*. Phytochemistry 19:483–84.

Left: The bird's eyes vine (*Rhynchosia*) in bloom. (Photographed in Palenque, Mexico)

Top right: The seeds of *Rhynchosia* resemble those of the poisonous jequirity (*Abrus precatorius* L.) to the smallest detail.

Bottom right: An Indian necklace made from the seeds of *Abrus precatorius*. Because the toxic alkaloids are supposedly absorbed through the skin, many tourists fear wearing such chains. Experience, however, has shown that the wearing of such a necklace will not result in any effects. Jequirity seeds contain the alkaloid abrin, which is lethal at a dosage of 0.01 mg/kg. This notwithstanding, particularly courageous Plains Indians appear to have used the seeds as "red medicine" for vision quests.

284 Pellagra (maidism) is a disease of malnutrition that results from a one-sided maize diet with deficiencies in niacin and nicotine amide.

"Just one single seed [of jequirity, *Abrus precatorius*]—which must however be chewed thoroughly before swallowing—can produce lethal effects in humans. Cattle and goats are more resistant, but 60–120 g of seeds can also kill a horse. A case of poisoning proceeds in the following sequence: First, stomach pains become noticeable, the urge to vomit appears, the patient falls into a coma, and the circulation collapses so that the patient dies. . . . In some tropical countries, the unripe seeds are used to murder a person; they are formed into sharp needles that are then stabbed into the victim. Because the toxin can directly enter the bloodstream in this manner, it is usually lethal. . . . In some places, the seeds are used to produce decorative objects. They are used, for example, to make necklaces, chains, and also rosaries."

FRANTISEK STARY
GIFTPFLANZEN [POISONOUS PLANTS]
(1983, 28*)

Ghosal, S., and S. K. Dutta. 1971. Alkaloids of *Abrus precatorius*. *Phytochemistry* 10:195–98.

Grear, J. W. 1978. A revision of the New World species of *Rhynchosia* (Leguminosae-Fabodeae). *Memoirs of the New York Botanical Garden* 31 suppl. (1): 1–168.

Nwodo, O. F. C. 1991. Studies on *Abrus precatorius* seeds. I: Uterotonic activity of seed oil. *Journal of Ethnopharmacology* 31 (3): 391–94.

Nwodo, O. F. C., and E. O. Alumanah. 1991. Studies on *Abrus precatorius* seeds. II: Antidiarrhoeal activity. *Journal of Ethnopharmacology* 31 (3): 395–98.

Ristic, S., and A. Thomas. 1962. Zur Kenntnis von *Rhynchosia pyramidalis* (Pega Palo). *Archiv für Pharmakologie* 295:510.

Santesson, C. G. 1937a. Notiz über *piule*, eine mexikanische Rauschdroge. *Etnologiska Studier* (Göteborg) 4:1–11.

———. 1937b. Piule, eine mexikanische Rauschdroge. *Archiv für Pharmazie*: 532–37.

———. 1938. Noch eine mexikanische "Piule"-Droge: Semina Rynchosiae phaseoloidis DC. [sic!]. *Etnologiska Studier* 6:179–83.

Wasson, R. Gordon, and Valentina P. Wasson. 1957. *Mushrooms, Russia, and history*. New York: Pantheon Books.

Salvia divinorum Epling et Játiva-M.

Ska María Pastora

Family
Labiatae (Lamiaceae; Mint Family); Subfamily Nepetoideae, Salvieae Tribe, Salviinae Subtribe, Dusenostachys Section

Forms and Subspecies
Only clones or races of varying bitter taste are known. The Wasson clone is very bitter and is derived from plants collected in 1962; the "palatable clone," which has hardly any bitter taste, was collected in Llano de Arnica, Oaxaca, by the American ethnobotanist Bret Blosser (Ott 1996, 33).

Synonyms
None

Folk Names
Aztekensalbei, blätter der hirtin, diviner's sage, foglie della pastora, hierba de la pastora, hierba de la virgen, hoja de la pastora (Spanish, "leaf of the shepherdess"), hojas de adivinación, hojas de maría pastora, la hembra, leaves of the Mary shepherdess, mazatekischer salbei, pipiltzitzintli, sage of the seers, salvia, salvia of the seers, ska maría pastora, ska pastora (Mazatec, "leaf of the shepherdess"), wahrsagesalbei, yerba de maría, yerba maría, zaubersalbei

History
The Aztecs knew and used a plant they called **pipiltzintzintli** (literally "the noblest little prince") in entheogenic rituals in a manner very similar to

the ways in which they used mushrooms (*Psilocybe* spp.). A number of authors have suggested that this plant was *Salvia divinorum* (Wasson 1962; Ott 1995, 1996).[285]

Gordon Wasson discovered the plant and its divinatory use in 1962. That same year, the plant was first botanically described by Carl Epling and Carlos D. Játiva-M., botanists from UCLA. In the 1960s, Albert Hofmann was unable to discover any active constituents in an initial analysis of juice pressed from the plant (Hofmann 1979, 151–68*; 1990). The chemistry and pharmacology was not clarified until the 1980s and 1990s, when **salvinorin A** was discovered (Ortega et al. 1982; Valdés 1994; Valdés et al. 1987; Siebert 1994).

Distribution
Salvia divinorum is endemic to the Mazatec region of the Sierra Madre Oriental in the Mexican state of Oaxaca. Apart from this, the plant is found only as a cultigen among "neo-shamans" and in botanical gardens. It occurs naturally in tropical rain and cloud forests at altitudes between 300 and 1,800 meters (Reisfield 1993). Because of its small original range, the plant is one of the rarest of all natural entheogens. It is now grown by plant enthusiasts around the world.

Cultivation
Propagation is performed with cuttings or layers/shoots. All leaves except the topmost pair are removed from an 8 to 12 cm long branch tip,

285 "To date, little that is certain is known about the age of the salvia ritual, as corresponding ethnographic or historical evidence is lacking, but Wasson attempted to identify a plant known in ancient Aztec as *pipizizintli* or *pepetzintle* with *Salvia divinorum*, basing his arguments on the Spanish chronicles. The monk Agustin de Vetancourt reported on an inebriating drug named *pepetichinque*, the root of which produced effects similar to those of the peyotl cactus [*Lophophora williamsii*] or ololiuqui seeds [*Turbina corymbosa*]. There was a male and a female form of this plant, the *macho* and the *hembra*. In the National Archives in Mexico City, Inquisition files from the years 1696, 1698, and 1706 refer to the plant *pipiltzintzin* and mention its inebriating effects, but the details are too vague to unequivocally recognize this as a species of salvia" (Mayer 1977, 779).

which is then placed in water. The cutting should develop roots in about two weeks. It can be planted in soil after about four weeks. *Salvia divinorum* requires a great deal of water and prefers high to very high humidity. If the edges of the leaves turn brown, this is a sure sign that the air is too dry. As a shade plant, it does not tolerate any direct sunlight, prefers dark soil, and needs copious amounts of water, i.e., it should be watered almost every day. Although the plant is sensitive to cold, cultivated *Salvia divinorum* can survive a mild frost.

Methods for cultivating the plant from seed are currently being investigated (cf. Reisfield 1993).

Appearance

The evergreen plant is an herbaceous perennial that can grow to over 1 meter in height. Its most characteristic feature is its completely four-sided, sometimes even square stem, which can grow as thick as 2 cm. Its edges are angular. Both the opposite leaves and the side branches develop from nodes on the stem. The light to dark green leaves are entirely covered in fine hairs and attain a length of over 20 cm and a width of some 10 cm. The leaves are lanceolate and tapered at both ends. The panicled inflorescences appear at the ends of the stalks and look exactly like those of *Coleus blumei*. The campanulate calyxes are bluish or purple in color, while the petals are always white (Reisfield 1993; cf. Brand 1994, 540). In Mexico, the plant blossoms between October and March but primarily in January. In cultivation, the plant seldom flowers, and fruits almost never develop. Recently, however, one clone has been discovered that develops fruits and seeds more frequently. A hummingbird has been observed as a pollinator (Reisfield 1993). The seeds germinate and begin to develop, but with our current gardening techniques, they all eventually die.

Salvia divinorum can be confused with a similar, closely related Central American species, *Salvia cyanea* Lamb. ex Benth. (Epling et Játiva-M. 1962; Mayer 1977, 777).

Psychoactive Material
— Fresh or dried leaves (salvia divinorum leaves, folia salviae divinorum, divination leaf)

Preparation and Dosage

The Mazatec take thirteen pairs of fresh leaves (twenty-six leaves in all) and roll them into a kind of cigar (quid) that they place in the mouth and suck on or chew while retaining it in the mouth. The juice is not swallowed, as the active constituents can be absorbed only through the mucous membranes of the mouth. At least six fresh leaves are needed to prepare one quid (threshold dosage), while more distinct effects will occur with eight to ten leaves. When consumed in the form of

a quid, the effects appear after almost exactly ten minutes and persist for some forty-five minutes. The dried leaves are best smoked by themselves. Here, as little as half an average-sized leaf (two or three deep inhalations) can be sufficent to elicit profound psychoactive effects. Usually, however, one or two leaves are smoked. The dried leaves can be soaked in a *Salvia divinorum* tincture, after which they should again be allowed to dry.

Dried *Salvia divinorum* leaves are becoming popular as an ingredient in **smoking blends** and even in the manufacture of psychoactive **incense** (Valdés 1994).

Tinctures are prepared from fresh or dried leaves by using an ethanol-water mixture (60% alcohol). The tincture can be either used to impregnate dried leaves, thereby potentiating their effects, or applied sublingually. Dosages appear to vary considerably in their effects among individuals. In addition, several experiments seem to be needed before the effects become apparent. Looking back, however, one realizes that there were noticeable effects before.

For information concerning the use and dosage of the primary constituent, see "Constituents" and also the discussion of **salvinorin A** (cf. also Ott 1995; Siebert 1994; Valdés 1994).

Ritual Use

The shamans and shamanesses of the Mazatec of Oaxaca use *Salvia divinorum* in divinatory and healing rituals, usually as a substitute for the preferred psychoactive mushrooms (cf. *Psilocybe mexicana, Psilocybe* spp.). Only a few shamans prefer to use this *Salvia*. The ritual use is very similar to that of the mushrooms (Hofmann 1990).

Salvia divinorum rituals almost always take place at night in complete darkness and silence. Either the healer is alone with the patient or other patients as well as healthy participants are present. Before the shaman and perhaps other people chew and suck the leaves in the form of a quid, the leaves are fumigated with copal (cf. **incense**) while prayers are spoken and the quids are consecrated to the higher powers. After chewing the leaves, the participants lie down and try not to make any sound. Both sounds and sources of light will greatly disturb the visionary experience. Because the effects of the leaves are much shorter in duration than those of the mushrooms, *Salvia* rituals rarely last more than one or two hours. If the visions are sufficiently pronounced, the shaman will have identified the cause of the illness or some other problem. He then reports to the patient, provides appropriate advice, and ends the nocturnal meeting (Hofmann 1990; Mayer 1977; Ott 1995; Valdés et al. 1987; Wasson 1962).

In Mazatec folk taxonomy, *Salvia divinorum* is related to two species of *Coleus. Salvia* is *la hembra*

The "sage of the seers" (*Salvia divinorum*), from the Mazatec region of Oaxaca, is known only from cultivation.

"Under the direction of Maria Sabina, one of the children, a girl of about ten years of age, prepared the pressed juice of five pairs of fresh leaves of hojas de la pastora for me. I wanted to make up the experience with this drug, which I had missed in San José Tenango. The drink is said to be especially powerful when it is prepared by an innocent child. The cup with the pressed juice was also fumigated and Maria Sabina and Don Aurelio spoke over it before it was handed to me. . . . Probably as an effect of the hojas, I found myself for a time in a state of increased sensitivity and intense experience that was not accompanied by any hallucinations."

ALBERT HOFMANN
LSD—MEIN SORGENKIND [LSD – MY PROBLEM CHILD]
(1979, 164 f.*)

"When I wish to heal a sick person during the time in which there are no mushrooms, then I must turn to the leaves of the pastora. When you rub and eat them, they work like the *nienn*. Of course, the pastora is not as powerful as the mushrooms."

Maria Sabina
in *Maria Sabina—Botinder heiligen pilze*
(Estrada 1980, 125**)

Salvinorin A

Salvinorin B

286 Loliolide is a known ant repellent (Brand 1994, 540).

287 Many Westerners have reported having experiences while under the influence of ketamine that are strongly reminiscent of shamanic experiences (dismemberment, death and rebirth, cosmic flight, body doubles, relocalization, clairvoyance, et cetera). Many users of *Salvia divinorum* who are also familiar with ketamine have shown a tendency to compare the two agents. To date, however, ketamine is known only as a synthetic substance related to PCP ("angel dust"), although it may soon be found in a natural form. The receptor to which the ketamine molecule binds is known. **Salvinorin A** does not attach to this receptor site.

("the mother), *Coleus pumila* (a species introduced from Europe) is *el macho* ("the father"), and *Coleus blumei* is both *el nene* ("the child") and *el ahijado* ("the godchild") (Wasson 1962, 79). It is this relationship that is responsible for the psychoactive reputation of *Coleus*.

In the region of Puebla, a similar and botanically as yet unidentified species of *Salvia* known as *xiwit* is cultivated for use in treating the folk ailment *susto* ("fright") and in rituals. The ritual is said to be very similar to that practiced by the Mazatec (Díaz 1979, 91*).

Artifacts

The botanist William Emboden has suggested that certain floral elements in the Mayan hieroglyphic manuscripts may represent *Salvia divinorum* (cf. *Nymphaea ampla*). This interpretation is difficult to imagine, for the plant is entirely unknown in the Yucatán Peninsula.

The American artist Brigid C. Meier has produced several paintings inspired by her own *Salvia divinorum* visions.

A riotous novel titled *Nice Guys Finish Dead* (Debin 1992) features *Salvia divinorum* and a "super drug" called NICE made from the plant.

Medicinal Use

Indians use nonpsychoactive preparations to treat defecation and urination disorders, headaches, rheumatism, and anemia and to reinvigorate the infirm, the aged, and the dying (Brand 1994, 541; Valdés 1994, 277).

Constituents

The leaves contain the neoclerodan **diterpenes** salvinorin A and salvinorin B (= divinorin A and divinorin B) as well as two other similar substances whose composition has not been completely determined (Brand 1994, 540; Siebert 1994; Valdés 1994). The main active constituent is **salvinorin A**, which can induce extreme effects in dosages as small as 150 to 500 [μ]g (Siebert 1994, Zubke 1997).

Loliolide,[286] a substance known from *Lolium perenne* L. (cf. *Lolium temulentum*), has also been detected (Valdés 1986).

Neither an **essential oil** nor thujone, which is known to occur in other *Salvia* species, has been discovered to date (Ott 1996, 35).

Effects

Most people who have ingested *Salvia divinorum* in the form of a quid or tincture or by smoking have reported very bizarre and unusual psychoactive effects that are difficult to compare to the known effects of euphoric or psychedelic substances. Space is often perceived as curved, and surging and rolling body sensations or out-of-body experiences are frequently described as typical.

Daniel Siebert has summarized the phenomenology of the effects of *Salvia divinorum* in the following way:

Certain themes are common to many of the visions and sensations described. The following is a listing of some of the more common themes:

1. Becoming objects (yellow plaid French fries, fresh paint, a drawer, a pant leg, a Ferris wheel, etc.).
2. Visions of various two dimensional surfaces, films and membranes.
3. Revisiting places from the past, especially childhood.
4. Loss of the body and/or identity.
5. Various sensations of motion, or being pulled or twisted by forces of some kind.
6. Uncontrollable hysterical laughter.
7. Overlapping realities. The perception that one is in several locations at once.

(Siebert 1994, 55)

These effects are strongly reminiscent of those that are experienced at subanesthetic dosages (50 to 100 mg) of ketamine (Ketanest®) (Bolle 1988; Jansen 1996).[287]

Commercial Forms and Regulations

Living plants are increasingly available from sources specializing in ethnobotanical products, especially in North America and Europe. The plant is not regulated in any way.

Literature

See also the entries for *Coleus blumei*, diterpenes, and **salvinorin A**.

Bolle, Ralf H. 1988. *Am Ursprung der Sehnsucht: Tiefenpsychologische Aspekte veränderter Wachbewußtseinszustände am Beispiel des Anästhetikums KETANEST*. Berlin: VWB.

Brand, Norbert. 1994. Salvia. In *Hagers Handbuch der pharmazeutischen Praxis*, 5th ed., 6:538–74. Berlin: Springer.

Clebsch, Betsy. 1997. *A book of salvias: Sages for every garden*. Cambridge, U.K.: Timber Press.

Debin, David. 1992. *Nice guys finish dead*. New York: Random House.

Epling, Carl, and Carlos D. Játiva-M. 1962. A new species of salvia from Mexico. *Botanical Museum Leaflets* 20 (3): 75–76.

Hofmann, Albert. 1990. Ride through the Sierra Mazateca in search for the magic plant "*Ska María Pastora*." In *The sacred mushroom seeker*, ed. Th. Riedlinger, 115–27. Portland, Ore.: Dioscorides Press.

Jansen, Karl L. R. 1996. Using ketamine to induce the near-death experience: Mechanism of action and therapeutic potential. *Yearbook for Ethnomedicine and the Study of Consciousness* 1995 (4): 55–79. Berlin: VWB.

Mayer, Karl Herbert. 1977. *Salvia divinorum*: Ein Halluzinogen der Mazateken von Oaxaca. *Ethnologia Americana* 14 (2): 776–79.

Ott, Jonathan. 1995. Ethnopharmacognosy and human pharmacology of *Salvia divinorum* and salvinorin A. *Curare* 18 (1): 103–29.

———. 1996. *Salvia divinorum* Epling et Játiva (foglie della pastora/leaves of the shepherdess). *Eleusis* 4:31–39. (Very good bibliography.)

Reisfield, Aaron S. 1993. The botany of *Salvia divinorum* (Labiatae). *Sida—Contributions to Botany* 15 (3): 349–66.

Siebert, Daniel J. 1994. *Salvia divinorum* and salvinorin A: New pharmacologic findings. *Journal of Ethnopharmacology* 43:53–56.

Valdés, Leander J., III. 1983. The pharmacology of *Salvia divinorum* Epling and Játiva-M. PhD thesis, University of Michigan, Ann Arbor.

———. 1986. Loliolide from *Salvia divinorum*. *Journal of Natural Products* 49 (1): 171.

———. 1994. *Salvia divinorum* and the unique diterpene hallucinogen, salvinorin (divinorin) A. *Journal of Psychoactive Drugs* 26 (3): 277–83.

Valdés, Leander J., José L. Díaz, and Ara G. Paul. 1983. Ethnopharmacology of *ska maría pastora* (*Salvia divinorum*, Epling and Játiva-M.). *Journal of Ethnopharmacology* 7:287–312.

Valdés, L. J., G. M. Hatfield, M. Koreeda, and A. G. Paul. 1987. Studies of *Salvia divinorum*. (Lamiaceae), an hallucinogenic mint from the Sierra Mazateca in Oaxaca, central Mexico. *Economic Botany* 41 (2): 283–91.

Wasson, R. Gordon. 1962. A new Mexican psychotropic drug from the Mint Family. *Botanical Museum Leaflets* 20 (3): 77–84.

Z[ubke], A[chim]. 1997. *Salvia divinorum*: Lieferant des stärksten aus dem Pflanzenreich bekannten Psychedelikums. *Hanfblatt* 4 (36): 15–19.

Sassafras albidum (Nutt.) Nees

Sassafras Tree

Family
Lauraceae (Laurel Family); Subfamily Lauroideae, Cinnamomeae Tribe, Cinnamominae Subtribe

Forms and Subspecies
The species is divided into two varieties whose appearance is very similar but whose geographical distribution is somewhat distinct:

Sassafras albidum (Nutt.) Nees var. *albidum*
Sassafras albidum (Nutt.) Nees var. *molle* (Raf.) Fern.

Synonyms
Laurus sassafras L.
Persea sassafras Spreng.
Sassafras officinale Th. Nees et Eberm.
Sassafras officinalis Nees et Eberm.
Sassafras sassafras (L.) Karst.
Sassafras variifolium (Salisbury) O. Kuntze
Sassafras variifolium (Salisbury) O. Kuntze var. *albidum* (Nutt.) Fern.

Folk Names
Ague tree, cinnamon wood, fenchelholz, fenchelholzbaum, laurus sassafras, nelkenzimtbaum, pavane, saloop, sassafrasbaum, sassafras tree, sassafrax, sassafrax tree, saxifrax, sommerlorbeer

History
North American Indians were already drinking a tea of sassafras root cortex in pre-Columbian times for stimulant, tonic, and medicinal purposes. In 1582, sassafras wood was included in lists of German medicines under the names *lignum pauamum*, *lignum floridum*, and *sassafrasbaum* (Schneider 1974, 3:230*).

The name *sassafras* is apparently a corruption of the Spanish word for the genus *Saxifraga*, which the Spanish botanist Monardes coined in the sixteenth century. Even into the twentieth century, a sassafras tea with milk and sugar known as saloop was sold on many London street corners in the early mornings (Grieve 1982, 715*).

During the American Civil War, the root cortex was used as a substitute for Chinese tea (**Camellia sinensis**) (Havard 1896, 45*). Until recently, it was also used in the United States as a flavoring agent in root beer, a nonalcoholic soft drink (Bremness 1995, 83*). In the southern states, dried, young leaves are used as a spice in gumbo, a Creole dish.

This illustration of a Mexican plant called *sasafrás*, presumably the oldest illustration of a plant of this name, may represent *Elaphrium pubescens* Schlecht. (From Hernández, 1942/46 [Orig. pub. 1615]*)

This leaf shape is characteristic of the sassafras tree (*Sassafras albidum* var. *albidum*).

This illustration of the American sassafras tree may in fact be the earliest European representation of the plant. (Woodcut from Gerard, *The Herball or General History of Plants*, 1633)

Distribution

Entire forests of the tree are occasionally found along the Atlantic Coast from northern Florida to Canada. The variety *albidum* occurs from Maine west to Michigan and Illinois and south to Virginia and Arkansas. The variety *molle* is found from Maine to New York; in Illinois, Iowa, and Kansas; and as far south as Florida and Texas (Zander 1994, 500*).

Cultivation

The tree can be propagated from ripe seeds that have not yet dried, from cuttings, or from root scions. The tree thrives in almost all types of soil but does best in good topsoil. It requires a temperate climate (Grubber 1991, 58 f.*).

The plant is cultivated for pharmaceutical purposes primarily in the states of New Jersey, Pennsylvania, and North Carolina and reportedly in northern Mexico and Taiwan as well (Bertram and Abel 1994, 611).

Appearance

This deciduous tree, which can grow as tall as 30 meters, bears foliage that is green in summer and golden red in autumn. The thick bark is deeply furrowed and has a different structure in each of the two varieties. The clusters of small yellow flowers appear before the new leaves. The small blue fruits (pea-sized drupes) are attached to red stalks.

The sassafras tree is particularly recognizable by the shape of its leaves. The tree develops three different forms of leaves, each of which appears on a separate branch. The smallest form is oval, while the somewhat larger form is oval with an indentation (two-lobed). The largest and most frequent (three-lobed) form is deeply digital with two indentations. The tree also can be identified by the typical scent of safrole, which is exuded when the leaves are rubbed or crushed.

The sassafras tree can be confused with the other two members of the genus, *Sassafras tzumu* (Hemsl.) Hemsl. and *Sassafras randaiensis* (Hay.) Rehd. (Bertram and Abel 1994, 610).

Psychoactive Material

— Root pith (sassafras lignum, lignum sassafras, sassafras wood, lignum pavanum, fenchelholz)
— **Essential oil**, obtained from the root pith through steam distillation (sassafras oil, sassafras aetheroleum, oleum sassafras, sassafrasöl, fenchelholzöl, essence de sassafras)
— Root cortex (= root bark; sassafras radix, sassafras cortex, fenchelholzrinde)

The very aromatic root cortex can be obtained from the living tree without killing it. A hole is dug to reveal a piece of the root (no more than one third). The root cortex is then carefully removed.

Other Species of Sassafras

Other trees are also known as sassafras in North America: **Magnolia virginiana** is known as swamp sassafras, *Massoja aromatica* is called *Sassafras goesianum*, and **Umbellularia californica** is known as California sassafras (Grieve 1982, 716*).

The tree *Mespilodaphne sassafras* Meister is called Brazilian sassafras and is also used as a counterfeit for true sassafras wood (Bertram and Abel 1994, 615).

The **essential oil** obtained from *Ocotea cymbarum* H.B.K. (Lauraceae) is permitted to be sold under the name *sassafras oil* or *Brazilian sassafras oil* (Bertram and Abel 1994, 611).

In Australia, the name *sassafras* is applied to trees from the Family Monimiaceae that "smell of sassafras" and also contain safrole: **Atherosperma moschatum** Labill (southern sassafras, black sassafras) and *Doryphora sassafras* Endl. (real sassafras, yellow sassafras, canary sassafras). Both trees are used to produce a "bush tea" with stimulant and tonic properties. In addition to the essential oil, the bark of *Doryphora* also contains the alkaloid dryphorine (Cribb 1984, 172, 174*).

Care must be taken not to damage the inner cortex so the tree is able to regrow the root cortex that has been removed (Grubber 1991, 59*).

Preparation and Dosage

Sassafras formerly was used as an additive to **beer** and to perfume tobacco (**Nicotiana tabacum**) (Schneider 1974, 3:231*).

To prepare as a tea (called sassafras tea or saloop), add 30 g of chopped root cortex to 0.5 liter of boiling water. A normal dosage for a blood depurative tea is regarded as 2.5 g of chopped root (sassafras wood). Pour boiling water over this amount and strain after ten minutes (Wichtl 1989). As a single dose, 5 g of sassafras wood can be ingested (Bertram and Abel 1994, 617).

One or two drops of sassafras oil, dissolved in **alcohol**, can be taken several times daily as a medicinal dosage. The *EB6* lists 0.1 g as a single dosage (Bertram and Abel 1994, 612). A good starting dosage for aphrodisiac and psychoactive purposes is regarded as 100 to 200 mg of the oil (Gottlieb 1973, 45*). This dosage should be increased only with great care, as overdoses can result in kidney irritation (Pahlow 1993, 418*). One teaspoon of sassafras oil can induce "vomiting, dilated pupils, stupor and collapse" (Grieve 1982, 716*). The safrole present in the oil is regarded as carcinogenic (Bertram and Abel

1994, 612 f.). In former times, sassafras oil was often mixed with opium (cf. *Papaver somniferum*) when the latter was administered to children in order to cover up its horribly bitter taste. For medicinal use, sassafras oil was usually mixed with *Guaiacum* spp. and sarsaparilla (*Smilax regellii* Kill. et C.V. Morton) (Grieve 1982, 716*).

In Louisiana, filé (dried, powdered young sassafras leaves) are used to bind soups and to prepare gumbo (Bremness 1995, 83*).

Ritual Use
To date, no ritual use of sassafras, especially for psychoactive purposes, has been documented. The use of sassafras oil as an inebriant did not become well known until MDMA was made illegal (cf. **herbal ecstasy**).

Artifacts
Gumbo (Atlantic Records, 1972), an album from Dr. John, the Night Tripper (the "high priest of voodoo rock"), is named after the Creole dish gumbo, which is prepared with sassafras leaves. Whether the ingestion of copious amounts of sassafras leaves contributed to the hallucinations immortalized on the album is unknown.

Medicinal Use
In Europe, sassafras is regarded as a panacea (Schneider 1974, 3:230*). Sassafras oil was administered internally in folk medicine to treat physical and mental debility, gout, menstrual complaints, urine retention, and inflammations of the urethra and bladder. It was applied externally to soothe the pains of rheumatism and insect stings (Bertram and Abel 1994, 612). The oil was used internally to alleviate the cramps and pains associated with menstruation (Grieve 1982, 716*). It also was used to induce abortions, and it should be avoided when pregnancy is desired (Bertram and Abel 1994, 612).

In central Europe, teas made from the leaves or the root cortex were especially popular as a blood depurative (Bremness 1995, 83*; Wichtl 1989, 425).

The mother tincture (Sassafras hom. *HAB34*, Sassafras officinale hom. *HPUS88*), obtained by extracting the dried root cortex in 90% ethyl alcohol, is used in homeopathic medicine (Bertram and Abel 1994, 617).

Constituents
The root cortex typically contains between 6 and 9% **essential oil** whose primary constituent is safrole (approximately 80%). Also present are safrole camphor (= camphor/D-camphor; cf. *Cinnamomum camphora*), tannins (sassafrid), red tannic acid (an orange dye), resin, wax, mucilage, sugar, and sitosterol (Bertram and Abel 1994, 611; Grieve 1982, 715*). A recent study has provided

further knowledge of the composition of the essential oil; it is 85% safrole, 3.25% camphor, and 1.1% methyleugenol. Each of the other components—including estragol, eugenol, elemicine, myristicin (cf. *Myristica fragrans*), 5-methoxyeugenol, and apiol—make up less than 1% of the mixture (Kamden and Gage 1995). According to a different analysis, the essential oil obtained through steam distillation of the root cortex consisted of 90% safrole, with the remaining 10% composed of 30% 5-methoxyeugenol, 18% asarone, 5% camphor, 7% coniferaldehyde, 11% piperonylacrolein, and traces of apiol, l-menthone, α-phellandrene, β-phellandrene, thujone, anethol, caryophyllene, copaene, elemicine, eugenol, myristicin, α-pinene, and syringaaldehyde (Sethi et al. 1976).

The root cortex also contains alkaloids (aporphine and benzylisoquinoline derivatives, boldine, isoboldine, norboldine, cinnamolaurin, norcinnamolaurin, reticuline) (Bertram and Abel 1994, 614; Chowdhury et al. 1976; Wichtl 1989, 425).

The root pith consists of some 1 to 2% essential oil (of which approximately 80% is safrole). The seeds are 60% fatty oil with linoleic and oleic acids (Bertram and Abel 1994, 611, 616).

Effects
There are few reports of psychoactive effects resulting from use of the tea: "Large dosages have stimulating and sudoriferous effects" (Schuldes 1995, 69*). Ingestion of high dosages of sassafras oil results in profound stimulation, erotic excitation, perceptual changes, and particularly a more profound sensitivity in the emotional domain. The effects are sometimes described as MDMA-like and empathogenic (cf. *Myristica fragrans*). Higher dosages also can result in unpleasant side effects (cold sweats, cramping of the chewing muscles, nervousness, unease).

Commercial Forms and Regulations
Sassafras oil has been illegal in the United States since 1960, when it was claimed that it was carcinogenic (Kamden and Gage 1995). Because sassafras oil is a suitable precursor for the illegal manufacture of MDMA, it is now a controlled substance worldwide and is almost never sold, even in small amounts. Even the raw plant material (root wood, root cortex) has practically disappeared from the market. In the United States, and especially in the southern states, the only product still available is powdered sassafras leaves, which are sold as gumbo filé.

Literature
See also the entry for **essential oil**.

Bertram, Barbara, and Gudrun Abel. 1994. *Sassafras*. In *Hagers Handbuch der pharmazeutischen Praxis*, 5th ed., 6:610–19. Berlin: Springer.

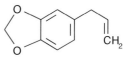

Safrole

Top: This bark, whose structure is typical for *Sassafras albidum* var. *albidum*, contains an essential oil that smells of safrole.

Bottom: The bark of *Sassafras albidum* var. *molle* has a very different structure.

"Sassafras may have been the first tree to be brought to Europe from the Americas."

LESLEY BREMNESS
KRÄUTER, GEWÜRZE UND HEILPFLANZEN [HERBS, SPICES, AND MEDICINAL PLANTS]
(1995, 83*)

Chowdhury, Bejoy K., Manohar L. Sethi, H. A. Lloyd, and Govind J. Kapadia. 1976. Aporphine and tetrahydrobenzylisoquinoline alkaloids in *Sassafras albidum*. *Phytochemistry* 15:1803–4.

Kamden, Donatien Pascal, and Douglas A. Gage. 1995. Chemical composition of essential oil from the root bark of *Sassafras albidum*. *Planta Medica* 61:574–75.

Sethi, Manohar L., G. Subbu Rao, B. K. Chowdhury, J. F. Morton, and Govind J. Kapadia. 1976. Identification of volatile constituents of *Sassafras albidum* root oil. *Phytochemistry* 15:1773–75.

Wichtl, Max. 1989. Sassafrasholz. In *Teedrogen*, ed. Max Wichtl, 424–25. Stuttgart: WVG.

Sceletium tortuosum (Linnaeus) N.E. Br.

Kougoed

The South African Hottentots (Khoikhoi) once chewed a variety of plants known as channa, kanna, or kougoed—including *Sceletium tortuosum*—as agents of pleasure and inebriation. (Copperplate engraving from Meister, *Der Orientalisch-Indianische Kunst- und Lustgärtner*, 1677) [The Oriental-Indian Art and Pleasure Gardener]

Family
Aizoaceae (Ice Plant Family) (Mesembryanthemaceae); Subfamily Mesembryanthemoideae (cf. Bittrich 1986)

Forms and Subspecies
None

Synonyms
Mesembryanthemum tortuosum L.

Folk Names
Canna, canna-root, channa, gunna, kanna, kauwgoed, kauwgoed, kon ("quid"), kou, kougoed, tortuose fig-marygold

History
Some three hundred years ago, it was reported that the Hottentots (Khoikhoi) of southern Africa chewed, sniffed, or smoked an inebriant that was said to be known as **kanna** or channa (Schleiffer 1979, 39 ff.*). The enthusiasm with which the Hottentots smoked was noted by all the early travelers to the region. Unfortunately, most of them neglected to provide any information about the botanical source of the "tobacco" (e.g., Meister 1677, 31 f.*). And so it was not until the end of the nineteenth century that it was suggested that the inebriant must have come from *Mesembryanthemum* spp., for these species were then still known by the name *kanna* in South Africa. The effects that were experienced at that time, however, were not nearly as dramatic and inebriating as had been hoped (Meiring 1898). Around the same time, Carl Hartwich was already suggesting that the species in question was *Mesembryanthemum tortuosum* (1911, 810*), which (following a taxonomic revision) is now known as *Sceletium tortuosum*. However, the first ethnobotanical evidence of the psychoactive use of *Sceletium tortuosum* as kougoed was obtained only a few years ago (Smith et al. 1996).

Distribution
The plant occurs only in South Africa, in the so-called kanna land. *Sceletium tortuosum* and other species (*Sceletium strictum*) have become rare in South Africa and are increasingly difficult to find (Smith et al. 1996, 128).

Cultivation
Propagation occurs through the seeds, which must be treated in the same manner as cactus seeds. The best method is to scatter them onto cactus or succulent soil, press them down slightly, and water (Schwantes 1953). Both the cultivation and care are similar to that for the Cactaceae, which is the most closely related family.

Appearance
This herbaceous plant, which resembles a leaf succulent, can grow as tall as 30 cm. It develops fleshy roots, a smooth and fleshy stalk, and low-growing branches that spread laterally. The thick, angular, fleshy leaves do not have stalks but are attached directly to the branches. The pale yellow flowers are 3 to 4 cm across and are attached to the ends of the branches. The plant produces angular-shaped fruits with small seeds.

Kougoed is easily confused with other members of the genus *Sceletium* (as well as with **Mesembryanthemum spp.**). Those species that not only look similar but also have similar effects and contain the same active constituent (mesembrine) were presumably also referred to as kougoed and used in the same manner (Arndt and Kruger 1970; Jeffs et al. 1970, 1974; D. McKenna 1995, 101*):

Sceletium anatomicum (Haw.) L. Bolus [syn. *Mesembryanthemum anatomicum* Haw.]
Sceletium expansum (L.) L. Bolus [syn. *Mesembryanthemum expansum* L.]
Sceletium joubertii L. Bolus[288]
Sceletium namaquense L. Bolus
Sceletium strictum L. Bolus

[288] This species also contains the β-**phenethylamine** hordenine, a characteristic constituent of the cacti (Arndt and Kruger 1970).

Psychoactive Material
— Entire plant with root

Preparation and Dosage
The method for preparing kougoed has only recently been discovered and described in great detail. The plant material—which should be collected in October, when the plant is most potent—is harvested, crushed between two rocks, and allowed to "ferment" for a few days in a closed container. At one time animal skins or hemp bags were used for this purpose, but plastic bags are now used in their place. The first step entails setting the bag containing the plant material in the sun. During the day, the plant will exude its juice, which condenses on the plastic and is later reabsorbed by the plant material. During the night, the material cools. After two or three days, the bag is opened and the contents are stirred well. Then the bag is sealed and placed outside again. On the eighth day after this procedure was started, the kougoed is taken from the bag and spread out to dry in the sun. It can be used as soon as it has dried. According to informants, the fresh leaves do not have any potency; only the "fermented" plant is psychoactive. The kougoed is now either chopped or powdered. This process presumably helps to substantially reduce the high content of oxalic acid that is characteristic of the genera *Sceletium* and *Mesembryanthemum*. Oxalic acid can produce severe irritation and allergies. A more hurried method involves simply toasting a fresh plant on glowing charcoals until it has completely dried and then powdering the result (Smith et al. 1996, 126).

The powder usually is taken orally with some **alcohol** and held in the mouth for about ten minutes. The saliva that collects can be swallowed. Two grams of the powder produces a "tranquil mellowness" in about thirty minutes; approximately 5 g of the powder is a dosage sufficient to relieve anxiety, and higher dosages can lead to more profound effects (euphoria, visions) (Smith et al. 1996, 126 f.).

The chopped plant material can be smoked alone or in combination with *Cannabis sativa* (cf. **smoking blends**). The finely ground powder purportedly also can be sniffed, either alone or mixed with tobacco (cf. **snuffs**).

This and other species were used as psychoactive additives to **beer** or to induce fermentation (Smith et al. 1996, 127).

Ritual Use
The South African Bushmen (San) use the same name for *Sceletium tortuosum* as they do for the eland antelope (*Taurotragus oryx* Pallas): *kanna*. The eland is regarded as the "trance animal" par excellence; since prehistoric times, it has played a central role as a magical ally in many ceremonies

and was closely associated both with the rainmakers and with divination, healing, and the communal trance dances (Lewis-Williams 1981). Kanna appears to have been used as part of these rituals (cf. also *Ferraria glutinosa*).

The Hottentots (Khoikhoi) apparently chewed *Sceletium* for their ritual and healing dances or smoked it together with *dagga* (*Cannabis sativa*). They also use the name *kanna* for the magical eland antelope (Smith et al. 1996, 120).

In contemporary South Africa, kougoed is now used primarily as an agent of pleasure; it is used as a party drug in the same way that *Cannabis sativa* is used in Western society.

Artifacts
It is possible that a great deal of the rock art of South Africa, some of which appears to be extremely visionary, was inspired by kougoed (Lewis-Williams 1981).

Medicinal Use
The natives of Namaqualand and Queenstown (southern Africa) drink a tea made from the leaves as an analgesic and to suppress hunger (Smith et al. 1996, 128).

Constituents
The leaves and stalks of the plant contain 0.3 to 0.86% mesembrine (empirical formula $C_{17}H_{23}NO_3$), along with some mesembrinine and tortuosamine (Smith et al. 1996). The leaves appear to also contain oxalic acid (Frohne and Jensen 1992, 125*). It is possible that tryptamines may occur in the plant as well. Methyltryptamine (MMT) and *N,N*-DMT have been detected in a *Delosperma* species, a close relative from the same family (Smith et al. 1996, 124).

Effects
The South African users describe the important effects of small dosages of kougoed as relief from anxiety and stress, deepening of social contact, increase in self-confidence, and dissolution of inhibitions and feelings of inferiority. "Some reported euphoria as well as a feeling of meditative tranquility. Several users felt that the relaxation

A cultivated South African kougoed (*Sceletium tortuosum*).

"And so it should be known how these clean mountain nymphs [the Hottentot women] are so unashamed that they even pass their maiden water in the presence of Europeans and are even accustomed to answering the call of nature.... Because they are also excellent lovers of the noble weed nicotianae or tobacco [a presumed reference to kougoed], then these brave women probably show to a curious, lewd lover everything that he asks of them before a pipe of tobacco."

GEORGE MEISTER
DER ORIENTALISCH-INDIANISCHE KUNST- UND LUSTGÄRTNER [THE ORIENTAL-INDIAN ART AND PLEASURE GARDENER]
(1677, CH. 4, P. 4*)

Mesembrine

Sceletium tortuosum, as shown in an eighteenth-century woodcut.

induced by 'kougoed' enabled one to focus on inner thoughts and feelings, if one wished, or to concentrate on the beauty of nature. Some informants reported heightened sensation of skin to fine touch, as well as sexual arousal" (Smith et al. 1996, 127f.).

Higher dosages, especially when combined with **Cannabis sativa** and **alcohol** (whiskey), produce mild visions. Chewing kougoed shortly after smoking *Cannabis* can considerably potentiate the effects of the hemp. Kougoed suppresses both the effects of tobacco (**Nicotiana tabacum**) and the desire for **nicotine**.

Commercial Forms and Regulations
Seeds of *Sceletium tortuosum* and other *Sceletium* species—usually under the synonym *Mesembryanthemum*—are occasionally available through flower shops and ethnobotanical specialty sources. Living members of the genus are sometimes offered by cactus dealers and shops.

Literature
See also the entries for **Mesembryanthemum spp.** and **kanna**.

Arndt, R. R., and P. E. J. Kruger. 1970. Alkaloids from *Sceletium joubertii* L. Bolus: the structure of joubertiamine, dihydrojoubertiamine and dehydrojoubertiamine. *Tetrahedron Letters* 37:3237–40.

Bittrich, V. 1986. Untersuchungen zu Merkmalbestand, Gliederung und Abgrenzung der Unterfamilie Mesembryanthemoideae (Mesembryanthemaceae Fenzl). *Mitteilungen aus dem Institut für Allgemeine Botanik* (Hamburg) 21:5–116.

Bodendorf, K., and K. Krieger. 1957. Über die Alkaloide von *Mesembryanthemum tortuosum* L. *Archiv für Pharmazie* 62:441–48.

Jeffs, P. W., G. Allmann, H. F. Campbell, D. S. Farrier, G. Ganguli, and R. L. Hawks. 1970. Alkaloids of *Sceletium* species III: The structures of four new alkaloids from *Sceletium strictum*. *Journal of Organic Chemistry* 35:3512–28.

Jeffs, P. W., T. Cappas, D. B. Johnson, J. M. Karle, N. H. Martin, and B. Rauckman. 1974. *Sceletium* alkaloid VI: Minor alkaloids from *Sceletium namaquense* and *Sceletium strictum*. *Journal of Organic Chemistry* 39:2703–9.

Laidler, P. W. 1928. The magic medicine of the Hottentots. *South African Journal of Science* 25:433–47.

Lewis-Williams, I. D. 1981. *Believing and seeing: Symbolic meanings in southern San rock paintings.* London: Academic Press.

Meiring, I. 1898. Notes on some experiments with the active principle of *Mesembryanthemum tortuosum*. *Transactions of the South African Philosophical Society* 9:48–50.

Schwantes, G. 1953. *The cultivation of the Mesembryanthemaceae.* London: Blandford Press.

Smith, Michael T., Neil R. Crouch, Nigel Gericke, and Manton Hirst. 1996. Psychoactive constituents of the genus *Sceletium* N.E. Br. and other Mesembryanthemaceae: A review. *Journal of Ethnopharmacology* 50:119–30. (Good bibliography.)

Scopolia carniolica Jacques
Scopolia

"Skopolie, indigenous to Carinthia and Ukraine, cultivated in gardens by the Lithuanians in eastern Prussia as 'crazy root.' The root serves them as a medicine to treat paralysis agitans [= Parkinson's disease]; also as an erotic inebriant, applied locally, as an abortifacient."

HERMANN FÜHNER
MEDIZINISCHE TOXIKOLOGIE
[MEDICAL TOXICOLOGY]
(1943, 195*)

Family
Solanaceae (Nightshade Family); Subfamily Solanoideae, Hyoscyameae Tribe, Hyoscyaminae Subtribe

Forms and Subspecies
Three to five species are now accepted botanically in the genus (D'Arcy 1991, 79*; Lu 1986, 6*). A number of varieties of *Scopolia carniolica* have been described:

Scopolia carniolica Jacq. var. *brevifolia* Dun.
Scopolia carniolica Jacq. var. *carniolica*
Scopolia carniolica Jacq. var. *concolor* Dun.
Scopolia carniolica Jacq. var. *hladnikiana* (Fleischm.) Fiori
Scopolia carniolica Jacq. var. *longifolia* Dun.

A new form that has pure yellow flowers and is found only in Slovenia (Hladnikov) has recently been described (Dakshobler 1996):

Scopolia carniolica Jacq. forma *hladnikiana*

Synonyms
Hyoscyamus scopolia L.
Scopolia hladnikiana Fleischm.
Scopolia longifolia Dun.
Scopolina atropoides Schultes

Scopolina hladnikiana Freyer
Scopolina viridiflora Freyer

Folk Names

Altsitzerkraut, deewa sales, durna rope (Lithuanian, "crazy root"), glockenbilsenkraut, gotteskraut, krainer tollkraut, matragun (Romanian),[289] mauda, maulda, pikt-rope ("evil root"),[290] pometis ropes ("pometis root"), Russian belladonna, scopolia (Italian), scopolie, skopolia, skopolie, tollkraut, tollrübe, volčič, walkenbaum

History

It is uncertain whether scopolia was known to the authors of antiquity. The "sleeping strychnos" (*Strychnos hypnoticos*) of Dioscorides (cf. **Solanum spp.**) has sometimes been interpreted as a *Scopolia* species (Fühner 1919, 223). The genus was named for the naturalist Antonio Scopoli (1723–1788), who was the first to study and describe the flora of Slovenia (Festi 1996, 35). In Slovenia, the plant may once have been used to prepare **witches' ointments**. In eastern Prussia, the root of the plant was used as a folk inebriant and aphrodisiac. It is said that women would use it to persuade young men to become their willing lovers. Sometimes some of the root was added to a person's coffee (*Coffea arabica*) as a practical joke so that others could amuse themselves on the seemingly nonsensical behavior of the inebriated victim (Fühner 1919).

In the history of pharmaceuticals, *Scopolia* has played only a minor role as a substitute or counterfeit for belladonna root (*Atropa belladonna*) and belladonna leaves (Schneider 1974, 3:240*). Today, the plant is used in the industrial manufacture of L-hyoscyamine and **atropine** (Wagner 1985, 172*).

Distribution

The plant occurs wild in the Alps, the Carpathian Mountains, and the Caucasus Mountains (Gelencir 1983, 217). It also grows in southeastern Europe (Slovenia), Lithuania, Latvia, and the Ukraine.

Cultivation

Cultivation is very simple. In spring, the seeds are sown into seedbeds to germinate. Later, the seedlings can be transplanted to the desired location. The plant does not tolerate a great deal of exposure to the sun (Festi 1996, 36), preferring dark, humid forests and calciferous humus soil. In Lithuania and Latvia, it has long been planted in gardens for use as a medicinal plant.

Appearance

This annual plant, which is typically 30 to 60 cm in height but can grow as tall as 80 cm, develops a fleshy, spindle-shaped root. The dull green leaves resemble those of the belladonna plant (*Atropa belladonna*)—hence the name *Russian belladonna*.

The small, pendulous, campanulate flowers are purple to pale yellow in color and are similar in shape to henbane flowers (*Hyoscyamus albus*)—hence the German name *glockenbilsenkraut* ("bell henbane"). The plant flowers from April to June. The fruit develops a capsule with a double partition and many small seeds.

Scopolia is easily confused with Chinese scopolia (*Scopolia carniolicoides* C.W. Wu et C. Chen) and Japanese scopolia (*Scopolia japonica* Maxim.). *Scopolia anomala* (Link et Otto) Airy Shaw [syn. *Scopolia lurida* Dun.], which is native to Nepal and Sikkim, is about twice as large as the European scopolia (Weinert 1972).

Psychoactive Material

— Root (rhizoma scopoliae, scopoliae radix, scopolia root, glockenbilsenkrautwurzel, europäische scopoliawurzel)
— Herbage (herba et radix scopoliae carniolicae)

Preparation and Dosage

The fresh root, when boiled and grated, can be eaten as a mush or taken in coffee (*Coffea arabica*). It also is added to **beer** or brewed with it in order to potentiate its effects (Fühner 1919, 224).

The root and the rootstock (scopoliae radix et rhizoma) are used. The root is dug up, dried, and used exactly as the belladonna root. Taste, color, and appearance are exactly like belladonna. Many plant collectors confuse the bell henbane with belladonna [*Atropa belladonna*], which is why one always finds scopolia mixed into belladonna, especially when the belladonna is from the Carpathians. (Gelencir 1983, 217)

The dried herbage, collected while the plant is in flower, can be smoked alone or in **smoking blends**.

Ritual Use

In eastern Prussia, Lithuania, and the Balkans, scopolia formerly was collected and used in magic in the same manner as mandrake (*Mandragora officinarum*). In the early twentieth century, only

Left: Scopolia (*Scopolia carniolica*) was used in eastern Europe much like the mandrake.

Right: The Asian scopolia *Scopolia anomala* also has psychoactive properties.

> "White thorn apple [*Datura metel* var. *alba*] and black scopolia [*Scopolia anomala*] heighten the sex drive, and together with henbane [*Hyoscyamus niger* var. *chinensis*], they alleviate illnesses caused by tiny little animals."
>
> "DER BLAUE BERYLL"
> (ARIS 1992, 67*)

289 In Romania, the same name is given to *Mandragora officinarum* (Fühner 1919).

290 In Lithuania, this name is also used for water hemlock (*Cicuta virosa*) (cf. **witches' ointments**).

"The incident occurred at the end of March 1901 in the church village of Lappienen, in the district of Niederung [eastern Prussia].... Here, some women added a decoction of scopolia to a man's afternoon coffee, supposedly as a joke. According to his own testimony, he developed a severe headache and a terrible burning soon after drinking the coffee. His tongue became completely rigid, and for a time he did not know where he was. Taken home, he lay in bed for a while and then vomited, whereupon the pains let up. But he remained lying in bed for three weeks, and during this time he suffered from severe headaches. A witness who took him home described his behavior in the following terms: 'Immediately after consuming the coffee, he complained of internal pains and talked all kinds of nonsense. On the way home, he acted like a madman: he ran, fell, claimed to see people, wood, a saw, and all manner of things. After arriving home, he did not recognize his wife, thought she was a young girl, lay down in bed, jumped back up, and was ill for quite a while."

HERMANN FÜHNER
"SCOPOLIAWURZEL ALS GIFT UND HEILMITTEL BEI LITAUEN UND LETTEN" [SCOPOLIA ROOT AS POISON AND MEDICINE AMONG THE LITHUANIANS AND LATVIANS] (1919, 225)

rudiments of this ritual use were still being practiced (Fühner 1919).

Artifacts

See *Mandragora officinarum.*

Scopolia anomala is depicted on Tibetan medical thangkas (Aris 1992, 67*).

Medicinal Use

Scopolia carniolica was used in eastern European folk medicine in the same manner as *Mandragora officinarum* (Schneider 1974, 3:240*). In Lithuania, the plant was used to treat rheumatism, gout, toothaches, colic, and Parkinson's disease; as a sedative for children and an aphrodisiac; and to induce abortions (Fühner 1919, 224).

In homeopathy, the essence obtained from the fresh-blooming herbage is known as Hyoscyamus scopolia and is used in accordance with the medical description (Schneider 1974, 3:240*).

Constituents

The entire plant contains hallucinogenic **tropane alkaloids** (Evans 1979, 249*). The total alkaloid content averages around 0.5% but can range from 0.3 to 0.8% (Fühner 1919, 223; Roth et al. 1994, 648*). The dried leaves contain 0.19% hyoscyamine and 0.13% **scopolamine** (Scholten et al. 1989). The root contains approximately 0.5% scopolamine (Gelencir 1983, 218). Also present are the alkaloids cuscohygrine, tropine, and 3α-tigloyloxytropane. Chemotaxonomically, *Scopolia* is thus closely related to henbane (*Hyoscyamus* spp.) (Evans 1979, 249*; Zito and Leary 1966). The alkaloid content of the dried roots can be as high as 1% (Wagner 1985, 172*).

In addition to the alkaloids, the entire plant also contains the **coumarins** scopoline and **scopoletin** as well as chlorogenic acid (Roth et al. 1994, 648*).

Effects

Few documents describing the actual effects of scopolia are available (Festi 1996). Depending upon dosages, all preparations are capable of producing psychoactive effects that are very similar to those produced by henbane. Low doses induce aphrodisiac sensations, whereas "larger quantities of the root are inebriating and produce a condition associated with unpredictable, comic actions" (Fühner 1919, 224). High doses have been observed to produce delirium, loss of awareness of reality, coma, severe pupillary dilation, headache, disturbances of coordination, and other symptoms typical of an overdose of *Atropa belladonna.*

Smoking the leaves produces only very mild psychoactive effects that are comparable with those resulting from smoking *Hyoscyamus niger* or *Datura stramonium.*

Commercial Forms and Regulations

The herbage and roots can sometimes be found in eastern European herb shops. The seeds are occasionally available from ethnobotanical specialty sources.

Literature

See also the entries for **coumarins, scopolamine, scopoletin,** and **tropane alkaloids.**

Dakskobler, Igor. 1996. Hladnikov volčič (*Scopolia carniolica* f. *hladnikiana*) tudi v Zelenem potoku. *Proteus* 58:102–3.

Festi, Francesco. 1996. *Scopolia carniolica* Jacq. *Eleusis* 5:34–45.

Fühner, Hermann. 1919. Scopoliawurzel als Gift und Heilmittel bei Litauen und Letten. *Therapeutische Monatshefte* 33:221–27.

Gelencir, Nikola. 1983. *Naturheilkunde des Balkans.* Steyr, Austria: Verlag Wilhem Ennsthaler.

Scholten, H. J., S. Batterman, and J. F. Visser. 1989. Formation of hyoscyamine in cell cultures of *Scopolia carniolica. Planta Medica* 55:230.

Weinert, E. 1972. Zur Taxonomie und Chorologie der Gattung *Scopolia* Jacq. *Feddes Repertorium* 82 (10): 617–28.

Zito, S. W., and J. D. Leary. 1966. Alkaloids of *Scopolia carniolica. Journal of Pharmaceutical Sciences* 55:1150–51.

Solandra spp.

Cup of Gold

Family

Solanaceae (Nightshade Family); Subfamily Solanoideae, Solandreae Tribe (formerly Datureae Tribe)

Species

Ten to twelve species are currently botanically recognized as belonging to the genus *Solandra* (D'Arcy 1991, 79*; Bärtels 1993, 207*; Schultes and Farnsworth 1982, 166*). However, the taxonomy of the genus is rather confusing or, as Schultes (1979b, 150*) expressed it, "very poorly understood."

The species of ethnopharmacological significance are:

Solandra brevicalyx Standl.—kieli, kieri, kiéri
Solandra guerrerensis Martinez—huipatli, hueypahtli, tecomaxochitl[291]
Solandra guttata D. Don ex Lindley (possibly identical to *Solandra brevicalyx*; Furst 1995, 55)
Solandra nitida Zucc. [syn. *Solandra maxima* P.S. Green, *Solandra hartwegii* N.E. Brown, *Swartzia nitida* Zucc.]—cutaquatzitziqui, copa de oro

To nonbotanists, these four species are difficult if not impossible to distinguish (Morton 1995, 20*). The Indians regard them as equivalent.

The following species, which occur in Mexico and are rich in alkaloids (Evans et al. 1972), have not been ethnobotanically described or investigated to date:

Solandra grandiflora Sw.
Solandra hirsuta Dun.
Solandra macrantha Dun.

Synonyms

Datura maxima Sessé et Mociña (= *Solandra* sp.)
Datura sarmentosa Lam. (= *Solandra grandiflora* Sw.)
Datura scandens Velloso (= *Solandra* sp.)

Solandra herbacea Mordant de Launay is a synonym for *Datura ceratocaula* (see **Datura spp.**).

Folk Names

In Mexico, these folk names are used for all of the species in the genus (cf. Martínez 1966): arbol del viento, bolsa de Judas (Spanish, "bag of Judas"), bolute, chalice vine, copa de oro (Spanish, "cup of gold"), cup of gold, cútacua (Tarascan), cutaquatzitziqui, floripondio del monte (Spanish, "angel's trumpet of the forest"), goldkelch, hueipatl, hueypatli, hueytlaca, itzucuatziqui, k'äni bäk'el (Lacandon, "yellow bone/scent"), kieli, kiéli, kieri, kiéri (Huichol, "tree of the wind"), lipa-ca-tu-hue (Chontal), ndari (Zapotec), perilla, tecomaxochitl (Aztec, "offering drink plant"), tetona, tima' wits (Huastec, "jicara decorated gourd flower"), tree of the wind, windbaum, wind tree, xochitecómatl (Nahuatl).

History

It is not known how ancient the ritual use of the potently hallucinogenic cup of gold in Mexico is, but it may have originated in prehistoric times. The Aztec plant *tecomaxochitl*, which is very likely to be interpreted as a *Solandra* species, was first described by Hernandez in the early colonial period. Maximino Martínez was the first to discuss the psychoactive use of *Solandra* species (1966). It is possible that the *Solandra* shamanism (also known as *kiéli* shamanism) of central Mexico may be older than the peyote cult, which arose in northern Mexico (cf. **Lophophora williamsii**) (Furst 1995).

The genus was named for the Swede D. C. Solander (1736–1786), a student of Linnaeus and a companion on the journeys of Captain Cook. To date, the ethnobotany of the genus has been only poorly studied, as the plants are often associated with witchcraft and harmful magic and their uses are consequently kept secret and suppressed. The plant (and its associated uses) was earlier often

Left: The flower of the cup of gold (*Solandra brevicalyx*) exudes a delicious perfume.

Center: The large cup of gold (*Solandra nitida*), cultivated by numerous Mexican Indians, develops very large flowers. (Photographed in Naha', Chiapas, Mexico)

Right: The pendulous flower of *Solandra guttata*.

291 This Nahuatl name is also used to refer to **Brugmansia arborea** (Díaz 1979, 84*).

292 At the time, the authors (Furst and Myerhoff 1966) still assumed that *kiéli* was either *Datura innoxia* or *Datura stramonium*, an interpretation that subsequently was corrected (Furst 1996).

confused with *Datura innoxia*. The Huichol refer to *Solandra brevicalyx* as the "true" *kiéli*, and to *Datura innoxia* as *kiélitsha*, "bad *kiéli*" (Knab 1977, 81).

Distribution

The genus *Solandra* is indigenous to Mexico (Schultes and Farnsworth 1982, 166*). Most of the species occur in central Mexico. The genus is represented to the south as far as the rain forests of Chiapas (Martínez 1966). Several species have spread into the Caribbean and to South America (Peru) (Furst 1995, 51).

Cultivation

Propagation is easily performed with cuttings. A piece of the stem (if possible from the end of the branch) approximately 20 cm long is placed in water. The plant can be placed in the ground as soon as its roots have started to develop. *Solandra* must be well watered and does not tolerate frost. In the rain forest, often all that is needed is to place a piece of the stem in the ground. Shoots will then quickly appear.

Solanda grandiflora and *Solandra nitida* are the most commonly cultivated species for garden and ornamental use (Bärtels 1993, 207*).

Appearance

The perennial, heavily branching, fast-growing climber develops oblong-elliptic leaves that are up to 15 cm in length and tapered at the end. The solitary, terminal, chalice-shaped yellow flowers exude a sweet scent, usually in the evening, that is intoxicating, delicious, and very fine. This scent is comparable to the perfume of *Brugmansia suaveolens* or *Brugmansia* x *insignis*. Because almost all of the plants are the product of cultivation, they only very rarely form fruits (spherical berries enclosed by the calyx). The flower of *Solandra nitida* can attain a length of 20 cm. Its fruits, known as *papaturra*, can weigh as much as 1 kg (Bärtels 1993, 207*).

Solandra species can be confused with the tropical dogbane *Allamandra cathartica* L., a potent laxative (Blohm 1962, 79 f.*).

Psychoactive Material

— Flowers
— Stalks
— Leaves

Preparation and Dosage

A tea can be made from the stalks (Schultes and Farnsworth 1982, 166*). The fresh stalks can be pressed to obtain a juice; "the shoot juice of *Solandra maxima* [= *S. nitida*] is an inebriant of the Mexican Indians" (Bremness 1995, 29*). Unfortunately, no information is available concerning dosages.

The fresh leaves (of *Solandra brevicalyx*) can be crushed and administered as an anal suppository or given as a decoction in the form of an **enema** (Knab 1977, 85). The dried flowers and leaves can be smoked alone or as a part of **smoking blends**.

A medicinal dosage is regarded as the tea prepared from one fresh flower (Yasumoto 1996, 247).

In colonial Mexico, Indians used the cup of gold to add zest to their cacao drinks (cf. *Theobroma cacao*) (Heffern 1974, 101*).

Ritual Use

The cup of gold is only rarely used as a shamanic trance drug, and the ethnographic reports are correspondingly few. The Huastec are said to still ingest the flowers of *Solandra nitida* ritually and to place the scented flowers on altars as an offering (Alcorn 1984, 320, 793*). The Mixtec also are reported to traditionally ingest *Solandra* as a hallucinogen for divination (Avila B. 1992*).

The most well-known use of the "plant of the gods" known as *kiéli* or *kiéri* occurs among the Huichol Indians who now live in the Mexican state of Jalisco. One of the plants they use has been botanically identified as *Solandra brevicalyx* (Knab 1977, 86). In the mythology of the Huichol, the plant was originally a god: Kiéli Tewiali, the god of wind and of magic. At the beginning of the world, he was born of the union of the cosmic serpent and the rain. Later, for the use and the blessing of humankind, he transformed himself into the enchantingly scented plant the "tree of the wind." An entire cycle of myths relates to this theme (Furst and Myerhoff 1966).[292] The *Solandra* is often identified with Kiéritáwe, the "drunken Kiéri" (Furst 1989; Yasumoto 1996).

This divine plant is regarded as very powerful and mighty and thus can be used for all types of magic ("kiéli shamanism"), including for dark purposes (harmful magic, death magic). Shamans-to-be must complete a five-year training period before they are allowed to use this potent magical plant. The leaves, which only experienced shamans (*mara'akame*) may remove from the tree, are later used as magical weapons for healing illnesses caused by magic or foreign, perfidious shamans (Knab 1977).

The divine plant must not be disturbed or offended lest one be punished with madness or death. The gifts offered to the plant are similar to those offered to the peyote (*Lophophora williamsii*): ceremonial pipes, tortillas, a homemade tequila known as *túche* (cf. *Agave* spp.), tobacco gourds (cf. *Nicotiana rustica*), coins, yarn paintings, jewelry, bead necklaces, et cetera. The Huichol sometimes approach the plant and offer it prayers, e.g., before they undertake a journey or make a pilgrimage to Wirikuta, the land of the peyote.

They also ask it for fertility, improvements in singing ability, and artistic creativity (Knab 1977, 83).

Shamans are able to receive sacred knowledge from the "tree of the wind." The Huichol artist José Bautista Corrillo provided the following explanation of such a ritual of knowledge portrayed in one of his yarn paintings:

> Kauyumari, the leader of the shamans in the shape of a deer, eats Kiéri, the tree of the wind, to learn about the legends of the past and the art of healing. He passes this knowledge on to the shaman who asks Kiéri to teach him everything while he sings throughout the entire night. The puma, who was once the fire, and the wolf, who was once a shaman, help the shaman to understand the teachings. (1996)

The plant is apparently used only extremely rarely as a hallucinogen. The leaves seem to be preferred for this purpose, although the fruits (which develop only infrequently) and the roots are thought to be more potent (Knab 1977, 85). It is said that the plant is able to help a person fly (Furst 1995, 53). Sometimes the hallucinogenic use of *Solandra* is regarded as a sure sign of sorcery, witchcraft, and black magic (Knab 1977, 85; Furst 1995). On the other hand, some Huichol say that this plant opens their mind for the "highest levels of enlightenment."

Some Huichol say that people are not allowed to ingest the plant but may only be exposed to its scent. Even the scent is capable of inducing trance, and the Huichol use it as a spiritual guide into mystical domains (Valadez 1992, 103 f.). They climb a steep mountain, upon which a *kiéli* plant is growing, for this purpose. They must fast (no food or beverages, including water) both before and while they are climbing, and they spend the night near the scented plant, inhaling its perfume and showing the bush their respect and attention (Meier 1996). While they sleep, they hope to receive meaningful visionary dreams in which they will be able to find messages.

Artifacts

Kiéri is sometimes depicted in the visionary yarn paintings of many Huichol artists (Valadez 1992). Although the plant can appear in varying degrees of abstraction, it usually is shown in a quite realistic and botanically correct manner (yellow flowers, leaf arrangement).

Many floral elements in the pre-Columbian wall paintings at Teotihuacán may symbolize *Solandra* vines (cf. **Turbina corymbosa**). Some of the illustrations resemble the typical iconography of the plant in modern Huichol yarn paintings (cf. **Lophophora williamsii**).

Medicinal Use

In Mexico, the cup of gold is used in folk medicine primarily as a love drink and aphrodisiac. Warnings against overdoses are common: one can dry out and die from an excessive sex drive. The Huastec use the rainwater or dew that has collected in the buds of *Solandra nitida* as eye-drops to improve sight (Alcorn 1984, 793*). A tea made from the flowers is drunk to treat coughing (Yasumoto 1996, 247).

Constituents

All of the Mexican species of *Solandra* contain potently hallucinogenic **tropane alkaloids**. The primary alkaloids are **atropine**, noratropine, and (–)-hyoscyamine (originally described as "solandrine"); the secondary alkaloids are littorine, hyoscine, norhyoscine, tigloidine, 3α-tigloyloxytropane, 3α-acetoxytropane, valtropine, norhyoscyamine, tropine, nortropine, χ-tropine, and cuscohygrine (Evans et al. 1972; Schultes and Farnsworth 1982, 166*). According to another source, scopolamine is the primary alkaloid, present at a concentration of 0.1 to 0.2% (Díaz 1979, 84*). The stalks of *Solandra guttata* have been found to contain norhyoscine. *Solandra* is chemotaxonomically closely related to the genera *Datura* and *Duboisia* (Evans 1979, 245*).

Most *Solandra* species contain approximately 0.15% alkaloids (Schultes 1979b, 150*). The highest concentration of alkaloids (calculated as **atropine**) was found in the roots of *Solandra grandiflora* (0.64%). The roots generally exhibit the highest alkaloid concentrations (Evans et al. 1972). However, in *Solandra nitida*, the alkaloid concentration is clearly highest in the fruits (Morton 1995, 20*).

Effects

The Huichol compare the visions produced by *Solandra brevicalyx* with the effects of **Lophophora williamsii** but warn against the former because they may frighten a person "to death" (Knab 1977).

In Mexico, *Solandra nitida* Zucc. (Perilla) is regarded as poisonous (Jiu 1966, 256*). A tea made from one flower induced a "toxic psychosis" in an adult, who required thirty-six hours to make

An illustration of the shamanic *kiéle* ritual, together with the corresponding *Solandra* plants, on a yarn painting by the Huichol artist José Bautista Corillo (1996).

The illustration to the left shows the *kiéle* plant (*Solandra* sp.) on a Huichol yarn painting; the blooming shrubs to the right are from wall paintings at Teotihuacán and may represent *Solandra* bushes. (From Rätsch 1994)

"Rue, leaves of henbane and thorn apple, dried Solandras and myrrh; these are the odors that are pleasing to Satan, our lord."

JORIS-KARL HUYSMANS
TIEF UNTEN [DEEP BELOW]
(1972, 186)

a complete recovery (Morton 1995, 20*). Internal administration of *Solandra* preparations can lead to severe hallucinations, delirium, delusions, et cetera. The spectrum of effects is very similar to that of *Brugmansia sanguinea.*

Smoking the flowers and/or leaves produces effects that are more subtle but still clearly psycho-active and aphrodisiac and generally very similar to the effects produced by smoking other night-shades (*Brugmansia, Datura, Latua pubiflora*).

It has been said that merely inhaling the scent can produce entheogenic states (Meier 1996). The Lacandon say that the scent has erotic effects and awakens sexual desire.

Commercial Forms and Regulations

Solandra species are not subject to any legal restrictions. In North America, young plants are occasionally available in nurseries.

Literature

See also the entries for **scopolamine** and **tropane alkaloids**.

Evans, W. C., A. Ghani, and Valerie A. Woolley. 1972. Alkaloids of *Solandra* species. *Phytochemistry* 11:470–72.

Furst, Peter T. 1989. The life and death of the crazy kiéri: Natural and cultural history of a Huichol myth. *Journal of Latin American Lore* 15 (2): 155–77.

———. 1995. The drunkard kiéri: New observations of an old problem in Huichol psychotropic ethnobotany. *Integration* 5:51–62.

———. 1996. Introduction to chapter 8. In *People of the peyote*, ed. Stacy Schaefer and Peter T. Furst, 232–34. Albuquerque: University of New Mexico Press.

Furst, Peter T., and Barbara G. Myerhoff. 1966. Myth as history: The jimson weed cycle of the Huichols of Mexico. *Antropológia* 17:3–39.

Huysmans, Joris-Karl. 1994. *Tief unten*. Stuttgart: Reclam. (Orig. pub. 1972.)

Knab, Tim. 1977. Notes concerning use of *Solandra* among the Huichol. *Economic Botany* 31:80–86.

Martínez, Maximino. 1966. Las solandras de México con una specie nueva. *Anales del Instituto de Biología* 37 (1/2): 97–106. Mexico City: UNAM.

Valadez, Mariano, and Susana Valadez. 1992. *Huichol Indian sacred rituals*. Oakland, Calif.: Dharma Enterprises.

Yasumoto, Masaya. 1996. The psychotropic kiéri in Huichol culture. In *People of the peyote*, ed. Stacy Schaefer and Peter T. Furst, 235–63. Albuquerque: University of New Mexico Press.

Solanum spp.

Nightshade Species

Family

Solanaceae (Nightshade Family); Subfamily Solanoideae, Solaneae Tribe, Solaninae Subtribe

Some one thousand to two thousand species are currently recognized in the genus *Solanum*

The Peruvian nightshade (*Solanum hispidum*), shown here with flowers and fruits, is found throughout South America.

(D'Arcy 1991, 79*; Schultes and Raffauf 1991, 43*; Teuscher 1994, 734). Many *Solanum* species are edible and are used as sources of food (eggplant, potato, kangaroo berry). A number of species have a long tradition of use as folk medicines. A few species are used as additives for coca chewing (see **Erythroxylum coca**). Some appear to be psycho-active or to be used for psychoactive purposes. At the turn of the twentieth century, the Zuni Indians were still using *Solanum elaeagnifolium* Cav. as a sedative **snuff** (von Reis and Lipp 1982, 271*).

Several members of the genus contain **tropane alkaloids** (Schultes and Raffauf 1991, 43*) and myricetin derivatives, substances that also are found in such plants as **Ledum palustre** (Kumari et al. 1984). The roots of many species have been found to contain steroidal alkaloids and sapogenins (Ripperger 1995).

Solanum dulcamara L. [syn. *Dulcamara flexuosa* Moench, *Solanum laxum* Royle, *Solanum lyratum* Thunb., *Solanum scandes* Lamk.]— bittersweet nightshade

This nightshade has been interpreted to be the "sleeping strychnos" (cf. *Strychnos nux-vomica*) of Dioscorides (Schneider 1974, 3:274*). In ancient times, the root cortex was drunk in **wine** as a sleeping agent.

The Germanic tribes used the plant as a narcotic and referred to it and *Solanum nigrum* as "night harm." Night harm is an illness induced by an elfish demon (nocturnal nightmare demon) during the night while one is sleeping and can be healed using *Solanum dulcamara*. The illness "should be fought off by the embodiment of another magically powerful elfish demon in the plant, i.e., the nocturnal unrest of the ill person is soothed by means of a narcotic agent" (Höfler 1990, 96*).

Bittersweet nightshade "was regarded as an elfen plant and is still known as alp [= elf] vine. It was placed in children's cradles to ward off enchantment and was placed around the necks of cattle to ward off 'hunsch,' or wheezing. Humans generally appear to have an aversion to this plant, for it has been called sow's vine, stinking devil, dog berry, choke plant, etc.; it is also regarded as a symbol for a treacherous person" (Perger 1864, 182*). During the Middle Ages, the berries were strung onto cords to make amulets or talismans that were worn around the neck to ward off evil gossip. The plant also played a role in other magical customs: "to initiate a vengeful magic, place the name of an enemy on the dry stem and lay this before the door of that person, [and] the berries can be used as a magical aid for all transformative magic, especially lycanthropic magic" (Magister Botanicus 1995, 193*). It may have been an ingredient in **witches' ointments**.

In Mexico, where the plant is known as *dulcamara* or *jazmincillo*, it is used in folk medicine as a sedative and narcotic agent. Mexican plant material has been found to contain solanine derivatives and **tropane alkaloids** (Díaz 1979, 85*).

The herbage contains between 0.3 and 3.0% and the roots approximately 1.4% steroid alkaloid glycosides. The alkaloid content of the fruits declines as they ripen, and ripe fruits are almost completely devoid of alkaloids (Teuscher 1994, 737). The alkaloid content and composition can exhibit considerable variation (Máthé and Máthé 1979). There may be chemical races with psychoactive properties.

Solanum hirtum Vahl

This neotropical nightshade species is known in Mayan as *put balam*, "papaya of the jaguar." To the Maya, the jaguar is the most important and most powerful shamanic animal (cf. *Nymphaea ampla*).

It is possible that this plant was or is still associated with shamanic practices.[293] Chewing the fresh leaves produces a narcotic and stimulating effect. The fruits also are used medicinally to treat angina (Pulido S. and Serralta P. 1993, 62*). In Mexico, the very similar species *Solanum rostratum* Dunal is known locally as *hierba del sapo*, "plant of the toads" (Martínez 1994, 434*).

Solanum hypomalacophyllum Bitter ex Pittier

This plant, which is known as *borrachera* in Venezuela, may contain **tropane alkaloids** (Schultes 1983a, 271*). Steroid alkaloids (solaphyllidin, solamaladin) and a steroid sapogenin (andesgenin) have been isolated from the plant (González et al. 1975; Schultes and Raffauf 1991, 44*). Whether the plant has psychoactive effects and has been used for this purpose is unknown, although possible. Why else would it be known as *borrachera*, "inebriator"?

Solanum leptopodum Van Heurck et Muell. Arg.

The Secoya Indians call this bush *oyo-ha'-o*, "bat leaf," and use the leaves for washings to treat and calm crying children. It may exert a sedative effect (Schultes and Raffauf 1991, 44 f.*).

Solanum ligustrinum Lodd.—natre

This bush is used in Chilean folk medicine to treat fever. It has mild analgesic properties and is known by the interesting name of *hierba de chavalongo*, usually translated as "typhus fever plant" (Hoffmann et al. 1992, 154*). But the similarity between this name and that of the still unidentified **cabalonga**, the psychoactive magical plant of the northern Andes region, is almost too striking. More detailed investigations into the ethnobotany of this plant would likely yield some very interesting findings. The plant is known to contain several alkaloids (natrine, huevine) as well as solanine (Hoffmann et al. 1992, 156*).

Solanum mammosum L.

In South America, the powdered fruits of this species are used as a cockroach poison. The plant reputedly is used in Colombia to "satisfy children," i.e., as a sedative narcotic (Schultes 1978a, 193*).

Left: The fruits of *Solanum dulcamara* have a bittersweet taste.

Top right: The bittersweet nightshade (*Solanum dulcamara*) in bloom.

Bottom right: The leaves of the tropical nightshade species *Solanum hirtum*, known in the Yucatán as "papaya of the jaguar," have narcotic effects. (Wild plant, photographed near Chichen Itzá, Yucatán, Mexico)

293 There is also an association between a nightshade species and this shamanic animal in South America: the Waorani of the Ecuadoran Amazon say that *Solanum pectinatum* Dunal in DC. was originally planted by a jaguar (Davis and Yost 1983, 204*).

Left: The black nightshade (*Solanum nigrum*) occurs in a number of varieties and forms throughout the world. (Wild plant, photographed in Hamburg)

Right: Originally from South America, the potato (*Solanum tuberosum*) has been found to contain trace amounts of a natural sedative agent.

Early illustration of the "garden nightshade," either *Solanum nigrum* or *Solanum dulcamara*. (Woodcut from Gerard, *The Herball or General History of Plants*, 1633)

"In the empire of Ethiopia, the 'criminal telepath' was an established institution until at least the time of the Second World War. Usually an as yet untouched boy, such a person was known as a *lebaschà* (concentrating searcher). He was called upon when a theft occurred. The lebaschà was required to imbibe a drink that contained the leaves of nightshade plants along with other things, there are also reports of drug smoking. The lebaschà would then enter an inebriated kind of state and follow 'the scent' to all of the places that might be connected with the theft, and would ultimately find the object and the person who had stolen it."

WERNER F. BONIN
NATURVÖLKER UND IHRE ÜBERSINNLICHEN FÄHIGKEITEN
[NATURE PEOPLE AND THEIR EXTRASENSORY ABILITIES]
(1986, 49)

Solanum nigrum L. [syn. *Solanum americanum* Mill., *Solanum caribaeum* Dun., *Solanum nodiflorum* Dun.; for other synonyms see Teuscher 1994, 744]—black nightshade

This nightshade has been interpreted as the "garden strychnos" (cf. *Strychnos nux-vomica*) of Dioscorides (Schneider 1974, 3:274*). It has often been attributed with psychoactive properties: "the nightshade was a true Germanic narcotic" (Höfler 1990, 96*). The plant was one of the ingredients in **witches' ointments**.

In Mexico, where the American variety of the black nightshade is known as *chichiquilitl* or *hierba mora*, the plant is used in folk medicine as a local analgesic, a sedative, and a stimulant and to treat Parkinson's disease and epilepsy (Díaz 1979, 85*). The plant is called *yocoyoco* in Venezuela, a name strikingly reminiscent of that of *Paullinia yoco* (see **Paullinia spp.**) (Blohm 1962, 97*). In Chile, the black nightshade is used as an antidote for overdoses of *Latua pubiflora*. The herbage contains chiefly solanine as well as related alkaloids. The unripe fruits can contain as much as 1.6% alkaloids, whereas ripe fruits are usually devoid of alkaloids (Teuscher 1994, 744). Further research is needed to determine whether the plant can be used for psychoactive purposes.

Solanum subinerme Jacq.—gujaco

The Witoto Indians add the ripe fruits of *gujaco* or *ujaca*, as they call the species, to their cassava **beer** to impart to it a special taste. It is not clear whether this affects only the taste or whether the fruits also contribute to the psychoactive effects (Schultes and Raffauf 1991, 46*).

Solanum topiro Humb. et Bonpl. [syn. *Solanum sessiliflorum* Dun.]—de-twa'

The Taiwano Indians dry and crush the small seeds from the edible fruits and add the powder to coca leaves when the oral mucosa have become irritated from too-frequent coca chewing (cf. **Erythroxylum coca**). The mixture is said to provide relief (Schultes 1978a, 194*; Schultes and Raffauf 1991, 46*).

Solanum tuberosum L.—potato

The potato is one of humankind's most important food plants. It originated in Peru, where there are numerous sorts, and is now planted worldwide. In the culture of inebriation, it is used both as a fermenting agent for **chicha** and **beer** and in the distillation of vodka (**alcohol**). The roots contain solanidine (Ripperger 1995). Recently, the potato was found to contain a "natural Valium" (see **diazepam**). Using very accurate methods of analysis, "traces of substances were found in the tubers that inhibit the binding of benzodiazepine to benzodiazepine receptors in the rat brain. To date, 8 of these have been identified as benzodiazepine derivatives, including diazepam and lormetazepam" (Teuscher 1994, 747). However, it is unlikely that one would notice any effects of these substances from eating potatoes, as one would probably need to consume an entire bushel for the effects to manifest.

Solanum verbascifolium L.

This plant, which has a pantropical distribution, is known as *toonpaap* ("hot tail"). The shamans of the Yucatec Maya (southern Mexico) appear to make use of the plant, although no precise information is available (Garza 1990, 189*). The Mayan name seems to suggest an aphrodisiac use of the plant.

Solanum villosum Mill. [syn. *Solanum nodiflorum* Jacq.]—witch's tomato

This species is very closely related to the black nightshade (Heiser et al. 1979). The plant is known as *tomate de la bruja* ("witch's tomato") in Spain, where it allegedly was once used for psychoactive purposes (J. M. Fericgla, pers. comm.; Fericgla 1996*). It may have been one of the ingredients in **witches' ointments**.

Literature

See also the entries for **Atropa belladonna** and **Datura stramonium**.

Bonin, Werner F. 1986. *Naturvölker und ihre übersinnlichen Fähigkeiten.* Munich: Goldmann.

Gonzáles, Antonio G., Cosme G. Francisco, Raimundo Freire, Rosendo Hernández, José A. Salazar, and Ernesto Suárez. 1975. [New sources of steroid sapogenins. 29:] Andesgenin, a new steroid sapogenin from *Solanum hypomalacophyllum*. *Phytochemistry* 14:2483–85.

Heiser, Charles B., Jr., Donald L. Burton, and Edward E. Schilling Jr. 1979. Biosystematic and taxonomic studies of the *Solanum nigrum* complex in eastern North America. In *The biology and taxonomy of the Solanaceae*, ed. J. G. Hawkes et al., 513–27. London: Academic Press.

Kumari, G. N. Krishna, L. Jagan Mohan Rao, and N. S. Prakasa Rao. 1984. Myricetin methyl esters from *Solanum pubecens*. *Phytochemistry* 23 (11): 2701–2.

Máthé, Imre, Jr., and Imre Máthé Sr. 1979. Variation in alkaloids in *Solanum dulcamara* L. In *The biology and taxonomy of the Solanaceae*, ed. J. G. Hawkes et al., 211–22. London: Academic Press.

Ripperger, Helmut. 1995. Steroidal alkaloids and sapogenins from roots of some *Solanum* species. *Planta Medica* 61:292.

Teuscher, Eberhard. 1994. *Solanum*. In *Hagers Handbuch der pharmazeutischen Praxis*, 5th ed., 6:734–52. Berlin: Springer.

Usubillaga, A. 1984. Alkaloids from *Solanum hypomalacophyllum*. *Journal of Natural Products* 47:52.

"Dulcamara. (Stipites.) One of the most excellent agents of the practice of the poor, powerful, and inexpensive. It is one of the most effective remedies for chronic rheumatism, catarrh, for beginning catarrhal and tubercular phthisis (the most common of them all), for chronic skin diseases, for whooping cough. The dosage is 2 to 4 drams daily, not as an infusion but rather as a decoction, because only the boiling brings out the power sufficiently.—The same thing is true of the extract that has been said about the plant."

C. W. HUFELAND
ARMEN-PHARMAKOPÖE
[PHARMACOPOEIA OF THE POOR]
(VIENNA; 1830, 19)

Sophora secundiflora (Gómez-Ortega) Lagasca ex DC.

Mescal Bean

Family
Leguminosae (Legume Family); Subfamily Lotoideae (Papilionoideae), Sophoreae Tribe

Forms and Subspecies
A form that occurs in Texas has exclusively yellow seeds; it has been described (Rudd 1968, 528) under the name *Sophora secundiflora* (Ort.) Lag. f. *xanthosoma* Render.

Synonyms
Agastianis secundiflora (Gómez-Ortega) Raf.
Broussonetia secundiflora Groussonetia
Calia erythrosperma Berlandier in Mier-Terán
Cladrastis secundiflora (Gómez-Ortega) Raf. ex Jacks.
Dermatophyllum speciosum Scheele
Sophora sempervirens Engelm. in A. Gray
Sophora speciosa (Scheele) Benth.
Virgilia secundiflora (Gómez-Ortega) Cav.

Folk Names
Big drunk bean, chilicote, colorín, colorines, coral bean tree, coral bean, frijolillo (Spanish, "little bean"), frijolillo of Texas, frijolito, frixolillo, k'awnk'odl (Comanche), mescal bean, meskalbohne, mountain laurel,[294] patiol, patol, red bean, red medicine, schnurbaum, Texas mountain laurel

History
In Texas, mescal beans have been found together with *Ungnadia speciosa* in archaeological contexts (ritual caves) that extend into strata dating as far back as eight thousand years (Adovasio and Fry 1976*). Mescal beans appear together with *Ungnadia* and peyote (**Lophophora williamsii**) in more recent layers. In the American Southwest, the use of the beans for ornamental purposes has been documented for eight thousand years (Merrill 1977).

Some anthropologists have assumed that the mescal bean cult (mescalism) represents a precursor to the peyote cult (peyotism). The mescal bean cult disappeared because the effects of the peyote cactus are much more pleasant and visionary (Campbell 1958; Howard 1957, 1960; La Barre 1957).

The use of the name *mescal* for this plant and its seeds has resulted in considerable confusion in both the ethnographic and the ethnobotanical literature. The name *mescal* is used in Mexico for an alcoholic beverage made from **Agave** spp., while *mescalito* is used in northern Mexico to refer to peyote, peyote buttons, and the peyote spirit. There also is the Mescalero Apache tribe, who were responsible for carrying the peyote cult into North America. The confusion is increased by the fact that necklaces of mescal beans have been used as ritual objects in both the historical and the modern peyote cult (cf. **Lophophora williamsii**).

In northern Mexico, mescal beans are used interchangeably with the seeds of **Erythrina flabelliformis** (Merrill 1977).

Distribution
The tree is found from Texas and New Mexico south into central Mexico (Rudd 1968, 528).

Cultivation
Propagation occurs from seeds, which should be pregerminated for best results. It also can be

---

294 This name is usually used for *Kalmia latifolia* (cf. kinnikinnick).

propagated from cuttings taken from the green wood (Grubber 1991, 49*). The bush requires a dry, warm climate.

Appearance

The shrub or small tree can attain a height of up to 12 meters. It has evergreen pinnate leaves with seven to eleven leaflets. The scented violet flowers are some 3 cm in length and form pendulous clusters. The siliquose, constricted fruits contain the actual mescal beans (seeds). The beans are 0.8 to 2 cm long and 0.5 to 1.5 cm wide. Although they are usually red, seeds that are dark red, light red, orange, and yellow also occur.

Mescal beans are difficult to distinguish from the similarly red seeds of *Erythrina flabelliformis* (cf. *Erythrina* spp.) and are often confused with them.

The closely related and similar species *Sophora conzatti* Standl. and *Sophora purpusii* Brandeg. are found in Mexico, where they too are known as *frijolillo* (Rudd 1968, 525 ff.).

Psychoactive Material

— Seeds (beans, mescal beans, meskal beans, colorines)

Preparation and Dosage

No more than a quarter of a bean is roasted on a fire until it turns yellow and then ground, chewed, and swallowed (Gottlieb 1973, 35*).

The Mescalero Apache add the seeds to the **beer** (*tiswin*, *tulbai*) they make from maize (***Zea mays***) in order to potentiate its effects (Bye 1979b, 38*).

Half of a bean is said to be sufficient to induce a state of delirium that can persist for two to three days (Havard 1896, 39*). In former times, up to 3 mg of cytisine (a component alkaloid) per day was used as a respiratory stimulant (Brown and Malone 1978, 9*).

It has been said that in the early twentieth century, Chinese immigrants in Oklahoma mixed mescal beans with sugar, vanilla, and musk to produce an aphrodisiac candy known as "red beans from China" (Reko 1938, 137*).[295]

Ritual Use

During the colonial period, the Coahuilteco Indians of southern Texas and northern Mexico were known to alternate eating mescal beans (the source calls them *frixolillo*) and eating peyote in their communal rituals (Merrill 1977). The Indians of San Antonio (Texas) formerly used the beans as a ritual inebriant (Havard 1896, 39*). Several Plains tribes also consumed the seeds.

The seeds were a common component of amulets. The seeds were stored in small leather medicine bags or carried on a person's body. Some of the Plains tribes had mescal bean secret societies, the members of which presumably used the beans in their vision quests. Unfortunately, we have few details about such use due to the secrecy that surrounded these groups (Merrill 1977).

Artifacts

Numerous necklaces of mescal beans are known and have been ethnographically described for North America. Ethnographic objects made with mescal beans have been documented for the following tribes: Apache, Arapaho, Arikara, Blackfoot, Caddo, Cheyenne, Coahuilteco, Comanche, Crow, Delaware, Hidatsa, Iowa, Kansa, Kickapoo, Kiowa, Kiowa-Apache, Mandan, Missouri, Ojibwa, Omaha, Osage, Oto, Pawnee, Ponca, Prairie Potawatomi, Pueblos, Sauk and Fox, Shawnee, Shoshone, northern Ute, Sioux, Tonkawa, Wichita, and Winnebago (Merrill 1977). Necklaces of mescal beans are still worn at peyote ceremonies today.

The red-black seeds of *Abrus precatorius* L. have also been found among the paraphernalia of the mescal bean secret societies and in several medicine bundles of the Iowa and Omaha (cf. ***Rhynchosia pyramidalis***).

A prehistoric medicine bundle containing seven mescal beans and the herbage of an *Ephedra* species (***Ephedra*** spp.) was discovered in southwestern Texas (Merrill 1977, 68). In general, the prehistoric art style (Pecos River style) of this region has been associated with the mescal cult (Wellmann 1981, 94*).

Left: The mescal bean tree (*Sophora secundiflora*) is indigenous to Texas.

Right: The seedpods and seeds of the mescal bean (*Sophora secundiflora*).

295 "Owners of various saloons in the port cities in China have long added a similar red bean, *Sophora tomentosa* L., to alcoholic beverages so that their patrons, typically sailors who would drink themselves into a heavy state of drunkenness after just a few draughts, could be duly robbed while in this condition" (Reko 1938, 138*).

Medicinal Use

The northern Mexican Kickapoo Indians use the seeds to treat ear ailments. They prepare a decoction from one ground seed and some tobacco (*Nicotiana* spp.), which is then dripped into the ear canal. A cold-water extract of crushed seeds is used as an ear wash. Ear drops also are produced by boiling a bean with a branch of juniper (*Juniperus* sp.) (Latorre and Latorre 1977, 352*).

Constituents

Mescal beans contain the alkaloids **cytisine** (= baptitoxine, sophorine, ulexine, laburnine, cytitone), *N*-methylcytisine, and sparteine (Keller 1975; Merrill 1977). Quinolizidine alkaloids have also been identified: *epi*-lupinine, Δ^5-dihydrolupanine, anagyrine, and thermopsine (Hatfield et al. 1977). At 0.25%, cytisine is the primary active constituent.

Effects

Depending upon the dosage, the seeds exhibit first psychoactive and then powerful toxic effects. The sequence of effects begins with a reddening of the face and inebriation and includes spasms, muscular rigidity, headache, nausea, vomiting, defecation, fainting, delirium, and, at the end, death. The reddening of the face is an effect that has been reported often (Howard 1957, 76).

Whether mescal beans are hallucinogenic remains an open question (Schultes and Hofmann 1992, 57*). To date, only a single case of poisoning has been reported in the toxicological literature (Hatfield et al. 1977, 374).

Commercial Forms and Regulations

Mescal beans are occasionally available on the international seed market.

Literature

See also the entries for *Lophophora williamsii* and **cytisine**.

Campbell, T. N. 1958. Origin of the mescal bean cult. *American Anthropologist* 60:156–60.

Hatfield, G. M., L. J. J. Valdes, W. J. Keller, W. L. Merrill, and V. H. Jones. 1977. An investigation of *Sophora secundiflora* seeds (mescalbeans). *Lloydia* 40 (4): 374–83. (Contains an extensive phytochemical bibliography.)

Howard, James H. 1957. The mescal bean cult of the central and southern Plains: An ancestor of the peyote cult? *American Anthropologist* 59:75–87.

———. 1960. Mescalism and peyotism once again. *Plains Anthropologist* 5:84–85.

———. 1962. Potawatomi mescalism and its relationship to the diffusion of the peyote cult. *Plains Anthropologist* 7:125–35.

Izaddoost, Mohamed. 1975. Alkaloid chemotaxonomy of the genus *Sophora*. *Phytochemistry* 14:203–4.

Keller, William J. 1975. Alkaloids from *Sophora secundiflora*. *Phytochemistry* 14:2305–6.

La Barre, Weston. 1957. Mescalism and peyotism. *American Anthropologist* 59:708–11.

Merrill, William L. 1977. An investigation of ethnographic specimens of mescalbeans (*Sophora secundiflora*) in American museums. Technical Reports 6, Research Reports in Ethnobotany 1. Ann Arbor: Museum of Anthropology, University of Michigan.

Rudd, Velva E. 1968. Leguminosae of Mexico—Faboideae. I: Sophoreae and Podalyrieae. *Rhodora* 70:492–532.

Troike, Rudolf C. 1962. The origin of Plains mescalism. *American Anthropologist* 64:946–63.

Sophora tomentosa, which is related to the mescal bean, also contains efficacious alkaloids. The seeds of the plant were once used as an inebriating beer additive.

Cytisine

"Those who have been poisoned or intoxicated with Sophora powder find themselves in a mildly sedated state (as if 'cheered up') without, however, exhibiting any disturbances in their intelligence. About one-half hour after ingesting the (raw, typically powdered) beans, they manifest a striking hypersensitivity of the skin. Every touch is perceived as a 'tickling' (hence the name 'laughing intoxication'), a gentle caress of the mouth provokes a flow of saliva, touching the umbilical region produces involuntary urination or wet dreams."

VICTOR A. REKO
MAGISCHE GIFTE [MAGICAL POISONS]
(1938, 135*)

Strychnos nux-vomica Linnaeus

Poison Nut

A botanical illustration of the flowers and fruits of the poison nut tree. (Engraving from Pereira, *De Beginselen der Materia Medica en der Therapie*, 1849)

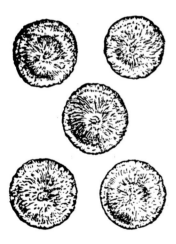

In Germany, the seeds of the poison nut tree (*Strychnos nux-vomica*) were once known as "crows' eyes" (*krähenaugen*). (Woodcut from Tabernaemontanus, *Neu Vollkommen Kräuter-Buch*, 1731)

The seeds of the poison nut tree (*Strychnos nux-vomica*) contain strychnine.

Family
Loganiaceae (Logania Family, Strychnos Family); Subfamily Strychneae

Forms and Subspecies
None

Synonyms
Strychnos colubrina Wight
Strychnos lucida R. Br.
Strychnos spireana Dop
Strychnos vomica St. Lag.

Folk Names
Azaraki, brauntaler, brechnußbaum, cilibucha, fuluz mahi (Persian), gemeines krähenauge, goda kaduru, kajara, kanchurai, krähenauglein, krähen-auge(n), krähenaugenbaum, kuchila (Hindi), kuchla, kuchla of India, kuchûlah, noce vomica, noix vomique, nux vomica, nux-vomica tree, poison nut, poison nut tree, Quaker buttons, rvotnyi orech (Russian), strychninbaum, strychnine, strychnine plant, visamusti

History
Probably the first person to describe the poison nut was Theophrastus, who discussed it under the name *strychnós manikós*, "strychnos that makes manic." It was once thought that this name referred to the thorn apple (cf. *Datura stramonium*), an interpretation that is now considered highly doubtful (Marzell 1922, 171*; Schneider 1974, 3:294*). The "sleeping strychnos" of Dioscorides is now interpreted as *Solanum dulcamara*, and the "garden strychnos" as *Solanum nigrum* (cf. *Solanum* spp.). Many very early Persian sources mention the poison nut as an agent that can induce paralysis (Hooper 1937, 175*). In Europe, the plant first became well known in the fifteenth century.

Distribution
The tree, which probably originated in the dry forests of Sri Lanka (Macmillan 1991, 416*), is native to India and Burma (Myanmar) but has now spread into all the tropical areas of the Indian Ocean and Southeast Asia (Bremness 1995, 131*). It is most commonly encountered in dry forests.

Cultivation
Propagation occurs through seeds or cuttings. The tree prefers sandy soils and a tropical but dry and hot climate. The seeds are collected primarily between August and November (Macmillan 1991, 417*). The most important areas of cultivation are in Southeast Asia, India, Pakistan, and tropical Africa (Teuscher 1994, 829).

Appearance
The projecting, shrublike tree can reach a height of up to 25 meters; the trunk can have a circumference of as much as 3 meters. The smooth, shiny, oval leaves are opposite and have five nerves. The greenish white umbels produce yellow fruits with gray, disk-shaped seeds that display a silky sheen through fine small hairs.

The poison nut tree is easily confused with the similar species *Strychnos nux-blanda* A.W. Hill. The latter, however, has larger flowers and fruits (Teuscher 1994, 829).

Psychoactive Material
— Ripe, dry seeds (brechnuß, krähenauge, strychni semen, nux metella, nux vomica, semen nucis vomicae, semen strychni)

Preparation and Dosage
After the fruits have been collected, the seeds are extracted and dried in the sun. The seeds must be stored in a cool, airtight location that is protected from the light. When seeds are stored properly, the raw constituents should remain stable for a long time. The seeds are used to produce extracts and tinctures for pharmaceutical purposes (Teuscher 1994, 832). To make Ayurvedic medicines, the seeds are boiled in milk or cow urine (this is known as the *sodhna* technique).

The largest therapeutic individual dosage is given as 0.1 g of dried poison nut (with a standardized alkaloid content of 2.4 to 2.6%); the largest total daily dosage is 0.2 g (Teuscher 1994, 836). The dosage of pure strychnine should never exceed 5 mg. In the "drug culture," strychnine is used as an adulterant to "stretch" **cocaine** and heroin (Teuscher 1994, 836).

Theophrastus described some rather drastic dosages (which should not be imitated!):

> If someone is simply playing a joke and wants to make the biggest fool out of oneself, you

take one dram [app. 3.4 g] of it, but 2 drams if he wants to become crazy and experience apparitions; persistent madness can be produced by three drams. Four drams are required to cause death.

In ancient times, the root (1 dram) was drunk in **wine** for psychoactive purposes. Poison nuts are an ingredient in bhang or *majun* (see *Cannabis indica*), **Oriental joy pills**, and similar aphrodisiacs. In Persia, an aphrodisiac tea was made from poison nuts, hemp (*Cannabis indica*), and poppy leaves (*Papaver somniferum*) (Most 1843, 570 f.*).

Ritual Use

It is only as an ingredient in other psychoactive products that poison nuts have acquired any ritual significance (cf. *Cannabis indica*, *Vitis vinifera*, Oriental joy pills, wine).

Artifacts

In India, poison nuts are used in amulets for magically protecting the house and farm (Jain 1991, 172*).

Medicinal Use

The Ayurvedic medical system regards the seed as a tonic and stimulant (Macmillan 1991, 417*) and especially as an aphrodisiac. In Indian folk medicine, the juice of the root cortex, along with cow's milk, is applied externally to treat snakebite (Bhandary et al. 1995, 154*). The bark also is used there to treat cholera. In Nepal, the seeds are used for palsy and rabies (Bremness 1995, 131*). In Iran, the seeds were still being used as a tonic in the twentieth century (Hooper 1937, 175*).

In Europe, poison nut seed was once seen as a remedy for the black death (Schneider 1974, 3:295*) and was long regarded as an "agent for strengthening the nerves" (Bremness 1995, 29*). The seeds are used in folk medicine to treat migraines, nervousness, and depression (Teuscher 1994, 835). Homeopathic preparations of poison nuts (Strychnos nux-vomica hom. *HAB1*, Nux vomica hom. *PFX*, Angustura spuria hom. *HAB34*) are used in accordance with the medical description to treat such ailments as ill moods, headaches, and nervous overstimulation (Teuscher 1994, 832). Nux-vomica D6 is said to be a good, dependable treatment for a hangover, even when accompanied by a severe headache (Olaf Rippe, pers. comm.). Combination preparations are also available (see *Claviceps purpurea*).

Constituents

The bark, the roots, and especially the seeds contain the **indole alkaloids strychnine** and brucine, as well as colubrine, pseudostrychnine, vomicine, and strychnicine (Bisset and Choudhury 1974).

The seeds contain an average of 2 to 3% alkaloids but, less frequently, can contain as little as 0.25% or as much as 5.3%. The strychnine content lies between 1.1 and 1.5% but can sometimes reach 2.3%. Also present are 1.1 to 2.1% brucine as well as the secondary alkaloids (comprising a total of no more than 1%) 12-hydroxystrychnine, 15-hydroxystrychnine, α-colubrine, β-colubrine, icajine, 11-methoxyicajine, novacine, vomicine, pseudostrychnine, pseudobrucine, pseudo-α-colubrine, pseudo-β-colubrine, *N*-methyl-*sec*-pseudo-β-colubrine, and isostrychnine (Teuscher 1994, 831).

The flesh and the shell of the fruit contain essentially the same alkaloids as the seeds. In addition, the alkaloid 4-hydroxystrychnine has been demonstrated to be present. The iridoids loganaine and secologanine have been detected as well (Bisset and Choudhury 1974).

The total alkaloid content of the leaves can range from 0.3 to 8.0% (Teuscher 1994, 829). The flowers also contain alkaloids. The bitter fruit pulp, which is sometimes characterized as edible, contains only 0.35% alkaloids.

The stem cortex contains up to 9.9% alkaloids, and the root cortex as much as 18%. The wood of the roots can contain as much as 1.8%, the bark of the branches up to 6.8%, the wood of the branches up to 1.4%, and the wood of the trunk only 0.3% alkaloids. Strychnine is always the primary alkaloid (Teuscher 1994, 829). The root cortex of a sample from Sri Lanka was found to contain a new alkaloid, which was named protostrychnine (Baser et al. 1979).

Effects

The effects of poison nuts are almost always the result of their **strychnine** content. With the exception of 12-hydroxystrychnine, none of the alkaloids exhibits any notable pharmacological activity. Strychnine and 12-hydroxystrychnine are specific antagonists of the neurotransmitter glycine and bind to the same receptors as glycine does. This results in a stimulation of the central nervous system. "The perception of sensory impressions is potentiated, differences in color and brightness are perceived better, the visual field becomes larger, and the sense of taste is improved" (Teuscher 1994, 835). Poison nuts have erotic/psychoactive effects similar to those produced by *Pausinystalia yohimba*, which is due primarily to the sharpening of sensory perception (vision, sense of smell, sense of taste). In addition, it is possible that men may experience "strong erections" (Roth et al. 1994, 684*).

Overdoses can result in ego dissolution associated with anxiety and severe spasms while remaining fully conscious and ultimately to death through respiratory paralysis. As little as 0.75 to 3 g can be lethal (Teuscher 1994, 836 f.).

Strychnine

Brucine

"In extracts and in powder form, the crows' eyes are occasionally used internally; but they always demand great care when they are used. Because of the narcotic properties, they must be stored using the necessary precautions."

DULK

IN *KOMMENTAR ZUR PREUSSISCHEN PHARMAKOPÖE* [COMMENTARY ON THE PRUSSIAN PHARMACOPOEIA] 1839

"The strychnos manikos, which some persons call persion ['round fruit'], others thryon [a plant from the magical gardens of Colchis], anhydron ['removed from water'], pentadryon ['five clusters'], enoron, orthogyion. . . . The root, in the amount of 1 dram drunk in wine, has the power to create not unpleasant imaginary images, 2 drams drunk persists for up to three days, 4 drams drunk can even kill. The antidote for this is honey mead, amply drunk and vomited up again."

DIOSCORIDES
DE MATERIA MEDICA
(4.74)

In India and Southeast Asia, the powdered seeds of *Syzygium cumini* (L.) Skeels [syn. *Myrtus cumini* L., *Eugenia cumini* (L.) Druce, *Eugenia jambolana* Lam., *Syzygium jambolana* (Lam.) DC.] are used as an antidote for overdoses of poison nuts (Macmillan 1991, 417*). In Oceania, *Piper methysticum* appears to have been used with success as an antidote. In ancient times, **mead** was considered an antidote. There have been reports of using curare to treat strychnine poisoning (Roth et al. 1994, 684*).

Commercial Forms and Regulations

Poison nuts (which are almost never available through retail sources) require a physician's prescription and can be obtained only from a pharmacy. The mother tincture as well as homeopathic dilutions up to and including D3 also require a prescription (Teuscher 1994, 838).

Literature

See also the entries for ***Strychnos* spp., indole alkaloids,** and **strychnine.**

Baser, Kemal H. C., Norman G. Bisset, and Peter J. Hylands. 1979. Protostrychnine, a new alkaloid from *Strychnos nux-vomica*. *Phytochemistry* 18:512–14.

Bisset, N. G., and A. K. Choudhury. 1974. Alkaloids and iridoids from *Strychnos nux-vomica* fruits. *Phytochemistry* 13:265–69.

Teuscher, Eberhard. 1994. *Strychnos*. In *Hagers Handbuch der pharmazeutischen Praxis*, 5th ed., 6:816–46. Berlin: Springer.

Strychnos spp.

Strychnos Species

Family

Loganiaceae (Logania Family, Strychnos Family); Subfamily Strychneae

The genus *Strychnos* is represented in both the Old and the New World by some two hundred species (Neuwinger 1994, 517*). Generally speaking, it can be said that the Old World species contain alkaloids of the **strychnine** type, while New World species contain substances from the curarine group (Macmillan 1991, 432*). Several New World *Strychnos* species were or still are used to produce curare and similar arrow poisons (Bauer 1965). In addition, several species are used for ethnomedicinal purposes. *Strychnos potatorum* L. f. is used in Ayurvedic medicine to treat hallucinations. Many New World species contain **indole alkaloids,** primarily in their bark. Norharman (see **harmaline and harmine**) was found in an extract of *Strychnos barnhartiana* leaves (Quetin-Leclercq et al. 1990). The mysterious **cabalonga** has been interpreted as a species from the genus *Strychnos* (*Strychnos cabalonga* hort. Lind.). Many species contain strychnine, which is why they are used as aphrodisiacs and also are potential sources of the raw materials for psychoactive substances.

Strychnos icaja L. (= *Strychnos ikaja*) [syn. *Strychnos dewevrei* Gilg, *Strychnos dundusanensis* De Wildeman, *Strychnos kipapa* Gilg, *Strychnos mildbraedii* Gilg, *Strychnos pusilliflora* S. Moore,

Strychnos venulosa Hutchinson et M.B. Moss]—
ikaja, bondes root
Known variously as *bondo, mbundu, mbondo,* or *icaja* (Fang), this plant is a 20- to 100-meter-long vine that can climb as high as 40 meters. Of all the African *Strychnos* species used to make hunting poison, this is the most important (Neuwinger 1994, 519*). In central Africa, it was used to make arrow poisons (Macmillan 1991, 432*). The red root cortex was boiled to obtain the poison. The plant is considered sacred because it was used in trials by ordeal. An accused person was given the plant to eat. If he survived he was considered innocent. In the Congo, "the macerate of the root is used in palm wine for very painful gastrointestinal complaints and broken bones. In low dosages, it is said to have diuretic and inebriating effects" (Neuwinger 1994, 521*). In Zaire, ashes of the root are used to treat insanity. The plant contains primarily **strychnine** as well as the related alkaloids icajine, vomicine, and novacine (Neuwinger 1994, 521 ff.*; Ohiri et al. 1983, 177). The missionary Alexander Le Roy (1854–1938) reported the following about this species:

Bwiti, which is the great fetish of the land, has its initiates in the area of Sette Cama [central Africa] and in other places. To be accepted into the secret society, the aspirant must first chew certain roots and drink a decoction of the bark of a tree which is known to botanists

"The rhinoceros bird is said to have such an immunity to strychnos seeds that they are his favorite food. . . . The poisonous seeds of *Strychnos cabalonga* are supposedly also eaten by several mammals, such as *Dasyprocta agouti*."

LOUIS LEWIN
GIFTE UND VERGIFTUNGEN
[POISONS AND POISONINGS]
(1992, 792F.*)

as *Strychnos icaja*. It does not take long for him to fall into a deep sleep and completely lose consciousness. Then a vine [*Ipomoea* **spp.**] is tied around his neck. Three days later, when he begins to recover, a magician will ask him to look into a piece of glass that is attached to the belly of *Bwiti*. He will see certain figures therein, about which he must report. If he says the correct things, he will be accepted; if not, this is taken as a sign that the fetish does not wish to reveal itself to him. (Le Roy 1922, 222)

This report is illuminating for a number of reasons. First, it provides evidence that the Bwiti cult was found in central Africa as well as West Africa; second, the Bwiti cult, which is now a syncretic, neomessianic movement, is characterized as a pure, typically African fetish cult (cf. Thiel et al. 1986); third, it establishes the use of *Strychnos icaja* as a psychoactive substance; and fourth, it supports the theory that *Strychnos icaja* is used as an iboga substitute (cf. **Tabernanthe iboga**). Of further interest is the description of how the aspirant is bound with a vine, another plant that can be included in the circle of psychoactive plants.

Strychnos ignatii Bergius [syn. *Ignatia amara* L. f., *Ignatia philippinensis* Blume, *Ignatiana phillippinica* Lour., *Strychnos balansae* A.W. Hill, *Strychnos beccarii* Gilg, *Strychnos blay-hitam* Dragendorff, *Strychnos cuspidata* A.W. Hill, *Strychnos hainanensis* Merr. et Chun., *Strychnos krabiensis* A.W. Hill, *Strychnos lanceolaris* Miq., *Strychnos ovalifolia* Wall. ex G. Don, *Strychnos philippensis* Blanco, *Strychnos pseudo-tieuté* A.W. Hill, *Strychnos tieuté* Lesch]—Ignatius bean
This creeping climber, which is also known as bitter fever nut and Saint-Ignatius's-bean, originated on the Sunda Islands and in the Philippines but is now found throughout Southeast Asia. In Malaysia, the Ignatius bean is regarded as *upas radja*, "royal poison," and was used both as an arrow poison and to commit murder (Lewin 1920, 556*). Reports of the plant's psychoactive properties had already appeared by the early modern period: "The Ignatius bean has a very energetic action upon the nervous system . . . acting in the very same manner as the poison nut [*Strchynos nux-vomica*]" (Meissner in Schneider 1974, 3:297*).

Today, the Ignatius bean is used by the pharmaceutical industry as a source of **strychnine**. It is used in folk medicine as an aphrodisiac and tonic. It also has acquired a certain significance in homeopathy (Strychnos ignatii hom. *HAB1*). Ignatius beans require a prescription and may be obtained only from a pharmacy. Homeopathic preparations (mother tincture up to and including

The Ignatius bean (*Strychnos ignatii*) contains the potent substance strychnine. (Photograph: Karl-Christian Lyncker)

D3) also require a prescription.

The seeds (faba febrifuga, faba indica, faba sancti ignatii, fabae St. Ignatii, semen ignatii, ignatii semen, Ignatius beans) contain 2.5 to 4% alkaloids (sometimes as much as 5.6%), of which some 45 to 60% is **strychnine**. Brucine, caffeic acid, and chlorogenic acid are also present. The therapeutic single dosage is given as 0.1 g and the total daily dosage as 0.3 g (Roth et al. 1994, 682*).

An aphrodisiac with psychoactive effects (single dosage) consists of 12.5 mg of yohimbé extract (*Pausinystalia yohimba*), 12.5 mg of Ignatius bean extract, 0.3 mg of **atropine** methonitrate, and 3.3 mg of **ephedrine** HCL (formerly registered as the medicine Tonaton®, to treat hypotonia of the bladder).

Strychnos usambarensis Gilg [syn. *Strychnos cooperi* Hutchinson et M.B. Moss, *Strychnos distichophylla* Gilg, *Strychnos micans* S. Moore]—little monkey orange, umuhoko
This species is one of the three most common members of the genus *Strychnos* in Africa. A tree form that can grow from 3 to 15 meters in height is found throughout eastern and southern Africa. A climbing bush form that can attain a length of over 70 meters (!) occurs in Zaire, the Congo, and West Africa. The Banyambo hunters of Rwanda use the roots and leaves of the tree form to make an arrow poison with curare-like effects (due to the presence of the alkaloids curarine, calebasin, dihydrotoxiferin, and afrocurarine). The leaves have been found to contain sixteen indole alkaloids of the usambarane type. Considerable amounts of harmane have been found in the stem bark of both forms (Quetin-Leclercq et al. 1991). It is possible that the bark of this *Strychnos* species may have psychoactive uses (cf. **ayahuasca analogs**).

Literature
See also the entries for *Strychnos nux-vomica* and **strychnine**.

Bauer, Wilhelm P. 1965. Der Curare-Giftkreis im Lichte neuer chemischer Untersuchungen. *Baessler-Archiv*, n.f., 13:207–53.

Botanical illustration of *Strychnos ignatii*. (From *Köhler's Medizinalpflanzen*, 1887/89)

Le Roy, Alexander. 1922. *The religion of the primitives.* New York: Macmillan.

Ohiri, F. C., R. Verpoorte, and A. Baerheim Svendsen. 1983. The African *Strychnos* species and their alkaloids: A review. *Journal of Ethnopharmacology* 9:167–223.

Quetin-Leclercq, Joëlle, Luc Angenot, and Norman G. Bisset. 1990. South American *Strychnos* species: Ethnobotany (except curare) and alkaloid screening. *Journal of Ethnopharmacology* 28:1–52.

Quetin-Leclercq, Joëlle, Monique Tits, Luc Angenot, and Norman G. Bisset. 1991. Alkaloids of *Strychnos usambarensis* stem bark. *Planta Medica* 57:501.

Richard, C., C. Delaude, L. Le Men-Olivier, J. Lévy, and J. Le Men. 1976. Alcaloides du *Strychnos variabilis. Phytochemistry* 15:1805–6.

Thiel, Josef F., Jürgen Frembgen, et al. 1986. *Was sind Fetische?* Frankfurt/M.: Museum für Völkerkunde. (Exhibit catalogue.)

Tabernaemontana spp.

Tabernaemontana Species

Left: The Indians refer to this *Tabernaemontana* species from southern Mexico as *u nek' tsimin,* "the genitals of the tapir." (Photographed in Yaxchilan, Chiapas, Mexico)

Center: A *Tabernaemontana* species, showing the typical flower of the genus.

Right: One *Tabernaemontana* species common in Belize is called "dog testicles" because of its fruit. (Wild plant, photographed in Belize)

Family
Apocynaceae (Dogbane Family); Subfamily Plumerioideae, Tabernaemontaneae Tribe, Tabernaemontaninae Subtribe

Synonyms
Ervatamia spp.
Peschiera spp.

Folk Names
Throughout the world, the suggestive appearance of the fruits of many *Tabernaemontana* species has led to them being named for the genitals of various mammals: dog's testicles, u nek' pek' ("the testicles of the dog"), u nek' tsimin ("the testicles of the tapir"), äh toon tsimin ("the penis of the tapir"), et cetera.

History
The genus *Tabernaemontana* is composed of some 120 tropical and several subtropical species (Sierra et al. 1991). Most occur in tropical rain forests, especially in Central and South America and in Africa (Schultes 1979). In Africa, many species are used for ethnomedicinal purposes (Omino and Kokwaro 1993*).

Linnaeus developed the genus name to honor the naturalist and "father of botany" Jakob Theodor, called Tabernaemontanus (1522–1590). Phytochemical studies of the genus have only recently taken place. **Indole alkaloids** dominate; several species have been found to contain **ibogaine** and voacangine (cf. *Tabernanthe iboga, Voacanga* **spp.**). As a result, this genus is of special interest in the search for new psychoactive plants. Several species with psychoactive effects and uses have already become known.

Appearance
Most species in the genus are bushy shrubs, shrubby herbs, climbers, or small trees. They have evergreen, lanceolate, more or less tapered leaves that frequently have a leathery upper surface. The pentacuspidate flowers often grow in clusters from the leaf axils. The fruits are always symmetrically bipartite with a more or less conspicuous constriction; they often appear remarkably similar to the scrotums of higher mammals. Some fruits turn luminously red as they ripen. The presence of whitish or yellowish latex in the bark is a characteristic of the genus.

Tabernaemontana coffeoides Bojer ex DC.

This plant is used in Madagascar as a stimulant. It contains voacangine and other alkaloids. Voacangine can be transformed in vitro to **ibogaine** (Ott 1993, 401*).

Tabernaemontana crassa Bentham

This midsize tree is from the rain forests of West Africa, where the local populace uses it for various folk medicinal purposes. The latex is applied externally to treat wounds and fleshworms. An extract of the leaves is ingested for fever. One especially popular application involves using the leaves as a local anesthetic, e.g., in the treatment of dislocations and broken bones (Agwu and Akah 1990). Whether the plant was or is also used for psychoactive purposes (as a narcotic) is still unknown.

Tabernaemontana dichotoma Roxburgh ex Wallich—divi kaduru

In India, the root and stem cortex of this species are used in folk medicine to treat wounds as well as snakebites and centipede bites (Perera et al. 1985, 2097). The bark is also regarded in India as a drug that can induce delirium—that is, have psychoactive effects (Ott 1993, 401*; Perera et al. 1983). In Sri Lanka, *divi kaduru* is regarded as a "forbidden fruit" and included in the same folk taxonomic category as *Strychnos nux-vomica* (*goda kaduru*); *kaduru* means "poisonous." Muslims call the fruit "the forbidden fruit of the Garden of Eden," while Europeans living in Sri Lanka call it "Eve's apple." The seeds are said to have potent narcotic and hallucinogenic effects, and folk healers consider them the equal of the seeds of *Datura metel* (Perera et al. 1984, 233 f.). The bark has been found to contain twenty-two alkaloids of the **ibogaine** type, including the stimulating vobasine and ibogamine (Perera et al. 1985).

Tabernaemontana heterophylla Vahl.—sanango

The Tukano Indians of the Brazilian Amazon give old people who are becoming slow and forgetful a tea made from the leaves twice a day for two weeks (Schultes 1993, 132*). *Sanango*, a word that essentially means "memory," is a name given to a number of Amazonian plants (Schultes 1979, 186). Whether this species is psychoactive or merely a brain tonic is an open question. It is possible that the leaves are used as an **ayahuasca** additive.

Tabernaemontana muricata Link ex Roemer et Schultes

The leaves and white flowers are dried in the sun and used as a stimulating additive to **chicha** made from *Manihot esculenta*. Such chicha is said to be especially good for the elderly. The leaves and flowers contain alkaloids (Schultes 1979, 186).

Tabernaemontana pandacaqui Poir. [syn. *Ervatamia pandacaqui* (Poir.) Pichon, *Tabernaemontana wallichiana* Steub.]

The root of this species, which is common in Thailand, is used in folk medicine to treat fever, pain, and dysentery. Pharmacological studies have demonstrated that an alcohol extract of the root, stem, leaves, and flowers has potent analgesic effects (Taesotikul et al. 1989b). To date, nothing is known about any psychoactive effects upon humans. The root has been found to contain 3*S*-hydroxyvoacangine, an **indole alkaloid** of the voacangine type that also occurs in *Voacanga* spp. Alkaloids of the **ibogaine** type are also present (Sierra et al. 1991).

Tabernaemontana rimulosa Woodson ex Schultes

In Venezuela, a few leaves of this species, boiled in milk, are drunk as a sleeping agent (Schultes 1979, 186).

Tabernaemontana sananho Ruíz et Pavón—sanango

In Amazonia, the *sanango* tree, which can grow as tall as 5 meters, is regarded as a cure-all; the leaves, the root, and the latex-rich bark are all used in folk medicine (Schultes 1979, 187 ff.). The leaves of the tree are used psychoactively as an additive to **ayahuasca** and also are combined with *Virola* spp. to produce an orally efficacious hallucinogen. The plant is also called *uch pa huasca sanango* and is known as a "memory plant," a reference to the fact that its inclusion in a psychoactive preparation causes a person to better remember the experiences he or she has had while under the influence

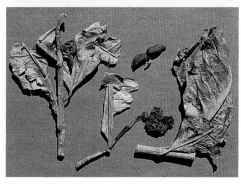

Top: The raw drug of *Tabernaemontana sananho* consists of stem pieces.

Bottom: Dried leaves, flowers, and fruits of *Tabernaemontana sananho*.

of that preparation. It is added to **ayahuasca** so that a person can, afterward, more clearly recall the visions he or she saw.

In Ecuador, the plant is known as *sikta* and is available in raw form (short branch pieces) at local markets.

The Jíbaro drip the freshly pressed juice into the nostrils of their dogs so that they may be better able to locate prey. The plant is also known as *yacu zanango*.[296]

It is rich in alkaloids (Schultes 1983a, 270*).

Tabernaemontana tetrastachys H.B.K.—uchu-sanango, saticu

The Makuna Indians call this plant *beé-e-ge* and use its latex as stimulating eyedrops (cf. ***Tabernanthe iboga***). A few drops is said to dispel tiredness and sleep (Schultes 1979, 189).

Constituents

Indole alkaloids are common in the Family Apocynaceae. To date, the approximately 120 species of the family have been found to contain 256 alkaloids, of which many are **ibogaine** analogs. Several species even contain pure ibogaine. Many *Tabernaemontana* species contain high concentrations of indole alkaloids, primarily tabernanthine, ibogaine, and ibogamine alkaloids (Achenbach and Raffelsberger 1980; van Beek et al. 1984). Other species, e.g., *Tabernaemontana campestris* (Rizz.) Leeuwenberg [syn. *Peschiera campestris* (Rizz.) Rizz.], contain voacangine, the main active constituent in **Voacanga spp.**, and similar alkaloids (Gower et al. 1986). Biochemical studies have shown that the genus's production of indoles is open to influence and can be modified (Dagnino et al. 1992). Many of the alkaloids have stimulant effects (van Beek et al. 1984).

Relatives

Some of the species that were originally assigned to the genus *Tabernaemontana* are now viewed as belonging to a genus of their own, *Pandaca*. **Indole alkaloids** of the ibogaine type also occur in the genus *Pandaca*.

Tabernaemontana van heurkii Muell. Arg. now bears the botanically valid name *Peschiera van heurkii* (Muell. Arg.) L. Allorge. The leaves and the stem bark contain twenty indole alkaloids, some of which have antibacterial properties (Muñoz et al. 1994).

Literature

See also the entry for **indole alkaloids**.

Achenbach, Hans, and Bernd Raffelsberger. 1980. 19-ethoxycoronaridine, a novel alkaloid from *Tabernaemontana glandulosa*. *Phytochemistry* 19:716–17.

Agwu, Ijere E., and Peter A. Akah. 1990. *Tabernaemontana crassa* as a traditional local

anesthetic agent. *Journal of Ethnopharmacology* 30:115–19.

Dagnino, D., J. Schripsema, and R. Verpoorte. 1992. Comparison of two cell lines of *Tabernaemontana divaricata* with respect to their indole alkaloid biosynthetic and transformation capacity. *Planta Medica* 58 suppl. (1): A608.

Delle Monache, G., et al. 1977. Studi sugli alcaloidi di *Tabernaemontana sananho* R. et P. *Atti Acc. Naz. Lincei* 62:221–26.

Gower, Adriana E., Benedito da S. Pereira, and Anita J. Marsaioli. 1986. Indole alkaloids from *Peschiera campestris*. *Phytochemistry* 25 (12): 2908–10.

Muñoz, V., C. Moretti, M. Sauvain, C. Caron, A. Porzel, G. Massiot, B. Richard, and L. Le Men-Olivier. 1994. Isolation of bis-indole alkaloids with antileishmanial and antibacterial activities from *Peschiera van heurkii* (syn. *Tabernaemontana van heurkii*). *Planta Medica* 60:455–59.

Perera, Premila, Duangta Kanjanapoothi, Finn Sandberg, and Robert Verpoorte. 1984. Screening for biological activity of different plant parts of *Tabernaemontana dichotoma*, known as *divi kaduru* in Sri Lanka. *Journal of Ethnopharmacology* 11:233–41.

———. 1985. Muscle relaxant activity and hypotensive activity of some *Tabernaemontana* alkaloids. *Journal of Ethnopharmacology* 13:165–73.

Perera, P., G. Samuelsson, T. A. van Beek, and R. Verpoorte. 1983. Tertiary indole alkaloids from leaves of *Tabernaemontana dichotoma*. *Planta Medica* 47:148–50.

Perera, P., F. Sanberg, T. A. van Beek, and R. Verpoorte. 1985. Alkaloids of stem and rootbark of *Tabernaemontana dichotoma*. *Phytochemistry* 24 (9): 2097–104.

Schultes, Richard Evans. 1979. De Plantis Toxicariis e Mundo Novo Tropicale Commentationes. XIX: Biodynamic Apocynaceous plants of the northwest Amazon. *Journal of Ethnopharmacology* 1:165–92.

Sierra, Marta I., Robert van der Heijden, Jan Schripsema, and Robert Verpoorte. 1991. Alkaloid production in relation to differentiation in cell and tissue cultures of *Tabernaemontana pandacaqui*. *Planta Medica* 57:543–47.

Taesotikul, T., A. Panthong, D. Kanjanapothi, R. Verpoorte, and J. J. C. Scheffer. 1989a. Cardiovascular effects of *Tabernaemontana pandacaqui*. *Journal of Ethnopharmacology* 27:107–19.

———. 1989b. Hippocratic screening of ethanolic extracts from two *Tabernaemontana* species. *Journal of Ethnopharmacology* 27:99–106.

[296] In Colombia, another alkaloid-rich plant from the Family Apocynaceae is also referred to as *sanango*: *Bonafousia tetrastachya* (H.B.K.) Markgraf (Schultes and Raffauf 1986, 277*).

van Beek, T. A., F. L. C. Kuijlaars, P. H. A. M.
Thomassen, R. Verpoorte, and A. Baerheim
Svendsen. 1984. Antimicrobially active alkaloids
from *Tabernaemontana pachysiphon*.
Phytochemistry 23 (8): 1771–78.

van Beek, T. A., and M. A. J. T. Van Gessel. 1988.
Alkaloids of *Tabernaemontana* species. In
Alkaloids: Chemical and biological perspectives, ed.
S. W. Pelletier, 6:75–226. New York: Wiley & Sons.

van Beek, T. A., R. Verpoorte, A. Baerheim Svendsen,
A. J. M. Leeuwenberg, and N. G. Bisset. 1984.
Tabernaemontana (Apocynaceae): A review of its
taxonomy, phytochemistry, ethnobotany and
pharmacology. *Journal of Ethnopharmacology*
10:1–156.

Tabernanthe iboga Baill.

Iboga Shrub

Family
Apocynaceae (Dogbane Family); Subfamily Plu-merioideae, Tabernaemontaneae Tribe

Forms and Subspecies
The synonyms are occasionally defined as distinct species. These may, however, merely represent varieties, forms, races, et cetera. The natives of Gabon make a distinction between two varieties on the basis of the shapes of the fruit (Brenneisen 1994, 890). The ethnographic literature sometimes distinguishes two varieties (Fernandez 1982):

Tabernanthe iboga var. *iboga* (iboga vrai, mabasoka)
Tabernanthe iboga var. *manii* (ñoké)

Synonyms
Iboga vateriana J. Br. et K. Schum.
Tabernanthe albiflora Stapf
Tabernanthe bocca Stapf
Tabernanthe mannii Stapf
Tabernanthe pubescens Pichon
Tabernanthe subsessilis Stapf
Tabernanthe tenuiflora Stapf
Tabernanthes eboka (incorrect spelling in the literature, e.g., Fernandez 1966, 46)

Folk Names
Abona, abonete, aboua, ahua (Pahuin), bocca, boccawurzel, boga, botola, bugensongo (Ngala), dibuga, dibugi, difuma (Eshira), eboga (Fang), eboga bush, ebôga, ébogé, eboghe, eboka ("miracle wood"), elahu (Mongo), eroga, gbana (Gbaya), gifuma, iboa, ibo'a, iboga (Galwa-Mpongwe/Miene), ibogakraut, ibogain-pflanze, iboga shrub, ibogastrauch, iboga typique (Congo), iboga vrai, ibogawortel (Dutch), ibogawurzel, ikuke (Mongo), inado a ebengabanga (Tshiluba), inaolo a ikakusa (Turumbu), inkomi (Mono), isangola, leboka, liboko (Vili/Yoombe), libuga, libuka, lofondja, lopundja, mabasoka, mbasaoka, mbasoka (Mitsogo), mbondo (Aka Pygmy), meboa (Bakwele), minkolongo (Fang), moabi, mungondo (Eshira), ñoké, nyokä (Mitsogo), obona, obuété, pandu (Mongo), sese (Fang), wunderholz

History
According to legend, the iboga bush arose from a person, just like many other psychoactive plants. According to the mythology of the West African Fang:

> Zame ye Mebege [the last of the creator gods] gave us Eboka. One day . . . he saw . . . the Pygmy Bitamu, high in an Atanga tree, gathering its fruit. He made him fall. He died, and Zame brought his spirit to him. Zame cut off the little fingers and the little toes of the cadaver of the Pygmy and planted them in various parts of the forest. They grew into the Eboka bush. (Schultes and Hofmann 1992, 112)

For many West African tribes, the plant became a "bridge to the ancestors," an instrument of initiation to the true world, a fetish in which the personal god dwells. The wisdom of the original ancestor, the Pygmy plant expert Bitamu, is embodied in the plant, and his sacrifice established the plant cult. The ingestion of iboga causes one to travel through time. Iboga is a sacrament and a symbol of the power of the forest (the Fang say that "Bwiti is a religion of the trees").

In West Africa, iboga has been used in fetish cults and in magic since ancient times (Bisset 1989, 21; Pope 1969). In the Congo, its psycho-active effects have been used to enable mediums to be possessed by fetishes (Schleiffer 1979, 49*). Hunters chewed the root so that they could stay

Botanical illustration of the West African iboga shrub (*Tabernanthe iboga*). (From Landrin, *Bull. Sc. Pharmacol.* 11, 1905)

awake and retain their strength during long hunting expeditions. It is said that the iboga root gives one the power to remain immobile for two full days while hunting the coveted lion trophies (Bouquet 1969).

Iboga found its greatest significance in the reformative cults that developed around the beginning of the twentieth century from the ancestor cults (*bieri*) of the Neo-Bantu peoples (Fang) and became known as Bwiti (Fernandez 1964; 1966, 44). Administrators in Gabon tried several times to suppress the Bwiti cult, sometimes using such spurious arguments as "Bwiti is a cult of cannibals and ritual murderers" (Schleiffer 1979, 54*). But Bwiti has remained vital in northern Gabon even into the present day and is, in fact, steadily gaining in popularity. The first white person ever to be initiated into the Bwiti cult and to undergo and survive the effects of the iboga root was the Italian ethnobotanist Giorgio Samorini (1993, 1996b).

The first report about the plant and its stimulant and aphrodisiac effects appeared in 1864 (Schultes 1970, 35*). The bush was botanically described in 1889 by Henri E. Baillon (1827–1895). The primary constituent of the plant, **ibogaine**, was isolated in 1901; pharmacological investigations of this substance have been conducted primarily in France.

Distribution
This tropical plant is found in Gabon and the surrounding areas of the Congo and from Cameroon to Angola. It also is planted in many parts of West Africa. It is a typical shade-loving underwood plant that thrives at altitudes from sea level to 1,500 meters. The plant is often encountered along rivers and in marshy areas (Vonk and Leeuwenberg 1989, 11).

Cultivation
Propagation usually occurs with root segments taken from the rootstock or from scions. Propagation from seed is very difficult, as the seeds remain viable only if they have not completely dried.

Tabernanthe iboga can be crossed with *Tabernanthe elliptica*. Even natural hybrids occur (Vonk and Leeuwenberg 1989, 12 f.). The so-called Kisantu hybrid was described under the name *Daturicarpa elliptica* x *Tabernanthe iboga*, but it may have been better interpreted as *Tabernanthe elliptica* x *Tabernaemontana* (*Pterotabera*) *inconspicua* (Bisset 1989, 24; Massiot et al. 1988).

Appearance
The evergreen, branching shrub can attain a height of 1.5 to, more rarely, 2 meters. The opposite, lanceolate leaves can grow 10 to 15 cm in length. The shrub develops strong, heavily branching roots that have a brownish rind and yellowish wood. The tiny yellow flowers have a corona 5 to 10 mm in length and often appear in clusters. The pendulous, orange-yellow fruits are ovoid and pointed at the end (18 to 24 mm long). The plant produces a white latex with a powerful scent. In the tropics, the shrub flowers from March to June/July (and sometimes even longer). The fruits ripen at the beginning of the dry period. Both flowers and fruits can appear simultaneously.

The iboga shrub can be easily confused with other members of the genus *Tabernanthe*. Following the last taxonomic revision of the genus, however, this confusion should be limited to the very similar *Tabernanthe elliptica* (Stapf) Leeuwenberg [syn. *Daturicarpa elliptica* Stapf, *Daturicarpa firmula* Stapf, *Daturicarpa lanceolata* Stapf]. *Tabernanthe elliptica* distinguishes itself by its copious white latex. The iboga shrub can also be confused with several *Tabernaemontana* species (*Tabernaemontana* **spp.**), regarded as its closest relatives (Vonk and Leeuwenberg 1989, 3).

Psychoactive Material
— Root (tabernanthe radix, tabernanthewurzel, boccawurzel, ibogawurzel, iboga root)
— Root cortex (tabernanthe radicis cortex, tabernanthewurzelrinde, tabernanthe root cortex)
— Leaves (tabernanthe folium, tabernantheblatt, tabernanthe leaves)

Above and center: The West African iboga shrub (*Tabernanthe iboga*)

Right: The tiny flowers of the iboga shrub (*Tabernanthe iboga*)

Preparation and Dosage

In Gabon, the roots are harvested from living plants. A small hole is dug in the ground near the rootstock. Part of the root is removed, but enough of the rootstock is left in the ground that the plant can remain alive and develop new roots.

The root or root cortex is dried and rasped or ground. The extremely bitter, repulsive-tasting root is either eaten and washed down with water or, less frequently, made into a tea.

In the Congo, an aphrodisiac wine is made by steeping the fresh or dried root in **palm wine** for a few hours and then removing it (Bouquet 1969, 67).

A heaping teaspoon of the root powder acts as a stimulant and produces an agreeable state of euphoria (Samorini 1993, 6). Six to 10 g of the dried root powder induces visions and psychedelic hallucinations. For initiation into the Bwiti cult, one eats 50 to 100 g, and sometimes probably even more (200 g).

A dosage corresponding to 2 to 10 mg per kilogram of body weight (calculated as ibogaine) produces a non-amphetamine-like stimulation of the central nervous system. With an amount corresponding to 40 mg of ibogaine per kilogram, the serotonin receptors are occupied, resulting in LSD-like effects (Brenneisen 1994, 892).

The bark and bark juice were used together with *Parquetina* and/or *Strophanthus* species to make arrow poisons (Bisset 1989, 21).

The fruits are edible and do not produce any psychoactive effects (Fernandez 1982, 474). Although leaf extracts contain different alkaloids, they are said to be more pharmacologically active (Bisset 1989, 25).

Iboga roots were or are sometimes prepared together with other plants, only a few of which have been botanically identified (see the table below) (Emboden 1979, 73; Schultes 1970, 36*).

Ritual Use

According to information provided by the Fang, the iboga plant was originally discovered in the

Psychoactive Eyedrops?

During Bwiti initiation, eyedrops (*ibama, ebama*) are sometimes dripped into the initiates' eyes so that they may receive more profound or clearer visions. It is possible that some of the eyedrop preparations may be psychoactive or have synergistic effects with iboga (Samorini 1996c). We have no precise recipes, but some of the ingredients are known:

Costus lucanusianus J. Braun et K. Schum. (Zingiberaceae), *Amorphophallus maculatus* N.E. Br. (Araceae), *Afromomum sanguineum* K. Schum. (Zingiberaceae), *Euphorbia hermentiana* Lem (Euphorbiaceae), *Mimosa pigra* L. (Leguminosae; cf. **Mimosa spp.**), *Buchholzia macrophylla* Pax (Capparidaceae), **Elaeophorbia drupifera** Stapf (Euphorbiaceae), and the juice of a large millipede (Fernandez 1972, 242 f.; Samorini 1996c).

rain forest by the Pygmies. The Apinji and Metsogo, who established the foundation of the initiatory use, learned the secret of the consciousness-expanding root from these small rain-forest people. Sometime around 1890, the Fang adopted the ancestor ritual (*bieri*) from them and fused it with Christian concepts and customs to produce the syncretic Bwiti cult (Samorini 1993). In the process, the iboga plant was occasionally identified with the cult god Bwiti himself (Fernandez 1966, 62 f.). In any case, iboga is regarded as the true tree of knowledge that came directly from the Garden of Eden so that people could use it to recognize God and the world and, initiated into paradisiacal secrets, to spend their life on the earth in joy (Samorini 1993). The Bwiti cult has certain parallels to the North American peyote cult (see **Lophophora williamsii**) and the Brazilian Santo Daime cult (see **ayahuasca**), both of which have members ingest a psychoactive

"The souls fish the pool of *eboka* to make us fertile. There is a mirror in the pool that reflects heaven and earth, and that is the one that god sees. The wind of creation will enable trees to spring from the ground, but not people, for their spiritual sources are in the water of creation—in the pool of *eboka*. For this reason, death has an inconsistency for us. The religion of *eboka* is like the rattan palm, although thorny, it offers a sweet repast to the wanderer. God, the source of all wisdom, has hidden himself and is as hidden as the heart of a palm, but we of the Bwiti cult learn directly from him."

BWITI SERMON BY EKANG ENGONO IN "UNBELIEVABLY SUBTLE WORDS" (FERNANDEZ 1966, 64)

Iboga Additives

Name	Stock Plant	Psychoactive Part	Active Principle
alan, niando	*Alchornea floribunda* Müll. Arg. (cf. *Alchornea* spp.)	root	alkaloids
ayañ beyem	*Elaeophorbia drupifera*	latex	alkaloids (?)
bangi	*Cannabis sativa*	flowers, leaves	THC
duna	unidentified mushroom[297]	fructification	?
ikaha	*Strychnos icaja* L. (cf. *Strychnos* spp.)	root cortex	strychnine, indoles
tava	*Nicotiana* spp.	leaves	nicotine
yohimbé	*Pausinystalia yohimba*	bark	yohimbine

[297] This mushroom was also used in other magical rites in West Africa but is now a component only of the Bwiti mythology, in which it serves as a symbol for the brain and for "the first man to die" (Samorini 1995, 111). It has been claimed that the powdered mushroom has psychedelic effects and was once used in connection with or as an additive to iboga (Fernandez 1972, 246).

"Soon the day will begin
the chief has left us
in the beginning, a rope came from
    the heavens
from our father Nzambe
Humans increased their numbers
And the earth received fertilizing
    rain
No one can pass the place
where the harp player sits
Except for the primordial ancestors,
who know all things
The heart of the new initiate
is full of bitterness
That of the old is full of wisdom
The young initiate is divided
By the light of the sun and of the
    moon."

TEXT OF THE BWITI RITUAL OF THE
METSOGO

The Fang use these fetishes, carved from wood, in the West African iboga cult.

298 In Gabon, some one thousand to two thousand temples are said to lie along the "streets of the iboga" (Samorini 1993, 6).

substance as a sacrament in a syncretic ritual. The Bwiti cult is first and foremost a rite of initiation:

The story of Bwiti is 150 years long; it arised from the influence of Christendom over the traditional cults in which *iboga* was used, widespread among different tribes of Gabon and of the neighboring countries. The Bwiti is differentiated in numerous sects, each one constituted by different communities, and the differences among them are particularly due to the degree of absorption of the christian symbols and practices. In all the sects *iboga* is used as the "true sacrament," in opposition to the ineffective christian host. . . . [T]he *ngozé*, or bwitist masses, are performed during three consecutive nights (from Thursday night to Saturday night), during which the faithful eat a "modest" quantity of powdered root of *iboga*, giving up to dances and songs until the coming of daybreak. . . . [T]he *tobe si* [is] the initiation rite, celebrated each time a person decides to enter into the religious community. In this case, the novice has to eat a huge quantity of *iboga*, comparable to hundreds of dosis as those used during the *ngozé*: a quantity that progressively carry him to a deep and long state of coma, during which his soul makes a trip into the "other world," while his body lies on the ground, watched by the officiants. Still today, sometimes someone does not awake from the state of unconsciousness, and dies" (Samorini 1995, 105)

The various cult communities, each of which is composed of around fifty people, usually have a temple[298] (*abeiñ*) that is used for initiations as well as special holidays (Easter, Christmas) and the weekly nocturnal masses. Before giving his sermon, the priest of the Bwiti cult will lie, under the influence of iboga, in a grave dug into the ground. He remains there until he has found the words for his sermon (*nkobo akyunge*, literally "amazing words"). He usually will lie for hours in the grave, coming out after midnight to proclaim his "amazing words" (Fernandez 1966, 46).

A ritual use of iboga root (often in combination with 50 μg of LSD) in ritual circles known as "vision circles" has developed in Europe and the United States as well. The ritual structure borrows from that of Indian mushroom circles (cf. *Psilocybe mexicana*) and peyote meetings (cf. *Lophophora williamsii*). Sometimes the vision circle will follow a psychedelic medicine wheel. Here, the participants who sit in the south will ingest *Trichocereus pachanoi*, those in the east *Psilocybe* spp., those in the west **ayahuasca**, and the participants in the north iboga root (Westerhout 1996).

Artifacts

In West Africa, the tradition of making and venerating fetishes for ancestor cults is very ancient and one of the characteristic features of the culture (Koloss 1980). The Fang were already carving anthropomorphic ancestor figures by the end of the nineteenth century. These were used as fetishes in the *bieri* cult and would later find use in the Bwiti cult as well.

A variety of paraphernalia is used in the iboga cult. The harp is especially important; it is made with great care and is played during the ritual (Swiderski 1970). The music and the texts that are sung with the harp represent the most important cultural artifacts of the Bwiti cult (Grebert 1928). The singer and harp player of the cult community sings Bwiti songs to accompany the soul of the initiate. For example:

So was the beginning. Spirit of earth, spirit of the sky. The place that we traverse. Father Zame, who is the gatekeeper. I come to a new land that is the cemetary. . . . Lightning and Thunder. Sun and Moon. Heaven and Earth. They are all of them twins. They are life and death. They are all twins. The yawning hole of the grave and the new life, they are all twins. . . . Joy, full of joy the ancestors greet you and hear the news. The anxious life of the born is at an end, at an end, at an end. And now come the disciples of death. I go to the dead. . . . Everything is pure, pure. Everything new, new. Everything is light, light. I have seen the dead and I am not afraid! (from Fernandez 1982)

Discography: Bwiti Music

Gabon: Les musicians de la forêt, vol. 1. Ocora 558569. Paris, 1981.

Gabon: Musica da un Microcosmo Equatoriale—Musica Fang Bwiti con esempi musicali Mbiri. Albatros VPA 8232/B. Milan, 1975.

Gabon: Musiques de Mitsogho et des Batéké. Ocora OCR 84. Paris, 1984.

Music from an Equatorial Microcosm: Fang Bwiti Music with Mbiri Selections. Folkway Records FE 4214. New York, 1973. (Recorded by James Fernandez)

Medicinal Use

Iboga root is used in West African folk medicine as a stimulant, tonic, and aphrodisiac; in cases of nervous weakness; and to treat fever and high blood pressure. Because of its anesthetic properties, it also is used for toothaches (Brennesien 1994, 892). The Metsogo also use iboga roots to divine and diagnose the causes of illness (Prins 1987). In the Congo, iboga is used to treat the tropical sleeping sickness (Hirschfeld and Linsert 1930, 202*).

The French in equatorial Africa once praised an iboga extract called *lambarence* as a cure-all and recommended it especially in the treatment of neurasthenia and syphilis (Miller 1985, 50*).

In homeopathic medicine, a mother tincture and various dilutions (Tabernanthe iboga hom.) obtained from the fresh root are used in accordance with the medical description.

Constituents

The dried root cortex can contain a total of up to 6% monoterpene **indole alkaloids** (Brennesien 1994, 892; Schultes 1970, 36*). The alkaloid concentration in the entire root is about 1% (Roth et al. 1994, 688*). The alkaloids can be divided into three types: the ibogaine type (**ibogaine**, tabernanthine, ibogamine, gabonine, ibogaline, et cetera), the voacangine type (voacangine, catharanthine, voacryptine, et cetera), and the voaphylline type (voaphylline) (Brennesien 1994, 890). The main active constituent is ibogaine. Voacangine can be regarded as another important component (cf. *Voacanga* spp.). The alkaloid mixture varies depending upon the race, location, et cetera. Many of the iboga alkaloids also occur in *Tabernaemontana* spp.

The seeds contain the alkaloids (–)-catharanthine, (+)-voaphylline, and (–)-coronaridine (Roth et al. 1994, 688*).

The form indigenous to Zaire, which was previously described under the name *Tabernanthe pubescens*, has been found to contain the following alkaloids: coronaridine, voaphylline, tetrahydroalstonine, voaphylline hydroxyindolenine, 11-hydroxytabersonine, ibogamine, ibogaine, ibogaline, iboxygaine, voacangine, voacangine hydroxyindoleine, voacristine, 3,6-oxido-iboxygaine, 10-hydroxycoronaridine, 10-hydroxyheyneanine, and 3,6-oxidoibogaine (Mulamba et al. 1981).

Effects

The Fang describe the visions that follow the ingestion of iboga as "wending through the forest." They experience carrying the entire wondrous world of the forest within themselves. Reports of visionary encounters with the ancestors are typical (Fernandez 1982, 476 ff.).

Although only a few white people have had the opportunity to use iboga root, their reports are also constant in some respects. They describe powerful yet peaceful visions and especially contact with deceased family members, unknown people, and animals ("ancestors" in the broadest sense). The following represents a typical report of an iboga experience:

A white light rose up from within me. First as an infinitely small point. The point grew in spite of all mathematical definitions. It became larger, but did not form a circle. . . . It became a triangle, or more precisely, a three-cornered crystal that glowed white. I knew that it was the center of the eternal circle. Its three crystalline faces were the past, the present, and the future. All three aspects of time were one, they touched one another and together they created the world. I had the cosmic jewel before me. Indeed, the etymology of *cosmos* is jewel. The Buddha also holds a magical jewel, from whose brilliance the world arises, in his enlightened hand. Many ochre brown layers lay around the three-cornered crystal. Each one housed another one. All of the layers penetrated into every direction. Every layer was a stage in the development of the universe, in the evolution of life, in the unfolding of consciousness. Every sequence of layers, whether it was that of the past, the present, or the future, transcended into the infinite. There the layers met once more. The infinite was the extreme outer limit of the crystal, and it lay precisely in its center. I saw a culture that is beyond all cultures and yet is inherent to all cultures. I saw gods that are beyond all known gods and yet are contained in all gods. I saw rows of ancestors, all of which are beyond humans and yet have effects upon them even into the present time. I saw the archetypes. They danced in rings around one another in all units of consciousness and led them securely through the universe. *Maya* is not the appearance of things. *Maya* is the masks of the archetypes. We need *Maya*, otherwise we would no longer understand the world. There was no stop to this swirl of insight, one face followed the next. And yet all of the faces remained in existence. I had never remembered faces so well before. Since then, they have never disappeared. I can remember everything clearly, nothing has gotten confused.

The extract of the root has a potent stimulating effect upon the brain that is not comparable to the stimulation produced by amphetamines (cf. **ephedrine**) (Bert et al. 1988). The effects of the root are also different from those of isolated or pure ibogaine, because the other alkaloids exhibit an affinity to certain receptors or represent antagonists (e.g., tabernanthine has an antagonistic activity toward benzodiazepine and GABA receptors).

Commercial Forms and Regulations

The plant is (still) legal, although there have been attempts to schedule the constituent **ibogaine** under current drug laws. The mother tincture of the root can be obtained in France and Switzerland. Dilutions (D3 and higher) can be purchased in the United States. Plant material is only rarely

Ibogaine

Voacangine

Voaphylline

"For this reason God left the *iboga*, so that men would see their bodies as God had made them, as He himself has hidden inside them. Therefore brothers take the *iboga*, the *iboga* plant that God gave to Adam and Eve, Obola and Biome."

A Bwiti initiate
in "Adam, Eve, and Iboga"
(Samorini 1993, 4)

available outside of West Africa. Iboga roots are sometimes offered by ethnobotanical specialty sources, but the material is often counterfeit.

Literature

See also the entries for *Tabernaemontana* spp., *Alchornea* spp., *Voacanga* spp., **ibogaine**, and **indole alkaloids**.

Bert, Maryse, René Marcy, Marie-Anne Quermonne, Michel Cotelle, and Michel Koch. 1988. Non-amphetaminenic central stimulation by alkaloids from ibogaine and vobasine series. *Planta Medica* 36:191–92.

Binet, J. 1974. Drugs and mysticism: The Bwiti cult of the Fang. *Diogenes* 86:31–54.

Bisset, N. G. 1989. *Tabernanthe*: Uses, phytochemistry, and pharmacology anatomy of *T. iboga. Wageningen Agricultural University Papers* 89 (4): 19–26.

Bouquet, Armand. 1969. *Féticheurs et médicines traditionelles du Congo (Brazzaville).* Mémoires, no. 36. Paris: ORSTOM.

Brenneisen, Rudolf. 1994. *Tabernanthe*. In *Hagers Handbuch der pharmazeutischen Praxis*, 5th ed., 6:890–93. Berlin: Springer.

Caignault, J. C., and J. Delourme-Houdé. 1977. Les alcaloïdes de l'iboga (*Tabernanthe iboga* H. Bn.). *Fitoterapia* 48:243–65.

Fernandez, James W. 1964. African religious movements: Types and dynamics. *Journal of Modern African Studies* 2 (4): 531–49.

———. 1966. Unbelievably subtle words: Representation and integration in the sermons of an African reformative cult. *Journal of the History of Religions* 6:53–69.

———. 1972. *Tabernanthe iboga*: Narcotic ecstasis and the work of the ancestors. In *Flesh of the gods*, ed. Peter T. Furst, 237–59. New York: Praeger.

———. 1982. *Bwiti: An ethnography of the religious imagination in Africa.* Princeton, N.J.: Princeton University Press.

Fromaget, M. 1986. Contribution du Bwiti mitsogho à l'anthropologie de l'imaginaire: A propos d'un cas de diagnostic divinatoire au Gabon. *Anthropos* 81:87–107.

Grebert, M. F. 1928. L'art musical chez les Fang du Gabon. *Archives Suisses d'Anthropologie Générale* 5:75–86.

Koloss, Hans Joachim. 1980. Götter und Ahnen, Hexen und Medizin. In *Zum Weltbild in Oku*, ed. Walter Raunig, 1–12. Frankfurt/M.: Pinguin-Verlag.

Massiot, Georges, Bernard Richard, Louisette Le Men-Olivier, Jean de Grave, and Clément Delaude. 1988. Alkaloids from leaves of *Pterotaberna inconspicua* and the Kisantu hybrid problem. *Phytochemistry* 27 (4): 1085–88.

Mulamba, T., C. Delaude, L. Le Men-Olivier, and J. Lévy. 1981. Alcaloïdes de *Tabernanthe pubescens. Journal of Natural Products* 44 (2): 184–89.

Naeher, Karl. 1996. Die Droge gegen Drogen. *Esotera* 8:57–59.

Pope, Harrison G., Jr. 1969. *Tabernanthe iboga*: An African narcotic plant of social importance. *Economic Botany* 23:174–84. (Very good bibliography.)

Prins, Marina. 1987. *Tabernanthe iboga*, die vielseitige Droge Äquatorial-Westafrikas: Divination, Initiation und Besessenheit bei den Mitsogho in Gabun. In *Ethnopsychotherapie*, ed. A. Dittrich and Ch. Scharfetter, 53–69. Stuttgart: Enke.

Samorini, Giorgio. 1993. Adam, Eve and iboga. *Integration* 4:4–10.

———. 1995. The Buiti religion and the psychoactive plant *Tabernanthe iboga* (equatorial Africa). *Integration* 5:105–14.

———. 1996a. Adam, Eva e iboga: Mi experiencia con los Bwitis del Gabón. *Takiwasi* 4 (2): 63–75.

———. 1996b. El rito de iniciación a la religión Buiti (Secta Ndea Narizanga, Gabon). Lecture given at the *II° Congrès International per l'Estudio dels Estats Modificats de Consciència*, 3–7 October 1994, Llèida.

———. 1996c. Visionary eye-drops. *Eleusis* 5:27–32.

———. 1997. Una bibliografia commentata sulla religione Buiti. *Eleusis* 7:3–16.

———. 1998. The initiation rite in the Bwiti religion (Ndea Narizanga sect, Gabon). *Yearbook for Ethnomedicine and the Study of Consciousness*, 1997/1998 (6–7): 39–55. Berlin: VWB.

Swiderski, Stanislaw. 1964. Symbol- und Kultwandel des Geheimbundes "Bwiti" in Gabun. *Anthropos* 59 (5/6).

———. 1965. Le Bwiti, société d'initiation chez les Apindji au Gabon. *Anthropos* 60: 541–76.

———. 1970. La harpe sacrée dans les cultes syncrétiques en Gabon. *Anthropos* 65:833–57.

———. 1972. Die sakrale Verzierung der Tempel in den synkretistischen Sekten in Gabun. *Mitteilungen der Anthropologischen Gesellschaft in Wien* 102:105–13.

———. 1981. Les visions d'iboga. *Anthropos* 76:393–429.

———. 1982. Le rite mortuaire pour un initié au Bouiti. *Anthropos* 77:741–54.

———. 1990. *La religion Bouiti.* 5 vols. New York, Ottawa, and Toronto: Legas.

Vonk, G. J. A., and A. J. M. Leeuwenberg. 1989. A taxonomic revision of the genus *Tabernanthe* and a study of wood anatomy of *T. iboga. Wageningen Agricultural University Papers* 89 (4): 1–18.

Westerhout, Hans. 1996. Een Vision Circle met ibogaine. *Pan* 3:9–12.

Tagetes spp.

Marigolds

Family

Compositae: Asteraceae (Aster Family); Helecieae Tribe

Species of Ethnobotanical Significance

Tagetes erecta L.—cempoalxóchitl, flor de los muertos (the subspecies *Tagetes erecta nana* is primarily used for cultivation purposes; *Tagetes erecta* hybrids)

Tagetes lucida Cav.—yauhtli

Tagetes minuta L. [syn. *Tagetes glandulifera* Schrank]—wild marigold

Tagetes patula L.—French marigold

Tagetes pusilla H.B.K. [syn. *Tagetes filifolia* Lag., *T. congesta* Hook et Arn., *T. multifida* DC.]—pampa anis

Folk Names

Most of these names are used for all species of *Tagetes*: anisillo, belbop (Nepali), cempoal, cempoalxóchitl, flor de los muertos (Spanish, "flower of the dead"), flor de tierradentro (Spanish, "flower of the underworld"), gainda, gendha, hierba anis, hierba de nubes (Spanish, "herb of the clouds"), marigold, marygold, pericirituela de muerto (Spanish, "rose of the dead"), sammetblume, Santa María,[299] sempoalxochitl, stinkende hofart (Switzerland), studentenblume, tagète, yerbanis

History

Tagetes species were bred and cultivated in pre-Columbian Mexico. They were first described in the Aztec Badianus manuscript of 1552 (Emmart 1940*). The Aztec name for those *Tagetes* species that were used as ritual incense is *yyauhtli* (also written *yyahitl*); the word is derived from *ujana*, "to offer incense in sacrifices" (Siegel et al. 1977, 20). The Spanish physician Fernando Hernandez wrote in his colonial-period work that *Tagetes* stimulates sexual desire and relieves the insane.

Most *Tagetes* species (especially *T. erecta* and *T. patula*) quickly spread throughout the world as ornamental plants. In India and Nepal, they even took on ritual significance as flowers for offering to the goddess Bhagwati and to Shiva (Majupuria and Joshi 1988, 221*). *Tagetes* has also acquired a certain significance as "American saffron," used to counterfeit the considerably more expensive true saffron (see **Crocus sativus**).

Distribution

All species of *Tagetes* originated in the Americas, where they occur from the North American Southwest to Argentina (Ferraro 1955). The main area of distribution and the area in which the greatest variety can be found is in southern Mexico (Neher 1968, 317). *Tagetes lucida* is very common in Nayarit and Jalisco at altitudes of up to 2,100 meters (Siegel et al. 1977, 20). *Tagetes*

"To dispel evil spirits, the Garinagu prepare a mixture of orange peels and marigolds. At burials, the helpers wash their hands in a water to which the flowers were added. The priests at the ceremonies of the Maya wash their hands and faces with a decoction of the leaves so that they will be better able to invoke the spirits."

ROSITA ARVIGO AND MICHAEL BALICK
DIE MEDIZIN DER REGENWALDES
[THE MEDICINE OF THE RAIN FOREST]
(1994, 147*)

Macuilxochitl ("five flowers"), a manifestation of Xochipilli, the Aztec god of inebriating plants (*Florentine Codex*, book 1; sixteenth century). One *Tagetes* species bore his name.

Top left: The typical flower of *Tagetes erecta*.

Bottom left: *Tagetes erecta* with filled flowers.

Top right: An interesting type of *Tagetes* aff. *erecta* from Nepal.

Bottom right: The fresh herbage of *Tagetes lucida* has a very pleasant scent.

299 The same name is used for **Piper auritum**.

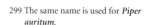

Left: The diminutive *Tagetes minuta*.

Right: One of the many cultivated sorts of *Tagetes* aff. *patula*.

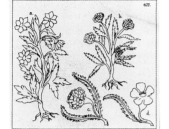

Illustration of the variety of *Tagetes erecta* known as *cempoalxóchitl* (b) in the Aztec-language work of Fra Bernardino de Sahagun. (Paso y Troncoso edition)

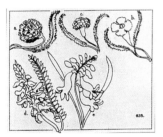

Illustration of the variety of *Tagetes erecta* known as *macuilxochitl* (a) in the Aztec-language work of Fra Bernardino de Sahagun (Paso y Troncoso edition); *macuilxochitl* is the flower of Xochipilli.

erecta is originally from Mexico (Dressler 1953, 147 f.*).

Cultivation
All *Tagetes* species are easy to grow from seed. The seeds are scattered onto the ground in March or April and covered only lightly with soil.

Appearance
Most of the species discussed here are available in numerous cultivated forms and strains (Kaplan 1960). They are often difficult to distinguish from one another. This difficulty is compounded by the fact that most plants have double flowers (Graf 1992, 330*). The plants, which grow from 20 to 50 cm in height, almost always have yellow flowers that either have five distinct petals or are filled to some degree, and most have pinnate leaves. All *Tagetes* species exude a strong, sometimes pungently "medicinal" scent.

Psychoactive Material
— Flowering herbage

Preparation and Dosage
The herbage can be infused, boiled, or ground to produce a paste. Bundles of fresh or dried flowering herbage of *Tagetes lucida* are sold in markets in Mexico; these may be used as a spice for preparing food (the aromatic herbage is used to flavor maize dishes), as a remedy, or as a ritual plant (Bye and Linares 1983, 6 f.*).

Two to three cups of an aromatic tea (infusion of one bundle) made from *Tagetes lucida* are sufficient to produce potent stimulant effects (Neher 1968, 321).

The leaves of *Tagetes pusilla*, which is known in southern Peru as *pampa anis* ("prairie anis"), are heated until they turn to ash and then added to coca quids (see *Erythroxylum coca*).[300]

In Lesotho (Africa), the leaves of *Tagetes minuta* are burned to ash and then finely ground with tobacco leaves (*Nicotiana tabacum*), an *Aloe* species, maize cobs, and millet stalks (*Sorghum* spp.) to produce a (medicinal?) **snuff** (Neher 1968, 320).

Unfortunately, no information is available concerning precise dosages.

Ritual Use
The Mexican Indians have attributed *Tagetes* species with magical properties since pre-Columbian times. One variety of *Tagetes erecta* with filled flowers is known in Aztec as *macuilxochitl* (according to Sahagun). *Macuilxochitl* (also spelled *macuilsuchitl*), "five flower," is a manifestation of Xochipilli, the god of psychoactive plants (Nicholson 1967*). The Maya used this flower as an additive to the sacred balche' drink. It is said that contemporary Mayan shamans still use the plant they call *xpuhuc* (*tagetes lucida*) as an inebriant (Ott 1993, 402*). The Mixe of Oaxaca drink a tea made from nine flowers for divination (Lipp 1991*).

In Mexico, the flowers of *Tagetes erecta* and *Tagetes patula* are known as *flores del muerto*, "flowers of the dead." They are offered to the dead the night of the All Saints' Day festival (November 1). The blossoms of these species are also used as flower offerings in many Hindu ceremonies in India and Nepal.

The Aztecs referred to *Tagetes lucida* as *yauhtli*, "plant of the clouds." They would sprinkle a powder of the plant into the faces of war prisoners who were to be burned as sacrifices so that they would be sedated during their ordeal. Even today, many Mexican Indians burn the dried herbage of *Tagetes lucida* as an **incense** on their house altars or at public ceremonies (Neher 1968, 322).

The Huichol Indians of the Sierra Madre (Mexico) call *Tagetes lucida* either *tumutsáli* or, more rarely, *yahutli*. They smoke the dried herbage alone or mixed with equal parts of the leaves of **Nicotiana rustica**. Although *Tagetes* is also smoked recreationally, the mixture does have a ceremonial character. The leaves and flowers are smoked in cigarettes made from corn husks. The **smoking blend** is often smoked in combination with the ingestion of peyote (**Lophophora williamsii**), *tesquino* or *nawa* (maize **beer**), or homemade *cí* or *soter* (cactus liquor; cf. **alcohol**). These combinations are said to produce lively hallucinations. Bundles of the dried herbage are placed as offerings in temples, administrative buildings, and sacred sites (Siegel et al. 1977, 20).[301]

300 The fresh leaves of this *Tagetes* species are also chewed alone as a cold remedy (Plowman 1980, 254).

301 The information that Siegel et al. published has not been verified by any other ethnographic research (Stacy Schaefer, pers. comm.).

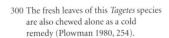

A number of herbs are used in Mexican *brujería* ("witchcraft") for *limpias*, "purifications," to dispel diseases. *Tagetes lucida* and *Tagetes erecta* are among these.

Artifacts

Representations of flowers having five petals are often found in pre-Columbian art. It is possible that some of these are depictions of *Tagetes* species. The Museo Carlos Pellicer Camara (Villahermosa, Tabasco) has on display a cylindrical polychrome ceramic vessel from the Classic Mayan period (300–900 C.E.) that depicts a yellow flower whose form and color suggest that it may represent *Tagetes lucida*.

In Mexican folk art, a wide variety of skulls, skeletons, et cetera, are produced for such purposes as the All Saints' Day festivals. These objects of wood, papier-mâché, or sugar are sometimes decorated with painted *Tagetes* flowers.

The album *IN Mixkoakali* (Cademac Records, 1996), by the Mexican music group Tribu, includes "Sempoalxochitl," a piece played with pre-Columbian instruments that is dedicated to the "flower of twenty scents" (*Tagetes erecta*).

Medicinal Use

The Aztecs used all species of *Tagetes* for medicinal purposes, e.g., to treat hiccups and diarrhea. People who had been struck by lightning were treated with extracts of *Tagetes lucida*.

Today, the fresh herbage of *Tagetes lucida* is made into a tea that is drunk for abdominal pains (Bye and Linares 1983, 8*). In Mexico, it is believed that the herbage promotes lactation (Jiu 1966, 252*). It is also used as a bath additive to treat rheumatism (Siegel et al. 1977, 20).

In Uttar Pradesh (India), juice that has been freshly pressed from *Tagetes erecta* leaves is administered to treat eczema (Siddiqui et al. 1989, 482*).

In Mexico, the crushed leaves or the juice that has been pressed from the herbage of *Tagetes erecta* is mixed with water or wine (pulque; cf. *Agave* spp.) and drunk as an aphrodisiac (Neher 1968, 318). A tea of the plant is used as a stimulant.

In Argentina, a decoction of the leaves of *Tagetes minuta* L. is drunk for coughs (Filipov 1994, 186*). It is also a well-known insect repellent.

Constituents

All *Tagetes* species contain potently aromatic essential oils. *Tagetes lucida* and *Tagetes erecta* contain salvinorin-like substances (see **salvinorin A**) whose structures have not yet been fully clarified. Also present are thiophene compounds, e.g., α-terthienyl (Roth et al. 1994, 689*). Benzofurans are present in *Tagetes patula* (Sütfeld et al. 1985).

The fresh, flowering herbage of *Tagetes minuta* contains mono- and sesquiterpenes (carvone, linalool, tagetone) as well as ocimenone. (5E)-ocimenone has been found to have lethal effects upon the larvae of the mosquito species *Aedes aegypti* (Maradufu et al. 1978).

Effects

Research is still needed to clarify the mechanism of activity of *Tagetes* species. When the eyes are closed, the Huichol **smoking blend** is said to be able to produce images and visions similar to those produced by peyote (***Lophophora williamsii***) (Siegel et al. 1977, 20 f.).

Commercial Forms and Regulations

The seeds of many *Tagetes* species, breeds, and cultivars as well as living plants can be obtained from any well-stocked flower store or nursery.

Literature

Ferraro, Matilde. 1955. Las species Argentinas del genero Tagetes. *Boletín de la Sociedad Argentina de Botánica* 6 (1): 30–39.

Kaplan, Lawrence. 1960. Historical and ethnobotanical aspects of domestication in *Tagetes. Economic Botany* 14:200–202.

Maradufu, Asafu, Richard Lubega, and Franz Dorn. 1978. Isolation of (5E)-ocimenone, a mosquito larvicide from *Tagetes minuta. Lloydia* 41:181–83.

Neher, Robert Trostle. 1968. The ethnobotany of Tagetes. *Economic Botany* 22:317–25.

Siegel, Ron K., P. R. Collings, and José L. Diaz. 1977. On the use of *Tagetes lucida* and *Nicotiana rustica* as a Huichol smoking mixture. *Economic Botany* 31:16–23.

Sütfeld, Rainer, Felipe Balza, and G. H. Neil Towers. 1985. A benzufuran from *Tagetes patula* seedlings. *Phytochemistry* 24 (4): 876–77.

"Yellow flowers open wide;
It is our mother, she with the masked face.
You, who have come from Tamoancan.
Yellow are your flowers."

AZTEC PRAYER
IN *LA LITERATURA DE LOS AZTECOS*
(A. GARIBAY, 1971)

The earliest European illustration of *Tagetes erecta*, which was introduced from Mexico. (Woodcut from Fuchs, *Kreüterbuch,* 1543)

Tanaecium nocturnum (Barb.-Rodr.) Bureau et K. Schum.

Koribo

The flowers of the tropical climber *Tanaecium nocturnum* open only at night and exude a scent like that of almonds. (Photographed in South America)

Family
Bignoniaceae (Bignonia Family)

Forms and Subspecies
None

Synonyms
None

Folk Names
Huangana huasca, hutkih (Lacandon),[302] koribo, koribó, koriboranke, pum-ap, puu tʰotʰo moki (Yanomamö), samedu-ap

History
It was only in the 1970s that any information about the ethnomedicinal and ritual uses of this plant by the Indians of the Americas became known.

Distribution
The tropical plant occurs in Amazonia, the West Indies, Central America, and southern Mexico (Yucatán).

Cultivation
The Paumarí Indians occasionally use cuttings to propagate the plant. This method is quite new, as the Paumarí were originally a nomadic tribe that has now been made sedentary (Prance et al. 1977, 131 f.). The plant presumably can be grown rather easily from seed.

Appearance
This climbing bush has cordate leaves and long, trumpet-shaped white flowers that curl up in the sun. In the evenings, the flowers exude a delicious scent, like that of almond oil.

The plant is easily confused with the white vine (*Ipomoea alba* L.; cf. **Ipomoea** spp.).

The closely related Colombian species *Tanaecium exitiosum* Dun. is dangerous to cattle. In Venezuela, a comparable species (*Tanaecium crucigerum* Seemann) is known, astonishingly enough, as *borrachera* (Blohm 1962, 97*)—the same name that is given to many other psychoactive plants (e.g., **Brugmansia** spp., **Iochroma fuchsioides**, **Pernettya** spp., et cetera).

Psychoactive Material
— Leaves and stems
— Root cortex

Preparation and Dosage
The green leaves are roasted, mixed with tobacco (*Nicotiana tabacum*), and ground for use as a **snuff** (Prance 1978, 72).

Ritual Use
The Paumarí (= Ija'ari) on the Purus River use the leaves to prepare a snuff known as *koribón-nafuni* that is used only in rituals. Only men use this snuff. Shamans snuff it when treating special cases, e.g., to extract magical objects (grasshoppers, pieces of wood, bones) from a patient's body. They also snuff it to enter a trance state as part of a ritual to protect children that is carried out whenever a child is learning to eat a new kind of food (e.g., an animal). This ritual is accompanied by sacred songs and begins with the snuffing of *koribo*. Men use bird bones to snuff the powder during puberty rites for girls (Prance et al. 1977, 131). Although women never use the snuff, they may use a tea that they prepare from the root cortex (two teaspoons per person) (Prance 1978, 72).

The Lacandon, who live in the rain forest of southern Mexico, use the latex of the stem of the plant, which they call *hutkih*, as a vulcanizing agent during the ritual manufacture of caoutchouc (rubber) figures known as *tulis k'ik'*, "full blood." These are offered to the gods and also are used in magical rites (Bruce 1974; Rätsch 1985, 128*).

Artifacts
None (apart from the rubber figures of the Lacandon)

Medicinal Use
The Karitiana Indians of Porto Velho (Brazil) mix the leaves with the leaves of a Leguminosae to treat diarrhea (Prance et al. 1977, 134). The Chocó Indians use the plant as an aphrodisiac. The Wayãpi Indians (Guyana) boil the bark and/or stems and use the decoction to bathe diseased areas. The Palikur make a decoction of the leaves and stalks that they use to bathe the head to treat migraines. The Brazilian Yanomamö (Yanomami) boil the leaves and rub the juice they press from them onto itchy areas of the skin (Milliken and Albert 1996, 18, 19).

The Colombian Creoles believe that the plant is efficacious in treating lung ailments. They also use leaf extracts to remove lice and fleas from their pets (Duke and Vasquez 1994, 165 f.*).

---

302 The name *hutkih* is given to all white-flowering scented vines, e.g., *Ipomoea alba* (cf. **Ipomoea** spp.).

Constituents

The leaves contain high concentrations of hydro-cyanic acid (HCN), which explains why they smell like bitter almonds (Duke and Vasquez 1994, 166*). They also contain toxic cyanoglycosides. Apart from this, the chemistry is essentially unknown. The cyanoglycosides appear to be destroyed when the leaves are roasted (D. McKenna 1995, 101*). Roasting the leaves also breaks down the hydrocyanic acid.

Effects

The **snuff** is said to induce somnambulant states entailing drowsiness, disturbances of concentration, and a clouding of consciousness (Müller 1995, 197*). The tea also produces "an inability to concentrate and reduces awareness" (Prance et al. 1977, 131). The Indians, on the other hand, describe the effects as identical to those of the *kawabó* snuff that is made from *Virola elongata*

(Benth.) Warb. (see *Virola* spp.) (Prance et al. 1977, 134).

Commercial Forms and Regulations

None

Literature

Bruce, Robert D. 1974. Figuras ceremoniales lacandones de hule. *Boletín* (INAH): 25–34.

Milliken, William, and Bruce Albert. 1996. The use of medicinal plants by the Yanomami Indians of Brazil. *Economic Botany* 50 (1): 10–25.

Prance, Ghillian T. 1978. The poisons and narcotics of the Dení, Paumarí, Jamamadí and Jarawara Indians of the Purus River region. *Revista Brasileira do Botanica* 1:71–82.

Prance, Ghillian T., David G. Campbell, and Bruce W. Nelson. 1977. The ethnobotany of the Paumarí Indians. *Economic Botany* 31:129–39.

Tulis k'ik', "the full blood," an offering figure of the Lacandon of Naha' made from rubber and the juice of the koribo vine.

Theobroma cacao Linnaeus

Cacao Tree

Family

Sterculiaceae (Cocoa Family); Byttnerieae Tribe

Forms and Subspecies

Like all cultivated plants with a long history, *Theobroma* is a quite variable plant, especially with regard to the color, shape, and size of its fruits and the seeds (Baumann and Seitz 1994, 943). A number of subspecies, varieties, and forms have been described:

Theobroma cacao ssp. *cacao* (L.) Cuatr.—criollo
Theobroma cacao ssp. *sphaerocarpum* (Chevalier) Cuatr.—forastero, calabacillo, amelonado
Theobroma cacao var. *catonga*
Theobroma cacao f. *lacandonense* Cuatr.—balamte' "jaguar tree") (a wild form)
Theobroma cacao f. *leiocarpum* (Bernoulli) Ducke—porcelaine java criollo, cacao calabacillo
Theobroma cacao f. *pentagonum* (Bernoulli) Cuatr.—alligator cacao, cacao lagarto

The form *lacandonense* is a semi-climbing wild shrub with relatively small fruits that occurs in the primary forest of the Selva Lacandona (Chiapas, Mexico). The form is regarded as the natural precursor of the cultivated cacao tree (Baumann and Seitz 1994, 943).

On commercial plantations, a distinction is made between essentially only two cultivated sorts:

forastero and *criollo*. The former is planted chiefly in Brazil and Africa, the latter in Central America. A number of hybrids have been produced from the two, and these are named after the places where they are cultivated, viz., Guayaquil, Caracas, Bahia, and Accra.

Synonyms

Cacao guianensis Aubl.
Cacao minus Gaertn.
Cacao sativa Aubl.
Theobroma caribaea Sweet
Theobroma interregima Stokes
Theobroma kalagua De Wild.
Theobroma leiocarpa Bernoulli
Theobroma pentagona Bernoulli
Theobroma saltzmanniana Bernoulli
Theobroma sapidum Pittier
Theobroma sativa (Aubl.) Lign. et Le Bey
Theobroma sphaerocarpa Chevalier

Folk Names

Äh kakaw (Lacandon), aka-'i (Ka'apor), aka-'iwa (Ka'apor), aka-'iwe-te (Ka'apor), ako'o-'i (Ka'apor), bana torampi (Shipibo), biziáa (Zapotec), bizoya, cacahoaquiahuit, cacahoatl, cacahua, cacahuatl, cacao, cacaocuáhuitl (Aztec), cacaotero, cacao tree, cacau, cacauatzaua (Zoque), cacauaxochitl (Aztec, "cacao flower"),[303] cacayoer, caco (Mixe), cágau (Popoluca), cajecua (Tarascan), chocolate, chudechú

This cacao tree (*Theobroma cacao*) is covered with an abundance of red fruits.

[303] This name is said to have been used for *Theobroma angustifolium* DC. as well (Furst 1995, 121*).

(Otomi), cocoa tree, haa (Maya), hach kakaw, kahau, kaka (Ka'apor), kakao, kakaobaum, ma-micha-moya (Chinantec), ma-mu-guía, mochá (Chinantec), palo de cacao, pizoya (Zapotec), quemitoqui, sarhuiminiqui, schokoladenbaum, sia (Cuna), si'e (Siona), tlapalcacauatl (Aztec, "colored cacao"), torampi (Shipibo-Conibo), turampi (Quechua), turanqui, tzon xua, xocoatl, yaga-bisoya (Zapotec), yaga-pi-zija, yau

History

The cacao tree was cultivated in Central America some four thousand years ago. There, it was venerated as a food of the gods and was consumed during rituals and offered to the gods. Linnaeus took this fact into consideration when he named the tropical plant *Theobroma cacao*. *Theobroma* means "gods' food," and *cacao* is a word borrowed from the Mayan language that refers to the tree, its fruit, and the drink prepared from the fruit. The word *chocolate* is derived from the Aztec *xocolatl*, a name for the beverage. Solid chocolate appears to have been a Swiss invention.

The Aztec held cacao beans in high regard. They used them as food, stimulants, medicine, and even currency (especially for paying prostitutes). They were revered as a food of the gods. The psychoactive effects of cacao were described in the Aztec-language texts of Bernardino de Sahagun (Ott 1985).

The conquistador Hernán Cortés brought the first cacao beans to Europe, where they were initially used almost exclusively in the production of love drinks. The first book about cacao, titled *Libro en el cual se trata del chocolate* [Book on the Preparation of Cacao], was published in New Spain (Mexico) in 1609. In 1639, a book appeared in Europe in which it was claimed that the sea god Neptune had brought chocolate from the New World to Europe (Morton 1986, 10, 14). Today, cocoa for drinking and the various kinds of chocolate are among the most commonly consumed foods and/or agents of pleasure in the world.

Distribution

The wild form of the plant is known only in southern Mexico. The cultivated cacao tree had

spread into all of the tropical rain forests of the Americas by prehistoric times. Today, there also are large occurrences of the plant in Africa and Southeast Asia, where it is grown as a crop.

Cultivation

Cacao is propagated from fresh seeds, which are pregerminated in plantations and allowed to develop into small trees before being planted into the ground. This tropical tree grows only in the tropics in areas that receive at least 130 cm of precipitation per year. The cacao tree is a shade-loving plant that does not tolerate exposure to direct sunlight. Because of this, modern cacao plantations plant groves of bananas (known as "cocoa mothers") next to the young trees (Morton 1986, 57–58). The first harvest occurs after the eighth year, and thereafter a rich harvest can be obtained at least twice a year.

In ancient Nicaragua, cacao farmers were required to abstain from sex for thirteen days before planting the seeds so that they would not make the god of chocolate (= moon god) angry, thereby protecting their harvest.

Appearance

This evergreen tree can attain a height of around 15 meters and can live for some sixty years. The tiny white, pink, or violet flowers grow directly from the trunk or the thicker primary branches, often at the same time as the pods (fruits), which hang from the trunk on short stems. A single tree can develop approximately one hundred thousand flowers annually. The pods are initially green and then turn yellow, red, or purple as they mature.

Psychoactive Material

— Cacao beans (cacao semen, avellanae mexicanae, faba cacao, fabae mexicanae, nuclei cacao, semen cacao, semen cacao tostum, semen theobromae, theobromatis semen, kakaosamen, cocoa beans)
— Cocoa shells (cacao cortex, cortex cacao, cortex cacao tostus, testae cacao, kakaotee)
— Cocoa butter (cacao oleum, butyrum cacao, oleum cacao, oleum theobromatis, kakaofett)
— Fresh fruit pulp (for brewing **beer** or **chicha**)

Preparation and Dosage

The Indians prepared a cacao mixture from roasted and ground cacao beans, cornmeal (cf. *Zea mays*), honey (from wild bees), vanilla, allspice, and chili pods. In former times, a variety of spices (cinnamon/canella, vanilla, almonds, pistachios, musk, nutmeg [cf. *Myristica fragrans*], cloves, allspice, anise) usually were added to the cocoa as well.

The Spanish conquistador Hernán Cortés is said to have brought the following cacao recipe to Spain in 1528 (from Montignac 1996, 27):

Top left: The flowers and fruits of the cacao tree grow directly from the trunk.

Bottom left: The ripe cacao fruit (*Theobroma cacao*).

Bottom right: The seeds of the cacao fruit are known as cacao beans.

700 g cacao

750 g white sugar

56 g (= 2 ounces) "cinnamon" (canella, perhaps *Canella winterana*)[304]

14 Mexican peppercorns (*Capsicum* spp.)

14 g "spice cloves" (*Pimenta dioica*)

3 vanilla pods

1 handful "anise" (probably *Tagetes lucida*)

1 hazelnut

musk, gray amber, and orange blossom water

One important ingredient in the traditional Indian preparation was the cacao flowers, which came from *Quararibea funebris* and not *Theobroma cacao*. Today, the cacao drink prepared with *Quararibea* flowers is called *tejate* (West 1992, 106). Generally speaking, cacao appears to have served an important function as a vehicle for administering other psychoactive plants and fungi (Ott 1985).

In the seventeenth and eighteenth centuries, a *succolade* made from powdered cacao beans, sugar, and **wine** was drunk in Germany, sometimes heavily fortified with cardamom (cf. **essential oils**) and saffron (*Crocus sativus*) (Root 1996, 364*).

At the beginning of the twentieth century, invigorating drinks made of cacao and *Catha edulis* were made in London and sold as Catha-Cocoa Milk. Preparations of cacao and *Cola* spp. or *Coffea arabica* are now popular. In Switzerland, a special chocolate with powdered *Psilocybe semilanceata* is being manufactured clandestinely.

Cacao shells can be brewed to make a tea. A normal dosage is 2 to 4 g per cup of water (Baumann and Seitz 1994, 946).

To date, no incidents of overdosing on cacao are known.

Ritual Use

Numerous archaeological finds demonstrate that the ritual use of cacao—as an offering, incense, or inebriant—must be very ancient in Mesoamerica. Among the prehistoric Toltecs, a cacao branch was placed in the hand of every person who made a public smoke offering to the gods as a sign of his religious respect.

The Aztecs viewed the cacao tree as a gift from their peace loving god Quetzalcoatl ("feathered serpent"). An Aztec text from the early colonial period provides a precise description of the tree and of the drink, which could be inebriating:

Cacaoaquavitl—Cacao Tree

It has broad branches. It is simply a round tree. Its fruit is like the ears of dried maize, like an ear of green maize. Its name is "cacao ear." Some are reddish brown, some whitish brown, some bluish brown. Its heart, that which is inside it, its filled insides, is like an ear of maize. The name of this when it grows is *cacao*. This is edible, is drinkable.

Aztec Additives to Cacao

(from Dressler 1953, 149*; Heffern 1974*; Navarro 1992, 124*; Ott 1993*; Reents-Budet 1994, 77–79; supplemented)

Aztec Name	Botanical Name	Active Constituent
cacahuaxochitl	*Quararibea funebris* (Llave) St. (see *Quararibea* spp.)	
teonanacatl	*Psilocybe mexicana* Heim *Psilocybe aztecorum* Heim *Psilocybe* spp.	psilocybin psilocybin psilocybin
achiotlín (matico)	*Piper angustifolium* Ruíz et Pav. (see *Piper* spp.)	**essential oil**, maticin, resins
mecaxochitl (mecaxuchitl)	*Piper* sp. [*amalago* L. ?]	**essential oil**
hueynacaztli/teonacaztli/ xochinacaztli	*Cymbopetalum penduliflorum* (Dun.) Bail. or *Enterolobium cyclocarpum* (Jacq.) Griseb.	alkaloids? tryptamines?
chili	*Capsicum annuum* L. **Capsicum** spp.	capsaicin
tlilxóchitl (vanilla)	*Vanilla planifolia* Andr. [syn. *V. fragrans* (Salisb.) Ames]	vanillin, **essential oil**
tecomaxochitl	**Solandra** spp.	**tropane alkaloids**
xocoxóchitl	*Pimenta dioica* (L.) Merr. [syn. *Pimenta officinalis* Lindl.]	**essential oil** (eugenol and others)
cempoalxóchitl / yauhtli	*Tagetes lucida* Cav. (see *Tagetes* spp.)	**essential oil**
tlacoxiloxochitl	*Calliandra anomala*	alkaloids

Top: The so-called cacao flowers come from the rare tree *Quararibea funebris* and are used to flavor cacao.

Bottom: Matico (*Piper angustifolium*) is used to flavor traditional Aztec cacao.

304 In Veracruz, the cacao flower tree (*Quararibea funebris*) is called *canela* (Martínez 1987, 1199*). *Calliandra anomala* is known as *canelo* in Mexico. For more on canella, *canelo*, et cetera, see **incense**.

"Quetzalcoatl ruled in Tula. Abundance and happiness were complete. No one paid money for food, nor for the things that were necessary for life. The gourds were so large and fat that a man could barely carry them under his arm. The ears of maize were as large and full as the grip of the stone that was used to grind maize and cacao. . . . Even jewels and gold were given without one having to give anything in return, in such abundance were they present. Cacao thrived magnificently. Cacao plants could be seen everywhere. All of the inhabitants of Tula were rich and happy, they suffered from no need, nothing was missing from their houses."

AZTEC TEXT
IN *LITERATURA DEL MÉXICO ANTIGUO*
(M. LEON-PORTILLA 1958)

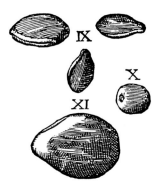

An early modern illustration of the cacao bean. (Woodcut from Tabernaemontanus, *Neu Vollkommen Kräuter-Buch*, 1731)

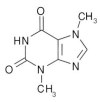

Theobromine

This cacao, when much is drunk, when one consumes much of it, especially that which is green, which is tender, makes one drunk, has an effect upon one, makes one ill, makes one confused. If a normal amount is drunk, it makes one happy, refreshes one, comforts one, strengthens one. Thus it is said: "I take cacao. I moisten my lips. I refresh myself." (Sahagun, 11)

The Aztecs ingested cacao or chocolate together with entheogenic mushrooms (*Psilocybe* spp.) in associated rituals, a practice that is still found among numerous tribes today (West 1992, 106).

The Yucatec Maya venerate a black god named Ek Chuah (= God M, presumably identical to Yacatecuhtli, the long-nosed Aztec god of merchants) as the cacao god. Cacao farmers held a festival in his honor during the month of Muan in the old Mayan calendar. During travel, incense (consisting perhaps of cacao beans and copal) was offered to effect a safe return. Ek Chuah was frequently depicted on incense vessels. The glyph of the god's name was a free-floating eye (Taube 1992, 88 ff.). The Maya and Lacandon use freshly whipped cacao as a ritual additive to **balche'**.

The shamans of the Cuna Indians of Panama (Darien) also use cacao beans as a ritual **incense**. The healers use it in their diagnoses. First, a clay incense vessel with two handles on its sides is filled with glowing charcoal. The shaman then scatters cacao beans onto the charcoal and peers into the ascending smoke. The shaman reads the patient's illness in the behavior and structure of the smoke. Cacao beans are burned as incense at almost every Cuna ritual occasion and tribal ceremony. The cacao smoke also finds medicinal use. The beans are mixed with chili pods (*Capsicum fructescens* L.; see *Capsicum* spp.) and then burned; this pungent smoke is said to promote healing for all types of fever diseases, including malaria (Duke 1975, 293*).

Artifacts

The Maya of the Classic Period (300–900 C.E.) left behind a rich trove of ritual drinking vessels. These polychrome ceramics are artistically decorated with hieroglyphic texts and varied depictions of visionary experiences and ritual activities. Many of the hieroglyphic texts found on such drinking vessels have now been deciphered. The owner of the vessel is frequently named, and the text then notes that "the vessel [was used] for cacao freshly picked from the tree," indicating that these drinking vessels were directly associated with the ritual ingestion of cacao (MacLeod and Reents-Budet 1994). The hieroglyph for cacao is a stylized monkey head.

Cacao fruits are frequently depicted in Aztec and related art. The tree, its fruits, and the drink prepared from the fruits are often depicted in the

illustrated manuscripts of numerous Meso-american peoples.

In the 1960s, an American psychedelic band took the name Chocolate Watch Band. Its music, however, was probably influenced by other types of drugs.

Medicinal Use

In ancient America, cacao was esteemed as a tonic and aphrodisiac. In Indian folk medicine, cacao is drunk to treat diarrhea and scorpion stings. Cuna women drink a decoction of the fruit pulp as a pregnancy tonic. Listless children are given a tea made from the leaves, and fresh, young leaves are applied externally as an antiseptic agent (Duke 1975, 293*). In Peru, cacao is drunk primarily as a diuretic and in cases of kidney infections (Chavez V. 1977, 322*).

In homeopathy, the mother tincture obtained by macerating the roasted seeds (Cacao hom. *HPUS88*) occasionally finds use (Baumann and Seitz 1994, 946).

The debate as to whether chocolate is harmful or beneficial to health continues today (Fuller 1994). A recently published book written by a physician argues that chocolate is very healthful for people:

Today, there is a moral obligation to disseminate its extraordinary nutritional properties everywhere, which together with its preventive effects especially in the area of cholesterol make it a beneficial and wholesome food source that should be used frequently and which, in the proper amounts, is to be recommended as a regular component of the diet. (Montignac 1996, 198)

Constituents

Cacao beans contain 18% protein, 56% lipids (fat), 13.5% carbohydrates, 1.45% theobromine, 0.05% **caffeine**, and 5% tannin (Montignac 1996, 203). Theobromine is also found in *Ilex cassine* and *Ilex guayusa*. Also present is theophylline, a compound with a similar structure, and β-phenethylamine, tyramine, tryptamine, serotonin, and catechin tanning agents (especially in the shell). Dried and roasted cocoa shells can contain up to 0.02% caffeine and 0.4 to 1.3% theobromine (Baumann and Seitz 1994, 946).

The leaves also contain the methylxanthines theobromine and caffeine. The concentrations vary depending upon the source but typically comprise less than 1% of the dry weight. They also contain chlorogenic acid and rutoside (Baumann and Seitz 1994, 944).

It has recently been discovered that cacao also contains anandamides (see **THC**).

Effects

The psychoactive effects described in the Aztec sources may be due to the cacao additives or to a synergism with the added substances. I have found the traditionally prepared Indian drink to be very stimulating and euphoric. These effects are not necessarily to be expected with commercial cocoa.

Chocolate is popularly referred to as "brain nourishment" or "nerve food." Moderate to high use clearly leads to an improvement in mood and other beneficial effects. This is attributed to the theobromine but may also result from the anandamide (cf. **THC**). Theobromine can apparently produce a kind of dependency (a so-called chocolate addiction).

It is unlikely that the smoke produced by burning cacao beans will have any pharmacological effects.

Commercial Forms and Regulations

Cacao is subject only to the laws governing foods and is freely available. Internationally, the *criollo* variety is especially prized, as it is regarded as superior in quality. In Europe, it is sometimes sold under the name *Mayan chocolate*.

Literature

See also the entries for **Theobroma spp.** and **caffeine**.

Baumann, Thomas, and Renate Seitz. 1994. *Theobroma.* In *Hagers Handbuch der pharmazeutischen Praxis*, 5th ed., 6:941–55. Berlin: Springer.

Bühler, Margrit. 1987. *Geliebte Schokolade*. Aarau and Stuttgart: AT Verlag.

Cuatrecasas, José. 1964. Cacao and its allies: A taxonomic revision of the genus *Theobroma*. *Contribution of the U.S. National Herbarium* 35 (6).

Fuller, Linda K. 1994. *Chocolate fads, folklore, and fantasies*. New York: The Haworth Press.

MacLeod, Barbara, and Dorie Reents-Budet. 1994. The art of calligraphy: Image and meaning. In *Painting the Maya universe: Royal ceramics of the Classic Period*, ed. Dorie Reents-Budet, 106–63. Durham, N.C. and London: Duke University Press.

Mitscherlich, A. 1859. *Der Cacao und die Chocolade*. Berlin: A. Hirschwald.

Montignac, Michel. 1996. *Gesund mit Schokolade*. Offenburg: Artulen-Verlag.

Morton, Marcia, and Frederic Morton. 1986. *Chocolate: An illustrated history*. New York: Crown Publishers.

Ott, Jonathan. 1985. *Chocolate addict*. Vashon Island, Wash.: Natural Products Co.

Reents-Budet, Dorie, ed. 1994. *Painting the Maya universe: Royal ceramics of the Classic Period*. Durham, N.C. and London: Duke University Press.

Schwarz, Aljoscha, and Ronald Schweppe. 1997. *Von der Heilkraft der Schokolade: Geniessen ist gesund*. Munich: Peter Erd.

Taube, Karl Andreas. 1992. *The major gods of ancient Yucatan*. Washington, D.C.: Dumbarton Oaks.

West, John A. 1992. A brief history and botany of cacao. In *Chilies to chocolate: Food the Americas gave the world*, ed. Nelson Foster and Linda S. Cordell, 105–21. Tucson and London: The University of Arizona Press.

Young, Allen M. 1994. *The chocolate tree: A natural history of cacao*. Washington, D.C., and London: Smithsonian Institution Press.

"O happie money [= cocoa beans], which provides the human race with a delightful and useful drink and exempts its immune possessor from the Tartaric plague of avarice, since it cannot be buried and cannot be stored for long!"

PETRUS MARTYR
IN *DIE MENSCHLICHEN GENUßMITTEL* [THE HUMAN AGENTS OF PLEASURE]
(HARTWICH 1911, 336*)

"Chocolate is divine, we all know that—divine as in delicious, delectable . . . a-a-ahhh. But chocolate really *is* divine—divine as in heaven-born. It came on this earth as the gift of a god. And earthly incarnations of divinity—emperors, kings, princesses—have always made it their own before passing it down to the rest of humanity. Today it still suggests luxury, opulence, and pleasure."

MARCIA AND FREDERIC MORTON
CHOCOLATE
(1986, 1)

Harvesting the flowers of the cacao flower tree (*Quararibea funebris*), which was added to cacao as an aromatic and possibly psychoactive spice. Other *Quararibea* species were used for psychoactive purposes (**espingo**) in Peru or were added to **ayahuasca** in Amazonia. (Colonial-period illustration from the Paso y Tronco edition of Sahagun)

Theobroma spp.

Wild Cacao

This unidentified Mexican plant, known in Aztecan as *cacahuaxochitl*, "cacao flower/plant," was used to flavor cacao. (From Hernández, 1942/46 [Orig. pub. 1615]*)

Family

Sterculiaceae (Cocoa Family); Byttnerieae Tribe

The genus *Theobroma* consists of some twenty neotropical, i.e., American, species. Several species of wild cacao that occur in the tropics of Central and South America have ethnopharmacological significance. To date, there have been few chemotaxonomic studies of the genus. For example, the presence of **caffeine** and theobromine has been demonstrated only for *Theobroma cacao* (Baumann and Seitz 1994, 942).

Theobroma bicolor Humb. et Bonpl.

This Central American cacao species is known in Mayan as *kakaw*, *xaw*, or *balamte'* ("jaguar tree"). In other parts of Mexico, it is called *cacao blanco*, *cacao malacayo*, *pataste*, or *pataxte* (Hernández 1987, 1228*). The Maya used it as a **balche'** additive. In Mexico, its fruits are used as a substitute for the cultivated cacao (*Theobroma cacao*). It is grown in plantations in the state of Guerrero, where the fruit juice is used to prepare a soft drink as well as a fermented, winelike beverage.

Theobroma grandiflorum (Willd. ex Spreng.) Schum.—cupuassú

The wild cacao species known as *cupuassú* occurs in the area of Manaus (Amazonia) and is esteemed especially for its fruits, which are rounder and flatter than those of *Theobroma cacao* (de Aguiar and Lleras 1983). The abundant fruit juice is used to make fermented drinks (cacao **wine**). The leaves are used to manufacture alkaline plant ashes for **snuffs** and coca quids (see *Erythroxylum coca*). This species of cacao contains neither caffeine nor theobromine; its only active constituent is the purine alkaloid theacrine (Baumann and Seitz 1994, 942).

Theobroma subincanum Martius—cacahuillo

A number of Amazonian tribes burn the bark of this cacao species and add the resulting ashes to **snuffs**, especially those from *Virola* **spp.** or tobacco (cf. *Nicotiana tabacum*) (Schultes 1978a, 187*). The powdered inner bark, mixed with tobacco, is said to be used as a hallucinogen (Duke and Vasquez 1994, 169*).

It has been suggested that *Theobroma cacao* may have been derived through the cultivation of this Amazonian species (Baumann and Seitz 1994, 942).

Literature

See also the entries for *Erythroxylum coca*, *Theobroma cacao*, **balche'**, **snuff**, and **caffeine**.

Baumann, Thomas, and Renate Seitz. 1994. *Theobroma*. In *Hagers Handbuch der pharmazeutischen Praxis*, 5th ed., 6:941–55. Berlin: Springer.

de Aguiar Falcão, Martha, and Eduardo Lleras. 1983. Aspectos fenológicos, ecológicos e de produtividade do cupuaçu—*Theobroma grandiflorum* (Willd. ex Spreng.) Schum. *Acta Amazônica* 13 (5–6): 725–735.

Left: Wild cacao (*Theobroma bicolor*)

Right: In Peru, the ashes of the leaves of a *Theobroma* species are used as an alkaline additive to coca. The black mass (*llipta de cacao*) has a pungent taste.

Trichocereus pachanoi Britton et Rose

San Pedro Cactus

Family
Cactaceae (Cactus Family); Cereus Subdivision

Forms and Subspecies
The Indians make a distinction between two forms of the cactus, a "male" (with long spines) and a "female" (with short or even no spines). These forms do not have botanical names.

Synonyms
Cereus peruvianus nom. nud.
Cereus giganteus
Echinopsis pachanoi (Britt. et Rose) Friedr. et Rowl. (cf. *Echinopsis* spp.)

Folk Names
Achuma, agua-colla, aguacolla, aguacolla-cactus alucinógena, cardo, cimarrón, cimora blanca, cuchuma, gigantón, huachuma, huachumo, huando hermoso, kachum, rauschgiftkaktus, sampedro, San Pedro, San Pedro cactus, San-Pedro-kaktus, San Pedrillo, símora

History
The San Pedro cactus was in use at the very beginning of Andean civilization (Burger 1992); it was the *materia prima* of the shamans of that time (Giese 1989a, 225). In Peru, the central Andes region, and neighboring desert areas, the cactus has been used ritually for at least two thousand years (Cabieses 1983). The oldest archaeological evidence of its ritual use was found in the early layers of the formative period of Chavín (Joralemon and Douglas 1993, 185). San Pedro was used both as a sacral drug and as a shamanic medicine (Andritzky 1989*; Donnan and Douglas 1977). The cactus has been cultivated on the Peruvian coast since 200 B.C.E. to 600 C.E. (Early Intermediate period) (Davis 1983, 368).

Amazingly, there are very few reports about the Indian use that date to the colonial period. Moreover, the Inquisition (which apparently did know about it) did not persecute the use of the cactus (Andritzky 1987*).

No one knows precisely how an Indian sacred plant received the name of a Catholic saint (Saint Peter). The cactus probably was associated with rain cults and pagan rain gods. Since San Pedro is the patron saint of rain, it seems likely that the cactus obtained its name as a result (perhaps in an attempt to save it from the pharmacratic Inquisition). In addition, Saint Peter is the keeper of the keys to heaven (cf. *Nicotiana rustica*).

A flowering *Trichocereus pachanoi*.

Distribution
The San Pedro cactus is originally from Peru, where it is found at altitudes between 2,000 and 3,000 meters (Giese 1989a; Polia and Bianchi 1991). It also is cultivated in many other parts of the Andes, e.g., Ecuador and Peru. The cactus can be found in most botanical gardens and cactus nurseries around the world. It thrives in both dry regions and moist zones.

Cultivation
The cactus can be grown from its tiny seeds or propagated from cuttings. The latter approach requires simply placing a piece of the cactus in the ground. One or two new stalks will grow from the place where the cactus was cut.

Today, large numbers of *Trichocereus pachanoi* are planted in particular in California, not only for decorative purposes but primarily for use as entheogens. The cactus thrives in the Californian climate and grows quickly when watered daily. Because it is not a desert inhabitant but comes from the moist-warm, rain-rich areas of the Andes, the cactus is accustomed to receiving large amounts of water. My own experiments with cultivation have demonstrated that with regular watering (daily!), one can almost watch the cactus shoot up from the ground.[305] At the same time, it is modest in its needs and can survive for months without water. Pieces cut from the cactus can survive for months or even years and can develop lateral shoots even without food and water. Anyone who plants this cactus at home will be impressed with its unbelievable vitality.

Trichocereus pachanoi is well suited for grafting with other cacti, e.g., peyote (*Lophophora williamsii*). The tip of the San Pedro cactus should be cut flat and the head of the other cactus placed on the cut surface and tied to it for a few days. However, the cactus that has been grafted onto San Pedro will not contain any **mescaline** unless it is a species that itself produces mescaline.

A mythical snail being with a cactus, which may represent a *Trichocereus* species. (Painting on a Mochica pottery vessel, Chimu, approximately 500 C.E.)

305 ". . . San Pedro is extremely fast growing. Specimens in my greenhouse have gained almost eighteen inches in one year. It is also uncommonly easy to cultivate, and will thrive almost anyplace where the winters are relatively mild" (DeKorne 1994, 86*).

Top left: The San Pedro cactus (*Trichocereus pachanoi*) has only a few soft, short spines. It is one of the most important shamanic plants of Peru.

Bottom left: A monstrous form of *Trichocereus pachanoi*.

Center: The dried powder of the rind of *Trichocereus pachanoi* contains mescaline. It is light green in color and tastes extremely bitter.

Right: The *Pernettya* species known as *toro-maique* is a fortifying additive to the San Pedro drink.

Appearance

The almost spineless columnar cactus can grow as tall as 6 meters (Britton and Rose 1963, 2:134 ff.*). It has several ribs—usually six, but often seven or eight, sometimes as many as twelve, and in very rare cases (or not at all?) only four (the Indians consider a four-ribbed cactus to be especially powerful, as it symbolizes the four cardinal points). The beautiful white flowers appear only at night. The very delicious red fruits, which are almost the size of a child's head, only very rarely develop.

Psychoactive Material
— Slices of the fresh cactus
— Powdered rind

The fresh cactus slices are sold at "witches' markets" in Peru (Anzeneder et al. 1993, 79*).

Preparation and Dosage

The San Pedro drink is prepared from fresh cactus stalks or pieces. The chopped stalks are boiled for a few hours in ample water (often with other plants added). The decoction is then poured off and boiled again for several hours until only about half of the original volume remains (Davis 1983; Dobkin de Rios 1968). Some *curanderos* ("healers") boil four thin stalks in 20 liters of water for seven hours (Sharon 1980, 66).

Usually, a piece of cactus approximately 25 cm long and 5 to 8 cm thick is sliced and boiled per person. Some lemon or lime juice may be added to improve the dissolution of the mescaline. A technique using a pressure cooker has also been developed (Torres and Torres 1996).

Traditional *curanderos* fortify the San Pedro drink with leaves of the angel's trumpet known as *misha* (*Brugmansia* spp.; Giese 1989b, 225) as well as with other plants. Some of these, such as *hornamo* and *condorillo*, have not yet been botanically identified with accuracy (Polia M. 1988*; Polia and Bianchi 1991, 66; Sharon 1980, 66). These plants clearly alter the qualities of the San Pedro drink (Dobkin de Rios 1968, 191; Giese 1989a, 228 ff.).

To harvest, the stalks are cut off some 5 to 10 cm above the ground. The remaining stumps will develop shoots again in just a short time. The stalks are cut into manageable pieces some 30 to 40 cm long. The ribs are then cut apart. The skin or rind is cut away at the place where the green coloration of the flesh disappears. The fresh skin is placed in the sun to dry. After a few hours, the pieces of skin will begin to roll up, and they should then be placed so that the inside faces the sun. The drying process can last from two to six days, depending on the amount of solar radiation. After the cactus skins have dried thoroughly, they are ground. This can be done with a mortar and pestle (very arduous), a Mexican metate (grinding stone), a coffee mill, or a professional device from a drug store for pulverizing raw drugs (i.e., raw plant material). The more finely the cactus material is ground, the more effective is the absorption of the mescaline. Because the cactus tastes extremely or even disgustingly bitter, many Californians pour the powder into gelatin capsules that hold 1 g each. This practice makes it easier to ingest the powder and also makes it easy to determine the dosage. The powder should be stored in a dry, dark location. Because mescaline is a relatively stable compound, the powder will remain active for a long time if stored properly. If the powder is dissolved in milk, water, apple juice, tea, or some other liquid, it should be consumed as quickly as possible, as otherwise it will congeal into a disgusting mass.

Ritual Use

In pre-Hispanic times, the cactus played a ritual role in oracles, sexual magic, and shamanism (Andritsky 1989; Burger 1992; Dobkin de Rios 1982). Although the ritual use appears to date back to very ancient times, no precise pre-Columbian rituals have been documented. Dobkin de Rios (1985) has suggested that the renowned earth drawings on the Nazca plain may represent a kind of sacred cartography (or visionary map) that Moche shamans used for their out-of-body flights.

Today, Peruvian *curanderos* still ingest the sacred cactus during their nocturnal *mesa* rituals and also give it to the other participants. The *mesa* (Spanish, "table") is an altar with numerous objects (sticks, shells, ceramics, images of saints, et cetera) whose structure dates back to pre-Hispanic times and represents a visionary map (Giese 1989a, 1989b; Joralemon 1985, 21; Joralemon and Sharon 1993, 167; Villoldo 1984). The drink is consumed primarily by shamans so that they can recognize the cause of an illness during their nocturnal ceremonies. Less frequently, the patient and other people who are present may also receive some of the drink. Prior to this, however, they must "drink" an alcohol extract of tobacco (*Nicotiana rustica*) through the nose from a snail or seashell to purify themselves and protect themselves from negative powers.

The use of the San Pedro drink among Peruvian folk healers is no longer truly shamanic but has taken on a more symbolic form. At the *mesa* rituals, the dosage that is now usually taken is not large enough to elicit psychoactive effects:

> Based on my observations of numerous *mesa* rituals in Lima and Huancabamba, the *achuma* drink does not have a hallucinogenic effect. Even in the *curandero* and his helpers (*rastreadores*), no signs of an altered state of consciousness could be noticed. According to their statements, the *achuma* strengthens the visionary/diagnostic sensitivity, animates the objects on the *mesa*, and allows the souls of the patients to "blossom." But true hallucinogenic visions were not reported at any *mesa*. (Andritzky 1989, vol. 1, 113 f.*)

Artifacts

There are numerous pre-Columbian artifacts from Nazca and from the Moche-Chimu period that depict columnar cacti that look exactly like *Trichocereus pachanoi* (and less like other species) (Dobkin de Rios 1977a*, 1980). An image engraved in a stele showing the oracle god of Chavín holding a cactus in his hand is particularly well known (Burger 1992; Cordy-Collins 1977, 1980; Mulvany de Peñaloza 1984*). The flowering cactus is also depicted on two-thousand-year-old shamanic textiles of the Chavín culture, although in an idealized form (with only four ribs). And while Peruvian shamans today still claim that four-ribbed cacti are the most potent, no such specimens have ever been observed in nature (Cordy-Collins 1982*). Many Mochican stirrup vessels have representations of cacti, either in three-dimensional relief or as drawings, that are clearly indicative of shamanic associations (Bourget 1990; Cordy-Collins 1977; Donnan and Sharon 1977; Kutscher 1997*; Sharon 1972, 1980, 1982). One especially interesting object is a vessel on which the magical cactus is shown growing out of a deer, i.e., here a connection is made between a cervine and a plant that contains mescaline, a connection that is also made in the Huichol peyote

cult (see *Lophophora williamsii*). A Mochica vessel with an image of an erotic scene shows a woman on her back and a man who is in the act of penetrating her while holding a slice of San Pedro in his hand (Furst 1996*).

The American artist Donna Torres has produced several paintings that were inspired by experiences with San Pedro. The Chilean museum illustrator and artist José Pérez de Arce Antoncich produced a lithograph, *Hume Adentro*, after his own first experience with the San Pedro cactus (signed copies are available at the National Museum for Pre-Columbian Cultures in Santiago de Chile).

A Peruvian postage stamp features a drawing of *Trichocereus pachanoi* in flower.

Medicinal Use

The cactus is used primarily by shamans in psychedelic rituals. In Peruvian folk medicine, preparations of the cactus flesh are used in a limited degree as aphrodisiacs and tonics (Dobkin de Rios 1968).

At the present time, the cactus does not find any use in homeopathic or Western medicine.

Constituents

The dried extract of *Trichocereus pachanoi* is said to contain 2% mescaline (Cabieses 1983, 138; Polia and Bianchi 1991, 66). Information in the literature about the concentrations of active ingredients often varies. According to Gottlieb (1978, 45*), 1 kg of fresh cactus contains 1.2 g of **mescaline**. The fresh cactus is said to have a mescaline content of 0.12% (Polia and Bianchi 1991, 66). Freeze-dried material has been found to have a mescaline content of 0.33% (Brown and Malone 1978, 14*). DeKorne (1994, 88*) has stated that 100 g of dried material contains 300 mg of mescaline. More recent chromatographic methods (HPLC) have achieved very precise results indicating that the mescaline content in six different samples of *Trichocereus pachanoi* ranged from 1.09 to 23.75 μg per mg of dried material. In other words, the mescaline concentration can vary substantially (Helmlin and Brenneisen 1992, 94). Human pharmacological experiments have clearly shown that the effects produced by cactus material from younger specimens are considerably more intense than those produced by older, woody individuals (Manuel Torres, pers. comm.).

In addition to **mescaline**, *Trichocereus pachanoi* contains tyramine and β-**phenethylamines** (Mata and McLaughlin 1976*). Also present are trichocerine (Polia and Bianchi 1991, 66), hordenine, 3,4-dimethoxy-β-phenethylamine, and anhalonidine (Brown and Malone 1978, 14*).

Even snails that live on the cactus are said to contain mescaline (Furst 1996*).

"According to the *curandero* teachings, there are different types of the San Pedro cactus that can be distinguished on the basis of the number of their longitudinal ribs. Four-ribbed cacti, like four-leaved clovers, are considered to be very rare and very lucky. They are attributed with special healing properties because they correspond to the 'four winds' and the 'four roads,' supernatural powers associated with the cardinal directions that are invoked during the healing rituals. The San Pedro species that are found in the foothills of the Andes are considered especially effective, regardless of the number of ribs, because the soil there has more minerals."

Douglas Sharon
Magier der vier Winde [Wizard of the Four Winds]
(1980, 66)

The feline oracle god of Chavín de Huantar (Peru) holding the psychedelic San Pedro cactus in his right hand, a clear indication that the cactus was already in use for divinations during the pre-Hispanic period.

Ritual snail collectors are shown with columnar cacti, most likely *Trichocereus pachanoi* or another *Trichocereus* species. The sacred snails actually do often sit between the ribs of *Trichocereus pachanoi*. (Painting on a Moche vessel, Chimu, approximately 500 C.E.)

Traditional Additives to the San Pedro Drink

(from Davis 1983; Dobkin de Rios 1968; Giese 1989a, 227 ff.*; Sharon 1980; supplemented by my own observations in Chiclayo, northern Peru; cf. **cimora**)

Indigenous Name	Stock Plant	Active Constituents
misha = floripondio	*Brugmansia* sp. (= "Datura arborea")	tropane alkaloids
misha curandera	*Brugmansia* sp.	tropane alkaloids
misha rastrera = misha colorada	*Brugmansia sanguinea*	tropane alkaloids
misha rastera blanca = cimora	*Brugmansia arborea* *Brugmansia* x *candida* Pers.	tropane alkaloids
cimora oso	*B.* x *candida* f.	
cimora galga	*B.* x *candida* f.	
cimora toto curandera	*B.* x *candida* f.	
chamico	*Datura stramonium* *D. stramonium* ssp. *ferox* *Datura innoxia*	tropane alkaloids
cóndor misha = hierba del cóndor	*Lycopodium saururus*	alkaloids[306]
cóndor purga = huaminga oso = trenza shimbe = huaminga misha	*Lycopodium* spp.	alkaloids
condorillo	*Lycopodium affine* Hook et Grev.	
condorillo de quatro filos	*Lycopodium tetragonum*	
condoro	*Lycopodium magellanicum* *Lycopodium reflexum*	
toro-maique	*Pernettya* sp.	?
contrahechizo	*Iochroma grandiflorum* (cf. *Iochroma fuchsioides*) *Fuchsia* sp.	?
piri-piri[307] = congona	*Peperomia galioides* H.B.K. (cf. *Peperomia* spp.) *Peperomia flavamenta* Trelease *Peperomia galioides* H.B.K. *Peperomia* sp.	essential oil
hornamo	*Senecio* spp. (?)	?
hornamo amarillo	*Senecio tephrosioides* Turz.	?
hornamo blanco	*Onoseris* sp. (?)	
hornamo caballero = hornamo caballo	*Pleurothallis* sp. or *Epidendron* sp.	
hornamo chancho	?	
hornamo cuti	?	
hornamo lírio	*Lycopodium* sp.	
hornamo loro	*Lycopodium* sp.	
hornamo morado	*Valeriana adscendens* Turz. (cf. *Valeriana officinalis*)	
hornamo toro	*Niphogeton scabra* (Wolff.) Macbr.	
hornamo verde	?	
ishpingo	(see **espingo**)	
marijuana	*Cannabis sativa*	THC
cimora = timora	*Iresine* spp. *Iresine celosia* L.	? ?
timora	*Euphorbia cotinifolia* L.	?
cimora misha = misha	*Pedilanthus tithymaloides* Poit. *Pedilanthus retusus* Benth. (cf. *Pedilanthus* spp.)	? ?
cimora toro = misha veneno	*Hippobroma longiflora* Don [syn. *Isotoma longiflora* L.; cf. **pituri**]	?
siempreviva	*Tillandsia* sp. (cf. *Lophophora williamsii*)	?

Effects

The effects of *Trichocereus pachanoi* are typically characterized as psychedelic or entheogenic. These effects make it appear to be the ideal shamanic drug for out-of-body journeys, et cetera (Giese 1989b, 83; Turner 1994, 32 f., 36*).

I have carried out experiments with varying dosages of the powder. With 1 g, I did not experience any effects. Two to 4 g produced a mild stimulation that persisted for approximately six to eight hours. This amount functions as a true tonic and restorative. I have also experimented with this dosage in the high mountains, where I noticed a distinct improvement in performance. If a person eats something during the time in which the effects are felt, the effects will increase as digestion begins.

With amounts of 5 to 6 g, empathogenic sensations appear alongside of the tonic qualities. Ten grams of the powder are unequivocally psychedelic, although few hallucinations occur. The psychedelic effects manifest more in the emotional domain. Very profound psychedelic effects can be achieved by taking some 50 µg of LSD with 10 g of San Pedro powder (cf. ergot alkaloids).

Recently, the use of cactus powder (sometimes in combination with *Peganum harmala* seeds) as a smoking substance has been on the rise. Whether psychoactive effects can be produced in this manner is questionable. I have not noticed any effects from such use.

Commercial Forms and Regulations

The cactus can be obtained through the international cactus trade (the seeds are also available from time to time). At present, there are no import restrictions.

Literature

See also the entries for *Trichocereus peruvianus*, *Trichocereus* spp., and **mescaline**.

Below, Till, and Anette Morvai. 1997. *Schamanistische Volksmedizin in Peru.* Beitrage zur Ethnomedizin, Kleine Reihe 13. Berlin: VWB.

Bourget, Steve. 1990. Caracoles sagrados en la iconografía moche. *Gaceta Arqueológiga Andina* 5 (20): 45–58.

Burger, Richard L. 1992. *Chavin and the origins of Andean civilization.* London: Thames and Hudson.

Cabieses, Fernando. 1983. Die magischen Pflanzen Perus. In *Peru durch die Jahrtausende* (exhition catalogue), 138–41. Niederosterreichische Landesausstellung, Schloß Schallaburg.

306 More than one hundred alkaloids have been found in the genus *Lycopodium* (club mosses) (Gerard and MacLean 1986). To date, it is not known whether any of these alkaloids are psychoactive. Six alkaloids have been detected in the Chilean species *Lycopodium magellanicum* (Loyola et al. 1979).

307 In Amazonia, this name is used for pleasantly scented *Cyperus* spp. (cf. **ayahuasca**).

Calderón, Richard Cowan, Douglas Sharon, and F. Kaye Sharon. 1982. *Eduardo el curandero: The words of a Peruvian healer.* Richmond, Calif.: North Atlantic Books.

Cordy-Collins, Alana. 1977. Chavín art: Its shamanic/hallucinogenic origins. In *Pre-Columbian art history: Selected readings,* ed. A. Cordy-Collins and Jean Stern, 353-61. Palo Alto, Calif.: Peek Publications.

———. 1980. An artistic record of the Chavín hallucinatory experience. *The Masterkey* 54 (3):84–93.

Crosby, D. M., and J. L. McLaughlin. 1973. Cactus alkaloids. XIX. Crystallization of mescaline HCL and 3-methoxytyramine HCL from *Trichocereus pachanoi. Lloydia* 36:417.

Davis, E. Wade 1983. Sacred plants of the San Pedro cult. *Botanical Museum Leaflets* 29 (4): 367–86.

Dobkin de Rios, Marlene. 1968. *Trichocereus pachanoi:* A mescaline cactus used in folk healing in Peru. *Economic Botany* 22:191–94.

———. 1969. Folk curing with a psychedelic cactus in north coast Peru. *International Journal of Social Psychiatry* 15:23–32.

———. 1980. Plant hallucinogens, shamanism and Nazca ceramics. *Journal of Ethnopharmacology* 2:233–46.

———. 1982. Plant hallucinogens, sexuality and shamanism in the ceramic art of ancient Peru. *Journal of Psychoactive Drugs* 14 (1–2): 81–90.

———. 1985. Schamanen, Halluzinogene and Erdaufschuttungen in der Neuen Welt. *Unter dem Pflaster liegt der Strand* 15:95–112. Berlin: Karin Kramer Verlag.

Donnan, Ch. B., and Douglas G. Sharon. 1977. The magic cactus: Ethnoarchaeological continuity in Peru. *Archaeology* 30:374–81.

Friedberg, Claudine. 1960. Utilisation d'un cactus à mescaline au nord du Pérou (*Trichocereus pachanoï* Britt. et Rose). *Actes du VIᵉ Congrès International des Sciences Anthropologiques et Ethnologiques* (Paris) 2:23–26.

Gerard, Robert V., and David B. MacLean. 1986. GC/MS examination of four *Lycopodium* species for alkaloid content. *Phytochemistry* 25 (5): 1143–50.

Giese, Claudius Cristobal. 1989a. "*Curanderos*": *Traditionelle Heiler in Nord-Peru (Küste und Hochland).* Münchner Beiträge zur Amerikanistik, vol. 20. Hohenschäftlarn: Klaus Renner Verlag.

———. 1989b. Die Diagnosemethode eines nordperuanischen Heilers. *Curare* 12 (2): 81–87.

Glass-Coffin, Bonnie. 1991. Discourse, Daño and healing in north coastal Peru. *Medical Anthropology* 13 (1–2): 33–55.

———. 1992. *Female healing and experience in northern Peru.* PhD dissertation, University of California at Los Angeles.

Gutiérrez-Noriega, C. 1950. Area de mescalinismo en el Perú. *América Indígena* 10:215–220.

Joralemon, Donald. 1985. Altar symbolism in Peruvian ritual healing. *Journal of Latin American Lore* 11:3–29.

Joralemon, Donald, and Douglas Sharon. 1993. *Sorcery and shamanism:* Curanderos *and clients in northern Peru.* Salt Lake City: University of Utah Press.

Kakuska, Rainer. 1994. San Pedro blues. *Connection* 12:29–32.

Loyola, Luis A., Glauco Morales, and Mariano Castillo. 1979. Alkaloids of *Lycopodium magellanicum. Phytochemistry* 18:1721–23.

Lundström, J. 1970. Biosynthesis of mescaline and 3,4-dimethoxyphenethylamine in *Trichocereus pachanoi* Britt. et Rose. *Acta Pharmaceutica Suecica* 7:651.

Polia, M., and A. Bianchi. 1991. Ethnological evidences and cultural patterns of the use of *Trichocereus pachanoi* Britt. et Rose among Peruvian *curanderos. Integration* 1:65–70.

Rätsch, Christian. 1994. Eine bisher nicht beschriebene Zubereitungsform von *Trichocereus pachanoi. Yearbook for Ethnomedicine and the Study of Consciousness,* 1995 (4): 267–81. Berlin: VWB.

Sharon, Douglas. 1972. The San Pedro cactus in Peruvian folk healing. In *Flesh of the gods,* ed. Peter T. Furst, 114–35. New York: Praeger.

———. 1980. *Magier der vier Winde: Der Weg eines peruanischen Schamanen.* Freiburg: Bauer.

———. 1981. San-Pedro-Kaktus: Botanik, Chemie and ritueller Gebrauch in den mittleren Anden. In *Rausch and Realitat,* ed. G. Volger, 2:785–800. Cologne: Rautenstrauch-Joest Museum für Völkerkunde.

———. 1987. Der gescheiterte Schamanenschüler. In *Heilung des Wissens,* ed. Amelie Schenk and Holger Kalweit, 187–211. Munich: Goldmann.

Torres, Donna, and Manuel Torres. 1996. San Pedro in the pressure pot. *Yearbook for Ethnomedicine and the Study of Consciousness,* 1995 (4): 283–84. Berlin: VWB.

Villoldo, Alberto. 1984. Die Mesa des Don Eduardo. *Sphinx* 26:10–17.

"It is certainly not pleasant to eat San Pedro. Most hard-core consumers can stand the taste. But it is difficult to consume the amount of cactus that is needed for an intensive trip. The taste of the various *Trichocereus* species varies from extremely bitter to neutral, but those species that have a less intense taste tend to have a slimy consistency. With these, this consistency is the greatest barrier to being able to consume large quantities. I chew the cactus to a paste and swallow it with liquid. It is also helpful to eat whole-grain bread to bind the liquid in the stomach. The dark green flesh directly under the skin contains the most active ingredients and should be eaten first. The V-shaped strips should be pressed flat so that the flesh can be scratched from the skin. The flesh of the protruding ribs can be eaten like an ear of corn. The middle part is woody and inedible. The effects begin some 45 minutes after ingestion, and since it takes a while to eat the cactus, one can feel the effects while still eating."

D. M. TURNER
DER PSYCHEDELISCHE REISEFÜHRER
[THE PSYCHEDELIC TRAVEL GUIDE]
(1997, 39*)

Trichocereus peruvianus Britton et Rose

Peruvian Columnar Cactus

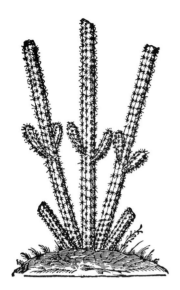

It is possible that *Trichocereus peruvianus* was known in Europe as early as the eighteenth century, for it may be identical with "*Cereus peruvianus.*" (Woodcut from Tabernaemontanus, *Neu Vollkommen Kräuter-Buch*, 1731)

Family
Cactaceae (Cactus Family); Cereus Subdivision

Forms and Subspecies
One geographically isolated variety has been referred to as *Trichocereus peruvianus* var. *truxilloensis*.

Synonyms
Echinopsis peruvianus (cf. **Echinopsis** spp.)

Folk Names
Cuchuma, peruanischer kaktus, Peruvian columnar cactus, San Pedro

History
This species was first described botanically in 1937 by Britton and Rose in their large monograph on cacti (2:136). Apart from this, nothing is known about the plant's history.

Distribution
This *Trichocereus* species is found almost exclusively in Peru at altitudes of around 2,000 meters.

Cultivation
Trichocereus peruvianus grows as fast or even faster than **Trichocereus pachanoi**, but only when it is watered daily. During arid periods, it requires considerably less water.

Appearance
Trichocereus peruvianus is distinguished from *Trichocereus pachanoi* primarily by its considerably longer, harder, and more pointed spines. It is very similar to *Trichocereus bridgesii* (cf. **Trichocereus** spp.) but does not grow as large, attaining a height of only 2 to 4 meters.

Psychoactive Material
— Fresh cactus flesh
— Dried rind powder

Preparation and Dosage
See **Trichocereus pachanoi**. Because of the long spines, care should be taken when preparing this cactus (wear gloves!). A piece of cactus approximately 10 cm long is sufficient for one person.

Ritual Use
As with **Trichocereus pachanoi**. *T. peruvianus* is sometimes regarded as the "male" counterpart of *T. pachanoi*.

Artifacts
It is possible that some pre-Columbian representations of cacti on Peruvian ceramics may be depictions of *T. peruvianus*.

Medicinal Use
As yet unknown

Left: The Peruvian columnar cactus *Trichocereus peruvianus* has deep ribs. It frequently has four ribs, regarded as symbolic of the four cardinal points. The cactus is rich in mescaline.

Right: The rare variety *Trichocereus peruvianus* var. *truxilloensis* occurs in northwestern Peru, the area in which *trujillo* coca (**Erythroxylum novogranatense** var. *truxillense*) is grown.

Constituents

Trichocereus peruvianus contains approximately three times as much **mescaline** as other *Trichocereus* species (Turner 1994, 31*). Its concentration of mescaline is said to be the same as or even higher than that of peyote (cf. *Lophophora williamsii*) (Pardanani et al. 1977, 585). On occasion, it may contain up to ten times as much mescaline as *Trichocereus pachanoi* (Rob Montgomery, pers. comm.).

Effects

See *Trichocereus pachanoi.*

Commercial Forms and Regulations

The cactus is available through the international cactus trade and is not subject to any regulations.

Literature

See also the entries for *Trichocereus pachanoi*, *Trichocereus* spp., and **mescaline.**

Pardanani, J. H., J. L. McLaughlin, R. W. Kondrat, and R. G. Cooks. 1977. Cactus alkaloids. XXXVI. Mescaline and related compounds from *Trichocereus peruvianus. Lloydia* 40 (6): 585–90.

Trichocereus spp.

Columnar Cacti

Family

Cactaceae (Cactus Family); Cereus Subdivision

Synonyms

Echinopsis spp.
Helianthocereus pasacana (Rümpl.) Backeb.

Folk Names

Achuma, cardón, cardón grande, columnar cactus, pasakana,[308] säulenkaktus, San Pedro

Psychoactive Species in the Genus *Trichocereus*

Trichocereus bridgesii (Salm-Dyck) Britt. et Rose
Trichocereus cuscoensis Britt. et Rose
Trichocereus fulvinanus Ritt.
Trichocereus macrogonus (Salm-Dyck) Ricc.
Trichocereus pachanoi Britt. et Rose
Trichocereus peruvianus Britt. et Rose
Trichocereus taquimbalensis Card.
Trichocereus terscheckii (Parmentier) Britt. et Rose
Trichocereus validus (Monv.) Backeb.
Trichocereus werdermannianus Backeb.

Mescaline has been detected in all of these species (Agurell 1969; Mata and McLaughlin 1976). Several of these species are known as San Pedro or are seen as substitutes for it (see *Trichocereus pachanoi*).

While the following species have been chemically investigated, no mescaline has (yet) been detected in them:

Trichocereus spachianus (Lem.) Ricc.[309]
Trichocereus candidans Britt. et Rose

Distribution

These species occur in Ecuador, Peru, Bolivia, northern Chile, and Argentina.

Cultivation

Trichocereus bridgesii can be grown from seed and requires conditions similar to those required by *T. pachanoi*. The other *Trichocereus* species can be grown from seed and also propagated from layers or cuttings.

Appearance

All species of *Trichocereus* develop long, ribbed columns with varying numbers of spines. Hybrids of different *Trichocereus* species now exist.

Ritual Use

In Bolivia, where *Trichocereus bridgesii* is commonly referred to as *achuma*, the cactus is said to be used both traditionally by the Indians and "for its stimulating effects upon the psyche" by young people in La Paz (Giese 1989b, 225*). Davis reported that this species has potent psychedelic effects (Giese 1983, 375*).

The *pasakana* cactus (*Trichocereus pasacana* [Webb.] Britt. et Rose) appears to have a long history of cultural use in South America. *Trichocereus pasacana* fruits have been found in a cave near Jujuy (Argentina) in layers that have been dated to 7670 to 6980 B.C.E. They were also found in all subsequent layers, i.e., continuously. They first begin to appear together with coca leaves (*Erythroxylum coca*) in layers from the fifteenth century (Inca period). It is unclear whether the cactus was used for psychoactive or simply culinary purposes in early times. The fruits (sans seeds) and flowers are still used today in the

A flowering *Trichocereus huanucensis* from Peru

308 This name is not used solely for *Trichocereus pasacana*. In northern Chile, the edible fruits of various columnar cacti are also referred to as *pasakana*, including those of *Trichocereus atacamensis, Oreocereus hendriksenianus* Backeb., and a *Soehrensia* species (Aldunate et al. 1981, 211, 213, 217*).

309 In only one sample of a *Trichocereus spachianus* cultivated in Indiana were traces of **mescaline** detected (Shulgin 1995*).

Top left: The Bolivian *Trichocereus bridgesii* is consumed as a psychedelic recreational drug in La Paz.

Top center: The border of Puna, the high plain between northern Chile and northwestern Argentina, is the area of natural occurrence of the *pasakana* cactus (*Trichocereus pasacana*), one of the oldest psychoactives in the plant world. Its fruits and flowers are burned to make ashes, which are added to coca quids for chewing.

Top right: A *Trichocereus* species with fruit.

Middle center: In the Chaco region of northwestern Argentina, a tall, red-blooming *Trichocereus* species is called San Pedro. It is said to produce psychoactive effects. (Photographed in its natural habitat)

Middle right: There are now numerous crosses and hybrids among *Trichocereus* species that can no longer be unequivocally assigned to a specific species. Whether these contain mescaline is unknown.

Bottom: *Trichocereus atacamensis*, indigenous to northern Chile (Atacama Desert), has stimulating powers and may possibly contain mescaline and similar constituents. This *Trichocereus* species is easily recognized by its extremely long spines (up to 20 cm in length). (Photographed in its natural habitat)

area of Jujuy to produce *llipta*, the alkaline coca additive (Fernández Distel 1984).

Northwestern Argentina is home to the red-blooming *Trichocereus tarijensis* [syn. *Trichocereus poco*] (Fernández Distel 1984) and the mescaline-containing *Trichocereus terscheckii* (Mata and McLaughlin 1982, 114*). There they are known as San Pedro or *cardón santo*, "sacred cactus." The Mataco have used the cactus flesh of these and perhaps other species to make *llipta* for coca chewing. Such coca quids are said not only to taste better but also to be much more potent.

Trichocereus atacamensis appears to have been used in the Atacama Desert in prehistoric times in connection with **snuffs**.

Constituents

Mescaline and occasionally other β-**phenethylamines** are present in *Trichocereus* species (Agurell 1969b). *Trichocereus terscheckii* contains 0.25 to 1.2% alkaloids (primarily trichocerine and mescaline; cf. Reti and Castrillo 1951 and Herrero-Ducloux 1932).

Trichocereus pasacana has been found to contain the alkaloid hordenine, which may have sympathomimetic effects (Agurell 1969a; Fernandez Distel 1984). The alkaloid candicine has also been detected (Meyer and McLaughlin 1980).

Effects

The cactus flesh of *Trichocereus atacamensis* (Phil.) Britt. et Rose [syn. *Helianthocereus atacamensis* (Phil.) Bckbg., *Cereus atacamensis* Phil.] (cf. Aldunate et al. 1981, 211*) has a very bitter taste (quite similar to that of *Trichocereus pachanoi*) and has distinctly stimulating properties.

Trichocereus terscheckii is said to produce the same effects as **Trichocereus pachanoi**.

Commercial Forms and Regulations

It is occasionally possible to purchase seeds of various *Trichocereus* species. The cacti are not subject to any special regulations.

Literature

See also the entries for **Trichocereus pachanoi**, **Trichocereus peruvianus**, and **mescaline**.

Agurell, Stig. 1969a. Cactaceae alkaloids. *Lloydia* 32 (2): 206–16.

———. 1969b. Identification of alkaloid intermediates by gas chromatography-mass spectometry. I. Potential mescaline precursors in *Trichocereus* species. *Lloydia* 32 (1): 40–45.

Agurell, Stig, J. G. Bruhn, J. Linundstrom, and U. Svensson. 1971. Cactaceae alkaloids. X: Alkaloids of *Trichocereus* and some other cacti. *Lloydia* 34 (2): 183–87.

Fernandez Distel, Alicia. 1984. Contemporary and archaeological evidence of llipta elaboration from the cactus *Trichocereus pasacana* in northwest Argentina. *Proceedings 44 International Congress of Americanists*, BAR International Series 194.

Gutiérrez-Noriega, C. 1950. Area de mescalinismo en el Perú. *América Indígena* 10:215–20.

Herrero-Ducloux, E. 1932. Datos quimicos sobre el *Trichocereus* sp. aff. *T. terscheckii*. *Revista Farmaceutica* 74:375.

Mata, Rachel, and Jerry L. McLaughlin. 1976. Cactus alkaloids. XXX. N-methylated tyramines from *Trichocereus spachianus, T. candicans*, and *Espostoa huanucensis*. *Lloydia* 39:461–63.

Meyer, B. N., and J. L. McLaughlin. 1980. Cactus alkaloids. XLI: Candicine from *Trichocereus pasacana*. *Planta Medica* 38:91.

Reti, L., and J. A. Castrillo. 1951. Cactus alkaloids. I: *Trichocereus terscheckii* (Parmentier) Britt. et Rose. *Journal of the American Chemical Society* 73:1767–69.

A variety of *Trichocereus* species helps make up the typical high mountain flora of the Andes and in some areas shapes the characteristic image of the landscape. (Illustrations from Mortimer, 1901)

Turbina corymbosa (Linnaeus) Rafinesque

Ololiuqui Vine

Family

Convolvulaceae (Morning Glory Family); Subfamily Convolvuloideae, Argyreieae Tribe

Forms and Subspecies

None

Synonyms

Convolvulus corymbosa L.
Convolvulus corymbosus L.
Convolvulus domingensis Desr.
Convolvulus sidaefolia H.B.K.
Convolvulus sidaefolius Kunth
Ipomoea antillana Millspaugh
Ipomoea burmanni Choisy
Ipomoea corymbosa (L.) Roth
Ipomoea dominguensis (Desr.) House
Ipomoea sidaefolia (H.B.K.) Choisy
Ipomoea sidaefolia (Kunth) Choisy
Rivea corymbosa (L.) Hall. f.

The ololiuqui vine (*Turbina corymbosa*) in blossom. (Photographed in Xalapa, Mexico)

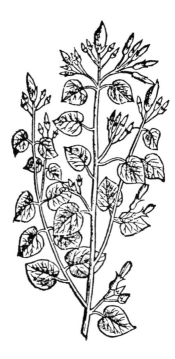

This is most likely the earliest non-Indian illustration of the ololiuqui vine (*Turbina corymbosa*). (From Francisco Hernandez, *Rerum medicarum Novae Hispania thesaurus*, Rome, 1651)

Folk Names

Aguinaldo (Cuba), a-mu-kia, angelito, badoh (Zapotec), badoh blanco ("white badoh"), badoh-shnaash, badoor, bejuco de San Pedro ("vine of St. Peter"), bidoh shnaash, bi-to, coatlxihuitl (Aztec, "snake plant"), coatlxoxouhqui (Aztec, "green snake" or "blue snake"), coatlxoxouqui, cuan-bodoa, cuetzpallin ("wall lizard"), cuexpalli, flor de la virgen (Spanish, "flower of the virgin"), flor de pascua (Spanish, "Easter flower"), grüne schlange, guana-lace, hierba de la virgen ("herb of the virgin"), hierba María, hoja del norte ("leaf of the north"), huan-mei (Chinantec), huan-men-ha-sey, loquetica, loquetico ("the crazy one"), manta, mantecón, manto ("coat"), ma:sungpahk (Mixe, "bones of the children"), m'+'oo quia' sée, mo-ho-quiot-mag, mo-so-lena (Mazatec), nicuana-laci, nocuana-laci, nosolena, ololiuhqui (Aztec, "that which causes turns"), ololiuqui, ololiuqui-ranke, ololiuqui vine, pamaxunk, pi-too (Zapotec), piule,[310] quahn shnaash, sachxoit (Tepehuan), santa ("the saint"), Santa Catarina, schlangen-pflanze, semillas de la virgen, señorita (Spanish, "lady"), tabentun, trepadora ("vine"), tumba caballo ("grave of the horse"), ua-men-hasey (Chinantec), weiße trichterwinde, xtabentum, xtabentun (Mayan, "jeweled cord"), xtabentún, yaga-bidoo, yerba de la serpientes, yerba de la virgen, yololique (Nahuat), yucu-yaha (Mixtec)

History

The ritual and medicinal use of the psychoactive seeds of the ololiuqui vine dates back far into the pre-Hispanic period. There has been evidence of the use of the "little gods" (= seeds) among the Aztecs and other Mesoamerican peoples since early colonial times. Sahagun documented the divinatory and medicinal uses. Hernandez was the first to discuss the plant and its properties. Ximenez described the plant and its use in *Cuatro-libros de la naturaleza* (1790). The most comprehensive report comes from Ruiz de Alarcón.

In the early twentieth century, it was thought that the use of ololiuqui had died out. The botanical source was also long uncertain. For example, ololiuqui had been interpreted as *Datura innoxia* or *Lophophora williamsii* (Safford 1915; Reko 1934). It was once even thought that ololiuqui must have been a poppy species (*Papaver* spp., *Argemone mexicana*) with narcotic properties (Cerna 1932, 305*).

The botanical identity of ololiuqui was first clarified in the 1940s by Richard Evans Schultes (Davis 1996, 94 ff.*; Schultes 1941). In the early 1960s Albert Hofmann isolated the active constituents, which he recognized to be **ergot alkaloids** that were closely related to the constituents of *Claviceps purpurea* and LSD (Hofmann 1961). From a chemotaxonomic perspective, this discovery was an absolute sensation. At the time,

no one had thought it possible that a primitive fungus was able to biosynthesize the same substances as a highly developed flowering plant. For this reason, Albert Hofmann was greeted with severe skepticism and disbelief when his report was first published and when he made his first presentations on the subject.

Distribution

The plant is very likely from tropical Mexico but is now very common in Cuba as well as on other islands of the West Indies and on the North American Gulf Coast. It also occurs in Central America; its southernmost occurrence is in the Amazon basin of southern Colombia (Richardson 1992, 69*). The plant was introduced into the Philippines at an early date and has now become wild there (Brenneisen 1994, 1014).

Cultivation

Propagation occurs via the seeds, which are preferably pregerminated or raised in growing pots until they are seedlings. Cultivation is not a particularly successful venture, as typically only a few seeds germinate. The plant requires a tropical climate and relatively large amounts of water and does not tolerate any frost (cf. also *Ipomoea violacea*).

Appearance

The large, woody, perennial creeper (up to 8 m long) has cordate leaves (5 to 9 cm long, 2.5 to 6 cm wide) and flowering branches that grow from the leaf axils. The branches bear the funnel-shaped flowers in clusters or umbels. The sepals are only about 1 cm wide, while the white petals can attain a length of 5 cm. The fruits contain only one light brown or ocher seed. In Mexico, the plant flowers between December and March. The flowers are the source of a great amount of (psychoactive) **honey**. Most of the honey produced in Cuba is from *Turbina corymbosa*.

The genus *Turbina* is composed of twelve to fifteen species, some of which are similar in appearance, but only a few of which are known (Brenneisen 1994, 1013). They occur in tropical Africa and on New Caledonia (Schultes 1941, 20). The seeds of other *Turbina* species may also one

310 For the name *piule* and its etymology, see *Rhynchosia pyramidalis*.

day be found to contain psychoactive **ergot alkaloids**, as does *Argyreia nervosa*. Ololiuqui is the only New World species in the genus.

Psychoactive Material
— Fresh or dried seeds (*Turbina corymbosa* seeds, pascua, piule,[311] badoh, ololiuqui)
— Leaves
— Roots

Preparation and Dosage
The fresh or dried seeds normally are added to such alcoholic drinks as mescal (cf. *Agave* spp.), *aguardiente* (sugarcane liquor; cf. **alcohol**), *tepache* (maize beer, **chicha**), and **balche'** (Schultes 1941, 37). The fresh seeds, when crushed, are added to pulque (cf. *Agave* spp.) and allowed to steep. This drink, known as *piule*, can be drunk to attain hypnotic states.

Fifteen or more seeds can be ground and allowed to soak in one-half cup of water (Gottlieb 1973, 39*). The Zapotec say that a shamanic dosage consists of thirteen pairs of seeds (Fields 1968, 206); traditional dosages also are said to consist of fourteen or twenty-two seeds (Wasson 1971, 343). Because such traditional dosages did not elicit any effects among Western test subjects, experiments were conducted using larger quantities:

Ingesting 60 to 100 seeds led to apathy, indifference, and increased sensitivity to optical stimuli. After some 4 hours, there followed a longer-lasting phase of relaxation and well-being. In contrast, in eight male subjects, dosages of up to 125 seeds did not elicit any effects except vomiting. (Brennesien 1994, 1015)

Dosages as high as three hundred to five hundred seeds have also been tested, usually with unsatisfactory results and such severe side effects as vomiting, diarrhea, et cetera (Brenneisen 1994, 1016).

Literature from the colonial period mentions an ointment known as "sacred flesh" that was prepared from the ashes of burned insects, tobacco (*Nicotiana tabacum, Nicotiana rustica*), and ololiuqui seeds (José de Acosta, *Historia natural y moral de las Indias . . .*, Seville 1590).

The root is purportedly used for divinatory purposes, but the manner of preparation is unknown (Fields 1968, 206).

Ritual Use
The plant and its psychoactive effects are discussed in some detail in the colonial literature:

Its leaves are slender, ropelike, small. Its name is ololiuhqui. It inebriates one; it makes one crazy, stirs one up, makes one mad, makes one

A comparison of the psychoactive seeds of the two vines traditionally used in Mexico: on the left are the light-colored seeds of *Turbina corymbosa* (= *badoh blanco*) and on the right the dark seeds of *Ipomoea violacea* (= *badoh negro*).

possessed. He who eats of it, he who drinks it, sees many things that will make him afraid to a high degree. He is truly terrified of the great snake that he sees for this reason.

He who hates people causes one to swallow it in drink and in food so as to make one mad. But it smells sour; it burns a little in the throat. It is applied on the surface alone to treat gout. (Sahagun, Florentine Codex 11.7*)

The Spanish physician Francisco Hernandez wrote about ololiuqui in his *Rerum medicarum Novae Hispaniae thesaurus*:

There is a plant in Mexico that is called snake-plant, a vine with arrow-shaped leaves, which is thus also called arrow plant. The seed is used in medicine. Ground and drunk with milk and Spanish pepper, it takes away pains, heals all manner of ailments, inflammations, and ulcers. When the priests of the Indians wish to commune with the spirits of the dead, they eat these seeds to induce a delirium and then see thousands of satanic figures and phantasms around them. (In Rätsch 1991a, 193*)

The Spanish missionary Hernando Ruiz de Alarcón has provided us with the most detailed reports of the Indian uses of magical plants (cf. ***Lophophora williamsii, Nicotiana rustica***). His writings were published in 1629 under the title *Treatise on the Heathen Superstitions That Today Live Among the Indians Native to This New Spain*. (This work became a kind of "witches' hammer," providing the juristic basis for persecuting witches in the New World.) He has the following to say about the use of ololiuqui, which is equated with the use of peyote:

The so-called *ololiuhqui* is a seed like lentils or lentil vetch which, when drunk, deprives one of judgment. And the faith that these unhappy natives have in this seed is amazing, since, by drinking it, they consult it like an oracle for everything whatever that they want to know, even those things which are beyond human knowledge, such as knowing the cause of

311 See *Rhynchosia pyramidalis* for more on the name *piule*.

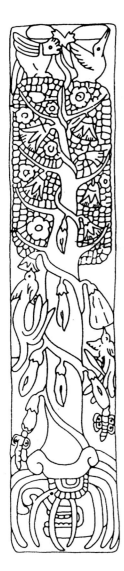

A depiction of the World Tree on a stele from the Classic Mayan period, showing how the tree grows out of the "earth monster" and is entwined by a vine that may be interpreted as *Turbina corymbosa*.

illnesses, because almost everyone among them who is consumptive, tubercular, with diarrhea, or with whatever other sickness of the persistent kind right away attributes it to sorcery. And in order to resolve this doubt and others like it, such as those about stolen things and of aggressors, they consult this seed by means of one of their deceitful doctors, some of whom have it as their job to drink this seed for such consultations, and this kind of doctor is called *Pàyni*, because of the job, for which he is paid very well, and they bribe him with meals and drinks in their fashion. If this doctor either does not have this function or wishes to excuse himself from that torment, he advises the patient himself to drink that seed, or another person for whose services they also pay as they do the doctor, but the doctor indicates to him the day and the hour in which he is to drink it, and he tells him for what purpose he will drink it.

Finally, whether it is the doctor or another person in his place, in order to drink the seed, or peyote, which is another small root and for which they have the same faith as for that other seed, he closes himself up alone in a room, which usually is his oratory, where no one is to enter throughout all the time that the consultation lasts, which is for as long as the consultant is out of his mind, for then they believe the *ololiuhqui* or peyote is revealing to them that which they want to know. As soon as the intoxication or deprivation of judgment passes from this person, he tells two thousand hoaxes, among which the Devil usually includes some truths, so that he has them deceived or duped absolutely. . . .

Also they make use of this drink to find things that have been stolen, lost, or misplaced and in order to know who took or stole them. . . .

When the wife leaves the husband or the husband the wife, they also take advantage of *ololiuhqui,* and in this case the imagination and fantasy work also, and even better than in the case of sicknesses, because in this second case conjectures follow that are the cause of more vehement suspicion, and thus it works with greater strength at the time of the intoxication, since it is easily seen that one person will be persuaded that another carried off his wife or stole his property. . . .

Finally these prophets make use of *ololiuhqui* or of peyote to solve these riddles, in the way already described. Then they say that a venerable old man appears to them who says that he is the *ololiuhqui* or the peyote and that he has come at their call in order to help them in whatever way might be necessary. Then, being asked about the theft or about the

absent wife, he answers where and how they will find it or her. . . .

Here it should be carefully noted how much these miserable people hide this superstition of the *ololiuhqui* from us, and the reason is that, as they confess, the very one they consult orders them not to reveal it to us. . . . And thus their excuse is *ipampa àmo nechtlahueliz,* which is as if to say "in order that the *ololiuhqui* will not declare himself to be my enemy." (Ruiz de Alarcón, treatise I, chapter 6; 1984, 59–60, 65, 67)

This use of ololiuqui seeds has continued in the same manner with only slight variations into the present day among the Zapotec, Mixtec, Mazatec, and Mixe. Among the Mixe, the plant and its parts (seeds, et cetera) are regarded as apotropaic. Witches can be kept away from the house if the appropriate parts of the plant are used. The seeds are used in the same manner as mushrooms (cf. **Psilocybe mexicana**). Twenty-six seeds is given as a dosage (Lipp 1991, 190*). The Zapotec consider the plant sacred, "like a little god"; they ingest thirteen individual seeds or thirteen pairs of seeds (= twenty-six) for divinatory, clairvoyant, and medicinal purposes (Fields 1968). The seeds are powdered (if possible by a virgin), soaked in water, and then ingested. In some areas, they are purportedly also made into a **snuff** (Furst 1976b, 155*). As the seeds are being collected, the patient should direct the following prayer to the plant (Fields 1968, 206): "I come here to purchase something from you. With your permission, you will heal my illness."

Several hours after the patient has ingested the seeds, things of importance are revealed to him in a dreamlike, hypnotic state. Two children (*niige*) or the *badoh* plant itself appear to the afflicted person and tell him the reasons for his illness. The patient usually stammers, and the *curandera* will interpret his sounds and words. After the session, she speaks with the patient about the messages from the plant (Fields 1968, 207).

In the area of the Maya (Yucatán), the medicinal and ritual use of the ololiuqui vine was documented during the colonial period. It was consistently listed under the name *xtabentum*, "jewel cord," in the Mayan lexica and was described in the *Libro dej Judío*:

The plant, *tabentun*, is a vine that gets white flowers. It is common in the gardens. Its quality is "moderate" [humoral doctrine] and it has many effects; the most well known is for those who cannot urinate. It can open the channels in which there is a stone. The bees take honey from its flowers. (Rätsch 1986a, 232*)

Honey plays an important role in the cult of the Maya during the production of the mildly alcoholic ritual drink known as balche', the use of which extends far back into the Classic Mayan period and is still found in the Yucatán today. *Xtabentum* honey is popular for making **balche'** because it improves its effects. The seeds also are ingested with balche'. One Mayan shaman (*h-mèn*) had the following to say about *xtabentum*:

> Especially when it is freshly harvested, ground, and taken as a drink; and when one drinks enough of it, one sees thousands of spirits, has contact with the devil and with hell. . . . When someone loses something valuable, we give him xtabentum to drink. Before he falls asleep, we repeatedly say into his ear: "Where is the lost object." And we describe it. He becomes clairvoyant in the ensuing xtabentum sleep and sees where the object is. And if it was stolen, he will recognize the thief. Because the sleep is not deep, we can speak with him by calling to him repeatedly, as with people under hypnosis. He will give clear answers, although slowly and haltingly. In the inebriation of xtabentum, one also becomes weak and regrets his sins. He admits to everything if he is asked. (Leuenberger 1979, 83 f.*)

Artifacts

Pre-Columbian art contains many illustrations of plants and floral elements that can be interpreted as vines (e.g., as *Ipomoea violacea*). Some of the images contained in the wall paintings at Tepantitla (Teotihuacán) are particularly fasci-

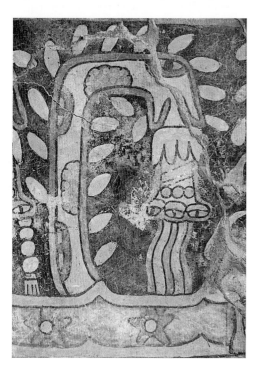

nating. Numerous vinelike plants bear white flowers whose petals contain disembodied eyes. Peter T. Furst views these as representations of the *Turbina* vine. They are in a direct iconographic association with a deity that was once interpreted as the (Aztec) rain god Tláloc (cf. *Argemone mexicana*) but probably represents a mother goddess (Furst 1974).

The Mayan codices also show vines that may represent ololiuqui (Rätsch 1986a, 232*). The *Codex Magliabecchi* contains a depiction of the plant as a climber on a field that is borne by rattlesnakes (cf. Guerra 1990, 177*).

The plant is depicted on a Cuban stamp that was issued at Christmas 1960–61 (Schultes and Hofmann 1992, 158*).

Medicinal Use

The Yucatec Maya use the plant medicinally as a diuretic and to treat wounds and bruises (Pulido S. and Serralta P. 1993, 20*). In Tecún Umán (Guatemala), the leaves are used to treat tumors (Fields 1968, 206). The plant finds use in Cuban folk medicine as an aid to parturition (Seoane Gallo 1984, 853*).

Constituents

The fresh seeds (only in the embryo) contain 0.012 to 0.07% indole or **ergot alkaloids** (ergoline alkaloids). The primary alkaloid is ergine (= 5*R*,8*R*-(+)-lysergic acid amide, LA-111 = ergobasin), which comprises 50% of the total amount. The most important secondary alkaloid is erginine (= isoergine, 5*R*,8*S*-(+)-isolysergic acid amide); also present are small amounts of chanoclavine, elymoclavine, and lysergol. Terpene glycosides (e.g., epicorymbosin) and galactomannanes have also been detected (Brenneisen 1994, 1014; Cook and Kealand 1962).

The leaves and stalks, but not the roots, also contain psychoactive **indole alkaloids**. The concentration in the dried leaves ranges from 0.016 to 0.027% and in the dried stems from 0.01 to 0.012% (primarily ergine and erginine = isoergine) (Brenneisen 1994, 1014).

Effects

Ololiuqui seeds do not produce psychedelic effects like those of *Psilocybe* spp., LSD, or *N,N*-DMT. Instead, they produce a hypnotic state similar to that induced by *Ipomoea violacea.* The Indians report powerful visions, even with very low doses. It is possible that the ololiuqui experience is particularly amenable to cultural conditioning. Or perhaps the seeds invoke visionary effects only when they are ingested by a qualified shaman.

While the main active constituent ergine has been demonstrated to produce psychoactive effects, these are not comparable to those produced by LSD. Rather, ergine induces a kind of

"With the studies of ololiuqui, my work in the area of hallucinogenic drugs nicely came full circle. It now formed a circle, one could say a magical circle; the starting point was the studies on the production of lysergic acid amides of the type of the naturally occurring ergot alkaloid ergobasin. These led to the synthesis of lysergic acid diethylamide, of LSD. The work with the hallucinogenic substance LSD then led to studies of the hallucinogenic magic mushroom teonanacatl, from which the active principles psilocybin and psilocin were isolated. The concern with the Mexican magic drug teonanacatl led to work on a second Mexican magic drug, ololiuqui. In ololiuqui, lysergic acid amides, including ergobasin, were once again found to be the hallucinogenic substances, and this closed the magical circle."

ALBERT HOFMANN
LSD—MEIN SORGENKIND [LSD— MY PROBLEM CHILD]
(1979, 149 f.*)

Lysergic acid amide (LSA, ergine)

This pre-Columbian representation of a plant on a wall painting in Tepantitla, Teotihuacán, can be interpreted as *Turbina corymbosa.*

A group of musicians who perform in the techno style known as "psychedelic trance" took their name from the Aztec magical plant ololiuqui. (CD cover, Spirit Zone Records, 1996)

trance or twilight sleep with dream images (Brenneisen 1994, 1015 f.).

Commercial Forms and Regulations

Although the plant is not subject to any legal restrictions, it is almost impossible to obtain. Plant material cannot be procured even in Mexico. Seeds are sometimes available from ethnobotanical suppliers (in very limited quantities). Unfortunately, many seeds that are sold as ololiuqui actually come from **Ipomoea spp.** (usually devoid of active constituents) or some other Convolvulaceae.

The liquor known as xtabentum is available for purchase in Mérida or Valladolid (Yucatán). But it is doubtful whether this is actually made with **honey** from *Turbina corymbosa*.

Literature

See also the entries for **Ipomoea violacea, Ipomoea spp.**, and **ergot alkaloids**.

Brenneisen, Rudolf. 1994. *Turbina*. In *Hagers Handbuch der pharmazeutischen Praxis*, 5th ed., 6:1013–16. Berlin: Springer.

Cook, W. B., and W. E. Kealand. 1962. Isolation and partial characterization of a glucoside from *Rivea corymbosa* (L.) Hall. f. *Journal of Organic Chemistry* 27:1061.

Der Marderosian, Ara. 1967. Psychotomimetic indoles in the Convolvulaceae. *American Journal of Pharmacology* 139:19–26.

Der Marderosian, Ara, and Heber W. Youngken, Jr. 1966. The distribution of indole alkaloids among certain species and varieties of *Ipomoea, Rivea* and *Convolvulus* (Convolvulacea). *Lloydia* 29 (1): 35.

Fields, F. Herbert. 1968. *Rivea corymbosa*: Notes on some Zapotecan customs. *Economic Botany* 23:206–9.

Furst, Peter T. 1974. Mother Goddess and morning glory at Tepantitla, Teotihuacan: Iconography and analogy in pre-Columbian art. In *Mesoamerican archaeology: New approaches*, ed. Norman Hammond. Austin: University of Texas Press.

Heim, E., H. Heimann, and G. Lukacs. 1968. Die psychische Wirkung der mexikanischen Droge "Ololiuqui" am Menschen. *Psychopharmacologia* (Berlin) 13:35–48.

Hofmann, Albert. 1961. Die Wirkstoffe der mexikanischen Zauberdroge Ololiuqui. *Planta Medica* 9:354–67.

———. 1963. The active principles of the seeds of *Rivea corymbosa* and *Ipomoea violacea*. *Botanical Museum Leaflets, Harvard University* 20:194–212.

———. 1964. Mexican witchcraft drugs and their active principles. *Planta Medica* 12:341–52.

———. 1971a. The active principles of the seeds of *Rivea corymbosa* (L.) Hall f. (ololiuhqui, badoh) and *Ipomoea tricolor* Cav. (badoh negro). In *Homenaje a Roberto J. Weitlaner*, 349–57. Mexico: UNAM.

———. 1971b. Teonanácatl and ololiuqui: Two ancient magic drugs of Mexico. *Bulletin on Narcotics* 23 (1): 3–14.

Hofmann, Albert, and A. Tscherter. 1960. Isolierung von Lysergsäure-Alkaloiden aus der mexikanischen Zauberdroge Ololiuqui (*Rivea corymbosa* [L.] Hall. f.). *Experientia* 16:414–16.

Isbell, H., and C. W. Gorodetzky. 1966. Effects of alkaloids of ololiuqui in man. *Psychopharmacologia* (Berlin) 8:331–39.

Osmond, Humphry. 1955. Ololiuhqui: The ancient Aztec narcotic. *Journal of Mental Science* 101:526–37.

Ott, Jonathan. 1996. *Turbina corymbosa (Linnaeus) Rafinesque*. Unpublished database

Reko, Blas Pablo. 1934. Das mexikanische Rauschgift Ololiuqui. *El México Antiguo* 3 (3/4): 1–7.

Safford, William E. 1915. An Aztec narcotic. *Journal of Heredity* 6 (7): 291–311.

Schultes, Richard Evans. 1941. *A contribution to our knowledge of* Rivea corymbosa: *The narcotic ololiuqui of the Aztecs*. Cambridge, Mass.: Botanical Museum of Harvard University.

Taber, W. A., et al. 1963. Ergot-type alkaloids in vegetative tissue of *Rivea corymbosa* (L.) Hall. f. *Phytochemistry* 2:99–101.

Wasson, R. Gordon. 1971. Ololiuqui and the other hallucinogens of Mexico. In *Homenaje a Roberto J. Weitlaner*, 329–48. Mexico: UNAM.

Wolff, Robert. 1966. Seeds of glory. *Psychedelic Review* 8:111–22.

Turnera diffusa Willd. ex Schultes

Damiana

Family
Turneraceae (Turnera Family)

Forms and Subspecies
The damiana that is found in Baja California has been described as a variety: *Turnera diffusa* Willd. var. *aphrodisiaca* (Ward) Urban

Synonyms
Turnera aphrodisiaca L.F. Ward
Turnera aphrodisiaca Willd.
Turnera humifusa Endl.
Turnera pringlei Rose

Folk Names
Ajkits, damiana, damiana amarilla, damiana americana, damiana de California, garañona, hierba de la mora, hierba de la pastora (Spanish, "plant of the shepardess"),[312] hierba del pastor (Spanish, "plant of the shepherd"), hierba del venado (Spanish, "plant of the deer"), itamo real, jícamo real, Mexican damiana, mezquitillo, miixkok, misibkok (Mayan, "asthma sweeper"), misibkook, mis kok (Mayan, "asthma broom"), old woman's broom, oreganillo ("little oregano"), oreja de venado (Spanish, "ear of the deer"), paraleña, pastorcita (Spanish, "little shepherdess"), pastorica, rosemary, salverreal, San Nicolás, shepherd's herb, stag's herb, xmisibkok, xmisibkook

Species Also Referred to and Sold as Damiana
(Martínez 1994, 120*)

Turnera pumilla L.	bruja (Spanish, "witch")
Turnera ulmifolia L.	clave de oro (Spanish, "gold clove")
Chrysactinia mexicana A. Gray	false damiana
Bigelowia veneta A. Gray [syn. *Haplopappus discoideus*]	false damiana
Haplopappus laricifolius	false damiana

History
It is very likely that damiana was in use as a medicine and love drink in northern Mexico and the area of the Maya during prehistoric times. The missionary Jesús María de Salvatierra first mentioned its aphrodisiac use among the Indians

of northern Mexico in his *Chronica* of 1699. The name *damiana* is derived from that of either Saint Damian, the patron saint of pharmacists, or Peter Damiani, who railed against the immorality of the clergy in the eleventh century.

The first botanical description of the plant was written by Austrian botanist Josef August Schultes (1773–1831) in 1820. In the nineteenth century, the plant was included in the U.S. (1874) and Mexican pharmacopoeias as a tonic and aphrodisiac (Martínez 1994, 121*). It was introduced into Europe in 1880 (Hirschfeld and Linsert 1930, 174*). Since the late 1960s, the plant has been regarded as a "legal high" and as a substitute for marijuana as well as tobacco (*Cannabis indica, Nicotiana tabacum*).

Prior to Prohibition in the United States, Dr. John S. Pemberton (1831–1888), the inventor of Coca-Cola, developed a tonic beverage inspired by Vin Mariani. Known as French Wine Coca, it contained extracts of coca, cola, Mediterranean sweet wine, and damiana (cf. *Erythroxylum novogranatense*).

Distribution
Damiana occurs from Southern California (Baja) to Argentina. The main region of the plant is in northern Mexico and Baja California.

Cultivation
The plant can be grown from seed (very difficult) as well as cuttings. It requires a warm or hot climate but does not have any great requirements in terms of soil type (Grubber 1991, 26 f.*). Damiana also grows well in desert zones (Miller 1985, 21*).

Appearance
The plant usually attains a height of about 30 cm but in rare cases can grow as tall as 2 m. The leaves, which are alternate, lanceolate, serrated, and covered with a few hairs, can grow up to 2 cm in length. The yellow flowers, which are only 12 mm

The yellow-blooming damiana (*Turnera ulmifolia*) occurs throughout the American tropics and has spread to Asia as well as various islands in the Indian Ocean. (Photographed on Oahu, Hawaii)

The damiana plant may be named after Saint Damian, the patron saint of pharmacists. (Colorized woodcut showing the brothers Cosmas and Damian; title page by Hans Wechtlin to *Feldbuch der Wundartzney,* by Hans von Gersdorf, Strassburg, 1517)

312 This name is used in Querétaro (Martínez 1994, 119*); in Oaxaca, it is used for *Salvia divinorum.*

Dried damiana herbage (*Turnera diffusa*).

in length and grow from the leaf axils, bloom from July to September. The round, three-chambered capsule fruit is just 2 to 4 mm in size and contains only one or two pear-shaped seeds.

Damiana can be easily confused with other species in the genus *Turnera* as well as with *Chrysactinia mexicana*. It can be distinguished from the rich green *Turnera ulmifolia* on the basis of the greenish blue color of its leaves and its distinctly smaller flowers.

Psychoactive Material
— Herbage sans roots (herba damianae, damiana herbage)

Preparation and Dosage
The dried herbage can be prepared as a tea or an alcohol extract, smoked, or burned as incense. For aphrodisiac purposes, one can either smoke a joint made from the leaves or drink a tea prepared from the herbage (Gottlieb 1974, 27 f.*).

Damiana herbage is an ingredient in some psychoactive **smoking blends** (Miller 1985, 23*). It is especially popular as a substitute for tobacco (**Nicotiana tabacum**) for smoking together with hashish (see **Cannabis indica, Cannabis sativa**).

Damiana tea can be prepared as an infusion, a decoction, or a cold-water extract. An infusion of damiana herbage, to which orange blossoms can be added if desired, should be allowed to steep for three to five minutes. Boil the herbage for up to an hour to produce a decoction that is more potently effective. Cold-water extracts should be allowed to sit for twenty-four hours. The dosage for teas is 4 g

Damiana Liqueur
1 bottle (0.7 l) **alcohol** (white rum [sugarcane spirits] or tequila [cf. **Agave spp.**])
approximately 10–20 g damiana herbage (*Turnera diffusa*)
approximately 20–25 g saw palmetto fruits (fructus sabalae serrulata tot.) (cf. **palm wine**)
2 vanilla pods (= 7–9 g) (*Vanilla planifolia*)
4 cinnamon sticks (approximately 15 g) (*Cinnamomum verum*)
approximately 2 g mace (**Myristica fragrans**)
approximately 0.5 g galangal root (*Alpinia galanga* [L.] Willd. [syn. *Maranta galangal* L.]) or *Alpinia officinarum* Hance) (cf. **Kaempheria galanga**)

Slice the vanilla pods lengthwise or in half. Add all of the ingredients to the spirits and allow the mixture to sit in a warm location for at least two weeks. Then either filter out the ingredients or leave them in the bottle. Drink one small glassful daily or one hour before an erotic encounter.

per cup or mug (Lowry 1984, 267). The dosages can be increased as desired, as side effects are unknown.

For aphrodisiac purposes, damiana is often combined with equal parts of saw palmetto fruits (cf. **palm wine, wine**) and occasionally also with cola nuts (*Cola* spp.). A preparation of former times known as *píldoras de damiana* consisted of 5.5 g of phosphorus, 9 g of *Strychnos nux-vomica*, and 10 g of damiana (Martínez 1994, 122*). Damiana also can be combined with pure **strychnine** (Lowry 1984).

The herbage is well suited for making liqueurs. In Mexico, it is used to manufacture a liqueur that is alleged to have aphrodisiac effects.

A Mexican Recipe for an Aphrodisiac Damiana Tea

30% damiana (*Chrysactinia mexicana* Gray)
10% gobernadora (*Larrea tridentata* [DC.] Cav.; also *Larrea divaricata* [?])
50% damiana californica (*Turnera diffusa* Willd.)
10% garañona (*Castilleja canescens* Benth. or *Castilleja arvensis* Schl. et Cham.)

Combine all the ingredients. Add 2 teaspoons of the mixture to 1 liter of water; drink 1 cup after every meal.

Ritual Use
No traditional use of damiana based upon its psychoactivity is known from Mexico.

In the voodoo cult practiced in the southern United States, damiana is consecrated to the love goddess Erzulie and is used in love magic (Müller-Ebeling and Rätsch 1986, 122 f.*; Riva 1974).

The use of damiana herbage as an **incense** is most likely a modern invention. Because it is also regarded as an aphrodisiac when burned, damiana herbage is often added to so-called Pan, Venus, or love incenses. When used as a fumigant, damiana produces a pleasantly herbal, sweet scent that is characteristic and easily recognizable when encountered again. It combines very well with copal (resin of *Protium copal* or *Bursera* spp.; cf. **Bursera bipinnata**).

"Damiana has mild diuretic and aphrodisiac effects, especially among women. Most specialists consider the libido-promoting effects to be induced psychogenically as a result of the mental state of expectation of the believed effectiveness."

BRUNO WOLTERS
AGAVE BIS ZAUBERNUSS [AGAVE TO MAGIC NUT]
(1996, 59*)

The Maká Indians once had a magic custom that used the roots of *Turnera ulmifolia* to improve the sound of the flutes they used in rituals (Arenas 1987, 287*).

Artifacts

In Mexico, a commercial damiana liqueur is bottled in containers in the shape of a female torso, an advertisement for the aphrodisiac effects of the drink (Rätsch 1990, 160*).

Medicinal Use

In Indian medicine, damiana is used primarily in the treatment of asthma. This use is reflected in the Mayan name *mis kok*, "asthma broom," for the plant "sweeps away" the illness. When used for this purpose, the herbage can be drunk as a tea, burned as a fumigant, or smoked. In Mexico, damiana's good reputation as an aphrodisiac has led to the nickname "shirt remover" (Argueta V. et al. 1994, 566*). In Mexican folk medicine, damiana tea is drunk twice a day for fifteen days as a diuretic and to regulate the menstrual cycle (Jiu 1966, 256*). The Indians of northern Mexico use the plant primarily to treat muscle weakness and nervousness and, of course, as an aphrodisiac (Martínez 1994, 121*). Damiana is also used in northern Mexico to treat stomach problems, rheumatism, headaches, and scorpion stings (Wolters 1996, 57*). A variety of different preparations are drunk to treat smoker's cough (cf. *Nicotiana tabacum*) (Argueta V. et al 1994, 566*).

In the Bahamas, the steam produced when damiana herbage is boiled in water is inhaled to treat headaches (Brown and Malone 1978, 12*). Bed wetters drink a damiana tea in the morning for three or four consecutive days "to strengthen their backs" (Eldridge 1975, 320*).

In phytotherapy, damiana has been found to be especially effective in treating menstrual pains and cramps, as it not only has antispasmodic properties but also improves the mood. When used for this purpose, cassia cinnamon (*Cinnamomum aromaticum* Nees [syn. *Cinnamomum cassia* Bl.]) can be added to the tea (Lowry 1984).

A tincture (mother tincture) of the dried leaves, known as Damiana, is used in homeopathy as an aphrodisiac and for other purposes (Schneider 1974, 3:362*):

> Should be of use for sexual neurasthenia. Sexual weakness as a result of nervous prostration. Incontinence in older persons. Chronic prostatorrhea. Kidney and bladder catarrh; frigidity in women. Helps to produce a normal menstrual flow in young girls.— Dosage. —Tincture and liquid extract, amounts of 10–40 drops. (Boericke 1992, 292*)

Damiana also is found in many homeopathic compounds used to treat such ailments as sexual weakness. For example, Damiana Pentarkan consists of damiana, ginseng (*Panax ginseng*), muira puama (*Liriosma ovata*), phosphoric acid, and ambergris.

Constituents

Damiana leaves are 0.2 to 0.9% **essential oil**, 6% hard brown resin, approximately 8% soft resin, 3.5% tannin, and 6% starch (Brown and Malone 1978, 12*). According to the Mexican pharmacopoeia, the herbage contains 8.06% chlorophyll, white resin, and essential oil; 6.39% hard brown resin; 3.46% tannin; and 7.08% yellow dye (Martínez 1994, 120*). According to a different analysis, the herbage contains 0.51% essential oil of a greenish color, two resins, 0.7% arbutin, the bitter substance damianin, tannin, sugar, and albuminoids (Steinmetz 1960). The essential oil consists of some twenty substances, of which 1,8-cineol, α-pinene, β-pinene, and *para*-cymene have been identified (Auterhoff and Hauffel 1968; Argueta V. et al. 1994, 566*). About half of the essential oil consists of sesquiterpenes (guajan derivatives and similar substances) and the other half of monoterpenes (pinene, thymol); cineol and *para*-cineol were detected in only a few of the samples (Wolters 1996, 59*). Although it has often been claimed that **caffeine** is present in the leaves, this is questionable (Lara Ochoa and Marquez Alonso 1996, 47*; Lowry 1984, 268). In contrast, the stems have been determined to contain caffeine (Argueta V. et al. 1994, 566*). The flavone 5-hydroxy-7,3',4'-trimethoxyflavone has been isolated from the herbage (Domínguez and Hinojosa 1976), as has tetraphylline B (Spencer and Seigler 1981). **Ephedrine** is not present.

The leaves of the related *Turnera ulmifolia* contain procyanidin, and the seeds and leaves higher concentrations of **caffeine** (Wolters 1996, 59*).

Effects

Smoking the herbage produces a pleasant state of euphoria and mild, marijuana-like effects (Miller 1993, 8*). The "high" lasts for some sixty minutes (Lowry 1984, 268). Drinking the tea or other preparations produces effects that are only subtly perceptible and in no way spectacular. Damiana has effects upon the lower abdomen, which can result in an increase in the flow of blood into the region. Women have repeatedly reported that damiana has a very relaxing effect on menstrual cramps or pains.

The herbage is generally considered to have tonic, diuretic, stimulant, and aphrodisiac effects.[313] Damiana received the best marks in a test of different plants and natural drugs said to have aphrodisiac properties (Radakovich 1992,

"The effects of damiana apply especially to the genital area. According to information provided by American physicians, its effects are exciting to the highest degree and are in this regard comparable to the effects of coca."

MAGNUS HIRSCHFELD AND RICHARD LINSERT
LIEBESMITTEL [LOVE AGENTS]
(1930, 174*)

[313] Unfortunately, no pharmacological studies of these effects have been conducted to date (Lowry 1984, 267).

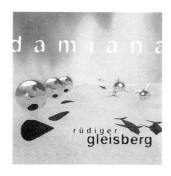

The love plant damiana inspired the German electronic musician Rüdiger Gleisberg to create this album. (BSC Music/Rough Trade 1997)

32). An ethanol extract of the plant had antibiotic effects upon *Staphylococcus aureus* and *Bacillus subtili* (Argueta V. et al. 1994, 566*). Jiu (1966, 257*) was able to measure an effect upon the central nervous system.

Commercial Forms and Regulations

Damiana is available without restriction from pharmacies and herb shops (damiana folium conc., herbae damianae). In the United States, tinctures and extracts of damiana can be purchased in health food stores and supermarkets. Damiana mother tincture is available in Europe. Damiana extracts and drops are sold in sex shops.

A "damiana essence" is occasionally offered for sale. However, this is actually davana oil, which is obtained from different stock plants (e.g., *Artemisia pallens*; cf. **Artemisia spp.**). Raw plant material from *Turnera ulmifolia*, *Haplopappus* spp., and *Chrysactinia mexicana* is also sold under the name *damiana*.

Literature

Auterhoff, H., and H. P. Hauffel. 1968. Inhaltsstoffe der Damiana-Droge. *Archiv für Pharmazie* 301:537–44.

Der Marderossian, Ara H., et al. 1977. Pharmacognosy: medicinal teas—boon or bane. *Drug Therapy* 7:178–86.

Domínguez, X. A., and M. Hinojosa. 1976. Mexican medicinal plants. XXVIII: Isolation of 5-hydroxy-7,3',4'-trimethoxy-flavone from *Turnera diffusa*. *Planta Medica* 30 (68): 68.

Fryer, F. A. 1965. A chemical investigation of damiana (*Turnera diffusa*). *Specialities* 1 (12): 21.

Lope, Vergara. 1906. Damiana. *Anales del Instituto Médico Nacionál* 8:238.

Lowry, Thomas P. 1984. Damiana. *Journal of Psychoactive Drugs* 16 (3): 267–68.

Radakovich, Anka. 1992. Love drugs. *Details* 8:32–33.

Ramírez, José. 1903. La damiana (*Turnera diffusa aphrodisiaca*). *Anales del Instituto Médico Nacionál* 5:238.

Riva, Anna. 1974. *Voodoo handbook of cult secrets.* Toluca Lake, Calif.: Occult Books.

Ruíz, Luis E. 1906. Damiana. *Anales del Instituto Médico Nacionál* 8:87.

Spencer, K. C., and D. S. Seigler. 1981. Tetraphyllin B from *Turnera diffusa*. *Planta Medica* 43:175–78.

Steinmetz, E. F. 1960. Damiana folia. *Acta Phyto Therapeutica* 7 (1): 1–2.

Zubke, Achim. 1998. Damiana, das sanfte Aphrodisiakum. *HanfBlatt* 5 (44): 8–10.

"Damiana is the means for loving thy neighbor!"

Max Amann (September 1997)

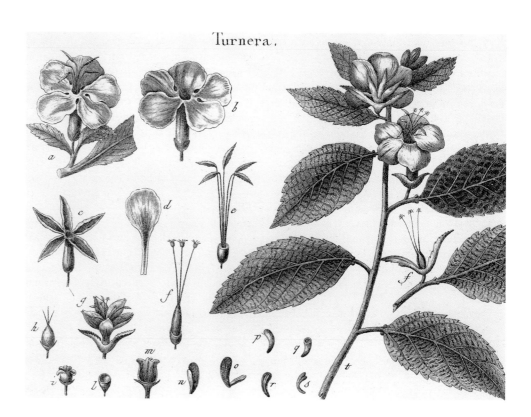

Vaccinium uliginosum Linnaeus

Bog Bilberry

Family
Ericaceae (Heath Family); Subfamily Vaccinioideae, Vaccinieae Tribe

Forms and Subspecies
The bog bilberry can vary considerably in appearance from one location to another. In low-lying areas, it can develop into an imposing bush, while in the high mountains it takes on a compact form (Hecker 1995, 288*).

Synonyms
None

Folk Names
Airelle uligineuse (French), bogbilberry, bog bilberry, bog whortleberry, lausbeere, mirtillo falso (Italian), moorbeere, moorheidelbeere, moosbeere, rausch, rauschbeere, rijsbes (Dutch), saftbeere, schwindelbeere, schwindelbeeri, sumpfheidelbeere, tollbeere, trunkelbeere

History
The renowned paleoanthropologist Björn Kurtén has reconstructed the prehistory of humans in northern Europe in novel form. He places the invention of an inebriating drink made from plants from the Heath Family there in the early Stone Age. During the Middle Ages, a wine made from bog bilberries was produced in Scandinavia. In Siberia, shamans used the berries together with fly agaric mushrooms (*Amanita muscaria*), a custom that may also have been known in Europe. In Tyrolia (Austria), it is still said that children will lose their minds if they eat bog bilberries (Engel 1982, 109*).

Distribution
The bog bilberry is a circumpolar plant that is at home in North America and in Siberia. It is also frequently encountered in dwarf shrub heaths and arolla pine forests in the Alps, e.g., on Bettmeralp and in other locations in Valais (Switzerland) (Hecker 1995, 288*).

Cultivation
Propagation is performed from seeds, which can be raised in seedbeds that should be kept moist. The seedlings can then be transplanted as desired.

Appearance
This undershrub can grow as tall as 1 meter. It has alternate, summer-green, deciduous leaves. The pinkish white flowers hang in clusters. The round, blue, pruinose fruits resemble blueberries; they

Other *Rauschbeeren* ("Inebriating Berries")
The German name *rauschbeere*, "inebriating berry," is also given to several other plants, especially the evergreen dwarf shrub *Empetrum nigrum* L. (Empetraceae/Crow Berry Family), known as the black crow berry. This inebriating berry occurs in two subspecies: ssp. *hermaphroditum* (Lange) Böcher [syn. *Empetrum hermaphroditum* (Lange) Hagerup] and ssp. *nigrum* (Zander 1994, 558*). This Scandinavian plant has a long history of use as an inebriant:

> In Norway, the juice of the drunken berry or inebriating berry (*Empetrum nigrum* L.) was used to make wine. King Sverre (12th century) attempted to use such native wines to drive out the foreign wine that German merchants were importing in. In 1203, Bishop Jon taught the Icelanders how to make a wine like the one that King Sverre had taught him about. It was apparently this wine or another that was prepared from berries that was involved when the Norwegian and Icelandic clergy asked Pope Gregory IX for permission to use domestic wine during mass because true wine was not available in the country. Although the Pope did not give his permission, tradition says that such inebriating berry wine was used in Iceland for Holy Communion. (Hartwich 1911, 761*)

Even today, the plant has a reputation of being a hallucinogen: "The inebriating berry of the North Sea coast, which is consumed raw and cooked, causes inebriated states and hallucinations but is not a [narcotic]" (Körner 1994, 1572*).

The entire plant contains quercetin, ursolic acid, rutin, isoquercitrin, ellagic acid, andromedotoxin, and alkaloids. **Honey** from this plant can be toxic (Roth et al. 1994, 319*).

The cowberry (*Vaccinium vitis-idaea* L.) is also known colloquially as "inebriating berry," perhaps because its berries are also used to produce inebriating beverages (e.g., Kroatzbeeren liquor).

Vaccinium floribundum H.B.K. var. *ramosissimum* (D. Don) Sleuner, a Bolivian species closely related to *Vaccinium uliginosum*, is known locally as *macha-macha* (cf. **macha**). The same name is given to the related *Pernettya* spp. (von Reis Altschul 1975, 215*), which also produce inebriating fruits (cf. **chicha**).

The bog bilberry (*Vaccinium uliginosum*), an alpine native, can have inebriating effects. (Wild plant, photographed near the Aletsch glacier, Bettmeralp, Switzerland)

The German name for the bog billberry, *rauschbeere* ("inebriating berry"), is appropriate, for the fruits can produce an inebriated state if enough are consumed. (Woodcut from Tabernaemontanus, *Neu Vollkommen Kräuter-Buch*, 1731)

produce a colorless juice and have a sour/sweet taste. The plant flowers in June and July, and the fruits mature in the fall (August to September).

The plant can be easily confused with the true blueberry (*Vaccinium myrtillus* L.) and other members of the Heath Family (e.g., *Vaccinium vitis-idaea* L., *Vaccinium oxycoccus* L.), especially before the fruits ripen.

Psychoactive Material
— Fruits (uliginosi fructus, fructus uliginosi, rauschbeeren, rauschbeerfrüchte, bog bilberries)
— Leaves (uliginosi folium, folia uliginosi, rausch-beerenblätter, bog bilberry leaves)

Preparation and Dosage
The fresh berries or the juice pressed from them is ingested. One handful of berries is said to be sufficient to induce inebriating effects. Bog bilberry has retained its reputation as an inebriant in part because a type of **wine** is made from the plant:

> In Norway, the juice of the fruits of *Vaccinium uliginosum* L., the bog bilberry, with some sugar, which is also a popular additive in the production of berry wine, is allowed to ferment into a wine. (Hartwich 1911, 761*)

In Siberia, the juice pressed from fresh bog bilberries is mixed with dried fly agaric mushrooms (*Amanita muscaria*) and drunk (Lewin 1980 [orig. pub. 1929], 168*; Schultes 1969, 246*). Water and yeast were sometimes added to the mixture to produce a kind of **beer**. It is possible that the berries were also used as an additional inebriating additive to Germanic **mead** and beer.

The berries can be preserved by drying. They should be collected when ripe and dried in the sun or in a warm location.

The dried leaves can be smoked (cf. **smoking blends, kinnikinnick**) or brewed into a tea known as *batum* (Lewin 1980 [orig. pub. 1929], 352*).

Ritual Use
In Old Germanic times, the **wine** produced from bog bilberries apparently was consumed as part of certain drinking rituals, for it was used as a wine in early Christian masses. Since the Catholic Church integrated traditional pagan customs into the local liturgies to ensure its position of power, it seems reasonable to assume that bog bilberry wine was once an offering drink to the Germanic gods, e.g., to Odin/Wotan, who the Eddas tell us was the "wine drinker" of the gods (cf. **mead**).

For information about the shamanic use of bog bilberries in Siberia, see ***Amanita muscaria***.

Artifacts
It is possible that some skaldic songs and head rhymes were inspired by bog bilberry wine (cf. **mead**).

Medicinal Use
Bog bilberry leaves were used in folk medicine in the same manner as bilberry or bearberry leaves (*Arctostaphylos uva-ursi*; cf. **kinnikinnick**). A tea (cold-water extract) of bog bilberry leaves and/or fruits is drunk for diarrhea and bladder ailments (Pahlow 1993, 245 f.*).

Constituents
The entire plant contains flavanols, flavonoid compounds, tanning agents, vitamins (especially C), minerals, a glycoside, and arbutin derivatives (Pahlow 1993, 254*).

The inebriating constituents of the berries are apparently a metabolic product or a constituent of a parasitic fungus (*Sclerotina megalospora* Wot.) that often infects the fruits (Frohne and Pfänder 1983, 111*). To date, however, no constituent has been isolated or identified (Roth et al. 1994, 718*). Because the inebriation is likely a metabolic product of a fungus, the active constituent may be some type of **ergot alkaloid**.

The leaves contain hyperoside, ursolic acid, α-amyrine, friedelin, oleanolic acid, (+)-catechin, and organic acids (Roth et al. 1994, 718*). The quercetin derivative found in the leaves, quercetin-3-glucuronide (Gerhardt et al. 1989), may be a narcotic compound (cf. *Psidium guajava*).

Effects
Consuming the fruits is reported to induce an inebriation characterized by excitation, pupillary dilation, feelings of dizziness, and vomiting and numbness (Frohne and Pfänder 1983, 111*; Roth et al 1994, 719*; Zipf 1944). Sometimes the only effect reported is "nausea" (Root 1996, 32 f.*).

Commercial Forms and Regulations
None

Literature

Gerhardt, G., V. Sinnwell, and Lj. Kraus. 1989. Isolierung von Quercetin-3-glucuronid aus Heidelbeer- und Rauschbeerblättern durch DCCC. *Planta Medica* 55:200 ff.

Moeck, Sabine. 1994. *Vacccinium*. In *Hagers Handbuch der pharmazeutischen Praxis*, 5th ed., 6:1051–1067. Berlin: Springer.

Zipf, K. 1944. Vergiftungen durch Rauschbeeren: Sammlung von Vergiftungsfällen. *Archiv für Toxikologie* 13:139–40.

Veratrum album Linnaeus

White Hellebore

Family
Liliaceae (Lily Family; previously Melanthiaceae)

Forms and Subspecies
The American hellebore is sometimes regarded as a distinct species but more recently has been considered a subspecies of white hellebore (Roth et al. 1994, 725*):

Veratrum album L. ssp. *viride* Ait.
Veratrum album L. var. *viride* Baker

Synonyms
Veratrum viride Baker

Folk Names
Brechwurz, condision, elabro bianco (Italian), elleborus albus, European white hellebore, false hellebore, fieberstellwurzel, gärwere, gentiana maior, germander, germar, germara, germâra, germaren, germarrun vel hemerun, germer, germerra, germerwurzel, gonos aetou (Greek, "eagle's chest" or "summer bird child"), heimwurz, helleboros leukos, helleborus albus, hemer, hemera, hemerum, kondochi, kundush (Persian), lagnion (Gaulish, "physician plant"), langwort, läusekraut, lüppwurzel, marsithila ("seat of a sea demon"), melampodium, nieskraut, nieswurz, politizon, rumex albus, scamphonie, schampanier-wurtzel, sichterwurtz alba, sitterwurz, sittirwurz, süttirwurz, somphia (Egyptian), veladro, vératre blanc (French), veratro bianco (Italian), weiß nießwurtz, weiße nieswurz, weißer germer, white-flowered veratrum, white hellebore, winterwurz, wis nisworz, wiswurz, witte nieswortel (Dutch)

History
According to Theophrastus, the two types of *helleboros*, black hellebore (*Helleborus niger* L.) and white hellebore (*Veratrum album*), were the most important of all medicinal plants in pre-historic Greece. They were the central medicines of the *rhizotomes*, diggers who nourished the magical plants with shamanic rituals. Hellebore was a sacred plant of the gods. The name *helleboros* may be derived from *hella-bora*, "food of Helle." Helle was a Pelasgian goddess for whom the Hellespont was named (Graves 1948, 440*). The most important mode of application of the root was nasally, as **snuff**. The artificially induced sneezing (the German name *nieswurz* means "sneezing root") was believed to cause the demons of sickness to leave the body. "The white sneezing plant, which very quickly produces sneezing, is best; but it is much more terrible than the black [*Helleborus niger* L.]" (Pliny 25.23, 56).

The plant's use as a sneezing powder has continued into the present day, although it has become increasingly profane. White hellebore is now mixed into Schneeberger Schnupftabak (Snow Mountain Snuff), which is used as an agent of pleasure (Höfler 2990, 85*; Schneider 1974, 3:386*). The root has been used as a sneezing

> "We ate poisonous fungi and
> *Veratrum album* as well; the
> herbage of the white hellebore.
> All of them the faces of night!
> I wanted to call it out loud and
> could not.
> I wanted to go to the side and could
> not.
> . . . .
> Albine Veretrine turned around to
> me once more
> And made an obscene gesture.
> I wanted to turn my eyes to the side
> or
> Close my eyelids
> And could not . . ."
>
> GUSTAV MEYRINK
> "BAL MACABRE"
> IN *DES DEUTSCHEN SPIESSERS
> WUNDERHORN 1* [THE GERMAN
> PHILISTINE'S MAGIC HORN]
> (1984, 63*)

Left: The typical wild form of white hellebore (*Veratrum album*) with white flowers. (Wild plant, photographed on Weissenstein, Jura, Switzerland)

Right: A green-flowering form of the alpine *Veratrum album*. (Wild plant, photographed in Valais, Switzerland)

Left: The rootstock of white hellebore (*Veratrum album*) is rich in highly active substances.

Right: Infructescence of the American hellebore (*Veratrum album* ssp. *viride* = *Veratrum viride*).

Botanical illustration of the white hellebore (*Veratrum album*). (Engraving from Pereira, *De Beginselen der Materia Medica en der Therapie*, 1849)

powder in pranks, e.g., on New Year's Eve (cf. *Calliandra anomala*). White hellebore has some significance as a psychoactive substance in occultism (Werner 1991, 469*).

Distribution
The plant occurs throughout Eurasia, especially in the Alps, the Pyrenees, central Asia, Scandinavia, Finland, Siberia, North America, and Alaska. The plant is frequently encountered among tall herbaceous vegetation, in mountain meadows, and in clearings. In Switzerland, white hellebore is part of the typical flora of the Jura.

Cultivation
Propagation can be performed with the seeds or with scions or root segments. The plant loves calciferous soils and also does well in humus- and nutrient-rich soil.

Appearance
This herbaceous perennial can attain a height of up to 1.5 meters. It has a straight, thick, round, fleshy stalk to which the broad, ovate, continuous, alternate leaves (25 to 30 cm long) are directly attached. The plant has a cylindrical rootstock from which numerous fleshy albeit thin roots as long as 20 cm develop. The green or white flowers are only 1 cm in size and are found in thick terminal panicles. The flowering period usually lasts from June to August. The fruits are small, brown, roundish capsules that are filled with seeds.

White hellebore can be easily confused with the closely related North American species or subspecies *Veratrum viride* Ait. The plant also can

be taken for the toxic yellow gentian (*Gentiana lutea* L.), although it can be distinguished by the arrangement of the leaves: white hellebore has three leaves at each stem attachment, while the gentian has decussate leaves that are streaked with numerous leaf nerves and whose footstalks become shorter toward the top (Pahlow 1993, 122*).

White hellebore is also easily confused with the West Indian sabadilla (*Schoenocaulon officinale* [Cham. Et Schlecht.] A. Gray [syn. *Veratrum sabadilla*]) (cf. Pereira 1849, 111*; Wolters 1996, 230*).

Psychoactive Material
— Rootstock with roots; rhizome (rhizoma veratri alba, radix veratri albi, veratri albi rhizome, germerwurzel, weiße nieswurzel, radix ellobori albi, radix campanica, white hellebore root)
— Leaves (folia veratri albi, germerblätter, nieswurzblätter, white hellebore leaves)

Preparation and Dosage
The roots of wild plants are collected in September or October, dried well, and powdered. Together with *Nicotiana tabacum*, the root powder is one of the ingredients in Schneeberger Schnupftabak, a modern European **snuff**.[314]

The dried leaves can be smoked alone or used as an ingredient in **kinnikinnick** or other **smoking blends**. For psychoactive purposes, white hellebore roots also appear to work well in combination with *Amanita muscaria* (Meyrink 1984*)

The root has been (accidentally) used to distill a "gentian liquor" (cf. **alcohol**) (Hruby et al. 1981), the effects of which can be devastating (Roth et al. 1994, 723*).

It has been suggested that white hellebore may have been an ingredient in **witches' ointments**. In early modern times, it was used as an inebriating additive to **beer**, and possibly to **mead** and **wine** as well.

Casual use of *Veratrum album* is very dangerous! One to 2 g of the dried root can be sufficient to cause death through respiratory and circulatory paralysis (Frohne and Pfänder 1983, 153*; Roth et al. 1994, 723*). No information is available about smoking dosages.

Ritual Use
The Greeks and Romans used white hellebore, which was dug up in a ritual similar to that used for *Mandragora officinarum*, as a ritual cleansing agent that was not only snuffed but also strewn about in houses and hearths (Höfler 1990, 82*).

The prophets and magicians of late antiquity called the plant the "seed of Hercules" (Dioscorides 4.148), thereby associating it with the semi-divine sperm. Unfortunately, only vague information about its magical uses has come down to us. Apparently, the "prophets and magicians" referred

---

314 "Small amounts of the Schneeberger Schnupftabak—which contains only a little hellebore—can be consumed without concern. Anyone who snuffs too frequently or too much may occasionally have nosebleeds" (Pahlow 1993, 242*).

to in the sources were Celtic Druids. In late antiquity, Herakles/Hercules enjoyed great popularity in Gaul, becoming an object of Celtic mythology and consecration rituals (Botheroyd 1992, 157). This may have been the reason why hellebore branches were hung on houses as an apotropaic protection. It is certain that the Gauls used the root as an arrow poison (Pliny 25.25, 61).

It is possible that the Celts used white hellebore for psychoactive purposes. The Celtic mother goddess Cerridwen, who was related to Demeter, or Ceres, was said to know the secret of the "drink of inspiration and all wisdom." Anyone who partook of it attained enlightenment and was able to experience the past, present, and future as one (cf. **mead**). The mythically and ritually significant "cauldron of Cerridwen"[315] contained "probably a mash of barley, acorns, honey, bull's blood, and such sacred herbs as ivy [*Hedera helix*], white hellebore and laurel [*Laurus nobilis*]" (Graves 1948, 439 f.*).

Honey mead was brewed in a cauldron for the "festival of the other world," which was strikingly shamanic (Botheroyd 1992, 182). It appears as though the intention of this ritual was to journey into the other world through the aid of the drink, for it is "the fount of all wisdom; here is where the heroes learned their magical arts, the poets obtained inspiration, and the Druids their magic with which they cast spells in the real world. Here too was where the wondrous treasure was guarded that the brave wished to raise: here dwelled absolute reality" (Botheroyd 1992, 18).

The Germanic name for the plant, *germâr*, is "probably the name of a old Germanic hero known for his use of the spear" (Höfler 1990, 84*). The Germanic tribes regarded the root as a *marrensitz*, i.e., as a place where elves would dwell (Höfler 1990, 85*). The ancient Germans may have used the plant to travel into other worlds where they could contact the elves, both the light elves in the heavens and the dark elves in the earth. It is possible that white hellebore was inhaled or smoked in the form of **incense** during the Germanic period, a use that continued into the late nineteenth century (Werner 1991, 468 f.*): "In Tyrolia, dried leaves of white hellebore are smoked from time to time" (Höfler 1990, 84*).

The Flathead Indians knew the North American hellebore (*Veratrum viride*) by the name *steso'o*, "sneeze," and used the powdered root as a **snuff** that would induce sneezing and thereby clear the respiratory tract (Hart 1979, 273*). The dried roots were smoked together with tobacco (*Nicotiana* spp.) or bearberry leaves (see **kinni-kinnick**). The Blackfeet call the closely related *Veratrum eschscholtzii* A. Gray *etarva-asi*, "that which makes you sneeze," and snuff the dried and powdered root as a headache remedy (Johnston 1970, 309*). In the Pacific Northwest, the Quinault

call the North American hellebore *tci'ai'nix*, while the Cowlitz Indians call it *mimu'n*,[316] and the plant appears to have had a certain significance in shamanism. It is said that a small piece of the root was chewed, and the resulting saliva was spat onto the water "to make sea monsters disappear" (Gunther 1988, 24*)!

Artifacts

Although *helleboros* was of great importance in ancient Greece, it does not appear in Greek art. Similarly, no snuff paraphernalia has been discovered or even described.

White hellebore inspired the author Gustav Meyrink (1868–1932) to several pieces of literature (Meyrink 1984*). In the French comic series *The Smurfs*, by Peyo, Papa Smurf is constantly preparing magical drinks from white hellebore. The drawings in which the effects of these alchemical brews are shown are both turbulent and amusing (cf. *Cannabis indica*).

Medicinal Use

In ancient times, white hellebore was used medicinally to treat a number of afflictions, especially those of a psychological nature:

> The body must be prepared beforehand for seven days by spicy food and abstention from wine, on the fourth and third day through vomiting, on the day preceding through fasting. White hellebore is also given in something sweet but is best in lentils or in a mush. . . . The emptying begins after about four hours; the entire treatment is over in seven hours. In this manner, white hellebore heals epilepsy, . . . dizziness, melancholy, insanity, possession, white elephantiasis, leprosy, tetanus, tremors, foot gout, dropsy, incipient tympanic water, stomach weakness, charley horse, hip pains, four-day fever, if this will not disappear in any other way, persistent coughing, flatulence, and recurrent stomachaches. (Pliny 25.24, 59 f.)

Hildegard of Bingen made similar use of white hellebore: "The white *sichterwurtz* dispels, when mixed with wild thyme [*Thymus pulegioides* L.] and fennel [*Foeniculum vulgare*] and fat . . . , even madness in a man" (*Physica* 1.130). The ancient Germans also used the root to induce abortions (Höfler 1990, 84*).

In Persia (Iran), the fresh root is used to produce a paste that is applied externally to relieve headaches and neuralgia (Hooper 1937, 183*). In Russian folk medicine, the root was administered to children in **honey** as an anthelmintic (Rowell 1978, 265*). White hellebore was used both internally and externally in the folk medicine of the Alpine countries, and for both humans and

White hellebore (*Veratrum album*) is also known as the "white sneezing plant," even though it is not related to the true sneezing plant (*Helleborus*). In earlier times, two forms were distinguished and were referred to as the "male" and the "female." (Woodcut from Tabernaemontanus, *Neu Vollkommen Kräuter-Buch*, 1731)

315 "Apart from the profane use, the cauldron, as archaeological and written sources have confirmed, was *the* sacred and ritual vessel of the Celts, comparable to the Christian chalice, which myth elevated from a household to a sacral object. It became a complex symbol, a true focal point of mythology" (Botheroyd 1992, 180). The Celts also used cauldrons to brew **beer** and **mead**.

316 Both names were also used for *Veratrum eschscholtzii* Gray [syn. *Veratrum eschscholtzianum* (R. et S.) Rydb.].

"Sneezing plant put in the nose cleanses the brain and makes one sneeze."

LEONARD FUCHS
KREÜTTERBUCH [HERBAL]
(1543, CH. 103)

"The white sneezing plant is important in magic as a narcotic agent much used for fumigations and witches' ointments."

HELMUT WERNER
LEXIKON DER ESOTERIK
[DICTIONARY OF THE ESOTERIC]
(1991, 469*)

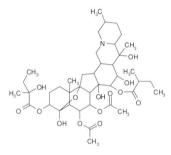

Protoveratrine A

A very closely related species (or subspecies, variety?) of the white hellebore (*Veratrum sabadilla*) grows in Mexico and in the Antilles. Research is needed to determine whether this plant has psychoactive effects and has ever been used for psychoactive purposes. (Engraving from Pereira, *De Beginselen der Materia Medica en der Therapie*, 1849)

animals. Ointments, poultices, and washings with extracts of white hellebore were used externally to treat scabies, lice, psoriasis, and other skin diseases. The root powder was used internally for melancholy (= depression), asthma, dropsy, paralysis, rheumatism, and fever (Pahlow 1993, 242*).

In homeopathic medicine, a tincture obtained from the dried rootstock (Veratrum—White Hellebore) is used in D3 and above in accordance with the medical description for such afflictions as mood disorders, depression, and the consequences of fright, anger, and migraines (Pahlow 1993, 242*; Schneider 1974, 3:386*).

Constituents

The entire plant contains steroid and steroidlike alkaloids with a C_{27} steroid framework (protoverine, jervine, protoveratrine, germerine, pseudojervine, veratrosine, *O*-acetyljervine, jervinone, 1-hydroxy-5,6-dihydrojervine) (Attar-ur-Rahman et al. 1993; Morton 1977, 63*). The root has an alkaloid content of 1.2 to 1.6% and the leaf bases 0.9 to 1.5%. Three ester alkaloids are considered to be the main active constituents: protoveratrine A, protoveratrine B, and germerine. The alkaloid content of the plant can vary considerably depending upon location and altitude. A basic rule is that the higher the altitude in which the plant occurs, the lower its alkaloid concentration (Roth et al. 1994, 723*).

In addition to the alkaloids, the plant contains the glycoside veratramine, chelidonic acid, veratrum acid, and fat (Morton 1977, 63*). White hellebore is very toxic. As little as 1 to 2 g of dried root (corresponding to approximately 20 mg of the ester alkaloids) can be lethal.

Effects

Because *Veratrum album* suppresses the sympathetic centers and induces a serious drop in blood pressure, use of the plant can easily result in coma (Attar-ur-Rahman et al. 1993; Frohne and Pfänder

1983, 152*). Typical symptoms include burning and tingling in the throat, followed by sensations of numbness and formication (similar to the symptoms produced by *Aconitum*; cf. **Aconitum napellus**). Patients remain conscious until they collapse and death occurs, although hallucinations also occur (Hruby et al. 1981). The toxicological literature contains a description of one case in which a thirteen-year-old boy smoked the dried leaves as "tobacco"; the only symptom reported was diarrhea, which persisted for a week (Roth et al. 1994, 723*).

Commercial Forms and Regulations

White hellebore can be obtained only in homeopathic dilutions from pharmacies. Pharmacologically active preparations are not available. The laws regulating games and practical joke articles prohibit preparations of sneezing powder that contain *Veratrum album* (Roth et al. 1994, 723*).

Literature

See also the entries for **beer**, **witches' ointments**, and **snuffs**.

Attar-ur-Rahman, Rahat Azhar Ali, Anwar-ul-Hassan Gilani, M. Iqbal Choudhary, Khalid Aftab, Bilge Sener, and Songol Turkoz 1993. Isolation of antihypertensive alkaloids from the rhizomes of *Veratrum album*. *Planta Medica* 59:569–71.

Botheroyd, Sylvia, and Paul F. 1992. *Lexikon der keltischen Mythologie*. Munich: Diederichs.

Hruby, K., K. Lenz, and J. Krausler. 1981. Vergiftungen mit *Veratrum album* (weißer Germer). *Wiener Klinische Wochenschrift* 93 (16): 517–19.

Kaneko, K., M. Wataname, S. Taira, and H. Mitsuhashi. 1972. Conversion of solanidin to jerveratrum alkaloids in *Veratrum grandiflorum*. *Phytochemistry* 11:3199–202.

Virola spp.

Epena, Parika Species

Family

Myristicaceae (Nutmeg Family); there are six sections

Species Used Psychoactively (and Synonyms)

The genus *Virola* encompasses some forty species, all of which are native to tropical (South) America (Plotkin and Schultes 1990, 357). Others recognize forty-five to sixty species (Holmstedt et al. 1982, 217).

Virola species that are made into psychoactive snuffs (Schultes 1979):

Virola calophylla Warb. [syn. *Myristica calophylla* Spruce, *Virola incolor* Warb., *Otoba incolor* Karsten ex Warb.]

Virola calophylloidea Markgraf [syn. *Virola lepidota* A.C. Smith]

Virola cuspidata (Benth.) Warb.

Virola elongata (Spruce ex Benth.) Warb. [syn. *Virola cuspidata* (Spruce) Warb., *Virola rufula* Warb.]

Virola loretensis A.C. Smith

Virola pavonis (DC.) A.C. Smith

Virola rufula (Mart. ex A. DC.) Warb. (questionable)

Virola surinamensis (Rol.) Warb.

Virola theiodora (Spruce ex Benth.) Warb. (some authors regard *V. calophylla* and *V. elongata* as synonyms for *V. theiodora* [Brenneisen and Hasler 1994, 1157])

Virola venosa (Benth.) Warb.

Virola species that are taken orally as hallucinogens:

Virola duckei A.C. Smith (huapa blanca)

Virola elongata (Spruce ex Benth.) Warb.

Virola loretensis A.C. Smith

Virola pavonis (DC.) A.C. Smith

Virola peruviana (A. DC.) Warb.

Virola surinamensis (Rol.) Warb.

Folk Names

Are-de-yé, camaticaro, cedrillo, cozoiba, cuajo, cudo rebalsero, cumala,[317] cumala caspi, ebene, epena, épena, huapa, isioma, jakuana, jeajeamadou, k-de'-ko, ko-gá, koó-na, krüdeeko, machfara-a, nyakwana, pa-ree-ká, paricá,[318] parika, parikabaum, parikana, parikaraná, rapá, ra-se-ñẽ-mee, rose-nameti, rose-nemee, sangerino, shomiá, talgmuskatnußbaum, tchkiana, trompillo, tsunem, ucuba, ucufe-ey, ucuúba preta, uucuba, vihó, yakee, yá-kee, yakoana, yakohana, yakohana-hi, yá-to, yeag aseiiñ

Top: *Virola* bark contains *N,N*-DMT, 5-MeO-DMT, and other tryptamines.

Bottom: A tree from the neotropical genus *Virola*.

History

The use of various *Virola* species as ritual **snuffs** was first discovered in the 1950s by the American ethnobotanist Richard Evans Schultes, who was astonished that this fact had not been noted before (Schultes 1954). The only previous information came from the Brazilian botanist Adolpho Ducke, who wrote that the Indians on the Rio Negro produced a snuff called *paricá* from the leaves of *Virola theiodora* and *Virola cuspidata* (Holmstedt et al. 1982, 216). Astonishingly, Richard Spruce had collected botanical material from a number of *Virola* species between 1851 and 1854 but did not notice the psychoactive use of the bark (Schultes 1983c*).

Distribution

Members of the genus are found primarily in Amazonia and adjacent tropical areas (Brazil, Colombia, Peru, Venezuela). Some species also occur in the tropical zones of Central America (Brenneisen and Hasler 1994, 1154; Schultes 1955, 79f.*). One species (*Virola guatemalensis* [Hemsl.] Ward) is found in southern Mexico and Guatemala; in Chiapas, this plant is known as *cacao volador*, "cacao flying device"[319] (Martínez 1987, 1238*).

Cultivation

Virola trees do not appear to be cultivated in Amazonia; no information about possible cultivated forms is available. To date, it does not

317 This name is also given to ***Osteophloeum platyspermum*** (DC.) Warburg and *Iryanthera macrophylla* (Benth.) Warburg (see ***Iryanthera juruensis***), both of which are used as oral hallucinogens (Schultes et al. 1977, 264).

318 In Amazonia, the following trees are also known by the name *paricá*: *Cassia fastuosa* Willd., *Cedrelinga catenaeformis* Ducke, *Parkia* spp., *Piptadenia* spp., *Pithecolobium* spp., *Schizolobium amazonicum* (Hub.) Ducke, *Schizolobium parahybum* (Vell.) Blake, *Senegalia* spp., and other Leguminosae (Schultes 1954, 257 f.).

319 This name may suggest a psychoactive use. The association with ***Theobroma cacao*** is also of interest.

Top: The seeds of *Virola surinamensis*, known as *ucuba*.

Bottom: The seeds of *Virola oleifera*.

Virola Species of Ethnobotanical Significance

(from Duke and Vasquez 1994, 174 ff.*; Beloz 1992; Schultes et al. 1977; modified and supplemented)

Botanical Name	Indian Name(s) (other than *cumala*)	Use
Species known as *cumala blanca*:		
Virola calophylla Warb.	epená	**snuff** (Bora, Huitoto)
Virola divergens Ducke		
Virola elongata (Benth.) Warb.	anya huapa, ko-de-ko	decoction of the branch tips for arthritic swelling (Barasana); snuff; oral hallucinogen (Bora)
Virola flexuosa A.C. Smith	caupuri de altura, huapa, pucuna huapa	protection insect
Virola loretensis A.C. Smith		hallucinogen (Huitoto)
Virola peruviana (DC.) A.C. Smith	sacha annona, sacha avio, ichilla muyu sebu	bark as hallucinogen
Virola sebifera Aubl. [syn. *Myristica sebifera* (Aubl.) S.W.]		folk medicine
Virola surinamensis (Rol.) Warb.	nyakwana, ucuba, diaru	**snuff** (Bora, Huitoto); latex for treating cheek ulcers (Warao)
	caupuri	**ayahausca** additive (in Iquitos)
	cumala colorada	oral hallucinogen (Bora)
Species known as *cumala negra*:		
Virola decorticans Ducke		leaf juice for cutting teeth (Jíbaro)
Virola multinerva Ducke	ila yura	timber
Species known as *aguano cumala*:		
Virola albidiflora Ducke		resin for treating wounds (Kumeo, Tukano)
Virola pavonis (DC.) A.C. Smith	caupuri del bajo, cedro ajua, puliu huapa, pucuna huapa huachig caspi	**snuff** (Bora, Huitoto) oral hallucinogen (Bora)

appear that anyone has had success in cultivating the plants (Rob Montgomery, pers. comm.).

Appearance

Virola species are large trees that can grow as tall as 30 meters. The leaves are undivided and pinnate and have continuous margins; there are no stipules. The tiny dioecious flowers grow in fascicles. The fruits are ellipsoid. The leaves can attain a length of more than 30 cm.

The different species are difficult to distinguish from one another.

Psychoactive Material

— Resin (= latex, exudates) and the inner bark (cambium)

Preparation and Dosage

The resin or latex (usually called *oom* or *yá-kee-oom*) of the *Virola* species can be obtained in a number of ways. An incision can be made in the bark, part of the bark can be removed, or the inner bark (cambium) can be warmed, which will cause the resin to exude. Because the pure resin is sticky, it usually is mixed with plant ashes, e.g., from the bark of a wild cacao tree (*Theobroma subincanum* Mart.; cf. **Theobroma spp.**), or with shell lime (from burned freshwater mussels) before being ground (Schultes 1954, 247 ff.). The **snuff** apparently will not have any effects unless the (alkaline) plant ashes are added.

The Indians say that the bark must be harvested in the early morning, before the sunlight has hit the trunk, lest the power of the powder dissipate. The rays of the sun are said to have a considerable impact upon the effects (Schultes 1954, 248).

For shamanic purposes, the dosage is given as a slightly heaping teaspoon of the powdered resin mixed with the plant ashes. This amount is usually snuffed three times in a row in short intervals (fifteen to twenty minutes) (Schultes 1954, 250).

Some Amazonian tribes make their snuff powder from the dried juice of the bark of different *Virola* species and the ashes of *Theobroma subincanum* Martius or dried leaves of **Justicia pectoralis** Jacquin (Schultes and Holmstedt 1968).

The Desana of the Colombian Vaupés use the inner bark of the species *Virola calophylla*, *V. calophylloidea*, and *V. theiodora* for their snuffs. Depending on the ritual occasion and the desired effects, powdered tobacco leaves (**Nicotiana tabacum**), powdered coca leaves (**Erythroxylum coca** var. *ipadu*), the ashes of *Cecropia* leaves, powdered pieces of bark from **Banisteriopsis** spp., or the lime scratched off stalactites may be added to the finely powdered bark (Reichel-Dolmatoff 1979, 32 f.).

Other recipes are used for oral ingestion. The Colombian Huitoto boil the juice until it has thickened into a syrup. The thickened juice is then rolled into bean-size balls and coated with the ashes of *Gustavia poeppigiana* Berg ex Martius. Three to six of these little balls are swallowed or dissolved in water and drunk (Schultes 1969). Because of the increasing pressures of acculturation, such oral use appears to be disappearing (Schultes et al. 1977, 259).

The Peruvian Bora and Huitoto also knew of oral use. They repeatedly boiled the inner bark (cambium) of a variety of species (especially *Virola elongata*) to yield a paste called *ko'do* that was then swallowed without any further preparation. In other areas, the paste was mixed with the ashes of a species from the genus *Carludovica* (Cyclanthaceae; cf. Bristol 1961) and the leaves of a palm of the genus *Scheelea* (Schultes et al. 1977, 262 f.). The ashes (called "salt") from the bark of the large tree *Eschweilera itayensis* Kunth (Lecythidiaceae) as well as the ashes from the buds and leaves of *Spathiphyllum cannaefolium* (Dryand.) Schott (Araceae) were added for the same purpose (Schultes 1979, 228).

Some species of *Virola* are used as **ayahuasca** additives. Some shamans in Iquitos add *Virola surinamensis* to the ayahuasca drink so that the ayahuasca will "teach medicine."

Ritual Use

The Bora and Huitoto of the Orinoco region use *Virola calophylla* as a snuff and also orally as a hallucinogen. They also make snuff from the cambium of *Virola elongata* and *Virola surinamensis* and use the species *Virola pavonis* as a hallucinogen. Usually only shamans use this snuff

(which appears to have quite potent effects); they use it to diagnose illnesses.

The use of *Virola* snuffs (*vihó*) is very common among the Desana. Here again, it is normally used only by shamans when they wish to diagnose illnesses. However, all boys who are being initiated into manhood must learn how to prepare the powder during the initiation celebrations, during which they must also use it for the first time (Reichel-Dolmatoff 1979). Many men take it along with **ayahuasca**.

The Quechua of Ecuador use the dried bark secretion of *Virola duckei* A.C. Smith as an oral hallucinogen (Bennett and Alarcón 1994). Unfortunately, no details about the precise nature of their ritual uses have become known.

The Yanomomö (= Waika) use *Virola theiodora* not only as a shamanic hallucinogen but also as an arrow poison (Soares Maia and Rodrigues 1974). They also use *Virola elongata* to manufacture arrow poisons (Macrae and Towers 1984).

The tribes that use *Virola* species to produce psychoactive compounds include the Puinave from the Río Inírida, Kuripako from Río Guainía, Kubeo, Tukano, Desana, Papurí, Barasana, Makuna from Río Piraparaná, Taiwano from Río Kananarí, Brazilian and Venezuelan Yanomamö/Waika, Mundurukú,[320] Huitoto (= Witoto), Bora, and various small tribes of the Rio Issana (Içana), and presumably other tribes or ethnic groups.

Artifacts

Apart from certain snuff tubes and other paraphernalia, to date no artifacts are known (cf. **snuffs**).

Medicinal Use

Venezuelan shamans smoke the dried inner bark of *Virola sebifera* at dances to treat fever diseases (Altschul 1973, 76*; Plotkin and Schultes 1990, 357). The bark, which is known by the names *wircawei-yek* and *erika-bai-yek*, is boiled to dispel evil spirits (Altschul 1973, 76*). One as yet unidentified *Virola* species is said to be used as a contraceptive (Plotkin and Schultes 1990, 357).

A number of *Virola* species are regarded as brain stimulants and are said to improve both memory and intelligence (Plotkin and Schultes 1990, 357).

Many *Virola* species (e.g., *V. elongata*, *V. melinonii* [Benoist] A.C. Smith, *V. sebifera*, *V. surinamensis*) are used in folk medicine to treat skin diseases (Brenneisen and Hasler 1994, 1158; Plotkin and Schultes 1990, 358 ff.).

The mother tincture of *Virola sebifera* (medicinal content 1/10) is known in homeopathy as *Myristica sebifera* hom. *HAB34* (also *HPUS88*) and is used for such conditions as suppuration (Brenneisen and Hasler 1994, 1157). The agent is regarded as a "medicine with great antiseptic

Two devices (made of bones and shells) from northwestern Brazil for ingesting *Virola* (parika) snuff powder. (From Koch-Grünberg *Zwei Jahre bei den Indianern Nordwest-Brasiliens*, 1921*)

320 Whether the Mundurukú actually make their parika from a *Virola* or from an **Anadenanthera** is unclear. One early report mentioned a Leguminosae (*Acacia angico*) as the stock plant, the seeds of which are mixed with the juice from the leaves of a moonseed known as *abuta* ("cocculus"; cf. **Anarmita cocculus**) (Schultes 1954, 257).

DMT

5-MeO-DMT

β-carboline

power" (Boericke 1992, 532*). It also is used in homeopathic compound medicines, e.g., Sulfur Pentarkan, which consists of sulfur, *Atropa belladonna*, mercury, "Myristica sebifera," and silicic acid.

Constituents

It was once thought that the active constituent of the parica drugs was myristicin (cf. *Myristica fragrans*) (Schultes 1954, 247). However, this conjecture has not been substantiated.

Many *Virola* species contain tryptamines (*N,N*-DMT, 5-MeO-DMT, and others) and β-carbolines. Some, e.g., *Virola cuspidentata*, even contain harmane derivatives (6-methoxyharmalane, 6-methoxyharmane, 6-methoxytetrahydroharmane) as well as diarylpropanes of the virolane and virolin types (Brenneisen and Hasler 1994, 1154). Those *Virola* species that have been the most intensively investigated contain tryptamines, most commonly DMT (Holmstedt et al. 1982).

Virola calophylla contains *N,N*-DMT, MMT, 5-MeO-DMT, 5-MeO-MMT, and β-carbolines (Duke and Vasquez 1994, 174). Astonishingly, no psychoactive indoles or tryptamines have been found in the latex of those species or individuals that produce a great deal of red resin (Schultes et al. 1977, 260). MMT, DMT, and 5-MeO-DMT have been detected in the bark (Farnsworth 1968, 1088*).

The resin of *Virola theiodora* contains 8% 5-MeO-DMT (Soares Maia and Rodrigues 1974).

The bark of *Virola elongata* contains resin as well as sesartemin and yangambin, substances that are said to inhibit aggressiveness.

Although no actual active constituent has yet been found in *Virola surinamensis* (the latex contains diarylpropanoids, neolignans, and long-chain esters: Barata et al. 1978; Gottlieb et al. 1973), its pharmacological activity has been experimentally demonstrated (Beloz 1992). Long-chain esters are found in many *Virola* species (Kawanishi and Hashimoto 1987).

The seeds contain an abundance of oil that is sold under the name *virola fat*, *ucuúba*, or *ucuúba butter* and is reminiscent of cocoa butter (it is even made into candles; Plotkin and Schultes 1990, 357).

Effects

The effects of *Virola* snuffs are described as very intense and not necessarily pleasant. Schultes reported almost only unpleasant side effects as a result of his own experiment (severe headache, ocular pressure, disturbances of coordination, et cetera). Shamans typically enter into a sleeplike trance state accompanied by dreams and hallucinations. It has even been reported that a shaman died while under the influence of *Virola* powder (Schultes 1954, 251).

The Colombian Desana characterize the effects

of the snuff as follows: "This *Virola* bark, this lucid dot, it penetrates into us and makes us dizzy/sedated" (Reichel-Dolmatoff 1979, 36).

The resins of various *Virola* species, especially *Virola elongata*, have antifungal properties (Duke and Vasquez 1994*).

It is uncertain whether oral ingestion can in fact produce hallucinogenic effects:

The efficaciousness of the oral application forms (pills, pastes, etc.) is controversial primarily with regard to the role of monoamine oxidase and therewith the metabolic deactivation of the β-carboline derivatives that inhibit the tryptamine derivatives, as these alkaloids as a rule are biogenically present in only small amounts. It is however conceivable that these alkaloids largely arise only during the preparation of the bark exudates as artifacts of tryptamine alkaloids. It has been conjectured that other constituents of Virola, e.g., flavonoids, neolignans, and diarylpropane, may as antioxidants nonspecifically inhibit the oxidative first-pass breakdown of the tryptamine alkaloids through MAO, mixed functioning oxigenases, and are thereby able to increase the oral efficaciousness. (Brenneisen and Hasler 1994, 1158)

Commercial Forms and Regulations

None, apart from the mother tincture known as Myristica sebifera (cf. Brenneisen and Hasler, 1994: 1157)

Literature

See also the entries for *Justicia pectoralis* and **snuffs**.

Agurell, S., B. Holmstedt, J.-E. Lindgren, and R. E. Schultes. 1969. Alkaloids in certain species of *Virola* and other South American plants of ethnopharmacologic interest. *Acta Chemica Scandinavica* 23:903–16.

Barata, L. E., P. M. Baker, O. R. Gottlieb, and E. A. Ruveda. 1978. Neolignans of *Virola surinamensis*. *Phytochemistry* 17:783–86.

Beloz, Alfredo. 1992. Brine shrimp bioassay screening of two medicinal plants used by the Warao: *Solanum straminifolium* and *Virola surinamensis*. *Journal of Ethnopharmacology* 37:225–27.

Bennet, B. C., and Rocío Alarcón. 1994. *Osteophloeum platyspermum* and *Virola duckei* (Myristicaceae): Newly reported as hallucinogens from Amazonian Ecuador. *Economic Botany* 48 (2): 152–58.

Brenneisen, Rudolf, and Felix Hasler. 1994. *Virola*. In *Hagers Handbuch der pharmazeutischen Praxis*, 5th ed., 6:1154–59. Berlin: Springer.

Bristol, Melvin Lee. 1961. *Carludovica palmata* in broommaking. *Botanical Museum Leaflets* 19 (9): 183–89.

Fernandes, João B., M. Nilce de S. Ribeiro, Otto R. Gottlieb, and Hugo E. Gottlieb. 1980. Eusiderins and 1,3-diarylpropanes from *Virola* species. *Phytochemistry* 19:1523–25.

Fernandes, João Batista, Paulo Cezar Vieira, and Regina Lúcia Fraga. 1988. Transformações químicas de liganas isoladas de *Virola sebifera* em análogos de podofilotoxina. *Supl. Acta Amazônica* 18 (1–2): 439–42.

Gottlieb, Otto R. 1979. Chemical studies on medicinal Myristicaceae from Amazonia. *Journal of Ethnopharmacology* 1:309–23.

Gottlieb, O. R., A. A. Loureiro, M. Dos Santos Carneiro, and A. Imbiriba Da Rocha. 1973. Distribution of diarylpropanoids in Amazonian *Virola* species. *Phytochemistry* 12:1830.

Holmstedt, B., J. E. Lindgren, T. Plowman, L. Rivier, R. E. Schultes, and O. Tovar. 1982. Indole alkaloids in Amazonian Myristicaceae: Field and laboratory research. *Botanical Museum Leaflets* 28 (3): 215–34.

Kawanishi, K., and Y. Hashimoto. 1987. Long chain esters of *Virola* species. *Phytochemistry* 26 (3): 749–52.

Lai, A., M. Tin-Wa, E. S. Mika, et al. 1973. Phytochemical investigation of *Virola peruviana*, a new hallucinogenic plant. *Journal of the Pharmaceutical Society* 62:1561–63.

Macrae, W. Donald, and G. H. Neil Towers. 1984. An ethnopharmacological examination of *Virola elongata* bark: A South American arrow poison. *Journal of Ethnopharmacology* 12:75–92.

Plotkin, Mark J., and Richard Evans Schultes. 1990. *Virola*: A promising genus for ethnopharmacological investigation. *Journal of Psychoactive Drugs* 22:357–61.

Reichel-Dolmatoff, Gerardo. 1979. Some source materials on Desana shamanistic initiation. *Antropológia* 51:27–61.

Rodrigues, William A. 1977. Novas espécies de *Virola* Aubl. (Myristicaceae) da Amazônia. *Acta Amazônica* 7 (4): 459–71.

———. 1980. Revisão taxonômica das especies de *Virola* Aublet (Myristicaceae) do Brasil. *Acta Amazônica* 10 (1) suppl.: 1–127.

Schultes, Richard Evans. 1954. A new narcotic snuff from the Northwest Amazon. *Botanical Museum Leaflets* 16 (9): 241–60.

———. 1969. De Plantis Toxicariis e Mundo Novo Tropicale Commentationes IV: *Virola* as an orally administered hallucinogen. *Botanical Museum Leaflets* 22:133–64.

———. 1979. Evolution of the identification of the Myristicaceous hallucinogens of South America. *Journal of Ethnopharmacology* 1 (2): 211–39.

Schultes, Richard Evans, and Bo Holmstedt. 1968. De Plantis Toxicariis e Mundo Novo Tropicale Commentationes II: The vegetable ingredients of the Myristicaceous snuffs of the Northwest Amazon. *Rhodora* 70:113–60.

———. 1971. De Plantis Toxicariis e Mundo Novo Tropicale Commentationes VIII: Miscellaneous notes on Myristicaceous plants of South America. *Lloydia* 34:61–78.

Schultes, Richard Evans, and Tony Swain. 1976. De Plantis Toxicariis e Mundo Novo Tropicale Commentationes XIII: Further notes on *Virola* as an orally administered hallucinogen. *Journal of Psychedelic Drugs* 8:317–24.

Schultes, Richard Evans, Tony Swain, and Timothy C. Plowman. 1977. De Plantis Toxicariis e Mundo Novo Tropicale Commentationes XVII: *Virola* as an oral hallucinogen among the Boras of Peru. *Botanical Museum Leaflets* 25 (9): 259–72.

Soares Maia, J. G., and William A. Rodrigues. 1974. *Virola theiodora* como alucinógena e tóxica. *Acta Amazônica* 4:21–23.

"In spite of its many miracle drugs, Western medicine does not yet have an effective treatment possibility for skin diseases caused by fungi. For many cancer and AIDS patients, this is often a serious problem, for they are often literally infested with fungi. Since fungal infections are common in the moist and warm rain forest, the natives have developed numerous methods of treatment. Of these, the juice of the nutmeg tree [*Virola* sp.] appears to be the most effective."

MARK J. PLOTKIN
DER SCHATZ DES WAYANA [THE TREASURE OF THE WAYANA] (1994, 253*)

Vitis vinifera Linnaeus

Grape

Family

Vitaceae (Grape Family; previously also Ampelideae)

Forms and Subspecies

A number of subspecies and varieties of grape have been described:

Vitis vinifera L. ssp. *caucasia* Vavilov
Vitis vinifera L. ssp. *sativa* DC. (cultivated form for producing fruit)
Vitis vinifera L. ssp. *sylvestris* (C.C. Gmel.) Berger (wild form)
Vitis vinifera L. ssp. *vinifera* (cultivated subspecies)
Vitis vinifera L. var. *apyrena* L.

There also are numerous cultivars (grape varietals) that are important primarily in viniculture because of their different tastes (Pabst 1887, 2:211*).

Synonyms

Vitis sylvestris C.C. Gmel.

Folk Names

Angur (Hindi), drakh, draksha (Sanskrit), duracina, grape, grapevine, grape vine, gvid (Celtic, "bush"), 'inab (Iraq), khamr (Arabic),[321] palmes, parra, reba, rebe, rebo, rebstock, vigne, vine, vitis sativa (Latin), weinranke, weinrebe, weinstock, wynreben, zame weinreben

Folk Names for Wine

Aqua vitae, oinos, sharab, vin, vinho, vino, vinum, wein, woinos

History

The grape plant originated in Asia and probably was used to prepare inebriating drinks from a very early date. Current research suggests that humans have been making beverages by fermenting grapes for at least some nine thousand years (McGovern 2003). In the summer of 1990, clay drinking vessels were found in Godin Tepe (Iran), chemical analyses of which have clearly shown that they were used for drinking wine. These sensational finds were dated from 3500 to 2900 B.C.E. (McGovern et al. 1995). Soon thereafter, the first well-documented viticulture (vineyards, wine cellars) flourished in Mesopotamia. It quickly spread from Asia Minor to Egypt, Crete, and Greece (Lesko 1978). In ancient times, the Romans spread viticulture into all the areas of their empire that had suitable climate and soil conditions.

In Egypt, the cultivation of wine did not become established until the New Kingdom, when grapes were pressed and poured into different containers with enthusiasm. Numerous wine flagons have been discovered that include details about the vintage, quality, and location and the name of the head vintner (Lesko 1978). In the Nile valley, wine was a drink of the upper classes; it was enjoyed at private feasts as well as at religious offering festivals (libations).

The spread of viticulture that occurred in the following years was very closely related to the Christianization that was decreed "from above" (Marzahn 1994, 90, 96*). Whereas the Dionysian religion had been a cult of ecstasy, Christianity degenerated into a religion of alcoholics (Daniélou 1992*).

Today, viticulture is practiced throughout the world and has become an industry of great economic significance. The plant itself is not psychoactive, but its most important products, **wine** and distilled **alcohol** (brandy, cognac, et cetera), are.

Distribution

Current evidence indicates that the grape and viticulture originated not in Greece but in Asia Minor. It may have begun in the region between the Caucasus and the Hindukush mountains, where wild grapes can still be found today (Pabst

Left: A grapevine (*Vitis vinifera*) growing on Crete.

Right: A wild grapevine (*Vitis* sp.) in the Mexican rain forest. (Photographed in Naha', Chiapas, Mexico)

321 It has often been claimed that the word *coffee* is derived from the Arabic word for wine, i.e., *khamr* or *gahwa* (cf. **Cata edulis, Coffea arabica**, **wine**). *Khamr* essentially means "inebriating."

1887, 2:212*). Today, cultivation has taken the grape into every part of the world (including North and South America, Australia, and South Africa).

Cultivation
Grapes are propagated primarily from cuttings. These are allowed sit in water until they develop roots and then planted. Grapes will grow well only in temperate climates where the average annual temperature does not exceed 17°C (63°F).

The "proper" method for growing grapes was a subject for many authors in ancient times (Hagenow 1982, 171 ff.).

Appearance
This twining climber can grow to a length of over 10 meters. It has a woody, often twisted stem and a woody, branched root that grows deep into the ground. The bush develops numerous climbing branches that divide like forks. The long-stemmed, heart-shaped leaves are retuse, with three to five lobes, and usually have a serrated or jagged margin. The flower panicles develop on the lower vines and have yellow-green, tiny, usually monoclinous flowers. These develop into the characteristic green, reddish, red, or blue clusters of fruits (grapes).

Vitis vinifera can be easily confused with wild *Vitis* species.

Psychoactive Material
— Fruits (grapes)
— Wine

Preparation and Dosage
The juice that is pressed from the grapes is fermented to produce **wine**. Over the course of history, numerous methods have been developed for making wine. Wine is a psychoactive compound in and of itself. But in ancient times, many psychoactive plants were also added to wine (see the table on the following page) to produce specific types of effects (Ruck 1995*). Such additives were known as "flowers of the wine" (Ruck 1982). There were essentially two methods: In the first, the additives were placed directly into the fermenting mixture, while in the second the finished wine served as a solvent for macerating specific substances. One famous concoction was mandrake wine, which was prepared by adding fresh or dried mandrake roots (*Mandragora officinarum*) to the grape must. Other recipes added root pieces to the finished wine. Because wine to which "flowers" had been added had much more potent effects, it was served with great care.

The ancient Greeks were very aware of the significance of dosing wine. The comedy *Dionysos or Semele*, by the poet Eubulos (fourth century B.C.E.), provides an indication:

For reasonable people I prepare only three mixing jars [with wine and water]: one for health (*hygíeia*), which they drink first; the second for love and pleasure, and the third for sleep. When that has been emptied, the people who are called wise will go home. The fourth mixing jar no longer belongs to me, but to immoderation. The fifth is full of howls; the sixth causes gushing and bawling; the seventh brings black eyes; the eighth calls to the servant of the court; the ninth is full of anger and disgust. The tenth leads to madness (*manía*) and stumbling. If you fill it into a small container, then this will easily knock the legs out from under he who empties it and throw him to the ground.

In general, the Greeks regarded their different wines as too inebriating to be consumed undiluted. Prior to consumption, wine was normally mixed with water in a 1:2 or 1:3 proportion. In addition, wine was rarely drunk unadulterated. Numerous aromatic, medicinal, and inebriating additives (*aromatites*) are known from ancient times (Weeber 1993, 35). Although the recipes for most preparations were kept secret, some have come down to us. These indicate that oleander (*Nerium oleander* L.; cf. **honey**), hemp ("wine of Democritus"; cf. **Cannabis sativa**), opium (cf. **Papaver somniferum**), and various nightshades, especially mandrake ("the grape of the field"; **Mandragora officinarum**) and henbane (**Hyoscyamus niger**), were all added to wine. In ancient Italy, *crapula*, "inebriating resin," was an important additive (Weeber 1993, 41).

During the Middle Ages, plants were added to wine to produce specific psychoactive effects:

A person in whom melancholy is growing, who has a dark mood and is always sad. And he should drink often the wine with the boiled arum root [*Aaron aculatum = Arum maculatum* L.], and it will lessen the melancholy in him, that is, it will disappear, as will the fever. (Hildegard von Bingen, *Physica* 1.49)

Spiced wine continued to enjoy great popularity during the fifteenth and sixteenth

Triumphal procession of the wine god Dionysos, who was originally of Asian origin. (Floor mosaic, Roman period, Cyprus)

"The wine has been transformed into nectar, thereby breaking the curse that has been attached to it since ancient times. It is now a magical drink that helps in crossing the threshold; it is the blood of the sun and of the moon."

MIGUEL SERRANO
EL/ELLA—DAS BUCH DER MAGISCHEN LIEBE [HE/SHE—THE BOOK OF MAGICAL LOVE] (1982, 32)

The grapevine originated in Asia, where it was derived from a wild form. The earliest Chinese literature describes wild grapevines (*Vitis* spp.) under the name *chien-sui-tzû*. (Illustration from *Nan-fang-ts'ao-mu chuang*)

"I saw Bacchus on distant crags
teaching songs—posterity,
    believe it—
and Nymphs learning and the keen
ears of the goat-footed Satyrs.

Euhoe, my mind quivers with
    strange fear,
and, breast full of Bacchus, rejoices
    in commotion,
euhoe, spare me, Father Freedom,
spare, fearsome with grievous stem-
    wand!

It is divine right for me to sing
the stubborn bacchants, the
    fountain of wine and rivers
of rich milk, the hollow
trunks honey-flow."

HORACE
ODE 2.19
(IN LOWRIE 1997, 206)

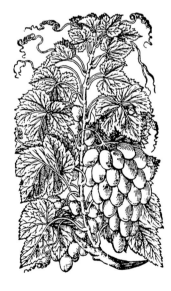

Grapes not only are the source of
the juice that can be fermented to
make wine but also are a very
healthful source of nourishment.
(Woodcut from
Tabernaemontanus, *Neu
Vollkommen Kräuter-Buch*, 1731)

Psychoactive Additives to Wine

(from Krug 1993; Macmillan 1991, 427*; Pabst 1887, 216*; Root 1996*; Ruck 1992; Weeber 1993; supplemented)

Name/Preparation	Stock Plant	Application/Era/Place
arum	*Arum maculatum* L. (cf. ***Arisaema dracontium***)	Middle Ages
cacao powder	*Theobroma cacao*	succolade (seventeenth-century Europe)
camphor	*Cinnamomum camphora*	vinum camphoratum (nineteenth century)
cloves	*Syzygium aromaticum* (cf. **essential oils**)	Clareth/Luter drink
coca leaves	*Erythroxylum coca*	Vin Mariani (nineteenth century)
cola nuts	*Cola acuminata* (cf. ***Cola* spp.**)	*kola-wine* (Africa)
coriander	*Coriandrum sativum* (cf. **essential oils**)	ancient Egypt
crapula ("inebriating resin")	?	Italy
cubeb	*Piper cubeba* (cf. ***Piper* spp.**)	antiquity, seventeenth-century East India
deadly nightshade	*Atropa belladonna*	Middle Ages
ergot	*Claviceps purpurea*	vinum ergotae (nineteenth century)
harmala seeds	*Peganum harmala*	harmel wine (Morocco)
hemp flowers (female)	*Cannabis indica*	modern era (India)
	Cannabis sativa	late antiquity
henbane	*Hyoscyamus albus*	antiquity
	Hyoscyamus muticus	Egypt
	Hyoscyamus niger	Middle Ages
horn poppy decoction	*Glaucium flavum* Crantz (cf. ***Papaver* spp.**)	late antiquity
ivy leaves	*Hedera helix*	antiquity/Dionysos cult
mandrake	*Mandragora officinarum*	antiquity (Greece, Rome, Egypt)
	Mandragora spp.	antiquity
monkshood root	*Aconitum napellus*	folk medicine
mushrooms	*Amanita muscaria*	antiquity
	Psilocybe spp.	
nightshade	*Solanum dulcamara* (cf. ***Solanum* spp.**)	antiquity
nutmeg, mace	*Myristica fragrans*	Clareth/Luter drink
olibanum	*Boswellia sacra*	antiquity (Orient)
opium	*Papaver somniferum*	antiquity/late antiquity (India); vinum opii (nineteenth century)
poison nut root	*Strychnos nux-vomica*	antiquity
saffron	*Crocus sativus*	Clareth/Luter drink
white hellebore	*Veratrum album*	uncertain
wormwood	*Artemisia absinthium*	wormwood wine (antiquity to the modern era), vinum de absinthio (nineteenth century)
yellow water lily root	*Nuphar lutea*	antiquity

centuries. Clareth, also known as Luter drink, was renowned in the city of Bremen and the rest of northern Germany. This was a heavily spiced wine to which honey, sugar, saffron (*Crocus sativa*), cloves, and nutmeg (*Myristica fragrans*) had been added. It was served in the Bremen *ratskeller* (the inn in the city hall) and in the pharmacy of the city hall (Marzahn 1994, 96*).

Ritual Use

Grape wine was at the center of the cult and mysteries of Dionysos (= Bacchus). Dionysos was both a god of fertility who was venerated as the lord of plants in rural festivals and a shamanic god of psychopharmaka who was celebrated in ecstatic cults and who revealed himself in secret mysteries (Merkelbach 1988).

Dionysos was the prototypical shaman of the ancient world (Emboden 1977). The mythology that surrounded him was involved with issues of life and death, healing, and ecstasy and frenzy. He was born two or even three times and was once dismembered and killed by the Titans. But since he is a god and by definition immortal, he was born again *knowing*. The experience of being dismembered gave him the knowledge of the infinity of life. The dismembered god demonstrates that no matter what happens, there is actually nothing to fear. Salvation awaits at the end of all horror. Dionysos acts like a shaman in other ways as well. He has animal spirits that aid him or animal identities (panther, lynx, lion, tiger, dolphin, snake, bull, goat), has ecstatic music (drums, tambourine, cymbals, flutes) at his command that bring him to rapture, and often garbs himself—like the Siberian shamans—in women's clothing and indulges in transsexual excesses. He is the wearer of a mask and a singer, and his goat song (*tragedy*) is renowned. He is the founder of the theater and of the bacchanal mysteries. He is also knowledgeable about plants and is a healer.

Orgiastic wine festivals were held in many places in Dionysos's honor, and these often escalated into wild Bacchanalias. The temple of Dionysos in Pompeii had a wine garden in which the symposiums of the inebriated god took place. Great quantities of wine flowed there; wine, the gift of Dionysos, was reverently known as "the blood of the earth" as well as the "blood of Dionysos" or was simply named after the god himself, *Dionysos*. It was hoped that he would make it possible for humans to partake of immortality. Thus did the blind prophet Tiresias speak:

> Young man, there are two things that come first for mankind: the goddess Demeter—she is the earth, but call her what you like—she nourishes mortals with food; the son of Semele came later, and he is Demeter's counterpart, since he discovered and gave to mortals a drink, the juice of the grape. It puts an end to the pain of suffering humans, when they are filled with the stream of the vine, and it gives sleep to forget the troubles of the day; there is no other cure for pain. Itself a god, to the gods it is poured as a libation, so that through Dionysus people may have good fortune. (Euripides, *The Bacchae*, 274 ff.; Eds. D. Franklin, J. Harrison, J. Affleck 2000, 15)

According to mythology, Dionysos carried the first vines to each of the places in the world where grapes still grow and planted them himself (Daniélou 1992a*). In the Greco-Roman world, a goat kid was sacrificed when grapevines were planted so that the grapes would burst from the vines and Dionysos could receive the blood of his favorite animal.

Wine was one of the agents used to induce the Dionysian ecstasy[322] (Detienne 1992; Emboden 1977; Evans 1988). Drinking festivals were held both in Greece and later in the Roman Empire (Murray 1990). Such a drinking festival was known as a *symposion* or *symposium*, "drinking together." The Latin name for the leader of the symposium was *magister*, "master." He was responsible for the dosages, for the ratio of wine to water, and for any psychoactive additives that may have been used. As Xenophon stated: "So is it best to come home from drinking: I am no longer sober, but also not too inebriated." A symposium was first and foremost a communal drinking fest, often of an intellectual nature. This was the place where the philosophy of the ancients (e.g., of Plato and Socrates) was born.

In India and the Himalayan region, wine is sacred to Shiva (cf. **Aconitum ferox, Cannabis indica, Papaver somniferum, alcohol**). Shiva and Dionysos were already being equated in ancient times (Daniélou 1992a*). Wine also plays a significant role in the tantric cult (Serrano 1982). Since one of the aims of the cult was to ritually break social taboos, wine drinking was an important means for doing so, as Hindus are normally forbidden to consume alcohol.

In the Catholic Church, the consumption of wine has remained central to the rite of communion into the present time:

> Although wine was already indispensable in the Church because of communion, it also played a role there for other reasons. For the German Pagan honored his gods and folk heroes by raising a glass to them, while the newly converted German Christian drank to commemorate those saints who had gained his admiration because their spiritual or bodily strength had been tested, and the Church in its tolerance incorporated these so-called *minnetrinken* into their rituals after having tried in vain for centuries to suppress them; the bishops were almost incapable of limiting the number of saints in whose honor, or *minne*, they would drink. (Schultze 1867, 104)

Many contemporary wine aficionados have developed a cult around wine that is known as "wine culture." But this activity is not so much concerned with the psychoactive or inebriating effects of wine as it is with a gourmet lifestyle, a mania for collecting, and a desire to acquire possessions.

Artifacts

There are numerous ancient representations of the grapevine, the harvest of its grapes, and the

"The Vine-Dionysus once had no father, either. His nativity appears to have been that of an earlier Dionysus, the Toadstool-god; for the Greeks believed that mushrooms and toadstools were engendered by lightning—not sprung from seed like all other plants. When the tyrants of Athens, Corinth and Sicyon legalized Dionysus-worship in their cities, they limited the orgies, it seems, by substituting wine for toadstools; thus the myth of the Toadstool-Dionysus became attached to the Vine-Dionysus."

ROBERT GRAVES
THE WHITE GODDESS
(1948, 159*)

From *Der Formenschatz* [Treasury of Forms], 1885, no. 105.

322 The dervishes, who have preserved certain elements of the orgiastic Orphic cult of Dionysos in their tradition, esteemed and "esteem wine (*sharab*) as a means to ecstasy" (Frembgen 1993, 198*).

"The vine bears three grapes,
The first brings the loss of the senses,
The second inebriation,
The third crime."

EPICTETUS
C.E. 55–C.E. 135

The shamanic god Dionysos offers a cup of his sacred wine. (Copperplate engraving, eighteenth century)

323 Henze's opera (a musical drama in one act by W. H. Auden and Chester Kallman) is a modern adaptation of Euripides' *The Bacchae.* "An—unfortunately lost—tragedy by the playwright Aischylos was called 'The Bassarids.' From what we know, this drama appears to have been a counterpart to 'The Bacchae' of Euripides. The name 'Bassarids' refers to Dionysos Bassareus, the 'Dionysos in a fox skin,' after whose example the Maenads would occasionally wrap themselves in the characteristic red fox furs that their name 'Bassarids' refers to" (Giani 1994, 98*). The "Bassariká" of an almost unknown poet named Dionysios provided the model for the "Dionysiaka" of Nonnos (Merkelbach 1988, 50).

consumption of wine as well as its effects. A marble sculpture from the first century C.E. shows a drunken Hercules, naked and holding his penis between his fingers as if urinating (Herkulaneum, House of the Deer). The Dionysian mysteries, which were celebrated with wine, are depicted on the murals of Pompeii (Grimal and Kossakowski 1993). Dionysos and his followers, festivals, and symposia have been artistic themes since antiquity (Hamdorf 1986). Some representations appear to make reference to *Amanita muscaria.*

Dionysos/Bacchus and his wine, festivals, and mysteries were described in numerous ancient writings (Brommer 1959; Merkelbach 1988; Preiser 1981a, 1981b; Weeber 1993).

The number of vessels for drinking wine (kraters, chalices, cups) that have been found is almost beyond count. Some ancient wine chalices have been interpreted as cryptic symbols for *Amanita muscaria* or other psychoactive fungi.

No other ancient deity survived as long as Dionysos/Bacchus. The mysterious god of inebriation continues to inflame people's feelings. He appears on wine and beer labels and in theaters—in *The Bacchae* of Euripides (ca. 480–406 B.C.E.), *The Bassarids* of Hans Werner Henze (born 1926),[323] and *Ariadne auf Naxos* by Richard Strauss (1864–1949). Richard Wagner (1813–1883) opulently immortalized Dionysos's wild and erotic festivals (Bacchanalias) in his opera *Tannhäuser.*

The "classic" horror novel *The Elixir of the Devil*, written in 1815 by E. T. A. Hoffmann (1776–1822), revolves around a "wine of Saint Anthony" (cf. *Claviceps purpurea*). The wine, which is preserved as a reliquary in a cloister, is described as a potent psychoactive agent. Anyone who partakes of it is cast into a schizophrenic world of hallucinations that he interprets as "the temptations of the devil." One monk dares to partake of the drink:

Among these [reliquaries] was a sealed flask, which contained within a seductive elixir that Saint Anthony was said to have taken from the devil. . . . I was overcome by an indescribable longing to find out what was in fact contained within the flask. I was able to take it aside, I opened it and found a wonderfully scented, sweet tasting, potent drink that I drank to the last drop.—How all of my senses then changed, how I experienced a burning thirst for the world, how the vices, in their seductive forms, appeared to be the highest pinnacle of life. (Hoffmann 1982, 123*)

In 1974, the American science-fiction author Robert Silverberg published "The Feast of St. Dionysus," a story that received the Jupiter Award. In this story, an astronaut who is traveling to Mars ends up in a utopian cult of Dionysos beyond

space and time and is initiated into the mysteries of the god with a special wine:

The rhythms are sharp and fierce. This is the music of the Bacchantes, this is an Orphic song, at first strange and terrifying, and then strangely soft and consoling. . . . Take, eat. This is my body. This is my blood. More wine. Figures move around him, other communicants step forward. He loses all sense of time and space. He separates from the physical dimension and glides over a swelling ocean, a large warm sea, a gently rolling sea that bears him lightly and merrily. He senses light, warmth, size, weightlessness, but he does not feel anything tangible. The wine. The host. Perhaps a drug in the wine? He slips out of the world and into the universe. This is my body. This is my blood. This is the experience of wholeness and unity. I take the cup of the god, and his wine dissolves me. . . . I call the name of the god, and his thunder sedates me. *Dionysos Dionysos!* (Silverberg 1984, 69)

Medicinal Use

Wine had approximately the same significance in Hippocratic medicine as **beer** did in the Babylonian and Egyptian healing arts. White, dark, red, sweet, tart, sweet-smelling, and heavy wines were all used as dietary vehicles for administering medicines. Numerous medicinal wines were made from wine and the appropriate herbs. Dioscorides names many of these.

The healing properties of wine grapes were mentioned in the founding works of Indian Ayurvedic medicine by Susrata and Charaka (Hopper 1937, 186*).

In 1753, a book titled *Der curieus- und offenherzige Wein-Artzt* [The Curious and Open-Hearted Wine Physician] appeared. Written by an anonymous "lover of the economic sciences," it revived the ancient tradition of healing wines and played an enormous role in popularizing the use of medicinal wines.

Today, wine—in moderate dosages—is still recommended as a healing agent and tonic, particularly for the elderly (Köhnlechner 1978).

Constituents

Wine grapes contain large amounts of grape sugar (= glucose), fructose, saccharose, citric acid, malic acid, tartaric acid, tannic acid, gallic acid, salicylic acid, succinic acid, oxalic acid, potassium salts, and traces of starch (Downton and Hawker 1973). An antifungal substance, α-viniferin, has also been described as a constituent in grapes (Pryce and Langcake 1977).

The alcohol content of wine can vary from 6 to 18%. White wines typically have between 10 and

12%, and red wines between 11 and 15%. It has been rumored that red wine also contains anandamide, a substance analogous to THC (cf. *Theobroma cacao*, THC).

Effects

During the eighteenth and nineteenth centuries, there were reports of a variety of peculiar psychoactive effects resulting from the consumption of wine; unfortunately, we have no information about these wines or the pharmacological manipulations that might have led to such effects. All that is known is that these were "old wines" that were prescribed by a physician and said to be healthful:

> But these wines also had other wondrous effects: "*I had only but tasted from these barrels*" wrote the theologian Johann Gottfried Hoche in the year 1800, "*and yet the stones on the street appeared to have grown when I came out.*" Wilhelm Hauff had his wonderful *Phantasien im Bremer Ratskeller* [Reveries in the Bremer Council Cellar][324] after tasting what may have been a [Bremen] rosé from 1615 or 1624, and Heinrich Heine saw drunken angels sitting on the rooftops and the spirit of the world with a red nose after a visit to a *ratskeller*. (Marzahn 1994, 109*)

Is it possible that chemical transformations in wines that have been stored for centuries might produce visionary substances? Unfortunately, it will be a difficult task to perform chemical and pharmacological studies of such old wines, which collectors safeguard as though they were precious jewels.

The effects of wine are very dependent upon dosage (cf. **alcohol**, **beer**). In moderate doses, wine uplifts the spirits, lowers inhibitions, and stimulates. If a person drinks wine and also water throughout the evening (in a ratio of 1:3, 1:2, or 1:1), the pleasantly stimulating and uplifting effects can be enjoyed the whole night. But if too great a quantity is consumed, the experience can end in a "blackout" or loss of consciousness. Individual reactions to wine can vary considerably and are much more difficult to control than, for example, dosages of *Cannabis indica*. Many people believe that the effects of sparkling wines (champagne, *sekt*) are different from those of wine. The former are more stimulating, have greater aphrodisiac effects, and also stimulate the circulation of the blood (of course, only when consumed in moderation).

In former times, the effects of wine ("drunkenness") were frequently compared to those of opium (cf. *Papaver somniferum*).

Commercial Forms and Regulations

Grapes and all types of wine are subject to the laws regulating food. Wine is legal in most countries of the world. In some Islamic countries, the use of wine is generally forbidden or frowned upon.

Literature

See also the entries for **alcohol** and **wine**.

Brommer, Frank. 1959. *Satyrspiele*. Berlin: de Gruyter.

Detienne, Marcel. 1992. *Dionysos: Göttliche Wildheit*. Frankfurt and New York: Campus.

Downton, W. John S., and John S. Hawker. 1973. Enzymes of starch metabolism in leaves and berries of *Vitis vinifera*. *Phytochemistry* 12:1557–63.

Emboden, William A. 1977. Dionysos as a shaman and wine as a magical drug. *Journal of Psychedelic Drugs* 9 (3): 187–92.

Evans, Arthur. 1988. *The god of ecstasy: Sex-roles and the madness of Dionysos*. New York: St. Martin's Press.

Grewening, Meinrad Maria, ed. 1996. *Mysterium Wein: Die Götter, der Wein und die Kunst*. Speyer: Verlag Gerd Hatje.

Grimal, P., and E. Kossakowski. 1993. *Pompeji: Ort der Mysterien*. Munich: Metamorphosis Verlag.

Hagenow, Gerd. 1982. *Aus dem Weingarten der Antike*. Mainz: Philipp von Zabern.

Hamdorf, Friedrich Wilhelm. 1986. *Dionysos-Bacchus: Kult und Wandlungen des Weingottes*. Munich: Callwey.

Hehn, Victor. 1992. *Olive, Wein und Feige: Kulturhistorische Skizzen*. Frankfurt/M.: Insel.

Köhnlechner, Manfred. 1978. *Heilkräfte des Weines*. Munich: Knaur.

Lesko, Leonard H. 1978. *King Tut's wine cellar*. Berkeley, Calif.: B. C. Scribe Publications.

Lowrie, Michèle. 1997. *Horace's narrative odes*. Oxford: Clarendon Press.

McGovern, Patrick E. 2003. *Ancient wine: The search for the origins of viniculture*. Princeton, N.J.: Princeton University Press.

McGovern, Patrick E., Stuart J. Fleming, and Solomon H. Katz, eds. 1995. *The origins and ancient history of wine*. Amsterdam: Gordon and Breach Publishers.

Merkelbach, Reinhold. 1988. *Die Hirten des Dionysos*. Stuttgart: Teubner.

Murray, Oswyn, ed. 1990. *Sympotica: A symposium on the symposion*. Oxford: Clarendon Press.

Otto, Walter E. 1933. *Dionysos: Mythos und Kultus*. Frankfurt/M.: Vittorio Klostermann.

Paris, Ginette. 1991. *Pagan grace: Dionysos, Hermes, and Goddess Memory in daily life*. Dallas: Spring Press.

"Wine, in which the fire of the sun joins with the liquid element, is seen everywhere as the counterpart to bread, the fruit of the earth. But wine has effects upon man that are more subtle than those of bread, for it becomes blood, life itself. Even more, because of its divine origin, it is a drink of immortality, embodying the presence of the supernatural light, the divine love in humans. It opens the spiritual inebriation, which produces 'a complete forgetting of all that exists in the world,' leaving room only for the burning desire to find once more the devoutly loved and unite with him."

JACQUES BROSSE
MAGIE DER PFLANZEN [MAGIC OF PLANTS]
(1992, 278*)

---

324 Hauff, who wrote a number of famous fairy tales, effusively described the Bremen *ratskeller* as the "seat of bliss."

Preiser, Gert. 1981a. Wein im Urteil der griechischen Antike. In *Rausch und Realität*, ed. G. Völger, 1:296–303. Cologne: Rautenstrauch-Joest-Museum für Völkerkunde.

———. 1981b. Wein im Urteil der Römer. In *Rausch und Realität*, ed. G. Völger, 1:304–8. Cologne: Rautenstrauch-Joest-Museum für Völkerkunde.

Pryce, R. J., and P. Langcake. 1977. α-viniferin: An antifungal resveratrol trimer from grapevines. *Phytochemistry* 16:1452–54.

Ruck, Carl A. P. 1982. The wild and the cultivated: Wine in Euripides' *Bacchae. Journal of Ethnopharmacology* 5:231–70.

Schultze, Rudolf. 1867. *Geschichte des Weins und der Trinkgelage*. Berlin. Repr. Sändig, 1984.

Serrano, Miguel. 1982. *EI/Ella—Das Buch der Magischen Liebe*. Basel: Sphinx.

Silverberg, Robert. 1987. The Feast of St. Dionysus. In *The Feast of St. Dionysus*. London: Hodder & Stoughton.

Smith, Huston. 1970. Psychedelic theophanies and the religious life. *Journal of Psychedelic Drugs* 3 (1): 87–91.

Weeber, Karl-Wilhelm. 1993. *Die Weinkultur der Römer*. Zurich: Artemis und Winkler.

Withania somnifera (Linnaeus) Dunal

Ashwagandha

Left: The fruits of *Withania somnifera*.

Right: The flowering herbage of ashwagandha (*Withania somnifera*).

Family
Solanaceae (Nightshade Family); Subfamily Solanoideae, Solaneae Tribe

Forms and Subspecies
Eight to ten species are currently recognized in the genus *Withania*, most of which occur in North Africa and the neighboring regions of Eurasia (D'Arcy 1991, 79*; Hepper 1991; Symon 1991, 146*).

Two morphologically and geographically distinct varieties or forms of *Withania somnifera* are distinguished in India. One form with a knotty rootstock occurs in the Indus valley. The other form, which has a fleshy root that can take on an anthropomorphic appearance, is found in Punjab and Rajasthan (Kumaraswamy 1985, 113).

Because of the various mixtures of **withanolides**, the species is sometimes subdivided into chemotypes (Eastwood et al. 1980).

Synonyms
Physalis somnifera L.
Solanum somniferum nom. nud.

Folk Names
Agol (Ethiopian), ambubi, amkuram kizhangu (Dravidian languages, "beautiful horse root"), amukkara, asgandh (Hindi), ashvaganda, ashwagandha, aswagandha, beautiful horse root, bûdîdân, hajarat el dib (Arabic, "wolf tree"), harhumbashir (Assyrian, "red coral"), henbane,[325] jangida, kakink (Pakistan), kuthmithi, marjân (modern Arabic, "coral"), rasbhari, salztiegel, schlafbeere, schlaffbeeren, schlafmachende schlute, sekran (Syrian, "inebriant"), slaepcruydt, solanum somniferum, timbutti eqli (Assyrian, "ring of the field" or "cantharides"), 'ubâd (Arabic/Yemen), winterkirsche

History
If the interpretation of the Assyrian name as ashwagandha is correct, the plant was in use for medicinal and narcotic purposes in Mesopotamia (Thompson 1949, 216*). It was well known in ancient Egypt (Germer 1985, 167*), and it was characterized and classified as a *sakrân*, "inebriant," in Old Arabic. The plant is regarded as a hypnotic and sedative throughout its entire range of distribution (Hooper 1937, 186*). It is possible that the plant was seen as a form of "sleeping strychnos" during late antiquity, e.g., by Pliny (21.180) (cf. **Datura stramonium, Solanum spp., Strychnos nux-vomica**). It may have been identical to the mysterious **haliacacabon**.

It has been claimed that the wondrous root *jangida*, whose praises were sung in the Vedic scriptures—especially the Atharva Veda—and which was regarded as a panacea, amulet, magical agent, and aphrodisiac, is identical to *Withania somnifera* (Kumaraswamy 1985).

The plant has been known in Europe since at

325 This English name is normally given to *Hyoscyamus niger*.

least the sixteenth century, for it is described and pictured in most of the herbals written by the fathers of botany.

Distribution

Ashwagandha is originally from North Africa and is very common throughout Iraq (Al-Hindawi et al. 1992). It also occurs in Pakistan and northern India. In China, the plant is a popular ornamental (Lu 1986, 81*).

Cultivation

Propagation is performed via the seeds, which are best pregerminated before planting. Water well at first, then water in moderation. Because the plant does not tolerate any frost, in cold climates it should be kept inside in a pot during the winter. As a houseplant, ashwagandha can blossom several times a year.

Appearance

This perennial, branchy, herbaceous plant can grow to more than 1 meter in height (or in rare cases up to 2.5 meters), but usually it remains a small bush. The small, oval leaves are alternate. The tiny flowers have greenish calyxes and white pistils. They are attached to the upper branches close to the main stem. The red berry fruits (hence the ancient Assyrian name *harhumbashir*, "red coral"; Thompson 1949, 215*) are surrounded by an inflated calyx (similar to *Physalis* spp.) so that they resemble small lanterns. The small, orange-yellow seeds are round, flat, and 1 to 2 mm in length. The thin, smooth roots can grow 30 to 40 cm long and 1 to 2 cm thick.

Ashwagandha can be confused with other members of the genus, in particular the Mediterranean *Withania frutescens* (L.) Pauq. and the Canary Island *Withania aristata* (Ait.) Pauq. It also can be confused with smaller members of the genus *Physalis* spp. (cf. **halicacabon**).

Psychoactive Material

— Root
— Aboveground herbage

Preparation and Dosage

The root is dried and left whole or finely ground. The powder can be poured into gelatin capsules for ingestion. A tonic and sedative tea can be prepared by boiling the root cortex for a few minutes. The root powder can be boiled in milk together with **honey** and pippali (*Piper longum*; cf. *Piper* spp.).

In Ayurvedic medicine, a single dosage consists of 250 mg to 1 g of the powdered root (Lad and Frawley 1987, 227*). Pronounced antistress effects occur at dosages of 100 mg of root powder per kg of body weight (Grandhi et al. 1994, 134*).

Tonic effects can be attained by chewing a piece of root the length of half a finger every day. The root has a not-unpleasant taste somewhat reminiscent of that of licorice.

Ritual Use

Ancient Arabs used the root as a tonic, aphrodisiac, and inebriant. Unfortunately, no information about their ritual use has come down to us.

Sushruta, the Indian physician and cofounder of Ayurveda, praised the root as a *rasayana* (an alchemical elixir) and as a *vajikarana* (aphrodisiac) with few equals. For this reason, ashwagandha (sometimes in combination with *Cannabis indica*) was used in sexual magic and tantric rituals as an aid in supporting the required duration of erections. The *vaidyas* (folk healers) still prepare a love drink from the root that is said to attract the opposite sex and make one ready for love (Kumaraswamy 1985, 114, 116, 119).

In Pakistan, the leaves of *panirbad*, the closely related *Withania coagulans* (Stocks) Dun. [syn. *Puneeria coagulans* Stocks], are used (presumably smoked) as an inebriant (Goodman and Ghafoor 1992, 43*). In India, the fruits are used to coagulate milk when rennet cannot be used in rituals and ceremonies for religious reasons (Macmillan 1991, 422*).

Artifacts

The remains of several Egyptian flower garlands have been found in El Faiyûm that date to late antiquity and incorporate ashwagandha fruits (Germer 1985, 167*).

The root was used as a substitute for mandrake (see *Mandragora officarum*).

Medicinal Use

The Assyrians used the root as a fumigant, directing the smoke onto painful teeth (cf. **incense**). This use is similar to the ways in which they used henbane (*Hyoscyamus niger*) (Thompson 1949, 216*). In Yemen, the root is still used for treating toothaches (Fleurentin and Pelt 1982, 102 f.*).

In the folk medicine of the inhabitants of the Golan Heights and the Negev Desert, the leaves, and less frequently the fruits, are applied externally as a paste and massaged into the skin to treat open wounds, swelling, rheumatism, and external inflammation (Dafni and Yaniv 1994, 16*).

In Africa, the root is given to children as a tranquilizer (Schuldes 1995, 77*). In Ethiopia, crushed leaves are smeared onto arthritic joints (Wilson and Mariam 1979, 33*).

In Baluchistan (Pakistan), the root cortex is powdered, mixed with water, and kneaded to a paste that is applied to treat wounds (Goodman and Gharfoor 1992, 42*). In India, the herbage is smoked to soothe coughing and asthma (cf. **smoking blends**) (Macmillan 1991, 425*).

An early illustration of the "sleep-inducing nightshade" (*Withania somnifera*). (Woodcut from Gerard, *The Herball or General History of Plants*, 1633*)

Ashwagandha has a significance in Ayurvedic medicine that is similar to that of ginseng (*Panax ginseng*) in Chinese herbalism. Ashwagandha is regarded as a "rejuvenative herb"; "*sattvic* in quality, it is one of the best herbs for the mind upon which it is nurturing and clarifying. It is calming and promotes deep, dreamless sleep" (Frawley and Lad 2001, 161).

Constituents

The plant contains steroid lactones, somniferin, withaferin A, and various steroids (Al-Hindawi et al. 1989, 167). The root contains approximately 2.8% steroid lactones, so-called **withanolides**, and also starch (Grandhi et al. 1994:134). The new **withanolides** withasomnilide, withasomniferanolide, somniferanolide, somniferawithanolide, and somniwithanolide were discovered in the stem bark of a sample from India (Ali et al. 1997).

Effects

The effects of the root are described as tranquilizing, sedative, and generally tonic. An aqueous extract of the root has antistress effects similar to those of ginseng (*Panax ginseng*) (Grandhi et al. 1994, 134). The antiserotinergic activity results in a stimulation of the appetite. An alcohol extract of the aboveground herbage has quite potent anti-inflammatory properties, primarily as a result of the steroids that are present, especially withaferin A (Al-Hindawi et al. 1989, 167; 1992). No toxic side effects have been reported to date, even when the plant was used during pregnancy (Grandhi et al. 1994, 132). The information concerning inebriating effects is sporadic and unreliable.

Commercial Forms and Regulations

The plant is not subject to any legal restrictions and is freely available. Although the root is difficult to obtain in Europe, it can be obtained in any herb shop in India.

Young plants can be purchased from sources offering ethnobotanical plants and from specialized nurseries.

Literature

See also the entry for **withanolides**.

Ali, Mohammed, Mohammed Shuaib, and Shahid Husain Ansari. 1997. Withanolides from the stem bark of *Withania somnifera*. *Phytochemistry* 44 (6): 1163–68.

Eastwood, Frank W., Isaac Kirson, David Lavie, and Arieh Abraham. 1980. New withanolides from a cross of South African chemotype by chemotype II (Israel) in *Withania somnifera*. *Phytochemistry* 19:1503–7.

Frawley, David, and Vasant Lad. 2001. *The yoga of herbs: An Ayurvedic guide to herbal medicine.* 2nd ed. Twin Lakes, Wis.: Lotus Press.

Grandhi, Anuradha, A. M. Mujumdar, and Bhushan Patwardhan. 1994. A comparative pharmacological investigation of ashwagandha and ginseng. *Journal of Ethnopharmacology* 44:131–35. (Lists additional literature.)

Hepper, F. Nigel. 1991. Old World *Withania* (Solanaceae): A taxonomic review and key to species. In *Solanaceae III: Taxonomy, chemistry, evolution*, ed. Hawkes, Lester, Nee, and Estrada, 211 ff. London: Royal Botanic Gardens Kew and Linnean Society.

Hindawi, Muhaned K. al-, Ishan H. S. Al-Deen, May H. A. Nabi, and Mudafar A. Ismail. 1989. Anti-inflammatory activity of some Iraqi plants using intact rats. *Journal of Ethnopharmacology* 26:163–68.

Hindawi, Muhaned K. al-, Saadia H. Al-Khafaji, and May H. Abdul-Nabi. 1992. Anti-granuloma activity of Iraqi *Withania somnifera*. *Journal of Ethnopharmacology* 37:113–16.

Kumaraswamy, R. 1985. Ethnopharmacognostical studies of the Vedic jangida and the siddha kattuchooti as the Indian mandrake of the ancient past. In Ethnobotanik, special issue, *Curare* 3:109–20.

Nittala, S. S., V. van den Velde et al. 1981. Chlorinated withanoloides from *Withania somnifera* and *Acnistus breviflorus*. *Phytochemistry* 20:2547.

Sour, K. Y. 1980. Phytochemical investigation of *Withania somnifera* grown in Iraq. MSc thesis, University of Baghdad.

Black henbane *Hyoscyamus niger* L. (from
Giftgewächse [Poisonous Plants], 1875).

Little-Studied Psychoactive Plants

Minor Monographs

There are a great number of plants about which almost no ethnobotanical or phytochemical research has been carried out but that are said to have or have been demonstrated to have psychoactive effects. Moreover, some of these plants' botanical identifications are questionable. In the case of some of the plants discussed in these minor monographs, information about their psychoactivity (e.g., in the case of *Cymbopogon densiflorus*) is contained only in notes found together with old herbarium specimens (Altschul 1975*; von Reis and Lipp 1982*). In other cases, very well-known plants that see occasional use for psychoactive purposes, such as ginger (*Zingiber officinale*), have had little real research into their psychoactive use, preparation, and application. For some plants we have no examples of preparations made with the psychoactive substances they yield (e.g., *Gomortega keule*). In other cases, plants whose psychoactive effects have been pharmacologically demonstrated have no known or reported traditional use (e.g., *Mikania cordata*). Many of the plants included in these minor monographs are used primarily as additives to other plant mixtures or products (e.g., *Alchornea* spp.). These plants do not exhibit psychoactive properties when used alone but have synergistic effects when combined with other substances.

This section will certainly provide many suggestions and impulses for future ethnopharmacological and phytochemical studies.

To the extent that it is possible, the minor monographs also include references to specialized literature. For many of the plants, however, the slim or sparse research situation means that there are no writings dedicated specifically to them.

The Genera Discussed in This Section

Ailanthus, Alchornea, Amaranthus, Anarmita, Archontophoenix, Armatocereus, Aspidosperma, Astragalus, Atherosperma

Benthamia, Bernoullia, Boophane, Brosimum, Bursera

Caesalpinia, Capsicum, Cardamine, Carissa, Castanopsis, Cecropia, Clematis, Comandra, Conium, Cordia, Cordyline, Coriaria, Crotalaria, Cymbopetalum, Cymbopogon, Cyperus, Cypripedium

Delphinium, Dictyoloma, Dictyonema, Dimorphandra, Dioscorea

Elaeophorbia

Ferraria

Gaultheria, Gelsemium, Gloeospermum, Gomortega, Goodenia

Hedera, Helichrysum, Helicostylis, Hieracium, Hipomosa, Homalomena, Huperzia

Iresine, Iryanthera

Jasminum, Jatropha, Juanulloa

Kaempferia

Lagochilus, Lancea, Leonotis, [lichen non ident.], Limonium, Lobelia, Lotus, Lucuma, Lupinus, Lycopodium

Macropiper, Magnolia, Malva, Manihot, Maquira, Matayba, Mentha, Metteniusa, Mikania, Mirabilis, Monadenium, Monodora, Mostuea

Neoraimondia, Nephelium

Ocimum, Osteophloeum, Oxytropis

Pancratium, Pandanus, Pedilanthus, Peperomia, Pernettya, Persea, Petunia, Peucedanum, Philodendron, Physalis, Pithecellobium, Polypodium, Pontederia, Pseuderanthemum

Quararibea

Ranunculus, Rauvolfia, Rhododendron

Sanango, Santalum, Scirpus, Sclerocarya, Scoparia, Securidaca, Senecio, Sida, Sloanea, Spiraea, Stephanomeria, Stipa

Teliostachys, Terminalia, Tetrapteris, Thamnosma, Thevetia, Tillandsia, Tribulus, Trichocline, Trichodesma

Umbellularia, Ungnadia, Urmenetea, Utricularia

Valeriana, Vanda, Voacanga

Zea

Left: The attractive devil's tobacco (*Lobelia tupa*) is from southern Chile and develops magnificent flowers. Studies about and tests of the plant are lacking.

Ailanthus altissima (Mill.) Swingle

[syn. *Ailanthus glandulosa* Desf., *Ailanthus peregrina* (Buc'hoz) Barkl.]

(Simaroubaceae)—tree-of-heaven

Originally from China, the tree of the gods has now been introduced into Europe and North America (Zander 1994, 98*). An herbarium specimen collected in Pennsylvania in 1937 bears the note, "Seedling 'of a possibly narcotic plant'" (von Reis and Lipp 1982, 146*). No research has been conducted to ascertain whether the tree actually does have psychoactive effects. The main agent (5%) is quassiine (= ailanthine), with the molecular formula $C_{36}H_{50}O_{10}$ (Reichert et al. 1949, 3:169*).

Alchornea spp.

(Euphorbiaceae; Acalyphoideae)

This tropical genus is composed of some seventy species, most of which are found in the Americas, although a few are native to Africa (Schneider 1992, 166). The South American *Alchornea castaneifolia* (Willd.) Juss. is used in Peru as an **ayahuasca** additive. The bark of this species, known as *pájaro arbol* (Spanish, "bird tree"), exhibits antifungal properties (Ott 1993, 403*). The Tikana Indians use the bark medicinally to treat diarrhea (Schultes and Raffauf 1986, 265*). It is not known whether the plant itself is psychoactive.

In Africa, the bushy species *Alchornea floribunda* Muell. Arg. is known as *niando* or *malande*. Found in the tropical regions between Sierra Leone and Zaire, it was formerly used ritually in the *bieri* ancestor cult (Smet 1996*). The fresh or dried root (which is also known as *niando*) is sometimes added to iboga preparations (see ***Tabernanthe iboga***). The plant is also used as a stimulant and inebriant in many parts of Africa (Raymond-Hamet 1952). It is regarded as a marijuana substitute (see ***Cannabis indica***) and smoked as an aphrodisiac as well (De Wildeman 1920; Schneider 1992, 171). "The natives of the Congo add ground roots of *A. floribunda* to palm wine for several days to prepare a stimulating drink, *niando*. The drink serves both as an

The African niando bush (*Alchornea floribunda*) is one of the sacred plants of the Bwiti cult. (Photograph: Giorgio Samorini)

aphrodisiac and to maintain vigor for wars and tribal festivals" (Scholz and Eigner 1983, 78*). The effects have been described as follows: "As a narcotic hallucinogen, the root drug added to palm wine initially induces a phase of stimulation, then a profound fatigue with isolated fatal results" (Schneider 1992, 71).

In contrast to an earlier report (Paris and Goutarel 1958), the plant does not contain **yohimbine**, although it does contain the alkaloids alchorneine, isoalchorneine, alchorneinone, and pyrimidine along with imidazole derivatives (Khuong-Huu et al. 1972; Ott 1993, 403*). The quantity of alkaloids can vary considerably. It is usually the highest in the roots, where it can range from 0.6 to 1.2%. As a free base or in the form of a simple derivative, alchornine has antidepressive, spasmolytic, and anticholinergic properties (Schneider 1992, 170 f.).

The closely related African species *Alchornea cordifolia* (Schum. et Thonn.) Muell. Arg. is easily confused with *A. floribunda*. It has a number of ethnomedical uses. Its large leaves are used to package cola nuts (***Cola* spp.**) for transport (Schneider 1992, 170). In West Africa, the dried leaves are used to brew a tonic (Assi and Guinko 1991, 26*).

Literature

De Wildeman, E. 1920. Le "Niando" succédané du chanvre au Congo belge. *Congo* 1:534–38.

Khuong-Huu, F., J.-P. Le Forestier, and R. Goutarel. 1972. Alchorneine, isoalchorneine et alchorneinone, produits isolés de l'*Alchornea floribunda* Muell. Arg. *Tetrahedron Letter* 28:5207–20.

Paris, R., and R. Goutarel. 1958. Les Alchornea africains. Présence de yohimbine chez l'*Alchornea floribunda* (Euphorbiaceae). *Ann. Pharm. Fr.* 16:15–20.

Raymond-Hamet. 1952. L'*Alchornea floribunda* Müller ou Niando. *Revue Internationale de Botanique Appliquée et d'Agriculture Tropicale* 32:427–42.

Schneider, Kurt. 1992. *Alchornea.* In *Hagers Handbuch der pharmazeutischen Praxis*, 5th ed., 4:166–73. Berlin: Springer.

Amaranthus spp.

(Amaranthaceae)—amaranths

In South America, a number of *Amaranthus* species are roasted into ashes and used as *llipta* for chewing coca (see ***Erythroxylum coca***). In Mexico, a **chicha** brewed from the seeds of *Amaranthus caudatus* L. is offered to Mother Earth and ritually consumed before the planting of a new field (Early 1992, 29). In Ecuador, the flowers of *Amaranthus* *hybridus* L. are decocted into a red liquid that is mixed with rum to produce a drink known as *draque*, which is used to purify the blood and regulate menstruation (Early 1992, 30). A species of *Amaranthus* was used by the Cherokee of North America for ethnogynecological and ceremonial purposes (Ott 1993, 403*).

The Lodha, a tribal people from West Bengal, are said to smoke the dried, pulverized roots of *Amaranthus spinosus* L. (prickly amaranth, *cauleyi, kateli, tanduliyah*) as a hallucinogen. A paste of the plant is said to produce "temporary insanity." In Ayurvedic medicine, the plant is considered a tonic and is used to treat hallucinations (Warrier et al. 1993, 1:107*) In Swaziland (Africa), the entire plant (*Amaranthus spinosus*) in burned to ashes and used as a **snuff**, either alone or mixed with tobacco (*Nicotiana tabacum*) (Ayensu 1978, 32*).

There is no information available about psychoactive substances in this genus (cf. also *Iresine* spp.).

Literature

Cole, John N. 1979. *Amaranth from the past for the future.* Emmous, Mich.: Reference Publications.

Early, Daniel K. 1992. The renaissance of amaranth. In *Chilies to chocolate: Food the Americas gave the world*, ed. Nelson Foster and Linda S. Cordell, 15–33. Tucson: University of Arizona Press.

Anarmita cocculus Wight et Arnott

[syn. *Anarmita paniculata* Colebrooke, *A. baueriana* Endl., *A. jucunda* Miers, *A. populifolia* (DC.) Miers, *A. toxifera* Miers, *Cissampelos cocculus* (L.) Miers, *Cocculus lacunosus* (Lam.) DC., *C. populifolius* DC., *C. suberosus* DC., *Menispermum cocculus* L., *M. heteroclitum* Roxb., *M. lacunosum* Lam.]
(Menispermaceae)—cocculus bush

This shrubby climber, a member of the Moonseed Family, is indigenous to eastern India. The round fruits, which are red when fresh and can be as large as 1 cm, are known as cocculus seeds (fructus cocculi), fish seeds, or crazy seeds (cf. **witches' ointments**) (Schneider 1974, 1:90*). The fruits contain 1.5 to 5% picrotoxin, consisting of picrotoxinine and pikrotin, as well as bases of the berberine and aporphinal alkaloid types. Picrotoxin has stimulating effects on the central nervous system but can lead to coma and delirium. It is considered one of the most effective antidotes for barbiturate poisoning (Roth et al. 1994, 122*). Cocculus seeds were used in early modern times as an inebriating additive to **beer** (Tabernaemontanus 1731*).

Cocculus seeds are now illegal in Germany under the Cosmetic Law of June 19, 1985. Their

only use today is in homeopathy. The closely related species *Cocculus leaeba* DC. and *Cocculus pendulus* are reputed to be psychoactive (Schultes and Farnsworth 1982, 187*; Schultes and Hofmann 1980, 368*).

Literature

Hänsel, Rudolf, and Renate Seitz. 1992. *Anarmita*. In *Hagers Handbuch der pharmazeutischen Praxis*, 5th ed., 4:267–72. Berlin: Springer.

Archontophoenix cunninghamiana (H. Wendl.) H. Wendl. et Drude

[syn. *Ptychosperma cunninghamiana* H. Wendl., *Seafortia elegans* Hook. non R. Br.]
(Palmae)—king palm

This frond palm, which can grow as tall as 25 meters, belongs to the Subfamily Arecoideae. Originally from Australia (Queensland, New South Wales), it bears round, red berries approximately 2 cm in size. The seeds inside are covered by wide fibers. The Papuas (Papua New

Top: The prickly amaranth (*Amaranthus spinosus*) is from India.

Bottom: Fruits of the Southeast Asian cocculus bush (*Cocculus* sp.) have toxic and psychoactive properties.

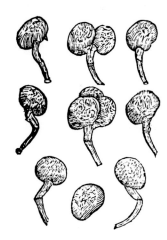

The exotic cocculus seeds (*Anarmita cocculus*) were used as an inebriating beer additive during the seventeenth and eighteenth centuries. (Woodcut from Gerard, *The Herball or General History of Plants*, 1633)

Botanical illustration of a Southeast Asian cocculus bush. (Engraving from Pereira, *De Beginselen der Materia Medica en der Therapie*, 1849)

Guinea) are said to chew the ripe seeds, which have an inebriating effect, as an **alcohol** substitute (Bärtels 1993, 43*). It is possible that the king palm was used to make **palm wine**.

Botanical illustration of the white quebracho tree (*Aspidosperma quebracho-blanco*), a member of the Dogbane Family. (From *Köhler's Medizinal-Pflanzen*, 1889)

Astragalus species are found throughout the world. In North America, some of them are known as locoweed. (Woodcut from Tabernaemontanus, *Neu Vollkommen Kräuter-Buch*, 1731)

Quebracho bark (*Aspidosperma quebracho-blanco*), which contains yohimbine, in its crude form.

Armatocereus laetus (H.B.K.) Backeb.

(Cactaceae)

This cactus, known in Peru as *pishicol*, is reputed to be psychoactive and is one of the plants of the San Pedro cult (see *Trichocereus pachanoi*). No chemical studies have been conducted (Ott 1993, 396*).

Aspidosperma quebracho-blanco Schlecht.

[syn. *A. chakensis* Speg., *A. crotalorum* Speg., *A. quebracho* Griseb., *Macaglia quebracho* O. Ktze., *Macaglia quebracho-blanco* (Schlect.) Lyons]

(Apocynaceae)—white quebracho

The white quebracho tree is found throughout Argentina (Chaco), Peru, and Bolivia. It grows up to 20 meters in height and has mythological, ritual, and ethnomedicinal significance. Among many South American Indian cultures, it is an important shaman's tree that shamans occasionally climb to communicate with animal spirits (cf. *Jatropha grossidentata*). The tree, known as *no'dik*, yields a bark that the Pilagá Indians (Chaco) decoct to treat stomach upsets, cough, headaches, and syphilis. It also finds use as an analgesic and an abortifacient (Filipov 1994, 185*). The Ayoreo Indians, who live in Paraguayan Chaco, consider the tree, which they call *ebedu*, to be a panacea and they use it to treat all manner of ailments (Schmeda-Hirschmann 1993, 107, 108*).

The common name *quebracho*, "ax breaker," is used for a number of species of hardwood trees: *horco quebracho* (*Schinopsis haenkeana* Engler), *quebracho colorado chaqueño* (*Schinopsis balansae* Engler), *quebracho colorado santiagueño* (*Schinopsis quebracho-colorado* [Schlecht.] Barkley et Meyer), *quebrachillo* (*Diatenopteryx sorbifolia* Radlk.), and *quebracho blanco chico* (*Aspidosperma triternatum* Rojas Acosta [syn. *Aspidosperma quebracho-blanco*

Schlecht. ssp. *brevifolium* Hassl]) (cf. Santos Biloni 1990, 35, 107, 109, 199, 239*). In Peru, Quechua speakers refer to *Aspidosperma quebracho-blanco* as *willca* (*kachakacha* in Aymara); the name *willca* is usually used as a name for **Anadenanthera colubrina** (Santos Biloni 1990, 118*).

Quebracho bark has often been attributed with aphrodisiac powers with psychoactive effects (Schuldes 1995, 20*). The bark of *Aspidosperma quebracho-blanco*, which finds phytotherapeutic use for asthma, contains some thirty **indole alkaloids**, including six principal alkaloids (totaling at least 1%): quebrachine (= **yohimbine**), aspidospermine, quebrachamine, hypoquebrachine, and aspidosamine (Santos Biloni 1990, 119*). The fruits contain the **indole alkaloid** aspidospermatine; the leaves also contain indoles. Pharmacological studies are lacking (Hoffmann-Bohm 1992, 401 f.).

Literature

Hoffmann-Bohm, Kerstin 1992. *Aspidosperma*. In *Hagers Handbuch der pharmazeutischen Praxis*, 5th ed., 4:400–405. Berlin: Springer.

Astragalus spp.

(Leguminosae: Fabaceae)—locoweeds

Of the more than five hundred species of this genus in North America, a few are referred to by the Spanish-English name *locoweed*, "crazy weed" (cf. **Oxytropis spp.**), and have toxic or psychotropic properties (Turner and Szczawinski 1992, 122*). Many *Astragalus* species of the North American prairies have cytotoxic properties, i.e., they kill cells and can thus be used in the treatment of cancer (McCracken et al. 1970). The common name of this plant refers to observations that grazing sheep, cattle, and horses "flip out" or "go crazy" after eating *Astragalus* or *Oxytropis*. In South Dakota, I once heard that the Dakota Indians used to and possibly still do eat (or perhaps smoke) locoweeds to produce visions. To date, no constituents with psychoactive or psychotropic effects have been found. The nitrogenous substance miserotoxin (or a derivative) may be responsible for these effects (Williams et al. 1975). Miserotoxin (= 3-nitro-1-propyl-β-D-gentiobioside) occurs especially in *Astragalus miser* Dougl. var. *serotinus* (Gray) Barneby and in at least ten additional species (Majak and Benn 1988). *Astragalus* species also possess unusually high concentrations of selenium (Emboden 1976, 160*; Turner and Szczawinski 1992, 123*).

The Navajo utilize a hallucinogenic locoweed, perhaps blue loco (known in Navajo as *dibéhaich'iidii*, "gray sheep scratcher"), together with **Datura innoxia** in magical rituals. They regard blue loco as a "life medicine" (Mayes and Lacy 1989, 59*).

The following species are said to be psychoactive: *Astragalus amphioxys* A. Gray, *A. besseyi* Rdb., *A. cagopus,* and *A. mollissimus* Torr. (Schultes and Farnsworth 1982, 187*; Schultes and Hofmann 1980, 367*). The Mexican species *Astragalus amphioxis* [1095] Gray is also attributed with such properties (Reko 1938, 187*).

The Old World species *Astragalus microcephalus* Willd. and *Astragalus gummifer* Labill. provide a gum (tragacanth) that is used in the production of **incense** (Scholz 1992).

Literature

Majak, Walter, and Michael H. Benn. 1988. 3-nitro-1-propyl-β-D-gentiobioside from *Astragalus miser* var. *serotinus. Phytochemistry* 27 (4): 1089–91.

McCracken, D. S., L. J. Schermeister, and W. H. Bhatti. 1970. phytochemical and cytotoxic evaluation of several *Astragalus* species of North Dakota. *Lloydia* 33 (1): 19–24.

Scholz, Eberhard. 1992. *Astragalus.* In *Hagers Handbuch der pharmazeutischen Praxis,* 5th ed., 4:405–17. Berlin: Springer.

Williams, M. Coburn, Frank R. Stermitz, and Richard D. Thomas. 1975. Nitro compounds in *Astragalus* species. *Phytochemistry* 14:2306–8.

Atherosperma moschatum Labill.

(Monimiaceae)—southern sassafras, black sassafras

This tree, which can grow as tall as 45 meters, is found in the cool and temperate rain forests of Tasmania (Collier 1992, 24; Kirkpatrick and Backhouse 1989, 59). It is usually refered to as sassafras[326] or, less frequently, as southern sassafras and should not be confused with the true *Sassafras albidum* of America.

Early European settlers prepared a tonic tea from the fresh or dried bark, which contains safrole and smells strongly of sassafras (Cribb and Cribb 1984, 172*). The settlers on the Australian mainland made a similar tea from the bark of what they called "real sassafras," the closely related tree *Doryphora sassafras* Endl. (Collier 1992, 24; Cribb and Cribb 1984, 174*). The black wattle tree (*Acacia decurrens*; cf. **Acacia spp.**) was used in the same manner. Apparently the settlers learned to use these tea plants through trial and error and not from the Aborigines (Low 1992a, 34*).

Tea made from the bark of the trunk (eight strips, 5 to 7 cm in length, steeped in 1/4 liter of water for five to eight minutes) has a strong safrole taste and leaves a slightly numb but invigorating sensation in the mouth and throat. The effects are clearly stimulating; higher dosages produce the typical safrole effects. Of particular interest is the report that early settlers used the leaves and bark in place of hops (*Humulus lupulus*) to brew **beer**

(Pettit 1989, 62). It is conceivable that the beer yeast metabolizes the safrole into an amphetamine derivative during the fermentation process. Home brewers in Tasmania allegedly still make a psychoactive sassafras beer.

In addition to the safrole-containing oil (cf. **essential oil**), the bark of *Atherosperma moschatum* contains several alkaloids: berbamine (main alkaloid), isotetradrine, isocorydine, atherospermidine, atherosperminine, spermatheridine, atheroline, moschatoline, and methoxyatherosperminine (Lassak and McCarthy 1987, 80*).

Literature

Collier, Phil 1992. *Rainforest plants of Tasmania.* Hobart, Australia: Society for Growing Australian Plants–Tasmania.

Kirkpatrick, J. B., and Sue Backhouse. 1989. *Native trees of Tasmania.* Hobart, Australia: Pandani Press.

Pettit, Rose. 1989. *Tasmanien: Reisen auf der urwüchsigen australischen Insel.* Alpers: SYRO.

Benthamia alyxifolia (Benth.) Tieghem

(Loranthaceae)—mistletoe
This Australian plant, which is similar to the true mistletoe (*Viscum album* L.; Loranthaceae), is a parasite on several plants, including *Duboisia myoporoides* (see **Duboisia spp.**). The leaves contain **scopolamine** and are smoked in Australia as an inebriant (Bock 1994, 85*). It is possible that the scopolamine is extracted from the host tree *Duboisia myoporoides* as a result of the mistletoe's parasitic activity and is then incorporated into the plant's own tissue.

Bernoullia flammea Oliver in Hook.

(Bombacaceae)—amapola blanca
This tree grows to a towering height of 40 meters in the tropical lowlands of Guatemala (El Petén), where it is known as *amapola blanca*, "white opium [tree]" (cf. **amapola**). It has a whitish bark and fire-red flowers, and its fruits are filled with winged seeds that are reminiscent of those of both *Banisteriopsis caapi* and the maple tree (*Acer* spp.;

Left: The Tasmanian tree *Atherosperma moschatum,* known as "southern sassafras," has a strong safrole scent. (Photographed in the cold rainforest of Tasmania)

Right: The rainforest tree *Bernoullia flammea,* shown here in front of a pyramid in Tikal (Guatemala), is also known as *amapola* ("opium").

---

326 In eastern Australia, the cinnamon wood tree (*Cinnamomum oliveri*) is also called sassafras (Pearson 1992, 62*).

Aceraceae) (Lanza Rosado 1996, 22 ff.). The seeds are smoked by the Guatemalan people living in the area around the spectacular Mayan ruins of Tikal and are said to have a strong, opium-like effect (Rob Montgomery, pers. comm.; Brett Blosser, pers. comm.). The tree is known in Itza Maya as *chunte'* and in Yucatec Maya as *wakut.*

Literature

Lanza Rosado, Felipe. 1996. *Manual de los árboles de Tikal.* Alicante and Barcelona: Agencia Española de Cooperación International.

Boophane disticha (L. f.) Herbert

[syn. *Amaryllis disticha* L., *Haemanthus toxicarius* Thumb., *Boophane toxicaria* (L. f.) Herb., *Brunsvigia toxica* Ker., *Bufane toxicaria* Herb., *Buphane toxicaria* Thunberg, *Haemanthus lemairei* (L. f.) Herbert; also *Buphane, Boöphone, Boophone*]

(Amaryllidaceae)—fan lily, cowbane

The bulb of this African species of amaryllis is used in folk medicine and to manufacture arrow poisons. The Bushmen use it as a hunting poison (Neuwinger 1994, 4 f.*). It is also used in (ritual) suicides (Lewin 1912). Another use occurred in the secret initiation ceremonies of the South African Basuto. The young boys ate the crushed bulb together with other ingredients so that they could contact their ancestors. The first signs of inebriation were interpreted to mean that the spirit of adulthood had entered into them. A dried powder of the bulb also serves as a ritual and psychoactive **incense**:

> The Sotho, among whom the plant enjoys a particularly high status (e.g., they call October, the month in which the plant flowers, *mphalane es leshoma = Boophane* stalk), utilize the "alcoholizing" properties of *Boophane disticha* during the preliminaries of initiation. A powder made from the bulb is mixed with other plants and heated, and the smoke is inhaled. It makes the initiates drunk as if from alcohol. When signs of intoxication appear, this taken as a sign that the spirit of manliness has entered into the youths. (Neuwinger 1994, 7*)

In West Africa, *Securidaca longepedunculata,* which contains **ergot alkaloids**, is used together with *Boophane* for psychoactive purposes (Neuwinger and Mels 1997).

Fresh *Boophane* bulbs contain 0.31% alkaloids (buphanidrine, undulatine, buphanisine, buphanamine, nerbowdine, lycorine), which exhibit the same bioactivity as the **tropane alkaloids** in *Datura* (Hauth and Stauffacher 1961; Rauwald and Kober 1992, 527; Tupin 1912; Roth et al. 1994,

In South Africa, the fan lily (*Boophane disticha*), also known as cowbane, is called *malgif,* "crazy poison." (From Neuwinger)

The strange and magical African root *Boophane disticha.* (From Lewin, "Untersuchungen über *Buphane disticha* [*Haemanthus toxicarius*]," 1912)

179*). Some of the alkaloids also have **morphine**like structures and effects (Neuwinger 1994, 6 f.*). Overdoses can be fatal. Both the ethnographic and the ethnopharmacological literature contain numerous indigenous reports of distinct hallucinogenic effects. In Zimbabwe, the bulb is used to help the ancestor spirits appear (de Smet 1996, 142 f.*). The bulb also has folk medical uses; it is taken internally to treat hysteria and insomnia (Rauwald and Kober 1992, 528).

Literature

Hauth, H., and D. Stauffacher. 1961. Die Alkaloide von *Buphane disticha* (L. f.) Herb. *Helvetia Chimica Acta* 44:491–502.

Lewin, Louis. 1912. Untersuchungen über *Buphane disticha* (*Haemanthus toxicarius*). *Archive für experimentelle Pathologie und Pharmakologie* 68:333–40.

Neuwinger, Hans Dieter, and Dietrich Mels. 1997. *Boöphane disticha*—Eine halluzinogene Pflanze Afrikas. *Deutsche Apotheker-Zeitung* 137 (14).

Rauwald, Hans-W., and Martin Kober. 1992. *Boophane.* In *Hagers Handbuch der pharmazeutischen Praxis,* 5th ed., 4:526–28. Berlin: Springer.

Tutin, F. 1912. Über die Bestandteile von *Buphane distacha. Archive für experimentelle Pathologie und Pharmakologie* 69:314.

Brosimum acutifolium ssp. obovatum (Ducke) C.C. Berg

(Moraceae)

The Amazonian Palikur and Wayãpi Indians use parts (which?) of this tree, known as *tamamuri* or *congona,* as hallucinogens, especially in initiation rites (Duke and Vasquez 1994, 32*). It is doubtful whether this vague information is correct. Coumarins (pyranocoumarins, furocoumarins) have been found in this species and in other Amazonian members of the genus (Gottlieb et al. 1972). Whether these have psychoactive effects, however, is unknown.

Literature

Gottlieb, O. R., M. Leão da Silva, and J. G. Soares Mia. 1972. Distribution of coumarins in Amazonian *Brosimum* species. *Phytochemistry* 11:3479–80.

Bursera bipinnata Engl.

[syn. *Elaphrium bipinnatum* (DC.) Schlecht.]

(Burseraceae)—sacred copal

Today, this balsam bush is known in Mexico as *copal amargo, copal cimarrón, copal chino, copal de santo, copal de la virgin, copalio, palo copal,* and *pom,* and it is used as a ritual **incense**. Emmar

Left: In Mexico, many species of the genus *Bursera* produce aromatic resins used as incense.

Right: The *yün-shih* fruit (*Caesalpinia decapetala*).

(1937*) suggested that the tree was known as *teuvetli* in pre-Spanish times and that it was used in the preparations for Aztec human sacrifice (cf. **Datura innoxia**). It is said that a decoction made from the resinous bark was administered to the victims prior to the ceremony. The extract was probably mixed with pulque (cf. **Agave spp.**), for the prisoners were required to drink four bowls of pulque prior to the ceremony (Davies 1983, 244). This beverage produced a subdued or sedated state of consciousness but did not impair muscle coordination or the ability to move. Both were needed, for the victims had to ascend the steep steps of the temple pyramids before their hearts were cut out of their chests while they were still alive (Tyler 1966, 291*). Unfortunately, it is uncertain whether the Aztec tree *teuvetli* is botanically identical to *Bursera bipinnata* (Emboden 1979, 4*).

Literature

Davies, Nigel. 1983. *Opfertod und Menschenopfer.* Frankfurt/M.: Ullstein.

Caesalpinia decapetala (Roth) Alston

[syn. *Caesalpinia sepiaria* Roxb.[327]]

(Leguminosae)—yün-shih, caesalpinia

This yellow-flowering climber is found in the Himalayas and in central Asia and China (Polunin and Stainton 1985, 89*). Known as *yün-shih*, the plant is used in traditional Chinese medicine to treat worm infestations, malaria, and inflammations. It is claimed that it was once used in China as a hallucinogen (Li 1978, 20*). According to ancient Chinese sources, the flowers "contain occult powers." The *Pên Ts'ao Ching*, the famous herbal of Li Shih-chên, states:

> [The flowers] make it possible for one to see spirits but make one idiotic if consumed in excess. When consumed over a long period of time, they produce levitation of the body and promote communication with the spirits.

In the herbal *Tao Hung-ching*, one can read:

> [The flowers] dispel the evil spirits. When added to water or burned [i.e., as an **incense**], spirits can be called.... The seeds are like those of *lang-tang* [see **Hyoscyamus niger**]; when they are burned, spirits can be called.

The Chinese botanist Hui-Lin Li has suggested that these statements are indicative of a psychoactive use as well as of psychoactive effects of the flowers and seeds (Li 1978, 20*). Whether these text passages actually do refer to psychoactivity has not yet been established. After all, there are hundreds of plants that are used in magical rituals and spirit conjurations, even though they do not exhibit the slightest psychoactive effects (cf. Gessmann n.d.*; Mercatante 1980*; Schöpf 1986*). Still, an alkaloid of unknown chemical makeup has been discovered in *Caesalpinia decapetala* (Schultes and Hofmann 1992, 37*). Its method of action is unknown.

An Indian relative, *Caesalpinia bonduc* (L.) Roxb. [syn. *Caesalpinia cristata* L.], was once proposed as a candidate for **soma**. The beautiful paradise flower (*Caesalpinia pulcherrima* (L.) Sw. [syn. *Poinciana pulcherrima* L.]) is a symbol of the divine phallus in the cosmic vulva and is thus sacred to the Hindu god Shiva; *C. decapetala* is often mistaken for the yellow-blooming variety of the cultivated *C. pulcherrima*. The related American species *Caesalpinia echinata* Lam. is used in South America as an **ayahuasca** additive.

Capsicum spp.

(Solanaceae)—chili pepper

In the tropical regions of the Americas, there are many species (approximately forty) and cultivated varieties of chilies or chili peppers, most of which are used as spices (Andrews 1992). Chilies also have ethnomedical and ritual significance (Long-Solís 1986). The pods are used as medicines to treat a variety of diseases and have bactericidal properties (Cichewicz and Thorpe 1996). In higher dosages (30 to 125 mg), chilies are considered to be aphrodisiac (Gottlieb 1974, 19*). It is possible that they are psychoactive under certain circumstances. Indeed, chilies are used as

The climbing *yün-shih* shrub (*Caesalpinia decapetala*) is reputed to have psychoactive effects. (Illustration from the *Chêng-lei pên-ts'ao*, 1249)

327 Some authors have suggested that the synonym refers to a different species. *Caesalpinia japonica* Sieb. et Zucc. is said to be another synonym (Bärtels 1993, 206*). Here again we find taxonomic confusion, an unfortunately frequent occurrence.

One of the many forms of chili pepper (*Capsicum* sp.), a plant of the Americas, that the Indians have cultivated.

additives to many different psychoactive products, including **ayahuasca**, **balche'**, **beer**, cacao (see *Theobroma cacao*), **incense**, kava-kava (see *Piper methysticum*), and **snuff** (cf. *Nicotiana tabacum*) (Weil 1976). The Kakusi Indians of Guyana use *Capsicum* species as a stimulant (Schultes 1967, 41*). The Waorani Indians of Ecuador cultivate *Capsicum chinensis* Jacq. and use the fruits as a stomachic. When the men are too strongly inebriated from ayahuasca, their women give them chilies to help them become sober (Schultes and Raffauf 1991, 35*).

In the "drug scene," the dried remnants of rotten green paprika pods (*Capsicum fructescens* var. *grossum*) are sometimes used as a marijuana substitute (see *Cannabis indica*).

All species of this genus contain the hot compound capsaicin (chemically related to vanillin) (Weil 1976). Some species also contain flavonoids. *Capsicum annuum* L. contains steroidal alkaloids and glycosides (Schultes and Raffauf 1991, 35*).

Literature (selection)

Andrews, Jean. 1992. The peripatetic chili pepper: Diffusion of the domesticated capsicums since Columbus. In *Chilies to chocolate: Food the Americas gave the world*, ed. Nelson Foster and Linda S. Cordell. 81–93. Tucson: The University of Arizona Press.

Cichewicz, Robert H., and Patrick A. Thorpe. 1996. The antimicrobial properties of chili peppers (*Capsicum* species) and their uses in Mayan medicine. *Journal of Ethnopharmacology* 52:61–70.

Long-Solís, Janet. 1986. *Capsicum y cultura: La historia del chilli*. Mexico City: Fondo de Cultura Económica.

Waldmann, Werner, and Marion Zerbst. 1995. *Chili, Mais und Kaktusfeigen*. Munich: Hugendubel.

Weil, Andrew. 1976. Hot! Hot!—I: Eating chilies. *Journal of Psychedelic Drugs* 8 (1): 83–86.

Cardamine sp.

[syn. *Dentaria* sp.]
(Cruciferae)—pepper root, tooth root

There are reports that this little-known plant (possibly *C. concatenata*) was used as a hallucinogen by the Iroquois (Moerman 1986, 100, 604*). Unfortunately, chemical and ethnobotanical information is lacking (Ott 1993, 405*). The "cuckoo flower" (*Cardamine pratensis* L.) was once used to treat epilepsy (Millspaugh 1974, 88*).

Carissa edulis (Forsk.) Vahl

(Apocynaceae)—dagams

The root of this member of the Dogbane Family is very popular in Kenyan folk medicine. A decoction of the root is drunk to treat headaches and as an aphrodisiac and stimulant. In Ghana, the root is used to treat diminishing virility. In South Africa, a stimulating and aphrodisiac tea is made from the stem. The plant contains **indole alkaloids** and may have psychoactive effects (Omino and Kokwaro 1993, 171, 176*).

Castanopsis acuminatissima

(Fagaceae)—kawang

People from the Banz region of Papua New Guinea steam and eat the seeds of this tree. When consumed in sufficient quantity, they are said to have sedative or psychoactive effects similar to those of certain mushrooms (*Russula, Boletus*) (Schleiffer 1979, 91*).

Cecropia spp.

(Moraceae)—ant tree

In Mexico (Veracruz), *Cecropia mexicana* Hemsl. [syn. *Cecropia obtusifolia* Bert., *Cecropia schiedeana* Klotzsch]) is known as *guaruma*. The large, dried leaves are smoked as a marijuana substitute (see *Cannabis indica*); the effects are said to be similar to those produced by cannabis (Ott 1993, 405*). In Palenque, severe delirium effects have been observed in *Cecropia* smokers (Chan K'in Tercero, pers. comm.). In Mexican folk medicine, the fresh leaves are used to prepare a bath for relieving pain (Argueta et al. 1994, 706*). The plant contains sterols and tannins (pirogalole). The leaves contain various sugars (rhamnose, glucose, xylose), stigmasterol and three isomers, and 4-ethyl-5-(*n*-3-valeroil)-6-hexahydrocumarin (cf. **coumarins**) and 1-(2-methyl-1-nonen-8-il)-aziridin (Argueta et al. 1994, 706*). Tests on rats have demonstrated antihypertensive effects (Vidrio et al. 1982).

The leaves of a very similar species, *Cecropia peltata* L. [syn. *Cecropia asperrima* Pitt.], known locally as *guarumbo* or *tzon ndue*, contain leucocyanidin; the bark contains sterin and ursolic acid; and the latex has been found to contain the alkaloid cowleyine (Wong 1976, 115*). In Yucatán, a tea made from the leaves is consumed for diabetes (Argueta et al. 1994, 708*).

In South America, the leaves of various species are burned to ash and used as *llipta* for coca chewing (*Erythroxylum coca*).

Literature

Vidrio, H., et al. 1982. Hypotensive activity of *Cecropia obtusifolia*. *Journal of Pharmaceutical Science* 71 (4): 475–76.

Clematis virginiana L.

(Ranunculaceae)—Virginia clematis

The stems of this plant were reputedly used by the Iroquois of North America as a bath or wash to produce "strange dreams." Unfortunately, no additional ethnographic or chemical information is available (Ott 1993, 406*). Other *Clematis* species are toxic and irritate the nervous system (Roth et al. 1994, 241*). In Bavaria, the sprouts of *Clematis vitalba* L. were once smoked as a tobacco substitute (cf. **Nicotiana tabacum**).

Comandra pallida

(Santalaceae)

The North American Kayenta Navajo allegedly used this plant as a narcotic. No other information is available (Ott 1993, 406*).

Conium maculatum L.

(Apiaceae)—poison hemlock

Since ancient times, hemlock has "enjoyed" an infamous reputation as a poisonous plant, known as a killing poison ("Socrates' cup") and sedative, but it is also known as an aphrodisiac. According to Germanic legend, hemlock was avoided by both man and beast. It was said that only the toad, which would "suck its poison," liked to live in its vicinity (Perger 1864, 184*). Hemlock thus stands in close association with the toad goddess, an earth deity venerated especially by Baltic tribes (the ancient Prussians) (Gimbutas 1983; cf. **bufotenine**).[328]

Saxo Grammaticus has described how Haddingus, who was raised by the giants but later became a favorite of Odin, was once visited at the hearth in the evening by an apparition of a woman who wore hemlock in her garb. This woman, interpreted as a shamanic helping spirit, placed her cloak around Haddingus and led him across a bridge, into the underworld, and to the realm of the dead (Lichtenberger 1986, 31 f.). In Scandinavia and England, hemlock was a magical plant closely associated with Wotan/Odin. This can be seen in the Norwegian name for the plant, *woden-dunk*, and the Anglo-Saxon *wôde-hwistle*, "rage reed/ Wotan's whistle." The English word *hemlock*, literally "hemp leek," still recalls the rune magic *lina laukaR*. In Old High German, the plant was known as *wuotscerling* or *wode-scerne*, "rage hemlock" (Höfler 1990, 95*). Hemlock was one of the primary ingredients in the **witches' ointments**. Andrés Laguna, a sixteenth-century physician from Lorraine, once found a vessel full of "witches' ointment." He analyzed its contents and found henbane (*Hyoscyamus niger*), mandrake root (*Mandragora officinarum*), and hemlock. He rubbed the ointment onto a woman, who then fell into trance.

In Bohemia, hemlock is reputed to have once been used as an additive to **beer**. Unfortunately, we know nothing about the effects of such a brew (Hansen 1981*). Perhaps all of the people who tried it died from overdoses . . .

The entire plant contains approximately 2% alkaloids. The fruits exhibit especially high concentrations, as much as 3.5%. The principal alkaloid is coniine (approximately 90% of the total alkaloid content); γ-coniceine, conhydrine, pseudoconhydrine, and methylconiine are also present (Teuscher 1992). Toxic effects include rapid paralysis, sensations of cold, lack of sensation, and, finally, death through respiratory paralysis (Roth et al. 1994, 259*).

Literature

Gimbutas, Marija. 1983. *The Balts*. London: Thames and Hudson.

Lichtenberger, Sigrid. 1986. Züge des Schamanentums in der germanischen Überlieferung. In *Schamanentum und Zaubermärchen*, ed. Heino Gehrts and Gabriele Lademann-Priemer, 28–41. Kassel: Röth.

Left: The Mexican ant tree, *Cecropia mexicana*, is a typical tree of the tropical rain forest. The dried leaves are used as a marijuana substitute, although they have a high tar content. (Wild plant, photographed in the Selva Lacandona, Chiapas, Mexico)

Center: Ashes of the leaves of the South American *Cecropia* are added to pinches of coca. (Photographed in Machu Picchu, Peru)

Right: Poison hemlock (*Conium maculatum*) in bloom.

328 "And inasmuch as they [the Prussians] did not know of [the Christian] God, it so happened that they worshipped the entire creature-world instead of God, namely: the sun, moon and stars, the thunder, birds, even the four-legged animals, including toads. They also had holy groves, sacred fields, and waters." (*Chronicon Prussiae*, by Peter Dusberg, 1326; in Gimbutas 1983, 179).

The *ti* plant (*Cordyline fruticosa*), which the Hawaiians venerate as sacred, was formerly used as a magical plant and inebriant. (Photographed on Oahu, Hawaii)

Bottom left: A tannin sumac (*Coriaria* sp.) used in South America for psychoactive purposes.

Bottom right: *Coriaria thymifolia*. (Photographed near Machu Picchu, Peru)

Teuscher, Eberhard. 1992. *Conium*. In *Hagers Handbuch der pharmazeutischen Praxis*, 5th ed., 4:970–75. Berlin: Springer.

Cordia boissieri DC.

(Boraginaceae)—nacahuita

It is said that the fruits of this Mexican plant have inebriating effects (von Reis Altschul 1975, 237*).

Cordyline fruticosa (L.) A. Chev.

[syn. *Cordyline terminalis* (L.) Kunth]
(Agavaceae, previously Liliaceae)—*ti*

This Polynesian plant, which is now found throughout the world as an ornamental and potted plant, is called *kî*, *tî*, or *ti* in Hawaii (Krauss 1993, 186*). There, the plant acquired a great significance in magical and religious rituals:

> *Tî* received heavy ceremonial use as well and frequently was planted around *heiau* [temples]. Priests wore leaves about their necks as an indication of high rank or divine power, and it was among the plants customary on the altar of the *hâlau hula*, representing Laka, the goddess of hula [a sacred dance]. It was also valued as a charm against evil spirits. (Abbott 1992, 115)

An inebriating beverage was brewed from this Hawaiian magical plant:

> On Hawaii and Samoa, the roots or rhizomes are used to prepare an inebriating beverage. Since the rhizomes contain sugar [saccharose], the drink may contain alcohol. The leaves of the plant are made into belts for dancing and loincloths. The root is also used to treat diarrhea and dysentery. *Cordyline ti* Schott [apparently a synonym] is also used on some South Seas islands to produce an inebriating drink. (Hartwich 1911, 811*)

It is unknown whether **alcohol** was the only active component of this drink, which was presumably similar to beer (cf. **beer**), or whether it also contained other psychoactive substances.

Literature

Abbott, Isabella Aiona. 1992. *Lâ'au Hawaii: Traditional Hawaiian uses of plants*. Honolulu: Bishop Museum Press.

Coriaria thymifolia H.B.K. ex Willd.

(Coriariaceae)—tannic sumac

This shrub is found in the high mountains from Colombia to Chile, where it is known as *shanshi*.

In Ecuador, its fruits are reportedly eaten to produce an inebriated state. The eater is said to experience "sensations of soaring through the air" (Alvear 1971, 22*; Schultes and Hofmann 1992, 40*). The effects are said to be similar to those produced by *Petunia violacea*. *Coriaria thymifolia* grows in Mexico as well; it has been suggested that it was the Aztec inebriant known as *tlacopétatl* (Díaz 1979, 93*). In the Las Huaringas region, a lake plateau in the northern Peruvian Andes, the local healers (*curanderos*) refer to *Coriaria thymifolia* as *contra-alergica*, "against allergies." They use the herbage to prepare a bath additive (*baño*) that they use to wash patients suffering from allergic reactions. I was unable to learn of any psychoactive use there. Astonishing numbers of *Coriaria thymifolia* can be found in the area around Macchu Picchu.

The fruits contain catechol derivatives and probably several sesquiterpenes (Schultes and Farnsworth 1982, 178*). Other sources state that a toxic substance named coriamyrtine has been isolated from the plant. The effects are described as initially stimulating but then becoming less pleasant. Death from nervous exhaustion can result (Blohm 1962, 62*). Coriamyrtine is a sesquiterpene; other sesquiterpenes coriatine, tutine, and pseudotutine have also been reported (Emboden 1979, 175*).

The closely related species *Coriaria ruscifolia* L., which is known in Chile as *deu*, *dewü*, *huique*, *huiqui*, and *matarratones* (literally "killer of rats"), is reputed to be a hallucinogen. It is definitely poisonous. The fruits are made into rat poison in Chile (Mösbach 1992, 89*) and are said to be lethal for small children (Donoso Zegers and Ramírez García 1994, 46*). The Mapuche use a tea made from the leaves as an emetic (Houghton and Manby 1985, 94*).

Crotalaria sagittalis L.

(Fabaceae, Leguminosae)—rattle box

The Delaware-Okl Indians of North America regarded the root of this papilionaceous plant as "a very strong narcotic" (Moerman 1986, 140*). Other species of the genus contain potent liver toxins (Ott 1993, 406*). Also present are flavonoids (Fleurentin and Pelt 1982, 96 f.*), amino

acids, and alkaloids, including neurotoxins (Pilbeam et at. 1979, 1983, 1983; Wong 1976, 126*). The dried seedpods of various *Crotalaria* species can be used as rattles, hence the English name "rattle box." In Argentina, the pods of *Crotalaria incana* L. are used to magically heal deafness (Schmeda-Hirschmann et al. 1987).

One related species, *Crotalaria juncea* L., is known as Bengali hemp, East Indian hemp, and Bombay hemp. These names suggest that the species may be psychoactive.

Literature

Pilbeam, David J., and E. Arthur Bell. 1979. Free amino acids in *Crotalaria* seeds. *Phytochemistry* 18:973–85.

Pilbeam, D. J., and A. J. Lyon-Joyce. 1983. Occurrence of the pyrrolizidine alkaloid monocrotaline in *Crotalaria* seeds. *Journal of Natural Products* 46:601–5.

Pilbeam, D. J., R. M. Pohlhill, and E. A. Bell. 1983. Free amino acids and alkaloids of South American, Asian and Australian *Crotalaria* species. *Botanical Journal of the Linnean Society* 79:259–66.

Schmeda-Hirschmann, Guillermo, Lucia Franco, and Estebán Ferro. 1987. A magic use of *Crotalaria incana* pods. *Journal of Ethnopharmacology* 21:187–88.

Cymbopetalum penduliflorum (Dunal) Baill.

(Annonaceae)

This plant was known among the Aztecs as *xochinacaztli*, "ear flower"; the aromatic flower was called *teonacaztli*,[329] "sacred ear," and it was said that "it inebriates like mushrooms" (Sahagún). The dried flowers were smoked together with tobacco (*Nicotiana tabacum*) (Díaz 1979, 94*). Today, the flowers are used in Mexico as a spice for cacao drinks (see *Theobroma cacao*) and are known as *hueynacaztli* (Ott 1993, 406*). It has been suggested that this plant is the still unidentified Aztec inebriant **poyomatli** (Díaz 1979, 94*).

Cymbopogon densiflorus (Stendl.) Stapf

(Gramineae)

This perennial plant is found in the Congo, Gabon, and Malawi (cf. **madzoka medicine**). Medicine men in Tanganyika used to smoke an extract of the leaves of this citronella, which have a lemon scent, either alone or mixed with tobacco (*Nicotiana tabacum*), in order to induce divinatory dreams so that they could predict the future (von Reis and Lipp 1982, 10*). The leaves were used by magicians in central Africa (Krause

1909, 4). The genus *Cymbopogon* is rich in essential oils (Schultes and Hofmann 1992, 41*). The **essential oil** of *C. densiflorus* has been chemically investigated, and it does not contain any psychoactive compounds (Da Cunha 1972; Koketsu et al. 1976).

Literature

Da Cunha, A. P. M. A. 1972. Estudio químico e cromatográfico de oleo essencial de *Cymbopogon densiflorus* (Stendl.) Stapf, de Angola. *Anais da Academia Brasileira de Ciencias* 44 suppl.: 285–88.

Koketsu, M., L. L. Moura, and M. T. Magalhaes. 1976. Essential oils of *Cymbopogon densiflorus* Stapf and *Tagetes minuta* L. grown in Brazil. *Anais da Academia Brasileira de Ciencias* 48:743–46.

Krause, M. 1909. Die Gifte der Zauberer im Herzen Afrikas. *Zeitschrift für experimentelle Pathologie und Therapie* 6:1–4.

Cyperus spp.

(Gramineae-Poaceae)—sedges, piripiri

The Amazonian Sharanahua Indians use a *Cyperus* species that has been infested with the fungus *Balansia cyperi* Edgerton as an **ayahuasca** additive. This fungus is related to ergot (*Claviceps purpurea*) and contains as yet unidentified **ergot alkaloids**. Many species of *Cyperus* that are used ethnogynecologically appear to be infested with this or another fungus of the genus *Balansia* (Ott 1993, 396*).

The Ecuadoran Shuar, Achuar, and Aguaruna use various *Cyperus* species that they call *piri-piri* (including *C. articulatus* L., *C. odoratus* L., and *C. prolixus* Humb. et Kunth) not just as ayahuasca additives but also as psychoactive substances in their own right. Some Shuar shamans (*uwishin*) drink a tea made from the root that serves them in place of ayahuasca when they are making diagnoses. The tea enables them to enter into trance and communicate with the dead (Bennett 1992, 490, 492*). The neighboring Secoya utilize *piripiri* to dispel evil spirits and to induce labor (Cipoletti 1988). The Jívaro add *Cyperus* extracts to their drinking tobacco (see *Nicotiana tabacum*).

Left: Many species of cypress grass (*Cyperus* spp.) are used in South America for ethnopharmacological purposes, including as ayahuasca additives. (Wild plant, photographed in northwestern Argentina)

Right: A grass from the genus *Cyperus* that is known in Amazonia as *piripiri* but has not yet been botanically identified is infested by a fungus that likely produces ergot alkaloids.

329 The Leguminosae *Enterolobium cyclocarpum* (Jacq.) Griseb. is listed in the Florentine Codex (Sahagun) under the same name (Díaz 1979, 94*).

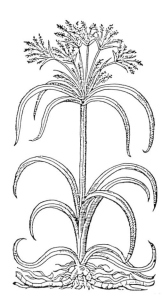

Sedges (*Cyperus* spp.) are occasionally infested by fungi that can introduce psychoactive metabolites into the plant material. (Woodcut from Gerard, *The Herball or General History of Plants*, 1633)

Left: Many species of larkspur (*Delphinium* spp.) are cultivated as ornamentals.

Right: The seeds of the South American tree *Dictyoloma incanescens*, in which 5-MeO-DMT was first demonstrated to occur naturally.

The women of the Achuar Jívaro use the rhizome of the species *Cyperus prolixus* H.B.K., infested with **Balansia cyperi** Edgerton, to induce labor (Lewis and Elvin-Lewis 1990). The species *Cyperus articulatus* and *Cyperus prolixus*, when infested with **Balansia cyperi**, contain as yet unidentified **ergot alkaloids** (Plowman et al. 1990).

In Venezuela, *Cyperus articulatus* L.—like other psychoactive plants (e.g., **Brugmansia spp.**, *Iochroma fuchsioides*)—is known as *borrachera*, "inebriating agent," because of its psychoactive effects. In El Salvador, the plant finds folk medicinal use as an analgesic for treating toothaches (von Reis and Lipp 1982, 15*).

Literature

Cipoletti, M. S. 1988. El piri-piri y su significado en el shamanismo Secoya. *Amazonia Peruana* 8:83–97.

Lewis, Walter H., and Memory Elvin-Lewis. 1990. Obstetrical use of the parasitic fungus *Balansia cyperi* by Amazonian Jivaro women. *Economic Botany* 44:131–33.

Plowman, Timothy C., Adrian Leuchtmann, Carol Blaney, and Keith Clay. 1990. Significance of the fungus *Balansia cyperi* infecting medicinal species of *Cyperus* (Cyperaceae) from Amazonia. *Economic Botany* 44:452–62. (Includes a list of additional literature.)

Cypripedium calceolus (Willd.) Correll var. *pubescens*

[syn. *Cypripedium calceolus* L., *Cypripedium luteum* Ait. var. *pubescens* Willd., *Cypripedium parviflorum* Willdenow, *Cypripedium pubescens* Willd.]

(Orchidaceae)—yellow lady's slipper

The North American Menominee Indians added this orchid to their sacred bundles to promote supernatural dreams (Moerman 1986, 604*). This possibly psychoactive plant was utilized by the neighboring Cherokee as a sedative and analgesic. The American colonists used the root as a substitute for the sedative valerian (*Valeriana officinalis*) to treat nervousness, hysteria, and insomnia (Emboden 1976, 166*; Millspaugh 1974, 683 f.*; Veit 1992, 1123). The aboveground parts of this orchid, which is now protected by law, probably contain cypripedine and similar chinones. To date, no investigations of the root have been carried out (Veit 1992, 1123).

Literature

Cribb, Philip. 1997. *The genus* Cypripedium. Cambridge, U.K.: Timber Press.

Veit, Markus. 1992. *Cypripedium*. In *Hagers Handbuch der pharmazeutischen Praxis*, 5th ed., 4:1122–24. Berlin: Springer.

Delphinium consolida L.

[syn. *Consolida regalis* S.F. Gray, *Delphinium nudicaule*]

(Ranunculaceae)—larkspur

The Mendocino Indians of California regarded this plant, a species introduced from Europe that is closely related to the common larkspur (*Delphinium elatum* L.), as a narcotic. It is questionable whether the plant does in fact have psychoactive properties. Only toxic glycosides and aconitine-like alkaloids (delphinium alkaloids) have been found in the genus (Ott 1993, 407*; Roth et al 1994, 296*). Other *Delphinium* species were used as ceremonial medicine (Moerman 1986, 150*). In the Himalayas, leaves of *Delphinium brunonianum* are mixed with **Nicotiana rustica** and smoked (Atkinson 1989, 756*).

The Cappella Indians of California used the root of *Delphinium consolida* L. as a sleeping aid for children (Emboden 1976, 160*). It contains aconitine-like diterpene alkaloids, delphinine, delphinedin, and ajacine (Emboden 1979, 176*; Wren 1988, 167*). The diterpenoid alkaloid tricornine occurs in a very rare Californian species, *Delphinium tricorne* Michx. (Pelletier and Bhattacharya 1977).

Literature

Pelletier, S. William, and J. Bhattacharya. 1977. Tricornine, a new diterpenoid alkaloid from *Delphinium tricorne*. *Phytochemistry* 16:1464.

Dictyoloma incanescens DC.

(Rutaceae)—tinqui

The bark of this tree contains *N,N*-dimethyl-5-methoxytryptamine (= **5-MeO-DMT**) (Pachter et al. 1959*). It may be suitable as an ingredient in **ayahuasca analogs**. 5-MeO-DMT was first discovered in this plant.

Dictyonema sp. nov.

(Basidiolichenes, Dictyonemataceae)—lichen

The Waorani, who live in the Ecuadoran region of Amazonia, appear to have a psychoactive use for this lichen. It is said that the Waorani shamans

formerly called the lichen *nenendape* and used it as a ritual entheogen (Davis and Yost 1983). Mixed with several unidentified mosses (Bryophyta) known as *quiguiwai*, the lichen was made into a tea that a shaman would drink when he wanted to cast a spell over or magically kill a person (Davis and Yost 1983, 163, 170, 209*). To date, only one other lichen has been reported to have psychoactive effects (cf. **Lichen non ident.**).

Literature

Davis, E. W., and J. A. Yost. 1983. Novel hallucinogens from Ecuador. *Botanical Museum Leaflets* 29 (3): 291–95.

Dimorphandra parviflora

(Leguminosae)

The seeds of this Brazilian tree are used to prepare *paricá* snuffs (cf. *Virola* spp.). The seeds likely contain alkaloids and may have psychoactive effects (Ott 1993, 407 f.*). It is possible that the botanical name is incorrect or out of date.

Dioscorea composita Hemsl.

(Dioscoreaceae)—barbasco, camotillo
This tuberous plant is alleged to have psychoactive effects:

> The camotillo induces not an inebriation but rather a latent semiconscious state that does not manifest itself until long after consumption. Affected persons become indifferent to their surroundings and external impressions. Thinking becomes restricted and is typically concerned with an event that lies far back in their life that they, in the manner of a monomaniac, continually attempt to reconstruct in their daydreams. (V. Reko 1938, 191*)

It is said that Empress Charlotte of Mexico was robbed of her wits by this agent (cf. **Datura innoxia**).

In Malaysia, the root tuber of a wild yam species (*Dioscorea triphylla* Lam. [syn. *Dioscorea daemona* Roxb., *Dioscorea hirsuta* Blume]), known locally as *gadong*, is attributed with narcotic powers. A hallucinogenic paste is mixed from the green shoots, opium (cf. *Papaver somniferum*), seeds of *Datura metel*, and the green inner bark of *Glycosmis citrifolia* (Rutaceae) (Gimlette 1981, 220, 222*).

Dioscorea tubers are occasionally used in **beer** brewing.

Elaeophorbia drupifera (Thonn.) Stapf

(Euphorbiaceae)—ayañ beyem

This little-known African spurge may have psychoactive effects (Schultes and Hofmann 1980, 367*). In West Africa, the herbage is used to treat animal bites and as a purgative (Ayensu 1978, 123*). The latex of the fresh plant is occasionally dripped into the eyes of initiates of the Bwiti cult in order to induce more powerful visions (see **Tabernanthe iboga**).

Ferraria glutinosa (Bak.) Rendle

(Iridaceae)—gaise noru noru

The San, or !Kung Bushmen, of the western Kalahari Desert conduct collective healing dances in which they strive to attain an ecstatic trance state. The dancers achieve the healing trance (*kia*) by dancing for hours, sometimes in combination with the consumption of *dagga* (**Cannabis sativa**) or the plant they call *gaise noru noru* (also called *!kaishe*) (Dobkin de Rios 1986). A drink concocted from the root was apparently used more often in earlier times. One Bushman provided the following description of its effects:

> Everybody at the dance, every man at the dance, drank it. . . . Anybody who danced would take it. The thing is that those who hadn't reached kia yet would drink more than those who had already reached kia. We elders who had experienced kia long ago would just drink a little of the preparation. . . . You start to feel something moving around in your stomach, in your chest, and in your back, a pulsating feeling in your back like a jabbing. . . . You feel your front spine starting to pulsate with your heartbeat and starting to tremble. . . . The reason I say this gaise noru noru is powerful is because you just didn't take it. You had to be washed, and you had to be fed certain foods. Certains [*sic*] foods were prohibited to you, and meat that was hunted, the blood of it, was rubbed on you. Then you were washed again with something else. All that is associated with this gaise noru noru. And that's why I say it was powerful. (In Katz 1982, 284, 286, 293)

It is possible that the plant used to be smoked or dropped in a tortoise shell onto hot coals and the smoke inhaled (Winkelman and Dobkin de Rios 1989, 56). The root contains substances that have been little studied but may possibly have hallucinogenic or psychoactive effects. Other species of the genus *Ferraria* and related plants have potent toxic properties (Winkelman and Dobkin de Rios 1989, 54).

The botanical identification of *gaise noru noru* as *Ferraria glutinosa* has been occasionally called into question but does appear to be correct (Dobkin de Rios 1984, 205, 208; 1986). The ethnomusicologist Richard Katz has reported

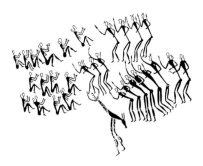

This rock art by an ancestor of the Bushmen in Nicosasanatal (South Africa) depicts the collective dance of ecstasy in which *Ferraria glutinosa* and other psychoactive plants were used. (From Müller-Ebeling 1991)

Botanical illustration of *Gelsemium sempervirens*. (From *Köhler's Medizinal-Pflanzen*, 1887)

Only the signs pointing toward a Mapuche settlement (southern Chile) recall the endangered *keule* or *queule* tree (*Gomortega keule*), the mysterious plant from which the locale took its name.

another plant, called *gwa*, that was used in a drumming dance to induce *kia*. Unfortunately, this plant has not been botanically identified (Dobkin de Rios 1984, 205 f.). The Bushmen may also have used other plants to induce *kia*: *Albizia anthelmintica* A. Brongn. (Leguminosae), *Cassia* spp. (Leguminosae), *Cissampelos mucronata* A. Rich (Menispermaceae; contains psychoactive compounds), *Loranthus oleaefolius* Cham. et Schlectend. (Loranthaceae; said to contain **scopolamine**), and *Plumbago zeylanica* L. (Plumbaginaceae). The ethnographic literature, however, contains references only to folk medicinal uses (Winkelman and de Rios 1989, 54 f.). It is hoped that additional research into these questions will be carried out.

Literature

Dobkin de Rios, Marlene. 1984. Review of *Boiling Energy* by Richard Katz. *Transcultural Psychiatric Research* 21 (3): 103–210.

———. 1986. Enigma of drug-induced altered states of consciousness among the !Kung Bushmen of the Kalahari desert. *Journal of Ethnopharmacology* 15:297–304.

Katz, Richard. 1982. *Boiling energy: Community healing among the Kalahari Kung.* Cambridge, Mass.: Harvard University Press.

Müller-Ebeling, Claudia. 1991. Die Ekstase-Tänze der Buschleute. In *Von den Wurzeln der Kultur*, ed. C. Rätsch, 189–204. Basel: Sphinx Verlag.

Winkelman, Michael, and Marlene Dobkin de Rios. 1989. Psychoactive properties of !Kung Bushmen medicine plants. *Journal of Psychoactive Drugs* 21 (1): 51–59.

Gaultheria sp.

(Ericaceae)—borrachera

An herbarium specimen of an unknown *Gaultheria* species collected by J. A. Steyermark in Venezuela in 1971 bears a note stating that this member of the Heath Family, closely related to *Pernettya* spp., is known as *borrachera*, "drunken maker" (von Reis and Lipp 1982, 227*). In Chile, a number of different *Gaultheria* species are used to manufacture **chicha**. In North America, *Gaultheria procumbens* L. (mountain tea, Canadian tea, Labrador tea) is used as a tea substitute (Lewin 1980 [orig. pub. 1929], 352*). The leaves contain **essential oils** and gaultherine (Roth et al. 1994, 367*).

Gelsemium sempervirens (L.) Jaume St.-Hil.

[syn. *Gelsemium nitidum* Michx.]
(Loganiaceae)—Carolina jessamine

This aromatic wild jasmine is from North America, where Indians have long used it as a

medicinal plant (Rätsch 1991a, 146*). The Aztec-speaking peoples of Mexico call this yellow-flowered climber *xomil-xihuite*, "paralyzing poison." The neighboring Otomí Indians know it as *beho-sito*, "glass coffin." The root was allegedly used as a poison in trials by ordeal. The effects are said to be quite drastic: "The person who has been poisoned enters a paralyzed state while remaining fully conscious, with open eyes; they cannot move, and yet they perceive with terrible clarity all of the things that are occurring around them" (V. Reko 1938, 168*). This pattern of effects is strongly reminiscent of that of the Haitian **zombie poison**. It is said that Mexicans used to add the root to schnapps (**alcohol**) to lend it a special effect (Reko 1938, 167*).

Profound psychic alterations with strong hallucinations occur even when *Gelsemium* is used for medicinal purposes (Lewin 1980 [orig. pub. 1929], 192*). It is for this reason that *Gelsemium* has repeatedly been characterized as a psychoactive plant or even a hallucinogen (Emboden 1976, 165*; Schultes and Hofmann 1980, 367*). The entire plant, but especially the root, contains **indole alkaloids**: gelsemicine, gelsemine, gelsedine, and sempervirine, some of the effects of which resemble those of **strychnine**. Gelsemine paralyzes the central nervous system (Blaw et al. 1979; Roth et al. 1994, 368 f.).

Literature

Blaw, M. E., M. A. Adkisson, D. Levin, J. C. Garriott, and R. S. A. Tindall. 1979. Poisoning with Carolina jessamine (*Gelsemium sempervirens*). *Journal of Pediatrics* 94:998–1001.

Gloeospermum sphaerocarpum Tr. et Pl.

(Violaceae)

The Amazonian Waunana Indians drink a cold-water extract of the leaves of this plant, which they call *tamarillo*, as a "ceremonial hallucinogen" (Duke and Vasquez 1994, 81*). Additional ethnopharmacological research is required.

Gomortega keule (Moldenke) I.M. Johnst.

[syn. *Gomortega nitida* Ruíz et Pavón]
(Gomortegaceae)—keule

This tall tree, the only species in the Family Gomortegaceae (related to the Family Lauraceae), is endemic to southern Chile. The Mapuche Indians call it *keule, queule, queuli* (pronounced kay-ulay), *linge*, or *hualhual* (literally "vicinity") and may once have used it as a psychoactive substance. The round fruits have inebriating effects, especially when fresh, and may even be hallucinogenic, possibly as a result of the essential oils they contain (Schultes and Farnsworth 1982, 180*). The opposite lanceolate leaves are said to contain an **essential oil** (Schultes and Hofmann 1980, 334*). A number of derivatives of methoxylated **coumarins** have been detected (D. McKenna 1995, 101*). Chemical studies of the fruits have not yet been conducted (Ott 1993, 408*).

Unfortunately, this fruit tree is very rare. It is found only in a single one-hundred-square-mile expanse (Schultes and Hofmann 1980, 334*) in the coastal area between Maule and Arauco, somewhat south of Concepción (Mösbach 1992, 79*). The tree appears to be on the brink of extinction.

The plant was first described by the Spanish botanist Don Hipólito Ruíz, who encountered it on a 1777–1788 expedition to Peru and Chile. He reported that the leaves have a sour-astringent taste and, when chewed, stick to the teeth as a result of their resin content. Rubbed between the fingers, they exude a scent reminiscent of that of rosemary and turpentine. "The beautiful fruits are as large as small chicken eggs, are shiny, have a yellow color, and invite one to eat them. But you will get a headache if you eat too many of them" (Schultes 1980, 97*).

The yellow fruits contain an extremely hard stone. They ripen toward the end of April and are used to make marmalade (Donoso Zegers 1995, 94*). They are regarded as culinary delicacies. A fishing village of the Mapuche ("earthlings"), on the coast some 50 km north of Valdivia, is named Queule after this mysterious tree. None of its residents seems to have any knowledge of the tree or of its purported psychoactive use. The fruits may formerly have been used in the preparation of **chicha**; for this reason, they are still reputed to be psychoactive.

Goodenia spp.

(Goodeniaceae)—goodenia

In the ethnobotany of the Aborigines, plants of the genus *Goodenia* enjoy a certain reputation as healing and food plants as well as **pituri** or pituri substitutes (O'Connell et al. 1983, 109). The Goodeniaceae are well represented in Australia. In addition to the genus that gave the family its name, *Scaevola taccada* (pipe tree) is especially important. This bushy shrub is a typical coastal plant that grows in sandy soil. The Aborigines use the juice of the fruits as an eye remedy and as an antidote for animal stings and bites. The ends of the branches are hollow; they are used as pipes for smoking pituri (Wightman and Mills 1991, 46 f.).

In the language of the Alyawara, *Goodenia lunata* J.M. Black is called *ngkulpa ankirriyngka*. The Alyawara mix the dried leaves with plant ashes (primarily from *Ventilago viminalis* Hook.; Rhamnaceae) and chew the result. They place it in the same folk taxonomic category as wild tobacco (*Nicotiana* **spp.**) (O'Connell et al. 1983, 109*). Macerations of the fresh leaves are used in a manner similar to those made from wild tobacco, namely for poisoning the watering holes of the emu (98*). When smoked or chewed, the leaves of *Goodenia lunata* appear to have mild psychoactive effects.

Literature

Wightman, Glenn, and Milton Andrews. 1991. *Bush Tucker Identikit: Common native food plants of Australia's top end*. Darwin: Conservation Commission of the Northern Territory.

Hedera helix L.

[syn. *Hedera caucasigena* Pojark, *Hedera chrysocarpa* Walsh, *Hedera helix* var. *chrysocarpa* Ten., *Hedera taurica* Carr., *Hedera helix* var. *taurica* Tobler]
(Araliaceae)—ivy

Ivy is an ancient sacred plant that was associated with the cult of Dionysos, the god of wine, inebriation, and ecstasy. Dioscorides described three types of ivy,[330] one of which bore the same name as the god, *Dionysos*. Plutarch, the philosopher, oracular priest, and disciple of Dionysos, wrote in his *Roman Questions* (112) that ivy contains "powerful spirits" that produce outbursts of madness and cramps. Ivy could induce an "inebriation without wine," or a type of possession in those who had a natural tendency to enter ecstatic states. When ivy leaves are added to wine (see *Vitis vinifera*), the resulting mixture is able to

Left: The little-studied Australian *Goodenia lunata* is used as a **pituri** substitute. (Wild plant, photographed in Tasmania)

Right: Many authors of antiquity characterized ivy (*Hedera helix*) as a psychoactive plant. (Wild plant, photographed in northern Germany)

330 This may correspond to the three subspecies: *Hedera helix* L. ssp. *canariensis* (Willd.) Cout. [syn. *H. canariensis* Willd., *H. algeriensis* Hibb.], ssp. *helix*, and ssp. *poetarum* Nym. It could also refer to *Hedera colchica* (K. Koch) K. Koch or *Hedera pastuchovii* Woron. of southwestern Asia.

produce a delirium, "a confusion that can otherwise be produced only by henbane" (see ***Hyoscyamus niger***). The Roman naturalist Pliny the Elder also wrote of the psychoactive effects:

> [Ivy] confuses the mind, cleanses, when drunk in excess, the head; taken internally, it damages the nerves, but is healthy for these same nerves when applied externally. . . . As a drink, [all species of ivy] are diuretic, soothe headaches, especially in the brain. . . . The berries, which are the color of saffron, provide certain protection against inebriation when they are taken beforehand as a drink. (Pliny 24.75/78)

Ivy has been linked to the Dionysian ecstasy of the maenads (= female bacchantes, bassarides; cf. ***Vitis vinifera***) primarily through the work of Robert Graves and his book *The White Goddess*. It has been rumored that Graves wrote this book while under the influence of **psilocybin**. Otherwise, the ancient sources would be difficult to interpret in this manner:

> October was the season of the Bacchanal revels of Thrace and Thessaly in which the intoxicated Bassarids rushed wildly about on the mountains, waving the fir-branches of Queen of Artemis (or Ariadne) spirally wreathed with ivy—the yellow-berried sort—in honour of Dionysus . . . , and with a roebuck tattooed on their right arms above the elbow. They tore fawns, children and even men to pieces in their ecstasy. The ivy was sacred to Osiris as well as to Dionysus. Vine and ivy come next to each other at the turn of the year, and are jointly dedicated to resurrection. . . . It is likely that the Bassarids' tipple was "spruce-ale," brewed from the sap of silver-fir [*Abies cephalonica* Loud.] and laced with ivy; they may also have chewed ivy-leaves for their toxic effect. Yet the main Maenad intoxicant will have been *amanita muscaria*. (Graves 1966, 183*)

The botanical identity of the inebriating ivy is a total mystery: "However, the Dionysian ivy was not the one native among us but the northern Indian with the yellow berries, of which it is said that it grows only on the mountain of Meros, near Indian Nysa" (Duerr 1978, 213*). This may refer to Himalayan ivy (*Hedera nepalensis* K. Koch [syn. *Hedera himalaica* Tobl.]), which bears orange-yellow fruits.

Some subjects who have smoked the dried leaves have reported them to be inebriating.

Ivy leaves contain glycosides, inositol, chlorogenic acid, hedera tannic acid, malic acid, formic acid, and triterpene hedera saponines (α-hedrine), as well as the trace elements arsenic, zinc, copper, manganese, iodine, lithium, and aluminum. The alkaloid emetine has been found in Egyptian specimens (Horz and Reichling 1993, 399). In the toxicological literature, it is noted that "a 3-year-old child ate a large amount and had hallucinations" (Roth et al. 1994, 391*). To date, however, no truly inebriating substances have been found in ivy. Francesco Festi has worked extensively on the botany and phytochemistry of ivy and has not found the slightest evidence of the presence of psychoactive compounds (F. Festi, pers. comm.).

It may be that the ancient word for *ivy* was a catchall phrase for climbing plants. There are vines (***Convolvulus tricolor***) in the Mediterranean region whose seeds contain lysergic acid derivatives. Or "ivy" may have been a designation for another plant that is no longer known or able to be identified but which had potent inebriating effects and contained psychoactive compounds.[331] Imagine that a medical historian of the future finds an article written in the present that notes, "Grass has potent effects when smoked." He might think that people were smoking grass from their lawns as an inebriant. If he were to smoke it himself, he would find that it produces no such effects. How would he know that "grass" is a common and generally understood name for hemp (***Cannabis sativa*** or ***Cannabis indica***) and for its female flowers?

Literature

Horz, Karl-Heinrich, and Jürgen Reichling. 1993. *Hedera*. In *Hagers Handbuch der pharmazeutischen Praxis*, 5th ed., 4:398–407. Berlin: Springer.

Helichrysum foetidum (L.) Moench

(Compositae: Asteraceae)—stinking strawflower

It is said that the magician-physicians of the African Zulu made a powder of this strawflower that they inhaled or smoked in order to induce a divinatory trance. This vague information comes only from a note belonging to an herbarium specimen of this plant and has no other ethnographic support (von Reis and Lipp 1982, 303*). *Helichrysum stenopterum* is said to have been used in the same manner (de Smet 1996, 142*; von Reis

One of the difficult-to-identify species of strawflower (*Helichrysum* sp.).

331 In Islamic literature, for example, hashish, the "inebriant par excellence for the wandering dervishes," is often referred to as "wine" (Frembgen 1993, 200*).

560

and Lipp 1982, 303*). Various derivatives of phloroglucinol have been detected in this plant (Jakupovic et al. 1986). Other members of the genus have yielded **coumarins** and **diterpenes** (D. McKenna 1995, 101*; Schultes and Hofmann 1992, 44*). *Helichrysum serpyllifolium*, which has been characterized as "Hottentot tea," is drunk as an infusion (Lewin 1980 [orig. pub. 1929], 352*).

Literature

Jakupovic, J., J. Kuhnke, A. Schuster, M. A. Metwally, and F. Bohlmann. 1986. Phloroglucinol derivatives and other constituents from South African *Helichrysum* species. *Phytochemistry* 25:1133–42.

Helicostylis tomentosa (Poepp et Endl.) Rusby

(Moraceae)—takini

In Amazonia, the inner bark of this tree, known locally as *misho chaqui*, is said to be used as a hallucinogen. Animal tests have shown that rats exhibit the same symptoms after ingesting this plant as they do when inebriated with *Cannabis* (Buckley et al. 1973; Duke and Vasquez 1994, 86*).

The magicians of the Caribs and Africans living in the wilds of Guyana used various species of the genus *Helicostylis* (*H. pedunculata* Benoist) to induce visions (Schultes and Farnsworth 1982, 184*). The reddish sap of this sacred tree is made into a raw drug known as *takini* (D. McKenna 1995, 101*; Schultes and Hofmann 1992, 45*).

Literature

Buckley, J. P., R. J. Theobald Jr., I. Cavero, et al. 1973. Preliminary pharmacological evaluation of extracts of takini: *Helicostylis tomentosa* and *Helicostylis pedunculata*. *Lloydia* 36:341–45.

Hieracium pilosella L.

(Cichoriaceae; Compositae/Asteraceae)—
   hawkweed, hairy hawkweed, long-haired
   hawkweed

This plant was originally native to Eurosiberia but is now common in Switzerland and has spread even to North America (Lauber and Wagner 196, 1204*).

In Denmark, the yellow-blossomed herbage,

known as *håret høgeurt*, is smoked in joints. One gram is said to produce good psychoactive or euphoriant effects (Larris 1980). The plant is known as hawkweed in the United States and was used by the Iroquois for ethnomedicinal purposes (Ott 1993, 409*). The leaf rosette (without the roots) of the blooming plant, which is common to meadows and moorlands, is collected and dried in the shade. It is sold in pharmacies and herb shops under the names *herba auriculae muris* and *hieracii pilosellae herba*. It contains tannins, flavonoids, and umbelliferone. It is used in folk medicine to treat and strengthen the eyes (as a tea or eyewash). The German name for the plant, *habichtskraut* ("goshawk's weed"), is derived from the belief that goshawks receive their excellent vision from this plant (Pahlow 1993, 146*). This belief may be rooted in Old Germanic shamanism. In Germany, the plant formerly was reputed to offer magical protection against witches and magic (Perger 1864, 133*). Also known as little mouse ears or nail weed, the plant is said to be harmful to sheep (Chamisso 1987, 228*).

Today, hawkweed is usually sold in German pharmacies under the name *pilosellae herba*. When I smoked about 1 g, I felt slightly euphoric and cannabis-like, but relatively weak, effects.

Literature

Larris, S. 1980. *Forbyde Hallucinogener? Forbyd Naturen at Gro!* 4th ed. Nimtoffe: Forlaget Indkøbstryk.

Hipomosa carnea

(Convolvulaceae)—Matacabra

This plant, which is supposedly also known as *flore* or *chalviande*, is said to be used as a hallucinogen in the coastal region of Ecuador (Alvear 1971, 23*). Whether the botanical identification given by Alvear is correct is more than doubtful. It may be that the plant being referred to is actually *Ipomoea carnea*, which is also called *matacabra* ("goat killer") (see *Ipomoea* spp.).

Homalomena sp.

(Araceae)—ereriba

This tropical araliaceous plant is said to have been used as a hallucinogen in Papua New Guinea. The

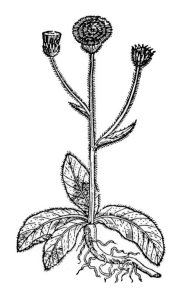

Top: In former times, hawkweed (*Hieracium pilosella*) was hung in many German houses as a protection against witchcraft (woodcut from Tabernaemontanus 1731)

Left: Dried hawkweed (*Hieracium pilosella*) herbage is used as a mild acting marijuana substitute.

Right: Many species of the genus *Homalomena*, which is found throughout Southeast Asia, have not yet been botanically described, even though they are likely used for ethnomedicinal purposes (a *Homalomena* sp. from Malaysia).

Some species of the genus *Iresine*, from the Family Amaranthaceae, are used in the manufacture of South American shamans' drinks (ayahuasca, cimora).

leaves of this plant (possibly *Homalomena belgraveana* Sprague) were ingested together with the bark of *Galbulimima belgraveana* (F. Muell.) Sprague[332] [syn. *Himantandra belgraveana* F. Muell.] and the root of *Zingiber zerumbet* (L.) Sm. [syn. *Alpinia speciosa*] (see **Zingiber officinale**). This allegedly produced strong visions followed by intense dreams (Barrau 1958). Since this plant is known locally as *maraba*, the same name given to **Kaempferia galanga** and *Galbulimima*, the botanical identity of this purported hallucinogen is still in question. Chemical studies are lacking (Schultes and Hofmann 1995, 45*).

The species *Homalomena cordata* Schott and *H. versteegii* Engler are used in New Guinea for rain and love magic, respectively (Ott 1993, 409*). Chemical studies of these species are also lacking (D. McKenna 1995, 101*). The ginger-scented rhizome of the East Indian species *Homalomena aromatica* was once used as an aphrodisiac (Hirschfeld and Linsert 1930, 180*). In Papua New Guinea, an ointment is made from the stem of a *Homalomena* species known as *iva iva* together with coconut oil (cf. **Cocos nucifera**) (von Reis and Lipp 1982, 10*).

Literature

Barrau, Jacques. 1958. Nouvelles observations au sujet des plantes hallucinogènes d'usage autochtone en Nouvelle-Guinée. *Journal d'Agriculture Tropicale et de Botanique Appliquée* 5:377–78.

Huperzia selago (L.) **Bernh. ex Schrank et Mart.**

[syn. *Lycopodium selago* L., *Urostachys selago* (L.) Herter]
(Lycopodiaceae)—devil's claw

This club moss (cf. **Lycopodium clavatum, Lycopodium spp.**), which is found in circumpolar and Antarctic regions and is also known as fir club moss, heckenysop, devil's clover, and selago, is an ancient Celtic-Germanic magical plant that was highly esteemed among the Druids:

It was gathered with great care, no iron instrument was allowed to touch it, even bare hands were unworthy of this honor. A special covering, or "sagus," was used with the right hand. This covering had to be consecrated and secretly received from a holy personage with the left hand. It could be collected only by a white-clad druid with bare feet that had been washed in clear water. Before he collected this plant, he had to make an offering of bread and wine; after this, the plant was carried away from the place in which it grew in a new, clean cloth. In the "Kadir Taliesin," selago is referred to as "the gift of god," and in modern Welsh as

Devil's claw (*Huperzia selago*) has certain psychoactive effects and is an ancient Celtic-Germanic ritual plant. (Woodcut from Tabernaemontanus, *Neu Vollkommen Kräuter-Buch*, 1731)

332 Two alkaloids and two lignans have been found in this plant, the pharmacology of which is unknown (D. McKenna 1995, 101*).

the "gras duw," or the "grace of god." This plant was viewed primarily as an amulet that protected its possessor against all harm. (Schöpf 1986, 58*)

The herbage contains 0.1 to 0.9% total alkaloids, which have been characterized as "selagine" and comprise lycopodines, arifoline, pseudoselagine (= isolycodoline), selagine, and lycodoline. The entire plant can induce vomiting, dizziness, wooziness, and unconsciousness in humans (Roth et al. 1994, 407*).

Iresine spp.

(Amaranthaceae)

Various species of this genus are used in South America as **ayahuasca** additives. Some species, under the name *cimora* or *timora*, are added to San Pedro drinks (see **Trichocereus pachanoi**). *Iresine* species also are said to be the main ingredient or at least one of the main ingredients in the mysterious South American magical drink **cimora** (Ott 1993, 409*). Unfortunately, chemical studies are lacking.

Betacyanin has been detected in the herbage of the Caribbean *Iresine herbstii* Hook. f. (Wong 1976, 119*).

Iryanthera juruensis Warb.

(Myristicaceae)—forest cacao

This small tree, also known as *cedro ajua, huapa, pucuna huapa* ("blowpipe huapa"), or *sacha cacao* ("forest cacao"), provides a resin that is used in the manufacture of **snuff**. It may be psychoactive. Active substances have not yet been detected (Ott 1993, 409*).

The Amazonian Indians (Venezuela, Colombia, Peru, Brazil) produce a **snuff** from *Iryanthera macrophylla* (Benth.) Warb. that may be psychoactive. One analysis found **5-MeO-DMT** in this plant material (Schultes 1985, 131); later studies have been unable to confirm this (Ott 1993, 409*).

The Bora and Witoto formerly used *Iryanthera ulei* Warb. as an oral hallucinogen. 5-MeO-DMT has also been found in the bark of this plant. *Iryanthera longiflora* Ducke is said to be a hallucinogen as well (Davis and Yost 1983, 186*).

Literature

Schultes, Richard Evans. 1985. De Plantis Toxicariis e Mundo Novo Tropicale Commentationes XXXV: Miscellaneous notes on biodynamic plants of the northwest Amazon. *Journal of Ethnopharmacology* 14:125–58.

Jasminum spp.

(Oleaceae)—jasmine

The flowers of several jasmine species are the source of the aromatic jasmine oil (oleum jasmini), which contains eugenol and plays an important role in the perfume industry (cf. **essential oils**). Moreover, two African species have been reported to have psychoactive effects. In Abyssinia, the leaves of *Jasminum floribundum* R. Br. (known as *hab el tsalim*) are used as an "inebriating agent"; the leaves of *Jasminum abyssinicum* R. Br. are used in Eritrea for the same purpose (Hartwich 1911, 811*). The constituents are unknown. *Gelsemium sempervirens* is also known as yellow jasmine.

Jatropha grossidentata Pax et Hoffm.

(Euphorbiaceae)—purgative nut

In the shamanism of the Ayoré Indians of Paraguay, the dried root of the plant, known as *caniroja*, is smoked in order to communicate with animal spirits and to initiate novices into shamanism. The shamans (*naijna*) occasionally climb into a quebracho tree (*Aspidosperma quebracho-blanca*) and sit in its crown, where they smoke the roots. In this way, they are able to speak directly to the animals (Schmeda-Hirschmann 1993, 108, 109*). During a self-experiment under the supervision of one of the last Ayoreo shamans, no psychotropic effects of any kind could be observed. However, rhamnofolane and **diterpenes** have been found in the roots (Jakupovic et al. 1988; Schmeda-Hirschmann et al. 1992), which require further investigation (the active principle in *Salvia divinorum* is also a diterpene). In South America, other *Jatropha* species are regarded as aphrodisiacs (Schultes 1980, 104*). In northern Peru, *Jatropha macrantha* Arg. is known locally as *huanarpo macho* and is one of the most famous aphrodisiacs for men. Further study is needed to ascertain whether this species has psychoactive effects.

Literature

Jakupovic, J., M. Grenz, and G. Schmeda-Hirschmann. 1988. Rhamnofolane derivatives from *Jatropha grossidentata*. *Phytochemistry* 27:2997–98.

Schmeda-Hirschmann, G., F. Tsichritzis, and J. Jakupovic. 1992. Further diterpenes and a lignan from *Jatropha grossidentata*. *Phytochemistry* 31:1731–35.

Juanulloa ochracea Cuatrecasas

(Solanaceae)—ayahuasca

This nightshade is known in Colombia as *ayahuasca* and may possibly be utilized as an **ayahuasca** additive (Ott 1993, 410*; Schultes and Raffauf 1991, 39*). It may also have been used alone for psychoactive purposes. In the region of Limón (Costa Rica), the leaves and stems are used to treat wounds (Schultes 1978a, 192*). The alkaloid parquine (cf. *Cestrum parqui*) has been detected in the genus, which is composed of some twelve species (Schultes 1979b, 151*; Schultes and Raffauf 1991, 39*).

Kaempferia galanga L.

(newly written as *Kempferia galanga*)

(Zingiberaceae)—galanga

A member of the ginger family, galanga is also known as the galgant-spice lily, resurrection lily, and *hinguru-piyali*. It is found in the tropical regions of Africa and in Southeast Asia. The very aromatic rootstock (rhizome), which often looks like a hand and is usually referred to as *maraba*, is used throughout the range of the plant as a spice and as a remedy for treating digestive ailments. *Kaempferia* has a strong, refreshing taste. In Malaysia, the root formerly was added to an arrow poison made with *Antiaris toxicaria* (Schultes and Hofmann 1995, 47*). *Kaempferia galanga* is one of the ingredients in the Indonesian spice mixture known as *jamu* (Rehm 1985) and is the main ingredient in those mixtures that are produced for tonic and aphrodisiac purposes (Macmillan 1991, 424*). In Japan, the root is sometimes used in the manufacture of **incense**. In Thailand, the root and young leaves are added to curries. The crushed root, mixed with whiskey (cf. **alcohol**), is applied as a paste to the forehead and scalp as a folk medicine for treating headaches (Jacquat 1990, 117).

The inhabitants of the area around Mount Hagen (Papua New Guinea) supposedly use or once used the rhizome as a hallucinogen, similar to *Homalomena* spp. (Barrau 1962). "The root is used as a spice and inebriant throughout all of Southeast Asia. . . . The rhizome induces hallucinations (without any side-effects)" (Bremness 1994, 180*). A European report states that ingestion of the powdered root produces "a surprising clarity of thought and alterations in vision" (Schuldes 1995, 46*).

The rootstock is rich in **essential oils**, the

The packaging of a traditional Indonesian *jamu* mixture, whose main component (20%) is *Kaempferia galanga* and which also includes gingerroot (*Zingiber officinale*), licorice root (*Glycyrrhiza glabra*), and *sindora* fruits. This *jamu* is recommended as a potent aphrodisiac and tonic for both men and women. The finely crushed herbal ingredients are mixed with some lukewarm water, freshly pressed curcuma juice, and a raw egg (if desired) and drunk in the morning, preferably before breakfast. The drink, of course, should be consumed daily!

Above: The most renowned aphrodisiac in Peru is *huanarpo macho* (*Jatropha macrantha*), which is reputed to have psychoactive effects. The raw plant material, seen here, is macerated with high-proof alcohol.

Left: Galanga (*Kaempferia galanga*), found throughout Southeast Asia, produces a very aromatic rhizome.

The true galanga root (from the stock plant *Kaempferia galanga*) is only rarely found in trade. It is usually confused with the rootstock of galgant (*Alpinia galanga* [syn. *Languas galanga*], *Alpinia officinarum* [syn. *Languas officinarum*]). In this illustration from an early modern herbal, the rhizome of "large galgan" is probably identical to *Kaempferia galanga*. (Woodcut from Tabernaemontanus, *Neu Vollkommen Kräuter-Buch*, 1731)

composition of which is unknown. It may contain psychoactive substances (Schultes and Hofmann 1992, 47*). Reports following ingestion of the powder often indicate mild or even no effects (Schuldes 1995, 95*). This may be due to the fact that the experimenters did not use genuine *Kaempferia* roots, for *galanga* is a name that has produced much confusion. Another member of the ginger family, galgant (*Alpinia officinarum* Hance [syn. *Languas officinarum*]), is known by the name *galanga* or *little galanga*. It too is used as a spice (Norman 1991, 64*). In Germany, *Alpinia galanga* (L.) Willd. [syn. *Galanga major* Rumpf., *Maranta galanga* L., *Languas galanga* Sw.] is known by the name *large galanga root*, and *Alpinia officinarum* by the name *little galanga root* (Jacquat 1990, 118; Norman 1991, 45*; Seidemann 1993, 180*).

Literature

Barrau, Jacques. 1962. Observations et travaux récents sur les végétaux hallucinogens de la Nouvelle-Guinée. *Journal d'Agriculture Tropicale et de Botanique Appliquée* 9:245–49.

Jacquat, Christiane. 1990. *Plants from the markets of Thailand.* Bangkok: Editions Duang Kamol.

Rehm, Klaus D. 1985. Jamu—die traditionellen Arzneimittel Indonesiens. *Curare*, Sonderband 3/85:403–10.

Lagochilus inebrians Bunge

(Labiatae)—intoxicating mint

This bushy mint is native to the central Asian steppes of Turkistan and Uzbekistan. It is gathered in autumn and hung on the rafters to dry over the winter. The leaves are used to make a tea that is sweetened with honey and induces a mild state of euphoria, although it also can be used as a sedative (D. McKenna 1995, 103*). In Russian folk medicine and phytotherapy, the plant is also used to treat allergies and skin diseases and to promote blood coagulation (Schultes and Hofmann 1992, 47*).

The dried material (leaves) contains up to 17% lagochiline, a diterpene alkaloid (the average is around 3%; Schultes 1970, 41*; Tyler 1966, 287*). Numerous studies are available in the Russian literature. The plant is, or at least was, listed as a natural tranquilizer in the Russian pharmacopoeia (D. McKenna 1995, 103*; Scholz and Eigner 1983, 78*).

Lancea tibetica Hook. f. et Thoms.

(Scrophulariaceae)—depgul

In Ladakh, India, the root of this plant (known locally as *depgul*) is roasted, crushed, and either smoked together with tobacco (*Nicotiana tabacum*) or drunk in milk. The product is called *berzeatsink* and is said to have potent stimulant and activating effects (Navchoo and Buth 1990, 320*).

Leonotis leonurus (L.) R. Br.

(Labiatae)—lion's tail, lion's ear

This South African bush bears orange blossoms and is purported to have hallucinogenic effects (Schultes and Hofmann 1980, 367*). In Africa, it is known by the names *dacha*, *daggha*, and *wild dagga*, "wild hemp" (cf. **Cannabis indica**). The Hottentots (Khoikhoi; Heusaquas) and Bushmen smoke the buds and leaves as inebriants (Schleiffer 1979, 93 ff.*; Schuldes 1995, 48*). This bush may be one of the inebriating plants subsumed under the name *kanna* (cf. **kanna**, *Mesembryanthemum* **spp.**, *Sceletium tortuosum*). The resinous leaves and the resin rubbed off or extracted from them are smoked either alone or mixed with tobacco (*Nicotiana tabacum*) (Grubber 1991, 44*). In northern California, many people now smoke the leaves and orange flowers. Chemical studies are lacking (Ott 1993, 411*). The rather bitter-tasting smoke of flowers grown in California has a mild psychoactive effect reminiscent of that of both **Cannabis** and **Datura**. In eastern South Africa, the closely related species *Leonotis ovata* is reportedly used for the same purpose (Schleiffer 1979, 93*).

Another closely related species, *Leonotis nepetaefolia* (L.) R. Br., is used in Caribbean folk medicine. The leaves and flowers of this species

Left: Seeds and herbage of *Lagochilus inebrians*.

Right: The South African wild dagga (*Leonotis leonurus*) in bloom.

have yielded bound oils, bitter principles, **diterpenes**, **coumarins**, and resins (Argueta V. et al. 1994, 229*; Puroshothaman et al. 1974a, 1974b; Wong 1976, 136*). In Mexico, this plant is known as *flor de mundo*, "world flower," or *mota*. The name *mota* is normally used to refer to marijuana (cf. *Cannabis indica*); this may indicate that the plant is used as a marijuana substitute. The extract of this plant has antispasmodic effects and appears to inhibit acetylcholine and histamine (Argueta V. et al. 1994, 229*).

Literature

Puroshothaman, K. K., et al. 1974a. 4,6,7-trimethoxy-5-methylchromon-2-one, a new coumarin from *Leonotis nepetaefolia*. *Journal of the Chemical Society, Perkin Transactions* 1 (1): 2594–95.

———. 1974b. Nepetaefolinol and two related diterpenoids from *Leonotis nepetaefolia*. *Journal of the Chemical Society, Perkin Transactions* 1 (1): 2661.

Lichen non ident.

(family non ident.)—jievut hiawsik
The Pima and O'odham (= Papago) Indians both use the name *jievut hiawsik*, "earth flower," to refer to lichens that live on rocks. One species, which unfortunately has not been identified botanically, exudes a strong scent, has an ashen gray color, and lives on rocks and old, dry wood. The lichen once had a religious significance. It was mixed with tobacco (*Nicotiana tabacum*) and smoked during the summer dances (cf. **kinnikinnick**). It is said to have an effect similar to that of marijuana (*Cannabis indica*) and to "make young men crazy." The Pima believe that a man can conquer any woman after he has smoked the lichen (Curtin 1984, 77). Until now, lichens have been almost completely unknown as psychoactive substances in ethnopharmacology (cf. *Dictyonema*). Recently, beard lichens have found use as **incense**.

Literature

Curtin, L. S. M. 1984. *By the prophet of the earth: Ethnobotany of the Pima*. Tucson: University of Arizona Press.

Limmonium macrorhabdos O. Kuntze

(Plumbaginaceae)—staspak
In Ladakh, the sun-dried leaves of this plant, known as *staspak*, are drunk in the form of a cold-water extract (the powder is left in the water for about one week). The drink, called *staspakchek*, is said to produce strong inebriating effects and even to be dangerous (Navchoo and Buth 1990, 320*).

Lobelia inflata L.

(Campanulaceae)—Indian tobacco, pukeweed
This delicate lobelia is native to North America, where it is known by the names *pukeweed* and *Indian tobacco*. The plant was used ceremonially by the North American Crow Indians and played a role in the love magic of the Pawnee and Mesquakie (Ott 1993, 411*). Lobelia is one of the ingredients in **kinnikinnick** and other **smoking blends**. The Indians also smoked the plant medicinally for asthma, bronchitis, irritations of the throat, and coughs. The herbage is finding increasing use as a tobacco substitute (see *Nicotiana tabacum*), especially among people who are trying to quit tobacco. Smoked by itself, lobelia is clearly psychoactive. It has both sedative and stimulating effects, which can surprise those who have no prior knowledge of the plant.

The herbage contains more than twenty piperidine alkaloids. The main alkaloid, α-lobeline, is a **nicotine** antagonist (Szôke et al. 1993) and is used as a nicotine substitute for medical withdrawal (Krochmal et al. 1972, 216). The α-lobeline content is almost twice as high in cultivated plants as in wild specimens (approximately 1.05 to 2.25% of dry weight; Krochmal et al. 1972, 216).

In Mexico, a closely related species, *Lobelia cliffordtiana* L., is numbered among the *hierbas locas*, the "herbs that make one crazy" (Martínez 1987, 427*; Reko 1938, 185*). It too may be suitable as an inebriating ingredient in **smoking blends**. An Asian species, *Lobelia nicotianaefolia*, is known as *rasni* or "wild tobacco." The long, tobacco-like leaves of this plant, which can grow up to 3 meters in height, are said to be poisonous but can be smoked (Macmillan 1991, 430*). The name *Lobelia longiflora* L. is an outdated synonym for *Hippobroma longiflora* (L.) G. Don, which is one of the ingredients in the South American **cimora** drink (Zander 1994, 312*).

Literature

Krochmal, Arnold, Leon Wilken, and Millie Chien. 1972. Plant and lobeline harvest of *Lobelia inflata* L. *Economic Botany* 26:216–20.

Szôke, É., A. Krajewska, and A. Nesmélyi. 1993. NMR characterization of alkaloids from *Lobelia inflata*. *Planta Medica* 59 suppl.: A704.

Lobelia tupa L.

(Campanulaceae)—tupa, devil's tobacco
This large lobelia, the flowers of which are fiery red, is found in the wild in South America in the Andes and their foothills. It is cultivated throughout the world as an ornamental. The most common name for this conspicuous plant is *tupa*, which can be translated as "spot," "point," "sunspot" or even "mark of disgrace." Many inhabitants

Indian tobacco (*Lobelia inflata*), with flowers and fruits.

The Chilean devil's tobacco (*Lobelia tupa*) can grow to as much as 3 meters in height. The long leaves, which are somewhat reminiscent of tobacco leaves, are smoked.

"Zapote. Tree belonging to the Sapatoceae. The Cora in the Mexican Sierra Madre occidental hang its branches on their huts to protect the newborn from evil demons."

SIEGFRIED SELIGMAN
DIE MAGISCHEN HEIL- UND SCHUTZMITTEL AUS DER BELEBTEN NATUR [THE MAGICAL HEALING AND PROTECTIVE AGENTS FROM ANIMATED NATURE]
(1996, 290*)

of the Andes consider this campanulaceous plant to be toxic and avoid it. Since it is often called *tabaco del diablo* ("devil's tobacco"), it was once thought that it might have psychoactive or even hallucinogenic effects (Schultes and Hofmann 1992, 47*). However, there is no ethnographic evidence indicating that devil's tobacco is or ever was used ritually for psychoactive purposes.

In Chile, *Lobelia tupa* and several other species are referred to as *trupa*, *tupa*, or *tabaco del diablo* (*Lobelia excelsa* Bonpl., *Lobelia polyphylla* H. et A.; cf. Mösbach 1992, 105*). The Mapuche use the name *tupa* to refer to a related species, *Lobelia salicifolia* Sweet, which they use as a medicinal plant to treat flu (they make a tea from the leaves). The latex is said to induce severe inflammation of the eye and of the digestive tract, together with vomiting and diarrhea (Houghton and Manby 1985, 100*).

Lobelia tupa has been shown to contain piperidine alkaloids; these do not, however, have any unequivocal psychoactive effects. As with **Lobelia inflata**, the principal alkaloid in the leaves is α-lobeline (Kaczmarek and Steinegger 1958). Lobelamidine and norlobelamidine are also present (Kaczmarek and Steinegger 1958; Schultes and Farnsworth 1982, 177*). Smoking the dried leaves strongly stimulates the production of saliva and an immediate stimulation occurs that is similar to that produced by *Lobelia inflata* and **Nicotiana tabacum**. The white smoke is relatively easy to inhale and produces almost no irritation (cf. **smoking blends**).

Literature

Kaczmarek, F., and E. Steinegger. 1958. Untersuchungen der Alkaloide von *Lobelia tupa* L. *Pharm. Helvetica Acta* 33:257–62.

———. 1959. Botanische Klassifizierung und Alkaloidvorkommen in der Gattung *Lobelia*. *Pharm. Helvetica Acta* 34:413–29.

Lotus wrightii (A. Gray) Greene
(Leguminosae)—deervetch, Wright's horn clover

The Navajo Indians regard this plant as a "life medicine" and use it ritually in hunting (Vestal 1958, 32*). The Apache used the roots as an inebriating additive for their homemade beer (see **beer**). It is possible that the root cortex contains alkaloids, e.g., tryptamines, which are also found in many other plants of the same family.

Lucuma salicifolia H.B.K.
(Sapotaceae)—zapote borracho

The Aztecs of Mexico called the fruits of this Sapodilla Family species, the flesh of which is bright yellow, *cozticzápotl* ("dizzy-making fruit"),

cochiz tzapotl, *zapote somnífero*, or *zapote blanco*. It is said that excessive consumption of the fruits can result in a peculiar state of inebriation not unlike that produced by alcohol. This is why the fruit is now known primarily as *zapote borracho*, "drunken zapote" (Martínez 1987, 1154*). Mexican farmers are said to use it to produce a state of inebriation. "In Oaxaca and Puebla, zapote fruits are purchased by barkeepers (as well as by housewives) and added to brandy (as we do with other fruits and rum). They impart a beautiful, cognaclike color to cheap alcohol and are said to make this 'stronger,' i.e., the inebriating effects that the drinker desires are manifested more rapidly than after the consumption of normal brandy" (V. Reko 1938, 151 f.*).

Both *zapote blanco*, "white zapote," and *zapote dormilte*, "sleep-inducing zapote" (*Casimiroa edulis* Llave et Lex. [syn. *Casimiroa sapota* Orst., *Fagara bombacifolium* Krug et Urbán, *Zanthoxylum bombacifolium* A. Rich, *Zanthoxylum aracifolium* Turcz.]; Rutaceae), have also been identified as *cozticzápotl* (Argueta V. et al. 1994, 1413*). They also are reputed to have sedative and hypnotic effects. The seeds, burned to ashes, were ingested by the Aztecs as a sleeping agent (Navarro 1992, 94*). Even today, a tea made from the leaves is used in Mexican folk medicine for sleep disorders and to regulate and stimulate dreaming (Argueta V. et al. 1994, 1413*). The seeds of *Casimiroa edulis* have been found to contain the alkaloids *N*-benzoyltyramine, methylhistamine, casimiroin, fagarine, and casimiroidin as well as **coumarins** (**scopoletin**) (Aebi 1956; Argueta V. et al. 1994, 1414*; Emboden 1979, 6, 173*). The leaves contain methylhistamine and dimethylhistamine as well as rutin (Argueta V. et al. 1994, 1414*). The Aztec name *cozticzápotl* has also been interpreted as referring to **Calea zacatechichi** (Lara Ochoa and Marquez Alonso 1996, 123*).

Literature

Aebi, A. 1956. The isolation of casimiroidin from the seeds of *Casimiroa edulis*. *Helvetica Chimica Acta* 39:1495.

Lupinus spp.
(Fabaceae)—lupines, wolf's beans

Several species of lupine (*Lupinus albus* L., *L. angustifolius* L., *L. luteus* L.) are found in the Mediterranean region. In ancient times, they were used for medicinal (described in Dioscorides 2.132), ritual, and apparently psychoactive purposes. The pilgrims who came to the Greek death oracle of Acheron (near Ephyra, Thesprotia, northern Greece)—the entrance to Hades—were required to eat large quantities of lupine seeds so that they could contact the souls of the dead (Dakaris 1989). "A strict diet was used to psycho-

Left: In ancient times, yellow lupine, also known as wolf's bean (*Lupinus luteus*), apparently was associated with magical practices and lycanthropy (transformation of a human into a wolf).

Center: Some species of the genus *Lupinus* appear to contain psychoactive substances.

Right: The condor plant, *Lycopodium magellanicum*. (Photographed at Laguna Shimbe, northern Peru)

logically prepare them for communicating with the underworld in the narrow passages of the labyrinthine shrine. . . . The consumption of the alkaloid-containing lupine seeds induced in the pilgrims the state of inebriation that the priests desired and diminished their faculties of perception, preconditions that were necessary for the initiated to be able to feign a genuine communication with the shadow figures of the deceased" (Baumann 1982, 146*). Since the oracular priests jealously guarded their secrets, we unfortunately know nothing precise about the ways in which the lupines were actually used (Vandenberg 1979*). It is likely that, in addition, sulfur was burned as a fumigant (Dakiris 1989, 160).

According to other sources, the pilgrims ate not lupine seeds when they visited the oracle but "pig beans," which were probably *Hyoscyamus.* These induced "states of dizziness, unreal sensory perceptions, and passivity" (Dakaris 1989, 162 f.). Lucian described a séance in which the sea onion was used as a magical plant (cf. **moly**).

Lupine seeds contain a number of toxic substances: lupanine, 13-hydroxylupanine, angustifoline, 13-tigloyloxylupanine, albine, multiflorine, α-isolupanine, 4-hydroxylupanine, ammodendrine, anagyrine, and sparteine (Roth et al. 1994, 473*). Lupanine is chemically related to **cytisine**. A new alkaloid, (–)-(*trans*-4'-β-D-glycopyranosyloxy-3'-methoxycinnamyl)-lupinine, has been detected in the yellow lupine (*Lupinus luteus*) (Murakoshi et al. 1979). However, nothing is known about the pharmacology of this substance.

Lupine seeds were once brewed as a coffee substitute (*Coffea arabica*). In Mexico, the lupine species *Lupinus elegans* H.B.K. is known as *hierba loca*, "crazy herb" (Martínez 1987, 427*). It may have inebriating effects (cf. *Astragalus* spp.).

Literature

Dakaris, Sotiris. 1989. Das Totenorakel am Acheron. In *Tempel und Stätten der Götter Griechenlands,* ed. Evi Melas, 157–64. Cologne: DuMont.

Murakoshi, Isamu, Kazuo Toriizuka, Joju Haginiwa, Shigeru Ohmiya, and Hirotaka Otomasu. 1979. (–)-(*Trans*-4'-β-D-glycopyranosyloxy-3'-methoxycinnamyl)-lupinine, a new lupin alkaloid in *Lupinus* seedlings. *Phytochemistry* 18:699–700.

Lycopodium clavatum L.

(Lycopodiaceae)—club moss

Lycopodium clavatum L. and other club mosses indigenous to Europe (*Lycopodium* **spp.**; cf. also *Huperzia selago*) are known by a variety of common names, including Druid's foot herb, Druid's foot, Druid's plant, Druid's flour, Druid's foot flour, witches' plant, witches' flour, witches' flour plant, witches' dust, snake moss, devil's claw, devil's claw flour, devil's rubbish, and disquiet. The spores are known as witches' flour, Druid's flour, lightning powder, dusting powder, and moss powder. These names suggest an ancient use in pagan rituals and strong associations with witchcraft. One German name, *bärlapp*, means "uterine ointment" (Beckmann and Beckmann 1990, 196*).

Lycopodium clavatum and similar species (*L. cernuum* L., *L. hamiltonii* Spreng., *L. serratum* Thunb., *L. subulifolium* Wall. ex Hook. et Grev.) are found also in Nepal. There, the club moss is sacred to the Hindu god Vishnu and is used in garlands and other objects at his festivals.

Club moss (*Lycopodium clavatum* L.) contains a toxic or psychoactively effective alkaloid complex that is generally referred to as "clavatine"

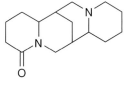

Lupinine

A bundle of condor plants (*Lycopodium* spp.) from the witches' market in Chiclayo, Peru.

which even includes **nicotine** (Roth et al. 1994, 477*).

Lycopodium spp.

(Lycopodiaceae)—condor plants, condoros

In northern Peruvian *curanderismo*, a variety of club mosses are used by folk healers as medicinal plants, amulets, and additives to the San Pedro drink (cf. ***Trichocereus pachanoi***). In the north-western lowlands, club mosses are normally subsumed under the name *cóndor, condor,* or *condor plant.* In the highlands of Huancabamba and Las Huaringas, they are known as *huaminga.* Only one as yet unidentified species is included among the magical plants of the category *hornamo* (cf. ***Senecio* spp.**). Club mosses are also used as bath additives and for magical defense during healing rituals (Giese 1989, 227 f.*).

Condor Plants Used in Northern Peruvian *Curanderismo*

Lycopodium spp.	cóndor purga
	condorillo
	hierba de condorillo
	hornamo lírio
	hornamo loro
	huaminga misha
	huaminga oso
	trenza amarilla
	trenza shimbe
Lycopodium affine Hook. et Grev.	condorillo
Lycopodium clavatum L.	trencilla verde
Lycopodium contigum Kltz.	trencilla blanca
Lycopodium crassum H.B.K.	trencilla
Lycopodium magellanicum	condoro
Lycopodium reflexum	condoro
Lycopodium saururus	hierba del cóndor
	cóndor misha
Lycopodium spurium	trencilla del lago
Lycopodium tetragonum	condorillo de quatro filos
Lycopodium vestitum	trencilla blanca

When *condorillo* or *cóndor misha* is added to the San Pedro drink, the plant spirit appears to the *curandero* as a condor. At the behest of the healer, the condor can go on astral journeys and fulfill small tasks. As a result, he can remedy harmful magic and bring the lost soul back to a patient suffering from *susto,* "fright" (Giese 1989, 249*). There may even be some club moss species with psychoactive effects:

It is possible that *Lycopodium* sp. also augments the hallucinogenic effects of the San Pedro drink. Manuel, a plant dealer from Trujillo, said that the plant that he called "trenza shimbe" and appears to be the same as "cóndor misha" serves to improve the "visionary sight." (Giese 1989, 228*)

One plant dealer at a "witches' market" in Chiclayo in July 1997 told me that *condoro,* which I was able to identify as *Lycopodium magellanicum,* has hallucinogenic effects, especially when combined with *Trichocereus pachanoi.*

More than one hundred alkaloids have been found in the genus *Lycopodium* (Gerard and MacLean 1986). To date, it is uncertain whether there are any psychoactive alkaloids among these. Six alkaloids have been detected in the Chilean species *Lycopodium magellanicum* (Loyola et al. 1979).

It is possible that a psychoactive use of club mosses is known in Chile or was practiced in former times. *Lycopodium paniculatum* A.N. Desv. is called *llanca-lahuén,* "precious medicine," in Mapuche and is known in the local Spanish as *licopodio, pimpinela,* and *palmita* (Mösbach 1992, 55*). The Mapuche use another species, *Lycopodium gayanum* Remy et Fée, which they call *ngalngal,* as a sedative medicine. In the local Spanish, it is known as *harina de los brujos,* "flour of the witches" (Mösbach 1992, 55*).

Literature

Gerard, Robert V., and David B. MacLean. 1986. GC/MS examination of four *Lycopodium* species for alkaloid content. *Phytochemistry* 25 (5): 1143–50.

Loyola, Luis A., Glauco Morales, and Mariano Castillo. 1979. Alkaloids of *Lycopodium magellanicum. Phytochemistry* 18:1721–23.

Macropiper excelsum (Forster) Miq.

(Piperaceae)—Maori kava

Since *Piper methysticum* does not grow in New Zealand, the Maoris looked for a substitute when they settled the islands. They found it in the form of an indigenous pepper species, which they used like *Piper methysticum* to produce a kavalike drink. The plant contains an **essential oil** with the active substances myristicin and elemicine (Bock 1994, 98*).

Magnolia virginiana L.

[syn. *Magnolia glauca* L.]
(Magnoliaceae)—Virginia magnolia, swamp sassafras

The Rappahannock Indians snuffed the leaves and

bark of this North American tree as a mild inebriant (cf. **snuff**). Chemical studies and additional ethnographic data are lacking (Ott 1993, 412*). Virginia magnolia, which has a safrole scent, is also known in the United States as swamp sassafras (Grieve 1982, 716*). It contains an **essential oil** that clearly includes a high quantity of safrole (cf. *Sassafras albidum*).

Another magnolia species has been associated with the Aztec inebriant **poyomatli**. Magnolias have been shown to contain alkaloids, e.g., magnoflorine (Roth et al. 1994, 479*).

Malva rotundifolia

(Malvaceae)—panirac

An herbarium specimen collected by E. Bacon in Afghanistan in 1939 bears the remark "the seeds make one drunk" (von Reis and Lipp 1982, 178*).

Manihot anomala Pohl ssp. *anomala*

(Euphorbiaceae)

The dried root of this manioc species, which is known as *sienejna*, is used by the Ayoreo Indians of Paraguay to initiate a shaman (*naijna*) so that he can communicate with the spirits (Schmeda-Hirschmann 1993, 108*). But not all of the Ayoreo believe that this plant works. It is said that the shaman feels as though he is drunk when he smokes *sienejna*. In this state, the spirits of the animals (especially those of iguanas, poisonous snakes, and birds) meet him in the shape of small people so that they may let him know of their whereabouts (Schmeda-Hirschmann 1993, 109*). Smoking experiments, however, have not revealed any type of hallucinogenic or other psychotropic effect (Schmeda-Hirschmann 1993, 111*). Chemical studies and other experiments are still needed.

Maquira sclerophylla Ducke

[syn. *Olmedioperebea sclerophylla*]
(Moraceae)—rapé dos indios

In the central Amazonian region (Xingu) of Brazil, the bark and perhaps the seeds of this tree, which can grow as tall as 30 meters and is known locally as *rapé dos indios*, are made into a **snuff** that is purported to have hallucinogenic effects and is consumed at religious festivals (Schultes and Raffauf 1990, 318*). This practice has apparently died out (D. McKenna 1995, 101*). The powder is said to stimulate the central nervous system and cause euphoria and visual hallucinations. Unfortunately, pharmacological studies of these effects using human subjects have not yet been conducted (Carlini and Gagliarid 1970; D. McKenna 1995, 101*). One experiment with rats and guinea pigs revealed—as is typical—little but amphetamine-like reactions (Carvalho and Lapa 1990). Earlier studies demonstrated the presence of **coumarins**. Later investigations revealed the presence of cardiac glycosides as well (Ott 1993, 412*).

Literature

Carlini, E. A., and R. J. Gagliarid. 1970. Comparação das acões farmacológicas de estratos brutos de *Olmedioperebea calophyllum* e *Cannabis sativa*. *Anais do Academia Brasileira dos Ciencies* 42:400–412.

Carvalho, João Ernesto de, and Antonio José Lapa. 1990. Pharmacology of an Indian-snuff obtained from Amazonian *Maquira sclerophylla*. *Journal of Ethnopharmacology* 30:43–54.

Matayba guianensis

(Sapindaceae)—papa-para

A note attached to an herbarium specimen collected by J. A. Steyermark in Venezuela in 1945 reads, "Fruit poisonous or makes one '*loco*' [= crazy] when eaten" (von Reis and Lipp 1982, 169*).

Mentha pulegium L.

(Labiatae)—pennyroyal

This mint species, once known as *blechon* or *glechon*, was apparently an ingredient of **kykeon**, the potion drunk by initiates to the Eleusinian mysteries (Ruck 1995, 142*). Aristophanes, in *Pax*, mentions a drink containing pennyroyal that was named *kykeon* and which Hermes, the messenger of the gods, recommended as a protection against

Pennyroyal (*Mentha pulegium*) plays a significant role in ethnogynecology and as a ritual incense.

Because of its aromatic leaves, the North American *Magnolia virginiana* is also known as sweet bay.

Fruits and seeds of the tree *Metteniusa edulis,* which the Kogi use to obtain a psychoactive product. (From Karsten, *Florae Columbiae,* 1858–61)

disease. Pennyroyal was also used to produce love drinks and was considered an obscene metaphor for a woman's pubic hair and a symbol of illicit sexuality. The herbalist Bodin (1591) identified it with Homer's **nepenthes**. Pennyroyal was one of the most renowned abortifacients of antiquity and was used medicinally to treat cramps in the lower abdomen (Rätsch 1995a, 237 f.*). In ancient times, it was also burned as an **incense**. In South America, the dried plant is still used as a ritual incense and is offered to the earth goddess Pachamama (Ott 1993, 412*).

A medicinal use of pennyroyal can still be found in the folk medicine of Cyprus, where fresh leaves are eaten in salad to treat male impotency. They are also consumed as a tea for stimulant and tonic purposes.

Pennyroyal has medicinal and ritual significance for the pagan Berber peoples of the Atlas Mountains (Morocco) that may have its roots in ancient ideas. A tea made from the herbage is drunk to treat abdominal pains, colic, rheumatism, and flatulence and as a tonic and digestive. During the summer solstice, the plant is burned as a ritual incense to protect humans and animals against misfortune. The feast consumed on the night of the summer solstice consists of snails that have been cooked in salt, pepper (*Piper nigrum*; cf. **Piper spp.**), pennyroyal, and thyme (*Thymus* spp.). Eating this preparation ensures good health throughout the coming year. The medicinal properties of the plant are said to be best when the plant is collected shortly before the solstice. It is used externally to treat wounds and is taken internally to treat coughs and colds (Venzlaff 1977*).

Hildegard von Bingen had the following to say about the psychoactive effects of pennyroyal: "He who has pains in the brain so that he is ill should add pennyroyal to wine and boil it, and he should lay it on his head while still warm, and he should tie a cloth over this, so that the brain is warm and suppresses the madness in him" (*Physica* 1.126).

Pennyroyal contains 1 to 2% **essential oil**, 80 to 94% of which is pulegone, a substance that can induce abortions in animals and humans (Boyd 1992). Piperitone and (-)-limonene also occur (Roth et al. 1994, 493*). Use of the plant as an abortifacient can be dangerous (Gunby 1979, Vallance 1955), and deaths have been reported (cf. *Focus* [1994] 32:95). Higher dosages of *oleum pulegii* can produce delirium and an anesthesia-like paralysis. Apart from the essential oil, no psychoactive compounds have been isolated (Ott 1993, 412*).

Literature

Boyd, E. L. 1992. *Hedeoma pulegioides* and *Mentha pulegium.* In *Adverse effects of herbal drugs,* ed. P. A. G. M. de Smet, K. Keller, R. Hänsel, and R. F. Chandler, 151–56. Berlin: Springer.

Gunby, P. 1997. Plant known for centuries still causes problems today. *Journal of the American Medical Association* 241 (21): 2246–47.

Vallance, W. B. 1955. Pennyroyal poisoning, a fatal case. *Lancet* (1955): 850-851.

Metteniusa edulis Karst.

[syn. *Pentandria monogynia* L., *Gamopetalae nuculiferae* Endl.]

(Metteniusaceae)—macagua, urupagua, canyí

The three species of the genus *Metteniusa* in Colombia thrive primarily in cloud forests. These trees bear large fruits (Gentry 1993, 474 f.*). The genus constitutes its own family but is also assigned to the families Alangiaceae and Icacinaceae (Brako and Zarucchi 1993, 573*). Karsten has even seen a certain resemblance to the Convolvulaceae:

> The fruit and seed formation of this tree— whose bitter-tasting seeds are a not unimportant source of nourishment for the tribe of Arguaco Indians who inhabit the peaks of the mountains of St. Marta—isolate it from its natural relatives, the Cordiceae and the Asperifoliae. (Karsten 1858, 1:80)

In the Sierra Madre de Santa Marta, the tree is known as *canyí.* It is said to have ritual significance for the Kogi of Colombia.[333] The priests (*mamas*) attribute strong psychoactive effects to the tree's chestnutlike fruits (Reichel-Dolmatoff 1977, 285*). Whether this plant truly is psychoactive is questionable, for it is eaten as a food in Venezuela, albeit after having been cooked (Lozano-C. and Lozano 1988, 26).

Another species, *Metteniusa tessmanniana* (Sleumer) Sleumer [syn. *Aveledoa tessmanniana* Sleumer], is found in Peru (Brako and Zarucchi 1993, 573*).

Literature

Karsten, Hermann. 1858–61. *Florae Columbiae,* I. Berlin.

Lozano-C., Gustavo, and Nulia B. de Lozano. 1988. *Metteniusaceae.* Vol. 11 of *Flora de Colombia.* Bogotá: Universidad de Colombia.

Mikania cordata (Burm) B.L. Robinson

(Compositae)

This shrub is common throughout the hot zones of India. The leaves are used in traditional medicine to treat itchiness and as a wound plaster. Neuropharmacological studies in animals (mice) have demonstrated that the root extract induces profound behavioral changes, particularly the disappearance of aggressive behavior. The root

333 The related species *Metteniusa nucifera* (Pittier) Sleumer is also called *canyí* and is also said to be used by the Kogi in ritual contexts. In other areas, it is regarded as a fruit tree (Lozano-C. and Lozano 1988, 26).

extract appears to have strong narcotic effects upon the central nervous system as well as analgesic properties (Bhattacharya et al. 1988).

In the medicine of the Andean Callawaya, a closely related species (*Mikania scandens* Willd., known as *guaco*) is used together with other plants (see *Erythroxylum coca*, *Cytisus* **spp.**) to treat rheumatism (Bastien 1987, 131*).

Literature

Bhattacharya, Siddhartha, Siddhartha Pal, and A. K. Nag Chaudhuri. 1988. Neuropharmacological studies on *Mikania cordata* root extract. *Planta Medica* 54:483–87.

Mirabilis multiflora (Torr.) Gray

[syn. *Quamoclidion multiflorum* Torr.; cf. Schultes and Farnsworth 1982, 189*]
(Nictaginaceae or Nyctaginaceae)—four-o'clock

The so-called Hopi hallucinogen belongs to the four-o'clocks, those amazing flowers whose blossoms always close at the same time each late afternoon. Known as *so:'ksi* or *so'kya*, the plant produces red flowers and a long, deeply penetrating root. Hopi medicine men chewed the root or drank the juice pressed from it in order to obtain diagnostic visions (Whiting 1939, 75). Twenty-eight to 57 g of the root is said to result in a "half-hour of gaiety." The Zuni Indians bake a bread using flour made from the root and, interestingly, use the bread as an appetite suppressant (Moerman 1986, 293*). The active principles are unknown (Ott 1993, 413*). The botanical name *Mirabilis nyctaginea* is also sometimes applied to this questionable hallucinogen (Moerman 1982, 81 f.*).

On the basis of this information from the older ethnographic literature and the superficial similarities between this genus and the nightshades, many closet shamans believe that another four-o'clock, *Mirabilis jalapa* L., is also psychoactive. The seeds of this plant, which is now cultivated throughout the world as an ornamental, are used ethnomedicinally as an antibacterial and anti-inflammatory agent (Kusamba et al. 1991). It is unknown whether the tuberous root has psychoactive effects. The Pima Indians of northern Mexico use the leaves to brew a tonic for the elderly (Pennington 1973, 221*).

Literature

Kusamba, Chifundera, Kizungu Byamana, and Wa Mpoyi Mbuyi. 1991. Antibacterial activity of *Mirabilis jalapa* seed powder. *Journal of Ethnopharmacology* 35:197–99.

Whiting, Alfred F. 1939. *Ethnobotany of the Hopi.* Museum of Northern Arizona Bulletin no. 15. Flagstaff.: Northern Arizona Society of Science and Art.

Monadenium lugardae N.E. Br.

(Euphorbiaceae)—tshulu, mhlebe

This South African spurge is used for folk medicinal purposes in Eastern Transvaal (Mpumalanga) and is reputedly psychoactive (Schultes and Hofmann 1980, 367*). Ingestion of a large piece of the root tuber (how much?) is said to cause hallucinations and delirium. The local diviners sometimes swallow pieces of the root in order to obtain prophetic visions. The plant contains bioactive alkaloids (Gundiza 1991; de Smet 1996, 143 f.*) and may contain methylamines (Emboden 1979, 184*).

Literature

Gundiza, M. 1991. Effect of methanol extract from *Monadenium lugardae* on contractile activity of guinea-pig ileum. *Central African Journal of Medicine* 37:141–44.

Monodora myristica Gaertner

(Curcubitaceae)—Jamaican nutmeg, calabash nutmeg tree

In West Africa, this treelike curcubit is known as *pebe*. Its seeds are reputedly used to establish contact with the water spirits (*mamiwata*). Apparently they are ingested and also smeared on the arms. The seeds are used by the Pygmies as a stimulant and to treat headaches. With a scent reminiscent of that of nutmeg, they are also used

Left: This four-o'clock (*Mirabilis* sp.) bears a strong resemblance to the nightshades and may be regarded as psychoactive for this reason. (Wild plant, photographed in southern Mexico)

Right: The treelike curcubitaceous *Monodora myristica*, which has a scent like that of nutmeg.

Catnip (*Nepeta cataria*).

as a substitute for the true nutmeg (*Myristica fragrans*), which may also represent a *pebe* for contacting the water spirits (Wagner 1991; cf. also Ott 1993, 413 f.*). The seeds contain an **essential oil** in which myristicin or safrole may be present; this would make them a useful psychoactive substance. African slaves introduced the plant into the Caribbean, where the seeds are used as a spice (Bärtels 1993, 69*).

Another curcubit species (*Echinocystis lobata* Torr. et Gray) is rumored to be psychoactive or even hallucinogenic (Schultes and Farnsworth 1982, 188*; Schultes and Hofmann 1980, 367*).

Literature

Wagner, Johanna. 1991. Das "dawa" der *mamiwata* (Ein möglicherweise pharmakologischer Aspekt des westafrikanischen Glaubens an Wassergeister.) *Integration* 1:61–63.

Mostuea spp.
(Loganiaceae)

In Gabon, the species *Mostuea gabonica* Baillon and *Mostuea stimulans* Chevalier, known locally as *sata mbwanda* or *sété mbwundè*, are regarded as potent aphrodisiacs. Their effects are said to be identical to those of *Tabernanthe iboga*. The roots of the plants, which have a taste like that of cola nuts (*Cola* spp.), produce euphoria and inebriation. The root was chewed extensively throughout the night and was swallowed by itself or mixed with iboga to produce sexual excitation (Chevalier 1946, 1947). Alkaloids are present in the genus *Mostuea*. The root cortex of *Mostuea stimulans* contains 0.33% alkaloids of an unknown structure. They have a pharmacological activity like sempervirine and gelsemine (cf. *Gelsemium sempervirens*) and effects similar to those of **strychnine** (de Smet 1996, 144*).

Literature

Chevalier, A. 1946. La *Sata mbwanda* racine stimulante et aphrodisiaque employée par les Noirs du Gabon et son identification botanique. *Comptes Rendus de l'Academie des Sciences* 223:767–69.

———. 1947. Les *Mostuea* africains et leurs propriétés stimulantes. *Revue de Botanique Appliquée* 27:104–9.

Neoraimondia arequipensis (Meyen) Backeb.
[syn. *Neoraimondia macrostibas* (Schum.) Britt. et Rose]
(Cactaceae)

This South American cactus from northern Peru has the reputation of being psychoactive or hal-lucinogenic (Schultes and Farnsworth 1982, 187*). It is one of the ingredients of the psychoactive **cimora** drink. It apparently contains β-**phenethylamines**.

Nepeta cataria L.
(Labiatae)—catnip

Cats seem to be magically attracted to this plant (and to its varietals), which is frequently grown as an ornamental, and they appear to feel a strong psychoactive effect—hence its name (Siegel 1991a*). The dried leaves can be smoked alone or in **smoking blends**. The extract can be sprayed onto other smoking herbs. A tea made from equal parts of catnip and damiana (**Turnera diffusa**) (add 2 tablespoons of each to ¹/₄ liter of water and allow to steep for five minutes) is said to have mild euphoriant effects (Schuldes 1995, 54*).

Catnip contains an aromatic **essential oil** composed of nepetalactones, dihydronepetalactone, and isodihydronepetalactone. It also contains the psychoactive alkaloid actinidine. There are many reports of the psychoactive efficacy of smoking catnip leaves, including some from sources that may be taken seriously (Ott 1993, 414 f.*; Schultes 1970, 42*).

Amazingly, the active substances in catnip (nepetalactones) are also found in the animal kingdom. They have been demonstrated to be present in the toxin of *Myrmacomecocystus*, a genus of ant from California. As part of their initiations, some California Indians swallowed these ants alive (wrapped in eagle down) to induce altered states of consciousness. The ants apparently bit into the stomach lining, thereby introducing the active principles into the blood. The ritual use of psychoactive ants was very similar to the use of *Datura wrightii* (Blackburn 1976*).

Literature

Jackson, B., and A. Reed. 1969. Catnip and the alteration of consciousness. *Journal of the American Medical Association* 207:1349–50.

Nephelium topengii (Merr.) H.S. Lo
(Sapindaceae)

This Southeast Asian tree may be identical to a plant that ancient Chinese sources refer to as *lung-li*, which was said to have hallucinogenic effects (Li 1978, 24 f.*). *Nephelium*, however, contains only toxic cyanogenic glycosides (Schultes and Hofmann 1995, 51*; Schultes and Farnsworth 1982, 187*). This botanical identity is questionable (Li 1978, 24*).

Ocimum micranthum Willd.

[syn. *Ocimum guatemalense* Gandoger]
(Lamiaceae, Labiatae)—small-flowered basil,
American basil

In Amazonia, where this plant is known as *alba-haca, iroro, pichana albaca,* or *pichana blanca,* it is said that this basil species is hallucinogenic (Duke and Vasquez 1994*). The leaves are used as an **ayahuasca** additive. The herbage has ethno-medicinal use as an analgesic in Mexico and Guatemala (Alcorn 1984, 715*; Ott 1993, 416*). The plant is known as *xkakaltun* in the Yucatán, where it is regarded as a **honey** plant (Barrera M. et al. 1976, 263*) and is used in an abortifacient medicine (Rätsch and Probst 1983). The Siona Indians call the aromatic plant *gõnõ ma'nya,* "chicha perfume," and it is similarly known as *kõnõ na'nya* among the Secoya (Vickers and Plowman 1984, 16*). It apparently was once used as a **chicha** additive. In Brazil, where the plant is known as *mangericão,* it is used in the Candomblé cult as an ingredient in the initiatory drink (see **madzoka medicine**). It has folk medicinal signi-ficance in the Caribbean. The plant contains an **essential oil** (Wong 1976, 137*) whose constitu-ents include camphene, cineol, linalool, myrcene, *cis-trans*-ocimene, α-pinene, β-pinene, α-terpineol, aromandrene, β-caryophyllene, β-elemene, Δ-ele-mene, γ-elemene, α-humulene, neriol, and euge-nol (Argueta V. et al. 1994, 89*; Maia et al. 1988).

Sacred basil (*Ocimum sanctum* L. [syn. *Ocimum tenuiflorum* L.]), a relative that is better known by the names *tulasi, tulsi,* and *madura-tala* (Knecht 1985), is not itself psychoactive,[334] although it is chewed as a substitute for **betel quids** (Macmillan 1991, 424*).

Literature

Knecht, Sigrid. 1985. Die heilige Heilpflanze Tulasi. In "Ethnobotanik," special issue, *Curare* 3/85:95–100.

Maia, J. G. S., et al. 1988. Uncommon Brazilian essential oils of the Labiatae and Compositae. *Dev. Food Science* 18:177–88.

Rätsch, Christian, and Heinz J. Probst. 1983. Kräuter zur Familienplanung. *Sexualmedizin* 12 (4): 173–76.

Osteophloeum platyspermum (DC.) Warburg

(Myristicaceae)—huapa

It has recently been discovered that the Ecuadoran Quecha use a tree that they call *anya huapa, huachig caspi, huapa, llauta caspi,* or *machin cara yura* ("monkey bark tree") as a hallucinogen. It is possible that this tree was already in use for this purpose in pre-Columbian times, for the infor-

mants explained that their ancestors used this plant to communicate with phantoms and spirits. The red sap from the trunk, which is ingested orally, must be boiled before use and is sometimes mixed with *guando* (**Brugmansia spp.**) and *tzicta* (*Tabernaemontana sananho* Ruíz et Pav.; see **Tabernaemontana spp.**). The Quechua drip some of the red sap into the nostrils of their dogs so that they are better able to hunt. A chemical quick test (Dragendorff test) has confirmed the presence of alkaloids (Bennett and Alarcón 1994). The Makú Indians drink the sap of the tree, which they call *tugnebānpe,* to treat colds (Prance 1972a, 20*). In the region around Manaus, the leaves are smoked as a treatment for asthma (Schultes 1978b, 230*; 1983b, 347*).

Literature

Bennett, B. C., and Rocío Alarcón. 1994. *Osteophloeum platyspermum* and *Virola duckei* (Myristicaceae): Newly reported as hallucinogens from Amazonian Ecuador. *Economic Botany* 48 (2): 152–58.

Oxytropis spp.

(Leguminosae: Fabaceae)—locoweeds

Several species from the genus *Oxytropis* are known in North America by the Spanish-English name *locoweed,* "crazy weed" (cf. **Astragalus spp.**), and have toxic or psychotropic properties (Turner and Szczawinski 1992, 122*). The Indians used some species for medicinal purposes (Johnston 1970, 314*). Several species were used as a ritual or medicinal wash during sweathouse ceremonies (Moerman 1986, 320 f.*). In Mexico, the species *Oxytropis lamberti* Pursh. is called *hierba loca,* "crazy herb" (Martínez 1987, 427*; Reko 1938, 185*).

Pancratium trianthum Herbert

(Amaryllidaceae)—pancrat lily, kwashi

This African sea lily, known as *kwashi,* is said to be a popular hallucinogen among the Bushmen of Botswana. Cut into slices, the bulb is rubbed into incisions made on the scalp (de Smet 1996, 142*).

Ocimum micranthum, a basil species found throughout tropical America, where it is used for ethnopharma-cological purposes.

334 This notwithstanding, in the Ayurvedic view *Ocimum sanctum* has the power to affect the mind: "Basil opens the heart and the mind and distributes the energy of love and devotion (*bhakti*). Basil is sacred to *Vishnu* and *Krishna* and strengthens faith, compassion, and clarity. *Tulsi* stalks are worn as garlands and strengthen the energy of attachment. Basil imparts divine protection by purifying the aura and invigorating the immune system. It contains natural mercury that, as the seed of Shiva, imparts the germinative power of pure consciousness" (Lad and Frawley 1987, 156*).

Right: The Mediterranean pancrat lily (*Pancratium maritimum*) is similar to the related African species and bears enchanting flowers.

Below: One of the many species of screw pine (*Pandanus* sp.) found throughout Southeast Asia.

Among the some fifteen species in the genus, this species is considered the most poisonous; it contains a variety of cardiac toxins (Schultes and Hofmann 1992, 52*). A Russian study isolated trisperidine, tacettin, hippeastrine, pancratine, galanthamine, lycorine, hordenine, and two unidentified bases from the bulb (Munvime and Muravjoza 1983). The main alkaloid in the bulb of the Mediterranean *Pancratium maritimum* L. is lycorine (Sener et al. 1993). Lycorine, which is present in many amaryllis species, causes paralysis of the central nervous system (Roth et al. 1994, 854*).

Literature

Munvime, F. D., and D. A. Muravjova. 1983. Alkaloids of *Pancratium trianthum* Herb. *Farmatsiya* 32:22–24.

Sener, B., S. Koenuekol, C. Krukl, and U. K. Pandit. 1993. Alkaloids of lycorine and lycorenine class from *Pancratium maritimum* L. *Archiv für Pharmazie* 326:61–62.

Pandanus sp.

(Pandanaceae)—screw tree

In Papua New Guinea, the fruit of an as yet unidentified *Pandanus* species is said to be used or to have been used as a hallucinogen. Unfortunately, we have no dependable ethnographic or ethnobotanical information about this. The fruits of several *Pandanus* species have been found to contain *N,N*-DMT (Schultes and Hofmann 1992, 52*). The species *Pandanus antaresensis* St. John is used in Papua New Guinea as an analgesic (Ott

1993, 401*). The Australian Aborigines make a **wine** from the fruits of *Pandanus spiralis* R. Br. (Bock 1994, 147*).

In Nepal, the screw pine (*Pandanus nepalensis* St. John [syn. *Pandanus furcatus* auct. non. Roxb.]) is considered sacred to Ganesha, the Hindu elephant-headed god. The leaves of the kevada or aromatic screw pine *Pandanus odoratissimus* L. [syn. *P. tectorius* auct. non. Soland. ex Parkinson], which is called *ketaka* in Sanskrit, are offered to his father Shiva (Majupuria and Joshi 1988, 170 f.*). In Ayurvedic medicine, the leaves are used as a tonic aphrodisiac, while in Thailand they are often used as a cooking spice (Norman 1991, 66*). In Hawaii, root tips that grow above the ground are used together with sugarcane juice to make a tonic (Krauss 1981, 6*). The flowers contain a stimulating **essential oil** composed of benzyl benzoate, benzyl acetate, benzyl alcohol, geranol, linalool, guiacol, phenethyl alcohol, and aldehydes (Majupuria and Joshi 1988, 171*). In India, the ripe spadix of *Pandanus tectorius* Parkins. ex Du Roi [syn. *Pandanus odoratissimus* L. f.] is the source of the so-called kewda perfume, one of whose uses is to aromatize smoking tobacco (**Nicotiana rustica**, **Nicotiana tabacum**) (Bärtels 1993, 122*).

In the Seychelles, a number of species known as *vacoa* are regarded as aphrodisiacs (Müller-Ebeling and Rätsch 1989, 72*).

Pedilanthus spp.

(Euphorbiaceae)—shoeflower

In Peru, *Pedilanthus retusus* Benth. is known locally as *misha*. This name is typically used as a folk taxonomic catchall for the various angel's trumpets (**Brugmansia** spp.) but is also occasionally given to other psychoactive plants (cf. **Lycopodium** spp., cimora). At the Institute for Traditional Medicine in Lima, it is claimed that *Pedilanthus retusus* is a potent hallucinogen, for it is said to contain a substance similar to **mescaline**.

Peperomia spp.

(Piperaceae)—tsemtsem, dwarf pepper

The Shuar of the Ecuadoran rain forest use this

Left: A South American peperomia (*Peperomia* sp.), also known as dwarf pepper.

Right: The master plant *piri-piri* (*Peperomia galioides*), which is used in the San Pedro cult.

Left: Peat myrtle (*Pernettya mucronata*).

Center: Found from Costa Rica to southern Chile, *Pernettya prostrata* is reputed to be a psychoactive plant.

Right: In South America, many species of the genus *Pernettya* were or still are used to prepare inebriating drinks. (Wild plant, photographed in Valdivia, southern Chile)

epiphytic, tropical pepper plant as a "mild hallucinogen." Parents given newborns who are just a few days old leaves that they have chewed. Older children are given the plant so that they may find their dream souls (*arutam*) (cf. ***Brugmansia suaveolens, Nicotiana tabacum***) (Bennett 1992, 492 f.*). The leaves apparently are also used as an **ayahuasca** additive.

Several *Peperomia* species contain alkaloids (Schultes and Raffauf 1990*). *Peperomia galioides* H.B.K., which is known in Peru as *piri-piri*,[335] is added to the San Pedro drink (cf. ***Trichocereus pachanoi***) to lend the psychoactive effects "more clarity, brightness, and distinctness" (Giese 1989, 252*).

In Trinidad, the dried leaves of *Peperomia emarginella* (Sw.) C. DC. are smoked for asthma. The **essential oil** has antispasmodic effects (Wong 1976, 114*).

Pernettya spp.

(Ericaceae)—peat myrtle

Several species of this South American member of the Heath Family are reputed to be psychoactive. The fruits of *macha-macha* (see **macha**), an Andean species (*Pernettya prostrata* [Cav.] Sleumer var. *pentlandii* [DC.] Sleumer) from Cochabamba (Bolivia), are said to cause dizziness when eaten in excess: "The fruit has a soporific property. A tame monkey who ate the berries of plants I had set aside to preserve became totally drunken" (Steinbach in Schultes 1967, 279; von Reis Altschul 1975, 215*) Some species and varieties are considered toxic (*Pernettya prostrata* var. *purpurea* [D. Don] Sleumer, *Pernettya mucronata* [L. f.] Gaudich. ex Spreng.). *Pernettya prostrata* (Cav.) DC. may be known as **macha** or *macha-macha*,

"drunk," in Quechua. This information, however, is questionable (Franquemont et al. 1990, 66*).

In Chile, *Pernettya furens* (Hook. ex DC.) Klotzch is known as *huedhued* or *hierba loca*, "crazy herb," and is said to cause mental confusion and possession (Schultes and Hofmann 1992, 53*). The fruits of *Pernettya parvifolia* Benth., known as *taglli*, are reputed to have toxic and hallucinogenic properties (Alvear 1971, 23*; Schultes and Farnsworth 1982, 179*). Andromedotoxins (= grayanotoxins) have been detected (Ott 1993, 417*). *Pernettya mucronata*, which is sometimes grown in Europe as an ornamental, also contains acetylandromedole (= andromedotoxin) (Roth et al. 1994, 549*). Sesquiterpenes have been demonstrated in *Pernettya furens* (Hosozawa et al. 1985).

Whether the fruits have psychoactive effects and were or are used culturally for psychoactive purposes is questionable. It is likely that the ripe fruits were used solely as a material for brewing **chicha**. Other species are used in Chile to prepare chicha (Mösbach 1992, 100*). In northern Peru, folk healers (*curanderos*) use a *Pernettya* species known as *toro-maique* as an inebriating additive to the San Pedro drink (cf. ***Trichocereus pachanoi***). The addition of this plant is said to give the drink "more power"; the spirit of the plant appears to the healer in the form of a bull (Giese 1989, 228*). In Venezuela, various species of the genus (perhaps including *P. prostrata*) are called *borracherita*, *borrachero, borrachera, borracherito*, or *chivacú* (Blohm 1962, 74*; von Reis and Lipp 1982, 228*). In South America, all plants with psychoactive or inebriating effects are typically subsumed under the name *borrachero*. For this reason, it is entirely possible that the Venezuelan peat myrtle exhibits some type of psychoactivity.

Probably the earliest depiction of an American peat myrtle (*Pernettya ciliata* Schlect. et Cham.). (From Hernández, 1942/46 [Orig. pub. 1615]*)

335 Not to be confused with the grass *Cyperus* spp., which is called *piripiri*.

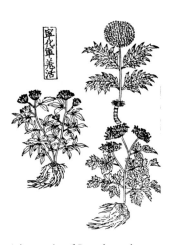

Asian species of *Peucedanum* have been used as aphrodisiacs and tonics in Chinese medicine since ancient times. (Illustration from the *Ch'ung-hsiu cheng-ho pen-ts'ao*)

Literature

Hosozawa, S., I. Miura, M. Kido, O. Munoz, and M. Castillo. 1985. Sesquiterpenes from *Pernettya furens*. *Phytochemistry* 24 (10): 2317–23.

Schultes, Richard Evans. 1967. De Plantis Toxicariis e Mundo Novo Tropicale Commentationes I. *Botanical Museum Leaflets* 21 (9): 265–84.

Persea indica (L.) Spreng.

(Lauraceae)—viñatigo

This tree, a relative of the avocado (*Persea americana* Mill.), belongs to the indigenous flora of the Canary Islands, where it is known as *viñatigo*. It is said that a person should not take a siesta or sleep under the tree because doing so will provoke a state of inebriation. Goats eat the foliage and branches with gusto, because it makes them "drunken." A psychoactive use has not yet been determined (vries 1993).

Literature

vries, herman de. 1993. Über die Wirkungen von *Persea indica* (L.) Spreng. *Integration* 4:57.

Petunia violacea Lindl.

(Solanaceae)—petunia, shanín

It is said that the Indians of the Ecuadoran highlands smoke this plant, known as *shanín*, the effects of which resemble those of **Coriaria thymifolia**. The plant is said to give its user "the feeling of rising into the air or floating weightlessly away" (Alvear 1971, 23*; Schultes and Hofmann 1992, 53*). Chemical studies to date have not detected the presence of any alkaloids (Butler at al. 1981; Ott 1993, 417*). It is possible, however, that new **diterpenes** may be present, as they have been found in a relative, *Petunia patagonica* (Speg.) Millan (Guerreiro et al. 1984). Ketones have also been found in the genus (Elliger et al. 1990).

Literature

Butler, Edward Grant, Trevor Robinson, and Richard Evans Schultes. 1981. *Petunia violacea*: Hallucinogen or not? *Journal of Ethnopharmacology* 4 (1): 111–14.

The violet-flowered petunia (*Petunia violacea*) is a common ornamental.

Elliger, Carl A., Anthony C. Waiss Jr., Marby Benson, and Rosalind Y. Wong. 1990. Ergostanoids from *Petunia parodii*. *Phytochemistry* 29 (9): 2853–63.

Guerreiro, Eduardo, J. de Fernandez, and O. S. Giordano. 1984. Beyerene derivatives and other constituents from *Petunia patagonica*. *Phytochemistry* 23 (12): 2871–73.

Peucedanum japonicum Thunb.

(Umbelliferae)—fang-k'uei

This Japanese species of master plant is also found in China. Known as *fang-k'uei*, the species is mentioned in the herbal *Tao Hung-ching*: "feverish persons should not take it, for it stupefies and enables spirits to appear." This passage has been interpreted as indicating a possible psychoactivity (Li 1978, 21*).

The roots (radix peucedani, qian hu, zenko) of several closely related species are used in traditional Chinese medicine to treat disorders of the lung and spleen (Paulus and Ding 1987, 376). The root of *Peucedanum decursivum* (Miq.) Maxim. is regarded as a nerve tonic and aphrodisiac (Stark 1984, 95*). The roots of the species known as *zenko* (*Peucedanum praeruptorum* Dunn) are used in Japanese kampo medicine to treat fever, shivering, and headaches (Tsumura 1991, 175*).

Alkaloid-like substances do occur in the genus *Peucedanum*. **Coumarins** have been detected in *Peucedanum japonicum* (Schultes and Hofmann 1992, 54*). Additional chemical studies are needed (Ott 1993, 417*).

Philodendron scandens K. Koch et Sello

(Araceae)—heartleaf philodendron

A note attached to an herbarium specimen collected in Peru in 1969 reads, "Narcotic, used to induce sleep" (von Reis and Lipp 1982, 20*). The principal agent of the allergologically efficacious plant is thought to be 5-heptadecatrin-8(Z),11(Z),14(Z)-benzylresorcinol (Roth et al. 1994, 559*). A related *Montrichardia* species is used as an **ayahuasca** additive.

Physalis spp.

(Solanaceae)—*Physalis* species

The genus *Physalis* is composed of some 120 species and is thus the largest genus of its family (Lu 1986, 80*). Several species are regarded as toxic, some are raised as ornamentals popular for their unusual flowers (Chinese lanterns), and others have ethnomedical significance. *Physalis pubescens* L. and *Physalis peruviana* L. (Cape gooseberry) are the two species most commonly grown for their fruit. Very mild toxic effects have been observed following consumption of a large number of berries of *Physalis peruviana* (Roth et al. 1994, 560*). The calyx, which surrounds the seeds like a lantern, can be smoked. It has definite psychoactive effects that tend to be narcotic in nature.

Physalis angulata L., a species from the northwestern Amazon, is said to be mildly narcotic. Its juice finds use in Brazilian folk medicine as a treatment for earaches (Schultes and Raffauf 1991, 43*). *Physalis minima*, a species from the Caroline Islands known as *poowa*, bears fruits that are said to have an inebriating effect when consumed in excess (von Reis Altschul 1975, 269*).

The roots of some species of the genus have yielded **tropane alkaloids** as well as alkaloids of the hygrine type (von Reis Altschul 1975, 269*). The Jew's cherry or lampion flower, *Physalis alkekengi* L. (cf. **halicacabon**), contains the mildly toxic bitter principles physaline A, B, and C (Roth et al. 1994, 560*).

Pithecellobium spp.

[sometimes written as *Pithecolobium*]
(Leguminosae)—jurema branca

The genus *Pithecellobium* encompasses some two hundred species. It is closely related to and is easily confused with the genus *Mimosa*. In the Brazilian Amazon, the name *jurema* is usually given to **Mimosa tenuiflora** [syn. *Mimosa hostilis*], which is used to prepare an **ayahuasca**-like drink. However, several species of the genus *Pithecellobium*, e.g., *P. diversifolium* (known as *jurema branca*, "white jurema"), also appear to be used for that purpose (Rätsch 1988, 83*).

In Brazil, the use of a *vinho do jurema* made from *Pithecellobium* has become established among the followers of various cults of West African origin. Apparently the use is connected with the Camdomblé god Ossain, who is regarded as a great magician, protector, and discoverer of healing herbs. Psychoactive and other constituents of the genus are largely unknown. Alkaloids and flavonoids have been detected (Schultes and Raffauf 1990, 251*).

The species *Pithecellobium laetum* Benth., which is known as *remo caspi*, *pashaquillo*, or *shimbillo*, contains alkaloids and is used as an **ayahuasca** additive. In Mexico, the species *Pithecellobium arboreum* (L.) Urb. and *P. donnell-smithii* Britt. et Rose are known as *frijolillo* (Martínez 1987, 1189 f.*). This name is also used for *Sophora secundiflora*.

Polypodium sp.

(Polypodiaceae)—polypody, tree fern

The Spanish botanist Hipólito Ruíz (1754–1816), who traveled through South America in the eighteenth century, provided the first descriptions of many plants and returned with extensive collections of botanical materials. In his *Relación*, he reported a number of experiences and provided ethnobotanical notes from Peru and Chile (Schultes 1980, 89*). Ruíz described a polypody known as *cucacuca*, *incapcocam*, or *coca del Inca* under the binomial *Polypodium incapcocam* (nomen nudum), whose botanical identity even Richard E. Schultes was unable to determine. Ruíz noted that the Indians had informed him that the Incas used the leaves of this plant instead of coca (see **Erythroxylum coca**). Moreover, the powdered plant was said to be used (how?) in place of tobacco (**Nicotiana** spp.) in order to "clear the head" (Schultes 1980, 89*).

Whether *Polypodium* does indeed have psychoactive effects remains to be seen, although it is possible. Interestingly, the root of one species of the genus *Polypodium* whose identity has not been specified is ingested orally together with the seeds of **Anadenanthera colubrina**. It may contain MAO-inhibiting β-**carbolines** (or other substances with the same effects). The root of the common polypody (*Polypodium vulgare* L.), also known as sweet angel, contains small amounts of an essential oil, tannin, bitter principles, and sweet-tasting saponins (Pahlow 1993, 119*).

It is quite possible that there are other psychoactive ferns; even German folklore contains tales of *filices*, "ferns" (e.g., the worm fern, *Aspidium filix-mas*)—also called *irrwurz* ("confused root")—that have magical powers, make the devil appear, and can make one invisible (Marzell 1964, 33 ff.*; Schöpf 1986, 84 f.*). There are also ferns that were added to **beer**. In Mexico, a fern known

This may be one of the earliest illustrations of an American species of *Physalis*. (From Hernández, 1942/46 [Orig. pub. 1615]*)

Left above: The papery bladder that surrounds the *Physalis* fruit is reminiscent of a Chinese lantern. When smoked, it produces mild narcotic effects.

Left below: The South American leguminous tree *Pithecellobium discolor*.

Below: A polypody (*Polypodium* sp.) growing on the trunk of a tree.

The fern is one of the Germanic magical plants. Whether it is in fact psychoactive is unknown, but possible. (Woodcut from Tabernaemontana, *Neu Vollkommen Kräuter-Buch*, 1731)

as *itamo real*[336] (*Pellaea cordata* J. Sm.) is said to have inebriating effects (Díaz 1979, 93*).

Pontederia cordata L.

(Pontederiaceae)

In the Colombian region of the Amazon, this plant is known as *amarrón borrachero* ("hazelwort inebriator"). It is used as an **ayahuasca** additive but may also possibly be used alone for psychoactive purposes (Schultes 1972, 141*). The plant is used ethnomedicinally to treat facial paralysis (Schultes 1981, 5*). There is also a North American variety: *Pontederia cordata* L. var. *lancifolia*.

Pseuderanthemum sp.

(Acanthaceae)—dormidero

A note on an herbarium specimen collected by Killip and Smith in Peru in 1929 describes this as a "narcotic plant" (von Reis and Lipp 1982, 281*). The Spanish name *dormidero* means "sleep inducer."

Quararibea spp.

(Bombacaceae)

The genus *Quararibea* consists of some twenty-nine species (Schultes 1957, 249). In Mexico, the aromatic leaves ("cacao flowers") of the small tree *Quararibea funebris* (La Llave) Vischer [syn. *Lexarza funebris* La Llave, *Myrodia funebris* Benth.] are used as a spice for cacao drinks (see ***Theobroma cacao***) (Rosengarten 1977; Schultes 1957). They have been described as a possible hallucinogen and also have been identified as the mys-

terious Aztec inebriant **poyomatli**. According to Jonathan Ott, who has experimented with these flowers as well as cacao preparations containing them, *Quararibea* flowers are not psychoactive (Ott 1993, 418*). However, several interesting substances are present (γ-butyrolactones, alkaloids) that may indeed have psychotropic effects (Raffauf and Zennie 1983).

The Peruvian inebriant **espingo**, which has not been clearly identified, has also been interpreted as the fruit of a *Quararibea* species. In Amazonia, *Quararibea* species are used as **ayahuasca** additives, and they are also added to Peruvian San Pedro preparations (see ***Trichocereus pachanoi***). The Kofán Indians use *Quararibea putumayensis* Cuatr. in the manufacture of arrow poisons (Ott 1993, 418*).

Literature

Raffauf, Robert F., and Thomas M. Zennie. 1983. The phytochemistry of *Quararibea funebris*. *Botanical Museum Leaflets* 29 (2): 151–58.

Rosengarten, Frederic, Jr. 1977. An unusual spice from Oaxaca: The flowers of *Quararibea funebris*. *Botanical Museum Leaflets* 25 (7): 183–202.

Schultes, Richard Evans. 1957. The genus *Quararibea* in Mexico and the use of its flowers as a spice for chocolate. *Botanical Museum Leaflets* 17 (9): 247–64.

Ranunculus sp.

(Ranunculaceae)—buttercup

In ancient China, a species of the genus *Ranunculus* (perhaps *Ranunculus acris* L. var. *japonicum* Maxim.) was known by the names *shui lang* and *maoken*. Unintentional consumption was said to result in a kind of delirium (Li 1978, 24*). The genus contains glycosides (protoanemonin) (Frohne and Pfänder 1983, 173*; Schultes and Hofmann 1995, 54*). Whether there really is a psychoactive species is questionable.

Left: The cacao flower tree *Quararibea funebris*, a native of Oaxaca (Mexico).

Right: The buttercup (*Ranunculus acris*), a Ranunculaceae, is associated with an ancient Chinese magical drug.

336 This name is also used to refer to puffballs (see ***Lycoperdon*** spp.) and to ***Ephedra*** spp.

Rauvolfia serpentina (L.) Bentham ex Kurz

[formerly spelled *Rauwolfia*]
(Apocynaceae)—snakeroot

Rauvolfia, or snakeroot, is occasionally regarded as a psychoactive plant. This is due primarily to theoretical considerations concerning the yohimbane-type alkaloids, here represented by corynanthine, isorauhimbine, and **yohimbine** (Kähler 1970). The principal active agent, however, is the alkaloid reserpine; its primary effect is hypotensive, although it also has sedative properties. *Rauvolfia* thus induces sleep (Hänsel and Henkler 1994, 369). Reserpine appears to work in a manner similar to the neuroleptica and played a significant role in the study of the function of the mono-amine transmitter in the nervous system (D. McKenna 1995, 103*). It is conceivable that certain as yet unknown methods of preparation could yield psychoactive effects. In addition, the some sixty species in the genus may very well include some that contain much higher concentrations of yohimbine and induce very different effects. Apart from *Rauvolfia serpentina*, the pharmacologically most important species are the African *Rauvolfia vomitoria* Afzel and the American *Rauvolfia tetraphylla* L. [syn. *Rauvolfia canescens* L., *R. hirsuta* Jacq., *R. heterophylla* Roem. and Schult.], which is also sometimes referred to as *borrachero* ("inebriator"; cf. **Brugmansia**) (Morton 1977, 243–57*). Most of the species are tropical and are found in both the Old and the New Worlds. Many species have ethnomedicinal significance. β-yohimbine has been detected in *Rauvolfia vomitoria* (Hofmann 1955; Stoll et al. 1955). In India, *Rauvolfia serpentina* has a long history as an antidote for snakebite (Jain 1991, 153*). Circumcised boys in Kenya use *Rauvolfia caffra* Sond. [syn. *R. natalensis* Sond.], known locally as *mwerere, rerendet, omomure,* or *mutu,* as a tea for inducing sleep. The stalks are used as a fermentation agent for making **beer** (Omino and Kokwaro 1993, 173*).

Literature (selection)

Hänsel, Rudolf, and Günter Henkler. 1994. *Rauvolfia.* In *Hagers Handbuch der pharmazeutischen Praxis,* 5th ed., 6:361–84. Berlin: Springer.

Hofmann, Albert. 1955. β-Yohimbin aus den Wurzeln von *Rauvolfia canescens* L. *Helvetica Chimica Acta* 38:536 ff.

Kähler, Hans Joachim, and coworkers. 1970. *Rauwolfia Alkaloide: Eine historische, pharmakologische und klinische Studie.* Mannheim: Boehringer.

Stoll, Arthur, Albert Hofmann, and R. Brunner. 1955. Alkaloide aus den Blättern von *Rauvolfia canescens* L. *Helvetica Chimica Acta* 38:270ff.

Rhododendron caucasicum Pallas and *Rhododendron* spp.

(Ericaceae)—Caucasus alpine rose, rhododendrons
The Osset live in the mountains of the northern Caucasus and are thought to be later descendants of the ancient Scythians. The Oriental scholar Julius Klaproth visited the Osset during the nineteenth century and returned with a description of a divination ritual in which the Caucasus rhododendron (whether the botanical

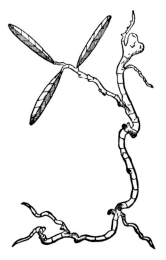

This illustration may be the earliest European representation of the Indian snakeroot (*Rauvolfia serpentina*). (Woodcut from Garcia da Orta, *Colloquies on the Simples and Drugs of India,* 1987 [orig. pub. 1563])

From top to bottom:
The leaves of the Caucasus rhododendron (*Rhododendron caucasicum*) have a resinous aroma and were once inhaled as an inebriating incense.

The aromatic high mountain rhododendron (*Rhododendron lepidotum*) is an important source of incense in the Himalayas. (Wild plant, photographed in Langtang, Nepal)

The yellow-blossomed *Rhododendron cinnabarinum* is found in the Himalayas. Burned as an incense, this plant has potent effects upon yaks.

The flowers of the Pontic alpine rose (*Rhododendron ponticum*) contain a nectar that bees transform into a psychoactive honey. (Photographed near the Aletsch glacier, Switzerland)

identification is correct remains open) was used as a psychoactive incense (Klaproth 1823, 2:223 f.):

> He described their ardent devotion to the prophet Elias, who was regarded as their greatest protector. In caves consecrated to him, they [the Osset] offered goats and consumed their flesh; after which they spread the skins out under a large tree and honored these in a special fashion on the prophet's feast day so that he would keep away the hail and grant them a bountiful harvest. The Osset would often go to these caves to inebriate themselves on the smoke of *Rhododendron caucasicum,* which would cause them to sleep deeply. The dreams that appeared to them under these circumstances were interpreted as prophecies. (Ginzburg 1990, 165)

The Caucasus rhododendron (section Pontica) is a broad bush that grows to only about 1 meter in height. The flowers are creamy or pale yellow, sometimes with pink spots. The plant typically blooms from April to May and is found primarily at an altitude between 1,800 and 2,700 meters. It is found across northeastern Turkey and the Caucasus Mountains (Cox 1985, 175). Its evergreen leaves are weakly aromatic. *Rhododendron caucasicum* is only rarely encountered in rhododendron gardens, for it is much more difficult to cultivate than are other species.

In Nepal, the closely related *Rhododendron lepidotum* Wall. ex Donn (in two forms: var. album Davidian and var. minutiforme Davidian; cf. Cox 1985, 113 f.) is still used today as a ritual and shamanic incense, the effects of which are very subtle (see **incense**). In Tibet and China, other rhododendron species are also used as incense. The yellow-flowered *Rhododendron cinnabarinum* Hook f. is found in the high mountains of Sikkim. Its smoke is said to have a profound effect on yaks, producing a strong inebriation and altering their behavior. It is possible that it also has psychoactive effects upon humans.

In Nepal, the leaves of a *Rhododendron* species are mixed with tobacco (**Nicotiana tabacum**) for smoking. A **snuff** is made from the bark of a *Rhododendron* species and tobacco leaves (Hartwich 1911, 108*).

Other rhododendron species, e.g., the rusty-leaved alpine rose (*Rhododendron ferrugineum* L.) and the Pontic rhododendron (*Rhododendron ponticum* L.), yield a psychoactive/toxic **honey**. The Tartars made a tea from the leaves (ten or more) of the gold-yellow alpine rose (*Rhododendron chrysanthum* Pall. [syn. *Rhododendron officinale* Salisb., *Rhododendron aureum* Georg]) that is said to have produced a state of inebriation (Roth et al 1994, 612*). There is also a cultivar, *Rhododendron* x *sochadzeae*, resulting from a cross

between *R. ponticum* and *R. caucasicum* (Cox 1985, 175 f.). This rare ornamental variety may have potent psychoactive effects.

The aromatic species of rhododendron contain relatively high concentrations of **essential oil.** Mongolian species contain primarily limonene, aromadendrene, caryophyllene, Δ-candinene, β-selinene, and gurjunene (Satar 1985).

It is would be an interesting task to investigate a possible cultural link between rhododendron forests and psilocybin mushrooms. Rhododendron groves are a preferred habitiat of some psychoactive mushrooms, e.g., *Psilocybe cyanescens.*

Literature

Cox, Peter A. 1985. *The smaller rhododendrons.* Portland, Ore.: Timber Press.

Ginzburg, Carlo. 1990. *Hexensabbat.* Berlin: Wagenbach.

Klaproth, Julius. 1823. *Voyage au Mont Caucase et en Géorgie.* 2 vols. Paris.

Satar, S. 1985. Analyse der ätherischen Öle aus drei Rhododendron-Arten der Mongolischen Volksrepublik. *Pharmazie* 40 (6): 432.

Sanango racemosum (Ruíz et Pav.) Barringer

[syn. *Gomara racemosa* Ruíz et Pav., *Gomaranthus racemosus* (R. et P.) Rauschert, *Sanango durum* Bunting et Duke]
(Loganiaceae)—sanango

This rare and mysterious bush is related to ***Desfontainia spinosa.*** It is found in the rain forests of Amazonia at altitudes up to 100 meters (Brako and Zarucchi 1993, 619*) and is endemic to Peru (Gentry 1993, 564*). The leaf is used as an inebriant. Reports circulate of its wondrous effects. However, there have been few ethnopharmacological investigations into this promising plant.

The leaves of the mysterious Amazonian magical plant *Sanango racemosum.*

Santalum murrayanum (Mitch.) Gardner

(Santalaceae)—bitter quandong

This Australian tree is a relative of sandalwood (*Santalum album* L.; cf. **incense**). The Aborigines of Lake Boga use the bark as a narcotic. From it, they produce an inebriating drink known as *cootha* (Bock 1994, 108*). The bark of the trunk contains 0.21% alkaloids, which have strong toxic effects above a certain dosage (2g/kg) (Collins et al. 1990, 65, 128*).

The leaves and wood of *gumamu*, the closely related species *Santalum lanceolatum* R. Br., were used by the Bardi during healing rituals as a medicinal **incense** (Lands 1987, 17). It is said that this treatment was "too strong" for children; this incense may be psychoactive. Alkaloids have been found in the leaves, trunk wood, and bark (Collins et al. 1990, 65*).

Literature

Lands, Merrilee, ed. 1987. *Mayi: Some bush fruits of Dampierland*. Broome, Australia: Magabala Books.

Scirpus spp.

(Cyperaceae)—bakana, simse

In Northern Mexico, the Tarahumara use a species from the genus *Scirpus* as a hallucinogen. They call the plant *bakánoa*, *bakánawa*, *bakánowa*, or *bakana*. The ethnobotanist Robert Bye has stated that this grass is the most important hallucinogen of the central and western Tarahumara (= Rarámuri), being even more important than peyote (*Lophophora williamsii*) (Bye 1979b, 35*). Little is known about the ritual use:

> Bakánowa is another medicinal plant used in rituals. A ceremony known as simse is associated with and named after the plant simse, bot. Scirpus sp. It is regarded as a source of vigor and is ritually venerated, especially by older women and men, who nourish it with offerings. Bakánowa is a kind of counterpart to híkuri [= peyote]. The plant is sought for in the western Sierra Tarahumara. The ceremonial circle with the offering altar also faces to the west, while the ritual semantics depict the híkuri to the east. The bakánowa root is clearly a potent drug that is not ingested in most cases but [is] merely ritually venerated. Here, some healers use a notched piece of wood, as in the híkuri rites. (Deimel 1996, 12)

Nourishing the plant with offerings is considered important for health. One Tarahumaran healer said, "If god onorúame, the goddess maria mechaka, or the dead or the sacred plants híkuri and bakánowa go hungry, humans will become ill" (Deimel 1996, 12).

The root is used in folk medicine as an analgesic and to treat the insane. The plant is regarded as a protective amulet and as a remedy for all mental illnesses. This is why it is periodically brought offerings. Anyone who treats the plant poorly will be punished with disease. Eating the root tuber is said to induce a deep sleep accompanied by visions and allows one to travel to other dimensions. Unfortunately, the species the Tarahumara use has not yet been identified.

Alkaloids have been found in one species of the genus *Scirpus* (Bye 1979b, 36*). These may be **ergot alkaloids** (cf. *Cyperus* spp.) that are deposited as metabolites of a parasitic fungus.

In South America, *Scirpus* species have been used since pre-Columbian times to produce mats and other woven goods, including some intended for ritual use (Towle 1952, 232 f.).

Literature

Deimel, Claus. 1996. *Híkuri ba—Peyoteriten der Tarahumara*. Ansichten der Ethnologie 1. Hannover: Niedersächsisches Landessmuseum.

Towle, Margaret Ashley. 1952. Plant remains from a Peruvian mummy bundle. *Botanical Museum Leaflets* 15 (9): 223–46.

Sclerocarya caffra Sond.

(Anacardiaceae)—marula

This tree, a relative of the cashew (*Anacardium occidentale* L.), can grow as tall as 18 meters. It supposedly is or was consumed in South Africa for its inebriating properties (Lewin 1980 [orig. pub. 1929], 297*). It is possible that *marula* was the **kanna** plant or one of the plants known as kanna (Schultes and Farnsworth 1982, 174*). The closely related species *Sclerocarya schweinfurthiana* Schinz. has also been characterized as psychoactive (Lewin 1980, 297*; Schultes and Farnsworth 1982, 187*). Constituents are unknown (Emboden 1979, 191*).

Scoparia dulcis L.

(Scrophulariaceae)—vacourinha, sweet broom

In Amazonia, this plant is referred to as *bati matoshi* or *piqui pichana*, and its dried leaves are smoked as a marijuana substitute (see **Cannabis indica**) (Duke and Vasquez 1994, 154*). In Brazil, the herbage finds folk medicinal use as an astringent and antispasmodic (Grieve 1982, 427*). The plant is known in central Africa as *osimmiseng*, and its leaves are used in a magical medicine for spells with "worms" (Akendengué 1992, 170*). Members of the Bastar tribe (India) make the leaves into pills that are swallowed to treat "weakness of the semen" (Jain 1965, 244*). The plant contains labdane (**diterpenes**).

Securidaca longepedunculata Fresenius
(Polygalaceae)

A native of western and tropical southern Africa, this tree is used in many cultures as a poison in trials by ordeal and in ordeals for uncovering witchcraft (Neuwinger 1994, 682 ff.*). In West Africa, it is used together with *Boophane disticha* for psychoactive purposes. This species is also venerated as a fetish tree in West Africa and is used to provide magical protection from the "evil eye" and the illnesses of the deceased. The plant is one of the most renowned and legendary medicinal plants and abortifacients in Africa. The Nigerian Haussa call it *uwar magunguna*, "mother of medicine."

Among the Kusase, who live in the extreme northeast of Ghana, the plant is used as a psychoactive substance when a new *baga* ("diviner") is being initiated. A **snuff** is made of *pelig* roots (*Securidaca longepedunculata*), *datin-vulin* roots (*Ipomoea mauritiana* Jacq. [syn. *Ipomoea digitata* auct. non. L., *Ipomoea paniculata* (L.) R. Br.]; cf. ***Ipomoea* spp.**) or *bailla/punung-buur* roots (*Tinospora bakis*), the root cortex of the *zurmuri* pepper (*Piper guineense* Schumach. et Thonn., also known as ashanti pepper; cf. ***Piper* spp.**), red *nansus* pepper (*Schinus molle* L.; cf. **chicha**), and the dried head of a bat. The snuff is blown into the initiate's nose, whereupon he falls into a trancelike state. The Ngindo of Tanzania use a flour made from the root as a snuff for headaches (Neuwinger 1994, 685*).

In Ethiopia, smoke from the root is inhaled as a medicinal **incense** to treat flatulence. In Gambia, "crazy" people burn the root cortex and inhale the smoke (Neuwinger 1994, 684*).

The root contains methylsalicylate, the bark contains the alkaloid securine, and the leaves have yielded tannins, saponins, terpenes, et cetera (Lenz 1913). Securine has stimulating effects upon the central nervous system and can produce effects like those of **strychnine** (Neuwinger 1994, 686). The root contains various **indole alkaloids**, particularly the psychoactive elymoclavine, from the family of the **ergot alkaloids** (Costa et al. 1992).

Literature

Costa, C., A. Bertazzo, G. Allegri, O. Curcuruto, and P. Traloli. 1992. Indole alkaloids from the roots of an African plant, *Securidaca longepedunculata*. *Journal of Heterocycl. Chem.* 29:1641–47.

Lenz, W. 1913. Untersuchungen der Wurzelrinde von *Securidaca longepedunculata*. *Arbeiten aus dem Pharm. Inst. d. Univ. Berlin* 10:177–80.

Senecio spp.
(Compositae)

Many species of this genus, which encompasses some 1,300 species and is found throughout the world, are said to be psychoactive (Schultes and Farnsworth 1982, 188*; Schultes and Hofmann 1992, 56*) or are at least associated with psychoactive plants or preparations (see ***Lophophora williamsii, Trichocereus pachanoi***). The Mexican species *Senecio cardiophyllus* Hemsl. is even referred to as *peyote* (Martínez 1994, 384*). Many *Senecio* species are used in South America as ritual **incense** (Aldunate et al. 1981*). In the Andes regions, they are known as *cundur-cundur* and appear to be mythologically associated with the condor, an animal sacred to the Indians. A *Senecio* species known as *chula-chula* is chewed together with coca (see ***Erythroxylum coca***). Many *Senecio* species contain alkaloids of the pyrrolizidine type (Röder and Wiedenfeld 1977; Schultes and Hofmann 1992, 56*). Cyanoglycosides have also been found (Schultes 1981, 43*). The alkaloid jacobine, together with other pyrrolizidines, passes into the **honey** that is produced from these plants (Frohne and Pfänder 1983, 66*).

In Nepal, various yellow-blooming crucifers are used as ritual offerings. A psychoactive use is unknown.

Literature

Röder, Erhard, and Helmut Wiedenfeld. 1977. Isolierung und Strukturaufklärung des Alkaloids Fuchsisenesionin aus *Senecio fuchsii*. *Phytochemistry* 16:1462–63.

A number of *Senecio* species are used in Mexico and South America to prepare psychoactive drinks and as ritual incense. (A *Senecio* species from Argentina)

Sida acuta Burm. f.
(Malvaceae)

This prostrate or sometimes bushy plant with yellow flowers is found in the tropical zones of Central and South America. The Cuna Indians (Darien, Panama) esteem the plant, which they call *kwala*, as a "mystical medicine" (Duke 1975, 292*). A tea made from the leaves is consumed in Bangladesh to help induce sleep (Ott 1993, 419*).

Among the Maya, this species is known as *chichibeh* ("the little one on the path"). On the Mexican Gulf Coast, its leaves, like those of the closely related *Sida rhombifolia* L., are smoked as a marijuana substitute (see *Cannabis indica*). The two species are known locally as *el macho*, "the male" (*S. rhombifolia*), and *la hembra*, "the female" (*S. acuta*) (Schultes and Hofmann 1980, 347*). Both of these *Sida* species apparently contain **ephedrine** (Schultes and Hofmann 1992, 56*). When drying, the herbage of *Sida acuta* exudes a distinct aroma of **coumarin**. The leaves are said to contain saponins. Several studies of Caribbean and Philippine specimens of *Sida acuta* have demonstrated the presence of asparagine and **ephedrine** in the roots (Wong 1976, 132*). The alkaloids choline, pseudoephedrine, β-**phenethylamine**, vasicine, vascicine, vasinole, and vasicinone have been found in all parts of *Sida rhombifolia*. The stem contains the **indole alkaloids** hipaphorine, hipaphorine methylester, and cryptolenine. The leaves contain traces of an **essential oil**. The seeds contain sesquiterpenes (gosipol, et cetera) (Argueta et al. 1994, 615*).

Sloanea laurifolia
(Elaeocarpaceae)—taque

An herbarium specimen collected by J. A. Steyermark in Venezuela in 1945 bears the note "Fresh fruit are said to produce a kind of *loco* [= crazy] feeling.... [T]he ground fruit is boiled" (von Reis and Lipp 1982, 175*). The fruit is also known as *arepa de maiz*, "maize breads" (cf. *Zea mays*).

Spiraea caespitosa Nutt. ex Torr. et A. Gray
[syn. *Petrophyton caespitosum* (Nutt. ex Torr. et Gray) Rydb., *Spirea caespitosum*]
(Rosaceae)—meadowsweet

The North American Navajo-Kayenta Indians are said to have used this bushlike plant as a narcotic (Moerman 1986, 466*). However, it contains only salicylic acid and should therefore be effective solely as an analgesic (Ott 1993, 420*; Grieve 1982, 525*).

Stephanomeria pauciflora (Torr.) A. Nels.
(Asteraceae)—blue gum

The root of this plant is said to have been used by the Navajo-Kayenta Indians as a narcotic (Moerman 1986, 469*). It is questionable whether this information is correct, for the root is also used as a kind of chewing gum (Vestal 1952, 53*). The chemistry of the plant is unknown (Ott 1993, 420*). It is, however, regarded as a "life medicine." Perhaps it is a psychoactive chewing gum.

Stipa spp.
(Gramineae)—sleepy grass

Various species of the genus *Stipa* are found from Texas to Guatemala. In the White Mountains of the region of the Rio Grande, *Stipa vaseyi* Scribn. [syn. *Stipa robusta* (Vasey) Scribn.] is known by the name *popoton sacaton*, which is probably of Aztec origin (Emboden 1979, 191*). This grass is said to have inebriating effects and in Guatemala supposedly is used as a sleeping agent. A related

Top left: *Sida acuta* herbage is thought to contain ephedrine. (Wild plant photographed in Belize)

Bottom left: The herbage of *Sida rhombifolia* is smoked as a marijuana substitute. (Wild plant photographed in northwestern Argentina)

Above: Meadowsweet (*Spirea* spp.) contains analgesic salicylic acid derivatives.

Left: Several American species from the genus *Stipa* are known by the name *sleeping grass* and consumed as sedative teas. (*Stipa ichu* from Chiapas, Mexico)

The seeds of bellerian myrobalan (*Terminalia bellirica*) purportedly are used as a hallucinogen in West Bengal. (Woodcut from Tabernaemontanus, *Neu Vollkommen Kräuter-Buch*, 1731)

species, *Stipa viridula*, is purported to have a narcotic effect (Emboden 1976, 161*). It has recently been claimed that another species native to the American Southwest, *Stipa robusta*, exhibits strong psychoactive effects. This sleepy grass lives in a symbiotic relationship with a fungus (*Acremonium*) that is thought to produce the **ergot alkaloid** D-lysergic acid amide (cf. *Turbina corymbosa*) in the seeds. It has been claimed that a dosage of nine seeds will produce LSD-like effects (DeKorne 1995, 127*). Ethnographic evidence for any psychoactive use is lacking.

Teliostachys lanceolata Nees var. *crispa* Nees ex Martius

(Acanthaceae)

This Acanthaceae, known as *toé negro* (cf. *Brugmansia suaveolens*), is used by the Colombian Kokama Indians both as an **ayahuasca** additive and as a psychoactive substance in its own right. For this purpose, ten leaves are boiled for seven hours over a low flame. The effects are said to be strong. A person loses his or her vision for three days, but during this time he or she can communicate with the spirit of the plant (Schultes 1972, 139*). Chemical studies of the plant have demonstrated that the leaves are devoid of alkaloids (Ott 1993, 402*). The plant is added to ayahuasca only when it is intended for use in "witchcraft" (Duke and Vasquez 1994, 167*).

Terminalia bellirica (Gaertner) Roxburgh

(Combretaceae)—bahera, bellerian myrobalan

The Lodha of West Bengal eat the dried seeds of this plant in order to induce hallucinations. In Southeast Asia, the seeds are known for their narcotic properties. They are used in traditional Chinese medicine as an anthelmintic agent, in Kerala to treat asthma, and in Nepal as a laxative (Ott 1993, 420*). In India, the tree (known as *vibhitika*) is associated with the goddess Kali and is used in black magic to kill enemies (Gupta 1991, 94*).

Bellerian myrobalan is closely related to black myrobalan (*Terminalia chebula* [Gaertn.] Retz.), which is venerated as a sacred tree in Nepal and India. It is said that the god Indra, inebriated on **soma**, was imbibing the drink of immortality (*amrita*, ambrosia) and let fall a drop from heaven to earth; from this drop the myrobalan tree arose. In Tantra, eating myrobalan is said to summon the goddess Shri, the erotic consort of Vishnu (Majupuria and Joshi 1988, 109*).

Myrobalan (Sanskrit *harîtaki*) is an attribute of the Tibetan Medicine Buddha (Bhaisajya-guru) and symbolizes the "elixir of long life." The second attribute of the Medicine Buddha is the begging bowl, carved out of lapis lazuli, that is filled with *amrita* (= ambrosia), the "divine nectar of enlightenment" (cf. **soma**) (Birnbaum 1982, 123 ff.).

Literature

Birnbaum, Raoul. 1982. *Der Heilende Buddha*. Bern: O. W. Barth/Scherz.

Tetrapteris methystica Schultes

[syn. *Tetrapteris styloptera* Jusseiu] (Malpighiaceae)—caapi, pinima

This little-known, yellow-flowered climbing bush was first described in 1954 by Richard E. Schultes. The Makú, who live on the Río Tikié (Amazonia), call it *caapi* (cf. *Banisteriopsis caapi*) and use it for ritual purposes (Schultes 1954, 204). They manufacture from the bark a psychoactive drink that apparently has effects like those of **ayahuasca** and is used in a similar manner. The bark may contain β-**carbolines** (Schultes and Hofmann 1992, 58*).

The Tanimukas call the closely related yellow-flowered *Tetrapteris styloptera* Jussieu—most likely a synonym—*weé-po-awk*. They use the powdered bark medicinally as a hemostatic (Schultes 1983, 137).

Tetrapteris mucronata Cav., known as *caapipinima*, also appears to be used for psychoactive purposes (Schultes and Hofmann 1992, 66 f.*).

Literature

Schultes, Richard Evans. 1954. Plantae Austro-Americanae IX: Plantarum Novarum vel Notabilium Notae Diversae. *Botanical Museum Leaflets* 16 (8): 179–228.

———. 1983. De Plantis Toxicariis e Mundo Novo Tropicale Commentationes XXXI: Further ethnopharmacological notes on malpighiaceous plants of the northwestern Amazon. *Botanical Museum Leaflets* 29 (2): 133–37.

Thamnosma montana Torr. et Frem.

(Rutaceae)—turpentine broom

Shamans of the North American Kawaiisu Indians were said to drink a tea from this plant so that they could become "crazy like coyotes," i.e., transform themselves into these animals (Moerman 1986, 481*). The plant contains numerous **coumarins** that may have psychoactive effects. It is also possible that it contains *N,N*-DMT (Ott 1993, 420*).

Thevetia spp.

(Apocynaceae)—rattle tree

The genus *Thevetia* is composed of nine species, known as rattle trees or tropical oleander

Left: The Peruvian rattle tree (*Thevetia peruviana*) is planted in tropical gardens around the world for its beautiful flowers.

Right: A number of tillandsias. (Photographed in northern Peru)

(Anzeneder et al. 1993, 61*). The peels of the fruits are hard and are used to make rattles and clappers for Indian dances.

One species whose identity has not been clarified is known in Colombia and the surrounding regions as *cabalonga blanca*. It is considered a weaker relative of the true **cabalonga**. *Cabalonga blanca* is said to have magical powers and psychoactive effects and is used as an **ayahuasca** additive.

In Mexico, *Thevetia thevetioides* (H.B.K.) K. Schum. is known as *yoyotl* and is used in folk medicine as a cardiac stimulant and analgesic (Jiu 1966, 252*). *Thevetia peruviana* (Pers.) Schum. [syn. *T. neriifolia* Juss.], known as yellow oleander, is originally from Peru but is now cultivated as an ornamental in all tropical zones of the world, and it is the species whose chemical makeup is best known. The seeds are rich in cardiac glycosides, e.g., peruvoside (Steinegger and Hänsel 1972, 193*). Eight to ten seeds is reportedly the lethal dose for an adult (Roth et al. 1994, 699*). In the Mexican state of San Luis Potosí, *Thevetia peruviana* is known as *palo de San Antonio*, "tree of St. Anthony" (Aguilar Contreras and Zolla 1982, 196*). The name may very well be derived from a psychoactive effect (see **Claviceps purpurea**). On the Gulf Coast of Mexico, in the territory of the Huastec, the plant is known as *cabalonga de la huasteca* (Aguilar Contreras and Zolla 1982, 196*).

Tillandsia spp.

(Bromeliaceae)—tillandsias

Tillandsias are epiphytes that grow on typical American flora. In pre-Columbian Peru, they were used as a stuffing in the false heads of mummies (Towle 1961, 31*). Tillandsias appear on Mochican ceramic paintings in connection with winged shamans (Andritzky 1989, 169 f.*). It may be that a psychoactive use was once known but has now been forgotten. The plant depicted in the paintings of the Mochica is sometimes interpreted as *Tillandsia purpurea* Ruíz et Pav. (Ott 1996, 108*). Flavonoids have been found in this species (Arslanian et al. 1986). The Tarahumara Indians refer to *Tillandsia mooreana* Smith as *waráruwi*, "peyote companion" (cf. **Lophophora williamsii**),

and presumably used it as a peyote substitute (Ott 1996, 108*). The Tarahumara use a related species, the ball moss (*Tillandsia recurvata* [L.] L.), which they call *muchiki chabóame*, as a cough medicine (Deimel 1989, 61). This plant was previously identified as *Tillandsia inflata* Mez. (Bye 1975).

In Brazilian ethnomedicine, *Tillandsia usneoides* (L.) L. (Spanish moss) is used as an analgesic. It is said that a watery extract of this plant induces "visions" (Ott 1996, 420*).

Literature

Arslanian, R. L., et al. 1986. 3-n=methoxy-5-hydroxyflavonols from *Tillandsia purpurea*. *Journal of Natural Products* 49 (6): 1177–78.

Bye, R. A. 1975. Plantas psicotrópicas de los Tarahumaras. *Cuadernos Científicos CEMEF* 4:49–72.

Deimel, Claus. 1989. Pflanzen zwischen den Kulturen: Tarahumaras und Mestizen der Sierra Madre im Noroeste de México. Ethnobotanische Vergleiche. *Curare* 12 (1): 41–64.

Tribulus terrestris L.

(Zygophyllaceae)—small caltrops

In Ayurvedic medicine, *Tribulus terrestris* is utilized as an aphrodisiac and as a geriatric agent. This plant, also known as *zama* or *zimpating*, produces fruits that are used in Ladakh to fortify **beer**. The young branches and ripe fruits also are crushed and consumed in milk. High dosages (how high?) are said to produce delirium (Navchoo and Buth 1990, 319, 320*). The plant has been shown to contain steroids and sapogenin along with some five alkaloids, including harmane, **harmine**, and harmol (Ott 1993, 426*; Festi and Samorini 1997, 26).

In Baluchistan (Pakistan), 10 to 20 g of the dried fruits (*ghur gan*) are ground, mixed with water, and drunk to increase the sexual abilities of men. The ground fruits (*gurgandako*) of the closely related species *Tribulus longipetalus* Viv. [syn. *Tribulus alatus* Del.] are used as a medicinal **snuff** for treating stuffy noses (Goodman and Ghafoor 1992, 55*).

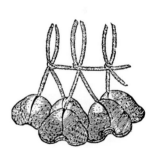

Ethnographic objects first brought the tropical rattle tree (*Thevetia* spp.) to the attention of Europe. The hard fruit husks were used as rattles for Indian rhythm instruments and dance belts and anklets. (Woodcut from Tabernaemontanus, *Neu Vollkommen Kräuter-Buch*, 1731)

Winged shamans with featherlike tillandsias. (Painting on a Moche ceramic vessel, from Kutschner 1954)

The caltrop (*Tribulus* spp.) is related to Syrian rue (*Peganum harmala*) and may be useful as a source of β-carbolines. (Woodcut from Tabernaemontanus, *Neu Vollkommen Kräuter-Buch*, 1731)

The psychoactive roots of various *Trichocline* species are known in South America as *contrayerba*. They may have served as the model for this European illustration. (Woodcut from Gerard, *The Herball or General History of Plants*, 1633)

The very pungent-smelling leaves of the California bay tree (*Umbellularia californica*) contain safrole. The bark of the tree is alleged to contain 5-MeO-DMT. (Wild plant photographed in Occidental, California)

337 Numerous medicinal plants, e.g., *Valeriana* sp., are known by these names in South America.

338 The Mataco whom I was able to interview had never heard anything about these plants.

Literature

Festi, Francesco, and Giorgio Samorini. 1997. *Tribulus terrestris* L. (tribolo/caltrop). *Eleusis* 7:24–32.

Trichocline spp.

(Compositae: Mutisieae)—coro

In the Chaco region of northern Argentina, a number of species of the genus *Trichocline* are utilized as psychoactive substances. Locals call them *coro* or *contrayerba*.[337] Jesuit reports from the eighteenth century describe how the Calchaqui Indians used the ground roots to strengthen their chicha (**beer** made from maize or other plants). The Mocovies, Toba, and Mataco[338] smoke the powdered root alone or mixed with tobacco (cf. **smoking blends**). The smoke is said to have medicinal effects upon stomachaches. Today, the root is also burned alone or with tobacco as an **incense**. The most commonly used species are *Trichocline reptans* (Webb.) Rob., *Trichocline exscapa* Griseb., and *Trichocline dealbata* (Hook. et Arn.) Griseb. (Zardini 1975, 649 f.; 1977). Unfortunately, no chemical studies of the root have been conducted to date. The roots supposedly are sold at herb stands in Argentinian markets in the Chaco region. In Salta, a seller from Germany offered imported calamus roots (**Acorus calamus**) as *coro*.

Literature

Zardini, Elsa M. 1975. Revision del genero *Trichocline* (Compositae). *Darwiniana* 19:618–733.

———. 1977. The identification of an Argentinian narcotic. *Botanical Museum Leaflets* 25 (3): 105–7.

Trichodesma zeylancium R. Br.

(Boraginaceae)—bush tobacco

Also known as cattle bush, this shrubby plant has blue flowers and lanceolate leaves and was once used as an inebriant (Webb 1969). In Australia (Arnhem Land), the dried leaves formerly were used as a substitute for tobacco (**Nicotiana tabacum**) (Low 1990, 190*). The entire plant contains 0.01 to 0.07% alkaloids (Collins et al. 1990, 31*).

Literature

Webb, I. J. 1969. The use of plant medicines and poisons by Australian Aborigines. *Mankind* 7:137–46.

Umbellularia californica (Hook. et Arn.) Nutt.

(Lauraceae)—California bay

This evergreen tree is known variously as California laurel, California bay, California olive, Oregon myrtle, pepperwood, headache tree, and California sassafras (Fuller and McClintock 1986, 184*). Its leaves are rich in an **essential oil** with a high concentration of safrole; the principal constituent, however, is umbellulone (Fuller and McClintock 1986, 184*). The bark of the trunk is said to contain 5-MeO-DMT (Rob Montgomery, pers. comm.). No traditional psychoactive use of this plant is known (cf. *Sassafras albidum*). The leaves are used in folk medicine for headaches (hence the common name headache tree), colic, and diarrhea (Grieve 1982, 716*). In California, the leaves are used as a spice in place of those of the true bay tree (*Laurus nobilis*).

Ungnadia speciosa Endl.

(Sapindaceae)—Mexican horse chestnut

It has been suggested that the seeds (sometimes called Texas buckeyes) of this small tree were once used for psychoactive purposes in northern Mexico and Texas (Schultes and Hofmann 1992, 59*). The black seeds, which are 1.5 cm in length, have been found in archaeological contexts together with peyote (*Lophophora williamsii*) and mescal beans (*Sophora secundiflora*) (Adovasio and Fry 1976*). *Ungnadia* seeds contain cyanogenetic compounds (Seigler et al. 1971).

Literature

Seigler, D., F. Seaman, and T. J. Mabry. 1971. New cyanogenetic lipids from *Ungnadia speciosa*. *Phytochemistry* 10:485–87.

Urmenetea atacamensis Phil.

[formerly also referred to as *Retanilla ephedra* (Vent.) Brogn.]
(Compositae)—coca del suri

This plant, which can grow up to 10 cm in height, is found only in the Atacama Desert of northern Chile—the driest desert in the world—where the local oasis dwellers know it as *coquilla*, "little coca," or *coca del suri*, "coca of the *suri* bird." Until recently, the whitish, downy leaves were chewed either alone or together with *llipta* as a coca substitute (see **Erythroxylum coca**). The inconspicuous plant is used as a source of nourishment by the ostrichlike cursorial birds known as *suri*. A tea of the leaves is used as a treatment for altitude sickness (*puna*) (Aldunate et al. 1981, 218*).

Chewing the leaves produces a slight sensation of numbness in the mouth. A mild psychoactivity (cocalike stimulation) was also observed. Smoking the dried leaves (a good dosage is 0.3 g) produces clear psychoactive effects that are initially somewhat narcotic and then more stimulating. The effects are similar to those produced from smoking dried coca leaves (see **Erythroxylum coca**). Chemical studies are lacking.

Utricularia minor L.

(Lentibulariaceae)—little bladderwort

In Ladakh, this insectivorous plant is known as *lingna*. There, the dried and powdered leaves are roasted on a flat stone, after which the powder is mixed with water and put in a bottle, which is then buried for ten to fifteen days. The Ladakhis enjoy this drink (*lingeatzish*) primarily in winter. It is said to have very inebriating effects and can cause death if consumed in excess (Navchoo and Buth 1990, 320*).

Valeriana officinalis L.

(Valerianaceae)—valerian

Valerian is an Old Germanic ritual and healing plant. It was sacred to the goddess Hertha, who rode upon the red deer. Wieland, the shamanic smith of the Germanic mythological world, used the root to heal diseases. For this reason, valerian was also known as *velandswurt*, "Wieland's root" (Weustenfeld 1995, 13*). In earlier times, valerian was hung on houses as a protection against witches and witchcraft, evil spirits, and devils. The root was also used as a fumigant to keep away the devil (cf. **incense**). In the early modern period, valerian root was regarded as an aphrodisiac and was used to treat the "sacred disease" (epilepsy) (Knoller 1996, 12 f.). It was also known as theriac root, for it was an important ingredient in the panacea **theriac** (Weustenfeld 1995, 15*).

Valerian (along with the variety *Valeriana officinalis* L. var. *sambucifolia* Mikan.) is also called cat weed and is renowned for the power it has to attract cats (cf. **Nepeta cataria**). The sedative effects that the root has upon the nervous system are quite well known (Pahlow 1993, 64*). Valerian roots are sometimes characterized as a "legal high" with psychoactive powers (Schultes and Hofmann 1980, 368*). In particular, a tea made of equal parts of valerian root and kava-kava (**Piper methysticum**) is said to produce "beautiful dreams" (Schuldes 1995, 76*). When mixed with hops (**Humulus lupulus**), valerian yields a potent tea for inducing sleep (cf. also **diazepam**).

In South America, *Valeriana longifolia* H.B.K. is regarded as a panacea and stimulant for elderly people suffering from infirmity. There, various *Valeriana* species are referred to as *contrayerba* (cf. **Trichocline spp.**). *Valeriana adscendens* Turz. is known as *hornamo morado* in Peru, where it is used as an additive to San Pedro drinks (cf. **Trichocereus pachanoi**). The North American Blackfeet Indians smoke the roots of *Valeriana sitchensis* Bong, known as tobacco root, either alone or mixed with tobacco (see **kinnikinnick**) (Johnston 1970, 320*). In India and Nepal, the aromatic root of *Valeriana jatamansi* (DC.) Jones [syn. *Valeriana wallichii* DC.], known as *samyo* or *muskbala*, is used as a fumigant or as an ingredient

Top: Valerian (*Valeriana officinalis*) is renowned for its folk medicinal uses as a sleeping and calming agent.

Bottom: The aromatic roots of the Himalayan valerian species *Valeriana jatamansi* are used as a ritual incense.

in **incense** for magical and religious rites (Shah 1982, 298*; Shah and Joshi 1971, 421*). The aromatic root of *jatamansi* or *masi*, the closely related species *Nardostachys jatamansi* (D. Don) DC., is even more highly regarded; it is used both as an incense and to treat epilepsy (Shah 1982, 297*). Whether these two incenses have psychoactive properties, as is sometimes asserted, remains an open question. The sesquiterpene ketone valeranon, which is present in *Valeriana officinalis*, *Valeriana jatamansi*, and *Nardostachys jatamansi*, is presumably responsible for the sedative (tranquilizing) effects (Hörster et al. 1977).

The alkaloid actinidine has been found in the genus (Schultes 1981, 42*). Of interest for further research into a possible psychoactivity beyond the sedative effects is the finding that an aqueous extract influences the central nervous system neurotransmitter GABA (γ-aminobutyric acid; see *Amanita muscaria*, ibotenic acid, **muscimole**) (Santos et al. 1994).

Literature

Gränicher, F., P. Christen, and I. Kapetanidis. 1992. Production of valepotoriates by hairy root cultures of *Valeriana officinalis* var. *sambucifolia*. *Planta Medica* 58 suppl. (1): A614.

Hörster, Heinz, Gerhard Rücker, and Joachim Tautges. 1977. Valeranon-Gehalt in den unterirdischen Teilen von *Nardostachys jatamansi* and *Valeriana officinalis*. *Phytochemistry* 16:1070–71.

Knoller, Rasso. 1996. *Baldrian*. Niederhausen/Ts.: Falken Taschenbuch Verlag.

Santos, Maria S., Fernanda Ferreira, António P. Cunha, Arsélio P. Carvalho, and Tice Macedo. 1994. An aqueous extract of *Valeriana* influences the transport of GABA in synaptosomes. *Planta Medica* 60:278–97.

Vanda roxburghii R. Por.

(Orchidaceae) [syn. *Vanda tesselata* (Roxb.) G. Don)—vanda

When bees sip nectar from these beautiful orchids, which are found in Sri Lanka, India, and Burma (Myanmar), they soon fall to the ground in a narcoticized stupor (cf. **honey**). In India, a psychoactive use of this plant is said to have been derived from observations of this occurrence: "Ayurvedic shamans used the flower in a decoction to achieve the hypnotic narcosis of their office, permitting them a transcendent state of being" (Emboden 1979, 17*).

No active principles are known (Emboden 1979, 194*).

Voacanga spp.

(Apocynaceae)—voacango bush

The bark and seeds of the African species *Voacanga africana* Stapf[339] contain up to 10% **indole alkaloids** of the iboga type (cf. *Tabernanthe iboga*, **ibogaine**) and reportedly induce stimulating and hallucinogenic effects (Bisset 1985b; Oliver-Bever 1982, 8). The principal alkaloid is voacamine. African sorcerers are said to use the seeds to produce visions. In West Africa, the bark is utilized as a hunting drug and stimulant (Schuldes 1995, 77*). It also is regarded as a potent aphrodisiac. The bark of *Voacanga bracteata* Stapf is used in Gabon to get "high" (most likely as a marijuana substitute; cf. *Cannabis indica*). It contains 2.46% alkaloids (voacamine, voacamine-*N*-oxide, 20-epi-voacorine, voacangine) that, although closely related to the compounds found in *Tabernanthe iboga*, apparently produce only mild depressant effects (de Smet 1996, 145*; Puiseux et al. 1965).

Voacanga dregei E. H. Mey. is also said to produce hallucinogenic effects (Schultes and

339 The taxonomy of this genus is rather chaotic. *Voacanga africana* alone is known under the following synonyms: *Voacanga glabra* Schum., *V. schweinfurthii* var. *parvifolia* Schum., *V. magnifolia* Wernham, *V. talbotti* Wern., *V. eketensis* Wern., *V. glaberrima* Wern., and *V. africana* var. *glabra* Schum. (Oliver-Bever 1982, 29).

Hofmann 1980, 366*). West African sorcerers ingest the seeds of *Voacanga grandiflora* (Miq.) Rolfe for visionary purposes. Unfortunately, the details of this use are still unknown, as the sorcerers keep their knowledge secret.

Literature

Bisset, N. G. 1985a. Phytochemistry and pharmacology of *Voacanga* species. *Agricultural University Wageningen Papers* 85 (3): 81–114.

———. 1985b. Uses of *Voacanga* species. *Agricultural University Wageningen Papers* 85 (3): 115–22.

Bombardelli, Ezio, Attilio Bonati, Bruno Gabetta, Ernseto Martinelli, Giuseppe Mustich, and Bruno Danieli. 1976. 17-O-acetyl-19,20-dihydrovoachalotine, a new alkaloid from *Voacanga chalotiana*. *Phytochemistry* 15:2021–22.

Oliver-Bever, B. 1982. Medicinal plants in tropical West Africa I: Plants acting on the cardiovascular system. *Journal of Ethnopharmacology* 5 (1): 1–71.

Puiseux, F., M. P. Patel, J. M. Rowson, and J. Poisson. 1965. Alcaloïdes des *Voacanga*: *Voacanga africana* Stapf. *Annales Pharmaceutiques Françaises* 23:33–39.

Zea mays L.

(Gramineae: Poaceae)—maize

First cultivated in Mexico some four thousand years ago, maize is the single most important source of nourishment for many Central and South American Indians. Grains of maize (ferment) are also used to brew numerous **beers** as well as **chicha** (Wedemeyer 1972).

Corn silk (stigmata maydis) plays a role in Indian medicine. It also has some significance in modern phytotherapy, where it is used as a diuretic (Czygan 1989; Rätsch 1991a, 174–78*). In addition, it is "smoked by the Indians in Peru as an inebriant" (Roth et al 1994, 742*; Czygan 1989, 326). In the "drug scene," corn silk is smoked for inebriating purposes, both alone and as an ingredient in **smoking blends**. In North America, corn silk is one of the ingredients in the ceremonial "tobacco" **kinnikinnick**. The silk (or,

more precisely, styles) contains up to 85% alkaloids of an as yet unknown structure (possibly from the family of the **ergot alkaloids** or tryptamine derivatives) that are able to induce states of excitation and delirium when inhaled (Roth et al. 1994, 742*).

Literature (selection)

Czygan, Franz-Christian. 1989. Maisgriffel. In *Teedrogen*, ed. Max Wichtl, 325–26. Stuttgart: WVG.

Wedemeyer, Inge von. 1972. Mais, Rausch- und Heilmittel im alten Peru. *Ethnomedizin* 2 (1/2): 99–112.

Zingiber officinale Roscoe

(Zingiberaceae)—ginger

Ginger comes from the tropical rain forests of Southeast Asia, but it has been planted throughout tropical Asia for at least three thousand years (Norman 1991, 62*) and is now planted in tropical areas around the world. It has acquired an ethnopharmacological significance among many American Indian peoples. It is frequently used both as a spice and as a medicine, e.g., for upset stomachs (Rätsch 1994b, 58*). In Ecuador, where ginger is known as *ajej*, the Shuar, Achuar, and Aguaruna all use it as a hallucinogen. The shamans ingest ginger to obtain magical power (Bennett 1992, 493*). The Cariña rub a mixture of gingerroot and tobacco (*Nicotiana tabacum*) onto the eyelids of apprentice shamans so that they may be able to see the spirits of the forest. Ginger is also

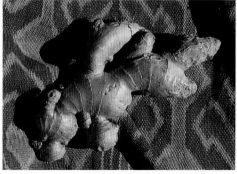

Left: *Voacanga grandiflora* produces fruits that resemble testicles. African sorcerers use the fruit as a visionary inebriant.

Top: Corn silk (*Zea mays*) contains a psychoactive alkaloid.

Bottom: Used throughout the world as a spice, ginger (*Zingiber officinale*) also has a certain significance as a magical shamanic drug.

Indians use ginger (*Zingiber officinale*), which they cultivate in the American tropics, for both medicinal and psychoactive purposes. (Photographed in Chiapas, Mexico)

one of the initiatory plants of the novice shamans on the Indonesian island of Siberut:

> Finally, every novice is given "seeing" eyes. He goes with the teacher to a secluded spot in the vicinity and must make a promise never to betray the secret. . . . Old shamans explain that the novice is called upon to massage a small disease stone out of a spear that he has brought with him and which the ancestors have placed inside as a test. After he has attempted this for a while without success, the master shows him how it is done. Afterward, burning ginger juice is dripped from a little bottle into the novice's eyes and he becomes able "to see." The master then asks him what he sees. (Schefold 1992, 116)

Ginger extracts have clear effects upon the central nervous system, but whether they are able to induce hallucinations (in what dosages?) is questionable (Bennett 1992, 490*). The use of ginger as an aphrodisiac is widespread. The Secoya number ginger among the *nuni*, the plants of supernatural origin (Vickers and Plowman 1984, 33*).

In Papua New Guinea, the roots of bitter ginger, a wild ginger species (*Zingiber zerumbet* [L.] Sm.), are known as *kaine*. They purportedly were used together with a *Homalomena* species (**Homalomena sp.**) as a hallucinogen (cf. also **Kaempferia galanga**). In the South Pacific, a ginger species is used for magical purposes. On the Gazelle Peninsula (formerly New Pomerania), ginger leaves and roots are used in all magical acts.

For these reasons, ethnologists have characterized ginger as the "mandrake root of the indigenous people" (cf. **Mandragora officinarum**) (Meier 1913).

Literature

Meier, P. Joseph. 1913. Die Zauberei bei den Küstenbewohner der Gazelle-Halbinsel, Neupommern, Südsee. *Anthropos* 8:1–11, 285–305, 688–713.

Rätsch, Christian. 1992. Nahrung für den Feuergott—Die Ingwergewächse. *Dao* 4/92:48–49.

Schefold, Reimar. 1992. Schamanen auf Siberut. In *Mentawai Schamane: Wächter des Regenwaldes*, Charles Lindsay, 105–17. Frankfurt/M.: Zweitausendeins.

Schulick, Paul. 1996. *Ginger: Common spice and wonderful drug.* 3rd ed. Brattleboro, Vt.: Herbal Free Press.

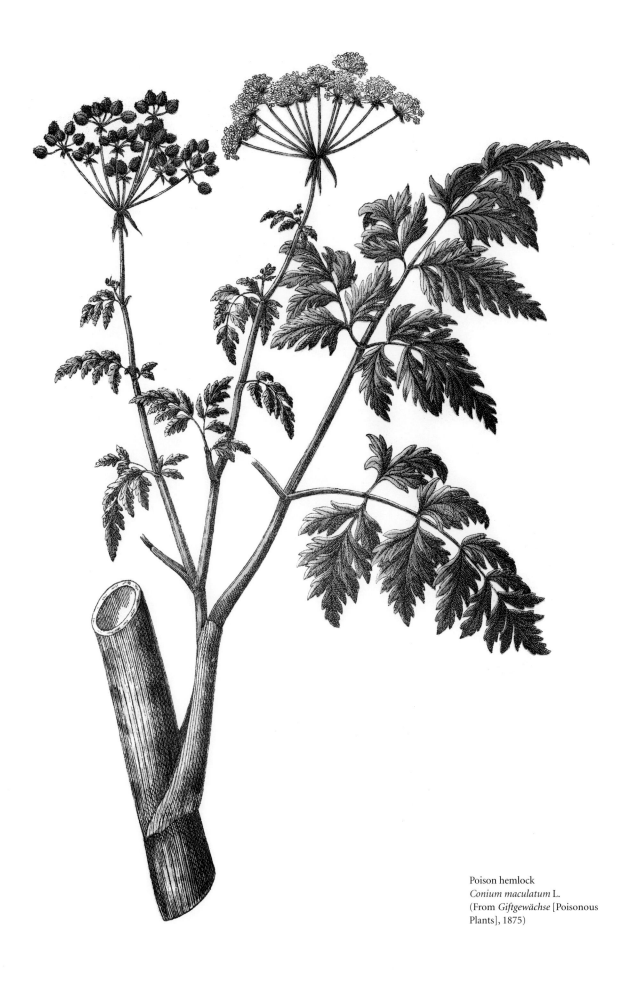

Poison hemlock
Conium maculatum L.
(From *Giftgewächse* [Poisonous
Plants], 1875)

Reputed Psychoactive Plants

"Legal Highs"

Because of the increasingly strict drug laws that began to be implemented in the 1960s, aficionados have looked for new possibilities for obtaining a legal high. In doing so, they have swallowed or smoked a large number of legally obtainable plant products. Some of these legal highs do in fact have psychoactive effects, others work for only some of their users (e.g., *Coleus blumei*), and still others appear to have not much of an effect upon anyone. At the beginning of the 1970s, brochures and booklets introducing these legal plants began to appear (Gottlieb 1973*; Grubber 1991*). Through these as well as the "grapevine," knowledge (or semi-knowledge) about exotic and legal psychoactive plants was disseminated (Brown and Malone 1978*). At the same time, numerous popular myths and legends about the efficacy of various plants also arose. This section provides a short introduction to those plants that are still regarded as legal highs but whose psychoactive properties are doubtful. It should be noted, however, that psychoactivity in these plants cannot be ruled out. Perhaps future research will uncover some interesting insights concerning, for example, appropriate methods of preparing and processing these plants, possible synergistic relationships with other ingredients, and other useful information.

The Genera Discussed in This Section

Actinidia, Anethum, Arisaema
Borago
Catharanthus, Cineraria
Daucus, Digitalis, Dioon
Equisetum, Evodia
Foeniculum
Hydrangea
Laurus, Liriosma
Matricaria (= *Chamomilla*), *Musa*
Panax, Phrygilanthus, Podophyllum, Polygala
Scutellaria, Sebastiana, Swainsonia
Ungernia
Wisteria

"'Legal high' means it does not work!"
Terence McKenna
(lecture, February 1996)

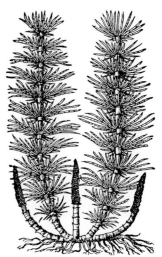

Many long-known medicinal plants now have a reputation of being psychoactive and are regarded as legal highs. It is highly questionable whether fennel, horsetail, and the others actually do produce anything apart from a placebo-induced state of inebriation. (Woodcut from Tabernaemontanus, *Neu Vollkommen Kräuter-Buch*, 1731.)

Left: From time to time, the leaves of the carrot (*Daucus carota*) have been smoked as a marijuana substitute. This "legal high" is sometimes attributed with a certain psychoactivity.

Many species of jack-in-the-pulpits (*Arisaema*), members of the Family Araceae, are capable of producing dangerous toxic effects. (Woodcut from Tabernaemontanus, *Neu Vollkommen Kräuter-Buch*, 1731)

Left: A species of Chinese cat powder (*Actinidia* sp.) has a psychoactive effect upon cats.

Right: Dill (*Anethum graveolens*) has a folk reputation as an aphrodisiac. It is used in the "drug scene" as a marijuana substitute.

Actinidia polygama (Sieb. et Zucc.) Planch. ex Maxim.

(Actinidiaceae)—Chinese cat powder

It is rumored that the dried leaf of this tree, which is closely related to the kiwi bush (*Actinida chinensis* Planch.) (Schneebeli-Graf 1992, 93*), is a potent inebriant or strong tranquilizer for animals; hence its common name *Chinese cat powder*. In Asian zoos, it purportedly is used to sedate large wild cats. The branches and young leaves of this plant are said to produce hallucinogenic effects (Grubber 1991, 60*). The bush is found in Manchuria, Korea, Japan, and western China and on the island of Sakhalin. Both metatabilacetone and actinidine have been found in this plant (Emboden 1979, 168*).

Anethum graveolens L.

(Apiaceae: Umbelliferae)—dill

Known primarily as a spice, dill has also long enjoyed a reputation as an aphrodisiac. It is also said to be an inebriant, for "the garden plant is numbered among the so-called legal highs; when dried and smoked, dill induces a mild euphoria" (Sahihi 1995, 153*). In the American "drug scene," dill leaves are sometimes smoked mixed with glutamate. Dill contains an **essential oil** (approximately 4%) consisting of carvone, limonene, phellandrene, terpinene, and myristicin. It is likely that dill is considered to be psychoactive because it contains myristicin (cf. *Myristica fragrans*) as well

as dillapiol, a nonaminated precursor for the synthesis of DMMDA-2 (cf. *Petroselinum crispum*) (Gottlieb 1973, 12*). However, as Hildegard von Bingen noted, "[N]o matter how [dill] is eaten, it makes people sad" (*Physica* 1.67).

Arisaema dracontium (L.) Schott

(Araceae)—green dragon, dragon root

Also known as memory root, dragon root is reputed to have hallucinogenic effects (Schultes and Farnsworth 1982, 187*; Schultes and Hofmann 1980, 366*). There are several potently toxic plants in the Family Araceae (e.g., *Arum*, *Dieffenbachia*, *Dracunculus*) as well as such questionable hallucinogens as sweet flag (*Acorus calamus*) (cf. Plowman 1969). The genus *Arisaema* is known for its allergenic effects when touched or eaten. Fruits and other parts of the plant contain microscopically small needles of crystallized calcium oxalate, contact with which can lead to heavy histamine secretion (Turner and Szczawinski 1992, 116*). The homeopathic agent Arum dracontium hom. is obtained from the flowers of this plant. The Ojibwa Indians are said to have used the root to counteract witchcraft (Moerman 1982, 101*). The related species *Arum maculatum* L. was used as a wine additive (cf. *Vitis vinifera*).

Literature

Plowman, Timothy. 1969. Folk uses of New World aroids. *Economic Botany* 23 (2): 97–122.

Borago officinalis L.

(Boraginaceae)—borage

This ancient cultigen and spice plant, which is common in Europe and North America, is purported to have psychoactive or hallucinogenic effects (Farnsworth 1972, 68*; Schultes and Hofmann 1980, 367*). Borage contains the slightly toxic pyrrolizidine alkaloids lycopsamine and intermedin, as well as their acetyl derivatives, amabiline and thesinine (Roth et al. 1994, 169*). In phytotherapy, borage is indicated for several conditions that are at least partially related to consciousness (Haas 1961): "An invigorating tea of leaves and flowers is ideal for stress, depression, or following a treatment with cortisone. Borage mitigates fever, dry coughs, and skin rash. The oil of the seeds is helpful for menstrual problems, nervous intestinal complaints, high blood pressure, and hangovers" (Bremness 1995, 233*). Borage pills are sold to dehydrate and "purify the blood." Flowers harvested during the time of blossom are ingested as a folk medicinal sedative (Ratka 1992).

Literature

Haas, H. 1961. Pflanzliche Heilmittel gegen Nerven- und Geisteskrankheiten. *Arzneimittel-Forschung* 4:49–59.

Ratka, Otto. 1992. *Borago*. In *Hagers Handbuch der pharmazeutischen Praxis*, 5th ed., 4:528–32. Berlin: Springer.

Catharanthus roseus (L.) G. Don

[syn. *Ammocallis rosea* Small, *Lochnera rosea* (L.) Reichb., *Vinca rosea* L.]
(Apocynaceae)—Madagascar periwinkle

The Madagascar periwinkle is apparently from the West Indies (Caribbean) but was first described for Madagascar (Morton 1977, 237*). It has pink flowers but also occurs in a pure white form (*Catharanthus roseus* f. *albus* [Sweet] Woodson). This evergreen is one of the truly well-investigated medicinal plants and is the subject of a rich monographic literature. In Caribbean folk medicine, periwinkle tea is used to treat diabetes. In Florida, the leaves are dried and smoked as a marijuana substitute (see *Cannabis indica*) (Morton 1977, 241*). It has often been claimed that the dried leaves are also smoked in Europe and are able to produce "euphoria and hallucinations" (Schuldes 1995, 30*). On the islands of Guadeloupe, the plant is known as *herba aux sorciers* ("plant of the sorcerers"); it may possibly be used in magical voodoo rites (see **zombie poison**).

The plant contains more than seventy alkaloids, most of them **indole alkaloids**, some of which are of the **ibogaine** type (e.g., catharanthine; Scott et al. 1980). The root cortex contains the sedative and antihypertensive alstonine (cf. *Alstonia scholaris*) (Morton 1977, 238*). Recent investigations have shown that different laboratory methods can influence the biosynthesis of indole alkaloids and may even be able to control it to produce a desired outcome (Schrisema and Verpoorte 1992). In the future, this may make it possible to breed strains that will in fact produce psychoactive indoles of the ibogaine or voacangine types (cf. *Tabernanthe iboga, Voacanga* spp.).

The use of *Catharanthus* is not without risk. Chronic use has been observed to result in severe damage to the central and peripheral nervous systems (Morton 1977, 241*; Roth et al. 1994, 204*).

The dwarf periwinkle (*Vinca minor* L.), also known as the sorcerer's violet (Emboden 1974, 66*), is occasionally characterized as having psychoactive properties (Schultes and Hofmann 1980, 366*). It contains a number of indole alkaloids (including vincamine) with antihypertensive effects (Roth et al. 1994, 730*; Wilms 1972). "It was believed to offer protection against witches and storms and was also used at séances. Periwinkle was a component of many love drinks" (Weustenfeld 1995, 45*).

Literature (selection)

Schrisema, J., and R. Verpoorte. 1992. Regulation of indole alkaloid biosynthesis in *Catharanthus roseus* cell suspension cultures, investigated with ^{1}H-NMR. *Planta Medica* 58 suppl. (1): A608.

Left: Today, borage (*Borago officinalis*) is usually planted as an ornamental.

Center: The white-blossomed variety of the Madagascar periwinkle (*Catharanthus roseus* f. *albus*).

Right: Madagascar periwinkle (*Catharanthus roseus*) was first encountered in the Caribbean. It is said to have psychoactive effects, although we know of no appropriate methods of preparation.

Red foxglove (*Digitalis purpurea* L.). (From *Giftgewächse*, 1875)

Scott, A. Ian, Hajime Mizukami, Toshifumi Hirata, and Siu-Leung Lee. 1980. Formation of catharanthine, akuammicine and vindoline in *Catharanthus roseus* suspension cells. *Phytochemistry* 19:488–89.

Wilms, K. 1972. Chemie und Wirkungsmechanismus von Vinca-Alkaloiden. *Planta Medica* 22:324–33.

Cineraria aspera Thunb.

(Compositae)—mohodu-wa-pela

It is rumored that this South African plant has psychoactive or hallucinogenic effects (Schultes and Farnsworth 1982, 187*; Schultes and Hofmann 1980, 367*). Unfortunately, details about its use or effects are lacking (Emboden 1979, 173*).

Daucus carota L. ssp. *sativus* (Hoffm.) Schübl. et G. Martens

(Umbelliferae)—carrot

It is difficult to believe that this popular vegetable could have any psychoactive effects, but the reports to this end are numerous (Schultes and Hofmann 1980, 367*). The aboveground leaves are dried and smoked and are said to produce marijuana-like effects (cf. **Cannabis indica**). Amazingly, no information is available about the components of the leaves (Roth et al. 1994, 295*). In former times, carrot root was used as a counterfeit mandrake (see **Mandragora officina-**

rum). The leaves and corolla of the Eurasian wild carrot (*Daucus carota* L. ssp. *carota*) are said to produce "better" effects. Mattiolus wrote that combining the seeds with **theriac** will incite "unchaste desires."

Digitalis purpurea L.

(Scrophulariaceae)—common foxglove
Common throughout the mountains of central Europe (Alps) and found in many gardens as an ornamental, foxglove is one of the most potent poisonous plants known. This notwithstanding, there are rumors that the herbage is used for psychoactive or even hallucinogenic purposes (Schultes and Hofmann 1980, 367*). Foxglove contains several cardiac glycosides that have medicinal value when used in low dosages (Withering 1776/1785); higher dosages can cause cardiac arrest (Luckner and Diettrich 1992). As little as 0.3 g of dried leaves can be dangerously toxic for adults (Roth et al. 1994, 307*).

Literature

Luckner, Martin, and Beate Diettrich. 1992. *Digitalis*. In *Hagers Handbuch der pharmazeutischen Praxis*, 5th ed., 4:1168–87. Berlin: Springer.

Withering, William. 1963. *Bericht über den Fingerhut und seine medizinische Anwendung mit praktischen Bemerkungen über Wassersucht und andere Krankheiten*. Mannheim: Boehringer.

Top left: In some circles, dwarf periwinkle (*Vinca minor*) is esteemed as a magical plant. It is said that the flowers are able to stimulate channeling.

Bottom left: When dried, the flowering top of wild carrot (*Daucus carota* ssp. *carota*) is claimed to be usable as a marijuana substitute.

Right: The pharmacologically highly active foxglove (*Digitalis purpurea*) is sometimes attributed with mind-altering powers.

Dioon edule Lindl.

(Cycadaceae)—Mexican cycad

This edible cycad is from Mexico, where it is known as *chamal*. It is reputed to have psychoactive or even hallucinogenic effects (Schultes and Farnsworth 1982, 187*; Schultes and Hofmann 1980, 367*). This assumption is apparently due to the fact that in Mexico the plant is also known as *hierba loca*, "crazy herb" or "crazy-making herb," and is said to cause animals to act strangely (Reko 1938, 185*). No other details suggesting any actual psychoactivity are known (Aguilar Contreras and Zolla 1982, 91*). The large seeds yield a good starch flour (Bärtels 1993, 59*). In Mexican folk medicine, the seeds are utilized to treat neuralgia (Martínez 1994, 409*). The plant contains the biflavones amentoflavone (main component), bilobetin, sesquioflavone, ginkgetin, sciadopitysin, 7,4',7",4"-tetra-*O*-methylamentoflavone, and diooflavone (Dossaji et al. 1973, 372).

Literature

Dossaji, S. F., E. A. Bell, and J. W. Wallace. 1973. Biflavones of *Dioon*. *Phytochemistry* 12:371–73.

Equisetum arvense L.

(Equisetaceae)—field horsetail

The horsetail is a well-known medicinal plant that is used in folk medicine around the world to treat diarrhea. It is uncertain why it is sometimes characterized as psychoactive (Schultes and Hofmann 1980, 367*). It contains primarily silicic acid (up to 10%), flavonoids, saponins, potassium salts, and other compounds (Pahlow 1993, 273*). The closely related species *Equisetum fluvatile* L. and *Equisetum hyemale* L. are regarded as mildly toxic. The marsh horsetail (*Equisetum palustre* L.) contains the alkaloids palustrine and palustridine, which can cause the "staggers" in animals. All species of *Equisetum* have been found to contain traces of **nicotine** (Roth et al. 1994, 321 f.*). In southern Mexico, *Equisetum myriochaetum* Schlecht. et Cham. is regarded as an aphrodisiac (Rätsch 1994b, 86*; cf. Pérez G. et al. 1985). Alison B. Kennedy has proposed that the **soma** plant was a Himalayan species of *Equisetum*.

Literature

Pérez Gutiérrez, R. M., G. Tesca Laguna, and Aleksander Walkowski. 1985. Diuretic activity of Mexican equisetum. *Journal of Ethnopharmacology* 14:269–72.

Evodia bonwickii F. Muell.

[= *Euodia*]
(Rutaceae)

This plant has a reputation of producing "psychotomimetic" effects (Farnsworth 1972, 71*; Schultes and Farnsworth 1982, 187*; Schultes and Hofmann 1980, 368*). The shrub is used in Papua New Guinean ethnomedicine for treating psychological complaints (Scott 1963). **Coumarins** have been detected in several species of the genus (*E. alata* F. Muell., *E. beleha* Baill., *E. hupehensis* Dode, *E. viteflora* F. Muell.).

Literature

Scott, K. 1963. Medicinal plants of the Mt. Hagen people in New Guinea. *Economic Botany* 17:16–22.

Foeniculum vulgare

[syn. *Foeniculum officinale*]
(Umbelliferae)—fennel

It has often been claimed that fennel or fennel oil can have psychoactive effects (Albert-Puleo 1980, 339). This psychoactivity has been suggested since at least the time of Hildegard von Bingen, who noted, "No matter how it is eaten, it makes people happy. . . . Even a person who is plagued by melancholy, he should pound fennel to a juice and rub this often on the forehead, temples, chest, and abdomen, and the melancholy in him will yield" (*Physica* 1.66).

Smoked, the herbage of this well-known spice is said to be psychoactive (Schultes and Hofmann 1980, 367*). Fennel contains a sweet-smelling **essential oil** (around 6%) consisting primarily of *trans*-anethol and fenchene (Brand 1993; Pahlow 1993, 132*). Fennel tea finds folk medicinal use as a calmative (which is probably how it acquired its psychoactive reputation). The seeds are said to contain the greatest amount of "psychotropic oil" in the plant (Grubber 1991, 32*). Estragole, found in the essential oil, is regarded as a precursor to 4-methoxy-amphetamine (Gottlieb 1973, 50*). Anethol has a primarily estrogenic effect (Albert-Puleo 1980). Fennel and anise are used in Greece in the production of ouzo (see **alcohol**).

Top: The Mexican cycad *Dioon edule* is edible and is said to have inebriating effects.

Bottom: The marsh horsetail (*Equisetum arvense*) is a very common wild plant in Europe.

Left: The blossoming herbage of wild fennel (*Foeniculum vulgare* var. *vulgare*) is said to be usable as a marijuana substitute.

The Asian *Hydrangea paniculata* in full bloom.

Literature

Albert-Puleo, Michael. 1980. Fennel and anise as estrogenic agents. *Journal of Ethnopharmacology* 2:337–44.

Brand, Norbert. 1993. *Foeniculum.* In *Hagers Handbuch der pharmazeutischen Praxis*, 5th ed., 4:156–81. Berlin: Springer

Hydrangea paniculata Sieb. var. *grandiflora*

(Saxifragaceae)—peegee hydrangea

This garden and ornamental plant, which is from China and Japan (Grubber 1991, 39*), has occasionally been described as a euphoriant, although its use is "strongly unadvised" (Schuldes 1995, 41*). When smoked, the dried leaves and flowers are said to have effects similar to those of marijuana (see *Cannabis indica*). The leaves contain the iso**coumarin** hydragenol, which has been linked to contact allergies (Roth et al. 1994, 411*). Other constituents include a substance known as hydrangin, saponins, and hydrocyanic acid compounds (Gottlieb 1973, 20*).

Literature

Takeda, Kosaku, Tomoko Yamashita, Akihisa Takahashi, and Colin F. Timberlake. 1990. Stable blue complexes of anthocyanin-aluminium-3-*p*-coumaroyl- or 3-caffeoyl-quinic acid involved in the blueing of *Hydrangea* flower. *Phytochemistry* 29 (4): 1089–91.

Laurus nobilis L.

(Lauraceae)—true laurel, sweet bay

The evergreen laurel tree was sacred to the ancient Greeks. It was consecrated especially to Apollo, the god of mental ecstasy. According to ancient mythology, the plant was originally an enchanting woman or nymph named Daphne. Hence, *daphne* was another name for this tree in ancient times.[340] The aromatic leaves were once used as an additive to **beer** and **wine**.

Laurel leaves were an important **incense** at Delphi. The Pythia, the oracular priestess of

Delphi, would chew fresh laurel leaves and inhale laurel smoke before falling into a trance, during which she would open up her body to the god Apollo and allow him to utter prophecies through her mouth (cf. *Hyoscyamus albus*). The ancient singers, poets, and seers who chewed laurel leaves or inhaled the smoke were known as *daphnephages* (Melas 1990, 54 ff.). The priest-physicians of Asclepius (cf. *Papaver somniferum*) would inhale laurel smoke so that they could diagnose the causes of diseases (so-called daphnomancy). Their practice entailed more than simply inhaling the smoke, however, for they also interpreted the crackling of the burning leaves as well as the shape of the rising smoke (Rätsch 1995a, 222 –27*).

In the ancient literature (Dioscorides, Pliny), the true laurel was characterized as a potent psychoactive plant. According to Proclus, laurel smoke was well suited for holding spirits that appeared so that they could be placed into service. To date, all attempts at using laurel for psychoactive purposes have failed. It is likely that other plants were also known as *daphne* in ancient times, and one of these species—whose botanical identity has remained unknown—may have been psychoactive:

> The leaves of other trees or bushes that we refer to as laurel are usually bitter and often even toxic. The snowball [*Viburnum* spp.] is known as bastard or stone laurel, while rose laurel refers to the oleander [*Nerium oleander;* cf. **honey**], summer laurel to the sassafras tree [*Sassafras albidum*], poison laurel to the false star anise [*Illicium anisatum* L. (syn. *Illicium religiosum* Sieb. et Zucc.)], wild laurel to the English holly [*Ilex aquifolium* L.; cf. *Ilex paraguariensis*], camphor laurel to the camphor tree [*Cinnamomum camphora*], mountain laurel to the laurel rose [*Kalmia* spp.; cf. **kinnikinnick**], and cherry laurel to the cherry laurel [*Laurus cerasi*]. The most poisonous are the leaves of the mountain laurel, which the Delaware Indians used to prepare their version of the hemlock drink [cf. *Conium maculatum,* **witches' ointments**] for purposes of committing suicide. (Root 1996, 240*)

Umbellularia californica is known as California laurel; its leaves are used as a substitute for those of the true laurel.

Laurel leaves (from *Laurus nobilis*) contain 2% **essential oil,** consisting of cineol, pinene, phellandrene, sesquiterpenes, eugenol, terpineol, linalool, geraniol, and bitter substances.

Literature

Melas, Evi. 1990. *Delphi: Die Orakelstätte des Apollon.* Cologne: DuMont.

A laurel tree (*Laurus nobilis*) at the temple of Apollo at Delphi.

340 The botanical genus name *Daphne* L. is not associated in any way with the laurels but rather belongs to the Family Thymelaeaceae.

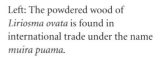

Left: The powdered wood of *Liriosma ovata* is found in international trade under the name *muira puama*.

Right: True chamomile (*Matricaria recutita*) is an ancient medicinal plant that is sometimes attributed with psychoactive powers.

Liriosma ovata Miers

[syn. *Dulacia inopiflora* (Miers) O. Kuntze, *D. ovata* (Miers) K., *Liriosma inopiflora* Miers, *L. micrantha* Spruce ex Engl.]
(Olacaceae)—potency wood, muira puama

This small tree, which grows to a height of only about 15 meters, comes from tropical South America (the Amazon basin). The wood of the trunk and roots is marketed internationally under the name *lignum muira puama* and is esteemed chiefly as an aphrodisiac (potency wood!) and nerve tonic (600 to 1,200 mg) (Gottlieb 1974, 54*; Stark 1984, 87*). Small pieces of the wood are added to a number of psychoactive **smoking blends**. It is sometimes claimed that the wood produces not just erotic but also psychoactive effects. The constituents are completely unknown (Schweins and Sonnenborn 1993, 706). The dried roots of a related tree, *Ptychopetalum olacoides* Benth., are also sold under the same name (lignum muira puama and also radix muira puama). These roots contain a mixture of esters composed of the behenic acid ester of lupeol (0.4 to 0.5%), phytosterols, and an **essential oil** consisting of camphene, camphor, β-caryophyllene, α-humulene, and α- and β-pinene. This root is said to have aphrodisiac effects. Experimental and pharmacological studies, however, are lacking (Brand 1994, 308 f.).

Literature

Brand, Norbert. 1994. *Ptychopetalum*. In *Hagers Handbuch der pharmazeutischen Praxis*, 5th ed., 6:307–10. Berlin: Springer.

Schweins, Sabine, and Ulrich Sonnenborn. 1993. *Liriosma*. In *Hagers Handbuch der pharmazeutischen Praxis*, 5th ed., 5:706–7. Berlin: Springer.

Matricaria recutita L.

[syn. *Matricaria chamomilla* L., *Chamomilla recutita* (L.) Rauschert]
(Asteraceae)—true chamomile

Chamomile was one of the most highly esteemed medicinal plants of Asclepius, the late ancient Greek god of healing (cf. **Laurus nobilis, Papaver somniferum**). It appears that the temple sleepers who came to seek help found that chamomile was frequently recommended during their therapeutic and visionary dreams (Rätsch 1995a, 194*). It is uncertain why chamomile acquired its reputation as a psychoactive plant (Schultes and Hofmann 1980, 367*). The entire plant contains an **essential oil** of complex composition whose main components are α-bisabolol and chamzulene. Also present are flavonoids and **coumarins**, which produce the well-known anti-inflammatory effects of chamomile only in connection with (synergistically with) the **essential oil** (Schilcher 1987). Apart from occasional allergic reactions, the toxicological literature contains no reports of truly interesting effects (Roth et al. 1994, 489*).

Literature

Schilcher, Heinz. 1987. *Die Kamille.* Stuttgart: WVG.

Musa x sapientum L.

(Musaceae)—banana

In the mid-1960s, a rumor surfaced espousing the idea that dried banana peels or scrapings of the inner peel could be smoked and that the effects were identical to those of marijuana (cf. **Cannabis indica**) (Schultes and Hofmann 1980, 367*). This rumor was based in no small part on the song "Mellow Yellow," by the folk/rock singer Donovan (DeRogatis 1996, 59; Krikorian 1968, 385). Many young hippies of the time truly believed that it was possible to "take a trip" using these dried or baked peels. *Time* magazine featured a lead article "Tripping on Banana Peels" (April 1967) that popularized the idea even more. In the United States, the government actually initiated studies to determine whether the banana should be scheduled as a dangerous drug (because of the tremendous "danger of misuse"). The rumor has remained alive up to the present day, and from time to time it appears anew. It has even been claimed that banana peels contain a highly efficacious alkaloid called "bananadine" (Krikorian 1968). The only substance in bananas that may possibly be active is serotonin. According to our current pharmacological understanding, however, serotonin is inactive when ingested orally (cf. **Panaeolus subbalteatus**).

Toward the end of the 1960s, smoking dried banana peels became a popular way to get "high." (Banana inflorescence with unripe fruit)

Top: The Asiatic ginseng root and especially the wild-grown Korean ginseng root are attributed with aphrodisiac, stimulating, and magical powers as well as with mind-altering effects. (Typical *Panax ginseng* roots from Korea)

Bottom: The mayapple (*Podophyllum peltatum*), which blooms in the month from which it takes its name, is also known as the American mandrake.

Literature

DeRogatis, Jim. 1996. *Kaleidoscope eyes*. Secaucus, N.J.: Citadel.

Krikorian, A. D. 1968. The psychedelic properties of banana peel: An appraisal. *Economic Botany* 22:385–89.

Panax ginseng C.A. Mey

[syn. *Panax schinseng* Th. Nees]
(Araliaceae)—ginseng (panacea)

Ginseng is the most renowned medicinal plant in Asia and has become the very symbol of traditional Asian or Chinese medicine and phytotherapy. The plant has also been called the mandrake of the East and the Chinese mandrake (cf. **Mandragora officinarum, Phytolacca acinosa**) (Kirchdorfer 1981, 30 ff.). These names may be the origin for ginseng's reputation as a psychoactive plant (Schultes and Hofmann 1980, 367*). Ginseng is one of the most well-known aphrodisiacs and is also regarded as a panacea (Kimmens 1975). The ginsenosides in the root have general tonic and stimulating ("harmonizing") effects on the body and mind (Fulder 1984, 1985). The Chinese say that ginseng "kindles the inner fire." Ginseng is used in homeopathy for a variety of ailments, including defects of memory and depression. It is found in numerous traditional and modern nerve tonics (Hu 1976). Ginseng is a harmonizing and somatensive medicine, i.e., it stimulates in a completely nontoxic manner and produces no stress. It increases the flow of oxygen to brain cells and can even alleviate the lack of oxygen in the brain that can be caused by amphetamines and other stimulants (Fulder 1995, 210). It lowers the amount of **alcohol** in the blood by about half, i.e., consuming ginseng can shield a person from becoming inebriated (Lee 1996, 47 ff.).

In the toxicological literature, the side effects of frequent use of ginseng are given as euphoria and sleeplessness (Roth et al. 1994, 532*). The dried leaves find use in **smoking blends**. It is doubtful whether these have psychoactive effects. The same is true for American ginseng, *Panax quinquefolium* L. (Emboden 1986, 165*; Pritts 1995).

Literature (selection)

Fulder, Stephen. 1984. *Über Ginseng*. Bonn: Hörnemann Verlag.

———. 1985. *Tao der Medizin*. Basel: Sphinx Verlag.

———. 1995. *Das Buch vom Ginseng*. Munich: Goldmann.

Hu, Shiu-Ying. 1976. The genus *Panax* (ginseng) in Chinese medicine. *Economic Botany* 30:11–28.

Kappstein, Stefan. 1980. *Das Buch vom Ginseng*. Bern: Morzsinay Verlag.

Kimmens, Andrew C., ed. 1975. *Tales of ginseng*. New York: William Morrow and Co.

Kirchdorfer, Anton Maria. 1981. *Ginseng: Legende und Wirklichkeit*. Munich: Droemer Knaur.

Lee, Florence C. 1996. *Facts about ginseng: The elixir of life*. Seoul: Hollym.

Pritts, Kim Derek. 1995. *Ginseng: How to find, grow, and use America's forest gold*. Mechanicsburg, Pa.: Stackpole Books.

Phrygilanthus eugenioides (L.) H.B.K.

(Loranthaceae)

A relative of mistletoe (*Viscum album* L.), *Phrygilanthus eugenioides* is used in the voodoo cult as a magical plant. It is said to have psychoactive or hallucinogenic powers (Schultes and Farnsworth 1982, 187*; Schultes and Hofmann 1980, 367*). Curiously, the ancient texts suggest that mistletoe may also produce psychoactive effects (cf. *Benthamia alyxifolia*).

Podophyllum peltatum L.

(Podophyllaceae)—mayapple

The mayapple is from North America, where it is also called mandrake, wild mandrake, American mandrake, Indian apple, devil's apple, et cetera (Morton 1977, 87*). The number of names can lead to some confusion. *Mandrake* is actually the English name for **Mandragora officinarum**. Settlers applied the name to the mayapple because North American Indians used its root as an amulet and as medicine (Emboden 1974, 149*). Because of this confusion, many people, especially English-speaking Americans, continue to believe that the mayapple is psychoactive. But the root contains no known psychoactive constituents, only toxic glycosides and podophyllin, a resin with cathartic effects (Meijer 1974; Morton 1977, 88*).

The Asian mayapple (*Podophyllum pleianthum* Hance [syn. *Dysosma pleiantha* (Hance) Woodson]), a native of China and Japan, is mixed with hemp (**Cannabis sativa**) and sweet flag (see **Acorus calamus**) to produce a psychoactive substance that "allows one to see spirits" (Li 1978, 23*). In the Kumaon region of India, the seeds of a species known as *bankakri* (*Podophyllum hexandrum* [syn. *P. emodi* Wall. ex Hook. f. et Th.]) are used to ferment an alcoholic beverage (**beer**) (Shah and Joshi 1971:417*).

In homeopathy, the extract Podophyllum is still used in various dilutions. While developing its symptom picture, some strong alterations of consciousness were observed:

> Podophyllum exhibits a bilious temperament.
> . . . Furthermore, there exists the delusion of a serious heart or liver disease, he [the patient]

believes that he is becoming seriously ill and will die. Everything makes him melancholy and sad, and nowhere does he see a ray of light. Sometimes, the delusion arises that through his own fault he has gambled away his own grace or endangered the well-being of his soul by a mortal sin. Still others feel as if the clouds in heaven were too dark or everything were running the wrong way. (Vonarburg 1996, 215)

This example provides a clear illustration of the ways in which the psychological patterns that arise when a medicine is administered can be influenced and shaped by a person's culture.

Literature

Meijer, Willem. 1974. *Podophyllum peltatum*—may apple: A potential new cash-crop plant of eastern North America. *Economic Botany* 28:68–72.

Vonarburg, Willem. 1996. Entenfuß—*Podophyllum peltatum* L. (Homöopathisches Pflanzenbrevier: Folge 11). *Naturheilpraxis* 49 (2): 212–16.

Polygala tenuifolia Willd.

(Polygalaceae)

This plant is native to China and Inner Mongolia and is the source of *yuan-zhi* (= *yuan-chih*, radix polygalae), which traditional Chinese medicine prescribes to "calm the mind" and "heal emotional disorders." It is used to treat nervousness, sleeplessness, forgetfulness, mood swings, and depression (Paulus and Ding 1987, 258*). Also known as *chodat* and *hsiao-ts'ao*, the plant was prescribed in Taoist medicine to increase brain activity and enhance memory. This may be the reason why it is occasionally regarded as psychoactive (Schuldes 1995, 63*). It is used in the manufacture of **herbal ecstasy.** The "active component" is said to be "senegine," which makes up 7% of the dry weight (Gottlieb 1973, 11*). The chemistry is well understood. The root contains primarily polygalitol, tetramethoxyanthones, and triterpenes (Paulus and Ding 1997, 259*) but not a trace of psychoactive compounds. The very similar *Polygala sibirica* L. is sometimes sold in place of this Chinese root drug.

Scutellaria lateriflora L.

(Labiatae)—mad-dog skullcap

Skullcap is a component of **smoking blends** with alleged psychoactive effects that are offered as a marijuana substitute (see *Cannabis indica*). The herbage formerly was used as a sedative and nerve tonic and was even prescribed for the treatment of epilepsy, neuralgia, and sleeplessness. The plant contains the flavonoid scutellarin, which has sedative and antispasmodic effects (Foster and Duke 1990, 186*). A species described under the name *Scutellaria arvense* is reputed to have psychoactive or hallucinogenic effects (Schultes and Hofmann 1980, 367*).

Sebastiana pavonia Muell. Arg.

(Euphorbiaceae)

This euphorbia has a questionable reputation as a hallucinogen (Schultes and Farnsworth 1982, 187*; Schultes and Hofmann 1980, 367*). This assumption apparently dates back to a report by Victor A. Reko in his book *Magische Gifte* [Magical Poisons] (1938, 176–81*), which claimed that North American Yaqui Indians use the crushed seeds of the plant as a tonic during strenuous exertion.

Swainsonia galegifolia R. Br.

(Leguminosae)

As with so many species from the Legume Family, it has been alleged that this plant may have psychoactive effects (Schultes and Farnsworth 1982, 187*; Schultes and Hofmann 1980, 367*).

Ungernia minor

(Amaryllidaceae)—ungernia

This little-known member of the Amaryllis Family is alleged to have psychoactive or even hallucinogenic effects (Farnsworth 1972, 68*). The biologically active alkaloid ungminorine was discovered in this plant (Abdymalikova et al. 1966). However, not even the botany of this plant is well understood (Schultes and Farnsworth 1982, 187*). It may have been confused with *Boophane disticha.*

Literature

Abdymalikova, N. V., Y. B. Zakirov, and I. K. Kamilov. 1966. Some pharmacological activities of the new alkaloid ungminorine. *Akad. Nauk. Uz. SSR, Khim. Biol. Otd.* Jg.:36–40.

Wisteria sinensis (Sims) Sweet

[syn. *Wisteria chinensis*]
(Leguminosae: Fabaceae)   Chinese wisteria

This climbing bush is originally from China. It produces clusters of wonderfully aromatic flowers and is now a popular ornamental. The plant has been alleged to be psychoactive, probably because of its close relationship with and similarity to *Sophora secundiflora* (Schultes and Hofmann 1980, 367*). The plant contains wistarine, a substance with effects similar to but more mild than those of **cytisine** (Roth et al. 1994, 736*). Wisteria is closely related to the soybean (*Glycine max* Merr.) (Keng 1974, 402*).

Whether there really is a psychoactive species of *Polygala* is doubtful. (Woodcut from Tabernaemontanus, *Neu Vollkommen Kräuter-Buch,* 1731)

Because of its purported inebriating effects, Chinese wisteria (*Wisteria sinensis*) is occasionally used as a tobacco substitute.

Psychoactive Plants That Have Not Yet Been Identified

Both the ancient and the ethnographic literature make mention of a number of psychoactive plants, the botanical identifications of which have not yet been determined.

Early Greek writings contain descriptions of plants with incredible effects and fantastic properties, such as **moly** and **nepenthes**. It is uncertain whether these names refer to real or to imaginary plants. There are many mythical plants, such as the tree of knowledge, the plant of life, the mushroom of immortality (cf. "*Polyporus mysticus*"), and the golden apples of Freia, that do not belong to the plant kingdom as we know it but thrive instead in the mythic beyond.

Some old names for magical plants have been adopted by modern botanical taxonomy and assigned to plants that have nothing in common with the earlier flora. The following genus names were all given by Carl von Linné (Carolus Linnaeus; 1707–1778), the great Swedish naturalist and the founder of binominal nomenclature. In choosing these names, he gave expression to his extraordinary love of classical mythology:

Silphium L. (Compositae)—cup plant, compass plant
Canna L. (Cannaceae)—canna lily
Nepenthes L. (Nepenthaceae)—pitcher plant
Daphne L. (Thymelaeaceae)—spurge laurel (cf. **Laurus nobilis**)
Strychnos L. (Loganiaceae)—nux-vomica (cf. **Strychnos spp.**)

In the ethnographic literature, magical plants are often not identified botanically, as the researchers did not possess sufficient botanical knowledge or they returned without the plant material that would have made an identification possible.

Some of the names for these unidentified plants appear to have been general terms used to refer to a variety of psychoactive plants or substances (e.g., **amapola**, **moly**, **soma**).

"In the meantime, the cunning scoundrel gave me . . . a poisonous drink of I know not what kind, sweet indeed and lovely scented, but also extremely treacherous and confusing to the senses; for immediately after I had partaken of it, everything seemed to spin around me, the entire cave stood upside down, in short, I was no longer my usual self, and I finally sank into a deep sleep."

LUCIAN
CONVERSATIONS WITH THE SEA GODS (II)

Left: The ancient magical plant ephemeron ("one-day plant") is often interpreted as the autumn crocus (*Colchicum autumnale*). Up until now, however, it has not been possible to produce the legendary effects of ephemeron using the crocus. The botanical identity of the psychoactive plant of Medea thus remains uncertain.

Achaemenidon

The name of this unidentified plant is derived from that of the ancient **achaimenis**:

> A plant that was used in the Orient for love magic, which it is not possible to identify with any of the plants we know of today. It is said to have had the appearance of "electrum" [= amber] and to have grown in Indian Tardistylis. Its root, consumed in the form of a lozenge, supposedly possessed the ability to evoke "terrible" visions. (Hirschfeld and Linsert 1930, 149*)

Achaimenis

Pliny described a psychoactive plant from *Cheirokmeta*, the lost book of Democritus:

> The *achaimenis*, which has the color of amber, grows without a leaf in India in the area of the Taradastyles [an unknown tribe]; when criminals consume this in wine, the torment of the appearances of different gods compels them to admit everything; he also calls [it] *hippophobos* ["horse's terror"] because the mares in particular watch out for it. (24.161)

This leafless plant may be a psychoactive fungus, such as **Panaeolus subbalteatus**, which grows wherever horses are kept. In the philological literature, the plant is often interpreted as *Euphorbia antiquorum* L.

Aglaophotis

"The Magnificent Shining One"

According to Pliny, Democritus was well versed in magic and knew a great number of magical plants:

> And so he told of the plant *aglaophotis*, which is said to have received its name as a result of the wonder humans had for its special color and which thrives in the marble quarries of Arabia on the Persian side, which is why it is also called *marmaritis* [marble plant]; the magicians made use of this when they wanted to summon the gods. (24.164)

This plant has often been interpreted as the peony (*Paeonia* sp.),[341] which is not psychoactive. **Mandragora officinarum** has also been suggested as a possibility.

Amapola

The word *amapola* is used in South America to refer to opium (cf. **Papaver somniferum**). In Mexico and Guatemala, many plants are referred to as *amapola* [*N.N.*], perhaps because they are able to exert a psychoactive effect or are used in folk

Plants Known as Amapola

Common Name	Botanical Name	Family
amapola	*Bernoullia flammea* Oliv.	Bombacaceae
amapola	*Hunnemannia fumariaefolia* Sweet	Papaveraceae
amapola	*Ipomoea fistulosa* Mart. [syn. *Ipomoea carnea*] (see **Ipomoea** spp.)	Convolvulaceae
amapola	*Kosteletzkya paniculata* Benth.	Malvaceae
amapola	*Papaver rhoeas* L. (see **Papaver** spp.)	Papaveraceae
amapola	*Papaver somniferum* L.	Papaveraceae
amapola	*Passiflora foetida* L. (see **Passiflora** spp.)	Passifloraceae
amapola	*Pseudobombax ellipticum* (H.B.K.) Dugand [syn. *Bombax ellipticum* H.B.K.]	Bombacaceae
amapola	*Tabebuia pentaphylla* Hemsl.	Bignoniaceae
amapola amarilla	*Chelidonium majus* Mill.	Papaveraceae
amapola amarilla	*Eschscholzia californica* Cham.	Papaveraceae
amapola blanca	*Bernoullia flammea* Oliv.	Bombacaceae
amapola blanca	*Pseudobombax ellipticum* (H.B.K.) Dugand	Bombacaceae
amapola colorada	*Pseudobombax ellipticum* (H.B.K.) Dugand	Bombacaceae
amapola de California	*Eschscholzia californica* Cham.	Papaveraceae
amapola de China	*Papaver rhoeas* L.	Papaveraceae
amapola del campo	*Argemone mexicana* L.	Papaveraceae
amapola de los indios	*Eschscholzia californica* Cham.	Papaveraceae
amapola grande	*Althaea rosea* Cav.	Malvaceae
amapola de opio	*Papaver somniferum* L.	Papaveraceae
amapola silvestre	*Pseudobombax ellipticum* (H.B.K.) Dugand	Bombacaceae

341 "We also find the peony in Hecate's magic garden, which suggests the plant has magical powers" (Baumann 1982, 100*).

Top left: The fruits of *Bernoullia flammea*, a tree that grows in the tropical lowlands, recall both the fruits of the genus *Banisteriopsis* and those of the maple (*Acer* spp.). Known in Guatemala as *amapola* ("opium"), the fruits shown here were collected near the ancient ceremonial center of Tikal.

Bottom left: The recipe for a medicine against "evil winds," consisting of espingo buds, *ashango* seeds, *pucho* seeds, cabalonga, and nutmeg.

Right: In Guatemala, the hollyhock (*Alcea rosea* L.) is known as *amapola*, "opium." Whether the plant produces narcotic effects has not yet been determined.

medicine as an opium substitute (Martínez 1987, 52 f.*' Argueta V. et al. 1994, 119*).

In Mexico, 4 g of the petals of *amapola* (*Passiflora foetida*?) brewed with 200 ml of water is drunk as a tea to treat lack of dreams, over-excitability, whooping cough, and asthma (de la Rosa 1995, 15). In the tropics, sedative teas made of the flowers or leaves of *Pseudobombax ellipticum* are drunk to treat coughs, asthma, and flu (Argueta V. et al. 1994, 120*). The chemistry is unknown.

Literature

de la Rosa, Francisco. 1995. *Ayúdese con las yerbas y plantas medicinales mexicanas.* Mexico City: Editores Mexicanos Unidos.

Cabalonga Negra
"Black Cabalonga"

In Colombia, the hard fruit of black cabalonga, said to be a large tree, is one of the most sought-after magical agents. It is purported to have potent psychoactive effects. The fruits are rare and are sold under the table at herb markets for exorbitant prices. For this reason, other fruits are often sold in their place:

> The true cabalonga can be recognized by placing it under the tongue for a little while. Only a few moments are sufficient to produce sensations of dizziness. . . . The Ashaninka of Atalaya, in whose medicine cabalonga plays a

major role, say that it grows in the land of the Amahuaca, in the area around the source of the Rio Inuya. (Faust and Bianchi 1997, 248 f.)

In shamanism, black cabalonga is used for healing as well as for harmful magic and also to prepare magical arrows for shamanic battles. *Ayahuasqueros* utilize the hallucinogenic effects of the fruit to learn how to use plants for healing.

At the "witches' market" (*mercado modelo*) in Chiclayo (northwestern Peru), cabalonga seeds are sold only when asked for and then at relatively steep prices. In the *curanderismo* of the folk healers (*curanderos*), they are an important ingredient in a medicine for treating "evil winds" (*aires*), i.e., ill-nesses of a psychosomatic or psychological nature. This treatment requires that cabalonga seeds, *ashango* seeds (possibly a Rubiaceae), nutmeg (***Myristica fragrans***), **espingo** buds, and *pucho* seeds (***Nectandra*** sp.) be ground; mixed with **chicha**, **wine**, hard liquor (cf. **alcohol**), or water; and ingested.

In northern Peru, the seeds of the Ignatius bean (*Strychnos ignatii*; cf. **Strychnos spp.**) are apparently also sold under the name *cabalonga*. These are imported from outside the country and are more prized than those of the Amazonian *cabalonga de la selva* (Giese 1989, 258*).

Black cabalonga has also been interpreted as *Strychnos cabalonga* hort. Lind. and as *Strychnos brachiata* Ruiz et Pav. (cf. **Strychnos spp.**).

The name *cabalonga blanca*, "white cabalonga" —considered to be a weaker relative—is used to refer to a *Thevetia* species (***Thevetia* spp.**) that is also used as an **ayahuasca** additive.

Above: The cabalonga fruit, obtained at the "witches' market" in Chiclayo, Peru.

In Mexico, several plants are known by the name *cabalonga*, some of which may have psychoactive properties (Martínez 1987, 119*):

In Spanish, the name *cabalonga* is generally used to refer to the Philippine Ignatius bean (*Strychnos ignatii* Bergius; cf. **Strychnos spp.**).

Plants Known as Cabalonga

cabalonga	*Jatropha multifida* L.	Euphorbiaceae
cabalonga	*Strychnos panamensis* Seem.	Loganiaceae
cabalonga	*Thevetia peruviana* (Pers.) Mer.	Apocynaceae
cabalonga de la huasteca	*Thevetia peruviana* (Pers.) Mer.	Apocynaceae
cabalonga de tabasco	*Thevetia peruviana* (Pers.) Mer. et Sandw.	Apocynaceae

Literature

Faust, Franz Xaver, and Antonio Bianchi. 1998. Die mysteriöse Cabalonga. *Yearbook for Ethnomedicine and the Study of Consciousness,* 1996 (5): 247–51. Berlin: VWB.

Channa

See **kanna**.

Charin Peco

This name is used to refer to an unidentified "parasitic plant" that formerly was used in the shamanism of the Peruvian Shipibo Indians as a psychoactive initiatory plant. Its use was apparently identical with the use of another uncertain epiphyte, **shahuan-peco**. However, the effects of charin peco were said to be weaker than those of **shahuan-peco** (see there).

Devil's Foot Root

In a 1966 article, Varro E. Tyler remarked upon the devil's foot root or *radix pedis diaboli*, a root that is said to grow on the Ubangi River between the Congo and former French Equatorial Africa. It supposedly has the form of a human foot and is red-brown in color. West African medicine men are said to have used it for trials by ordeal. The root is purported to have psychoactive effects when it is burned as an **incense** and the smoke inhaled. Tyler's statements were based not upon his own studies but on an article in a book from A. Conan Doyle (*The Complete Sherlock Holmes*) (Tyler 1966, 292*). It is possible that the entire story is nothing more than a late Victorian fiction.

Dionysonymphas

"Bride of Dionysos"

Dionysonymphas is another name for **hestiateris**.

Dodecatheon

"Twelve Gods' Plant"

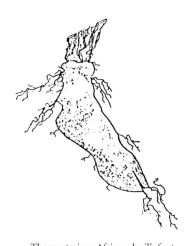

The mysterious African devil's foot root is said to resemble a foot—half human, half goat's hoof—and to produce profound psychoactive effects when burned as an incense. The plant may be nothing more than a literary joke. (Illustration from 1910)

Pliny discussed this wondrous and as yet not clearly identified "plant of the gods" directly after **moly**:

> After [*moly*], the most highly regarded plant is one that is known as the twelve gods' plant, recommending it [as a symbol] for the power of all the gods. Drunk in water, it is said to heal all ailments. It has seven leaves that are very similar to those of lettuce [*Lactuca*] and that arise from a yellow root. (*Natural History* 25.28)

The plant may have been a species of *Lactuca* (cf. **Lactuca virosa**). The Gaul Marcellus (ca. 400 A.D.) wrote in his Latin recipe book *De medicamentis* that dodecatheon was also called *donax* (27.7). This may be a reference to **Arundo donax**. The twelve gods' plant has also been identified with primrose (*Primula elatior* [L.] Hill or *Primula veris* L.) and butterwort (*Pinguicula vulgaris* L.) (Dierbach 1833, 176*). Other sources suggest that dodecatheon was a medicine compounded from twelve plants that was said to be effective against all diseases (Baumann 1982, 115*). It may have been an alchemical product made of the twelve sacred plants that were consecrated to the twelve Olympian gods. This divine preparation may have borne those who used it up to Olympus (cf. Müller-Ebeling et al. 1998).

Ephemeron

"One-Day Flower"

Since ancient times, the magical plant that Medea used to induce the dragon that guarded the Golden Fleece to fall asleep has been known as *kolchikon*. Linnaeus subsequently chose this name as the genus name for the crocus (*Colchicum autumnale* L.; Baumann 1982, 111*). But ephemeron and the crocus are almost certainly not identical (Engel 1978, 18*). Virgil described how Moeris, a herdsman versed in magic, used this plant to transform himself into a wolf:

> This plant here, this poison, gathered beforehand
> Moeris himself gave to me; it grows in abundance on the Pontus.
> With it—often did I see it—Moeris transformed himself into a wolf
> and hid in the forest; in this form, he lured the souls
> from the grave; in this form, he could even move the seeds in the ground.

The shifting of forms from human to wolf was well known in ancient times and was usually associated with the consumption of human flesh (Burkert 1997, 98 ff.), and sometimes also with the use of *Aconitum napellus* or *Lupinus* spp. (cf. witches' ointments).

Literature

Burkert, Walter. 1997. *Homo Necans: Interpretationen altgriechischer Opferriten and Mythen.* 2nd ed. Berlin: de Gruyter.

Espingo

The literature from the colonial period, and especially Arriaga's report from Peru (1621), makes frequent mention of a fruit named *espingo, ishpingo, ispinku, yspincu,* or *ispincu.* It was said to be rather small and roundish and somewhat similar to an almond. It was used as a medicine and as an inebriating additive to **beer**:

> In the plains, from Chancay and downriver, the chicha that the huacas [= temples] offer is known as yale. It is made from zora [= malted maize] mixed with chewed maize and has espingo powder added to it. They prepare the chicha so that it is very potent and syrupy. After they have poured as much of this over the temple as they think proper, the magicians drink the rest. The chicha makes them completely crazy. (Arriaga 1992, 40*)

Espingo had a magical and religious significance similar to that of villca (see **Anadenanthera colubrina**):

> The espingo is a dry, small fruit similar to the round almond, with a penetrating but not very pleasant scent. They bring it from the Chechepoyas [i.e., from the tropical lowlands]. They say that when taken in the form of a powder, espingo is a very effective medicine against stomachaches, bloody diarrhea, and other diseases, and they purchase it for a good price. People are wont to sell espingo for these purposes. Not too many years ago, in Jaén de Bracamoros, the Indios paid their tribute in espingo. The deceased Lord Archbishop forbade the sale of espingo to the Indios under pain of excommunication, for he knew that it was a common offering at the huacas. In the flatlands in particular, there is not a one who has conopas [= images of the gods, sacred objects] who does not also have espingo, no matter how often they have been searched. (Arriaga 1992, 43*)

The Incas offered a **chicha** with an espingo additive at the festivals of Raimi, Citua, and Aymoray (Wassén 1973, 40). One source states that those who consume espingo, e.g., the Incan magicians (*omo*), would "go crazy" because of the fruit (Wassén 1973, 38). In the seventeenth century, the chronicler and missionary Bernabé Cobo claimed that the pleasantly scented fruit of the espingo tree was called *vainilla* ("vanilla") and was traded over long distances, and that the Indians esteemed it highly for its medicinal properties. Henry Wassén, who has worked on the botanical identification of the stock plant, has assumed that a number of plants with similar properties were subsumed under the name *espingo.* Today, there is still a plant known as *asango-espingo* (perhaps a species from the Rubiaceae or the Lauraceae Family). The following plants have all been suggested as possible candidates for the espingo of the colonial period (cf. Wassén 1979, 59 ff.; 1973, 40):

Quararibea **spp.**: possibly psychoactive (Ott 1993, 418*)
Gnaphalium dysodes Spreng.: unknown[342]
Trifolium sp. (*trébol*): botanical identity unknown
nambicuara (Bombacaeae): unknown effectiveness

A botanical and anatomical study compared espingo seeds with the seeds of *Quararibea* and concluded that the two may be identical (Bondeson 1973). However, if no additional ethnohistorical sources or archaeological or ethnobotanical samples are discovered, the knowledge of the identity of the espingo tree may be lost.

In Ecuador and Peru, the American cinnamon tree (*Ocotea quixos* Lam. [syn. *Nectandra cinnamomoides* Nees]; Lauraceae) is still referred to as *ishpino* or *espingo.* A chemical analysis of a botanically unidentified sample of espingo found substances that have the scent of cinnamon: cinnamaldehyde, *O*-methoxycinnamaldehyde, cinnamic acid, and methylcinnamate. The same substances are found in *Ocotea quixos* (Naranjo et al. 1981). *Ocotea* continues to find favor as a cinnamon-like spice and is used in Ecuadoran folk medicine as an appetite stimulant, disinfectant, and diarrhea remedy. The scented calyxes are used as an aromatic additive to a **chicha** made from maize and sugarcane juice that is known as *alajua.* This mildly alcoholic drink is offered in Ecuador to the ancestors of one's own family (Naranjo et al. 1981, 234).

Today, the following plants are known in Peru as *ishpingo* or *espingo*:

Ajouea tambillensis Mez (Lauraceae)
Jacaranda copaia (Aublet) D. Don (Bignoniaceae)
Ocotea jelskij Mez (Lauraceae)
Guarea trichilioides L. (Meliaceae): among ayahuasqueros (cf. **ayahuasca**)

The wood of this tree is used to make magical wands for the northern Peruvian *mesa* rituals (Giese 1989, 259*; cf. **Trichocereus pachanoi**).

In northern Peru, espingo buds are still used in folk medicine, sometimes in combination with

Above: An espingo necklace from the "witches' market" in Chiclayo, Peru.

342 In Ecuador, *Gnaphalium dysodes* Spreng. (Compositae) is still known as *ispingo* (Naranjo et al. 1981).

Left: The South American coca bush *Erythroxylum hondense*.

Right: The coca plant *Erythroxylum cumanense* with its typically small, very round leaves. Many species of *Erythroxylum* contain alkaloids, but most contain only traces of cocaine.

Halicacabon, a mysterious psychoactive plant of antiquity, was identified in early modern times as the Chinese lantern plant (*Physalis alkekengi* L.). (Woodcut from Fuchs, *Läebliche abbildung und contrafaytung aller kreüter*, 1545)

cabalonga and nutmeg (*Myristica fragrans*). The precise identification of these buds is not known.

In the Ucayali region of Peru, a leguminous tree, *Amburana cearensis*, is known in the local Spanish as *ishpingo*. Healers and shamans esteem it as a "master plant." The bark is used to prepare medicinal baths to treat illnesses that are the result of harmful magic (Arévalo V. 1994, 251*).

Literature

Bondeson, Wolmar E. 1973. Anatomical notes on espingo and seeds of *Quararibea*. *Göteborgs Etnografiska Museum Årstryck* 1972:48–52.

Brenner, F. 1975. El ishpingo: su oso pre-columbino y actual. *Folklore Amer.* 19:101–4.

Naranjo, Plutarco, Anake Kijjoa, Astréa M. Giesbrecht, and Otto R. Gottlieb. 1981. *Ocotea quixos*, American cinnamon. *Journal of Ethnopharmacology* 4:233–36.

Wassén, S. Henry. 1973. Ethnobotanical follow-up of Bolivian Tihuanacoid tomb material, and of Peruvian shamanism, psychotropic plant constituents, and espingo seeds. *Göteborgs Etnografiska Museum Årstryck* 1972:35–47.

———. 1979. Was *espingo* (*ispincu*) of psychotropic and intoxicating importance for the shamans in Peru? In *Spirits, shamans, and stars*, ed. D. L. Browman and R. A. Schwarz, 55–62. The Hague: Mouton.

Gelotophyllis
"The Leaf That Evokes Laughter"

Pliny has provided us with a description of this plant that is based upon the writings of Democritus:

> The *gelotophyllis* grows in Bactria [= Afghanistan] and on the Borysthenes [= Dnieper]; when drunk together with myrrh and wine, various figures will float before one's eyes, and the urge to laugh will cease only if one eats pine nuts with pepper [*Piper* spp.] and drinks honey in palm wine. (24.164)

This "plant that incites laughter" has often been interpreted as *Cannabis indica*. Indeed, Indian hemp does grow in Afghanistan, and the laughter that the use of hemp can evoke is legendary.

Guanguára
The Kogi Indians of the Sierra Nevada de Santa Marta (Colombia) call their coca *hayo*, a word that comes from the Tairona language. They distinguish three "species" of coca, each of which belongs to one of the tribes of the region. The Kamkuama tribe, which has now either disappeared or become extinct, cultivated a type of coca that had long leaves. The Kogi plant a variety with small leaves, and the Ika a type with tiny leaves (cf. *Erythroxylum novogranatense*). It is said that the ancestors, i.e., the Tairona, planted or harvested a tree named *guanguára* or *guanguála* in the *páramos* (high marshes) that had cocalike leaves with similar effects (Reichel-Dolmatoff 1985, 1:87*; Uscategui M. 1959, 281 f.*). Unfortunately, the botanical identification of this tree is unknown. It may simply have been another wild species of the genus *Erythroxylum*.

Haoma
The original haoma plant has not yet been identified with certainty. See the section on **haoma** on pages 747 f.

Halicacabon
"Salt Jar"

Also known as *halikákabon* or *halicacabum*, this plant belonged to a group of psychoactive plants known as *trychnos* or *strychnos* (cf. **strychnos manikos**), the botanical identity of which is unknown:

> There is yet another kind [of *strychnos*], called "salt jar," that provokes sleep and can cause death even more quickly than opium; others call it "fool's herb" [*morion*], still others moly. But it was praised by Diocles and Euenor, and Timaristos even lauded it in a poem, wherein she strangely forgot about its harmlessness, for her, [the plant] is a fast-acting agent for firming up wobbly teeth by rinsing these with halicacabon in wine; at least they added the qualification that this should not be done for very long, lest madness be the result. . . . The root of halicacabon is imbibed as a drink by those who wish to utter prophecies and who want to be seen as truly enthusiastic about

God in order to strengthen superstitious conceptions. As an antidote—which I am that much happier to be able to mention—water to which an ample quantity of mead has been added should be drunk warm. And I do not want to neglect to say that halicacabon is so contrary to the nature of the adder that its root, when placed very close to it, will quiet the very deadly power that its stupefaction can bring. This is why grinding it up and adding it to oil it can aid a person that has been bitten. (Pliny 21.180–82)

This entheogenic plant has been interpreted as *Atropa belladonna* or *Withania somnifera*. Dioscorides described *morion* as a species of *Mandragora* (cf. *Mandragora* spp.).

Leonhard Fuchs depicted two plants under the name *halicacabum* (Fuchs 1545). *Halicacabum vulgare* is the Chinese lantern (*Physalis alkekengi* L.; cf. *Physalis* spp.). The other species (*halicacabum peregrinum*, welsch schlutten) is probably *Cardiospermum halicacabum* L. Linnaeus selected the ancient name as a species name for the American balloon vine (*Cardiospermum halicacabum* L.; Sapindaceae), which is not known to have any psychoactive effects.

Hestiateris
"Plant That Gives a Meal"

Pliny's information about this plant is based upon the statements of Democritus, who was well versed in the magical arts:

> In Persia, the *hestiateris* has its name from the guest meal, for it produces gaiety there; it is also called *protomedia* ["preferred by the Medians"] because it can be used to ingratiate oneself with their kings; *kasignete* ["the fraternal"] because it grows only with its own kind and not with any other plants; it is also called dionysonymphas ["bride of Dionysos"] because it goes wonderfully with wine. (24.165)

This plant has been variously interpreted as *Areca catechu*, pimpinella (*Pimpinella major* [L.] Huds.; Umbelliferae), and *Sanguisorba minor* Scop. [syn. *Poterium sanguisorba* L., *Pimpinella minor* (Scop.) Lam.] (Rosaceae).

Hippophobos
The name means "refused by horses" and is a synonym for **achaimenis**.

Jénen-Joni-Rau
"Water People Medicine"

This as yet unidentified little plant with lanceolate leaves is said to be similar to the **shahuan peco** plant, which is also botanically unknown. The Peruvian Shipibo-Conibo Indians regard it as a "master of the healing arts." A shaman will consume parts of this plant with **ayahuasca** when he wishes to undertake a spirit or astral journey into the water world (Gebhart-Sayer 1987, 337*).

Kanna
Some 250 years ago, the Hottentots (Khoikhoi) chewed kanna, also known as *canna* or *channa*, in order to induce visual hallucinations:

> About 200 years ago, Kolbe used the name Kanna (Channa) to refer to a plant whose root he supposedly saw the Hottentots use as an agent of pleasure. They chewed it and retained it in their mouths for a long time. It made them drunken and stimulated. Their "animal spirits became animated," their eyes sparkled, and their faces were marked by laughter and happiness. A thousand charming ideas arose in them, a gentle joy that amused itself over the simplest of jests. When they used too much of the agent, they would ultimately lose consciousness and fall into horrible deliriums. (L. Lewin 1980 [orig. pub. 1929], 296*)

Because the name was used at the end of the nineteenth century to refer to one or more *Mesembryanthemum* species (*Mesembryanthemum* spp.), it was assumed that the originally described inebriant was in fact *Mesembryanthemum tortuosum* (Hartwich and Zwicky 1914). Today, the valid botanical name for this plant is *Sceletium tortuosum*. Louis Lewin was of the opinion that the plant may also have been *Sclerocarya caffra* or *Sclerocarya schweinfurthiana* (1980, 297*).

In the seventeenth century, white settlers in South Africa applied the name *kanna* to *Aureliana canadensis*, the root of which was used as a substitute for the true mandrake (*Mandragora officinarum*).

The word *kanna* is almost too reminiscent of *cannabis* (cf. *Cannabis indica*). Perhaps the original kanna was actually a mixture of *Cannabis sativa* and *Sceletium tortuosum*.

In its genus name, the canna lily carries on the name *kanna*, spelled as *canna*. Canna seeds are sometimes strung together to make bracelets and necklaces. (*Canna indica*, photographed in Chiapas, Mexico)

The Andean inebriant *macha-macha* may be a pseudoberry. (*Gaultheria acuminata*, from Latin America)

Apart from the name, kanna has nothing to do with the lilylike plant *Canna indica* L. (Cannaceae), a native of tropical America whose seeds the Indians string into bracelets and necklaces. Kanna is sometimes confused with *Salsola dealata* Botsch., known as *ganna*.

Literature

Hartwich, Carl, and E. Zwicky. 1914. Über Channa, ein Genußmittel der Hottentotten. *Apotheker-Zeitung* 29:925 f., 937–39, 949 f., 961 f.

Keu

In 1875, the ethnologist Miklucho-Maclay described an inebriating drink named *keu*, which the Papuas of the Maclay Coast (Papua New Guinea) prepared in a manner similar to the Polynesian kava-kava. Although the men, young men, and boys produced the drink on the occasion of tribal festivals, only the older people were allowed to consume it. It is possible that this bush, the leaves, stalk, and roots of which were used, was *Piper methysticum*, a different *Piper* species (*Piper* spp.), or an as yet unknown plant. This theory is not unlikely, because we do know that *Piper methysticum* was used on Papua New Guinea (Haddon 1916). But the drink that was prepared from this plant is known as *wati*. More recently, a previously undescribed *Piper* species was discovered on Papua New Guinea. Chemical investigations isolated 2-methoxyangonin, kavalactones, and flavonoids from the material (Hänsel et al. 1966; Sauer and Hänsel 1967).

Literature

Haddon, A. C. 1916. Kava drinking in New Guinea. *Man* 16:145–52.

Hänsel, R., H. von Sauer, and H. Rimpler. 1966. II-methoxyangonin aus einer botanisch nicht beschriebenen Piperart Neu-Guineas. *Archiv der Pharmazie* 299:507–12.

Miklucho-Maclay. 1875. Ethnologische Bemerkungen über die Papuas der Maclayküste. *Naturkundig Tijdschrift* (Batavia) 1:8.

Sauer, H. von, and R. Hänsel. 1967. Kawalaktone und Flavonoide aus einer endemischen *Piper*-Art Neu-Guineas. *Planta Medica* 15:443–58.

Macha

In South America, a number of plants of the Family Ericaceae are given the name *macha* or *macha-macha* and are regarded as psychoactive. The ripe fruits are sweet and delicious and invoke inebriating effects when consumed in great quantities. These effects are described as "drunkenness." The botanist Hippolito Ruíz identified macha as *Arbutus parviflora* [nom. nud.!] and characterized the Ericaceae known as *macha-*

macha as *Thibaudia* (Schultes 1980, 111*). It is possible that this name refers to various species of the genus *Vaccinium* (cf. **Vaccinium uliginosum**) or **Pernettya**. In Peru, the wandering healers of the Callawaya refer to a *Gaultheria* species as *macha-macha*; it contains methylsalicylate (Bastien 1987, 128*). There is also said to be *macha-macha* in the Las Huaringas region of northern Peru, especially at Laguna Shimbe. Unfortunately, nothing is known about its botanical identity. The Ericaceae *Befaria glauca* var. *coarctata* has fruits that cause dizziness and have a stimulating effect. It is known as *macha-macha* in Bolivia (von Reis Altschul 1975, 214*).

Marari

The renowned ethnologist Alfred Métraux wrote that the shamans of the Mojo Indians (a Brazilian Arawak tribe) consumed a drink called *marari*, which they made from a plant of the same name, when they wished to consult the spirits. The effects of the drink were said to last for some twenty-four hours and were described as excitation, sleeplessness, and pains—nothing particularly pleasant. Marari was said to compare to "our verbena." It is not clear whether Métraux was comparing this plant to the European verbena (*Verbena officinalis* L.)[343] or to another plant from South America that has long been cultivated in the Mediterranean region, lemon verbena (*Lippia citriodora* H.B.K. [syn. *Aloysia triphylla* (L'Herit.) Britt.]). There are a number of plants (*Verbena* spp., *Lippia* spp.) in South America that are very similar to these two verbenas (and which the literature on **essential oils** and **incense** usually confuses). Only careful ethnobotanical research will be able to cast light upon these questions (Schultes 1966, 298*).

The wonderfully scented lemon verbena (*Aloysia triphylla* = *Lippia citriodora*) is also called the lemon bush and is often confused with the true verbena (*Verbena officinalis*).

343 If the ancient sources are correct, a verbena known as *hiera botane*, "sacred plant," was a magical and psychoactive plant (cf. Rätsch 1995a, 154 ff. *).

Literature

Métraux, Alfred. 1943. The social organization and religion of the Mojo and Manasi. *Primitive Man* 16:1–30.

Marmaritis

See **aglaophotis**.

Medeaka

This tropical rain forest plant is found in West Africa, where the Pygmies used it as an alternative agent of pleasure and inebriant to *bangi* (***Cannabis sativa***) and *tava* (***Nicotiana tabacum***):

> In an emergency, when there is no tobacco or hemp to be found, the Pygmies turn to the leaves of a forest plant, *medeaka*, which they smoke. The effects are said to be stronger than those of *bangi* [= *Cannabis sativa*]. But it does not seem correct that the leaves of the poisonous *tava* tree are smoked as well. (Schebesta 1941, 179)

The root, which was known as *masili*, was chewed as an aphrodisiac (Schebesta 1941, 236). This plant may be one of the many West African *Mitragyna* species (cf. **Mitragyna speciosa**).

Literature

Schebesta, Paul. 1941. *Die Bambuti-Pygmäen von Ituri, vol. 2: Ethnographie der Ituri-Bambuti.* Brussels: George van Campenhout.

Moly

Homer provided only a brief description of moly, the divine magical plant of Hermes. Odysseus used moly, which was classified as a *pharmakon*, "remedy/poison," to protect himself from the sorceress Circe, who had transformed his men into swine (Schmiedeberg 1981). An important piece of information is that Hermes (= Mercury) harvested the plant in Circe's garden, and thus the messenger of the gods transformed one of Circe's own magical plants into an almost homeopathic type of antidote:

> [Hermes] pulled from the ground a
>   poisonous plant,
> Presented it to me and showed me what it
>   was and how it grew.
> Black was the root, white as milk were the
>   flowers, the gods
> call it moly. It is very difficult for mortal men
> to find; the gods, of course, can do anything.
>
> (*Odyssey* 10.302–6)

Generations of alchemists, Greek scholars, philologists, pharmacologists, and ethnobotanists have attempted to determine the botanical identity of Homer's magical plant (cf. Baumann 1982, 110*; Dierbach 1833, 192*; Rahner 1957*). Even Theophrastus, the father of botany, made an effort to ascertain the botanical identity of the stock plant:

> Panakeia, the plant that heals all, thrives in large quantities and prefers rocky ground near Psophis, Moly by Pheneos, and on the mountain of Cyllene. They say that this plant would be like the moly that Homer mentions; that it has a root like an onion and leaves like the squill; and that it is used against magical formulae and magic, but it is not, as Homer stated, difficult to dig up. (Theophrastus, *History of Plants*, 9.15)

Dioscorides (3.21) wrote that the sea holly or sea hulver *Eryngium maritimum* L. (Umbelliferae), prized as an aphrodisiac, was also known as moly (Rätsch 1995a, 228 ff.*).

Moly was already being construed as a psychoactive plant in ancient times. Remarkably, both moly and the magical plant of Circe were interpreted as the mandrake (***Mandragora officinarum***) (Dierbach 1833, 204*; Kreuter 1982, 29*). Dioscorides passed down the name *circeon* for the mandrake (cf. Rahner 1957, 201*) and *mandragora circaea* for the plant that Circe used to transform the men of Odysseus into "swine" (= sexually aroused men):

> The Mandragora. Some call it *antimelon* [= "in the apples' place"], others *dirkaia*, also *kirkaia* [= "plant of Circe"], as the root appears to be effective as a love charm. (Dioscorides 4.76)

According to Apollodorus (second century B.C.), the most important scholar of his time, mandrake (which he called *kirkaia riza*) was used as an amulet against the harmful magic of Pasiphae, daughter of Helios (Circe was also a daughter of Helios), wife of King Minos, and mother of Ariadne and the Minotaur (frag. 2.15).

In late ancient times, it was assumed that the mandrake had been a gift from the Greek/Egyptian god Hermes Trismegistus—the god of alchemy (Fowden 1993)—and that it was suitable for conjuring up spirits and for use in alchemical practices. Because Hermes/Mercury is depicted in some ancient works of art with an opium capsule in his hand, moly could be interpreted as ***Papaver somniferum***.

Bauhinius identified the renowned Hermetic magical plant moly, which Odysseus used to restore the men who had been turned into swine back into men, as rue (*Ruta graveolens* L.; cf.

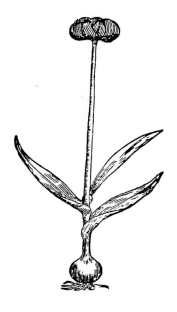

The so-called fathers of botany went to great lengths in their efforts to botanically identify moly, the magical plant of Homer. The squill was long considered to be the legendary plant. However, with the discovery of new plants in the New World, many of the species that were returned to Europe were also regarded as candidates. The botanical illustrations, however, suggest a plant that grew from the imagination of the engraver and his client. (Woodcut from Tabernaemontanus, *Neu Vollkommen Kräuter-Buch*, 1731)

The beautiful Helena offering the wondrous nepenthes. (Illustration: John Flaxman)

In late ancient times, the squill (*Urginea maritima* [syn. *Scilla maritima*]) was already being interpreted as the Hermetic magical plant moly. However, any information about the psychoactive effects squill may have has not come down to us. (Wild plant, photographed on Naxos, Greece)

soma). Stannard (1962) identified the Hermetic moly with *Peganum harmala* and explicitly referred to a passage in Dioscorides:

> Wild rue. Some also call wild peganon the plant that is known as *moly* in Cappadocia and in Asian Galatia. It is a bush that develops several branches from a single root, it has leaves, much larger and more tender than the other peganon [= rue] and with a penetrating scent, a white flower, small capitula on the tips, larger than with cultivated peganon [= rue], mostly consisting of three parts, in which is found a trigonal, light yellow seed that is useful. In late autumn, the seed becomes ripe and can be used, with honey, wine, chicken gall, saffron [*Crocus sativa*], and fennel juice [*Foeniculum vulgare*], finely ground, to treat dull vision. Some also call this same plant *harmala*, the Syrians *besasa*, the Egyptians *epnubu*, the Africans *churma*, but the Cappadocians *moly*, because it is essentially similar to the *moly*, for it has a black root and white flowers. It grows in hilly and fertile soil. (Dioscorides 3.46)

Moly has long been interpreted to be the squill (*Urginea maritima* [L.] Baker [syn. *Scilla maritima* L.]) (Rahner 1957*):

> According to the testimony of Homer, the most renowned of all plants is *moly*, about which he believed that it received its name from the gods; he attributes Hermes with its discovery and use against the strongest enchantment. It is said to grow today in the region of Pheneos and on Cyllene in Arcadia. According to Homer, it is a plant with a round, black root the size of an onion and with one leaf like the squill [*Scilla*], but it is (not) difficult to dig up. The Greek writers have reported its flowers to be yellow, whereas Homer wrote that they are white. I found a physician wise in herbal lore who told me that it also grows in Italy, and after a few days (in autumn) he had one brought to me from Campania that was dug up under difficult circumstances from craggy rocks. It had a 30-foot-long root that was not complete but had been ripped out. (Pliny 25.26–27)

Nevertheless, Lucian describes in a dialogue (Menippus) how the priests or magicians, dressed in the style of the Persians, used the squill as a magical plant for conjuring up the dead for oracular purposes (cf. *Lupinus* spp.). The squill contains highly effective cardiac glycosides (cf. *Digitalis purpurea*) (Roth et al. 1994, 714*).

Other plants with bulbs (Liliaceae) have also been interpreted as moly, including *Nectaroscordum siculum* (Ucria) Lindl. [syn. *Allium dios-*

coridis Sibthorp, *Allium siculum* Ucria], *Allium magicum* L. (magic leek), *Allium moly* L. (Linnaeus actually named this species after the magical plant), *Allium nigrum* L., and *Allium sativum* L. (Dierbach 1833, 192; Schöpf 1986, 117*). *Allium victorialis* L. is known as false mandrake, for its root was often used in place of that of the true mandrake (cf. *Mandragora officinarum*).

Wedel, a pharmacologist, wrote in one of his dissertations (*De Mythologia Moly Homeri*, Jena, 1713) that moly was a *Nymphaea* (cf. **Nymphaea caerulea**). In his work *De Moly Homerico et fibula Circaea* (Lipsiae 1716), D. W. Triller identified moly as black hellebore (*Helleborus niger* L.; cf. **Veratrum album**, snuff).

It appears that in ancient times, the word *moly* was more or less a catchall term that meant "magical plant" or "entheogen." It was used to refer to plants that were psychoactive and employed for magical purposes (in a manner quite similar to the use of the words **haoma** and **soma**).

Literature

Fowden, Garth. 1993. *The Egyptian Hermes: A historical approach to the late pagan mind.* Princeton, N.J.: Princeton University Press.

Schmiedeberg, O. 1918. Über die Pharmaka in der Illias und Odyssee. *Schriften der wissenschaftlichen Gesellschaft in Straßburg* 36.

Stannard, J. 1962. The plant called moly. *Osiris* 14:254–307.

Nepenthes

In the *Odyssey*, Homer sang the praises of the psychoactive effects of a wondrous *pharmakon* called *nepenthes* (literally "consoling plant"):

> But Helen, the daughter of Zeus, remembered something:
> and immediately put a magical agent into the wine which they drank,
> good against sorrow and bilious nature: for all evils
> it created forgetfulness. It was in the mixing jug: anyone who then drank of it,
> on that day would no tears flow down his cheeks,
> even when both his father and mother would die,
> yea even when his son, the beloved, or his brother
> would be slain with swords by the enemy directly before him,
> so that he would see it with his own eyes. Now the daughter of Zeus could avail herself of agents with such skillful effects."

(*Odyssey* 4.219–28)

Theophrastus himself attempted to fathom the origins of this *pharmakon*:

> The places outside of Hellas that yield special healing herbs are regions in Tyrrhene and Latium (where, it is said, lives Circe), and several regions in Egypt, about which Homer said: Helen brought from there the useful things that Polydamna, the wife of Thon, had given her. There the fruit-bearing earth produces the greatest number of *pharmaka*, many of these are beneficial and others harmful. Among these, so he tells us, was *nepenthes*, the famed *pharmakon* that heals worries and afflictions, for it allows one to forget diseases. These regions have been named by the poets. (*History of Plants*, 9.15)

Because of the effects attributed to it, the Homeric nepenthes has often been interpreted as opium (cf. **Papaver somniferum**) as well as certain other plants:

> In contrast, another assumption is more likely, that the agent of Helena was the juice of *Cannabis indica*, hashish. The use of hemp and of pills made from it was . . . known in Egypt at an early date and was very popular, and because of the connections between Greece and Egypt, Homer may . . . have had very good knowledge of this. . . . Moreover, the condition produced by its use, which corresponds to the Oriental names for it, "stimulator of happiness, soul cheerer" [cf. **Oriental joy pills**], accords with that which Homer described, namely a cheerfulness that cannot be dampened by the hardest strokes of fate. (Berendes 1891, 131 f.*)

Since the stock plant was said to be indigenous to Egypt, it has also been identified as Egyptian henbane (*Hyoscyamus muticus*):

> The Egyptian priests used this plant in many ways during their religious practices, especially to placate that hostile god whom they called Typhon. The sedative powers of this Egyptian henbane appear to be even more energetic than those of the one we are used to [*Hyoscyamus niger*]; for when someone receives some of the powder of the plant either accidentally or on purpose, there follows a condition of madness that continues for several days, after which clear consciousness returns once more. This plant has been interpreted as the *nepenthes* of Homer, a statement that should not be entirely disregarded, but that in fact has much that speaks for it. (Dierbach 1833, 189*)

Nepenthes has also been identified as **Mentha pulegium**, **Mandragora officinarum**, and even **Catha edulis**.

In modern times, the name *nepenthes* is used to refer to a genus of carnivorous plants (Family Nepenthaceae; pitcher plants) that encompasses a number of species (e.g., *Nepenthes maxima* Reinw., *Nepenthes sanguinea* Lindley). These plants, however, are not psychoactive (Emboden 1974, 135f.*).

Ophiussa
"Snake Plant"

Pliny provided us with a description of this plant, which comes from the island of Elephantine on the Nile, renowned as the place where the mandrake (**Mandragora officinarum**) was found:

> The *ophiussa* on Elephantine in the same Ethiopia has a lead-colored and unsightly appearance and is said, when drunk, to produce such dread and such terror of snakes that people would kill themselves out of fear; this is why the desecrators of temples are forced to drink of it. Palm wine is an antidote. (24.163)

This information inspired the film *Young Sherlock Holmes* (from the studio of Steven Spielberg). In this film, an Egyptian sect of murderers kills its victims by shooting an arrow that has been treated with an unidentified hallucinogen into the carotid artery. The victims experience such terrible visions, including snakes, that they commit suicide. Today, many people continue to believe that people will throw themselves out of windows because they have been driven to madness by psychedelics. This motif, which is quite popular in the yellow press, appeared in a novel that Leo Perutz (1882–1957) wrote in 1923, *Der Meister des jüngsten Tages* [The Master of Judgment Day] (cf. **Claviceps purpurea**).

Literature

Perutz, Leo. 1990. *Der Meister des jüngsten Tages*. Reinbek: Rowohlt.

Pipiltzintzintli

This Aztec inebriant, also written as *pipizizintli*, *pepetzintle*, *pepetichinque*, and *pipiltzintzin*, is now usually interpreted as **Salvia divinorum**. This interpretation, however, is by no means certain.

Potamaugis
"River Sheen"

Potamaugis is another name for **thalassaigle**.

Poyomatli

This Aztec inebriant is mentioned in the *Florentine Codex*, written in colonial times by

The relatively unknown Mexican *Magnolia dealbata* may have been the source of the Aztec inebriant poyomatli.

Sahagún. It has not yet been clearly identified. One suggestion is *Quararibea funebris* (see **Quararibea spp.**, **Theobroma cacao**). Another possibility is that the name refers to the flowers of *Cymbopetalum penduliflorum* (Dunal) Baill. *Magnolia dealbata* Zucc. (see **Magnolia virginiana**), now known as *elexuch wu*, has also been suggested as a candidate (Diaz 1979, 94*; Ott 1993, 412*). If no additional ethnohistorical sources are discovered, the secret of poyomatli may have been lost with the destruction of the Aztec culture.

Sakaka

The Desana Indians of the Tukano tribe, who live in the Vaupés region of Colombia and Brazil, have different classes of shamans. The sakaka shamans—the word is from the *língua-geral*—specialize in journeying to the underworld, especially the underwater worlds. In order to travel there, a shaman will chew the root of a plant known by the name *sakaka*. This enables him to traverse great distances underwater. Because they are said to be related to malevolent water beings, these shamans are regarded as rather dangerous. To date, the sakaka plant has not been clearly identified; it may be a species from the Family Connaraceae (Buchillet 1992, 212). Whether this genus contains psychoactive species is unknown; several species do have toxic properties (Buchillet 1992, 228). The Indians use some species of the genus *Connarus* as fish poisons (Schultes and Raffauf 1990, 141*).

Literature

Buchillet, Dominique. 1992. Nobody is there to hear: Desana therapeutic incantations. In *Portals of power: Shamanism in South America*, ed. E. J. M. Langdon and Gerhard Baer, 211–30. Albuquerque: University of New Mexico Press.

Semnios

"Sacred Plant"

Also written as *semnos*, *semnios* is another name for **theombrontion**.

Shahuan Peco

"Moult of the Ara Parrots"

The Shipibo-Conibo, who live in the Ucayali region of Peru, use this name to refer to an as yet botanically unidentified "parasitic plant that grows on tall trees" (Illius 1991, 122). It is said to be very rare, is considered to be difficult to find, and cannot be recognized by everyone. This hallucinogenic parasite nests in the forks of branches of various jungle trees in the area around Pucallpa (Peru). It has three-fingered leaves, and the

Shipibo-Conibo shamans use only the longest, middle leaflet. The freshly crushed leaflet is either mixed with tobacco (*Nicotiana tabacum*) and smoked or mixed with tobacco juice and drunk. In addition, the diluted juice is rubbed over the entire body (Gebhart-Sayer 1987, 211*).

In former times, anyone who wished to become a *merayabo*, a priest-shaman, was required to use this ominous plant. The novice needed to consume a drink made of the small leaflets and "bathe" in it just once during his training. The visionary effects began some twelve hours later and lasted for a similar period; they were culturally stereotyped:

> You hear loud thunder, even when the sun is shining and not a cloud can be seen, you begin to tremble all over your body and you hear voices. The two "lords" of the *shahuan pecó* appear in human form. They threaten the drinker and attempt to talk him out of wanting to make use of the power of the plant. When he has become almost unconscious (*shiná-oma*) as a result of the continued trembling, the two *yoshinbo* [= plant spirits] throw him back and forth many times like a ball. A jaguar appears, bites the adept in the back of the neck, and sucks out all of his blood "so that he becomes light." Then the jaguar bears him upward for several hours. Suddenly the *meraya*-to-be notices that he can move through the air himself and that he "sees everything." In the clouds, the *yoshinbo* provide him with information about all manner of diseases. Finally, the jaguar leads him to *anta yoshin*. The *anta yoshin* is a great physician who reveals to the shaman how he can heal himself and protect himself from the attacks of "brujos" ("sorcerers"). (Illius 1991, 122 f.)

Unfortunately, the use of this plant is disappearing among the Shipibo. Shahuan peco is an initiatory plant that should be ingested by shamans-in-training after they have been consecrated with **ayahuasca**. It is regarded as the most powerful of the visionary plants (Gebhart Sayer 1987, 340*), and a shaman typically will take it only once during his lifetime. The plant spirit (*shahuan-anta-jonibo*) gives him the ability to undertake spirit or astral journeys into far distant places. One of the few Shipibo shamans actually to have had an experience with this rare plant described its effects:

> I shook. Then the two lords of the plant appeared. I also heard the sounds of many other spirits, who threatened me and wanted to deny me the power of the *shahuán-peco*. When I became unconscious, the four lords of the plant threw me around the four corners of

the world like a ball. I no longer felt my weight. When I was close to death, the large, shining jaguar *ani ino* came and sucked out my blood and carried me far away. In this way, the shaman learns to fly through the air on his own and to see things from above. In the cloud villages, he meets with the spirits that support him in therapy. During the *shahuán-peco* visions, the shaman sees the people in their true nature, he sees their true intentions, and he sees them naked. (Gebhart-Sayer 1987, 212*)

Another Shipibo shaman clearly recalled his initiation with the plant:

Shahuán peco is stronger than *charin pecó*. Oh, how strong it is! You can hardly stand it—it is much stronger than ayahuasca! Once you have drunk *shahuán peco*, you see all people naked. Then a man comes, a *nai yoshin* (heavenly spirit); he holds an empty book in his hand. He holds it before me. A *pino* (hummingbird) comes and draws *quené* (designs) in the book with its beak. I still possess this book! I open it up and look at the *quené*. The *pino* comes, and I sing with him. I open the book every night—this is why I sing so well! (in Illius 1991, 123)

The statement that shahuan peco is considered more highly effective than ayahuasca is worth noting, for the Shipibo are renowned for preparing an especially potent ayahuasca.

Unfortunately, we have no information at all about the botanical classification of this "parasitic plant with small leaves." It may possibly be a Bromeliaceae. There are reports that one species of *Tillandsia* is used as a psychoactive substitute for peyote (*Lophophora williamsii*).

Literature

Illius, Bruno. 1991. *Ani Shinan: Schamanismus bei den Shipibo-Conibo (Ost-Peru). Ethnologische Studien,* vol. 12. Münster: Lit Verlag.

Shlain

In Cambodia, *shlain* is a name for both a tree and its product. The very hard wood of the shlain tree is light colored, almost white. It is shaved, mixed with hemp (*Cannabis indica, Cannabis sativa*), and smoked. The addition of shlain to the hemp clearly increases the effects of the **THC** and makes it somewhat more psychedelic. The wood has been tested many times with success, but unfortunately its botanical identification is still uncertain. It is, however, possible that it is *Strychnos nux-vomica.* In Cambodia, a traditional cough medicine is made from dried *kancha* (hemp herbage), which is chopped up on a board made from nux-vomica (*Strychnos nux-vomica*) wood. Mixed with small

pieces of the wood (which contains **strychnine**), the resulting product is then smoked (Martin 1975, 67).

Literature

Martin, Marie Alexandrine. 1975. Ethnobotanical aspects of cannabis in Southeast Asia. In *Cannabis and culture,* ed. V. Rubin, 63–75. The Hague: Mouton.

Silphion

In ancient times, silphion was a renowned medicinal plant, apparently with psychoactive effects. Its botanical identity has been completely lost (Dioscorides 3.84). In modern times, it has been proposed that it might have been food of the gods or devil's dung (*Ferula asafoetida* L.); it may also have been *Ferula moschata* (Reinisch) Kozo-Polj. [syn. *Ferula sumbul* (Kauffm.) Hook. f., *Euryangium sumbul* Kauffm.], a purportedly hallucinogenic plant (Schultes and Hofmann 1980, 368*; cf. **incense**). It may even have been a species of bracket fungus (*Laricifomes officinalis* [Vill. ex Fr.] Kotl. et Pouz., *Polyporus officinalis* Fr., *Boletus laricis* Jacq.; cf. "*Polyporus mysticus*"):

Bracket fungus. *Agarikon* is regarded as a root, similar to that of *silphion,* however it is not close to the surface like *silphion,* but completely lax. (Dioscorides 3.1)

A remark by Dioscorides concerning silphion's effects upon glaucoma is reminiscent of the effect of hemp (*Cannabis indica, Cannabis sativa*): "It induces sharp-sightedness and dispels the beginnings of glaucoma when it is rubbed in with honey. (3.84)

It has been suggested that the ancient *silphion* plant is now extinct. The genus that bears its name, *Silphium* L. (Compositae), is with certainty not identical to the ancient healing plant, nor is *Thapsia silphium* [syn. *Thapsia garganica* L. var. *silphium*] (Umbelliferae) (cf. Dierbach 1833, 213*).

Soma

The original soma plant has not yet been identified with certainty. Cf. **soma** on pages 792 ff.

Strychnos Manikos

The ancient literature contains descriptions of several species of strychnos, of which the fourth, strychnos manikos ("the strychnos that makes one wild"), must have been a potent psychoactive plant (Theophrastus, Dioscorides, Pliny). Many philologists and botanists have attempted to uncover the botanical identification of strychnos manikos, interpreting it variously as *Atropa*

A piece of wood from the trunk of the Cambodian shlain tree. When ground, it is said to potentiate the effects of hemp.

belladonna, *Datura stramonium*, *Solanum* spp., *Strychnos nux-vomica*, *Withania somnifera*, and *Physalis* spp. (Baumann 1982, 111*). It is possible that the name *strychnos* was a catchall term for psychoactive plants.

Suíja

This hallucinogenic plant, found in the Amazonian lowlands of Peru, is a vine "whose leaves are black on top, yellow underneath." It is said to have effects similar to those of **ayahuasca** but also to produce visions by itself. The shamans of the Shipibo-Conibo Indians regard it as a "master of the healing arts" (Gebhart-Sayer 1987, 340*). The shaman ingests two grated roots with water each night for ten nights. The visionary experience follows the tenth administration. The few known reports of experiences suggest that the effects are very much like the visions produced by ayahuasca (Gebhart-Sayer 1987, 206 ff.*) as well as the effects of **shahuan peco**.

Thalassaigle

"Sheen of the Ocean"

This Indian plant with visionary powers was described by Democritus in a lost work. Pliny paraphrases:

> The *thalassaigle* is found on the River Indus and is thus called by another name, *potamaugis* ["river sheen"]; whoever drinks of it becomes insane, whereby strange faces hover in front of him. (24.164)

It is possible that this information refers to **soma**, which was drunk in the Indus Valley.

Theangelis

"Messenger of the Gods"

Pliny wrote the following about this plant, which has been neither identified nor even interpreted:

> The *theangelis* comes from the mountains of Lebanon in Syria, from the Dikte Mountains of Crete, and from Babylon and Susa in Persia; when drunk, it imparts magicians with the ability to prophesy. (24.164)

Theombrontion

Pliny, basing his account upon the report of his informant Democritus, notes:

> The *theombrontion* grows 30 schoine [app. 166.5 km] away from Choaspes, is the color of the peacock, and has a wonderful scent; the

kings of Persia drink it for all physical ailments and to protect themselves from inconstancy of the mind and the sense of justice; because of its powerful effects, it is also called *semnios* ["the exalted" or "the sacred plant"]. (24.162)

This description suggests that this plant had a tonic effect upon the brain. Theombrontion was also an ingredient in a fertility-promoting mixture known as *hermesias* ("drink of Hermes") that also included pine nuts, **honey**, myrrh (cf. **incense**), saffron (*Crocus sativus*), and **palm wine** (24.166). The name *theombrontion* may have been a corruption of *theos*, "god," and *bromios*, "food."

The *theombrontion* plant has sometimes been interpreted as *Aeonium arboreum* (L.) Webb et Berth. [syn. *Sempervivum arboreum* L.] (Crassulaceae).

Tila

Tila is actually the Spanish name for the linden (*Tilia* spp.), which has mild sedative effects (cf. **diazepam**). Since the linden is not indigenous to the Americas, the European name was applied to American plants. Thus, in Cuba, *Justicia pectoralis* is called either *tilo* or *tila*. Numerous plants are known in Latin America as *tila*. In southern Mexico, an as yet unidentified plant is also called *tila*. Its fruits, which resemble the large fruit capsules of **Turbina corymbosa**, are referred to as *flor de tila*, "linden flowers." A tea made from these flowers is used in folk medicine as a tranquilizer and nerve tonic. This plant may be identical with or related to *Tilia mexicana* Schlechtend. (Tiliaceae) (Argueta et al. 1994, 1337*).

Tobo Tree

There is a sacred mountain in Iran known as Kohi-Gabr, "mountain of the fire worshippers." In ancient Persia, before the advent of Islam, the fire worshippers were the people who would ritually drink **haoma** (cf. *Peganum harmala*). A ruin at the peak of the mountain is thought to once have been a Zoroastrian fire temple (cf. *Ephedra* spp.):

> Concentrated magical essence has remained in this place, and an army of especially gifted genies dwells there. It is said that the "power" causes people who wish to approach the spot to shrink back. . . . Stories relate how those who climbed Kohi returned as madmen or cripples or wasted away. (I. Shah 1994, 153*)

We do not know what really transpired on this mountain. But it was associated both with other places of power and with certain psychoactive substances:

Not far from this place, there are other mountains that are also linked to magical concepts. Here the fire-worshipping magicians once made offerings of fruit to placate certain spirits and to entice them into captivity so that they would obey their commands. Anyone who wanted to have a wish fulfilled would write it down and place it in a bowl filled with fruit, which the magicians would then take up the mountain. At the peak of one such mountain grew the tobo tree, the tree of eternal bliss. It is said to resemble the tree that stands at the right side of Allah in paradise. Good fairies carry the great sufferings and fears to this place, where they are purified so that the sufferers will be freed from their afflictions. (I. Shah 1994, 153 f.*)

The tobo tree is presumably some type of psychoactive plant that was identified with the tree of knowledge.

"Trees with Special Fruits"

Herodotus (ca. 500–424 B.C.) was a Greek historian and the father of ethnography. In his *Histories*, he provides us with the following story from Assyria:

> Of the River Araxes [= Jaxartes/Syr Darja], some say that it is larger than the Istros, others that it is smaller. Numerous islands are said to lie within it, about the size of the island of Lesbos, and these are inhabited by people who dig out all kinds of roots in the summer and nourish themselves from these, but in winter they live from the fruits of trees, which they collect and store after they have ripened. They also have other trees that bear very special fruits. When many people have come together, they light a fire, sit around it in a circle, and throw these fruits into the fire. When the smell of the burning fruits enters their noses, they become inebriated like the Greeks from their wine. They throw more and more fruits into the fire, so that they become more and more inebriated and finally jump up to dance and to sing. This is what is said about the way they live. (2.202)

The "fruits" being burned here as an **incense** sound suspiciously like female hemp flowers (*Canabis indica, Cannabis sativa*) and the ritual like that of the associated Scythian purification ritual. Their true identity, however, cannot be proved.

Woi

The Yecuana Indians of southern Venezuela are said to have used a magical plant called woi, apparently for psychoactive purposes. The botanical identity of this plant is completely unknown (Schultes 1966, 298*).

These fruits of an as yet unidentified plant are sold in Mexico under the name *flores de tila*, "linden flowers." They are said to have narcotic effects.

PSYCHOACTIVE
FUNGI

"The mushrooms give me the power to see everything completely. I can look back to the beginning. I can go to where the world originated. The sick become healthy, and the relatives then come and visit me to tell me that relief has come. They give me thanks and bring me alcohol, cigarettes, and a little money."

María Sabina
in *Maria Sabina*
(Estrada 1980, 72 f.**)

For many musicians (e.g., The Allman Brothers), the universe was created in a divine mushroom trip. Consequently, the psychedelic paradise in which these musicians (the techno-pop band Deeelite) are shown frolicking is a garden of gigantic mushrooms. (CD covers: 1994, Sony Music; 1994, Elektra Records)

* = literature listed on pages 878–907

** = literature listed on pages 689–93

344 Ambrosia consisted of **honey**, water, fruits, olive oil, cheese, and barley. The first letters of the Greek names of these ingredients yield the word *myceta*, the accusative form of *mykes*, "fungus." Graves sees this as a secret message indicating that the ambrosia of the gods actually consisted of (psychoactive) fungi (Graves 1957**).

In ancient times, the dictum that "mushrooms are the food of the gods" was a familiar one (Graves 1957**). Probably the oldest written record of fungi in general was provided by Euripides (480–406 B.C.). The first written mention of inebriating mushrooms and the rituals associated with them comes from the historical document *Historia General de las Cosas de Neuva España*, which was compiled between 1529 and 1590 by the Franciscan missionary Bernardino de Sahagún. This work contains what is probably the oldest graphic depiction of the ritual consumption of mushrooms (*teonanacatl*). The very earliest representation of mushrooms is a rock painting in the Tassili plain of the southern Sahara (Algeria), which has been dated to the late Neolithic period. The conjectures that Euripides wrote about psychedelic mushrooms and that the rock art in the desert are depictions of early mushroom shamans appear in a new light as a result of the knowledge and discoveries of modern ethnomycology.

The Neoplatonist Porphyrios (third century A.D.) referred to mushrooms as "the children of the gods." The poets spoke lovingly of the "children of the earth" (Lonicerus 1679, 160*). Many researchers have suggested that the divine "drink of immortality"—whether known as **soma**, **haoma**, *amrita*, ambrosia,[344] or *nektar*—was a fungus or, more precisely, a psychoactive mushroom (Graves 1957**). Even the tree of knowledge has been interpreted as a fly agaric mushroom, and Christianity as originally a secret mushroom cult. Similarly, the Sufis are said to have employed mushrooms, which they called the "bread of crows," so that they could know God (Shah 1980, 104 ff.*). Terence McKenna (1988*, 1996*) has advanced the hypothesis that psilocybin mushrooms of the species *Psilocybe cubensis* were the catalyst in primate evolution that led our apelike ancestors to become human.

We are only beginning to understand the immensely important role psychoactive fungi have played in human cultures. The branch of science that investigates these questions was founded by an American banker, R. Gordon Wasson (1898–1986), and is known as ethnomycology (Riedlinger 1990**). Those psychedelic or entheogenic mushrooms that contain **psilocybin** and psilocin have had the greatest impact upon the history of culture (Metzner 1970**).

Amazingly, psilocybin mushrooms, first "discovered" in Mexico, are now found on all continents (and are being consumed by increasing numbers of mycophile psychonauts). New species are constantly being described (e.g., *Psilocybe samuensis* Guzmán, Bandala et Allen and *Psilocybe azurescens* Stamets et Gartz) and new ranges and occurrences of known species are being discovered.

The Flesh of the Gods

Although ancient Europe almost certainly had its own indigenous mushroom cults (Graves 1957**; Samorini and Camilla 1995**), the first mention of magic mushrooms and their associated rituals is found in (ethnohistorical) sources from the early colonial period of the Americas. Because the Catholic Church and the Mexican and Peruvian Inquisitions forbade the religious use of psychoactive plants and severely punished any transgressions (Andritzky 1987*), the sacred use of mushrooms was forced underground. It was not rediscovered until the 1950s, when the Mazatec shaman María Sabina initiated the Wassons into the nocturnal cult (Wasson 1957**). Today, María Sabina, now deceased, is regarded as a kind of "saint" of the psychedelic movement. The botanical identity of the mushroom that the colonial sources referred to as *teonanacatl* (Nahuatl: "flesh of the gods" or "wondrous mushrooms") was determined in the late 1930s (Johnson 1940**; Reko 1940**; Schultes 1939,** 1940,** 1978**; Singer 1958**; Wasson 1963**).

Because of their enormous powers, these mushrooms (*Psilocybe mexicana, Psilocybe* spp.) are often regarded as something sacred or divine or even as deities in their own right. Those who consume the "mushrooms of the gods" in a ritual context undertake a journey into other realities where they might enter the realm of divine beings, retrieve the souls of the sick, and foretell the future. One Mazatec mushroom shaman had the following to say about such divination:

The words come only when the mushroom is in my body. A wise man does not learn by heart that which he must say in his ceremonies. It is the sacred mushroom that speaks. The wise man simply lends it his voice. (Estrada 1980, 144**)

The *velada*, "night watch," is a nocturnal meeting in the house of a female or male shaman, during most of which the participants sit in a circle. After **incense** is burned and offerings and prayers are made, the participants are given the mushrooms, which they consume in pairs while prayers are being offered. The shaman then begins a series of songs that make it possible for her and the group to enter trance and explore the psychedelic universe of the entheogenic mushroom (Wasson et al. 1974**; Estrada 1980**; Hofmann 1979*; Liggenstorfer and Rätsch 1996**).

Setting a date for the nocturnal event already implies that the patient, the healer, and perhaps even the group are going to deal with the situation at hand. The problems that are to be solved (the illness or other reasons for holding the meeting) are placed at center stage. Doña Julieta explained that each and every one of the participants must clearly understand the questions that he wishes to ask the "santitos." It often happens that the "velada" will not focus on the healing of one single, clearly "defined" patient, but that the various participants also announce their interest in a cleansing ("limpia") or healing. (Donati 1991, 86 f.**)

In Europe, the American Indian model for entheogenic mushroom rituals has been adapted for use with mushrooms (*Psilocybe cubensis, Psilocybe cyanescens, Psilocybe semilanceata*) and other psychoactive substances (e.g., LSD, MDMA; cf. **herbal ecstasy**) in group settings (Müller-Ebeling and Rätsch n.d.**). These sessions are known variously as ritual circles, ceremonial circles, healing circles, and mushroom circles. This ritualized use has led to the emergence of a new movement (Rätsch 1996**). Today's entheogenic mushroom culture is decentralized, both anarchic and partnership oriented, and it transcends religious, cognitive, and political boundaries. It resembles, in fact, the manners in which the mushrooms themselves grow: an underground network of interlaced roots. The fruiting bodies appear at the proper time and the proper place, often in a circle ("fairy ring"), at which time they disperse their spores throughout the world. In order to be introduced or initiated into the mushroom culture, one needs only to ritually consume some mushrooms in a circle of like-minded individuals and "be accepted" by them. Many people have experienced the ritual circle as an initiation into the mysteries of the mushroom. In the Western world, entheogenic mushrooms have brought forth a spiritual cult with its own rituals not unlike those of the peyote cult (cf. *Lophophora williamsii*) of the Native American Church and other, similar movements (La Barre 1979b*).

The World of Fairies and Witches
Toward the end of the Middle Ages, it was widely believed that witches and demons caused mushrooms to grow. "Fairy rings," the circular fruiting characteristic of a number of fungi (e.g., the edible wood blewit *Lepista nuda* [Bull.: Fr.] Cooke and *Amanita muscaria*), were regarded as especially dangerous, for they were thought to be the nocturnal meeting places of witches, female magicians, elves, and mushroom spirits (Englbrecht 1994, 8, 56**):

In days gone by, and especially in the Middle Ages, mushrooms stimulated the human imagination. To superstitious individuals, they have always appeared to be strange beings. The most amazing tales have been told about fungi. Various meanings have been attributed to the "elf rings," also known as "elf dancing places" and "fairy rings." They can still be found today; in fields, meadows, and forest clearings, mushrooms often grow in a closed

circle. Folk tales about such rings have been passed down in songs, stories, and poems. It was believed that they were the sites where delicate winged and happy elves or lusty groups of goblins would assemble during the night and perform their graceful or turbulent dances in the forest. These folk beliefs were not limited to the enchanting forms of the fairy rings alone, for they also attributed the mushrooms themselves with mystical significance. As a result, those mushrooms that were shaped like umbrellas became the umbrellas and parasols that dwarves and elves used to protect themselves from the rain or the sun, and they also became known as "elf stools." Nymphs were said to sip the cool morning dew from mushrooms that looked like small goblets, and the little magical beings of the forest were said to take refreshing baths in the larger ones. It was also believed that evil witches would use poisonous mushrooms when they were preparing their dangerous magical drinks. (Rippchen 1993, 13**)

Such conceptions are not limited to medieval and modern Europe but are also found in many other parts of the world. In Japan, mushrooms, especially those of the genus *Amanita* (cf. *Amanita muscaria, Amanita pantherina*), are thought to be the food of the long-nosed goblins known as *tengus*, or the places where tengus dallied. It is not unlikely that these stories of elves, goblins, and fairies arose as a result of the visions that certain mushrooms can evoke (Golowin 1973*). Some authors have even recommended that people

"This mushroom is a transdimensional doorway which sly fairies have left slightly ajar for anyone to enter into who can find the key and who wishes to use this power—the power of vision—to explore this peculiar and naturally occurring psychoactive complex."

TERENCE MCKENNA
TRUE HALLUCINATIONS
(1993, 41*)

Even the truffle (*Tuber* spp.), that culinary delicacy, has a type of psychoactive effect. It produces an odoriferous substance that is an analog of human pheromones, thereby eliciting a specific sexual search behavior in humans. For this reason, truffles enjoy a reputation as an aphrodisiac. (Woodcut from Gerard, *The Herball or General History of Plants*, 1633)

Common Names for Psychoactive Mushrooms in Mexico

Aztec	*nanacatl*	"flesh"
[Molina dictionary]	*teonanacatl*	"flesh of the gods"
	xochinanacatl	"flower flesh"
	teyhuinti-nanacatl	"inebriating mushroom"
Purépecha (= Tarascan)	*cauigua-terékua*	"mushrooms that make one drunk"
Mazatec	*dishitu*	"born of the earth"
Modern Mayan	*lol lú'um*	"flowers of the earth"
Modern Nahuatl	*nanakatsisten*	"flesh of the earth"
	tlakatsitsin	"little people"
	a-pipil-tzin	"little child of the water"
Mixe	*naax wiin mux*	"mushrooms of the earth [goddess]"
	pi:tpi	"spindle staff"
Zapotec	*beyo-zoo*	
Spanish	*hongos*	"mushrooms"
	hongos maravillosos	"wondrous mushrooms"
	hongitos	"little mushrooms"
	niños	"children"
	niños santos	"sacred children"
	santitos	"little saints"
	cositas	"little things"

"It is reported that Nero said that 'mushrooms are the food of the gods'; perhaps the joke of a madman who wished to express that it is possible to at least enter paradise by being conveyed by the mushrooms into the beyond."

ANTHONY MERCATANTE
DER MAGISCHE GARTEN [THE MAGICAL GARDEN]
(1980, 254*)

"The Mazateco Indians identify the hallucinogenic mushroom with Jesus Christ. They believe that while on earth, Jesus spat onto the ground. The mushrooms grew out of his saliva. They are his speaking tube, through which he speaks when their chemical constituents produce hallucinations."

EUNICE PIKE AND FLORENCE COWAN
MUSHROOM RITUAL VERSUS CHRISTIANITY
(1959, 145**)

"He [Gwyddyon] then resorted to his skills and began to demonstrate the power of his magic. He made appear a dozen stallions, twelve black hunting dogs . . . twelve golden bucklers. These shields were mushrooms that he had transformed."

MABINOGION (CELTIC MYTH)
IN THE DRUIDS
(MARKALE 1999, 143*)

---

desiring to gain access to the fairy world consume mushrooms (Morris 1992, 5**).

In Celtic mythology, at least as it is represented in modern times, mushrooms are closely related to the "other world" and to the realms of fairies and elves. In Celtic Wales, tradition tells of a mushroom, known as *bwyd-ellyllon* ("elf mushroom"), that "is one of the delicacies of the elves but is feared by man and beast" (Perger 1864, 210*). It is very likely that the pagan Celts partook of the "elf mushroom":

We can assume with certainty that they used in their rituals various mushrooms, such as the fly agaric, which when dosed carefully evoke visions and trance states. (Markale 1989, 203*)

Today, in such Celtic areas of Great Britain as Wales and Cornwall, neopagans and underground Druids continue to consume "liberty caps" (*Psilocybe semilanceata*) as a sacrament to invoke the elves. Megalithic dolmens, such as the one found at Chûn Quoit (which looks like a gigantic mushroom), are preferred locales. The British geomancer Paul Devereux has called these dolmens "the dreaming stones" and suggested that at such dream-stimulating[345] sites, appropriate earth energies can have inspiring effects upon human consciousness—especially when said consciousness has been expanded by mushrooms (Devereux 1990*, 1992a*, 1992b*; Devereux et al. 1989, 117, 180 ff.*). A traditional Celtic belief holds that these dolmens are entrances to the fairy world (Evans-Wentz 1994*). The keys, of course, are the mushrooms.

The Archaeology of Entheogenic Mushroom Cults

Since fungi are very poorly or not able to be preserved as organic material over long periods of time, they and their fruiting bodies are almost never found during archaeological excavations (Gartz 1992). For this reason, it is practically impossible to state definitively whether fungi had any role to play as ritual offerings or grave goods or to find them as remnants of shamanic activities. From an archaeological perspective, both the shamanic use of fungi and religious mushroom cults can be demonstrated only by means of artifacts and pictorial representations. Often, petroglyphs are the only evidence of a lost culture. It was through rock carvings in the Sahara that the "oldest psychedelic mushroom culture in the world" was discovered (Samorini 1992). Scandinavian petroglyphs have been taken as evidence of mushroom use among the Vikings (Kaplan 1975). Similarly, many Indian petroglyphs have been interpreted as indicating the shamanic use of psychoactive plants (cf. *Datura wrightii*) and as representations of these plants (Wellmann 1978*, 1981*).

Mushrooms and Petroglyphs

In the Tassili plain of northern Africa (Algeria), a now barren Saharan landscape, rock paintings and petroglyphs dated to between 9000 and 7000 B.C. have been discovered. The illustrations portray a flourishing hunting and pastoral culture (Mori 1974). One group of rock drawings appears to be related to entheogenic mushrooms:

The polychromic scenes of harvest, adoration, and the offering of mushrooms, and large masked "gods" covered with mushrooms . . . lead us to suppose we are dealing with an ancient hallucinogenic mushroom cult. . . . [T]hese mushrooms can be distinguished by means of a complex system; every type having its own mythological representation. . . . [T]hey could indeed reflect the most ancient human culture as yet documented in which the ritual use of hallucinogenic mushrooms is explicitly represented. (Samorini 1992, 69)

In addition to *Psilocybe cubensis*, the rock images apparently depict other psychoactive fungi, including *Psilocybe cyanescens* and *Panaeolus* spp. (Gartz 1992). The most impressive of the Tassili images feature running or dancing figures that look like anthropomorphized mushrooms and are shown holding a mushroom in their hands. In the entheogenic mushroom culture, a bee-shaped god, out of whose body mushrooms are sprouting, is now regarded as a kind of "primordial deity" (Rätsch 1996**).

A large number of petroglyphs are located in the Southwest of North America. Both the Indians who now live in this region and the archaeologists and ethnologists who study these artifacts interpret them as shamanic scenes (Schaafsma 1992). Some petroglyphs portray "shamans" holding mushrooms or mushroomlike objects in their hands (Rätsch 1994; Samorini 1995a*). While no mushrooms are used for ritual or shamanic purposes in the Southwest today, it is possible that they were used in prehistoric times, when there was a lively cultural exchange between Mexico and the Southwest. Trade routes connected Mexico to the Southwest, and ritual objects (e.g., the sacred *Spondylus princeps* shells) were known to move north. Dried psilocybin mushrooms may also have traveled along these routes to the cultures of the north (cf. McGuire 1982**).

Some of the "mushroom" petroglyphs may represent the fly agaric mushroom, which was used in shamanic contexts by other North American Indian peoples. The fly agaric (*Amanita muscaria*) and a subspecies or variety (*A. muscaria* var. *formosa*), as well as the even more potently hallucinogenic *Amanita pantherina*, do occur in the Southwest and are not considered rare (States 1990, 57 f.**). Perhaps the shamans of the prehistoric Pueblo Indians ritually consumed fly agarics in

---

345 "Whether invoked in a natural manner or through special techniques, through hallucinogenic essences, or through inebriating drinks—in Druidism as in shamanism, the dream is always an elementary component of the world" (Markale 1989, 208*).

remembrance of their Asian origins (Rätsch 1996).

Entheogenic Mushrooms in Antiquity

The Hittites were known to have possessed idols that had the appearance of anthropomorphized mushrooms (Morgan 1995, 112**). The Mycenae-ans were originally of Asian descent, and they etymologized their "new home" Mycenae from *mykes*, "fungus," to remind themselves of the entheogenic cult that stood at the beginning of their history (Ruck 1995, 32*). When seen from inside, the so-called Treasury of Atreus in Mycenae is a perfect reproduction of a *Psilocybe semilanceata* cap. According to legend, Perseus, the founder of Mycenae, discovered and apparently ate a mushroom at the site, whereupon he received a vision of his city.

The fact that mushrooms had a special religious or ritual importance in ancient Greek cultures can also be seen in the mushroom chains, strings of beads that are clearly shaped like mushrooms, that were were produced in both Minoan and Hellenistic times (one such example of the jeweler's art can be seen in the museum of Iráklion, Crete). The Bronze Age has yielded characteristic ceramic figures that are now known as *kourotrophos* ("wet nurses"). Both the wet nurses and the children they suckled have the shape of mushrooms (Ripinsky-Naxon 1993*). An Etruscan bronze mirror features an image of the Thessalian hero Ixion, an uncle of the legendary Asclepius, as he dances over a mushroom with eagle wings. An amphora from the fourth century B.C. shows Perseus and the decapitated Medusa, she "who has lost her head." Above the hero float long-stemmed mushrooms with small, roundish heads. If something looks like a mushroom, it is a mushroom. And all of these images bear strong resemblances to psilocybin mushrooms.

The Greek scholar Carl Ruck interpreted a very enigmatic scene in Aristophanes' *Birds* as an allusion to a mushroom cult surrounding the philosopher Socrates:

> Amidst the shade-foots,
> there is a certain swamp
> where Socrates, unwashed,
> summons up souls.
> Amongst his clients came Peisander,
> who begged to see a spirit that had forsaken
>   him
> while he remained alive. (1553 ff.)

"Shade-foot," or *monocoli*, was a paraphrase for anthropomorphic mushrooms.[346] The "pond" was the sacred swamp of Dionysos in Athens. The "unwashed" Socrates was impure because he had profaned the lesser Eleusinian mysteries, i.e., had carried them out in his own home. This is the reason why he was regarded as a summoner of

souls, because he had induced the youth, represented here by Peisander, to consume the sacred drug. That the latter would search for the soul that had slipped out of his living body (a thoroughly shamanic motif) during an appropriate mushroom ritual can be understood only in this way. According to Ruck, the main mysteries at Eleusis, during which **kykeon** was administered, were contrasted by these "lesser rites" in which entheogenic mushrooms were cultically consumed (Ruck 1981**).

Support for this assertion comes from a late ancient relief found in Eleusis that dates to the fourth century A.D. On the relief are Demeter and Persephone; the Great Goddess is shown holding a mushroom in her hand, which she is presenting to Persephone (Stamets 1996, 14**).

The Mesoamerican Mushroom Stones

A number of stone artifacts have been found in southern Mesoamerica, primarily in Chiapas and Oaxaca (Mexico), Guatemala, and El Salvador. Around 30 cm in height, they have become known by the name *mushroom stones* in both the academic and the popular literature (Ohi and Torres 1994; Rose 1977; Trebes 1997). Some of these artifacts are well over two thousand years old. Most are small objets d'art (stone sculptures) depicting animals or humans, from whose backs or heads mushroom-shaped objects emerge. Many of these objects have been found and documented (Mayer 1977, 1979). Recently, a mushroom stone was discovered in the region occupied by the Maya (Trebes 1997).

The German geographer Carl Sapper (1898) regarded these stones as "mushroom-shaped images of idols." The American scholar Daniel G. Brinton (1898) was of the opinion that these objects are moon symbols. In contrast, the American archaeologist Thomas W. Gann (1911) viewed them as phallic symbols (cf. also Bruder 1978). R. Gordon Wasson (1961**) saw the mushroom stones as symbols of an archaic entheogenic mushroom religion. The German pre-Columbian scholar Ulrich Köhler (1976) argued that the mushroom stones are forms for making pottery (see the criticism in Lowy 1981). The Mexican mycologist Gastón Guzmán (1984)

The ancient "mushroom stone" of Delphi looks like a naturalistic representation of a *Psilocybe semilanceata* cap. Classical archaeologists, however, have interpreted it as a model of the *omphalos*, "navel of the world." It does make sense that the navel of the world should be sought in the cap of a psychedelic mushroom.

A Central American mushroom stone (Maya, El Salvador, ca. 300 B.C. to 200 A.D.), in which the spirit of the mushroom appears to be emerging from the stem. (Illustration: Sebastian Rätsch)

A *monocoli* or one-foot, which R. Gordon Wasson sees as a symbol of the sacred mushroom. (From *Schedels Weltchronik*, 1493)

346 For Theophrastus, *monocoli*, literally "one-footed," was a metaphor for plants that have only one stem (Ruck 1981, 181**).

"Then, indeed, as you lie there bemushroomed, listening to the music and seeing the visions, you know a soul-shattering experience, recalling as you do the belief of some early peoples that mushrooms, the sacred mushrooms, are divinely engendered by Parjanya, the Aryan God of the Lightning-bolt, in the Soft Mother Earth."

R. GORDON WASSON
THE ROAD TO ELEUSIS
(1984, 34)

A running mushroom shaman from Tassili. (Illustration: Christian Rätsch)

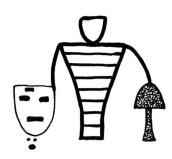

A petroglyph from the American Southwest (Dry Fork Valley, Utah) may be a depiction of a shaman with a mask and a sacred mushroom. (Illustration: Christian Rätsch.)

347 In Sanskrit, the word for mushroom is *chattra*, a word derived from *chad*, "to cover," which originally meant "parasol" (Samorini 1995, 35).

has argued that the mushroom stones are representations of culinary mushrooms, e.g., the edible "porcini mushroom," *Boletus edulis* Buillard ex Fries (cf. *Boletus* spp.). The most commonly held view is that the mushroom stones are ritual objects that were associated with the ingestion of psilocybin mushrooms.

The discovery of nine small mushroom stones (14 to 18 cm in height) together with small, easy-to-use grinding stones (*metate* and *mano*) in a ritual depot at Kaminaljuyú (Guatemala) suggests that the mushroom stones were venerated as idols and the other stones were used to grind dried mushrooms as part of a mushroom cult (Borhegyi 1961; Lowy 1971). Ethnohistorical sources (Lowy 1980**) from the area where the stones were recovered bear witness to the ritual use of the mushroom stones:

> The "Titulo de Totonicapán," which dates to the colonial period, confirms that during their enthronement, the rulers of the Quiché received the insignia of their kingly power as well as *nanakat abaj beleje*, "nine mushroom stones." Although these text passages refer to practices from the post-Classic period, they clearly demonstrate the continuation of a sculptural tradition that is rooted in pre-Classic times. Mushroom stones may have already been a part of the insignia of kingly power in the pre-Classic period, because only the princes enjoyed the special privilege of consuming mushrooms with hallucinogenic effects. (*Die Welt der Maya*, 1992, 314)

In western Mexico (Nayarit), a number of ceramic objects have been found that may be representations of ritual mushroom consumption (Furst 1965; Kan et al. 1989, 82 f., 91). Ceramic figures discovered in Jalisco depict drumming men with strange, mushroomlike growths on their heads (Kan et al. 1989, 126 f.). Are these representations of shamans in a mushroom-induced trance state?

The contemporary Maya regard artifacts from pre-Columbian times as divine objects or gifts of the gods. These artifacts are considered to be especially powerful and are used as amulets and images of gods. The Lacandon Maya occasionally find artifacts while working their fields or in the forest. They call these ceramics *u pat k'uh*, "pottery of the gods." Ceremonial axes are known as *u baat k'uh*, "axes of the gods." Artifacts and stones of spherical shape are referred to as *kuxun tun*, "mushroom stones" or "living stones." The usual word for stone is *tunich*, while *tun* tends to be used to refer to precious stones. Perhaps the term *kuxun tun* is a last linguistic reminder of the pre-Columbian mushroom stones.

Today, *Psilocybe* mushroom fans from around the world collect reproductions of these mushroom stones. Both reproductions and forgeries

have been on the market since the 1960s (Mayer 1977, 2). Those tourists who visit Mexico in search of mushrooms have led enterprising souvenir makers to produce small mushroom stones (*hongitos*) and amulet-like pendants. These are made from stone (onyx, marble, chalcedony, obsidian), black coral, amber, ceramic, and even papier-mâché. Many of these souvenirs have landed on the house altars of mushroom lovers throughout the world (Rätsch 1996**).

The Golden Mushrooms of the Tairona

In pre-Hispanic Colombia, a number of cultures flourished that were well versed in the manufacture of gold jewelry. Of these, the goldwork of the Tairona incorporates obvious shamanic elements (Reichel-Dolmatoff 1981, 1988). It is probable that the Tairona knew of and ritually used psychoactive mushrooms. A large number of the figured gold objects known as the Darien pectorals display unequivocal mushroom ornamentations (Schultes and Bright 1979). Reichel-Dolmatoff has described the ritual use of psychoactive mushrooms among the Kogi, the descendents of the Tairona (Reichel-Dolmatoff 1977, 285*). From a mycological perspective, the "blue puffball" (cf. *Lycoperdon*) is completely unknown; the other species, which is not described, may be a *Psilocybe*. After all, *Psilocybe yungensis* Sing. et Smith [syn. *Psilocybe acutissima* Heim, *P. isauri* Singer; cf. *Psilocybe* spp.] is found throughout South America (Singer 1978, 59**). There also are reports of visionary experiences produced by *Psilocybe* species collected near the village of San Agustín (Mandel 1992).

The Dolmens of India

In Kerala (southern India), there are a number of prehistoric stone structures (dolmens) that belong to the megalithic culture and date to the period between 1400 B.C. and 100 A.D. In the Malayalam language of Kerala, they are known as *kuda-kallu*, "umbrella stones"[347]; the interiors are called *garbba-gripa*, "uterus chambers" (Rippchen 1993, 99**). The structures are 1.5 to 2 meters in height and are built of five stones that are put together in such a way that they resemble a mushroom: "In the indigenous tradition, they are typically interpreted as parasols and regarded as archaic symbols of power, authority, and sacredness" (Samorini 1995, 33). In modern India, it is thought that these dolmens, known as umbrella stones, are a symbol for *Psilocybe* spp. (Jain 1991, 151*). It has also been suggested that these stones are related to an archaic cult of the dead or to the Vedic **soma** cult:

> If the *kuda-kallu* represented mushrooms, then these were psychoactive mushrooms, i.e., mushrooms that were much more suitable than others—e.g., culinary mushrooms—for associating with a cult of the dead. There appears to be no direct connection between

the *kuda-kallu* and the Vedic soma, in the sense that these monuments do not appear to be emblems of a cult which may have emerged because of or been influenced by the soma cult, for the cult associated with the *kuda-kallu* developed in a period which certainly predated the contact with the Aryans in South India. (Samorini 1995, 33)

The appearance of the Kerala mushroom stones does not so much recall species from the genera *Psilocybe* and *Paneolus* as it does those from *Amanita* and *Boletus* (cf. **nonda**). Both the fly agaric (*Amanita muscaria*) and the panther cap (*Amanita pantherina*) are found in Kerala. It is possible that the *kuda-kallu* were associated with an *Amanita* cult. In the village of Ambalathara, near Thiruvananthapuram in modern Kerala, there is a temple to Devi, the radiant goddess of the plant devas (Storl 1997, 21 ff.*), that appears to have been built to resemble the *kuda-kallu* and which looks like a gigantic mushroom (Samorini 1995, 36). The Naga of this region are said to still consume fly agaric mushrooms for psychoactive purposes.

There are several megalithic dolmens in Cornwall (e.g., Chûn Quoit) that strongly resemble the Indian *kuda-kallu*. In the barren landscape, these giant mushroomlike objects are visible over great distances (Devereux 1990, 154 f.*; 1992, 182 f.).

Literature

Borhegyi, Stephan F. de. 1961. Miniature mushroom stones from Guatemala. *American Antiquity* 26 (4): 498–504.

———. 1963. Pre-Columbian pottery mushrooms from Mesoamerica. *American Antiquity* 28 (3): 328–38.

———. 1965. Some unusual Mesoamerican portable stone sculptures in the Museum für Völkerkunde Berlin. *Baessler-Archiv*, n.f., 13:171–206.

———. 1969a. "Miniature" and small stone artifacts from Mesoamerica. *Baessler-Archiv*, n.f., 17:245–64.

———. 1969b. Stone, bone, and shell objects from Lake Amatitlan, Guatemala. *Baessler-Archiv*, n.f., 17:265–302.

Brinton, Daniel G. 1898. Mushroom-shaped images. *Science*, n.s., 8 (187): 127.

Bruder, Claus J. 1978. Die Phallus-Darstellung bei den Maya: Ein Fruchtbarkeits-Symbol. *Ethnologia Americana* 14 (5): 809–15.

Devereux, Paul. 1992. *Secrets of ancient and sacred places*. London: Blandford.

Die Welt der Maya. 1992. Mainz: Philipp von Zabern. (An exhibition catalogue.)

Furst, Peter T. 1965. West Mexican tomb art as evidence for shamanism in pre-Hispanic Mesoamerica. *Antropologica* 15:29–80.

Gann, Thomas W. 1911. Exploration carried on in British Honduras during 1908/1909. *Annals of Archaeology and Anthropology* 4:72–87.

Gartz, Jochen. 1992. Der älteste bekannte Pilzkult— ein mykologischer Vergleich. *Jahrbuch des Europäischen Collegiums fürBewußtseinsstudien* (ECBS) 1992: 91–94. Berlin: VWB.

Guzmán, Gastón. 1984. El uso de los hongos en Mesoamérica. *Ciencia y Desarrollo* 59:17–27.

Kan, Michael, Clement Meighan, and H. B. Nicholson. 1989. *Sculpture of ancient west Mexico*. Los Angeles: County Museum of Art.

Kaplan, Reid W. 1975. The sacred mushroom in Scandinavia. *Man*, n.s., 10:72–79.

Köhler, Ulrich. 1976. Mushrooms, drugs, and potters: A new approach to the function of pre-Columbian Mesoamerican mushroom stones. *American Antiquity* 41 (2): 145–53.

Lajoux, Jean-Dominique. 1963. *The rock paintings of the Tassili*. New York: World Publishing.

Lowy, Bernard. 1971. New records of mushroom stones from Guatemala. *Mycologia* 63 (5): 983–93.

———. 1981. Were mushroom stones potter's molds? *Revista/Review Interamericana* 11:231–37.

Mandel, Michael. 1992. Eine sonderbare Begegnung. *Integration* 2/3:132–33. (A report on an experience with a *Psilocybe* species in San Augustín, Colombia.)

Mayer, Karl Herbert. 1977. *The mushroom stones of Mesoamerica*. Ramona, Calif.: Acoma Books. (Contains an excellent bibliography.)

———. 1979. Pilzsteine und Pilzkulte Mesoamerikas. *Das Altertum* 25 (1): 40–48.

Mori, Fabrizio. 1974. The earliest Saharan rock-engravings. *Antiquity* 48 (197): 87–92.

Ohi, Kuniaki, and Miguel R. Torres, eds. 1994. *Piedras-Hongo*. Tokyo: Museo de Tabaco y Sal. (Japanese and Spanish.)

Rätsch, Christian. 1994. Pilze und Petroglyphen im Südwesten der USA. *Yearbook for Ethnomedicine and the Study of Consciousness*, 1994 (3): 199–206. Berlin: VWB.

———. 1996. Addendum zu "Pilze und Petroglyphen im Südwesten Nordamerikas." *Yearbook for Ethnomedicine and the Study of Consciousness*, 1995 (4): 307. Berlin: VWB. (An addendum to the preceding article.)

Reichel-Dolmatoff, Gerardo. 1981. Things of beauty replete with meaning: Metals and crystals in Colombian Indian cosmology. In *Sweat of the sun, tears of the moon*, 17–33. Los Angeles: Natural History Museum of Los Angeles County.

———. 1988. *Goldwork and shamanism: An iconographic study of the Gold Museum*. Medellín: Editorial Colina.

These Mesoamerican seals from prehistoric times look like mushroom-covered mandalas.

Mushroom-shaped petroglyphs on a stone from Stonehenge— evidence of a psychoactive mushroom cult? (From Samorini)

"The mushrooms were moist, looked greenish, and were very dirty. As I bit into the first one, I had to choke. . . . I noticed that my husband's red-checked shirt was intensely colorful. I stared at the rough wooden furniture. The cracks and knotholes in the wood seemed to change form. Suddenly Masha cried: 'I am a chicken!' We both broke into resounding laughter. This comment was just too funny. . . . Then the walls shrank back, and I was carried off—up and away—on swaying waves of light turquoise. I do not know how long I was under way. I came to France, to the caves of Lascaux in Dordogne. We had been to France once, and I immediately recognized the giant stone vault above me and the beautiful primitive paintings of the former cave dwellers, the horses, bison, and deer on the walls. The paintings were even more beautiful than in reality. They appeared to be covered in a crystal-clear light. . . . Now I knew what the shamans meant when they say: 'The mushroom carries you to a divine place.'"

VALENTINA P. WASSON (1901–1957)
"ICH Aß DIE HEILIGEN PILZE" ["I ATE THE SACRED MUSHROOM"]
IN *ZAUBERPILZE*
(RIPPCHEN 1993, 128 f.**)

Top: A homegrown culture of *Psilocybe cyanescens* with mature fruiting bodies.

Bottom: A culture of *Psilocybe azurescens* that is just beginning to fruit.

Rose, Richard Maurice. 1977. Mushroom stones of Mesoamerica. PhD thesis, Harvard University, Cambridge, Mass.

Samorini, Giorgio. 1989. Etnomicologia nell'arte rupestre sahariana (Periodo delle "Teste Rotonde"). *B. C. Notizie* 6 (2): 18–22.

———. 1992. The oldest representations of hallucinogenic mushrooms in the world (Sahara Desert, 9000–7000 B.P.). *Integration* 2/3:69–78.

———. 1995. Umbrella-stones or mushroom-stones? (Kerala, southern India). *Integration* 6:33–40.

Samorini, Giorgio, and Gilberto Camilla. 1995. Rappresentazioni fungine nell'arte greca. *Annali dei Musei Civici di Rovereto* 10 (1994): 307–25.

Sapper, Carl. 1898. Pilzförmige Götzenbilder aus Guatemala und San Salvador. *Globus* 73 (20): 327.

Schaafsma, Polly. 1992. *Indian rock art of the Southwest.* Santa Fe: A School of American Research Book; Albuquerque: University of New Mexico Press.

Schultes, Richard Evans, and Alec Bright. 1979. Ancient gold pectorals from Colombia: Mushroom effigies? *Botanical Museum Leaflets* 27 (5–6): 113–41.

Trebes, Stefan. 1997. Ein Pilzstein aus dem mesoamerikanischen Maya-Gebiet. *Yearbook for Ethnomedicine and the Study of Consciousness,* 1996 (5): 241–46. Berlin: VWB.

Cultivating Mushrooms

In a comic story by Gilbert Sheldon, his heroes, the Fabulous Furry Freak Brothers, decide that they would like to grow psilocybin mushrooms. Fat Freddy collects a wheelbarrowful of cow dung from a nearby pasture, which he then tips into the bathtub. Because the mushrooms sprout every time it rains, he sprays the dung thoroughly with water. The following day, Fat Freddy goes in to gather his harvest. He finds no sign of mushrooms but does find millions of cockroaches. Disturbed in the midst of their feast, they quickly flee for other parts. . . .

Growing mushrooms at home can have results such as these. Although it appears simple in principle, it can entail almost insurmountable difficulties. Some methods for growing mushrooms for culinary or medicinal purposes have long been in use, e.g., growing mushrooms on straw, a practice that comes from ancient China (Chang 1977).

The simplest way to grow mushrooms is outdoors, e.g., in one's own garden. The psychoactive mushroom species most suited for this purpose are *Psilocybe azurescens* and *Psilocybe cyanescens.* Both species grow wild in forests and are excellent for the climate of central Europe.

First dig a pit 20 to 25 cm deep. Fill the bottom with a layer of small pieces of wood (approxi-mately 5 to 10 cm thick). Mix each mycelium with wood chips or mulch, moisten the mixture thoroughly, and spread it over the bottom layer. Spread a layer of wood shavings over this. Cover the site with leaves and small branches. The bed should never dry out, lest the mycelia stop growing. To prevent moisture loss, foil or wooden boards may be laid over the bed. When the mycelia feel at home and have spread, a rich harvest will erupt when the weather conditions are right (in central Europe, this usually occurs in October, after a period of some ten days with much precipitation and temperatures under 10°C). Since the mushrooms require nutrients after the fruiting bodies have been harvested, the upper layer of mycelium-filled wood should be replaced with fresh wood chips or mulch. Sometimes the mushrooms will spread outside of the bed and may suddenly appear in neighbors' gardens, much to their delight.

Some methods of laboratory cultivation were developed for the mycological and chemical investigations of *Psilocybe mexicana.* In the mid-1970s, mushroom hobbyists discovered a method for growing *Psilocybe (Stropharia) cubensis* ("the mushroom that comes from the stars") at home and published their findings (Oss and Oeric 1976). This book has been translated into a number of languages and has been banned more than once. To date, more than five hundred thousand copies have been published (not to mention the illegal copies). Many people have successfully followed these cultivation methods. With time, other books on growing mushrooms have appeared, some of which make no mention of magic mushrooms or refer to them only cryptically. Nevertheless, these too have provided home growers with useful information (Harris 1989; Meixner 1989; Pollock 1977; Stamets 1993; Stamets and Chilton 1983). In recent years, the species *Psilocybe cyanescens* and *Panaeolus cyanescens* have also been successfully grown in home laboratories.

Ethnobotanical specialty stores are increasingly offering spore prints and mycelia of psilocybin mushrooms (the trade in spores is still legal in many countries). Kits for growing mushrooms, together with detailed instructions, are also available. Growing spores on a substrate is actually quite easy and usually fails only when the petri dishes are not sterilized properly and other wild molds settle on the substrate. Substrates such as agar and rye are available in specialty shops. It is also worthwhile to consult the specialized literature on the subject. The books by Paul Stamets and the articles by Jochen Gartz are especially useful.

The methods of mushroom cultivation are continually being improved and simplified for home use (Gartz 1993a, 1993b; Stamets 1993). Some methods that can increase the psilocybin content have been discovered (Badham 1985).

When tryptamine is added to the substrate, the mushrooms are able to quickly metabolize it into **psilocybin** and psilocin. Phosphates can also aid in this process (Gartz 1991).

Literature

Chang, Shu-Ting. 1977. The origin and early development of straw mushroom cultivation. *Economic Botany* 31:374–76.

Dittmer, Werner. 1994. *Frische Pilze selbst gezogen.* 2nd ed. Munich: BLV.

Englbrecht, Jolanda. 1994. *Pilzanbau in Haus und Garten.* Stuttgart: Ullmer.

Gartz, Jochen. 1991. Einflüsse von Phosphat auf Fruktifikation und Sekundärmechanismen der Myzelien von *Psilocybe cubensis, Psilocybe semilanceata* und *Gymnopilus purpuratus. Zeitschrift für Mykologie* 57:149–54.

———. 1993a. Eine neuere Methode der Pilzzucht aus Nordamerika. *Integration* 4:37-38.

———. 1993b. New aspects of the occurrence, chemistry and cultivation of European hallucinogenic mushrooms. In *Atti del 2° Convegno Nazionale sugli Avvelenamenti da Funghi, Annali dei Musei Civici di Rovereto* suppl. 8 (1992): 107–23.

———. 1994. Extraction and analysis of indole derivatives from fungal biomass. *Journal of Basic Microbiology* 34 (1): 17–22.

———. 1995. Cultivation and analysis of *Psilocybe* species and an investigation of *Galerina steglichii. Annali dei Musei Civici di Rovereto* 10 (1994): 297–305.

Gottlieb, Adam. 1976. *Psilocybin producer's guide.* Hermosa Beach, Calif.: Kistone Press.

———. 1997. *Psilocybin production.* Berkeley, Calif.: Ronin.

Hadeler, Hajo. 1995. *Medicinal mushrooms you can grow.* Sechelt, British Columbia: Cariaga.

Harris, Bob. 1989. *Growing wild mushrooms. A complete guide to cultivating edible and hallucinogenic mushrooms.* Rev. ed. Seattle: Homestead Book Co.

Meixner, Axel. 1989. *Pilze selber züchten.* Aarau: AT Verlag.

Oss, O. T., and O. N. Oeric (= Terence McKenna and Dennis McKenna). 1976. *Psilocybin: Magic mushroom grower's guide.* Berkeley, Calif.: And/Or Press.

Pollock, Steven Hayden. 1977. *Magic mushroom cultivation.* San Antonio: Herbal Medicine Research Foundation.

Stamets, Paul. 1995. *Growing gourmet and medicinal mushrooms.* Rev. ed. Berkeley, Calif.: Ten Speed Press.

———. 1998. *Gardening with gourmet and medicinal mushrooms.* Berkeley, Calif.: Ten Speed Press.

Stamets, Paul, and J. S. Chilton. 1983. *The mushroom cultivar.* Olympia, Wash.: Agarikon Press.

Stevens, Jule, and Rich Gee. 1977. *How to identify and grow psilocybin mushrooms.* Rev. ed. Seattle: Sun Magic Publishing.

Mushrooms and the Law

Psilocybin and psilocin, the active compounds in many psychoactive mushrooms (but not the mushrooms themselves!), are listed in Section I of the German drug laws of 1982 (*Betäubungsmittelgesetz*) as drugs, trafficking in which is prohibited (Geschwinde 1990, 110*; Körner 1994, 40*):

> As "plants," mushrooms are not subject to the German drug laws, for they simply contain psilocybin. The cultivation of mushrooms is not punishable, so it is not a question of the growth stage. The harvesting and preparation of mushrooms, however, can constitute a manufacturing offense. Consequently, trafficking in spores is not illegal. In the same way, a setup for cultivating mushrooms is not illegal. (Böhm 1993, 174)

Psilocybin mushrooms thus thrive in a legal limbo. If the active constituents psilocybin and psilocin were not proscribed as "narcotics," dried mushrooms and/or Galenic preparations made from them would be available in German pharmacies as medicines requiring a prescription.

The legal situation in the United States is similar, although the laws are applied much more frequently and with much greater severity than in Germany (Boire 1995; Shulgin 1992*). In Mexico, mushrooms of the genera *Panaeolus* and *Psilocybe* are illegal (the Conquest and the Inquisition have not yet ended!).

In the Netherlands, it is permitted to sell freshly cultivated mushrooms (*Psilocybe cubensis, Psilocybe cyanescens, Panaeolus cyanescens*).

Fly agaric (*Amanita muscaria*) and panther cap mushrooms (*Amanita pantherina*) are legal in Germany (Körner 1994, 1572*).

Literature

Böhm, Rüdiger. 1993. Zauberpilze im Recht. In *Zauberpilze,* ed. R. Rippchen, 173–74. Löhrbach: MedienXperimente.

Boire, Richard Glen. 1995. *Sacred mushrooms and the law.* Davis, Calif.: Spectral Mindustries (P.O. Box 73401, Davis, CA 95617-3401).

"'MEXICAN MUSHROOMS!' Meanwhile, psiloc(yb)in mushrooms are being sold openly at some Goa open-air parties. Outside of these 'temporary autonomous zones,' the uncertain legal situation means that people are not quite as courageous as in the Netherlands. In the summer of 1994, the owner of the Amsterdam smart drug shop 'Conscious Dreams' was the first to dare to sell cultivated psiloc(yb)in mushrooms of the species *Psilocybe cubensis* openly across the counter. It did not take long for the police to show up. The case went before the court. There, as here, the almost identically functioning constituents of the mushroom, psilocin and psilocybin, are forbidden by the opium and narcotic laws. The mushrooms, however, are not explicitly forbidden. The court's decision noted that the amount of active constituents becomes highly concentrated only when the mushrooms are dried (namely, by about a factor of 10 compared to fresh mushrooms), and only then could one speak of an illegal substance. In no time, only fresh mushrooms were being offered."

Achim Zubke
"Spass Attacks" ["Fun Attacks"]
(1997, 21 f.**)

The Genera and Species from A to Z

Psychoactive fungi are found not only in Mexico—as is so often assumed—but throughout the world, on all continents and almost all of the islands of Oceania (Gartz 1995**; 1996**; Stamets 1996**; Stijve 1995**), as well. On the basis of the active compounds they contain, psychoactive fungi can be divided into three groups:

Psilocybin type:
with species from the genera *Conocybe, Copelandia, Galerina, Gymnopilus, Inocybe, Panaeolus, Pluteus,* and *Psilocybe*; possibly also *Agrocybe, Gerronema, Hygrocybe, Mycena, Naematoloma, Panaeolina, Pholiotina, Psathyrella,* and *Stropharia*

Ibotenic acid/muscimol (isoxazole) type:
with species from the genus *Amanita*; possibly also *Boletus*

Ergot alkaloid (ergoline) type:
with species from the genera *Balansia* and *Claviceps*; possibly also *Aspergillus* and *Cordyceps*

In addition, there are many fungi that are purported to or may be psychoactive but whose constituents are largely unknown. These genera include *Boletus, Heimiella, Laetiporus, Lycoperdon, Piptoporus, Polyporus, Russula,* and *Scleroderma*.

Of all the psychoactive fungi, those species of the genus *Psilocybe* that contain **psilocybin** have had the greatest cultural significance. Moreover, of all naturally occurring psychoactive substances,

these fungi are generally considered to be the best, safest, and most free of side effects. *Psilocybe* mushrooms that contain psilocybin generally require no extensive preparation techniques and can be eaten fresh ("straight from the meadow," so to speak) or dried and ingested at a later time. The dosage depends upon the concentration of psilocybin (plus psilocin and baeocystin) present in the mushroom material. Between 20 to 30 mg of psilocybin suffices for a psychedelically effective dosage. The classic "psychedelic," "visionary," or "entheogenic" effects begin some ten to sixty minutes after ingestion and continue for almost exactly four hours. The first report of an experience with a *Psilocybe* species that appeared in Western literature came from R. Gordon Wasson, the founder of ethnomycology:

> [T]he visions came whether our eyes were opened or closed. . . . They were in vivid color, always harmonious. They began with art motifs. . . . Then they evolved into palaces with courts, arcades, gardens—resplendent palaces all laid over with semiprecious stones. Then I saw a mythological beast drawing a regal chariot. Later it was as though the walls of our house had dissolved, and my spirit had flown forth, and I was suspended in mid-air viewing landscapes of mountains, with camel caravans advancing slowly across the slopes, the mountains rising tier above tier to the very heavens. . . . The visions were not blurred or uncertain. They were sharply focused, the lines and colors being so sharp that they seemed more real to me than anything I had ever seen with my own eyes. . . . I was seeing the archetypes, the Platonic ideals, that underlie the imperfect images of everyday life. (Wasson 1957, 102, 109**)

Few people who have tried *Psilocybe* mushrooms will be surprised to learn that both ancient peoples and the American Indians lovingly referred to the species of this genus as divine mushrooms, food of the gods, flesh of the gods, sacred mushrooms, ambrosia, *amrita*, and magic mushrooms. Only people who have no knowledge of them may describe them as the "bread of the devil" or the "devil's mushroom" (Graves 1957**).

Magic mushrooms—whether *Psilocybe* spp. or *Panaeolus* spp.—cannot be consumed on a daily basis, for they would quickly cease to be effective. According to one common aphorism, "Marijuana is our daily bread, mushrooms the uncommon feast." The pharmacodynamics of the mushroom

Rock art from Rio Chinchipe, Peru, which may depict anthropomorphic mushrooms, mushroom spirits, or mushroom shamans.

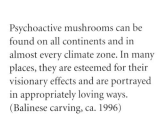

Psychoactive mushrooms can be found on all continents and in almost every climate zone. In many places, they are esteemed for their visionary effects and are portrayed in appropriately loving ways. (Balinese carving, ca. 1996)

make it impossible to use it on a daily, recreational basis:

> If psilocybin is ingested more frequently than once a week, this can quickly lead to the development of a considerable tolerance. As with LSD-25, a dosage twice as high will be needed to produce the same effects. Abstention from the drug, as with LSD-25, will just as quickly result in a disappearance of the tolerance effects. (Geschwinde 1990, 110*)

Identifying Mushrooms That Contain Psilocybin

Many people in Western cultures are afraid of mushrooms (an attitude known popularly as mycophobia). They believe that all mushrooms are toxic and that many of them are lethally poisonous. In fact, there are really very few truly dangerous mushrooms, the most lethal of which is the destroying angel (*Amanita phalloides* [Vaillant ex Fries] Secretan) (Graves 1957**). It contains amatoxins that can lead to death within three days. To date, no true antidote has been found for this mushroom. While the destroying angel is very easy to recognize, it can under some circumstances be confused with **Amanita muscaria** or **Amanita pantherina.**

Species of the genus *Psilocybe* can be mistaken for dangerously poisonous *Inocybe* spp. (cf. Gartz 1996):

> Since psychoactive mushrooms are typically self-collected, the danger of confusing these with "true" poisonous mushrooms should not be underestimated. There has been a report of a case of mistaken identity between *P. semilanceata* and the muscarinic *Inocybe geophylla*, which looks quite similar at first glance. (Bresinsky and Besl 1985, 115**)

In order to avoid such mistakes, it is advisable to prepare a spore print when an identification is in doubt. Every mushroom that produces a brown-violet spore print and whose stem turns blue when squeezed is a psilocybin-containing mushroom of the genus *Psilocybe* or *Panaeolus.*

However, the blueing discoloration alone is not a sure sign of mushrooms of the genus *Psilocybe* or *Panaeolus* (cf. Stamets 1996**). The possibility of mistaken identity is especially acute with species from the genus *Galerina*, all of which, however, produce an orange-colored spore print (cf. *Galerina steglichii*).

In order to avoid danger when consuming mushrooms, it is advisable to carefully consult the literature on mushroom identification (e.g., Winkler 1996).

Literature

Gartz, Jochen. 1994. Fuchsbandwurm und Pilze. *Yearbook for Ethnomedicine and the Study of Consciousness*, 1993 (2): 165–66. Berlin: VWB.

———. 1996. Das Hauptrisiko bei Verwendung psilocybin-haltiger Pilze—Verwechslung mit anderen Arten. *Jahrbuch für Transkulturelle Medizin und Psychotherapie* 6 (1995): 287–97.

Winkler, Rudolf. 1996. *2000 Pilze einfach bestimmen*. Aarau: AT Verlag.

Top: Psilocybin-containing mushrooms of the genus *Psilocybe* are often confused with species from the closely related genus *Hyphaloma*. (*Hyphaloma fasciculare*, photographed in Olympia, Washington)

Bottom: Spore prints from freshly collected mushrooms often help in identification. The mushroom cap is placed on a piece of paper with the lamellae (gills) facing down. Covering the mushroom with a plastic bowl or a glass may help. After a short time, the mushroom spores will accumulate on the paper. Most of the species that contain psilocybin produce a radial print that is brownish violet in color. The spores give the lamellae their characteristic color when dried. (Cap of *Psilocybe cubensis*)

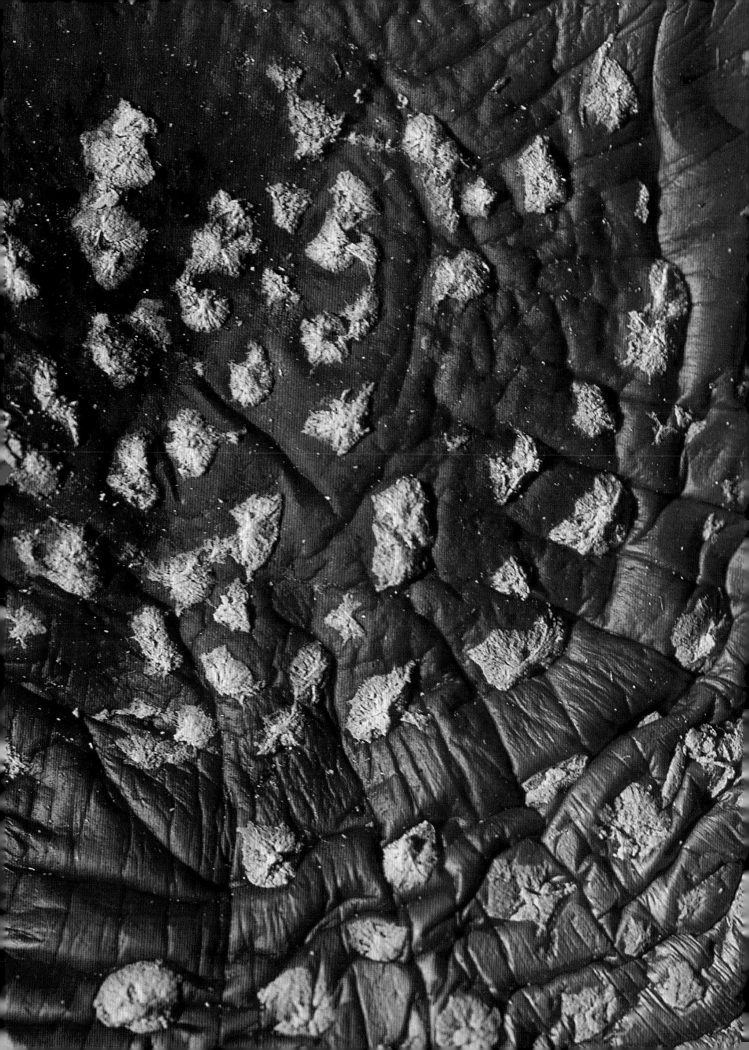

Amanita muscaria (L. ex Fr.) Persoon ex Hooker

Fly Agaric

Family

Agaricaceae; Subfamily Amanitaceae, Amanita Section

Forms and Subspecies

The following varieties of fly agaric are generally accepted (Ott 1996):[348]

Amanita muscaria var. *alba* (Peck) Saccardo—is completely white and is found primarily in Idaho

Amanita muscaria var. *aureola* Kalchbr.—rare!

Amanita muscaria var. *formosa* (Fr.) Saccardo—has a yellow cap

Amanita muscaria var. *muscaria*

In addition, a subspecies with an orange-yellow cap has been described for Central America (Mexico):

Amanita muscaria ssp. *flavivolvata* Singer

Synonyms

Agaricus muscarius L.
Agaricus muscarius Pers.
Amanita formosa Gom. et Rab.
Amanita mexicana Reko nom. nud.
Amanita muscaria var. *mexicana* Reko nom. nud.

Folk Names

Agaric au mouches (French), agaric moucheté (French), äh kib lu'um (Lacandon Mayan, "the light of the earth"), aka-haetori (Japanese, "red flycatcher"), amanite tue-mouches, amoroto (Basque, "the toadlike thing"), ampakʰaw (Igorot), ashitaka-beni-take (Japanese, "long-legged mushroom"), beni-tengu-take (Japanese, "red tengu mushroom"), bolg losgainn (Irish, "toad mushroom"), bolond gomba (Hungarian, "fool's mushroom"), bunte poggenstool, caws llyffant (Welsh), crapaudin (French, from *crapaud* "toad"), düwelsbrûet, escula, fanká'am (Tawgi), fausse-oronge, fleugenschwamm, fliegenkredling, fliegenpilz, fliegenschwamm, fliegenschwemme, fliegenteufel, fluesop (Norwegian), fluesvamp (Danish), flugsvamp (Swedish), flugswampt, fly agaric, fly amanite, flybane, fly fungus, fungus muscarius, giftblaume, grapudin, grzyb muszy (Polish), ha-ma chün (Chinese, "toad mushroom"), hango (Celtic), how k'an c/uh (Chuj, "poisonous yellow squash skin"), itzel ocox (Quiché, "diabolical mushroom"), kabell tousec (Breton), kaqualjá (Quiché), kaquljá (Quiché), kaquljá okox (Quiché, "thunderbolt mushroom"), kärbseseene (Estonian), kärpässieni (Finnish), kässchwamm, krötenpilz, krötenstuhl, matamosques (Catalan, "fly killer"), migeschwamb, miggeschamm, miskwedo (Ojibwa), moucheté, mousseron, muchomor (Polish), mucho-more, muchumor, mückenpfeffer, muckenschwamm, mückenschwamm, muckenschwemme, muhamor, muhovna goba (Slovenian), mukamor, muk-homor (Russian, "fly death"), mukkenswam, muscinery, musmira (Latvian), mussiomiré (Lithuanian), narrenschwamm, oriol foll (Catalan, "crazy Loriot"), oronja (Spanish), paddehat (Danish), paddockstool, pain de crapault, panga (Ostyak), panx (Vogul), pfifferling, pin d'crapâ (French, "toad bread"), pinka, poddehût (Frisian), ponx (Ostyak), premate-it, puddockstool, rabenbrot, reig bord (Catalan, "untrue mushroom"), rocox aj tza (Kekchi, "devil's mushroom"), röd flugsvamp, roter fliegenschwamm, rote tüfus-beeri, ruk'awach q'uatzu:y (Cakchiquel), shtantalok, shtantilok, skabell tousec, soma, sunneschirmche, tignosa dorata (Italian), toad-cheese, toad's bread, toad's cap, toadskep, toad's meat, toadstool, todestoll, tshashm baskon (Afghanistan, "eye opener"), tue-mouche, tzajal yuy chauk (Tzeltal, "red thunderbolt mushroom"), vliegenpaddestoel, vliegenzwam (Dutch), wapaq (Koryak), wliachenschbomm, wliagenschbamm, yuy chauk (Tzeltal), yuyo de rayo (Mexican Spanish)

History

There is no debate about the fact that the psychoactive fly agaric mushroom is associated with shamanism. Over the past decades, it has become ever clearer that this mushroom was or is used throughout the world. In spite of all the efforts that have been made to prove it, Wasson's thesis that the fly agaric was the renowned **soma** of the Aryans is still unproved (Wasson 1968, 1972), as is the question of whether the fly agaric was the tree of knowledge. The suggestion that the fly agaric was in fact a secret means for Buddhist monks to induce states of enlightenment remains speculation as well (Hajicek-Dobberstein 1995). Moreover, it is uncertain when the mushroom was first used for shamanic purposes. However, it has been possible to confirm that it had a shamanic significance in the Germanic regions. And it is possible that the fly agaric found ritual use among the prehistoric Beaker people, who used Stonehenge as a ritual site (Burl 1987, 106 f.).

Although the shamanic use of the fly agaric in Siberia was discovered only in the eighteenth century, it has been suggested that its use is rooted in the Stone Age and that it was used throughout Europe. Wasson (1961**; 1986, 78 f.**) has suggested that the fly agaric and its effects were well known and its shamanic usage was common

"If we do not want to admit that the reason why this poisonous mushroom was used was to produce a sacred inebriation, then we can ask what other purpose it was used for? And here there is only one possible answer, namely that when the grain failed, and under the pressure of the desire for stimulating agents, the natives turned to every kind of plant material. . . . And certainly nothing merited the religious veneration with which such plants were regarded more than this amazing poisonous plant—*Amanita muscaria*."

JOHN GREGORY BOURKE
DER UNRAT IN SITTE, BRAUCH, GLAUBEN, UND GEWOHNHEITSRECHT DER VÖLKER [FILTH IN THE CUSTOMS, TRADITIONS, BELIEF, AND COMMON LAW OF PEOPLES] (1913) (1996, 63**)

Left: The skin of the fly agaric mushroom (*Amanita muscaria*) acquires a velvety sheen when dried. It can be smoked either alone or mixed with other plants (e.g., hemp, thorn apple, tobacco, belladonna, peppermint, angel's trumpet).

348 In 1809, G. H. von Langsdorf suggested referring to the fly agaric of Kamchatka as a distinct variety: *Amanita muscaria* var. *camtschatica*. This name (nom. nud.), however, has never been accepted.

The fly agaric (*Amanita muscaria*) in the three stages of its development, each of which can be interpreted mythologically: the world egg, the divine phallus, and Eros with wings spread. (Photographed near Seattle, Washington)

throughout Asia before the Bering Straits were crossed. When the Paleoindians migrated into North America, they brought the fly agaric cult with them and continued it in the Americas. However, because of the availability of psilocybin mushrooms (*Psilocybe* spp.), which are more easily tolerated and produce more intense visions, the cult was largely forgotten. Traces of the fly agaric cult have been found in Mesoamerica (Rätsch 1995b), and it still exists (at least in part) in North America (Navet 1993; Wasson 1979). The Siberian use of the fly agaric bears many similarities to the North American fly agaric cult of the Ojibwa Indians (Kutalek 1995).

In his book *The Sacred Mushroom and the Cross* (1970), John Allegro, a former Jesuit who apparently had access to certain ancient writings that are preserved in the Vatican but unavailable to the public, advanced the theory that Jesus was actually a fly agaric mushroom and that the so-called original Christianity was a secret fly agaric cult. The fly agaric was the flesh of Christ that was consumed at the evening meal—a nocturnal cult circle—together with the blood of Christ, red wine (cf. *Vitis vinifera*). If Allegro is correct, then the original Christianity would have been a direct continuation of the cult of Dionysos, in which the adherents apparently consumed a **wine** that contained mushrooms (Allegro 1971).

Robert Graves (1957**, 1960**) has argued that ambrosia, which the centaurs used to honor Dionysos in the autumn, was in fact fly agaric.[349] He has also suggested that the maenads did not merely consume **beer** or **wine** (cf. *Vitis vinifera*) to which ivy had been added (cf. *Hedera helix*), but that they also were inebriated from fly agarics.

Wohlberg (1990) views the Thracian Dionysos Sabazios as the analog of the Indian **soma** and the Persian **haoma** (cf. *Peganum harmala*) and has propounded the theory that the Thracian god is identical to the fly agaric mushroom.[350] Carl Ruck has suggested that the secret offering of the Hyperboreans to the Delian Apollo was a fly agaric mushroom and was thus the last reminder of the Indo-Germanic soma (Ruck 1983**). He views the leopard, the sacred animal of Dionysos, as a symbol for the fly agaric, which was consumed ritually and used for entheogenic purposes, because the marks on the leopard's coat resemble those on a dried fly agaric cap (Ruck 1995, 133**). In general, Ruck regards the fly agaric as the original entheogen of the Greek culture(s), which over the course of time was replaced by a variety of (placebo) agents (olives, *Viola odorata* L., *Consolida ajacis* [L.] Schur. [syn. *Delphinium ajacis* L.; cf. ***Delphinium consolida***], *Apium graveolens* L., hippomanes,[351] ***Aconitum napellus***, ***Crocus sativus***, ***Conium maculatum***) and ultimately forgotten (Ruck 1995*). The pine (*Pinus pinea* L.) and spruce (*Pinus pinaster* L., *Pinus nigra* Arnold, *Pinus* spp.)

were sacred to Dionysos because they are the trees with which the fly agaric lives in a symbiotic relationship (Ruck 1995, 137*). The Golden Fleece and the golden apples of the Hesperides[352] have also been interpreted as fly agarics (Allegro 1971; Hajicek-Dobberstein 1995). Vestiges of this ancient or archaic fly agaric cult may have been preserved among the Basques and in Catalonia (Fericgla 1992, 1994).

It is possible that the fly agaric was known in Egypt by the name *raven's bread*.[353] Because some legends say that Saint Anthony (cf. *Claviceps purpurea*) nourished himself in the wilderness on bread that had been brought to him by ravens or similar birds, it has been suggested that fly agarics produced the visions that tempted Saint Anthony (Klapp 1985). It has also been proposed that the fly agaric was the "elixir" of the alchemists of the late ancient and subsequent periods; it has even been interpreted as the Grail (Heinrich 1995).

Pliny was apparently aware of the fly agaric. But like so many after him, he erroneously thought it to be a deadly poison:

> With certain [mushrooms], the toxicity can be easily seen in the pale red color; the disgusting appearance, the bluish coloration of the inside, the furrowed lammellae, and the pallid border that surrounds it. These features are not found on some; dry, and similar to the truffle [cf. *Lycoperdon*], these bear more or less whitish drops from their skin on the cap. (22.93)

The name *fungus muscarius* appeared in 1256 in the work *De Vegetabilibus*, written by the monk Albertus Magnus (Neukom 1996, 390). One of the oldest sources to mention the fly agaric by name is

349 Fly agaric mushrooms still grow in Greece today (Baumann 1982, 140 f.*).

350 Dionysos was usually worshipped in the form of a phallus. Also called *mykes*, "mushroom," this phallus was eaten by the frenzied maenads (Danielou 1992*)—an obvious metaphor for the mycomorphosed god as a fly agaric. In general, the fly agaric is regarded as the "penis of god" (Heinrich 1992).

351 *Hippomanes* (literally, "breast secretion of mares in heat") was originally the term for a "tough body obtained from the pineal glands (epiphyses) of newborn foals that (because of its melatonin content) was used primarily as an aphrodisiac" (Genaust 1996, 290*). Ruck has argued that *hippomanes* means "horse madness" and was a name for *Datura stramonium* (Ruck 1995, 135*).

352 The mythical apples of the Hesperides have also been interpreted as mandrake fruits (cf. ***Mandragora officinarum***).

353 In Egypt, mushrooms in general are known as *'eisch-al-ghorâb*, "raven's bread" (Gholam M. and Geerken 1979, 64).

the *Kräuterbuch* [Herbal] of the physician Johannes Hartlieb, published in 1440:

> There is also the sort of fungus that is impure, broad and thick and red with white spots on the top, when mixed with milk, it will kill the flies that are around, this is why it is known as the fly fungi, *muscinery* in Latin. (folio 16)

After Hartlieb, the fly agaric received only sporadic mention or description in later herbals, including those of Gerard (1633) and Lonicerus (1679). Those that did remark on the fly agaric noted its use as a fly poison, but they made no mention of its psychoactive effects! These psychoactive effects were first described in modern times by travelers to Siberia (Bauer et al. 1991, 121–64; Rosenbohm 1991, 26–60). Around 1880, during a period of widespread **wine** shortages, an Italian physician suggested that the populace turn to fly agaric as an inebriant (Samorini 1996).

In some areas, fly agarics are eaten for food. In the region around Hamburg, which is very rich in fly agarics, mushrooms with their red skin removed were made into soups. In some Alpine valleys, fresh fly agarics are still sliced and made into an appetizer with vinegar, oil, salt, and pepper (cf. *Piper* spp.). In Japan, fly agarics are one of the culinary specialties of the rural population. In Russia, fresh fly agarics are added to vodka (cf. **alcohol**) to improve its effects.

Distribution

The fly agaric mushroom grows only in symbiosis with birch (*Betula* spp.) and/or pine (*Pinus* spp.) trees. But in those places in which these trees occur, it is found throughout the world. It can be found in arctic, temperate, and even tropical climate zones (Alaska, Siberia, Scandinavia, central Europe, North America, Australia, Mexico, the Philippines). It sometimes occurs in the form of fairy rings.[354]

Cultivation

To date, attempts to grow or cultivate fly agaric mushrooms have been unsuccessful.

Appearance

The mycelium is white in color. The fruiting body can grow as tall as 25 cm and form caps as large as 20 cm in diameter. The remnants of the velum remain on the cap in the form of white spots. The fly agaric fruits in central Europe from August until the beginning of November and in North America usually in October.

Although the fly agaric is the most easily recognizable of all the fungi, confusion can sometimes occur. The fly agaric is most commonly confused with a related species, *Amanita regalis* (Fr.) Michael [syn. *Amanita muscaria* var. *umbrina* Fr.],

which is found primarily in mountains at altitudes above 400 meters. The fly agaric can also be confused with the panther mushroom (*Amanita pantherina*) or Caesar's amanita (*Amanita caesarea* [Scop. ex Fr.] Pers. ex Schw.), a delicious culinary mushroom (cf. Roth et al. 1990, 42**). In the button stage, the fly agaric bears a certain resemblance to puffballs: "Very young fly agarics that are not yet showing any red on the outside can also be confused with the common puffball (*Lycoperdon perlatum*) [cf. **Lycoperdon**]" (Bresinsky and Besl 1985, 104**).

Psychoactive Material

— Fruiting body (fungus muscarius)

Preparation and Dosage

The fruiting bodies can be used both fresh and dried. When the fresh mushrooms are to be used for culinary purposes, they must be soaked in cold water for at least one hour before they are prepared (this dissolves the active substances). The soaking water can be drunk to produce psychoactive effects. Fresh mushrooms are well suited for use with **alcohol**. One to three specimens can be added to a bottle of vodka (or any other type of alcohol) and placed in a warm location or, preferably, on a windowsill in the sun. After one week, the fly agaric schnapps will be ready for use. Often, one glass is sufficient to produce psychoactive effects. Fresh mushrooms can also be sautéed in butter and eaten (they are delicious).

To dry the mushrooms, place them in the sun or on a rack in an oven set at low heat (30° to 40° Celsius). The dried results can be smoked (either

Top: The rare yellow North American subspecies of the fly agaric, *Amanita muscaria* var. *formosa*. (Photographed near Olympia, Washington)

Bottom: The dried cap of the fly agaric (*Amanita muscaria*) has been used as a shamanic inebriant since the Stone Age.

354 "Certain species of mushrooms grow in 'witches' rings'—also known as fairy rings or elf dance sites—small circles in which, it was assumed, the witches—or the fairies—would dance at night. These fairy rings appear when the decaying fruiting bodies from the mushrooms of the previous year fertilize the mycelia that are growing out at the border. The tender, threadlike fibers—which have the same function as a root system— then cause new mushrooms to grow around" (Mercatante 1980, 254 f.*).

Petroglyphs from Yenisei (Russia), apparently depicting shamans covered with fly agaric or panther mushrooms.

355 Today, one still finds repeated in the literature, where the idea is perpetuated and passed on without reflection, the thesis that the berserkers acquired their enormous power and their legendary courage, the so-called berserker madness or *furor teutonica*, through their use of fly agaric (e.g., Roth et al. 1990, 274**). But the psychoactive effects of the fly agaric more closely resemble those of opium, and the user will seldom experience anything more than a euphoric state with certain synaesthetic visions. The fly agaric only rarely reveals its entheogenic effects (Cosack 1995). The fly agaric is most certainly *not* the drug of the berserkers. The only psychoactive that is able to produce real aggression, raving madness, and rage is **alcohol**. The berserker madness was also induced by a **beer** to which *Ledum palustre* had been added.

alone or in **smoking blends**), crumbled into a drink (e.g., **beer** or **wine**), or simply eaten as they are. Synergistic effects, typically with aphrodisiac sensations, can be produced by smoking mixtures that include, e.g., henbane (*Hyoscyamus niger*), thorn apple (*Datura stramonium*), and hemp (*Cannabis indica*). Fly agaric prevents the mucous membranes from drying out, a side effect that can occur with both nightshade plants and hemp.

In Siberia, dried fly agaric mushrooms are mixed for consumption or use with freshly pressed bog bilberries (*Vaccinium uliginosum*) or rose bay willow herb (*Epilobium angustifolium* L.) (Hartwich 1911, 257*; Saar 1991, 168; Schultes 1969, 246*). The mixture is either used as is or diluted with water, fermented, and made into a kind of fly agaric **beer**. The urine from people inebriated by fly agarics also can be drunk (Bourke 1996, 55 ff.*).

It is also possible to drink rainwater that has collected in fly agaric caps when they are in the goblet stage (in which the caps are turned upward). This is essentially a cold-water extract, and it has distinct psychoactive effects. This naturally occurring extract has been called "dwarves' wine" (Bauer 1995).

It has been suggested that fly agarics were an ingredient in **witches' ointments**.

The literature provides varying information about dosages, ranging from one mushroom to more than ten (Festi and Biachi 1992):

> The lethal dose lies above 100 grams of fresh mushrooms. In some locations, however, the quantity of toxins can be very low and the fly agaric may be eaten without side effects. Fly agaric intoxications make up 1–2% of all cases of mushroom poisoning. (Roth et al. 1990, 42**)

Fly agarics are usually not as effective fresh as when dried. The fresh material can produce mild sensations of nausea. The effective dosage of fly agaric is individually extremely variable, and probably more so than with any other entheogen. Approach with care!

Ritual Use

Siberian shamans eat dried fly agarics in order to enter a clairvoyant trance state and mobilize their shamanic powers of healing. According to Koryak tradition, the fly agaric grew from the saliva of the highest god; for this reason, it is regarded as a sacred plant (Bauer et al. 1991, 147 f.). The shamans ingest the mushroom especially when they wish to communicate with the souls of the ancestors or to contact spirits, when a newborn is to be given a name, to find a way out of threatening situations, to see into the future and peer into the past, and to be able to journey or fly to other worlds. Among the Khanty (Siberian Ostyak), shamans in training are tested with high dosages

of fly agaric to see whether they can master the mushrooms and are fit for their future office. In Siberia, fly agaric mushrooms are consumed fresh, cooked, and dried (Saar 1991).

The Siberian usage provided the basis for Wasson's proposal that fly agaric mushrooms were the **soma** of the Aryans (Wasson 1968, 1972, 1995). In the Vedic tradition, however, it is said that soma grows in the high mountains, that is, the Himalayas. No evidence of *Amanita muscaria* has yet been found anywhere in the Himalayan region. According to the current state of ethnopharmacological knowledge, the identification of soma with the fly agaric is untenable. However, there are remnants of a ritual consumption of fly agaric in the Hindu Kush, where the mushroom is known as *tshashm baskon*, "eye opener" (Gholam M. and Geerken 1979). While it was thought that the Siberian use of fly agaric had vanished, it was recently discovered that the mushroom is still used for shamanic and divinatory purposes on the Kamchatka peninsula (Salzman et al. 1996).

In Germanic mythology (as recorded in modern times), several stories associate Wotan (also known as Wodan or Odin), the shamanic god of ecstasy and knowledge, with the fly agaric. According to legend, the fly agaric would appear after the Wild Chase, when Wotan rode through the clouds on his horse at the winter solstice with his followers. The following autumn—exactly nine months later—fly agarics would sprout from the impregnated earth in those very spots where the foam from the mouth of Wotan's horse fell onto the ground (Haseneier 1992**). In the common parlance, the fly agaric is known by the name *raven's bread* (Klapp 1985). Ravens not only are ancient shamanic and power animals but also are the messengers of Wotan, who is also known as the raven god. In pagan times, it is entirely possible that the fly agaric also found use in ecstatic rituals (cf. **mead**). It has been suggested that the berserkers ("bear skinners"), warriors who were sacred to Wotan, may have used fly agaric mushrooms in the rituals of their secret society (Thorsen 1948).[355]

In Styria (Austria), a tradition has been passed down that illustrates the mushroom's relationship to the fertility-bringing wild storm god Donar, the son of Wotan. Fly agarics are sought out at the beginning of the "mushroom season" (which of course takes place on a Donnerstag (= Thursday), after the first *donner* (= thunder). The seeker will hold the mushroom first out toward the forest, then against him- or herself, and address it: "If you do not show me the good mushrooms, then I will cast you onto the ground, so that you will decay into dust and ashes!" (*Handwörterbuch des Deutschen Aberglaubens* [Hand Dictionary of German Superstitions], 7:30).

It is also entirely within reason that Santa

Claus, who always appears dressed in red and white and flies through the air with his team of reindeer, is simply an anthropomorphized fly agaric mushroom or a fly agaric shaman (van Renterghem 1995).

In contemporary neopagan circles, the fly agaric is now used as a psychoactive sacrament:

> Another pagan custom that has come down to us is that of drinking on Samhein [November 1] a special tea brewed from the peeled-off skin of a fly agaric that was picked during the full moon. This is probably based on traditions of Siberian and Norwegian shamans, in which the fly agaric was repeatedly referred to as a psychoactive working plant. (Magister Botanicus 1995, 186*)

One family described a traditional ritual that they still conduct on this ancient Celtic holiday (cf. *Atropa belladonna*):

> For this—and only on this night—we prepare a fly agaric tea according to the following recipe from my grandmother: On the full moon preceding Samhain, the heads of the family go into the forest and search for a few fly agarics, with whom they establish contact. The healthy mushrooms (those that have not been infested by worms or eaten by snails!) are cut off at the stipe and placed in a wicker basket; at the place where we harvested the mushrooms, we usually leave some tobacco [*Nicotiana tabacum*] and an apple as an offering. After this, the red skin of the cap is pulled off and quickly dried; the dried skin is kept in a red linen cloth in a dark and cool place until Samhain. During that night a cold [water] extract is made that all of the members of the family drink before they go to bed. The next morning, the resulting dreams are described and interpreted in the family circle. (Magister Botanicus 1995, 197*)

In the pre-Columbian fly agaric cult of the Americas, the fly agaric (known as the light of the earth, the flower of the earth, the underworld mushroom, or the thunderbolt/lightning mushroom) was regarded as a being that was in contact with the underworld (*xibalba, metlan*) (Rätsch 1995b). It was symbolically associated with toads and flies (helping spirits) and formed a door to the realm of the dead. It was also associated with the *bolon ti kuh*, the nine gods of the underworld, which were represented in the form of mushroom stones. It was a ritual inebriant with unique effects that shamans, oracular priests, and healers consumed (either eating it or smoking it together with tobacco [*Nicotiana rustica, Nicotiana tabacum*]) in order to enter a desired altered state

of consciousness in which they would carry out necromantic rites (*uay xibalba*),[356] liberate the souls of sick people from the underworld, and generally improve their visionary abilities.

Caves were typically regarded as entrances to the underworld and were often used for necromantic rituals and sacrificial offerings. The mushroom was used for shamanic initiation (the journey into the underworld). Only a few initiates were granted knowledge of the uses of the fly agaric. In order to protect (monopolize) this knowledge, the mushroom was publically portrayed as poisonous or dangerous. Since fly agarics were not available in all places and at all times, they were collected in pine groves and air- or fire-dried, which improved their effects. The mushrooms were sold by special merchants together with other ritual paraphernalia ("Transcendental Interaction Model"). There were three important centers of the ancient American fly agaric cult: the northeastern forests of North America (Algonquian, Ojibwa, Dogrib; cf. Keewaydinoquay 1979; Larsen 1977), central Mesoamerica (Mayan peoples, Aztecs, Purépecha; cf. Rätsch 1995b), and western Peru (Mochica).

The Tzeltal still make use of the fly agaric mushroom, which they call *tzajal yuy chauk*, "red thunderbolt mushroom." They remove the reddish skins from the fresh caps, dry the skins, and smoke them together with *may*, wild tobacco (*Nicotiana rustica*). It is said that smoking this mixture helps the Tzeltal shamans become clearsighted so that they may recognize diseases in their patients, track down lost or stolen objects, and utter prophecies. The shamans of the Chuj, a Mayan people of the southern Selva Lacandona and northern Guatemala, smoke dried pieces of fly agarics mixed with wild tobacco (Müller-Ebeling and Rätsch 1986, 96–97*; Rätsch 1992, 78*). In the high valley of Puebla, Timothy Knab found an indigenous *curandero* (= healer) who smoked dried fly agarics mixed with tobacco (*Nicotiana tabacum*) in order to make ritual diagnoses (Díaz 1979, 86*). Among the Tzutujil, a figure of a deity named Maximon that is made from the wood of *palo de pito* (*Erythrina rubrinervia*; cf. *Erythrina* spp.) is associated with the fly agaric (Lowy 1980, 102**).

Japan is rich in mushrooms that were or still are used for culinary or shamanic purposes. The genus *Amanita* is represented by a number of species, many of them endemic (Imazeki et al. 1988, 140–71**). All three psychoactive *Amanita* species (cf. *Amanita pantherina*) of the Japanese mycoflora are included in the taxon *tengu take*, "*tengu* mushroom" (Wasson 1973, 15**). The *tengu* is the spirit of the fly agaric and is one of the most popular figures in Japanese mythology and folklore (De Visser 1908; Fister 1985). *Tengus* appear sometimes as birdlike demons and at other

"In the birch forest
The wind whispers and runs softly
Through the heart-shaped leaves.
The trunks glow as white as snow.

"From the grass
sprouts a magical egg,
its grows round redly, spots itself
        white
and jumps asunder.

"I wished for a witch,
fourteen years old.
I rode with her
through the fly agaric forest.

"Smeared with ointment, she flies
swiftly like the swallow
through the green forest.
Tender skin, smooth fur.

"That which she lightly
touches with magical power,
loses its weight and is
carried off into the air.

"Stick, stem, or ledge,
it is all the same. She lifts, lifts
tenderly around the body.
She flies gently and floats.

"I can see that the forest is
        enchanted.
He who eats of the fly agaric,
becomes mad, he dances
sings, flies, and forgets."

FRIEDRICH GEORG JÜNGER
FLIEGENPILZE [FLY AGARIC
MUSHROOMS]

Toad and stool, an ancient mythological pair. *Toadstool* is another name for the fly agaric. (Illustration from a nineteenth-century children's book)

356 In Yucatec Maya dictionaries from the colonial period, *uay xibalba* is translated as "necromancer" or "conjurerer of the dead"; my own knowledge of Mayan, however, suggests that the name may mean "changer of the other world" or "shaman of the parallel world."

The folk literature in particular has helped make the fly agaric a "dangerous fellow." (From Usteri, *Pflanzenmärchen und -sagen*, 1926)

times as wild and reclusive monks of the mountains. Occasionally *tengus* are regarded as transformed shamans. At times it is said that there is only one *tengu*; other reports speak of many *tengus*, who even have a king. *Tengus* are seen sometimes as gods and at other times as demons, but they are usually regarded as *kami*.[357] *Tengus* can change their form; they can be humans or birds,[358] fly through the air, make themselves invisible, and create phantoms. The male *tengus*, which have a bright red skin and a phallic nose, are regarded as tricksters and sexual demons, but also as benefactors. Mountain shrines were erected in their honor. Fossil shark teeth are thought to be visible reminders of their passing. The teeth are known as the "claws of *tengu*" and are sold as talismans. They are even venerated in temples and shrines and guarded as religious relics (Rätsch 1995). *Tengus* have a magical leaf or an enchanted fan that they use to carry out their tricks and magical acts. In some traditional illustrations, this leaf clearly resembles a hemp leaf (*Cannabis indica*).

Although normally invisible, *tengus* reveal themselves as spooks or speak through the mouths of people whom they have possessed. Japanese who are praying on mountain peaks and in mountain shrines fall into possession states particularly often; while possessed, they lend the *tengus* their voices and utter prophecies (Lowell 1894, 1–15). *Tengus* are known for their unlimited thirst for **sake**, which is why people are advised to offer them the beverage. *Tengus* also are excellent with the sword. They sometimes abduct children or youngsters and teach them sword fighting or impart some other knowledge. People make offerings to the *tengus* so that they will protect them and teach them the wisdom of nature (Rätsch 1995a).

Artifacts

A number of petroglyphs discovered in Asia appear to be connected with a shamanic fly agaric cult.

The Italian ethnomycologists Giorgio Samorini and Gilberto Camilla recently proposed the theory that certain Greek depictions of wine grapes (*Vitus vinifera*) are actually epithetic representations of the fly agaric (or other psychoactive mushrooms) that were kept secret, and that these were related to the cult of Dionysos (Samorini and Camilla 1995**).

There are a large number of anthropomorphic mushroom representations in the pre-Columbian ceramics of the Peruvian Moche. The reference to shamanism is especially apparent in depictions that show mushrooms growing directly out of the head of a seer (Furst 1976a, 82*). A Mochican stirrup vessel (ca. 500 A.D.) at the Peabody Museum at Harvard University has the shape of a human head. In the middle, above the forehead, is

a very realistic depiction of a fly agaric mushroom that is more or less growing out of the hat. Other examples of Mochican ceramic work also appear to contain representations of fly agaric mushrooms (Ripinsky-Naxon 1993, 180 f.*).

In Nayarit (western Mexico), a number of small ceramic objects have been found that depict fly agaric–like figures, beneath which a person is sitting (Furst 1974*; Schultes and Hofmann 1992, 82*; Wasson 1986, 51**). A ceramic piece in the Remojada style from Tenenexpan (Veracruz, ca. 300 A.D.) depicts an oversized object, which looks like a fly agaric, together with a human figure in the throes of ecstasy. The person is shown touching the mushroom or mushroom stone (?) with the left hand while pointing the right hand toward the sky (Heim et al. 1966, plate 2**). In Michoacán, a small stone figure was recovered from a pre-Spanish site of the Purépecha culture that Guzmán has interpreted in the following manner: "The one side resembles a fly agaric cap, the other a death skull" (1990, 100 f.**).

The fly agaric is a popular figure in German literature, appearing in fairy tales, legends, songs, and poems (Bauer 1992). One German folk song[359] about the fly agaric, "Ein Männlein steht im Walde" [A Little Man Stands in the Forest], is quite well known. Engelbert Humperdinck (1854–1921) even included the song in his children's opera *Hänsel und Gretel* (1893).

Fly agaric mushrooms are common in illustrated children's books (e.g., *Wir fahren ins Zwergenland!* [We're Going to the Land of the Dwarves!], *Hochzeit im Walde* [A Forest Wedding], *Wichtelmanns Reise* [Wichtelmann's Journey], and *Alice in Wonderland*), where they are usually the dwellings of dwarfs or elves. In one children's book, *Mecki bei den 7 Zwergen* [Mecki and the Seven Dwarfs], the hero of the story smokes dried fly agarics together with his friends. While under the influence of the mushrooms, they realize that the seven dwarfs are actually fly agaric spirits:

The dwarfs shook their heads back and forth and back and forth. As they did so, their funny hats slowly turned into red caps with white spots. Their legs and little tummies grew together or sank into the ground. At the same time, their white necks grew longer and longer, so that they finally stood there like large, strange fly agarics and stared at me. (Rhein n.d., 45)

The life story of a Siberian shaman is told in the novel *Der Herr des Feuers* [The Lord of the Fire] (Braem 1994). During his training, he must have several experiences with fly agaric mushrooms. The author, who has also written nonfiction works on shamanism (Braem 1994*),

357 In Japan, the word *kami* is usually used to refer to ancient Shinto deities or sacred objects (Lowell 1894). The word itself appears to be related to the Turko-Tatar word *kam* and the Mongolian word *kami*, both of which mean "shaman" (Couliana 1995, 63**; Eliade 1975, 14**).

358 The Japanese red kite (*Accipiter gularis*) is usually identified with *tengus*.

359 This was actually an art song that became popularized among the people. The text was written by August Heinrich Hoffmann von Fallersleben (1798–1874) (Bauer 1992, 46).

describes these as though he himself had had such experiences:

Then there was the new seeing, an overly clear vision, that had been brought about by the frequent use of raw fly agarics. Shorting after eating the aromatic flesh of the mushroom, the effects began: Initially, the accelerated pulse and heartbeat were the sure messengers of the coming changes. Then the colors began to light up, a burning green, longingly deep blue, rich brown, and gleaming silvery gray of the rock. The room, the forest, expanded every time he exhaled, and contracted with every inhalation. Breath and forest and vision became one single act. The power of the helping spirits in the mushroom revealed how nature vibrated in a rhythm, illuminated the big pattern, imparted the smallest detail with an unimagined importance. . . . If the fly agaric was good and the helping spirit within it powerful, then his hearing would also change. He perceived the slightest snap in the woods, the rustle of a mouse, he heard crackling in the moss, the movement of the leaves in the wind. And the most astonishing thing was that he could sometimes also understand the language of the animals. (Braem 1994, 149)

Fly agarics often appear in comic books. In the book *Asterix at the Olympic Games* (Uderzo and Goscinny, 1968), Methusalix, the oldest of the unruly Gauls, collects fly agarics for his soup. The Druid Miraculix tells him that the fly agarics should be "sautéed in butter, for only in that way will they retain their typical flavor." Moebius, in his comic story *A True Wonder of the Universe*, transports the fly agaric to a planet a million light-years away. It speaks to a cosmonaut, who then tries a piece of it, whereupon he becomes a supernova. The volume *Soluna* (1996), from the series John Difool vor dem Incal [John Difool Before the Incal] by Janjetov and Jodorowsky, features a utopian city, the center of which is a gigantic mushroom-shaped temple. The entire fantasy comic cycle Alef-Thau, by Jodorowsky and Arno (1986–1991), takes place in a forest of fly agaric mushrooms. In the three volumes of the cycle Die Gefährten der Dämmerung [The Fellowship of the Dawn] (1984–1990), François Bourgeon artistically portrays a fly agaric trip. The second volume, *Die drei Augen der blaugrünen Stadt* [The Three Eyes of the Blue-Green City], contains detailed instructions for using fly agarics and tips about their effects:

This red and white cap contains more colors than your poor human eyes have ever seen! . . . If you dry this and chew it, you will be able to uncover terrible secrets in your dreams. You can go to the earliest times in the world . . . in the times before your gods . . . before my gods. (p. 13)

The comic *Fliegenpilz* [Fly Agaric], by Christian Farner (1993), depicts a fly agaric trip that is both haunting and oppressive.

Fly agarics are not featured as frequently in paintings, perhaps because they are too popular a symbol. The pen-and-ink drawing *Die Hexe* [The Witch] (ca. 1900) by Heinrich Vogeler (1872–1942) depicts a witch passing through the forest, fly agarics sprouting at her feet. In *Brekkek-kwakkwak*, an oil painting by Johan Fabricius (1899–1981) from 1926, fly agaric mushrooms are shown growing alongside a fantastic pond. The illustrator Alan Lee contributed several pictures of fly agarics and elves to *Das große Buch der Geister* [The Big Book of Spirits] (Frond and Lee 1979).

Because fly agarics are symbols of good luck in Europe, they are often portrayed on greeting and congratulatory cards. Fly agaric spirits appear on many decorative items and are common motifs on Easter eggs and Christmas ornaments (perhaps because Santa Claus himself is simply an anthropomorphized mushroom). Countless images of fly agarics are used for decorative purposes, ranging from plastic Smurf figures with fly agarics (cf. **Veratrum album**) to fireworks for New Year's Eve parties (these fireworks are known as *glückspilzen*, which is usually translated as "lucky dogs" but which literally means "lucky mushrooms"). These pretty mushrooms are also commonly reproduced in the form of Easter cakes, chocolate (cf. **Theobroma cacao**), and marzipan (Bauer et al. 1991).

In the 1990s, fly agaric mushrooms were frequently featured as emblems on handbills advertising raves (Rätsch 1995d; 1995, 12**). They are also found on the covers of CDs of psychedelic trance music (e.g., *Holy Mushroom, Ironic Beat*) and of other musical styles. The cover of the LP *Granny Takes a Trip*, by the Purple Gang (1968), shows fly agarics in an alchemical context. Another album, *Early One Morning*, by the band Mushroom (1973), has a picture of a mushroom whose colors are reversed: it has a white cap with red spots. The German combo Witthüser and Westrupp gave expression to the theories of John Allegro (1970) on the cover of the record *Der Jesuspilz—Musik vom Evangelium* [The Jesus Mushroom—Music from the Gospels] (1971) (cf. **Cannabis indica**). Mani Neumeier, founder of the band Guru Guru and the godfather of "kraut rock," is depicted on the cover of his solo CD *Privat* (ATM Records, 1993) as a six-armed Shiva holding a fly agaric in one hand. OM Records, a San Francisco–based company, selected the fly agaric mushroom as the symbol for its series *Mushroom Jazz*. And the cover of the avant-garde/psychedelic CD *Venus Square Mars*, by Mark

"According to the view of the natives, however, the fly agaric mushroom, in contrast to alcohol, has an innate power of revealing the future to those who consume it; namely when the wish to be able to look into the future is uttered over the mushroom in a specific form before it is eaten, whereupon the wish will come true in a dream."

J. ENDERLI
"ZWEI JAHREN BEI DEN TSCHUKTCHEN UND KORJAKEN" ["TWO YEARS AMONG THE CHUKCHEE AND KORYAK"] (1903, 185)

The cosmopolitan fly agaric on stamps from the Southeast Asian countries of Laos and Cambodia.

"Because a feeling of flying often occurs after the consumption [of fly agaric mushrooms], this may be the origin of the Scandinavian and English version of Santa Claus, in which he flies through the air on a sleigh drawn by reindeer."

LESLEY BREMNESS
HERBS
(1994, 286*)

Ibotenic acid

Muscimol

This flyer for a psychedelic trance party features a fly agaric mushroom. (Switzerland, ca. 1995)

360 "Russian forest workers have reported chewing dried pieces of fly agarics so that they could deal more easily with the physical exertion" (Neukom 1996, 392).

Naussef and Dave Philipson, features a *tengu* "fractalizing out of" a fly agaric (M·A Records, 1995).

In Japan, masks of long-nosed *tengus*, the fly agaric spirits, are still made out of wood and other materials and sold in the paraphernalia shops located near shrines. The masks are hung on houses during the Japanese New Year's festival.

Medicinal Use

It is likely that the fly agaric was originally a ritual medicine (Rosenbohm 1995). In Siberia, it was ingested to treat psychophysiological states of exhaustion.[360] For snakebites, a fly agaric tea (a cold-water extract made from dried fruiting bodies) was massaged into the affected area of the body (usually the legs). This was said to neutralize the toxins (Saar 1991, 177**).

In the nineteenth century, the fly agaric was used as a home remedy and was also prescribed by physicians as a medicine. It was used internally, for example, to treat epilepsy and fever and externally to treat ulcerated fistulae:

> It is officinal under the name *fungus muscarius*. Only the lower portion of the stalk is chosen. . . . The fly agaric, in powder form (for which it must be dried as quickly as possible without destroying it), is administered internally (with care) in small doses (10 to 30 grains) against falling sickness, et cetera, and is sprinkled externally onto malignant tumors, gangrene, et cetera. Meinhard gives a tincture to treat favus and other persistent eruptions. (W. Schneider 1974, 1:80*).

In homeopathy, the preparation Agaricus muscarius is an "agent to treat complaints of the entire nervous system" (Bremness 1995, 286*). Depending upon the symptom picture, it is used in homeopathic dilutions (D4, D6, D30, D200) for problems associated with menopause, overexcitability, and bladder and intestinal cramps. One physician who frequently utilizes the mother tincture in his practice reported:

> One portion (15–20%) of the patients I have treated with Agaricus muscarius had altered dreams during or after their therapy. Especially: dreams of flying with positive contents, dreams reminiscent of Alice in Wonderland, and other pleasant dream experiences. In no case did nightmares occur, although one must consider that the majority of dosages used in therapy are small. Even with higher dosages, on the following day the patients were normally found to be well and to exhibit a strong eagerness to work, without negative side effects or symptoms of a hangover. . . . Following the prescription of fly agaric,

almost all of the patients exhibited increased motivation, improved mood, and improved mental and physical well-being. Here again it is the dosage that determines that something is not a poison! (Waldschmidt 1992, 67)

Constituents

Fresh fly agaric mushrooms contain choline, acetylcholine, muscarine, muscaridine, muscazone (empirical formula $C_5H_6O_2N_2$), large amounts of **ibotenic acid** (= prämuscimol, "pilzatropin"; Ott 1996), very little **muscimol**, and the rare trace elements selenium and vanadium. As a result of the decarboxylation of ibotenic acid, dried fly agarics contain high amounts of muscimol, which is responsible for the psychoactive effects (Festi and Bianchi 1992). The pigment is a derivative of ibotenic acid (Talbot and Vining 1963). The amount of muscarine is 0.0003% at most (Roth et al. 1990, 42**). The amount of ibotenic acid in fresh specimens from Germany and Switzerland averaged 0.03% but can run as high as 0.1% (Eugster 1969). Traces of **bufotenine** (Schultes and Farnsworth 1982, 155*) and the **tropane alkaloid** L-hyoscyamine (Salemink et al. 1963) have also been reported.

Muscimol is regarded as the actual psychoactive constituent, although this view (from Eugster 1967a and 1967b) is not without controversy (Cosack 1995). Nevertheless, muscimol can be detected in the urine of people who have ingested fly agarics. Several experiments have demonstrated that the urine from a person who has consumed fly agaric will produce psychoactive effects in other people (McDonald 1978; Ott 1976). Theodore Schurr, an anthropologist and molecular biologist, has summarized our current pharmacological understanding of Siberian fly agaric shamanism:

> Once ingested, the psychoactive alkaloids and substances in [*Amanita muscaria*] acted as agonists of normal brain neurotransmitter function, disrupting the coordinate action between the catecholaminergic and serotoninergic systems and producing hallucinogenic effects similar to those generated by LSD and harmine. (Schurr 1995, 31)

Effects

Most people in the German- and English-speaking countries—regardless of their level of education—believe that fly agaric mushrooms are deadly poisonous and should be avoided at all costs. Paracelsus's rule, which says that the dosage is the factor that determines whether something is a poison or a medicine, has apparently not become widely known. If the fly agaric was evaluated according to this criterion, then people would have to leave their cherished black-and-white thinking behind—for many, a very painful or difficult task.

In recent years, many reports and descriptions of the effects of fly agaric have appeared (cf. Cosack 1995; Festi and Bianchi 1992; Ott 1976, 1993). In general, the symptoms of fly agaric inebriation include a strong parasympatholytic stimulation, wavelike shifts between sleepiness and wakefulness, illusions, hallucinations, and delirium. (Leuner 1981, 54*). The effects are often described as unpleasant, and inexperienced users may easily interpret them as signs of a "toxic ecstasy" (Leuner). In the older literature, fly agaric is portrayed as a deadly poison, and people were urgently warned against its use. This notwithstanding, the toxicological literature does not contain a single case of lethal fly agaric poisoning: "there is no evidence of fatalities" (Garnweidner 1993, 41**). The more recent literature also notes: "If the mushroom is consumed with the expectation of hallucinogenic effects, then it tends to produce a pleasant outcome" (Roth et al. 1990, 42**). Once again we see that the expectations a person brings into the experience (the set) exert a powerful influence upon the experience of an altered state. If the fly agaric is believed to be poisonous, horrible apparitions may arise; if it is regarded as a pleasurable inebriant, enjoyable visions and feelings will result.

Some people report temporary sensations of nausea after ingesting fly agaric mushrooms, after which most are overwhelmed by sleep. The visionary effects, which are often characterized by synaesthesia, begin after they awaken and can last for several hours. Reports of visions of giants in the world of dwarfs are remarkably common (Cosack 1995). The effects are more subtle when fly agarics are smoked and are manifested primarily as a heightened perception and increased sensitivity in the musculature. Regardless of the manner in which the mushrooms are consumed, auditory perception is heightened, refined, or altered.

Commercial Forms and Regulations

All homeopathic preparations (including the mother tincture) may be sold only by pharmacists but do not require a physician's prescription. Neither collecting the mushroom nor consuming it is illegal.

Literature

See also the entries for *Amanita pantherina*, **soma**, **ibotenic acid**, and **muscimol**.

Allegro, John M. 1970. *The sacred mushroom and the cross: A study of the nature and origins of Christianity within the fertility cults of the ancient Near East.* Garden City, N.Y.: Doubleday & Company.

Bauer, Wolfgang. 1992. Der Fliegenpilz in Zaubermärchen, Märchenbildern, Sagen, Liedern und Gedichten. *Integration* 2/3:39–54.

———. 1995. Ein Versuch mit "Zwergenwein." *Integration* 6:45–46.

Bauer, Wolfgang, Edzard Klapp, and Alexandra Rosenbohm. 1991. *Der Fliegenpilz. Ein kulturhistorisches Museum.* Cologne: Wienand-Verlag.

Benedict, R. G., V. E. Tyler Jr., and L. R. Brady. 1966. Chemotaxonomic significance of isoxazole derivatives in *Amanita* species. *Lloydia* 29 (4): 333–42.

Braem, Harald. 1994. *Der Herr des Feuers: Roman eines Schamanen.* Munich: Piper.

Buck, R. W. 1963. Toxicity of *Amanita muscaria*. *Journal of the American Medical Association* 185 (8): 663-664.

Burl, Aubrey. 1987. *The Stonehenge people.* London: Dent & Sons.

Cosack, Ralph. 1995. Die anspruchsvolle Droge: Erfahrungen mit dem Fliegenpilz. *Yearbook for Ethnomedicine and the Study of Consciousness,* 1994 (3): 209–44. Berlin: VWB.

De Visser, M. M. 1908. The Tengu. *Transactions of the Asiatic Society of Japan* 36 (2): 27–32.

Enderli, J. 1903. Zwei Jahre bei den Tschuktschen und Korjaken. *Petermanns Mitteilungen* 49 (8): 183 ff.

Eugster, Conrad Hans. 1956. Über Muscarin aus Fliegenpilzen. *Helvetica Chimica Acta* 39 (4): 1002.

———. 1967a. Isolation, structure and synthesis of central-active compounds from *Amanita muscaria* (L. ex Fr.) Hooker. In *Ethnopharmacological search for psychoactive drugs,* ed. D. H. Efron, 416–18. Washington, D.C.: U.S. Government Printing Office.

———. 1967b. *Uber den Fliegenpilz.* Zurich: Naturforschende Gesellschaft (Neujahrsblatt).

———. 1968. Wirkstoffe aus dem Fliegenpilz. *Die Naturwissenschaften* 55 (7).

———. 1969. Chemie der Wirkstoffe aus dem Fliegenpilz (*Amanita muscaria*). In *Fortschritte der Chemie organischer Naturstoffe,* vol. 27. Berlin: Springer.

Fabing, H. D. 1956. On going berserk: A neurochemical inquiry. *Scientific Monthly* 83:232–37.

Fericgla, Josep Maria. 1992. *Amanita muscaria* usage in Catalunya. *Integration* 2/3:63–65.

———. 1993. Las supervivencias culturales y el consumo actual de *Amanita muscaria* en Cataluña. In *Atti del 2° Convegno Nazionale sugli Avvelenamenti da Funghi, Annali dei Musei Civici di Rovereto* suppl. 8 (1992): 245–56.

"Sometimes [in Kamchatka] people also eat the fresh [fly agaric] mushroom in soups and broths, and then it loses much of its inebriating properties. When dipped into the juice of berries of *Vaccinium uliginosum*, it acts like strong wine. One large or two small mushrooms will usually suffice to induce a proper state of inebriation for the entire day, especially if water is consumed as well, which increases the narcotic properties.

"The desired effects begin one or two hours after the mushroom is ingested. Dizzy spells and drunkenness manifest themselves in the same way as with wine or schnapps; at first one feels a happy mental stimulation, the face turns red, then follow involuntary words and movements, and finally complete unconsciousness sometimes occurs."

DR. GEORG HEINRICH VON LANGSDORF (1809)
IN *DER UNRAT IN SITTE, BRAUCH, GLAUBEN, UND GEWOHNHEITSRECHT DER VÖLKER* [FILTH IN THE CUSTOMS, TRADITIONS, BELIEF, AND COMMON LAW OF PEOPLES] (BOURKE 1996, 57**)

"Once a year, the tradition of these nomads [the Koryak of Kamchatka] takes them back to the beginnings of time. This occurs when they slaughter the reindeer. The festival is celebrated according to ancient custom, and playing a special role is a mushroom: the fly agaric. Dried and ground to a powder, it is mixed, so I was told, with fresh reindeer blood that has been filled into vessels. The Koryak who have come to dance drink this. Along with the rhythms, which sound strange to our ears, the dancing becomes ever more intense, the stamping of the feet grows harder and faster. It almost has no end. The inebriation of drink and rhythm can last for three or four days. . . . Only those who have tried the fly agaric themselves can understand this custom of the Koryak."

Markus Wolf
Geheimnisse der russischen Küche [Secrets of the Russian Kitchen]
(1995, 178)

———. 1994. *El Hongo y la génesis de las culturas.* Barcelona: Los Libros de la Liebre de Marzo.

Festi, Francesco, and Antonio Bianchi. 1992. *Amanita muscaria. Integration* 2/3:79–89.

Fister, Pat. 1985. *Tengu,* the mountain goblin. In *Japanese ghosts and demons: Art of the supernatural,* ed. Stephen Addiss, 103–12. New York: George Braziller.

Frond, Brian, and Alan Lee. 1979. *Das große Buch der Geister.* Oldenburg: Stalling.

Hajicek-Dobberstein, Scott. 1995. Soma siddhas and alchemical enlightenment: Psychedelic mushrooms in Buddhist tradition. *Journal of Ethnopharmacology* 48:99–118.

Heinrich, Clark. 1992. *Amanita muscaria* and the penis of God. *Integration* 2/3:55–62.

———. 1995. *Strange fruit: Alchemy and religion— the hidden truth.* London: Bloomsbury.

Keewaydinoquay. 1979. The legend of Miskwedo. *Journal of Psychedelic Drugs* 11 (1–2): 29–31.

Klapp, Edzard. 1985. Rabenbrot. *Curare,* special issue, 3:67–72.

Kutalek, Ruth. 1995. Ethnomykologie des Fliegenpilzes am Beispiel Nordamerikas und Sibiriens. *Curare* 18 (1): 25–30.

Langsdorf, G. H. von. 1924. Einige Bemerkungen, die Eigenschaften des Kamtchadalischen Fliegenschwammes betreffend. *Wetterauische Gesellschaft für die gesamte Naturkunde, Annalen* 1 (2) (Frankfurt/M.).

Larsen, Stephen. 1977. *The shaman's doorway.* New York: Harper Colophon Books.

Lowell, Percival. 1894. *Occult Japan: Shinto, shamanism and the way of the gods.* Boston: Houghton-Mifflin.

Lowy, Bernard. 1972. Mushroom symbolism in Maya codices. *Mycologia* 64:816–21.

———. 1974. *Amanita muscaria* and the thunderbolt legend in Guatemala and Mexico. *Mycologia* 66 (1): 188–91.

McDonald, A. 1978. The present status of soma: The effects of California *Amanita muscaria* on normal volunteers. In *Mushroom poisoning: Diagnosis and treatment,* eds. B. H. Rumack and E. Salzman, 215–23. West Palm Beach, Fla.: CRC-Press.

Mochtar, Said Gholam, and Hartmut Geerken. 1979. Die Halluzinogene Muscarin und Ibotensäure im Mittleren Hindukush: Ein Beitrag zur volksheil-praktischen Mykologie in Afghanistan. *Afghanistan Journal* 6 (2): 63–65.

Navet, Eric. 1993. Die Ojibway und der Fliegenpilz. *Integration* 4:45–54.

Neukom, Hans-Peter. 1996. Geheimnisvoller Fliegenpilz. *Schweizer Apothekenzeitung* 134 (16): 390–92.

Ott, Jonathan. 1975. *Amanita muscaria:* Usos y química. *Cuadernos Científicos CEMEF* 4:203–21.

———. 1976. Psycho-mycological studies of *Amanita*—from ancient sacrament to modern phobia. *Journal of Psychedelic Drugs* 8 (1): 27–35.

———. 1977. *Amanita muscaria:* Mushroom of the gods. *Head* (March/April): 55–62.

———. 1996. *Amanita muscaria.* Unpublished electronic file. (Cited 1998.)

Pollock, S. H. 1975. The Alaska *Amanita* quest. *Journal of Psychedelic Drugs* 7 (4): 397–99.

Rätsch, Christian. 1995a. Die Klauen des Tengu. *Dao* 1/95:18–20.

———. 1995b. Äh kib lu'um: "Das Licht der Erde"—Der Fliegenpilz bei den Lakandonen und im alten Amerika. *Curare* 18 (1): 67–93.

Rhein, Eduard. no date. *Mecki und die 7 Zwerge.* Cologne: Lingen Verlag.

Römer, Stefan. 1992. *Amanita muscaria. Integration* 2/3:133–34.

Rosenbohm, Alexandra. 1991. Der Fliegenpilz in Nordasien. In *Der Fliegenpilz: Ein kulturhistorisches Museum,* ed. Wolfgang Bauer et al., 121–64. Cologne: Wienand-Verlag.

———. 1995. Zwischen Mythologie und Mykologie: Der Fliegenpilz als Heilmittel. *Curare* 18 (1): 15–23.

Saar, Maret. 1991. Ethnomycological data from Siberia and north-east Asia on the effect of *Amanita muscaria. Journal of Ethnopharmacology* 31 (2): 157–73.

Salemink, C. A., J. W. ten Broeke, P. L. Schuller, and E. Veen. 1963. Über die basischen Inhaltsstoffe des Fliegen Pilzes. XIII. Mitteilung: Über die Anwesenheit von L-Hyoscyamin. *Planta Medica* 11:139–44.

Salzman, Emanuel, Jason Salzman, Joanne Salzman, and Gary Lincoff. 1996. In search of *mukhomor,* the mushroom of immortality. *Shaman's Drum* 41:36–47.

Samorini, Giorgio. 1996. Un singolare documento storico inerente l'agarico muscaria. *Eleusis* 4:3–16.

Schurr, Theodore G. 1995. Aboriginal Siberian use of *Amanita muscaria* in shamanistic practices: Neuropharmacological effects of fungal alkaloids ingested during trance induction, and the cultural patterning of visionary experience. *Curare* 18 (1): 31–65.

Talbot, G., and L. Vining. 1963. Pigments and other extractives from carpophores of *Amanita muscaria. Canadian Journal of Botany* 41:639.

Thorsen, P. 1948. *Amanita muscaria* and the fury of the berserks. *Friesia* 3:333–51.

van Renterghem, Tony. 1995. *When Santa was a shaman: The ancient origins of Santa Claus and the Christmas tree.* St. Paul, Minn.: Llewellyn.

Waldschmidt, Eberhard. 1992. Der Fliegenpilz als Heilmittel. *Integration* 2/3:67–68.

Waser, Peter G. 1967. The pharmacology of *Amanita muscaria*. In *Ethnopharmacological search for psychoactive drugs*, ed. D. H. Efron, 419–39. Washington, D.C.: U.S. Government Printing Office.

Wasson, R. Gordon. 1967. Fly agaric and man. In *Ethnopharmacological search for psychoactive drugs*, ed. D. H. Efron, 405–14. Washington, D.C.: U.S. Government Printing Office.

———. 1968. *Soma—divine mushroom of immortality.* New York: Harcourt Brace Jovanovich.

———. 1972. *Soma and the fly-agaric: Mr. Wasson's rejoinder to Professor Brough.* Ethno-mycological Studies, no. 2. Cambridge, Mass.: Botanical Museum of Harvard University.

———. 1979. Traditional use in North America of *Amanita muscaria* for divinatory purposes. *Journal of Psychedelic Drugs* 11 (1–2): 25–27.

———. 1995. Ethnomycology: Discoveries about *Amanita muscaria* point to fresh perspectives. In *Ethnobotany: Evolution of a discipline*, ed. Richard Evans Schultes and Siri von Reis, 385–91. Portland, Ore.: Dioscorides Press.

Wohlberg, Joseph. 1990. Haoma-soma in the world of ancient Greece. *Journal of Psychoactive Drugs* 22 (3): 333–42.

Wolf, Markus. 1995. *Geheimnisse der russischen Küche.* Hamburg: Rotbuch-Verlag.

Amanita pantherina (DC. ex Fr.) Krombh.
Panther Cap

Family
Agaricaceae; Subfamily Amanitaceae, Amanita Section

Forms and Subspecies
Three varieties are usually distinguished:

Amanita pantherina var. *abietinum* (Gilb.) Ves.—fir panther cap
Amanita pantherina var. *multisquamosa* (Pk.) Jenkins
Amanita pantherina var. *pantherina*

Synonyms
Agaricus pantherinus Fr.
Amanita cothurnata Atkinson

Folk Names
Agarico panterino, amanite panthère, crapaudin gris, fausse golmelle, fausse golmotte, fongo rospèr (Treviso, "toad mushroom"), haitori (Japanese, "fly catcher"), haitori-goke, haitori-kinoko, haitori-take, hyô-take (Japanese, "panther mushroom"), panther cap, panther fungus, panther mushroom, pantherpilz, tengudake (Japanese, "*tengu* mushroom"), tengutake,[361] tignosa bigia, tignosa bruna

The panther cap is similar in appearance to the fly agaric (**Amanita muscaria**) but has a brownish or brown cap. It yields a white spore print and is easily confused with the nonpoisonous and non-psychoactive pearl mushroom (*Amanita rubescens* [Pers. ex. Fr.] S.F. Gray [syn. *Amanita rubens* Scopo. ex Fr.]), also called the blusher because of its tendency to turn red when bruised (Roth et al. 1990, 48**). The panther cap is found almost exclusively in deciduous and fir woods. In Europe, it fruits from July to October, while in North America it usually fruits in early spring. Although many people avoid the panther cap as poisonous, it actually has a long tradition of culinary use:

> It is remarkable that the panther cap is characterized as inedible in some books on mushrooms. Many of our patients who were relatively knowledgeable about mushrooms told us with conviction that they have been eating these mushrooms for years without any harm. (Leonhardt 1992, 127)

And yet the panther cap is clearly psychoactive and normally even more potent than **Amanita muscaria**. In Russia, the panther cap is preferred over the fly agaric because the effects of the former are thought to be more pleasant. The shamans of central Asia and Siberia apparently consumed it ritually as an alternative to the fly agaric. It is said that the Russian panther cap induces beautiful visions. The dosage is given as one to four mushrooms.

Today, the panther cap is used for psychoactive purposes wherever it is found:

> *Amanita pantherina* is also a widespread "recreational and party drug" whose effects are stronger than those of *Amanita muscaria*. Pleasant sensations are more likely to arise when the mushroom is ingested with an expectation of hallucinogenic effects. In many areas of the U.S., Russia, France, and Italy, the panther cap is also consumed for culinary purposes. These may be varieties that are relatively devoid of toxins. The psychotropic

This old English illustration of *fungi lethales*, "deadly mushrooms," may depict a group of panther cap, fly agaric, or other amanita mushrooms (*Amanita* spp.). (Woodcut from Gerard, *The Herball or General History of Plants*, 1633)

In northern Asia, dried panther caps (*Amanita pantherina*) are popular inebriants.

361 The very poisonous species *Amanita verna* (Fr.) Quél. is known in Japanese as *shiro-tamago-tengu-take*, "white egg *tengu* mushroom" (Imazeki 1973, 47**).

The North American variety of the panther cap (*Amanita pantherina*). (Photograph: Paul Stamets)

A *Cyperus* species grass known as *piripiri*. (Photographed in Peru)

Amanita Species

Species of the genus *Amanita* that are regarded as psychoactive or are used as psychoactive substances (from Beutler and Der Marderosian 1981; Ott 1978**; Weil 1977**):

Amanita citrina Schaeff. ex St. Gray [syn. *Amanita mappa* (Batsch. ex Lasch) Quél.]—false death cap; contains up to 7.5 mg **bufotenine** in 1 gram of dry mass (Beutler and Der Marderosian 1981, 423)

Amanita cothurnata Atkinson [syn. *Amanita pantherina* var. *multisquamosa* (Pk.) Jenkins]—booted amanita

Amanita gemmata (Fr.) Gill—contains **ibotenic acid/muscimol**

Amanita gemmata (Fr.) Gill x *A. pantherina* hybrids—contain **ibotenic acid/muscimol**

Amanita parcivolvata Pk.—contains traces of isoxazole

Amanita porphyria (Alb. et Schw. ex Fr.) Secretan—contains **bufotenine** (trace amounts)

Amanita strobiliformis (Paul) Quélet—ibo-tengu-take (Japanese, "warty *tengu* mushroom"); contains **ibotenic acid/muscimol**

Amanita tomentella Krombholz—contains some **bufotenine**

effects can also be elicited by smoking the dried skins of the caps or mushroom bodies. For adults, the lethal toxic dose is contained in more than 100 grams of fresh mushrooms. (Roth et al. 1990, 43 ff.**)

The effects subside no later than ten to fifteen hours after consumption (Roth et al. 1990, 44**). In contrast, the older literature speaks of euphoric

and psychotic states lasting as long as eight days (Leonhardt 1992).

Panther caps, at least those of North American origin, have been found to contain **ibotenic acid** and **muscimol**. When dried and stored, the quantity of ibotenic acid declines proportionally as the quantity of muscimol increases (Beutler and Der Marderosian 1981, 423, 427; cf. Benedict et al. 1966). Panther caps also contain stizolobic acid and stizolobinic acid (amino acids), which are also found in *Stizolobium* and *Mucuna* species (cf. **Mucuna pruriens**) (Bresinsky and Besl 1985, 107 f.**).

Literature

See also the entry for **Amanita muscaria**.

Benedict, R. G., V. E. Tyler Jr., and L. R. Brady. 1966. Chemotaxonomic significance of isoxazole derivatives in *Amanita* species. *Lloydia* 29 (4): 333–42.

Beutler, John A., and Ara H. Der Marderosian. 1981. Chemical variation in *Amanita*. *Journal of Natural Products* 44 (4): 422–31.

Beutler, John A., and Paul P. Verger. 1980. Amatoxins in American mushrooms: Evaluation of the Maixner test. *Mycologia* 72 (6): 1142–49.

Chilton, W. S., and J. Ott. 1976. Toxic metabolites of *Amanita pantherina*, *A. cothurnata*, *A. muscaria* and other *Amanita* species. *Lloydia* 39 (2/3): 150–57.

Kendrick, Bryce, and Arthur Shimizu. 1984. Mushroom poisoning—analysis of two cases, and a possible new treatment, plasmapheresis. *Mycologia* 76 (3): 448–53.

Leonhardt, Wolfram. 1992. Über Rauschzustände bei Pantherpilzvergiftungen. *Integration* 2/3:119–28.

Yocum, R. R., and D. M. Simons. 1977. Amatoxins and phallotoxins in *Amanita* species of the northeastern United States. *Lloydia* 40:178–90.

Balansia cyperi Edgerton

Cyprus Grass Fungus

Family

Clavicipitaceae; Subfamily Balansiae

The *Balansia cyperi* fungus is closely related to *Claviceps* and is parasitic exclusively on Cyprus grasses (Cyperacea; Edgerton 1919). In Ecuador, it infests chiefly *Cyperus prolixus* H.B.K., a grass known locally as *piripiri* that the women of the Jívaro use to aid in birthing (Lewis and Elvin-Lewis 1990). *Piripiri* is also used as an **ayahuasca** additive (cf. **Cyperus spp.**) and as a folk medicine for treating snakebites (Plowman et al. 1990).

Balansia cyperi is a parasite that preferentially

infests the following species of Cyprus grasses (Clay 1986, Lewis and Elvin-Lewis 1990, Plowman et al. 1990):

Cyperus articulatus L.
Cyperus articulatus L. var. *nodosus* (H. et B. ex Willd.) Kuek.
Cyprus prolixus H.B.K.—piripiri
Cyperus pseudovegetus Steudel
Cyperus rotundus L.—nut grass
Cyperus surinamensis Rottb.
Cyperus virens Michx.

Two of the **ergot alkaloids** found in Cyprus grasses infested with *Balansia* (Plowman et al. 1990), ergobalansine and ergobalansinin, were recently detected in the vine *Ipomoea piureusis* (cf. *Ipomoea* spp.) (Jnett-Siems et al. 1994).

Ergot alkaloids are also present in the following *Balansia* species: *Balansia claviceps* Speg., *Balansia epichloë* (Weese) Diehl, *Balansia henningsiana* (Moell.) Diehl, *Balansia strangulans* (Mont.) Diehl (Plowman et al. 1990, 459).

Literature

See also the entries for ***Ipomoea* spp.**, ***Cyperus* spp.**, ***Scirpus* spp.**, and **ergot alkaloids**.

Clay, Keith. 1986. Induced vivipary in the sedge *Cyperus virens* and the transmission of the fungus *Balansia cyperi* (Clavicipitaceae). *Canadian Journal of Botany* 64:2984–88.

Edgerton, C. W. 1919. A new *Balansia* on *Cyperus*. *Mycologia* 11:259–61.

Jnett-Siems, K., M. Kaloga, and E. Eich. 1994. Ergobalansine/ergobalansinin, a proline-free peptide-alkaloid of the genus *Balansia*, is a constituent of *Ipomoea piurensis*. *Journal of Natural Products* 57:1304–6.

Leuchtmann, Adrian, and Keith Clay. 1988a. *Atkinsonella hypoxylon* and *Balansia cyperi*, epiphytic members of the Balansiae. *Mycologia* 80 (2): 192–99.

———. 1988b. Experimental infection of host grasses and sedges with *Atkinsonella hypoxylon* and *Balansia cyperi* (Balansiae, Clavicipitaceae). *Mycologia* 80 (3): 291–97.

Lewis, Walter H., and Memory Elvin-Lewis. 1990. Obstetrical use of the parasitic fungus *Balansia cyperi* by Amazonian Jivaro women. *Economic Botany* 44:131–33.

Plowman, Timothy C., Adrian Leuchtmann, Carol Blaney, and Keith Clay. 1990. Significance of the fungus *Balansia cyperi* infecting medicinal species of *Cyperus* (Cyperaceae) from Amazonia. *Economic Botany* 44:452–62.

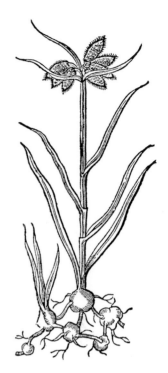

Claviceps paspali Stevens et Hall

Paspalum Ergot

Family

Class Ascomycetes, Order Clavicipitales, Family Clavicipitaceae

Synonyms

Claviceps fusiformis-paspali (also refers to a type group)
Claviceps rolfsii Stev. et Hall.

Folk Names

Ergot of paspalum, paspalum ergot, paspalummutterkorn, paspalum staggers

Paspulum ergot is a parasitic fungus that infests and forms its sclerotia exclusively in the spikes of wild grasses of the cosmopolitan genus *Paspalum* (approximately two hundred species; Gramineae; Poaceae) (cf. *Claviceps purpurea*, *Claviceps* spp.). Infected paspalum grains contain psychoactive or hallucinogenic **ergot alkaloids** (lysergic acid amide and its derivatives; Aaronson 1988; Acramone et al. 1960, 1961; Petroski and Kelleher 1978). The bioactive, antibiotic, and nontoxic substances paspaline, paspalinine, and paspalicine are also present (Gallagher et al. 1980; Springer and Clardy 1980). For cattle breeders, *Paspalum* grasses that have been infected with ergot represent a certain threat for their animals (Hindmarsh and Hart 1939). Ergot-infected grains of the North American Dallis grass (*Paspalum dilatatum* Poir.), which are dangerous to cattle, have also been found to contain paspalinine (Cole et al. 1977).

The ergot fungi of the following *Paspalum* species have acquired a certain significance as psychoactive substances:

***Paspalum distichum* L. [syn. *Paspalum paspaloides* (Michx.) Scribn.]—knotgrass**
Common in North America, knotgrass is now a common wild grass in the Mediterranean region[362] as well (Wasson et al. 1985). Ergot infesting this grass produces the **ergot alkaloids** lysergic acid amide (= ergine, LSA) and lysergic acid hydroxyethylamide, both of which also occur in ***Ipomoea violacea*** and ***Turbina corymbosa***. Since it contains only hallucinogenic alkaloids, Wasson et al. advanced the hypothesis that *Paspalum* ergot was the secret ingredient in **kykeon**, the Eleusinian initiatory drink (cf. *Claviceps purpurea*):

> "Early Man in ancient Greece could have arrived at an hallucinogen from ergot. He might have done this from ergot growing on wheat or barley [*Claviceps purpurea*]. An easier way would have been to use the ergot growing on the common wild grass Paspalum. This is based on the assumption that the herbalists of ancient Greece were as intelligent and resourceful as the herbalists of pre-Conquest Mexico. (Wasson et al. 1998, 44)

The cosmopolitan Cyprus grasses (*Cyperus* spp.) are often infected with the fungus *Balansia cyperi*, which produces psychoactive alkaloids as metabolites. (Woodcut from Gerard, *The Herball or General History of Plants*, 1633)

362 According to Zander (1994, 420*), the grass is originally from North America and became naturalized in the Mediterranean region only after the conquest of the New World! It is not mentioned by a single ancient author (cf. *Arundo donax*, *Phalaris arundinacea*, *Phalaris* spp., *Phragmites australis*).

If this grass was in fact not introduced into Europe until the modern era—as all of the sources suggest—then *Paspalum* ergot must be eliminated as a possible candidate for the active constituent in **kykeon**.

Paspalum plicatulum Michx. and *Paspalum unispicatum* (Scribn. et Merr.) Nash

In Paraguay, these two grasses are apparently often infested with *Claviceps paspali*. As a result, a sweet secretion (= honeydew) forms on the spikes and is consumed by wasps and bees. Bees that have visited these grasses inoculate their honey with this secretion. In the language of the Makai Indians, the resulting **honey** is called *fic'e* and is recognizable by its pungent aroma. Consumption of large amounts is dangerous, for it can produce dizziness, headaches, and drunkenness; it is said that it can be lethal. This honey is used to brew a **beer** or **mead**, certainly a beverage with potent effects. The grass *Elionurus muticus* (Spreng.) Kunth is used as an antidote for poisoning from this honey or drinks made from it. A cold-water extract made from the lower portions of five plants is drunk both to counteract the effects of the honey and to treat poisoning caused by manioc (*Manihot esculenta* Crantz) (Arenas 1987, 289 f.*).

Paspalum scrobiculatum L. [syn. *Paspalum commersonii* Lam.]—koda millet, kodrava, kodo

The indigenous people of India, where this grass is common, used it as a type of millet (*kodo*) for food (cf. DeWet et al. 1983). It is mentioned in the Vedic scriptures, the Puranas, and the Ayurvedic classics (e.g., *Sushruta Samhita*) under the name *kodrava*. Ayurvedic writings note that the seeds are "sweetish and bitter, tonic, and antidotal to poisons, useful in the treatment of ulcers; it caused constipation and flatulence, upset the physiological balance of the body and led to hallucinations and dysuria" [painful urge to urinate] (Aaronson 1988, 346). The effects and symptoms described are produced not by the grass itself but rather by *Paspalum* grains that have been infected with ergot[363] (Bhide and Aimen 1959). In India, ergot-infected *Paspalum* was used to treat postpartum pain (Aaronson 1988, 347).

In ancient India, the sclerotia were also used in religious contexts; unfortunately, little is known about such use:

It is said that the monks ate the spikes with the husks, whereupon they displayed symptoms of inebriation and were no longer able to stand. It is said that these effects lasted for several days. (Chauduri and Pal 1978)

During the rice shortage of 1946, many Indians collected and ate spikes of kodo millet. This resulted in numerous cases of LSD-like inebriation. Ergot-infested grains of this *Paspalum* species were found to contain 0.003% **ergot alkaloids** (dry weight): D-lysergic acid, methylcarbinol amide, D-lysergic acid amide (= ergine), and ergometrine (= ergobasin, ergotocine, ergostetetrine) (Aaronson 1988, 345 f.; Arcamone et al. 1960).

The Lodha of West Bengal either collect the wild grass or harvest the cultivated grass. The seed coats are used as a "ritual hallucinogen" (Pal and Jain 1989, 468). Unfortunately, we have no details about this ritual. Seeds that have had their coats removed are used to distill a type of **alcohol** (Aaronson 1988, 346).

Literature

See also the entries for ***Claviceps purpurea***, ***Claviceps* spp.**, **kykeon**, and **ergot alkaloids**.

Aaronson, S. 1988. *Paspalum* spp. and *Claviceps paspali* in ancient and modern India. *Journal of Ethnopharmacology* 24 (2, 3): 345–48.

Arcamone, F., C. Bonino, E. B. Chain, A. Ferretti, P. Pennella, A. Tonolo, and L. Vero. 1960. Production of lysergic acid derivatives by a strain of *Claviceps paspali* Stevens et Hall in submerged culture. *Nature* 187:238–39.

Arcamone, F., E. B. Chain, A. Ferretti, A. Minghetti, P. Pennella, A. Tonolo, and L. Vero. 1961. Production of a new lysergic acid derivative in submerged culture by a strain of *Claviceps paspali* Stevens et Hall. *Proceedings of the Royal Society* (London), series B, 155:26–54.

Bhide, N. K., and R. A. Aimen. 1959. Pharmacology of a tranquilizing principle in *Paspalum scrobiculatum* grain. *Nature* 183:1735–36.

Chaudhuri, R. H. N., and D. C. Pal. 1978. Less known uses of some grasses of India. *Bulletin of the Botanical Society of Bengal* 32:48–53.

Cole, Richard J., Joe W. Dorner, John A. Lansden, Richard H. Cox, Courtney Pape, Barry Cunfer, Stephen S. Nicholson, and David M. Bedell. 1977. Paspalum staggers: Isolation and identification of tremorgenic metabolites from sclerotia of *Claviceps paspali*. *Journal of Agric. Food Chem.* 25 (5): 1197–1201.

DeWet, J. M. J., et al. 1983. Diversity in kodo millet, *Paspalum scrobiculatum*. *Economic Botany* 37 (2): 159–63.

Gallagher, Rex T., Janet Finer, and Jon Clardy. 1980. Paspaline, a tremorgenic metabolite from *Claviceps paspali* Stevens et Hall. *Tetrahedron Letters* 21:235–36.

Hindmarsh, W. L., and L. Hart. 1939. Poisoning of cattle by ergotized paspalum. *Veterinary Report of New South Wales* 1938:78–88.

Pal, D. C., and S. K. Jain. 1989. Notes on Lodha medicine in Midnapur District, West Bengal, India. *Economic Botany* 43 (4): 464–70.

The ergot fungus (*Claviceps*) can assume quite different forms on different host plants. Left: *Claviceps purpurea* from rye (*Secale cornutum*); middle: *Claviceps paspali* from a Mediterranean *Paspalum* grass; right: a spike from a *Paspalum* species. (Photograph: F. Ratek, with permission of LEK, Slovenia)

363 In India, the closely related grass *Paspalum dilatatum* Poir. may also be infected with *Claviceps paspali* (Aaronson 1988, 346).

Petroski, Richard J., and William J. Kelleher. 1978. Biosynthesis of ergot alkaloids. Cell-free formation of three products from L-tryptophan and isopentenylpyrophosphate and their incorporation into lysergic acid amide. *Lloydia* 41:332–41.

Springer, James P., and Jon Clardy. 1980. Paspaline and paspalicine, two indole-mevalonate metabolites from *Claviceps paspali*. *Tetrahedron Letters* 21:231–34.

Wasson, R. Gordon, Albert Hofmann, and Carl A. P. Ruck. 1998. *The road to Eleusis*. 20th-anniversary ed. Los Angeles: Hermes Press.

Claviceps purpurea (Fries) Tulasne

Ergot

Family

Class Ascomycetes (Sac Fungi), Order Clavicipitales (Ergot Fungi), Family Clavicipitaceae

Forms and Subspecies

This species is composed of numerous races, which are distinguished primarily by their constituents and the differing composition of their alkaloids (Hofmann 1964, 5). Varieties, e.g., *Claviceps purpurea* var. *glyceriae*, are occasionally listed; these are not, however, botanically accepted (Teuscher 1992, 912).

Synonyms

Clavarius clavus nom. nud.
Claviceps microcephala (Wallr.) Tul.
Claviceps sesleriae Staeger
Claviceps setulosa Quél.
Clavis secalinus nom. nud.
Cordyceps purpurea Fries
Fusarium heterosporum nom. nud.
Sclerotium clavus DC.
Secale cornutum nom. nud.
Spacelia segetum Leveillé

Folk Names

Achterkorn, acinula davus, afterkorn, blé, bléavorté, blé cornu, blé noire, bockshorn, brandkorn, calvi siliginis, centeio espigado (Portuguese), charbon de seigle, chiodo segalino, clavaria clavus, clavus, clavus secalinus, cockspur, cockspur rye, conichuelo, cornadillo (Spanish), cornezuelo del centeno, cornichos, cravagem de centeio, cuernecillo de centeno, dürrkorn, ergot, ergota, ergot de seigle, ergot of rye, ergotum secale, esparo de centeio, espolón de centeno, esporão de centeio, faux seigle, fungus secalis, giftkorn, grano alloglia to, grano cornuto, grano speranota, grano sperone, hahnensporn, horned rye, horn seed, hungerkorn, kindesmord, kornmuhme, kornmutter, kornzapfen, krähenkopf, kriebelkorn, kriekelkorn, madre segal, mascarello, mater secalis, mehlmutter, meldrøje, mjyöldrya, mother of rye, mutterkorn, mutterkornpilz, mutterzapfen, rey ergot, rockenmutter, roggenbrand, roggenmutter, roter keulenkopf, rye smut, schwarzkopf, schwarzkorn, sclerotium clavus, secale clavatum, secale corniculatum, secale cornutum, secale luxurians, secale maternum, secale temulentum, secale turgidum, secalis mater, segala alloglioto, segale cornuta, seigle ergoté, seigle ergotisé, seigle ivre (French, "drunken grain"), spawn, spermoedia clavus, sperone di gallo, spiked rye, spur, spurred rye, tizón de centeno, todtenkorn, tollkorn, wolf(s)zahn, zapfenkorn

History

The history of ergot is presumably as old as that of rye (*Secale cereale* L.), which is the primary host for this parasitic fungus. Rye seems to have been largely unknown in ancient times (Germer 1985*). Apparently it was cultivated in central Europe in the first millennium B.C., during the Hallstatt (early) Iron Age. It first arrived in Greece in the fourth century A.D. (Renfrew 1973, 83*). For this reason, neither rye nor ergot is mentioned by the authors of antiquity (cf. the discussion in Rätsch 1995a, 250 ff.*).

The ancient Hebrews knew of a *son* or *sonin*, a "degenerate wheat of black color and bitter taste." Consuming this was said to have both inebriating and lethal effects, depending upon the dosage. Berendes (1891, 108*) has interpreted this as grain infected with *Claviceps purpurea*. The Bible makes mention of "rust-infected grain"; some (e.g., Moldenke 1986*) have suggested that this "rust" was the uredinial species *Puccinia*, while others have pointed to ergot.

Shelley (1995) has attempted to demonstrate that ergot was the **soma** of the Aryans, the **haoma** of the Parsis, the initiatory drug of the Mithraic mysteries (cf. **Peganum harmala**), and the elixir or philosopher's stone of the alchemists. Gordon Wasson believed that ergot was the secret psychoactive ingredient in **kykeon** (cf. *Claviceps paspali*).

In his work *On Nature*, the Roman poet Lucretius (ca. 94–55 B.C.) described a disease that can be interpreted as ergotism:

Suddenly this new and devastating air of plague
    descended

Grains on an ear of rye (*Secale cereale*) are infected with ergot (*Claviceps purpurea*).

Saint Anthony, the patron saint of those afflicted with ergotism, with his attributes: the *ignis sacer* ("holy fire"), the swine, and the tau-shaped walking stick. (Woodcut, fifteenth century)

Botanical illustration of ergot sclerotia. (Engraving from Pereira, *De Beginselen der Materia Medica en der Therapie*, 1849)

Ergot, or *secale cornutum*, is an ancient agent for inducing labor and still finds use today in homeopathy.

364 *Lolium temulentum* is also known as *tollkorn*.

down upon the water, or it nested in the fruits of the field . . .
the entire body was reddened by burning sores,
as when the "sacred fire" [*ignis sacer*] spread over the limbs.
Throughout the inside of a person, so that it burned all the way down to the bones,
burned in the stomach as brightly as the fire in the interior of the earth. . . .
Completely confused condition with fear and melancholia,
darkened brow and a sharp, even angry look in the eyes;
moreover, a fearfully excited hearing and buzzing in the ears.

(6.1125, 1166 ff., 1183 ff.)

During the Middle Ages, a devastating epidemic called *ignis sacer*, "holy fire," struck. The French physician Gay Didier, who worked in the service of the Antonite hospital in Saint-Antoine du Viennois, wrote about Saint Anthony's "fire" in his *Epitome Chirugieae*:

The fire consists of a mortifying gangrene of a limb, it is also known as Saint Antonious's or St. Martialis's fire. It is remarkable to note that such pain and such heat arise with this disease that it is tantamount to a real burn. (In Bauer 1973, 22)

Saint Anthony's fire (ergotism) was an epidemic disease that swept through central and southern Europe in the sixteenth century (Ruffie and Sournia 1993). The etiology of this devastating disease, which is the result of ergot poisoning, was not discovered until the seventeenth century. Since *Claviceps purpurea* infests primarily rye, the disease usually appeared in those areas where the populace lived primarily on rye and were constantly forced to eat the "bread of dreams" or "devil's bread" (Camporesi 1990; Gebelein 1991, 298 f.*; Matossian 1989). It was not until the seventeenth century that the connection between Saint Anthony's fire and ergotism was recognized. At the time, it was believed that ergot was a creature of the devil. Those afflicted by ergot regarded Saint Anthony as their patron saint, for he himself had once successfully resisted the hallucinatory temptations of the devil—which the victims now faced themselves—in the Egyptian desert (Athanasius 1987). Saint Anthony was long venerated as the healer of ergotism (Kolta 1987; Müller-Ebeling 1985).

The German common names *rockenmutter* [rye mother], *afterkorn* [anal grain], *todtenkorn* [death grain], *tollkorn* [mad grain],[364] and *mutterkorn* [mother grain] are all indicative of the effects and usages of this fungus. Since the Middle Ages, midwives have used ergot to induce labor. The first written source that directly referred to ergot in this

context was Lonicerus's *Kräuterbuch* (seventeenth century). During the late modern era, ergot was the most important agent for inducing contractions (Schneider 1974, 1:335*). The first scientific report on ergot as a uterotonic agent appeared in 1808. In 1824, the American physician David Hosack recommended ergot to accelerate childbirth.

In 1918, Arthur Stoll isolated and described ergotamine, the first **ergot alkaloid** to be identified. Later analyses of ergot by Albert Hofmann uncovered the structures of numerous other ergot alkaloids and also resulted in the "accidental" synthesis of LSD. These studies also produced several preparations that are still in use today (e.g., methergine and hydergine).

Distribution

Claviceps purpurea occurs worldwide as a parasite on grasses (June grass [*Poa pratensis* L.], orchard grass [*Dactylis glomerata* L.], meadow foxtail [*Alopecurus pratensis* L.]), and cereal grains (rye, barley, wheat). The fungus is found as a parasite on four hundred species of the Family Gramineae (= Poaceae) (Teuscher 1992, 912).

Cultivation

Ergot is reproduced through a process known as rye inoculation:

To effect this, the blossoming rye spikes are infected using a conidial [spore] suspension obtained through in-vitro culture. This can occur by spraying or by injection, the latter being more effective. Today, inoculating machines that make it possible to efficiently infect large fields using the injection method are used for large-scale industrial production. (Hofmann 1964, 7)

To produce ergot for pharmaceutical purposes, large fields of rye are now inoculated in Chechnya, Hungary, and Portugal (Teuscher 1992, 914).

Appearance

The sclerotium of *Claviceps purpurea* is dark purple. On rye (*Secale cereale* L.), the conelike sclerotium is dark violet to black, can grow up to 6 cm in length, and resembles a long, slender tooth.

Psychoactive Material

— *Secale cornutum* (ergot)

Only the sclerotia (fruiting bodies) of ergot infecting rye (*Secale cereale* L.) are used (Teuscher 1992, 912).

Preparation and Dosage

The sclerotia are collected when the rye is ripe and dry and are then dried and crushed. The resulting powder can be used to produce alcoholic extracts and cold-water infusions. Because ergot preparations are almost impossible to standardize and exhibit considerable variations in alkaloid concentration and in the proportions of the constituent alkaloids, one seldom finds information about dosages. The older literature often notes that four ergot grains are sufficient to speed up the process of labor. Such statements, however, should not necessarily be taken at face value. Alcoholic extracts can be very dangerous, as they dissolve the toxic alkaloids. It is only with cold-water extracts that the poisonous alkaloids remain undissolved. *There is no data available concerning cold-water extracts. Self-experimentation is absolutely not recommended!*

It has been conjectured that ergot was an ingredient in the Delphic **incense** (cf. *Hyoscyamus albus, Laurus nobilis*) and in **witches' ointments**. The recipe for a "lotion for prophetic dreams" (from Piobb) contains ergot (as the primary agent) as well as turpentine, the yolk from the egg of a wild duck, diascordium (?), red roses, goat's or mare's milk, ivy (*Hedera helix*), alchemilla, ironweed (*Verbena*?), shavings from deer antlers, monkshood (*Aconitum* spp.), and whale blubber. All of the ingredients are boiled in alcohol with camphor (cf. *Cinnamomum camphora*); mixed with coral syrup, black salisify (radix consolid.), balsam, and ammoniac; and finally dissolved in Malvasian wine. Three drops are added to a liter of water. The resulting solution is smeared onto the hands, feet, head, and abdomen before sleep (Spilmont 1984, 142 f.).

Ergot is also a component of many homeopathic preparations, e.g., Secale Pentarkan, which consists of secale cornutum, glonoine, Ignatius bean (see *Strychnos* spp.), magnesium phosphate, and nux-vomica (*Strychnos nux-vomica*).

A variety of plants were utilized to treat Saint Anthony's fire, including mandrake (*Mandragora officinarum*; cf. Gebelin 1991, 299*). The precious spice saffron (*Crocus sativus*) was also regarded as a magical antidote. Plants used to treat the epidemic are depicted on the Isenheim Altar together with Saint Anthony (Seidl and Bauer 1983, 64). They are easily identified as common plantain (*Plantago major* L.) and aconite (*Aconitum napellus*). During the Middle Ages, plantain was known as *herba proserpinacia*, "plant of Proserpina [= Persephone]" (Storl 1996, 101*). Persephone, the daughter of Demeter, was kidnapped by Pluto/Hades, as a result of which the Great Goddess ultimately established her mysteries (cf. **kykeon**).

Most (1842, 457*) lists rhubarb (*Rheum* spp.) as a remedy for ergot poisoning. Rhubarb is one of the many plants associated with **soma**.

Ritual Use

If ergot was in fact an ingredient in **haoma, soma,** or **kykeon,** then its ritual use as a psychoactive substance is rooted in prehistoric times.

In ancient central Europe, it may have been used in the ritual transformation of a person into an animal: "The werewolf crouches amidst the grain," it was said, and "amidst the grain" is where ergot grows. Here again, the best we can do is to speculate.

Ergot or preparations containing ergot appear to have been used for divinatory purposes in French occultism (Spilmont 1984).

Today, ergot-infected grains are used in Peru by Indian soothsayers in conjunction with the coca oracle (cf. *Erythroxylum coca*).

Artifacts

An ancient Celtic coin features what may be the oldest representation of an ergot-infected grain (Lengyel 1976). A woodcut in the *Kräuterbuch* of Lonicerus (1679) is usually regarded as the earliest illustration of a grain of ergot-infected rye.

It has frequently been suggested that the pictures of Hieronymus Bosch (ca. 1450–1516) were inspired by ergot-induced hallucinations (Brod 1991; Dixon 1984; cf. *Datura stramonium*). In particular, those pictures that have the temptation of Saint Anthony as their theme have been interpreted as representations of the ghastly visions induced by Saint Anthony's fire (Müller-Ebeling 1983). The famous Isenheim Altarpiece of Matthias Grünewald (ca. 1470–1528) is directly related to ergot:

On the contrary, this altar stood in an isolated place, before a strange and isolated community: in a hospital church in Isenheim (Alsace) belonging to a cloister of the order of Saint Anthony. Saint Anthony was the patron saint of the lepers, those unfortunates who were afflicted by an epidemic known at the time as "hell fire" or "burning sickness." This devastating epidemic, from which thousands died, had been moving through the West since the tenth century. The cloister at Isenheim was a house for the afflicted, a hospital for people suffering from this disease. The altar was intended for those wretched individuals who were doomed to this slow, suppurating death. It was the great shrine of their church. (Fraenger 1983, 11 f.)

The right wing of the second opening of the

"The 'sacred fire' ate into the flesh of the people, so that it fell from the bones in burning shreds, made the earth delirious, and turned humans into beasts."

Stanislav Przybyszewski
Die Synagoge Satans [The Synagogue of Satan]
(1979, 43)

This illustration of an ear of grain infected with ergot is from an antique Celtic coin and may be the oldest representation of its kind. (From Lengyel, *Das geheime Wissen der Kelten*, 1976)

"On the grain cone, Latin Clavi Siliginis: One often finds in the heads of rye or grain long, black, hard, narrow cones growing out betwixt and between the grain, that is in the heads, and which stand out and look like long little nails; they are white on the inside like the grain, and they are entirely harmless to the grain. Such grain cones are regarded by the women as a great aid and proven medicine for the rise and cramping of mothers when one ingests and uses three of these several times."

LONICERUS
KRÄUTERBUCH ["HERBAL"] (1679)

retable depicts the temptation of Saint Anthony. A "human figure, upon whose body fester all of the devastations of the epidemic" (Fraenger 1983, 46), is visible on the bottom left (Seidel and Bauer 1983). The afflicted person is unconscious and appears to be hallucinating the wild scene. On the bottom right of the same panel is a tree stump, from which polypores (cf. "*Polyporus mysticus*") are growing. Paul Hindemith (1895–1963) dramatized the creation of the altar in his opera *Mathis der Maler* [Matthias the Painter] (1935) by having his painter experience visions like those that had once tempted Saint Anthony.[365] The scene was also set to music in "The Temptation of St. Anthony" by the psychedelic rock group The Sensational Alex Harvey Band (1972).

In later centuries, the pictorial motif of the temptation of Saint Anthony was a popular device for depicting hallucinatory states or visionary experiences (Müller-Ebeling 1989, 1997). A poem of the same name by Gustave Flaubert (1821–1880), which appears very "trippy" from today's perspective, should also be mentioned in this context. In his horror novel *Die Elixiere des Teufels* [The Elixirs of the Devil] (1815), E. T. A. Hofmann (1776–1822) described a "wine of Saint Anthony," the consumption of which produced extreme hallucinations (Hofmann 1982*; cf. *Vitis vinifera*). The novel *The Day of St. Anthony's Fire* (Fuller 1968) is based upon a purported epidemic of ergotism in Pont-St.-Esprit (France) that produced mass hysteria and collective hallucinations (Latimer 1981, 119).

In his novel *St.-Petri-Schnee* [Saint Peter's Snow] (1933), Leo Perutz (1882–1957) strangely anticipated the discovery of LSD from ergot (cf. **ophiussa**). He describes an experiment in which a chemist is attempting to isolate from a cereal fungus a drug of the gods that would make mystical experiences possible:

This Dionysos [Areopagita] relates in his writings that he imposed a two-day fast on the members of his community who longed for the true presence of God, after which he served them "bread prepared from sacred flour."—"For this bread"—he wrote—"leads to union with God and enables one to comprehend the infinite."— . . . Like all priests, the Roman priests of the fields knew the secrets of the drug that placed humans into that state of ecstasy in which they "became able to see" and "recognized the power of God." The white frost—this was not a type of grain, but a disease of the grain, a parasite, a fungus that penetrates into the grain plant and nourishes itself on its substance. . . . In Spain, it is known as "the Magdalena lichen," in Alsace as "the dew of poor souls." The "doctor's book" of Adam of Cremona described it under the name

"misericordia grain"; it was known in the Alps as "St. Peter's snow." In the area around St. Gallen, it was called the "mendicant," and in northern Bohemia "St. John's rot." Among us here in Westphalia, where it occurs particularly often, the farmers call it "the fire of the mother of God." (Perutz 1960, 119, 120, 121)

The American Beat poet Dale Pendell (b. 1944) immortalized the discovery of LSD from ergot in a poem (Pendell 1997; cf. Pendell 1995*).

The internationally famous American author Marion Zimmer Bradley (b. 1930) is known especially for her romantic fantasy and science-fiction novels. Psychoactive drugs are featured remarkably often in her novels and stories. In a recent novel, Bradley demonstrates that she has carefully considered Gordon Wasson's thesis that ergot was used in Eleusis.[366] This historical novel, titled *The Firebrand*, tells the tale of the Trojan War from the (feminist) perspective of the young seeress Cassandra (Bradley 1987). She was consecrated to Apollo, raised in Colchis, and trained by Penthesilea, the queen of the Amazons. At the conclusion of her training among the Amazons, Cassandra was initiated into the mysteries of the Great Goddess. The ritual, presided over by older women, is conducted in a temple. The female initiates are given a drink made of ergot:

She thought, till she touched her lips to the brew, that it was wine.

She tasted a curious slimy bitterness which made her think of the smell of the blighted rye Penthesilea had bidden her remember; as she drank she thought her stomach would rebel, but with a fierce effort she controlled the queasiness and brought her attention back to the drums. (1987, 135)

The drink quickly manifests its effects, which are described as follows:

Strange colors crawled before her eyes, and it seemed for a moment that she was walking through a great dark tunnel. (135)

After this typical psychedelic tunnel experience, Cassandra perceives the voice of the Great Goddess:

Share with Earth's Daughter the descent into darkness, a voice guided her from afar; whether a real voice or not she never knew. *One by one you must leave behind all the things of this world which are dear to you, for now you have no part in them. . . . This is the first of the gates of the Underworld; here you must give up that which binds you to Earth and the realms of Light.* (135, 136)

[365] In 1934, Hindemith composed a symphony entitled *Mathis der Maler*, which also set the temptations to music.

[366] Bradley had already mentioned the "dancing madness" that ergot induced during the Middle Ages in her novel *Darkover Landfall*, which was first published in 1972 (Bradley 1972, 60).

Cassandra passes the test and walks through the gate. A second psychedelic tunnel now opens up to her:

This is the second Gate of the Underworld, where you must give up your fears or whatever holds you back from traveling this realm as one of those whose feet know and tread the Path in My very footprints. (136)

After experiencing her death, she reaches the third gate, whereupon she is reborn. The culmination is the discovery of the Great Goddess:

She hung senseless in the darkness, shot through with fire, surrounded by the rushing sound of wings.
    Goddess, If I am to die for You, at least let me once behold Your face!
    There was a little lightening of the darkness; before her eyes she saw a swirling paleness, from which gradually emerged a pair of dark eyes, a pallid face. She had seen the face before, reflected in a stream . . . it was her own. A voice very close to her whispered through the drumming and the whining flutes:
    Do you not yet know that you are I, and I am you?
    Then the rushing wings took her, blotting out everything. Wings and dark hurricane winds, thrusting her upward, upward toward the light. (137)

At the conclusion of the ritual, Penthesilea warns her:

Hush; it is forbidden to speak of the Mystery. (138)

As of 1994, 440,000 copies of the German translation of *Firebrand* had been sold. Assuming that every copy sold is read by at least two people, this means that the book has been read by almost one million people in Germany alone. Certainly, more people have learned of the ergot mysteries from this book than from all of the scientific or nonfiction discussions in the academic literature (Rätsch 1996).

The experimental music group Psychic TV released a song titled "Eleusis" on its album *Dreams Less Sweet*. It was clearly influenced by Wasson's theory (Sony/Some Bizarre, SBZ CD 011, 1992).

Medicinal Use

Since medieval times, ergot-infested grains have been used in folk gynecology as a means of inducing contractions, hastening childbirth, and treating postpartum complications (Mühle and Breuel 1977). In the eighteenth century, ergot was used in Thuringia (Germany) as a hemostatic

agent. In the nineteenth century, ground grains of ergot were known as *pulvis parturiens*; they were administered to treat palsy and to induce abortions (Most 1843, 457*).

Today, ergot is no longer used in allopathic medicine, as it is almost impossible to standardize the drug (Teuscher 1992, 918). Preparations made from it are used only in homeopathy (Secale cornutum hom. *HAB1*, Secale cornutum hom. *HPUS78*) to treat particular symptom pictures, such as uterine cramps, spasm disorders, and migraine headaches (Teuscher 1992, 921).

Some of the substances that have been derived from **ergot alkaloids** (e.g., methergine, dihydergot, and hydergine) are still prescribed frequently.

Constituents and Effects

Depending upon the host plant, climate, and location, the ergot fungus can produce alkaloids (the **ergot alkaloids** ergotamine, ergotine, ergocristine, ergokryptine,[367] ergocornine, ergometrine; also ergoclavine, histamine, tyramine, choline, acetylcholine) that have varying effects (cf. Horwell and Verge 1979). Some of these are toxic; others are psychedelic. The harmful alkaloids produce two different forms of ergotism (ergot poisoning, Saint Anthony's fire):

Gangrenous ergotism began with vomiting and diarrhea, with itching in the fingers and inflamed appearances that were accompanied by severe burning pains. After a few days, the symptoms of gangrene then appeared. Starting with the fingers and toes, the limbs began to take on a blue-blackish color and to mummify. In cases of severe poisoning, it could even occur that the arms and legs would fall from the body without any loss of blood whatsoever. The gangrenous form of ergotism was referred to by such names as "mal des ardents," "ignis sacer," and "heiliges feuer" [sacred fire]. The convulsive form, which began with symptoms similar to those of the gangrenous, was characterized primarily by severe nervous disturbances. Painful muscle contractions, especially in the extremities, occurred, which ultimately led to epileptic-like spasms. (Hofmann 1964, 8)

Ergot from wheat, barley, and rye contains essentially the same alkaloids (ergotamine and ergotoxine groups, ergonovine, and occasionally traces of lysergic acid amide). In contrast to the dangerous toxins, the hallucinogenic substances lysergic acid amide, lysergic acid hydroxyethylamide, and ergonovine are water soluble and can therefore be separated from one another. Unfortunately, reports about the effects of ergonovine and methylergonovine are quite unsatisfactory (Ripinsky-Naxon 1993*). Further ethnopharmacological research is required.

Plantain (*Plantago*), the "plant of Persephone," was used to treat Saint Anthony's fire (ergotism). (Woodcut from Tabernaemontanus, *Neu Vollkommen Kräuter-Buch*, 1731)

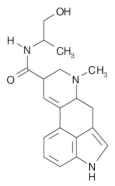

Ergometrine

367 Ergot infesting the Argentinian grass *Spartina alternifolia* (Gramineae) produces primarily α-ergokriptine (Ferraro et al. 1978).

"'I will bring you the strong tea. Drink it quickly. The ergot has overwintered. It is dissolved.' Taking small, hasty sips, Medea hurried to wash down the fungus. The pustules on the ears of rye were dark violet. The color of her flowers. Pain and happiness. The parasite rushed into her body and worked its way through her. Not a convolution was neglected. Medea lay with her legs and arms spread out, so that the poison could quickly find its way."

<small>Ursula Haas
Freispruch für Medea
[Acquittal for Medea]
(1991, 154)</small>

Commercial Forms and Regulations

In Germany, pharmaceutical preparations of ergot may be obtained only from a pharmacy. All preparations of **ergot alkaloids** require a prescription from a physician (Teuscher 1992, 920). Only homeopathic dilutions of D3 and below are freely available.

Literature

See also the entries for *Claviceps paspali*, **kykeon**, and **ergot alkaloids**.

Athanasius. 1987. *Vita Antonii*. Graz: Verlag Styria.

Bauer, Veit Harold. 1973. *Das Antonius-Feuer in Kunst und Medizin*. Historische Schriftenreihe, vol. 2. Basel: Sandoz.

Bove, Frank James. 1970. *The story of ergot*. Basel and New York: S. Karger.

Bradley, Marion Zimmer. 1987. *The firebrand*. New York: Simon and Schuster.

———. 1972. *Darkover landfall*. New York: Daw Books.

Brod, Thomas M. 1991. Hieronymus Bosch and ergot hallucinations. Paper presented at the 144th annual meeting of the American Psychiatric Association, May 11–16, 1991.

Camporesi, Piero. 1990. *Das Brot der Träume*. Frankfurt/M.: Campus.

Dixon, Laurinda S. 1984. Bosch's "St. Anthony Triptych"—An apothecary's apotheosis. *Art Journal*, summer: 119–131.

Ferraro, G. E., S. L. Debenedetti, and J. D. Coussio. 1978. Isolation of α-ergokriptine from an Argentine ergot. *Lloydia* 41:179–80.

Fraenger, Wilhelm. 1983. *Matthias Grünewald*. Munich: C. H. Beck.

Fuller, John G. 1968. *The day of St. Anthony's fire*. New York: Macmillan.

Haas, Ursula. 1991. *Freispruch für Medea*. Frankfurt/M.: Ullstein.

Hofmann, Albert. 1964. *Die Mutterkornalkaloide*. Stuttgart: Enke.

Horwell, David C., and John P. Verge. 1979. Isolation and identification of 6,7-seco-agroclavine from *Claviceps purpurea*. *Phytochemistry* 18:519.

Kolta, K. S. 1987. Der heilige Antonius als Heiler im Spätmittelalter. *Beiträge zur Geschichte der Medizin* 31 (38): 97–101.

Latimer, Dean. 1981. Mutterkorn und Roggenbrot. In *Der Hexengarten*, ed. H. A. Hansen, 109–46. Munich: Trikont-Dianus.

Lengyel, Lancelot. 1976. *Das geheime Wissen der Kelten*. Freiburg: Bauer.

Mannhardt, Wilhelm. 1865. *Roggenwolf und Roggenhund. Ein Beitrag zur germanischen Sittenkunde*. Danzig: Ziemssen.

———. 1868. *Die Korndämonen: Ein Beitrag zur germanischen Sittengeschichte*. Berlin: Dümmler's.

Matossian, Mary Kilbourne. 1989. *Poisons of the past: Molds, epidemics, and history*. New Haven: Yale University Press.

Mühle, Erich. 1953. *Vom Mutterkorn*. Leipzig: Akademische Verlagsgesellschaft.

Mühle, Erich, and Klaus Breuel. 1977. *Das Mutterkorn, ein Gräserparasit als Gift- und Heilpflanze*. Lutherstadt: Wittenberg.

Müller-Ebeling, Claudia. 1983. "Die Versuchung des Heiligen Antonius" als Identifikationsmodell der Maler des Fin De Siècle. Master's thesis, Hamburg.

———. 1985. Was hat der hl. Antonius mit dem Wilden Mann zu tun? In *Namaste Yeti: Geschichten vom Wilden Mann*, ed. C. Rätsch and H. J. Probst, 89–98. Munich: Knaur.

———. 1989. The return to matter—the temptations of Odilon Redon. In *Gateway to inner space*, ed. C. Rätsch, 167–78. Bridport, Dorset: Prism Press.

———. 1997. *Die "Versuchung des hl. Antonius" als "Mikrobenepos": Eine motivgeschichtliche Studie zu den drei Lithographiefolgen Odilon Redons zu Gustave Flauberts Roman*. Berlin: VWB.

Pendell, Dale. 1997. Das Mutterkorn: The making of Delysid. In *Entheogens and the future of religion*, ed. R. Forte, 23–29, San Francisco: Council of Spiritual Practices.

Perutz, Leo. 1960. *St.-Petri-Schnee*. Vienna: Paul Zsolnay Verlag.

Przybyszewski, Stanislaw. 1979. *Die Synagoge Satans*. Berlin: Zerling. (Orig. pub. 1900.)

Rätsch, Christian. 1996. Die Mutterkornmysterien im Roman von Marion Zimmer Bradley. *Yearbook for Ethnomedicine and the Study of Consciousness*, 1995 (4): 331–34. Berlin: VWB.

———. 1997. Eine kurze Bibliographie zum Mutterkorn. In *Schutzschrift für Das Mutterkorn, als einer angeblichen Ursache der sogenannten Kriebelkrankheit*, by Rudolph Augustin Vogel, 67–73. Berlin: VWB.

Ruffie, Jacques, and Jean-Charles Sournia. 1993. *Die Seuchen in der Geschichte der Menschheit*. Munich: dtv/Klett-Cotta.

Seidel, Max, and Christian Bauer. 1983. *Grünewald: Der Isenheimer Altar*. Stuttgart: Belser Verlag.

Shelley, William Scott. 1995. *The elixir: An alchemical study of the ergot mushrooms*. Notre Dame, Indiana: Cross Cultural Publications Inc.

Siemens, Fritz. 1880. Psychosen bei Ergotismus. *Archiv für Psychiatrie und Nervenkrankheiten* 11 (1–2): 366–90.

Spilmont, Jean-Pierre. 1984. *Magie*. Munich: Heyne.

Stoll, Arthur. 1943. Altes und Neues über Mutterkorn. *Mitteilungen der Naturforschenden Gesellschaft*, Bern 1942: 45–80.

———. 1951. *Die spezifischen Wirkstoffe des Mutterkorns und ihre therapeutische Anwendung.* Aulendorf: Editio Cantor.

Teuscher, Eberhard. 1992. *Claviceps.* In *Hagers Handbuch der Pharmazeutischen Praxis,* 5th ed., 4:911–22. Berlin: Springer.

Vogel, Rudolph Augustin. 1997. *Schutzschrift für Das Mutterkorn, als einer angeblichen Ursache der sogenannten Kriebelkrankheit.* Reihe Ethnomedizin und Bewußtseinsforschung—Historische Materialien, vol. 9. Berlin: VWB. (Orig. pub. 1771.)

Claviceps spp.

Ergot Fungi

Family

Class Ascomycetes (Sac Fungi), Order Clavicipitales (Ergot Fungi), Family Clavicipitaceae

There are thirty-five to fifty species of ergot fungi (*Claviceps* spp.). The name *ergot* refers to the overwintering stage (sclerotium) of sac fungi that parasitically infest various cereal grains (rye, wheat, barley, millet), wild sweet grasses (Gramineae = Poaceae; e.g., bearded darnel, *Lolium temulentum*; *Paspalum* species), and rushes (Juncaceae) and sedges (Cyperaceae; cf. *Cyperus* spp.) (Dörfelt 1989, 92**).

It is possible that sleepy grass (*Stipa* spp.) is also infested with *Claviceps* species and consequently may be psychoactive. All ergot fungi produce psychoactive and/or toxic alkaloids (**ergot alkaloids**, **indole alkaloids**).

Claviceps gigantea Fuentes—diente de caballo ("horse's tooth"), giant ergot

This fungus apparently infests only maize (*Zea mays*), a New World grass (Moreno and Fucikovsky 1972). It contains **ergot alkaloids** of unknown composition and is regarded in Mexico as poisonous (Guzmán 1994, 1438**). Many Indians regard this fungal infestation as a disease of the maize plant (Rätsch 1989).

When the Yucatán peninsula of Mexico was conquered by the Spanish in the sixteenth century, Franciscans came into the country, compiling various reports and dictionaries. One of the three most important lexicons that have been preserved is now known as the *Wiener Wörterbuch* [Vienna Lexicon] (Andrews Heath de Zapata 1978). Compiled around 1625, it contains several extremely unusual and interesting comments related to Saint Anthony and to Saint Anthony's fire, the malady named after him (cf. *Claviceps purpurea*):

— *metnalil kak: fuego de San Antón,* " 'Hell' fire: Saint Anthony's fire"
— *he metnalil kake humpati uinic lic yulel: Esta enfermedad mata sin remedio,* "This hell fire burns all people in the same manner: This affliction kills without remedy." (WW 306)

These entries clearly indicate that the author of the *Wiener Wörterbuch* interpreted certain manifestations and terminology of the Indians as Saint Anthony's fire (ergotism), which he knew from Europe. Since the link between the disease and ergot had not yet been discovered, and because it is possible that the ergot fungus was not present in the Yucatán at that time, *fuego de San Antón* may have referred to another, similar phenomenon (cf. **Datura innoxia**). However, the entries in the

Left: Giant ergot (*Claviceps gigantea*) infests chiefly ears of maize. (Photographed in southern Mexico)

Right: The spikes of the dune grass *Ammophila maritima*, which lives in a symbiotic relationship with *Psilocybe azurescens*, are often infested with ergot (*Claviceps* sp.).

Wiener Wörterbuch suggest that in the Yucatán there may also have been epidemics of some types of affliction that resembled gangrenous ergotism. The reports of two Spaniards, Diego Lopez de Cogolludo and Fra Diego de Landa, demonstrate that Saint Anthony was being venerated in Yucatán during the early colonial period (Hermanns and Probst 1994). It may be that *fuego de San Antón* was exactly that which it refers to, a form of poisoning due to the giant ergot that infests maize, *Claviceps gigantea*.

Claviceps glabra Langdon
Infests numerous wild grasses.

Claviceps nirgicans Tul.
Infests numerous wild grasses.

Claviceps paspali
Infests only grasses of the genus *Paspalum*.

Claviceps purpurea
Infests preferentially rye but also wheat and barley.

Claviceps sp.
Infests the dune grass *Ammophila maritima*, which lives in a symbiotic relationship with *Psilocybe azurescens* (Stamets 1996, 95**).

Literature

See also the listings for ***Claviceps paspali***, ***Claviceps purpurea***, and **ergot alkaloids**.

Andrews Heath de Zapata, Dorothy. 1978. *Vocabulario de Mayathan por sus abecedarios*. Mérida: Area Maya.

Hermanns, Barbara, and Heinz Jürgen Probst. 1994. Bericht über die Dinge von Yucatán (1572). In *Chactun—Die Götter der Maya*, ed. C. Rätsch, 2nd ed., 175–211. Munich: Diederichs.

Moreno, M., and L. Fucikovsky. 1972. Effect of position and number of sclerotia of *Claviceps gigantea* on maize germination. *Fitopatologia* 5/6:7–9.

Rätsch, Christian. 1989. St. Anthony's fire in Yucatán. In *Gateway to inner space*, ed. C. Rätsch, 161–65. Bridport and Dorset: Prism Press.

———. n.d. *Das Antoniusfeuer in Yucatán: Eine ethnopharmakologische Spekulation*. Unpublished manuscript, Hamburg.

Conocybe spp.
Cone Caps

The tiny fruiting bodies of *Conocybe cyanopus* contain 0.33 to 0.93% psilocybin. (Photograph: Paul Stamets)

Family
Agaricaceae: Bolbitiaceae (cone caps)

The cone caps are small, thin mushrooms of pale color and with conic to bell-shaped caps. The lamellae turn rusty brown as the spores form. Cone caps prefer open forest areas. Some species of the genus contain **psilocybin** and are psychoactive. A rudimentary cult based around *tamu*, a *Conocybe* species known as the mushroom of knowledge, was recently discovered in the Ivory Coast (Samorini 1995). The Mazatec used the species *Conocybe siligineoides* Heim, which they referred to as *ta'a'ya*, as an entheogen. Chemical analyses are lacking (Ott 1993, 313*; Stamets 1996, 176**).

Conocybe cyanopus (Atkins) Kühner (= *Conocybe cyanopus* (Atkins) Sing.) [syn. *Pholiotina cyanopus*]—blue-footed cone cap
This mushroom grows in grassy areas and mossy locales from summer to fall. It forms an obtusely convex cap 0.5 to 2.5 cm in diameter. The whitish stem is blue-green at its base and turns blue when squeezed. This mushroom is found in Germany and Switzerland as well as in North America. Generally regarded as poisonous, it contains 0.93% psilocybin (dry weight) and some baeocystin but no psilocin (Gartz 1985, 1992).

The following species also contain **psilocybin** (Allen et al. 1992, 93**; Gartz 1985):

Conocybe kuehneriana Singer
Conocybe siligineoides Heim—ta'a'ya
Conocybe smithii Watling (also contains baeocystin; Repke et al. 1977)

Literature
See also the entry for **psilocybin**.

Benedict, R. G., L. R. Brady, A. H. Smith, and V. E. Tyler. 1962. Occurrence of psilocybin and psilocin in certain *Conocybe* and *Psilocybe* species. *Lloydia* 25:156–59.

Benedict, R. G., V. E. Tyler Jr., and R. Watling. 1967. Blueing in *Conocybe*, *Psilocybe* and a *Stropharia* species and the detection of psilocybin. *Lloydia* 30:150–57.

Christensen, A. L., K. E. Rasmussen, and K. Høiland. 1984. Detection of psilocybin and psilocin in Norwegian species of *Pluteus* and *Conocybe*. *Planta Medica* 45:341–43.

Gartz, Jochen. 1985. Zur Analytik der Inhaltsstoffe zweier Pilzarten der Gattung *Conocybe*. *Pharmazie* 40 (5): 366.

————. 1992. Further investigations on psychoactive mushrooms of the genera *Psilocybe*, *Gymnopilus* and *Conocybe*. *Annali dei Musei Civici di Rovereto* 7 (1991): 265–74.

Repke, David B., Dale Thomas Leslie, and Gastón Guzmán. 1977. Baeocystin in *Psilocybe*, *Conocybe* and *Panaeolus*. *Lloydia* 40 (6): 566–78.

Samorini, Giorgio. 1995. Traditional use of psychoactive mushrooms in Ivory Coast? *Eleusis* 1:22–27.

Copelandia spp.

Ink Caps

Family
Coprinaceae (Ink Caps)

Some of the psychoactive mushrooms originally assigned to this genus are now classified in the genus *Panaeolus*: *Panaeolus cambodginiensis*, **Panaeolus cyanescens**, *Panaeolus tropicales* (cf. **Panaeolus spp.**).

The following species of the genus *Copelandia* contain **psilocybin** (Allen et al. 1992, 93**; Ott 1993, 309*):

Copelandia anomalus (Murr.) Sacc. et Trott.

Copelandia bispora (Malencon et Bertault) Sing. et Weeks
Copelandia chlorocystis Sing., Weeks et Hearns
Copelandia mexicana Guzmán
Copelandia westii (Murr.) Sing.

None of these species is known to have or to have had a traditional use (cf. Weeks et al. 1979).

Literature
See also the entries for **Panaeolus cyanescens**, **Panaeolus spp.**, and **psilocybin**.

Weeks, R. A., R. Singer, and W. Hearns. 1979. A new psilocybian species of *Copelandia*. *Journal of Natural Products* 42:71–74.

Galerina steglichii Besl

Steglich's Galerina

Family
Cortinariaceae

This tiny mushroom, which suddenly appeared in Regensburg, Germany, in 1993, develops a bluish color when pressed or squeezed. An extract of the mushroom was found to contain **psilocybin**, psilocin, and baeocystin. This was the first discovery of these psychoactive alkaloids in a species of the genus *Galerina* (Besl 1993). The dried fruiting bodies contain 0.21 to 0.51% psilocybin, 0.02 to 0.07% baeocystin, and 0.08 to 0.21% psilocin (Gartz 1995, 304). This species is relatively easy to cultivate on an agar substrate.

Beware! Many other species of the genus *Galerina* are very poisonous and may be lethal.

Literature
Besl, H. 1993. *Galerina steglichii* spec. nov., ein halluzinogener Häubling. *Zeitschrift für Mykologie* 59:215–18.

Gartz, Jochen. 1995. Cultivation and analysis of *Psilocybe* species and an investigation of *Galerina steglichii*. *Annali dei Musei Civici di Rovereto* 10 (1994): 297–305.

"The genus *Galerina* contains dangerously poisonous mushrooms in which the same deadly amatoxins are present as in the death cap [*Amanita phalloides* (Vaill.) Secr.]. These toxins manifest their effects after a latency period of 12 hours and are typically lethal in spite of therapy. *Galerina autumnalis* (Peck) Singer et Smith is a common North American species that, like some *Psilocybe* species, grows on remnants of wood in parks and forests. At first glance, it is very similar to *Psilocybe stuntzii* Guzman et Ott and can grow directly next to this. But the *Galerina* species do not blue."

JOCHEN GARTZ
NARRENSCHWÄMME [FOOLS' FUNGI] (1993, 67 f.**)

Gymnopilus spp.

Gyms

Family

Cortinariaceae

Gyms are typically small to medium in size and have convex to plane caps and yellow- or orange-colored lamellae. They thrive on trees from summer to fall. Most species have a bitter taste and are usually regarded as poisonous or inedible. Several species are said to have hallucinogenic effects.

The following species have been demonstrated to contain **psilocybin** (Allen et al. 1992, 93**; Hatfield et al. 1978):

Gymnopilus aeruginosus (Peck) Singer [syn.
   Pholiota aeruginosa Peck]
Gymnopilus braendlei (Peck) Hesler
Gymnopilus intermedius (Sing.) Singer
Gymnopilus leteoviridis Thiers
Gymnopilus liquiritae (Fr.) Karst.
Gymnopilus luteus (Peck) Hesler
Gymnopilus purpuratus (Cooke et Mass.) Sing.
   [syn. *Flammula purpurata* Cooke et Massee]—
   purple gym

   The purple gym apparently originated in South America but also occurs in Germany. It contains **psilocybin** as well as psilocin (Gartz 1993, 55**). This mushroom may once have been used for inebriating purposes (cf. "*Polyporus mysticus*").
Gymnopilus validipes (Peck) Hesler (cf. Hatfield et
   al. 1977)
Gymnopilus viridans Murrill

The following species may contain psychoactive substances:

Gymnopilus luteofolius (Peck) Singer [syn.
   Pholiota luteofolia (Peck) Saccardo]
Gymnopilus ventricosus

Only one species of the genus *Gymnopilus* has attained any cultural significance:

Gymnopilus spectabilis (Fries) Singer [syn.
   Gymnopilus junonius (Fr.: Fr.) P.D. Orton,
   Pholiota spectabilis (Fries) Gillet]—big gym,
   giant laughing mushroom

Gymnopilus spectabilis grows on deciduous wood and, less frequently, coniferous wood. Its cap can attain a diameter of 5 to 15 cm. It has a very bad taste and is low in active components. It is commonly regarded as poisonous.
   The yellow variety *Gymnopilus spectabilis* var. *junonius* (Fr.) J.E. Lange is one of the largest mushrooms known. It can develop stems as tall as 60 cm (Gartz 193, 55 f.**).

The big gym also occurs in North America and Japan; psychoactive effects have been reported from both regions (Walters 1965). An experiential report is now available from Canada:

> I tried three grams of *G. spectabilis*. Slight optical changes, but (together with my meditation techniques) enough to drive me into an intense convulsive trembling. When I gave in to this, I found that this was a memory of the room in which I had found myself after my birth. I was very cold there, I was afraid and needed my mother. The culmination came 3 to $3^1/2$ hours after ingestion. . . . *Psilocybe* always takes away my clarity; this mushroom increases it. . . . The effects are markedly different. (In *Entheogene* no. 2 [1994/1995]: 23 f.)

In Japan, this mushroom is known as *o-waraitake*, "giant laughing mushroom," a name with a long history. A story recorded in the *Konjaku monogatarishu* [Stories from Ancient Times] (eleventh century) indicates that psychoactive fungi must have been known in Japan at a very early date. In this story, woodcutters meet a group of nuns who are singing and dancing. The woodcutters regard the nuns as manifestations of demons or goblins. The nuns explain that they had eaten certain mushrooms, and that these had caused them to sing and dance. The woodcutters eat some of the same mushrooms, and they too then dance and sing. After this event became known, the mushroom was given the name *maitake*, "dancing mushroom." One old dictionary identified this mushroom as *Grifola frondosa* (Dicks. ex Fr.) S.F. Gray [syn. *Boletus frondosus* Vahl., *Cladomeris frondosa* Quél., *Polyporus frondosus*] (cf. "*Polyporus mysticus*"); it was noted, however, that the *maitake* of the *Konjaku* story was actually known as *waraitake*, "laughing mushroom."[368] In the *Daijiten*, on the other hand, *waraitake* is identified as *Panaeolus papilionaceus* (cf. **Panaeolus spp.**) or *Gymnopilus spectabilis* (Sanford 1972, 174**). The *Daijiten* has the following to say about this mushroom:

> People who eat this mushroom become inebriated. They can become extremely excited, dance and sing, and see different visions. Other names for it are *odoritake* ["jumping mushroom"] and *maitake*. (In Sanford 1972, 175**)

A Taoist source from the eleventh or twelfth century states that an elixir of life (known as "earth drink") was obtained from the "laughing mushroom" (Sanford 1972, 178**).

It is possible that *Gymnopilus spectabilis* contains indole derivatives; the analysis of Hatfield et al. (1978, 142) found that it contains **psilocybin** (cf. Benjamin 1995, 326**). Styrylpyrone, *bis*-noryangonine, and bitter principles (gymnopilins) have also been detected (Aoyagi et al. 1983; Tanaka et al. 1993). The psychoactivity, on the other hand, is unquestioned (Buck 1967; Romagnesi 1964; Sanford 1972**).

Literature

See also the entry for **psilocybin**.

Aoyagi, F., et al. 1983. Gymnopilins, bitter principles of the big-laughter mushroom *Gymnopilus spectabilis*. *Tetrahedron Letters* (1983): 1991–93.

Buck, R. W. 1967. Psychedelic effect of *Pholiota spectabilis*. *New England Journal of Medicine* 267:391–92.

Gartz, Jochen. 1989. Occurrence of psilocybin, psilocin and baeocystin in *Gymnopilus purpuratus*. *Persoonia* 14:19–22.

———. 1992. Further investigations on psychoactive mushrooms of the genera *Psilocybe, Gymnopilus* and *Conocybe*. *Annali dei Musei Civici di Rovereto* 7 (1991): 265–74.

Guzmán-Davalos, L., and Gastón Guzmán. 1991. Additions to the genus *Gymnopilus* (Agaricales) from Mexico. *Mycotaxon* 40 (1): 43–56.

Hatfield, G. M., L. J. Valdes, and A. H. Smith. 1977. Proceedings—isolation of psilocybin from the hallucinogenic mushroom *Gymnopilus validipes*. *Lloydia* 40:619.

———. 1978. The occurrence of psilocybin in *Gymnopilus* species. *Lloydia* 41 (2): 140–44.

Romagnesi, M. H. 1964. Champignons toxiques au Japon. *Bulletin de la Société Mycologique de France* 80 (1): iv–v.

Tanaka, Masayasu, Kimiko Hashimoto, Toshikatsu Okuno, and Haruhisa Shirahama. 1993. Neurotoxic oligoisoprenoids of the hallucinogenic mushroom *Gymnopilus spectabilis*. *Phytochemistry* 34 (3): 661–64.

Walters, Maurice B. 1965. *Pholiota spectabilis*, a hallucinogenic fungus. *Mycologia* 57:837–38.

"You should try to give King Emma several of the 'laughing mushrooms.'"

HAIKU BY KOBAYASHI ISSA IN "MUSHROOMS IN JAPANESE VERSE" (BLYTH 1973, 11**)

Inocybe spp.

Inocybe Mushrooms

Family
Cortinariaceae

These small- to medium-size mushrooms form first conic and later slightly convex caps with a slightly incurved margin. Their German name, *risspilze*, means "ripped or cracked fungi," referring to their typically fissured surface. They grow from summer to fall in forests, on pastures and moors, and in alpine settings. Of the approximately 160 species, several are feared as poisonous. The brick red *Inocybe erubescens* Blytt. [syn. *Inocybe patouillardii* Bres.] contains muscarine and can produce serious and even deadly poisonings. The few species that color blue-green (*I. aeruginascens, I. corydalina, I. haemacta*) contain psilocybin and baeocystin (Stivje et al. 1985). These are nonpoisonous and occur in central Europe (Germany, Switzerland) (Gartz and Drewitz 1985). They have no traditional use.

Inocybe aeruginascens Babos
This mushroom was first found in 1965 in Hungary, spread from there, and suddenly appeared in 1975 in Berlin. In 1980, it came to Holland, and in 1984 it even made it to the Rhône valley of Switzerland. It can be assumed that this species is a new one that arose only a few years ago (Gartz 1992).

The cap is only 2 to 3 cm across; the stem colors heavily bluish green all the way to the swelling base. This mushroom grows from spring

The blueing mushroom *Inocybe calamistrata* is psychoactive and may contain psilocybin.

368 In Japan, *Psilocybe venenata* (Imai) Hongo [syn. *Stropharia venenata, Stropharia caerulescens* Imai] (cf. **Psilocybe spp.**) is referred to as *waraitakemodoki*, "false laughing mushroom," or *shibiretake*, "inebriating mushroom."

"The mushrooms [*Inocybe aeruginascens*] taste like ordinary culinary mushrooms. After some 30 minutes, while lying relaxed with no other somatic effects, there gradually appeared an extremely pleasant neutralization of the sense of weight. Abstract hallucinations in the form of sparkling colors and lights slowly developed. When the sense of weight had completely disappeared, there arose a very lively perception of a flight of the soul with corresponding euphoric feelings."

JOCHEN GARTZ
NARRENSCHWÄMME [FOOL'S FUNGI]
(1993, 51**)

The mushroom *Inocybe patouillardii* on a Cuban stamp

to fall near deciduous trees in the grassy areas of parks. It contains **psilocybin** and has unequivocal psychoactive effects (Gartz 1986a). To produce visions, 2.4 g of dry weight suffice (Gartz 1995). None of the samples tested has yielded the toxic muscarine, which does appear elsewhere in the genus (Gartz 1986b).

Inocybe coelestium Kuyper

Inocybe corydalina Quélet
This species occurs in two varieties:

Inocybe corydalina var. *corydalina* Quélet
Inocybe corydalina var. *erinaceomorpha* (Stangl et Veselsky) Kuyper—brown cap, more than 5 cm across

This mushroom grows from summer to fall, primarily in deciduous forests.

Inocybe haemacta (Berkeley et Cooke) Saccardo
This mushroom grows in the fall in deciduous forests and parks.

Literature
See also the entry for **psilocybin**.

Gartz, Jochen. 1986a. Psilocybin in Mycelkulturen von *Inocybe aeruginascens*. Biochem. Physiol. *Pflanzen* 181:511–17.

———. 1986b. Untersuchungen zum Vorkommen des Muscarins in *Inocybe aeruginascens* Babos. *Zeitschrift für Mykologie* 52 (2): 359–61.

———. 1992. *Inocybe aeruginascens*, ein 'neuer' Pilz Europas mit halluzinogener Wirkung. *Yearbook for Ethnomedicine and the Study of Consciousness*, 1992 (1): 89–98. Berlin: VWB.

———. 1995. *Inocbye aeruginascens* Babos. *Eleusis* 3:31–34. (Additional literature.)

Gartz, J[ochen], and G. Drewitz. 1985. Der erste Nachweis von Psilocybin in Rißpilzen. *Zeitschrift für Mykologie* 51 (2): 199–203.

Semerdzieva, Marta, M. Wurst, T. Koza, and Jochen Gartz. 1986. Psilocybin in Fruchtkörpern von *Inocybe aeruginascens*. *Planta Medica* 47:83–85.

Stijve, T., J. Klan, and Th. Kuyper. 1985. Occurrence of psilocybin and baeocystin in the genus *Inocybe* (Fr.) Fr. *Persoonia* 12:469–73.

Panaeolus cyanescens Berkeley et Broome
Jambur

Family
Coprinaceae (Ink Caps); Subfamily Panaeoloideae

Synonyms
Campanularius anomalus Murr.
Campanularius westii Murr.
Copelandia cyanescens (Berk. et Br.) Boedjin
Copelandia cyanescens (Berk. et Br.) Sacc.
Copelandia cyanescens (Berk. et Br.) Sing.
Copelandia papilionacea (Bull. ex Fr.) Bres.
Copelandia papilionacea (Bull.) Bres. non Fr.
Copelandia westii (Murr.) Sing.
Panaeolus anomalus (Murr.) Sacc. et Torr.
Panaeolus westii (Murr.)

The tropical mushroom *Panaeolus cyanescens* [syn. *Copelandia cyanescens*] prefers to grow on cow or horse dung. (Photographed in Palenque, Chiapas, Mexico.)

Folk Names
Blauender düngerling, blue meanies, faleaitu (Samoan, "spirit house" or "comedy"), falterdüngerling, Hawaiian copelandia, jambur, jamur, pulouaitu (Samoan, "spirit hat"), taepovi (Samoan, "cow patty"), tenkech (Chol)

In the early 1960s, reports emerged from southern France of strange "intoxications" produced by mushrooms that grew on horse dung. The mushrooms were identified as the tropical species *Copelandia cyanescens* and were analyzed by Albert Hofmann. He found high concentrations of psilocin in the fruiting bodies and only slight quantities of **psilocybin** (Heim et al. 1966). The reason these mushrooms had so suddenly appeared in France was also discovered. The mushroom grows on horse dung, i.e., in a kind of symbiotic relationship with horses. When horses from Indonesia were brought to southern France to take part in a horse race, the mushroom became established in the wild via their feces (Gerhardt 1987). This species clearly comes from Southeast Asia and occurs in Indonesia, Australia (Low 1990, 206*), and, since ancient times, Samoa (Cox 1981).

Panaeolus cyanescens is easily confused with *Panaeolus tropicales* and *Panaeolus cambodginiensis* (cf. *Panaeolus* spp.). It is possible that the latter species are merely varieties or races and are in fact synonymous with *Panaeolus cyanescens*.

The mushroom is cultivated in Bali and purportedly is used both in native festivals and in the tourist trade (Cox 1981, 115). In Java, it may possibly have a long tradition of use as a ritual drug. The Javanese batik artists in Yogyarkata eat jambur mushrooms to obtain inspiration for their artistic endeavors. Not surprisingly, the mushroom is often featured in their art.

In Samoa, the caps are boiled in water for a long period of time until a black juice is produced. This juice is mixed with coffee (cf. *Coffea arabica*) and drunk. Sometimes the caps are eaten raw and washed down with Coca-Cola. Occasionally, they may be dried and smoked (Cox 1981).

The effects of the mushroom are manifested quite rapidly, as they usually contain a preponderance of psilocin, i.e., the actual active component. It produces strong feelings of euphoria with visual and auditory hallucinations that may last as long as seven hours. Very high dosages can result in loss of muscle control. In Samoa, it is said that regular use of the mushroom will produce a painful red rash around the neck (Cox 1981). This may be due to the presence of urea (Stivje 1987, 1992).

Literature

See also the entries for *Panaeolus* spp. and **psilocybin**.

Cox, Paul Allen. 1981. Use of a hallucinogenic mushroom, *Copelandia cyanescens*, in Samoa. *Journal of Ethnopharmacology* 4 (1): 115–16.

Gerhardt, E. 1987. *Panaeolus cyanescens* (Berk. et Br.) Sacc. und *Panaeolus antillarum* (Fr.) Dennis, zwei Adventivarten in Mitteleuropa. *Beiträge zur Kenntnis der Pilze Mitteleuropas* 3:223–27.

Heim, Roger, Albert Hofmann, and H. Tscherter. 1966. Sur une intoxication collective à syndrome psilocybien causée en France par un *Copelandia*. *Comptes rendus de l'Académie des Sciences* (Paris) 262:519–23.

Stijve, T. 1987. Vorkommen von Serotonin, Psilocybin und Harnstoff in Panaeoloideae. *Beiträge zur Kenntnis der Pilze Mitteleuropas* 3:229–34.

———. 1992. Psilocin, psilocybin, serotonin and urea in *Panaeolus cyanescens* from various origins. *Persoonia* 15:117–21.

"The blue meanie is the perfect alchemist: it transforms dung into gold, into the golden light of enlightenment."

GALAN O. SEID
DIE NEUE ALCHEMIE [THE NEW ALCHEMY]

This illustration, inspired by the use of mushrooms, clearly depicts jambur mushrooms (*Panaeolus cyanescens*) at work inside the figure's head. (Indonesian batik, twentieth century)

Panaeolus subbalteatus Berkeley et Broome

Dark-rimmed Mottlegill

Family

Coprinaceae (Ink Caps); Subfamily Panaeoloideae

Synonyms

Panaeolus cinctulus Bolt.
Panaeolus subalteatus (misspelling!)
Panaeolus subbalteatus (Berk. et Br.) Sacc.
Panaeolus venenosus Murr.

Folk Names

Dunkelrandiger düngerling, gezoneerde vlek plaat (Dutch), gezonter düngerling, magusotake (Japanese, "horse pasture mushroom")

This fungus is common throughout Europe and is also found in the subtropics and tropics (Asia, the Americas). It thrives in fields fertilized with manure, in grassy soil, and especially in horse pastures and in connection with horse manure. Its somewhat convex cap quickly becomes plane and is 2 to 6 cm in diameter. It is initially moist and brown but fades in the center as it dries, so that the margin often appears much darker (which accounts for its German name *dunkelrandiger düngerling* ("dark-banded dung mushroom"). The

reddish brown lamellae are emarginate and later turn black because of the spores. This species is easily confused with the changing pholiota (*Kuehneromyces mutabilis* [Schaef. ex Fr.] Sing. et Smith) (Roth et al. 1990, 95**).

No traditional uses of this mushroom are known. It is possible that it was used as an additive to the **mead** or **beer** of the Germanic peoples. The mushroom does have a symbiotic connection to the horse, the sacred animal of Wotan, the Germanic god of ecstasy.

Panaeolus subbalteatus contains approximately 0.7% **psilocybin** and 0.46% baeocystin along with

Panaeolus subbalteatus is found chiefly in the immediate vicinity of horse stud farms. (Photographed near the Externsteine, a series of standing stones)

The cover of this CD by Shaw Blades, with the telling title *Hallucination*, features specimens of *Panaeolus subbalteatus* in the right foreground. (Warner Bros. Records, 1995)

large amounts of serotonin and 5-hydroxy-tryptophan, but it does not contain psilocin (Gartz 1989). It is questionable whether serotonin can in fact reach the brain when the mushrooms are ingested. Experimental pharmacology has demonstrated that serotonin is not absorbed by the brain when ingested orally. Nevertheless, according to all reported experiences, the effects of *Panaeolus subbalteatus* differ from the effects of mushrooms that contain only **psilocybin**; they are more empathogenic and aphrodisiac and yet still visionary. The individual visions can be observed for longer periods of time and contemplated at a leisurely pace. Psychoactive effects are produced by as little as 1.5 g dry weight (Stein 1959); a visionary dosage is 2.7 g. The psychoactivity of this mushroom was discovered following its accidental ingestion (Bergner and Oettel 1971).

Literature

See also the entries for *Panaeolus* **spp.** and **psilocybin**.

Bergner, H., and R. Oettel. 1971. Vergiftungen durch Düngerlinge. *Mykologisches Mitteilungsblatt* 15:61–63.

Brodie, H. J. 1935. The heterothallism *of Panaeolus subbalteatus* Berk., a sclerotium-producing agaric. *Canadian Journal of Research* 12:657–60.

Gartz, Jochen. 1989. Analyse der Indolderivate in Fruchtkörpern und Mycelien von *Panaeolus subbalteatus* (Berk. et Br.) Sacc. *Biochemie und Physiologie der Pflanzen* 184:171–78.

Stein, Sam I. 1959. Clinical observations on the effect of *Panaeolus venenosus* versus *Psilocybe caerulescens* mushrooms. *Mycologia* 51:49–50.

Panaeolus spp.

Panaeolus Mushrooms

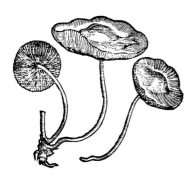

This old English illustration of "poisonous mushrooms or those that are usually not eaten" may represent a *Panaeolus* species with a wavy cap. (Woodcut from Gerard, *The Herball or General History of Plants*, 1633*)

Family

Coprinaceae (Ink Caps); Subfamily Panaeoloideae The cosmopolitan genus *Panaeolus*, with more than twenty species, forms fragile fruiting bodies that are small to medium in size. The caps are usually hemispheric to campanulate. The pale lamellae become increasingly dark as the black spores develop. Panaeolus mushrooms grow on nutrient-rich, grassy soils or dung.

Panaeolus acuminatus (Schaeffer) Quélet sensu Ricken [syn. *Panaeolus rickenii* Hora]
Found in North America; said to be psychoactive, although no analyses have detected **psilocybin** or psilocin.

Panaeolus africanus Ola'h—African panaeolus
Found from central Africa to Sudan; thrives in rhinoceros and elephant dung. It contains various quantities of **psilocybin** and psilocin.

Panaeolus antillarum (Fries) Dennis sensu Dennis [syn. *Panaeolus phalaenarum* (Fr.) Quélet, *Panaeolus sepulcralis* Berk., *Anellaria sepulchralis* (Berk.) Singer]—Antilles panaeolus
While this mushroom is regarded as psychoactive, it does not always contain active substances (Merlin and Allen 1993**).

Panaeolus ater (Lange) Kühner et Romagnesi—black panaeolus
Now considered to be a synonym for *Panaeolus fimicola*.

Panaeolus cambodginiensis Ola'h et Heim [syn. *Copelandia cambodginiensis* (Ola'h et Heim) Singer—gold top
Contains 0.55 to 0.6% **psilocybin** and psilocin (Merlin and Allen 1993).

Panaeolus castaneifolius (Murrill) Ola'h [syn. *Panaeolina castaneifolius* (Murr.) Smith]
Contains traces of active compounds.

Panaeolus cinctulus Bolt.
Regarded as a synonym for *Panaeolus subbalteatus*.

Panaeolus fimicola (Fries) Gillet [syn. *Panaeolus ater* (Lange) Kühner et Romagnesi]
Found in Africa, the Americas, and Europe; contains only trace amounts of **psilocybin** and psilocin (Roth et al, 1990, 95**). This small, reddish to brown-black mushroom (cap 2 to 4 cm across) thrives in grassy forest areas.

Panaeolus foenisecii (Fries) Kühner [syn. *Panaeolina foenisecii* (Pers.: Fr.) Maire = *Panaeolus foeniseci* (Pers.: Fr.) Schroeter]—haymaker's panaeolus
This cosmopolitan mushroom grows in central Europe from spring until fall on freshly mown meadows, along roadsides, and in pastures. Not all samples have been found to contain **psilocybin** (Allen and Merlin 1992; Gartz 1985a).

Panaeous olivaceus Moller
A Finnish sample was found to contain **psilocybin**.

Panaeolus papilionaceus (Bull. ex Fries) Quélet [syn. *Agaricus callosus* Fr., *Agaricus* (*Panaeolus*) *sphinctrinus* Fries, *Panaeolus campanulatus* (Fries) Quélet, *Panaeolus retirugis* (Fries) Quélet, *Panaeolus sphinctrinus* (Fries) Quélet]

This mushroom is quite variable, which is why it was formerly divided into different species that are now regarded as synonymous. It apparently occurs in different chemical races; some of these contain **psilocybin**, while others are lacking in psychoactive substances. Serotonin has also been detected (Gartz 1985b). It grows in pastures, in nutrient-rich meadows with dung deposits, and directly on dung. It is found throughout the world, including central Europe.

In Japan, this mushroom is known as *warai-take*, "laughing mushroom" (cf. *Gymnopilus* spp.). In ancient China, it was called *hsiao-ch'ün*, which has the same meaning. Consumption of the mushroom was known to result in "excessive laughter" (Li 1975, 175*).

During his attempts to find the Mexican magic mushroom, Richard Evans Schultes identified as *teonanacatl* a variety of this species: *Panaeolus campanulatus* L. var. *sphinctrinus* (Fries) Bres. [syn. *Panaeolus papilionaceus*] (Schultes 1939**). The psychoactivity of this species, however, is doubtful.

In addition to the fly agaric mushroom (*Amanita muscaria*), Graves regarded *Panaeolus papilionaceus*, which is "still used by Portuguese witches," as an additional candidate for the divine ambrosia and nectar (1966, 45*). To support his hypothesis, he cited a number of myths and works of art, including an Attic vase that depicts Nessus, the centaur. A mushroom can be seen sprouting from between his hooves. This fungal ambrosia later became the sacrament of the Eleusinian and Orphic mysteries. Graves even etymologically associated the word *kekyon* (= **kykeon**; cf. **Claviceps purpurea**) with the word *mykon* ("mushroom"). In Greek folklore, mushrooms are still referred to as the "food of the gods" (Ripinisky-Naxon 1988, 5*).

Panaeolus semiovatus Fries (Lundell) [syn. **Panaeolus separatus** Gillet, *Anellaria separata* Karst.]

Found throughout North America; may contain **psilocybin**.

Panaeolus tropicales Ola'h [syn. *Copelandia tropicales* (Ola'h) Sing. et Weeks]—tropical panaeolus

Found in tropical regions of Hawaii, central Africa, and Cambodia (cf. **Panaeolus cyanescens**).

Literature

See also the entries for **Panaeolus cyanescens** and **Panaeouls subbalteatus**.

Allen, John W., and Mark D. Merlin. 1992. Observations regarding the suspected psychoactive properties of *Panaeolina foenisecii* Maire. *Yearbook for Ethnomedicine and the Study of Consciousness,* 1992 (1): 99–115. Berlin: VWB.

Breitfeld, Matthias. 1996. Der falsche Pilz der Götter. *Der Tintling* 4:4–5.

Gartz, Jochen. 1985a. Zum Nachweis der Inhaltsstoffe einer Pilzart der Gattung *Panaeolus. Pharmazie* 40 (6): 431–32.

———. 1985b. Zur Analyse von *Panaeolus campanulatus* (Fr.) Quél. *Pharmazie* 40 (6): 432.

Gurevich, L. S. 1993. Indole derivatives in certain *Panaeolus* species from east Europe and Siberia. *Mycological Research* 97: 251–54.

Moser, M. 1984. *Panaeolus alcidis*, a new species from Scandinavia and Canada. *Mycologia* 76 (3): 551–54.

Ola'h, G. M. 1968. Étude chromataxinomique sur les *Panaeolus*, recherches sur les présences des corps indoliques psychotropes dans ces champignons. *Comptes Rendus de l'Académie des Sciences* 267:1369–72.

———. 1969. A taxonomic and physiological study of the genus *Panaeolus* with the Latin descriptions of the new species. *Review of Mycology* 33:284–90.

———. 1970. Le Genre *Panaeolus. Revue de Mycologie, Mémoire, Hors-Série* 10:1–273.

Pollock, Steven H. 1974. A novel experience with *Panaeolus*: A case study from Hawaii. *Journal of Psychedelic Drugs* 6 (1): 85–89.

———. 1976. Psilocybian mycetismus with special reference to *Panaeolus. Journal of Psychedelic Drugs* 8 (1): 43–57.

Robbers, J. E., V. E. Tyler, and G. M. Ola'h. 1969. Additional evidence supporting the occurrence of psilocybin in *Panaeolus foenisecii. Lloydia* 32 (3): 399–400.

Weeks, R. Arnold, Rolf Singer, and William Lee Hearn. 1979. A new psilocybian species of *Copelandia. Journal of Natural Products* 42 (5): 469–74.

Left: A tropical *Panaeolus* species that thrives on cow dung and has psychoactive effects. (Photographed in Belize)

Below: It is uncertain whether *Panaeolus papilionaceus* [syn. *Panaeolus campanulatus, Panaeolus sphinctrinus*] is psychoactive. (Photographed near the Externsteine, a series of standing stones)

Panaeolus papilionaceus is a common species found throughout the world. (From Winkler, *2000 Pilze selber bestimmen* [Identify 2000 Fungi Yourself], 1996**)

Pluteus spp.
Pluteus Species

The fawn pluteus (*Pluteus cervinus*). (From Winkler, *2000 Pilze selber bestimmen*, 1996**)

Family
Pluteaceae (Pluteus Mushrooms)

The small- to medium-size pluteus mushrooms have more or less plane to convex caps. They are well represented in Europe (Singer 1956). Many species are considered edible. Some have been found to contain **psilocybin** and baeocystin (Allen et al. 1992**; Gartz 1986; Stamets 1996**).

Pluteus cervinus (Schaeff.) P. Kumm.
This species, the fawn pluteus, is now seen as synonymous with *Pluteus atricapillus* Singer. In Europe, the fawn pluteus is regarded as edible. It is doubtful whether the fruiting bodies of *Pluteus cervinus* contain psychoactive substances. According to Allen et al. (1992, 93**), the species does contain **psilocybin**.

Pluteus cyanopus Quélet
This small species is found throughout central Europe. It develops a greenish blue color at the base of its stem. The mushroom is regarded as "somewhat poisonous" because it contains **psilocybin**.

Pluteus nigriviridis Babos
This pluteus species has been found to contain **psilocybin** (Stijve and Bonnard 1986).

Pluteus salicinus (Persoon ex Fries) Kummer—willow pluteus
This somewhat uncommon grayish mushroom, which usually appears alone, has a convex cap that later becomes plane. The center of the cap is scaly.

In Europe, it grows from early summer to fall on deciduous wood. It contains the psychoactive substances **psilocybin** and baeocystin (Gartz 1987). North American samples have yielded psilocybin and psilocin (Saupe 1981).

A nonblueing variety is known by the name *Pluteus salicinus* var. *achloes* Sing. (Saupe 1981, 783).

Pluteus villosus Bull. [syn. *Pluteus ephebeus* (Fr.: Fr.) Gillet, *Pluteus murinus* Bres., *Pluteus lepiotoides* A. Pearson, *Pluteus pearsonii* P.D. Orton]
In central Europe, this species thrives on decaying deciduous wood from early summer to fall. It purportedly contains **psilocybin**.

Literature

Christansen, A. L., K. E. Rasmussen, and K. Høiland. 1984. Detection of psilocybin and psilocin in Norwegian species of *Pluteus* and *Conocybe*. *Planta Medica* 45:341–43.

Gartz, Jochen. 1987. Vorkommen von Psilocybin und Baeocystin in Fruchtkörpern von *Pluteus salicinus*. *Planta Medica* (1987), no. 3: 290–91.

Saupe, Stephen G. 1981. Occurrence of psilocybin/psilocin in *Pluteus salicinus* (Pluteaceae). *Mycologia* 73 (4): 781–84.

Singer, R. 1956. Contributions towards a monograph of the genus *Pluteus*. *Transactions of the British Mycological Society* 39:145–232.

Stivje, T., and J. Bonnard. 1986. Psilocybine et urée dans le genre *Pluteus*. *Mycologia Helvetica* 2:123–30.

"*Polyporus mysticus*"
Polypore Species

"The delicious aroma of the polyporus of the birch tree (*Polyporus betulinus*) increases the smoker's enjoyment, which is why it is mixed into the tobacco."

Lippincotts Magazine (Philadelphia, 1888) in Der Unrat in Sitte, Brauch, Glauben und Gewohnheitsrecht der Völker [Filth in the Customs, Traditions, Belief, and Common Law of Peoples] (Bourke 1996, 68**)

Classification
Order Poriales (Aphyllophorales):
— Family Polyporaceae (Pore Fungi)
— Family Ganodermataceae (Lacquer Fungi)
— Family Poriaceae

A variety of evidence and numerous hypotheses suggest that there may be at least one as yet unidentified psychoactive polypore. Here, I have given this polypore the nickname *Polyporus mysticus*. But there is also evidence that well-known polypore species, such as the birch polypore and the larch polypore, have psychoactive

effects and have found use in shamanic contexts.

The sensational and still somewhat controversial discovery of Ötzi, a 5,300-year-old mummy found in a glacier (known popularly as Glacier Man, Adam of the Alps, Frozen Fritz, and Homo tyrolensis), indicates that polypores may have played a ritual or shamanic role among pre-Germanic peoples (Heim and Nosko 1993; Rätsch 1994). With him, Ötzi carried two dried mushrooms attached to a string. The mushrooms have been identified as either the birch polypore (*Piptoporus betulinus* [Bull. ex Fr.] Karst. [syn. *Polyporus betulinus* Karst., *Ungulina betulina* Pat.]) or the larch polypore (*Laricifomes officinalis* [Vill. ex Fr.] Kotl. et Pouz. [syn. *Polyporus officinalis* Fr., *Boletus laricis* Jacq.]) (cf. **silphion**). They were initially interpreted as "tinder fungi":[369]

> The two specimens from the Hauslabjoch were cut from the fruiting body of the birch polypore [*Piptoporus betulinus*]. The result was very surprising to us, as we had actually assumed that this was tinder material. The birch polypore, however, is not suitable for this purpose, as its tissue is difficult to set afire. We must thus look for the solution of this problem on another level. (Spindler 1993, 132)

Then a sensational press release made its way through the newspapers. A mycologist from Innsbruck, Reinhold Pöder, told the science editor of the weekly newsmagazine *Stern* that he had found LSD-like substances, i.e., **indole alkaloids**, in the first of the mushrooms he had studied:

> The newest evidence for the shamanic thesis was discovered by Dr. Reinhold Pöder of the University of Innsbruck. . . . The microbiological analysis . . . revealed: the two tree fungi that Ötzi was carrying were not for making fire at all. "The fungi are hallucinogenic," Dr. Pöder stated over the marble-size pieces from the genera of the larch and birch polypores. Among Ötzi experts, the discovery of this Stone Age relative of LSD is regarded as a minor sensation. (Scheppach 1992, 22)

Following this news, the yellow press (e.g., the *Hamburger Morgenpost*) stamped Ötzi as a junkie. The psychedelic scene celebrated him as a primordial shaman (Rätsch 1994):

> There are several things that suggest that he may have been a shaman . . . , especially the dried pieces of birch polypore that, strung onto a strip of leather, he also carried with him. The latter is a tree fungus that is used for medicinal and spiritual purposes ("spirit bread") in a number of shamanistic cultures, e.g., among the Haida Indians. The assertion that this tree fungus could contain a hallucinogen, i.e., an inebriating substance (which would fit the shamanic hypothesis without a problem), is plausible but is avoided in the scientific report on the fungal finds. (Scheidt 1992, 100)

The official scientific report on the mushrooms ignored the sensational news of the presence of a "Stone Age LSD" and degenerated into a trivial discussion:

> What could have led a person 5,000 years ago

to collect birch polypores and tie these to a strip of leather? At this time, it is hardly possible to suggest an appropriate answer to this question. The . . . evidence for the "medicinal and spiritual use of polypores," however, suggests that the discussions about such a use may be merited. The investigation of the second piece of "tree fungus" may provide additional information. (Pöder et al. 1992, 318)

Later, the head of the Innsbruck research project cautiously wrote:

> A few apocryphal places in the literature on this matter have drawn attention to a purported hallucinogenic effect of the birch polypore. To date, however, this supposition has not received either medical or pharmaceutical support. Therefore, at the present time, they should not yet be taken into consideration in discussions pertaining to the meaning that the tree fungi had for the man in the ice. (Spindler 1993, 133)

Among the Indians of the Pacific Northwest, polypores were directly related to shamanism. They are said to have been used as a **snuff** for healing rituals, and shamans are said to have carved the fruiting bodies of *Fomitopsis officinalis* into the shapes of protective spirits to watch over graves (Blanchette et al. 1992). The Indians of the northwestern coast used birch polypores, which they called "spirit breads," in shamanic contexts as a trance-inducing snuff (Andrew Weil, pers. comm.).

The larch polypore (*Laricifomes officinalis* = *Fomitopsis officinalis* [Vill. ex Fr.] Bond. et Sing.) formerly was made into an *elixier de longue vie*, "elixir of long life," together with aloe (*Aloe* sp.), gentian (*Gentiana lutea* L.), Levantine saffron (***Crocus sativus***), rhubarb (*Rheum officinale* Baill.), and **theriac** (Chapuis 1985, 111).

In the seventeenth century, Jesuits reported that the Yarimagua Indians of the Peruvian Amazon prepared an inebriating drink from "fungi that grow on fallen trees":

> These appear as reddish, pungent-tasting

Left: The Indians of the Pacific Northwest (e.g., the Haida) used the polypore *Fomitopsis pinicola* for medicinal and shamanic purposes. (Photographed near Seattle, Washington)

Right: The birch polypore (*Piptoporus betulinus*) was once used as a medicinal drug. It was one of the few objects found with the glacier mummy known as Ötzi. According to a statement by a mycologist from Innsbruck, it was a type of "Stone Age LSD." The statement was later retracted. (Photographed at Duvenstedter Brook, Hamburg)

369 The true tinder fungus is *Fomes fomentarius* (Fr. ex L.) Kichx. (Cetto 1987, 1:377**). "No fire can be made using a birch fungus, as its tissue is too thick. But the birch polypore does contain two interesting substances: polyporic acid C and ugulinic acid. Polyporic acid has antibacterial effects. This species of fungus can be used as an antibiotic, i.e., a medicinal agent. The glacier man was thus carrying with him a kind of first-aid kit" (Barfield et al. 1992, 96).

"The earth came forth from a tree fungus, 'as from an egg.' More precisely: The egg-shaped tree fungus parted in the middle: the upper part ascended and became the heavens, the lower became the earth. From the two halves of *alonkok*, as the tree fungus was called, came all the things that are visible: the stars, the sun, mountains, rivers, plants, animals, and the 'primordial mother,' which was also called *alonkok*. In a separate egg lay the lightning; this is how the primordial mother came to have fire. She bore two pairs of twins, the first were called 'Mud Hill' (Nkombodo) and 'Great Mountain' (Odangemeko), the second was Mebere ['Highest Being'] and his sister."

MYTH OF THE PANGWE OF THE AFRICAN CONGO
IN *DIE GÖTTER SCHWARZAFRIKAS* [THE GODS OF BLACK AFRICA] (BONIN 1979, 193 f.*)

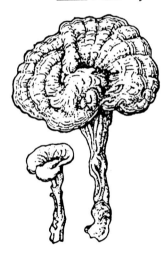

The polypore *Ganoderma lucidum* has been identified with the legendary Chinese mushroom of immortality. (From an old Chinese herbal)

370 *Fomes fomentarius* is depicted in the scene of the temptation of Saint Anthony on Matthias Grünewald's Isenheim Altar (cf. *Claviceps purpurea*). It is a symbol for the transitory nature of life. (Claudia Müller-Ebeling, pers. comm.)

371 The *ling chih* is usually interpreted as *Ganoderma lucidum* but has occasionally been seen as a fly agaric (*Amanita muscaria*) (Mackenzie 1994, 107).

growths on fallen trees. It was so strong that no man who took three draughts of the brew was able to resist its effects. The fungus was considered to be *Psilocybe yungensis* Singer and Smith [cf. *Psilocybe* spp.]. But because gym species are reddish and also settle on compact trunks, they are more likely candidates for this ominous tree fungus. . . . The red tree species was probably a close relative of *Gymnopilus purpuratus*. (Gartz 1993, 57 f.**)

Ott (1993, 316*) also considers it possible that this tree fungus was a *Gymnopilus purpuratus* that thrived on wood (cf. *Gymnopilus* spp.). But it may also have been a polypore. Ian Watson (1983) has written an outstanding science-fiction novel about an Indian mushroom cult in Amazonia. His description is based upon the Jesuit accounts.

Many travelers and researchers have reported psychoactive mushroom use in the Amazon, often in connection with **ayahuasca** (Leginger 1981*; McKenna 1989*; McKenna and McKenna 1994*; Ott 1993, 316*). One indication that these reports may indeed be based in truth comes from the Paumarí Indians of central Amazonia, who refer to polypores as *badiadimurobuni*, "the ear of the spirit" (Pance et al. 1977, 129*).

It was recently discovered that in Amazonia (Ecuador), the powdered fruiting body of a *Ganoderma* species is mixed with tobacco (*Nicotiana tabacum*) and smoked. According to the descriptions of indigenous shamans, this **smoking blend** has an **ayahuasca**-like effect (Paul Stamets, pers. comm.).

The Makai Indians of Paraguay use the fruiting bodies of various polypores (*Dedalea elegans* Fries, *Polyporus guaraniticum* Speg., *Pycnoporus sanguineus* [Fries.] Murr.) as **incense** to pacify crying children (Arenas 1987, 287*). The Siberian Khanty (= Ostyak) used a mixture of tinder fungus (*Fomes fomentarius* [Fr. ex L.] Kichx.) and the bark of the silver pine (*Picea obovata* Ledeb. or *Abies sibirica* Ledeb.) as a ritual **incense** when a person had died, burning the incense until the corpse had been removed from the house (Saar 1991, 177**).[370] The Khanty also used the polypore *Inonotus obliquus* (Fr.) Pilát as an incense. They burned the fruiting bodies of the polypore *Phellinus nigricans* (Fr.) Karst., which thrives on birches (*Betula* spp.), and mixed the ashes with ground wild tobacco (*Nicotiana rustica*) to produce a narcotic chewing tobacco (Saar 1991, 178**).

In Alaska, the polypores *Fomes igniarius* (Fr. ex L.) Kichx. and *Fomes fomentarius* (Fr. ex L.) Kichx. allegedly are smoked or sniffed in combination with *Nicotiana* spp. for their narcotic properties (Ott 1978, 234**).

In Chignahuapan (Puebla, Mexico) is a church known as the Iglesia de Neustro Señor del

Honguito (Church of Our Lord, the Mushroom). Built especially for the *honguito*, this church preserves as a relic the fruiting body of a mushroom that is said to bear the mark of the hand of Jesus. This mushroom, which has been identified as *Ganoderma lobatum* (Schw.) Atk., is believed to have miraculous and healing powers. It has been suggested that this folk church cult may preserve elements of entheogenic mushroom rituals from the pre-Spanish period (Guzmán et al. 1975).

Paul Stamets has suggested that **soma** may have been a psychoactive polypore that has not yet been discovered and described.

Many ancient Chinese myths and traditions concerning the *ling-shih* (also *ling chih*, *lingzhi*), or the mushroom of immortality, now usually identified mycologically as the lacquer fungus (*Ganoderma lucidum* [Fr.] Karst.),[371] suggest that this mushroom has potent psychoactive effects (Camilla 1995a, 1995b; Rätsch 1996). It is said to grow on a mysterious island in the East (probably the South Korean shaman's island of Chejudo). The effects of this magic mushroom have been described as fantastic:

The Isle of Tsu lies close by, in the eastern sea. It is there that the plant that does not die grows, shaped like a piece of seaweed with leaves that are up to four feet in length. A man who has already been dead for three days will immediately awaken to a new life if he is touched by it. If you eat of it, it will extend your life. . . . When dead bodies lay around on the side of the road, birds like ravens or crows would fly by with the miraculous plant in their beaks, which they laid upon the faces of the corpses. These immediately stood up and became alive once more. (*Shih chou chi*)

The Taoist alchemists later numbered the *lingzhi* among a group of "five wondrous fungi of immortality"—corresponding to the five signs of happiness (Imazeki and Wasson 1973, 6**). The mysterious elixir of immortality was prepared from these fungi, which probably included the fly agaric (*Amanita muscaria*) and other psychedelic magic mushrooms (such as the "nine stemmed, purple-red mushroom"), together with cinnabar and jade. Those recipes that have come down to us have not yet been tested (Michel Strickman, pers. comm.; cf. **han-shi**).

Ganoderma lucidum has been very well studied but has not yet yielded any evidence indicating any possible psychoactive effects or psychoactive constituents (Laatsch 1992). It does contain bioactive triterpenes and ganodermic acids (Lin et al. 1988).

The Japanese furry polypore *Inonotus hispidus* (Bull. ex Fr.) Karst. [syn. *Polyporus hispidus* Bull. ex Fr.], also known as the Japanese laughing mushroom (cf. *Gymnopilus* spp.), contains hispidin, a substance that is chemically closely related to the kavains (see *Piper methysticum*) and the longistylines (see *Lonchocarpus violaceus*) (H. Laatsch, pers. comm., July 1, 1986). *Polyporus berkeleyi* Fries has been shown to contain the β-phenethylamine hordenine, which is also found in many psychoactive cacti (*Ariocarpus fissuratus, Coryphantha* spp., *Pelecyphora aselliformis, Trichocereus* spp.) (Ott 1978, 234**).

One species of polypore, the sulfur fungus (*Laetiporus sulphureus* [Bull. ex Fr.] Bond et Sing. [syn. *Boletus sulphureus* Bull., *Boletus caudicinus* Scop., *Cladomeris sulphurea* Quél., *Polyporus caudicinus* Köhl, *Polyporus sulphureus* Fr.]),[372] has become known for having induced strong tryptamine-like hallucinations in at least one clinically documented case (Appleton 1988). Otherwise, it has been said that "used as a snuff, [it] can be described as a weak disinfectant" (Chapius 1985, 111).

Literature

Appleton, Richard Edward. 1988. *Laetiporus sulphureus* causing visual hallucinations and ataxia in a child. *Canadian Medical Association Journal* 39:48–49.

Barfield, Lawrence, Ebba Koller, and Andreas Lippert. 1992. *Der Zeuge aus dem Gletscher: Das Rätsel der frühen Alpen-Europäer*. Ed. Alfred Payrleitner. Vienna: Verlag Carl Ueberreuter (Edition Universum).

Blanchette, Robert, Brian D. Compton, Nancy Turner, and Robert L. Gilbertson. 1992. Nineteenth century shaman grave guardians are carved *Fomitopsis officinalis* sporophores. *Mycologia* 84 (1): 119–24.

Camilla, Gilberto. 1995a. I funghi allucinogeni in China e in Giappone: Sopravvivenze mitologiche, folkloriche e linguistiche. I. *Eleusis* 2:10–13.

———. 1995b. I funghi allucinogeni in China e in Giappone: Sopravvivenze mitologiche, folkloriche e linguistiche. II. *Eleusis* 3:25–28.

Chapuis, Jean-Robert. 1985. Die Verwendung von Pilzen als Arzneimittel (I). *Schweizerische Zeitschrift für Pilzkunde* 63 (5/6): 110–14.

Gartz, Jochen. 1994. Das Letzte von Ötzi. *Yearbook for Ethnomedicine and the Study of Consciousness,* 1993 (2): 157–63. Berlin: VWB.

Guzmán, Gastón, R. Gordon Wasson, and Teófilo Herrera. 1975. Una iglesia dedicada al culto de un hongo, "Nuestro Señor del Honguito," en Chignahuapan, Puebla. *Bol. Soc. Mex. Mic.* 9:137–47.

Heim, Michael, and Werner Nosko. 1993. *Die Öztal-Fälschung: Anatomie einer archäologischen Groteske*. Reinbek: Rowohlt.

Laatsch, Hartmut. 1992. Polysaccharide mit Antitumor-Aktivität aus Pilzen. *Pharmazie in unserer Zeit* 21 (4): 159–66.

Lewinsky-Sträubli, Marianne. 1989. *Japanische Dämonen und Gespenster*. Munich: Diederichs.

Lin, Lee-Juian, Ming-Shi Shiao, and Sheau-Farn Yeh. 1988. Triterpenes from *Ganoderma lucidum*. *Phytochemistry* 27 (7): 2269–71.

Mackenzie, Donald. 1994. *China and Japan: Myths and legends*. London: Studio Editions.

Matsumoto, Kosai. 1979. *The mysterious reishi mushroom*. Santa Barbara, Calif.: Woodbridge Press.

Pöder, Reinhold, Ursula Peintner, and Thomas Pümpel. 1992. Mykologische Untersuchungen an den Pilz-Beifunden der Gletschermumie vom Hauslabjoch. In *Bericht über das Internationale Symposion 1992 in Innsbruck*, vol. 1 of Der Mann im Eis (2nd ed.), 313–20. Innsbruck: Veröffentlichungen der Universität Innsbruck 187.

Rätsch, Christian. 1992. Nachwort: Sternstunde der Entheogeneologie? In *Das Tor zum inneren Raum*, ed. C. Rätsch, 257–60. Südergellersen: Verlag Bruno Martin.

———. 1994. Ötzis Pilze in Literaturzitaten. *Yearbook for Ethnomedicine and the Study of Consciousness,* 1993 (2): 157–62. Berlin: VWB

———. 1996. Lingzhi: Der Pilz der Unsterblichkeit. *Natürlich* 16 (3): 22–24.

Scheidt, Jürgen V. 1992. Der Zufall setzt ein Zeichen. *Esotera* 12/92: 96–101.

Scheppach, Joseph. 1992. Was uns der Gletschermann erzählt. *Stern*, no. 29 (July 9, 1992): 10–22.

Spindler, Konrad, et al. 1993. *Der Mann im Eis: Die Ötztaler Mumie verrät die Geheimnisse der Steinzeit*. Munich: C. Bertelsmann Verlag.

Watson, Ian. 1983. *Das Babel-Syndrom*. Munich: Heyne.

Willard, Terry. 1990. *Reishi mushroom: Herb of spiritual potency and medical wonder*. Issaquah, Wash.: Sylvan Press.

"Once an ancient peach tree stood for many years in a marsh in the deep mountains. It eventually became rotten and died. I am a mushroom that sprouted from its trunk. For many years I grew on a tree trunk, becoming larger and larger, until I had truly become an important mushroom. One day a beautiful small bird flew by and sat in the branches of the tree, where it sang with its beautiful chirping voice. Watching it, I at first found it to be simply pretty, but then I felt an urge to eat it. I had such a terribly great desire to eat, eat, eat it that while I was looking at it, I suddenly had eyes and a mouth and gulped it down and swallowed it. . . . With time the tree decayed and crumbled. I broke off from its stump and crawled around, then I grew arms and legs and I could walk."

DAS PILZUNGEHEUER [THE MUSHROOM MONSTER] (LEWINSKY-STRÄUBLI 1989, 52)

---

372 This species was used in Japan as a tinder fungus (Wasson 1973, 23**).

Psilocybe azurescens Stamets et Gartz

Indigo Psilocybe

Family
Agaricaceae: Strophariaceae; Stropharioideae Tribe, Caerulescentes or Cyanescens Section

Synonyms
Psilocybe astoriensis nom. nud.
Psilocybe cyanescens Ossip nom. nud.

Folk Names
Astoriensis, azureus-kahlkopf, blue runner, flying saucer mushroom, indigo psilocybe

The history of *Psilocybe azurescens* is very new and mysterious. This mushroom was only recently discovered:

> In 1979, Boy Scouts found an unusually large and strongly blueing Stropharia near the city of Astoria in the state of Oregon, close to the mouth of the majestic Columbia River. Stem lengths of up to 20 cm (!) and caps as broad as 10 cm were no rarity. By 1981, the mushroom was being grown outdoors on wood chips or cow mulch, methods that were originally developed for *Psilocybe cyanescens* and have been described in detail elsewhere (Gartz 1994). Soon these "new" mushrooms, which proved to be potently psychoactive, were called "*Psilocybe astoriensis*" (Gartz, in Rippchen 1993[**]), although no mycological description with a valid Latin diagnosis was published. (Gartz 1996, 189)

The initial description by Paul Stamets and Jochen Gartz was not published until 1995. This largest species of the genus *Psilocybe* grows on the remains of wood in the coastal region of Oregon and Washington, where it fruits in the fall (from the end of September to the beginning of November). Of all the *Psilocybe* species, this one spreads the most aggressively. "The mushroom even grew spontaneously on wooden clothespins that were lying about" (Gartz, in Rippchen 1993, 72**).

This mushroom is quite similar to *Psilocybe cyanofibrillosa* Stamets et Guzmán (cf. **Psilocybe spp.**) but is also easily confused with the sulfur tuft (*Hypholoma fasciculare* [Hus.: Fr.] Kummer).

The spore print can germinate on an agar surface within three days. Cultivating the mycelium is best done on a substrate of rye. Unfortunately, in Germany the cultivated mushrooms commonly do not fruit (B. Schuldes 1995).

Psilocybe azurescens can develop caps that are 10 cm in diameter and stems that are as long as 20 cm. Because it was discovered only at the end of the 1970s, it has been speculated that this is a new species that only recently evolved. It grows in a habitat that is very untypical for members of the *Psilocybe* genus. It thrives on sandy soil close to the ocean, usually in association with the grass *Ammophila maritima* (cf. **Claviceps spp.**). The mushroom quickly spread from Astoria and will certainly soon be common in the entire Pacific Northwest. It is the most potent member of the genus *Psilocybe*.

The indigo psilocybe (dry weight) contains 1.29 to 1.78% **psilocybin**, 0.18 to 0.37% baeocystin, and 0.27 to 0.5% psilocin. The alkaloid concentration is almost consistently the same among mushrooms that have been collected in the wild, those that have been cultivated in Washington state, and those that have become feral in Germany (Stamets and Gartz 1995, 23).

Typical Values as Analyzed in Individual Dried Mushrooms of *Psilocybe azurescens* (from Gartz 1996)

Mushroom	Psilocybin	Psilocin	Baeocystin
1	1.71	0.34	0.41
2	1.68	0.28	0.38
3	1.56	0.30	0.32
4	1.40	0.31	0.28

An experiential report written by a mycologist clearly demonstrates the potent effects of *Psilocybe azurescens*:

> To my surprise, the mushroom powder (1 gram) in the mixture of orange juice was tasteless, which I found positive in comparison to earlier experiments with *Psilocybe semilanceata* and *Psilocybe cubensis*. After some 20 minutes, the effects suddenly began in such a way that my body immediately dissolved into pure energy. This sensation of a remaining, isolated soul without a Christian/

Psilocybe azurescens, which resembles a flying saucer (photographed in its type locality [natural habitat] near Astoria, Oregon)

church context that existed somewhere and sometime was extremely impressive. There began a journey through historical periods that was without precedent for me. The coarse structure of the white ceiling of the room dissolved completely, as when cobwebs are brushed aside, and a stage opened up. A multitude of historical events and experiences that my soul experienced as totally real alternated in quick succession. Once, the room was transformed entirely into the style of an ancient Egyptian tomb, and I lay in the center, which provoked a moment of sudden terror, for it had an absolute character of being real. But generally this flight through historical times took place in an essentially peaceful and meditative mood with many experiences of a transpersonal nature, as [Stanislav] Grof has described in detail. Personal problems no longer existed.

After some 5 hours, this journey beyond time and space ended with the rather abruptly developing sensation that my body and its union with the free-floating soul had been created anew. Aside from a certain tiredness, no other aftereffects were observed. During the following day, there was a very pleasant sensation of a special freshness of spirit that then slowly dissipated. (Gartz 1996, 91)

Literature

See also the entries for the other *Psilocybe* species and for **psilocybin**.

Gartz, Jochen. 1994. Ethnopharmakologie psilocybinhaltiger Pilze im pazifischen Nordwesten der USA. In *Jahrbuch des Europäischen Collegiums für Bewußtseinsstudien* 1993/1994: 159–64.

———. 1996. Ein neuer psilocybinhaltiger Pilz. In *María Sabina—Botin der heiligen Pilze*, ed. Roger Liggenstorfer and Christian Rätsch, 189–92. Solothurn: Nachtschatten Verlag.

Schuldes, Bert Marco. 1995. Erfahrungen mit *Psilocybe astoriensis*. *Entheogene* 4:30–31.

Stamets, Paul, and Jochen Gartz. 1995. A new caerulescent *Psilocybe* from the Pacific Coast of northwestern America. *Integration* 6:21–27.

"A cold weather–tolerant species, *P. azurescens* is one of the most potent species in the world and exhibits one of the strongest bluing reactions I have seen."

PAUL STAMETS
PSILOCYBIN MUSHROOMS OF THE WORLD
(1996, 96**)

Psilocybe (*Stropharia*) *cubensis* (Earle) Singer
Magic Mushroom, Divine Dung Mushroom, Golden Cap

Family

Agaricaceae: Strophariaceae; Stropharioideae Tribe, Cubensae Section

Forms and Varieties

The recently described species *Psilocybe subcubensis* may possibly be merely a subspecies or variety of *P. cubensis* (cf. **Psilocybe spp.**). Three varieties have been described:

Psilocybe cubensis var. *caerulescens* (Murr.) Singer et Smith
Psilocybe cubensis var. *cubensis*
Psilocybe cubensis var. *cyanescens* (Murr.) Singer et Smith

Synonyms

Hypholoma caerulescens (Pat.) Sacc. et Trott.
Naematoloma caerulescens Pat.
Psilocybe cubensis var. *caerulescens* (Murr.) Singer et Smith
Stropharia cubensis Earle
Stropharia caerulescens (Pat.) Sing.
Stropharia cyanescens Murr.
Stropharia subcyanescens Rick.

Folk Names

Champiñon, derrumbe de estiércol de vaca (Spanish, "abyss of the cow patties"), di-ki-sho-lerraja, dishitjolerraja (Mazatec, "divine dung mushroom"), divine dung mushroom, golden top, gold top, göttlicher düngerpilz, hed keequai (Thai), hongo de San Isidro, hongo maravilloso, honguillos de San Isidro Labrador ("mushroom of Saint Isidro the Farmer" [= the saint of agriculture]), hysteria toadstool, kubanischer kahlkopf, kubanischer träuschling, lòl lú'um (Yucatec Mayan, "flowers of the earth"), magic mushroom, nocuana-be-neeche (Zapotec), nti-xi-tjolencha-ja (Mazatec, "mushroom like that which grows on cow patties"), San Isidro, San Isidro Labrador, tenkech (Chol), tenkech (Chol: Panlencano), teotlaquilnanácatl (modern Nahuatl, "the sacred mushroom that paints in colors"), zauberpilz

The tropical *Psilocybe cubensis* (= *Stropharia cubenis*) prefers to live on cow dung. (Photographed in Palenque, Chiapas, Mexico)

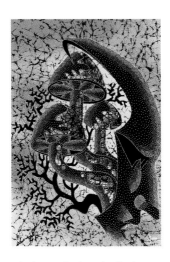

The fruiting bodies of *Psilocybe cubensis*, growing from the head of a mushroom user who has been enlightened from within. (Indonesian batik, twentieth century)

History

Psilocybe cubensis (Earle) Sing. (= *Stropharia cubensis* Earle), known internationally by the names magic mushroom and golden cap, is originally from Africa. It thrives on cattle dung and in meadows with deposits of dung. In symbiosis with African cattle, it has spread around the world, although it grows only in tropical or subtropical areas. Terence McKenna believed that this psychoactive mushroom exerted an important influence upon human evolution. According to his theory, consuming these mushrooms resulted in a "mental quantum leap" that transformed our apelike ancestors into "intelligent beasts" with a greater ability to survive. This psychedelic "primordial experience" led to the development of the first mystical mushroom rituals, which formed the basis for shamanism, mythologies, and religions (McKenna 1996*). It has even been suggested that this mushroom was the original **soma**.

This mushroom was first found in Cuba (hence its species name *cubensis*, "Cuban"). The Englishman S. Baker provided the first description of its traditional use in Ceylon (Sri Lanka) (*Eight Years in Ceylon*, London, 1855 [1884]). The shamanic use of *Psilocybe cubensis* in Mexico was discovered during research into the magic mushrooms of Mexico (cf. **Psilocybe mexicana**). There, it is known as *hongo de San Isidro*, "mushroom of Saint Isidro." Among the Mazatec Indians, Saint Isidro is the patron saint of fields and meadows, the same locales in which this mushroom—which is exclusively coprophilous—is found (Heim and Hofmann 1958a).

Because this mushroom is frequently found in Palenque (Mexico), it has been suggested that the ancient Maya may have used it as an entheogen. Before the Spanish, however, there were no cattle in the Americas, and the mushroom requires their dung to grow. All of the evidence suggests that *Psilocybe cubensis* was introduced into Mexico during the late colonial period (Coe 1990).

In Thailand, *Psilocybe cubensis* is now the most commonly offered mushroom on the vacation islands of Koh Samui and Koh Pha-Ngan (Allen 1991; Allen and Merlin 1992a, 1992b). The omelets made with this mushroom are renowned. It is also common in Bali (Wälty 1981).

Appearance

The mushroom forms relatively large fruiting bodies with slightly convex caps that can grow as large as 8 cm in diameter. The caps usually have a yellow or golden color at their center.

Psilocybe cubensis can be distinguished from *Psilocybe subcubensis*, a Central American species known as *suntiama*, only on the basis of the size of its spores (Guzmán 1994, 1472**).

Distribution

Psilocybe cubensis is found throughout the tropics wherever there is cattle or water buffalo breeding or ranching, including Mexico (Oaxaca, Chiapas), Cuba, Guatemala, Colombia, Bolivia, Brazil, Argentina, Florida, Thailand, Vietnam, Cambodia, Indonesia, the Philippines, and Australia. In tropical areas, the mushroom can fruit throughout the year. The mushrooms usually sprout from cow dung after it has rained.

Cultivation

Of all species of *Psilocybe*, this is the easiest to grow. The mushroom produces more **psilocybin** when grown on malt agar (Gartz 1987). Fruiting occurs most readily when air humidity is high and the temperatures are tropically warm (24 to 34°C).

Harvest, Storage, and Consumption

In the tropics, the fruiting bodies of *Psilocybe cubensis* are easy to collect. During the harvest, however, certain things should be taken into consideration:

Although many people eat the fresh mushrooms right from the field, this unhygienic practice is to be discouraged. Some mushrooms grow directly on the dung and may possibly have particles of dung adhering to their flesh. For reasons of safety, the wise user should select only fresh, healthy specimens that are free of insects and avoid those that are rotting. Before consumption, [the mushrooms] should be washed thoroughly with water; the conscientious consumer will also cut off the lower end of the stem.

To store, the mushrooms are dried in the air at more or less room temperature (devices for drying food are also suitable; they can also be dried on a grate near a source of warmth). Overly long drying processes and high temperatures should absolutely be avoided. When the mushrooms are crispy, they should be filled into airtight containers. They can then be placed in a freezer. In this way, they can be stored for months with only a very slight loss of effectiveness. The mushrooms should not be frozen until they are completely dry (otherwise, they will quickly lose their effectiveness). They also should not be preserved in honey while fresh (this will result in a disgusting fermented mass). If the mushrooms are going to be stored for only a few days, it will suffice to place them in the refrigerator....

The dried mushrooms are clearly not very easy to digest, especially when they have not been sufficiently mixed with saliva. Mixing the mushrooms with juice or chocolate breaks up the tissue and allows the psilocybin to more easily enter into solution. It goes without saying that the mushrooms should be mixed with these carrier substances only immediately prior to consumption. Some users prefer mushrooms that have been sautéed in butter and are eaten with toast or potato chips. Lightly sautéing them over a low flame will not seriously lower the psilocybin content (it may be better to fry fresh mushrooms, so that any toxic components that may be present, e.g., gyromitrine and other methylhydrazines, will be destroyed). (Ott 1996, 191 f.)

An effective dosage of *Psilocybe cubensis* is regarded as 3 to 5 g of dried mushrooms. The user may want to use different dosages for different purposes, ranging from mild psychostimulation produced by a small mushroom to a "full blast" or a psychedelic breakthrough (Terence McKenna's famous "heroic" recipe calls for 5 g "on an empty stomach in total silent darkness"). *Psilocybe cubensis* is the psilocybin mushroom that is most commonly available on the black market (Turner 1994, 27*).

Magic mushrooms are usually consumed in fresh or dried form. With time, certain specific forms of ingesting the mushrooms have been developed: Dipped in **honey** or powdered, the mushrooms may be drunk with cacao (cf. *Theobroma cacao*). From time to time, the mushrooms are also eaten with chocolate (cf. Remann 1989, 248*).

In Thailand, the mushroom is dried and then smoked or baked into cookies together with hemp (*Cannabis indica*) (Allen and Merlin 1992b, 213). The fresh mushrooms are incorporated into dishes in the same way that normal culinary mushrooms are used.

Ritual Use

In central Europe, cultivated mushrooms are used in ritual circles in the same manner as *Psilocybe semilanceata*. In Mexico, wild mushrooms growing on cow dung are used in shamanic rituals in the same manner as *Psilocybe mexicana*.

In central Europe, this mushroom has also been used with success in private healing rituals (Strassmann 1996).

Artifacts

On the Thai "mushroom island" of Koh Samui, an entire T-shirt industry has arisen that offers tourists hand-painted T-shirts with mushroom designs (Allen 1991; Allen and Merlin 1992a). The mushroom is also frequently depicted on Indonesian batiks (cf. *Panaeolus cyanescens*).

Constituents

The fruiting body contains a maximum of 1% **psilocybin** by dry weight. An analysis by Gartz (1994, 19**) found an average of approximately 0.6% psilocybin, 0.15% psilocin, and 0.02% baeocystin by dry weight. The quantity of active constituents is greater in the caps than in the stems (Gartz 1987).

Effects

As with all psilocybin mushrooms, *Psilocybe cubensis* produces strong visions that often feature shamanic characteristics:

> The effects of the mushrooms [*Psilocybe cubensis*] began by manifesting themselves as waves of energy that ran through my body. I found the beauty that was proffered to my eyes to be even more valuable.
>
> Suddenly a large snake glided toward me from the desert that surrounded us and slipped into my body. The next thing I noticed was that I myself had become the snake. No sooner had I gotten used to this condition than a large eagle descended and snatched me with its talons. My body shook from the blow, but I did not feel any pain. The eagle held me firmly in its clutches, ascended again, and flew directly into the sky until it had become one with the sunlight. My personal identity as a separate consciousness dissolved. The only thing that remained was the unity with the light. (Pinkson 1992, 144)

Literature

See also the entries for the other *Psilocybe* species and for **psilocybin**.

Allen, John W. 1991. Commercial activities related to psychoactive fungi in Thailand. *Boston Mycological Club Bulletin* 46 (1): 11–14.

Allen, John W., and Mark D. Merlin. 1992a. Psychoactive mushrooms in Thailand: Some aspects of their relationship to human use, law and art. *Integration* 2/3:98–108.

———. 1992b. Psychoactive mushroom use in Koh Samui and Koh Pha-Ngan, Thailand. *Journal of Ethnopharmacology* 35 (3): 205–28.

Bigwood, Jeremy, and Michael W. Beug. 1982. Variation of psilocybin and psilocin levels with repeated flushes (harvests) of mature sporocarps of *Psilocybe cubensis* (Earle) Singer. *Journal of Ethnopharmacology* 5 (3): 287–91.

Coe, Michael D. 1990. A vote for Gordon Wasson. In *The sacred mushroom seeker*, ed. T. Riedlinger, 43–45. Portland, Ore.: Dioscorides Press.

Gartz, Jochen. 1987. Variation der Indolalkaloide von *Psilocybe cubensis* durch unterschiedliche Kultivierungsbedingungen. *Beiträge zur Kenntnis der Pilze Mitteleuropas* 3:275–81.

———. 1989. Bildung und Verteilung der Indolalkaloide in Fruchtkörpern, Mycelien und Sklerotien von *Psilocybe cubensis*. *Beiträge zur Kenntnis der Pilze Mitteleuropas* 5:167–74.

Heim, Roger, and Albert Hofmann. 1958a. Isolement de la Psilocybine à partir de *Stropharia cubensis* Earle et d'autres espèces de champignons hallucinogènes mexicains appartenant au genre *Psilocybe*. *Comptes rendus de l'Académie des sciences, Paris* 247:557–61.

"Did you know that it has been suggested that some ancient Indian brahman was unable to properly get his metaphor together, so that instead of declaring the mushroom that grows on the dung holy, he proclaimed holy the cow that produces the dung? Did you know that such misunderstandings also make it clear why it is indeed idolatrous to dance around the golden calf?"

MICKY REMANN
GLÜCKSPILZE [HAPPY MUSHROOMS] (1989, 260 f.*)

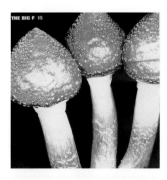

The American heavy-metal band The Big F was obviously inspired by the visionary powers of *Psilocybe* (*Stropharia*) *cubensis* when it named its album *Is*. Or is there another way to interpret this apparently mycological picture of magic mushrooms? (CD cover, 1993, Chrysalis Records)

————. 1958b. La psilocybine et la psilocine chez les psilocybes et strophaires hallucinogènes. In *Les champignons hallucinogènes du Mexique*, by Roger Heim and R. Gordon Wasson, 258–62**. Paris: Muséum National d'Histoire Naturelle.

Katerfeld, Raoul. 1995. A glimpse into heaven—a meeting with Thailand mushroom spirits. *Integration* 6:47–49.

Ott, Jonathan. 1996. Zum modernen Gebrauch des Teonanácatl. In *María Sabina—Botin der heiligen Pilze*, ed. Roger Liggenstorfer and Christian Rätsch, 161–63. Solothurn: Nachtschatten Verlag.

Pinkson, Tom. 1992. Reinigung, Tod und Wiedergeburt: Der klinische Gebrauch von Entheogenen in einem schamanischen Kontext. In *Das Tor zu inneren Räumen*, ed. C. Rätsch, 141–66. Südergellersen: Verlag Bruno Martin.

Strassmann, René. 1996. Sarahs Stimmen—ein traditionelles europäischen Pilzritual. In *María Sabina—Botin der heiligen Pilze*, ed. Roger Liggenstorfer and Christian Rätsch, 183–88. Solothurn: Nachtschatten Verlag.

Wälty, Samuel. 1981. Einfluß des Tourismus auf den Drogenbrauch in Kuta, Bali. In *Rausch und Realität*, ed. G. Völger, 2:572–75. Cologne: Rautenstrauch-Joest Museum für Völkerkunde.

Psilocybe cyanescens Wakefield emend. Kriegelsteiner

Wavy Caps

Family
Agaricaceae: Strophariaceae; Stropharioideae Tribe, Semilanceata = Cyanescens Section

Synonyms
Geophila cyanescens (Maire) Kühner et Rom.
Hypholoma coprinifacies (Rolland ex Herink) Pouzar
Hypholoma cyanescens Maire
Psilocybe bohemica Sebek (cf. ***Psilocybe* spp.**)
Psilocybe mairei Sing.
Psilocybe serbica Moser et Horak

Folk Names
Blaufärbender kahlkopf, blauwwordend kaalkopje (Dutch), blue halos, böhmischer kahlkopf, oink, wavy caps, zauberpilz, zyanescens

This mushroom is most easily recognized on the basis of its peculiar wavy cap. It lives not on dung but on the remains of plants, rotting wood, and humus-rich soils. In older mushroom guides, it often appears under the name *Hyphaloma cyanescens* (Cooper 1980, 18**). It is native to North America and central Europe and is found even in Hamburg (Findeisen 1982):

> This species settles on wood chips, which are often lying directly on the ground, so that the mushroom appears to sprout directly out of the earth. It is found primarily in parks in the Pacific Northwest, often in fairy circles, in amounts of up to 100 pounds. These mushrooms are one of the most potent species known and contain psilocybin and psilocin in amounts of up to 2% dry weight. (Gartz in Rippchen 1993, 70*)

Right: This painting by the Swiss artist Fred Weidmann was inspired by *Psilocybe cyanescens*.

Below: The very potent *Psilocybe cyanescens* is easily recognized by its undulating cap.

Gartz (1994, 19**) found that specimens grown in Germany contain approximately 0.3% **psilocybin**, 0.5% psilocin, and 0.01% baeocystin by dry weight.

In central Europe, *Psilocybe cyanescens* is used in rituals in precisely the same manner as ***Psilocybe semilanceata*** (Liggenstorfer 1996). Here, cultivated mushrooms that contain a very high concentration of psilocybin are consumed. A visionary dosage is regarded as 1 g dry weight.

Literature

See also the entries for the other *Psilocybe* species and for **psilocybin**.

Findeisen, Lotte. 1982. *Psilocybe serbica* Moser et Horak, ein blauender Kahlkopf. *Berichte des Botanischen Vereins zu Hamburg*, no. 4: 27–29.

Kriegelsteiner, G. J. 1984. Studien zum *Psilocybe-cyanescens*-Komplex in Europa. *Beiträge zur Kenntnis der Pilze in Mitteleuropa* 1:61–94.

———. 1986. Studien zum *Psilocybe-cyanescens-callosasemilanceata*-Komplex in Europa. *Beiträge zur Kenntnis der Pilze in Mitteleuropa* 2:57–72.

Liggenstorfer, Roger. 1996. Oink, der kosmische Kicherfaktor. In *María Sabina—Botin der heiligen Pilze*, ed. Roger Liggenstorfer and Christian Rätsch, 179–82. Solothurn: Nachtschatten Verlag.

Moser, M., and E. Horak. 1968. *Psilocybe serbica* spec. nov., eine neue Psilocybin und Psilocin bildende Art aus Serbien. *Zeitschrift für Pilzkunde* 34:137–44.

Müller, G. K., and Jochen Gartz. 1986. *Psilocybe cyanescens*—eine weitere halluzinogene Kahlkopfart in der DDR. *Mykologisches Mitteilungsblatt* 29:33–35.

Tjallingii-Beukers, D. 1976. Een blauwwordernde *Psilocybe* (*Psilocybe cyanescens* Wakefield 1946). *Coolia* 19:38–43.

"Studying the surface structure of the mushroom, this wave-shaped cap with the skin of an elephant, I know that this must be an Oink. Of course, an Oink, it could not be anything else. Explaining to the 'tour guide' what these mushrooms are really called, he agrees that this could only be an Oink! Since that time, this name for *Psilocybe cyanescens*, which was conceived of in a bemushroomed state, has been spreading like mycelia."

ROGER LIGGENSTORFER
OINK, DER KOSMISCHE KICHERFAKTOR [OINK, THE COSMIC LAUGH FACTOR] (1996, 181)

Psilocybe mexicana Heim

Mexican Magic Mushroom, Teonanacatl

Family
Agaricaceae: Strophariaceae; Stropharioideae Tribe, Mexicanae Section

Forms and Varieties
The following forms have been named (all nom. nud.!; Ott 1996):

Psilocybe mexicana f. *angulata-olivacea* Heim et Cailleux

Psilocybe mexicana f. *distorta-intermedia* Heim et Cailleux

Psilocybe mexicana f. *galericulata-convexa* Heim et Cailleux

Psilocybe mexicana f. *galericulata-viscosa* Heim et Cailleux

Psilocybe mexicana f. *grandis-gibbosa* Heim et Cailleux

Psilocybe mexicana f. *navicula-viscosa* Heim et Cailleux

Psilocybe mexicana f. *reflexa-conica* Heim et Cailleux

The variety *Psilocybe mexicana* var. *longispora* Heim, first proposed by Roger Heim, is now regarded as a synonym for *Psilocybe aztecorum* Heim (cf. **Psilocybe spp.**).

Folk Names
Alcalde, amokia, a-mo-kid (Chinantec), amokya, angelito (Spanish, "little angel"), a-ni, atkat, atka:t (Mixe), chamaquillo (Spanish, "little boy"), cui-ya-jo-to-ki (Chatino), di-chi-to-nize (Mazatec), di-nize, hongo sagrado, kong, kongk (Mixe), konk, little bird, mbey-san (Zapotec), mexikanischer kahlkopf, mexikanischer zauberpilz, Mexican liberty cap, Mexican magic mushroom, nashwinmush (Mixe, "earth mushroom" or "world mushroom"), ndi-shi-tjo-ni-se (Mazatec), nize (Mazatec, "little bird"), pajarito (Spanish, "little bird"), piitpa, pi-tpa (Mixe), pi-tpi, pi:tpi, piule de churis,[373] teonanacatl, teonanácatl (Aztec), teo-tlaquilnanácatl (Nahuatl)

History
Ethnohistorical sources indicate that *teonanacatl*, the "divine mushroom" or "flesh of the gods" (*Psilocybe mexicana* and other species of the genus *Psilocybe*), was being ritually consumed and used in religious ceremonies in Mexico before the arrival of the Spanish. During the colonial period, the indigenous use of the mushroom was forbidden and brutally suppressed by the Spanish Inquisition. In spite of this, the mushroom cult has survived underground even into the present day. The psychoactive use of *Psilocybe mexicana* in Indian shamanism was rediscovered at the end of the 1930s. In the late 1950s, it was found that the Mixe Indians of Coatlan, Oaxaca, also used *Psilocybe mexicana* for shamanic purposes (Hoogshagen 1959).

Psilocybe mexicana was the first mushroom in which Albert Hofmann discovered the LSD-like substances **psilocybin** and psilocin (Heim et al. 1958; Hofmann 1958, 1959).

Distribution
Psilocybe mexicana is found exclusively in Mexico (Michoacán, Morelos, Jalisco, Oaxaca, Puebla, Xalapa, Veracruz) and Guatemala (Stamets 1996, 129 f.**). It grows in subtropical forests at altitudes of 1,000 to 1,800 meters and is found in the vicinity of liquidambar (*Liquidambar styraciflua* L.), oak (*Quercus* spp.), alder (*Alnus* spp.), and plane (*Platanus lindeniana* Mart. et Gall.) trees.

The historic culture of the Mexican magic mushroom (*Psilocybe mexicana*), photographed in the laboratory by Albert Hofmann. (Photograph: Brack)

373 See *Rhynchosia pyramidalis* for more on this name, its etymology, and its meaning.

Nanacatl, the Mexican magic mushroom (*Psilocybe mexicana* and *Psilocybe aztecorum*), shown here in the Florentine Codex, the Aztec-language chronicle of Sahagun (Paso y Troncoso edition).

374 In his first three books, Castaneda reported that entheogenic mushrooms were being used as a "little smoke." This claim has been severely criticized by ethnomycologists. No one would believe him, although many people tried smoking the mushrooms—but without success (cf. Clare 1988**; Siegel 1988*). Gordon Wasson apparently wrote a very critical letter to Castaneda; Castaneda later stated his own position in an interview: "After the publication of *The Teachings of Don Juan*, I received a thoughtful letter from Gordon Wasson, the founder of ethnomycology, who has studied the human use of mushrooms and fungi. . . . And then Dr. Wasson [*sic*] asked me to clarify certain aspects of Don Juan's use of psychotropic mushrooms. I happily sent him several pages of my field notes that were relevant to his area of interest, after which we met two more times. From that point on, he spoke of me as an 'honest and serious young man' or in a similar manner." (From "Carlos Castaneda: Portrait eines Zauberers—1994," in *Energy* 6/94: 4–7, 14–18)

375 Mixe soothsayers also use other entheogens besides the mushrooms: *ma"zhun paHk* (*Turbina corymbosa*/*Ipomoea violacea*), *ama'y mushtak* (*Datura stramonium*), *po:b piH* (*Brugmansia* x *candida*), and *piH* (*Tagetes erecta*; see *Tagetes* spp.) (Lipp 1990, 151 f.).

Appearance

Albert Hofmann noted that *Psilocybe mexicana* can be recognized by its cap, which resembles a typical Mexican sombrero. It can grow up to 10 cm tall and has small bell- or hat-shaped caps (3 to 5 cm in diameter). In Mexico, it fruits from June to September.

Psilocybe mexicana can be confused with poisonous muscarinic *Inocybe* mushrooms, e.g., *I. geophylla* (Sow. ex Fr.) Kummer (cf. *Inocybe* spp.). It is also very similar to the species *Psilocybe semilanceata* and *Psilocybe pelliculosa* and is often confused with them (cf. *Psilocybe* spp.).

Psilocybe mexicana can be easily grown on a substrate of a *Lolium* species (cf. *Lolium temulentum*).

The fruiting bodies can be consumed fresh or dried. Mexican Indians often ingest the mushroom together with **honey** or chocolate (cf. *Theobroma cacao*). In former times, the mushrooms were steeped in pulque and drunk (cf. *Agave* spp.).

Carlos Castaneda's claim (1973*, 1975*) that these mushrooms are dried and then smoked for their psychedelic effects has been the subject of considerable controversy and is highly doubtful (Clare 1988**; Siegel 1981, 330*).[374]

Ritual Use

The literature from the colonial period contains numerous texts that provide information concerning the mushrooms, their effects, and their ritual and/or medicinal uses. The Florentine Codex, an early colonial chronicle by the Franciscan missionary Fra Bernardino de Sahagun, written in Aztec, reports:

> Nanacatl. They are called teonanacatl, "flesh of the gods." They grow in the flatlands, in grass. The head is small and round, the stem long and thin. It is bitter and scratches, it burns in the throat. It makes one foolish; it confuses one, it distresses one. It is a remedy for fever, for gout. Only two, three are eaten. It makes sad, depressed, distressed; it makes one run away, become afraid, hide. He who eats many of them sees many things that scare him and that make him happy. He runs away, hangs himself, throws himself from a cliff, screams, is afraid. It is eaten with honey. I eat mushrooms; I take mushrooms. It is said of one who is haughty, impertinent, vain that: "He has bemushroomed himself." (Sahagun, Florentine Codex 11.7*)

Another Aztec text by Sahagun provides a rudimentary description of the mushroom ritual:

> The first thing that one ate at such meetings was a black mushroom that they called nanacatl. It has inebriating effects, produces visions,

and incites to obscene acts. They already take the thing early on the morning of the festival day and drink cacao before they arise. They eat the mushrooms with honey. When they have made themselves drunk with these, they begin to become excited. Some sing, others cry, others sit in their rooms as if they were deep in sorrow. They have visions in which they see themselves die, and this hurts them bitterly. Others see scenes in which they are attacked by wild animals and believe that they are being eaten up. Some have beautiful dreams in which they believe they are very rich and possess many slaves. But others have quite embarrassing dreams: they have the feeling of being caught while committing adultery or of being wicked forgers or thieves who are now facing their punishment. They all have their visions. When the inebriation that the mushrooms produce is over, they speak of that which they have dreamed, and one tells the other about his visions. (Sahagun 9)

In his *Historia de las Indias de Nueva Espane*, the missionary Diego Duran noted several times that mushrooms were ingested at festivals and were "drunk like wine" (= pulque; cf. *Agave* spp.), although they were mixed with chocolate (cf. *Theobroma cacao*) (Wasson 1980**). Today, *Psilocybe mexicana* is still used by shamans of the Mazatec, Mixe, Zapotec, and Cuitlatec in a manner that is quite similar to its pre-Spanish use (Hoogshagen 1959; Lipp 1990; Miller 1966; Ravicz 1961).

Among the Mixe, the most important deity is the Earth Mother Naaxwin or Na:shwin (literally, "the eye of the earth"). The earth is regarded as the source of wisdom; the Earth Mother is omniscient and can see the past, present, and future. Since the mushrooms grow from the earth, they are regarded as extremely wise and full of knowledge. The Mixe originally believed that the mushrooms were born from the bones of primordial shamans and prophets. According to a different version of that belief, which was influenced by Christianity, the mushrooms are regarded as soothsayers because they are equated with the blood of Christ. It is said that as Jesus hung on the cross, blood flowed from his heart to the ground. Numerous flowers and edible mushrooms grew from this blood. Finally, the magic mushrooms emerged and supplanted the plants that had previously turned green. For this reason they are called *na:shwin mux*, "mushrooms of Mother Earth" (Lipp 1991, 187*). Accordingly, the messages of the mushrooms are known as the "voice of the Earth" (Mayer 1975, 604**).

Magic mushrooms are used primarily in ritual contexts by the mostly female shamans. They are eaten for divinatory purposes.[375] They are used to recognize the causes of diseases, to predict the death and loss of family members, to localize lost objects, to uncover thieves and magicians, and to search for answers to familial problems. The mushrooms can also help in finding hidden treasures, discovering ruins, and experiencing ritual knowledge. The mushrooms normally speak Mixe, although they occasionally speak Zapotec as well (Lipp 1991, 187). Among the Mixe, the old pre-Spanish *tonalámatl* divination calendar is still in use. Some shamans use the mushrooms in conjunction with the calendar divination (Miller 1966).

Magic mushrooms[376] can be harvested only in summer. It is said that they grow only on sacred ground. When a person encounters a mushroom, he should offer it three candles, kneel before it, and speak the following prayer:

Tum'Uh. Thou who art the queen of all there is and who was placed here as the healer of all sicknesses. I say to you that I will carry you from this place to heal the sickness I have in my house, for you were named as a great being of the earth. Forgive this molestation, for I am carrying you to the place where the sick person is, so that you make clear what the suffering is that has come to pass. I respect you. You are the master of all and you reveal all to the sick. (Lipp 1991, 189*)

The collected mushrooms are carefully placed on the house altar or stored in the village church for three days. Incense (copal; cf. **incense**) is offered to them. They are consumed either fresh or sun-dried. For three days before ingesting the mushrooms, a person must remain abstinent from sex and refrain from eating poultry, pork, eggs, and vegetables. It also is forbidden to drink **alcohol** (mescal; cf. *Agave* spp.) or to use other drugs or medicines. During this time, a person should also refrain from agricultural activities. On the morning of the fourth day, he or she takes a bath and eats a light breakfast (of only foods made from maize; cf. *Zea mays*). He or she fasts for the rest of the day. On the morning after the session, the person must eat a large quantity of chili peppers (*Capsicum* spp.); he or she should abstain from meat and alcohol for the following month.

The mushrooms are always eaten in pairs and also dosed in pairs: three pair for children, seven pair for women, nine pair for men (Lipp 1991, 189 f.*). Sometimes only the caps are eaten (Mayer 1975, 604**). In each session, a person should eat mushrooms of just a single species, because mixing the species can result in unpleasant, i.e., threatening, visions. Two eggs are laid next to the mushrooms before they are eaten. At the same time, "copal"[377] (incense; the resin of the palm *Acrocomia mexicana* Karw., from which **palm wine** is

also obtained) is burned and a candle is lit. A prayer is offered to the mushrooms before they are eaten:

Thou who art blessed. I am now going to swallow you so that you heal me of the illness I have. Please give me the knowledge I need, thou, who knows all of what I need and of what I have, of my problem. I ask of you the favor that you only tell me and divine what I need to know but do nothing bad to me. I do not wish an evil heart and wickedness. I only wish to know of my problems and illness and other things that you can do for me. But I ask you, please do not frighten me, do not show me evil things but only tell all. This is for the person with a pure heart. You can do many things, and I ask you to do them for me. I now ask your forgiveness for being in my stomach this night. (Lipp 1991, 190*)

After the mushrooms are swallowed whole with water, one should be quiet. It is said that the mushrooms, like all other magical plants, do not like noise and will not speak if they feel disturbed. Normally the person who has eaten the mushrooms is accompanied by one or two friends or family members. They should pay attention to the things that the "bemushroomed" person says and fumigate him or her with copal smoke if problems arise. The visions that appear are shaped by culture. First one sees snakes and jaguars. After these have disappeared, the sun and the moon appear as a boy and girl, the children of the wind and the Earth Mother. Often, the "bemushroomed" person only hears voices that give advice, provide diagnoses, or ask about the reasons for ingesting the mushrooms. In these visions, most people obtain profound insights into their state of health and learn how they may become healthy and complete (Rätsch 1996).

Artifacts
Some pre-Columbian Aztec manuscript illustrations (*tlacuilolli*) depict scenes that are usually interpreted as mushroom rituals (Caso 1963). In particular, several pages in a manuscript that has become known as the *Codex Vindobonensis Mexicanus I* give the impression of an entheogenic ceremony. A number of figures, each holding two mushrooms (pairs!) in their hand, are shown sitting in a ritual arrangement (cf. Rätsch 1988a, 174 f.*; Wasson 1983**).

In the comic *Azteken* [Aztecs], by Andreas (1992), Mexican magic mushrooms are ingested in order to solve problems.

Medicinal Use
The Aztecs used *teonanacatl* as a medicine to treat fever and gout (Rätsch 1991a, 267*). Today, Mexican magic mushrooms are still used as a remedy

"No one said that anyone had partaken of any kind of wine, not to mention had become inebriated; they spoke only of mushrooms of the forest that they ate raw and that made them become happy and beside themselves, but not wine. The only thing they spoke of was the tremendous quantities of chocolate that were drunk during these festivals."

THE MISSIONARY DIEGO DURAN, REPORTING ON THE FESTIVITIES SURROUNDING THE CORONATION OF THE AZTEC EMPEROR AHUITZOTL (CA. 1496)

376 Like the Mazatec, the Mixe use the species *Psilocybe caerulescens*, *Psilocybe cordispora*, *Psilocybe hoogshagenii*, and *Psilocybe yungensis* (cf. **Psilocybe spp.**).

377 In modern Mexico, the word *copal* is used to refer to all aromatic incenses. The true copal is obtained from a tropical deciduous tree (*Protium copal* [Schl. et Cham.] Engl. [Burseraceae]; cf. **Bursera bipinnata**.

Psilocybin

Psilocin

"I saw Mexican scenes. Although I attempted to see things in the normal manner, this did not happen, everything was simply Mexican. I had the feeling that the physician who was supervising this experiment was a Mexican priest who had come to remove my heart. I thought that I was imagining this only because I knew that these mushrooms were from Mexico."

ALBERT HOFMANN, REPORTING ON A SELF-EXPERIMENT[378]

378 From an interview in Maurizio Venturini and Claudio Vannini, *Zur Geschichte der Halluzinogenforschung: Schwerpunkt Schweiz (Teil I: 1938–1965)*, licentiate thesis at the Philosophischen Fakultät I of the Universität Zürich, 1995, page 101.

for a number of illnesses, including stomach and intestinal disturbances, migraines and headaches, swelling, broken bones, epileptic seizures, and acute and chronic ailments. Most Indians who are not shamans avoid the mushrooms and ingest them in low (subpsychedelic) dosages only in cases of illness. They fear confrontation with the mushrooms, which speak to them and can reveal things that may be unpleasant (Lipp 1991, 187 f.*).

Constituents

In his "classic" analysis, Albert Hofmann found concentrations of 0.25% **psilocybin** and 0.15% psilocin by dry weight (Heim and Hofmann 1958; Hofmann 1960b). Fresh mushrooms contain more psilocin (Stamets 1996, 130**).

Effects

Timothy Leary (1920–1996), a consciousness researcher and former Harvard professor, took his first "trip" with the magic mushrooms of Mexico. He encountered the "divine mushroom" while he was staying in Cuernavaca, Mexico, in 1960. This event did not simply change his life and thought but also led to profound changes in society and in the ways that science looks at the world. One of the first effects that Leary noted during his historical experience was that famous "cosmic laughter," especially about one's self and science:

> I laughed about my daily pomposity, that narrow-minded arrogance of the scientist, the impertinence of the rational, the glib naïveté of words in contrast to the unadulterated, rich, eternally changing panoramas that flooded my brain. . . . I surrendered to the joy, as mystics have done for centuries when they looked through the veil and discovered that the world—as plastic as it might appear—is actually a small stage setting that is constructed by our mind. There was a flood of possibilities out there (in here?), other realities, an infinite arrangement of programs for other scenarios of the future. (Leary 1986, 33 f.*)

At the peak of his mushroom encounter, Leary had a profound and mystical experience of the world:

> Then I was gone, off to the department for fantastic optical effects. The palaces of the Nile, the temples of the Bedouins, shimmering jewels, finely woven silk garments that breathed colors, of muso-emerald glistening mosaics, Burmese rubies, sapphires from Ceylon. There were jewel-encrusted snakes, Moorish reptiles whose tongues flickered, turned and reeled down into the drain in the center of my retina. Next there followed a journey through evolution that everyone who

travels through their brain is guaranteed to experience. I slipped down the channel of recapitulation into the ancient production rooms of the midbrain: snake time, fish time, big-jungle-palm time, green time of the ferns.

> Peacefully I observed how the first ocean creature crawled onto the land. I lay next to it, the sand crunching under my neck, then it fled back into the deep green ocean. Hello, I am the first living creature. (34)

This initiatory experience permanently changed the academically trained scientist:

> The trip lasted somewhat more than four hours. Like most everyone who has the veil lifted, I came back a changed person. . . . In four hours at the pool in Cuernavaca, I learned more about the mind, the brain, and its structures than I was able to during the preceding fifteen years as a busy psychologist. (35)

As they did for so many people before and after him, the mushrooms taught Leary something important (or would it be more appropriate to say that he discovered it through the mushrooms?):

> I experienced that the brain is an unused biocomputer that contains billions of unexplored neurons. I learned that normal waking consciousness is a drop in the ocean of intelligence. That the brain can be programmed anew. The knowledge about the functioning of our brain is the most pressing scientific task of our time. I was beside myself with enthusiasm, convinced that we had found the key we had been looking for. (35)

For many scientists and psychonauts, the Mexican mushrooms—and later the European and North American species as well—became keys to other worlds, realities, and conceptions of life that opened the normally locked doors to an expanded, visionary, or cosmic consciousness. Since that time, many have passed through these "doors of perception" and allowed the overwhelming adventures of consciousness to flow into their thoughts and actions, their scientific theories and philosophical treatises.

Literature

See also the entries for the other *Psilocybe* species and for **psilocybin**.

Andreas. 1992. *Azteken*. Hamburg: Carlsen.

Caso, Alfonso. 1963. Representaciones de hongos en los códices. *Estudios de Cultura Náhuatl* 4:27–38.

Heim, Roger, Arthur Brack, Hans Kobel, Albert Hofmann, and Roger Cailleux. 1958. Déterminisme de la formation des carpophores et des sclérotes dans la culture du "*Psilocybe mexicana*" Heim, agaric hallucinogène du Mexique, et mise en évidence de la psilocybine et dans de la psilocine. *Comptes Rendus des Séances de l'Académie des Sciences* (Paris) 246: 1346–51.

Hofmann, Albert. 1958. La psilocybine sur une auto-expérience avec le *Psilocybe mexicana* Heim. In *Les champignons hallucinogènes du Mexique*, by Roger Heim and R. Gordon Wasson, 278–80**. Paris: Muséum National d'Histoire Naturelle.

———. 1959. Chemical aspects of psilocybin, the psychotropic principle from the Mexican fungus, *Psilocybe mexicana* Heim. In *Neuropsychophar-macology*, ed. Bradley et al., 446–48. Amsterdam: Elsevier.

———. 1960a. Das Geheimnis der mexikanischen Zauberpilze gelüftet. *Radio + Fernsehen, Schweizer Radiozeitung* (1960), no. 4: 8–9.

———. 1960b. Die psychotropen Wirkstoffe der mexikanischen Zauberpilze. *Chimia* 14:309–18.

———. 1960c. Die psychotropen Wirkstoffe der mexikanischen Zauberpilze. *Verhandlungen der Naturforschenden Gesellschaft in Basel* 71:239–56.

———. 1961. Die Erforschung der mexikanischen Zauberpilze. *Schweizerische Zeitschrift für Pilzkunde* 1:1–10.

———. 1964. Die Erforschung der mexikanischen Zauberpilze und das Problem ihrer Wirkstoffe. *Basler Stadtbuch* (1964): 141–56.

———. 1969. Investigaciones sobre los hongos alucinogenos mexicanos y la importancia que tienen en la medicina sus substancias activas. *Artes de México* 16 (124): 23–31.

Hoogshagen, Searle. 1959. Notes on the sacred (narcotic) mushrooms from Coatlán, Oaxaca, Mexico. *Oklahoma Anthopological Society Bulletin* 7:71–74.

Lipp, Frank J. 1990. Mixe concepts and uses of entheogenic mushrooms. In *The sacred mushroom seeker: Essays for R. Gordon Wasson*, ed. Thomas J. Riedlinger, 151–59. Portland, Ore.: Dioscorides Press.

Miller, Walter S. 1956. *Cuentos Mixes*. Introduction by Alfonso Villa Rojas. Mexico City: INI.

———. 1966. El tonalamtl mixe y los hongos sagrados. In *Homenaje a Roberto J. Weitlaner*, 349–57. Mexico: UNAM.

Ott, Jonathan. 1996. *Psilocybe mexicana* Heim. Unpublished computer file. (Cited 1998.)

Rätsch, Christian. 1996. Das Pilzritual der Mixe. In *María Sabina—Botin der heiligen Pilze*, ed. Roger Liggenstorfer and C. Rätsch, 139–41. Solothurn: Nachtschatten Verlag.

Ravicz, Robert. 1961. La mixteca en et estudio comparativo del hongo alucinante. *Anales del Instituto Nacional de Antropología e Historia* 13 (1960): 73–92.

An ithyphallic shaman with magical staff and mushroom, above which a soul bird flies. (Petroglyph in the Petrified Forest National Park, Arizona)

Psilocybe semilanceata (Fries) Quélet

Liberty Cap

Family
Agaricaceae: Strophariaceae; Stropharioideae Tribe, Semilanceatae = Cyanescens Section

Forms and Subspecies
There are color variants that have white, brown, and bluish caps (cf. Dähncke 1993, 614 f.**):

Psilocybe semilanceata (Fr.) Quélet f.—brown caps
Psilocybe semilanceata (Fr.) Quélet var. *semilanceata*
Psilocybe semilanceata var. *caerulescens* (Cke.) Sacc.—the margin of the cap and the base of the stem turn blue

Two varieties that have been described are now regarded as constituting a separate species (*Psilocybe strictipes* Singer et Smith; cf. **Psilocybe spp.**):

Psilocybe semilanceata var. *obtusa* Bon.
Psilocybe semilanceata var. *microspora* Singer

Synonyms
Agaricus glutinosus Curtis
Agaricus semilanceatus Fr.
Coprinarius semilanceatus Fr.
Geophila semilanceata Quél.
Panaeolus semilanceatus (Fr.) Lge.
Psilocybe semilanceata Fr.
Psilocybe semilanceata (Fr.: Secretan) Kummer

Folk Names
Blue leg, halluzipilz, kaalkopje (Dutch), kleiner prinz, kleines zwergenmützchen, lanzenförmiger düngerling, liberty cap, magic mushroom, meditationspilz, narrenschwamm, paddlestool, pilzli, pixie cap, psilo, psilocybinpilz, puntig kaalkopje, sandy sagerose, schwammerl, spitzkegeliger kahlkopf, traumpilz, witch cap, zauberpilz, zuckerpuppe von der wasserkuppe, zwergenhut, zwergenmützchen

History
The liberty cap is the most common psychedelic mushroom in Italy. It is thought to have grown there for ten thousand to twelve thousand years. A number of late Neolithic rock paintings in northern Italy (Monte Bego, Valcamonica) depict

The liberty cap (*Psilocybe semilanceata*) can be recognized by its cap, which resembles a dwarf's hat. (From Winkler, *2000 Pilze selber bestimmen*, 1996**)

Dried fruiting bodies of the liberty cap (*Psilocybe semilanceata*); this amount corresponds to approximately one psychedelic dose.

mushrooms in shamanic contexts (Ripinsky-Naxon 1993, 154*).

Toward the end of the Middle Ages, women in Spain who were accused of being witches apparently used liberty caps as a visionary inebriant (Fericgla 1996*).

After numerous Mexican species (*Psilocybe spp.*) were collected, described, and chemically analyzed (Heim and Wasson 1958; Wasson 1961**), the Swiss chemist Albert Hofmann, who had first discovered the active substances **psilocybin** and psilocin in *Psilocybe mexicana*, was informed by an inhabitant of the Alps that there were mushrooms in the Alps that had effects like those from Mexico. He said that he had often eaten these mushrooms and had a very precise knowledge of their effects. Hofmann then received a sample of the mushrooms, which belonged to the species *Psilocybe semilanceata*, and was able to detect psilocybin in them as well. The original work was published in a small scientific journal (Hofmann et al. 1963). Nevertheless, knowledge of the indigenous magic mushroom, which Alpine nomads had apparently once consumed in ritual contexts (Golowin 1991*), spread very quickly (Gartz 1986):

> Today, it can be said that *Psilocybe semilanceata* is **the** psychotropic mushroom of Europe with regard to its distribution, study, and use. (Gartz 1993, 23**)

In Switzerland, collecting and ingesting *Psilocybe semilanceata* has been an established tradition for over twenty years (Venturini and Vanini 1995, 38 f.*). In Germany, the collecting and eating of *Psilocybe semilanceata* began somewhat later. The ritual consumption of the indigenous magic mushroom was first described by Linder (1981).

The mushroom can be collected from the end of August until the middle of January (Leistenfels n.d., 22**). It is either eaten while fresh or dried and stored. Occasionally, dried mushrooms are powdered and then ingested together with fruit juices, cacao, or chocolate (cf. *Theobroma cacao*). A handful of fresh mushrooms (30 to 40 g) or 2 to 3 g of dried mushrooms is regarded as a high dosage. In Switzerland, chocolate and 0.5 g of powdered liberty caps are mixed together to make

cookies ("the genuine Swiss chocolate"). It is possible that the mushroom was once used as a **beer** additive.

Distribution

Liberty caps do not occur in Europe and the Americas alone but instead are now found throughout the world (even in Australia) (Gartz 1986; Jokiranta et al. 1984). Although the mushroom is now cosmopolitan, it has not yet been found in Mexico. This has led to the suggestion that **Psilocybe mexicana** may be simply a subspecies or variety of *Psilocybe semilanceata*, which is regarded as the most common and most widely distributed of all members of the genus *Psilocybe*.

Liberty caps prefer to grow in meadows with old dung deposits and in grassy, nutrient-rich locales (pastures). It is equally at home in the flatlands of northern Germany, the meadows of the low mountains of central Germany, and the pastures of the Alpine countries. It has not yet been found in forests. It thus appears to be an anthropophilous species that has spread through human activity. Its fruiting bodies mature by late summer and early autumn. In the United States, temperate zones in the Northwest (Oregon, Washington) are especially good areas to collect the mushroom (Weil 1975**); similarly good European areas include the Swiss Alps, Valcamonica (Fest and Aliotta 1990**), the Rhône valley, and Wales (cf. Remann 1989, 247, 262*).

Appearance

The cap (1 to 2 cm in diameter) is campanulate and acutely umbonate and often has a distinct papilla; it usually feels moist or slimy to the touch. The pellicle is easily removed. The narrow lamallae are olive to red-brown in color, while the spores are dark brown or purple-brown.

Liberty caps can be confused with some muscarinic fungi, e.g., *Inocybe geophylla* (Sow. ex Fr.) Kummer (cf. **Inocybe spp.**). It is very similar to the closely related species **Psilocybe mexicana** and *Psilocybe pelliculosa* (cf. **Psilocybe spp.**) and is often confused with them.

Ritual Use

Alpine nomads are said to have referred to *Psilocybe semilanceata* as the "dream mushroom" and to have used it traditionally as a psychoactive substance. Unfortunately, we have no details concerning this use (Golowin 1991, 63*).

The first description of a modern European mushroom cult appeared in 1981 in the catalogue *Rausch und Realität—Drogen im Kulturvergleich* [Inebriation and Reality—Drugs in Cross-Cultural Perspective], which was published on the occasion of a museum exhibition of the same name:

> I [was able to take part] in a solstice ceremony in Canton Bern from Dec. 21–23, 1979, in

An English postcard depicting the native magic mushroom (*Psilocybe semilanceata*).

which the little mushroom that I identified as *Psilocybe semilanceata* was used in the context of a cult that had existed for some seven years and that featured complicated sweat-bath rituals, prayers, pipe ceremonies (without psychoactive substances), mandatory fasting, fumigations, offerings, and music in a specially furnished room with a central altar. All of the people who were present (5 women and 6 men) were required to strictly avoid all drugs, including alcohol, sexual contact, eating meat, and "bad thoughts" for four days prior to and following the ceremony. During the meeting, strict fasting was required, although only two of the men ate twenty mushrooms each on the second evening following rituals of purification. This use clearly served an oracular function for the group. It was supported by intensive, hours-long drumming of all the participants. . . . For the group, whose ideology was shaped by a broad "pagan"-Christian-Buddhist-Hindu syncretism . . . the mushroom does not appear to fall into the category of "drugs" but was said to be a component of "the primordial religion." (Linder 1981, 727)

Most participants in these modern mushroom rituals regard them as a form of "psychedelic shamanism" (DeKorne 1994*) related to American Indian rituals. However, they feel that they are practicing a revived primeval form of an entheogenic ritual that is accessible to all people as a result of the "collective unconscious" or the "morphogenetic field."

Harvesting of the mushrooms is often preceded by a prayer to the Earth Goddess Gaia or an ominous mushroom deity; offerings such as small crystals may be placed on the borders of the pasture or meadow to give thanks to the spirit of the mushroom. The collector should eat the first two mushrooms so that he or she will then be able to recognize the right species and find it everywhere. Some mushroom collectors told me that the mushroom can be found only when the collector is in a "good mood"; people who are "weird or in a bad mood" will not be able to locate them. In Switzerland, a ritual method of collecting the mushrooms was observed in the 1970s:

The mushrooms are collected on native meadows while adhering to avoidance taboos, they are purified with sage smoke, dried, and stored in vessels that are also ritually purified. They are regarded as a "gift from God" or "from nature," and only a limited number are collected. The largest specimens in each group are allowed to remain as the "leaders," which are thanked and offered flour and other gifts. Songs are sung for the mushrooms so that the mushrooms that are hidden in the grass will show themselves. (Linder 1981, 727)

In Italy, a small booklet that provides easy-to-understand instructions for the sacramental use of the indigenous *Psilocybe semilanceata* has been published in an enormously large press run (Pagani 1993).

In modern rituals in central Europe, the mushroom is used in groups consisting of six to twenty participants. The rituals take place outdoors in especially beautiful settings or power places, in special rooms, or in tepees. Sweat baths, meditations, walks in the forest, and other such preparations precede the communal ingestion. Most rituals begin in the evening and last for about four hours, corresponding to the duration of the mushroom's effects.

By far the most important ritual device is the talking stick, which is modeled after that of the North American peyote cult (cf. *Lophophora williamsii*) and plays an extremely important role in the rituals. It is a stafflike object that may be individually chosen and modified. After he or she receives the talking stick, each person is urged to communicate to the circle (by singing, speaking, silence, rattling). The other participants remain silent and offer their complete attention to the person holding the stick. The stick is passed around during all three phases of the ritual (always in a clockwise direction). Because each person can hold the stick as long as he or she pleases, individuals are able to ensure the ritual attention they desire. By using the talking stick, the communal circle, and the power of the mushroom, collective visions, ecstatic laughter, and individual insights occur. After the circle has ended, an evening meal is shared.

The next morning, the group meets for a communal breakfast. Most of the participants are hungry and have a healthy appetite. Much joking and laughter accompanies the meal, and sometimes people describe and discuss the dreams of the preceding night. After breakfast, the participants gather in the ritual room and assume their places in the circle. Sage (*Artemisia* spp.) is burned as an **incense**. The follow-up or integration of the experience is actually the most important part of the ritual. It is said that visions are valuable only when they have been communicated. The visions are to be taken seriously, for they provide guidelines for the future. The talking stick is then passed around for the final time, and the participants are urged to speak of their experiences. Often, it is only at this time that they realize that their question has been answered and that the mushroom has taught them much. Strong emotional reactions and expressions of thanks are common. Almost all of the participants leave the ritual with deep feelings of gratitude and with the

"I . . . have come to celebrating with mushrooms the presence of all the Gods and Goddesses. Gods and Goddesses are my name for forces that move me. . . . Teo-Nanacatl was yesterday's visitor, and having entered this abode of soul called Jeannine Parvati, is now written as first in my heart on the list of plant allies for healing."

JEANNINE PARVATI
HYGIEIA
(1978, 85*)

An English group known as the Magic Mushroom Band took its name from *Psilocybe semilanceata*, which is common in Great Britain. The free-floating eye in the cosmos depicted on the cover illustrates the visionary look into the infinite universe provided by magic mushrooms. It appears that the lettering can be deciphered only while under the influence of the mushroom (the album is titled *RU Spaced Out 2*). (CD cover 1993, Magick Eye Records)

Top: *Psilocybe semilanceata* in the heart chakra. (Embroidery on a T-shirt from Kathmandu, Nepal)

Bottom: A computer-generated picture of liberty caps in the modern design of the techno/rave culture. (Detail from a postcard, ca. 1997)

sense that they have been initiated into the mysteries of the entheogenic mushroom and have recognized their own place in the cosmos. A ritual leader once noted:

> You can always depend on the mushrooms. No matter what happened while the effects were being felt, whether people completely flipped out, experienced pure horror, were shamanically dismembered, or fell into paranoia, in the end the mushroom illuminates and disseminates its legendary healing power. (In Rätsch 1996**)

Artifacts

For several years, naturalistic wooden models of *Psilocybe semilanceata* have been produced in Switzerland on lathes.

In the United States and England, it is especially common to see T-shirts decorated with mushrooms of the genus *Psilocybe*, and often with *Psilocybe semilanceata*. One T-shirt even features a guide to identifying *Psilocybe semilanceata* (Rätsch 1996**).

Liberty caps also appear in the iconography of the techno and rave cultures. They are sometimes found on the album covers, posters, and admission tickets of psychedelic rock groups (e.g., Grateful Dead, *Aoxomoxoa*, 1971; The Golden Dawn, *Power Plant*, 1988; Merrell Fankhauser and H.M.S. Bounty, *Things Goin' Round In My Mind*, 1985; Phish). There is even a Magic Mushroom Band, which took its name from the native fungus.

In the Jugendstil and art deco movements, numerous lamps that look like naturalistic representations of the liberty cap were manufactured (Uecker 1992).

Constituents

Psilocybe semilanceata often contains high concentrations of **psilocybin**, some psilocin, and baeocystin. This species is one of the most potent *Psilocybe* species. According to an analysis by Gartz (1994, 19**), German specimens contained 0.97% **psilocybin**, no psilocin, and 0.33% baeocystin by dry weight. Mushrooms collected in the wild usually exhibit a higher concentration of psilocybin (up to 1.34% by dry weight has been measured) than is found in cultivated mushrooms. Total indole concentrations as high as 1.9% have been found (Gartz 1986). Swiss collections have been found to contain as much as 2.02% total alkaloids (Brenneisen and Borner 1988). In dried material, the active constituents can be preserved for quite some time:

> The length of time in which the psilocybin remains in the mushroom material is astonishing. A mushroom exsiccation from 1869 that was in a Finnish herbarium was found to still contain 0.014% psilocybin. On

the other hand, a sample from 1843 was found to not contain any more alkaloids. However, it is of course no longer possible to determine the manner in which drying was performed at that time. Temperatures above 50° C. effect the breakdown of psilocybin and its derivatives. In laboratory experiments, dried mushrooms and freeze-dried fruiting bodies were examined at room temperature. Here it should be pointed out that because of the porous structure of the freeze-dried mushrooms, a relatively rapid breakdown of the alkaloids will occur when they are stored for an extended period of time (months) at 20°C. For this reason, exsiccates that are produced for the analysis of natural products should be stored in a dry state at -10°C until extraction and chromatography are performed. (Gartz 1993, 31**)

Literature

See also the entries for the other *Psilocybe* species and for **psilocybin**.

Brenneisen, Rudolf, and Stefan Borner. 1988. The occurrence of tryptamine derivatives in *Psilocybe semilanceata*. *Zeitschrift für Naturforschung* 43c: 511–14.

Christansen, A. L., K. E. Rasmussen, and K. Høiland. 1981. The content of psilocybin in Norwegian *Psilocybe semilanceata*. *Planta Medica* 42:229–35.

Dawson, P. 1975. *A guide to the major psilocybin mushrooms of British Columbia* (Psilocybe semilanceata). Vancouver, B.C.: self-published.

Gartz, Jochen. 1986. Quantitative Bestimmung der Indolderivate von *Psilocybe semilanceata* (Fr.) Kumm. *Biochemie und Physiologie der Pflanzen* 181:117–24.

Hausner, Milan, and Marta Semerdzieva. 1991. "Acid Heads" and "Kahlköpfe" in Forschung und Therapie—Zum Stand der Psycholyse in der Tschechoslowakei. In *Jahrbuch des Europäischen Collegiums für Bewußtseinsstudien* (ECBS) (1991), 109–18. Berlin: VWB.

Hofmann, Albert, Roger Heim, and Hans Tscherter. 1963. Présence de la psilocybine dans une espèce européenne d'agaric, le *Psilocybe semilanceata* Fr. Note (*) de MM. In *Comptes rendus des séances de l'Académie des Sciences* (Paris) 257:10–12.

Jokiranta, J., et al. 1984. Psilocybin in Finnish *Psilocybe semilanceata*. *Planta Medica* 50:277–78.

Linder, Adrian. 1981. Kultischer Gebrauch psychoaktiver Pflanzen in Industriegesellschaften – kulturhistorische Interpretation. In *Rausch und Realität*, ed. G. Völger, 2:724–29. Cologne: Rautenstrauch-Joest-Museum für Völkerkunde.

Pagani, Silvio (pseudonym of a well-known mycologist). 1993. *Funghetti*. Turin: Nautilus.

Schwaiger, Saskia. 1994. Schwammerlrausch. *Profil*, no. 42 (Oct. 17, 1994): 88–89.

Schwester Krötenstuhl. 1992. Eine Reise im Herbst. *Integration* 2/3:129–30.

Stijve, T. 1984. *Psilocybe semilanceata* als hallucinogene paddestoll. *Coolia* 27:36–43.

Uecker, Wolf. 1992. *Licht-Kunst: Lampen des Art Nouveau und Art Deco.* Rastatt: Neff.

Young, R. E., R. Milroy, S. Hutchison, and C. M. Kesson. 1982. The rising price of mushrooms. *The Lancet* 8265 (1): 213–15.

Psilocybe spp.

Psilocybin Mushrooms

Family

Agaricaceae: Strophariaceae; Stropharioideae Tribe

The genus *Psilocybe* is divided into eighteen sections and includes at least 150 species (Brenneisen and Stalder 1994; Guzmán 1983 and 1995). The genus is represented in all areas of the world. Most species are rather small and have thin stems and more or less conic caps. In all species, the spore print ranges from purplish to violet to dark violet/blackish. Most species live on dung or prefer nutrient-rich soils with old dung deposits.

Some species have attained great cultural significance as traditional entheogens (see *Psilocybe cubensis, Psilocybe mexicana*). Some species play a role primarily in the modern Western mushroom cult (*Psilocybe azurescens, Psilocybe cyanescens, Psilocybe semilanceata*). In contrast, most psychoactive species are not used in traditional contexts. New species, some of which represent very potent entheogens, are being discovered and described at an increasing rate (Gartz 1995; Gartz et al. 1994, 1995; Guzmán 1995; Guzmán et al. 1993, n.d.; Marcano et al. 1994).

The following species of the genus *Psilocybe* contain **psilocybin**. Most also contain psilocin, and some contain baeocystin (Allen et al. 1992**). Only a few of the potent species have any ethnopharmacological significance.

Psilocybe acutipilea (Speg.)

Psilocybe aeruginosa (Curtis: Fr.) Noordeloos [syn. *Stropharia aeruginosa* (Curtis: Fries) Quélet]—verdigris cap

Psilocybe angustispora Smith

Psilocybe argentipes Yokoyama

Psilocybe armadii Guzmán et Pollock

Psilocybe atrobrunnea (Lasch) Gillet
This small species is found in central European marshlands and grows on peat and peat moss (*Sphagnum*).

Psilocybe aucklandii Guzmán, King et Bandala

Psilocybe augustipleurocystidiata Guzmán

Psilocybe australiana Guzmán et Watling—Australian psilocybe

Psilocybe aztecorum Heim emend. Guzmán—Aztec psilocybe
This mushroom occurs in at least two varieties:

Psilocybe aztecorum var. *aztecorum* (Guz.) Guzmán
Psilocybe aztecorum var. *bonetti* (Guz.) Guzmán

It is used in Mexico in the same manner as *Psilocybe mexicana* (Guzmán 1994, 1462**).

Folk names: Nahua, apipiltzin, tejuinti, teunanácatl, teyhuinti nanácatl; Spanish, dormilón ("long sleeper"), niño de las aguas ("child of the waters"), niños ("boys")

Psilocybe baeocystis Singer et Smith emend. Guzmán

Psilocybe banderiliensis Guzmán

Psilocybe barrerae Cifuentes et Guzmán

Psilocybe bohemica Sebek [syn. *Hypholoma coprinifacies* (Roll.) Herink]—Bohemian psilocybe
Now regarded as a synonym for *Psilocybe cyanescens*, as the species is confused with *Psilocybe mairei* (Sebek 1983).

Psilocybe brasiliensis Guzmán—Brazilian psilocybe

Psilocybe brunneocystidia Guzmán

Top: *Psilocybe baeocystis*, common in the Pacific Northwest, contains psilocybin and the psychoactive constituent it is named for, baeocystin (approximately 0.1%). (Photograph: Paul Stamets)

Bottom: This small and strongly blueing mushroom from the genus *Psilocybe* is found near Astoria, Oregon. According to reports of local mushroom collectors, its effects are primarily aphrodisiac.

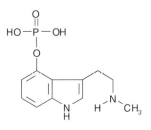

Baeocystin

The Mixtec god Seven Flowers is depicted here holding entheogenic mushrooms in his hand. In cross section, these look exactly like *Psilocybe caerulescens*. He is listening to the music of the wind god Nine Wind (= Ehecatl), a manifestation of Quetzalcoatl, the feathered serpent. (Codex Vindobonensis, p. 24)

A wooden shamanic staff in the form of a *Psilocybe* species, found in a prehistoric context of the Hopewell culture (Mound City, Ohio), approximately 40 cm long. (From Devereux)

Psilocybe caerulea (Kriesel) Noordeloos [syn. *Stropharia caerulea* Kriesel, *Stropharia cyanea* (Bolt. ex Secr.) Tuomikoski]—blue psilocybe

Psilocybe caeruleoannulata Sing.: Guzmán

Psilocybe caerulescens Murrill [syn. *Stropharia caerulescens*]—derrumbe
This mushroom occurs in several varieties:

Psilocybe caerulescens var. *albida* Heim
Psilocybe caerulescens var. *caerulescens* Heim—derrumbe (Spanish, "abyss"), di-chi-te-ki-sho (Mazatec), cañadas (Spanish, "canyons"), razón-bei, teotlaquilnanácatl (modern Nahuatl, "the sacred mushroom that paints in colors")
Psilocybe caerulescens vat. *mazatecorum* Heim—ntixitho ntikixo (Mazatec, "abyss")
Psilocybe caerulescens var. *nigripes* Heim—cui-ya'-jo'-o-su (Chatina, "mushroom of great understanding"), derrumbe negro (Spanish, "the black abyss"), kong (Mixe, "lord/ruler"), ko:ng-mus (Mixe, "ruler mushroom"), ndi-ki-sho (Mazatec), ndi-shi-tjo-ki-sho (Mazatec, "the small, dear things that shoot out")
Psilocybe caerulescens var. *ombrophila* (Heim) Guzmán
This species is used in Mexico (Oaxaca) in the same manner as *Psilocybe mexicana* (= *P. mixaeensis*) (Guzmán 1994, 1441**), known in the Mixe language as *atkat*.

Psilocybe caerulipes (Peck) Saccardo

Psilocybe carbonaria Singer

Psilocybe collybioides Singer et Smith

Psilocybe columbiana Guzmán—Colombian psilocybe

Psilocybe coprinifacies (Roll.) Pouz. [syn. *Hypholoma coprinifacies* (Roll.) Herink]

Psilocybe coprophila (Bulliard ex Fries) Kummer [syn. *Psilocybe mutans* McKnight]—dung psilocybe
This mushroom is very similar to *Psilocybe subcoprophila* (Britz.) Sacc., although the latter has larger spores. It grows on dung throughout most of the year in central Europe.

Psilocybe cordispora Heim—atka:t (Mixe/Coatlán), dulces clavitos del señor (Spanish, "sweet cloves of the lord"), enedi:z (Mixe, "thunder teeth"), pitpi (Mixe), pi:tpimus (Mixe)
This species is used in Mexico in the same manner as *Psilocybe mexicana*.

Psilocybe crobula (Fries) Kühner et Romagnesi [syn. *Geophila crobula* (Fr.) Kühner et Romagnesi,

Psilocybe inquilina var. *crobula* (Fr.) Holland]
Sometimes regarded as a synonym for *Psilocybe inquilina*.

Psilocybe cyanofibrillosa Stamets et Guzmán [syn. *Psilocybe rhododendronensis* Stamets nom. prov.]

Psilocybe dumontii Sing.: Guzmán

Psilocybe eucalypta Guzmán et Watling—eucalyptus psilocybe

Psilocybe fagicola Heim et Callieux
This mushroom occurs in at least two varieties:

Psilocybe fagicola Heim et Callieux var. *fagicola* Guzmán
Psilocybe fagicola Heim et Callieux var. *mesocystidiata* Guzmán

Both are known as *señores principales* (Spanish, "the principal lords")

Psilocybe farinacea Rick.

Psilocybe fimetaria (Orton) Watling [syn. *Psilocybe fimetaria* (Orton) Singer, *Psilocybe caesioannulata* Singer, *Stropharia fimetaria* Orton]

Psilocybe fuliginosa (Murr.) Smith

Psilocybe furtadoana Guzmán

Psilocybe galindii Guzmán

Psilocybe gastoni Sing. (?)—di-nizé-te-aya (Mazatec)

Psilocybe goniospora (B. et Br.) Singer

Psilocybe graveolens Peck

Psilocybe heimii Guzmán [syn. *Psilocybe hoogshagenii* Heim var. *hooghagenii*]—atkadmus (Mixe, "mushroom judge"), atka:t, cihuatsinsintle (Nahuatl), Heim's psilocybe, los chamaquitos (Spanish, "little boys"), los niños (Spanish, "children"), pajarito de monte (Spanish, "little bird of the forest")

Psilocybe herrerae Guzmán

Psilocybe hoogshagenii Heim sensu lato [syn. *Psilocybe caerulipes* var. *gastonii* Singer, *Psilocybe zapotecorum* Heim sensu Singer, *Psilocybe semperviva* Heim et Callieux]

Psilocybe hoogshagenii Heim var. *convexa* Guzmán [syn. *P. semperviva* Heim et Callieux]

Psilocybe hoogshagenii Heim var. *hoogshagenii* Guzmán

Psilocybe hoogshagenii Heim [var. *hoogshagenii*] [syn. *Psilocybe gastoni* Sing. (?)]—atkadmus (Mixe, "mushroom judge"), atka:t (Mixe), cihuatsinsintle (Nahuatl), di-nizé-te-aya (Mazatec), los chamaquitos (Spanish, "little boys"), los niños (Spanish, "children"), teotla-quilnanácatl (modern Nahuatl, "the sacred mushroom that paints in colors")
This species is used in Mexico in the same manner as *Psilocybe mexicana.*

Psilocybe inconsicua Guzmán et Horak

Psilocybe inquilina (Fries ex Fries) Bresadola [syn. *Psilocybe ecbola* (Fries) Singer]
This species, which is common in Europe and grows on small branches, decaying wood, and sawdust, is best recognized by its pellicle, which is sticky and easy to remove. A variety of this species, *Psilocybe inquilina* var. *crobula* Fr. [syn. *Psilocybe crobula* (Fr.) Lange ex Sing.], also occurs in central Europe.

Psilocybe jacobsii Guzmán—Jacob's psilocybe

Psilocybe kashmeriensis Abraham—Kashmiri psilocybe

Psilocybe kumaenorum Heim

Psilocybe liniformans Guzmán et Bas.
There are two or more varieties:

Psilocybe liniformans Guzmán et Bas. var. *americana* Guzmán et Stamets
Psilocybe liniformans Guzmán et Bas. var. *liniformans*

Psilocybe lonchopharus (Berk. et Br.) Horak: Guzmán

Psilocybe luteonitens (Peck) Saccardo [syn. *Stropharia umbonatescens* (Peck) Saccardo]

Psilocybe magnivelaris (Peck apud Haariman) Noordeloos [syn. *Psilocybe percevalii* (Berkeley et Broome) Orton, *Stropharia percevalii* (Berkeley et Broome) Saccardo, *Stropharia magnivelaris* Peck apud Harriman]

Psilocybe mairei Singer [syn. *Psilocybe maire* Singer sensu Guzmán, *Hypholoma cyanescens* Maire]
This species is now regarded as a synonym for *Psilocybe cyanescens.*

Psilocybe makarorae Johnston et Buchanan

Psilocybe mammillata (Murrill) Smith

Psilocybe merdaria (Fries) Ricken—dung psilocybe

This little mushroom, which thrives on dung, has a cap that is 1 to 4 cm in diameter. In central Europe, it grows from spring until fall.

Psilocybe moellerii Guzmán [syn. *Stropharia merdaria* Fr. sensu Rea, *Stropharia merdaria* var. *macrospora* (Moller) Singer]

Psilocybe montana (Fries) Quélet [syn. *Psilocybe atrorufa* (Schaeffer ex Fries) Quélet]
This small mushroom feels dry to the touch. It lives in sandy soil, among short mosses, and at elevations well above the tree line. In Europe, it is found in the Alps.

Psilocybe muliericula Singer et Smith [syn. *Psilocybe wassonii* Heim, *Psilocybe mexicana* var. *brevispora* Heim]—cihuatsinsintle (modern Nahuatl), mujercita, mujercitas (Spanish, "girls"), nano-catsintli (modern Nahuatl), netochhuatata (modern Nahuatl), ne-to-chutáta (Matlazinca, "[dear] little sacred lord"), niñas (Spanish, "daughters"), niño (Spanish, "son"), quauhtan-nanácatl (?) (modern Nahuatl), siwatsitsíntli (Nahua, "little girls")
This species is used in Mexico in the same manner as *Psilocybe mexicana.*

Psilocybe natalensis Gartz, Reid, Smith et Eicker—Natal psilocybe
The discovery of this potently psychoactive African species in Natal is of great ethno-pharmacological significance:

> It is the first blueing and entheogenic species that has been demonstrated to exist in this land. The comparatively large and entirely white mushrooms grow in pastures in the summer, although not directly on dung. . . . Germinating the spores on agar yielded a fast growing mycelium that also turned blue. (Gartz et al. 1995, 29)

The fruiting bodies contain up to 0.6% psilocybin, up to 0.04% baeocystin, and up to 0.21% psilocin by dry weight (Gartz et al. 1995).

Psilocybe ochreata (Berk. et Br.) Horak

Psilocybe papuana Guzmán et Horak—Papuan psilocybe

Psilocybe pelliculosa (Smith) Singer et Smith [syn. *Psathyra pelliculosa* A.H. Smith]—liberty cap

Psilocybe physaloides (Bull. ex Merat) Quélet [syn. *Psilocybe caespitosa* Murrill]
Found throughout central Europe, where it lives on nutrient-rich soils.

Psilocybe pelliculosa, a North American species, is easily confused with *Psilocybe semilanceata* and can have the same effects. It contains, however, considerably less psilocybin. (Photograph: Paul Stamets)

"True overdoses with psilocybin mushrooms appear to be almost impossible. Psilocybe intoxications are usually considered harmless. Between 1978 and 1981, 318 cases of psilocybe intoxication were registered in England, but none of them were fatal. The amount ingested ranged between a few mushrooms and 900 to 1360 grams and did not normally correlate with the symptoms of intoxication."

RUDOLF BRENNEISEN AND ANNA-BARBARA STALDER
PSILOCYBE
(1994, 293)

"The contact with Christianity and with modern ideas has had little influence upon the reverence the mushroom ritual is accorded in Mexico. The mushroom ceremony lasts the entire night and often includes a healing ritual. Most of the celebration is accompanied by songs. Depending upon the species, between 2 and 30 mushrooms are eaten. These may be eaten fresh or crushed and drunk as a hot-water infusion. The choice of the species of mushroom depends upon the personal taste of the shaman, the purpose it is being used for, and what is seasonally available."

RUDOLF BRENNEISEN AND ANNA-BARBARA STALDER
PSILOCYBE
(1994, 293)

Like most species of *Psilocybe*, the dung psilocybe (*Psilocybe coprophila*) "loves" to live on dung, which it alchemically transforms into the so-called food of the gods. (From Winkler, *2000 Pilze selber bestimmen* [Identify 2000 Fungi Yourself], 1996)

Psilocybe pintonii Guzmán

Psilocybe pleurocystidiosa Guzmán

Psilocybe plutonia (B. et C.) Sacc.

Psilocybe pseudobullacea (Petch) Pegler

Psilocybe pseudocyanea (Desmazieres: Fries) Noordeloos [syn. *Stropharia pseudocyanea* (Desm.) Morgan, *Stropharia albocyanea* (Des.) Quélet]

Psilocybe quebecensis Ola'h et Heim—Quebec psilocybe

Psilocybe rzedowski Guzmán

Psilocybe samuensis Guzmán, Allen et Merlin—Samoan psilocybe

Psilocybe sanctorum Guzmán—sacred psilocybe

Psilocybe schultesii Guzmán et Pollock—Schultes's psilocybe

Psilocybe semiglobata (Batsch: Fries) Noordeloos [syn. *Stropharia semiglobata* (Fries) Quélet]

Psilocybe serbica Moser et Horak—Serbian psilocybe
Now regarded as a synonym for *Psilocybe cyanescens*.

Psilocybe silvatica (Peck) Singer et Smith

Psilocybe singeri Guzmán—Singer's psilocybe

Psilocybe squamosa (Persoon ex Fries) Orton

Psilocybe strictipes Singer et Smith [syn. *Psilocybe callosa* (Fries ex Fries) Quélet sensu auct., sensu Guzmán (1983), *Psilocybe semilanceata* var. *obtusa* Bon., *Psilocybe semilanceata* var. *microspora* Singer]

Psilocybe stuntzii Guzmán et Ott [syn. *Psilocybe pugetensis* Harris]—Stuntz's psilocybe

Psilocybe subaeruginascens Hohnel [syn. *Psilocybe aerugineomaculans* (Hohnel) Singer et Smith; *Psilocybe subaeruginosa* Cleland]
Two varieties have been described:

Psilocybe subaeruginascens Hohnel var. *septentrionalis* Guzmán
Psilocybe subaeruginascens Hohnel var. *subaeruginascens*

Psilocybe subcaerulipes Hongo

Psilocybe subcubensis Guzmán (see *Psilocybe* (*Stropharia*) *cubensis*, soma)

Psilocybe subfimetaria Guzmán et Smith

Psilocybe subviscida (Peck) Kauffman

Psilocybe subyungensis Guzmán

Psilocybe tampanensis Guzmán

Psilocybe tasmaniana Guzmán et Watling—Tasmanian psilocybe

Psilocybe thrausta (Schulzer ex Kalchbremer) Orton [syn. *Psilocybe squamosa* var. *thrausta* (Schulzer ex Kalchbremer) Guzmán, *Stropharia thrausta* (Schulzer et Kalchbremer) Bon]

Psilocybe uruguayensis Sing.: Guzmán—Uruguayan psilocybe

Psilocybe uzpanapensis Guzmán

Psilocybe venenata (Imai) Imazecki et Hongo [syn. *Psilocybe fasciata* Hongo, *Stropharia caerulescens* Imai, *Stropharia venenata* Imai]
In Japan, this psilocybin mushroom is known as *waraitakemodoki*, "false laughing mushroom" (Wasson 1973, 14**), or *shibiretake*, "narcotic mushroom" (cf. *Gymnopilus* spp.).

Psilocybe veraecrucis Guzmán et Perez-Ortiz—Veracruz psilocybe

Psilocybe washingtonensis Smith—Washington psilocybe

Psilocybe wassonii Heim [probably synonymous with *Psilocybe muliericula*; cf. Ott 1993, 312*]—siwatsitsíntli (Nahua, "little girls"), mujercitas (Spanish, "girls")

Psilocybe wassoniorum Guzmán et Pollock—Wasson's psilocybe

Psilocybe weilii Guzmán, Stamets et Tapia—Weil's psilocybe

Psilocybe weldenii Guzmán—Welden's psilocybe

Psilocybe wrightii Guzmán—Wright's psilocybe

Psilocybe xalapensis Guzmán et Lopez

Psilocybe yungensis Singer et Smith [syn. *Psilocybe acutissima* Heim, *Psilocybe isauri* Singer]—atkad (Mixe, "judge"), derrumbe negro (Spanish, "the black abyss"), di-nezé-ta-a-ya (Mazatec), di-shi-to-ta-a-ya (Mazatec), hongo genio (Spanish, "genious mushroom"), pajarito

de monte (Spanish, "little bird of the forest"), piitpa (Mixe), si-shi-tjo-leta ja (Mazatec)
This species is used in Mexico in the same manner as *Psilocybe mexicana*.

Psilocybe zapotecorum Heim emend. Guzmán [syn. *Psilocybe bolivari, P. candidipes* Singer et Smith, *P. zapotecorum* forma *elongata*]—crown of thorns

This mushroom is known in Mexico as *hongo de la corona de cristo, badaoo, piule de barda, hongo santo*, et cetera, and is used in the same manner as *Psilocybe mexicana* (Guzmán 1994, 1450**).

Folk names: Zapotec, badao zoo, badoo, bei, be-meeche, beya-zoo, beneechi, mbey san, njte-jé, patao-zoo, paya-zoo, peacho, pea-zoo; Chatina, cui-ya-jo-otnu ("the great mushroom saint"); Mazatec, di-nizé-ta-a-ya, nche-je; Spanish, corona de cristo ("crown of Christ"), derrumbe de agua ("abyss of the water"), derrumbe negro ("the black abyss"), hongos de la razón ("mushrooms of reason"), piule de barda ("inebriating plant of the crown of thorns" [cf. **Rhynchosia pyramidalis**]), razón guiol ("the guiding reason"), razón viejo ("old reason")

Literature

See also the entries for the other *Psilocybe* species and for **psilocybin**.

Beck, J. E., and D. V. Bordon. 1982. Psilocybian mushrooms. *The PharmChem Newsletter* 11 (1): 1–4.

Benjamin, C. 1979. Persistent psychiatric symptoms after eating psilocybin mushrooms. *British Medical Journal* 6174:1319–20.

Beug, Michael W., and Jeremy Bigwood. 1982. Psilocybin and psilocin levels in twenty species from seven genera of wild mushrooms in the Pacific Northwest, U.S.A. *Journal of Ethnopharmacology* 5 (3): 271–85.

Brenneisen, Rudolf, and Anna-Barbara Stalder. 1993. *Psilocybe*. In *Hagers Handbuch der Pharmazeutischen Praxis*, 5th ed., 287–95. Berlin: Springer Verlag.

Gartz, Jochen. 1986. Ethnopharmakologie und Entdeckungsgeschichte der halluzinogenen Wirkstoffe von europäischen Pilzen der Gattung *Psilocybe. Zeitschrift für ärztliche Fortbildung* 80:803–5.

———. 1995. Psychotrope Pilze in Ozeanien. *Curare* 18 (1): 95–101.

Gartz, Jochen, John W. Allen, and Mark D. Merlin. 1994. Ethnomycology, biochemistry, and cultivation of *Psilocybe samuiensis* Guzmán, Bandala and Allen, a new psychoactive fungus from Koh Samui, Thailand. *Journal of Ethnopharmacology* 43:73–80.

Gartz, Jochen, Derek A. Reid, Michael T. Smith, and Albert Eicker. 1995. *Psilocybe natalensis* sp. nov.—the first indigenous blueing member of the Agaricales of South Africa. *Integration* 6:29–32.

Guzmán, Gastón. 1978. Further investigations of the Mexican hallucinogenic mushrooms with descriptions of new taxa and critical observations on additional taxa. *Nova Hedwigia* 29:625–64.

———. 1983. *The genus Psilocybe*. Nova Hedwigia, no. 74. Vaduz, Liechtenstein: Beihefte. Supplemental volume to Nova Hedwigia.

———. 1995. Supplement to the monograph of the genus *Psilocybe*. *Taxonomic Monographs of Agaricales, Bibliotheca Mycologica* 159:91–141.

Guzmán, Gastón, Victor M. Bandala, and John W. Allen. 1993. A new blueing psilocybe from Thailand. *Mycotaxon* 46:155–60.

Guzmán, Gastón, Victor M. Bandala, and Chris C. King. 1993. Further observations on the genus *Psilocybe* from New Zealand. *Mycotaxon* 46:161–70.

Guzmán, Gastón, and Jonathan Ott. 1976. Description and chemical analysis of a new species of hallucinogenic *Psilocybe* from the Pacific Northwest. *Mycologia* 68 (6): 1261–67.

Guzmán, Gastón, Jonathan Ott, Jerry Boydston, and Steven H. Pollock. 1976. Psychotropic mycoflora of Washington, Idaho, Oregon, California and British Columbia. *Mycologia* 68 (6): 1267–72.

Guzmán, Gastón, Fidel Tapia, and Paul Stamets. n.d.. A new blueing *Psilocybe* from U.S.A. *Mycotaxon* 65 (Oct–Dec.) 191–96.

Høiland, K. 1978. The genus *Psilocybe* in Norway. *Norwegian Journal of Botany* 25:111–22.

Imai, S. 1932. On *Stropharia caerulescens*, a new species of poisonous toadstool. *Transactions of the Sapporo Natural History Society* 13 (3): 148–51.

Koike, Yutaka, Kohko Wada, Genjiro Kusano, Shigeo Nozoe, and Kazumasa Yokoyama. 1981. Isolation of psilocybin from *Psilocybe argentipes* and its determination in specimens of some mushrooms. *Journal of Natural Products* 44 (3): 362–65.

Marcano, V., A. Morales Méndez, F. Castellano, F. J. Salazar, and L. Martinez. 1994. Occurrence of psilocybin and psilocin in *Psilocybe pseudobullacea* (Petch) Pegler from the Venezuelan Andes. *Journal of Ethnopharmacology* 43:157–59.

Matsuda, I. 1960. Hallucination caused by *Psilocybe venenata* (Imai) Hongo. *Transactions of the Mycological Society of Japan* 2 (4): 16–17.

Sebek, Svatopluk. 1983. Lysohlávka ceská—*Psilocybe bohemica*. *Ceská Mykologie* 37:177–81.

Semerdzieva, Marta, and F. Nerud. 1973. Halluzinogene Pilze in der Tschechoslowakei. *Ceská Mykologie* 27:42–47.

Semerdzieva, Marta, and M. Wurst. 1986. Psychotrope Inhaltsstoffe zweier Psilocybearten (Kahlköpfe) aus der CSSR. *Mykologisches Mitteilungsblatt* 29:65–70.

"It should be mentioned that the rich fungal flora of Japan also includes a number of psychotropic mushroom species. Reports from the eleventh century already spoke of the renowned 'laughing mushroom,' and cases of unintended ingestion are noted. In our century as well, many involuntary intoxications are known from Japan, resulting from confusing psychotropic species with edible mushrooms. In addition to dung mushrooms, *Stropharia venenata* Imai, a close relative of *Psilocybe cubensis* that can also be classified in the genus *Psilocybe*, also grows there. Other mushrooms that contain the active constituent psilocybin are *Psilocybe argentipes* Yokoyama, *Psilocybe subcaerulipes* Hongo, and *Psilocybe subaeruginascens* Höhnel . . . the demarcations between the species are the subject of debate. To date, however, there are no definitive reports about a possible subcultural use of such species in Japan."

JOCHEN GARTZ
PSYCHOTROPE PILZE IN OZEANIEN
[PSYCHOTROPIC FUNGI IN OCEANIA]
(1995, 100)

"I was haunted by legends of ancient beers, obsessed with rumors of fanciful brews like the Guatemalan millet beer, spiked with psilocybin."

T. CORAGHESSAN BOYLE
"QUETZALCOATL LITE"
(1979, 170)

The cover of this CD of psychedelic/trance music features a Mexican mushroom stone in the middle of a computer-generated image of psychedelic perception. (Spirit Zone Records, 1996)

A CD of psychedelic/trance/techno music, under the sign of sacred psychoactive mushrooms. (EFA Medien, ca. 1997)

Singer, Rolf, and Alexander H. Smith. 1958a. Mycological investigation on teonanácatl, the Mexican hallucinogenic mushroom. Part II. A taxonomic monograph of psilocybe, Section Caerulescentes. *Mycologia* 50:262–303.

———. 1958b. New species of psilocybe. *Mycologia* 50:141–42.

Stijve, T., and T. W. Kuyper. 1985. Occurrence of psilocybin in various higher fungi from several European countries. *Planta Medica* 5:385–87.

Yokoyama, Kazumasa. 1973. Poisoning by hallucinogenic mushroom, *Psilocybe subcaerulipes* Hongo. *Transactions of the Mycological Society of Japan* 14:317–20.

———. 1976. A new hallucinogenic mushroom, *Psilocybe argentipes* K. Yokoyama sp. nov. from Japan. *Transactions of the Mycological Society of Japan* 17:349–54.

Discography of "Bemushroomed" Music

Allman Brothers Band, *Where It All Begins* (Sony Music, 1994)

Awkana, *Earth's Call* (Wergo, 1993)—the cover features a Huichol yarn painting with a glowing mushroom

The Big F, *Is* (Chrysalis, 1993)—features a photograph of laboratory-raised *Psilocybe cubensis*

Braindub, *In Your Brain* (Sun Records, 1995)—a techno-mushroom-trip

Deee-Lite, *Dewdrops in the Garden* (Elektra, 1994)—the band is shown sitting in a garden with an oversized *Psilocybe cubensis* and fly agaric mushrooms

Merrill Fankenhauser & H.M.S. Bounty, *Things Goin' Round My Mind* (1968)

The Golden Dawn, *Power Plant* (Independent Artists, 1967)

Harald Grosskopf, *World of Quetzal* (CMS Music, 1992)—musical version of an Aztec myth in which the god Quetzalcoatl ingests the magic mushroom

Hans Hass, Jr., *Magic Mushroom—Strong* (Aquarius Records, 1996)

Holy Mushroom (Efa Records, 1997)—a psychedelic trance sampler

Ironic Beat, *Move on Groove On* (Rough Trade Records, 1995)

Jefferson Airplane, *Surrealistic Pillow* (RCA Records, 1967)

Dr. Timothy Leary, *Turn On, Tune In, Drop Out* (The Original Motion Picture Soundtrack, 1967)

Magic Mushroom Band, *RU Spaced Out 2* (Magic Eye Records, 1993)

Ministry, *Filth Pig* (Warner Bros. Records, 1996)

Mushroom, *Early One Morning* (1973)

Mushroom Trail, *My Medicine* (LSD/A&M Records, 1993)—opening the cover reveals textbook illustrations of a fly agaric and *Psilocybe semilanceata*

Mark Nauseef & Dave Philipson, *Venus Square Mars* (M•A Recordings, 1995)

Nevermore, *The Politics of Ecstasy* (Century Media Records, 1996)—mushrooms are praised as a sacrament

Phunk Junkeez (Naked Language Records, 1992)—the cover art includes psychedelic mushrooms

Porno For Pyros, *Good God's Urge* (Warner Bros. Records, 1996)—the lyrics refer to Balinese mushrooms

Robbie Robertson & The Red Road Ensemble, *Music for the Native Americans* (Capitol Records, 1994)—"We ate the sacred mushroom / And waded in the water / Howling like coyotes / At the naked moon"

Sacred Mushroom, *same* (Parallax, 1969)

Shamen, *Boss Drum* (Rough Trade, 1992)

Shaw & Blades, *Hallucinations* (Warner Bros. Records, 1995)—the cover shows the musicians sitting in a field of psychoactive mushrooms

Space Time Continuum, *Alien Dreamtime* (Caroline Records, 1993)

Space Tribe, *Sonic Mandala* (Spirit Zone Records, Hamburg, 1996)

Stereo MC's, *Connected* (Island Records, 1992)—the cover depicts several psychoactive mushrooms

The Tassili Players, *Outer Space* (Universal Egg Records, 1996)

Tiamat, *Wildhoney* (Magic Arts, 1994)—"Honey tea, psilocybe larvae / Honeymoon, silver spoon / Psilocybe tea"

Tribu, *IN Mixkoakali* (Cademac Records, 1996)—includes a song titled "Teonanakatl (Hongo Divino)"

Rick Wakeman, *Journey to the Centre of the Earth* (A& M Records, 1974)

Yo La Tengo, *May I Sing With Me* (Slang 017/EFA, 1992)—includes "Mushroom Cloud of His"

Zuvuya & Terence McKenna, *Dream Matrix Telemetry* (Delirium Records, 1993)

Zuvuya & Terence McKenna, *Shamania* (Delirium Records, 1994)

"Bemushroomed" Literature

Boyle, T. Coraghessan. 1979. *Descent of Man*. New York: Penguin Books. Includes the psilocybin-inspired story "Quetzalcoatl Lite."

Bradley, Marion Zimmer. 1982. *The mists of Avalon*. New York: Alfred A. Knopf. In this international best seller, psychedelic mushrooms are said to produce a different state of consciousness.

Braem, Harald. 1994. *Der Herr des Feuers: Roman eines Schamanen*. Munich: Piper. See **Amanita muscaria**.

Carroll, Lewis. 1946. *Alice in Wonderland*. New York: Grosset & Dunlap. See Carmichael 1996.**

Geerken, Hatmut. 1988. *mappa*. Spenge: Klaus Kramm. A literary mushroom trip; see the review by Martin Hanslmeier in *Integration* 2/3 (1992): 137–40.

———. 1992. fliegen pilze? merkungen und anmerkungen zum schamanismus in sibirien und andechs. *Integration* 2/3:109–14. According to the author's own statement, the text was written while he was under the influence of fly agaric.

Huxley, Aldous. 1962. *Island*. New York: Harper & Row. In his novel *Island*, Huxley helped psychedelic mushrooms gain literary renown as "*moksha*-medicine,"[379] the "reality revealer," and the "truth-and-beauty pill" (p. 157). He wrote the following about the value of using mushrooms once a year in a ritual context: "The fact remains that the experience can open one's eyes and make one blessed and transform one's whole life" (p. 160). For Huxley, this ritual "mycomysticism" was a way to use the power of the mushrooms for a positive purpose: "And all that the *moksha*-medicine can do is to give you a succession of beatific glimpses, an hour or two, every now and then, of enlightening and liberating grace. It remains for you to decide whether you'll co-operate with the grace and take those opportunities" (p. 197).

Lloyd, John Uri. 1895. *Etidorpha or the end of the Earth: The strange history of a mysterious being and the account of a remarkable journey*. Cincinnati: self-published. The mycologist John Uri Lloyd (1849–1916), after whom the journal *Lloydia* is named, worked his experiences with English mushrooms into this fantasy story, which had a considerable influence on *Alice in Wonderland*: "In any case, it seems clear that John Uri Lloyd's bizarre hollow-earth novel *Etidorpha* was for him a kind of labyrinth at whose center he wished to place the apotheosis he had personally experienced in his own peregrinations in the realm of gigantic fungi" (McKenna 1990, 169**).

Moers, Walter. 1992. *Schöner Leben mit dem Kleinen Arschloch: Sex, Drogen und Alkohol*. Frankfurt/M.: Eichborn Verlag. The author provides clear instructions for using mushrooms: "Mushrooms. The mildest results are achieved with champignons, the wildest with fly agarics. Psilocybin mushrooms, small, inconspicuous fellows that sometimes have magical effects, lie somewhere in between. But don't be afraid—if

LSD is the Porsche among consciousness-expanding drugs, then psilocybin is the bicycle. This means that the limits of perception are never opened so drastically that you believe that you have five lips—three lips is the maximum. If you have taken the proper dosage, you will soon feel a love for all forms of life that you have never felt before: for people, for animals, for plants, and especially for mushrooms" (p. 32 f.).

Remann, Micky. 1989. *SolarPerplexus: Achterbahn für die Neunziger*. Basel: Sphinx. Contains a chapter titled "Glücks-Pilze" [Happiness Mushrooms], a report on an experience with *Psilocybe*.

Shea, Robert. 1991. *Shaman*. New York: Ballantine Books. This best seller, which has appeared in numerous editions, begins with a mushroom trip. A young man destined to be a shaman goes to a cave and eats magic mushrooms. In this way, he is initiated into the secrets of the shamanic universe and obtains great visions that are full of both significance and consequences.

Wells, Herbert George. 1904. *The food of the gods and how it came to Earth*. London: Macmillan and Co. In this novel, scientists develop a drug, "the food of the gods," that causes enormous growth.

Widmer, Urs. 1995. Meine Jahre im Koka-Wald. *NZZ-Folio* (June 1995): 64–65. A satirical tale about Castaneda and his experiences with mushrooms.

"Bemushroomed" Comics

Mushrooms appear frequently in comics, usually as *Psilocybe* species or fly agarics (see **Amanita muscaria**). They also take on fantastic shapes. Sometimes mushroom trips are depicted (e.g., Jürgen Mick, *Träume*; Seyfried + Ziska, *Space Bastards*; Travel/Aoumri, *Das Volk der Wurzeln*; Jodorowsky/Arno, *Alef-Thau*). Some of the stories illustrate the use and effects of the mushrooms (Andreas, *Azteken*; Gilbert Shelton, *The Fabulous Furry Freak Brothers*; Howard Cruse, in *Dope Comix* no. 2, 1978). Magic mushrooms even appear in the *Abenteuern des kleinen Spirou* [Adventures of the Little Spirou]. The volume *Der geheimnisvolle Stern* from Hergé's Tim und Struppi series (1947) deals with strange mushrooms that are both red and white. Similar mushrooms appear in *Gazoline und der Rote Planet* (1991) by Jano. At the beginning of *Die Zeit der Asche* (1987), by Chevalier and Segur, the hero swallows a mushroom egg and goes on a fantastic trip. Mali and Werner, in *Der grüne Planet*, show the "path to the troll oracle": fly agaric mushrooms. In *Gon 3*, by Tanaka (1995), a small dinosaur eats psycho-active mushrooms and almost laughs itself to death.

"He knew about the mushrooms. Carlos also offered me some, and since I was starving, I ate quite a few as well. They tasted better than the coca, and I gained a little weight again. We often sat in the stubbly grass of an Indian savanna with a basket of hallucinogenic mushrooms, munching and looking up into the infinite starry sky. We both had visions. Carlos floated and conversed with dead magicians, while I always saw a woman as large as the heavens who was wearing a white doctor's coat and who longingly held out her arms toward me."

Urs Widmer
Meine Jahre im Koka-Wald [My Years in the Coca Forest]
(1995, 65)

On the cover of the album of the film music to *Turn On, Tune In, Drop Out* (1967), Dr. Timothy Leary (1920–1996) is depicted as a psychoactive mushroom. Leary's life changed dramatically after his first encounter with Mexican magic mushrooms in 1960.

379 In northern India, the meadow champignon (*Agaricus campestris* L.) is still known as *mokshai*, "the liberator" (Morgan 1995, 149**).

Purported Psychoactive Fungi

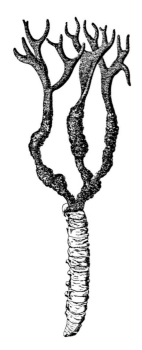

The caterpillar fungus *Cordyceps taylori*, shown adhering to a larva. (From Kerner and Oliver, *The Natural History of Plants*, 1897)

In Asia, dried larvae that live in symbiosis with the fungus *Cordyceps sinensis* are sought as aphrodisiacs and tonics. Some *Cordyceps* species are also attributed with psychoactive effects. (Dried larvae, collected in Tibet)

380 Since ancient times, the deer truffle (*Elaphomyces*) has been esteemed as an aphrodisiac and magical agent (Brøndegaard 1975).

A number of fungi are alleged to be psychoactive. Some species or genera are thought to contain the active constituent **psilocybin** (see the box below); the information contained in the literature, however, is contradictory. Other genera or species have a reputation of being psychoactive. Here again, the reports and other information contained in the literature are decidedly contradictory or incomplete.

In Australia, species from the genera *Aspergillus, Hypomyces, Hygrocybe,* and *Psathyrella* (Bock 1994*) may be psychoactive. The mold fungus *Aspergillus fumigatus* Fres. is known to produce **indole alkaloids** (Moreau 1982; Roth et al. 1990, 174**).

Genera and Species That May Contain Psilocybin

(from Ott 1993, 309, 313, 317*)
Agrocybe farinacea Hongo
Hygrocybe psittacina (Schaeffer ex Fr.) Wunsche—parrot mushroom
Hygrocybe psittacina var. *californica* Hesler et Smith
Hygrocybe psittacina var. *psittacina* Hesler et Smith
Mycena amicta (Fries) Quélet
Mycena cyanescens Velenovsky
Mycena cyanorrhiza Quélet
Naematoloma popperianum Singer
Panaeolina spp. (cf. **Panaeolus spp.**)
Pholiotina spp.
Stropharia *coronilla* (Bull. ex Fr.) Quél. (cf. Roth et al. 1990, 112**)

Cordyceps spp.

Clavicipitaceae
This parasitic genus of the Order Hypocreales (Class Ascomycetes) is found throughout the world (Jones 1997). In Mexico, two species are reputed to be psychoactive:

Cordyceps capitata (Holmskjold) Link— hombrecitos, soldaditos
Cordyceps ophioglossoides (Fries) Link— hombrecitos, club head fungus

Cordyceps ophioglossoides and *C. capitata* infest the fruiting bodies of truffles (*Elaphomyces* spp.),[380] which are found in oak and pine forests. In the area around Toluca (Mexico), these parasitic fungi were or are still used in nocturnal healing rituals. They are mixed with *Psilocybe muliericula* (cf. **Psilocybe spp.**), *Elaphomyces granulatus* Fr. [syn. *Elaphomyces cervinus* (Pers.)

Schroeter, *Hypogaeum cervinum* Pers.], and *Elaphomyces muricatus* f. *variegatus* and then powdered and ingested (Guzmán 1994b, 1446**). This fungus has been found to contain an **indole alkaloid**, the structure of which has not yet been clarified (Hobbs 1995, 86**; Ott 1993, 397*). *Cordyceps ophioglossoides* contains the antibiotic substance ophiocordine.

The two Mexican species are closely related to and hardly distinguishable from the species *Cordyceps sinensis* (Berk.) Sacc., which occurs in Tibet and southwestern China—especially in the highlands of Tibet—and is used for medicinal purposes. Rutting yaks sniff out the fungus and eat it as an aphrodisiac. The substance was praised as an aphrodisiac in the very oldest Chinese herbals, a reputation that continues to echo in the Chinatowns of our time (Davis 1983, 62–64). Attempts to cultivate the fungus, which is difficult to collect, have only recently proved successful. Extracts (alcoholic tinctures) are drunk as tonics and aphrodisiacs and are ingested even by athletes (not yet forbidden!) as doping agents. The effects are similar to those of ginseng (**Panax ginseng**). The tincture is also used as an antidote for opium overdoses and to treat opium addiction (Hobbs 1995, 82**). The coveted Chinese aphrodisiac *dong chong xia cao* is composed of the dried larvae of a moth (*Hepialus armoricanus* Oberthür) in which *Cordyceps sinensis* has grown. Six to 12 g is powdered and ingested as a tea (or mixed with other substances) in order to strengthen the body for erotic adventures and to overcome impotence. The drug contains cordycepinic acid, quinic acid, cordycepin, proteins, saturated and unsaturated fatty acids, D-mannitol, and vitamin B_{12}. Cordycepin has been demonstrated to have antibiotic effects (Paulus and Ding 1987, 114 f.*). Tinctures made from the powdered raw fungus are reported to have strong aphrodisiac effects as well as mild psychedelic effects that are primarily mood-improving in nature.

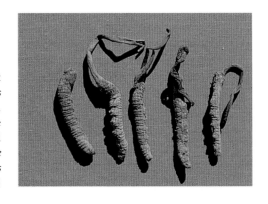

Dictyophora

Phallaceae (Stinkhorns)

The Chinantec use *Dictyophora indusiata* (Vent.: Pers.) Desv. [syn. *Dictyophora phalloides* Desv.], a species distributed widely throughout the tropics (Americas, Southeast Asia, Seychelles, et cetera), together with ***Psilocybe mexicana*** for ritual purposes. The shamans of Chinantla use the mushroom for divinatory purposes. They dry the mushroom and grind it. Two to three hours before the planned divination, they drink the powder with water (Guzmán 1994b, 1478**; McGuire 1982, 231**). The Lacandon Maya call this ritually important mushroom *u ba'ay äh och*, "the carrying net of the opossum." The Lacandon and other peoples of Mexico consider the opossum[381] a rascally trickster that is also associated with fertility and birth. The mushroom is regarded as inedible; its Spanish name is *vela de novia*, "bridal veil." In Thailand, this mushroom or another species of the same genus is used in magical rites. Chemical studies of the genus are lacking (Ott 1993, 399*).

Hydnum repandum L.: Fr.

Hydnaceae (Hedgehog Mushroom)

Some Swiss mushroom enthusiasts have reported that the hedgehog mushroom, which is quite common and well known in Jura, is psychoactive. The mushroom is generally regarded as edible (Dähncke 1993, 1036**). The red-yellow hedgehog mushroom (*Hydnum repandum* var. *rufescens* [Fr.] Barla [syn. *Hydnum refescens* Fries]) is also considered edible (Dähncke 1993, 1036**). It is interesting to note that in Japan, mushrooms of the genus *Hydnum* are known as *yamabushi-take*, "mushrooms of the mountain priests" (Imazeki 1973, 37**).

Gerronema

Tricholomataceae

An analysis by Gartz (1986) detected the presence of **psilocybin** in two East German species of this genus:

Gerronema fibula (Bull.: Fr.) Singer [syn. *Rickenella fibula* (Bull. ex Fr.) Raith., *Omphalia fibula* (Fr.) Kummer]
Gerronema solipes (Fr.) Singer [syn. *Gerronema swartzii* (Fries ex Fries) Kreisel]

A subsequent analysis found neither psilocybin nor any other active substances (Stijve and Kuyper 1988).

Psathyrella

Coprinaceae

This genus, in the Order Agaricales (Class Basidiomycetes), encompasses approximately one hundred species. It is closely related to ***Panaeolus***. **Psilocybin** was reportedly found in a Japanese sample of *Psathyrella candolleana* (Fr.) Maire [syn. *Agaricus violaceus-lamellatus* DC., *Agaricus appendiculatus* Bull.], which is found throughout the world. This mushroom is attributed with psychoactive effects (Gartz 1986; Ott 1993, 310*).

Literature

Brøndegaard, V. J. 1975. Die Hirschtrüffel. *Ethnomedizin* 3 (1/2): 169–76.

Davis, E. Wade. 1983. Notes on the ethnomycology of Boston's Chinatown. *Botanical Museum Leaflets* 29 (1): 59–67.

Furuya, Tsutomu, Masao Hirotani, and Masayuki Matsuzawa. 1983. N⁶-(2-hydroxyethyl)adenosine, a biologically active compound from cultured mycelia of *Cordyceps* and *Isaria* species. *Phytochemistry* 22:2509–12.

Gartz, Jochen. 1986. Nachweis von Tryptaminderivaten in Pilzen der Gattungen *Gerronema*, *Hygrocybe*, *Psathyrella* und *Inocybe*. *Biochemie und Physiologie der Pflanzen* 181:275–78.

Ginns, J. 1988. Typification of *Cordyceps canadensis* and *C. capitata,* and a new species, *C. longisegmentis*. *Mycologia* 80 (2): 217–22.

Jones, Kenneth. 1997. *Cordyceps: Tonic food of ancient China.* Rochester, Vt.: Healing Arts Press.

Moreau, Claude. 1982. Les mycotoxines neurotropes de l'*Aspergillus fumigatus,* une hypothèse sur le "pain maudit" de Pont-Saint-Esprit. *Bulletin, Société Mycologique Française* 98 (3): 261 ff.

Stijve, T., and Th. Kuyper. 1988. Absence of psilocybin in species of fungi previously reported to contain psilocybin and related tryptamine derivatives. *Persoonia* 13:463–65.

Left: The tropical stinkhorn *Dictyophora phalloides* has acquired a certain significance as a shamanic magical agent. (Photographed in the Seychelles)

Right: Some species of the genus *Psathyrella* are attributed with psychoactive properties. (A *Psathyrella* species, photographed in Washington state)

In one old English herbal, the fruiting bodies of a *Psathyrella* species were described as "deadly poisonous fungi." (Woodcut from Gerard, *The Herball or General History of Plants,* 1633*)

381 The opossum (*Didelphis marsupialis*) will eat fly agaric mushrooms (*Amanita muscaria*), but it learns to recognize their inebriating qualities and then avoids the mushroom (Ott 1993, 335*).

Lycoperdon

Puffballs

Family

Lycoperdaceae (Puffballs)

In 1962, during research into the Mexican magic mushrooms (cf. **Psilocybe mexicana**), it was discovered that Indians used two species of this genus that are attributed with psychoactive properties:

Lycoperdon marginatum Vitt. ex Morris et De Not. [syn. *Lycoperdon candidum* Persoon; cf. Guzmán 1994b, 1452**]

Lycoperdon mixtecorum Heim [syn. *Lycoperdon qudenii* Bottom., *Vascellum qudenii* (Bottom.) Ponce de Leon]—Mixtec puffball

Mixtec	*gi'-i-wa*	"fungus of first quality" (*L. mixtecorum*)
	gi'-i-sa-wa	"fungus of second quality" (*L. marginatum*)
Spanish	*bolita*	"little ball"
	bolita de lagartija	"little ball of the wall lizard"
	pata de perro	"dog's paw"
	pedo del diablo	"fart of the devil"
	hongo adivinador	"fungus of divination"
	jitamo real	"royal puffball"[382]
	jitamo real de venado	"royal puffball of the stag"

These fungi appear to have narcotic or dream-inducing effects. People under their influence are said to hear voices. One Indian reported, "I fell asleep for an hour or an hour and a half and the puffball spoke to me then, saying that I would become ill but would recover from the sickness" (in Schultes and Hofmann 1980, 41 f.*).

The Mixtec apparently use these fungi for divinatory purposes. In the region of the Tarahumara, the fungi are known as *kalamota* and are associated with witchcraft (Bye 1976*; 1979b, 39*). Additional research into their ritual use is needed. It is possible that they were used as peyote substitutes (see **Lophophora williamsii**).

Most puffballs are edible when young. The Mexican species *Lycoperdon umbrium* Pers. is known among the Tepehuan Indians as *ju'ba'pbich*

Some allegedly psychoactive Mexican species of puffball (*Lycoperdon* spp.) are used in Mazatec shamanism.

382 In Mexico, this name is also used to refer to *Ephedra* spp. as well as the fern *Pellaea cordata* J. Sm. (see **Polypodium spp.**) (Díaz 1979, 93*).

383 The literature occasionally claims that *Lycoperdon mixtecorum* and *L. marginatum* contain tryptamine derivatives (Benjamin 1995, 326**).

384 Under certain circumstances, this puffball can be mistaken for very young specimens of *Amanita muscaria*.

nakai, "excrement of the stars fungus," and it is a popular edible fungus when it is young (Gonzalez E. 1991, 170). It may be that only older specimens are psychoactive.

A number of species (including some from the related genera *Astraeus*, *Scleroderma*, and *Vascellum*) have been studied to determine whether they have any psychoactive effectiveness. No psychoactive effects have been observed, nor has any psychedelic component[383] been detected (Díaz 1979, 93*; Ott et al. 1975). However, it is possible that ingesting a puffball may result in subtle effects upon dreaming. Because most Indians have been trained to consciously experience their dreams and to analyze them for divinatory contents, it is entirely possible that non-Indian researchers have not been able to notice these effects.

In North America, puffballs possess a certain ritual and medicinal importance that may indicate a possible psychoactivity. The Blackfeet call them *ka-ka-toos*, "falling stars" or "dusty stars." They used the pleasantly scented species to make necklaces and decorated their tepees with representations of these fungi, considering them symbols of the life that arose from the earth. They sniffed the spores to treat nosebleeds (Johnston 1970, 303 f.*). They burned *Lycoperdon* species as magical **incense** to drive off spirits (Burk 1983, 55). The Cherokee used the common puffball (*Lycoperdon perlatum* Pers. ex Pers.)[384] as a remedy for treating chafed areas. The amino acid lycoperdic acid has been isolated from *Lycoperdon perlatum* (Rhugenda-Banga et al. 1979). The North American stump puffball (*Lycoperdon pyriforme* Schaeff. ex Pers.) is reputed to have sleep-inducing effects (Morgan 1995, 127**).

The following North American puffballs have also been attributed with psychoactive effects (Burk 1983, 60; Guzmán 1994, 1452**):

Lycoperdon pedicellatum Peck [syn. *Lycoperdon candidum* Pers. ex Pers.]

Scleroderma verrucosum Bull. ex Pers.—warted devil's snuffbox

Vascellum pratense (Pers.: Pers.) Kreisel [syn. *Lycoperdon hiemale*]

The Mapuche Indians of Chile refer to various *Lycoperdon* species as *pëtremquilquil*, "tobacco of Chuncho" or "powder of the devil" (Mösbach 1992, 52*). Chuncho or Chonchon is a bird with a human head that is sometimes regarded as an incarnation of a sorcerer, that is, a shaman (Brech 1985, 7). It is entirely possible that the fungus known as the tobacco of Chuncho, which has unfortunately not yet been identified, was smoked to effect the transformation into an animal and the ability to fly.

Reichel-Dolmatoff wrote that the *mamas* (shaman priests) of the Colombian Kogi used a number of psychoactive fungi, including a "bluish puffball," in ritual contexts. Unfortunately, the identity of this "bluish puffball" is unknown (Reichel-Dolmatoff 1977, 285*; 1996b, 167, 297*).

Before leaving his body for all time, the Buddha ate a piece of a fungus known as *pûtika* (*Scleroderma* sp. or *Lycoperdon pusillum*). *Pûtika* is said to have been a substitute for **soma** (Wasson 1983). The Santala, a Dravidian group of India, refer to *Lycoperdon pusillum* as "toad" and believe that the fungus possesses a soul (Morgan 1995, 148**). This belief may be a relic of an earlier psychoactive use.

Literature

Brech, Martha. 1985. *Kultrún—Zur Schamanentrommel der Mapuche*. Berlin: Peter Oberhofer.

Burk, William R. 1983. Puffball usages among North American Indians. *Journal of Ethnobiology* 3 (1): 55–62.

Gonzalez Elizondo, Martha. 1991. Ethnobotany of the southern Tepehuan of Durango, Mexico: I. Edible mushrooms. *Journal of Ethnobiology* 11 (2): 165–73.

Ott, Jonathan, Gastón Guzmán, J. Romano, and J. Luis Díaz. 1975. Nuevos datos sobre los supuestos Licoperdaceos psicotropicos y dos casos de intoxicación provocadas por hongos del genero *Scleroderma* en México. *Boletín de la Sociedad Mexicana de Micología* 9:67–76.

Rhugenda-Ganga, Nziraboba, André Welter, Joseph Jadott, and Jean Casimi. 1979. Un nouvel acide amide isole de *Lycoperdon perlatum*. *Phytochemistry* 18:482–84.

Wasson, R. Gordon. 1983. The last meal of the Buddha. *Botanical Museum Leaflets* 29 (3): 219–49.

> "One view that is widely held geographically but is inaccurate claims that the spore dust of *Lycoperdon* is harmful to the eyes and can even lead to blindness. This may be connected to the mistrust of the puffballs and of fungi in general and to the general fear of getting dust in one's eyes."
>
> V. J. Brøndegaard
> *Ethnobotanik* [Ethobotany]
> (1985, 241*)

Nonda

In Papua New Guinea, a number of mushrooms from the genera *Boletus*, *Russula*, and *Heimiella* are generally referred to by the name *nonda*. They reportedly are consumed by the Kuma tribe and produce a temporary state of "mushroom madness"[385] that is characterized by manic, wild behavior (Reay 1960). The mushrooms are eaten for culinary purposes throughout the year; it is only at a particular time (during moderate rainfall) that they are said to elicit these psychoactive effects. The fruits (*nong'n*) of a ***Pandanus*** **sp.** are said to be able to produce the same effects (Reay 1959, 188).

The Kuma once used the effects of nonda mushrooms to induce a wild and uninhibited aggressiveness before undertaking acts of warfare (Heim 1972, 170). Occasionally, the "mushroom madness" was also said to produce hallucinations of a terrible or pleasantly cheerful nature. Some Papuans describe the condition as a "bad trip" (Nelson 1970, 10).

This "mushroom madness" is strongly reminiscent of the "wild man" behavior that is so well known on Papua New Guinea (Newman 1964) as well as of the Balinese phenomenon of amok. Running amok, however, is not induced by any psychoactive substances but, rather, appears to be a traditional pattern of behavior within the culture (cf. Kertonegoro 1991, 61–102). In the same way, the "wild man" behavior, which is known as *long-long*, also appears to be learned and culturally patterned. It can appear without any pharmacological stimuli, at least among the Gururumba:

It begins when a person simply stops reacting to words. Because of this, he is also unable to understand anything, and his speech consists solely of inarticulate babbling or shrieking. Often, this condition leads to a violent shaking of the body, shortness of breath, and uncontrolled movements. In this state, the afflicted person may then take up his weapon and run through the village. (Rätsch and Probst 1985, 305 f.)

The French mycologist Roger Heim identified the following fungi as nonda (Heim 1972; Heim and Wasson 1965):

Boletus (**Boletaceae**)
This genus includes the delicious porcini mushroom or king bolete (*Boletus edulis* Bull.: Fr.), as well as *Boletus luridus* Schaeff.: Fr., which is toxic when taken in combination with **alcohol**, and the very toxic Satan's mushroom (*Boletus satanas* Lenz).

Boletus (*Tubiporus*) *flammeus* Heim	nondo ulné kobi
Boletus (*Tubiporus*) *kumaeus* Heim	nonda ngamp kindjkants
Boletus (*Tubiporus*) *manicus* Heim	nonda gegwants ngimbigl
Boletus (*Tubiporus*) *nigerrimus* Heim	nondo kermaipip
Boletus (*Tubiporus*) *nigroviolaceus* Heim	nonda tua-rua
Boletus (*Tubiporus*) *reayi* Heim	nonda ngam ngam

Boletus manicus, the largest and supposedly most effective species of nonda, closely resembles the Satan's mushroom (*Boletus satanas* Lenz) of Europe. A powder produced by grinding the dried

In Papua New Guinea, species of mushrooms that are closely related to the coveted king bolete (*Boletus edulis*) are said to induce a temporary state of insanity.

385 In the psychiatric literature, the effects produced by mushrooms that contain psilocybin are often also characterized as "mushroom madness" (McDonald 1980).

fruiting bodies is said to be able to induce colorful visions. *Boletus manicus* has been found to contain traces of **indole alkaloids** (Heim 1972, 173; Ott 1993, 422*). One *Boletus* species from the New Guinea highlands that is known as *namanama* was found to contain only amino acids and steroids, none of which is known to have any psychoactive effects (Gellert et al. 1973).

Heimiella (Boletaceae)

This genus is composed of just two or three species and is found only in Asia. It is characterized by long, fleshy stems and small caps.

Heimiella anguiformis Heim   nonda mbolbe
Heimiella retispora Heim

To date, no psychoactive compounds have been discovered in the genus *Heimiella* (Schultes and Hofmann 1992, 44*).

Russula (Russulaceae)—brittle caps

Brittle caps are found throughout the world. Some species are coveted as culinary mushrooms, some are regarded as inedible, and some are attributed with a certain degree of toxicity. The taste can be used to estimate the toxicity of a specimen. Species that have a mild taste are edible, while pungent varieties tend to be inedible or poisonous. Because the pungent taste is often not immediately apparent, a sample should be retained in the mouth for at least two minutes. Two species of *Russula* have been found to contain stearic acid. Some varieties also contain **ibotenic acid** and **muscimol**, both of which are also present in the fly agaric (**Amanita muscaria**) (Schultes and Hofmann 1992, 55*).

The following brittle caps have been described for Papua New Guinea and are classified as nonda:

Russula agglutinata Heim	nonda mos
Russula kirinea Heim	kirin
Russula maenadum Heim	nonda mos
Russula nondorbingi Singer	nonda bingi
Russula pseudomaenadum Heim	nonda wam

Nonda mushrooms from the genus *Russula* are said to induce the mushroom madness (*ndaadl*) in

Some brittle caps (*Russula* spp.) are purported to have psychoactive effects.

women but not in men (Heim 1972, 177).

Self-experiments with nonda mushrooms (ingestion) conducted by various ethnographers and mushroom enthusiasts have not detected any type of psychoactive effects (D. McKenna 1995, 102*). It is of course possible that the nonda mushrooms contain substances that react only in the context of some specific chemical properties of the Kuma (cf. Nelson 1970). All of the information that is available indicates that the mushroom madness represents a traditional and learned pattern of behavior that is integrated into the Kuma culture in a complex manner. Mushroom madness is a cultural institution that makes it possible for individuals to "flip out" on a temporary basis, thereby enabling them to undergo a social catharsis and enact a ritual drama (Heim and Wasson 1965).

Literature

Gellert, E., B. Halpern, and R. Rudzats. 1973. Amino acids and steroids of a New Guinea boletus. *Phytochemistry* 12:689–92.

Heim, Roger. 1972. Mushroom madness in the Kuma. *Human Biology in Oceania* 1 (3): 170–78.

Heim, Roger, and R. Gordon Wasson. 1964. Note préliminaire sur la folie fongique des Kuma. *Comptes Rendus des Séances de l'Académie des Sciences* (Paris) 258:1593–98.

———. 1965. The "mushroom madness" of the Kuma. *Botanical Museum Leaflets* 21 (1): 1–36.

Kertonegoro, Madi. 1991. *Flug des Geistes: Eine Reise in das andere Bali*. Basel: Sphinx.

McDonald, A. 1980. Mushrooms and madness: Hallucinogenic mushrooms and some psychopharmacological implications. *Canadian Journal of Psychiatry* 25:586–94.

Nelson, Hal. 1970. On the etiology of "mushroom madness" in highland New Guinea: Kaimbi culture and psychotropism. Paper presented at 69th annual meeting of the American Anthropological Association, San Diego, Calif., Nov. 18–20, 1970.

Newmann, Philip. 1964. "Wild man" behavior in a New Guinea highland community. *American Anthropologist* 66 (1): 1–19.

Rätsch, Christian, and Heinz J. Probst. 1985. *Namaste Yeti—Geschichten vom Wilden Mann*. Munich: Knaur.

Reay, Marie. 1959. *The Kuma: Freedom and conformity in the New Guinea highlands*. [Carlton]: Melbourne University Press.

———. 1960. "Mushroom madness" in the New Guinea highlands. *Oceania* 21 (2): 137–39.

General Literature on Psychoactive Fungi

This bibliography contains those sources that are cited in this section on psychoactive fungi and are marked with **.

Aaronson, Sheldon. 1989. Fungal parasites of grasses and cereals: Their role as food or medicine, now and in the past. *Antiquity* 63:247–57.

Adelaars, Arno. 1997. *Alles over Paddo's.* Amsterdam: Prometheus.

Allen, John W. 1993. Iconae plantarum inebriantium—2. *Integration* 4:81–87.

Allen, John W., Jochen Gartz, and Gastón Guzmán. 1992. Index to the botanical identification and chemical analysis of the known species of the hallucinogenic mushrooms. *Integration* 2/3:91–97.

Allen, John W., Mark D. Merlin, and Karl L. R. Jansen. 1991. An ethnomycological review of psychoactive agarics in Australia and New Zealand. *Journal of Psychoactive Drugs* 23 (1): 39–69.

Avila B., Alejandro de, A. L. Welden, and Gastón Guzmán. 1980. Notes on the ethnomycology of Hueyapan, Morelos, Mexico. *Journal of Ethnopharmaçology* 2 (4): 311–21.

Benítez, Fernando. 1964. *Los hongos alucinantes.* Mexico City: Ediciones Era. Repr. 1992 (7th ed.).

Benjamin, Denis R. 1995. *Mushrooms: Poisons and panaceas.* New York: Freeman and Company.

Birkfeld, Alfred. 1954. *Pilze in der Heilkunde.* Die Neue Brehm-Bücherei, no. 135 Wittenberg Lutherstadt: A. Ziemsen.

Blyth, R. H. 1973. Mushrooms in Japanese verse. *The Transactions of the Asiatic Society of Japan*, 3rd series, 11:1–14.

Böttcher, Helmuth M. 1959. *Wunderdrogen: Die abenteuerliche Geschichte der Heilpilze.* Cologne: Kiepenheuer und Witsch.

Bourke, John Gregory. 1996. *Der Unrat in Sitte, Brauch, Glauben und Gewohnheitsrecht der Völker.* Frankfurt/M.: Eichborn. (Orig. pub. 1913.)

Bresinsky, Andreas, and Helmut Besl. 1985. *Giftpilze.* Stuttgart: WVG.

Brown, Christopher. 1987. R. Gordon Wasson: 22 September 1893–23 December 1986. *Economic Botany* 41 (4): 469–73.

Camilla, Gilberto. 1997. Reminiscenze enteogeniche nella tradizione giudaico-cristiana. *Eleusis* 7:18–23.

Carmichael, Michael. 1996. Wonderland revisited. *The London Miscellany* 28:19–28.

Cerletti, A., and Albert Hofmann. 1963. Mushrooms and toadstools. *The Lancet* (January 5, 1963): 58–59.

Cetto, Bruno. 1987/88. *Enzyklopädie der Pilze.* 4 vols. Munich: BLV.

Clare, Ray. 1984. The Mayan magic mushroom dust of Palenque. Unpublished manuscript, Los Angeles. (Cited 1998.)

———. 1988. The breaching of Don Juan's teaching. Unpublished manuscript, Los Angeles. (Cited 1998.)

Conover, Philip. 1994. *Teonanacatl: The food of the gods.* Mexico City: The Huautla Press.

Cooper, Richard. 1980. *A guide to British psilocybin mushrooms.* London: Red Shift Books.

Cortés, Jesús. 1979. La medicina traditional en la Sierra Mazateca. *Actes du XL11e Congrès des Américanistes* 6:349–56. Paris: Société de Américanistes.

Dähncke, Rose Marie. 1993. *1200 Pilze in Farbfotos.* Aarau: AT Verlag.

Donati, Dario. 1991. *Doña Julieta, die kranke Heilerin: Dimensionen im Umgang mit Krankheit und Heilung einer mazatekischen Frau.* Zurich: Lizentiatsarbeit.

Dörfelt, Heinrich, ed. 1989. *Lexikon der Mykologie.* Stuttgart: Gustav Fischer Verlag.

Dörner, Gerd. 1963. Die "heiligen Pilze" Mexikos, ihre Zeremonie und ihre Wirkung. *Deutsche Apotheker-Zeitung* 103 (51): 1699–1702.

Eisner, Betty. 1996a. Ein Abenteuer in Huautla. In *María Sabina—Botin der heiligen Pilze*, ed. Roger Liggenstorfer and C. Rätsch, 117–31. Solothurn: Nachtschatten Verlag.

———. 1996b. Huautla—place where eagles are born. *Yearbook for Ethnomedicine and the Study of Consciousness*, 1995 (4): 13–33. Berlin: VWB.

Emboden, William A. 1990. Whence ethnomycology? *The Albert Hofmann Foundation Bulletin* 1 (4): 8.

Enos, L. 1970. *A key to the American psilocybin mushroom.* Lemon Grove, Calif.: Youniverse.

Escalante H., Roberto, and Antonio Lopez G. 1974. Hongos sagrados de los Matlazincas. In *Atti del XL Congresso Internazionale degli Americanisti* 2:245.

Estrada, Alvaro. 1980. *María Sabina—Botin der heiligen Pilze.* Munich: Trikont. (See also Liggenstorfer and Rätsch 1996.)

Festi, Francesco. 1985. *Funghi allucinogeni: Aspetti psichofisiologici e storici.* Pubblicazione 86. Rovereto: Musei Civici di Rovereto.

Findlay, W. P. K. 1982. *Fungi: Folklore, fiction, and fact.* England: Richmond Publishing.

Furst, Peter T. 1992. *Mushrooms: Psychedelic fungi.* New York: Chelsea House.

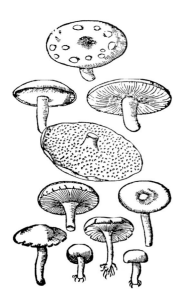

In ancient times, the word *boletus* was the name for the savory Caesar's mushroom (*Amanita caesarea*); it was also used for other species of the genus *Amanita*, e.g., the fly agaric. (Woodcut from Matthiolus, *Kräuterbuch*, 1627)

Garnweidner, Edmund. 1993. *Pilze: Bestimmen, Kennenlernen, Sammeln*. Munich: Gräfe und Unzer.

Gartz, Jochen. 1991. Psychotrope Inhaltsstoffe in verschiedenen einheimischen Pilzarten. *Jahrbuch des Europäischen Collegiums für Bewußtseinsstudien*, 1991: 101–8.

———. 1993. *Narrenschwämme: Psychotrope Pilze in Europa*. Geneva: Editions Heuwinkel.

———. 1994. Ethnopharmakologie psilocybinhaltiger Pilze im pazifischen Nordwesten der USA. *Jahrbuch des Europäischen Collegiums für Bewußtseinsstudien*, 1993/1994: 159–64.

———. 1996. *Magic mushrooms around the world*. Los Angeles: Lis Publications.

Ghouled, F. C. 1972. *Field guide to the psilocybin mushroom: Species common to North America*. New Orleans: Guidance Publications.

Gillman, Linnea, Art Goodtimes, Gary Lincoff, Emanuel Salzman, and Jason Salzman, eds. 1996. *Wild mushrooms of Telluride*. Denver: Fungophile.

Gonzalez Torres, Yolotl. 1989. Altered states of consciousness and ancient Mexican ritual techniques. In *Shamanism: Past and present*, ed. M. Hoppal and O. J. von Sadovsky, 349–53. Budapest: ISTOR Books.

Graves, Robert [= Robert Ranke-Graves]. 1957. Mushrooms, food of the gods. *Atlantic Monthly* 200 (2): 73–77.

———. 1960. *Food for Centaurs*. New York: Doubleday.

———. 1992. *The Greek myths*. Complete ed. London: Penguin Books.

Gurevich, Luydmila. 1995. Study of Russian psilocybine-containing basidiomycetes. *Integration* 6:11–20.

Guzmán, Gastón. 1978a. Further investigations of the Mexican hallucinogenic mushrooms with descriptions of new taxa and critical observations on additional taxa. *Nova Hedwigia* 29:625–64.

———. 1978b. *Hongos*. Mexico City: Editorial Limusa.

———. 1980. *Identificación de los hongos: comestibles, venenosos y alucinantes*. Mexico City: Editorial Limusa.

———. 1983a. *The genus Psilocybe*. Beihefte zur Nova Hedwigia, no. 74. Vaduz, Liechtenstein: J. Cramer.

———. 1983b. Los hongos de la Península de Yucatán. *Biotica* 8:71–100.

———. 1990. Wasson and the development of mycology in Mexico. In *The sacred mushroom seeker*, ed. Thomas J. Riedlinger, 83–110. Portland, Ore.: Dioscorides Press.

———. 1994a. Los hongos en la medicina tradicionál de Mesoamérica y de México. *Revista Iberoamericana de Micología* 11:81–85.

———. 1994b. Los hongos y liquenes en la medicina traditional. In *Atlas de las plantas de la medicina traditional mexicana*, ed. A. Argueta V. et al., 3:1427–78. Mexico City: INI.

Guzmán, Gastón, Jonathan Ott, Jerry Boydston, and Steven H. Pollock. 1976. Psychotropic mycoflora of Washington, Idaho, Oregon, California and British Columbia. *Mycologia* 68 (6): 1267–72.

Haard, Richard, and Karen Haard. 1980. *Poisonous and hallucinogenic mushrooms*. 2nd ed. Seattle: Homestead.

Haseneier, Martin. 1992. Der Kahlkopf und das kollektive Unbewußte: Einige Anmerkungen zur archetypischen Dimension des Pilzes. *Integration* 2/3:5–38.

Heim, Roger. 1959. *Les investigations anciennes et récentes propos aux agarics hallucinogènes du Mexique, à leur action et aux substances qui en sont responsables*. Paris: Masson.

———. 1963. *Les Champignons toxiques et hallucinogènes*. Paris: Editions N. Boubee. Repr. 1978 (2nd ed.).

Heim, Roger, et al. 1966. Nouvelles investigations sur les champignons hallucinogènes. *Archives du Muséum National d'Histoire Naturelle* 9 (1965–1966): 111–220, plus eleven plates.

Heim, Roger, and R. Gordon Wasson. 1958. Les champignons hallucinogènes du Mexique. In *Archives du Muséum National d'Histoire Naturelle*, 7th series, vol. 6. Paris: Muséum National d'Histoire Naturelle.

Hobbs, Christopher. 1995. *Medicinal mushrooms: An exploration of tradition, healing, and culture*. 2nd ed. Santa Cruz, Calif.: Botanica Press.

Hofmann, Albert. 1964. Die Erforschung der mexikanischen Zauberpilze und das Problem ihrer Wirkstoffe. *Basler Stadtbuch* 1964:141–56.

———. 1965. Pilzgifte als Halluzinogene. *Selecta* 7, 2146 (no. 49).

———. 1969. Investigaciones sobre los hongos alucinogenos mexicanos y la importancia que tienen en la medicina sus substancias activas. *Artes de México* 16 (124): 23–31.

———. 1971. Teonanácatl and ololiuqui, two ancient magic drugs of Mexico. *Bulletin on Narcotics* 23 (1): 3–14. (Also available in French.)

———. 1987a. Die heiligen Pilze in der Heilbehandlung der Maria Sabina. In *Ethnopsychotherapie*, ed. Adolf Dittrich and Christian Scharfetter, 45–52. Stuttgart: Enke.

———. 1987b. Pilzliche Halluzinogene vom Mutterkorn bis zu den mexikanischen Zauberpilzen. *Der Champignon* 310:22–28.

———. 1990. The discovery of the psychoactive components of the magic mushrooms of Mexico. *The Albert Hofmann Foundation Bulletin* 1 (4): 6–7.

———. 1993a. Chemistry and pharmacology of the "sacred mushrooms" of Mexico. In *Atti del 2° Convegno Nazionale sugli Avvelenamenti da Funghi, Annali dei Musei Civici di Rovereto* 8 suppl. (1992): 97–106.

———. 1993b. Maria Sabina und die heiligen Pilze. In *Naturverehrung und Heilkunst*, ed. C. Rätsch, 213–22. Südergellersen: Verlag Bruno Martin.

Hyde, C., G. Glancy, P. Omerod, D. Hall, and G. S. Taylor. 1978. Abuse of indigenous psilocybin mushrooms: A new fashion and some psychiatric complications. *British Journal of Psychiatry* 132:602–4.

Imazeki, Rokuya. 1973. Japanese mushroom names. *The Transactions of the Asiatic Society of Japan*, 3rd series, 11:26–80.

Imazeki, Rokuya, Yoshio Otani, and Tsuguo Hongo. 1993. *Fungi of Japan*. Tokyo: Yam-kei Publishers. (In Japanese.)

Imazeki, Rokuya, and R. Gordon Wasson. 1973. *Kinpu*, mushroom books of the Toku-Gawa period. *The Transactions of the Asiatic Society of Japan*, 3rd series, 11:1–12.

Inchaustegui, Carlos. 1977. *Relatos del mundo mágico mazateco*. Mexico City: INAH.

———. 1983. *Figuras en la niebla: relatos y creencias de los mazatecos*. Mexico City: Premia Editora.

Johnson, Jean Basset. 1939a. The elements of Mazatec witchcraft. *Ethnological Studies* 9:128–50.

———. 1939b. Some notes on the Mazatec. *Revista Mexicana de Estudios Antropológicos* 3:142–56. (This and the 1939a entry above are two of the earliest references on ritual mushroom use among the Mazatec.)

———. 1940. Note on the discovery of teonanacatl. *American Anthropologist*, n.s., 42:549–50.

Jordan, Michael. 1989. *Mushroom magic*. London: Elm Tree Books.

Kell, Volkbert. 1991. *Giftpilze und Pilzgifte*. Die Neue Brehm-Bücherei, vol. 612. Wittenberg Lutherstadt: Ziemsen Verlag.

Knecht, Sigrid (later Sigrid Lechner-Knecht). 1961. Magische Pilze. *Neue Wissenschaft* 10 (2).

Laatsch, Hartmut. 1994. Das Fleisch der Götter— Von den Rauschpilzen zur Neurotransmission. In *Welten des Bewußtseins*, ed. Adolf Dittrich, Albert Hofmann, and Hanscarl Leuner, 3:181–95. Berlin: VWB.

Leistenfels, H. v. n.d. *Pilze*. Der Grüne Zweig 65c. Löhrbach: Werner Pieper's MedienXperimente. (Includes "Glücks-Pilze," by Micky Remann.)

Lévi-Strauss, Claude. 1970. Les Champignons dans la culture. *L'Homme* 10 (1): 5–16.

———. 1992. *Strukturale Anthropologie II*. Frankfurt/M.: Suhrkamp. (Contains the chapter "Die Pilze in der Kultur," by Levi-Strauss [1970].)

Liggenstorfer, Roger, and Christian Rätsch, eds. 1996. *María Sabina—Botin der heiligen Pilze: Vom tra-ditionellen Schamanentum zur weltweiten Pilzkultur*. Edition Rauschkunde. Solothurn: Nachtschatten Verlag.

———. 1998. *Pilze der Götter*. Aarau: AT Verlag. (A new edition of Liggenstorfer and Rätsch 1996, with a CD of María Sabina's songs.)

Lowy, Bernard. 1975. Notes on mushrooms and religion. *Revista/Review Interamericana* 5 (1): 110–17.

———. 1977. Hallucinogenic mushrooms in Guatemala. *Journal of Psychedelic Drugs* 9 (2): 123–25.

———. 1980. Ethnomycological inferences from mushroom stones, Maya codices, and Tzutuhil legend. *Revista/Review Interamericana* 10 (1): 94–103.

Mayer, Karl Herbert. 1975. Die heiligen Pilze Mexikos. *Ethnologia Americana* 11 (5): 594–96; 11 (6): 603–8.

McGuire, Thomas. 1982. Ancient Maya mushroom connection: A transcendental interaction model. *Journal of Psychoactive Drugs* 14 (3): 221–38.

McKenna, Terence. 1988. Hallucinogenic mushrooms and evolution. *ReVision* 10 (4): 51–57.

———. 1990. Wasson's literary precursors. In *The sacred mushroom seeker: Essays for R. Gordon Wasson*, ed. Thomas J. Riedlinger, 165–75. Portland, Ore.: Dioscorides Press.

Menser, Gary P. 1997. *Hallucinogenic and poisonous mushroom field guide*. Berkeley, Calif.: Ronin.

Merlin, Mark D., and John W. Allen. 1993. Species identification and chemical analysis of psychoactive fungi in the Hawaiian Islands. *Journal of Ethnopharmacology* 40:21–40.

Metzner, Ralph. 1970. Mushrooms and the mind. In *Psychedelics*, ed. Bernard Aaronson and Humphry Osmond, 90–107. Garden City, N.Y.: Anchor Books (Doubleday).

Mills, P., D. Lesinskas, and G. Watkinson. 1979. The danger of hallucinogenic mushrooms. *Scottish Medical Journal* 24 (4): 316–17.

Morgan, Adrian. 1995. *Toads and toadstools: The natural history, folklore, and cultural oddities of a strange association*. Berkeley, Calif.: Celestial Arts.

Morris, Brian. 1992. Mushrooms: For medicine, magic and munching. *Nyala* 16 (1): 1–8.

Munn, Henry. 1973. The mushrooms of language. In *Hallucinogens and shamanism*, ed. M. Harner, 86–122. London: Oxford University Press.

Norland, Richard Hans. 1976. *What's in a mushroom?* Psychoactive Mushrooms, part 3. Ashland, Ore.: Pear Tree Publications.

Ohenoja, E., et al. 1987. The occurrence of psilocybin and psilocin in Finnish fungi. *Journal of Natural Products* 50:741–44.

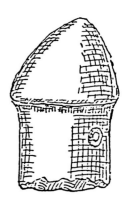

In the area inhabited by the Maya, objects have been found that have been interpreted as phallus stones but are nevertheless associated with mushroom stones. (From Theobert Maler, ca. 1895)

Psychoactive Molds

A number of species of mold fungi produce **indole alkaloids** (lysergic acid, elymoclavine, agroclavine, fumigaclavine, fumitremorgine, ergokryptine, ergosine, costaclavine, ergosinine, noragroclavine, and others), some of which have psychoactive as well as highly toxic effects (from Samorini 1997a, 41):

Ascochyta imperfecta Peck
Aspergillus clavatus Desmaz.
Aspergillus conicus Bloch.
Aspergillus flavus Link.
Aspergillus fumigatus Fres.
Aspergillus nidulans (Eidam) Wint
Aspergillus versicolor (Vuill.) Tirab.
Dematium chodati Nechitsch
Geotrichum candidum Link.
Isariopsis grieseola Sacc.
Mucor subtilissimus Berk.
Penicillum chermesinum Biourge
Penicillum expansum Link.
Penicillum granulatum Bain.
Penicillum roqueforti
Penicillum rugulosum Thom.
Rizopus arrhizus Fischer
Rizopus nigricans Ehramb.
Streptomyces rimosus Finlay
Trichochoma paradoxa Jungh

"While researching *Psilocybe*, I became accustomed to meeting great resistance from professional mycologists, many of whom had an instant distrust of anyone expressing a passion for *Psilocybe*. There were some mycologists who stated publicly that it would be better for people to die from mistakes in identification than to provide them with the tools for recognizing a *Psilocybe* mushroom. This bizarre attitude towards *Psilocybe* mushrooms and the people who used them reflected a chasm between generations."

PAUL STAMETS
PSILOCYBIN MUSHROOMS OF THE WORLD
(1996, 3**)

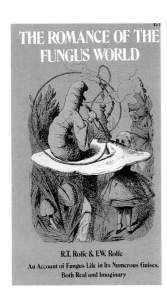

THE ROMANCE OF THE FUNGUS WORLD

R.T. Rolfe & F.W. Rolfe
An Account of Fungus Life in Its Numerous Guises,
Both Real and Imaginary

Ott, Jonathan. 1975. Notes on recreational use of hallucinogenic mushrooms. *Boletín de la Sociedad Mexicana de Micología* 9:131–35.

———. 1978. Recreational use of hallucinogenic mushrooms in the United States. In *Mushroom poisoning: Diagnosis and treatment*, ed. B. H. Rumack and E. Salzman, 231–43. West Palm Beach, Fla.: CRC Press.

Ott, Jonathan, and Jeremy Bigwood, eds. 1978. *Teonanácatl: Hallucinogenic mushrooms of North America*. Seattle: Madrona.

Pike, Eunice, and Florence Cowan. 1959. Mushroom ritual versus Christianity. *Practical Anthropology* 6 (4): 145–50.

Puharich, Andrija. 1959. *The sacred mushroom: Key to the door of eternity*. Garden City, N.Y.: Doubleday & Co. Repr. 1974 (paperback issue).

Rätsch, Christian. 1993. Halluzinogene Pilze und unsere Ahnen. In *Zauberpilze*, ed. R. Rippchen, 21–24. Löhrbach, Werner Pieper's MedienXperimente.

———. 1995. Pilze, Schamanen und die Facetten des Bewußtseins. *Curare* 18 (1): 3–14.

———. 1996. Die Rückkehr zur Kultur: Heilige Pilze in modernen Ritualen. *Jahrbuch für Transkulturelle Medizin und Psychotherapie* 6 (1995): 299–339.

Reko, Blas Pablo. 1940. Teonanacatl, the narcotic mushroom. *American Anthropologist*, n.s., 42:368–69.

Reyes G., Luis. 1970. Una relación sobre los hongos alucinantes. *Tlalocan* 6 (2): 140–45.

Ricks, David F. 1963. Mushrooms and mystics: A caveat. *The Harvard Review* 1 (4): 51–55.

Riedlinger, Thomas J., ed. 1990. *The sacred mushroom seeker: Essays for R. Gordon Wasson*. Portland, Ore.: Dioscorides Press.

Rippchen, Ronald, ed. [1993]. *Zauberpilze*. Der Grüne Zweig 155. Löhrbach: Werner Pieper's MedienXperimente; Solothurn: Nachtschatten Verlag.

Roldan, Dolores. 1975. *Teonanácatl (Carnita Divina): Cuentos antropológicos*. Mexico City: Editorial Orion.

Rolfe, R. T., and F. W. Rolfe. 1974. *The romance of the fungus world*. New York: Dover. (Orig. pub. 1925.)

Roth, Lutz, Hans Frank, and Kurt Kormann. 1990. *Giftpilze—Pilzgifte, Schimmelpilze—Mykotoxine*. Munich: ecomed.

Rubel, Arthur, and Jean Gettelfinder-Krejci. 1976. The use of hallucinogenic mushrooms for diagnostic purposes among some highland Chinantecs. *Economic Botany* 30:235–48.

Ruck, Carl A. P. 1981. Mushrooms and philosophers. *Journal of Ethnopharmacology* 4:179–205.

———. 1983. The offerings from the Hyperboreans. *Journal of Ethnopharmacology* 8:177–207.

Rumack, Barry H., and Emanuel Salzman, eds. 1978. *Mushroom poisoning: Diagnosis and treatment*. West Palm Beach, Fla.: CRC Press.

Saar, Maret. 1991. Fungi in Khanty folk medicine. *Journal of Ethnopharmacology* 31 (2): 175–79.

Samorini, Giorgio. 1990a. Sciamanismo, funghi psicotropi e stati alterati di coscienza: Un rapporto da chiarire. *Boll. Camuno Studi Preistorici* 25/26:147–50.

———. 1990b. Sullo stato attuale della conoscenza dei Basidiomiceti psicotropi Italiani. *Annali dei Musei civici di Rovereto* 5 (1989): 167–84.

———. 1993. Funghi allucinogeni italiani. In *Atti del 2° Convegno Nazionale sugli Avvelenamenti da Funghi, Annali dei Musei Civici di Rovereto* 8 suppl. (1992): 125–49. (Very good bibliography!)

———. 1996. New frontiers of ethnomycology. Lecture to the Entheobotany Conference, San Francisco, Oct. 18–20, 1996.

———. 1997a. *Aspergillus fumigatus* Fres. *Eleusis* 8:38–43.

———. 1997b. L'albero-fungo die Plaincourault / The mushroom-tree of Plaincourault. *Eleusis* 8:29–37.

Samorini, Giorgio, and Francisco Festi. 1989. Le micotossicosi psicotrope volontarie in Europa: Osservazioni sui casi clinici. *Annali dei Musei civici di Rovereto* 4 (1988): 251–57.

Sandford, Jeremy. 1973. *In search of the magic mushroom: A journey through Mexico*. New York: Potter.

Sanford, James H. 1972. Japan's "laughing mushrooms." *Economic Botany* 26:174–81.

Schlichting, Michael. 1996. Reise nach Oaxaca. In *María Sabina—Botin der heiligen Pilze*, ed. Roger Liggenstorfer and C. Rätsch, 133–38. Solothurn: Nachtschatten Verlag.

Schultes, Richard Evans. 1939. Plantae mexicanae II: The identification of teonanacatl, a narcotic basidiomycete of the Aztecs. *Botanical Musuem Leaflets* 7 (3): 37–54.

———. 1940. Teonanacatl: The narcotic mushroom of the Aztecs. *American Anthropologist*, n.s., 42:429–43.

———. 1978. Evolution of the identification of the sacred hallucinogenic mushrooms of Mexico. In *Teonanácatl: Hallucinogenic mushrooms of North America*, ed. J. Ott and J. Bigwood, 25–43. Seattle: Madrona.

Semerdzieva, Marta, and F. Nerud. 1973. Halluzinogene Pilze in der Tschechoslowakei. *Ceska Mykologie* 27:42–47.

Shepherd, C. J., and C. J. Totterdell. 1988. *Mushrooms and toadstools of Australia*. Melbourne and Sydney: Inkata Press.

Singer, Rolf. 1958. Mycological investigation on teonanácatl, the Mexican hallucinogenic mushroom. Part I. The history of teonanácatl, field work and culture work. *Mykologia* 50:239–61.

———. 1978. Interesting and new species of basidiomycetes from Ecuador II. *Nova Hedwigia* 29:1–98.

Singer, Rolf, and Alexander H. Smith. 1960. Hongos psicotópicos. *Lilloa* 30:124–26.

———. 1982. *A correction*. Ethnomycological Studies, no. 8. Cambridge, Mass.: Botanical Museum of Harvard University.

Smith, Alexander H. 1977. Comments on hallucinogenic agarics and the hallucinations of those who study them. *Mycologia* 69:1196–200.

Solier, René de. 1965. *Curandera, les champignons hallucinogènes*. [Paris]: Chez Jean-Jacques Pauvert.

Stafford, Peter. 1980. *Psilocybin und andere Pilze*. Markt Erlbach: Raymond Martin Verlag.

Stamets, Paul. 1978. *Psilocybe mushrooms and their allies*. Seattle: Homestead.

———. 1996. *Psilocybin mushrooms of the world*. Berkeley, Calif.: Ten Speed Press.

States, Jack S. 1990. *Mushrooms and truffles of the Southwest*. Tucson: The University of Arizona Press.

Stijve, Tjacco. 1995. Worldwide occurrence of psychoactive mushrooms: An update. *Czech. Mycol.* 48:11–19.

Stijve, Tjakko. 1997. Boleti allucinogeni in China? *Eleusis* 7:33.

Thompson, John P., M. Douglas Anglin, William Emboden, and Dennis Gene Fischer. 1985. Mushroom use by college students. *Journal of Drug Education* 15 (2): 111–24.

Tibón, Gutierre. 1983. *La ciudad de los hongos alucinantes*. Mexico City: Panorama.

Viola, Severino. 1972. *Die Pilze*. Munich: Hirmer.

Walleyn, R., and J. Rammeloo. 1995. *The poisonous and useful fungi of Africa south of the Sahara: A literature survey*. Scripta Botanica Belgica, vol. 10. Meise: National Botanic Garden of Belgium.

Walters, Bill. 1995. Hallelujah! Praise the mushrooms. *Psychedelic Illuminations* 8:38–40.

Wasson, R. Gordon. 1957. Seeking the magic mushroom. *Life* (May 13, 1957) 42 (19): 100 ff.

———. 1961. The hallucinogenic fungi of Mexico: An inquiry into the origins of the religious idea among primitive peoples. *Botanical Museum Leaflets, Harvard University* 19 (7): 137–62.

———. 1963a. The hallucinogenic mushrooms of Mexico and psilocybin: A bibliography. *Botanical Museum Leaflets, Harvard University* 20 (2a): 25–73c. (Corrected and expanded second version.)

———. 1963b. The mushroom rites of Mexico. *The Harvard Review* 1 (4): 7–17.

———. 1973. Mushrooms in Japanese culture. *The Transactions of the Asiatic Society of Japan*, 3rd series, 11:5–25.

———. 1978. The hallucinogenic fungi of Mexico. In *Teonanacatl: Hallucinogenic mushrooms of North America*, ed. J. Ott and J. Bigwood, 63–84. Seattle: Madrona.

———. 1980. *The wondrous mushroom: Mycolatry in Mesoamerica*. New York: McGraw-Hill.

———. 1982. *R. Gordon Wasson's rejoinder to Dr. Rolf Singer*. Ethnomycological Studies, no. 9. Cambridge, Mass.: Botanical Museum of Harvard University.

———. 1986. Lightningbolt and mushroom. In *Persephone's quest: Entheogens and the origins of religion*, by R. G. Wasson et al., 83–94. New Haven, Conn.: Yale University Press.

Wasson, R. Gordon, George Cowan, Florence Cowan, and Willard Rhodes. 1974. *Maria Sabina and her Mazatec mushroom velada*. New York and London: Harcourt Brace Jovanovich.

Wasson, R. Gordon, and Valentina P. Wasson. 1957. *Mushrooms, Russia, and history*. New York: Pantheon Books.

Weil, Andrew. 1975. Mushroom hunting in Oregon. *Journal of Psychedelic Drugs* 7 (1): 89–102.

———. 1977. The use of psychoactive mushrooms in the Pacific Northwest: An ethnopharmacologic report. *Botanical Museum Leaflets, Harvard University* 25 (5): 131–49.

Winkler, Rudolf. 1996. *2000 Pilze selber bestimmen*. Aarau: AT Verlag.

Ying Jianzhe, Mao Xiaolan, Ma Qiming, Zong Yichen, and Wen Huaan. 1989. *Icons of medicinal fungi from China*. Beijing: Science Press.

Zeitlmayer, Linus. 1976. *Knaurs Pilzbuch: Leben, Erkennen, Verwerten, Sammeln*. Munich: Droemer-Knaur.

Z[ubke], A[chim]. 1997. Spass Attacks: Die Invasion der lachenden Pilze. *HanfBlatt* 4 (34): 18–25.

"And thus, as is still told here and there in Bavarian and Austrian states, their [the mushrooms'] growth depends upon higher and lower gods of growth. Once it depended upon Donar and Wodan, later upon their Christian equivalents, upon God and his saints, especially St. Peter, who controls the weather, and St. Veit, the successor to the Slavic sun god Svantevit, and upon St. Procopius and Anthony the hermit. On the other hand, some claimed that it was the devils, witches, elves, and good or evil fungi spirits that enabled the race of fungi to grow, or mushroom souls, mushroom dwarves, or perhaps a dwarflike mushroom prophet. All of them are offered small gifts and 'mushroom prayers.'"

LINUS ZEITLMAYER
KNAURS PILZBUCH [KNAUER'S BOOK OF FUNGI]
(1976, 10**).

"The secret of the castle ghosts has been solved! In Great Britain, which is rich in castles, microbiologists investigated moist, mildewed, and moldy cellars, in which thrive tiny little fungi with psychoactive effects. Breathing in their spores induces hallucinations—in other words, people get 'high' and believe that they are seeing things that in actuality do not exist."

FRANZ AUF DER MAUR
(IN *NATÜRLICH* 16/11:30, 1996)

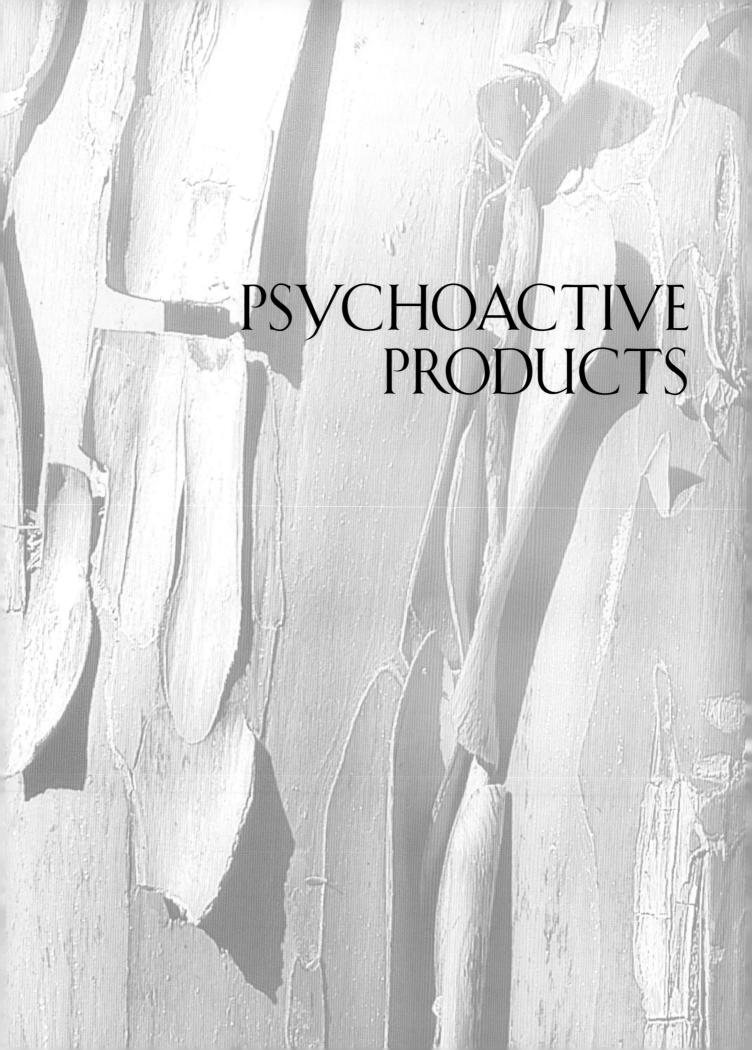

PSYCHOACTIVE PRODUCTS

The psychoactive effects of many plants can be improved, modified, or even simply made possible by preparing or processing the plants using both traditional and pharmaceutical methods or by combining them with other plants and substances. The resulting products typically are culturally significant, whether as agents of pleasure (**betel quids, chicha, palm wine, sake, wine**), shamanic tools (**ayahuasca, cimora, snuff**), or sacred drugs (**balche', mead**). These products bear witness to humanity's amazing powers of invention and creativity. Some (**beer, pituri, incense, honey**) have been produced and used since the Stone Age. However, precise details about many of the recipes for or ingredients in some ancient products (**han-shi, haoma, witches' ointments, kykeon, soma**) are not well known, whether because of a desire for secrecy, because that knowledge was suppressed, or because the ingredients were simply forgotten over time. Other products (**ayahuasca analogs, energy drinks, herbal ecstasy**) are more recent developments. Today, more and more people are searching for new psychoactive substances and products. Countless closet shamans in North America and Europe have been experimenting with new combinations, extracts, and possibilities for preparing such products. Considerable research into **ayahuasca analogs** and **smoking blends** is currently under way.

Combination Preparations

Some products (such as **alcohol**) can be produced only through elaborate techniques, while others are distinguished by a skillful and deliberate combination of various substances. At times, plants or certain parts of plants become psychoactive only when combined with other ingredients. Sometimes various admixtures can be used in combination to produce a synergistic effect, that is, the two effects shape one another, resulting in a new effect that is different from that of the individual substances (as is the case for **madzoka medicine** and **zombie poison**). In some cases, a particular substance is more easily tolerated when mixed with another. At other times, combining substances can potentiate a constituent's primary effects or steer them in a particular direction (as can be seen in **Oriental joy pills, kinnikinnick, enemas, soporific sponges**, and **theriac**).

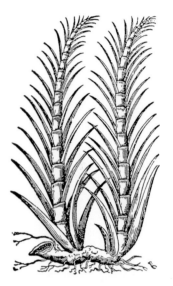

The sugarcane plant (*Saccharum officinarum* L.) originated in Melanesia but is now found in all the tropical regions of the world as a result of human activity. The fresh juice of the plant is used to brew beerlike beverages and to produce potent alcoholic spirits (firewater). Many plants can be used to produce psychoactive products such as beer and alcohol. (Woodcut from Tabernaemontanus, *Neu Vollkommen Kräuter-Buch*, 1731)

Left: An Amazonian shaman prepares the psychoactive and purgative ayahuasca drink from the leaves of *Psychotria viridis* and the stems of *Banisteriopsis caapi*. Drinking ayahuasca enables the astral body, which is composed of colorful lights, to leave the shaman's body and travel to the stars. (Painting by Pablo Amaringo, detail, ca. 1994)

Alcohol (Distilled)

Ethanol

"You can easily take this literally. Alcoholic thrushes really do exist. So do other animal drinkers, who are absolutely wild about alcohol and other drugs. They are all of them all too human: it appears that inebriation also produces feelings of happiness in animals."

LOTHAR FRANZ
DIE GRÖSSTEN TRUNKENBOLDE DES TIERREICHS [THE GREATEST DRINKERS OF THE ANIMAL KINGDOM]
(1995, 14)

The medieval distillation of alcoholic spirits. (Woodcut from *Von allen geprannten wassern* [Of All Distilled Waters], 1498)

An apparatus for distilling palm wine on Amboina Island. (From Hartwich, *Die menschlichen Genußmittel*, 1911)

Other Names

Alcohol, alk, aqua vitae, bourbon, brandy, branntwein, cañaza, dharu, ethanol, ethyl alcohol, gola, moonshine, pox, rakshi, schnapps, schnaps, spirits, spiritus, sprit, soju, whiskey, whisky

Since humans first became acquainted with sweet substances and the various sugars, we have been making them into alcohol by using yeast as a fermenting agent (Bush 1974). The resulting products can be either consumed as **wine** or distilled into alcoholic spirits. Because alcohol evaporates more readily than water, it can be distilled from a fermented product by heating it with care. And because alcohol is strongly hydroscopic, i.e., it attracts water, a portion of the water in which it is contained will be carried into the distillate during distillation. The final distillate product contains

Plants Used for Distilling Alcohol

(after Bärtels 1993, 21, 28, 29, 34*; Havard 1896*; Höschen n.d.; Jain and Dam 1979*; modified and expanded)

Plant/Plant Part	Botanical Name	Name of Alcohol
agave juice	*Agave* spp.	tequila, mescal
anise and other herbs	*Pimpinella anisum* L.	anisado, ouzo, raki, pastis, Pernod
apples	*Malus sylvestris* Mill.	Calvados
apricots	*Prunus armeniaca* L.	barack
bearded darnel	*Lolium temulentum*	korn schnapps
belladonna	*Atropa belladonna*	Tollkirsch
coconut milk	*Cocos nucifera*	brandy
corn/maize	*Zea mays*	(bourbon) whiskey
fruit	various	fruit schnapps
gentian root	*Gentiana lutea* L.	enzian, Swedish bitters
grain	*Triticum* spp.	whiskey, brandy, rakshi
grapes	*Vitis vinifera*	pisco
juniper berries	*Juniperus communis* L.	gin, genever, juniper schnapps (Häger)
marthuarong	*Croton roxburghii* Balak [syn. *C. oblongifolius* Roxb.]	daru
millet	various species	rakshi
palm honey	*Jubaea chilensis* (Mol.) Baill. (honey palm)	aguardiente
palm syrup	*Copernicia prunifera* (Mill.) H.E. Moore (carnauba wax palm)	arrack
	Nypa fruticans Wurmb. (nipa palm)	nipa brandy, nipa whiskey
palm wine	*Borassus flabellifer* L.	arrack
	Hyphaene coriacea	arrack
	Hyphaene thebaica (L.) Mart.	arrack
palm wine (toddy)	*Cocos nucifera*	arrack, rak, kolwater
plums	*Prunus domestica* L.	slivovitz
potatoes	*Solanum tuberosum* L. (cf. *Solanum* spp.)	vodka
rice	*Oryza sativa* L.	rakshi, soju
star anise	*Illicium anisatum* L.	pastis (old)
sugarcane	*Saccharum officinarum* L.	rum, ron, pox, aguardiente, pitú
white hellebore root	*Veratrum album*	enzian
wine	*Vitis vinifera*	brandy, cognac
remnants from pressing wine		trester, grappa, marc
wormwood	*Artemisia absinthium*	absinthe
yucca fruits	*Yucca baccata* Torr.	aguardiente
	Yucca macrocarpa Coville	
	Yucca treculeana Carr.	

approximately 38% alcohol, along with the water and the **essential oils** that are also distilled during the process.

The origin of the distilling arts lies far back in the mists of time. Distilling equipment has been discovered at the temple of Memphis. The ancient Egyptians had allegedly been distilling wine and apple wine since around 4000 B.C.E. (Bosi 1994, 11). Distilling alcohol to manufacture cosmetics has been practiced in Egypt since the eighth century B.C.E. Whether or not a distillate with a high percentage of alcohol was also being distilled during this period, however, remains uncertain. In Wales, various distillation procedures were experimented with during the fourth century C.E. The Saracens brought the Arabian art of distillation to Spain in the eighth century C.E., and from there it quickly spread across Europe (Höschen n.d.). The word *alcohol* is derived from Arabic (cf. **Catha edulis, Coffea arabica**). The Arabs believed that the distillate of wine was a "medicine that could soothe both physical and spiritual pain" (Bosi 1995, 13). The Arabian art of distillation exerted a powerful influence on the alchemical practices of medieval Europe. In Germany and Italy, a rich tradition emerged in which not only fermented alcohol but also practically every herb and every animal was distilled (Braunschweig 1610). For these reasons, distilled alcohol also became known by the names *spiritual drink*, *spirits*, and *alchemical elixir*.

The number of plants that contain starch or sugar constituents that can be mashed and fermented with yeast is beyond count. Numerous other ingredients may be combined with these plants before, during, or after the fermentation period. Most of these additives are aromatic plants whose active constituents are distilled along with the alcohol. For this reason, herbs such as wormwood (**Artemisia absinthium**) and juniper are often added to the mash. During the production of some palm alcohols (*arrack, kolwater*), the bitter bark of the large *muna-mal* or *mukalai* tree (*Mimusops elengi* L.) may be added to the **palm wine** either before or during the distillation process (Macmillan 1991, 424*).

Alcohol is a very effective solvent for herbs. The active constituents or extracts not only are absorbed into the solution but also are simultaneously preserved by the high alcohol content (cf. **theriac**). Many different kinds of alcohol can be aromatized by adding either herbal extracts or **essential oils** (Mayr 1984). For example, Scandinavian aquavit, "the water of life," is actually a grain alcohol to which the essential oil of caraway (*Carum carvi* L.) has been added. An absinthe substitute can be made by adding wormwood (**Artemisia absinthium**) to alcohol:

> Place the upper parts of the flowering herb in alcohol and allow the mixture to sit in a sunny location for two weeks. Shake often. Allow to sit undisturbed for another two weeks, strain, and store for a long time before drinking. (Mayr 1984, 96)

Many psychoactive plants are suitable for adding to alcohol. **Cannabis** may be added to tequila, **Mandragora officinarum** to brandy, **Ephedra** to brandy, **Brugmansia** to light rum, **Datura innoxia** to tequila, peyote (**Lophophora williamsii**) to mescal, and fly agaric (**Amanita muscaria**) to vodka.

Left: The root of yellow gentian (*Gentiana lutea*), which grows in the Alps, is used to distill a schnapps known locally as *enzian* (= gentian). Unfortunately, the plant is sometimes confused with white hellebore (*Veratrum album*), a plant capable of producing a powerful and toxic state of inebriation.

Middle: The starchy root-balls of the yucca plant are used to distill alcohol. (*Yucca* sp. from North America)

Right: The sugarcane plant (*Saccharum officinarum* L.) is originally from Melanesia. From there, humans have carried it to all the tropical regions of the world. The freshly pressed juice of the plant can be used to brew beerlike beverages and can also be distilled into potent alcoholic beverages (firewater).

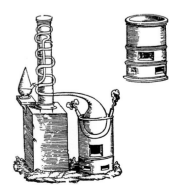

Distilling equipment from the early modern era. (Woodcut from Lonicerus, *Kreuterbuch*, 1679)

Top: This traditional *rakshi* container features the face of Bhairab, the god responsible for alcohol inebriation. (Nepal, twentieth century)

Bottom: A Mexican bottle of cheap sugarcane alcohol. Shamans drink copious quantities of this during their healing ceremonies.

Postcard, Ireland, ca. 1993

Ritual Use

Like all other alcoholic beverages (**beer, balche', palm wine, wine,** pulque [cf. *Agave* spp.]), distilled alcohol is used throughout the world for ritual purposes (Babor 1988). Surprisingly, some kinds of alcohol are used as shamanic drugs to induce trance and are offered to mountain spirits, gods, and Mama Coca (cf. *Erythroxylum coca*). Alcohol is also consumed ritually in many modern societies, such as when one shares a drink with a guest. Similar modern rituals are initiating communal drinking by clinking glasses or proposing a toast, the drinking binges of fraternities, et cetera.

In Nepal, Parvati, the divine wife and shakti of the Hindu god Shiva, is usually regarded as the creator of *rakshi* (a distilled alcohol typically made from a mash of millet). *Rakshi*, which is typically made at home, is both offered and drunk at the Buddhist sacrificial rites of the Newari and other Nepalese ethnic groups. The Newari say that it is good to drink some *rakshi*, but never so much that one becomes drunk.

Shiva, in his terrifying form as Bhairab, is a great lover of **beer** and alcohol. Because of this, his followers often drink (great amounts of) alcohol so that they may better identify with him (Fouce and Tomecko 1990, 19). In Darjeeling and Sikkim, millet alcohol is sometimes made more potent by adding the seeds of **Datura metel**, a plant that is also sacred to Shiva/Bhairab (Biswas 1956, 70*).

The Aghoris, Tantrists who follow the left-hand path, are able to consume enormous amounts of distilled alcohol without becoming drunk. They use the alcohol to train and sharpen their minds, which they can then use to transform the inebriating effects of the alcohol (Svoboda 1993, 173; cf. **Aconitum ferox, Cannabis indica**).

The North American Iroquois have long allowed unusual dreams to guide them in their lives. In later times, they used the "firewater" that the Europeans introduced to produce such dream states (Carpenter 1959).

The Indians of Mexico offer and drink great amounts of alcohol (*aguardiente, refino, yolixpa, pox*) during many of their rituals and prayer ceremonies. They frequently consume such large quantities that the sacred events often culminate in a state of collective drunkenness (cf. Knab 1995, 160*; Loyola 1986).

The shamans (*nahualli*) of the modern Nahuat offer *aguardiente* to the soul eaters of Talocan, the underworld, who hold captive the lost soul-parts (*tonalli*) of humans. The *aguardiente* is intended to make these otherworldly beings drunk, as it is then easier for the shamans to pull the lost *tonal* away from the soul eaters (Knab 1995*). A close connection between alcohol and witchcraft is found in other Mexican Indian cultures as well (Viqueira and Palerm 1954).

In the highlands of Chiapas, *aguardiente* is one of the most important shamanic drugs (Siverts 1973). In Zinacatán (Chiapas), a special *pox* prepared for the feast of San Lorenzo consists of sugarcane alcohol, sugarcane, pineapple juice, and an extract of **Ipomoea violacea** (Dawn Delo, pers. comm., 1996). The renowned shamans (healers) of the village of Masao, near Cuzco (Peru), drink large quantities of *cañazo* (homemade sugarcane alcohol) during their healing rituals, offering ceremonies (*t'inkupas*), and coca oracles. They usually also ingest equally substantial amounts of coca (cf. **Erythroxylum coca**).

A nineteenth-century observer provided the following description of the use of alcohol by the Siberian Samoyed and eastern Yakut shamans:

The shaman is familiar with his own kind, and yet he lives in enmity with the evil shamans. The shaman practices his art without compensation. Shamanism is passed down from father to son, although there are also female shamans. All of these shamans have their own specialty. One may be able to find lost objects, while another knows how to find the best places to fish. Others know how to reveal the site where a disease is located in the human body (worms on the heart for example!) or to locate stolen property. The shaman requires a glass of cognac, a knife, and a cross for this purpose. The latter object is needed especially by Christian, i.e., baptized shamans, for the power of the shaman has in no way been limited by Christianity and the cross. During the conjuration, the thief is struck in the eye with the knife. (Brutsgi 1987, 215)

When the Siberian people were forbidden to use the fly agaric mushroom for shamanic and hedonistic purposes during the Soviet era, many turned to vodka as a substitute. The shamans were able to utilize the alcohol, but many others became alcoholics.

Alcohol, and especially the associated problem of addiction, has been the subject of numerous autobiographical novels (famous examples include *John Barleycorn: Alcoholic Memoirs*, by Jack London, and *Der Trinker* [The Drinker], by Hans Fallada).

The Effects of Alcohol

The euphoric effects of certain amounts of alcohol may result from the release of endorphins or the activation of the endorphinergic system induced by alcohol (Verebey and Blum 1979). It has been speculated that acetaldehyde, the initial metabolite of alcohol, reacts with dopamine and enzymes to produce **morphine**like substances, and that it is these that lead to alcoholism (Davis and Walsh 1970). It is possible that alcohol consumption may cause psychoactive β-**carbolines** (tetrahydroharmane, harmane; cf. **harmaline and harmine**) to be created in the body, and that these in turn are responsible for certain euphoric effects of the alcohol. Recent experiments have demonstrated that acetaldehyde and tryptamine are converted in vivo into tetrahydroharmane via an enzymatic process (Callaway et al. 1996). Elevated concentrations of harmane have been detected in the bodies of alcoholics (Susilo 1994).

The effects of alcohol can be changed, suppressed, or intensified by combining it with other substances. The effects may be suppressed by coca (*Erythroxylum coca, Erythroxylum novogranatense*), *Ephedra* species, **ephedrine**, **mescaline**, **cocaine**, henbane (*Hyoscyamus niger*), **nicotine**, LSD, and **psilocybin**. Both *Ledum palustre* and *Piper methysticum* can potentiate the effects of alcohol. Synergistic effects (interactions) can result when alcohol is used in combination with MAO inhibitors (β-**carbolines**), **diazepam**, and numerous medicines (psychopharmaceuticals).

Hallucinogenic Salamander Brandy

In the mountains northwest of Ljubljana, Slovenia, a distinctly hallucinogenic alcohol is still being distilled today according to ancient (alchemical) recipes. Following the distillation of a fruit mash, living European fire salamanders (*Salamandra salamandra*) are placed in the distillation vessel. The salamanders are then slowly heated to a very high temperature. It is said that the more the animals suffer, the more the desired alkaloids are exuded into the distillate. The salamanders' skin contains the psychoactive steroidal alkaloids samandarin[386] and samandaridin. Samandenon is also present. The effects of salamander brandy are said to be similar to those of **ibogaine** or **strychnine**. In Slovenia, it is legal to distill live salamanders (Ogorevc 1995). An alternative method consists of simply adding living salamanders to high-percentage spirits (Valenčič 1998).

Literature

Babor, Thomas. 1988. *Alcohol: Customs and rituals.* London: Burke Publishing.

Bosi, Roberto. 1995. *I Distillati—Edle Brände: Von der Kunst des Destillierens.* Edition Spangenberg. Munich: Droemer Knaur.

Bourke, John G. 1893. Primitive distillation among the Tarascos. *American Anthropologist,* o.s., 6:65–69.

———. 1894. Distillation by early American Indians. *American Anthropologist,* o.s., 7:297–302.

Braun, Stephen. 1996. *Buzz.* New York: Oxford University Press.

Braunschweig, Hieronymus. 1610. *Ars destillandi oder die rechte Kunst zu destillieren.* Strasbourg.

Brutsgi, Franz Georg, ed. 1987. *Forschungsreisen des Grafen Karl von Waldburg-Zeil nach Spitzbergen und Sibirien 1870, 1876, 1881.* Constance: Rosengarten Verlag.

Bunzel, Ruth. 1940. The role of alcoholism in two Central American cultures. *Psychiatry Journal of the Biology and Pathology of Interpersonal Relations* 3:361–87.

Carpenter, E. S. 1959. Alcohol in the Iroquois dream quest. *American Journal of Psychiatry* 116:148–51.

Callaway, James C., Malmo M. Airaksinen, Katja S. Salmela, and Mikko Salaspuro. 1996. Formation of tetrahydroharman (1-methyl-1,2,3,4-tetrahydro-beta-carboline) by *Helicobacter pylori* in the presence of ethanol and tryptamine. *Life Sciences* 58 (21): 1817–21.

Davis, Virginia, and Michael J. Walsh. 1970. Alcohol, amines, and alkaloids: A possible biochemical basis for alcohol addiction. *Science* 167:1005–7.

Dennis, P. A. 1975. The role of the drunk in an Oaxacan village. *American Anthropologist,* n.s., 77 (4): 856–63.

Douglas, Mary, ed. 1987. *Constructive drinking: Perspectives on drink from anthropology.* New York: Cambridge University Press.

Fallada, Hans. 1959. *Der Trinker.* Hamburg: Rowohlt.

Fouce, Paula, and Denise Tomecko. 1990. *Shiva.* Bangkok: The Tamarind Press.

Frence, Lothar. 1995. Die größten Trunkenbolde des Tierreichs. *Das Tier* 2/95:14–17.

Gast, Arbo. 1986. *Liköre, Schnäpse und Wein selbstgemacht aus Früchten, Beeren und Kräutern.* Munich: Heyne.

Habermehl, Gerhard G. 1987. *Gift-Tiere und ihre Waffen.* 4th ed. Berlin: Springer Verlag.

Höschen, Ulrich. n.d.. *Das große Buch der feinen Spirituosen.* Cologne: Naumann und Göbel.

Lall, Kesar. 1993. *The origin of alcohol and other stories.* Kathmandu: Ratna Pustak Bhandar.

London, Jack. 1981. *John Barleycorn: Alcoholic memoirs.* Santa Cruz, Calif.: Western Tanager Press. (Orig. pub. 1913.)

Loyola, Luis J. 1986. The use of alcohol among Indians and Ladinos in Chiapas, Mexico. In *Drugs in Latin America,* ed. Edmundo Morales, 125–48. Studies in Third World Societies, no. 37. Williamsburg, Va.: College of William and Mary.

Marshall, Mac, ed. 1979. *Beliefs, behaviors, and alcoholic beverages: A cross-cultural survey.* Ann Arbor: University of Michigan Press.

Marsteller, Phyllis, and Karen Karnchanapee. 1980. The use of women in the advertising of distilled spirits. *Journal of Psychedelic Drugs* 12 (1): 1–12.

Mayr, Christoph. 1984. *Schnapsfibel: Kräutergeist für Gesunde und Kranke.* Bozen: Athesia.

"I would say that the drug that gets you knocked up, blindly and unconsciously, is alcohol. Alcohol does reduce inhibitions—people become aggressive, indiscriminately loving *or* hostile, weeply self-pitying or self-expansive. Alcohol stimulates the social emotions."

TIMOTHY LEARY
THE POLITICS OF ECSTASY
(1998, 213*)

"Salamanders have been known to be poisonous since the most ancient of times. Like toads, for thousands of years salamanders have played an important role as animals with magical powers. In ancient Persian mythology, the salamander was the animal that extinguishes fire, just as it did for the alchemists of the Middle Ages."

GERHARD G. HABERMEHL
GIFT-TIERE UND IHRE WAFFEN
[POISONOUS ANIMALS AND THEIR WEAPONS]
(1987, 123)

A Chinese illustration of an alcoholic, who can be recognized by his red drinker's nose.

386 "Samandarin is a convulsive poison; it affects the central nervous system but also has hypertensive and local anesthetic effects. Externally, it is extremely irritating to the mucous membranes" (Altmann 1980, 130*).

"The other type of drinker has imagination, vision. Even when most pleasantly jingled he walks straight and naturally, never staggers nor falls, and knows just where he is and what he is doing. It is not his body but his brain that is drunken. He may bubble with wit, or expand with good fellowship. Or he may see intellectual specters and phantoms that are cosmic and logical and that take the forms of syllogisms. It is when in this condition that he strips away the husks of life's healthiest illusions and gravely considers the iron collar of necessity to be welded about the neck of his soul. This is the hour of John Barleycorn's subtlest power."

JACK LONDON
JOHN BARLEYCORN: ALCOHOLIC
MEMOIRS
(1981, 12)

McKenna, Terence, and Werner Pieper. [1993]. *Die süßeste Sucht. Ist Zucker eine Killer-Droge?* Löhrbach: Werner Pieper's MedienXperimente; Solothurn: Nachtschatten Verlag.

McDonald, Maryon, ed. 1994. *Gender, drink and drugs.* Oxford: Berg Publisher.

Ogorevc, Blaž. 1995. Halluzinogene Droge, gemacht in Slovenia: Salamander brandy. *Mladina* 23:26–32. (In Slovakian.)

Pischl, Josef. 1996. *Schnapsbrennen.* Munich: Heyne.

Rose, A. H., ed. 1977. *Alcoholic beverages.* New York: Academic Press.

Siverts, Henning, ed. 1973. *Drinking patterns in highland Chiapas.* Bergen, Oslo, and Trom: Norwegian Research Council for Science and the Humanities.

Spode, Hasso. 1993. *Die Macht der Trunkenheit: Kultur- und Sozialgeschichte des Alkohols in Deutschland.* Opladen: Leske + Budrich.

———. 1994. Vom Archaischen des Gelages. *NZZ-Folio* (August): 18–21.

Susilo, Rudy. 1994. Metaboliten der Indolaminneurotransmitter: Schlüsselsubstanzen zum Alkoholismus? *Pharmazie in unserer Zeit* 23 (5): 303–11.

Svoboda, Robert E. 1993. *Aghora: At the left hand of God.* New Delhi: Rupa.

Valenčič, Ivan. 1998. Salamander brandy: A psychedelic drink made in Slovenia. *Yearbook for Ethnomedicine and the Study of Consciousness,* 1996 (5): 213–25. Berlin: VWB.

Verebey, Karl, and Kenneth Blum. 1979. Alcohol euphoria: Possible mediation via endorphinergic mechanisms. *Journal of Psychedelic Drugs* 11 (4): 305–11.

Viqueira, C., and Angel Palerm. 1954. Alcoholismo, brujería y homicidio en dos communidades rurales de México. *América Indígena* 14 (1): 7–36.

Ayahuasca

Other Names

Ambihuasca, ambiwáska, ayawáska, biaxíi, brew ("the brew"), caapi, cají, calawaya, camaramti (Shipibo), chahua (Shipibo), cipó, daime, dapa, dapá, djunglehuasca, djungle tea, doctor, dschungel-ambrosia, el remedio, hoasca, honi, iyaona (Zapara), jungle ambrosia, jungle-huasca, jungle tea, kaapi, kahi, kahpi, la droge (Spanish, "the drug"), la purga (Spanish, "the purgative"), la soga, masha (Shipibo), metí, mihi, mii (Huaorani), moca jene (Shipibo, "bitter brew"), muka dau (Cashinahua, "bitter medicine"), natem (Achuar), natema, natemá, natemä, nepe, nepi, nichi cubin (Shipibo, "boiled liana"), nishi sheati (Shipibo, "liana drink"), nixi honi, nixi paé, notema, ohoasca, ondi (Yaminahua), pilde, pildé, pinde, pindé, rao (Shipibo, "medicinal plant"), remedio, sachahuasca, santo diame, uni (Conibo), vegetal, yagé, yajé, yaxé

The psychoactive drink known as ayahuasca has been used by shamans and medicine men in the Amazon region for healing rituals and shamanic experiences since pre-Columbian times (Naranjo 1979). Such use is probably as ancient as South American civilization itself and apparently was first discovered in the western regions of the Amazon basin (now Ecuador) (Naranjo 1979). In the coastal regions of Ecuador, so-called witches' pots—used for brewing ayahuasca—have been discovered in archaeological excavations. These pots are estimated to be about 3,500 years old (Andritzsky 1989, 57*).

How the beverage was first discovered is still a mystery. But it certainly was more than just an accidental discovery made by primitive Indians:

A long time ago, a skilled hunter lived in the rain forest. One day, when he was far from his home, he heard a liana speaking to him. The hunter, who knew many things about making arrow poisons from roots, barks, and seeds for hunting, understood the power of plants. He returned to his house with his new find.

Left: *Banisteriopsis caapi* stems are the basis of all ayahuasca preparations.

Right: The leaves of *Psychotria viridis,* which contain DMT, are the most popular ayahuasca admixture.

During the following night, he had a dream in which the spirit of the liana explained to him how to prepare it into a brew that one could use to heal many diseases.

Today, shamans still use the "drink of true reality" to determine the origins of diseases, travel into the normally invisible world of the forest, communicate with the lords of the animals and plants, and accompany the participants of tribal rituals into the world of myths.

The beverage is a unique pharmacological combination of the *Banisteriopsis caapi* liana, which contains harmaline, and chacruna leaves (*Psychotria viridis*), which contain DMT. Harmaline is an MAO inhibitor—it inhibits the excretion of the endogenous substance monoamine oxidase, which breaks down the psychoactive substance *N,N*-DMT. It is only through this combination of active ingredients that the drink is able to produce its consciousness-expanding effects (Rivier and Lindgren 1972). Because of the powerful and often three-dimensional visions, ayahuasca is sometimes jokingly referred to as "Amazonian television" ("the Nature Channel") or "jungle cinema."

The shamans attribute the effects of ayahuasca not to any active constituent but to the plant spirits who reveal themselves as master-teachers to humans when they are under the influence of ayahuasca. The plant spirits make it possible for humans to discover the origins of a disease, obtain the recipe for a medicine, and find out where the wild game is hiding deep in the forest. The shamans have been using their magical drink for a long time, and clearly with great success. As former rain-forest regions have become increasingly urbanized, more and more non-Indians have come into contact with the magical drink, which has led to the development of an urban shamanism. Now, many Catholic mestizos have set themselves up as urban shamans and use the drink to treat the afflictions of city dwellers. Their rituals are a colorful mix of Indian and Catholic customs during which Christian songs are sung and the spirits of the forest invoked (Dobkin de Rios 1970, 1972, 1989, 1992; Luna 1986). Numerous ayahuasca churches and sects have emerged, as has a vigorous ayahuasca tourism.

Recipes

In the past, methods for preparing ayahuasca were well-protected secrets of the shamans. Only they knew the ingenious recipes. Only they knew which plants to use, where to find the lianas and herbs, which protective spirits needed to be invoked, and how to prepare the brew.

Banisteriopsis caapi stems are the basis for all ayahuasca recipes. To prepare ayahuasca, manageable-size stems of this liana must be boiled, after which chacruna leaves (*Psychotria viridis*) are

added. The mixture is allowed to sit on the fire until a black, thick, horrible-tasting liquid results. The drink should never be prepared in aluminum pots, as it will corrode the aluminum and may in some cases produce inedible aluminium salts. Although cold-water extracts of *Banisteriopsis caapi* and *Psychotria viridis* will also produce the desired effects, they are only rarely made.

In the recipes of the Amazonian Indians, the liana itself is typically the main ingredient. Tests of different samples have found 20 to 40 mg, 144 to 158 mg, and even 401 mg of β-carbolines as well as 25 to 36 mg of *N,N*-DMT per dose. The ayahuasca prepared by the urban mestizos contains consistently higher concentrations of alkaloids (especially *N,N*-DMT) than are found in the Indian preparations. The highest concentrations are said to be found in the preparations of the Barquinha Santo Daime church (Luis Eduarda Luna, pers. comm., 1996).

Natema Recipe of the Shuar

The Shuar shamans (*uwishin*) split a 1- to 2-meter-long piece of *Banisteriopsis caapi* stem into small strips. They place the strips in a pot along with several liters of water. They then add leaves of *Diplopterys cabrerana*, a *Herrania* species, *Ilex guayusa*, *Heliconia stricta*, and an unidentified Malpighiaceae known as *mukuyasku*. The resulting mixture is boiled until most of the water has evaporated and a syrupy fluid remains (Bennett 1992, 486*). The Kamsá, Inga, and Secoya make similar preparations (Bristol 1965, 207 ff.*).

Ecuadoran Recipe

The bark of the *Banisteriopsis caapi* liana is peeled off and placed beneath a certain tree in the forest.

A Shipibo man inspects his chacruna plant (*Psychotria viridis*).

A Tukano ayahuasca vessel and drinking bowls. (From Koch-Grünberg, *Zwei Jahre bei den Indianern Nordwest-Brasiliens* [Two Years among the Indians of Northwest Brazil], 1921)

These paintings on the community house (*maloca*) were inspired by the use of ayahuasca. (From Koch-Grünberg, *Zwei Jahre bei den Indianern Nordwest-Brasiliens* [Two Years among the Indians of Northwest Brazil], 1921)

Ayahuasca prepared from *Banisteriopsis caapi* and *Psychotria viridis*.

A very ancient, presumably pre-Columbian petroglyph found on a granite cliff in Nyi on the lower Río Piraparaná (Colombia). The Tukano believe that it was at this sacred place that the Sun-Father wed the first earth woman, thereby creating the Tukano people. The Desana (one of the Tukano tribes) interpret the triangular face as the cosmic vagina and the stylized human figure below it as a winged phallus. They say that ayahuasca was created at the beginning of history when the two poles were united. (Redrawn by C. Rätsch)

387 Amazonian Indians claim that the fruits of this tree produce an unusual effect. They are eaten by certain birds (*Nothocrax urumutum* Spix), which are in turn hunted and eaten by the Indians. When their dogs eat the bones of these birds, they immediately exhibit severe toxic symptoms (Schultes 1960).

388 Older texts claim that *Prestonia* contains *N,N*-DMT; this information is unfortunately incorrect. The common name *yagé* probably refers solely to the fact that the plant is used as an ayahuasca admixture (Schultes and Raffauf 1960).

The bare stems are then split into four to six strips and boiled together with fresh or dried *Psychotria viridis* leaves. A piece of liana approximately 180 cm long and forty *Psychotria* leaves represent a single dosage, although a piece of stem just 40 cm long and 3 cm thick is also said to be sufficient. In general, the less vine that is used, the easier the ayahuasca is on the stomach.

Preparation of the União do Vegetal (UDV), Brazil

Pieces from the *Banisteriopsis caapi* vine are pounded, mixed with leaves from *Psychotria viridis*, and boiled for 10 to 12 hours in rust-free steel pots until all that remains is a thick liquid with globules of fat on the surface that shimmer in all colors of the spectrum.

Recipe of the Shipibo of San Francisco/Yarinacocha

A fresh piece of *Banisteriopsis caapi* bark is boiled together with a handful of chacruna leaves (*Psychotria viridis*) and a *flor de toé* (*Brugmansia suavolens* flower) until a thick liquid decoction is produced. This preparation is said to have especially strong effects and to produce many visions.

Indigenous ayahuasca preparations exhibit considerable variation. Numerous plant admixtures can be used to induce psychoactive effects, and stimulating or medicinal drinks can also be produced. An Ecuadoran preparation of *Banisteriopsis caapi* and *Ilex guayusa* is purported to be a strong purgative. Recipes that cause delirium often contain tobacco and angel's trumpets (*Brugmansia*). Experienced ayahuasca shamans possess a vast wealth of knowledge about the effects of many plants and may utilize more than one hundred different admixtures in order to achieve the effects they desire.

These traditional preparations are often devoid of *N,N*-DMT. However, it is precisely those drinks that do contain high concentrations of DMT and that do produce visionary effects that have exerted such a powerful attraction on legions of Western enthobotanists, psychedelic cognoscenti, artists, New Age tourists, and seekers of the esoteric (Leginger 1981*; McKenna 1989*; McKenna and McKenna 1994*; Perkins 1995). For most outsiders, experiences with Amazonian ayahuasca have tended to be rather disappointing (McKenna 1993). Westerners seeking "highs" or healing experiences are often duped by the pranks of *curanderos* or self-proclaimed shamans. As early as 1953, William Burroughs reported, ". . . I had been conned by medicine men" (Burroughs and Ginsberg 1963, 15). But there are also examples of more positive experiences (Pinkson 1993; Wolf 1992).

Traditional Ayahuasca Admixtures

(from Ayala Flores and Lewis 1978; Bennett 1992*; Bianchi and Samorini 1993; Faust and Bianchi 1996; Luna 1984b, 1986; Ott 1993, 269 ff.*; Ott 1995; Pinkley 1969; Schultes 1972; modified and expanded)

Botanical Name	Common Name	Active Constituents
ACANTHACEAE		
Teliostachys lanceolata var. *crispa* Nees	toé negro	
AMARANTHACEAE		
Alternanthera lehmannii Hieronymus	picurullana quina, borrachera	
***Iresine* sp.**		
Pfaffia iresinoides	marosa	
APOCYNACEAE		
Himatanthus sucuuba (Spruce) Woodson	bellaco-caspi, sucuuba, platanote	fulvoplumieron
Malouetia tamaquarina (Aubl.) DC.[387]	cuchura-caspi, chicle	**indole alkaloids,** conessine, dihydrokurchessine, kurchessine, tetramethylholarhimine
Mandevilla scabra Schumann		
Prestonia amazonica (Benth.) Macbride [syn. *Haemadyction amazonicum*]	yajé	?[388]
Tabernaemontana sananho Ruíz et Pav.	tzicta	
***Tabernaemontana* sp.**	uchu-sanango	alkaloids
Thevetia sp.	cabalonga blanca	cardiac glycosides
AQUIFOLIACEAE		
***Ilex guayusa* Loes.**	guayusa, wais	**caffeine**

Botanical Name	Common Name	Active Constituents
ARACEAE		
Montrichardia aborescens Schott	raya balsa, camotillo	
BIGNONIACEAE		
Mansoa alliacea (Lam.) A. Gentry	ajo sacha	
Tabebuia heteropoda (DC.) Sandwith		
Tabebuia incana A. Gentry	tahuarí	
Tabebuia sp.		
Tynanthas panurensis (Burman) Sandwith	clavohuasca	
BOMBACACEAE		
Cavanillesia hylogeiton Ulbrich	puca lupuna, embirana	
Cavanillesia umbellata Ruíz. et Pav.		
Ceiba pentandra (L.) Gaertn.	lupuna, kapok, ceiba	
Chorisia insignis H.B.K.	lopuna, yuchán, palo borracho	resin
Chorisia speciosa St.-Hil.	samohú, ceiba	
Quararibea sp.	ishpingo	(see **espingo**)
BORAGINACEAE		
Tournefortia angustiflora Ruíz et Pav.		
CACTACEAE		
Epiphyllum sp.	pokere, wamapanako	
Opuntia sp.	thai	**mescaline**
CARYOCARACEAE		
Anthodiscus pilosus Ducke		
CELASTRACEAE		
Maytenus ebenifolia Reiss.	chuchuhuasi	
Maytenus laevis Reiss.	chuchuasca	**caffeine (?)**
CLUSIACEAE		
Tovomita sp.	chullachaqui caspi	
CONVOLVULACEAE		
Ipomoea carnea (cf. ***Ipomoea* spp.**)	toé	**ergot alkaloids**
CYCLANTHACEAE		
Carludovica divergens Ducke	tamshi	
CYPERACEAE		
Cyperus digitatus Roxb.	chicorro	
Cyperus prolixus H.B.K.		
***Cyperus* spp.**[389]	piripiri	**ergot alkaloids**
DRYOPTERIDACEAE		
Lomariopsis japurensis (Martius) J. Sm.	shoka, dsuiitetetseperi	
ERYTHROXYLACEAE		
Erythroxylum coca var. *ipadú* Plowman	ipadú	**cocaine**
EUPHORBIACEAE		
Alchornea castaneifolia (Willd.) Just. (cf. ***Alchornea* spp.**)	hiporuru	alkaloids (?)
Croton sp. (?)	tipu, tipuru	**morphine**
Euphorbia sp.	ai curo	
Hura crepitans L.	catahua, assacu	piscidides, lectins
GNETACEAE		
Gnetum nodiflorum Brongn.	tap-kam', hoo-roo', itua	
GRAMINEAE		
Arundo donax	carrizo	tryptamines, DMT
GUTTIFERAE		
Clusia sp.	miya, tara	

389 In the Afro-Brazilian Candomblé cult, the aromatic roots of *dandá da costa* (*Cyperus rotundus*) are chewed in order to give one the power to influence other people and to acquire personal power (Voeks 1989, 122, 123*).

"Gradually, faint lines and forms began to appear in the darkness, and the shrill music of the *tsentsak*, the spirit helpers, arose around him. The power of the drink fed them. He called, and they came. First, *pangi*, the anaconda, coiled about his head, transmuted into a crown of gold. Then *wampang*, the giant butterfly, hovered above his shoulder and sang to him with its wings. Snakes, spiders, birds and bats danced in the air above him. On his arms appeared a thousand eyes as his demon helpers emerged to search the night for enemies. The sound of rushing water filled his ears, and listening to its roar, he knew he possessed the power of *Tsungi*, the first shaman. Now he could see. Now he could find the truth. He stared at the stomach of the sick man. Slowly, it became transparent like a shallow mountain stream, and he saw within it, coiling and uncoiling, *makanchi*, the poisonous serpent, who had been sent by the enemy shaman. The real cause of the illness had been found."

MICHAEL HARNER
THE SOUND OF RUSHING WATER
(1973, 15 f.)

Botanical Name	Common Name	Active Constituents
HELICONIACEAE		
Heliconia stricta Huber		
Heliconia sp.	winchu	
LABIATAE		
Ocimum micranthum Willd.	pichana, abaca	**essential oil**
LECYTHIDACEAE		
Couroupita guianensis Aubl.	ayahuma	**indole alkaloids** (couroupitine A, B), stigmasterol, campesterol
LEGUMINOSAE		
Bauhinia guianensis Aubl.		
Caesalpinia echinata Lam.	cumaseba	
Calliandra angustifolia Spruce ex Benth.	bobinsana, quinilla blanca, chipero	alkaloids (harmane)
Calliandra pentandra (cf. **Calliandra anomala**)		harmane, DMT (?)
Campsiandra laurifolia Benth.	huacapurana	
Cedrelinga catenaeformis Ducke	huairacaspi, cedrorana	
Erythrina fusca Lour.	amasisa, gachica	erythraline, erythramine, erythratine
Erythrina glauca Willd.	amasisa	
Erythrina poeppigiana (Walpers) Cook (cf. **Erythrina** spp.)	amaciza, oropel	alkaloids
Pithecellobium laetum Benth.	remo caspi, pashaquillo, shimbillo	alkaloids
Sclerobium setiferum Ducke	palisangre, palisanto	
Vouacapoua americana Aubl.	huacapo, hucapù	
LORANTHACEAE		
Phrygilanthus eugenioides (L.) H.B.K.	miya, ho-ho-ho	
Phrygilanthus eugenioides var. *robustus* Galz.		
Phtirusa pyrifolia (H.B.K.) Eichler	suelda con suelda	
MALPIGHIACEAE		
Banisteriopsis rusbyana (Niedenzu) Morton	oco-yagé	DMT, β-carbolines
Diplopterys cabrerana (Cuatr.) Gates	yaco-ayahuasca, yajé, yaji	DMT
Diplopterys involuta (Turcz.) Niedenzu [syn. *Mezia includens* (Benth.) Cuatr.]		
Mascagnia psilophylla var. *antifebrilis* Niedenzu [syn. *Cabi paraensis* (Juss.) Griseb., syn. *Callaeum antifebrile* (Grisb.) Johnson]		
Stygmaphyllon fulgens (Lam.) Jussieu	ki-ria, kairia	
MARANTHACEAE		
Calathea veitchiana Veitch ex Hook. fil.	pulma	
MELIACEAE		
Trichilia tocacheana C. DC.	lupuna	latex
MENISPERMACEAE		
Abuta grandifolia (Martius) Sandwith	abuta, trompetero, sanango	palmatine
MORACEAE		
Coussapoa tessmannii Mildbread	renaco	
Ficus insipida Willd.	renaco, hojé, huito, bamba	
Ficus ruiziana Standl.		
Ficus sp.		
MYRISTICACEAE		
Virola surinamensis (Roland) Warb.	caupuri, cumala blanca	neolignans
Virola spp.	cumala	DMT

Botanical Name	Common Name	Active Constituents
NYMPHAEACEAE		
Cabomba aquatica Aubl.	mureru, murere	
PHYTOLACCACEAE		
Petiveria alliacea L.[390]	muckra, mucura, chanviro	**coumarins** (nineteen), isoarboriol, trithiolan, trithiolaniacine
PIPERACEAE		
Peperomia sp.	tsemtsem	**essential oil**
Piper sp.		**essential oil**
POLYGONACEAE		
Triplaris surinamensis Chamisso	tangarana	
Triplaris surinamensis var. *chamissoana* Meissner	tangarana	
PONTEDERIACEAE		
Pontederia cordata L.	amarrón borrachero	
RUBIACEAE		
Calycophyllum spruceanum (Benth.) Hook. fil.	capirona negro	
Capirona decorticans Spruce	capirona negro, kashi muna	
Guettarda ferox Standl.	garabata	
Psychotria carthaginensis Jacq.	yage-chacruna, rami appani, sameruca	DMT
Psychotria psychotriaefolia (Seem.) Standl.	chacruna	DMT
Psychotria viridis Ruíz et Pav.	chacruna	
Psychotria spp.	batsikawa, kawa kui, nai kawa, pishikawa, rami appane	
Rudgea refifolia Standl.		
Sabicea amazonensis Wernham	chà-dê-kê-na, kana, koti-kana-ma	
Uncaria guianensis (Aubl.) Gmelin	garabata	indoles: angustine, isorhynchophylline, rhynchophylline-*N*-oxide, dihydrocory-nantheine hirsutine, hirsutein
Uncaria tomentosa (?)	uña de gato	**indole alkaloids**
SAPINDACEAE		
Paullinia yoco Schultes et Killip (cf. **Paullinia** spp.)	yoco	**caffeine**
SCHIZAEACEAE		
Lygodium venustum Swartz	tchai del monte, rami	
SCROPHULARIACEAE		
Scoparia dulcis L.[391]		amellin, triterpenes, 6-methoxybenzoxozolinone
SOLANACEAE		
Brugmansia insignis	toa-toé, sacha-toé, danta borrachera	**tropane alkaloids**
Brugmansia suaveolens	(flor de) toé, tsuak, borrachero, floripondio	**tropane alkaloids**
Brunfelsia chiricaspi Plowman	chiricaspi, chiricsanango	**scopoletin**
Brunfelsia grandiflora D. Don	chiricaspi, chiricsanango	**scopoletin**
Brunfelsia grandiflora ssp. *schultesii* Plowman (cf. **Brunfelsia** spp.)	sanango, chiricsanango	**scopoletin**
Capsicum sp.	catsi, aji	capsaicin
Iochroma fuchsioides (H.B.K.) Miers	borrachero, guatillo, paguando, campanitas	alkaloids (tropane derivatives)

"I came to the realization, that all plants which are called the *doctores*, or *vegetales que enseñan* (= plants, the teachers) either 1) elicit hallucinations when they are taken alone, 2) in some way influence the effect of the *ayahuasca* drink, 3) cause dizziness, 4) have strongly emetic and/or cathartic qualities, or 5) elicit very vivid dreams. Often a plant has all of these characteristics or at least some of them."

LUIS EDUARDO LUNA
"THE CONCEPT OF PLANTS AS TEACHERS"
(1984a, 135–56)

390 The Shipibo-Conibo call *Petiveria alliacea* L., a member of the Family Phytolaccaceae (cf. *Phytolacca acinosa*), *vuén mucura* or *mucura sacha*. They claim that it has "hallucinogenic properties" and esteem it as a "master teacher of the healing arts" (Gebhart-Sayer 1987, 336).

391 In West Africa, this plant is used in combination with *Combretum micranthum* to treat fever (Assi and Guinko 1991, 122*).

Botanical Name	Common Name	Active Constituents
Juanulloa ochracea Cuatre.	ayahuasca, bi-ti-ka-oo-k, na-ka-te-pê	parquine (?)
Markea formicarium Dammer	ree-ko-pa	scopoletin (?)
Nicotiana rustica L.	tabaco	nicotine
Nicotiana tabacum L.	mapacho	nicotine
STERCULIACEAE		
Herrania sp.	kushiniap	alkaloids (?)
VERBENACEAE		
Cornutia odorata (P. et Endlicher) Poeppig	shinguarana, ulape, tal	
Vitex triflora Vahl.	tahuari, taruma	
VIOLACEAE		
Rinorea viridiflora Rusby	chacruna, amanga, capinuri, ayahuasca	

Botanically Unidentified Ayahuasca Admixtures
(from Bennett 1992*; Schultes 1966*, 1972*; also Ott 1993, 418*)

Indigenous Name	Part Used	Culture/Tribe
caapi-pinima	leaves ?	Brazil
doxké-mo-reri-dá	leaves	Tukano (Brazil)
duxtú-sarēnō-da	leaves	Tukano (Brazil)
ishpingo	fruit (?)	Shipibo-Conibo (Peru)
jénen-joni-rau	"parts"	Shipibo-Conibo (Peru)
kāna-puri	crushed fresh leaves	Tukano (Brazil)
kaxpi-puri	crushed fresh leaves	Tukano (Brazil)
ma-kaxpi-dá	?	Tariano (Vaupés)
muchípu-gahpí-dá ("sun yagé")	?	Tukano (Vaupés)
mukuyasku (a Malpighiaceae)	leaves	Shuar (Ecuador)
para-para (an unspecified Violaceae)	?	Shuar (Ecuador)
tipuru (*Croton* sp.?)[392]	?	Shuar (Ecuador)
vaí-gahpí ("fish-yagé")	?	Tukano (Vaupés)

In the Brazilian Santo Daime cult, certain religious songs are sung at worship ceremonies while worshippers are under the influence of ayahuasca. The most important of these songs were released privately on a CD in 1996. (CD cover, ©Richard Yensen, Orenda Institute)

Traditional Uses

To the shaman, ayahuasca is inseparable from the rain forest. The power of the drink enables him to see the spirits that are present in the plants and animals of the forest. He communicates with them and acquires a knowledge of their innermost being. In this way, he comes to know the significance of every individual animal and every single plant and to understand why each species has its necessary place in the "circle of life." With the aid of ayahuasca, he seeks out the lords of the animals. In the "true reality," these appear to him in human form, and from them he learns, for example, why the hunters are no longer able to find their children, the animals. The reason might be that an unknown hunter has killed too many animals, leaving the bodies behind in the jungle, unused and rotting. This enrages the lord of the animals, and he demands compensation. The shaman must impregnate the soul of a female animal so that there will again be enough offspring. The shaman returns to everyday life, reports about his experiences, and warns the hunters that they will be able to hunt these animals again only if they give them

time to recover. Any hunter who violates this law will be punished by the lord of the animals, perhaps with an invisible magical arrow.

The shamans utilize the visionary effects of ayahuasca to travel to the true reality, often referred to as the "blue zone," where they are able to explore the secrets of the past, present, and future; to heal sick members of the tribe; or to fight against a harmful sorcerer, a "black shaman" (Reichel-Dolmatoff 1996b*), who typically sends invisible objects (such as arrows, thorns, or crystals) into the body of his victim. In the mundane world, nothing about the victim appears to be different. It is only when the shaman drinks the ayahuasca that he is truly able to *see*. If he can recognize the alien magical objects, he can suck them out and remove them. To make the invisible healing process manifest to the external, visible world, he proudly shows the patient and the others in attendance a bloody thorn that he had previously placed in his mouth with a clever sleight of hand. This small "deception"—which serves only to make the invisible visible—is essential to the success of the treatment (Ott 1979).

392 Several South American *Croton* species contain morphinelike alkaloids (see **morphine**) (Bennett 1992, 490; Schultes and Raffauf 1990*).

Left from top to bottom:
Fiery hot chili pods (*Capsicum* sp.) may be added to the ayahuasca drink.

In Mayan culture, the *kapok* tree (*Ceiba pentandra*) is regarded as the World Tree, which links the different levels of the universe with one another. The shaman can use the tree to climb up to the heavens or to descend to the underworld. The *kapok* is also viewed as a World Tree and shamanic tree in South America, where it is used as an ayahuasca admixture. Although the tree does have a symbolic significance, it probably has no psychoactive effects. (Photographed at Palenque in Chiapas, Mexico)

Chorisia insignis, the sacred Argentinean tree known as *palo borracho* ("drunken tree"), looks like a "pregnant" *Ceiba pentandra*, and the two trees are very easily confused. Its bark is sometimes used as an ayahuasca additive.

Many South American peoples venerate the pink-flowering *Chorisia speciosa* as a shamanic tree or World Tree. The bark of this tree is used as an ayahuasca additive.

Right from top to bottom:
The fruit of this rattletree (*Thevetia* sp.) is added to ayahuasca as a powerful magical substance; such use, however, appears to be very dangerous.

In Amazonia, the tropical tree *Couroupita guianensis* is regarded as a medicine that stimulates dreams; for this reason it is added to ayahuasca. Because of the flowers' unusual beauty and the unusual way in which they grow directly out of the trunk, the tree has spread into all tropical regions as a decorative plant. In Thailand, it is even religiously venerated.

This *Tournefortia* species is used as an ayahuasca additive. (Photographed in northwestern Peru)

Ipomoea carnea, known locally as *toé*, is used as an ayahuasca admixture in the Pucallpa region. (Photographed in Peru)

"*Nixi honi*
vision vine
boding spirit of the forest
origin of our understanding
give up your magic power
to our potion
illuminate our mind
bring us foresight
show us the designs of our enemies
expand our knowledge
expand our understanding
of our forest."

AYAHUASCA SONG OF THE
AMAHUACA INDIANS
IN *THE NATURAL MIND*
(WEIL 1972, 106)

Amazonian shamans usually travel to the other reality while in a different form. The *payé* has "turned his stomach inside out"—this is how the Tukano describe the condition in which the shaman's body lies as if dead while his consciousness has taken off into another reality. The shamanic soul has transformed itself into a jaguar and now flies over a rainbow to the Milky Way. The most fantastic colors and forms unfold before the shaman's inner eye. Honeycomb patterns dance by and change into crystals filled with an other-worldly light. Wavy lines flow out and back together into colorful swirls. The jaguar shaman is irresistibly sucked in. The swirls open into a tunnel made of circling skulls, at the end of which shines a warm, blue light. The jaguar shaman has reached the Milky Way, where he meets the ayahuasca woman who revealed the true reality to humans at the dawn of creation and gave them the secret of the "drink of true reality" (Reichel-Dolmatoff 1971*, 1975*, 1978*).

Although the true reality lies beyond the Milky Way, it is nevertheless a mirror image of the rain forest. To the shamans, the mundane visible world is only a world of appearances; behind its manifestations, the world of myths, gods, and spirits operates. All that happens in the mundane world finds its cause in this true reality. Here is where the causes of diseases, absent wild game, and droughts and floods can be found. The true reality is hidden from the normal eye (Baer 1987; Reichel-Dolmatoff 1996*), and one can attain a glimpse of it only through dreams and visions. The shaman is a specialist in dreams and visions, and he functions as a kind of tour guide in the other reality. His most important tool for this task is the magical drink ayahuasca (Deltgen 1993). Because the contents of the ayahuasca visions are culturally interpreted, they often become standardized patterns that make it possible for the Indians to swiftly and purposefully arrive in the realm of the all-important visionary world (Langdon 1979).

Shamans often prepare for an ayahuasca session by remaining sexually abstinent for a time (from three days to six months), adhering to special dietary rules, and using purgative and laxative substances, **enemas**, libations, and so on. The ayahuasca diet prohibits the consumption of salt, chili (*Capsicum* spp.), spices, and fat. The Jíbaro drink guayusa (*Ilex guayusa*) to make themselves vomit. The Siona and Secoya drink a cold-water extract of the liana they call *hetu bisi* (*Tournefortia angustiflora* Ruíz et Pav.; Boraginaceae) to purify themselves before the ritual (Vickers and Plowman 1984, 8*).

Ayahuasca rituals are often accompanied by the continuous smoking of tobacco (*Nicotiana rustica* or *Nicotiana tabacum*). The tobacco is said to banish evil spirits, that is, unpleasant and threatening visions. In addition, participants may also drink great quantities of **chicha** or some other type of **alcohol** (Bennett 1992, 486*) and chew *ipadú* (*Erythroxylum coca*). Sometimes, *Brugmansia* leaves will be soaked in rum and drunk as a tonic. The mestizo shamans burn camphor (*Cinnamomum camphora*) as an **incense** during their ayahuasca healing ceremonies, possibly to enhance the psychoactivity (Luna 1992, 246 f.). If a shaman wants to learn about an unknown plant and its healing properties, he may add the plant to an ayahuasca mixture before he drinks it.

Artifacts

A great number of different artifacts are related to ayahuasca (Mallol de Recasens 1963). The Tukanos interpret many Amazonian petroglyphs as ayahuasca images (Reichel-Dolmatoff 1967*). They decorate their houses with the patterns and figures they see during their ayahuasca journeys. These patterns often take on symbolic dimensions and can tell initiates entire stories from the other reality (Reichel-Dolmatoff 1978*). The patterns seen under the influence of ayahuasca are also woven into such everyday objects as baskets (Reichel-Dolmatoff 1985*, 1987*).

Jíbaro artists make many images depicting their ayahuasca experiences (Münzel n.d., 204, 205, 207, 212, 213). Yando Rios, a Peruvian artist who now lives in California, made numerous paintings of his early ayahuasca visions (Andritzky 1989, 191). The paintings of the former *ayahuasquero* Pablo Amaringo are particularly impressive and have become quite renowned. His visionary paintings portray the entire ayahuasca mythology (including the syncretic elements), the jungle pharmacopoeia, and the other reality of the shamanic universe (Luna 1991; Luna and Amaringo 1991). Many Latin American and Western artists, e.g., Alexandre Segrégio, have also painted their ayahuasca visions (Weiskopf 1995).

In 1953, William Burroughs (1914–1997) traveled to the Amazon basin to learn about the ethnopharmacology of the mysterious magical drink of the Amazon Indians and to try it himself. His experiences were published in the book *The Yage Letters* (Burroughs and Ginsberg 1963).

The best-selling novel *The Incas* contains numerous descriptions of ayahuasca experiences (Peters 1991*). The Peruvian poet César Calvo published a magnificent poem, inspired directly by ayahuasca and the story of Bruce Lamb (1974), in his book *The Three Halves of Ino Moxo* (1995).

The Shipibo-Conibo Indians have developed an astonishing method for encoding and decoding the patterns they see under the influence of ayahuasca in songs. When they want to use a certain pattern to paint their face, an article of clothing, or a piece of pottery, they sing the pattern. The person who is doing the painting, usually a woman, will then translate the song into a design (Gebhardt-Sayer 1985, 1987; Illius 1991).[393]

Ayahuasca songs also play a role in traditional shamanism, where they are used primarily as a map for the journey into the other reality. Whistled melodies for creating certain standardized visions have an important part in the ayahuasca shamanism practiced by the urban healers of Iquitos (Katz and Dobkin de Rios 1971). The so-called *icaros* (shamanic power songs) are also used for this purpose (Luna 1986, 1992).

Ayahuasca Music—A Discography

Ethnic recordings

Brazil—The Bororo World of Sound (Auvidis-Unesco D 8201, 1989)

Brésil Central—Chants et danses des Indiens Kaiapó (VDE-Gallo, 1989)

Indian Music of the Upper Amazon: Cocama, Shipibo, Campa, Conibo (Folkways Records FE 4458, 1954)

Indiens d'Amazonie (Le Chant du Monde LDX 74501, n.d.)

Music of the Jivaro of Ecuador (Ethnic Folkways Records FE 4386, 1972)

Santo Diame—Sacred Music from the 1930s to the 1990s (Orenda Institute, Baltimore, 1996)

Waorani Waaponi—Archaic Chanting in the Amazon Rainforest (Tumi Records CD043, 1994)

Music inspired by ayahuasca

Tori Amos, *Under the Pink* (WEA/Warner, 1994)

Greg White Hunt, *Enter the Orienté* (All Is Well Records, 1997)

Inti César Malasquez, *Earth Incarnation* (Meistersinger Musik NGH-CD-453, 1996)

Ayahuasca Churches

In addition to the true shamanic use of ayahuasca, in recent decades various syncretic churches that use ayahuasca as a sacrament have emerged in Amazonia. Some ayahuasca churches have integrated African orishas into their cult. In Brazil, there are also non-Christian sects that use ayahuasca ritually to venerate spirits (Prance 1970, 67*).

Both in the Santo Daime cult and in the ayahuasca church known as União do Vegetal, regular meetings are held at which the followers—overwhelmingly mestizos from the lower and middle classes—drink ayahuasca communally and sing devotional songs. Guided by a priest, the congregation travels to both the spirits of the forest and to the Christian saints. Many of the cult participants discover a new purpose in life and find healing for their souls. Like the shamans of the forest, the adherents of these Brazilian churches (which have now gained a foothold in Europe as well) are legally permitted to use the magical and ritual drink, which they prepare from ***Banisteriopsis caapi*** and ***Psychotria viridis*** (Lowy 1987). The DMT content of their brew is typically higher than that of the more traditional preparations (Liwszyc et al. 1992). At the meetings, pleasant-smelling *breuzinho* (**incense** made from *Protium heptaphyllum* and *Protium* spp.) is burned. Sometimes ***Cannabis indica*** (Santa María) is consumed as well.

Barquinha Daniel Periera de Matos, the founder of one branch of the Santo Daime cult, was once in the navy. As a result, his congregations are structured along military lines and the congregants all wear uniforms. Barquinha saw his church as an institution open to the entire world, and the cult has been carried to the Western world with missionary zeal (Bogers 1995). Men, women,

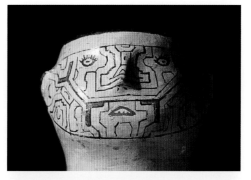

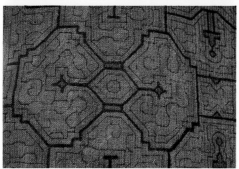

Top: A Shipibo ceramic jug (*chomo*) painted with ayahuasca designs.

Middle: Ayahuasca designs—originally seen under the influence of the magical drink—on a piece of Shipibo fabric.

Bottom: The preparation of the ayahuasca drink, as depicted on a painting by Pablo Amaringo, a former *ayahuasquero*.

393 In the Ecuadoran Amazon, a black "paint" is prepared from *ko-pi,* the fruits of the nightshade *Cyphomandra hartwegii* (Mier) Sendtner ex Walper. This preparation is used to paint on pottery the geometric patterns that represent the visual hallucinations of ayahuasca (Schultes and Raffauf 1991, 37*).

and children are all allowed to take part in the ceremony. Even pregnant women drink *daime*. It is said that this has never resulted in problems of any kind!

In the Daime cult, the *Banisteriopsis caapi* liana is regarded as the embodiment of Jesus and the leaves of *Psychotria viridis* as Mary. The finished brew is regarded as the flesh and blood of Christ. During the ceremony, *Cannabis* (which they call Santa María) is smoked as a sacrament. Europeans often report having profound spiritual experiences, during which Jesus, Mary, and the spirits of the forest reveal themselves (Luczyn 1994; Weigle 1995). Often consumed daily, *daime* is considered both healthy and therapeutic (Groisman and Snell 1996), and various types of addictions are now being treated within the context of the Santo Daime cult (Yatra 1995). In Peru, the Takiwasi project is investigating the effects of shamanic therapy with ayahuasca on drug addicts. Initial reports indicate that this method has great promise (Mabit et al. 1986).

The other large ayahuasca church, which was founded by Gabriel da Acosta, is known as the União do Vegetal (UDV). In recent years, this church has attracted increasing numbers of Brazilians, most of whom have been culturally uprooted. Because the effects of the ayahuasca drink are clearly regarded as positive, the Brazilian government has made this psychoactive agent legal when used within the framework of cult and religious ceremonies. This unusual sociopolitical situation has made it possible to carry out interdisciplinary research. Under the direction of Charles Grob, Dennis McKenna, and James Callaway, universities in the United States (California), Finland, and Brazil joined together to investigate the medicinal, pharmacological, and health effects of regular ayahuasca use on the members of the church. The research was carried out on location, that is, in the temples of the UDV. Standardized psychiatric tests accepted throughout the world were administered to a test group before, during, and after the ingestion of ayahuasca. The test subjects were also given standard medical tests while they were under the influence of ayahuasca. The data that was obtained was analyzed at the University of California at Los Angeles. A control group of Brazilian workers who had never taken ayahuasca were given the same tests for purposes of comparison. The pilot study for this large-scale project revealed that the cult members who regularly ingested ayahuasca were on average far more healthy in both body and mind than the group that had never had contact with ayahuasca. This research project will certainly provoke a radical rethinking of the medicinal and psychiatric attitudes about psychoactive substances that are currently accepted in the West (Callaway et al. 1994; Grob et al. 1996; cf. Dobkin de Rios 1996).

Ayahuasca Tourism

The many reports of travelers' experiences with Amazonian shamans have led to a kind of pilgrimage of Westerners to the rain forest. For years, tourists from around the world have been attracted to the magical realm of ayahuasca. Many of these seekers have promised themselves personal insights and mystical experiences. In turn, numerous greedy, self-proclaimed pseudo-shamans have exploited the hopes of these tourists. They openly offer ayahuasca rituals, but the desired effects typically do not occur. These entrepreneurs are not initiated shamans who know what their recipes can do. In the best cases, the brew they offer simply is ineffective. Sometimes, however, even highly toxic plants may be tossed into the brew, which is then offered at great cost as "genuine Amazonian ayahuasca." The journey, which should take one to mystical domains, may actually end with the tourist in a coma or at the hospital (Dobkin de Rios 1995).

But those travelers who have the good fortune to encounter real shamans typically report having spiritual, mystical, and healing experiences (Ayala Flores and Lewis 1978; Pinkson 1993; Wolf 1992).

Pharmacology

Unbelievable rumors and amazing reports of the wondrous effects of ayahuasca have been making their way to the West since the nineteenth century. It was said that people under the influence of ayahuasca were able to go through walls, locate hidden treasures, see through mountains, know the future, and take part in events happening in far distant places. Both missionaries and physicians have claimed that the drink could awaken and even increase telepathic abilities.

It was only in recent times that the neurochemical secrets of the visionary effects of ayahuasca could be clarified (Rivier and Lindgren 1972). The two main active constituents of ayahuasca are harmaline (= telepathine) and *N,N*-DMT. When ingested, DMT cannot reach the brain because it is broken down by the enzyme monoamine oxidase (MAO). **Harmaline** (as well as harmine and several other β-**carbolines**) inhibits the release of MAO. (It was recently discovered that ayahuasca inhibits only MAO-A [D. McKenna 1996*].) This makes it possible for DMT to pass unhindered through the blood–brain barrier, occupy specific receptor sites, and cause the nervous system to enter into an extraordinary state characterized by brilliant and overwhelming visions (McKenna et al. 1994a, 1994b; McKenna and Towers 1985*).

The effects last for some four hours. The **harmaline** first induces a sedation that sometimes results in immobility. During this phase, in which the effects first become manifest, harmaline can cause profound nausea and frequently induces vomiting. The psychedelic DMT effects set in some forty-five minutes after ingestion of the drink. The main visionary effects last for about an hour and then suddenly cease. The nausea usually disappears when the DMT effects begin. When ayahuasca is used on a regular basis, the body becomes accustomed to the pharmacological action of the harmaline, and many chronic users find that they no longer experience nausea. Because the body does not build up tolerance to *N,N*-DMT, a person can consume ayahuasca several times a day.

The Legal Situation

At the present time, ayahuasca is completely legal in Brazil. The legal situation in other Amazonian countries is less clear. Because the drink contains the prohibited substance *N,N*-DMT, the legal situation is very difficult in Western countries.

Patent No. 5751, which the International Medicine Corporation (represented by Loren Miller) registered at the United States Marks and Patents Office in June 1996, created a perversely odd situation (Fericgla 1996a). In taking the step, this corporation hoped to secure for itself a patent and copyright on ayahuasca, that is, to monopolize the chemical and pharmacological principles of ayahuasca. If the patent had come into force, the Indians—the inventors and guardians of the ayahuasca brew—would have been forbidden to brew their own drink or, more precisely, would have been allowed to make it only if they paid licensing fees to the corporation. In Ecuador, for example, which recognizes American patent laws, the few remaining traditional Indians would have had to pay money to the owners of the copyright on ayahuasca every time they prepared it. As of 2004, however, this patent has not gone into effect.

The heads of some four hundred Amazonian tribes wrote an open letter to then president Bill Clinton protesting this incredible impudence. As Valerio Grefa, the speaker of the Confederation of Indian Organizations of the Amazon Basin, exclaimed: "Patenting our medicine, which we have inherited through many generations, is an attack on the culture of our peoples and on all of humankind."

Literature

See also the entries for **ayahuasca analogs,** *Banisteriopsis caapi,* *Psychotria viridis,* and **harmaline and harmine.**

Andritzky, Walter. 1989a. Ethnopsychologische Betrachtung des Heilrituals mit Ayahuasca (*Banisteriopsis caapi*) unter besonderer Berücksichtigung der Piros (Ostperu). *Anthropos* 84:177–201.

———. 1989b. Sociopsychotherapeutic functions of ayahuasca healing in Amazonia. *Journal of Psychoactive Drugs* 21 (1): 77–89.

Arévalo V[alera], Guillermo. 1986. Al ayahuasca y el curandero Shipibo-Conibo del Ucayali (Perú). *América Indígena* 46 (1): 147–61.

Ayala Flores, Franklin, and Walter H. Lewis. 1978. Drinking the South American hallucinogenic ayahuasca. *Economic Botany* 32:154–56.

Baer, Gerhard. 1969. Eine Ayahuasca-Sitzung unter den Piro (Ost-Peru). *Bulletin de la Société Suisse des Américanistes* 33:5–8.

———. 1987. Peruanische Ayahuasca-Sitzungen. In *Ethnopsychotherapie*, ed. A. Dittrich and C. Scharfetter, 70–80. Stuttgart: Enke.

Baer, Gerhard, and Wayne W. Snell. 1974. An ayahuasca ceremony among the Matsigenka (eastern Peru). *Zeitschrift für Ethnologie* 99 (1/2): 63–80.

Bianchi, Antonio, and Giorgio Samorini. 1993. Plants in association with ayahuasca. *Yearbook for Ethnomedicine and the Study of Consciousness* 2:21–42. Berlin: VWB. (Contains an excellent bibliography.)

Bogers, Hans. 1995. De Santa Daime Leer: Ayahuascagebruik in een religieuze setting. *Pan* 1:2–10.

Burroughs, William, and Allen Ginsberg. 1963. *The Yage Letters.* San Francisco: City Lights Books.

Califano, M., et al. 1987. Schamanismus und andere rituelle Heilungen bei indianischen Völkern Südamerikas. In *Ethnopsychotherapie*, ed. A. Dittrich and C. Scharfetter, 114–34. Stuttgart: Enke.

Callaway, James. 1995a. Ayahuasca: A correction. *Eleusis* 2:26–27.

———. 1995b. Ayahuasca, now and then. *Eleusis* 1:4–10.

———. 1995c. *Pharmahuasca* and contemporary ethnopharmacology. *Curare* 18 (2): 395–98.

———. 1995d. Some chemistry and pharmacology of ayahuasca. *Yearbook for Ethnomedicine and the Study of Consciousness* 3 (1994): 295–98. Berlin: VWB.

Callaway, James, Charles Grob, and Dennis McKenna. 1994. Platelet serotonin uptake sites increased in drinkers of *Ayahuasca*. *Psychopharmacology* 116:385–87.

Calvo, César. 1995. *The three halves of Ino Moxo: Teachings of the wizard of the upper Amazon.* Rochester, Vt.: Inner Traditions International.

Chango, Alfonso. 1984. *Yachaj sami yachachina.* Quito: Ediciones Abya-yala.

Deltgen, Florian. 1993. *Gelenkte Ekstase: Die halluzinogene Droge Cají der Yebámasa-Indianer.* Acta Humboldtiana 14. Stuttgart: Franz Steiner Verlag.

Dobkin de Rios, Marlene. 1970. A note on the use of ayahuasca among urban mestizo populations in the Peruvian Amazon. *American Anthropologist* 72 (6): 1419–22.

"Because the Yebamasa have a theory of reality that differs from ours, they also have a different conception of true consciousness and of the true recognition of reality. The ability that enables the Yebamasa to recognize reality is neither intellectual nor emotional nor intuitive. It encompasses all these kinds of awareness and adds another—the mystical experience, through hallucinations and visions in the state of toxic ecstasy, of a realm of reality that cannot normally be experienced. According to the Yebamasa conception, humans are able to penetrate the foreground of ordinary reality with cají [= ayahuasca]. To those who are knowledgeable, this ordinary reality appears as the world of effects, whereas the world of myths is that of causes. . . . The mythological reality experienced in the cají visions is just as real and natural as the ordinary world—perhaps even more real."

FLORIAN DELTGEN
GELENKTE EKSTASE [GUIDED ECSTASY]
(1993, 125 f.)

"The songs have a special importance during the ayahuasca séances, for through them all the figures of the ayahuasca world manifest themselves as visions. The songs are the activating and formative stimulus that structures and controls the sequence of the mentally preexisting cultural-mythic pattern of the visions."

WALTER ANDRITZKY
"ETHNOPSYCHOLOGISCHE BETRACHTUNG DES HEILRITUALS MIT AYAHUASCA" ["AN ETHNO-PSYCHOLOGICAL CONSIDERATION OF THE AYAHUASCA HEALING RITUAL"] (1989a, 186)

———. 1972. *Visionary vine: Hallucinogenic healing in the Peruvian Amazon.* San Francisco: Chandler.

———. 1981. Socio-economic characteristics of an Amazon urban healer's clientele. *Social Sciences and Medicine* 15B:51–63.

———. 1989. A modern-day shamanistic healer in the Peruvian Amazon: Pharmacopoeia and trance. *Journal of Psychoactive Drugs* 21 (1): 91–99.

———. 1992. *Amazon healer: The life and times of an urban shaman.* Bridport, Dorset: Prism Press.

———. 1994. Drug tourism in the Amazon. *Anthropology of Consciousness* 5 (1): 16–19.

———. 1995. Drug tourism in the Amazon. *Yearbook for Ethnomedicine and the Study of Consciousness* 3 (1994): 307–14. Berlin: VWB.

———. 1996. Commentary on "human psychopharmacology of hoasca": A medical anthropology perspective. *The Journal of Nervous and Mental Disease* 181 (2): 95–98.

Fericgla, Josep Ma. 1994. *Los Jíbaros, cazadores de suental.* Barcelona: Integral.

———. 1996a. Ayahuasca patented! *Eleusis* 5:19–20.

———. 1996b. Theory and applications of ayahuasca-generated imagery. *Eleusis* 5:3–18.

Fischer-Fackelmann, Ruth. 1996. *Fliegender Pfeil.* Munich: Heyne. (On Santo Daime.)

Gebhardt-Sayer, Angelika. 1985. The geometric designs of the Shipibo-Conibo in ritual context. *Journal of Latin American Lore* 11 (2): 143–75.

———. 1987. *Die Spitze des Bewußtseins: Unter-suchungen zu Weltbild und Kunst der Shipibo-Conibo.* Münchner Beiträge zur Amerikanistik. Hohenschäftlarn: Klaus Renner Verlag.

Giove, Rosa. 1992. Madre ayahuasca. *Takiwasi* 1 (1): 7–10.

Grob, Charles S., et al. 1996. Human psychopharma-cology of hoasca, a plant hallucinogen in ritual context in Brazil. *The Journal of Nervous and Mental Disease* 181 (2): 86–94.

Groisman, Alberto, and Ari Bertoldo Snell. 1996. Heal-ing power: Cultural-neurophenomenological therapy with Santo Daime. *Jahrbuch für Transkul-turelle Medizin und Psychotherapie* 6 (1995): 241–55.

Harner, Michael. 1973. The sound of rushing water. In *Hallucinogens and shamanism,* ed. Michael Harner, 15–27. London: Oxford University Press.

Illius, Bruno. 1991. *Ani Shinan: Schamanismus bei den Shipibo-Conibo.* Münster and Hamburg: Lit., Ethnologische Studien Bd. 12.

Junquera, Carlos. 1989. Botanik und Schamanismus bei den Harakmbet-Indianern im südwestlichen Amazonasgebiet von Peru. *Ethnologia Americana* 25/1 (114): 1232–38.

Katz, Fred, and Marlene Dobkin de Rios. 1971. Healing sessions. *Journal of American Folklore* 84 (333): 320–27.

Kusel, Heinz. 1965. Ayahuasca drinkers among the Chama Indians of northeast Peru. *Psychedelic Review* 6:58–66.

Lamb, F. Bruce 1974. *Wizard of the upper Amazon.* Boston: Houghton Mifflin.

———. 1985. *Rio Tigre and beyond: The Amazon jungle medicine of Manuel Córdova.* Berkeley, Calif.: North Atlantic Books.

Lamb, F. Bruce, and Manuel Córdova-Rios. 1994. *Kidnapped in the Amazon jungle.* Berkeley, Calif.: North Atlantic Books.

Langdon, E. Jean. 1979. *Yagé* among the Siona: Cultural patterns in visions. In *Spirit, shamans, and stars,* ed. D. L. Browman and R. A. Schwarz, 63–80. The Hague: Mouton.

Liwszyc, G. E., E. Vuori, I. Rasanen, and J. Issakainen. 1992. Daime—a ritual herbal potion. *Journal of Ethnopharmacology* 36:91–92.

Lowy, B. 1987. Caapi revisited—in Christianity. *Economic Botany* 41:450–52.

Luczyn, David. 1994. Reise zum Geist des Waldes. *Esotera* 5/94:30–35. (On Santo Daime.)

Luna, Luis Eduardo. 1984a. The concept of plants as teachers among four mestizo shamans of Iquitos, northeast Peru. *Journal of Ethnopharmacology* 11 (2): 135–56.

———. 1984b. The healing practices of a Peruvian shaman. *Journal of Ethnopharmacology* 11 (2): 123–33.

———. 1986. *Vegetalismo: Shamanism among the mestizo population of the Peruvian Amazon.* Acta Universitatis Stockholmiensis, Stockholm Studies in Comparative Religion 27. Stockholm: Almqvist und Wiskell International.

———. 1991. Plant spirits in ayahuasca visions by Peruvian painter, Pablo Amaringo: An iconographic analysis. *Integration* 1:18–29.

———. 1992. Icaros: The magic melodies among the mestizo shamans of the Peruvian Amazon. In *Portals of power: Shamanism in South America,* ed. E. Jean M. Langdon and Gerhard Baer, 231–53. Albuquerque: University of New Mexico Press.

Luna, Luis Eduardo, and Pablo Amaringo. 1991. *Ayahuasca visions.* Berkeley, Calif.: North Atlantic Books.

Mabit, Jacques, Rosa Giove, and Joaquín Vega. 1996. Takiwasi: The use of Amazonian shamanism to rehabilitate drug addicts. *Jahrbuch für Transkulturelle Medizin und Psychotherapie* 6 (1995): 257–85.

Mac Rae, Edward. 1995. El uso religioso de la ayahuasca en el Brasil contemporáneo. *Takiwasi* 3:17–23.

Mallol de Recasens, Maria Rosa. 1963. Cuatro representaciones de las imágenes alucinatorias originadas por la toma del yagé. *Revista Colombiana de Folklore* 8 (3): 61–81.

McKenna, Dennis J. 1996. Ayahuasca: An overview of its chemistry, botany and pharmacology. Lecture given at the Entheobotany Conference, San Francisco, October 18–20, 1996.

McKenna, Dennis J., Luis Eduardo Luna, and G. N. Towers. 1995. Biodynamic constituents in ayahuasca admixture plants: An uninvestigated folk pharmacopeia. In *Ethnobotany: Evolution of a discipline*, ed. Richard Evans Schultes and Siri von Reis, 349–61. Portland, Ore.: Dioscorides Press.

McKenna, Dennis J., G. H. N. Towers, and F. Abbott. 1994a. Monoamine oxydase inhibitors in South American hallucinogenic plants: Tryptamine and β-carboline constituents of ayahuasca. *Journal of Ethnopharmacology* 10:195–223.

———. 1994b. Monoamine oxydase inhibitors in South American hallucinogenic plants. Part 2: Constituents of orally active myristicaceous hallucinogens. *Journal of Ethnopharmacology* 12:179–211.

McKenna, Terence. 1993. Bei den Ayahuasqueros. In *Das Tor zu inneren Räumen*, ed. C. Rätsch, 105–39. Südergellersen: Verlag Bruno Martin.

Münzel, Mark. n.d. *Schrumpfkopf-Macher? Jíbaro-Indianer in Südamerika*. Frankfurt/M.: Museum für Völkerkunde.

Naranjo, Plutarco. 1979. Hallucinogenic plant use and related indigenous belief systems in the Ecuadoran Amazon. *Journal of Ethnopharmacology* 1:121–45.

———. 1983. *Ayahuasca: Etnomedicina y mitología*. Quito: Ediciones Libri Mundi.

———. 1986. El ayahuasca en la arqueologia ecuatoriana. *América Indígena* 46 (1): 117–27.

Narby, Jeremy. 1999. *The cosmic serpent: DNA and the origins of knowledge*. Los Angeles: Jeremy P. Tarcher.

Ott, Jonathan. 1995. Ayahuasca—ethnobotany, phytochemistry and human pharmacology. *Integration* 5:73–97.

Ott, Theo. 1979. *Der magische Pfeil: Magie und Medizin*. Zurich: Atlantis.

Payaguaje, Fernando. 1990. *El bebedor de yajé*. Shushufindi: Ediciones CICAME, Vicariato Apostolico de Aguarico.

Paymal, Noemi, and Catalina Sosa, eds. 1993. *Mundos amazonicos: Pueblos y culturas de la Amazonia Ecuatoriana*. Quitó: Fundación Sinchi Sacha.

Perkins, John. 1995. *Und der Traum wird Welt: Schamanische Impulse zur Aussöhnung mit der Natur*. Wessobrunn: Integral Volkar-Magnum.

Pinkley, Homer V. 1969. Plant admixtures to *ayahuasca*, the South American hallucinogenic drink. *Lloydia* 32 (3): 305–14.

Pinkson, Thomas. 1993. Amazonian shamanism: The ayahuasca experience. *Psychedelic Monographs and Essays* 6:12–19.

Rätsch, Christian. 1994. Ayahuasca: Der Zaubertrank. *Geo Special: Amazonien* 5/94:62–65.

———. 1997. Ayahuasca, der Schamanentrunk von Amazonien. *Naturheilpraxis* 50 (10): 1581–85.

Reichel-Dolmatoff, Gerardo. 1969. El contexto cultural de un alucinogeno aborigen: *Banisteriopsis caapi. Revista de la Academia Colombiana de Ciencias Exactas, Físicas y Naturales* 13 (51): 327–45.

———. 1970. Notes on the cultural context of the use of yagé (*Banisteriopsis caapi*) among the Indians of the Vaupés, Colombia. *Economic Botany* 24 (1): 32–33.

———. 1972. The cultural context of an aboriginal hallucinogen: *Banisteriopsis caapi*. In *Flesh of the gods: The ritual use of hallucinogens*, ed. Peter T. Furst, 84–113. New York: Praeger. Repr. rev. ed., Prospect Heights, Ill.: Waveland Press, 1990.

Rivas, Agustin. 1989. Meisterpflanze Ayahuasca. In *Amazonas: Mae Mañota*, ed. C. Kobau, 182–83. Graz: Leykam.

Rivier, Laurent, and Jan-Erik Lindgren. 1972. "Ayahuasca," the South American hallucinogenic drink: An ethnobotanical and chemical investigation. *Economic Botany* 26:101–29.

Schultes, Richard Evans. 1960. A reputedly toxic *Malouetia* from the Amazon. *Botanical Museum Leaflets* 19 (5): 123–24.

Schultes, Richard Evans, and Robert F. Raffauf. 1960. Prestonia: An Amazon narcotic or not? *Botanical Museum Leaflets* 19 (5): 109–22.

Shoemaker, Alan. 1997. The magic of *curanderismo*: Lessons in mestizo ayahuasca healing. *Shaman's Drum* 46:28–39.

Taussig, Michael. 1987. *Shamanism, colonialism, and the wild man*. Chicago and London: The University of Chicago Press.

Weigle, Ewald. 1995. Die wunderbare Heilkraft des Ayahuasca. *Yearbook for Ethnomedicine and the Study of Consciousness* 3 (1994): 299–305. Berlin: VWB.

Weil, Andrew. 1972. *The natural mind: A new way of looking at drugs and higher consciousness*. Boston: Houghton Mifflin.

Weiskopf, Jimmy. 1995. From agony to ecstasy: The transformative spirit of yajé. *Shaman's Drum* (Fall 1994): 41–47. (With images by Alexandre Segrégio.)

Wolf, Fred Alan. 1992. *The eagle's quest*. New York: A Touchstone Book (Simon & Schuster).

Yatra, Atmo. 1995. *From addiction to health with the magic of ayahuasca*. Unpublished manuscript, Amsterdam. (Cited 1998.)

"The narcotics produced a miracle! My performance is greeted with appreciative comments and deemed to consist of true *natem* songs. But soon my intoxication takes a new turn. Against the serene glow of the night, phosphorescent circles begin to whirl, then merge and separate, forming constantly changing kaleidoscopic designs. One after another all the symmetrical patterns invented by nature pass before me in a subtle continuum: lozenges first red, then yellow, then indigo, delicate traceries, crystalline prisms, iridescent scales, the eyes of butterfly wings, feline pelt markings, reticular carapaces."

THE ETHNOLOGIST PHILIPPE DESCOLA, WRITING ABOUT HIS PERSONAL EXPERIENCE WITH AYAHUASCA IN *THE SPEARS OF TWILIGHT* (DESCOLA 1996, 207*)

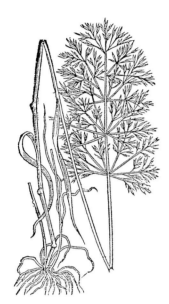

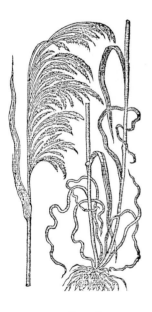

Many grasses (from the genera *Arundo*, *Phalaris*, and *Phragmites*) contain DMT and are increasingly being tested for use as ingredients in ayahuasca analogs. (Woodcut from Gerard, *The Herball or General History of Plants*, 1633)

394 In Nepal, the Himalayan species *Hypericum choisianum* Choisy [syn. *Hypericum cernuum* Roxb.] is one of the plants sacred to Kali (the wife of Shiva; cf. *Cannabis indica*). Its veneration as a plant of the gods suggests that it may have some kind of psychoactivity.

Other Names
Anahuasca, ayahuasca borealis

The effects of the pharmacological principle that was discovered during the investigations of traditional **ayahuasca** can be imitated with other plants that contain the same constituents (**harmaline**, **harmine**, *N*,*N*-DMT/5-MeO-DMT). Today, nontraditional combinations of plants with these ingredients are known as ayahuasca analogs or anahuasca. Combinations composed of isolated or synthesized constituents are referred to as pharmahuasca:

> Paradoxically, psychonautic research on *pharmahuasca . . .*, which is so far out of the scientific mainstream that nearly three decades had to pass before unfunded and independent scientists working underground and in secrecy put the enzyme-inhibitor theory of *ayahuasca* pharmacology to the test, may turn out to be at the center of research on the biochemistry of consciousness and the genetics of pathological brain function! Not only is *ayahuasca* research now at the neuroscientific cutting edge, but the reversible MAO-inhibitors in *ayahuasca* may prove to be viable, less toxic alternatives to the noxious compounds currently in use! (Ott 1994, 69)

The term *ayahuasca analog* appears have been coined by Dennis McKenna. The American ethnobotanist Jeremy Bigwood was probably the first person to test pharmahuasca (100 mg each of **harmaline** hydrochloride and *N*,*N*-DMT) on himself; he reported "DMT-like hallucinations" (Ott 1994, 52). The chemist and chaos theorist Mario Markus used the Heffter technique (self-experimentation) to perform extensive experiments into the optimal proportions for mixing the alkaloids:

> Markus reported on the studies that he had carried out several years earlier in which the plant combinations used by Indian peoples in the Amazon region for ritual purposes were simulated experimentally. In those experiments, Markus mixed one sample each of a representative of the β-carbolines (harmine, harmaline, or 6-MeO-harmalane) with a tryptamine (5-MeO-DMT). This yielded a domain of different optimal mixture ratios within which a marked psychoactive productivity with hallucinatory effects occurred. Within quite specific dosage limits, there was a

good overall tolerance without serious side effects. (Leuner and Schlichting 1986, 170*)

For Jonathan Ott, the value of the ayahuasca analogs lies in their entheogenic effects, which can help one attain a more profound spiritual ecology and a mystical perspective. **Ayahuasca** and its analogs can induce a state of shamanic ecstasy, but only when used at the proper dosage:

> Shamanic ecstasy is the *real* "Old Time Religion," of which modern churches are but pallid evocations. Our forebears discovered in many times and places that in the ecstatic, entheogenic experience, suffering humankind could reconcile the cultivated braininess, which isolated each individual human being from all other creatures and even from other human beings, with the wild and feral, beastly magnificent bodies that we also are. . . . There is no need for faith, it is the ecstatic experience itself that *gives* one faith in the intrinsic unity and integrity of the universe, in ourselves as integral parts of the whole; that reveals to us the sublime majesty of our universe, and the fluctuant, scintillant, alchemical miracle that is quotidian consciousness. . . .
>
> Entheogens like *ayahuasca* may be just the right medicine for hypermaterialistic humankind on the threshold of a new millennium which will determine whether our species continues to grow and prosper, or destroys itself in a massive biological Holocaust unlike anything the planet has experienced in the last 65 million years. . . .
>
> The Entheogenic Reformation is our best hope for healing Our Lady Gaea, while fostering a *genuine* religious revival for a new millennium. (Ott 1994, 89–90)

Recipes
All recipes must contain an MAO inhibitor as well as a source of **DMT**. To date, experiments have been conducted only with *Banisteriopsis caapi*, *Banisteriopsis* spp., *Peganum harmala*, and synthetic (pharmaceutical) MAO inhibitors. But there are other MAO inhibitors in nature, such as *Tribulus terrestris*. The ongoing investigations into St. John's wort (*Hypericum perforatum* L.) and other *Hypericum* species[394] as possible MAO-inhibiting admixtures are very interesting. Hypericin, the primary active constituent in *Hypericum* spp., "has been proven to be a monoamine oxidase inhibitor" (Becker 1994, 48*). *Psychotria viridis* and *Mimosa tenuiflora* have been looked at as sources of DMT, but numerous other possibilities

also exist (see the tables on the following pages). The dosages are determined by the alkaloid concentrations in the various admixtures (DeKorne 1996; Ott 1994).

As with traditional **ayahuasca**, most ayahuasca analogs have a thoroughly disgusting taste and are therefore generally difficult to force down (because they are forced up again from below). Chewing sliced ginger (*Zingiber officinale*) can help counteract the often repulsive taste (DeKorne 1994, 98*).

The following recipes are formulated to yield a single dose.

Classic Ayahuasca Analog
25 g *Psychotria viridis* leaves, dried and ground
3 g *Peganum harmala* seeds, crushed
Juice of one lemon
Enough water to boil all the ingredients (approximately 200–350 ml)

Place all the ingredients in a steel pot. Slowly bring to a boil, then boil rapidly for two to three minutes. Reduce the heat and simmer for approximately five more minutes. Pour off the decoction. Add some water to the herbs remaining in the pot and boil again. Pour the first decoction back into the pot. After a while, pour out the liquid once more. Add fresh water to the remaining herbs and bring to a boil again. Remove the plant remnants and compost them, if possible. Mix together the three extracts. Carefully heat the mixture to reduce the total volume. The tea should be drunk as fresh as possible (allow to cool first), although it can be stored in the refrigerator for a few days. The effects begin about forty-five minutes after ingestion. The visionary phase lasts about an hour.

Juremahuasca or Mimohuasca
Connoisseurs consider this ayahuasca analog to be both the most easily tolerated and the most psychoactive preparation.
3 g *Peganum harmala* seeds, finely ground
9 g *Mimosa tenuiflora* root cortex
Juice of one lime or lemon

The crushed Syrian rue (*P. harmala*) seeds may be either swallowed in a gelatin capsule or mixed in water and drunk. The decoction of lemon juice and mimosa root cortex should be drunk fifteen minutes later.

Prairie Ayahuasca
This blend is especially popular in North America. Predominantly pleasant experiences have been reported (Ott 1994, 63; cf. DeKorne 1994, 97*).
3–4 g *Peganum harmala* seeds, finely ground
30 g *Desmanthus illinoensis* root cortex (prairie mimosa, Illinois bundleweed, Illinois bundleflower)
Juice of one lemon or lime

Prepare in the same manner as juremahuasca (above).

Acaciahuasca
This blend is especially popular in Australia and has been used with good success.
3 g *Peganum harmala* seeds, finely ground
20 g *Acacia phlebophylla* leaves, ground (cf. **Acacia spp.**)
Juice of one lemon or lime

Prepare in the same manner as juremahuasca (above).

Phalahuasca
In Europe, various combinations of **Phalaris arundinacea** or *Phalaris aquatica* (see **Phalaris spp.**) and **Peganum harmala** have been investigated. Unfortunately, the experiments have met with little success to date as far as pleasant visionary experiences are concerned. Because of the toxic alkaloid (gramine) that occurs in the reed grasses, these preparations can be very dangerous (Festi and Samorini 1994).

Peyohuasca
This preparation is a combination of *Peganum harmala* and *Lophophora williamsii*. It may be pharmacologically very dangerous.

San Pedro Ayahuasca
The following amounts and ingredients have been reported to produce pleasant effects (in *Entheogene* 5 [1995], 53).
1–3 g Syrian rue (*Peganum harmala*)
20–25 g San Pedro cactus powder (see *Trichocereus pachanoi*)

This blend may be pharmacologically dangerous.

Top: Precisely measured ingredients for mimohuasca, an ayahuasca analog made with 9 g of *Mimosa tenuiflora* root cortex and 3 g of Syrian rue (*Peganum harmala*).

Bottom: The root of the American shrub known as prairie mimosa (*Desmanthus illinoensis*) is rich in DMT and is a popular ingredient in ayahuasca analogs.

"May the Entheogenic Reformation prevail over the Pharmacratic Inquisition, leading to a spiritual rebirth of humankind at Our Lady Gaea's breasts, from which may ever copiously flow the *amrta*, the *ambrosia*, the *ayahuasca* of eternal life!"

JONATHAN OTT
AYAHUASCA ANALOGUES
(1994, 12)

Plants That Contain MAO-Inhibiting β-Carbolines and May Be Useful for Ayahuasca Analogs

(from Ott 1994; also Fleurentin and Pelt 1982*; Schultes and Farnsworth 1982*; Shulgin 1996; expanded)

AGARICAEAE
Coriolus maximus (Mont.) Murrill — harmane

APOCYNACEAE
Amsonia tabernaemontana Walt. — harmine and others
Apocynum cannabinum L. — harmalol
Ochrosia nakaiana Koidz — harmane

ARACAEAE
Pinellia pedatisecta — norharmane

BIGNONIACEAE
Newbouldoia laevis Benth. et Hook. f. — harmane

CALYCANTHACEAE
Calycanthus occidentalis Hook. et Arnot — harmine

CHENOPODIACEAE[395]
Hammada leptoclada (Pop) Iljin — tetrahydroharmane and others

Kochia scoparia (L.) Schrad. — harmine, harmane, triterpene glycosides
  [syn. *Bassia scoparia* (L.) A.J. Scott]
K. scoparia var. *childsii* Kraus
K. scoparia var. *trichophylla* (Voss) Boom

COMBRETACEAE
Guiera senegalensis Lam. — harmane and others

CYPERACEAE
Carex brevicollis DC. — harmine and others

ELAEAGNACEAE
Elaeagnus angustifolia L. — harmane and others
Elaeagnus hortensis M.B. — tetrahydroharmane and others

Elaeagnus orientalis L. — tetrahydroharmane
Elaeagnus spinosa L. — tetrahydroharmane
Hippophae rhamnoides L. — harmane and others
Shepherdia argentea Nutt. — tetrahydroharmol
Shepherdia canadensis Nutt. — tetrahydroharmol

GRAMINEAE
Arundo donax L. — tetrahydroharmane and others

Festuca arundinacea Schreber — harmane and others
Lolium perenne L. — harmane and others

LEGUMINOSAE
Acacia baileyana F. v. Muell — tetrahydroharmane
Acacia complanata A. Cunn. — tetrahydroharmane
Burkea africana Hook. — harmane and others
Calliandra pentandra — tetrahydroharmine
Desmodium pulchellum Benth. ex Bak. — harmane and others
Mucuna pruriens DC. — 6-methoxyharmane
Petalostylis labicheoides R. Brown — tetrahydroharmane
Petalostylis labicheoides var. *cassioides* — tetrahydroharmane, **N,N-DMT**

Prosopis nigra (Griseb.) Heironymus — harmane and others

LOGANIACEAE
Strychnos usambarensis Gilg. — harmane
  (cf. **Strychnos** spp.)

MALPIGHIACEAE
Banisteriopsis spp. — harmine
Cabi paraensis Ducke [syn. *Callaeum antifebrile* (Griseb.) Johnson] — harmine

MYRISTICACEAE
Virola cuspidata (Benth.) Warb. — 6-methoxyharmane

PASSIFLORACEAE
Passiflora actinea Hook. — harmane
Passiflora alata Aiton — harmane
Passiflora alba Link et Otto — harmane
Passiflora bryonoides H.B.K. — harmane
Passiflora caerulea L. — harmane
Passiflora capsularis L. — harmane
Passiflora decaisneana Nichol — harmane
Passiflora edulis L. — harmane, harmol, harmaline, harmine

Passiflora eichleriana Mast. — harmane
Passiflora foetida L. — harmane
Passiflora incarnata L. — harmane, harmine, harmaline

Passiflora involucrata (Mast.) Gentry — β-carbolines
Passiflora quadrangularis L. — harmane
Passiflora aff. *ruberosa* — harmane
Passiflora subpeltata Ortega — harmane
Passiflora warmingii Mast. — harmane
  (cf. **Passiflora** spp.)

POLYGONACEAE
Calligonum minimum Lipski — harmane and others

RUBIACEAE
Leptactinia densiflora Hook. fil. — tetrahydroharmine (= leptaflorin)

Nauclea diderrichii — harmane and others
Ophiorrhiza japonica Blume — harmane
Pauridiantha callicarpoides Bremek — harmane
Pauridiantha dewevrei Bremek — harmane
Pauridiantha lyalli Bremek — harmane
Pauridiantha viridiflora Hepper — harmane
Simira klugii Standl. — harmane
Simira rubra K. Schum. — harmane
Uncaria attenuata Korth. — harmane
Uncaria canescens Korth. — harmane
Uncaria orientalis Guillemin — harmane

SAPOTACEAE
Chrysophyllum lacourtianum De Wild. — norharmane and others

SYMPLOCACEAE
Symplocos racemosus Roxb. — harmane

ZYGOPHYLLACEAE
Fagonia cretica L. — harmane
Fagonia indica Burm.[396] — harmine
Peganum harmala L. — harmine, tetrahydroharmane, dihydroharmane, harmane, isoharmine, tetrahydroharmol, harmalol, harmol, norharmine, harmalicin, tetrahydroharmine, harmaline

Tribulus terrestris L. — harmine and others
Zygophyllum fabago L. — harmine and others

---

395 Tetrahydro-β-carbolines are common in this family, although often only in trace amounts (Drost-Karbowska et al. 1978, 289).

396 In Yemen, the shoots of this plant are used as an antispasmodic agent (Fleurentin and Pelt 1982, 92 f.*).

Psilohuasca

This mixture, which is also known as mushroom ayahuasca or soma ayahuasca, consists of:

3 g *Peganum harmala* and 3 g mushrooms (*Psilocybe cubensis*)

or

2 g *Peganum harmala* and 1.5 g *Psilocybe semilanceata* in sage tea

Because the effects of these blends can be extremely unpleasant, people are generally warned against using them (Kent 1995; Malima 1995).

LSA/Desmanthus Ayahuasca

Although the report (in *Entheogene* 5 [1995]: 40 f.) spoke of a quite pleasant experience, this mixture appears to be potentially dangerous.

3 g *Peganum harmala*

1 *Argyreia nervosa* seed

3–4 g *Desmanthus illinoensis* root cortex

Mayahuasca

For several years, there has been considerable speculation that the pre-Columbian Maya may have used a psychoactive ritual drink that was an ayahuasca analog. It has been conjectured that the Mayans used a *Banisteriopsis* species (*Banisteriopsis* spp.) that grows in the Mesoamerican lowlands in combination with a source of DMT to make a "Mayahuasca" (Hyman 1994). It is entirely possible that *Banisteriopsis muricata* was used for this purpose, as its stems contain harmine and its leaves DMT. In other words, it is possible that an ayahuasca analog was made from just one plant.

Pharmahuasca

For pharmahuasca, 100 mg *N,N*-DMT and 50 mg **harmaline** is usually the recommended dosage per person. However, combinations of 50 mg harmaline, 50 mg harmine, and 50 mg *N,N*-DMT have also been tested with success. As a rule, the fewer β-**carbolines**, the less nausea; the more DMT, the more spectacular the visions. The constituents are put into separate gelatin capsules. The capsule with harmaline/harmine is swallowed first and the capsule containing the DMT is taken some fifteen to twenty minutes later. The purely synthetic MAO inhibitor Marplan is also suitable in place of **harmaline and harmine** (Ott 1996, 34).

Plants That Contain DMT and May Be Used for Making Ayahuasca Analogs

(from Montgomery, pers. comm.; Ott 1993*, 1994; supplemented)

Stock Plant	Part(s) Used	Tryptamine
GRAMINEAE (POACEAE)		
Arundo donax L.	rhizome	DMT
Phalaris arundinacea L.	grass, root	DMT
Phalaris tuberosa L. (Italian race)	leaf	DMT
Phragmites australis (Cav.) Tr. ex St.	rhizome	DMT, 5-MeO-DMT
LEGUMINOSAE (FABACEAE)		
Acacia maidenii F. v. Muel.	bark	DMT (0.36%)
Acacia phlebophylla F. v. Muel.	leaf	DMT (0.3%)
Acacia simplicifolia Druce	leaf, bark	DMT (0.81%)
Anadenanthera peregrina (L.) Speg.	bark	DMT, 5-MeO-DMT
Desmanthus illinoensis (Michx.) MacM. (Illinois bundleflower; cf. Kindscher 1992, 239–40*)	root cortex	DMT (up to 0.34%), indole
Desmanthus leptolobus	root cortex	DMT
Desmodium pulchellum Benth. ex Baker [syn. *Phyllodium pulchellum*] (lodrum)	root cortex[397]	DMT
Desmodium adscendens (SW.) DC. var. *adscendens* (cf. N'Gouemo et al. 1996)[398]		DMT (?)
Lespedeza capitata Michx. (cf. Kindscher 1992, 257 f.*)	?	DMT
Mimosa scabrella Benth.	bark	DMT
Mimosa tenuiflora (Willd.) Poir.	root cortex	DMT (0.57%)
MALPIGHIACEAE		
Diplopterys cabrerana (Cuatr.) Gates	leaf	DMT, 5-MeO-DMT
MYRISTICACEAE		
Virola sebifera Aubl.	bark	DMT
Virola theiodora (Spruce ex Benth.) Warb.	flower	DMT (0.44%)
Virola spp.	bark/resin	DMT, 5-MeO-DMT
RUBIACEAE		
Psychotria carthaginensis Jacquin	leaf	DMT
Psychotria poeppigiana Muell. Arg.	leaf	DMT
Psychotria viridis	leaf	DMT
RUTACEAE		
Dictyoloma incanescens DC.	bark	5-MeO-DMT (0.04%)
Limonia acidissima L.		DMT (traces)
Melicope leptococca (Baillon) Guillaumin	leaf/branch	DMT (0.21%)
Pilocarpus organensis Rizzini et Occhioni		alkaloids (1.06%), primarily 5-MeO-DMT
Vepris ampody H. Perr.	leaf/branch	*N,N*-DMT (0.22%)
Zanthoxylum arborescens Rose		DMT (traces)

The summer cypress, *Kochia scoparia* [syn. *Bassia scoparia*], indigenous to Asia, contains substantial amounts of harmala alkaloids and may be useful in ayahuasca analogs as an MAO inhibitor.

397 Many *Desmodium* species have ethnomedicinal and folk medicinal significance (e.g., Akendengué 1992:169*); to date, no psychoactive use among traditional peoples has been observed.

398 The leaves of this plant are used in African folk medicine to treat asthma (N'Gouemo et al. 1996).

The root cortex of *Desmanthus pulchellus* is rich in DMT and is thus suitable for use as an ingredient in ayahuasca analogs.

Tribulus terrestris (known as *burra gokhru*) may be of psychopharmacological interest and urgently merits further research. (Woodcut from Gerard, *The Herball or General History of Plants*, 1633)

5-MeO-DMT can be used instead of *N,N*-DMT, or a mixture of both DMTs can be used.

Endohuasca

The pharmacologist James Callaway has hypothesized that under certain circumstances a kind of pharmahuasca (which he calls endohuasca) is produced in the brain when both endogenous β-**carbolines** and endogenous **DMT** are excreted. This endohuasca produces dreams in a neurochemical manner (Callaway 1995; cf. also Ott 1996).

Literature

See also the entries for *Peganum harmala*, *Phalaris arundinacea*, *Phalaris* **spp.**, **ayahuasca**, *N,N*-**DMT**, **harmaline and harmine**, and β-**phenethylamine**.

Appleseed, Johnny. 1993. Ayahuasca analog plant complexes of the temperate zone. *Integration* 4:59–62.

Callaway, James. 1995. *Pharmahuasca* and contemporary ethnopharmacology. *Curare* 18 (2): 395–98.

DeKorne, Jim, ed. 1996. *Ayahuasca analogs and plant-based tryptamines.* E. R. Monograph Series, no. 1. El Rito, N.M.: The Entheogen Review.

Drost-Karbowska, K., Z. Kowalewski, and J. David Phillipson. 1978. Isolation of harmane and harmine from *Kochia scoparia. Lloydia* 41:289–90.

Festi, Francesco, and Giorgio Samorini. 1994. "Ayahuasca-like" effects obtained with Italian plants. Lecture at the II° Congrés Internacional per a l'Estudio dels Estats Modificats de Consciencis, October 3–7, 1994, Llèida, Catalonia (manuscript).

Hyman, Richard. 1994. *Speculations on the ritual use of* Banisteriopsis *by the ancient Maya.* Unpublished manuscript, London. (Cited 1998.)

Kent, James. 1995. Mushroom ayahuasca. *Psychedelic Illuminations* 8:74–75.

Malima. 1995. Psilocybin und Harmala—Psilohuasca. *Entheogene* 5:6–12.

N'Gouemo, P., M. Baldy-Moulinier, and C. Nguemby-Bima. 1996. Effects of an ethanolic extract of *Desmodium adscendens* on central nervous system in rodents. *Journal of Ethnopharmacology* 52:77–83.

Ott, Jonathan. 1994. *Ayahuasca analogues: Pangaean entheogens.* Kennewick, Wash.: Natural Products Co. [See book review by Christian Rätsch in *Curare* 18 (1) (1995): 246–48.]

———. 1995. Ayahuasca and ayahuasca analogues: Pan-gaean entheogens for the new millennium. *Yearbook for Ethnomedicine and the Study of Consciousness* 3 (1994): 285–293.

———. 1996. Pharmahuasca: On phenethylamines and potentiation. *Maps* 6 (3): 32–35.

Shulgin, Alexander T. 1996. *Carbolines.* Unpublished manuscript. (Cited 1998.)

Thompson, Alonzo C., Gilles F. Nicollier, and Daniel F. Pope. 1987. Indolealkylamines of *Desmanthus illinoensis* and their growth inhibition activity. *Journal of Agriculture and Food Chemistry* 35 (3): 361–65.

Ye Wen, Yingje Chen, Zhiping Cui, Jiahe Li, and Zhixue Wang. 1995. Triterpenoid glycosides from the fruits of *Kochia scoparia. Planta Medica* 61:450–52.

Balche'

Other Names
Ba'che', ba'alche, balché, pitarilla

The name *balche'* refers to three things: a tree (see *Lonchocarpus violaceus*), a drink prepared from this tree through fermentation (with *Saccharomyces cerevisiae*), and a religious ritual during which the drink is collectively consumed. The balche' beverage is a ritual drink used by the pre-Hispanic Maya, the Yucatec Maya, and the Lacandon. It is a kind of **mead** made from water, **honey** (or more rarely sugarcane syrup, pineapple juice, or refined sugar), and the fresh or previously used bark of *Lonchocarpus violaceus*. The drink is mildly alcoholic (2 to 5% alcohol), and sometimes other ingredients are added to it.

Numerous so-called *chultunes* (Mayan, "moist stone holes") have been discovered in archaeological excavations at Petén (Guatemala), e.g., at the ceremonial center of Tikal. Some archaeologists have suggested that these underground, shoe-shaped hollows in the limestone may have been used as storage spaces. Others have speculated that the *chultunes* were used during the Classic Mayan period (300 to 900 C.E.) as vessels for brewing balche' (Dahlin and Litzinger 1986). Chacmol, a reclining god of the post-Classic period, has been interpreted as the god of the balche' drink (Cuéllar 1981).

Both the tree and the ritual drink made from it are mentioned in all of the ethnohistorical sources about the Yucatec Maya (Blom 1928; Roys 1967; Rätsch 1986*). The Motul dictionary, prepared during the early colonial period, describes the balche' tree as the tree from which "the ancient wine" was brewed (MS 45r). The translator Gaspar Antonio Chi remarked that balche' was the most important ritual drink of the Maya prior to the Spanish conquest (Blom 1928, 260). The Spanish Franciscan and book burner Fray Diego de Landa wrote about it in his sixteenth-century *Relación de las cosas de Yucatán*:

> The Indians were very dissolute, drinking until they were intoxicated. . . . They make wine from honey, water and the root of a certain tree that they cultivated for this purpose. The drink thus produced was strong and smelled foul. (de Landa 2000, 67)

The drink and its ritual and medicinal significance were described in de Landa's early colonial text:

> A further reason why these Indians are decreasing in number is that they have been

The Lacandon, a Mayan tribe, prepare the divine balche' drink in a ritual canoe. (Photograph taken in Naha', Chiapas, 1990)

prevented from drinking a wine that they were accustomed to preparing and of which they said it was healthy for them and that they called *balche*. They made it out of honey, water, and a root called *balche*. They filled this into large containers that were like big tubs and held fifty *arrobas* [= 200 gallons] or more of water. There it fermented and frothed for two days, became very strong, and then smelled very bad. At their dances and songs, and when they were dancing and singing, they gave everyone who was dancing or singing a small bowl to drink from. They gave them so much until they became very inebriated from it so that they did strange things and made such faces that their condition did not remain hidden from the spectators. When they were drunk they vomited and emptied themselves. This purified them and made them so hungry that they ate with a great appetite. Some of the elder men said that this was very good for them, that it was a medicine for them, that it healed them, because it worked like a very good laxative. With it they remained healthy and strong and many grew to be very old because of it. (2.188)

Both the balche' tree and the drink of the same name are mentioned repeatedly in the esoteric texts of the *chilam balam*, the shamanic divining priest. In these texts, the drink is metaphorically called *u ci maya* "the exquisite of the Maya." In a

Freshly pressed pineapple juice (*Ananas comosus*) was once used as a fermenting agent for balche'. The use of pineapple to prepare fermented drinks is widespread throughout Central and South America. (Copperplate engraving from Meister, *Der Orientalisch-Indianische Kunst- und Lustgärtner* [The Oriental-Indian Art and Pleasure Gardener], 1677)

Depiction of the traditional balche' ceremony. (Drawing by K'ayum Ma'ax, ca. 1980)

The magical incantation for brewing balche' invokes Chäk Xok ("red shark"), a water dweller, and asks him for assistance in starting the fermentation. (Drawing by a Lacandon child, Naha', ca. 1982)

399 *Xuchit* is the Mayan version of the Aztec *xochitl*, "flower" (Helfrich 1972, 145*).

passage dealing with the symbolic (secret) Yucatec Maya language of Zuyua, the balche' ritual is metaphorically described as "little woman":

This is the green blood of the little woman, who is called for, it is the Mayan's exquisite.
These are the entrails of the little woman, they are beehives.
This is the head of the little woman, it is the untouched vessel of the exquisite that is being prepared.
This is the stool of the little woman, it is the honeycomb [?] of the bees.
This is the left ear of Ah Bol [probably the god of inebriation], it is the small *zul* cup [a small tree gourd vessel used in the preparation of the drink] of the exquisite.
These are the bones of the little woman, they are the bark strips of the balche' tree.
These are the thighs—so it is said—, they are the trunk of the balche' tree.
These are the arms of the little woman, they are the branches of the balche' tree.
This—so it is said—is her crying, it is the language of inebriation.
(Chumayel MS 37c; in Rätsch 1986, 216*)

The balche' drink is associated with the origin of the world. The first god or gods were Ah Muzencab, "the honey collector(s)" (bee deities). During the creation of the world, the balche' drink was (re)born with the rain gods. In the Mayan creation myth, as related in *The Book of Chilam Balam of Chumayel*, the divine origins of the drink and the plants associated with it are noted:

Then the Lord (of Katun) 11 Ahau released his bees. Then came the word of the Bolon Dzacab (nine descendant) from the tip of his tongue. Then the burden of Katun was searched for. Ninefold is his burden. Then it came from heaven. *Kan* is the day to which his burden is bound. Then water came out, it came out of the heart of heaven, for the rebirth, nine years old is his house. And with it came Bolon Mayel ["nine *mayel*"; from Mayahuel, the Aztec goddess of pulque inebriation; see *Agave* **spp.**]; sweet was his mouth and the tip of his tongue. Sweet too was his brain [i.e., they were sexually aroused]. Then came the four Chacs [the rain gods of the four cardinal points], they have pots full of blessings [filled with the balche' drink]. There is the honey of the flowers [of the balche' drink]. Then emerged the red opening vessel and the white opening vessel and the black opening vessel and the yellow opening vessel and the opened water lily and the closed water lily. And with it emerged the five-petaled flower, the five-petaled flower, the toothed cocoa [= *ninichh cacao*] and the *chabil tok*

plant [?] and the *bac* flower [probably *Polianthes tuberosa*] and the *macuil xuchit* flower [*macuilxochitl* "five flower,"[399] a manifestion of Xochipilli, the Aztec god of psychedelic plants], the flower with the hollow interior and the laurel flower and the paralyzing (crooked flower). (Smailus 1986, 132 f.)

Both this text and various archaeological sources and other ethnohistorical reports (such as the 1710 report from Thomas Gage [1969, 225]) suggest that a number of ingredients were added to the balche' drink of the pre-Hispanic Maya period (see the table on page 725). The balche' drink of that time thus seems to have been a kind of "witch's cauldron" with quite potent, synergistic effects (Gonçalves de Lima et al. 1977).

There are only a few sparse reports about balche' from the colonial era (Blom 1928). The most interesting of these is contained in the travel journal of the English clergyman Thomas Gage, who visited the New World in the sixteenth century. The tenth chapter contains a "description of an unusual drink of the Indians":

They [the Pokomchi] make among other things a special drink / that is much stronger than wine / in large earthen jugs or pots / that are brought from Spain / in this way: They first add a little sugarcane / or a little honey / so that the drink will be sweet / as well as other roots / that grow in that land / and of which they know / that they have the same effect. I have myself seen at different locations / that they have thrown a living toad into it.

Then the vessel is closed / and they let this all sit together fermenting for fifteen days or a month / until everything is well worked through / the frog totally decayed / and the drink has reached its desired potency.

And then they open the vessel up again / and invite their friends to the feast / which usually takes place at night / so that they are not caught by the village priests / and they do not stop drinking / until they are totally crazy and full.

They call this drink, which smells foul as it comes from the vessel, *chicha* / and it often causes many to die / particularly in those places / where they add toads. (Gage 1710, 307 f.)

The Maya who live in the Yucatán today still consume the balche' drink (Steggerda 1943, 209*; Redfield and Villa Rojas 1962, 38), which they brew from water, **honey** from native nonstinging bees, and four pieces of balche' tree bark per person. There appear to be a number of magical sayings that the Mayan shamans (*h-mènó'ob*) use when they are preparing the drink (Bolles 1982).

One Mayan from Campeche informed me that there is a balche' song (*u k'àyil bà'lché*) that invokes many poisonous animals and the bees of the forest.

Balché is brewed only for non-Catholic ceremonies. The *h-mèn* (shaman) uses the drink, which is stored in gourd containers called *homa*, during pagan ceremonies for divinatory purposes (Redfield and Villa Rojas 1962, 36, 125). Drops of balche' are offered during prayers to Yuntsilob (the lord of the animals) (Redfield and Villa Rojas 1962, 128). Before the *h-mèn* gazes into his *sastun* (quartz crystal; cf. *Turbina corymbosa*) so that he may divine, he dips the *sastun* into the balche', thereby awakening its powers (Redfield and Villa Rojas 1962, 171). The *h-mènò'ob* also ingest seeds of *Datura innoxia* and *Turbina corymbosa* with the balche'.

In Valladolid (Yucatán), balche' is also prepared for wedding ceremonies (Aguilera 1985, 131*; Barerra M. et al. 1976, 302*). The Yucatec Maya drink balche' at rain and agricultural ceremonies to honor the rain god Chakó'b (Friedel et al. 1993). They believe that a man who owns chickens that are currently incubating their eggs should not drink balche', lest the chicks die in the eggs (Steggerda 1943, 209*).

There is also evidence for the use of balche' among the Chol-Lacandon, who lived in Chiapas during colonial times and have since disappeared. They also used sugarcane and pineapple juice for fermentation (Tozzer 1984, 15).

It is remarkable that the classic balche' ritual of the Mayan-speaking Lacandon of Chiapas has been preserved. The earliest ethnographic reports noted that "honey, mixed with the bark of the balche' tree, was fermented into an inebriating drink (balche')" (Sapper 1891, 892). At that time, it was said that the Lacandon still performed their rituals in Yaxchilan, a ruin from the Classic Mayan period:

> It seems likely to me that the Lacandon have always had a connection to Manché Tinamit [= Yaxchilan], for otherwise we could not explain why they (until recently) would travel every year from their residences in imprecisely known places in Lacanjá and other areas to Manché in order to celebrate their festivals there with balche' carousals and unusual customs, and to make offerings to their gods in the different buildings, in particular an exceptional three-storied structure (apparently the main temple of the former city). (Sapper 1891, 894)

The balche' ceremony is a ritual circle that is still performed by the Lacandon of Naha'. The gods gave this ritual to the Lacandon primordial ancestors, the "great, ancient people, who still found the paths to heaven." The god of creation himself originally conceived of the ritual, during which the inebriating balche' drink is communally consumed. The brewing of the drink, on the other hand, was the job of Bol, the god of inebriation:

> Bol made a balche' drink for Hachäkyum, our true lord. He tried some of it. Bol stood up and gave it to the gods. How many of them were soon lying there! Mensäbäk lay there, and even Hachäkyum; they did not move, they were completely drunk. All of the gods were stretched out there on the ground. They were drunk all at once. They were very drunk, but they were happy and sang. (Ma'ax and Rätsch 1984, 127)

From that time on, the primordial ancestors of the Lacandon as well as their descendants were to imitate the inebriation of the gods in their ritual circle in order to continually relive the drunken state of consciousness that creates harmony between heaven and earth. The drink is prepared in a special ceremonial mahogany canoe (called *u chemi balche'*) from honey (or sugar), water, and the bark of the balche' tree, which is cultivated specifically for this purpose. The solution will ferment, and when the fermentation is finished—usually after two to three days—the drink is ready. It contains 1 to 5% ethanol, which dissolves various other substances from the bark (Rätsch 1985b).

Brewing the balche' is a well-established magical act. Magical incantations and prayers are recited during the brewing process. The brewer identifies himself with Bol, the god of inebriation. Uttering a long incantation (*u t'ani balche'*), he calls the invisible spirits of all the poisonous animals and plants of the forest to him and asks them to add the essences of their poisons to the drink so that it will be especially potent. One after another, the following animals are called forth and asked to add the "juice of their stings/teeth/nettles" to the brew: wasps, caterpillars, ants, spider killers, bird spiders, scorpions, poisonous lizards, beetles, and snakes. Then the brewer invokes other beings, such as Chäk Xok ("red shark"), the helpful water spirit, and these stir the drink, heat it, and cause it to froth (cf. Ma'ax and Rätsch 1984, 270–82; also Boremanse 1981).

One or two days later, the drink is done fermenting. Now it is time to offer the "soul" of the balche' drink to the gods. The "head of the balche'" is carried in a ceramic jug to the house of the gods. The gods, in the form of ceramic incense bowls (*u läkil k'uh*), are placed on the ground. As prayers are continuously offered, a palm leaf is used to give every god and every goddess a mouthful of the drink. From the bowls of the gods on the ground, the soul of the drink rises into the heavens, where it manifests as a drink that will

> "Stab through, bore through my foot
> stab through, bore through my hand
> I, were I Bol
> I, were I Istal
> Here is the green chili pepper
> It passes through my canoe
> Here is the green *chawa'* chili
> Here is the raging chili pepper
> With it the heat comes into my canoe
> It comes in and bites into my vessel
> Here is the pot of my solution
> Here are the wasps
> That is the root of my canoe
> Here is the nose of my canoe
> Here are the wasps
> They pass by on the flanks of my canoe
> Come, give me the juice of your stingers for my balche' drink
> Come and give it to me
> Here are the raging wasps
> Here are the hard red drinking wasps
> Here are the wasps
> Here are the branch wasps
> Here are the hornets
> Here are the bumblebees
> Here are the white bumblebees
> Here are the big bumblebees
> Here come here, give me the juice of your stingers for my canoe
> In my vessel
> Here is my balche' drink
> Come, give me the juice of your stingers . . .
> Here is the ground chili in my canoe
> There they come
> Here the bubbles rise
> The air bubbles are rising up in my vessel
> Here is Chäk Xok, the water man
> He stands in the middle of my canoe
> He stands at the root of my canoe
> Here is the nose of my canoe
> It goes to the ground of the vessel of the solution
> He enters and stirs the juice of my balche' drink
> He enters and awakens the fermentation
> He makes the bubbles rise
> He, Chäk Xok, the water man
> So come then . . .
> Here are the ch'el birds [*Penelope nigra*]
> So come then
> Come and light the fire
> Beneath my vessel
> Come, kindle the fire
> With your tails
> You ch'el birds . . .
> There the ants rise up and pass by
> Through my vessel

There the tree toads croak
That is their croaking
The tree toads croak in my vessel
the *wo'* toads croak in my vessel
Here are the white salmon
Here are the *bayok'* fish
They swim through my canoe
They dive into it
These are the things in my canoe . . .
It bubbles in my vessel
Here are my tree toads
The crocodiles make noise
Here are the crocodiles of the sea . . .
The turtles pass by . . .
There the ants are passing by on the
     air bubbles
There are the things of my vessel
They rumble in my vessel
In my canoe
All came and gave of themselves for
     my vessel
I, were I Bol
I, were I Itsal."

A MAGICAL INCANTATION FOR
MAKING BALCHE'
IN *EIN KOSMOS IM REGENWALD*
[A COSMOS IN THE RAIN FOREST]
(MA'AX AND RÄTSCH 1984, 270–82)

Depiction of a toad (*uo*-glyph) on
lintel 48 at Yaxchilan (from the
Classic Mayan period).

inebriate the gods. The gods love to be inebriated, and they sing and dance and amuse themselves with delight for the ritual offerings the humans have provided.

After the balche' drink has been offered to the gods, the humans may drink it. This occurs in a ritual, the form of which can be characterized as a ritual circle. The communal drinking, in which the *tu wolol winik,* "whole circle of humans," takes part, usually begins just before dawn. When the first offering has been made, the brewer blows into a conch trumpet (from *Strombus gigas,* the queen conch) to sound the call and invitation to the ritual. The people come. No one is required to participate in the ritual, but all who have been initiated are allowed to come. The men go into the house of the gods and the women meet in the ceremonial kitchen. The brewer assigns each person a place in the circle around the ceramic jug filled with balche'. Using a measure (*u p'iis*), the brewer fills special drinking glasses (*luch* or *hama',* made from *Crescentia cujete* L.; cf. Morton 1968). Each participant receives the same amount. Everyone drinks at the same time, and all the participants are supposed to drink the same amount. Within just a few hours, each of the participants will consume approximately 17 liters. The communal inebriation and the corresponding alteration of consciousness rather quickly becomes apparent. It is said that consciousness gets lost or turned around and is exposed to the adventure of inebriation. The Lacandon say that the effects of the drink are determined by the quality of the bark used and the magical abilities of the brewer. They attribute the least part of the effects to the alcohol.

The effects of balche' are not like those of **beer** or **wine** or any other inebriant known to us. Balche' is not hallucinogenic. Its effects may be more precisely described as empathogenic. Consciousness becomes euphoric, perception more acute, the muscles relaxed, and the stomach and intestines emptied, and the heart is brought to laughter. Very strong doses (20 liters) have narcotic and analgesic effects. The effects on mood are especially pronounced. The inebriated people are prone to cramps from laughing and sentimental or affable feelings. As the effects increase, aggressive feelings disappear. Sometimes the drink may also produce a mildly psychedelic effect, particularly when a lot of fresh bark has been used (Rätsch 1985b).

The main purpose of the ritual circle is to heal the sick or to improve the state of the "consciousness of heaven" (as Hachäkyum, the main god, is referred to ceremonially). When humans come together into the circle and imitate the gods' inebriation, the inebriated gods in heaven develop such a good mood that their magical healing powers are able to heal the sick, counteract ecological catastrophes, and make the rain fall and the corn grow.

One important function of the balche' ritual is to promote social contact among the Lacandon community, the communal alteration of consciousness, and often a kind of social therapy that appears to be reinforced by the specific effects of the drink. When two men who have been in an aggressive mood toward each other take part in the ritual, one is able to say to the other, "Come, let us drink, for there is something between us." Then the two quickly and simultaneously drink such an enormous quantity that they can barely keep it down. In this way, they shift their state of consciousness together and, because of the large amounts of liquid, must soon vomit and empty their bladders and intestines together as well. The drink flushes out their bodies completely, alters their consciousness with its strong inebriation, and cleanses them both of their problem.

The ritual is over when all of the balche' has been consumed. The inebriated participants then typically fall into a dreamless sleep. Several hours later, they awaken with a clear consciousness and a purified body (the drink has diuretic, purgative, and laxative properties). No unpleasant side effects or aftereffects are known.

Balche' Additives

Balche' appears to have had meaning and function in Mesoamerican and Mayan culture similar to that of **ayahuasca** in Amazonia. The archaeological, ethnohistorical, and ethnographic information makes it clear that balche' has long been a drink to which various other, often more potent psychoactive substances were added. There has been considerable speculation about which other substances may have been used.

Older Lacandon still remember that the Lacandon who once lived near Piedras Negras

(Guatemala) brewed an extremely potent balche'. One small bowlful was said to have induced powerful inebriation and visions. We can only conjecture as to whether these effects were achieved by the addition of turtles, frogs (e.g., *Dendrobates* spp. or *Phyllobates* spp.; cf. Daly and Myers 1967 and Myers et al. 1978), or plants. I assume that the plants and the animals referred to in the magical spell for balche' brewing were once used to flavor or strengthen the drink. Among the plants that are mentioned are some (e.g., **Acacia spp.**) that contain *N,N*-DMT or other tryptamines. It is entirely possible that the longistylines that are present in balche' bark may have MAO-inhibiting properties and could thus have orally activated any ingredients that contained DMT (cf. **ayahuasca**).

It has been asserted in the literature that the Lacandon add psychedelic mushrooms to their balche' (Furst 1976; Greene Robertson 1972). Unfortunately, there is no ethnographic information of any kind to support this conjecture. Today, the Lacandon use only aromatic additives (*Polianthes tuberosa, Plumeria* spp., cocoa beans, vanilla pods, balche' flowers).

It is very likely that the Maya of the Classic period added ***Nymphaea ampla*** to their balche'; it is probable that ***Tagetes* spp.** and other as yet unidentified plants were also added. A report from the colonial period states that the roots of a maguey agave (perhaps *Agave americana* var. *expansa*; cf. ***Agave* spp.**) were used as a balche' additive in the northern Yucatán (*Relacíon de Mérida*, Col. Doc. inéd., 11:49).

Known and Purported (*) Balche Additives

Mayan Name	Identification	Active Constituent(s)
akunte'	*Acacia cornigera*	tryptamines, DMT?
bab/äh bäb	*Bufo marinus*	bufotenine, tryptamines, glycosides
	*Bufo alvarius**	**5-MeO-DMT**
bac nicte/bäk nikte'	*Polianthes tuberosa*	**essential oil**
	Solandra spp.*	**tropane alkaloids**
bukluch	*Vanilla planifolia*	vanillin, coumarins
hach käkaw	*Theobroma cacao* L.	theobromine, phenethylamine
ik	*Capsicum* spp.	capsaicin
kih	*Agave americana* var. *expansa*	sugar, enzymes
	Agave spp.	
k'uts	*Nicotiana rustica*	**nicotine**
	Nicotiana tabacum	**nicotine**
	*Nicotiana ondulata**	**nicotine, harmine**
kuxum lu'um	*Panaeolus venenosus** [syn. *Panaeolus subbalteatus*]	**psilocybin, psilocin, serotonin**
lol lú'um	*Psilocybe* spp.*	**psilocybin**
	Psilocybe (Str.) *cubensis**	**psilocybin, psilocin**
macuil xuchit	*Tagetes lucida, T. erecta*	**essential oil**
naab	*Nymphaea ampla*	aporphine, nuciferin
nicte[400]	*Plumeria alba, P. rubra*	**essential oil**
ninichh cacao (= balamte')	*Theobroma bicolor* Humb. et Bonpl. (cf. *Theobroma* spp.)	theobromine, phenethylamine
poch, pochil-ak	*Passiflora* spp.*	**harmine**
wi' ("root")	*Lophophora williamsii**	**mescaline, phenethylamines**
wo'	*Bufo* sp. (?)	(tryptamine derivatives)
	Physalaemus natereri	
xtabentum	*Turbina corymbosa*	lysergic acid amide
xtohk'ùh	*Datura innoxia*	**tropane alkaloids**
	Datura stramonium	
xut'	*Dendrobates* sp.*	steroidal alkaloids

Opposite page: In the Lacandon creation myth, the gods and goddesses were born enveloped in the sweet scent of night hyacinth (*Polianthes tuberosa*) blossoms. These enchantingly scented flowers are sometimes used to perfume the balche' drink. (Photographed in Naha', Chiapas)

Left: Vanilla, a native of the southern Mexican rain forest, is used to flavor the balche' drink. (Wild plant, photographed in the Selva Lacandona region of Chiapas)

Right: The balche' drink was once flavored with chili pods. (A variety cultivated by the Lacandon, photographed in Naha', Chiapas)

400 The name *nicte* is used both for the vulva and for *Plumeria* (Helfrich 1972, 145*).

Top: The exquisitely aromatic *Plumeria rubra* flowers are used to perfume the balche' drink.

Bottom: The Maya use frangipani or temple tree (*Plumeria alba*) flowers for love magic and also occasionally add them to the balche' drink.

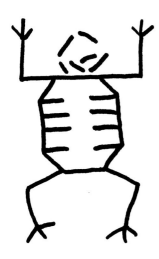

This image of a turtle (*äh bäb*) was carved into a Lacandon balche' drinking vessel. (Drawing: C. Rätsch)

A Mayan god (God F) with the hands of a tree toad (*Hyla eximia*) or a tree frog (*Dendrobates* sp.). The secretions of such tree-dwelling amphibians were likely added to the balche' drink. (Codex Tro-Cortesianus 26b, 26a)

Literature

See also the entries for ***Lonchocarpus violaceus***, ***Nymphaea ampla***, **bufotenine**, and **5-MeO-DMT**.

Blom, Frans. 1928. Gaspar Antonio Chi, interpreter. *American Anthropologist* 30:250–62.

———. 1956. On Slotkin's "fermented drinks in Mexico." *American Anthropologist* 58:185–86.

Bolles, David. 1982. Two Yucatec Maya ritual chants. *Mexicon* 4 (4): 65–68.

Boremanse, Didier. 1981. Una forma de clasificación simbólica: Los encantamientos al balche' entre los lacandones. *Journal of Latin American Lore* 7 (2): 191–214.

Cuéllar, Alfredo. 1981. *Tezcatzoncatl escultorico—el "Chac-Mool"—(El dios mesoamericano del vino)*. Mexico City: Avangrafica, S.A.

Dahlin, Bruce H., and William J. Litzinger. 1986. Old bottle, new wine: The function of chultuns in the Maya lowlands. *American Antiquity* 51 (4): 721–36.

Daly, John W., and Charles W. Myers. 1967. Toxicity of Panamanian poison frogs (Dendrobates): Some biological and chemical aspects. *Science* 156:970–73.

Freidel, David, Linda Schele, and Joy Parker. 1993. *Maya cosmos: Three thousand years on the shaman's path*. New York: William Morrow and Co.

Furst, Peter T. 1976. Fertility, vision quest and auto-sacrifice. In *Segunda Mesa Redonda de Palenque* 3, 181–93. Pebble Beach, Calif.: Pre-Columbian Art Research.

Gage, Thomas. 1710. *Neue, merkwürdige Reise-Beschreibung nach Neu-Spanien*. Gotha: Verlegts Johann Herbordt Kloß.

———. 1969. *Thomas Gage's travels in the New World*. Ed. J. E. S. Thompson. Norman: University of Oklahoma Press.

Gonçalves de Lima, O., J. F. de Mello, I. L. d'Albuquerque, F. delle Monache, G. B. Marini-Bettolo, and M. Sousa. 1977. Contribution to the knowledge of the Maya ritual wine: Balche. *Lloydia* 40:195–200.

Greene Robertson, Merle 1972. The ritual bundles of Yaxchilan. Lecture given at the Tulane University Symposia on the Art of Latin America, April 15, New Orleans.

Landa, Diego de. 2000. *An account of the things of Yucatán*. Mexico City: Monclem Ediciones.

Ma'ax, K'ayum, and Christian Rätsch. 1984. *Ein Kosmos im Regenwald: Mythen und Visionen der Lakandonen-Indianer*. Cologne: Diederichs. Repr. Munich, 1994.

McGee, R. Jon. 1984. The influence of pre-Hispanic Maya religion in contemporary Lacandon Maya ritual. *Journal of Latin American Lore* 10 (2): 175–187.

———. 1985. *Sacrifice and cannibalism: An analysis of myth and ritual among the Lacandon Maya of Chiapas, Mexico*. Ann Arbor, Michigan: University Microfilms International.

———. 1988. *The Lacandon Maya balche ritual*. VHS. Berkeley: The Extension Media Center, University of California.

———. 1990. *Life, ritual, and religion among the Lacandon Maya*. Belmont, Calif.: Wadsworth Publishing Co.

Metzner, Ralph. 1996. The true, original first world and the fourth—a visit to the Lacandon Maya in Chiapas. *Yearbook for Ethnomedicine and the Study of Consciousness* 4 (1995): 231–44.

Miyanishi, Teruo. 1992. La cultura de trance en los grupos mayas. In *Memoria de Primer Simposium Internacional de Medicina Maya—the ancient Maya and hallucinogens*, ed. Teruo Miyanishi, 107–38. Wakayama, Japan: Wakayama University.

Morton, Julia F. 1968. The calabash (*Crescentia cujete*) in folk medicine. *Economic Botany* 22:273–80.

Myers, Charles W., John W. Daly, and Borys Malkin. 1978. A dangerously toxic new frog (*Phyllobates*) used by Emberá Indians of western Colombia, with discussion of blowgun fabrication and dart poisoning. *Bulletin of the American Museum of Natural History,* vol. 161, art. 2. New York: American Museum of Natural History.

Rätsch, Christian. 1985a. Der Rausch der Götter: Zum kulturellen Gebrauch von Datura und Balche' in Mexico. Lecture given at the symposium "Über den derzeitigen Stand der Forschung auf dem Gebiet der psychoaktiven Substanzen," Nov. 29 to Dec. 1, 1985, Burg Hirschhorn. (See Leuner and Schlichting 1986*.)

———. 1985b. Eine Hamburger *balche'*-Zeremonie. *Trickster* 12/13:50–58.

———. 1986. Balche'—der Rausch der Götter. In *Rausch und Erkenntnis—Das Wilde in der Kultur,* ed. Sigi Höhle et al., 90–94. Munich: Knaur Taschenbuch.

———. 1987. Alchemie im Regenwald—Dichtung, Zauberei und Heilung. *Salix* 2 (2): 44–64.

———. 1988. Das Bewußtsein von der Welt: Mensch und "Umwelt" im lakandonischen Kosmos. In *Die neuen "Wilden,"* ed. Peter E. Stuben, 166–71 (*Ökozid* 4). Giessen: Focus. Repr. in *Politische Ökologie* 24, 51–53 (1991).

———. 1992. Their word for world is forest: Cultural ecology and religion among the Lacandon Maya Indians of southern Mexico. *Yearbook for Ethnomedicine and the Study of Consciousness* 1:17–32. Berlin: VWB.

———. [1993]. *Kinder des Regenwaldes.* Löhrbach: Werner Pieper's MedienXperimente.

———. 1994. Der Stamm der Anarchisten. *Esotera* 3/94:88–94.

Redfield, Robert, and Alfonso Villa Rojas. 1962. *Chan Kom: A Maya village.* Chicago and London: The University of Chicago Press.

Roys, Ralph L. 1967. *The book of Chilam Balam of Chumayel.* Norman: University of Oklahoma Press.

Sapper, Carl. 1891. Ein Besuch bei den östlichen Lacandones. *Ausland* 64:892–95.

Slotkin, J. S. 1954. Fermented drinks in Mexico. *American Anthropologist* 56:1089–90.

Smailus, Ortwin. 1986. Die Bücher des Jaguarpriesters—Darstellung und Texte. In *Chactun—Die Götter der Maya,* ed. C. Rätsch, 107–36. Cologne: Diederichs.

Tozzer, Alfred M. 1907. *A comparative study of the Mayas and the Lacandones.* New York: Macmillan.

———. 1984. *A Spanish manuscript letter on the Lacandones.* Culver City, Calif.: Labyrinthos.

Beer

Other Names

Acca, acupe, ahai, akka, ale, asua, badek, bakhar, bier, binburam, birra, biru, bosa, bouza, burukutu, busaa, cangüi, cashirí, cauim, caxiri, caysuma, cerveza, chang, chhang, chica, **chicha**, darassun, dolo, huicú, ikigage, kaffir, kalya, kiwa, kufa, kwass, lugri, masato, mazamorro, mekzu, merissa, mqombothi, munkoya, murcha, nawá, øl, pachwai, paiva, paiwariu, pajuarú, pissioina, pito, sende, sendechó, talla, taroba, tesvino, tesgüino, tizwin, to, toach, torani, tulapi, tulbai, tulpi, utywala, yale

Beer and beerlike drinks are found throughout the world (Bücheler 1934; Fairley 1992; Hürlimann 1984). Beer consists chiefly of water in which starch or sugar has been dissolved (see the table on page 729; cf. **chicha**). By adding cultivated or wild yeast (see the table on page 728), the mixture begins to ferment (for more on fermentation, cf. Hlavacek 1961). This yields a brew with an **alcohol** content of generally between 2 and 5%. However, modern brewing techniques make it possible to increase the alcohol content to around 10% (as in *bock* and strong beer). Today, most industrial beer is made with malted barley (Delos 1994; Jackson 1988); in former times, almost all the various kinds of grain known to man (many of which were also made into bread) were fermented (Gastineau et al. 1979; Lazzarini and Lonardoni 1984; Ziehr and Bührer 1984).

Throughout the world, beer was originally a ritual drink that was drunk during shamanic or religious ceremonies to honor the gods (libations) and in order to establish contact with other realities (Huber 1929). Psychoactive plants were usually added to these ritual beers (see the table on page 730). Some fifty known psychoactive plants have been added to beer at some time and place in the world. Such beers were consecrated to the gods or goddesses (e.g., Thor, Dionysos, Bacchus, Hathor, Bhairab/Shiva, Isis; cf. Golowin n.d.). Some of the more famous of these beers are the mandrake beer of the Egyptians (see *Mandragora officinarum*), the maize beer of the South American Indians—which is flavored with seeds from angel's trumpets or thorn apple (see *Brugmansia sanguinea, Datura innoxia, Datura stramonium*)—and the "true pilsner," the Germanic henbane beer (see *Hyoscyamus niger*). The *porst* beer of the Vikings, which had potent inebriating

Today, barley (*Hordeum vulgare*) is the most commonly used brewing grain in the world. (Woodcut from Tabernaemontanus, *Neu Vollkommen Kräuter-Buch,* 1731)

Top: Barley (*Hordeum vulgare*) is the most important source of fermentable material for beer brewing.

Middle: The root of the cassava bush (*Jatropha multifida*) is used to brew beer in South America.

Bottom: In Africa and Asia, beer is brewed primarily from true millet (*Panicum miliaceum*).

"The art of brewing beer was revealed to the humans out of a special goodness and grace. When no one yet knew how barley could be used, Dionysos conceived of the drink and taught it to those who had no vineyards, so that they would not have to drink water like the geese and ducks."

CHRISTOPH WEIGE
(REGENSBURG, 1698)

properties, was brewed with marsh rosemary (**Ledum palustre**) or bog myrtle (*Myrica gale*) (Simpson et al. 1996).

In the Middle Ages, beer brewing was associated with alchemy and witchcraft and was consequently a sometimes disreputable practice. This was often due to the "secret ingredients" (Eckstein 1927). Not only strong, inebriating beers but also aphrodisiac and medicinal beers were made. In Germany, "the land of beer," beer was brewed with strong psychoactive additives, some of which cannot be botanically identified, well into the early modern era. In 1720, Paul Hönn, a royal councilor of Saxon, wrote in the *Betrugslexicon* [Dictionary of Deception]:

Brewers deceive, when they throw so-called cat brains [?], valerian [**Valeriana officinalis** L.], and other such mind-boggling things into the pot, so that they will make the beer strong and the people who drink it stagger, even more so when hops becomes expensive and they instead use wormwood [**Artemisia absinthium**], ox bile, and such things in the beer simply to make it bitter. (In Mathäser 1996, 57)

The so-called Bavarian Purity Law of 1516 was the first German drug law; it specifically forbade the use of henbane (**Hyoscyamus niger**) as a beer admixture. The use of hops (see **Humulus lupulus**) as a flavoring agent in beer is an invention of Christian monks. It has been suggested that the

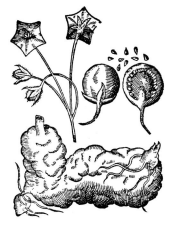

The potato (*Solanum tuberosum* L.) was cultivated in South America; it is one of the most important sources of highly nutritive carbohydrates in the Andes. The entire plant is highly toxic, and only the tubers are suitable for consumption. The native peoples of the Andean region use the tubers to brew beer. Potatoes are also used to make vodka. (Woodcut from Tabernaemontanus, *Neu Vollkommen Kräuter-Buch*, 1731)

The Most Important Beer Yeasts
(from Lappe and Ulloa 1989; Litzinger 1983)

Species	Distribution
Candida famata [syn. *Torulopsis candida*]	southeastern Europe
Candida guilliermondi	Nigeria, South Africa
Candida krusei	Kenya, South Africa
Candida pseudotropicales	Nigeria
Candida tropicales	South Africa, South America
Candida valida [syn. *Candida mycoderma*]	
Candida vini [syn. *Mycoderma vini*]	South America
Candida spp.	Nigeria, India
Hansenula anomala	Nigeria, South Africa
Hansenula anomala var. *scheggi*	Asia
Hansenula sp.	India
Pichia burtonii [syn. *Endomycopsis burtonii*]	Asia
Pichia membranaefaciens	Mexico, Nigeria
Saccharomyces apiculata	South America
Saccharomyces cerevisiae	cosmopolitan
Saccharomyces pastorianus	Nigeria, South America
Saccharomyces uvarum	Himalayas
Saccharomyces spp.	Sudan
Saccharomycopsis fibuligera [syn. *Endomycopsis fibuliger*]	Asia
Trichosporon cutaneum	Zambia

The Most Important Fermenting Agents

(from Cutler and Cardenas; Hartmann 1958; Havard 1896*; La Barre 1938; Low 1990, 189*; Mowat 1989; Nicholson 1960; Rätsch 1996; supplemented)

Name	Botanical Name	Culture/Region
agave	*Agave* spp.	northern Mexico
airampu	*Opuntia sulphurea* G. Don	Peru
	Opuntia soerensii Britt. et Rose	Peru (Cochabamba)
algarrobo	*Prosopis* spp.	Peru
arrowroot	*Maranta arundinacea* L.	South America, Caribbean
assai	*Euterpe* spp.	Tupí/South America
banana	*Musa* x *sapientum*	South America, Asia
banksia	*Banksia* spp.	Australia
barley	*Hordeum distichon* L.	
	Hordeum hexastichon L.	
	Hordeum vulgare L.	cosmopolitan
bataua palms	*Oenocarpus* spp.	South America
bearded darnel	*Lolium temulentum*	Gaul
bi palm	*Mauritia* sp.	South America
birch juice	*Betula* spp.	North America
bread wheat	*Triticum aestivum* L.	ancient Greece
bristle grasses	*Setaria* spp.	Asia, Africa
cashew tree	*Anacardium occidentale* L.	Aruak/South America
chañar	*Gourleia spinosa* (Mol.) Skeels	Chaco/northern Chile
emmer	*Triticum dicoccoides* Körn	ancient Greece
	Triticum dicoccum Schübl.	Egypt, Mesopotamia
	Triticum monococcum L.	ancient Greece
foambark tree	*Jagera pseudorhus* (A. Rich.) Radlk.	Australia
frutilla	*Fragaria chilensis* Ehrh.	Peru/Chile
hard wheat	*Triticum durum* Desf.	ancient Greece
honey	*mel*	cosmopolitan
gourd	*Cucurbita pepo* L.	South America
maize	*Zea mays* L.	Mexico to Peru
makrozamia	*Macrozamia spiralis* (Sal.) Miq.	Australia
mangareto	*Xanthosoma sagittifolium* Schott	South America
manioc	*Manihot esculenta* Crantz [syn. *Manihot utilissima* Pohl]	Central and South America
manioc, sweet	*Manihot dulcis* (Gmel.) Pax [syn. *Manihot aypi* Pohl]	Mesoamerica, South America
molle	*Schinus molle* L.	Peru
mwerere	*Rauvolfia caffra* Sond. (cf. *Rauvolfia* spp.)	Kenya
oaks	*Quercus* spp.	North America
oats	*Avena sativa*	Thrace
oca	*Oxalis tuberosa* Mol.	Peru
panic grasses	*Panicum* spp.	Asia, Africa
peanut (mani)	*Arachis hypogaea* L.	Tuparí/Peru
pearl millet	*Pennisetum* spp.	Asia
potato	*Solanum tuberosum* L. (cf. *Solanum* spp.)	South America
pupunha palm	*Bactris* spp.	Tupí/South America
quinoa	*Chenopodium quinoa* Willd.	Peru
rice	*Oryza sativa* L.	Asia (cf. **sake**)
reed roots	*Phragmites australis*	New World
rye	*Secale cereale* L.	Europe
schinus fruit	*Schinus latifolius* (Gill.) Engl.	Chile (cf. **chicha**)
	Schinus polygamus Cav. Cabr.	Chile (cf. **chicha**)
sorghum	*Sorghum* spp.	Asia, Africa
sotol	*Dasylirion* spp.	northern Mexico
spelt	*Triticum spelta* L.	Germania
sugarcane	*Saccharum officinarum* L.	Central and South America
sweet potato	*Ipomoea batatas* Pore.] (cf. *Ipomoea* spp.)	Central and South America
ti	*Cordyline fruticosa*	Hawaii, Samoa
tusca	*Acacia aroma* Gill. (cf. *Acacia* spp.)	Chaco
wheat	*Triticum* spp.	cosmopolitan
wild einkorn	*Triticum boeoticum* Boiss.	ancient Greece
yams	*Dioscorea sativa* div. ssp.	South America
yucca	*Yucca* spp.	Jíbaro/Peru

"Of the herb margosae [neem tree]. This herb, which the Dutch use instead of hops for flavoring sweet beer, is dried because of its bitterness. Its leaves are almost like those of hemp but grow somewhat smaller, or like cinquefoil, it vines up trees."

GEORGE MEISTER
DER ORIENTALISCH-INDIANISCHE KUNST-UND LUSTGÄRTNER [THE ORIENTAL-INDIAN ART AND PLEASURE GARDENER]
(1677)

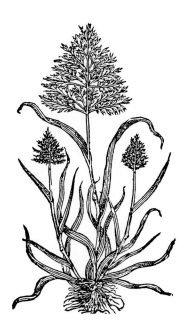

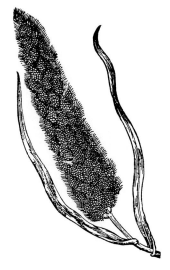

In Asia and Africa, most beer is brewed from many different kinds of millet. (Woodcut from Tabernaemontanus, *Neu Vollkommen Kräuter-Buch*, 1731)

"In any case, the Chinook of the northwestern coast of North America report an unusual form of alcohol fermentation for the purposes of inebriation. They consume as delicacies acorns that have been allowed to soak in human urine for over five weeks, which causes them to ferment. These *Chinook olives* produce a pleasant state of inebriation."

Andrea Blätter
"Drogen im präkolumbischen Nordamerika" ["Drugs in Pre-Columbian North America"]
(1996a, 178*)

Eine Wirthin vom Teufel geholt.
Altenglische Sculptur.

In the early modern era, beer and beer drinking were often demonized and associated with pagan witches' cults. Indeed, beer, especially that brewed with henbane (*Hyoscyamus niger*), is an ancient, heathen ritual drink. The caption reads: "A female innkeeper being taken by the devil." (Illustration of an old English sculpture)

401 Although it is uncertain whether the bog myrtle (*Myrica gale* L.; Myricaceae) induces psychoactive effects, it is possible; the fruits contain substances (chalcones, flavonoids) with preservative properties (Mathiesen et al. 1995). The **essential oil** has exhibited sedative properties in animal studies (Simpson et al. 1996, 127).

Psychoactive Beer Additives

(from La Barre 1938; Lappe and Ulloa 1989; Maurizio 1933; Navchoo and Buth 1990; Rätsch 1996; supplemented)

Name	Part(s)	Culture/Region
Acacia campylacantha (cf. *Acacia* spp.)	bark	Africa (*dolo* beer)
Acorus calamus	rhizome	Middle Ages
Amanita muscaria	fructification	Siberia
Anadenanthera colubrina	seeds	Inca/Peru
Anarmita cocculus	seeds	early modern period
Ariocarpus fissuratus	buttons	Tarahumara/Mexico
Artemisia absinthium	herbage	Germany
Artemisia tournefortiana	herbage	Ladakh
Artemisia vulgaris	herbage	Germania
Artemisia spp.	herbage	Peru, Sherpa/Nepal
Atherosperma moschatum	leaves, bark	Tasmania
Atropa belladonna	fruits, leaves	Slav, Middle Ages
Brugmansia sanguinea	seeds	Quechua/Peru
Brugmansia spp.	seeds, leaves	South America
Cannabis indica	female flowers	United States, Europe, ancient Orient, Scythia
Capsicum spp.	pods	Chile, United States
Claviceps paspali	**honey**	Paraguay
Conium maculatum	herbage	Bohemia
Corynanthe pachyceras (cf. *Corynanthe* spp.)	bark	West Africa
Coryphantha compacta (cf. *Coryphantha* spp.)	cactus flesh	Tarahumara/Mexico
Crocus sativus	pistils	Egypt, Middle Ages, modern period
Datura innoxia	seeds	Huichol/Mexico, Apache/United States
	roots, leaves	Tarahumara/Mexico
Datura metel	seeds	Africa, India
Datura stramonium	seeds	Araucana/Chile, Africa (*dolo* beer)
Datura spp.	seeds	Chile
Ephedra spp.	herbage	ancient Orient
Erythrina flabelliformis	seeds	northern Mexico
Erythroxylum coca	leaves	Peru
Filices (ferns)	?	Germany (?)
Humulus lupulus	female flowers	now cosmopolitan
Hedera helix	leaves	late antiquity
Hyoscyamus niger	herbage	Germania
Hyoscyamus physaloides (cf. *Hyoscyamus* spp.)	roots	Siberia
Juniperus recurva	branch tips	Himalayas
Ledum palustre	herbage	Germania
Lolium temulentum	infructescence	Middle Ages/Gaul
Lophophora williamsii	buttons	northern Mexico
Lotus wrightii	roots	Apache/United States
Lupinus spp.	seeds	Gaul, Babylon
Mandragora officinarum	roots	ancient Egypt
Mesembryanthemum mahonii N.E. Br. (cf. *Mesembryanthemum* spp.)	roots	Bantu/South Africa
Myrica cerifera	herbage	North America
Myrica gale L.[401]	herbage	Vikings/northern Europe
Myristica fragrans	seeds	Middle Ages
Nicotiana tabacum	leaves	Jíbaro/Ecuador
Pachycereus pecten-aboriginum	cactus flesh	Tarahumara/Mexico
Panaeolus subbalteatus	fructification	Germania ?
Papaver somniferum	opium	Babylon, Egypt, Middle Ages, modern era

Name	Part(s)	Culture/Region
Paullinia cupana	extract of seeds	Brazil
Petroselinum crispum	roots	Egypt, Germania
Phaseolus sp.	roots	Tarahumara/Mexico
Physalis peruviana (cf. *Physalis* spp.)	leaves	Australia
Piper spp.	leaves/fruits	ancient Orient
Psilocybe cubensis	fructification	"underground"
Psilocybe spp. (?)	fructification	Germania
Quararibea sp.	fruit	coastal culture/Peru
Salvia sclarea L.	herbage	England (nineteenth century)
Sarothamnus scoparius (cf. *Cytisus* spp.)	herbage	modern era
Scopolia carniolica	herbage, roots	eastern Europe, Lithuania
Solanum subinerme (cf. *Solanum* spp.)	fruits	South America
Sophora secundiflora	seeds	Tarahumara/Mexico
Theobroma cacao	chocolate	Belgium
Trichocline spp.	roots	Chaco/Argentina
Trichodiadema stellatum (Mill.) Schw.	roots	South Africa
Tribulus terrestris	fruits	Ladakh
Turbina corymbosa	seeds	Mexico
Vaccinium uliginosum	fruits	Siberia, modern era
Veratrum album	herbage/roots	modern era
Bufo marinus (cf. **balche'**, **bufotenine**)	all	Guatemala (*chicha*)
Oil beetles[402]	all	Aymara/Bolivia

Bavarian Purity Law was intended primarily to suppress the use of pagan ritual plants and thus to finish the work of the Inquisition (Rätsch 1996).

In recent years, home brewing has been gaining in popularity again. At home, of course, people can put anything they want in their beer. In the United States, spices (cinnamon, coriander, ginger, paradise grains, nutmeg and mace, cardamom, pepper, chilies, cumin, turmeric, vanilla) are often added. In Germany, henbane beer (*pilsner*) is being brewed once again, and a Swiss brewery introduced a hemp beer to the market in 1996. A mixture of wheat beer and a guaraná extract (*Paullinia cupana*) is popular in Brazil and is now available in Europe as well. In Belgium, a wheat beer known as Floris Chocolat is made with chocolate (cf. *Theobroma cacao*).

In Central and South America, ritual beers (see **chicha**) with psychoactive ingredients were very common in pre-Hispanic times (Arriaga 1992*; Cobo 1990*). Some tribes, such as the Tarahumara, Huichol, and Quechua, still use varied admixtures to increase the strength of their maize beer. Many of the same plants are also added to **ayahuasca**, **cimora**, and San Pedro drinks (cf. *Trichocereus pachanoi*).

Literature

See also the entry for **chicha**.

Appun, Carl Ferdinand. 1870. Die Getränke der Indianer Guayanas. *Globus* 18.

Ardussi, John A. 1977. Brewing and drinking the beer of enlightenment in Tibetan Buddhism: The doha tradition in Tibet. *Journal of the American Oriental Society* 97 (2): 115–24.

Baldus, Herbert. 1950. Bebidas e narcoticos dos indios do Brasil. *Sociologia* (São Paulo) 12.

Behre, K. E. 1983. Aspects of the history of beer flavouring agents based on fruit finds and written sources. In *Plants and ancient man: Studies in palaeoethnobotany*, ed. W. van Zeist and W. Casparie, 115–22. Rotterdam: A. A. Balkema.

Bücheler, Walther. 1934. *Bier und Bierbereitung in den frühen Kulturen und bei den Primitiven*. Berlin: VGGB.

Delos, Gilbert. 1994. *Biere aus aller Welt*. Erlangen: Karl Müller Verlag.

Eckstein, F. 1927. Bier. In *Handwörterbuch des Deutschen Aberglaubens*, ed. Bächtold-Stäubli, 1:1255–82. Berlin: De Gruyter.

Fairley, Peter. 1992. Probably the oldest lager in the world . . . *New Scientist* 16 (May): 6.

Feest, C. E. 1983. New wines and beers of North America. *Journal of Ethnopharmacology* 9 (2/3): 329–35.

Gaessner, Heinz. 1941. *Bier und bierartige Getränke im germanischen Kulturkreis*. Berlin: GGBB.

Gastineau, C., W. Darb, and T. Turner, eds. 1979. *Fermented foods in nutrition*. New York: Academic Press.

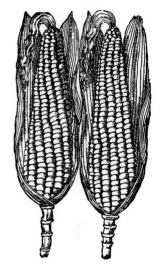

Maize/corn (*Zea mays* L.) was first cultivated in Mexico, and it is the most important source of nourishment for many Indians. In all of those places where maize was adopted into Indian culture, it is known that it can be used to make beer. The Indians usually improve their maize beer (*chicha*) with such powerful psychoactive substances as peyote, angel's trumpets, thorn apple, tobacco, and coca. (Woodcut from Tabernaemontanus, *Neu Vollkommen Kräuter-Buch*, 1731)

402 Unfortunately, no further details about this species are known. There is considerable evidence for the existence of psychoactive beetles, e.g., *Phromnia marginella* Oliv. This insect from the Family Fulgoridae occurs in Uttar Pradesh (India). It is purportedly used in the Garwhal district for its alleged psychoactive effects (Reichel-Dolmatoff 1975, 200, n. 248*).

Golowin, Sergius. n.d. *Die weisen Frauen und ihr Bier.* Brauerei Hürlimann.

Hartman, Louis Francis, and A. Leo Oppenheim. 1950. On beer and brewing techniques in ancient Mesopotamia. *Journal of the American Oriental Society* suppl. (10) (Baltimore).

Helck, Wolfgang. 1971. *Das Bier im Alten Ägypten.* Berlin: GGBB.

Hlavacek, Frantisek. 1961. *Brauereihefen.* Leipzig: Fachbuchverlag.

Huber, E. 1926. Bier und Bierbereitung im alten Babylon . . . im alten Ägypten. In *Bier und Bierbereitung bei den Völkern der Urzeit,* 9–28, 33–46. Berlin: VGGBB.

———. 1929. *Das Trankopfer im Kulte der Völker.* Hannover-Kirchrode: Oppermann.

Hürlimann, Martin. 1984. *Das Buch vom Bier.* Zurich: Brauerei Hürlimann.

Jackson, Michael. 1988. *Das große Buch vom Bier.* Bern, Stuttgart: Hallwag.

Kistemaker, R. E., and V. T. van Volsteren. 1994. *Bier! Geschiedenis van een volksdrank.* Amsterdam: De Bataafsche Leeuw.

La Barre, Weston. 1938. Native American beers. *American Anthropologist* 40 (2): 224–34.

Lappe, Patricia, and Miguel Ulloa. 1989. *Estudios étnicos, microbianos y químicos del tesgüino tarahumara.* Mexico City: UNAM.

Lazzarini, Ennio, and Anna Rota Lonardoni. 1983. *Gesundheit aus Halm und Korn: Heilsame Kräfte aus Gräsern und Getreide.* Freiburg i. Br.: Bauer.

Litzinger, William J. 1983. The ethnobiology of alcoholic beverage production by the Lacandon, Tarahumara, and other aboriginal Mesoamerican peoples. PhD diss., Department of Biology, University of Colorado.

Lohberg, Rolf, et al. 1984. *Das große Lexikon vom Bier.* 3rd ed. Stuttgart: Scripta.

Mathäser, Willibal. 1996. *Flüssiges Brot: Andechs und sein Klosterbier.* 2nd ed. Munich: Hugendubel.

Mathiesen, Liv, Karl Egil Malterud, and Reidar Bredo Sund. 1995. Antioxidant activity of fruit exudate and C-methylated dihydrochalcones from *Myrica gale. Planta Medica* 61:515–18.

Maurizio, A. 1933. *Geschichte der gegorenen Getränke.* Berlin and Hamburg: Paul Parey. Repr. Wiesbaden: Sändig, 1970.

Navchoo, Irshad A., and G. M. Ruth. 1990. Ethnobotany of Ladakh, India: Beverages, narcotics, foods. *Economic Botany* 44 (3): 318–21.

Rasanen, Matti. 1975. *Vom Halm zum Faß: Die volkstümlichen alkoholarmen Getreidegetränke in Finnland.* Kansatieteelinen Arkisto 25. Helsinki: Suomen Muinaismuistoyhdistys.

Rätsch, Christian. 1996a. *Urbock: Bier jenseits von Hopfen und Malz.* Aarau: AT Verlag.

———. 1996b. Vom Bilsenkraut zum Pils. *Natürlich* 16 (7–8): 50–3.

Röllig, Wolfgang. 1970. *Das Bier im alten Mesopotamien.* Berlin: GGBB.

Rose, A. H., ed. 1977. *Alcoholic beverages.* New York: Academic Press.

Rosenthal, Ed. 1984. *Marijuana beer.* Berkeley, Calif.: And/Or Press.

Simpson, Michael J. A., Donald F. Macintosh, John B. Cloughley, and Angus E. Stuart. 1966. Past, present and future utilisation of *Myrica gale* (Myricaceae). *Economic Botany* 50 (1): 122–29.

Ziehr, Wilhelm, and Emil Bührer. 1984. *Le pain à travers les âges.* Tielt, Belgium: Editions Lannoo.

A betel palm plantation. (Copperplate engraving, nineteenth century)

Betel Quids

Other Names

Asia's chewing gum, betel, betelbissen, betele, betelpriem, betelpriemchen, bulath, paan, pan, pán, pan masala, pin-lang, pynan, sirih, supari, tambul, tembul

Betel quids consist of essentially three ingredients: betel nuts (**Areca catechu**), betel leaves (**Piper betle**), and slaked lime, to which other ingredients (*masala*) are almost always added. About half of betel quid preparations call for mixing in specially treated (e.g., limed or fermented) tobacco (**Nicotiana tabacum**) (Gowda 1951, 196), as well as an assortment of spices and other psychoactive substances (see the table on page 735). These mixtures can be varied to change the taste or to produce specific effects.

The oldest recipe that is known to us from the literature was passed down by Sushrata, the first-century founder (or at least one of the founders) of the Ayurvedic system of medicine. He lists as ingredients betel leaves filled with broken betel nuts, camphor (**Cinnamomum camphora**), nutmeg (**Myristica fragrans**), and cloves and adds that intelligent people chew betel after a meal.

In India, there are veritable gardens of betel, in

which betel palms are cultivated together with betel pepper. The pepper plants climb up the stems of the palms, and the space between the palms is used to grow the spices added to the quids.

Slaked lime, which is indispensable for increasing the effectiveness and absorption of the active constituents, is obtained by burning mussel shells, conch shells, river snail shells, coral, or limestone and then dousing the ashes with water.

Today, an estimated 450 million people worldwide chew betel (Roth et al. 1994, 141*). Betel chewing is found throughout India, Nepal, Sri Lanka (Ceylon), the Maldives, the Nicobar Islands, Myanmar (Burma), Thailand, southern China, Malaysia, Singapore, Indonesia, Taiwan, the Philippines, Papua New Guinea, and Melanesia.

There is evidence suggesting that people have practiced betel chewing for more than twelve thousand years. Archaeologists excavating the Spirit Cave site (in northwestern Thailand) have unearthed fragments of betel nuts, traces of a *Piper* species, and gourds, all of which suggest the use of betel. Radiocarbon dating has indicated that these finds are between 12,000 and 750 years old (Gorman 1972; Seyfarth 1981, 562).

Betel chewing has a long tradition in India, perhaps one of the longest in the world. In addition to its ritual use, betel is chewed hedonistically for its stimulating effects. Mentions of betel merchants date back to the late Vedic period (Moser-Schmitt 1981).

Great quantities of betel are chewed in Varanasi (Benares). Small betel shops and the stands of betel sellers (*pan vala*) can be found wherever one goes. Both freshly made betel (*pan*)—typically spiced according to the customer's particular wishes—and hygienically packed mixtures (*pan masala*), of which there are many brands and makers, are available. These packages cost between 0.75 and 1.50 rupees each. While it is difficult to predict whether these factory-made betel blends might displace the traditional betel, it is rather unlikely.

Although the effects of betel were formerly described as narcotic, the opposite is actually the case; betel has stimulating effects (Charpentier 1977) and, in general, a primarily parasympatho-mimetic action (a "muscarinic character"). It promotes salivation, suppresses hunger and thirst, and can also act as a laxative. Betel has stimulating effects upon the central nervous system. The strongest action (on the central and peripheral nervous systems) begins six to eight minutes after chewing a quid (Chu 1995, 183). The Trobriand Islanders say that betel produces sensations of heat, increased perspiration, and feelings of happiness. These euphoric feelings are more powerful when unripe betel nuts are used in the blend (Jüptner 1969, 371). Internationally, betel is considered not to be "addictive" but to "structure

Fresh betel leaves sit on the table of a betel merchant, while the bowls and tins in the foreground contain the other ingredients. (Photographed in Varanasi, India, 1995)

social behavior" (Charpentier 1977, 117). "Betel makes the ears hot, the face red, [and] the eyes blurry and produces a mood like drunkenness, at least that is what the Chinese texts state. It is believed that betel is a remedy for malaria" (Eberhard 1983, 39).

In the eighteenth century, European observers came up with the idea that betel chewing causes cancer. In Sri Lanka, a disease known as "betel chewer's cancer" was recorded in the literature (Charpentier 1977, 110). Even today, it is often claimed that extensive and regular chewing of betel over years or even decades can increase the risks of or even cause mouth or tongue cancer. Sen et al. (1989) summarized the results of the studies that have been carried out to date. These indicate that only those betel quids that contain tobacco exhibit these characteristics. Among betel chewers who have never used tobacco as an additive, the carcinostatic action of the betel leaves (*Piper betle*) appears sufficient to protect the mucous membranes of the mouth from cell damage, an effect that presumably is due to the formation of cytotoxic *N*-nitrosamines when the leaves are chewed (Sen et al. 1989). The slaked lime and the catechu have also been attributed with carcinogenic effects, although these views are based solely upon animal experiments.

The notorious red coloration that betel chewing imparts to saliva is thought to be a product of areca red, a phlobatannin. This phenol-like substance is a constituent of *Areca catechu* and turns red as a result of the slaked lime (Heubner 1952, 17*; Roth et al. 1994, 140*).

Slaked Lime

(lime, quick lime, calcium hydroxide)

In Sri Lanka, slaked lime is known as *huna* or *chunam* and is produced by burning mussel shells, freshwater snails, coral (*hirigal/hunugal*), and, in rare cases, limestone in large kilns. This lime is used not only for betel chewing but also to build houses, in agriculture, and as a paint pigment. The slaked lime used for betel chewing is sometimes colored with powdered yellow turmeric (*Curcuma longa* L. [syn. *Curcuma domestica*]; Zingiberaceae); coconut oil is added to keep the lime from

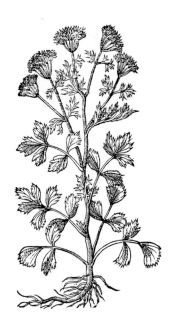

In Nepal and India, pieces of betel nuts (*Arecha catechu*) and anise seeds (*Pimpinella anisum*) may be chewed together in place of betel quids. This mixture is used mainly after meals as a digestive aid. (Anise plant; woodcut from Gerard, *The Herball or General History of Plants*, 1633)

drying out and becoming hard (Charpentier 1977, 111).

In Melanesia, the third ingredient in betel chewing, lime, is produced primarily by burning conches or mineral limestone. For the coastal inhabitants, mussels and coral species naturally offer themselves, while inland groups use chiefly limestone and, when available, river snails; the Yimar in the southern Sepik region provide an example. They first cook the mussels until the shells can be easily opened and the mussel flesh can be removed [and eaten]. The shells are then air-dried, wrapped in sago leaves, bound with *rotang* [palm fibers], and burned. The burned shells are crumbled by hand and dissolved in water, and the coarse lime powder is wrapped again in leaves to dry, until it is finally put in the lime container. (Seyfarth 1981, 563)

In Taiwan, a so-called red lime, a mixture of lime and catechu, is available and used in betel (Chu 1995, 183).

Paraphernalia

The paraphernalia associated with betel are expressions of specific cultural characteristics and the cultural significance of betel (Beran 1988), and they are often artistically and splendidly produced. Siamese and Thai kings used only betel instruments made of gold.

The following tools are traditionally used to prepare and consume betel (Brownrigg 1992):

- Betel scissors (*tong, giraya, girri*).
- Betel chopper (*bulath wangediya, wanggedi moolah kaimili*). The betel chopper is a kind of tube with a built-in chopping knife to cut the betel nuts. It is often used by older people without teeth.
- Betel presentation plate (*ilah thattuwa, heppuwa*). Presentation plates are used primarily on ceremonial occasions.
- Tobacco box (*dumkola heppuwa*). These small, right-angled boxes or chests are made from a variety of materials.
- Lime box (*hunu killotaya, yaguma, sunnadu dabbi*). Lime is usually carried in a small bottle with a stopper to which a spatula or spoon is fastened with a band or a chain. Very large lime boxes also exist, but they are used only for ritual occasions. In Melanesia and other areas, lime is stored in bottle gourds (*Lagenaria* spp.).
- Lime spatula (*kaiaku*). In Melanesia, the lime spatula is sometimes the only betel-chewing tool. It is usually made of an animal or bird bone (e.g., from a pig or cassowary), but it can also be made from the leg bone of a deceased family member. The spatula is used especially

when chopped betel nuts and pieces of betel leaves, rather than a betel quid, are being consumed. In this case the spatula is used to continuously add lime to the mouth (Seyfarth 1981, 564).

On the Trobriand Islands, which are part of Papua New Guinea, the making of lime spatulas has developed into a high art. Here, they are usually carved out of hardwood (ebony) or, less frequently, from tortoise shells, and they bear depictions of snakes, birds, and crocodiles or of ancestors (Jüptner 1969). Betel utensils were once also made from the bones of ancestors:

In former times, it was common for the son to remove individual bones from his dead father's corpse, to make this into a kind of "reliquary," and to save it: the skullcap became a lime container, pieces of the shin or arm bones were used as lime spatulas. Such reliquaries were passed down in the family for a time and then finally placed on a rock overlooking the sea. (Jüptner 1969, 375)

Symbolic and Ritual Use

The Dayak of central Borneo, like nearly all the peoples of the Malaysian archipelago and the islands of Southeast Asia, use betel as an agent of pleasure and as an important element in ritual activities. The Dayak have a detailed conception of what happens to the soul of the dead in the afterlife. After dying, the soul travels on a ghost ship into the next world, which is connected to the world of the gods and ghosts. This ghost canoe is filled with all sorts of status symbols and other valuable objects. The most important of these are the betel utensils (betel scissors) and betel leaves and nuts. If the deceased person chewed betel constantly during his lifetime, then his soul will be able to indulge in this pleasure in the afterlife for all eternity. The betel-laden ghost ship is frequently depicted in the cult drawings of the Dayak (Seyfarth 1981, 560 f.).

Betel chewing often plays a great role in maintaining social dynamics (quite similar to the roles played by *Camellia sinensis, Cannabis indica, Catha edulis, Coffea arabica, Erythroxylum coca, Nicotiana tabacum, Piper methysticum,* **alcohol, beer,** and **wine**):

The offering of and the communal chewing of betel can strengthen partnerships, seal negotiations, and end conflicts and is not infrequently a firm component of peace treaties following feuds and wars. The areca nut itself is the very symbol of friendship and peace. (Seyfarth 1981, 566)

Ingredients in Betel Quids
(from Charpentier 1977; Chinnery 1922; Chu 1995; Gowda 1951; Hartwich 1911*; Jain and Dam 1979*; Krenger 1942c; Seyfarth 1981; modified)

The betel pepper vine (*Piper betle*) prefers to climb shade trees. (Woodcut from Gerard, *The Herball or General History of Plants*, 1633)

Name	Stock Plant/Source	Active Constituent(s)
amber	*Succinium*	resins
anise	*Pimpinella anisum* L.	**essential oil** (*trans*-anethol)
beets, red	*Beta vulgaris* L.	sugar
betel leaf	*Piper betle* L.	**essential oil**
	Piper **spp.** (as a surrogate)	**essential oil**
betel nut	*Areca catechu* L.	arecoline
	Areca macrocalyx Zipp.[403]	alkaloids
camphor	*Cinnamomum camphora*	camphor
	Dryobalanops aromatica Gaertn.	camphor
caraway	*Carum carvi* L.	**essential oil**
	Carum bulbocastanum Koch	**essential oil**
cardamom	*Elettaria cardamomum* (L.) Mat.	**essential oil**
	Amomum subulatum Roxb.	**essential oil**
catechu	*Acacia catechu* L.	catechins
	Acacia chundra	catechins
	Acacia catechuoides	catechins
	Acacia polyantha Willd.	
cinnamon	*Cinnamomum verum* Presl	**essential oil**
	Cinnamomum cassia Nees	**essential oil**
cloves	*Syzygium aromaticum* (L.) Merr. et Perry	eugenol
cocaine	*Erythroxylum coca* Lam.	cocaine[404]
coconut shells (copra)	*Cocos nucifera* L.	carbohydrates
coriander	*Coriandrum sativum* L.	**essential oil**
crow's eyes	*Strychnos nux-vomica*	**strychnine**
cumin	*Cuminum cyminum* L.	**essential oil**
	Nigella sativa L. (small garden fennel)	**essential oil**
dill	*Anethum graveolens* L.	**essential oil**
fennel	*Foeniculum vulgare* Miller	**essential oil**
foliis syryboae	*Piper* **sp.**	**essential oil**
gambir[405]	*Uncaria gambir* (Hunt.) Roxb.	flavonols, tannin
garn-garn (aloe wood)	*Aquilaria agallocha* Roxb.	resins
ginger	*Zingiber officinale* Rosc.	**essential oil**
hashish	*Cannabis indica*	THC
heroin	from **morphine**	heroin[406]
kava-kava	*Piper methysticum* G. Forst.	kavains
kratom	*Mitragyna speciosa*	**indole alkaloids**
melon seeds	*Cucumis melo* L.	
menthol	*Mentha* sp.	**essential oil**
nutmeg	*Myristica fragrans* Hout.	**essential oil**
opium	*Papaver somniferum* L.	**opium alkaloids**
potentilla	*Potentilla fulgens* Hook.	
ratabulath	*Vitis* sp. (cf. *Vitis vinifera*)	?
saffron	*Carthamus tinctorius* L.	coloring agent
	Crocus sativus L.	**essential oil**
sago leaves	*Metroxylon sagu* Rottb.	
sandalwood bark	*Pterocarpus santalinus* L. f.	coloring agent
sandalwood oil	*Santalum album* L.	**essential oil**
smilax roots	*Smilax calophylla* Wall.	
speed	synthetic	amphetamine[407]
tamarind leaves	*Tamarindus indica* L.	
thorn apple seeds (*kecubong*)	*Datura metel*	**tropane alkaloids**
	Datura innoxia	**tropane alkaloids**
	Datura stramonium	**tropane alkaloids**
tobacco	*Nicotiana tabacum* L.	nicotine
	Nicotiana rustica L.	nicotine

403 This wild relative of the cultivated betel palm is indigenous to the mountain forests of New Guinea (Seyfarth 1981, 562).

404 The use of cocaine as an additive to betel quids has been previously described for the region from Iran to Pakistan (Krenger 1942c, 2929).

405 This extract of the Rubiaceae is sometimes sold under the name *catechu* or *pale catechu*.

406 The use of heroin is a new invention that appears to be found primarily in Taiwan (Chu 1995, 183).

407 The use of amphetamines as an (illegal) betel additive is common in Taiwan (Chu 1995, 183).

Name	Stock Plant/Source	Active Constituent(s)
turmeric	*Curcuma longa* L.	**essential oil**, curcumin
ylang-ylang flowers	*Cananga odorata* (Lam.) Hook.	**essential oil**
ashes	pearls	
lime, slaked lime	limestone	Ca(OH)$_2$
	corals	
	mussel and snail shells	
	river shells (Sepik)	
	river snails (Ceylon)	
lime soil	slaked lime	
marmalade	various fruits	fructose
perfumes, various		
sugar (sugarcane)	*Saccharum officinarum* L.	saccharose
sugar syrup		
tree bark	various species	slaked lime ashes

In Ceylon, it was customary for rulers to have a special "royal betel," which was made by a special betel preparer and was constantly offered to the king (Charpentier 1977, 109). The king had the privilege of enjoying betel quids with pearls that had been powdered or burned to ashes.

The natives of the Trobriand Islands have a betel ceremony called *kakaui* in which many people gather together and, within the span of one to three hours, consume large quantities of betel quids. The amount of betel they share is calculated based on the number of people in attendance and the amount of betel to be consumed by each. Eight, ten, or twelve betel nuts are used per person (Jüptner 1969, 371).

Betel quids, as well the objects used and needed for their preparation and consumption, often have symbolic and ritual significance. In Ceylon, it was customary to carry around a betel presentation plate during wedding ceremonies. The hairdresser who shaved and bathed the groom before the ceremony was rewarded with a roll of seven betel leaves (*Piper betle*), seven silver coins, and seven slices of betel nut (Charpentier 1977, 110). Betel is also a ritual wedding gift among the minorities of southern China (Eberhard 1983, 39). On the Trobriand Islands, a man typically brings a woman he is pursuing or his lover betel nuts or tobacco when he meets her (Jüptner 1969, 376).

Betel quids often carry sexual connotations. In Melanesia, the betel quid or the betel nut alone is

Left: Nepalese betel scissors for chopping the hard betel nuts.

Right: A lime container, one of the traditional betel utensils, from Timor.

given as a sign of sexual desire and is used for love magic. The ground betel mixture is painted onto arrows to enhance their accuracy, rubbed onto fishing lines in order to improve the catch, and smeared onto hunting fetishes in order to utilize the spirits that live in them. When sprayed onto the belly of a pregnant woman, the betel mixture is supposed to induce labor and make the delivery easier. Betel saliva, spat into the wind, is supposed to dispel rain and storms; spat onto the grain, it is supposed to encourage the growth of the soil.

Ready-Made Mixtures (*Pan Masala*)

In India, ready-made betel mixtures are increasingly competing with traditional betel quids. None of these industrially packaged betel mixtures contains betel leaf (*Piper betle*). The precise ingredients are listed on the back of each package. Such mixtures contain the following ingredients:

betel nut	*Areca catechu*
catechu	*Acacia catechu* (cf. *Acacia* spp.)
lime	calcium hydroxide
cardamom	*Elettaria cardamomum*
camphor	presumably synthetic (cf. *Cinnamomum camphora*)
menthol	
sandalwood oil	from *Santalum album*
saffron	presumably *Carthamus tinctorius*
tobacco	*Nicotiana tabacum*
perfumes	no precise specification
"permitted spices"	no precise specification[408]

About half of the available products contain tobacco.

Some of these products are exported (primarily to Nepal). Apart from a mild stimulation, a suppression of the effects of hemp, and the promotion of digestion, I have not been able to perceive any particular psychoactive effects (Rätsch 1996).

The packages feature the caution "Betel chewing may be injurious to your health." A traditional Ayurvedic medicine to cure "betel addiction"[409] recommends chewing a few leaves of *tulsi* (= holy basil; *Ocimum sanctum*) after a meal instead of the customary betel quid (cf. *Ocimum micranthum*). These too promote digestion while simultaneously alleviating withdrawal symptoms (Rai 1988, 117). In Southeast Asia, *tulsi* leaves are sometimes chewed as a betel substitute (Macmillan 1991, 424*).

Substitutes

The seeds of other *Areca* species are occasionally chewed in place of betel quids (see *Areca catechu*). Sometimes the bark, leaves, and roots of entirely different plants (which the literature unfortunately does not botanically specify) may be used as surrogates (Charpentier 1977, 115).

Literature

See also the entries for *Areca catechu* and *Piper betle*.

Beran, Harry. 1988. *Betel-chewing equipment of East New Guinea.* Shire Ethnography, no. 8. Aylesbury, U.K.: Shire Publications.

Brownrigg, Henry. 1992. *Betel cutters, from the Samuel Eilenberg Collection.* London: Thames and Hudson.

Charpentier, C.-J. 1977. The use of betel in Ceylon. *Anthropos* 72:107–18.

Chinnery, E. W. Person. 1922. *Piper methysticum* in betel chewing. *Man* 22:24–27.

Chu, Nai-Shin. 1995. Sympathetic response to betel chewing. *Journal of Psychoactive Drugs* 27 (2): 183–86.

Eberhard, Wolfram. 1983. *Lexikon der chinesischen Symbole.* Cologne: Diederichs.

Gorman, C. F. 1972. Excavations at Spirit Cave, North Thailand: Some interim interpretations. *Asian Perspectives* 13:79–107.

Gowda, M. 1951. The story of pan chewing in India. *Botanical Museum Leaflets* 14 (8): 181–214.

Grabowsky, F. 1888. Das Betelkauen bei den Malaiischen Völkern, besonders auf Java und Borneo. *Internationales Archiv für Ethnographie* (Leiden) 1:188–91.

Hartwich, Carl. 1905. Beiträge zur Kenntnis des Betelkauens. *Bulletin va het Koloniaal Museum te Haarlem* 32:49–97.

Huu, Tien, ed. 1985. *Augen lachen, Lippen blühen: Erotische Lyrik aus Vietnam.* Munich: Simon & Magiera.

Jüptner, Horst. 1968. Klinisch-experimentelle Beobachtungen über intensives Betelkauen bei den Eingeborenen der Trobriand-Inseln. *Zeitschrift für Tropenmedizin und Parasitologie* 19:245–57.

———. 1969. Über das Betelnusskauen auf den Trobriand-Inseln (Neuguinea) und den Versuch einer Klassifizierung der Kalkspatel. *Baessler-Archiv,* n.f., 17:371–86.

Krenger, W. 1942a. Kulturgeschichtliches zum Betelkauen. *Ciba-Zeitschrift* 7 (84): 2922–28.

———. 1942b. Über die Wirkung des Betels. *Ciba-Zeitschrift* 7 (84): 2942–47.

———. 1942c. Zusammensetzung und Zubereitung des Betels. *Ciba Zeitschrift* 7 (84): 2929–41.

Lewin, Louis. 1889. *Über* Areca catechu, Chavica betle *und das Betelkauen.* Stuttgart: Enke.

———. 1890. Über das Betelkauen. *Internationales Archiv für Ethnographie* 3:61–65, Leiden.

Millot, J. 1966. Le bétel au Népal. *Objets et Mondes* (Paris) 6:153–68.

"Betel illuminates the spirit and dispels worries. . . . Whoever uses it will be filled with joy; he will have a perfumed breath and a sound sleep. . . . Betel takes the place of wine for the Indians, who use it often."

Sheriff
(cited by Abd Allah Ibn Ahmad)

408 What these "permitted spices" are is unclear. I presume that the ingredients that are not permitted are datura seeds, crow's eyes (*Strychnos nux-vomica*), opium, hashish (*Cannabis indica*), and possibly arsenic and *Aconitum*—all ingredients found in Tantric inebriating mixtures. Such ingredients, however, are sometimes available under the counter: Prince Pan, a betel shop in Delhi, sells special betel quids with such rare and illegal ingredients as gold leaf, opium, and even **cocaine**!

409 Interestingly, the World Health Organization classifies betel as a nonaddictive drug (cf. *Areca catechu*). In contrast, the literature frequently uses such terms as *habit, addiction,* and *betel nut psychoses* (cf. Seyfarth 1981, 565).

"The groom, the bride, and a woman of the barber caste participate in a certain ritual procedure. The barber woman cuts in the little finger of the left hand of the bride with her nail clippers so that blood comes out. This blood is put on betel, which is then put into the mouth of the groom. In this instance betel is the pleasant packaging for another important thing, the blood of the woman, with whom he enters into a fictitious blood relationship."

ERIKA MOSER-SCHMITT
"SOZIOKULTURELLER GEBRAUCH VON BETEL IN INDIEN" ["SOCIO-CULTURAL USE OF BETEL IN INDIA"] (1981, 549)

Moser-Schmitt, Erika. 1981. Sozio-kultureller Gebrauch von Betel in Indien. In *Rausch und Realität*, ed. G. Völger, 2: 546–51. Cologne: Rautenstrauch-Joest-Museum für Völkerkunde.

Rai, Yash. 1988. *Holy basil: Tulsi (a herb)*. Ahmedabad, Bombay: GALA Publ.

Rätsch, Christian. 1996. Pan Masala: Betel aus der Tüte. *Yearbook for Ethnomedicine and the Study of Consciousness* 4 (1995): 289–92. Berlin: VWB.

Rooney, Dawn F. 1993. *Betel chewing traditions in South-East Asia*. Images of Asia series. Kuala Lumpur: Oxford University Press.

Schomburgk, R. 1868. Die Arekanuß und das Betelblatt als Reizmittel in Siam. *Globus* 14:120–21.

Sen, Soumitra, Geeta Talukder, and Archana Sharma. 1989. Betel cytotoxicity. *Journal of Ethnopharmacology* 26:217–47. (Contains an extensive bibliography on the pharmacology.)

Seyfarth, Siegfried. 1981. Betelkauen in Melanesien. In *Rausch und Realität*, ed. G. Völger, 2:560–66. Cologne: Rautenstrauch-Joest-Museum für Völkerkunde.

Stöhr, Waldemar. 1981. Betel in Südost- und Südasien. In *Rausch und Realität*, ed. G. Völger, 2:552–59. Cologne: Rautenstrauch-Joest-Museum für Völkerkunde.

True, R. H. 1896. Betel chewing. *Pharmaceutical Review* 14 (6): 130–33.

Chicha

Other Names

Akha (Quechua, colonial period), asua (Quichua), cachir, cachiri, catchir, cono (Secoya), corn beer, kashiri, kasuma (Zapara), maisbier, nijiamanch (Achuar/Shuar), tepae (Huaorani), tesguino, tesvino, tizwin, tsetsepa (Kofán), tulpi

Chicha is produced primarily from maize/corn (*Zea mays*), tubers, and fruits. The substance to be fermented is always mashed with water; in other words, chicha is actually a **beer**.

Maize beer is brewed in the American Southwest, for example by the Apaches (Hrdlicka 1904). In Central and South America, chicha made from maize has been known from time immemorial and has long been valued as both a food and an inebriant. Other plants believed to improve fermentation or to protect the grains of maize are sometimes added during the preparation of the drink. In Peru, the leaves of certain ferns (*Thelypteris glandulosolanosa* [C. Chr.] Tyron, *Thelypteris rufa* [Poiret] A.R. Smith) are used for this purpose (Franquemont et al. 1990, 40*). There, chicha is made primarily from the fruit pods of *Prosopis pallida*.

In the Amazon region of Ecuador, the fruits of the chonta palm (*Bactris gasipaës*) are boiled and fermented to make chicha. The tree's extremely hard wood is used to make bows, spears, arrowheads, and the small, magical shamanic arrows that are used mainly for black magic and during shamanic wars.

In Colombia, chicha is made from a type of corn known as *maíz blanda*. The kernels of maize are ground with a stone pestle, soaked in diluted sugarcane molasses (*aguamiel*, "honey water"), and allowed to ferment for twelve days. Magical additives such as ground bones, rat skulls, or cow's skin are often added to the brew (cf. **zombie poison**). During fermentation, the corn gluten can produce ptomaine, a toxic substance that can produce undesired side effects (cf. *Zea mays*).

In Central and South America, chicha is also prepared from various palms (cf. **palm wine**). The fruits of the following palms are used in the preparation of chicha (after Vickers and Plowman 1984*):

Bactris gasipaës H.B.K.
Jessenia bataua (Mart.) Burret
Mauritia flexuosa L.
Mauritia minor Burret (*chicha de canaguche*)

The palm fruits are first boiled and then usually pressed, soaked in water, and allowed to ferment.

Chile is a veritable land of chicha. In Santiago de Chile, the name *chicha* is now used for freshly fermented apple cider (Seeler 1994, 247). In rural and Indian regions, *chicha* is a catchall term given to all fermented beverages, especially aqueous fruit solutions (see the table on page 740). The *chicha de algarrobo* is very popular. Interestingly, the *algarrobo* tree (*Prosopis chilensis* [Mol.] Stuntz), the sweet fruit pods of which are fermented to prepare the drink, is also called *tacu*, *huancu*, and *huilca* in Peru. These same names are given to *Anadenanthera colubrina* (Mösbach 1992, 84*).

Some Chilean types of chicha are attributed with medicinal qualities; those from *huighan* or *huighnan* (*Schinus dependens* Orteg.), for example, have strong diuretic properties and are consumed

An Indian chicha merchant (drawing from Mortimer, *History of Coca*, 1901).

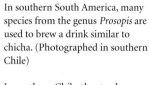

Left, from top to bottom:
The hard fruit pods of the Chilean *algarrobo* (*Prosopis chilensis*) contain a large amount of sugar and are thus suitable for preparing fermented drinks.

In southern South America, many species from the genus *Prosopis* are used to brew a drink similar to chicha. (Photographed in southern Chile)

In southern Chile, the starchy, sugary sprouts of the primordial monkey puzzle (*araucarie*) tree (*Araucaria araucana*) are fermented to produce chicha. (Photographed in southern Chile)

The people who live in the oases of the Atacama Desert use the ripe fruits of the unusual *chañar* tree (*Geoffrea decorticans*), which contains large amounts of chlorophyll in its bark, as an ingredient in brewing. (Wild plant, photographed in San Pedro de Atacama, northern Chile)

Right, from top to bottom:
The fruits of *Gaultheria phillyreifolia* are used on the southern Chilean island of Chiloé for brewing chicha. (Wild plant, photographed near Ancud, Chiloé)

Known as red pepper, the seeds of the molle tree (*Schinus molle*) are used as a fermentation substance and as a spicy beer additive. (Wild plant, photographed in San Pedro de Atacama, northern Chile)

The fruits of the Chilean barberry (*Berberis darwinii*) are used to brew chicha.

The people of Chiloé use the sweet fruits of the Chilean guava (*Ugni molinae*) to make chicha. (Wild plant, photographed near Ancud, Chiloé)

Chilean Chicha Plants

(from Donoso Zegers and Ramírez García 1994*; Franquemont et al. 1990*; Gómez Parra and Siarez Flores 1995; Mösbach 1992*; modified and supplemented)

Chilean Name	Botanical Identity	Part(s) Used
algarrobo	*Prosopis alba* Griseb. var. *alba*	
	Prosopis chilensis (Mol.) Stuntz	fruit pods
	Prosopis chilensis (Mol.) Stuntz var. *chilensis*	
araucarie	*Araucaria araucana* (Mol.) Koch [syn. *Araucaria imbricata*]	shoots
calafate	*Berberis linearifolia* Phil.	fruits
chañar	*Geoffrea decorticans* (Gill. ex H. et A.) Burk	fruits
chaura	*Gaultheria* spp.[410]	
	Gaultheria phillyreifolia (Pers.) Sleumer	
	Pernettya spp.	
	Pernettya mucronata (L. f.) Gaud.	
	Pernettya mucronata var. *angustifolia* (Lindl.) Reiche	
	Pernettya mucronata var. *mucronata*	
	Pernettya myrtilloides Zucc. ex Steud.	
chaura común	*Gaultheria phillyreifolia* (Pers.) Sleumer	fruits
cüd-cüd	*Pernettya insana* (Mol.) Gunckel	fruits
huingán	*Schinus dependens* Orteg.	fruits
	Schinus polygamus (Cav.) Cabr.	fruits
keule	**Gomortega keule**	fruits
litre	*Lithrea caustica* (Mol.) H. et A.	fruits
luma	*Amomyrtus luma* (Mol.) Legr. et Kaus.	fruits
maíz	**Zea mays** L.	grains
michay	*Berberis darwinii* Hook.	fruits
michay blanco	*Berberis congestiflora* Gay	fruits
molle	*Schinus molle* L.	fruits
muchi, müchü	*Schinus montanus* (Phil.) Engler	fruits
murta, üñü	*Ugni molinae* Turcz.	fruits
	Ugni philippii Berg.	
	Ugni poeppigii Berg.	
quëlón	*Aristotelia chilensis* (Mol.) Stuntz	fruits[411]
tamarugo	*Prosopis tamarugo* Phil.	fruit pods
trautrau[412]	*Ugni candollei* (Barn.) Berg.	fruits

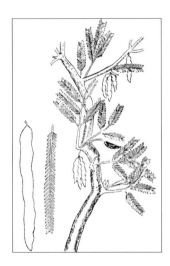

In South America, several trees from the Family Leguminosae that are widely referred to as *algarrobo* yield fruit pods that are used to make fermented drinks (chicha). (From Pedro de Montenegro, *Materia médica misionera,* seventeenth century)

410 Some *Gaultheria* species are reputed to be psychoactive (Schultes and Hofmann 1995, 43*). Wintergreen (*Gaultheria procumbens* L.), however, is drunk only as a tea substitute (called Salvador tea or mountain tea) and does not have any psychoactive effects. It contains only methylsalicylate (Frohne and Pfänder 1983, 106*).

411 The chicha prepared from these is usually known as *tecu* (Mösbach 1992,91*).

412 This name is also used for **Desfontainia spinosa**.

by people suffering from dropsy (Schultes 1980, 106*). A chicha prepared from the pepper tree (*Schinus molle*) has potent stimulating and possibly other psychoactive effects, as the **essential oils** of the fruit include β-phellandrene, α-pinene, carvacrol, o-ethylphenol, β-pinene, camphene, myrcene, α-phellandrene, limonene, π-cymene, and β-spathulene (Terhune et al. 1974). The chicha prepared from *Schinus molle* is also used to treat dropsy.

Sometimes admixtures are used to alter the taste or the effect of chicha. In the Colombian Vaupés region, the dried greenish yellow flowers of *Duguetia odorata* (Diels) Macbride (Annonaceae) are used to perfume (and strengthen?) chicha. The plant is rich in alkaloids (Schultes and Raffauf 1986, 259*). In the same region, the powdered flowers of *Heterostemon mimosoides* Desf. are also used as an aromatic additive to the drink (Schultes 1978b, 231*). The Barasa Indians add the powdered bark of *Vochysia lomatophylla* Standl., known locally as *ka-kwee'-gaw-ya*, as an aborti-

facient (Schultes 1977b, 117*). A wide variety of other psychoactive plants may also be added to chicha (see the table on page 741).

Literature

See also the entries for **beer.**

Caspar, Franz. 1952. Die Tupari, ihre Chicha-Braumethode und ihre Gemeinschaftsarbeit. *Zeitschrift für Ethnologie* 77 (2): 245–60.

Cutler, Hugh C., and Martin Cardenas. 1947. Chicha, a native South American beer. *Botanical Museum Leaflets* 13 (3): 33–60.

Gómez Parra, Domingo, and Eva Siarez Flores. 1995. *Alirrrerunción tradicionál atacameña.* Antofagasta, Chile: Fondart.

Hartmann, Günther. 1958. Alkoholische Getränke bei den Naturvölkern Südamerikas. Diss., Berlin.

———. 1960. Alkoholische Getränke bei den südamerikanischen Naturvölkern. *Baessler-Archiv* 8 (1).

Psychoactive Chicha Additives

Stock Plant	Part(s) Used	Culture/Region
Anadenanthera colubrina	seeds	Inca/Andes, Mataco/Argentina
Ariocarpus fissuratus	cactus flesh	Tarahumara/Mexico
Brugmansia arborea	seeds	Peru
Brugmansia aurea	seeds	Peru
Brugmansia sanguinea	seeds	Andes region
Coryphantha spp.	cactus flesh	Tarahumara/Mexico
Datura innoxia	roots	Tarahumara/Mexico
Lolium temulentum	seeds	Peru
Lophophora williamsii	buttons, powder	Tarahumara/Mexico, Huichol/Mexico
Mammillaria spp.	cactus flesh	Tarahumara/Mexico
Nicotiana glauca (see *Nicotiana* spp.)	herbage	America
Pachycereus pecten-aboriginum	cactus flesh	Tarahumara/Mexico
Paullinia yoco (see *Paullinia* spp.)	bark, latex	Putumayo
Tabernaemontana muricata (see *Tabernaemontana* spp.)	leaves/flowers	Amazonia

———. 1981. Alkoholische Getränke bei den südamerikanischen Indianern. In *Rausch und Realität*, ed. G. Völger, 1:152–62. Cologne: Rautenstrauch-Joest-Museum für Völkerkunde.

Hrdlicka, A. 1904. Method of preparing tesvino among the White River Apaches. *American Anthropologist*, n.s., 6:190–91.

Karsten, Rafael. 1920. Berauschende und narkotische Getränke unter den Indianern Südamerikas. *Acta Acad. Åboensis.*

Lomnitz, L. 1973. Influencia de los cambios políticos y económicos en la ingestión de alcohol: el caso Mapuche. *América Indígena* 33 (1): 133–50.

Moore, Jerry D. 1989. Pre-Hispanic beer in coastal Peru: Technology and social context of prehistoric production. *American Anthropologist*, n.s., 91:682–95.

Mowat, Linda. 1989. *Cassava and chicha: Bread and beer of the Amazonian Indians.* Aylesbury, U.K.: Shire Ethnography.

Nachtigall, Horst. 1954. Koka und Chicha. *Kosmos* 50 (9): 423–27.

Nicholson, G. Edward. 1960. Chicha maize types and chicha manufacture in Peru. *Economic Botany* 14 (4): 290–99.

Scheffer, Karl-Georg. 1981. Chicha in Südamerika. In *Rausch und Realität*, ed. G. Völger, 1: 146–51. Cologne: Rautenstrauch-Joest-Museum für Völkerkunde.

Seeler, Rolf. 1994. *Chile mit Osterinsel.* Cologne: DuMont.

Terhune, Stuart J., James W. Hogg, and Brian M. Lawrence. 1974. β-spathulene: A new sesquiterpene in *Schinus molle* oil. *Phytochemistry* 13:865–66.

Vásquez, Mario. 1967. La chicha en los países andinos. *América Indígena* 27 (2): 265–82.

"When the inhabitants of the river-filled forests and meadows no longer felt like hunting and gathering, they lived from their reserves, drank home-brewed stimulants, and indulged in polygamy. Hate and jealousy began to creep around the Rucas, and depravity raised its ugly face. Then the giant serpent Caicaivilu awoke from its thousand-year sleep and roused up the wind and the sea so that they would destroy the land. Hurricanes laid waste to the forest, earthquakes in the ocean caused tidal waves. Humans and animals fled to the mountaintops and fought each other for a safe haven. The giant serpent Tentenvilu, who slumbered in the ancient cordilleras, was also awoken by the chaotic din. It realized the distress of the besieged and pushed the mountains upward with all its strength as the ocean rose ever higher. The battle of the enemy serpents lasted for moons upon moons. Then the god Ngenechen intervened, neutralized the battle, and brought the wrathful nature to a standstill. The gulfs, fjords, canals, and islands remained from the waters of the evil serpent and the earth masses of the good serpent."

A FLOOD LEGEND FROM CHILOÉ, SOUTHERN CHILE IN *CHILE MIT OSTERINSEL* [CHILE AND EASTER ISLAND] (SEELER 1994, 174)

Cimora

Other Names
Timora

In Peru, the name *cimora* or *timora* is given to a psychoactive drink used for shamanic purposes. This drink consists primarily of *Iresine* species (*Iresine celosia* L. and others; see *Iresine* spp.), *Brugmansia* species, or a mixture of the following plants (Ott 1993, 409*; Schultes and Farnsworth 1982, 159*; Schultes 1966, 302*):

Trichocereus pachanoi Br. et R.
Neoraimondia arequipensis (Meyen) Backeb.
    [syn. *Neoraimondia macrostibas* (K. Schum.)
    Br. et R., *Neoraimondia roseiflora* (Werderm. et
    Bckbg.) Bckbg., *Pilocereus macrostibas* K.
    Schum.]
Hippobroma longiflora (L.) G. Don[413] [syn.
    Isotoma longiflora Ducke or (L.) Presl,
    Laurentia longiflora (L.) Peterm., *Lobelia
    longiflora* L.]
Pedilanthus tithymaloides (L.) Poit. [syn.
    Pedilanthus carinatus Spreng.]
Brugmansia spp. [syn. *Datura*]

Iresine does not appear to contain any alkaloids and presumably does not induce any psychoactive effects. In Peru, *Euphorbia cotinifolia* L. is known as *timora* (cf. *Trichocereus pachanoi*). Although a related euphorbia, *Pedilanthus tithymaloides* Poit. (cf. *Pedilanthus* spp.), is known in Peru by the folk name *cimora misha*, it does not appear to have any psychoactive effects (Müller-Ebeling and Rätsch 1989, 32 f.*).[414]

In Peru, different cultivars of *Brugmansia* x *candida* as well as *Brugmansia arborea* are known by the name *cimora*, and it is these, along with *Trichocereus pachanoi*, that presumably represent the actual psychoactive components of the cimora drink. More precise recipes for preparing cimora or timora are lacking, as are precise pharmacological investigations of the purported blend.

Literature

See also the entry for *Trichocereus pachanoi*.

Davis, E. Wade 1983. Sacred plants of the San Pedro cult. *Botanical Museum Leaflets* 29 (4): 367–86.

413 This plant, originally from the Caribbean region, was taken to Sri Lanka, where it is now feared as a deadly horse poison (Macmillan 1991, 430*).

414 "Lampe describes especially the eye injuries caused by the [milk] latex. The seeds or latex induce long-lasting vomiting and diarrhea, leading to electrolyte imbalances" (Roth et al. 1994, 547*).

Enemas

Other Names

Clistere, clysma, cluster, clysterium, clystiere, eingießungen, einläufe, klistiere, klystier, lavement

The term *enema* refers to a liquid that is administered rectally for medicinal/therapeutic, hedonistic, or ritual/psychoactive purposes. Enemas often consist of nothing more than lukewarm water, but for medicinal purposes, a decoction or infusion of certain plants may be used, e.g., as a laxative. The fluids may also take the form of alcoholic beverages (beer,[415] **wine, chicha, balche'**, pulque [cf. *Agave* spp.]). Many medicines that can cause discomfort to the stomach are often administered in enema form (e.g., opium; cf. *Papaver somniferum*). Cleansing enemas often play a role during ritual preparation for entheogenic rituals.

It has often been asserted that the ancient Egyptians invented enemas after observing the behavior of the ibis. This bird supposedly uses its long, curved, tubelike beak to administer enemas to itself. Because the ibis is the symbolic animal or even the embodiment of the shamanic god Thoth, Thoth became known as the god of enemas (Degenhard 1985, 13). In fact, enema devices have been found in various places around the world (Hallowell 1935; Heizer 1944; Lieberman 1944). Of particular significance are the caoutchouc (rubber) balls that were used to administer enemas that have been found in South America. South American shamans also made enema devices from jaguar bladders and bird bones (Nordenskiöld 1930, 188). Such tools are sometimes also used to ingest dry **snuffs** through the nose.

In Europe, special enema syringes were once popular. Devices known as blowing tubes and also specially designed machines were used to blow smoke, especially from tobacco (*Nicotiana tabacum*), into the rectum (Schäffer 1772; Degenhard 1985, 22 ff.). Enemas of tobacco smoke were administered not only to humans but also to horses (Degenhard 1985, 171).

During antiquity and the Middle Ages, people believed that enemas had been discovered by observing nature: "the stork that purges itself." (Woodcut, late Middle Ages)

Enema scene on a ritual vessel (drinking glass?) of the Classic Mayan period, showing a diviner or shaman administering either a tobacco decoction or balche'.

Ingredients in Psychoactive Enemas
(from de Smet 1983, 1985; Hovorka and Kronfeld 1908*; Rätsch 1987; supplemented)

Stock Plant	Part(s) Used	Culture/Region
Agave spp.	pulque	Mesoamerica
Anadenanthera colubrina	seeds	Maué/Brazil, Inca
Anadenanthera peregrina	seeds	South America
Banisteriopsis spp.	bark, stems	Amazonia
Boswellia sacra	olibanum	antiquity
Brugmansia arborea		South America
Brugmansia x *insignis*	bark decoction	Huachipaire, Amazonia (Peru)
Brugmansia suaveolens		
Cannabis indica	flowers/oil	antiquity/Assyria
Capsicum spp.	fruits	Central and South America
Coffea arabica	decoction	United States
Datura ceratocaula (see *Datura* spp.)		Mesoamerica
Datura innoxia	leaves, seeds	North America
Datura stramonium	leaves, seeds	North America
Erythroxylum coca	cocaine	gay culture/San Francisco
Hyoscyamus niger	decoction	antiquity into the present
Ilex guayusa	decoction	Tiahuanaco
Ipomoea violacea	seeds	Maya
Lonchocarpus violaceus	**balche'**	Maya
Lophophora williamsii	pressed juice, powder, decoction	Huichol, Aztec
Mandragora officinarum	wine	antiquity
Nicotiana rustica	leaves	Mexico
Nicotiana tabacum	leaves	Jíbaro/Ecuador
Nicotiana undulata	leaves	Maya
Nicotiana spp.	leaves	Mesoamerica, South America
Nymphaea caerulea	flowers, roots	Egypt
Papaver somniferum	opium	Orient
Solandra brevicalyx (cf. *Solandra* spp.)	leaves	Huichol/Mexico
Virola spp.	seeds	Brazil

415 Even wheat beer (*weizenbier*) has been administered as an enema (Degenhard 1985, 333).

"The Indians who live in the interior of Brazil, such as the Caripuna, Murás, Maukés, Pouporó, and the Catanixi, had already hit upon the idea of using an enema to affect consciousness, and they used parica enemas for inebriation. 'Parica' is the name of the seeds of the angico tree, which are powdered and mixed with the ashes of the imbauwa tree and then, added to water, were administered as an enema in a rubber syringe with a very long tip made from a hollow bird bone."

Armin von Degenhard
Das Klistier [The Enema]
(1985, 334)

This early-modern-era apparatus from England was used to administer enemas of concentrated tobacco smoke. (Copperplate engraving from Johann Andreas Stisser, *De machini fumiductoriis curiosis*, Hamburg, 1686)

In Japan, enema blowing tubes were used to administer different psychoactive substances, especially green tea and opium, for hedonistic and other purposes. (Woodcut, nineteenth century)

In ancient Mexico, sexual rituals included enemas containing pulque, the fermented juice of *Agave* spp., to which other psychoactive substances (e.g., *Lophophora williamsii*) were usually added (de Smet 1985, 20). The Mayans appear to have used enemas made from *Nicotiana* **spp.** and **balche'** in ritual contexts (de Smet 1981; Furst and Coe 1977). There is also evidence suggesting that the pre-Columbian Mochica administered enemas containing aphrodisiacs to men during ritual anal coitus (cf. Dobkin de Rios 1982*). The ritual use of enemas for purification and/or to administer psychoactive preparations was especially important in South American shamanism.

In the modern period, enemas are often used as part of erotic activities (Degenhard 1985). In the Orient, and especially in the world of the harem, opium was given in enemas that were aphrodisiac and also intended to induce forgetfulness. In contemporary anal-erotic circles, a variety of psychoactive substances (PCP, ketamine, **cocaine**, **scopolamine**) are used in aphrodisiac enemas in orgies (Rätsch 1987).

Different preparations were administered as narcotic enemas in ancient times (cf. **soporific sponges**). In the nineteenth century, narcotic enemas consisting of olive oil and ether were administered for surgical purposes (Degenhard 1985, 333 f.). Enemas have also been abused in the context of Catholic exorcisms and in the persecution and humiliation of women accused of being witches.

Literature

Degenhard, Armin von. 1985. *Das Klistier.* Flensburg: Carl Stephenson Verlag.

de Smet, Peter A. G. M. 1981. Enema scenes on ancient Maya pottery. *Pharmacy International* 2:217–19.

———. 1983. A multidisciplinary overview to intoxicating enema rituals in the Western Hemisphere. *Journal of Ethnopharmacology* 9:129–66. (Very good bibliography.)

———. 1985. *Ritual enemas and snuffs in the Americas.* Latin America Studies 33. Amsterdam: CEDLA.

de Smet, Peter, and Nicholas M. Hellmuth. 1986. A multidisciplinary approach to ritual enema scenes on ancient Maya pottery. *Journal of Ethnopharmacology* 16 (1–2): 213–62.

Furst, Peter T., and Michael D. Coe. 1977. Ritual enemas. *Natural History* 86 (3): 88–91.

———. 1989. Ritual enemas. In *Magic, witchcraft, and religion*, ed. Arthur C. Lehmann and James E. Myers, 127–31. Mountain View, Calif.: Mayfield. (Reprint of 1977.)

Hallowell, A. Irving. 1935. The bulbed enema syringe in North America. *American Anthropologist* 37:708–10.

Heizer, R. F. 1944. The use of the enema among the aboriginal American Indians. *Ciba Symposia* 5 (11): 1686–93.

Lieberman, William. 1944. The history of the enema. *Ciba Symposia* 5 (11): 1694–1708.

Nordenskiöld, Erland. 1930. Modifications in Indian culture through inventions and loans: Appendix 1: The use of enema tubes and enema syringes among Indians. *Comparative Ethnographical Studies* 8:184–95.

Rätsch, Christian. [1987]. Das Zepter der heroischen Medizin. In *Das Scheiß Buch*, 80–83. Löhrbach: Werner Pieper's MedienXperimente.

Schäffer, Johann Gottlieb. 1772. *Der Gebrauch und Nutzen des Tabackrauchclystiers.* Regensburg: Montag und Gruner.

Energy Drinks

During the past several years, new types of beverages have been developed in connection with the rave and techno culture. Marketed as "energy drinks," they are offered to ravers as stimulating, healthy alternatives to **alcohol**, which in the rave culture is becoming increasingly frowned upon as a party drug (Ahrens 1994, Millman and Beeder 1994). Many of the names of these drinks—for example, Mystery (an "official Michael Jackson Product"), Fit for Fun, Flying Horse, Warp 4 Space Drink, Cult Energy Activator, Magic Man, Taurus, and XTC (= ecstasy = MDMA)—suggest improbable psychoactive effects.

The basis of most of these products is guaraná (see *Paullinia cupana*). They usually also contain various vitamins, DHA (polyunsaturated fatty acids), taurine (a substance that appears to be pharmacologically inactive), propolis, and pure **caffeine**. However, the caffeine concentration typically is not as high as that found in a cup of coffee (cf. *Coffea arabica*). In other words, these products are as frustrating as **herbal ecstasy**.

A collection of so-called energy drinks. Advertisements for such beverages often claim that they are more stimulating than cocaine and more psychedelic than LSD.

Literature

See also the entry for **herbal ecstasy**.

Ahrens, Helmut. 1994. *Partydrogen—safer-use-info zu: Ecstasy, Speed, LSD, Kokain*. Berlin: Arbeitgruppe "Eve and Rave."

Die Gestalten Berlin and Chromapark, eds. 1995. *Localizer 1.0: The techno house book*. Berlin: Die-Gestalten-Verlag.

Millman, Robert B., and Ann Bordwine Beeder. 1994. The new psychedelic culture: LSD, ecstasy, "rave" parties and The Grateful Dead. *Psychiatric Annals* 24 (3): 148–50.

Han-Shi

Other Names

Cold food powder, cold mineral powder, five-mineral powder, han-shih, han-shih san, han-shi powder, medicinal powder made from the five minerals, wu-shi

The Chinese politician He Yan (in office 240–249 C.E.) was one of the most important philosophers of the Wei dynasty. After trying the han-shi powder, he reported enthusiastically, "When one takes the five mineral powder, not only are diseases cured, but the mind is awakened and opened to clarity" (in Wagner 1981, 321).

The purported inventor or discoverer of the drug, Huangfu Mi (215–282 C.E.), commented:

In recent times, He Yan has devoted his time to music and esteemed sex, and when he took the drug for the first time, he attained an additional clarity of consciousness, and his physical strength gradually grew stronger. [Because of this], everyone was soon passing the drug around in the capital city. . . . After his death, the number of those who took it grew even larger, and this did not change over time. (In Wagner 1981, 321)

The poet Su Shi (1036–1101) listed the main ingredients of the drug:

It began with He Yan, that the people took stalactites with aconite and uninhibitedly abandoned themselves to wine [= **sake**] and sex in order to extend their lives. In his youth, He Yan was rich and respected, why should it be such a surprise that he took the *han-shi* powder in order to satisfy his desires? (In Wagner 1981, 321)

Although there is some connection between the efficacious powder and the recipes of Taoist alchemy, han-shi was used chiefly as an agent of pleasure (Strickman 1979, 168). It was generously consumed in circles that were already interested in inebriants:

From the Wei period on [after 220 C.E.], one encounters wine [pressed from grapes] in a totally new context. It was consumed by the feudal class with a consciousness-expanding and potency-promoting drug—the han-shi powder. According to the instructions of the inventor Huang-fu Mi, the drug needed to be

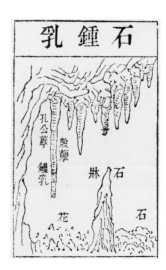

Stalactites in a limestone cave; since ancient times, stalactites have been used in traditional Chinese medicine to strengthen *yang* and to move *chi* (= life energy) in the lower abdomen. (Woodcut from *Ben cao gang mu*, 1596)

The Chinese drug called *fu tzu* is made from the roots of different species of monkshood (*Aconitum* spp.). It was once an important ingredient in the psychoactive han-shi powder.

taken with hot, high-quality wine in order for its effects to be released. The literature of the time contains various reports of wine societies, which were actually drug parties. The combined effects of wine and drugs sometimes caused things to get out of control. For example, it was said that the wealthy Shi Ch'ung would use so-called beautiful women to encourage his guests to drink wine at his banquets. In the event the guest did not drink to complete excess, the woman would be executed. (Majlis 1981, 318)

Many han-shi consumers—and not only Taoists and/or alchemists—also experimented with such other drugs as **sake, wine,** brandy (**alcohol**), and psychoactive mushrooms (Wagner 1973; Wagner 1981, 322; cf. Strickman 1979 and Cooper 1984, 23, 54, 62*). Unfortunately, we do not yet know the identity of these psychoactive mushrooms (cf. "*Polyporus mysticus*"). It also appears that the han-shi powder was often used in the context of Taoist sexual practices and sexual magic exercises.

Yü Chia-hsi (1938) has conducted research into the recipe or recipes for making han-shi powder, but only imprecise details are known:

The recipe for the drug is known. In addition to various ingredients containing calcium (stalactites [*é guan shí*], oyster shells [*mu lì*], both ground) and numerous herbs, it contains above all the poisonous aconite. Unfortunately, no pharmacologist has studied this complex drug to date, so that we provide no information about experiments or theoretical effects. (Wagner 1981, 321)

Unfortunately, Yü Chia-hsi did not indicate whether the *é guan shí* (literally "gooseneck stones" = stalactites) and oyster shells (most likely *Crassostrea gigas* [Thunberg 1793])[416] were pulverized or burned/slaked. But it seems likely that this was a slaked lime, for the recipes for all known psychoactive products that are mixed with lime require slaked lime (i.e., calcium hydroxide); cf. *Areca catechu, Erythroxylum coca, Nicotiana tabacum,* and **betel quids**. In addition to stalactites, Shen Kuo names an additional plant ingredient: *Atractylodes macrocephala* Koidz. (cf. **sake**). He con-

siders the effects of the powder to be a product of synergistic interactions between its components:

When a person ingests many minerals in one medicine, then the minerals must be able to work in synergy, and when in addition a person further stimulates them with medicinal plants, then the effect must be very strong. Thus, when the powder of the five minerals is combined with different medicinal plants, one should use only an extremely small amount of mineral powder, for one must produce the effects with only small amounts of additives. (Shen Kuo 1997, 127 f.*)

Shen Kuo then mentions Sun Simajao, who claimed that poison sumac (*Rhus toxicondendron* L. [syn *Toxicondendron quercifolium* (Michx.) Greene]) and kudzu vine (*Pueraria lobata* [Willd.] Ohwi [syn. *Dolichos lobatus* Willd., *Pueraria thunbergiana* (Sieb. et Zucc.) Benth., *Pueraria hirsuta* (Thunb.) Scheid. non Kurz]) were effective substitutes for the dangerous five mineral powder (Shen Kuo 1997, 129*).

Aconite is the only psychoactive ingredient (cf. *Aconitum* spp.) known to be present in the powder. It is possible that it reacts with the lime and the other herbs in a synergistic manner, although Su Shi claims that the other plants are not significant. In addition, the inebriating effects of alcohol, which was used as a carrier substance, should not be underestimated or forgotten.

It would be truly interesting to reconstruct the recipes and perform pharmacological tests on humans with them. However, caution is advised, for the Chinese literature also contains descriptions of unpleasant side effects, emaciation resulting from chronic use, and death from overdose (Wagner 1981, 322 f.).

Literature

Mailis, Brigitte. 1981. Alkoholische Getränke im Alten China. In *Rausch und Realität*, ed. G. Völger, 1: 314–19. Cologne: Rautenstrauch-Joest-Museum für Völkerkunde.

Needham, Joseph, and He Ping-Yü. 1959. Elixir poisoning in mediaeval China. *Janus* 48.

Strickman, Michel. 1979. On the alchemy of T'ao Hung-ching. In *Facets in Taoism*, ed. Holmes Welch and Anna Seidel, 123–92. New Haven and London: Yale University Press.

Wagner, Rudolf G. 1973. Lebensstil und Drogen im chinesischen Mittelalter. *T'oung Pao* 59:79–178.

———. 1981. Das Han-shi Pulver—eine "moderne" Droge im mittelalterlichen China. In *Rausch und Realität*, ed. G. Völger, 1:320–23. Cologne: Rautenstrauch-Joest-Museum für Völkerkunde.

Yü Chia-hsi. 1938. Han-shih san k'ao. In *Fu-jen hsüeh-chih* 7:29–63. (In Chinese.)

416 In traditional Chinese medicine, it is said that oyster shells (from *Crassostrea gigas* Thunb., *Ostrea rivularis* Gould, *Ostrea talienwhanensis* Crosse) "stabilize and calm the mind: used for pounding heart accompanied by anxiety, restlessness, and insomnia" (Bensky and Gamble 1986, 571*).

Haoma

Other Names
Chaoma, hauma, hom, homa, sauma

The ancient Parsis had a sacred drink known by the name *haoma* (also *hauma*, corresponding to the Indian **soma**[417]), which is reputed to have had inebriating effects and to have been a source of divine inspiration. This inebriating drink was consumed during the communal bull sacrifice. It was venerated as a god but was condemned by Zarathustra (= Zoroaster), the founder of the religion that bears his name. According to Pliny, Zarathustra was "the creator of magic" who rejected[418] both haoma and the ancient (Indo-Iranian) gods, who were the personifications of the stars, waters, and natural occurrences (fire) (Gaube 1992, 108, 114).[419] These *daiwas* ("demons, idols") were primordially related to the devas, the plant spirits of the Indians (cf. Storl 1997*). The god of the inebriating drink was also known as Hauma or Haoma. Today, Iranians still refer to Syrian rue (**Peganum harmala**) as *hom* or *homa*.

> In order to extract the sacred juice from the plant, Haoma as a god must in a certain sense be killed, and this happens during the pressing of the juice. During the main ceremony of the Parsis—the sacrifice—not only is haoma drunk, that is, one god is offered to another dying god, but sacred bread is also consumed. By doing this, the priests and the faithful desired to partake in the immortality of the gods and therewith the resurrection of eternal life. (von Prónay 1989, 27)

Not only did the Parsis regard haoma as the primordial plant out of which all other medicinal plants came, but they also viewed it as a powerful medicine in itself:

> The haoma inebriation is invigorating. Any mortal that praises haoma like a young son: to him will haoma make itself available and heal his body. Since that time have you been growing on these mountains, the multifarious, milky, gold-colored haoma; your medicines are tied to the blisses of Vohu Manah. (Avesta, Yasna 10)

The Persians considered the haoma plant to be a "miracle tree" or an "all-seed tree" from which the seeds of all trees descended. During the Hellenic period, the ancient Iranian god Mithra became the god Mithras, who was cultically worshipped in a secret male organization. The veneration of the Parsi haoma lived on in the Mithraic mysteries (Cumont 1981; Ulansey 1991). Some cult images depict Mithras as a young god who is grabbing a bull by its nostrils with one hand and stabbing him with the other:[420]

> It is then that the miracle takes place, that blessings flow from the body of the bull as it is collapsing in death. All of the nourishing and healing plants issue forth from it. This is suggested by the ears of grain that grow out of the end of his tail; the most important is the generative seed that gushes from the bull, and from which comes future life. Diabolic animals, snake, scorpion, crab, attempt to steal this source of life, but the seed is caught in a vessel and brought to the moon. Purified in the light of the moon, from there this seed produces a pair of cattle, and with this pair, from which the entire earthly race of cattle are descended, arise all the useful animals. As so it is that all plant and animal life on earth is created from the death of the bull. This bull was the first living being to be created, and the brutal and gruesome deed that Mithras was prepared to do against his will upon the command of the highest god, to kill the primordial life, brought forth all that is good in the world, increased life in infinite fashion, the multifaceted all-life of nature comes from a mythical, unified living being that had to be killed for this very purpose . . . this bull is haoma. (Lommel 1949, 212)[421]

Haoma was stirred together with the fat of a bull to make the "drink of immortality" (cf. "**Polyporus mysticus**"); the psychoactive plant "wards off death" and symbolizes the energy of life:

> This sacred plant is the embodiment or paragon of the plant world or the primordial plant; it encompasses the entire plant world within itself, and its juice represents all of the nutritional and medicinal powers contained in the plant world. It is the symbol of nourishment and healing. . . . Soma-haoma is thus the all-life, which comes from heaven and pulses through all of nature and is given form in all living beings. . . . During the full moon, when the vessel is full with the bright life potion, the gods drink from it. It is from this that they derive their immortality, for the contents of the moon is the drink of immortality, *amrta*, a word related to ambrosia. (Lommel 1949, 213)

Carl Ruck believes that the Parsis remembered the fly agaric mushroom (**Amanita muscaria**)

417 Despite the linguistic connection between *soma* and *haoma*, there is no evidence supporting the assumption that the two terms refer to the same plant, much less the same preparation.

418 "In his divine zeal, the prophet [Zarathustra] characterized the inebriating beverage (hauma), which was drunk during meals, as 'urine' (Yasna 48, 10)" (Merkelbach 1984, 11). It is possible that a urine rich in active constituents was actually consumed (cf. *Amanita muscaria*).

419 It has been suggested that "the pre-Zoroastrian Iranian religion was ruled by communities of ecstatic warriors who cultivated shamanic ecstasies and journeyed in the realms beyond. Inebriated by *haoma*, these warriors would enter into a dangerous state of murderous frenzy (*aeshma*). Zarathustra's reforms were directed against these male, shamanic brotherhoods of warriors" (Couliano 1995, 136*).

420 It is here that the Spanish bullfight has its origins.

421 The following description of the first sacrifice of a bull is found in the Persian book on the "creation" (*Bundahishin*): "The good god Ohrmazd (the later form of Ahura Mazda) created the bull, 'white and brilliant like the moon' (1, 49), but later the evil one (Ahriman) slipped into the world, and Ohrmazd saw in advance that he would slaughter the bull. And so he gave the bull hemp to eat, thereby inducing in the animal a hashish sedation, 'so that the injustice of its killing and the affliction of its pain would be lessened' (chapter 4). From the bull were all the small animals then created, as well as 55 kinds of grain and 12 kinds of healing plants. The bull's semen was borne to the moon, where it was filtered; it contains the seeds of all life. The soul of the bull will nourish all earthly creatures and be recreated as a beneficial animal in the material life" (Merkelbach 1984, 12). "The death of a cosmic primordial being . . . , which is sacrificed and cut into pieces and, in dying, brings forth the universe out of itself" (Gina 1994, 11*), is an archetypal image and is, in fact, the most primordial experience of life. It is only by killing our food that we are able to preserve our lives!

(which has often been construed as ambrosia) as haoma (Ruck 1995, 132*). Unfortunately, the identity of the true haoma plant remains undetermined. The limited sources also make it difficult to reconstruct the method or methods used to prepare the drink. However, it is very likely that haoma, like **soma**, was a plant or preparation that produced potent psychoactive effects. It may have been a kind of **ayahuasca analog**, such as a preparation made from *Peganum harmala* and *Phragmites australis* or *Phalaris arundinacea*. Archaeological finds suggest that *Ephedra* species (*Ephedra* spp.) were consumed ritually in the haoma cult as part of a beerlike preparation.

The psychedelic or visionary effects of hamoa were described in the Persian text called the Book of Arda Viraf (fourth century C.E.). A holy man named Viraz was inebriated on haoma—his haoma being a drink called *mang* made from "wine and henbane" (cf. *Hyoscyamus niger, Vitis vinifera*)—and fell asleep. His soul was led across the bridge that spans the world mountain and binds this world with the one beyond and into heaven. The holy man passed beyond the sphere of the stars and into the realm of the wise lord of the heavens, Ahura Mazda or Ohrmuzd, where he was initiated into the secrets of life after death. After seven days, he descended back to the earth with instructions to tell the people what he had seen (Couliano 1995, 140 f.*):

In Persia, vision into the spirit world was not thought to come about simply by divine grace nor as a reward for saintliness. From the apparent role of sauma [= haoma] in initiation rites, experience of the effects of sauma, which is to say of *menog* existence, must have at one time been required of all priests (or the shamans antecedent to them). (Flattery and Schwartz 1989, 31)

Some rudiments of the ancient haoma cult have been preserved in modern Iran. Today, the ritual drink is brewed either from pomegranate juice (*Punica granatum* L.) and ephedra (*Ephedra* spp.) or from rue (*Ruta graveolens* L.) and milk (Flattery and Schwartz 1989, 80). The fire ritual of the haoma cult has been integrated into the rites of tantric Buddhism and has survived into the present day; it is still practiced in Japan (Saso 1991).

Literature

See also the entries for *Ephedra gerardiana, Mandragora* spp., *Peganum harmala*, and **soma**.

Clauss, Manfred. 1990. *Mithras: Kult und Mysterien.* Munich: C. H. Beck.

Cumont, Franz. 1981. *Die Mysterien des Mithra.* Stuttgart: Teubner.

Flattery, David S., and Martin Schwartz. 1989. *Haoma and harmaline.* Near Eastern Studies, vol. 21. Berkeley: University of California Press.

Gaube, Heinz. 1992. Zoroastrismus (Die Religion des Zarathustra). In *Die großen Religionen des Alten Orients und der Antike,* ed. Emma Brunner-Traut, 95–121. Stuttgart: Kohlhammer.

Merkelbach, Reinhold. 1984. *Mithras.* Königstein/Ts.: Hain.

Lindner, Paul. 1933. Das Geheimnis um Soma, das Getränk der alten Inder und Perser. *Forschungen und Fortschritte* 9 (5): 65–66.

Lommel, Herman. 1949. Mithra und das Stieropfer. *Paideuma* 3 (6/7): 207–18.

Saso, Michael. 1991. *Homa rites and mandala meditation in Tendai Buddhism.* New Delhi: Aditya Prakashan/International Academy of Indian Culture.

Ulansey, David. 1991. *The origins of the Mithraic mysteries.* New York: Oxford University Press.

von Prónay, Alexander. 1989. *Mithras und die geheimen Kulte der Römer.* Braunschweig: Aurum.

Wolf, Fritz. 1910. *Avesta: Die Heiligen Bücher der Parsen.* Strasbourg: Trübner.

Mithras slaying the bull—a symbol of the pressing of haoma juice? (bronze medallion of Emperor Gordian III, 238–244)

Haoma Candidates

(from Couliano 1995*; Flattery and Schwartz 1989; Lindner 1933; Ruck 1995*; supplemented)

Stock Plant	Persian Name	Active Constituent(s)
Amanita muscaria	haoma	**ibotenic acid/muscimol**
Cannabis indica	beng, bang	**THC**
Ephedra spp.	hôm	**ephedrine**
Ephedra ciliata F. et M.	hum-i-bandak	**ephedrine**
Ephedra intermedia Schr. et Mey.	hôm, hum, huma	**ephedrine**
Ephedra nebrodensis Tineo.	omah, umah	**ephedrine**
Ephedra pachyclada Boiss.	hôm, hum, huma	**ephedrine**
Hyoscyamus niger	bhanga, bang	**tropane alkaloids**
Mandragora turcomanica (cf. *Mandragora* spp.)	?	**tropane alkaloids**
Peganum harmala	hom	**harmaline, harmine, etc.**
Punica granatum L.	hadânaêpatâ	**N,N-DMT (?), alkaloids**
Ruta graveolens L.	sudâb, sadâb	**essential oil, harmaline**
Vitis vinifera	hom	**alcohol**

Herbal Ecstasy

Other Names

Cyberorganic ecstasy, herbal XTC, natural ecstasy, nature XTC, thrill pills

The modern techno parties or raves of the 1990s were like a reflowering of the ancient bacchanalias and recalled the medieval dance frenzies. To non-participating observers, they seem like witches' sabbaths, Haitian voodoo dances, Indian powwows, or the trance dances of the San (!Kung Bushmen) of the Kalahari Desert (cf. *Ferraria glutinosa*). In particular, the so-called goa parties, which usually take place outdoors and are attended mainly by aging hippies and neo-hippies, resemble the ecstatic dance rituals of archaic peoples (Saunders and Doblin 1996).

These weekend parties, which usually begin around midnight and last until the following afternoon, seem to be an imitation of what Aldous Huxley described as "experiences of heaven or paradise" in his classic book *Heaven and Hell* (Huxley 1963*). Light shows reveal the mystical "otherworldly" light, and burning sticks of incense (cf. **incense**) evoke the delightful scents of heaven, carried on the "breezes of paradise." The DJs (disc jockeys) function as the high priests of a cult community, which clothes itself in special ("techno-style") garb. The DJs have the task of leading their congregation into an altered, ecstatic state—a task similar to that performed by tribal shamans. For this reason, many DJs like to refer to themselves as techno-shamans. These modern shamans employ the latest technology to take themselves and others into different realities. The powerful rhythmic music, called techno, trance, or psychedelic trance music, plays a central role (Cousto 1995).

Most techno-shamans and most of the partygoers agree that the rhythm of the music, especially in combination with psychoactive substances, helps induce trance experiences. A number of studies of the techno scene have shown that the partygoers selectively seek out the sound and rhythm that best functions for them. In general, a longing for ecstatic experience seems to provide the impetus for such party and dance activities (Krollpfeiffer 1995; Böpple and Knüpfer 1996; Rabes and Harm 1997).

Most partygoers ingest psychoactive substances, such as ecstasy (= MDMA; cf. *Myristica fragrans*), LSD, psilocybin mushrooms (*Psilocybe semilanceata, Psilocybe cyanescens*), hashish and marijuana (*Cannabis sativa*), **cocaine**, amphetamines, and guaraná (*Paulliania cupana*). The problems usually

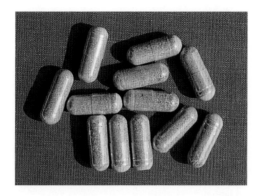

associated with MDMA began when the substance was banned internationally (1987). As a result, the black market was flooded with poor-quality preparations containing ingredients that the users often did not know about (Ahrens 1995). Uncertainties about the purity and quality of these products, combined with the development of a tolerance toward MDMA and the need for a "natural" alternative, have led to increasing numbers of plant products (so-called natural drugs) appearing on the party scene (cf. **energy drinks**). Distributors and producers advertise these products, usually sold under the name *herbal ecstasy*, as natural surrogates for MDMA, promising "very similar effects" (Leitner 1995; Saunders and Wright 1995).

The recipes for herbal ecstasy are based on the so-called brain foods and smart drugs coming from the United States. These compounds are made from plant stimulants (**ephedrine**, **caffeine**), vitamins, amino acids, and hormones (Pelton and Pelton 1989; Potter and Orfali 1993) and are reminiscent of traditional Chinese tonics (Teeguarden 1984). Many of them contain *fo-ti-tieng* (*Centella asiatica* [L.] Urban [syn. *Hydrocotyle asiatica* L.]), a tonic and sometimes mildly psychoactive plant (Emboden 1985; Storl 1995). They often include an extract of *Ginkgo biloba* L. [syn. *Salisburia adiantifolia* Sm.] (Gingkoaceae) as well; it is said to stimulate the brain and to strengthen memory[422] (Schmid and Schmoll 1994). But the main ingredient is (was) usually ma-huang (*Ephedra sinica*). Sometimes yohimbé bark or extract (*Pausinystalia yohimba*) may also be added (Saunders and Doblin 1996, 157). It is doubtful whether these herbal ecstasy preparations actually do induce psychoactive, not to mention empathogenic, experiences. Even when ephedrine is present, the dosages would typically be too small.

One recipe that is considered to be effective in

One of the many products—sold mainly at raves and techno parties—circulating under the name *herbal ecstasy*.

"Gingko biloba
This tree's leaf, which from the east
in my garden propagates,
On its secret sensations we feast,
As the wise ones it elevates.

Is it a living being and single,
which in itself divides?
Is it two, which choose to mingle
So that one the other hides?

To answer such a question,
I found a sense that's right and true;
Do you see by my songs and
    suggestions,
That I am both one and two?"

JOHANN WOLFGANG VON GOETHE
WEST-ÖSTLICHER DIVAN [WEST-EASTERN DIVAN]
(1819/20)

422 A recent double-blind study found that regular administration of ginkgo extracts led to substantial and significant improvements in memory performance in people affected by senile dementia of the Alzheimer's type (Hofferberth 1994).

The packaging of the American product Herbal Ecstacy. The "psychedelic" butterfly is intended to symbolize the purported effects ("soul flight").

The ginkgo tree (*Ginkgo biloba*) is an ancient Asian medicinal plant. Since its positive effects on brain function have become known in the West, there has been renewed interest in this traditional herbal medicine. Because ginkgo is now regarded as a brain tonic, the extract is often added to so-called brain foods and to herbal ecstasy. (Copperplate engraving from Kaempfer, *Amoenitates Exoticae*, 1712)

423 A pharmacognostic study conducted at the Pharmaceutical Institute in Bern (Switzerland) was unable to identify this component as a raw drug obtained from *Myristica* spp. (Leitner 1995, 6).

the European scene contains the following ingredients:

Angelica dahurica (Fisch. ex Hoffm.) Benth. et Hook. f.	furanocoumarins
Carthamus tinctorius L.	?
Epimedium grandiflorum C. Morr.	
Syzygium aromaticum (L.) Merr. et M. Perry [syn. *Eugenia caryophyllus* (Spreng.) Bull et Harr.]	eugenol
Glycyrrhiza uralensis Fisch.	glycyrrhizine and others
Inula japonica Thunb. [syn. *I. britannica* L. var. *japonica* (Thunb.) Franch. et Savat.]	
Ephedra sinica Stapf	**ephedrine**
Paeonia veitchii Lynch	paeoniflorine and others
Panax notoginseng (Burk.) F.H. Chen [syn. *P. pseudoginseng* Wall. var. *notoginseng* (Burk.) Hoo et Tseng]	ginsenosides
Polygala tenuifolia Willd.	polygalitol and others
Salvia miltiorrhiza Bge.	tanshinones and others
Zizyphus vulgaris Lam. var. *spinosus* Bge. [syn. *Zizyphus jujuba* Mill.]	betuline and others

The recommended dosage is one to three capsules (0.8 g per capsule) per person. Unfortunately, the recipe does not provide any information as to the relative proportions of ingredients. I perceived mild stimulant and aphrodisiac effects after taking three capsules but did not notice any similarities to the effects of MDMA.

Herbal Ecstasy, a trademarked product marketed in the United States as 100% natural, consists of the following ingredients:

Tibetan ma huang	*Ephedra intermedia* var. *tibetica* or *E. monosperma* (see **Ephedra** spp.)
wild Brazilian guaraná	**Paullinia cupana**
Chinese black ginseng	*Panax* sp. (see **Panax ginseng**)
wild ginkgo biloba	*Ginkgo biloba* L.
African raw cola nut	**Cola** spp.
gotu kola	*Hydrocotyle* sp. ?
fo-ti-tieng	*Centella asiatica*
green tea extract	**Camellia sinensis**
rou gui (a rare form of Chinese nutmeg) [423]	*Myristica* sp. ? (cf. **Myristica fragrans**)

The effects of this product are comparable to those of the first (disappointing).

The marketing of Herbal Ecstasy and similar products (Ultimate Euphoria, et cetera) is big business ($300 million in annual sales; Jolly 1996). In the United States, this has resulted in an FDA investigation into the ingredients and the component that is typically the only one with any activity, namely the ephedrine-containing *Ephedra* extracts. Ultimately, this and other problems led to a ban on ephedra and on all products that contain ephedra (cf. Saunders and Doblin 1996, 160). Since the summer of 1996, the makers of herbal ecstasy (and other preparations) have been advertising ephedra-free herbal ecstasy. The result is a kind of "decaffeinated coffee" or, in other words, "ecstasy-free ecstasy," a harmless but expensive placebo.

Literature

See also the entries for **Myristica fragrans** and **energy drinks**.

Ahrens, Helmut. 1995. Safer Use von Partydrogen. In *Risiko mindern beim Drogengebrauch*, ed. J.-H. Heudtlass, H. Stöver, and P. Winkler, 37:129–38. Frankfurt/M.: Fachhochschulverlag.

Böpple, Friedhelm, and Ralf Knüfer. 1996. *Generation XTC: Techno und Ekstase*. Berlin: Verlag Volk und Welt.

Cousto, Hans. 1995. *Vom Urkult zur Kultur: Drogen und Techno*. Solothurn: Nachtschatten Verlag.

Emboden, William A. 1985. The ethnopharmacology of *Centella asiatica* (L.) Urban (Apiaceae). *Journal of Ethnobiology* 5 (2): 101–7.

Hofferberth, B. 1994. The efficacy of EGb 761 in patients with senile dementia of the Alzheimer type, a double-blind, placebo-controlled study on different levels of investigation. *Human Psychopharmacology* 9:215–22.

Jolly, Mark. 1996. King of the thrill pill cult. *Details* (December): 170–76, 208.

Krollpfeiffer, Katrin. 1995. *Auf der Suche nach ekstatischer Erfahrung: Erfahrungen mit Ecstasy*. Berlin: VWB.

Leitner, Simone. 1995. Herbal Ecstasy. *4U—Das Jugendmagazin der Berner Zeitung* BZ (no. 34; July 7, 1996): 5–7.

Pelton, Ross, and Taffy Clarke Pelton. 1989. *Mind food and smart pills*. New York: Doubleday.

Potter, Beverly, and Sebastian Orfali. 1993. *Brain boosters: Foods and drugs that make you smarter*. Berkeley, Calif.: Ronin Publishing.

Rabes, Manfred, and Wolfgang Harm, eds. 1997. *XTC und XXL—Ecstasy*. Reinbek: Rowohlt.

Saunders, Nicholas. 1996. *Ecstasy: Dance, trance and transformation*. With Rick Doblin. Oakland, Calif.: Quick American Archives.

Saunders, Nicholas. 1995. *Ecstasy and the dance culture.* With Mary Anna Wright. London: self-published.

Schmid, Maria, and Helga Schmoll [Helga Eisenwerth], eds. 1994. *Ginkgo: Ur-Baum und Arzneipflanze—Mythos, Dichtung und Kunst.* Stuttgart: WVG.

Storl, Wolf-Dieter. 1995. An ethnobotanical portrait of the Indian pennywort. *Yearbook for*

Ethnomedicine and the Study of Consciousness 3 (1994): 267–82

Teeguarden, Ron. 1984. *Chinese tonic herbs.* Tokyo and New York: Japan Publications.

Trebes, Stefan. 1996. MDMA—Eine aktuelle Übersicht. *Jahrbuch des Europäischen Collegiums für Bewußtseinstudien* (1995): 209–19.

Honey

Other Names

Cab, honig, kab, ksandra (Sanskrit), mel, mella, miel

Honey is a substance produced by the domestic honeybee (*Apis mellifica*) and by wild bees (*Melipona* spp., *Trigona* spp.) from the nectar and pollen of various plants. "Honey is perhaps the only predigested food that humans know" (Root 1996, 127*).

Honey has been used to make **mead** since the Stone Age. The fact that honey can be toxic and/or psychoactive—in other words, inebriating—has long been known and has been demonstrated throughout the world (Palmer-Jones 1965). Honey also has a long history of use as a healing remedy or a "heavenly medicine." In Hippocratic medicine, honey was used as "a kind of psychopharmacological agent to treat depression and melancholia, and as a geriatric medicine." It was also used as an antidote for opium overdoses (Uccusic 1987, 38 f.; see *Papaver somniferum*).

There are three categories of plants that are associated with toxic honey: 1) plants whose nectar or pollen kills the bees before they can transform it into honey (e.g., locoweed [*Astragalus lentiginosus*], *Veratrum californicum*, *Vernonia* spp.); 2) plants whose nectar is harmless to bees but when turned into honey becomes toxic/inebriating to humans (e.g., oleander [*Nerium oleander*], thorn apple [*Datura* spp.], angel's trumpet [*Brugmansia* spp.], mountain laurel [*Kalmia* spp.], false jasmine [*Gelsemium sempervirens*], *Euphorbia marginata*, *Serjania lethalis*); and 3) known poisonous plants that are harmless to bees and yield edible and often exquisite honey (e.g., *Rhus toxicodendron*, *Metopium toxiferum*, *Jatropha curcas*, *Baccharis halimifolia*, *Ricinus communis*) (Morton 1964, 415).

Xenophon (ca. 430–355 B.C.E.) reported in his *Anabasis* that soldiers became inebriated and poisoned by the honey that had been produced from the Pontic rhododendron (*Rhododendron ponticum* L.) and apparently from a red-flowering

oleander (*Nerium oleander* L.; cf. Rätsch 1995, 267 f.*) (Roth et al. 1994, 615*). "In modern terms, they 'got high.' . . . This condition did not last long amongst the Greeks and quickly abated" (Rüdiger 1974, 93). The toxicological literature refers to this Pontic (Turkish) honey as "mad honey" or "toxic honey of Asia Minor" (Führer 1943, 203*). This inebriating honey was well known in ancient times (Krause 1926; Plugge 1891), and it may have had been involved with the Dionysian frenzies:

> In the district on the Pontus, among the people of the Sanni, there is a kind of honey that is known as *maenomenon* ["mad maker"] because of the insanity it induces. It is thought that this is caused by the flowers of the oleanders [*Rhododendron*], which abound in the woods. (Pliny 21.77)

In ancient times, it was believed that oleander first came from the land of Colchis (on the Black Sea); it was regarded as a plant of the "witch" Medea (who may have been a Scythian shamaness). Apparently, oleander also had something to do with the wines that were drunk during the Dionysian orgies. Oleander was a popular subject in the wall frescoes of Pompeii, a city known for its Bacchic mysteries. Oleander leaves contain the powerful cardiac poison oleandrin, which can be life-threatening for humans and animals because it can paralyze the heart. Also present are digitalis-like

Left: On Cypress, the island of the great love goddess, honey is still regarded as "Aphrodite's secret."

Right: Oleander (*Nerium oleander*) was once used as an inebriating ingredient in wine. The honey yielded by its flowers is also said to have inebriating effects. (Photographed in Delphi, Greece)

Left, from top to bottom:
In the Yucatán, a liquor made with honey from *Turbina corymbosa* is known by the Mayan name for the vine: *xtabentun*.

The Pontic rhododendron or Pontic oleander (*Rhododendron ponticum* ssp. *ponticum*) is the source of an inebriating "mad honey."

A violet-flowering *Rhododendron ponticum* var. *variegatum*.

Right, from top to bottom:
The Japanese azalea (*Rhododendron simsii*) produces a honey that is feared as poisonous.

The honey collected from the flowering alpine rose (*Rhododendron ferrugineum*) is said to have toxic or inebriating effects. (Wild plant, photographed near the Aletsch glacier, Switzerland)

The honey of European ragwort (*Senecio jacobaea*) appears to contain pyrrolizidine alkaloids that can have toxic or inebriating effects.

"The first ceremonial house (*alnáua*, *arhuínese*) was created by the 'Great Mother' in the ocean, at the same time as the levels of the universe and according to the same ninefold pattern. The ceremonial house had the shape of a beehive, but the upper structure was complemented below by another beehive, placed on its head and invisible, which it gave it the same egg-shaped form as that of the universe."

GERARDO REICHEL-DOLMATOFF
"THE LOOM OF LIFE"
(1978)

"According to ancient beliefs, honey falls on the plants (trees and flowers) as dew from heaven or out of the air and thus is considered to be a kind of food of the heavens."

WILHELM HEINRICH ROSCHER
NEKTAR UND AMBROSIA [NECTAR AND AMBROSIA]
(1883, 1)

glycosides (neriine, nerianthine, adynerin, cortenerin). The milky latex contains salacin and other alkaloids. Although oleander is frequently said to be toxic, the toxicological literature contains no observations of dangerous intoxications resulting from the consumption of the flowers and leaves (Frohne and Pfänder 1983, 47*).

An alchemical papyrus dated from late ancient times contains a puzzling recipe made with "mad honey": "Preparation of emerald. 1 part burned copper, 2 parts verdigris, and a corresponding amount of Pontic honey, cook for one hour" (in Hengstl 1978, 272).

Like all later alchemical recipes, this recipe appears to contain secret instructions for a consciousness process associated with the transmutation of matter. The inclusion of psychoactive honey is particularly interesting.

In ancient times, honey was mixed with ground medicinal herbs (such as wormwood [cf. *Artemisia absinthium*]) and other pharmaceutical substances to produce what were known as "lick agents," a kind of pharmaceutical "hard candy." Some of these may have had psychoactive effects, for example: "A remedy to cool the uterus: hemp is pounded in honey and administered to the vagina.

This is a contraction [of the uterus]" (*Papyrus Ebers* 821 [1550 B.C.E.]; in Manniche 1989, 82*).

The Mayans regard honey (*cab*) as a gift of the bee gods (*ah muzen cab*), a food from the heavens (Tozzer and Allen 1910, 298 ff.). An indigenous form of apiculture was practiced in the Yucatán in pre-Columbian times (Brunius 1995). In the Yucatán and Selva Lacandona regions (Chiapas), several species of native stingless bees (Family Meliponidae) make their honey from the nectar of specific flowers. The Lacandon know that at certain times of the year (the flowering periods), bees produce types of honey that have psychoactive or inebriating effects, even when consumed in small amounts. As little as one tablespoon is sufficient to produce noticeable effects. I once tried two tablespoons of such a honey dissolved in *atole* (a maize drink) and experienced rather strong feelings of inebriation and extreme good cheer.

The Yucatec Maya have domesticated *Melipona beecheii* and now keep these bees in special hives (hollowed-out tree trunks) to produce honey (Buchmann and Nabhan 1996). The significance of this honey is more religious and ritual than culinary. It is offered at various planting rites and also is fermented to make **balche'**, which is thus a type of **mead** (Brunius 1995). In the Yucatán, the honey made from certain vines (*Ipomoea* spp. and *Turbina corymbosa*) is called *xtabentum* or *xtabentun* (Souza Novelo et al. 1981, 32). Such honey has psychoactive effects and is preferred for making balche'. A liquor

known by the same name is produced in the region of Valladolid. This honey is usually harvested in November and December (Brunius 1995, 20).

Certain active constituents in plants can pass into the nectar of the flowers, and the bees metabolize these either not at all or only a little when they produce honey. For example, the toxic grayanotoxins present in rhododendrons and the tropane alkaloids (especially **atropine**) found in belladonna flowers can both pass into the honey that is derived from their flowers.

Some species of rhododendron, for example azaleas, contain the toxic terpene andromedotoxin (= grayanotoxin, rhodotoxin).[424]

Literature

See also the entries for **balche'** and **mead**.

Brunius, Staffan. 1995. Facts and thoughts about past and present Maya traditional apiculture. *Acta Americana* 3 (1): 5–30.

Buchmann, Stephen L., and Gary Paul Nabhan. 1996. The survival of Mayan beekeeping. *The Seedhead News* 54:1–3.

Charlton, Jane, and Jane Newdick. 1996. *Honig.* Munich: Irisiana.

Glock, Joh. Ph. 1897. *Die Symbolik der Bienen und ihrer Produkte in Sage, Dichtung, Kultus, Kunst und Brauchen der Völker.* Heidelberg: Th. Groos.

Hazslinsky, B. 1956. Toxische Wirkung eines Honigs der Tollkirsche (*Atropa belladonna* L.). *Zeitschrift für Bienenforschung* 3 (5): 93–96; 3 (10): 240.

The long-nosed Mayan god Ek Chuah, with honeycombs and the bee-shaped god Ah Muzencab ("the honey gatherer"). This black god is also the protector of cacao and appears to have numerous other associations with psychoactive plants, mushrooms, and brews. (Codex Tro-Cortesianus 109c)

An early-modern-era apiary. (Woodcut from Tabernaemontanus, *Neu Vollkommen Kräuter-Buch,* 1731)

Some Plants Known to Produce Psychoactive/Toxic Honey

Name	Botanical Name	Reference
aconite	*Aconitum napellus*	
alpine rose	*Rhododendron ferrugineum* L.	Roth et al. 1994, 613*
azalea	*Rhododendron simsii* Planch.	Roth et al. 1994, 614 f.*
belladona	*Atropa belladonna* L.	Hazslinsky 1956
euphorbia (Africa)	*Euphorbia* spp.	Rüdiger 1974, 93
grasses infected with *Claviceps*	*Paspalum plicatulum* Michx. *Paspalum unispicatum* (Sm.) Nash	Arenas 1987, 289*
Greenland tea	*Ledum groenlandicum* L.	Palmer-Jones 1965
hemp	*Cannabis*	reports by hemp growers
oleander	*Nerium oleander* L.	Rätsch 1995a, 267*; Roth et al. 1994, 511*
paullinia	*Paullinia australis*	Millspaugh 1974, 167*
ragwort	*Senecio jacobaea* L.	Frohne and Pfänder 1983, 66*
rhododendron (Pontic alpine rose)	*Rhododendron ponticum* L. [syn. *Azalea pontica, Heraclea pontica*]	Fühner 1943, 203*; Plugge 1891
	Rhododendron flavum Don	Krause 1926, 978
toé	*Brugmansia sanguinea*	
tutu	*Coriaria arborea* Lindsay (cf. *Coriaria thymifolia*)	Palmer-Jones and White 1949
water hemlock	*Cicuta virosa* L.	Rüdiger 1974, 93
wild rosemary	*Ledum palustre* L.	Ott 1993, 404*
xtabentun	*Turbina corymbosa*	
	Ipomoea triloba L.	Souza Novelo et al. 1981, 32
	Ipomoea spp.	
yew	*Taxus baccata* L.	Rüdiger 1974, 93

424 This terpene is also present in many other plants, e.g., bog rosemary (*Andromeda polifolia* L.), mountain laurel (*Kalmia latifolia* L.), prickly heath (*Pernettya mucronata* [L. f.] Gaudich ex Spreng.), and the Siberian alpine rhododendron (*Rhododendron chrysanthum* Pall.) (Roth et al. 1994, 758*). It is conceivable that bees also visit these plants.

Humans have gathered and prized honey since the Stone Age. This Stone Age painting showing honey being gathered was discovered in the caves of Araña, near Valencia, Spain.

Hengstl, Joachim, G. Häge, and H. Kühnert, eds. 1978. *Griechische Papyri aus Ägypten: Zeugnisse des öffentlichen und privaten Lebens.* Munich: Heimeran.

Huber, Ludwig. 1905. *Die neue, nützlichste Bienenzucht.* 14th ed. Lahr: M. Schauenburg.

Krause, K. 1926. Über den giftigen Honig des pontischen Kleinasien. *Die Naturwissenschaften* 44 (29, 10): 976–78.

Morton, Julia F. 1964. Honeybee plants of South Florida. *Proceedings of the Florida State Horticultural Society* 77:415–36.

Palmer-Jones, T. 1965. Poisonous honey overseas and in New Zealand. *New Zealand Medical Journal* 64:631–37.

Palmer-Jones, T., and E. P. White. 1949. A recent outbreak of honey poisoning. *New Zealand Journal of Science and Technology* 31:246–56.

Plugge, P. C. 1891. Giftiger Honig von *Rhododendron ponticum. Archiv der Pharmazie* 229:554–58.

Ransome, Hilda M. 1937. *The sacred bee in ancient times and folklore.* Boston and New York: Houghton Mifflin Co.

Reichel-Dolmatoff, Gerardo. 1978. The loom of life: A Kogi principle of integration. *Journal of Latin American Lore* 4 (1): 5–27.

Roscher, Wilhelm Heinr. 1883. *Nektar und Ambrosia. Mit einem Anhang über die Grundbedeutung der Aphrodite und Athene.* Leipzig: B. G. Teubner.

Rüdiger, Wilhelm. 1974. *Ihr Name ist Apis: Kleine Kulturgeschichte der Bienen.* Illertissen: Mack.

Schwarz, H. F. 1948. Stingless bees (Meliponidae) of the Western Hemisphere. *Bulletin of the American Museum of Natural History* 90:1–536.

Souza Novelo, Narciso. 1940. *Plantas melíferas y políniferas que viven en Yucatán.* Mérida: El Povenir.

Souza Novelo, Narciso, Victor M. Suarez Molina, and Alfredo Barrera Vasquez. 1981. *Plantas melíferas y políniferas de Yucatán.* Mexico City: Fondo Editorial de Yucatán.

Tozzer, Alfred M., and Glover M. Allen. 1910. Animal figures in the Maya codices. *Papers of the Peabody Museum* (Cambridge) 4 (3): 277–372.

Uccusic, Paul. 1987. *Doktor Biene.* Munich: Heyne.

Valli, Eric, and Diane Summers. 1988. *Honey hunters of Nepal.* London: Thames & Hudson.

White, J. W., Jr. 1966. Honey. In *The hive and the honey bee,* ed. Roy A. Grout, 369–406. Hamilton, Ill.: Dadant and Sons.

Incense

In every climate zone, humans have discovered plants that can be used as incenses for ritual, medicinal, and psychoactive purposes. The inhabitants of the oases in the extremely arid Atacama Desert (northern Chile) use the aromatic, resinous herbage of *Fabiana bryoides* as a shamanic incense. (Watercolor: Donna Torres)

Other Names

Dhoop, dhup, fumigium, incensio, räucherwerk, sahumerio, saumerio, weihrauch

The use of incenses for ritual, religious, magical, medicinal, hygienic, and other purposes is found throughout the world. Incense is a transcultural phenomenon. In most cultures, incense is regarded as a "food of the gods." Some incenses are used for their unusually pleasant scents. Others are burned for their pharmacological effects. The latter typically have an unpleasant smell but induce psychoactive effects. Many parts of plants (such as laurel leaves; cf. ***Laurus nobilis***) are used as incense during esoteric rituals for their allegedly magical effects. From the form that the smoke takes on as it rises, the will of the gods or the malicious intent of demons can be read. Some incenses stimulate erotic feelings, while others are said to protect house and farm from diseases, spirits, and thieves. The use of incense gave rise to the practice of smoking (cf. **smoking blends**).

There are three main cultural centers for the use of incense: the ancient world (including the ancient Orient), the Indian subcontinent (includ-ing the Himalayan region), and Mesoamerica. The Native American incense culture developed separately and independently, while strong reciprocal influences existed between the ancient world, the Orient, and India. Extensive trade relationships between India and Egypt, resulting in the exchange of spices, medicinal plants, inebriants, and incenses, existed in ancient times. Egyptian graves dating to late antiquity have yielded containers filled with aromatic substances and inscribed with Indian and Chinese writing (Wollner 1995, 19).

The use of incense was very widespread during antiquity. Greek, Egyptian, and Roman gods all had their respective incenses (cf. Rätsch 1995a, 312 ff.*). In the ancient literature, e.g., the writings of Dioscorides and Pliny, the laurel (***Laurus nobilis***) was known as *daphne* and was attributed with powerful mind-altering properties. So far, however, all attempts to use laurel for psychoactive purposes have failed (see Rätsch 1996b), and no psychoactive constituents have been found to date (Hogg et al. 1974). This suggests that *daphne* was likely a name given to at least one or more other plants in antiquity. Perhaps one of these *daphne*,

The Uses of Incense
(from Rätsch 1996b)

Incense is used to:

— make offerings to the gods and goddesses
— establish contact with deities, demons, and spirits
— make contact with the ancestors
— escort the dead into the afterlife
— banish or ward off negative spirits
— support meditation
— intensify prayers
— kindle love and the readiness for love
— increase one's own attractiveness
— honor guests
— conduct magical rituals
— improve conditions of hygiene
— disinfect rooms
— heal illnesses and dispel their causes
— induce specific spiritual experiences
— effect changes in consciousness
— effect changes in mood
— achieve specific medicinal or therapeutic effects
— disinfect or serve as an insecticide
— preserve food
— perfume clothes and hair
— spread joy and create amusement
— poison witches and offerings

whose botanical identity is now unknown, was psychoactive.

It is very likely that white henbane (*Hyoscyamus albus*), known as *apollinaris*, or the plant of Apollo, was the inebriant that the Pythia of Delphi, the famous oracular priestess of antiquity, used to enter a state of ecstasy (cf. also *Catha edulis*). The mysterious smoke of Delphi that the Pythia inhaled before sitting on the tripod and uttering her prophecies was most probably produced by burning henbane seeds.

During the Middle Ages, "sleep-bringing" incenses were used in medicine. These almost certainly had powerful psychoactive effects. A fourteenth-century Roman codex contains a recipe for such an incense; it consisted of equal parts of arsenic,[425] mandrake root (*Mandragora officinarum*), and opium (see *Papaver somniferum*). Storax (the gum resin of *Liquidambar officinalis*) and olibanum (= frankincense; the resin from *Boswellia sacra*) were also added to this already potent blend, which was then strewn over glowing coals. This recipe was used as an alternative to **soporific sponges** and is also somewhat reminiscent of **witches' ointments**.

During the Renaissance, the burning of incense was construed as an alchemical process (Krumm-Heller 1934, 1955). Matter is transformed through fire and then affects the mind, whether pharma-

cologically or psychologically. The element of fire creates smoke out of the element of earth, and the smoke joins with the element of air to "transform the mind." That Agrippa's recipe for incense (see below) has the ability to transform the mind is clear: he names many psychoactive plants (mandrake, henbane, hemp, poppy) whose smoke one should inhale. It was said that one could also conjure demons—in the ancient sense of the word—with incense:

> For the purpose of prophecy, incenses for stimulating the imagination shall be preferred. These, in correspondence with certain higher spirits, make us skilled to receive the divine inspiration. . . . Thus, when a smoke is made from coriander and celery, or henbane along with hemlock, the demons should gather momentarily, and this is why these plants are called spirit herbs. (1.43) [1387]

In Asia and Arabia, the most valuable incense was considered to be the dark brown, resinous lignum aloe (*Aquilaria agallocha*), also known as aloe wood, agar wood, agar-agar, garugaru, and lignum aspalathi. It was used primarily for ritual purposes. There have been repeated reports of psychoactive effects resulting from burning lignum aloe or inhaling the very expensive **essential oil** extracted from it. Lignum aloe contains sesquiterpenes, chromone derivatives, a **coumarin** lignan derivative, and an alkaloid (Kletter 1992, 308).

Today, incenses continue to play a significant ceremonial and symbolic role during the commencement of entheogenic rituals around the world. *Breuzinho* (from *Protium heptaphyllum* and *Protium* spp.) is the main incense used by the Santo Daime cult (cf. **ayahuasca**). In Afro-American possession cults, various aromatic herbs and mints are burned (cf. **madzoka medicine**) to greet and attract the *orishas* (the Yoruba deities) (Voeks 1989, 123*). The wood of *Juniperus* species, known as cedar, is used as a fumigant in the North American peyote cult (the Native American Church; cf. *Lophophora williamsii*). Copal (from *Protium copal, Bursera* spp.) or pine resin (*Pinus* spp.) is burned during the Mesoamerican balche' ritual (cf. **balche'**). During the Japanese tea ceremony, incense sticks made of lignum aloe or special tea ceremony blends made of several ingredients are burned (cf. *Camellia sinensis*).

The burning of certain plants during healing ceremonies is a widespread phenomenon. In the Vaupés region of Colombia, the leaves of a Rubiaceae (*Retiniphyllum concolor* [Spruce ex Benth.] Muell. Arg.) are used for this purpose (Schultes 1978a, 196*). In Switzerland, juniper branches (*Juniperus communis* L.) are used.

Since ancient times, incenses have played a central role in shamanism. There is hardly any shamanic activity that does not incorporate the

> "In Mexico, they know of a very unusual juniper bush, *Juniperus thurifera* (L.), and in Tibet we even find a species of juniper that grows at an altitude of 4,000 meters [and] that the Chinese call 'hyiang ching' (scented green). *Juniperus scopulorum* (Sarg.) and *Cupressus benthamii* (Endl.) are found in Mexico, while one encounters *Juniperus squamata* (Lamb.) and *recurva* in Nepal and Kashmir and *Juniperus sabina* (L.; sade tree) in Siberia. All of these juniper plants are different from one another and are also used by the indigenous medicine men for the most varied of treatments for diseases."
>
> ARNOLD KRUMM-HELLER
> *OSMOLOGISCHE HEILKUNDE* [THE OSMOLOGICAL ART OF HEALING] (1955, 28)

425 Arsenic not only is one of the most famous poisons of criminal history but also is a tonic and inebriant (cf. Lewin 1980 [orig. pub. 1929], 417 ff.*).

Incense and Shamanism

Culture/Region	Most Commonly Used Incense
Amazonia	breuzinho; resin from *Protium heptaphyllum*
ancient Greece	henbane seeds (*Hyoscyamus albus*)
Atacama (Chile)	fabiana herbage (*Fabiana imbricata*)
Burjat/Mongolia	thyme (*Thymus* spp.)
Germania	mugwort herbage (*Artemisia vulgaris*)
Himalayas	mountain juniper branch tips (*Juniperus recurva*, *Juniperus pseudosabina*)
Korea	sandalwood (*Santalum album* L.), either pieces or in the form of incense sticks
Malaya (= Siamese benzoin tree)	benzoin[426]; resin gum of *Styrax tonkinensis* (Pierre) Craib ex Hartwich
Mapuche/Chile	canelo bark (*Drimys winteri* Forst.)
Mesoamerica	copal; resin of *Protium copal* (Schl. et Cham.) Engl. or *Pinus* spp.
Plains tribes/North America	sage (*Artemisia ludoviciana*, *Artemisia* spp.)
Pueblo/North America	piñon pine, resin, needles (*Pinus edulis* Engelm. [syn. *Pinus cembroides* Zucc.])
	American juniper, cedar, red cedar (*Juniperus virginiana* L.)
Scythia (antiquity)	hemp (*Cannabis sativa*)
Scythia (Mongolia)	hemp (*Cannabis ruderalis*)
Siberia	wild rosemary herbage (*Ledum palustre*), juniper/sade tree (*Juniperus sabina*)

burning or fumigation of precisely defined plants or substances. Various species of juniper (cf. ***Juniperus recurva***) grow throughout the world, particularly in Europe, Asia (the Himalayan regions and Mongolia), and North America. They are used for ritual, magical, and medicinal purposes almost wherever they grow. In most cultures familiar with shamanism, juniper has a reputation for being a shamanic incense. Juniper is probably one of the oldest—it may even be the oldest—incense used by humans. This is certainly due to the fact that its leaves exude an exquisite and spicy aroma, even when burned fresh (Rätsch 1996a, 1996b).

Shamans in the Colombian Vaupés region inhale the aromatic balsam of *Styrax tessmannii* Perk., which is purported to be psychoactive (Schultes and Raffauf 1986, 277*).

In Peru and northern Chile (Atacama region), a number of incenses (from *Fabiana* spp., *Mentha* spp., *Senecio* spp.) are referred to by the names *koa*, *k'oa*, *khoa*, et cetera. They are used primarily during the ceremonies known as *señaladas*, when they are burned as an offering to the mother-goddess Pachamama (cf. ***Erythoxylum coca***).

During shamanic healings and rituals, the modern Nahuat (Mexico) burn a blend of tobacco leaves (*Nicotiana tabacum*) and pine resin (from a *Pinus* species). Such an incense is said to keep away the malicious soul-eaters that live in Talocan ("the great flower of darkness"), the realm of dreams (Knab 1995, 29*). Mazatec shamans inhale large amounts of a blend of copal (resin from a *Pinus* species) and chili pods (*Capsicum* spp.) before they begin their divinations. This incense blend is said to have psychoactive effects (Jonathan Ott,

pers. comm.). The Cuna of the San Blas Islands (Panama) burn generous amounts of chili pods, often mixed with cocoa beans (*Theobroma cacao*), in order to dispel malicious spirits (Duke 1975, 286*).

Among the North American Plains Indians, the most important ritual and shamanic incense is sage (***Artemisia* spp.**). The Flathead use an incense of *qepqepte* (*Artemisia ludoviciana* Nutt.) and *cqelshp* (*Pseudotsuga menziesii* [Mirbel.] Franco) in their sweat lodge ceremonies (Hart 1979, 278*).

The Indians of the Southwest prefer the piñon pine (*Pinus edulis*), which has played an important role in native culture as a food, medicine, and incense for at least six thousand years. The seeds are edible and high in nutritional value. The needles and resin are a component of many medicines. The Navajo believe that a squirrel planted the piñon pine (*cá'ol*) at the beginning of creation and that the first humans nourished themselves exclusively on pine nuts (*nictc'íi pináa'*). They use the resin as an incense in the

An Indian depiction of a shaman burning cocoa beans (Cuna-Mola Panama, ca. 1990)

426 The resin known as benzoin is derived from two plant sources. The Siamese benzoin tree grows in Thailand and Malaysia, while in Indonesia the source is *Styrax benzoin* Dryander (Sumatran benzoin tree, benzoin storax tree) [syn. *Laurus benzoin* Houtt., *Benzoin officinale* Hayne, *Lithocarpus benzoin* Blume].

Psychoactive Incenses

The psychoactivity of some of the incenses listed here is doubtful.
(from Fischer 1971; Krumm-Heller 1934; Krumm-Heller 1955; Ludwig 1982, 134 f.*; Rätsch 1995; Rätsch 1996a; Rätsch 1996b; Vinci 1980; supplemented)

Incense/Part Used	Stock Plant(s)	Culture/Region of Use
aloe wood[427] (lignum aloe)	*Aquilaria agallocha* Roxb. [syn. *A. malaccensis* Lam.]	Orient, Asia
artemisia	*Artemisia mexicana*	Mexico
	Artemisia spp.	North America, Asia
asafoetida	*Ferula asafoetida* L.	Asia
	Ferula narthex Boiss.	
	Euphorbia spp. (see *Papaver somniferum*)	modern era
ashwagandha root	*Withania somnifera*	Assyria
bejuco de la vibora ("snake vine")	unknown	Mexico[428]
belladonna	*Atropa belladonna*	Renaissance, alchemy
boldo leaves, folo	*Peumus boldus* Mol.	Chile
boophane	*Boophane disticha*	Africa
broom	*Cytisus* spp.	Peru
cacao beans	*Theobroma cacao*	Cuna/Panama
calamus root	*Acorus calamus*	Asia
camphor	*Cinnamomum camphora*	India
canelo	*Drimys winteri* Forst.	southern Chile
coca leaves	*Erythroxylum coca*	Andes
copal	*Bursera bipinnata*	Mexico
coriander seeds[429]	*Coriandrum sativum* L.	Egypt
coro	*Trichocline* spp. (+ *Nicotiana* spp.)	Chaco/Argentina
cundur-cundur	*Senecio* spp.	Andes
damiana herbage	*Turnera diffusa*	esoteric[430]
ginger lily	*Kaempferia galanga*	Japan
hemlock root	*Conium maculatum*	antiquity, Renaissance
hemp flowers	*Cannabis indica*	Asia
	Cannabis ruderalis	Scythia
	Cannabis sativa	Europe
henbane leaves ("mad dill")	*Hyoscyamus* spp.	modern era
henbane seeds	*Hyoscyamus niger*	antiquity, Middle Ages
	H. niger var. *chinensis*	China
juniper branches	*Juniperus recurva*	Himalayas
khat leaves	*Catha edulis*	Yemen
khoa/khoba	*Mentha pulegium* L.	Peru
koa	*Fabiana* spp.	Atacama (Chile)
latúe	*Latua pubiflora*	Mapuche/southern Chile
laurel leaves	*Laurus nobilis*	ancient Greece, Rome
lobelia	*Lobelia inflata*	esoteric
magic mushrooms	*Psilocybe cubensis*	esoteric
	Psilocybe semilanceata	
mandrake	*Mandragora officinarum*	antiquity, Renaissance
olibanum	*Boswellia sacra*	antiquity, present
	Boswellia spp.	cosmopolitan
palqui leaves	*Cestrum parqui*	Chile
parsley root	*Petroselinum crispum*	Renaissance, occultism
pelig	*Securidara longepedunculata*	West Africa
peyote	*Lophophora williamsii*	Tarahumara/Mexico
pichi-pichi	*Fabiana imbricata*	Mapuche/southern Chile
poppy (seeds, opium)	*Papaver somniferum*	antiquity, modern era
rhododendron leaves	*Rhododendron caucasicum*	Caucasus
	R. lepidotum	Himalayas
	R. cinnabarinum	Sikkim
	Rhododendron spp.	China
saffron	*Crocus sativus*	antiquity

Many manufacturers of incense sticks offer varieties that hint at psychoactive effects. However, these products do not contain any true psychoactive substances. (Incense stick packaging)

427 The popular and historical literature often confuses the rare and very valuable lignum aloe with *Aloe vera*.

428 This incense is said to hold witches (*nahuallis*) at bay. The ascending smoke is said to be deadly (for them) (Knab 1995, 156*).

429 "Sorcerers burned coriander seeds to dispel evil spirits and to induce hallucinations. The narcotic effect has been confirmed by modern science—as long as large quantities of coriander are consumed. Perhaps it is because of this narcotic effect that the seeds are still used to make gin" (Drury 1989, 55).

430 The modern, worldwide esoteric movement, especially in the United States (California) and Europe, has developed numerous recipes for a variety of types of incense. These have become known chiefly through esoteric publications (e.g., Caland 1992, Lee 1993, Rose 1979, Vinci 1980, Wollner 1995).

Botanical illustration of the Southeast Asian benzoin tree (*Styrax benzoin*), the source of the pleasant-smelling incense ingredient benzoin. (Engraving from Periera, *De Beginselen der Materia Medica en der Therapie*, 1849)

Incense/Part Used	Stock Plant(s)	Culture/Region of Use
ska María pastora	*Salvia divinorum*	esoteric
somalata	*Ephedra gerardiana*	Tamang/Himalayas
storax	*Styrax tessmannii* Perkins	Río Vaupés
sumbul	*Ferula sumbul* Hook. f.	Asia[431]
Syrian rue seeds	*Peganum harmala*	Morocco, Near East
thorn apple leaves	*Datura wrightii*	Chumash/California
thorn apple seeds	*Datura* spp.	cosmopolitan
thyme	*Thymus* spp.	Mongolia
tobacco leaves	*Nicotiana rustica*	Mexico
	Nicotiana tabacum	America
	Nicotiana sp.	Hopi
valerian root	*Valeriana officinalis*	central Europe
white hellebore leaves	*Veratrum album*	Europe
wild rosemary	*Ledum palustre*	Eurasia
	Ledum spp.	North America
yauhtli	*Tagetes lucida*	Mexico
yün-shih	*Caesalpinia decapetala*	China

night chant, their most important religious healing ceremony. The Tewa and Santa Clara Pueblo regard the pine as the first of all trees and its seeds as the first food (*tô*). The pine is the most important incense of the Pueblo Indians. The Zuni call the tree *he'sho tsi'tonne*, "gum branch." The Hopi use primarily the needles of the pine as incense. Occasionally they also crumble the needles, mix them with wild tobacco (***Nicotiana* sp.**), and use the product as an incense powder. After a burial ceremony, pine resin is tossed into the fire at the house of the surviving relatives so that they can "smudge" and purify themselves. The resin is also used to protect against magic. For this purpose, the Hopi rub a drop of the resin onto their foreheads (Lanner 1981).

Shamans often use the leaves, branches, or bark of sacred trees as ritual and even psychoactive incenses. In the Himalayan region, shamans inhale copious amounts of juniper smoke (see ***Juniperus recurva***). The Chilean Mapuche shamans (*machi*), most of whom are female, use the cinnamon-scented bark of the sacred canelo tree (*Drimys winteri* Forst.[432]; Magnoliaceae), known locally as *voique*, *foique*, or *foye*, as an incense at all of their tribal and healing ceremonies (Mösbach 1992, 79*). Whether the incense actually does have psychoactive effects remains an open question; in any case, the Mapuche use an infusion of the leaves as an inebriating narcotic (Houghton and Manby 1985, 93*). The *machi* regard the tree as a panacea and use the bark as a tonic and stimulant (Mösbach 1992, 79*). The bark contains an essential oil, a pungent resin, and tannin. The

Top left: The Mapuche shamans of southern Chile use the sacred canelo tree (*Drimys winteri*) as a ritual incense. (Photographed in Valdivia, southern Chile)

Bottom left: The aromatic leaves of the Chilean boldo tree (*Peumus boldus*) are used as a ritual incense and also are purported to have psychoactive effects.

Right: Poison ivy (*Toxicodendron radicans* [syn. *Rhus radicans*]) may have been used as a psychoactive incense in certain "witches' cults."

431 This species has a reputation for being hallucinogenic (Schultes and Farnsworth 1982, 189*; Schultes and Hofmann 1980, 367*).

432 This species occurs in two varieties: *Drimys winteri* var. *winteri* and *D. winteri* var. *chilensis* (DC) A. Gray. Some botanists assign it to its own family: Winteraceae (Mösbach 1992, 78*). [1389]

leaves contain sesquiterpenes, such as drimendiol (Brown 1994; Wren 1988, 284*). Canelo is often combined with *Latua pubiflora, Fabiana imbricata,* and *Cestrum parqui* and burned as a psychedelic incense.

Another ritual incense attributed with psychoactive effects is composed of the leaves of the Chilean boldo tree (*Peumus boldus*), which contain alkaloids (1% boldine in the leaves; also norboldine) that have stimulating and possibly psychoactive effects (Mösbach 1992, 80*). Boldine is an aporphine alkaloid (cf. *Nymphaea ampla*) that increases the secretion of gastric juices and thus aids in digestion; it also promotes bile production and is antispasmodic. Overdoses and chronic use are said to have toxic effects. Psychotropic effects and even hallucinations have been mentioned (Pahlow 1993, 365*). The flowery-fruity-scented essential oil contains ascaridol, cineol, eucalyptol, and p-cymol and has anthelmintic effects.

Recipes for Psychoactive Incenses

Countless recipes for incense have been created, tested, discarded, or passed down through tradition. The following selection includes recipes that were developed over time and either were or are used for psychoactive purposes (Rätsch 1991b, 230–36*).

Incense of Hecate (late antiquity)
Equal parts of:
laurel leaf (*Laurus nobilis*)
myrrh (*Commiphora* spp.)
olibanum (*Boswellia sacra*)
storax (*Styrax officinalis*)
Syrian rue seed (*Peganum harmala*)

Incense of Delphi (reconstruction of original recipe)
Equal parts of:
henbane seeds (*Hyoscyamus albus*)
laurel leaf (*Laurus nobilis*)
myrrh (*Commiphora* spp.)

Incense for Conjuring Lesser Devils (sixteenth century)
Equal parts of:
parsley root (*Petroselinum crispum*)
belladonna (*Atropa belladonna*)

coriander seed (*Coriandrum sativum*)
hemlock root (*Conium maculatum*)
henbane (*Hyoscyamus niger*)
opium (*Papaver somniferum*)
sandalwood (*Santalum album*)
Mix together and sprinkle onto glowing charcoals.

Spirit-Herb Incense (from Agrippa of Nettesheim)
Equal parts of:
celery (*Apium graveolens*)
coriander seed (*Coriandrum sativum*)
hemlock root (*Conium maculatum*)
henbane (*Hyoscyamus niger*)
Mix together and sprinkle onto glowing charcoals.

Incense for Leaving That Which Is Hidden Unknown (from Porphyrius)
Equal parts of:
coriander seed (*Coriandrum sativum*)
celery seed (*Apium graveolens*)
henbane (*Hyoscyamus niger*)
opium/poppy capsule (*Papaver somniferum*)
saffron (*Crocus sativus*)
Chop and mix all of the ingredients and bind with freshly pressed hemlock juice. Sprinkle the dried mixture onto glowing charcoals.

Roman Incense (from Pliny)
laurel leaf (*Laurus nobilis*)
juniper branch (*Juniperus* spp.; cf. *Juniperus recurva*)
vervain (*Verbena officinalis* L. [?])
sage (*Salvia officinalis* L.)
thyme (*Thymus* sp.)

The four components of the psychoactive incense of the Mapuche shamans: *Fabiana* herbage, *Cestrum parqui* leaves, canelo bark, and a branch tip of *Latua pubiflora.*

A traditional European recipe for a psychoactive incense used to conjure spirits: The main ingredient is henbane, to which is added olibanum, fennel seeds, cassia cinnamon, and coriander.

Incense for Divining the Future (from Jeannine Rose)

Equal parts of:
olibanum (*Boswellia sacra*)
magic mushroom (*Psilocybe* [*Stropharia*] *cubensis* or *Psilocybe semilanceata*)
ska María pastora (*Salvia divinorum*)

Mix all of the ingredients with a pinch of parsley root (*Petroselinum crispum*) and sprinkle onto glowing charcoals.

Incense for Seeing Visions (from Jeannine Rose)

Equal parts of:
hemp flower, female (*Cannabis sativa*)
sandalwood (*Santalum album*)
thorn apple seed (*Datura innoxia* or *Datura* spp.)

Add a pinch of violet root (*Viola odorata* L.) and perfume with sandalwood oil, benzoin, and tolu balsam. Sprinkle the finished blend onto glowing charcoals.

Incense for Conjuring Spirits

4 parts henbane (*Hyoscyamus niger*)
1 part cassia cinnamon bark (*Cinnamomum cassia*)
1 part coriander seed
1 part fennel root (or seed)
1 part olibanum (*Boswellia sacra*)

Ground the ingredients in a mortar to produce an incense powder: This incense should be taken into a ghostly, dark forest. Light a black candle on a tree stump and heat the incense censer. Burn the powder until the candle suddenly goes out. In the darkness, the spirits of the night will then appear from the smoke. To disperse them, burn a blend of equal parts of asafoetida and olibanum (Hyslop and Ratcliffe 1989, 15*).

Ritual Incense of the Tarahumara

copal (*Bursera* spp. or *Protium copal* resin)
peyote (*Lophophora williamsii*)

Mongolian Shamanic Incense[433]

Equal parts of:
juniper branch (*Juniperus* sp.; cf. *Juniperus recurva*)
wild thyme herbage (*Thymus* spp.)
rabbit droppings (as desired)

Mongolian Shamanic Incense

Equal parts of (from Tschubinow 1914, 44*):
juniper/sade tree (*Juniperus sabina* L.)
silver fir bark (*Abies alba* Mill. [syn. *Pinus picea* L.])
wild thyme (*Thymus serpyllum* L.)

Incense for Bodnath (Boudha)

Equal parts of:
balu (*Rhododendron lepidotum* Wall. ex Don)
pama (pamo) (*Juniperus indica* Bertol.; Indian juniper)
shupa (*Juniperus recurva*)

Mix all ingredients and powder. Strew the powder (*sang*) onto glowing charcoals.

"Pressant"—Asthma Fumigant (1904)

40% fol. stramonii (*Datura stramonium*)
10% herba cannabis indic. (*Cannabis indica*)
2.5% herba hyoscyami (*Hyoscyamus niger*)
30% kalium nitricum (potassium nitrate)
2% anethol (from *Anethum graveolens* or others)
15.5% binding agent (e.g., gum arabic)

The smoke produced by burning this mixture is inhaled for asthma attacks.

"Hadra"—Asthma Incense Powder (ca. 1920)

This incense powder was once available in the pharmacies of central Europe. It was intended to be burned and inhaled to treat asthma attacks. It may also have been used for "other purposes." Unfortunately, while the ingredients are known, the proportions of each are not.
herb. cannabis ind. (*Cannabis indica*), herbage
fol. stramoni (*Datura stramonium*), leaf
herb. hyoscyami (*Hyoscyamus niger*), herbage
herb. lobelia (*Lobelia inflata*), herbage
fol. eucalypti (*Eucalyptus* sp.), leaf
kal. nitric (potassium nitrate)
menthol, **essential oil**

There are a number of other, similar preparations, all of which are no longer available. Many contemporary purveyors of incenses offer blends with psychoactive effects (based on their own self-experimentation). Pharmacognostic research has demonstrated that these often contain the branch tips of *Fabiana imbricata* and various resins (olibanum and others).

Some incenses are used to aromatize other psychoactive substances, such as betel nut (*Areca catechu*), ipadú (*Erythroxylum coca*), opium (*Papaver somniferum*), and tobacco (*Nicotiana tabacum*).

Activity and Pharmacology of Incenses

Today, there are a variety of scientific models for explaining the psychoactive effects of smoke and scent on human consciousness (cf. Laatsch 1991). There are essentially three mechanisms of action

433 From personal communication with Amelie Schenk (September 1996).

that can play a role, both alone and in combination with one another (from Rätsch 1996b):

1. The smoke may contain pharmacologically active substances that, at appropriate dosages, can act like neurotransmitters or their antagonists in the nervous system.
2. The smoke may give off a characteristic scent with a demonstrably strong psychological activity.

Nearly all of the plants or raw extracts used as incenses contain **essential oils**, which are responsible for their scent. Experiments have demonstrated that certain odors induce powerful changes in the brain's activity and thus unequivocal alterations in consciousness (Steele 1991*, 1992*, 1993*). It is assumed that these aromatic substances elicit primarily psychotogenic effects, i.e., although the substance does not act pharmacologically, the experience of the scent can alter a person's state of consciousness (scent as a catalyst for memory!). Some essential oils have been observed to have both pharmacological and psychological effects. When inhaled or taken internally, higher dosages of certain essential oils can result in powerful states of inebriation that from a neurophysiological perspective cannot yet be fully explained. The following incenses have been observed to produce the strongest psychoactive effects as a result of their essential oils: camphor, cedar (*Cedrus* spp.), cinnamon (*Cinnamomum verum*), cloves (*Syzygium aromaticum* (L.) Merr. et Perry [syn. *Eugenia caryophyllata* Thunberg]), copal, coriander, damiana, juniper (*Juniperus* spp.), laurel, lignum aloe, mugwort (*Artemisia* spp.), rosemary (*Rosmarinus officinalis* L.), sade tree (*Juniperus sabina*), various sages (*Artemisia* spp., *Salvia* spp.), and wild rosemary (*Ledum palustre*). In addition, some of the constituents in **essential oils**, including thujone, eugenol, myristicin, safrole, and ledol, also have powerful psychoactive effects.

3. The smoke contains pheromones, which carry messages to the brain via the sensory organs.

Pheromones are rather simple chemical compounds that are related to hormones and function as sexual attractors in the plant and animal kingdoms (Jaenicke 1972). Although often odorless, they are as a result that much more effective. Animals and humans exude pheromones when it is time to mate. When a potential partner breathes in these molecules, they induce in him or her an irresistible urge to copulate. Male and female pheromones often differ in their chemical makeup. Certain aromatic substances that are produced in the fungal, plant, and animal kingdoms (e.g., the scent of the truffle, *Tuber* spp.) are chemically or structurally analogous or even identical to human pheromones. When these substances are inhaled in an incense, they can inflame a person's amorous desires. Vanillin, the main aromatic substance in vanilla (*Vanilla planiflora*; cf. **balche'**), which is also found in many balsams and resins, is very closely related to human pheromones and appears to exert a corresponding effect upon the nervous system. Almost all of the plants that contain vanillin are traditionally regarded as aphrodisiacs. The following incenses contain or create substances that are pheromone analogs: ambergris, benzoin, cloves, copal, ladanum (the exudate of *Cistus ladaniferus* L. [syn. *Cistus ladanifer* L.]), Peruvian balsam (*Myroxylon balsamum* [L.] Harms var. *pereirae* [Royle] Harms [syn. *Myroxylon pereira* (Royle) Baill.]), rockrose and gray-haired rockrose (*Cistus creticus* L. [syn. *Cistus incanus* L. ssp. *creticus*] = Cretian rockrose), storax (*Liquidambar* spp.), tolu balsam (*Myroxylon balsamum* [L.] Harms var. *balsamum*), and white sandalwood.

The effects of incenses on humans result from a complex series of psychological, pharmacological, and hormonal events. Unfortunately, there has been almost no research in this area. In addition, other factors that can play a role include hyperventilation, the "choking fits" that may occur as a result of deep inhalation, oxygen deprivation (which can also lead to hyperventilation), combination with other activities (such as drumming, rattling, body postures, songs), and cognitive structures. For example, some shamans inhale incense to the beat of a drum. In this manner, they are able to precisely control the rate at which they hyperventilate and the depth to which they inhale the incense. This enables them to purposefully induce and control different altered states of consciousness.

Literature

See also entries for *Boswellia sacra, Cinnamomum camphora,* **smoking blends,** and **essential oils.**

Brown, Geoffrey D. 1994. Drimendiol, a sesquiterpene from *Drymis winterii* (sic!). *Phytochemistry* 35 (4): 975–77.

Caland, Marianne, and Patrick Caland. 1992. *Weihrauch und Räucherwerk.* Aitrang: Windpferd.

Droesbeke, Erna. 1998. *Weihrauch.* Amsterdam: Iris Bücher.

Drury, Nevill, and Susan Drury. 1989. *Handbuch der heilenden Öle, Aromen und Essenzen.* Durach: Windpferd.

Fischer, L. 1917. Ein "Hexenrauch": Eine volkskundlich-liturgiegeschichtliche Studie. *Bayerische Hefte für Volkskunde* 4:93–212.

Fischer-Rizzi, Susanne. 1996. *Botschaft an den Himmel: Anwendung, Wirkung und Geschichten von duftendem Räucherwerk.* Munich: Irisiana.

Gardner, Gerald B. 1965. *Ursprung und Wirklichkeit der Hexen.* Weilheim: O. W. Barth.

Hinrichsen, Torkild. 1994. *Erzgebirge: "Der Duft des Himmels."* Hamburg: Altonaer Museum.

"I was told that in earlier times the witches knew an herb called *kat*.[434] Used with incense, it opened the inner eye, the unconscious. In combination with another herb, sumac,[435] it produced hallucinations and was therefore not to be taken too frequently. When both herbs were used in the proper way, one could acquire the ability to leave one's body. Unfortunately, the witches of today no longer know which herbs they were, although both are said to grow in England. It is also said that a woman will appear more beautiful if a man inhales incense mixed with kat. Magicians use something similar for the same purpose. Their mixture contains hemp [*Cannabis sativa*] and many other ingredients with tonic effects."

GERALD B. GARDNER
URSPRUNG UND WIRKLICHKEIT DER HEXEN [ORIGIN AND REALITY OF THE WITCHES]
(1965, 109)

434 It is very unlikely that the plant being referred to is khat (*Catha edulis*), as this does not grow in England.

435 This may be one of the sumac species that were introduced from North America to Europe, such as fragrant sumac (*Rhus aromatica* Ait. [syn. *Rhus canadensis* Marsh. non Mill.]), staghorn sumac (*Rhus typhina* L.), and poison sumac (*Toxicodendron quercifolium* [Michx.] Greene [syn. *Rhus toxicodendron* L., *Toxicodendron radicans* (L.) O. Kuntze]). Poison sumac has severe toxic effects, including dizziness, stupor, and states of excitation, and may indeed be experienced as psychoactive (cf. Roth et al. 1994, 704 f.*).

"Aromatic substances have been used in the form of incense in all confessions or earlier forms of religion, from the Mexican or the Egyptian mysteries to the Catholic mass of today. According to the religious beliefs, such incense is suitable for invoking beings from the invisible world, who can then exert a beneficial effect upon us according to their kind."

ARNOLD KRUMM-HELLER
VOM WEIHRAUCH ZUR OSMOTHERAPIE [FROM INCENSE TO OSMOSIS THERAPY]
(1934, 55)

Hogg, James W., Stuart J. Terhune, and Brian Lawrence. 1974. Dehydro-1,8-cineole: A new monoterpene oxide in *Laurus nobilis* oil. *Phytochemistry* 13:868–69.

Jaenicke, Lothar. 1972. *Sexuallockstoffe im Pflanzen-reich.* Lecture no. 217. Opladen: Westdeutscher Verlag.

Kletter, Christa. 1992. *Aquilaria.* In *Hagers Handbuch der pharmazeutischen Praxis,* 5th ed., 4:306–11. Berlin: Springer.

Krumm-Heller, Arnold. 1934. *Vom Weihrauch zur Osmotherapie.* Berlin-Steglitz: Astrologischer Verlag W. Becker.

———. 1955. *Osmologische Heilkunde: Die Magie der Duftstoffe.* Berlin: Verlag Richard Schikowski.

Laatsch, Hartmut. 1991. Wirkung von Geruch und Geschmack auf die Psyche. *Jahrbuch des Europäischen Collegiums für Bewußtseinsstudien* (1991): 119–33. Berlin: VWB.

Lanner, Ronald M. 1981. *The piñon pine: A natural and cultural history.* Reno: University of Nevada Press.

Lee, Dave. 1993. *Magische Räucherungen.* Soltendieck: Boheimer Verlag.

Rätsch, Christian. 1995. Nahrung für die Götter. *Esotera* 11/95:70–74.

———. 1996a. Einige Räucherstoffe der Tamang. *Yearbook for Ethnomedicine and the Study of Consciousness* 4 (1995): 153–61.

———. 1996b. *Räucherstoffe—Der Atem des Drachen: Ethnobotanik, Rituale und praktische Anwendungen.* Aarau and Stuttgart: AT Verlag.

Rose, Jeannine. 1979. *Hygieia: A woman's herbal.* Berkeley, Calif.: Freestone Collective Book.

Vinci, Leo. 1980. *Incense: Its ritual significance, use and preparation.* New York: Samuel Weiser.

Wollner, Fred. 1995. *Räucherwerk und Ritual: Die vergessene Kunst des Räucherns.* Kempten: Buchverlag Fred Wollner.

Kinnikinnick

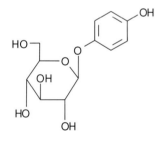

Arbutin

Other Names

Atamaoya, heiliger tabak, holy tobacco, Indianer-tabak, Indian tobacco, killikinnick, kinikinnik, kinnecanick, kinnickkinnick, k'nickk'neck, larb, native blend, ninnegahe, Uakan tobacco

Kinnikinnick is an Algonquian word that means "the mixed" or "something that is mixed." It is used to refer to various ingredients and **smoking blends** that are ritually smoked by Native Americans, for example, in the proverbial peace pipe. The ingredients of such smoking blends are often called *larb,* a corruption of the French word *l'herbe* (Johnston 1970, 317*).

Smoking was and continues to be a part of all Native American ceremonies, including shamanic healings, powwows, meetings of the tribal councils, ratification of treaties, and vision quests. The Kiowa, for example, smoke sumac leaves (*Rhus glabra* L.) before they ingest peyote (**Lophophora**

williamsii) to purify themselves for the ceremony (Kindscher 1992, 185*).

Early reports about the smoking habits of North American Indians often attributed kinni-kinnick with a variety of psychoactive effects. One document described the effects as "narcotic," while another claimed that it was "like opium" or that it made one drunk (Ott 1993*).

The main ingredient of the various mixtures is uva-ursi (*Arctostaphylos uva-ursi* [L.] Sprengel [syn. *Arbutus uva-ursi* L., *Arctostaphylos media* Greene, *A. officinalis* Wimm., *A. procumbens* Patzke, *Mairania uva-ursi* Desv., *Uva-ursi buxifolia* S.F. Gray, *Uva-ursi procumbens* Moench]), a member of the Ericaceae (Heath) Family. In North America, uva-ursi is also known by the names *ka-sin, ka-sixie, kaya'nl, kwicá, sklêwat, kinnikinnik,* or *smoking weed.* This prostrate-growing plant can easily be confused (and adulterated) with bog bilberry (**Vaccinium uliginosum** L.) and winter-

Left: Many Native American Indians know bearberry (*Arctostaphylos uva-ursi*) by the name *kinnikinnick* and use it as a basic ingredient in their ritual smoking mixtures. (Photographed in Colorado)

Right: A kinnikinnick blend that is smoked in a peace pipe at ritual events. (Photograph: Karl-Christian Lyncker)

green (*Gaultheria procumbens* L.) (Hoffmann-Bohm and Simon 1992, 331). Its leaves are added as an antiseptic ingredient to many diuretic teas (Paper et al. 1993). The Flathead Indians smoked uva-ursi in pipes and blew the smoke they inhaled into the ears of those suffering from earaches because of its numbing effects (Hart 1979, 281*). Prior to the introduction of tobacco, uva-ursi was smoked throughout the American Northwest. Later, the leaves were often mixed with tobacco (*Nicotiana tabacum*). The Chehali say that uva-ursi smoke causes a "drunken feeling" when inhaled. A S'Klallam man even warned against mixing yew needles (*Taxus brevifolia* Nutt.) and uva-ursi because the blend would have "too strong of an effect" (Gunther 1988, 44*).

Dried uva-ursi leaves contain 5 to 12%, sometimes even as much as 15%, arbutin and occasionally up to 2.5% methylarbutin (Hoffmann-Bohm and Simon 1992, 331). The leaves have antibacterial properties. High doses can induce labor (Hoffmann-Bohm and Simon 1992, 335). The leaves of the American plant contain the flavones myricetin and quercetin (cf. **Psidium guajava**, **Vaccinium uliginosum**) as well as arbutin, hydrochinone, and gallic acid (Veit et al. 1992). A closely related Mexican species, *Arctostaphylos arguta* (Zucc.) DC., is known in the local vernacular as *madroño borracho* ("drunken strawberry tree"), which may indicate a possible psychoactive activity (Martínez 1994, 205*).

One very popular admixture is the inner bark of *Cornus stolonifera* Michx., which is also often smoked by itself and is also known by the name *kinnikinnik* (Johnston 1970, 317*).

To aromatize the mixture, musk glands as well as various animal fats (such as buffalo fat) were sometimes used (Kindscher 1992, 226*).

Some kinnikinnick ingredients are clearly psychoactive, including the roots of *Veratrum viride*, the leaves and seeds of the thorn apple (**Datura stramonium**, **Datura innoxia**), the herbage of **Lobelia inflata**, the various tobacco species (**Nicotiana** spp.), sassafras bark (**Sassafras albidum**), and others (Hart 1979, 281*). Many of the ingredients in eastern kinnikinnick mixtures are commonly considered to be poisonous, and some are especially dangerous: *Datura stramonium*, *Euonymus* spp. (alkaloids; cf. Bishay et al. 1973), *Kalmia latifolia*, *Prunus serotina*, *Taxus* spp., *Veratrum viride* (de Wolf 1974). The different mixtures can have very different effects, which are likely due to a wide range of possible synergisms and pyro-chemical modifications.

Recipes

It is quite possible that the psychoactive effects of some kinnikinnick recipes are produced by a skillful combination of the various ingredients and the resulting synergistic effects, even though the individual ingredients might not themselves

be psychoactive. On the other hand, there are also recipes that do contain powerful hallucinogenic ingredients (e.g., *Datura stramonium*). Unfortunately, information about the relative proportions of ingredients is not always provided.

The smoking blend called *sagackhomi* consists of equal parts of (Emboden 1986, 162*):

| kinnikinnick | *Arctostaphylos uva-ursi* (L.) Spreng. |
| tobacco | ***Nicotiana tabacum*** L. |

The ceremonial pipe tobacco of the Blackfeet contains equal parts of (Johnston 1970*):

| pistacan (wild tobacco[436]) | *Nicotiana attenuata* Torr. (cf. **Nicotiana** spp.) |
| siputsimo (sweetgrass) | *Hierochloë odorata* (L.) Beauv. |

Another mixture produced by the Blackfeet consists of equal parts of (Johnston 1970, 317 f.*):

| kuk-see | *Arctostaphylos uva-ursi* (L.) Spreng. |
| pistacan (wild tobacco) | *Nicotiana attenuata* Torr. |

The smoking blend of the Omaha consists of equal parts of (Kindscher 1992, 184*):

chanzi	*Rhus glabra* L.
tobacco	***Nicotiana*** spp.
or	
arrowroot	*Viburnum* sp.
or	
red willow	*Cornus stolonifera* Michx.

The Cheyenne prepared a particularly "potent" smoking blend from (Kindscher 1992, 185*):

aromatic sumac	*Rhus aromatica* Ait.
bearberry, uva-ursi	*Arctostaphylos uva-ursi* (L.) Spreng.
dogwood	*Cornus stolonifera* Michx.
tobacco	***Nicotiana*** sp.

One popular modern ritual blend that can also be bought as a finished product consists of:

blackberry leaf	*Rubus* sp.
California poppy flower	***Eschscholzia californica*** Cham.
catnip	***Nepeta cataria*** L.
comfrey leaf	*Cynoglossum virginianum* L.
mullein	*Verbascum thapsus* L.
raspberry leaf	*Rubus idaeus* L.
spearmint	*Mentha spicata* L.
Virginia strawberry leaf	*Fragaria virginiana* Duchesne[437]

A different Native American smoking blend consists of equal parts of the following ingredients (Rätsch 1991, 168*):

damiana herbage	***Turnera diffusa***
lobelia herbage	***Lobelia inflata*** L.
passionflower herbage	***Passiflora*** incarnata L.
water mint	*Mentha aquatica* L. (cf. **kykeon**)

Pre-Columbian pipe heads from North America, used for the ritual smoking of kinnikinnick. (From Hartwich, *Die menschlichen Genußmittel*, 1911)

436 *Lobelia nicotianaefolia* is also known as wild tobacco (cf. *Lobelia inflata*).

437 Settlers drank a tea made from the leaves of this strawberry as a nerve tonic (Foster and Duke 1990, 38).

Kinnikinnick Ingredients

(from Foster and Duke 1990; Hart 1979*; Hartwich 1911, 32 f.*; Johnston 1970*; Kindscher 1992*;
Ott 1993*; Rutsch 1973; Schroeter 1989; modified)

Botanical Name	[Indian] Name(s)/Part Used	Active Constituent(s)
Acorus calamus	muskrat root	essential oil
Amorpha fruticosa L.	leadplant	amorphastibol[438]
Antennaria microphylla Rydb.	kinnikinnick	essential oil
Antennaria rosea Greene	pussytoes	essential oil
Arctostaphylos alpina (L.) Spreng.	kinnikinnick	arbutin (2%)
Arctostaphylos glauca L.		
Arctostaphylos pungens H.B.K.	kinnikinnick	
Arctostaphylos uva-ursi (L.) Spreng.	kinnikinnick, sagackhomi, inkashapack	arbutin
Arenaria spp.	sandwort	
Artemisia ludoviciana Nutt. (cf. *Artemisia mexicana*, *Artemisia* spp.)	sage	essential oil
Berberis spp.	bearberries	berberine
Betula lenta L.	sweet birch	methylsalicylate
Cannabis sp.[439]	hemp	THC
Carpinus caroliniana Welt.	ironwood	
Chimaphila umbellata (L.) Nutt.	pipsissewa[440]	arbutin, sitosterol
Chimaphila umbellata var. *occidentalis* (Rydb.) Blake		
Cornus amomum Du Roi	kinnikinnick	tannin
Cornus rugosa Lam.		
Cornus sanguinea L. [syn. *Thelycrania sanguinea* (L.) Fourr., *Swida sanguinea* (L.) Opiz] *Cornus sericea* L. [syn. *Cornus alba* Wangenh. non L., *Swida sericea* (L.) Holub, *Cornus stolonifera* Michaux]	kinnikinnick, dogwood, mekotsipis, pl'likinick	tannin
Cynoglossum virginianum L.	wild comfrey[441]	pyrrolizidines
Datura innoxia L.	jimsonweed	tropane alkaloids
Datura stramonium L.	jimsonweed	tropane alkaloids
Elaeagnus sp.	mistletoe	
Eriodictyon californicum Greene	yerba santa[442]	essential oil
Eriogonum sp.	desert trumpet	hordenine
Eschscholzia californica Cham.	California poppy	alkaloids
Euonymus atropurpurea Jacq.	wahoo	glycosides, alkaloids
Eupatorium berlandieri DC.	water dost	
Fragaria virginiana Duchesne	strawberry	flavonoids
Hierochloë odorata (L.) P. Beauv.	sweetgrass	coumarins
Kalmia angustifolia L.	sheep laurel	toxines
Kalmia angustifolia var. *angustifolia*[443]	narrow-leafed laurel	
Kalmia latifolia L.	calico, mountain laurel	arbutin, andromedotoxin
Ledum palustre [syn. *Ledum groenlandicum* L.]	Greenland tea	ericoline, ledol
Lobelia inflata L.	Indian tobacco	alkaloids (lobeline)
Mentha aquatica L.	wild water mint	essential oil
Mentha spicata L.	spearmint	essential oil
Nicotiana attenuata Torr.	wild tobacco	alkaloids
Nicotiana bigelovii (Torr.) Watson	wild tobacco	alkaloids
Nicotiana multivalvis Gray	wild tobacco	alkaloids
Nicotiana quadrivalvis Pursh.	nonchaw	alkaloids
Nicotiana rustica L.	oyenkwa honne	nicotine
Nicotiana tabacum L.	tobacco	nicotine
Nicotiana trigonophylla Dunal ex DC.	wild tobacco	alkaloids
Osmorhiza occidentale (Nutt.) Torr.	cicely	
Passiflora incarnata L.	passionflower	alkaloids

Botanical Name	[Indian] Name(s)/Part Used	Active Constituent(s)
Pinus sp.	pine bark	resin
Prunus serotina Ehrh. [syn. *Prunus virginiana* L. p.p.]	wild cherry, capulin	cyanoglycosides, scopoletin
Rhus aromatica Ait.	aromatic sumac	**essential oil**
Rhus glabra L.	sumac, mokola	tannin
Rubus idaeus L.	red raspberry	
Salix lasiolepis Benth.	willow bark	
Salix nigra Marsh.	willow bark	
Salix purpurea L.	purple willow bark	
Salix spp.	willow bark	
Sassafras albidum (Nutt.) Nees	sassafras	safrole
Taxus brevifolia Nutt.	yew	taxol, taxane
Taxus spp.	yew	taxol, taxane
Turnera diffusa	damiana	**essential oil**
Vaccinium stamineum L. [syn. *Polycodium stamineum*]	blueberry	arbutin
Vaccinium uliginosum	drunken berry	arbutin
Valeriana sitchensis Bong.	tobacco root	
Veratrum viride Ait. [syn. *Veratrum eschscholtzii*] (cf. *Veratrum album*)	Indian poke	alkaloids
Verbascum thapsus L.	mullein	rotenone, **coumarins**
Verbascum spp.		
Viburnum acerifolium	haw	viburnine
Viburnum sp.	arrowwood	
Zea mays L.	corn silk	alkaloids

Literature

See also the entries for **smoking blends**.

Bishay, D. W., Z. Kowalewski, and J. D. Phillipson. 1973. Peptide and tetrahydroisoquinoline alkaloids from *Euonymus europaeus*. *Phytochemistry* 12:693–98.

Black Elk. 1989. *The sacred pipe*. Norman: University of Oklahoma Press.

de Wolf, Gordon R. 1974. Guide to potentially dangerous plants. *Arnoldia* 34 (2): 45–91.

Foster, Steven, and James A. Duke. 1990. *Eastern/central medicinal plants*. A Peterson Field Guide. Boston: Houghton Mifflin Co.

Hoffmann-Bohm, Kerstin, and Peter Simon. 1992. *Arctostaphylos*. In *Hagers Handbuch der pharmazeutischen Praxis*, 5th ed., 4:328–38. Berlin: Springer.

McGuire, Joseph D. 1897. *Pipes and smoking customs of the American aborigines*. Washington, D.C.: U.S. National Museum.

"With this wakan tobacco, we place You in the pipe, O winged Power of the west. We are about to send our voices to Wakan-Tanka, and we wish You to help us! This day is wakan because a soul is about to be released. All over the universe there will be happiness and rejoicing! O You sacred Power of the place where the sun goes down, it is a great thing we are doing in placing You in the pipe. Give to us for our rites one of the two sacred red and blue days which You control!"

PRAYER TO THE SACRED KINNIKINNICK IN *THE SACRED PIPE* (BLACK ELK 1971, 19)

Top left: The Plains Indians smoke many species from the genus *Rhus*.

Bottom left: The American cranberry bush (*Viburnum trilobum*) is one of the plants traditionally smoked by the forest Indians.

Right: The North American black cherry (*Prunus serotina*) is an ingredient in traditional smoking blends.

Left, from top to bottom:
The leaves of the alpine bearberry (*Arctostaphylos alpina*) are smoked by Native Americans.

The silky dogwood (*Cornus amomum*) is known as kinnikinnick in the northeastern region of North America, where it is an important ingredient in smoking blends.

The bark of the dogwood variety *Cornus sericea* var. *sericea* is smoked by the forest Indians of North America.

Right, from top to bottom:
The dried leaves of the North American wild strawberry (*Fragaria virginiana*) are an ingredient in kinnikinnick blends.

The leaves of the narrow-leafed laurel (*Kalmia angustifolia*), which are regarded as toxic, certainly belong among the pharmacologically active smoking herbs.

The mountain laurel (*Kalmia latifolia*) is a traditional ingredient in kinnikinnick.

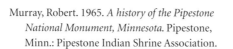

Murray, Robert. 1965. *A history of the Pipestone National Monument, Minnesota.* Pipestone, Minn.: Pipestone Indian Shrine Association.

———. 1983. *Pipes on the Plains.* Pipestone, Minn.: Pipestone Indian Shrine Association.

Paper, D. H., J. Koehler, and G. Franz. 1993. Bioavailability of drug preparations containing a leaf extract of *Arctostaphylos uva-ursi* (uvae ursi folium). *Planta Medica* 59 suppl.: A589.

Paper, Jordan. 1988. *Offering smoke: The sacred pipe and Native American religion.* Moscow: The University of Idaho Press.

Reagan, Albert. 1934. Plants used by the Hoh and Quileute Indians. *Transactions of the Kansas Academy of Sciences* 37.

Rutsch, Edward S. 1973. *Smoking technology of the aborigines of the Iroquois area of New York State.* Rutherford, N.J.: Fairleigh Dickinson University Press.

Schroeter, Willy. 1989. *Calumet: Der heilige Rauch— Pfeifen und Pfeifenkulte bei den nordamerikanischen Indianern.* Wyk auf Föhr: Verlag für Amerikanistik.

Veit, M., I. Van Rensen, J. Kirch, H. Geiger, and F.-C. Czygan. 1992. HPLC analysis of phenolics and flavonoids in *Arctostaphylos uvae-ursi. Planta Medica* 58 suppl. (1): A687.

West, George A. 1934. *Tobacco, pipes and smoking customs of the American Indians.* Bulletin 17:1–994. Milwaukee, Wis.: Milwaukee Public Museum.

Kykeon

Other Names

Ciceone, cyceon, einweihungstrank, initiation drink, kekyon, mischtrank, mixed drink

The word *kykeon* refers to a "mixed drink," in particular the one that was used as the initiatory drink in the Eleusinian mysteries. According to the myth of Eleusis, the grieving Demeter, the Great Goddess and mother of grain, was wandering in search of her daughter Persephone (Kore, Prosperina), whom Hades had abducted and taken to the underworld. Because of her sorrow, the earth's fertility had disappeared and its grain had withered. It was only when she encountered Metaneira that Demeter became happy once more:

> Metaneira offered her a cup filled with wine, as sweet as honey, but she refused it, telling her the red wine would be a sacrilege.[444] She asked instead for barley and water to drink mixed with tender leaves of *glechon*. Metaneira made the potion and gave it to the goddess as she had asked; and great Deo received the potion as the precedent for the Mystery. (*Homeric Hymn to Demeter,* line 207 ff.; in Wasson et al. 1998, 74)[445]

This drink, known as kykeon, the production of which was the well-guarded secret of two Eleusinian families, became the sacrament of initiation in the Telesterion.[446] All of the mystics who wished to participate in the initiatory ceremonies had to drink it:

> I fasted; I drank the mixed drink [*kykeon*]; I took it from the chest; after I had fulfilled my task, I placed it into the basket and from the basket into the chest. (Clemens of Alexandria, *Exhortation to the Heathen* 2.21, 2)

We can no longer reconstruct what really occurred inside the Telesterion.[447] Yet the extremely sparse details that are known to us—for as in all mystery cults, there was an absolute rule of silence—indicate that initiates experienced collective psychedelic visions (Eyer 1993). One of the last lines in the *Hymn to Demeter* is "whoever among men who walk the earth has seen these Mysteries is blessed." (*Homeric Hymn to Demeter,* line 479 ff.; in Wasson et al. 1998, 83)

Eleusis was founded around 2000 B.C.E., and the Telesterion was built around 600 B.C.E. In the beginning, the Eleusinian mysteries were probably more of a private cult, but they soon took on a local character and ultimately captivated the entire ancient world. The Lesser Mysteries took place

during the month of Anthesterión (March), while the Greater Mysteries occurred in the month of Boëdromión (September). Because initiates were forbidden to speak about the secrets of the mysteries under penalty of death, scarcely anything about them is known (Travlos 1989).

But what was the kykeon made from? The Homeric hymns name the most important ingredients: water, barley (presumably malted), and a type of mint, probably water mint (*Mentha aquatica* L.) or polei mint (= pennyroyal; ***Mentha pulegium*** L.).[448]

The **essential oil** in water mint consists primarily of limonene, caryophyllene, and menthol, as well as some psychoactive α-thujone (Malingré and Maarse 1974). High dosages of limonene may produce psychoactive effects. But a drink that consisted solely of these ingredients would never have been capable of inducing profound entheogenic experiences, even if it was a kind of **beer**. It can be assumed that the barley drink fermented during the time between its preparation and its use and, thus, had a low alcohol content. Karl Kerényi (1961) and Robert Graves (1957) appear to have been the first to speculate that the drink also contained an additional secret ingredient, namely a substance with visionary effects.

Robert de Ropp, in his classic book *Drugs and the Mind* (de Ropp 1961*), suggested that the mystery ingredient was opium, obtained from the poppy (***Papaver somniferum***), one of the plants sacred to Demeter. Could the many ancient depictions of Demeter holding barley and poppy capsules in her hands have been iconographic recipes for kykeon?

Robert Graves was the first to suggest that psychoactive mushrooms were the secret ingredient (Graves 1957). He subsequently provided a more precise formulation:

> Tantalus's crime, the mythographers explain, was that, having been privileged to eat ambrosia, the food of the gods, with the Olympians, he later invited commoners to try it. *Ambrosia*

The water mint (*Mentha aquatica*), a shallow-water plant, may have been an ingredient in the initiatory drink used in the Eleusinian mysteries.

444 Apparently, Demeter had to refuse the sacred sacrament of the wild Dionysos and provided a **beer** recipe in its place (cf. Ruck 1982*).

445 Unfortunately, the next lines (22 to 26 of them) have been lost. They may have contained further instructions for the drink and may have been destroyed intentionally.

446 The grain-mother Demeter was well versed in the world of plants and said of herself: "For I know the powerful great herbs that are gathered and the talisman plant that wards off possession" (*Homeric Hymn to Demeter,* line 229 ff.; in Wasson et al. 1998, 75).

447 The Telesterion was quite unsuited for use as a theater (Eyer 1993).

448 The name *polei* was once used to refer to various plants, including species from the genus *Nepeta* (cf. **Nepeta cataria**) (Fuchs 1545, 245*). *Polion* (= polium) has also been construed as golden germander (*Teucrium polium* L.) (Baumann 1982, 114*).

"It must have been very difficult for the initiated priests of Eleusis to procure the required doses for the approximately two thousand mystics with the necessary regularity."

Wolfgang Schmidbauer
"Halluzinogene in Eleusis?"
["Hallucinogens in Eleusis?"]
(1969, 34)

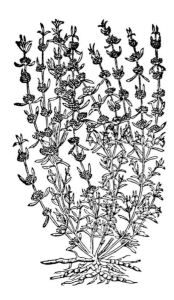

In the early modern era, the plant known in the ancient world as *glechon* was interpreted as pennyroyal (*Mentha pulegium*), and a Cretan variety was distinguished from a German variety. Whether or not these two varieties are in fact the same species is difficult to determine. (Woodcut from Tabernaemontanus, *Neu Vollkommen Kräuter-Buch*, 1731)

was the name of Dionysus's autumnal feast in which, I suggest, the intoxicant toadstool once inspired his votaries to a divine frenzy; and in my *What Food the Centaurs Ate*, I show that the ingredients given by Classical grammarians for ambrosia, nectar, and *kekyon* (Demeter's drink at Eleusis) represent a food-oghan—their initial letters all spell out forms of a Greek word for "mushroom." The story of Tantalus's crime may have been told when wine displaced toadstools at the Maenad revels, and a toadstool—perhaps not *Amanita muscaria*, but the milder, more entrancing *Panaeolus papilionaceus* [cf. **Panaeolus** spp.]—was eaten by adepts at the Eleusinian, Samothracian, and Cretan Mysteries, who became as gods by virtue of the transcendental visions it supplied. (Graves 1966, 333 f.*)

The hypothesis advanced by Wasson et al. (1984), namely that the kykeon was mixed with ergot (***Claviceps paspali, Claviceps purpurea***), has found support among some authors (Illmaier 1995; Ott 1978), while others have rejected the idea (McKenna 1996*; cf. the discussion in Valenčič 1995).

The following psychoactive plants have been considered as possible additives to kykeon (McKenna 1996*; Ruck 1995, 142*; Schmidbauer 1969; Wohlberg 1990):

Amanita muscaria
Claviceps paspali
Claviceps purpurea
Convolvulus tricolor
Lolium temulentum
Panaeolus papilionaceus (cf. **Panaeolus** spp.)
Papaver somniferum
Peganum harmala
Psilocybe cubensis
Psilocybe semilanceata
Psilocybe spp.

One central problem in attempting to find a solution to this riddle has to do with the questions concerning the availability and reliable entheogenic activity of the psychoactive substance, for every year thousands of initiates would have drunk kykeon and had wonderful experiences because of it. Of all the possible ingredients, only the easy-to-grow mushrooms of the genus *Psilocybe* withstand such scrutiny. First, they are native to Greece and were known at the time (in contrast to ergot). Second, they can easily be grown, harvested, and dried throughout the year. Third, of all the candidates, they are the only substances that are free of side effects. Finally, there is no plant that is as dependable in its use as these mushrooms. Toxic reactions are unknown. In addition, none of the other candidates even

comes close to inducing such magnificent visions. Because their effects very quickly manifest (in contrast to the fly agaric), especially when they are dissolved, the mystics could have entered those worlds that would make them "blissful" only a short time after having entered the Telesterion. It is also worth noting that Gordon Wasson (Wasson et al. 1998) became convinced that it was psychoactive mushrooms that had been used in Eleusis while he was having his own experiences with the Mexican magic mushrooms (*Psilocybe mexicana*).

Most Greek scholars, by the way, have called into question, ignored, or even dismissed this "mystery ingredient" hypothesis as a pipe dream (Burkert 1990; Foley 1994; Giebel 1990; Lauenstein 1987). Others have suggested that the kykeon of Eleusis was identical to the Persian **haoma** and the Indian **soma** (Wohlberg 1990).

Literature
See also the entries for ***Claviceps paspali, Claviceps purpurea***, and **ergot alkaloids**.

Burkert, Walter. 1990. *Antike Mysterien*. Munich: Beck.

Eyer, Shawn. 1993. Psychedelic effects and the Eleusinian Mysteries. *Alexandria: The Journal for the Western Cosmological Traditions* 2:63–95.

Foley, Helene P., ed. 1994. *The Homeric Hymn to Demeter: Translation, commentary, and interpretive essays*. Princeton, N.J.: Princeton University Press.

Giani, Leo Maria. 1994. *In heiliger Leidenschaft: Mythen, Kulte und Mysterien*. Munich: Kösel.

Giebel, Marion. 1990. *Das Geheimnis der Mysterien: Antike Kulte in Griechenland, Rom und Ägypten*. Zurich and Munich: Artemis.

Graves, Robert. 1957. Mushrooms, food of the gods. *Atlantic Monthly* 200 (2): 73–77.

———. 1960. *Food for centaurs*. New York: Doubleday.

———. 1992. *The Greek myths*. London: Penguin Books.

Hofmann, Albert. 1993. Die Botschaft der Mysterien von Eleusis an die heutige Welt. In *Welten des Bewußtseins*, ed. Adolf Dittrich, Albert Hofmann, and Hanscarl Leuner, 1:9–19. Berlin: VWB.

Illmaier, Thomas. 1995. Die Mysterien von Eleusis. *Esoterik und Wissenschaft* 1/95:36–38.

Jensen, Ad. E. 1944. Das Weltbild einer frühen Kultur. *Paideuma* 3 (1/2): 1–83.

Kerényi, Carl [= Karl]. 1962. *Die Mysterien von Eleusis*. Zurich.

———. 1991. *Eleusis*. Princeton, N.J.: Princeton University Press.

Lauenstein, Diether. 1987. *Die Mysterien von Eleusis*. Stuttgart: Urachhaus.

Malingré, Theo M., and Henk Maarse. 1974. Composition of the essential oil of *Mentha aquatica*. *Phytochemistry* 13:1531–35.

Meyer, Marvin W., ed. 1987. *The ancient mysteries.* San Francisco: Harper & Row.

Ott, Jonathan. 1978. Review: The road to Eleusis. *Journal of Psychedelic Drugs* 10 (2): 163–64.

Reitzenstein, Richard. 1956. *Die hellenistischen Mysterienreligionen.* Darmstadt: WBG.

Riedweg, Christoph. 1987. *Mysterienterminologie bei Platon, Philo und Klemens von Alexandrien.* Berlin and New York: Walter de Gruyter.

Ripinsky-Naxon, Michael. 1988. Systematic knowledge of herbal use in ancient Egypt and Greece: From the divine origins to *De Materia Medica.* Paper delivered at the 11th Annual Conference of the Society of Ethnobiology, Universidad Nacional Autonoma de Mexico, Mexico City, March 9–13, 1988.

Scheffer, Thassilo von. 1940. *Hellenische Mysterien und Orakel.* Stuttgart: Spemann.

Schmidbauer, Wolfgang. 1996. Halluzinogene in Eleusis? *Antaios* 10:18–37.

Travlos, J. 1989. Die Anfänge des Heiligtums von Eleusis. In *Tempel und Stätten der Götter Griechenlands*, ed. Evi Melas, 55–70. Cologne: DuMont.

Valenčič, Ivan. 1993. Misterij elevzinskih misterijev. *Razgledi* 18 (1001): 30–31.

———. 1995. Has the mystery of the Eleusinian Mysteries been solved? *Yearbook for Ethnomedicine and the Study of Consciousness,* 1994 (3): 325–36. Berlin: VWB.

Wasson, R. Gordon, Albert Hofmann, and Carl A. P. Ruck. 1998. *The road to Eleusis: Unveiling the secret of the mysteries.* 20th anniversary ed. Los Angeles: William Dailey Rare Books Ltd.

Wohlberg, Joseph. 1990. Haoma-soma in the world of ancient Greece. *Journal of Psychoactive Drugs* 22 (3): 333–42.

Madzoka Medicine

Among many African tribes, spirit possession is both known and culturally encouraged as a sacred or magical act. There are numerous possession cults in which special mediums—often or even primarily women—enter into a state of trance or ecstasy and allow their bodies to be possessed by a spirit being. The spirit—whether a deity, demon, bush spirit, animal spirit, ancestor, spirit of a deceased person, or something else—speaks through the body of the enraptured person, who shouts out oracles and prophecies, can perform magical healings, and so on (Lewis 1978). The African possession cults have become established in the New World in the form of Santería, Umbanda, Candomblé, voodoo, et cetera. From an anthropological point of view, the African possession cults are related to shamanism but must be regarded as a separate phenomenon (Goodman 1991). Nevertheless, there are a number of parallels and overlaps, particularly with the cults of Southeast Asia (van Quekelberghe and Eigner 1996). "Also included in spirit possession are such spectacular practices as dervish dancing, walking on hot coals, sword swallowing, and transvestitism, not to mention such mysterious phenomena as 'automatic writing'" (Lewis 1989, 42).

In the literature on possession, it is often claimed that the state of possession occurs "on its own" or, at best, in the context of magical rituals, sacrificial ceremonies, ecstatic drumming ("voodoo drumming"), and dancing. The literature on possession has a very similar tone to the early literature on shamanism in that it ignores the significance of pharmacological stimuli. However, the use of **incense**, for example, has been documented in most possession cults. And psychoactive plants are clearly used during the initiation ceremonies of the African voodoo cult in Benin (Verger 1995). Substantial amounts of the psychoactive pennyroyal (*Mentha pulegium*; cf. *Fabiana imbricata*, kykeon) are used in the Brazilian Candomblé cult (Voeks 1989, 123, 126*). In Haitian voodoo, hemp (*Cannabis sativa*) is said to play a specific role in triggering possession, and there are also reports of excessive rum drinking (see **alcohol**). *Justicia pectoralis* and *Cola acuminata* are used in the Afro-Cuban Santería cult (González-Wippler 1981, 95). It is quite possible that the use of certain psychoactive plants or products from Indian ethnoflora was adopted by the Afro-American possession cults. The following plants are used to prepare the initiation drink of the Candomblé cult: *Ipomoea pescaprae* Sweet [1402] (see **Ipomoea spp.**), *Mimosa pudica* L. and *Mimosa pudica* L. var. *acerba* Benth. (see **Mimosa spp.**), *Vernonia bahiensis* Tol., *Hibiscus* sp., *Hibiscus rosa-sinensis* L., *Mentha sativa* L., **Ocimum micranthum** Willd., **Camellia sinensis**, *Vismia guinensis* Pers., *Vismia cayennensis* Pers., *Urostigma doliarium* Miq., *Eugenia* sp., and *Eugenia jambosa* L. (Fichte 1985, 248).

It was once believed that no use of psychoactive

"There are hardly any publications about the inebriants of the Afro-American religions. It is practically unknown that in Candomblé, when the conventional initiation drink is not sufficient to subdue the consciousness of the novice, a type of brainwashing is carried out . . . so that the conditioning of the initially wild-appearing trance can take place. If the orisha, the god, does not appear because the novice is consciously following the ritual, then the novice is given a drink of cold water in which a few leaves have been soaked. Witnesses attest that immediately and without exception, the novice falls into a deep stupor, the god possesses him in trance, and the initiation can be completed."

HUBERT FICHTE
"PSYCHOLEPTICA DER 'OBRIGAÇÂO DA CONSCIENCIA'"
(1985, 247)

Madzoka medicine consists of four plants:

Malawi Name	Botanical Identification	Part Used
chiwanga azimu ("spirit banishing")	Chenopodium ambrosioides L.	leaf
bwazi ("cord")	**Securidaca longepedunculata** Fresen.	root
ampoza	Annona senegalensis Pers.	root
kachachi mkazukwa	Asparagus africanus Lam.	leaf

The African annona is a component of the possession medicine that has psychoactive effects. To date, nothing is known about any possible psychoactive activity for this plant. (Copperplate engraving from Meister, *Der Orientalisch-Indianische Kunst- und Lust-Gärtner* [The Oriental-Indian Art and Pleasure Gardener], 1677)

Mexican wormseed (*Chenopodium ambrosioides*) is highly esteemed in folk medicine, especially for its anthelmintic and abortifacient properties. In Africa, it is one of the ingredients of an allegedly psychoactive preparation said to induce possession states. The plant has not been reported to have any psychoactive effects. It is possible that it reacts with the other ingredients or exerts a synergistic effect.

or hallucinogenic plants occurred in Africa or in its cultures. Only in the past two decades has this area of ethnobotany come under greater scrutiny (de Smet 1996*). It can be expected that a great deal of interesting information will come to light.

The possession cult that serves for divination and healing in Malawi uses an herbal mixture, a madzoka medicine, to induce the trance that is required for spirit possession (*madzoka*). The fresh ingredients (presumably in equal parts) are crushed together, and the resulting paste is rubbed on the face, arms, and legs and sniffed into the nose. The trance is said to begin immediately. The mixture may be sniffed again during the trance (Hargreaves 1986, 27).

In South America, wormseed (*Chenopodium ambrosioides*), a plant introduced from North America, is used as an additive to coca (see **Erythroxylum coca**). **Securidaca longipedunculata** is drunk in Mozambique by those who are "possessed by evil spirits." The powdered root acts as a potent sneezing powder when inhaled (cf. **Veratrum album, snuffs**). The Karanga people chew the root cortex to treat impotence. During their religious rites, the Balanta (Guinea-Bissau) use an aqueous extract from the root (which they

call *tchúnfki*) because of its alleged psychoactive effects (Samorini 1996). The root, which contains 4% saponins, tannin, steroid glycosides, and gaultherine, numbs the mucous membranes. The root was recently found to contain three **ergot alkaloids**: elymoclavine, dehydroelymoclavine, and a new ergoline derivative, called compound A (Samorini 1996).

The bark of *Annona senegalensis* contains substantial amounts of tannin; mixed with palm oil, it is used as an antidote for poisoning (Assi and Guinko 1991, 30*). *Asparagus africanus*, the African asparagus, is used in Sotholand during circumcision rituals, when it is rubbed into artificially created wounds to give an initiate strength (Hargreaves 1986, 30 f.). It is possible that mixing the four components together may result in synergistic effects that are psychoactive.

Literature

Fichte, Hubert. 1985. Psycholeptica der "Obrigaçáo da Consciencia." *Curare*, Sonderband 3/85:247–48.

Goodman, Felicitas D. 1991. *Ekstase, Besessenheit, Dämonen: Die geheimnisvolle Seite der Religion.* Gütersloh: Gütersloher Verlagshaus.

González-Wippler, Migene. 1981. *Santería: African magic in Latin America.* Bronx, N.Y.: Original Products.

Hargreaves, Bruce J. 1986. Plant induced "spirit possession" in Malawi. *The Society of Malawi Journal* 39 (1): 26–35.

Lewis, Ioan M. 1978. *Ecstatic religion: An anthropological study of spirit possession and shamanism.* Harmondsworth, U.K.: Penguin Books.

———. 1989. *Schamanen, Hexer, Kannibalen: Die Realität des Religiösen.* Frankfurt/M.: Athenäum.

Samorini, Giorgio. 1996. An African kykeon? *Eleusis* 4:40–41.

van Quekelberghe, Renaud, and Dagmar Eigner, eds. 1996. Trance, Besessenheit, Heilrituale und Psychotherapie. In *Jahrbuch für Transkulturelle Medizin und Psychotherapie* (1994). Berlin: VWB.

Verger, Pierre. 1995. Del papel de las plantas psicoactivas durante la inición a ciertas religiones africanas. *Takiwasi* 3:80–87.

Mead

Other Names

Aqua mulla, **balche'**, cashirí, honey beer, honey kwass, honey mead, honey wine, honigbier, honigmeth, honigwasser, honigwein, hydromel, hydromeli, kaschiri (Arawak), madhu, melicraton, met, meth, metu, mid, mydromel, t'ädj

Mead is an alcoholic drink that is brewed from water, **honey**, other additives ("bitter herbs"), and wild or cultivated yeasts (*Saccharomyces cerevisiae*). Traditional mead has a very low alcoholic content (approximately 2 to 4%) and is not at all sweet, because the sugar in the honey is completely transformed into **alcohol**. The mead that is most popular today is a sweet, sticky drink with 14% alcohol that is brewed by fermenting a saturated solution of honey. In former times, honey was often fermented together with malt. As a result, the ancient literature often did not make a distinction between mead and **beer**. In recent years, an increasing number of drinks have come on the market that are reminiscent of mead (honey beer).

Mead, which probably was invented during the Stone Age, was found in many regions around the world. It was sacred in all ancient pagan cultures and was used ritually as a libation and for collective inebriation (Maurizio 1933). It was also considered sacred in ancient India and is sometimes associated with **soma**. The Indian gods were referred to as *madhava*, those who "sprang from the mead." The beverage was also known to all Indo-European peoples. In ancient times, it was used primarily for medicinal purposes. The Celtic and Germanic tribes—both enthusiastic drinking peoples—considered mead sacred (Markale 1989, 203*) and were aware of the divine origins of the inebriating drink: "Among the Germanic peoples, mead itself was the symbol of the drink of the gods, which had fallen from the world tree like a heavenly dew" (Delorez 1963, 23*).

During Germanic libation ceremonies, the sacred mead (and/or **beer**) that was specially brewed for the festival was passed around the circle of participants in drinking horns decorated with mythical motifs. The priest or chief took the horn and drank to the gods, offered some to the earth, and sprinkled a few drops into the air. He thanked Wotan (Odin, Woden), the god of ecstasy and the lord of magical drinks. He called to the ancestors and to the heroes who had founded human culture, and he wished his tribe peace, well-being, and health. Then he passed the horn to the next participant, who once again drank to the gods, to friends, or to specific ancestors. The horn was passed on around the circle until it was empty.

Then another would immediately be brought to the circle, passed around, and emptied, until everyone in the circle was communally and simultaneously inebriated and the gods were present among the people (Gaessner 1941). As the effects of the alcohol became apparent, the door to the world of the gods and goddesses opened:

> Mead was attributed with the power to enthuse humans and open to them the entrance to the supernatural world. It was thus to a certain extent the source of wisdom and artistic inspiration. (Fischer-Fabian 1975, 196)

It is likely that the Germanic peoples prepared their mead with inebriating berries (*Empetrum nigrum* and *Vaccinium uliginosum*) and possibly also the root of white hellebore (*Veratrum album*).

The earliest sources on Germanic beer and mead brewing indicate that a variety of psychoactive plants were added to mead, including henbane (*Hyoscyamus niger*), wild rosemary (*Ledum palustre*), bog myrtle (*Myrica gale*), and bearded darnel (*Lolium temulentum*) (cf. Maurizio 1933).

There have been some suggestions that mead or beer was brewed with the addition of mushrooms. But why would mushrooms be added to what was an only mildly alcoholic drink? The only sensible explanation is to improve the effects. Could the Germanic peoples have enriched their mead with such psychedelic mushrooms as liberty caps (*Psilocybe semilanceata*) or dark-rimmed mottlegills (*Panaeolus subbalteatus*)? After all,

Odin garbed as an eagle—in this shape, the Germanic god of ecstasy stole the mead of knowledge and inspiration from the giants. (Stone carving from Stora Hammars III, ca. 700 C.E.; reproduction drawing by C. Rätsch)

Left: The Maya use the honey produced in the Selva Lacandona (Chiapas, Mexico) as a fermenting substance for making mead.

Right: In recent years, fermented honey beverages have appeared on the market. Although these are usually regarded as honey beers, they are not allowed to be sold in Germany under this name because of the Bavarian Purity Law. The drinks are more similar to the mildly alcoholic (and not sweet) mead drinks.

"At the entrance to the enclosure is
a tree
From whose branches there comes
beautiful and harmonious
music.
It is a tree of silver, which illumines;
It glistens like gold.

There are thrice fifty trees.
At times their leaves mingle, at
times, not.
Each tree feeds three hundred
people
With abundant food, without rind.

...

There is a cauldron of invigorating
mead,
For the use of the inmates house.
It never grows less; it is a custom
That it should be full forever."

"The Sickbed of Cu Chulainn"
(an Irish epic)
in Ancient Irish Tales
(Cross and Slover 1936, 189)

"In ancient times there was no
mead. An old man had tried to
make it out of honey. He mixed the
honey with water and left the
mixture to ferment overnight. The
next day, he tried it and found it to
be very good. The other people did
not want to try the drink because
they thought it would be
poisonous. The old man said, 'I will
drink it. Because I am very old and
it doesn't matter if I die from it.'
The old man drank a lot of the
mixture and fell down as if he were
dead. He awoke that night and said
that the brew was not poisonous.
Then the men carved a large
fermenting vat and drank all of the
mead that they brewed. It was a
bird who carved the first drum [=
fermenting vat]. It drank all night
long and in the morning, it
transformed itself into a human."

myth of the Mataco Indians
in Folk Literature of the
Mataco Indians
(Wilbert and Simoneau 1982,
119*)

mead was a ritual drink that was consumed at communal gatherings so that the gods might come down and stay awhile among the inebriated people. A last memory of this practice was documented in the late Middle Ages, when Johannes Hartlieb wrote that a man died in Vienna because he had drunk a mead that contained mushrooms (chanterelles!) (cf. **witches' ointments**). The fact that mead was being brewed with the addition of plant products can also be seen in the herbals written by the "fathers of botany." For example, Tabernaemontanus wrote:

To a measure of good honey / take eight measures of water / mix together in a wide kettle / allow to simmer over a gentle fire without smoke / and continually remove the foam / until it becomes entirely clear: and the longer one wishes for the mead to keep / the longer it should simmer: afterward when it cools / pour it into a small cask / leaving three fingers / that he pours out.

If one wants it to be stronger and more powerful / then put ginger / cinnamon / cloves / galanga root / nutmeg [*Myristica fragrans*] and such herbs in it / one can also add a little saffron [*Crocus sativus*]; when it has been poured / one should store it for three months / and thereafter use it. (Tabernaemontanus 1731, 1526*)

In medieval England and Ireland, it was said that mead could increase a man's virility. For this reason, a newlywed couple were given a great amount of mead at their wedding in order to ensure the continuation of the clan. This practice is the source of the term *honeymoon*.

Mead also was and is still prized among some Native American tribes, who use it as a ritual drink (cf. **balche'**). The South American Mataco Indians brew their mead from honey, dried and ground *tusca* fruits (?), and water. They use the thick, hollowed-out stem of the bottle tree (*Chorisia insignis* H.B.K.; cf. **ayahuasca**) as a fermenting vat; as a result, the tree is known in Argentina as *palo borracho*, "drunken tree" (Wilbert and Simoneau 1982, 120 f.*). **Mead** was also known in North America. A note included with a North American herbarium specimen of the honey locust tree

(*Gleditsia triacanthos* L.; Leguminosae) reads: "[T]he sweet pith of the pods is used as a remedy for catarrh, a mead is also simmered from it" (von Reis and Lipp 1982, 126*).

Africa, in addition to its ever-popular barley beer, also has mead and honey beers that are ascribed with magical protective powers. Because of this, people often sprinkle a few drops of the drink. In Ethiopia, the chopped branches of a buckthorn known as *gescho* (*Rhamnus prinoides*; Rhamnaceae) are added to brewing mead (Haberland 1981, 172). The honey collected from the mimosa (*Mimosa* spp.) is preferred for brewing there. Mead brewed from a mixture of honey and water (1:5) is distilled to make a kind of schnapps (**alcohol**) (Haberland 1981, 173). Mead was also administered as an antidote for *Strychnos nux-vomica* poisoning.

In the summer of 1997, a "hemp mead" was introduced to the German market; the drink, however, contains no **THC**. Recipes for making mead with psychoactive mushrooms have recently been making their way around the underground (Kelly 1995).

Literature
See also the entries for **honey**.

Cross, Tom Peete, and Clark Harris Slover, eds. 1936. *Ancient Irish tales*. New York: Henry Holt and Co.

Fischer-Fabian, S. 1975. *Die ersten Deutschen*. Munich: Knaur.

Gaessner, Heinz. 1941. *Bier und bierartige Getränke im germanischen Kulturkreis*. Berlin: Veröffentlichungen der Gesellschaft für die Geschichte und Bibliographie des Brauwesens.

Haberland, Eike. 1981. Honigbier in Äthiopien. In *Rausch und Realität*, ed. G. Völger, 1:170–73. Cologne: Rautenstrauch-Joest-Museum für Völkerkunde.

Kelly, I. 1995. Mushroom mead. *Psychedelic Illuminations* 8:84.

Maurizio, A. 1933. *Geschichte der gegorenen Getränke*. Berlin: Verlag Paul Parey.

Rätsch, Christian. 1994. Der Met der Begeisterung und die Zauberpflanzen der Germanen. In *Der Brunnen der Erinnerung*, ed. Ralph Metzner, 231–49. Braunschweig: Aurum.

Oriental Joy Pills

Other Names

Fröhlichkeitspillen, gandshakini (Sanskrit), gods-chaki, hab-i nishad (Arabic, "joy pills"), happy pills, joy pills, madgiun, madjnun, madshun, majoon, majun, ma'jun, mojun, nepenthe, orientalische fröhlichkeitspillen

The term *joy pills* refers to combination preparations consisting of four basic ingredients: opium (see *Papaver somniferum*), *Cannabis* products, *Datura* seeds, and spices. These combinations are efficacious psychoactive aphrodisiacs that activate the nervous system simultaneously in numerous places. The recipes are from the Orient and are thought to be very ancient.

In ancient India, the most important of the *vajikarana* (= aphrodisiacs) were those that exerted a psychoactive effect; these were made chiefly of opium, hashish, **wine**, et cetera (Bose 1981). According to the early Ayurvedic literature (e.g., the *Bavasita*) [1408], many recipes for aphrodisiacs contained opium, *Datura metel*, camphor, nutmeg, long pepper (*Piper longum*), ginger (*Zingiber officinale*), and bhang (Chaturvedi et al. 1981). During the Moghul period, inebriants composed of opium, hemp, *Datura*, and other substances (spices, **alcohol**) were common (Saleh 1981; Sangar 1980).

Joy pills were also known in Arabian lands, where they were used mainly by the dervishes:

They take opium dissolved in wine, milk, or water, and ingest it as "joy pills" (*hab-i nishad*), and since the 17th century they have also smoked it as a syruplike substance with various ingredients. . . . Opium leads the mystic within, transports him from the here and now, and inspires his contemplation of God. (Frembgen 1993, 202*)

In Europe, a variety of recipes for joy pills became known at the beginning of the nineteenth century. The *Encyclopädie der gesammten Volksmedizin* [Encyclopedia of All Folk Medicine] of 1843 states that "hemp is the main ingredient in the joy pills of the Orient" (Most 1943, 225*). The name given the pills in this reference is interesting, as it makes reference to the famous Homeric **nepenthe**:

In order to dispel bad moods and hypochondria, the Orientals, who are also known to take delight in opium smoking and opium eating, take their refuge in a mixture: *nepenthe* by name, which consists of the powder of the dried, uppermost leaves and flowers of hemp, in combination with opium, areca nut, spices, and sugar, which they swallow in pill form. (Most 1843, 194*)

A book about poisonous plants contains the following entry under *Datura metel*:

The seeds are also a component of the Oriental joy pills, which also contain poppy juice, hemp, and several spices; to the Orientals, for whom wine is forbidden, these represent a surrogate for the same and are said to incite an indescribable sense of well-being. They have recently also strayed into Europe, and in Marseilles they produced veritable symptoms of poisoning (Berge and Riecke 1845, 101).

Freiherr von Bibra was also aware of this agent of pleasure:

Mojun is a very strong preparation of hemp, poppy, thorn apple, crow's eyes, milk, and sugar, in other words, *gondschaki*, or *joy pills*, which were already mentioned by the ancient Sanskrit writers, appear to be entirely identical with one of the lighter Oriental preparations. (Bibra 1855, 271*)

We can certainly presume that joy pills were received with enthusiasm in certain European circles. They have always been renowned as powerful inebriants and extraordinarily effective aphrodisiacs. Reports of their effects often take on an effusive and poetic expression:

The joy pills are a flying carpet that bears one to the pearly strands of pleasurable sensuality. All of the senses are immeasurably heightened in the most exquisite manner. The inner happiness radiates throughout the body with the smile of bliss, just as the light of the sun allows the tears of heaven to appear as wonderful rainbows. The enjoyment of one's own body, of one's own being, and of existence, has a cultivated and subtle quality that sweetens life with the sense of divine eternity. The soul kisses the body, dances with it, and rides on the dragon of wisdom to the stars, which sparkle like the jeweled twinkling of the eyes of the immortals. Just as the blood flows through the body, the peace of the heart streams through the universe, which is illuminated in love by the breath of the gods. The

"Bangue is also common in India, almost like amfion [= opium]; the seeds are like hemp seeds, the same is true for its leaves although they are a little smaller. The Indians eat these seeds and its leaves, but somewhat crushed, they say that it provides a good appetite for eating. Sometimes the leaves and seeds are mixed together, combined with areca or nutmeg leaves or massa, and sold on the corners in order to attract the good graces of the women. The wealthy enrich this bangue with cloves, camphor, ambergris, musk, and amfion, it dispels all of one's worries and makes them forget their misery, while at the same time it makes them happy and finally puts them to sleep, which some women make diligent use of when conducting the business of marital love, and themselves approach the male customers. It is also used by those burdened with much work and by slaves kept under hard conditions so that they might forget their burden from time to time. Even among those who are most sorrowful, or who are given to melancholy, it awakens an unnatural, happy mood and is thus a kind of remedy for melancholy, but only in the proper proportion, not too much and not too little, and as so long as one experiences its effects."

GEORGE MEISTER
DER ORIENTALISCH–INDIANISCHE KUNST– UND LUSTGÄRTNER [THE ORIENTAL-INDIAN ART AND PLEASURE GARDENER]
(1677, CH. 9, 30*)

Oriental joy pills are the ultimate, specific aphrodisiac.

Even today, Oriental joy pills are a part of the Ayurvedic pharmacopoeia:

Majun or sweetness from hemp (Cannabis), consists of *ghee* and water as well as *bhang*, *ganja*, *caras*, *opium*, poppy seeds, *dhatura* (*Datura innoxia*) leaves and seeds, cloves, resin, anise, caraway, sugar, butter, flour, milk, cardamom, and *tabasir*. A dosage of one and a half to one *drachm* is sufficient for someone who takes this drug frequently. It tastes sweet and has a decidedly pleasant scent. Occasionally, *Stramonium* seeds are also added, but never *Nux-vomica*. The effects are astonishing: ecstasy, a sense of elation, the feeling of flying, heightened appetite, and strong sexual desires. (Thakkur 1977, 317)

It is certainly no coincidence that this recipe is reminiscent of the composition of **witches' ointments**.

The following ingredients are required for each person: 0.3 g opium, ten datura seeds, 0.3 to 0.5 g

Ingredients in Oriental Joy Pills and Related Preparations

(from Chaturvedi et al. 1981; Meister 1677, 94*; Rätsch 1990; Thakkur 1977; supplemented)

Name	Botanical Name	Active Constituent(s)
almonds	*Prunus dulcis* (Mill.) D.A. Webb	vitamins
aloe	*Aloe vera* (L.) Burm.	anthracenes
ambergris	*Physeter macrocephalus* L.	pheromones
anise	*Pimpinella anisum* L.	**essential oil**
betel nuts	**Areca catechu**	arecoline
betel pepper	**Piper betle**	**essential oil**
bhang	**Cannabis indica**	THC
bonduc	*Caesalpinia bonduc* (L.) Roxb. (cf. **Caesalpinia decapetala**)	alkaloids
camphor	**Cinnamomum camphora**	camphor
cardamom	*Elettaria cardamomum* (L.) Maton	**essential oil**
cinnamon	*Cinnamomum verum* Presl	**essential oil**
cloves	*Syzygium aromaticum* (L.) M. et P.	**essential oil**
coconut flakes	**Cocos nucifera**	vitamins
coriander	*Coriandrum sativum* L.	**essential oil**
crow's eyes	**Strychnos nux-vomica**	strychnine
cubeb	*Piper cubeba* L. (cf. **Piper** spp.)	cubebin
cumin	*Cuminum cyminum* L.	**essential oil**
datura	*Datura innoxia*	tropane alkaloids
	Datura metel	tropane alkaloids
	Datura stramonium	tropane alkaloids
fennel seeds	**Foeniculum vulgare**	**essential oil**
galangan	**Kaempferia galanga**	**essential oil**
galgant	*Alpinia officinarum* Hance	**essential oil**
ghee	clarified butter	fat
ginger	**Zingiber officinale**	**essential oil**
henbane	**Hyoscyamus niger**	tropane alkaloids
honey		sugar, enzymes
long pepper	*Piper longum* L. (cf. **Piper** spp.)	**essential oil**
musk	*Moschus moschiferus*	pheromones
myrrh	*Commiphora molmol* Engl.	resin, **essential oil**
nutmeg flowers	**Myristica fragrans**	**essential oil**
oleander	*Nerium oleander* L.	oleandrine, alkaloids
olibanum	**Boswellia sacra**	**essential oil**, resin
opium	*Papaver somniferum*	**opium alkaloids**
poppy seeds	*Papaver somniferum*	fat, traces of alkaloids
pumpkin seeds	*Cucurbita pepo* L.	vitamin E
saffron	**Crocus sativus**	crocine
Spanish fly	*Lytta vesicatoria*	cantharidine
sugar	*Saccharum officinarum* L.	saccharose
turmeric	*Curcuma longa* L.	**essential oil**

hashish, and spices, resins, et cetera as desired. The ingredients are mixed together and added to clarified butter. The mixture should be poured out as soon as it has melted into a mass (cf. Rätsch 1990).

Green Chinese or Japanese tea (cf. *Camellia sinensis*) should be drunk as the pills are ingested to counteract the somniferous effects of the opium. The full effects usually manifest about four hours later and last for at least twelve hours.

Literature

See also the entries for *Cannabis indica, Datura metel,* and *Papaver somniferum.*

Abel, Ernest L. 1984. Opiates and sex. *Journal of Psychoactive Drugs* 16 (3): 205–16.

Anwari-Alhosseyni, Schams. 1981. Über Haschisch und Opium im Iran. In *Rausch und Realität,* ed. G. Völger, 2:482–87. Cologne: Rautenstrauch-Joest Museum für Völkerkunde.

Berge, Fr., and W. A. Riecke. 1985. *Giftpflanzen-Buch.* Stuttgart: Hoffmann'sche Verlags-Buchhandlung.

Bose, A. K. 1981. Aphrodisiacs—a psychosocial perspective. *Indian Journal of History of Science* 16 (1): 100–103.

Chaturvedi, G. N., S. K. Tiwari, and N. P. Rai. 1981. Medicinal use of opium and cannabis in medieval India. *Indian Journal of History of Science* 16 (1): 31–35.

Gawin, Frank H. 1978. Drugs and eros: Reflections on aphrodisiacs. *Journal of Psychedelic Drugs* 10 (3): 227–36.

Rätsch, Christian. 1990. *Die "Orientalischen Fröhlichkeitspillen" und verwandte psychoaktive Aphrodisiaka.* Berlin: VWB.

Saleh, Ahmed. 1981. Alkohol und Haschisch im heutigen Orient. In *Rausch und Realität,* ed. G. Völger, 2:488–91. Cologne: Rautenstrauch-Joest-Museum für Völkerkunde.

Sangar, S. P. 1980. Intoxicants in Mughal India. *Indian Journal of History of Science* 16 (2): 202–14.

Thakkur, Ch. G. 1977. *Ayurveda: Die indische Heil- und Lebenskunst.* Freiburg: Bauer.

Vetschera, Traude, and Alfonso Pillai. 1979. The use of hemp and opium in India. *Ethnomedizin* 5 (1/2; 1978/79): 11–23.

Wilson, Robert Anton. 1990. *Sex and drugs.* Phoenix: New Falcon Publications.

Palm Wine

Other Names

Bourdon, cachiry, chica de caanguche, coroxo, maboca, mimbo, palmenwein, palmwein, salap, sura, suri, toddy, vino palmeo

In many parts of the world, winelike drinks are produced by fermenting palms (Palmae; formerly Arecaceae); these preparations are generally referred to as palm wine (cf. **wine**). Either the juice pressed from the sweet fruits or the sap that flows through the trunk and the leaf ribs (bleeding sap) is used; both are fermented undiluted. Some palm fruits may also be mixed with water and used to make **beer** and **chicha**.

Usually it is the bleeding sap of various palms that is fermented to make wine. The young male inflorescences are often used; incisions are made into them after they are pressed or squeezed to stimulate the secretions. The sugary sap often starts fermenting as it flows from the stem (cf. *Cocos nucifera*). Sometimes a tapping hole may also be made in the upper trunk. Fermentation is usually initiated with yeast: *Saccharomyces* spp., *Candida* spp., or *Endomycopsis* sp. (Ofakor 1972).

Palm wine is a much-loved drink in Southeast Asia, where it is made from either the coconut palm (*Cocos nucifera*) or the sugar palm (*Arenga pinnata* [Wurmb.] Merr.):

Since the earliest times, it [the sugar palm, *Arenga pinnata*] has been tapped to obtain its sugar sap. To do this, the young, male inflorescences are cut off. From the site of the incision, 2 to 7 liters of sap flows daily over a period of two to five months. When the first incision site has been exhausted, an inflorescence below it will be tapped. One palm is said to provide up to 1,800 liters of sap, which is drained off into bamboo tubes. The sap contains approximately 15% sucrose and is either fermented into palm wine or boiled and made into a brown cane sugar that, pressed into disks, is then sold. (Bärtels 1993, 56*)

Pliny noted that the Egyptians used the date palm (*Phoenix dactylifera* L.) to produce wine. This Old World palm, which grows as tall as 30 meters, is found throughout Africa, the Near East, Arabia, and India. It has been cultivated for its fruits since antiquity (Stewart 1994, 151*). The sap tapped from the stems of older date palms begins to ferment immediately. This fermentation was known as palm wine (*vino palmeo*), and it was used as a ritual inebriant drink but was especially esteemed for its aphrodisiac qualities. It was often fortified with various other magical plants, presumably henbane (*Hyoscyamus niger, Hyoscyamus*

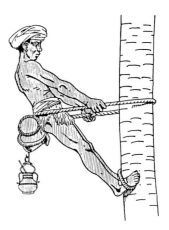

A man climbs a palm in order to tap its bleeding sap. (From Hartwich, *Die menschlichen Genußmittel* [The Human Agents of Pleasure], 1911)

775

"Every year the Shuar, who live in the rain forest in the eastern Andean foothills of Venezuela, celebrate a festival in which *B.[actris] gasipaës*, the chonta palm, plays a great role. When the strong winds of spring herald the approaching rain season and therewith the return of Uwi, the lord of fertility, the Shuar celebrate a sacred festival. Palm fruits are collected and the women chew the flesh of the fruit to prepare it for fermentation. Later, the beer is drunk by the men. Various sacred songs are also sung, all of which describe the life and uses of the palm."

ANDREAS BÄRTELS
FARBATLAS TROPENPFLANZEN
[COLOR ATLAS OF TROPICAL PLANTS]
(1993, 52*)

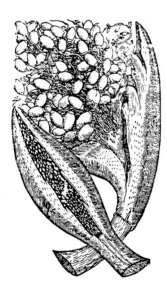

Since ancient times, the date palm and its fruits have been used to produce palm wine. (Woodcut from Tabernaemontanus, *Neu Vollkommen Kräuter-Buch*, 1731)

Palms Used to Make Wine

(from Allen 1947; Allen 1965; Bartels 1993*; Balick and Beck 1990*; Ferguson 1851; Hartwich 1911, 632*; Hawkes 1946; Lévi-Strauss 1950, 470; Lévi-Strauss 1952; Plotkin and Balick 1984; Rehm and Espig 1996, 75*; supplemented)

Palm Species	Part(s) Used	Culture/Location
Acrocomia aculeata (macaw palm)	fruits, bleeding sap	South America, Caribbean
Acrocomia mexicana Karw. (taberna/coyol palm)	bleeding sap	Honduras, Mexico
Acrocomia vinifera Oerst. (coyol palm)	sap	Honduras
***Areca catechu* L.**	fruits	India
Arenga pinnata (Wurmb.) Merr. [syn. *A. saccharifera* Labill.] (sugar palm)	bleeding sap	Southeast Asia, Cambodia
Attalea cohune Mart.		South America
Attalea speciosa Mart.		South America
Bactris gasipaës H.B.K. [syn. *Guilielma gasipaës* Bai.] (peach palm, pupuña)	fruits	Ecuador, Colombia, Bolivia, Venezuela (Yanomamö)
Bactris major Jacq.	fruits	Brazil
Bactris sp.	fruits	Central and South America
Borassus aethiopium Mart.	sap	Ivory Coast
Borassus flabellifer L. (palmyra palm)	bleeding sap	India, Ceylon
Caryota urens L. (fishtail wine palm, kitul palm)	bleeding sap	Southeast Asia
Cocos butyracea L. (palma de vino)	bleeding sap	Brazil
Cocos eriospatha Mart.	bleeding sap	Brazil
Cocos nucifera L.	sap	tropics
Copernicia prunifera (M.) Moore [syn. *Copernicia cerifera* Mart.] (carnauba wax palm)	fruits, seeds	Brazil, Argentina
Corypha silvestris Blume		Moluccas
Elaeis guinensis Jacq. [syn. *Elaeis melanocca* Gaertn.]	sap	Nigeria, Brazil
Euterpe edulis Mart.	fruits	Brazil
Euterpe oleracea Mart. (acai palm)	fruits	Brazil
Euterpe precatoria Mart.	fruits	Brazil
Euterpe spp.	fruits	Bolivia
Hyphaene natalensis (ilala palm)	bleeding sap	Tongaland
Hyphaene thebaica (L.) Mart. (dhoum palm)	bleeding sap	Africa
Hyphaene ventricosa	seed coat	Namibia
Jubaea chilensis (Mol.) Baill.	bleeding sap from the trunk	Chile
Jubaea spectabilis H.B.K.	bleeding sap	South America
Mauritia flexuosa L. f. (mirity palm)	sap	Guaraon/Orinoco (Venezuela)
Mauritia minor Burret (canbanguche palm)	fruits	Colombia (Amazonia)
Mauritia vinifera Mart.	sap	Warrau/South America, Brazil
Nypa fruticans Wurmb. (nipa palm, atap palm)	bleeding sap	Indochina, Philippines
Orbignya cohune (Mart.) Dahl. [syn. *Attalea cohune* Mart.]	sap	Honduras
Orbignya spp. (babassu palm)	seeds	Brazil
Phoenix dactylifera L. (date palm)	fruits/bleeding sap from the trunk	antiquity, Near East
Phoenix spinosa Schumach. [syn. *Phoenix reclinata* Jacq.]	fruits	tropical Africa
Phoenix sylvestris (L.) Roxb. (forest date)	bleeding sap from the trunk	India
Pholidocarpus ihur Blume		Sunda Islands
Phytelephas macrocarpa Ruíz et Pav.		neotropics
Raphia hookeri Mann et Wendl.	bleeding sap	Nigeria
Raphia vinifera P. Beauv. (wine palm)	bleeding sap	tropical Africa
Roystonea regia (H.B.K.) Cook (royal palm)	bleeding sap	Haiti
Roystonea venezuelana Bai. et Moore	bleeding sap	Venezuela
Sabal bermudana	fruits	Bermuda
Scheelea princeps Karst.		Brazil
Serenoa repens (Bartr.) Small [syn. *S. serrulata* (Michx.) Nichols.] (saw palmetto; cf. **wine**)	fruits	coastal tribes/southeastern North America

spp.), mandrake (*Mandragora officinarum*, *Mandragora* spp.), or hemp (*Cannabis indica*). The inebriating effects of such a wine were described in a cuneiform text:

When a person has drunk the inebriating drink and his head has been seized by it, he forgets his words, they disappear as he speaks, his reason does not hold firm, the eyes of such a man are fixed, to help him recover, you should grate together liquorice juice . . . , beans, oleander . . . , he should drink it with oil and the inebriating drink before the descent of the *gula* [= "in the evening before the stars rise"], in the morning before the sun rises, and before anyone has kissed him, and so he will recover. (Sigerist 1963, 30)[449]

The fruits of the date palm were also made into an inebriating drink, which the Egyptians called *srm.t*. This preparation may have been a beer to which date must was added (Cranach 1981*). This drink was often added to medicines. Palm wine was also used to wash corpses as part of the process of embalming mummies. Palm wine was also drunk as a remedy for hallucinations (Pliny 24.165 f.). In ancient times, saffron (*Crocus sativus*) and palm wine were mixed together with many magical plants.

There are seventeen species in the genus *Phoenix*, many of which are easily confused with the true date palm. Some species (e.g., *Phoenix reclinata* Jacq.) develop edible fruits that are also called dates. The sap of the Indian forest date palm (*Phoenix sylvestris* [L.] Roxb.) is also fermented into palm wine.

Palm wine is very popular in Africa, where it is used as a refreshing drink, a solvent for medicine, and an offering drink (libation). In West Africa, palm wine with cola nuts (*Cola* spp.) is an important offering in the orisha rites. In the land of the Yoruba, Ogun is the shamanic god of iron and the smithy, of war, of the hunt, and of stones, and he is considered a mighty tamer of snakes. Because of his power, offerings made to him have great importance:

A woman wants to address Ogun; she comes and brings a calabash with cola nuts, for the sacrifice she has a dog and roasted yams in addition to palm oil and palm wine. The priest rises and turns to face the shrine. He begins with a libation of water or palm wine, then he takes a hammer made entirely from metal and touches the emblem of the god so that it makes a sound. As he does this, he says: "Hear us, O Ogun, *Awo*, controller of the world, chief of the gods, whose pupils man never sees, supporter of orphans, lord of the countless heavenly palaces!" . . . Then he pours the palm wine and the palm oil on or in front of the shrine and questions the cola nut. If the answer is favorable, then he places a piece of the nut on the shrine. Then the dog is sacrificed. (Bonin 1979, 251*)

In India, palm wine (called *salap*) is also used for shamanic purposes. The tribes who live in the jungles of Orissa, in particular the Sora or Saura, have for the most part retained their pre-Hindu, prehistoric religion, which consists primarily of establishing contact with the underworld and the spirits and ancestors who dwell in the realm beyond. It is said that the spirits and ancestors live

"Yesterday's drunkenness will not relieve today's thirst."

Egyptian saying (late antiquity)

Top left: The saw palmetto (*Serenoa repens*), which grows throughout the southern regions of North America, was once used by Indians of the region to prepare inebriating drinks.

Bottom left: Saw palmetto fruits, which come from the saw palmetto (*Serenoa repens*), are fermented into a palm wine. In phytotherapy, the fruits are regarded as aphrodisiac.

Right: The fruits of the cohune palm (*Orbignya cohune*) are used in the Central and South American tropics to make palm wine. (Photographed in the Maya Mountains, Belize)

449 The symptoms described here suggest the effects of nightshade plants.

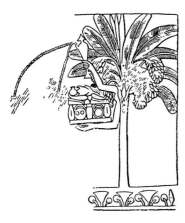

The goddess Hathor, shown here personified as a date palm, pours the inebriating palm wine. (Egyptian relief, eighteenth dynasty, sixteenth to fourth centuries B.C.E., from Luschan)

in the underworld directly beneath the wine palms. Copious amounts of palm wine are consumed during the shamanic underworld ceremonies. The *kunan*, or shaman, consumes the inebriating drink from a special calabash. He uses the palm wine as a "fuel for traveling to the underworld" (Gerhard Heller, pers. comm.).

In some places, palm wine may be distilled into *arrak* or palm schnapps (cf. **alcohol**).

Palm Wine Additives

As with all other alcoholic drinks, plants or other substances are sometimes added to palm wine to alter its effects (cf. *Vitis vinifera*, **alcohol**, **balche'**, **beer**, **chicha**, **wine**).

In the Congo, roots of *Alchornea floribunda* are added to palm wine to make a beverage called *niando* (cf. *Alchornea* **spp.**). It has psychoactive and aphrodisiac effects (Scholz and Eigner 1983, 78*). Another palm wine from the Congo is made by adding iboga roots (see *Tabernanthe iboga*). In Ghana, palm wine is mixed with the leaves of *Vernonia conferta*, known as *flakwa*, to produce aphrodisiac effects (Bremness 1995, 29*). In West Africa, the bark of *Mitragyna stipulosa* (DC.) O. Kuntze, which may contain alkaloids of the **yohimbine** type, is drunk with palm wine (cf. *Mitragyna speciosa*). In central Africa, the root bark of *Strychnos icaja* L. (cf. *Strychnos* **spp.**) is added, and in West Africa the bark of *Corynanthe pachyceras* (cf. *Corynanthe* **spp.**).

In India, palm wine was once fortified with the seeds of *Datura metel*. The Sora, a tribal people of Orissa, add an as yet undetermined root to their palm wine, known as *salap*, to give it a cannabis-like effect (Gerhard Heller, pers. comm.).

Literature

See also the entries for *Cocos nucifera*, **alcohol**, and **wine**.

Allen, P. H. 1947. Indians of southeastern Colombia. *Geographical Review* 37 (4): 567–82.

———. 1965. Miscellaneous notes: Coyol wine. *Principes* 9 (2): 66.

Balick, Michael J. 1979a. Amazonian oil palms of promise: A survey. *Economic Botany* 33 (1): 11–28.

———. 1979b. Economic botany of Guahibo. I. Palmae. *Economic Botany* 33 (4): 361–76.

———. 1980. Wallace, Spruce, and Palm Trees of the Amazon: An historical perspective. *Botanical Museum Leaflets* 28 (3): 263–69.

Dahlgren, B. E. 1944. Economic products of palms. *Tropical Woods* 78:10–34.

Davis, T. A. 1972. Tapping the wild date. *Principes* 16 (1): 12–15.

Duke, James A. 1977. Palms as energy sources: A solicitation. *Principes* 21 (2): 60–62.

Faparusi, S. I. 1981. Sugars identified in *Raphia* palm wine. *Food Chemistry* 7:81–86.

Ferguson, W. 1851. Description of the palmyra palm of Ceylon. *Hooker's Journal of Botany* 3:63–64.

Fox, James J. 1981. Der Gebrauch von Palmwein und Palmschnaps in Süd- und Südostasien. In *Rausch und Realität*, ed. G. Völger, 1:182–87. Cologne: Rautenstrauch-Joest-Museum für Völkerkunde.

Freytag, G. F. 1953. The coyol palm as a beverage tree. *Missouri Botanical Garden Bulletin* 41 (3): 47–49.

Hawkes, A. 1946. The mirity palm. *Fairchild Tropical Garden Bulletin* 2 (3): 4–7.

Johnson, D. 1972. The carnauba wax palm (*Copernicia prunifera*). IV. Economic use. *Principes* 16 (4): 128–31.

Lévi-Strauss, Claude. 1950. The use of wild plants in tropical South America. In *Handbook of South American Indians* (B.A.E. Bulletin 143), ed. J. Steward, 465–86. Washington, D.C.: Smithsonian Institution.

———. 1952. The use of wild plants in tropical South America. *Economic Botany* 6 (3): 252–70.

Miller, R. H. 1964. The versatile sugar palm. *Principes* 8 (4): 115–47.

Molisch, H. 1898. Botanische Beobachtungen auf Java. III: Die Sekretion des Palmweines und ihre Ursachen. *Sitzungsberichte der Kaiserlichen Akademie der Wissenschaften in Wien* 107.

Nash, L. J., and C. H. Bornman. 1973. Constituents of ilala wine. *South African Journal of Science* 69:89–90.

Ofakor, N. 1972. Palm-wine yeasts from parts of Nigeria. *Journal of the Science of Food and Agriculture* 23:1399–1407.

Plotkin, Mark J., and Michael J. Balick. 1984. Medicinal uses of South American palms. *Journal of Ethnopharmacology* 10:157–79.

Schaareman, Danker H. 1981. Palmwein im rituellen Gebrauch auf Bali. In *Rausch und Realität*, ed. G. Völger, 1:188–93. Cologne: Rautenstrauch-Joest-Museum für Völkerkunde.

Sigerist, Henry E. 1963. *Der Arzt in der mesopotamischen Kultur*. Esslingen: Robugen.

Vasaniya, P. C. 1966. Palm sugar: A plantation industry in India. *Economic Botany* 20:40–45.

Pituri

Other Names

Bedgery, pedgery, pitchery, pitchuri, pitjuri, pituri-bissen, pituripriem, pituri quid

Pituri, in the broadest sense, is a word given to all plants and all products obtained from these plants (with additives) that the Australian Aborigines chewed or chew for hedonistic and magical purposes. The more recent literature uses *pituri* only to refer to the nightshade *Duboisia hopwoodii* (Horton 1994).

Usually pituri leaves are mixed with alkaline plant ashes and chewed as a quid. In this practice, the chewing of various wild tobacco species (*Nicotiana ingulba*, *N. gossei*, *N. stimulans*, *N. benthamiana*, *N. velutina*, *N. megalosiphon* [cf. *Nicotiana* spp.], and *Goodenia lunata*) has more of a hedonistic character, whereas the chewing of *Duboisia hopwoodii* and *Datura* species[450] is more magical and religious in nature. The smoking of pituri may have developed as a result of exposure to the smoking habits of Europeans (Emboden 1979, 146*).

Pituri takes away hunger and thirst, has inebriating effects, and induces ardent dreams. This is presumably the reason why the Aborigines use(d) pituri as a magical agent. In Aboriginal magic, entering into what is known as the Dreaming, the transcendent and primordial state of being, is of primary significance. The Dreaming is an altered or different state of consciousness:

> Everything in nature is a symbolic footprint of the metaphysical world, through whose actions our world was created. As with a seed, the inherent power of a place is also coupled with the memory of its origin. The Aborigines call this power the *Dreaming*, the dream of the place, and this dream is the foundation for the sacredness of the earth. Only in extraordinary levels of consciousness can one perceive the inner dream of the earth or attune oneself to it. (Lawlor 1993, 1)

It is in the Dreaming that all magical actions that affect the normal state (which is understood to be unreal) are determined and carried out. It appears that there were different types of pituri for different purposes and that the different types were associated with different songs, totems, and the corresponding dream paths or songlines. Several songlines were sung as "pituri paths." There were even pituri clans (Watson et al. 1983, 308). Pituri was considered to be charged with the place of the land; it carries the dream of the place in which it grows and passes this on to humans.

The ritual and hedonistic use of pituri may be the longest continuous cultural use of a psycho-active substance in the history of all humankind. The culture of the Aborigines is the longest continuous culture in the world. It is possible that the Dreaming ancestors of the Aborigines were chewing pituri some forty thousand to sixty thousand years ago (Lawlor 1993).

The Gathering and Preparation of Pituri

Although *Duboisia hopwoodii* and *Nicotiana* spp. are both widespread in Australia, there are nevertheless certain areas that are preferred for gathering the plants. In the ethnographic and ethnobotanical literature, the authors repeatedly express their surprise that the Aborigines of the northern desert regions, where *Duboisia* bushes grow in abundance, nevertheless prefer leaves imported from the distant east. Unfortunately, no sources have been preserved that provide information as to why the Aborigines prefer the leaves from the eastern regions. There were presumably magical reasons, as the gathering of pituri always takes place according to certain songlines. The leaves are charged with the energy of the place or the land on which they grew. (It may be that the Aborigines were psychoactive gourmets who, like other gourmets, preferred cognac over all other brandies.) Before the Aborigines had contact with Europeans, there was an extensive network of trade in the central desert, and the so-called pituri roads on which the prized

Top: Fermented pituri leaves (*Duboisia hopwoodii*) are the basis of pituri quids.

Bottom: This painting by the Aboriginal artist Walangari Karntawarra Jakamarra (Collin McCormick) depicts the pituri plants as round, gray dots. (Detail, oil painting, ca. 1993)

"The natives also stick the chewed [pituri] in their ear, which causes their eyes to take on an unusual shine and their pupils to become very dilated."

CARL HARTWICH
DIE MENSCHLICHEN GENUßMITTEL
[THE HUMAN AGENTS OF PLEASURE]
(1911, 834 f.*)

450 There are six species of *Datura* in Australia, of which only one is native. All the others were introduced either inadvertently or purposely as medicinal and decorative plants. Both the native thorn apple (*Datura leichhardtii* F. Muell.; cf. *Datura* ssp.) and the imported species (*Datura stramonium* L., *Datura stramonium* ssp. *ferox*) were (are?) used as pituri or pituri substitutes (Dowling and McKenzie 1993, 126 ff.*).

Plants Whose Ashes Are Added to Pituri

PROTEACEAE
    Grevillea striata R. BR. (ijinyja)   Meggitt 1966, 126
     O'Connell et al. 1983

MIMOSACEAE (LEGUMINOSAE)
    Acacia aneura F. Muell. ex Benth. (mulga)   Meggitt 1966, 126
     O'Connell et al. 1983
    Acacia coriacea DC. (awintha)   Meggitt 1966, 126
    Acacia kempeana F. Muell. (witchetty bush)   Meggitt 1966, 126
    Acacia lingulata A. Cunn. ex Benth.   Meggitt 1966, 126
    Acacia pruinocarpa   Meggitt 1966, 126
    Acacia salicina Lindl. (cf. *Acacia* spp.)   Aiston 1930, 49

CAESALPINIACEAE (LEGUMINOSAE)
    Cassia spp.   Peterson 1979, 179

RHAMNACEAE
    Ventilago viminalis Hook. (atnyira)   O'Connell et al. 1983
     Lassak and McCarthy 1987, 43*

MYRTACEAE
    Eucalyptus microtheca F. Muell. (angkirra)   O'Connell et al. 1983
    Eucalyptus spp. (gums)[451]   Peterson 1979, 179
    Eucalyptus sp. (red gum)
    Melaleuca sp.

The resins of various species of eucalyptus are added to pituri quids.

451 There are more than seven hundred species of *Eucalyptus* in Australia. Eucalyptus gum (crystallized resin), which is exuded from the bark when the wood is injured, can be obtained from almost all of them. The Aborigines use the gum (= *mijilypa, munuun, arrkiypira, wokalba, jior*) of numerous species primarily for medicinal purposes (Barr 1990, 122–25*; Macpherson 1939).

452 I have discussed this problem extensively with Jonathan Ott, Rob Montgomery, and Manuel Torres, who share the views given here.

453 South American **snuffs** made from *Anadenanthera peregrina* act in a similar manner. The self-experimentation by and reports of C. Manuel Torres indicate that pure *Anadenanthera* powder has almost no effects, while the full DMT effects immediately become apparent when the powder is mixed with ash (C. M. Torres, pers. comm.).

pituri was traded were a part of this (Emboden 1979, 145*).

The very elaborate, complicated, and sophisticated processing methods used in the preparation of botanical raw products, e.g., to detoxify or enhance a plant's activity, are a typical feature of Australian ethnopharmacology (Beck 1992). Simple methods for preparing food and products play only a subordinate role. The Aborigines appear to have possessed a great skillfulness in the processing of medicinal and culinary products, and they spent a great deal of their time practicing these techniques (Isaacs 1987*).

Various ingredients are added to the dried or fermented pituri leaves to produce a quid. Some of these ingredients are plant ashes, while others are binding agents, such as animal hair (from wallabies, euros or wallaroos [a small kangaroo species], or rabbits), plant fibers (*Linum marginale*), yellow ocher, eucalyptus resin, and, in recent times, sugar (Peterson 1979, 179). *Duboisia* leaves can also be chewed by themselves, but the effects are not

considered to be particularly powerful. *Nicotiana* species are always chewed in combination with plant ashes (O'Connell et al. 1983, 108).

All of the plants that are used as sources of ashes contain active constituents. The Aborigines have long used acacia wood, especially that from the mulga acacia, to make boomerangs, spear tips, digging sticks, and shields (Low 1992b, 181*). Acacias (*Acacia* spp.) have been found to contain alkaloids (tryptamines, e.g., *N,N*-DMT). Several acacias, such as *Acacia georginae*, contain the toxic substance fluoroacetate (Dowling and McKenna 1993, 146*). The ashes of both the oceanic acacia (*Acacia manguim*) and *Melaleuca* species contain salts and minerals and are rich in sodium (Ohtsuka et al. 1987). Some acacias, e.g., *Acacia aneura*, also produce a gum that can be added to pituri (O'Connell et al. 1983, 105). Unfortunately, we do not know how the plant ashes were produced. If acacia wood is simply burned, then it can be assumed that the DMT is destroyed by the fire. But if a special process is used to turn the wood into an ashlike substance, such as smoldering the wood at a lower temperature or a similar process, then the DMT may have been preserved, and the concentration may actually have been increased[452] (cf. *Erythoxylum coca*).

The gum-producing genus *Ventilago* has also been found to contain alkaloids (Collins et al. 1990, 61*).

Rock isotome (*Isotoma petraea* Muell.; Campanulaceae) was added to pituri to increase its potency (Low 1990, 192*). Members of the genus *Isotoma* are also used for psychoactive purposes in South America, primarily as additives to **ayahuasca** and **cimora**. Some additives were also used as substitutes for *Duboisia* and tobacco (see the table on page 781).

Actual reports of experiences with pituri are extremely rare. The effects of the various pituri species can differ considerably, and some can be quite powerful, as Gary Thomas has experienced (G. Thomas, pers. comm.). Some appear to have potently and others only weakly stimulating effects; the effects of some are euphoriant and of others visionary. According to the reports of the painter Collin McCormick, the effects of pituri leaves, whether from *Duboisia* or *Nicotiana*, are not very good when they are used alone. Only when combined with ashes are the desired effects brought out.[453] He said that "the ash functions as an amplifier to the pituri." Because pituri quids are often only briefly chewed and then are placed behind or even into the ear, it is possible that the DMT may be absorbed into the bloodstream from this location, which is known to be highly permeable to alkaloids (**scopolamine** patches can be placed behind the ear as a remedy for motion sickness). Perhaps the leaves also contain MAO inhibitors, so the DMT could be orally active (cf. **ayahuasca**). However, it is also likely that the DMT is able to enter the brain directly via the mucous membranes of the mouth, just as it can through those of the nose. In other words, pituri could be a highly effective psychoactive combination drug whose significance has not yet been fully recognized. Future research into pituri may provide some ethnopharmacological sensations.

Literature

See also the entries for *Duboisia hopwoodii*, *Duboisia* spp., and *Nicotiana* spp.

Aiston, Georg. 1930. Magic stones of the tribes east and north-east of Lake Eyre. *Papers and Proceedings of the Royal Society of Tasmania for the Year 1929*: 47–50.

———. 1937. The Aboriginal narcotic pitcheri. *Oceania* 7 (3): 372–77.

Aplin, T. E. H., and J. R. Cannon. 1971. Distribution of alkaloids in some western Australian plants. *Economic Botany* 25:366–80.

Bancroft, J. 1879. *Pituri and tobacco.* Brisbane: Gov. Printer.

Beck, Wendy. 1992. Aboriginal preparation of cycad seeds in Australia. *Economic Botany* 46 (2): 133–47.

Burnum Burnum. 1988. *Burnum Burnum's Aboriginal Australia.* Ed. David Stewart. North Ryde, New South Wales: Angus & Robertson.

Chatwin, Bruce. 1988. *The songlines.* New York: Penguin.

Cleland, J. Burton, and T. Harvey Johnston. 1933/34. The history of the Aboriginal narcotic pituri. *Oceania* 4 (2): 201–23.

Dingle, Tony. 1988. *Aboriginal economy: Patterns of experience.* Ringwood, Victoria: McPhee Gribble/Penguin Book.

Glowczewski, Barbara. 1991. *Träumer der Wüste: Leben mit den Ureinwohnern Australiens.* Vienna: Promedia.

Hamlyn-Harris, R., and F. Smith. 1916. On fish poisoning and poisons employed among the Aborigines of Queensland. *Memoirs of the Queensland Museum* 5:1–22.

Hartwich, Carl. 1910. Über Pituri. *Apotheker-Zeitung.*

Hicks, C. S. 1963. Climatic adaptation and drug habituation of the central Australian Aborigine. *Perspectives in Biology and Medicine* 7:39–57.

Higgin, J. A. 1903. An analysis of the ash of *Acacia salicina. Transactions of the Royal Society of South Australia* 17:202–4.

Horton, David, ed. 1994. *The encyclopaedia of Aboriginal Australia: Aboriginal and Torres Strait Islander history, society and culture.* Canberra: Aboriginal Studies Press for the Australian Institute of Aboriginal and Torres Strait Islander Studies.

Pituri Substitutes

(from Bock 1994, 59*; Low 1990*; supplemented)

Name	Constituent(s)
Centipeda spp. (sneezeweed)	alkaloids
Datura leichhardtii F. v. Muell. ex Benth. (cheeky bugger; cf. *Datura* spp.)	**tropane alkaloids**
Dendrocnide sp. (stinging tree)	
Duboisia myoporoides	**scopolamine** and others
Evolvulus alsinoides L. (speedwell)	alkaloids (?)
Goodenia lunata	alkaloids
Hippobroma longiflora (L.) G. Don[454] [syn. *Isotoma longiflora* (L.) Presl, *Lobelia longiflora* L., *Laurentia longiflora* (L.) Peterm.]	lobeline (cf. *Lobelia inflata*)
Isotoma anethifolia (Summerh.) F.E. Wimm.	lobeline
Isotoma axillaris Lindl. [syn. *Isotoma senecioides* A. DC., *Laurentia axillaris* (Lindl.) F.E. Wimm.]	lobeline
Isotoma petraea F. v. Muell. (rock isotome)	lobeline
Nicotiana spp.	**nicotine**, nornicotine, anabasine
Pterocaulon serrulatus (Montr.) Guill.	alkaloids
Pterocaulon sphacelatum (Labill.) F. v. Muell. (ragwort)	alkaloids
Solanum ellipticum R. Br. (wild tomato; cf. *Solanum* spp.)	?
Trichodesma zeylanicum (bush tobacco, cattle bush)	alkaloids

Johnston, T. H., and J. B. Clelland. 1933. The history of the Aborigine narcotic, pituri. *Oceania* 4 (2): 201–23, 268, 289.

Lawlor, Robert. 1993. *Am Anfang war der Traum: Die Kulturgeschichte der Aborigines.* Munich: Droemer Knaur.

Löffler, Anneliese, ed. 1994. *Australische Märchen: Traumzeitmythen der Aborigines.* Reinbek: Rowohlt.

Macpherson, J. 1939. The *Eucalyptus* in the daily life and medical practice of the Australian Aborigines. *Mankind* 2 (6): 175–80.

Mathews, Janet. 1994. *Opal that turned into fire.* Broome, Wash.: Magabala Books.

Meggit, M. J. 1966. Gadjari among Walpiri Aborigines of central Australia. *Oceania* 37:124–47.

O'Connell, James F., Peter K. Latz, and Peggy Barnett. 1983. Traditional and modern plant use among the Alyawara of central Australia. *Economic Botany* 27 (1): 80–109.

Ohtsuka, Ryutaro, Tsuguyoshi Suzuki, and Masatoshi Morita. 1987. Sodium-rich tree ash as a native salt source. *Economic Botany* 41 (1): 55–59.

Peeters, Alice. 1968. Les plantes masticatoires d'Australie. *Journal d'Agriculture Tropicale et de Botanique Appliquée* 15 (4/5/6): 157–71.

Peterson, Nicolas. 1979. Aboriginal uses of Australian Solanaceae. In *The biology and taxonomy of the Solanaceae*, ed. J. G. Hawkes et al., 171–89. London: Academic Press.

454 This species is originally from the West Indies (Zander 1994, 312*).

Spencer, B., and F. J. Gillen. 1899. *Native tribes of central Australia.* London: Macmillan.

Thomson, D. F. 1939. Notes on the smoking pipes of North Queensland and the Northern Territory of Australia. *Man* 39:81–91.

Watson, Pamela. 1983. *This precious foliage: A study of the Aboriginal psychoactive drug pituri.* Oceania Monograph 26. Sydney: University of Sydney Press.

Watson, P.[amela], L. O. Luanratana, and W. J. Griffin. 1983. The ethnopharmacology of pituri. *Journal of Ethnopharmacology* 8 (3): 303–11.

Sake

"The storax-scented wine . . . is very well suited for bringing the five organs [heart, liver, spleen, lungs, kidneys] into harmony with one another and for dispelling all diseases of the stomach. If one awakens in the morning with a cold, one should always drink a cup of this wine."

SHEN KUO
PINSELUNTERHALTUNGEN AM TRAUMBACH [BRUSH DISCUSSIONS BY THE DREAM STREAM] (1997, 68*)

Traditional sake vessels in Kamakura, Japan.

455 The chrysanthemum flower is a symbol of long life. A "liquor of longevity" is prepared by adding the flowers (flos chrysanthemi), mixed with *Polygonum multiflorum* Thunb., *Lycium chinense* Mill., and red dates (*Zizyphus jujuba* Mill.), to sake or rice schnapps. Chrysanthemum flowers contain flavone derivatives, sesquiterpenoids, amino acids, borneol, camphor (cf. *Cinnamomum camphora*), chrysantheone, and other substances (Paulus and Ding 1987, 183 f.*).

Other Names

Chongha, chongjung, ju, kukhuaju, makoli, reisbier, reiswein, rice beer, rice wine, saké, saki, taenju, tong dong ju, "wine"

Sake is brewed from water, rice (*Oryza sativa* L.), yeast (*Saccharomyces cerevisiae*), and the koji fungus (*Aspergillus oryzae*). A variety of methods may be used to prepare the hulled rice, after which it is mashed. Koji is then added, which transforms the rice's starch into sugar. The rice is then mixed with water and fermented with the help of yeast. The alcohol content of the finished drink depends upon how much of the starch has been transformed into sugar. If 40% of the starch becomes sugar, the sake will contain 20% ethanol (= **alcohol**) (Kondo 1992, 42 f.). In former times, a *kuchikami no sake*, "sake chewed in the mouth," was made using a very archaic method: Rice, chestnuts (*Castanea sativa* Mill.), and millet were chewed thoroughly and spat into a trough of water. After a few days, the mix would finish fermenting and be ready to drink. Sake is actually more like **beer** than **wine**. In Korea, sakelike drinks and their mashes have been distilled into high-proof rice schnapps for some five hundred years.

In ancient China, alcoholic drinks (*chiu*) made from rice were already known during the Neolithic period (over four thousand years ago). During the Chou era, rice wine was drunk as a gift of offering in the ancester cult. This wine was brewed with the addition of wormwood (*Artemisia absinthium*) or dogwood (*Cornus* spp.) (Majlis 1981, 314). Some-

times saffron (*Crocus sativa*), ginger (*Zingiber officinale*), or the seeds of *Datura metel* were also used.

The art of sake brewing originated in China, rapidly spread to Korea, and was introduced into Japan around the seventh century. There, it has become a national drink and has retained its position in the world of inebriating drinks as a typically Japanese specialty.

In Japan, sake brewing falls under the protection of Matsuno'o, the god of sake (Kondo 1992, 26). The origins of the drink are traced back to the god Susanoonomikata, the brother of the sun goddess Amaterasu. He is said to have invented the inebriating drink in order to sedate the great snake of Lake Yamata so that he could kill it with his magical sword. Originally, sake was brewed as a "drink of the gods," only for festivals and rituals of the Shinto cult. Great quantities were offered so that the gods would look with favor upon humans. When people drank it, they felt "like gods." In contemporary Japan, sake is still also known as "nirvana wine," because drinking it is supposed to lead one to nirvana. Over the course of history, certain sake-drinking rituals have evolved that bear a strong resemblance to the tea ceremony (cf. *Camellia sinensis*). But today, sake is enjoyed primarily as a part of profane life.

There are many varieties of sake, from dry to sweet, of which some are drunk ice cold and others lukewarm, warm, or hot. Dry, high-quality sake is always enjoyed cold or ice cold. Lower-quality sakes are heated so that the poor taste will not be apparent.

Sake was sometimes brewed with additives. A "black sake" was made with the addition of ashes from aromatic woods (presumably lignum aloe; cf. **incense**) and a *kikuzaké* ("chrysanthemum sake") with the addition of chrysanthemum blossoms (*Chrysanthemum morifolium* Ramat.)[455] (Kondo 1992, 17). Sake also may have once been brewed with *Phytolacca acinosa* (pokeweed). In tenth- and eleventh-century China, a "medicinal wine" was made from sake and *sû hé xîang* (= storax, the resin of *Liquidambar orientalis* Mill.; cf. **incense**), to which the dried rhizome of *bai zhu* (*Atractylodes macrocephala* Koidz.) was also added. One of the uses of this storax wine was to restore consciousness to a person who had passed out (Shen Kuo 1997, 68, 261*; cf. **han-shi**).

In Japan sake is often added to green tea (*Camellia sinensis*) during the cold months. In Korea, all sakelike drinks, and in particular the milky, mildly alcoholic *makoli*, play a significant role in the native shamanism, which continues to thrive in spite of thousands of years of suppression. *Taenju*, "wine," is a central offering (libation) in all Korean shamanic ceremonies (Cho 1982, 107, 117). Because of the copious consumption of *makoli* and ecstatic dancing, powerful altered states of consciousness are often experienced during these offering ceremonies. A document from the thirteenth century notes:

> During *Wei* (Korean *Ye*), people made offerings to the heavens at the October ceremony and drank, sang, and danced day and night. They called their ceremony *wutian* [Korean *much'on* = dance (in honor) of the heavens]. (Cho 1982, 12)

It is possible that in the past, a psychoactive mushroom (e.g., **Amanita pantherina**, **Panaeolus subbalteatus**, or **Psilocybe** spp.) may have been added to the *makoli*. After all, the "mushroom of

A view of a Japanese sake brewery (woodcut from *Nihon Sankai Meisan Zu-e*, ca. 1800).

immortality" is regarded as a gift of the *tengu*, the shamanic mountain deity.

Literature

See also the entries for **beer**.

Cho, Hung-Youn. 1982. *Koreanischer Schamanismus: Eine Einführung.* Hamburg: Hamburgisches Museum für Völkerkunde.

Kondo, Hiroshi. 1992. *Saké: A drinker's guide.* Tokyo, New York, and London: Kodansha International.

Majlis, Brigitte. 1981. Alkoholische Getränke im alten China. In *Rausch und Realität*, ed. G. Völger, 1:314–319. Cologne: Rautenstrauch-Joest-Museum für Völkerkunde.

Smoking Blends

Other Names

Blends, rauchkräuter, rauchmischungen, smoking mixtures

Almost any plant can be smoked after it has been dried. Many psychoactive plants are smoked alone and unblended. Usually, however, one or more other herbs or extracts are combined with them. Herbs are very often mixed to produce specific psychoactive effects. Many smoking blends are important components of shamanic rituals or social situations.

In Amazonia, mixtures of resins from different *Virola* species (*Virola* spp.) and tobacco (*Nicotiana tabacum*) are ritually smoked. Mataco shamans mix the seeds of *Anadenanthera colubrina* var. *cebil* with tobacco (*Nicotiana* spp.) and an *Amaranthus* species so that they can diagnose and heal illnesses. In Mexico, Mayan shamans smoke blends of tobacco (*Nicotiana tabacum*) and thorn apple (*Datura innoxia*). The Huichol use a combination of wild tobacco (*Nicotiana rustica*) and marigold (*Tagetes lucida*). The Mam and Tzeltal Indians smoke wild tobacco together with fly agaric skins (*Amanita muscaria*) for divinatory purposes. Countless blends are used in North America (see

kinnikinnick). The sadhus in India and Nepal are fond of blending hemp products (*Cannabis indica*) with *Datura metel*, *Aconitum ferox*, and cobra venom. In central Asia, henbane (*Hyoscyamus niger*, **H. spp.**) is mixed with tobacco or hemp (*Cannabis sativa*, *Cannabis ruderalis*). In Pakistan and North Africa, it is common to smoke hashish with tobacco (*Nicotiana tabacum*). In Southeast Asia, cloves (*Syzygium aromaticum*) are added to tobacco, and clove cigarettes are even manufactured in factories. In Australia, **pituri** is occasionally smoked. Sometimes hemp leaves are soaked in opium (*Papaver somniferum*) and a tincture of lobelia (*Lobelia inflata*) for smoking.

Recipes

In fact, of course, most herbs can be combined with one another for smoking. However, one should nevertheless proceed with caution when experimenting and begin with small dosages (cf. **kinnikinnick**).

Shiva/Shakti Blend

Equal parts of:
ganja (hemp leaves; *Cannabis indica*)
dhatura (thorn apple leaves; *Datura metel*)

"The smoke of dried coltsfoot with root, sucked in by means of a tube, is said to heal an old cough, but one must take a sip of raisin wine after each inhalation."

PLINY
(26.36)

"Each of the different smoking blends allows one to strike a different key on the piano of consciousness states!"

A CONSUMER
IN *ISOLDENS LIEBESTRANK* [ISOLDE'S LOVE DRINK]
(MÜLLER-EBELING AND RÄTSCH 1986, 166*)

Psychoactive Smoking Herbs

Name	Botanical Name	Active Constituent(s)
aconite herbage	*Aconitum ferox*	aconitine
angel's trumpet leaves	*Brugmansia* spp.	tropanes
ayahuasca leaves	*Banisteriopsis caapi*	**harmaline, harmine**
bearberry leaves	*Arctostaphylos uva-ursi*	arbutin
belladonna	*Atropa belladonna*	atropine
catnip	*Nepeta cataria*	essential oil
cebil seeds	*Anadenanthera colubrina*	bufotenine
coca del suri	*Urmenetea atacamensis*	?
coca leaves	*Erythroxylum coca*	cocaine
coleus	*Coleus* spp.	diterpenes
coro	*Trichocline* spp.	?
cup of gold leaves	*Solandra* spp.	**tropane alkaloids**
damiana	*Turnera diffusa*	essential oil
desfontainia leaves	*Desfontainia spinosa*	?
devil's tobacco	*Lobelia tupa*	lobeline and others
ephedra herbage	*Ephedra* spp.	ephedrine
fly agaric skins	*Amanita muscaria*	**muscimol**
hawkweed	*Hieracium pilosella*	umbelliferone
henbane	*Hyoscyamus* spp.	**tropane alkaloids**
justicia leaves	*Justicia pectoralis*	**coumarin**, tryptamines
khat leaves	*Catha edulis*	[cathinone], cathine
kougoed	*Sceletium tortuosum*	mesembrine and others
latua leaves	*Latua pubiflora*	**atropine, scopolamine**
lobelia herbage	*Lobelia inflata*	lobeline
magic mushrooms	*Psilocybe* spp.	**psilocybin**
mandrake leaves	*Mandragora officinarum*	**tropane alkaloids**
marigold herbage	*Tagetes* spp.	essential oil
mugwort	*Artemisia vulgaris*	essential oil
passionflower herbage	*Passiflora incarnata*	β-carbolines
peyote pieces	*Lophophora williamsii*	mescaline
pituri	*Duboisia hopwoodii*	nornicotine
poppy capsules	*Papaver somniferum*	**opium alkaloids**
prairie mugwort (sage)	*Artemisia* spp.	essential oil
reed	*Phalaris arundinacea*	*N,N*-DMT
San Pedro cactus	*Trichocereus pachanoi*	mescaline
scopolia leaves	*Scopolia carniolica*	scopolamine
Scotch broom	*Cytisus scoparius*	sparteine
ska maría pastora	*Salvia divinorum*	**salvinorin A**
Syrian rue seeds	*Peganum harmala*	**harmaline, harmine**
thorn apple leaves	*Datura* spp.	tropanes
toad foam	*Bufo alvarius*	5-MeO-DMT
tobacco	*Nicotiana rustica*	nicotine
	Nicotiana tabacum	nicotine
	Nicotiana spp.	anabasine, **nicotine**, and others
white hellebore leaves	*Veratrum album*	steroid alkaloids
wild lettuce	*Lactuca virosa*	lactucarium
wormwood herbage	*Artemisia absinthium*	thujone
	Artemisia mexicana	essential oil

Smoking Herbs Whose Psychoactive Effects Are Questionable

Name	Botanical Name	Active Constituent(s)
balm	*Melissa officinalis*	essential oil
basil	*Ocimum basilicum*	essential oil
black tea	*Camellia sinensis*	theine (= caffeine)
bog bilberry leaves	*Vaccinium uliginosum*	arbutin
cinnamon bark	*Cinnamomum verum*	essential oil
coltsfoot	*Tussilago farfara*	mucins
mints	*Mentha* spp.	essential oil
oregano	*Origanum vulgare*	essential oil
sage	*Salvia officinalis*	essential oil
stinging nettle	*Urtica dioica*	histamine

Indian Cigarettes (nineteenth century)

In the nineteenth century, pharmacies sold numerous pharmaceutical smoking blends that were prerolled into cigarettes. This recipe of the Parisian company Grimault et Cie. from 1870 is strongly reminiscent of **witches' ointments.** Per cigarette:

0.3 g belladonna leaves (*Atropa belladonna*)
0.15 g henbane leaves (*Hyoscyamus niger*)
0.15 g thorn apple leaves (*Datura stramonium*)
0.1 g Indian hemp leaves (*Cannabis indica*), soaked in opium extract and cherry laurel schnapps

Smoke one cigarette as needed.

Neumeier's Cigarillos (1913)

These pharmaceutical cigars were smoked as a remedy for asthma. Unfortunately, no amounts were given with the ingredients. They consisted of:

herba and radix brachycladi (*Trichocline argentea* Grisebach [syn. *Brachyclados stuckeri* Speg.]; cf. *Trichocline* spp.)
Cannabis indica
Grindelia robusta Nutt.
folia eucalypti (eucalyptus leaves)
folia stramoni (thorn apple leaves)
an outer leaf of *Nicotiana tabacum*

Smoke as needed.

Yuba Gold

In the international herb trade, one can even find finished herbal blends based upon hemp products. One popular product is Yuba Gold, which consists of (Miller 1985, 23*):

4 parts damiana leaf (*Turnera diffusa*)
4 parts skullcap herb (*Scutellaria lateriflora* L.)
$\frac{1}{4}$ part lobelia herb (*Lobelia inflata*)
4 parts passionflower herb (*Passiflora incarnata*; cf. *Passiflora* spp.)
1 part spearmint leaf (*Mentha spicata*)

Legal Grass

This mixture is sold as a marijuana substitute (Brown and Malone 1978, 23*). It consists of:

Korean ginseng leaves (*Panax ginseng*)
damiana (*Turnera diffusa/ Turnera aphrodisiaca*)
high-grade lobelia herb (*Lobelia inflata*)
African yohimbe bark (*Pausinystalia yohimba*)
hops (*Humulus lupulus*)

Creative Euphoria

Equal parts of:

marijuana (*Cannabis indica, Cannabis sativa*)
damiana (*Turnera diffusa*)
fly agaric mushroom (*Amanita muscaria*)
ska maría pastora (*Salvia divinorum*)
yohimbe bark (*Pausinystalia yohimba*)

"Legal High" Blend

Equal parts of:

leonotis herbage (*Leonotis leonurus*)
bog bilberry leaves (*Vaccinium uliginosum*)
aristolochia herbage (*Aristolochia triangularis*)
papaya leaves (*Carica papaya* L.)
marsh marigold herbage (*Caltha palustris* L.)

Hottentot Tobacco

Equal parts of:

kougoed (*Sceletium tortuosum*)
dagga (*Cannabis sativa*)

Aphrodisiac Smoking Blend I

Equal parts of:

hashish (*Cannabis indica, Cannabis sativa*)
fly agaric mushroom, dried (*Amanita muscaria*)
coca leaves, toasted (*Erythroxylum coca* var. *coca*)

Aphrodisiac Smoking Blend II

Equal parts of:

hashish (*Cannabis indica, Cannabis sativa*)
fly agaric mushroom, dried (*Amanita muscaria*)
thorn apple leaves (*Datura innoxia, D. stramonium, D.* spp.)

Mixture for Peyote Rituals

(cf. *Lophophora williamsii*, kinnikinnick)

Bull Durham tobacco (*Nicotiana tabacum*)
mokola leaves (*Rhus glabra* L.)

Hataj Mixture

5–8 hataj seeds (*Anadenanthera colubrina* var. *cebil*)
1 cigarette (*Nicotiana tabacum*)
1 pinch of aroma (*Amaranthus* sp.)
some aromo (*Mimosa* spp.)

"Mazatec Blend"

Equal parts of:

coleus (*Coleus blumei*)
ska maría pastora (*Salvia divinorum*)

Blending smoking herbs can clearly produce synergistic effects that may be desirable to a smoker (cf. Siegel 1976*). Some consumers have even begun to practice what may be termed "chemical engineering," in which certain blends, combinations, and proportions of mixtures are designed to steer the specific effects of hemp in a particular direction. For example, combinations with tropane-rich nightshades (henbane, thorn apple, angel's trumpet) are smoked for aphrodisiac purposes. To produce more profound psychedelic effects, hemp products may be combined with belladonna (*Atropa belladonna*) and fly agaric mushrooms (*Amanita muscaria*) (belladonna and

Coltsfoot leaves (*Tussilago farfara*) were once smoked as a tobacco substitute. Today, they are often mixed with hashish and used in medicinal and psychoactive smoking blends. (Woodcut from Tabernaemontanus, *Neu Vollkommen Kräuter-Buch*, 1731)

fly agaric mushrooms have a positive synergy). Kitchen spices, coca leaves (*Erythroxylum coca*), and *Ephedra* herbage can be used for refreshing and invigorating effects. Some ingredients, e.g., *Psilocybe* mushrooms, are thought to have no pharmacological effect when smoked.[456] Others, such as toad foam (cf. **bufotenine, 5-MeO-DMT**), are extremely potent and can clearly obscure the effects of THC. The effects of many *Artemisia* spp. (*Artemisia* **spp.**) are antiasthmatic and muscle relaxing and are suitable for use in smoking blends, and it is possible that they may improve the absorption of the active constituents of the other ingredients in the blends. Continuing ethnopharmacological research into smoking herbs and mixtures will certainly provide many stimulating insights.

Literature

Golowin, Sergius, ed. 1982. *Kult und Brauch der Kräuterpfeife in Europa.* Dokumente zur einheimischen Ethnologie. Allmendingen: Verlag der Melusine.

Ohsawa, George, Herman Aihara, and Fred Pulver. 1985. *Rauchen, Marihuana und Drogen.* Holthausen and Münster: Verlag Mahajiva.

Snuffs

"I felt how every single grain shot up through my nostrils and then exploded in my skull. Gradually, a wonderful weariness spread through my body. I turned my gaze to the river and nearly expected to see a mythical creature there rising from the depths."

FLORINDA DONNER
SHABONO
(1985)

Other Names

Polvo alucinogeno, polvo psicoactivo, rapé, rapé dos Indios, rapé halucinogênico, schnupfdrogen, schnupfpulver, sternutators

A number of substances are traditionally ingested via the mucous membranes of the nasal passages by snuffing or inhaling for medicinal, ritual, shamanic, or hedonistic purposes. It is difficult to determine when humans first began the practice of snuffing. It probably originated around the time that milling stones or other milling techniques were being invented. Usually, very finely ground powders are used.

In the Old World, we know of only the occasional use of "sneezing powders"[457] (sternutators) (cf. *Veratrum album*). In India, various plants, including hemp (see *Cannabis indica*), are used as medicinal snuffs. The same is true in Africa (see *Mesembryanthemum* spp., *Sceletium tortuosum*).

In North America, a shamanic-ritual use has been demonstrated for only a few snuffs. In the Pacific Northwest, powders from polypores were snuffed during shamanic healing ceremonies (cf. "*Polyporus mysticus*"). In the forest regions of the Northeast, the ritual use of a snuff made of calamus roots was widespread (cf. *Acorus calamus*). We have no evidence indicating that tobacco was snuffed in North America. Although the practice of snuffing tobacco is now known throughout the world, it is actually seldom practiced.

South America is the center of the use of psychoactive snuffs, although such use has also been confirmed for many Caribbean islands (Hispaniola, Greater Antilles). Christopher Columbus was the first to describe the use of *cohoba* among the Taino Indians. Taino healers placed the powder on the heads of the *cenis*, the wooden figures of their deities. From there, they would use a tube to draw it into their noses, after which they would question the gods about the origins of diseases (Torres 1998; Torres n.d.). The first mention of the use of snuff powders in the Amazon came from the missionary Fray Pedro de Aguado in 1560 (Torres et al. 1991, 645).

Numerous objects that were used as snuff trays are known from the South America Andes region (Wassén 1985). Most of these are carved from wood, but some are made of such exotic materials as (fossilized or partially fossilized) whale bones. Most of these two-thousand-year-old snuff trays were found in San Pedro de Atacama (Chile), located at an altitude of 2,450 meters, and in neighboring areas. To date, the excavations of approximately five thousand graves have yielded 612 snuff kits. These usually consist of a woolen bag with a four-sided snuff tray, a snuffing tube of wood or bone,[458] a small spoon, a small mortar with a pestle, and one or more leather bags containing snuff (Torres et al. 1991, 641; Cornejo B. 1994; Núñez A. 1969). In addition to the snuff, many of the leather bags contained a tiny bag of crushed malachite. The snuff has usually been preserved as an amorphous mass, although one bag yielded seeds that clearly belong to the *Anadenanthera* family. Chemical analyses of two samples of the amorphous masses verified the presence of the tryptamines *N,N*-DMT, 5-MeO-DMT, and **bufotenine** (Torres et al. 1991, 643). These findings suggest that the snuff was produced from *Anadenathera colubrina* var. *cebil*.

The ritual use of psychoactive snuffs appears to have spread north from South America, a fact attested to by both the age and the geographic concentration of the archaeological finds of snuff tools (Torres et al. 1991). The oldest known snuff device dates to 3000 B.C.E., while the oldest snuff tray found thus far has been dated to

456 Nevertheless, some experimenters have repeatedly reported experiencing mild psychoactive effects from smoking *Psilocybe* mushrooms.

457 The medicinal use of sneezing powder is also found in South America, where the powdered bark of *Myrica stornatatoria* (nom. nud.!) is snuffed to clear the mind and to treat headaches (Schultes 1980, 92*).

458 Thorns from *Trichocereus atacamensis*, which are more than 20 cm long, were used to clean the tubes and prevent them from becoming clogged (cf. *Trichocereus* spp.).

1200 B.C.E. The iconography of this snuff paraphernalia incorporates all the elements of a shamanic worldview, including animal spirits, chimeras, erotic scenes, deities, winged beings, et cetera (Torres 1987a, 1987b, 1988). The iconography is strongly reminiscent of that of the later Tiahuanaco (= Tiwanaku) culture, where the use of snuffs is well documented (Berenguer 1987; Torres et al. 1991, 646; Wassén 1972). Snuff paraphernalia in the Tiahuanaco style appears in the region of the southern Andes and has been dated from 300 B.C.E. It can thus be assumed that the use of psychoactive snuffs was closely interwoven with the cultural development of this region (Torres 1993; cf. also Boetzkes et al. 1986, 62). The snuff trays from the late phase (Incan period) are nothing more than crude imitations of the tools from the archaic period. However, in the eighteenth century the art of the snuff tray flowered in Brazil once again (Wassén 1983).

The manner in which these snuff tools are used is clear. The powder is laid out in lines on the tray and snuffed into the nose through a tube. In general, the snuff powders are used to make contact with the higher reality and with beings that are ordinarily invisible. Sometimes hunters will take snuff in order to *see* where the game is hiding. The hunter will transform himself into an eagle, a condor, a jaguar; in this form, he will fly through the air or pace through the primordial forest and *see* the game.

The shamanic use of psychoactive snuffs is also widespread among the Indians of the Amazon and Orinoco regions. Here, snuff trays are not used; instead, many different kinds of tubes are used through which a person blows the powder into his own nose or into another's nose. There are also Y-shaped snuffing tubes made from hollow plant stalks that a person can use to snuff powder through both nostrils at the same time. Snuff tubes made from snails are a special invention. These are made from the shells of large land or freshwater snails (e.g., *Ampullaria* sp., *Strophocheilus* spp., *Poulimus gallina* Sultana, *Helix terrestris*) (Zerries 1980, 174, table 86; Wassén 1965, 62 ff.; Wassén 1967, 119). The apex (= tip, embryonic spire) of the snail shell is removed. The resulting opening is extended with a bone tube or hollow plant stalk. The powder is placed in the orifice of the snail shell, and the tube is then placed in a nostril and the snuff inhaled. Snail shells are also used to store snuff. Numerous snails from the genus *Strophocheilus* have been found in graves at San Pedro de Atacama (Llagostera et al. 1988, 93). Since these snails are from the tropical Chaco region, this may indicate that the shells were imported along with the seeds of **Anadenanthera colubrina**.

The Desana, a Tukano-speaking tribe in Colombia, call their snuff powder *vihó*. It is usually

Top: This pre-Columbian snuff tray from San Pedro de Atacama is decorated with a jaguar, the most important shamanic animal, and fits nicely in the hand for snuffing. (Universidad del Norte Museum, San Pedro de Atacama)

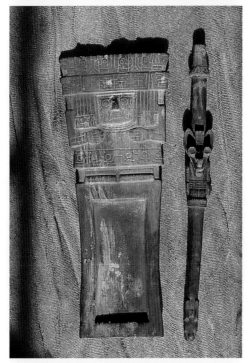

Center: Snuff tools, consisting of a tray and a snuff tube decorated with a "psychedelic" llama. (Universidad del Norte Museum, San Pedro de Atacama)

Bottom: The Atacameños (Kunza) used the stems of the plant called *sorona, peril,* or *brea* (*Tessaria absinthioides* [H. et A.] DC.; Compositae) as snuff tubes. (Wild plant, photographed in the San Pedro de Atacama, northern Chile)

In ancient times, the black hellebore or Christmas rose (*Helleborus niger* L.; Ranunculaceae) was used to prepare a medicinal sneezing powder. However, these ancient *Helleborus* snuffs were likely not psychoactive. (Woodcut from Fuchs, *Kreüterbuch,* 1543)

"The ingestion of the hallucinogens had begun: gray-, green-, and khaki-colored powders waltz from one nostril to the next. Several people stand up, suddenly gripped by nausea, and vomit in the central plaza; emaciated dogs then come over and lick up this unexpected manna. Others have tears in their eyes, they spit out thick spittle that they can expel from their mouths only with great effort. Kremoanawe is dizzy. He sees fantastic landscapes dipped in orange, red, carmine, or scarlet. The round house, shaken by spasms, bends grotesquely. All at once a tornado of blood rises and floods everything, beings and things. Strange and horrifying people appear and dissolve again."

Jacques Lizot
Im Kreis der Feuer [In the Circle of Fire]
(1982, 118)

459 2-CB (4-bromo-2,5-dimethoxyphenethylamine), which produces effects somewhere between those of **mescaline** and MDMA, was discovered by Alexander Shulgin (cf. Shulgin and Shulgin 1991, 503 ff.).

produced from various species of *Virola*. According to the mythology of the Desana, the sun created Vihó-mahse, "snuff-powder being," at the beginning of time so that it could establish contact between humans and the creator (the sun god) through hallucinations. The actual snuff was the property of the sun god himself, who kept it hidden in his navel (from whom was his umbilical cord separated?), until his daughter scratched him there and discovered the powder. Vihó-mahse normally dwells in the Milky Way, the "blue zone" of hallucinations and visions, and is in a permanent trance. Using the earthly snuff, the shaman (*payé*) is able to reach this place and contact Vihó-mahse (Reichel-Dolmatoff 1971*).

The use of powerful psychedelic snuff powders is widespread among the Yanomamö (= Waika) of the Orinoco region and northern Brazil (Brewer-Carias and Steyermark 1976; Chagnon 1977; Chagnon 1994; Donner 1985; Lizot 1982). Most of the men—and not only the shamans—take *epena* or *ebene* daily. Even five- to six-year-old boys are allowed to snuff the powder, in accordance with the motto: Practice early if you wish to become a master. Women are not allowed to use snuff. The Yanomamö believe that spirit beings (*hekura* or *hekula*) dwell in their chests, as well as under cliffs and in mountains, and that the snuff aids them in contacting these beings (Brewer-Carias and Steyermark 1976, 63; Goetz 1970, 45; Henley 1995).

The shamans of the rain forest often take snuff before they treat a patient; this enables them to better *see* the origins of a disease. Sometimes (for example, among the Sanama), snuff is ingested collectively during funerals (Prance 1970, 62*).

An Amazonian snuff plays a central role in the novel *The Emerald Forest*, which was made into a motion picture by John Boorman. Both the novel and the film show the cultural significance of the visions or journeys to other realities facilitated by the psychoactive snuff (Holdstock 1986, 117 f., 152, 163 ff., 190 f.).

Peter T. Furst has proposed the theory that psychoactive snuffs were used in pre-Hispanic Mexico, but that the knowledge of their use had disappeared by the time of the Conquest (Furst 1974). The actual functions of the archaeological objects Furst draws upon to support his theory are disputed. The snuff expert Manuel Torres has called Furst's theory into question on the grounds that, for example, the size of the objects in question made it impossible for them to fit them into a person's nostril. But according to Torres (pers. comm.), an object from Guerrero that is now housed in the Museo Nacionál de Antropolgía e Historia (Mexico City) may be a snuff tray. The discovery of a hollow ceramic figure from Colima first established proof that snuffing (of whatever substances) must have been known in Mesoamerica. In addition, certain Olmec pieces pos-

sibly were used as snuff trays (Furst 1996, 77, 78*). Furst regards the following plants as candidates for Mesoamerican snuffs: peyote (*Lophophora williamsii*), ololiuqui (*Turbina corymbosa*), Piptadenia flava (Spreng.) Benth., *Piptadenia constricta* (Mich. et Rose) Macbride, **Mimosa spp.**, **Acacia spp.**, **Psychotria spp.**, and *Justicia pectoralis* (Furst 1974, 3 f.). It is also conceivable that *Virola guatemalensis* was used.

Today, the snuffing of more or less pure **cocaine** is a worldwide phenomenon. Other psychoactive substances are also snuffed, including the synthetic phenethylamine 2-CB,[459] MDMA (ecstasy or XTC), **DMT**, **scopolamine**, and crystallized ketamine (cf. Höhle et al. 1986, 65*; de Smet 1985, 102). As part of the "back to nature" movement, herb sellers have even begun blending psychoactive snuffs from legal ingredients. For instance, a mixture sold under the name Storm's Breath consists of kava-kava (*Piper methysticum*), cola nut (*Cola* spp.), guaraná (*Paullinia cupana*), nutmeg (*Myristica fragrans*), and cinnamon (*Cinnamomum verum*); its effects are mildly stimulating and, surprisingly, not especially irritating.

Traditional Snuff Recipes

Amazonian Snuff Powder
Roasted seeds of **Anadenanthera peregrina** are ground very fine with tobacco and ash. Most people who try this powder experience extremely strong (allergic) reactions or pain. No one who has tried it has expressed a desire to try it again.

Shinā or Tsinā
This snuff powder is produced by the Jamamadis and Denís of the Brazilian Amazon. It consists of equal parts of roasted **Nicotiana tabacum** leaves and ashes from the bark of *Theobroma subincanum* Mart. and other *Theobroma* species (known as *cacau*). Both are ground fine and snuffed in the evening (Prance 1972b, 221*).

Baduhu-tsinā
The word means "deer snuff." This snuff is made from a lichen (*Pyrenocarpus*) that grows on trees. The effects are more like those of a sneezing powder than those of a psychoactive shamanic drug (Prance 1972a, 16*; 1972b, 227*).

Yanomamö/Waika Snuff

Epena, Ebena, Ebene
The Yanomamö make a potent psychoactive snuff from the bark of *Virola theiodora* or *Virola elongata* and the leaves of **Justicia pectoralis**. The active constituent is the *Virola*; the *Justicia* leaves impart a more pleasant aroma to the powder and also appear to make it easier for the snuff to be absorbed through the nose (Prance 1972, 234 f.*). The ashes of the magnificent *Elizabetha princeps*

Plants Used in Snuffs

Family	Botanical Name	Part(s) Used
Acanthaceae	*Justicia pectoralis* var. *stenophylla*	leaves
Araceae	*Acorus calamus* L.	rhizome
Bignoniaceae	*Tanaecium nocturnum* (B.-R.) Bureau et Schum.	leaves
Convolvulaceae	*Ipomoea guineense*	root bark
	Ipomoea mauritiana Jacq. (see *Securidaca longepedunculata*)	roots
Ephedraceae	*Ephedra gerardiana* var. *saxatilis*	ashes
Ericaceae	*Rhododendron* spp.	bark
Erythroxylaceae	*Erythroxylum coca*	leaves
	Erythroxylum novogranatense	leaves
	Erythroxylum spp.	leaves
Euphorbiaceae	*Manihot esculenta* Crantz	flour from roots
Lecythidaceae	Spec. non id.	bark
Leguminosae	*Anadenanthera colubrina* var. *cebil*	seeds
	Anadenanthera colubrina var. *colubrina*	seeds
	Anadenanthera peregrina var. *falcata*	seeds[460]
	Anadenanthera peregrina var. *peregrina*	seeds
	Anadenanthera spp.	seeds, bark
	Calliandra anomala	resin
	Elizabetha leiogyne	bark[461]
	Elizabetha princeps Schomb. ex Benth.	bark
	Erythrina falcata Benth.	seeds (?)
	Mimosa acacioides	seeds, bark
	Piptadenia excelsa (Gris.) Lillo[462]	pods, seeds
	Piptadenia macrocarpa Benth.[463]	bark, seeds, pods
Lichenes	*Pyrenocarpus* sp.	entire plant (lichen)
Liliaceae	*Veratrum album*	roots
Magnoliaceae	*Magnolia virginiana* L.	leaves, bark
Malpighiaceae	*Banisteriopsis caapi*	bark[464]
Meliaceae	*Trichilia* sp.	juice (?)[465]
Mesembryanthemaceae	*Mesembryanthemum* spp.	roots
	Rabaiea albinota (Haw.) N.E. Br. [syn. *Nananthus albinotus* N.E. Br.]	herbage
	Sceletium tortuosum	herbage/roots
	Sceletium spp.	
Moraceae	*Cecropia* spp.	ashes
	Maquira sclerophylla Ducke	seeds, bark (fruits)
Myristicaceae	*Iryanthera juruensis* Warb.	resin
	Myristica fragrans	nutmeg
	Virola calophylla Warb.	resin
	Virola calophylloidea Markgr.	resin
	Virola elongata (Spruce ex Benth.) Warb. [syn. *Virola cuspidata* (Spruce ex Benth.) Warb., *Virola rufula* Warb.]	resin
	Virola loretensis A.C. Smith	resin
	Virola pavonis (DC.) A.C. Smith	resin
	Virola surinamensis (Rol.) Warb.	cambium
	Virola theiodora (Spruce ex Benth.) Warb.	resin
Piperaceae	*Piper interitum* Trel.	leaves, roots
	Piper sp.	leaves
Polygalaceae	*Securidaca longepedunculata*	roots
Rubiaceae	*Pagamea macrophylla* Spruce ex Benth.	leaves[466]
Solanaceae	*Brugmansia* spp.	leaves, seeds
	Brunfelsia hopeana	roots
	Datura spp.	leaves
	Nicotiana rustica	leaves
	Nicotiana tabacum	leaves
	Nicotiana spp.	leaves
	Solanum elaeagnifolium Cav. (cf. *Solanum* spp.)	?
Sterculiaceae	*Cola* spp.	nuts
	Theobroma subincanum Mart.	bark
	Theobroma spp.	bark
Zygophyllaceae	*Peganum harmala*	seeds

460 Whether this variety was indeed used traditionally to prepare snuffs is questionable (C. Manuel Torres, pers. comm.).

461 A note on an herbarium specimen collected by J. A. Steyermark in Brazil in 1970 includes the remark: "The bark of the tree is used here by the Guaica Indians [= Waika/Yanomamö] as an ingredient in *yopo*" (von Reis and Lipp 1982, 126*).

462 This species, which has been described for Argentina, is synonymous with *Anadenanthera colubrina*. Its seeds and pods contain DMT and other tryptamines (Holmstedt and Lindgren 1967, 362).

463 It is very likely that this Argentinean species is synonymous with a variety of *Anadenanthera colubrina*.

464 It is rather unlikely that a snuff made from *Banisteriopsis* would have psychopharmacological effects.

465 The Peruvian mestizo *ayahuasqueros* use the species *Trichilia tocacheana*, a tree known as *lupana*, to make **ayahuasca**. They fill ayahuasca, mixed with tobacco juice, into a hollow in the trunk of the tree and allow it to be enriched by the toxic resin that is exuded (Schultes and Raffauf 1990, 435*).

466 The shamans of the Barasana and Makuna used the powdered leaves as a ritual snuff for divination. However, there are no reports demonstrating that this powder exerts any psychoactive effects (Schultes 1980).

Indole Alkaloids in South American Snuff Powders

(from Holmstedt and Lindgren 1967, 361; cf. Bernauer 1964 and de Budowski et al. 1975; modified)

Name of Snuff	Source	Alkaloids
epéna	Waika (Brazil)	DMT, 5-OH-DMT,[467] **5-MeO-DMT**
epéna	Waika	DMT, MMT, **5-MeO-DMT**
epéna	Yanomamö	DMT, DMT-*N*-oxide, 5-OH-DMT, 5-OH-DMT-*N*-oxide
epéna	Surára	**harmine**, tetrahydroharmine
epéna	Surára	**harmine**, tetrahydroharmine
epéna	Tukano	DMT, **5-MeO-DMT**, 5-MeO-MMT
epéna	Araraibo	DMT, **5-MeO-DMT**
paricà	Venezuela	5-OH-DMT
paricà	Colombia	5-OH-DMT
paricà	Tukano	**harmine, harmaline**, tetrahydroharmine
paricà	Piaroa	DMT, 5-OH-DMT, **5-MeO-DMT, harmine**
yopo	Colombia	DMT, 5-OH-DMT, **5-MeO-DMT**
yopo	Venezuela	**bufotenine**, methylbufotenine

tree are sometimes added to this blend (Brewer-Carias and Steyermark 1976, 60).

The Yanomamö (Waika) of northern Brazil use the bark of the *Virola theiodora* tree (also known as *epena*) to make a snuff and add ashes from the bark of *Elizabetha princeps*, which they call *amá, ama-asita,* or *chopó* (Brewer-Carias and Steyermark 1976, 63; Schultes and Raffauf 1990, 239*; cf. also Chagnon et al. 1970).

The Yanomamö do not consider *Elizabetha princeps* to be in and of itself hallucinogenic, but they believe that it potentiates the effects of the active constituents (*Virola, Anadenanthera*) (Brewer-Carias and Steyermark 1976, 63).

They also make a snuff from the ground, roasted seeds of ***Anadenanthera peregrina*** (Prance 1972, 234 f.*).

Active Constituents

The main constituents of most South American snuffs used for shamanic purposes are the tryptamine derivatives ***N,N*-DMT, 5-MeO-DMT,** and **bufotenine**. Some powders contain all three substances, while others contain only two or even just one (see the table on page 789). The sources of these tryptamines are species from the genera *Anadenanthera* and *Virola* (Holmstedt 1965). All other plants (e.g., *Elizabetha, Justicia, Manihot, Piper, Theobroma*) are only additives or substitutes and often have no psychoactive effects of their own. However, it is entirely possible that certain, as yet unknown synergistic activities may play an important role. Only the nightshade plants (*Brugmansia, Brunfelsia, Datura, Nicotiana*) contain potent psychoactive alkaloids. Chemical studies of some of the plants used as snuffs (*Maquira, Pagamea*), to the extent that they have been carried out at all, have not yet uncovered any definitive active constituents (Schultes 1980, 274). They may serve only as symbolic elements in the

rituals (Schultes and Raffauf 1990, 389*). It would be interesting to conduct further research into the use of *Banisteriopsis* as a snuff (ingredient?).

Literature

Berenguer, José. 1987. Consumo nasal de alucinógenos en Tiwanaku: Una aproximación iconográfica. *Boletín del Museo Chileno de Arte Precolumbino* 2:33–53.

Bernauer, K. 1964. Notiz über die Isolierung von Harmin und (+)–1,2,3,4-Tetrahydro-harmin aus einer indianischen Schnupfdroge. *Helvetica Chimica Acta* 47 (4): 1075–77.

Boetzkes, Manfred, Wolfgang Gockel, and Manfred Höhl. 1986. *Alt-Peru: Auf den Spuren der Zivilisation.* Hildesheim: Roemer-Museum.

Brewer-Carias, Charles, and Julian A. Steyermark. 1976. Hallucinogenic snuff drugs of the Yanomamo Caburiwe-Teri in the Cauaburi River, Brazil. *Economic Botany* 30:57–66.

Chagnon, Napoleon. 1977. *Yanomamö: The fierce people.* 2nd ed. New York: Holt, Rinehart & Winston.

———. 1994. *Die Yanomamö: Leben und Sterben der Indianer am Orinoko.* Berlin: Byblos Verlag.

Chagnon, Napoleon A., Philip Le Quesne, and James M. Cook. 1970. Algunos aspectos de uso de drogas comercio y domesticación de plantas entre los indígenas yanomamö de Venezuela y Brazil. *Acta Científica Venezolano* 21:186–93.

Cornejo B., Luis E. 1994. San Pedro de Atacama: Desmasiado mundo terrenal (DMT). *Mundo Precolombino—Revista del Museo Chileno de Arte Precolombino* 1:14–24.

de Budowski, J., G. B. Marini-Bettolo, E. Delle Monache, and F. Ferrari. 1975. On the alkaloid composition of the snuff drug yopo from upper Orinoco (Venezuela). *Il Farmaco* 29 (8): 574–78.

467 5-OH-DMT = bufotenine.

de Smet, Peter A. G. M. 1985a. A multidisciplinary overview of intoxicating snuff rituals in the Western Hemisphere. *Journal of Ethnopharmacology* 13 (1): 3–49.

———. 1985b. *Ritual enemas and snuffs in the Americas.* Latin America Studies 33. Amsterdam: CEDLA.

de Smet, Peter A. G. M., and Laurent Rivier. 1974. Intoxicating snuffs of the Venezuelan Piaroa Indians. *Journal of Psychoactive Drugs* 17:93–103.

Donner, Florinda. 1985. *Shabono.* Munich: Knaur.

Furst, Peter T. 1974. Archaeological evidence for snuffing in prehispanic Mexico. *Botanical Museum Leaflets, Harvard University* 24:1–28.

Goetz, Inga Steinvorth. 1970. *Uriji jami! Die Waika-Indianer in den Urwäldern des Oberen Orinoko.* Caracas: Asociación Cultural Humboldt.

Henley, Paul. 1995. *Yanomami: Masters of the spirit world.* Tribal Wisdom Series. San Francisco: Chronicle Books.

Holdstock, Robert. 1986. *Der Smaragdwald.* Munich: Goldmann.

Holmstedt, Bo. 1965. Tryptamine derivatives in epená, an intoxicating snuff used by some South American Indian tribes. *Archives internationales de Pharmacodynamie et de Thérapie* 156 (2): 285–305.

Holmstedt, Bo, and Jan-Erik Lindgren. 1967. Chemical constituents and pharmacology of South American snuff. In *Ethnopharmacologic search for psychoactive drugs*, ed. Daniel H. Efron, 339–73. Washington, D.C.: U.S. Government Printing Office.

Lizot, Jacques. 1982. *Im Kreis der Feuer: Aus dem Leben der Yanomami-Indianer.* Frankfurt/M.: Syndikat.

Llagostera, Agustín, Manuel C. Torres, and María Antonietta Costa. 1988. El complejo psicotrópico en Solcor-3 (San Pedro de Atacama). *Estudios Atacameños* 9:61–98.

Núñez A., Lautaro. 1969. Informe arqueologico sobre una muestra de posible narcotico, del sitio Patillos-1 (Provincia de Tarapaca, Norte de Chile). *Etnografiska Museet Göteborg Årstryck* 1967–1968: 83–95.

Rätsch, Christian. [1990]. Und Moleküle bohren sich durch die Nase: Von psychedelischen Schnupfpulvern. In *Ene Mene Mopel—Die Nase und der Popel* (Der Grüne Zweig 139), ed. Werner Pieper, 111–18. Löhrbach: Werner Pieper's MedienXperimente.

Ritchie, Mark Andrew. 1996. *Spirit of the rainforest: A Yanomamö shaman's story.* Chicago: Island Lake Press.

Schultes, Richard Evans. 1967. The botanical origins of South American snuffs. In *Ethnopharmacologic search for psychoactive drugs*, ed. Daniel H. Efron,

291–306. Washington, D.C.: U.S. Government Printing Office.

———. 1980. De plantis toxicariis e mundo novo tropicale commentationes XXIX: A suspected new Amazonian hallucinogen. *Botanical Museum Leaflets* 28 (3): 271–75.

———. 1984. Fifteen years of study of psychoactive snuffs of South America: 1967–1982, a review. *Journal of Ethnopharmacology* 11 (1): 17–32.

Seitz, George J. 1965. Einige Bemerkungen zur Anwendung und Wirkungsweise des Epena-Schnupfpulvers der Waika-Indianer. *Etnologiska Studier* 28:117–32.

———. 1967. Epena, the intoxicating snuff powder of the Waika Indians and the Tucano medicine men, agostino. In *Ethnopharmacologic search for psychoactive drugs*, ed. Daniel H. Efron, 315–38. Washington, D.C.: U.S. Government Printing Office.

Torres, Constantino Manuel. 1981. Evidence for snuffing in the prehispanic stone sculpture of San Agustín, Colombia. *Journal of Psychoactive Drugs* 13 (1): 53–60.

———. 1987a. *The iconography of South American snuff trays and related paraphernalia.* Etnologiska Studier 37. Göteborg, Sweden: Göteborgs Etnografiska Museum.

———. 1987b. The iconography of the prehispanic snuff trays from San Pedro de Atacama, northern Chile. *Andean Past* 1:191–245.

———. 1988. Tabletas para alucinogenos de San Pedro de Atacama: Estilo e iconografia. In *Tesoros de San Pedro de Atacama*, 23–36. Santiago: Museo Chileno de Arte Precolombino.

———. 1993. Snuff trails of Atacama: Psychedelics and iconography in prehispanic San Pedro de Atacama. *Integration* 4:17–28.

———. 1995. Archaeological evidence for the antiquity of psychoactive plant use in the central Andes. *Annali dei Musei Civici di Rovereto* 11 (1995): 291–326.

———. 1998. Status of research on psychoactive snuff powders: A review of the literature. *Yearbook for Ethnomedicine and the Study of Consciousness*, 1996 (5): 15–39. Berlin: VWB.

Torres, Constantino Manuel, David B. Repke, Kelvin Chan, Dennis McKenna, Agustín Llagostera, and Richard Evans Schultes. 1991. Snuff powders from pre-Hispanic San Pedro de Atacama: Chemical and contextual analysis. *Current Anthropology* 32 (5): 640–49.

Wassén, S. Henry. 1965. *The use of some specific kinds of South American Indian snuff and related paraphernalia.* Etnologiska Studier 28. Göteborg, Sweden: Göteborgs Etnografiska Museum.

———. 1967a. Anthropological survey of the use of South American snuffs. In *Ethnopharmacologic*

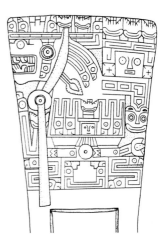

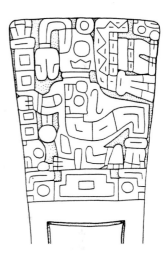

These depictions of gods and mythical beings were found on pre-Columbian snuff trays from San Pedro de Atacama (Chile). (Drawing: Donna Torres; reproduced with kind permission)

search for psychoactive drugs, ed. Daniel H. Efron, 233–89. Washington, D.C.: U.S. Government Printing Office.

———. 1967b. Om några indianska droger och speciellt om snus samt tillbehör. *Etnografiska Museet Göteborg Årstryck* 1963–1966: 97–140.

———. 1972. A medicine-man's implements and plants in a Tiahuanacoid tomb in highland Bolivia. *Etnologiska Studier* 32:8–114.

———. 1983. Revival in coimbra of 18th century Brazilian snuff trays. *Göteborgs Etnografiska Museum Årstryck* 1981/82:39–42.

———. 1985. Convergent approches to the analysis of hallucinogenic snuff trays. *Göteborgs Etnografiska Museum Årstryck* 1983/84:26–37.

Zearies, Otto. 1960. Medizinmannwesen und Geisterglaube der Waiká-Indianer des oberen Orinoco. *Ethnologica*, n.f., 2:485–507.

———. 1980. *Unter Indianern Brasiliens: Sammlung Spix und Martius 1817–1820.* Innsbruck: Pinguin; Frankfurt/M.: Umschau.

Soma

"Then the gods commanded us to gather all of the plants that could be dipped in the ocean and were suitable for obtaining the nectar of immortality; and in those days we were very strong."

VALMIKI
RAMAYANA
(1983, 165)

"These are the plants, these highly gifted ones, that shall free the sick from their suffering! Truly, I confirm, you herbs, that your lord is Soma and that you were created by none other than [the seven-faced] Brishaspati [planetary god Jupiter]! The shadow that lies over us, that threatens us, shall be overcome."

MAGICAL SAYING FROM THE
ATHARVA VEDA
IN *MAGIE DES OSTENS* [MAGIC OF
THE EAST]
(I. SHAH 1994, 159*)

Some authors believe that the original soma drink was pressed from rhubarb stems (*Rheum palmatum*) and fermented into a winelike drink.

468 The urine of people inebriated on psilocybin mushrooms (cf. *Psilocybe* spp.) can also be used as an entheogen. Twenty-five percent of orally ingested psilocybin ends up in the urine one to two hours after the mushrooms are eaten (J. Ott, pers. comm.).

Other Names

Ambrosia, amrita, bolud rtzi (Tibetan), haoma, homa, nectar, nektar, sauma, saumya, som

Soma is the earthly counterpart of *amrita*, the drink of immortality that is reserved for the gods in the heavens. The name *soma* was given to a deity, to a plant, and to the sacrificial drink prepared from the plant. The Aryans of the Indus Valley venerated and drank soma as part of a cult that existed some three thousand years ago (Aguilar I Matas 1991). Soma is the Indian counterpart of the Persian **haoma**. Today, the definitive botanical identity of soma is unknown.

In the Hindu tradition, the moon (originally called *soma*) is the ambrosia-filled drinking vessel of the gods. When the moon is full, the vessel is full; by the time the new moon appears, it has been emptied. It fills up again as the moon waxes. The moon is divided into sixteen sections. Every day, the gods drink up one of these sections. The moon is the lord of the plants, the deity who protects all plant life. For this reason, the soma plant must be gathered by moonlight and brought to the place of offering on a platform pulled by two goats. The sacrificial altar itself was made solely of *kusa* grass (Gupta 1991, 85*).

There were three soma preparations, called *asir*: the one made with milk was called *go*, the one with sour milk *dadhi*, and the one with barley *yava* (Gupta 1991, 85*). To prepare the drink, the stems of the soma plant were pressed between two stones. The "soma juice, which dissolves all sins" (Valmiki 1983, 21), was then mixed with water, milk, and other ingredients. It has sometimes been assumed that the soma prepared with barley was a kind of **beer** (*surā*).

The priest at the fire altar offered soma (as a libation) to and consumed the drink in honor of Indra, the god of thunder, who is believed to be eternally inebriated on soma. The drink was also consumed by singers and poets because it inspired them to their art (Gonda 1978; Hauschild 1954).

The Vedas state that the urine of people inebriated on soma could be drunk, producing the same effects. Because the urine of people inebriated on fly agaric mushrooms is drunk in Siberia to induce further inebriation (Bourke 1996, 54 ff.**), Gordon Wasson put forth the hypothesis that the original soma plant must have been *Amanita muscaria*.[468] However, Wasson's hypothesis has been strongly contested (McKenna 1996, 135 ff.*). The fact that there are no fly agaric mushrooms in the Himalayas is especially problematic. How were the Vedic religious communities supplied with the inebriant?

According to the Rig Veda, the soma plant grows only in the mountains; for this reason, all lowland plants, such as *Peganum harmala*, can be eliminated as potential soma candidates (cf. Ott n.d.). It is also known that the Aryans acquired the plant by trading with the indigenous mountain tribes (Gupta 1991, 84*).

Some authors have speculated that soma was a **mead** made from **honey** (Hermanns 1954, 75), a kind of **wine** (rhubarb wine; Hummel 1959), or even a hopped **beer**. Others believe that soma was a fermented drink made from **honey** and the pressed juice of ephedra (*Ephedra* spp.) soaked in milk (Tyler 1966, 285*). However, the Rig Veda makes a clear distinction between soma and alcoholic drinks (*surā*) (Stutley 1980, 74).

Other authors, such as Terence McKenna (1996*), have speculated that the original soma plant was the magic mushroom *Psilocybe cubensis* (or another *Psilocybe* species), which grows abundantly on cow dung—and that this may be why cows are sacred in India. It has also been suggested that soma was made from a combination of *Peganum harmala* and *Amanita muscaria*

Soma Plants

The original soma plant—hypothetical candidates

Amanita muscaria (L. ex Fr.) Pers. ex Hooker

Argyreia nervosa (Burm. f.) Boj.

Bacopa monnieri (L.) Pennell [syn. *Bacopa monniera* Wettst., *Monniera cuneifolia* Michx., *Herpestris monniera* (L.) H.B.K.]—sarasvati (Sanskrit)[469]

Calonyction muricatum (L.) Don [syn. *Ipomoea turbinata* Lag., *Ipomoea muricata*] (cf. **Ipomoea spp.**)

Claviceps paspali Stevens et Hall (parasitic on the koda grass *Paspalum scrobiculatum* L.)

Claviceps purpurea

***Ephedra* spp.:**

    Ephedra ciliata F. et M. [syn. *Ephedra foliata* Boiss. ex C.A. Mey.]

    Ephedra distachya L.

    Ephedra intermedia Schrenk et C.A. Mey.

    Ephedra pachyclada Boiss.—hum (Hindi), huma (Hindi) (Gupta 1991, 84*)

    Ephedra vulgaris Rich. (= *Ephedra gerardiana*)

Equisetum sp. (cf. **Equisetum arvense**)

***Humulus lupulus* L.**

Mandragora turcomanica Mizgireva (cf. **Mandragora spp.**)

***Peganum harmala* L.**

Peganum harmala in combination with *Psilocybe* sp. ("somahuasca"; cf. **ayahuasca analogs**)

Polyporus sp. (see "*Polyporus mysticus*")

***Psilocybe* (*Stropharia*) *cubensis* (Earle) Singer**

Psilocybe semilanceata (Fr.) Kummer

Psilocybe (*Stropharia*) *subcubensis* Guzmán

Rheum spp.[470]—wild rhubarb:

    Rheum emodi Wall.

    Rheum officinale Baill.

    Rheum palmatum L.

    Rheum rhaponticum L.

Succulent(s) (cf. Ajaya 1980, 273 ff.)

Vitis sp., ***Vitis vinifera* ssp.** *sylvestris*—wild Afghani wine

Post-Vedic soma substitutes (O'Flaherty 1968)

adara (not identified; cf. Gupta 1991, 84*)

Andropogon sp. (arjunnâni) [syn. *Cymbopogon* sp.] (cf. **Cymbopogon densiflorus**)

Basella cordifolia Lam.—pûtîkâ

***Cannabis indica* Lam.**

Ceropegia decaisneana

Ceropegia elegans

***Ephedra gerardiana* Wall. ex Stapf**

Ficus religiosa L.

Periploca aphylla Dene.

putika (not identified with certainty, but perhaps *Basella cordifolia* or a fungus; cf. Kramrisch 1986 and Heim and Wasson 1970)

Sarcostemma brevistigma W. et A. [syn. *Asclepias acida, Cynanchium viminale, Sarcostemma acidum* Voigt, *Sarcostemma viminale* R. Br.[471]]

Setaria italica (L.) Beauv. (according to the *Satapatha Brahmana*)

Vitex negundo L.—indrasura ("Indra's inebriating drink")[472]

Plants that are referred to as soma in Sanskrit or another language

soma (Sanskrit):

    Eleusine coracana (L.) Gärtn. [syn. *Cynosurus coracanus* L.]—African millet, finger millet

    Setaria glauca (L.) Beauv.

somlata (Nepali, "soma plant/moon plant"):

    Ephedra gerardiana

somalata (Sanskrit):

    Caesalpinia bonduc (L.) Roxb.[473]

    Calotropis gigantea (L.) Dryander[474]

    Periploca aphylla Dene.

    Sarcostemma brevistigma W. et Arn. (also called soma)

It has been suggested that an Asian species of horsetail (*Equisetum* sp.) may have been the soma plant. Horsetail, generally called *mu-ts'ê,* "wood thief," was used in ancient China for polishing wood, rhinoceros horn, and ivory. (Illustration from Chi Han, *Nan-fang-ts'ao-mu chuang,* A.D. 304)

469 To date, this plant has been found to contain triterpenes and sapogenins with no known psychoactive effects (Kulshreshtha and Rastogi 1973). However, the leaves are the primary ingredient of Brahmi Rasayan, an Ayurvedic nerve tonic that does exert an effect upon the central nervous system (Shukia et al. 1987).

470 In Tibet, dried rhubarb leaves were mixed with tobacco (**Nicotiana tabacum**) or smoked as a tobacco substitute (Hartwich 1911, 108*).

471 This plant is from Africa. Its latex is used as a fish poison (Tyler 1966, 284*). Other species contain triterpenes (Domínguez et al. 1974).

472 The aromatic bush *Vitex negundo* L. (Verbenaceae) contains large amounts of flavonoids with antiandrogenous effects (Bhargava 1989) as well as an essential oil whose primary constituents are sabinene, terpinene-4-ol, β-caryophyllene, globulol, and bis[1,1-dimethyl]-methylphenol. In Ayurvedic medicine, the plant is used as a remedy for feverish and rheumatic complaints (Mallavarapu et al. 1994).

473 This plant, which is also known as *kumburu-wel,* is used in Southeast Asian folk medicine to treat toothaches and parasitic worms (Macmillan 1991, 423*).

474 The Bodos and Kacharis (tribal peoples of Assam, India) regard this plant as sacred and make offerings to it. The flowers are strung together to make garlands for offerings (Boissya et al. 1981, 221*). The tribal peoples of Bastar (India) believe that *Dendrophtoe falcata* (L. f.) Etting (Loranthaceae), which grows parasitically on *Calotropis gigantea,* is able to increase "brain power" (intelligence) (Jain 1965, 245*). In Southeast Asia, where *Calotropis gigantea* is known as *wara,* its roots are used as a tonic (Macmillan 1991, 423*).

"One day Indra, the chief of the gods, felt a great thirst. The gods of his court asked the goddess Gayatri to go to the celestial mountain Mujavana where the Soma creeper grew and bring it back so that Indra would then have an uninterrupted supply of Soma forever after.

"Gayatri disguised herself as an eagle. She flew to the mountain and found it guarded by the sentries of the Moon. She swooped down and, in a trice, seized the creeper in her beak. Before the startled sentries could do anything she flew away, screeching triumphantly.

"One of the sentries, Krishanu, let fly an arrow at the bird. The arrow missed Gayatri but struck the vine. One of the leaves fell off and it fell to Earth and grew into the Palasa tree."

AN EARLY INDO-ARYAN TALE
IN *BRAHMA'S HAIR*
(GANDHI AND SINGH 1989, 42*)

somalutâ (Sanskrit):
  Ruta graveolens L.
somaraj (Hindi):
  Paederia scandens (Lour.) Merr. [syn. *Paederia foetida* L.]
somarâjî:
  Vernonia anthelmintica (L.) Willd. [syn. *Serratula anthelmintica*, *Conyza anthelmintica*, *Centrathera anthelmintica*]
somatvak (Sanskrit):
  Acacia catechu (cf. **Acacia spp.**)
somavalka (Sanskrit):
  Acacia polyantha (cf. **Acacia spp.**)
somavalli (Sanskrit; Bengali: amrtavallî):
  Tinospora cordifolia (L.) Merr.[475] [syn. *Menispermum glabrum*]—guduchi
  Cocculus cordifolius DC.
saumya/amsúmat ("rich in soma juice"):
  Desmodium gangeticum DC.[476]

Plants that are cultically or mythologically associated with soma

"brother of the great soma" (Shah 1994, 198*):
  Mucuna pruriens
  Limonia acidissima L. [syn. *Feronia limonia* (L.) Swingle, *Feronia elephantum* Corrêa]—elephant apple
palasha (flame of the forest, parrot tree):
  Butea monosperma (Lam.) Kuntze (Leguminosae) [syn. *Butea frondosa* Koen. ex Roxb.]
The palasa tree appears to have been a sacred tree even in the Vedic period. It is consecrated to the moon. Today, the tree is still associated in folklore with the divine soma drink (Gandhi and Singh 1991, 41*).

Top left: In post-Vedic times, the sacred *peepal*, or bodhi tree (*Ficus religiosa*), was used as a soma substitute.

Bottom left: Finger millet (*Eleusine coracana*) is still known by the name *soma* today; in Nepal, it is used to make beer and schnapps. (Millet field on the edge of the Kathmandu Valley, Nepal)

Top right: The shrub *Calotropis gigantea* is called *somalata*, "soma plant," in Sanskrit; whether it has psychoactive effects is unknown.

Bottom right: Rue (*Ruta graveolens*) is called *somalutâ* in Sanskrit (and in several other languages) and is regarded as a possible ingredient in soma and perhaps haoma. Although the herb contains harmala alkaloids, no clear evidence of a psychoactive effect has been found.

475 Known as *guduchi*, this Himalayan plant was referred to in the Sanskrit literature as *amrita*, "nectar of immortality." *Amrita* is also translated as "life-giving drink."

476 The roots contain desmodium alkaloids and possibly *N,N*-DMT and other tryptamines (Ghosal and Bhattacharya 1972).

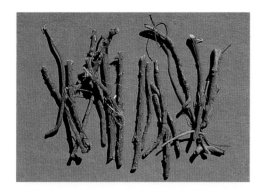

Left: The roots of the shrub *Desmodium gangeticum*, known in India by the name *salparni*, contain the psychoactive substances DMT and bufotenine. In Sanskrit, the plant is called *saumya*, "rich in soma juice."

Right: The sacred palasa tree, also known as "flame of the forest" (*Butea monosperma*), is closely connected to soma in mythology. (Wild plant, photographed in the Terai, Nepal)

or *Psilocybe cubensis* (cf. **ayahuasca analogs**). Of all of the candidates for soma that have been proposed, psilocybin mushrooms are the only psychoactive plants that produce effects like those in the fantastic descriptions of the Rig Veda:

> Your juices, o purified Soma, all penetrating, as quick as thought, move of themselves like the descendents of quickly hurrying steeds; the heavenly, winged sweet-tasting juices, inciter of great cheer, radiant in the vessel. (Rig Veda 9)

Jonathan Ott has speculated that an Indian pharmacratic inquisition may have taken place during the post-Vedic period, with the result that the original visionary soma plant was replaced with a substitute that had only mildly stimulating or even placebo-like properties. We do know that in the post-Vedic period, the soma ritual was carried out with *Cannabis indica* and *Ephedra gerardiana* (O'Flaherty 1968). Other, nonactive plants were also used as substitutes (see table, page 793).

A relatively recent theory suggests that ergot (*Claviceps* spp.) from an *Eleusine* species of grass was used as a substitute for soma (Greene 1993). Indeed, in western Bengal (India), a psychoactive use of *Paspalum* ergot (see *Claviceps paspali*) has continued into the present day.

In western India, the soma ritual in honor of Indra has been preserved in certain tribal rites (such as the *babo* ritual) into the present day (Jain n.d.). The Vedic fire ritual that was connected with the soma sacrifice has survived in India, eastern Asia, and even Japan (Staal 1983). The soma ritual is also thought to have served as the prototype for the kava ceremony of the South Pacific (see *Piper methysticum*).

Soma was likely nothing more than a catchall word, similar to the current words *drug, entheogen, psychedelic*, et cetera. Soma has become a symbol for the "perfect drug" (Huxley 1958).

The aura-soma therapy popular in esoteric circles has nothing to do with the Aryan soma but is a modern invention in which no psychoactive substances are used (Dalichow and Booth 1994).

Literature

See also the entries for *Ephedra gerardiana, Amanita muscaria, Terminalia bellirica*, and **haoma**.

Aguilar I Matas, Enric. 1991. *Rgvedic society*. Leiden: E. J. Brill.

Ayaya, Swami. 1980. *Living with the Himalayan masters: Spiritual experiences of Swami Rama*. Honesdale, Penn.: Himalayan International Institute. (Pages 273–77 are on soma.)

Bhargava, S. K. 1989. Antiandrogenic effects of a flavonoid-rich fraction of *Vitex negundo* seeds: A histological and biochemical study in dogs. *Journal of Ethnopharmacology* 27:327–39.

Dalichow, Irena, and Mike Booth. 1994. *Aura-Soma: Heilung durch Farbe, Pflanzen und Edelsteinenergie*. Munich: Knaur.

Daniélou, Alain. 1964. *Hindu polytheism*. London: Routledge and Kegan Paul. (See pp. 98 ff.)

Domínguez, Xorge A., Jorge Marroquín, Luz Ma. Olguín, Francisco Morales, and Victoria Valdez. 1974. β-amyrin juarezate, a novel ester from *Marsdenia pringlei* and triterpenes from *Asclepias linaria*. *Phytochemistry* 13:2617–18.

Gershevitch, Ilya. 1974. An Iranianist's view of the soma controversy. *Mémorial: Jean de Menasce* 1985:45–75.

Ghosal, S., and S. K. Bhattacharya. 1972. Desmodium alkaloids II: Chemical and pharmacological evaluation of *Desmodium gangeticum*. *Planta Medica* 22:434.

Gonda, Jan. 1978. *Die Religionen Indiens I: Veda und älterer Hinduismus*. 2nd ed. Stuttgart: Kohlhammer.

Greene, Mott. 1993. *Natural knowledge in preclassical antiquity*. Baltimore: Johns Hopkins University Press.

Hauschild, Richard. 1954. Das Selbstlob (*Âtmastuti*) des somaberauschten Gottes Agni. In *Asiatica— Festschrift Friedrich Weller*, 247–88. Leipzig: Otto Harrassowitz.

Heim, Roger, and R. Gordon Wasson. 1970. Les Putka des Santals: Champignons doués d'une âme. *Cahiers du Pacific* 14:59–85.

"Inebriation is one of the inevitable effects of *mythos*. Cultures that no longer possess this are sober and burned out. The longing for a mythology is the drive for an inebriating drink, which excites the imbiber like soma, the divine inebriating drink of the Indian war and thunder god, who ingests it three times a day at the Brahmans' sacrifices, and is consequently able to do his world-conquering deeds and is capable of presiding over the lightning bolt of the ancient god of heaven to open the path of victory for the campaigns of conquest of his chosen people, the Vedic Aryans."

HEINRICH ZIMMER
ABENTEUER UND FAHRTEN DER SEELE
[ADVENTURES AND TRAVELS OF THE SOUL]
(1987, 310)

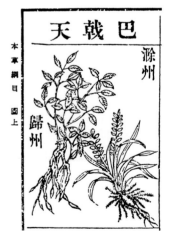

The Asian medicinal plant *Bacopa monnieri* has been interpreted as soma. In Sanskrit, the plant is known as *sarasvati* ("that which flows") and is named after the goddess of writers, artists, and poets. (Chinese illustration from the *Pen-t'sao-kang-Mu*, sixteenth century)

This Nepalese depiction of the soma-inebriated Indra, the god of thunder and of the heavens, shows him holding his thunderbolt (*vajra/dorje*) in his right hand. (Woodcut print from Shakya, *The Iconography of Nepalese Buddhism*, 1994, 63)

Hermanns, Matthias. 1954. *Mythen und Mysterien der Tibeter*. Stuttgart: Magnus.

Hummel, K. 1959. Aus welcher Pflanze stellten die arischen Inder den Somatrank her? *Mitteilungen der Deutschen Pharmazeutischen Gesellschaft* 29:57–61.

Huxley, Aldous. 1958. *Brave new world revisited*. New York: Harper & Row.

Jain, Jyotindra. n.d. *Painted myths of creation: Art and ritual of an Indian tribe*. New Delhi: Lalit Kata Akademi.

Kashikar, C. G. 1990. *Identification of soma*. Research Series, no. 7. Pune, India: Tilak Maharashtra Vidyapeeth.

Kramrisch, Stella. 1986. The mahâvira vessel and the plant pûtika. In *Persephone's quest*, ed. R. G.Wasson et al., 95–116. New Haven, Conn., and London: Yale University Press.

Kulshreshtha, D. K., and R. P. Rastogi. 1973. Bacogenin-al: A novel dammarane triterpene sapogenin from *Bacopa monniera*. *Phytochemistry* 12:887–92.

La Barre, Weston. 1970. Soma: The three-and-one-half millennia mystery. *American Anthropologist* 72:368–73.

Mallavarapu, Gopal R., Srinivasaiyer Ramesh, Pran N. Kaul, Arun K. Bhattacharya, and Bhaskaruni Rajeswara Rao. 1994. Composition of the essential oil of the leaves of *Vitex negundo*. *Planta Medica* 60:583–84.

Müller, Reinhold F. G. 1954. Soma in der altindischen Heilkunde. In *Asiatica—Festschrift Friedrich Weller*, 428–41. Leipzig: Otto Harrassowitz.

Napier, A. David 1986. *Masks, transformation, and paradox*. Berkeley: University of California Press.

O'Flaherty, Wendy Doniger. 1968. The post-Vedic history of the soma plant. In *Soma—divine mushroom of immortality*, ed. R. G. Wasson, 95–147. New York: Harcourt Brace Jovanovich.

Ott, Jonathan. 1994. La historia de la planta 'soma' después de R. Gordon Wasson. In *Plantas, chamanismo y estados de consciencia*, ed. Josep Maria Fericgla, 117–50. Barcelona: Los Libros de la Liebre de Marzo.

———. n.d. *The post-Wasson history of the soma plant*. Unpublished manuscript, Jalapa, Mexico.

Riedlinger, Thomas J. 1993. Wasson's alternate candidates for soma. *Journal of Psychoactive Drugs* 25 (2): 149–56.

Schneider, Ulrich. 1971. *Der Somaraub des Manu: Mythus und Ritual*. Freiburger Beiträge zur Indologie, vol. 4. Wiesbaden: Otto Harrassowitz.

Schroeder, R. F., and Gastón Guzmán. 1981. A new psychotropic fungus in Nepal. *Mycotaxon* 13 (2): 346–48. (On *Psilocybe* [*Stropharia*] *cubensis*, *Psilocybe subcubensis*.)

Shakya, Min Bahadur. 1994. *The iconography of Nepalese Buddhism*. Kathmandu: Handicraft Association of Nepal (HAN).

Shukia, Bina, N. K. Khanna, and J. L. Godhwani. 1987. Effect of brahmi rasayan on the central nervous system. *Journal of Ethnopharmacology* 21:65–74.

Staal, Frits, ed. 1983. *Agni: The Vedic ritual of the fire altar*. Berkeley, Calif.: Asian Humanities Press.

Stutley, Margaret. 1980. *Ancient Indian magic and folklore*. Boulder, Colo.: Great Eastern.

Thomas, P. 1983. *Secrets of sorcery spells and pleasure cults of India*. Bombay: D. B. Taraporevala Sons.

Valmiki. 1983. *Ramayana*. Cologne: Diederichs.

Wasson, Gordon. See entries for *Amanita muscaria*

Wilson, Peter Lamborn. 1995. Irish soma. *Psychedelic Illuminations* 8:42–48.

Zimmer, Heinrich. 1984. *Indische Mythen und Symbole*. Cologne: Diederichs.

———. 1987. *Abenteuer und Fahrten der Seele: Ein Schlüssel zu indogermanischen Mythen*. Cologne: Diederichs.

"Various drinks with somniferous and anodyne effects are mentioned in legends and reports from the past as well as from antiquity, and when one takes into account the ingredients used—in addition to mandrake, also opium, hemlock, Indian hemp, gall-nuts, as well as wine and other substances—one can believe that effects were indeed occasionally achieved with these. Indeed, we find confirmation for this assumption, for we hear of fatal incidents resulting from overdose when physicians used such mixtures."

DETLEF RÜSTLER
ALTE CHIRUGIE [ANCIENT SURGERY] (1991, 77)

Soporific Sponge

Other Names

Spongia somnifera

In ancient times, herbalists and physicians searched for anesthetic agents that could be used during operations and in the treatment of wounds. Numerous psychoactive plants and their products were used in antiquity to anesthetize patients, including *Cannabis indica*, *Cannabis sativa*, *Conium maculatum*, *Hyoscyamus albus*, *Hyoscyamus muticus*, *Mandragora officinarum*, and *Papaver somniferum* (Grover 1965; Rüster 1991, 77 f.; Schmitz and Kuhlen 1989):

The use of narcotics during antiquity, for which henbane, Indian hemp, mandragora, opium, hemlock, and wine were the ones most often recommended, did not always revolve around the alleviation of pain but was also from time to time related to ritual customs and the attainment of states of inebriation. (Amberger-Lahrmann 1988, 1)

As the early modern era began, the anesthetics used in medicine and surgery continued to be based primarily on opium (see *Papaver somniferum*) and henbane (*Hyoscyamus niger*) (Rüster 1991). *Atropa belladonna* was also used (Grover 1965). Henbane was apparently also used to sedate convicted criminals, for the oil that was pressed from it was known as "delinquent oil" (Arends 1935, 58*).

In the late Middle Ages and the early modern era, the most commonly used sedative that was also used as an anesthetic was the so-called soporific sponge. The recipes for soporific sponges tended to be relatively uniform (Brunn 1928; Kuhlen 1983) and were based upon the preparations of ninth- and tenth-century Islamic physicians (e.g., Rhazes). They were especially popular in the thirteenth, fourteenth, and fifteenth centuries. The primary ingredient was opium, to which mandrake roots (*Mandragora officinarum*) and henbane seeds (*Hyoscyamus niger*) were added. This mixture was kneaded in rose-hip juice (*Rosa canina* L.) and mixed with **wine** (cf. the fourteenth-century Roman Codex). The recipe for this narcotic is strongly reminiscent of that of the **witches' ointments** of the early modern era as well as that of **theriac**. One recipe called for opium, juice pressed from mandrake leaves, hemlock, and henbane (Schmitz and Kuhlen 1989, 12). A twelfth-century recipe from Salerno used opium, henbane, poppy, mandrake, ivy (*Hedera helix*), mulberries, lettuce (*Lactuca virosa*), and hemlock (Brandt 1997, 41 ff.).

Soaked in wine, these mixtures were dripped onto a bath sponge (*Euspongia officinalis* L.), which was then inserted into the nostrils of the patient. The patient would then fall into a sleep filled with wild fantasies.

A number of authors have speculated that such soporific sponges were in use in ancient Jerusalem, and that the sponge dipped in vinegar that was offered to Jesus on the cross was actually one of these.

In the fifteenth and sixteenth centuries, there were still a number of *sedativa* and *anodyna specifica*, which are strongly reminiscent of the mixtures used to make soporific sponges. The physician and chemist Paracelsus (1493–1541) left such a recipe (cf. Schneider 1981):

2 drachmas opium thebaicum
1 half ounce cinnamon (*Cinnamomum verum* Presl)
1 pinch musk and ambergris
1 half ounce poppy seeds (*Papaver somniferum*)
1 half drachma mandrake roots (*Mandragora* sp.)
3 drachmas mastic resin (from *Pistacia lentiscus* L.)
1 drachma henbane juice (*Hyoscyamus niger*)

This mixture was later supplanted by laudanum, in particular *laudanum liquidum sydenhami*, which consisted of the following ingredients:

2 ounces opium
1 ounce saffron (*Crocus sativus*)
1 drachma cinnamon (*Cinnamomum verum*)
1 drachma cloves (*Syzygium aromaticum*)

These ingredients were digested in a pound of Malaga wine (Schmitz and Kuhlen 1989, 15). This agent was more of a psychoactive agent of pleasure than an anesthetic.

Literature

See also the entries for **theriac**.

Amberger-Lahrmann, M. 1988. Narkotika. In *Gifte: Geschichte der Toxikologie*, ed. M. Amberger-Lahrmann and D. Schmähl, 1–46. Berlin: Springer.

Brandt, Ludwig. 1997. *Illustrierte Geschichte der Anästhesie*. Stuttgart: WVG.

Brunn, Walter von. 1928. Von den Schlafschwämmen. *Schmerz* 1.

Grover, Norman. 1965. Man and plants against pain. *Economic Botany* 19:99–111.

Kuhlen, Franz-Josef. 1983. *Zur Geschichte der Schmerz-, Schlaf- und Betäubungsmittel in Mittelalter und früher Neuzeit*. Stuttgart: Deutscher Apotheker-Verlag.

Rüster, Detlef. 1991. *Alte Chirurgie*. 3rd ed.. Berlin: Verlag Gesundheit.

Schmitz, Rudolf, and Franz-Josef Kuhlen. 1989. Schmerz- und Betäubungsmittel vor 1600. *Pharmazie in unserer Zeit* 18 (1): 11–19.

Schneider, Wolfgang. 1981. Mittelalterliche Arzneidrogen und Paracelsus. In *Rausch und Realität*, ed. G. Völger, 1:368–72. Cologne: Rautenstrauch-Joest-Museum für Völkerkunde.

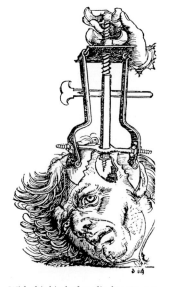

With this kind of medical treatment, it is easy to understand why powerful narcotics are needed. (Woodcut from Hans von Gersdorf, *Feldbuch der Wundartzney*, 1517)

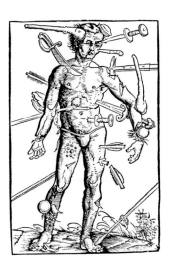

The history of Western surgery is directly connected to the tradition of the narcotic sleeping sponge. (Woodcut, "The Wounded Man," from Hans von Gersdorf, *Feldbuch der Wundartzney*, 1517)

Theriac

Because it was one of the main aromatic ingredients in theriac, angelica (*Angelica archangelica* L.) was once known as theriac spice or theriac herb. Angelica has no psychoactive activity. (Woodcut from the herbal of Matthiolus, 1627)

"I drink images:
Theriac against death.
How much longer?"

Uwe Dick
Theriak: 13 Fügungen [Theriak: 13 Strokes of Fate]
(1986, 47)

Vipers' flesh was one of the basic ingredients in the ancient theriac mixtures. (Woodcut from Gesner, *Historia Animalium*, 1670)

477 In Persian, *taryāk* means both "poison" and "antidote" and is also a name for opium. The opium-eating Turks were known as *theriakis*.

Other Names
Deridek, electarium theriacale, electuarium theriaca, mithridatium, taryāk, teryāk (Iranian), theriaca, theriacum, theriak, theriakos, tiriaque, tyriacke

Toxicology, the science of poisons and the toxicity of substances, was established in ancient times. Because of the many murders by poisoning, at that time toxicology was focused primarily on the *antidoton*, "antidote."

> Antidotes are what physicians call those medicines that are not laid externally onto the body but are applied inside the body. Three general types of these can be distinguished: some are administered to counteract fatal poisons, others for so-called poisonous animals, the third is for ailments that occur as a result of a bad diet. Some, such as the so-called theriac, promise help in all three cases. (Galen, *De Antidotis* 1)

Theriac is the most renowned antidote of antiquity and was regarded as the "wonder drug of all wonder drugs" (Watson 1966). Theriac was developed by Andromachus, the personal physician of Emperor Nero (37–68 C.E.), and contained opium[477] (*Papaver somniferum*) and vipers' flesh along with various spices, roots, **honey**, and **wine**. At first, theriac was a further development of the so-called *mithridatium*, the antidote of the tyrannical king Mithridates of Pontos (132–63 B.C.E.). *Mithridatium* was formulated by the king himself, who, as a result of his public atrocities, was in constant fear of being poisoned.

According to Celsus (5.23, 3), *mithridatium* consisted of costus (*Costus* sp.), calamus (***Acorus calamus***), St. John's wort, gum arabic, *Sapapenum*, acacia juice (***Acacia* spp.**), Illyric iris, cardamom, anise, Gallic spikenard, gentian root, dried rose leaves, poppy juice (= opium), parsley (***Petroselinum crispum***), cassia cinnamon (*Cinnamomum cassia*), sil, bearded darnel (***Lolium temulentum***), long pepper (see ***Piper* spp.**), storax, castoreum, turis, *Hypocistis* juice, myrrh, opopanax, malabathron leaves, flowers of the round rush, terebinth resin, galbanum, Cretan carrot seeds, spikenard, opobalsam, shepherd's purse, rhubarb root (*Rheum* sp.), saffron (***Crocus sativus***), ginger (***Zingiber officinale***), and cinnamon (*Cinnamomum verum*). In other words, it contained a number of psychoactive plants.

The recipes for theriac and *mithridatium* were later refined by Arabic physicians (Steinschneider 1971). But alongside the approximately sixty additional ingredients, the primary constituent was always opium. Among the other important

ingredients were theriac spice (*Angelica archangelica* L. [syn. *Archangelica officinalis* Hoffm.]) and theriac root (***Valeriana officinalis***), as well as carrot seeds (***Daucus carota***).

Theriacs were increasingly employed (with quite differing degrees of success) to treat all manner of diseases, from the pervasive syphilis to the plague. The opium that was needed to prepare these panaceas was imported mainly from Egypt, whereby Venice acquired a central significance as the primary harbor of transshipment. As the demand for the drug grew, the price rose as well, so that opium was often "stretched" with foreign admixtures. Since the Venetian dealers generally received the drug in an as yet unsullied state, the theriacs of the Serenissima soon became the most preferred. In Germany, the city of Nuremberg gained a similar reputation, and it continued to be one of the European market leaders in the production of theriac into the eighteenth century. In order to demonstrate the good quality of the ingredients, theriacs were often mixed together in the marketplace during a folk festival. (Kupfer 1996a, 27*)

After the work of Andromachus, Galen, and the Arabic physicians of the Middle Ages, there were numerous recipes for preparing theriac. All of them listed **honey**, **wine**, bread (?), vipers' flesh, opium, and spices as their most important ingredients. These mixtures even found their way into modern pharmacopoeias (*electuarium theriaca con opii*). Ultimately, theriac gave rise to the so-called elixir of long life and to Swedish bitters (Treben 1980, 60). At first, both of these products also contained opium. It was only during the time of the "drug wars" that opium was banished from these recipes. Cynical tongues have suggested that this step resulted in the removal of the only truly active constituent in these elixirs.

Literature
See also the entries for ***Papaver somniferum*** and **soporific sponge**.

Dick, Uwe. 1986. *Theriak: 13 Fügungen*. Munich: Piper.

Treben, Maria. 1980. *Gesundheit aus der Apotheke Gottes*. Steyr: Ennsthaler.

Steinschneider, Moritz. 1971. *Die toxikologischen Schriften der Araber*. Hildesheim: Gerstenberg.

Watson, Gilbert. 1966. *Theriac and mithridatium: A study in therapeutics*. London: The Wellcome Historical Medical Library.

Wine

Other Names

Ju, khamr (Arabic, "inebriating"), oinos, sdh, vin, vinho, vino, wein

The term *wine* is generally used to refer to alcoholic products that through the action of yeast are fermented from undiluted fruit juices or, less frequently, bleeding sap (**palm wine**). The alcohol content is between 8 and 14% by volume and is thus significantly higher than in other fermented drinks (**balche', beer, chicha**). The term *wine* is often construed as the product made from pressed grapes (*Vitis vinifera*), while the products fermented from garden or wild fruit juices are usually referred to by the name *fruit wines*. Wines can also be distilled, thereby yielding a corresponding variety of spirits (cf. **alcohol**). Winemaking was invented in many places around the world. All wines are well suited for use as solvents for other psychoactive ingredients.

It is possible that the Egyptians knew of a winelike drink containing mandrake that was called *sdh* (cf. *Mandragora officinarum*). Apparently, vintners produced it not from grapes (*Vitis vinifera*) but from pomegranate juice (*Punica granatum* L.). The texts describe the *sdh* drink as more inebriating than wine. It was praised in love songs as an aphrodisiac and was a popular libation (Cranach 1981, 266*). Many Egyptian drinking vessels were patterned after the lotus flower. In the pyramid texts, the lotus is mentioned together with the *sdh* drink. The two most important symbolic plants in Egyptian art and iconography are the lotus (*Nymphaea caerulea*, *Nymphaea lotus*) and the mandrake. In order to maintain the harmonic balance between the symbolic pair of lotus and mandrake, any drink that contained mandrake would theoretically need to be consumed from a lotus-shaped vessel.

In Scandinavia, bilberries (*Empetrum nigrum* L., *Vaccinium uliginosum*) were used to make inebriating wines. In northern Eurasia, birch sap (the bleeding sap that is produced when the bark is injured, usually from *Betula alba* L.) was fermented to produce alcoholic beverages (Hartwich 1911, 764 ff.*).

The stems of various rhubarb species can also be pressed to make wine. In fact, it has been speculated that **soma** was a kind of rhubarb wine.

Often, pressed fruit juices (e.g., from *Berberis vulgaris* L.) were mixed with **honey** to produce wines with a higher alcohol content. Honey was added to quince juice (*Cydonia vulgaris* L.; cf. *Erythroxylum coca*) in ancient times (Hartwich 1911, 760*).

In England, many people make fruit wine at home. Although wild fruits are preferred, almost

any kind of fruit can be used for this purpose. A "counterculture" psychedelic wine is made from the fresh-pressed juice of forest berries (e.g., blackberries) and *Psilocybe semilanceata*.

In Chihuahua (Mexico), wines are made from the fruits of various yucca species (Havard 1896, 37*). Mexican Indians also ferment the juice of the fruits of *Opuntia tuna* Mill. and *Opuntia ficus-indica* Haw. to produce a pink-colored wine known as *colonche*, the taste of which is similar to that of cider or apple wine (Havard 1986, 36 f.*). Because **mescaline** is present in *Opuntia*, it is possible that wines fermented from these fruits contain traces of the alkaloid. Pineapple juice (*Ananas comosus* [L.] Merr., *Ananas nanus* [L.B. Sm.] L.B. Sm.) can be made into a wine that is called *matzaoctli* in Nahuatl; it is produced and consumed primarily in Mazatlán, the "land of pineapples" (Bruman 1940, 148*).

In South America, a so-called *vino de cebil* ("cebil wine") with presumably psychedelic activity was or is brewed from the seeds/fruits of **Anadenanthera colubrina** var. *cebil*. Unfortunately, no recipes are known. In Chile, the Indians used maqui fruits (*Aristotelia maqui* l'Herit.) to prepare a "pleasant-tasting wine" called *tecu* (Hartwich 1911, 762*).

Wine is pressed from many different palm fruits, for example, from the fruits of the betel nut palm (**Areca catechu**) and the fruits of the saw palmetto (*Serenoa repens* [Bartr.] Small [syn. *Sabal serrulata* Michx., *Serenoa serrulata* (Michx.) Nichol.]; cf. **palm wine**). Saw palmetto wine has aphrodisiac effects in addition to the inebriating effects of alcohol. In phytotherapy and homeopathy, saw palmetto fruits are considered to be aphrodisiacs (cf. **Turnera diffusa**). They also have beneficial effects on benign prostate hyperplasia (Metzker et al. 1996).

The Australian Aborigines make fruit wines from *Pandanus spiralis* R. Br. (cf. **Pandanus spp.**), *Banksia* species, *Hakea* species, and a *Xanthorrhoea* species that are known respectively as pandanus wine, banksia wine, hakea wine, and grass tree wine (Bock 1994, 147*). The banksia wine is actually a kind of **beer** (Low 1990, 189*). The sap of *Eucalyptus gunnii* (cider gum), which collects in

A wine can be made simply by allowing the fruits of the prickly pear (*Opuntia phaecantha*) to ferment.

"This morning how grand is the space!
Without bridle or spurs, in our haste
Let us set out by horseback on wine,
For the heavens—enchanted, divine!"

CHARLES BAUDELAIRE
"THE LOVERS' WINE"
IN *LES FLEURS DU MAL* [THE FLOWERS OF EVIL] (1857)
(1993, 225*)

"Drunkenness, however, is that state of ecstasy in which one can leave reality and open oneself up to the supernatural."

JEAN MARKALE
THE DRUIDS
(1999, 174*)

The stems of rhubarb (*Rheum officinale* Baill., *Rheum palmatum* L.), a plant originally from Tibet, not only are eaten as a gruel but also can be fermented into wine. The roots (rhei radix) have medicinal qualities and are used in laxative and diet teas. (Woodcut from Tabernaemontanus, *Neu Vollkommen Kräuter-Buch*, 1731)

Brazilian Indians esteem the fruits of the cashew tree (*Anacardium occidentale*), especially for their role in the production of inebriating winelike drinks. (Copperplate engraving from Meister, *Der Orientalisch-Indianische Kunst- und Lustgärtner*, [The Oriental-Indian Art and Pleasure Gardener] 1677)

For many connoisseurs, wine is a presentiment of paradisiacal joys. And sometimes even more: this Swiss label portrays the grape vine as the tree of knowledge.

hollows of the trunk when the tree is injured, ferments more or less on its own, yielding a potently inebriating wine (Low 1990, 189*).

Literature

See also the entries for *Vitis vinifera* and **palm wine**.

Feest, Christian F. 1983. New wines and beers of Native North America. *Journal of Ethnopharmacology* 9:329–35.

Lorey, Elmar M. 1997. *Die Wein-Apotheke.* 2nd., suppl. ed. Bern, Stuttgart: Hallwag.

Metzker, H., M. Kieser, and U. Hölscher. 1996. Wirksamkeit eines Sabal-Urtica-Kombinationspräparats bei der Behandlung der benignen Prostatahyperplasie (BPH). *Der Urologe* B 36:292–300.

Witches' Ointments

The famous witches' or flying brooms are said to have been made from birch brush. The witches' ointments were supposedly smeared onto the handle, which was then used as a dildo. The use of birch is interesting; in northern Eurasia, birch (*Betula* spp.) represents the shamanic world tree and is both culturally and biologically associated with the fly agaric mushroom. (Woodcut from Bock, *Kreutterbuch*, 1577)

One of the few depictions of a witch smearing herself with the flying ointment, after which she immediately flies out of the chimney. (Woodcut, early sixteenth century)

Other Names

Buhlsalbe, demon salve, flugsalbe, flying ointment, hexensalbe, hexenschmiere, oyntment, schlafsalbe, sleeping ointment, unguenta somnifera, unguenti sabbati, unguentum pharelis, unguentum populi

The famous "witches' ointments," that is, the substances alleged witches used to undertake their nocturnal "excursions," were not an invention of the Inquisition, for they had already been mentioned in ancient texts (Luck 1962). The first mention of a "flying ointment" comes from Homer, the father of poetry: Hera smears herself with ambrosia in order to reach Zeus on Mount Ida by flying down from Mount Olympus and over Thrace's snowcapped mountains, "over the highest peak, without ever touching the ground." Zeus is profoundly amazed at how quickly she was able to make the journey without horse and wagon (*Iliad* 2.14.169 ff.). The most famous picaresque novel of late antiquity, the *Metamorphosis* (= *The Golden Ass*) of Apuleius (second century C.E.), includes a well-known reference to a witches' ointment. The hero, Lucius, reports on the magical practices and witchcraft of the inhabitants of Thessalia, "the world-renowned home of magic" (Apuleius 2). According to his reports, the Thessalian witches knew how to bring mandrake manikins (cf. ***Mandragora officinarum***) to life so that they could send them out and cause harm according to their wishes. They were also able to change their shapes and travel wherever they desired:

First Pamphile completely stripped herself; then she opened a chest and took out a number of small boxes. From one of these she removed the lid and scooped out some ointment, which she rubbed between her hands for a long time before smearing herself with it all over from head to foot. Then there was a long muttered address to the lamp during which she shook her arms with a fluttering motion. As they gently flapped up and down there appeared on them a soft fluff, then a growth of strong feathers; her nose hardened into a hooked beak, her feet contracted into talons—and Pamphile was an owl. (Apuleius 3.21)

Unfortunately, none of the ancient recipes has come down to us.

The medieval sources are silent on the topic. Only toward the end of the late Middle Ages was there speculation about witches' ointments (Haage 1984), which were used variously to enable witches to fly or to transform them into animals (e.g., werewolves; cf. Leubscher 1950, Völker 1977). The Renaissance not only rekindled an interest in the ancient world but also witnessed a revival in which all sorts of narcotic ointments with clear classical roots began to be used again in folk medicine and surgery (cf. **soporific sponge**; Piomelli and Pollio 1994).

Dr. Johannes Hartlieb (ca. 1400–1468) was the personal physician of the Wittelsbach dukes, writers, and diplomats. He not only left us one of the earliest German herbals (written around 1440; Werneck and Franz 1980) but also published the most important medieval source on the vestiges of the pagan world. Because he was a devoted Christian, he depicted the magical practices presented in his book *Das Buch aller verbotenen Künste* [The Book of All Forbidden Arts] (1456, originally titled *Das puoch aller verpoten kunst, ungelaubens und der zaubrey*) as reprehensible and dangerous.

Hartlieb was the first physician to write down a recipe for a witches' ointment, which can be found in chapter 32 of his book:

How the journey through the air occurs
In order to undertake such a journey, men and women, and in particular the fiends, use an ointment called *Unguentum pharelis*. It is made from seven herbs. Each of these herbs is picked exactly on the day to which it is assigned. Thus, they pick or dig up *solsequium*

Hartlieb's Name	Translation	Possible Interpretation(s)[479]
solsequium	"following the sun"	dandelion (*Taraxacum officinale* Web. s.l.)
		marigold/turning to the sun (*Calendula officinalis* L.)
		chicory (*Cichorium intybus* L.)
		English marigold
lunaria	"moon [plant]"	carline thistle (*Carlina* sp.?)
		pointed thistle, royal fern, "Lunaria maior"
verbena	"iron plant"	vervain[480]
		Verbena officinalis L.
mercurialis	"Mercury [plant]"	annual mercury (*Mercurialis annua* L.)
		mercury (*Mercurialis* sp.)
barba jovis	"Jupiter's beard"	house leek/sengreen (*Sempervivum tectorium* L.)
capillus veneris	"hair of Venus"	true maidenhair (various fern species, including *Adiantum capillus-veneris* L.)

on Sunday, *lunaria* on Monday, *verbena* on Tuesday, *mercurialis* on Wednesday, *barba jovis* on Thursday, *capillus veneris* on Friday. From these they then prepare ointments by mixing in bird's blood and animal lard. But I will not describe this in detail, so that none will be corrupted by it. When they then wish, they smear this onto benches or chairs, rakes, or pitchforks, and fly away (on these). This is nothing other than nigromantie [= "black divination"[478]] and is strictly forbidden. (Hartlieb 1989a, 45)

It is not possible to make a definite botanical identification of the herbs that were associated with the days of the week (and the planetary gods). However, some of the names can be identified with plants that were called by the same name in the late Middle Ages.

It seems that this recipe from Johannes Hartlieb is more an agent of sympathetic magic than a psychoactive substance (cf. Biedermann 1974). That is, unless there are psychoactive ferns that have not yet been identified.

It appears that in the entire history of the witch hunts, only once was an ointment actually found that was successfully able to produce the effects attributed to it. When, in the year 1545, the duke of Lothringen lay in bed critically ill, a married couple was arrested and charged with having bewitched the duke. Both "confessed" to witchcraft while being stretched on the rack. During a subsequent search of their house, a container of ointment was found that was examined by Andrés de Laguna (1499–1560), the personal physician to the pope (cf. Rothman 1972). He recognized in the ointment *un cierto unguento verde como el del populeon* ("a certain green unguent like the poplar ointment"; cf. Vries 1991). Laguna speculated that the ointment contained *Cicuta* (hemlock), *Solanum* (?), *Hyoscyamus*, and *Mandragora* and tested it on the executioner's wife. She fell into a kind of coma or deep sleep that lasted for three days and complained irritably when she was pulled from her sleep, which had been filled with sweet dreams and erotic adventures.

For some Inquisitors, such as Pedro Ciruelo, it was clear that the alleged witches did not really fly to the sabbat but rather had hallucinatory experiences because of the ointment (Dinzelbacher 1995, 209).[481]

From the beginning, those who speculated about the nature of the witches' ointments—usually physicians of the early modern era (Vries 1991)—attributed the effects of the ointment to the presence of plants from the Nightshade Family (Evans 1978; Führer 1919; Harner 1973; cf. also Duerr 1978).

Johannes Wier (1515–1588), the personal physician of Duke Wilhelm von Jülich, discussed witches' ointments in his work (Weyer 1563) and cited a recipe for a witches' ointment from the book *De Subtilitate Rerum* by Hieronimus Cardanus (= Girolamo Cardano):

Ointments, which are said to have the power and the effect that one is able to see wondrous things because of them. These are prepared from children's fat, as they say, and celery juice, wolf's plant, tormentill, solano (nightshade), and soot. However, they are considered to be asleep because they see such things. But the things they see that widen their eyes are most often houses of pleasure, green gardens of delight, splendid meals, many and various decorations, pretty clothes, handsome youths, kings, lords, yes everything that they are anxious about and long for, they also mean nothing more than to enjoy such leisure and pleasure and to be made happy. But they also see the devil, ravens, dungeons, wastelands, and the hangman's or prosecutor's bag of tricks. . . . For this reason, the more strong is the conclusion that they are imagining that they travel through many far and strange lands, therein experiencing

478 *Nigromantie* is a neologism from the Middle Ages that can be traced back to Isidor (ca. 530–636), the archbishop of Seville. It is derived from the word *necromancy* ("divination of the dead" or "divination by the dead"—i.e., conjuring the dead).

479 From Heinrich Marzell, *Wörterbuch der deutschen Pflanzennamen* [Dictionary of German Plant Names], 5 vols. (Leipzig, 1943–1979); Hanns Bächtold-Stäubli, ed., *Handwörterbuch des deutschen Aberglaubens* [Hand Dictionary of German Superstitions], 10 vols. (Berlin, 1927–1942).

480 The "iron plant" of antiquity was known as *hiera botane*, "sacred plant," and was presumably not identical to *Verbena officinalis*, because the ancient literature (Pliny 25.105 f.) ascribed it with powerful magical and mind-altering properties (cf. Rätsch 1995, 154–57*). The iron plant is also known by the folk names *druidenkraut* ("druid plant"), *merkurblutkraut* ("Mercury's blood plant"), and *sagenkraut* ("plant of legend"). *Verbena officinalis* has no psychoactive properties of any kind.

481 "In the sixteenth century, scholars such as Cardano or Della Porta formulated a different view: animal transformations, flying, and manifestations of the devil were the effects of malnutrition or the use of hallucinogenic substances such as found in plant decoctions or in ointments. These explanations have not yet lost any of their fascinating effects. Yet in and of itself, no form of deprivation, no substance, and no ecstatic technique, is able to provoke the repeated appearance of such complex experiences. In the face of this biological determinism, it is emphatically important to remember that the key to this codified repetition can only be a cultural one. Naturally, the intentional ingestion of psychotropic or hallucinogenic substances, even when they do not explain the ecstasy of the followers of the nocturnal goddess, the werewolves, etc., nevertheless places them in a dimension that is not exclusively mythical. Would it be possible to prove the existence of such ritual frameworks?" (Ginzburg 1990, 296 f.)

In pagan times, the yew (*Taxus baccata*) was a sacred tree. Today, it is feared as poisonous. However, its red berries can be eaten without harm. According to folk tradition, one should not sleep beneath a yew tree, lest this lead to insanity and hallucinations.

many different things, and also that the ointments described are not evil. (in Hauschild et al. 1979, 37)

Wier also described an "oil" that he invented himself that allegedly induces the exact same effects as those attributed to the witches' ointments. It consisted of:

lolium	bearded darnel (*Lolium temulentum*)
hyoscyamus	black henbane (*Hyoscyamus niger*)
cicuta	hemlock (*Cicuta virosa* or *Conium maculatum*)
papaver ruber	"red poppy"; possibly *Papaver rhoeas*
papaver niger	"black poppy"; possibly *Papaver somniferum*
lactuca	"lettuce"; possibly *Lactuca virosa*
portulacca	purslane (*Portulaca* sp.)
solanum somniferum fructus	fruits of ashwaganda (*Withania somnifera*); also possibly belladonna (*Atropa belladonna*)
opium thebaicum	opium from *Papaver somniferum*

This mixture was said to induce a long-lasting sleep with hallucinatory dreams. The English politician, philosopher, and author[482] Francis Bacon (1561–1626) discussed the recipe of the Italian Cardano in his text *The Oyntment That Witches Use*. However, he misinterpreted the nature of the "soot," replacing it with wheat flour. But soot almost certainly referred to grain smut (also known as grain soot) or even rye ergot (*Claviceps purpurea*). In addition, Bacon conjectured that the New World plants tobacco (cf. **Nicotiana tabacum** and **Hyoscyamus spp.**) and thorn apple (*Datura stramonium*) were suitable ingredients.

Paracelsus was also said to be knowledgeable about witches' ointments and other magical substances (cf. **soporific sponge,** *Papaver somniferum*). Johannes Praetorius (1630–1680) mentioned this in his book, which first appeared in 1668:

Paracelsus reported that the witches made their witches' ointments from the flesh of young, newborn children. They cooked them like a mush, together with herbs that cause sleep, such as poppy, nightshades, chicory, hemlock, and such. When the witches then smeared themselves with the ointment and uttered the following words: up and out, and

off to nowhere, then in his opinion they believed that they could travel away through the fire walls, the windows, and through other small holes with the help of the devil. (Praetorius 1979, 40)

Hoffmann (1660–1742), a chemist from Halle (Germany), placed the yew (*Taxus baccata* L.),[483] which is considered to be poisonous, alongside the nightshades and opium as ingredients in "sleeping ointments." The Germanic peoples regarded the yew as a magical tree, and it was one of the rune names (*eihwaz*). It is entirely possible that the yew does have psychoactive properties (cf. **kinni-kinnick**). During a witchcraft trial in 1758, a witches' ointment was said to contain the following ingredients: "mandrake root, henbane seeds, nightshade berries, opium juice" (Grünther 1992, 24). That this document lists other ingredients than the fat of a child is a complete exception to the standard recipes that were presented at witch trials.

Speculative recipes continued to appear in the literature of the nineteenth century. In one of his tales of witches, Ludwig Bechstein (1801–1860), the Romantic compiler of fairy tales, mentions the ointment of a weather witch that consisted of marsh violet (*Pinguicula vulgaris* L.), bewitching herb (possibly *Conyza* sp.), agrimony (*Agrimonia eupatoria* L.), black horehound (*Ballota nigra* L.), and devil's bit scabious (*Scabiosa succisa* L. [syn. *Succisa pratensis* Moench]); the mixture was rubbed onto the arms (Bechstein 1985, 255).

The idea that the fly agaric mushroom (see **Amanita muscaria**) might have been an ingredient in witches' ointments is a twentieth-century notion. Another recent assumption is that the witches' ointments were applied vaginally or rectally using a "witches' broom," a kind of dildo.

It was only toward the close of the nineteenth century that several brave individuals actually began to prepare and try the recipes that had been passed down. One of these was the esoteric scholar Carl Kiesewetter (1854–1895). After a number of seemingly successful experiments, he appears to have died as a result of a careless overdose. The spectacular and often cited report of Wilhelm Mrsich is probably more a literary product than an actual human pharmacological experiment, for he intentionally divulged no recipes (Mrsich 1978).

The only self-experiment that appears to have been authentic (using Portas's recipe) seems to have been that of the German folklorist Will-Erich Peukert (1895–1969):

We had wild dreams. At first, horribly distorted faces danced before my eyes. Then I suddenly had the feeling [that] I was flying through the air for miles. The flight was repeatedly interrupted by deep dives. In the

---

482 It is likely that Bacon wrote most, if not all, of the plays attributed to William Shakespeare. This is the reason Shakespeare's pharmacological details concerning poisons and medicinal plants are so correct (cf. Tabor 1970*).

483 Although the yew tree is regarded almost everywhere as very poisonous, it also finds use in ethnomedicine. In India, a beverage called *jya* is cooked from a few pieces of the bark and some salt and ghee (clarified butter). It is said to bestow mental power and vitality (Shah and Joshi 1971, 419*).

closing phase, there was an image of an orgiastic party with grotesque sensual excesses. (Peukert 1960)

The experiences of Siegbert Ferkel, cited in Marzell's book (1964, 48*), also appear to be authentic. Hanscarl Leuner undertook a self-experiment but did not experience any effects (Leuner 1981, 67*). Most of the authors who have published works on witches' ointments in recent years (e.g., Duerr, Grünther, Hansen, Harner, Kuhlen, vom Scheidt, Vries, Yilmaz) have been unable to report on any personal experiences. In contrast, the historian Walter Ulreich tried a homemade ointment of belladonna, poppy, hem-

lock, "other herbs," and pig fat; he too, however, did not meet with any success. The ride to the

Above left: The Romans called the white water lily (*Nymphaea alba*) Hercules' club. It was regarded as an anaphrodisiac in late antiquity, while in the early modern era it was considered to be an ingredient in witches' ointments. Whether it has psychoactive effects like its Mexican and Egyptian relatives is unknown. It is conceivable that the plant could produce synergistic effects when combined with other substances.

Above right: The black hellebore or Christmas rose (*Helleborus niger*) has been associated with black magic and the preparation of witches' ointments since the Middle Ages.

Ingredients in the So-called Witches' Ointments

plant products	presumed stock plant
aconite	*Aconitum napellus*
agrimony	*Agrimonia eupatoria* L.
asafoetida	*Ferula asafoetida* L.
bearded darnel, ryegrass	*Lolium temulentum*
betel nut (black)	*Areca catechu*
bewitching herb	*Conyza* sp.
black horehound	*Ballota nigra* L.
botrychium lunaria	*Botrychium lunaria* L. Sw.
calamus	*Acorus calamus*
celandine	*Chelidonium majus* L.
celery (wild)	*Apium graveolens* L.
celery juice	*Apium* or *Aethusa* (?)
cicuta[484]	*Cicuta virosa* L.
	Conium maculatum L. [syn. *Cicuta maculata* Gaertn.]
devil's bit scabious	*Scabiosa succisa* L.
dragon's blood	*Dracaena cinnabari* Balf. f.
fly agaric mushroom	*Amanita muscaria*
fool's parsley	*Aethusa cynapium* L.
hemp	*Cannabis sativa*
hyoscyamus	*Hyoscyamus niger*
	Hyoscyamus spp.
lily	*Iris* sp.
magsaamen	*Papaver somniferum*
mandragora	*Mandragora* spp.
nasturium	*Nasturtium* sp.
nightshades	*Solanum* spp.
olibanum, frankincense	resin of *Boswellia sacra*, *Boswellia* spp.
opium thebaicum	opium from *Papaver somniferum*
papaver niger	*Papaver somniferum*
papaver ruber	*Papaver rhoeas*
pastinaca	*Pastinaca* sp.
pentaphyllum	*Potentilla* sp. (?)
pepper	*Piper nigrum* L.
populi	*Populus nigra* L.
portulaca	*Portulaca* sp.
saffron	*Crocus sativus*
smallage	*Apium graveolens* (?)
Smyrna paste	opium (?)
solano, solanum	*Solanum* spp.
	Datura spp.
solanum somniferum	*Withania somnifera*
	Atropa belladonna
soot	cereal blight, smut (*Ustomycetes*)
	Claviceps purpurea
spurge	*Euphorbia* spp.
stramonii	*Datura stramonium*
sweet sedge	*Acorus calamus*
thebaicum	= opium
tobacco	*Nicotiana tabacum*
tollkraut	*Scopolia carniolica*
	Atropa belladonna
tormentil	*Potentilla* spp.
vervain	*Verbena officinalis* L. (?)
water lily	*Nymphaea alba* L.
	Nuphar lutea
water merck	*Apium graveolens* or *Sium* sp.
white hellebore	*Veratrum album*
	Helleborus spp.
wild lettuce	*Lactuca virosa*, *Lactuca sativa* L.
wolf's plant	*Aconitum* spp.
yew	*Taxus baccata* L.

Animal products

badger's lard, bat's blood, fox's lard, cat's brain, child's fat, child's blood, infant's blood, wolf's blood, wolf's fat, wolf's lard, toad poison (*Bufo bufo*), Spanish fly (*Lytta vesicatoria*), bird's blood, hoopoe's blood, owl's blood, tawny owl's blood, vulture's fat

Other

oil, salt, rust (?), communion hosts, wine

484 In the literature, one occasionally encounters the claim that hemlock is psychoactive in and of itself. Since both *Conium maculatum* and *Cicuta virosa* are among the most well studied of all medicinal and poisonous plants, it can be presumed that any purported psychoactivity would not have escaped the pharmacologists of the past three thousand years. Both hemlocks contain highly toxic alkaloids, furanocoumarins, and bioactive polyacetylenes (Teuscher 1992; Wittstock et al. 1992; Wittstock et al. 1995).

Witches preparing an ointment to apply to a young novice (illustration from the magazine *Jugend* [Youth], 1926)

"Tonight I rode far on my staff, and I now know things I did not know before."

Fóstbroedhara saga
(late thirteenth century)

"She [the witch] knows no greater delight than mixing the body of God [= the host] into her filthy salves, to stuff it into her sexual organs and to season the rotten carrion of a defiled corpse with it."

Stanislaw Przybyszewki
The Synagogue of Satan
(2002, 30)

In the Middle Ages, vervain (*Verbena officinalis* L.) was a renowned magical plant that was used for magical protection and as an aphrodisiac. In the early modern era, it was thought to be one of the ingredients in witches' ointments. Vervain has no known psychoactive effects. The magical plant that classical texts refer to as *heira botane* ("sacred plant") has not yet been botanically identified. (Woodcut from Brunfels, *Contrafayt Kreuterbuch*, 1532)

Blocksberg (a famous witches' gathering place) did not occur ("On the Belladonna," in *People in Motion*, Summer 1996, 66 f.).

The pharmacognosy of witches' ointments is far from understood and represents a worthwhile area for human pharmacological research. Future researchers, however, must be willing to conduct self-experiments. Experimental research is also needed to determine whether the active constituents contained in the ingredients are actually able to be absorbed through the skin (cf. Grünther 1992, Waldvogel 1979).

The notion of hallucinogenic witches' ointments has produced a rich trove of prose literature. Here are just a few stories and novels of particular interest: Bechstein 1986; Delaney 1994; Görres 1948; Meyrink 1984, 179–186*; and Tieck 1988.

Literature

Apuleius. 1998. *The golden ass or metamorphoses*. Translation, notes, preface by E. J. Kenney. London: Penguin.

Barnett, Bernard. 1965. Witchcraft, psychopathology and hallucinations. *British Journal of Psychiatry* 3:439–45.

Bechstein, Ludwig. 1986. *Hexengeschichten*. Frankfurt/M.: Insel.

Biedermann, Hans. 1974. *Hexen—Auf den Spuren eines Phänomens*. Graz: Verlag für Sammler.

Caro Baroja, Julio. 1967. *Die Hexen und ihre Welt*. Intro. by Will-Erich Peuckert. Stuttgart: Klett.

Daxelmüller, Christoph. 1996. *Aberglaube, Hexenzauber, Höllenängste*. Munich: dtv.

Delany, Daniel. 1994. *Der Hexentrank*. Munich: Ehrenwirth.

Dinzelbacher, Peter. 1995. *Heilige oder Hexen?* Zurich: Artemis und Winkler.

Dross, Annemarie. 1978. *Die erste Walpurgisnacht*. Reinbek: Rowohlt.

Duerr, Hans-Peter. 1976. Können Hexen fliegen? *Unter dem Pflaster liegt der Strand* 3:55–82.

———. 1978. *Traumzeit*. Frankfurt/M.: Syndikat.

Ehrenreich, Barbara, and Deike English. 1981. *Hexen, Hebammen und Krankenschwestern*. Munich: Frauenoffensive.

Evans, Arthur. 1978. *Witchcraft and the gay counterculture*. Boston: Fag Rag Books.

Ferkel, Siegbert. 1954. "Hexensalben" und ihre Wirkung. *Kosmos* 50:414–15.

Fühner, Hermann. 1919. Solanazeen als Berauschungsmittel: eine historische Studie. *Archiv für experimentelle Pathologie und Pharmakologie* 111:281–94.

Gardner, Gerald B. 1965. *Ursprung und Wirklichkeit der Hexen*. Weilheim: O. W. Barth.

Ginzburg, Carlo. 1980. *Die Benandanti*. Frankfurt/M.: Syndikat.

———. 1990. *Hexensabbat*. Berlin: Wagenbach.

Görres, Josef von. 1948. *Das nachtländische Reich*. Villach: Moritz Stadler.

Grünther, Ralf-Achim. 1992. Hexensalbe— Geschichte und Pharmakologie. *Jahrbuch des Europäischen Collegiums für Bewußtseinsstudien* (ECBS) 1992:21–32.

Haage, Bernhard. 1984. Dichter, Drogen und Hexen im Hoch- und Spätmittelalter. *Würzburger medizinhistorische Mitteilungen* 4:63–83.

Hansen, Harold A. 1981. *Der Hexengarten*. Munich: Trikont-dianus.

Harner, Michael. 1973. The role of hallucinogenic plants in European witchcraft. In *Hallucinogens and shamanism*, ed. Michael Harner, 125–50. London: University of Oxford Press.

Hartlieb, Johannes. 1989a. *Das Buch aller verbotenen Künste*. Ed., trans., commentary by Falk Eisermann and Eckhard Graf. Ahlerstedt: Param.

———. 1989b. *Das Buch aller verbotenen Künste*. Ed., trans. Frank Fürbeth. Frankfurt/M.: Insel.

Hauschild, Thomas. 1981. Hexen und Drogen. In *Rausch und Realität*, ed. G. Völger, 1:360–66. Cologne: Rautenstrauch-Joest-Museum.

Hauschild, Thomas, Heidi Staschen, and Regina Troschke. 1979. *Hexen: Katalog zur Ausstellung*. Hamburg: Hochschule für bildende Künste.

Howard, Michael. 1994. Flying witches: The *unguenti sabbati* in traditional witchcraft. In *Witchcraft and shamanism* (Witchcraft Today, vol. 3), ed. Chas. S. Clifton, 35–55. Saint Paul: Llewellyn.

Kiesewetter, Carl. 1902. *Die Geheimwissenschaften*. 2nd ed. Leipzig: Wilhelm Friedrich.

Kuhlen, F.-J. 1980. Hexenwesen—Hexendrogen. *Pharmaziegeschichtliche Rundschau* 9:29–31, 41–43.

———. 1984. Von Hexen und Drogenträumen. *Deutsche Apotheker-Zeitung* 124:2195–2202.

Labouvie, Eva. 1991. *Zauberei und Hexenwerk: Ländlicher Hexenglaube in der frühen Neuzeit*. Frankfurt/M.: Fischer.

Leubuscher, Rud. 1850. *Ueber die Wehrwölfe und Thierverwandlungen im Mittelalter*. Berlin: G. Reimer.

Luck, Georg. 1962. *Hexen und Zauberei in der römischen Dichtung*. Zurich: Artemis.

Madejsky, Margret. 1997. Hexenpflanzen—oder: Über die Zauberkünste der weisen Frauen. *Naturheilpraxis* 50 (10): 1552–63.

Mrsich, Wilhelm. 1978. Erfahrungen mit Hexen und Hexensalben. *Unter dem Pflaster liegt der Strand* 5:109–19.

Peuckert, Will-Erich. 1960. Hexensalben. *Medizinischer Monatsspiegel* 8:169–74.

Piomelli, Daniele, and Antonio Pollio. 1994. *In upupa o strige:* A study in Renaissance psychotropic plant ointments. *Hist. Phil. Life Sciences* 16:241–73.

Porta, Giambattista della. 1589. *Magia naturalis.* Lugduni. (Original Latin edition.)

———. 1680. *Haus-Kunst- und Wunderbuch.* (German edition).

———. 1957. *Natural magick.* New York: Basic Books. (English version.)

Praetorius, Johannes. 1979. *Hexen-, Zauber- und Spukgeschichten aus dem Blocksberg.* Frankfurt/M.: Insel.

Przybyszewski, Stanislaw. 2002. *The synagogue of Satan.* Trans. Istvan Sarkady. Smithville, Texas: Rûna-Raven Press. (Orig. pub. 1900.)

Quayle, Eric, and Michael Foreman. 1986. *The magic ointment and other Cornish legends.* London: Macmillan.

Richter, E. 1960. Der nacherlebte Hexensabbat. *Forschungsfragen unserer Zeit* 7:97 ff.

Robbins, Rossell Hope. 1959. *The encyclopedia of witchcraft and demonology.* New York: Crown Publ.

Rothman, T. 1972. De Laguna's commentaries on hallucinogenic drugs and witchcraft in Dioscorides' materia medica. *Bulletin of the History of Medicine* 46:562–67.

Schmitt, Jean-Claude. 1993. *Heidenspaß und Höllenangst: Aberglaube im Mittelalter.* Frankfurt/M. and New York: Campus.

Sebald, Hans. 1990. *Hexen: Damals—und heute?* Frankfurt/M. and Berlin: Ullstein.

Spilmont, Jean-Pierre. 1984. *Magie.* Munich: Heyne.

Stramberg, Chr. von, et al. 1986. Hexenfahrten. In *Spuk- und Hexengeschichten,* ed. Hermann Hesse, 26–62. Frankfurt/M.: Insel. (Orig. pub. in *Rheinischen Antiquarius* 2 (4), 334–61, ca. 1851.)

Tieck, Ludwig. 1988. *Hexen-Sabbat.* Frankfurt/M.: Insel.

Van Dülmen, Richard, ed. 1987. *Hexenwelten: Magie und Imagination.* Frankfurt/M.: Fischer.

Völker, Klaus, ed. 1977. *Von Werwölfen und anderen Tiermenschen.* Munich: dtv.

Vom Scheidt, Jürgen. 1984. Hexensalben. In *Handbuch der Rauschdrogen,* by W. Schmidbauer and J. vom Scheidt, Frankfurt/M.: Fischer.

Vries, Herman de. 1986. Über die sogenannten Hexensalben. *Salix* 2 (2).

———. 1991. Über die sogenannten Hexensalben. *Integration* 1:31–42. (Revised version of von Vries 1986.)

Waldvogel, Ruth. 1979. *Pharmakopsychologie der Hexensalben.* Manuscript from an independent project at the Psychologischen Institut der Universität Zurich, Zurich.

Werneck, Heinrich L., and Speta Franz. 1980. *Das Kräuterbuch des Johannes Hartlieb.* Graz: Akademische Druck- und Verlagsanstalt (ADEVA).

Weyer, Johann. 1575. *Von den Teuffeln/Zaubrern/Schwarzkünstlern/Hexen . . .* Frankfurt/M.

Wittstock, Ute, Franz Hadacek, Gerald Wurz, Eberhard Teuscher, and Harald Greger. 1992. Bioactive polyacetylenes from *Cicuta virosa. Planta Medica* 58 suppl. (1): A722–23.

———. 1995. Polyacetylenes from water hemlock, *Cicuta virosa. Planta Medica* 61:439–45.

Yilmaz, Martina. 1985. *Zauberkräuter Hexengrün.* Berlin: Johanna Bohmeier.

"In order to investigate the so-called witches' ointments, we must remind ourselves of the actual meaning of the natural sciences as an expansion of human consciousness, as a deepened insight into the nature of reality, into the unity of all living creatures, and the integration of humans in the bio-universe. When seen in this manner, the insights of the natural sciences into the microcosm and the macrocosm of objective reality, together with our subjective perceptions and our own mystical experience, could open the door to the spiritual world."

PATRICIA OCHSNER
ÜBER HEXENKULT, FLUGSALBEN AND HEXENKRÄUTER [ON THE WITCHES' CULT, FLYING OINTMENTS, AND WITCHES' HERBS]
(1997)

"Faces floated toward me from the darkness, first blurry and then taking on form. . . . I floated upward at great speed. It became light and through a pink veil I hazily recognized that I was floating over the city. The figures that had already oppressed me in my room accompanied me on this flight through the clouds. More and more of them joined me, and every minute lasted an eternity. The next morning, as the first light came into my room, I thought I had awakened to a new life."

SIEGBERT FERKEL, ON AN EXPERIENCE WITH A WITCHES' OINTMENT IN *ZAUBERPFLANZEN— HEXENTRÄNKE* [SPELL PLANTS— WITCHES' DRINKS]
(MARZELL 1964, 48*)

Zombie Poison

"My first impression of the three 'zombies,' who apathetically continued with their work, was that there was something unnatural and strange about them. They appeared to do their work completely automatically, like robots, and held their heads so low that I would have had to bend down to see into their faces. . . . The worst thing was the eyes. No, it had nothing to do with my imagination. They were in truth the eyes of a corpse, not blind, but gazing rigidly into emptiness, without expression, lifeless, nonseeing. The entire face was like that, saying nothing, slack and empty. It appeared not just to be without expression, but to not even be capable of any expression at all. . . . These 'zombies' were nothing more than pitiable insane people, idiots, that were forced to perform heavy work in the fields."

W. B. SEABROOK
GEHEIMNISVOLLE HAITI [THE SECRET ISLAND]
(1931, 107, 108)

Other Names

Zombie powder, zombi poison

The voodoo cult (also called *vodou, vodun,* or *wodu*) is a syncretic system composed of elements of traditional Yoruba religion, early-modern-era Catholicism, and various influences from the magic of India, Native American concepts, and occult practices. The voodoo cult is practiced almost exclusively by the descendants of slaves brought to the New World from Africa and is centered in Haiti. Voodoo and other similar cults are also practiced on other Caribbean islands, in the southeastern United States, and in the northern areas of South America. Voodoo is a possession cult in which the participants individually or collectively enter into trance in order to have spiritual experiences, to heal, or to perform divinations (Planson 1975). Whether psychoactive substances play a significant part in this remains an open question (cf. **madzoka medicine**).

Although the zombie phenomenon is concentrated in Haiti, it is also known in Guadeloupe. Zombies, the "living dead," were long regarded as the stuff of legend and dismissed as mere folklore (Metraux 1972). The fact that zombies have nothing to do with folklore but are a real phenomenon with a sociocultural background was first reported by Seabrook (1931), who even claimed to have encountered zombies.[485] Zombies are not the resurrected dead but people who have been turned into the living dead (the process is known as zombification) by a poison known as zombie poison. Zombies are "in reality certainly individuals who have been artificially placed into a state

Ingredients in the Zombie Poison

(from Davis 1983b, 1988)

Haitian Name	Botanical Name	Active Constituent
PLANTS		
bois piné	*Zanthoxylum martinicense* (Lam.) DC.	
bresillet	*Comocladia glabra* Spreng.	
calmador	*Dieffenbachia sequine* (Jacq.) Schott.	calcium oxalate
concombre zombi	***Datura stramonium*** L.	**tropane alkaloids**
consigne	*Trichilia hirta* L.	
desmembre	unidentified	
maman guêpes	*Urera baccifera* (L.) Gaud.	
mashamasha	*Dalechampia scandens* L.	
pois gratter	***Mucuna pruriens*** (L.) DC	indoles, DMT
pomme cajou	*Anacardium occidentale* L.	
tcha-tcha	*Albizia lebbeck* (L.) Benth. [syn. *Mimosa lebbeck* L., *Acacia lebbeck* (L.) Willd.]	?[486]
tremblador	unidentified	
ANIMALS		
bango	*Bufo marinus* L.	**bufotenine**, glycosides
bilan	*Diodon holacanthus*	tetrodotoxin
centipedes	Spirobolida and Polydesmida Orders	alkaloids, glomerines
crabe araignée	Therphosidae (tarantula)	
crapaud blanc	*Osteopilus dominicensis* Tschudi	
crapaud de mer	*Sphoeroides testudineus* L.	tetrodotoxin
fou-fou	*Diodon hystrix* L.	tetrodotoxin
lèzard	*Ameiva chrysolaema* Cope / *Leiocephalus schreibersi* Graven.	
mabouya	*Epicrates striatus* Fischer	
miti verde	*Anolis coelestinus* Cope	
"serpente"	*Hermodice carunculata* Pallas	toxines (?)
zanolite	*Anolis cybotes* Cope	
OTHER		
human bones	*Homo sapiens sapiens*	

485 Zombies and "voodoo people" number among the figures that Westerners often hallucinate while under the influence of a psychoactive substance, e.g., *Datura* (Siegel 1981, 325*).

486 In West Africa, a drug known as *ibok usiak owo* is obtained from the closely related species *Albizia zygia*. It was used in ordeals and as a truth serum (Davis 1983b, 144). The first part of the drug's name is reminiscent of that of iboga (see *Tabernanthe iboga*).

Ingredients in the Antidotes for Zombie Poison

(from Davis 1983b)

PLANTS (FRESH OR DRIED LEAVES ONLY)

aloe	*Aloe vera* L.
guaiac	*Guaiacum officinale* L.
cedre	*Cedrela odorata* L.
bois ca-ca	*Capparis cynophyllophora* L.
bois chandelle	*Amyris maritima* Jacq.
cadavre gaté	*Capparis* sp.
bayahond	*Prosopis juliflora* (Sw.) DC. (cf. ***Agave*** spp., **chicha**)
ave	*Petiveria alliacea* L.

OTHER

mineral salt[487]
mothballs (naphthalene)
seawater
clairin (inexpensive perfume)
human bones
dog skulls
mule tibia
talcum (= talc, talcum powder; magnesium silicate)
sulfur powder

of apparent death, buried, and then awakened and dug up, and are consequently as obedient as beasts of burden, as they must assume in good faith that they are dead" (Leiris 1978, 9). The earliest assumption was that the *bokors*,[488] the voodoo sorcerers, used ***Datura stramonium*** to poison their victims. In Haiti and other Caribbean islands (e.g., Guadeloupe), *Datura* is called *concombre zombi*, "zombie cucumber." In the Dominican Republic, the fruit of *Passiflora rubra* (cf. ***Passiflora*** spp.) is known as *pomme de liane zombie*, "the potato of the zombie liana" (von Reis and Lipp 1982, 197*). Other psychoactive substances have also been associated with zombification. One of the popular street names for PCP (= "angel dust") is *zombie weed* (Linder et al. 1981, 10).

In the early 1980s, the American ethnobiologist Wade Davis was able to obtain in Haiti numerous samples of as well as the corresponding recipes for zombie poisons that were purportedly used for the zombification of humans (Davis 1983a, 1983b). The *bokors* prepare the poison and use it when a payment has been made. Surprisingly, the main ingredients of the recipes are plants and animals that produce psychoactive effects (see table, page 806). If possible, the zombie poison must be administered to the chosen victim through skin contact. Moreover, one "treatment" is usually not enough. Instead, the victim must come into repeated contact with the poison, often over a period of weeks, so that he will fall into a state of

apparent death and can be buried and then dug up (Davis 1986). When the apparently "dead" victim is dug up, he is given an antidote that, so to speak, restores him to life. The antidote can also be used to protect a person from the poison (Davis 1983b).

The Effects of Zombie Poison

Davis has assumed that the primary active constituent in zombie poison is tetrodotoxin, a substance derived from the porcupine fish (*Diodon hystrix*). Although this substance can elicit states of apparent death, in higher concentrations it can induce actual death (Davis 1988). Tetrodotoxin ($C_{11}H_{17}N_3O_3$) is a neurotoxin and is one of the most poisonous, nonproteinaceous substances known, inducing complete neuromuscular paralysis. It is sixty times more potent than **strychnine** or D-tubocurarine (the active constituent in curare), and some five hundred to one thousand times more potent than hydrocyanic acid. For a man weighing 70 kilos, 0.5 mg of the pure substance can be fatal, and 20 g of the fish's skin is lethal (Davis 1988, 145; Gage 1971). One piquant effect of tetrodotoxin is the occasional induction of apparent death with full consciousness, a condition that cannot be physiologically distinguished from true death, or only with extreme difficulty. As a result, many victims have been buried alive (Davis 1986, 1988).

However, Davis's assumption that tetrodotoxin is the primary active constituent has been the subject of some contention (Anderson 1988). Though tetrodotoxin has been found to be present in the liver of the Caribbean porcupine fish, it was not detected in any of the samples of zombie poison that Davis was able to obtain (Yasumoto and Kao 1986).

Tetrodotoxin, a Potentially Psychoactive Substance

Japanese waters are home to a genus of pufferfish (*Fugu* spp.), known as fugu, that is prized in Japanese cuisine. Fugu, however, has one small drawback: it contains tetrodotoxin. Fugu was mentioned in the oldest Chinese herbal, the *Pên-ts'ao Ching*. As early as the Han dynasty (202 B.C.E.–220 C.E.), the poison was known to be

Guaiac wood (*Guaiacum officinale*) is regarded as an ingredient in the antidotes for zombie poison. Because of its resin content, the very hard and heavy wood is also used as an incense.

"If work was beautiful, then the rich would not have left it for the poor to do."

HAITIAN SAYING

"The plantation and landowners had more or less smiled about it and claimed that zombies had simply been invented by the Danes, so that the fieldworkers would remain in their huts after the break of night and in this way would preserve the nocturnal peace and also minimize the damage from plundering the ripening harvest."

HENRY S. WHITEHEAD
DER ZOMBIE [THE ZOMBIE]
(1986, 8)

487 Seabrook noted that the food the zombies eat cannot contain salt of any kind; otherwise, the effects of the zombification will be counteracted (1931, 103, 105).

488 The sorcerers, who are able to transform themselves into wolves or other animals, are called *loups garous* in Haiti, and they are greatly feared by the people (Simpson 1942). It is possible that these sorcerers use a preparation made with cathedral bells (*Kalanchoe pinnata* [Lam.] Pers.; Crassulaceae), which is known locally as *loup garou*.

present in fugu liver. The poison is also present in the fish's skin, ovaries, and viscera. The poison immediately attacks the tongue and the viscera. Fugu poisoning was regarded as a disease for which there is no cure. Nevertheless, since at least 1596, these fish have been regarded as one of the most refined, expensive, and prized delicacies of Japanese cuisine. Fugu is "considered one of those rare delicacies that straddles the border between food and drug" (Davis 1986, 162). The following species are eaten in Japan:

Fugu rubripes rubripes Temminck et Schlegel
Fugu paradalis Temminck et Schlegel
Fugu vermicularis prophyreus Temminck et
    Schlegel
Fugu vermicularis vermicularis Temminck et
    Schlegel

The art of preparing fugu, which in Japan requires an examination and a license, does not entail removing the poison from the fish but, rather, leaving a trace of the poison in the food. Usually only well-to-do businessmen and the women they have hired for the night are present in fugu restaurants. There are approximately two thousand licensed chefs and a similar number of specialty restaurants. A fugu dinner is expensive and can cost as much as $300 to $600 per person. The erotic meal is also risky. For fugu patrons, playing with death has an almost erotic allure, and a meal of fugu is the ultimate aesthetic experience. Fugu is extremely delicious and produces an incredible effect!

Chiri is fugu meat that has been cooked in a soup containing the poisonous viscera. As a result, it is impregnated with the poison and has become an inebriating, euphoric, and aphrodisiac drug (Davis 1988, 152). The effects are phenomenal. At first, a pleasant tingling runs up one's back and head and hairs seem to vibrate where they meet the skin. One becomes mentally alert, awake, and libidinous. The effects of the alcohol in **beer** or **sake** are suppressed (in a manner similar to the effect of **cocaine**). Slowly, all of one's muscles become filled with an enormous tension and erotic prickling. Streams of energy swirl around the spinal column. One feels completely electrified. The effects on the sacral ganglia are very clear. In a male, this stimulation quickly becomes noticeable as a hard erection. Sexual desires are amazingly heightened. After my own experience, I understand all too well that it is best to have a fugu dinner with a romantic companion. I have personally found fugu to be a powerful psychoactive aphrodisiac.

And thus, in the repressive and otherwise "drug-hostile" society of Japan (where possession of 1 g of hashish can be punishable by five years of prison), fugu is the only legal psychotropic substance apart from tea (*Camellia sinensis*), coffee (*Coffea arabica*), **alcohol**, and **nicotine**. Although fugu is clearly an inebriating drug, it has been spared the negative connotations normally associated with drugs because it is regarded as a food (and a delicacy).

Literature

Anderson, William H. 1988. Tetrodotoxin and the zombi phenomenon. *Journal of Ethnopharmacology* 23:121–26.

Davis, E. Wade. 1983a. The ethnobiology of the Haitian zombi. *Journal of Ethnopharmacology* 9:85–104.

———. 1983b. Preparation of the Haitian zombi poison. *Botanical Museum Leaflets* 29 (2): 139–49.

———. 1986. *Die Toten kommen zurück: Die Erforschung der Voodoo-Kultur und ihrer geheimen Drogen.* Munich: Droemer Knaur.

———. 1988. *Passages of darkness: The ethnobiology of the Haitian zombie.* Chapel Hill, N.C., and London: University of North Carolina Press.

Gage, P. W. 1971. Tetrodotoxin and saxitoxin as pharmaceutical tools. In *Neuropoisons: Their pharmacological actions,* ed. L. L. Simpson, 187–212. New York and London: Plenum Press.

Leiris, Michel. 1978. *Das Auge des Ethnographen.* Frankfurt/M.: Syndikat.

Linder, Ronald L., Steven E. Lerner, and R. Stanley Burns. 1981. *PCP: The Devil's Dust.* Belmont, Calif.: Wadsworth Publishing Co.

Metraux, Alfred. 1972. *Voodoo in Haiti.* New York: Schocken Books Edition.

Planson, Claude. 1975. *Vaudou: rituels et possessions.* Paris: Pierre Horay Editeur.

Seabrook, W. B. 1929. *The magic island.* New York: Harcourt Brace and Co.

———. 1931. *Geheimnisvolles Haiti: Rätsel und Symbolik des Wodu-Kultes.* Berlin: R. Mosse.

Simpson, George Eaton. 1942. Loup garou and loa tales from northern Haiti. *Journal of American Folklore* 55:219–27.

Whitehead, Henry S. 1986. *Der Zombie.* Frankfurt/M.: Suhrkamp.

Yasumoto, Takashi, and C. Y. Kao. 1986. Tetrodotoxin and the Haitian zombie. *Toxicon* 24:747–49.

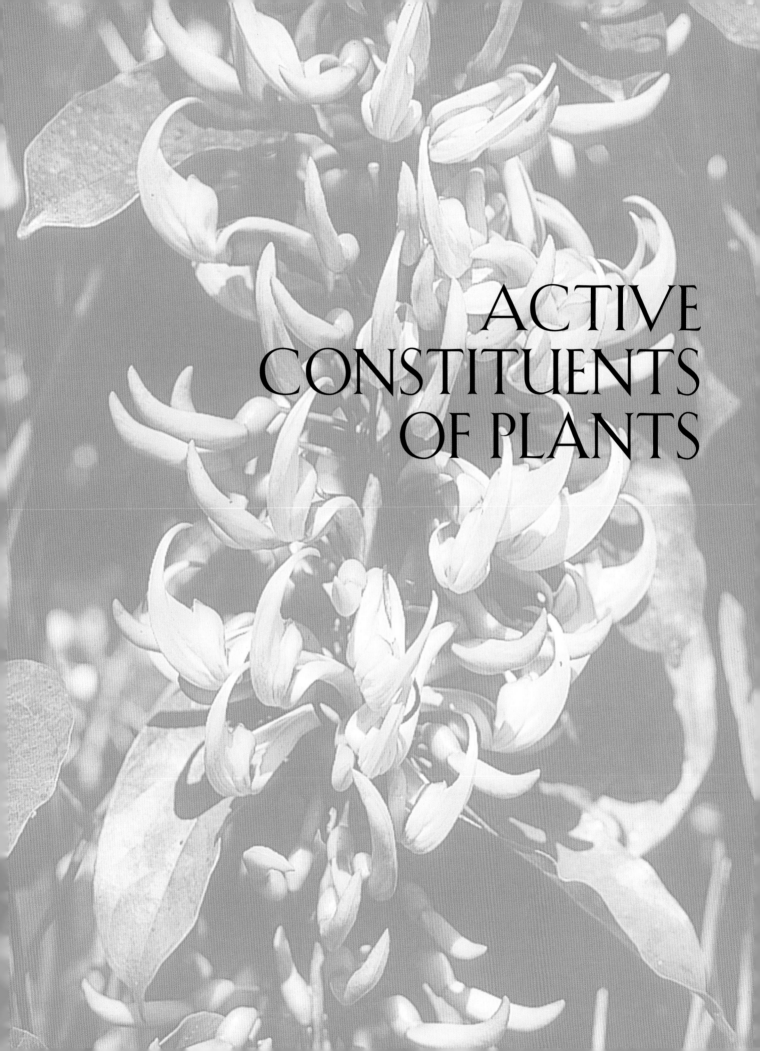

ACTIVE
CONSTITUENTS
OF PLANTS

"Chemistry is applied theology."

TIMOTHY LEARY
POLITICS OF ECSTASY
(1982, 271*)

"Life is the convergence of mind and matter."

GALAN O. SEID
(11/1996)

"It is in fact a miracle, a secret of the world in the full light of the day, to experience how the prosaic alcoholic matter in a glass of wine dispels sadness and worries, how ether fumes or chloroform can cause a person to temporarily lose consciousness, how morphine can dull even the most severe pain, and how Veronal, Luminal, and other preparations help a sleepless person to find slumber. No philosopher, no medical doctor, and no chemist has ever solved these riddles, and it is not likely that they will be solved in the future either. Today, and in the centuries to come, we will have to be satisfied with collecting the most comprehensive, certain, and reliable knowledge possible that stems from experience with these substances. And indeed, man will investigate ever more precisely how the countless natural and synthetic substances affect human beings, which concentrations and methods of administration are the most useful, which aftereffects appear, and how such factors as age, physical constitution, race, sex, occupation, and state of health have to be taken into consideration, etc. We should not look down upon such experience-based knowledge, for it leads to healing success or at least to the elimination of the anguishing symptoms of disease— and for a compassionate human being, this is certainly a matter of greater importance than the most wonderful theory about the physical and mental processes."

HERMANN RÖMPP
CHEMISCHE ZAUBERTRÄNKE
[CHEMICAL POTIONS]
(1950, 7*)

Over the past two centuries, pharmaceutical and pharmacological research has recognized that it is not the plants themselves that produce the particular effects but the principles or active constituents that reside within them. But what *is* an active constituent? Active constituents are chemically uniform substances (molecules) that can be extracted from plants with the help of solvents and which cause an effect when ingested. They either occur in the form of oily substances (bases) or can be crystallized out as salts.

Jonathan Ott, a chemist of natural substances, once discussed the effects of plants and plant spirits with a shaman in the Amazon. Ott used the metaphor that modern chemistry had discovered that the spirit of a plant exists in the form of a crystal. This picture made sense to the traditional shaman, for whom crystals are gateways to another reality, a kind of crystallized consciousness.

However, experiences with plants and their active constituents have made it evident that the effects of the molecule or the so-called main active constituent are not necessarily identical to the effects of the plant that contains the substance (cf. Storl 1996a*, 1996b*). The pharmacological explanation for this fact is that most plants contain a mixture of active constituents, and it is the synergy among these that is responsible for the characteristic effects of that plant. A plant usually induces a broader spectrum of effects than is produced by its isolated components. In other words, a pure constituent is more specific in its action.

There are people who believe that only natural substances are capable of producing good or tolerable effects, and that synthesizing those same molecules alters their pharmacological behavior so they will not work as well as the natural molecules. However, neither the chemical nor the pharmacological perspectives provide support for this notion.

Many people believe that "new" molecules that have been artificially produced by chemists, such as LSD, MDMA, and ketamine, are not as good as compounds found in the natural world. It would be wise to remember, however, that a chemist is not really able to create artificial molecules. All he or she can do is to make use of the properties of substances so that a specific molecule can form. Moreover, simply because a molecule that was first produced in a laboratory was at the time unknown in nature does not mean that the molecule does not occur in some plant that has not yet been discovered or studied. For example, *N,N*-DMT was first synthesized in a laboratory and described as an artificial molecule. Later, it was discovered that this same molecule occurs naturally in plants, animals, and even humans. Valium (= **diazepam**) is regarded as the artificial drug par excellence and feared as an addictive poison. Yet this substance, which was initially synthesized in a laboratory, has since been found to occur naturally in potatoes and cereal grains. It is quite likely that LSD, MDMA, and ketamine will soon be detected in psychoactive plants as well. Chemists are merely transformers of matter; they are not gods. All they do is apply a "divine law."

Active Plant Constituents and Neurotransmitters

Neurotransmitters, also known as transmitters or chemical messengers, are substances that are released at presynaptic nerve endings (or axon terminals), cross the synaptic cleft, and induce changes at the postsynaptic membrane, i.e., the next neuron (Black 1993*; Snyder 1989*; Spitzer 1996*). In other words, they are a chemical message that is sent out at the end of one neuron and read by the next. The first neurotransmitters to be discovered were the endorphins, substances that are produced in the body itself (= endogenous) and act within the nervous system in the same manner as **morphine** and other related opiates. We now know that the purposeful activation of particular neurotransmitters can induce psychoactive experiences or altered states of consciousness that resemble or even correspond to those induced by active plant constituents:

> Man is his own drug producer; he simply needs to relearn how to stimulate his endogenous drugs as he needs and desires. . . . Until now, the deliberate and targeted stimulation of endogenous drugs has been an unexplored territory in the field of biomedicine. But ritual healing cults and archaic methods of healing (shamanism, the voodoo cult, healing dances, yoga, meditation) include numerous elements for stimulating the production of endogenous drugs, although the participants of course are typically unaware of the biochemical background. (Zehentbauer 1992*)

In other words, the perspective offered by neurochemistry tells us that an alteration in a person's state of consciousness is always triggered by some type of drug, whether this is a neurotransmitter produced by one's own body or one that comes from outside in the form of an active plant constituent: "Endogenous drugs can produce effects similar to those produced by exogenous 'miracle drugs'" (Zehentbauer 1992, 113*).

Most shamans, however, prefer to use "plant spirits" to produce the states they desire, for the ingestion of a psychoactive substance is the most reliable method for achieving a desired altered state of consciousness. Mushrooms, for example, are always dependable. Shamans have no time for techniques that work only on occasion.

Research into migraines has shed some light onto the relationship between neurotransmitters and hallucinations or visions. Migraines are frequently accompanied by hallucinations (phosphenes, abstract patterns, strange shapes) whose form and content are often indistinguishable from the hallucinations or perceptual changes induced by the active constituents of plants. The only real difference is that migraines are very painful, whereas the states of consciousness produced by plant constituents are usually euphoric and pleasant in nature.

Analogies between Exogenous and Endogenous Neurotransmitters
(from Perrine 1996*; Snyder 1989*; Zehentbauer 1992*; supplemented)

Exogenous Neurotransmitter	Endogenous Neurotransmitter
TRYPTAMINES/INDOLES	
bufotenine	bufotenine
5-MeO-DMT	5-MeO-DMT
harmaline/harmine	β-carbolines: (harmane)
ibogaine	β-carbolines
lysergic acid derivates	endopsychedelics
N,N-DMT	*N,N*-DMT
psilocybin/psilocin	serotonin
strychnine	glycine
yohimbine	β-carbolines
PHENETHYLAMINES	
amphetamine and its derivatives (ephedrine, MDMA, etc.)	adrenaline
β-phenethylamine	β-phenethylamine
cocaine	noradrenaline
mescaline	dopamine
MORPHINE/OPIUM ALKALOIDS	
codeine	codeine
morphine	morphine
opiates/heroin	endorphins/ encephalins
TROPANE ALKALOIDS	
atropine	acetylcholine
hyoscyamine	acetylcholine
scopolamine	acetylcholine
VARIOUS GROUPS	
diazepam (Valium)	endovalium (= diazepam)
ibotenic acid	glutamate
muscimol	GABA
nicotine	acetylcholine
PCP/ketamine	angeldustine (= ketamine)
THC/cannabinoids	anandamide

"Molecules are alive!—I believe that chemical compounds can teach us just as well as plants."

DR. ALEXANDER T. SHULGIN
ASK DR. SHULGIN ONLINE,
DRUG INFORMATION WEB SITE
OF THE CENTER FOR COGNITIVE
LIBERTY AND ETHICS
(1/1996)

There is evidence that *all* natural neurotransmitters may show significant changes during an attack of migraine. . . . We see changes in adrenalin, nor-adrenalin, acetylcholine, histamine; and, often prominently, in 5-hydroxy-tryptamine (or serotonin). (Sacks 1985, 193*)

All of these neurotransmitters have their analogs in psychoactive plants: **cocaine**, **scopolamine/atropine**, **mescaline**, histamine, **muscimol/ibotenic acid**, **morphine**, **psilocybin**, and **psilocin**. This means that the ability to produce hallucinations and visions is a normal property of our nervous system. It apparently makes no difference whether these states are caused by endogenous neurotransmitters or by exogenous neurotransmitters, i.e., active plant constituents.

Several neurotransmitters that are vitally important in the human nervous system also occur in plants. Acetylcholine can be found in the scarlet runner bean (*Phaseolus coccineus* L. [syn. *Phaseolus multiflorus* Lam.]), in various mimosas (*Mimosa* spp.), in *Albizia julibrissin* Durazz., and in peas (*Pisum sativum* L.), all members of the Family Leguminosae. Serotonin occurs in numerous plants and fungi (*Panaeolus subbalteatus*). Bananas (*Musa* x *sapientum*) and the orange root (*Hydrastis canadensis* L.; Ranunculaceae) contain norepinephrine.

Why plants produce and contain human neurotransmitters is unknown to the scientific community (Applewhite 1973*). For the shaman, on the other hand, the answer is clear: The active constituent, the plant spirit, is a messenger in the neural network of nature. Every human, like every plant and animal, is one of infinitely many neurons in the nervous system of Gaia. The active constituents found in plants are the neurotransmitters, the communication system, of living nature.

Some psychoactive constituents can be found in humans as well as in other animals and plants. **Morphine**, for example, occurs in cow's milk, in the human brain, and in poppy juice. **Bufotenine** has been detected in human urine, in the secretions of certain toads, and in many plants and fungi. In those cases in which an active constituent of a plant is not identical to an endogenous substance, it is analogous. In other words, the active constituent behaves within the human nervous system in the same way as the endogenous neurotransmitter, occupying the same specific receptors at the neural terminals. This is the only reason that plant active constituents can do what they do. Substances that are not identical or analogous to neurotransmitters appear to be unable to induce any psychoactive effects (laughing gas is an exception[489]). Alcohol appears to affect neurotransmission in a variety of ways.

For the shaman, it is of no importance whether one of his plant spirits affects serotonin transmission, the adrenergic system, or some other aspect of the nervous system. He can achieve the same effects—trance, ecstasy, and journeying to another world—with pharmacologically quite different substances. Adolf Dittrich (1996*) came to a similar conclusion when he was developing a phenomenology of altered states of consciousness based upon his experimental and empirical research. He found that states with very different etiologies can have the same content.

Neurotransmission provides but one perspective for explaining the occurrence of altered states of consciousness. And all pharmacological explanations are ultimately nothing more than models for understanding the mysterious play of our consciousness.

489 Laughing gas (nitrous oxide, nitrous monoxide, N$_2$O) was first produced in 1776 by Joseph Priestley (1733–1804), the theologian and chemist who discovered oxygen. By 1799, Humphrey Davy was suggesting that laughing gas could be used as an anesthesia, although its first successful application for this purpose did not occur until 1844. Today, laughing gas is back in fashion, for it is inhaled at techno-parties and raves (cf. **herbal ecstasy**). Despite the fact that the substance has been known for a long time, its mechanism of activity has not yet been clarified (see *Laughing Gas*, ed. Michael Sheldin and Davis Wallechinski [Berkeley, Calif: Ronin Publishing, 1992]).

The Active Plant Constituents from A to Z

The following section discusses the most important active constituents and types of substances found in psychoactive plants. Particular attention is given to those substances that have acquired a significance as a psychoactive drug in a cultural setting or have played a key role in the history of pharmacology. This section will also enable the reader to locate plants that contain a particular agent or substances of a particular type (see the various tables).

The names of the substances are spelled as they are most commonly found in the popular literature and consequently will not always conform to the spelling accepted in the chemical literature. Indeed, inconsistencies in both spelling and nomenclature are not uncommon in this field.

A primary focus of this section is on the cultural significance of the different active plant constituents. For other information, the reader is referred to the chemical, pharmacological, and neurochemical literature (e.g., Du Quesne and Reeves 1982*; Ebel and Roth 1987*; *Hagers Handbuch der pharmazeutischen Praxis* [Hagers Handbook of Pharmaceutical Practice]; Hunnius 1975*; Inaba and Cohen 1993*; Lenson 1995*; Lin and Glennon 1994*; Ott 1993*; Perrine 1996*; *Römpp Chemielexikon* [Roempp's Dictionary of Chemistry]; Roth et al. 1994*; Seymour and Smith 1987*; Shulgin 1992*; and Wagner 1985*).

The groups or types of substances discussed below include: β-carbolines, β-phenethylamines, coumarins, diterpenes, ergot alkaloids, essential oils, indole alkaloids, opium alkaloids, tropane alkaloids, and withanolides. The following individual substances are treated in their own sections: atropine, bufotenine, caffeine, cocaine, codeine, cytisine, diazepam, ephedrine, 5-MeO-DMT, harmaline and harmine, ibogaine, ibotenic acid, mescaline, morphine, muscimol, nicotine, *N,N*-DMT, papaverine, psilocybin/psilocin, salvinorin A, scopolamine, scopoletin, strychnine, THC, and yohimbine.

The flowers of the linden tree contain sedative substances that bind to benzodiazepine receptors. (Woodcut from Fuchs, *Läebliche abbildung und contrafaytung aller kreüter*, 1545)

The milky sap that oozes from scored poppy capsules thickens into raw opium when exposed to air. In 1805, German pharmacist Friedrich Sertürner became the first person in the history of pharmacology to extract a pure chemical constituent from opium. He named the alkaloid morphium (= morphine) after Morpheus, the ancient god of sleep.

Atropine

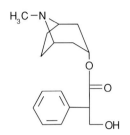

Atropine

Other Names

Atropin, atropina, atropinum, atropium, DL-hyoscyamine, d,l-hyoscyaminum, DL-tropyltropate, (±)-hyoscyamine, $3\alpha(1\alpha H,5\alpha H)$-tropanyl-(RS)-tropate, tropintropate

Empirical formula: $C_{17}H_{23}NO_3$

Substance type: **tropane alkaloid**

Atropine was first isolated from the deadly nightshade (*Atropa belladonna*) in 1820 by Rudolph Brandes, who named the compound after the genus. Atropine is found in many plants of the Nightshade Family (including the genera *Atropa, Brugmansia, Datura, Hyoscyamus, Latua, Mandragora*). Atropine is chemically related to **cocaine** (Willstaedter 1889). It also is closely related to **scopolamine** and hyoscyamine. Hyoscyamine, which is present in many living plants, quickly racemizes into atropine when the raw drugs are dried or stored.

A therapeutic dosage is usually considered to be 1 mg. It is possible that 10 mg is lethal for children and babies, but not for adults:

> Relatively high doses (10 mg atropine sulfate and above) have a stimulating effect on the central nervous system, affecting especially the cerebrum, diencephalon, and medulla oblongata. The arousal is followed by an anesthesia-like paralysis that can lead to coma and a fatal respiratory paralysis. (Roth et al. 1994, 945*)

The lethal oral dosage for an adult is approximately 100 mg (Roth et al. 1994, 765*).

The range of atropine's effects includes psychomotor agitation, excitation, constant repetition of a particular activity pattern, a need to talk, euphoria, crying spells, confused speech, hallucinations, spasms, delirium, flushing of the skin, drying of the mucous membranes, coma, unconsciousness, and heart arrhythmia (Roth et al. 1994, 945*). One particularly characteristic effect is a long-lasting dilation of the pupils (mydriasis). It is because of this effect that atropine has long been used in ophthalmology (Jürgens 1930). Atropine is also utilized as a component in certain anesthetics (in combination with **morphine**). Injections of atropine are often administered prior to surgery so that the mucous membranes will remain dry during the procedure and the patient will not choke on his or her own saliva. Atropine has also been used to treat asthma (Terray 1909).

When it is given orally, the typical effects of atropine (dry mouth, pupillary dilation, increased pulse rate) manifest about twice as strong as compared to intramuscular injection (Mirakhur 1978). Some of the atropine is excreted in the urine unchanged (Roth et al. 1994, 945*).

Atropine is an important antidote in cases of poisoning (overdoses) caused by the fungal toxin muscarine (cf. *Inocybe* **spp.**), *Digitalis purpurea*, hydrogen cyanide, opium (cf. *Papaver somniferum*), and **morphine** (Römpp 1995, 298*). Overdoses of atropine can be successfully treated with morphine.

Because of its unpleasant side effects (dryness of the mouth, difficulties in swallowing, disturbances in vision, confusion), atropine as a pure alkaloid has never acquired any cultural significance as a psychoactive substance. However, the medical literature does contain a few reports of "atropine addiction" (Flincker 1932).

Commercial Forms and Regulations

Atropine is available both as a pure substance and as atropine sulfate. Although regulated as a dangerous substance, it can be obtained with a prescription and is not included on any list of "narcotic drugs" (Koerner 1994, 1573*).

Literature

See also the entries for *Atropa belladonna, Latua pubiflora*, **cocaine**, and **tropane alkaloids**.

Brandes, Rudolph. 1920. Über das Atropium, ein neues Alkaloid in den Blättern der Belladonna (*Atropa belladonna* L.). *Journal für Chemie und Physik* 28:9–31.

Flincker, R. 1932. Über Abstinenz-Erscheinungen bei Atropin. *Münchner Medizinische Wochenschrift* 17:540–41.

Jürgensen, E. 1930. Atropin im Wandel der Zeiten. *Ärztliche Rundschau* (Munich; 1930): 5–8.

Ketchum, J. S., F. R. Sidell, E. B. Crowell, G. K. Aghajanian, and A. H. Hayes. 1973. Atropine, scopolamine and ditran: Comparative pharmacology and antagonists in man. *Psychopharmacology* 28:121–45.

Mirakhur, R. K. 1978. Comparative study of the effects of oral and i.m. atropine and hyoscine in volunteers. *British Journal of Anaesthesia* 50 (6): 591–98.

Terray, Paul von. 1909. Über Asthma Bronchiale und dessen Behandlung mit Atropin. *Medizinische Klinik* 1 (5): 79–83.

Willstätter, R. 1898. Über die Constitution der Spaltungsprodukte von Atropin und Cocain. *Berichte der Deutschen Chemischen Gesellschaft* 31:1534–53.

The deadly nightshade (*Atropa belladonna* L.). (From *Giftgewächse* [Poisonous Plants], 1875)

β-Carbolines

Other Names

Beta-carbolines, β-carbolines, βCs

β-carbolines are derived from the actual β-carbo-line (norharmane). They belong to the group of **indole alkaloids** and are closely related to trypt-amines. They consist of an indole skeleton and various side chains.

The psychoactive effects of β-carbolines are due primarily to the harmala alkaloids **harmaline**, **harmine**, harmalol, harmane (1-methyl-β-carbo-line), and norharmane (β-carboline) (Naranjo 1967). The simpler (β-carboline) alkaloids occur in numerous plants (Allen and Holmstedt 1980).

Many plants that produce psychoactive effects or are utilized for psychoactive purposes contain β-carbolines (including *Acacia* spp., *Arundo donax*, *Banisteriopsis caapi*, *Banisteriopsis* spp., *Mucuna pruriens*, *Papaver* spp., *Passiflora* spp., *Peganum harmala*, *Phalaris arundinacea*, *Phalaris* spp., *Psychotria* spp., *Strychnos* spp., *Virola* spp., *Tribulus terrestris*, and *Amanita muscaria*). These compounds are also present in tobacco smoke (cf. *Nicotiana tabacum*) and in many plants that are used traditionally to make **aya-huasca** or are now used as **ayahuasca analogs** (Schultes 1982).

Many β-carbolines occur as endogenous sub-stances in animals and in humans, where they serve important functions in the nervous system (Bringmann et al. 1991). They appear to influence both moods and dreaming. It is likely that norharmane (β-carboline) occupies a specific β-carboline receptor. Harmane is the endogenous MAO (monoamine oxidase) inhibitor, suppress-ing MAO-A (Rommelspacher et al. 1991). This allows the endogenous *N,N*-DMT to persist for a longer duration and trigger visionary perceptions that manifest either as spontaneous visions during the waking state or as dreams while sleeping (Callaway et al. 1995).

The harmala alkaloids harmaline, harmine, harmane, and tetrahydroharmane are all MAO inhibitors that inhibit primarily MAO-A (Buckholtz and Bogan 1977; McIsaac and Estévez 1966).

In the presence of certain foods, MAO inhibitors are considered to be dangerous or even very dangerous. Tyramine, which is found in such foods as aged cheese, is especially hazardous. If it is not broken down by MAO, it can cause severe toxic effects to an organism. More-recent studies, however, have shown that the dangers have been greatly exaggerated in both the literature and "on the street." Moreover, the amount of tyramine contained in most "dangerous" foods tends to be rather low (Berlin and Lecrubier 1996).

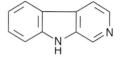

β-carboline

Literature

Allen, J. R. F., and Bo Holmstedt. 1980. The simple β-carboline alkaloids. *Phytochemistry* 19:1573–82.

Berlin, Ivan, and Yves Lecrubier. 1996. Food and drug interactions with monoamine oxidase inhibitors: How safe are the newer agents? *CNS Drugs* 5 (6): 403–13.

Bringmann, Gerhard, Doris Feineis, Heike Friedrich, and Anette Hille. 1991. Endogenous alkaloids in man—synthesis, analytics, *in vivo* identification, and medicinal importance. *Planta Medica* 57 suppl. (1): 73–84.

Buckholtz, N. S., and W. O. Bogan. 1977. Monoaminooxydase inhibition in brain and liver produced by β-carbolines: Structure-activity relationships and substrate specificity. *Biochemical Pharmacology* 26:1991–96.

Callaway, James C., M. M. Airaksinen, and J. Gynther. 1995. Endogenous β-carbolines and other indole alkaloids in mammals. *Integration* 5:19–33. (Includes a very comprehensive bibliography.)

McIsaac, W. M., and V. Estévez. 1966. Structure-activity relationship of β-carbolines monoamine oxidase inhibitors. *Biochemical Pharmacology* 15:1625–27.

Naranjo, Claudio. 1967. Psychotropic properties of the harmala alkaloids. In *Ethnopharmacologic search for psychoactive drugs*, ed. D. H. Efron, 385–91. Washington, D.C.: U.S. Department of Health, Education, and Welfare.

Rommelspacher, Hans, Torsten May, and Rudi Susilo. 1991. β-carbolines and tetrahydroisoquinolines: Detection and function in mammals. *Planta Medica* 57 suppl. (1): 93 ff.

Schultes, Richard Evans. 1982. The beta-carboline hallucinogens of South America. *Journal of Psychoactive Drugs* 14 (3): 205–20.

Stohler, R., H. Rommelspacher, D. Ladewig, and G. Dammann. 1993. Beta-carboline (Harman/Norharman) sind bei Heroinabhängigen erhöht. *Therapeutische Umschau* 50:178–81.

β-Phenethylamines

PEA

Hordenine

MDMA

Notocactus ottonis, a cactus originally from eastern South America, contains the β-phenethylamine hordenine.

Numerous cacti, including this *Melocactus* species, contain phenethylamines, some of which have psychoactive properties. (Woodcut from Tabernaemontanus, *Neu Vollkommen Kräuter-Buch*, 1731)

Other Names
β-phenethylamines, PEAs, 2-phenethylamines

β-phenethylamines are derivatives of phenethylamine (Shulgin 1979). The biogenic 2-phenethylamine (PEA) dilates the blood vessels in the brain and consequently can, under certain circumstances, cause headaches or migraines (cf. *Theobroma cacao*). The most well-known psychoactive β-phenethylamine is **mescaline**, a component of numerous cacti.

Many cacti (including *Gymnocactus* spp.[490] and *Opuntia* spp.) contain phenethylamines that are structurally very similar to mescaline but whose effects are practically unknown (West et al. 1974). It is quite possible that such substances as candicine (*Trichocereus* **spp.**), hordenine (*Ariocarpus* **spp.**, *Opuntia clavata* Eng.; cf. Meyer et al. 1980 and Vanderveen et al. 1974), and macromerine (*Coryphantha* spp.) produce psychoactive effects when used at the appropriate dosages. This area still offers many opportunities for experimental human pharmacology (Heffter technique). Such experimentation could, for example, lead to a psychoactive use of the South American *Notocactus ottonis* (Lehm.) Berg. [syn. *Parodia ottonis*] (cf. Hecht 1995, 82), a cactus that is often found at places where cacti are sold, is very easy to grow, and contains hordenine (Shulgin 1995, 16*). The genus *Lobivia* also contains hordenine (Follas et al. 1977).

Hordenine and related substances (occasionally in high concentrations) are also found in other plants, such as the Himalayan Leguminosae

Desmodium tiliaefolium G. Don (Ghosal and Srivastava 1973).

Numerous phenethylamines that have psychoactive effects (both empathogenic and/or psychedelic) have been synthesized (e.g., MDMA, MDA, MMDA, MDE, 2-CB, et cetera; cf. Shulgin and Shulgin 1991*).

Literature
See also the entries for **mescaline**.

Follas, W. D., J. M. Cassidy, and J. L. McLaughlin. 1977. β-phenethylamines from the cactus genus *Lobivia*. *Phytochemistry* 16:1459–60.

Ghosal, S., and R. S. Srivastava. 1973. β-phenethylamine, tetrahydroisoquinoline and indole alkaloids of *Desmodium tilaefolium*. *Phytochemistry* 12:193–97.

Meyer, Brian N., Yehia A. H. Mohamed, and Jerry L. McLaughlin. 1980. β-phenethylamines from the cactus genus *Opuntia*. *Phytochemistry* 19:719–20.

Shulgin, Alexander T. 1979. Chemistry of phenethylamines related to mescaline. *Journal of Psychedelic Drugs* 11 (1–2): 41–52.

Vanderveen, Randall L., Leslie C. West, and Jerry L. McLaughlin. 1974. *N*-methyltryramine from *Opuntia clavata*. *Phytochemistry* 13:866–67.

West, Leslie G., Randell L. Vanderveen, and Jerry L. McLaughlin. 1974. β-phenethylamines from the genus *Gymnocactus*. *Phytochemistry* 3:665–66.

490 Despite its name, *Gymnocactus mandragora* (Berger) Backeb. [syn. *Neolloydia mandragora* (Berger) Anderson] should not be confused with the mandrake (*Mandragora officinarum*).

Bufotenine

Other Names

Bufotenin, 5-hydroxy-*N,N*-dimethyltryptamine, 5-OH-DMT, mappin, *N,N*-dimethylserotonin, 3-[2-(dimethylamino)ethyl]-1H-indol-5-ol

Empirical formula: $C_{12}H_{16}ON_2$

Substance type: tryptamine (**indole alkaloid**)

Bufotenine was first isolated in 1893 from secretions of the common toad (*Bufo vulgaris* L.) (Shulgin 1981). In 1954, it was found in *Anadenanthera peregrina*. Bufotenine also occurs in the false death cap (*Amanita citrina* [Schaeff.] S.F. Gray) (Keup 1995, 11; Wieland and Motzel 1953) and in other members of the genus (cf. *Amanita pantherina*). Indeed, the symbolic relationship between toads and mushrooms is very interesting in this context (see *Amanita muscaria*). Bufotenine has also been found in *Anandenanthera colubrina* (*Piptadenia* spp.), *Arundo donax*, *Banisteriopsis* spp., *Mucuna pruriens*, and *Phragmites australis*. Bufotenine is a tryptamine derivative and is closely related to *N,N*-DMT, 5-MeO-DMT, and **psilocybin** and psilocin. Chemically, it is almost identical to melatonin (for more on melatonin, see Reiter and Robinson 1996).

Bufotenine has been detected in human urine on numerous occasions (Räisänen 1985) and thus we know that it is a natural substance that is metabolized in the human body. The molecule is very stable. Approximately 16 mg is considered to be an effective dosage. The pharmacology of the substance has been little studied.

The first report of the hallucinogenic effects of bufotenine was published by Fabing and Hawkins (1956), who tested the substance on prison inmates (probably against their will). This was followed by additional research with humans, including some studies in which the substance was tested under highly unethical conditions by being injected into patients in a closed psychiatric institution without their permission, or even against their will. The patients were administered overly high doses and also subjected to electroshocks and other procedures. In this setting, no visions were reported. It was concluded that bufotenine does not produce visions but has only toxic effects (Turner and Merlin 1959). Subsequent studies strengthened the theory that bufotenine should not be classified as a hallucinogen (Mandell and Morgan 1971). A more recent study using only one subject found no hallucinogenic effects, although changes in the emotional domain were observed (McLeod and Sitram 185). Almost all of the reports have noted that the faces of the test subjects turned red or even purple (Fabing and Hawkins 1956). The belief that bufotenine is not a true psychedelic drug has persisted into the present day (e.g., Lyttle et al. 1996; this study, however, is not based on any personal experiments). In its pure form, bufotenine has never acquired any cultural significance as a psychoactive substance.

Bufotenine and *Bufo marinus*

Since ancient times, there have been numerous reports of toads being used to prepare love drinks and other witches' brews, and even **witches' ointments** (Degraaff 1991; Hirschberg 1988). Researchers (prematurely) dismissed such reports as fantasy. In China and Mesoamerica, there is good evidence for the use of toads in magical brews. Chinese toad secretions (*ch'an su*) contain large amounts of bufotenine (Chen and Jensen 1929). In China and Japan, preparations containing bufotenine are sold as aphrodisiacs (Lewis 1989, 70).[491]

In Mesoamerica, the toad was regarded as a manifestation of the Earth Mother, for example, in the form of the Aztec earth goddess Tlatecuhtli (Furst 1972; 1974, 88*). In the region, toads (and frogs) are associated with the rain gods (*chac*) and rainmaking. The Tarahumara refer to toads as "powerful rainmakers." The Olmecs—whose culture is thought to have been the first Mesoamerican civilization—depicted toads in their sacred art and probably used them as hallucinogens. An Olmec object made of green jade and shaped like a toad has been interpreted as a tray for **snuff powder** (Peterson 1990, 46*). In general, the toad was probably the most important Olmec deity (Furst 1981; Furst 1996*; Kennedy 1982; Taylor n.d.).

A cylindrical ceramic container (late Classic period) containing a *Bufo marinus* (cane toad) skeleton was found in Seibal, a Mayan ceremonial

The secretions of *Bufo melanostictus* contain bufotenine. In traditional Chinese medicine, the crystallized raw drug obtained from these secretions is regarded as an excellent aphrodisiac.

491 A study of secretions of the Vietnamese *Bufo melanostictus* Schneider found only four sterols (Verpoorte et al. 1979).

817

In traditional Chinese medicine, the secretions of various toads (*Bufo bufo gargarizans* Cantor, *Bufo melanostictus* Schneider) have long been used as remedies, tonics, and aphrodisiacs. Many of these secretions (*secretio bufonis*) contain the hallucinogenic constituent bufotenine. (Illustration from the *Ch'ung-hsiu cheng-ho pen-ts'ao*.)

The common toad (*Bufo bufo*) was sacred to the Germanic and Baltic peoples. Only with the introduction of Christianity was the toad demonized and declared to be an animal of the devil. The hallucinogenic substance bufotenine was isolated from its secretions. (Woodcut from Gessner, *Historia Animalium*, sixteenth century.)

center. It may have been used as a vessel for drinking **balche'**. Hundreds of ritually interred cane toad skeletons were discovered in post-Classic Mayan ritual depots on the Caribbean island of Cozumel (Hamblin 1981; 1984, 53 ff.). A report from the colonial period indicates that toads were an ingredient in **balche'** or chicha. *Bufo marinus* is also an ingredient in **zombie poison**.

In Mesoamerica, *Bufo marinus* is known variously as *henhen* (Tzeltal; cf. Hunn 1977, 247), *bab* (Mayan), *äh bäb* (Lacandon), and *tamazolin* (Aztec). Numerous stone sculptures of toads as well as some toad bones were discovered in the main temple (Templo Mayor) of the Aztecs (Ofrenda 23; Alvarez and Ocaña 1991, 117, 128). All of the finds suggest that *Bufo marinus* was used in rituals or had a cosmological significance. Today, some Mexican Indians still eat skinned toads, while their secretions are sold at Mexican *brujería* markets as a love powder. The toad itself is invoked as a love magic in magical prayers (*oración del sapo*). The toad's mucus is rolled into little balls that are then rubbed behind the ear as an aphrodisiac. Many people in Mexico still wear toad-shaped amulets today; they may, for example, be made from amber or obsidian from Chiapas.

In the southern part of Veracruz, *curanderos* ("healers") or *brujos* ("sorcerers") still use a preparation of *Bufo marinus*. To prepare it, they capture and kill ten toads. The glands are removed and crushed to produce a paste, to which lime (probably slaked lime) and ashes from a botanically unidentified plant called *tamtwili* are added. The combination is then mixed in water and boiled until there is no more "bad smell" (usually all night long). The solution is then mixed with **chicha** (maize beer) and filtered. The remaining fluid is kneaded into maize dough, lime brine, and five kernels of sprouted maize, and the mixture is laid out in the sun for a few days so that it can ferment, after which it is dried over a fire. This product (*piedrecita*, "little rocks") is then stored far away from any human dwellings. In earlier times, special huts were used to store this magical substance. For consumption, a few pieces are cut off, ground, and soaked in water. After the insoluble ingredients have settled, the solution is poured off and boiled for a considerable time until it gives off a certain odor. Today, the drink is no longer ingested collectively but is used by only one person at a time under the supervision of a *curandero*. The effects begin after approximately thirty minutes and are first manifested as an increase in the pulse rate and a shaking of the muscles and limbs, followed by headaches and delirium. This state lasts three to five hours. In former times, the drinking of this brew was an important part of the initiation of boys into adulthood. Sacred songs were sung to the initiate

while he was delirious. The initiate was told to allow the visions he was going to experience to impress themselves well upon him (Knab n.d.).

It appears that the hallucinogenic effects of *Bufo marinus* were also known in Argentina, for it is considered there to be one of the "temptations" of Saint Anthony (see *Claviceps purpurea*) (Rosemberg 1951).

Originally from the Americas, *Bufo marinus* was introduced to Australia, where it is now known as the cane toad and its secretions are allegedly used as a psychoactive drug. Under Queensland's Drug Misuse Act, bufotenine is an illegal substance in Australia (Ingram 1988, 66). In recent years, the press has been reporting increasing cases of toad lickin' (Lyttle 1993), a practice in which the secretions of *Bufo marinus* are licked off the toad:

When licking the expressed secretions (abusers reported that it is possible to "milk" twice a day), a furry sensation on the lips and tongue quickly becomes manifest. Five to ten (up to 30) minutes later, nausea is common, and only 20 to 30 minutes after ingestion, sometimes earlier, hallucinations of various kinds set in, beginning more rapidly and not lasting as long as with LSD. (Keup 1995, 12)

The thickened juice of boiled animals is also ingested in Australia (Keup 1995, 14). A decoction of the dried skin (known as cane skin tea) is also used (*Der Spiegel* 32 [1994]: 92).

The secretions from *Bufo marinus* contain catecholamines (dopamine, *N*-methyldopamine, adrenaline, noradrenaline) and tryptamines (serotonin, *N*-methyl-serotonin, bufotenine, bufotenidine, dehydrobufotenine), as well as glycoside-like toad toxins (Deulofeu and Rúveda 1971; Lyttle 1993, 523 f.). The skin has been found to contain **morphine**. The toad toxins (bufotoxine, bufogenine, and bufadienolides) are cardiotoxic and are similar to digitalis in their effects: nausea, vomiting, increase in blood pressure, confusion, and psychotic states (Keup 1995, 12). Smoking is probably the safest method to ingest *Bufo marinus* secretions, as the burning process apparently destroys the toxic components while leaving the bufotenine intact (Alexander Shulgin, pers. comm.). The ethnobotanist Brett Blosser smoked dried *Bufo marinus* secretions (approximately 1 mg every few minutes) and reported experiencing tryptamine-like hallucinations similar to those induced by the secretions of *Bufo alvarius* (cf. **5-MeO-DMT**) (B. Blosser, pers. comm.).

The few reports of the effects of smoked toad skin indicate that they are hallucinogenic. One Australian user stated, "I am seeing the world through the consciousness of a toad" (Lewis 1989, 71).

The following species of toads contain significant amounts of bufotenine: *Bufo alvarius* (cf. **5-MeO-DMT**), *B. americanus, b. arenarum, b. bufo bufo, b. calamita, b. chilensis, b. crucifer, b. formosus, b. fowleri, b. paracnemis, b. viridis* (Deulofeu and Rúveda 1971, 483).

Commercial Forms and Regulations

Bufotenine is marketed as bufotenine hydrogen oxalate. In the United States, bufotenine is classified as a Schedule I drug (Shulgin 1981). In contrast, in Germany it is not considered a narcotic and is not illegal (Körner 1994, 1572*).

Literature

See also the entries for *Anandenanthera colubrina* and **5-MeO-DMT**.

Allen, E. R., and W. T. Neill. 1956. Effects of marine toad toxins on man. *Herpetologica* 12:150–51.

Alvarez, Ticul, and Aurelio Ocaña. 1991. Restos óseos de vertebrados terrestres de las ofrendas del Templo Mayor, ciudad de México. In *La fauna en el Templo Mayor*, ed. B. Quintanar, 105–46. Mexico City: INAH.

Chen, K. K., and H. Jensen. 1929. A pharmacognostic study of ch'an su, the dried venom of the Chinese toad. *Journal of the American Pharmaceutical Association* 23:244–51.

Davis, Wade. 1988. *Bufo marinus*: New perspectives on an old enigma. *Revista de la Academia Columbiana de las Ciencies Exactas, Fisicas y Naturales* 14 (63): 151–56.

Degraaff, Robert M. 1991. *The book of the toad*. Rochester, Vt.: Park Street Press.

Deulofeu, Venancio, and Edmundo A. Rúveda. 1971. The basic constituents of toad venoms. In *Venomous animals and their venoms*, ed. Wolfgang Bücherl and Eleanor E. Buckley, 475–556. New York and London: Academic Press.

Fabing, Howard D., and J. Robert Hawkins. 1956. Intravenous bufotenine injection in the human being. *Science* 123:886–87.

Furst, Peter T. 1972. Symbolism and psychopharmacology: The toad as earth mother in Indian America. In *Religión en Mesoamérica, XII Mesa Redondo*, 37–46. Mexico City: S.M.A.

———. 1981. Jaguar baby or toad mother: A new look at an old problem in Olmec iconography. In *The Olmec and their neighbors*, ed. E. Benson, 149–62. Washington, D.C.: Dumbarton Oaks.

Hamblin, Nancy L. 1981. The magic toads of Cozumel. *Mexicon* 3 (1): 10–13.

———. 1984. *Animal use by the Cozumel Maya*. Tucson: The University of Arizona Press.

Hirschberg, Walter. 1988. *Frosch und Kröte in Mythos und Brauch*. Vienna: Böhlau.

Hunn, Eugene S. 1977. *Tzeltal folk zoology*. New York: Academic Press.

Ingram, Glen. 1988. The "Australian" cane toad. In *Venoms and victims*, ed. John Pearn and Jeanette Covacevich, 59–66. Brisbane: The Queensland Museum and Amphion Press.

Kennedy, Alison B. 1982. *Ecce Bufo*: The toad in nature and Olmec iconography. *Current Anthropology* 23 (2): 273–90.

Keup, Wolfram 1995. Die Aga-Kröte und ihr Sekret: Inhaltsstoffe und Mißbrauch. *Pharmazeutische Zeitung* 140 (42): 9–14.

Knab, Tim. n.d. Narcotic use of toad toxins in southern Veracruz. Unpublished manuscript. (Ten typewritten pages.)

Lewis, Stephanie. 1989. *Cane toads: An unnatural history*. New York: Dolphin/Doubleday.

Lyttle, Thomas. 1993. Misuse and legend in the "toad licking" phenomenon. *The International Journal of the Addictions* 28 (6): 521–38.

Lyttle, Thomas, David Goldstein, and Jochen Gartz. 1996. Bufo toads and bufotenine: Fact and fiction surrounding an alleged psychedelic. *Journal of Psychoactive Drugs* 28 (3): 267–90. (Contains an excellent bibliography.)

Mandell, A. J., and M. Morgan. 1971. Indole(ethyl)amine N-methyltransferase in human brain. *Nature* 230:85–87.

McLeod, W. R., and B. R. Sitram. 1985. Bufotenine reconsidered. *Psychiatria Scandinavia* 72:447–50.

Räisänen, Martti. 1985. *Studies on the synthesis and excretion of bufotenine and N,N-dimethyltryptamine in man*. Academic dissertation, Helsinki, University of Helsinki.

Reiter, Russel J., and Jo Robinson. 1996. *Melatonin*. Munich: Droemer Knaur.

Rosenberg, Tobias. 1951. *El sapo en el folklore y en la medicina*. Buenos Aires: Editorial Periplo.

Shulgin, Alexander T. 1981. Bufotenine. *Journal of Psychoactive Drugs* 13 (4): 389.

Taylor, Michael. 1993. The use of the *Bufo marinus* toad in ancient Mesoamerica. *Crash Collusion* 4: 53–55.

Turner, W. J., and S. Merlis. 1959. Effects of some indolalkylamines on man. *Archives of Neurology and Psychiatry* 81:121–29.

Verpoorte, R., Phan-Quôc-Kinh, and A. Baerheim Svendsen. 1979. Chemical constituents of Vietnamese toad venom collected from *Bufo melanostictus* Schneider. *Journal of Ethnopharmacology* 1:197–202.

Wieland, Theodor, and Werner Motzel. 1953. Über das Vorkommen von Bufotenin im gelben Knollenblätterpilz. *Justus Liebigs Annalen der Chemie* 581:10–16.

"SMOKE IT! At the Police Department of Drugs in Brisbane [Australia] is a Heinz baby food can labeled 'Venom cane toad, hallucinogenic, bufotenine.' Inside the can is a dry, flaky, crystalline, brown substance with an unpleasant odor. This substance comes from dried toad skin, which contains natural bufotenine. It can be broken into small pieces and smoked in a water pipe or a simple pipe. It is said to cause an intense hallucinogenic effect which lasts for several hours."

STEPHANIE LEWIS
CANE TOADS: AN UNNATURAL HISTORY
(1989, 70)

The active constituent bufotenine was first isolated from secretions of the common toad. The heathen tribes of the Baltic region venerated the toad as a goddess. It is possible that they were aware of the psychoactive effect of the toad's secretions. (Illustration from 1912)

The title and cover art for the Australian rock music sampler *Toad Lickin'* alludes to a practice known in northeastern Australia in which hallucinogenic secretions are licked from toads. (CD cover, 1990)

Caffeine

Caffeine

Other Names

Cafeina, caféina, caffeina, coffein, coffeinum, guaranin, koffein, methyltheobromin, methyltheobromine, 1,3,7-trimethyl-2,6(1H,3H)-purindion, 1,3,7-trimethylxanthine, thein, theine

Empirical formula: $C_8H_{10}N_4O_2$

Substance type: purine

Caffeine was first isolated from coffee beans (*Coffea arabica*) and named after the genus. However, this stimulating substance actually occurs in many plants (see the table on page 821). Caffeine has stimulating effects upon the central nervous system because it inhibits the enzyme phosphodiesterase, which inhibits the conversion of endogenous substances (cAMP into AMP). This stimulation is usually accompanied by an increase in heart activity, increased urgency of micturition, sensations of heat, and a rise in body temperature. Vasodilation in the brain dispels tiredness and perception becomes more acute. A normal effective dosage is about 100 mg (equal to approximately one cup of strongly brewed coffee). Undesirable side effects begin to appear at 300 mg unless one is used to this level of consumption. There have been repeated reports of "caffeine addiction" in the United States (Weil 1974). Overdoses of caffeine tend to be unpleasant in nature (cf. *Ilex guayusa*):

> Acute caffeine poisoning causes inebriation-like states of excitation accompanied by ringing in the ears, headaches, dizziness, racing heart, muscle tension, insomnia, restlessness, confusion, deliria, cramps, urge to vomit, diarrhea, urgency of micturition. (Roth et al. 1994, 786*)

It has occasionally been suggested that *Catha edulis* contains caffeine, but this assumption has never been confirmed and can now actually be ruled out.

Caffeine is used medicinally to treat heart weakness, neuralgia, headaches, asthma, and hay fever and is also administered in homeopathic preparations. It is used as an antidote for poisonings and overdoses of **alcohol**, **nicotine**, **morphine**, and THC.

Commercial Forms and Regulations

Caffeine is available as a pure substance and as caffeine monohydrate. It is a legal substance.

Literature

See also the entries for *Camellia sinensis*, *Coffea arabica*, *Paullinia cupana*, and **energy drinks**.

Blanchard, J., and S. J. A. Sawers. 1983. The absolute bioavailability of caffeine in man. *European Journal of Clinical Pharmacology* 24:93–98.

Bohinc, P., J. Korbar-Smid, and A. Marinsek. 1977. Xanthine alkaloids in *Ilex ambigua* leaves. *Farmacevtski Vestnik* 28:89–96.

Braun, Stephen. 1996. *Buzz: The science and lore of alcohol and caffeine.* New York and Oxford: Oxford University Press.

Dews, Peter B., ed. 1984. *Caffeine.* Berlin: Springer.

Freise, F. W. 1935. Vorkommen von Koffein in brasilianischen Heilpflanzen. *Pharmazeutische Zentralhalle Deutschlands* 76:704 ff.

Gilbert, Richard J. 1981. Koffein—Forschungsergehnisse im Überblick. In *Rausch und Realität*, ed. G. Völger, 2:770–75. Cologne: Rautenstrauch-Joest-Museum für Völkerkunde.

———. 1988. *Caffeine: The most popular stimulant.* The Encyclopedia of Psychoactive Drugs. London: Burke Publishing.

Goulart, Frances Sheridan. 1984. *The caffeine book: A user's and abuser's guide.* New York: Dodd, Mead & Co.

Graham, D. M. 1978. Caffeine—its identity, dietary sources, intake and biological effects. *Nutrition Reviews* 36:97–102.

James, J. E. 1991. *Caffeine and health.* London: Academic Press.

Lee, Richard S., and Mary Price Lee. 1994. *Caffeine and nicotine.* New York: The Rosen Publishing Group.

Mosher, Beverly A. 1981. *The health effects of caffeine.* New York: The American Council on Science and Health.

Partington, David. 1996. *Pills, poppers & caffeine.* London: Hodder & Stoughton.

Spiller, Gene A., ed. 1984. *The methylxantine beverages and foods: Chemistry, consumption, and health effects.* New York: Alan R. Liss.

Weil, Andrew. 1974. Caffeine. *Journal of Psychedelic Drugs* 6 (3): 361–64.

Plants Containing Caffeine

(from Bohinc et al. 1977; Freise 1935; Gilbert 1988; Hartwich 1911*; Mata and McLaughlin 1982*; Schultes 1977b, 123*; Spiller 1984; supplemented)

Family/Name	Distribution	Average Caffeine Content
AQUIFOLIACEAE		
Ilex ambigua (Michx.) Torrey	North America	traces
Ilex cassine	southeastern North America	0–0.05%
Ilex guayusa	Ecuador	4–7.6%
Ilex paraguariensis	Paraguay, Chaco	0.4–1.6%
Ilex vomitoria	southeastern North America	0.09%
Ilex spp.	South America, Asia	traces
CACTACEAE		
Cereus jamacaru DC.	Brazil	
Harrisia adscendens (Gürke) Br. et R.	Bahia (Brazil)	
Leocereus bahiensis Br. et R.	Bahia (Brazil)	
Pilocereus gounellei (Web.) Byl. et Rowl	Pernambuco (Brazil)	
COMBRETACEAE		
Combretum spp.[492]	Brazil	
NYCTAGINACEAE		
Neea theifera (doubtful; cf. Hartwich 1911, 264, 266*)		
RUBIACEAE		
Coffea arabica	Arabia, Africa	1.16%
Coffea canephora Pierre ex Froehn. [syn. *Coffea robusta* Lindl.]	Arabia	2.15%
Coffea liberica Bull ex Hiern	Liberia	
Coffea spp.	Arabia, Africa	
SAPINDACEAE		
Paullinia cupana	Amazonia	6%
Paullinia yoco	Amazonia	2.73%
Paullinia spp.	Amazonia	
STERCULIACEAE		
Brachychiton diversifolius R. Br.[493]	Australia	
Cola acuminata	West Africa	up to 2.2%
Cola nitida	West Africa	up to 3.6%
Cola spp.	West Africa	traces
Firmiana simplex (L.) W.F. Wight [syn. *Sterculia platanifolia* L.]	East Asia	
Theobroma cacao	Central, South America	0.05%
Theobroma spp.	South America (Amazonia)	
THEACEAE		
Camellia sinensis	Asia; now cosmopolitan	1–4.5%
TURNERACEAE		
Turnera diffusa	Mexico	?
Turnera ulmifolia	Mexico, South America	

This French engraving from 1688 depicts the most important caffeinated beverages together with representatives of the cultures of their origin. The South American (on the left) is drinking maté, the Chinese man is drinking tea, and the Muslim sitting to the right is drinking coffee.

492 In Africa, *Combretum mucronatum* (Schum. et Thonn.) is used as an anthelmintic (Ayensu 1978, 90*).

493 The seeds of the Australian tree *kurrajong* (*Brachychiton diversifolius* R. Br.) contain 1.8% caffeine. The seeds are traditionally roasted, brewed, and drunk as a stimulant (Bock 1994).

Cocaine

In the history of human culture, no other psychoactive plant constituent has had such an impact as cocaine. Because of its enormous cost, however, cocaine is consumed primarily in wealthier circles.

Cocaine

Other Names
Benzoylecgoninmethylester, cocain, cocaïn, cocaina, d-cocain, erythroxylin, kokain, methylbenzoylec-gonine, methylbenzylekgonin, (±)-methyl-[3β-benzoyloxy-2α(1αH,5αH)-tropancarboxylate], O-benzoyl-[(−)-ekgonin]-methylester, 3-benzoyloxy-8-methyl-8-azabicyclo[3.2.1]octan-2-carboxylica-cidmethylester, 3β-benzoyloxy-2β-tropancarboxy-licacid-methylester

Street Names
Autobahn, blow, C, candy, charlie, coca, coca pura (Spanish, "pure coca"), coco, coke, cousin, do-nuts, doppelter espresso, flake, koks, la blanca, lady snow, la rubiecita, line, linie, mama coca, nasen-puder, nose candy, peach, perica, puro (Spanish, "pure"), schnee, schneewittchen, schniefe, schnupf-schnee, sniff, snow, snowwhite, strasse, strässchen, Ziggy's stardust

Empirical formula: $C_{17}H_{21}NO_4$

Substance type: coca alkaloid

The cocaine molecule is structurally related to tropine and other **tropane alkaloids** (Roth and Fenner 1988, 311*). Today, cocaine is the most consumed psychoactive plant constituent in the world. Pure cocaine (as a base) is not water soluble but can be dissolved in **alcohol**, chloroform, turpentine oil, olive oil, or acetone. Cocaine salts are water soluble.

History
In 1860, the German chemist Albert Niemann first isolated cocaine from the leaves of the Peruvian coca bush (***Erythroxylum coca***). The German pharmacist Friedrich Gaedeke (1855) may have represented the alkaloid before this. By around 1870, cocaine was being used as an agent of pleasure, and it was employed at this time to treat alcohol and morphine withdrawal as well as melancholy. The ophthalmologist Karl Koller, a friend of Sigmund Freud, introduced cocaine as a local anesthetic for eye surgery in 1884. Hermann Göring's use of cocaine was famous, and Adolf Hitler, who also used other stimulants (cf. **strychnine**), is thought to have consumed cocaine as well (Phillips and Wynne 1980, 112).

Later, other substances derived from cocaine, including eucaine, procaine (= Novocaine), tetra-caine (= Pantocaine) (1930), lidocaine (= Xylo-caine) (1944), mepivacaine (= Scandicaine) (1957), prilocain (= Xylonest) (1960), bupivacaine (1963), and etidocain (= Duranest) (1972), were also used as local anesthetics (Büsch and Rummel 1990;

Schneider 1993, 19*). Holocaine was also regarded as a substitute.

The goal of chemists and pharmacologists to carve out the effective core of the cocaine molecule and retain the desirable and remove the undesirable effects was achieved in an exemplary manner with the synthesis of pro-caine (1905). (Büsch and Rummel 1990, 490)

In 1923, Willstädter and his coworkers worked out the complete synthesis of cocaine. The pre-cursors are succindialdehyde, methylamine, and mono-methyl-β-keto-glutarate. However, this syn-thesis has never achieved pharmaceutical impor-tance. Practically speaking, all of the cocaine used in the pharmaceutical industry is derived from the coca plant. In 1976, 410 kg of cocaine were legally extracted for this purpose (Täschner and Richtberg 1982, 64).

Production and Use
An analysis of thirteen South American *Eryth-roxylum* species found that cocaine is present only in ***Erythroxylum coca*** and ***Erythroxylum novo-granatense*** (Holmstedt et al. 1977). Hair analysis of Egyptian mummies has revealed the presence of ecgonin, the first metabolite of cocaine, which indi-cates that the ancient Egyptians either consumed cocaine or an unknown African plant that metabo-lizes to ecgonin (Balabanova et al. 1992*).

The coca plantations that are the source of cocaine are known as *cocales*. Bolivian *huanaco* leaves (*Erythroxylum coca* var. *coca*) are preferred for cocaine production because they are the highest yielding. With good chemicals and chemists, it is possible to produce 1 kg of pure cocaine from 100 kg of coca leaves. In the early 1980s, some 100 tons of pure cocaine were exported from Colombia alone.

The entire process of cocaine production, as well as the smuggling routes, the cartels, and every-thing from the connections between politicians and the cartels to the consumption of cocaine even by politicians in the White House, has been documented in countless reports on the radio and television and in magazines and well-researched books (Morales 1989). It is difficult to escape the impression that the cocaine saga is one of the best-known stories of our times but one that is offi-cially ignored. Our leaders still act as though the Mafia is using the white powder to corrupt and dominate the world. In reality, the chief benefac-tors of the billion-dollar business are the banks and the countless politicians and law-enforcement personnel involved in the trade (Sauloy and Le Bonniec 1994).

The snuffing of crystallized cocaine appears to have been discovered in North America at the beginning of the twentieth century and spread from there. Shortly after 1900, pure cocaine was being ingested together with **betel** and lime in India, Ceylon (Sri Lanka), and Java. The use of cocaine as an athletic doping agent began in the 1940s (Fühner 1943, 195*). Little has changed since that time. Cocaine dealers still find some of their best customers in the soccer stars of the German first league and sports heroes in the United States.

Basuko is dried cocaine base (an intermediate step in the production of the pure alkaloid). *Sucito*, or joints made of *basuko*, have been smoked in Colombia since about 1930 (Siegel l982b, 274). Cocaine is usually produced as a hydrochloride but sometimes also as an oxalate or hypochloride (HCL). Street cocaine is almost exclusively cocaine HCL. Most of the illicit cocaine available in Europe is only about 30% pure, as the expensive pure drug is usually "cut." The substances that are most commonly used to "cut" cocaine are:

- Inactive additives: milk sugar (lactose), grape sugar (glucose), baking powder, talc (talcum), borax, cornstarch, innosital, mannitol

- Active additives: speed (amphetamine, fenetyllin, ritalin) and "freeze" (novocaine, benzocaine), PCP ("angel dust"), methedrine, pemoline, **yohimbine**, lidocaine, procaine, tetracaine, **caffeine**, quinine, heroin (Täschner and Richtberg 1982, 65; Voigt 1982, 84)

Dosage

A "line" of cocaine typically contains between 20 and 100 mg of cocaine, depending on the purity of the substance and the consumer's preference. Many users consume between 2 and 3 g in a day or night. It is said that "the first line of the day is the best."

Ritual Use

Cocaine has been called the champagne of drugs, the drug of high society, the drug of the rich, et cetera, and it is certainly most often associated with the wealthier classes. As a result, consumption of the drug has taken on a strong social character. Cocaine is rarely used by one person alone. When it is taken with others, the consumption follows a rather well-defined ritual. The person providing the costly substance lays out several lines (preferably on a mirror), then takes a currency note (often of high value) and rolls it up. One end of the rolled bill is placed in a nostril and held with one hand, while the other hand is used to press the other nostril closed. Half of one line, or a small line, is then snuffed into the nostril. The person then switches nostrils and snuffs the remaining powder, after which the mirror is passed to the next person. This circle may be repeated several time, and it is customary for each of several participants to prepare lines from their own supply.

Artifacts

The cultural significance of cocaine in the modern world cannot be overlooked. Artists, musicians, and writers use it as a stimulant, while highly paid computer experts, software engineers, and programmers would hardly be able to keep up with the demands of their jobs without their "coke." Stockbrokers, financial gurus, and election staffers may use cocaine until they are ready to collapse. Even some of the soccer stars who jog into the stadium sporting T-shirts with such incongruous imprints as "Keine Macht den Drogen" ("No Power to Drugs") are high as a kite on cocaine. According to several estimates, the highest per capita consumption of cocaine is found in Silicon Valley and on Wall Street.

The first literary treatment of cocaine is found in the Sherlock Holmes novel *A Scandal in Bohemia,* by Sir Arthur Conan Doyle, published only two years after Koller's discovery (Phillips and Wynne 1980, 45). In this book, the astonishing abilities of this brilliant detective are attributed in part to his use of cocaine. By the time of the following novel, *The Sign of the Four,* Sherlock Holmes is injecting the pure alkaloid intravenously (Voigt 1982, 38).

The most famous novel of the British writer Robert Louis Stevenson, *Dr. Jekyll and Mr. Hyde,* was written in only four or six days and nights—with the assistance of the magic powder, of course (Springer 1989, 8; Voigt 1982, 38).

The novellas of the expressionist poet Walter Rheiner (1895–1925), in which he referred to the drug as "the eternal poison" and "the loved and hated poison," played a great role in shaping the image of demonic seduction by pharmaceutical cocaine (Rheiner 1979).

At the beginning of the twentieth century, the physician Gottfried Benn (1886–1956) wrote and published numerous poems about cocaine (of which he was very fond) that at the time were deemed rather shocking (Benn 1982; vom Scheidt 1981, 401). Many other authors have also been inspired by cocaine, including Georg Trakl, Thomas Zweifel, Josef Maria Frank Fritz von Ostini, Klaus Mann, and Jean Cocteau (Springer 1989).

Cocaine is also the subject of many novels. The classic cocaine novel, *Cocaine,* was written by Pitigrilli (= Dino Sergè, 1927). The drug has often been treated within its current criminal context (Bädekerl 1983; Fauser 1983), while other novels have been written from a futuristic perspective

"Cocaine-using policemen chase cocaine-using pimps away from cocaine-using whores visited by cocaine-using johns, while cocaine-using journalists report about it for their cocaine-using target audiences. I despise all of this: every time I find myself bending over a plate or a mirror with a bill in my nose, I quietly despise myself. When the line is in the brain, this contempt is washed away. When the line is in the brain, the mind becomes deaf and mute. When the line is in the brain, the character turns around."

HELGE TIMMERBERG
"KALTMACHER KOKAIN" [KILLER COCAINE]
(1996, 36)

The famous novel *Cocaine,* penned by the Italian author Pitigrilli (a pseudonym), is a literary cocaine hallucination that continues to influence the literature inspired by cocaine in the present day. (Cover of an American edition)

"Now look! The shivering stars stand still again, just for a moment.—Sacred poison! Sacred poison!—Tobias sensed and saw the demon standing far above the night sky, as familiar as it was terrifying. He knew and whispered up to the sky: *You are death, mercy and life. You have no god beside you.*"

WALTER RHEINER
KOKAIN [COCAINE]
(1918)

"Cocaine"
"The decomposition of the Self,—
    how sweet, how deeply desired.
You grant it to me: already my
    throat is raw,
already the foreign sound has
    reached
unspoken structures at the
    foundation of my Self.

"No longer bearing the sword that
    was born from the mother's
    womb.
to fulfill a task here and there
To hit with steely might—: sunken
    into the heath,
where hills give rest to barely hidden
    forms!

"A tranquil smoothness, a little
    something.
There—
And now for the space of a breath
the primordial rises, condensed.
Not-his quake brain showers of
    mellow impermanence.

"Blasted Self—oh quenched ulcer—
dispersed fevers—sweetly destroyed
    defense:—
Flow out, flow out!—give birth
with bloodied womb
to the deformed."

GOTTFRIED BENN
DER PSYCHIATER [THE
PSYCHIATRIST]
(1917)

494 This opera, the libretto of which was written by Hugo von Hoffmannsthal, includes a character named Mandryka. Could the author have been making an allusion to *Mandragora*?

495 Kavapyrones have a similar anesthetic effect to cocaine and its derivatives procaine and lidocaine (cf. *Piper methysticum*).

(Boye 1986). The "coke scene" has also provided a rich source of literary inspiration (McInerney 1984; Ellis 1986).

The composer Richard Strauss (1864–1949) wrote his opera *Arabella* while under the influence of cocaine (Springer 1989, 8; Timmerberg 1996).[494] Countless compositions have had cocaine as their subject, including *Cocaine Lil*, for a mezzo-soprano and four female jazz singers, by the contemporary composer Nancy van de Vate (CD Ensemble Belcanto, Koch, 1994). From the 1920s to the 1940s, the white powder fueled the work of especially jazz and blues musicians, and Chick Webb, Luke Jordan, and Dick Justice even gave it a musical treatment ("Cocaine Blues").

Veritable blizzards of cocaine have passed through the brains of many of rock music's greats, who then set their experiences with the "fuel" to music. A few examples are Country Joe McDonald ("Cocaine"), Black Sabbath ("Snowblind"), Little Feat ("Sailin' Shoes"), the Rolling Stones ("Let It Bleed"), Jackson Browne ("Cocaine"), and David Bowie ("Ziggy Stardust").

The "hippie" band known as the Grateful Dead sang about the white powder in their song "Truckin'," one of their few hits to make it onto the charts. Eric Clapton's interpretation of J. J. Cale's song "Cocaine" became a worldwide success and has been played millions of times over. The reggae artist Dillinger released an album named *Cocaine*. The drug also left its mark on the German music scene, influencing or even appearing in the music of Hannes Wader, Konstantin Wecker, Abi Ofarim, and T'MA a.k.a. Falco ("Mutter, der Mann mit dem Koks ist da" ["Mother, the Man with the Coke Is Here"]; BMG Records 1995).

Cocaine has been the subject of at least one theater work: The American playwright Pendleton King wrote a piece entitled *Cocaine* that was produced for the stage in 1917 (Phillips and Wynne 1980, 93 ff.).

Medicinal Use

The medicinal applications of cocaine were discovered only a short time after the isolation of the molecule itself. Cocaine was initially used for local anesthesia[495] in ophthalmology and dentistry, and infiltration anesthesia was developed just a few years later (Custer 1898). Because analogs (e.g., procaine) were developed that produce specific effects with no psychoactive side effects, cocaine is rarely used as an anesthetic today.

Pharmacology and Effects

Cocaine stimulates the central nervous system, especially the autonomic (sympathetic) system, where it inhibits the reuptake of the neurotransmitters noradrenaline, dopamine, and serotonin and increases the time in which they remain in the synaptic cleft. Cocaine has a powerful effect upon the peripheral nervous system, which explains its efficaciousness as a local anesthetic. It has strong stimulant and vasoconstricting properties. Very high dosages of cocaine are said to be able to induce hallucinations, an effect that is frequently noted in the neurological literature (Pulvirenti and Koob 1996, 49) as well as in prose and poetry (Rheiner 1979, 27). Hallucinations (of nonexistent people, images, flickering lights) often occur during nights in which dosages of 2 to 3 g have been taken. For many people, cocaine also dispels fear. It stimulates a need for alcoholic beverages at the same time that it strongly suppresses the effects of **alcohol**. A similar dynamic applies to **nicotine**.

In a certain sense, there is something unsatisfying about the effects of cocaine. A person may sense that satisfaction could be achieved if the effects could possibly be increased. However, using more cocaine does not produce an enhancement of its effects.

Just as coca was and is employed in South America as an aphrodisiac, cocaine has a similar use in the West. Cocaine's reputation as an aphrodisiac can be traced back to Sigmund Freud (1884) and has been repeatedly confirmed in the pharmacological literature:

At a high level of intoxication, central excitation sets in with characteristic shivering, an initial state of euphoria that turns into delirium and hallucinations. For women, the stimulation . . . not infrequently has an erotic character and has resulted in later accusations of sexual misconduct against the operating physician. (Führer 1943, 196*)

Some psychiatrists believe that cocaine stimulates the "sexual center" of the brain (Siegel 1982a). For many users, cocaine is inevitably associated with sexuality (MacDonald et al. 1988; Phillips and Wynne 1980, 221).

Cocaine relaxes and opens the sphincter muscles, which makes anal penetration easier as well as substantially more pleasurable. However, cocaine (much like **ephedrine**) often has an adverse effect on erectile function and consequently leads to temporary impotence (cf. Siegel 1982a).

The addictive potential of cocaine has been the subject of much debate. This issue does not appear to be oriented toward the user as much as it reflects the current legal situation. In recent years, there have been efforts to develop a vaccination against "cocaine addiction." Of course, the research in this area is conducted on rats (Hellwig 1996). The effect of cocaine on the brain is also an object of much research, since studies that confirm the adverse effects of cocaine are likely to receive financial support from the government.

Studies that do not have a political agenda are the exception rather than the rule (Volkow and Swann 1990).

People who use cocaine frequently suffer from a runny nose ("coke sniffles") the following day. Users may counteract this undesirable and unpleasant aftereffect by rinsing their nose with a saline solution (e.g., with medicinal salts). Many users rub vitamin E oil in their nose, a practice said to regenerate the highly irritated mucous membranes in the nose (Voigt 1982, 72). Although cocaine can be very helpful in dealing with an acute attack of hay fever, chronic use can actually contribute to the condition.

Crack or Free-Base Cocaine

In the German press, crack has been portrayed as "death for a few dollars," "the devil's drug from the U.S.A.," et cetera. The general idea seems to be that "cocaine was a miracle, but crack, crack was better than sex" or "cocaine was purgatory—but crack is hell" (in *Wiener* 6 [1986]: 65, 66).

Crack, which is also known as base, free base, baseball, rocks, Roxanne, and supercoke, is nothing more than smokeable free-base cocaine (Siegel 1982b). In other words, crack is cocaine in the form of a free base (Pulvirenti and Koob 1996, 48). It can be obtained from an aqueous solution of cocaine hydrochloride to which an alkaline substance (such as sodium carbonate) is added. The cocaine salt is transformed into the pure base, or, in other words, the pure substance. It can then be purified with ether, causing the cocaine to crystallize out. Crack is usually "smoked" (i.e., vaporized and inhaled) in glass pipes. A typical dosage ranges from 0.05 to 0.1 g. The effect is very similar to that of snuffed cocaine but is much more intense:

Although crack is a derivative of cocaine, there is little comparison between the mild and mostly stimulating cocaine inebriation and the effects of the short-term crack high, which can literally bowl one over. Whereas cocaine produces a euphoric sensation of great concentration and razor-sharp intelligence for about 20 to 60 minutes, crack lasts for only three to five minutes while giving the consumer an incredibly strong kick with regard to physical sensations as well as the euphoria of absolute omnipotence. Of course, this has resulted in many myths, including one that crack is particularly pure. (Sahihi 1995, 37*)

Ethnologists have begun using the field methods typical of the discipline to study the "crack phenomenon," which appears to be a typically American product (Holden 1989). "Crack life" is a reflection of the problems in American society and reveals deep social fissures and cultural anomalies. For users, the "crack way" is an important form of identity formation. Crack is frequently found together with prostitution, as "addicts" may accept it as a form of payment for sexual services (Carlson and Siegal 1991).

On the street, the following substances may be used as substitutes for cocaine or crack in times of shortage: procaine, **caffeine**, benzocaine, phenyl-propanolamine, lidocaine, and **ephedrine** (Siegel 1980).

Commercial Forms and Regulations

Cocaine hydrochloride is available through the pharmacy trade. The German Drug Law lists cocaine as a "narcotic drug in which trafficking is allowed but which may not be prescribed" (Körner 1994, 42). In the United States, the Controlled Substances Act classifies cocaine as a Schedule II substance.

Literature

See also the entries for *Erythroxylum coca*, *Erythroxylum novogranatense*, atropine, and **tropane alkaloids**.

Ashley, Richard. 1975. *Cocaine: Its history, use and effects*. New York: St. Martin's Press.

Aurep, B. von. 1880. Über die physiologische Wirkung des Cocaïn. *Archiv für Physiologie* 21:38–77.

Bädekerl, Klaus. 1983. *Ein Kilo Schnee von Gestern*. Munich and Zurich: Piper.

Benn, Gottfried. 1982. *Gedichte, in der Fassung der Erstdrucke*. Frankfurt/M.: Fischer.

Boye, Karin. 1986. *Kallocain: Roman aus dem 21. Jahrhundert*. Kiel: Neuer Malik Verlag.

Büsch, H. P., and W. Rummel. 1990. Lokalanästhetika, Lokalanästhesie. In *Allgemeine und spezielle Pharmakologie und Toxikologie* (5th ed.), ed. W. Forth, D. Heuschler, and W. Rummel, 490–96. Mannheim, Vienna, and Zurich: B. I. Wissenschaftsverlag.

Carlson, Robert G., and Harvey A. Siegal. 1991. The crack life: An ethnographic overview of crack use and sexual behavior among African-Americans in a Midwest metropolitan city. *Journal of Psychoactive Drugs* 23 (1): 11–20.

Crowley, Aleister. 1973. *Cocaine*. San Francisco: And/Or Press.

Custer, Julius, Jr. 1898. *Cocain und Infiltrationanästhesie*. Basel: Benno Schwabe.

Ellis, Bret Easton. 1987. *Less Than Zero*. New York: Random House.

Fauser, Jörg. 1983. *Der Schneemann*. Reinbek: Rowohlt.

Fischer S., A. Raskin, and E. Uhlenhuth, eds. 1987. *Cocaine: Clinical and biobehavioral aspects*. New York: Oxford University Press.

"Look at this shining heap of crystals! They are Hydrochloride of Cocaine. . . . [T]here was never any elixir so instant magic as cocaine. Give it to no matter whom. Choose me the last loser on the earth; take hope, take faith, take love away from him. Then look, see the back of that worn hand, its skin discolored and wrinkled. . . . He places on it that shimmering snow, a few grains only, a little pile of starry dust. The wasted arm is slowly raised to the head that is little more than a skull; the feeble breath draws in that radiant powder. . . . Then happens the miracle of miracles, as sure as death, and yet as masterful as life . . . at least faith, hope and love throng very eagerly to the dance; all that was lost is found."

ALEISTER CROWLEY
COCAINE
(1918)

The reggae album *Cocaine*, by the Jamaican artist Dillinger, glorifies not only the white powder but also hemp smoke. (CD cover 1986, Charly Records)

"Mother, the Man with the Coke Is Here," a German cocaine hit of the Golden Twenties, was recently reissued in a new version with updated lyrics by Falco, a veteran of the Neue Deutsche Welle ("German New Wave"). (CD cover 1995, Sing Sing)

Cocaine was used widely in Germany and Italy during the Golden Twenties. Many cartoonists poked fun at the subject. (Cartoon, Germany, circa 1920)

Freud, Sigmund. 1884. *Über Coca. Centralblatt für die gesamte Therapie* 2:289–314. Repr. in Täschner and Richtberg 1982, 206–31 (see below).

———. 1885. Über die Allgemeinwirkung des Cocains. *Medizinisch-chirurgisches Centralblatt* 20:374–75.

———. 1887. Bemerkungen über Cocainsucht und Cocainfurcht, mit Beziehung auf einen Vortrag von W. A. Hammonds. *Wiener medizinische Wochenschrift* 37:927–32.

———. 1996. *Schriften über Kokain*. Frankfurt/M.: Fischer. (Orig. pub. 1884.)

Gay, George R. 1981. You've come a long way, baby! Coke time for the new American lady of the eighties. *Journal of Psychoactive Drugs* 13 (4): 297–318.

Gottlieb, Adam. 1979. *The pleasures of cocaine*. San Francisco: And/Or Press.

Grinspoon, Lester, and James B. Bakalar. 1985. *Cocaine: A drug and its social evolution*. Rev. ed. New York: Basic Books.

Hartmann, Walter. 1990. *Informationsreihe Drogen: Kokain*. Markt Erlbach: Raymond Martin Verlag.

Hellwig, Bettina. 1996. Impfung gegen Cocain? *Deutsche Apotheker-Zeitung* 136 (4): 46/270.

Holden, Constance. 1989. Streetwise crack research. *Science* 246:1376–81.

Holmstedt, Bo, Eva Jäätmaa, Kurt Leander, and Timothy Plowman. 1977. Determination of cocaine in some South American species of *Erythroxylum* using mass fragmentography. *Phytochemistry* 16:1753–55.

Kennedy, J. 1985. *Coca exotics: The illustrated story of cocaine*. New York: Cornwall Books.

Koller, Carl [= Karl]. 1884. Über die Verwendung des Cocaïn zur Anästhetisierung am Auge. *Wiener medizinische Wochenschrift* 34:1276–1278, 1309–11.

———. 1935. Nachträgliche Bemerkungen über die ersten Anfänge der Lokalanästhesie. *Wiener medizinische Wochenschrift* 85:7.

———. 1941. History of cocaine as a local anesthetic. *Journal of the American Medical Association* 117:1284.

Lindgren, J.-E. 1981. Guide to the analysis of cocaine and its metabolites in biological material. *Journal of Ethnopharmacology* 3:337–51.

Lossen, W. 1865. Über das Cocain. *Liebig's Annalen* 133:351–71.

MacDonald, P. T., V. Waldorf, C. Reinarman, and S. Murphy. 1988. Heavy cocaine use and sexual behavior. *Journal of Drug Issues* 18 (3): 437–55.

Maier, Hans Wolfgang. 1926. *Der Kokainismus*. Leipzig: Thieme.

McInerney, Jay. 1984. *Bright Lights, Big City*. New York: Knopf.

Morales, Edmundo. 1989. *Cocaine: White gold rush in Peru*. Tucson and London: The University of Arizona Press.

Niemann, Albert. 1860. Über eine neue organische Base in den Cocablättern. Dissertation, Göttingen University.

Pernice, Ludwig. 1890. Über Cocainanaesthesie. *Deutsche medizinische Wochenschrift* 16:287.

Phillips, Joel L., and Ronald D. Wynne. 1980. *Cocaine: The mystique and the reality*. New York: Avon Books.

Plasket, B., and E. Quillen. 1985. *The white stuff*. New York: Dell Publishing Co.

Pulvirenti, Luigi, and George F. Koob. 1996. Die Neurobiologie der Kokainabhängigkeit. *Spektrum der Wissenschaft* 2:48–55. (An unethical and nauseating study on animals.)

Rheiner, Walter. 1979. *Kokain: Eine Novelle und andere Prosa*. Berlin and Darmstadt: Agora Verlag. Repr. 2nd ed., 1982.

Richards, Eugene. 1994. *Cocaine true, cocaine blue*. New York: Aperture.

Roles, R., M. Goldberg, and R. G. Sharrar. 1990. Risk factors for syphilis: Cocaine use and prostitution. *American Journal of Public Health* 80 (7): 853–57.

Sabbag, Robert. 1976. *Snowblind: A brief career in the cocaine trade*. Indianapolis and New York: The Bobbs-Merrill Co.

Sauloy, Mylène, and Yves Le Bonniec. 1994. *Tropenschnee—Kokain: Die Kartelle, ihre Banken, ihre Gewinne. Ein Wirtschaftsreport*. Reinbek bei Hamburg: Rowohlt.

Siegel, Ronald K. 1978. Cocaine hallucinations. *American Journal of Psychiatry* 135:309–14.

———. 1980. Cocaine substitutes. *New England Journal of Medicine* 302:817–18.

———. 1982a. Cocaine and sexual dysfunction: The curse of Mama Coca. *Journal of Psychoactive Drugs* 14 (1–2): 71–74.

———. 1982b. Cocaine smoking. *Journal of Psychoactive Drugs* 14 (4): 271–359.

Smith, David E., and Donald R. Wesson. 1978. Cocaine. *Journal of Psychedelic Drugs* 10 (4): 351–60.

Springer, Alfred, ed. 1989. *Kokain: Mythos und Realität—Eine kritisch dokumentierte Anthologie*. Vienna and Munich: Verlag Christian Brandstätter.

Täschner, Karl-Ludwig, and Werner Richtberg. 1982. *Kokain-Report*. Wiesbaden: Akademische Verlagsgesellschaft.

Thamm, Berndt Georg. 1985. *Das Kartell: Von Drogen und Märkten—ein modernes Märchen*. Basel: Sphinx.

———. 1986. *Andenschnee: Die lange Linie des Kokain.* Basel: Sphinx.

Timmerberg, Helge. 1996. Kaltmacher Kokain. *Tempo* 3:34–42.

Turner, Canton E., Beverly S. Urbanek, G. Michael Wall, and Coy W. Waller. 1988. *Cocaine: An annotated bibliography.* 2 vols. Jackson and London: Research Institute of Pharmaceutical Sciences/University Press of Mississippi.

Voigt, Hermann P. 1982. *Zum Thema: Kokain.* Basel: Sphinx.

Volkow, Nora V., and Alan C. Swann, eds. 1990. *Cocaine in the brain.* New Brunswick, N.J.: Rutgers University Press. (See book review by Ronald Siegel in *Journal of Psychoactive Drugs* 23 (1; 1991): 93 f.)

vom Scheidt, Jürgen. 1973. Freud und das Kokain. *Psyche* (Munich) 27:385–430.

———. 1981. Kokain. In *Rausch und Realität,* ed. G. Volger, 1:398–402. Cologne: Rautenstrauch-Joest Museum für Völkerkunde.

Wesson, Donald R. 1982. Cocaine use by masseuses. *Journal of Psychoactive Drugs* 14 (1–2): 75–76.

Wolfer, P. 1922. Das Cocain, seine Bedeutung und seine Geschichte. *Schweizerische medizinische Wochenschrift* 3:674–79.

Codeine

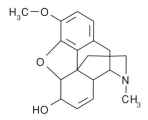

Codeine

Other Names

Codein, codeina, codéine, codeinum, 4,5α-epoxy-3-methoxy-17-methyl-7-morphinen-6α-ol, kodein

Empirical formula: $C_{18}H_{21}NO_3H_2O$

Substance type: **opium alkaloid**

In 1832, codeine was isolated from opium, which has a codeine content of 2 to 3% (see *Papaver somniferum*). Codeine is also biosynthesized in the roots of *Papaver somniferum* L. cv. Marianne (Tam et al. 1980). It is possible that trace amounts of codeine can also be found in other *Papaver* species (*Papaver bracteatum, Papaver decaisnei*; cf. *Papaver* **spp.**) (Theuns et al. 1986). Codeine is also an endogenous neurotransmitter in humans (cf. **morphine**).

A dosage of 20 to 50 mg produces "general mental stimulation, warmth in the head, and an increase in the pulse rate, as also appear after the consumption of alcohol" (Römpp 1950*). Codeine does not appear to be metabolized in the body and is excreted unchanged.

Because codeine suppresses the urge to cough, its most important pharmaceutical use is in cough syrups. The dosage when codeine is used as a cough suppressant is 50 mg three times a day. A dosage of 100 to 200 mg results in sleep and sedation. Higher dosages elicit effects comparable to those of morphine. The medical literature contains repeated mentions of "codeine addiction." Codeine "addicts" are said to ingest up to 2 g of codeine daily (Römpp 1950, 115*). Today, codeine is gaining increasing medicinal importance as a substitution therapy for heroin addicts (Gerlach and Schneider 1994). The pharmaceutical industry synthesizes codeine primarily from thebaine, the main active constituent in *Papaver bracteatum* Lindl. (cf. *Papaver* **spp.**) (Morton 1977, 125*; Theuns et al. 1986).

Codeine has acquired a certain significance in the music scene (jazz, rock, psychedelia), primarily as a substitute for heroin or **morphine**. Buffy Saint-Marie sang about the anguish of her codeine dependence in the song "Cod'ine" (LP *It's My Way!* Vanguard Records 1964). Quicksilver Messenger Service later covered the song and made it famous. In the 1990s, the wave band Codeine had several albums out through Sub Pop. Cough syrups[496] with a high codeine content were often consumed as inebriants at concerts, festivals, et cetera (usually in combination with **alcohol** and **cannabis**) (Bangs 1978, 158).

Commercial Forms and Regulations

Codeine is available as a pure substance and as codeine hydrochloride, codeine phosphate, and codeine phosphate hemihydrate. Codeine is on Schedule III in the United States. Preparations containing codeine (tinctures, cough syrups, et cetera) require a special prescription (i.e., with no refills allowed and/or on special prescription forms). But in other countries, including France, Spain, Nepal, and India, a prescription is still unnecessary and the medicine can be obtained over the counter from any pharmacy.

Literature

See also the entries for *Papaver somniferum*, **morphine**, and **opium alkaloids**.

Bangs, Lester. 1978. Ich sah Gott und/oder Tangerine Dream. *Rocksession* 2:155–58. Reinbek: Rowohlt.

Esser, Barbara. 1998. Vom Regen in die Traufe: Das Verbot des Ersatzstoffs Codein . . . *Focus* 26 (6): 58–60.

The album *Barely Real*, by the American underground band Codeine, provides a clear demonstration of the effects of the opium alkaloid codeine on the artists. The drawn-out melodies, slow rhythms, and restrained monotony make it difficult for the listener to remain awake. (CD cover 1992, Sub Pop)

496 In addition to codeine, many cough syrups also contain **ephedrine**, camphor (*Cinnamomum camphora*), essential oils, and sometimes even extracts of *Atropa belladonna* and *Aconitum napellus*. They are thus somewhat reminiscent of **witches' ointments**, **soporific sponges**, and **theriac**.

Gerlach, Ralf, and Wolfgang Schneider. 1994. *Methadon- und Codeinensubstitution: Erfahrungen, Forschungsergebnisse, Praxiskonsequenzen.* Berlin: VWB.

Tam, W. H. John, Friedrich Constabel, and Wolfgang G. W. Kurz. 1980. Codeine from cell suspension

cultures of *Papaver somniferum. Phytochemistry* 19:486–87.

Theuns, Hubert G., H. Leo Theuns, and Robert J. J. Ch. Lousberg. 1986. Search for new natural sources of morphinians. *Economic Botany* 40 (4): 485–97.

Coumarins

The black seeds (tonka beans) of the South American tonka tree (*Dipteryx odorata*) exude crystalline, almost pure white coumarin. In Venezuela, the coumarin that is rubbed from the seeds is utilized to aromatize tobacco and snuffs.

Coumarin

Extracts from ripe and wonderfully fragrant vanilla beans are rich in coumarins. Vanilla has a pheromone-like effect on humans, inducing courtship behavior. (Illustration from Hernández, *Rerum medicarum Novae Hispaniae*, Rome, 1651)

Other Names

Benzopyrones, coumarines, cumarines, kumarine

Empirical formula: $C_9H_6O_2$ (= 1,2-benzopyrone)

Substance type: benzopyrone

Coumarin (= chromen-2-on, kumarin, 2H-1-benzopyran-2-on, o-cumar[in]acid lactone), which has a scent like that of vanilla, crystallizes into colorless prisms and is easily soluble in **alcohol**, ether, and **essential oils**. Pure coumarin is exuded from what are known as tonka beans, and for this reason it is also called tonka bean camphor. Coumarin is biosynthesized by the hydroxylation of cinnamic acid or coumarin glycoside. Even plants that do not actually contain any coumarin

often produce it when they wilt (giving off the smell of hay) or dry (e.g., *Anthoxanthum odoratum, Galium odoratum, Sida acuta*).

Coumarins in Psychoactive Plants
(from Gray and Waterman 1978; Römpp 1995*; Shoeb et al. 1973; supplemented)
Coumarins (e.g., benzofuran) have been found in the following plants with demonstrated or purported psychoactivity:

Aegle marmelos Corr.	coumarin
Anthoxanthum odoratum L. (sweet vernal grass)	coumarin
Dipteryx (*Coumarouna*) *odorata* (Aubl.) Willd.	coumarin
Dipteryx oppositifolia (Aubl.) Willd. (tonka bean)	coumarin
Evodia spp. (cf. **Evodia bonwickii**)	
Galium odoratum (L.) Scop. (woodruff) [syn. *Asperula odorata* L.]	coumarin
Hierochloë australis (L.) P. Beauv. (buffalo grass; vodka additive)	coumarin
Hierochloë odorata (L.) P. Beauv. (sweet grass, vanilla grass; cf. **incense**)	coumarin
Justicia pectoralis	unidentified
Lavandula angustifolia Mill. [syn. *Lavandula officinalis* Chaix] (cf. **essential oils**)	coumarin and others
Melilotus officinalis (L.) Pall.	coumarin and others
Melilotus spp. (sweet clover)	various
Petroselinum crispum	furanocoumarins
Ruta graveolens L. (cf. **haoma, soma**)	rutin, gravolenic acid
Sida acuta	coumarin
Sida spp.	coumarin
Tagetes spp.	various
Thamnosma montana	various

For plants containing the coumarin derivative scopoletin, see **scopoletin**.

Umbelliferone, aesculine, and furocoumarin are all coumarin derivatives. More than six hundred natural coumarins are now known. About two hundred coumarins occur in the Family Rutaceae (including the genera *Zanthoxylum*, *Evodia*, *Ruta*, *Thamnosma*, *Dictamnus*, *Eriostemon*, *Citrus*, and *Aegle*), where they appear to have great chemotaxonomic importance (Gray and Waterman 1978; Tatum and Berry 1979).

Coumarins occur in some plants that are used for psychoactive purposes (see **scopoletin**). Coumarin is the substance responsible for the specific taste of woodruff punch, and it is also present in fahan tea (*Angraecum fragrans* Du Petit-Thouars), which Bibra (1855*) described as psychoactive. Fahan was once used as a substitute for green tea (*Camellia sinensis*) and was mixed with tobacco (*Nicotiana tabacum*) and rolled into cigars (Frerichs et al. 1938, 1234*).

High dosages of pure coumarin can cause headaches, dizziness, lethargy, stupor, and even respiratory paralysis (Roth et al. 1994, 796*). Coumarin is said to be toxic to the liver and for this reason was banned as a component or ingredient in food. However, the toxicity is very doubtful, and the alleged carcinogenic effects are also questionable (Marles et al. 1987).

Commercial Forms and Regulations
In the United States, coumarin has been banned as a food additive since 1954. It has been placed in Class 3 of the Swiss Poison List. In Germany, drinking brandies (38% **alcohol**) are allowed to contain a maximum of 10 mg of coumarin per liter (Roth et al. 1994, 402*).

Literature
See also the entries for **scopoletin**.

Gray, Alexander I., and Peter Waterman. 1978. Coumarins in the Rutaceae. *Phytochemistry* 17:845–64. (Contains a rich bibliography.)

Marles, R. J., C. M. Compadre, and N. R. Farnsworth. 1987. Coumarin in vanilla extracts: Its detection and significance. *Economic Botany* 41:41–47.

Mendez, R. D. H., J. Murray, and S. A. Brown. 1982. *The natural coumarins*. Chichester, U.K.: John Wiley.

Reisch, J., et al. 1968. Über weitere C$_3$-substituierte Cumarin-Derivate aus *Ruta graveolens*: Daphnoretin und Daphnoretin-methyläther. *Planta Medica* 15:372–76.

———. 1969. Über die Cumarine der Wurzel von *Ruta graveolens*. *Planta Medica* 17:116–19.

Shoeb, Aboo, Rhandhir S. Kapil, and Satya P. Popli. 1973. Coumarins and alkaloids of *Aegle marmelos*. *Phytochemistry* 12:2071–72.

Tatum, James H., and Robert E. Berry. 1979. Coumarins and psoralkens in grapefruit peel oil. *Phytochemistry* 18:500–502.

"Coumarins, which Vogel recently discovered in tonka beans (*Diperix* [sic] *odorata*) and Fontana then found in sweet clover (*Melilotus officinalis*), also occur in woodruff (*Asperula odorata*). Everyone knows that woodruff is added to wine in many places in Germany and in France because, as it is said, this improves the taste, but in reality this is probably done to make it more stimulating. . . . Nevertheless, it is doubtlessly worthy of notice that both in Africa [the fahan tea] and in Europe, two plants are used as stimulating or exhilarating agents that contain the very same substance that they very likely thank for precisely this stimulating effect."

ERNST FREIHERR VON BIBRA
DIE NARKOTISCHEN GENUßMITTEL [THE NARCOTIC AGENTS OF PLEASURE]
(1855, 128 f.*)

Cytisine

Other Names
Baptitoxin, cytiton, laburnin, 1,2,3,4,5,6-hexahydro-8*H*-1,5-methano-pyrido[1,2α][1,5]diazocin-8-ol, sophorin, ulexin

Empirical formula: C$_{11}$H$_{14}$N$_2$O

Substance type: quinolizidine alkaloid, lupine alkaloid

Cytisine is found in numerous legumes (Leguminosae) (Plugge 1895), such as the rain shower tree (*Laburnum anagyroides* Medikus [syn. *Cytisus laburnum* L.]).[497]

> Since cytisine acts to stimulate the central nervous system, states of excitation and confusion (with hallucinations, delirium), muscle spasms, as well as general clonic-tonic spasms in the extremities not infrequently occur. (Roth et al. 1994, 443*)

Cytisine docks to the acetylcholine (ACh) receptors of the central nervous system, the ganglia, and the neuromuscular endplate. The ganglia-blocking effects of cytisine are similar to those of **nicotine** and can induce **strychnine**like spasms, especially hallucinations, and even unconsciousness and ultimately death. However, the lethal dosage for humans is unknown (Roth et al. 1994, 801 f.*). The nicotine-like effects also explain the ethnopharmacological use of plants containing cytisine as tobacco substitutes.

Other lupine alkaloids and cytisine derivatives have been found in many plants from the Family Leguminosae, including **Lupinus spp.** and *Echinosophora koreensis* Nakai (a close relative of the genus *Sophora*) (Murakoshi et al. 1977).

Commercial Forms and Regulations
Cytisine is sold in its pure form and is not subject to any regulations (Roth et al. 1994, 802).

Cytisine

---

497 The derivative hydroxynorcytisine is also present in the fruit pods of the golden chain tree (Hayman and Gray 1989).

Left: The ripe seeds of the golden chain tree (*Laburnum anagyroides* = *Cytisus laburnum*) contain up to 3% cytisine. In Germany, the dried leaves of the tree were smoked during the years following World War II as a stimulating substitute for tobacco.

Right: The magnificent flowers of the Philippine jade vine or emerald creeper (*Strongylodon macrobotrys* A. Gray) contain cytisine and other alkaloids.

"During the World War, the leaves of plants containing cytisine were a popular substitute for tobacco in many countries in which tobacco was not readily available. The toxicity of these plants can be seen in the fact that a cytisus cigarette causes the same nausea to nonsmokers as a real tobacco cigarette does, while those who are used to tobacco do not notice any discomfort."

VICTOR REKO
MAGISCHE GIFTE [MAGICAL POISONS]
(1938, 134)

Literature

See also the entries for *Cytisus* spp. and *Sophora secundiflora*.

Hayman, Alison R., and David O. Gray. 1989. Hydroxynorcytisine, a quinolizidine alkaloid from *Laburnum anagyroides*. *Phytochemistry* 28 (2): 673–75.

Murakoshi, Isamu, Kyoko Fukuchi, Joju Haginiwa, Shigeru Ohmiya, and Hirotaka Otomasu. 1977. *N*-(3-oxobutyl)cytisine: A new lupin alkaloid from *Echinosophora koreensis*. *Phytochemistry* 16:1460–61.

Plugge, P. C. 1895. Uber das Vorkommen von Cytisin in verschiedenen Papilionaceae. *Archiv für Pharmazie* 233:430 ff.

Plugge, P. C., and A. Rauwerda. 1896. Fortgesetzte Untersuchungen Über das Vorkommen von Cytisin in verschiedenen Papilionaceae. *Archiv für Pharmazie* 234:685 ff.

Seeger, R., and H. G. Neumann. 1992. Cytisin. *Deutsche Apotheker Zeitung* 132:303–6.

Plants Containing Cytisine

(from Bock 1994, 75 ff.*; Römpp 1995*; Roth et al. 1994*; supplemented)

Stock Plant	Distribution
Ammodendron spp.	
Anagyris spp.	southern Europe
Baptisia spp.	North America
Colutea arborescens L.	Mediterranean region
Colutea spp.	southeastern Europe, Asia Minor
Cytisus canariensis	Canary Islands, Mexico
***Cytisus* spp.**	Europe
Eucresta spp.	Australia
Genista germanica L.	central Europe
Genista tinctoria L.	Europe
Laburnum alpinum (Mill.) Bercht. et Presl [syn. *Cytisus alpinus* Mill.]	Alps, southern Europe
Laburnum anagyroides Medik. [syn. *Laburnum vulgare* Bercht. et Presl, *Cytisus laburnum* L.]	central and southern Europe
Lamprolobium fruticosum Benth.	Australia
Lamprolobium grandiflorum Everist	Australia
Hovea acutifolia Cunn.	Australia
Hovea spp.	Australia
Plagiocarpus axillaris Benth.	Australia
Sophora secundiflora	Mexico, Texas
Sophora tomentosa L.	Australia, Oceania
Spartium junceum L.	Spain, southern Europe
Strongylodon macrobotrys A. Gray	Philippines
Templetonia spp.	Australia
Thermopsis spp.	Australia
Ulex europaeus L.	central Europe

Diazepam

Other Names

7-chlor-1,3-dihydro-1-methyl-5-phenyl-2H-1,4-benzodiazepin-2-on, sleeping pill, tranquilizer, Valium

Empirical formula: $C_{16}H_{13}ClN_2O$

Substance type: benzodiazepine

Diazepam, better known as Valium, was originally synthesized in the laboratory and introduced as a therapeutic drug (psychopharmaca, tranquilizer) in the 1960s. The substance produces sedative, euphoric, and especially anxiolytic (anxiety-reducing) effects (Henningfield 1988, 17, 35*).

During the investigation of diazepam's pharmacology, it was discovered that the human nervous system has a special receptor for this molecule, known as the benzodiazepine receptor or the [^3H]-diazepam receptor. Luk et al. (1983) found three isoflans in the urine of cattle that may possibly dock (as neurotransmitters) in the benzodiazepine receptor. It is known that the kavapyrones (cf. *Piper methysticum*) bind to the [^3H]-diazepam receptor. Recently, flavonoids in the buds of the South American linden tree (*Tilia tomentosa* Moench; Tiliaceae; cf. **tila**) were found to bind to the benzodiazepine receptor. A substance found in *Passiflora caerulea* L. (cf. **Passiflora spp.**), 5,7-dihydroxyflavone, also docks to the same location (Viola et al. 1994).

The benzodiazepine receptor has been shown to be present in all vertebrates, suggesting that it appeared at a very early date in the evolution of the nervous system and has been preserved into the present. This indicates that it plays an important function in the nervous system and that there are endogenous substances that bind to it in order to transmit certain messages (Müller 1988).

But what do these substances look like? At first they were thought to be polypeptides, but then traces of diazepam and desmethyldiazepam were discovered in the brains of humans and other animals. Because diazepam and its initial metabolite appear in breast milk and the placenta after the ingestion of Valium (Wessen et al. 1985), it was first believed that the diazepam must have been introduced into the body from outside. But when diazepam was subsequently also found to be present in brains that dated to a time before the discovery of Valium synthesis, it was concluded that diazepam was not a synthetic chemical at all but a naturally occurring neurotransmitter in the nervous system (Müller 1988). Thus it was demonstrated that "Valium, the very symbol of chemical psychopharmaca" (Zehentauer 1992, 121*), is actually a natural substance.

Pharmacologists were surprised when subsequent research demonstrated the presence of diazepam and desmethyldiazepam in potatoes (*Solanum tuberosum* L.; cf. **Solanum spp.**) and in such diverse grains as wheat (*Triticum aestivum* L.; cf. **beer**), corn/maize (*Zea mays*), and rice (*Oryza sativa* L.; cf. **sake**) (Müller 1988). Valium, in other words, is a natural active constituent in plants. However, the concentration in these plants is so low that a person would likely not notice any Valium effects even after consuming a whole sack of potatoes.

Valium is one of the most widely used sedative drugs in modern society and is normally prescribed for the treatment of anxiety and sleeping disorders.[498] Not surprisingly, Valium also finds use as a recreational drug in some circles, particularly in combination with other substances. Its euphoric properties can be greatly affected by **alcohol**, which can at times counteract the sedative properties, resulting in powerful stimulating effects.

Valium is one of the more commonly used psychopharmaca in the music scene. Several rock bands, including the classic "space rock" band Hawkwind ("Valium 10," 1978), have dedicated titles to the substance.

Commercial Forms and Regulations

Valium is available by prescription only. In the United States, it is listed as a Schedule IV drug under the Controlled Substances Act.

Literature

Flesch, Peter. 1996. Schlafstörungen bei älteren Patienten: Auf Benzodiazepine kann meist verzichtet werden. *Jatros Neurologie* 12:6–7 (interview).

Henningsfield, Jack E. 1988. *Barbiturates: Sleeping potion or intoxicant.* The Encyclopedia of Psychoactive Drugs. London, Toronto, and New York: Burke Publishing Company.

Luk, Kin-Chun, Lorraine Stern, Manfred Weigele, Robert A. O'Brien, and Nena Sprit. 1983. Isolation and identification of "diazepam-like" compounds from bovine urine. *Journal of Natural Products* 46 (6): 852–61.

Müller, Walter E. 1988. Sind Benzodiazipine 100% Natur? *Deutsche Apotheker Zeitung* 126 (13): 672–74.

Viola, H., C. Wolfman, M. Levi de Stein, C. Wasowski, C. Pena, J. H. Medina, and A. C. Paladini. 1994. Isolation of pharmacologically active benzodiazepine receptor ligands from *Tilia tomentosa* (Tiliaceae). *Journal of Psychopharmacology* 44:47–53.

Wesson, Donald R., Susan Camber, Martha Harkey, and David E. Smith. 1985. Diazepam and desmethyldiazepam in breast milk. *Journal of Psychoactive Drugs* 17 (1): 55–56.

Diazepam

"The fact that Valium is ultimately 100% natural is not without a certain irony. For many ideologically tinged critics of psychopharmaca, benzodiazepines have become the very symbol of the evils of chemistry for the mind."

WALTER MÜLLER
"SIND BENZODIAZIPINE 100% NATUR?" [ARE BENZODIAZEPINES 100% NATURAL?]
(1988, 674)

The potato (*Solanum tuberosum* L.), a member of the Nightshade Family, contains highly toxic solanum alkaloids, primarily α-solanine, in all parts of the plant, including the green potatoes and the young sprouts. The tubers have been discovered to contain traces of naturally occurring diazepam. (Woodcut from Tabernaemontanus, *Neu Vollkommen Kräuter-Buch*, 1731)

498 It has been discovered that people who suffer from sleep disorders can do without diazepam, and that a combination of hops (*Humulus lupulus*) and valerian (*Valeriana officinalis*) can be used as a substitute (Flesch 1996).

Diterpenes

Other Names

Diterpene, diterpènes, diterpenoids, diterpenos

Diterpenes are not alkaloids but non-nitrogenous natural substances composed of four isoprene groups. They are related to the monoterpenes and sesquiterpenes and belong to the terpene group. Diterpenes occur in numerous plants and several **essential oils.**

Some diterpenes regulate plant growth. Termites, sponges (*Spongia spongens* L.), and coelenterates contain bioactive diterpenes that have inhibiting effects upon certain bacteria (Buchbauer et al. 1990, 28). There are even sweet-tasting diterpenes, such as the natural sweetening agents in *Stevia rebaudiana* (Bert.) Hemsl., the leaves of which are used to sweeten maté (cf. *Ilex paraguariensis*).

The first psychoactive diterpene to be discovered was **salvinorin A**. It is very likely that there are other psychoactive diterpenes that have not yet been isolated, pharmacologically tested, or chemically described. Some psychoactive alkaloids are diterpene derivatives. Aconitine, the primary active constituent in monkshood (cf. *Aconitum ferox*, *Aconitum napellus*), is a diterpene alkaloid. Diterpene alkaloids also occur in *Delphinium* and *Spiraea.*

Literature

See also the entries for *Coleus blumei*, *Salvia divinorum*, and **salvinorin A**.

Buchbauer, Gerhard, Helmut Spreizer, and Gabriele Kiener. 1991. Biologische Wirkungen von Diterpenen. *Pharmazie in unserer Zeit* 19 (1): 28–37.

Reid, W. W. 1979. The diterpenes of the *Nicotiana* species and *N. tabacum* cultivars. In *The biology and taxonomy of the Solanaceae*, ed. J. G. Hawkes, R. N. Lester, and A. Skelding, 273–78. London: Academic Press.

Diterpenes in Psychoactive Plants

(from Buchbauer et al. 1990; Rein 1979; supplemented)

Stock Plant	Known Diterpenes
Coleus blumei	bicyclic diterpenes
Coleus spp.	forskolin, labdanes, coleones
Crocus sativus	crocetine (as glycoside)
Helichrysum spp. (cf. *Helichrysum foetidum*)	various
Jatropha grossidentata	jatrophone
Lagochilus inebrians	lagochiline (diterpene alcohol)
Leonotis leonurus	various
Leonurus sibiricus	various
Nicotiana sylvestris Spegazz.	2,7,11-duvatrien-4,6-diol
Nicotiana tabacum	labdanes or duvanes
Nicotiana tomentosiformis Goodspeed	labdanes
Nicotiana spp.	
Petunia patagonica (Spegazz.) Millan (cf. *Petunia violacea*)	various
Piper auritum	*trans*-phytol
Salvia divinorum	**salvinorin A**, salvinorin B (clerodanes)
Scoparia dulcis	labdanediterpenes
Taxus baccata L. (cf. **witches' ointments**)	taxines, taxol
Taxus brevifolia Nutt.	taxines
Taxus canadensis Marsh.	taxines
Taxus cuspidata Sieb. et Zucc.	taxines
Taxus wallichiana	taxines

Ephedrine

Other Names

Aphetonin, efedrina, ephedrin, ephédrine, ephedrinum, ephetonin, erythro-2-methylamino-1-hydroxyl-1-phenylpropane, (1R,2S)-2-methyl-amino-1-phenyl-1-propanol

Empirical formula: $C_{10}H_{15}NO$

Substance type: ephedra alkaloid

Ephedrine was first isolated in 1887 by Nagai from *Ephedra distachya* (cf. *Ephedra* spp.) and was first introduced into ophthalmology as Mydriaticum (cf. **atropine**). Since around 1925, the alkaloid also been an important asthma medication (Schneider 1974, 2:54*).

Ephedrine occurs in almost all species of ephedra (cf. *Ephedra gerardiana, Ephedra sinensis, Ephedra* spp.). Two Malvaceae, *Sida acuta* Burm. and *Sida rhombifolia* L. (*Sida* spp.), which are smoked along the Mexican Gulf Coast as a marijuana substitute (cf. *Cannabis indica*), also contain ephedrine (Schultes and Hoffmann 1992, 56*). Ephedrine is probably present in other species of *Sida* as well. Ephedrine has also been found in *Aconitum* spp., yew (*Taxus bacata* L.; cf. **witches' ointments**), and khat (*Catha edulis*) (Römpp 1995, 1191*; Roth et al. 1994, 695*).

Ephedrine has sympathomimetic effects and causes an increased excretion of the endogenous neurotransmitter noradrenaline, which is responsible for the stimulant effects (Kalix 1991). Ephedrine hydrochloride has potent stimulant effects; it improves the general mood and may even induce euphoria. These effects can last up to eight hours. It is known that "therapeutic overdoses of ephedrine (Aphetonin) can also cause pronounced states of excitation combined with sexual arousal" (Fühner 1943, 199*). In men, however, ephedrine induces a temporary state of impotence. Ephedrine is a popular doping agent for athletes but is prohibited for this purpose (Körner 1994, 1483*). There have been reports of "ephedrine addiction" (Prokop 1968).

Because ephedrine helps reduce swelling of the mucous membranes, it is a component of many cough syrups (see **codeine**). Ephedrine suppresses the effects of **alcohol** and is administered subcutaneously to prevent hypotension during anesthesia (Morton 1977, 35*). Between 55 and 75% of ephedrine is excreted in the urine unchanged (Roth et al. 1994, 812*). The effective oral dosage is 5 to 10 mg.

The closely related ephedra alkaloids have similar effects but vary in their potency (Reti

1953). Pseudoephedrine is significantly weaker, while the related ephedroxanes tend to have depressant effects (Hikino et al. 1985). Pseudoephedrine can be used to produce methcathinone, which in the United States is smoked as "speed" or snuffed like cocaine (it is also used as a substitute for cocaine) (Glennon et al. 1987).

Although *Catha edulis* does contain *d*-norisoephedrine, it is not the plant's primary active constituent, as was previously assumed (Wolfes 1930). However, cathinone, the psychoactive constituent in khat leaves, is metabolized into ephedrinene (Brenneisen et al. 1986; Kalix 1991). Norephedrine, the *nor*-form (a *threo*-isomer) of ephedrine, lacks a methyl group on the side chain. Up to 90% of norephedrine is excreted unchanged (Cho and Segal 1994, 58).

Removing the hydroxyl group from the ephedrine molecule by either reduction or β-hydroxylation yields amphetamine (Cho and Segal 1994, 57). Amphetamine is one of the most highly effective stimulants known. Numerous derivatives have been developed from amphetamine (e.g., Ritalin, methamphetamine, MDMA; cf. **herbal ecstasy**). In addition to their stimulating effects, several of these substances also induce empathogenic and even hallucinogenic effects (Cho and Segal 1994). Amphetamine has not yet been found to occur in nature.

Commercial Forms and Regulations

Ephedrine is available as anhydrous ephedrine (ephedrinum anhydricum), ephedrine hemihydrate, or (most often) ephedrine hydrochloride ([+]-ephedrine-HCL). Ephedrine and ephedrine preparations (medicinal drugs) require a prescription. Because ephedrine is now regarded as a precursor substance for the illegal synthesis of MDMA, it is only rarely prescribed and is strictly controlled. In Germany, only combination preparations (cough medicines) in which a single dosage may not exceed 10 mg of ephedrine can be purchased in a pharmacy without prescription (Roth et al. 1994, 812*). A number of high-profile cases, including one in which a young professional athlete died after ingesting ephedrine before training, resulted in the banning of most ephedrine preparations in the United States in 2004.

Literature

See also the entries for *Catha edulis, Ephedra gerardiana, Ephedra sinica,* and *Ephedra* spp.

Brenneisen, R., S. Geisshüsler, and X. Schorno. 1986. Metabolism of cathinone to (–)-norephedrine and (–)-norpseudoephedrine. *Journal of Pharmacy and Pharmacology* 38:298–300.

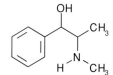

Ephedrine

Swiss ephedra (*Ephedra helvetica*) contains ephedrine.

Cho, Arthur K., and David S. Segal, eds. 1994. *Amphetamine and its analogs: Psychopharmacology, toxicology and abuse*. San Diego: Academic Press.

Costa, E., and S. Garattini, eds. 1970. *Amphetamine and related compounds*. New York: Raven Press.

Glennon, R., M. Yousif, N. Naiman, and P. Kalix. 1987. Methcathinone, a new and potent amphetamine-like agent. *Pharmacol. Biochem. Behav.* 26:547–51.

Hikino, Hiroshi, Kuniaki Ogata, Yoshimasa Kasahara, and Chohachi Konno. 1984. Pharmacology of ephedroxanes. *Journal of Ethnopharmacology* 13:175–91.

Hofmann, H., K. Opitz, and H. J. Schnelle. 1955. Die Wirkung des nor-c-Ephedrins. *Arzneimittel-Forshung* 5:367–70.

Kalix, P. 1991. The pharmacology of psychoactive alkaloids from *Ephedra* and *Catha*. *Journal of Ethnopharmacology* 32:201–8.

Panse, F., and W. Klages. 1964. Klinisch-pathologische Beobachtungen bei chronischem Mißbrauch von Ephedrin. *Archiv für Psychiatrie und Neurologie* 206:69 ff.

Prokop, H. 1968. Halluzinose bei Ephedrinsucht. *Der Nervenarzt* 1968:71 ff.

Reti, L. 1953. Ephedra bases. In *The alkaloids: Chemistry and physiology*, ed. R. H. F. Manske and H. L. Holmes, 339–62. New York: Academic Press.

Wolfes, O. 1930. Über das Vorkommen von *d*-Norisoephedrin in *Catha edulis*. *Archiv der Pharmazie* 268:81–83.

Lysergol

Lysergic acid amide (LSA)

Lysergic acid diethylamide (LSD)

Ergot Alkaloids

Other Names
Ergoline, ergoline alkaloids, ergot alkaloids, ergotalkaloide, mutterkornalkaloide

Ergot alkaloids are derivatives of lysergic acid or clavine derivatives and belong to the group of **indole alkaloids.** They are found in many climbing plants (Convolvulaceae) and fungi (*Claviceps purpurea, Claviceps paspali, Claviceps* spp.). The ergot alkaloids can be divided into two groups that exhibit stark pharmacological differences. One group is composed of alkaloids that are highly toxic and cause gangrenous ergotism, while the other group consists of psychoactive alkaloids with hallucinogenic effects. Both types may be present in the same plant (Hofmann 1964).

The following ergot alkaloids have been found in the Convolvulaceae: agroclavine, ergine, ergonovine, isoergine (= isolysergic acid amide), chanoclavine I and II, racemic chanoclavine II, elymoclavine, festuclavine, lysergene, lysergol, isolysergol, molliclavine, penniclavine, cycloclavine, stetoclavine, isostetoclavine, ergometrinine, lysergic acid-α-hydroxyethylamide (= lysergic acid methylcarbinolamide), isolysergic acid-α-hydroxyethylamide (= isolysergic acid methylcarbinolamide), ergosine, and ergosinine (cf. *Argyreia nervosa, Convolvulus tricolor, Ipomoea violacea, Ipomoea* **spp.,** *Turbina corymbosa*).

One hallucinogenic ergot alkaloid is ergonovine (ergometrine, D-lysergic acid-L-2-propanolamide, ergobasin, ergotocine, ergostetrine, ergotrate, syntometrine, *N*-[α-(hydroxymethyl)ethyl]-D-lysergic amide). Ergonovine maleate is psychoactive at dosages between 3 and 10 mg (Bigwood et al. 1979). The semi-synthetic methylergonovine has also been reported to induce psychoactive effects (Ott and Neely 1980).

Ergine (= lysergic acid amide, LSA, lysergic amide, 9,10-didehydro-6-methylergoline-8β-carboxamide) induces psychoactive effects reminiscent of those produced by LSD (lysergic acid diethylamide). LSD is a slight chemical variant of lysergic acid amide that can be produced from ergot (*Claviceps purpurea*). LSD is a psychopharmaca, a "remedy for the soul" (Albert Hofmann) whose entheogenic effects are very well known (Hofmann 1979*).

The ergot alkaloids dihydroergotaminemesilate, dihydroergotamintartrate, ergometrine hydrogenmaleate, and ergotamine tartrate have a variety of uses in medicine, including as treatments for labor contractions and migraines.

Commercial Forms and Regulations
Ergonovine requires a prescription. In the United States, ergine is a controlled substance (Ott 1993, 437*). LSD is illegal throughout the world.

Literature

See also the entries for *Claviceps paspali*, *Claviceps purpurea*, and indole alkaloids.

Bigwood, Jeremy, Jonathan Ott, Catherine Thompson, and Patricia Neely. 1979. Entheogenic effects of ergonovine. *Journal of Psychedelic Drugs* 11 (1–2): 147–49.

Hofmann, Albert. 1964. *Die Mutterkorn-Alkaloide.* Stuttgart: Enke.

Ott, Jonathan, and Patricia Neely. 1980. Entheogenic (hallucinogenic) effects of methylergonovine. *Journal of Psychedelic Drugs* 12 (2): 165–66.

Rivier, L. 1984. Ethnopharmacology of LSD and related compounds. In *50 years of LSD: Current status and perspectives of hallucinogens*, ed. A. Pletscher and D. Ladewig, 43–55. New York and London: Parthenon Publishing.

Yui, T., and Y. Takeo. 1958. Neuropharmacological studies on a new series of ergot alkaloids. *Japanese Journal of Pharmacology* 7:157.

Saint Anthony is the patron saint of people afflicted with Saint Anthony's fire, a severe condition caused by the ingestion of ergot alkaloids. (Statue in Cloister Unterlinden, Colmar, Elsass)

Essential Oils

Other Names

Aroma, ätherische öle, ätherischöl, essence, essenz, etherisches öl, volatile oil

Essential oils are complex mixtures of carbohydrates, alcohols, ketones, acids and esters, ethers, aldehydes, and sulfur compounds that are volatile and evaporate at low temperatures. Essential oils can exhibit tremendous variation in their composition. Each specific mixture produces its own characteristic scent. For the most part, essential oils are distilled from raw drugs or stock plants through a variety of techniques. Essential oils are used medicinally in what has become known as aromatherapy. This healing system was founded by René-Maurice Gattefossé (1881–1950) and is gaining increasing recognition throughout the world (Carle 1993; Henglein 1985; Kraus 1990; Strassmann 1991).

Many psychoactive plants contain essential oils. They are sometimes the only active constituents, while sometimes they occur only in trace amounts. Several components are present in the essential oils of most plants that have unequivocal psychoactive effects.

Eugenol

Eugenol is known to be stimulating, anesthetic, and psychoactive (Sensch et al. 1993; Toda et al. 1994). High concentrations of eugenol occur in the essential oil of clove (*Syzygium aromaticum*).

Myristicin

Myristicin is regarded as the hallucinogenic component of many essential oils (Wulf et al. 1978, 271). Myristicin is present in dill (*Anethum*), lovage (*Levisticum officinale*), parsnips (*Pastinaca* sp.), and parsley (**Petroselinum crispum**). The essential oil of the Australian *Zieria* species (Rutaceae) contains up to 23.4% myristicin. It is thought that myristicin is metabolized into an amphetamine derivative (MDA) (cf. ***Myristica fragrans***).

Safrole

Safrole is found in cloves (*Syzygium aromaticum*) and in the sassafras tree (*Sassafras albidum*). Safrole is one of the most important precursors for the synthesis of MDMA and other similar substances (MMDA, MDE, MDA). The halogen derivatives of safrole, the closely related piperonal and isosafrole, are also suitable for this purpose (Yourspigs 1995). In the body, safrole is thought to be metabolized into amphetamine derivatives.

Thujone

In nature, thujone exists in two forms: α-thujone and β-thujone. The common tansy (*Tanacetum vulgare*), whose name is derived from the Greek word *athanaton* ("immortal"), is very rich in thujone (= tanacetone; cf. Semmler 1900). According to myth, Ganymede became immortal because he had eaten tansy (Albert-Puleo 1978, 65).

Eugenol

Myristicin

Safrole

Thujone

Top: The Romans held rosemary (*Rosmarinus officinalis*) in high esteem as an incense; it contains an essential oil with psychoactive constituents.

Middle: Cloves are the flower buds of a tropical tree (*Syzygium aromaticum* [L.] Merr. et L.M. Perry). They contain a great quantity of an essential oil with anesthetic and stimulating properties; the oil consists primarily of eugenol.

Bottom left: The Western or giant red cedar (*Thuja plicata*), also known as the giant arborvitae, contains an essential oil that is rich in thujone and can be potently inebriating.

Bottom right: The Moroccan cedar (*Cedrus atlantica*) has been found to contain thujone.

Clary sage (muscatel sage) (*Salvia sclarea* L.) has a high thujone content. During the nineteenth century, this plant was used in England instead of hops (**Humulus lupulus**) to produce a more potent type of **beer**. Other plants that contain thujone (**Artemisia absinthium**, *Artemisia vulgaris*) were used for the same purpose (Albert-Puleo 1978, 69).

Thujone kills the common roundworm *Ascaris lumbricoides* (Albert-Puleo 1978, 65). The pharmacological effects of thujone are very similar to those of THC (cf. **Artemisia absinthium**).

Ud Oil

It has often been reported that *ud* oil, the essential oil in lignum aloe or aloe wood (*Aquillaria agallocha* Roxb. [syn. *Aquillaria malaccensis* Lam.]; Thymeleaceae), can induce psychoactive effects:

> As an incense or aromatic oil, it is used to treat mental and psychological disturbances as well as emotional instability, particularly in cases when this has been produced by negative mental energies. Our experience indicates that aloe wood has unusually relaxing and mood-enhancing properties. It produces a state of trance and introspection and lifts the mind to higher plains of perception. It facilitates the attainment of high levels in meditation. For this reason, it should not necessarily be used prior to a work-filled day in which concentration and quick reactions are required. (Ashisha and Mahahradanatha 1994, 10)

Sufis utilize the precious lignum aloe or distilled *ud* oil (essence) for advanced stages of Islamic mysticism:

> One could say that only those individuals whose minds are more highly developed

Plants Containing Psychoactive Essential Oils

(from Albert-Puleo 1978; Bock 1994*; supplemented)

Stock Plant	Primary Constituent(s) of Essential Oil
ANNONACEAE	
Cananga odorata (Lam.) Hook. f. et Thoms.	safrole, eugenol
APIACEAE (= UMBELLIFERAE)	
Anethum graveolens	anethol, myristicin, and others
Coriandrum sativum L.	coriandrol
Foeniculum vulgare Mill. ssp. *vulgare*	*trans*-anethol
Levisticum officinale Koch	myristicin and others
Pastinaca sativa L.	myristicin and others
Petroselinum crispum	apiol/myristicin
apiol race	apiol (58–80%)
myristicin race	myristicin (49–77%)
ssp. *tuberosum*	apiol
ARACEAE	
Acorus calamus	safrole, asarone (not in all strains), eugenol
Acorus gramineus	safrole, eugenol, and others
ARISTOLOCHIACEAE	
Asarum europaeum L.	asarone
BURSERACEAE	
Commiphora spp. (myrrh)	eugenol and others
CANELLACEAE	
Canella winterana (L.) Gaertn.	eugenol
CANNABACEAE	
Humulus lupulus[499]	
CISTACEAE	
Cistus ladaniferus L.	eugenol, ledol
COMPOSITAE (= ASTERACEAE)	
Achillea millefolium L.	thujone and others
Artemisia absinthium	β-thujone
Artemisia mexicana	β-thujone
Artemisia tilesii Ledeb.	thujone, isothujone
Artemisia tridentata ssp. *vasyana* (Rydb.) Beetle	thujone, isothujone
Artemisia vulgaris L.	β-thujone
***Artemisia* spp.**	β-thujone and others
Salvia officinalis L.	α-thujone
Salvia sclarea L.	α-thujone
Tanacetum vulgare L.	β-thujone
CUPRESSACEAE	
Juniperus recurva	limonene (23.6%), α-thujone
Juniperus sabina L.	thujone and others
Thuja occidentalis L.	α-thujone, thujone isomers, thujone acid
Thuja orientalis L.	α-thujone, thujone isomers, thujone acid
Thuja plicata D. Don	α-thujone, thujone isomers, thujone acid
CURCUBITACEAE	
Monodora myristica	myristicin, safrole, and others
ERICACEAE	
Ledum groenlandicum Oed.	ledol
Ledum palustre	ledol

The Southeast Asian ylang-ylang tree (*Cananga odorata*) is the source of an essential oil with a sweet, pungent aroma that is used in tantric rituals as an aphrodisiac and has also been shown to have psychoactive effects.

Early European illustration of the American tuberose (*Polianthes tuberosa*), which produces a valuable essential oil with an enchanting scent. (Woodcut from Gerard, *The Herball or General History of Plants*, 1633)

499 2-methyl-3-buten-2-ol, found in the essential oil of hops (***Humulus lupulus***), has strong sedative properties.

The essential oil of sage (*Salvia officinalis*) contains constituents with inebriating properties.

"Coriander was one of the first spices used by man. Its seeds have been found in Bronze Age ruins on Thera (Santorini) and Therasia as well as in the graves of pharaohs, and we know that it was cultivated in Assyria and Babylon. The Egyptians added it to wine to potentiate its inebriating effects. It used in India during religious and magical ceremonies. The Hebrews called it *gad* and were very fond of it. In Mycenae, one of the oldest cities in Greece, it was used as a spice to improve bland meals. The Romans believed that the coriander from Egypt was the best and they used it to spice their bread and stews and knew it as an ingredient in their Roman bouquet garni."

WAVERLY ROOT
WACHTEL, TRÜFFEL, SCHOKOLADE
[GAME HENS, TRUFFLES, CHOCOLATE]
(1996, 195 f.*)

"Let my soul inhabit various flowers,
let it become inebriated on their
    scent,
only too soon will I have to depart
    crying,
to stand before the face our mother in
    the kingdom of the dead."

AZTEC SONG
IN *ANCIENT NAHUATL POETRY*
(BRINTON 1887, 79*)

Stock Plant	Primary Constituent(s) of Essential Oil
ILLICIACEAE	
Illicium verum Hook. f.	anethol, safrole
IRIDACEAE	
Crocus sativus	?
LAMIACEAE (LABIATAE)	
Hyssopus officinalis L.	thujone
Mentha aquatica L.	limonene, caryophyllene, α-thujone
Mentha pulegium	pulegon (80–94%)
Orthodon sp.	myristicin
Thymus spp.	thymol, thujone
LAURACEAE	
Cinnamomum camphora	safrole, eugenol
Cinnamomum glanduliferum	myristicin
Cinnamomum verum Presl.	eugenol, cinnamic aldehyde
Laurus nobilis	eugenol and others
Ocotea cymbarum H.B.K.	safrole (90–93%)
Sassafras albidum	safrole (80–90%)
Umbellularia californica (H. et A.) Nutt.	umbellulone, safrole
MAGNOLIACEAE	
Magnolia virginiana	safrole and others
MONIMIACEAE	
Atherosperma moschatum	methyleugenol (60%), safrole (10%)
Doryphora sassafras Endl.	safrole
MYOPORACEAE	
Eremophila longifolia (R. Br.) Muell.	methyleugenol
MYRISTICACEAE	
Myristica fragrans	myristicin, safrole
MYRTACEAE	
Backhousia myrtifolia Hook.	methyleugenol
Pimenta dioica (L.) Merr.	eugenol and others
Syzygium aromaticum (L.) Merr. et Perry	eugenol, acetyleugenol
OLEACEAE	
Jasminum officinale L. (cf. ***Jasminum spp.***)	eugenol and others
PINACEAE	
Cedrus atlantica (Endl.) Manetti	thujone and others
PIPERACEAE	
Macropiper excelsum (Forster) Miq.	myristicin, elemicine
Piper amalago L.	safrole
Piper auritum	safrole (70%)
Piper betle	eugenol, isoeugenol
Piper elongatum	apiol, asarone
Piper sanctum Schl.	safrole
Piper spp.	safrole and others
RUTACEAE	
Zieria spp.	myristicin
WINTERACEAE	
Tasmannia glancifolia Williams	safrole (17%), myristicin (5.3%)
ZINGIBERACEAE	
Alpinia officinarum Hance	eugenol

experience the benefits of *ud*. In fact, it is only used to treat imbalances during the last three states of mental development. (Moinuddin Chishti 1991, 118*)

Aromatic lignum aloe (lignum aquillariae resinatum) contains p-methoxycinnamic acid, agarotetrol, and the sesquiterpenoids agarol, agarospirole, α- and β-agarofurane, dihydroagarofurane, 4-hydrodioxydihydroagarofurane, and oxo-nor-agarofuran, among other constituents.

Essential Oils as Aphrodisiacs

Some essential oils are attributed with aphrodisiac effects. Perfume manufacturers consider the aroma of the Mexican tuberose (*Polianthes tuberosa* L.; Agavaceae) (cf. Dressler 1953, 144*) to be aphrodisiac.

The evergreen ylang-ylang tree (*Cananga odorata* [Lam.] Hook. f. et Thoms. [f. *genuina*] [syn. *Canangium oderatum* Baill.]), which thrives only in tropical regions, is the source of the oil of the same name. In India, ylang-ylang oil is regarded as the favorite oil for tantric rituals because it is believed to have potent aphrodisiac effects and to stimulate and refine erotic sensations. Today, individuals throughout the world use ylang-ylang oil to ritualize their own eroticism (Huron 1994; Kraus 1990; Strassmann 1991). The flowers contain 1.5 to 2.5% of an essential oil composed of linalool, safrole, eugenol, geraniol, pinene, cadinene, and sesquiterpenes. There have been frequent reports of ylang-ylang having mind-altering powers. Pharmacologically, this effect is probably due to the safrole component of the essential oil (Rätsch 1996). Above a certain concentration, safrole appears to produce psychoactive effects that are quite similar to those of MDMA (see **herbal ecstasy**).

Literature

See also the entries for *Artemisia absinthium*, *Artemisia* spp., *Myristica fragrans*, **herbal ecstasy**, and **incense**.

Albert-Puleo, Michael. 1978. Mythobotany, pharmacology, and chemistry of thujone-containing plants and derivatives. *Economic Botany* 32:65–74.

Ashisha, Ma Deva, and Mahahradantha. 1994. *Duftkräuter und ätherische Öle in der ayurvedischen Heilkunst.* Tostedt: Yogini Verlag.

Carle, Reinhold. 1993. *Ätherische Öle—Anspruch and Wirklichkeit.* Stuttgart: WVG.

Chandler, R. F., S. N. Hooper, and M. J. Harvey. 1982. Ethnobotany and phytochemistry of yarrow, *Achillea millefolium*, Compositae. *Economic Botany* 36 (2): 203–23.

Cipolla, Carlo M. 1992. *Allegro ma non troppo.* Frankfurt/M.: Fischer.

Dandiya, P. C., and M. K. Menon. 1963. Effects of asarone and β-asarone on conditioned responses, fighting behaviour and convulsions. *British Journal of Pharmacology* 20:436–42.

———. 1964. Actions of asarone on behaviour, stress hyperpyrexia and its interaction with central stimulants. *Journal of Pharmacology and Experimental Therapeutics* 145:42–46.

Gattefossé, René-Maurice. 1994. *Aromatherapie.* Aarau: AT Verlag.

Harnishfeger, Götz. 1994. Thuja. In *Hagers Handbuch der pharmazeutischcn Praxis*, 5th ed., 6:955–66. Berlin: Springer.

Hengelein, Martin. 1985. *Die heilende Kraft der Wohlgerüche und Essenzen.* Munich: Schönbergers.

Hurton, Andrea. 1994. *Erotik des Parfums: Geschichte und Praxis der schönen Düfte.* Frankfurt/M.: Fischer.

Kraus, Michael. 1991. *Ätherische Öle für Körper, Geist and Seele.* Gaimersheim: Simon und Wahl.

Kremer, Bruno P. 1988. *Duft und Aromapflanzen.* Stuttgart: Franckh-Kosmos.

Laatsch, Hartmut. 1991. Wirkung von Geruch und Geschmack auf die Psyche. In *Jahrbuch des Europäischen Collegiums für Bewußtseinsstudien* (1991), 119–33. Berlin: VWB.

Miller, Richard Alan, and Iona Miller. 1990. *Das magische Parfum.* Braunschweig: Aurum.

Morwyn. 1995. *Witch's brew: Secrets of scents.* Arglen, Penn.: Whitford Press/Schiffer Publishing.

Rätsch, Christian. 1996. Ylang-Ylang, "die Blume der Blumen." *Dao* 6:68.

Richter, Dieter. 1984. *Schlaraffenland.* Cologne: Diederichs.

Rimmel, Eugene. 1985. *Das Buch des Parfums.* Dreieich: Hesse und Becker. (Orig. pub. 1864.)

Schivelbusch, Wolfgang. 1983. *Das Paradies, der Geschmack und die Vernunft.* Frankfurt/M.: Ullstein.

Semmler, F. W. 1900. Über Tanaceton und seine Derivate. *Berichte der Deutschen Chemischen Gesellschaft* 33:275–77

Sensch, O., W. Vierling, W. Brandt, and M. Reiter. 1993. Calcium-channel blocking effect of constituents of clove oil. *Planta Medica* 59 suppl.: A687.

Strassmann, René A. 1991. *Duftheilkunde.* Aarau: AT Verlag.

Toda, Shizuo, Motoyo Ohnishi, Michio Kimura, and Tomoko Toda. 1994. Inhibitory effects of eugenol and related compounds on lipid peroxidation induced by reactive oxygen. *Planta Medica* 60:282.

Wieshammner, Rainer-Maria. 1995. *Der 5. Sinn: Düfte als unheimliche Verführer.* Rott am Inn: F/O/L/T/Y/S Edition.

Wulf, Larry W., Charles W. Nagel, and Larry Branen. 1978. High-pressure liquid chromatographic separation of the naturally occurring toxicants myristicin, related aromatic ethers and falcarinol. *Journal of Chromatography* 161:271–78.

Yourspigs, U. P. 1995. *The complete book of ecstasy.* 2nd ed. N.p.: Synthesis Books.

"It has been said that a Sufi *attar*, a maker of aromatic essences, one day reached the *Fardaws* region, the highest of the heavens, while performing his mystic exercises. When he arrived there, he smelled a certain scent. After returning to his normal state of mind, he concocted a similar aroma, and this is the origin of the name: 'gate to the highest heavens' [*jannat al-Fardaws*]."

SHAYKH HAKIM MOINUDDIN CHISHTI
THE BOOK OF SUFI HEALING
(1991, 118*)

"And on that meadow is a tree which is of such beauty to behold. Its roots are galanga and ginger, its branches are pure white turmeric, and its flowers are of delicate nutmeg, the bark is sweet aromatic cinnamon, the fruits are wonderfully scented cloves.
You also find plenty of cubeb."

"THE BALLAD OF THE LAND OF PLENTY" (IRELAND, FOURTEENTH CENTURY)
IN *SCHLARAFFENLAND* [LAND OF MILK AND HONEY]
(RICHTER 1984, 137)

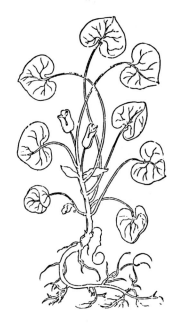

Asarabacca (*Asarum europaeum* L.; Aristolochiaceae) is rich in the psychoactive asarone. (Woodcut from Fuchs, *Läebliche abbildung und contrafaytung aller kreüter*, 1545)

5-MeO-DMT

5-MeO-DMT

"We sat in a circle with some friends and smoked a joint of toad secretions. I inhaled deeply and passed the joint on. Immediately, a mandala unfolded before me. A dragon was sitting in each corner, and in the circle in the center was a cyclone. No sooner had I recognized the cyclone than I was pulled into it. The cyclone was turning in a clockwise direction—and yet time was going backward. Dragons, amphibians, and dinosaurs appeared in the cyclone, and they too were pulled into eternity. At first, I was surprised that Tibetan dragons and dinosaurs were appearing together in the waves, but then I realized that they represented two metaphors for the same principle. I was drawn further and further along by the deluge of images until I finally arrived at the journey's end. I sat like a toad or a newt in a Permian bog. The water around me was black, and in the murky fog I saw gigantic ferns and horsetails. Somehow, I communicated with the strange amphibians and realized that I was able to communicate not only with beings of another kind but also across the barriers of time, across millions of years."

AN EXPERIENCE WITH *BUFO ALVARIUS* SECRETIONS [1495]

The Colorado River toad (*Bufo alvarius*) lives only in Arizona. Its secretions contain high concentrations (up to 15% of dry weight) of 5-MeO-DMT.

Other Names
Dimethyl-5-methoxytryptamine, 5-methoxy-DMT, 5-methoxy-*N,N*-dimethyl-tryptamine, *O*-methyl-bufotenine, 3-[2-(dimethylamino)ethyl]-5-methoxy-indole, toad foam

Empirical formula: $C_{13}H_{18}N_2O$

Substance type: tryptamine (**indole alkaloid**)

5-MeO-DMT was first discovered in *Dictyoloma incanescens* DC. and later was isolated from *Anadenanthera peregrina* as well. It occurs in a very large number of plants, often in association with *N,N*-**DMT** (see the table on pages 853–854). Its effects are somewhat more potent than those of *N,N*-DMT. When the two are administered simultaneously, 5-MeO-DMT more quickly occupies the specific receptors. 5-MeO-DMT is a natural neurotransmitter in the human nervous system. When 5-MeO-DMT (10 to 20 mg) is smoked or vaporized and inhaled, the effects are almost immediately apparent, are incredibly extreme, and last about ten minutes. Many people report having shamanic experiences with this substance as well as experiencing states of enlightenment and the clear light of nirvana (Metzner 1988).

The Colorado River toad (*Bufo alvarius*) is native to the area around Tucson, Arizona. These toads spend nine months of the year underground, buried in the mud that keeps them protected from the burning desert sun. The toads emerge from their hiding places with the first rains and begin their courtship (Smith 1982, 97–100). They remain visible for only three months. Like all toads, *Bufo alvarius* develops mucous secretions in two glands that are located on the neck The secretions of the Colorado River toad, however, do not contain bufotoxine, the toxic substance that is found in the secretions of most other toads. Instead, the dried mass contains 15% 5-MeO-DMT (Erspamer et al. 1965, 1967).

The native tribes that lived in the North American Southwest made fetishes of this *Bufo alvarius*. However, it was only in recent times that the toad's cultural importance and its psychedelic use were discovered, or more likely *re*discovered (cf. Davis and Weil 1992). The toad is "milked" by being held firmly without being crushed. Both glands are then massaged gently until a fat stream of the secretion squirts out. The secretion is caught on a piece of glass, where it is allowed to dry and crystallize. The yellowish crystalline mass then can be scraped off, mixed with different herbs (such as damiana [*Turnera diffusa*]), and smoked. The toad, which is released unharmed, is quickly able to replenish the loss in its secretions.

When taken orally, *Bufo alvarius* secretions are apparently toxic, whereas they are not poisonous when smoked (Weil and Davis 1994). Davis and Weil have suggested that the dried secretions of *Bufo alvarius* were traded to Mexico in pre-Columbian times and that the priests and shamans there smoked or used it in some other manner (Davis and Weil 1992; cf. **balche'**, **bufotenine**).

In Arizona, there is now a Church of the Toad of Light, which uses the secretions of *Bufo alvarius* as a sacrament (Most 1984; Ott 1993, 396*).

Commercial Forms and Regulations
Pure 5-MeO-DMT is available from chemical suppliers. While the substance is not explicitly mentioned in the narcotics laws, the fact that it could be interpreted as a DMT analog may result in problems with the law.

Literature
See also the entries for **bufotenine**.

Davis, Wade, and Andrew T. Weil. 1992. Identity of a New World psychoactive toad. *Ancient Mesoamerica* 3:51–59.

Erspamer, V., T. Vitali, M. Roseghini, and J. M. Cei. 1965. 5-methoxy and 5-hydroxy-indolalkylamines in the skin of *Bufo alvarius*. *Experientia* 21:504.

———. 1967. 5-methoxy- and 5-hydroxyindoles in the skin of *Bufo alvarius*. *Biochemical Pharmacology* 16:1149–64.

Metzner, Ralph. 1988. Hallucinogens in contemporary North American shamanic practice. In *Proceedings of the Fourth International Conference on the Study of Shamanism and Alternate Modes of Healing* (Independent Scholars of Asia), 170–75.

Most, A. 1984. *Bufo alvarius: The psychedelic toad of the Sonoran Desert*. Denton, Texas: Venom Press.

Rätsch, Christian. 1993. Die Krötenmutter. In *Naturverehrung und Heilkunst*, ed. C. Rätsch, 125–28. Südergellersen: Bruno Martin.

Smith, Robert L. 1982. *Venomous animals of Arizona.* Bulletin 8245. Tucson: The University of Arizona.

Weil, Andrew T., and Wade Davis. 1994. *Bufo alvarius*: A potent hallucinogen of animal origin. *Journal of Ethnopharmacology* 41:1–8.

This drawing depicts the experience of a volunteer subject after smoking 5-MeO-DMT.

Harmaline and Harmine

Other Names

Harmaline: 4,9-dihydro-7-methoxy-1-methyl-3*H*-pyriol[3,4-*b*] indole, harmalin, harmalolmethyl-ester, harmidin, harmidin, 3,4-dihydroharmin

Harmine: banisterin, banisterine, harmin, 7-methoxy-1-methyl-β-carboline, telepathin, tele-pathine, yageine

Empirical formula: $C_{13}H_{14}N_2O$ (harmaline), $C_{13}H_{12}N_2O$ (harmine)

Substance type: β-**carbolines**, harmala alkaloids (**indole alkaloids**)

Harmaline and harmine are found in *Banisteriopsis caapi* and *Peganum harmala* (Beringer 1928; Beringer 1929; Chen and Chen 1939). Harmine also occurs in numerous other plants (see **ayahuasca analogs**). Harmaline and harmine not only are strong MAO inhibitors (Pletscher et al. 1959; cf. β-**carbolines**) but also have antibacterial properties (Ahmad et al. 1992). Harmine was an early treatment for Parkinson's disease (Halpern 1930b):

> Harmine lessens the exaggerated excitability of the parasympathetic system in Parkinson's patients, increases the low excitability of the sympathetic system, also promotes the excitability of the vestibular apparatus, and puts the patient in a state of euphoria, helping them to better accept their affliction. (Roth et al. 1994, 548*)

In the 1960s, the Chilean psychiatrist Claudio Naranjo (1969*) introduced harmaline and harmine to psychotherapy as "fantasy-increasing drugs" (cf. **ibogaine**). The extent to which these substances are psychoactive is questionable. To

investigate the supposed "psychedelic" effect of harmine,

Maurer (together with Lamparter and Dittrich) tested the hypothesis that harmine is a hallucinogen in 11 self-experiments with a sublingual dosage between 25 and 750 mg. Contrary to expectations, however, harmine did not prove to be a substance that exhibited many similarities to classic hallucinogens such as mescaline or psilocybin. Maurer characterized the state that harmine induced as more of a retreat from one's surroundings and as a pleasant relaxation with a mildly reduced ability to concentrate. Short-term and elementary optic hallucinatory phenomena were observed only to the degree that they would otherwise also appear naturally during reduced contact with one's surroundings. With dosages above 300 mg, such undesirable vegetative and neurological symptoms as dizziness, nausea, and ataxia became more apparent, precluding any increase in dosage above 750 mg. (Leuner and Schlichting 1986, 170*)

Crystallized harmaline has a characteristic light yellow color. (Photo: Karl-Christian Lyncker)

Harmaline

Harmine

Most experimenters have called into question the reports that Naranjo (1979*) published from his psychotherapeutic practice. It may be that he administered **ayahuasca**, and not any pure substances, to his patients.

Today, harmaline and harmine are primarily used in the production of pharmahuasca (**ayahuasca analogs**).

Commercial Forms and Regulations

Both substances are available through chemical suppliers. They may be purchased without restriction and are not subject to any legal regulations (Ott 1993, 438*).

Literature

See also the entries for *Banisteriopsis caapi*, *Peganum harmala*, ayahuasca, ayahuasca analogs, β-carbolines, and indole alkaloids.

Ahmad, Aqeel, Kursheed Ali Khan, Sabiha Sultana, Bina S. Siddiqui, Sabira Begum, Shaheen Faizi, and Salimuzzaman Siddiqui. 1992. Study of *in vitro* antimicrobial activity of harmine, harmaline and their derivatives. *Journal of Ethnopharmacology* 35:289–94.

Beringer, Kurt. 1928. Über ein neues, auf das extrapyramidal-motorische System wirkendes Alkaloid (Banisterin). *Der Nervenarzt* 1:265–75.

———. 1929. Zur Banisterin und Harminfrage. *Der Nervenarzt* 2:545–49.

Beringer, Kurt, and K. Willmanns. 1929. Zur Harmin-Banisterin Frage. *Deutsche medizinische Wochenschrift* 55:2081–86.

Chen, A. L., and K. K. Chen. 1939. Harmine: The alkaloid of caapi. *Quarterly Journal of Pharmacy and Pharmacology* 12:30–38.

Halpern, L. 1930a. Über die Harminwirkung im Selbstversuch. *Deutsche medizinische Wochenschrift* 56:1252–54.

———. 1930b. Der Wirkungsmechanismus des Harmins und die Pathophysiologie der Parkinsonschen Krankheit. *Deutsche medizinische Wochenschrift* 56:651–55.

Manske, R. H. F., et al. 1927. Harmine and harmaline: Part IX: A synthesis of harmaline. *Journal of the Chemical Society (Organic)* (1927): 1–15.

Pennes, H. H., and P. H. Hoch. 1957. Psychotomimetics, clinical and theoretical considerations: Harmine, WIN-2299 and nalline. *American Journal for Psychiatry* 113:887–92.

Pletscher, A., et al. 1959. Über die pharmakologische Beeinflussung des Zentralnervensystems durch kurzwirkende Monoaminooxydasehemmer aus der Gruppe der Harmala-Alkaloide. *Helvetica Physiologica et Pharmacologica Acta* 17:202–14.

Späth, E., and F. Lederer. 1930. Synthese der Harma Alkaloide: Harmalin, Harmin und Harman. *Berichte der deutschen chemischen Gesellschaft* 63:120–25.

Ibogaine

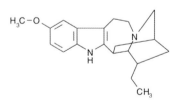

Ibogaine

Other Names

Endabuse, ibogain, ibogaina, ibogaïne, NIH 10567, 12-methoxy-ibogamin, 12-methoxy-ibogamine

Empirical formula: $C_{20}H_{26}N_2O$

Substance type: **indole alkaloid**, indole alkylamine, ibogane type

Chemically, ibogaine is closely related to the β-carbolines, and particularly to **harmaline and harmine**. It belongs to the group of cyclic tryptamine derivatives.

Ibogaine was first isolated from the root cortex of *Tabernanthe iboga* in France in 1901 (Dybowsky and Landgren 1901). Ibogaine and analogous alkaloids (ibogane type) also occur in *Pandaca retusa* (Lam.) Mgf. [syn. *Tabernaemontana retusa* (Lam.) Pichon] (cf. **Tabernaemontana spp.**), a dogbane species native to Madagscar (Le Men-Olivier et al. 1974). Many genera in the Family Apocynaceae, including *Tabernaemontana*, **Voacanga spp.**, *Stemmadenia*, *Ervatamia*, and *Gabunea*, contain ibogaine-type **indole alkaloids** (ibogamine, tabernanthine, voacangine, ibogaline) (Prins 1988, 5).

Between 1940 and 1950, most research into ibogaine was conducted in France. Because it exhibited potent stimulating properties, the initial pharmacological research focused on ibogaine's neuropharmacological effects. Only later were the hallucinogenic effects more precisely studied (Sanchez-Ramos and Mash 1996, 357).

In the 1960s, the Chilean psychiatrist Claudio Naranjo introduced ibogaine into psychotherapy as a "fantasy-enhancing drug" (Naranjo 1969*). One subject provided the following account of a shamanic experience during a psychotherapeutic session with the "stomach drug" ibogaine:

I am a panther! A black panther! I defend myself, I stand up. I snort powerfully, with the breath of a panther, predator breath! I move like a panther, my eyes are those of a panther, I see my whiskers. I roar, and I bite. I react like a panther, offense is the best defense.

Now I hear drums. I dance. My joints are gears, hinges, hubs. I can be a knee, a bolt, could do something, indeed almost anything. And I can loose [sic] myself again in this chaos of nonexistence and the perception of vague, abstract ideas of changing forms, where there exists a sense of the truth of all things and an order that one should set out to discover. (Naranjo 1979, 188*)

In Europe, the Swiss psychiatrist Peter Baumann provided the main impetus for the use of ibogaine in psychotherapy:

Baumann reported about experiments with completely synthetic ibogaine, which he used on only a few patients with whom a long and positive therapeutic relationship existed. The dosage was usually 5 mg/kg of body weight. At this dosage level, the effects lasted for approximately 5 to 8 hours and diminished only very slowly. In his experiments with ibogaine, the author found that it was not the substance as such that triggered a specific effect but that it induced an unspecific psychological and physical stimulus that was then responded to in the language that patient was accustomed to using with this therapist. (Leuner and Schlichting 1986, 162)

Unfortunately, an accident led to this initially promising research being halted. Marina Prins (1988) subsequently compared Baumann's results with those reported by Naranjo.

Today, ibogaine is in the spotlight of neuropharmacological research because it has been shown that this alkaloid can be used to reduce and cure the addictive behavior of people dependent on other drugs (heroine, cocaine) (Sanchez-Ramos and Mash 1996; cf. *Maps* 6 [2; 1996]: 4–6). For example, ibogaine has been found to suppress the motor activity that occurs during opiate withdrawal. It has been proposed that ibogaine, when

ingested by opiate addicts in a single high dosage, dramatically reduces withdrawal symptoms while simultaneously causing a trip that provides the patient with such deep insights into the personal causes of the addiction that a majority of the individuals who receive such therapy can live for months without relapse. However, it should be noted that several additional sessions may be

necessary before a persistent stabilization occurs. (Naeher 1996, 12)

Experiments with primates have shown that ibogaine reduces opiate addiction and partially blocks withdrawal symptoms. Although the neuropharmacological mechanism behind these effects has not yet been discovered, Deborah Mash and her team in Miami (Mash 1993; Mash et al. 1995) are researching this question. Ibogaine has been demonstrated to interact with numerous different receptors, and it has been concluded that this breadth of interaction is the reason for ibogaine's effectiveness in addiction therapy (Sweetman et al. 1995).

In the United States, the use of ibogaine to treat addiction has been patented as the clinical Lotsof procedure (Lotsof 1995). Whether this procedure will receive endorsement from the medical community remains to be seen (Touchette 1995). A novel about this facet of ibogaine (which incorporates such actual people as Howard Lotsof) was published in Slovenia (Knut 1994).

Ibogaine enjoys a reputation for being an exceptionally potent and stimulating aphrodisiac (Naranjo 1969*).[500] The research to date has entirely neglected this aspect.

Another substance of pharmacological and therapeutic interest is noribogaine, which is chemically and pharmacologically very similar to Prozac (fluoxetine). In the United States, Prozac is one of the most frequently prescribed psychopharmaca for depression, and it is celebrated as the "happy drug" in the popular press (Kramer 1995; Rufer 1995*).

Dosage and Application

Two to four tablets containing up to 8 mg ibogaine per tablet may be given daily as a stimulant for states of exhaustion, debility, et cetera. Nausea, vomiting, and ataxia are possible side effects. When used for psychotherapeutic purposes (Baumann), dosages of 3 to 6 mg of ibogaine hydrochloride per kg of body weight were administered. For psychoactive purposes, dosages of around 200 mg are recommended (Prins 1988, 47).

Commercial Forms and Regulations

Ibogaine was formerly available as a medicine under the trade name Bogadin (Schneider and McArthur 1956). In the United States, ibogaine is considered a Schedule I drug and has been prohibited since 1970. However, ibogaine hydrochloride is marketed under the trade name Endabuse and can be used with the appropriate special permit. In Germany, ibogaine is not considered a narcotic under the guidelines of the narcotic laws and is therefore legal (Körner 1994, 1573*).

"We are assuming that ibogaine triggers a fractal time structure that resembles the architecture of the REM dream phase during sleep by activating certain amygdaloid–cortico-thalamic circuits in the brain stem."

JULIE STALEY
LECTURE AT THE ENTHEOBOTANY
CONFERENCE (SAN FRANCISCO, 1996)

The Slovenian novel *Iboga*, by Amon Knut Jr., has as its subject the therapeutic effects of ibogaine for alcoholism, cocaine abuse, heroin addiction, and nicotine dependency. (Book cover; Maribor: Skupina Zrcalo Publishers, 1994)

500 In West Africa, *Tabernanthe iboga* is supposedly preferred over yohimbe (*Pausinystalia yohimba*) as an aphrodisiac (Prins 1988, 6).

Literature

See also the entries for *Tabernaemontana* spp., *Tabernanthe iboga*, *Voacanga* spp., and **indole alkaloids**.

Baumann, Peter. 1986. "Halluzinogen"-unterstützte Psychotherapie heute. *Schweizerische Ärztezeitung* 67 (47): 2202–5.

Dybowski, J., and E. Landrin. 1901. Sur l'iboga, sur ses propriétés excitantes, sa composition et sur l'alcaloïde nouveau qu'il renferme. *Comptes Rendues* 133:748.

Fromberg, Eric. 1996. Ibogaine. *Pan* 3:2–8. (Includes a very good bibliography.)

Knut, Amon Jr. 1994. *Iboga*. Maribor: Skupina Zrcalo. (Cf. *Curare* 18 (1; 1995): 245–46.)

Kramer, Peter D. 1995. *Glück auf Rezept: Der unheimliche Erfolg der Glückspille Fluctin*. Munich: Kösel.

Le Men-Olivier, L., B. Richards, and Jean Le Men. 1974. Alcaloïdes des graines du *Pandaca retusa*. *Phytochemistry* 13:280–81.

Lotsof, Howard S. 1995. Ibogaine in the treatment of chemical dependence disorders: Clinical perspectives. *Maps* 5 (3): 15–27.

Mash, Deborah C. 1995. Development of ibogaine as an anti-addictive drug: A progress report from the University of Miami School of Medicine. *Maps* 6 (1): 29–30.

Mash, Deborah C., Julie K. Staley, M. H. Baumann, R. B. Rothman, and W. L. Hearn. 1995. Identification of a primary metabolite of ibogaïne that targets serotonin transporters and elevates serotonin. *Life Sciences* 57 (3): 45–50.

Naeher, Karl. 1996. Ibogain: Eine Droge gegen Drogenahhängigkeit? *Hanfblatt* 3 (21): 12–15 (interview).

Prins, Marina. 1988. "Von Iboga zu Ibogain: Über eine vielseitige Droge Westafrikas und ihre Anwendung in der Psychotherapie." Unpublished licentiate thesis, Zurich. (Very rich bibliography.)

Sanchez-Ramos, Juan R., and Deborah Mash. 1996. Pharmacotherapy of drug-dependence with ibogain. *Jahrbuch für Transkulturelle Medizin und Psychotherapie* 6 (1995): 353–67.

Schneider, J., and M. McArthur. 1956. Potentiation action of ibogain (Bogadin™) on morphin analgesia. *Experimenta* 8:323–24.

Sweetman, P. M., J. Lancaster, Adele Snowman, J. L. Collins, S. Perschke, C. Bauer, and J. Ferkany. 1995. Receptor binding profile suggests multiple mechanisms of action are responsible for ibogaine's putative anti-addiction activity. *Psychopharmacology* 118:369–76.

Touchette, Nancy. 1995. Anti-addiction drug ibogain on trial. *Nature Medicine* 1 (4): 288–89.

Ibotenic Acid

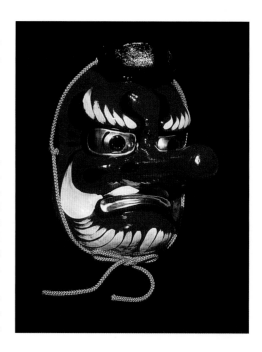

Other Names

α-amino (3-hydroxy-5-isoxazolyl)acetic acid, α-amino-2,3-dihydro-3-oxo-5-isoxazole-acetic acid, ibotenic acid, "pilzatropin," prämuscimol

Empirical formula: $C_5H_6O_4N_2$

Substance type: amino acid

Ibotenic acid was first isolated in 1964 from the Japanese mushroom *Amanita strobiliformis* (Paul) Quél. The Japanese name for this mushroom is *ibo-tengu-take* ("warty *tengu* mushroom"), and ibotenic acid was named after it (Ott 1993, 341*; Takemoto et al. 1964). Ibotenic acid is also found in **Amanita muscaria** and **Amanita pantherina** (Eugster et al. 1965). It may also be present in members of the genus **Botelus** (porcini mushrooms).

Ibotenic acid is structurally related to the neurotransmitter glutamate and may behave similarly in the nervous system. A psychoactive

Ibotenic acid

A traditional mask of *tengu*, the Japanese fly agaric mushroom spirits. Ibotenic acid was named after these spirits and their mushroom.

dose is regarded as 50 to 100 mg. Ibotenic acid is converted into **muscimol** when stored (Good et al. 1965).

Commercial Forms and Regulations

Ibotenic acid is available from chemical suppliers and is a legal substance (Ott 1993, 440*).

Literature

See also the entries for *Amanita muscaria*, *Amanita pantherina*, and **muscimol**.

Eugster, C. H., G. F. R. Müller, and R. Good. 1965. Wirkstoffe aus *Amanita muscaria*: Ibotensäure und Muscazon. *Tetrahedron Letters* 23:1813–15.

Gagneux, A. R., et al. 1965. Synthesis of ibotenic acid. *Tetrahedron Letters* 965:2081–84.

Good, R., et al. 1965. Isolierung und Charakterisierung von Prämuscimol und Muscazon aus *Amanita muscaria* (L. ex Fr.) Hooker. *Helvetica Chimica Acta* 48 (4): 927–30.

Romagnesi, M. H. 1964. Champignons toxiques au Japon. *Bulletin de la Société Mycologique de France* 80 (1): iv–v.

Takemoto, T., T. Nakajima, and R. Sakuma. 1964. Structure of ibotenic acid. *Yakugaku Zasshi* 84:1233.

Indole Alkaloids

Other Names

Indolalkaloide, indolamine alkaloids, indole, indoles

Indole alkaloids are derived from the indole ring system and appear almost exclusively in the families Apocynaceae[501] (*Alchornea* spp., *Alstonia scholaris*, *Aspidosperma quebracho-blanco*, *Catharanthus roseus*, *Rauvolfia* spp., *Tabernaemontana* spp., *Tabernanthe iboga*, *Vinca* spp., *Voacanga* spp.), Loganiaceae (*Gelsemium sempervirens*, *Strychnos nux-vomica*, *Strychnos* spp.), and Rubiaceae (*Corynanthe* spp., *Mitragyna speciosa*, *Pausinystalia yohimba*). Indole alkaloids also occur in certain ascomycetes (*Balansia cyperii*, *Claviceps paspali*, *Claviceps purpurea*, *Claviceps* spp.), other fungi (Tyler 1961), and several climbing vines (*Ipomoea violacea*, *Turbina corymbosa*) (Hofmann 1966; cf. **ergot alkaloids**).

Included among the large group of indole alkaloids (Trojánek and Blaha 1966) are the β-carbolines with **harmaline and harmine**; the tryptamine derivates **bufotenine**, *N,N*-**DMT**, **5-MeO-DMT**, **psilocybin**, and psilocin; the ergot alkaloids; and the alkaloids of the ibogane type (ibogaine, voacangine), yohimbane type (**yohimbine**), and strychnane type (**strychnine**). Indoles are also found in the genus *Uncaria*, several species of which are used as **ayahuasca** additives (Phillipson and Hemingway 1973).

Many indole alkaloids are psychoactive or occur in plants that are utilized for traditional psychoactive purposes (Lindgren 1995; Rivier and Pilet 1971; Schultes 1976).

Literature

See also the entries for β-**carbolines**, **ergot alkaloids**, and **yohimbine**.

Gershon, S., and W. J. Lang. 1962. A psychopharmacological study of some indole alkaloids. *Archives Internationales de Pharmacodynamie et de Thérapie* 135 (1–2): 31–56.

Hesse, M. 1968. *Indolalkaloide in Tabellen*. Berlin: Springer.

Hofmann, Albert. 1966. Alcaloïdes indoliques isolés de plantes hallucinogènes et narcotiques du Mexique. In *Colloques internationaux du Centre National de la Recherche Scientifique: Phytochimie et plantes médicinales des terres du Pacifique, Noméa (Nouvelle Calédonie) 28.4–5.5.1964*, 223–41. Paris: Centre National de la Recherche Scientifique.

Lindgren, Jan-Erik. 1995. Amazonian psychoactive indols: A review. In *Ethnobothany: Evolution of a discipline*, ed. Richard Evans Schultes and Siri von Reis, 343–48. Portland, Ore.: Dioscorides Press.

Phillipson, John David, and Sarah Rose Hemingway. 1973. Indole and oxindol alkaloids from *Uncaria bernaysia*. *Phytochemistry* 12:1481–87.

Rivier, Laurent, and Paul-Emile Pilet. 1971. Composés hallucinogènes indoliques naturels. *Année Biol.* 3:129–49.

Schultes, Richard Evans. 1976. Indole alkaloids in plant hallucinogens. *Journal of Psychedelic Drugs* 80 (1): 7–25.

Trojánek, J., and K. Bláha. 1966. A proposal for the nomenclature of indole alkaloids. *Lloydia* 29 (3): 149–55.

Tyler, Varro E. 1961. Indole derivatives in certain North American mushrooms. *Lloydia* 24:71–74.

Indole

501 The Subfamily Plumerioideae (Apocynaceae) is rich in indole alkaloids (Omino and Kokwaro 1993, 174*).

Mescaline

Mescaline

"Peyote is always on mescaline.
Humans are sometimes on
    mescaline.
But no peyote could stand to
    always be on human.
We are but a side-effect of god."

GALAN O. SEID
UNPUBLISHED POEM (CA. 1985)

"Mescaline is the final, ultimate
cult that does away with every
cult."

GÜNTER WALLRAFF
MESKALIN
(1968, 13)

The cover of a very rare publication
by the author Günter Wallraff
regarding his experiences with
mescaline.

Opuntia cylindrica is found in the
Atacama Desert (northern Chile). It
is easily identifiable by its very long,
reddish thorns. The cactus contains
mescaline, and the former
inhabitants of oases in the region
may have once used it for
psychoactive purposes.

Other Names

Mescalin, meskalin, mezcalin, mezkalin, 3,4,5-tri-methoxy-benzolmethanamine, 3,4,5-trimethoxy-β-phenethylamine, 3,4,5-trimethoxyethyl- phenyl-amine, TMPFA, 2-(3,4,5-trimethoxy-phenyl)-ethylamine

Empirical formula: $C_{11}H_{17}NO_3$

Substance type: lophophora alkaloid, **β-phen-ethylamine**

Mescaline was first isolated in 1886 from "mescal buttons," the aboveground parts of the peyote cactus (*Lophophora williamsii*), and was named after them. Mescaline is the most thoroughly studied of all psychoactive plant constituents. In the period between 1886 and 1950, more than one hundred mescaline research studies were published in the German language alone (Passie 1994). This alkaloid was found to be a component of numerous cacti (see the table on page 847). And it is possible that mescaline is produced from dopamine in vitro (Paul et al. 1969; Rosenberg et al. 1969).

Arthur Heffter was the first person to initially test an isolated plant constituent on himself (Heffter 1894). The classic Heffter dosage consisted of 150 mg mescaline hydrochloride (HCL). A psychedelic dosage is now considered to be 178 to 256 mg of mescaline HCL or 200 to 400 mg of mescaline sulfate. The highest measured dosage reported in the literature was 1,500 mg. Taken orally, 5 mg/kg of pure mescaline is regarded as a hallucinogenic dosage. In the toxicological literature, there is no known lethal dosage of mescaline when it is ingested orally (Brown and Malone 1978, 14).

Western psychiatry has been aware of consciousness-altering drugs since the nineteenth century. Mescaline was the first substance to be tested and applied in psychiatry. At the time, researchers regarded the effects of mescaline on a

healthy subject as inducing a state that was otherwise known only in psychopathic patients. This led to the idea of pharmacologically induced "model psychoses" (Leuner 1962*). The effects of mescaline (and also of **psilocybin**) were described as "intoxication, toxic ecstasy, clouding of consciousness, hallucinosis, model psychosis, drug intoxication, emphasis, daydream," et cetera (Passie 1994). Only in recent years has there been a shift in thinking away from the model psychosis concept and a recognition that psychedelic states and psychoses do not have a common origin (Hermle et al. 1988*, 1992*, 1993*).

The predominant effects of mescaline are a "reveling of the individual senses and primarily visual orgies" (Ellis 1971, 21). The mescaline inebriation was first systematically described by Kurt Beringer in 1927. To date, there have been many encounters with the substance, and the most commonly reported experiences are ecstatic and visionary in nature:

> My awareness of subject and object disappeared, and I felt dissolved, rising in an orchestra of sounds. This ecstatic state was accompanied by an indescribable sensation of happiness. (Ammon and Götte 1971, 32)

It has often been suggested that pure mescaline can be taken in place of *Lophophora williamsii*. "However, most peyote users are of the opinion that synthetic mescaline cannot be compared with the effects of peyote" (Harp 1996, 16).

On the Cultural History of Mescaline

Aldous Huxley (1894–1963) made the psychedelic effects of mescaline famous in his two essays "The Doors of Perception" and "Heaven and Hell."

> Usually the person taking mescaline will discover an inner world that is so obviously something given, so enlighteningly eternal and sacred, as the transformed outer world that I had perceived with my eyes open. (Huxley 1970, 32*)

It is very likely that Hermann Hesse also had contact with mescaline, and that it may have inspired his novel *Steppenwolf,* one of the cult books of the hippie generation. The psychedelic rock band Steppenwolf took its name from the book, and the novel also became a motion picture starring Max von Sydow (USA 1974).

Nationalgalerie, a German New Wave band, sings on its album *Mescaline,* "To be transformed

Cacti Containing Mescaline

(from Doetsch et al. 1980; La Barre 1979; Mata and McLaughlin 1982*; Shulgin 1995*; Lundström 1971; Pardanini et al. 1978; Ott 1993*; Turner and Heyman 1960)

Species	Occurrence	Use
Gymnocalycium gibbosum (Haw.) Pfeiffer	Argentina	
Gymnocalycium leeanum (Hook.) Br. et R.	Argentina, Uruguay	
Islaya minor Backeb.	southern Peru	
Lophophora diffusa (Croizat) Bravo	Mexico	peyote substitute
[syn. *Lophophora echinata*]		
Lophophora jourdaniana [nom. nud.]		
Lophophora williamsii (Lem.) Coult.	Mexico	entheogen
[syn. *Lophophora fricii* Habermann]		
Myrtillocactus geometrizans (Mart.) Cons.	Mexico	
Opuntia acanthocarpa Engelm. et Bigel.		
Opuntia basilaria Engelm. et Bigel.		
Opuntia cylindrica (Lam.) S.-D.	Chile	inebriant[502]
Opuntia echinocarpa Engelm. et Bigel.		
Opuntia ficus-indica (L.) Mill.	Mexico, Egypt[503]	food
Opuntia imbricata (Haw.) DC.	Arizona	
Opuntia spinosior (Engelm.) Toumey	Arizona	
Pelecyphora aselliformis Ehrenb.	Mexico	peyote substitute
Pereskia corrugata Cutak	Florida	
Pereskia tampicana Web.	Mexico	
Pereskiopsis scandens Br. et R.	Yucatán	
Polaskia chende (Gossel.) Gibs.	California	
Polaskia sp.	California	
Pterocereus gaumeri (Br. et R.) Mac-Doug. et Mir.	California	
Pterocereus sp.	California	
Stenocereus beneckei (Ehrenb.) Buxbaum	California	
Stenocereus eruca (Brand.) Gibs. et Horak	Baja California	
Stenocereus stellatus (Pfeiffer) Rice	California	
Stenocereus treleasei (Br. et R.) Backeb.	California	
Stenocereus sp.		
Stetsonia coryne (SD.) Br. et R.	Argentina	
Trichocereus bridgesii (SD.) Br. et R.	Peru, Bolivia	entheogen
Trichocereus cuscoensis Br. et R.	Peru	
Trichocereus fulvinanus Ritt.	Chile	
Trichocereus macrogonus (SD.) Ricc.	Peru	
Trichocereus pachanoi Br. et R.	Peru, Ecuador	entheogen
Trichocereus peruvianus Br. et R.	Peru	entheogen
Trichocereus spachianus (Lem.) Ricc.	Indiana (cultivated)	
Trichocereus strigosus (SD.) Br. et R.	Argentina	
Trichocereus taquimbalensis Card.	Peru	
Trichocereus terscheckii (Parm.) Br. et R.	Peru, northwestern Argentina	
Trichocereus validus (Monv.) Backbg.	Peru, Bolivia	
Trichocereus werdermannianus Backbg.	Peru, Bolivia	
Trichocereus spp. (cf. **Echinopsis spp.**)	South America	

Top: Some species from the genus *Gymnocalycium* have been found to contain mescaline.

Bottom: Although mescaline has been detected in some members of the genus *Opuntia*, no traditional use of these plants as psychoactive substances is known.

502 This cactus has a very high mescaline content of 0.9% (Schultes and Farnsworth 1982, 159*).

503 Although the prickly pear doubtlessly originated in Mexico, it was introduced to Europe and North Africa at an early date (Dressler 1953, 140*).

by a trickster fairy. My lawyer advised me to take some mescaline" (Sony Records, 1995).

The French novelist and artist Henri Michaux (1899–1984) studied mescaline during the 1960s and ingested it to see what effects it might have upon his creativity. Like many other Frenchmen, however, he summarized his experience as an "accursed miracle" and scribbled his experiences of inner turmoil on paper (Michaux 1986). Today, these "drawings" are still reproduced in publications as an example of the "psychosis-like" effects of mescaline.

Commercial Forms and Regulations

Mescaline is available primarily as a hydrochloride or sulfate. In Germany, it is considered a "narcotic

Many species of the Mexican prickly pear (*Opuntia* spp.) contain mescaline and other phenethylamines. To date, we know of no *Opuntia* that was used traditionally as an entheogen. (Woodcut from Tabernaemontanus, *Neu Vollkommen Kräuter-Buch*, 1731)

in which trafficking is prohibited." In the United States, the Controlled Substances Act lists mescaline as a Schedule I substance (Körner 1994, 38*).

Literature

See also the entries for *Lophophora williamsii*, *Trichocereus pachanoi*, *Trichocereus* spp., and β-phenethylamines.

Ammon, Günter, and Jürgen Götte. 1971. Ergebnisse früher Meskalin-Forschung. In Bewußtseinserweiternde Drogen aus psychoanalytischer Sicht, special issue, *Dynamische Psychiatrie*, 23–45.

Beringer, Kurt. 1927. *Der Meskalinrausch*. Berlin: Springer. Repr. 1969.

Blofeld, John. 1966. A high yogic experience achieved with meskalin. *Psychedelic Review* 7:27–32.

Doetsch, P. W., J. M. Cassidy, and J. L. McLaughlin. 1980. Cactus alkaloids. XL: Identification of mescaline and other phenethylamines in *Pereskia*, *Pereskiopsis* and *Islaya* by use of fluorescamine conjugates. *Journal of Chromotography* 189:79.

Ellis, Havelock. 1971. Zum Phänomen der Meskalin-Intoxikation, Bemerkungen zum Problem der Meskalin-Intoxikation. In Bewußtseinserweiternde Drogen aus psychoanalytischer Sicht, special issue, *Dynamische Psychiatrie*, 17–22.

Frederking, W. 1954. Meskalin in der Psychotherapie. *Medizinischer Monatsspiegel*, 3:5–7.

Harf, Jürgen C. 1996. Meskalin und Peyote. *Grow!* 6/96:15–16.

Heffter, Arthur. 1894. Über zwei Kakteenalkaloide. *Berichte der deutschen Chemischen Gesellschaft* 27:2975.

Klüver, Heinrich. 1926. Mescal vision and eidetic vision. *American Journal of Psychology* 37:502–15.

———. 1969. *Mescal and mechanisms of hallucinations*. Chicago: The University of Chicago Press.

La Barre, Weston. 1979. Peyotl and mescaline. *Journal of Psychedelic Drugs* 11 (1–2): 33–39.

Lundström, Jan. 1971. Biosynthetic studies on mescaline and related cactus alkaloids. *Acta Pharm. Suecica* 8:275–302.

Michaux, Henri. 1986. *Unseliges Wunder: Das Meskalin*. Munich and Vienna: Carl Hanser.

Pardanani, J. H., B. N. Meyer, and J. L. McLaughlin. 1978. Cactus alkaloids. XXXVII. Mescaline and related compounds from *Opuntia spinosior*. *Lloydia* 41 (3): 286–88.

Passie, Torsten. 1994. Ausrichtungen, Methoden und Ergebnisse früher Meskalinforschungen im deutschsprachigen Raum (bis 1950). In *Jahrbuch des Europäischen Collegiums für Bewußtseinsstudien* (1993/1994), 103–11. Berlin: VWB.

Paul., A.G., H. Rosenberg, and K. L. Khanna. 1969. The roles of 3,4,5-trihydroxy-β-phenethylamine and 3,4-dimethoxy-β-phenethylamine in their biosynthesis of mescaline. *Lloydia* 32 (1): 36–39.

Rosenberg, H., K. L. Khanna, M. Takido, and A. G. Paul. 1969. The biosynthesis of mescaline in *Lophophora williamsii*. *Lloydia* 32 (3): 334–38.

Turner, W. J., and J. J. Heyman. 1960. The presence of mescaline in *Opuntia cylindrica*. *Journal of Organic Chemistry* 25:2250.

Wallraff, Günter. 1968. *Meskalin—Ein Selbstversuch*. Berlin: Verlag Peter-Paul Zahl.

Morphine

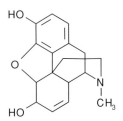

Morphine

"Morphine is to opium as alcohol is to wine."

Jean-Marie Pelt
Pflanzenmedizin [Plant Medicine]
(1983, 70*)

Other Names

4,5α-epoxy-17-methyl-7-morphinen-3,6α-diol, morfina, morphin, morphinium, morphium

Empirical formula: $C_{17}H_{19}NO_3$

Substance type: **opium alkaloid**

Sometime around 1803 and 1804, Friedrich Wilhelm Adam Sertürner (1783–1841), a pharmacist's assistant, first isolated morphine as the "sleep-inducing principle" in opium (cf. *Papaver somniferum*, **opium alkaloids**). This achievement was the most important "quantum leap" in the history of pharmacology and repre-

sents the beginning of the true chemical investigation of plants. Today, the Sertürner Medal is still awarded for exceptional work in pharmaceutics.

Morphine may also be present in *Papaver decaisnei* Hochst., *Papaver dubium* L. [syn. *Papaver modestum* Jordan, *Papaver obtusifolium* Desf.], and *Papaver hybridum* L. (Slavík and Slavíková 1980). Whether morphine occurs in **Argemone mexicana** and other *Papaver* species (**Papaver spp.**) is doubtful, while the idea that hops (**Humulus lupulus**) contains morphine is a figment of someone's imagination. Tiny traces of the substance have been found in hay and lettuce (cf. **Lactuca virosa**) (Amann and Zenk 1996, 19). Morphine has also been detected in the skin of

Bufo marinus toads (cf. **bufotenine**) (Amann and Zenk 1996, 18).

Since the time when morphine was first detected in breast milk, cow's milk, and human cerebrospinal fluid, it has been known that it is a natural endogenous neurotransmitter in higher vertebrates, including humans (Amann and Zenk 1996; Cardinale et al. 1987; Hazum et al. 1981). Morphine does not bind well to the encephalin receptors (to which the endorphins dock) but docks at the specific morphine (m) receptors (Hazum et al. 1981). It is most likely biosynthesized in the body from dopamine (Brossi 1991). Another closely related substance, **codeine**, is also endogenous in humans (Cardinale et al. 1987).

Morphine is the best and strongest natural painkiller known. Its efficaciousness is surpassed only by that of the synthetic morphine analogs (heroin, fentanyl). Morphine is particularly well suited for treating chronic pain, such as in cancer therapy (Amann and Zenk 1996; Melzack 1991). Endogenous morphine constitutes the body's own pain medication:

> Studies on rats have shown that among animals who were suffering from arthritis, morphine concentrations in the spinal cord and urine were significantly elevated. Because of this, it is assumed today that the organism produces increased amounts of morphine in certain disease states. Consequently, endogenous morphine may serve to regulate pain in the organism. Morphine exists in animal and human tissue and is excreted in significant amounts in the urine. (Amann and Zenk 1996, 24)

About 30 mg orally represents an effective dosage. Habitual morphine users may use as much as 1 g per day (Hirschfeld and Linsert 1930, 255*):

> It is known that opium eaters experience a significant increase in sexual functions during the initial period of opium consumption. During opium inebriation, erotic images appear and may even include extraordinary sexual fantasies. . . . The effects of morphine are similar, where an increase in sexual excitability was observed following several weeks of taking 0.03 to 0.06 g per day. (Max Marcuse, 1923, *Handwörterbuch der Sexualwissenschaft* [Handbook of Sexual Sciences])

When used for sedation, in anesthesia, and for calming and antispasmodic purposes, pharmaceutical preparations of morphine hydrochloride and **atropine** sulfate or morphine hydrochloride and **scopolamine** hydrobromide are used—the final reminders of the recipes for the former **soporific sponges.**

During the Golden Twenties, the use of morphine in Berlin society circles was depicted in numerous pictures and illustrations (e.g., by Paul Kamm) that appeared in magazines. These illustrations played a great role in creating the stereotype of the "Morphinist" (cf. *Papaver somniferum*), who also became the object of literary treatments (Bulgaka 1971; Mac From 1931). Even the life story of the man who first discovered the substance, Friedrich Wilhelm Sertürner, became the subject of a novel (Schumann-Ingolstadt n.d.). Heroin, a derivative of morphine, has also inspired a rich body of literature. One of the first of this was the novel *Heroin*, by Rudolf Brunngraber (1952*), which dealt with the role of heroin in Egypt during the Golden Twenties.

Morphine was and still is a popular inebriant in the music scene (particularly that of jazz and rock). "Sister Morphine," a song by the Rolling Stones (*Sticky Fingers*, Virgin Records, 1971), is arguably the most famous hymn to the drug. Morphine, a crossover band that mixes elements of cool jazz and modern rock, took its name from the alkaloid, and one of its albums is titled *Cure for Pain* (Rykodisc, 1993).

Commercial Forms and Regulations

Morphine is available from pharmacies in the form of morphine hydrochloride. Although morphine is covered by narcotics laws, it can be obtained with a special prescription. In the United States, morphine is a Schedule II substance.

Literature

See also the entries for *Papaver somniferum* and *Papaver* spp.

Amann, Tobias, and Meinhart H. Zenk. 1996. Endogenes Morphin: Schmerzmittelsynthese in Mensch und Tier. *Deutsche Apotheker Zeitung* 136 (7): 17–25. (Contains a very good bibliography.)

Brossi, Arnold. 1991. Mammalian alkaloids: Conversions of tetrahydroisoquinoline-1-carboxylic acid derived from dopamin. *Planta Medica* 57 suppl. (1): 93 ff.

Bulgaka, M. 1971. *Morphium Erzählungen.* Zurich: Arche Verlag.

Cardinale, George J., Josef Donnerer, A. Donald Finck, Joel D. Kantrowitz, Kazuhiro Oka, and Sydney Spector. 1987. Morphine and codeine are endogenous compounds of human cerebrospinal fluid. *Life Sciences* 40:301–6.

Fairbain, J. W., S. S. Handa, E. Gürkan, and J. D. Phillipson. 1978. *In vitro* conversion of morphine to its *N*-oxide in *Papaver somniferum* latex. *Phytochemistry* 172:261–62.

Ferres, H. 1926. Gefährliche Betäubungsmittel: Morphium und Kokain. In *Bibliothek der Unterhaltung und des Wissens*, 5:136–44. Stuttgart: Union Deutsche Verlagsgesellschaft.

"Illusionary ecstasies travel down confused nerves,
nervous trembling moves through pale hands,
the lips, red as a poppy, narrow as a knife, foam,
As if sensing blood in cups of gold,
A fragrance of orchids in the room,
pale yellow, heavy azaleas,
An arm bends in crazed desire,
A mouth grimaces quietly,
A gasp presses from the narrow chest: Morphium!"

Ed. Golland Breslau
in *Der Junggeselle* [The Bachelor]
(1921)

In the early twentieth century, morphine (also known as morphium) was used as a cure for alcoholism. (Advertisements from German magazines 1907/1908)

The discovery of morphine or morphium led to some of the most significant innovations in the history of pharmacology and neurophysiology. The discovery has been described many times, even in novels. (Book cover, no date; novelized version of the life of German researcher F. W. Sertürner)

"Please, Sister Morphine, turn my nightmares into dreams.
Oh, can't you see I'm fading fast and that this shot will be my last . . ."

Mick Jagger and Keith Richards
"Sister Morphine"
(*Sticky Fingers*, Virgin Records, 1971)

The American jazz-rock fusion trio Morphine has the best "cure for pain": one of their songs is called "Let's Take a Trip Together." (CD cover 1993, Rykodisc)

Hazum, Eli, Julie J. Sabatka, Kwen-Jen Chang, David A. Brent, John W. A. Findlay, and Pedro Cuatrecasas. 1981. Morphine in cow and human milk: Could dietary morphine constitute a ligand for specific morphine (m) receptors? *Science* 213:1010–12.

Kramer, John C. 1980. The opiates: Two centuries of scientific study. *Journal of Psychedelic Drugs* 12 (2): 89–103.

Mac From, ed. 1931. *Täglich 5 Gramm Morphium—Aufzeichnungen eines Morphinisten.* Berlin-Pankow: A. H. Müller.

Melzack, Ronald. 1991. Morphium und schwere chronische Schmerzen. Offprint of *Spektrum der Wissenschaft.* Heidelberg: Spektrum der Wissenschaft Verlag.

Schmitz, Rudolf. 1983. Friedrich Wilhelm A. Sertürner und die Morphinentdeckung. *Pharmazeutische Zeitung* 128:1350–59.

Schuhmann-Ingolstadt, Otto. n.d. *Morphium: Lebensroman des Entdeckers.* Berlin and Frankfurt/M.: Deutscher Apothekerverlag.

Slavík. J., and L. Slavíková. 1976. Occurrence of morphin as a minor alkaloid in *Papaver decaisnei* Hochst. *Collection Czechoslov. Chem. Commun.* 45:2706–9.

Muscimol

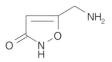

Muscimol

The panther cap (*Amanita pantherina*) is found throughout Eurasia and North America. Its effects are similar to, but more pronounced than, those of the fly agaric. (Photograph: Paul Stamets)

Other Names

Agarin, 5-(aminomethyl)-3-[2*H*]-isoxazolone, pyro-ibotenic acid, 3-hydroxy-5-aminomethyl-isoxazole

Empirical formula: $C_4H_6O_2N$

Substance type: amino acid, isoxazole derivative

Muscimol was first described in 1964 as a constituent of *Amanita pantherina.* Muscimol is the decarboxylated product of **ibotenic acid** and is considered to be more psychoactive than it. Some 15 to 20 mg is regarded as a psychoactive dosage (Müller and Eugster 1965; Ott 1993, 446*; Scotti et al. 1969).

Muscimol is an analog of the neurotransmitter GABA (gamma amino butyric acid) and docks to its receptor (Johnston 1971). Kavapyrones (cf. **Piper methysticum**) also bind to the same receptor.

Ibotenic acid as well as muscimol were detected in the urine of people who had eaten fly agaric mushrooms (*Amanita muscaria*) about an hour earlier. An experiment with mice conducted by the same team of researchers found that the amount of active constituents in the urine passed by one inebriated animal was not sufficient to inebriate another animal (Ott et al. 1975).

Commercial Forms and Regulations

Muscimol can be purchased from chemical suppliers. The substance is legal and is not subject to any specific regulations.

Literature

See also the entries for *Amanita muscaria* and **ibotenic acid**.

Johnston, G. A. R. 1971. Muscimol and the uptake of γ-aminobutyric acid by rat brain slices. *Psychopharmacologia* 22:230.

Müller, G. F. R., and C. H. Eugster. 1965. Muscimol, ein pharmakodynamisch wirksamer Stoff aus *Amanita muscaria. Helvetica Chemica Acta* 48:910–26.

Ott, Jonathan, Preston S. Wheaton, and William Scott Chilton. 1975. Fate of muscimol in the mouse. *Physiol. Chem. and Physics* 7:381–84.

Scotti de Carolis, A., et al. 1969. Neuropharmacological investigations on muscimol, a psychotropic drug extracted from *Amanita muscaria. Psychopharmacologia* 15:186–95.

Nicotine

Other Names

Nicotin, nikotin, (–)-nikotin , 1-methyl-2(3-pyridyl)-pyrrolidine, 3-(1-methyl-2-pyrrolidinyl)pyridine

Empirical formula: $C_{10}H_{14}N_2$

Substance type: pyrrolidine alkaloid, pyridine alkaloid, tobacco alkaloid

Nicotine was first discovered in tobacco (*Nicotiana tabacum*) and was named for the genus. It occurs in numerous species of *Nicotiana* as well as other members of the Nightshade Family. It has also been found in club moss (*Lycopodium clavatum*).

Nicotine is very easily absorbed through the mucous membranes and even through the skin. Consequently, plants that contain nicotine can be smoked or administered as **enemas**. Nicotine is broken down through oxidation, and about 10% is excreted unchanged. It has stimulating effects upon the central nervous system and has paralyzing effects at very high dosages (cf. **cytisine**). In the peripheral nervous system, nicotine behaves in a similar manner as the neurotransmitter acetylcholine. High dosages can result in sudden death due to respiratory paralysis or cardiac arrest within five minutes of ingestion (Roth et al. 1994, 864*). From 40 to 60 mg represents a lethal dosage for humans (Frerichs, Arends, Zörnig, *Hagers Handbuch*). **Diazepam** can be effective as an antidote for nicotine poisoning (Roth et al. 1994, 865*). Nicotine is now generally regarded as highly "addictive" (Schiffman 1981). Although it is commonly assumed that nicotine causes cancer, uncertainty has been expressed about this theory (Schievelbein 1972).

Nicotine has been detected in Egyptian mummies (New Kingdom) (Balabanova et al. 1992*). However, this discovery should not be taken as evidence that the Egyptians knew of wild tobacco (*Nicotiana rustica*), as the Balabanova team in Munich has suggested, for some Old World plants also contain nicotine (see the table at right).

Commercial Forms and Regulations

Nicotine can be obtained in its pure form from chemical suppliers. It is subject to the laws covering the transport of dangerous substances and is classified as a Category 1 substance on the Swiss Poison List. In the United States, pure nicotine is available only with a prescription (Ott 1993, 447*). In Germany, it is subject to the laws regarding dangerous substances but is not regarded as a "narcotic."

Plants Containing Nicotine

(from Bock 1994, 93*; Römpp 1995, 2995*; Schultes and Raffauf 1991, 37*; supplemented)

Nicotine

Stock Plant	Plant Part(s)
ARACEAE	
Arum maculatum L. (cuckoo pint)	herbage
ASCLEPIADACEAE	
Asclepias syriaca L. (Syrian milkweed)	
EQUISETACEAE	
Equisetum palustre L. (marsh horsetail; cf. ***Equisetum arvense***)	herbage
ERYTHROXYLACEAE	
Erythroxylum coca	roots, stems
Erythroxylum spp.	
LEGUMINOSAE	
Acacia retinodes Schlechtend. (cf. **Acacia spp.**)	leaves
Mucuna pruriens	leaves
LYCOPODIACEAE	
Lycopodium spp. (club moss; cf. *Trichocereus pachanoi*)	herbage
SOLANACEAE (NIGHTSHADES)	
Cestrum spp. (cf. **Cestrum**, nocturnum Cestrum parqui)	
Cyphomandra spp.	
Datura metel	herbage
Duboisia hopwoodii	leaves
Duboisia spp.	leaves, bark
Nicotiana rustica	entire plant
Nicotiana tabacum	entire plant
Nicotiana spp.	

Literature

See also the entries for *Nicotiana rustica*, *Nicotiana tabacum*, and *Nicotiana* spp.

Lee, Richard S., and Mary Price Lee. 1994. *Caffeine and nicotine.* New York: The Rosen Publishing Group.

Schievelbein, H. 1972. Biochemischer Wirkungsmechanismus des Nikotins oder seiner Abbauprodukte hinsichtlich eines eventuellen carcinogenen, mutagenen oder teratogenen Effektes. *Planta Medica* 22:293–305.

Shiffman, Saul. 1981. Tabakkonsum und Nikotinabhängigkeit. In *Rausch und Realität*, ed. G. Völger, 2:780–83. Cologne: Rautenstrauch-Joest-Museum für Völkerkunde.

A smoking Indian inspects his tobacco plants. Native Americans cultivate tobacco strains with especially high nicotine contents.

N,N-DMT

N,N-DMT

The seeds of the Leguminosae *Desmodium tiliaefolium* contain detectable amounts of *N,N*-DMT and bufotenine.

Other Names

Dimethyltryptamin, dimethyltryptamine, DMT, nigerin, nigerina, nigerine (1946), *N,N*-dimethyl-tryptamine, 3-[2-(dimethyl-amino)ethyl]-indole

Empirical formula: $C_{12}H_{16}N_2$

Substance type: tryptamine (**indole alkaloid**)

R. H. F. Manske first created DMT as a synthetic substance in the laboratory in 1931. It was not until 1955 that it was isolated as a natural compound in the seeds of *Anadenanthera peregrina*. It is found in a great number of plants and also occurs naturally in humans and other mammals (see table, pages 853–854). *N,N*-DMT is closely related to **5-MeO-DMT** and **psilocybin**/psilocin.

N,N-DMT and 5-MeO-DMT are among the psychedelics whose effects are short in duration. They are not orally effective in an isolated form (as a salt or a base) because the MAO enzyme breaks them down before they can pass through the blood–brain barrier (cf. **ayahuasca**, β-**carbolines**). They reveal their awesome effects only when injected by syringe (Strassman et al. 1994), snuffed, or smoked. When *N,N*-DMT or 5-MeO-DMT is injected intravenously, the effects last some forty-five minutes; when the substance is smoked or snuffed, the effects last only ten minutes. Subjectively, however, those few minutes may seem to have spanned centuries. People who have had experiences with DMT unanimously agree that it is easily the most powerful psychedelic known (cf. McKenna 1992; Meyer 1992). Some people have described DMT as "crystallized consciousness." When DMT is smoked, it is said that the "scent of enlightenment" fills the air. Only "a few seconds after taking it, DMT acts like the trumpets of Jericho upon the gates of perception" (Kraemer 1995, 98). DMT experiences can be so extraordinarily alien that most subjects find it extremely difficult or even impossible to describe them in words. Many people speak of contact with strange beings (aliens, fairies, machine elves, et cetera) (Bigwood and Ott 1977; Leary 1966; McKenna 1992; Meyer 1992).

When pure DMT is smoked or vaporized and inhaled, the effective dosage lies around 20 mg (although amounts as high as 100 mg are sometimes smoked). The dosage for **ayahuasca** and **ayahuasca analogs** ranges from 50 to 100 mg. When DMT is injected, the typical dose is 1 mg/kg of body weight (Ott 1993, 433*).

DMT is also produced in the human nervous system, where it appears to serve an important function as a neurotransmitter (Barker et al. 1981; Callaway 1996; Siegel 1995b). Neurobiologists are as yet uncertain about the role DMT might play in the nervous system. Hyperventilating causes the concentration of DMT in the lungs to increase (Callaway 1996). One physician has reported that the release of endogenous DMT is highest at the moment of death. It is my opinion that this chemical messenger is responsible for the ultimate shamanistic ecstasy, for enlightenment, and for the merging into the "clear light of death." An experiment in which DMT was given to practicing Buddhists found that the subjects had experiences and visions that corresponded to the Buddhist teachings (Strassman 1996).

DMT has clearly inspired numerous novels in the fantasy and science-fiction genres. The novel *Kalimantan* deals with the search for a fictitious hallucinogenic drug called *seribu aso*. The descriptions of the effects of this drug agree perfectly with the descriptions of DMT trips (Shepard 1993). Several novels, including the *Valis* trilogy of the science-fiction master Philip K. Dick (1928–1982), also appear to represent a literary attempt to understand the hyperdimensionality of DMT experiences (Dick 1981a, 1981b, 1982).

Commercial Forms and Regulations

DMT occurs as a free base, an HCL, and a fumarate. Although the fumarate crystallizes out easily, it contains only 60% of the pure substance. DMT is classified as a Schedule I drug in the United States and is a "narcotic drug in which trafficking is not allowed" in Germany as well as in Switzerland (Körner 1994, 38*).

Literature

See also the entries for **5-MeO-DMT**.

Arnold, O. H., and G. Hofman. 1957. Zur Psychopathologie des Dimethyltryptamine. *Wiener Zeitschrift für Nervenheilkunde* 13:438–45.

Barker, S., J. Monti, and S. Christian. 1981. *N,N*-dimethyltryptamine: An endogenous hallucinogen. *International Review of Neurobiology* 22:83–110.

Plants Containing DMT

(from Block 1994*; Smith 1977; Montgomery, pers. comm.; Ott 1993*; Schultes and Hofman 1980, 155*; supplemented)

Species	Demonstrated Tryptamines
AGARICACEAE (FUNGI)	
Amanita citrina Gray	DMT, 5-MeO-DMT
Amanita porphyria (Fries) Secretan	**5-MeO-DMT**
Amanita spp.	DMT, **bufotenine**
AIZOACEAE/MESEMBRYANTHEMACEAE	
Delosperma sp.	DMT, MMT
***Mesembryanthemum* spp.**	DMT (?)
GRAMINEAE (POACEAE)	
***Arundo donax* L.**	DMT, **bufotenine**, and others
***Phalaris arundinacea* L.**	DMT, **bufotenine**, and others
Phalaris tuberosa L.	DMT, **bufotenine**, and others
***Phragmites australis* (Cav.) Trin. ex Steud.**	DMT
LAURACEAE	
Umbellularia californica (Hook. et A.) Nutt.	**5-MeO-DMT**
LEGUMINOSAE	
Acacia confusa Merr.	DMT
Acacia maidenii F. von Muell.	DMT (0.36%)
Acacia nubica Benth.	DMT
Acacia phlebophylla F. von Muell.	0.3% DMT
Acacia simplicifolia Druce	0.81% DMT
***Acacia* spp.**	DMT
***Anadenanthera colubrina* (Vell.) Bren.**	DMT, 5-MeO-DMT, **bufotenine**
***Anadenanthera peregrina* (L.) Spag.**	DMT, 5-MeO-DMT, **bufotenine**
Desmanthus illinoensis (Michx.) MacMillan	DMT (to 0.34%)
Desmodium adscendens (Sw.) DC. var. *adscendens*	DMT (?)
Desmodium caudatum DC.	DMT
Desmodium gangeticum DC.	DMT, **bufotenine**, and others
Desmodium gyrans DC.	DMT, **bufotenine**, and others
Desmodium pulchellum Benth. ex Bak.	DMT, **bufotenine**, and others
Desmodium racemosum Thunb.	**5-MeO-DMT**
Desmodium tiliaefolium G. Don	DMT, **bufotenine**, and others
Desmodium triflorum DC.	DMT, **bufotenine**, and others
Lespedeza bicolor Turcz.	DMT, 5-MeO-DMT
Lespedeza bicolor var. *japonica* Nakai	DMT, 5-MeO-DMT
Lespedeza capitata Michx.	DMT
Mimosa scabrella Benth.	DMT
***Mimosa tenuiflora* (Willd.) Poir. [syn. *Mimosa hostilis* Benth., *Mimosa nigra*]**	0.57% DMT
Mimosa verrucosa	DMT
***Mimosa* spp.**	DMT and others
***Mucuna pruriens* DC.**	DMT, 5-MeO-DMT, **bufotenine**
Mucuna spp.	DMT and others
Petalostylis cassioides Pritzel	DMT, tetrahydroharmane
Petalostylis labicheoides R. Brown	DMT, tryptamine
Phyllodium pulchellum (L.) Desv.	DMT
MALPIGHIACEAE	
Banisteriopsis argentea Spring. [syn. *B. muricata* (Cav.) Cuatr.]	DMT, DMT-*N*-oxide
***Diplopterys cabrerana* (Cuatr.) Gates [syn. *Banisteriopsis rusbyana*]**	DMT, 5-MeO-DMT
MYRISTICACEAE	
Iryanthera ulei Warb.	**5-MeO-DMT**
Osteophloeum platyspermum (DC.) Warb.	DMT, 5-MeO-DMT
Virola calophylla Warb.	DMT, 5-MeO-DMT
Virola calophylloidea Markgr.	DMT, 5-MeO-DMT

DMT has been detected in several species of *Desmodium*. (From Hernández, 1942/46 [Orig. pub. 1615]*)

"The first puff immediately took me into a completely different reality, a reality immanent to the normal one but surpassing it by leaps and bounds. Dripping and pearling off my companions were blue and violet dabs of color that extended and flowed together through the room. The small drops were overlapping and pushing into each other, arranging themselves into the most magnificent patterns and forming a circle.

"When the room had become a cathedral of dancing patterns, rays of light shot through the room, jumping from one person to the next. A glowing ring of bodiless consciousness and high-caliber energy shot through the circle. The chakras were blooming. I was able to see and feel them clearly. The kundalini energy shot to the top and sprayed from the heads, to join together with the eternity of that which can be experienced. I sensed the power of the ring and was initiated into an ancient cult. The person across from me looked like a priest from an earlier culture.

"The bustling activity of the patterns of magnificent color and thousandfold patterns became wilder and ever more intense. I was able to perceive absolutely the same things whether I kept my eyes open or closed. The flying lights and swinging colors became pure light that condensed into cosmic laughter. A tryptamine initiation."

A DMT EXPERIENCE

Species	Demonstrated Tryptamines
Virola carinata (Spruce ex Benth.) Warb.	DMT, **5-MeO-DMT**
Virola divergens Ducke	DMT
Virola elongata (Spruce ex Benth.) Warb.	DMT, 5-MeO-DMT
Virola melinonii (Benoist) A.C. Smith	DMT, 5-MeO-DMT
Virola multinerva Ducke	DMT, 5-MeO-DMT
Virola pavonis (DC.) A.C. Smith	DMT
Virola peruviana (DC.) Warb.	DMT, 5-MeO-DMT
Virola rufula (DC.) Warb.	DMT, 5-MeO-DMT
Virola sebifera Aubl.	DMT
Virola theiodora (Spruce ex Benth.) Warb.	DMT, 5-MeO-DMT
Virola venosa (Benth.) Warb.	DMT, 5-MeO-DMT
Virola spp.	DMT, **5-MeO-DMT**, and others
OCHNACEAE	
Testulea gabonensis Pellegr.	DMT
POLYGONACEAE	
Eriogonum sp.	DMT
RUBIACEAE	
Psychotria carthaginensis Jacq.	DMT
Psychotria poeppigiana Muell. Arg.	DMT
Psychotria viridis Ruíz et Pav. [syn. *P. psychotriaefolia* Standl.]	DMT
RUTACEAE	
Dictyoloma incanescens DC.	**5-MeO-DMT**
Dutaillyea drupacea (Baill.) Hartley	**5-MeO-DMT**
Dutaillyea oreophila (Baill.) Sévenet-Pusset	**5-MeO-DMT**
Evodia rutaecarpa Benth.	**5-MeO-DMT**
Limonia acidissima L.	DMT traces
Melicope leptococca (Baill.) Guill.	0.21% DMT
Pilocarpus organensis Rizzini et Occhioni	**5-MeO-DMT**
Vepris ampody H. Perr.	DMT
Zanthoxylum arborescens Rose	DMT traces
Zanthoxylum procerum Donn. Sm.	DMT

Bigwood, Jeremy, and Jonathan Ott. 1977. DMT: The fifteen minute trip. *Head* 11:56 ff.

Callaway, James. 1996. DMTs in the human brain. In *Yearbook for Ethnomedicine and the Study of Consciousness* (1995), 4:45–54. Berlin: VWB.

Dick, Philip K. 1981a. *The divine invasion*, New York: Vintage Books.

———. 1981b. *Valis*. New York: Vintage Books.

———. 1982. *The transmigration of Timothy Archer*. New York: Vintage Books.

Kraemer, Olaf. 1995. Die Trompeten Jerichos. *Wiener* 9:97–99.

Lamparter, Daniel, and Adolf Dittrich. 1996. Intraindividuelle Stabilität von ABZ unter sensorischer Deprivation, N,N-Dimethyltryptamin (DMT) und Stickoxydul. In *Jahrbuch des Europäischen Collegiums für Bewußtseinsforschung* (1995), 33–43. Berlin: VWB.

Leary, Timothy. 1966. Programmed communication during experience with DMT. *Psychedelic Review* 8:83–95.

Manske, R. H. F. 1931. A synthesis of the methyltryptamines and some derivatives. *Canadian Journal of Research* 5:592–600.

McKenna, Terence. 1992. Tryptamin hallucinogens and consciousness. In *Yearbook for Ethnomedicine and the Study of Consciousness* (1992), 1:133–48. Berlin: VWB.

Meyer, Peter. 1992. Apparent communication with discarnate entities induced by dimethyltryptamine DMT. In *Yearbook for Ethnomedicine and the Study of Consciousness* (1992), 1:149–74. Berlin: VWB.

Shepard, Lucius. 1993. *Kalimantan*. New York: Tom Doherty Associates.

Shulgin, Alexander T. 1976. Profiles of psychedelic drugs. 1: DMT. *Journal of Psychedelic Drugs* 8 (2): 167–68.

Smith, Terence A. 1977. Tryptamine and related compounds in plants. *Phytochemistry* 16:171–75.

Strassman, Rick J. 1996. Sitting for sessions: Dharma and DMT research. *Tricycle* 6 (1): 81–88.

Strassman, Rick J., Clifford R. Qualls, Eberhard H. Uhlenhuth, and Robert Kellner. 1994. Dose-response study of *N,N*-dimethyltryptamin in humans. *Archive of General Psychiatry* 51:85–97, 98–108.

Szára, S. I. 1956. Dimethyltryptamin: Its metabolism in man; the relation of its psychotic effect to the serotonin metabolism. *Experientia* 15 (6): 441–42.

Opium Alkaloids

Other Names
Opiate, opiates, opium compounds

The study of opium and the isolation of its constituents ranks among the most important achievements in the history of pharmacology (cf. *Papaver somniferum*). In ancient times, opium was already known as the best of all analgesics (cf. **soporific sponge**). The isolation of **morphine** from opium revolutionized pain therapy in Europe. No other component of opium has a comparably powerful effect. The potency of morphine would not be exceeded until heroin (diacetylmorphine) was synthesized (Snyder 1989). Subsequent pharmacological research has led to the creation of numerous morphine analogs (fentanyls), some of which are as much as 7,500 times more potent than morphine (Sahihi 1995, 31 ff.*).

The opium alkaloids **codeine** and **morphine** have become culturally significant as psychoactive substances. **Papaverine** is used in medicine as a treatment for impotence.

Some opium alkaloids are also found in other *Papaver* species (*Papaver* spp.), although these species usually contain only traces of these substances (Khanna and Sharma 1977; Küppers et al. 1976; Phillipson et al. 1973; Phillipson et al. 1976).

Aporphines, whose structures are analogous to those of the opium alkaloids, occur in *Papaver fugax* Poir. [syn. *Papaver caucasicum* M.-B., *Papaver floribundum* Desf.] and *Nymphaea ampla*. Other substances related to the opium alkaloids are present in *Argemone mexicana*, *Eschscholzia californica*, *Nuphar lutea*, and *Papaver* spp.

Literature
See also the entries for *Argemone mexicana*, *Papaver somniferum*, *Papaver* spp., **codeine**, **morphine**, and **papaverine**.

Khanna, P., and G. L. Sharma. 1977. Production of opium alkaloids from in vitro tissue culture of *Papaver rhoeas* L. *Indian Journal of Experimental Biology* 15:951–52.

Krikorian, A. D., and M. C. Ledbetter. 1975. Some observations on the cultivation of opium poppy (*Papaver somniferum* L.) for its latex. *Botanical Review* 41:30–103.

Küppers, F. J. E. M., C. A. Salmink, M. Bastart, and M. Paris. 1976. Alkaloids of *Papaver bracteatum*: Presence of codeine, neopine and alpinine. *Phytochemistry* 15:444–45.

Phillipson, J. D., S. S. Handa, and S. W. El-Dabbas. 1976. *N*-oxides of morphine, codeine and thebaine and their occurrence in *Papaver* species. *Phytochemistry* 15:1297–1301.

Phillipson, J. D., G. Sariyar, and T. Baytop. 1973. Alkaloids from *Papaver fugax* of Turkish origin. *Phytochemistry* 12:2431–34.

Scully, Rock, with David Dalton. 1996. *An American odyssey: Die legendäre Reise von Jerry Garcia und den Grateful Dead*. St.Andrä-Wördern: Hannibal Verlag.

Snyder, Solomon H. 1989. *Brainstorming: The science and politics of opiate research*. Cambridge and London: Harvard University Press.

The Constituents of Opium
The composition of the alkaloid mixture can vary greatly, depending upon the strain of poppy, the location of cultivation, and the processing technique (Krikorian and Ledbetter 1975).

I. Isocholine derivates

codamine*	
cryptopine*	
gnoscopine*	
hydrocotarnine*	
laudanindine (= tritopine)*	
dl-laudanine*	
laudanosine*	
norlaudanosine*	
narceine (= narceinum)	0.1–0.2%
l-narcotine (= narcotine = noscapine)	1–11%
oxynarcotine*	
papaverine	0.5–1%
protopine* (also in *Papaver rhoeas*, *Argemone mexicana*, and *Eschscholzia californica*)	
reticuline*	
xanthaline (= papaveraldine)*	

II. Bases that yield phenanthrene derivatives when broken down

codeine	0.2–4%
morphine	2.8–23%
neopine*	
porphyroxine*	
pseudomorphine*	
thebaine (also in *Papaver bracteatum*)	0.1–4%

III. Other bases

lanthopine*	
meconine*	
oripavine* (also in *Papaver orientale*)	
papaveramine*	
rhoeadine*	

Alkaloids marked with * are present in only trace amounts.

"The hall of wonder drugs [at the German Museum in Munich] is huge, a curious demonstration of German pharmacology. Morphine, methamphetamine, adolphine (a synthetic heroin named after Hitler), etc. Every drug on display comes with a complete little biography of its own. The little sign next to methamphetamine says that it was given to Stuka pilots in World War II and that soldiers at the eastern front took it in order to stay awake. And that it was Hitler's favorite drug. We learn about the various nicknames: 'coffee substitute,' 'lightning powder,' etc."

ROCK SCULLY
AN AMERICAN ODYSSEY
(1996, 15 f.)

The most powerful tool of the American heavy metal band Tool is a strong *Opiate* (CD cover 1992, BMG Music)

Papaverine

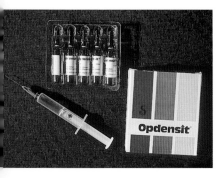

Papaverine

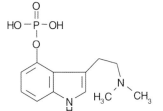

Papaverine, an active constituent isolated from opium, is prescribed for impotence and is injected into the penis.

Other Names

1-(3,4-dimethoxybenzyl)-6,7-dimethoxyisochinoline, papaverin, papaverina, papavérine

Empirical formula: $C_{20}H_{21}NO_4$

Substance type: **opium alkaloid**

Papaverine is a component of opium (0.3 to 0.8%) and was named after the genus *Papaver* (cf. ***Papaver somniferum***). Papaverine has very weak psychoactive properties but is a powerful vasodilator. Effective dosages start at 200 mg. An extract of *Nuphar lutea* has similar effects.

In recent years, papaverine has been used to treat impotence, often with good success (Mellinger et al. 1987). When used for this purpose, the substance is injected directly into the corpus cavernosum when the penis is flaccid (so-called SKAT therapy; cf. Ernst et al. 1993). Among the problems that this method may cause are painful priapism (persistent erections for up to thirty-six hours without sexual arousal!) and inflammation of the penis (Sanders 1985).

Commercial Forms and Regulations

The substance, available as papaverine hydrochloride, is sold in suppository form and in solution for injection. Papaverine is available only with a prescription.

Literature

See also the entries for ***Papaver somniferum*** and **opium alkaloids**.

Ernst, Günter, Hans Finck, and Dieter Weinert. 1993. *Dem Manne kann geholfen werden*. Munich: Ehrenwirth.

Mellinger, Brett C., E. Darracott Vaughan, Stephen L. Thompson, and Marc Goldstein. 1987. Correlation between intracavernous papaverine injection and Doppler analysis in impotent men. *Urology* 30 (5): 416–19.

Porst, H. 1996. Orale und intracavernöse Pharmakotherapie. *TW Urologie Nephrologie* 8 (2): 88–94.

Sanders, Kevin. 1985. 30-Stunden Erektion. *Penthouse* 4/85:65–68, 196, 200.

Schnyder von Wartensee, M., A. Sieber, and U. E. Studer. 1988. Therapie der erektilen Dysfunktion mit Papaverin—$2^{1}/_{2}$ Jahre Erfahrung. *Schweizer medizinische Wochenschrift* 118 (30): 1099–1103.

Psilocybin/Psilocin

Psilocybin

Psilocin

504 Because psilocybin mushrooms are sometimes difficult to digest, more time (up to one and a half hours) may pass before the effects become manifest.

Other Names

Psilocybin: CY-39, indocybin, *O*-phosphoryl-4-hydroxy-*N,N*-dimethyltryptamine, 3(2-dimethylamino)ethylindol-4-ol dihydrogenphosphatester

Psilocin: 4-hydroxy-*N,N*-dimethyltryptamine, psilocine, psilocyn (misspelling in the legal literature), 3-[2-(dimethylamino)ethyl]-1*H*—indole-4-ol

Empirical formula: $C_{12}H_{17}N_2O_4P$ (psilocybin), $C_{12}H_{16}N_2O$ (psilocin)

Substance type: tryptamines, indole amines (**indole alkaloids**)

Psilocybin was first isolated from ***Psilocybe mexicana*** and identified by Albert Hofmann in 1955 (Hofmann et al. 1958, 1959). The phosphorylated indole amine psilocybin is transformed into psilocin by splitting off the phosphoric acid group (Hofmann and Troxler 1959). Because the protection the phosphoric acid would provide is lacking, psilocin easily oxidizes with the phenolic hydroxyl group, resulting in blue quinonoid products. This explains why psilocybin mushrooms turn blue after they have been squeezed and harvested (cf. ***Panaeolus cyanescens***, ***Psilocybe cyanescens***). In the body, psilocybin is immediately metabolized into psilocin, which is the actual psychoactive constituent.

Psilocybin and psilocin are closely related to baeocystin (= *O*-phosphoryl-4-hydroxy-*N*-methyltryptamine, norpsilocybin), which probably represents the biogenic precursor of psilocybin (Repke et al. 1977; cf. also Brack et al. 1961 and Chilton et al. 1979). Baeocystin may be a derivative of tryptophan (Brack et al. 1961).

The usual psychedelic dosage of psilocybin is 10 mg. When psilocybin is taken orally, the effects typically become apparent in about twenty minutes[504] (Shulgin 1980). Rudolf Gelpke (1928–1972) took between 6 and 20 mg during his self-experiments; with 10 mg, he made his historic "journey to the outer space of the soul":

This inebriation was a space flight not into the outer realm, but into the inner person, and for a moment I experienced reality from a position located somewhere beyond the gravity of time. (Gelpke 1962, 395)

With very high dosages, it is common to perceive voices (Beach 1997). This could explain why Indians say that the mushroom talks to them. Toxic dosages are unknown!

Walter Pahnke's "Good Friday experiment," in which theology students were administered psilocybin in a church on Good Friday, has become renowned. Pahnke applied the theory of dosage, set, and setting as part of the test to see whether mystical revelations would occur, which was indeed the case (Pahnke 1972; Pahnke and Richards 1970; cf. Doblin 1991).

Timothy Leary and his colleagues at Harvard experimented with psilocybin on prisoners. Their experiments were aimed at determining whether the psychedelic constituent was suitable for use in therapy with inmates. It was hoped that the drug experience would enable the prisoners to attain insights into their behavior that would then enable them to change themselves on their own. Although these experiments showed great promise, they had to be terminated (Clark 1970; Forcier and Doblin 1994; Riedlinger and Leary 1994).

Both psilocybin and two synthetic derivatives (CZ-74, CY-19) have been used with success in psychedelic and psycholytic therapy (Leuner 1963; Leuner and Baer 1965; Passie 1995, 1996). Psilocybin can release, stimulate, and inspire creativity (Fischer et al. 1972), as an increasing number of studies have shown (Baggott 1997; Spitzer et al. 1996), and "archetypal art therapy" is making use of this effect (Allen 1995).

Today, psilocybin is playing a central role in neurochemical research into brain activity, in which it is being studied with the very elaborate and costly positron-emission tomography (PET) method (Vollenweider 1996).

Jochen Gartz has discovered that fungal enzymes synthesize the "synthetic" psilocin analog CZ-74 (diethyl-4-hydroxytryptamine, 4-OH-DET) from diethyltryptamine when it is added to a *Psilocybe* spp. substrate (J. Gartz, pers. comm.). It is possible that the "synthetic" CY 19 (= diethyl-4-phosphoryloxytryptamine) can be produced in the same fashion.

Commercial Forms and Regulations

Both psilocybin and psilocin are classified as Schedule I drugs in the United States (Shulgin 1980). They are internationally regarded as illegal "narcotics." The analog substances psilocin-(eth) and psilocybin-(eth) are also illegal (Körner 1994, 40*).

PSILOCYBIN   PSILOCIN

Microscopic views of psilocybin and psilocin, the active constituents in mushrooms, after crystallization from methanol. (Photograph: Albert Hofmann)

Literature

See also the entries for *Psilocybe mexicana* and *Psilocybe* spp.

Allen, Tamara D. 1994. Research in archetypal art therapy with psilocybin. *Maps* 5 (1): 39–40.

———. 1995. Archetypal art therapy: Hearing psilocybin in the art & metaphor work of volunteer no. 31. *Maps* 6 (1): 23–26.

Baggot, Matthew. 1997. Psilocybin's effects on cognition: Recent research and its implications for enhancing creativity. *Maps* 7 (1): 10–11.

Beach, Horace. 1997. Listening for the logos: Study of reports of audible voices at high doses of psilocybin. *Maps* 7 (1): 12–17.

Bocks, S. M. 1968. The metabolism of psilocin and psilocybin by fungal enzymes. *Biochemical Journal* 106:12–13.

Borner, Stefan, and Rudolf Brenneisen. 1987. Determination of tryptamines in hallucinogenic mushrooms using high-performance liquid chromatography with photodiode array detection. *Journal of Chromatography* 408:402–8.

Brack, A., Albert Hofmann. F. Kalberer, H. Kobel, and J. Rutschmann. 1961. Tryptophan als biogenetische Vorstufe des Psilocybins. *Archiv der Pharmazie* 294/66 (4): 230–34.

Chilton, W. Scott, Jeremy Bigwood, and Robert E. Jensen. 1979. Psilocin, bufotenine, and serotonin: Historical and biosynthetic observations. *Journal of Psychedelic Drugs* 11 (1–2): 61–69.

Clark, Jonathan. 1970. Psilocybin: The use of psilocybin in a prison. In *Psychedelics*, ed. Bernard Aaronson and Humphry Osmond, 40–44. Garden City, N.Y.: Anchor Books.

Doblin, Rick. 1991. Pahnke's 'Good Friday experiment': A long-term follow-up and methodological critique. *The Journal of Transpersonal Psychology* 23 (1): 1–28.

Fischer, Roland, Ronald Fox, and Mary Ralstin. 1972. Creative performance and the hallucinogenic drug-induced creative experience. *Journal of Psychedelic Drugs* 5 (1): 29–36. (On psilocybin and creativity research.)

Forcier, Michael W., and Rick Doblin. 1994. Long-term follow-up to Leary's Concord Prison psilocybin study. *Maps* 4 (4): 20–21.

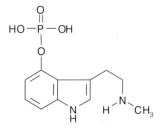

Baeocystin

Tryptophan

The structural formulas of the two active mushroom constituents psilocybin and psilocin, shown here in the handwriting of their discoverer, Albert Hofmann.

"The chemical structures of psilocybin and psilocin were new and important in a number of ways. These two mushroom constituents were the first naturally occurring indole compounds with a hydroxyl function in the 4th position of the indole system. Psilocybin is the only known indole alkaloid [apart from its analog baeocystin] in which there is a substituted phosphoryloxy group. The chemical relationship between psilocin and the neurotransmitter serotonin is of particular importance."

ALBERT HOFMANN

"PILZLICHE HALLUZINOGENE" ["FUNGAL HALLUCINOGENS"]

(DER CHAMPIGNON, JUNE 1987, P. 24)

Gelpke, Rudolf. 1962. Von Fahrten in den Weltraum der Seele: Berichte über Selbstversuche mit Delysid (LSD) und Psilocybin (CY). *Antaios* 3:393–411.

———. [1997]. *Von Fahrten in den Weltraum der Seele: Berichte über Selbstversuche mit LSD und Psilocybin*. Löhrbach: Werner Pieper's MedienXperimente and Edition Rauschkunde.

Gnirss, Fritz. 1959. Untersuchung mit Psilocybin, einem Phantastikum aus dem mexikanischen Rauschpilz *Psilocybe mexicana*. *Schweizer Archiv für Neurologie, Neurochirurgie und Psychiatrie* 84:346–48.

Hofmann, Albert, A. Frey, H. Ott, Th. Petrzilka, and F. Troxler. 1958. Konstitutionsaufklärung und Synthese von Psilocybin. *Experientia* 14 (11): 397–401.

Hofmann, Albert, Roger Heim, A. Brack, and H. Kobel. 1958. Psilocybin, ein psychotroper Wirkstoff aus dem mexikanischen Rauschpilz *Psilocybe mexicana* Heim. *Experientia* 14 (3): 107–12.

Hofmann, Albert, Roger Heim, A. Brack, H. Kobel, A. Frey, H. Ott, T. Petrzilka, and F. Troxler. 1959. Psilocybin und Psilocin, zwei psychotrope Wirkstoffe aus mexikanischen Rauschpilzen. *Helvetica Chimica Acta* 42 (162): 1557–72.

Hofmann, Albert, and F. Troxler. 1959. Identifizierung von Psilocin. *Experientia* 15 (3): 101–4.

Jones, Richard. 1963. "Up" on Psilocybin. *The Harvard Review* 1 (4): 38–43.

Krippner, Stanley. 1970. Psilocybin: An adventure in psilocybin. In *Psychedelics*, ed. Bernard Aaronson and Humphry Osmond, 35–39, Garden City, N.Y.: Anchor Books.

Laatsch, Hartmut. 1994. Das Fleisch der Götter— Von den Rauschpilzen zur Neurotransmission. In *Welten des Bewußtseins*, ed. A. Dittrich et al., 3:181–95. Berlin: VWB.

———. 1996. Zur Pharmakologie von Psilocybin und Psilocin. In *Maria Sabina—Botin der heiligen Pilze*, ed. Roger Liggenstorfer and Christian Rätsch, 193–202. Solothurn: Nachtschatten Verlag.

Leuner, Hanscarl. 1963. Die Psycholytische Therapie: Klinische Psychotherapie mit Hilfe von LSD-25 und verwandten Substanzen. *Zeitschrift für Psychotherapie und medizinische Psychologie* 13:57 ff.

Leuner, Hanscarl, and G. Baer. 1965. Two short-acting hallucinogens of the psilocybin-group. In *Neuro-pharmacology*, ed. D. Bente and P. B. Bradley. Amsterdam: Elsevier.

Ott, Jonathan, and Gastón Guzmán. 1976. Detection of psilocybin in species of *Psilocybe, Panaeolus* and *Psathyrella*. *Lloydia* 39:258–60.

Pahnke, Walter N. 1972. Drogen und Mystik. In Josuttis and Leuner, 54–76*.

Pahnke, Walter N., and William A. Richards. 1970. Implications of LSD and experimental mysticism. *Journal of Psychedelic Drugs* 3 (1): 92–108.

Passie, Torsten. 1995. Psilocybin in der westlichen Psychotherapie. *Curare* 18 (1): 131–52.

———. 1996. Psilocybin in der westlichen Psychotherapie. In *María Sabina—Botin der heiligen Pilze*, ed. Roger Liggenstorfer and Christian Rätsch, 211–25. Solothurn: Nachtschatten Verlag.

Repke, David B., Dale Thomas Leslie, and Gastón Guzmán. 1977. Baeocystin in *Psilocybe, Conocybe* and *Panaeolus*. *Lloydia* 40 (6): 566–78.

Riedlinger, Thomas, and Timothy Leary. 1994. Strong medicine for prisoner reform: The Concord Prison experiments. *Maps* 4 (4): 22–25.

Shulgin, Alexander T. 1980. Psilocybin. *Journal of Psychedelic Drugs* 12 (1): 79.

Spitzer, M., M. Thimm, L. Hermle, P. Holzmann, K. A. Kovar, H. Heimann, E. Gouzoulis-Mayfrank, U. Kischka, and F. Schneider. 1996. Increased activation of indirect semantic associations under psilocybin. *Biological Psychiatry* 39:1055–57.

Strassmann, Rick. 1992. DMT and psilocybin research. *Maps* 3 (4): 8–9.

———. 1995. University of New Mexico DMT and psilocybin studies. *Maps* 5 (3): 14–15.

Troxler, F., F. Seemann, and Albert Hofmann. 1959. Abwandlungsprodukte von Psilocybin und Psilocin. *Helvetica Chimica Acta* 42 (226): 2073–103.

Vollenweider, Franz. 1996. Perspektiven der Bewußtseinsforschung mit Halluzinogenen. In *Maria Sabina—Botin der heiligen Pilze*, ed. Roger Liggenstorfer and Christian Rätsch, 203–10. Solothurn: Nachtschatten Verlag.

Salvinorin A

Other Names
Divinorin A

Empirical formula: $C_{23}H_{28}C_8$

Substance type: **diterpene** (clerodane)

Salvinorin A is the active constituent in *Salvia divinorum*. Apart from **THC** and the constituents in **essential oils**, it is the only known non-nitrogenous psychoactive plant constituent. Salvinorin is not an alkaloid.

The substance was first described by Ortega et al., who named it salvinorin (1982). The same substance was subsequently described under the name divinorin A (Valdes et al. 1984). Salvinorin A is extracted from fresh plant material. The effective dosage is between 200 and 500 μg.

Salvinorin A can be smoked in a glass pipe; a better technique involves vaporizing the plant and then inhaling the fumes. It can also be taken in solution under the tongue. When the substance is smoked or inhaled, the effects are immediately apparent, and the primary effects last from five to ten minutes. When it is administered sublingually, the effects become manifest after about ninety seconds and reach their peak some ten to fifteen minutes later, after which they gradually diminish (Turner 1996).

The potent and strange psychoactive effects of salvinorin A were probably discovered by Daniel Siebert:

> Salvinorin A is an extremely powerful compound for altering consciousness. In fact, it is **the most potent naturally occurring hallucinogen** that has been isolated to date. But before potential experimenters become too interested, it must be clearly stated that the effects are often *extremely* unnerving and that there is a very real risk that persons may physically harm themselves when using it. . . .
>
> I have seen people get up and jump across the room, thereby falling over the furniture, babbling incomprehensible nonsense, and hitting their heads against the wall. Several people tried to leave the house. When the experience was over, they did not remember

what had happened. In fact, they actually believed that they remembered entirely different events. To an outside observer, it appears as though these people have an empty expression in their eyes, as though they are not present (and perhaps they really are not). (Siebert 1995, 4)

This description is strongly reminiscent of phenomena that occur with high dosages (overdoses) of nightshades (*Atropa belladonna, Brugmansia* spp., *Hyoscyamus niger, Datura* spp.) and the tropane alkaloids **atropine** and **scopolamine**. Most subjects have no desire at all to repeat an experiment with salvinorin.

The neurochemistry of salvinorin A is still unresolved. In spite of extensive receptor testing (NovaScreen method), salvinorin A has not been found to bind to any known neurotransmitter receptors, including the receptor that ketamine occupies (David Nichols, pers. comm.). The daring and extreme experiments of D. M. Turner suggest that salvinorin A does not have any negative cross-tolerance with other psychoactive substances (such as LSD, *N,N*-**DMT**, ketamine) (Turner 1996).

Commercial Forms and Regulations
None

Literature
See also the entries for *Coleus blumei, Salvia divinorum*, and **diterpenes**.

Ortega, A., J. F. Blount, and P. S. Marchand. 1982. Salvinorin, a new trans-neoclerodane diterpene from *Salvia divinorum* (Labiatae). *Journal of the Chemical Society, Perkin Transactions* I:2505–8.

Siebert, Daniel J. 1995. Salvinorin A: Vorsicht geboten. *Entheogene* 3:4–5.

Turner, D. M. 1996. *Salvinorin: The psychedelic essence of* Salvia divinorum. San Francisco: Panther Press.

Valdes, Leander, William M. Butler, George M. Hatfield, Ara G. Paul, and Masato Koreeda. 1984. Divinorin A, a psychotropic terpenoid, and divinorin B from the hallucinogenic Mexican mint *Salvia divinorum*. *Journal of Organic Chemistry* 49 (24): 4716–20.

"Salvinorin A is the end of everything!"

ANDREW WEIL
(1/1995)

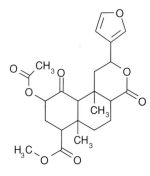

Salvinorin A

Salvinorin, the active constituent of the sage known as ska maría pastora (*Salvia divinorum*), has the reputation of being the most extreme of all psychedelics. (Book cover, 1996)

Scopolamine

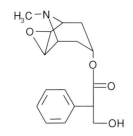

Scopolamine

Other Names

Hyoscin, (–)-hyoscin, hyoscine, hyoszin, L6(–),7-epoxytropin-tropate, *l*-hyoscine, scopolamin, [7(S)-(1α,2β,4β,5α,7β)]-α-(hydroxymethyl)benzene-acetic acid 9-methyl-3-oxa-9-azatricyclo-[3.3.1.0²,⁴] non-7-ylester, skopolamin, tropane acid ester of skopolin

Empirical formula: $C_{17}H_{21}ON_4$

Substance type: **tropane alkaloid**

Scopolamine was first isolated in 1888 by E. Schmidt from the roots of "*Scopolia atropoides*" (= **Scopolia carniolica**). It is very closely related to **atropine** and is a characteristic component of plants from the Nightshade Family (Solanaceae), especially the psychoactive species. For the pharmaceutical industry, the most important sources of scopolamine are the Australian duboisias (**Duboisia spp.**), the dried leaves of which can contain up to 7% alkaloids. Scopolamine is also produced by the recrystallization of hyoscyamine.

For medicinal purposes, scopolamine is administered at dosages of 0.5 to 1 mg, with a total daily maximum dosage of 3 mg. The lowest lethal dosage for humans is about 14 mg (Roth et al. 1994:921*).

Scopolamine is a very potent hallucinogen that

Plants Containing Scopolamine

(from Festi 1995*; Hagemann et al. 1992; Ripperger 1995; supplemented)

LORANTHACEAE	
Benthamia alyxifolia	leaves
SOLANACEAE	
Anthoceris ilicifolia Hook.	root
Atropa belladonna (L-scopolamine)	root
Atropanthe sinensis (Hemsl.) Pascher	fruits, root
Brugmansia (all species)	entire plant
Datura stramonium	entire plant
Datura spp.	entire plant
Duboisia hopwoodii	leaves
Duboisia spp.	leaves, bark
Hyoscyamus niger	entire plant
Hyoscyamus spp.	entire plant
Iochroma fuchsioides	leaves
Iochroma spp.	?
Latua pubiflora	entire plant
Lycium barbarum L. [syn. *Lycium halimifolium* Mill.]	entire plant
Mandragora officinarum	root
Mandragora chinghaiensis Kuang et Lu (cf. *Mandragora* spp.)	root
Scopolia carniolica (= *Scopolia atropoides*)	root
Solandra spp.	entire plant

Leuner (1981*) classified as a "Class II" hallucinogen because of its simultaneously hallucinogenic and narcotic/consciousness-clouding effects (cf. also Dittrich 1996*).

According to Hunnius, scopolamine is utilized in medicine as a hypnotic agent, especially for cases of "agitated states in the mentally ill, for Parkinson's and paralysis agitans, and for treating withdrawal of morphine users" (Hunnius 1975, 609*).

> In contrast to atropine, which initially stimulates the central nervous system, scopolamine induces predominantly a narcotic paralysis from the beginning, which is why it serves as a "chemical straitjacket" for agitated mental patients. Delirium and hallucinations are not infrequently seen . . . with therapeutic application. . . . Chronic scopolamine poisoning with gradually increasing dosages leads to psychoses with hallucinations. (Fühner 1943, 202 f.*)

In the former East Germany, scopolamine was still being used as a "chemical straightjacket" in the 1980s (Ludwig 1982, 148*; Schwarz 1984). Scopolamine may be combined with **morphine** for the same purpose (Römpp 1950, 264*). A combination of scopolamine hydrobromide and morphine hydrochloride is used as a preoperative anesthesia (cf. **soporific sponge**). Recent tests with mice found that scopolamine hydrobromide causes a marked increase in anxiety as compared to scopolamine methylbromide (Rodgers and Cole 1995).

To treat motion sickness—a use to which scopolamine has long been put (Römpp 1950, 265*)—an adhesive patch was developed that contains 1.5 mg scopolamine and can be adhered behind the ear as needed. The active component is absorbed through the skin into the blood vessels of the ear region and affects the organs of balance in the ear. This property of scopolamine supports the idea that the constituents found in **witches' ointments** could be absorbed when the mixtures were rubbed onto the skin.

Scopolamine was a popular inebriant in the Munich jazz scene of the 1950s. Because the dosages used were often so high, many of the concerts had to end early.

Commercial Forms and Regulations

The alkaloid is available as scopolamine hydrobromide and scopolamine hydrochloride. Pharmacies usually carry these substances in the form of small flasks for use in injections. Scopolamine requires a prescription.

Literature

See also the entries for *Atropa belladonna*, **cocaine**, and **tropane alkaloids**.

Flicker, C., M. Serby, and S. H. Ferris. 1990. Scopolamine effects on memory, language, visuospatial praxis and psychomotor speed. *Psychopharmacology* 100:243–50.

Hagemann, K., K. Piek, J. Stöckigt, and E. W. Weiler. 1992. Monoclonal antibody-based enzyme immunoessay for the quantitative determination of the tropane alkaloid, scopolamine. *Planta Medica* 58:68–72.

Heimann, Hans. 1952. *Die Skopolaminwirkung*. Basel and New York: S. Karger.

Keeler, M. H., and F. J. Kane. 1968. The use of hyoscyamine as a hallucinogen and intoxicant. *American Journal of Psychiatry* 124:852–54.

Ripperger, Helmut. 1995. (*S*)-scopolamine and (*S*)-norscopolamine from *Atropanthe sinensis. Planta Medica* 61:292–93.

Rodgers, R. J., and J. C. Cole. 1995. Effects of scopolamine and its quaternary analogue in the murine elevated plus-maze test of anxiety. *Behavioral Pharmacology* 6:283–89.

Schwarz, H.-D. 1984. Hyoscin (= Scolpolamin) statt Zwangsjacke. *Zeitschrift für Phytotherapie* 5 (3): 840–41.

Scopoletin

Other Names

Chrysatropasäure, gelseminsäure, scopoletina, scopolétine, 7-hydroxy-6-methoxycumarine, 6-methoxyumbelliferon, skopoletin, 3-methylesculetin

Empirical formula: $C_{10}H_{18}O_4$

Substance type: **coumarin**

The coumarin-derivative scopoletin was first isolated from the genus *Scopolia* and is named for the genus (Chaubal and Iyer 1977). Scopoletin is found in numerous plants that are utilized for medicinal or psychoactive purposes. It is the characteristic constituent in *Brunfelsia* spp. (Mors and Ribeiro 1957).

Scopoletin is known to inhibit plant growth. It may possibly have a certain psychoactive effect on humans, although there is no data at present to support this assertion. Scopoletin is a substance that clearly merits additional research.

Commercial Forms and Regulations

None

Literature

See also the entries for *Fabiana imbricata* and **coumarins**.

Chaubal, M., and R. P. Iyer. 1977. Carbon-13 NMR spectrum of scopoletin. *Lloydia* 40:618.

Mors, W. B., and O. Ribeiro. 1957. Occurrence of scopoletin in the genus *Brunfelsia. Journal of Organic Chemistry* 22:978–79.

Schilcher, H., and R. St. Effenberger. 1986. Scopoletin und β-Sitosterol—zwei geeignete Leitsubstanzen fur Urtica radix. *Deutsche Apotheker-Zeitung* 126:79–81.

Plants Containing Scopoletin

ACANTHACEAE
Justicia pectoralis

APOCYNACEAE
Nerium oleander L. (see **honey**)

CONVOLVULACEAE
Convolvulus scammonia L. (cf. **Convolvulus tricolor**)

LOGANIACEAE
Gelsemium sempervirens

ROSACEAE
Prunus serotina Ehrh. (see **kinnikinnick**)

RUTACEAE
Casimiroa edulis Llave ex Lex. (see *Lucuma salicifolia*)

SOLANACEAE
Atropa belladonna
Atropa spp. (see *Atropa belladonna*)
Brugmansia arborea
Brunfelsia brasiliensis (see *Brunfelsia* spp.)
Brunfelsia chiricaspi (see *Brunfelsia* spp.)
Brunfelsia grandiflora (see *Brunfelsia* spp.)
Brunfelsia pauciflora (see *Brunfelsia* spp.)
Fabiana imbricata
Mandragora officinarum
Markea formicarium Dammer (see **ayahuasca**)
Nicotiana tabacum
Scopolia carniolica
Scopolia spp. (see *Scopolia carniolica*)

URTICACEAE
Urtica dioica L.

Scopoletin

Strychnine

Strychnine

Poison nuts or crow's eyes (the seeds of *Strychnos nux-vomica*) contain high concentrations of strychnine. (Engraving from Pereira, *De Beginselen der Materia Medica en der Therapie*, 1849)

"The symptoms of strychnine are marked by spontaneous convulsions and states of nervousness in animals."

HANS HAAS AND HANS FRIEDRICH ZIPF

"ÜBER DIE ERREGENDE WIRKUNG VON BARBITURSÄUREABKÖMMLINGEN UND IHRE BEEINFLUSSUNG DURCH STRYCHNIN, PERVITIN UND CARDIAZOL" [ON THE STIMULATING EFFECT OF BARBITURIC ACID DERIVATIVES AND THE INFLUENCE OF STRYCHNINE, PERVITIN, AND CARDIAZOL UPON THEM]

(1949, 685)

Other Names

Estricnina, stricnina, strychnidin-10-on, strychnin, 2,4*a*,5,5*a*,8,15*a*,15*b*,15*c*-decahydro-4,6-methano-14*H*,16*H*-indolo[3,2,1,*ij*]oxepino-[2,3,4-de]-pyrrolo[2,3-*h*]chinolin-14-on

Empirical formula: $C_{21}H_{22}N_2O_2$

Substance type: **indole alkaloid**, strychnos alkaloid

Strychnine was first isolated in 1818 by Caventou and Pelletier from the Philippine Ignatius bean (*Strychnos ignatii* Berg.; cf. **Strychnos spp.**). Strychnine occurs in many *Strychnos* species (Loganiaceae); the primary sources are **Strychnos nux-vomica** and *Strychnos ignatii*. Contrary to widespread belief, the hairs of *Lophophora williamsii* do not contain any strychnine!

Strychnine is an analeptic, a substance that at low dosages activates certain parts of the central nervous system and in higher dosages acts as a convulsive poison:

> Milligram dosages of strychnine nitrate administered internally or subcutaneously cause an increased sensitivity of the senses (the feeling that vision, hearing, taste, smell are more acute) and faster reflex response. (Fühner 1943, 221*)

Strychnine docks to the glycine receptor. At lower dosages it is clearly psychoactive, in a manner very similar to **yohimbine**. The therapeutic dosage for tonic purposes is listed as 1 to 3 mg; a dosage of 5 mg produces aphrodisiac and psychoactive effects. However, 10 mg can cause convulsions, and dosages above 30 mg can lead to difficulty in breathing and severe anxiety (Neuwinger 1994, 527*). From 100 to 300 mg is normally regarded as a lethal dosage for adults, while dosages as low as 1 to 5 mg can prove fatal to small children (Roth et al. 1994, 935*). Strychnine is an extremely stable molecule and could still be detected in corpses that were exhumed as much as four years after burial (Roth et al. 1994, 935*). **Diazepam** is recommended as an antidote in cases of strychnine poisoning or overdose (Moeschlin 1980). Kavapyrones and kava-kava can also be used as antidotes for strychnine poisoning (cf. *Piper methysticum*).

Strychnine is also an effective aphrodisiac, but the dosage must be very precise:

> The literature contains numerous references to the stimulating effects of strychnine on the sexual apparatus. Many experienced immedi-ate erections. But the extraordinary toxicity of the substance makes it an especially dangerous aphrodisiac. Strychnine has for this reason always played a dangerous role in criminality in this regard as well. (Hirschfeld and Linsert 1930, 210*)

A very effective recipe for a "firm erection" can be prepared with strychnine and other substances (from Gotttlieb 1974, 81*):

5 mg	**yohimbine** hydrochloride
5 mg	methyltestosterone
25 mg	pemoline
2 mg	strychnine sulfate

Strychnine is said to have been the favorite drug of Adolf Hitler, who also appears to have used cocaine (Schmidbauer and vom Scheidt 1984, 260*):

> Moreover, we will never know if and how Hitler's strategy and war leadership might have changed if he had not been making his decisions while in a euphoric trance state induced by high dosages of strychnine. . . . (Irving 1980, 135)

Strychnine has also had an impact on sports because of its prominent role as a doping agent (Schmidbauer and vom Scheidt 1984, 289*).

Strychnine is a popular rat poison and is still used for this purpose today. In the United States, the members of some rather extreme Christian sects drink such rat poisons as an ordeal and an inebriant during their worship services. It is said that the Holy Spirit will protect the true believers from dying from the poison. Surprisingly, these cults have not yet become extinct.

Commercial Forms and Regulations

The substance is available in the form of a base and as strychnine hydrochloride, strychnine nitrate, strychnine phosphate, and strychnine sulfate. All forms of the substance are subject to the regulations concerning dangerous substances. Strychnine is listed in Class 1 of the Swiss Poison List. In principle, however, strychnine is legal.

Literature

See also the entries for *Strychnos nux-vomica* and *Strychnos* spp.

Haas, Hans, and Hans Friedrich Zipf. 1949. Über die erregende Wirkung von Barbitursäureabkömmlingen und ihre Beeinflussung durch Strychnin, Pervitin und Cardiazol. *Archiv für experimentelle Pathologie und Pharmakologie* 206 (5/6): 683–97.

Irving, David. 1980. *Wie krank war Hitler wirklich?* Munich: Wilhelm Heyne Verlag.

Moeschlin, S. 1980. *Klinik und Therapie der Vergiftung.* 6th ed. Stuttgart: Thieme.

Seeger, R., and H. G. Neumann. 1986. Strychnin/Brucin. *Deutsche Apotheker Zeitung* 126 (26): 1386–88.

THC

Other Names

Δ^9-tetrahydrocannabinol, Δ^9-THC, delta-9-THC, Δ^1-3,4-*trans*-tetrahydrocannabinol, tetrahydro-6,6,9-trimethyl-3-pentyl-*6H*-di-benzo[*b,d*]pyran-1-ol, *trans*-THC

Empirical formula: $C_{21}H_{30}O_2$

Substance type: cannabinoid, pyrane derivative, pyranol derivative

THC is the main active constituent of the three hemp species *Cannabis indica*, *Cannabis ruderalis*, and *Cannabis sativa*. THC has not yet been found in any other plants. The information suggesting that THC is pyrochemically synthesized when olibanum (the resin of *Boswellia sacra*) is burned is contradictory. Similarly, no trace of THC or its analogs has yet been found in hops (*Humulus lupulus*). THC and its metabolites have been found in Egyptian mummies (Balabanova et al. 1992*).

While *trans*-THC is psychoactive, the isomer *cis*-THC is not (Kempfert 1977):

> The effective dosage of THC when smoked is between 2 and 22 mg and when taken orally is between 20 and 90 mg. When smoked under normal conditions, 16 to 19% of the THC is consumed and the rest is pyrolized. No lethal dosage is known. However, experiments with animals indicate that the ratio between an effective and a lethal dosage can be estimated to be 4,000 to 40,000. In comparison, this ratio for alcohol is 4 to 10. (Fromberg 1996, 37)

In the blood, THC is transformed into the active metabolite 11-hydroxy-Δ^9-THC. This substance is absorbed by fatty tissues after about thirty minutes and is then released back into the blood, metabolized, and excreted. After only a few days, all of the substance has been excreted by the body. With chronic use, 11-hydroxy-THC accumulates in the fatty tissues and in the liver and can be detected for a longer period of time (urine tests!; cf. Rippchen 1996).

THC receptors have been discovered both in the central nervous system and in the peripheral pathways (Compton 1993; Devane et al. 1989;

Matsuda et al. 1990). The THC or cannabinoid receptor in the nervous system has now been studied extensively and is very well understood (Pertwee 1995). Normally, endogenous neurotransmitters known as anandamides bind to these receptors (Devane et al. 1992; Devane and Axelrod 1994; Kruszka and Gross 1994). Nerve diseases (such as multiple sclerosis) can result if the body does not produce sufficient amounts of anandamides. If anandamide deficiencies are responsible for these diseases, it is possible that they could be successfully treated with THC (Mechoulam et al. 1994).

Anandamide (= arachidonylethanolamide)—the name is derived from the Sanskrit word *ananda*, "bliss"—binds to THC receptors in the brain and is the endogenous THC analog, even though the inner structures of the two are quite different. Recently, anandamide has been discovered in chocolate and cocoa beans (*Theobroma cacao*) as well as in red wine (cf. *Vitis vinifera*) (Grotenhermen 1996).

Since 1971, cannabis products have been tested experimentally as medicines for treating alcoholism, heroin and amphetamine addiction, emotional disturbances, muscle spasms, and glaucoma. In 1990, the microbiologist Gerald Lancs of the University of South Florida discovered that marijuana kills the herpes virus (AFP announcement on May 16, 1990), providing scientific validation of an old Roman remedy for herpes. The traditional use of hemp products for asthma has also received scientific support: "THC dilates the bronchial passages. Like other medicines, it can be inhaled as an aerosol to treat bronchial asthma and produces equally positive effects" (Maurer 1989, 48).

The medicinal use of THC and its analogs for the treatment of glaucoma has become an established practice. No other substance has been demonstrated to be better tolerated or more effective than THC (Maurer 1989). A Swiss group of researchers was able to show that THC relaxes the muscular cramping associated with central nervous system spasticity (e.g., due to multiple sclerosis or spinal cord injury) (Maurer et al. 1990). The researchers found that THC (at a dosage of 5 mg) produces effects that are similar to

THC

"While cannabis and chocolate do not contain the same substances, they do produce similar effects. Moreover, the body itself also produces the substance in chocolate. These substances were named anandamides, after the Sanskrit word for bliss."

Franjo Grotenhermen "Schokolade, Haschisch und Anandamide" [Chocolate, Hashish, and Anandamide] (1996, 14)

THC is the main active constituent in the resin produced by the hemp plant. (Engraving from Pereira, *De Beginselen der Materia Medica en der Therapie*, 1849)

"While a person who smokes cannabis can at most smoke [himself or herself] to sleep and an overdose is not possible, a handful of THC capsules is sufficient to induce unconsciousness in a person for a long time. In addition to this flaw, the occasional indications of psychosis as well as such side effects as sleep disorders that last for weeks, irritability, and diarrhea speak against the use of the synthetic substance. It appears as though the gentle drug has been turned into a bitter pill in the laboratory."

Sebastian Schmidt
"Die THC-Pille auf Rezept"
[The THC Pill on Prescription]
(1996, 31)

The cocoa tree produces anandamide (or a precursor) in its fruits. Anandamide is the substance that binds to the THC or cannabinoid receptors in the human brain. (Engraving from Peireira, *De Beginselen der Materia Medica en der Therapie*, 1849)

those of **codeine** but more effective and that THC is also more easily tolerated. There have also been encouraging attempts to utilize THC in the clinical treatment of spasticity and the associated pain (Hagenbach 1996).

> The potential applications [of synthetic THC] range from the treatment of epilepsy, chronic pain, multiple sclerosis, and lack of appetite to a reduction in the "addictive pressure" associated with opiate addiction. (Schmidt 1996, 30)

Synthetic THC is better known by the trade name Marinol. A dosage of 20 to 45 mg of Marinol induces a "high" that lasts for only sixty to ninety minutes. Many patients in the United States who take Marinol complain that the expensive medicine is ineffective compared to marijuana when either smoked or eaten (Jack Herer, pers. comm.).

Pharmacological research is now under way to develop synthetic THC analogs that could be marketed as medicines. The goal is to isolate the medically useful properties of THC while removing the psychoactive ones (Evans 1991). One of the products that has been synthesized as a result of this research is the cannabinoid analog HU-210, chemically known as $(-)$1l-OH-Δ^8-THC-dimethyl-heptyl. This substance not only is psychoactive but is some one hundred to eight hundred times more potent than natural THC (Ovadia et al. 1995). However, government health departments and pharmaceutical companies are more interested in THC analogs that are devoid of psychoactive effects. Some critics of this research take a different position, arguing that the therapeutic effects of THC are directly related to its psychoactivity.

Commercial Forms and Regulations

In principle, THC is an illegal substance throughout the world (cf. *Cannabis indica*). However, for the past several years certain prescription drugs containing THC have been available in the United States under the trade names Canasol and Marinol. Physicians may prescribe these for glaucoma and cancer patients. In Europe, these drugs can be obtained only from pharmacies that sell foreign medicines, and they are extremely expensive. Recently, there have been efforts in several states in the United States as well as in several European nations to make THC and/or *Cannabis* products more readily available to patients suffering from a variety of conditions. There is, however, considerable resistance to such liberalization efforts. In spite of the very long history of use of THC and *Cannabis* in numerous cultures and for a wide variety of purposes (see Rätsch 2001*), it remains to be seen whether these substances will ever become widely accepted and legitimately used.

Literature
See also the entries for *Cannabis indica* and *Cannabis sativa*.

Compton, David R., Kenner C. Rice, Brian R. de Costa, Raj K. Razdan, Lawrence S. Melvin, M. Ross Johnson, and Billy R. Martin. 1993. Cannabinoid structure-activity relationships: Correlation of receptor binding and *in vivo* activities. *The Journal of Pharmacology and Experimental Therapeutics* 265:218–26.

Devane, William A., and Julius Axelrod. 1994. Enzymatic synthesis of anandamide, an endogenous ligand for the cannabinoid receptor, by brain membranes. *Proceedings of the National Academy of Science, USA* 91:6698–701.

Devane, William A., Francis A. Dysarz III, M. Ross Johnson, Lawrence S. Melvin, and Alynn C. Howlett. 1988. Determination and characterization of a cannabinoid receptor in rat brain. *Molecular Pharmacology* 34:605–13.

Devane, William A., Lumir Hanus, Aviva Breuer, Roger G. Pertwee, Lesley A. Stevenson, Graeme Griffin, Dan Gibson, Asher Mandelbaum, Alexander Etinger, and Raphael Mechoulam. 1992. Isolation and structure of a brain constituent that binds to the cannabinoid receptor. *Science* 258:1946–49.

Evans, Fred J. 1991. Cannabinoids: The separation of central from peripheral effects on a structural basis. *Planta Medica* 57 suppl. (1): 60–67.

Fromberg, Erik. 1996. Die Pharmakologie von Cannabis. In *Cannabis*, ed. Jürgen Neumeyer, 36–42. [Munich]: Packeispresse Verlag Hans Schickert.

Grotenhermen, Franjo. 1996. Schokolade, Haschisch und Anandamide. *Hanf!* 12/96:14–15.

Hagenbach, Ulrike. 1996. Spinale Spastik und Spasmolyse: Ist die Therapie mit THC eine unerwartete Bereicherung? In *Jahrbuch des Europäischen Collegiums für Bewußtseinsstudien* (1995), 199–207. Berlin: VWB.

Iversen, Leslie L. 1993. Medical uses of marijuana? *Nature* 365:12–13.

Kettenes-van den Bosch, J. J., and C. A. Salemink. 1980. Biological activity of the tetrahydrocannabinols. *Journal of Ethnopharmacology* 2:197–231. (Very good bibliography.)

Kruszka, Kelly K., and Richard W. Gross. 1994. The ATP- and coA-independent synthesis of arachidonoylethanolamide: A novel mechanism underlying the synthesis of the endogenous ligand of the cannabinoid receptor. *The Journal of Biological Chemistry* 269 (20): 14345–48.

Matsuda, Lisa A., Stephen J. Lolait, Michael J. Brownstein, Alice C. Young, and Tom I. Bonner. 1991. Structure of a cannabinoid receptor and functional expression of the cloned cDNA. *Nature* 346:561–64.

Maurer, Maja. 1989. Therapeutische Aspekte von Cannabis in der westlichen Medizin. In *3. Symposion über psychoaktive Substanzen und veränderte Bewußtseinszustände in Forschung und Therapie*, ed. M. Schlichting and H. Leuner, 46–49. Göttingen: ECBS.

Maurer, M., V. Henn, A. Dittrich, and A. Hofmann. 1990. Delta-9-tetrahydrocannabinol shows antispastic and analgesic effects in a single case double-blind trial. *European Archives of Psychiatry and Clinical Neuroscience* 240:1–4.

Mechoulam, Raphael, Zvi Vogel, and Jacob Barg. 1994. CNS cannabinoid receptors: Role and therapeutic implications for CNS disorders. *CNS Drugs* 2 (4): 255–60.

Mestel, Rosie. 1993. Cannabis: The brain's other supplier. *New Scientist* 7/93:21–23.

Ovadia, H., A. Wohlman, R. Mechoulam, and J. Weidenfeld. 1995. Characterization of the hypothermic effect of the synthetic cannabinoid HU-210 in the rat: Relation to the adrenergic system and endogenous pyrogens. *Neuropharmacology* 34 (2): 175–80.

Pertwee, Roger, ed. 1995. *Cannabinoid receptors.* New York: Harcourt Brace Jovanovich.

Rippchen, Ronald, ed. [1996]. *Mein Urin gehört mir.* Lörbach: Edition Rauschkunde.

Schmidt, Sebastian. 1996. Die THC-Pille auf Rezept. *Hanfblatt* 3 (20): 30–31.

Smith, R. Martin, and Kenneth D. Kempfert. 1977. Δ^1-3,4-*cis*-tetrahydrocannabinol in *Cannabis sativa. Phytochemistry* 16:1088–89.

Zeeuw, Rokus A. de, and Jaap Wijsbeek. 1972. Cannabinoids with a propyl side chain in cannabis: Occurrence and chromatographic behavior. *Science* 175:778–79

Tropane Alkaloids

Other Names

Tropanalkaloide, tropanes, tropeine

Tropane alkaloids are esters of tropanal combined with various acids. They occur primarily in nightshades (Solanaceae), especially the psychoactive ones. The most important psychoactive tropane alkaloids are **atropine, scopolamine**, and hyoscyamine. These substances are "quickly absorbed through the mucous membranes but also through the intact skin" (Roth et al. 1994, 944*). For this reason, plant preparations in the form of ointments with these tropane alkaloids can induce psychoactive effects (cf. **Datura innoxia**, witches' ointments). Atropine, scopolamine, and hyoscyamine are found in the genera *Atropa, Brugmansia, Datura, Hyoscyamus, Iochroma, Juanulloa, Mandragora, Solandra*, and *Scopolia*.

The psychoactive tropane alkaloid hyoscyamine (cf. *Hyoscyamus niger*) occurs in the following nightshades in concentrations that appear to make them useful for psychoactive purposes (Festi 1995, 132 f.*): *Anthoceris littorea* Labill. (herbage), *Crenedium spinescens* Haegi (leaves), *Cyphanthera anthocercidea* (F.v. Muell.) Haegi (leaves), *Mandragora caulescens* C.B. Clarke (entire plant; cf. **Mandragora spp.**), *Physochlaina praealta* (Decne.) Miers (entire plant), and *Scopolia lurida* Dunal (roots; cf. **Scopolia carniolica**). As a plant dries, the hyoscyamine it contains is usually transformed into its analog **scopolamine**. The profile of effects of hyoscyamine is essentially the same as that of scopolamine.

Tropanes and **cocaine** are chemically related and can under certain conditions elicit similar pharmacological effects (Sauerwein et al. 1993). The tropane alkaloid 2-tropanone is a metabolic product of cocaine. Tropane alkaloids occur in most if not all *Erythroxylum* species (Al-Said et al. 1989). The bark of *Erythroxylum zambesiacum* N. Robson has been found to contain various tropanes (Christen et al. 1993). The root bark of *Erythroxylum hypericifolium* Lam., a species indigenous to Mauritius that is used in folk medicine to treat kidney problems, contains large amounts of hygrine as well as other tropanes (e.g., cuscohygrine) (Al-Said et al. 1989). Both hygrine and cuscohygrine are also found in the leaves and bark of the two coca species **Erythroxylum coca** and **Erythroxylum novogranatense** (Al-Said et al. 1989, 672). The leaves of the Southeast Asian species *Erythroxylum cuneatum* (Wall.) Kurz, which is used in Malaysia as a tonic, were found to contain

The tropane alkaloids are typical constituents of almost all members of the Nightshade Family (Solanaceae). (A *Lycianthes* species from South America)

Tropane skeleton

"But how does it happen that people ill with fevers or who are destroying their nervous system with narcotics—in other words, making it more sensitive— 'imagine' that they see gruesome faces when their consciousness is dulled? Why do mice, after having been injected with hyoscyamine, raise themselves up on their hind legs—something they would normally never do when they see an enemy—and with gestures of utmost terror betray to us that they are perceiving something that is hidden from our senses?"

GUSTAV MEYRINK
SÜDSEE-MASKEN/DAS HAUS DER LETZTEN LATERN 1 [SOUTH SEA MASKS/THE HOUSE OF THE LAST LANTERN 1] (REISSUE 1973)

as their primary alkaloid (±)-3α,6β-dibenzoyloxytropane; another major constituent in the leaves is **nicotine**. The main alkaloid in the leaves of another ethnomedicinally useful Southeast Asian species, *Erythroxylum ecarinatum* Burck., is tropacacaine. The root bark of the Australian species *Erythroxylum australe* F.v. Muell. also contain numerous tropanes (meteloidine) (El-Imam et al. 1988).

Tropane alkaloids also appear to be present in the Proteaceae Family, e.g., in the species *Knightia strobolina* (El-Imam et al. 1988:2182). In Australia, several members of the genera *Hakea* and *Banksia* are used to produce **wine**.

The recent discovery of tropane alkaloids (tropine, tropinone, cuscohygrine, hygrine) in field bindweed (*Convolvulus arvensis* L.; cf. **Convolvulus tricolor**) is very interesting; the species also contains **ergot alkaloids** (Todd et al. 1995). Tropane alkaloids have also been found in the hedge bindweed *Calystegia sepium* (L.) R. Br. [syn. *Convolvulus sepium*] (Goldmann et al. 1990).

Literature

See also the entries for **atropine** and **scopolamine**.

Bauer, Eduard. 1919. *Studium über die Bedeutung der Alkaloide in pharmakognostisch wichtigen Solanaceen, besonders in Atropa Belladonna und Datura Stramonium.* Bern: Hallwag.

Christen, P., M. F. Roberts, J. D. Phillipson, and W. C. Evans. 1993. Recent aspects of tropane alkaloid biosynthesis in *Erythroxylum zambesiacum* stem bark. *Planta Medica* 59 suppl.: A583–84.

Goldmann, Arlette, Marie-Louise Milat, Paul-Henri Ducrot, Jean-Yves Lallemand, Monique Maille, Andree Lepingle, Isabelle Charpin, and David Tepfer. 1990. Tropane derivates from *Calistegia sepium. Phytochemistry* 29 (7): 2125–27.

Imam, Yahia M. A. el-, William C. Evans, and Raymond J. Grout. 1988. Alkaloids of *Erythroxylum cuneatum, E. ecarinatum* and *E. australe. Phytochemistry* 27 (7): 2181–84.

Said, Mansour S. al-, William C. Evans, and Raymond J. Grout. 1989. Alkaloids of *Erythroxylum hypericifolium* stem bark. *Phytochemistry* 28 (2): 671–73.

Sauerwein, M., F. Sporer, and M. Wink. 1993. Allelochemical properties of derivatives from tropane and ecgonine. *Planta Medica* 59 suppl.: A662

Todd, G. Fred, F. R. Stermitz, P. Schultheiss, A. P. Traub-Dargatz, and J. Traub-Dargatz. 1995. Tropane alkaloids and toxicity of *Convolvulus arvensis. Phytochemistry* 39:301–3.

Xiao, P., and L. Y. He. 1983. Ethnopharmacologic investigation on tropane-containing drugs in Chinese Solanaceous plants. *Journal of Ethnopharmacology* 8:1–18.

Withanolides

Other Names

None

Withanolides are not alkaloids but C_{28} steroidal lactones. More than one hundred withanolides have been isolated and described to date (Christen 1989). They occur only (or primarily) in members of the Nightshade Family (Solanaceae) (Christen 1989; Evans et al. 1984; Lavie 1986).

The withanolides withaferine A and withanolide E have very interesting biological and pharmacological effects: they are antinflammatory, stimulate the immune system, and inhibit the formation of tumors (Christen 1989). Despite the fact that many plants with psychoactive properties contain only or primarily withanolides, not a single psychoactive constituent from this group has been isolated or described to date.

Literature

See also the entries for **Withania somnifera.**

Buddhiraja, R. D., and S. Sudhir. 1987. Review of biological activity of withanolides. *Journal of Scientific and Industrial Research* 46:488–91.

Christen, P. 1989. Withanolide: Naturstoffe mit vielversprechendem Wirkungsspektrum. *Pharmazie in unserer Zeit* 18 (5): 129–39.

Evans, William C., Raymond J. Grout, and Merlin L. K. Mensah. 1984. Withanolides of *Datura* spp. and hybrids. *Phytochemistry* 23 (8): 1717–20.

Lavie, David. 1986. The withanolides as a model in plant genetics: Chemistry, biosynthesis, and distribution. In *Solanaceae: Biology and systematics,* ed. William G. D'Arcy, 187–200. New York: Columbia University Press

Top: The shoo-fly or apple-of-Peru (*Nicandra physalodes*; Solananceae), a plant from Peru, contains primarily withanolides and produces an effect similar to that of hyoscyamine. This may be the reason behind the French name for the plant: *belladonne de pays*, "belladonna of the countryside."

Bottom: The central European hedge bindweed (*Calystegia sepium*) contains psychoactive tropane alkaloids. (Photographed in Schönbrühl, near Bern, Switzerland)

Plants Containing Withanolides

Acnistus arborescens (L.) Schlechtend.	withaferine A
Acnistus (*Dunalia*) spp.	acnistines
Datura metel	daturiline
Datura quercifolia H.B.K. (cf. **Datura** spp.)	withaferoxolide
Datura stramonium var. *violacea*	withaferoxolide
Datura stramonium ssp. *ferox*	withaferoxolide
Datura ferox x *D. quercifolia* F_1 hybrid	withaferoxolide
Dunalia australis Griseb.	dunawithanine A and B
Iochroma coccineum Scheidw. (cf. **Iochroma fuchsioides**)	
Jaborosa spp.	jaborosa lactones, jaborosa latoles
Lycium spp.	withanolides
Nicandra physalodes (L.) Gaertn.	nicandrenone
Nicandra spp.	
Physalis ixocarpa Brot. ex Hornem. [syn. *Physalis edulis* hort. non Sims]	ixocarpalactones
Physalis peruviana L. [syn. *Physalis edulis* Sims]	withanolides, withaperuvines, perulactones, physalolactone B-3-*O*-glucoside
Physalis peruviana var. *varanasi*	perulactone
Physalis spp.	withaphysalines, physalines, ixocarpalactones, physalolactones, withaperuvines
Trechonaetes laciniata Miers.	trechonolide A
Trechonaetes sativa Miers.	trechonolides
Trechonaetes spp.	trechonolides
Withania frutescens Pauq.	withanolides
Withania somnifera	withaferine A, withanolide
Withania spp.	withaferines
Witheringia spp.	

Yohimbine

Yohimbine

Yohimbine is a characteristic constituent in the bark of many *Alstonia* species, some of which have been traditionally utilized as aphrodisiacs. (*Alstonia macrophylla*, photographed in Hawaii)

505 These results have occasionally been called into question: "Nevertheless, in animal studies the author was able to demonstrate an increase in frequency of copulation and in blood flow through the penis after premedication with yohimbine" (Weyers 1982, 64).

Other Names

Aphrodin, corymbin, corynin, hydroergotocin, johimbin, quebrachin, quebrachina, yohimbenin, yohimbin, yohimbina, yohimbinum, yohimvetol

Empirical formula: $C_{21}H_{26}N_2O_3$

Substance type: aspidosperma alkaloid, **indole alkaloid**

Yohimbine was first extracted from the bark of *Pausinystalia yohimba* and described in the nineteenth century. It is a typical alkaloid in plants from the Apocynaceae Family and is related to the Rauvolfia alkaloids, and it constitutes the primary alkaloid (1%) in *Alstonia angustifolia*. It is also present in some species of *Rauvolfia*, especially the African species *Rauvolfia macrophylla* Stapf (Timmins and Court 1974).

Yohimbine was once regarded as an MAO inhibitor, a view that is no longer considered accurate. Rather, it is simply an α-adrenergic blocker that consequently stimulates the release of noradrenaline at the nerve endings. This makes noradrenaline available in the corpus cavernosum and results in an erection (Roth et al. 1994, 955*; Wren 1988, 292*).

> As a sympatholytic agent, [yohimbine] dilates the peripheral blood vessels and reduces blood pressure. The aphrodisiac effect is explained through a vasodilatation of the genital organs and an increased excitability of the reflexes in the sacral medulla. (Roth et al. 1994, 545*)

Yohimbine's aphrodisiac and virility-enhancing effects, and its therapeutic efficaciousness in treating impotence, have been demonstrated in a number of clinical double-blind studies (Buffum 1982; Miller 1968; Sobotka 1969).[505]

Consequently, yohimbine hydrochloride has been approved as a specific medicine for the treatment of impotence (sexual neurasthenia). The recommended dosage is 5 to 10 mg taken three times daily as a short-term treatment over three to four weeks. Higher individual dosages (15 to 25 mg) result in psychoactive effects that are somewhat reminiscent of those of LSD, but with much less emotional content and an emphasis on physical phenomena (sexual desire, erotic enjoyment, and increased sensations of pleasure). Overdoses can be unpleasant but do not appear to be particularly dangerous (cf. Lewin 1992, 750*):

> A chemist had taken an almost 1000-fold dosage (1.8 g). He became unconscious for a few hours (during which time a pronounced priapism was observed) but was able to be discharged from the hospital within a day. (Roth et al. 1994, 956*)

Plants Containing Yohimbine

(from Geschwinde 1996, 145 f.*; Hofmann 1954; Lewin 1992*; Römpp 1995, 5093*; Roth et al. 1994*; supplemented)

Stock Plant	Distribution
Alstonia spp.	
Alstonia angustifolia	Old World
Alstonia scholaris	Southeast Asia
Aspidosperma quebracho-blanco	South America
Catharanthus lanceus	North America
Corynanthe spp.	Africa
Mitragyna stipulosa (cf. **palm wine**)	Africa
Pausinystalia yohimba	West Africa
Pausinystalia macroceras	
Pausinystalia trillesii	
Rauvolfia spp.	
Rauvolfia macrophylla Stapf	Africa
Rauvolfia serpentina Benth.	
Rauvolfia volkensii	Africa
Vinca spp. (cf. **Catharanthus roseus**)	Africa

Commercial Forms and Regulations

The alkaloid is available as yohimbine hydrochloride. Yohimbine is a prescription medication.

Literature

See also the entries for *Alstonia scholaris*, *Corynanthe* spp., and *Pausinystalia yohimba*.

Buffum, John. 1982. Pharmacosexology: The effects of drugs on sexual function—a review. *Journal of Psychoactive Drugs* 14 (1–2): 5–44.

Finch, N., and W. I. Taylor. 1962. Oxidative transformation of indole alkaloids. 1: Preparation of oxindoles from yohimbine. *Journal of the American Chemical Society* 84:3871–77.

Hofmann, Albert. 1954. Die Isolierung weiterer Alkaloide aus *Rauwolfia serpentina* Benth. *Helvetica Chimica Acta* 37:849–65.

Lambert, G. A., W. J. Lang, E. Friedman, E. Meller, and S. Gershon. 1978. Pharmacological and biological properties of isomeric yohimbine alkaloids. *European Journal of Pharmacology* 49:39–48.

Leary, Timothy. 1985. Auf der Suche nach dem wahren Aphrodisiakum und elektronischer Sex. *Sphinx Magazin* 35.

Miller, W. W. 1968. Afrodex in the treatment of male impotence: A double-blind cross-over study. *Current Therapeutic Research* 10:354–59.

Poisson, J. 1964. Recherches récentes sur les alcaloïdes du pseudocinchona et du yohimbine. *Ann. Chim.* 9:99–121.

Porst, H. 1996. Orale und intracavernöse Pharmakotherapie. *TW Urologie Nephrologie* 8 (2): 88–94.

Sobotka, J. J. 1969. An evaluation of Afrodex in the management of male impotency: A double-blind cross-over study. *Current Therapeutic Research* 11:87–94.

Timmins, Peter, and William E. Court. 1974. Alkaloids of *Rauwolfia macrophylla*. *Phytochemistry* 13:281–82.

Weyers, Wolfgang. 1982. *Die Empfehlung in der Selbstmedikation*. Heusenstamm: Keppler Verlag.

"Yohimbe acts both as a central nervous-system stimulant and as a mild hallucinogen.... The first effects are a lethargic weakness of the limbs and a vague restlessness, similar to the initial effects of LSD. Chills and warm spinal shivers may also be felt, along with slight dizziness and nausea.... Then the effects produce a relaxed, somewhat inebriated mental and physical feeling accompanied by slight auditory/visual hallucinations. Spinal ganglia are then affected, causing erection of the sex organs. These effects last from two to four hours."

RICHARD ALAN MILLER
THE MAGICAL AND RITUAL USE OF APHRODISIACS
(1985, 116–17*)

Botanical Taxonomy of Psychoactive Plants and Fungi

The following list contains all of the plants and fungi that are discussed in this encyclopedia in both the major and the minor monographs. The psychoactive properties of some of the species are uncertain or only purported and may even have been incorrectly described; these species are nevertheless included here for the sake of comprehensiveness (and further research can often bring astonishing discoveries).

This taxonomy follows that found in Dörfelt 1989,* Frohne and Jensen 1992,* Roth et al. 1990,** and Zander 1994.* It is based on the taxonomic order established by the late botanist Richard Evans Schultes of Harvard.

Mushrooms—Fungi

Class Deuteromycetes—conidial fungi
  mold fungi
    Aspergillus fumigatus

Class Ascomycetes—ascomycota
Order Clavicipitales
  Balansiae
    Balansia cyperi
  Hypocreaceae (= Clavicipitaceae)
    Claviceps paspali
    Claviceps purpurea
    Claviceps spp.
    Cordyceps spp.

Class Basidiomycetes
Order Lycoperdales
  Lycoperdaceae
    Lycoperdon marginatum
    Lycoperdon mixtecorum
    Scleroderma verrucosum
    Vascellum pratense
Order Phallales
  Phallaceae
    Dictyophora spp.
Order Russulales
  Russulaceae
    Russula spp.
Order Agaricales
  Amanitaceae
    Amanita muscaria
    Amanita pantherina
    Amanita spp.
  Bolbitiaceae
    Agrocybe farinacea
    Conocybe cyanopus
    Conocybe siligineoides
    Conocybe spp.
    Pholiotina spp.

Boletaceae
  Boletus flammeus
  Boletus kumaeus
  Boletus manicus
  Boletus nigroviolaceus
  Boletus reayi
  Heimiella anguiformis
  Heimiella retispora
Coprinaceae
  Copelandia spp.
  Naematoloma popperianum
  Panaeolina spp.
  Panaeolus (Copelandia) cyanescens
  Panaeolus subbalteatus
  Panaeolus spp.
  Psathyrella candolleana
Cortinariaceae
  Galerina steglichii
  Gymnopilus spectabilis (= *Pholiota spectabilis*)
  Gymnopilus spp.
  Inocybe aeruginascens
  Inocybe spp.
Hygrophoraceae
  Hygrocybe psittacina
Pluteaceae
  Pluteus spp.
Strophariaceae
  Psilocybe azurescens
  Psilocybe (Stropharia) cubensis
  Psilocybe cyanescens
  Psilocybe mexicana
  Psilocybe semilanceata
  Psilocybe spp.
  Stropharia coronilla
  Stropharia spp.
Tricholomataceae
  Gerronema spp.
  Mycena spp.
Order Polyporales/Poriales
  Polyporaceae/Ganodermataceae
    Ganoderma sp.
    Laetiporus sulphureus
    Polyporus spp.
Order Cantharellales
  Hydnaceae
    Hydnum repandum

Lichens
Basidiolichenes
  Dictyonemataceae
    Dictyonema sp.
  [not yet assigned to a family]
    Lichen non ident.

Pteridophyta

Class Lycopsida
Lycopodiaceae
Huperzia selago
Lycopodium clavatum
Lycopodium spp.

Class Filicopsida—Ferns
Polypodiaceae
Polypodium spp.

Class Equisetatae
Equisetaceae
Equisetum arvense

Gymnospermae

Class Cycadopsida
Cycadaceae
Dioon edule

Class Coniferopsida
Cupressaceae
Juniperus recurva
Juniperus spp.

Class Taxopsida
Taxaceae
Taxus baccata
Taxus spp.

Class Chlamydospermae (= Gnetopsida)
Ephedraceae
Ephedra alata
Ephedra altissima
Ephedra americana
Ephedra andina
Ephedra breana
Ephedra campylopoda
Ephedra distachya
Ephedra equisetina
Ephedra gerardiana
Ephedra helvetica
Ephedra intermedia
Ephedra major
Ephedra nevadensis
Ephedra sinica
Ephedra torreyana
Ephedra trifurca
Ephedra spp.

Angiospermae—Flowering Plants

Class Dicotyledoneae
Subclass Archichlamydeae
Fagaceae
Castanopsis acuminatissima
Moraceae
Brosimum acutifolium ssp. *obovatum*

Cecropia mexicana
Cecropia spp.
Helicostylis tomentosa
Maquira sclerophylla
Cannabaceae
Cannabis indica
Cannabis ruderalis
Cannabis sativa
Humulus lupulus
Olacaceae
Liriosma ovata
Ptychopetalum olacoides
Santalaceae
Comandra pallida
Santalum murrayanum
Metteniusaceae
Metteniusa edulis
Loranthaceae
Benthamia alyxifolia
Phrygilanthus eugenioides
Phytolaccaceae
Phytolacca acinosa
Nyctaginaceae (= Nictaginaceae)
Mirabilis multiflora
Aizoaceae
Mesembryanthemum spp.
Sceletium expansum
Sceletium tortuosum
Sceletium spp.
Amaranthaceae
Amaranthus caudatus
Amaranthus spinosus
Amaranthus spp.
Iresine spp.
Cactaceae
Ariocarpus fissuratus
Ariocarpus retusus
Ariocarpus spp.
Armatocereus laetus
Carnegia gigantea
Coryphantha compacta
Coryphantha macromeris
Coryphantha macromeris var. *runyonii*
Coryphantha palmeri
Coryphantha spp.
Echinocereus triglochidiatus
Echinopsis spp.
Epithelantha micromeris
Lophophora diffusa
Lophophora echinata
Lophophora williamsii
Mammillaria craigii
Mammillaria grahamii var. *oliviae*
Mammillaria spp.
Neoraimondia arequipensis
Pachycereus pecten-aboriginum
Pelecyphora aselliformis
Trichocereus atacamensis
Trichocereus bridgesii
Trichocereus cuscoensis

Trichocereus fulvinanus
Trichocereus macrogonus
Trichocereus pachanoi
Trichocereus pasacana
Trichocereus peruvianus
Trichocereus taquimbalensis
Trichocereus terscheckii
Trichocereus validus
Trichocereus werdermannianus
Trichocereus spp.
Magnoliaceae
Drimys winteri
Magnolia dealbata
Magnolia virginiana
Himantandraceae
Galbulimima belgraveana
Annonaceae
Cymbopetalum penduliflorum
Myristicaceae
Iryanthera juruensis
Iryanthera longiflora
Iryanthera macrophylla
Iryanthera ulei
Myristica fragrans
Myristica spp.
Osteophloeum platyspermum
Virola calophylla
Virola calophylloidea
Virola cuspidata
Virola duckei
Virola elongata
Virola loretensis
Virola pavonis
Virola peruviana
Virola rufula
Virola surinamensis
Virola theiodora
Virola venosa
Virola spp.
Monimiaceae
Atherosperma moschatum
Gomortegaceae
Gomortega keule
Lauraceae
Cinnamomum camphora
Laurus nobilis
Persea indica
Peumus boldus
Sassafras albidum
Umbellularia californica
Ranunculaceae
Aconitum ferox
Aconitum napellus
Aconitum spp.
Clematis virginiana
Delphinium consolida
Ranunculus acris var. *japonicum*
Ranunculus sp.

Podophyllaceae
Podophyllum peltatum
Menispermaceae
Anarmita cocculus
Cocullus spp.
Nymphaeaceae
Nuphar lutea
Nymphaea ampla
Nymphaea caerulea
Piperaceae
Macropiper excelsum
Peperomia spp.
Piper amalago
Piper angustifolium
Piper auritum
Piper betle
Piper cubeba
Piper elongatum
Piper interitum
Piper longum
Piper methysticum
Piper plantagineum
Piper spp.
Actinidiaceae
Actinidia polygama
Theaceae
Camellia sinensis
Papaveraceae
Argemone mexicana
Argemone spp.
Eschscholzia californica
Glaucium flavum
Papaver bracteatum
Papaver orientale
Papaver somniferum
Papaver rhoeas
Papaver spp.
Cruciferae (previously: Brassicaceae)
Cardamine sp.
Saxifragaceae
Hydrangea paniculata
Rosaceae
Spiraea caespitosa
Leguminosae (previously: Fabaceae)
Acacia angustifolia
Acacia campylacantha
Acacia catechu
Acacia confusa
Acacia cornigera
Acacia maidenii
Acacia phlebophylla
Acacia retinodes
Acacia simplicifolia
Acacia spp.
Anadenanthera colubrina
Anadenanthera colubrina var. *cebil*
Anadenanthera peregrina
Anadenanthera peregrina var. *falcata*
Astragalus amphioxys

Astragalus besseyi
Astragalus cagopus
Astragalus mollissimus
Astragalus spp.
Caesalpinia decapetala [syn. *C. sepiaria*]
Calliandra anomala
Calliandra spp.
Canavalia maritima
Crotalaria sagittalis
Cytisus canariensis
Cytisus scoparius
Cytisus spp.
Dimorphandra parviflora
Erythrina americana
Erythrina berteroana
Erythrina corallodendron
Erythrina falcata
Erythrina flabelliformis
Erythrina fusca
Erythrina glauca
Erythrina indica
Erythrina mulungu
Erythrina poeppigiana
Erythrina standleyana
Erythrina vespertilio
Erythrina spp.
Lonchocarpus violaceus
Lotus wrightii
Lupinus albus
Lupinus angustifolius
Lupinus luteus
Mimosa pudica
Mimosa scabrella
Mimosa tenuiflora [syn. *M. hostilis*]
Mimosa verrucosa
Mimosa spp.
Mucuna pruriens
Mucuna spp.
Oxytropis lamberti
Oxytropis spp.
Pithecellobium diversifolium
Pithecellobium laetum
Pithecellobium spp.
Rhynchosia longeracemosa
Rhynchosia phaseoloides
Rhynchosia pyramidalis
Rhynchosia spp.
Sophora secundiflora
Sophora spp.
Spartium junceum
Swainsonia galegifolia
Wisteria sinensis
Zornia latifolia
Zygophyllaceae
Peganum harmala
Tribulus terrestris
Erythroxylaceae
Erythroxylum coca
Erythroxylum novogranatense
Erythroxylum spp.

Euphorbiaceae
Alchornea castaneifolia
Alchornea floribunda
Elaeophorbia drupifera
Jatropha grossidentata
Manihot anomala ssp. *anomala*
Monadenium lugardae
Pedilanthus retusus
Pedilanthus tithymaloides
Pedilanthus spp.
Sebastiania pavonia
Rutaceae
Dictyoloma incanescens
Evodia bonwickii
Thamnosma montana
Simaroubaceae
Ailanthus altissima
Burseraceae
Boswellia sacra
Boswellia spp.
Bursera bipinnata
Malpighiaceae
Banisteriopsis argentea
Banisteriopsis caapi [syn. *B. quitensis*]
Banisteriopsis inebrians (= *B. caapi*)
Banisteriopsis maritiniana var. *laevis*
Banisteriopsis muricata
Banisteriopsis spp.
Diplopterys cabrerana [syn. *Banisteriopsis rusbyana*]
Mascagnia psilophylla var. *antifebrilis*
Tetrapteris methystica
Tetrapteris mucronata
Tetrapteris spp.
Coriariaceae
Coriaria ruscifolia
Coriaria thymifolia
Coriaria spp.
Anacardiaceae
Sclerocarya caffra
Sclerocarya schweinfurthiana
Sapindaceae
Matayba guianensis
Nephelium topengii
Paullinia cupana
Paullinia yoco
Paullinia spp.
Ungnadia speciosa
Polygalaceae
Polygala tenuifolia
Aquifoliaceae
Ilex cassine
Ilex guayusa
Ilex paraguariensis
Ilex vomitoria
Ilex spp.
Celastraceae
Catha edulis

Vitaceae
Vitis vinifera
Vitis sp.
Elaeocarpaceae
Sloanea laurifolia
Malvaceae
Malva rotundifolia
Sida acuta
Sida rhombifolia
Bombacaceae
Bernoullia flammea
Pseudobombax ellipticum
Quararibea funebris
Quararibea spp.
Sterculiaceae
Brachychiton diversifolius
Cola acuminata
Cola nitida
Cola spp.
Theobroma bicolor
Theobroma cacao
Theobroma grandiflorum
Theobroma subincanum
Theobroma spp.
Violaceae
Gloeospermum sphaerocarpum
Turneraceae
Turnera diffusa
Turnera pumilla
Turnera ulmifolia
Passifloraceae
Passiflora involucrata
Passiflora spp.
Cucurbitaceae
Echinocystis lobata
Monodora myristica
Lythraceae
Heimia salicifolia
Heimia spp.
Myrtaceae
Backhousia myrtifolia
Psidium guajava
Combretaceae
Combretum mucronatum
Terminalia bellirica
Araliaceae
Hedera helix
Panax ginseng
Panax spp.
Umbelliferae (previously: Apiaceae)
Anethum graveolens
Conium maculatum
Daucus carota
Ferula asafoetida
Ferula sumbul
Foeniculum vulgare
Petroselinum crispum
Peucedanum japonicum
Siler divaricatum

Subclass Sympetalae
Ericaceae
Arctostaphylos uva-ursi
Arctostaphylos spp.
Gaultheria anastomosans
Gaultheria phillyreifolia
Gaultheria spp.
Ledum palustre
Ledum spp.
Pernettya furens
Pernettya mucronata
Pernettya parvifolia
Pernettya prostrata
Pernettya spp.
Rhododendron caucasicum
Rhododendron lepidotum
Rhododendron ponticum
Rhododendron spp.
Vaccinium floribundum
Vaccinium uliginosum
Plumbaginaceae
Limmonium macrorhabdos
Sapotaceae
Lucuma salicifolia
Oleaceae
Jasminum spp.
Loganiaceae
Gelsemium sempervirens
Mostuea gabonica
Mostuea stimulans
Mostuea spp.
Sanango racemosum
Strychnos ignatii
Strychnos nux-vomica
Strychnos spp.
Desfontainiaceae
Desfontainia spinosa
Apocynaceae
Alstonia scholaris
Alstonia venenata
Alstonia spp.
Aspidosperma quebracho-blanco
Aspidosperma spp.
Carissa edulis
Catharanthus roseus
Catharanthus roseus f. *albus*
Malouetia tamaquarina
Rauvolfia serpentina
Rauvolfia tetraphylla
Rauvolfia vomitoria
Rauvolfia spp.
Tabernaemontana coffeoides
Tabernaemontana crassa
Tabernaemontana dichotoma
Tabernaemontana heterophylla
Tabernaemontana muricata
Tabernaemontana sananho
Tabernaemontana spp.
Tabernanthe iboga

Tabernanthe spp.
Thevetia sp.
Vinca minor
Voacanga africana
Voacanga bracteata
Voacanga dregei
Voacanga grandiflora
Voacanga spp.
Rubiaceae
Coffea arabica
Coffea spp.
Corynanthe spp.
Mitragyna speciosa
Mitragyna spp.
Pausinystalia yohimba
Pausinystalia spp.
Psychotria brachypoda
Psychotria carthaginensis
Psychotria colorata
Psychotria poeppigiana
Psychotria psychotriaefolia
Psychotria viridis
Psychotria spp.
Convolvulaceae
Argyreia acuta
Argyreia barnesii
Argyreia cuneata
Argyreia hainanensis
Argyreia luzonensis
Argyreia mollis
Argyreia nervosa
Argyreia obtusifolia
Argyreia philippinensis
Argyreia speciosa
Argyreia splendens
Argyreia wallichi
Argyreia spp.
Calonyction muricatum
Convolvulus arvensis
Convolvulus tricolor
Ipomoea batatas
Ipomoea calobra
Ipomoea carnea
Ipomoea crassicaulis
Ipomoea hederacea
Ipomoea involucrata
Ipomoea muricata
Ipomoea murucoides
Ipomoea nil
Ipomoea pes-caprae
Ipomoea purpurea
Ipomoea rubrocaerulea
Ipomoea tricolor
Ipomoea violacea
Ipomoea spp.
Merremia tuberosa
Stictocardia titiaefolia
Turbina corymbosa

Boraginaceae
Borago officinalis
Cordia boissieri
Trichodesma zeylanicum
Labiatae (previously: Lamiaceae)
Coleus blumei
Coleus pumila
Lagochilus inebrians
Leonotis leonurus
Leonurus sibiricus
Mentha pulegium
Nepeta cataria
Ocimum micranthum
Salvia divinorum
Salvia sp.
Scutellaria arvense
Scutellaria lateriflora
Solanaceae
Atropa belladonna
Atropa spp.
Brugmansia arborea
Brugmansia aurea
Brugmansia x *candida*
Brugmansia x *insignis*
Brugmansia sanguinea
Brugmansia suaveolens
Brugmansia versicolor
Brugmansia vulcanicola
Brugmansia spp.
Brunfelsia australis
Brunfelsia chiricaspi
Brunfelsia grandiflora
Brunfelsia grandiflora ssp. *grandiflora*
Brunfelsia grandiflora ssp. *schultesii*
Brunfelsia maritima
Brunfelsia mire
Brunfelsia uniflora [syn. *B. hopeana*]
Brunfelsia spp.
Capsicum spp.
Cestrum laevigatum
Cestrum nocturnum
Cestrum parqui
Cestrum spp.
Datura ceratocaula
Datura discolor
Datura ferox
Datura innoxia [syn. *D. meteloides*]
Datura kymatocarpa
Datura leichhardtii
Datura metel
Datura pruinosa
Datura quercifolia
Datura reburra
Datura stramonium
Datura velutinosa
Datura villosa
Datura wrightii
Duboisia hopwoodii
Duboisia leichhardtii
Duboisia myoporoides

Fabiana imbricata
Fabiana spp.
Hyoscyamus albus
Hyoscyamus aureus
Hyoscyamus bohemicus
Hyoscyamus boveanus
Hyoscyamus desertorum
Hyoscyamus x *györffyi*
Hyoscyamus muticus
Hyoscyamus niger
Hyoscyamus niger var. *chinensis*
Hyoscyamus pallidus
Hyoscyamus physaloides
Hyoscyamus pusillus
Hyoscyamus reticulatus
Hyoscyamus spp.
Iochroma fuchsioides
Iochroma spp.
Juanulloa ochracea
Latua pubiflora
Mandragora autumnalis
Mandragora officinarum
Mandragora turcomanica
Mandragora spp.
Methysticodendron amesianum
Nicotiana acuminata
Nicotiana attenuata
Nicotiana benthamiana
Nicotiana bigelovii
Nicotiana clevelandii
Nicotiana glauca
Nicotiana gossei
Nicotiana ingulba
Nicotiana megalosiphon
Nicotiana palmeri
Nicotiana plumbaginifolia
Nicotiana quadrivalvis
Nicotiana rustica
Nicotiana stimulans
Nicotiana sylvestris
Nicotiana tabacum
Nicotiana trigonophylla
Nicotiana undulata
Nicotiana velutina
Nicotiana spp.
Petunia violacea
Petunia sp.
Physalis spp.
Scopolia carniolica
Scopolia spp.
Solandra brevicalyx
Solandra guerrerensis
Solandra guttata
Solandra nitida
Solandra spp.
Solanum dulcamara
Solanum hirtum
Solanum nigrum
Solanum villosum

Solanum spp.
Withania somnifera
Scrophulariaceae
Digitalis purpurea
Lancea tibetica
Scoparia dulcis
Bignoniaceae
Tanaecium nocturnum
Acanthaceae
Justicia pectoralis var. *stenophylla*
Pseuderanthemum sp.
Teliostachys lanceolata var. *crispa*
Lentibulariaceae
Utricularia minor
Valerianaceae
Valeriana officinalis
Valeriana spp.
Campanulaceae
Isotoma longiflora
Lobelia inflata
Lobelia tupa
Lobelia spp.
Goodeniaceae
Goodenia lunata
Goodenia spp.
Compositae (previously: Asteraceae)
Artemisia absinthium
Artemisia mexicana
Artemisia spp.
Cacalia cordifolia
Calea zacatechichi
Cineraria aspera
Helichrysum foetidum
Helichrysum stenopterum
Hieracium pilosella
Lactuca sativa
Lactuca virosa
Lactuca spp.
Matricaria recutita
Mikania cordata
Senecio hartwegii
Senecio spp.
Stephanomeria pauciflora
Tagetes erecta
Tagetes lucida
Tagetes minuta
Tagetes patula
Tagetes pusilla
Tagetes spp.
Trichocline dealbata
Trichocline exscapa
Trichocline reptans
Trichocline spp.
Urmenetea atacamensis

Class Monocotyledoneae
Liliaceae
Veratrum album
Veratrum viride

Agavaceae
Agave americana
Agave atrovirens
Agave cerulata spp. *dentiens*
Agave latissima
Agave mapisaga
Agave palmeri
Agave parryi
Agave tequilana
Agave spp.
Cordyline fruticosa
Amaryllidaceae
Boophane disticha
Pancratium trianthum
Ungernia minor
Dioscoreaceae
Dioscorea composita
Pontederiaceae
Pontederia cordata
Iridaceae
Crocus sativus
Ferraria glutinosa
Gramineae (previously: Poaceae)
Arundo donax
Cymbopogon densiflorus
Lolium temulentum
Phalaris aquatica
Phalaris arundinacea
Phalaris spp.
Phragmites australis
Stipa robusta
Stipa vaseyi
Stipa viridula

Stipa spp.
Zea mays
Palmae (previously: Arecaceae)
Archontophoenix cunninghamiana
Areca catechu
Areca spp.
Cocos nucifera
Phoenix dactylifera
Araceae
Acorus calamus
Acorus gramineus
Arisaema dracontium
Homalomena belgraveana
Homalomena sp.
Philodendron scandens
Pandanaceae
Pandanus spp.
Cyperaceae
Cyperus spp.
Scirpus spp.
Musaceae
Musa x *sapientum*
Zingiberaceae
Kaempferia galanga
Zingiber officinale
Cannaceae
Canna sp.
Orchidaceae
Cypripedium calceolus
Oncidium cebolleta
Vanda roxburghii
Vanilla planifolia

General Bibliography

"The mysterious and essential center of a culture is its books. If we do not have a living culture before us, then we can still experience it through its books. Books are the soul of a culture."

TERENCE MCKENNA
(1995)

Internet addresses of discussion forums, news groups, and Web sites that provide information about psychoactive plants:
www.alb2c3.com/drugs/index.htm
www.csp.org
www.erowid.com/entheo.shtml
www.fungi.com
www.hanfblatt.de
www.heffter.org
www.hempBC.com
www.hightimes.com
www.hofmann.org
www.island.org
www.lycaeum.org
www.maps.org
www.med.uni-
muenchen.de/medpsy/ethno

This bibliography lists only works of an introductory or general nature. Specialized literature and monographs are listed under the specific plants, products, and active components. See also the general literature on psychoactive fungi on pages 689–693.

Those sources that are marked in the monographs with a single * may be found in this bibliography. Those marked with a double ** can be found in the general literature on psychoactive fungi.

Bibliographies

Beifuss, Will. 1996. *Psychedelic sourcebook.* Berkeley, Calif.: Rosetta/Flowers Joint Production.

Hanna, Jon. 1996. *Psychedelic resource list.* Sacramento: Soma Graphics.

Hefele, Bernhard. 1988. *Drogenbibliographie: Verzeichnis der deutschsprachigen Literatur über Rauschmittel und Drogen von 1800 bis 1984.* 2 vols. Munich, London, New York, and Paris: K. G. Saur.

Passie, Torsten. 1992. *Schamanismus, eine kommentierte Auswahlbibliographie.* Hannover: Laurentius.

Rajbhandari, Keshab R. 1994. *A bibliography of the plant science of Nepal.* Kathmandu: Nepal Press.

SISSC [= Società Italiana per lo Studio degli Stati di Coscienza]. 1994. *Bibliographia Italiana su allucinogeni e cannabis: Edizione commentata.* Bologna: Grafton 9 edizioni.

Periodicals

The following journals, magazines, yearbooks, and series contain numerous articles about psychoactive plants.

Altrove. Has been published since 1993 and is the yearbook of the Italian Society for the Study of the States of Consciousness (= SISSC).

Anthropology of Consciousness. Has been published since 1989 by the Society for the Anthropology of Consciousness, a section of the American Anthropological Association. However, most of the contributions focus on the anthropology of consciousness, with only an occasional article on psychoactive substances.

Botanical Museum Leaflets, Harvard University. This journal was published between 1932 and 1986; it contains numerous important articles on the ethnopharmacology of psychoactive plants and fungi.

Curare—Zeitschrift für Ethnomedizin. This interdisciplinary journal is the organ of the German Society Arbeitsgemeinschaft Ethnomedizin e.V.; it frequently contains articles on psychoactive

plants, shamanism, traditional healers; reports on congresses; book reviews; et cetera.

Economic Botany. Published by the New York Botanical Garden, this journal is probably the most important ethnobotanical periodical.

Eleusis—A Quarterly Bulletin of the Italian Society for the Study of the States of Consciousness. This additional organ of the Italian Society for the Study of the States of Consciousness (= SISSC) has been published since April 1995. The bulletin is published in two languages (Italian and English). Address: SISSC, c/o Museo Civico di Rovereto, Largo S. Catarina 43, 1-38068 Rovereto (TN), Italy.

Entheogene. This journal, which is edited by Jim DeKorne and Bert Marco Schuldes, has been published in Germany since the fall of 1994. The full title is *Entheogene: Forum für entheogene Forschungen, Verfahren und Erfahrungen* [Entheogens: A forum for entheogenic studies, techniques, and experiences]. In this journal, the themes from Jim DeKorne's book *Psychedelic Shamanism* (Port Townsend, Wash.: Loompanics Unlimited, 1994) are continued and discussed in greater depth. The publication is an adaptation of the *Entheogen Newsletter,* published in the United States by Jim DeKorne. Address: B. M. Schuldes, Hauptstr. 70, D-99759 Rehungen, Germany.

Grow! Marihuana Magazin. The first issue of this quarterly journal appeared in the summer of 1995.

HanfBlatt. This monthly journal has been published in northern Germany since 1994.

Hanf! Das zeitkritische Journal. This bimonthy journal has been published since April/May 1995.

Hemplife Magazine. This bimonthly magazine has been published since December 1998. It is produced in Holland.

Integration—Zeitschrift für geistbewegende Pflanzen und Kultur. Published at irregular intervals since 1991, this journal consists almost exclusively of articles about psychoactive substances.

Jahrbuch des Europäischen Collegiums für Bewußtseinsstudien / Yearbook of the European College for the Study of Consciousness. Edited by Michael Schlichting and Hanscarl Leuner, this series has been published since 1992; almost all of the articles deal with psychoactive substances.

Jahrbuch für Ethnomedizin und Bewußtseinsforschung / Yearbook for Ethnomedicine and the Study of Consciousness. Edited by Christian Rätsch and John R. Baker, this series was published from 1992 until 1998; almost all of the articles deal with ethnobotany and psychoactive substances.

Jahrbuch für Transkulturelle Medizin und Psychotherapie / Yearbook of Cross-Cultural Medicine and Psychotherapy. Edited by Walter Andritzky, this series has been published since 1990; all of the volumes have contained articles on psychoactive substances.

Journal of Ethnobiology. Published since 1980, this journal occasionally contains articles on psychoactive substances.

Journal of Ethnopharmacology. Published by Elsevier since 1978, the first volumes contained almost exclusively ethnopharmacological articles about psychoactive plants. It has now become more of a clearinghouse for pharmacognostic and purely pharmacological works by authors from Third World countries.

Journal of Psychedelic Drugs. Vol. 1–12, 1967–1980; renamed *Journal of Psychoactive Drugs* as of vol. 13 in 1981.

Journal of the International Hemp Association. This scientific journal has been published in Amsterdam since 1994 (two issues per year).

Lloydia/Journal of Natural Products. This journal, which focuses on the chemistry of naturally occurring substances, was named for the Lloyd brothers. It contains numerous articles on the chemistry of psychoactive plants and fungi.

MAPS Bulletin. Published by the Multidisciplinary Association for Psychedelic Studies, this bulletin contains primarily reports on current research projects, conferences, and publications. The emphasis is on MDMA, psilocybin, DMT, medicinal cannabis, and therapy.

Mreža Drog. From Slovenia, this journal (ISSN 1318-2609) is devoted primarily to the classic psychedelics and entheogens. Many of the articles it contains are translations of German, Italian, or English literature. Published since 1994; the editors are Bojan Dekleva (editor in chief) and Ivan Valenčič (associate editor).

Pan. A "psychoactive network" has been producing a quarterly publication in the Netherlands since May 1995 titled *PAN FORUM* (ISSN 1382-4538). All of the articles are in Dutch.

Phytochemistry. Published by Pergamom Press since 1962; contains numerous reports on chemical studies of psychoactive plants.

Psychedelic Monographs and Essays. Published since 1985 by Thomas Lyttle. Numerous volumes have appeared to date. Lyttle published the best articles in book form in 1994.

Psychedelic Review. Published from 1964 until 1972. Weil et al. published the best articles as a book (*The Psychedelic Reader*) in 1973.

Shaman's Drum. Published in the United States; contains articles discussing all aspects of shamanism, including many that deal with the shamanic use of psychoactive substances.

Takiwasi. Has appeared at irregular intervals since 1991; contains only articles about psychoactive substances and altered states of consciousness. Many of the articles are translations of previously published material.

Books and Articles

A

Aaronson, Bernard, and Humphrey Osmond, eds. 1970. *Psychedelics.* New York: Anchor Books.

Abraham, Hartwig, and Inge Thinnes. 1995. *Hexenkraut und Zaubertrank.* Greifenberg: Urs Freund.

Abraham, Henry David, Andrew M. Aldridge, and Prashant Gogia. 1996. The psychopharmacology of hallucinogens. *Neuropsychopharmacology* 14 (4): 285–98.

Adly, Abdallah, ed. 1982. *The history of medicinal and aromatic plants.* Karachi, Pakistan: Hamdard Foundation Press.

Adovasio, J. M., and G. F. Fry. 1976. Prehistoric psychotropic drug use in northeastern Mexico and Trans-Pecos Texas. *Economic Botany* 30:94–96.

Agrippa von Nettesheim, and Heinrich Cornelius. 1982. *Die magischen Werke.* Wiesbaden: Fourier.

Aguilar, Abigail, Arturo Argueta, and Leticia Cano, eds. 1994. *Flora medicinal indígena de México.* 3 vols. Mexico City: INI.

Aguilar Contreras, Abigail, and Carlos Zolla. 1982. *Plantas tóxicas de México.* Mexico City: IMSS.

Aguilera, Carmen. 1985. *Flora y fauna mexicana: Mitología y tradiciones.* Mexico City: Editorial Everest Mexicana.

Aguirre Beltrán, Gonzalo. 1963. *Medicina y magía.* Mexico City: INI.

Aichele, Dietmar, and Reinhild Hofmann. 1991. *Unsere Gräser.* 10th ed. Stuttgart: Franckh-Kosmos.

Akendengué, B. 1992. Medicinal plants used by the Fang traditional healers in Equatorial Guinea. *Journal of Ethnopharmacology* 37:165–73.

Alcorn, Janis B. 1984. *Huastec Mayan ethnobotany.* Austin: University of Texas Press.

Aldunate, Carlos, Juan J. Armesto, Victoria Castro, and Carolina Villagrán. 1981. Estudio etnobotanico en una communidad precordillerana de Antofagasta: Toconce. *Boletín del Museo Nacional de Historia Natural de Chile* 38:183–223.

———. 1983. Ethnobotany of pre-altiplanic community in the Andes of northern Chile. *Economic Botany* 37 (1): 120–35.

Alexiades, Miguel N., and Jennie Wood Sheldon. 1996. *Selected guidelines for ethnobotanical research: A field manual.* New York: The New York Botanical Garden.

Aliotta, Giovanni, Danielle Piomelli, and Antonio Pollio. 1994. Le piante narcotiche e psicotrope in Plinio e Dioscoride. *Annali dei Musei Civici de Rovereto* 9 (1993): 99–114.

Allen, O. N., and Ethel K. Allen. 1981. *The Leguminosae: A source book of characteristics, uses, and nodulation.* Madison: The University of Wisconsin Press.

Allen, Timothy F. 1975. *The encyclopedia of pure materia medica.* New York: Boericke & Tafel.

Altmann, Horst. 1980. *Giftpflanzen—Gifttiere.* Munich: BLV.

Altschul, Siri von Reis (see also von Reis and Lipp). 1973. *Drugs and foods from little-known plants: Notes in Harvard University herbaria.* Cambridge, Mass.: Harvard University Press.

Alvear, Silvio Luis Haro. 1971. *Shamanismo y farmacopea en el reino de Quito.* Quito, Ecuador: Instituto Ecuatoriana de Ciencias Naturales (Contribución).

Ambasta, Shri S. P., ed. 1994. *The useful plants of India.* New Delhi: Publications and Information Directorate.

Amberger-Lahrmann, M., and D. Schmähl, eds. 1993. *Gifte: Geschichte der Toxikologie.* Wiesbaden: Fourier.

Amorín, J. L. 1974. *Plantas de la flora argentina relacionadas con alucinógenos americanos.* Publicaciones de la Academia Argentina de Farmacia y Bioquímica, no. 1. Buenos Aires: La Academia Argentina de Farmacia y Bioquímica.

Anderson, Edward F. 1993. *Plants and people of the Golden Triangle: Ethnobotany of the hill tribes of northern Thailand.* Portland, Ore.: Dioscorides Press.

Andoh, Anthony. 1986. *The science and romance of selected herbs used in medicine and religious ceremony.* San Francisco: North Scale Institute.

Andrews, J. Richard, and Ross Hassig. 1984. *Treatise on the heathen superstitions by Hernando Ruiz de Alarcón.* Norman and London: University of Oklahoma Press.

Andritzky, Walter. 1987. Die Volksheiler in Peru während der spanisch-kolonialen Inquisition. *Anthropos* 82:543–66.

———. 1989. *Schamanismus und rituelles Heilen im Alten Peru.* 2 vols. Berlin: Clemens Zerling.

———. 1995. Sakrale Heilpflanze, Kreativität und Kultur: indigene Malerei, Gold- und Keramikkunst in Peru und Kolumbien. *Curare* 18 (2): 373–93.

Andritzky, Walter, and Stefan Trebes. 1996. Vision, Kreativität, Heilung: Das konstruktive Potential sakraler Heilpflanzen in der Industriegesellschaft. *Jahrbuch für Transkulturelle Medizin und Psychotherapie*, 1995 (6): 381–408.

Anzeneder, Robert, Mario Miyagawa, and Gisela Rödl-Linder. 1993. *Pflanzenführer Tropisches Lateinamerika.* Pforzheim: Goldstadtverlag.

Applewhite, P. B. 1973. Serotonin and norepinephrine in plant tissues. *Phytochemistry* 12:191–92.

Arevalo Valera, Guillermo. 1994. *Medicina indígena: las plantas medicinales y su beneficio en la salud—Shipibo-Conibo.* Lima: Edicion Aidesep.

Arenas, P. 1987. Medicine and magic among the Maká Indians of the Paraguayan Chaco. *Journal of Ethnopharmacology* 21:279–95.

Arends, G. 1935. *Volkstümliche Namen der Arzneimittel, Drogen, Heilkräuter und Chemikalien.* 12th ed. Berlin: Julius Springer.

Argueta Villamar, Arturo, Leticia M. Cano Asseleih, and María Elena Rodarte, eds. 1994. *Atlas de las plantas de la medicina traditional mexicana.* 3 vols. Mexico City: INI.

Aris, Anthony, ed. 1992. *Tibetan medical paintings.* 2 vols. London: Serindia Publications.

Arnau, Frank. 1967a. *Flucht in den Sex: Vom Liebestrank zu den Hormonen.* Munich: Rütten & Loening Verlag.

———. 1967b. *Rauschgift: Träume auf dem Regenbogen.* Lucerne and Frankfurt/M.: C. J. Bucher.

Arriaga, Pablo José de. 1992. *Eure Götter werden getötet: 'Ausrottung des Götzendienstes in Peru' (1621).* Ed. Karl A. Wipf. Darmstadt: Wissenschaftliche Buchgesellschaft.

Arvigo, Rosita, and Michael Balick. 1994. *Die Medizin des Regenwaldes: Heilkraft der Maya-Medizin—die 100 heilenden Kräuter von Belize.* Aitrang: Windpferd.

Asolka, L. V., K. K. Kakkar, and O. J. Chakre. 1992. *Second supplement to glossary of Indian medicinal plants with active principles. Part I (A–K) (1965–1981).* New Delhi: CSIR.

Assi, Laurent Aké, and Sita Guinko. 1991. *Plants used in traditional medicine in West Africa.* Basel: Editiones Roche.

Atkinson, E. T. 1989. *Economic botany of the Himalayan regions.* New Delhi: Cosmo Publications. (Contains very detailed descriptions of manners of using and preparing hemp, opium, betel, alcohol, and tobacco.)

Avila B., Alejandro de. 1992. Plants in contemporary Mixtec ritual: *Juncus, Nicotiana, Datura,* and *Solandra. Journal of Ethnobiology* 12 (2): 237–38.

Axton, Joe E. 1984. *Hallucinogens: A comprehensive guide for laymen and professionals.* 2nd ed. Ed. Jeremy Bigwood and Jonathan Ott. Tempe, Ariz.: Do It Now Foundation.

Ayensu, Edward S. 1978. *Medicinal plants of West Africa.* Algonac, Mich.: Reference Publications.

GUILLERMO AREVALO V.

MEDICINA INDIGENA

LAS PLANTAS MEDICINALES Y SU BENEFICIO EN LA SALUD

SHIPIBO - CONIBO

EDICION AIDESEP

B

Baer, Gerhard. 1984. *Die Religion der Matsigenka, Ost-Peru*. Basel: Wepf.

———. 1986. 'Der vom Tabak Berauschte'—Zum Verhältnis von Rausch, Ekstase und Wirklichkeit. *Verhandlungen Naturforschende Gesellschaft Basel* 96:41–84.

———. 1987. Peruanische ayahuasca-Sitzungen. In *Ethnopsychotherapie*, ed. A. Dittrich and Ch. Scharfetter, 70–80. Stuttgart: Enke.

Baker, John R. 1989. The emergence of culture: A general anthropological approach to the relationship between the individual and his external world. MS dissertation, Hamburg.

———. 1995. Nachtschattengewächse in der Behandlung von Asthma: Physiologische und psychologische Aspekte. *Jahrbuch des Europäischen Collegiums für Bewußtseinsstudien*, 1993/1994: 137–52.

Balabanova, S., F. Parsche, and W. Pirsig. 1992. First identification of drugs in Egyptian mummies. *Naturwissenschaften* 79:358.

Balée, William. 1994. *Footprints of the forest: Ka'apor ethnobotany—the historical ecology of plant utilization by an Amazonian people*. New York: Columbia University Press.

Balick, Michael J., and Hans T. Beck, eds. 1990. *Useful palms of the world: A synoptic bibliography*. New York: Columbia University Press.

Balick, Michael J., and Paul Alan Cox. 1996. *Plants, people, and culture: The science of ethnobotany*. New York: Scientific American Library.

Bangen, Hans G. 1992. *Geschichte der medikamentösen Therapie der Schizophrenie*. Berlin: VWB.

Barber, Theodore X. 1970. *LSD, marihuana, yoga, and hypnosis*. Chicago: Aldine.

Barr, Andy, ed. 1990. *Traditional bush medicines: An aboriginal pharmacopoeia*. Northern Territory of Australia [Publishing].

Barrera Marin, Alfredo, Alfredo Barrera Vazquez, and Rosa María Lopez Franco. 1976. *Nomenclatura etnobotanica maya: Una interpretación*. Collección Científica 36. Mexico City: INAH/SEP.

Barrows, David Prescott. 1967. *The ethno-botany of the Coahuilla Indians of Southern California*. Banning, Calif.: Malki Museum Press. (Orig. pub. 1900.)

Bärtels, Andreas. 1993. *Farbatlas Tropenpflanzen: Zier- und Nutzpflanzen*. 3rd ed. Stuttgart: Ulmer.

Bartels, Max. 1893. *Die Medicin der Naturvölker*. Leipzig: Th. Grieben's Verlag.

Bastien, Joseph W. 1987. *Healers of the Andes: Kallawaya herbalists and their medicinal plants*. Salt Lake City: University of Utah Press.

Baudelaire, Charles. 1993. *The flowers of evil*. Trans. James McGowan. Oxford: Oxford University Press. (Orig. pub. 1857.)

Bauereiss, Erwin, ed. 1995. *Heimische Pflanzen der Götter: Ein Handbuch für Hexen und Zauberer*. Markt Erlbach: Raymond Martin Verlag.

Baumann, B. B. 1960. The botanical aspects of ancient Egyptian embalming and burial. *Economic Botany* 14:84–104.

Baumann, Hellmut. 1982. *Die Griechische Pflanzenwelt in Mythos, Kunst und Literatur*. Munich: Hirmer.

Baumgartner, Daniela. 1994. Das Priesterwesen der Kogi. *Yearbook for Ethnomedicine and the Study of Consciousness*, 1994 (3): 171–98. Berlin: VWB.

Bayrle-Sick, Norbert. 1984. *Drogensubkultur*. Linden: Volksverlag.

Becker, Stefan R. 1994. Das Johanniskraut (*Hypericum perforatum*)—Antidepressivum aus der Natur: Möglichkeiten einer Therapie leichter bis mittelschwerer Depressionen. *Schweizerische Zeitschrift für Ganzheitsmedizin* 1:46–49; 2:92–94.

Beckmann, Dieter, and Barbara Beckmann. 1990. *Alraune, Beifuß und Andere Hexenkräuter*. Frankfurt/M. and New York: Campus.

Belledame, ed. 1990. *Die Persönliche Magie der Pflanzen: Traditionelle Grundlagen der Aromatherapie*. Bad Münstereifel: Edition Tramontane.

Bennett, B. C. 1992. Hallucinogenic plants of the Shuar and related indigenous groups in Amazonian Ecuador and Peru. *Brittonia* 44: 483–93.

Bensky, Dan, and Andrew Gamble. 1986. *Chinese herbal medicine: Materia medica*. Seattle: Eastland Press.

Beredonk, Brigitte. 1992. *Doping—Von der Forschung zum Betrug*. Reinbek: Rowohlt.

Berendes, Julius. 1891. *Die Pharmacie bei den alten Culturvölkern*. Halle: Tausch & Grosse.

Berlin, Brent, Dennis E. Breedlove, and Peter H. Raven. 1974. *Principles of Tzeltal plant classification*. New York and London: Academic Press.

Bhandary, M. J., K. R. Chandrashekar, and K. M. Kaveriappa. 1995. Medical ethnobotany of the Siddis of Uttara Kannada District, Karnataka, India. *Journal of Ethnopharmacology* 47:149–58.

Bhattarai, N. K. 1989. Traditional phytotherapy among the Sherpas of Helambu, central Nepal. *Journal of Ethnopharmacology* 27:45–54.

Bibra, Baron Ernst von. 1855. *Die narkotischen Genußmittel und der Mensch*. Nuremberg: Verlag von Wilhelm Schmid. Repr. Leipzig: Zentralantiquariat der DDR; Wiesbaden: Fourier-Verlag, 1983. Repr. Leipzig: Reprint-Verlag, 1995. (Beautiful hardcover edition.)

———. 1995. *Plant intoxicants: A classic text on the use of mind-altering plants*. Trans. Hedwig Schleiffer. Foreword by Martin Haseneier. Tech. notes by Jonathan Ott. Rochester, Vt.: Healing Arts Press. (Orig. pub. 1855; see previous entry.)

Biedermann, Hans. 1972. *Medicina Magica*. Graz: Akademische Druck und Verlagsanstalt.

———. 1984. *Höhlenkunst der Eiszeit*. Cologne: Dumont.

Biswas, K. 1956. *Common medicinal plants of Darjeeling and the Sikkim Himalayas*. Alipore: West Bengal Government Press.

Black, Ira B. 1993. *Symbole, Synapsen und Systeme: Die molekulare Biologie des Geistes*. Heidelberg: Spektrum.

———. 1994. *Information in the Brain*. Cambridge, Mass.: Bradford Books.

Blackburn, Thomas. 1976. A query regarding the possible hallucinogenic effects of ant ingestion in south-central California. *The Journal of California Anthropology* 3 (2): 78–81.

Blätter, Andrea. 1995. Die Funktionen des Drogengebrauchs und ihre kulturspezifische Nutzung. *Curare* 18 (2): 279–90.

———. 1996a. Drogen im präkolumbischen Nord-amerika. *Yearbook for Ethnomedicine and the Study of Consciousness*, 1995 (4): 163–83. Berlin: VWB.

———. 1996b. High for health: Über die Ethnomedizinische Nutzung von halluzinogenen Drogen. *Infoemagazin* 11:24–26.

Blohm, Henrik. 1962. *Poisonous plants of Venezuela*. Cambridge, Mass.: Harvard University Press.

Bock, Hieronymus. 1577. *Kreutterbuch*. Strasbourg: Rihel.

Bock, Michael. 1994. The psychoactive flora and fauna of Australia. Unpublished manuscript.

Boericke, William. 1992. *Handbuch der homö-opathischen Materia medica*. Heidelberg: Haug.

Bogers, Hans, Stephen Snelders, and Hans Plomp. 1994. *De Psychedelische (R)evolutie*. Amsterdam: Bres.

Boissya, C. L., R. Majumder, and A. K. Majumder. 1981. Some medicinal plants from Darring District of Assam, India. *Anthropos* 76:220–22.

Bonin, Werner F. 1979. *Die Götter Schwarzafrikas*. Graz: Verlag für Sammler.

Bourguignon, Erika, ed. 1973. *Religion, altered states of consciousness, and social change*. Columbus: Ohio State University Press.

Bourke, John Gregory. 1996. *Der Unrat in Sitte, Brauch, Glauben und Gewohnheitsrecht der Völker*. Frankfurt/M.: Eichborn. (Orig. pub. 1913.)

Boyd, Carolyn E., and J. Philip Dering. 1996. Medicinal and hallucinogenic plants identified in the sediments and pictographs of the lower Pecos, Texas Archaic. *Antiquity* 70 (268): 256–75.

Braem, Harald. 1994. *Die magische Welt der Schamanen und Höhlenmaler*. Cologne: Dumont.

Brako, Lois, and James L. Zarucchi. 1993. *Catalogue of the flowering plants and gymnosperms of Peru*. Saint Louis: Missouri Botanical Garden.

Bramley, Serge. 1977. *Im Reiche des Wakan: Das magische Universum der nordamerikanischen Indianer*. Basel: Sphinx.

Brandenburg, Dietrich. 1973. *Medizinisches in Tausendundeiner Nacht*. Stuttgart: Fink.

Brau, Jean-Louis. 1969. *Vom Haschisch zum LSD*. Frankfurt/M.: Insel.

Brautigan, Richard. 1968. *The pill versus the Springhill Mine disaster*. Boston: Houghton Mifflin.

Bremness, Lesley. 1994. *Kräuter, Gewürz und Heilpflanzen, aus dem Englischen von Barbara Schmidt und Michael Conrad*. Ravensburg: Ravensburger Buchverlag. (Originally published in 1994 as an Eyewitness Handbook. *Herbs*. London: Dorling Kindersly.)

Breton, André. 1968. *Die Manifeste des Surrealismus*. Reinbek: Rowohlt.

Brinton, Daniel G. 1887. *Ancient Nahuatl poetry*. Library of American Aboriginal Literature, vol. 7. Philadelphia: D. G. Brinton.

Bristol, Melvin Lee. 1965. Sibundoy ethnobotany. PhD diss., Harvard University, Cambridge, Mass.

Britton, N. L., and J. N. Rose. 1963. *The Cactaceae: Descriptions and illustrations of plants of the Cactus Family*. 2 vols. New York: Dover. (Orig. pub. 1937 in 4 vols.)

Brøndegaard, V. J. 1985. *Ethnobotanik*. Berlin: Mensch und Leben.

Brosse, Jacques. 1990. *Mythologie der Bäume*. Olten and Freiburg: Walter-Verlag.

———. 1992. *Magie der Pflanzen*. Olten and Freiburg: Walter-Verlag.

Brown, John K., and Marvin H. Malone. 1978. "Legal Highs"—constituents, activity, toxicology, and herbal folklore. *Clinical Toxicology* 12 (1): 1–31.

Brown, Stanley D., John L. Massingill Jr., and Joe E. Hodgkins. 1968. Cactus alkaloids. *Phytochemistry* 7:2031–36.

Bruman, Henry John. 1940. Pre-Columbian brewing. PhD thesis, Los Angeles.

———. 2000. *Alcohol in Ancient Mexico*. Salt Lake City: University of Utah Press.

Brunngraber, Rudolf. 1952. *Heroin: Roman der Rauschgifte*. Hamburg: Rowohlt Verlag.

Brunton, Paul. 1983. *Von Yogis, Magiern und Fakiren: Begegnungen in Indien*. Munich: Knaur.

Buchmann, Werner. 1983. *Hahnemanns Reine Arznei-mittellehre: Die Grundlinien*. Heidelberg: Haug.

Buhner, Stephen Harrod. 1996. *Sacred plant medicine: Explorations in the practice of indigenous herbalism*. Boulder, Colo.: Robert Rinehart Publishers.

Burn, Harold. 1963. *Drogen, Medikamente und wir.* Zurich: Orell Füssli Verlag.

Bye, Robert A. 1979a. An 1878 ethnobotanical collection from San Luis Potosí: Dr. Edward Palmer's first major Mexican collection. *Economic Botany* 33 (2): 135–62.

———. 1979b. Hallucinogenic plants of the Tarahumara. *Journal of Ethnopharmacology* 1:23–48.

Bye, Robert A., and Edelmira Linaris. 1983. The role of plants found in the Mexican markets and their importance in ethnobotanical studies. *Journal of Ethnobiology* 3 (1): 1–13.

C

Cabieses, Fernando. 1983. Die magischen Pflanzen Perus. In *Peru durch die Jahrtausende*, 138–41. Exhibit catalogue, Niederösterreichische Landesausstellung, Schloß Schallaburg.

Califano, Mario. 1995. Los rostros del chamán: Nombres y estados. In *Chamanismo en Latinoamérica*, ed. Isabel Lagarriga, Jacques Galinier, and Michel Perrin, 103–42. Mexico City: Plaza y Valdés Editores.

Callaway, James Clayton Jr. 1994. Pinoline and other tryptamine derivatives: Formations and functions. PhD dissertation, Department of Pharmaceutical Chemistry at the University of Kuopio, Finland.

Camilla, Gilberto. 1995. Le erbe del diavolo 1: Aspetti antropologici. *Altrove* 2:105–15.

Camporesi, Piero. 1991. *Geheimnisse der Venus: Aphrodisiaka vergangener Zeiten.* Frankfurt and New York: Campus.

Candre, Hipólito. 1996. *Cool tobacco, sweet coca: Teachings of an Indian sage from the Colombian Amazon.* Totnes and Devon: Themis Books.

Carneiro Martins, José Evandro. 1989. *Plantas medicinais de uso na Amazônia.* 2nd ed. Belém-Pará: Cultural CEJUP.

Carnochan, F. G., and Hans Christian Adamson. 1986. *Das Kaiserreich der Schlangen.* Berlin: Verlag Clemens Zerling.

Carstairs, G. M. 1954. Daru and Bhang: Cultural factors in the choice of intoxicants. *Quarterly Journal for the Study of Alcohol* 15:220–37. (Repr. in *Psychedelic Review* 6 [1965]: 67–83.)

Carvajal, P. A. 1980. *Plantas que curan y plantas que matan.* Mexico City: Editores Mexicanos Unidos.

Castaneda, Carlos. 1968. *The teachings of Don Juan: A Yaqui way of knowledge.* Berkeley: University of California Press.

———. 1971. *A separate reality.* New York: Simon & Schuster. (See book review by Gordon Wasson in *Economic Botany* 26 [1972]: 98–99.)

———. 1972. *Journey to Ixtlan.* New York: Simon & Schuster. (See book review by Gordon Wasson in *Economic Botany* 27 [1973]: 151–52.)

Cawte, J. 1985. Psychoactive substances of the South Seas: Betel, kava and pituri. *Australian and New Zealand Journal of Psychiatry* 19 (1): 83–87.

Cerna, David. 1932. The pharmacology of the ancient Mexicans. *Annals of Medical History*, n.s., 4 (3): 298–305.

Chamisso, Adelbert von. 1987. *Illustriertes Heil-, Gift- und Nutzpflanzenbuch.* Berlin: Dietrich Reimer.

Chavez Velasquez, Nancy A. 1977. *La materia medica en el incanato.* Lima: Editorial Mejia Baca.

Cherikoff, Vic. 1993. *The bushfood handbook.* Boronioa Park, Australia: Bush Tucker Supply Australia Pty. Ltd. (Orig. pub. by Vic Cherikoff and Jennifer Isaacs, Ti Tea Press, 1989.)

Chopra, R. N., I. C. Chopra, and B. S. Varma. 1992. *Supplement to glossary of Indian medicinal plants.* New Delhi: CSIR.

Christiansen, Morgens Skytte, and Verner Hancke. 1993. *Gräser.* Munich: BLV.

CIORO [= Centro de Investigaciones de Quintana Roo, A.C.]. 1982. *Imagenes de la flora Quintanarroense.* Puerto Morelos, Mexico: CIORO.

Clarke, Charlotte Bringle. 1977. *Edible and useful plants of California.* Berkeley: University of California Press.

Clarke, Robert C. 1997. *Hanf—Botanik, Anbau, Vermehrung und Züchtung.* Aarau: AT Verlag. (Orig. pub. as *Marijuana botany*, Berkeley, Calif.: Ronin Publ., 1981.)

Clifton, Chas S., ed. 1994. *Witchcraft today: Witchcraft and shamanism.* Saint Paul, Minn.: Llewellyn Publications.

Clottes, Jean, and David Lewis-Williams. 1997. *Schamanen: Trance und Magie in der Höhlenkunst der Steinzeit.* Sigmaringen: Thorbecke Verlag.

Cobo, Bernabé. 1990. *Inca religion and customs.* Trans., ed. Roland Hamilton. Austin: University of Texas Press.

Collins, D. J., C. C. J. Culvenor, J. A. Lamberton, J. W. Loder, and J. R. Price. 1990. *Plants for medicine: A chemical and pharmacological survey of plants in the Australian region.* East Melbourne: CSIRO Publication.

Cooke, Mordecai C. 1989. *The seven sisters of sleep.* Lincoln, Mass.: Quarterman Publications. (Orig. pub. 1860. See book review by Jonathan Ott in *Journal of Ethnobiology* 12 [2; 1992]: 278 f.)

———. 1997. *The seven sisters of sleep.* Foreword by Rowan Robinson. Rochester, Vt.: Park Street Press.

Cooper, J. C. 1984. *Chinese alchemy: The Taoist quest for immortality.* Wellingborough, U.K.: The Aquarian Press.

Cooper, J. M. 1949. Stimulants and narcotics. In *The Comparative Ethnology of South American Indians*, vol. 5 of *Handbook of South American Indians*,

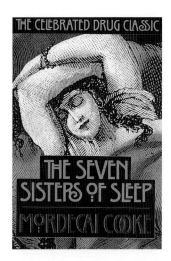

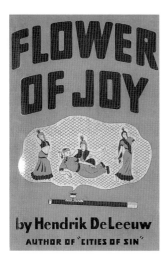

Bureau of American Ethnology Bulletin 143, ed. J. H. Stewart: 525–58.

Cordy-Collins, Alana. 1982. Psychoactive painted Peruvian plants: The shamanism textile. *Journal of Ethnobiology* 2 (2): 144–53.

Couliano, Ioan P. 1995. *Jenseits dieser Welt: Außerweltliche Reisen von Gilgamesch bis Albert Einstein.* Munich: Diederichs Verlag.

Coult, Allan D. 1977. *Psychedelic anthropology.* Philadelphia and Ardmore, Penn.: Dorrance & Co.

Cowan, Eliot. 1994. *Pflanzengeist-Medizin: Der schamanistische Weg mit Heilkräutern.* Munich: Knaur.

Cranach, Diana von. 1981. Drogen im alten Ägypten. In *Rausch und Realität,* ed. G. Völger, 1:266–69. Cologne: Rautenstrauch-Joest-Museum für Völkerkunde.

Cribb, A. B., and J. W. Cribb. 1981. *Wild medicine in Australia.* Sydney: Collins.

———. 1984. *Wild food in Australia.* Sydney: Fontana/Collins.

Crowley, Aleister. 1977. *Diary of a drug fiend.* York Beach, Maine: Samuel Weiser. (Orig. pub. 1922.)

Cunningham, Sue, and Ghillian T. Prance. 1992. *Out of the Amazon.* London: HMSO.

D

Dafni, A., and Z. Yaniv. 1994. Solanaceae as medicinal plants in Israel. *Journal of Ethnopharmacology* 44:11–18.

Danielou, Alain. 1992a. Las divinidades alucinógenas. *Takiwasi* 1 (1): 25–29.

———. 1992b. *Gods of love and ecstasy: The traditions of Shiva and Dionysus.* Rochester, Vt.: Inner Traditions.

———. 1994. *The complete Kama Sutra.* Rochester, Vt.: Park Street Press.

D'Arcy, William G. 1991. The Solanaceae since 1976, with a review of its bibliography. In *Solanaceae III: Taxonomy, chemistry, evolution,* ed. Hawkes, Lester, Nee, and Estrada, 75–138. London: Royal Botanic Gardens Kew and Linnean Society.

da Orta, Garcia. 1987. *Colloquies on the simples and drugs of India.* Delhi: Sri Satguru Publications. (Orig. pub. 1563.)

Dastur, J. F. 1985. *Medicinal plants of India and Pakistan.* Bombay: Taraporevala.

Daumal, René. 1960. *Mount Analogue: An authentic narrative.* New York: Pantheon. (On experiments with hashish, mescaline, and other substances.)

Davis, Wade. 1985. Hallucinogenic plants and their use in traditional societies. *Cultural Survival Quarterly* 9 (4): 2–5.

———. 1992. The traditional use of psychoactive plants among America's native peoples. *Canadian Journal of Herbalism* 13 (1): 7–14, 30.

———. 1996. *One river: Explorations and discoveries in the Amazon rain forest.* New York: Simon & Schuster.

Davis, E. Wade, and James A. Yost 1983a. The ethnobotany of the Waorani of eastern Ecuador. *Botanical Museum Leaflets* 29 (3): 159–217.

———. 1983b. The ethnomedicine of the Waorani of Amazonian Ecuador. *Journal of Ethnopharmacology* 9:273–97.

de Cuveland, Helga. 1989. Der Gottorfer Codex. Worms: Wernersche Verlagsgesellschaft.

DeKorne, Jim. 1994. *Psychedelic shamanism.* Port Townsend, Wash.: Loompanics Unlimited.

———. [1995]. *Psychedelischer neo-Schamanismus.* Löhrbach: Werner Pieper's MedienXperimente (Edition Rauschkunde).

de Landa, Fray Diego. 2000. *An account of the things of Yucatán.* Mexico City: Monclem Ediciones. (Orig. pub. in Spanish, sixteenth century.)

DeLeeuw, Hendrik. 1939. *Flower of joy.* New York: Lee Furman.

Deltgen, Florian. 1978. Culture, drug and personality: A preliminary report about the results of a field research among the Yebámasa Indians of Río Piraparaná in the Colombian Comisaría del Vaupés. *Ethnomedizin* 5 (1–2): 57–81.

———. 1979. *Mit Flinte und Blasrohr: Urwaldindianer in Kolumbien.* Cologne: Rautenstrauch-Joest-Museum für Völkerkunde.

———. 1993. *Gelenkte Ekstase.* Stuttgart: Franz Steiner Verlag.

Denkow, Wesselin. 1992. *Gifte der Natur.* Steyr: Ennsthaler.

Derolez, R. L. M. 1963. *Götter und Mythen der Germanen.* Einsiedeln: Benziger.

de Ropp, Robert. 1961. *Drugs and the mind.* New York: Grove.

———. 1985. *Das Meisterspiel.* Munich: Knaur.

Descola, Philippe. 1996. *The spears of twilight: Life and death in the Amazon jungle.* Trans. Janet Lloyd. New York: The New Press.

de Smet, Peter A. G. M. 1995. Considerations in the multidisciplinary approach to the study of ritual hallucinogenic plants. In *Ethnobotany: Evolution of a discipline,* ed. Richard Evans Schultes and Siri von Reis, 369–83. Portland, Ore.: Dioscorides Press.

Devereux, Paul. 1990. *Places of power.* London: Blandford.

———. 1992a. *Shamanism and the mystery lines.* London: Quantum.

———. 1992b. *Symbolic landscapes.* Glastonbury, U.K.: Gothic Image Publications.

———. 1993. Schamanische Landschaften. In *Naturverehrung und Heilkunst,* ed. C. Rätsch, 165–88. Südergellersen: Verlag Bruno Martin.

———. 1997. *The long trip: A prehistory of psychedelia.* Harmondsworth and New York: Penguin (Arkana).

Devereux, Paul, John Steele, and David Kubrin. 1992. *Earthmind: Communicating with the living world of gaia.* Rochester, Vt.: Inner Traditions.

Diaz, José Luis, ed. 1975. *Etnofarmacología de plantas alucinógenas latinoamericanas.* Mexico City: Cuadernos Cientificos CEMEF 4.

———. 1977. Ethnopharmacology of sacred psycho-active plants used by the Indians of Mexico. *Annual Review of Pharmacology and Toxicology* 17:647–75.

———. 1979. Ethnopharmacology and taxonomy of Mexican psychodysleptic plants. *Journal of Psychedelic Drugs* 11 (1–2): 71–101.

Dibble, Charles E. 1966. La base científica para el estudio de las yerbas medicinales de los aztecas. *Actas y Memorias del XXXVI Congreso Internacionál de Americanistas* 2:63–67.

Dibble, Charles E., and Arthur J. O. Anderson. 1963. Florentine Codex, Book 11—*Earthly Things.* Santa Fe, N.M.: School of American Research; Salt Lake City: University of Utah Press.

Diener, Harry. 1989. *Fachlexikon abc: Arzneipflanzen und Drogen.* 2nd ed. Thun and Frankfurt/M.: Verlag Harri Deutsch.

Dierbach, Johann Heinrich. 1833. *Flora Mythologica.* Frankfurt. (Repr. Schaan, Liechtenstein: Sändig Reprint, 1981.)

Dioscorides. 1610. *Kreutterbuch.* Frankfurt/M.: Conrad Corthons.

———. 1902. *Arzneimittellehre.* Stuttgart: Enke.

Dittrich, Adolf. 1985. *Ätiologie-unabhängige Strukturen veränderter Wachbewußtseinszustände.* Stuttgart: Enke.

———. 1996. *Ätiologie-unabhängige Strukturen veränderter Wachbewußtseinszustände.* 2nd ed. Berlin: VWB.

Dittrich, Adolf, and Christian Scharfetter, eds. 1987. *Ethnopsychotherapie.* Stuttgart: Enke.

Dittrich, Bernd. 1988. *Duftpflanzen.* Munich: BLV.

Dobkin de Rios, Marlene. 1972. The anthropology of drug-induced altered states of consciousness. *Sociologus* 22 (1/2): 147–51.

———. 1974a. Cultural persona in drug-induced altered states of consciousness. In *Social and cultural identity,* ed. Thomas K. Fitzgerald, 16–23. Athens, Ga.: Southern Anthropological Society.

———. 1974b. The influence of psychotropic flora and fauna on Maya religion. *Current Anthropology* 15 (2): 147–64.

———. 1977a. Hallucinogenic ritual as theatre. *Journal of Psychedelic Drugs* 9 (3): 265–68.

———. 1977b. Plant hallucinogens and the religion of the Mochica. *Economic Botany* 31 (2): 189–203.

———. 1982. Plant hallucinogens, sexuality and shamanism in the ceramic art of ancient Peru. *Journal of Psychoactive Drugs* 14 (1–2): 81–90.

———. 1984. *Hallucinogens: Cross-cultural perspectives.* Albuquerque: University of New Mexico Press.

———. 1985. Schamanen, Halluzinogene and Erdaufschüttungen in der Neuen Welt. *Unter dem Pflaster liegt der Strand* 15:95–112.

———. 1995. Hallucinogens in cross-cultural perspective. In *Abstracts and selected papers,* vol. 5 of *Worlds of Consciousness,* ed. Michael Schlichting and Hanscarl Leuner, 81–96. Berlin: VWB.

Dobkin de Rios, Marlene, and David E. Smith. 1976. Using or abusing? An anthropological approach to the study of psychoactive drugs. *Journal of Psychedelic Drugs* 8 (3): 263–66.

———. 1977. The function of drug rituals in human society: Continuities and changes. *Journal of Psychedelic Drugs* 9 (3): 269–75.

Donoso Zegers, Claudio. 1995. *Arboles nativos de Chile.* Valdivia, Chile: Marisa Cúneo Ediciones.

Donoso Zegers, Claudio, and Carlos Ramírez García. 1994. *Arbustos nativos de Chile.* Valdivia, Chile: Marisa Cúneo Ediciones.

Dörfelt, Heinrich, ed. 1989. *Lexikon der Mykologie.* Stuttgart and New York: G. Fischer.

Dowling, Ralph M., and Ross A. McKenzie. 1993. *Poisonous plants: A field guide.* Brisbane, Queensland: Dept. of Primary Industry.

Dressler, Robert L. 1953. The pre-Columbian cultivated plants of Mexico. *Botanical Museum Leaflets* 16 (6): 115–72.

Drury, Nevill. 1989. *Vision quest.* Bridport, U.K.: Prism Press.

———. 1991. *The visionary human.* Shaftesbury, U.K.: Element Books.

———. 1996. *Shamanism.* Shaftesbury, U.K.: Element Books.

Duerr, Hans Peter. 1978. *Traumzeit.* Frankfurt/M.: Syndikat.

Duke, James A. 1970. Ethnobotanical observations on the Chocó Indians. *Economic Botany* 24:344–66.

———. 1975. Ethnobotanical observations on the Cuna Indians. *Economic Botany* 29:278–93.

———. 1987. *Living Liqueurs.* Lincoln, Mass.: Quarterman Publications, Inc.

Duke, James A., and Rodolfo Vasquez. 1994. *Amazonian ethnobotanical dictionary.* Boca Raton, Fla.: CRC Press.

DuQuesne, Terence, and Julian Reeves. 1982. *A handbook of psychoactive medicines.* London: Quartet Books.

DuToit, Brian M. 1977. *Drugs, rituals and altered states of consciousness.* Rotterdam: Balkema.

HALLUCINOGENS:
Cross-Cultural Perspectives

Marlene Dobkin de Rios

E

Ebel, S., and H. J. Roth. 1987. *Lexikon der Pharmazie.* Stuttgart and New York: Thieme.

Efron, Daniel H., Bo Holmstedt, and Nathan S. Kline, eds. 1967. *Ethnopharmacologic search for psychoactive drugs.* Washington, D.C.: U.S. Department of Health, Education, and Welfare.

Eldridge, Joan. 1975. Bush medicine in the Exumas and Long Island, Bahamas: A field study. *Economic Botany* 29:307–32.

Eliade, Mircea. 1964. *Shamanism: Archaic techniques of ecstasy.* Princeton, N.J.: Princeton University Press.

———. 1975. *Schamanismus und archaische Ekstasetechnik.* Frankfurt/M.: Suhrkamp.

———. 1992. *Schamanen, Götter und Mysterien: Die Weit der alten Griechen.* Freiburg: Herder.

Emboden, William A. 1972. *Narcotic plants.* London: Studio Vista.

———. 1974. *Bizarre plants: Magical, monstrous, mythical.* New York: Macmillan.

———. 1976. Plant hypnotics among the North American Indians. In *American folk medicine: A symposium,* ed. Wayland D. Hand, 159–67. Berkeley: University of California Press.

———. 1979. *Narcotic plants.* Rev. ed. New York: Macmillan.

———. 1995. Art and artifact as ethnobotanical tools in the ancient Near East with emphasis on psychoactive plants. In *Ethnobotany: Evolution of a discipline,* ed. Richard Evans Schultes and Siri von Reis, 93–107. Portland, Ore.: Dioscorides Press.

Emmart, E. W. 1937. Herb medicine of the Aztecs. *Journal of the American Pharmaceutical Association* 26:42–45.

———. 1940. *The Badianus manuscript.* Baltimore: The Johns Hopkins Press.

Engel, Fritz-Martin. 1978. *Zauberpflanzen— Pflanzenzauber.* Hannover: Landbuch-Verlag.

———. 1982. *Die Giftküche der Natur.* Hannover: Landbuch-Verlag.

Epstein, Mark, and Lobsang Rabgay. 1982. Mind and mental disorders in Tibetan medicine. *Tibetan Medicine* 5:66–82.

Erichsen-Brown, Charlotte. 1989. *Medicinal and other uses of North American plants.* New York: Dover.

Escohotado, Antonio. 1990. *Historia de las drogas.* 3 vols. Madrid: Alianza Editorial.

———. 1994. *Las drogas: De los orígenes a la prohibición.* Madrid: Alianza Cien.

Estrada Lugo, Erin Ingrid Jane. 1989. *El Códice Florentino: su información etnobotánica.* Chapingo, Edo. de Mexico: Colegio de Postgraduados (CP).

Evans, Hilary. 1989. *Alternate states of consciousness.* Wellingborough, U.K.: The Aquarian Press.

Evans, W. C. 1979. Tropane alkaloids of the Solanaceae. In *The biology and taxonomy of the Solanaceae,* ed. J. G. Hawkes, R. N. Lester, and A. D. Skelding, 241–54. London: Academic Press.

Evans-Wentz, W. Y. 1994. *The fairy faith in Celtic countries.* Intro. by Terence McKenna. New York: Library of Mystic Arts/Citadel Press.

Everist, Selwyn L. 1974. *Poisonous plants of Australia.* Sydney: Angus and Robertson.

F

Fabrega, Horacio, and Daniel B. Silver. 1973. *Illness and shamanistic curing in Zinacantan.* Stanford, Calif.: Stanford University Press.

Farnsworth, Norman R. 1968. Hallucinogenic plants. *Science* 162:1086–92.

———. 1972. Psychotomimetic and related higher plants. *Journal of Psychedelic Drugs* 5 (1): 67–74.

———. 1974. Psychotomimetic plants II. *Journal of Psychedelic Drugs* 6 (1): 83–84.

Fässler, Benjamin. 1997. *Drogen zwischen Herrschaft und Herrlichkeit.* Solothurn: Nachtschatten Verlag.

Faure, Paul. 1990. *Magie der Düfte: Eine Kulturgeschichte der Wohlgerüche von den Pharaonen zu den Römern.* Munich, Zurich: Artemis.

Faust, Franz X. 1983. *Medizinische Anschauungen und Praktiken der Landbevölkerung im andinen Kolumbien.* Hohenschäftlarn: Renner.

———. 1989. *Medizin und Weltbild.* Munich: Trickster.

Faust, Volker. 1994. *Psychopharmaka: Arzneimittel mit Wirkung auf das Seelenleben.* Stuttgart: Trias.

Felger, Richard S., and Mary Beck Moser. 1974. Seri Indian pharmacopoeia. *Economic Botany* 28:414–36.

———. 1991. *People of the desert and sea: Ethnobotany of the Seri Indians.* Tucson: University of Arizona Press.

Félice, Philippe de. 1936. *Poisons sacrés, ivresses divines.* Paris: Albin Michel.

———. 1990. *Le droghe degli dei: Veleni sacri, estasi divine.* Genoa, Italy: ECIG.

Fericgla, Josep Mª. 1994a. Etnopsiquiatría: Delirios; cultura y pruebas de realidad. *Rev. Psiquiatría Fac. Med. Barna.* 21 (4): 92–99.

———, ed. 1994b. *Plantas, chamanismo y estados de consciencia.* Barcelona: Los Libros de la Liebre de Marzo (Colección Cogniciones).

———. 1996. Traditional entheogens in the Mediterranean basin. Lecture presented at the Entheobotany Conference, San Francisco, October 18–20, 1996.

Fernández Chiti, Jorge. 1995. *Hierbas y plantas curativas.* Buenos Aires: Ediciones Condorhuasi.

Festi, Francesco. 1995. Le erbe del diavolo. 2: Botanica, chimica e farmacologia. *Altrove* 2:117–45.

Festi, Francesco, and Giovanni Aliotta. 1990. Piante psicotrope spontanee o coltivate in Italia. *Annali dei Musei Civici di Rovereto* 5 (1989): 135–66.

Filipov, A. 1994. Medicinal plants of the Pilagá of central Chaco. *Journal of Ethnopharmacology* 44:181–93.

Fischer, Georg, and Erich Krug. 1984. *Heilkrauter und Arzneipflanzen.* 7th ed. Heidelberg: Haug.

Flaubert, Gustave. 1979. *Die Versuchung des heiligen Antonius.* Frankfurt/M.: Insel. [English edition: Flaubert, Gustave. 2001. *The Temptation of Saint Anthony.* Lafcadio Hearn, trans. New York: Random House, Modern Library Edition.]

Fleurentin, Jacques, and Jean-Marie Pelt. 1982. Repertory of drugs and medicinal plants of Yemen. *Journal of Ethnopharmacology* 6:85–108.

Fontquer, Pio. 1993. *Plantas medicinales: El Dioscórides renovado.* 3 vols. Barcelona: Editorial Labor, S.A.

Ford, Joel. 1970. *Künstliche Freuden.* Bergisch-Gladbach: Lübbe.

Forte, Robert, ed. 1997. *Entheogens and the future of religion.* San Francisco: Council on Spiritual Practices/Promind Services (Sebastopol).

Foster, Steven, and James A. Duke. 1990. *Eastern/central medicinal plants.* A Peterson Field Guide. Boston and New York: Houghton Mifflin Co.

Franke, Gunther, ed. 1994. *Nutzpflanzen der Tropen únd Subtropen. Bd. 3: Spezieller Pflanzenbau.* Stuttgart: Ulmer (UTB). (On plants that provide agents of pleasure.)

Franquemont, Christine, Edward Franquemont, Wade Davis, Timothy Plowman, Steven R. King, Calvin R. Sperling, and Christine Niezgoda. 1990. The ethnobotany of Chinchero, an Andean community in southern Peru. *Fieldiana* (Botany), n.s., 24:1–126.

Frembgen, Jürgen. 1993. *Derwische: Gelebter Sufismus.* Cologne: DuMont.

Frerichs, G., G. Arends, and H. Zörnig, eds. 1938. *Hagers Handbuch der pharmazeutischen Praxis.* Berlin: J. Springer. [Note: *Hagers Handbuch* receives annual updates.]

Friedreich, J. B. 1966. *Zur Bibel: Naturhistorische, anthropologische und medicinische Fragmente.* Bad Reichenhall: Antiquariat Rudolf Kleinert. (Orig. pub. 1848.)

Frohne, Dietrich, and Uwe Jensen. 1992. *Systematik des Pflanzenreichs.* 4th ed. Stuttgart and New York: G. Fischer.

Frohne, Dietrich, and Hans Jürgen Pfander. 1997. *Giftpflanzen.* 4th ed. Stuttgart: WVG.

Fuchs, Leonhart. 1543. *Kreütterbuch.* Basel: Michael Isingrin.

———. 1545. *Läebliche abbildung und contrafaytung aller kreüter.* Basel: Michael Isingrin.

Fühner, Hermann. 1925. Solanazeen als Berauschungsmittel: Eine historisch-ethnologische Studie. *Archiv für experimentelle Pathologie und Pharmakologie* 111:281–94.

———. 1943. *Medizinische Toxikologie.* Leipzig: Georg Thieme.

Fuller, Thomas C., and Elizabeth McClintock. 1986. *Poisonous plants of California.* Berkeley: University of California Press.

Furst, Peter T., ed. 1972a. *Flesh of the gods.* New York: Praeger.

———. 1972b. Ritual use of hallucinogens in Mesoamerica. In *Religión en Mesoamérica, XII Mesa Redonda,* ed. K. J. Litvak and T. N. Castillo, 61–68. Mexico City.

———. 1974. Hallucinogens in precolumbian art. In *Art and environment in Native America* (Special Publication no. 7), ed. Mary Elizabeth King and Idris R. Traylor Jr. Lubbock: The Museum of Texas Tech, Texas Tech University.

———. 1976a. *Hallucinogens and culture.* Novato, Calif.: Chandler & Sharp.

———. 1976b. Shamanistic survivals in Meso-american religion. *XLI ICA (1974)* 3:149–57.

———. 1981. Pflanzenhalluzinogene in frühen amerikanischen Kulturen—Mesoamerika und die Anden. In *Rausch und Realität,* ed. G. Völger, 1:330–39. Cologne: Rautenstrauch-Joest Museum für Völkerkunde.

———. 1986. *Mushrooms: Psychedelic fungi.* New York: Chelsea House Publishers. (Rev. ed. in 1992.)

———. 1990. Schamanische Ekstase und botanische Halluzinogene: Phantasie und Realität. In *Der Gesang des Schamanen,* ed. G. Guntern, 211–43. Brig: ISO-Stiftung.

———. 1995. "This little book of herbs": Psychoactive plants as therapeutic agents in the Badianus manuscript of 1552. In *Ethnobotany: Evolution of a discipline,* ed. Richard Evans Schultes and Siri von Reis, 108–30. Portland, Ore.: Dioscorides Press.

———. 1996a. Intoxicating treasures: Native American entheogens in art and archaeology. Paper presented at the Entheobotany Conference, San Francisco, October 18–20, 1996.

———. 1996b. Shamanism, transformation, and Olmec art. In *The Olmec world: Ritual and ruler-ship,* 69–81. Princeton, N.J.: The Art Museum, Princeton University; New York: Harry N. Abrams.

G

Gandhi, Maneka, and Yasmin Singh. 1989. *Brahma's hair: Mythology of Indian plants.* Calcutta: Rupa.

Garza, Mercedes de la. 1990. *Sueños y alucinación en el mundo náhuatl y maya.* Mexico City: UNAM.

Peter T. Furst
ALUCINOGENOS
Y CULTURA
COLECCION POPULAR

Gawlik, Willibald. 1994. *Götter, Zauber und Arznei.* Schäftlarn: Barthel & Barthel.

Gebelein, Helmut. 1991. *Alchemie.* Munich: Diederichs.

Gebhart-Sayer, Angelika. 1987. *Die Spitze des Bewußt-seins: Untersuchungen zu Weltbild und Kunst der Shipibo-Conibo.* Münchener Beiträge zur Amerikanistik. Hohenschäftlarn: Klaus Renner Verlag.

Gelpke, Rudolf. 1995. *Vom Rausch im Orient und Okzident.* 2nd ed. Afterword by Michael Klett. Stuttgart: Klett-Cotta.

Genaust, Helmut. 1996. *Etymologisches Wörterbuch der botanischen Pflanzennamen.* 3rd ed. Basel: Birkhäuser.

Gentry, Alwyn H. 1993. *A field guide to the families and genera of woody plants of northwest South America (Colombia, Ecuador, Peru) with supplementary notes on herbaceous taxa.* Washington, D.C.: Conservation International.

Gentry, Johnnie L., and William G. D'Arcy. 1986. Solanaceae of Mesoamerica. In *Solanaceae: Biology and systematics,* ed. W. G. D'Arcy, 15–26. New York: Columbia University Press.

Georgiades, Christos Ch. 1987. *Flowers of Cyprus. Plants of medicine.* 2 vols. Nicosia: Cosmos Press.

Gerard, John. 1633. *The herball or general history of plants.* Ed. Thomas Johnson. London: Norton & Whitaker.

Germer, Renate. 1985. *Flora des pharaonischen Ägypten.* Mainz: Philipp von Zabern.

———. 1988. *Katalog der altägyptischen Pflanzenreste der Berliner Museen.* Ägyptologische Abhandlungen, vol. 47. Wiesbaden: Otto Harrassowitz.

———. 1991. *Mumien: Zeugen des Pharaonenreiches.* Zurich and Munich: Artemis & Winkler.

Geschwinde, Thomas. 1990. *Rauschdrogen: Marktformen und Wirkungsweisen.* Berlin: Springer Verlag.

———. 1996. *Rauschdrogen: Marktformen und Wirkungsweisen.* 3rd ed. Berlin: Springer Verlag.

Gessmann, G. W. [ca. 1899]. *Die Pflanze im Zauberglauben.* Den Haag: Couver.

Giant, Leo Maria. 1994. *In heiliger Leidenschaft: Mythen, Kulte und Mysterien.* Munich: Kösel.

Giese, Claudius Cristobal. 1989a. *"Curanderos": Traditionelle Heiler in Nord-Peru (Küste und Hochland).* Münchner Beiträge zur Amerikanistik, vol. 20. Hohenschäftlarn: Klaus Renner Verlag.

———. 1989b. Die Diagnosemethode eines nordperuanischen Heilers. *Curare* 12 (2): 81–87.

Gimlette, John D. 1981. *Malay poisons and charm cures.* Kuala Lumpur: Oxford in Asia Paperbacks.

Golowin, Sergius. 1971. Psychedelische Volkskunde. *Antaios* 12:590–604.

———. 1973. *Die Magie der verbotenen Märchen.* Gifkendorf: Merlin.

———. 1991. Psychedelische Volkskunde. In *Der Fliegenpilz,* ed. W. Bauer et al., 43–65. Cologne: Wienand Verlag. (This is an expanded and updated edition of Golowin 1971.)

Gonçalves de Lima, Oswaldo. 1986. *El maguey y el pulque en los códices mexicanos.* Mexico: Fondo de Cultura Económica. (See book review by John B. Tompkins in *American Anthropologist* 59 [1957]: 170–71.)

———. 1990. *Pulque, balché y pajauru, en la etnobiología de las bebidas y de los alimentos fermentados.* Mexico: Fondo de Cultura Económica.

Goodman, Felicitas D. 1992. *Trance—der uralte Weg zum religiösen Erleben.* Gütersloh: GTB.

Goodman, Jordan, Paul E. Lovejoy, and Andrew Sherratt, eds. 1995. *Consuming habits: Drugs in history and anthropology.* London and New York: Routledge.

Goodman, Steven M., and Abdul Ghafoor. 1992. The ethnobotany of southern Balochistan, Pakistan, with particular reference to medicinal plants. *Fieldiana* (Botany), n.s., 31:1–84.

Gottlieb, Adam. 1973. *Legal highs.* Manhattan Beach, Calif.: 20th Century Alchemist.

———. 1974. *Sex, drugs, and aphrodisiacs.* Manhattan Beach, Calif.: 20th Century Alchemist.

Graf, Alfred Byrd. 1992. *Tropica: Color cyclopedia of exotic plants and trees.* 4th ed. East Rutherford, N.J.: Roehrs Co.

Graupner, Heinz. 1966. *Dämon Rausch.* Hamburg: Hapus Verlag.

Graves, Robert. 1966. *The white goddess.* Amended, enlarged ed. New York: Farrar, Straus and Giroux.

Grey-Wilson, Christopher. 1995. *Poppies: The poppy family in the wild and in cultivation.* Portland, Ore.: Timber Press.

Grieve, Maude. 1982. *A modern herbal.* New York: Dover. (Orig. pub. 1931.)

Griffith, F. L., and Herbert Thompson. 1974. *The Leyden papyrus: An Egyptian magical book.* New York: Dover.

Grimm, Gorm. 1992. *Drogen gegen Drogen: Eine Bilanz.* Kiel: Veris Verlag.

Grinspoon, Lester, and James B. Bakalar. 1981. *Psychedelic drugs reconsidered.* New York: Basic Books.

———. 1983, eds. *Psychedelic reflections.* New York: Human Sciences Press.

———. 1987. Medical uses of illicit drugs. In *Dealing with drugs,* ed. Ronald Hamowy. Lexington, Mass.: Lexington Books.

Grob, Charles S. 1995. Psychiatric research with hallucinogens: What have we learned? *Yearbook* for *Ethnomedicine and the Study of Consciousness*, 1994 (3): 91–112. Berlin: VWB.

Grob, Charles S., and Marlene Dobkin de Rios. 1992. Adolescent drug use in cross-cultural perspective. *The Journal of Drug Issues* 22 (1): 121–38.

Grob, Charles S., and Willis Harman. 1995. Making sense of the psychedelic issue. *Noetic Sciences Review* (Summer 1995): 1–10.

Grof, Stanislav. 1976. *Realms of the human unconscious: Observations from LSD research.* New York: E. P. Dutton.

———. 1978. *Topographie des Unbewußten.* Stuttgart: Klett-Cotta.

———. 1985. *Geburt, Tod und Transzendens.* Munich: Kösel.

———. 1996. Technologien des Heiligen. *Esotera* 11/97:16–21.

———. 1997. *Kosmos und Psyche: An den Grenzen des menschlichen Bewußtseins.* Frankfurt/M.: Wolfgang Krüger Verlag.

Grubber, Hudson. 1991. *Growing the hallucinogens.* Berkeley, Calif.: 20th Century Alchemist. (Orig. pub. 1973.)

Guerra, Francisco. 1967. Mexican phantastica: A study of the early ethnobotanical sources on hallucinogenic drugs. *British Journal of Addiction* 62:171–87.

———. 1971. *The pre-Columbian mind.* London: Seminar Press.

———. 1990. *La medicina precolombiana.* [Spain]: Instituto de Cooperacion Iberoamericana.

Gunther, Erna. 1988. *Ethnobotany of western Washington: The knowledge and use of indigenous plants by Native Americans.* Rev. ed. Seattle and London: University of Washington Press.

Gupta, Shakti M. 1991. *Plant myths and traditions in India.* 2nd ed. New Delhi: Munshiram Manoharlal Publishers.

H

Haerkötter, Gerd, and Marlene Haerkötter. 1986. *Hexenfurz und Teufelsdreck.* Frankfurt/M.: Eichborn.

———. 1991. *Wüterich und Hexenmilch: Giftpflanzen.* Frankfurt/M.: Eichborn.

Haerkötter, Gerd, and Thomas Lasinski. 1989. *Das Geheimnis der Pimpernuß: Liebeskräuter aus Gottes Garten.* Frankfurt/M.: Eichborn.

Halifax, Joan, ed. 1981. *Die andere Wirklichkeit der Schamanen.* Bern and Munich: O.W. Barth/Scherz.

———. 1983. *Schamanen.* Frankfurt/M.: Insel.

———. 1994. *The fruitful darkness.* San Francisco: HarperCollins.

Hansen, Harold A. 1981. *Der Hexengarten.* Munich: Trikont-Dianus.

Hargous, Sabine. 1976. *Beschwörer der Seelen: Das*

magische Universum der südamerikanischen Indianer. Basel: Sphinx.

Harner, Michael, ed. 1973. *Hallucinogens and shamanism.* London: Oxford University Press.

———. 1984. *The Jíbaro: People of the sacred waterfalls.* Berkeley: University of California Press.

———. 1989. Was ist ein Schamane? In *Opfer und Ekstase,* ed. Gary Doore, 20–31. Freiburg: Bauer.

———. 1990. *The way of the shaman.* Rev. ed. San Francisco: HarperSanFrancisco.

———. 1994. *Der Weg des Schamanen.* Genf: Ariston.

Hart, Jeffrey A. 1979. The ethnobotany of the Flathead Indians of western Montana. *Botanical Museum Leaflets* 27 (10): 261–307.

Hartwich, Carl. 1911. *Die menschlichen Genußmittel: Ihre Herkunft, Verbreitung, Geschichte, Anwendung und Wirkung.* Leipzig: Tauchnitz.

Hasenfratz, Hans-Peter. 1992. *Die religiöse Welt der Germanen.* Freiburg: Herder.

Hasterlik, Alfred. 1918. *Von Reiz- und Rauschmitteln.* Stuttgart: Kosmos-Franckh'sche.

Havard, V. 1896. Drink plants of the North American Indians. *Bulletin of the Torrey Botanical Club* 23 (2): 33–46.

Hecht, Hans. 1995. *Kakteen und andere Sukkulenten.* 7th ed. Munich: BLV.

Hecker, Ulrich. 1995. *Bäume und Sträucher.* Munich: BLV.

Heffern, Richard. 1974. *Secrets of mind-altering plants of Mexico.* New York: Pyramid.

Heilmann, Peter. 1984. *Das Kräuterbuch der Elisabeth Blackwell.* Dortmund: Harenberg.

Heiser, Charles B. 1987. *The fascinating world of the nightshades.* New York: Dover.

———. 1990. *Seed to civilization: The story of food.* Cambridge, Mass., and London: Harvard University Press.

Helfrich, Klaus. 1972. Sexualität und Repression in der Kultur der Maya. *Baessler-Archiv,* n.f., 20:139–71.

Hellinga, Gerben, and Hans Plomp. 1994. *Uit je bol: Over XTC, paddestoelen, wiet en andere middelen.* Amsterdam: Prometheus.

Helmlin, Hans-Jörg, and Rudolf Brenneisen. 1992. Determination of psychotropic phenylalkylamine derivatives in biological matrices by high-performance liquid chromatography with photo-diode-array detection. *Journal of Chromatography* 593:87–94. (On peyote, San Pedro cactus.)

Henman, Anthony R., R. Lewis, and T. Maylon, eds. 1985. *Big deal: The politics of the illicit drugs business.* London: Pluto.

Hepper, F. Nigel 1992. *Pflanzenwelt der Bibel.* Stuttgart: Deutsche Bibelgesellschaft.

Hermle, Leo, et al. 1992. Untersuchungen über Arylalkylamin-induzierte Wirkungen bei gesunden Probanden. *Jahrbuch des Europäischen Collegiums für Bewußtseinsstudien*, 1992: 53–62. Berlin: VWB.

Hermle, Leo, E. Gouzoulis, G. Oepen, M. Spitzer, K. A. Kovar, D. Borchardt, M. Fünfgeld, and M. Berger. 1993. Zur Bedeutung der historischen und aktuellen Halluzinogenforschung in der Psychiatrie. *Der Nervenarzt* 64:562–71.

Hermle, Leo, G. Oepen, and M. Spitzer. 1988. Zur Bedeutung der Modellpsychosen. *Fortschritte der Neurologie, Psychiatrie* 56 (2): 35–68.

Hernandez, Francisco. 1942/46. *Historia de las plantas de Nueva España*. 3 vols. Mexico City: Imprenta Universitaria. (Orig. pub. 1615.)

Herodotus. 1987. *The history*. Trans. David Grene. Chicago: The University of Chicago Press. (Orig. pub. 440 B.C.E.)

Hesse, Hermann, ed. 1986. *Spuk- und Hexengeschichten*. Frankfurt/M.: Insel.

Heubner, Wolfgang. 1952. *Genuss und Betäubung durch chemische Mittel*. Wiesbaden: Verlag für angewandte Wissenschaften.

Heyden, Doris. 1979. Flores, creencias y el control social. *Actes du XLIIe Congrès International des Américanistes* (Paris) 6:85–97.

———. 1985. *Mitologia y simbolismo de la flora en el México prehispanico*. Mexico City: UNAM.

Hiller, Karl, and Günter Bickerich. 1988. *Giftpflanzen*. Stuttgart: Enke.

Hilton, M. G., and P. D. G. Wilson. 1995. Growth and the uptake of sucrose and mineral ions by transformed root cultures of *Datura stramonium*, *Datura candida x aurea*, *Datura wrightii*, *Hyoscyamus muticus* and *Atropa belladonna*. *Planta Medica* 61:345–50.

Hirschfeld, Magnus, and Richard Linsert. 1930. *Liebesmittel*. Berlin: Man Verlag.

Hlava, Bohumir, and Dagmar Lanska. 1977. *Lexikon der Küchen- und Gewürzkräuter*. Herrsching: Pawlak.

Hobson, J. Allen. 1994. *The chemistry of conscious states: How the brain changes its mind*. Boston: Little, Brown & Co.

Hoffer, Abraham, and Humphry Osmond. 1967. *The hallucinogens*. New York and London: Academic Press.

Hoffmann, Adriana, Cristina Farga, Jorge Lastra, and Esteban Veghazi. 1992. *Plantas medicinales de use común en Chile*. Santiago: Ediciones Fundación Claudio Gay.

Hoffmann, E. T. A. 1982. *Die Elixiere des Teufels*. Stuttgart: Reclam.

Höfler, Max. 1990. *Volksmedizinische Botanik der Germanen*. Berlin: VWB. (Orig. pub. 1908.)

Hofmann, Albert. 1980. *LSD: My problem child*. New York: McGraw-Hill.

———. 1986. *Einsichten—Ausblicke*. Basel: Sphinx.

———. 1995. Medicinal chemistry's debt to ethnobotany. In *Ethnobotany: Evolution of a discipline*, ed. Richard Evans Schultes and Siri von Reis, 311–19. Portland, Ore.: Dioscorides Press.

———. 1996. *Lob des Schauens*. Private printing (limited edition of 150).

Höhle, Sigi, Claudia Müller-Ebeling, Christian Rätsch, and Ossi Urchs. 1986. *Rausch und Erkenntnis: Das wilde in der kultur*. Munich: Knaur.

Honeychurch, Penelope N. 1986. *Caribbean wild plants and their uses*. London: Macmillan.

Hooper, David. 1937. *Useful plants and drugs of Iran and Iraq*. Botanical Series IX, 3. Chicago: Field Museum of Natural History.

Horman, Richard E., and Allan M. Fox, eds. 1970. *Drug awareness*. New York: Avon Books.

Houghton, P. J., and J. Manby. 1985. Medicinal plants of the Mapuche. *Journal of Ethnopharmacology* 13 (1): 89–103.

Hovorka, O. von, and A. Kronfeld. 1908. *Vergleichende Volksmedizin*. Stuttgart: Strecker & Schröder.

Hubinger Tokarnia, Carlos, Jürgen Döbereiner, and Marlene Freitas da Silva. 1979. *Plantas tóxicas da Amazônia a Bovinos e outros herbívoros*. Manaus, Brazil: Instituto Nacional de Pesquisas da Amazônia (INPA).

Human, Dawn. 1988. *How to get high: The use of drugs as a spiritual path*. Portland, Ore.: self-published.

Hunnius, Curt. 1975. *Pharmazeutisches Wörterbuch*. 5th ed. Berlin and New York: Walter de Gruyter.

Hutchens, Alma R. 1992. *A handbook of Native American herbs*. Boston and London: Shambhala.

Huxley, Aldous. 1963. *The doors of perception and heaven and hell*. New York: Perennial.

———. 1983. *Moksha*. Munich: Piper.

Huxley, Anthony. 1984. *Green inheritance*. London: Gaia Books.

Hyslop, Jon, and Paul Ratcliffe. 1989. *A folk herbal*. Oxford: Radiation Publications.

I

Illmaier, Thomas. 1997. *Rauschzeit*. Berlin: VWB.

Imam, Y. M. A. el, and W. C. Evans. 1990. Alkaloids of a *Datura candida* cultivar, *D. aurea* and various hybrids. *Fitoterapía* 61 (2): 148–52.

Inaba, Darryl S., and William E. Cohen. 1993. *Uppers, downers, all arounders: Physical and mental effects of psychoactive drugs*. 2nd ed. Ashland, Ore.: CNS Productions.

Inchaustegui, Carlos. 1977. *Relatos del mundo mágico mazateco*. Mexico City: INAH.

Inciardi, James N. 1992. *The war on drugs II: The continuing epic of heroin, cocaine, crack, crime, AIDS, and public policy.* Mountain View, Calif., London, and Toronto: Mayfield Publishing Co.

Isaacs, Jennifer. 1987. *Bush food: Aboriginal food and herbal medicine.* MacMahons Point, Australia: Weldons.

Isbell, H., and C. W. Gorodetzky. 1966. Effect of alkaloids of oloiuqui in man. *Psychopharmacologia* 8 (5): 331–39.

J

Jackes, Betsy R. 1992. *Poisonous plants in northern Australian gardens.* Townsville, Australia: James Cook University of North Queensland.

Jacob, Irene, and Walter Jacob, eds. 1993. *The healing past: Pharmaceuticals in the biblical and rabbinic world.* Leiden: Brill.

Jain, S. K. 1965. Medicinal plant lore of the tribals of Bastar. *Economic Botany* 19:236–50.

———. 1991. *Dictionary of Indian folk medicine and ethnobotany.* New Delhi: Deep Publications.

Jain, S. K., and S. K. Borthakur. 1986. Solanaceae in Indian tradition, folklore, and medicine. In *Solanaceae: Biology and systematics,* ed. William G. D'Arcy, 577–83. New York: Columbia University Press.

Jain, S. K., and Namita Dam [nee Goon]. 1979. Some ethnobotanical notes from northeastern India. *Economic Botany* 33 (1): 52–56.

Jain, S. K., V. Ranjan, E. L. S. Sikarwar, and A. Saklani. 1994. Botanical distribution of psychoactive plants in India. *Ethnobotany* 6:65–75.

Jiu, James. 1966. A survey of some medicinal plants of Mexico for selected biological activities. *Lloydia* 29 (3): 250–59.

Johnston, Alex. 1970. Blackfoot Indian utilization of the flora of the northwestern Great Plains. *Economic Botany* 24:301–24.

Johnston, James F. 1853. The narcotics we indulge in. *Blackwood's Edinburgh Magazine* 74:129–39, 605–28.

———. 1855. *The chemistry of common life. Vol. II: The narcotics we indulge in.* New York: D. Appleton & Co.

———. 1869. *Die Chemie des täglichen Lebens.* 2 vols. Berlin. (German edition of Johnston 1855.)

———. 1985. The narcotics we indulge in. Part I. *Journal of Psychoactive Drugs* 17 (3): 191–99. (Reprint of Johnston 1853.)

———. 1986. The narcotics we indulge in. Part II. *Journal of Psychoactive Drugs* 18 (2): 131–50. (Reprint of Johnston 1853.)

Jones, Hardin B., and Helen C. Jones. 1978. *Sensual drugs: Deprivation and rehabilitation of the mind.* Cambridge, U.K.: Cambridge University Press.

Joralemon, Donald, and Douglas Sharon. 1993. *Sorcery and shamanism: Curanderos and clients in northern Peru.* Salt Lake City: University of Utah Press.

Josuttis, Manfred, and Hanscarl Leuner, eds. 1972. *Religion und die Droge.* Stuttgart: Kohlhammer.

Jované, Ana, ed. 1994. *De México al mundo: Plantas.* Mexico City: Grupo Azabache.

Joyce, Christopher. 1994. *Earthly goods: Medicine-hunting in the rainforest.* Boston: Little, Brown & Company.

Jünger, Ernst. 1980. *Annäherungen—Drogen und Rausch.* Frankfurt/M.: Ullstein.

K

Kakar, Sudhir. 1984. *Schamanen, Heilige und Ärzte.* Munich: Biederstein.

Kalweit, Holger. 1984. *Traumzeit und innerer Raum: Die Welt der Schamanen.* Bern: Scherz.

———. 1992. *Urheiler, Medizinleute und Schamanen.* Munich: Heyne.

Karlinger, Felix, and Elisabeth Zacherl. 1976. *Südamerikanische Indianermärchen.* Cologne: Diederichs.

Kaufmann, Richard. 1985. *Die Krankheit erspüren: Tibets Heilkunst und der Westen.* Munich and Zurich: Piper.

Keng, Hsuan. 1974. Economic plants of ancient north China as mentioned in *Shih Ching* (Book of Poetry). *Economic Botany* 28:391–410.

Kindscher, Kelly. 1992. *Medicinal wild plants of the prairie: An ethnobotanical guide.* Lawrence: University of Kansas Press. (See book review by Jonathan Ott in *Journal of Ethnobiology* 12 [2; 1992]: 279 ff.)

Klüver, Heinrich. 1966. *Mescal and mechanisms of hallucinations.* Chicago: University of Chicago Press.

Knab, Timothy J. 1995. *A war of witches: A journey into the underworld of the contemporary Aztecs.* San Francisco: HarperCollins.

Koch-Grünberg, Theodor. 1921. *Zwei Jahre bei den Indianern Nordwest-Brasiliens.* Stuttgart: Strecker & Schröder.

———. 1956. *Geister am Roraima.* Eisenach and Kassel: Röth-Verlag.

Kohn, Marek. 1992. *Dope girls: The birth of the British drug underground.* London: Lawrence & Wishart.

Kondo, Norio, Hiroshi Yuasa, and Fumio Maekawa. 1987. *Resource-handbook of legumes.* Tokyo: The Japan Science Society. (In Japanese.)

Körner, Harald Hans. 1994. *Beck'sche Kurz-Kommentare,* vol. 37 of *Betäubungsmittelgesetz—Arzneimittelgesetz.* 4th ed. Munich: C. H. Beck.

Körner, Wolfgang. 1980. *Drogenreader.* Frankfurt/M.: Fischer.

Kotschenreuther, Hellmut. 1978. *Das Reich der Drogen und Gifte.* Frankfurt/ M.: Ullstein.

Kottek, Samuel S. 1994. *Medicine and hygiene in the works of Flavius Josephus.* Leiden: E. J. Brill.

Kraemer, Olaf. 1997. *Luzifers Lichtgarten: Expeditionen ins Reich der Halluzinogene.* Munich: Hugendubel/Sphinx. (With music CD.)

Kraepelin, Emil. 1892. *Über die Beeinflussung einfacher psychologischer Vorgänge durch einige Arzneimittel.* Jena: Fischer Verlag.

Kraft, Hartmut. 1995. *Über innere Grenzen: Initiation in Schamanismus, Religion und Psychoanalyse.* Munich: Diederichs Verlag.

Krause, M. 1909. Die Gifte der Zauberer im Herzen Afrikas. *Zeitschrift für experimentelle Pathologie und Therapie* 6.

Krauss, Beatrice H. 1981. *Native plants used as medicine in Hawaii.* Honolulu: Lyon Arboretum, University of Hawaii at Manoa.

———. 1993. *Plants in Hawaiian culture.* Honolulu: University of Hawaii Press.

Kreuter, Marie-Luise. 1982. *Wunderkräfte der Natur: Von Alraunen, Ginseng und anderen Wunderwurzeln.* Munich: Heyne.

Krochmal, Arnold, and Connie Krochmal. 1984. *A field guide to medicinal plants.* New York: Times Books.

Kronfeld, Moritz. 1981. *Donnerwurz und Mäuseaugen.* Berlin: Zerling. (Orig. pub. 1898.)

Kruedener, Stephanie von, Isolde Hagemann, and Bernhard Zepernick. 1993. *Arzneipflanzen—altbekannt und neu entdeckt.* Berlin: Botanischer Garten und Botanisches Museum Berlin-Dahlem.

Krug, Antje. 1993. *Heilkunst und Heilkult: Medizin in der Antike.* Munich: C. H. Beck.

Kupfer, Alexander. 1996a. *Göttliche Gifte: Kleine Kulturgeschichte des Rausches seit dem Garten Eden.* Stuttgart and Weimar: Verlag J. B. Metzler.

———. 1996b. *Die künstlichen Paradiese: Rausch und Realität seit der Romantik.* Stuttgart and Weimar: Verlag J. B. Metzler.

Kutscher, Gerdt. 1977. *Chimu: Eine altindianische Hochkultur.* Hildesheim: Gerstenberg.

Küttner, Michael. 1995. *Psychedelische Handluneselemente in den Märchen der Brüder Grimm.* Wetzlar: Schriftenreihe und Materialien der Phantastischen Bibliothek Wetzlar, vol. 4.

L

La Barre, Weston. 1964. Le complexe narcotique de l'Amérique autochthone. *Diogène* 48:120–34.

———. 1970. Old and New World narcotics. *Economic Botany* 24 (1): 73–80.

———. 1972. Hallucinogens and the shamanic origins of religion. In *Flesh of the gods,* ed. Peter T. Furst, 261–94. New York: Praeger.

———. 1979a. *The peyote cult.* Norman: University of Oklahoma Press.

———. 1979b. Shamanic origins of religion and medicine. *Journal of Psychedelic Drugs* 11 (1–2):7–11.

———. 1980. *Culture in context: Selected writings.* Durham, N.C.: Duke University Press.

———. 1995. The importance of ethnobotany in American anthropology. In *Ethnobotany: Evolution of a discipline,* ed. Richard Evans Schultes and Siri von Reis, 226–34. Portland, Ore.: Dioscorides Press.

Labrousse, Alain. 1993. Coca, peyotl, ayahuasca y opio—guerra a la droga y represión de las minorias nacionales. *Takiwasi* 2 (1): 121–33.

Lad, Vasant, and David Frawley. 1987. *Die Ayurveda-Pflanzen-Heilkunde.* Haldenwang: Edition Shangrila.

Lame Deer, Archie Fire, and Richard Erdoes. 1992. *Gift of power: The life and teachings of a Lakota medicine man.* Santa Fe, N.M.: Bear & Co. Publishing.

Landy, Eugene E. 1971. *The underground dictionary.* New York: Simon & Schuster.

Langdon, E. Jean Matteson, and Gerhard Baer, eds. 1992. *Portals of power: Shamanism in South America.* Albuquerque: University of New Mexico Press.

Lanteri-Laura, G. 1994. *Las alucinaciones.* Mexico City: Fondo de Cultura Economica.

Lara Ochoa, Francisco, and Carmen Marquez Alonso. 1996. *Plantas medicinales de México: Composición, usos y actividad biológica.* Mexico City: UNAM.

Lassak, Erich V., and Tara McCarthy. 1987. *Australian medicinal plants.* Port Melbourne, Australia: Mandarin. [Reprinted in paperback in 2001 by Australian publisher Reed/New Holland.]

Latorre, Dolores L., and Felipe A. Latorre. 1977. Plants used by the Mexican Kickapoo Indians. *Economic Botany* 31 (3): 340–57.

Lauber, Konrad, and Gerhart Wagner. 1996. *Flora Helvetica.* Bern, Stuttgart, and Vienna: Haupt.

Leary, Timothy. 1982. *Politik der Ekstase.* Linden: Volksverlag.

———. 1986. *Denn sie wussten, was sie tun: Eine Rückblende.* Basel: Sphinx.

———. 1990. *The politics of ecstasy.* Rev. ed. Berkeley, Calif.: Ronin Publishing.

———. n.d. *Über die Kriminalisierung des Natürlichen.* Der Grüne Zweig 138. Löhrbach: Werner Pieper's MedienXperimente.

Leary, Timothy, Ralph Metzner, and Richard Alpert.

1964. *The psychedelic experience.* New York: University Books.

Leginger, Thomas. 1981. *Urwald: Eine Reise zu den Schamanen des Amazonas.* Munich: Trikont-dianus.

Leibrock-Plehn, Larissa. 1992. *Hexenkräuter oder Arznei: Die Abtreibungsmittel im 16. und 17. Jahrhundert.* Stuttgart: WVG.

Leippe, Peter. 1997. *Gegenwelt Rauschgift: Kulturen und ihre Drogen.* Cologne: vgs (ZDF).

Lenson, David. 1995. *On drugs.* Minneapolis and London: University of Minnesota Press.

Lenz, Harald Othmar. 1966. *Botanik der Griechen und Römer.* Vaduz: Sändig. (Orig. pub. 1859.)

Leu, Daniel. 1984. *Drogen: Sucht oder Genuß.* Basel: Lenos Verlag.

Leuenberger, Hans. 1969. *Zauberdrogen: Reisen ins Weltall der Seele.* Stuttgart: Henry Goverts Verlag.

———. 1970. *Im Rausch der Drogen.* Munich: Humboldt.

———. 1979. *Mexiko—Land links vom Kolobri.* Frankfurt/M.: Fischer.

Leuner, Hanscarl. 1962. *Die experimentelle Psychose.* Berlin: Springer.

———. 1981. *Halluzinogene.* Bern: Huber.

———, ed. 1996. *Psychotherapie und religiöses Erleben: Ein Symposion über religiöse Erfahrungen unter Einflug von Halluzinogenen.* Berlin: VWB.

Leuner, Hanscarl, and Michael Schlichting. 1986. *Symposion "Über den derzeitigen Stand der Forschung auf dem Gebiet der psychoaktiven Substanzen" vom 29.11. bis 1.12.1985, Burg, Hirschhorn.* Berlin: EXpress Edition.

Leung, Albert Y. 1995. *Chinesische Heilkräuter.* Munich: Diederichs Verlag.

Lewin, Louis. 1920. *Die Gifte in der Weltgeschichte.* Berlin: Julius Springer.

———. 1998. *Phantastica.* Rochester, Vt.: Park Street Press. (Orig. pub. 1929.)

———. 1984. *Die Pfeilgifte.* Hildesheim: Gerstenberg. (Orig. pub. 1923.)

———. 1992. *Gifte und Vergiftungen: Lehrbuch der Toxikologie.* 6th ed. Heidelberg: Haug.

Lewin, Roger. 1991. Stone Age psychedelia. *New Scientist* 8/91:30–34.

Lewington, Anna. 1990. *Plants for the people.* London: The Natural History Museum.

Lewis-Williams, J. D., and T. A. Dowson. 1988. The signs of all times: Entoptic phenomena in upper Palaeolithic art. *Current Anthropology* 29 (2): 201–45.

———. 1993. On vision and power in the Neolithic: Evidence from the decorated monuments. *Current Anthropology* 34 (1): 55–65.

Li, Hui-Lin. 1975. Hallucinogenic plants in Chinese herbals. *Botanical Museum Leaflets* 25 (6): 161–81.

———. 1979. *Nan-fang ts'ao-mu chuang: A fourth century flora of Southeast Asia.* Hong Kong: The Chinese University Press.

Lin, Geraline C., and Richard A. Glennon, eds. 1994. *Hallucinogens: An update.* NIDA Research Monograph 146. Rockville, Md.: National Institute on Drug Abuse.

Lindequist, Ulrike. 1992. Datura. In *Hagers Handbuch der pharmazeutischen Praxis,* 5th ed., 4:1138–54. Berlin: Springer.

Lindstrom, Lamont, ed. 1987. *Drugs in western Pacific societies.* ASAO Monograph 11. Lanham, Md.: University Press of America.

Lingeman, Richard R. 1969. *Drugs from A to Z: A dictionary.* New York: McGraw-Hill.

Lipp, Frank J. 1971. Ethnobotany of the Chinantec Indians, Oaxaca, Mexico. *Economic Botany* 25:234–44.

———. 1991. *The Mixe of Oaxaca: Religion, ritual, and healing.* Austin: University of Texas Press.

———. 1996. *Herbalism.* London and Boston: Little, Brown & Co.

Lippert, Herbert. 1972. *Einführung in die Pharmakopsychologie.* Munich: Kindler.

Löbsack, Theo. 1979. *Die manipulierte Seele.* Dusseldorf and Vienna: Econ.

Lohs, Karlheinz, and Dieter Martinetz. 1986. *Gift: Magie und Realität.* Munich: Callway.

Lommel, Andreas. 1980. *Schamanen und Medizinmänner.* 2nd ed. Munich: Callway.

Lonicerus, Adamus. 1679. *Kreuterbuch.* Frankfurt: Matthius Wagner.

Low, Tim. 1990. *Bush medicine: A pharmacopoeia of natural remedies.* North Ride, Australia: Angus & Robertson.

———. 1992a. *Bush tucker: Australia's wild food harvest.* Pymble, Australia: Angus & Robertson.

———. 1992b. *Wild food plants of Australia.* Pymble, Australia: Angus & Robertson.

———. 1993. *Wild herbs of Australia and New Zealand.* Pymble, Australia: Angus & Robertson.

Lu, An-ming. 1986. Solanaceae in China. In *Solanaceae: Biology and systematics,* ed. William G. D'Arcy, 79–85. New York: Columbia University Press.

Lu, Henry C. 1991. *Legendary Chinese healing herbs.* New York: Sterling Publishing Co.

Luck, Georg. 1962. *Hexen und Zauberei in der Römischen Dichtung.* Zurich: Artemis.

———. 1990. *Magie und andere Geheimlehren in der Antike.* Stuttgart: Kröner.

Ludwig, Otto. 1982. *Im Thüringer Kräutergarten: Von Heilkräutern, Hexen und Buckelapothekern.* Gütersloh: Prisma Verlag.

Lyttle, Thomas, ed. 1994. *Psychedelics: A collection of the most exciting new material on psychedelic drugs.* New York: Barricade Books.

M

MacFarlane, Ruth B. [Alford]. 1994. *Collecting and preserving plants.* New York: Dover.

Macmillan, H. F. 1991. *Tropical planting and gardening.* 6th ed. Kuala Lumpur: Malayan Nature Society.

Madhihassan, S. 1991. *Indian alchemy or rasayana.* Delhi: Motilal Banarsidass Publ.

Madsen, William, and Claudia Madsen. 1972. *A guide to Mexican witchcraft.* Mexico City: Minutae Mexicana.

Magister Botanicus. 1995. *Magisches Kreutherkompendium.* 2nd ed. Runkel, Germany: Verlag Die Sanduhr—Fachverlag für altes Wissen.

Majupuria, Trilok Chandra, and D. P. Joshi. 1988. *Religious and useful plants of Nepal and India.* Lautpur, Nepal: M. Gupta.

Malcolm, Andrew I. 1972. *The pursuit of intoxication.* New York: Washington Square Press.

Manandhar, N. P. 1980. *Medicinal plants of Nepali Himalaya.* Kathmandu: Rama Pustak Bhandar. (On *Aconitum, Acorus, Alstonia,* and *Ephedra.*)

Mandl, Elisabeth. 1985. *Arzneipflanzen in der Homöopathie.* Vienna, Munich, and Bern: Verlag Wilhem Maudrich.

Mann, John. 1994. *Murder, magic and medicine.* Oxford, New York, and Tokyo: Oxford University Press.

Manniche, Lise. 1988. *Liebe und Sexualität im alten Ägypten.* Zurich and Munich: Artemis.

———. 1989. *An ancient Egyptian herbal.* London: British Museum.

Mantegazza, Paolo. 1858. Sulle virtù igieniche e medicinali della coca e sugh alimenti nervosi in generale. *Ann. Univ. Med.* 167:449–519.

———. 1871. *Quadri della natura umana: Feste ed ebbrezze.* 2 vols. Milan: Brigola.

———. 1887. *Le estasi umane.* Milan: Dumolard.

Marhenke, Dorit, and Ekkehard May. 1995. *Shunga: Erotic art in Japan.* Heidelberg: Edition Braus.

Markale, Jean. 1999. *The Druids: Celtic Priests of Nature.* Rochester, Vt.: Inner Traditions.

Marquardt, Hans, and Siegfried G. Schäfer, eds. 1994. *Lehrbuch der Toxikologie.* Mannheim: B.I. Wissenschaftsverlag. (Includes a short chapter titled "Stoffe mit hypnotischen und psychotropen Wirkungen" ["Substances with Hypnotic and Psychotropic Effects"] 679 ff.)

Martinetz, Dieter. 1994. *Rauschdrogen und Stimulantien: Geschichte—Fakten—Trends.* Leipzig, Jena, and Berlin: Urania.

Martínez, Maximino. 1987. *Catálogo de nombres vulgares y científicos de plantas mexicanas.* Mexico City: Fondo de Cultura Económica.

———. 1994. *Las plantas medicinales de México.* 6th ed. Mexico City: Ediciones Botas.

Martini, F. C. 1977. *Pianti medicamentosi e rituali magico-religiosi in Plinio.* Rome: Bulzoni.

Marzahn, Christian. 1994. *Bene Tibi—Über Genuß und Geist.* Bremen: Edition Temmen.

Marzell, Heinrich. 1922. *Die heimische Pflanzenwelt im Volksbrauch und Volksglauben.* Leipzig: Hirzel.

———. 1926. *Alte Heilkräuter.* Jena: Eugen Diederichs.

———. 1935. *Die Pflanze im Deutschen Brauchtum.* Berlin: Enckehaus.

———. 1943–77. *Wörterbuch der deutschen Pflanzennamen.* Leipzig: S. Hirzel Verlagsbuchhandlung.

———. 1964. *Zauberpflanzen—Hexentränke.* Stuttgart: Kosmos.

Mata, Rachel, and Jerry L. McLaughlin. 1982. Cactus alkaloids. 50: A comprehensive tabular summary. *Revista Latinoamerica de Quimica* 12:95–117.

Matthiolus, Pierandrea. 1627. *Kreutterbuch.* Franckfurt am Mayn: Jacob Fischers Erben.

Mautner, Uli, and Bernd Küllenberg. 1989. *Arzneigewürze.* Wiesbaden: Jopp.

Mayes, Vernon O., and Barbara Bayless Lacy. 1989. *Nanise': A Navajo herbal.* Tsaile, Ariz.: Navajo Community College Press.

McClure, Susan A., and W. Hardy Eshbaugh. 1983. Love potions of Andros Island, Bahamas. *Journal of Ethnobiology* 3 (2): 149–56.

McGlothlin, William H. 1965. Hallucinogenic drugs—a perspective with special reference to peyote and cannabis. *Psychedelic Review* 6:16–57.

McKenna, Dennis. 1995. Bitter brews and other abominations: The uses and abuses of some little-known hallucinogenic plants. *Integration* 5:99–104.

McKenna, Dennis J., and Terence K. McKenna. 1975. *The invisible landscape: Mind, hallucinogens, and the I Ching.* New York: The Seabury Press.

———. 1994. *The invisible landscape: Mind, hallucinogens, and the I Ching.* Rev. ed. San Francisco: HarperSanFrancisco.

McKenna, Dennis J., and G. H. N. Towers. 1985. On the comparative ethnopharmacology of Malpighiaceous and Myristicaceous hallucinogens. *Journal of Psychoactive Drugs* 17 (1): 35–39.

McKenna, Terence. 1989. *Wahre Halluzinationen.* Basel: Sphinx. [English edition: McKenna, Terence. 1993. *True hallucinations.* San Francisco: HarperSanFrancisco.

———. 1991. *The archaic revival.* San Francisco: HarperCollins.

———. [1996]. *Die Speisen der Götter: Die Suche nach dem Baum der Erkenntnis.* Löhrbach: Werner Pieper's MedienXperimente (Edition Rauschkunde).

Mehra, K. L. 1979. Ethnobotany of Old World Solanaceae. In *The biology and taxonomy of the Solanaceae,* ed. J. G. Hawkes et al., 161–70. London: Academic Press.

Meister, George. n.d. *Der Orientalisch-Indianische Kunst- und Lust-Gärtner.* Weimar: Kiepenheuer. (Orig. pub. 1677.)

Melas, Evi. 1990. *Delphi: Die Orakelstätte des Apollon.* Cologne: DuMont.

Mercatante, Anthony. 1980. *Der magische Garten.* Zurich: Schweizer Verlagshaus.

Metha, Ashvin, and P. V. Bole. 1991. *100 Himalayan flowers.* Ahmadabad, India: Mapin Publishing.

Metzner, Ralph, ed. 1968. *The ecstatic adventure.* New York: Macmillan.

———. 1989. States of consciousness and transpersonal psychology. In *Existential-phenomenological perspectives in psychology,* ed. R. Valle and S. Halling. New York: Plenum Press.

———. 1991. Shamanism, animism and ecological awareness. *The City* 2 (7): 52–58.

———. [1993]. *Sucht und Transzendenz als Zustände veränderten Bewußtseins.* Der Grüne Zweig 158. Löhrbach: Werner Pieper's MedienXperimente; Solothurn: Nachtschatten Verlag.

———. 1994a. Addiction and transcendence as altered states of consciousness. *The Journal of Transpersonal Psychology* 26 (1): 1–17.

———. 1994b. *Der Brunnen der Erinnerung.* Braunschweig: Aurum.

Meyrink, Gustav. 1984. *Des deutschen Spießers Wunderhorn 1: Das Wachsfigurenkabinett.* Rastalt: Moewig.

Mierow, Dorothy, and Tirtha Bahadur Shrestha. 1987. *Himalayan flowers and trees.* Kathmandu: Sahayogi Press.

Mildner, Theodor. 1956. *Giftpflanzen in Wald und Flur.* Wittenberg Lutherstadt: Ziemsen (Die Neue Brehm-Bücherei).

Miller, Arthur G. 1973. *The mural paintings of Teotihuacán.* Washington, D.C.: Dumbarton Oaks.

Miller, Richard Alan. 1985. *The magical and ritual use of aphrodisiacs.* New York: Destiny Books.

———. 1988. *Liebestrank und Ritual: Aphrodisiaka und die Kunst des Liebens.* Basel: Sphinx.

———. 1993. *The magical and ritual use of herbs.* Rochester, Vt.: Destiny Books.

Millspaugh, Charles F. 1974. *American medicinal plants.* New York: Dover. (Orig. pub. as *Medicinal Plants,* 1892.)

Miner, H. 1939. Parallelism in alkaloid-alkali quids. *American Anthropologist,* n.s., 41:617–19.

Miranda, Faustino. 1975. *La vegetación de Chiapas.* 2nd ed. Chiapas: Tuxtla Gutiérrez.

Mitsuhashi, Hiroshi. 1976. Medicinal plants of the Ainu. *Economic Botany* 30:209–17.

Moerman, Daniel E. 1982. *Geraniums for the Iroquois: A field guide to American Indian medicinal plants.* Algonac, Mich.: Reference Publications.

———. 1986. *Medicinal plants of Native America.* 2 vols. Technical Reports, no. 19; Research Reports in Ethnobotany, contribution 2. Ann Arbor: University of Michigan Museum of Anthropology.

Moinuddin Chishti, Shaykh Hakim. 1991. *The Book of Sufi Healing.* Rochester, Vt.: Inner Traditions.

Moldenke, Harold N., and Alma L. Moldenke. 1986. *Plants of the Bible.* New York: Dover.

Møller, Knud O. 1951. *Rauschgifte und Genußmittel.* Basel: Benno Schwabe.

Montes, Marco, and Tatiana Wilkomirsky. 1987. *Medicina tradicional chilena.* Concepción: Editorial de la Universidad de Concepción.

Montgomery, Rob. 1989. Ethnobotanical research field kit. *Whole Earth Review* (Fall 1989): 30–31.

———. 1991. *Botanical preservation corps field training manual.* Sebastopol, Calif.: BPC.

———. n.d. *Cultivation details for exotic plants with plans for a green house and propagation unit.* Sebastopol, Calif.: . . . of the jungle.

Morton, Julia F. 1977. *Major medicinal plants: Botany, culture and uses.* Springfield, Ill.: Charles C. Thomas.

———. 1981. *Atlas of medicinal plants of middle America.* Springfield, Ill.: Charles C. Thomas.

———. 1995. *Plants poisonous to people in Florida and other warm areas.* 3rd ed. Miami: Hallmark Press.

Mösbach, Ernesto Wilhelm de. 1992. *Botanica indígena de Chile.* Ed. Carlos Aldunate and Carolina Villagran. Santiago: Museo Chileno de Arte Precolombino.

Moscher, Richi. [1994]. *Too Much: Erste Hilfe bei Drogenvergiftungen.* Der Grüne Zweig 172. Löhrbach: Werner Pieper's MedienXperimente; Solothurn: Nachtschatten Verlag. (2nd ed. published in 1995.)

Most, Georg Friedrich. 1843. *Encyclopädie der Volksmedizin.* Leipzig: F. A. Brockhaus. (Repr. Graz: Akademische Druck und Verlagsanstalt, 1984.)

Motolinia, Fray Toribio [de Benevente o]. 1979. *Historia de los indios de la Nueva España.* Edmundo O'Gorman, ed. 3rd ed. Mexico, D.F.: Editorial Porrua [written in 1541].

Alte Heilkräuter

Müller, Irmgard. 1982. *Die pflanzlichen Heilmittel bei Hildegard von Bingen.* Salzburg: Otto Müller.

Müller, R. K., and O. Prokop. 1988. Geschichte der Genußgifte. In *Gifte: Geschichte der Toxikologie,* ed. M. Amberger-Lahrmann and D. Schmähl, 253–91. Berlin: Springer.

Müller, Wolfgang. 1988. *Kleine Geschichte der altamerikanischen Kunst.* Cologne: DuMont.

———. 1995. *Die Indianer Amazoniens.* Munich: C. H. Beck.

Müller-Ebeling, Claudia. 1986. Malerei im Labyrinth des Innenraumes. In *Rausch und Erkenntnis,* ed. Sigi Höhle et al., 280–304. Munich: Knaur.

———. 1992. Psychedelische und visionäre Malerei. In *Das Tor zu inneren Räumen,* ed. C. Rätsch, 183–96. Südergellersen: Bruno Martin.

Müller-Ebeling, Claudia, and Christian Rätsch. 1986. *Isoldens Liebestrank.* Munich: Kindler. (Repr. in paperback by Knaur TB, 1989.)

———. 1989. *Heilpflanzen der Seychellen.* Berlin: VWB.

Müller-Ebeling, Claudia, Christian Rätsch, and Wolf-Dieter Storl. 1998. *Hexenmedizin.* Aarau: AT Verlag.

Mulvany De Peñaloza, Eleonora. 1984. Motivos fitomorfos de alucinógenos en Chavin. *Revista Chungará* 12:57–80.

Murr, Josef. 1890. *Die Pflanzenwelt in der griechischen Mythologie.* Innsbruck.

N

Nadler, Kurt H. 1991. *Drogen: Rauschgift und Medizin.* Munich: Quintessenz.

Namba, Tsuneo. 1980. *Colored illustrations of wakan-yaku (the crude drugs in Japan, China and the neighbouring countries.* 2 vols. Osaka: Hoikusha Publishing Co. (In Japanese.)

Naranjo, Claudio. 1969. Psychotherapeutic possibilities of new fantasy-enhancing drugs. *Clinical Toxicology* 2 (2): 209–24.

———. 1979. *Die Reise zum Ich: Psychotherapie mit heilenden Drogen.* Frankfurt/M.: Fischer TB.

Naranjo, Plutarco. 1995. Archaeology and psychoactive plants. In *Ethnobotany: Evolution of a discipline,* ed. Richard Evans Schultes and Siri von Reis, 393–99. Portland, Ore.: Dioscorides Press.

Navarro, Fray Juan. 1992. *Historia natural o Jardín Americano (manuscrito de 1801).* Mexico City: UNAM.

Navchoo, Irshad A., and G. M. Buth. 1989. Medicinal system of Ladakh, India. *Journal of Ethnopharmacology* 26:137–46.

———. 1990. Ethnobotany of Ladakh, India: Beverages, narcotics, foods. *Economic Botany* 44 (3): 318–21.

Neuwinger, Hans Dieter. 1994. *Afrikanische Arzneipflanzen und Jagdgifte: Chemie, Pharmakologie, Toxikologie.* Stuttgart: Wissenschaftliche Verlagsgesellschaft.

———. 1997. *Afrikanische Arzneipflanzen und Jagdgifte.* 2nd ed. Stuttgart: WVG.

Nicholson, Irene. 1967. *Mexikanische Mythologie.* Wiesbaden: Emil Vollmer Verlag.

Norman, Jill. 1991. *Das grosse Buch der Gewürze.* Aarau: AT Verlag.

O

Ocaña Fernandez, Enrique. 1993. *El Dioniso moderno y la farmacia utópica.* Barcelona: Editorial Anagram.

O'Connell, James F., Peter K. Latz, and Peggy Barnett. 1983. Traditional and modern plant use among the Alyawara of central Australia. *Economic Botany* 27 (1): 80–109.

Ortiz de Montellano, Bernard R. 1981. Entheogens: The interaction of biology and culture. *Reviews of Anthropology* 8 (4): 339–65.

Ott, Jonathan. 1979. *Hallucinogenic plants of North America.* Rev. ed. Berkeley, Calif.: Wingbow Press.

———. 1985. *Chocolate addict.* Vashon, Wash.: Natural Products Co.

———. 1993. *Pharmacotheon.* Kennewick, Wash.: Natural Products Co.

———. 1994. *Ayahuasca analogues: Pangæn entheogens.* Kennewick, Wash.: Natural Products Co.

———. 1995. *The age of entheogens and the angels' dictionary.* Kennewick, Wash.: Natural Products Co.

———. 1996a. Enteobotánica: Embriagantes chamánicos. Unpublished manuscript.

———. 1996b. Entheogens II: On entheology and entheobotany. *Journal of Psychoactive Drugs* 28 (2): 205–9.

———. 1996c. El estado actual de los embriagantes chamánicos mexicanos. Unpublished manuscript.

———. 1996d. *Pharmacotheon.* 2nd ed. Kennewick, Wash.: Natural Products Co. (Rev. ed. of Ott 1993.)

Ott, Jonathan, and R. Gordon Wasson. 1983. Carved "disembodied eyes" of Teotihuacan. *Botanical Museum Leaflets* 29 (4): 387–400.

P

Pabst, G., ed. 1887/89. *Köhler's Medicinal-Pfanzen.* Gera-Untermhaus: Eugen Köhler.

Pachter, Irwin J., David E. Zacharias, and Oscar Ribeiro. 1959. Indole alkaloids of *Acer saccharinum* (the silver maple), *Dictyoloma incanescens, Piptadenia colubrina,* and *Mimosa hostilis. Journal of Organic Chemistry* 24:1285–87.

Pahlow, Mannfried. 1993. *Das große Buch der Heilpflanzen*. Munich: Gräfe und Unzer.

———. 1995. *Gewürze: Genuß und Arznei*. 2nd ed. Stuttgart: Edition Medpharm.

Palmer, Cynthia, Michael Horowitz, and Ronald Rippchen, eds. n.d. *Tänzerinnen zwischen Himmel und Hölle*. Der Grüne Zweig 136. Löhrbach: Werner Pieper's MedienXperimente.

Papajorgis, Kostis. 1993. *Der Rausch: Ein philosophischer Aperitif*. Stuttgart: Klett-Cotta.

Parnefjord, Ralph. 1997. *Das Drogentaschenbuch*. Stuttgart: Ferdinand Enke Verlag.

Parvati, Jeannine. 1978. *Hygiea: A woman's herbal*. Berkeley, Calif.: Freestone Collective.

Pasztory, Esther. 1974. *The iconography of the Teotihuacan Tlaloc*. Studies in Pre-Columbian Art and Archaeology 15. Washington, D.C.: Dumbarton Oaks.

Patnaik, Naveen. 1993. *The garden of life: An introduction to the healing plants of India*. New York: Doubleday.

Patzelt, Erwin. 1996. *Flora del Ecuador*. 2nd ed. Quito: Banco Central del Ecuador.

Paulus, Ernst, and Yu-he Ding. 1987. *Handbuch der traditionellen chinesischen Heilpflanzen*. Heidelberg: Haug.

Pavia, Fabienne. 1995. *Der Amazonas*. Amsterdam: Time-Life Bücher.

Pearson, Steve, and Alison Pearson. 1992. *Rainforest plants of eastern Australia*. Kenthurst, Australia: Kangaroo Press.

Pelt, Jean-Marie. 1983a. *Drogues et plantes magiques*. Paris: Fayard.

———. 1983b. *Pflanzenmedizin*. Dusseldorf and Vienna: Econ.

Pendell, Dale. 1995. *Pharmako/poeia: Plant powers, poisons, and herbcraft*. San Francisco: Mercury House.

Pennington, Campbell W. 1973. Plantas medicinales utilizadas por el pima montañes de Chihuahua. *América Indígena* 33 (1): 213–32.

Pereirra, Jonathan, and L. C. E. E. Fock. 1849. *De Beginselen der Materia Medica en der Therapie*. Amersfoort, Netherlands: W. J. van Bommel van Vloten.

Perez de Barradas, José. 1957. *Plantas magicas americanas*. Madrid: Inst. "Bernardino de Sahagún."

Perger, K. Ritter von. 1864. *Deutsche Pflanzensagen*. Stuttgart and Oehringen: Schaber.

Perrin, Michel. 1992. Enfoque antropológico sobre las drogas. *Takiwasi* 1 (1): 31–51.

Perrine, Daniel M. 1996. *The chemistry of mind-altering drugs: History, pharmacology, and cultural context*. Washington, D.C.: American Chemical Society.

Perry, Lily M., and Judith Metzger. 1980. *Medicinal plants of East and Southeast Asia*. Cambridge, Mass., and London: MIT Press.

Peters, Daniel. 1991. *The Incas*. New York: Random House.

Peterson, Jeanette Favrot. 1990. *Precolumbian flora and fauna: Continuity of plant and animal themes in Mesoamerican art*. San Diego, Calif.: Mingei International Museum.

Pfeiffer, Wolfgang M. 1988. Zustände veränderten Bewußtseins in kulturvergleichender Sicht. *Salix* 4 (1.88): 8–22.

Pfeiffer, W. M., and W. Schoene, eds. 1980. *Psychopathologie im Kulturvergleich*. Stuttgart: Enke.

Phillips, Roger. 1992. *Kosmos-Atlas Bäume*. Stuttgart: Franckh-Kosmos.

Plotkin, Mark J. 1994. *Der Schatz der Wayana: Abenteuer bei den Schamanen im Amazonas-Regenwald*. Bern, Munich, and Vienna: Scherz Verlag. (See book review by Daniela Baumgartner in *Curare* 18 [2; 1995]: 573–74.)

Polia Meconi, Mario. 1988. *Las lagunas de los encantos: Medicina tradicional andina del Perú septentrional*. Piura, Peru: Central Peruana de Servicios—CEPESER/Club Grau de Piura.

Pollio, Antonino, Giovanne Aliotta, and E. Giuliano. 1988. Etnobotanica delle Solanaceae allucinogene europee. *Atti del Congresso Internazionale di Storia della Farmacia* (Piacenza): 217–19.

Polunin, Miriam, and Christopher Robbins. 1992. *The natural pharmacy*. New York: Collier Books/Macmillan.

Polunin, Oleg, and Adam Stainton. 1985. *Flowers of the Himalaya*. Delhi: Oxford University Press. (See also Stainton 1988.)

Posner, Michael I., and Marcus E. Raichle. 1996. *Bilder des Geistes: Hirnforscher auf den Spuren des Denkens*. Heidelberg: Spektrum-Akademischer Verlag.

Pozo, Efrén C. del. 1965. La botánica medicinal indígena de Mexico. *Estudios de Cultura Náhuatl* 5:57–73.

———. 1967. Empiricism and magic in Aztec pharmacology. In *Ethnopharmacologic search for psychoactive drugs*, ed. Daniel H. Efron, 59–76. Washington, D.C.: U.S. Dept. of Health, Education, and Welfare.

Prance, Ghillian T. 1970. Notes on the use of plant hallucinogens in Amazonian Brazil. *Economic Botany* 24:62–68.

———. 1972a. An ethnobotanical comparison of four tribes of Amazonian Indians. *Acta Amazônica* 2 (2): 7–27.

———. 1972b. Ethnobotanical notes from Amazonian Brazil. *Economic Botany* 26:221–37.

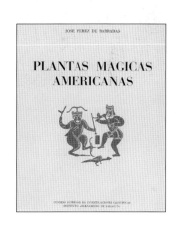

———. 1991. What is ethnobotany today? *Journal of Ethnopharmacology* 32:209–16.

Prance, Ghillian T., David G. Campbell, and Bruce W. Nelson. 1977. The ethnobotany of the Paumarí Indians. *Economic Botany* 31:129–39.

Prance, Ghillian T., and J. A. Kallunki, eds. 1984. *Ethnobotany in the neotropics.* Advances in Economic Botany 1. New York: The New York Botanical Garden.

Preisendanz, Karl. 1973. *Papyri Graecae magicae: Die griechischen Zauberpapyri.* Stuttgart: Teubner.

Preston-Mafham, Rod, and Ken Preston-Mafham. 1995. *Kakteen-Atlas.* 2nd ed. Stuttgart: Ulmer.

Pulido Salas, Ma. Teresa, and Lidia Serralta Peraza. 1993. *Lista anotada de las plantas medicinales de use actual en el estado de Quintana Roo, México.* Chetumal, Mexico: Centro de Investigaciones de Quintana Roo (CIORO).

Püschel, Klaus. 1995. Drogen—ihre Wirkungen, Nebenwirkungen, Wechselwirkungen. In *Risiko mindern beim Drogengebrauch*, vol. 37, ed. J.-H. Heudtlass, H. Stöver, and P. Winkler, 14–67. Frankfurt/M.: Fachhochschul-Verlag.

Q

Quezada, Noemí. 1989. *Amor y magia amorosa entre los aztecas.* Mexico City: UNAM.

R

Raffauf, Robert F. 1970. *A handbook of alkaloids and alkaloid-containing plants.* New York: Wiley-Interscience.

Rahner, Hugo. 1957. *Griechische Mythen in christlicher Deutung.* Zurich: Rhein-Verlag.

Rätsch, Christian. 1985. *Bilder aus der unsichtbaren Welt.* Munich: Kindler.

———. 1986. Heilige Bäume und halluzinogene Pflanzen. In *Chactun—Die Götter der Maya*, ed. C. Rätsch, 213–36. Cologne: Diederichs.

———. 1987. Mexikanische Prophetien—Träume und Visionen. *Grenzgebiete der Wissenschaft* 36 (2): 116–34.

———. 1988a. *Lexikon der Zauberpflanzen aus ethnologischer Sicht.* Graz: ADEVA.

———. 1988b. Tarot und die Maya. *Ethnologia Americana* 241 (112): 1188–90.

———. 1989a. Die Pflanzen der Götter auf der Erde. *Imagination* 4 (1): 18–20.

———. 1989b. St. Anthony's fire in Yucatán. In *Gateway to inner space*, ed. C. Rätsch, 161–65. Bridport, U.K.: Prism Press.

———. 1990. *Pflanzen der Liebe.* Bern: Hallwag. (2nd ed., Aarau: AT Verlag, 1995.)

———. 1991a. *Indianische Heilkräuter.* 2nd ed. Munich: Diederichs.

———. 1991b. *Von den Wurzeln der Kultur: Die Pflanzen der Propheten.* Basel: Sphinx.

———. 1992. *The dictionary of sacred and magical plants.* Santa Barbara, Calif.: ABC-Clio.

———. 1993. Zur Ethnologie veränderter Bewußtseinszustände. In *Welten des Bewußtseins*, ed. A. Dittrich et al., 1:21–45. Berlin: VWB.

———. 1994a. Die Pflanzen der blühenden Träume: Trancedrogen mexikanischer Schamanen. *Curare* 17 (2): 277–314.

———. 1994b. Schamanismus versus halluzinogen-unterstützte Psychotherapie. In *Das Wendepunkt-buch* 1, ed. Ralph Cosack, 15–42. Hamburg: Wendepunkt.

———. 1994c. Ts'ak: Die Heilpflanzen der Lakandonen von Naha'. *Yearbook for Ethnomedicine and the Study of Consciousness*, 1994 (2): 43–93. Berlin: VWB.

———. 1995a. *Heilkräuter der Antike in Ägypten, Griechenland und Rom.* Munich: Diederichs Verlag (DG).

———. 1995b. Naturverehrung und Heilkunst oder: Das Nervensystem der Gaia. In *Lichtensteiner Exkurse II Was wäre Natur?*, ed. Norbert Hass et al., 45–62. Eggingen: Edition Isele.

———. 1995c. *Pflanzen der Venus: Aphrodisiaka und Liebestränke.* Hamburg: Ellert und Richter.

———. 1995d. Ritueller Gebrauch von psychoaktiven Substanzen im modernen Mitteleuropa: Eine ethnographische Skizze. *Curare* 18 (2): 297–324.

———. 1996a. Aus der ethnopharmakologischen Praxis. *Infoemagazin* 11:6–7.

———. 1996b. Die wichtigsten Schamanendrogen Kolumbiens. In *Das schamanische Universum*, ed. G. Reichel-Dolmatoff, 280–304. Munich: Diederichs.

———. 1997a. *Medizin aus dem Regenwald.* Neckarsulm: Hampp/Natura Med.

———. 1997b. Schamanische Bewußtseinszustände und religiöse Erfahrungen. In *Paranormologie und Religion*, ed. Andreas Resch, 39–70. Innsbruck: Resch Verlag.

———. 1997c. *Die Steine der Schamanen: Kristalle, Fossilien und die Landschaften des Bewußtseins.* Munich: Diederichs.

———. 2001. *Marijuana medicine: A world tour of the healing and visionary powers of cannabis.* Rochester, Vt.: Healing Arts Press.

Ray, Oakley, and Charles Ksir. 1996. *Drugs, society, and human behavior.* 7th ed. Saint Louis, Mo.: William C. Brown.

Read, Bernard E. 1977. *Chinese materia medica.* Taipei: Southern Materials Center.

Rehm, Sigmund, and Gustav Espig. 1996. *Die Kulturpflanzen der Tropen und Subtropen.* 3rd ed. Stuttgart: Ulmer.

Reichel-Dolmatoff, Gerardo. 1950/51. *Los Kogi: Una tribu indigéna de la Sierra Nevada de Santa Marta, Colombia.* 2 vols. Bogotá: Ervista del Instituto Etnológico National/Editorial Iqueima.

———. 1967. Rock-paintings of the Vaupés: An essay of interpretation. *Folklore Americas* 26 (2): 107–33.

———. 1971. *Amazonian cosmos: The sexual and religious symbolism of the Tukano Indians.* Chicago and London: University of Chicago Press. (See book review by Richard Evans Schultes in *Economic Botany* 26 [1972]: 197.)

———. 1972. *San Agustin: A culture of Colombia.* Art and Civilization of Indian America. New York: Praeger. (See book review by Erward B. Dwyer in *American Anthropologist,* n.s., 76 [1974]: 127–28.)

———. 1975. *The shaman and the jaguar: A study of narcotic drugs among the Indians of Colombia.* Philadelphia: Temple University Press.

———. 1977. Training for the priesthood among the Kogi of Colombia. In *Enculturation in Latin America: An anthology,* ed. Johannes Wilbert. Los Angeles: UCLA Latin American Center Publications.

———. 1978. *Beyond the Milky Way: Hallucinatory imagery of the Tukano Indians.* Los Angeles: UCLA Latin American Center Publications.

———. 1981. Brain and mind in Desana shamanism. *Journal of Latin American Lore* 7 (1): 73–98.

———. 1985a. *Basketry as metaphor: Arts and crafts of the Desana Indians of the northwest Amazon.* Los Angeles: Los Angeles Museum of Cultural History.

———. 1985b. *Los Kogi: Una tribu indígena de la Sierra Nevada de Santa Marta, Colombia.* 2 vols., 2nd ed. Bogotá: PROCULTURA, Presidencia de la República.

———. 1987. *Shamanism and art of the eastern Tukanoan Indians.* Iconography of Religions 9 (1). Leiden: Brill.

———. 1996a. *The forest within: The world-view of the Tukano Amazonian Indians.* Totnes, U.K.: Green Books.

———. 1996b. *Das schamanische Universum: Schamanismus, Bewußtsein und Ökologie in Südamerika.* Munich: Diederichs.

Reid, Daniel E. 1988. *Chinesische Naturheilkunde.* Vienna: Orac.

Reko, Blas Pablo [= Blasius Paul]. 1919. De los nombres botánicos aztecas. *El México Antigua* 1 (5): 113–57.

———. 1996. *On Aztec botanical names.* Trans. Jonathan Ott. Berlin: VWB. (Orig. pub. 1919.)

Reko, Victor A. 1936. *Magische Gifte: Rausch- und Betäubungsmittel der neuen Welt.* Stuttgart: Enke.

———. 1938. *Magische Gifte: Rausch- und Betäubungsmittel der neuen Welt.* 2nd ed. Stuttgart: Enke.

(Repr., Berlin: EXpress Edition, 1987; VWB, 1996.)

Remann, Micky. 1989. *SolarPerplexus.* Basel: Sphinx.

Renfrew, Jane M. 1973. *Paleoethnobotany.* New York: Columbia University Press.

Richardson, P. Mick. 1992. *Flowering plants: Magic in bloom.* The Encyclopedia of Psychoactive Drugs. Rev. ed. New York and Philadelphia: Chelsea House Publications.

Ripinsky-Naxon, Michael. 1989. Hallucinogens, shamanism, and the cultural process. *Anthropos* 84:219–24.

———. 1992. Shamanism: Religion or rite? *Journal of Prehistoric Religion* 6:37–44.

———. 1993. *The nature of shamanism: Substance and function of a religious metaphor.* Albany: State University of New York Press. (See book review by John Baker in *Yearbook for Ethnomedicine and the Study of Consciousness,* 1995 [4]: 309–11.)

———. 1995. Cognition, symbolization, and the beginnings of shamanism. *Journal of Prehistoric Religion* 9:43–54.

———. 1996. Psychoactivity and shamanic states of consciousness. *Yearbook for Ethnomedicine and the Study of Consciousness,* 1995 (4): 5–43. Berlin: VWB.

Rivas, Augustin. 1989. Meisterpflanze Ayahuasca. In *Amazonas: Mae mañota,* ed. C. Kobau, 182–83. Graz: Leykam.

Robinson, D. 1994. Plants and Vikings: Everyday life in Viking-age Denmark. *Botanical Journal of Scotland* 46 (4): 542–51.

Robson, Philip. 1994. *Forbidden drugs: Understanding drugs and why people take them.* Oxford, New York, and Tokyo: Oxford University Press.

Roddick, James G. 1991. The importance of the Solanaceae in medicine and drug therapy. In *Solanaceae III: Taxonomy, chemistry, evolution,* ed. Hawkes, Lester, Nee, and Estrada, 7–23. London: Royal Botanic Gardens Kew and Linnaean Society.

Rodriguez, Eloy, Jan Clymer Cavin, and Jan E. West. 1982. The possible role of Amazonian psychoactive plants in the chemotherapy of parasitic worms—a hypothesis. *Journal of Ethnopharmacology* 6:303–9.

Rogers, Dilwyn S. 1980. *Edible, medicinal, useful, and poisonous wild plants of the northern Great Plains—South Dakota region.* Sioux Falls, S.D.: Augustana College.

Römpp, [Hermann]. 1995. *Chemie Lexikon.* 9th ed. 6 vols. Ed. Jürgen Falbe and Manfred Regitz. Stuttgart, New York: Thieme.

———. 1950. *Chemische Zaubertränke.* 5th ed. Stuttgart: Kosmos-Franckh'sche Verlagshandlung. (Orig. pub. 1939.)

Root, Waverley. 1996. *Wachtel, Trüffel, Schokolade: Die Enzyklopädie der kulinarischen Köstlichkeiten.* Munich: Goldmann.

Rosenbohm, Alexandra. 1991. *Halluzinogene Drogen im Schamanismus.* Berlin: Reimer.

Rosengarten, Frederic, Jr. 1977. An unusual spice from Oaxaca: The flowers of *Quararibea funebris. Botanical Museum Leaflets* 25 (7): 183–202.

Rossi-Wilcox, Susan M. 1993. Henry Hurd Rusby: A biographical sketch and selectively annotated bibliography. *Harvard Papers* 4:1–30.

Roth, Hermann J., and Helmut Fenner. 1988. *Pharmazeutische Chemie III: Arzneistoffe.* Stuttgart, New York: Thieme.

Roth, Lutz, Max Daunderer, and Kurt Kormann. 1994. *Giftpflanzen—Pflanzengifte.* 4th ed. Munich: Ecomed.

Rouhier, Alexandre. 1996. *Die Hellsehen hervorrufenden Pflanzen.* Berlin: VWB. (Orig. pub. 1927.)

Rowell, Margery. 1978. Plants of Russian folk medicine. *Janus* 65:259–82.

Roys, Ralph L. 1976. *The ethno-botany of the Maya.* Intro. by Sheila Cosminsky. Philadelphia: ISHI Reprint. (Orig. pub. 1931.)

Ruben, Walter. 1952. *Tiahuanaco, Atacama und Araukaner.* Leipzig: Otto Harrassowitz.

Rublowsky, John. 1974. *The stoned age: A history of drugs in America.* New York: G. P. Putnam's Sons (Capricorn Books).

Ruck, Carl A. P. 1995. Gods and plants in the classical world. In *Ethnobotany: Evolution of a discipline,* ed. Richard Evans Schultes and Siri von Reis, 131–43. Portland, Ore.: Dioscorides Press.

Ruck, Carl A. P., et al. 1979. Entheogens. *Journal of Psychedelic Drugs* 11 (1–2): 145–46.

Ruck, Carl A. P., and Danny Staples. 1994. *The world of classical myth: Gods and goddesses, heroines and heroes.* Durham, N.C.: Carolina Academic Press.

Rudgley, Richard. 1994. *Essential substances: A cultural history of intoxicants in society.* Foreword by William Emboden. New York, Tokyo, and London: Kodansha International. (Orig. pub. as *The alchemy of culture,* British Museum Press, 1993).

———. 1995. The archaic use of hallucinogens in Europe: An archaeology of altered states. *Addiction* 90:163–64.

Rufer, Marc. 1995. *Glückspillen: Ecstasy, Prozac und das Comeback der Psychopharmaka.* Munich: Knaur. (Contains a chapter on the history of the discovery of psychoactive plant constituents.)

Ruiz de Alarcón, Hernando (see Andrews and Hassig 1984). 1984. *Treatise on the heathen superstitions that today live among the Indians native to this New Spain, 1629.* Trans., ed. J. Richard Andrews and Ross Hassig. Norman: University of Oklahoma Press.

Rutter, Richard A. 1990. *Catalogo de plantas utiles de la Amazona peruana.* Yarinacocha, Peru: Ministerio de Educación, Instituto Lingüístico de Verano.

S

Sacks, Oliver. 1985. *Migraine: Understanding a common disorder.* Berkeley: University of California Press.

Safford, William E. 1917. Narcotic plants and stimulants of the ancient Americans. *Annual Report of the Smithsonian Institution,* 1916: 387–424.

———. 1922. Daturas of the Old World and New. *Annual Report of the Smithsonian Institution,* 1920: 537–67.

Sahagun, Fray Bernardino de. 1989. *Aus der Welt der Azteken: Die Chronik des Fray Bernardino de Sahagun.* Frankfurt/M.: Insel. (Selections.)

———. 1978. Florentine Codex. J. O. Anderson and C. E. Dibble, trans. 12 volumes. Santa Fe, N.M.: The School of American Research and the University of Utah. (Written in the sixteenth century.)

Sahihi, Arman. 1995. *Designer-Drogen: Gifte, Sucht und Szene.* 3rd ed. Munich: Heyne.

Sala. 1993. See Warrier et al. 1993 ff.

Salomon, Frank, and George L. Urioste. 1991. *The Huarochirí manuscript: A testament of ancient and colonial Andean religion.* Austin: University of Texas Press.

Samorini, Giorgio. 1995a. *Gli allucinogeni nee mito: Racconti sull'origine delle piante psicoattive.* Turin: Nautilus.

———. 1995b. Paolo Mantegazza (1831–1910): Pioniere italiano degli studi sulle droghe. *Eleusis* 2:14–20.

———. 1998. *Halluzinogene in Mythos: Vom Ursprung psychoaktiver Pflanzen.* Foreword by Christian Rätsch. Solothurn: Nachtschatten Verlag.

Santos Biloni, José. 1990. *Arboles autoctonos argentinos.* Buenos Aires: Tipográfica Editora Argentina.

Schaffner, Willi. 1992. *Heilpflanzen und ihre Drogen.* Munich: Mosaik Verlag.

Schaffner, Willi, Barbara Häfelfinger, and Beat Ernst. 1992. *Phytopharmakokompendium.* Hinterkappelen: Arboris-Verlag.

Schall, Paul. 1965. *Zaubermedizin im Alten China?* Stuttgart: J. Fink Verlag.

Scheerer, Sebastian, and Irmgard Vogt, eds. 1989. *Drogen und Drogenpolitik.* Frankfurt/M. and New York: Campus.

Scheffler, Lilian. 1983. *Magía y brujería en México.* Mexico City: Panorama Editorial.

Scheiblich, Wolfgang, ed. 1987. *Rausch, Ekstase, Kreativität.* Freiburg: Lambertus-Verlag.

Schenk, Amelie. 1994. *Schamanen auf dem Dach der Welt.* Graz: ADEVA.

———. [1996]. *Was ist Schamanentum?* Der Grüne Zweig 192. Löhrbach: Werner Pieper's MedienXperimente.

Schenk, Gustav. 1948. *Schatten der Nacht.* Hannover: Sponholtz.

———. 1954. *Das Buch der Gifte.* Berlin: Safari.

Scheuch, Erwin K. 1970. *Haschisch und LSD als Modedrogen.* Osnabrück: Verlag A. Fromm.

Schivelbusch, Wolfgang. 1983. *Das Paradies, der Geschmack und die Vernunft: Eine Geschichte der Genußmittel.* Frankfurt/M.: Ullstein.

Schleiffer, Hedwig, ed. 1973. *Narcotic plants of the New World Indians: An anthology of texts from the 16th century to date.* New York: Hafner Press (Macmillan).

———. 1979. *Narcotic plants of the Old World: An anthology of texts from ancient times to the present.* Monticello, N.Y.: Lubrecht & Cramer.

Schmeda-Hirschmann, Guillermo. 1993. Magic and medicinal plants of the Ayoreos of the Chaco Boreal (Paraguay). *Journal of Ethnopharmacology* 39:105–11.

Schmidbauer, Wolfgang, and Jürgen vom Scheidt. 1984. *Handbuch der Rauschdrogen.* Frankfurt/M.: Fischer.

———. 1997. *Handbuch der Rauschdrogen.* 8th ed. Munich: Nymphenburger.

Schneebeli-Graf, Ruth. 1992. *Nutz- und Heilpflanzen Chinas.* Frankfurt/M.: Umschau.

Schneider, Ernst. 1993. Arzneipflanzen der Neuen Welt. *Pharmazie in unserer Zeit* 22 (1): 15–24.

Schneider, Wolfgang. 1974. *Pflanzliche Drogen.* Vol. 5 (1–3) of *Lexikon der Arzneimittelgeschichte.* Frankfurt/M.: Govi-Verlag/Pharmazeutischer Verlag.

Schoen, M. 1909. Alter und Entwickelung der Berauschungsmittel. *Globus* 96:277–81.

Scholz, Dieter, and Dagmar Eigner. 1983. Zur Kenntnis der natürlichen Halluzinogene. *Pharmazie in unserer Zeit* 12 (3): 74–79.

Schönfelder, Ingrid, and Peter Schönfelder. 1994. *Kosmos-Atlas Mittelmeer- und Kanarenflora.* Stuttgart: Franckh-Kosmos.

Schöpf, Hans. 1986. *Zauberkräuter.* Graz: ADEVA.

Schopen, Armin. 1983. *Traditionelle Heilmittel in Jemen.* Wiesbaden: Franz Steiner Verlag.

Schröder, D. Johann. 1685. *Höchstkostbarer Artzeney-Schatz.* Jena: Johann Hoffmann. (Repr. Munich: Konrad Kölbl, 1963.)

Schröder, Rudolf. 1991. *Kaffee, Tee und Kardamom: Tropische Genußmittel und Gewürze.* Stuttgart: Ulmer.

Schuldes, Bert Marco. [1993]. *Psychoaktive Pflanzen.* Löhrbach: Werner Pieper's MedienXperimente; Solothurn: Nachtschatten Verlag.

———. [1995]. *Psychoaktive Pflanzen.* 2nd ed. Der Grüne Zweig 164. Löhrbach: Werner Pieper's MedienXperiment; Solothurn: Nachtschatten Verlag.

Schultes, Richard E. 1955. Plantae Colombianae XIII: De plantis principaliter Colombiae Amazonicae notae diversae significantes. *Botanical Museum Leaflets* 17 (3): 65–100.

———. 1960. Trapping our heritage of ethnobotanical lore. *Economic Botany* 14 (4): 257–62.

———. 1963. Hallucinogenic plants of the New World. *The Harvard Review* 1 (4): 18–32.

———. 1965. Ein halbes Jahrhundert Ethnobotanik amerikanischer Halluzinogene. *Planta Medica* 13:125–57.

———. 1966. The search for new natural hallucinogens. *Lloydia* 29 (4): 293–308.

———. 1967. The place of ethnobotany in the ethnopharmacologic search for psychotomimetic drugs. In *Ethnopharmacologic search for psychoactive drugs,* ed. Daniel H. Efron, 33–57. Washington, D.C.: U.S. Dept. of Health, Education, and Welfare.

———. 1969. Hallucinogens of plant origin. *Science* 163:245–54.

———. 1970a. The botanical and chemical distribution of hallucinogens. *Annual Review of Plant Physiology* 21:571–94.

———. 1970b. The New World Indians and their hallucinogenic plants. *Bulletin of the Morris Arboretum* 21:3–14.

———. 1970c. The plant kingdom and hallucinogens. *Bulletin on Narcotics* 22 (1): 25–51.

———. 1972a. De plantis toxicariis e mundo novo tropicale commentationes X: New data on the Malpighiaceous narcotics of South America. *Botanical Museum Leaflets* 23 (3): 137–47.

———. 1972b. The utilization of hallucinogens in primitive societies—use, misuse or abuse? In *Drug abuse: Current concepts and research,* ed. W. Keup, 17–26. Springfield, Ill.: Charles C. Thomas.

———. 1976. *Hallucinogenic plants.* Racine, Wis.: Western.

———. 1977a. De plantis toxicariis e mundo novo tropicale commentationes XVI: Miscellaneous notes on biodynamic plants of South America. *Botanical Museum Leaflets* 25 (4): 109–30.

———. 1977b. Mexico and Colombia: Two major centres of aboriginal use of hallucinogens. *Journal of Psychedelic Drugs* 9 (2): 173–76.

———. 1978a. De plantis toxicariis e mundo novo tropicale commentationes XXIII: Ethnopharmacological notes from northern South America. *Botanical Museum Leaflets* 26 (6): 225–36.

———. 1978b. De plantis toxicariis e mundo novo tropicale commentationes XXIII: Notes on biodynamic plants of aboriginal use in the northwestern Amazonia. *Botanical Museum Leaflets* 26 (5): 177–97.

———. 1979a. Evolution of the identification of the major South American narcotic plants. *Journal of Psychedelic Drugs* 11 (1–2): 119–34.

———. 1979b. Hallucinogenic plants: Their earliest botanical descriptions. *Journal of Psychedelic Drugs* 11 (1–2): 13–24.

———. 1979c. Solanaceous hallucinogens and their role in the development of New World cultures. In *The biology and taxonomy of the Solanaceae*, ed. J. G. Hawkes, R. N. Lester, and A. D. Skelding, 137–60. London: Academic Press.

———. 1980. Ruiz as an ethnopharmacologist in Peru and Chile. *Botanical Museum Leaflets* 28 (1): 87–122.

———. 1981. De plantis toxicariis e mundo novo tropicale commentationes XXVI: Ethnopharmacological notes on the flora of northwestern South America. *Botanical Museum Leaflets* 28 (1): 1–45.

———. 1983a. De plantis toxicariis e mundo novo tropicale commentationes XXXIII: Ethnobotanical, floristic and nomenclatural notes on plants of the northwest Amazon. *Botanical Museum Leaflets* 29 (4): 343–65.

———. 1983b. De plantis toxicariis e mundo novo tropicale commentationes XXXII: Notes, primarily of field tests and native nomenclature, on biodynamic plants of the northwest Amazon. *Botanical Museum Leaflets* 29 (3): 251–72.

———. 1983c. Richard Spruce: An early ethnobotanist and explorer of the northwest Amazon and northern Andes. *Journal of Ethnobiology* 3 (2): 139–47.

———. 1988. *Where the gods reign: Plants and peoples of the Colombian Amazon*. Oracle, Ariz.: Synergetic Press.

———. 1993. Plants in treating senile dementia in the northwest Amazon. *Journal of Ethnopharmacology* 38:129–35.

———. 1995. Antiquity of the use of New World hallucinogens. *Integration* 5:9–18.

Schultes, Richard E., and Norman R. Farnsworth. 1982. Ethnomedical, botanical and phytochemical aspects of natural hallucinogens. *Botanical Museum Leaflets* 28 (2): 123–214.

Schultes, Richard E., and Albert Hofmann. 1980. *The botany and chemistry of hallucinogens*. Springfield, Ill.: Charles C. Thomas.

———. 1992. *Plants of the gods: Their sacred, healing and hallucinogenic powers*. Rochester, Vt.: Healing Arts Press.

———. 1995. *Pflanzen der Götter*. Aarau: AT Verlag.

Schultes, Richard Evans, Albert Hofmann, and Christian Rätsch. 2001. *Plants of the gods: Their sacred, healing and hallucinogenic powers*. Rev. ed. Rochester, Vt.: Healing Arts Press.

Schultes, Richard Evans, and Robert F. Raffauf. 1986. De plantis toxicariis e mundo novo tropicale commentationes XXXVII: Miscellaneous notes on medicinal and toxic plants of the northwest Amazon. *Botanical Museum Leaflets* 30 (4): 255–85.

———. 1990. *The healing forest. Medicinal and toxic plants of the northwest Amazonia*. Portland, Ore.: Dioscorides Press.

———. 1991. 3. De plantis toxicariis e mundo novo tropicale commentationes XXXVI: Phytochemical and ethnopharmacological notes on the Solanaceae of the northwest Amazon. In *Solanaceae III: Taxonomy, chemistry, evolution*, ed. Hawkes, Lester, Nee, and Estrada, 25–49. London: Royal Botanic Gardens Kew and Linnaean Society.

———. 1992. *Vine of the soul: Medicine men, their plants and rituals in the Colombian Amazonia*. Oracle, Ariz.: Synergetic Press.

Schultes, Richard E., and Siri von Reis, eds. 1995. *Ethnobotany: Evolution of a discipline*. Portland, Ore.: Dioscorides Press.

Schultes, Richard Evans, and Michael Winkelman. 1996. The principal American hallucinogenic plants and their bioactive and therapeutic properties. *Jahrbuch für Transkulturelle Medizin und Psychotherapie*, 1995 (6): 205–39.

Schurz, Josef. 1969. *Vom Bilsenkraut zum LSD: Giftsuchten und Suchtgifte*. Stuttgart: Kosmos.

Schütz, Harald, Björn Ahrens, Freidoon Erdmann, and Gertrud Rochholz. 1993. Nachweis von Arznei- und anderen Fremdstoffen in Haaren. *Pharmazie in unserer Zeit* 22 (2): 65–78.

SFA ISPA [Schweizerische Fachstelle für Alkohol- und andere Drogenprobleme]. 1993. *Zahlen und Fakten zu Alkohol und anderen Drogen*. Lausanne: SFA.

Seaforth, C. E. 1991. *Natural products in Caribbean folk medicine*. Rev. ed. Saint Augustine, Trinidad: The University of the West Indies.

Seaman, Gary, and Jane S. Day, eds. 1994. *Ancient traditions: Shamanism in central Asia and the Americas*. Niwot, Colo.: Denver Museum of Natural History and University Press of Colorado.

Seidemann, Johannes. 1993. *Würzmittel-Lexikon*. Hamburg: Behr's Verlag.

Seler, Eduard. 1927. *Einige ausgewählte Kapitel aus dem Geschichtswerke des Fray Bernardino de Sahagun*. Stuttgart: Strecker & Schröder.

Seligmann, Siegfried. 1996. *Die magischen Heil- und Schutzmittel aus der belebten Natur: Das Pflanzenreich*. Berlin: Reimer. (Prepared from notes and fragments found in Seligmann's estate by Jürgen Zwernemann.)

Seoane Gallo, José. 1984. *El folclor medico de Cuba*. Havana: Editorial de Ciencias Sociales.

Sepulveda, Maria Teresa. 1983. *Magia, brujería y superticiones en México*. Mexico City: Editorial Everest Mexicana.

Seymour, Richard, and David E. Smith. 1987. *Guide to psychoactive drugs: An up-to-the-minute reference to mind-altering substances*. New York and London: Harrington Park Press.

———. 1993. *The psychedelic resurgence: Treatment, support, and recovery options*. Center City, Minn.: Hazelden.

Sfikas, Georg. 1990. *Wild flowers of Cyprus*. Athens: Efstathiadis.

Shah, Indies. 1980. *Die Sufis*. Dusseldorf and Cologne: Diederichs.

———. 1994. *Magie des Ostens*. Munich: Piper.

Shah, N. C. 1982. Herbal folk medicines in northern India. *Journal of Ethnopharmacology* 6:293–301.

Shah, N. C., and M. C. Joshi. 1971. An ethnobotanical study of the Kumaon region of India. *Economic Botany* 25:414–22.

Sheldrake, Rupert, Terence McKenna, and Ralph Abraham. 2001. *Chaos, creativity, and cosmic consciousness*. Rochester, Vt.: Park Street Press. (Orig. pub. as *Trialogues at the edge of the West*, Santa Fe, N.M.: Bear & Co., 1992.)

Shen Kuo. 1997. *Pinselunterhaltungen am Traumbach: Das gesamte Wissen des alten China*. Munich: Diederichs.

Sherratt, Andrew. 1991. Sacred and profane substances: The ritual use of narcotics in later neolithic Europe. In *Sacred and profane*, ed. Paul Garwood et al., 50–64. Oxford, U.K.: Oxford University Committee for Archaeology, monograph no. 32.

Shulgin, Alexander T. 1969. Psychotomimetic agents related to the catecholamines. *Journal of Psychedelic Drugs* 2 (2): 14–19.

———. 1992. *Controlled substances: Chemical and legal guide to federal drug laws*. 2nd ed. Berkeley, Calif.: Ronin.

———. [1993]. *Drogenpolitik: Zur schleichenden Entmündigung des Bürgers*. Der Grüne Zweig 160. Löhrbach: Werner Pieper's MedienXperimente.

———. 1995. Cactus species tabulation. Unpublished electronic file. (Cited 1998.)

Shulgin, Alexander, and Ann Shulgin. 1991. *PIHKAL: A chemical love story*. Berkeley, Calif.: Transform Press.

———. 1997. *TIHKAL: The continuation*. Berkeley, Calif.: Transform Press.

Sibly, E. 1988. *A herbal of foreign plants being a supplement to Culpeper's British herbal*. Lampeter, Wales: Llanerch Enterprises. (Orig. pub. 1821.)

Siddiqui, M. Badruzzaman, M. Mashkoor Alam, and Wazahat Husain. 1989. Traditional treatment of skin disease in Uttar Pradesh, India. *Economic Botany* 43 (4): 480–86.

Siegel, Ronald K. 1981. Inside Castaneda's pharmacy. *Journal of Psychoactive Drugs* 13 (4): 325–32.

———. 1992. *Fire in the brain: Clinical tales of hallucination*. New York: Dutton.

———. 1995a. *Halluzinationen: Expedition in eine andere Wirklichkeit*. Frankfurt/M.: Eichborn.

———. 1995b. *Rauschdrogen: Sehnsucht nach dem Künstlichen Paradies*. Frankfurt/M.: Eichborn.

Siegel, Ronald K., and Louis Joylon West, eds. 1975. *Hallucinations*. New York: John Wiley & Co.

Siegmund, Georg, and Anton Christian Hofmann. 1962. *Der Mensch im Rausch*. Würzburg: Echter-Verlag; Zurich: NZN.

Singh, M. P, S. B. Malla, S. B. Rajbhandari, and A. Manandhar. 1979. Medicinal plants of Nepal—retrospects and prospects. *Economic Botany* 33 (2): 185–98.

Slocum, Perry D., Peter Robinson, and Frances Perry. 1996. *Water gardening: Water lilies and lotuses*. Portland, Ore.: Timber Press.

Smith, Michael Valentine. 1981. *Psychedelic chemistry*. Port Townsend, Wash.: Loompanics.

Snyder, Solomon H. 1989. *Chemie der Psyche: Drogenwirkungen im Gehirn*. Heidelberg: Spektrum.

Spitta, Heinrich. 1892. *Die Schlaf- und Traumzustände der menschlichen Seele mit besonderer Berücksichtigung ihres Verhältnisses zu den psychischen Alienationen*. 2nd ed. Freiburg i.B.: J.C.B. Mohr. (Orig. pub. 1877.)

Spitzer, Manfred. 1996. *Geist im Netz: Modelle für Lernen, Denken und Handeln*. Heidelberg: Spektrum Akademischer Verlag.

Spivak, Leonid I. 1991. Psychoactive drug research in the Soviet scientific tradition. *Journal of Psychoactive Drugs* 23 (3): 271–81.

Spode, Hasso. 1994. Die Entstehung der Suchtgesellschaft. *Traverse* 1/94:23–37.

Spruce, Richard. 1970. *Notes of a botanist on the Amazon and Andes*. Foreword by R. E. Schultes. New York: Johnson Reprint Corporation. (Orig. pub. 1908.)

Stafford, Peter. 1980. *Enzyklopädie der psychedelischen Drogen*. Linden: Volksverlag.

———. 1992. *Psychedelics encyclopedia*. 3rd ed. Berkeley, Calif.: Ronin.

Stainton, Adam (see also Polunin and Stainton 1985). 1988. *Flowers of the Himalaya: A supplement*. Delhi: Oxford University Press.

Stark, Raymond. 1984. *Aphrodisiaka und ihre Wirkung*. Munich: Heyne.

Starks, Michael. 1982. *Cocaine fiends and reefer madness: An illustrated history of drugs in the movies*. New York and London: Cornwall Books.

Stary, Frantisek. 1983. *Giftpflanzen*. Hanau: Dausien.

Steckel, Ronald. 1969. *Bewußtseinserweiternde Drogen*. Berlin: Edition Voltaire.

Steele, John J. 1991. The transformational use of fragrance in pharaonic and shamanic cultures: The anthropology of smell and scent in ancient Egypt and South American shamanism. Paper presented at the Second International Conference on the Psychology of Perfumery, University of Warwick, Coventry, U.K., July 22–26, 1991.

———. 1992. The anthropology of smell and scents. In *Fragrance: The psychology and biology of perfume*, ed. S. Van Toller and G. H. Dodd. London: Elsevier.

———. 1993. The fragrant hospital: Environmental fragrancing in health care design. Paper presented at Aroma 1993, University of Sussex, U.K., July 2–4, 1993.

Steggerda, Morris. 1943. Some ethnological data concerning one hundred Yucatecan plants. *Anthropological Papers*, no. 29: 189–266. (BAE Bulletin 136.)

Steinegger, E., and R. Hänsel. 1972. *Lehrbuch der Pharmakognosie*. 3rd ed. Berlin: Springer.

Sterneck, Wolfgang, ed. 1996. *Cybertribe-Visionen: Rhythmus und Widerstand, Liebe und Bewußtsein*. Hanau: KomistA.

Stevenson, Matilda Coxe. 1993. *The Zuñi Indians and their uses of plants*. New York: Dover.

Stewart, Lynette. 1994. *A guide to palms and cycads of the world*. Sydney: Angus & Robertson.

Stoffler, Hans-Dieter. 1978. *Der Hortulus des Walahfried Strabo: Aus dem Kräutergarten des Klosters Reichenau*. Sigmaringen: Jan Thorbecke Verlag.

Storl, Wolf-Dieter. 1988. *Feuer und Asche, Dunkel und Licht: Shiva—Urbild des Menschen*. Freiburg i.B.: Bauer.

———. 1993. *Von Heilkräutern und Pflanzengottheiten*. Braunschweig: Aurum.

———. 1996a. Heilkräuter: Komplexität des Lebendigen. *Natürlich* 16 (5): 6–14.

———. 1996b. *Heilkräuter und Zauberpflanzen zwischen Haustür und Gartentor*. Aarau: AT Verlag.

———. 1996c. *Kräuterkunde*. Braunschweig: Aurum.

———. 1997. *Pflanzendevas—Die Göttin und ihre Pflanzenengel*. Aarau: AT Verlag.

———. 2004. *Shiva: The Wild God of Power and Ecstasy*. Rochester, Vt.: Inner Traditions.

Storrs, Adrian, and Jimmie Storrs. 1987. *Enjoy trees*. Kathmandu: Sahayogi Press.

———. 1990. *Trees and shrubs of Nepal and the Himalayas*. Kathmandu: Pilgrims Book House.

Strabo, Walahfried. See Stoffler 1978.

Strassman, Rick J. 1995. Hallucinogenic drugs in psychiatric research and treatment: Perspectives and prospects. *The Journal of Nervous and Mental Disease* 1983 (3): 127–38.

Suwal, P. N., et al. 1993. Medicinal plants of Nepal. *Bulletin of the Department of Medicinal Plants*, no. 3. Kathmandu: Department of Medicinal Plants.

Svoboda, Robert E. 1993. *Aghora: At the left hand of God*. New Delhi: Rupa.

Symon, David E. 1991. Gondwanan elements of the Solanaceae. In *Solanaceae III: Taxonomy, chemistry, evolution*, ed. Hawkes, Lester, Nee, and Estrada, 139–50. London: Royal Botanic Gardens Kew and Linnaean Society.

Szasz, Thomas. 1996. *Our right to drugs: The case for a free market*. Syracuse, N.Y.: Syracuse University Press.

T

Tabernaemontanus, Jacobus Theodorus. 1731. *Neu Vollkommen Kräuter-Buch*. Supplemented by Casparum and Hieronymum Bauhinium. Basel: Verlag Johann Ludwig König.

Tabor, Edward. 1970. Plant poisons in Shakespeare. *Economic Botany* 24 (1): 81–94.

Taeger, Hans-Hinrich. 1988. *Spiritualität und Drogen*. Markt Erlbach: Raymond Martin Verlag.

Taylor, Norman. 1966. *Narcotics: Nature's dangerous gifts*. New York: Laurel Edition.

Télles Valdez, Oswaldo, Edgar F. Cabreracano, Edelmira Linares Mazari, and Robert Bye. 1989. *Las plantas de Cozumel*. Mexico City: Instituto de Biología, UNAM.

Thamm, Berndt Georg. 1994. *Stichwort: Drogen*. Munich: Heyne.

Thomas, Klaus. 1970. *Die künstlich gesteuerte Seele*. Stuttgart: Enke.

Thompson, Eric. 1970. *Maya history and religion*. Norman: University of Oklahoma Press.

———. 1977. Hallucinatory drugs and hobgoblins in the Maya lowlands. *Tlalocan* 7:295–308.

Thompson, R. Campbell. 1949. *A dictionary of Assyrian botany*. London: British Academy.

Timbrook, Jan. 1990. Ethnobotany of Chumash Indians, California, based on collections by John P. Harrington. *Economic Botany* 44 (2): 236–53.

Towle, Margaret A. 1961. *The ethnobotany of pre-Columbian Peru*. Viking Fund Publications in Anthropology, no. 30. Chicago: Aldine.

Trebes, Stefan. 1995. Psychedelische Heilbehandlungen im Kulturvergleich. *Jahrbuch des Europäischen Collegiums fürBewußtseinsstudien*, 1993/1994: 165–80.

Trew, Christoph Jakob. 1981. *Erlesene Pflanzen*. Dortmund: Harenberg.

Trupp, Fritz. 1984. *Die letzten Indianer: Kulturen Südamerikas*. Wörgl: Perlinger.

Tsarong, Tsewang Jigme. 1986. *Handbook of traditional Tibetan drugs*. Kalimpong: Tibetan Medical Publications.

———. 1991. Tibetan psychopharmacology. *Integration* 1:43–60.

Tschubinow, Georg. 1914. *Beiträge zum psychologischen Verständnis des sibirischen Zauberers*. Halle: Inaugural-Dissertation.

Turner, D. M. [pseud.]. 1994. *The essential psychedelic guide*. San Francisco: Panther Press.

———. 1997. *Der psychedelische Reiseführer*. Solothurn: Nachtschatten Verlag.

Turner, Nancy J., and Adam F. Szczawinski. 1992. *Common poisonous plants and mushrooms of North America*. Portland, Ore.: Timber Press. (See book review by Richard Evans Schultes in *Journal of Ethnopharmacology* 35 [1991]: 201.)

Tyler, Varro E. 1966. The physiological properties and chemical constituents of some habit-forming plants. *Lloydia* 29 (4): 275–92.

———. 1979. The case for Victor A. Reko: An unrecognized pioneer writer on New World hallucinogens. *Journal of Natural Products* 42 (5): 489–95.

———. 1993. *The honest herbal*. 3rd ed. New York: Pharmaceutical Products Press.

U

Uhe, George. 1974. Medicinal plants of Samoa. *Economic Botany* 28:1–30.

Unschuld, Paul Ulrich. 1973. *Pen-Ts'ao: 2000 Jahre traditionelle pharmazeutische Literatur Chinas*. Munich: Heinz Moos.

Uscátegui M., Nestor. 1959. The present distribution of narcotics and stimulants amongst the Indian tribes of Colombia. *Botanical Museum Leaflets* 18 (6): 273–304.

Usteri, A. 1926. *Pflanzenmärchen und -sagen*. Basel: Rudolf Seering.

V

Valette, Simone. 1990. Die Pharmakologie im alten Ägypten. In *Illustrierte Geschichte der Medizin*, ed. R. Toellner, 1:463–79. Salzburg: Andreas & Andreas.

Vandenberg, Philip. 1979. *Das Geheimnis der Orakel*. Munich: Orbis.

van Treek, Bernhard. 1997. *Partydrogen*. Berlin: Schwarzkopf & Schwarzkopf.

Venturini, Maurizio, and Claudio Vannini. 1995. *Zur Geschichte der Halluzinogenforschung: Schwerpunkt Schweiz (Teil I: 1938–1965)*. Zurich: Private printing. (Licentiate's thesis at the Philosophischen Fakultät I der Universität, Zurich.)

Venzlaff, Helga. 1977. *Der marokkanische Drogenhändler und seine Ware*. Wiesbaden: Franz Steiner.

Vestal, Paul A. 1952. Ethnobotany of the Ramah Navaho. *Papers of the Peabody Museum of American Archaeology and Ethnology, Harvard University* 40 (4).

Vickers, William T., and Timothy Plowman. 1984. Useful plants of the Siona and Secoya Indians of eastern Ecuador. *Fieldiana* (Botany), n.s., 15.

Villoldo, Alberto, and Erik Jendresen. 1993. *Die Macht der vier Winde: Eine Reise ins Reich der Schamanen*. Reinbek: Rowohlt.

Viola, Severino. 1979. *Piante medicinali e velenose della flora italiana*. Milan: Edizioni Artistiche Maestretti.

Vitebsky, Piers. 1995. *The shaman: Voyages of the soul—trance, ecstasy and healing from Siberia to the Amazon*. London: Macmillan.

Voeks, Robert. 1989. Sacred leaves of Brazilian Candomblé. *Geographical Review* 80 (2): 118–31.

Volkan, Kevin. 1994. *Dancing among the maenads: The psychology of compulsive drug use*. New York: Peter Lang.

von Hagen, Victor W. 1979. *Die Wüstenkönigreiche Perus*. Bergisch-Gladbach: Bastei-Lübbe.

von Reis, Siri, and Frank J. Lipp Jr. 1982. *New plant sources for drugs and foods from the New York Botanical Garden Herbarium*. Cambridge and London: Harvard University Press.

Vries, Herman de. 1984. *Natural relations I—die marokkanische sammlung*. Nuremberg: Institut für moderne Kunst; Stuttgart: Galerie d+c mueller-roth.

———. 1989. *Natural relations*. Nuremberg: Verlag für moderne Kunst.

W

Waal, M. de 1988. *Medicinal herbs in the Bible*. York Beach, Maine: Samuel Weiser.

Wagner, Hildebert. 1970. *Rauschgift-Drogen*. 2nd ed. Berlin: Springer.

———. 1985. *Pharmazeutische Biologie 2: Drogen und ihre Inhaltsstoffe*. Stuttgart, New York: G. Fischer.

Waldman, Carl. 1985. *Atlas of the North American Indian*. New York: Facts On File Publications.

Waldmann, Helmut. 1970. *Phantastika im Untergrund*. Vol. 32 of *Akademische Vorträge und Abhandlungen*. Bonn: H. Bouvier & Co. Verlag.

Walker, Winifred. 1964. *All the plants of the Bible*. London: Lutterworth.

Wallnöfer, Heinrich. 1991. *Die vergessene Heilkunst der Azteken*. Stuttgart: Naglschmid.

Walter, Heinz. 1956. Der Jaguar in der Vorstellungswelt der südamerikanischen Naturvölker. Dissertation, Hamburg.

herman de vries
natural relations
eine skizze

karl ernst osthaus-museum hagen

verlag für moderne kunst, nürnberg

Warrier, P. K., V. P. K. Nambar, and C. Ramankutty, eds. 1993 ff. *Indian medicinal plants: A compendium of 500 species.* 5 vols. Madras: Orient Longman.

Wasson, R. Gordon. 1971. Ololiuqui and the other hallucinogens of Mexico. In *Homenaje a Roberto J. Weitlaner*, 329–48. Mexico City: UNAM.

———. 1973. The role of "flowers" in Nahuatl culture: A suggested interpretation. *Botanical Museum Leaflets* 23 (8): 305–24.

———. 1974. The role of "flowers" in Nahuatl culture. *Journal of Psychedelic Drugs* 6 (3): 351–60. (Reprint of Wasson 1973.)

———. 1986. Persephone's quest. In *Persephone's quest: Entheogens and the origins of religion*, ed. R. G. Wasson et al., 17–81. New Haven and London: Yale University Press.

Weatherford, Jack. 1995. *Das Erbe der Indianer: Wie die Neue Welt Europa verändert hat.* Munich: Diederichs.

Wedeck, Harry. 1961. *A dictionary of aphrodisiacs.* New York: Philosophical Society.

Wee, Yeow-Chin, and Hsuan Keng. 1992. *An illustrated dictionary of Chinese medicinal herbs.* Sebastopol, Calif.: CRCS Publications.

Wegmann, Donald H. 1994. *Gift und Helvetia.* Zurich: Virus Verlag.

Weil, Andrew. 1980. *The marriage of the sun and moon: A quest for unity in consciousness.* Boston: Houghton Mifflin.

———. 1986. *The natural mind: An investigation of drugs and the higher consciousness.* Rev. ed. Boston: Houghton Mifflin.

———. 1993. Was uns gesund macht. In *Naturverehrung und Heilkunst*, ed. C. Rätsch, 223–40. Südergellersen: Bruno Martin.

———. 1995. *Spontanheilung.* Foreword by Rüdiger Dahlke. Munich: C. Bertelsmann.

Weil, Andrew, and Winifred Rosen. 1983. *Chocolate to morphine: Understanding mind-active drugs.* Boston: Houghton Mifflin.

———. 1993. *From chocolate to morphine.* Rev. ed. Boston and New York: Houghton Mifflin.

Weil, Gunther M., Ralph Metzner, and Timothy Leary, eds. 1973. *The psychedelic reader.* Secaucus, N.J.: The Citadel Press.

Wellmann, Klaus F. 1978. North American Indian rock art and hallucinogenic drugs. *Journal of the American Medical Association* 239:1524–27.

———. 1981. Rock art, shamans, phosphenes and hallucinogens in North America. *Bollettino del Centro Camuno di Studi Preistorici* 18:89–103.

Wells, Brian. 1974. *Psychedelic drugs: Psychological, medical and social issues.* Baltimore: Penguin.

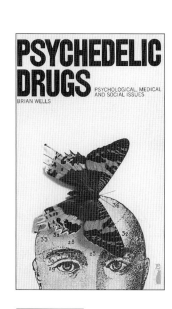

Werner, Helmut. 1991. *Lexikon der Esoterik.* Wiesbaden: Fourier.

———. 1993. *Die Magie der Zauberpflanzen, Edelsteine, Duftstoffe und Farben.* Munich: Droemer-Knaur.

Westbrooks, Randy G., and James W. Preacher. 1986. *Poisonous plants of eastern North America.* Columbia: University of South Carolina Press.

Westendorf, Wolfhart. 1992. *Erwachen der Heilkunst: Die Medizin im Alten Ägypten.* Zurich: Artemis & Winkler.

Westermeyer, Joseph. 1988. The pursuit of intoxication: Our 100-century-old romance with psychoactive substances. *American Journal of Drug and Alcohol Abuse* 14 (2): 175–87.

Westrich, LoLo. 1989. *California herbal remedies.* Houston: Gulf Publishing Co.

Weustenfeld, Wilfried. 1995. *Zauberkräuter von A bis Z: Heilende und mystische Wirkung.* Munich: Verlag Peter Erd.

Wheelwright, Edith Grey. 1974. *Medicinal plants and their history.* New York: Dover.

Whistler, W. Arthur. 1996. *Samoan herbal medicine.* Honolulu: Isle Botanica.

Wichtl, Max, ed. 1989. *Teedrogen.* 2nd ed. Stuttgart: VWG.

Wiesner, Christian. 1964. Aus der Geschichte der indianischen Zauberpflanzen. *Die Grünenthal Waage* 5 (3): 197–200.

Wilbert, Johannes. 1959. *Zur Kenntnis der Yabarana.* Antropologica Supplementband, no. 1. Cologne: Naturwissenschaftliche Gesellschaft.

———. 1993. *Mystic endowment: Religious ethnography of the Wirao Indians.* Cambridge, Mass.: Harvard University Press.

Wilbert, Johannes, and Karin Simoneau, eds. 1982. *Folk literature of the Mataco Indians.* Los Angeles: UCLA Latin American Center Publications.

Willerding, Ulrich. 1970. Vor- und frühgeschichtliche Kulturpflanzenfunde in Mitteleuropa. *Neue Ausgrabungen und Forschungen in Niedersachsen* 5:288–375.

Wilson, R. T., and Woldo Gebre Mariam. 1979. Medicine and magic in central Tigre: A contribution to the ethnobotany of the Ethiopian plateau. *Economic Botany* 33 (1): 29–34.

Winkelman, Michael J. 1992. *Shamans, priests, and witches: A cross-cultural study of magico-religious practitioners.* Tempe: Arizona State University.

———. 1996. Psychointegrator plants: Their roles in human culture, consciousness and health. *Jahrbuch für Transkulturelle Medizin und Psychotherapie*, 1995 (6): 9–53.

Winkelman, Michael, and Walter Andritzky, eds. 1996. *Sakrale Heilpflanzen, Bewußtsein und Heilung: Transkulturelle und Interdisziplinäre Perspektiven/Jahrbuch für Transkulturelle Medizin und Psychotherapie*, 1995 (6). Berlin: VWB.

Wlislocki, Heinrich von. 1891. *Volksglaube und religiöser Brauch der Zigeuner*. Münster: Aschendorffsche Buchhandlung.

Wolke, William, and Laurence Cherniak. 1997. *Opium, Morphin und Heroin*. Markt Erlbach: Raymond Martin Verlag.

Wolters, Bruno. 1994. *Drogen, Pfeilgift und Indianermedizin: Arzneipflanzen aus Südamerika*. Greifenberg: Verlag Urs Freund.

———. 1996. *Agave bis Zaubernuß: Heilpflanzen der Indianer Nord- und Mittelamerikas*. Greifenberg: Verlag Urs Freund.

Wong, Wesley. 1976. Some folk medicinal plants from Trinidad. *Economic Botany* 30:103–42.

Wren, R. C. 1988. *Potter's new cyclopaedia of botanical drugs and preparations*. Rev. ed. Saffron Walden, U.K.: The C. W. Daniel Co.

Y

Yensen, Rich. 1992a. Towards a psychedelic medicine. *Yearbook for Ethnomedicine and the Study of Consciousness*, 1992 (1): 51–69. Berlin: VWB.

———. 1992b. Vom Mysterium zum Paradigma: Die Reise des Menschen von heiligen Pflanzen zu psychedelischen Drogen. In *Das Tor zu inneren Räumen*, ed. C. Ratsch, 17–61. Südergellersen: Verlag Bruno Martin.

Z

Zaehner, R. C. 1974. *Zen, drugs & mysticism*. New York: Vintage Books.

Zander, Robert. 1994. *Handwörterbuch der Pflanzennamen*. 15th ed. Stuttgart: Ulmer.

Zehentbauer, Josef. 1991. *Chemie für die Seele*. Frankfurt/M.: Zweitausendeins.

———. 1992. *Körpereigene Drogen*. Munich and Zurich: Artemis & Winkler.

Zehentbauer, Josef, and Wolfgang Steck. 1986. *Chemie für die Seele*. Königstein/Ts.: Athenäum.

Zimmer, Heinrich. 1973. *Philosophie und Religion Indiens*. Frankfurt/M.: Suhrkamp.

Zimmerer, E. W. 1896. *Kräutersegen*. Donauwörth: Auer.

Zinberg, Norman E. 1984. *Drug, set, and setting: The basis for controlled intoxicant use*. New Haven, Conn., and London: Yale University Press.

Zinser, Hartmut. 1991. Zur Faszination des Schamanismus. In *Hungrige Geister und rastlose Seelen: Texte zur Schamanismusforschung*, ed. Michael Kuper, 17–26. Berlin: Dietrich Reimer.

Zohary, Michael. 1986. *Pflanzen der Bibel*. 2nd ed. Stuttgart: Calwer.

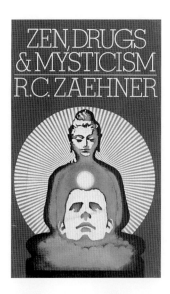

Acknowledgments for the English Edition

I am very gratified that my *Encyclopedia of Psycho-active Plants* has received so much international recognition that it is now being translated into English and published by Inner Traditions and the courageous Ehud Sperling.

I would like to give special thanks to Dr. John R. Baker for his outstanding translation of this *Encyclopedia* into English and for his meticulous efforts to eliminate typographical and printing errors and bibliographic inconsistencies! I would also like to thank Annabel Lee and Cornelia Ballent for their excellent translation work.

Translating a work of this size into another language is an enormous undertaking. For the translators, such a task poses many challenges; for the editorial staff, it requires a painstaking attention to detail and a continuous awareness of the big picture. And it is a risk for the publisher. For these reasons, I would also like to thank all of the readers of the English edition for their support of this project!

Christian Rätsch

From the Translator

This is an extraordinarily detailed and complex book, and translating it into English has posed many challenges. Thanks to the people that have been involved in the preparation and production of this edition, these challenges have all been met. It gives me great pleasure to acknowledge the people that have played a part.

I would like to thank the entire staff at Inner Traditions for their professionalism and expertise. Jon Graham (acquisitions editor) got the project off the ground. Nancy Ringer (copy editor) repeatedly amazed me with her sharp eyes and uncanny ability to point out that the spelling of a term on page 854 deviated from its spelling on page 532. The quality and accuracy of the finished text is a direct reflection of her attention to detail. Doris Troy (proofreader) was indispensable in pointing out and helping to clarify ambiguities and inconsistencies in the text. Priscilla Baker (typesetter) is responsible for the wonderful look of the book. Special thanks go to the two people that I worked with the closest. Laura Schlivek (project editor) provided much expert guidance and good-humored support. She kept the leash loose and still managed to keep things under control. Jeanie Levitan (managing editor) has hovered over this project like an angel. From the beginning, her flexibility, can-do attitude, and unusual working hours have made my side of things much easier.

Annabel Lee and Cornelia Ballent provided excellent translations of parts of the manuscript, and I cannot thank them enough for their assistance. Without their help, I'd still be somewhere in the jungle.

Finally, a word of thanks to Big George and Fatima. Their acceptance of my irregular schedule and support for this project have made things the easiest where I needed it the most: at home with the pack. You guys are the best!

John R. Baker

Index